COMPUTER POWER

NAVIGATION AND COMMUNICATION SYSTEMS

Enjoy sailing to the full with the computing power of Navico marine electronics. Why? because at the heart of every Wheelpilot 4000, System 200 integrated instruments and the new RT6500 series of marine VHF radio telephones, there is an immensely powerful microprocessor. Designed to ensure extreme ease of use, and never lose concentration the latest in 8 bit processor technology has enabled the design of a fully featured Wheelpilot with just two simple controls. A reassuring compass dial to select or change course and an ingenious rubber keypad with LCD display. The System 200 range of thinking instruments is no exception. Can you name just one other ship's log that concurrently displays speed, trip or log and even trend on one large LCD display, with quartz accuracy and also calculates average speed, countdown and elapsed time, even processes data for your SatNav or Decca Navigator. Impressed? You will be for sure when you discover all of the facilities provided by the RT6500. This all British marine transceiver packs real communication power into an amazingly slim, pressure diecast aluminium case. Insist unpon the computer power of Navico marine electronics, backed by a worldwide customer support network that knows no boundaries.

Navico, Star Lane, Margate, Kent UK. (0843) 290290

NAVICO®

© Macmillan Press Ltd 1980, 1981, 1982, 1983, 1986, 1987, 1988; Macmillan London Ltd 1984, 1985

All rights reserved. No reproduction, copy or transmission of this publication may be made without written permission. No paragraph of this publication may be reproduced, copied or transmitted save with written permission or in accordance with the provisions of the Copyright Act 1956 (as amended). Any person who does any unauthorized act in relation to this publication may be liable to criminal prosecution and civil claims for damages.

First published 1980
This edition published 1988 by Macmillan Press Ltd
a division of MACMILLAN PUBLISHERS LTD
London and Basingstoke
Associated companies throughout the world

This book is sold subject to the standard conditions of the Net Book Agreement.

British Library Cataloguing in Publication Data
The Macmillan & Silk Cut nautical almanac.—
    1989–
    1. Coastwise navigation—Great Britain—
    Periodicals 2. Yachts and yachting—Great
    Britain—Periodicals
    623.89′29′41   VK555

    ISBN 0–333–46054–5
    ISSN 0260–2709

MACMILLAN CONSULTANT EDITOR
Klaus Boehm

IMPORTANT NOTE
Whilst every care has been taken in compiling the information contained in this Almanac, the Publishers and Editors accept no responsibility for any errors or omissions, or for any accidents or mishaps which may arise from its use.

CORRESPONDENCE
Letters on editorial matters should be addressed to:
The Editors, The Macmillan & Silk Cut Nautical Almanac,
Macmillan Press Ltd, Houndmills, Basingstoke,
Hampshire RG21 2XS

Enquiries about despatch, invoicing or commercial matters should be addressed to:

Customer Services Department,
Macmillan Press Ltd, Houndmills, Basingstoke,
Hampshire RG21 2XS

Enquiries about advertising space should be addressed to:
Communications Management International,
Chiltern House, 120 Eskdale Avenue, Chesham,
Buckinghamshire HP5 3BD

ACKNOWLEDGMENT
The Publishers gratefully acknowledge the loan of the following equipment used in the photograph on the front cover:

Sextant   – Carl Zeiss (Jena Ltd)
Compass   – Henry Browne Ltd
Dividers – W & H China Ltd

Chapters 1–9   Page preparation by
               Wyvern Typesetting Limited, Bristol
Chapter 10   Cartography and page preparation by
               Lovell Johns Ltd, Oxford

Colour artwork by Dick Vine
Editorial colour separations by Excel Photolit Ltd, Slough

Printed and bound in Hong Kong

THE MACMILLAN & SILK CUT NAUTICAL ALMANAC

1989

EDITORS   Commander R.L. Hewitt LVO RN
                     MIMechE, MRINA, FRIN

                Rear Admiral I.J. Lees-Spalding CB

CONSULTANT EDITORS   M.W. Richey MBE Hon MRIN

                W. Nicholson BSc FRAS

                K.E. Best

                Wing Commander B. D'Oliveira OBE
                                        MRIN

Our support team is the best in the business

With an Autohelm, you've invested in an autopilot that's designed and built to give years of outstanding service.

However, should it ever need attention we want you to know that there is a network of Service Centres to get you back in action quickly – if you're on a holiday cruise that's a really important point.

Each Service Centre has factory trained staff and comprehensive stocks of parts. Service Centres are located conveniently in every major sailing area and each one is a leading marine electronics specialist.

As part of the best team in the business they are ready to give you their help – when you need it – wherever you sail.

Autohelm Service Centres - U.K., Eire and Channel Islands

Factory Service
Nautech Ltd, Portsmouth
(0705) 693611

South Coast and Channel Islands

Brighton, E Sussex
D M S Seatronics
(0273) 605166

Chichester, W Sussex
Pennant Marine
(0243) 511070

Emsworth, Hants
Greenham Marine Ltd
(0243) 378314

Hamble, Hants
Hudson Marine Electronics
(0703) 455129

Hamble, Hants
Marine Technology Ltd
(0703) 455743

Warsash, Hants
B K Electro Marine
(048 95) 2170

Cowes, Isle of Wight
Lecmar Marine Electronics
(0983) 293996

Lymington, Hants
Greenham Marine Ltd
(0590) 75771

Poole, Dorset
Danlea Electronics Ltd
(0202) 673880

Poole, Dorset
Greenham Marine Ltd
(0202) 676363

Brixham, S Devon
Quay Electrics (Teignmough) Ltd
(080 45) 3030

Dartmouth, S Devon
Burwin Marine Electronics
(080 43) 5417

Salcombe, S Devon
Burwin Marine Electronics
(054 884) 3321

Plymouth, S Devon
Ocean Marine Services
(0752) 223922

Plymouth, S Devon
Sutton Marine Electronics Ltd
(0752) 662129

Falmouth, Cornwall
Mylor Marine Electronics
(0326) 74001

Guernsey
Boatworks +
(0481) 26071

Jersey
Jersey Marine Electronics
(0534) 21603

Alderney
Mainbrayce ltd
(048 182) 2772

West Coast, Isle of Man and Ireland

Bideford, N Devon
Marine Electronic Systems
(0805) 22870

Avonmouth, Bristol
A N D Electronics
(0272) 821441

Swansea, W Glam
Caxios Instrumentation
(0792) 797898

Dale, Dyfed
Dale Sailing Co Ltd
(064 65) 349

Pwllheli, Gwynedd
Rowlands Marine Electronics Ltd
(0758) 613193

Colwyn Bay, Clwyd
Sailtronic Marine
(0492) 68536

Liverpool
Robbins marine Radio Services
(051 709) 5431

Fleetwood, Lancs
John N Jones Ltd
(039 17) 5241

Co Cork, Eire
Rider Services
(010 353) 2184 1176

Dublin, Eire
A E Brunker
(0001) 342590

Carrickfergus, N Ireland
Belfast Lough Marine Electronics
(09603) 51565

Isle of Man
Bevan ltd
(0624) 812583

Scotland

Troon, Ayrshire
Boat Electrics & Electronics Ltd
(0292) 315355

Largs, Ayrshire
Yacht Electrical & Electronic Services
(0475) 686091

Oban, Argyll
Camus Marine Ltd
(085 22) 248/279

Shetland Isles
H Williamson & Sons Ltd
(059588) 645

Aberdeen, Grampian
B & P Instrumentation
(0244) 874003

Edinburgh, Lothian
Forth Area Marine Electronics
(031 331) 4343

East Coast

Hull, Humberside
Electronics Marine Ltd
(0482) 25163

Wroxham, Norfolk
Greenham marine Ltd
(060 53) 2238

Ipswich, Suffolk
R & J Marine Electronics
(047 388) 737

Harwich, Essex
R & J Marine Electronics
(0255) 502849

Maldon, Essex
Mantsbrite Marine Electronics
(0621) 53003

Herne Bay, Kent
Heron Marine Services
(0227) 361255

Autohelm Service Centres - France

Lille
Chantiers Navals du Nord:
59000 Lille
(20) 55.17.29/(20) 04.94.60

Le Havre
Heilmann SA: 76057 Le Havre
(35) 42.42.49

Deauville-Trouville
Manche Electronique: 14800
Deauville. (31) 88.63.07

Ouistreham
Nauti Plaisance: 14150
Ouistreham. (31) 97.03.08

Cherbourg
Ets Ergelin: 50130
Octeville. (33) 53.20.26

Granville
Hérélec: 50400
Granville. (33) 50.33.58
Lécoulant Marine:
50400 Granville (33) 50.20.34

Saint-Malo
Electrotechnique Malouine:
35400 Saint-Malo.
(99) 81.52.01

Paimpol
Le Lionnais Marine:
22500 Paimpol. (96) 20.85.18

Brest
Service Electronique de Navigation:
29281 Brest (98) 42.10.35

Morlaix
S.E.N.: 29210 Morlaix
(98) 88.42.42

Le Guilvinec
S.E.A.: 29115 Le Guilvinec
(98) 58.23.52

Concarneau
St. Gué Electronique: 29110
Concarneau (98) 97.50.86

La Trinite-Sur-Mer
S.E.E.M.A.: 56470
La Trinite-sur-Mer. (97) 55.78.06

Noirmoutier
S.E.V.: 85330. Noirmoutier
(51) 39.36.60

Saint Gilles Croix de Vie
Transnav: 85800.
St Gilles Croix de Vie (51.31.60)

Les Sables D'Olonne
Masson: 85100
Les Sables d'Olonne (51) 32.01.07

La Rochelle
Pochon: 17000 La Rochelle
(46) 41.30.53

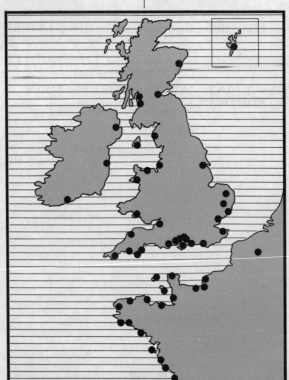

Head straight for the best
Autohelm™

Contents

CHAPTER 1	Page 1

ABOUT THIS ALMANAC

Numbering system, index, acknowledgments, standard terms and abbreviations, suggestions for improvements, free supplements.

CHAPTER 2	7

GENERAL INFORMATION

Rule of the road, traffic schemes, conversion factors, documentation of yachts, HM Customs, regulations in foreign countries, useful addresses.

CHAPTER 3	23

COASTAL NAVIGATION

Compass variation and deviation, true bearing of Sun at sunrise/sunset, distance off by Vertical Sextant Angle, distance of horizon for various heights of eye, distance of lights when rising/dipping, time and distance table, measured mile table, navigation by electronic calculator, chart abbreviations, light characters, IALA buoyage.

CHAPTER 4	37

RADIO NAVIGATIONAL AIDS

Radiobeacons, position fixing systems, VHF direction finding, VHF Radio Lighthouses, radar, radar beacons (Racons).

CHAPTER 5	67

ASTRO-NAVIGATION

The use of calculators, astronomical data (ephemeris), star diagrams, azimuth diagrams, conversion tables.

CHAPTER 6	123

COMMUNICATIONS

International Code, miscellaneous signals, radiotelephony, navigation warnings, coast radio stations, notes on flag etiquette.

CHAPTER 7	145

WEATHER

Beaufort scale, barometer and thermometer conversion scales, terms used in forecasts, sources of weather information.

CHAPTER 8	161

SAFETY

Safety equipment, distress signals, Search and Rescue procedures.

CHAPTER 9	167

TIDES

Explanation, definitions, calculations of times of high and low water, depths of water at specific times, times at which tide reaches certain height, clearance under bridges etc, tidal streams, meteorological conditions.

CHAPTER 10	177

HARBOUR, COASTAL AND TIDAL INFORMATION

Introduction.

Area			
	1	South-West England, including Scilly Islands	185
	2	Central Southern England	217
	3	South-East England	265
	4	Eastern England	307
	5	North-East England	335
	6	South-East Scotland	363
	7	North-East Scotland	389
	8	North-West Scotland	417
	9	South-West Scotland	443
	10	North-West England, Isle of Man and North Wales	465
	11	South Wales and Bristol Channel	493
	12	South Ireland	525
	13	North Ireland	555
	14	South Biscay	583
	15	South Brittany	615
	16	North Brittany	639
	17	Channel Islands and Adjacent Coast of France	661
	18	North-East France	691
	19	Belgium and Netherlands	725
	20	Federal Republic of Germany	761

ADVERTISERS INDEX 785

LATE CORRECTIONS 786

INDEX 789

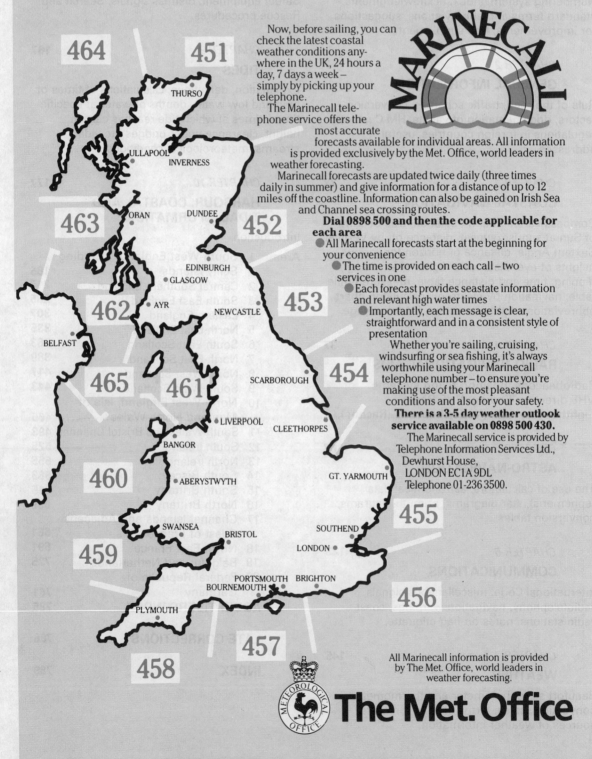

QUICK PAGE REFERENCE
(for main index see page 789)

Abandon ship	166	Link calls (RT)	130-131
Beaufort scale	146	Map of areas (Chap 10)	183
Buoyage	36, 119	MAYDAY (RT distress signal)	132
Chart abbreviations	34-35	Metres/feet conversion	18
Chart symbols	116-117	Morse code	120-121
Customs, HM	20	Navigation lights	115, 118
Distress messages (RT)	132	Phonetic tables	120-121
Distress signals	163-164	Radiobeacons	42-61
Emergency signals	125, 163	Rule of the road	8, 16-17
Ephemeris	86-97	Shipping forecasts	152-154
Feet/metres conversion	18	Single letter signals	120-121
Helicopter rescue	166	Sound signals	17
High Water, Dover	284-286	Tidal calculations	169-173
International Code flags	120-121	Time signals	127
International Code, selected groups	125-126	Traffic schemes	9-15
Late corrections	786-787	Traffic signals	122, 128

PERSONAL INDEX OF IMPORTANT PAGES

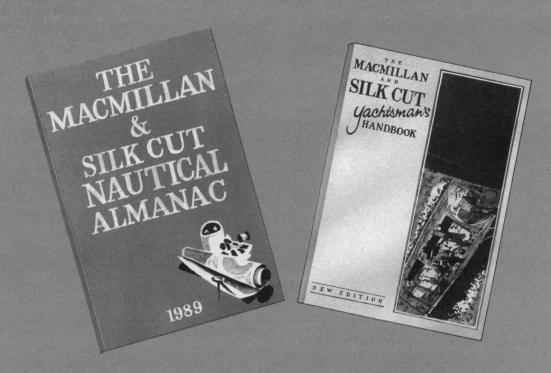

Macmillan and Silk Cut Nautical Almanac and The Yachtsman's Handbook

"Don't drop anchor without it." The Macmillan and Silk Cut Nautical Almanac, once again crammed with important navigational data, is an essential sailing companion.

1989 is *the* year! Not only has the Almanac been as usual up-dated, but so too has its mate, the Macmillan and Silk Cut Yachtsman's Handbook. The Handbook, comprising detailed standing information on owning, maintaining and operating a yacht, has been re-styled and re-edited.

There is no doubt that these two publications, making up what has become popularly known as "the yachtsman's bible," should be alongside every yachtsman's chart table.

— ATTENTION

SILK CUT SEAMANSHIP

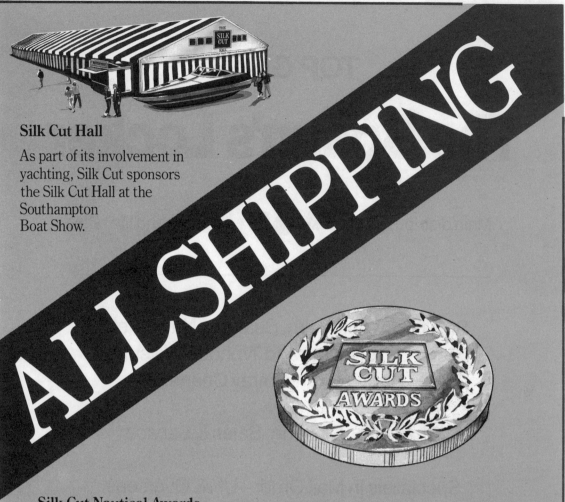

Silk Cut Hall

As part of its involvement in yachting, Silk Cut sponsors the Silk Cut Hall at the Southampton Boat Show.

Silk Cut Nautical Awards

The Silk Cut Nautical Awards were launched at the Southampton Boat Show in September 1983 as a comprehensive Awards scheme designed to reward those who have made an outstanding contribution to the nautical and marine world.

The Awards are given on an annual basis in five key areas:— Seamanship, Rescue, Design, Club/Yachting Service and Yacht Racing, and it is the first time that such a range of awards have been brought together under one umbrella.

Nominations for the Awards are considered by a distinguished judging committee, chaired by Chay Blyth and consisting of leading representatives from the following prestigious organisations — The R.N.L.I., R.Y.A. Seamanship Foundation, H.M. Coastguard, Royal Ocean Racing Club and the British Marine Industries Federation.

The Committee considers candidates via:— The monitoring of press, radio and television coverage — The organisations represented by the Judging Committee — Nominations by you, the public.

These Awards are now established as one of the most important and prestigious in the sailing world. Award winners are presented with their specially struck Gold Commemorative Plaques at a gala luncheon held in London during November of each year.

If you require further information on the Awards, or wish to put in a nomination please contact the Awards Secretary, 11a West Halkin Street, London, SW1X 8JL. Tel. 01-235 7040.

TOP GEAR
The Bosun's Locker

Main distributors of **AVON** dinghies and life rafts

Stockist for Autohelm, Seafarer, Navico, etc.
Henri Lloyd, Musto and Typhoon Wetsuits.
Admiralty and Imray Charts.

Chandlery, Clothing, Sales & Service

Specialists in Mail Order – UK & Overseas

Repairs undertaken for
Marine Electronics and Sails.

Hours of opening 7 days a week
9.00 a.m. to 5.30 p.m.
High Season: 8.00 a.m. to 8.00 p.m.

Ring or write for our free catalogue

Military Road, Royal Harbour, Ramsgate, Kent
Telephone: Thanet (0843) 597158 Telex: 946240 CWEASY G

Quick Reference Marine Products and Services Guide

The following guide gives readers a quick reference to companies and organisations currently offering supplies or services to boat owners and the boating industry. Where possible, each entry carries a concise description and, while every care is taken to ensure accuracy, the publisher does not necessarily endorse the information supplied.

- BOATBUILDERS AND REPAIRS
- BOOKS AND CHARTS
- CHANDLERS
- CLOTHING
- COMMUNICATIONS EQUIPMENT
- ELECTRONIC EQUIPMENT
- ELECTRONIC AND ELECTRICAL ENGINEERS
- ENGINES
- GENERAL MARINE EQUIPMENT AND MACHINERY
- INSTRUMENTATION AND POSITION FIXING
- LUBRICANTS
- MAIL ORDER
- MARINAS
- MARINE FINANCE
- MARINE INSURANCE
- MARINE TRANSPORT AND YACHT DELIVERIES
- PAINT
- SAFETY EQUIPMENT
- SAILING SCHOOLS
- SECURITY SYSTEMS
- SPARS, ROPES AND RIGGING
- YACHT BROKERS

BOATBUILDERS AND REPAIRS

ARDFERN YACHT CENTRE LTD
Ardfern By Lochgilphead,
Argyll, Scotland PA31 8QN
Tel: (08525) 247/636

Boatyard with full repair and maintenance facilities. Timber and GRP repairs, painting and engineering. Sheltered moorings and pontoon berthing. Winter storage, chandlery, showers, fuel calor, brokerage. Hotel, bars and restaurant.

ARDORAN MARINE
Lerags, By Oban, Argyll
PA34 4SE
Tel: (0631) 66123

Swinging moorings, pontoons, diesel, gas, water, toilets, hot showers, slip and cranage for winter storage, engine sales and repairs, Suzuki/mariner outboards, Zodiac inflatables, Orkney boats, chain/shackes in stock.

BERTHON BOAT COMPANY LTD
The Shipyard, Lymington,
Hampshire SO41 9YL
Tel: (0590) 73312

Berthon Boat Company offers a total service of repair and new construction in superb facilities, plus brokerage, new boat sales of Baltic and Trintella yachts and marine berthing.

BOATWORKS + LTD
Castle Emplacement, St Peter Port
Tel: (0481) 26071
Telex: 4191576 Postiv G

Boatworks + provides a comprehensive range of services including electronics, chandlery, boatbuilding and repairs, engine sales and services, yacht brokerage, clothing, fuel supplies.

COBB'S QUAY LTD
Hamworthy, Poole, Dorset
Tel: (0202) 674299

Marina and boatyard with ample storage facilities. All amenities on site.

CUXTON MARINA LTD
Station Road, Cuxton, nr
Rochester, Kent ME2 1AB
Tel: (0634) 721941

Situated on the River Medway the marina offers Pontoon Berthing. 12-ton hoist, on-site security, fuel, dry storage, brokerage and visitors welcome.

EVERSON AND SONS LTD
Phoenix Works, Riverside,
Woodbridge, Suffolk
Tel: (03943) 4358

Boatbuilding and marine engineers; chandlers and riggers; insurance and brokerage arranged. Full yard facilities.

FALMOUTH BOAT CONSTRUCTION LTD
Little Falmouth Yacht Yard,
Flushing, Falmouth, Cornwall
Tel: (0326) 74309

Repair, refit and new building Boat Yard, slipways to 75′ craft. Commercial and pleasure; moorings, re-engining, full engineering facilities, storage, lay-up undercover or hard standing.

A H MOODY & SON LTD
Swanwick Shore Road,
Swanwick, Southampton
SO3 7ZL
Tel: (04895) 6116

Sail and motor new boat sales. New construction. Brokerage. Major refits, repair facilities. Winter lay-up, marina. Chandlery, bunkering, valeting. Insurance.

L H MORGAN & SONS (Marine) LTD
The Boat Centre, 32/42
Waterside, Brightlingsea,
Essex
Tel: (020630) 2003

Marine engineers, chandlers, boat and engine repairers. Yacht brokers, marine insurance, tools and hardware, sailing and fashion clothing.

SEAWARD MARINE ENTERPRISES
Southdown Quay, Millbrook,
Torpoint, Cornwall PL10 1EZ
Tel: (0752) 823084

Extensive quayside berthing with small marina in quiet scenic lake on edge of River Tamar just 3/4 miles to sea. New amenity block and deep water basin now available.

BOOKS AND CHARTS

ADLARD COLES
8 Grafton Street, London
W1X 3LA
Tel: 01-493 7070

Publishers of the best sailing books in the world. Complete range of titles for power and sail; pilotage, instruction, navigation, cruising, sails, rigging, building, design, racing, travel and adventure.

DUBOIS PHILLIPS & McCALLUM LTD
Oriel Chambers, Covent
Garden, Liverpool L2 8UD
Tel: (051) 236 2776

Admiralty chart agents and nautical booksellers. Chart correction service.

HYDROGRAPHER OF THE NAVY
Hydrographic Department, MOD, Taunton, Somerset TA1 2DN
Tel: (0823) 337900

Admiralty charts and hydrographic publications – worldwide coverage corrected to date of issue. Available from appointed admiralty chart agents together with notices to mariners.

THE SOLENT TRADING COMPANY
Port Hamble Marina, Hamble, Southampton SO3 5QD
Tel: (0703) 454858

Comprehensive ranges of chandlery, marine and leisure clothing. Admiralty chart correction, liferaft hire and service inflatables, outboard sales and service, charts and nautical instruments.

WARSASH NAUTICAL BOOKSHOP
31 Newtown Road, Warsash, Southampton
Tel: (04895) 2384

Nautical bookseller & chart agent. Callers and mail order. New & secondhand books. Free lists. Credit cards taken. Publishers of the bibliography of nautical books.

CHANDLERS

ABINGDON BOAT CENTRE
The Bridge, Abingdon, Oxon OX14 3HX
Tel: (0235) 21125

Chandlers.

ARDFERN YACHT CENTRE LTD
Ardfern By Lochgilphead, Argyll, Scotland PA31 8QN
Tel: (08525) 247/636

Boatyard with full repair and maintenance facilities. Timber and GRP repairs, painting and engineering. Sheltered moorings and pontoon berthing. Winter storage, chandlery, showers, fuel calor, brokerage. Hotel, bars and restaurant.

BOATWORKS + LTD
Castle Emplacement, St Peter Port
Tel: (0481) 26071
Telex: 4191576 Postiv G

Boatworks + provides a comprehensive range of services including electronics, chandlery, boatbuilding and repairs, engine sales and services, yacht brokerage, clothing, fuel supplies.

BOSUN'S LOCKER/TOPGEAR
4 Military Road, Royal Harbour, Ramsgate, Kent
Tel: (0843) 597158/602343

Expert mail order chandler for both UK and overseas – also retail outlet with knowledgeable assistants at the harbour in Ramsgate.

BRIXHAM YACHT SUPPLIES LTD
72 Middle Street, Brixham, Devon
Tel: (0803) 882290

COBB'S QUAY LTD
Hamworthy, Poole, Dorset
Tel: (0202) 674299

Marina and boatyard with ample storage facilities. All amenities on site.

CUXTON MARINA LTD
Station Road, Cuxton, nr Rochester, Kent ME2 1AB
Tel: (0634) 721941

Situated on the River Medway the marina offers Pontoon Berthing. 12-ton hoist, on-site security, fuel, dry storage, brokerage and visitors welcome.

EVERSON AND SONS LTD
Phoenix Works, Riverside, Woodbridge, Suffolk
Tel: (03943) 4358

Boatbuilding and marine engineers; chandlers and riggers; insurance and brokerage arranged. Full yard facilities.

MAYFLOWER INTERNATIONAL MARINA (Sailport) plc
Ocean Quay, Richmond Walk, Plymouth PL1 4LS
Tel: (0752) 556633

Marina operators with boat-hoist (25-ton) facility. Restaurant, clubroom, launderette, fuel, chandlery, shop and off-licence. Winter storage. Owned by berth holders and operated to a very high standard.

A H MOODY & SON LTD
Swanwick Shore Road, Swanwick, Southampton SO3 7ZL
Tel: (04895) 6116

Sail and motor new boat sales. New construction. Brokerage. Major re-fits, repair facilities. Winter lay-up, marina. Chandlery, bunkering, valeting. Insurance.

L H MORGAN & SONS (Marine) LTD
The Boat Centre, 32/42 Waterside, Brightlingsea, Essex
Tel: (020630) 2003

Marine engineers, chandlers, boat and engine repairers. Yacht brokers, marine insurance, tools and hardware, sailing and fashion clothing.

ROCHFORD MARINE ENTERPRISES
Unit 3, Orlando Court, Vicarage Lane, Walton-on-the-Naze, Essex
Tel: (0255) 672036

Manufacturers of 'Rochford' telescopic gangways. Sole importers and distributors of Guest (USA) range of lighting. Marinaspec navigation lights. Exclusive distributors SA Equipment Safety Torches, LAGO headlights etc.

SEAWARD MARINE ENTERPRISES
Southdown Quay, Millbrook, Torpoint, Cornwall PL10 1EZ
Tel: (0752) 823084

Extensive quayside berthing with small marina in quiet scenic lake on edge of River Tamar just 3/4 miles to sea. New amenity block and deep water basin now available.

CLOTHING

SHIPSIDES MARINE LTD
5 New Hall Lane, Preston, Lancs
Tel: (0772) 797079

Yacht and Dinghy Chandlery Store close to M6 motorway. Also worldwide mail-order service (catalogue £1.75). Chart agents. International yacht paint centre, Liferaft hire. Yacht clothing department.

THE SOLENT TRADING COMPANY
Port Hamble Marina, Hamble, Southampton SO3 5QD
Tel: (0703) 454858

Comprehensive ranges of chandlery, marine and leisure clothing. Admiralty chart correction, liferaft hire and service inflatables, outboard sales and service, charts and nautical instruments.

COMMUNICATIONS EQUIPMENT

ICOM (UK) LTD
Sea Street, Herne Bay, Kent CT6 8LD
Tel: (0227) 363859
Telex: 965179 Icom G

Importers and suppliers of ICOM marine radio equipment, all frequencies in all situations. Suppliers to Ministry of Defence.

ROWLANDS MARINE ELECTRONICS LTD
The Outer Harbour, Pwllheli, Gwynedd LL53 5HD
Tel: (0758) 613193

Marine Electronic Trade Association member, dealer for Autohelm, Brookes & Gatehouse, Cetrek, ICOM, Kelvin Hughes, Marconi, Nasa, Navico, Naustar, Neco, Seafarer, Racal-Decca, V-Tronix, Ampro, Walker. Equipment supplied installed and serviced.

ELECTRONIC EQUIPMENT

EMTRAD LTD
William Wright Dock, Hull
HU3 4PG
Tel: (0482) 25163

Suppliers to the leisure, fishing and professional marine market of Locat Radio Distress Beacons, Mohawk TV Antenna, Microguard Burglar alarm system and SP-5 solar panels.

NAVICO
Star Lane, Margate, Kent
CT9 4NP
Tel: (0843) 290290
Fax: + 44 843 290471

Manufacturers of marine electronics; VHF radios, cockpit instruments, automatic wheel and tillerpilots, echosounders.

ORWELL AUTO ELECTRONICS LTD
37 Boss Hall Industrial Estate, Sproughton Road, Ipswich
Tel: (0473) 49624/5

Specialists in the supply and maintenance of all marine electrical and electronic equipment

ROWLANDS MARINE ELECTRONICS LTD
The Outer Harbour, Pwllheli, Gwynedd LL53 5HD
Tel: (0758) 613193

Marine Electronic Trade Association member, dealer for Autohelm, Brookes & Gatehouse, Cetrek, ICOM, Kelvin Hughes, Marconi, Nasa, Navico, Naustar, Neco, Seafarer, Racal-Decca, V-Tronix, Ampro, Walker. Equipment supplied installed and serviced.

SEAWARD MARINE ENTERPRISES
Southdown Quay, Millbrook, Torpoint, Cornwall PL10 1EZ
Tel: (0752) 823084

Extensive quayside berthing with small marina in quiet scenic lake on edge of River Tamar just 3/4 miles to sea. New amenity block and deep water basin now available.

W & H CHINA
Howley Properties Ltd, Howley Tannery, Howley Lane, Warrington, Cheshire
WA1 2DN
Tel: (0925) 34621
Telex: 94013565 Chin G

Manufacturers of China chart dividers. Agents for LINEX navigational aids, slide rules etc.

ELECTRONIC AND ELECTRICAL ENGINEERS

DMS SEATRONICS LTD
Unit 14, Brighton Marina, Brighton, East Sussex
BN2 5UD
Tel: (0273) 605166
Telex: 878210

Marine electronics/electrical. Sales, installations, repairs and advice. Agents for all leading manufacturers. Switch panel design, build and installation, battery systems, etc.

EMTRAD LTD
William Wright Docks, Hull
HU3 4PG
Tel: (0482) 25163

Suppliers to the leisure, fishing and professional marine market of Locat Radio Distress Beacons, Mohawk TV Antenna, Microguard Burglar alarm system and SP-5 solar panels.

ENGINES

ARDORAN MARINE
Lerags, By Oban, Argyll
PA34 4SE
Tel: (0631) 66123

Swinging moorings, pontoons, diesel, gas, water, toilets, hot showers, slip and cranage for winter storage, engine sales and repairs, Suzuki/mariner outboards, Zodiac inflatables, Orkney boats, chain/shackes in stock.

VOLVO PENTA UK LTD
Otterspool Way, Watford
WD2 8HW
Tel: (0923) 28544

Volvo Penta's leading marine power – petrol and diesel for leisure craft and workboats – is supported by an extensive network of parts and service dealers.

GENERAL MARINE EQUIPMENT AND MACHINERY

AQUA MARINE MANUFACTURING (UK) LTD
216 Fair Oak Road, Bishopstoke, Eastleigh, Hants
Tel: (0703) 694949
Fax: (0703) 601381

Manufacturers and distributors of a comprehensive range of marine chandlery including Aqua Flow water systems, Danforth anchors, Rule pumps, Engel refrigerators, Dutton Lanson winches and TWC regulators.

BOATWORKS + LTD
Castle Emplacement, St Peter Port
Tel: (0481) 26071
Telex: 4191576 Postiv G

Boatworks + provides a comprehensive range of services including electronics, chandlery, boatbuilding and repairs, engine sales and services, yacht brokerage, clothing, fuel supplies.

CAMUS MARINE LTD
Kilmelford, Oban, Argyll
Tel: (08522) 248/279
Telex: 779828 Camus G

Yacht repairs, Fitting out. Engineering. Moorings. Storage ashore. Painting. Osmosis treatment. Rigging. Chandlery, brokerage. Insurance. Also at Craobh Haven Marina tel: (08525) 225.

FALMOUTH BOAT CONSTRUCTION LTD
Little Falmouth Yacht Yard, Flushing, Falmouth, Cornwall
Tel: (0326) 74309

Repair, refit and new building Boat Yard, slipways to 75' craft. Commercial and pleasure; moorings, re-engining, full engineering facilities, storage, lay-up undercover or hard standing.

GLASPLIES
2 Crowland Street, Southport, Lancs
Tel: (0704) 40626

GREENHAM MARINE LTD
King's Saltern Road, Lymington, Hampshire
Tel: (0590) 71144

With branches nationwide, Greenham Marine can offer yachtsmen one of the most comprehensive selections of marine electronic equipment currently available.

C T HARWOOD LTD
Ashley House, Hurlands Close, Farnham, Surrey GU9 9JF
Tel: (0252) 733312

Manufacturers of water separators and

filters to remove harmful water from diesel fuel. Also a range of cold starting aids for diesel engines.

E C SMITH & SONS (MARINE FACTORS) LTD
Unit H & J Kingsway Industrial Estate, Kingsway, Luton, Beds
Tel: (0582) 29721

Manufacturers and distributors of a comprehensive range of marine equipment including Lofrans Anchor Windbasses. PAG solar panels. Firemast extinguishers, ETA circuit breakers and ARLO sealants.

SOUTH WESTERN MARINE FACTORS LTD
43 Pottery Road, Parkstone, Poole, Dorset BH14 8RE
Tel: (0202) 745414

Sowesters engineering division distributes Mercury outboards, Mercruiser stern drives. BUKH diesels and Teleflex remote controls, the chandlery division distributes Bruce anchors, Maxwell winches, Achilles inflatables, Pains Wessex flares and most hardware.

WYN LIMITED
Plas Paradwys, Bodorgan, Isle of Anglesey, Gwynedd LL62 5PE
Tel: (0407) 840199

Wyn Ltd imports and distributes Sealand's Vacuflush, a luxury marine toilet which is quietly efficient. Force 10's stainless steel cookers, barbecues and cabin heaters are fuelled by a choice of diesel/paraffin or gas.

INSTRUMENTATION AND POSITION FIXING

AUTOHELM
Nautech Ltd, Anchorage Park, Portsmouth, Hants PO3 5TD
Tel: (0705) 693611
Telex: 86384 Nautec G
Fax: (0705) 694642

Manufacturers of the Autohelm range of microprocessor-based Tepad autopilots. Tiller, wheel and inboard installations for sail and power craft from 28' to 125'.

GREENHAM MARINE LTD
King's Saltern Road, Lymington, Hampshire
Tel: (0590) 71144

With branches nationwide, Greenham Marine can offer yachtsmen one of the most comprehensive selections of marine electronic equipment currently available.

KELVIN HUGHES BOAT ELECTRONICS
New North Road, Hainault, Ilford, Essex IG6 2UR
Tel: 01-500 1020

Kelvin Hughes Boat Electronics markets the Kingfisher range of small radars, Husun VHFs, a range of echo sounders/fish finders, 600 series professional navigators and Dolphin chart and track plotters.

RIGEL COMPASSES LTD
Shamrock Quay, William Street, Southampton SO1 1QL
Tel: (0703) 632967/638663

Designers and manufacturers of magnetic compasses for all craft. Electronic compass systems and repeaters. Also clocks, barometers, etc. Full instrument repair and compass adjusting service.

ROWLANDS MARINE ELECTRONICS LTD
The Outer Harbour, Pwllheli, Gwynedd LL53 5HD
Tel: (0758) 613193

Marine Electronic Trade Association member, dealer for Autohelm, Brookes & Gatehouse, Cetrek, ICOM, Kelvin Hughes, Marconi, Nasa, Navico, Naustar, Neco, Seafarer, Racal-Decca, V-Tronix, Ampro, Walker. Equipment supplied installed and serviced.

SESTREL COMPASSES
36/44 Tabernacle Street, London EC2A 4DT
Tel: 01-253 4517

Sestrel Compasses have been manufacturing compasses for over 100 years. Their products can be found on great ocean liners; on ships of the Royal Navy and commercial fleets; as well as leisure crafts – whatever the craft simply the best.

THOMAS WALKER & SON LTD
37–41 Bissell Street, Birmingham B5 7HR
Tel: (021) 622 4475

Manufacturers of the world renowned Walker towed and paddle wheel logs, wind instruments, mentor instrumentation system, watchman and sentinel boat security products and satellite navigators.

WANSBROUGH-WHITE & CO LTD
Chiswick Mall, London W4 2PW
Tel: 01-994 0964

Designers and manufacturer's of Gnav 'Nautrack' plotting system. 'Affix' chart cover, bearing plotter/pelorus/sun compass. DME range finder. Gnav is stable and accurate in any sea state.

LUBRICANTS

AMSOIL UK
184 Watford Road, St Albans, Herts AL2 3EB
Tel: (0727) 66971

Manufacturers of high performance lubricating oils and greases for marine use. Including winch and stern gland grease, diesel and petrol engine oils, two stroke oils and gearbox oils.

MAIL ORDER

BOSUN'S LOCKER/ TOPGEAR
4 Military Road, Royal Harbour, Ramsgate, Kent
Tel: (0843) 597158/602343

Expert mail order chandler for both UK and overseas – also retail outlet with knowledgeable assistants at the harbour in Ramsgate.

MARINAS

ARDFERN YACHT CENTRE LTD
Ardfern By Lochgilphead, Argyll, Scotland PA31 8QN
Tel: (08525) 247/636

Boatyard with full repair and maintenance facilities. Timber and GRP repairs, painting and engineering. Sheltered moorings and pontoon berthing. Winter storage, chandlery, showers, fuel calor, brokerage. Hotel, bars and restaurant.

ARDORAN MARINE
Lerags, by Oban, Argyll PA34 4SE
Tel: (0631) 66123

Swinging moorings, pontoons, diesel, gas, water, toilets, hot showers, slip and cranage for winter storage, engine sales and repairs, Suzuki/mariner outboards, Zodiac inflatables, Orkney boats, chain/ shackes in stock.

BRIGHTON MARINA CO LTD
Brighton Marina BN2 5VF
Tel: (0273) 693636

Britain's largest marine with 1800 pontoon berths offering comprehensive facilities including boatyard, security staff, restaurants, boat sales and yacht club.

BERTHON BOAT COMPANY LTD
The Shipyard, Lymington, Hampshire SO41 9YL
Tel: (0590) 73312

Berthon Boat Company offers a total

service of repair and new construction in superb facilities, plus brokerage, new boat sales of Baltic and Trintella yachts and marina berthing.

CALEY MARINA AND CHANDLERY
Canal Road, Inverness
IV3 6NF
Tel: (0463) 236539

Pontoon berths, boatbuilding, repairs, chandlery, brokerage, slipway, lifting, storage, inboard and outboard service centre (Volvo and Yamaha). Holiday hire cruisers, launderette, showers. Visitors always very welcome.

CAMPER & NICHOLSONS MARINAS LTD
Portway Village Marina,
Penarth, Glamorgan, South Wales.
Tel: (0222) 705021

500 pontoon berths, 50 visitors pontoons, rates on application.

CAMPER & NICHOLSON MARINAS LTD
Brunel Quay,
Neyland, Milford Haven,
Dyfed SA73 1PY
Tel: (0646) 601601

350 pontoon berths, 50 visitors pontoons, rates on application.

CAMPER & NICHOLSONS MARINAS LTD
Mumby Road, Gosport, Hants
PO12 1AH
Tel: (0705) 524811

350 Pontoon berths. Visitors 40 pontoons. Rates on application.

CAMUS MARINE LTD
Kilmelford, Oban, Argyll
Tel: (08522) 248/279
Telex: 779828 Camus G

Yacht repairs. Fitting out. Engineering. Moorings. Storage ashore. Painting. Osmosis treatment. Rigging. Chandlery, brokerage. Insurance. Also at Craobh Haven marina tel: (08525) 225.

COBB'S QUAY LTD
Hamworthy, Poole, Dorset
Tel: (0202) 674299

Marina and boatyard with ample storage facilities. All amenities on site.

CHICHESTER YACHT BASIN
Birdham, Chichester, Sussex
PO20 7EJ
Tel: (0243) 512731

A very attractive marina offering superb services and shore installations, welcoming long and short-term visitors. Use our VHF radio to secure your berth.

CUXTON MARINA LTD
Station Road, Cuxton, nr
Rochester, Kent ME2 1AB
Tel: (0634) 721941

Situated on the River Medway the marina offers Pontoon Berthing. 12-ton hoist, on-site security, fuel, dry storage, brokerage and visitors welcome.

ELMHAVEN MARINA
Rochester Road, Halling, Kent
Tel: (0634) 240489

Tidal berths available including pontoons with electricity and water, walkway berths mud and hard standing. Good toilet and shower facility. John Hawkins Volvo Penta Dealer on site.

ESSEX MARINA LTD
Wallasea Island, nr Rochford,
Essex SS4 2HQ
Tel: ((03706) 531

Yacht marina, Pontoon and offshore moorings. Fuel sales. Brokerage. Night security. Hauling and launching. Engineering and shipwrights. Bars and restaurant. One hour from London.

FALMOUTH YACHT MARINA
North Parade, Falmouth,
Cornwall TR11 2TD
Tel: (0326) 316620

The most westerly marina in England, strategically placed for transatlantic departures and arrivals. Fully serviced permanent and visitor berths. Diesel fuel. Chandlery. 50-ton hoist. Famous friendly service.

GRANARY YACHT HARBOUR SERVICES
The Granary, Dock Lane,
Melton, Woodbridge, Suffolk
IP12 1PE
Tel: (03943) 6327

KIP MARINA
Holt Leisure Parks Ltd,
Inverkip, Renfrewshire,
Scotland PA16 0AS
Tel: (0475) 521485

Scotland's premier marine with over 700 berths – visitors always welcome. Full facility yard, 40-ton boat hoist, chandlery, yacht sales and superb new club house.

LARGS YACHT HAVEN LTD
Irvine Road, Largs, Ayrshire
KA30 8EZ
Tel: (0475) 675333

Perfectly situated in the Firth of Clyde for the best cruising grounds in Britain. 450 berths with first class shore facilities. 45-ton hoist. Regular racing off the marina. Visitors always welcome.

LITTLEHAMPTON MARINA LTD
Ferry Road, Littlehampton,
Sussex
Tel: (0903) 713553

Marina and boatyard with storing ashore for all types of craft. Full marina amenities on site. Compressed air for diving cylinders with large slipway for visitors.

LYMINGTON YACHT HAVEN LTD
King's Saltern Road,
Lymington, Hants SO4 9XY
Tel: (0590) 77071

Established 750 berth marina with all the required boatyard and shore facilities required by visiting yachtsmen. 50-ton hoist and emergency call out engineer.

MAYFLOWER INTERNATIONAL MARINA (Sailport) plc
Ocean Quay, Richmond Walk,
Plymouth PL1 4LS
Tel: (0752) 556633

Marina operators with boat-hoist (25-ton) facility. Restaurant, clubroom, launderette, fuel, chandlery, shop and off-licence. Winter storage. Owned by berth holders and operated to a very high standard.

THE MELFORT MARITIME CO LTD
Melfort Pier, Kilmelford, By
Oban, Argyll, Scotland
Tel: (08522) 333

We offer facilities to fulfil needs of visiting and cruising yachtsmen. Ideally located in one of the most attractive anchorages in the area – 2 hours from Glasgow and you are in the centre of some of the most beautiful scenery on the West coast of Scotland.

A H MOODY & SON LTD
Swanwick Shore Road,
Swanwick, Southampton
SO3 7ZL
Tel: (04895) 6116

Sail and motor new boat sales. New construction. Brokerage. Major re-fits, repair facilities. Winter lay-up, marina. Chandlery, bunkering, valeting. Insurance.

NEWHAVEN MARINA LTD
The Yacht Harbour,
Newhaven, East Sussex
BN9 9BY
Tel: (0273) 513881

Awarded National Yacht Harbour Association FOUR GOLD ANCHORS for comprehensive facilities including boat yard, boat hoist, superb yacht club with accommodation and restaurant. Visiting craft always welcome.

PORT FLAIR LTD
Bradwell Marina, Waterside,
Bradwell-on-Sea, Essex
CM0 7RB
Tel: (0621) 76235/76391

280 pontoon berths with water and electricity, petrol and diesel, chandlery, marine slip/hoistage to 16 tons, repairs, winter lay-ups, licensed club, yacht brokerage.

PORT SOLENT LIMITED
Port Solent, North Harbour,
Portsmouth, Hants PO6 4SX
Tel: (0705) 210765

Port Solent is a developing community of houses, apartments, restaurants and retail encompassing 900 berths. This includes 500 public berths with facilities for visitors, including travel hoist and yacht club.

SEAWARD MARINE ENTERPRISES
Southdown Quay, Millbrook,
Torpoint, Cornwall PL10 1EZ
Tel: (0752) 823084

Extensive quayside berthing with small marina in quiet scenic lake on edge of River Tamar just 3/4 miles to sea. New amenity block and deep water basin now available.

SWANSEA YACHT HAVEN LTD
Lockside, Maritime Quarter,
Swansea, West Glamorgan
SA1 1WN
Tel: (0792) 470310

Centre piece of multi-million development. Locked marina with 360 berths and good facilities. Tidal access HW +/− 3 hrs. Excellent motorway/rail communications. Ideal for day-sailing, cruising, racing.

ST KATHARINE'S YACHT HAVEN
Ivory House, St Katherine By
The Tower, London E1 9AT
Tel: 01-488 2400

150 berths, water, electrics, showers, sewerage disposal, yacht club. Entry via a lock. Operational HW − 2 hrs to HW + 1½ hrs London Bridge, Oct–March 0800–1800. April–August 0600–2030.

THE SOLENT TRADING COMPANY
Port Hamble Marina, Hamble,
Southampton SO3 5QD
Tel: (0703) 454858

Comprehensive ranges of chandlery, marine and leisure clothing. Admiralty chart correction, liferaft hire and service inflatables, outboard sales and service, charts and nautical instruments.

WOOLVERSTONE MARINE MDL (Marinas) LTD
Woolverstone, Ipswich, Suffolk
Tel: (047384) 206/354

Pontoon berths accessible any time, moorings, VHF channel 37, workshops, marine engineers, mobile crane, slipway, large chandlery, pantry off-licence, launderette, brokerage. Visitors welcome. NYHA 4 Gold Anchors

MARINE FINANCE

HOUSEMANS INSURANCE CONSULTANTS LTD
38a West Street, Marlow,
Bucks SL7 2NB
Tel: (0628) 890888

Marine insurance specialists covering all aspects of yacht and motor boat insurance throughout the world but especially United Kingdom and Europe. Also finance arranged.

SOUTH EASTERN CREDIT
206 South Coast Road,
Peacehaven, East Sussex
Tel: (07914) 81911/2

Marine finance and insurance, marine finance up to 10 years to repay loan. Capital and interest 80 per cent of purchase price. No brokerage fee. No penalties – early settlement.

MARINE INSURANCE

FALMOUTH BOAT CONSTRUCTION LTD
Little Falmouth Yacht Yard,
Flushing, Falmouth, Cornwall
Tel: (0326) 74309

Repair, refit and new bulding Boat Yard, slipways to 75′ craft. Commercial and pleasure; moorings, re-engining, full engineering facilities, storage, lay-up undercover or hard standing.

HOUSEMANS INSURANCE CONSULTANTS LTD
38a West Street, Marlow,
Bucks SL7 2NB
Tel: (0628) 890888

Marine insurance specialists covering all aspects of yacht and motor boat insurance throughout the world but especially United Kingdom and Europe. Also finance arranged.

SOUTH EASTERN CREDIT
206 South Coast Road,
Peacehaven, East Sussex
Tel: (07914) 81911/2

Marine finance and insurance, marine finance up to 10 years to repay loan. Capital and interest 80 per cent of purchase price. No brokerage fee. No penalties – early settlement.

MARINE TRANSPORT AND YACHT DELIVERIES

IAIN KERR HUNTER YACHT DELIVERIES
61 Belwood Road, Milton
Bridge, Midlothian EH26 0QN
Tel: (0968) 74486

Specialists in the delivery of yachts and commercial craft by sea. Coastal, offshore and ocean deliveries to owners requirements. Phone for a brochure or quotation.

PAINT

INTERNATIONAL YACHT PAINTS
24/30 Canute Road,
Southampton, Hants
Tel: (0703) 226722

Manufacturers of yacht enamels, varnishes, antifoulings and anti-osmosis treatments. Also operators of the 'Yacht Paint Centre' and 'Interspray' franchises.

SAFETY EQUIPMENT

CREWSAVER LTD
Mumby Road, Gosport,
Hampshire PO12 1AQ
Tel: (0705) 528621

Crewsaver – leading UK manufacturers of lifejackets, buoyancy aids and personal safety equipment. Crewsaver is able to offer a complete range designed to meet your individual water sport needs.
mm

E C SMITH & SONS (MARINE FACTORS) LTD
Unit H & J Kingsway
Industrial Estate, Kingsway,
Luton, Beds
Tel: (0582) 29721

Manufacturers and distributors of a comprehensive range of marine equipment including Lofrans Anchor Windbasses. PAG solar panels. Firemast extinguishers, ETA circuit breakers and ARLO sealants.

EMTRAD LTD
William Wright Docks, Hull
HU3 4PG
Tel: (0482) 25163

Suppliers to the leisure, fishing and professional marine market of Locat Radio Distress Beacons, Mohawk TV Antenna, Microguard Burglar alarm system and SP-5 solar panels.

SAILING SCHOOLS

FOWEY CRUISING SCHOOL
32 Fore Street, Fowey,
Cornwall PL23 1AQ
Tel: (072683) 2129

RYA courses; Practical Cruising March to Christmas, dinghy Easter to October, shorebased intensive in winter with John Myatt author of *The Shorebased Sailor* Skippered charter and dingy hire also available.

WHEELHOUSE SCHOOL OF NAVIGATION
Rudley Mill, Hambledon,
Hants PO7 6QZ
Tel: (070132) 467

RYA shorebased certificate courses by correspondence for day skipper. Yachtmaster ofshore and yachtmaster ocean.

SECURITY SYSTEMS

EC SMITH & SONS (MARINE FACTORS) LTD
Unit H & J Kingsway
Industrial Estate, Kingsway,
Luton, Beds
Tel: (0582) 29721

Manufacturers and distributors of a comprehensive range of marine equipment including Lofrans Anchor Windbasses. PAG solar panels. Firemast extinguishers, ETA circuit breakers and ARLO sealants.

SPARS, ROPE AND RIGGING

SS SPARS (1982) LTD
Unit 4, Parsons Hall Industrial Estate, High Street, Irchester,
Northants NN9 7AB
Tel: (0933) 317143

Manufacturers of alloy yacht masts, alloy toe-rails, stanchions, stanchion bases, fairleads, turning blocks and ancillary equipment.

YACHT BROKERS

ARDFERN YACHT CENTRE LTD
Ardfern By Lochgilphead,
Argyll, Scotland PA31 8QN
Tel: (08525) 247/636

Boatyard with full repair and maintenance facilities. Timber and GRP repairs, painting and engineering. Sheltered moorings and pontoon berthing. Winter storage, chandlery, showers, fuel calor, brokerage. Hotel, bars and restaurant.

BERTHON BOAT COMPANY LTD
The Shipyard, Lymington,
Hampshire SO41 9YL
Tel: (0590) 73312

Berthon Boat Company offers a total service of repair and new construction in superb facilities, plus brokerage, new boat sales of Baltic and Trintella yachts and marina berthing.

BOATWORKS + LTD
Castle Emplacement, St Peter Port
Tel: (0481) 26071
Telex: 4191576 Postiv G

Boatworks + provides a comprehensive range of services including electronics, chandlery, boatbuilding and repairs, engine sales and services, yacht brokerage, clothing, fuel supplies.

CHANNEL YACHT BROKERS LTD
Concordia, Les Vardes, St Peter Port, Guernsey, Channel Islands
Tel: (0481) 22282
Fax: (0481) 20127

Practical offshore yacht brokers specialising in VAT free boats. Conditions and commissions as per BMIF Code of Practice but with no VAT added. No sale no commission.

CUXTON MARINA LTD
Station Road, Cuxton, nr Rochester, Kent ME2 1AB
Tel: (0634) 721941

Situated on the River Medway the marina offers Pontoon Berthing. 12-ton hoist, on-site security, fuel, dry storage, brokerage and visitors welcome.

ESSEX MARINA LTD
Wallasea Island, nr Rochford,
Essex SS4 2HQ
Tel: (03706) 531

Yacht marina, Pontoon and offshore moorings. Fuel sales. Brokerage. Night security. Hauling and launching. Engineering and shipwrights. Bars and restaurant. One hour from London.

L H MORGAN & SONS (Marine) LTD
The Boat Centre, 32/42 Waterside, Brightlingsea,
Essex
Tel: (020630) 2003

Marine engineers, chandlers, boat and engine repairers. Yacht brokers, marine insurance, tools and hardware, sailing and fashion clothing.

YACHT DELIVERIES

IAIN KERR HUNTER YACHT DELIVERIES
61 Belwood Road, Milton Bridge, Midlothian EH26 0QN
Tel: (0968) 74486

Specialists in the delivery of yachts and commercial craft by sea. Coastal, offshore and ocean deliveries to owners requirements. Phone for a brochure or quotation.

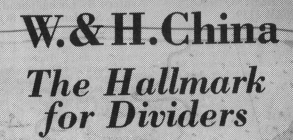

Chapter 1

About this Almanac

Contents

1.1	**INTRODUCTION**	**Page 2**
1.1.1	Numbering system	
1.1.2	Index	
1.1.3	General acknowledgments	
1.1.4	Acknowledgments — tidal information	
1.1.5	Standard terms	
1.1.6	Abbreviations	
1.2	**IMPROVING THE ALMANAC**	**4**
1.2.1	Suggestions for improvements	
1.2.2	Notification of errors	
1.3	**KEEPING IT UP TO DATE**	**4**
1.3.1	Late corrections	
1.3.2	Sources of amendments	
1.3.3	Our free supplements	
1.3.4	Record of amendments	
1.3.5	Supplement application form	

Explanation

The 1989 edition of *The Macmillan & Silk Cut Nautical Almanac* follows the pattern established in 1985, when much of the standing information which does not alter from year to year was transferred to a new companion volume — *The Macmillan & Silk Cut Yachtsman's Handbook*, which will not be republished on an annual basis, although a Revised Edition is now on sale.

The 1989 *Almanac* contains the essential navigational data needed by yachtsmen for the waters round the United Kingdom, Ireland, and the coast of Europe from the border of Spain and France on the Atlantic coast to the North Sea border of Germany and Denmark.

Chapters 2–9 of the *Almanac* deal with the same subjects as Chapters 2–9 of the *Handbook*. Matters which are not likely to change, or which are of a permanent nature, are in the *Handbook*; things which are liable to alter from year to year (or which change completely, such as the tide tables and the ephemeris) are in this *Almanac*. Chapters 2–9 of the *Almanac* and the *Handbook* are cross-referenced where this is helpful to the user.

Chapter 10 — the bulk of the *Almanac* — contains harbour, coastal and tidal information, arranged area by area. A map of the twenty areas is shown on page 183.

1.1 INTRODUCTION

1.1.1 Numbering system

There are ten chapters. For ease of reference each chapter is divided into numbered sections, prefaced by the number of the chapter. Thus the sections in Chapter 7, for example, are numbered 7.1, 7.2 etc.

Within each section the key paragraphs are numbered. Thus in section 7.2 (say) the main paragraphs are numbered 7.2.1, 7.2.2, 7.2.3 etc.

Diagrams carry the chapter number and a figure in brackets, thus: Fig. 7(1), Fig. 7(2), Fig. 7(3) etc.

Tables carry the chapter number and a figure in brackets, thus Table 3(1), Table 3(2) etc.

1.1.2 Index

The main paragraph headings and the page number of each section are listed on the contents page at the start of each chapter. At the back of the book is a full page index, while at the front is a quick reference and personal index for important items.

1.1.3 General acknowledgments

The Editors wish to record their thanks to the many individuals and official bodies who have assisted by providing essential information and much advice in the preparation of this almanac. They include the Hydrographic Department of the Ministry of Defence (Navy) at Taunton, the Proudman Oceanographic Laboratory at Bidston, HM Nautical Almanac Office, HM Stationery Office, HM Customs, the Meteorological Office, HM Coastguard, British Telecom, Trinity House, the National Maritime Museum, the BBC and IBA, the Department of Transport, the Royal National Lifeboat Institution, the Port of London Authority, Associated British Ports, countless Harbour Masters, and our many individual agents.

Chartlets, tidal stream diagrams and tidal curves are produced from British Admiralty Charts and from Hydrographic Publications with the permission of the Controller of HM Stationery Office and of the Hydrographer of the Navy, and from French publications by permission of the Service Hydrographique et Océanographique de la Marine.

Information from the *Admiralty Lists of Lights*, *Admiralty Sailing Directions* and from the *Admiralty List of Radio Signals* is reproduced with the sanction of the Controller HM Stationery Office, and of the Hydrographer of the Navy.

Extracts from the following are published by permission of the Controller of HM Stationery Office: *International Code of Signals, 1969*; *Meteorological Office Leaflets Met 0.1* and *Met 0.3*.

Astronomical data is derived from the current edition of *The Nautical Almanac*, and is included by permission of HM Nautical Almanac Office and of the Controller of HM Stationery Office.

Material from the *Handbook for Radio Operators* is by permission of British Telecom.

1.1.4 Acknowledgments — tidal information

Tidal predictions for Southampton, Dover, Sheerness, London Bridge, Harwich, Lowestoft, Immingham, River Tees, Leith, Aberdeen, Liverpool, Holyhead, Milford Haven, Avonmouth, Shoreham, Galway, Cobh, Belfast and St Helier are computed by the Proudman Oceanographic Laboratory, copyright reserved. Predictions for Dublin are prepared by the Proudman Oceanographic Laboratory for the Dublin Port and Docks Board, copyright reserved. Phases of the Moon are supplied by the Science and Engineering Research Council.

Tidal predictions for Devonport, Dartmouth, Poole, Portsmouth, Lerwick, Ullapool, Oban and Greenock are Crown Copyright and are supplied by permission of the Controller of HM Stationery Office, and the Hydrographer of the Navy.

Acknowledgment is made to the following authorities for permission to use the tidal predictions stated. Service Hydrographique et Océanographique de la Marine, France: Pointe de Grave, Brest, Cherbourg, Le Havre and Dieppe. Department van Waterstaat, Netherlands: Flushing and Hook of Holland. Deutsches Hydrographisches Institut: Helgoland.

1.1.5 Standard terms

All bearings given in this almanac are 'True', from seaward. For example, the sector of a light shown as G (Green) from 090°–180° is visible over an arc of 90° from the moment that the observer is due west of the light until he is due north of it.

Dimensions, in general, are stated in metric terms, the Imperial equivalent being included where appropriate. Distances, unless otherwise stated, are in nautical (sea) miles, abbreviated as nm or M. All depths and heights are shown in metres (m) unless otherwise indicated.

Times are given in GMT, unless stated otherwise (e.g. LT or local time), and are reckoned from 0000 (midnight) to 2400. DST refers to Daylight Saving Time (e.g. BST — British Summer Time). For BST and for standard time (in winter) in France, Belgium, Netherlands and West Germany — add one hour to GMT. For Daylight Saving (DST) from last Sunday in March to last Saturday in September in France, Belgium, Netherlands and West Germany — add two hours to GMT. Tidal predictions are given in GMT for the United Kingdom, and Zone −0100 for the Continent. (See 9.1.2.)

VHF frequencies are identified throughout by their International Maritime VHF series channel (Ch) designator. Frequencies used for calling and working may be separated thus Ch 16; 12.

1.1.6 Abbreviations

The following abbreviations are in general use, while others are explained where they appear in the Almanac. Abbreviations which appear on charts are shown in 3.1.4 on page 34, while those which refer to harbour facilities, lights, fog signals etc are given in 10.0.2, 10.0.3 and 10.0.4 at the start of Chapter 10.

Abbreviations for harbour facilities are also shown on the Dover tide table bookmark.

Alt	Altitude
anch	Anchor, anchorage
App	Apparent
approx	Approximate
Az	Azimuth
B	Bay
Bcst	Broadcast
Bn	Beacon
BFO	Beat frequency oscillator
brg	Bearing
BS	British Standard
BST	British Summer Time
°C	Degrees Celsius (Centigrade)
ca	Cable
CG	Coastguard
Ch	Channel (VHF)
chan	Channel (navigational)
conspic	Conspicuous
cont	Continuous
Corr	Correction
CS	Calibration Station
Dec	Declination
DF	Direction finding
DST	Daylight Saving Time
DW	Deep Water Route
E	East
ext	Extension
°F	Degrees Fahrenheit
Fcst	Forecast
Freq	Frequency
GHA	Greenwich Hour Angle
GMT	Greenwich Mean Time
h	Hours
H + ..	Commencing .. minutes past hour
H24	Continuous
harb	Harbour
Hd	Head
HF	High Frequency
HMSO	Her Majesty's Stationery Office
ht	Height
HW	High water
Hz	Hertz
I	Island
IALA	International Association of Lighthouse Authorities
Ident	Identification signal
in	Inch
inop	Inoperative
ITZ	Inshore traffic zone
kHz	Kilohertz
km	Kilometres
kn	Knot(s)
kW	Kilowatts
Lat	Latitude
LAT	Lowest Astronomical Tide
LANBY	Large Navigational Buoy
Ldg	Leading
LHA	Local Hour Angle
Long	Longitude
LT	Local time
Lt	Light
Lt F	Light float
Lt Ho	Lighthouse
Lt V	Light-vessel
LW	Low water
MF	Medium Frequency
MHWN	Mean High Water Neaps
MHWS	Mean High Water Springs
MHz	Megahertz
MLWN	Mean Low Water Neaps
MLWS	Mean Low Water Springs
MRCC	Maritime Rescue Co-ordination Centre
MRSC	Maritime Rescue Sub-Centre
N	North
n mile	International nautical mile
np	Neaps
Occas	Occasional
(P)	Provisional
PA	Position approximate
Pt	Point
RC	Non-directional Radiobeacon
RD	Directional Radiobeacon
RG	Radio Direction Finding Station
rk(s)	Rock(s)
rky	Rocky
RT	Radio telephony
RNLI	Royal National Lifeboat Institution
S	South
sec, s	Seconds
Seq	Sequence
SHA	Sidereal Hour Angle
Sig	Signal
sp	Springs
SSB	Single sideband
Stn	Station
(T)	Temporary
Tel	Telephone
temp inop	Temporarily inoperative
Tr	Tower
TSS	Traffic Separation Scheme
Twi	Twilight
ufn	Until further notice
VHF	Very High Frequency
W	West (longitude), White
wef	With effect from
WT	Radio telegraphy

About this Almanac

1.2 IMPROVING IT

1.2.1 Suggestions for improvements
The Editors would be particularly grateful for suggestions for improving the content of the almanac. Ideas based on experience with its practical use afloat would be specially welcome. It is not always feasible to implement suggestions received, but all will be carefully considered. Even minor ideas are welcome. Please send any ideas or comments to:

The Editors, *The Macmillan & Silk Cut Nautical Almanac*, Macmillan Press Limited, Houndmills, Basingstoke, Hampshire, RG21 2XS.

1.2.2 Notification of errors
Although very great care has been taken in compiling all the information from innumerable sources, it is recognised that in a publication of this nature some errors may occur. The Editors would be extremely grateful if their attention could be called to any such lapses, by writing to them at the address in 1.2.1 above.

1.3 KEEPING IT UP TO DATE

1.3.1 Late corrections
Late corrections are at the back of the almanac, in front of the index.

1.3.2 Sources of amendments
It is most important that charts and other navigational publications — such as this almanac — are kept up to date. Corrections to Admiralty charts and publications are issued weekly in *Admiralty Notices to Mariners*. These are obtainable from Admiralty Chart Agents (by post if required), or they can be sighted at Customs Houses or Mercantile Marine Offices.

An alternative, but less frequent, service is given by the *Admiralty Notices to Mariners, Small Craft Edition*. This contains reprinted Notices for the British Isles and the European coast from the Gironde to the Elbe. Notices concerning depths greater in general than 7 metres (23ft), or which do not affect small craft for some other reason, are not included. They are available from Admiralty Chart Agents, or through the Royal Yachting Association.

1.3.3 Our free supplements
Important navigational information in this almanac is corrected up to and including *Admiralty Notices to Mariners*, Weekly Edition No. 13 of 1988. Later amendments are at back of almanac, before index. On page 5 are application forms for two free supplements.

The first, to be published in January 1989, will contain corrections up to November 1988. The second, to be published in May 1989, will contain corrections up to March 1989.

Please enter your name and address clearly in block capitals on each form, which will then become a label for the return of the supplement to you.

Place the completed form(s) in a stamped envelope, and post to:

The Editors, *The Macmillan & Silk Cut Nautical Almanac*, Macmillan Press Limited, Houndmills, Basingstoke, Hampshire, RG21 2XS.

At the same time we would be grateful for any comments or suggestions, on the form provided.

It is not necessary to stamp the application forms for the supplements; postage will be paid by the publishers.

Supplements also include important corrections to *The Macmillan & Silk Cut Yachtsman's Handbook*.

1.3.4 Record of amendments
The amendment sheet below is intended to assist you in keeping the almanac up to date, although it can also be used to record corrections to charts or other publications. Tick where indicated when the appropriate amendments have been made.

Weekly Notices to Mariners		Small Craft Editions	
1	27	1 May 1988	....
2	28		
3	29	1 July 1988	....
4	30		
5	31	1 Sept 1988	....
6	32		
7	33		
8	34	1 March 1989	....
9	35		
10	36	1 May 1989	....
11	37		
12	38	1 July 1989	....
13	39		
14	40	1 Sept 1989	....
15	41		
16	42	**Late corrections**	
17	43	(see back of	
18	44	almanac, before	
19	45	index)	
20	46		
21	47	**Macmillan Supplements**	
22	48	First	....
23	49	(Jan 1989)	
24	50		
25	51	Second	....
26	52	(May 1989)	

Keeping it up to date 5

APPLICATION FORMS FOR FREE SUPPLEMENTS

Please follow the directions in 1.3.3 carefully. Write your name and address clearly in BLOCK CAPITALS. No stamps are needed on the forms. For our address see over.

― *Cut here* ― ― ―

Type of boat .... Things I like about the almanac

....

Where kept ....

....

Main cruising areas .... Things I dislike ....

....

....

....

Second supplement (1989)

Name ....

Address ....

....

....

First supplement (1989)

Name ....

Address ....

....

....

Cut here

Please post the application forms overleaf to:

The Editors
Macmillan & Silk Cut Nautical Almanac
Macmillan Press Ltd.
Houndmills
Basingstoke
Hampshire, RG21 2XS
ENGLAND

Please stamp the envelope enclosing the forms, but there is no need to affix postage stamps to the application forms themselves — postage will be paid by the publishers.

Supplements will be posted in January and May.

Please apply as early as possible.

FOR THE BEST MARINE EQUIPMENT

There's only one name you need to know....

AQUA-MARINE Manufacturing (UK) Ltd

216 Fair Oak Road, Bishopstoke, Eastleigh, Hants SO5 6NJ
Tel: (0703) 694949. Fax: (0703) 601381. Telex: 47220 Aquama G. Cable: Aqua-England Southampton

---

**DEDICATED MARINE POWER
9 to 422hp**

VOLVO PENTA

**DEDICATED MARINE SERVICE
OVER 1300 LOCATIONS ACROSS EUROPE**

BOAT LOANS
Marine Finance

We offer Marine Mortgages ...... secured on vessels new or secondhand from £5000 upwards.

Personal Loans up to £10,000, unsecured 100% boat finance secured on property at the markets' lowest rates.

Mortgage and re-mortgage finance ...... the cheapest way to finance a boat. Interest only loans up to 10 years.

Southern Marine has helped thousands of people to buy boats for over ten years, but is not tied to any bank or finance company.

We offer free, independent advice on all aspects of boat purchase, plus insurance and registration services.

We arrange finance fast, after taking details over the phone.

Our rates are always competitive, there are no brokerage fees to pay.

We are available 9am to 9pm, 7 days a week.

**Talk to Tony Tawell or John Mathews
phone 0860-392221**

Southern Marine, Brighton Marina BN2 5US
0273-690999
Licenced Credit Brokers

PAG Daylight Panels

**AN EVEN MORE POWERFUL
SOLUTION TO FLAT BATTERIES**

E.C. Smith & Sons Ltd.,
Dept R900, Units H&J,
Kingsway Ind. Estate,
Kingsway, Luton, Beds LU1 1LP.
Tel: 0582 29721 Telex: 825473G
Fax: 0582 458893

**DEDICATED MARINE POWER
9 to 422hp**

VOLVO PENTA

**DEDICATED MARINE SERVICE
OVER 1300 LOCATIONS ACROSS EUROPE**

CHICHESTER
Yacht Basin

BIRDHAM, CHICHESTER, SUSSEX PO20 7EJ
TELEPHONE BIRDHAM (0243) 512731

As a result of our on-going refurbishment programme, we can now offer new berths at one of the most attractive marinas in the country. The Basin, located in the sheltered conservation area of Chichester Harbour, offers superb sailing and cruising in beautiful surroundings with good access to the Solent and Isle of Wight.

Our excellent amenities include –

- ***Superb new pontoons***
- ***Electricity and finger piers***
- ***Walkaway lighting***
- ***Good sheltered moorings***
- ***Twenty-four hour security***
- ***VHF Radio – Channel 37***
- ***Full boatyard services including lift out and lay up***
- ***Yacht brokerage, insurance and surveys***
- ***Sailing school***
- ***Luxurious shower facilities***
- ***Supermarket, shops & chandlery***
- ***Excellent road communications (London 65 miles)***
- ***Short term visitors welcome***
- ***Limited facilities for trailer sailors***

For further details or an appointment to view the marina, please contact our Berthing Manager, John Haffenden.

Chapter 2
General Information

Contents

2.1 INTERNATIONAL REGULATIONS FOR PREVENTING COLLISIONS AT SEA — Page 8
- 2.1.1 General
- 2.1.2 Traffic separation schemes
- 2.1.3 Vessels in sight of each other
- 2.1.4 Restricted visibility
- 2.1.5 Lights and shapes
- 2.1.6 Distress signals
- 2.1.7 Sound signals

2.2 CONVERSION FACTORS — 18
- 2.2.1 Conversion factors
- 2.2.2 Feet to metres, metres to feet

2.3 DOCUMENTATION — 19
- 2.3.1 Registration
- 2.3.2 International Certificate for Pleasure Navigation
- 2.3.3 Helmsman's (Overseas) Certificate of Competence
- 2.3.4 Licences
- 2.3.5 Insurance
- 2.3.6 Classification
- 2.3.7 Cruising formalities

2.4 HM CUSTOMS — 20
- 2.4.1 General information
- 2.4.2 Notice of departure
- 2.4.3 Stores
- 2.4.4 Immigration
- 2.4.5 Arrivals
- 2.4.6 Full report
- 2.4.7 Quick report
- 2.4.8 Arrival after 2300
- 2.4.9 Customs offices — telephone numbers

2.5 FOREIGN CUSTOMS — PROCEDURES — 21
- 2.5.1 General
- 2.5.2 Irish Republic
- 2.5.3 France
- 2.5.4 Belgium
- 2.5.5 Netherlands
- 2.5.6 Federal Republic of Germany

2.6 USEFUL ADDRESSES — 22

General information — introduction

The following subjects are described in detail in Chapter 2 of *The Macmillan & Silk Cut Yachtsman's Handbook*:

Limits and dangers — eg territorial waters; fishing limits; measured distances; hovercraft; warships on exercises; practice and exercise areas; submarines; minefields; wrecks; offshore oil and gas fields; power cables; traffic schemes. HM Customs — notice of departure; immigration; full and quick reports. Customs regulations in European countries. Yacht tonnage measurement — Net and Gross Tonnages; Lloyd's Register Tonnage; Deadweight Tonnage; One Ton Cup etc. Units and conversions. Glossaries of nautical terms. Yachting organisations — Royal Yachting Association; Seamanship Foundation; British Marine Industries Federation; Trinity House; useful addresses.

Here in the Almanac are given brief notes on the *International Regulations for Preventing Collisions at Sea*, traffic separation schemes, useful conversion factors, a summary of documentation and Customs procedures, and some useful addresses. For further details of these items, and of the subjects listed above, reference should be made to *The Macmillan & Silk Cut Yachtsman's Handbook*.

2.1 INTERNATIONAL REGULATIONS FOR PREVENTING COLLISIONS AT SEA

2.1.1 General
The regulations are stated in full, with diagrams and explanatory notes, in *The Macmillan & Silk Cut Yachtsman's Handbook* (2.1). The following are notes on provisions of special concern to yachtsmen. The numbers of the rules quoted are given for reference.

The rules must be interpreted in a seamanlike way if collisions are to be avoided (Rule 2). A vessel does not have right of way over another regardless of special factors — such as other vessels under way or at anchor, shallow water or other hazards, poor visibility, traffic schemes, fishing boats etc — or the handling characteristics of the vessels concerned in the prevailing conditions. Sometimes vessels must depart from the rules to avoid a collision.

A sailing vessel is so defined (Rule 3) when she is under sail only. When under power she must show the lights for a power-driven vessel, and when under sail and power a cone point down forward (Rule 25).

Keep a good lookout, using eyes and ears, particularly at night or in poor visibility (Rule 5).

Safe speed is dictated by visibility, traffic, depth of water, navigational dangers, and the manoeuvrability of the boat (Rule 6). Excess speed gives less time to appreciate the situation, less time to take avoiding action, and produces a worse collision if such action fails.

Risk of collision must be assessed by all available means (Rule 7). A yacht should take a series of compass bearings of a converging ship. Unless the bearings change appreciably, there is risk of collision. Take special care with large ships.

Take early and positive action to avoid collision (Rule 8). Large alterations of course and/or speed are more evident to the other skipper, particularly at night or on radar. Do not hesitate to slow down, stop (or even go astern, under power). While keeping clear of one vessel, watch out for others.

In narrow channels, keep to starboard whether under power or sail (Rule 9). A yacht under 20m in length must not impede larger vessels confined to a channel.

2.1.2 Traffic separation schemes
Yachts, like other vessels, must conform to traffic schemes (Rule 10). Traffic schemes are essential for the safety of larger vessels and, while inconvenient for yachtsmen, must be accepted as another element of passage planning, and be avoided where possible. They are shown on charts, and those around the British Isles are summarised in Figs 2(1)–2(8) on pages 9–15.

Proceed in the correct lane, and in the general direction of traffic. Normally join or leave a lane at its extremity, but when joining or leaving at the side, do so at as small an angle as possible. Boats under 20m in length, and any sailing yacht, may use inshore traffic zones — often the most sensible action for a yacht. If essential to cross a traffic lane, do so heading at as near right angles as possible to the lane, and do not impede vessels using the lane. If under sail, start the engine if speed falls below about three knots or if a reasonable course cannot be maintained.

Rule 10 does not modify the Collision Regulations when two vessels meet or converge in a traffic scheme and are in risk of collision. Some traffic schemes are under surveillance by radar, aircraft or patrol vessels. There are heavy penalties for breaking the rules. 'YG' in the International Code means 'You appear not to be complying with the traffic separation scheme'.

2.1.3 Vessels in sight of each other
When two sailing vessels are in risk of collision and on opposite tacks, the port tack one keeps clear. If on the same tack, the windward one keeps clear (Rule 12).

Any overtaking vessel, whether power or sail, keeps clear of a vessel she is overtaking (Rule 13). Overtaking means approaching the other vessel from a direction more than $22\frac{1}{2}°$ abaft her beam (in the sector of her sternlight by night). An overtaken vessel must not hamper one overtaking: always look astern before altering course.

When two power-driven vessels approach head-on, each must alter course to starboard, to pass port to port (Rule 14). A substantial alteration may be needed, with the appropriate sound signal (see page 17 and Rule 34), to make intentions clear.

When two power-driven vessels are crossing and in risk of collision, the one with the other on her starboard side must keep clear and, if possible, avoid passing ahead of the other (Rule 15). The give-way vessel should normally alter to starboard; exceptionally, an alteration to port may be justified, in which case a large alteration may be needed to avoid crossing ahead of the other.

When one vessel has to keep clear, the other shall hold her course and speed. But if she realises that the give-way vessel is failing to keep clear, she must take independent action to avoid collision (Rule 17).

Under Rule 18, except where Rules 9 (Narrow Channels), 10 (Traffic Schemes) and 13 (Overtaking) otherwise require:

continued on page 16

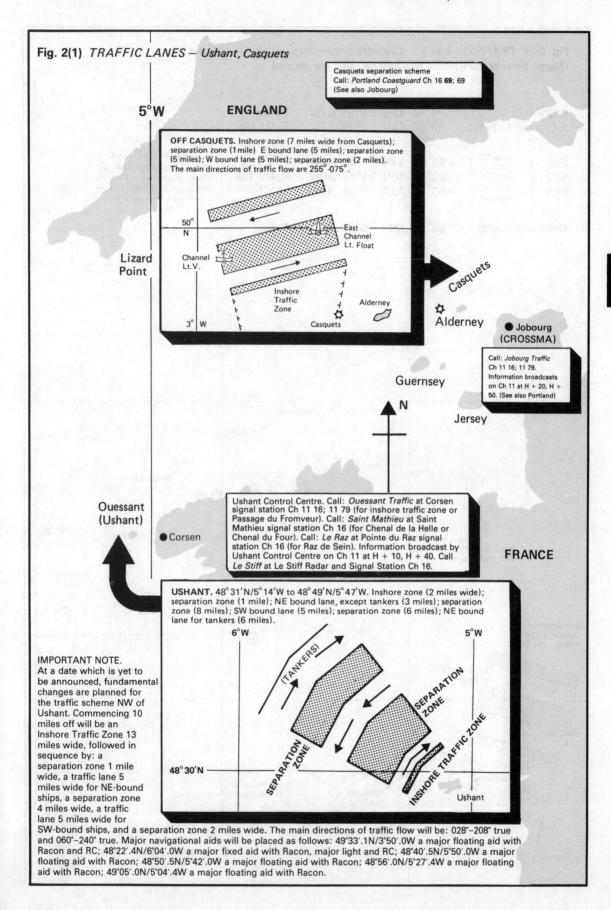

Fig. 2(1) TRAFFIC LANES – Ushant, Casquets

10   General Information

Fig. 2(2) *TRAFFIC LANES – Casquets, Dover Strait, Sandettie, Noord Hinder*
(Note: Only certain major navigational aids are shown)

Channel Lt V	49°54′.42N	2°53′.67W	Fl 15s
E Channel Lt Float	49°58′.67N	2°28′.87W	Fl (2) 10s
EC1 } Large Y	50°05′.90N	1°48′.35W	Fl Y 2.5s
EC2 } pillar buoys,	50°12′.10N	1°12′.40W	Fl(4)Y 15s
EC3 } X topmarks Racons	50°18′.30N	0°36′.10W	Fl Y 5s
Greenwich Lanby	50°24′.5N	0°00′.00	Fl 5s

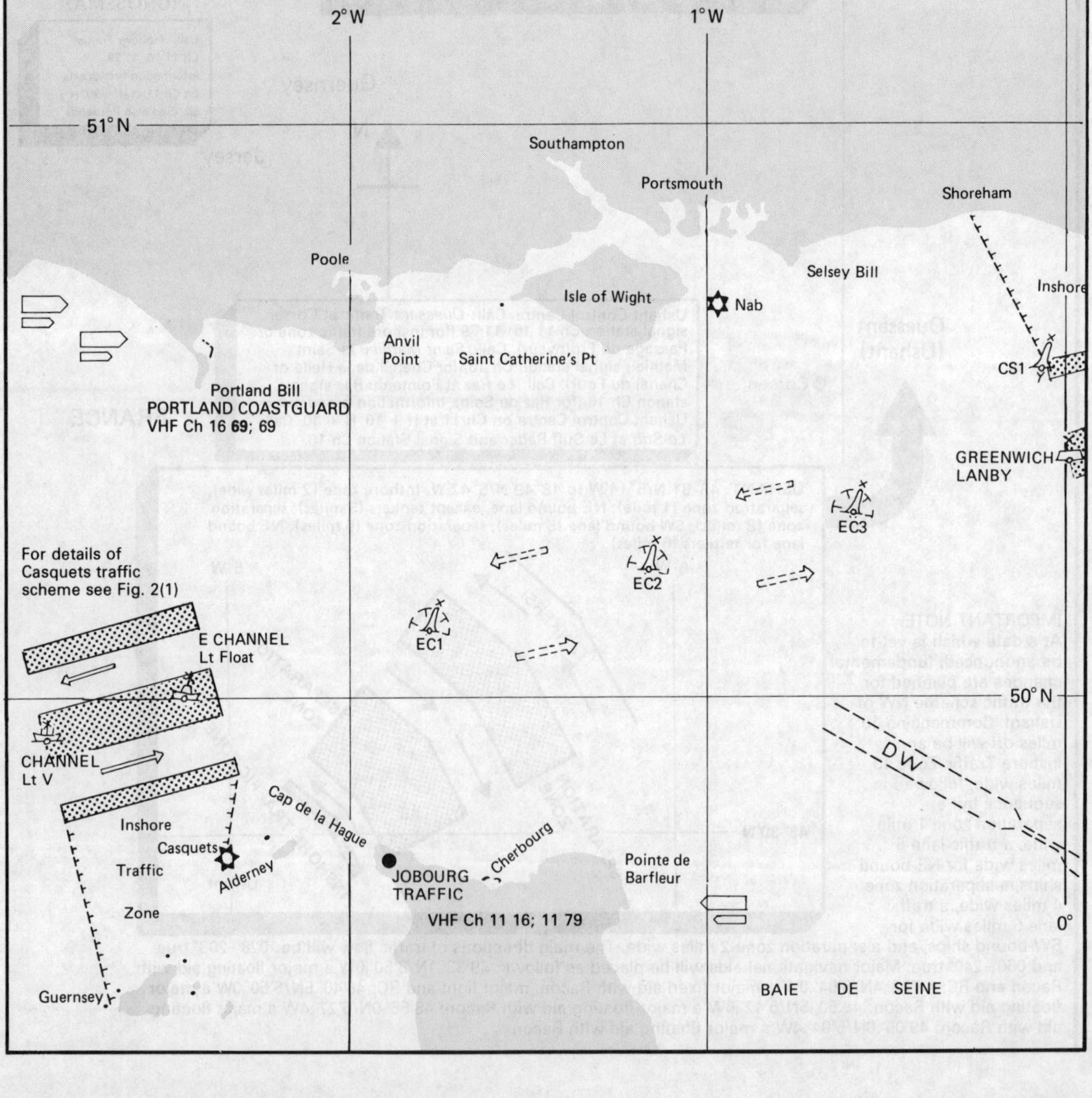

Around each buoy is a circle 2M radius — area to be avoided. Vessels going between Casquets and Dover Strait traffic schemes should leave these areas to port

Portland Bill
PORTLAND COASTGUARD
VHF Ch 16 **69**; 69

For details of Casquets traffic scheme see Fig. 2(1)

E CHANNEL Lt Float

CHANNEL Lt V

Inshore Casquets Traffic Zone

Guernsey

Alderney
Cap de la Hague
JOBOURG TRAFFIC
VHF Ch 11 16; 11 79
Cherbourg
Pointe de Barfleur

Southampton
Poole
Anvil Point
Isle of Wight
Saint Catherine's Pt
Portsmouth
Selsey Bill
Nab
Shoreham
Inshore
CS1
GREENWICH LANBY

EC1 EC2 EC3

BAIE DE SEINE

DW

Traffic Separation Schemes

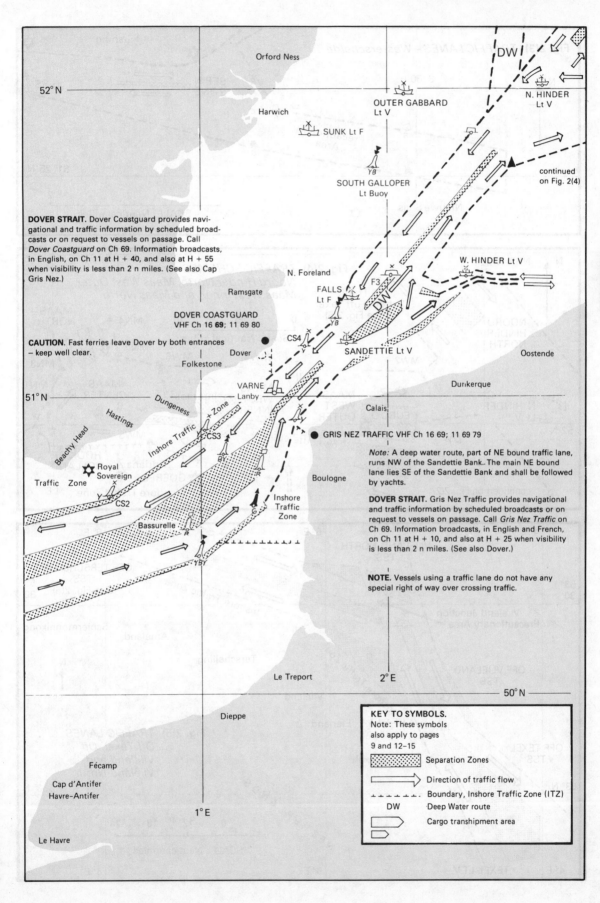

General Information

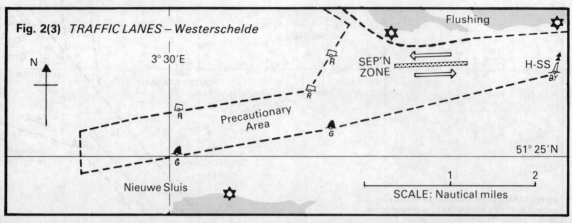

Fig. 2(3) TRAFFIC LANES – Westerschelde

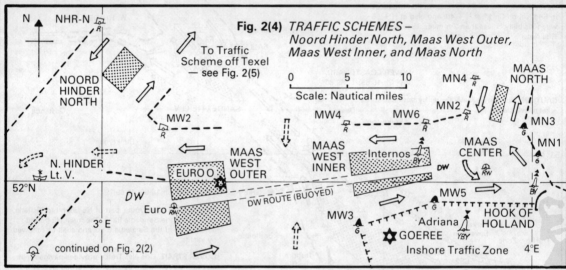

Fig. 2(4) TRAFFIC SCHEMES – Noord Hinder North, Maas West Outer, Maas West Inner, and Maas North

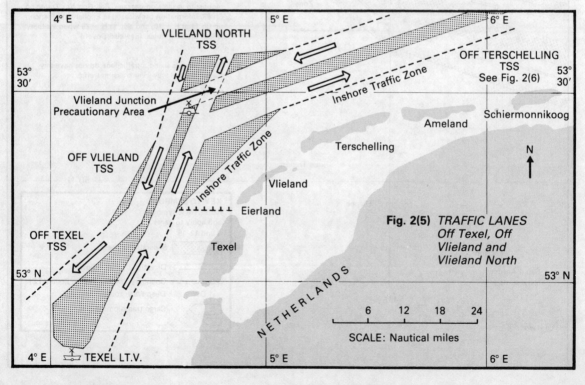

Fig. 2(5) TRAFFIC LANES Off Texel, Off Vlieland and Vlieland North

Traffic Separation Schemes

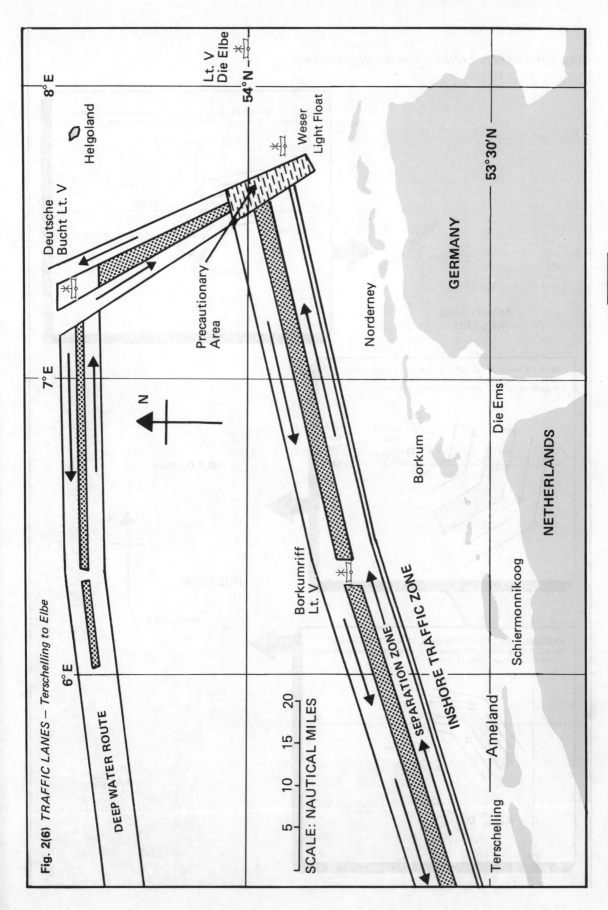

Fig. 2(6) *TRAFFIC LANES* – Terschelling to Elbe

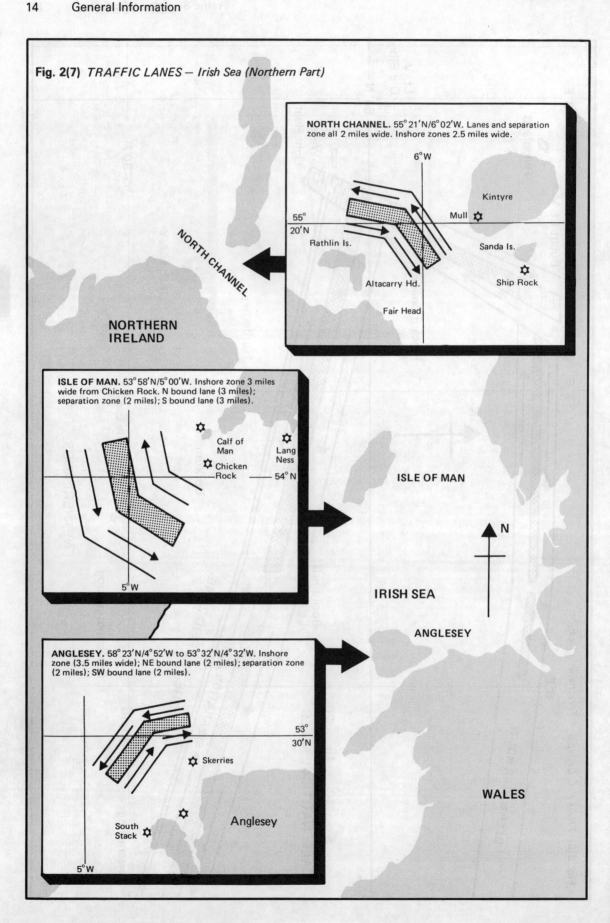

Fig. 2(7) TRAFFIC LANES – Irish Sea (Northern Part)

Traffic Separation Schemes 15

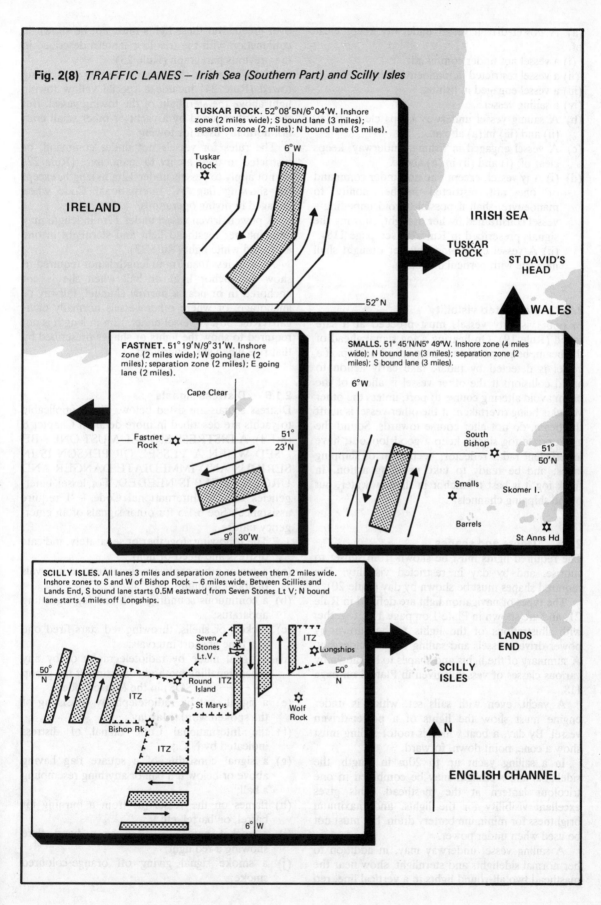

Fig. 2(8) TRAFFIC LANES – Irish Sea (Southern Part) and Scilly Isles

(a) A power-driven vessel underway keeps clear of:
 (i) a vessel not under command;
 (ii) a vessel restricted in manoeuvrability;
 (iii) a vessel engaged in fishing;
 (iv) a sailing vessel.
(b) A sailing vessel underway keeps clear of: (i), (ii) and (iii) in (a) above.
(c) A vessel engaged in fishing, underway, keeps clear of: (i) and (ii) in (a) above.
(d) (i) Any vessel, except one not under command or one not restricted in her ability to manoeuvre, shall if possible avoid impeding a vessel constrained by her draught, showing the signals prescribed in Rule 28 (see page 118).
 (ii) A vessel constrained by her draught shall navigate with particular caution.

2.1.4 Restricted visibility

In poor visibility vessels must proceed at a safe speed (Rule 19). On hearing a fog signal ahead of the beam, be prepared to reduce speed or stop. If a vessel is detected by radar, take early action to avoid collision: if the other vessel is ahead of the beam avoid altering course to port, unless the other vessel is being overtaken: if the other vessel is abaft the beam, do not alter course towards. Sound the appropriate fog signal; keep a good lookout; have an efficient radar reflector; keep clear of shipping lanes; and be ready to take avoiding action. In thick fog it is best to anchor in shallow water, out of the shipping channels.

2.1.5 Lights and shapes

The required lights must be shown from sunset to sunrise, and by day in restricted visibility. The required shapes must be shown by day (Rule 20).

The types of navigation light are defined in Rule 21, and are shown in Plate 1 on page 115, together with illustrations of the lights to be shown by power-driven vessels and sailing vessels underway. A summary of the lights and shapes to be shown by various classes of vessel is given in Plate 4 on page 118.

A yacht, even with sails set, which is under engine must show the lights of a power-driven vessel. By day, a boat which is motor-sailing must show a cone, point down, forward.

In a sailing yacht up to 20m in length, the sidelights and sternlight may be combined in one tricolour lantern at the masthead. This gives excellent visibility for the lights, and maximum brightness for minimum battery drain, but must not be used when under power.

A sailing vessel underway may, in addition to her normal sidelights and sternlight, show near the masthead two all-round lights in a vertical line, red over green. But these lights must not be shown in conjunction with the tricolour lantern described in the previous paragraph (Rule 25).

Lights required for vessels towing and being towed (Rule 24) include a special yellow towing light above the sternlight of the towing vessel. But this is not required by a yacht or other small craft not normally used for towing.

The rules for vessels not under command, or restricted in their ability to manoeuvre (Rule 27) do not apply to vessels under 12m in length, except for showing flag 'A' International Code when engaged in diving operations.

A power-driven vessel under 12m in length may combine her masthead light and sternlight in one all-round white light (Rule 23).

A yacht less than 7m in length is not required to show an anchor light or ball when she is not anchored in or near a narrow channel, fairway or anchorage, or where other vessels normally navigate (Rule 30). A vessel under 12m in length is not required to show the lights or shapes prescribed by that rule when she is aground.

2.1.6 Distress signals

Distress signals are listed below. Those applicable to yachts are described in more detail in Chapter 8 (8.1.4). A DISTRESS SIGNAL MUST ONLY BE USED WHEN A VESSEL OR PERSON IS IN SERIOUS AND IMMEDIATE DANGER AND URGENT HELP IS NEEDED. For lesser emergencies use 'V' International Code – 'I require assistance'. See 6.1.5 for other signals of an emergency nature.

1. The following, together or separately, indicate distress and need of help:
(a) a gun or other explosive signal fired at intervals of about a minute;
(b) a continuous sounding with any fog-signalling apparatus;
(c) rockets or shells, throwing red stars fired one at a time at short intervals;
(d) a signal made by radiotelegraphy or by any other signalling method consisting of the group ··· ——— ··· (SOS) in the Morse Code;
(e) a signal sent by radiotelephony consisting of the spoken word 'Mayday';
(f) the International Code Signal of distress indicated by N.C.;
(g) a signal consisting of a square flag having above or below it a ball or anything resembling a ball;
(h) flames on the vessel (as from a burning tar barrel, oil barrel, etc.);
(i) a rocket parachute flare or a hand flare showing a red light;
(j) a smoke signal giving off orange-coloured smoke;

(k) slowly and repeatedly raising and lowering arms outstretched to each side;
(l) the radiotelegraph alarm signal;
(m) the radiotelephone alarm signal;
(n) signals transmitted by emergency position-indicating radio beacons.
2. The use or exhibition of any of the foregoing signals except for the purpose of indicating distress and need of assistance and the use of other signals which may be confused with any of the above signals is prohibited.
3. Attention is drawn to the relevant sections of the *International Code of Signals*, the *Merchant Ship Search and Rescue Manual* and the following signals:
(a) a piece of orange-coloured canvas with either a black square and circle or other appropriate symbol (for identification from the air);
(b) a dye marker.

2.1.7 Sound signals

Sound signals required (by Rules 34 and 35) are summarised in the table below. Vessels over 12m in length must be provided with a whistle (foghorn) and a bell. A boat under 12m is not obliged to carry these sound signalling appliances, but must have some means of making an efficient sound signal. The effectiveness of a yacht's sound signal should be judged against its audibility from the bridge of a large ship, with conflicting noises from other sources.

Note that a short blast is about one second, and a prolonged blast four to six seconds in duration. A sailing vessel underway in fog sounds one prolonged blast, followed by two short blasts ('D'). The maximum intervals between sound signals for vessels underway in restricted visibility is two minutes, but they should be sounded more frequently if other craft are near.

Summary of important sound signals — Rules 34 and 35

Note: • indicates a short blast of foghorn, of about one second's duration.
— indicates a prolonged blast of foghorn, of four to six seconds' duration.

Vessels in sight of each other (Rule 34)
• I am altering course to starboard (power-driven vessel)
•• I am altering course to port (power-driven vessel)
••• I am operating astern propulsion (power-driven vessel)
—— • (In a narrow channel) I intend to overtake you on your starboard side
—— •• (In a narrow channel) I intend to overtake you on your port side
— • — • Agreement with the overtaking signal above
••••• I fail to understand your intentions or actions/I doubt if you are taking sufficient action to avoid collision
— Warning signal by vessel(s) approaching a bend in channel

Sound signals in restricted visibility (Rule 35)
— Power-driven vessel making way through the water
— — Power-driven vessel under way, but stopped and not making way through the water
— •• Vessel not under command, or restricted in her ability to manoeuvre, or constrained by her draught, or engaged in fishing, or towing or pushing, or a sailing vessel
— ••• Vessel being towed, or if more than one vessel is towed the last vessel in the tow
•••• Pilot vessel engaged on pilotage duties

Bell rung rapidly for about 5 seconds, every minute	Vessel at anchor
Gong rung rapidly for about 5 seconds following above signal, every minute	Vessel of 100 metres or more in length at anchor: the bell being sounded in the fore part of the vessel and the gong aft
•—•	Vessel at anchor (optional additional signal)
Bell rung rapidly for about 5 seconds, with three separate and distinct strokes before and after	Vessel aground

2.2.1 Conversion factors

To convert	Multiply by	To convert	Multiply by
sq in to sq mm	645.16	sq mm to sq in	0.00155
sq ft to sq m	0.0929	sq m to sq ft	10.76
lb/hp/hr to gram/hp/hr	447.4	gram/hp/hr to lb/hp/hr	0.0022
in to mm	25.40	mm to in	0.0394
ft to m	0.3048	m to ft	3.2808
fathoms to m	1.8288	m to fathoms	0.5468
naut miles to statute	1.1515	statute miles to naut	0.8684
lbf to N	4.4482	N to lbf	0.2248
kgf to N	9.8066	N to kgf	0.101972
lb to kg	0.4536	kg to lb	2.205
tons to tonnes (1000 kg)	1.016	tonnes to tons (2240 lb)	0.9842
horsepower to kW	0.7457	kW to hp	1.341
hp to metric hp	1.014	metric hp to hp	0.9862
metric hp to kW	0.735	kW to metric hp	1.359
lb/sq in to kg/sq cm	0.0703	kg/sq cm to lb/sq in	14.22
lb/sq in to ft of water	2.31	ft of water to lb/sq in	0.433
lb/sq in to atmospheres	0.0680	atmospheres to lb/sq in	14.7
ft/sec to m/sec	0.3048	m/sec to ft/sec	3.281
ft/sec to miles/hr	0.682	miles/hr to ft/sec	1.467
ft/min to m/sec	0.0051	m/sec to ft/min	196.8
knots to miles/hr	1.1515	miles/hr to knots	0.868
knots to km/hr	1.8520	km/hr to knots	0.5400
lbf ft to Nm	1.3558	Nm to lbf ft	0.7376
kgf m to Nm	9.8066	Nm to kgf m	0.1020
lbf ft to kgf m	0.1383	kgf m to lbf ft	7.2330
cu ft to galls	6.25	galls to cu ft	0.16
cu ft to litres	28.33	litres to cu ft	0.035
pints to litres	0.568	litres to pints	1.76
galls to litres	4.546	litres to galls	0.22
Imp galls to US galls	1.2	US galls to Imp galls	0.833

2.2.2 Feet to metres, metres to feet

Explanation: The central columns of figures in bold type can be referred in either direction. To the left to convert metres into feet, or to the right to convert feet into metres. For example, five lines down: 5 feet = 1.52 metres, and 5 metres = 16.40 feet.

Feet		Metres	Feet		Metres	Feet		Metres	Feet		Metres
3.28	1	0.30	45.93	14	4.27	88.58	27	8.23	131.23	40	12.19
6.56	2	0.61	49.21	15	4.57	91.86	28	8.53	134.51	41	12.50
9.84	3	0.91	52.49	16	4.88	95.14	29	8.84	137.80	42	12.80
13.12	4	1.22	55.77	17	5.18	98.43	30	9.14	141.08	43	13.11
16.40	5	1.52	59.06	18	5.49	101.71	31	9.45	144.36	44	13.41
19.69	6	1.83	62.34	19	5.79	104.99	32	9.75	147.64	45	13.72
22.97	7	2.13	65.62	20	6.10	108.27	33	10.06	150.92	46	14.02
26.25	8	2.44	68.90	21	6.40	111.55	34	10.36	154.20	47	14.33
29.53	9	2.74	72.18	22	6.71	114.83	35	10.67	157.48	48	14.63
32.81	10	3.05	75.46	23	7.01	118.11	36	10.97	160.76	49	14.94
36.09	11	3.55	78.74	24	7.32	121.39	37	11.28	164.04	50	15.24
39.37	12	3.66	82.02	25	7.62	124.67	38	11.58			
42.65	13	3.96	85.30	26	7.92	127.95	39	11.89			

2.3 DOCUMENTATION

2.3.1 Registration

Two forms of registration are available for British owned yachts. Both are described more fully in section 1.5 of *The Macmillan & Silk Cut Yachtsman's Handbook*.

Full registration, under the Merchant Shipping Act of 1894, is a relatively complex and expensive business since a yacht has to follow the same procedure as a large merchant vessel. It does however have the advantage of establishing title (ownership), and is also required for recording a marine mortgage. The procedure is described in Customs Notice No. 382, obtainable from any Registrar of British Ships, who can normally be located at the Custom House in major ports.

The Small Ships Register, established in 1983, is sufficient for most purposes. It satisfies the law that a British yacht proceeding abroad must be registered, and it also meets the registration requirement for a privileged ensign. The cost is only £10 for a five year period, and measurement is a simple matter of taking the overall length of the boat — well described in the instructions which accompany the application form, obtainable from the Royal Yachting Association, RYA House, Romsey Road, Eastleigh, Hants, SO5 4YA.

2.3.2 International Certificate for Pleasure Navigation

With the introduction of the Small Ships Register, this certificate has no relevance for yachts owned by UK or Commonwealth citizens resident in the UK. It is however available for other persons.

2.3.3 Helmsman's (Overseas) Certificate of Competence

Some countries require a skipper to have a certificate of competence. If a suitable RYA Certificate (e.g. Yachtmaster Offshore) is not held, the above certificate can be issued by the RYA to UK residents on receipt of the appropriate application form, endorsed by the secretary or flag officer of an RYA club or the principal of an RYA recognised teaching establishment.

2.3.4 Licences

A licence is required to operate a boat on most inland waterways (e.g. the River Thames above Teddington Lock, the Norfolk Broads, Yorkshire Ouse above Naburn Lock, and canals or rivers controlled by the British Waterways Board).

2.3.5 Insurance

Any cruising boat represents a large capital investment, which should be protected against possible loss or damage by adequate insurance. It is also essential to insure against third-party risks, and cover for at least £500,000 is recommended.

The value for which a boat is insured should be the replacement cost of the boat and all equipment. Read the proposal form carefully, and fill in the details required as accurately as possible. Take care to abide by the nominated period in commission and cruising area. Note the various warranties which are implied or expressed in the policy. For example, the owner is required to keep the boat in good, seaworthy condition; insurance does not cover charter, unless specially arranged; prompt notice must be given of any claim; a reduction may be made for fair wear and tear for items such as outboards, sails and rigging; theft is only covered if forcible entry or removal can be shown; engines and other mechanical items are only covered in exceptional circumstances; personal effects are not covered, unless specially arranged; and motor boats with speeds of 17 knots or more are subject to special clauses and often to extra premiums.

2.3.6 Classification

Lloyd's Register of Shipping provides an advisory and consultancy service to owners, builders, moulders and designers, and publishes rules for the construction of yachts in various materials. Experienced surveyors approve drawings, supervise moulding, inspect fitting out, check the machinery, and certify the completed yacht. To remain 'in class' a yacht must be subjected to periodical surveys.

As an alternative to full classification, Lloyd's Register Building Certificate is provided to newly-built yachts which have been constructed of any approved material in accordance with the Society's rules, and under the supervision of its surveyors, without the requirement for periodical survey.

2.3.7 Cruising formalities

Before, or while cruising abroad, certain formalities are necessary, as summarised below:
(1) Conform to HM Customs regulations, see 2.4.
(2) The yacht must be registered, see 2.3.1.
(3) Take valid passports for all the crew, and conform to health regulations (e.g. by reporting any infectious disease): exceptionally vaccination certificates may be needed.
(4) Conform to Customs regulations in countries visited. Brief notes are given in section 2.5, but if in doubt about specific items or procedures, ask. All

countries are sensitive to the importation (including carriage on board) of illegal quantities of alcohol and tobacco, and drugs of any kind.

(5) Make sure the yacht is covered by insurance for the intended cruising area, including third party cover.

(6) It is wise for the skipper to carry a Certificate of Competence or similar document (e.g. Yachtmaster Certificate).

(7) The yacht should wear the Red Ensign (or a Special Ensign, if so authorised), and fly a courtesy ensign of the country concerned at starboard crosstree. Flag 'Q' must be carried to comply with Customs procedures.

(8) In most countries it is illegal to use a visiting cruising yacht for any commercial purpose (e.g. charter).

2.4 HM CUSTOMS

2.4.1 General information

Yachts arriving in or departing from the UK from or to places abroad must conform to the regulations in Customs Notice No. 8 (December 1985) summarised below. This notice, the forms mentioned, and further information may be obtained from any Customs and Excise office or from HM Customs and Excise, CDE 1, Dorset House, Stamford Street, London SE1 9PS (Tel: 01-928 0533). Foreign yachts are subject to Customs Notice No. 8A.

Yachtsmen are warned that a boat may be searched at any time; everything obtained abroad must be declared; nothing must be landed or transferred to another vessel until permission is given or the reporting formalities completed; all animals or birds must be securely confined below and out of contact with any other animals, and not landed except when licensed by an officer representing the Ministry of Agriculture, Fisheries and Food; the penalties for smuggling are severe and may include forfeiture of the boat.

2.4.2 Notice of departure

Each intended departure from the UK must be notified to HM Customs on Part I of Form C1328, copies of which are available at Customs offices and from most yacht clubs etc. The form is in three parts. When completed Part I should be posted or delivered to the Customs office nearest the place of departure, so as to arrive before departure. It is valid for up to 48 hours after the stated time of departure. Should the voyage be abandoned, Parts II and III should be posted or delivered to the same office marked 'voyage abandoned'. Failure to give notice of departure may result in delay and inconvenience on return, and possible prosecution.

2.4.3 Stores

In general there is no restriction on reasonable quantities of food, fuel and other stores on which all duties and VAT have been paid. Duty free stores are normally restricted to vessels of 40 tons net register or more, but may be allowed for smaller yachts proceeding south of Brest or north of the Eider — details can be obtained from any Customs office. No other goods may be shipped unless all customs and export procedures are met.

2.4.4 Immigration

The skipper must ensure that the presence of any non-patrials (people not having the right to live in the UK) is reported to the immigration authorities, usually represented by the Customs officer, unless only bound for the Channel Islands, Isle of Man, or Republic of Ireland.

2.4.5 Arrivals

On arrival from abroad (including the Channel Islands and the Republic of Ireland) yachts are subject to Customs control, and the skipper must make a written report on Form C1328 (C1329 for foreign yachts). The skipper must also conform to public and animal health and immigration requirements. Health regulations are stated in Form Port 38, Appendix A to Customs Notice No. 8, whereby any death, illness, infectious disease, animal, bird etc on board must be reported. The signal 'QQ' means 'I require health clearance'.

Entering UK waters flag 'Q' must be flown (illuminated at night) until formalities are completed. The skipper of a British yacht must notify the nearest Customs office by telephone (which may be by link call) or in person within two hours of arrival. Where an answering machine operates, pass this information: Names of vessel and person notifying arrival; time and date of arrival; where the vessel is lying; whether the vessel is eligible for full or quick report procedure (see below); whether or not any animals or birds on board. If there is no reply, phone the nearest alternative office. If that fails, notify arrival by completed Part II of Form C1328 in Customs office letter box.

2.4.6 Full report

The full report procedure is applicable if duty and/or VAT is payable on the vessel; or if the vessel has goods on which duty and/or VAT is payable; or has had repairs/modifications done abroad; or has on board any animal or bird, or goods which are prohibited or restricted; or non-patrials or patrial without a passport (except from Channel Isles or Ireland); or the vessel left the UK more than one year previously; or was not cleared outwards on Part I of Form C1328; or has any death or notifiable illness on board. Under the above circumstances the skipper must make a full report by completing Parts II and III of Form C1328 and awaiting the arrival of a Customs Officer.

2.4.7 Quick report
When none of the conditions in the previous paragraph apply, and animal and health clearance is not required, the skipper is to complete Parts II and III of Form C1328. If the yacht is not visited by a Customs officer within two hours of notifying arrival, deliver Part II to the Customs office concerned (or post it) and retain Part III for reference. The crew may then proceed ashore. This procedure is not available to foreign yachts.

2.4.8 Arrival after 2300
If arriving between 2300 and 0600 and you do not wish to leave the yacht then, providing that there are no animals or birds on board, notification of arrival and making a full or quick report may be delayed until 0800. Flag 'Q' must remain flying.

2.4.9 Customs offices — telephone numbers
The telephone number of the appropriate Customs office (HMC) is given under the heading 'Telephone' for each British harbour in Chapter 10. Note that special numbers must be used if reporting by RT, in which case Freefone does not apply.

2.5 FOREIGN CUSTOMS PROCEDURES

2.5.1 General
It is the skipper's responsibility to find out and to observe the Customs formalities which vary from country to country. In some cases it is obligatory to fly flag 'Q' on arrival, but in others only when carrying dutiable stores above the normal 'tourist' allowances. It is useful to have duplicated lists of those people on board, with their passport numbers.

2.5.2 Irish Republic
It is preferable for yachts to clear Customs at a harbour where there are Customs facilities – Dublin, Dun Laoghaire, Dunmore East, Waterford, New Ross, Dungarvan, Cobh, Cork, Kinsale, Baltimore (summer only), Crookhaven, Bantry, Castletownbere, Cahirciveen, Fenit, Kilrush, Foynes, Limerick, Galway, Westport, Sligo or Killybegs. But yachts may arrive at any port, and if the correct procedure is followed may report to the Civil Guard if there is no Customs officer present.

On arrival therefore, all yachts from abroad (whether carrying dutiable stores or not) must fly flag 'Q' by day, or show a red light over a white light by night, and report to the local Customs (or Civil Guard). Dutiable, prohibited or restricted goods must be declared. United Kingdom citizens do not require passports or visas.

2.5.3 France
Provided that a boat which is on a temporary visit from Great Britain, Northern Ireland or the Channel Islands is only carrying effects of the crew, authorised goods intended for personal use which are within duty-free allowances (and not for commercial purposes) and foreign currency which does not exceed 5000 French francs, there are no Customs formalities on arriving in French waters.

If, for whatever reason, it is required to clear Customs, a yacht should berth in a port which has Customs facilities (*Douanes*), and must fly flag 'Q' by day, or show a red light over a white light by night. It is a requirement that a boat is registered, either under the full registration procedure or on the Small Ships Register (see 2.3.1), or equivalent documentation for the Channel Islands.

A private owner taking his yacht to France qualifies for six months *Importation en Franchise Temporaire* (IFT), with no duty payable. To qualify for IFT the yacht should normally be skippered by the owner: while in France the boat must not be chartered, or even lent to somebody else, and the owner and crew must not pursue any paid activity. There should not be any crew changes. Nor should the crew contribute to running expenses, which might be construed by the authorities as a charter operation.

The six month period may be unbroken, or it may be split into two or more visits within one year. After that, the boat becomes liable to TVA (French VAT) and to an annual tax called *Droit de Passeport*. Details should be obtained from the French Customs. Special regulations apply to yachts chartered from outside France, and to yacht deliveries in and out of the country.

2.5.4 Belgium
Foreign yachts are temporarily admitted duty-free, and there are no arduous formalities unless a stay of more than two months is contemplated, in which case it is necessary to apply for a *plaque d'immatriculation*.

In normal circumstances, fly flag 'Q' at the port of entry, where the yacht will be boarded by Customs officers (in light blue uniform) and possibly by marine police (in blue uniform) for immigration control.

2.5.5 Netherlands
On arrival the skipper must report to Customs with documentation of the yacht and the passports of the crew, to obtain an entry certificate. Flag 'Q' need only be flown if dutiable stores are carried. Yachts are expected to carry, and to obey, the *Binnenvaart Politie Reglement* (collision regulations etc), and also, where applicable, the special regulations which apply to the Rivers Rhine, Lek and Waal. It is possible that on these rivers and other inland waterways a Certificate of Competence or equivalent documentation may have to be produced.

2.5.6 Federal Republic of Germany
On arrival it is necessary to clear Customs at one of the main ports of entry — Borkum, Norderney, Norddeich, Wilhelmshaven, Bremerhaven or Cuxhaven. It is not possible to clear Customs at Helgoland. Yachts arriving from EEC or Scandinavian harbours need not fly flag 'Q'. If the yacht is not boarded by Customs officers, go ashore and report to the Customs office (*Zoll*).

There are special arrangements for yachts arriving in the Elbe and which only intend to transit the Kiel Canal: they are not required to clear Customs, but should fly the Third Substitute flag. Special regulations governing the canal and giving details of signals etc are available at the locks.

It is probable that a Certificate of Competence or equivalent document will have to be produced for navigation on the rivers and inland waterways.

2.6 SOME USEFUL ADDRESSES

Amateur Yacht Research Society, 10 Boringdon Terrace, Turnchapel, Plymouth, PL9 9TQ. Tel: Plymouth (0752) 492646.

Association of Brokers and Yacht Agents, The Wheelhouse, 5 Station Road, Liphook, Hants, GU30 7DW. Tel: Liphook (0428) 722322.

British Marine Industries Federation, Boating Industry House, Vale Road, Oatlands, Weybridge, Surrey. Tel: Weybridge (0932) 54511.

British Waterways Board, Melbury House, Melbury Terrace, London, NW1 6JX. Tel: 01-262 6711.

Clyde Cruising Club, S.V. *Carrick*, Clyde Street, Glasgow, G1 4LN. Tel: 041-552 2183.

Cowes Combined Clubs, Secretary, 18/19 Bath Road, Cowes, I.O.W. Tel: Cowes (0983) 295744.

Cruising Association, Ivory House, St Katharine's Dock, World Trade Centre, London, E1 9AT. Tel: 01-481 0881.

HM Coastguard, Department of Transport, Sunley House, 80–84 High Holborn, London, WC1V 6LP. Tel: 01-405 6911

HM Customs and Excise, Dorset House, Stamford Street, London SE1 9PS. Tel: 01-028 0533.

International Maritime Organization, 4 Albert Embankment, London, SE1. Tel: 01-735 7611.

Junior Offshore Group, 59 Queen's Road, Cowes, I.O.W. Tel: Cowes (0983) 291572.

Little Ship Club, at The Naval Club, 38 Hill Street, London, W1X 8DP.

Lloyd's Register of Shipping, Yacht and Small Craft Services, 71 Fenchurch Street, London, EC3M 4BS. Tel: 01-709 9166.

Maritime Trust, 16 Ebury Street, London, SW1H 0LH. Tel: 01-730 0096.

Meteorological Office, London Road, Bracknell, Berkshire, RG12 2SZ. Tel: Bracknell (0344) 420242.

Royal Cruising Club, 42 Half Moon Street, London, W1. Tel: 01-499 2103.

Royal Institute of Navigation, 1 Kensington Gore, London, SW7 2AT. Tel: 01-589 5021.

Royal National Lifeboat Institution, West Quay Road, Poole, Dorset, BH15 1HZ. Tel: Poole (02013) 71133.

Royal Naval Sailing Association, c/o Royal Naval Club, Pembroke Road, Portsmouth, Hants, PO1 2NT. Tel: Portsmouth (0705) 23524.

Royal Ocean Racing Club, 20 St James Place, London, SW1A 1NN. Tel: 01-493 5252.

Royal Thames Yacht Club, 60 Knightsbridge, London SW1X 7LF. Tel: 01-235 2121.

Royal Yachting Association, RYA House, Romsey Road, Eastleigh, Hants, SO5 4YA. Tel: Eastleigh (0703) 629962.

Royal Yachting Association (Scotland), 18 Ainslie Place, Edinburgh, EH3 GAU. Tel: Edinburgh (031) 226 4401.

Solent Cruising and Racing Association, 18/19 Bath Road, Cowes, I.O.W. Tel: Cowes (0983) 295744.

Sports Council, 70 Brompton Road, London, SW3 1HE. Tel: 01-589 3411.

Trinity House, Corporation of, Trinity House, Tower Hill, London, EC3N 4DH. Tel: 01-480 6601.

UK Offshore Boating Association, Burn's House, 144 Holdenhurst Road, BH8 8AS. Tel: Bournemouth (0202) 298555.

Yacht Designers and Surveyors Association, The Wheelhouse, 5 Station Road, Liphook, Hants, GU30 7DW. Tel: Liphook (0428) 722322.

ROYAL OCEAN RACING CLUB

WARNING

IYR RULE 22.2 SHIPPING, UNSHIPPING OR SHIFTING BALLAST; WATER
(AMENDED IN RORC SAILING INSTRUCTIONS PART 1)

From the warning signal until she is no longer racing, a yacht shall not ship, unship or shift ballast, whether moveable or fixed, or take in or discharge water, except for ordinary ship's use and the removal of bilge water.

MOVING BALLAST TO IMPROVE PERFORMANCE IS CHEATING

**MOVABLE GEAR IS BALLAST
THIS INCLUDES:
ANCHORS,
SAILS, BEER,
WATER, TOOLS,
LIFERAFT.**

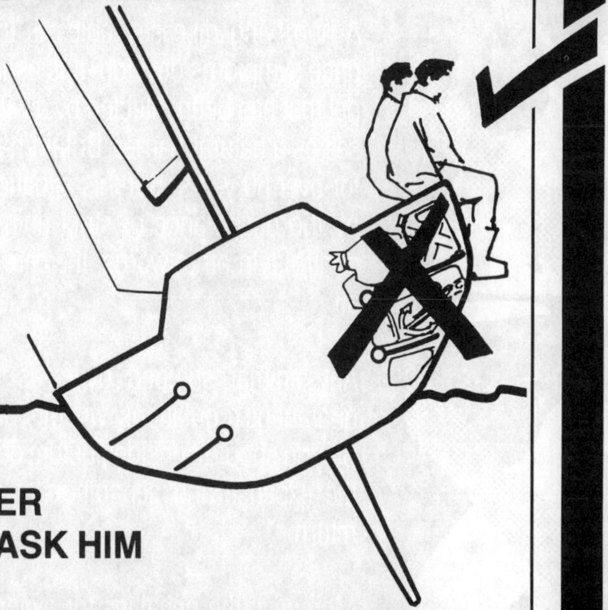

READ THE RULES

THE OWNER(S)/SKIPPER
IS RESPONSIBLE — ASK HIM

SAIL FAST BUT FAIR!

Issued by the RORC in co-operation with the ORC

The Royal Ocean Racing Club offers a unique range of offshore racing activities.
**For full details contact the RORC, 20 St. James's Place, London SW1A 1NN.
Telephone: 01-493 2248   Fax: 01-493 5252**

RABIES
PREVENTION

NEVER SAIL ABROAD WITH ANIMALS ON BOARD

Sailing overseas? If you're travelling to any place outside territorial waters, always leave pets at home and never bring back any animals. The reason, in a word, is rabies.

A horrifying disease, rabies is invariably fatal in humans, once symptoms develop. It is usually caused by the bite of a pet that has been infected by a wild or stray animal. With the sea as a natural barrier, Britain has been free of the disease for over 60 years. But this situation can only be maintained by the observance of strict animal import and quarantine controls. In other words, the care and common-sense of us all.

Any animal shipped overseas is automatically liable to quarantine regulations, whether or not it has stepped ashore. So you should consider, before deciding to take pets abroad, the expense and separation of six-months quarantine on return.

Moreover, the penalties for evading regulations are severe. So-called 'pet-lovers' attempting to break the rules face an unlimited fine or up to a year's imprisonment. In addition, the animal may be destroyed, and at best will have to undergo the full six-months quarantine.

Just one infected animal would endanger humans and animals throughout Britain.

IT'S JUST NOT WORTH THE RISK.

Contents

Coastal Navigation

Chapter 3

3.1.1 Compass variation and deviation
3.1.2 Explanation of the following tables:
   3(1) True bearing of Sun at sunrise/sunset
   3(2) Distance off by Vertical Sextant Angle
   3(3) Distance of horizon for various heights of eye
   3(4) Lights — distance off when rising/dipping
   3(5) Time, speed and distance table
   3(6) Measured mile table
3.1.3 Navigation by electronic calculator
3.1.4 Chart abbreviations. For chart symbols see Plates 2 and 3, on pages 116–117
3.1.5 Light characters (Fathoms and Metric Charts)
3.1.6 IALA Buoyage System (Region A). See also Plate 5 on page 119

Coastal navigation — introduction
The following subjects are described in detail in Chapter 3 of *The Macmillan & Silk Cut Yachtsman's Handbook*:
   The terms and definitions used in coastal navigation; magnetic variation and deviation; compass checks; compass adjusting and compass swinging; charts and their symbols; lights and fog signals; methods of laying off courses and position fixing; time, speed and distance; measured mile table; pilotage; IALA buoyage system; passage planning; the use of calculators; practical passage making; sailing directions.
Here in the Almanac is provided basic information, in the form of tables and other data, needed for coastal navigation. Some brief descriptions on how to use the tables etc are included, but for a fuller explanation of these matters or of the subjects listed above, reference should be made to Chapter 3 of *The Macmillan & Silk Cut Yachtsman's Handbook*.

3.1.1 Compass variation and deviation

The compass is the most vital navigational instrument in a cruising boat. It is affected by variation (the amount magnetic North is displaced from True North, and which varies from place to place) and by deviation (caused by the boat's local magnetic field). Variation, which alters slightly from year to year, is shown on the chart — normally at the compass rose. Deviation varies according to the boat's heading: it should be shown, for different headings, on a deviation card — produced as a result of adjusting and swinging the compass. With a properly adjusted compass, deviation should not be more than about 2° on any heading — in which case it can often be ignored except on long passages.

When converting a True course or a True bearing to Magnetic: add Westerly variation or deviation and subtract Easterly.

When converting a Magnetic course or bearing to True: subtract Westerly variation or deviation, and add Easterly.

Bearings given on charts and in Sailing Directions are normally True bearings, from seaward.

3.1.2 Tables — explanations

The tables listed below are given on pages 25–32. Here are brief explanations of their use. For greater detail see Chapter 3 of *The Macmillan & Silk Cut Yachtsman's Handbook*.

Table 3(1) — True bearing of Sun at sunrise and sunset

Enter with approximate latitude and with declination (extracted from the ephemeris in Chapter 5). The tabulated figure is the True bearing, measured from North if declination is North or from South if declination is South, towards the East if rising or towards the West if setting. Having extracted the True bearing, apply variation before comparing with compass to determine deviation on course steered. The bearing of the Sun should be taken when its lower limb is a little over half a diameter above the horizon.

Table 3(2) — Distance off by Vertical Sextant Angle

Enter with the height of the body (in metres or feet) and read across the page until the required sextant angle (corrected for index error) is met. Take out the distance of the object (in nautical miles) at the head of the column. Caution is needed when the base of the object (e.g. a lighthouse) is below the horizon. For precise ranges the distance that sea level is below MHWS must be added to the height of the object (above MHWS) before entering the table.

Table 3(3) — Distance of horizon for various heights of eye

Enter with height of eye (in metres or feet), and extract distance of horizon (in nautical miles). The actual distance may be affected by abnormal refraction.

Table 3(4) — Lights — distance off when rising or dipping

This table combines selected heights of eye with selected heights of lights, to give the range at which a light dips below or rises above the horizon.

Table 3(5) — Time, speed and distance table

Enter with time (in decimals of an hour, or in minutes) and speed (in knots) to determine distance run (in nautical miles).

Table 3(6) — Measured mile table

Enter with time (in minutes at the head of columns, and in seconds down the side of the table) to extract speed (in knots).

3.1.3 Navigation by electronic calculator

Speed, time and distance

$$\text{Speed (in knots)} = \frac{\text{Distance (n miles)} \times 60}{\text{Time (mins)}}$$

$$\text{Time (in mins)} = \frac{\text{Distance (n miles)} \times 60}{\text{Speed (knots)}}$$

$$\text{Distance (in n miles)} = \frac{\text{Speed (knots)} \times \text{Time (mins)}}{60}$$

Distances and speed

Distance of horizon (in n miles) =
$$1.144 \times \sqrt{\text{Ht of eye}} \text{ (in ft)}$$
or
$$2.072 \times \sqrt{\text{Ht of eye}} \text{ (in m)}$$

Distance a light is visible (in n miles) =
$$1.144 \times (\sqrt{h_o} + \sqrt{h_e})$$
where h_o and h_e are heights of light and eye (ft)
or
$$= 2.072 \times (\sqrt{h_o} + \sqrt{h_e})$$
where heights are in metres.

Distance of mountains etc beyond horizon, in n miles = $\sqrt{1.13(h_o - h_e) + (a - .972 \times \sqrt{h_e})^2} - (a - .972 \times \sqrt{h_e})$ where h_o is height of mountain (ft), h_e is height of eye (ft), and a is the sextant angle in minutes; or $= \sqrt{3.71(h_o - h_e) + (a - 1.76 \times \sqrt{h_e})^2} - (a - 1.76 \times \sqrt{h_e})$ where the heights are in metres.

(Continued on page 33)

TABLE 3(1) True bearing of sun at sunrise and sunset.

LAT	0°	1°	2°	3°	4°	DECLINATION 5°	6°	7°	8°	9°	10°	11°	LAT
30°	90	88.8	87.7	86.5	85.4	84.2	83.1	81.9	80.7	79.6	78.4	77.3	30°
31°	90	88.8	87.7	86.5	85.3	84.2	83.0	81.9	80.6	79.5	78.3	77.1	31°
32°	90	88.8	87.6	86.5	85.3	84.1	82.9	81.7	80.5	79.4	78.2	77.0	32°
33°	90	88.8	87.6	86.4	85.2	84.0	82.8	81.6	80.4	79.2	78.0	76.8	33°
34°	90	88.8	87.6	86.4	85.2	84.0	82.7	81.5	80.3	79.1	77.9	76.7	34°
35°	90	88.8	87.5	86.3	85.1	83.9	82.7	81.4	80.2	79.0	77.8	76.5	35°
36°	90	88.8	87.5	86.3	85.0	83.8	82.6	81.3	80.1	78.8	77.6	76.3	36°
37°	90	88.7	87.5	86.2	85.0	83.7	82.5	81.2	80.0	78.7	77.4	76.2	37°
38°	90	88.7	87.5	86.2	84.9	83.6	82.4	81.1	79.8	78.5	77.3	76.0	38°
39°	90	88.7	87.4	86.1	84.8	83.6	82.3	81.0	79.7	78.4	77.1	75.8	39°
40°	90	88.7	87.4	86.1	84.8	83.5	82.1	80.8	79.5	78.2	76.9	75.6	40°
41°	90	88.7	87.3	86.0	84.7	83.4	82.0	80.7	79.4	78.0	76.7	75.3	41°
42°	90	88.6	87.3	86.0	84.6	83.3	81.9	80.6	79.2	77.8	76.5	75.1	42°
43°	90	88.6	87.3	85.9	84.5	83.1	81.8	80.4	79.0	77.6	76.3	74.9	43°
44°	90	88.6	87.2	85.8	84.4	83.0	81.6	80.2	78.8	77.4	76.0	74.6	44°
45°	90	88.6	87.2	85.7	84.3	82.9	81.5	80.1	78.6	77.2	75.8	74.3	45°
46°	90	88.6	87.1	85.7	84.2	82.8	81.3	79.9	78.4	77.0	75.5	74.0	46°
47°	90	88.5	87.1	85.6	84.1	82.6	81.2	79.7	78.2	76.7	75.2	73.7	47°
48°	90	88.5	87.0	85.5	84.0	82.5	81.0	79.5	78.0	76.5	75.0	73.4	48°
49°	90	88.5	86.9	85.4	83.9	82.4	80.8	79.3	77.7	76.2	74.6	73.1	49°
50°	90	88.4	86.9	85.3	83.8	82.2	80.6	79.1	77.5	75.9	74.3	72.7	50°
51°	90	88.4	86.8	85.2	83.6	82.0	80.4	78.8	77.2	75.6	74.0	72.4	51°
52°	90	88.4	86.7	85.1	83.5	81.9	80.2	78.6	76.9	75.3	73.6	71.9	52°
53°	90	88.3	86.7	85.0	83.3	81.7	80.0	78.3	76.6	74.9	73.2	71.5	53°
54°	90	88.3	86.6	84.9	83.2	81.5	79.8	78.0	76.3	74.6	72.8	71.1	54°
55°	90	88.2	86.5	84.8	83.0	81.3	79.5	77.7	76.0	74.2	72.4	70.6	55°
56°	90	88.2	86.4	84.6	82.8	81.0	79.2	77.4	75.6	73.8	71.9	70.0	56°
57°	90	88.2	86.3	84.5	82.6	80.8	78.9	77.0	75.2	73.3	71.4	69.5	57°
58°	90	88.1	86.2	84.3	82.4	80.5	78.6	76.7	74.8	72.8	70.9	68.9	58°
59°	90	88.1	86.1	84.2	82.2	80.3	78.3	76.3	74.3	72.3	70.3	68.3	59°
60°	90	88.0	86.0	84.0	82.0	80.0	77.9	75.9	73.8	71.8	69.7	67.6	60°

LAT	12°	13°	14°	15°	16°	DECLINATION 17°	18°	19°	20°	21°	22°	23°	LAT
30°	76.1	74.9	73.8	72.6	71.4	70.3	69.1	67.9	66.7	65.5	64.4	63.2	30°
31°	76.0	74.8	73.6	72.4	71.2	70.0	68.9	67.7	66.5	65.3	64.1	62.9	31°
32°	75.8	74.6	73.4	72.2	71.0	69.8	68.6	67.4	66.2	65.0	63.8	62.6	32°
33°	75.6	74.4	73.2	72.1	70.8	69.6	68.4	67.1	65.9	64.7	63.5	62.2	33°
34°	75.5	74.2	73.0	71.8	70.6	69.3	68.1	66.9	65.6	64.4	63.1	61.9	34°
35°	75.3	74.1	72.8	71.6	70.3	69.1	67.8	66.6	65.3	64.1	62.8	61.5	35°
36°	75.1	73.8	72.6	71.3	70.1	68.8	67.5	66.3	65.0	63.7	62.4	61.1	36°
37°	74.9	73.6	72.4	71.1	69.8	68.5	67.2	65.9	64.6	63.3	62.0	60.7	37°
38°	74.7	73.4	72.0	70.8	69.5	68.2	66.9	65.6	64.3	62.9	61.6	60.3	38°
39°	74.5	73.2	71.9	70.5	69.2	67.9	66.6	65.2	63.9	62.5	61.2	59.8	39°
40°	74.2	72.9	71.6	70.2	68.9	67.6	66.2	64.8	63.5	62.1	60.7	59.3	40°
41°	74.0	72.7	71.3	69.9	68.6	67.2	65.8	64.4	63.0	61.6	60.2	58.8	41°
42°	73.7	72.4	71.0	69.6	68.2	66.8	65.4	64.0	62.6	61.2	59.7	58.3	42°
43°	73.5	72.1	70.7	69.3	67.9	66.4	65.0	63.6	62.1	60.7	59.2	57.7	43°
44°	73.2	71.8	70.3	68.9	67.5	66.0	64.6	63.1	61.6	60.1	58.6	57.1	44°
45°	72.9	71.4	70.0	68.5	67.0	65.6	64.1	62.6	61.1	59.5	58.0	56.4	45°
46°	72.6	71.1	69.6	68.1	66.6	65.1	63.6	62.0	60.5	58.9	57.4	55.8	46°
47°	72.2	70.7	69.2	67.7	66.2	64.6	63.1	61.5	59.9	58.3	56.7	55.0	47°
48°	71.9	70.3	68.8	67.2	65.7	64.1	62.5	60.9	59.3	57.6	55.9	54.3	48°
49°	71.5	69.9	68.4	66.8	65.1	63.5	61.9	60.2	58.6	56.9	55.2	53.4	49°
50°	71.1	69.5	67.9	66.2	64.6	62.9	61.3	59.6	57.8	56.1	54.3	52.6	50°
51°	70.7	69.1	67.4	65.7	64.0	62.3	60.6	58.8	57.1	55.3	53.5	51.6	51°
52°	70.3	68.6	66.9	65.1	63.4	61.6	59.9	58.1	56.3	54.4	52.5	50.6	52°
53°	69.8	68.1	66.3	64.5	62.7	60.9	59.1	57.3	55.4	53.5	51.5	49.5	53°
54°	69.3	67.5	65.7	63.9	62.0	60.2	58.3	56.4	54.4	52.4	50.4	48.3	54°
55°	68.7	66.9	65.1	63.2	61.3	59.4	57.4	55.4	53.4	51.3	49.2	47.1	55°
56°	68.2	66.3	64.4	62.4	60.5	58.5	56.5	54.4	52.3	50.1	47.9	45.7	56°
57°	67.6	65.6	63.6	61.6	59.6	57.5	55.4	53.3	51.1	48.9	46.5	44.2	57°
58°	66.9	64.9	62.8	60.8	58.7	56.5	54.3	52.1	49.8	47.4	45.0	42.5	58°
59°	66.2	64.1	62.0	59.8	57.6	55.4	53.1	50.8	48.4	45.9	43.3	40.7	59°
60°	65.4	63.3	61.1	58.8	56.5	54.2	51.8	49.4	46.8	44.2	41.5	38.6	60°

Chartwork 25

TABLE 3(2) Distance off by Vertical Sextant Angle

Height of object ft	m	\\ Distance of object (nautical miles) \\ 0.1	0.2	0.3	0.4	0.5	0.6	0.7	0.8	0.9	1.0	1.1	1.2	1.3	1.4	1.5	Height of object m	ft
		° ′	° ′	° ′	° ′	° ′	° ′	° ′	° ′	° ′	° ′	° ′	° ′	° ′	° ′	° ′		
33	10	3 05	1 33	1 02	0 46	0 37	0 31	0 27	0 23	0 21	0 19	0 17	0 15	0 14	0 13	0 12	10	33
39	12	3 42	1 51	1 14	0 56	0 45	0 37	0 32	0 28	0 25	0 22	0 20	0 19	0 17	0 16	0 15	12	39
46	14	4 19	2 10	1 27	1 05	0 52	0 43	0 37	0 32	0 29	0 26	0 24	0 22	0 20	0 19	0 17	14	46
53	16	4 56	2 28	1 39	1 14	0 59	0 49	0 42	0 37	0 33	0 30	0 27	0 25	0 23	0 21	0 20	16	53
59	18	5 33	2 47	1 51	1 24	1 07	0 56	0 48	0 42	0 37	0 33	0 30	0 28	0 26	0 24	0 22	18	59
66	20	6 10	3 05	2 04	1 33	1 14	1 02	0 53	0 46	0 41	0 37	0 34	0 31	0 29	0 27	0 25	20	66
72	22	6 46	3 24	2 16	1 42	1 22	1 08	0 58	0 51	0 45	0 41	0 37	0 34	0 31	0 29	0 27	22	72
79	24	7 23	3 42	2 28	1 51	1 29	1 14	1 04	0 56	0 49	0 45	0 40	0 37	0 34	0 32	0 30	24	79
85	26	7 59	4 01	2 41	2 01	1 36	1 20	1 09	1 00	0 54	0 48	0 44	0 40	0 37	0 34	0 32	26	85
92	28	8 36	4 19	2 53	2 10	1 44	1 27	1 14	1 05	0 58	0 52	0 47	0 43	0 40	0 37	0 35	28	92
98	30	9 12	4 38	3 05	2 19	1 51	1 33	1 20	1 10	1 02	0 56	0 51	0 46	0 43	0 40	0 37	30	98
105	32	9 48	4 56	3 18	2 28	1 58	1 39	1 25	1 14	1 06	0 59	0 54	0 49	0 46	0 42	0 40	32	105
112	34	10 24	5 15	3 30	2 38	2 06	1 45	1 30	1 19	1 10	1 03	0 57	0 53	0 49	0 45	0 42	34	112
118	36	11 00	5 33	3 42	2 47	2 14	1 51	1 35	1 24	1 14	1 07	1 01	0 56	0 51	0 48	0 45	36	118
125	38	11 36	5 41	3 55	2 56	2 21	1 58	1 41	1 28	1 18	1 11	1 04	0 59	0 54	0 50	0 47	38	125
131	40	12 11	6 10	4 07	3 05	2 28	2 04	1 46	1 33	1 22	1 14	1 07	1 02	0 57	0 53	0 49	40	131
138	42	12 47	6 28	4 19	3 15	2 36	2 10	1 51	1 37	1 27	1 18	1 11	1 05	1 00	0 56	0 52	42	138
144	44	13 22	6 46	4 32	3 24	2 43	2 16	1 57	1 42	1 31	1 22	1 14	1 08	1 03	0 58	0 54	44	144
151	46	13 57	7 05	4 44	3 33	2 51	2 22	2 02	1 47	1 35	1 25	1 18	1 11	1 06	1 01	0 57	46	151
157	48	14 32	7 23	4 56	3 42	2 58	2 28	2 07	1 51	1 39	1 29	1 21	1 14	1 09	1 04	0 59	48	157
164	50	15 07	7 41	5 09	3 52	3 05	2 35	2 13	1 56	1 43	1 33	1 24	1 17	1 11	1 06	1 02	50	164
171	52	15 41	7 59	5 21	4 01	3 13	2 41	2 18	2 01	1 47	1 36	1 28	1 20	1 14	1 09	1 04	52	171
177	54	16 15	8 18	5 33	4 10	3 20	2 47	2 23	2 05	1 51	1 40	1 31	1 23	1 17	1 12	1 07	54	177
184	56	16 49	8 36	5 45	4 19	3 28	2 53	2 28	2 10	1 55	1 44	1 34	1 27	1 20	1 14	1 09	56	184
190	58	17 23	8 54	5 58	4 29	3 35	2 59	2 34	2 15	2 00	1 48	1 38	1 30	1 23	1 17	1 12	58	190
197	60	17 57	9 12	6 10	4 38	3 42	3 05	2 39	2 19	2 04	1 51	1 41	1 33	1 26	1 20	1 14	60	197
203	62	18 31	9 30	6 22	4 47	3 50	3 12	2 44	2 24	2 08	1 55	1 45	1 36	1 29	1 22	1 17	62	203
210	64	19 04	9 48	6 34	4 56	3 57	3 18	2 50	2 28	2 12	1 59	1 48	1 39	1 31	1 25	1 19	64	210
217	66	19 37	10 06	6 46	5 05	4 05	3 24	2 53	2 33	2 16	2 02	1 51	1 42	1 34	1 27	1 22	66	217
223	68	20 10	10 24	6 59	5 15	4 12	3 30	3 00	2 38	2 20	2 06	1 55	1 45	1 37	1 30	1 24	68	223
230	70	20 42	10 42	7 11	5 24	4 19	3 36	3 05	2 42	2 24	2 09	1 58	1 48	1 40	1 33	1 27	70	230
236	72	21 15	11 00	7 23	5 33	4 27	3 42	3 11	2 47	2 48	2 14	2 01	1 51	1 43	1 35	1 29	72	236
246	75	22 03	11 27	7 41	5 47	4 38	3 52	3 19	2 54	2 35	2 19	2 07	1 56	1 47	1 39	1 33	75	246
256	78	22 50	11 54	7 59	6 01	4 49	4 01	3 27	3 01	2 41	2 24	2 12	2 01	1 51	1 43	1 36	78	256
266	81	23 37	12 20	8 18	6 14	5 00	4 10	3 35	3 08	2 47	2 30	2 17	2 05	1 56	1 47	1 40	81	266
276	84	24 24	12 47	8 36	6 28	5 11	4 19	3 42	3 15	2 53	2 36	2 22	2 10	2 00	1 51	1 44	84	276
289	88	25 25	13 22	9 00	6 46	5 26	4 32	3 53	3 24	3 01	2 43	2 28	2 16	2 06	1 57	1 49	88	289
302	92	26 25	13 57	9 24	7 05	5 40	4 44	4 04	3 33	3 10	2 51	2 35	2 22	2 11	2 02	1 54	92	302
315	96	27 24	14 32	9 48	7 23	5 55	4 56	4 14	3 42	3 18	2 58	2 42	2 28	2 17	2 07	1 59	96	315
328	100	28 22	15 07	10 12	7 41	6 10	5 09	4 25	3 52	3 26	3 05	2 49	2 35	2 23	2 13	2 04	100	328
341	104	29 19	15 41	10 36	7 59	6 24	5 21	4 35	4 01	3 34	3 13	2 55	2 41	2 28	2 18	2 09	104	341
358	109	30 29	16 24	11 06	8 22	6 43	5 36	4 48	4 12	3 44	3 22	3 04	2 48	2 36	2 24	2 15	109	358
374	114	31 37	17 06	11 36	8 45	7 01	5 51	5 02	4 24	3 55	3 31	3 12	2 56	2 43	2 31	2 21	114	374
394	120	32 56	17 57	12 11	9 12	7 23	6 10	5 17	4 38	4 07	3 42	3 22	3 05	2 51	2 39	2 28	120	394
427	130	35 04	19 20	13 10	9 57	8 00	6 40	5 44	5 01	4 28	4 01	3 39	3 21	3 05	2 52	2 41	130	427
459	140	37 05	20 42	14 09	10 42	8 36	7 11	6 10	5 24	4 48	4 19	3 56	3 36	3 20	3 05	2 53	140	459
492	150	39 00	22 03	15 07	11 27	9 12	7 41	6 36	5 47	5 09	4 38	4 13	3 52	3 34	3 19	3 05	150	492
574	175		25 17	17 29	13 17	10 42	8 57	7 41	6 44	6 00	5 24	4 55	4 30	4 09	3 52	3 36	175	574
656	200		28 22	19 48	15 07	12 11	10 12	8 46	7 41	6 51	6 10	5 36	5 09	4 45	4 25	4 07	200	656
738	225			22 03	16 54	13 39	11 27	9 51	8 38	7 41	6 56	6 18	5 47	5 20	4 58	4 38	225	738
820	250			24 14	18 39	15 07	12 41	10 55	9 35	8 32	7 41	7 00	6 25	5 56	5 30	5 09	250	820
902	275			26 20	20 22	16 32	13 54	11 59	10 31	9 22	8 27	7 41	7 03	6 31	6 03	5 39	275	902
984	300				22 03	17 57	15 07	13 02	11 27	10 12	9 12	8 23	7 41	7 06	6 36	6 10	300	984
1148	350					20 42	17 29	15 07	13 17	11 51	10 42	9 45	8 57	8 16	7 41	7 11	350	1148
1312	400						19 48	17 09	15 07	13 30	12 11	11 07	10 12	9 26	8 46	8 12	400	1312
ft m Height of object		0.1	0.2	0.3	0.4	0.5	0.6	0.7	0.8	0.9	1.0	1.1	1.2	1.3	1.4	1.5	m ft Height of object	
							Distance of object (nautical miles)											

TABLE 3(2) Distance off by Vertical Sextant Angle (continued)

Height of object ft	m	1.6	1.7	1.8	1.9	2.0	2.1	2.2	2.3	2.4	2.5	2.6	2.7	2.8	2.9	3.0	m	ft
		° ′	° ′	° ′	° ′	° ′	° ′	° ′	° ′	° ′	° ′	° ′	° ′	° ′	° ′	° ′		
33	10	0 12	0 11	0 10	0 10												10	33
39	12	0 14	0 13	0 12	0 12	0 11	0 11	0 10	0 10	0 10							12	39
46	14	0 16	0 15	0 14	0 14	0 13	0 12	0 12	0 11	0 11	0 10	0 10	0 10				14	46
53	16	0 19	0 17	0 16	0 16	0 15	0 14	0 13	0 13	0 12	0 12	0 11	0 11	0 11	0 10	0 10	16	53
59	18	0 21	0 20	0 19	0 18	0 17	0 16	0 15	0 15	0 14	0 13	0 13	0 12	0 12	0 12	0 11	18	59
66	20	0 23	0 22	0 21	0 20	0 19	0 18	0 17	0 16	0 15	0 15	0 14	0 14	0 13	0 13	0 12	20	66
72	22	0 26	0 24	0 23	0 21	0 20	0 19	0 19	0 18	0 17	0 16	0 16	0 15	0 15	0 14	0 14	22	72
79	24	0 28	0 26	0 25	0 23	0 22	0 21	0 20	0 19	0 19	0 18	0 17	0 16	0 16	0 15	0 15	24	79
85	26	0 30	0 28	0 27	0 25	0 24	0 23	0 22	0 21	0 20	0 19	0 19	0 18	0 17	0 17	0 16	26	85
92	28	0 32	0 31	0 29	0 27	0 26	0 25	0 24	0 23	0 22	0 21	0 20	0 19	0 19	0 18	0 17	28	92
98	30	0 35	0 33	0 31	0 29	0 28	0 27	0 25	0 24	0 23	0 22	0 21	0 21	0 20	0 19	0 19	30	98
105	32	0 37	0 35	0 33	0 31	0 30	0 28	0 27	0 26	0 25	0 24	0 23	0 22	0 21	0 20	0 20	32	105
112	34	0 39	0 37	0 35	0 33	0 31	0 30	0 29	0 27	0 26	0 25	0 24	0 23	0 23	0 22	0 21	34	112
118	36	0 42	0 39	0 37	0 35	0 33	0 32	0 30	0 29	0 28	0 27	0 26	0 25	0 24	0 23	0 22	36	118
125	38	0 44	0 41	0 39	0 37	0 35	0 34	0 32	0 31	0 29	0 28	0 27	0 26	0 25	0 24	0 24	38	125
131	40	0 46	0 44	0 41	0 39	0 37	0 35	0 34	0 32	0 31	0 30	0 29	0 27	0 27	0 26	0 25	40	131
138	42	0 49	0 46	0 43	0 41	0 40	0 37	0 35	0 34	0 32	0 31	0 30	0 29	0 28	0 27	0 26	42	138
144	44	0 51	0 48	0 45	0 43	0 41	0 39	0 37	0 36	0 34	0 33	0 31	0 30	0 29	0 28	0 27	44	144
151	46	0 53	0 50	0 47	0 45	0 43	0 41	0 39	0 37	0 36	0 34	0 33	0 32	0 30	0 29	0 28	46	151
157	48	0 56	0 52	0 49	0 47	0 45	0 42	0 40	0 39	0 37	0 36	0 34	0 33	0 32	0 31	0 30	48	157
164	50	0 58	0 55	0 52	0 49	0 46	0 44	0 42	0 40	0 39	0 37	0 36	0 34	0 33	0 32	0 31	50	164
171	52	1 00	0 57	0 54	0 51	0 48	0 46	0 44	0 42	0 40	0 39	0 37	0 36	0 34	0 33	0 32	52	171
177	54	1 03	0 59	0 56	0 53	0 50	0 48	0 46	0 44	0 42	0 40	0 39	0 37	0 36	0 35	0 33	54	177
184	56	1 05	1 01	0 58	0 55	0 52	0 49	0 47	0 45	0 43	0 42	0 40	0 38	0 37	0 36	0 35	56	184
190	58	1 07	1 03	1 00	0 57	0 54	0 51	0 49	0 47	0 45	0 43	0 41	0 40	0 38	0 37	0 36	58	190
197	60	1 10	1 06	1 02	0 59	0 56	0 53	0 51	0 48	0 46	0 45	0 43	0 41	0 40	0 38	0 37	60	197
203	62	1 12	1 08	1 04	1 01	0 58	0 55	0 52	0 50	0 48	0 46	0 44	0 43	0 41	0 40	0 38	62	203
210	64	1 14	1 10	1 06	1 03	0 59	0 57	0 54	0 52	0 49	0 48	0 46	0 44	0 42	0 41	0 40	64	210
217	66	1 17	1 12	1 08	1 05	1 01	0 58	0 56	0 53	0 51	0 49	0 47	0 45	0 44	0 42	0 41	66	217
223	68	1 19	1 14	1 10	1 06	1 03	1 00	0 57	0 55	0 53	0 50	0 49	0 47	0 45	0 44	0 42	68	223
230	70	1 21	1 16	1 12	1 08	1 05	1 02	0 59	0 56	0 54	0 52	0 50	0 48	0 46	0 45	0 43	70	230
236	72	1 24	1 19	1 14	1 10	1 07	1 04	1 01	0 58	0 56	0 53	0 51	0 49	0 48	0 46	0 45	72	236
246	75	1 27	1 22	1 17	1 13	1 10	1 06	1 03	1 01	0 58	0 56	0 54	0 51	0 50	0 48	0 46	75	246
256	78	1 30	1 25	1 20	1 16	1 12	1 09	1 06	1 03	1 00	0 58	0 56	0 54	0 52	0 50	0 48	78	256
266	81	1 34	1 28	1 23	1 19	1 15	1 12	1 08	1 05	1 03	1 00	0 58	0 56	0 54	0 52	0 50	81	266
276	84	1 37	1 32	1 27	1 22	1 18	1 14	1 11	1 08	1 05	1 02	1 00	0 58	0 56	0 54	0 52	84	276
289	88	1 42	1 36	1 31	1 26	1 22	1 18	1 14	1 11	1 08	1 05	1 03	1 00	0 58	0 56	0 54	88	289
302	92	1 47	1 40	1 35	1 30	1 25	1 21	1 18	1 14	1 11	1 08	1 06	1 03	1 01	0 59	0 57	92	302
315	96	1 51	1 45	1 39	1 34	1 29	1 25	1 21	1 17	1 14	1 11	1 09	1 06	1 04	1 01	0 59	96	315
328	100	1 56	1 49	1 43	1 38	1 33	1 28	1 24	1 21	1 17	1 14	1 11	1 09	1 06	1 04	1 02	100	328
341	104	2 01	1 54	1 47	1 42	1 36	1 32	1 28	1 24	1 20	1 17	1 14	1 11	1 09	1 07	1 04	104	341
358	109	2 06	1 59	1 52	1 46	1 41	1 36	1 32	1 28	1 24	1 21	1 18	1 15	1 12	1 10	1 07	109	358
374	114	2 12	2 04	1 58	1 51	1 46	1 41	1 36	1 32	1 28	1 25	1 21	1 18	1 16	1 13	1 11	114	374
394	120	2 19	2 11	2 04	1 57	1 51	1 46	1 41	1 37	1 33	1 29	1 26	1 22	1 20	1 17	1 14	120	394
427	130	2 31	2 22	2 14	2 07	2 01	1 55	1 50	1 45	1 41	1 36	1 33	1 29	1 26	1 23	1 20	130	427
459	140	2 42	2 33	2 24	2 17	2 10	2 04	1 58	1 53	1 48	1 44	1 40	1 36	1 33	1 30	1 27	140	459
492	150	2 54	2 44	2 35	2 26	2 19	2 13	2 07	2 01	1 56	1 51	1 47	1 43	1 39	1 36	1 33	150	492
574	175	3 23	3 11	3 00	2 51	2 42	2 35	2 28	2 21	2 15	2 10	2 05	2 00	1 56	1 52	1 48	175	574
656	200	3 52	3 38	3 26	3 15	3 05	2 57	2 49	2 41	2 35	2 28	2 23	2 17	2 13	2 08	2 04	200	656
738	225	4 21	4 05	3 52	3 40	3 29	3 19	3 10	3 01	2 54	2 47	2 41	2 36	2 29	2 24	2 19	225	738
820	250	4 49	4 32	4 17	4 04	3 52	3 41	3 31	3 22	3 13	3 05	2 58	2 52	2 46	2 40	2 35	250	820
902	275	5 18	5 00	4 43	4 28	4 15	4 03	3 52	3 42	3 32	3 24	3 16	3 09	3 02	2 56	2 50	275	902
984	300	5 47	5 27	5 09	4 52	4 38	4 25	4 13	4 02	3 52	3 42	3 34	3 26	3 19	3 12	3 05	300	984
1148	350	6 44	6 21	6 00	5 41	5 24	5 09	4 55	4 42	4 30	4 19	4 09	4 00	3 52	3 44	3 36	350	1148
1312	400	7 41	7 14	6 51	6 29	6 10	5 52	5 36	5 22	5 09	4 56	4 45	4 34	4 25	4 16	4 07	400	1312
ft	m	1.6	1.7	1.8	1.9	2.0	2.1	2.2	2.3	2.4	2.5	2.6	2.7	2.8	2.9	3.0	m	ft

Height of object   Distance of object (nautical miles)   Height of object

TABLE 3(2) Distance off by Vertical Sextant Angle (continued)

Height of object ft	m	\multicolumn{14}{c	}{Distance of object (nautical miles)}	Height of object m	ft
		3.1	3.2	3.3	3.4	3.5	3.6	3.7	3.8	3.9	4.0	4.2	4.4	4.6	4.8	5.0		
		° ′	° ′	° ′	° ′	° ′	° ′	° ′	° ′	° ′	° ′	° ′	° ′	° ′	° ′	° ′		
33	10																10	33
39	12																12	39
46	14																14	46
53	16	0 10															16	53
59	18	0 11	0 10	0 10	0 10	0 10											18	59
66	20	0 12	0 12	0 11	0 11	0 11	0 10	0 10	0 10	0 10							20	66
72	22	0 13	0 13	0 12	0 12	0 12	0 11	0 11	0 11	0 10	0 10						22	72
79	24	0 14	0 14	0 13	0 13	0 13	0 12	0 12	0 12	0 11	0 11	0 11	0 10				24	79
85	26	0 16	0 15	0 15	0 14	0 14	0 13	0 13	0 13	0 12	0 12	0 11	0 11	0 10	0 10		26	85
92	28	0 17	0 16	0 16	0 15	0 15	0 14	0 14	0 14	0 13	0 13	0 12	0 12	0 11	0 11	0 10	28	92
98	30	0 18	0 17	0 17	0 16	0 16	0 15	0 15	0 15	0 14	0 14	0 13	0 13	0 12	0 12	0 11	30	98
105	32	0 19	0 19	0 18	0 17	0 17	0 16	0 16	0 16	0 15	0 15	0 14	0 13	0 13	0 12	0 12	32	105
112	34	0 20	0 20	0 19	0 19	0 18	0 17	0 17	0 17	0 16	0 16	0 15	0 14	0 14	0 13	0 13	34	112
118	36	0 22	0 21	0 20	0 20	0 19	0 19	0 18	0 18	0 17	0 17	0 16	0 15	0 14	0 14	0 13	36	118
125	38	0 23	0 22	0 21	0 21	0 20	0 20	0 19	0 19	0 18	0 18	0 17	0 16	0 15	0 15	0 14	38	125
131	40	0 24	0 23	0 22	0 22	0 21	0 21	0 20	0 20	0 19	0 19	0 18	0 17	0 16	0 15	0 15	40	131
138	42	0 25	0 24	0 24	0 23	0 22	0 22	0 21	0 21	0 20	0 19	0 19	0 18	0 17	0 16	0 16	42	138
144	44	0 26	0 25	0 25	0 24	0 23	0 23	0 22	0 22	0 21	0 20	0 19	0 19	0 18	0 17	0 16	44	144
151	46	0 28	0 27	0 26	0 25	0 24	0 24	0 23	0 22	0 22	0 21	0 20	0 19	0 19	0 18	0 17	46	151
157	48	0 29	0 28	0 27	0 26	0 25	0 25	0 24	0 23	0 23	0 22	0 21	0 20	0 19	0 19	0 18	48	157
164	50	0 30	0 29	0 28	0 27	0 27	0 26	0 25	0 24	0 24	0 23	0 22	0 21	0 20	0 19	0 19	50	164
171	52	0 31	0 30	0 29	0 28	0 28	0 27	0 26	0 25	0 25	0 24	0 23	0 22	0 21	0 20	0 19	52	171
177	54	0 32	0 31	0 30	0 29	0 29	0 28	0 27	0 26	0 26	0 25	0 24	0 23	0 22	0 21	0 20	54	177
184	56	0 34	0 32	0 31	0 31	0 30	0 29	0 28	0 27	0 27	0 26	0 25	0 24	0 23	0 22	0 21	56	184
190	58	0 35	0 34	0 33	0 32	0 31	0 30	0 29	0 28	0 28	0 27	0 26	0 24	0 23	0 22	0 21	58	190
197	60	0 36	0 35	0 34	0 33	0 32	0 31	0 30	0 29	0 29	0 28	0 26	0 25	0 24	0 23	0 22	60	197
203	62	0 37	0 36	0 35	0 34	0 33	0 32	0 31	0 30	0 30	0 29	0 27	0 26	0 25	0 24	0 23	62	203
210	64	0 38	0 37	0 36	0 35	0 34	0 33	0 32	0 31	0 30	0 30	0 28	0 27	0 26	0 25	0 24	64	210
217	66	0 40	0 38	0 37	0 36	0 35	0 34	0 33	0 32	0 31	0 31	0 29	0 28	0 27	0 26	0 25	66	217
223	68	0 41	0 39	0 38	0 37	0 36	0 35	0 34	0 33	0 32	0 32	0 30	0 29	0 27	0 26	0 25	68	223
230	70	0 42	0 41	0 39	0 38	0 37	0 36	0 35	0 34	0 33	0 32	0 31	0 29	0 28	0 27	0 26	70	230
236	72	0 43	0 42	0 40	0 39	0 38	0 37	0 36	0 35	0 34	0 33	0 32	0 30	0 29	0 28	0 27	72	236
246	75	0 45	0 44	0 42	0 41	0 40	0 39	0 38	0 37	0 36	0 35	0 33	0 32	0 30	0 29	0 28	75	246
256	78	0 47	0 45	0 44	0 43	0 41	0 40	0 39	0 38	0 37	0 36	0 34	0 33	0 31	0 30	0 29	78	256
266	81	0 48	0 47	0 46	0 44	0 43	0 42	0 41	0 40	0 39	0 38	0 36	0 34	0 33	0 31	0 30	81	266
276	84	0 50	0 49	0 47	0 46	0 45	0 43	0 42	0 41	0 40	0 39	0 37	0 35	0 34	0 32	0 31	84	276
289	88	0 53	0 51	0 49	0 48	0 47	0 45	0 44	0 43	0 42	0 41	0 39	0 37	0 36	0 34	0 33	88	289
302	92	0 55	0 53	0 52	0 50	0 49	0 47	0 46	0 45	0 44	0 43	0 41	0 39	0 37	0 36	0 34	92	302
315	96	0 57	0 56	0 54	0 52	0 51	0 49	0 48	0 47	0 46	0 45	0 42	0 41	0 39	0 37	0 36	96	315
328	100	1 00	0 58	0 56	0 55	0 53	0 52	0 50	0 49	0 48	0 46	0 44	0 42	0 40	0 39	0 37	100	328
341	104	1 02	1 00	0 58	0 57	0 55	0 54	0 52	0 51	0 49	0 48	0 46	0 44	0 42	0 40	0 39	104	341
358	109	1 05	1 03	1 01	1 00	0 58	0 56	0 55	0 53	0 52	0 51	0 48	0 46	0 44	0 42	0 40	109	358
374	114	1 08	1 06	1 04	1 02	1 00	0 59	0 57	0 56	0 54	0 53	0 50	0 48	0 46	0 44	0 42	114	374
394	120	1 12	1 10	1 07	1 06	1 04	1 02	1 00	0 59	0 57	0 56	0 53	0 51	0 48	0 46	0 45	120	394
427	130	1 18	1 15	1 13	1 11	1 09	1 07	1 05	1 03	1 02	1 00	0 57	0 55	0 52	0 50	0 48	130	427
459	140	1 24	1 21	1 19	1 16	1 14	1 12	1 10	1 08	1 07	1 05	1 02	0 59	0 56	0 54	0 52	140	459
492	150	1 30	1 27	1 24	1 22	1 20	1 17	1 15	1 13	1 11	1 10	1 06	1 03	1 01	0 58	0 56	150	492
574	175	1 45	1 41	1 38	1 36	1 33	1 30	1 28	1 25	1 23	1 21	1 17	1 14	1 11	1 08	1 05	175	574
656	200	2 00	1 56	1 52	1 49	1 46	1 43	1 40	1 38	1 35	1 33	1 28	1 24	1 21	1 17	1 14	200	656
738	225	2 15	2 10	2 06	2 03	1 59	1 56	1 53	1 50	1 47	1 44	1 39	1 35	1 31	1 27	1 24	225	738
820	250	2 30	2 25	2 20	2 16	2 13	2 09	2 05	2 02	1 59	1 56	1 50	1 45	1 41	1 37	1 33	250	820
902	275	2 45	2 39	2 34	2 30	2 26	2 22	2 18	2 14	2 11	2 08	2 01	1 56	1 51	1 46	1 42	275	902
984	300	2 59	2 54	2 48	2 44	2 39	2 35	2 30	2 26	2 23	2 19	2 13	2 07	2 01	1 56	1 51	300	984
1148	350	3 29	3 23	3 16	3 11	3 05	3 00	2 55	2 51	2 46	2 42	2 35	2 28	2 21	2 15	2 10	350	1148
1312	400	3 59	3 52	3 44	3 38	3 32	3 26	3 20	3 15	3 10	3 05	2 57	2 49	2 41	2 35	2 28	400	1312
ft m Height of object		3.1	3.2	3.3	3.4	3.5	3.6	3.7	3.8	3.9	4.0	4.2	4.4	4.6	4.8	5.0	m ft Height of object	

Distance of object (nautical miles)

Chartwork 29

TABLE 3(3) Distance of horizon for various heights of eye

Height of eye		Horizon distance	Height of eye		Horizon distance	Height of eye		Horizon distance
metres	feet	n. miles	metres	feet	n. miles	metres	feet	n. miles
1	3.3	2.1	21	68.9	9.5	41	134.5	13.3
2	6.6	2.9	22	72.2	9.8	42	137.8	13.5
3	9.8	3.6	23	75.5	10.0	43	141.1	13.7
4	13.1	4.1	24	78.7	10.2	44	144.4	13.8
5	16.4	4.7	25	82.0	10.4	45	147.6	14.0
6	19.7	5.1	26	85.3	10.6	46	150.9	14.1
7	23.0	5.5	27	88.6	10.8	47	154.2	14.3
8	26.2	5.9	28	91.9	11.0	48	157.5	14.4
9	29.6	6.2	29	95.1	11.2	49	160.8	14.6
10	32.8	6.6	30	98.4	11.4	50	164.0	14.7
11	36.1	6.9	31	101.7	11.6	51	167.3	14.9
12	39.4	7.2	32	105.0	11.8	52	170.6	15.0
13	42.7	7.5	33	108.3	12.0	53	173.9	15.2
14	45.9	7.8	34	111.6	12.1	54	177.2	15.3
15	49.2	8.1	35	114.8	12.3	55	180.4	15.4
16	52.5	8.3	36	118.1	12.5	56	183.7	15.6
17	55.8	8.6	37	121.4	12.7	57	187.0	15.7
18	59.1	8.8	38	124.7	12.8	58	190.3	15.9
19	62.3	9.1	39	128.0	13.0	59	193.6	16.0
20	65.6	9.3	40	131.2	13.2	60	196.9	16.1

TABLE 3(4) Lights — distance off when rising or dipping (n. miles)

Height of light		Height of eye									
		metres	1	2	3	4	5	6	7	8	9	10
metres	feet	feet	3	7	10	13	16	20	23	26	30	33
10	33		8.7	9.5	10.2	10.8	11.3	11.7	12.1	12.5	12.8	13.2
12	39		9.3	10.1	10.8	11.4	11.9	12.3	12.7	13.1	13.4	13.8
14	46		9.9	10.7	11.4	12.0	12.5	12.9	13.3	13.7	14.0	14.4
16	53		10.4	11.2	11.9	12.5	13.0	13.4	13.8	14.2	14.5	14.9
18	59		10.9	11.7	12.4	13.0	13.5	13.9	14.3	14.7	15.0	15.4
20	66		11.4	12.2	12.9	13.5	14.0	14.4	14.8	15.2	15.5	15.9
22	72		11.9	12.7	13.4	14.0	14.5	14.9	15.3	15.7	16.0	16.4
24	79		12.3	13.1	13.8	14.4	14.9	15.3	15.7	16.1	16.4	17.0
26	85		12.7	13.5	14.2	14.8	15.3	15.7	16.1	16.5	16.8	17.2
28	92		13.1	13.9	14.6	15.2	15.7	16.1	16.5	16.9	17.2	17.6
30	98		13.5	14.3	15.0	15.6	16.1	16.5	16.9	17.3	17.6	18.0
32	105		13.9	14.7	15.4	16.0	16.5	16.9	17.3	17.7	18.0	18.4
34	112		14.2	15.0	15.7	16.3	16.8	17.2	17.6	18.0	18.3	18.7
36	118		14.6	15.4	16.1	16.7	17.2	17.6	18.0	18.4	18.7	19.1
38	125		14.9	15.7	16.4	17.0	17.5	17.9	18.3	18.7	19.0	19.4
40	131		15.3	16.1	16.8	17.4	17.9	18.3	18.7	19.1	19.4	19.8
42	138		15.6	16.4	17.1	17.7	18.2	18.6	19.0	19.4	19.7	20.1
44	144		15.9	16.7	17.4	18.0	18.5	18.9	19.3	19.7	20.0	20.4
46	151		16.2	17.0	17.7	18.3	18.8	19.2	19.6	20.0	20.3	20.7
48	157		16.5	17.3	18.0	18.6	19.1	19.5	19.9	20.3	20.6	21.0
50	164		16.8	17.6	18.3	18.9	19.4	19.8	20.2	20.6	20.9	21.3
55	180		17.5	18.3	19.0	19.6	20.1	20.5	20.9	21.3	21.6	22.0
60	197		18.2	19.0	19.7	20.3	20.8	21.2	21.6	22.0	22.3	22.7
65	213		18.9	19.7	20.4	21.0	21.5	21.9	22.3	22.7	23.0	23.4
70	230		19.5	20.3	21.0	21.6	22.1	22.5	22.9	23.2	23.6	24.0
75	246		20.1	20.9	21.6	22.2	22.7	23.1	23.5	23.9	24.2	24.6
80	262		20.7	21.5	22.2	22.8	23.3	23.7	24.1	24.5	24.8	25.2
85	279		21.3	22.1	22.8	23.4	23.9	24.3	24.7	25.1	25.4	25.8
90	295		21.8	22.6	23.3	23.9	24.4	24.8	25.2	25.6	25.9	26.3
95	312		22.4	23.2	23.9	24.5	25.0	25.4	25.8	26.2	26.5	26.9
metres	feet	metres	1	2	3	4	5	6	7	8	9	10
Height of light		feet	3	7	10	13	16	20	23	26	30	33
		Height of eye										

TABLE 3(5)   Time, Speed and Distance Table

Time Decimal of hr.	Mins	2.5	3	3.5	4	4.5	5	5.5	6	6.5	7	7.5	8	8.5	9	9.5	10	Mins	Time Decimal of hr.
								Speed in knots											
.0167	1					0.1	0.1	0.1	0.1	0.1	0.1	0.1	0.1	0.1	0.2	0.2	0.2	1	.0167
.0333	2	0.1	0.1	0.1	0.1	0.1	0.2	0.2	0.2	0.2	0.2	0.2	0.3	0.3	0.3	0.3	0.3	2	.0333
.0500	3	0.1	0.1	0.2	0.2	0.2	0.2	0.3	0.3	0.3	0.3	0.4	0.4	0.4	0.4	0.5	0.5	3	.0500
.0667	4	0.1	0.2	0.2	0.3	0.3	0.3	0.4	0.4	0.4	0.5	0.5	0.5	0.6	0.6	0.6	0.7	4	.0667
.0833	5	0.2	0.2	0.3	0.3	0.4	0.4	0.5	0.5	0.5	0.6	0.6	0.7	0.7	0.7	0.8	0.8	5	.0833
.1000	6	0.2	0.3	0.3	0.4	0.4	0.5	0.5	0.6	0.6	0.7	0.7	0.8	0.8	0.9	0.9	1.0	6	.1000
.1167	7	0.3	0.4	0.4	0.5	0.5	0.6	0.6	0.7	0.8	0.8	0.9	0.9	1.0	1.1	1.1	1.2	7	.1167
.1333	8	0.3	0.4	0.5	0.5	0.6	0.7	0.7	0.8	0.9	0.9	1.0	1.1	1.1	1.2	1.3	1.3	8	.1333
.1500	9	0.4	0.4	0.5	0.6	0.7	0.7	0.8	0.9	1.0	1.0	1.1	1.2	1.3	1.3	1.4	1.5	9	.1500
.1667	10	0.4	0.5	0.6	0.7	0.8	0.8	0.9	1.0	1.1	1.2	1.3	1.3	1.4	1.5	1.6	1.7	10	.1667
.1833	11	0.5	0.5	0.6	0.7	0.8	0.9	1.0	1.1	1.2	1.3	1.4	1.5	1.6	1.6	1.7	1.8	11	.1833
.2000	12	0.5	0.6	0.7	0.8	0.9	1.0	1.1	1.2	1.3	1.4	1.5	1.6	1.7	1.8	1.9	2.0	12	.2000
.2167	13	0.5	0.6	0.8	0.9	1.0	1.1	1.2	1.3	1.4	1.5	1.6	1.7	1.8	2.0	2.0	2.2	13	.2167
.2333	14	0.6	0.7	0.8	0.9	1.1	1.2	1.3	1.4	1.5	1.6	1.7	1.9	2.0	2.1	2.2	2.3	14	.2333
.2500	15	0.6	0.7	0.9	1.0	1.1	1.2	1.4	1.5	1.6	1.8	1.9	2.0	2.1	2.2	2.4	2.5	15	.2500
.2667	16	0.7	0.8	0.9	1.1	1.2	1.3	1.5	1.6	1.7	1.9	2.0	2.1	2.3	2.4	2.5	2.7	16	.2667
.2833	17	0.7	0.8	1.0	1.1	1.3	1.4	1.6	1.7	1.8	2.0	2.1	2.3	2.4	2.5	2.7	2.8	17	.2833
.3000	18	0.7	0.9	1.0	1.2	1.3	1.5	1.6	1.8	1.9	2.1	2.2	2.4	2.5	2.7	2.8	3.0	18	.3000
.3167	19	0.8	1.0	1.1	1.3	1.4	1.6	1.7	1.9	2.1	2.1	2.4	2.5	2.7	2.9	3.0	3.2	19	.3167
.3333	20	0.8	1.0	1.2	1.3	1.5	1.7	1.8	2.0	2.2	2.3	2.5	2.7	2.8	3.0	3.2	3.3	20	.3333
.3500	21	0.9	1.0	1.2	1.4	1.6	1.7	1.9	2.1	2.3	2.4	2.6	2.8	3.0	3.1	3.3	3.5	21	.3500
.3667	22	0.9	1.1	1.3	1.5	1.7	1.8	2.1	2.2	2.4	2.6	2.8	2.9	3.1	3.3	3.5	3.7	22	.3667
.3833	23	1.0	1.1	1.3	1.5	1.7	1.9	2.1	2.3	2.5	2.7	2.9	3.1	3.3	3.4	3.6	3.8	23	.3833
.4000	24	1.0	1.2	1.4	1.6	1.8	2.0	2.2	2.4	2.6	2.8	3.0	3.2	3.4	3.6	3.8	4.0	24	.4000
.4167	25	1.0	1.3	1.5	1.7	1.9	2.1	2.3	2.5	2.7	2.9	3.1	3.3	3.5	3.8	4.0	4.2	25	.4167
.4333	26	1.1	1.3	1.5	1.7	1.9	2.2	2.4	2.6	2.8	3.0	3.2	3.5	3.7	3.9	4.1	4.3	26	.4333
.4500	27	1.1	1.3	1.6	1.8	2.0	2.2	2.5	2.7	2.9	3.1	3.4	3.6	3.8	4.0	4.3	4.5	27	.4500
.4667	28	1.2	1.4	1.6	1.9	2.1	2.3	2.6	2.8	3.0	3.3	3.5	3.7	4.0	4.2	4.4	4.7	28	.4667
.4833	29	1.2	1.5	1.7	1.9	2.2	2.4	2.7	2.9	3.1	3.4	3.6	3.9	4.1	4.3	4.6	4.8	29	.4833
.5000	30	1.2	1.5	1.7	2.0	2.2	2.5	2.7	3.0	3.2	3.5	3.7	4.0	4.2	4.5	4.7	5.0	30	.5000
.5167	31	1.3	1.6	1.8	2.1	2.3	2.6	2.8	3.1	3.4	3.6	3.9	4.1	4.4	4.7	4.9	5.2	31	.5167
.5333	32	1.3	1.6	1.9	2.1	2.4	2.7	2.9	3.2	3.5	3.7	4.0	4.3	4.5	4.8	5.1	5.3	32	.5333
.5500	33	1.4	1.6	1.9	2.2	2.5	2.7	3.0	3.3	3.6	3.8	4.1	4.4	4.7	4.9	5.2	5.5	33	.5500
.5667	34	1.4	1.7	2.0	2.3	2.6	2.8	3.1	3.4	3.7	4.0	4.3	4.5	4.8	5.1	5.4	5.7	34	.5667
.5833	35	1.5	1.7	2.0	2.3	2.6	2.9	3.2	3.5	3.8	4.1	4.4	4.7	5.0	5.2	5.5	5.8	35	.5833
.6000	36	1.5	1.8	2.1	2.4	2.7	3.0	3.3	3.6	3.9	4.2	4.5	4.8	5.1	5.4	5.7	6.0	36	.6000
.6117	37	1.6	1.8	2.1	2.4	2.8	3.1	3.4	3.7	4.0	4.3	4.6	4.9	5.2	5.5	5.8	6.1	37	.6117
.6333	38	1.6	1.9	2.2	2.5	2.8	3.2	3.5	3.8	4.1	4.4	4.7	5.1	5.4	5.7	6.0	6.3	38	.6333
.6500	39	1.6	1.9	2.3	2.6	2.9	3.2	3.6	3.9	4.2	4.5	4.9	5.2	5.5	5.8	6.2	6.5	39	.6500
.6667	40	1.7	2.0	2.3	2.7	3.0	3.3	3.7	4.0	4.3	4.7	5.0	5.3	5.7	6.0	6.3	6.7	40	.6667
.6833	41	1.7	2.0	2.4	2.7	3.1	3.4	3.8	4.1	4.4	4.8	5.1	5.5	5.8	6.1	6.5	6.8	41	.6833
.7000	42	1.7	2.1	2.4	2.8	3.1	3.5	3.8	4.2	4.5	4.9	5.2	5.6	5.9	6.3	6.6	7.0	42	.7000
.7167	43	1.8	2.2	2.5	2.9	3.2	3.6	3.9	4.3	4.7	5.0	5.4	5.7	6.1	6.5	6.8	7.2	43	.7167
.7333	44	1.8	2.2	2.6	2.9	3.3	3.7	4.0	4.4	4.8	5.1	5.5	5.9	6.2	6.6	7.0	7.3	44	.7333
.7500	45	1.9	2.2	2.6	3.0	3.4	3.7	4.1	4.5	4.9	5.2	5.6	6.0	6.4	6.7	7.1	7.5	45	.7500
.7667	46	1.9	2.3	2.7	3.1	3.5	3.8	4.2	4.6	5.0	5.4	5.8	6.1	6.5	6.9	7.3	7.7	46	.7667
.7833	47	2.0	2.3	2.7	3.1	3.5	3.9	4.3	4.7	5.1	5.5	5.9	6.3	6.7	7.0	7.4	7.8	47	.7833
.8000	48	2.0	2.4	2.8	3.2	3.6	4.0	4.4	4.8	5.2	5.6	6.0	6.4	6.8	7.2	7.6	8.0	48	.8000
.8167	49	2.0	2.5	2.9	3.3	3.7	4.1	4.5	4.9	5.3	5.7	6.1	6.5	6.9	7.4	7.8	8.2	49	.8167
.8333	50	2.1	2.5	2.9	3.3	3.7	4.2	4.6	5.0	5.4	5.8	6.2	6.7	7.1	7.5	7.9	8.3	50	.8333
.8500	51	2.1	2.5	3.0	3.4	3.8	4.2	4.7	5.1	5.5	5.9	6.4	6.8	7.2	7.6	8.1	8.5	51	.8500
.8667	52	2.2	2.6	3.0	3.5	3.9	4.3	4.8	5.2	5.6	6.1	6.5	6.9	7.4	7.8	8.2	8.7	52	.8667
.8833	53	2.2	2.6	3.1	3.5	4.0	4.4	4.9	5.3	5.7	6.2	6.6	7.1	7.5	7.9	8.4	8.8	53	.8833
.9000	54	2.2	2.7	3.1	3.6	4.0	4.5	4.9	5.4	5.8	6.3	6.7	7.2	7.6	8.1	8.5	9.0	54	.9000
.9167	55	2.3	2.8	3.2	3.7	4.1	4.6	5.0	5.5	6.0	6.4	6.9	7.3	7.8	8.3	8.7	9.2	55	.9167
.9333	56	2.3	2.8	3.3	3.7	4.2	4.7	5.1	5.6	6.1	6.5	7.0	7.5	7.9	8.4	8.9	9.3	56	.9333
.9500	57	2.4	2.8	3.3	3.8	4.3	4.7	5.2	5.7	6.2	6.6	7.1	7.6	8.1	8.5	9.0	9.5	57	.9500
.9667	58	2.4	2.9	3.4	3.9	4.4	4.8	5.3	5.8	6.3	6.8	7.3	7.7	8.2	8.7	9.2	9.7	58	.9667
.9833	59	2.5	2.9	3.4	3.9	4.4	4.9	5.4	5.9	6.4	6.9	7.4	7.9	8.4	8.8	9.3	9.8	59	.9833
1.0000	60	2.5	3.0	3.5	4.0	4.5	5.0	5.5	6.0	6.5	7.0	7.5	8.0	8.5	9.0	9.5	10.0	60	1.0000
Decimal of hr.	Mins.	2.5	3	3.5	4	4.5	5	5.5	6	6.5	7	7.5	8	8.5	9	9.5	10	Mins.	Decimal of hr.
Time								**Speed in knots**											Time

Chartwork 31

TABLE 3(5) Time, Speed and Distance Table (continued)

Time Decimal of hr.	Mins	\multicolumn{15}{c}{Speed in knots}	Mins	Time Decimal of hr.
		10.5	11.0	11.5	12.0	12.5	13.0	13.5	14.0	14.5	15.0	15.5	16.0	17.0	18.0	19.0	20.0		
.0167	1	0.2	0.2	0.2	0.2	0.2	0.2	0.2	0.2	0.2	0.3	0.3	0.3	0.3	0.3	0.3	0.3	1	.0167
.0333	2	0.3	0.4	0.4	0.4	0.4	0.4	0.4	0.5	0.5	0.5	0.5	0.5	0.6	0.6	0.6	0.7	2	.0333
.0500	3	0.5	0.5	0.6	0.6	0.6	0.6	0.7	0.7	0.7	0.7	0.8	0.8	0.8	0.8	0.9	1.0	3	.0500
.0667	4	0.7	0.7	0.8	0.8	0.8	0.9	0.9	0.9	1.0	1.0	1.0	1.1	1.1	1.2	1.3	1.3	4	.0667
.0833	5	0.9	0.9	1.0	1.0	1.0	1.1	1.1	1.2	1.2	1.2	1.3	1.3	1.4	1.5	1.6	1.7	5	.0833
.1000	6	1.0	1.1	1.1	1.2	1.2	1.3	1.3	1.4	1.4	1.5	1.5	1.6	1.7	1.8	1.9	2.0	6	.1000
.1167	7	1.2	1.3	1.3	1.4	1.5	1.5	1.6	1.6	1.7	1.8	1.8	1.9	2.0	2.1	2.2	2.3	7	.1167
.1333	8	1.4	1.5	1.5	1.6	1.7	1.7	1.8	1.9	1.9	2.0	2.1	2.1	2.3	2.4	2.5	2.7	8	.1333
.1500	9	1.6	1.6	1.7	1.8	1.9	1.9	2.0	2.1	2.1	2.2	2.3	2.4	2.5	2.7	2.8	3.0	9	.1500
.1667	10	1.8	1.8	1.9	2.0	2.1	2.2	2.3	2.3	2.4	2.5	2.6	2.7	2.8	3.0	3.2	3.3	10	.1667
.1833	11	1.9	2.0	2.1	2.2	2.3	2.4	2.5	2.6	2.7	2.7	2.8	2.9	3.1	3.3	3.5	3.7	11	.1833
.2000	12	2.1	2.2	2.3	2.4	2.5	2.6	2.7	2.8	2.9	3.0	3.1	3.2	3.4	3.6	3.8	4.0	12	.2000
.2167	13	2.3	2.4	2.5	2.6	2.7	2.8	2.9	3.0	3.1	3.2	3.3	3.5	3.7	3.9	4.1	4.3	13	.2167
.2333	14	2.4	2.6	2.7	2.8	2.9	3.0	3.1	3.3	3.4	3.5	3.6	3.7	4.0	4.2	4.4	4.7	14	.2333
.2500	15	2.6	2.7	2.9	3.0	3.1	3.2	3.4	3.5	3.6	3.7	3.9	4.0	4.2	4.5	4.7	5.0	15	.2500
.2667	16	2.8	2.9	3.1	3.2	3.3	3.5	3.6	3.7	3.9	4.0	4.1	4.3	4.5	4.8	5.1	5.3	16	.2667
.2833	17	3.0	3.1	3.3	3.4	3.5	3.7	3.8	4.0	4.1	4.2	4.4	4.5	4.8	5.1	5.4	5.7	17	.2833
.3000	18	3.1	3.3	3.4	3.6	3.7	3.9	4.0	4.2	4.3	4.5	4.6	4.8	5.1	5.4	5.7	6.0	18	.3000
.3167	19	3.3	3.5	3.6	3.8	4.0	4.1	4.3	4.4	4.6	4.8	4.9	5.1	5.4	5.7	6.0	6.3	19	.3167
.3333	20	3.5	3.7	3.8	4.0	4.2	4.3	4.5	4.7	4.8	5.0	5.2	5.3	5.7	6.0	6.3	6.7	20	.3333
.3500	21	3.7	3.8	4.0	4.2	4.4	4.5	4.7	4.9	5.1	5.2	5.4	5.6	5.9	6.3	6.6	7.0	21	.3500
.3667	22	3.9	4.0	4.2	4.4	4.6	4.8	5.0	5.1	5.3	5.5	5.7	5.9	6.2	6.6	7.0	7.3	22	.3667
.3833	23	4.0	4.2	4.4	4.6	4.8	5.0	5.2	5.4	5.6	5.7	5.9	6.1	6.5	6.9	7.3	7.7	23	.3833
.4000	24	4.2	4.4	4.6	4.8	5.0	5.2	5.4	5.6	5.8	6.0	6.2	6.4	6.8	7.2	7.6	8.0	24	.4000
.4167	25	4.4	4.6	4.8	5.0	5.2	5.4	5.6	5.8	6.0	6.3	6.5	6.7	7.1	7.5	7.9	8.3	25	.4167
.4333	26	4.5	4.8	5.0	5.2	5.4	5.6	5.8	6.1	6.3	6.5	6.7	6.9	7.4	7.8	8.2	8.7	26	.4333
.4500	27	4.7	4.9	5.2	5.4	5.6	5.8	6.1	6.3	6.5	6.7	7.0	7.2	7.6	8.1	8.5	9.0	27	.4500
.4667	28	4.9	5.1	5.4	5.6	5.8	6.1	6.3	6.5	6.8	7.0	7.2	7.5	7.9	8.4	8.9	9.3	28	.4667
.4833	29	5.1	5.3	5.6	5.8	6.0	6.3	6.5	6.8	7.0	7.2	7.5	7.7	8.2	8.7	9.2	9.7	29	.4833
.5000	30	5.2	5.5	5.7	6.0	6.2	6.5	6.7	7.0	7.2	7.5	7.7	8.0	8.5	9.0	9.5	10.0	30	.5000
.5167	31	5.4	5.7	5.9	6.2	6.5	6.7	7.0	7.2	7.5	7.8	8.0	8.3	8.8	9.3	9.8	10.3	31	.5167
.5333	32	5.6	5.9	6.1	6.4	6.7	6.9	7.2	7.5	7.7	8.0	8.3	8.5	9.1	9.6	10.1	10.7	32	.5333
.5500	33	5.8	6.0	6.3	6.6	6.9	7.1	7.4	7.7	8.0	8.2	8.5	8.8	9.3	9.9	10.4	11.0	33	.5500
.5667	34	6.0	6.2	6.5	6.8	7.1	7.4	7.7	7.9	8.2	8.5	8.8	9.1	9.6	10.2	10.8	11.3	34	.5667
.5833	35	6.1	6.4	6.7	7.0	7.3	7.6	7.9	8.2	8.5	8.7	9.0	9.3	9.9	10.5	11.1	11.7	35	.5833
.6000	36	6.3	6.6	6.9	7.2	7.5	7.8	8.1	8.4	8.7	9.0	9.3	9.6	10.2	10.8	11.4	12.0	36	.6000
.6117	37	6.4	6.7	7.0	7.3	7.6	8.0	8.3	8.6	8.9	9.2	9.5	9.8	10.4	11.0	11.6	12.2	37	.6117
.6333	38	6.6	7.0	7.3	7.6	7.9	8.2	8.5	8.9	9.2	9.5	9.8	10.1	10.8	11.4	12.0	12.7	38	.6333
.6500	39	6.8	7.1	7.5	7.8	8.1	8.4	8.8	9.1	9.4	9.7	10.1	10.4	11.0	11.7	12.3	13.0	39	.6500
.6667	40	7.0	7.3	7.7	8.0	8.3	8.7	9.0	9.3	9.7	10.0	10.3	10.7	11.3	12.0	12.7	13.3	40	.6667
.6833	41	7.2	7.5	7.9	8.2	8.5	8.9	9.2	9.6	9.9	10.2	10.6	10.9	11.6	12.3	13.0	13.7	41	.6833
.7000	42	7.3	7.7	8.0	8.4	8.7	9.1	9.4	9.8	10.1	10.5	10.8	11.2	11.9	12.6	13.3	14.0	42	.7000
.7167	43	7.5	7.9	8.2	8.6	9.0	9.3	9.7	10.0	10.4	10.8	11.1	11.5	12.2	12.9	13.6	14.3	43	.7167
.7333	44	7.7	8.1	8.4	8.8	9.2	9.5	10.0	10.3	10.6	11.0	11.4	11.7	12.5	13.2	13.9	14.7	44	.7333
.7500	45	7.9	8.2	8.6	9.0	9.4	9.7	10.1	10.5	10.9	11.2	11.6	12.0	12.7	13.5	14.2	15.0	45	.7500
.7667	46	8.1	8.4	8.8	9.2	9.6	10.0	10.4	10.7	11.1	11.5	11.9	12.3	13.0	13.8	14.6	15.3	46	.7667
.7833	47	8.2	8.6	9.0	9.4	9.8	10.2	10.6	11.0	11.4	11.7	12.1	12.5	13.3	14.1	14.9	15.7	47	.7833
.8000	48	8.4	8.8	9.2	9.6	10.0	10.4	10.8	11.2	11.6	12.0	12.4	12.8	13.6	14.4	15.2	16.0	48	.8000
.8167	49	8.6	9.0	9.4	9.8	10.2	10.6	11.0	11.4	11.8	12.2	12.7	13.1	13.9	14.7	15.5	16.3	49	.8167
.8333	50	8.7	9.2	9.6	10.0	10.4	10.8	11.2	11.7	12.1	12.5	12.9	13.3	14.2	15.0	15.8	16.7	50	.8333
.8500	51	8.9	9.3	9.8	10.2	10.6	11.0	11.5	11.9	12.3	12.7	13.2	13.6	14.4	15.3	16.1	17.0	51	.8500
.8667	52	9.1	9.5	10.0	10.4	10.8	11.3	11.7	12.1	12.6	13.0	13.4	13.9	14.7	15.6	16.5	17.3	52	.8667
.8833	53	9.3	9.7	10.2	10.6	11.0	11.5	11.9	12.4	12.8	13.2	13.7	14.1	15.0	15.9	16.8	17.7	53	.8833
.9000	54	9.4	9.9	10.3	10.8	11.2	11.7	12.1	12.6	13.0	13.5	13.9	14.4	15.3	16.2	17.1	18.0	54	.9000
.9167	55	9.6	10.1	10.5	11.0	11.5	11.9	12.4	12.8	13.3	13.8	14.2	14.7	15.6	16.5	17.4	18.3	55	.9167
.9333	56	9.8	10.3	10.7	11.2	11.7	12.1	12.6	13.1	13.5	14.0	14.5	14.9	15.9	16.8	17.7	18.7	56	.9333
.9500	57	10.0	10.4	10.9	11.4	11.9	12.3	12.8	13.3	13.8	14.2	14.7	15.2	16.1	17.1	18.0	19.0	57	.9500
.9667	58	10.2	10.6	11.1	11.6	12.1	12.6	13.1	13.5	14.0	14.5	15.0	15.5	16.4	17.4	18.4	19.3	58	.9667
.9833	59	10.3	10.8	11.3	11.8	12.3	12.8	13.3	13.8	14.3	14.7	15.2	15.7	16.7	17.7	18.7	19.7	59	.9833
1.0000	60	10.5	11.0	11.5	12.0	12.5	13.0	13.5	14.0	14.5	15.0	15.5	16.0	17.0	18.0	19.0	20.0	60	1.0000
Decimal of hr.	Mins	10.5	11.0	11.5	12.0	12.5	13.0	13.5	14.0	14.5	15.0	15.5	16.0	17.0	18.0	19.0	20.0	Mins	Decimal of hr.
Time		\multicolumn{16}{c}{Speed in knots}		Time

TABLE 3(6) Measured Mile Table — Knots related to time over one nautical mile

Secs	1 min	2 min	3 min	4 min	5 min	6 min	7 min	8 min	9 min	10 min	11 min
0	60.00	30.00	20.00	15.00	12.00	10.00	8.57	7.50	6.67	6.00	5.45
1	59.02	29.75	19.89	14.94	11.96	9.97	8.55	7.48	6.66	5.99	5.45
2	58.06	29.51	19.78	14.88	11.92	9.94	8.53	7.47	6.64	5.98	5.44
3	57.14	29.27	19.67	14.81	11.88	9.92	8.51	7.45	6.63	5.97	5.43
4	56.25	29.03	19.57	14.75	11.84	9.89	8.49	7.44	6.62	5.96	5.42
5	55.38	28.80	19.46	14.69	11.80	9.86	8.47	7.42	6.61	5.95	5.41
6	54.55	28.57	19.35	14.63	11.76	9.84	8.45	7.41	6.59	5.94	5.41
7	53.73	28.35	19.25	14.57	11.73	9.81	8.43	7.39	6.58	5.93	5.40
8	52.94	28.12	19.15	14.52	11.69	9.78	8.41	7.38	6.57	5.92	5.39
9	52.17	27.91	19.05	14.46	11.65	9.76	8.39	7.36	6.56	5.91	5.38
10	51.43	27.69	18.95	14.40	11.61	9.73	8.37	7.35	6.55	5.90	5.37
11	50.70	27.48	18.85	14.34	11.58	9.70	8.35	7.33	6.53	5.89	5.37
12	50.00	27.27	18.75	14.29	11.54	9.68	8.33	7.32	6.52	5.88	5.36
13	49.32	27.07	18.65	14.23	11.50	9.65	8.31	7.30	6.51	5.87	5.35
14	48.65	26.87	18.56	14.17	11.46	9.63	8.29	7.29	6.50	5.86	5.34
15	48.00	26.67	18.46	14.12	11.43	9.60	8.28	7.27	6.49	5.85	5.33
16	47.37	26.47	18.37	14.06	11.39	9.58	8.26	7.26	6.47	5.84	5.33
17	46.75	26.28	18.27	14.01	11.36	9.55	8.24	7.24	6.46	5.83	5.32
18	46.15	26.09	18.18	13.95	11.32	9.52	8.22	7.23	6.45	5.83	5.31
19	45.57	25.90	18.09	13.90	11.29	9.50	8.20	7.21	6.44	5.82	5.30
20	45.00	25.71	18.00	13.85	11.25	9.47	8.18	7.20	6.43	5.81	5.29
21	44.44	25.53	17.91	13.79	11.21	9.45	8.16	7.19	6.42	5.80	5.29
22	43.90	25.35	17.82	13.74	11.18	9.42	8.14	7.17	6.41	5.79	5.28
23	43.37	25.17	17.73	13.69	11.15	9.40	8.13	7.16	6.39	5.78	5.27
24	42.86	25.00	17.65	13.64	11.11	9.37	8.11	7.14	6.38	5.77	5.26
25	42.35	24.83	17.56	13.58	11.08	9.35	8.09	7.13	6.37	5.76	5.26
26	41.86	24.66	17.48	13.53	11.04	9.33	8.07	7.11	6.36	5.75	5.25
27	41.38	24.49	17.39	13.48	11.01	9.30	8.05	7.10	6.35	5.74	5.24
28	40.91	24.32	17.31	13.43	10.98	9.28	8.04	7.09	6.34	5.73	5.23
29	40.45	24.16	17.22	13.38	10.94	9.25	8.02	7.07	6.33	5.72	5.22
30	40.00	24.00	17.14	13.33	10.91	9.23	8.00	7.06	6.32	5.71	5.22
31	39.56	23.84	17.06	13.28	10.88	9.21	7.98	7.04	6.30	5.71	5.21
32	39.13	23.68	16.98	13.24	10.84	9.18	7.96	7.03	6.29	5.70	5.20
33	38.71	23.53	16.90	13.19	10.81	9.16	7.95	7.02	6.28	5.69	5.19
34	38.30	23.38	16.82	13.14	10.78	9.14	7.93	7.00	6.27	5.68	5.19
35	37.89	23.23	16.74	13.09	10.75	9.11	7.91	6.99	6.26	5.67	5.18
36	37.50	23.08	16.67	13.04	10.71	9.09	7.89	6.98	6.25	5.66	5.17
37	37.11	22.93	16.59	13.00	10.68	9.07	7.88	6.96	6.24	5.65	5.16
38	36.73	22.78	16.51	12.95	10.65	9.05	7.86	6.95	6.23	5.64	5.16
39	36.36	22.64	16.44	12.90	10.62	9.02	7.84	6.94	6.22	5.63	5.15
40	36.00	22.50	16.36	12.86	10.59	9.00	7.83	6.92	6.21	5.62	5.14
41	35.64	22.36	16.29	12.81	10.56	8.98	7.81	6.91	6.20	5.62	5.13
42	35.29	22.22	16.22	12.77	10.53	8.96	7.79	6.90	6.19	5.61	5.13
43	34.95	22.09	16.14	12.72	10.50	8.93	7.78	6.89	6.17	5.60	5.12
44	34.62	21.95	16.07	12.68	10.47	8.91	7.76	6.87	6.16	5.59	5.11
45	34.29	21.82	16.00	12.63	10.43	8.89	7.74	6.86	6.15	5.58	5.10
46	33.96	21.69	15.93	12.59	10.40	8.87	7.72	6.84	6.14	5.57	5.10
47	33.64	21.56	15.86	12.54	10.37	8.85	7.71	6.83	6.13	5.56	5.09
48	33.33	21.43	15.79	12.50	10.34	8.82	7.69	6.82	6.12	5.56	5.08
49	33.03	21.30	15.72	12.46	10.32	8.80	7.68	6.80	6.11	5.55	5.08
50	32.73	21.18	15.65	12.41	10.29	8.78	7.66	6.79	6.10	5.54	5.07
51	32.43	21.05	15.58	12.37	10.26	8.76	7.64	6.78	6.09	5.53	5.06
52	32.14	20.93	15.52	12.33	10.23	8.74	7.63	6.77	6.08	5.52	5.06
53	31.86	20.81	15.45	12.29	10.20	8.72	7.61	6.75	6.07	5.51	5.05
54	31.58	20.69	15.38	12.24	19.17	8.70	7.59	6.74	6.06	5.50	5.04
55	31.30	20.57	15.32	12.20	10.14	8.67	7.58	6.73	6.05	5.50	5.04
56	31.03	20.45	15.25	12.16	10.11	8.65	7.56	6.72	6.04	5.49	5.03
57	30.77	20.34	15.19	12.12	10.08	8.63	7.55	6.70	6.03	5.48	5.02
58	30.51	20.22	15.13	12.08	10.06	8.61	7.53	6.69	6.02	5.47	5.01
59	30.25	20.11	15.06	12.04	10.03	8.59	7.52	6.68	6.01	5.46	5.00
Secs	1 min	2 min	3 min	4 min	5 min	6 min	7 min	8 min	9 min	10 min	11 min

(Continued from page 24)

Distance to radar horizon (in n miles) =

$1.22 \times \sqrt{\text{Ht of scanner}}$ (in ft),

or $2.21 \times \sqrt{\text{Ht of scanner}}$ (in m)

Boat speed over measured distance of 1 n mile =

$$\frac{3600}{\text{time in seconds}} \text{ (knots)}$$

Horizontal sextant angle
Radius of position circle (in n miles) =

$$\frac{D}{2 \times \sin A}$$

where D is distance between objects in n miles, and A is the angle between them in degrees.

Vertical sextant angles

$$\frac{\text{Distance off}}{\text{(in n miles)}} = \frac{\text{Ht of object (above MHWS, in ft)}}{6076 \times \tan (\text{sextant angle})}$$

or $= \dfrac{\text{Ht of object (above MHWS, in m)}}{1852 \times \tan (\text{sextant angle})}$

Note: sextant angle above is in degrees and minutes, and must be corrected for index error.

An approximate distance off, in n miles, adequate for most purposes is given by:

$$\frac{\text{Distance}}{\text{(n miles)}} = \frac{\text{Ht of object (in feet)} \times 0.565}{\text{Sextant angle (in minutes)}}$$

or $= \dfrac{\text{Ht of object (in metres)} \times 1.854}{\text{Sextant angle (in minutes)}}$

Coastal navigation
To find the DR/EP, as bearing and distance from start position; example using TI.57. Key in:

1st distance run $\boxed{5.2}$ x⇌t
1st course (°T) $\boxed{230}$ →R STO 0 x⇌t STO 6
2nd distance run $\boxed{1.9}$ x⇌t
2nd course (°T) $\boxed{255}$ →R SUM 0 x⇌t SUM 6

Repeat for each subsequent Co(°T) and distance.
For EP, treat Set/Drift as for Co(°T) and distance.

To display bearing and distance from start:
RCL 6 x⇌t RCL 0 → P $\boxed{236.6}$ (°T)
x⇌t $\boxed{6.97}$ (nm)

Note: To find EP, treat Set/Drift as for Co(°T) and distance run.
Example: Using RPN calculator. Key in:

1st course (°T) $\boxed{230}$ ENTER
1st distance run $\boxed{5.2}$ →R Σ+
2nd course (°T) $\boxed{255}$ ENTER
2nd distance run $\boxed{1.9}$ →R Σ+

Repeat for each subsequent Co (°T) and distance.

To display distance and bearing from start:
RCL 13[1]   RCL 11[1]   → P $\boxed{6.97}$ nm
x ⇌ y $\boxed{236.6^{[2]}}$ (°T)

Notes: (1) Check the actual stores used for vector summation in your calculator.
(2) If display negative (−), add 360

To find EP, treat Set/Drift as for Co(°T) and distance run.

Distance (D, in n miles) of
object at second bearing $= \dfrac{R \times \sin A}{\sin (B - A)}$

Predicted distance (in n miles) object will be off when abeam = $D \times \sin B$, where R is distance run (nm) between two relative bearings of an object, first A degrees and then B degrees.

Course to steer and speed made good
Co(°T) = Tr(°T) − $\sin^{-1}$ (Drift ÷ Speed) × sin (Set − Track)
SMG = Speed × cos (Co.T − Track) + Drift × cos (Set − Track)
Note: The Drift must be less than yacht's speed.

Conversion angle (half convergency)
Radio bearings follow great circles, and become curved lines when plotted on a Mercator chart, see 4.1.10. A correction may be needed for bearings of beacons more than about 60 n mile away, and can be calculated from the formula:
Conversion angle = ½ d.Long × sin mid Latitude

A great circle always lies on the polar side of the rhumb line, and conversion angle is applied towards the equator.

Short distance sailing
(Note: These formulae should not be used for distances over 600 n miles).

Departure = Distance × sin Course
          = d.Long × cos Mean Latitude
          = tan Course × d.Lat
d.Lat   = Distance × cos Course
d.Long   = Departure ÷ cos Mean Latitude
Distance = Departure ÷ sin Course
          = d.Lat × sec Course
sin Course = Departure ÷ Distance
cos Course = d.Lat ÷ Distance
tan Course = Departure ÷ d.Lat

Further explanation of the use of calculators, and formulae for the calculation of tracks and distances for distances over 600 n miles are contained in Chapter 3 of *The Macmillan & Silk Cut Yachtsman's Handbook*.

3.1.4 Chart abbreviations. (For symbols see Plates 2 and 3, pages 116–117)

Coastal Features
G. Gulf
B. Bay
L. Loch, Lough, Lake
Cr. Creek
Str. Strait
Sd. Sound
Pass. Passage
Chan. Channel
Appr. Approaches
Ent. Entrance
R. River
Est. Estuary
Mth. Mouth
Rds. Roads, Roadstead
Anch. Anchorage
Hr. Harbour
Hn. Haven
P. Port
I. Island
It Islet
C. Cape
Prom. Promotory
Pt. Point
Mt. Mountain, Mount
Lndg. Landing place
Rk. Rock

Units
m Metre(s)
dm Decimetre(s)
cm Centimetre(s)
mm Millimetre(s)
km Kilometre(s)
ft Foot, feet
M Sea Mile(s)
kn knot(s)
Lat Latitude
Long Longitude
Ht Height
No Number

Adjectives etc
Gt Grt Great
Lit Little
Mid Middle
Anct Ancient
S St Saint
conspic Conspicuous
dest Destroyed
proj Projected
dist Distant
abt About
illum Illuminated
Aero Aeronautical
Hr Higher
Lr Lower
exper Experimental

discont Discontinued
prohib Prohibited
explos Explosive
priv Private
prom Prominent
subm Submerged
approx Approximate
NM Notices to Mariners
(P) Preliminary (NM)
(T) Temporary (NM)
SD Sailing Directions
LL List of Lights

Buildings etc
Cas Castle
Ho House
Va Villa
Fm Farm
Ch Church, Chapel
Cath Cathedral
Cemy Cemetery
Ft Fort
Baty Battery
St Street
Ave Avenue
Tel Telegraph
PO Post Office
Hosp Hospital
Mon Monument, Memorial
Cup Cupola
Ru Ruin
Tr Tower
Chy Chimney
Sch School
Bldg Building
Tel Telephone
Col Column, Obelisk
Sta Station
CG Coastguard
LB Lifeboat
SS Signal Station
Sem Semaphore
SS (Storm) Storm signal station
FS Flagstaff
Sig Signal
Obsy Observatory
Off Office
NB Notice Board

Dangers
Bk. Bank
Sh. Shoal
Rf. Reef
Le. Ledge
Obstn Obstruction
Wk Wreck
dr Dries
cov Covers

uncov Uncovers
PA Position approximate
PD Position doubtful
ED Existence doubtful
pos Position
unexam Unexamined
Rep Reported

Quality of the Bottom
Gd Ground
S Sand
M Mud
Oz Ooze
Ml Marl
Cy Clay
G Gravel
Sn Shingle
P Pebbles
St Stones
R Rock
Bo Boulders
Ck Chalk
Qz Quartz
Sh Shells
Oy Oysters
Ms Mussels
Wd Weed
f Fine
c Coarse
so Soft
h Hard
sf Stiff
sm Small
l Large
sy Sticky
bk Broken
ga Glacial

Tides and Currents
HW/LW High Water/Low Water
MTL Mean Tide Level
MSL Mean Sea Level
Sp/Np Spring Tides/Neap Tides
MHWS Mean High Water Springs
MHWN Mean High Water Neaps
MLWS Mean Low Water Springs
MLWN Mean Low Water Neaps
HAT Highest Astronomical Tide
LAT Lowest Astronomical Tide
Vel Velocity
kn Knots
Dir Direction
OD Ordnance Datum

Compass
Mag Magnetic
Var Variation
annly Annually

3.1.5 LIGHT CHARACTERS (Fathoms and Metric Charts)

Reproduced by kind permission of H.M. Stationery Office and the Hydrographer of the Navy

CLASS OF LIGHT	International abbreviations	Older form (where different)	Illustration — Period shown
Fixed *(steady light)*	F		
Occulting *(total duration of light more than dark)*			
Single-occulting	Oc	Occ	
Group-occulting e.g.	Oc(2)	Gp Occ(2)	
Composite group-occulting e.g.	Oc(2+3)	Gp Occ(2+3)	
Isophase *(light and dark equal)*	Iso		
Flashing *(total duration of light less than dark)*			
Single-flashing	Fl		
Long-flashing *(flash 2s or longer)*	L Fl		
Group-flashing e.g.	Fl(3)	Gp Fl(3)	
Composite group-flashing e.g.	Fl(2+1)	Gp Fl(2+1)	
Quick *(50 to 79—usually either 50 or 60—flashes per minute)*			
Continuous quick	Q	Qk Fl	
Group quick e.g.	Q(3)	Qk Fl(3)	
Interrupted quick	IQ	Int Qk Fl	
Very Quick *(80 to 159—usually either 100 or 120—flashes per minute)*			
Continuous very quick	V Q	V Qk Fl	
Group very quick e.g.	V Q(3)	V Qk Fl(3)	
Interrupted very quick	IV Q	Int V Qk Fl	
Ultra Quick *(160 or more—usually 240 to 300—flashes per minute)*			
Continuous ultra quick	UQ		
Interrupted ultra quick	IUQ		
Morse Code e.g.	Mo(K)		
Fixed and Flashing	F Fl		
Alternating e.g.	Al.WR	Alt.WR	R W R W R W

COLOUR	International abbreviations	Older form (where different)	RANGE in sea miles	International abbreviations	Older form
White	W *(may be omitted)*		*Single range* e.g.	15M	
Red	R				
Green	G		*2 ranges* e.g.	14/12M	14.12M
Yellow	Y				
Orange	Y	Or	*3 or more ranges* e.g.	22-18M	22,20,18M
Blue	Bu	Bl			
Violet	Vi				
ELEVATION is given in metres (m) or feet (ft)			**PERIOD** in seconds e.g.	5s	5sec

3.1.6 IALA Buoyage System (Region A)
(See also Plate 5 on page 119.)

International buoyage is harmonized into a single system which, applied to Regions A and B, differs only in the use of red and green lateral marks. In Region A (which includes all Europe) lateral marks are red on the port hand, and in Region B red on the starboard hand, related to direction of buoyage. Five types of marks are used, as illustrated in Plate 5, on page 119.

(1) *Lateral marks* are used in conjunction with a direction of buoyage, shown by a special arrow on the chart. In and around the British Isles its general direction is from SW to NE in open waters, but from seaward when approaching a harbour, river or estuary. Where port or starboard lateral marks do not rely on can or conical buoy shapes for identification they carry, where practicable, the appropriate topmarks. Any numbering or lettering follows the direction of buoyage.

In Region A, port hand marks are coloured red, and port hand buoys are can or spar shaped. Any topmark fitted is a single red can. Any light fitted is red, any rhythm. Starboard hand marks are coloured green, and starboard hand buoys are conical or spar shaped. Any topmark fitted is a single green cone, point up. Any light fitted is green, any rhythm. In exceptional cases starboard hand marks may be coloured black.

At a division, the preferred channel may be shown by lateral marks with red or green stripes:

Preferred channel	Indicated by	Light (if any)
To starboard	Port lateral mark with green stripe	Flashing red (2 + 1)
To port	Starboard lateral mark with red stripe	Flashing green (2 + 1)

(2) *Cardinal marks* are used in conjunction with a compass to show where dangers exist or where the mariner may find navigable water. They are named after the quadrant in which the mark is placed, in relation to the danger or point indicated. The four quadrants (North, East, South and West) are bounded by the true bearings NW–NE, NE–SE, SE–SW and SW–NW, taken from the point of interest. The name of a cardinal mark indicates that it should be passed on the named side.

A cardinal mark may indicate the safe side on which to pass a danger, or that the deepest water is on the named side of the mark, or it may draw attention to a feature in a channel such as a bend, junction or fork, or the end of a shoal.

Cardinal marks are pillar or spar shaped, painted black and yellow, and always carry black double cone topmarks, one cone above the other. Their lights are white, either very quick flashing (VQ or VQkFl) 120 to 100 flashes per minute, or quick flashing (Q or QkFl) 60 to 50 flashes per minute. A long flash is one of not less than two seconds duration.

North cardinal mark
Two black cones — Points up
Colour — Black above yellow
Light (if fitted) — White; VQ or Q

East cardinal mark
Two black cones — Base to base
Colour — Black, with horizontal yellow band
Light (if fitted) — White; VQ(3) 5 sec or Q(3) 10 sec

South cardinal mark
Two black cones — Points down
Colour — Yellow above black
Light (if fitted) — White; VQ(6) plus long flash 10 sec or Q(6) plus long flash 15 sec

West cardinal mark
Two black cones — Point to point
Colour — Yellow, with horizontal black band
Light (if fitted) — White; VQ(9) 10 sec or Q(9) 15 sec

(3) *Isolated danger marks* are placed on or above an isolated danger such as a rock or a wreck which has navigable water all around it. The marks are black, with one or more broad horizontal red bands. Buoys are pillar or spar shaped. Any light is white, flashing (2). Topmark—two black spheres.

(4) *Safe water marks* indicate that there is navigable water all round the mark, and are used for mid-channel or landfall marks. Buoys are spherical, or pillar with spherical topmark, or spar, and are coloured with red and white vertical stripes. Any topmark fitted is a single red sphere. Any light fitted is white — either isophase, occulting or long flash every 10 seconds.

(5) *Special marks* do not primarily assist navigation, but indicate a special area or feature (e.g. spoil grounds, exercise areas, water ski areas, cable or pipeline marks, outfalls, Ocean Data Acquisition Systems (ODAS), or traffic separation marks where conventional channel marks may cause confusion). Special marks are yellow, and any shape not conflicting with lateral or safe water marks. If can, spherical or conical are used they indicate the side on which to pass. Any topmark fitted is a yellow X. Any light fitted is yellow, and may have any rhythm not used for white lights.

New dangers (which may be natural obstructions such as a sandbank, or a wreck for example) are marked in accordance with the rules above, and lit accordingly. For a very grave danger one of the marks may be duplicated.

Camper & Nicholsons Marinas Amble

Tele No: 0665 712168
Radio: VHF Channel 37(M)
Call Sign — Camper Base
Contact: Marina Master
Address: Camper & Nicholsons Marinas Ltd, The Braid, Amble, Northumberland
Total no of berths: 200 pontoon
Visitors berths: 30 pontoon
Hours of access: 8 hrs in 12

The Marina is situated within the River Coquet in an area of outstanding natural beauty. The entrance to the marina is approx 1000 metres inside the harbour on the portside.

Facilities include:
Fuel (diesel), Brokerage, Chandlery, Repairs etc. The town centre is a 3 min walk and is well served with shops and restaurants.

Camper & Nicholsons Marinas Ltd.

Also operate marinas at:— Gosport, Hampshire (0705) 524811
Neyland, South Wales (0646) 601601   Penarth, South Wales (0222) 705021

Marine Electronics AND A First Class Service From
GREENHAM MARINE

With branches nationwide Greenham Marine can offer yachtsmen one of the most comprehensive selections of Marine Electronic Equipment currently available.

Our experienced sales staff will provide advice on ● **Equipment Selection** ● **Tuition** ● **After Sales Care and Maintenance.**

Strategically placed throughout the U.K. our team of highly skilled engineers can install and service your electronic equipment quickly and efficiently.

Estimates and quotations are free of charge so why not contact us today by writing to:

Greenham Marine Ltd, Kings Salterns Road, Lymington, Hampshire. Tel: Lymington (0590) 71144 for more details on the range of services and advice available, plus a free copy of our latest Marine Electronic Equipment Catalogue.

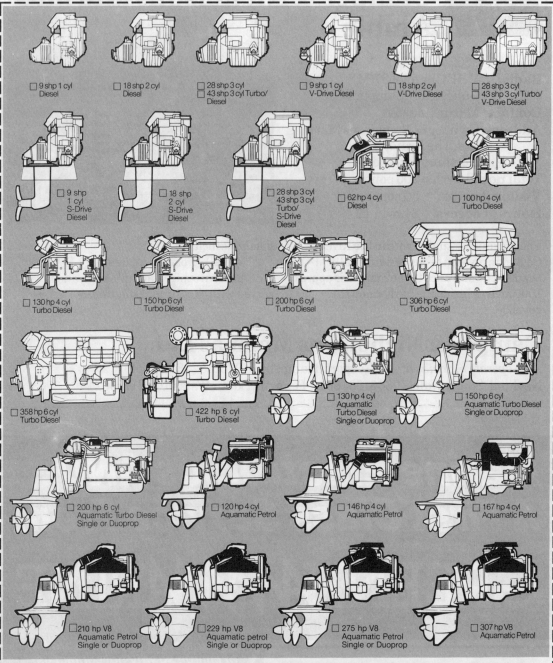

CHOOSING PERFORMANCE STARTS WITH A TICK ✓

To get the facts on Volvo Penta performance choose your engine, tick the box alongside and send in the *whole* advertisement. In return you will receive the features, specifications and dimensions.

Volvo Penta has the greatest choice in marine power. Totally dedicated petrol and diesel engines spanning the range of 9 to 656 hp.

Choose Volvo Penta and you choose power, performance, reliability and worldwide service.

To: Volvo Penta UK Ltd Marine Pleasure Sales, FREEPOST, Watford WD1 8FP. Tel: (0923) 228544

Please send me details on the engines I have indicated.

Name
Address

Boat Type

VOLVO PENTA

Chapter 4
Radio Navigational Aids

Contents

4.1 RADIOBEACONS	**Page 38**	**4.4 RADAR**	**63**
4.1.1 Marine radiobeacons – general		4.4.1 Radar in yachts	
4.1.2 DF receiving sets		4.4.2 Radar for collision avoidance	
4.1.3 Grouping and sequence of beacons		4.4.3 Radar as a navigational aid	
4.1.4 Beacons incorporating distance finding		4.4.4 Radar beacons (Racons)	
4.1.5 Aero Radiobeacons			
4.1.6 Types of emission			
4.1.7 Directional Radiobeacons (RD)			
4.1.8 Errors in radio bearings			
4.1.9 Radiobeacons – calibration			
4.1.10 Half convergency			

4.2 POSITION FIXING SYSTEMS 40
4.2.1 Types of systems
4.2.2 Waypoints

4.3 RADIO DIRECTION FINDING 40
4.3.1 Principle of operation
4.3.2 VHF emergency direction finding service
4.3.3 VHF Radio Lighthouses

Radio navigational aids – introduction
The following subjects are described in detail in Chapter 4 of *The Macmillan & Silk Cut Yachtsman's Handbook*:
   Marine radiobeacons; DF receiving sets; grouping and sequence of beacons; directional radiobeacons; beacons incorporating distance finding; Aero radiobeacons; types of emissions; operating procedures; errors in radio bearings; calibration; QTG service from Coast Radio Stations; VHF emergency direction finding service; half convergency; Consol; Decca Navigator; Loran-C; Omega; satellite navigation; radar; radar beacons (Racons).
Here in the Almanac is provided the necessary detail about individual radiobeacons and Racons for navigational use, but for information on procedures and possible errors, together with the other subjects listed above, reference should be made to Chapter 4 of *The Macmillan & Silk Cut Yachtsman's Handbook*.

4.1 RADIOBEACONS

4.1.1 Marine radiobeacons — general

This widely-used system of direction finding allows a yachtsman to obtain a bearing from a non-directional radiobeacon, using a receiving set with an aerial which has directional qualities. This is useful in poor visibility, or when out of sight of land, but the effective range is limited, and few beacons give bearings of reasonable accuracy at distances of more than 50 miles.

Beacons transmit on certain known frequencies and at fixed intervals. The navigator simply tunes to the listed frequency, identifies the beacon by its Morse call sign, and rotates the aerial of the set so that it registers the minimum signal (or null).

Often up to six beacons covering a certain area work on the same frequency, transmitting for one minute in turn.

On Admiralty charts marine radiobeacons are shown by a magenta circle with the letters 'RC'.

The details of radiobeacons in the area covered by this Almanac are given in Table 4(2), commencing on page 42.

4.1.2 DF receiving sets

Most sets have ferrite rod aerials, which give a minimum signal when in line with the direction of the beacon. Often the aerial incorporates a small compass, so that the operator can read the bearing at the moment that he identifies the null in his earphones. Such compasses are just as liable to error as any other magnetic compass, and must not be held close to magnetic objects.

With some sets the receiver itself is fixed, and the aerial rotates against a graduated scale, which shows the bearing relative to the ship's head. In this case it is necessary to record the course steered at the same moment as the null and the relative bearing.

An Automatic Direction Finding (ADF) set is tuned to the required frequency, and will then lock on to any transmission automatically, indicating the bearing. If there is a sequence of beacons on one frequency it will point to each in turn. Some sets even assess the reliability of individual bearings taken.

A quartz clock may be incorporated, to allow beacons to be identified by their time sequence, but this should only be considered an aid. A beacon should always be positively identified by its call (ident) sign in order to avoid possible errors.

4.1.3 Grouping and sequence of beacons

Where two or more beacons share a common frequency, they transmit on a strict time schedule, for one minute in turn. The minutes past each hour at which a beacon transmits, depending on its sequence number, is shown in the table at the foot of each relevant page of Table 4(2).

4.1.4 Beacons incorporating distance finding

A few beacons allow radio and sound signals to be synchronised for distance finding. The two signals are synchronised at an easily identifiable point in the cycle, e.g. at the start or end of a long dash. Using a stop watch, the difference between the two times in seconds, multiplied by 0.18, gives the distance in nautical miles.

Another system involves the transmission of a number of measuring signals, started when the fog signal is sounded. The number of measuring signals received before the fog signal is heard indicates the distance. An intervals of 5.5 seconds between measuring signals is equivalent to a unit distance of one nautical mile.

4.1.5 Aero Radiobeacons

Although intended for aircraft, some of those near the coast are useful to yachtsmen. They are shown on a chart by a small magenta circle, and the letters 'Aero RC'. These beacons transmit continuously during operational hours. They must be used with care, because the land effect (see 4.1.8) is unpredictable. Details of those beacons which may be of use to yachtsmen, in the area covered by this Almanac, are included in Table 4(2).

4.1.6 Types of emission

Intelligence is impressed upon a radio emission in various ways. Some understanding of the different modes is helpful, in order to get good results from radio equipment. Where a separate Beat Frequency Oscillator (BFO) control is provided, it should be adjusted as follows when taking radio bearings:

Emission		BFO setting	
Old style	New style	For DF use	For ident.
A1	A1A	ON	ON
AOA1	NON A1A	ON	ON
AOA2	NON A2A	ON	OFF[2]
A2	A2A	ON or OFF[1]	OFF[2]
A2*		ON or OFF[1]	OFF[2]
A3	A3E	ON or OFF[1]	OFF[2]

Notes: (1) For best performance consult the maker's handbook.
(2) If BFO cannot be switched off, it may be difficult to hear the Morse identification.

4.1.7 Directional Radiobeacons (RD)
In a very few places a directional signal is transmitted, the signal varying on the boat's position relative to the required bearing line. Automatic gain control should be switched off when using such devices, and a direction finding aerial should be turned to the position for maximum reception.

4.1.8 Errors in radio bearings
The various errors to which radio bearings are subject are described in Chapter 4 of *The Macmillan & Silk Cut Yachtsman's Handbook* (4.2.9). They fall into two categories — signal errors (caused by distance from the beacon, night or sky-wave effect particularly near sunset and sunrise, land effect or coastal refraction where the beam passes over high ground or along the coast, and synchronised transmissions of two beacons), and errors on board the boat (caused by quadrantal error due to magnetic objects re-radiating the incoming signal, compass error, the possibility of inadvertently taking a reciprocal bearing, and operating error due to inexperience or bad weather). Do not rely on radio bearings exclusively, unless three or more give an acceptable cocked hat.

4.1.9 Radiobeacons — calibration
A DF set can be calibrated for quadrantal error by taking simultaneous radio and visual bearings of a beacon on different headings. Alternatively radio bearings may be taken from a known position, and the bearing of the beacon taken from the chart. To save time it is helpful if the beacon transmits continuously. Beacons which provide a calibration service are shown in Table 4(2).

4.1.10 Half convergency
Unless taking radio bearings of a distant station, this is unlikely to worry the average yachtsman. But, since a radio wave follows a great circle, and meridians (depicted as parallel straight lines on a Mercator chart) in reality converge towards the poles, an allowance called half convergency has to be made when plotting bearings if the difference in longitude between station and vessel is more than about 3°. For practical purposes the values of half convergency in Table 4(1) can be used for distances up to 1000 n miles.

In north latitudes, for bearings taken from a boat — if the boat is East of the station, half convergency must be subtracted; if West of the station it must be added. For bearings provided by a DF station — if the boat is East of the station, half convergency must be added; if West of the station it must be subtracted. The opposite applies in the southern hemisphere.

TABLE 4(1) — HALF CONVERGENCY
Enter with difference in longitude between the station and the boat (along the top), and mid latitude between the station and the boat (down the side). The figures extracted are half convergency, in degrees.

Mid Lat	Difference in longitude (degrees)									
	3°	6°	9°	12°	15°	18°	21°	24°	27°	30°
5°	0.1	0.3	0.4	0.5	0.7	0.8	0.9	1.0	1.2	1.3
10°	0.3	0.5	0.8	1.0	1.3	1.6	1.8	2.1	2.3	2.6
15°	0.4	0.8	1.2	1.6	1.9	2.3	2.7	3.1	3.5	3.9
20°	0.5	1.0	1.5	2.1	2.6	3.1	3.6	4.1	4.6	5.1
25°	0.6	1.3	1.9	2.5	3.2	3.8	4.4	5.1	5.7	6.3
30°	0.7	1.5	2.2	3.0	3.7	4.5	5.2	6.0	6.7	7.5
35°	0.9	1.7	2.6	3.4	4.3	5.2	6.0	6.9	7.7	8.6
40°	1.0	1.9	2.9	3.9	4.8	5.8	6.7	7.7	8.7	9.6
45°	1.1	2.1	3.2	4.2	5.3	6.4	7.4	8.5	9.5	10.6
50°	1.1	2.3	3.4	4.6	5.7	6.9	8.0	9.2	10.3	11.5
55°	1.2	2.5	3.7	4.9	6.1	7.4	8.6	9.8	11.0	12.3
60°	1.3	2.6	3.9	5.2	6.5	7.8	9.1	10.4	11.7	13.0

Example. A boat in DR position 42°N, 2°W obtains a radio bearing from a powerful station in position 58°N, 7°E. The difference in longitude between the two positions is 9°. The mid latitude is 50°. From inspection, the half convergency is 3°.4. Because the boat is to the west of the station, this figure should be added to the bearing taken before it is plotted on the chart.

4.2 POSITION FIXING SYSTEMS

4.2.1 Types of systems
Four systems are available for yachtsmen: Decca, Loran-C and Omega (all hyperbolic systems) and satellite navigation from the American Transit satellites. Each has its merits for particular applications, and they are described in *The Macmillan & Silk Cut Yachtsman's Handbook* (4.5).

The Decca Yacht Navigator uses the established chains of Decca Navigator transmitting stations, and provides good coverage in NW Europe and in some other parts of the world (but not the Mediterranean). Loran-C covers the United States and much of the northern hemisphere, but coverage of European waters is poor. Omega gives extensive global coverage, but the accuracy is not so good as Decca or Loran-C. Transit satellites give worldwide coverage, but a fix can only be obtained when there is a suitable satellite pass – about every 1½ hours in British waters. Navstar GPS will give better coverage and accuracy.

4.2.2 Waypoints
For use with position fixing systems, over 2000 waypoints are given in Chapter 10 – for individual harbours (at the start of 'Navigation' in each case), in section 4 of each Area (where they are underlined in the lists of 'Lights, Fog Signals and Waypoints'), for the Solent area (10.2.9), and for cross-Channel passages (10.1.7).

Latitudes and longitudes are normally stated to one-hundredth of a minute, as taken from a large scale chart. But it should be realised that a chart using a different datum or based on another survey may give a slightly different position. Charts may contain small errors, just like the read-out from an electronic instrument.

Electronic systems are only aids to navigation and are subject to fixed and variable errors – or sometimes total failure. It is essential to maintain a DR plot, not only as a stand-by but to make sure that the boat's track is well clear of all dangers. Take great care when using waypoints on shore.

4.3 RADIO DIRECTION FINDING AND RADIO LIGHTHOUSES

4.3.1 Principle of operation — radio direction finding
Radio direction finding stations are shown on charts by the letters 'RG', and are equipped with apparatus to determine the direction of signals transmitted by a vessel. The vessel calls the station, and is requested to transmit a series of long dashes, followed by her call sign. There are no stations in the United Kingdom except as shown in 4.3.2, where the procedure is modified.

4.3.2 VHF emergency DF service
The stations below operate a VHF DF service for emergency use only. They are controlled by a Coastguard MRCC or MRSC as shown in brackets. On request from a yacht in distress, the station transmits her bearing *from the DF site*. Watch is kept on Ch 16. A yacht should transmit on Ch 16 (distress only) or on Ch 67 (Ch 82 for Jersey, and Ch 11 for French stations) to allow the station to obtain the bearing, which is transmitted on the same frequency.

Station	Control	Position
St Mary's	(Falmouth)	49°55'.7N 6°18'.2W
Pendeen	(Falmouth)	50°08'.1N 5°38'.2W
Pendennis	(Falmouth)	50°08'.7N 5°02'.7W
Rame Head	(Brixham)	50°19'.0N 4°13'.1W
Berry Head	(Brixham)	50°23'.9N 3°29'.0W
Grove Point	(Portland)	50°32'.9N 2°25'.2W
Stenbury Down	(Solent)	50°36'.8N 1°14'.5W
Selsey Bill	(Solent)	50°43'.8N 0°48'.1W
Newhaven	(Solent)	50°46'.9N 0°03'.1E
Fairlight	(Dover)	50°52'.2N 0°38'.8E
South Foreland	(Dover)	51°08'.4N 1°22'.4E
North Foreland	(Dover)	51°22'.5N 1°26'.8E
Bawdsey	(Thames)	51°59'.5N 1°24'.6E
Trimingham	(Yarmouth)	52°54'.5N 1°20'.7E
Easington	(Humber)	53°39'.1N 0°05'.9E
Flamborough	(Humber)	54°07'.0N 0°05'.0W
Whitby	(Humber)	54°29'.4N 0°36'.2W
Tynemouth	(Tyne/Tees)	55°01'.1N 1°24'.9W
Fife Ness	(Forth)	56°16'.7N 2°35'.2W
Inverbervie	(Aberdeen)	56°51'.1N 2°15'.7W
Windyheads Hill	(Moray)	57°38'.9N 2°14'.5W
Compass Head	(Shetland)	59°52'.0N 1°16'.3W
Thrumster	(Pentland)	58°23'.5N 3°07'.2W
Dunnet Head	(Pentland)	58°40'.3N 3°22'.5W
Sandwick	(Stornoway)	58°12'.6N 6°21'.2W
Tiree	(Oban)	56°30'.3N 6°57'.8W
Kilchiaran	(Clyde)	56°46'.0N 6°27'.1W
Snaefell	(Ramsey)	54°15'.8N 4°27'.6W
Great Ormes Hd	(Holyhead)	53°20'.0N 3°51'.2W
Mynydd Rhiw	(Holyhead)	52°50'.0N 4°37'.7W
St Anns Head	(Milford Haven)	51°41'.0N 5°10'.5W
Hartland	(Hartland)	51°01'.1N 4°31'.3W
Trevose Head	(Falmouth)	50°32'.9N 5°01'.9W
Orlock Point	(Belfast)	54°40'.4N 5°35'.0W
West Torr	(Belfast)	55°11'.9N 6°05'.6W
Guernsey	—	49°26'.3N 2°35'.8W
Jersey	—	49°10'.8N 2°14'.3W
Etel	(CROSS)	47°39'.8N 3°12'.0W
Créac'h	(CROSS)	48°27'.6N 5°07'.7W
Roches Douvres	(CROSS)	49°06'.5N 2°48'.8W
Jobourg	(CROSS)	49°41'.1N 1°54'.6W
Gris-Nez	(CROSS)	50°52'.1N 1°35'.0E

For 4.3.3, details of VHF Radio Lighthouses, see p. 61–62.

Position fixing systems 41

FIG. 4(1) DECCA CHAINS — WESTERN EUROPE

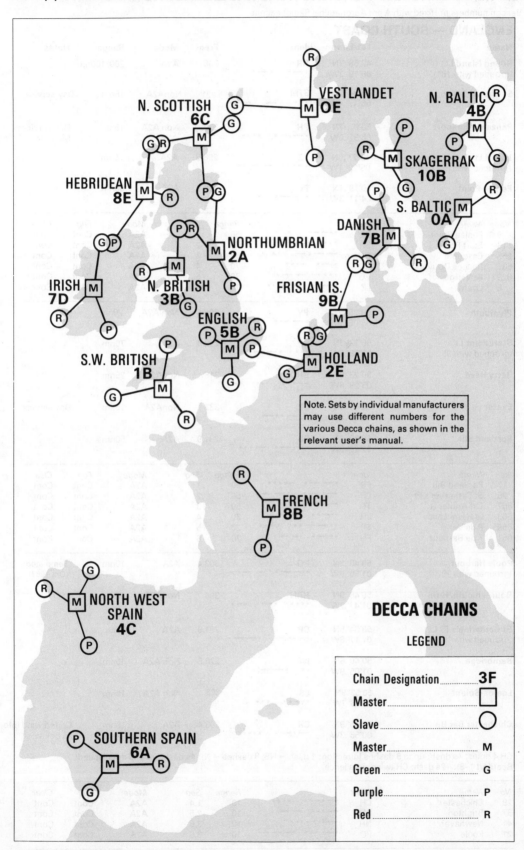

TABLE 4(2) MARINE AND AERONAUTICAL RADIOBEACONS

Note: Beacon numbers prefixed with A are Aeronautical Radiobeacons

ENGLAND — SOUTH COAST

No	Name	Lat/Long	Ident	Freq	Mode	Range	Notes
1	Round Island Lt (grouped with 181)	49°58'.7N 06°19'.3W	RR	308	A2A	200/100nm	
A1	St Mary's, Scilly	49°54'.8N 06°17'.4W	STM	321	NonA2A	15nm	Day service
A2	Penzance Heliport	50°07'.7N 05°31'.0W	PH	333	Non A2A	15nm	Day service: Mon–Sat
5	Lizard Lt (grouped with 9)	49°57'.6N 05°12'.1W	LZ	298.8	A2A	70nm	
9	Penlee Point	50°19'.1N 04°11'.3W	PE	298.8	A2A	50nm	

No	Name	Ident	Range	Seq	Mode	Fog	Clear
9	Penlee Point	PE	50	1	A2A	Cont	Cont
13	Start Point	SP	70	2	A2A	Cont	Cont
255	Casquets	QS	50	3	A2A	Cont	Cont
635	Roches Douvres	RD	70	4	A2A	Cont	Cont
623	Ile Vierge	VG	70	5	A2A	Cont	Cont
5	Lizard	LZ	70	6	A2A	Cont	Cont

No	Name	Lat/Long	Ident	Freq	Mode	Range	Notes
A6	Plymouth	50°25'.4N 04°06'.7W	PY	396.5	Non A2A	20nm	
13	Start Point Lt (grouped with 9)	50°13'.3N 03°38'.5W	SP	298.8	A2A	70nm	
A10	Berry Head	50°23'.9N 03°29'.6W	BHD	318	NonA2A	25nm	
A14	Exeter	50°45'.1N 03°17'.6W	EX	337	NonA2A	15nm	Day service
17	Portland Bill	50°30'.8N 02°27'.3W	PB	291.9	A2A	50nm	

No	Name	Ident	Range	Seq	Mode	Fog	Clear
17	Portland Bill	PB	50	1	A2A	Cont	Cont
25	St Catherine's Pt	CP	50	2	A2A	Cont	Cont
667	C d'Antifer	TI	50	3	A2A	Cont	Cont
663	Le Havre Lanby	LH	30	4	A2A	Cont	Cont
659	P de Ver	ÉR	20	5	A2A	Cont	Cont
651	P de Barfleur	FG	70	6	A2A	Cont	Cont

No	Name	Lat/Long	Ident	Freq	Mode	Range	Notes
21	Poole Harbour (grouped with 29)	50°40'.9N 01°56'.7W	PO	303.4	A2A	10nm	Temp inop (Oct 1987)
A18	Bournemouth/Hurn	50°48'.0N 01°43'.7W	HRN	394	NonA2A	35nm	
25	St Catherine's Pt Lt (grouped with 17)	50°34'.5N 01°17'.8W	CP	291.9	A2A	50nm	
A26	Bembridge	50°40'.6N 01°05'.9W	IW	276.5	Non A2A	15nm	
A30	Lee-on-Solent	50°52'.9N 01°06'.7W	LS	323	Non A2A	10nm	
29	Chichester Bar Bn	50°45'.9N 00°56'.3W	CH	303.4	A2A	10nm	Coded wind info. as below

CH 4 times: 4s dash: up to 8 dashes (direction; 1 dash — NE, 8 dashes — N): 4s dash: up to 8 dots (speed: Beaufort 1–8): 12s dash: CH twice: 5s silence.

No	Name	Ident	Range	Seq	Mode	Fog	Clear
29	Chichester	CH	10	1,4	A2A	Cont	Cont
37	Brighton	BM	10	2,5	A2A	Cont	Cont
41	Newhaven	NH	10	3,6	A2A	Cont	Cont
21	Poole	PO	10	3,6	A2A	Cont	Cont

Marine and Aeronautical Radiobeacons

No	Name	Lat/Long	Ident		Freq	Mode	Range	Notes
33	Nab Tower Lt	50°40'.0N 00°57'.1W	NB	—· —···	312.6	A2A	10nm	

No	Name	Ident		Range	Seq	Mode	Fog	Clear
33	Nab	NB	—· —···	10	1,3,5	A2A	Cont	Cont
647	Cherbourg	RB	·—· —···	20	2,4,6	A2A	Cont	Cont

No	Name	Lat/Long	Ident		Freq	Mode	Range	Notes
A38	Shoreham	50°49'.9N 00°17'.6W	SHM	··· ···· ——	332	Non A2A	10nm	Day service
37	Brighton Marina (grouped with 29)	50°48'.7N 00°05'.9W	BM	—··· ——	303.4	A2A	10nm	
41	Newhaven (grouped with 29)	50°46'.9N 00°03'.5E	NH	—· ····	303.4	A2A	10nm	
45	Royal Sovereign Lt (grouped with 687)	50°43'.4N 00°26'.1E	RY	·—· —·——	310.3	A2A	50nm	
A42	Lydd	50°58'.2N 00°57'.3E	LYX	·—·· —·—— —··—	397	Non A2A	15nm	
49	Dungeness Lt (grouped with 687)	50°54'.8N 00°58'.7E	DU	—·· ··—	310.3	A2A	30nm	
53	South Foreland Lt (grouped with 61)	51°08'.4N 01°22'.4E	SD	··· —··	305.7	A2A	30nm	
57	North Foreland Lt	51°22'.5N 01°26'.8E	NF	—· ··—·	301.1	A2A	50nm	
61	Falls Lt V	51°18'.1N 01°48'.5E	FS	··—· ···	305.7	A2A	50nm	

No	Name	Ident		Range	Seq	Mode	Fog	Clear
61	Falls Lt V	FS	··—· ···	50	1	A2A	Cont	Cont
701	W Hinder Lt V	WH	·—— ····	20	3	A2A	Cont	Cont
713	Oostende	OE	——— ·	30	4	A2A	Cont	Cont
691	Calais Main Lt	CS	—·—· ···	20	5	A2A	Cont	Cont
53	S Foreland Lt	SD	··· —··	30	6	A2A	Cont	Cont

ENGLAND — EAST COAST

No	Name	Lat/Long	Ident		Freq	Mode	Range	Notes
A46	Southend	51°34'.5N 00°42'.1E	SND	··· —·· —··	362.5	Non A2A	20nm	
69	Sunk Lt F	51°51'.0N 01°35'.0E	UK	··— —·—	312.6	A2A	10nm	
73	Outer Gabbard Lt V (grouped with 77)	51°59'.4N 02°04'.6E	GA	——· ·—	287.3	A2A	50nm	
77	Smith's Knoll Lt V	52°43'.5N 02°18'.0E	SK	··· —·—	287.3	A2A	50nm	

No	Name	Ident		Range	Seq	Mode	Fog	Clear
77	Smith's Knoll Lt V	SK	··· —·—	50	1	A2A	Cont	Cont
721	Goeree Lt	GR	——· ·—·	50	2	A2A	Cont	Cont
85	Dudgeon Lt V	LV	·—·· ···—	50	3	A2A	Cont	Cont
73	Outer Gabbard Lt V	GA	——· ·—	50	4	A2A	Cont	Cont
81	Cromer	CM	—·—· ——	50	5	A2A	Cont	Cont
709	N Hinder Lt V	NR	—· ·—·	50	6	A2A	Cont	Cont

BEACON SEQUENCE NUMBERS   Commence transmission at the following minutes past the hour:

1 00 06 12 18 24 30 36 42 48 54	**2** 01 07 13 19 25 31 37 43 49 55	**3** 02 08 14 20 26 32 38 44 50 56
4 03 09 15 21 27 33 39 45 51 57	**5** 04 10 16 22 28 34 40 46 52 58	**6** 05 11 17 23 29 35 41 47 53 59

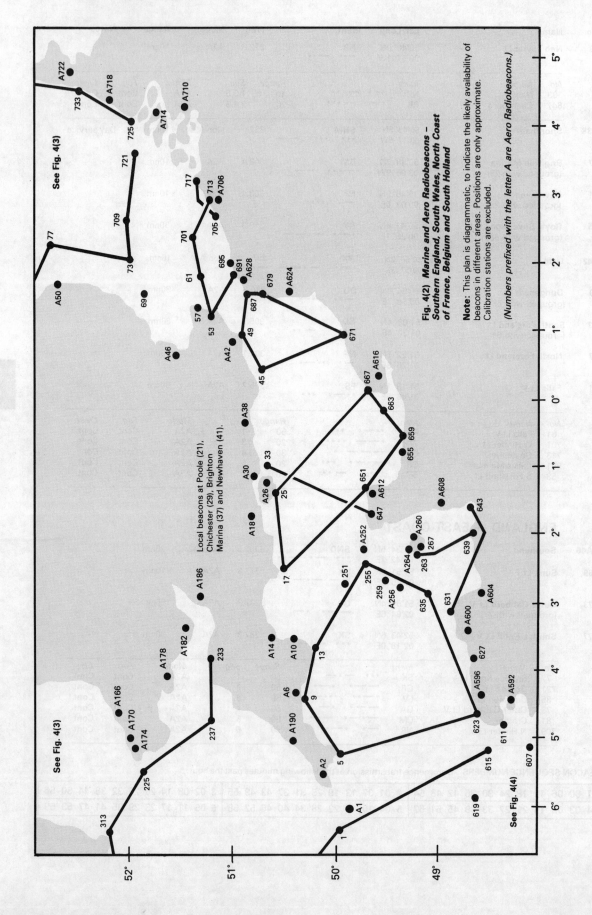

Fig. 4(2) *Marine and Aero Radiobeacons – Southern England, South Wales, North Coast of France, Belgium and South Holland*

Note: This plan is diagrammatic, to indicate the likely availability of beacons in different areas. Positions are only approximate. Calibration stations are excluded.

(Numbers prefixed with the letter A are Aero Radiobeacons.)

Marine and Aeronautical Radiobeacons

No	Name	Lat/Long	Ident		Freq	Mode	Range	Notes
A50	Great Yarmouth/North Denes	52°38'.2N 01°43'.5E	ND	— · — · ·	397	Non A2A	15nm	
81	Cromer Lt (grouped with 77)	52°55'.5N 01°19'.1E	CM	— · — · — —	287.3	A2A	50nm	
85	Dudgeon Lt V (grouped with 77)	53°16'.6N 01°17'.0E	LV	· — · · · · · —	287.3	A2A	50nm	
A54	Ottringham	53°41'.9N 00°06'.1W	OTR	— — — — · — ·	335	Non A2A	50nm	
93	Humber Lt V (calibration station)	53°36'.7N 00°21'.6E	MB	— — — · · ·	312.6	A2A	5nm	Cont from 1 hour after sunrise to 1 hr before sunset
97	Flamborough Head Lt (grouped with 109)	54°07'.0N 00°04'.9W	FB	· · — · — · · ·	303.4	A2A	70nm	
A58	Teeside	54°33'.6N 01°20'.0W	TD	— — · ·	347.5	Non A2A	25nm	
101	Heugh Lt	54°41'.8N 01°10'.5W	HS	· · · · · · ·	294.2	A2A	20nm	
109	Souter Lt	54°58'.2N 01°21'.8W	SJ	· · · · — — —	303.4	A2A	70nm	
	Calibration station		PT	· — — · —	312.6	A2A	5nm	Cont from 1 hour after sunrise to 1 hr before sunset

No	Name	Ident		Range	Seq	Mode	Fog	Clear
109	Souter Lt	SJ	· · · · — — —	70	1,4	A2A	Cont	Cont
97	Flamborough Hd	FB	· · — · — · · ·	70	2	A2A	Cont	Cont
113	Longstone	LT	· — · · —	20	5	A2A	Cont	Cont
121	Isle of May	LM	· — · · — —	100/70	6	A2A	Cont	Cont

No	Name	Lat/Long	Ident		Freq	Mode	Range	Notes
113	Longstone Lt (Farne I) (grouped with 109)	55°38'.6N 01°36'.5W	LT	· — · · —	303.4	A2A	20nm	

SCOTLAND — EAST COAST

No	Name	Lat/Long	Ident		Freq	Mode	Range	Notes
A70	Edinburgh	55°58'.7N 03°17'.0W	EDN	· — · · — ·	341	Non A2A	35nm	
117	Inchkeith Lt	56°02'.0N 03°08'.1W	NK	— · — · —	296.5	A2A	10nm	
121	Isle of May (grouped with 109)	56°11'.1N 02°33'.3W	LM	· — · · — —	303.4	A2A	100/70nm	
125	Fife Ness Lt (grouped with 133)	56°16'.7N 02°35'.1W	FP	· · — · · — — ·	310.3	A2A	50nm	
A74	Leuchars	56°22'.3N 02°51'.4W	LU	· — · · · · —	255.5	Non A2A	100nm	Subject to night interference
A78	Dundee	56°27'.3N 03°06'.6W	DND	— · · — · — · ·	394	Non A2A	25nm	
133	Girdle Ness Lt	57°08'.3N 02°02'.8W	GD	— — · — · ·	310.3	A2A	50nm	

No	Name	Ident		Range	Seq	Mode	Fog	Clear
133	Girdle Ness	GD	— — · — · ·	50	1,4	A2A	Cont	Cont
125	Fife Ness	FP	· · — · · — — ·	50	2,5	A2A	Cont	Cont

BEACON SEQUENCE NUMBERS Commence transmission at the following minutes past the hour:

1	00 06 12 18 24 30 36 42 48 54	2	01 07 13 19 25 31 37 43 49 55	3	02 08 14 20 26 32 38 44 50 56
4	03 09 15 21 27 33 39 45 51 57	5	04 10 16 22 28 34 40 46 52 58	6	05 11 17 23 29 35 41 47 53 59

Radio Navigational Aids

No	Name	Lat/Long	Ident	Freq	Mode	Range	Notes
A82	Aberdeen	57°16'.1N 02°14'.7W	ADN	377	Non A2A	50nm	
A86	Scotstown Head	57°33'.6N 01°48'.9W	SHD	383	Non A2A	80nm	
137	Kinnairds Head	57°41'.9N 02°00'.1W	KD	291.9	A2A	100/70nm	

No	Name	Ident		Range	Seq	Mode	Fog	Clear
137	Kinnairds Hd	KD		100/70	1,4	A2A	Cont	Cont
145	N Ronaldsay	NR		100/70	2	A2A	Cont	Cont
141	Stroma (Swilkie Pt)	OM		50	3,6	A2A	Cont	Cont
149	Sumburgh Hd	SB		70	5	A2A	Cont	Cont

No	Name	Lat/Long	Ident	Freq	Mode	Range	Notes
A90	Kinloss	57°39'.0N 03°35'.0W	KS	370	Non A2A	50nm	
A94	Wick	58°26'.8N 03°03'.7W	WIK	344	Non A2A	40nm	
141	Stroma, Swilkie Pt Lt (grouped with 137)	58°41'.8N 03°06'.9W	OM	291.9	A2A	50nm	
A98	Kirkwall, Orkney	58°57'.6N 02°54'.6W	KW	395	Non A2A	30nm	
145	North Ronaldsay Lt (grouped with 137)	59°23'.4N 02°22'.8W	NR	291.9	A2A	100/70nm	
A102	Sumburgh	59°52'.1N 01°16'.3W	SUM	351	Non A2A	75nm	
149	Sumburgh Hd Lt (grouped with 137)	59°51'.3N 01°16'.4W	SB	291.9	A2A	70nm	
153	Bressay Lt, Shetland I	60°07'.2N 01°07'.2W	BY	287.3	A2A	30nm	
A106	Lerwick/Tingwall	60°11'.3N 01°14'.7W	TL	376	Non A2A	25nm	
A110	Scatsa	60°27'.7N 01°12'.8W	SS	315.5	Non A2A	25nm	
A114	Unst	60°44'.3N 00°49'.2W	UT	258	Non A2A	20nm	
157	Muckle Flugga (N Unst Lt) (grouped with 165)	60°51'.3N 00°53'.0W	MF	298.8	A2A	150/70nm	
161	Sule Skerry Lt (grouped with 165)	59°05'.1N 04°24'.3W	LK	298.8	A2A	100/70nm	

SCOTLAND — WEST COAST

No	Name	Lat/Long	Ident	Freq	Mode	Range	Notes
A118	Dounreay/Thurso	58°34'.9N 03°43'.6W	DO	364.5	Non A2A	15nm	
165	Cape Wrath Lt	58°37'.5N 04°59'.9W	CW	298.8	A2A	50nm	

No	Name	Ident		Range	Seq	Mode	Fog	Clear
165	Cape Wrath	CW		50	1	A2A	Cont	Cont
169	Butt of Lewis	BL		150	2,5	A2A	Cont	Cont
157	Muckle Flugga	MF		150/70	3,6	A2A	Cont	Cont
161	Sule Skerry	LK		100/70	4	A2A	Cont	Cont

Marine and Aeronautical Radiobeacons

No	Name	Lat/Long	Ident	Freq	Mode	Range	Notes
169	Butt of Lewis Lt (grouped with 165)	58°30'.9N 06°15'.7W	BL —··· ·—··	299.8	A2A	150nm	
A122	Stornoway	58°17'.2N 06°20'.6W	STN ··· — —·	669.5	Non A2A	60nm	
173	Eilean Glas Lt (grouped with 189)	57°51'.4N 06°38'.4W	LG ·—·· ——·	294.2	A2A	50nm	
177	Hyskeir Lt, Oigh Sgeir (grouped with 189)	56°58'.1N 06°40'.8W	OR ——— ·—·	294.2	A2A	50nm	
A124	Barra	57°01'.4N 07°26'.4W	BRR —··· ·—· ·—·	316	Non A2A	20nm	Occas
181	Barra Head Lt, Berneray	56°47'.1N 07°39'.2W	BD —··· —··	308	A2A	200/70nm	

No	Name	Ident		Range	Seq	Mode	Fog	Clear
181	Barra Hd	BD	—··· —··	200/70	1	A2A	Cont	Cont
337	Tory I	TY	— —·——	100/70	2	A2A	Cont	Cont
341	Eagle I	GL	——· ·—··	200/100	3	A2A	Cont	Cont
301	Mizen Hd	MZ	—— ——··	200/100	4	A2A	Cont	Cont
1	Round I	RR	·—· ·—·	200/100	5	A2A	Cont	Cont
615	Pte de Creac'h	CA	—·—· ·—	100	6	A2A	Cont	Cont

No	Name	Lat/Long	Ident	Freq	Mode	Range	Notes
A126	Connel/Oban	56°27'.0N 05°24'.0W	CNL —·—· —· ·—··	404	Non A2A	15nm	
185	Rhinns of Islay Lt (grouped with 189)	55°40'.4N 06°30'.7W	RN ·—· —·	294.2	A2A	70nm	
189	Pladda Lt, Arran I	55°25'.5N 05°07'.1W	DA —·· ·—	294.2	A2A	30nm	

No	Name	Ident		Range	Seq	Mode	Fog	Clear
189	Pladda	DA	—·· ·—	30	1	A2A	Cont	Cont
329	Mew I	MW	—— ·——	50	2	A2A	Cont	Cont
333	Altacarry Hd	AH	·— ····	50	3	A2A	Cont	Cont
185	Rhinns of Islay	RN	·—· —·	70	4	A2A	Cont	Cont
177	Hyskeir	OR	——— ·—·	50	5	A2A	Cont	Cont
173	Eilean Glas	LG	·—·· ——·	50	6	A2A	Cont	Cont

No	Name	Lat/Long	Ident	Freq	Mode	Range	Notes
193	Cloch Pt Lt (calibration station)	55°56'.5N 04°52'.7W	CL —·—· ·—··	308		8nm	On request Tel: (0475) 26221 (6 hrs notice)
A138	Turnberry	55°18'.8N 04°47'.0W	TRN — ·—· —·	355	Non A2A	25nm	
A142	New Galloway	55°10'.6N 04°10'.0W	NGY —· ——· —·——	399	Non A2A	35nm	

ENGLAND (WEST COAST), ISLE OF MAN, WALES

No	Name	Lat/Long	Ident	Freq	Mode	Range	Notes
197	Point of Ayre High Light (grouped with 217)	54°24'.9N 04°22'.0W	PY ·——· —·——	301.1	A2A	50nm	
201	Douglas, Victoria Pier Lt (grouped with 209)	54°08'.8N 04°28'.0W	DG —·· ——·	287.3	A2A	50nm	
A146	Carnane	54°08'.5N 04°29'.4W	CAR —·—· ·— ·—·	366.5	Non A2A	25nm	
A150	IOM/Ronaldsway	54°05'.1N 04°36'.4W	RWY ·—· ·—— —·——	359	Non A2A	20nm	Day service

BEACON SEQUENCE NUMBERS Commence transmission at the following minutes past the hour:

1	00 06 12 18 24 30 36 42 48 54	2	01 07 13 19 25 31 37 43 49 55	3	02 08 14 20 26 32 38 44 50 56
4	03 09 15 21 27 33 39 45 51 57	5	04 10 16 22 28 34 40 46 52 58	6	05 11 17 23 29 35 41 47 53 59

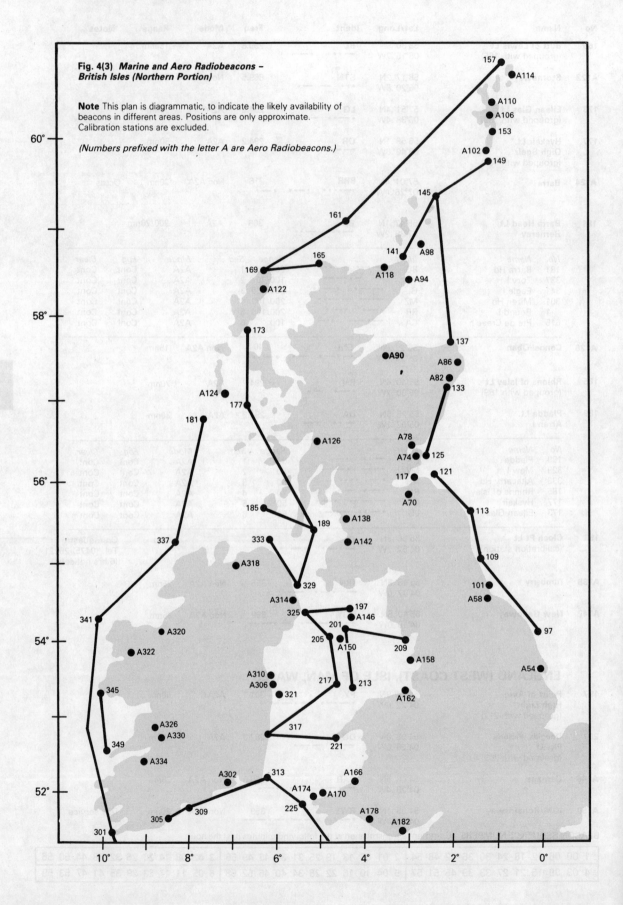

Marine and Aeronautical Radiobeacons

No	Name	Lat/Long	Ident		Freq	Mode	Range	Notes
205	Cregneish, IOM (grouped with 217)	54°03'.9N 04°45'.9W	CN	—·—· —·	301.1	A2A	50nm	
209	Walney Island Lt	54°02'.9N 03°10'.5W	FN	··—· —·	287.3	A2A	30nm	

No	Name	Ident		Range	Seq	Mode	Fog	Clear
209	Walney I	FN	··—· —·	30	1,4	A2A	Cont	Cont
213	Point Lynas	PS	·——· ···	40	2,5	A2A	Cont	Cont
201	Douglas	DG	—·· ——·	50	3,6	A2A	Cont	Cont

No	Name	Lat/Long	Ident		Freq	Mode	Range	Notes
A158	Blackpool	53°46'.2N 02°59'.3W	BPL	—··· ·—·· ·—··	278.5	Non A2A	15nm	Day service
A162	Wallasey	53°23'.4N 03°08'.0W	WAL	·—— ·— ·—··	331.5	Non A2A	50nm	
213	Point Lynas Lt (grouped with 209) Calibration station	53°25'.0N 04°17'.3W	PS	·——· ···	287.3 310.3	A2A A2A	40nm 5nm	H24
217	Skerries Lt	53°25'.3N 04°36'.5W	SR	··· ·—·	301.1	A2A	50nm	

No	Name	Ident		Range	Seq	Mode	Fog	Clear
217	Skerries	SR	··· ·—·	50	1	A2A	Cont	Cont
221	Bardsey Lt	IB	·· —···	30	2	A2A	Cont	Cont
317	Wicklow Hd	WK	·—— —·—	70	3	A2A	Cont	Cont
205	Cregneish	CN	—·—· —·	50	4	A2A	Cont	Cont
197	Point of Ayre	PY	·—. —·——	50	5	A2A	Cont	Cont
325	South Rock Lt F	SU	··· ··—	50	6	A2A	Cont	Cont

No	Name	Lat/Long	Ident		Freq	Mode	Range	Notes
221	Bardsey Lt (grouped with 217)	52°45'.0N 04°47'.9W	IB	·· —···	301.1	A2A	30nm	
A166	Aberporth	52°07'.0N 04°33'.6W	AP	·— ·—··	370.5	Non A2A	20nm	
A170	Strumble	52°00'.5N 05°01'.0W	STU	··· — ··—	400	Non A2A	40nm	
A174	Brawdy	51°53'.4N 05°07'.3W	BY	—··· —·——	414.5	Non A2A	30nm	
225	South Bishop Lt (grouped with 309)	51°51'.1N 05°24'.6W	SB	··· —···	296.5	A2A	50nm	
A178	Swansea	51°36'.1N 04°03'.9W	SWN	··· ·—— —·	320.5	Non A2A	15nm	Day service
A182	Cardiff/Rhoose	51°23'.6N 03°20'.2W	CDF	—·—· —·· ··—·	363.5	Non A2A	20nm	
233	Nash Point Lt (grouped with 309)	51°24'.0N 03°33'.1W	NP	—· ·——·	296.5	A2A	50nm	
A186	Weston	51°20'.3N 02°56'.3W	WS	·—— ···	390.5	Non A2A	15nm	Day service
235	Lynmouth Foreland Lt Calibration station	51°14'.7N 03°47'.1W	FP	··—· ·——·	312.6	A2A	5nm	Cont from 1 hr after sunrise to 1 hr before sunset
237	Lundy I, South Lt (grouped with 309)	51°09'.7N 04°39'.3W	LS	·—·· ···	296.5	A2A	50nm	Reliable 234°–216°

BEACON SEQUENCE NUMBERS Commence transmission at the following minutes past the hour:

1	00 06 12 18 24 30 36 42 48 54	2	01 07 13 19 25 31 37 43 49 55	3	02 08 14 20 26 32 38 44 50 56
4	03 09 15 21 27 33 39 45 51 57	5	04 10 16 22 28 34 40 46 52 58	6	05 11 17 23 29 35 41 47 53 59

No	Name	Lat/Long	Ident	Freq	Mode	Range	Notes
A190	St Mawgan	50°26'.8N 04°59'.6W	SM ··· — —	356.5	Non A2A	50nm	

CHANNEL ISLANDS

No	Name	Lat/Long	Ident	Freq	Mode	Range	Notes
251	Channel Lt V	49°54'.4N 02°53'.7W	CR —·—· ·—·	287.3	A2A	10nm	
255	Casquets Lt (grouped with 9)	49°43'.4N 02°22'.5W	QS ——·— ···	298.8	A2A	50nm	
A252	Alderney	49°42'.6N 02°11'.9W	ALD ·— ·—·· —··	383	Non A2A	50nm	Day service
259	Castle Breakwater, St Peter Port[1]	49°27'.4N 02°31'.4W	GY ——· —·——	285	A2A	10nm	

[1] Synchronised with horn for distance finding. Horn begins simultaneously with 27 sec long dash after the four GY ident signals. Time in secs from start of long dash until horn is heard, multiplied by 0.18, gives distance in nm

A256	Guernsey	49°26'.1N 02°38'.3W	GUR ——· ··— ·—·	361	Non A2A	30nm	Day service
263	La Corbière[1,2] (grouped with 639)	49°10'.9N 02°14'.9W	CB —·—· —···	305.7	A2A	20nm	

[1] For details of coded wind information see 7.2.12. [2] Synchronised for distance finding. Horn blast (Morse 'C') begins with end of 18 sec long dash. Each pip heard before the blast corresponds to a distance of 335m from the light.

267	St Helier Harbour	49°10'.6N 02°07'.5W	EC · —·—·	287.3	A2A	10nm	
A260	Jersey East	49°13'.2N 02°02'.1W	JEY ·——— · —·——	367	Non A2A	75nm	
A264	Jersey West	49°12'.4N 02°13'.3W	JW ·——— ·——	329	Non A2A	25nm	

IRELAND — SOUTH AND EAST COASTS

No	Name	Lat/Long	Ident	Freq	Mode	Range	Notes
301	Mizen Head Lt (grouped with 181)	51°27'.0N 09°48'.8W	MZ —— ——··	308	A2A	200/100nm	070° 475m from Lt.
305	Old Head of Kinsale Lt (grouped with 309)	51°36'.3N 08°32'.0W	OH ——— ····	296.5	A2A	50nm	
309	Ballycotton Lt	51°49'.5N 07°59'.0W	BN —··· —·	296.5	A2A	50nm	
	Calibration station		BC —··· —·—·	312.6	A2A	5nm	On request

No	Name	Ident		Range	Seq	Mode	Fog	Clear
309	Ballycotton Lt	BN	—··· —·	50	1	A2A	Cont	Cont
305	Kinsale	OH	——— ····	50	2	A2A	Cont	Cont
237	Lundy	LS	·—·· ···	50	3	A2A	Cont	Cont
233	Nash Point	NP	—· ·——·	50	4	A2A	Cont	Cont
225	South Bishop	SB	··· —···	50	5	A2A	Cont	Cont
313	Tuskar Rock	TR	— ·—·	50	6	A2A	Cont	Cont

A302	Waterford	52°11'.8N 07°05'.3W	WTD ·—— — —··	368	Non A2A	25nm	
313	Tuskar Rock Lt (grouped with 309)	52°12'.1N 06°12'.4W	TR — ·—·	296.5	A2A	50nm	
317	Wicklow Hd Lt (grouped with 217)	52°57'.9N 05°59'.8W	WK ·—— —·—	301.1	A2A	70nm	
A306	Killiney	53°16'.2N 06°06'.3W	KLY —·— ·—·· —·——	378	Non A2A	50nm	
A310	Dublin/Rush	53°30'.6N 06°06'.8W	RSH ·—· ··· ····	326	Non A2A	30nm	
321	Kish Bank Lt	53°18'.7N 05°55'.4W	KH —·— ····	312.6	A2A	20nm	
	Calibration station		KH —·— ····	312.6	A2A	5nm	On request

Marine and Aeronautical Radiobeacons

No	Name	Lat/Long	Ident	Freq	Mode	Range	Notes
325	South Rock Lt F (grouped with 217)	54°24'.5N 05°21'.9W	SU ••• ••—	301.1	A2A	50nm	
A314	Belfast Harbour	54°37'.0N 05°52'.9W	HB •••• —•••	275	Non A2A	15nm	
329	Mew Island Lt (grouped with 189)	54°41'.9N 05°30'.7W	MW —— •——	294.2	A2A	50nm	
	Calibration station		MC —— —•—•	312.6	A2A	5nm	On request
333	Altacarry Head Lt (grouped with 189)	55°18'.1N 06°10'.2W	AH •— ••••	294.2	A2A	50nm	

IRELAND — WEST COAST

No	Name	Lat/Long	Ident	Freq	Mode	Range	Notes
A318	Eglinton/Londonderry	55°02'.7N 07°09'.2W	EGT •—• ——• —	328.5	Non A2A	25nm	Occasional
337	Tory Island Lt (grouped with 181)	55°16'.3N 08°14'.9W	TY — —•——	308	A2A	100/70nm	
341	Eagle Island Lt (grouped with 181)	54°17'.0N 10°05'.5W	GL ——• •—••	308	A2A	200/100nm	
A320	Sligo	54°16'.5N 8°36'.0W	SL ••• •—••	384	Non A2A	25 nm	
345	Slyne Head Lt	53°24'.0N 10°14'.0W	SN ••• —•	289.6	A2A	50nm	

No	Name	Ident		Range	Seq	Mode	Fog	Clear
345	Slyne Head	SN	••• —•	50	1,3,5	A2A	Cont	Cont
349	Loop Head	LP	•—•• •—•••	50	2,4,6	A2A	Cont	Cont

No	Name	Lat/Long	Ident	Freq	Mode	Range	Notes
A322	Carnmore	53°18'.0N 08°57'.0W	CRN —•—• •—• —•	321	Non A2A	——	
A326	Ennis	52°54'.3N 08°55'.6W	ENS • —• •••	371	Non A2A	25nm	
A330	Bunratty	52°41'.8N 08°49'.3W	BNY —••• —• —•——	352	Non A2A	100nm	
A334	Foynes	52°34'.0N 09°11'.7W	FOY ••—• ——— —•——	395	Non A2A	50nm	
349	Loop Head Lt (grouped with 345)	52°33'.6N 09°55'.9W	LP •—•• •—•••	289.6	A2A	50nm	

SPAIN — NORTH COAST

No	Name	Lat/Long	Ident	Freq	Mode	Range	Notes
543	Cabo Mayor Lt (grouped with 551)	43°29'.5N 03°47'.4W	MY —— —•——	296.5	A2A	50nm	
A548	Bilbao	43°19'.4N 02°58'.4W	BLO —••• •—•• ———	370	Non A2A	70nm	
547	Cabo Machichaco Lt (grouped with 551)	43°27'.4N 02°45'.1W	MA —— •—	296.5	A2A	100nm	Reliable sector 110°–220°
A552	San Sebastian	43°23'.3N 01°47'.7W	HIG •••• •• ——•	328	Non A2A	50nm	

BEACON SEQUENCE NUMBERS Commence transmission at the following minutes past the hour:

1	00 06 12 18 24 30 36 42 48 54	**2**	01 07 13 19 25 31 37 43 49 55	**3**	02 08 14 20 26 32 38 44 50 56
4	03 09 15 21 27 33 39 45 51 57	**5**	04 10 16 22 28 34 40 46 52 58	**6**	05 11 17 23 29 35 41 47 53 59

FRANCE

No	Name	Lat/Long	Ident		Freq	Mode	Range	Notes
A556	Biarritz	43°28'.2N 01°24'.2W	BZ	—··· —·— —···	341	A1A	35nm	
A560	Cazaux	44°33'.1N 01°07'.1W	CAA	—·—· ·— ·—	382	A1A	80nm	
551	Cap Ferret Lt	44°38'.8N 01°15'.0W	FT	··—· —	296.5	A2A	100nm	Temp inop (July 1987)

No	Name	Ident		Range	Seq	Mode	Fog	Clear
551	Cap Ferret	FT	··—· —	100	1,2	A2A	Cont	Cont
547	Cabo Machichaco	MA	—— ·—	100	3,4	A2A	Cont	Cont
543	Cabo Mayor	MY	—— —·——	50	5,6	A2A	Cont	Cont

No	Name	Lat/Long	Ident		Freq	Mode	Range	Notes
A564	Bordeaux/Merignac	44°55'.9N 00°33'.9W	BD	—··· —··	393	A1A	30nm	
A568	Cognac/Chateaubernard	45°40'.1N 00°18'.5W	CGC	—·—· ——· —·—·	354	A1A	75nm	
559	Pointe de la Coubre (grouped with 591)	45°41'.9N 01°13'.9W	LK	·—·· —·—	303.4	A2A	100nm	
563	La Rochelle, Tour Richelieu Lt	46°09'.0N 1°10'.3W	RE	·—· ·	291.9	A2A	5nm	Temp inop (Feb 1988)
571	Les Baleines Lt (Ile de Ré) (grouped with 591)	46°14'.7N 01°33'.6W	BN	—··· —·	303.4	A2A	50nm	
575	Les Sables d'Olonne Tour de la Chaume Lt	46°29'.6N 01°47'.8W	SO	··· ———	291.9	A2A	5nm	Temp inop (Mar 1987)
579	Ile d'Yeu Main Lt	46°43'.1N 02°22'.9W	YE	—·—— ·	312.6	A2A	70nm	Sequence 3
583	Ile du Pilier Lt	47°02'.6N 02°21'.5W	PR	·——· ·—·	298.8	A2A	10nm	Temp inop (Mar 1987)
587	St Nazaire, Pointe de St Gildas Lt (grouped with 603)	47°08'.1N 02°14'.7W	NZ	—· ——··	289.6	A2A	35nm	
A572	St Nazaire/Montoir	47°20'.0N 02°02'.6W	MT	—— —	398	A1A	50nm	
591	Belle Ile, Goulphar Lt	47°18'.7N 03°13'.6W	BT	—··· —	303.4	A2A	100nm	Temp inop (March 1987)

No	Name	Ident		Range	Seq	Mode	Fog	Clear
591	Belle Ile	BT	—··· —	100	2	A2A	Cont	Cont
607	Ile de Sein	SN	··· —·	70	3	A2A	Cont	Cont
559	Pte de la Coubre	LK	·—·· —·—	100	4	A2A	Cont	Cont
571	Les Baleines	BN	—··· —·	50	6	A2A	Cont	Cont

No	Name	Lat/Long	Ident		Freq	Mode	Range	Notes
A576	Vannes/Meucon	47°46'.2N 02°39'.6W	VA	···— ·—	342.5	A1A	25nm	
A580	Lorient/Lann-Bihoué	47°45'.7N 03°26'.4W	LOR	·—·· ——— ·—·	294.2	A1A	80nm	

BEACON SEQUENCE NUMBERS Commence transmission at the following minutes past the hour:

1 00 06 12 18 24 30 36 42 48 54	**2** 01 07 13 19 25 31 37 43 49 55	**3** 02 08 14 20 26 32 38 44 50 56
4 03 09 15 21 27 33 39 45 51 57	**5** 04 10 16 22 28 34 40 46 52 58	**6** 05 11 17 23 29 35 41 47 53 59

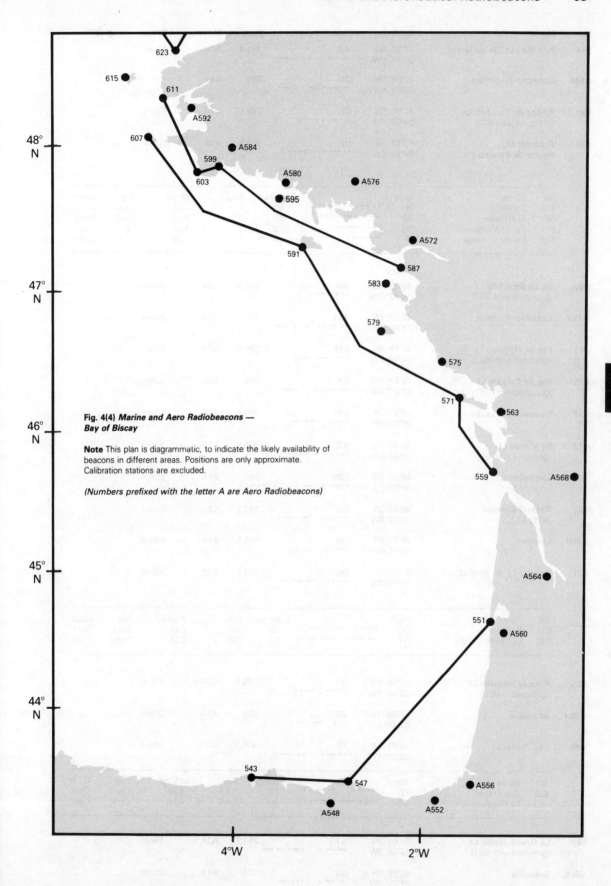

Fig. 4(4) *Marine and Aero Radiobeacons —
Bay of Biscay*

Note This plan is diagrammatic, to indicate the likely availability of beacons in different areas. Positions are only approximate. Calibration stations are excluded.

(Numbers prefixed with the letter A are Aero Radiobeacons)

54 Radio Navigational Aids

No	Name	Lat/Long	Ident		Freq	Mode	Range	Notes
595	Pen Men Lt, Ile de Groix	47°38'.9N 03°30'.5W	GX	— —・ —・・—	301.1	A1A	10nm	
A584	Quimper/Pluguffan	47°58'.1N 03°59'.8W	QR	— —・— ・—・	380	A1A		
599	Pointe de Combrit Lt (grouped with 603)	47°51'.9N 04°06'.7W	CT	—・—・ —	289.6	A2A	20nm	
603	Eckmühl Lt, Pointe de Penmarc'h	47°47'.9N 04°22'.4W	ÜH	・・— — ・・・・	289.6	A2A	50nm	

No	Name	Ident		Range	Seq	Mode	Fog	Clear
603	Eckmühl	ÜH	・・— — ・・・・	50	1	A2A	Cont	Cont
587	St Nazaire	NZ	—・ ——・・	35	2	A2A	Cont	Cont
611	Pte St Mathieu	SM	・・・ — —	20	3	A2A	Cont	Cont
599	Pte de Combrit	CT	—・—・ —	20	4	A2A	Cont	Cont

No	Name	Lat/Long	Ident		Freq	Mode	Range	Notes
607	Ile de Sein NW Lt (grouped with 591)	48°02'.7N 04°52'.0W	SN	・・・ —・	303.4	A2A	70nm	
A592	Lanvéoc, Poulmic	48°17'.1N 04°26'.0W	BST	—・・・ ・—・ —	316	A1A	80nm	
611	Pte St Mathieu Lt (grouped with 603)	48°19'.8N 04°46'.2W	SM	・・・ — —	289.6	A2A	20nm	
615	Pte de Créac'h Lt (grouped with 181)	48°27'.6N 05°07'.6W	CA	—・—・ ・—	308	A2A	100nm	
619	Ouessant SW Lanby	48°31'.7N 05°49'.1W	SW	・・・ ・——	294.2	A2A	10nm	
623	Ile Vierge Lt (grouped with 9)	48°38'.4N 04°34'.0W	VG	・・・— ——・	298.8	A2A	70nm	
A596	Landivisiau	48°32'.8N 04°08'.2W	LDV	・—・・ —・・ ・・・—	324	A1A	60nm	
627	Roscoff-Bloscon Jetty Lt	48°43'.3N 03°57'.6W	BC	—・・・ —・—・	287.3	A2A	10nm	
A600	Lannion	48°43'.3N 03°18'.4W	LN	・—・・ —・	345.5	A1A	50nm	
631	Rosédo Lt, Ile Bréhat	48°51'.5N 03°00'.3W	DO	—・・ ———	294.2	A2A	10nm	

No	Name	Ident		Range	Seq	Mode	Fog	Clear
631	Rosédo	DO	—・・ ———	10	1,5	A2A	Cont	Cont
643	Le Grand Jardin	GJ	——・ ・———	10	2,6	A2A	Cont	Cont

No	Name	Lat/Long	Ident		Freq	Mode	Range	Notes
635	Roches-Douvres Lt (grouped with 9)	49°06'.5N 02°48'.8W	RD	・—・ —・・	298.8	A2A	70nm	
A604	St Brieuc	48°34'.1N 02°46'.9W	SB	・・・ —・・・	353	A1A	25nm	
639	Cap Fréhel Lt	48°41'.1N 02°19'.1W	FÉ	・・—・ ・・—・・	305.7	A2A	20nm	

No	Name	Ident		Range	Seq	Mode	Fog	Clear
639	Cap Fréhel	FÉ	・・—・ ・・—・・	20	1,3,5	A2A	Cont	Cont
263	La Corbière	CB	—・—・ —・・・	20	2,4,6	A2A	Cont	Cont

No	Name	Lat/Long	Ident		Freq	Mode	Range	Notes
643	Le Grand Jardin Lt (grouped with 631)	48°40'.3N 02°04'.9W	GJ	——・ ・———	294.2	A2A	10nm	
A608	Granville	48°55'.1N 01°28'.9W	GV	——・ ・・・—	321	A1A	25nm	

Marine and Aeronautical Radiobeacons

No	Name	Lat/Long	Ident		Freq	Mode	Range	Notes
647	Cherbourg W Fort Lt (grouped with 33)	49°40'.5N 01°38'.9W	RB	•–• –•••	312.6	A2A	20nm	
651	Pte de Barfleur Lt (grouped with 17)	49°41'.9N 01°15'.9W	FG	••–• ––•	291.9	A2A	70nm	
655	Port en Bessin Rear Lt	49°21'.0N 00°45'.6W	BS	–••• •••	313.5	A2A	5nm	
A612	Cherbourg	49°38'.3N 01°22'.3W	MP	–– •––•	373	A1A	—	
659	Pte de Ver Lt (grouped with 17)	49°20'.5N 00°31'.1W	ÉR	••–•• •–•	291.9	A2A	20nm	
663	Le Havre Lanby (grouped with 17)	49°31'.7N 00°09'.8W	LH	•–•• ••••	291.9	A2A	30nm	Temp inop (Jan 1987)
A616	Le Havre/Octeville	49°35'.7N 00°11'.0E	LHO	•–•• •••• –––	346	A2A	15nm	
667	Cap d'Antifer Lt (grouped with 17)	49°41'.1N 00°10'.0E	TI	– ••	291.9	A2A	50nm	
671	Pte d'Ailly Lt (grouped with 687)	49°55'.1N 00°57'.6E	AL	•– •–••	310.3	A2A	50nm	
A620	Eu/Le Tréport	50°04'.1N 01°25'.9E	EU	• ••–	330	A1A	20nm	
A624	Le Touquet/Paris Plage	50°32'.2N 01°35'.4E	LT	•–•• –	358	A2A	20nm	
679	Cap d'Alprech Lt (grouped with 687)	50°41'.9N 01°33'.8E	PH	•––• ••••	310.3	A2A	20nm	
687	Cap Gris Nez Lt	50°52'.1N 01°35'.1E	GN	––• –•	310.3	A2A	30nm	

No	Name	Ident		Range	Seq	Mode	Fog	Clear
687	Cap Gris Nez	GN	––• –•	30	1,4	A2A	Cont	Cont
45	Royal Sovereign	RY	•–• –•––	50	2	A2A	Cont	Cont
671	Pte d'Ailly	AL	•– •–••	50	3	A2A	Cont	Cont
679	Cap d'Alprech	PH	•––• ••••	20	5	A2A	Cont	Cont
49	Dungeness	DU	–•• ••–	30	6	A2A	Cont	Cont

No	Name	Lat/Long	Ident		Freq	Mode	Range	Notes
691	Calais Main Lt (grouped with 61)	50°57'.7N 01°51'.3E	CS	–•–• •••	305.7	A2A	20nm	
A628	Saint Inglevert	50°53'.0N 01°44'.6E	ING	•• –• ––•	387.5	A1A	50nm	
695	Dunkerque Lanby	51°03'.1N 01°51'.8E	DK	–•• –•–	294.2	A2A	10nm	Temp inop
A632	Calais/Dunkerque	50°59'.8N 02°03'.3E	MK	–– –•–	275	A1A	15nm	

BEACON SEQUENCE NUMBERS Commence transmission at the following minutes past the hour:

1	00 06 12 18 24 30 36 42 48 54	**2**	01 07 13 19 25 31 37 43 49 55	**3**	02 08 14 20 26 32 38 44 50 56
4	03 09 15 21 27 33 39 45 51 57	**5**	04 10 16 22 28 34 40 46 52 58	**6**	05 11 17 23 29 35 41 47 53 59

56 Radio Navigational Aids

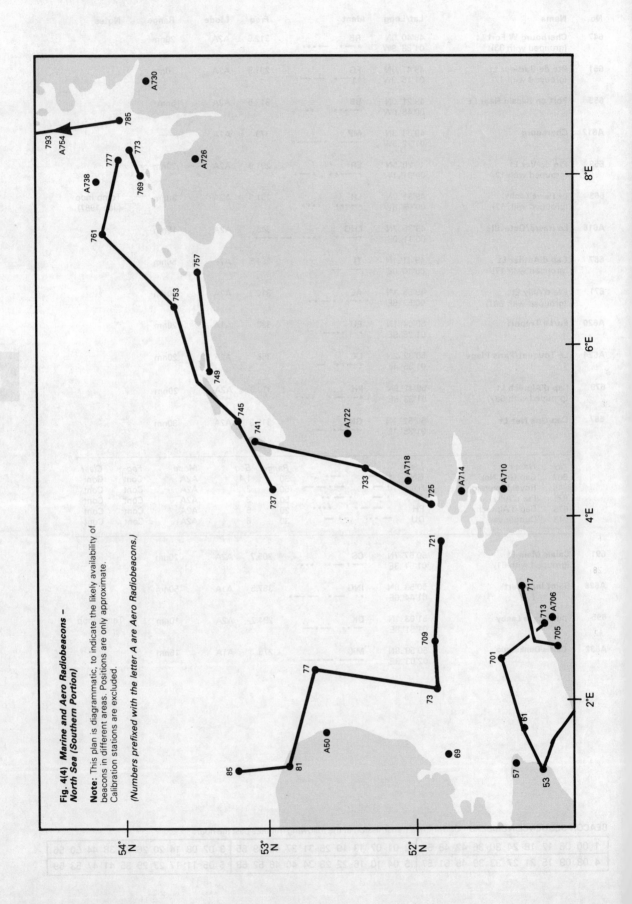

Fig. 4(4) *Marine and Aero Radiobeacons – North Sea (Southern Portion)*

Note: This plan is diagrammatic, to indicate the likely availability of beacons in different areas. Positions are only approximate. Calibration stations are excluded.

(Numbers prefixed with the letter A are Aero Radiobeacons.)

Marine and Aeronautical Radiobeacons 57

No	Name	Lat/Long	Ident	Freq	Mode	Range	Notes

BELGIUM, NETHERLANDS AND FEDERAL REPUBLIC OF GERMANY

No	Name	Lat/Long	Ident	Freq	Mode	Range	Notes
701	West Hinder Lt V (grouped with 61)	51°23'.0N 20°26'.3E	WH	305.7	A2A	20nm	
705	Nieuwpoort W Pier Lt (grouped with 717)	51°09'.4N 02°43'.1E	NP	296.5	A2A	5nm	
709	Noord Hinder Lt V (grouped with 77)	52°00'.2N 02°51'.2E	NR	287.3	A2A	50nm	
713	Oostende Rear Lt (grouped with 61)	51°14'.2N 02°55'.9E	OE	305.7	A2A	30nm	
A706	Oostende	51°13'.1N 02°59'.9E	ONO	399.5	A2A	50nm	
717	Zeebrugge Mole Lt	51°20'.9N 03°12'.2E	ZB	296.5	A2A	5nm	

No	Name	Ident	Range	Seq	Mode	Fog	Clear
717	Zeebrugge	ZB	5	1,2	A2A	Cont	Cont
705	Nieuwpoort	NP	5	4,5	A2A	Cont	Cont

No	Name	Lat/Long	Ident	Freq	Mode	Range	Notes
721	Goeree Lt (grouped with 77)	51°55'.5N 03°40'.2E	GR	287.3	A2A	50nm	
725	Hoek van Holland (grouped with 733)	51°58'.9N 04°06'.8E	HH	294.2	A2A	20nm	
A710	Woensdrecht	51°26'.4N 04°20'.9E	WDT	345	Non A2A	25nm	
A714	Stad	51°44'.5N 04°14'.6E	STD	386	Non A2A	15nm	
729	Scheveningen Lt Calibration station	52°06'.3N 04°16'.2E	HO	312.6	A2A		On request Tel 01747 4840 or call Scheveningen
A718	Valkenberg/Scheveningen	52°05'.6N 04°15'.2E	GV	364	Non A2A	25nm	
733	IJmuiden Front Lt	52°27'.8N 04°34'.6E	YM	294.2	A2A	20nm	
	Calibration station		YC	314.5	A2A	5nm	On request Tel: 02550 19027 or call Scheveningen

No	Name	Ident	Range	Seq	Mode	Fog	Clear
733	IJmuiden	YM	20	1,4	A2A	Cont	Cont
725	Hoek van Holland	HH	20	2,5	A2A	Cont	Cont
741	Eierland	ER	20	3,6	A2A	Cont	Nil

No	Name	Lat/Long	Ident	Freq	Mode	Range	Notes
A722	Amsterdam/Spijkerboor	52°32'.5N 04°50'.5E	SPY	381	Non A2A	75nm	
737	Texel Lt V (grouped with 745)	52°47'.1N 04°06'.6E	HK	308	A2A	50nm	
741	Eierland Lt (grouped with 733)	53°11'.0N 04°51'.4E	ER	294.2	A2A	20nm	
745	Vlieland Lt	53°17'.8N 05°03'.6E	VL	308	A2A	70nm	

No	Name	Ident	Range	Seq	Mode	Fog	Clear
745	Vlieland	VL	70	1	A2A	Cont	Cont
753	Borkumriff Lt V	BF	10	3	A2A	Cont	Cont
761	Deutsche Bucht Lt V	DB	50	4	A2A	Cont	Cont
737	Texel Lt V	HK	50	5	A2A	Cont	Cont
777	Elbe No 1 Lt V	EL	50	6	A2A	Cont	Cont

Radio Navigational Aids

No	Name	Lat/Long	Ident	Freq	Mode	Range	Notes
749	Ameland Lt	53°27'.0N 05°37'.6E	AD	298.8	A2A	20nm	

No	Name	Ident	Range	Seq	Mode	Fog	Clear
749	Ameland	AD	20	1,3,5	A2A	Cont	Nil
757	Borkum Little	BE	20	2,4,6	A2A	Cont	Cont

No	Name	Lat/Long	Ident	Freq	Mode	Range	Notes
753	Borkumriff Lt V (grouped with 745)	53°47'.5N 06°22'.1E	BF	308	A2A	10nm	
757	Borkum Little Lt (grouped with 749)	53°34'.8N 06°40'.1E	BE	298.8	A2A	20nm	
761	Deutsche Bucht Lt F (grouped with 745)	54°10'.7N 07°26'.1E	DB	308	A2A	50nm	
765	Weser Lt Float FS1 Calibration station	53°54'.3N 07°50'.0E	WB	313.5	A2A		On request, call Jade Revier Ch 16 20 63
769	Wangerooge Lt	53°47'.4N 07°51'.5E	WE	291.9	A2A	30nm	

No	Name	Ident	Range	Seq	Mode	Fog	Clear
769	Wangerooge	WE	30	1,3,5	A2A	Cont	Cont
773	Alte Weser	AR	10	2,4,6	A2A	Cont	Cont

No	Name	Lat/Long	Ident	Freq	Mode	Range	Notes
773	Alte Weser Lt (grouped with 769)	53°51'.9N 08°07'.7E	AR	291.9	A2A	10nm	
777	Elbe No 1 Lt V (grouped with 745) Calibration station	54°00'.0N 08°06'.6E	EL LB	308 312.6	A2A A2A	50nm	On request, Tel 04721 01221 or call Lt V Ch 11 16
A726	Jever	53°31'.2N 08°00'.9E	JEV	390	Non A2A	50nm	
779	Wilhelmshaven Calibration station	53°31'.0N 08°08'.8E	JB	312.6	A2A		On request, call Jade Revier Ch 16 20
781	Bremerhaven Calibration station	53°32'.4N 08°34'.9E	GB	312.6	A2A		On request, call West Revier Ch 05 07
785	Grosser Vogelsand Lt (grouped with 793)	53°59'.8N 08°28'.7E	VS	301.1	A2A	10nm	
A730	Nordholz	53°47'.2N 08°48'.5E	NDO	372	Non A2A	30nm	
789	Hollerwettern Lt Calibration station	53°50'.4N 09°21'.3E	HB	313.5	A2A		On request, Tel 04852 8011–16
791	Krautsand Calibration station	53°45'.6N 09°22'.6E	KB	312.6	A2A		On request, Tel 04852 8400
A734	Glukstadt	53°51'.1N 09°27'.3E	GLX	365	Non A2A	30nm	
A738	Helgoland Lt	54°11'.0N 07°53'.0E	DHE	397.2	A1A	100nm	Aeromarine beacon
	Calibration station		NB	313.5	A2A		On request, Eider Lock Radio Ch 14
A754	Westerland/Sylt	54°51'.4N 08°24'.7E	SLT	286	Non A2A	25nm	
793	Kampen/Rote Kliff Lt, Sylt	54°56'.9N 08°20'.5E	RF	301.1	A2A	20nm	

No	Name	Ident	Range	Seq	Mode	Fog	Clear
793	Kampen	RF	20	1,3,5	A2A	Cont	Cont
785	Gr Vogelsand	VS	10	2,4,6	A2A	Cont	Cont

BEACON SEQUENCE NUMBERS Commence transmission at the following minutes past the hour:

1	00 06 12 18 24 30 36 42 48 54	2	01 07 13 19 25 31 37 43 49 55	3	02 08 14 20 26 32 38 44 50 56
4	03 09 15 21 27 33 39 45 51 57	5	04 10 16 22 28 34 40 46 52 58	6	05 11 17 23 29 35 41 47 53 59

RADIOBEACONS — CALL SIGNS AND STATION NUMBERS

(Numbers prefixed with the letter A are Aero Radiobeacons)
(CS) = Calibration Station

Call Sign	Station	No
AD	Ameland Lt	749
ADN	Aberdeen	A82
AH	Altacarry Hd Lt	333
AL	Pte d'Ailly Lt	671
ALD	Alderney	A252
AP	Aberporth	A166
AR	Alte Weser Lt	773
BC	Bloscon	627
BC	Ballycotton (CS)	309
BD	Barra Hd Lt	181
BD	Bordeaux/Merignac	A564
BE	Borkum Little	757
BF	Borkumriff Lt V	753
BHD	Berry Hd	A10
BL	Butt of Lewis Lt	169
BLO	Bilbao	A548
BM	Brighton Marina	37
BN	Ballycotton Lt	309
BN	Les Baleines Lt	571
BNY	Bunratty	A330
BPL	Blackpool	A158
BRR	Barra	A124
BS	Port en Bessin	655
BST	Lanvéoc/Poulmic	A592
BT	Belle Ile, Goulphar Lt	591
BY	Bressay Lt	153
BY	Brawdy	A174
BZ	Biarritz	A556
CA	Pte de Creac'h Lt	615
CAA	Cazaux	A560
CAR	Isle of Man/Carnane	A146
CB	La Corbière	263
CDF	Cardiff/Rhoose	A182
CGC	Cognac/Chateaubernard	A568
CH	Chichester Bar	29
CL	Cloch Pt Lt (CS)	193
CM	Cromer Lt	81
CN	Cregneish, IOM	205
CNL	Connel/Oban	A126
CP	St Catherine's Pt Lt	25
CR	Channel Lt V	251
CRN	Carnmore	A322
CS	Calais Main Lt	691
CT	Pte de Combrit Lt	599
CW	Cape Wrath Lt	165
DA	Pladda Lt	189
DB	Deutsche Bucht Lt V	761
DG	Douglas	201

Call Sign	Station	No
DHE	Helgoland	A738
DK	Dunkerque Lt Buoy	695
DND	Dundee	A78
DO	Rosédo Lt (Brehat)	631
DO	Dounreay	A118
DU	Dungeness Lt	49
EC	St Helier	267
EDN	Edinburgh	A70
EGT	Eglington	A318
EL	Elbe No 1 Lt V	777
ENS	Ennis	A326
ER	Eierland Lt	741
ÉR	Pte der Ver Lt	659
EU	Eu/LeTréport	A620
EX	Exeter	A14
FB	Flamborough Hd	97
FÉ	Cap Fréhel	639
FG	Pte de Barfleur Lt	651
FN	Walney Island Lt	209
FOY	Foynes	A334
FP	Fife Ness Lt	125
FS	Falls Lt V	61
FT	Cap Ferret Lt	551
GA	Outer Gabbard Lt V	73
GB	Bremerhaven (CS)	781
GD	Girdle Ness Lt	133
GJ	Grand Jardin Lt	643
GL	Eagle Island Lt	341
GLX	Gluckstadt	A734
GN	Cap Gris Nez Lt	687
GR	Goeree Lt	721
GUR	Guernsey	A256
GV	Granville	A608
GV	Valkenburg	A718
GX	Ile de Groix, Pen Men	595
GY	St Peter Port	259
HB	Hollerwettern (CS)	789
HB	Belfast Harbour	A314
HH	Hoek van Holland	725
HIG	San Sebastian	A552
HK	Texel Lt V	737
HO	Scheveningen (CS)	729
HRN	Hurn	A18
HS	Heugh Lt	101
IB	Bardsey Lt	221
ING	St Inglevert	A628

Radio Navigational Aids

(Numbers prefixed with the letter A are Aero Radiobeacons)
(CS) = Calibration Station

Call Sign	Station	No
IW	Bembridge	A26
JB	Wilhelmshaven (CS)	779
JEV	Jever	A726
JEY	Jersey East	A260
JW	Jersey West	A264
KB	Krautsand (CS)	791
KD	Kinnairds Hd	137
KH	Kish Bank Lt	321
KLY	Killiney	A306
KS	Kinloss	A90
KW	Kirkwall	A98
LB	Elbe No 1 Lt V (CS)	777
LDV	Landivisiau	A596
LG	Eilean Glas Lt	173
LH	Le Havre Lanby	663
LHO	Le Havre/Octeville	A616
LK	Sule Skerry Lt	161
LK	Pointe de la Coubre	559
LM	Isle of May	121
LN	Lannion	A600
LOR	Lorient/Lann-Bihoué	A580
LP	Loop Head Lt	349
LS	Lundy Island, S Lt	237
LS	Lee-on-Solent	A30
LT	Longstone Lt	113
LT	Le Touquet	A624
LU	Leuchars	A74
LV	Dudgeon Lt V	85
LYX	Lydd	A42
LZ	Lizard Lt	5
MA	Cabo Machichaco Lt	547
MB	Humber Lt V (CS)	93
MC	Mew Island (CS)	329
MF	Muckle Flugga	157
MK	Calais/Dunkerque	A632
MP	Cherbourg	A612
MT	St Nazaire/Montoir	A572
MW	Mew Island Lt	329
MY	Cabo Mayor Lt	543
MZ	Mizen Head Lt	301
NB	Nab Tower Lt	33
NB	Helgoland Lt (CS)	A738
ND	Gt Yarmouth	A50
NDO	Nordholz	A730
NF	N Foreland Lt	57
NGY	New Galloway	A142
NH	Newhaven	41
NK	Inchkeith Lt	117
NP	Nash Point Lt	233
NP	Nieuwpoort	705
NR	N Ronaldsay	145
NR	Noord Hinder Lt V	709
NZ	St Nazaire	587
OE	Oostende Rear Lt	713
OH	Kinsale	305
OM	Stroma	141
ONO	Oostende	A706
OR	Hyskeir Lt	177
OTR	Ottringham	A54
PB	Portland Bill	17
PE	Penlee Point	9
PH	Penzance	A2
PH	Cap d'Alprech Lt	679
PO	Poole Harbour	21
PR	Ile du Pilier Lt	583
PS	Point Lynas Lt	213
PT	Souter Lt (CS)	109
PY	Point of Ayre	197
PY	Plymouth	A6
QR	Quimper/Pluguffan	A584
QS	Casquets Lt	255
RB	Cherbourg	647
RD	Roches Douvres Lt	635
RE	La Rochelle	563
RF	Kampen/Rote Kliff Lt	793
RN	Rhinns of Islay Lt	185
RR	Round Island Lt	1
RSH	Dublin (Rush)	A310
RWY	Ronaldsway, IOM	A150
RY	Royal Sovereign Lt	45
SB	Sumburgh Hd Lt	149
SB	South Bishop Lt	225
SD	S Foreland Lt	53
SHD	Scotstown Hd	A86
SHM	Shoreham	A38
SJ	Souter Lt	109
SK	Smiths Knoll Lt V	77
SL	Sligo	A320
SLT	Westerland/Sylt	A754
SM	St Mawgan	A190
SM	Pt de St Mathieu Lt	611
SN	Slyne Head Lt	345
SN	Ile de Sein	607
SND	Southend	A46

(Numbers prefixed with the letter A are Aero Radiobeacons)
(CS) = Calibration Station

Call Sign	Station	No	Call Sign	Station	No
SO	Les Sables d'Olonne	575	UK	Sunk Lt F	69
SP	Start Point Lt	13	UT	Unst	A114
SPY	Amsterdam	A722			
SR	Skerries Lt	217	VA	Vannes/Meucon	A576
SS	Scatsa	A110	VG	Ile Vierge Lt	623
STD	Stad	A714	VL	Vlieland Lt	745
STM	St Mary's Scilly	A1	VS	Gr Vogelsand Lt	785
STN	Stornoway	A122			
STU	Strumble	A170	WAL	Wallasey	A162
SU	South Rock Lt F	325	WB	Weser Lt Float (CS)	765
SUM	Sumburgh	A102	WDT	Woensdrecht	A710
SW	Ouessant SW Lanby	619	WE	Wangerooge Lt	769
SWN	Swansea	A178	WH	West Hinder Lt V	701
			WIK	Wick	A94
			WK	Wicklow Head Lt	317
TD	Teeside	A58	WS	Weston	A186
TI	Cap d'Antifer Lt	667	WTD	Waterford	A302
TL	Lerwick	A106			
TR	Tuskar Rock Lt	313	YC	IJmuiden (CS)	733
TRN	Turnberry	A138	YE	Ile d'Yeu Main Lt	579
TY	Tory Island Lt	337	YM	IJmuiden	733
ÜH	Eckmühl Lt, Penmarc'h	603	ZB	Zeebrugge Mole Lt	717

4.3.3 VHF Radio Lighthouse

This is a beacon which transmits a rotating directional signal that can be received by a VHF set capable of receiving frequency modulated signals. The signals are modulated with an audio tone varying between a maximum and a null. The 'null radial' rotates at a uniform speed of 4° per second, and the bearing of the beacon is determined by measuring the time that the null radial takes to reach the observer from a known starting point. To measure the time interval, the tone is broken into a number of half-second beats, each beat being the equivalent of 2° of bearing. The observer simply counts the number of beats from the start of each transmission, until the tone disappears, and then refers to Table 4(3).

The signal composition is as follows: Pause 0.1s; Morse ident. 3.2s; Pause 1.0s; Digital data 0.3s; Pause 1.0s; 70 Nav. beats 35.0s; Pause 1.0s; Morse ident. 3.2s; Pause 1.0s; Digital data 0.1s; Pause 12.1s; Static gap 2.0s. Total 60.0s.

The digital data is of no navigational significance, and is transmitted for the use of test equipment.

Beat numbers 10, 20, 30, 40, 50 and 60 are marked by a change in audio tone frequency.

All the VHF Radio Lighthouses currently in service transmit on VHF Ch 88 (162.025 MHz). The mode of emission used is designated as FXX, in which the main carrier is frequency modulated.

In the event of a transmitting equipment fault, a warbling tone will be transmitted, rendering the preceding beats invalid.

In the absence of a table, the bearing can be found from the formula:

Bearing = $A + 2(N - 7)$

where N is the number of beats counted, and A is the start bearing for 7 beats.

Alternatively, the time (T) in seconds can be taken with a stopwatch between beat 7 and the null. Then:

Bearing = $A + 4T$

The overall bearing accuracy is estimated as $\pm 2°$ (root mean square error) or better.

The stations are experimental, and transmissions may be altered or discontinued without warning.

TABLE 4(3) VHF RADIO LIGHTHOUSES

ANVIL POINT LT — 50°35'.5N 1°57'.5W

Morse ident: AL ·— ·—··
Frequency: Ch 88 (162.025 MHz)
Range: 14 n miles
Times: H24 (transmits alternately with Scratchell's Bay)

Count of beats	Bearing of Lt Ho from seaward (degrees)									
	0	1	2	3	4	5	6	7	8	9
0	—	—	—	—	—	—	—	247	249	251
10	253	255	257	259	261	263	265	267	269	271
20	273	275	277	279	281	283	285	287	289	291
30	293	295	297	299	301	303	305	307	309	311
40	313	315	317	319	321	323	325	327	329	331
50	333	335	337	339	341	343	345	347	349	351
60	353	355	357	359	001	003	005	007	—	—

SCRATCHELL'S BAY — 50°39'.7N 1°34'.6W

Morse ident: HD ···· —··
Frequency: Ch 88 (162.025 MHz)
Range: 30 n miles
Times: H24 (transmits alternately with Anvil Point Lt)

Count of beats	Bearing of Lt Ho from seaward (degrees)									
	0	1	2	3	4	5	6	7	8	9
0	—	—	—	—	—	—	—	337	339	341
10	343	345	347	349	351	353	355	357	359	001
20	003	005	007	009	011	013	015	017	019	021
30	023	025	027	029	031	033	035	037	039	041
40	043	045	047	049	051	053	055	057	059	061
50	063	065	067	069	071	073	075	077	079	081
60	083	085	087	089	091	093	095	097	—	—

NORTH FORELAND LT — 51°22'.5N 1°26'.8E

Morse ident: ND —·· —·
Frequency: Ch 88 (162.025 MHz)
Range: 20 n miles
Times: H24 (transmits alternately with Calais Main Lt)

Note: Temp inop (Aug 1986)

Count of beats	Bearing of Lt Ho from seaward (degrees)									
	0	1	2	3	4	5	6	7	8	9
0	—	—	—	—	—	—	—	240	242	244
10	246	248	250	252	254	256	258	260	262	264
20	266	268	270	272	274	276	278	280	282	284
30	286	288	290	292	294	296	298	300	302	304
40	306	308	310	312	314	316	318	320	322	324
50	326	328	330	332	334	336	338	340	342	344
60	346	348	350	352	354	356	358	360	—	—

CALAIS MAIN LT — 50°57'.7N 1°51'.3E

Morse ident: CL —·—· ·—··
Frequency: Ch 88 (162.025 MHz)
Range: 20 n miles
Times: H24 (transmits alternately with North Foreland Lt)

Count of beats	Bearing of Lt Ho from seaward (degrees)									
	0	1	2	3	4	5	6	7	8	9
0	—	—	—	—	—	—	—	090	092	094
10	096	098	100	102	104	106	108	110	112	114
20	116	118	120	122	124	126	128	130	132	134
30	136	138	140	142	144	146	148	150	152	154
40	156	158	160	162	164	166	168	170	172	174
50	176	178	180	182	184	186	188	190	192	194
60	196	198	200	202	204	206	208	210	—	—

4.4 RADAR

4.4.1 Radar in yachts

Radar is useful both for navigation and for collision avoidance, but to take full advantage of it and to use it in safety demands a proper understanding of its operation and of its limitations. Read the instruction book carefully, and practice using and adjusting the set so as to get optimum performance in different conditions. It is important to learn how to interpret what is actually seen on the display.

Radar beams do not discriminate so well in bearing as they do in range — so an accurate fix is sometimes best obtained by a radar range and a visual bearing of the same object.

The effective range of radar is approximately line of sight, but this can be decreased or increased by abnormal conditions. It is necessary to be aware that radar will not detect a low-lying coastline which is over the radar horizon.

The details of radar, and its use in yachts, are described in *The Macmillan & Silk Cut Yachtsman's Handbook* (4.6).

4.4.2 Radar for collision avoidance

Yacht radars invariably have a "ship's head up display", with the boat at the centre, apparently stationary. If a target is moving in the same direction and at the same speed, it is stationary relative to own ship, and its echo should be sharp and well defined. If it is on a reciprocal course, it paints an echo with a long tail.

If an echo is on a steady bearing, and the range is decreasing, there is risk of collision. But to determine the proper action to take it is necessary to plot an approaching echo three or four times, in order to determine her actual course and speed, and how close she will actually approach.

4.4.3 Radar as a navigation aid

Radar cannot see behind other objects, or round corners; it may not pick up small objects, or differentiate between two targets that are close together. As already stated, radar ranges are more accurate than radar bearings. Objects with sharp features such as buildings give a better reflection than those with curved or sloping surfaces. High cliffs make a good target, but approaching a low coastline the first thing to show on the radar display may be hills some distance inland.

4.4.4 Radar beacons (Racons)

A Racon is a transponder beacon, triggered by the emissions of a vessel's radar set, and sending out a distinctive signal, which appears on the display of the set. They are fitted to some light-vessels, buoys and lighthouses, and are marked on the chart.

In most cases the Racon flash on the display is a line extending radially outwards from a point slightly beyond the actual position of the Racon, due to the slight delay in the response of the beacon apparatus. Thus the distance to the mark of the Racon flash is a little more than the vessel's real distance from the Racon. Some Racons give a flash composed of a Morse identification signal, often with a tail to it, the length of the tail depending on the number of Morse characters.

The maximum range of a radar beacon is usually about 10 nautical miles, but may be more. In practice, picking up a Racon at greater distances depends also on the effective range of the boat's radar. With abnormal radio propagation, a spurious Racon flash may be seen at much greater distances than the beacon's normal range, appearing at any random position along the correct bearing on the display. Only rely on a Racon flash if its appearance is consistent, and the boat is believed to be within its range. At short range a Racon sometimes causes unwelcome interference on the radar display, and this may be reduced by adjusting the rain clutter control on the set.

The characteristics of radar beacons around the coasts of Great Britain, and adjacent coasts of Europe, are given in Table 4(4). Details are arranged in the following columns:
(1) Reference number.
(2) The type of radar beacon. Unless otherwise stated, all radar beacons sweep the frequency range of marine 3cm (X-band) radar emissions. A few, where indicated, respond to 10cm (S-band) emissions. Some in-band Racons (marked by an asterisk) are called frequency agile: their response is within the bandwidth of a yacht's radar, and they may cease to respond for a short period to allow echoes otherwise obscured by the Racon signal to be seen — otherwise the response appears automatically on the display. The term F Racon indicates a fixed frequency Racon transmitting outside the marine radar band. By tuning to that frequency the Racon flash is selected to the exclusion of normal echoes.
(3) Name of the station.
(4) Latitude and longitude.
(5) The time, in seconds, for a slow-sweep radar beacon to sweep the frequency range of the marine radar band.
(6) The sector within which signals may be received, bearings being towards the beacon, clockwise from 000° to 359°. 360° indicates all round operation.
(7) Approximate range, in nautical miles. This also depends on the range of the yacht's radar set.
(8) The form of the beacon's flash on the radar display. Morse signals are shown alphabetically, and are often followed by a "tail". Racons coded 'D' are used to mark new dangers.

Table 4(4) LIST OF RADAR BEACONS
(For heading details see 4.4.4)

(1) No.	(2) Type	(3) Name	(4) Lat	(4) Long	(5) Sweep time	(6) Sector	(7) Approx range	(8) Form of flash
		GREAT BRITAIN						
1	3 cm	Bishop Rock Lighthouse	49°52'.3N	6°26'.7W	90s	360°	10 nm	T
3	3 & 10 cm*	Seven Stones Lt V	50°03'.6N	6°04'.3W	30s	360°	15 nm	O
5	3 & 10 cm*	Wolf Rock Lt	49°56'.7N	5°48'.5W	30s	360°	10 nm	T
7	3 cm	Eddystone Lt	50°10'.8N	4°15'.9W	90s	360°	10 nm	T
9	3 cm	West Bramble Buoy	50°47'.2N	1°18'.6W	45s	360°	3 nm	T
11	3 cm	Nab Tower Lt	50°40'.1N	0°57'.1W	90s	360°	10 nm	T
13	3 & 10 cm*	Greenwich Lanby	50°24'.5N	0°00'.0	30s	360°	15 nm	T
15	3 & 10 cm*	Varne Lanby	51°01'.2N	1°24'.0E	30s	360°	10 nm	T
17	3 & 10 cm*	East Goodwin Lt V	51°13'.0N	1°36'.3E	30s	360°	10 nm	T
19	3 cm	Falls Lt V	51°18'.1N	1°48'.5E	90s	360°	10 nm	O
21	3 & 10 cm*	NE Goodwin Lt Buoy	51°20'.3N	1°34'.3E	30s	360°	10 nm	G
23	3 cm	F3 Lanby	51°23'.8N	2°00'.6E	90s	360°	10 nm	T
25	3 cm	Sea Reach Lt Buoy No. 1	51°29'.4N	0°52'.7E	90s	360°	10 nm	T
27	3 cm	Sea Reach Lt Buoy No. 7	51°30'.1N	0°37'.1E	90s	360°	10 nm	T
29	3 & 10 cm*	Outer Tongue Lt Buoy	51°30'.7N	1°56'.5E	30s	360°	10 nm	T
31	3 & 10 cm*	Barrow No 3 Lt Buoy	51°42'.0N	1°19'.9E	30s	360°	10 nm	B
33	3 & 10 cm	S Galloper Lt Buoy	51°43'.9N	1°57'.8E	30s	360°	10 nm	T
35	3 cm	Sunk Lt F	51°51'.0N	1°35'.0E	90s	360°	10 nm	T
37	3 & 10 cm*	Harwich Channel Lt Buoy No 1	51°55'.9N	1°26'.8E	30s	360°	10 nm	T
39	3 & 10 cm*	Orfordness Lt	52°05'.0N	1°34'.6E				
41	3 cm	Cross Sand Lt Buoy	52°37'.0N	1°59'.2E	90s	360°	10 nm	T
43	3 cm	Winterton Old Lighthouse	52°42'.8N	1°41'.8E	90s	360°	10 nm	T
45	3 cm	Smiths Knoll Lt V	52°43'.5N	2°18'.0E	90s	360°	10 nm	T
47	3 & 10 cm	Newarp Lt F	52°48'.4N	1°55'.8E		360°	10 nm	O
49	3 & 10 cm	Cromer Lt	52°55'.4N	1°19'.1E	30s	360°	25 nm	C
51	3 cm*	North Well Lt Buoy	53°03'.0N	0°28'.0E	30s	360°	10 nm	T
53	3 & 10 cm*	Dudgeon Lt V	53°16'.6N	1°17'.0E	30s	360°	10 nm	O
55	3 & 10 cm*	Inner Dowsing Lt	53°19'.7N	0°34'.0E	30s	360°	25 nm	T
57	3 & 10 cm	Spurn Lt F	53°33'.5N	0°14'.3E		360°	5 nm	S
59	3 cm	Humber Lt V	53°36'.7N	0°21'.6E	90s	360°	10 nm	T
61	3 cm	Tees Fairway Buoy	54°40'.9N	1°06'.4W	72s	360°		B
63	3 cm	Souter Lt	54°58'.2N	1°21'.8W	90s	135°-350°	10 nm	T
65	3 cm	St Abb's Head Lt	55°55'.0N	2°08'.2W	30s	360°	18 nm	T
67	3 cm	Inchkeith Fairway Buoy	56°03'.5N	3°00'.0W	75s	360°	5 nm	T
69	3 cm	Forth N Channel Lt Buoy No 7	56°02'.8N	3°10'.9W	60s	360°	5 nm	T
71	3 & 10 cm	Bell Rock Lt	56°26'.0N	2°23'.1W	60s	360°	18 nm	M
73	3 cm	Abertay Lt Buoy	56°27'.4N	2°40'.6W	70s	360°	8 nm	T
75	3 cm	Scurdie Ness Lt	56°42'.1N	2°26'.1W	70s	360°	15 nm	T
77	3 cm	Girdle Ness Lt	57°08'.3N	2°02'.8W	72s	165°-055°[1]	25 nm	G
		1) Reduced coverage within sector 055°-165°.						
79	3 cm	Aberdeen Fairway Buoy	57°09'.3N	2°01'.8W	70s	360°	7nm	T
81	3 cm	Buchan Ness Lt	57°28'.2N	1°46'.4W	72s	155°-045°[1]	25 nm	O
		1) Reduced coverage within sector 045°-155°.						
83	3 cm	Cromarty Firth Fairway Buoy	57°40'.0N	3°54'.1W	75s	360°	5 nm	T
85	3 cm	Tarbat Ness Lt	57°51'.9N	3°46'.5W	70s	360°	12 nm	T
87	3 cm	Duncansby Head Lt	58°38'.7N	3°01'.4W	72s	360°	20 nm	T
89	3 cm	Lother Rock Lt	58°43'.8N	2°58'.6W	70s	360°	10 nm	M
91	3 cm	North Ronaldsay Lt	59°23'.4N	2°22'.8W	75s	360°	10 nm	T
93	3 cm	Rumble Rock Bn	60°28'.2N	1°07'.1W	70s	360°	10 nm	O

Radar Beacons 65

(1) No.	(2) Type	(3) Name	(4) Lat	Long	(5) Sweep time	(6) Sector	(7) Approx range	(8) Form of flash
95	3 cm	Gruney Island Lt	60°39'.2N	1°18'.0W	70s	360°	18 nm	T
97	3 cm	Ve Skerries Lt	60°22'.4N	1°48'.7W	70s	360°	15 nm	T
99	3 cm	Sule Skerry Lt	59°05'.1N	4°24'.3W	120s	360°	25 nm	T
101	3 cm	Eilean Glas Lt	57°51'.4N	6°38'.5W	70s	360°	12 nm	T
103	3 cm	Castlebay South Buoy	56°56'.1N	7°27'.2W	70s	360°	7 nm	T
105	3 cm	Monach Lt Ho	57°31'.5N	7°41'.6W	70s	360°	16 nm	T
107	3 cm	Kyleakin Lt Ho	57°16'.7N	5°44'.5W	70s	360°	16 nm	T
109	3 cm	Skerryvore Lt	56°19'.4N	7°06'.7W	120s	360°	25 nm	M
111	3 cm	Sanda Lt	55°16'.5N	5°34'.9W	70s	360°	20 nm	T
113	3 cm	Point of Ayre Lt	54°24'.9N	4°22'.0W	30s	360°	15 nm	M
115	3 & 10 cm*	Lune Deep Lt Buoy	53°55'.8N	3°11'.0W	30s	360°	10 nm	T
117	3 cm	Bar Lanby	53°32'.0N	3°20'.9W	90s	360°	10 nm	T
119	3 & 10 cm*	Skerries Lt	53°25'.3N	4°36'.4W	30s	360°	25 nm	T
121	3 & 10 cm*	The Smalls Lt	51°43'.2N	5°40'.1W	30s	360°	25 nm	T
123	3 & 10 cm*	St Gowan Lt V	51°30'.5N	4°59'.8W	30s	360°	15 nm	T
125	3 & 10 cm*	W Helwick Lt Buoy	51°31'.4N	4°23'.6W	30s	360°	10 nm	T
127	3 & 10 cm*	W Scar. Lt Buoy	51°28'.3N	3°55'.5W	30s	360°	10 nm	T
129	3 cm	English and Welsh Grounds Lt Float	51°26'.9N	3°00'.1W	90s	360°	10 nm	O
131	3 cm	Breaksea Lt Float	51°19'.9N	3°19'.0W	90s	360°	10 nm	T

CHANNEL/CHANNEL ISLANDS (see also No. 13)

151	3 cm	Lt Buoy EC1	50°05'.9N	1°48'.3W	90s	360°	10 nm	T
153	3 cm	Lt Buoy EC2	50°12'.1N	1°12'.4W	90s	360°	10 nm	T
155	3 cm	Lt Buoy EC3	50°18'.3N	0°36'.1W	90s	360°	10 nm	T
157	3 cm	East Channel Lt Float	49°58'.7N	2°28'.9W	90s	360°	10 nm	T
159	3 & 10 cm*	Channel Lt V	49°54'.4N	2°53'.7W	30s	360°	15 nm	O
161	3 & 10 cm*	Casquets Lt Ho	49°43'.4N	2°22'.5W	30s	360°	25 nm	T
163	3 cm	Platte Fougère Lt	49°30'.9N	2°29'.0W				P
165	3 cm	St Helier — Demie de Pas Lt	49°09'.1N	2°06'.0W	120s	360°	10 nm	T
167	3 cm	St Helier — Mount Ubé Leading Lt	49°10'.3N	2°03'.5W	60s	360°	14 nm	T

IRELAND

201	3 cm	Mizen Head Lt	51°27'.0N	9°49'.2W		360°	24 nm	T
203	3 cm	Cork Lt Buoy	51°42'.9N	8°15'.5W	72s	360°	7 nm	T
205	3 cm	Hook Head Lt	52°07'.4N	6°55'.7W	120s	237°-177°	10 nm	K
207	3 cm	Coningbeg Lt F	52°02'.4N	6°39'.4W		360°	13 nm	M
209	3 cm	Tuskar Rock Lt	52°12'.2N	6°12'.4W	120s	360°	18 nm	T
211	3 cm	Arklow Lanby	52°39'.5N	5°58'.1W		360°	10 nm	O
213	3 cm	Codling Lanby	53°03'.0N	5°40'.7W		360°	10 nm	G
215	3 cm	Kish Bank Lt	53°18'.7N	5°55'.4W	120s	360°	15 nm	T
217	3 cm	South Rock Lt F (H + 02½ to H + 00)	54°24'.5N	5°21'.9W		360°	13 nm	T
219	3 cm	Inishtrahull Lt	55°25'.9N	7°14'.6W		060°-310°[1]	24 nm	T

1) Reduced or no signal 310°-060°.

Radio Navigational Aids

(1) No.	(2) Type	(3) Name	(4) Lat	Long	(5) Sweep time	(6) Sector	(7) Approx range	(8) Form of flash
FRANCE (Atlantic Coast)								
301	3 cm	BXA Lanby	45°37'.6N	1°28'.6W	120-150s	360°		B
305	3 cm	St Nazaire Lt Buoy No 10[1]	47°13'.7N	2°16'.1W		360°	3-5 nm	See[1]

1) Signal appears as a series of dots, the distance between each dot corresponding to 0.2 nm.

306	3 cm	St Nazaire Lt Buoy SN1	47°00'.0N	2°39'.9W		360°	3–8 nm	Z
307	3 cm	Chausée de Sein Lt Buoy	48°03'.8N	5°07'.7W	120-150s	360°	10 nm	O
311	3 cm	Pointe de Créac'h Lt	48°27'.6N	5°07'.6W	120-150s	030°-248°	20 nm	C
313	3 & 10 cm	Ouessant SW Lanby	48°31'.7N	5°49'.3W		360°	20 nm	M
315	3 cm	Ouessant NE, Lt Buoy	48°45'.9N	5°11'.6W		360°	20 nm	B
325	3 & 10 cm	Le Havre Lanby	49°31'.7N	0°09'.8W		360°	8-10 nm	See[1]

1) Signal appears as a series of 8 dots or 8 groups of dots, the distance between each dot or group of dots corresponding to 0.3 nm.

327	3 cm	Bassurelle Lt Buoy	50°32'.7N	0°57'.8E	120-150s	360°	6-10 nm	B
328	3 & 10 cm	Vergoyer Lt Buoy N	50°39'.7N	1°22'.3E		360°	5–8 nm	C
329	3 cm	Sangatte	50°57'.2N	1°46'.6E		360°	11 nm	See[1]

1) Signal appears as 3 successions of 3 dots, the distance between each dot corresponding to 0.3 nm.

331	3 cm	Dunkerque Lt Buoy[1]	51°03'.1N	1°51'.8E		360°		See[1]

1) The signal appears as a succession of 8 dots, the distance between each dot corresponding to 0.3 nm.

335	3 cm	Sandettié Lt V	51°09'.4N	1°47'.2E		360°	4-10 nm	See[1]

1) Signal appears as a succession of 8 dots, the distance between each dot corresponding to 0.3 nm.

(1) No.	(2) Type	(3) Name	(4) Lat	Long	(5) Sweep time	(6) Sector	(7) Approx range	(8) Form of flash
BELGIUM AND NETHERLANDS								
375	3 & 10 cm	Wandelaar Lt MOW 0	51°23'.7N	3°02'.8E		360°		W
377	3 & 10 cm	Bol Van Heist Lt MOW 3	51°23'.4N	3°12'.0E		360°		H
399	3 cm	Noord Hinder Lt V	52°00'.2N	2°51'.2E	120s	360°	6-10 nm	T
401	3 cm	Noord Hinder Buoy NHR-SE	51°45'.5N	2°40'.0E	120s	360°	10 nm	N
403	3 cm	Noord Hinder Buoy NHR-N	52°13'.3N	2°59'.5E	120s	360°	10 nm	K
405	3 cm	Schouwenbank Lt Buoy	51°45'.0N	3°14'.4E	120s	360°	10 nm	O
407	3 cm	Zuid Vlije Lt Buoy ZVl5	51°38'.2N	4°14'.5E				K
409	3 & 10 cm*	Goeree Lt	51°55'.5N	3°40'.2E	100s	360°	14-20 nm	T
411	3 cm	Maas Center Lt Buoy	52°01'.2N	3°53'.6E	120s	360°	8 nm	M
413	3 cm	IJmuiden Lt Buoy	52°28'.7N	4°23'.9E		360°		Y
417	3 cm	Texel Lt V	52°47'.1N	4°06'.6E	120s	360°	6-10 nm	T
421	3 cm	Logger Platform	53°00'.9N	4°13'.0E	120s	60°–270°	10 nm	X
425	3 cm	Vlieland Lanby VL-CENTER	53°27'.0N	4°40'.0E	120s	360°	12 nm	C
431	3 cm	Vlieland N Lt Buoy VL–N	53°35'.5N	4°37'.2E	120s	360°	10 nm	G
433	3 cm	Petroland Platform L4-B	53°40'.6N	4°00'.2E		360°	15 nm	B
435	3 cm	DW Lt Buoy	54°02'.4N	4°44'.3E	120s	360°	6-10 nm	M
FEDERAL REPUBLIC OF GERMANY (North Sea Coast)								
501	3 cm	Ems, Hubertgat Lt Buoy H2	53°35'.0N	6°16'.4E	48s	360°	8 nm	B
503	3 cm	Borkumriff Lt V	53°47'.5N	6°22'.1E	48s	360°	8 nm	T
505	3 cm	Westerems Lt Buoy	53°37'.1N	6°19'.5E	48s	360°	8 nm	G
507	3 cm	TW/EMS Lt Buoy	54°10'.0N	6°20'.8E	55s	360°	6-10 nm	T
509	3 cm	Deutsche Bucht Lt V	54°10'.7N	7°26'.1E	48s	360°	8 nm	T
511	3 cm	DB/Weser Lt Buoy	54°02'.4N	7°43'.0E	48s	360°	6 nm	K
513	3 cm	Weser Lt Float FS1	53°54'.2N	7°50'.0E	95s	360°	8 nm	T
515	3 cm	Elbe No 1 Lt V	54°00'.0N	8°06'.6E	48s	360°	8 nm	T

You don't have to be a bilge rat to find a little luxury below decks. Wyn Ltd provide practical below decks comfort for everyone

WITH

The SEALAND VACU-FLUSH Sanitation System
— a simple to operate luxury toilet system with low water usage at a realistic price

— for all below decks applications
The FORCE 10 range of cookers
— of which Yachting Monthly said: "possibly the best cooker we have seen on a yacht..." Also available: barbeques and cabin heaters.

All of Wyn Ltd's products are carefully selected from the best of North American and Canadian manufacturers. They were chosen because they give the highest quality, practicality and value for money available.

Wyn's high quality marine products provide affordable luxury below decks — the kind that even bilge rats can't find.

WYN LTD, Plas Paradwys, Paradwys, Bodorgan, Isle of Anglesey LL62 5PE. Tel: 0407 840199

If you think all Aerials are the same... ...you haven't got the message.

Whether it is business, Pleasure, or a real emergency, the aerial is the vital link in your VHF communication.
To ensure the best result you should insist on

Communication Aerials
We have a wide range to suit every kind of vessel. Some of our aerials for sailing boats include Wind Direction indicators. Good performance does not end with a good aerial. You need good cable and connectors that will continue to perform well for many years in our harsh environment. Again we offer a wide choice. Ask to see the complete range of our VHF aerials together with the New 4 in 1 receiving aerial and 24v-12v convertor.

COMMUNICATION AERIALS LIMITED
Woodland Industrial Estate, Eden Vale Road, Westbury, Wiltshire BA13 3QS, England
Telephone: Westbury (0373) 822835

Don't blow your investment, protect it with a Harwood Filter Separator

You won't need reminding of the cost and dangerous consequences of a damaged or blown up engine. Now you can protect your engine, cruiser and crew simply and effectively by fitting a Harwood filter separator – thus eliminating 99.9% of water in the fuel and filtering out particles down to 2 microns.
Harwoods are fitted as standard equipment by many leading boatbuilders, so for peace of mind fit one now and protect your investment.

Engine-uity at its best.

For full details write to:
C.T. Harwood Ltd.,
Ashley House, Hurlands Close, Farnham, Surrey GU9 9JF Tel: 0252 733312 Telex: 859641

IAIN KERR HUNTER YACHT DELIVERIES

IAIN KERR HUNTER YACHT DELIVERIES are quality specialists in the business of delivering yachts and commercial craft by sea. Our professional service has been carefully developed over several years of experience in navigating a wide range of craft to a variety of locations worldwide.

Our portfolio of regular and repeated custom is founded on our fundamental principle of supplying prompt, responsible and safe delivery of the craft in our care.

We pay special attention to the maintenance and upkeep of the vessels in our charge and our number one priority at all times is to meet the personal needs of our customer in each individual assignment.

Call us now on:
0968 - 74486

FOR THE BEST QUOTATIONS...

For competitive quotations with the best security consistent with economy, and a secure and prompt claims service,

contact the marine specialists

WE ANTICIPATE THE UNEXPECTED
Housemans

38a West Street, Marlow, Bucks. SL7 2NB.
Telex: 846835 HOUINS G
Fax: Marlow (0628) 890636

0628 890888

All our policies are through approved BIA and Lloyds insurers.

MARINE RADIO

ICOM proudly present their 1988 range of marine VHF and HF radio telephones. Shown here are the IC-M12 VHF handheld, the IC-M55 and NEW IC-M100c VHF transceivers plus 3 or 5 watt versions of the keypad operated IC-M5. Also shown is 'big brother' namely the ICOM IC-M700-UK HF transceiver for longer range communications. Complete the FREEPOST coupon today for more information.

AS USED ON VIRGIN ATLANTIC CHALLENGER

ICOM Dept. MA, FREEPOST, Herne Bay, Kent CT6 8BR
Tel: 0227 363859. Telex: 965179 ICOM G

Admiralty Charts and Hydrographic Publications

The Hydrographic Department has been publishing charts and navigational publications for almost 200 years. To-day, a range of over 3,000 charts provide world-wide coverage for all vessels from yachts to supertankers. Admiralty Notices to Mariners are issued weekly to allow chart users to incorporate the latest navigational information so vitally important for safety at sea.

● ADMIRALTY CHARTS

- SAILING DIRECTIONS (PILOTS)
  Fully updated with supplements
- RADIO PUBLICATIONS
- MARINER'S HANDBOOK
- ADMIRALTY NOTICES TO MARINERS
  – SMALL CRAFT EDITION

- TIDAL INFORMATION
  Admiralty Tide Tables – Tidal Stream Atlases
- LIST OF LIGHTS & FOG SIGNALS
- OCEAN PASSAGES FOR THE WORLD
  (NEW EDITION)

Admiralty charts and hydrographic publications are obtainable through an extensive network of appointed Admiralty Chart Agents in the British Isles and overseas.

New Catalogues available January 1989 including FREE Home Waters Catalogue from Admiralty Chart Agents.

Small Craft Facilities

Additional symbols showing the location of facilities of interest to small craft users are being introduced on certain coastal and harbour charts of the British Isles.

- Visitors' mooring
- Visitors' berth
- Slipway
- Public landing
- Water tap
- Fuel
- Public telephone
- Post box
- Public house or inn
- Restaurant
- Yacht or sailing club
- Public toilets
- Public car park
- Parking for boats/trailers
- Laundrette
- Caravan site
- Camping site
- Nature reserve

**HYDROGRAPHER OF THE NAVY
TAUNTON, SOMERSET
TELEPHONE: TAUNTON 87900**

YACHTING MONTHLY
Britain's top cruising magazine

Cruising under sail – that's the cover to cover theme of Yachting Monthly. From beginner to expert, creek crawler to blue water yachtsman there's something for everyone in each issue.

The *Practical Seamanship* section discusses and analyses seamanship problems in a down to earth and easily understood fashion. Then there's cruising narrative near and far to whet the appetite and technical features on navigation and pilotage. Each month there are authoritative test reports on new boats from our experienced team and also reviews of the latest equipment for boats of all types and sizes. And rounded off by our Around the Coast regional cruising news section it makes a complete package for the cruising yachtsman.

Editor: Andrew Bray
01-261 6040

Advertisement Manager: David Williams
01-261 5582

IPC Magazines Ltd, Kings Reach Tower, Stamford Street, London SE1 9LS

New pre-wired mounting pod.
Perfect protection for today's top instruments

The Voyager mounting pod is a handsome and practical housing which presents the vital speed, depth and wind data you need to back your cruising decisions and racing tactics. It is weatherproof, so you can install it on the steering pedestal or over the companionway. But its compact dimensions are equally at home in the wheelhouse or by the chart table. Supplied complete with junction box and 3 metres of cable. No fuss. No costly installation. Simply fit, and go voyaging.

VOYAGER SPEED – 7 functions
- Speed measurement to 32 knots (paddle wheel), 12 knots (trailing) and 50 knots (electro-magnetic)
- Distance to 999.9 nautical miles
- Trim to ± 2 knots
- Average trip speed
- 10-5-0 minute alarm
- Elapsed trip time to 100 hours
- Distance alarm 0.1 to 255 nautical miles
- All key data retained in memory

VOYAGER WIND
- Apparent windspeed and direction
- Windspeed in knots, metres per second and Beaufort Scale. Displayed in digital or analogue
- Wind direction displayed 360° analogue and digitally
- Magnified readout to 60° off bows and quarters
- Mast head unit incorporates opto-electronic sensor
- True windspeed/direction and V.M.G. when interfaced with Voyager Log

VOYAGER DEPTH
- Digital readings up to 300 metres
- Analogue presentation to 100 units
- Shallow depth resolution to 0.1' (digital) or 6" (analogue)
- Computer controlled gain eliminates spurious echoes and reading errors
- Audible and visual deep and shallow alarms

ELECTRONIC COMPASS
– 6 functions
- Remote located sensor unit avoids errors caused by proximity of electrical circuits and steel
- Analogue and digital display
- Functions comprise Compass, Heading, Rate of Turn, Header/Lifter, Audible Off Course Alarm and Dead Reckoning

NEW

All Voyager models are suitable for cockpit, bulkhead or chart table installation. For further data or advice, please ask your nearest Seafarer Distributor or contact Seafarer direct.

International SEAFARER RANGE

Seafarer Navigation International Limited, Fleets Lane, Poole, Dorset, BH15 3BW. Telephone: 0202-674641. Telex: 41169.

Crown Copyright. Reproduced from Admiralty Charts with the permission of Her Majesty's Stationery Office.

THREE OF A KIND

All of our Marinas around UK operate with the same high standards of efficient service, modern facilities and friendliness. Whatever your sailing area, base your yacht with us

LARGS YACHT HAVEN

Perfectly situated, the Clyde's newest Marina provides safe and sheltered access to the beautiful sailing waters of the Firth. The spectacular Kyles of Bute are within a couple of hours sail, or you can cruise North, South and West from the Marina to visit Arran or Ireland, or to explore the superb 3000 mile coastline of Scotland's West Coast and the Hebrides

SWANSEA YACHT HAVEN

The focal point of the City's multi-million pound Maritime Quarter, linked by excellent motorways to the Midlands and the South of England

LYMINGTON YACHT HAVEN

The finest base for top class racing or for cruising the South Coast or across the Channel

LARGS YACHT HAVEN
Irvine Road
Largs
Ayrshire
KA30 8EZ
Tel: 0475 675333

SWANSEA YACHT HAVEN
Lockside
Maritime Quarter
Swansea
SA1 1WN
Tel: 0792 470310

LYMINGTON YACHT HAVEN
King's Saltern Road
Lymington
Hants
SO41 9QD
Tel: 0590 77071

Chapter 5
Astro-Navigation

Contents

5.1 THE USE OF CALCULATORS — Page 68
5.1.1 General
5.1.2 Dip
5.1.3 Refraction
5.1.4 Amplitudes
5.1.5 Sight reduction

5.2 ASTRONOMICAL DATA (EPHEMERIS) — EXPLANATION — 69
5.2.1 Summary of contents
5.2.2 Main data
5.2.3 Rising and setting phenomena
5.2.4 Altitude correction tables
5.2.5 Polaris table
5.2.6 Auxiliary and planning data
5.2.7 Illustrations
5.2.8 The use of a calculator for interpolation
5.2.9 Standard Times

5.3 EPHEMERIS — 75
5.3.1 Calendar and festivals
5.3.2 Altitude correction tables
5.3.3 Selected stars and Sun (1989) (for calculator use)
5.3.4 Planet planning data, phases of the Moon and eclipses
5.3.5 Stars, 1989 — SHA and Dec
5.3.6 Polaris table
5.3.7 Sunrise, sunset and twilight — 1989
5.3.8 Moonrise and moonset — 1989
5.3.9 Ephemeris, Aries, planets, Sun and Moon — 1989
5.3.10 Interpolation tables
5.3.11 Star diagrams
5.3.12 Azimuth diagrams
5.3.13 Conversion tables: Arc to time; hours, minutes and seconds to decimals of a day; decimals of a degree to minutes of arc; decimals of a minute to seconds of arc

Astro-Navigation — introduction
The following subjects are described in detail in Chapter 5 of *The Macmillan & Silk Cut Yachtsman's Handbook*

A general introduction to astro-navigation; the sextant; sextant errors and adjustments; sextant handling. The principles of nautical astronomy — angular and geographical distances; time and hour angle; glossary of terms. Altitude corrections — dip; refraction; semi-diameter; parallax. Theory into practice — the astronomical triangle; sight reduction methods. Practical sight taking — notes on observation; the accuracy of sights; Sun sights; Moon sights; star sights. Sight reduction — tables; latitude by meridian altitude; latitude by Polaris; plotting the Sun's geographical position; equal altitudes. The use of calculators — dip; refraction; amplitudes; sight reduction. The position line — the use of a single position line; angle of cut; rate of change of bearing and altitude. Plotting and evaluating the sight. Sight reduction table — description; notation, method, equations and rules; instructions and illustrative examples; table. Bibliography.

Here in the Almanac are given the necessary astronomical data for determining position at sea — the Greenwich Hour Angle of the first point of Aries, the Greenwich Hour Angle and the Declination of the Sun, Moon, planets and selected stars for any instant of Greenwich Mean Time, together with auxiliary data comprising corrections to observed altitude, times of rising and setting of the Sun and Moon, times of twilight, planning data, star diagrams and azimuth diagrams.

5.1 THE USE OF CALCULATORS

5.1.1 General
For sight reduction either a 'dedicated' calculator (designed for the purpose) or a calculator offering program capability, and adequate memory storage together with polar-rectangular conversion and statistical functions should be used. Further information on this subject, and on the formulae below, is given in Chapter 5 (section 5.7) of *The Macmillan & Silk Cut Yachtsman's Handbook*.

5.1.2 Dip
Dip corrections to the observed altitude are always negative ($-$), and may be calculated from:

$$\text{Dip (in mins)} = 1'.76 \times \sqrt{\text{HE}} \text{ (in metres)}$$
$$\text{or} = 0'.97 \times \sqrt{\text{HE}} \text{ (in feet)}$$

where HE is height of eye above sea level.

5.1.3 Refraction
The correction for refraction is always subtractive ($-$), and may be calculated from the formula:

$$R_M = \left(\frac{74}{h+2}\right) - 0.7$$

where R_M is mean refraction in minutes of arc
h is observed altitude in degrees at sea level air temperature 10°C and atmospheric pressure 1010 millibars

5.1.4 Amplitudes
Amplitude is the bearing of a body measured from true east or west when it is on the horizon (i.e. rising or setting), and is useful for checking the accuracy of the compass.

$$\sin \text{Amplitude} = \frac{\sin \text{Declination}}{\cos \text{Latitude}}$$

Amplitudes should not be used in high latitudes.

5.1.5 Sight reduction
Several different formulae can be used for basic sight reduction. Three different methods are given below:

(1) $\sin \text{Alt} = (\sin \text{Lat} \times \sin \text{Dec}) +$
$(\cos \text{Lat} \times \cos \text{Dec} \times \cos \text{LHA})$

where: Dec is positive if the SAME name as Lat, and negative if CONTRARY name.

For the Azimuth Angle (Z): $\tan Z =$

$$\frac{\sin \text{LHA}}{(\cos \text{Lat} \times \tan \text{Dec}) - (\sin \text{Lat} \times \cos \text{LHA})}$$

where the sign conventions are:

(a) If Lat or Dec are SOUTH change their sign to minus ($-$).
(b) If Z and sin LHA have the same sign, Zn = Z + 180°; if Z is minus ($-$) and LHA is less than 180°, Zn = Z + 360°; if Z is plus (+) and LHA is greater than 180°, Zn = Z.

(2) Using a different method, the procedure with a simple scientific calculator begins with loading the variable information into the memory stores.

(a) Key in LHA, convert if necessary to (D.d), and put into STO 1.
(b) Key in Dec, convert if necessary to (D.d). If it is CONTRARY name to latitude, change its sign in the display to negative: put into STO 2.
(c) Key in DR Lat, convert if necessary to (D.d), and put into STO 3.

Then proceed as follows:

RCL 1 cos ÷ RCL 2 tan = inv tan SUM 3 sin ×
RCL 1 tan ÷ RCL 3 cos = inv tan

The display will now show the Azimuth Angle (Z). Note Z and its sign (+) or ($-$), and continue:

cos × RCL 3 tan = inv tan
inv 2nd D.MS

The display now shows the calculated altitude (Hc). Note the sign of the final display, and establish the true azimuth (Zn) from Z according to these rules:

(a) If the final display is (+), in NORTH latitudes Zn = 360° $-$ Z, and in SOUTH latitudes Zn = 180° + Z.
(b) If the final display is ($-$), in NORTH latitudes Zn = 180° $-$ Z, and in SOUTH latitudes Zn = 360° + Z. Note: Remember that if (Z) was displayed as negative, you must apply the usual arithmetic rules when adding or subtracting.

(3) Calculators with the facility of rectangular to polar conversion can use the following method and sign convention. If Lat or Dec are SOUTH change their sign to minus ($-$).

$Z = R \rightarrow P (\cos \text{Dec} \times \sin \text{Lat} \times \cos \text{LHA} - \cos \text{Lat} \times \sin \text{Dec}, \sin \text{LHA} \times \cos \text{Dec})$

$Zn = Z + 180°$

$Hc = \sin^{-1}(\sin \text{Dec} \times \sin \text{Lat} + \cos \text{Dec} \times \cos \text{LHA} \times \cos \text{Lat})$

5.2 ASTRONOMICAL DATA (EPHEMERIS) — EXPLANATION

5.2.1 Summary of contents

The object of this Almanac is to provide, in a condensed form, the data needed to determine position at sea to an accuracy of one or two minutes of arc. It is designed for European waters between latitudes N.30° and N.60°.

The main contents consist of data from which the *Greenwich Hour Angle* (GHA) and the *Declination* (Dec) of the Sun, Moon and planets (Venus, Mars, Jupiter and Saturn) can be determined for any instant of *Greenwich Mean Time* (GMT). The *Local Hour Angle* (LHA) can then be obtained from:

$$\text{LHA} = \text{GHA} \begin{matrix} -\text{west} \\ +\text{east} \end{matrix} \text{ longitude}$$

Auxiliary data consist of: corrections to observed altitude, calendarial data, times of rising and setting of the Sun and Moon, times of twilight, planning data, star diagrams and azimuth diagrams.

For the Sun and planets, the GHA and Dec are tabulated directly for 00^h of each day throughout the year. For the Moon the tabulations are for 00^h, 08^h and 16^h each day. For the stars the *Sidereal Hour Angle* (SHA) is given, and the GHA is obtained from:

$$\text{GHA Star} = \text{GHA Aries} + \text{SHA Star}$$

The SHA and Dec of the stars change so slowly that, to the accuracy required, mean values are adequate and are given for use throughout the year. GHA Aries, or the Greenwich Hour Angle of the first point of Aries, is tabulated for 00^h each day. Permanent tables give the increments and corrections to the tabulated daily values of GHA and Dec for hours, minutes and seconds of GMT.

The tabular accuracy of GHA and Dec of the Sun, Moon, planets, GHA Aries and of the SHA and Dec of the stars is $0°.01$; for the stars the equivalent of the decimals of a degree of the SHA and Dec is given in brackets alongside in minutes of arc.

The ephemeral data for two months are given on an opening of two pages; the left-hand page contains data for Aries and the planets and the right-hand page data for the Sun and Moon. Separate tabulations give Local Mean Times (LMT), for every third day, of Morning Nautical Twilight, Sunrise, Sunset and Evening Civil Twilight for latitude N.50°, together with variations (v) which, with the aid of auxiliary tables, enable the LMT for other latitudes in the range N.30° to N.60° to be found. Tabulations for Moonrise and Moonset for each day for latitude N.50° are given together with variations (v) which, with the auxiliary tables, enable the LMT of Moonrise and Moonset to be found for latitudes N.30° to N.60°.

5.2.2 Main data

Monthly pages. These give on each left-hand page the GHA of Aries and the GHA and Dec of the four navigational planets at 00^h GMT each day. On the right-hand page, the GHA and Dec of the Sun are given each day for 00^h GMT, and the mean semi-diameter (SD) is shown each month; for the Moon the GHA and Dec are given each day for 00^h, 08^h and 16^h GMT, and horizontal parallax (HP) for 12^h. Values of the variations (v) in GHA and (d) in Declination allow the GHA and Dec of Sun, Moon or planets to be found at any instant of GMT. South declinations are treated as negative.

Stars. The SHA and Dec of 48 selected stars are tabulated in numerical order in 5.3.3 on page 77 and are intended for use throughout the year; the tabulations are given in degrees with two decimals (the equivalent of the decimal part is also given in minutes of arc, in brackets, alongside); an alphabetical list of star names and numbers is also given. A further list of the SHA and Dec of 154 stars is tabulated in 5.3.5 on pages 80–81; this list includes the selected stars and is tabulated to the same accuracy, with minutes of arc alongside.

Interpolation tables. These tables provide the increments for hours, minutes and seconds to be applied to the tabular values of GHA and Dec; the tables for Aries, Sun and planets comprise four pages, and those for the Moon six pages. The increment for GHA Aries is taken out in two parts, the first with argument GMT in units of 12 minutes and the second part, with argument the remainder, in minutes and seconds, is given on the right-hand side of the opening. For the GHA of the Sun and planets the increment is found in three parts: the first with argument GMT in units of 12 minutes, the second is for the remainder in minutes and seconds, and the third with vertical argument GMT (in units of 12 minutes) and horizontal argument v. The increments for Dec are found with vertical argument GMT (in units of 12 minutes) and horizontal argument d.

For the Moon, the increment of GHA is found in three parts. First with argument GMT in units of 4 minutes, second with argument the remainder of GMT in minutes and seconds, and third with vertical argument GMT (in units of 4 minutes) and horizontal argument v. The increments for Dec are found with vertical argument GMT (in units of 4 minutes) and horizontal argument d.

The increments are based on the following adopted rates of increase: Aries $15°.041$; Sun and planets $15°$ precisely; Moon $14°.31$. The values of v are the excesses in units of $0°.001$ of the actual hourly motions over the adopted values. The values of d are the hourly rates of change of Dec, in units of $0°.001$.

5.2.3 Rising and setting phenomena

These tabulations give for every third day of the year, Local Mean Times (LMT) of Morning Nautical Twilight, Sunrise, Sunset and Evening Civil Twilight for latitude N.50°, together with variations (v) which, with the aid of auxiliary tables, enable the LMT for other latitudes (in the range N.30° to N.60°) to be found. The tabulations for Moonrise and Moonset are given for each day for latitude N.50°, together with variations (v) which, together with the auxiliary tables, enable the LMT to be found for latitudes N.30° to N.60°. The tabular values are for the Greenwich Meridian, but are approximately the Local Mean Times of the corresponding phenomena for the other meridians; only for the Moon phenomena need a correction for longitude be applied and a note of this correction is given in 5.3.8 on page 85. The GMT of a phenomenon is obtained from the LMT by

$$\text{GMT} = \text{LMT} \begin{array}{c} +\text{west} \\ -\text{east} \end{array} \text{longitude}$$

in which the longitude must first be converted to time by the table in 5.3.13 on page 114.

At Sunrise and Sunset 16' is allowed for semidiameter and 34' for horizontal refraction, so that at the times given the Sun's upper limb is on the visible horizon; all times refer to phenomena as seen from sea level with a clear horizon.

At the times given for the beginning and end of twilight, the Sun's zenith distance is 96° for Civil, and 102° for Nautical Twilight. The degree of illumination at the times given for Civil Twilight (in good conditions and in the absence of other illumination) is such that the brightest stars are visible and the horizon is clearly defined. At the times given for Nautical Twilight the horizon is in general not visible, and it is too dark for observation with a marine sextant.

At Moonrise and Moonset allowance is made for semi-diameter, parallax and refraction (34'), so that at the times given the Moon's upper limb is on the visible horizon as seen from sea level.

5.2.4 Altitude correction tables

Tables for the correction of dip of the horizon, due to height of eye above sea level, are given in 5.3.2 on page 76; this correction should be applied to the sextant altitude to give apparent altitude, which is the correct argument for the other tables.

The dip table is arranged as a critical table, where an interval of height of eye corresponds to a single value of dip. No interpolation is required. At a 'critical' entry the upper of the two possible values of the correction is to be used.

Separate tables for refraction are given for the Sun, stars and planets, and Moon. With the Sun and Moon, to allow for semi-diameter, separate corrections are given for the lower and upper limbs. For the Moon, corrections are also given for parallax.

The altitude correction tables are given to one decimal of a minute of arc so that the addition of several corrections does not lead to any significant error.

5.2.5 Polaris table

The table in 5.3.6 on page 81 provides means by which the latitude can be deduced from an observed altitude of *Polaris*; the azimuth of *Polaris* is also given. The correction Q, to be applied to the apparent altitude, corrected for refraction, is given corresponding to an argument of LHA Aries (LHA♈).

5.2.6 Auxiliary and planning data

Pages 78 and 79 (5.3.4) give the times of meridian passage (mer pass) of the Sun; symbols, approximate SHA and Dec, magnitude (mag) and meridian passage of the planets; an explanation of the symbols used and planet notes are given on page 79. Phases of the Moon, with brief eclipse notes, are also given on page 79.

Star diagrams. Star diagrams are given on pages 108–110; these have been drawn for latitude N.45° and show the approximate positions in altitude and azimuth, of stars in the Selected list for each range of 15° of LHA Aries; indications are also given of the horizons for latitude N.30° (shown as ··· ···) and N.60° (shown as ·—·—); these will enable the user to estimate the approximate altitude and azimuth of any particular star and will indicate stars suitable for observation at a given value of LHA Aries.

Symbols indicate the magnitude of the stars as follows:

- • Mag 2.1–3.1
- ○ Mag 1.1–2.0   note that the larger symbols
- ● Mag 0.0–1.0   indicate greater brilliance
- ⊙ Mag −1.6

The symbol for any star indicates its position at the beginning of the range of LHA Aries and the arrow point the end of the range.

Azimuth diagrams are given on pages 111–113; they are used as follows: a point is plotted on the diagrams corresponding to the LHA and Dec of the object observed; the plotted point is then moved horizontally to the *left* corresponding to the co-latitude (90° − latitude) of the observer; the background graticule and the scale along the top of the diagram will assist in this movement; at the new position the corresponding value of LHA is read and is converted to azimuth according to the rules given below the diagrams.

5.2.7 Illustrations

Example (a) Required the GHA and Dec of the Sun on 1989 January 22 at GMT 15^h 47^m 13^s.

				GHA	v	Dec	d
Monthly page,	Jan 22			177.°12	−3	S.19.°73	+10
Interpolation	GMT	15^h 36^m		234			
Tables	GMT	11^m 13^s		2.80			
Pages 100, 101	GMT	15^h 36^m,	$v = -3, d = +10$	− .05		+0.16	
				413.87		S.19.57	
				−360			
				53.87			

Note that, in this case, the declination is south (i.e. negative) and the correction for d is positive, so that the correction must be subtracted.

Example (b) Required the GHA and Dec of the Moon on 1989 January 25 at GMT 15^h 47^m 13^s.

				GHA	v	Dec	d
Monthly page,	Jan 25	8^h-16^h		79.°92	290	N.5.°68	−233
Interpolation	GMT	7^h 44^m		110.66			
Tables	GMT	3^m 13^s		0.76			
Page 107	GMT	7^h 44^m,	$v = 290, d = -233$	2.24		−1.80	
				193.58		N.3.88	

Example (c) Required the LHA and Dec of Venus on 1989 February 22 at GMT 07^h 47^m 13^s in longitude E.5°15′.

				GHA	v	Dec	d
Monthly page,	Feb 22			186.°27	−10	S.14.°99	+16
Interpolation	GMT	07^h 36^m		114			
Tables	GMT	11^m 13^s		2.80			
Pages 98, 99	GMT	07^h 36^m,	$v = -10, d = +16$	−0.08		+0.12	
				302.99		S.14.87	
Longitude East				+5.25			
	LHA			308.24			

The first correction for the GHA of the planets is taken from the column headed Sun/Planet.

Here, the declination is south (i.e. negative) and the correction for d is positive, so that the correction must be subtracted.

Example (d) Required the GHA and Dec of *Aldebaran* on 1989 January 22 at GMT $15^h\ 55^m\ 13^s$.

		GHA	Dec
Page 77,	SHA and Dec	291°.17	N.16°.49
Monthly page,	GHA Aries	121.32	
Interpolation	GMT $15^h\ 48^m$	237.65	
Tables	GMT $\quad\ \ 7^m\ 13^s$	1.80	
Page 101			
		651.94	
		−360	
		291.94	

Example (e) Required the GMT of the beginning of nautical twilight and sunrise on 1989 January 22 for latitude N.30°50′, longitude W.7°25′.

		Naut twi h m	v	Sunrise h m	v
Page 82,	Jan 22	06 31	+31	07 46	+51
Page 83,	Corrections for N.30°50′	−30		−49	
	LMT	06 01		06 57	
Page 114,	Long. equivalent in time	+30		+30	
	GMT	06 31		07 27	

Example (f) Required the GMT of sunset and evening civil twilight on 1989 January 22 for latitude N.36°07′, longitude E. 18°20′.

		Sunset h m	v	Civil Twi h m	v
Page 82,	Jan 22	16 37	−51	17 14	−40
Page 83,	Corrections for N.36°07′	+38		+30	
	LMT	17 15		17 44	
Page 114,	Long. equivalent in time	−1 13		−1 13	
	GMT	16 02		16 31	

Example (g) Required the GMT of moonrise and moonset on 1989 April 8 for latitude N.36°07′, longitude W.6°30′.

		Moonrise h m	v	Moonset h m	v
Page 84,	Apr 8	05 57	−62	22 43	+75
Page 85,	Corrections for N.36°07′	+ 47		−57	
	LMT	06 44		21 46	
Page 114,	Long. equivalent in time	+ 26		+ 26	
	GMT	07 10		22 12	

Ephemeris

Example (h) Required the correction to an observed altitude of *Polaris*, made on 1989 January 22 at $16^h 45^m$ in DR longitude E.18°20' the sextant angle being 37°05'.5, height of eye 15ft.

Monthly page,	Jan 22	GHA Aries	121°.32
Page 100,	Correction for $16^h 36^m$		249.68
Page 101,	Correction for 9^m		2.25
			373.25
	Correction for Long E.18°20'		+ 18.33
			391.58
			−360
	LHA Aries		31.58

Sextant altitude			37°05'.5
Page 76,	Dip, for 15ft		− 3.8
Page 76,	Refraction, stars		− 1.3
Corrected observed altitude			37 00.4
Page 81,	Q for LHA Aries 32°		− 47
	Latitude		36 13

From page 81 the azimuth of *Polaris* at LHA Aries of 32° is seen to be 0°.

The following examples illustrate the use of the altitude correction tables on page 76. The sextant altitudes are assumed to be taken on 1989 January 22 with a marine sextant at height 5.4 metres (18ft).

	SUN lower limb	SUN upper limb	MOON lower limb	MOON upper limb	Venus	*Polaris*
Sextant altitude	21°19'.7	3°20'.2	33°27'.6	26°06'.7	4°32'.6	49°36'.5
Dip, height 5.4 metres (18ft)	− 4.1	− 4.1	− 4.1	− 4.1	− 4.1	− 4.1
Main correction	+ 13.7	− 29.4	+ 58.5	+ 31.7	− 10.8	− 0.8
Parallax correction for Moon (HP = 55'.3, from page 87)			+ 1.4	+ 0.8		
Corrected sextant altitude	21 29.3	2 46.7	34 23.4	26 35.1	4 17.7	49 31.6

The following example illustrates the use of the Star Diagrams: on 1989 January 22 an observer in an approximate position of latitude N.51°, longitude W.7° wishes to determine which stars are suitable for observation at about GMT $7^h 20^m$, and determines LHA Aries as follows.

Monthly page,	Jan 22	GHA Aries	121°.32
Interpolation	⎰ GMT	$7^h 12^m$	108.30
Tables,	⎱ GMT	8^m	2.00
Page 99.			
GHA Aries at	GMT	$7^h 20^m$	231.62
	Longitude W.7°		−7.00
	LHA Aries		224.62

Inspection of the Star Diagram for LHA Aries 210° to 225° indicates that stars 53, 49, and 46 are visible towards the east, 39, 37 and 33 are towards the south, 28 and 26 are westerly and 3 (rather low) and 40 (rather high) are northerly; *Polaris* will also be available. The names of the stars corresponding to the numbers, can be found on page 77 and at the foot of the second star diagram.

5.2.8 The use of a calculator for interpolation

When using a calculator the tabular values of GHA, v, Dec and d should be extracted from the monthly pages for the date of observation (and in the case of the Moon, the nearest 8^h before the GMT of observation). The GMT must be converted to hours and decimals, i.e. hours + (minutes ÷ 60) + (seconds ÷ 3600), then the required values are found as follows. Southern declinations are treated as negative.

Stars
GHA star = Tabular GHA Aries + (15°.041 × GMT) + SHA star
Dec star = Tabular Dec (from Star list)

Sun
GHA Sun = Tabular GHA Sun + (15° + (v ÷ 1000)) × GMT (See also page 77).
Dec Sun = Tabular Dec Sun + (d ÷ 1000) × GMT

Planets
GHA Planet = Tabular GHA Planet + (15° + (v ÷ 1000)) × GMT
Dec Planet = Tabular Dec Planet + (d ÷ 1000) × GMT

Moon
GHA Moon = Tabular GHA Moon + (14°.310 + (v ÷ 1000)) × GMT*
Dec Moon = Tabular Dec Moon + (d ÷ 1000) × GMT*

* For the Moon, multiples of 8 hours must be subtracted from the GMT.

5.2.9 Standard Times

List 1 — Places fast on GMT

The times given below should be { *added* to GMT to give Standard Time / *subtracted* from Standard Time to give GMT

	h		h		h
Albania*	01	France*	01	Netherlands, The*	01
Algeria	01	Germany, East*	01	Norway*	01
Austria*	01	Germany, West[2]*	01	Poland*	01
Balearic Islands*	01	Gibraltar*	01	Romania*	02
Belgium*	01	Greece*	02	Sardinia*	01
Bulgaria*	02	Hungary*	01	Sicily*	01
Corsica*	01	Israel*	02	Spain*	01
Crete*	02	Italy*	01	Sweden*	01
Cyprus, Ercan*	02	Latvia	03	Switzerland*	01
Cyprus, Larnaca*	02	Lebanon*	02	Syria*	02
Czechoslovakia*	01	Libya*	01	Tunisia*	01
Denmark*	01	Liechtenstein	01	Turkey*	02
Egypt	02	Luxembourg*	01	USSR west of 40°E*	03
Estonia	03	Malta*	01	Yugoslavia*	01
Finland*	02	Monaco*	01		

List 2 — Places normally keeping GMT

Canary Islands*	Great Britain[1]	Irish Republic[1]
Channel Islands[1]	Iceland	Morocco*
Faeroes, The*	Ireland, Northern[1]	Portugal*

* Summer time may be kept in these countries.

[1] Summer time, one hour in advance of GMT, is kept from March 26 01^h to October 29 01^h GMT (1989).

[2] Including West Berlin.

5.3.1 CALENDAR, 1989

DAYS OF THE WEEK AND DAYS OF THE YEAR

Day of month	JAN		FEB		MAR		APR		MAY		JUN		JUL		AUG		SEP		OCT		NOV		DEC	
1	Su	1	W	32	W	60	S	91	M	121	Th	152	S	182	Tu	213	F	244	Su	274	W	305	F	335
2	M	2	Th	33	Th	61	Su	92	Tu	122	F	153	Su	183	W	214	S	245	M	275	Th	306	S	336
3	Tu	3	F	34	F	62	M	93	W	123	S	154	M	184	Th	215	Su	246	Tu	276	F	307	Su	337
4	W	4	S	35	S	63	Tu	94	Th	124	Su	155	Tu	185	F	216	M	247	W	277	S	308	M	338
5	Th	5	Su	36	Su	64	W	95	F	125	M	156	W	186	S	217	Tu	248	Th	278	Su	309	Tu	339
6	F	6	M	37	M	65	Th	96	S	126	Tu	157	Th	187	Su	218	W	249	F	279	M	310	W	340
7	S	7	Tu	38	Tu	66	F	97	Su	127	W	158	F	188	M	219	Th	250	S	280	Tu	311	Th	341
8	Su	8	W	39	W	67	S	98	M	128	Th	159	S	189	Tu	220	F	251	Su	281	W	312	F	342
9	M	9	Th	40	Th	68	Su	99	Tu	129	F	160	Su	190	W	221	S	252	M	282	Th	313	S	343
10	Tu	10	F	41	F	69	M	100	W	130	S	161	M	191	Th	222	Su	253	Tu	283	F	314	Su	344
11	W	11	S	42	S	70	Tu	101	Th	131	Su	162	Tu	192	F	223	M	254	W	284	S	315	M	345
12	Th	12	Su	43	Su	71	W	102	F	132	M	163	W	193	S	224	Tu	255	Th	285	Su	316	Tu	346
13	F	13	M	44	M	72	Th	103	S	133	Tu	164	Th	194	Su	225	W	256	F	286	M	317	W	347
14	S	14	Tu	45	Tu	73	F	104	Su	134	W	165	F	195	M	226	Th	257	S	287	Tu	318	Th	348
15	Su	15	W	46	W	74	S	105	M	135	Th	166	S	196	Tu	227	F	258	Su	288	W	319	F	349
16	M	16	Th	47	Th	75	Su	106	Tu	136	F	167	Su	197	W	228	S	259	M	289	Th	320	S	350
17	Tu	17	F	48	F	76	M	107	W	137	S	168	M	198	Th	229	Su	260	Tu	290	F	321	Su	351
18	W	18	S	49	S	77	Tu	108	Th	138	Su	169	Tu	199	F	230	M	261	W	291	S	322	M	352
19	Th	19	Su	50	Su	78	W	109	F	139	M	170	W	200	S	231	Tu	262	Th	292	Su	323	Tu	353
20	F	20	M	51	M	79	Th	110	S	140	Tu	171	Th	201	Su	232	W	263	F	293	M	324	W	354
21	S	21	Tu	52	Tu	80	F	111	Su	141	W	172	F	202	M	233	Th	264	S	294	Tu	325	Th	355
22	Su	22	W	53	W	81	S	112	M	142	Th	173	S	203	Tu	234	F	265	Su	295	W	326	F	356
23	M	23	Th	54	Th	82	Su	113	Tu	143	F	174	Su	204	W	235	S	266	M	296	Th	327	S	357
24	Tu	24	F	55	F	83	M	114	W	144	S	175	M	205	Th	236	Su	267	Tu	297	F	328	Su	358
25	W	25	S	56	S	84	Tu	115	Th	145	Su	176	Tu	206	F	237	M	268	W	298	S	329	M	359
26	Th	26	Su	57	Su	85	W	116	F	146	M	177	W	207	S	238	Tu	269	Th	299	Su	330	Tu	360
27	F	27	M	58	M	86	Th	117	S	147	Tu	178	Th	208	Su	239	W	270	F	300	M	331	W	361
28	S	28	Tu	59	Tu	87	F	118	Su	148	W	179	F	209	M	240	Th	271	S	301	Tu	332	Th	362
29	Su	29			W	88	S	119	M	149	Th	180	S	210	Tu	241	F	272	Su	302	W	333	F	363
30	M	30			Th	89	Su	120	Tu	150	F	181	Su	211	W	242	S	273	M	303	Th	334	S	364
31	Tu	31			F	90			W	151			M	212	Th	243			Tu	304			Su	365

RELIGIOUS CALENDARS

Epiphany	Jan 6
Septuagesima Sunday	Jan 22
Ash Wednesday	Feb 8
Good Friday	Mar 24
Easter Day	Mar 26
Low Sunday	Apr 2
Ascension Day	May 4
Whit Sunday	May 14
Christmas Day	Dec 25
Jewish New Year (5750)	Sep 30
Islamic New Year (1410)	Aug 4

CIVIL CALENDAR

Accession of the Queen	Feb 6
St David (Wales)	Mar 1
Commonwealth Day	Mar 13
St Patrick (Ireland)	Mar 17
Birthday of the Queen	Apr 21
St George (England)	Apr 23
Coronation Day	Jun 2
Birthday of Prince Philip	Jun 10
The Queen's Official Birthday	Jun 17
Remembrance Sunday	Nov 12
Birthday of the Prince of Wales	Nov 14
St Andrew (Scotland)	Nov 30

BANK HOLIDAYS

ENGLAND & WALES
Jan 2, Mar 27, May 1, May 29, Aug 28, Dec 26

NORTHERN IRELAND
Jan 2, Mar 17, Mar 27, May 1, May 29, July 12, Aug 28, Dec 26

SCOTLAND
Jan 2, Jan 3, Mar 24, May 1, May 29, Aug 7, Dec 25, Dec 26

5.3.2 ALTITUDE CORRECTION TABLES

DIP

Ht of eye (m)	Corr	Ht of eye (ft)
1.6	−2.3	5.3
1.7	−2.4	5.8
1.9	−2.5	6.3
2.1	−2.6	6.9
2.2	−2.7	7.4
2.4	−2.8	8.0
2.6	−2.9	8.6
2.8	−3.0	9.2
3.0	−3.1	9.8
3.2	−3.2	10.5
3.4	−3.3	11.2
3.6	−3.4	11.9
3.8	−3.5	12.6
4.0	−3.6	13.3
4.3	−3.7	14.1
4.5	−3.8	14.9
4.7	−3.9	15.7
5.0	−4.0	16.5
5.2	−4.1	17.4
5.5	−4.2	18.3
5.8	−4.3	19.1
6.1	−4.4	20.1
6.3	−4.5	21.0
6.6	−4.6	22.0
6.9	−4.7	22.9
7.2	−4.8	23.9
7.5	−4.9	24.9
7.9	−5.0	26.0
8.2	−5.1	27.1
8.5	−5.2	28.1
8.8	−5.3	29.2
9.2	−5.4	30.4
9.5	−5.5	31.5
9.9	−5.6	32.7
10.3	−5.7	33.9
10.6	−5.8	35.1
11.0	−5.9	36.3
11.4	−6.0	37.6
11.8	−6.1	38.9
12.2	−6.2	40.1
12.6	−6.3	41.5
13.0	−6.4	42.8
13.4	−6.5	44.2
13.8	−6.6	45.5
14.2	−6.7	46.9
14.7	−6.8	48.4
15.1	−6.9	49.8
15.5	−7.0	51.3
16.0	−7.1	52.8
16.5	−7.2	54.3
16.9	−7.3	55.8
17.4	−7.4	57.4
17.9	−7.5	58.9
18.4	−7.6	60.5
18.8		62.1

SUN / STARS and PLANETS

App Alt	Sun Lower limb	Sun Upper limb	Stars and Planets
2.5	+ 0.1	−32.0	−16.1
3.0	1.8	30.2	14.4
3.5	3.2	28.8	13.0
4.0	4.4	27.7	11.8
4.5	5.4	26.6	10.7
5	+ 6.3	−25.8	− 9.9
6	7.7	24.3	8.5
7	8.7	23.3	7.4
8	9.6	22.4	6.6
9	10.3	21.8	5.9
10	+ 10.8	−21.2	− 5.3
12	11.7	20.4	4.5
14	12.4	19.7	3.8
16	12.8	19.2	3.3
18	13.2	18.8	3.0
20	+ 13.5	−18.5	− 2.6
25	14.1	18.0	2.1
30	14.5	17.6	1.7
35	14.8	17.3	1.4
40	15.0	17.0	1.2
45	+ 15.2	−16.9	− 1.0
50	15.3	16.8	0.8
55	15.5	16.6	0.7
60	15.6	16.5	0.6
70	15.7	16.3	0.4
80	+ 15.8	−16.1	− 0.2

MOON

App Alt	Lower limb	Upper limb	App Alt	Lower limb	Upper limb
2.5	+ 52.5	+ 23.1	40	+ 55.1	+ 25.3
3.0	54.2	24.8	42	53.9	24.2
3.5	55.7	26.2	44	52.7	23.0
4.0	56.8	27.4	46	51.5	21.7
4.5	57.8	28.4	48	50.2	20.4
5	+ 58.6	+ 29.2	50	+ 48.8	+ 19.0
6	60.0	30.5	52	47.4	17.6
7	60.9	31.5	54	46.0	16.1
8	61.7	32.2	56	44.5	14.6
9	62.2	32.7	58	42.9	13.1
10	+ 62.6	+ 33.1	60	+ 41.4	+ 11.5
12	63.1	33.6	62	39.8	9.9
14	63.3	33.8	64	38.1	8.3
16	63.3	33.7	66	36.5	6.6
18	63.2	33.6	68	34.8	4.9
20	+ 62.9	+ 33.3	70	+ 33.1	+ 3.2
22	62.5	32.9	71	32.2	2.3
24	62.0	32.3	72	31.3	1.4
26	61.4	31.7	73	30.4	+ 0.5
28	60.7	31.0	74	29.6	− 0.3
30	+ 59.9	+ 30.3	75	+ 28.7	− 1.2
32	59.1	29.4	76	27.8	− 2.1
34	58.2	28.5	77	26.9	− 3.0
36	57.2	27.5	78	26.0	− 3.9
38	56.2	26.4	79	25.1	− 4.8
40	+ 55.1	+ 25.3	80	+ 24.2	− 5.7

CORRECTION FOR MOON'S PARALLAX

App Alt HP	0° L	0° U	20° L	20° U	40° L	40° U	50° L	50° U	60° L	60° U	70° L	70° U	80° L	80° U
54.0	+0.0	+0.0	+0.0	+0.0	+0.0	+0.0	+0.0	+0.0	+0.0	+0.0	+0.0	+0.0	+0.0	0.0
54.5	0.6	0.4	0.6	0.3	0.5	0.2	0.4	0.2	0.4	0.1	0.3	0.0	0.2	0.0
55.0	1.3	0.7	1.2	0.7	1.0	0.5	0.9	0.4	0.8	0.2	0.6	0.0	0.4	−0.1
55.5	1.9	1.1	1.8	1.0	1.5	0.7	1.3	0.6	1.1	0.3	0.9	0.1	0.6	−0.1
56.0	2.5	1.5	2.4	1.3	2.1	1.0	1.8	0.7	1.5	0.5	1.2	0.1	0.9	−0.2
56.5	3.1	1.8	3.0	1.7	2.6	1.2	2.2	0.9	1.9	0.6	1.5	0.2	1.1	−0.2
57.0	3.8	2.2	3.6	2.0	3.1	1.5	2.7	1.1	2.3	0.7	1.8	0.2	1.3	−0.3
57.5	4.5	2.5	4.2	2.3	3.6	1.7	3.2	1.3	2.7	0.8	2.1	0.2	1.5	−0.3
58.0	5.1	2.9	4.8	2.7	4.1	2.0	3.6	1.5	3.1	1.0	2.4	0.3	1.8	−0.4
58.5	5.7	3.3	5.4	3.0	4.7	2.2	4.1	1.7	3.4	1.1	2.8	0.3	2.0	−0.4
59.0	6.3	3.6	6.0	3.3	5.2	2.5	4.6	1.8	3.8	1.2	3.1	0.3	2.3	−0.5
59.5	7.0	4.0	6.6	3.7	5.7	2.7	5.0	2.0	4.2	1.3	3.4	0.4	2.5	−0.5
60.0	7.6	4.4	7.2	4.0	6.2	3.0	5.5	2.2	4.6	1.4	3.7	0.4	2.7	−0.6
60.5	8.2	4.8	7.8	4.3	6.7	3.2	6.0	2.4	4.9	1.5	4.0	0.4	2.9	−0.6
61.0	8.9	5.1	8.4	4.7	7.2	3.5	6.4	2.6	5.2	1.6	4.3	0.5	3.1	−0.7
61.5	9.5	5.5	9.0	5.0	7.7	3.7	6.8	2.7	5.5	1.7	4.6	0.5	3.4	−0.7

ALTITUDE CORRECTION TABLES

App Alt = Apparent altitude
    = Sextant altitude corrected for index error and dip.

The correction for the Sun's lower or upper limbs can be taken out directly as also can the corrections for the stars and planets.

The correction for the Moon is in two parts; the first correction is taken from the upper table with argument apparent altitude, and the second correction from the lower table with arguments apparent altitude and HP.

Separate corrections are given for lower (L) and upper (U) limbs.

5.3.3 SELECTED STARS, 1989

No	Name	Mag	SHA ° ′	Dec ° ′	Name	No
1	Alpheratz	2.2	358.04 (02)	N29.04 (02)	Acamar	7
2	Ankaa	2.4	353.55 (33)	S42.36 (22)	Adhara	19
3	Schedar	2.5	350.02 (01)	N56.48 (29)	Aldebaran	10
4	Diphda	2.2	349.23 (14)	S18.04 (03)	Alioth	32
6	Hamal	2.2	328.35 (21)	N23.42 (25)	Alkaid	34
7	Acamar	3.1	315.53 (32)	S40.34 (21)	Al Na'ir	55
8	Menkar	2.8	314.56 (34)	N 4.05 (03)	Alnilam	15
9	Mirfak	1.9	309.10 (06)	N49.83 (50)	Alphard	25
10	Aldebaran	1.1	291.17 (10)	N16.49 (30)	Alphecca	41
11	Rigel	0.3	281.49 (29)	S 8.21 (13)	Alpheratz	1
12	Capella	0.2	281.02 (01)	N45.99 (59)	Altair	51
13	Bellatrix	1.7	278.85 (51)	N 6.34 (21)	Ankaa	2
14	Elnath	1.8	278.59 (35)	N28.60 (36)	Antares	42
15	Alnilam	1.8	276.08 (05)	S 1.21 (12)	Arcturus	37
16	Betelgeuse	*	271.34 (21)	N 7.41 (25)	Bellatrix	13
17	Canopus	-0.9	264.07 (04)	S52.69 (41)	Betelgeuse	16
18	Sirius	-1.6	258.82 (49)	S16.70 (42)	Canopus	17
19	Adhara	1.6	255.44 (27)	S28.96 (57)	Capella	12
20	Procyon	0.5	245.31 (18)	N 5.25 (15)	Deneb	53
21	Pollux	1.2	243.83 (50)	N28.05 (03)	Denebola	28
23	Suhail	2.2	223.10 (06)	S43.39 (23)	Diphda	4
25	Alphard	2.2	218.23 (14)	S 8.61 (37)	Dubhe	27
26	Regulus	1.3	208.04 (02)	N12.02 (01)	Elnath	14
27	Dubhe	2.0	194.22 (13)	N61.81 (48)	Eltanin	47
28	Denebola	2.2	182.87 (52)	N14.63 (38)	Enif	54
29	Gienah	2.8	176.18 (11)	S17.48 (29)	Fomalhaut	56
32	Alioth	1.7	166.60 (36)	N56.01 (01)	Gienah	29
33	Spica	1.2	158.84 (50)	S11.11 (06)	Hamal	6
34	Alkaid	1.9	153.21 (13)	N49.36 (22)	Kaus Aust	48
36	Menkent	2.3	148.49 (29)	S36.32 (19)	Kochab	40
37	Arcturus	0.2	146.20 (12)	N19.23 (14)	Markab	57
39	Zuben'ubi	2.9	137.43 (26)	S16.00 (00)	Menkar	8
40	Kochab	2.2	137.31 (19)	N74.20 (12)	Menkent	36
41	Alphecca	2.3	126.44 (26)	N26.75 (45)	Mirfak	9
42	Antares	1.2	112.81 (49)	S26.41 (25)	Nunki	50
44	Sabik	2.6	102.56 (33)	S15.72 (43)	Pollux	21
45	Shaula	1.7	96.78 (47)	S37.10 (06)	Procyon	20
46	Rasalhague	2.1	96.39 (23)	N12.57 (34)	Rasalhague	46
47	Eltanin	2.4	90.91 (55)	N51.49 (29)	Regulus	26
48	Kaus Aust	2.0	84.13 (08)	S34.39 (24)	Rigel	11
49	Vega	0.1	80.85 (51)	N38.77 (46)	Sabik	44
50	Nunki	2.1	76.35 (21)	S26.31 (19)	Schedar	3
51	Altair	0.9	62.43 (26)	N 8.84 (50)	Shaula	45
53	Deneb	1.3	49.73 (44)	N45.24 (15)	Sirius	18
54	Enif	2.5	34.08 (05)	N 9.83 (50)	Spica	33
55	Al Na'ir	2.2	28.10 (06)	S47.01 (01)	Suhail	23
56	Fomalhaut	1.3	15.73 (44)	S29.68 (41)	Vega	49
57	Markab	2.6	13.94 (56)	N15.15 (09)	Zuben'ubi	39

* Variable 0.1 – 1.2

SUN 1989—Interval 10 days
(For use with calculator)

Date (0ʰ GMT)	GHA °	A °	B °	Dec °	C °	D °
Jan 1	179.15	359.880	+0.0009	S23.02	+0.082	+0.0037
11	178.04	359.901	.0014	21.84	.155	.0033
21	177.19	359.929	.0016	19.96	.223	.0029
31	176.64	359.962	.0017	17.45	.279	.0023
Feb 10	176.43	359.997	.0015	14.43	.326	.0017
20	176.55	360.027	+0.0012	S11.00	+0.359	+0.0012
Mar 2	176.94	360.051	.0008	7.30	.383	.0006
12	177.53	360.067	+ .0004	S 3.41	.395	+ .0001
22	178.24	360.076	.0000	N 0.54	.395	− .0004
Apr 1	179.00	360.075	− .0004	4.45	.387	.0009
11	179.71	360.068	−0.0008	N 8.23	+0.369	−0.0014
21	180.31	360.051	.0010	11.78	.342	.0019
May 1	180.72	360.031	.0011	15.01	.304	.0023
11	180.92	360.008	.0012	17.82	.259	.0027
21	180.88	359.984	.0010	20.14	.204	.0030
31	180.62	359.963	−0.0007	N21.88	+0.145	−0.0033
June 10	180.18	359.949	− .0002	23.00	.078	.0034
20	179.65	359.945	+ .0002	23.44	+ .009	.0034
30	179.12	359.949	.0007	23.19	− .059	.0033
July 10	178.68	359.965	.0010	22.27	.126	.0031
20	178.43	359.986	+0.0012	N20.71	−0.187	−0.0027
30	178.40	360.009	.0013	18.57	.241	.0023
Aug 9	178.61	360.036	.0011	15.93	.288	.0019
19	179.08	360.058	.0009	12.86	.326	.0015
29	179.74	360.076	.0006	9.45	.357	.0010
Sept 8	180.55	360.087	+0.0002	N 5.78	−0.377	−0.0006
18	181.43	360.090	− .0003	N 1.96	.388	− .0001
28	182.30	360.085	.0007	S 1.93	.390	+ .0004
Oct 8	183.08	360.072	.0011	5.79	.383	.0009
18	183.69	360.049	.0014	9.53	.365	.0014
28	184.04	360.020	−0.0016	S13.04	−0.337	+0.0020
Nov 7	184.08	359.987	.0017	16.21	.298	.0026
17	183.78	359.951	.0016	18.93	.246	.0031
27	183.13	359.918	.0013	21.08	.185	.0035
Dec 7	182.18	359.891	− .0007	22.58	.115	.0038
17	181.02	359.876	0.0000	S23.35	−0.038	+0.0039
27	179.78	359.877	+0.0006	S23.34	+0.040	+0.0038

GHA and Dec of the Sun can be determined by:

$GHA(t) = GHA(tab) + A.t + B.t^2$
$Dec(t) = Dec(tab) + C.t + D.t^2$

where t is the time difference in days and decimals of a day (see p. 114) between the time of observation and the nearest tabular date before that time; t must be determined to five decimal places, but t^2 is only required to the nearest whole number.

In evaluating GHA Sun (and GHA Aries, below) multiples of 360 degrees should be removed as required.

GHA Aries

GHA Aries can be determined by
GHA Aries = 99.64 + 360.98565 × d

where d is the day of the year (p. 75) plus the decimal of a day corresponding to the GMT of observation.

5.3.4 PLANET PLANNING DATA, 1989

Date	SUN Mer Pass	Symbol	VENUS SHA	Dec	Mag	Mer Pass	Symbol	MARS SHA	Dec	Mag	Mer Pass	Symbol	JUPITER SHA	Dec	Mag	Mer Pass	Symbol	SATURN SHA	Dec	Mag	Mer Pass
	h m		°	°		h m		°	°		h m		°	°		h m		°	°		h m
Jan 1	12 04	Eb	103	S22	−3.9	10 25	WA	342	N 8	0.0	18 30	EA	305	N19	−2.7	20 52	☉	84	S23	+0.5	11 40
6	12 06		96	23	3.9	10 33		339	9	+0.1	18 20	MA	306	19	2.7	20 32	☉	83	23	0.5	11 23
11	12 08		90	23	3.9	10 40		337	11	0.2	18 10		306	18	2.6	20 11	Eb	83	23	0.5	11 06
16	12 10		83	23	3.9	10 48		334	12	0.3	18 00		306	18	2.6	19 51		82	23	0.5	10 49
21	12 11		76	23	3.9	10 55		332	13	0.4	17 52		306	18	2.6	19 31		81	23	0.5	10 31
26	12 13	Eb	69	S22	−3.9	11 02	WA	329	N14	+0.5	17 42	MA	306	N19	−2.5	19 12	Eb	81	S22	+0.6	10 14
31	12 13		63	22	3.9	11 09		326	15	0.5	17 34	MA	306	19	2.5	18 53	Eb	80	22	0.6	09 57
Feb 5	12 14		56	20	3.9	11 16		323	16	0.6	17 25	WA	306	19	2.4	18 34	EB	80	22	0.6	09 39
10	12 14	Eb	50	19	3.9	11 22		320	17	0.7	17 17		305	19	2.4	18 16		79	22	0.6	09 22
15	12 14	☉	43	18	3.9	11 28		317	18	0.8	17 09		305	19	2.4	17 58		79	22	0.6	09 04
20	12 14	☉	37	S16	−3.9	11 33	WA	314	N18	+0.9	17 01	WA	304	N19	−2.3	17 40	EB	78	S22	+0.6	08 47
25	12 13		31	14	3.9	11 38		311	19	0.9	16 54		304	19	2.3	17 22		78	22	0.6	08 29
Mar 2	12 12		25	12	3.9	11 43		308	20	1.0	16 46		303	19	2.3	17 05		77	22	0.6	08 11
7	12 11		19	9	3.9	11 47		305	21	1.1	16 39		303	19	2.2	16 48		77	22	0.6	07 53
12	12 10		13	7	3.9	11 50		302	22	1.1	16 32		302	20	2.2	16 32		76	22	0.6	07 35
17	12 08	☉	7	S 5	−3.9	11 53	WA	299	N22	+1.2	16 25	WA	301	N20	−2.2	16 15	EB	76	S22	+0.6	07 16
22	12 07		1	S 2	3.9	11 57		296	23	1.2	16 19		300	20	2.1	15 59		76	22	0.6	06 58
27	12 05		356	0	3.9	12 00		292	23	1.3	16 12		299	20	2.1	15 43		76	22	0.5	06 39
Apr 1	12 04		350	N 3	3.9	12 03		289	24	1.3	16 06		298	20	2.1	15 27	EB	75	22	0.5	06 20
6	12 02		344	5	3.9	12 06		286	24	1.4	15 59		297	20	2.1	15 11	MB	75	22	0.5	06 01
11	12 01	☉	339	N 8	−3.9	12 09	WA	282	N24	+1.4	15 53	WA	296	N21	−2.0	14 56	MB	75	S22	+0.5	05 42
16	12 00		333	10	3.9	12 12		279	25	1.5	15 47	WA	295	21	2.0	14 40		75	22	0.5	05 23
21	11 59		327	12	3.9	12 16		276	25	1.5	15 41	Wa	294	21	2.0	14 25		75	22	0.5	05 03
26	11 58		321	15	3.9	12 20		272	25	1.6	15 35		293	21	2.0	14 10		75	22	0.4	04 43
May 1	11 57		315	17	3.9	12 25		269	25	1.6	15 29		292	21	2.0	13 54		75	22	0.4	04 24
6	11 57	☉	309	N18	−3.9	12 30	WA	265	N25	+1.6	15 22	Wa	291	N22	−2.0	13 39	MB	75	S22	+0.4	04 04
11	11 56		303	20	3.9	12 36		262	25	1.7	15 16		290	22	2.0	13 25	WB	75	22	0.3	03 43
16	11 56	☉	296	21	3.9	12 41		259	24	1.7	15 10		288	22	2.0	13 10		75	22	0.3	03 23
21	11 57	Wa	290	23	3.9	12 48		255	24	1.7	15 04	Wa	287	22	2.0	12 55		76	22	0.3	03 03
26	11 57		283	23	3.9	12 55		252	24	1.7	14 58	☉	286	22	1.9	12 40		76	22	0.2	02 42
31	11 58	Wa	276	N24	−3.9	13 02	WA	248	N23	+1.7	14 51	☉	285	N22	−1.9	12 25	WB	76	S22	+0.2	02 21
June 5	11 58		270	24	3.9	13 09		245	23	1.8	14 45		284	22	1.9	12 11	WB	76	22	0.2	02 00
10	11 59		263	24	3.9	13 16		242	22	1.8	14 39		282	23	1.9	11 56	V	77	22	0.2	01 39
15	12 00		256	24	3.9	13 23	WA	239	22	1.8	14 32		281	23	1.9	11 41		77	22	0.1	01 18
20	12 01		250	23	3.9	13 30	Wa	235	21	1.8	14 25	☉	280	23	1.9	11 27		78	22	0.1	00 57
25	12 03	Wa	243	N23	−3.9	13 36	Wa	232	N20	+1.8	14 18	Eb	279	N23	−1.9	11 12	V	78	S22	+0.1	00 36
30	12 04		237	21	3.9	13 42		229	19	1.8	14 11		277	23	1.9	10 57		78	22	0.0	00 15
Jul 5	12 05		230	20	3.9	13 48		226	19	1.8	14 04		276	23	1.9	10 42		79	22	0.0	23 49
10	12 05		224	18	3.9	13 53		223	18	1.8	13 57		275	23	1.9	10 28		79	22	+0.1	23 28
15	12 06		218	17	3.9	13 57		219	17	1.8	13 50		274	23	2.0	10 13		79	22	0.1	23 07
20	12 06	Wa	212	N15	−3.9	14 01	Wa	216	N16	+1.8	13 43	Eb	273	N23	−2.0	09 58	V	80	S23	+0.1	22 46
25	12 06		206	13	3.9	14 05		213	15	1.8	13 35	Eb	271	23	2.0	09 43	V	80	23	0.1	22 24
30	12 06		201	10	3.9	14 08		210	14	1.8	13 28	EB	270	23	2.0	09 27	EA	81	23	0.2	22 03
Aug 4	12 06		195	8	3.9	14 10		207	12	1.8	13 20		269	23	2.0	09 12		81	23	0.2	21 43
9	12 05		190	5	3.9	14 12		204	11	1.8	13 12		268	23	2.0	08 57		81	23	0.2	21 22
14	12 05	Wa	184	N 3	−4.0	14 14	Wa	201	N10	+1.8	13 04	EB	267	N23	−2.0	08 41	EA	81	S23	+0.3	21 01
19	12 04		179	0	4.0	14 16		198	9	1.8	12 57		266	23	2.1	08 26	EA	81	23	0.3	20 41
24	12 02		173	S 2	4.0	14 18	☉	195	8	1.8	12 49		265	23	2.1	08 10	MA	82	23	0.3	20 20
29	12 01		168	5	4.0	14 19		192	6	1.8	12 41		264	23	2.1	07 54		82	23	0.3	20 00
Sep 3	11 59		163	7	4.0	14 21		189	5	1.8	12 33		263	23	2.1	07 37		82	23	0.4	19 40
8	11 58	Wa	157	S10	−4.0	14 23	☉	187	N 4	+1.8	12 25	EB	262	N23	−2.1	07 21	MA	82	S23	+0.4	19 20
13	11 56	Wa	152	12	4.0	14 25		184	3	1.7	12 17		262	23	2.2	07 04		82	23	0.4	19 01
18	11 54	WA	146	15	4.1	14 28		181	N 1	1.7	12 09		261	23	2.2	06 47	MA	82	23	0.4	18 41
23	11 52		141	17	4.1	14 30		178	0	1.7	12 01		260	23	2.2	06 30	WA	82	23	0.5	18 22
28	11 51		135	19	4.1	14 33		175	S 1	1.7	11 53		260	23	2.3	06 13		82	23	0.5	18 03
Oct 3	11 49	WA	129	S21	−4.1	14 36	☉	172	S 3	+1.7	11 46	EB	259	N23	−2.3	05 55	WA	82	S23	+0.5	17 44

Ephemeris

PLANET PLANNING DATA — PHASES OF THE MOON, 1989

Date	SUN Mer Pass		VENUS SHA	Dec	Mag	Mer Pass		MARS SHA	Dec	Mag	Mer Pass		JUPITER SHA	Dec	Mag	Mer Pass		SATURN SHA	Dec	Mag	Mer Pass
	h m	Symbol	°	°		h m	Symbol	°	°		h m	Symbol	°	°		h m	Symbol	°	°		h m
Oct 3	11 49	WA	129	S21	−4.1	14 36	☉	172	S 3	+1.7	11 46	EB	259	N23	−2.3	05 55	WA	82	S23	+0.5	17 44
8	11 48		123	22	4.2	14 40		169	4	1.7	11 38	MB	259	23	2.3	05 37		81	23	0.5	17 25
13	11 46		118	24	4.2	14 44		166	5	1.7	11 30		259	23	2.4	05 18		81	23	0.5	17 06
18	11 45		112	25	4.2	14 48		163	7	1.7	11 23		258	23	2.4	05 00		81	23	0.5	16 48
23	11 44		106	26	4.3	14 52		160	8	1.7	11 15		258	23	2.4	04 41		81	23	0.6	16 30
28	11 44	WA	100	S26	−4.3	14 56	☉	156	S 9	+1.7	11 08	MB	258	N23	−2.5	04 21	WA	80	S23	+0.6	16 11
Nov 2	11 44		94	27	4.4	15 00		153	10	1.7	11 01	MB	258	23	2.5	04 01		80	23	0.6	15 53
7	11 44		88	27	4.4	15 03	☉	150	12	1.7	10 54	WB	258	23	2.5	03 41		79	23	0.6	15 36
12	11 44		83	27	4.4	15 05	Eb	147	13	1.7	10 47		259	23	2.6	03 21		79	23	0.6	15 18
17	11 45		77	26	4.5	15 07		144	14	1.7	10 41		259	23	2.6	03 00		78	23	0.6	15 00
22	11 46	WA	72	S26	−4.5	15 07	Eb	140	S15	+1.7	10 34	WB	259	N23	−2.6	02 39	WA	78	S23	+0.6	14 42
27	11 48		67	25	4.6	15 06		137	16	1.7	10 28		260	23	2.6	02 17	WA	77	23	0.6	14 25
Dec 2	11 49		63	24	4.6	15 03		134	17	1.6	10 22	WB	260	23	2.7	01 56	Wa	77	23	0.6	14 08
7	11 51		59	23	4.6	14 59		130	18	1.6	10 16	V	261	23	2.7	01 34		76	23	0.6	13 50
12	11 54		56	22	4.7	14 51		127	19	1.6	10 10		261	23	2.7	01 11		76	22	0.5	13 33
17	11 56	WA	54	S21	−4.7	14 41	Eb	123	S20	+1.6	10 05	V	262	N23	−2.7	00 49	Wa	75	S22	+0.5	13 16
22	11 59	Wa	52	19	4.7	14 27		119	21	1.6	10 00		263	23	2.7	00 26	Wa	74	22	0.5	12 59
27	12 01		51	18	4.6	14 10		116	21	1.6	09 55		264	23	2.7	00 04	☉	74	22	0.5	12 41
32	12 04	Wa	52	S17	−4.6	13 49	Eb	112	S22	+1.5	09 50	V	264	N23	−2.7	23 37	☉	73	S22	+0.5	12 24

The symbols indicate the position of the planet as follows:

☉	Too close to the Sun for observation.		
Eb	Low in the east before sunrise.	Wa	Low in the west after sunset.
EB	Well placed in the east before sunrise.	WA	Well placed in the west after sunset.
MB	Well placed near the meridian before sunrise.	MA	Well placed near the meridian after sunset.
WB	Well placed in the west before sunrise.	EA	Well placed in the east after sunset.
V	Visible all night or most of the night, in the east after sunset or in the west before sunrise.		

PLANET NOTES

The approximate SHA and Dec of the planets as tabulated above can be used for plotting the positions of the planets on a star chart, or for determining their positions amongst the stars in a star list.

Venus, the most brilliant of the planets, is visible in the morning sky until late February and then in the evening sky from mid-May to the end of the year. It is in conjunction with Saturn on January 16 and November 15, with Mercury on February 1, Jupiter on May 23 and with Mars on July 12.

Mars, which can be distinguished by its reddish colour, is too close to the Sun from mid-August to early November. Before that it is visible in the evening and afterwards in the morning. It is in conjunction with Jupiter on March 12, with Venus on July 12 and with Mercury on August 5.

Jupiter, the second brightest planet, is too close to the Sun from late May to late June. Before that it is visible all night/evening and afterwards morning/all night. It is in conjunction with Mars on March 12, Venus on May 23, and with Mercury on July 2.

Saturn is visible morning/all night/evening except at the beginning and end of the year. It is in conjunction with Venus on January 16 and November 15, and with Mercury on December 16.

Do not confuse — Venus (the brighter) with Saturn in mid-January and mid-November;
with Jupiter in late May; with Mars in mid-July.
— Jupiter (slightly brighter) with Mars (reddish) in March

PHASES OF THE MOON

	New Moon	First Quarter	Full Moon	Last Quarter
	d h m	d h m	d h m	d h m
Jan	7 19 22	14 13 58	21 21 33	30 02 02
Feb	6 07 37	12 23 15	20 15 32	28 20 08
Mar	7 18 19	14 10 11	22 09 58	30 10 21
Apr	6 03 33	12 23 13	21 03 13	28 20 46
May	5 11 46	12 14 19	20 18 16	28 04 01
Jun	3 19 53	11 06 59	19 06 57	26 09 09
Jul	3 04 59	11 00 19	18 17 42	25 13 31
Aug	1 16 06	9 17 28	17 03 07	23 18 40
Aug	31 05 44			
Sep	29 21 47	8 09 49	15 11 51	22 02 10
Oct	29 15 27	8 00 52	14 20 32	21 13 19
Nov	28 09 41	6 14 11	13 05 51	20 04 44
Dec	28 03 20	6 01 26	12 16 30	19 23 54

ECLIPSES

1. Total eclipse of the Moon, February 20. Visible in north-east Europe.

2. Partial eclipse of the Sun, March 7. Not visible in Europe.

3. Total eclipse of the Moon, August 17. Visible in Europe (except northern Scandinavia).

4. Partial eclipse of the Sun, August 31. Not visible in Europe.

5.3.5 STARS, 1989 — SHA AND DEC

Mag	Name and Number	SHA ° '	Dec ° '	Mag	Name and Number	SHA ° '	Dec ° '
3.4	γ Cephei	5.28 (17)	N77.58 (35)	3.0	γ Lupi	126.39 (24)	S41.13 (08)
2.6	α Peg Markab 57	13.94 (56)	N15.15 (09)	2.3	α CrB Alphecca 41	126.44 (26)	N26.75 (45)
2.6	β Peg Scheat	14.18 (11)	N28.03 (02)	3.1	γ Ursæ Minoris	129.81 (49)	N71.87 (52)
1.3	α PsA Fomalhaut 56	15.73 (44)	S29.68 (41)	2.7	β Libræ	130.89 (53)	S 9.35 (21)
2.2	β Gruis	19.49 (29)	S46.94 (56)	2.8	β Lupi	135.54 (33)	S43.09 (06)
2.2	α Gru Al Na'ir 55	28.10 (06)	S47.01 (01)	2.2	β UMi Kochab 40	137.31 (19)	N74.20 (12)
3.0	δ Capricorni	33.38 (23)	S16.18 (11)	2.9	α Lib Zuben'ubi 39	137.43 (26)	S16.00 (00)
2.5	ε Peg Enif 54	34.08 (05)	N 9.83 (50)	2.6	ε Bootis	138.87 (52)	N27.12 (07)
3.1	β Aquarii	37.25 (15)	S 5.62 (37)	2.9	α Lupi	139.70 (42)	S47.34 (21)
2.6	α Cep Alderamin	40.42 (25)	N62.54 (32)	2.6	η Centauri	141.29 (18)	S42.11 (07)
2.6	ε Cygni	48.55 (33)	N33.93 (56)	3.0	γ Bootis	142.08 (05)	N38.35 (21)
1.3	α Cyg Deneb 53	49.73 (44)	N45.24 (15)	0.2	α Boo Arcturus 37	146.20 (12)	N19.23 (14)
3.2	α Indi	50.79 (47)	S47.33 (20)	2.3	θ Cen Menkent 36	148.49 (29)	S36.32 (19)
2.3	γ Cygni	54.54 (32)	N40.22 (13)	3.1	ζ Centauri	151.28 (17)	S47.24 (14)
0.9	α Aql Altair 51	62.43 (26)	N 8.84 (50)	2.8	η Bootis	151.45 (27)	N18.45 (27)
2.8	γ Aquilæ	63.56 (34)	N10.58 (35)	1.9	η UMa Alkaid 34	153.21 (13)	N49.36 (22)
3.0	δ Cygni	63.84 (50)	N45.10 (06)	2.6	ε Centauri	155.20 (12)	S53.41 (25)
3.2	β Cyg Albireo	67.43 (26)	N27.94 (56)	1.2	α Vir Spica 33	158.84 (50)	S11.11 (06)
3.0	π Sagittarii	72.72 (43)	S21.04 (03)	2.2	ζ UMa Mizar	159.12 (07)	N54.98 (59)
3.0	ζ Aquilæ	73.77 (46)	N13.85 (51)	2.9	ι Centauri	160.00 (00)	S36.66 (39)
2.7	ζ Sagittarii	74.51 (31)	S29.90 (54)	3.0	ε Virginis	164.58 (35)	N11.01 (01)
2.1	σ Sgr Nunki 50	76.35 (21)	S26.31 (19)	2.9	α CVn Cor Caroli	166.11 (07)	N38.37 (22)
0.1	α Lyr Vega 49	80.85 (51)	N38.77 (46)	1.7	ε UMa Alioth 32	166.60 (36)	N56.01 (01)
2.9	λ Sagittarii	83.17 (10)	S25.43 (26)	2.9	γ Virginis	169.72 (43)	S 1.39 (24)
2.0	ε Sgr Kaus Aust. 48	84.13 (08)	S34.39 (24)	2.4	γ Cen Muhlifain	169.77 (46)	S48.90 (54)
2.8	δ Sagittarii	84.92 (55)	S29.84 (50)	2.8	β Corvi	171.54 (32)	S23.34 (20)
3.1	γ Sagittarii	88.72 (43)	S30.43 (26)	2.8	γ Crv Gienah 29	176.18 (11)	S17.48 (29)
2.4	γ Dra Eltanin 47	90.91 (55)	N51.49 (29)	2.9	δ Centauri	178.05 (03)	S50.66 (40)
2.9	β Ophiuchi	94.26 (16)	N 4.57 (34)	2.5	γ UMa Phecda	181.68 (41)	N53.75 (45)
2.5	κ Scorpii	94.56 (34)	S39.03 (02)	2.2	β Leo Denebola 28	182.87 (52)	N14.63 (38)
2.0	θ Scorpii	95.86 (52)	S42.99 (60)	2.6	δ Leonis	191.61 (36)	N20.58 (35)
2.1	α Oph Rasalhague 46	96.39 (23)	N12.57 (34)	3.2	ψ Ursæ Majoris	192.73 (44)	N44.55 (33)
1.7	λ Sco Shaula 45	96.78 (47)	S37.10 (06)	2.0	α UMa Dubhe 27	194.22 (13)	N61.81 (48)
3.0	α Aræ	97.24 (15)	S49.87 (52)	2.4	β UMa Merak	194.69 (41)	N56.44 (26)
3.0	β Draconis	97.45 (27)	N52.31 (18)	2.8	μ Velorum	198.42 (25)	S49.36 (22)
2.8	υ Scorpii	97.49 (29)	S37.29 (17)	2.3	γ Leo Algeiba	205.15 (09)	N19.89 (54)
2.8	β Aræ	98.89 (54)	S55.52 (31)	1.3	α Leo Regulus 26	208.04 (02)	N12.02 (01)
*	α Herculis	101.46 (28)	N14.40 (24)	3.1	ε Leonis	213.68 (41)	N23.82 (49)
2.6	η Oph Sabik 44	102.56 (33)	S15.72 (43)	2.2	α Hya Alphard 25	218.23 (14)	S 8.61 (37)
3.1	ζ Aræ	105.57 (34)	S55.98 (59)	2.6	κ Velorum	219.55 (33)	S54.97 (58)
2.4	ε Scorpii	107.63 (38)	S34.28 (17)	2.2	λ Vel Suhail 23	223.10 (06)	S43.39 (23)
3.0	ζ Herculis	109.78 (47)	N31.62 (37)	3.1	ι Ursæ Majoris	225.37 (22)	N48.08 (05)
2.7	ζ Ophiuchi	110.86 (51)	S10.55 (33)	2.0	δ Velorum	228.90 (54)	S54.67 (40)
2.9	τ Scorpii	111.19 (12)	S28.20 (12)	1.9	γ Velorum	237.70 (42)	S47.30 (18)
2.8	β Herculis	112.56 (33)	N21.51 (31)	2.9	ρ Puppis	238.22 (13)	S24.27 (16)
1.2	α Sco Antares 42	112.81 (49)	S26.41 (25)	2.3	ζ Puppis	239.19 (12)	S39.97 (58)
2.9	η Draconis	114.04 (02)	N61.53 (32)	1.2	β Gem Pollux 21	243.83 (50)	N28.05 (03)
3.0	δ Ophiuchi	116.55 (33)	S 3.67 (40)	0.5	α CMi Procyon 20	245.31 (18)	N 5.25 (15)
2.8	β Scorpii	118.79 (48)	S19.78 (47)	1.6	α Gem Castor	246.51 (31)	N31.91 (55)
2.5	δ Sco Dschubba	120.07 (04)	S22.59 (36)	3.3	σ Puppis	247.77 (46)	S43.28 (17)
3.0	π Scorpii	120.45 (27)	S26.09 (05)	3.1	β Canis Minoris	248.35 (21)	N 8.31 (19)
2.8	α Serpentis	124.06 (04)	N 6.46 (27)	2.4	η Canis Majoris	249.08 (05)	S29.28 (17)

* Variable 3.0–3.7

STARS, 1989

Mag	Name and Number			SHA	Dec
				° ′	° ′
2.7	π	Puppis		250.80 (48)	S37.08 (05)
2.0	δ	CMa	Wezen	253.00 (00)	S26.37 (22)
3.1	o	Canis Majoris		254.35 (21)	S23.82 (49)
1.6	ε	CMa	Adhara 19	255.44 (27)	S28.96 (57)
2.8	τ	Puppis		257.58 (35)	S50.60 (36)
−1.6	α	CMa	Sirius 18	258.82 (49)	S16.70 (42)
1.9	γ	Gem	Alhena	260.72 (43)	N16.41 (25)
−0.9	α	Car	Canopus 17	264.07 (04)	S52.69 (41)
2.0	β	CMa	Mirzam	264.44 (26)	S17.95 (57)
2.7	θ	Aurigæ		270.24 (15)	N37.21 (13)
2.1	β	Aur	Menkalinan	270.30 (18)	N44.95 (57)
*	α	Ori	Betelgeuse 16	271.34 (21)	N 7.41 (25)
2.2	κ	Orionis		273.18 (11)	S 9.67 (40)
1.9	ζ	Ori	Alnitak	274.94 (56)	S 1.95 (57)
2.8	α	Col	Phact	275.18 (11)	S34.08 (05)
3.0	ζ	Tauri		275.74 (44)	N21.14 (08)
1.8	ε	Ori	Alnilam 15	276.08 (05)	S 1.21 (12)
2.9	ι	Orionis		276.27 (16)	S 5.91 (55)
2.7	α	Leporis		276.93 (56)	S17.83 (50)
2.5	δ	Orionis		277.13 (08)	S 0.30 (18)
3.0	β	Leporis		278.05 (03)	S20.77 (46)
1.8	β	Tau	Elnath 14	278.59 (35)	N28.60 (36)
1.7	γ	Ori	Bellatrix 13	278.85 (51)	N 6.34 (21)
0.2	α	Aur	Capella 12	281.02 (01)	N45.99 (59)
0.3	β	Ori	Rigel 11	281.49 (29)	S 8.21 (13)
2.9	β	Eridani		283.16 (10)	S 5.10 (06)
2.9	ι	Aurigæ		285.92 (55)	N33.15 (09)
1.1	α	Tau	Aldebaran 10	291.17 (10)	N16.49 (30)
3.2	γ	Eridani		300.61 (37)	S13.54 (32)
3.0	ε	Persei		300.71 (43)	N39.98 (59)
2.9	ζ	Persei		301.63 (38)	N31.86 (51)
3.0	η	Tau	Alcyone	303.28 (17)	N24.08 (05)
1.9	α	Per	Mirfak 9	309.10 (06)	N49.83 (50)
†	β	Per	Algol	313.13 (08)	N40.92 (55)
2.8	α	Cet	Menkar 8	314.56 (34)	N 4.05 (03)
3.1	θ	Eri	Acamar 7	315.53 (32)	S40.34 (21)
2.1	α	UMi	Polaris	324.90 (54)	N89.22 (13)
3.1	β	Trianguli		327.77 (46)	N34.94 (56)
2.2	α	Ari	Hamal 6	328.35 (21)	N23.42 (25)
2.2	γ	And	Almak	329.19 (11)	N42.28 (17)
2.7	β	Ari	Sheratan	331.48 (29)	N20.76 (46)
2.8	δ	Cas	Ruchbah	338.72 (43)	N60.18 (11)
2.4	β	And	Mirach	342.71 (43)	N35.57 (34)
‡	γ	Cassiopeiæ		345.98 (59)	N60.66 (40)
2.2	β	Cet	Diphda 4	349.23 (14)	S18.04 (03)
2.5	α	Cas	Schedar 3	350.02 (01)	N56.48 (29)
2.4	α	Phe	Ankaa 2	353.55 (33)	S42.36 (22)
2.9	γ	Peg	Algenib	356.83 (50)	N15.13 (08)
2.4	β	Cas	Caph	357.85 (51)	N59.09 (06)
2.2	α	And	Alpheratz 1	358.04 (02)	N29.04 (02)

* Variable 0.1–1.2
† Variable 2.3–3.5
‡ Irregular var. 1987 Mag. 2.2

5.3.6 POLARIS TABLE, 1989

LHA ♈	Q	LHA ♈	Q	LHA ♈	Q
°	′	°	′	°	′
0	−38	120	− 4	240	+43
3	40	123	− 1	243	41
6	41	126	+ 1	246	40
9	42	129	4	249	39
12	43	132	6	252	38
15	−44	135	+ 8	255	+36
18	45	138	11	258	34
21	45	141	13	261	33
24	46	144	15	264	31
27	46	147	18	267	29
30	−47	150	+20	270	+27
33	47	153	22	273	25
36	47	156	24	276	23
39	47	159	26	279	21
42	46	162	28	282	19
45	−46	165	+30	285	+16
48	46	168	32	288	14
51	45	171	34	291	12
54	44	174	35	294	9
57	43	177	37	297	7
60	−42	180	+38	300	+ 5
63	41	183	40	303	+ 2
66	40	186	41	306	0
69	39	189	42	309	− 3
72	37	192	43	312	− 5
75	−36	195	+44	315	− 8
78	34	198	45	318	10
81	32	201	45	321	12
84	31	204	46	324	15
87	29	207	46	327	17
90	−27	210	+47	330	−19
93	25	213	47	333	22
96	22	216	47	336	24
99	20	219	47	339	26
102	18	222	46	342	28
105	−16	225	+46	345	−30
108	13	228	46	348	32
111	11	231	45	351	33
114	9	234	44	354	35
117	6	237	43	357	37
120	− 4	240	+43	360	−38

Azimuth of Polaris

LHA ♈	0°	11°	59°	190°
Az	1°	0°		359°

LHA ♈	190°	240°	360°
Az	0°		1°

Latitude = Apparent altitude (corrected for refraction)+Q

5.3.7 SUNRISE, SUNSET AND TWILIGHT — 1989

Date	Naut Twi	ν	Sun-rise	ν	Sun-set	Civil Twi	ν	Date	Naut Twi	ν	Sun-rise	ν	Sun-set	Civil Twi	ν
	h m		h m		h m	h m			h m		h m		h m	h m	
Jan 1	06 39	+39	07 59	+63−	16 09	16 47	−51	Jul 3	02 08	−115	03 56	−67+	20 12	20 56	+84
4	06 39	38	07 58	61	16 12	16 50	50	6	02 12	112	03 59	65	20 10	20 54	82
7	06 39	38	07 57	60	16 16	16 53	49	9	02 16	110	04 01	65	20 09	20 52	81
10	06 38	36	07 56	58	16 20	16 57	47	12	02 21	107	04 04	63	20 06	20 49	78
13	06 37	35	07 54	57	16 24	17 01	46	15	02 26	104	04 07	62	20 04	20 46	77
16	06 35	+34	07 52	+55−	16 28	17 05	−44	18	02 32	−100	04 11	−60+	20 01	20 42	+74
19	06 33	32	07 49	53	16 33	17 09	43	21	02 37	97	04 14	58	19 58	20 38	71
22	06 31	31	07 46	51	16 37	17 14	40	24	02 43	93	04 18	56	19 54	20 34	69
25	06 28	28	07 43	49	16 42	17 18	39	27	02 49	89	04 22	54	19 50	20 29	66
28	06 25	26	07 39	46	16 47	17 23	36	30	02 55	85	04 26	52	19 46	20 25	64
31	06 21	+24	07 35	+44−	16 52	17 27	−34	Aug 2	03 01	−82	04 30	−49+	19 41	20 19	+61
Feb 3	06 17	21	07 31	42	16 58	17 32	32	5	03 07	78	04 35	46	19 36	20 14	58
6	06 13	19	07 26	39	17 03	17 37	29	8	03 14	73	04 39	44	19 31	20 08	55
9	06 09	17	07 21	36	17 08	17 42	27	11	03 20	69	04 43	42	19 26	20 02	52
12	06 04	14	07 16	33	17 13	17 47	24	14	03 26	65	04 48	39	19 20	19 56	49
15	06 00	+13	07 11	+31−	17 18	17 52	−21	17	03 32	−62	04 52	−36+	19 15	19 50	+46
18	05 54	9	07 05	28	17 23	17 57	18	20	03 37	59	04 57	33	19 09	19 44	43
21	05 49	7	07 00	26	17 29	18 02	16	23	03 43	55	05 01	31	19 03	19 38	40
24	05 44	5	06 54	23	17 34	18 06	14	26	03 49	51	05 06	28	18 57	19 31	37
27	05 38	+2	06 48	20	17 39	18 11	11	29	03 54	48	05 10	25	18 51	19 24	33
Mar 2	05 32	−1	06 42	+17−	17 44	18 16	−8	Sep 1	04 00	−44	05 15	−22+	18 44	19 18	+31
5	05 26	4	06 35	13	17 49	18 21	5	4	04 05	41	05 19	19	18 38	19 11	28
8	05 19	8	06 29	11	17 54	18 26	−2	7	04 10	38	05 24	16	18 31	19 04	24
11	05 13	10	06 23	8	17 58	18 31	+1	10	04 16	33	05 28	13	18 25	18 58	22
14	05 06	14	06 16	5	18 03	18 35	3	13	04 21	30	05 33	10	18 18	18 51	19
17	05 00	−16	06 10	+2−	18 08	18 40	+7	16	04 26	−27	05 37	−8+	18 12	18 44	+16
20	04 53	19	06 03	−1+	18 13	18 45	10	19	04 31	24	05 42	5	18 05	18 37	13
23	04 46	23	05 57	3	18 18	18 50	13	22	04 36	20	05 46	−2+	17 58	18 31	10
26	04 39	26	05 50	7	18 22	18 55	16	25	04 40	18	05 51	+1−	17 52	18 24	7
29	04 32	29	05 44	9	18 27	19 00	19	28	04 45	15	05 55	3	17 45	18 17	4
Apr 1	04 25	−32	05 37	−13+	18 32	19 05	+22	Oct 1	04 50	−12	06 00	+7−	17 39	18 11	+2
4	04 18	36	05 31	15	18 36	19 10	25	4	04 55	8	06 04	9	17 32	18 04	−2
7	04 11	39	05 24	18	18 41	19 15	28	7	04 59	6	06 09	12	17 26	17 58	4
10	04 03	43	05 18	21	18 46	19 20	32	10	05 04	−3	06 14	15	17 19	17 52	7
13	03 56	46	05 12	24	18 51	19 25	35	13	05 09	0	06 18	17	17 13	17 46	10
16	03 49	−50	05 05	−27+	18 55	19 30	+38	16	05 13	+3	06 23	+21−	17 07	17 40	−12
19	03 42	53	04 59	30	19 00	19 35	41	19	05 18	6	06 28	24	17 01	17 34	15
22	03 35	57	04 53	33	19 05	19 40	44	22	05 22	8	06 33	27	16 55	17 28	18
25	03 27	61	04 48	35	19 09	19 45	47	25	05 27	11	06 38	29	16 50	17 23	21
28	03 20	65	04 42	38	19 14	19 50	49	28	05 31	13	06 43	32	16 44	17 18	23
May 1	03 13	−69	04 37	−40+	19 19	19 55	+52	31	05 36	+16	06 48	+35−	16 39	17 13	−26
4	03 06	73	04 31	44	19 23	20 00	55	Nov 3	05 40	18	06 53	38	16 34	17 08	28
7	03 00	76	04 26	46	19 28	20 06	59	6	05 45	21	06 58	40	16 29	17 03	31
10	02 53	80	04 22	48	19 32	20 11	62	9	05 49	23	07 03	43	16 24	16 59	33
13	02 47	84	04 17	51	19 36	20 16	65	12	05 53	25	07 08	45	16 20	16 55	36
16	02 40	−88	04 13	−53+	19 41	20 20	+67	15	05 58	+27	07 13	+48−	16 16	16 52	−37
19	02 35	91	04 09	55	19 45	20 25	70	18	06 02	29	07 17	50	16 12	16 48	40
22	02 29	95	04 05	58	19 49	20 30	73	21	06 06	31	07 22	52	16 09	16 45	42
25	02 24	99	04 02	60	19 52	20 34	75	24	06 10	33	07 27	55	16 06	16 43	43
28	02 19	102	03 59	61	19 56	20 38	77	27	06 14	34	07 31	56	16 04	16 41	45
31	02 14	−106	03 57	−63+	19 59	20 42	+79	30	06 17	+35	07 35	+58−	16 02	16 39	−47
Jun 3	02 10	109	03 54	65	20 02	20 45	80	Dec 3	06 21	37	07 39	59	16 00	16 38	48
6	02 07	111	03 53	66	20 05	20 49	83	6	06 24	38	07 43	61	15 59	16 37	49
9	02 04	114	03 51	67	20 07	20 51	83	9	06 27	39	07 46	62	15 58	16 36	51
12	02 02	116	03 51	68	20 09	20 54	85	12	06 30	40	07 49	63	15 58	16 36	51
15	02 01	−117	03 50	−68+	20 11	20 55	+85	15	06 32	+40	07 52	+64−	15 58	16 37	−51
18	02 00	118	03 50	69	20 12	20 57	86	18	06 34	40	07 54	64	15 59	16 38	52
21	02 00	118	03 51	68	20 13	20 58	86	21	06 36	41	07 56	64	16 00	16 39	52
24	02 01	118	03 51	68	20 13	20 58	86	24	06 37	40	07 57	64	16 02	16 41	52
27	02 03	117	03 53	68	20 13	20 58	86	27	06 38	40	07 58	64	16 04	16 43	51
30	02 05	−116	03 54	−68+	20 13	20 57	+84	30	06 39	+40	07 59	+64−	16 07	16 45	−51
Jul 3	02 08	−115	03 56	−67+	20 12	20 56	+84	Jan 2	06 39	+39	08 00	+63−	16 09	16 47	−51

SUNRISE, SUNSET AND TWILIGHT — 1989

Corrections to Sunrise or Sunset

N.Lat	30°	35°	40°	45°	50°	52°	54°	56°	58°	60°
v	m	m	m	m	m	m	m	m	m	m
0	0	0	0	0	0	0	0	0	0	0
2	−2	−1	−1	0	0	+1	+1	+1	+2	+2
4	4	3	2	−1	0	1	1	1	2	3
6	6	5	3	2	0	1	2	3	4	5
8	8	6	5	2	0	1	2	3	5	6
10	−10	−8	−6	−3	0	+1	+3	+4	+6	+8
12	12	10	7	4	0	2	4	6	8	10
14	14	11	8	5	0	2	4	6	9	12
16	16	13	9	5	0	2	5	8	11	14
18	18	14	10	6	0	2	5	9	12	16
20	−20	−16	−11	−6	0	+3	+6	+10	+14	+18
22	22	17	12	7	0	3	7	11	15	20
24	24	19	14	8	0	3	7	12	16	22
26	26	21	15	8	0	4	8	13	18	24
28	28	22	16	9	0	4	9	14	19	26
30	−30	−24	−17	−9	0	+5	+10	+15	+21	+28
32	32	26	19	10	0	5	10	16	22	30
34	34	27	20	10	0	5	11	17	24	32
36	36	29	21	11	0	5	11	17	25	33
38	38	31	22	12	0	5	11	18	26	35
40	−40	−32	−23	−13	0	+6	+12	+20	+28	+37
42	42	34	24	13	0	6	13	21	30	40
44	44	36	26	14	0	6	13	22	31	42
46	46	37	27	15	0	7	14	23	33	44
48	48	39	28	15	0	7	15	25	35	46
50	−50	−40	−29	−16	0	+7	+16	+26	+36	+48
52	52	42	30	16	0	8	17	27	38	51
54	54	43	31	17	0	8	17	27	39	53
56	56	45	33	18	0	8	17	28	40	55
58	58	47	34	19	0	9	18	30	42	57
60	−60	−48	−35	−20	0	+9	+19	+31	+45	+61
62	62	50	37	21	0	9	20	32	47	64
64	64	52	38	21	0	10	21	34	49	67
66	66	53	39	21	0	11	22	35	51	70
68	68	55	41	22	0	11	23	37	54	73
70	−70	−57	−42	−23	0	+11	+24	+38	+56	+76

Corrections to Nautical Twilight

N.Lat	30°	35°	40°	45°	50°	52°	54°	56°	58°	60°
v	m	m	m	m	m	m	m	m	m	m
+40	−40	−31	−22	−12	0	+5	+11	+17	+23	+30
30	30	23	16	9	0	3	7	12	16	21
20	20	15	10	6	0	2	4	7	10	13
+10	−10	−7	−5	−2	0	+1	+2	+3	+4	+5
0	0	+1	+1	0	0	0	−1	−2	−3	−5
−10	+10	+8	+6	+4	0	−2	−4	−6	−9	−13
20	20	17	13	7	0	3	7	12	17	23
30	30	25	19	11	0	4	10	16	24	32
40	40	33	24	14	0	6	14	22	32	44
50	50	41	30	17	0	9	18	29	42	58
−60	+60	+49	+37	+21	0	−11	−22	−36	−53	−74
70	70	58	43	24	0	13	27	45	67	96
80	80	66	50	29	0	14	32	53	83	−134
90	90	75	56	32	0	18	39	68	−118	TAN
100	100	84	63	37	0	20	46	−87	TAN	TAN
−110	+110	+92	+70	+41	0	−25	−60	TAN	TAN	TAN
−120	+120	+100	+78	+46	0	−29	−83	TAN	TAN	TAN

Corrections to Civil Twilight

N.Lat	30°	35°	40°	45°	50°	52°	54°	56°	58°	60°
v	m	m	m	m	m	m	m	m	m	m
−50	+50	+40	+29	+16	0	−7	−15	−23	−33	−44
40	40	32	22	12	0	6	12	19	26	35
30	30	24	17	10	0	4	8	13	18	24
20	20	16	11	7	0	3	6	9	12	16
−10	+10	+8	+5	+3	0	−1	−3	−4	−5	−7
0	0	0	0	0	0	0	0	+1	+1	+2
+10	−10	−8	−6	−4	0	+1	+3	5	8	11
20	20	16	12	7	0	3	7	11	15	20
30	30	24	18	10	0	5	10	16	22	30
40	40	32	24	13	0	6	13	21	30	41
+50	−50	−41	−30	−16	0	+8	+17	+27	+39	+53
60	60	49	36	20	0	9	20	33	47	65
70	70	57	42	23	0	12	25	41	60	85
80	80	65	48	27	0	14	30	50	75	111
83	83	68	50	28	0	15	32	52	80	121
+86	−86	−71	−52	−30	0	+15	+33	+55	+85	+135

The times given on the opposite page are the local mean times (LMT) of nautical twilight, sunrise, sunset and civil twilight for latitude N.50°, together with variations v. The variations are the differences in time between the times for latitude N.50° and those for latitude N.30°. The sign on the left-hand side of v (between sunrise and sunset) applies to sunrise, and the right-hand sign applies to sunset. The LMT for any latitude from N.30° to N.60° can be found by applying corrections, taken from the tables above, to the tabulated times.

To determine the LMT of sunrise or sunset, take out the tabulated time and v corresponding to the required date and, using v and latitude as arguments in the table of "Corrections to Sunrise or Sunset", extract the correction and apply it to the tabulated time as follows. If v is positive, apply the correction with the sign as tabulated. If v is negative, apply the correction with the opposite sign to that which is tabulated.

To determine LMT of nautical twilight the correction to the tabulated time must be taken from the table of "Corrections to Nautical Twilight" using as argument latitude and v; in entering the correction table the correct sign of v must be used, and in applying the correction to the tabulated time its sign must be used as given in the table. When TAN (Twilight All Night) is given, the Sun does not reach a zenith distance of 102°.

The corrections to civil twilight must be found in the same manner as those for nautical twilight.

LMT can be converted to GMT by adding the longitude (in time) if west or subtracting if east.

Examples of the use of these tables are given on page 72.

5.3.8 MOONRISE AND MOONSET — 1989

Date	JANUARY Rise	v	Set	v	MARCH Rise	v	Set	v	MAY Rise	v	Set	v	JULY Rise	v	Set	v
	h m		h m		h m		h m		h m		h m		h m		h m	
1	01 19	+29	11 37	−34	02 46	+88	09 48	−90	02 45	+19	14 11	−11	01 32	−84	19 19	+88
2	02 31	45	11 52	50	03 48	91	10 44	90	03 00	+ 1	15 37	+ 8	02 27	88	20 15	85
3	03 45	60	12 13	65	04 38	86	11 55	83	03 16	−16	17 05	27	03 36	83	20 56	75
4	05 01	76	12 41	79	05 16	74	13 18	69	03 33	35	18 36	47	04 53	71	21 24	60
5	06 15	86	13 22	88	05 43	56	14 47	50	03 55	53	20 08	66	06 12	56	21 45	44
6	07 21	+90	14 19	−90	06 04	+38	16 17	−31	04 24	−70	21 35	+81	07 29	−39	22 01	+29
7	08 14	86	15 32	83	06 22	20	17 47	−11	05 04	83	22 51	88	08 42	23	22 15	+15
8	08 52	72	16 57	68	06 38	+ 2	19 16	+ 8	05 58	89	23 50	87	09 52	− 8	22 27	0
9	09 20	56	18 25	51	06 53	−17	20 46	28	07 06	84	24 31	76	11 01	+ 8	22 39	−14
10	09 41	38	19 53	31	07 10	36	22 16	48	08 21	73	00 31	76	12 09	22	22 51	28
11	09 58	+20	21 19	−12	07 31	−54	23 45	+66	09 38	−58	01 00	+62	13 18	+38	23 06	−42
12	10 13	+ 3	22 43	+ 7	07 58	71	25 10	82	10 53	42	01 21	47	14 28	52	23 24	56
13	10 27	−15	24 06	25	08 35	84	01 10	82	12 04	27	01 38	33	15 41	67	23 47	71
14	10 43	32	00 06	25	09 24	90	02 23	90	13 13	−12	01 51	18	16 53	80	24 19	82
15	11 01	49	01 30	43	10 26	87	03 20	89	14 21	+ 3	02 03	+ 4	18 00	88	00 19	82
16	11 25	−65	02 54	+61	11 36	−78	04 02	+80	15 28	+18	02 14	−10	18 58	+88	01 04	−88
17	11 56	80	04 16	76	12 50	64	04 32	68	16 37	33	02 27	24	19 43	81	02 03	88
18	12 40	88	05 32	87	14 04	48	04 54	53	17 48	49	02 40	39	20 17	67	03 17	78
19	13 37	89	06 35	90	15 16	32	05 10	38	19 00	63	02 57	53	20 43	52	04 38	64
20	14 45	82	07 23	84	16 25	18	05 24	24	20 13	77	03 19	67	21 02	34	06 03	46
21	16 00	−69	07 57	+72	17 33	− 3	05 36	+10	21 22	+86	03 49	−79	21 19	+17	07 28	−27
22	17 15	54	08 22	58	18 41	+13	05 47	− 5	22 22	89	04 30	87	21 34	0	08 52	− 9
23	18 29	37	08 40	42	19 49	27	05 58	19	23 10	84	05 24	89	21 49	−17	10 15	+ 9
24	19 40	22	08 55	28	20 59	43	06 11	34	23 46	72	06 32	81	22 06	34	11 39	28
25	20 48	− 7	09 07	+13	22 11	58	06 26	48	24 12	57	07 49	68	22 26	51	13 04	46
26	21 56	+ 8	09 19	− 1	23 24	+73	06 45	−63	00 12	+57	09 10	−52	22 52	−68	14 30	+64
27	23 04	23	09 30	15	24 34	84	07 11	76	00 33	41	10 32	35	23 28	81	15 53	79
28	24 14	39	09 42	29	00 34	84	07 46	86	00 50	24	11 54	−17	24 17	88	17 08	87
29	00 14	39	09 56	44	01 38	90	08 34	91	01 05	+ 7	13 17	+ 2	00 17	88	18 09	88
30	01 26	55	10 14	59	02 31	88	09 37	87	01 20	−10	14 41	20	01 20	86	18 54	79
31	02 40	+70	10 38	−73	03 12	+79	10 53	−76	01 36	−28	16 08	+39	02 34	−76	19 26	+66
	FEBRUARY				APRIL				JUNE				AUGUST			
1	03 54	+83	11 12	−85	03 43	+64	12 17	−59	01 56	−45	17 37	+58	03 52	−62	19 49	+51
2	05 03	90	12 00	91	04 06	47	13 44	40	02 21	62	19 05	74	05 10	45	20 07	36
3	06 02	90	13 05	88	04 24	29	15 11	22	02 55	77	20 27	86	06 24	30	20 21	20
4	06 47	81	14 24	77	04 41	+11	16 40	− 2	03 42	87	21 34	88	07 36	−14	20 34	+ 6
5	07 19	64	15 53	60	04 56	− 8	18 09	+17	04 44	88	22 24	81	08 45	+ 1	20 46	− 8
6	07 43	+46	17 24	−40	05 13	−26	19 40	+37	05 58	−79	22 59	+69	09 54	+17	20 58	−22
7	08 02	28	18 53	21	05 32	45	21 13	58	07 17	64	23 23	53	11 02	31	21 11	37
8	08 18	+10	20 21	− 1	05 57	62	22 43	75	08 34	49	23 42	38	12 12	46	21 27	52
9	08 33	− 8	21 48	+18	06 30	78	24 04	86	09 48	33	23 56	23	13 23	61	21 48	65
10	08 49	26	23 14	37	07 15	88	00 04	86	10 59	17	24 09	9	14 35	75	22 16	77
11	09 07	−43	24 41	+56	08 14	−89	01 10	+89	12 08	− 2	00 09	+ 9	15 44	+85	22 54	−86
12	09 29	60	00 41	56	09 24	81	01 59	83	13 15	+12	00 21	− 5	16 46	90	23 46	89
13	09 58	76	02 05	72	10 39	68	02 34	72	14 23	27	00 33	19	17 36	85	24 53	83
14	10 37	87	03 24	85	11 53	53	02 59	58	15 33	43	00 46	33	18 15	74	00 53	83
15	11 29	91	04 31	91	13 06	37	03 17	43	16 45	58	01 02	47	18 44	59	02 11	71
16	12 34	−85	05 22	+87	14 16	−22	03 32	+28	17 58	+72	01 22	−62	19 06	+42	03 36	−54
17	13 46	74	06 00	77	15 24	− 7	03 44	+14	19 09	83	01 48	75	19 24	24	05 03	35
18	15 01	59	06 27	63	16 31	+ 8	03 55	− 1	20 13	88	02 25	85	19 40	+ 7	06 29	−17
19	16 15	43	06 47	48	17 39	23	04 07	14	21 06	86	03 15	89	19 55	−11	07 55	+ 2
20	17 26	28	07 02	33	18 49	38	04 19	29	21 46	76	04 20	85	20 12	28	09 21	21
21	18 36	−12	07 15	+18	20 00	+53	04 34	−43	22 16	+61	05 36	−73	20 31	−46	10 48	+40
22	19 44	+ 3	07 27	+ 5	21 13	68	04 52	57	22 39	46	06 58	57	20 56	63	12 16	58
23	20 51	17	07 38	−10	22 24	80	05 15	72	22 56	28	08 21	39	21 28	78	13 41	74
24	22 00	33	07 50	24	23 31	89	05 47	83	23 12	+12	09 43	22	22 13	87	15 00	86
25	23 11	49	08 03	38	24 27	89	06 32	89	23 27	− 5	11 05	− 4	23 11	88	16 04	88
26	24 23	+64	08 19	−53	00 27	+89	07 30	−88	23 42	−22	12 27	+14	24 21	−81	16 54	+83
27	00 23	64	08 40	67	01 11	82	08 40	79	24 00	39	13 51	33	00 21	81	17 29	71
28	01 36	+77	09 08	−81	01 44	69	09 59	65	00 00	39	15 17	51	01 37	68	17 54	56
29					02 08	53	11 22	48	00 22	56	16 43	68	02 54	52	18 13	41
30					02 28	+36	12 46	−30	00 51	−72	18 06	+86	04 09	36	18 29	26
31													05 22	−19	18 42	+12

MOONRISE AND MOONSET — 1989

Date	SEPTEMBER Rise	v	Set	v	NOVEMBER Rise	v	Set	v	N Lat	| 30°	35°	40°	45°	50°	52°	54°	56°	58°	60°
	h m		h m		h m		h m		v	m	m	m	m	m	m	m	m	m	m
1	06 31	− 5	18 54	− 2	10 12	+82	17 30	−84	0	0	0	0	0	0	0	0	0	0	0
2	07 40	+11	19 06	17	11 12	87	18 18	87	2	− 2	− 1	− 1	0	0	0	+ 1	+ 1	+ 2	+ 2
3	08 48	25	19 18	32	12 01	84	19 18	83	4	4	3	2	− 1	0	1	2	2	3	3
4	09 58	41	19 33	46	12 39	75	20 29	72	6	6	5	3	2	0	+ 1	2	3	4	5
5	11 08	56	19 52	60	13 09	63	21 45	58	8	8	6	5	2	0	1	2	3	5	6
6	12 19	+70	20 16	−73	13 31	+47	23 05	−41	10	−10	− 8	− 6	− 3	0	+ 1	+ 3	+ 4	+ 6	+ 8
7	13 28	81	20 49	83	13 50	31	24 25	25	12	12	10	7	4	0	2	4	5	8	10
8	14 32	88	21 34	89	14 06	+15	00 25	25	14	14	11	8	5	0	2	4	6	9	12
9	15 27	88	22 33	87	14 22	− 2	01 48	− 6	16	16	13	9	5	0	2	5	8	11	14
10	16 10	80	23 45	77	14 38	20	03 12	+11	18	18	14	10	6	0	2	5	9	12	16
11	16 42	+66	25 06	−62	14 58	−37	04 40	+31	20	−20	−16	−11	− 6	0	+ 3	+ 6	+10	+14	+18
12	17 07	50	01 06	62	15 22	55	06 11	49	22	22	17	12	7	0	3	7	11	15	20
13	17 27	33	02 31	45	15 54	73	07 44	68	24	24	19	14	8	0	3	7	12	16	22
14	17 44	+15	03 57	27	16 40	84	09 12	82	26	26	21	15	8	0	4	8	13	18	24
15	18 00	− 2	05 25	− 7	17 41	86	10 26	87	28	28	22	16	9	0	4	9	14	19	26
16	18 17	−20	06 53	+13	18 55	−79	11 21	+81	30	−30	−24	−17	− 9	0	+ 5	+10	+15	+21	+28
17	18 35	39	08 22	32	20 14	66	12 00	70	32	32	26	19	10	0	5	10	16	22	30
18	18 58	57	09 53	51	21 34	50	12 28	55	34	34	27	20	10	0	5	11	17	24	32
19	19 29	72	11 23	69	22 49	34	12 48	40	36	36	29	21	11	0	5	11	17	25	33
20	20 10	84	12 46	82	24 01	19	13 04	25	38	38	31	22	12	0	5	11	18	26	35
21	21 05	−88	13 58	+89	00 01	−19	13 17	+10	40	−40	−32	−23	−13	0	+ 6	+12	+20	+28	+37
22	22 12	83	14 52	85	01 10	− 4	13 30	− 3	42	42	34	24	13	0	6	13	21	30	40
23	23 27	71	15 31	74	02 18	+11	13 42	17	44	44	36	26	14	0	6	13	22	31	42
24	24 44	56	15 59	60	03 26	26	13 56	31	46	46	37	27	15	0	7	14	23	33	44
25	00 44	56	16 20	46	04 35	41	14 11	46	48	48	39	28	15	0	7	15	25	35	46
26	01 59	−40	16 36	+31	05 45	+55	14 31	−59	50	−50	−40	−29	−16	0	+ 7	+16	+26	+36	+48
27	03 11	26	16 50	16	06 55	69	14 56	72	52	52	42	30	16	0	8	17	27	38	51
28	04 21	− 9	17 02	+ 2	08 03	79	15 30	81	54	54	43	31	17	0	8	17	27	39	53
29	05 29	+ 5	17 14	−12	09 06	86	16 15	86	56	56	45	33	18	0	8	17	28	40	55
30	06 37	+20	17 27	−26	09 58	+85	17 12	−85	58	58	47	34	19	0	9	18	30	42	57
									60	−60	−48	−35	−20	0	+ 9	+19	+31	+45	+61
									62	62	50	37	21	0	9	20	32	47	64
									64	64	52	38	21	0	10	21	34	49	67
	OCTOBER				DECEMBER				66	66	53	39	21	0	11	22	35	51	70
1	07 46	+35	17 41	−40	10 40	+78	18 21	−75	68	68	55	41	22	0	11	23	37	54	73
2	08 56	50	17 58	55	11 11	66	19 35	62	70	−70	−57	−42	−23	0	+11	+24	+38	+56	+76
3	10 06	64	18 20	68	11 36	52	20 53	46	72	72	58	43	24	0	12	25	40	58	80
4	11 16	77	18 50	79	11 55	36	22 11	30	74	74	60	44	24	0	12	26	41	60	84
5	12 21	86	19 29	87	12 11	20	23 30	−13	76	76	62	45	25	0	12	27	43	62	87
6	13 18	+88	20 22	−88	12 27	+ 4	24 51	+ 6	78	78	63	46	26	0	12	27	44	64	90
7	14 05	84	21 27	81	12 42	−13	00 51	6	80	−80	−65	−48	−27	0	+12	+27	+45	+67	+94
8	14 40	72	22 42	69	12 59	30	02 13	22	82	82	67	49	28	0	13	28	47	70	100
9	15 07	57	24 03	52	13 20	47	03 40	41	84	84	68	50	28	0	14	30	49	73	106
10	15 29	42	00 03	52	13 47	65	05 09	59	86	86	70	51	29	0	14	31	50	76	112
11	15 47	+25	01 26	−35	14 26	−78	06 38	+75	88	88	72	53	30	0	15	32	52	80	119
12	16 03	+ 7	02 51	−17	15 19	85	07 59	85	90	−90	−73	−54	−30	0	+16	+33	+55	+84	+127
13	16 19	−11	04 17	+ 2	16 27	84	09 05	85											
14	16 37	29	05 46	21	17 47	72	09 52	75											
15	16 58	48	07 18	41	19 09	57	10 26	62											
16	17 26	−65	08 51	+61	20 29	−40	10 49	+46											
17	18 04	79	10 21	77	21 44	25	11 08	31											
18	18 55	87	11 41	86	22 56	− 9	11 22	15											
19	20 00	85	12 44	86	24 05	+ 6	11 36	+ 2											
20	21 15	75	13 30	78	00 05	6	11 48	−13											
21	22 33	−60	14 02	+64	01 14	+21	12 01	−27											
22	23 49	45	14 25	49	02 22	35	12 16	41											
23	25 02	29	14 43	35	03 32	51	12 34	55											
24	01 02	29	14 58	21	04 42	64	12 57	68											
25	02 12	−14	15 10	+ 6	05 51	76	13 28	79											
26	03 20	+ 1	15 22	− 8	06 57	+85	14 09	−86											
27	04 28	16	15 35	22	07 53	86	15 03	86											
28	05 36	31	15 49	35	08 39	81	16 09	79											
29	06 45	45	16 05	50	09 14	70	17 24	66											
30	07 56	60	16 26	63	09 40	55	18 42	51											
31	09 06	+73	16 53	−76	10 01	+40	20 01	−34											

The times given on the opposite page and alongside are the Local Mean Times (LMT) of moonrise and moonset for latitude N.50° and their variations v from the times for N.30°.
To find the times for any latitude between N.30° and N.60° enter the table with arguments latitude and v, and apply the correction found to the tabulated time, *using the tabulated sign of the correction if v is positive, and the opposite sign of the correction if v is negative*. A very small extra correction due to the daily difference of the times has a mean value of $+3^m$ in longitude W.20° and -3^m in longitude E.20°.
The LMT can be converted to GMT by adding the longitude (in time) if west or subtracting if east.
Examples of the use of this table are given on page 72.

5.3.9 1989 JANUARY, FEBRUARY — ARIES AND PLANETS

0h GMT	ARIES GHA	VENUS GHA	v	Dec	d	MARS GHA	v	Dec	d	JUPITER GHA	v	Dec	d	SATURN GHA	v	Dec	d
Jan								**JANUARY**									
1	100.62	203.81	−15	S 22.07	− 6	82.18	+21	N 8.40	+ 9	45.98	+44	N 18.55	0	184.56	+36	S 22.61	0
2	101.61	203.45	15	22.22	6	82.69	21	8.61	9	47.03	44	18.54	0	185.42	36	22.60	0
3	102.60	203.10	15	22.35	5	83.21	21	8.83	9	48.08	44	18.53	0	186.28	36	22.60	0
4	103.58	202.74	15	22.48	5	83.71	21	9.04	9	49.12	43	18.52	0	187.14	36	22.60	0
5	104.57	202.37	15	22.60	4	84.22	21	9.25	9	50.17	43	18.51	0	188.00	36	22.59	0
6	105.55	202.01	−15	S 22.70	− 4	84.72	+21	N 9.47	+ 9	51.21	+43	N 18.51	0	188.85	+36	S 22.59	0
7	106.54	201.64	15	22.79	3	85.22	21	9.68	9	52.24	43	18.50	0	189.71	36	22.58	0
8	107.52	201.27	15	22.87	3	85.71	20	9.89	9	53.27	43	18.50	0	190.57	36	22.58	0
9	108.51	200.90	16	22.94	2	86.21	20	10.10	9	54.30	43	18.49	0	191.43	36	22.57	0
10	109.50	200.52	16	23.00	2	86.69	20	10.32	9	55.33	42	18.48	0	192.29	36	22.57	0
11	110.48	200.15	−16	S 23.05	− 1	87.18	+20	N 10.53	+ 9	56.35	+42	N 18.48	0	193.15	+36	S 22.56	0
12	111.47	199.77	16	23.08	− 1	87.66	20	10.74	9	57.36	42	18.48	0	194.01	36	22.56	0
13	112.45	199.40	16	23.10	0	88.14	20	10.95	9	58.38	42	18.48	0	194.87	36	22.55	0
14	113.44	199.02	16	23.12	0	88.62	20	11.17	9	59.39	42	18.47	0	195.73	36	22.55	0
15	114.42	198.64	16	23.11	0	89.09	20	11.38	9	60.40	42	18.47	0	196.60	36	22.54	0
16	115.41	198.27	−16	S 23.10	+ 1	89.56	+19	N 11.59	+ 9	61.40	+42	N 18.47	0	197.46	+36	S 22.54	0
17	116.40	197.89	16	23.08	2	90.02	19	11.80	9	62.40	41	18.48	0	198.32	36	22.53	0
18	117.38	197.52	16	23.04	2	90.49	19	12.01	9	63.40	41	18.48	0	199.18	36	22.53	0
19	118.37	197.14	16	22.99	3	90.95	19	12.22	9	64.39	41	18.48	0	200.04	36	22.52	0
20	119.35	196.77	15	22.93	3	91.41	19	12.43	9	65.38	41	18.48	0	200.91	36	22.51	0
21	120.34	196.40	−15	S 22.86	+ 3	91.86	+19	N 12.64	+ 9	66.37	+41	N 18.49	0	201.77	+36	S 22.51	0
22	121.32	196.03	15	22.78	4	92.32	19	12.84	9	67.35	41	18.49	0	202.64	36	22.50	0
23	122.31	195.66	15	22.68	4	92.77	19	13.05	9	68.33	41	18.50	0	203.50	36	22.50	0
24	123.29	195.29	15	22.58	5	93.22	19	13.26	9	69.30	40	18.51	0	204.37	36	22.49	0
25	124.28	194.93	15	22.46	5	93.66	18	13.47	9	70.27	40	18.52	0	205.23	36	22.48	0
26	125.27	194.57	−15	S 22.33	+ 6	94.10	+18	N 13.67	+ 8	71.24	+40	N 18.52	0	206.10	+36	S 22.48	0
27	126.25	194.21	15	22.19	6	94.54	18	13.87	8	72.21	40	18.53	0	206.97	36	22.47	0
28	127.24	193.85	15	22.04	7	94.98	18	14.08	8	73.17	40	18.54	0	207.84	36	22.47	0
29	128.22	193.50	15	21.87	7	95.42	18	14.28	8	74.13	40	18.56	0	208.71	36	22.46	0
30	129.21	193.15	14	21.70	8	95.85	18	14.48	8	75.08	40	18.57	0	209.58	36	22.45	0
31	130.19	192.80	−14	S 21.51	+ 8	96.28	+18	N 14.68	+ 8	76.03	+39	N 18.58	+ 1	210.45	+36	S 22.45	0
Feb								**FEBRUARY**									
1	131.18	192.46	−14	S 21.32	+ 9	96.71	+18	N 14.88	+ 8	76.98	+39	N 18.59	+ 1	211.32	+36	S 22.44	0
2	132.17	192.12	14	21.11	9	97.13	18	15.08	8	77.92	39	18.60	1	212.19	36	22.43	0
3	133.15	191.78	14	20.90	10	97.55	18	15.28	8	78.86	39	18.62	1	213.07	36	22.43	0
4	134.14	191.45	14	20.67	10	97.97	17	15.48	8	79.80	39	18.63	1	213.94	36	22.42	0
5	135.12	191.12	13	20.43	10	98.39	17	15.67	8	80.74	39	18.65	1	214.81	36	22.41	0
6	136.11	190.80	−13	S 20.18	+11	98.81	+17	N 15.87	+ 8	81.67	+39	N 18.67	+ 1	215.69	+37	S 22.41	0
7	137.09	190.48	13	19.93	11	99.22	17	16.06	8	82.60	38	18.68	1	216.56	37	22.40	0
8	138.08	190.17	13	19.66	12	99.63	17	16.25	8	83.52	38	18.70	1	217.44	37	22.39	0
9	139.07	189.86	13	19.38	12	100.04	17	16.44	8	84.44	38	18.72	1	218.32	37	22.39	0
10	140.05	189.55	13	19.09	12	100.45	17	16.63	8	85.36	38	18.74	1	219.20	37	22.38	0
11	141.04	189.25	−12	S 18.80	+13	100.85	+17	N 16.82	+ 8	86.27	+38	N 18.76	+ 1	220.08	+37	S 22.37	0
12	142.02	188.95	12	18.49	13	101.25	17	17.00	8	87.18	38	18.78	1	220.96	37	22.37	0
13	143.01	188.66	12	18.18	13	101.65	17	17.19	8	88.09	38	18.80	1	221.84	37	22.36	0
14	143.99	188.38	12	17.86	14	102.05	16	17.37	8	89.00	38	18.83	1	222.72	37	22.35	0
15	144.98	188.10	12	17.53	14	102.45	16	17.55	7	89.90	38	18.85	1	223.61	37	22.34	0
16	145.96	187.82	−11	S 17.19	+14	102.84	+16	N 17.73	+ 7	90.80	+37	N 18.87	+ 1	224.49	+37	S 22.34	0
17	146.95	187.55	11	16.84	15	103.23	16	17.91	7	91.69	37	18.89	1	225.38	37	22.33	0
18	147.94	187.28	11	16.49	15	103.62	16	18.09	7	92.59	37	18.92	1	226.26	37	22.32	0
19	148.92	187.02	11	16.12	15	104.01	16	18.26	7	93.48	37	18.94	1	227.15	37	22.32	0
20	149.91	186.77	10	15.75	16	104.39	16	18.44	7	94.36	37	18.97	1	228.04	37	22.31	0
21	150.89	186.52	−10	S 15.38	+16	104.77	+16	N 18.61	+ 7	95.25	+37	N 19.00	+ 1	228.93	+37	S 22.30	0
22	151.88	186.27	10	14.99	16	105.15	16	18.78	7	96.13	37	19.02	1	229.82	37	22.30	0
23	152.86	186.03	10	14.60	17	105.53	16	18.95	7	97.01	36	19.05	1	230.71	37	22.29	0
24	153.85	185.79	10	14.20	17	105.91	16	19.11	7	97.88	36	19.08	1	231.61	37	22.28	0
25	154.84	185.56	9	13.80	17	106.29	16	19.28	7	98.76	36	19.11	1	232.50	37	22.28	0
26	155.82	185.33	− 9	S 13.39	+17	106.66	+15	N 19.44	+ 7	99.63	+36	N 19.14	+ 1	233.40	+37	S 22.27	0
27	156.81	185.11	9	12.97	18	107.03	15	19.60	7	100.49	36	19.16	1	234.30	37	22.26	0
28	157.79	184.89	− 9	S 12.55	+18	107.40	+15	N 19.76	+ 7	101.36	+36	N 19.19	+ 1	235.20	+38	S 22.25	0

1989 JANUARY, FEBRUARY — SUN AND MOON

0h GMT	SUN GHA	v	Dec	d	MOON 0h–8h GHA	v	Dec	d	MOON 8h–16h GHA	v	Dec	d	MOON 16h–24h GHA	v	Dec	d	HP
Jan	(SD 16'.3)						JANUARY										
1	179.15	−5	S 23.02	+4	264.26	273	S 10.72	−218	20.92	264	S 12.47	−213	137.52	254	S 14.17	−207	54.7
2	179.03	5	22.94	4	254.03	244	15.82	199	10.46	232	17.42	191	126.80	219	18.95	182	55.2
3	178.91	5	22.85	4	243.03	205	20.40	171	359.15	191	21.76	158	115.16	176	23.03	144	55.8
4	178.80	5	22.75	4	231.05	161	24.18	129	346.82	147	25.22	112	102.47	133	26.12	94	56.5
5	178.68	5	22.64	5	218.01	119	26.87	74	333.45	108	27.47	−53	88.79	98	27.89	−31	57.3
6	178.57	−5	S 22.53	+5	204.06	90	S 28.14	−7	319.26	85	S 28.20	+17	74.42	82	S 28.06	+41	58.1
7	178.46	5	22.40	5	189.56	82	27.73	+66	304.69	85	27.20	91	59.85	89	26.47	114	58.7
8	178.35	4	22.27	6	175.05	96	25.56	137	290.30	105	24.46	159	45.62	115	23.18	180	59.3
9	178.25	4	22.14	6	161.01	125	21.75	198	276.49	136	20.16	215	32.06	147	18.44	230	59.7
10	178.15	4	21.99	6	147.72	157	16.60	243	263.46	168	14.65	255	19.28	176	12.61	264	59.8
11	178.04	−4	S 21.84	+7	135.17	185	S 10.50	+272	251.13	191	S 8.32	+277	7.14	197	S 6.11	+281	59.8
12	177.95	4	21.68	7	123.19	201	S 3.86	283	239.28	204	S 1.60	283	355.39	205	N 0.67	282	59.7
13	177.85	4	21.52	7	111.51	205	N 2.92	279	227.63	203	N 5.15	274	343.74	200	7.34	268	59.4
14	177.76	4	21.35	8	99.82	196	9.49	261	215.87	191	11.57	251	331.88	184	13.58	241	59.0
15	177.67	4	21.17	8	87.83	177	15.51	229	203.72	169	17.34	216	319.55	160	19.07	200	58.6
16	177.58	−4	S 20.98	+8	75.31	151	N 20.67	+184	191.00	142	N 22.15	+167	306.61	134	N 23.48	+148	58.1
17	177.50	3	20.79	8	62.16	126	24.67	128	177.65	119	25.69	107	293.08	114	26.55	86	57.7
18	177.41	3	20.59	9	48.47	110	27.24	+64	163.83	108	27.75	+41	279.18	109	28.08	+19	57.2
19	177.34	3	20.39	9	34.53	111	28.23	−4	149.89	116	28.20	−26	265.30	122	27.99	−48	56.7
20	177.26	3	20.17	9	20.75	130	27.61	68	136.28	140	27.06	88	251.88	151	26.36	107	56.2
21	177.19	−3	S 19.96	+9	7.57	163	N 25.50	−124	123.35	175	N 24.51	−140	239.23	188	N 23.39	−154	55.7
22	177.12	3	19.73	10	355.22	201	22.16	168	111.31	214	20.81	179	227.49	226	19.38	190	55.3
23	177.05	3	19.50	10	343.78	237	17.86	199	100.15	248	16.27	207	216.62	257	14.61	214	54.9
24	176.99	2	19.26	10	333.15	266	12.90	219	89.76	274	11.14	224	206.43	280	9.35	228	54.5
25	176.93	2	19.02	10	323.16	286	7.52	231	79.92	290	5.68	233	196.73	293	N 3.82	234	54.3
26	176.88	−2	S 18.77	+11	313.55	296	N 1.95	−234	70.40	297	N 0.08	−234	187.25	296	S 1.79	−233	54.1
27	176.82	2	18.52	11	304.10	295	S 3.65	231	60.94	292	S 5.50	228	177.75	288	7.33	225	54.1
28	176.77	2	18.26	11	294.54	283	9.13	221	51.28	277	10.89	216	167.98	269	12.62	210	54.3
29	176.73	2	17.99	11	284.61	261	14.31	204	41.18	251	15.94	197	157.66	240	17.51	188	54.6
30	176.68	2	17.72	11	274.07	228	19.02	179	30.37	216	20.45	168	146.58	203	21.79	156	55.1
31	176.64	−1	S 17.45	+12	262.68	189	S 23.04	−143	18.67	175	S 24.18	−128	134.55	160	S 25.21	−112	55.8
Feb	(SD 16'.2)						FEBRUARY										
1	176.61	−1	S 17.17	+12	250.31	146	S 26.11	−95	5.96	133	S 26.87	−76	121.50	121	S 27.48	−56	56.6
2	176.58	1	16.88	12	236.95	110	27.93	−35	352.31	101	28.20	−12	107.60	94	28.30	+11	57.5
3	176.55	1	16.59	12	222.83	89	28.21	+36	338.02	86	27.92	+60	93.19	86	27.45	84	58.4
4	176.52	1	16.30	13	208.36	88	26.77	109	323.54	92	25.90	133	78.76	98	24.84	155	59.2
5	176.50	1	15.99	13	194.03	106	23.60	177	309.35	114	22.18	197	64.75	123	20.60	216	60.0
6	176.48	−1	S 15.69	+13	180.21	132	S 18.87	+233	295.75	142	S 17.01	+248	51.37	151	S 15.03	+261	60.5
7	176.46	−1	15.38	13	167.06	159	12.94	272	282.81	167	10.77	280	38.62	173	8.52	287	60.7
8	176.45	0	15.07	13	154.49	178	S 6.23	291	270.40	182	S 3.90	294	26.34	185	S 1.55	294	60.7
9	176.44	0	14.75	13	142.30	187	N 0.80	292	258.27	187	N 3.14	289	14.24	185	N 5.45	283	60.4
10	176.43	0	14.43	14	130.20	183	7.72	276	246.15	179	9.93	267	2.06	174	12.07	257	59.9
11	176.43	0	S 14.10	+14	117.94	169	N 14.12	+244	233.77	162	N 16.08	+231	349.55	155	N 17.92	+215	59.3
12	176.43	0	13.77	14	105.27	148	19.65	199	220.93	141	21.24	181	336.54	134	22.69	162	58.6
13	176.44	0	13.44	14	92.09	127	23.99	142	207.58	122	25.13	122	323.04	117	26.10	100	57.9
14	176.44	0	13.10	14	78.45	114	26.90	78	193.84	111	27.53	+56	309.22	113	27.97	+33	57.2
15	176.45	+1	12.76	14	64.60	115	28.24	+11	180.00	119	28.33	−11	295.43	125	28.24	−33	56.6
16	176.47	+1	S 12.41	+15	50.91	132	N 27.98	−53	166.44	141	N 27.55	−73	282.05	151	N 26.96	−92	56.1
17	176.48	1	12.07	15	37.74	162	26.23	110	153.51	174	25.35	127	269.38	186	24.34	142	55.6
18	176.50	1	11.72	15	25.34	197	23.20	156	141.40	209	21.96	169	257.56	221	20.61	180	55.1
19	176.52	1	11.36	15	13.81	232	19.17	190	130.15	243	17.64	199	246.57	252	16.05	207	54.8
20	176.55	1	11.00	15	3.07	261	14.39	214	119.64	269	12.68	220	236.27	276	10.92	225	54.5
21	176.58	+1	S 10.64	+15	352.96	282	N 9.12	−228	109.69	287	N 7.29	−231	226.47	291	N 5.44	−233	54.2
22	176.61	1	10.28	15	343.28	294	N 3.58	234	100.10	295	N 1.70	235	216.95	296	S 0.18	234	54.1
23	176.64	1	9.92	15	333.80	295	S 2.05	233	90.64	294	S 3.91	231	207.47	292	5.76	228	54.0
24	176.67	2	9.55	15	324.28	288	7.59	225	81.07	283	9.38	220	197.81	277	11.14	215	54.1
25	176.71	2	9.18	15	314.51	270	12.86	209	71.15	263	14.54	202	187.73	254	16.15	194	54.2
26	176.75	+2	S 8.81	+16	304.25	244	S 17.71	−185	60.68	234	S 19.19	−175	177.03	222	S 20.59	−165	54.6
27	176.80	2	8.43	16	293.29	211	21.91	153	49.45	198	23.14	140	165.52	186	24.25	125	55.0
28	176.84	+2	S 8.06	+16	281.48	173	S 25.25	−110	37.35	161	S 26.13	−93	153.11	149	S 26.88	−75	55.6

1989 MARCH, APRIL — ARIES AND PLANETS

0ʰ GMT	ARIES GHA	VENUS GHA	v	Dec	d	MARS GHA	v	Dec	d	JUPITER GHA	v	Dec	d	SATURN GHA	v	Dec	d
Mar								**MARCH**									
1	158.78	184.67	−9	S 12.12	+18	107.77	+15	N 19.92	+6	102.22	+36	N 19.23	+1	236.10	+38	S 22.25	0
2	159.76	184.46	9	11.69	18	108.14	15	20.07	6	103.08	36	19.25	1	237.00	38	22.24	0
3	160.75	184.26	8	11.25	19	108.50	15	20.23	6	103.94	36	19.28	1	237.90	38	22.23	0
4	161.74	184.05	8	10.80	19	108.86	15	20.38	6	104.79	35	19.32	1	238.80	38	22.23	0
5	162.72	183.85	8	10.35	19	109.22	15	20.53	6	105.64	35	19.35	1	239.71	38	22.22	0
6	163.71	183.66	−8	S 9.90	+19	109.58	+15	N 20.67	+6	106.49	+35	N 19.38	+1	240.61	+38	S 22.22	0
7	164.69	183.47	8	9.44	19	109.94	15	20.82	6	107.34	35	19.41	1	241.52	38	22.21	0
8	165.68	183.28	8	8.98	19	110.29	15	20.96	6	108.18	35	19.45	1	242.43	38	22.20	0
9	166.66	183.09	8	8.51	20	110.65	15	21.10	6	109.02	35	19.48	1	243.34	38	22.20	0
10	167.65	182.91	8	8.04	20	111.00	15	21.24	6	109.86	35	19.51	1	244.25	38	22.19	0
11	168.63	182.73	−7	S 7.57	+20	111.35	+15	N 21.37	+6	110.70	+35	N 19.55	+1	245.17	+38	S 22.18	0
12	169.62	182.56	7	7.09	20	111.70	14	21.51	5	111.53	35	19.58	1	246.08	38	22.18	0
13	170.61	182.38	7	6.61	20	112.04	14	21.64	5	112.37	35	19.61	1	247.00	38	22.17	0
14	171.59	182.21	7	6.13	20	112.39	14	21.77	5	113.20	35	19.65	1	247.92	38	22.17	0
15	172.58	182.05	7	5.65	20	112.73	14	21.89	5	114.02	34	19.69	1	248.84	38	22.16	0
16	173.56	181.88	−7	S 5.16	+20	113.07	+14	N 22.02	+5	114.85	+34	N 19.72	+1	249.76	+38	S 22.16	0
17	174.55	181.72	7	4.67	21	113.41	14	22.14	5	115.67	34	19.76	1	250.68	39	22.15	0
18	175.53	181.56	7	4.17	21	113.75	14	22.26	5	116.49	34	19.79	1	251.60	39	22.15	0
19	176.52	181.40	7	3.68	21	114.09	14	22.37	5	117.31	34	19.83	2	252.53	39	22.14	0
20	177.51	181.24	6	3.18	21	114.43	14	22.49	5	118.13	34	19.86	2	253.46	39	22.13	0
21	178.49	181.09	−6	S 2.69	+21	114.76	+14	N 22.60	+5	118.94	+34	N 19.90	+2	254.38	+39	S 22.13	0
22	179.48	180.93	6	2.19	21	115.10	14	22.71	4	119.76	34	19.94	2	255.32	39	22.13	0
23	180.46	180.78	6	1.69	21	115.43	14	22.82	4	120.57	34	19.97	2	256.25	39	22.12	0
24	181.45	180.62	6	1.19	21	115.76	14	22.92	4	121.38	34	20.01	2	257.18	39	22.11	0
25	182.43	180.47	6	0.68	21	116.09	14	23.02	4	122.19	34	20.05	2	258.12	39	22.11	0
26	183.42	180.32	−6	S 0.18	+21	116.42	+14	N 23.12	+4	122.99	+33	N 20.08	+2	259.05	+39	S 22.11	0
27	184.41	180.17	6	N 0.32	21	116.74	14	23.22	4	123.79	33	20.12	2	259.99	39	22.10	0
28	185.39	180.02	6	0.83	21	117.07	14	23.31	4	124.60	33	20.16	2	260.93	39	22.10	0
29	186.38	179.87	6	1.33	21	117.40	14	23.40	4	125.40	33	20.20	2	261.87	39	22.09	0
30	187.36	179.72	6	1.83	21	117.72	13	23.49	4	126.19	33	20.23	2	262.81	39	22.09	0
31	188.35	179.57	−6	N 2.33	+21	118.04	+13	N 23.57	+3	126.99	+33	N 20.27	+2	263.76	+39	S 22.09	0
Apr								**APRIL**									
1	189.33	179.42	−6	N 2.83	+21	118.36	+13	N 23.66	+3	127.78	+33	N 20.31	+2	264.71	+40	S 22.08	0
2	190.32	179.27	6	3.33	21	118.69	13	23.74	3	128.58	33	20.35	2	265.65	40	22.08	0
3	191.30	179.11	6	3.83	21	119.00	13	23.81	3	129.37	33	20.38	2	266.60	40	22.08	0
4	192.29	178.96	6	4.33	21	119.32	13	23.89	3	130.16	33	20.42	2	267.55	40	22.07	0
5	193.28	178.80	7	4.83	21	119.64	13	23.96	3	130.94	33	20.46	2	268.51	40	22.07	0
6	194.26	178.65	−7	N 5.32	+21	119.96	+13	N 24.03	+3	131.73	+33	N 20.50	+2	269.46	+40	S 22.07	0
7	195.25	178.49	7	5.82	20	120.27	13	24.10	3	132.51	33	20.54	2	270.42	40	22.07	0
8	196.23	178.33	7	6.31	20	120.59	13	24.16	3	133.30	33	20.57	2	271.38	40	22.06	0
9	197.22	178.17	7	6.80	20	120.90	13	24.22	2	134.08	33	20.61	2	272.34	40	22.06	0
10	198.20	178.00	7	7.28	20	121.21	13	24.28	2	134.86	32	20.65	2	273.30	40	22.06	0
11	199.19	177.84	−7	N 7.77	+20	121.53	+13	N 24.33	+2	135.64	+32	N 20.68	+2	274.26	+40	S 22.06	0
12	200.18	177.67	7	8.25	20	121.84	13	24.38	2	136.41	32	20.72	2	275.23	40	22.06	0
13	201.16	177.50	7	8.72	20	122.15	13	24.43	2	137.19	32	20.76	2	276.20	40	22.05	0
14	202.15	177.33	7	9.20	20	122.46	13	24.48	2	137.96	32	20.80	2	277.17	40	22.05	0
15	203.13	177.15	7	9.67	19	122.77	13	24.52	2	138.74	32	20.84	2	278.14	41	22.05	0
16	204.12	176.98	−8	N 10.13	+19	123.08	+13	N 24.56	+2	139.51	+32	N 20.87	+2	279.11	+41	S 22.05	0
17	205.10	176.80	8	10.60	19	123.39	13	24.60	1	140.28	32	20.91	2	280.08	41	22.05	0
18	206.09	176.61	8	11.06	19	123.69	13	24.63	1	141.05	32	20.95	2	281.06	41	22.05	0
19	207.08	176.42	8	11.51	19	124.00	13	24.66	1	141.82	32	20.98	2	282.04	41	22.05	0
20	208.06	176.23	8	11.96	19	124.31	13	24.69	1	142.58	32	21.02	2	283.02	41	22.05	0
21	209.05	176.04	−8	N 12.41	+18	124.62	+13	N 24.72	+1	143.35	+32	N 21.06	+2	284.00	+41	S 22.05	0
22	210.03	175.84	8	12.85	18	124.92	13	24.74	1	144.11	32	21.09	2	284.98	41	22.05	0
23	211.02	175.64	9	13.28	18	125.23	13	24.76	1	144.88	32	21.13	2	285.97	41	22.05	0
24	212.00	175.43	9	13.71	18	125.54	13	24.78	+1	145.64	32	21.17	2	286.95	41	22.05	0
25	212.99	175.22	9	14.13	17	125.84	13	24.79	0	146.40	32	21.20	2	287.94	41	22.05	0
26	213.97	175.01	−9	N 14.55	+17	126.15	+13	N 24.80	0	147.16	+32	N 21.24	+1	288.93	+41	S 22.05	0
27	214.96	174.79	9	14.96	17	126.45	13	24.81	0	147.92	32	21.27	1	289.92	41	22.05	0
28	215.95	174.57	9	15.37	17	126.76	13	24.81	0	148.68	32	21.31	1	290.92	41	22.05	0
29	216.93	174.34	10	15.77	16	127.06	13	24.81	0	149.43	31	21.35	1	291.91	42	22.05	0
30	217.92	174.11	−10	N 16.16	+16	127.37	+13	N 24.81	0	150.19	+31	N 21.38	+1	292.91	+42	S 22.05	0

Ephemeris

1989 MARCH, APRIL — SUN AND MOON

0h GMT	SUN GHA	v	Dec	d	MOON 0h–8h GHA	v	Dec	d	MOON 8h–16h GHA	v	Dec	d	MOON 16h–24h GHA	v	Dec	d	HP
Mar	(SD 16'.1)					MARCH											
1	176.89	+2	S 7.68	+16	268.79	138	S 27.48	−56	24.37	128	S 27.93	−36	139.87	120	S 28.21	−14	56.4
2	176.94	2	7.30	16	255.31	113	28.33	+7	10.69	108	28.27	+30	126.03	104	28.03	+53	57.3
3	176.99	2	6.92	16	241.35	103	27.60	77	356.65	103	26.99	100	111.96	106	26.19	123	58.3
4	177.04	2	6.53	16	227.28	110	25.21	145	342.64	115	24.04	167	98.04	121	22.71	188	59.2
5	177.10	2	6.15	16	213.48	127	21.21	207	328.98	134	19.55	225	84.53	141	17.75	241	60.1
6	177.15	+2	S 5.76	+16	200.14	148	S 15.82	+256	315.81	154	S 13.77	+269	71.52	160	S 11.62	+280	60.8
7	177.21	3	5.37	16	187.28	164	9.39	288	303.07	168	S 7.08	295	58.90	170	S 4.73	299	61.2
8	177.27	3	4.98	16	174.74	172	S 2.34	301	290.59	172	N 0.07	301	46.45	170	N 2.48	299	61.3
9	177.33	3	4.59	16	162.29	168	N 4.87	294	278.12	164	7.22	287	33.91	160	9.52	278	61.1
10	177.40	3	4.20	16	149.67	154	11.75	268	265.39	148	13.89	255	21.05	141	15.93	240	60.6
11	177.46	+3	S 3.81	+16	136.66	134	N 17.85	+224	252.21	126	N 19.64	+206	7.70	119	N 21.29	+186	59.9
12	177.53	3	3.41	16	123.13	113	22.78	166	238.52	107	24.10	144	353.86	103	25.26	122	59.0
13	177.59	3	3.02	16	109.16	100	26.23	99	224.44	99	27.02	75	339.71	99	27.62	+52	58.1
14	177.66	3	2.63	16	94.99	102	28.03	+28	210.29	107	28.26	+5	325.62	113	28.30	−17	57.3
15	177.73	3	2.23	16	81.01	121	28.16	−39	196.46	131	27.85	−60	311.99	142	27.37	79	56.5
16	177.80	+3	S 1.84	+16	67.60	154	N 26.74	−98	183.31	166	N 25.96	−115	299.12	179	N 25.04	−131	55.8
17	177.87	3	1.44	16	55.03	191	23.99	145	171.04	204	22.83	159	287.15	216	21.56	170	55.2
18	177.95	3	1.05	16	43.36	228	20.20	182	159.66	239	18.75	191	276.06	249	17.22	200	54.8
19	178.02	3	0.65	16	32.53	258	15.62	207	149.07	267	13.96	214	265.68	274	12.25	219	54.4
20	178.09	3	S 0.26	16	22.36	280	10.50	224	139.08	285	8.71	228	255.84	290	6.89	230	54.2
21	178.17	+3	N 0.14	+16	12.64	293	N 5.04	−232	129.46	295	N 3.19	−233	246.30	296	N 1.32	−234	54.0
22	178.24	3	0.54	16	3.15	296	S 0.55	233	119.99	295	S 2.42	232	236.83	293	S 4.28	230	54.0
23	178.32	3	0.93	16	353.66	290	6.12	227	110.46	286	7.94	224	227.22	281	9.73	219	54.0
24	178.40	3	1.32	16	343.95	275	11.48	214	100.63	268	13.19	207	217.25	260	14.85	200	54.1
25	178.47	3	1.72	16	333.81	251	16.45	192	90.30	242	17.98	183	206.72	232	19.45	172	54.3
26	178.55	+3	N 2.11	+16	323.05	221	S 20.82	−161	79.30	211	S 22.11	−149	195.47	199	S 23.30	−135	54.6
27	178.62	3	2.50	16	311.54	188	24.39	121	67.53	177	25.35	105	183.43	167	26.19	88	55.1
28	178.70	3	2.89	16	299.24	157	26.90	70	54.98	148	27.46	−52	170.64	140	27.87	−32	55.6
29	178.78	3	3.28	16	286.24	133	28.13	−12	41.78	128	28.23	+9	157.29	125	28.15	+31	56.3
30	178.85	3	3.67	16	272.77	123	27.91	+52	28.23	122	27.49	74	143.69	124	26.90	96	57.1
31	178.93	+3	N 4.06	+16	259.16	126	S 26.13	+117	14.65	130	S 25.19	+138	130.17	135	S 24.09	+158	58.0
Apr	(SD 16'.0)					APRIL											
1	179.00	+3	N 4.45	+16	245.73	140	S 22.82	+178	1.33	145	S 21.40	+196	116.97	151	S 19.83	+213	58.9
2	179.08	3	4.83	16	232.66	157	18.12	229	348.39	162	16.29	244	104.17	166	14.33	257	59.8
3	179.15	3	5.22	16	219.98	170	12.28	269	335.82	173	10.12	279	91.69	175	7.89	287	60.5
4	179.22	3	5.60	16	207.57	176	S 5.60	293	323.45	175	S 3.26	297	79.34	173	S 0.88	299	61.1
5	179.30	3	5.98	16	195.20	171	N 1.51	298	311.05	166	N 3.90	296	66.86	160	N 6.27	291	61.4
6	179.37	+3	N 6.36	+16	182.62	154	N 8.60	+285	298.33	146	N 10.88	+275	53.98	138	N 13.08	+263	61.3
7	179.44	3	6.74	16	169.57	129	15.19	250	285.08	120	17.19	234	40.52	111	19.06	216	60.9
8	179.51	3	7.12	16	155.88	102	20.78	196	271.18	94	22.35	175	26.41	87	23.75	152	60.3
9	179.58	3	7.49	15	141.58	82	24.97	128	256.72	78	26.00	104	11.82	77	26.83	79	59.4
10	179.64	3	7.86	15	126.92	78	27.46	+54	242.03	82	27.89	+29	357.16	88	28.12	+4	58.5
11	179.71	+3	N 8.23	+15	112.35	96	N 28.15	−20	227.59	106	N 27.99	−42	342.92	118	N 27.66	−64	57.5
12	179.78	3	8.60	15	98.34	130	27.15	84	213.86	144	26.48	102	329.50	159	25.66	120	56.6
13	179.84	3	8.96	15	85.25	173	24.70	135	201.11	188	23.62	150	317.09	202	22.42	163	55.8
14	179.90	3	9.32	15	73.18	215	21.12	174	189.38	228	19.73	184	305.68	239	18.25	193	55.2
15	179.97	3	9.68	15	62.08	250	16.71	201	178.56	260	15.10	208	295.12	268	13.43	214	54.7
16	180.03	+3	N 10.04	+15	51.74	276	N 11.72	−219	168.43	282	N 9.96	−223	285.17	287	N 8.17	−227	54.3
17	180.09	2	10.39	15	41.94	291	6.36	229	158.76	294	N 4.53	231	275.59	296	N 2.68	232	54.1
18	180.14	2	10.74	14	32.44	296	N 0.82	232	149.29	296	S 1.04	232	266.14	294	S 2.89	230	54.0
19	180.20	2	11.09	14	22.97	292	S 4.73	228	139.79	289	6.56	225	256.57	283	8.36	222	54.0
20	180.26	2	11.44	14	13.32	278	10.13	217	130.02	271	11.87	212	246.67	264	13.56	205	54.1
21	180.31	+2	N 11.78	+14	3.26	255	S 15.20	−198	119.78	246	S 16.78	−189	236.23	236	S 18.30	−180	54.3
22	180.36	2	12.12	14	352.61	226	19.73	169	108.89	216	21.09	157	225.10	205	22.35	145	54.6
23	180.41	2	12.45	14	341.22	194	23.50	131	97.25	183	24.55	115	213.19	173	25.47	99	54.9
24	180.45	2	12.78	14	329.05	163	26.27	82	84.84	155	26.92	64	200.55	147	27.43	−45	55.3
25	180.50	2	13.11	14	316.21	141	27.80	−25	71.81	136	28.00	−5	187.38	133	28.04	+16	55.8
26	180.54	+2	N 13.44	+13	302.93	131	S 27.91	+36	58.46	132	S 27.62	+58	173.99	133	S 27.16	+78	56.4
27	180.58	2	13.76	13	289.54	136	26.53	99	45.11	140	25.74	119	160.71	146	24.79	138	57.1
28	180.62	1	14.08	13	276.36	151	23.68	157	32.05	157	22.43	175	147.79	164	21.03	192	57.8
29	180.66	1	14.39	13	263.58	170	19.49	207	19.42	176	17.84	222	135.30	181	16.06	235	58.6
30	180.69	+1	N 14.70	+13	251.23	185	S 14.18	+247	7.19	189	S 12.20	+258	123.18	192	S 10.14	+267	59.3

1989 MAY, JUNE — ARIES AND PLANETS

0h GMT	ARIES GHA	VENUS GHA	v	Dec	d	MARS GHA	v	Dec	d	JUPITER GHA	v	Dec	d	SATURN GHA	v	Dec	d
May	°	°		°		°		°		°		°		°		°	
							MAY										
1	218.90	173.88	−10	N 16.55	+16	127.67	+13	N 24.81	0	150.94	+31	N 21.42	+1	293.91	+42	S 22.05	0
2	219.89	173.64	10	16.93	15	127.98	13	24.80	0	151.70	31	21.45	1	294.91	42	22.05	0
3	220.87	173.39	10	17.30	15	128.28	13	24.79	−1	152.45	31	21.48	1	295.91	42	22.06	0
4	221.86	173.14	11	17.67	15	128.58	13	24.78	1	153.20	31	21.52	1	296.91	42	22.06	0
5	222.85	172.89	11	18.02	15	128.89	13	24.76	1	153.95	31	21.55	1	297.92	42	22.06	0
6	223.83	172.63	−11	N 18.37	+14	129.20	+13	N 24.74	−1	154.71	+31	N 21.59	+1	298.93	+42	S 22.06	0
7	224.82	172.37	11	18.71	14	129.50	13	24.72	1	155.46	31	21.62	1	299.94	42	22.07	0
8	225.80	172.10	11	19.05	13	129.81	13	24.69	1	156.20	31	21.65	1	300.95	42	22.07	0
9	226.79	171.82	12	19.37	13	130.11	13	24.66	1	156.95	31	21.68	1	301.96	42	22.07	0
10	227.77	171.55	12	19.69	13	130.42	13	24.63	1	157.70	31	21.72	1	302.97	42	22.08	0
11	228.76	171.27	−12	N 19.99	+12	130.72	+13	N 24.60	−2	158.45	+31	N 21.75	+1	303.99	+42	S 22.08	0
12	229.75	170.98	12	20.29	12	131.03	13	24.56	2	159.19	31	21.78	1	305.00	42	22.08	0
13	230.73	170.69	12	20.58	12	131.33	13	24.52	2	159.94	31	21.81	1	306.02	43	22.08	0
14	231.72	170.39	13	20.86	11	131.64	13	24.48	2	160.68	31	21.84	1	307.04	43	22.09	0
15	232.70	170.09	13	21.13	11	131.95	13	24.43	2	161.43	31	21.88	1	308.07	43	22.09	0
16	233.69	169.79	−13	N 21.39	+10	132.26	+13	N 24.38	−2	162.17	+31	N 21.91	+1	309.09	+43	S 22.10	0
17	234.67	169.48	13	21.64	11	132.57	13	24.33	2	162.91	31	21.94	1	310.11	43	22.10	0
18	235.66	169.17	13	21.88	10	132.88	13	24.28	2	163.66	31	21.97	1	311.14	43	22.11	0
19	236.64	168.85	13	22.11	9	133.18	13	24.22	3	164.40	31	22.00	1	312.17	43	22.11	0
20	237.63	168.53	14	22.33	9	133.49	13	24.16	3	165.14	31	22.03	1	313.20	43	22.11	0
21	238.62	168.21	−14	N 22.54	+8	133.81	+13	N 24.10	−3	165.88	+31	N 22.05	+1	314.23	+43	S 22.12	0
22	239.60	167.88	14	22.74	8	134.12	13	24.03	3	166.62	31	22.08	1	315.26	43	22.12	0
23	240.59	167.55	14	22.92	7	134.43	13	23.96	3	167.36	31	22.11	1	316.29	43	22.13	0
24	241.57	167.22	14	23.10	7	134.74	13	23.89	3	168.10	31	22.14	1	317.33	43	22.13	0
25	242.56	166.88	14	23.27	6	135.05	13	23.82	3	168.84	31	22.17	1	318.37	43	22.14	0
26	243.54	166.54	−14	N 23.42	+6	135.37	+13	N 23.74	−3	169.58	+31	N 22.19	+1	319.40	+43	S 22.14	0
27	244.53	166.20	14	23.57	5	135.68	13	23.66	3	170.32	31	22.22	1	320.44	43	22.15	0
28	245.52	165.85	14	23.70	5	136.00	13	23.57	4	171.05	31	22.25	1	321.48	43	22.15	0
29	246.50	165.50	15	23.82	5	136.31	13	23.49	4	171.79	31	22.27	1	322.53	43	22.16	0
30	247.49	165.15	15	23.93	4	136.63	13	23.40	4	172.53	31	22.30	1	323.57	43	22.17	0
31	248.47	164.80	−15	N 24.02	+4	136.95	+13	N 23.31	−4	173.27	+31	N 22.32	+1	324.61	+44	S 22.17	0
Jun							**JUNE**										
1	249.46	164.45	−15	N 24.11	+3	137.27	+13	N 23.21	−4	174.00	+31	N 22.35	+1	325.66	+44	S 22.18	0
2	250.44	164.09	15	24.18	3	137.58	13	23.12	4	174.74	31	22.37	1	326.70	44	22.19	0
3	251.43	163.74	15	24.25	2	137.90	13	23.02	4	175.48	31	22.40	1	327.75	44	22.19	0
4	252.42	163.38	15	24.30	2	138.23	13	22.92	4	176.21	31	22.42	1	328.80	44	22.20	0
5	253.40	163.03	15	24.33	1	138.55	13	22.81	4	176.95	31	22.45	1	329.85	44	22.20	0
6	254.39	162.67	−15	N 24.36	+1	138.87	+13	N 22.70	−5	177.68	+31	N 22.47	+1	330.90	+44	S 22.21	0
7	255.37	162.31	15	24.38	0	139.19	14	22.59	5	178.42	31	22.49	1	331.95	44	22.22	0
8	256.36	161.95	15	24.38	0	139.52	14	22.48	5	179.16	31	22.51	1	333.01	44	22.22	0
9	257.34	161.60	15	24.37	−1	139.84	14	22.37	5	179.89	31	22.53	1	334.06	44	22.23	0
10	258.33	161.24	15	24.35	1	140.17	14	22.25	5	180.63	31	22.56	1	335.11	44	22.24	0
11	259.31	160.88	−15	N 24.31	−2	140.50	+14	N 22.13	−5	181.36	+31	N 22.58	+1	336.17	+44	S 22.24	0
12	260.30	160.53	15	24.27	2	140.83	14	22.00	5	182.10	31	22.60	1	337.23	44	22.25	0
13	261.29	160.17	15	24.21	3	141.16	14	21.88	5	182.83	31	22.62	1	338.28	44	22.26	0
14	262.27	159.82	15	24.14	3	141.49	14	21.75	5	183.57	31	22.64	1	339.34	44	22.27	0
15	263.26	159.47	14	24.06	4	141.82	14	21.62	6	184.31	31	22.66	1	340.40	44	22.27	0
16	264.24	159.13	−14	N 23.97	−4	142.15	+14	N 21.49	−6	185.04	+31	N 22.68	+1	341.46	+44	S 22.28	0
17	265.23	158.78	14	23.86	5	142.49	14	21.35	6	185.78	31	22.69	1	342.52	44	22.29	0
18	266.21	158.44	14	23.75	5	142.82	14	21.21	6	186.51	31	22.71	1	343.58	44	22.29	0
19	267.20	158.10	14	23.62	6	143.16	14	21.07	6	187.25	31	22.73	1	344.64	44	22.30	0
20	268.19	157.77	14	23.48	6	143.49	14	20.93	6	187.99	31	22.75	1	345.70	44	22.31	0
21	269.17	157.43	−14	N 23.33	−7	143.83	+14	N 20.79	−6	188.72	+31	N 22.76	+1	346.77	+44	S 22.31	0
22	270.16	157.11	14	23.17	7	144.17	14	20.64	6	189.46	31	22.78	1	347.83	44	22.32	0
23	271.14	156.78	13	22.99	8	144.51	14	20.49	6	190.20	31	22.80	1	348.89	44	22.33	0
24	272.13	156.46	13	22.81	8	144.86	14	20.34	6	190.93	31	22.81	1	349.95	44	22.34	0
25	273.11	156.14	13	22.61	9	145.20	14	20.18	7	191.67	31	22.83	1	351.02	44	22.35	0
26	274.10	155.83	−13	N 22.41	−9	145.54	+14	N 20.03	−7	192.41	+31	N 22.84	+1	352.08	+44	S 22.35	0
27	275.08	155.52	13	22.19	9	145.89	14	19.87	7	193.15	31	22.85	1	353.15	44	22.36	0
28	276.07	155.22	13	21.97	10	146.23	14	19.71	7	193.88	31	22.87	1	354.21	44	22.37	0
29	277.06	154.92	12	21.73	10	146.58	15	19.54	7	194.62	31	22.88	1	355.28	44	22.37	0
30	278.04	154.62	−12	N 21.48	−11	146.93	+15	N 19.38	−7	195.36	+31	N 22.90	+1	356.34	+44	S 22.38	0

1989 MAY, JUNE — SUN AND MOON

0ʰ GMT	SUN GHA	v	Dec	d	MOON 0ʰ–8ʰ GHA	v	Dec	d	8ʰ–16ʰ GHA	v	Dec	d	16ʰ–24ʰ GHA	v	Dec	d	HP
May		(SD 15'.8)					MAY										
1	180.72	+1	N 15.01	+13	239.20	193	S 8.00	+275	355.22	193	S 5.80	+281	111.24	192	S 3.55	+286	60.0
2	180.75	1	15.31	12	227.26	189	S 1.26	288	343.25	185	N 1.05	289	99.21	180	N 3.36	288	60.6
3	180.78	1	15.60	12	215.13	173	N 5.67	285	330.99	165	7.95	280	86.79	156	10.19	273	60.9
4	180.81	1	15.90	12	202.52	146	12.37	263	318.16	135	14.47	251	73.72	124	16.48	237	60.9
5	180.83	1	16.19	12	189.19	112	18.38	221	304.56	101	20.15	202	59.85	90	21.76	182	60.7
6	180.85	+1	N 16.47	+12	175.05	81	N 23.22	+160	290.18	73	N 24.50	+136	45.24	67	N 25.59	+111	60.2
7	180.87	1	16.75	11	160.25	64	26.48	86	275.24	63	27.16	+60	30.23	65	27.64	+34	59.4
8	180.88	+1	17.02	11	145.23	70	27.91	+8	260.26	77	27.97	−17	15.36	87	27.84	−41	58.6
9	180.90	0	17.29	11	130.54	99	27.51	−64	245.81	113	27.00	85	1.20	128	26.32	105	57.7
10	180.91	0	17.56	11	116.70	144	25.48	123	232.33	159	24.50	139	348.09	175	23.38	154	56.8
11	180.92	0	N 17.82	+11	103.97	191	N 22.15	−167	219.97	205	N 20.82	−178	336.10	220	N 19.40	−188	55.9
12	180.92	0	18.08	10	92.34	233	17.89	197	208.68	244	16.31	205	325.11	255	14.67	211	55.3
13	180.93	0	18.33	10	81.63	265	12.98	217	198.23	273	11.25	221	314.89	280	9.48	225	54.7
14	180.93	0	18.57	10	71.61	285	7.69	228	188.38	290	5.87	229	305.18	293	N 4.03	231	54.4
15	180.93	0	18.81	10	62.00	295	N 2.19	231	178.84	296	N 0.34	231	295.68	295	S 1.51	230	54.1
16	180.93	0	N 19.05	+10	52.53	294	S 3.35	−229	169.36	291	S 5.18	−226	286.16	287	S 6.99	−223	54.1
17	180.92	0	19.28	9	42.94	282	8.78	219	159.67	276	10.53	215	276.36	269	12.25	209	54.2
18	180.91	0	19.50	9	32.99	260	13.92	203	149.55	252	15.54	195	266.04	242	17.10	186	54.4
19	180.90	0	19.72	9	22.46	231	18.59	177	138.79	220	20.01	166	255.03	209	21.33	154	54.7
20	180.89	−1	19.93	9	11.19	198	22.57	141	127.25	186	23.69	127	243.22	175	24.71	111	55.0
21	180.88	−1	N 20.14	+8	359.10	165	S 25.60	−94	114.90	155	S 26.35	−77	230.62	147	S 26.96	−58	55.4
22	180.86	1	20.34	8	346.27	139	27.43	−38	101.87	134	27.73	−18	217.42	130	27.88	+3	55.9
23	180.84	1	20.54	8	332.94	128	27.86	+24	88.44	128	27.67	+45	203.94	129	27.31	66	56.4
24	180.82	1	20.72	8	319.45	132	26.78	86	74.99	137	26.09	107	190.57	143	25.24	126	56.9
25	180.80	1	20.91	7	306.19	149	24.23	145	61.87	157	23.07	163	177.60	164	21.77	179	57.4
26	180.77	−1	N 21.09	+7	293.40	172	S 20.33	+195	49.25	179	S 18.78	+209	165.17	186	S 17.10	+222	58.0
27	180.74	1	21.26	7	281.14	193	15.33	234	37.16	198	13.45	245	153.22	202	11.50	254	58.5
28	180.72	1	21.42	7	269.32	206	9.47	262	25.44	207	7.37	268	141.58	208	S 5.23	273	59.0
29	180.68	1	21.58	6	257.73	207	S 3.05	277	13.86	205	S 0.84	278	129.99	202	N 1.39	279	59.5
30	180.65	1	21.74	6	246.08	196	N 3.62	278	2.13	190	N 5.85	275	118.13	182	8.04	270	59.9
31	180.62	−2	N 21.88	+6	234.06	173	N 10.20	+263	349.92	163	N 12.31	+255	105.70	151	N 14.35	+244	60.1
Jun		(SD 15'.8)					JUNE										
1	180.58	−2	N 22.02	+6	221.39	139	N 16.30	+231	336.99	127	N 18.15	+217	92.48	114	N 19.88	+200	60.2
2	180.54	2	22.16	5	207.87	102	21.48	181	323.17	91	22.93	161	78.38	81	24.21	138	60.0
3	180.50	2	22.29	5	193.50	73	25.32	114	308.56	66	26.23	90	63.57	63	26.95	+64	59.6
4	180.46	2	22.41	5	178.56	63	27.46	+38	293.54	65	27.77	+12	48.54	70	27.86	−13	59.0
5	180.41	2	22.52	4	163.58	78	27.76	−38	278.69	89	27.46	−61	33.88	101	26.97	84	58.3
6	180.37	−2	N 22.63	+4	149.17	115	N 26.30	−104	264.57	131	N 25.46	−123	20.10	147	N 24.48	−141	57.5
7	180.32	2	22.73	4	135.75	163	23.35	156	251.53	179	22.11	169	7.44	194	20.75	182	56.7
8	180.28	2	22.83	4	123.47	209	19.30	192	239.63	223	17.76	201	355.89	235	16.15	209	56.0
9	180.23	2	22.91	3	112.25	247	14.48	215	228.71	257	12.77	220	345.25	266	11.00	224	55.3
10	180.18	2	23.00	3	101.86	274	9.21	228	218.53	281	7.39	230	335.26	286	N 5.55	232	54.8
11	180.13	−2	N 23.07	+3	92.02	289	N 3.70	−232	208.82	292	N 1.84	−232	325.64	293	S 0.02	−231	54.4
12	180.08	2	23.14	3	82.46	293	S 1.87	230	199.29	292	S 3.71	228	316.11	289	5.53	225	54.2
13	180.03	2	23.20	2	72.90	286	7.34	222	189.67	281	9.11	218	306.40	275	10.85	213	54.2
14	179.97	2	23.25	2	63.08	268	12.55	207	179.70	260	14.21	200	296.26	251	15.81	193	54.4
15	179.92	2	23.30	2	52.75	241	17.35	184	169.16	230	18.83	174	285.48	219	20.22	163	54.6
16	179.87	−2	N 23.34	+1	41.71	207	S 21.53	−152	157.84	195	S 22.75	−138	273.88	183	S 23.85	−124	55.0
17	179.81	2	23.38	1	29.82	171	24.84	108	145.67	160	25.71	91	261.42	149	26.44	73	55.5
18	179.76	2	23.40	+1	17.10	140	27.03	−54	132.70	132	27.47	−34	248.23	126	27.74	−14	56.1
19	179.71	2	23.42	0	3.72	122	27.85	+8	119.17	119	27.79	+29	234.61	119	27.55	+51	56.6
20	179.65	2	23.44	0	350.04	121	27.14	73	105.48	124	26.56	94	220.96	129	25.81	115	57.1
21	179.60	−2	N 23.44	0	336.47	136	S 24.89	+134	92.03	143	S 23.82	+153	207.66	151	S 22.59	+171	57.6
22	179.54	2	23.44	0	323.35	160	21.22	187	79.10	168	19.73	203	194.93	176	18.10	217	58.1
23	179.49	2	23.43	−1	310.82	184	16.37	229	66.78	192	14.54	240	182.79	198	12.62	249	58.5
24	179.44	2	23.42	1	298.85	203	10.63	257	54.96	207	8.57	264	171.10	211	S 6.47	269	58.8
25	179.38	2	23.40	1	287.26	212	S 4.32	272	43.44	212	S 2.14	274	159.61	211	N 0.06	275	59.1
26	179.33	−2	N 23.37	−1	275.78	208	N 2.25	+274	31.93	204	N 4.45	+272	148.04	199	N 6.62	+267	59.3
27	179.28	2	23.34	2	264.11	192	8.76	262	20.12	184	10.85	254	136.07	174	12.89	246	59.4
28	179.22	2	23.29	2	251.95	164	14.85	235	7.74	153	16.73	222	123.44	141	18.51	208	59.4
29	179.17	2	23.25	2	239.05	129	20.17	192	354.56	117	21.70	174	109.98	106	23.09	154	59.3
30	179.12	−2	N 23.19	−3	225.31	96	N 24.32	+133	340.56	87	N 25.39	+110	95.74	80	N 26.27	+87	59.1

1989 JULY, AUGUST — ARIES AND PLANETS

0ʰ GMT	ARIES GHA	VENUS GHA	v	Dec	d	MARS GHA	v	Dec	d	JUPITER GHA	v	Dec	d	SATURN GHA	v	Dec	d
Jul								JULY									
1	279.03	154.33	−12	N 21.22	−11	147.28	+15	N 19.21	−7	196.10	+31	N 22.91	0	357.41	+44	S 22.39	0
2	280.01	154.04	12	20.96	12	147.63	15	19.04	7	196.84	31	22.92	0	358.47	44	22.39	0
3	281.00	153.76	11	20.68	12	147.98	15	18.87	7	197.58	31	22.93	0	359.54	44	22.40	0
4	281.98	153.49	11	20.39	12	148.33	15	18.70	7	198.32	31	22.94	0	0.60	44	22.41	0
5	282.97	153.22	11	20.10	13	148.69	15	18.52	7	199.07	31	22.96	0	1.67	44	22.42	0
6	283.96	152.96	−11	N 19.79	−13	149.04	+15	N 18.34	−8	199.81	+31	N 22.97	0	2.73	+44	S 22.43	0
7	284.94	152.70	11	19.48	13	149.40	15	18.16	8	200.55	31	22.98	0	3.80	44	22.43	0
8	285.93	152.44	10	19.15	14	149.76	15	17.98	8	201.29	31	22.99	0	4.86	44	22.44	0
9	286.91	152.19	10	18.82	14	150.11	15	17.80	8	202.04	31	23.00	0	5.93	44	22.45	0
10	287.90	151.95	10	18.48	15	150.47	15	17.61	8	202.78	31	23.00	0	6.99	44	22.45	0
11	288.88	151.71	−10	N 18.14	−15	150.83	+15	N 17.42	−8	203.53	+31	N 23.01	0	8.05	+44	S 22.46	0
12	289.87	151.48	9	17.78	15	151.20	15	17.24	8	204.27	31	23.02	0	9.12	44	22.47	0
13	290.86	151.26	9	17.42	15	151.56	15	17.04	8	205.02	31	23.03	0	10.18	44	22.47	0
14	291.84	151.04	9	17.04	16	151.92	15	16.85	8	205.77	31	23.04	0	11.24	44	22.48	0
15	292.83	150.82	9	16.66	16	152.29	15	16.66	8	206.51	31	23.04	0	12.31	44	22.49	0
16	293.81	150.61	−9	N 16.28	−16	152.65	+15	N 16.46	−8	207.26	+31	N 23.05	0	13.37	+44	S 22.50	0
17	294.80	150.41	8	15.89	17	153.02	15	16.26	8	208.01	31	23.06	0	14.43	44	22.50	0
18	295.78	150.21	8	15.48	17	153.39	15	16.06	8	208.76	31	23.06	0	15.49	44	22.51	0
19	296.77	150.01	8	15.08	17	153.76	15	15.86	8	209.51	31	23.07	0	16.55	44	22.52	0
20	297.75	149.83	8	14.66	18	154.13	15	15.65	9	210.26	31	23.07	0	17.61	44	22.52	0
21	298.74	149.64	−7	N 14.24	−18	154.50	+15	N 15.45	−9	211.01	+31	N 23.08	0	18.67	+44	S 22.53	0
22	299.73	149.46	7	13.82	18	154.87	16	15.24	9	211.77	31	23.08	0	19.73	44	22.53	0
23	300.71	149.29	7	13.39	18	155.24	16	15.03	9	212.52	31	23.09	0	20.79	44	22.54	0
24	301.70	149.12	7	12.95	18	155.62	16	14.82	9	213.28	32	23.09	0	21.84	44	22.55	0
25	302.68	148.96	7	12.51	19	155.99	16	14.61	9	214.03	32	23.09	0	22.90	44	22.55	0
26	303.67	148.80	−6	N 12.06	−19	156.37	+16	N 14.40	−9	214.79	+32	N 23.09	0	23.95	+44	S 22.56	0
27	304.65	148.64	6	11.61	19	156.75	16	14.18	9	215.55	32	23.10	0	25.01	44	22.57	0
28	305.64	148.49	6	11.15	19	157.13	16	13.96	9	216.31	32	23.10	0	26.06	44	22.57	0
29	306.63	148.35	6	10.69	19	157.50	16	13.75	9	217.07	32	23.10	0	27.11	44	22.58	0
30	307.61	148.20	6	10.22	20	157.88	16	13.53	9	217.83	32	23.10	0	28.16	44	22.58	0
31	308.60	148.07	−6	N 9.75	−20	158.26	+16	N 13.31	−9	218.59	+32	N 23.10	0	29.21	+44	S 22.59	0
Aug								AUGUST									
1	309.58	147.93	−5	N 9.27	−20	158.64	+16	N 13.08	−9	219.36	+32	N 23.11	0	30.26	+44	S 22.59	0
2	310.57	147.80	5	8.79	20	159.02	16	12.86	9	220.12	32	23.11	0	31.31	44	22.60	0
3	311.55	147.67	5	8.31	20	159.41	16	12.63	9	220.89	32	23.11	0	32.36	44	22.60	0
4	312.54	147.55	5	7.82	20	159.79	16	12.41	10	221.66	32	23.11	0	33.40	44	22.61	0
5	313.53	147.43	5	7.33	20	160.18	16	12.18	10	222.43	32	23.11	0	34.45	43	22.62	0
6	314.51	147.32	−5	N 6.84	−21	160.56	+16	N 11.95	−10	223.20	+32	N 23.10	0	35.49	+43	S 22.62	0
7	315.50	147.20	5	6.34	21	160.94	16	11.72	10	223.97	32	23.10	0	36.54	43	22.63	0
8	316.48	147.09	4	5.85	21	161.33	16	11.49	10	224.74	32	23.10	0	37.58	43	22.63	0
9	317.47	146.98	4	5.35	21	161.72	16	11.26	10	225.51	32	23.10	0	38.62	43	22.64	0
10	318.45	146.88	4	4.84	21	162.11	16	11.02	10	226.29	32	23.10	0	39.65	43	22.64	0
11	319.44	146.78	−4	N 4.34	−21	162.49	+16	N 10.79	−10	227.07	+32	N 23.10	0	40.69	+43	S 22.65	0
12	320.42	146.68	4	3.83	21	162.88	16	10.55	10	227.84	33	23.10	0	41.73	43	22.65	0
13	321.41	146.58	4	3.32	21	163.27	16	10.31	10	228.62	33	23.09	0	42.76	43	22.66	0
14	322.40	146.48	4	2.81	21	163.66	16	10.07	10	229.41	33	23.09	0	43.79	43	22.66	0
15	323.38	146.39	4	2.30	21	164.05	16	9.83	10	230.19	33	23.09	0	44.83	43	22.66	0
16	324.37	146.30	−4	N 1.78	−21	164.44	+16	N 9.59	−10	230.97	+33	N 23.08	0	45.86	+43	S 22.67	0
17	325.35	146.21	4	1.27	21	164.84	16	9.35	10	231.76	33	23.08	0	46.88	43	22.68	0
18	326.34	146.12	4	0.75	21	165.23	16	9.11	10	232.55	33	23.08	0	47.91	43	22.68	0
19	327.32	146.03	4	N 0.24	22	165.62	16	8.86	10	233.34	33	23.07	0	48.94	43	22.68	0
20	328.31	145.94	4	S 0.28	22	166.01	16	8.62	10	234.13	33	23.07	0	49.96	43	22.69	0
21	329.30	145.86	−4	S 0.80	−22	166.41	+16	N 8.37	−10	234.92	+33	N 23.06	0	50.98	+43	S 22.69	0
22	330.28	145.77	4	1.31	22	166.80	16	8.13	10	235.71	33	23.06	0	52.00	43	22.70	0
23	331.27	145.69	4	1.83	22	167.20	16	7.88	10	236.51	33	23.06	0	53.02	43	22.70	0
24	332.25	145.60	3	2.34	21	167.59	16	7.63	10	237.31	33	23.05	0	54.04	42	22.70	0
25	333.24	145.52	3	2.86	21	167.99	16	7.38	10	238.11	33	23.05	0	55.06	42	22.71	0
26	334.22	145.44	−3	S 3.37	−21	168.38	+16	N 7.13	−10	238.91	+33	N 23.04	0	56.07	+42	S 22.71	0
27	335.21	145.35	3	3.89	21	168.78	17	6.88	10	239.71	34	23.04	0	57.08	42	22.72	0
28	336.20	145.27	3	4.40	21	169.18	17	6.63	10	240.52	34	23.03	0	58.10	42	22.72	0
29	337.18	145.18	4	4.91	21	169.57	17	6.37	11	241.33	34	23.03	0	59.11	42	22.72	0
30	338.17	145.10	4	5.42	21	169.97	17	6.12	11	242.13	34	23.02	0	60.11	42	22.73	0
31	339.15	145.01	−4	S 5.93	−21	170.36	+17	N 5.87	−11	242.95	+34	N 23.01	0	61.12	+42	S 22.73	0

1989 JULY, AUGUST — SUN AND MOON

0h GMT	SUN GHA	v	Dec	d	MOON 0h–8h GHA	v	Dec	d	MOON 8h–16h GHA	v	Dec	d	MOON 16h–24h GHA	v	Dec	d	HP
Jul	(SD 15'.8)						JULY										
1	179.07	−2	N 23.13	−3	210.86	76	N 26.97	+63	325.95	74	N 27.47	+38	81.02	74	N 27.77	+13	58.8
2	179.02	2	23.06	3	196.09	78	27.87	−12	311.19	84	27.78	−36	66.34	92	27.48	−60	58.3
3	178.98	2	22.98	3	181.56	103	27.01	82	296.86	115	26.35	103	52.26	129	25.53	122	57.7
4	178.93	2	22.90	4	167.77	143	24.55	140	283.40	158	23.43	156	39.14	174	22.18	170	57.0
5	178.89	2	22.81	4	155.01	188	20.82	183	271.00	203	19.36	194	27.10	217	17.81	203	56.4
6	178.84	−2	N 22.72	−4	143.31	229	N 16.18	−211	259.62	241	N 14.49	−218	16.03	251	N 12.75	−223	55.7
7	178.80	2	22.61	5	132.52	261	10.96	227	249.09	269	9.14	231	5.72	276	7.30	233	55.2
8	178.76	2	22.51	5	122.40	281	N 5.43	234	239.13	286	N 3.56	235	355.90	289	N 1.68	234	54.7
9	178.72	2	22.39	5	112.68	290	S 0.19	233	229.49	291	S 2.06	231	346.29	290	S 3.91	229	54.4
10	178.68	1	22.27	5	103.10	288	5.74	226	219.88	285	7.55	222	336.64	281	9.32	217	54.3
11	178.65	−1	N 22.14	−6	93.37	275	S 11.06	−212	210.06	269	S 12.75	−206	326.69	261	S 14.40	−199	54.3
12	178.62	1	22.01	6	83.26	253	15.99	191	199.76	243	17.52	182	316.19	233	18.98	173	54.5
13	178.59	1	21.87	6	72.53	221	20.36	162	188.78	210	21.66	150	304.94	198	22.86	137	54.8
14	178.56	1	21.72	6	61.00	186	23.95	123	176.96	173	24.93	107	292.83	161	25.79	91	55.3
15	178.53	1	21.57	7	48.60	151	26.52	73	164.28	140	27.10	−54	279.89	132	27.53	−34	55.9
16	178.51	−1	N 21.41	−7	35.42	124	S 27.80	−13	150.89	119	S 27.91	+8	266.32	116	S 27.84	+30	56.6
17	178.48	1	21.24	7	21.73	114	27.60	+52	137.12	114	27.18	75	252.51	117	26.58	97	57.3
18	178.46	1	21.07	7	7.93	121	25.81	118	123.37	126	24.87	138	238.86	133	23.76	158	57.9
19	178.44	1	20.89	8	354.41	141	22.50	177	110.01	149	21.08	194	225.68	157	19.53	210	58.5
20	178.43	−1	20.71	8	341.42	165	17.85	224	97.22	174	16.06	237	213.09	181	14.16	248	59.0
21	178.41	0	N 20.52	−8	329.02	188	S 12.18	+257	85.00	193	S 10.12	+265	201.03	198	S 8.00	+271	59.3
22	178.40	0	20.33	8	317.09	201	S 5.83	276	73.18	203	S 3.63	278	189.29	204	S 1.40	279	59.5
23	178.39	0	20.13	9	305.40	204	N 0.83	279	61.51	202	N 3.06	276	177.61	199	N 5.28	273	59.5
24	178.39	0	19.92	9	293.67	194	7.46	267	49.70	188	9.60	260	165.69	181	11.68	251	59.4
25	178.38	0	19.71	9	281.62	173	13.69	241	37.48	164	15.61	229	153.27	154	17.45	216	59.3
26	178.38	0	N 19.49	−9	268.99	144	N 19.17	+200	24.62	134	N 20.77	+183	140.17	124	N 22.24	+165	59.0
27	178.38	0	19.27	10	255.64	114	23.56	145	11.04	106	24.72	124	126.36	98	25.72	102	58.7
28	178.38	0	19.04	10	241.63	93	26.53	80	356.85	89	27.17	+56	112.04	87	27.62	+32	58.3
29	178.39	0	18.81	10	227.22	88	27.88	+8	342.40	92	27.94	−16	97.62	97	27.81	−39	57.9
30	178.40	0	18.57	10	212.87	105	27.50	−61	328.19	114	27.01	83	83.58	126	26.35	103	57.4
31	178.41	+1	N 18.33	−10	199.07	138	N 25.52	−122	314.65	151	N 24.55	−140	70.33	165	N 23.43	−155	56.9
Aug	(SD 15'.8)						AUGUST										
1	178.42	+1	N 18.08	−11	186.13	178	N 22.19	−170	302.04	192	N 20.83	−183	58.05	205	N 19.37	−194	56.4
2	178.44	1	17.83	11	174.17	218	17.82	203	290.40	230	16.19	212	46.71	240	14.50	219	55.8
3	178.45	1	17.57	11	163.12	250	12.75	224	279.60	259	10.96	229	36.15	267	9.12	232	55.3
4	178.47	1	17.31	11	152.77	274	7.27	234	269.44	279	N 5.39	236	26.15	283	N 3.51	236	54.9
5	178.50	1	17.04	11	142.90	287	N 1.61	236	259.67	288	S 0.27	235	16.45	289	S 2.15	233	54.5
6	178.52	+1	N 16.77	−12	133.25	289	S 4.01	−230	250.04	287	S 5.85	−227	6.81	284	S 7.66	−222	54.3
7	178.55	1	16.49	12	123.57	281	9.44	218	240.30	270	11.18	211	356.98	270	12.88	205	54.2
8	178.58	1	16.21	12	113.62	263	14.52	198	230.21	255	16.10	190	346.73	246	17.62	181	54.2
9	178.61	2	15.93	12	103.18	237	19.07	171	219.56	226	20.44	160	335.85	216	21.72	149	54.5
10	178.65	2	15.64	12	92.05	204	22.91	135	208.17	193	23.99	122	324.19	181	24.97	106	54.9
11	178.69	+2	N 15.34	−12	80.12	170	S 25.82	−90	195.96	159	S 26.54	−73	311.71	149	S 27.12	−55	55.4
12	178.73	2	15.05	13	67.38	140	27.56	−35	182.98	132	27.85	−15	298.52	126	27.97	+6	56.1
13	178.77	2	14.74	13	54.01	122	27.92	+27	169.46	119	27.70	+49	284.89	118	27.31	71	56.9
14	178.82	2	14.44	13	40.31	118	26.74	93	155.74	121	26.00	115	271.19	125	25.08	136	57.7
15	178.86	2	14.13	13	26.66	130	24.00	156	142.18	136	22.75	175	257.75	142	21.34	194	58.6
16	178.91	+2	N 13.82	−13	13.36	149	S 19.79	+211	129.03	156	S 18.11	+226	244.76	163	S 16.30	+240	59.3
17	178.97	2	13.50	13	0.55	169	14.38	252	116.38	175	12.36	263	232.26	180	10.25	272	59.8
18	179.02	2	13.18	13	348.18	184	8.08	278	104.13	187	S 5.85	284	220.10	188	S 3.58	287	60.2
19	179.08	2	12.86	14	336.09	189	S 1.29	288	92.08	188	N 1.01	287	208.07	186	N 3.31	284	60.3
20	179.13	3	12.53	14	324.04	183	N 5.59	280	79.98	178	7.83	274	195.89	173	10.01	265	60.2
21	179.19	+3	N 12.20	−14	311.75	166	N 12.14	+255	67.56	159	N 14.18	+243	183.31	150	N 16.13	+230	59.9
22	179.26	3	11.86	14	298.99	142	17.97	215	54.61	133	19.69	198	170.15	124	21.27	180	59.5
23	179.32	3	11.53	14	285.62	116	22.71	160	41.03	108	23.99	139	156.37	102	25.10	117	59.0
24	179.39	3	11.19	14	271.66	97	26.04	95	26.92	93	26.80	71	142.14	92	27.37	+48	58.4
25	179.45	3	10.84	14	257.36	92	27.76	+24	12.57	95	27.95	+1	127.82	100	27.95	−23	57.8
26	179.52	+3	N 10.50	−15	243.10	107	N 27.77	−45	358.44	116	N 27.41	−67	113.84	126	N 26.88	−87	57.2
27	179.59	3	10.15	15	229.33	138	26.18	106	344.91	150	25.33	124	100.59	163	24.34	141	56.7
28	179.67	3	9.80	15	216.37	176	23.21	156	332.26	189	21.96	170	88.25	201	20.60	182	56.2
29	179.74	3	9.45	15	204.34	214	19.15	193	320.53	225	17.61	202	76.81	236	15.99	211	55.7
30	179.81	3	9.09	15	193.18	246	14.30	217	309.63	255	12.56	223	66.15	263	10.78	228	55.2
31	179.89	+3	N 8.73	−15	182.74	270	N 8.95	−231	299.37	276	N 7.10	−234	56.06	280	N 5.23	−235	54.8

1989 SEPTEMBER, OCTOBER — ARIES AND PLANETS

0h GMT	ARIES GHA	VENUS GHA	v	Dec	d	MARS GHA	v	Dec	d	JUPITER GHA	v	Dec	d	SATURN GHA	v	Dec	d
Sep							**SEPTEMBER**										
1	340.14	144.92	−4	S 6.44	−21	170.76	+17	N 5.61	−11	243.76	+34	N 23.01	0	62.12	+42	S 22.73	0
2	341.12	144.84	4	6.94	21	171.16	17	5.36	11	244.57	34	23.00	0	63.13	42	22.73	0
3	342.11	144.75	4	7.45	21	171.55	17	5.10	11	245.39	34	23.00	0	64.13	42	22.74	0
4	343.09	144.66	4	7.95	21	171.95	17	4.84	11	246.21	34	22.99	0	65.13	42	22.74	0
5	344.08	144.56	4	8.45	21	172.35	17	4.59	11	247.03	34	22.98	0	66.12	42	22.74	0
6	345.07	144.47	−4	S 8.94	−20	172.75	+17	N 4.33	−11	247.86	+34	N 22.98	0	67.12	+41	S 22.75	0
7	346.05	144.37	4	9.43	20	173.14	17	4.07	11	248.68	35	22.97	0	68.11	41	22.75	0
8	347.04	144.28	4	9.92	20	173.54	17	3.81	11	249.51	35	22.96	0	69.11	41	22.75	0
9	348.02	144.18	4	10.41	20	173.94	17	3.55	11	250.34	35	22.96	0	70.10	41	22.75	0
10	349.01	144.08	4	10.89	20	174.33	17	3.29	11	251.18	35	22.95	0	71.09	41	22.76	0
11	349.99	143.98	−4	S 11.37	−20	174.73	+17	N 3.03	−11	252.01	+35	N 22.94	0	72.07	+41	S 22.76	0
12	350.98	143.87	5	11.85	20	175.13	17	2.77	11	252.85	35	22.94	0	73.06	41	22.76	0
13	351.97	143.76	5	12.32	19	175.53	17	2.51	11	253.69	35	22.93	0	74.04	41	22.76	0
14	352.95	143.65	5	12.78	19	175.92	17	2.25	11	254.53	35	22.92	0	75.02	41	22.76	0
15	353.94	143.54	5	13.25	19	176.32	17	1.99	11	255.38	35	22.92	0	76.00	41	22.77	0
16	354.92	143.43	−5	S 13.70	−19	176.71	+17	N 1.73	−11	256.23	+35	N 22.91	0	76.98	+41	S 22.77	0
17	355.91	143.31	5	14.16	19	177.11	16	1.46	11	257.08	36	22.90	0	77.96	41	22.77	0
18	356.89	143.19	5	14.61	18	177.51	16	1.20	11	257.93	36	22.90	0	78.93	41	22.77	0
19	357.88	143.06	5	15.05	18	177.90	16	0.94	11	258.78	36	22.89	0	79.91	40	22.77	0
20	358.87	142.94	5	15.49	18	178.30	16	0.68	11	259.64	36	22.88	0	80.88	40	22.77	0
21	359.85	142.81	−6	S 15.92	−18	178.69	+16	N 0.41	−11	260.50	+36	N 22.88	0	81.85	+40	S 22.77	0
22	0.84	142.67	6	16.34	17	179.09	16	N 0.15	11	261.37	36	22.87	0	82.81	40	22.78	0
23	1.82	142.54	6	16.76	17	179.48	16	S 0.12	11	262.23	36	22.86	0	83.78	40	22.78	0
24	2.81	142.40	6	17.18	17	179.87	16	0.38	11	263.10	36	22.86	0	84.74	40	22.78	0
25	3.79	142.26	6	17.59	17	180.27	16	0.64	11	263.97	36	22.85	0	85.70	40	22.78	0
26	4.78	142.11	−6	S 17.99	−16	180.66	+16	S 0.91	−11	264.85	+37	N 22.85	0	86.67	+40	S 22.78	0
27	5.76	141.96	6	18.38	16	181.05	16	1.17	11	265.73	37	22.84	0	87.62	40	22.78	0
28	6.75	141.81	7	18.77	16	181.44	16	1.43	11	266.61	37	22.84	0	88.58	40	22.78	0
29	7.74	141.65	7	19.15	16	181.83	16	1.70	11	267.49	37	22.83	0	89.54	40	22.78	0
30	8.72	141.49	−7	S 19.52	−15	182.22	+16	S 1.96	−11	268.38	+37	N 22.82	0	90.49	+40	S 22.78	0
Oct							**OCTOBER**										
1	9.71	141.33	−7	S 19.89	−15	182.61	+16	S 2.23	−11	269.27	+37	N 22.82	0	91.44	+40	S 22.78	0
2	10.69	141.16	7	20.25	15	183.00	16	2.49	11	270.16	37	22.81	0	92.39	40	22.78	0
3	11.68	141.00	7	20.60	14	183.39	16	2.75	11	271.05	37	22.81	0	93.34	39	22.78	0
4	12.66	140.82	7	20.94	14	183.77	16	3.01	11	271.95	38	22.80	0	94.29	39	22.78	0
5	13.65	140.65	7	21.28	14	184.16	16	3.28	11	272.85	38	22.80	0	95.23	39	22.78	0
6	14.64	140.47	−7	S 21.60	−13	184.55	+16	S 3.54	−11	273.76	+38	N 22.80	0	96.17	+39	S 22.78	0
7	15.62	140.29	8	21.92	13	184.93	16	3.80	11	274.67	38	22.79	0	97.12	39	22.78	0
8	16.61	140.11	8	22.23	13	185.32	16	4.07	11	275.58	38	22.79	0	98.06	39	22.78	0
9	17.59	139.92	8	22.53	12	185.70	16	4.33	11	276.49	38	22.78	0	98.99	39	22.78	0
10	18.58	139.73	8	22.82	12	186.08	16	4.59	11	277.41	38	22.78	0	99.93	39	22.78	0
11	19.56	139.54	−8	S 23.11	−11	186.46	+16	S 4.85	−11	278.33	+38	N 22.77	0	100.87	+39	S 22.78	0
12	20.55	139.35	8	23.38	11	186.84	16	5.11	11	279.26	39	22.77	0	101.80	39	22.78	0
13	21.54	139.15	8	23.64	11	187.22	16	5.37	11	280.18	39	22.77	0	102.73	39	22.77	0
14	22.52	138.95	8	23.90	10	187.60	16	5.63	11	281.12	39	22.77	0	103.66	39	22.77	0
15	23.51	138.76	8	24.14	10	187.98	16	5.89	11	282.05	39	22.77	0	104.59	39	22.77	0
16	24.49	138.56	−8	S 24.38	−9	188.36	+16	S 6.15	−11	282.99	+39	N 22.76	0	105.52	+39	S 22.77	0
17	25.48	138.36	8	24.60	9	188.73	16	6.41	11	283.93	39	22.76	0	106.44	38	22.77	0
18	26.46	138.15	8	24.82	9	189.11	16	6.66	11	284.87	40	22.76	0	107.37	38	22.77	0
19	27.45	137.95	8	25.03	8	189.48	15	6.92	11	285.82	40	22.76	0	108.29	38	22.77	0
20	28.43	137.75	8	25.22	8	189.85	15	7.18	11	286.77	40	22.75	0	109.21	38	22.76	0
21	29.42	137.55	−8	S 25.41	−7	190.22	+15	S 7.43	−11	287.73	+40	N 22.75	0	110.13	+38	S 22.76	0
22	30.41	137.34	8	25.58	7	190.59	15	7.69	11	288.69	40	22.75	0	111.05	38	22.76	0
23	31.39	137.14	8	25.75	6	190.96	15	7.94	11	289.65	40	22.75	0	111.96	38	22.76	0
24	32.38	136.94	8	25.90	6	191.33	15	8.20	11	290.61	40	22.75	0	112.88	38	22.76	0
25	33.36	136.74	8	26.04	6	191.69	15	8.45	10	291.58	41	22.75	0	113.79	38	22.75	0
26	34.35	136.54	−8	S 26.18	−5	192.05	+15	S 8.70	−10	292.56	+41	N 22.75	0	114.70	+38	S 22.75	0
27	35.33	136.34	8	26.30	5	192.42	15	8.95	10	293.53	41	22.75	0	115.61	38	22.75	0
28	36.32	136.14	8	26.41	4	192.78	15	9.20	10	294.51	41	22.76	0	116.52	38	22.75	0
29	37.31	135.95	8	26.51	4	193.14	15	9.45	10	295.50	41	22.76	0	117.43	38	22.74	0
30	38.29	135.76	8	26.60	3	193.50	15	9.70	10	296.48	41	22.76	0	118.34	38	22.74	0
31	39.28	135.57	−8	S 26.68	−3	193.85	+15	S 9.95	−10	297.47	+41	N 22.76	0	119.24	+38	S 22.74	0

1989 SEPTEMBER, OCTOBER — SUN AND MOON

0h GMT	SUN GHA	v	Dec	d	MOON 0h–8h GHA	v	Dec	d	MOON 8h–16h GHA	v	Dec	d	MOON 16h–24h GHA	v	Dec	d	HP
Sep	(SD 15'.9)						SEPTEMBER										
1	179.97	+3	N 8.37	−15	172.78	284	N 3.35	−236	289.54	287	N 1.46	−236	46.31	288	S 0.42	−235	54.5
2	180.05	3	8.01	15	163.09	289	S 2.30	233	279.88	288	S 4.16	230	36.67	286	6.00	226	54.3
3	180.13	3	7.64	15	153.44	284	7.81	222	270.19	280	9.59	217	26.91	275	11.32	211	54.1
4	180.21	3	7.27	15	143.59	270	13.01	204	260.23	263	14.65	197	16.81	256	16.22	188	54.0
5	180.29	3	6.90	15	133.34	248	17.73	179	249.81	239	19.16	169	6.20	230	20.51	158	54.1
6	180.38	+3	N 6.53	−16	122.52	220	S 21.78	−146	238.76	210	S 22.95	−133	354.92	200	S 24.01	−119	54.4
7	180.46	4	6.16	16	111.00	189	24.96	104	226.99	179	25.80	88	342.90	170	26.51	72	54.8
8	180.55	4	5.78	16	98.74	161	27.08	−54	214.51	153	27.51	−35	330.21	146	27.80	−16	55.4
9	180.63	4	5.41	16	85.85	140	27.93	+4	201.46	136	27.90	+24	317.02	133	27.71	+45	56.1
10	180.72	4	5.03	16	72.56	131	27.35	66	188.09	131	26.82	87	303.62	133	26.12	108	56.9
11	180.81	+4	N 4.65	−16	59.17	135	S 25.26	+128	174.73	139	S 24.23	+149	290.32	143	S 23.04	+168	57.8
12	180.90	4	4.27	16	45.94	148	21.70	186	161.60	153	20.21	204	277.31	158	18.58	220	58.8
13	180.98	4	3.89	16	33.05	163	16.83	235	148.84	168	14.95	248	264.66	172	12.96	260	59.7
14	181.07	4	3.51	16	20.51	175	10.88	271	136.39	177	8.71	279	252.29	178	S 6.48	286	60.4
15	181.16	4	3.12	16	8.20	179	S 4.20	290	124.11	178	S 1.88	293	240.01	176	N 0.47	293	60.9
16	181.25	+4	N 2.74	−16	355.90	172	N 2.81	+292	111.75	168	N 5.15	+288	227.57	162	N 7.46	+282	61.1
17	181.34	4	2.35	16	343.35	155	9.71	274	99.07	148	11.91	264	214.74	139	14.02	252	61.0
18	181.43	4	1.96	16	330.33	130	16.03	237	85.85	121	17.93	221	201.30	112	19.70	203	60.6
19	181.52	4	1.58	16	316.68	104	21.32	183	71.99	96	22.79	162	187.24	89	24.08	139	60.0
20	181.61	4	1.19	16	302.43	84	25.20	116	57.58	80	26.13	92	172.71	79	26.86	+67	59.2
21	181.70	+4	N 0.80	−16	287.82	80	N 27.40	+42	42.94	83	N 27.74	+18	158.09	89	N 27.88	−6	58.5
22	181.79	4	0.41	16	273.28	96	27.83	−30	28.52	106	27.60	−52	143.85	117	27.18	73	57.7
23	181.87	4	N 0.02	16	259.26	129	26.59	93	14.77	142	25.85	112	130.39	156	24.95	129	56.9
24	181.96	4	S 0.37	16	246.11	169	23.92	145	1.94	183	22.77	159	117.89	197	21.50	172	56.2
25	182.05	4	0.76	16	233.94	209	20.12	183	350.10	222	18.66	193	106.35	233	17.11	202	55.7
26	182.13	+4	S 1.15	−16	222.70	244	N 15.49	−210	339.13	253	N 13.82	−216	95.63	261	N 12.09	−222	55.2
27	182.22	4	1.54	16	212.20	269	10.32	226	328.83	275	8.51	229	85.51	280	6.68	232	54.8
28	182.30	3	1.93	16	202.23	284	N 4.82	233	318.98	287	N 2.96	234	75.75	289	N 1.09	234	54.4
29	182.39	3	2.32	16	192.54	290	S 0.78	233	309.34	289	S 2.64	230	66.14	288	S 4.48	228	54.2
30	182.47	+3	S 2.71	−16	182.92	286	S 6.31	−224	299.69	283	S 8.10	−220	56.43	279	S 9.86	−215	54.0
Oct	(SD 16'.1)						OCTOBER										
1	182.55	+3	S 3.09	−16	173.14	274	S 11.58	−209	289.81	268	S 13.25	−202	46.44	261	S 14.87	−194	53.9
2	182.63	3	3.48	16	163.01	254	16.43	186	279.53	246	17.91	176	35.98	238	19.32	166	54.0
3	182.71	3	3.87	16	152.36	229	20.65	155	268.67	220	21.89	143	24.91	211	23.03	129	54.1
4	182.79	3	4.26	16	141.07	201	24.06	115	257.17	192	24.98	100	13.18	184	25.79	84	54.3
5	182.86	3	4.64	16	129.13	176	26.46	68	245.02	168	27.00	−50	0.85	162	27.41	−32	54.7
6	182.94	+3	S 5.03	−16	116.62	157	S 27.66	−14	232.35	152	S 27.77	+6	348.05	150	S 27.73	+25	55.3
7	183.01	3	5.41	16	103.73	148	27.53	+45	219.39	148	27.17	65	335.05	148	26.66	84	55.9
8	183.08	3	5.79	16	90.72	150	25.98	104	206.40	153	25.15	123	322.11	156	24.17	141	56.7
9	183.15	3	6.17	16	77.84	160	23.04	160	193.60	165	21.76	177	309.40	169	20.35	194	57.7
10	183.22	3	6.55	16	65.23	173	18.80	209	181.10	177	17.12	224	297.00	180	15.33	237	58.6
11	183.29	+3	S 6.93	−16	52.92	183	S 13.43	+250	168.87	185	S 11.43	+261	284.82	186	S 9.35	+270	59.6
12	183.35	3	7.31	16	40.79	186	7.19	278	156.76	184	S 4.96	284	272.71	182	S 2.69	288	60.4
13	183.41	2	7.68	15	28.64	178	S 0.38	291	144.55	173	N 1.94	291	260.41	166	N 4.27	290	61.1
14	183.47	2	8.06	15	16.22	159	N 6.59	286	131.96	150	8.87	279	247.64	140	11.10	270	61.4
15	183.53	2	8.43	15	3.24	130	13.27	259	118.76	119	15.34	246	234.19	107	17.31	230	61.4
16	183.59	+2	S 8.80	−15	349.52	96	N 19.15	+212	104.77	86	N 20.85	+192	219.94	76	N 22.38	+170	61.1
17	183.64	2	9.16	15	335.02	68	23.74	147	90.05	62	24.91	122	205.02	58	25.89	96	60.5
18	183.69	2	9.53	15	319.96	57	26.65	+70	74.89	58	27.21	+43	189.83	62	27.56	+17	59.7
19	183.74	2	9.89	15	304.81	69	27.70	−8	59.84	78	27.63	−33	174.95	89	27.36	−56	58.8
20	183.78	2	10.25	15	290.14	103	26.91	78	45.44	117	26.28	99	160.86	132	25.49	118	57.8
21	183.83	+2	S 10.61	−15	276.40	148	N 24.55	−135	32.07	164	N 23.48	−150	147.86	180	N 22.27	−164	56.9
22	183.87	2	10.97	15	263.78	195	20.96	176	19.82	209	19.56	187	135.97	222	18.06	196	56.1
23	183.90	1	11.32	15	252.22	234	16.49	204	8.58	245	14.86	211	125.02	255	13.17	217	55.5
24	183.94	1	11.67	14	241.54	264	11.44	221	358.13	271	9.67	225	114.78	278	7.86	228	54.9
25	183.97	1	12.02	14	231.48	283	6.04	230	348.23	286	N 4.20	231	105.00	289	N 2.36	231	54.5
26	184.00	+1	S 12.36	−14	221.79	291	N 0.51	−231	338.60	291	S 1.34	−229	95.40	290	S 3.18	−228	54.2
27	184.02	1	12.70	14	212.21	289	S 5.00	225	328.99	286	6.79	221	85.76	282	8.56	217	54.0
28	184.04	1	13.04	14	202.50	278	10.30	212	319.20	272	11.99	206	75.86	266	13.64	199	53.9
29	184.06	+1	13.37	15	192.46	259	15.22	191	309.01	251	16.75	182	65.50	243	18.21	173	53.9
30	184.08	0	13.70	14	181.92	234	19.59	162	298.27	225	20.89	150	54.54	216	22.09	138	54.0
31	184.09	0	S 14.03	−13	170.75	206	S 23.19	−124	286.88	198	S 24.19	−110	42.94	189	S 25.07	−95	54.2

1989 NOVEMBER, DECEMBER — ARIES AND PLANETS

0ʰ GMT	ARIES GHA	VENUS GHA	v	Dec	d	MARS GHA	v	Dec	d	JUPITER GHA	v	Dec	d	SATURN GHA	v	Dec	d
Nov									NOVEMBER								
1	40.26	135.39	−8	S 26.75	−2	194.21	+15	S 10.20	−10	298.47	+42	N 22.76	0	120.15	+38	S 22.73	0
2	41.25	135.20	7	26.81	2	194.56	15	10.44	10	299.47	42	22.76	0	121.05	38	22.73	0
3	42.23	135.03	7	26.86	2	194.92	15	10.69	10	300.47	42	22.77	0	121.95	38	22.72	0
4	43.22	134.86	7	26.89	1	195.27	15	10.93	10	301.47	42	22.77	0	122.85	38	22.72	0
5	44.21	134.69	7	26.92	−1	195.62	15	11.17	10	302.48	42	22.77	0	123.75	37	22.72	0
6	45.19	134.53	−6	S 26.94	0	195.97	+14	S 11.41	−10	303.50	+42	N 22.78	0	124.65	+37	S 22.71	0
7	46.18	134.38	6	26.94	0	196.31	14	11.65	10	304.51	43	22.78	0	125.54	37	22.71	0
8	47.16	134.23	6	26.94	+1	196.66	14	11.89	10	305.53	43	22.78	0	126.44	37	22.70	0
9	48.15	134.09	5	26.92	1	197.00	14	12.13	10	306.56	43	22.79	0	127.33	37	22.70	0
10	49.13	133.96	5	26.90	1	197.34	14	12.37	10	307.58	43	22.79	0	128.22	37	22.69	0
11	50.12	133.84	−5	S 26.86	+2	197.68	+14	S 12.60	−10	308.61	+43	N 22.80	0	129.11	+37	S 22.69	0
12	51.10	133.73	4	26.82	2	198.02	14	12.83	10	309.65	43	22.80	0	130.00	37	22.68	0
13	52.09	133.62	4	26.76	3	198.36	14	13.07	10	310.68	43	22.81	0	130.89	37	22.68	0
14	53.08	133.52	4	26.70	3	198.69	14	13.30	10	311.73	44	22.81	0	131.78	37	22.67	0
15	54.06	133.44	3	26.63	3	199.02	14	13.52	10	312.77	44	22.82	0	132.67	37	22.67	0
16	55.05	133.36	−3	S 26.54	+4	199.35	+14	S 13.75	−9	313.82	+44	N 22.83	0	133.55	+37	S 22.66	0
17	56.03	133.30	2	26.45	4	199.68	14	13.98	9	314.87	44	22.83	0	134.44	37	22.66	0
18	57.02	133.25	2	26.35	5	200.01	14	14.20	9	315.92	44	22.84	0	135.32	37	22.65	0
19	58.00	133.21	1	26.24	5	200.33	13	14.43	9	316.98	44	22.85	0	136.21	37	22.64	0
20	58.99	133.18	−1	26.12	5	200.66	13	14.65	9	318.04	44	22.85	0	137.09	37	22.64	0
21	59.98	133.17	0	S 25.99	+6	200.98	+13	S 14.87	−9	319.11	+44	N 22.86	0	137.97	+37	S 22.63	0
22	60.96	133.17	+1	25.86	6	201.30	13	15.08	9	320.18	45	22.87	0	138.85	37	22.63	0
23	61.95	133.18	1	25.71	6	201.61	13	15.30	9	321.25	45	22.87	0	139.73	37	22.62	0
24	62.93	133.21	2	25.56	7	201.93	13	15.51	9	322.32	45	22.88	0	140.61	37	22.61	0
25	63.92	133.25	2	25.40	7	202.24	13	15.72	9	323.40	45	22.89	0	141.48	37	22.60	0
26	64.90	133.31	+3	S 25.23	+7	202.56	+13	S 15.93	−9	324.48	+45	N 22.90	0	142.36	+36	S 22.60	0
27	65.89	133.38	4	25.06	7	202.86	13	16.14	9	325.56	45	22.91	0	143.23	36	22.59	0
28	66.88	133.48	5	24.88	8	203.17	13	16.35	9	326.65	45	22.92	0	144.11	36	22.58	0
29	67.86	133.59	5	24.69	8	203.48	13	16.55	8	327.74	45	22.93	0	144.98	36	22.58	0
30	68.85	133.72	+6	24.50	+8	203.78	+13	S 16.76	−8	328.83	+46	N 22.93	0	145.86	+36	S 22.57	0
Dec									DECEMBER								
1	69.83	133.86	+7	S 24.30	+9	204.08	+13	S 16.96	−8	329.92	+46	N 22.94	0	146.73	+36	S 22.56	0
2	70.82	134.03	8	24.09	9	204.38	12	17.15	8	331.02	46	22.95	0	147.60	36	22.55	0
3	71.80	134.22	9	23.88	9	204.68	12	17.35	8	332.12	46	22.96	0	148.47	36	22.54	0
4	72.79	134.43	10	23.67	9	204.97	12	17.54	8	333.22	46	22.97	0	149.34	36	22.53	0
5	73.77	134.66	11	23.45	9	205.27	12	17.74	8	334.33	46	22.98	0	150.21	36	22.53	0
6	74.76	134.91	+12	S 23.22	+10	205.56	+12	S 17.92	−8	335.43	+46	N 22.99	0	151.08	+36	S 22.52	0
7	75.75	135.19	13	22.99	10	205.85	12	18.11	8	336.54	46	23.00	0	151.95	36	22.51	0
8	76.73	135.49	14	22.76	10	206.14	12	18.30	8	337.65	46	23.01	0	152.81	36	22.50	0
9	77.72	135.81	15	22.52	10	206.42	12	18.48	8	338.77	47	23.02	0	153.68	36	22.49	0
10	78.70	136.16	16	22.28	10	206.70	12	18.66	7	339.88	47	23.03	0	154.55	36	22.48	0
11	79.69	136.54	+17	S 22.04	+10	206.98	+12	S 18.84	−7	341.00	+47	N 23.04	0	155.41	+36	S 22.47	0
12	80.67	136.94	18	21.80	10	207.26	12	19.01	7	342.12	47	23.05	0	156.28	36	22.46	0
13	81.66	137.38	19	21.55	10	207.54	11	19.18	7	343.24	47	23.05	0	157.14	36	22.45	0
14	82.65	137.84	21	21.30	10	207.81	11	19.35	7	344.36	47	23.06	0	158.01	36	22.44	0
15	83.63	138.33	22	21.05	10	208.09	11	19.52	7	345.48	47	23.07	0	158.87	36	22.43	0
16	84.62	138.86	+23	S 20.80	+10	208.36	+11	S 19.68	−7	346.61	+47	N 23.08	0	159.73	+36	S 22.42	0
17	85.60	139.41	25	20.55	10	208.63	11	19.85	7	347.73	47	23.09	0	160.60	36	22.41	0
18	86.59	140.00	26	20.30	10	208.89	11	20.01	7	348.86	47	23.10	0	161.46	36	22.40	0
19	87.57	140.62	27	20.05	10	209.16	10	20.16	6	349.99	47	23.11	0	162.32	36	22.39	0
20	88.56	141.28	29	19.80	10	209.42	11	20.32	6	351.12	47	23.12	0	163.18	36	22.38	0
21	89.55	141.97	+30	S 19.55	+10	209.68	+11	S 20.47	−6	352.25	+47	N 23.13	0	164.04	+36	S 22.37	0
22	90.53	142.70	32	19.30	10	209.94	11	20.61	6	353.38	47	23.14	0	164.90	36	22.36	0
23	91.52	143.47	34	19.05	10	210.19	11	20.76	6	354.51	47	23.15	0	165.77	36	22.35	0
24	92.50	144.28	35	18.81	10	210.45	10	20.90	6	355.65	47	23.16	0	166.63	36	22.33	0
25	93.49	145.12	37	18.57	10	210.70	10	21.04	6	356.78	47	23.16	0	167.49	36	22.32	0
26	94.47	146.00	+39	S 18.33	+10	210.95	+10	S 21.18	−6	357.91	+47	N 23.17	0	168.35	+36	S 22.31	0
27	95.46	146.93	40	18.09	10	211.20	10	21.31	5	359.05	47	23.18	0	169.20	36	22.30	0
28	96.44	147.90	42	17.86	10	211.44	10	21.44	5	0.18	47	23.19	0	170.06	36	22.29	0
29	97.43	148.90	44	17.63	9	211.69	10	21.57	5	1.31	47	23.20	0	170.92	36	22.28	0
30	98.42	149.96	46	17.41	9	211.93	10	21.69	5	2.44	47	23.20	0	171.78	36	22.26	0
31	99.40	151.05	+49	S 17.19	+9	212.17	+10	S 21.81	−5	3.58	+47	N 23.21	0	172.64	+36	S 22.25	0

1989 NOVEMBER, DECEMBER — SUN AND MOON

0h GMT	SUN GHA	v	Dec	d	MOON 0h-8h GHA	v	Dec	d	MOON 8h-16h GHA	v	Dec	d	MOON 16h-24h GHA	v	Dec	d	HP
Nov			(SD 16'.2)				**NOVEMBER**										
1	184.10	0	S 14.35	−13	158.93	181	S 25.82	−79	274.86	174	S 26.45	−62	30.74	168	S 26.95	−44	54.5
2	184.10	0	14.67	13	146.56	163	27.30	−26	262.34	159	27.51	−8	18.10	157	27.57	+11	54.8
3	184.10	0	14.99	13	133.83	156	27.48	+31	249.55	156	27.23	+50	5.28	157	26.83	68	55.3
4	184.10	0	15.30	13	121.02	160	26.29	87	236.77	163	25.59	106	352.55	167	24.74	123	55.9
5	184.10	0	15.61	13	108.37	172	23.76	141	224.23	177	22.63	157	340.12	182	21.37	173	56.6
6	184.09	0	S 15.91	−12	96.06	187	S 19.98	+188	212.03	192	S 18.48	+203	328.05	196	S 16.86	+216	57.4
7	184.08	−1	16.21	12	84.10	199	15.13	228	200.17	202	13.31	239	316.27	204	11.40	249	58.3
8	184.07	1	16.50	12	72.38	205	9.40	258	188.50	204	7.34	266	304.61	202	S 5.21	272	59.1
9	184.05	1	16.79	12	60.71	199	S 3.04	277	176.79	195	S 0.82	280	292.83	189	N 1.41	281	60.0
10	184.03	1	17.08	12	48.82	182	N 3.67	+281	164.75	173	N 5.91	279	280.61	163	8.14	274	60.7
11	184.00	−1	S 17.36	−11	36.39	152	N 10.34	+268	152.08	139	N 12.48	+259	267.68	126	N 14.55	+247	61.2
12	183.97	1	17.63	11	23.17	112	16.53	234	138.55	99	18.40	218	253.81	85	20.14	199	61.3
13	183.94	2	17.90	11	8.98	72	21.73	178	124.04	61	23.16	156	239.00	51	24.40	131	61.2
14	183.91	2	18.17	11	353.90	45	25.45	105	108.73	40	26.29	+78	223.54	39	26.91	+50	60.7
15	183.87	2	18.42	11	338.33	41	27.31	+23	93.14	47	27.49	−4	208.00	55	27.46	−30	60.0
16	183.82	−2	S 18.68	−10	322.92	67	N 27.22	−56	77.94	80	N 26.77	−79	193.06	96	N 26.14	−101	59.1
17	183.78	2	18.93	10	308.31	112	25.33	121	63.68	130	24.37	139	179.20	147	23.25	155	58.1
18	183.73	2	19.17	10	294.86	164	22.01	169	50.65	181	20.66	182	166.58	197	19.21	192	57.2
19	183.68	2	19.41	10	282.64	212	17.67	202	38.81	226	16.06	209	155.09	238	14.38	215	56.3
20	183.62	3	19.64	9	271.48	249	12.66	220	27.95	259	10.90	224	144.50	267	9.10	227	55.5
21	183.56	−3	S 19.86	−9	261.12	274	N 7.28	−230	17.79	280	N 5.45	−231	134.51	284	N 3.60	−231	54.9
22	183.50	3	20.08	9	251.27	287	N 1.75	231	8.05	289	S 0.10	230	124.84	290	S 1.93	228	54.5
23	183.43	3	20.29	9	241.64	289	S 3.76	226	358.44	288	5.56	222	115.22	285	7.34	218	54.2
24	183.36	3	20.50	8	231.98	281	9.09	214	348.71	276	10.80	208	105.40	270	12.46	202	54.0
25	183.28	3	20.70	8	222.04	264	14.08	195	338.63	256	15.64	187	95.16	248	17.14	178	54.0
26	183.21	−3	S 20.90	−8	211.62	239	S 18.57	−169	328.02	230	S 19.92	−158	84.34	220	S 21.18	−146	54.1
27	183.13	3	21.08	7	200.58	211	22.35	134	316.75	201	23.42	120	72.84	192	24.38	105	54.3
28	183.05	4	21.26	7	188.85	183	25.22	89	304.80	175	25.93	73	60.67	168	26.51	−56	54.5
29	182.96	4	21.44	7	176.50	162	26.96	−38	292.27	157	27.26	−19	48.01	154	27.41	0	54.9
30	182.87	−4	S 21.61	−7	163.73	152	S 27.41	+19	279.42	152	S 27.26	+38	35.12	153	S 26.96	+57	55.3
Dec			(SD 16'.3)				**DECEMBER**										
1	182.78	−4	S 21.77	−6	150.83	156	S 26.50	+76	266.56	160	S 25.89	+95	22.32	165	S 25.13	+113	55.7
2	182.68	4	21.92	6	138.12	170	24.23	130	253.96	176	23.19	147	9.85	183	22.02	162	56.2
3	182.59	4	22.07	6	125.80	189	20.72	177	241.79	196	19.31	191	357.84	202	17.78	203	56.8
4	182.49	4	22.21	6	113.93	207	16.15	215	230.07	212	14.43	226	346.24	215	12.62	236	57.4
5	182.39	4	22.34	5	102.45	218	10.74	244	218.67	220	8.79	252	334.91	220	6.77	258	58.1
6	182.28	−4	S 22.47	−5	91.15	219	S 4.71	+263	207.39	217	S 2.61	+266	323.60	213	S 0.48	+269	58.7
7	182.18	5	22.58	5	79.79	208	N 1.67	270	195.94	202	N 3.83	269	312.03	193	N 5.99	267	59.4
8	182.07	5	22.69	4	68.06	184	8.13	263	184.01	173	10.23	258	299.87	161	12.29	250	60.0
9	181.96	5	22.80	4	55.64	148	14.29	240	171.30	133	16.21	228	286.84	119	18.04	214	60.4
10	181.85	5	22.89	4	42.27	104	19.75	198	157.58	90	21.33	179	272.78	76	22.76	158	60.7
11	181.73	−5	S 22.98	−3	27.87	64	N 24.02	+135	142.85	53	N 25.10	+111	257.76	46	N 25.99	+85	60.6
12	181.62	5	23.06	3	12.60	40	26.67	+58	127.41	39	27.14	+31	242.20	41	27.39	+3	60.4
13	181.50	5	23.13	3	357.01	46	27.41	−23	111.86	54	27.23	−49	226.77	66	26.83	−74	59.8
14	181.38	5	23.20	2	341.78	79	26.23	98	96.89	95	25.45	119	212.12	111	24.50	139	59.1
15	181.27	5	23.26	2	327.49	128	23.39	156	83.00	146	22.14	172	198.65	163	20.76	185	58.2
16	181.14	−5	S 23.31	−2	314.43	180	N 19.28	−197	70.35	196	N 17.71	−207	186.39	210	N 16.05	−215	57.3
17	181.02	5	23.35	1	302.56	224	14.34	221	58.83	236	12.57	226	175.20	247	10.76	230	56.4
18	180.90	5	23.38	1	291.65	257	8.92	233	48.19	265	7.06	234	164.79	272	N 5.18	235	55.7
19	180.78	5	23.41	−1	281.44	277	N 3.30	235	38.13	281	N 1.42	234	154.86	284	S 0.45	232	55.0
20	180.65	5	23.43	0	271.61	285	S 2.30	230	28.37	285	S 4.14	227	145.13	284	5.95	223	54.6
21	180.53	−5	S 23.44	0	261.88	282	S 7.74	−218	18.62	279	S 9.48	−213	135.33	274	S 11.18	−207	54.3
22	180.41	5	23.44	0	252.00	269	12.84	200	8.63	262	14.45	193	125.21	255	15.99	185	54.1
23	180.28	5	23.44	+1	241.73	247	17.47	176	358.19	238	18.88	166	114.57	229	20.20	155	54.1
24	180.16	5	23.43	1	230.88	219	21.44	143	347.11	209	22.59	130	103.26	199	23.63	117	54.3
25	180.03	5	23.41	1	219.33	189	24.56	102	335.33	180	25.38	86	91.25	171	26.07	69	54.6
26	179.91	−5	S 23.38	+2	207.10	164	S 26.62	−52	322.89	157	S 27.04	−34	78.62	152	S 27.31	−15	55.0
27	179.78	5	23.34	2	194.32	148	27.42	+4	309.98	146	27.39	+24	65.63	146	27.20	+44	55.4
28	179.66	5	23.30	2	181.27	146	26.85	63	296.93	149	26.34	83	52.60	153	25.68	102	55.9
29	179.54	5	23.24	2	168.30	158	24.86	120	284.04	164	23.91	138	39.83	170	22.80	154	56.4
30	179.41	5	23.19	3	155.68	178	21.57	170	271.58	185	20.21	185	27.54	192	18.73	198	56.9
31	179.29	−5	S 23.12	+3	143.55	199	S 17.15	+211	259.62	205	S 15.47	+222	15.74	210	S 13.69	+231	57.4

5.3.10 INTERPOLATION TABLES — ARIES, SUN, PLANETS 0ʰ–12ʰ

GMT	Corr. to GHA Aries	Sun Planet	\multicolumn{19}{c	}{Correction to GHA or Declination for v or d — UNIT $0°.01$}																			
h m	°	°	1	2	3	4	5	6	7	8	9	10	11	12	13	14	15	16	17	18	19	20	22	24	26	28	30
0 00	0.00	0	0	0	0	0	0	0	0	0	0	0	0	0	0	0	0	0	0	0	0	0	0	0	0	0	0
12	3.01	3	0	0	0	0	0	0	0	0	0	0	0	0	0	0	0	0	0	0	0	0	0	0	1	1	1
24	6.02	6	0	0	0	0	0	0	0	0	0	0	0	1	1	1	1	1	1	1	1	1	1	1	1	1	1
36	9.02	9	0	0	0	0	0	0	0	1	1	1	1	1	1	1	1	1	1	1	1	1	1	1	2	2	2
0 48	12.03	12	0	0	0	0	0	0	1	1	1	1	1	1	1	1	1	1	1	1	2	2	2	2	2	2	2
1 00	15.04	15	0	0	0	0	1	1	1	1	1	1	1	1	1	2	2	2	2	2	2	2	2	2	3	3	3
12	18.05	18	0	0	0	0	1	1	1	1	1	1	1	2	2	2	2	2	2	2	2	3	3	3	3	3	4
24	21.06	21	0	0	0	1	1	1	1	1	1	2	2	2	2	2	2	2	3	3	3	3	3	3	4	4	4
36	24.07	24	0	0	0	1	1	1	1	1	2	2	2	2	2	2	3	3	3	3	3	3	4	4	4	4	5
1 48	27.07	27	0	0	1	1	1	1	1	1	2	2	2	2	2	3	3	3	3	3	3	4	4	4	5	5	5
2 00	30.08	30	0	0	1	1	1	1	2	2	2	2	2	3	3	3	3	4	4	4	4	4	5	5	6	6
12	33.09	33	0	0	1	1	1	2	2	2	2	2	3	3	3	3	4	4	4	4	4	5	5	6	6	7
24	36.10	36	0	1	1	1	1	2	2	2	2	3	3	3	3	4	4	4	5	5	5	5	6	6	7	7
36	39.11	39	0	1	1	1	1	2	2	2	3	3	3	3	4	4	4	5	5	5	5	6	6	7	7	8
2 48	42.12	42	0	1	1	1	1	2	2	2	3	3	3	4	4	4	5	5	5	5	6	6	7	7	8	8
3 00	45.12	45	0	1	1	1	2	2	2	3	3	3	4	4	5	5	5	6	6	7	7	8	8	9
12	48.13	48	0	1	1	1	2	2	2	3	3	3	4	4	5	5	5	6	6	6	7	8	8	9	10
24	51.14	51	0	1	1	1	2	2	2	3	3	3	4	4	5	5	6	6	6	7	7	8	9	9	10	10
36	54.15	54	0	1	1	1	2	2	3	3	3	4	4	5	5	5	6	6	7	7	7	8	9	9	10	11
3 48	57.16	57	0	1	1	2	2	2	3	3	3	4	4	5	5	5	6	6	6	7	7	8	9	10	10	11	11
4 00	60.16	60	0	1	1	2	2	2	3	3	4	4	4	5	5	6	6	6	7	7	8	8	9	10	10	11	12
12	63.17	63	0	1	1	2	2	3	3	3	4	4	5	5	5	6	6	7	7	8	8	8	9	10	11	12	13
24	66.18	66	0	1	1	2	2	3	3	4	4	4	5	5	6	6	7	7	7	8	8	9	10	11	11	12	13
36	69.19	69	0	1	1	2	2	3	3	4	4	5	5	6	6	6	7	7	8	8	9	9	10	11	12	13	14
4 48	72.20	72	0	1	1	2	2	3	3	4	4	5	5	6	6	7	7	8	8	9	9	10	11	12	12	13	14
5 00	75.21	75	1	1	2	2	3	3	4	4	5	5	6	6	7	7	8	8	9	9	10	10	11	12	13	14	15
12	78.21	78	1	1	2	2	3	3	4	4	5	5	6	6	7	7	8	8	9	9	10	10	11	12	14	15	16
24	81.22	81	1	1	2	2	3	3	4	4	5	5	6	6	7	8	8	9	9	10	10	11	12	13	14	15	16
36	84.23	84	1	1	2	2	3	3	4	4	5	6	6	7	7	8	8	9	10	10	11	11	12	13	15	16	17
5 48	87.24	87	1	1	2	2	3	3	4	5	5	6	6	7	8	8	9	9	10	10	11	12	13	14	15	16	17
6 00	90.25	90	1	1	2	2	3	4	4	5	5	6	7	7	8	8	9	10	10	11	11	12	13	14	16	17	18
12	93.25	93	1	1	2	2	3	4	4	5	6	6	7	7	8	9	9	10	11	11	12	12	14	15	16	17	19
24	96.26	96	1	1	2	3	3	4	4	5	6	6	7	8	8	9	10	10	11	12	12	13	14	15	17	18	19
36	99.27	99	1	2	2	3	3	4	5	5	6	7	7	8	9	9	10	11	11	12	13	13	15	16	17	18	20
6 48	102.28	102	1	1	2	3	3	4	5	5	6	7	7	8	9	10	10	11	12	12	13	14	15	16	18	19	20
7 00	105.29	105	1	1	2	3	4	4	5	6	6	7	8	8	9	10	11	11	12	13	13	14	15	17	18	20	21
12	108.30	108	1	1	2	3	4	4	5	6	6	7	8	9	9	10	11	12	12	13	14	14	16	17	19	20	22
24	111.30	111	1	1	2	3	4	4	5	6	7	7	8	9	10	11	11	12	13	13	14	15	16	18	19	21	22
36	114.31	114	1	2	2	3	4	5	5	6	7	8	8	9	10	11	11	12	13	14	14	15	17	18	20	21	23
7 48	117.32	117	1	2	2	3	4	5	5	6	7	8	9	9	10	11	12	12	13	14	15	15	17	19	20	22	23
8 00	120.33	120	1	2	2	3	4	5	6	6	7	8	9	10	10	11	12	13	14	14	15	16	18	19	21	22	24
12	123.34	123	1	2	2	3	4	5	6	7	7	8	9	10	11	12	13	13	14	15	16	16	18	20	21	23	25
24	126.35	126	1	2	3	3	4	5	6	7	8	8	9	10	11	12	13	14	15	16	16	17	18	20	22	24	25
36	129.35	129	1	2	3	3	4	5	6	7	8	9	9	10	11	12	13	14	15	15	16	17	19	21	22	24	26
8 48	132.36	132	1	2	3	4	4	5	6	7	8	9	10	11	11	12	13	14	15	16	17	18	19	21	23	25	26
9 00	135.37	135	1	2	3	4	5	5	6	7	8	9	10	11	12	13	14	14	15	16	17	18	20	22	23	25	27
12	138.38	138	1	2	3	4	5	6	7	8	8	9	10	11	12	13	14	15	16	17	17	18	20	22	24	26	28
24	141.39	141	1	2	3	4	5	6	7	8	8	9	10	11	12	13	14	15	16	17	18	19	21	23	24	26	28
36	144.39	144	1	2	3	4	5	6	7	8	9	10	11	12	13	14	15	16	17	18	19	21	23	25	27	29
9 48	147.40	147	1	2	3	4	5	6	7	8	9	10	11	12	13	14	15	16	17	18	19	20	22	24	25	27	29
10 00	150.41	150	1	2	3	4	5	6	7	8	9	11	12	13	14	15	16	17	18	19	20	22	24	26	28	30
12	153.42	153	1	2	3	4	5	6	7	8	9	11	12	13	14	15	16	17	18	19	20	22	24	27	29	31
24	156.43	156	1	2	3	4	5	6	7	8	10	11	12	14	15	16	17	18	19	20	21	23	25	27	29	31
36	159.44	159	1	2	3	4	5	6	7	8	10	11	12	13	14	15	16	17	18	19	20	21	23	25	28	30	32
10 48	162.44	162	1	2	3	4	5	6	8	9	10	11	12	13	14	15	16	17	18	19	21	22	24	26	28	30	32
11 00	165.45	165	1	2	3	4	6	7	8	9	10	11	12	13	14	15	17	18	19	20	21	22	24	26	29	31	33
12	168.46	168	1	2	3	4	6	7	8	9	10	11	12	13	15	16	17	18	19	20	21	22	25	27	29	31	34
24	171.47	171	1	2	3	5	6	7	8	9	10	12	13	14	15	16	17	18	19	21	22	23	25	27	30	32	34
36	174.48	174	1	2	3	5	6	7	8	9	10	12	13	14	15	16	17	19	20	21	22	23	26	28	30	32	35
11 48	177.48	177	1	2	4	5	6	7	8	9	11	12	13	14	15	16	18	19	20	21	22	24	26	28	31	33	35
12 00	180.49	180	1	2	4	5	6	7	8	10	11	12	13	14	16	17	18	19	20	22	23	24	26	29	31	34	36

INTERPOLATION TABLES — ARIES, SUN, PLANETS 0ʰ –12ʰ

GMT	Corr. to GHA Aries	Corr. to GHA Sun Planet	\-\- Correction to GHA or Declination for v or d \-\-																Correction to GHA of Aries, Sun, Planets				
			30	32	34	36	38	40	42	44	46	48	50	52	54	56	58	60					
h m	°	°						UNIT 0°.01											GMT	0ᵐ	1ᵐ	2ᵐ	3ᵐ
0 00	0.00	0	0	0	0	0	0	0	0	0	0	0	0	0	0	0	0	0	s	°	°	°	°
12	3.01	3	1	1	1	1	1	1	1	1	1	1	1	1	1	1	1	1	00	0.00	0.25	0.50	0.75
24	6.02	6	1	1	1	1	2	2	2	2	2	2	2	2	2	2	2	2	04	.02	.27	.52	.77
36	9.02	9	2	2	2	2	2	3	3	3	3	3	3	3	3	3	4	08	.03	.28	.53	.78
0 48	12.03	12	2	3	3	3	3	3	4	4	4	4	4	4	4	5	5	12	.05	.30	.55	.80
																		16	.07	.32	.57	.82
1 00	15.04	15	3	3	3	4	4	4	4	5	5	5	5	5	6	6	6					
12	18.05	18	4	4	4	4	5	5	5	6	6	6	6	7	7	7	7	20	0.08	0.33	0.58	0.83
24	21.06	21	4	4	5	5	5	6	6	6	7	7	7	8	8	8	8	24	.10	.35	.60	.85
36	24.07	24	5	5	5	6	6	6	7	7	8	8	8	9	9	9	10	28	.12	.37	.62	.87
1 48	27.07	27	5	6	6	7	7	8	8	8	9	9	9	10	10	10	11	32	.13	.38	.63	.88
																		36	.15	.40	.65	.90
2 00	30.08	30	6	6	7	7	8	8	9	9	10	10	10	11	11	12	12					
12	33.09	33	7	7	7	8	8	9	9	10	10	11	11	12	12	13	13	40	0.17	0.42	0.67	0.92
24	36.10	36	7	8	8	9	9	10	10	11	11	12	12	13	13	14	14	44	.18	.43	.68	.93
36	39.11	39	8	8	9	9	10	11	11	12	12	13	14	14	15	15	16	48	.20	.45	.70	.95
2 48	42.12	42	8	9	10	10	11	11	12	12	13	13	14	15	15	16	16	17	52	.22	.47	.72	.97
																		56	.23	.48	.73	0.98
3 00	45.12	45	9	10	10	11	11	12	13	13	14	14	15	16	16	17	17	18					
12	48.13	48	10	10	11	12	12	13	13	14	15	15	16	17	17	18	19	19	60	0.25	0.50	0.75	1.00
24	51.14	51	10	11	12	12	13	14	14	15	16	16	17	18	18	19	20	20					
36	54.15	54	11	12	12	13	14	14	15	16	17	17	18	19	19	20	21	22		Correction to GHA of Aries, Sun, Planets			
3 48	57.16	57	11	12	13	14	14	15	16	17	17	18	19	20	21	21	22	23					
4 00	60.16	60	12	13	14	14	15	16	17	18	18	19	20	21	22	22	23	24					
12	63.17	63	13	13	14	15	16	17	18	18	19	20	21	22	23	24	24	25	GMT	4ᵐ	5ᵐ	6ᵐ	7ᵐ
24	66.18	66	13	14	15	16	17	18	19	20	21	22	23	24	25	26	26	s	°	°	°	°
36	69.19	69	14	15	16	17	17	18	19	20	21	22	23	24	25	26	27	28	00	1.00	1.25	1.50	1.75
4 48	72.20	72	14	15	16	17	18	19	20	21	22	23	24	25	26	27	28	29	04	.02	.27	.52	.77
																		08	.03	.28	.53	.78
5 00	75.21	75	15	16	17	18	19	20	21	22	23	24	25	26	27	28	29	30	12	.05	.30	.55	.80
12	78.21	78	16	17	18	19	20	21	22	23	24	25	26	27	28	29	30	31	16	.07	.32	.57	.82
24	81.22	81	16	17	18	19	21	22	23	24	25	26	27	28	29	30	31	32					
36	84.23	84	17	18	19	20	21	22	24	25	26	27	28	29	30	31	32	34	20	1.08	1.33	1.58	1.83
5 48	87.24	87	17	19	20	21	22	23	24	26	27	28	29	30	31	32	34	35	24	.10	.35	.60	.85
																		28	.12	.37	.62	.87
6 00	90.25	90	18	19	20	22	23	24	25	26	28	29	30	31	32	34	35	36	32	.13	.38	.63	.88
12	93.25	93	19	20	21	22	24	25	26	27	29	30	31	32	33	35	36	37	36	.15	.40	.65	.90
24	96.26	96	19	20	22	23	24	26	27	28	29	31	32	33	35	36	37	38					
36	99.27	99	20	21	22	24	25	26	28	29	30	32	33	34	36	37	38	40	40	1.17	1.42	1.67	1.92
6 48	102.28	102	20	22	23	24	26	27	29	30	31	33	34	35	37	38	39	41	44	.18	.43	.68	.93
																		48	.20	.45	.70	.95
7 00	105.29	105	21	22	24	25	27	28	29	31	32	34	35	36	38	39	41	42	52	.22	.47	.72	.97
12	108.30	108	22	23	24	26	27	29	30	32	33	35	36	37	39	40	42	43	56	.23	.48	.73	1.98
24	111.30	111	22	24	25	27	28	30	31	33	34	36	37	38	40	41	43	44					
36	114.31	114	23	24	26	27	29	30	32	33	35	36	38	40	41	43	44	46	60	1.25	1.50	1.75	2.00
7 48	117.32	117	23	25	27	28	30	31	33	34	36	37	39	41	42	44	45	47					
8 00	120.33	120	24	26	27	29	30	32	34	35	37	38	40	42	43	45	46	48		Correction to GHA of Aries, Sun, Planets			
12	123.34	123	25	26	28	30	31	33	34	36	38	39	41	43	44	46	48	49					
24	126.35	126	25	27	29	30	32	34	35	37	39	40	42	44	45	47	49	50					
36	129.35	129	26	28	29	31	33	34	36	38	40	41	43	45	46	48	50	52					
8 48	132.36	132	26	28	30	32	33	35	37	39	40	42	44	46	48	49	51	53	GMT	8ᵐ	9ᵐ	10ᵐ	11ᵐ
																		s	°	°	°	°
9 00	135.37	135	27	29	31	32	34	36	38	40	41	43	45	47	49	50	52	54	00	2.00	2.25	2.50	2.75
12	138.38	138	28	29	31	33	35	37	39	40	42	44	46	48	50	52	53	55	04	.02	.27	.52	.77
24	141.39	141	28	30	32	34	36	38	39	41	43	45	47	49	51	53	55	56	08	.03	.28	.53	.78
36	144.39	144	29	31	33	35	36	38	40	42	44	46	48	50	52	54	56	58	12	.05	.30	.55	.80
9 48	147.40	147	29	31	33	35	37	39	41	43	45	47	49	51	53	55	57	59	16	.07	.32	.57	.82
10 00	150.41	150	30	32	34	36	38	40	42	44	46	48	50	52	54	56	58	60					
12	153.42	153	31	33	35	37	39	41	43	45	47	49	51	53	55	57	59	61	20	2.08	2.33	2.58	2.83
24	156.43	156	31	33	35	37	40	42	44	46	48	50	52	54	56	58	60	62	24	.10	.35	.60	.85
36	159.44	159	32	34	36	38	40	42	45	47	49	51	53	55	57	59	61	64	28	.12	.37	.62	.87
10 48	162.44	162	32	35	37	39	41	43	45	48	50	52	54	56	58	60	63	65	32	.13	.38	.63	.88
																		36	.15	.40	.65	.90
11 00	165.45	165	33	35	37	40	42	44	46	48	51	53	55	57	59	62	64	66					
12	168.46	168	34	36	38	40	43	45	47	49	52	54	56	58	60	63	65	67	40	2.17	2.42	2.67	2.92
24	171.47	171	34	36	39	41	43	46	48	50	52	55	57	59	62	64	66	68	44	.18	.43	.68	.93
36	174.48	174	35	37	39	42	44	46	49	51	53	56	58	60	63	65	67	70	48	.20	.45	.70	.95
11 48	177.48	177	35	38	40	42	45	47	50	52	54	57	59	61	64	66	68	71	52	.22	.47	.72	.97
																		56	.23	.48	.73	2.98
12 00	180.49	180	36	38	41	43	46	48	50	53	55	58	60	62	65	67	70	72	60	2.25	2.50	2.75	3.00

INTERPOLATION TABLES — ARIES, SUN, PLANETS 12ʰ–24ʰ

GMT	Corr. to GHA Aries	Corr. to GHA Sun Planet	1	2	3	4	5	6	7	8	9	10	11	12	13	14	15	16	17	18	19	20	22	24	26	28	30
h m	°	°												UNIT 0°.01													
12 00	180.49	180	1	2	4	5	6	7	8	10	11	12	13	14	16	17	18	19	20	22	23	24	26	29	31	34	36
12	183.50	183	1	2	4	5	6	7	9	10	11	12	13	15	16	17	18	20	21	22	23	24	27	29	32	34	37
24	186.51	186	1	2	4	5	6	7	9	10	11	12	14	15	16	17	19	20	21	22	24	25	27	30	32	35	37
36	189.52	189	1	3	4	5	6	8	9	10	11	13	14	15	16	18	19	20	21	23	24	25	28	30	33	35	38
12 48	192.53	192	1	3	4	5	6	8	9	10	12	13	14	15	17	18	19	20	22	23	24	26	28	31	33	36	38
13 00	195.53	195	1	3	4	5	7	8	9	10	12	13	14	16	17	18	20	21	22	23	25	26	29	31	34	36	39
12	198.54	198	1	3	4	5	7	8	9	11	12	13	15	16	17	18	20	21	22	24	25	26	29	32	34	37	40
24	201.55	201	1	3	4	5	7	8	9	11	12	13	15	16	17	19	20	21	23	24	25	27	29	32	35	38	40
36	204.56	204	1	3	4	5	7	8	10	11	12	14	15	16	18	19	20	22	23	24	26	27	30	33	35	38	41
13 48	207.57	207	1	3	4	6	7	8	10	11	12	14	15	17	18	19	21	22	23	25	26	28	30	33	36	39	41
14 00	210.58	210	1	3	4	6	7	8	10	11	13	14	15	17	18	20	21	22	24	25	27	28	31	34	36	39	42
12	213.58	213	1	3	4	6	7	9	10	11	13	14	16	17	18	20	21	23	24	26	27	28	31	34	37	40	43
24	216.59	216	1	3	4	6	7	9	10	12	13	14	16	17	19	20	22	23	24	26	27	29	32	35	37	40	43
36	219.60	219	1	3	4	6	7	9	10	12	13	15	16	18	19	20	22	23	25	26	28	29	32	35	38	41	44
14 48	222.61	222	1	3	4	6	7	9	10	12	13	15	16	18	19	21	22	24	25	27	28	30	33	36	38	41	44
15 00	225.62	225	2	3	5	6	8	9	11	12	14	15	17	18	20	21	23	24	26	27	29	30	33	36	39	42	45
12	228.62	228	2	3	5	6	8	9	11	12	14	15	17	18	20	21	23	24	26	27	29	30	33	36	40	43	46
24	231.63	231	2	3	5	6	8	9	11	12	14	15	17	18	20	22	23	25	26	28	29	31	34	37	40	43	46
36	234.64	234	2	3	5	6	8	9	11	12	14	16	17	19	20	22	23	25	27	28	30	31	34	37	41	44	47
15 48	237.65	237	2	3	5	6	8	9	11	13	14	16	17	19	21	22	24	25	27	28	30	32	35	38	41	44	47
16 00	240.66	240	2	3	5	6	8	10	11	13	14	16	18	19	21	22	24	26	27	29	30	32	35	38	42	45	48
12	243.67	243	2	3	5	6	8	10	11	13	15	16	18	19	21	23	24	26	28	29	31	32	36	39	42	45	49
24	246.67	246	2	3	5	7	8	10	11	13	15	16	18	20	21	23	25	26	28	30	31	33	36	39	43	46	49
36	249.68	249	2	3	5	7	8	10	12	13	15	17	18	20	22	23	25	27	28	30	32	33	37	40	43	46	50
16 48	252.69	252	2	3	5	7	8	10	12	13	15	17	18	20	22	23	25	27	29	30	32	34	37	40	44	47	50
17 00	255.70	255	2	3	5	7	9	10	12	14	15	17	19	20	22	24	26	27	29	31	32	34	37	41	44	48	51
12	258.71	258	2	3	5	7	9	10	12	14	15	17	19	21	22	24	26	28	29	31	33	34	38	41	45	48	52
24	261.71	261	2	3	5	7	9	10	12	14	16	17	19	21	23	24	26	28	30	31	33	35	38	42	45	49	52
36	264.72	264	2	4	5	7	9	11	12	14	16	18	19	21	23	25	26	28	30	32	33	35	39	42	46	49	53
17 48	267.73	267	2	4	5	7	9	11	12	14	16	18	20	21	23	25	27	28	30	32	34	36	39	43	46	50	53
18 00	270.74	270	2	4	5	7	9	11	13	14	16	18	20	22	23	25	27	29	31	32	34	36	40	43	47	50	54
12	273.75	273	2	4	5	7	9	11	13	15	16	18	20	22	24	25	27	29	31	33	35	36	40	44	47	51	55
24	276.76	276	2	4	6	7	9	11	13	15	17	18	20	22	24	26	28	29	31	33	35	37	40	44	48	52	55
36	279.76	279	2	4	6	7	9	11	13	15	17	19	20	22	24	26	28	30	32	33	35	37	41	45	48	52	56
18 48	282.77	282	2	4	6	8	9	11	13	15	17	19	21	23	24	26	28	30	32	34	36	38	41	45	49	53	56
19 00	285.78	285	2	4	6	8	10	11	13	15	17	19	21	23	25	27	29	30	32	34	36	38	42	46	49	53	57
12	288.79	288	2	4	6	8	10	12	13	15	17	19	21	23	25	27	29	31	33	35	36	38	42	46	50	54	58
24	291.80	291	2	4	6	8	10	12	14	16	17	19	21	23	25	27	29	31	33	35	37	39	43	47	50	54	58
36	294.81	294	2	4	6	8	10	12	14	16	18	20	22	24	25	27	29	31	33	35	37	39	43	47	51	55	59
19 48	297.81	297	2	4	6	8	10	12	14	16	18	20	22	24	26	28	30	32	34	36	38	40	44	48	51	55	59
20 00	300.82	300	2	4	6	8	10	12	14	16	18	20	22	24	26	28	30	32	34	36	38	40	44	48	52	56	60
12	303.83	303	2	4	6	8	10	12	14	16	18	20	22	24	26	28	30	32	34	36	38	40	44	48	53	57	61
24	306.84	306	2	4	6	8	10	12	14	16	18	20	22	24	27	29	31	33	35	37	39	41	45	49	53	57	61
36	309.85	309	2	4	6	8	10	12	14	16	19	21	23	25	27	29	31	33	35	37	39	41	45	49	54	58	62
20 48	312.85	312	2	4	6	8	10	12	15	17	19	21	23	25	27	29	31	33	35	37	40	42	46	50	54	58	62
21 00	315.86	315	2	4	6	8	11	13	15	17	19	21	23	25	27	29	32	34	36	38	40	42	46	50	55	59	63
12	318.87	318	2	4	6	8	11	13	15	17	19	21	23	25	28	30	32	34	36	38	40	42	47	51	55	59	64
24	321.88	321	2	4	6	9	11	13	15	17	19	21	24	26	28	30	32	34	36	39	41	43	47	51	56	60	64
36	324.89	324	2	4	6	9	11	13	15	17	19	22	24	26	28	30	32	35	37	39	41	43	48	52	56	60	65
21 48	327.90	327	2	4	7	9	11	13	15	17	20	22	24	26	28	31	33	35	37	39	41	44	48	52	57	61	65
22 00	330.90	330	2	4	7	9	11	13	15	18	20	22	24	26	29	31	33	35	37	40	42	44	48	53	57	62	66
12	333.91	333	2	4	7	9	11	13	16	18	20	22	24	27	29	31	33	36	38	40	42	44	49	53	58	62	67
24	336.92	336	2	4	7	9	11	13	16	18	20	22	25	27	29	31	34	36	38	40	43	45	49	54	58	63	67
36	339.93	339	2	5	7	9	11	14	16	18	20	23	25	27	29	32	34	36	38	41	43	45	50	54	59	63	68
22 48	342.94	342	2	5	7	9	11	14	16	18	21	23	25	27	30	32	34	36	39	41	43	46	50	55	59	64	68
23 00	345.94	345	2	5	7	9	12	14	16	18	21	23	25	28	30	32	35	37	39	41	44	46	51	55	60	64	69
12	348.95	348	2	5	7	9	12	14	16	19	21	23	26	28	30	32	35	37	39	42	44	46	51	56	60	65	70
24	351.96	351	2	5	7	9	12	14	16	19	21	23	26	28	30	33	35	37	40	42	44	47	51	56	61	66	70
36	354.97	354	2	5	7	9	12	14	17	19	21	24	26	28	31	33	35	38	40	42	45	47	52	57	61	66	71
23 48	357.98	357	2	5	7	10	12	14	17	19	21	24	26	29	31	33	36	38	40	43	45	48	52	57	62	67	71
24 00	360.99	360	2	5	7	10	12	14	17	19	22	24	26	29	31	34	36	38	41	43	46	48	53	58	62	67	72

INTERPOLATION TABLES — ARIES, SUN, PLANETS 12^h—24^h

GMT	Corr. to GHA Aries	Corr. to GHA Sun Planet	\multicolumn{16}{c}{Correction to GHA or Declination for v or d}															
			30	32	34	36	38	40	42	44	46	48	50	52	54	56	58	60
h m	°	°	\multicolumn{16}{c}{UNIT 0°.01}															
12 00	180.49	180	36	38	41	43	46	48	50	53	55	58	60	62	65	67	70	72
12	183.50	183	37	39	41	44	46	49	51	54	56	59	61	63	66	68	71	73
24	186.51	186	37	40	42	45	47	50	52	55	57	60	62	64	67	69	72	74
36	189.52	189	38	40	43	45	48	50	53	55	58	60	63	66	68	71	73	76
12 48	192.53	192	38	41	44	46	49	51	54	56	59	61	64	67	69	72	74	77
13 00	195.53	195	39	42	44	47	49	52	55	57	60	62	65	68	70	73	75	78
12	198.54	198	40	42	45	48	50	53	55	58	61	63	66	69	71	74	77	79
24	201.55	201	40	43	46	48	51	54	56	59	62	64	67	70	72	75	78	80
36	204.56	204	41	44	46	49	52	54	57	60	63	65	68	71	73	76	79	82
13 48	207.57	207	41	44	47	50	52	55	58	61	63	66	69	72	75	77	80	83
14 00	210.58	210	42	45	48	50	53	56	59	62	64	67	70	73	76	78	81	84
12	213.58	213	43	45	48	51	54	57	60	62	65	68	71	74	77	80	82	85
24	216.59	216	43	46	49	52	55	58	60	63	66	69	72	75	78	81	84	86
36	219.60	219	44	47	50	53	55	58	61	64	67	70	73	76	79	82	85	88
14 48	222.61	222	44	47	50	53	56	59	62	65	68	71	74	77	80	83	86	89
15 00	225.62	225	45	48	51	54	57	60	63	66	69	72	75	78	81	84	87	90
12	228.62	228	46	49	52	55	58	61	64	67	70	73	76	79	82	85	88	91
24	231.63	231	46	49	52	55	59	62	65	68	71	74	77	80	83	86	89	92
36	234.64	234	47	50	53	56	59	62	66	69	72	75	78	81	84	87	90	94
15 48	237.65	237	47	51	54	57	60	63	66	70	73	76	79	82	85	88	92	95
16 00	240.66	240	48	51	54	58	61	64	67	70	74	77	80	83	86	90	93	96
12	243.67	243	49	52	55	58	62	65	68	71	75	78	81	84	87	91	94	97
24	246.67	246	49	52	56	59	62	66	69	72	75	79	82	85	89	92	95	98
36	249.68	249	50	53	56	60	63	66	70	73	76	80	83	86	90	93	96	100
16 48	252.69	252	50	54	57	60	64	67	71	74	77	81	84	87	91	94	97	101
17 00	255.70	255	51	54	58	61	65	68	71	75	78	82	85	88	92	95	99	102
12	258.71	258	52	55	58	62	65	69	72	76	79	83	86	89	93	96	100	103
24	261.71	261	52	56	59	63	66	70	73	77	80	84	87	90	94	97	101	104
36	264.72	264	53	56	60	63	67	70	74	77	81	84	88	92	95	99	102	106
17 48	267.73	267	53	57	61	64	68	71	75	78	82	85	89	93	96	100	103	107
18 00	270.74	270	54	58	61	65	68	72	76	79	83	86	90	94	97	101	104	108
12	273.75	273	55	58	62	66	69	73	76	80	84	87	91	95	98	102	106	109
24	276.76	276	55	59	63	66	70	74	77	81	85	88	92	96	99	103	107	110
36	279.76	279	56	60	63	67	71	74	78	82	86	89	93	97	100	104	108	112
18 48	282.77	282	56	60	64	68	71	75	79	83	86	90	94	98	102	105	109	113
19 00	285.78	285	57	61	65	68	72	76	80	84	87	91	95	99	103	106	110	114
12	288.79	288	58	61	65	69	73	77	81	84	88	92	96	100	104	108	111	115
24	291.80	291	58	62	66	70	74	78	81	85	89	93	97	101	105	109	113	116
36	294.81	294	59	63	67	71	74	78	82	86	90	94	98	102	106	110	114	118
19 48	297.81	297	59	63	67	71	75	79	83	87	91	95	99	103	107	111	115	119
20 00	300.82	300	60	64	68	72	76	80	84	88	92	96	100	104	108	112	116	120
12	303.83	303	61	65	69	73	77	81	85	89	93	97	101	105	109	113	117	121
24	306.84	306	61	65	69	73	78	82	86	90	94	98	102	106	110	114	118	122
36	309.85	309	62	66	70	74	78	83	87	91	95	99	103	107	111	115	119	124
20 48	312.85	312	62	67	71	75	79	83	87	92	96	100	104	108	112	116	121	125
21 00	315.86	315	63	67	71	76	80	84	88	92	97	101	105	109	113	118	122	126
12	318.87	318	64	68	72	76	81	85	89	93	98	102	106	110	114	119	123	127
24	321.88	321	64	68	73	77	81	86	90	94	98	103	107	111	116	120	124	128
36	324.89	324	65	69	73	78	82	86	91	95	99	104	108	112	117	121	125	130
21 48	327.90	327	65	70	74	78	83	87	92	96	100	105	109	113	118	122	126	131
22 00	330.90	330	66	70	75	79	84	88	92	97	101	106	110	114	119	123	128	132
12	333.91	333	67	71	75	80	84	89	93	98	102	107	111	115	120	124	129	133
24	336.92	336	67	72	76	81	85	90	94	99	103	108	112	116	121	125	130	134
36	339.93	339	68	72	77	81	86	90	95	99	104	108	113	118	122	127	131	136
22 48	342.94	342	68	73	78	82	87	91	96	100	105	109	114	119	123	128	132	137
23 00	345.94	345	69	74	78	83	87	92	97	101	106	110	115	120	124	129	133	138
12	348.95	348	70	74	79	84	88	93	97	102	107	111	116	121	125	130	135	139
24	351.96	351	70	75	80	84	89	94	98	103	108	112	117	122	126	131	136	140
36	354.97	354	71	76	80	85	90	94	99	104	109	113	118	123	127	132	137	142
23 48	357.98	357	71	76	81	86	90	95	100	105	109	114	119	124	129	133	138	143
24 00	360.99	360	72	77	82	86	91	96	101	106	110	115	120	125	130	134	139	144

Correction to GHA of Aries, Sun, Planets

GMT S	0^m °	1^m °	2^m °	3^m °
00	0.00	0.25	0.50	0.75
04	.02	.27	.52	.77
08	.03	.28	.53	.78
12	.05	.30	.55	.80
16	.07	.32	.57	.82
20	0.08	0.33	0.58	0.83
24	.10	.35	.60	.85
28	.12	.37	.62	.87
32	.13	.38	.63	.88
36	.15	.40	.65	.90
40	0.17	0.42	0.67	0.92
44	.18	.43	.68	.93
48	.20	.45	.70	.95
52	.22	.47	.72	.97
56	.23	.48	.73	0.98
60	0.25	0.50	0.75	1.00

Correction to GHA of Aries, Sun, Planets

GMT S	4^m °	5^m °	6^m °	7^m °
00	1.00	1.25	1.50	1.75
04	.02	.27	.52	.77
08	.03	.28	.53	.78
12	.05	.30	.55	.80
16	.07	.32	.57	.82
20	1.08	1.33	1.58	1.83
24	.10	.35	.60	.85
28	.12	.37	.62	.87
32	.13	.38	.63	.88
36	.15	.40	.65	.90
40	1.17	1.42	1.67	1.92
44	.18	.43	.68	.93
48	.20	.45	.70	.95
52	.22	.47	.72	.97
56	.23	.48	.73	1.98
60	1.25	1.50	1.75	2.00

Correction to GHA of Aries, Sun, Planets

GMT S	8^m °	9^m °	10^m °	11^m °
00	2.00	2.25	2.50	2.75
04	.02	.27	.52	.77
08	.03	.28	.53	.78
12	.05	.30	.55	.80
16	.07	.32	.57	.82
20	2.08	2.33	2.58	2.83
24	.10	.35	.60	.85
28	.12	.37	.62	.87
32	.13	.38	.63	.88
36	.15	.40	.65	.90
40	2.17	2.42	2.67	2.92
44	.18	.43	.68	.93
48	.20	.45	.70	.95
52	.22	.47	.72	.97
56	.23	.48	.73	2.98
60	2.25	2.50	2.75	3.00

INTERPOLATION TABLES — MOON 0^h–4^h, 8^h–12^h, 16^h–20^h

GMT	Corr. to GHA	Correction to GHA or Declination for v or d UNIT $0°.01$																				GMT	Corr. to GHA
		5	10	15	20	25	30	35	40	45	50	55	60	65	70	75	80	85	90	95	100		
h m	°																						m s	°
0 00	0.00	0	0	0	0	0	0	0	0	0	0	0	0	0	0	0	0	0	0	0	0	0 00	0.00
(8 04	0.95	0	0	0	0	0	0	0	0	0	0	0	0	0	0	1	1	1	1	1	1	04	.02
or 08	1.91	0	0	0	0	0	0	0	1	1	1	1	1	1	1	1	1	1	1	1	1	08	.03
16) 12	2.86	0	0	0	0	1	1	1	1	1	1	1	1	1	1	2	2	2	2	2	2	12	.05
16	3.82	0	0	0	1	1	1	1	1	1	1	1	2	2	2	2	2	2	2	3	3	16	.06
0 20	4.77	0	0	1	1	1	1	1	1	2	2	2	2	2	2	3	3	3	3	3	3	0 20	0.08
24	5.72	0	0	1	1	1	1	1	2	2	2	2	2	3	3	3	3	3	4	4	4	24	.10
28	6.68	0	0	1	1	1	1	2	2	2	2	3	3	3	3	4	4	4	4	4	5	28	.11
32	7.63	0	1	1	1	1	2	2	2	2	3	3	3	3	4	4	4	5	5	5	5	32	.13
36	8.59	0	1	1	1	2	2	2	2	3	3	3	3	4	4	5	5	5	5	6	6	36	.14
0 40	9.54	0	1	1	1	2	2	2	3	3	3	4	4	4	5	5	5	6	6	6	7	0 40	0.16
44	10.49	0	1	1	1	2	2	3	3	3	4	4	4	5	5	5	6	6	6	7	7	44	.17
48	11.45	0	1	1	2	2	2	3	3	4	4	4	5	5	6	6	6	7	7	8	8	48	.19
52	12.40	0	1	1	2	2	3	3	3	4	4	5	5	6	6	7	7	7	8	8	9	52	.21
0 56	13.36	0	1	1	2	2	3	3	4	4	5	5	6	6	7	7	7	8	8	9	9	0 56	.22
1 00	14.31	1	1	2	2	3	3	4	4	5	5	6	6	7	7	8	8	9	9	10	10	1 00	0.24
(9 04	15.26	1	1	2	2	3	3	4	4	5	5	6	6	7	7	8	9	9	10	10	11	04	.25
or 08	16.22	1	1	2	2	3	3	4	5	5	6	6	7	7	8	9	9	10	10	11	11	08	.27
17) 12	17.17	1	1	2	2	3	4	4	5	5	6	7	7	8	8	9	10	10	11	11	12	12	.29
16	18.13	1	1	2	3	3	4	4	5	6	6	7	8	8	9	10	10	11	11	12	13	16	.30
1 20	19.08	1	1	2	3	4	4	5	5	6	7	7	8	9	9	10	11	11	12	13	13	1 20	0.32
24	20.03	1	1	2	3	4	4	5	6	6	7	8	8	9	10	11	11	12	13	14	14	24	.33
28	20.99	1	1	2	3	4	4	5	6	7	7	8	9	10	10	11	12	12	13	14	15	28	.35
32	21.94	1	2	2	3	4	5	5	6	7	8	8	9	10	11	12	12	13	14	15	15	32	.37
36	22.90	1	2	2	3	4	5	6	6	7	8	9	10	10	11	12	13	14	14	15	16	36	.38
1 40	23.85	1	2	3	3	4	5	6	7	8	8	9	10	11	12	13	13	14	15	16	17	1 40	0.40
44	24.80	1	2	3	3	4	5	6	7	8	9	10	10	11	12	13	14	15	16	16	17	44	.41
48	25.76	1	2	3	4	5	5	6	7	8	9	10	11	12	13	14	14	15	16	17	18	48	.43
52	26.71	1	2	3	4	5	6	7	7	8	9	10	11	12	13	14	15	16	17	18	19	52	.45
1 56	27.67	1	2	3	4	5	6	7	8	9	10	11	12	13	14	15	15	16	17	18	19	1 56	.46
2 00	28.62	1	2	3	4	5	6	7	8	9	10	11	12	13	14	15	16	17	18	19	20	2 00	0.48
(10 04	29.57	1	2	3	4	5	6	7	8	9	10	11	12	13	14	16	17	18	19	20	21	04	.49
or 08	30.53	1	2	3	4	5	6	7	9	10	11	12	13	14	15	16	17	18	19	20	21	08	.51
18) 12	31.48	1	2	3	4	6	7	8	9	10	11	12	13	14	15	17	18	19	20	21	22	12	.52
16	32.44	1	2	3	5	6	7	8	9	10	11	12	14	15	16	17	18	19	20	22	23	16	.54
2 20	33.39	1	2	4	5	6	7	8	9	11	12	13	14	15	16	18	19	20	21	22	23	2 20	0.56
24	34.34	1	2	4	5	6	7	8	10	11	12	13	14	16	17	18	19	20	22	23	24	24	.57
28	35.30	1	2	4	5	6	7	9	10	11	12	14	15	16	17	19	20	21	22	23	25	28	.59
32	36.25	1	3	4	5	6	8	9	10	11	13	14	15	16	18	19	20	22	23	24	25	32	.60
36	37.21	1	3	4	5	7	8	9	10	12	13	14	16	17	18	20	21	22	23	25	26	36	.62
2 40	38.16	1	3	4	5	7	8	9	11	12	13	15	16	17	19	20	21	23	24	25	27	2 40	0.64
44	39.11	1	3	4	5	7	8	10	11	12	14	15	16	18	19	21	22	23	25	26	27	44	.65
48	40.07	1	3	4	6	7	8	10	11	13	14	15	17	18	20	21	22	24	25	27	28	48	.67
52	41.02	1	3	4	6	7	9	10	11	13	14	16	17	19	20	22	23	24	26	27	29	52	.68
2 56	41.98	1	3	4	6	7	9	10	12	13	15	16	18	19	21	22	23	25	26	28	29	2 56	.70
3 00	42.93	2	3	5	6	8	9	11	12	14	15	17	18	20	21	23	24	26	27	29	30	3 00	0.72
(11 04	43.88	2	3	5	6	8	9	11	12	14	15	17	18	20	21	23	25	26	28	29	31	04	.73
or 08	44.84	2	3	5	6	8	9	11	13	14	16	17	19	20	22	24	25	27	28	30	31	08	.75
19) 12	45.79	2	3	5	6	8	10	11	13	14	16	18	19	21	22	24	26	27	29	30	32	12	.76
16	46.75	2	3	5	7	8	10	11	13	15	16	18	20	21	23	25	26	28	29	31	33	16	.78
3 20	47.70	2	3	5	7	8	10	12	13	15	17	18	20	22	23	25	27	28	30	32	33	3 20	0.80
24	48.65	2	3	5	7	9	10	12	14	15	17	19	20	22	24	26	27	29	31	32	34	24	.81
28	49.61	2	3	5	7	9	10	12	14	16	17	19	21	23	24	26	28	29	31	33	35	28	.83
32	50.56	2	4	5	7	9	11	12	14	16	18	19	21	23	25	27	28	30	32	34	35	32	.84
36	51.52	2	4	5	7	9	11	13	14	16	18	20	22	23	25	27	29	31	32	34	36	36	.86
3 40	52.47	2	4	6	7	9	11	13	15	17	18	20	22	24	26	28	29	31	33	35	37	3 40	0.87
44	53.42	2	4	6	7	9	11	13	15	17	19	21	22	24	26	28	30	32	34	35	37	44	.89
48	54.38	2	4	6	8	10	11	13	15	17	19	21	23	25	27	29	30	32	34	36	38	48	.91
52	55.33	2	4	6	8	10	12	14	15	17	19	21	23	25	27	29	31	33	35	37	39	52	.92
3 56	56.29	2	4	6	8	10	12	14	16	18	20	22	24	26	28	30	31	33	35	37	39	3 56	.94
4 00	57.24	2	4	6	8	10	12	14	16	18	20	22	24	26	28	30	32	34	36	38	40	4 00	0.95

Ephemeris

INTERPOLATION TABLES – MOON 4ʰ–8ʰ, 12ʰ–16ʰ, 20ʰ–24ʰ

GMT	Corr. to GHA	Correction to GHA or Declination for *v* or *d* UNIT 0°.01																			GMT	Corr. to GHA
		5	10	15	20	25	30	35	40	45	50	55	60	65	70	75	80	85	90	95	100		
h m	°																					m s	°
4 00	57.24	2	4	6	8	10	12	14	16	18	20	22	24	26	28	30	32	34	36	38	40	0 00	0.00
(12) 04	58.19	2	4	6	8	10	12	14	16	18	20	22	24	26	28	31	33	35	37	39	41	04	.02
or 08	59.15	2	4	6	8	10	12	14	17	19	21	23	25	27	29	31	33	35	37	39	41	08	.03
(20) 12	60.10	2	4	6	8	11	13	15	17	19	21	23	25	27	29	32	34	36	38	40	42	12	.05
16	61.06	2	4	6	9	11	13	15	17	19	21	23	26	28	30	32	34	36	38	41	43	16	.06
4 20	62.01	2	4	7	9	11	13	15	17	20	22	24	26	28	30	33	35	37	39	41	43	0 20	0.08
24	62.96	2	4	7	9	11	13	15	18	20	22	24	26	29	31	33	35	37	40	42	44	24	.10
28	63.92	2	4	7	9	11	13	16	18	20	22	25	27	29	31	34	36	38	40	42	45	28	.11
32	64.87	2	5	7	9	11	14	16	18	20	23	25	27	29	32	34	36	39	41	43	45	32	.13
36	65.83	2	5	7	9	12	14	16	18	21	23	25	28	30	32	35	37	39	41	44	46	36	.14
4 40	66.78	2	5	7	9	12	14	16	19	21	23	26	28	30	33	35	37	40	42	44	47	0 40	0.16
44	67.73	2	5	7	9	12	14	17	19	21	24	26	28	31	33	36	38	40	43	45	47	44	.17
48	68.69	2	5	7	10	12	14	17	19	22	24	26	29	31	34	36	38	41	43	46	48	48	.19
52	69.64	2	5	7	10	12	15	17	19	22	24	27	29	32	34	37	39	41	44	46	49	52	.21
4 56	70.60	2	5	7	10	12	15	17	20	22	25	27	30	32	35	37	39	42	44	47	49	0 56	.22
5 00	71.55	3	5	8	10	13	15	18	20	23	25	28	30	33	35	38	40	43	45	48	50	1 00	0.24
(13) 04	72.50	3	5	8	10	13	15	18	20	23	25	28	30	33	35	38	41	43	46	48	51	04	.25
or 08	73.46	3	5	8	10	13	15	18	21	23	26	28	31	33	36	39	41	44	46	49	51	08	.27
(21) 12	74.41	3	5	8	10	13	16	18	21	23	26	29	31	34	36	39	42	44	47	49	52	12	.29
16	75.37	3	5	8	11	13	16	18	21	24	26	29	32	34	37	40	42	45	47	50	53	16	.30
5 20	76.32	3	5	8	11	13	16	19	21	24	27	29	32	35	37	40	43	45	48	51	53	1 20	0.32
24	77.27	3	5	8	11	14	16	19	22	24	27	30	32	35	38	41	43	46	49	51	54	24	.33
28	78.23	3	5	8	11	14	16	19	22	25	27	30	33	36	38	41	44	46	49	52	55	28	.35
32	79.18	3	6	8	11	14	17	19	22	25	28	30	33	36	39	42	44	47	50	53	55	32	.37
36	80.14	3	6	8	11	14	17	20	22	25	28	31	34	36	39	42	45	48	50	53	56	36	.38
5 40	81.09	3	6	9	11	14	17	20	23	26	28	31	34	37	40	43	45	48	51	54	57	1 40	0.40
44	82.04	3	6	9	11	14	17	20	23	26	29	32	34	37	40	43	46	49	52	54	57	44	.41
48	83.00	3	6	9	12	15	17	20	23	26	29	32	35	38	41	44	46	49	52	55	58	48	.43
52	83.95	3	6	9	12	15	18	21	23	26	29	32	35	38	41	44	47	50	53	56	59	52	.45
5 56	84.91	3	6	9	12	15	18	21	24	27	30	33	36	39	42	45	47	50	53	56	59	1 56	.46
6 00	85.86	3	6	9	12	15	18	21	24	27	30	33	36	39	42	45	48	51	54	57	60	2 00	0.48
(14) 04	86.81	3	6	9	12	15	18	21	24	27	30	33	36	39	42	46	49	52	55	58	61	04	.49
or 08	87.77	3	6	9	12	15	18	21	25	28	31	34	37	40	43	46	49	52	55	58	61	08	.51
(22) 12	88.72	3	6	9	12	16	19	22	25	28	31	34	37	40	43	47	50	53	56	59	62	12	.52
16	89.68	3	6	9	13	16	19	22	25	28	31	34	38	41	44	47	50	53	56	60	63	16	.54
6 20	90.63	3	6	10	13	16	19	22	25	29	32	35	38	41	44	48	51	54	57	60	63	2 20	0.56
24	91.58	3	6	10	13	16	19	22	26	29	32	35	38	42	45	48	51	54	58	61	64	24	.57
28	92.54	3	6	10	13	16	19	23	26	29	32	36	39	42	45	49	52	55	58	61	65	28	.59
32	93.49	3	7	10	13	16	20	23	26	29	33	36	39	42	46	49	52	56	59	62	65	32	.60
36	94.45	3	7	10	13	17	20	23	26	30	33	36	40	43	46	50	53	56	59	63	66	36	.62
6 40	95.40	3	7	10	13	17	20	23	27	30	33	37	40	43	47	50	53	57	60	63	67	2 40	0.64
44	96.35	3	7	10	13	17	20	24	27	30	34	37	40	44	47	51	54	57	61	64	67	44	.65
48	97.31	3	7	10	14	17	20	24	27	31	34	37	41	44	48	51	54	58	61	65	68	48	.67
52	98.26	3	7	10	14	17	21	24	27	31	34	38	41	45	48	52	55	58	62	65	69	52	.68
6 56	99.22	3	7	10	14	17	21	24	28	31	35	38	42	45	49	52	55	59	62	66	69	2 56	.70
7 00	100.17	4	7	11	14	18	21	25	28	32	35	39	42	46	49	53	56	60	63	67	70	3 00	0.72
(15) 04	101.12	4	7	11	14	18	21	25	28	32	35	39	42	46	49	53	57	60	64	67	71	04	.73
or 08	102.08	4	7	11	14	18	21	25	29	32	36	39	43	46	50	54	57	61	64	68	71	08	.75
(23) 12	103.03	4	7	11	14	18	22	25	29	32	36	40	43	47	50	54	58	61	65	68	72	12	.76
16	103.99	4	7	11	15	18	22	25	29	33	36	40	44	47	51	55	58	62	65	69	73	16	.78
7 20	104.94	4	7	11	15	18	22	26	29	33	37	40	44	48	51	55	59	62	66	70	73	3 20	0.80
24	105.89	4	7	11	15	19	22	26	30	33	37	41	44	48	52	56	59	63	67	70	74	24	.81
28	106.85	4	7	11	15	19	22	26	30	34	37	41	45	49	52	56	60	63	67	71	75	28	.83
32	107.80	4	8	11	15	19	23	26	30	34	38	41	45	49	53	57	60	64	68	72	75	32	.84
36	108.76	4	8	11	15	19	23	27	30	34	38	42	46	49	53	57	61	65	68	72	76	36	.86
7 40	109.71	4	8	12	15	19	23	27	31	35	38	42	46	50	54	58	61	65	69	73	77	3 40	0.87
44	110.66	4	8	12	15	23	27	31	35	39	43	46	50	54	58	62	66	70	73	77		44	.89
48	111.62	4	8	12	16	20	23	27	31	35	39	43	47	51	55	59	62	66	70	74	78	48	.91
52	112.57	4	8	12	16	20	24	28	31	35	39	43	47	51	55	59	63	67	71	75	79	52	.92
7 56	113.53	4	8	12	16	20	24	28	32	36	40	44	48	52	56	60	63	67	71	75	79	3 56	.94
8 00	114.48	4	8	12	16	20	24	28	32	36	40	44	48	52	56	60	64	68	72	76	80	4 00	0.95

INTERPOLATION TABLES — MOON 0^h–4^h, 8^h–12^h, 16^h–20^h

GMT h m	Corr. to GHA °	Correction to GHA or Declination for v or d — UNIT $0°.01$																				GMT m s	Corr. to GHA °
		100	105	110	115	120	125	130	135	140	145	150	155	160	165	170	175	180	185	190	195	200		
0 00	0.00	0	0	0	0	0	0	0	0	0	0	0	0	0	0	0	0	0	0	0	0	0	0 00	0.00
(8 or 16) 04	0.95	1	1	1	1	1	1	1	1	1	1	1	1	1	1	1	1	1	1	1	1	1	04	.02
08	1.91	1	1	1	2	2	2	2	2	2	2	2	2	2	2	2	2	2	2	3	3	3	08	.03
12	2.86	2	2	2	2	2	3	3	3	3	3	3	3	3	3	3	4	4	4	4	4	4	12	.05
16	3.82	3	3	3	3	3	3	3	4	4	4	4	4	4	4	5	5	5	5	5	5	5	16	.06
0 20	4.77	3	4	4	4	4	4	4	5	5	5	5	5	6	6	6	6	6	6	6	7	7	0 20	0.08
24	5.72	4	4	4	5	5	5	5	5	6	6	6	6	7	7	7	7	7	7	8	8	8	24	.10
28	6.68	5	5	5	5	6	6	6	6	7	7	7	7	8	8	8	9	9	9	9	9	9	28	.11
32	7.63	5	6	6	6	6	7	7	7	7	8	8	8	9	9	9	9	10	10	10	10	11	32	.13
36	8.59	6	6	7	7	7	8	8	8	8	9	9	9	10	10	10	11	11	11	11	12	12	36	.14
0 40	9.54	7	7	7	8	8	8	9	9	9	10	10	10	11	11	11	12	12	12	13	13	13	0 40	0.16
44	10.49	7	8	8	8	9	9	10	10	10	11	11	11	12	12	13	13	14	14	14	14	15	44	.17
48	11.45	8	8	9	9	10	10	10	11	11	12	12	12	13	13	14	14	15	15	15	16	16	48	.19
52	12.40	9	9	10	10	10	11	11	12	12	13	13	13	14	14	15	15	16	16	16	17	17	52	.21
0 56	13.36	9	10	10	11	11	12	12	13	13	14	14	14	15	15	16	16	17	17	18	18	19	0 56	.22
1 00	14.31	10	11	11	12	12	13	13	14	14	15	15	16	16	17	17	18	18	19	19	20	20	1 00	0.24
(9 or 17) 04	15.26	11	11	12	12	13	13	14	14	15	15	16	17	17	18	18	19	19	20	20	21	21	04	.25
08	16.22	11	12	12	13	14	14	15	15	16	16	17	17	18	19	19	20	20	21	22	22	23	08	.27
12	17.17	12	13	13	14	14	15	16	16	17	17	18	19	19	20	20	21	22	22	23	23	24	12	.29
16	18.13	13	13	14	15	15	16	16	17	18	18	19	20	20	21	22	22	23	23	24	25	25	16	.30
1 20	19.08	13	14	15	15	16	17	17	18	19	19	20	21	21	22	23	23	24	25	25	26	27	1 20	0.32
24	20.03	14	15	15	16	17	18	18	19	20	20	21	22	22	23	24	25	25	26	27	27	28	24	.33
28	20.99	15	15	16	17	18	18	19	20	21	21	22	23	23	24	25	26	26	27	28	29	29	28	.35
32	21.94	15	16	17	18	18	19	20	21	21	22	23	24	25	25	26	27	28	28	29	30	31	32	.37
36	22.90	16	17	18	18	19	20	21	22	22	23	24	25	26	26	27	28	29	30	30	31	32	36	.38
1 40	23.85	17	18	18	19	20	21	22	23	23	24	25	26	27	28	28	29	30	31	32	33	33	1 40	0.40
44	24.80	17	18	19	20	21	22	23	23	24	25	26	27	28	29	29	30	31	32	33	34	35	44	.41
48	25.76	18	19	20	21	22	23	23	24	25	26	27	28	29	30	31	32	32	33	34	35	36	48	.43
52	26.71	19	20	21	21	22	23	24	25	26	27	28	29	30	31	32	33	34	35	35	36	37	52	.45
1 56	27.67	19	20	21	22	23	24	25	26	27	28	29	30	31	32	33	34	35	36	37	38	39	1 56	.46
2 00	28.62	20	21	22	23	24	25	26	27	28	29	30	31	32	33	34	35	36	37	38	39	40	2 00	0.48
(10 or 18) 04	29.57	21	22	23	24	25	26	27	28	29	30	31	32	33	34	35	36	37	38	39	40	41	04	.49
08	30.53	21	22	23	25	26	27	28	29	30	31	32	33	34	35	36	37	38	39	41	42	43	08	.51
12	31.48	22	23	24	25	26	28	29	30	31	32	33	34	35	36	37	39	40	41	42	43	44	12	.52
16	32.44	23	24	25	26	27	28	29	31	32	33	34	35	36	37	39	40	41	42	43	44	45	16	.54
2 20	33.39	23	25	26	27	28	29	30	32	33	34	35	36	37	39	40	41	42	43	44	46	47	2 20	0.56
24	34.34	24	25	26	28	29	30	31	32	34	35	36	37	38	40	41	42	43	44	46	47	48	24	.57
28	35.30	25	26	27	28	30	31	32	33	35	36	37	38	39	41	42	43	44	46	47	48	49	28	.59
32	36.25	25	27	28	29	30	32	33	34	35	37	38	39	41	42	43	44	46	47	48	49	51	32	.60
36	37.21	26	27	29	30	31	33	34	35	36	38	39	40	42	43	44	46	47	48	49	51	52	36	.62
2 40	38.16	27	28	29	31	32	33	35	36	37	39	40	41	43	44	45	47	48	49	51	52	53	2 40	0.64
44	39.11	27	29	30	31	33	34	36	37	38	40	41	42	44	45	46	48	49	51	52	53	55	44	.65
48	40.07	28	29	31	32	34	35	36	38	39	41	42	43	45	46	48	49	50	52	53	55	56	48	.67
52	41.02	29	30	32	33	34	36	37	39	40	42	43	44	46	47	49	50	52	53	54	56	57	52	.68
2 56	41.98	29	31	32	34	35	37	38	40	41	43	44	45	47	48	50	51	53	54	56	57	59	2 56	.70
3 00	42.93	30	32	33	35	36	38	39	41	42	44	45	47	48	50	51	53	54	56	57	59	60	3 00	0.72
(11 or 19) 04	43.88	31	32	34	35	37	38	40	41	43	44	46	48	49	51	52	54	55	57	58	60	61	04	.73
08	44.84	31	33	34	36	38	39	41	42	44	45	47	49	50	52	53	55	56	58	60	61	63	08	.75
12	45.79	32	34	35	37	38	40	42	43	45	46	48	50	51	53	54	56	58	59	61	62	64	12	.76
16	46.75	33	34	36	38	39	41	42	44	46	47	49	51	52	54	56	57	59	60	62	64	65	16	.78
3 20	47.70	33	35	37	38	40	42	43	45	47	48	50	52	53	55	57	58	60	62	63	65	67	3 20	0.80
24	48.65	34	36	37	39	41	43	44	46	48	49	51	53	54	56	58	60	61	63	65	66	68	24	.81
28	49.61	35	36	38	40	42	43	45	47	49	50	52	54	55	57	59	61	62	64	66	68	69	28	.83
32	50.56	35	37	39	41	42	44	46	48	49	51	53	55	57	58	60	62	64	65	67	69	71	32	.84
36	51.52	36	38	40	41	43	45	47	49	50	52	54	56	58	59	61	63	65	67	68	70	72	36	.86
3 40	52.47	37	39	40	42	44	46	48	50	51	53	55	57	59	61	62	64	66	68	70	72	73	3 40	0.87
44	53.42	37	39	41	43	45	47	49	50	52	54	56	58	60	62	63	65	67	69	71	73	75	44	.89
48	54.38	38	40	42	44	46	48	49	51	53	55	57	59	61	63	65	66	68	70	72	74	76	48	.91
52	55.33	39	41	43	44	46	48	50	52	54	56	58	60	62	64	66	68	70	72	73	75	77	52	.92
3 56	56.29	39	41	43	45	47	49	51	53	55	57	59	61	63	65	67	69	71	73	75	77	79	3 56	.94
4 00	57.24	40	42	44	46	48	50	52	54	56	58	60	62	64	66	68	70	72	74	76	78	80	4 00	0.95

Ephemeris 105

INTERPOLATION TABLES — MOON 4ʰ–8ʰ, 12ʰ–16ʰ, 20ʰ–24ʰ

GMT	Corr. to GHA	Correction to GHA or Declination for *v* or *d*																				GMT	Corr. to GHA
		100	105	110	115	120	125	130	135	140	145	150	155	160	165	170	175	180	185	190	195	200		
h m	°							UNIT 0°.01															m s	°
4 00	57.24	40	42	44	46	48	50	52	54	56	58	60	62	64	66	68	70	72	74	76	78	80	0 00	0.00
(12 04	58.19	41	43	45	47	49	51	53	55	57	59	61	63	65	67	69	71	73	75	77	79	81	04	.02
or 08	59.15	41	43	45	48	50	52	54	56	58	60	62	64	66	68	70	72	74	76	79	81	83	08	.03
20) 12	60.10	42	44	46	48	50	53	55	57	59	61	63	65	67	69	71	74	76	78	80	82	84	12	.05
16	61.06	43	45	47	49	51	53	55	58	60	62	64	66	68	70	73	75	77	79	81	83	85	16	.06
4 20	62.01	43	46	48	50	52	54	56	59	61	63	65	67	69	72	74	76	78	80	82	85	87	0 20	0.08
24	62.96	44	46	48	51	53	55	57	59	62	64	66	68	70	73	75	77	79	81	84	86	88	24	.10
28	63.92	45	47	49	51	54	56	58	60	63	65	67	69	71	74	76	78	80	83	85	87	89	28	.11
32	64.87	45	48	50	52	54	57	59	61	63	66	68	70	73	75	77	79	82	84	86	88	91	32	.13
36	65.83	46	48	51	53	55	58	60	62	64	67	69	71	74	76	78	81	83	85	87	90	92	36	.14
4 40	66.78	47	49	51	54	56	58	61	63	65	68	70	72	75	77	79	82	84	86	89	91	93	0 40	0.16
44	67.73	47	50	52	54	57	59	62	64	66	69	71	73	76	78	80	83	85	88	90	92	95	44	.17
48	68.69	48	50	53	55	58	60	62	65	67	70	72	74	77	79	82	84	86	89	91	94	96	48	.19
52	69.64	49	51	54	56	58	61	63	66	68	71	73	75	78	80	83	85	88	90	92	95	97	52	.21
4 56	70.60	49	52	54	57	59	62	64	67	69	72	74	76	79	81	84	86	89	91	94	96	99	0 56	.22
5 00	71.55	50	53	55	58	60	63	65	68	70	73	75	78	80	83	85	88	90	93	95	98	100	1 00	0.24
(13 04	72.50	51	53	56	58	61	63	66	68	71	73	76	79	81	84	86	89	91	94	96	99	101	04	.25
or 08	73.46	51	54	56	59	62	64	67	69	72	74	77	80	82	85	87	90	92	95	98	100	103	08	.27
21) 12	74.41	52	55	57	60	62	65	68	70	73	75	78	81	83	86	88	91	94	96	99	101	104	12	.29
16	75.37	53	55	58	61	63	66	68	71	74	76	79	82	84	87	90	92	95	97	100	103	105	16	.30
5 20	76.32	53	56	59	61	64	67	69	72	75	77	80	83	85	88	91	93	96	99	101	104	107	1 20	0.32
24	77.27	54	57	59	62	65	68	70	73	76	78	81	84	86	89	92	95	97	100	103	105	108	24	.33
28	78.23	55	57	60	63	66	68	71	74	77	79	82	85	87	90	93	96	98	101	104	107	109	28	.35
32	79.18	55	58	61	64	66	69	72	75	77	80	83	86	89	91	94	97	100	102	105	108	111	32	.37
36	80.14	56	59	62	64	67	70	73	76	78	81	84	87	90	92	95	98	101	104	106	109	112	36	.38
5 40	81.09	57	60	62	65	68	71	74	77	79	82	85	88	91	94	96	99	102	105	108	111	113	1 40	0.40
44	82.04	57	60	63	66	69	72	75	77	80	83	86	89	92	95	97	100	103	106	109	112	115	44	.41
48	83.00	58	61	64	67	70	73	75	78	81	84	87	90	93	96	99	102	104	107	110	113	116	48	.43
52	83.95	59	62	65	67	70	73	76	79	82	85	88	91	94	97	100	103	106	109	111	114	117	52	.45
5 56	84.91	59	62	65	68	71	74	77	80	83	86	89	92	95	98	101	104	107	110	113	116	119	1 56	.46
6 00	85.86	60	63	66	69	72	75	78	81	84	87	90	93	96	99	102	105	108	111	114	117	120	2 00	0.48
(14 04	86.81	61	64	67	70	73	76	79	82	85	88	91	94	97	100	103	106	109	112	115	118	121	04	.49
or 08	87.77	61	64	67	71	74	77	80	83	86	89	92	95	98	101	104	107	110	113	117	120	123	08	.51
22) 12	88.72	62	65	68	71	74	78	81	84	87	90	93	96	99	102	105	109	112	115	118	121	124	12	.52
16	89.68	63	66	69	72	75	78	81	85	88	91	94	97	100	103	107	110	113	116	119	122	125	16	.54
6 20	90.63	63	67	70	73	76	79	82	86	89	92	95	98	101	105	108	111	114	117	120	124	127	2 20	0.56
24	91.58	64	67	70	74	77	80	83	86	90	93	96	99	102	106	109	112	115	118	122	125	128	24	.57
28	92.54	65	68	71	74	78	81	84	87	91	94	97	100	103	107	110	113	116	120	123	126	129	28	.59
32	93.49	65	69	72	75	78	82	85	88	91	95	98	101	105	108	111	114	118	121	124	127	131	32	.60
36	94.45	66	69	73	76	79	83	86	89	92	96	99	102	106	109	112	116	119	122	125	129	132	36	.62
6 40	95.40	67	70	73	77	80	83	87	90	93	97	100	103	107	110	113	117	120	123	127	130	133	2 40	0.64
44	96.35	67	71	74	77	81	84	¦88	91	94	98	101	104	108	111	114	118	121	125	128	131	135	44	.65
48	97.31	68	71	75	78	82	85	88	92	95	99	102	105	109	112	116	119	122	126	129	133	136	48	.67
52	98.26	69	72	76	79	82	86	89	93	96	100	103	106	110	113	117	120	124	127	130	134	137	52	.68
6 56	99.22	69	73	76	80	83	87	90	94	97	101	104	107	111	114	118	121	125	128	132	135	139	2 56	.70
7 00	100.17	70	74	77	81	84	88	91	95	98	102	105	109	112	116	119	123	126	130	133	137	140	3 00	0.72
(15 04	101.12	71	74	78	81	85	88	92	95	99	102	106	110	113	117	120	124	127	131	134	138	141	04	.73
or 08	102.08	71	75	78	82	86	89	93	96	100	103	107	111	114	118	121	125	128	132	136	139	143	08	.75
23) 12	103.03	72	76	79	83	86	90	94	97	101	104	108	112	115	119	122	126	130	133	137	140	144	12	.76
16	103.99	73	76	80	84	87	91	94	98	102	105	109	113	116	120	124	127	131	134	138	142	145	16	.78
7 20	104.94	73	77	81	84	88	92	95	99	103	106	110	114	117	121	125	128	132	136	139	143	147	3 20	0.80
24	105.89	74	78	81	85	89	93	96	100	104	107	111	115	118	122	126	130	133	137	141	144	148	24	.81
28	106.85	75	78	82	86	90	93	97	101	105	108	112	116	119	123	127	131	134	138	142	146	149	28	.83
32	107.80	75	79	83	87	90	94	98	102	105	109	113	117	121	124	128	132	136	139	143	147	151	32	.84
36	108.76	76	80	84	87	91	95	99	103	106	110	114	118	122	125	129	133	137	141	144	148	152	36	.86
7 40	109.71	77	81	84	88	92	96	100	104	107	111	115	119	123	127	130	134	138	142	146	150	153	3 40	0.87
44	110.66	77	81	85	89	93	97	101	104	108	112	116	120	124	128	131	135	139	143	147	151	155	44	.89
48	111.62	78	82	86	90	94	98	101	105	109	113	117	121	125	129	133	137	140	144	148	152	156	48	.91
52	112.57	79	83	87	90	94	98	102	106	110	114	118	122	126	130	134	138	142	146	149	153	157	52	.92
7 56	113.53	79	83	87	91	95	99	103	107	111	115	119	123	127	131	135	139	143	147	151	155	159	3 56	.94
8 00	114.48	80	84	88	92	96	100	104	108	112	116	120	124	128	132	136	140	144	148	152	156	160	4 00	0.95

INTERPOLATION TABLES — MOON 0^h-4^h, 8^h-12^h, 16^h-20^h

GMT	Corr. to GHA	\multicolumn{21}{c}{Correction to GHA or Declination for v or d — UNIT $0°.01$}	GMT	Corr. to GHA
h m	°	200	205	210	215	220	225	230	235	240	245	250	255	260	265	270	275	280	285	290	295	300	m s	°
0 00	0.00	0	0	0	0	0	0	0	0	0	0	0	0	0	0	0	0	0	0	0	0	0	0 00	0.00
(8 04	0.95	1	1	1	1	1	2	2	2	2	2	2	2	2	2	2	2	2	2	2	2	2	04	.02
or 08	1.91	3	3	3	3	3	3	3	3	3	3	3	4	4	4	4	4	4	4	4	4	4	08	.03
16) 12	2.86	4	4	4	4	4	5	5	5	5	5	5	5	5	5	5	6	6	6	6	6	6	12	.05
16	3.82	5	5	6	6	6	6	6	6	6	7	7	7	7	7	7	7	7	8	8	8	8	16	.06
0 20	4.77	7	7	7	7	7	8	8	8	8	8	8	9	9	9	9	9	9	10	10	10	10	0 20	.08
24	5.72	8	8	8	9	9	9	9	9	10	10	10	10	10	11	11	11	11	11	12	12	12	24	.10
28	6.68	9	10	10	10	10	11	11	11	11	11	12	12	12	12	13	13	13	13	14	14	14	28	.11
32	7.63	11	11	11	11	12	12	12	13	13	13	13	14	14	14	14	15	15	15	15	16	16	32	.13
36	8.59	12	12	13	13	13	14	14	14	14	15	15	15	16	16	16	17	17	17	17	18	18	36	.14
0 40	9.54	13	14	14	14	15	15	15	16	16	16	17	17	17	18	18	18	19	19	19	20	20	0 40	0.16
44	10.49	15	15	15	16	16	17	17	17	18	18	18	19	19	20	20	20	21	21	21	22	22	44	.17
48	11.45	16	16	17	17	18	18	18	19	19	20	20	20	21	21	22	22	23	23	23	24	24	48	.19
52	12.40	17	18	18	19	19	20	20	20	21	21	22	22	23	23	23	24	24	25	25	26	26	52	.21
0 56	13.36	19	19	20	20	21	21	21	22	22	23	23	24	24	25	25	26	26	27	27	28	28	0 56	.22
1 00	14.31	20	21	21	22	22	23	23	24	24	25	25	26	26	27	27	28	28	29	29	30	30	1 00	0.24
(9 04	15.26	21	22	22	23	23	24	25	25	26	26	27	27	28	28	29	29	30	30	31	31	32	04	.25
or 08	16.22	23	23	24	24	25	26	26	27	27	28	28	29	29	30	31	31	32	32	33	33	34	08	.27
17) 12	17.17	24	25	25	26	26	27	28	28	29	29	30	31	31	32	32	33	34	34	35	35	36	12	.29
16	18.13	25	26	27	27	28	29	29	30	30	31	32	32	33	34	34	35	35	36	37	37	38	16	.30
1 20	19.08	27	27	28	29	29	30	31	31	32	33	33	34	35	35	36	37	37	38	39	39	40	1 20	0.32
24	20.03	28	29	29	30	31	32	32	33	34	34	35	36	36	37	38	39	39	40	41	41	42	24	.33
28	20.99	29	30	31	32	32	33	34	34	35	36	37	37	38	39	40	40	41	42	43	43	44	28	.35
32	21.94	31	31	32	33	34	35	35	36	37	38	38	39	40	41	41	42	43	44	44	45	46	32	.37
36	22.90	32	33	34	34	35	36	37	38	38	39	40	41	42	42	43	44	45	46	46	47	48	36	.38
1 40	23.85	33	34	35	36	37	38	38	39	40	41	42	43	43	44	45	46	47	48	48	49	50	1 40	0.40
44	24.80	35	36	36	37	38	39	40	41	42	42	43	44	45	46	47	48	49	49	50	51	52	44	.41
48	25.76	36	37	38	39	40	41	41	42	43	44	45	46	47	48	49	50	50	51	52	53	54	48	.43
52	26.71	37	38	39	40	41	42	43	44	45	46	47	48	49	49	50	51	52	53	54	55	56	52	.45
1 56	27.67	39	40	41	42	43	44	44	45	46	47	48	49	50	51	52	53	54	55	56	57	58	1 56	.46
2 00	28.62	40	41	42	43	44	45	46	47	48	49	50	51	52	53	54	55	56	57	58	59	60	2 00	0.48
(10 04	29.57	41	42	43	44	45	47	48	49	50	51	52	53	54	55	56	57	58	59	60	61	62	04	.49
or 08	30.53	43	44	45	46	47	48	49	50	51	52	53	54	55	57	58	59	60	61	62	63	64	08	.51
18) 12	31.48	44	45	46	47	48	50	51	52	53	54	55	56	57	58	59	61	62	63	64	65	66	12	.52
16	32.44	45	46	48	49	50	51	52	53	54	56	57	58	59	60	61	62	63	65	66	67	68	16	.54
2 20	33.39	47	48	49	50	51	53	54	55	56	57	58	60	61	62	63	64	65	67	68	69	70	2 20	0.56
24	34.34	48	49	50	52	53	54	55	56	58	59	60	61	62	64	65	66	67	68	70	71	72	24	.57
28	35.30	49	51	52	53	54	56	57	58	59	60	62	63	64	65	67	68	69	70	72	73	74	28	.59
32	36.25	51	52	53	54	56	57	58	60	61	62	63	65	66	67	68	70	71	72	73	75	76	32	.60
36	37.21	52	53	55	56	57	59	60	61	62	64	65	66	68	69	70	72	73	74	75	77	78	36	.62
2 40	38.16	53	55	56	57	59	60	61	63	64	65	67	68	69	71	72	73	75	76	77	79	80	2 40	0.64
44	39.11	55	56	57	59	60	62	63	64	66	67	68	70	71	72	74	75	77	78	79	81	82	44	.65
48	40.07	56	57	59	60	62	63	64	66	67	69	70	71	73	74	76	77	78	80	81	83	84	48	.67
52	41.02	57	59	60	62	63	65	66	67	69	70	72	73	75	76	77	79	80	82	83	85	86	52	.68
2 56	41.98	59	60	62	63	65	66	67	69	70	72	73	75	76	78	79	81	82	84	85	87	88	2 56	.70
3 00	42.93	60	62	63	65	66	68	69	71	72	74	75	77	78	80	81	83	84	86	87	89	90	3 00	0.72
(11 04	43.88	61	63	64	66	67	69	71	72	74	75	77	78	80	81	83	84	86	87	89	90	92	04	.73
or 08	44.84	63	64	66	67	69	71	72	74	75	77	78	80	81	83	85	86	88	89	91	92	94	08	.75
19) 12	45.79	64	65	67	69	70	72	74	75	77	78	80	82	83	85	86	88	90	91	93	94	96	12	.76
16	46.75	65	67	69	70	72	74	75	77	78	80	82	83	85	87	88	90	91	93	95	96	98	16	.78
3 20	47.70	67	68	70	72	73	75	77	78	80	82	83	85	87	88	90	92	93	95	97	98	100	3 20	0.80
24	48.65	68	70	71	73	75	77	78	80	82	83	85	87	88	90	92	94	95	97	99	100	102	24	.81
28	49.61	69	71	73	75	76	78	80	81	83	85	87	88	90	92	94	95	97	99	101	102	104	28	.83
32	50.56	71	72	74	76	78	80	81	83	85	87	88	90	92	94	95	97	99	101	102	104	106	32	.84
36	51.52	72	74	76	77	79	81	83	85	86	88	90	92	94	95	97	99	101	103	104	106	108	36	.86
3 40	52.47	73	75	77	79	81	83	84	86	88	90	92	94	95	97	99	101	103	105	106	108	110	3 40	0.87
44	53.42	75	77	78	80	82	84	86	88	90	91	93	95	97	99	101	103	105	106	108	110	112	44	.89
48	54.38	76	78	80	82	84	86	87	89	91	93	95	97	99	101	103	105	106	108	110	112	114	48	.91
52	55.33	77	79	81	83	85	87	89	91	93	95	97	99	101	102	104	106	108	110	112	114	116	52	.92
3 56	56.29	79	81	83	85	87	89	90	92	94	96	98	100	102	104	106	108	110	112	114	116	118	3 56	.94
4 00	57.24	80	82	84	86	88	90	92	94	96	98	100	102	104	106	108	110	112	114	116	118	120	4 00	0.95

Ephemeris

INTERPOLATION TABLES – MOON 4ʰ–8ʰ, 12ʰ–16ʰ, 20ʰ–24ʰ

GMT	Corr. to GHA	\multicolumn{7}{c}{Correction to GHA or Declination for *v* or *d*}	GMT	Corr. to GHA
h m	°	200 205 210	215 220 225	230 235 240	245 250 255	260 265 270	275 280 285	290 295 300	m s	°
				UNIT 0°.01						
4 00	57.24	80 82 84	86 88 90	92 94 96	98 100 102	104 106 108	110 112 114	116 118 120	0 00	0.00
(12) 04	58.19	81 83 85	87 89 92	94 96 98	100 102 104	106 108 110	112 114 116	118 120 122	04	.02
or 08	59.15	83 85 87	89 91 93	95 97 99	101 103 105	107 110 112	114 116 118	120 122 124	08	.03
(20) 12	60.10	84 86 88	90 92 95	97 99 101	103 105 107	109 111 113	116 118 120	122 124 126	12	.05
16	61.06	85 87 90	92 94 96	98 100 102	105 107 109	111 113 115	117 119 122	124 126 128	16	.06
4 20	62.01	87 89 91	93 95 98	100 102 104	106 108 111	113 115 117	119 121 124	126 128 130	0 20	0.08
24	62.96	88 90 92	95 97 99	101 103 106	108 110 112	114 117 119	121 123 125	128 130 132	24	.10
28	63.92	89 92 94	96 98 101	103 105 107	109 112 114	116 118 121	123 125 127	130 132 134	28	.11
32	64.87	91 93 95	97 100 102	104 107 109	111 113 116	118 120 122	125 127 129	131 134 136	32	.13
36	65.83	92 94 97	99 101 104	106 108 110	113 115 117	120 122 124	127 129 131	133 136 138	36	.14
4 40	66.78	93 96 98	100 103 105	107 110 112	114 117 119	121 124 126	128 131 133	135 138 140	0 40	0.16
44	67.73	95 97 99	102 104 107	109 111 114	116 118 121	123 125 128	130 133 135	137 140 142	44	.17
48	68.69	96 98 101	103 106 108	110 113 115	118 120 122	125 127 130	132 134 137	139 142 144	48	.19
52	69.64	97 100 102	105 107 110	112 114 117	119 122 124	127 129 131	134 136 139	141 144 146	52	.21
4 56	70.60	99 101 104	106 109 111	113 116 118	121 123 126	128 131 133	136 138 141	143 146 148	0 56	.22
5 00	71.55	100 103 105	108 110 113	115 118 120	123 125 128	130 133 135	138 140 143	145 148 150	1 00	0.24
(13) 04	72.50	101 104 106	109 111 114	117 119 122	124 127 129	132 134 137	139 142 144	147 149 152	04	.25
or 08	73.46	103 105 108	110 113 116	118 121 123	126 128 131	133 136 139	141 144 146	149 151 154	08	.27
(21) 12	74.41	104 107 109	112 114 117	120 122 125	127 130 133	135 138 140	143 146 148	151 153 156	12	.29
16	75.37	105 108 111	113 116 119	121 124 126	129 132 134	137 140 142	145 147 150	153 155 158	16	.30
5 20	76.32	107 109 112	115 117 120	123 125 128	131 133 136	139 141 144	147 149 152	155 157 160	1 20	0.32
24	77.27	108 111 113	116 119 122	124 127 130	132 135 138	140 143 146	149 151 154	157 159 162	24	.33
28	78.23	109 112 115	118 120 123	126 128 131	134 137 139	142 145 148	150 153 156	159 161 164	28	.35
32	79.18	111 113 116	119 122 125	127 130 133	136 138 141	144 147 149	152 155 158	160 163 166	32	.37
36	80.14	112 115 118	120 123 126	129 132 134	137 140 143	146 148 151	154 157 160	162 165 168	36	.38
5 40	81.09	113 116 119	122 125 128	130 133 136	139 142 145	147 150 153	156 159 162	164 167 170	1 40	0.40
44	82.04	115 118 120	123 126 129	132 135 138	140 143 146	149 152 155	158 161 163	166 169 172	44	.41
48	83.00	116 119 122	125 128 131	133 136 139	142 145 148	151 154 157	160 162 165	168 171 174	48	.43
52	83.95	117 120 123	126 129 132	135 138 141	144 147 150	153 155 158	161 164 167	170 173 176	52	.45
5 56	84.91	119 122 125	128 131 134	136 139 142	145 148 151	154 157 160	163 166 169	172 175 178	1 56	.46
6 00	85.86	120 123 126	129 132 135	138 141 144	147 150 153	156 159 162	165 168 171	174 177 180	2 00	0.48
(14) 04	86.81	121 124 127	130 133 137	140 143 146	149 152 155	158 161 164	167 170 173	176 179 182	04	.49
or 08	87.77	123 126 129	132 135 138	141 144 147	150 153 156	159 163 166	169 172 175	178 181 184	08	.51
(22) 12	88.72	124 127 130	133 136 140	143 146 149	152 155 158	161 164 167	171 174 177	180 183 186	12	.52
16	89.68	125 128 132	135 138 141	144 147 150	154 157 160	163 166 169	172 175 179	182 185 188	16	.54
6 20	90.63	127 130 133	136 139 143	146 149 152	155 158 162	165 168 171	174 177 181	184 187 190	2 20	0.56
24	91.58	128 131 134	138 141 144	147 150 154	157 160 163	166 170 173	176 179 182	186 189 192	24	.57
28	92.54	129 133 136	139 142 146	149 152 155	158 162 165	168 171 175	178 181 184	188 191 194	28	.59
32	93.49	131 134 137	140 144 147	150 154 157	160 163 167	170 173 176	180 183 186	189 193 196	32	.60
36	94.45	132 135 139	142 145 149	152 155 158	162 165 168	172 175 178	182 185 188	191 195 198	36	.62
6 40	95.40	133 137 140	143 147 150	153 157 160	163 167 170	173 177 180	183 187 190	193 197 200	2 40	0.64
44	96.35	135 138 141	145 148 152	155 158 162	165 168 172	175 178 182	185 189 192	195 199 202	44	.65
48	97.31	136 139 143	146 150 153	156 160 163	167 170 173	177 180 184	187 190 194	197 201 204	48	.67
52	98.26	137 141 144	148 151 155	158 161 165	168 172 175	179 182 185	189 192 196	199 203 206	52	.68
6 56	99.22	139 142 146	149 153 156	159 163 166	170 173 177	180 184 187	191 194 198	201 205 208	2 56	.70
7 00	100.17	140 144 147	151 154 158	161 165 168	172 175 179	182 186 189	193 196 200	203 207 210	3 00	0.72
(15) 04	101.12	141 145 148	152 155 159	163 166 170	173 177 180	184 187 191	194 198 201	205 208 212	04	.73
or 08	102.08	143 146 150	153 157 161	164 168 171	175 178 182	185 189 193	196 200 203	207 210 214	08	.75
(23) 12	103.03	144 148 151	155 158 162	166 169 173	176 180 184	187 191 194	198 202 205	209 212 216	12	.76
16	103.99	145 149 153	156 160 164	167 171 174	178 182 185	189 193 196	200 203 207	211 214 218	16	.78
7 20	104.94	147 150 154	158 161 165	169 172 176	180 183 187	191 194 198	202 205 209	213 216 220	3 20	0.80
24	105.89	148 152 155	159 163 167	170 174 178	181 185 189	192 196 200	204 207 211	215 218 222	24	.81
28	106.85	149 153 157	161 164 168	172 175 179	183 187 190	194 198 202	205 209 213	217 220 224	28	.83
32	107.80	151 154 158	162 166 170	173 177 181	185 188 192	196 200 203	207 211 215	218 222 226	32	.84
36	108.76	152 156 160	163 167 171	175 179 182	186 190 194	198 201 205	209 213 217	220 224 228	36	.86
7 40	109.71	153 157 161	165 169 173	176 180 184	188 192 196	199 203 207	211 215 219	222 226 230	3 40	0.87
44	110.66	155 159 162	166 170 174	178 182 186	189 193 197	201 205 209	213 217 220	224 228 232	44	.89
48	111.62	156 160 164	168 172 176	179 183 187	191 195 199	203 207 211	215 218 222	226 230 234	48	.91
52	112.57	157 161 165	169 173 177	181 185 189	193 197 201	205 208 212	216 220 224	228 232 236	52	.92
7 56	113.53	159 163 167	171 175 179	182 186 190	194 198 202	206 210 214	218 222 226	230 234 238	3 56	.94
8 00	114.48	160 164 168	172 176 180	184 188 192	196 200 204	208 212 216	220 224 228	232 236 240	4 00	0.95

5.3.11 STAR DIAGRAM (LHA ARIES 0°–120°)

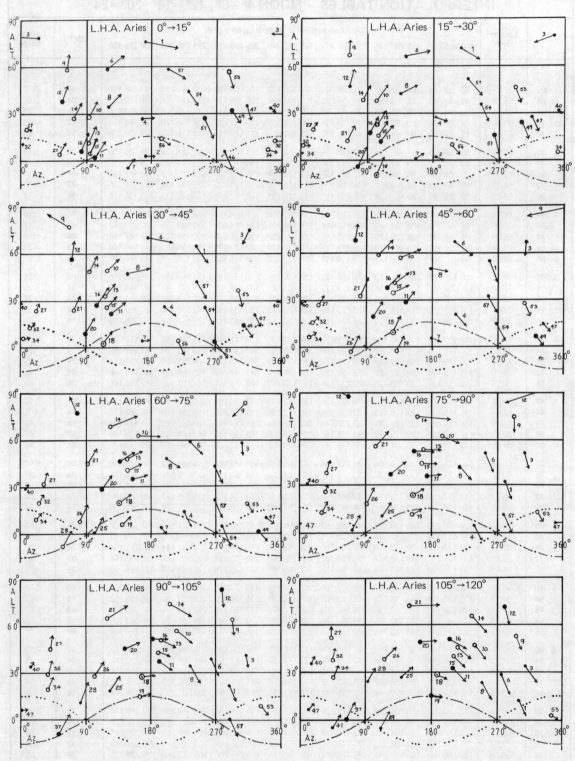

The diagrams are explained in 5.2.6 on page 70. Note that when a star is near the zenith its azimuth is liable to a considerable change for a small change in latitude. The magnitude corresponding to the star symbols is shown alongside.

- • Mag 2.1–3.1
- ○ Mag 1.1–2.0
- ● Mag 0.0–1.0
- ⊙ Mag −1.6

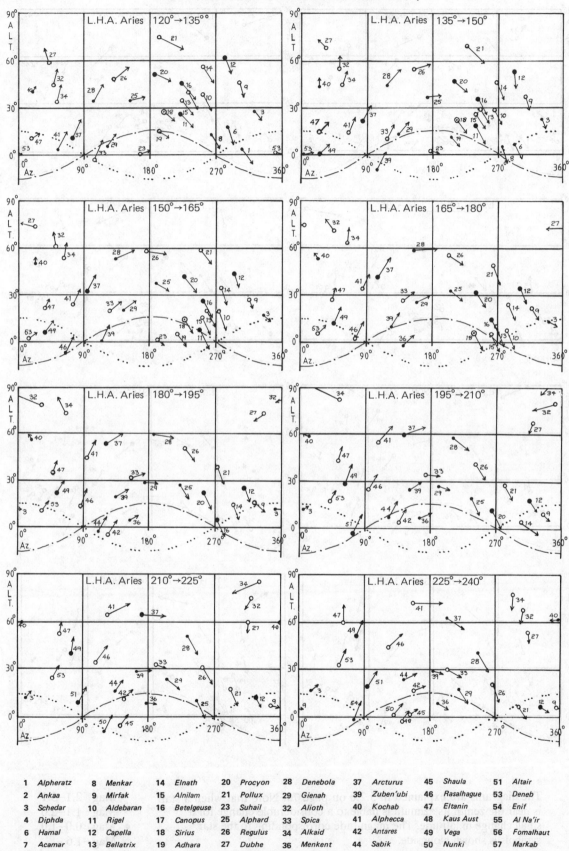

STAR DIAGRAM (LHA ARIES 240°-360°)

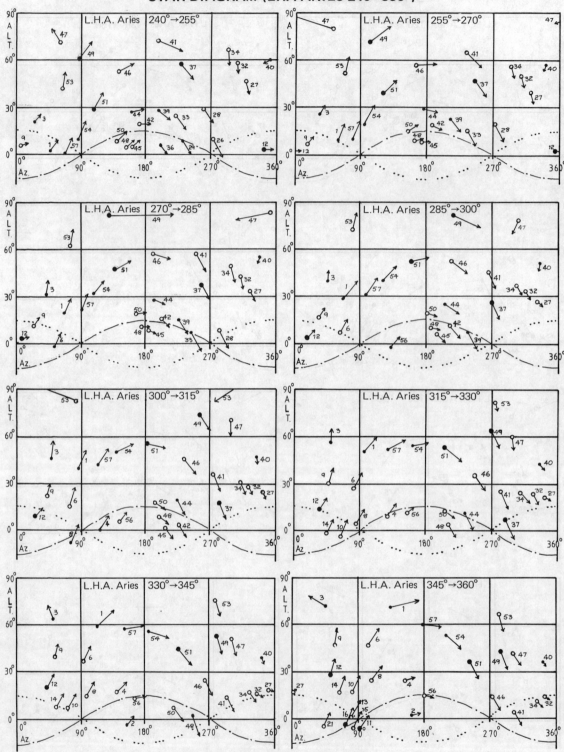

The diagrams are explained in 5.2.6 on page 70. Note that when a star is near the zenith its azimuth is liable to a considerable change for a small change in latitude. The magnitude corresponding to the star symbols is shown alongside.

- · Mag 2.1–3.1
- ○ Mag 1.1–2.0
- ● Mag 0.0–1.0
- ⊙ Mag −1.6

5.3.12 AZIMUTH DIAGRAM A

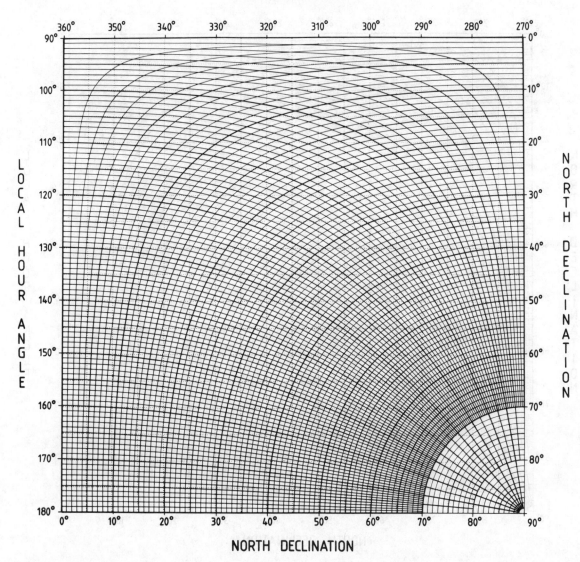

Instructions — To determine the Azimuth of an observed body given the latitude of the observer (Lat) and the LHA and Dec of the body.
(1) Plot on the diagrams a position corresponding to the LHA and Dec; if the LHA is greater than 180° use 360° — LHA when plotting.
(2) Read the position of the plotted point on the rectangular graticule, using the scale at the top for the horizontal reading, and the declination scale for the vertical reading.
(3) To the horizontal reading in (2) add the co-latitude (i.e. add 90° and subtract Lat). The vertical co-ordinate is unchanged.
(4) At the point corresponding to the co-ordinates in (3) read the corresponding LHA (= Z).
(5) Z is converted to Azimuth as follows:

LHA (body)	Azimuth
0° to 180°	180° + Z
180° to 360°	180° − Z

Note that in step (4) a reading of the Dec at the same point gives an approximate value of the altitude, and so gives a check on the plotting.

The diagrams A, B and C are portions of a continuous diagram (see p. 113), so that if the movement for co-latitude goes off the diagram on which step (2) is plotted, then the overlap can be found on the previous diagram.

AZIMUTH DIAGRAM B

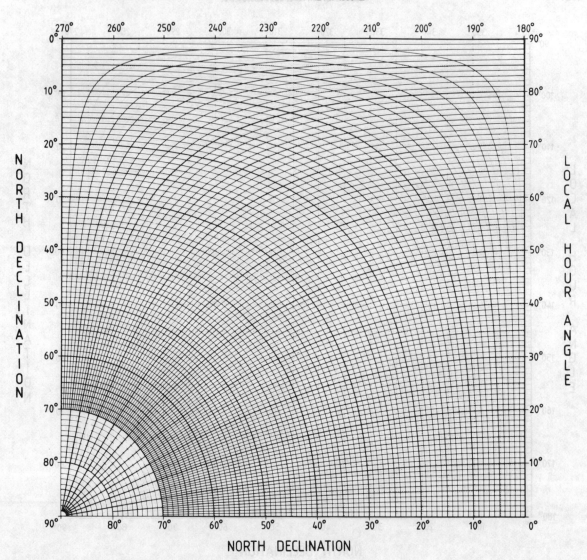

Illustration — On 1989 January 5 at GMT 14ʰ 40ᵐ 18ˢ, the Sun is observed from a position N.44°36' W.2°40' and it is required to determine the azimuth.

Page 87,	Jan 5,	00ʰ		GHA	178.68	$v-5$	Dec S.22.64	$d+5$
Interpolation	GMT	14ʰ 36ᵐ			219			
Tables	GMT		4ᵐ 18ˢ		1.08			
Pages 100, 101	v correction				− .07			+ .07
	GMT	14ʰ 40ᵐ 18ˢ		GHA	398.69			
	Longitude W.2°40'				− 2.67			
				LHA	396.02			
					− 360			
				LHA	36.02		Dec S.22.57	

Plot on diagram C, LHA = 36°, Dec = S.22°.6; the corresponding graticule co-ordinates are found to be 153° and 57° (down); to the horizontal co-ordinate add 90° and subtract 44°.6; the new co-ordinates being 198°.4 and 57° the corresponding value of LHA (= Z) is 34°. The Dec is 15°. The Azimuth is therefore 180° + 34° = 214°, and the approximate altitude is 15° which can be compared with the observed altitude as a check on the plotting.

AZIMUTH DIAGRAM C

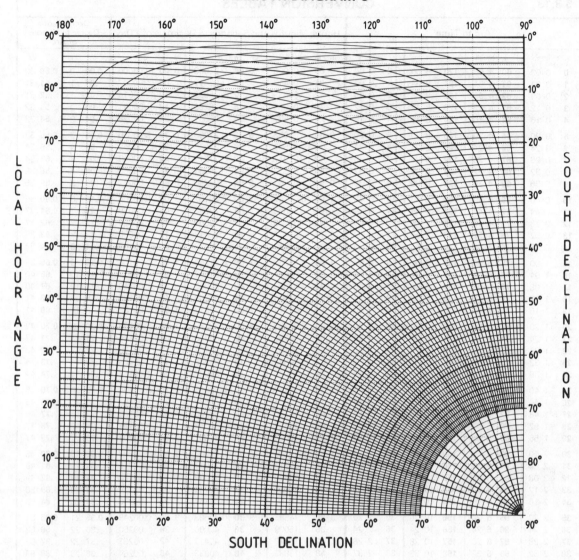

The illustration below shows how A, B and C are portions of a continuous diagram. The points P and Q show the plotting of the example opposite.

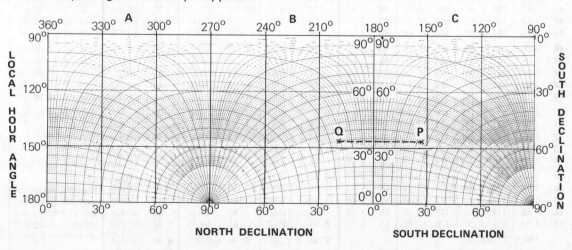

5.3.13 CONVERSION TABLES

Arc to Time								Hours, Minutes and Seconds to Decimals of a Day							Decimals of a Degree to Minutes of Arc			
°	h m	°	h m	°	h m	′	m s	h m	d	h m	d	m s	d	°	′	°	′
0	0 00	60	4 00	120	8 00	0	0 00	0 00	0.00000	6 00	0.25000	0 00	0.00000	0.00	0	0.50	30
1	0 04	61	4 04	121	8 04	1	0 04	06	.00417	06	.25417	06	.00007	.01	1	.51	31
2	0 08	62	4 08	122	8 08	2	0 08	12	.00833	12	.25833	12	.00014	.02	1	.52	31
3	0 12	63	4 12	123	8 12	3	0 12	18	.01250	18	.26250	18	.00021	.03	2	.53	32
4	0 16	64	4 16	124	8 16	4	0 16	24	.01667	24	.26667	24	.00028	.04	2	.54	32
5	0 20	65	4 20	125	8 20	5	0 20	0 30	.02083	6 30	.27083	0 30	.00035	0.05	3	0.55	33
6	0 24	66	4 24	126	8 24	6	0 24	36	.02500	36	.27500	36	.00042	.06	4	.56	34
7	0 28	67	4 28	127	8 28	7	0 28	42	.02917	42	.27917	42	.00049	.07	4	.57	34
8	0 32	68	4 32	128	8 32	8	0 32	48	.03333	48	.28333	48	.00056	.08	5	.58	35
9	0 36	69	4 36	129	8 36	9	0 36	0 54	.03750	6 54	.28750	0 54	.00062	.09	5	.59	35
10	0 40	70	4 40	130	8 40	10	0 40	1 00	.04167	7 00	.29167	1 00	.00069	0.10	6	0.60	36
11	0 44	71	4 44	131	8 44	11	0 44	06	.04583	06	.29583	06	.00076	.11	7	.61	37
12	0 48	72	4 48	132	8 48	12	0 48	12	.05000	12	.30000	12	.00083	.12	7	.62	37
13	0 52	73	4 52	133	8 52	13	0 52	18	.05417	18	.30417	18	.00090	.13	8	.63	38
14	0 56	74	4 56	134	8 56	14	0 56	24	.05833	24	.30833	24	.00097	.14	8	.64	38
15	1 00	75	5 00	135	9 00	15	1 00	1 30	.06250	7 30	.31250	1 30	.00104	0.15	9	0.65	39
16	1 04	76	5 04	136	9 04	16	1 04	36	.06667	36	.31667	36	.00111	.16	10	.66	40
17	1 08	77	5 08	137	9 08	17	1 08	42	.07083	42	.32083	42	.00118	.17	10	.67	40
18	1 12	78	5 12	138	9 12	18	1 12	48	.07500	48	.32500	48	.00125	.18	11	.68	41
19	1 16	79	5 16	139	9 16	19	1 16	1 54	.07917	7 54	.32917	1 54	.00132	.19	11	.69	41
20	1 20	80	5 20	140	9 20	20	1 20	2 00	.08333	8 00	.33333	2 00	.00139	0.20	12	0.70	42
21	1 24	81	5 24	141	9 24	21	1 24	06	.08750	06	.33750	06	.00146	.21	13	.71	43
22	1 28	82	5 28	142	9 28	22	1 28	12	.09167	12	.34167	12	.00153	.22	13	.72	43
23	1 32	83	5 32	143	9 32	23	1 32	18	.09583	18	.34583	18	.00160	.23	14	.73	44
24	1 36	84	5 36	144	9 36	24	1 36	24	.10000	24	.35000	24	.00167	.24	14	.74	44
25	1 40	85	5 40	145	9 40	25	1 40	2 30	.10417	8 30	.35417	2 30	.00174	0.25	15	0.75	45
26	1 44	86	5 44	146	9 44	26	1 44	36	.10833	36	.35833	36	.00181	.26	16	.76	46
27	1 48	87	5 48	147	9 48	27	1 48	42	.11250	42	.36250	42	.00188	.27	16	.77	46
28	1 52	88	5 52	148	9 52	28	1 52	48	.11667	48	.36667	48	.00194	.28	17	.78	47
29	1 56	89	5 56	149	9 56	29	1 56	2 54	.12083	8 54	.37083	2 54	.00201	.29	17	.79	47
30	2 00	90	6 00	150	10 00	30	2 00	3 00	.12500	9 00	.37500	3 00	.00208	0.30	18	0.80	48
31	2 04	91	6 04	151	10 04	31	2 04	06	.12917	06	.37917	06	.00215	.31	19	.81	49
32	2 08	92	6 08	152	10 08	32	2 08	12	.13333	12	.38333	12	.00222	.32	19	.82	49
33	2 12	93	6 12	153	10 12	33	2 12	18	.13750	18	.38750	18	.00229	.33	20	.83	50
34	2 16	94	6 16	154	10 16	34	2 16	24	.14167	24	.39167	24	.00236	.34	20	.84	50
35	2 20	95	6 20	155	10 20	35	2 20	3 30	.14583	9 30	.39583	3 30	.00243	0.35	21	0.85	51
36	2 24	96	6 24	156	10 24	36	2 24	36	.15000	36	.40000	36	.00250	.36	22	.86	52
37	2 28	97	6 28	157	10 28	37	2 28	42	.15417	42	.40417	42	.00257	.37	22	.87	52
38	2 32	98	6 32	158	10 32	38	2 32	48	.15833	48	.40833	48	.00264	.38	23	.88	53
39	2 36	99	6 36	159	10 36	39	2 36	3 54	.16250	9 54	.41250	3 54	.00271	.39	23	.89	53
40	2 40	100	6 40	160	10 40	40	2 40	4 00	.16667	10 00	.41667	4 00	.00278	0.40	24	0.90	54
41	2 44	101	6 44	161	10 44	41	2 44	06	.17083	06	.42083	06	.00285	.41	25	.91	55
42	2 48	102	6 48	162	10 48	42	2 48	12	.17500	12	.42500	12	.00292	.42	25	.92	55
43	2 52	103	6 52	163	10 52	43	2 52	18	.17917	18	.42917	18	.00299	.43	26	.93	56
44	2 56	104	6 56	164	10 56	44	2 56	24	.18333	24	.43333	24	.00306	.44	26	.94	56
45	3 00	105	7 00	165	11 00	45	3 00	4 30	.18750	10 30	.43750	4 30	.00312	0.45	27	0.95	57
46	3 04	106	7 04	166	11 04	46	3 04	36	.19167	36	.44167	36	.00319	.46	28	.96	58
47	3 08	107	7 08	167	11 08	47	3 08	42	.19583	42	.44583	42	.00326	.47	28	.97	58
48	3 12	108	7 12	168	11 12	48	3 12	48	.20000	48	.45000	48	.00333	.48	29	.98	59
49	3 16	109	7 16	169	11 16	49	3 16	4 54	.20417	10 54	.45417	4 54	.00340	.49	29	0.99	59
50	3 20	110	7 20	170	11 20	50	3 20	5 00	.20833	11 00	.45833	5 00	.00347	0.50	30	1.00	60
51	3 24	111	7 24	171	11 24	51	3 24	06	.21250	06	.46250	06	.00354				
52	3 28	112	7 28	172	11 28	52	3 28	12	.21667	12	.46667	12	.00361				
53	3 32	113	7 32	173	11 32	53	3 32	18	.22083	18	.47083	18	.00368				
54	3 36	114	7 36	174	11 36	54	3 36	24	.22500	24	.47500	24	.00375				
55	3 40	115	7 40	175	11 40	55	3 40	5 30	.22917	11 30	.47917	5 30	.00382				
56	3 44	116	7 44	176	11 44	56	3 44	36	.23333	36	.48333	36	.00389				
57	3 48	117	7 48	177	11 48	57	3 48	42	.23750	42	.48750	42	.00396				
58	3 52	118	7 52	178	11 52	58	3 52	48	.24167	48	.49167	48	.00403				
59	3 56	119	7 56	179	11 56	59	3 56	5 54	.24583	11 54	.49583	5 54	.00410				
	180° = 12^h 00^m							6 00	0.25000	12 00	0.50000	6 00	0.00417				

This table can also be used for the conversion of decimals of a minute to seconds of arc and vice versa.

PLATE 1 – NAVIGATION LIGHTS (SEE ALSO PLATE 4)

Port sidelight (red) shows from ahead to 22½° abaft the beam

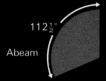

For yachts 12–50m overall, visibility – 2 miles. For yachts under 12m – 1 mile

(May be combined with starboard sidelight in one centreline lantern in boats under 20m overall)

White masthead light shows over arc of 225° – from ahead to 22½° abaft the beam each side. Shown by vessels under power only

Ahead 225°

(Masthead light and sternlight may be combined in one all-round white light in boats under 12m overall)

White sternlight shows over arc of 135°, 67½° on each side of vessel

For yachts 20–50m overall, visibility – 5 miles. For yachts 12–20m – 3 miles. For yachts under 12m – 2 miles

Astern 135°

For yachts under 50m overall, visibility – 2 miles

Starboard sidelight (green) shows from ahead to 22½° abaft the beam

For yachts 12–50m overall, visibility – 2 miles. For yachts under 12m – 1 mile

(May be combined with port side light in one centreline lantern in boats under 20m overall)

Lights for power-driven vessels underway (plan views)

Note: Also apply to sailing yachts or other sailing craft when under power

Motor boat under 7m, less than 7 knots

Motor boat under 12m (combined masthead & sternlight)

Motor yacht under 20m (combined lantern for sidelights)

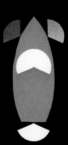
Motor yacht over 20m

Larger vessel, over 50m, with two masthead lights – the aft one higher

Lights for sailing vessels underway (plan views)

Note: These lights apply to sailing craft when under sail ONLY. If motor-sailing the appropriate lights for a power-driven vessel must be shown, as above

Sailing boat under 7m shows white light to prevent collision. If practicable she should show sidelights and sternlight

Combined sidelights plus sternlight

Masthead tricolour lantern

or

Tricolour lantern at masthead

Separate sidelights and sternlight for sailing vessel over 20m

Sailing yacht under 20m

Bow view

If *not* using tri-colour masthead lantern, a sailing yacht may show (in addition to other lights) two all-round lights near masthead, the upper red and the lower green

PLATE 2 – ADMIRALTY METRIC CHART SYMBOLS

A selection of the more common symbols from Admiralty Publication 5011

Reproduced by kind permission of H.M. Stationery Office and the Hydrographer of the Navy.

THE COASTLINE	ARTIFICIAL FEATURES	RADIO AND RADAR
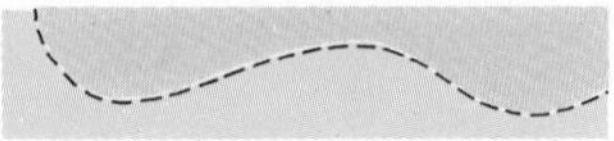 Coast imperfectly known or shoreline unsurveyed.	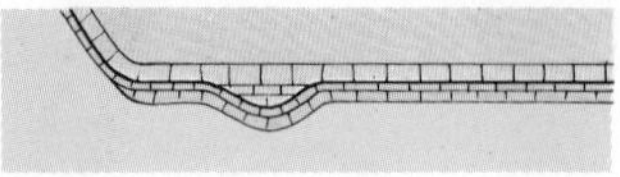 Sea wall	 RC Non-directional Radiobeacon
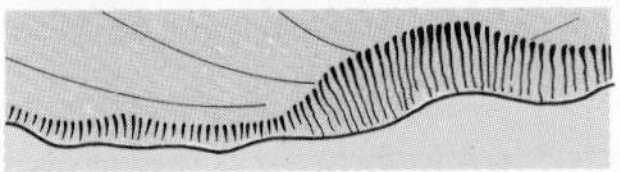 Steep coast	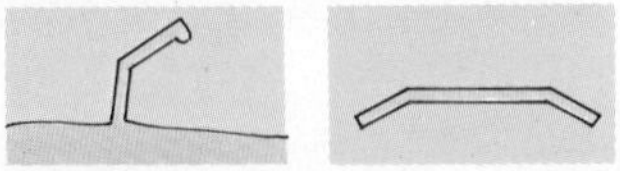 Breakwater	 RD RD 269°30′ Directional Radiobeacon
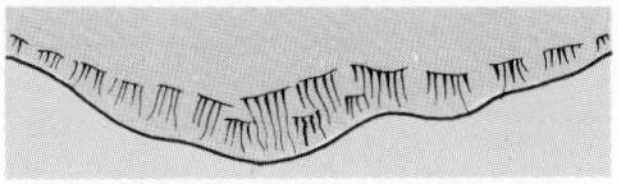 Cliffy coast	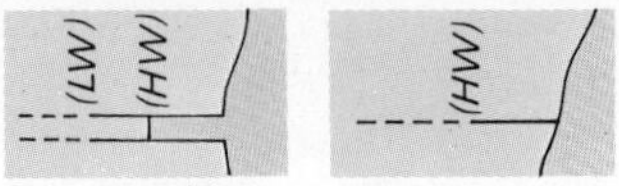 Submerged jetty	 RW Rotating Pattern Radiobeacon
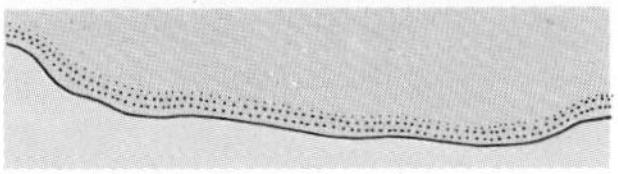 Sandy shore	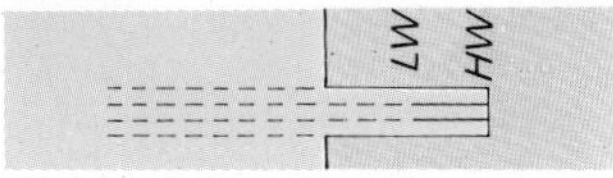 Patent slip	 RG Radio Direction Finding Station
Low Water Line	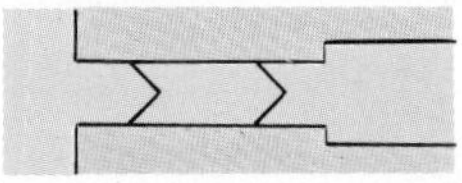 Lock	Radio Mast — Radio mast or tower / Radio Tr — Radio tower or scanner / Radar Tr / Radar Sc — Landmarks for visual fixing only
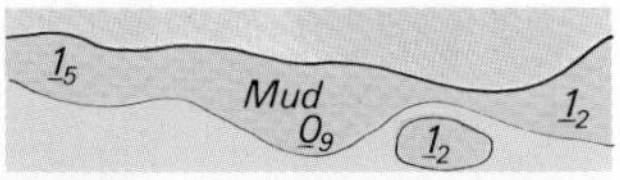 Foreshore, Mud	 Hulk	TV Mast / TV Tr Television mast or tower
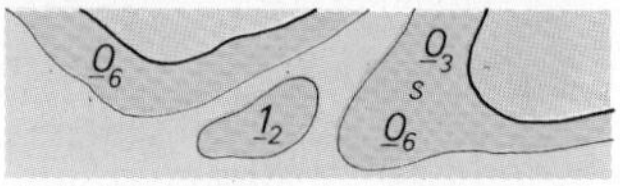 Foreshore, Sand	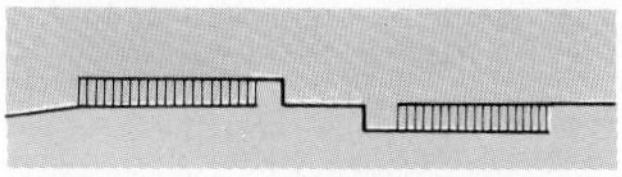 Steps	 R Coast Radio Station providing QTG service
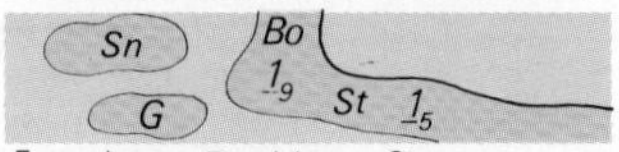 Foreshore, Boulders, Stones, Gravel and Shingle	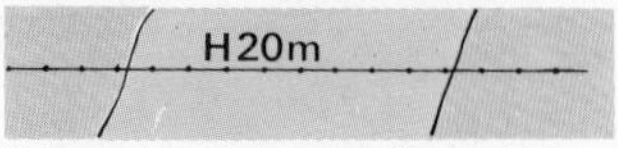 Telegraph or telephone line, with vertical clearance (above HW) H20m	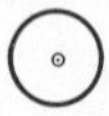 Ra Coast Radar Station
 Foreshore, Rock	 Sewer, Outfall pipe	 Racon Radar Responder Beacon
 Foreshore, Sand and Mud	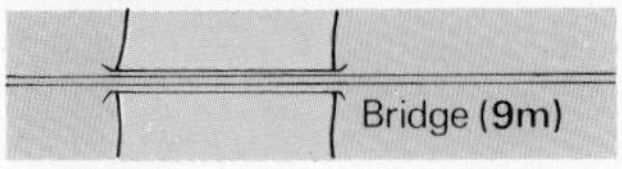 Bridge (9m)	 Radar Reflector
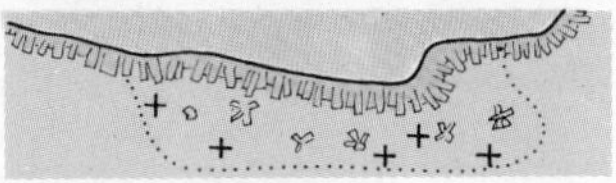 Limiting danger line	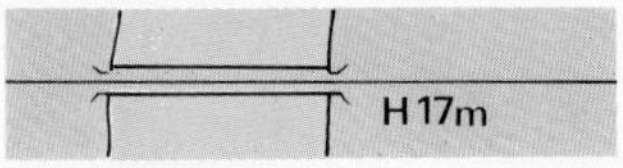 Fixed bridge with vertical clearance (above HW) H17m	Ra (conspic) Radar conspicuous object.
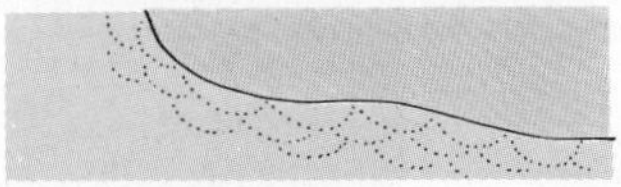 Breakers along a shore	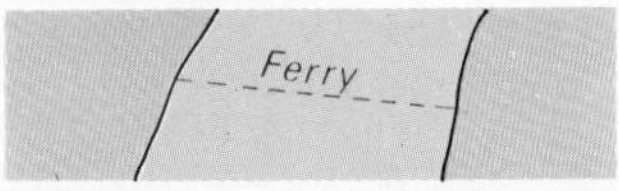 Ferry	 Aero RC Aeronautical radiobeacon.
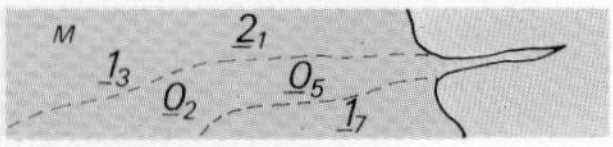 Half-tide channel (on intertidal ground)	 Training wall (covers)	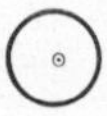 Consol Bn Consol beacon.

PLATE 3 – ADMIRALTY METRIC CHART SYMBOLS

A selection of the more common symbols from Admiralty Publication 5011

Reproduced by kind permission of H.M. Stationery Office and the Hydrographer of the Navy.

DANGERS	DANGERS	LIMITS
Rock which does not cover (with elevation above MHWS or MHHW, or where there is no tide, above MSL).	Wreck over which the depth has been obtained by sounding, but not by wire sweep.	Leading Line
Rock which covers and uncovers (with elevation above chart datum).	Wreck over which the exact depth is unknown but thought to be more than 28 metres, or a wreck over which the depth is thought to be 28 metres or less, but which is not considered dangerous to surface vessels capable of navigating in the vicinity.	Limit of sector
Rock awash at the level of chart datum.		
Submerged rock with 2 metres or less water over it at chart datum, or rock ledge on which depths are known to be 2 metres or less, or a rock or rock ledge over which the exact depth is unknown but which is considered to be dangerous to surface navigation.	The remains of a wreck, or other foul area no longer dangerous to surface navigation, but to be avoided by vessels anchoring, trawling, etc.	Traffic separation scheme: one-way traffic lanes (separated by zone).
		Submarine cable (telegraph & telephone).
Shoal sounding on isolated rock.		Submarine cable (power).
Submerged rock not dangerous to surface navigation.	Overfalls and tide-rips	Limits of national fishing zones.
Submerged danger with depth cleared by wire drag.	Eddies	Anchorage Area. Type of anchorage is usually indicated by legend, e.g. Small Craft Anchorage, Naval Anchorage, Quarantine Anchorage, etc.
Wreck showing any portion of hull or superstructure at the level of chart datum. Large scale charts	Kelp	Recommended track for deep draught vessels (track not defined by fixed mark(s)). Where the minimum safe depth along the recommended track (or section thereof) is guaranteed by the competent harbour regional or national authority, the depth is indicated thus:
Wreck of which the masts only are visible. Large scale charts	Breakers	
Wreck over which the exact depth of water is unknown but is thought to be 28 metres or less, and which is considered dangerous to surface navigation.	Limiting danger line	DW 27m DW 25m

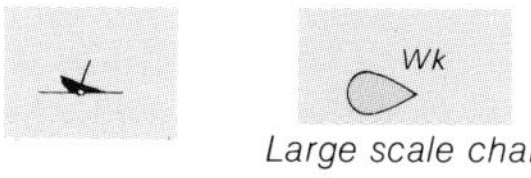

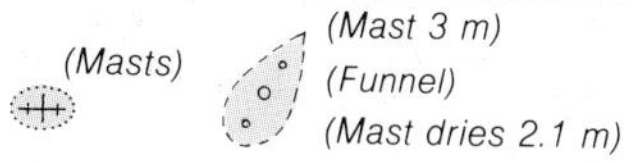

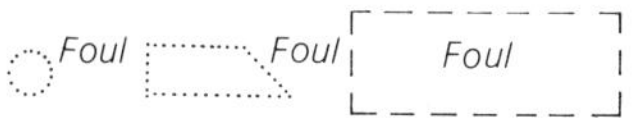

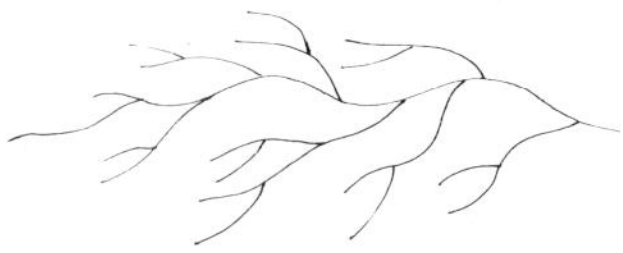

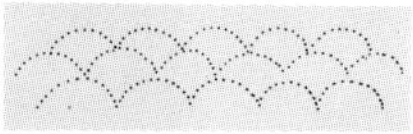

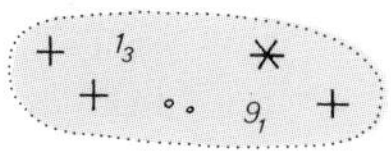

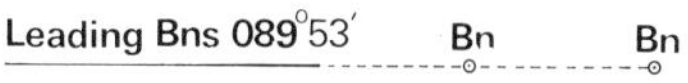

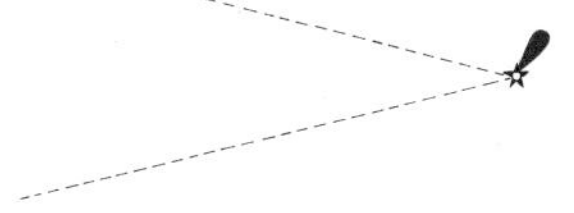

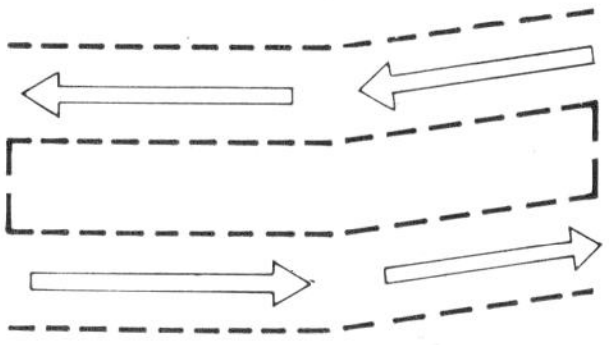

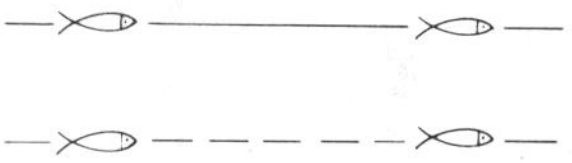

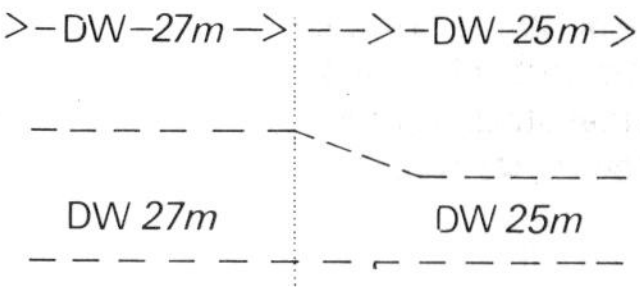

PLATE 4 – PRINCIPAL NAVIGATION LIGHTS AND SHAPES

(*Note*: All vessels seen from starboard beam)

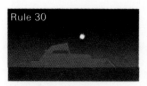

Vessel at anchor
All-round white light: if over 50m, a second light aft and lower

Black ball forward

Not under command
Two all-round red lights, plus sidelights and sternlight when making way

Two black balls vertically

Motor sailing
Cone point down, forward

Divers down
Letter 'A' International Code

Vessel aground
Anchor light(s), plus two all-round red lights in a vertical line

Three black balls in a vertical line

Vessels being towed and towing
Vessel towed shows sidelights (forward) and sternlight
Tug shows two masthead lights, sidelights, sternlight, yellow towing light

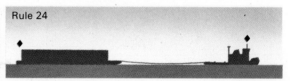

Towing by day – Length of tow more than 200m
Towing vessel and tow display diamond shapes. By night, the towing vessel shows three masthead lights instead of two as for shorter tows

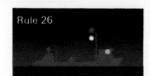

Vessel fishing
All-round red light over all-round white, plus sidelights and sternlight when underway

Fishing/trawling
Two cones point to point, or a basket if fishing vessel is less than 20m

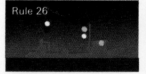

Vessel trawling
All-round green light over all-round white, plus sidelights and sternlight when underway

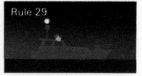

Pilot boat
All-round white light over all-round red; plus sidelights and sternlight when underway, or anchor light

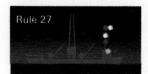

Vessel restricted in her ability to manoeuvre
All-round red, white, red lights vertically; plus normal steaming lights when under way

Three shapes in a vertical line – ball, diamond, ball

Dredger
As left, plus two all-round red lights (or two balls) on foul side, and two all-round green (or two diamonds) on clear side

Constrained by draught
Three all-round red lights in a vertical line, plus normal steaming lights. By day – a cylinder

PLATE 5 – IALA BUOYAGE

Lateral marks

Used generally to mark the sides of well defined navigable channels.

Port Hand marks

Light:
Colour – red
Rhythm – any

Navigable channel

Direction of buoyage

Starboard Hand marks

Light:
Colour – green
Rhythm – any

Cardinal marks

Used to indicate the direction from the mark in which the best navigable water lies, or to draw attention to a bend, junction or fork in a channel, or to mark the end of a shoal.

Lights: Always white

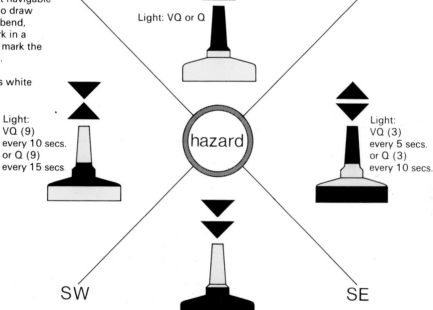

NW — Light: VQ or Q — NE

Light: VQ (9) every 10 secs. or Q (9) every 15 secs.

hazard

Light: VQ (3) every 5 secs. or Q (3) every 10 secs.

SW — SE

Light: VQ (6) + LFl every 10 secs. or Q (6) + LFl every 15 secs.

Other marks

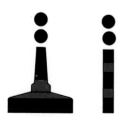

Isolated danger marks

Use: To mark a small isolated danger with navigable water all round.
Light: Colour – white
Rhythm – group flashing (2)

Safe water marks

Use: Mid-channel or landfall.
Light: Colour – white
Rhythm – Isophase, occulting or 1 long flash every 10 seconds.

Special marks

Any shape not conflicting with lateral or safe water marks.
Light: Colour – yellow
Rhythm – different from other white lights used on buoys.

PLATES 6-7 – INTERNATIONAL CODE OF SIGNALS
CODE FLAGS, PHONETIC ALPHABET, MORSE SYMBOLS AND SINGLE-LETTER SIGNALS

Notes:
1. Single letter signals may be made by any method of signalling. Those marked * when made by sound must comply with the *International Regulations for Preventing Collisions at Sea*, Rules 34 and 35.
2. Signals 'K' and 'S' have special meanings as landing signals for small boats with persons in distress.
3. In the phonetic alphabet, the syllables to be emphasised are in italics.

A Alfa (*AL* FAH)

I have a diver down; keep well clear at slow speed

***B Bravo** (*BRAH* VOH)

I am taking in, or discharging, or carrying dangerous goods

***C Charlie** (*CHAR* LEE)

Yes (affirmative or 'The significance of the previous group should be read in the affirmative)

***D Delta** (*DELL* TAH)

Keep clear of me; I am manoeuvring with difficulty

***E Echo** (*ECK* OH)

I am altering my course to starboard

F Foxtrot (*FOKS* TROT)

I am disabled; communicate with me

***G Golf** (*GOLF*)

I require a pilot. When made by fishing vessels operating in close proximity on the fishing grounds it means: I am hauling nets

***H Hotel** (*HOH* TELL)

I have a pilot on board

Code and Answering Pendant

***I India** (*IN* DEE AH)

I am altering my course to port

J Juliett (*JEW* LEE *ETT*)

I am on fire and have dangerous cargo on board: keep well clear of me

K Kilo (*KEY* LOH)

I wish to communicate with you

L Lima (*LEE* MAH)

You should stop your vessel instantly

***M Mike** (*MIKE*)

My vessel is stopped and making no way through the water

N November (NO *VEM* BER)

No (negative or 'The significance of the previous group should be read in the negative'). This signal may be given only visually or by sound

O Oscar (*OSS* CAH)

Man overboard

P Papa (PAH *PAH*)

In harbour: all persons should report on board as the vessel is about to proceed to sea. **At sea**: it may be used by fishing vessels to mean 'My nets have come fast upon an obstruction'

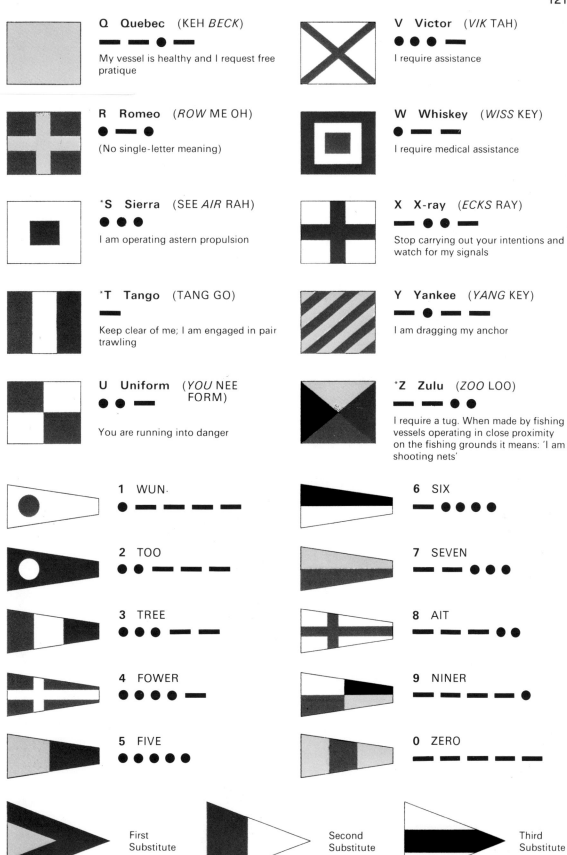

PLATE 8 – INTERNATIONAL PORT TRAFFIC SIGNALS

No	Lights		Main message
1	 ● ● ● (flashing)	Flashing	Serious emergency – all vessels to stop or divert according to instructions
2	 ● ● ●	Fixed or Slow Occulting	Vessels shall not proceed (*Note*: Some ports may use an exemption signal, as in 2a below)
3	● ● ●		Vessels may proceed. One way traffic
4	● ● ○		Vessels may proceed. Two way traffic
5	 ● ○ ●		A vessel may proceed only when she has received specific orders to do so (*Note*: Some ports may use an exemption signal, as in 5a below)
			Exemption signals and messages
2a	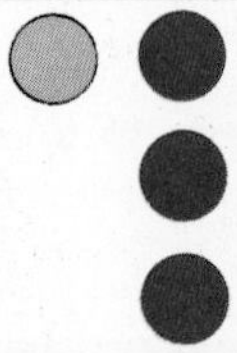 (yellow) ● / ● ●	Fixed or Slow Occulting	Vessels shall not proceed, except that vessels which navigate outside the main channel need not comply with the main message
5a	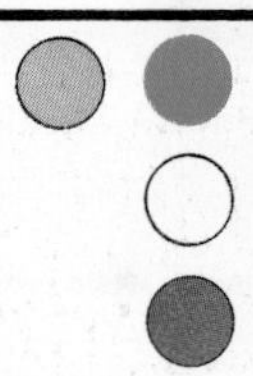 (yellow) ● / ● ○ ●		A vessel may proceed only when she has received specific orders to do so, except that vessels which navigate outside the main channel need not comply with the main message
			Auxiliary signals and messages
	White and/or yellow lights, displayed to the right of the main lights		Local meanings, as promulgated in local port orders

This new system is gradually being introduced, but its general adoption is likely to take many years.

Notes on use:

(1) The main movement message given by a port traffic signal shall always comprise three lights, disposed vertically. No additional light shall be added to the column carrying the main message. (The fact that the main message always consists of three vertical lights allows the mariner to recognise it as a traffic signal, and not lights of navigational significance). The signals may also be used to control traffic at locks and bridges.

(2) Red lights indicate 'Do not proceed'.

(3) Green lights indicate 'Proceed, subject to the conditions stipulated'.
 (For examples, see opposite).
 Note that, to avoid confusion, red and green lights are never displayed together.

(4) A single yellow light, displayed to the left of the column carrying main messages Nos 2 or 5, at the level of the upper light, may be used to indicate that 'Vessels which can safely navigate outside the main channel need not comply with the main message'. This signal is of obvious significance to yachtsmen.

(5) Signals which are auxiliary to the main message may be devised by local authorities. Such auxiliary signals should employ only white and/or yellow lights, and should be displayed to the right of the column carrying the main message. Ports with complex entrances and much traffic may need many auxiliary signals, which will have to be documented; but smaller harbours with less traffic may only need one or two of the basic signals, such as 'Vessels shall not proceed' and 'Vessels may proceed, two way traffic'.

Some signals may be omni-directional – exhibited to all vessels simultaneously: others must be directional, and be shown either to vessels entering or to vessels leaving harbour.

Signal No 5 is based on the assumption that some other means of communication such as VHF radio, signal lamp, loud hailer, or auxiliary signal will be used to inform a vessel that she may specifically proceed.

The 'Serious Emergency' signal must be flashing, at least 60 flashes per minute. All other signals must be either fixed or slow occulting (the latter useful when background glare is a problem). A mixture of fixed and occulting lights must not be used.

Chapter 6
Communications

Contents

6.1 INTERNATIONAL CODE   Page 124
- 6.1.1 International Code of Signals
- 6.1.2 Using the International Code
- 6.1.3 General instructions
- 6.1.4 Procedure signals
- 6.1.5 Selected groups from the International Code
- 6.1.6 Morse code
- 6.1.7 Flag signalling
- 6.1.8 Radiotelephony (RT)
- 6.1.9 Morse code by hand flags, or arms
- 6.1.10 Sound signals

6.2 MISCELLANEOUS SIGNALS   127
- 6.2.1 Radio time signals
- 6.2.2 Pratique messages
- 6.2.3 Distress signals
- 6.2.4 Signals between shore and ships in distress
- 6.2.5 Signals used by SAR aircraft
- 6.2.6 Directing signals by aircraft
- 6.2.7 Port signals
- 6.2.8 Visual storm signals
- 6.2.9 Tide signals

6.3 RADIO COMMUNICATIONS   129
- 6.3.1 Radiotelephones — general
- 6.3.2 Radiotelephones — general provisions
- 6.3.3 RT procedures
- 6.3.4 Coast Radio Stations
- 6.3.5 Link calls
- 6.3.6 Weather information by RT
- 6.3.7 Medical help by RT
- 6.3.8 Flotilla reporting
- 6.3.9 Port operations
- 6.3.10 Distress messages and procedures
- 6.3.11 INMARSAT
- 6.3.12 Navigational Warnings — general
- 6.3.13 Navigational Warnings — Europe
- 6.3.14 NAVTEX
- 6.3.15 UK Coast Radio Stations
- 6.3.16 Irish and Continental Coast Radio Stations

6.4 SHORT NOTES ON FLAG ETIQUETTE   144
- 6.4.1 Ensign
- 6.4.2 Special ensigns
- 6.4.3 Burgee
- 6.4.4 Flag officer's flag
- 6.4.5 Choice of burgee
- 6.4.6 Courtesy ensign
- 6.4.7 House flag
- 6.4.8 Salutes

Communications — introduction

It is useful — and sometimes essential — for yachts to be able to communicate with other vessels and with shore stations. A full description of the various methods and procedures is given in Chapter 6 of *The Macmillan & Silk Cut Yachtsman's Handbook*, including:

The International Code; Morse code; phonetic tables; procedure signals; complements; single-letter signals; flag signalling; Morse code by light; sound signals; radio time signals; pratique messages; port signals; tide signals; radiotelephony; flag etiquette. There are also included colour plates showing national maritime flags, flag etiquette, and burgees of selected yacht clubs.

Here in this Almanac explanations are kept to a minimum, and only sufficient information is given to allow limited use of the International Code in cases of emergency, and to permit yachtsmen to pass simple messages. Particular emphasis is placed on the use of VHF radiotelephones. For further information on all aspects of signalling, including the subjects listed above, reference should be made to Chapter 6 of *The Macmillan & Silk Cut Yachtsman's Handbook*.

6.1 INTERNATIONAL CODE

6.1.1 International Code of Signals

Most communication afloat is based on the *International Code of Signals* (HMSO), which provides particularly where there are language problems.

The Code can be used by flags, flashing light, sound signalling, voice (e.g. loud hailer), radiotelegraphy, radiotelephony, or by hand flags.

Signals consist of: single-letter signals which are very urgent or very common; two-letter signals in the General Section; and three-letter signals starting with 'M' in the Medical Section.

Complements

In principle each signal has a complete meaning, but complements (which are numerals from 0 to 9, added to the letters of a signal) are used to supplement the available groups — expressing variations in the meaning of a basic signal, asking or answering questions on the same subject, or giving more detailed information. For example:

YD   What is the wind expected to do?
YD1   The wind is expected to back.
YD2   The wind is expected to veer.
YD3   The wind is expected to increase.
etc.

Complements should only be used as and when specified in the Code.

6.1.2 Using the International Code

To make full and proper use of the International Code, whereby seamen of nine nations can communicate with each other on matters concerning navigation and safety without knowing any foreign language, it is necessary to hold a copy of the *International Code of Signals*, published by HMSO. It is however possible for yachtsmen to make limited use of the Code so as to pass and receive simple messages, using the information contained in this chapter.

In Plates 6 and 7 on pages 120–121 are shown the International Code flags for the alphabet and for numerals, the phonetic alphabet, the phonetic figure-spelling table, the Morse symbols for letters and numerals, and the single-letter signals of the letters of the alphabet.

6.1.3 General instructions

(1) *Numbers* are signalled: by flags — by numeral pendants; by light — by Morse numerals, but spelt if important; by voice — by phonetic code words, see Plate 7.
(2) *Decimal point* is indicated: by flags — by Answering Pendant; by Light — by AAA; by voice — by the word 'Decimal'.
(3) *Depths* are signalled in figures followed by 'F' for feet, or 'M' for metres.
(4) *Bearings* are signalled in three figures, denoting degrees (True) from 000 to 359.
(5) *Courses* are signalled as for bearings, with figures prefixed by the letter 'C'.
(6) *Dates* are signalled by two, four or six figures preceded by 'D'. The first two figures show day of month, the next two the month of the year, and the final two (when six are signalled) the year.
(7) *Latitude* is signalled by four figures preceded by 'L', and followed where necessary by 'N' or 'S'. The first two figures indicate degrees, and the second two indicate minutes.
(8) *Longitude* is signalled by four (or five) figures preceded by 'G', and followed by 'E' or 'W'. The last two figures show minutes, and the first two (or three) show degrees.
(9) *Distance* is signalled by figures (in nautical miles) preceded by 'R'.
(10) *Speed* is signalled by figures preceded by 'S' (for knots) or by 'V' (kilometres per hour).
(11) *Time* is signalled by 24 hour clock, preceded by 'T' for local time or by 'Z' for GMT.

6.1.4 Procedure signals

Note: A bar over the letters of a signal means that the letters are joined together and made as one symbol.

(1) Signals for voice transmissions (R/T or loud hailer)

Signal	Pronunciation	Meaning
INTERCO	IN-TER-CO	International Code group(s) follow(s).
CORRECTION	KOR-REK-SHUN	Cancel my last word or group. The correct word or group follows.
DECIMAL	DAY-SEE-MAL	Decimal point
STOP	STOP	Full stop

(2) Signals for Morse transmissions by light

$\overline{AA}$ $\overline{AA}$ $\overline{AA}$ etc	Call for unknown station or general call
$\overline{EEEEEE}$ etc	Erase signal.
$\overline{AAA}$	Full stop or decimal point.
TTTT etc	Answering signal.
T	Word or group received.

(3) Signals for flags, radiotelephony and radiotelegraphy transmissions

CQ — Call for unknown station(s) or general call to all stations.

Note: When this signal is used in voice transmission, it should be pronounced in accordance with the letter-spelling table (i.e. Charlie Quebec).

(4) Signals for use where appropriate in all forms of transmission

- **AA** 'all after ...' (used after the 'Repeat' signal (RPT) — see below — means 'Repeat all after ...').
- **AB** 'All before ...' (used after the 'Repeat' signal (RPT) — see below — means 'Repeat all before ...').
- **AR** Ending signal, or End of transmission or signal.
- **AS** Waiting signal or period.
- **BN** 'All between ... and ...' (used after the 'Repeat' signal (RPT) — means 'Repeat all between ... and ...').
- **C** Affirmative — YES or 'The significance of the previous group should be read in the affirmative'.
- **CS** 'What is the name or identity signal of your vessel (or station)?'
- **DE** 'From ...' (used to precede the name or identity signal of the calling station).
- **K** 'I wish to communicate with you' or 'Invitation to transmit'.
- **NO** Negative — NO or 'The significance of the previous group should be read in the negative'. When used in voice transmission the pronunciation should be 'NO'.
- **OK** Acknowledging a correct repetition or 'It is correct'.
- **RQ** Interrogative or 'The significance of the previous group should be read as a question'.
- **R** 'Received' or 'I have received your last signal'.
- **RPT** Repeat signal 'I repeat' or 'Repeat what you have sent' or 'Repeat what you have received'.
- **WA** 'Word or group after ...' (used after the 'Repeat' signal (RPT) means 'Repeat word or group after ...').
- **WB** 'Word or group before ...' (used after the 'Repeat' signal (RPT) means 'Repeat word or group before ...').

Notes:
1. The procedure signals 'C', 'NO' and 'RQ' cannot be used in conjunction with single-letter signals.
2. When these signals are used by voice transmission the letters should be pronounced in accordance with the letter-spelling table, except that 'NO' is pronounced 'NO'.

6.1.5 Selected groups from the International Code

- **AC** I am abandoning my vessel
- **AE** I must abandon my vessel
- **AF** I do not intend to abandon my vessel
- **AN** I need a doctor
- **CB** I require immediate assistance
- **CB4** I require immediate assistance; I am aground
- **CB5** I require immediate assistance; I am drifting
- **CB6** I require immediate assistance; I am on fire
- **CB7** I require immediate assistance; I have sprung a leak
- **CJ** Do you require assistance?
- **CK** Assistance is not (or is no longer) required by me (or vessel indicated).
- **CV** I am unable to give assistance.
- **DX** I am sinking.
- **ED** Your distress signals are understood.
- **EF** SOS/MAYDAY has been cancelled.
- **FA** Will you give me my position?
- **IL** I can only proceed at slow speed.
- **IM** I request to be escorted until further notice.
- **IT** I am on fire.
- **IW** Fire is under control.
- **IX** Fire is gaining.
- **IZ** Fire has been extinguished.
- **JG** I am aground. I am in a dangerous situation.
- **JH** I am aground. I am not in danger.
- **JI** Are you aground?
- **JL** You are running the risk of going aground.
- **JO** I am afloat.
- **JW** I have sprung a leak.
- **JX** Leak is gaining rapidly.
- **KM** I can take you in tow.
- **KN** I cannot take you in tow.
- **LO** I am not in my correct position (to be used by a lightvessel).
- **MG** You should steer course ...
- **NC** I am in distress and require immediate assistance.
- **NG** You are in a dangerous position.
- **NH** You are clear of all dangers.
- **PD** Your navigation light(s) is (are) not visible.
- **PH** You should steer as indicated.
- **PI** You should maintain your present course.
- **PP** Keep well clear of me.
- **QO** You should not come alongside.
- **QP** I will come alongside.
- **QR** I cannot come alongside.
- **QT** You should not anchor. You are going to foul my anchor.
- **RA** My anchor is foul.
- **RB** I am dragging my anchor.
- **RN** My engines are out of action.
- **RY** You should proceed at slow speed when passing me (or vessels making signal).
- **SC** I am under way.
- **SD** I am not ready to get under way.
- **SQ** You should stop or heave to.

UM	The harbour or port is closed to traffic.
UN	You may enter harbour immediately.
UO	You must not enter harbour.
UW	I wish you a pleasant voyage.
YU	I am going to communicate with your station by International Code.
YV	The groups which follow are from the International Code of Signals.
ZD2	Please report me to Lloyd's London.
ZL	Your signal has been received but not understood.
ZM	You should send (or speak) more slowly.

6.1.6 Morse code

When making Morse, it is most important to get the right rhythm and spacing. If a dot is taken as the unit of time, the correct spacing is as follows:

Dot	1 unit
Dash	3 units
Space between each dot/dash in a letter	1 unit
Space between each letter or symbol	3 units
Space between each word or group	7 units

The most likely method whereby a yachtsman will use the Morse code is by light. It helps to have a good light, with a proper flashing key, or trigger.

6.1.7 Flag signalling

The International Code flags (see Plates 6 and 7) consist of 26 alphabetical flags, 11 pendants (numerals 0–9 plus the Answering Pendant or Code Flag), and three triangular flags — the First, Second and Third Substitutes.

Some definitions are important. *Group* — one or more continuous letters and/or numerals comprising a signal. *Hoist* — one or more groups on one halyard. *At the dip* — a signal half-hoisted. *Close up* — a signal fully hoisted. *Tackline* — a line separating two groups. *Superior* — a flag or group above another. *Inferior* — a flag or group below another. *Class* — whether a flag is alphabetical or numeral.

The Answering Pendant is used to answer or acknowledge signals from another vessel. It may also be used as a decimal point.

Substitutes allow for the repetition of one or more letters (or numerals) within a group. The First Sub. repeats the first flag of the group in the class immediately superior to it. The Second Sub. repeats the second flag of that class, and the Third Sub. repeats the third.

A substitute can only repeat a flag of the same class as that immediately preceding it. The Answering Pendant used as a decimal point is disregarded in deciding which substitute to use.

Procedure
The sending ship hoists the identity signal of the ship she is calling: or she hoists the group 'VF' ('You should hoist your identity signal') or 'CS' ('What is the name or identity signal of your vessel/station') — at the same time hoisting her own identity signal.

The sending ship then hoists her message, and when sighted the receiving ship hoists her Answering Pendant at the dip — and close up when the signal has been understood. The procedure is repeated for subsequent hoists.

6.1.8 Radiotelephony (RT)

Plain language is normally used for RT (see 6.3.3), but in the event of language difficulties use the International Code and the following procedure. Letters and figures are spelt in accordance with the spelling tables, see Plates 6 and 7.

(1) *Method of calling.* Call sign or name of station called, not more than three times; the group 'DELTA ECHO'; and the call sign or name of the calling station, not more than three times. After contact is made, the call sign or name need not be sent more than once.

(2) *Reply to call.* Call sign or name of calling station; the group 'DELTA ECHO'; and the call sign or name of the station called.

(3) *Code groups.* The word 'INTERCO' indicates that International Code groups follow. Plain language may be used for names. 'YANKEE ZULU' means 'The words which follow are in plain language'.

(4) For other procedure signals see 6.1.4 (1), (3) and (4).

6.1.9 Morse Code by hand flags, or arms

Exceptionally, it may be useful to signal Morse by hand flags or arms. A dot is made by extending both flags (arms) above the head, and a dash by extending them horizontally at shoulder level. Between dots and dashes the flags (arms) are brought in front of the chest. To separate letters, groups or words the flags (arms) are extended downwards 45° away from the body. Circular motion of the flags (arms) indicates the erase signal if made by the transmitting station, or a request for repetition if made by the receiving station.

A station wishing to communicate by this method sends 'K2' or makes the general Morse call 'AA AA AA'. The station called should make the answering signal (TTTTTT etc) or, if unable to communicate by this method the signal 'YS2' by any means. Other procedure signals as in 6.1.4(4).

6.1.10 Sound signals

The International Code may be sent by sound signal (e.g. whistle, siren, foghorn) but the method is slow, and if misused can cause confusion. In poor visibility it should be reduced to a minimum. Signals other than the single-letter ones should only be used in emergency, and never where there is other traffic around. The signals should be made slowly and distinctly. They may be repeated, if necessary, but at sufficiently long intervals to ensure that no confusion can arise. The single-letter signals of the Code marked by an asterisk, when made by sound, must conform with the *International Regulations for Preventing Collisions at Sea* (Rules 34 and 35), for which see 2.1.7.

6.2 MISCELLANEOUS SIGNALS

6.2.1 Radio time signals
The BBC broadcast time signals at the local times and on the frequencies indicated in the table below. The start of the final, longer pulse marks the minute.
BBC Radio 1 1053 1089 kHz
BBC Radio 2 693 909 kHz, 88–92 MHz
BBC Radio 3 1215 kHz, 90–97 MHz
BBC Radio 4 198 603(Tyneside) 720(N Ireland & London) 1449(Aberdeen) 1485(Carlisle) 756 (Redruth) 774(Plymouth) kHz, 92.7–94.7 MHz

Local time	Mon–Fri Radio	Sat Radio	Sun Radio
0000	1,2	1,2	1,2
0500	2	2	
0600	1,2,4	1,4	1,4
0700	2,3,4	2,3,4	2,4
0800	1,2,3,4	2,4	2,4
0900	3,4	3,4	3,4
1000	4		
1100	4		
1200	4	4	
1300	2,4	1,2,4	4
1400	4	4	4
1500	4	4	
1600	4 (not Mon)		4
1700	2,4		1,2,4
1800	2	2	
1900	2,4		
1930		1,2	
2100			4
2200	1,2,4		

6.2.2 Pratique messages
All yachts, whether carrying dutiable stores or not, arriving in the United Kingdom from abroad (including the Channel Islands) are subject to Customs, Health and Immigration requirements — see Customs Notice No 8, summarised in Chapter 2 (2.4). The following Health Clearance Messages (International Code) apply:

Q or ZS   My vessel is 'healthy' and I request free pratique. (Note: Flag 'Q' to be flown on entering UK waters, illuminated at night).
QQ   (Flag 'Q' over First Substitute, or by night a red light over a white light). I require health clearance.
ZT   My Maritime Declaration of Health has negative answers to the six health questions.
ZU   My Maritime Declaration of Health has a positive answer to question(s) ... indicated.
ZW   I require Port Medical Officer.
ZY   You have health clearance.
ZZ   You should proceed to anchorage for health clearance.
AM   Have you a doctor?

6.2.3 Distress signals
A complete list of distress signals, as in Annex IV of the *International Regulations for Preventing Collisions at Sea*, is given in Chapter 2 (2.1.6). Those which are more appropriate for yachts to use are described in 8.1.4. Full details of RT distress messages and procedures, control of distress traffic on RT, the Urgency Signal and the Safety Signal are given in 6.3.10.

Distress signals must only be used when the boat is in serious and immediate danger, and help is urgently required. For lesser emergencies use 'V' (Victor) International Code — 'I require assistance', or one of the groups shown in 6.1.5.

6.2.4 Signals between shore and ships in distress
The following are used if a vessel is in distress or stranded off the coast of the United Kingdom:

(1) Acknowledgment of distress signal
By day: Orange smoke signal, or combined light and sound signal consisting of three single signals fired at about one minute intervals.
By night: White star rocket consisting of three single signals at about one minute intervals.

(2) Landing signals for small boats
'This is the best place to land' — Vertical motion of a white flag or arms (or white light or flare by night), or signalling 'K' (— · —) by light or sound. An indication of direction may be given by placing a steady white light or flare at a lower level.

'Landing here is highly dangerous' — Horizontal motion of a white flag or arms extended horizontally (or white light or flare by night), or signalling 'S' (· · ·). In addition, a better landing place may be signalled by carrying a white flag (or flare, or light), or by firing a white star signal in the direction indicated; or by signalling 'R' (· — ·) if the better landing is to the right in the direction of approach, or 'L' (· — · ·) if it is to the left.

(3) Signals for shore life-saving apparatus
'Affirmative' or specifically 'Rocket line is held', 'Tail block is made fast', 'Hawser is made fast', 'Man is in breeches buoy' or 'Haul away' — Vertical motion of a white flag or the arms (or of a white light or flare).

'Negative' or specifically 'Slack away' or 'Avast hauling' — Horizontal motion of a white flag or the arms (or of a white light or flare).

(4) Warning signal
'You are running into danger' — International Code signal 'U' (· · —) or 'NF'.

Note: Attention may be called to the above signals by a white flare, a white star rocket, or an explosive signal.

6.2.5 Signals used by SAR aircraft
A searching aircraft normally flies at about 3000–5000ft (900–1500m), or below cloud, firing a green Very light every five or ten minutes and at each turning point. On seeing a green flare, a yacht in distress should take the following action:
(1) Wait for the green flare to die out.
(2) Fire one red flare.
(3) Fire another red flare after about 20 seconds (this enables the aircraft to line up on the bearing).
(4) Fire a third red flare when the aircraft is overhead, or appears to be going badly off course.

6.2.6 Directing signals by aircraft
(1) To direct a yacht towards a ship or aircraft in distress — The aircraft circles the yacht at least once; it then crosses low ahead of the yacht, opening and closing the throttle or changing the propeller pitch. Finally it heads in the direction of the casualty.
(2) To indicate that assistance of the yacht is no longer required — The aircraft passes low, astern of the yacht, opening and closing the throttle or changing the propeller pitch.

6.2.7 Port signals
For signals for individual harbours, see Chapter 10. On the Continent traffic signals are to some extent standardised — see 10.14.7 for France, 10.19.7 for the Netherlands, and 10.20.7 for the Federal Republic of Germany.

A new system of International Port Traffic Signals was introduced in 1983, and is illustrated in Plate 8 on page 122. Although this new system has been adopted in a few ports, it is likely to be some time before it is used widely.

VHF channels for port operations are shown for harbours in Chapter 10. Call on the indicated working channel. The following International Code groups refer to port operations.

UH Can you lead me into port?
UL All vessels should proceed to sea as soon as possible owing to danger in port.
UM The harbour (or port indicated) is closed to traffic.
UN You may enter harbour immediately (or at time indicated).
UO You must not enter harbour.
UP Permission to enter harbour is urgently requested. I have an emergency case.
UQ You should wait outside the harbour (or river mouth). UQ 1 You should wait outside the harbour until daylight.
UR My estimated time of arrival (at place indicated) is (time indicated). UR 1 What is your estimated time of arrival (at place indicated)?
RZ 1 You should not proceed out of harbour/anchorage.
RV 2 You should proceed into port.

The following notes apply to individual countries:

British Isles
In an emergency, an Examination Service might be instituted for certain ports, when the following signals would apply:

By day	By night	Meaning
Three red balls shown vertically	Three flashing red lights vertically	Entry to port prohibited
—	Three green lights shown vertically	Entry to port permitted
A blue flag	Red, green, red lights shown vertically	Movement of shipping within port or anchorage prohibited

Vessels of the Examination Service wear a special flag, with a blue border and a square in the centre — the top half white and the bottom half red.

France
See 10.14.7 for simplified and full codes of traffic signals. The simplified code is used in some ports where there is less traffic.

Netherlands
In the event of government control of entry to Dutch harbours, the following signals indicate that

entry is prohibited, and that a yacht should proceed towards the vessel flying the same signal.

By day
Three red balls, disposed vertically, or
Two cones points together over a ball

By night
Three red lights, disposed vertically, or
Three lights, disposed vertically, green over red over white

For other signals see 10.19.7.

Federal Republic of Germany
See 10.20.7.

6.2.8 Visual storm signals
Official storm signals, as previously displayed by Coastguard stations etc, are now discontinued in the British Isles, but signals may be shown in a few places by private arrangement. These should be treated with caution, in case they are not up to date. Signals are shown when a gale is expected within 12 hours, or is already blowing, in the adjacent sea area. The signal is lowered when the wind is below gale force, if a renewal is not expected within six hours.

The signals consist of black cones. The North cone, point upwards, indicates gales from a point north of the east–west line. The South cone, point downwards, indicates gales from a point south of the east–west line. A few stations display night signals consisting of a triangle of lights.

International System
An International System of visual storm signals is used in France, the Netherlands, and the Federal Republic of Germany. This is illustrated in 10.14.7.

6.2.9 Tidal signals
Tidal signals are shown for individual harbours in Chapter 10. There is no standard system for British ports, nor for the Netherlands and West Germany. France uses a system for indicating whether the tide is rising or falling, and to show the height of the tide above chart datum by a combination of cones, cylinders and balls by day, and by green, red and white lights at night. This is illustrated in 10.14.7.

6.3 RADIOTELEPHONY (RT)

6.3.1 Radiotelephones — general
Many yachts are fitted with radiotelephones, mostly Very High Frequency (VHF) sets with ranges of about 20 miles. Medium Frequency (MF) sets give much greater ranges, but must be Single Sideband (SSB). Double Sideband (DSB) transmissions are prohibited except for emergency transmissions on 2182 kHz, the international MF distress frequency. Full details of the various licences and operating procedures are given in Chapter 6 of *The Macmillan & Silk Cut Yachtsman's Handbook*. The brief notes which follow cover only the most important points.

A Ship Licence is required for the set. Apply to the Department of Trade and Industry, Marine Licensing Section, Room 613(c), Waterloo Bridge House, Waterloo Road, London, SE1 8UA. The same authority will give approval for the use of VHF Ch M, for communication with yacht clubs and marinas on this frequency (157.85 MHz).

A Certificate of Competence and an Authority to Operate are required by the person in charge of a set. For yachtsmen this is likely to be the Certificate of Competence, Restricted VHF Only. The Royal Yachting Association (RYA) is responsible for the conduct of this examination. Details of the syllabus and examination are in RYA booklet G26, available (with application form for the examination) from the RYA.

6.3.2 Radiotelephones — general provisions
For details see the *Handbook for Radio Operators* (Lloyds of London Press). Briefly: operators must not divulge the contents of messages heard; stations must identify themselves when transmitting; except in cases of distress, coast radio stations control communications in their areas; at sea a yacht may call other vessels or shore stations, but messages must not be sent to an address ashore except through a coast radio station; in harbour a yacht may not use inter-ship channels except for safety, and may only communicate with the local Port Operations Service, British Telecom coast stations, or with stations on Ch M; operators must not interfere with the working of other stations — before transmitting, listen to see that the channel is free; it is forbidden to transmit unnecessary or superfluous signals; priority must be given to distress calls; the transmission of bad language is forbidden; a log must be kept, recording all transmissions, etc.

VHF radio
Very High Frequency (VHF) radio has a range slightly better than the line of sight between the aerials. It pays to fit a good aerial, as high as possible. Maximum power output is 25 watts, and a lower power (usually one watt) is used for short ranges. Most UK coast stations and many harbours now have VHF. So do the principal Coastguard stations and other rescue services.

Marine VHF frequencies are in the band 156.00–174.00 MHz. Ch 16 (156.80 MHz) is for distress and safety purposes, and for calling and answering. Once contact has been made the stations concerned must switch to a working channel, except for safety matters. Yachts at sea are encouraged to listen on Ch 16.

Basic VHF sets are 'simplex', transmitting and receiving on the same frequency, so that it is not possible to speak and listen simultaneously. 'Semi-duplex' sets transmit and receive on different frequencies, while fully 'duplex' sets can do this simultaneously so that conversation is normal.

There are three main groups of frequencies, but certain channels can be used for more than one purpose. They are shown in order of preference.
(1) *Public correspondence* (through coast radio stations). All can be used for duplex. Ch 26, 27, 25, 24 23, 28, 04, 01, 03, 02, 07, 05, 84, 87, 86, 83, 85, 88, 61, 64, 65, 62, 66, 63, 60, 82.
(2) *Inter-ship*. These are all simplex. Ch 06, 08, 10, 13, 09, 72, 73, 69, 67, 77, 15, 17.
(3) *Port operations*. Simplex: Ch 12, 14, 11, 13, 09, 68, 71, 74, 10, 67, 69, 73, 17, 15. Duplex: Ch 20, 22, 18, 19, 21, 05, 07, 02, 03, 01, 04, 78, 82, 79, 81, 80, 60, 63, 66, 62, 65, 64, 61, 84.

Ch M (157.85 MHz) may be specially authorised for communication with yacht clubs and marinas. Ch 67 (156.375 MHz) is operated in the UK by principal Coastguard stations as the Small Craft Safety Channel, accessed via Ch 16 (see 8.2.2).

Ch 70 (previously an inter-ship channel) is now reserved exclusively for digital selective calling for distress and safety purposes.

MF and HF radio
Medium Frequency (MF) radiotelephones can have ranges of 200 miles or more, while with High Frequency (HF) communication can be worldwide. But MF/HF sets are more complex and more expensive than VHF and are subject to more regulations. Except for distress calls, they must be single sideband (SSB).

The international MF distress frequency is 2182 kHz. Silence periods are observed on this frequency for three minutes commencing every hour and half-hour. During these silence periods only Distress and Urgency messages may be transmitted (see 6.3.10 below).

6.3.3   RT procedures
Except for distress, urgency or safety messages, communications between a ship and a coast station are controlled by the latter. Between two ship stations, the station called controls the working. A calling station must use a frequency on which the station called is keeping watch. After making contact, comunication can continue on an agreed working channel. For VHF the name of the station called need normally only be given once, and that of the calling station twice. Once contact is made, each name need only be transmitted once. If a station does not reply, check the settings on the transmitter and repeat the call at three-minute intervals (if the channel is clear).

Prowords
It is important to understand the following:

ACKNOWLEDGE	'Have you received and understood?'
CONFIRM	'My version is ... is that correct?'
CORRECTION	'An error has been made; the correct version is ...'
I SAY AGAIN	'I repeat ... (e.g. important words)'
I SPELL	'What follows is spelt phonetically'
OUT	End of work
OVER	'I have completed this part of my message, and am inviting you to reply'
RECEIVED	'Receipt acknowledged'
SAY AGAIN	'Repeat your message (or part indicated)'
STATION CALLING	Used when a station is uncertain of the identity of a station which is calling

Attention is also called to other procedure signals which can be used for RT, as shown in 6.1.4.

Before making a call, decide exactly what needs to be said. It may help to write the message down. Speak clearly and distinctly. Names or important words can be repeated or spelt phonetically.

For a position, give latitude and longitude, or the yacht's bearing and distance from a charted object. For bearings use 360° True notation. For times use 24 hour notation, and specify GMT, BST etc.

6.3.4   Coast Radio Stations
In the United Kingdom these are operated by British Telecom, and details are shown in 6.3.15. They control communications, and link ship stations with the telephone network ashore. At fixed times they transmit traffic lists, navigational warnings and weather bulletins. They play an important role in distress, urgency, safety and medical messages — see 6.3.10.

The major UK coast radio stations operate on both MF and VHF, and they have remotely controlled VHF stations to extend their VHF coverage.

MF calls to UK coast stations are made and answered on paired working frequencies. Stations listen on 2182 kHz for Distress, Urgency and Safety purposes, and in case of difficulties in calling.

Except in emergency, VHF calls to all UK coast stations should be made on a working channel. Monitor each channel to locate a free one — with no carrier noise, speech or engaged signal (a series of pips). When possible avoid the broadcast channel, particularly at scheduled broadcast times. Having located a free channel, the initial call must last at least five seconds to activate equipment at the coast station. The engaged signal will then be heard, meaning that the call has been accepted. Wait for the operator to speak. If you do not activate the station's transmitter, you may be out of range. Try another station, or call when closer.

French and German coast stations are called on VHF on the appropriate working channel. In Belgium and the Netherlands this depends on the position of the yacht. VHF calls to Scheveningen Radio should last several seconds and state the calling channel. A four-tone signal indicates temporary delay.

6.3.5 Link calls

A yacht can be connected, via a coast radio station, into the telephone network. It is possible to make personal calls to certain countries and collect (transferred charge) calls to the UK. World-wide accounting for calls is arranged by quoting the 'Accounting Authority Indicator Code' (AAIC). For British yachts this is 'GB 14', followed by the yacht's call sign. Yacht owners making VHF link calls to the UK through a British coast station can have the charge transferred to their home telephone bill by quoting the AAIC 'YTD', followed by the UK telephone number to which the call is to be charged. Or payment may be by British Telecom credit card, or transferred charge.

To make a link call proceed as described in 6.3.4. Have the following details ready for the operator: yacht's name, call sign, number required, accounting (AAIC) code, type of traffic (e.g. transfer charge call, YTD call, personal call, credit card call, telegram etc). Call the station by name (e.g. NITON RADIO) not more than three times, THIS IS . . . (the name of the yacht or her radio call sign, not more than three times). Here is an example when reception is good:

'NITON RADIO. THIS IS YACHT SEABIRD, YACHT SEABIRD (GOLF OSCAR ROMEO INDIA). LINK CALL PLEASE'.

Once contact is established names need not be repeated. If, exceptionally, the initial call is made on Ch 16, you will be told the working channel.

The operator makes the connection. When channels are congested calls may be limited to six minutes. Timing is automatic.

Data may be exchanged on VHF over the telephone network by yachts suitably equipped with V22bis specification modem units. Procedures and charges are as for an ordinary VHF call except that data modems are switched into circuit.

Link call to a yacht

For a link call to a yacht, telephone the appropriate coast station; the number is shown against each station in 6.3.15, or dial 100 and ask for 'Ships Telephone Service' followed by the name of the required station. The information needed by the coast station is; the vessel's name (and call sign if known), the Selcal number (if known and if the yacht is so fitted), the caller's telephone number, and the voyage particulars. Replace the receiver and await a call from the coast station when communication is established.

6.3.6 Weather information by RT

Sources of weather information are given in 7 (7.2) and it is possible, for example, to make a link call to a Weather Centre through the nearest coast radio station. Reports on actual weather can be obtained by link calls to Coastguard Stations and lighthouses, as listed in 7.2.16.

MRCCs and MRSCs broadcast strong wind warnings on receipt, and local forecasts about every four hours — on Ch 67 after announcement on Ch 16, see 7.2.15. They will also give weather or other safety information on request.

6.3.7 Medical help by RT

Medical advice can be obtained through any UK coast radio station. If medical assistance is required (a doctor, or a casualty to be off-lifted) the request will be passed to the Coastguard. In either case the messages are passed free of charge. Where appropriate, the Urgency Signal PAN PAN (see 6.3.10) may be used.

6.3.8 Flotilla reporting schemes

Where numbers of yachts are required to report on a regular basis to a control centre, BTI offer two schemes. (a) Where vessels are given a window during which to report, and (b) Where vessels report at fixed times. Details from BTI Maritime Radio, 43 Bartholomew Close, London, EC1A 7HP.

6.3.9 Port operations

Many harbours now have VHF RT facilities, while a few also operate on MF (mostly for pilotage). The nominated channels must only be used for messages concerning port operations or (in emergency) the safety of persons, and not for public correspondence. For details of harbour RT information in Chapter 10, see 10.0.1. In a busy port a yacht can learn a great deal by merely listening on the right channel.

6.3.10 Distress messages and procedures

Anybody in the crew should be able to pass a distress message, should this be necessary. But this must only be done if the yacht or a person is in serious danger and requires immediate assistance. The procedure is:

Check main battery switch ON
Switch set ON, and turn power selector to HIGH
Tune to VHF Ch 16 (or 2182 kHz for MF)
If alarm signal generator fitted, operate for at least 30 seconds
Press 'transmit' button, and say slowly and distinctly:

    MAYDAY MAYDAY MAYDAY
    THIS IS (name of boat, spoken three times)
    MAYDAY (name of boat, spoken once)
    MY POSITION IS (latitude and longitude, or true bearing and distance from a known point)
    Nature of distress (whether sinking, on fire etc)
    Aid required
    Number of persons on board
    Any other important, helpful information (e.g. if the yacht is drifting, whether distress rockets are being fired)
    OVER

The yacht's position is of vital importance, and should be repeated if time allows. On completion of the distress message, release the 'transmit' button and listen. In coastal waters an immediate acknowledgment should be expected, in the following form:

    MAYDAY (name of station sending the distress message, spoken three times)
    THIS IS ... (name of the station acknowledging, spoken three times)
    RECEIVED MAYDAY

If an acknowledgment is not received, check the set and repeat the distress call. For 2182 kHz the call should be repeated during the three-minute silence periods which commence every hour and half-hour.

A yacht which hears a distress message from a vessel in her immediate vicinity and which is able to give assistance should acknowledge accordingly, but only after giving an opportunity for the nearest shore station or some larger vessel to do so.

If a yacht hears a distress message from a vessel further away, and it is not acknowledged, she must try to pass on the message in this form:

    MAYDAY RELAY, MAYDAY RELAY, MAYDAY RELAY.
    This is (name of vessel re-transmitting the distress message, spoken three times)
    Followed by the intercepted message.

Control of distress traffic

A distress (MAYDAY) call imposes general radio silence, until the vessel concerned or some other authority (e.g. the nearest MRCC, MRSC or coast station) cancels the distress.

If necessary the station controlling distress traffic may impose radio silence in this form:

    SEELONCE MAYDAY, followed by its name or other identification, on the distress frequency.

If some other station nearby believes it necessary to do likewise, it may transmit:

    SEELONCE DISTRESS, followed by its name or other identification.

When appropriate the station controlling distress traffic may relax radio silence as follows:

    MAYDAY
    HELLO ALL STATIONS (spoken three times)
    THIS IS (name or identification)
    The time
    The name of the station in distress
    PRU-DONCE

When all distress traffic has ceased, normal working is authorised as follows:

    MAYDAY
    HELLO ALL STATIONS (spoken three times)
    THIS IS (name or identification)
    The time
    The name of station which was in distress
    SEELONCE FEENEE

Urgency Signal

This consists of the words PAN PAN, spoken three times, and indicates that the station has a very urgent message concerning the safety of a ship or person. Messages prefixed by PAN PAN take priority over all traffic except distress, and are sent on VHF Ch 16 or on 2182 kHz. The Urgency Signal is appropriate when urgent medical advice or attention is needed, or when someone is lost overboard. It should be cancelled when the urgency is over.

Safety Signal

This consists of the word SÉCURITÉ (pronounced SAY-CURE-E-TAY) spoken three times, and indicates that the station is about to transmit an important navigational or meteorological warning. Such messages usually originate from a coast station, and are transmitted on a working frequency after an announcement on the distress frequency.

6.3.11 INMARSAT

World-wide satellite communication (Satcom) is provided by the International Maritime Satellite Organisation (INMARSAT), through satellites in geostationary orbits above the equator over the Atlantic, Pacific and Indian Oceans. These satellites are the links between the Coast Earth Stations (CESs), operated by organisations such as British Telecom, and the on-board terminals called Ship Earth Stations (SESs). Satcom is more reliable and gives better reception than HF SSB radio.

A CES connects the satellite system with the landbased communication network. A message to a ship originating on land is transmitted by the CES to one of the satellites, and thence to the ship. A call from a ship is received via the satellite by the CES, which then transmits it to its destination over landbased networks.

Standard 'A' terminals have a large antenna about 4ft (1.2m) in diameter inside a radome, and are only suitable for larger vessels. They provide direct dialling facilities, Telex, data and facsimile transmission, and speedy connection to HM Coastguard's MRCC at Falmouth.

New, Standard 'C' terminals are very much smaller and provide Satcom facilities for transmitting and receiving data or text (but not voice) in even a small yacht.

6.3.12 Navigational Warnings — general

The world-wide Navigational Warning Service covers 16 sea areas (NAVAREAS) numbered I–XVI, each with a country nominated as Area Co-ordinator responsible for issuing long range warnings. These are numbered consecutively through the year and are transmitted in English and in one or more other languages at scheduled times by WT. Other forms of transmission (RT, radiotelex and facsimile) may also be used. Warnings cover items such as failures or changes to navigational aids, wrecks and navigational dangers of all kinds, SAR operations, cable or pipe laying, naval exercises etc.

Within each NAVAREA, Coastal Warnings and Local Warnings may also be issued. Coastal Warnings, up to 100 or 200 n miles offshore, are broadcast in English and in the national language by coast radio stations. Local Warnings are issued by harbour authorities in the national language.

6.3.13 Navigational Warnings — Europe

United Kingdom

The United Kingdom (together with Northern Europe and Scandinavia) come within NAVAREA I, with Britain as the Area Co-ordinator for long range navigational warnings which are broadcast by Portishead Radio (GKA). Messages are numbered in sequence, and the text is published in the weekly *Notices to Mariners* together with a list of warnings still in force.

NAVTEX warnings (se 6.3.14) are broadcast by Niton, Cullercoats and Portpatrick Radio for the areas shown in Fig.6(1).

Coastal Warnings are broadcast by RT at scheduled times from coast radio stations, for the Sea Regions lettered A–N, in Fig.6(2). Important warnings are broadcast at any time on the distress frequencies of 500kHz, 2182kH and VHF Ch 16.

Local warnings from harbour authorities are broadcast by nearby coast radio stations. HM Coastguard broadcasts local warnings for inshore waters outside of harbour limits, on VHF Ch 67 after an announcement on Ch 16.

Vessels which encounter dangers to navigation should notify other craft and the nearest coast radio station, prefacing the message by the safety signal (see 6.3.10).

France

Long range warnings are broadcast in English and French for NAVAREA II, which includes the west coast of France, by St Lys Radio. The north coast comes within NAVAREA I.

AVURNAVS (AVis URgents aux NAVigateurs) are coastal and local warnings issued by regional authorities:

(1) Avurnavs Brest for the west coast of France and the western Channel (Spanish frontier to Mont St Michel).

(2) Avurnavs Cherbourg for the eastern Channel and the North Sea (Mont St Michel to the Belgian frontier).

Avurnavs are broadcast by the appropriate coast radio station, urgent ones on receipt and at the end of the next silence period as well as at scheduled times. RT warnings are prefixed by 'Sécurité Avurnav', followed by the name of the station.

Belgium

Navigational warnings are broadcast by Oostende Radio at scheduled times (see 6.3.16) on 2761kHz MF, 518kHz (NAVTEX) and on Ch 27.

Netherlands

Navigational warnings are broadcast by Scheveningen Radio at scheduled times (see 6.3.16) on 1862kHz (Nes), 1890kHz and 2824kHz MF, and on 518kHz (NAVTEX).

Federal Republic of Germany

Navigational warnings are broadcast by Norddeich Radio on 2614kHz on receipt, after the next silence period, and at scheduled times. Broadcasts commence with the safety signal, the words *Nautische Warnnachricht*, and the serial number; they are in English and German. Decca Warnings for the German and Frisian Islands chains (*Decca Warnnachricht*) are included. Dangers to navigation should be reported to Seewarn Cuxhaven.

6.3.14 NAVTEX

NAVTEX provides navigational and meteorological warnings and other safety information by automatic print-outs from a dedicated receiver, and is a component of the International Maritime Organization (IMO) Global Maritime Distress and Safety System (GMDSS) due for world-wide implementation by August 1991.

At present broadcasts are all in English on a single frequency of 518kHz, with excellent coverage of coastal waters in areas such as NW Europe where the scheme originated. Interference between stations is avoided by time sharing and by limiting the range of transmitters to about 300 n miles, so that three stations cover the United Kingdom. IMO is expected to make a second NAVTEX channel available for broadcasts in the local national language.

The use of a single frequency allows a simple receiver with a printer using cash-roll paper, although some yacht receivers have a video screen. The user programmes the receiver for the vessel's area and for the types of messages required. For example, if Decca is not carried Decca Warnings can be rejected. The receiver automatically rejects messages which are corrupt or ones that have already been printed.

Each message is pre-fixed by a four character group. The first character is the code letter of the transmitting station (in the United Kingdom S for Niton, G for Cullercoats and O for Portpatrick). The second character indicates the category of the message as in the code opposite. The third and fourth are message serial numbers, from 01 to 99. The serial number 00 denotes urgent traffic such as gale warnings and SAR alerts, and messages with this prefix are always printed.

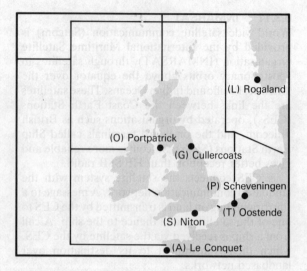

Fig.6(1) NAVTEX areas around British Isles.

Message categories
A Navigational warnings
B Gale warnings
C Ice reports (unlikely to apply in UK)
D Search and Rescue information
E Weather forecasts
F Pilot Service messages
G Decca messages
H Loran-C messages
I Omega messages
J Satnav messages
L Oil and gas rig information (UK trial)
Z No messages on hand at scheduled time

Information in a NAVTEX broadcast applies only to the area for which the broadcast station is responsible, as shown in Fig.6(1). A user may accept messages from one or more stations.

NAVTEX stations

The table opposite shows the NAVTEX stations in NAVAREAS I and II, with their station identity codes and transmission times (GMT). The times of weather bulletins for the British stations are shown in italics. (P) indicates that the station is provisional or projected.

The diagram above shows the stations and their areas around the British Isles.

NAVAREA I

R – **Reykjavik,** Iceland	0318	0718	1118	1518	1918	2318
B – **Bodø,** Norway	0018	0418	0900	1218	1618	2100
J – **Stockholm,** Sweden	0330	0730	1130	1530	1930	2330
H – **Härnösand,** Sweden	0000	0400	0800	1200	1600	2000
U – **Tallin,** USSR	0030	0430	0830	1230	1630	2030
P – **Scheveningen,** Netherlands	0348	0748	1148	1548	1948	2348
T – **Oostende,** Belgium	0248	0648	1048	1448	1848	2248
G – **Cullercoats,** UK	0048	0448	*0848*	1248	1648	*2048*
S – **Niton,** UK	0018	0418	*0818*	1218	1618	*2018*
O – **Portpatrick,** UK	0130	0530	*0930*	1330	1730	*2130*
V – **Vardø,** Norway	0300	0700	1100	1500	1900	2300
L – **Rogaland,** Norway	0148	0548	0948	1348	1748	2148

NAVAREA II

A – **Le Conquet,** France (P)	0000	0400	0800	1200	1600	2000
R – **Lisboa,** Portugal (P)	0250	0650	1050	1450	1850	2250
F – **Azores** (P)	0050	0450	0850	1250	1650	2050
D – **Finisterre,** Spain (P)	0030	0430	0830	1230	1630	2030
I – **Islas Canarias,** Spain (P)	0100	0500	0900	1300	1700	2100

6.3.15 UK Coast Radio Stations

Coast Station: Name Position	MF (Medium Range) RT				VHF Ch	Traffic List Times (GMT)	Navigation Warning Times (GMT)	Weather Bulletin Times (GMT)	Gale Warning Times (GMT) *	Decca Warning Times (GMT) for Chains
	Distress, Urgency, Safety (kHz)	Calling and Working Paired Frequencies (kHz)								
		Channel	Coast Station							
			Transmit	Receive						
Land's End Radio 50°07'N 05°40'W Tel: 0736 87 1363	2182	W X	2782 3610	2002 3373	16 27 88 85 64	Broadcasts on 2670 kHz and Ch 27 and Ch 64 at: 0303 1503 0033 0303 0703 1703 0433 0803 0903 0903 1903 0833 1503 1103 2103 1233 2103 1303 2303 1633 2003 2033				As for Navigation Warnings 1B 7D
	Note: VHF Ch 64 directed to Scillies. Selcal (3204): 2170.5 kHz Ch 16									
Start Point Radio 50°21'N 03°43'W	VHF station remotely controlled from Land's End Radio Selcal (3224): Ch 16				16 26 65 60	Traffic Lists, Weather Bulletins and Warnings broadcast on Ch 26 at same times as for Land's End Radio above				
Pendennis Radio 50°09'N 05°03'W	VHF station remotely controlled from Land's End Radio Selcal (3238): Ch 16				16 62 66	Traffic Lists, Weather Bulletins and Warnings broadcast on Ch 62 at same times as for Land's End Radio above				
Niton Radio 50°35'N 01°18'W (For Navtex see 6.3.14 p. 134) Tel: 0983 730495	2182	U V	2628 2810	2009 2562	16 04 28 81 85 64 87	Broadcasts on 1834 kHz and Ch 28 at: 0133 1533 0233 0533 1733 0633 0833 0303 0733 1933 1033 0903 0933 2133 1433 1503 1133 2333 1833 2033 2103 1333 2233				As for Navigation Warnings 1B 5B
	Note: VHF Ch 04 directed to Brighton. Selcal (3203): 2170.5 kHz Ch 16									
Weymouth Bay Radio 50°33'N 02°26'W	VHF station remotely controlled from Niton Radio Selcal (3242): Ch 16				16 05	Traffic Lists, Weather Bulletins and Warnings broadcast on Ch 05 at same times as for Niton Radio above				
St Peter Port Radio 49°27'N 02°32'W Note: Commercial calls only. Link calls available on VHF Ch 62 only. Tel: 0481 20085	Ship calls on 2182. Station replies on 1810	1662.5[1] 2182 **1810**		1662.5[1] 2182[2] 2381[3]	16 62[4] 78 67[5] (for notes see right)	Broadcasts on 1810 kHz and Ch 62 and 78 at:				
		(1) Trinity House and SAR only. (2) H24. (3) When 2182 is distress working.				After navigational warnings. (Ships also called individually on 2182 kHz and Ch 16)	Every 4h from 0133 for local waters	Notes for VHF services: Direct calling on Ch 62 and 78. (4) Link calls available Ch 62. (5) Ch 67 available for yacht safety messages.		
Jersey Radio 49°11'N 02°32'W Note: Link calls only available on VHF Ch 25. Tel: 0534 41121	Ship calls on 2182. Station replies on 1726	**1726** 2182		2381 2182 2049[1] 2104[2] 2534[2]	16 82 25	Broadcasts on 1726 kHz and Ch 25 and 82 at:				
		(1) Foreign vessels (2) UK reg. vessels				After weather messages. (Ships also called individually on 2182 kHz and Ch 16)	0433 0645 0745 0833 1245 1633 1845 2033 2245	0645 0745 1245 1845 2245	On receipt and at 0307 0907 1507 2107	(1) on receipt (2) at three minutes past the next two hours 1B
North Foreland Radio 51°22'N 01°25'E Tel: 0843 220592	2182	T	2698	2016	16 05 26 66 65	Broadcasts on 1848 kHz and Ch 26 at: 0103 1503 0033 0303 0503 1703 0433 0803 0903 0903 1903 0833 1503 1103 2103 1233 2103 1303 2303 1633 2003 2033				As for Naviation Warnings 1B 5B 2E
	Selcal (3201): 2170.5 kHz Ch 16									
Hastings Radio 50°53'N 00°37'E	VHF station remotely controlled from North Foreland Radio Selcal (3225): Ch 16				16 07 63	Traffic Lists, Weather Bulletins and Warnings broadcast on Ch 07 at same times as for Northforeland Radio above				
Thames Radio 51°20'N 00°20'E	VHF station remotely controlled from North Foreland Radio Selcal (3202): Ch 16				16 02 83	Traffic Lists, Weather Bulletins and Warnings broadcast on Ch 02 at same times as for Northforeland Radio above				
Orfordness Radio 52°00'N 01°25'E	VHF station remotely controlled from North Foreland Radio Selcal (3235): Ch 16				16 62 82	Traffic Lists, Weather Bulletins and Warnings broadcast on Ch 62 at same times as for Northforeland Radio above				

* Gale warnings — see note at foot of page 137

continued

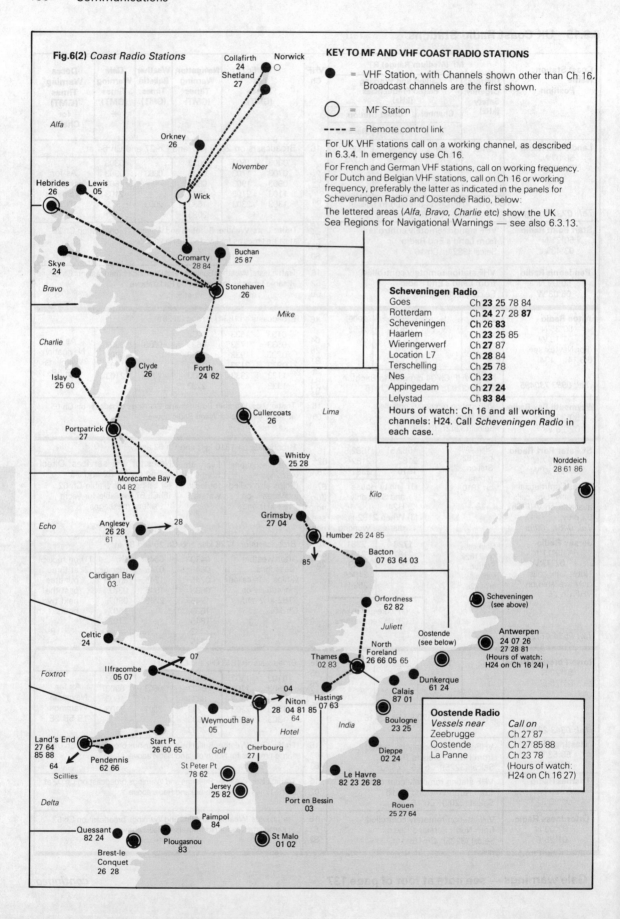

Fig.6(2) Coast Radio Stations

6.3.15 UK Coast Radio Stations continued

Coast Station: Name Position	MF (Medium Range) RT				VHF Ch	Traffic List Times (GMT)	Navigation Warning Times (GMT)	Weather Bulletin Times (GMT)	Gale Warning Times (GMT) *	Decca Warning Times (GMT) for chains
	Distress, Urgency, Safety (kHz)	Calling and Working Paired Frequencies (kHz)								
		Channel	Coast Station							
			Transmit	Receive						
Humber Radio 53°20'N 00°17'E Tel: 0521 73447	2182	Q R S	1925 2684 3778	2569 2111 3179	16 24 26 85	Broadcasts on 1869 kHz and Ch 26 at:				
						0333 1533 0733 1733 0933 1933 1133 2133 1333 2333	0233 0633 1033 1433 1833 2233	0833 2033	0303 0903 1503 2103	As for Navigation Warnings 5B 2A 2E 9B
	Note: VHF Ch 85 directed to The Wash. Selcal (3212): 2170.5 kHz Ch 16									
Bacton Radio 52°51'N 01°28'E	VHF station remotely controlled from Humber Radio Selcal (3214): Ch 16				16 07 63 64 03	Traffic Lists, Weather Bulletins and Warnings broadcast on Ch 07 same times as for Humber Radio above				
Grimsby Radio 53°34'N 00°05'W	VHF station remotely controlled from Humber Radio Selcal (3239): Ch 16				16 04 27	Traffic Lists, Weather Bulletins and Warnings broadcast on Ch 27 at sames times as for Humber Radio above				
Cullercoats Radio 55°04'N 01°28'W (For Navtex see 6.3.14 p. 134) Tel: 091 253 1318	2182	N O P	1838 2828 3750	2527 1953 2559	16 26	Broadcasts on 2719 kHz and Ch 26 at:				
						0303 1503 0703 1703 0903 1903 1103 2103 1303 2303	0033 0433 0833 1233 1633 2033	0803 2003	0303 0903 1503 2103	As for Navigation Warnings 5B 2A 6C 2E 9B 0E
	Selcal (3211): 2170.5 kHz Ch 16									
Whitby Radio 54°29'N 00°36'W	VHF station remotely controlled from Cullercoats Radio Selcal (3231): Ch 16				16 25 28	Traffic Lists, Weather Bulletins and Warnings broadcast on Ch 25 at same times as for Cullercoats Radio above				
Stonehaven Radio 56°57'N 02°13'W Tel: 0569 62917	2182	I J K L M	1856 1866 1946 2779 3617	2555 2552 2566 2146 3249	16 26	Broadcast on 2691 kHz and Ch 26 at:				
						0133 1533 0533 1733 0733 1933 0933 2133 1133 2333 1333	0233 0633 1033 1433 1833 2233	0833 2033	0303 0903 1503 2103	As for Navigation Warnings 2A 6C 0E
	Selcal (3222): 2170.5 kHz Ch 16									
Forth Radio 55°56'N 02°27'W	VHF station remotely controlled from Stonehaven Radio Selcal (3228): Ch 16				16 24 62	Traffic Lists, Weather Bulletins and Warnings broadcast on Ch 24 at same times as for Stonehaven Radio above				
Buchan Radio 57°36'N 02°02'W	VHF station remotely controlled from Stonehaven Radio Selcal (3237): Ch 16				16 25 87	Traffic Lists, Weather Bulletins and Warnings broadcast on Ch 25 at same times as for Stonehaven Radio above				
Hebrides Radio 58°14'N 07°02'W (remotely controlled from Stonehaven Radio) Tel: 0569 62917	2182	Z	1715	2534	16 26	Broadcasts on 2667 kHz and Ch 26 at:				
						0103 1503 0503 1703 0703 1903 0903 2103 1103 2303 1303	0033 0433 0833 1233 1633 2033	0803 2003	0303 0903 1503 2103	As for Navigation Warnings 6C 8E
	Selcal (3234): 2170.5 kHz Ch 16									
Lewis Radio 58°28'N 06°14'W	VHF station remotely controlled from Stonehaven Radio Selcal (3216): Ch 16				16 05	Traffic Lists, Weather Bulletins and Warnings broadcast on Ch 05 at same times as for Hebrides Radio above				
Skye Radio 57°28'N 06°41'W	VHF station remotely controlled from Stonehaven Radio Selcal (3232): Ch 16				16 24	Traffic Lists, Weather Bulletins and Warnings broadcast on Ch 24 at same times as for Hebrides Radio above				

***GALE WARNINGS**

Gale warnings are broadcast at the end of the first silence period after receipt and at the next of the times shown or if the first broadcast is on a scheduled time the warning is repeated at the end of the next silence period.

continued

6.3.15 UK Coast Radio Stations continued

Coast Station: Name Position	MF (Medium Range) RT				VHF Ch	Traffic List Times (GMT)	Navigation Warning Times (GMT)	Weather Bulletin Times (GMT)	Gale Warning Times (GMT) *	Decca Warning Times (GMT) for Chains
	Distress, Urgency, Safety (kHz)	Calling and Working Paired Frequencies (kHz)								
		Channel	Coast Station							
			Transmit	Receive						
Wick Radio 58°26'N 03°06'W	2182	E F G H	1706 1827 2604 2625	2524 2548 2013 2381		Broadcasts on 1792 kHz (Wick) and 1824 kHz (N Shetland) at:				
At Norwick, Shetland Is.	2182	A B C D	2751 2840.6 3525 3538	2006 2277 3335 3328		0503 0703 0903 1103 1303	1503 1703 1903 2103 2303	0033 0433 0833 1233 1633 2033	0803 2003	0303 0903 1503 2103	As for Navigation Warnings 6C 8E 0E
Tel: 0955 2271	Selcal (3221): 2170.5 kHz									
Cromarty Radio 57°36'N 02°58'W	VHF station remotely controlled from Wick Radio Selcal (3227): Ch 16				16 28 84	Traffic Lists, Weather Bulletins and Warnings broadcast on Ch 28 at same times as for Wick Radio above				
Orkney Radio 58°47'N 02°57'W	VHF station remotely controlled from Wick Radio Selcal (3226): Ch 16				16 26	Traffic Lists, Weather Bulletins and Warnings broadcast on Ch 26 at same times as for Wick Radio above				
Shetland Radio 60°09'N 01°12'W	VHF station remotely controlled from Wick Radio Selcal (3215): Ch 16				16 27	Traffic Lists, Weather Bulletins and Warnings broadcast on Ch 27 at same times as for Wick Radio above				
Collafirth Radio 60°32'N 01°23'W	VHF station remotely controlled from Wick Radio Selcal (3230): Ch 16				16 24	Traffic Lists, Weather Bulletins and Warnings broadcast on Ch 24 at same times as for Wick Radio above				
Portpatrick Radio 54°51'N 05°07'W (For Navtex see 6.3.14 p. 134)	2182	Y	2607	2104	16 27	Broadcasts on 1883 kHz and Ch 27 at:				
						0133 0533 **0733** 0933 1133 1333	1533 1733 1933 2133 2333	0233 0633 1033 1433 1833 2233	0833 2033	0303 0903 1503 2103	As for Navigation Warnings 8E 3B 7D
Tel: 077 681 311	Selcal (3207): 2170.5 kHz Ch 16									
Clyde Radio 55°55'N 04°48'W	VHF station remotely controlled from Portpatrick Radio Selcal (3213): Ch 16				16 26	Traffic Lists, Weather Bulletins and Warnings broadcast on Ch 26 at same times as for Portpatrick Radio above				
Islay Radio 55°46'N 06°27'W	VHF station remotely controlled from Portpatrick Radio Selcal (3233): Ch 16				16 25 60	Traffic Lists, Weather Bulletins and Warnings broadcast on Ch 25 at same times as for Portpatrick Radio above				
Anglesey Radio 53°20'N 04°18'W	VHF station remotely controlled from Portpatrick Radio Note: VHF Ch 28 directed to R Mersey. Selcal (3206): Ch 16				16 26 28 61	Broadcasts on Ch 26 at:				
						0103 0503 0903 1103 1303	1503 1703 1903 2103 2303	0033 0433 0833 1233 1633 2033	0803 2003	0303 0903 1503 2103	As for Navigation Warnings 3B 1B 7D
Morecambe Bay Radio 54°10'N 03°12'W	VHF station remotely controlled from Portpatrick Radio Selcal (3240): Ch 16				16 04 82	Traffic Lists, Weather Bulletins and Warnings broadcast on Ch 04 at same times as for Anglesey Radio above				
Cardigan Bay Radio 52°50'N 04°37'W	VHF station remotely controlled from Portpatrick Radio Selcal (3241): Ch 16				16 03	Traffic Lists, Weather Bulletins and Warnings broadcast on Ch 03 at same times as for Anglesey Radio above				
Ilfracombe Radio 51°11'N 04°07'W	VHF station remotely controlled from Niton Radio Note: Ch 07 directed to Severn Bridge Selcal (3205): Ch 16				16 05 07	Broadcasts on Ch 05 at:				
						0333 0733 0933 1133 1333	1533 1733 1933 2133 2333	0233 0633 1033 1433 1833 2233	0833 2033	0303 0903 1503 2103	As for Navigation Warnings 1B 7D
Celtic Radio 51°41'N 05°11'W	VHF station remotely controlled from Niton Radio Selcal (3218): Ch 16				16 24	Traffic Lists, Weather Bulletins and Warnings broadcast on Ch 24 at same times as for Ilfracombe Radio				

* **Gale warnings — see note at foot of page 137**

Radiotelephony 139

6.3.16 Irish and Continental Coast Radio Stations

Coast Station Name Position	Medium Frequency RT Service (kHz) Transmits / Receives	VHF Ch Tx Rx	Traffic List Times (GMT)	Navigation Warnings Times (GMT)	Weather Bulletin Times (LT)	Gale Warning Times (LT)	Decca Warning Times (GMT)
Malin Head Radio 55°22'N 7°21'W Tel: (077) 70103	1841[1] 2182 2182 2049[1] 2593 (1) When 2182 distress working	16 16 67 67 (safety only)	1841 kHz every odd H+03 (not 0303 0703)	1841 kHz every 4h from 0033	No service	No service	1841 kHz O/R & H+03 for next 2h 6C 3B 7D 8E
Crockalough Radio 55°21'N 7°16'W	VHF station remotely controlled from Malin Head (*Ch 67 safety messages only)	16 16 23 23 85 85 67* 67*	Ch 23 every odd H+03 (not 0303 0703)	Ch 23 every 4h from 0033	Ch 23 every 3h from 0100	Ch 23 O/R and 0030 0630 1230 1830	Ch 23 as for Malin Head
Ballybunnion Radio 52°31'N 9°36'W	VHF station remotely controlled from Valentia (*Ch 67 safety messages only)	16 16 28 28 24 24 67* 67*	Ch 28 every odd H+33 (not 0133 0533)	Ch 28 every 4h from 0233	Ch 28 every 3h from 0100	Ch 28 O/R and 0030 0630 1230 1830	Ch 28 as for Valentia
Valentia Radio 51°56'N 10°21'W Tel: (0667) 6109	1827[1] 2182 2182 2049[1] 2590 (1) When 2182 2614 distress working	16 16 67 67 (safety only)	1827 kHz every odd H+33 (not 0133 0533)	1827 kHz every 4h from 0233	1827 kHz 0833 2033 (GMT)	1827 kHz S/P1 0303 0903 1503 2103(GMT)	1827 kHz O/R & H+03 for next 2h 1B 7D
Kilkeaveragh Radio 51°52'N 10°20'W	VHF station remotely controlled from Valentia (*Ch 67 safety messages only)	16 16 24 24 28 28 67* 67*	Ch 24 every odd H+33 (not 0133 0533)	Ch 24 every 4h from 0233	Ch 24 every 3h from 0100	Ch 24 O/R and 0030 0630 1230 1830	Ch 24 as for Valentia
Knockgour Radio 51°38'N 10°00'W	VHF station remotely controlled from Valentia (*Ch 67 safety messages only)	16 16 23 23 85 85 67* 67*	Ch 23 every odd H+33 (not 0133 0533)	Ch 23 every 4h from 0233	Ch 23 every 3h from 0100	Ch 23 O/R and 0030 0630 1230 1830	Ch 23 as for Valentia
Dublin Radio 53°26'N 6°15'W Tel: (01) 379900	VHF station established primarily for safety. No link calls (*Ch 67 safety messages only)	16 16 65 65 67* 67*	Ch 65 every odd H+33 (not 0133 0533)	Ch 65 every 4h from 0233	Ch 65 every 3h from 0100	Ch 65 O/R and 0030 0630 1230 1830	Ch 65 as for Valentia

FRANCE

Bayonne Radio 43°16'N 1°24'W	VHF station remotely controlled from Bordeaux-Arcachon (0700–2200 local time)	16 **24** 24			Ch 24 0733 1233 LT (Fr)	Ch 24 O/R & S/P2 (Fr)	
Bordeaux-Arcachon Radio 44°39'N 1°10'W (MF remotely controlled from Brest-Le Conquet 2200–0700 local time. VHF operates 0700–2200 local time) Tel: 56.83.40.50	1820 **1862** 2182 2775 3722	At Royan and Cap Ferret: 2182 (0700–2200 LT) 2049 2056 2069 2153 2167 2506 2541 At Royan: 2449 3161 At Cap Ferret: 2037 2321 2421 2527 3168	16 **28** 28 **82** 82	1862 kHz every even H + 07	1820 kHz 0703 1703 (in French and English)	1820 kHz 0703 1703 (in French for Areas 23, 24) Ch 82 0733 1233 LT (Fr) (Sables d'Olonne to Spanish border)	1862 kHz every even H + 07 (in French for Areas 23, 24) 1820 kHz & Ch 82 O/R & S/P2 (Fr)	1820 kHz 0703 1703 (in French and English) 1B 8B
Bordeaux Radio 44°53'N 0°30'W	VHF station remotely controlled from Bordeaux-Arcachon (0700–2200 local time)	16 **27** 27					
Royan Radio 45°34'N 0°58'W	VHF station remotely controlled from Bordeaux-Arcachon (0700–2200 local time)	16 **23** 23 **25** 25			Ch 23 0733 1233 LT (Fr)	Ch 23 O/R & S/P2 (Fr)	
La Rochelle Radio 46°14'N 1°33'W	VHF station remotely controlled from Bordeaux-Arcachon (0700–2200 local time)	16 **21** 21 **26** 26			Ch 21 0733 1233 LT (Fr)	Ch 21 O/R & S/P2 (Fr)	

continued

NOTES:
1. H + ... means commencing at ... minutes past the hour.
2. O/R means on receipt.
3. Ch 16 is reserved for calls from French Coast Radio Stations. Vessels should call on a working frequency. For MF, call on 2182 or (if 2182 is busy) on 2321 kHz.
4. LT means Local Time.
5. (Fr) means 'in French'.
6. S/P1 means after first silence period after receipt.
7. S/P2 means after first and second silence periods after receipt.
8. French VHF Bcsts of weather bulletins and gale warnings refer to coastal waters.
9. For French Fcst Areas (MF Bcsts) see Fig. 7(1), page 159.

6.3.16 Continental Coast Radio Stations continued (for notes see opposite page)

Coast Station Name Position	Medium Frequency RT Service (kHz)		VHF Ch		Traffic List Times (GMT)	Navigation Warning Times (GMT)	Weather Bulletin Times (GMT)	Gale Warning Times (GMT)	Decca Warning Times (GMT)
	Transmits	Receives	Tx	Rx					
St Gilles-Croix-de-Vie Radio 46°44'N 1°59'W	VHF station remotely controlled from Saint Nazaire		16 **27** 85	27 85			Ch 27 0733 1233 LT (Fr)	Ch 27 O/R & S/P2 (Fr)	
Nantes St Herblain Radio 47°13'N 1°37'W	VHF station remotely controlled from Saint Nazaire		16 **28**	28			Ch 28 0733 1233 LT (Fr)	Ch 28 O/R & S/P2 (Fr)	
Saint Nazaire Radio 47°21'N 2°06'W 47°17'N 2°14'W } VHF (Remotely controlled from Brest-Le Conquet 2200–0700 local time) Tel: 40.22.39.04	**1687** 2182 2740 3795 1722	2182[1] Also at Batz-sur Mer: 1995 2049 2056 2153 2167 2321 2491 2506 3168 Note: (1) 2182 kHz: H24.	16 **23** **24**	23 24	1687kHz every odd H + 07	1687 kHz O/R 1722 kHz 0803 1803 (in French)	1722 kHz 0803 1803 (in French) for Areas 14–24 Ch 23 0733 1233 LT (Fr) (Sables d'Olonne to Cap de la Hague)	1687 kHz every odd H + 07 (in French for Areas 14–16) 1722 kHz & Ch 23 O/R & S/P2 (Fr)	1722kHz O/R 0803 1803 (in French) Chains 1B 8B
Belle Ile Radio 47°21'N 3°09'W	VHF station remotely controlled from Saint Nazaire		16 **25** **87**	25 87			Ch 87 0733 1233 LT (Fr)	Ch 87 O/R & S/P2 (Fr)	
Pont-l'Abbé Radio 47°53'N 4°13'W	VHF station remotely controlled from Brest-Le Conquet		16 **27**	27			Ch 27 0733 1233 LT (Fr)	Ch 27 O/R & S/P2 (Fr)	
Brest-Le Conquet Radio 48°20'N 4°44'W Located at St Malo 48°38'N 2°02'W Located at Quimperlé 47°53'N 3°30'W (For Navtex see 6.3.14) Tel: 98.80.40.26	1673 **1806** 2182 2726 3722 { 1673 2182 **2691** { 1673 1806 **1876** 2182	2182 (H24) 2182[1] (H24) 2182[1,2] (H24) Notes: 1) Also receives on 2049 2956 2097 2153 2160 2167 2463 2506 3168 kHz. 2) Located at Treffiagat 47°49'N 4°16'W.	16 **26** **28**	26 28	1806 kHz (Le Conquet) 2691 kHz (St Malo) every even H + 03	1673 kHz (Le Conquet) 2691 kHz (St Malo) 0333 0733 1133 1533 1933 2333 1876 kHz (Quimperlé) 0733 1633 (in English and French)	1673 kHz (Le Conquet) 2691 kHz (St Malo) 1876 kHz 0733 1633 2153 and on request (in French for Areas 14–22) Ch 26 0733 1233 LT (Fr)	1673 kHz 1876 kHz 2691 kHz O/R & S/P2 1806 kHz 2691 kHz every even H + 03 (in French for Areas 14–22) Ch 26 O/R & S/P2 (Fr)	
Ouessant Radio 48°27'N 5°05'W	VHF station remotely controlled from Brest-Le Conquet		16 **24** **82**	24 82			Ch 82 0733 1233 LT (Fr)	Ch 82 O/R & S/P2 (Fr)	
Plougasnou Radio 48°42'N 3°48'W	VHF station remotely controlled from Brest-Le Conquet		16 **83**	83			Ch 83 0733 1233 LT (Fr)	Ch 83 O/R & S/P2 (Fr)	
Paimpol Radio 48°45'N 2°59'W	VHF station remotely controlled from Brest-Le Conquet		16 **84**	84			Ch 84 0733 1233 LT (Fr)	Ch 84 O/R & S/P2 (Fr)	

6.3.16 Continental Coast Radio Stations continued

Coast Station Name Position	Medium Frequency RT Service (kHz) Transmits	Medium Frequency RT Service (kHz) Receives	VHF Ch Tx	VHF Ch Rx	Traffic List Times (GMT)	Navigation Warning Times (GMT)	Weather Bulletin Times (GMT)	Gale Warning Times (GMT)	Decca Warning Times (GMT)
Cherbourg Radio 49°38'N 1°36'W	VHF station remotely controlled from Boulogne		16 **27**	27			Ch 27 0733 1233 LT (Fr)	Ch 27 O/R & S/P2 (Fr)	
Port en Bessin Radio 49°20'N 0°42'W	VHF station remotely controlled from Boulogne		**03** 16	03			Ch 03 0733 1233 LT (Fr)	Ch 03 O/R & S/P2 (Fr)	
Rouen Radio 49°27'N 1°02'E	VHF station remotely controlled from Boulogne		16 **25** 27 64	25 27 64					
Le Havre Radio 49°31'N 0°05'E Tel: 35.70.90.04	VHF station remotely controlled from Boulogne		16 **23** **26** **28** 82	23 26 28 82			Ch 82 0733 1233 LT (Fr)	Ch 82 O/R & S/P2 (Fr)	
Dieppe Radio 49°55'N 1°04'E	VHF station remotely controlled from Boulogne		16 **02** **24**	02 24			Ch 02 0733 1233 LT (Fr)	Ch 02 O/R & S/P2 (Fr)	
Boulogne Radio 50°43'N 1°37'E Escalles: 50°55'N 1°43'E Tel: 21.31.44.00	1694 **1771** 2182 2747 3795	2182[1,2] Also receives on following frequencies at Escalles: 2049 2056 2097 2153 2167 2321 2506 2576 3161 3168 Notes: (1) At Boulogne and Escalles. (2) 2182 kHz: H24.	16 **23** **25**	23 25	1771 kHz every odd H + 03	1694 kHz 0133 0533 0933 1333 1733 2133 (in French and English)	1694 kHz 0703 1733 (in French for Areas 1–14) Ch 23 0733 1233 LT (Fr)	1694 kHz O/R & S/P2 1771 kHz every odd H + 03 (Fr) Ch 23 O/R & S/P2 (Fr)	1694 kHz (as Navigation Warning times, in French and English) Chain 5B
Calais Radio 50°55'N 1°43'E	VHF station remotely controlled from Boulogne		**01** 16 **87**	01 87			Ch 87 0733 1233 LT (Fr)	Ch 87 O/R & S/P2 (Fr)	
Dunkerque Radio 51°02'N 2°24'E	VHF station remotely controlled from Boulogne		16 **24** **61**	24 61			Ch 61 0733 1233 LT (Fr)	Ch 61 O/R & S/P2 (Fr)	

BELGIUM (for Oostende Radio see next page)

Antwerpen Radio VHF facilities at following positions. The call in each case is Antwerpen Radio									
Antwerpen	Ch 07 16 **24** 26 27 **28** 83	51°17'N 4°20'E			Ch 24 every H+05	Ch 24 O/R and every H+03 H+48 (in English and Dutch for the Schelde)	Ch 24 Fog warnings O/R and every H+03 H+48 (in English and Dutch)	Ch 24 O/R and every H+03 H+48 (strong breeze warnings in English)	
Kortrijk	Ch 16 **24** 83	50°50'N 3°17'E							
Gent	Ch 16 **24** 26 81	51°02'N 3°44'E							
Vilvoorde	Ch 16 **24** 28	50°56'N 4°25'E							
Ronquières	Ch 16 **24** 27	50°37'N 4°13'E							
Mol	Ch 16 **24** 87	51°11'N 5°07'E							
Liège	Ch 16 **24** 27	50°34'N 5°33'E							

continued

NOTES:
1. H + ... means commencing at ... minutes past the hour.
2. O/R means on receipt.
3. Ch 16 is reserved for calls from French Coast Radio Stations. Vessels should call on a working frequency. For MF, call on 2182 or (if 2182 is busy) on 2321 kHz.
4. LT means Local Time.
5. (Fr) means 'in French'.
6. S/P1 means after first silence period after receipt.
7. S/P2 means after first and second silence periods after receipt.
8. French VHF Bcsts of weather bulletins and gale warnings refer to coastal waters.
9. For French Fcst Areas (MF Bcsts) see Fig. 7(1), page 159.

6.3.16 Continental Coast Radio Stations continued

Coast Station Name Position	Medium Frequency RT Service (kHz)		VHF Ch		Traffic List Times (GMT)	Navigation Warning Times (GMT)	Weather Bulletin Times (GMT)	Gale Warning Times (GMT)	Decca Warning Times (GMT)
	Transmits	Receives	Tx	Rx					
Oostende Radio 51°11'N 2°48'E (See note below VHF column concerning which channels to be used in different areas) (For Navtex see 6.3.14)	**1817** **1820**[1] 1905 1908 2087 2090 2170.5 2182 2253 2256 2373 2376 2481 2484 2758 **2761**[1] 2814 **2817**[2] 3629 **3632**[2] 3681 **3684**[1]	2182[3] 2484[4] 3178[4] 2191[5]	16 (24) **23**[1] **27**[2,3] (H24) 78[1] 85[3] 87[2] 88[3] Used for calls in vicinity of: 1) La Panne 2) Zeebrugge 3) Oostende Hours of watch Ch 16 27: H24		2761 kHz every even H + 20, Ch 27 every H + 20	2761 kHz and Ch 27 O/R S/P2 0233 0633 1033 1433 1833 2233 (in Dutch and English)	2761 kHz and Ch 27 0820 1720 (in Dutch and English for Dover and Thames) Ice Reports on Ch 27 every 4h from 0103 Fog Warnings on 2761 kHz O/R and S/P2 in English and Dutch for the Schelde	2761 kHz and Ch 27 O/R and S/P2 (in Dutch and English for Dover and Thames)	2761 kHz and Ch 27 S/P2 (in English) Chains 5B 2A 6C 8E 2E 9B
Notes: 1) Working frequencies for Belgian vessels. 2) Working frequencies for foreign vessels. 3) 2182 kHz: H24. 4) 2484 3178 kHz: H24 for Belgian vessels. 5) 2191 kHz: when 2182 kHz is distress working.									

NETHERLANDS

Coast Station Name Position	Medium Frequency RT Service (kHz)			VHF Ch		Traffic List Times (GMT)	Navigation Warning Times (GMT)	Weather Bulletin Times (GMT)	Gale Warning Times (GMT)	Decca Warning Times (GMT)
	Transmits	Receives		Tx	Rx					
Scheveningen Radio 50°06'N 4°16'E (For Navtex see 6.3.14) Tel: 550.19104	1764 **1862**[1] 1890 1939 2182[3] **2600** 2824 **3673**	2030 2160[1] 2049[2] 2056 2513 2182[3] (H24) 1995 2520[4] (H24) 3191		See details of VHF in table below Stations remotely controlled (all H24). In each case call Scheveningen Radio Goes Rotterdam Scheveningen Haarlem Wieringerwerf Location L7 Terschelling Ness Appingedam Lelystad		1862 kHz 1890 kHz every odd H + 05 VHF (not Ch 16) every H + 05 2824 kHz 0105 0305 0505 2305 Ch 16 **23 25** 78 84 Ch 16 **24 27 28 87** Ch 16 26 **83** Ch 16 **23 25** 85 Ch 16 **27** 87 Ch 16 **28** 84 Ch 16 **25** 78 Ch 16 **23** Ch 16 **27 24** Ch 16 **83 84**	1862 kHz 1890 kHz O/R S/P1 and every 4h from 0333 2824 kHz 0333 2333 (all in Dutch and English) 51°31'N 51°56'N 52°06'N 52°23'N 52°55'N 53°32'N 53°22'N 53°24'N 53°18'N 52°32'N	1862 kHz 1890 kHz every 6h from 0340 (in Dutch and English) VHF at (LT) 0005 0705 1305 1905 (in Dutch) 3°54'E 4°28'E 4°16'E 4°38'E 5°04'E 4°13'E 5°13'E 6°04'E 6°52'E 5°26'E	1862 kHz 1890 kHz O/R S/P1 (in Dutch and English) VHF O/R and every H + 05 (in Dutch)	As for Navigation Warnings (in Dutch and English) Decca Chains 2E 9B
1) Located at Nes 53°24'N/6°04'E. 2) Keeps watch on 2049 when distress working on 2182 kHz. 3) Located at Nes and at Scheveningen. 4) Calling frequency; alternative is 2182 kHz. Hours of service: 1764 2182 2824 kHz: H24 1862 1939 kHz: Mon–Sat: 0700–2200[5] Sun: 0800–2200[5] 1890 kHz: 0700–2300[5] 2600 kHz: Mon–Sat: 0700–2200[5] 3673 kHz: Occas 5) All 1h earlier when DST is in force.										

NOTES:
1. H + ... means commencing at ... minutes past the hour.
2. LT means Local Time.
3. O/R means on receipt.
4. S/P1 means after first silence period after receipt.
5. S/P2 means after first and second silence periods after receipt.

6.3.16 Continental Coast Radio Stations *continued*

Coast Station Name Position	Medium Frequency RT Service (kHz)		VHF Ch	Traffic List Times (GMT)	Navigation Warning Times (GMT)	Weather Bulletin Times (GMT)	Gale Warning Times (GMT)	Decca Warning Times (GMT)
	Transmits	Receives						
FEDERAL REPUBLIC OF GERMANY								
Norddeich Radio 53°38'N 7°12'E	1799 1911 2182 2614[2] 2799 2848	2491 2541 2182[1] 2023[3] 2049[4] 2153 3161	16 **28** 61 86	2614 kHz and Ch 28 every H + 45 Also on 2614 kHz at 0810 2010 after weather report	2614 kHz O/R S/P1 0133 0533 0933 1333 1733 2133 and on request (In English and German)	2614 kHz 0810 2010 (in German for North Sea areas and Skagerrak)	2614 kHz O/R S/P1 and every 4h from 0133 0533 0933 1333 1733 2133 and on request. In English for German Bight)	2614 kHz O/R S/P1 0133 0533 0933 1333 1733 2133 and on request (in English and German) 9B
	Notes: 1) 2182 kHz: H24. 2) During Bcsts station replies on 2848 kHz. 3) 2023 kHz 0700–2300 LT 4) 2049 kHz when 2182 is distress working. Only available 2300–0700 LT to German vessels.							
Bremen Radio 53°05'N 8°48'E	VHF station remotely controlled from Elbe-Weser		16 25 **28**	Ch 28 every H + 40				
Helgoland Radio 54°11'N 7°53'E	VHF station remotely controlled from Elbe-Weser		03 16 **27** 88	Ch 27 every H + 20				
Elbe-Weser Radio 53°50'N 8°39'E	Note: Ch 23 28 and 62 are for communication with vessels on the Nord-Ostsee Canal		01 16 **23** 24 **26** 28 62	Ch 23 every H + 20 Ch 26 every H + 50				
Hamburg Radio 53°33'N 9°58'E	VHF station remotely controlled from Elbe-Weser		16 25 **27** 82 83	Ch 27 every H + 40				
Eiderstedt Radio 54°20'N 8°47'E	VHF station remotely controlled from Elbe-Weser		16 **25** 64	Ch 25 every H + 40				
Nordfriesland Radio 54°55'N 8°18'E	VHF station remotely controlled from Elbe-Weser		05 16 **26**	Ch 26 every H + 50				

NOTES:
1. H + ... means commencing at ... minutes past the hour.
2. LT means Local Time.
3. O/R means on receipt.
4. S/P1 means after first silence period after receipt.
5. S/P2 means after first and second silence periods after receipt.

6.4 SHORT NOTES ON FLAG ETIQUETTE

6.4.1 Ensign

A yacht's ensign is the national maritime flag corresponding to the nationality of her owner. Thus a British yacht should wear the Red Ensign, unless she qualifies for a special ensign (see 6.4.2). At sea the ensign must be worn when meeting other vessels, when entering or leaving foreign ports, or when approaching forts, signal and coastguard stations etc. In British harbours the ensign should be hoisted at 0800 (0900 between 1 November and 14 February) and lowered at sunset (or 2100 local time if earlier). The ensign should normally be worn at the stern, but if this is not possible the nearest position should be used, e.g. at the peak in a gaff-rigged boat, at the mizzen masthead in a ketch or yawl, or about two-thirds up the leech of the mainsail.

6.4.2 Special ensigns

Members of certain clubs may apply for permission to wear a special ensign (e.g. Blue Ensign, defaced Blue Ensign, or defaced Red Ensign). For this purpose the yacht must either be a registered ship under Part I of the Merchant Shipping Act 1894 and of at least 2 tons gross tonnage, or be registered under the Merchant Shipping Act 1983 (Small Ships Register) and of at least 7 metres overall length. The owner or owners must be British subjects, and the yacht must not be used for any professional, business or commercial purpose. Full details can be obtained from Secretaries of Clubs concerned.

A special ensign must only be worn when the owner is on board or ashore in the vicinity, and only when the yacht is flying the burgee (or a Flag Officer's flag) of the club concerned. The permit must be carried on board. When the yacht is sold, or the owner ceases to be a member of the Club, the permit must be returned to the Secretary of the Club.

6.4.3 Burgee

A burgee shows that a yacht is in the charge of a member of the club indicated, and does not necessarily indicate ownership. It should be flown at the masthead. If the yacht is on loan, or is chartered, it is correct to use the burgee of the skipper or charterer — not that of the absent owner. Normal practice has been to lower the burgee at night, at the same time as the ensign, but nowadays many owners leave the burgee flying if they are on board or ashore in the vicinity.

6.4.4 Flag officer's flag

Clubs authorise their flag officers to fly special swallow-tailed flags, with the same design as the club burgee and in place of it. The flags of a vice-commodore and a rear-commodore carry one and two balls respectively. A flag officer's flag is flown day and night while he is on board, or ashore nearby. A flag officer should fly his flag with the Red Ensign (or special ensign, where authorised) in preference to the burgee of some other club.

6.4.5 Choice of burgee

An owner who is not a flag officer, and who belongs to more than one club, should normally fly the burgee (and if authorised the special ensign) of the senior club in the harbour where the yacht is lying. An exception may be if another club is staging a regatta or similar function.

6.4.6 Courtesy ensign

It is customary when abroad to fly a small maritime ensign of the country concerned at the starboard crosstrees. The correct courtesy flag for a foreign yacht in British waters is the Red Ensign (not the Union Flag). British yachts do not fly a courtesy flag in the Channel Islands since these are part of the British Isles.

6.4.7 House flag

An owner may fly his personal flag when he is on board in harbour, provided it does not conflict with the design of some existing flag. A house flag is normally rectangular, and is flown at the crosstrees in a sloop or cutter, at the mizzen masthead in a ketch or yawl, or at the foremast head in a schooner.

6.4.8 Salutes

Yachts should salute all Royal Yachts, and all warships of whatever nationality. A salute is made by dipping the ensign (only). The vessel saluted responds by dipping her ensign, and then re-hoisting it, whereupon the vessel saluting re-hoists hers.

SELL'S marine market

FIRST CHOICE FOR THE BOAT OWNER AND BUILDER

£12

- Comprehensive Boat Section includes full specifications of Sailing Cruisers, Sailing Dinghies, Motor Cruisers, Sportsboats and Workboats, followed by Boatbuilders & Distributors.
- The latest technical information available appears in the Engine and Electronics sections.
- For enthusiasts a complete UK County Guide to Local Services is included... AND MUCH MUCH MORE.

Marine Market 1989 edition available December 1988

MARINA GUIDE

Coastal and Inland Marinas, Yacht Harbours and Moorings

- Expanded Coastal and Inland Marina Guide now covers the UK, Holland, Belgium and France, with full details of all facilities and charges.

Marina Guide 1989 edition available December 1988

£5

ORDER YOUR COPIES NOW

Please send me .... copy/copies of Sell's Marine Market 1989 edition at £12.00 per copy, postage & packing included (UK only).

Please send me .... copy/copies of Sell's Marina Guide 1989 edition at £5.00 per copy, postage & packing included (UK only).

Please debit my Access/Visa Card No. ☐☐☐☐☐☐☐☐☐☐☐☐☐☐

Name .... Address ....

.... 89NA

DON'T SAIL WITHOUT THEM!

Cheques and Postal Orders should be made payable to: **Sell's Publications Ltd,
55 High Street, Epsom, Surrey KT19 8DW. Tel: (03727) 26376. Fax: (03727) 29241**

Also available from leading Chandlers and Branches of W.H. Smith

Working for Britain's marine industry

Written to serve the needs of the British leisure marine industry, Boating Business and Marine Trade News provides a medium through which suppliers to the industry can establish and maintain contact with their sales outlets: the boatbuilders, chandlers, manufacturers, distributors and dealers of marine equipment throughout the UK plus key buyers across the EEC countries. Boating Business and Marine Trade News publishes reviews of new products and developments within the industry, plus a wide range of features covering many of the important aspects of an industry which serves one of the largest participation sports in this country.

Boating Business and Marine Trade News is published monthly and has a circulation in excess of 6,500 copies which are distributed free of charge across the UK and relevant EEC countries as a service to Britain's marine trade.

Published by
RUSHTON MARINE PRESS, **Woodside, Burnhams Road, Little Bookham, Leatherhead, Surrey KT23 3BA. Telephone 0372 53316**

Chapter 7
Weather

Contents

7.1	**GENERAL WEATHER INFORMATION**	**Page 146**
7.1.1	Beaufort scale	
7.1.2	Barometer and thermometer conversion scales	
7.1.3	Meanings of terms used in weather bulletins	
7.2	**SOURCES OF WEATHER INFORMATION**	**148**
7.2.1	BBC Radio Shipping Forecasts	
7.2.2	BBC Inshore Waters Forecasts	
7.2.3	BBC General Forecasts	
7.2.4	Special forecasts for sea areas	
7.2.5	Automatic Telephone Weather Services — Marinecall	
7.2.6	Facsimile broadcasts	
7.2.7	NAVTEX	
7.2.8	VOLMET	
7.2.9	W/T transmissions	
7.2.10	Press forecasts	
7.2.11	Television forecasts and Prestel	
7.2.12	Visual storm signals	
7.2.13	Wind information — La Corbière Lt Ho	
7.2.14	British and Irish Coast Radio Stations — Weather Bulletins by R/T	
7.2.15	HM Coastguard — VHF Ch 67	
7.2.16	Reports of present weather	
7.2.17	Local Radio Stations — coastal forecasts	
Table 7(1)	British and Irish Coast Stations — Weather Bulletins by R/T	**150**
Table 7(2)	British Isles — Daily Shipping Forecasts and Forecasts for Coastal Waters	**154**
Table 7(3)	Gale Warnings	**155**
Table 7(4)	Local Radio Stations — Coastal Weather Forecasts	**156**
Table 7(5)	Western Europe — Shipping Forecasts, Coastal Waters Forecasts, Gale Warnings	**158**

Weather — introduction

The following subjects are described in detail in Chapter 7 of *The Macmillan & Silk Cut Yachtsman's Handbook*:

Transfer of heat; world weather; air masses; atmospheric pressure; wind; humidity; clouds; depressions and fronts; the passage of a depression; anticyclones; fog; sea and land breezes; thunderstorms; tropical storms; glossary of meteorological terms; forecasting your own weather; bibliography.

Here in the Almanac, emphasis is given to obtaining and interpreting forecasts. For more general information on the weather, including the subjects listed above, reference should be made to Chapter 7 of *The Macmillan & Silk Cut Yachtsman's Handbook*.

7.1 GENERAL WEATHER INFORMATION

7.1.1 Beaufort scale

Force	Wind speed (knots)	Description	State of sea	Probable wave height (m)	Probable max. wave height (m)
0	0–1	Calm	Like a mirror	0	0
1	1–3	Light air	Ripples like scales are formed	0	0
2	4–6	Light breeze	Small wavelets, still short but more pronounced, not breaking	0.1	0.3
3	7–10	Gentle breeze	Large wavelets, crests begin to break; a few white horses	0.4	1
4	11–16	Moderate breeze	Small waves growing longer; fairly frequent white horses	1	1.5
5	17–21	Fresh breeze	Moderate waves, taking more pronounced form; many white horses, perhaps some spray	2	2.5
6	22–27	Strong breeze	Large waves forming; white foam crests more extensive; probably some spray	3	4
7	28–33	Near gale	Sea heaps up; white foam from breaking waves begins to blow in streaks	4	5.5
8	34–40	Gale	Moderately high waves of greater length; edges of crests break into spindrift; foam blown in well marked streaks	5.5	7.5
9	41–47	Severe gale	High waves with tumbling crests; dense streaks of foam; spray may affect visibility	7	10
10	48–55	Storm	Very high waves with long overhanging crests; dense streams of foam make surface of sea white. Heavy tumbling sea; visibility affected	9	12.5
11	56–63	Violent storm	Exceptionally high waves; sea completely covered with long white patches of foam; edges of wave crests blown into froth. Visibility affected	11	16
12	64 plus	Hurricane	Air filled with foam and spray; sea completely white with driving spray; visibility very seriously affected	14	—

Notes: (1) The state of sea and probable wave heights are a guide to what may be expected in the open sea, away from land. In enclosed waters, or near land with an off-shore wind, wave heights will be less but possibly steeper — particularly with wind against tide.
(2) It should be remembered that the height of sea for a given wind strength depends upon the fetch and the duration for which the wind has been blowing. For further information on sea state, see Chapter 9 of *The Macmillan & Silk Cut Yachtsman's Handbook*.

7.1.2 Barometer and thermometer conversion scales

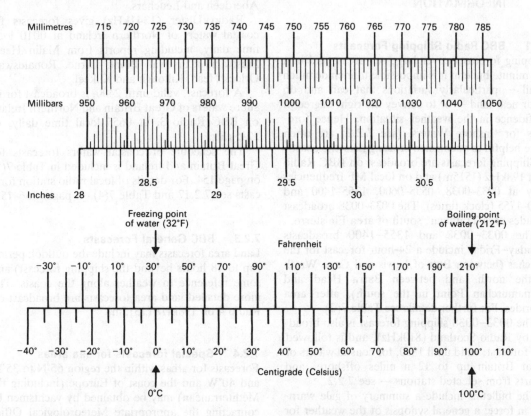

7.1.3 Meaning of terms used in weather bulletins

Visibility
Good   More than 5 miles
Moderate   5–2 miles
Poor   2 miles — 1100 yards (1000 metres)
Fog   Less than 1100 yards (1000 metres)

Timing (of gale warnings)
Imminent   Within 6 hours of time of issue
Soon   6–12 hours
Later   12–24 hours

Speeds (of weather systems)
Slowly   0–15 knots
Steadily   15–25 knots
Rather quickly 25–35 knots
Rapidly   35–45 knots
Very rapidly   Over 45 knots

Barometric pressure changes (tendency)
Now falling   ) pressure
Rising more slowly (or now steady)) higher than
Rising   ) 3 hours ago
Steady
Now rising (pressure lower or same as 3 hours ago)
Falling more slowly) Pressure lower than
Falling   ) 3 hours ago

'Rising' and 'Falling' may be qualified by:
Slowly   Change less than 1.6mb in 3 hours
Quickly   3.5–6mb in 3 hours
Very rapidly More than 6mb in 3 hours

Gale warnings
Indicate that winds of at least force 8 or gusts reaching 43 knots are expected somewhere in the area. 'Severe gale' implies winds of at least force 9 or gusts of 52 knots. 'Storm' implies winds of force 10 or above, or gusts of 61 knots. Gale warnings remain in force until amended or cancelled ('gales now ceased'). If a gale persists for more than 24 hours the warning is re-issued.

Land area forecasts — wind strength
In land area forecasts winds are given in the following terms, which relate to Beaufort forces as indicated:
Calm   0   Fresh   5
Light   1–3   Strong   6–7
Moderate 4   Gale   8

Land area forecasts — visibility
The following definitions are used in land area forecasts:
Mist   200–1100 yards (183–1000m)
Fog   Less than 200 yards (183m)
Dense fog Less than 50 yards (46m)

7.2 SOURCES OF WEATHER INFORMATION

7.2.1 BBC Radio Shipping Forecasts
Shipping forecasts cover large sea areas, and in a five minute bulletin it is impossible to include much detail — particularly variations that can and do occur near land — or to convey the right degree of confidence in the weather situation. Hence forecasts for 'Inshore Waters' (see 7.2.2) are often more helpful to yachtsmen cruising near the coast.

Shipping forecasts are broadcast on BBC Radio 4 on 198kHz (1515m) and on local MF frequencies daily at 0033–0038, 0555–0600, 1355–1400 and 1750–1755 (clock times). The 0033–0038 broadcast includes area Trafalgar, south of area Finisterre.

The 0033–0038 and 1355–1400 broadcasts Monday–Friday include a 24-hour forecast for the Minches (between Butt of Lewis and Cape Wrath in the north, and between Barra Head and Ardnamurchan Point in the south), after area Hebrides. For further details see Table 7(2).

The 0033–0038 shipping forecast is also broadcast by Radio Scotland (810kHz), and is followed by a forecast, valid until 1800, for coastal waters of Great Britain up to 12 n miles offshore, and reports from selected stations — see 7.2.2.

The bulletins include a summary of gale warnings in force; a general synopsis of the weather for the next 24 hours and expected changes within that period; forecasts for each sea area for the next 24 hours, giving wind direction and speed, weather and visibility; and the latest reports from selected stations from those shown at the foot of page 153. For each station are given wind direction and Beaufort force, present weather, visibility, and (if available) sea-level pressure and tendency. These stations are marked by their initial letters on the chart for forecast areas on page 152.

Apart from being included in shipping forecasts, gale warnings are broadcast at the earliest juncture in the BBC Radio 4 programme after receipt, and also after the next news bulletin.

Instructions on recording and interpreting the shipping forecasts are given in 7.6.4 of *The Macmillan & Silk Cut Yachtsman's Handbook*.

7.2.2 BBC Inshore Waters Forecasts
Forecasts are given for inshore waters (up to 12 miles offshore) of Great Britain until 1800 next day at the end of English, Welsh and Scottish Radio 4 programmes, and on Radio Scotland, at about 0038 local (or clock) time. The forecast of wind, weather and visibility is followed by the 2200 reports from the following stations: Boulmer, Bridlington, Walton-on-the-Naze, St Catherine's Point, Land's End, Mumbles, Valley, Blackpool, Prestwick, Benbecula, Stornoway, Lerwick, Wick, Aberdeen and Leuchars.

Radio Ulster (1341kHz) gives forecasts for coastal waters of Northern Ireland at 0010 local time daily, including reports from Malin Head, Prestwick, Corsewall Point, Larne, Ronaldsway, Orlock Head, Killough and Kilkeel.

A forecast, valid until 2400, is broadcast for inshore waters of Great Britain and Northern Ireland on BBC Radio 3, at 0655 local time daily, on 1215kHz.

The schedule of coastal waters forecasts for Great Britain and Ireland is included in Table 7(2) on page 154. For details of local radio station forecasts see 7.2.17 and Table 7(4) on pages 156–157.

7.2.3 BBC General Forecasts
Land area forecasts may include the outlook period (up to 48 hours beyond the shipping forecast) and some reference to weather along the coasts. The more detailed land area forecasts are broadcast on Radio 4 on 198kHz (1515m).

7.2.4 Special forecasts for sea areas
Forecasts for areas within the region 65°N to 35°N, and 40°W and the coast of Europe (including the Mediterranean) may be obtained by yachtsmen by contacting the appropriate Meteorological Office forecasting centre, listed below. This may be done by telephone, which may be by link call. No charge is made by the Met. Office.

Alternatively, the request may be addressed to the nearest UK Coast Radio Station. Such a call might be in the form: 'North Foreland Radio. Request weather forecast next 24 hours for sea areas Dover and Wight on passage Ramsgate to Cherbourg.'

When telephoning a forecast centre it must be realised that during busy periods, such as occasions of bad weather, the staff may be fully occupied and there is likely to be a delay.

If a forecast is required for some future occasion or period, or if the forecast is to be kept under review and up-dated, the request should be addressed to The Director-General, Meteorological Office, Met.02a, London Road, Bracknell, Berks, RG12 2SZ, or sent by telex 849801 WEABKA G, giving full details of the service required and the address to which the account is to be forwarded. Cheques should be crossed and made payable to 'Met. Office HQ. Public a/c'.

Weather Centres (marked by asterisk*) and Weather Information Offices

Plymouth*	(0752) 402534
Southampton*	(0703) 228844
London*	01-836 4311

Norwich*	(0603) 660779
Nottingham*	(0602) 384092
Leeds*	(0532) 451990
Newcastle*	091-232 6453
Aberdeen Airport	(0224) 722334
Kirkwall Airport	(0856) 3802
Sella Ness, Sullom Voe	(0806) 242069
Glasgow*	041-248 3451
Manchester*	061-477 1060
Cardiff*	(0222) 397020
Bristol*	(0272) 279298
Belfast Airport, Crumlin	(08494) 52339
Jersey	(0534) 23660

Irish Republic

Central Forecast Office, Dublin (H24)	(01) 424655
Cork Airport Met. (0900–2000)	(021) 965974
Shannon Airport Met. (H24)	(061) 61333

The numbers of overseas forecast offices and recorded weather messages are shown under 'Telephone' for individual harbours in Areas 14–20 of Chapter 10.

7.2.5 Automatic Telephone Weather Service – Marinecall

Recorded Met Office forecasts are updated twice daily (three times in summer) and cover up to 12 miles offshore including Channel and Irish Sea crossings, Isles of Scilly, Channel Islands, Orkney and the Isle of Man. There are 15 areas, as below. In each case dial 0898 500, and then the three figure number of the area required. The charge is 38p per minute (peak and standard rates) and 25p per minute (evenings and weekends), including VAT. Calls can be made through British Telecom coast stations when the VHF per minute tariff is 95p.

458 South West (Hartland Point to Lyme Regis)
457 Mid-Channel (Lyme Regis to Selsey Bill)
456 Channel East (Selsey Bill to North Foreland)
455 Anglia (North Foreland to The Wash)
454 East (The Wash to Whitby)
453 North East (Whitby to Berwick)
452 Scotland East (Berwick to Rattray Head)
451 Scotland North (Rattray Head to Cape Wrath)
464 Minch (Cape Wrath to Ardnamurchan Point)
463 Caledonia (Ardnamurchan Point to Mull of Kintyre)
462 Clyde (Mull of Kintyre to Mull of Galloway)
461 North West (Mull of Galloway to Colwyn Bay)
460 Wales (Colwyn Bay to St David's Head)
459 Bristol Channel (St David's Head to Hartland Point)
465 Ulster (Carlingford Lough to Lough Foyle)

7.2.6 Facsimile broadcasts

Facsimile receivers are available of a type and price suitable for use in yachts. They will receive and reproduce the various charts (isobaric, isothermal, wind direction etc.) which are broadcast at specific times each day.

Information from many meteorological stations is processed at major centres such as Bracknell (England), Paris (France) and Offenbach (Germany), and is then codified and transmitted as a radio signal.

For further details see 7.6.12 of *The Macmillan & Silk Cut Yachtsman's Handbook* and for times and frequencies of transmissions refer to the *Admiralty List of Radio Signals, Vol 3*.

7.2.7 Navtex

Gale warnings are broadcast on receipt and are repeated at scheduled times. Weather messages are broadcast at scheduled times. Details of NAVTEX stations, identify codes, message categories and transmission times for NAVAREAS I and II are given in Chapter 6 (6.3.14).

7.2.8 Volmet

Volmet broadcasts of weather information for aircraft in flight can be useful with respect to airfields close to the UK and European coasts. The most useful is the HF SSB (H3E) broadcast by Shannon which operates H24 as follows:

3413 kHz	Sunset to sunrise
5640 kHz	H24
8957 kHz	H24
13264 kHz	Sunrise to sunset

The schedule starts at H+00 at five minute intervals until all airfields are covered. UK and Eire airfield broadcasts are at H+05 and H+35.

Royal Air Force SSB VOLMET broadcasts actual weather (H24) for a number of military and civil airfields, mostly in the UK, on 4722 kHz and 11200 kHz.

7.2.9 W/T transmissions

W/T transmissions of weather information in considerable detail are intended for ocean going vessels. The average yachtsman has neither the equipment nor the ability to receive the coded signals at the speed at which they are transmitted by W/T. However, equipment is available which will decode (or encode) Morse signals, and print out the message in English.

Portishead Radio broadcasts Morse transmissions for the Atlantic Weather Bulletin Areas, see *Admiralty List of Radio Signals Vol 3*.

7.2.10 Press forecasts

The delay between the time of issue and the time at which they are available next day make press forecasts of limited value to yachtsmen. However, the better papers publish forecasts which include a synoptic chart which, in the absence of any other chart, can be helpful when interpreting the shipping forecast on first putting to sea.

7.2.11 Television forecasts and Prestel

Some TV forecasts show a synoptic chart which, with the satellite pictures, can be a useful guide to the weather situation at the start of a passage.

Weather information is given on teletext by Ceefax (BBC) and Oracle (ITV). Oracle shows the shipping forecast (updated three times daily) on page 162. Antiope is the French system.

Prestel (operated over a telephone link by British Telecom) has a great deal of weather data supplied by the Met. Office, including shipping forecasts and synopsis for all British sea areas. Main index page, key 209: land areas 2091: shipping and sailing 2093: actual weather in UK 20940: actual weather world-wide 2094: aviation Fcsts 20971: European Fcsts 20915: UK weather index 20904: What's New 209091.

Shipping Fcst, 4 times/day 20930 (20p). Gale warnings since last Fcst 20931 (4p).

Sea crossings. 3 times/day (4p). Southern North Sea 209330. Dover Strait 209331. English Channel (East) 209332. St Georges Channel 209333. Irish Sea 209334.

Channel East (North Foreland – Selsey Bill and Le Havre – Calais) 209352 (50p). Mid-Channel (Selsey Bill – Lyme Bay plus Channel Isles/Cherbourg Peninsular) 209353 (50p).

UK land Fcsts. Caption chart. 3 times/day 20911 (5p). Text, 4 times/day 20910 (5p). 3 day outlook, once a day 20913 (as shown). 7 day outlook, issued Sun 20914 (as shown).

7.2.12 Visual storm signals

Visual storm signals used on the Continent are summarised in 10.14.7, 10.19.7 and 10.20.7.

7.2.13 Wind information — La Corbière Lt Ho

Wind information at La Corbière Lt radiobeacon (49°10′.8N/2°14′.9W) is transmitted in sequence 6 on carrier frequency 305.7 kHz modulated as follows: Call sign CB 4 times at 500 Hz. One to eight dots at 1000 kHz for wind direction by eight cardinal points (one dot = NE, two dots = E, clockwise to eight dots = N). Up to eight dots at 500 kHz for average Beaufort scale (eight dots = Force 8 or more). One or more dots at 1000 kHz for maximum gust above average Beaufort scale.

7.2.14 British and Irish Coast Radio Stations — Weather Bulletins by R/T

Forecasts originating from the Meteorological Office are broadcast by British Telecom Coast Radio Stations, as indicated in Table 7(1) below, by radiotelephone on Medium Frequency, and simultaneously on VHF (where available) after an initial announcement on VHF Ch 16. These stations also give weather information on request — see 7.2.4. Weather messages comprise gale warnings, synopsis and 24-hour forecast for the areas stated.

TABLE 7(1)

Station (*VHF only)	Frequency MF — kHz	VHF Ch	Forecast areas
At 0803 and 2003 (GMT)			
Land's End	2670	27	Wight, Portland,
Start Point*	—	26	Plymouth, Biscay
Pendennis*	—	62	Finisterre, Sole, Lundy, Fastnet, Shannon.
North Foreland	1848	26	Humber, Thames,
Thames*	—	02	Dover, Wight.
Hastings*	—	07	
Orfordness*	—	62	
Cullercoats	2719	26	Tyne, Dogger,
Whitby*	—	25	Fisher, German Bight, Humber.
Wick	1792 1824	—	Viking, North
Shetland*	—	27	Utsire, South
Orkney*	—	26	Utsire, Forties,
Cromarty*	—	28	Cromarty, Fair
Collafirth*	—	24	Isle, Faeroes, SE Iceland.
Hebrides	2667	26	Malin, Hebrides,
Skye*	—	24	Bailey, Rockall.
Lewis*	—	05	
Anglesey		26	Irish Sea.
Morecambe Bay*		04	
Cardigan Bay*		03	
At 0833 and 2033 (GMT)			
Niton	1834	28	Dover, Wight,
Weymouth Bay*		05	Portland
Humber	1869	26	Tyne, Dogger,
Bacton*	—	07	German Bight,
Grimsby*	—	27	Humber, Thames.
Stonehaven	2691	26	South Utsire,
Forth*	—	24	Forties, Cromarty,
Buchan*	—	25	Forth, Fisher.

At 0833 and 2033 (GMT) (continued)

Portpatrick	1883	27	Lundy, Irish Sea,
Clyde*	—	26	Malin.
Islay*	—	25	
Ilfracombe	—	05	Lundy, Fastnet.
Celtic*	—	24	
Valentia	1827	—	Shannon, Fastnet

Note: Irish VHF coast stations (see 6.3.16) broadcast weather bulletins every three hours from 0100 local time, and gale warnings on receipt and at the next of the following times: 0030 0630 1230 1830.

At 0645, 0745, 1245, 1845, and 2245 (GMT)

Jersey	1726	25 Channel Islands
		82 waters, south of 50°N and east of 3°W.

Note: Strong wind warnings for Channel Islands waters are broadcast on receipt, and repeated at next silence period and at 0307, 0907, 1507, 2107.

Gale warnings

British Telecom Coast Radio Stations transmit gale warnings at the end of the next R/T Medium Frequency silence period after receipt. These silence periods are from 00 to 03 and from 30 to 33 minutes past each hour. Gale warnings are repeated at the next of the following times: 0303, 0903, 1503, 2103 GMT. Gale warnings are preceded by the R/T Safety Signal 'SÉCURITÉ' (pronounced 'SAY-CURE-E-TAY').

Gale warnings remain in force unless amended or cancelled. If the gale persists for more than 24 hours from the time of origin, the gale warning is re-issued.

7.2.15 HM Coastguard — VHF Ch 67

Each MRCC and MRSC keeps watch on VHF Ch 16 and operates Ch 67 – see 8.2.2. They broadcast strong wind warnings for their local area (only) on receipt on Ch 67 after an announcement on Ch 16; also forecasts for their local area on Ch 67 after an announcement on Ch 16 normally every four hours (every two hours if strong wind or gale warning in force) commencing from the local (clock) times shown below. They will also respond to telephone enquires (for the number see each harbour in Chap 10).

Forecasts every four (or two) hours from: Swansea 0005, Moray 0010, Clyde 0020, Yarmouth 0040, Solent 0040, Brixham 0050, Shetland 0105, Dover 0105, Stornoway 0110, Hartland 0120, Pentland 0135, Falmouth 0140, Tyne/Tees 0150, Forth 0205, Liverpool 0210, Portland 0220, Holyhead 0235, Oban 0240, Thames 0250, Belfast 0305, Aberdeen 0320, Milford Haven 0335, Humber 0340, Ramsey 0350.

7.2.16 Reports of present weather

Reports of actual local weather can be obtained from Meteorological Office stations at the following telephone numbers. Shoeburyness (office hours) (03708) 2271. Kinloss, Forres (0309) 72161, Ext 674. Wick (0955) 2216. Sella Ness, Sullom Voe (0806) 242069. Kirkwall (0856) 3802. Stornoway (0851) 2256 (night 2282). Benbecula (0870) 351. Tiree, Scarinish (08792) 456. Prestwick (0292) 78475. Carlisle (0228) 23422, Ext 440. Ronaldsway, Castletown (0624) 823311 (night 823313). Blackpool (0253) 43061 (night 43063).

HM Coastguard MRCCs/MRSCs

Each MRCC/MRSC (manned continuously) can give information on local weather. In Chapter 10, the telephone number of the nearest centre is given for each harbour in Areas 1–11, and in Area 13 (for Northern Ireland). At sea, if RT contact cannot be made with the local Coast Radio Station, call the MRCC/MRSC on Ch 16 and work Ch 67. See 7.2.15 and 8.2.2.

Lighthouses, CG Sector Bases etc

The following lighthouses and other stations may be able to give information on actual weather locally. Those shown in *italics* are not manned continuously. Round Island LH (Scilly), Sennen (073687) 559. *Lizard*, Falmouth (0326) 290444. Brixham, Brixham (08045) 2156. *Portland Bill*, Portland (0305) 820400. St Catherine's LH, Niton (0983) 730284. *Eastbourne*, Eastbourne (0323) 20634. *Fairlight*, Hastings (0424) 813171. *Aldeburgh*, Aldeburgh (072885) 2779. *Cromer LH*, Cromer (0263) 512123, *Cromer*, Cromer (0263) 512507. *Whitby*, Whitby (0947) 602107. St Abb's Head LH, Coldingham (03903) 287. Bass Rock LH, Dunbar (0368) 63640. *Kinnaird Head LH*, Fraserburgh (03462) 3044. Strathy Point LH, Strathy (06414) 210. Cape Wrath LH, Durness (097181) 230. Butt of Lewis LH, Stornoway (0851) 81201. Hyskeir LH, Tobermory (0688) 2423. Ardnamurchan LH, Kilchoan (09723) 210. Neist Point LH, Glendale (047081) 200. Rhinns of Islay LH, Bowmore (049681) 233. Pladda LH, Turnberry (06553) 657. Corsewall Point LH, Kirkholm (077685) 220. Mull of Galloway LH, Drummore (077684) 211. St Bees LH, Whitehaven (0946) 2635. Point of Ayre LH, Kirkandreas (062488) 238. *Rhyl*, Rhyl (0745) 53284. *Porth Dinllaen*, Nefyn (0758) 720204. *Aberdovey*, Aberdovey (065472) 327. Lundy South LH, Clovelly (02373) 455. *Killough*, Ardglass (0396) 203. Bangor, Donaghadee (0247) 882982. *Ballycastle*, Ballycastle (02657) 63519. *Portrush*, Portrush (0265) 823356.

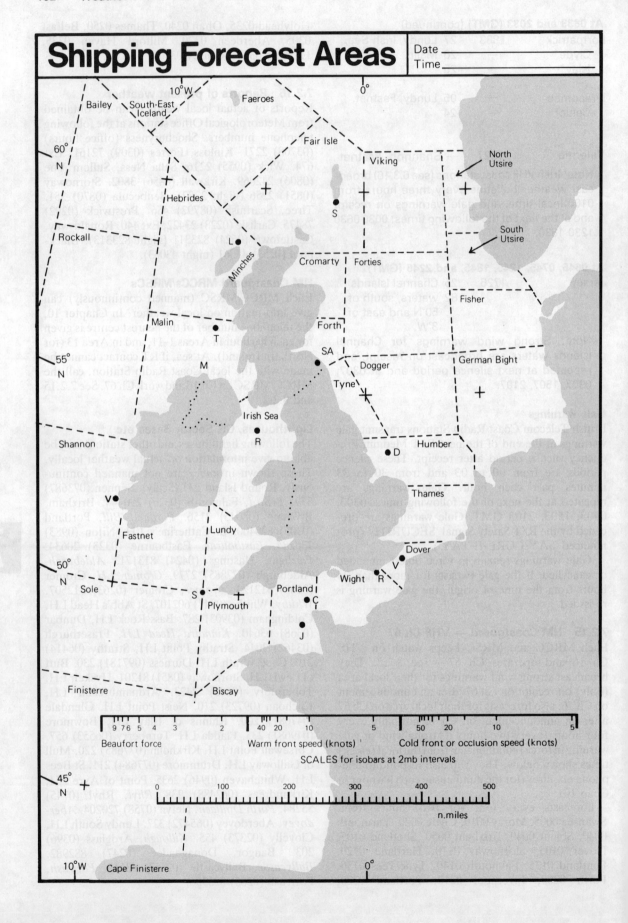

Shipping Forecast Record

GENERAL SYNOPSIS at _____ GMT/BST

System	Present position	Movement	Forecast position	at

Gales	SEA AREA FORECAST	Wind (At first)	Wind (Later)	Weather	Visibility
	VIKING				
	NORTH UTSIRE				
	SOUTH UTSIRE				
	FORTIES				
	CROMARTY				
	FORTH				
	TYNE				
	DOGGER				
	FISHER				
	GERMAN BIGHT				
	HUMBER				
	THAMES				
	DOVER				
	WIGHT				
	PORTLAND				
	PLYMOUTH				
	BISCAY				
	FINISTERRE				
	SOLE				
	LUNDY				
	FASTNET				
	IRISH SEA				
	SHANNON				
	ROCKALL				
	MALIN				
	HEBRIDES				
	BAILEY				
	FAIR ISLE				
	FAEROES				
	SE ICELAND				

COASTAL REPORTS at ____ BST/GMT	Wind Direction	Force	Weather	Visibility	Pressure	Change
Tiree						
Butt of Lewis						
Sumburgh						
St Abb's Head						
Dowsing						
Varne/Dover						

COASTAL REPORTS	Wind Direction	Force	Weather	Visibility	Pressure	Change
Royal Sovereign						
Channel Lt V						
Scilly						
Valentia						
Ronaldsway						
Malin Head						
Jersey						

TABLE 7(2) British Isles — Daily Shipping Forecasts and Forecasts for Coastal Waters
as broadcast by British Broadcasting Corporation (BBC) and Radio Telefis Eireann (RTE)

Note: All times are local or 'clock' times, unless otherwise stated

Time	Forecast	Contents	Stations and Frequencies
0010	Coastal Waters (Northern Ireland)	Fcst, valid until 1800 for coastal waters of N Ireland up to 12 nm offshore, and reports from selected stations.	*BBC Radio Ulster:* 1341kHz.
0033	Shipping Forecast — Home Waters Fcst Areas	Gale warnings in force, synopsis, 24h Fcst for Home Waters Fcst Areas (including Trafalgar, and Mon-Fri a 24h Fcst for area Minch); reports from selected stations.	*BBC Radio 4:* 198kHz Tyneside 603kHz, London 720kHz, N Ireland 720kHz, Aberdeen 1449kHz, Carlisle 1485kHz, Plymouth 774kHz, Redruth 756kHz. *BBC Radio Scotland:* 810kHz.
0038	Coastal Waters (Great Britain)	Fcst, valid until 1800, for coastal waters of Great Britain up to 12 nm offshore; reports from selected stations.	*BBC Radio 4 and Radio Scotland:* as for 0033 Shipping Forecast.
0555	Shipping Forecast — Home Waters Fcst Areas	Gale warnings in force, synopsis, 24h Fcst for Home Waters Fcst Areas; reports from selected stations.	*BBC Radio 4:* as for 0033 Shipping Forecast.
0655	Coastal Waters (Great Britain)	Fcst, valid until 2400, for coastal waters of Great Britain up to 12 nm offshore.	*BBC Radio 3:* 1215kHz.
0633	Coastal Waters (Ireland)	Gale warnings in force; 24h Fcst for Irish coastal waters up to 30 nm offshore and the Irish Sea.	*RTE – Radio 1:* Tullamore 567kHz, Cork 729kHz.
1155 (1255 on Sun)	Coastal Waters (Ireland)	Gale warnings in force; 24h Fcst for Irish coastal waters up to 30 nm offshore and the Irish Sea.	*RTE – Radio 1:* Tullamore 567kHz, Cork 729kHz.
1355	Shipping Forecast — Home Waters Fcst Areas	Gale warnings in force, synopsis, 24h Fcst for Home Waters Fcst Areas; 24h Fcst for area Minch (Mon–Fri); reports from selected stations.	*BBC Radio 4:* as for 0033 Shipping Forecast.
1750	Shipping Forecast — Home Waters Fcst Areas	Gale warnings in force, synopsis, 24h Fcst for Home Waters Fcst Areas; reports from selected stations.	*BBC Radio 4:* as for 0033 Shipping Forecast.
1755	Coastal Waters (Ireland)	Gale warnings in force; 24h Fcst for Irish coastal waters up to 30 nm offshore and the Irish Sea.	*RTE – Radio 1:* Tullamore 567kHz, Cork 729kHz.
2352	Coastal Waters (Ireland)	Gale warnings in force; 24h Fcst for Irish coastal waters up to 30 nm offshore and the Irish Sea.	*RTE – Radio 1:* Tullamore 567kHz, Cork 729kHz.

Forecasts 155

TABLE 7(3) Gale Warnings		
Stations	**Areas covered**	**Times**
BBC Radio 4: 198kHz, Tyneside 603 kHz, London 720kHz, N Ireland 720kHz, Aberdeen 1449kHz, Carlisle 1485kHz, Plymouth 774kHz, Redruth 756kHz.	Broadcast gale warnings for all Home Waters Fcst Areas, including Trafalgar	At the first available programme junction after receipt and after the first news bulletin after receipt
RTE – Radio 1: Tullamore 567kHz, Cork 729kHz.	Broadcast gale warnings for Irish coastal waters up to 30 nm offshore and the Irish Sea	At first programme junction after receipt and with news bulletins (0630–2352)
RTE – Radio 2: Athlone 612kHz, Dublin and Cork 1278kHz.		At first programme junction after receipt and with news bulletins (0630–0150).

British Coast Radio Stations broadcast gale warnings for adjacent areas at the end of the first silence period after receipt (i.e. at H + 03 or H + 33) and subsequently at the next of the following times: 0303, 0903, 1503, 2103 GMT. For further details see 7.2.14. For Irish coast stations see 6.3.16 on page 139.

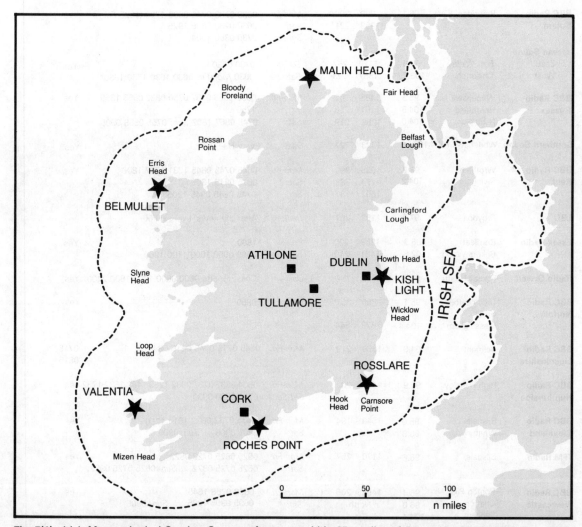

Fig. 7(1) Irish Meteorological Service. Sea area forecasts within 25 n miles of the coast, plus the Irish Sea. Also shown are the stations (starred) from which reports are included in the first and last broadcasts each day and the headlands used to divide up the coastline into smaller areas, depending on the expected weather situation. RTE broadcast stations are also indicated.

Table 7(4) Local Radio Stations — Coastal Weather Forecasts

Station	VHF Transmitter(s)	VHF (MHz)	MF (kHz)	(m)		Coastal Waters Forecasts (local times) (*summer months only)	Small Craft Warnings
BBC Radio Cornwall							
Redruth	Redruth	103.9	630	476	Mon-Fri:	0605 0745 0845 1245 1710 1740	Yes
Bodmin	Caradon Hill	95.2	657	457	Sat:	0715 0745 0815 0845	
Scilly	Scilly	96.0	—	—	Sun:	0915 0945 1015	
Plymouth Sound	Plympton	97.0	1152	261	Daily:	Approx 0603 0615	Yes
Devonair Radio (Exeter/Torquay)	St Thomas	97.0	666	450	Daily:	Every H+00 after the news Brixham Coastguard 0830	Yes
	Beacon Hill	96.4	954	314			
BBC Radio Devon							
Exeter	Exeter	95.8	990	303	Mon-Fri:	0633 0733 0833 1310 1740	Yes
Plymouth	N Hessary Tor	103.4	855	351	Sat:	0633 0833 1310	
North Devon	Huntshaw Cross	94.8	801	375	Sun:	0833 1310	
Torbay	N Hessary Tor	103.4	1458	206			
Okehampton	Okehampton	96.0	—	—			
2CR Bournemouth	Poole	97.2	828	362	Daily:	0710 0810 0906 1310 1710	Yes
BBC Radio Solent	Rowridge, IOW	96.1	999	300	Mon-Fri:	0633 0733 0745 0833 1725 (1825 Fri)	Yes
			1359	221	Sat:	0630 0730 0835 1925	
					Sun:	0730 0850 1504	
Ocean Sound							
East	Fort Widley	97.5	1170	257	Daily:	0730 0830	Yes
West	Chillerton	103.2	1557	193	Sat:	0630 Mon-Fri: 0630 1630 1730 1830*	
BBC Radio Sussex	Whitehawk Hill	95.3	1485	202	Mon-Fri:	0632 0653 0732 0750 0832 0855 1355 1750	Yes
	Heathfield	104.5	1161	258			
	N Sussex	104.0	1368	219	Sat:	0748 0857 1802 Sun: 0755 0858 0905	
Southern Sound	Whitehawk Hill	103.5	1323	227	Daily:	Every H+59	Yes
BBC Radio Kent	Wrotham	96.7	1035	290	Mon-Fri:	0645 0745 0845 1231 1707 1806	Yes
	Swingate	104.2	774	388	Sat:	0815 0915 1305	
			1602	187	Sun:	0745 0845 0945 1305	
LBC	Croydon	97.3	1152	261	Daily:	After the news, every H+00	—
Essex Radio	Benfleet	96.3	1359	220	Fri:	1800	Yes
	Bakers Wood	102.6	1431	210	Sat-Sun:	0800 0900 1000 1100 1200	
Radio Orwell	Foxhall Heath	97.1	1170	257	Daily:	Coast reports 0600 0800 1200 1600 2400	Yes*
BBC Radio Norfolk	Tacolneston	95.1	855	351	Daily:	0850	Yes
	Great Massingham	104.4	873	344			
BBC Radio Lincolnshire	Belmont	94.9	1368	219	Mon-Fri:	0645 0715 0745 0815 0845	0715 0815
BBC Radio Humberside	High Hunsley	95.9	1485	202	Mon-Fri:	0604 0632 0732 0832 1259 1432 1632 1732	Yes
					Sat-Sun:	0730 0830 0930	
BBC Radio Cleveland	Bilsdale	95.0	1548	194	Mon-Fri:	0635 0732 0832 1314 1750	Yes
	Whitby	95.8			Sat:	0805 0905 Sun: 0905	
TFM Radio	Bilsdale	96.6	1170	257	Mon-Fri:	0525 0625 0722 0822	Yes
					Sat:	0625 0725 0822 Sun: 0625 0725 0922	
BBC Radio Newcastle	Pontop Pike	95.4	1458	206	Mon-Fri:	0658 0758 1645	Yes
	Chatton	96.0	in North		Sat:	0805 0905 1005 Sun: 0905	
	Fenham	104.4					
Metro Radio	Burnhope	97.1	1152	261	Mon-Fri:	2400 Sun: 0600 0900	Yes

Sources of Weather Information 157

Table 7(4) Local Radio Stations — Coastal Weather Forecasts Continued

Station	VHF Transmitter(s)	Frequencies VHF (MHz)	MF (kHz)	(m)	Coastal Waters Forecasts (local times) (*summer months only)		Small Craft Warnings
Radio Forth	Craigkelly	97.3	1548	194	Mon-Sun:	0615 0715 (October-March)	Yes
					Sat-Sun*:	0600 0700 0900 1200 1500 1800 2100	
Radio Tay	Dundee	102.8	1161	258	Daily:	At regular intervals	Yes
	Perth	96.4	1584	189			
Radio Clyde	Black Hill	102.5	1152	261	Daily:	At regular intervals (0600-1800)	Yes
West Sound, Ayr	Symington	96.7	1035	290	Daily:	Every H+00 and H+30 after news (0600-0900). Then every H+00	Yes
	Girvan	97.5	—	—			
BBC Radio Cumbria	Sandale	95.6	756	397	Mon-Fri:	0658 0758 1258	—
			1458	206	Sat:	0805 1030 Sun: 0840	
	Morecambe Bay	96.1	837	358			—
Manx Radio	Snaefell	89.0	1368	219	Mon-Fri:	0700 0800 0900 1310 1740	
	Richmond Hill	97.2			Sat:	0700 0800 0900 1200 1700	
					Sun:	0800 0859 1300	
BBC Radio Merseyside	Allerton	95.8	1485	202	Mon-Fri:	0633 0733 1145 1309 1744	
					Sat:	0725 1304 1804	
					Sun:	0904 1404 1804	
Radio City	Allerton	96.7	1548	194	Mon-Fri:	1315 Sat-Sun: 0715	
Swansea Sound	Kilvey Hill	96.4	1170	257	Mon-Fri:	0725 0825 0925 1725 Sat-Sun: 0825 1003	Yes
Red Dragon Radio	Wenallt	103.2	1359	221	Daily:	0630 0730 0830 1630 1730 H+00 (0600-2400)	Yes
	Christchurch	97.4	1305	230			
BBC Radio Bristol	Bristol	94.9	1548	194	Mon-Fri:	0605 0632 0659 0733 0759 0833 0859 every H+05 (0905-1705) 1739	Yes
	Wells	95.5	1323	227	Sat:	0805 0905 1105 1302	
	Bath	104.6			Sun:	0905 1105 1305	
GWR Radio	Dundry Hill	96.3	1260	238	Daily:	0645	Yes
Downtown Radio	Limavady	96.4			Mon-Fri:	0705 0805 0905 1005 1312 1403 1503 1710 2315	Yes
	Black Mountain	97.4					
	Sheriff's Mountain	102.4			Sat:	0705 0805 1215	
	Brougher	96.6			Sun:	1105 2303 (0003 Mon)	
BBC Radio Guernsey	—	93.2	1116	269	Mon-Fri:	0732* 0832* 1232* 1715	Yes
					Sat:	0810 0910 1005 Sun: 1005 1205	
BBC Radio Jersey	—	88.8	1026	292	Mon-Fri:	0700 0735 0815 0829 0900 1315 1740	Yes
					Sat:	0800 0815 0830 0903	
					Sun:	0805 0815 0912 1259	

7.2.17 Local Radio Stations — coastal forecasts

The details and usefulness of forecasts broadcast by local radio stations vary considerably. Some give no more than an indication of the present weather conditions, while others provide more responsible forecasts in conjunction with the local Weather Centre. The timings of weather information from local radio stations most likely to be of interest to yachtsmen in local coastal waters are shown in Table 7(4). Note that many stations are due to change their VHF frequencies shortly.

Many local radio stations in coastal areas participate in a scheme for broadcasting 'Small Craft Warnings' when winds of Force 6 or more are expected within the next 12 hours on the coast or up to five miles offshore. These warnings are handled in much the same way as gale warnings on Radio 4, being broadcast at the first programme junction or at the end of the first news bulletin after receipt. The stations which participate in this scheme are indicated in Table 7(4). In most cases the services operates from Good Friday until 31 October.

Table 7(5) Western Europe — Shipping Forecasts, Coastal Waters Forecasts, Gale Warnings

Notes:
1. All times local or 'clock' times unless otherwise stated.
2. Unless otherwise described, forecasts include gale or near gale warnings, synopsis, 12h Fcst and outlook for further 12h.
3. Forecasts read in English are printed in **bold type** in columns 2 and 3.
4. On receipt and at end of next two silence periods.

Time	Forecast areas	Contents	Stations and Frequencies	Gale Warnings
FRANCE 0655 2005	French Fcst areas 1-24	See note 2; in French	*France Inter Allouis* 164kHz; **Bayonne, Brest & Lille** 1071kHz	
0733 1233	Spanish border to Sables d'Olonne	See note 2; in French	**Bayonne** Ch 24, **Bordeaux-Arcachon** Ch 82, **Royan** Ch 23, **La Rochelle** Ch 21	See note 4 (0700-2200); in French for coastal waters, Spanish border to Sables d'Olonne
0800 1430 1100 1800	Spanish border to Sables d'Olonne	See note 2; in French	**Soulac**, CROSS: Ch 13	See note 4.
0703 GMT 1703 GMT	French Fcst areas 23-24	See note 2; in French	**Bordeaux-Arcachon** 1820kHz (and on request 0700-2200)	See note 4 (0700-2200); in French for areas 23-24. Also 1862kHz every even H+07 (0607-1807) GMT.
0803 GMT 1803 GMT	French Fcst areas 14-24	See note 2; in French	**St Nazaire** 1722kHz	See note 4; in French for areas 14-16. Also 1687 kHz every odd H+07 GMT
0400 1410 0830 1910	Sables d'Olonne to Penmarc'h	See note 2; in French	**Etel**, CROSS: Ch 13	
0733 1233	Sables d'Olonne to Cap de la Hague	See note 2; in French	**St Gilles Croix de Vie** Ch 27, **Nantes St Herblain** Ch 28, **St Nazaire** Ch 23, **Belle Ile** Ch 87, **Pont-l'Abbé** Ch 27, **Brest-Le Conquet** Ch 26, **Ouessant** Ch 82, **Plougasnou** Ch 83, **Paimpol** Ch 84, **St Malo** Ch 02	See note 4; in French for coastal waters, Sables d'Olonne to Cap de la Hague
0733 GMT 1633 GMT 2153 GMT	French Fcst areas 14-22	See note 2; in French	**Brest-Le Conquet**: 1673kHz, **Quimperlé**: 1876kHz, **St Malo**: 2691kHz	See note 4; in French for areas 14-22. 1806 & 2691kHz every even H+03 GMT
Every 3 hours from 0150 GMT	French Fcst areas 14–16	See note 2; **in English** (or French)	**Ouessant Traffic** (Ushant Control Centre, at Corsen): Ch 11 after announcement on Ch 16. General bulletin Ch 11, H+10, H+40	
0900 1900 1600	Penmarc'h to Granville	See note 2; in French	**CROSS**, Ile de Sein Ch 13	
0930 1930 1630	Penmarc'h to Granville	See note 2; in French	**CROSS**, Ile de Batz Ch 13	
0733 1233	Cap de la Hague to Belgian border	See note 2; in French	**Cherbourg** Ch 27, **Port en Bessin** Ch 03, **Le Havre** Ch 82, **Dieppe** Ch 02, **Boulogne** Ch 23, **Calais** Ch 87, **Dunkerque** Ch 61	See note 4; in French for coastal waters Cap de la Hague to Belgian border
0703 GMT 1733 GMT	French Fcst areas 1–14	See note 2; in French	**Boulogne** 1694kHz	See note 4; in French for areas 1-14. 1771kHz every odd H+03 GMT

continued on page 160

Forecasts — Western Europe 159

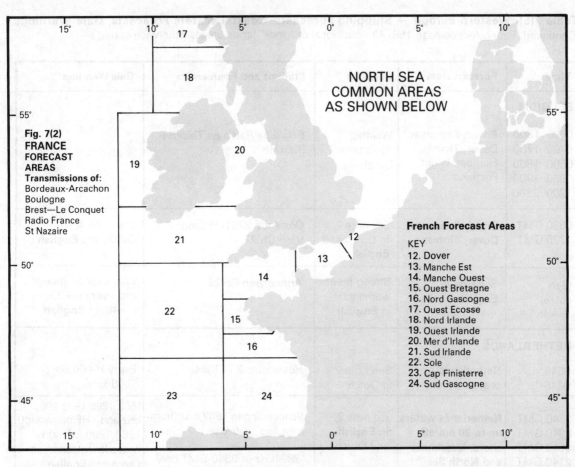

**Fig. 7(2)
FRANCE
FORECAST
AREAS
Transmissions of:**
Bordeaux-Arcachon
Boulogne
Brest—Le Conquet
Radio France
St Nazaire

French Forecast Areas
KEY
12. Dover
13. Manche Est
14. Manche Ouest
15. Ouest Bretagne
16. Nord Gascogne
17. Ouest Ecosse
18. Nord Irlande
19. Ouest Irlande
20. Mer d'Irlande
21. Sud Irlande
22. Sole
23. Cap Finisterre
24. Sud Gascogne

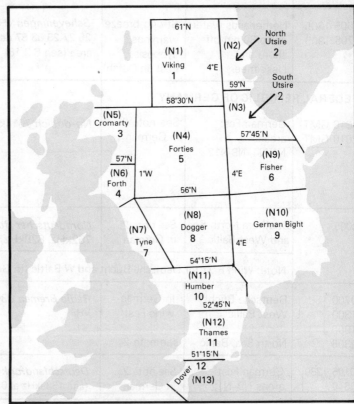

**Fig. 7(3)
NORTH SEA COMMON SHIPPING
FORECAST AREAS**
As used for forecasts from:
 Belgium
 Netherlands
 West Germany
 Denmark
 Norway
 United Kingdom

Notes:
(1) Numbers are as for
French Forecast Areas.
(2) Numbers in brackets
with prefix N are as
for German Forecast Areas.

Table 7(5) Western Europe — Shipping Forecasts, Coastal Waters Forecasts, Gale Warnings
(Continued. For notes see page 158. All times local or 'clock' times unless otherwise stated.)

Time	Forecast areas	Contents	Stations and Frequencies	Gale Warnings
BELGIUM				
0600 1300 0700 1700 0800 1800 0900 1900 1200 2300	British Fcst areas Dover, Thames, Humber, Wight, Portland	Weather reports; in Dutch	*Belgische Radio en Televisie* 926kHz	
0820 GMT 1720 GMT	**British Fcst areas Dover, Thames**	See note 2; in Dutch and **English**	*Oostende* 2761kHz and VHF Ch 27	See note 4; in Dutch and **English**
(See col. 5)	**Schelde** in **English**	Strong breeze warnings in **English**	*Antwerpen* Ch 24	Warnings on receipt and every H+03, H+48; in **English**
NETHERLANDS				
0545 0645	Netherlands coastal waters	See note 2; in Dutch	*Hilversum 2* 747 kHz	Every H+00 on 1008 kHz
0340 GMT 0940 GMT 1540 GMT 2140 GMT	**Netherlands waters up to 30 nm offshore, IJsselmeer and North Sea**	see note 2; in **English** and Dutch	*Scheveningen* 1862 kHz (from Nes) and 1890 kHz. 2824kHz at 0340 GMT only	1862 1890 1939 2600 kHz and VHF on receipt. 1862 1890 kHz after next silence period; in Dutch and **English**
0005 1305 0705 1905	Netherlands waters up to 30 nm offshore and IJsselmeer	Strong breeze warnings, synopsis, Fcst (in Dutch)	*Scheveningen* VHF Ch 23 25 27 28 83 87 depending on area (see 6.3.16)	Strong breeze warnings on receipt and every H+05 on VHF; in Dutch
FEDERAL REPUBLIC OF GERMANY				
0810 GMT 2010 GMT	German Fcst areas N1-N4, N8-N12	See note 2; in German	*Norddeich* 2614kHz	In English for N10. In German for N1-N4, N8-N12. On receipt, after next silence period, and every 4h from 0133 GMT
0005	Southern North Sea and West Baltic	See note 2; in German	*Norddeutscher Rundfunk:* 702kHz, 828kHz, 972kHz	
	Note: Wind Fcst for Deutsche Bucht and W Baltic (in German) in weather reports after news			
0700 1300 1900 2305	Deutsche Bucht, West Baltic North Sea, Baltic	In German, wind Fcst see note 2;	*Radio Bremen* 936kHz and VHF.	
0105 1230 0640	German Fcst areas N9–N12	See note 2; in German	*Deutschlandfunk* 1269kHz (and 1539kHz at 0105 only)	Every H+00 (except 2100) in German

Chapter 8
Safety

Contents

8.1 SAFETY EQUIPMENT   Page 162
8.1.1   Safety equipment — general
8.1.2   Safety equipment — legal requirements
8.1.3   Safety equipment — recommendations for seagoing yachts 18–45ft (5.5–13.7m) overall length
8.1.4   Distress signals

8.2 SEARCH AND RESCUE (SAR)   164
8.2.1   SAR — general
8.2.2   HM Coastguard
8.2.3   Royal National Lifeboat Institution (RNLI)
8.2.4   SAR — communications
8.2.5   Helicopter rescue
8.2.6   Abandon ship

Safety — introduction

The following subjects are described in detail in Chapter 8 of *The Macmillan & Silk Cut Yachtsman's Handbook*:

Safety equipment — legal requirements and recommended outfits; radar reflectors; bilge pumps; guardrails; fire prevention and fire fighting; lifejackets; safety harnesses; man overboard gear and drill; liferafts; distress signals; SAR organisation; response to distress calls; abandoning ship; liferafts; helicopter rescues; first aid afloat.

Here in the Almanac is given basic information about safety equipment, distress signals and SAR operations. For further information on these subjects, together with those listed above, reference should be made to Chapter 8 of *The Macmillan & Silk Cut Yachtsman's Handbook*.

8.1 SAFETY EQUIPMENT

8.1.1 Safety equipment — general

The skipper is responsible for the safety of the boat and all on board. He must ensure that:

(1) The boat is suitable in design and in construction for her intended purpose.
(2) The boat is maintained in good condition.
(3) The crew is competent and sufficiently strong.
(4) The necessary safety and emergency equipment is carried, is in good condition, and the crew know how to use it.

Individual crew members are responsible for their personal gear. Non-slip shoes or boots are essential. So is foul-weather clothing with close fastenings at neck, wrists and ankles. At least two changes of sailing clothing should be carried, including warm sweaters and towelling strips as neck scarves. Other personal items include a sailor's knife and spike on a lanyard, a waterproof torch, and a supply of anti-seasick pills. Lifejackets and safety harnesses are usually supplied on board, but if individuals bring their own the skipper should make certain they are up to standard.

8.1.2 Safety equipment — legal requirements

Yachts more than 45ft (13.7m) in length are required to carry safety equipment as in the *Merchant Shipping (Life Saving Appliances)* and *Merchant Shipping (Fire Appliances) Rules* (HMSO).

All yachts must carry navigation lights and sound signals which comply with the *International Regulations for Preventing Collisions at Sea*.

Racing yachts are required to carry the safety equipment specified for the class/event concerned.

8.1.3 Safety equipment — recommendations for sea-going yachts 18–45ft (5.5–13.7m) overall length

Full details of recommended safety equipment are given in *The Macmillan & Silk Cut Yachtsman's Handbook*. Below are brief reminders of the minimum equipment which should be carried for (a) coastal and (b) offshore cruising.

But prevention is better than cure, and simple precautions can eliminate accidents. Be particularly careful with bottled gas and petrol. Fit a gas detector. Turn off the gas at the bottle after use. If gas or petrol is smelt — no naked lights, and do not run electrical equipment. Test systems regularly. Insist that crew wear lifejackets and harnesses when necessary, and that they do clip on. Make sure a good look-out is maintained at all times. Listen to every forecast. Double-check all navigational calculations. Take nothing for granted.

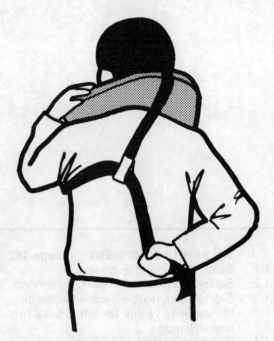

Fig. 8(1) *Putting on a lifejacket — first read the instructions. Hold the jacket up in front of you, put your head through the hole, and secure the waistband at the side or front as appropriate.*

Safety equipment list

	(a) Coastal	(b) Offshore
Safety		
Lifejackets, BS3595, per person	1	1
Harnesses, BS4224, per person	1	1
Navigation		
Charts, almanac, pilot	Yes	Yes
Compass with deviation card	1	1
Hand bearing compass	1	1
Chart table instruments	Yes	Yes
Watch/clock	1	2
Echo sounder	1	1
Leadline	1	1
Radio direction finding set	1	1
Radio receiver (forecasts)	1	1
Barometer	1	1
Navigation lights	Yes	Yes
Radar reflector	1	1
Foghorn	1	1
Powerful waterproof torch	1	1
Anchor with warp or chain	2	2
Towline	1	1
Man overboard		
Lifebuoy, with drogue and light	2	2
Buoyant heaving line	1	1
Dan buoy	–	1
Rope (or boarding) ladder	1	1

Distress Signals

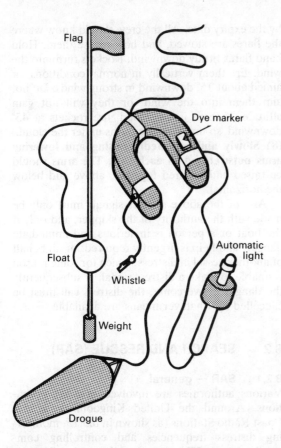

Fig. 8(2) *Lifebuoy with dan buoy and flag, automatic light, whistle, dye marker, and drogue.*

	(a) Coastal	(b) Offshore
Fire		
Fire extinguishers	2	3
Fire blanket	1	1
Sinking		
Bilge pumps	2	2
Buckets with lanyards	2	2
Leak stopping gear	Yes	Yes
Distress signals		
Hand flares, red	2	4
Hand flares, white (warning)	4	4
Red parachute rockets	2	4
Hand smoke signals	2	–
Buoyant orange smoke signals	–	2
Emergency radio transmitter	–	1
Abandon ship		
Liferaft for whole crew	1	1
or		
Dinghy with buoyancy, or inflated inflatable	1	–
Panic bag, extra water etc	–	1
Miscellaneous	(a) Coastal	(b) Offshore
First aid kit	1	1
Engine tool kit	1	1
Name/number prominently displayed	Yes	Yes
Storm canvas	Yes	Yes
Emergency steering arrangements	Yes	Yes

8.1.4 Distress signals

A distress signal must only be made if the yacht or a person is in serious and immediate danger, and help is urgently required.

For a lesser emergency, an urgency signal (PAN PAN) as described in 6.3.10 may be appropriate. If help is needed, but the boat is not in immediate danger, the proper signal is 'V' (Victor) International Code, meaning 'I require assistance'. This can be sent as a flag signal (a white flag with a red St Andrew's cross), or by light or sound in Morse code ($\cdots -$). Other signals of an emergency nature in the International Code are given in 6.1.5.

If medical help is required, the proper signal is 'W' (Whiskey) International Code, meaning 'I require medical assistance'.

A full list of the recognised distress signals is given in Annex IV of the *International Regulations for Preventing Collisions at Sea*, and is reproduced in Chapter 2 (2.1.6). The following are those most appropriate for yachts and small craft, together with notes on their use.

(1) **Continuous sounding with any fog signalling apparatus.** In order to avoid confusion, this is best done by a succession of letters SOS in Morse ($\cdots - - - \cdots$).

(2) **A signal made by radiotelegraphy or by any other signalling method consisting of the group** $\cdots - - - \cdots$ **(SOS) in the Morse Code.** For a yacht the most likely methods are by sound signal as in (1) above, or by flashing light.

(3) **A signal sent by radiotelephony consisting of the spoken word MAYDAY.** This procedure is fully described in 6.3.10, but for ease of reference the basic rules are repeated briefly below.

Check the battery switch is ON. Switch the set ON, and select HIGH power. Tune to VHF Ch 16 (or 2182 kHz for MF). If an alarm signal generator is fitted, operate it for at least 30 seconds. Press the 'transmit' button, and say slowly and distinctly:

'MAYDAY MAYDAY MAYDAY — THIS IS (name of boat spoken three times) — MAYDAY — (name of boat spoken once) — MY POSITION IS (latitude and longitude, or true bearing and distance from a known object) — (nature of distress) — (aid required) — (number of persons on board) — (any other important information) — OVER'

Release the transmit button, and listen. If no acknowledgment is heard, repeat the message. If transmitting on 2182 kHz, repeat the call during the silence periods commencing each hour and half-hour.

The Urgency Signal (PAN PAN, spoken three times) indicates a lesser emergency than a distress signal, and should be used when there is a very urgent message concerning the safety of a ship or person. For further detail see 6.3.10.

Portable emergency radiotelephones, on 2182 kHz or Ch 16, can be used from the yacht or the liferaft. Also available are Personal Locator Beacons (PLBs) and Emergency Position Indicating Radio Beacons (EPIRBs) operating on 121.5 and 243 MHz. These are primarily for aeronautical purposes and since maritime SAR authorities do not keep watch on these frequencies they are of limited use in coastal waters. More useful is an EPIRB transmitting on 406 MHz, the dedicated COSPAS/SARSAT satellite frequency with full global coverage. Like other radio transmitters, all such equipment must be licensed.

(4) **The International Code signal of distress 'NC'.** This can be made by flag hoist, N being a blue and white chequered flag and C one which is horizontally striped blue, white, red, white, blue.

(5) **A signal consisting of a square flag having above or below it a ball or anything resembling a ball.** This is not too difficult to contrive from any square flag, and a round fender or anchor ball.

(6) **A rocket parachute flare or a hand flare showing a red light.** A red flare is the most effective distress signal at night. Flares serve two purposes — first to raise the alarm, and then to pinpoint the boat's position. Within about three miles from land a hand flare will do both. At greater distances a red parachute rocket (which projects a suspended flare to a height of more than 1000ft, or 300m, and which burns for more than 40 seconds) is needed to raise the alarm, but hand flares are useful to indicate the boat's position. See further comments under (7) below.

(7) **A smoke signal giving off orange-coloured smoke.** By day orange smoke signals (hand held for short distances, or the larger buoyant type for greater ranges) are more effective than flares, although the smoke disperses quickly in a strong wind.

White flares are not distress signals, but are used to indicate the presence of a boat — to another vessel on a collision course for example. An outfit of four is suggested for boats which make night passages. Shield your eyes when using them, to protect night vision.

Pyrotechnics must be stowed where they are accessible, but protected from damp. In good storage conditions they should have a life of three years. Examine them regularly, and replace them by the expiry date. All the crew should know where the flares are stowed, and how to use them. Hold hand flares firmly downwind. Rockets turn into the wind: fire them vertically in normal conditions, or aimed about 15° downwind in strong winds. Do not aim them into the wind, or they will not gain altitude. If there is low cloud, fire rockets at 45° downwind, so that the flare burns under the cloud.

(8) **Slowly and repeatedly raising and lowering arms outstretched to each side.** The arms should be raised and lowered together, above and below the horizontal.

Any of the above distress signals must only be made with the authority of the skipper, and only if the boat or a person is in serious and immediate danger, and help is urgently required: or on behalf of another vessel in distress, which for some reason is unable to make a distress signal. If subsequently the danger is overcome, the distress call must be cancelled by whatever means are available.

8.2 SEARCH AND RESCUE (SAR)

8.2.1 SAR — general

Various authorities are involved in SAR operations. Around the United Kingdom these are: Coast Radio Stations (as shown in 6.3.15, monitoring distress frequencies and controlling communications); HM Coastguard (responsible for initiating and controlling all civil maritime SAR — see below); the RNLI (which supplies and mans the lifeboats — see below); the Royal Navy (which assists with ships and aircraft, including helicopters); the Royal Air Force (operating through Rescue Co-ordination Centres at Edinburgh and Plymouth); and Air Traffic Control Centres. Similar organisations to the above exist in other countries in Western Europe — see 10.12.7 (Republic of Ireland), 10.14.7 (France), 10.19.7 (Belgium, Netherlands), 10.20.7 (West Germany).

8.2.2 HM Coastguard

HM Coastguard initiates and controls SAR around the United Kingdom and over a large part of the eastern Atlantic. The area is divided into six Maritime Search and Rescue Regions (SRRs), supervised by Maritime Rescue Co-ordination Centres (MRCCs) at Falmouth, Dover, Great Yarmouth, Aberdeen, the Clyde and Swansea. It also includes 'Shannon' area which is the responsibility of the Republic of Ireland. Each SRR is divided into Districts, each with a Maritime Rescue Sub-Centre (MRSC). Their boundaries are shown in the maps at the start of Areas 1–11 in Chapter 10. The telephone number of the nearest MRCC or MRSC (or the equivalent in other countries) is shown for each harbour.

Within each of the 24 districts thus formed there is an organisation of Auxiliary Coastguard watch and rescue stations, grouped within Sectors under the management of Regular Coastguard Officers.

All MRCCs and MRSCs keep watch on VHF Ch 16 at 99 radio sites, and are connected to telex and telephone. A visual lookout is maintained when necessary. Some stations have a radar watch facility, and the Channel Navigation Information Service (CNIS) keeps a constant radar watch on the Dover Strait, with broadcasts of navigational and traffic information on Ch 10 at 10 and 40 minutes past each hour. There are about 100 Auxiliary Watch Stations, where watch is set in bad weather. In all there are about 560 Regular Coastguard Officers, backed up by more than 8000 Auxiliaries on call for emergencies. HM Coastguard also have a cliff and beach rescue role.

A VHF emergency direction finding service, controlled by HM Coastguard, is operated from various stations round Britain. See 4.3.2.

The voice callsign of an MRCC or MRSC is the geographical name, followed by 'Coastguard' — for example 'SOLENT COASTGUARD'.

HM Coastguard operates a Local Warning Service relating to hazards which may affect craft in inshore waters, but outside Port and Harbour Authority limits. These local warnings are broadcast on Ch 67, after an announcement on Ch 16. There is no numerical sequence, and no specific broadcast schedule; any repetition of the broadcast is at the discretion of the originating Coastguard station. Strong wind warnings are broadcast on receipt, and forecasts for local sea areas about every four hours or on request. See also Chapter 7.

Yacht and Boat Safety Scheme
This free scheme provides useful information for the Coastguard to mount a successful SAR operation. Owners can obtain a post-paid card (Form CG66) from their local Coastguard station, harbour master or marina. This should be completed with details of the boat and her equipment, and then posted to the local Coastguard Rescue Centre. There is a tear-off section which can be given to a friend or relative so that they know the Coastguard station to contact if they are concerned for the boat's safety.

It is not the function of HM Coastguard to maintain watch for boats on passage, but they will record information by phone before departure or from intermediate ports, or while on passage by visual signals or VHF Ch 67 (the Small Craft Safety Channel). When using Ch 67 for safety messages it is requested that yachts give the name of the Coastguard Rescue Centre holding the boat's Safety Scheme card. In these circumstances the Coastguard must be told of any change to the planned movements of the boat, and it is important that they are informed of the boat's safe arrival at her ultimate destination.

Raising the alarm
If an accident afloat is seen from shore, dial 999 and ask for the Coastguard. You will be asked to report on the incident, and possibly to stay near the telephone for further communications.

If at sea you receive a distress signal and you are in a position to give assistance, you are obliged to do so with all speed, unless or until you are specifically released.

When alerted, the Coastguard summons the most appropriate help. They might direct a Coastguard Shore Boat or an Auxiliary (Afloat), if in the vicinity; they might request the launch of a RNLI lifeboat or inflatable lifeboat; the Royal Navy or the Royal Air Force might be asked for a SAR helicopter; other Coastguard stations might be contacted; or shipping might be alerted through nearby Coast Radio Stations.

8.2.3 Royal National Lifeboat Institution (RNLI)

The RNLI, which is supported entirely by voluntary contributions, has about 200 stations around the United Kingdom, the Republic of Ireland, the Isle of Man, and the Channel Islands. From them are deployed about 130 lifeboats over 10m in length and a similar number of smaller lifeboats. Some of the latter only operate in summer.

When launched on service lifeboats over 10m keep watch on 2182 kHz and Ch 16. They can also use other frequencies (including VHF Ch 0, which is reserved exclusively for SAR) to contact HM Coastguard or Coast Radio Stations. Smaller lifeboats are fitted with VHF. All lifeboats show a quick-flashing blue light.

Similar organisations to the RNLI exist in other countries in Western Europe. The positions of lifeboat stations are indicated on the maps at the start of each Area in Chapter 10.

Yachtsmen can help support the RNLI by joining Shoreline. Details from RNLI, West Quay Road, Poole, Dorset, BH15 1HZ.

8.2.4 SAR — communications
For details of signals between shore and ships in distress, signals used by SAR aircraft, and directing signals used by aircraft — see 6.2.4, 6.2.5 and 6.2.6 respectively.

8.2.5 Helicopter rescue
SAR helicopters in the UK are based at Chivenor, Culdrose, Lee-on-Solent, Manston, Woodbridge, Coltishall, Leconfield, Boulmer, Leuchars,

Lossiemouth, Sumburgh, Prestwick, Valley and Brawdy.

Wessex helicopters can carry up to ten survivors, but do not usually operate at night, or when the wind exceeds 45 knots. Sea King helicopters can operate to a distance of 300 miles, and at night, and can rescue up to 18 survivors. All can communicate with lifeboats etc on VHF, and some on MF.

When the helicopter is sighted by a boat in distress, a flare, smoke signal, dye marker or a well trained Aldis' lamp will assist recognition (very important if there are other vessels in the vicinity). Dodgers with the boat's name or sail number are useful aids to identification.

While hovering the pilot has limited vision beneath him, and relies on instructions from the winch operator. Survivors from a yacht with a mast must be picked up from a dinghy or liferaft streamed at least 100ft (30m) away. It is helpful if the drift of the yacht can be reduced by a sea anchor. In a small yacht with no dinghy, survivors (wearing lifejackets) may have to be picked up from the water, at the end of a long warp. Sails should be lowered.

If a crewman descends from the helicopter, he will take charge. Obey his instructions quickly. Never secure the winch wire to the yacht, and beware that it may carry a lethal static charge if it is not dipped (earthed) in the sea before handling.

Survivors may be lifted by double lift in a strop, accompanied by the crewman in a canvas seat. Or it may be necessary, with no crewman, for a survivor to position himself in the strop. Put your head and shoulders through the strop so that the padded part is in the small of the back and the toggle is in front of the face. Pull the toggle down, as close to the chest as possible. When ready, give a thumbs up sign with an extended arm, and place both arms close down by the side of the body (resist the temptation to hang onto the strop). On reaching the helicopter, do exactly as instructed by the crew.

In some circumstances a 'Hi-line' may be used. This is a rope tail, attached to the winch wire by a weak link, and weighted at its lower end. When it is lowered to the yacht do not make it fast, but coil it down carefully. The helicopter pays out the winch wire and descends, while the yacht takes in the slack (keeping it outboard and clear of all obstructions) until the winch hook and strop are on board. When ready to lift, the helicopter ascends and takes in the wire. Pay out the tail, and cast it off well clear of the yacht. But if a further lift is to be made the tail should be retained on board (not made fast) to facilitate recovery of the strop for the next lift.

Injured persons can be lifted, strapped into a special stretcher carried in the helicopter.

When alighting from a helicopter, beware of the tail rotor.

8.2.6 Abandon ship

Although preparations must be made, do not abandon a yacht until she is doomed. She is a better target for rescue craft than a liferaft, and while she is still afloat it is possible to use her resources (such as RT, for distress calls) and to select what extra equipment is put in the liferaft or lashed into the dinghy (which should be taken too, if possible).

Before entering the raft, and cutting it adrift:
(1) Send a distress message (Mayday call), saying that yacht is being abandoned, and position.
(2) Dress warmly with sweaters etc under oilskins, and lifejackets on top. Take extra clothes.
(3) Fill any available containers with tops about $\frac{3}{4}$ full with fresh water, so that they will float.
(4) Collect additional food — tins and tin opener.
(5) Collect navigational gear, torch, extra flares, bucket, length of line, first aid kit, knife etc.

Once in the liferaft, plan for the worst. If there has not been time to collect items listed above, collect whatever flotsam is available.
(1) Keep the inside of the raft as dry as possible. Huddle together for warmth. Close the opening as necessary, but keep a good lookout for shipping.
(2) Stream the drogue if necessary for stability, or if it is required to stay near the position.
(3) Ration fresh water to $\frac{3}{4}$ pint ($\frac{1}{2}$ litre) per person per day. Do not drink sea water or urine. Collect rain water.
(4) Use flares sparingly, on the skipper's orders.
(5) Issue and commence anti-seasick pills.

Chapter 9
Tides

Contents

9.1 GENERAL — Page 168
9.1.1 Explanation
9.1.2 Times
9.1.3 Predicted heights

9.2 DEFINITIONS — 168
9.2.1 Chart datum
9.2.2 Charted depth
9.2.3 Drying height
9.2.4 Duration
9.2.5 Height of tide
9.2.6 Interval
9.2.7 Mean Level
9.2.8 Range

9.3 CALCULATIONS OF TIMES AND HEIGHTS OF HIGH AND LOW WATER — 169
9.3.1 Standard Ports
9.3.2 Secondary Ports — times of HW and LW
9.3.3 Secondary Ports — heights of HW and LW
9.3.4 Graphical method for interpolating time and height differences

9.4 TIDAL CURVES — CALCULATIONS OF DEPTHS AT SPECIFIC TIMES, AND OF TIMES AT WHICH TIDE REACHES CERTAIN HEIGHTS — 170
9.4.1 Standard Ports
9.4.2 Secondary Ports
9.4.3 The use of factors

9.5 CALCULATIONS OF CLEARANCES UNDER BRIDGES, ETC. — 174

9.6 TIDAL STREAMS — 174
9.6.1 General
9.6.2 Computation of tidal stream rates

9.7 METEOROLOGICAL CONDITIONS — 175

9.8 CONVERSION — FEET TO METRES, AND METRES TO FEET — 175

Tides — introduction

The following subjects are described in detail in Chapter 9 of *The Macmillan & Silk Cut Yachtsman's Handbook*:
  The theory of tides; definitions of terms; calculations of times and heights of HW and LW; calculations of depths of water at specific times; calculations of times at which tide reaches certain heights; Twelfths Rule; tidal calculations by pocket calculator; French tidal coefficients; co-tidal and co-range charts; harmonic constituents; establishment of a port; tidal stream diamonds; tidal stream information on charts and in Sailing Directions ... plus *general information on the sea — how waves are formed; freak waves; wind against tide; bars; overfalls and tide races; refraction of waves; reflected waves; ocean currents* etc.
Here in the Almanac is given sufficient information for the use of the tidal data provided in Chapter 10, but for fuller details of matters concerning tides and the sea reference should be made to Chapter 9 of *The Macmillan & Silk Cut Yachtsman's Handbook*.

9.1 GENERAL

9.1.1 Explanation
This chapter explains how to use the tidal information contained in Chapter 10, where are given the daily times and heights of High Water (HW) and Low Water (LW) for Standard Ports, together with time and height differences for many other places. Tidal predictions are for average meteorological conditions. In abnormal weather the times and heights of HW and LW may differ considerably.

9.1.2 Times
Clock (or local) time in the United Kingdom is Greenwich Mean Time (GMT) in the winter, and British Summer Time (BST) in the summer. Clock time in France, Belgium, Netherlands and the Federal Republic of Germany is Zone −0100 (one hour ahead of GMT) in the winter, and Zone −0200 (two hours ahead of GMT) during the period in which Daylight Saving Time (DST) operates — from the last Sunday in March to the last Saturday in September.

The Zone Time in which predictions are given is stated at the head of each tide table in Chapter 10, and takes no account of BST or DST. For Standard Ports in the United Kingdom, the Channel Islands and Ireland it is Zone 0 (GMT). For Standard Ports in France, Netherlands and the Federal Republic of Germany it is Zone −0100.

As an example, for British ports one hour must be added to the predicted times for a Standard Port while BST is being kept. For convenience the dates where BST does and does not apply are shown by non-tinted and tinted areas (Similarly for DST in Standard Ports on the Continent).

Time differences for a Secondary Port, when applied to the printed times of HW and LW at the Standard Port, give the times of HW and LW at the Secondary Port in the Zone Time as indicated under 'Tides' for the Secondary Port.

Care must be taken in a few instances, as in Area 17 where certain French harbours which are Zone −0100 (e.g. St Malo) have as their Standard Port St Helier, for which the tidal predictions are expressed in GMT. Applying the rule in the previous paragraph, if HW St Helier is 0806 (GMT) and the time correction for St Malo is +0044, then HW St Malo is 0850 (Zone −0100), or 0750 GMT.

9.1.3 Predicted heights
Predicted heights are given in metres and tenths of metres above Chart Datum (see 9.2.1). Care must be taken when using charts which show depths in fathoms/feet.

9.2 DEFINITIONS

Certain definitions are given below and in fig. 9(1). For further details see Chapter 9 of *The Macmillan & Silk Cut Yachtsman's Handbook*.

9.2.1 Chart datum
Chart datum (CD) is the reference level above which heights of tide are predicted, and below which charted depths are measured. Hence the actual depth of water is the charted depth (at that place) plus the height of tide (at that time).

Tidal predictions for British ports use as their datum Lowest Astronomical Tide (LAT), which is the lowest sea level predicted under average meteorological conditions. All Admiralty charts of the British Isles use LAT as chart datum. But most fathom charts do not use LAT, and are drawn to a datum approximating to Mean Low Water Springs (MLWS). Since MLWS is slightly higher than LAT, when tide tables with LAT datum are used in conjunction with charts which have MLWS datum, there will be slightly less depth than is calculated — possibly as much as 0.5 metre.

9.2.2 Charted depth
Charted depth is the distance of the sea bed below chart datum, and is shown in metres and tenths of metres on metric charts, or in fathoms and/or feet on older charts. Make sure which units are used.

9.2.3 Drying height
Drying height is the height above chart datum of the top of any feature occasionally covered by water. The figures are underlined on the chart — in metres and tenths of metres on metric charts, and in feet on older charts. The depth is the height of tide (at the time) minus the drying height. If the result is negative, then that place is above sea level.

9.2.4 Duration
Duration is the time between LW and HW, normally slightly more than six hours, and can be used to calculate the time of LW when only the time of HW is known.

9.2.5 Height of tide
The height of tide is the distance of sea level above (or very occasionally below) chart datum, as defined in 9.2.1.

9.2.6 Interval
The interval is the time between a given time and HW, expressed in hours and minutes before (−) or after (+) High Water.

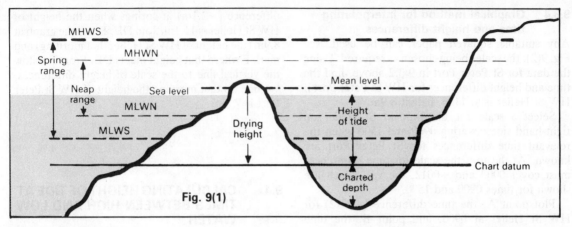

Fig. 9(1)

9.2.7 Mean Level
Mean Level (ML) is the average between the heights of Mean High Water Springs (MHWS), Mean High Water Neaps (MHWN), mean Low Water Springs (MLWS) and Mean Low Water Neaps (MLWN).

9.2.8 Range
The range of a tide is the difference between the heights of successive High and Low Waters. Spring range is MHWS–MLWS. Neap range is MHWN–MLWN.

9.3 CALCULATIONS OF TIMES AND HEIGHTS OF HIGH AND LOW WATER

9.3.1 Standard Ports
The times and heights of HW and LW are given for Standard Ports in Chapter 10. Zone times are shown at the head of each tide table, and the predicted times must be adjusted for BST or DST as explained in 9.1.2.

9.3.2 Secondary Ports — times of HW and LW
For Secondary Ports, the approximate times of HW and LW are calculated by adding (when +) or subtracting (when −) the time difference shown, to or from the time of HW or LW at the Standard Port indicated. This example is for St Peter Port:

TIDES
 −0450 Dover; ML 5.0; Duration 0500; Zone 0 (GMT)

Standard Port ST HELIER

Times				Height (metres)			
HW		LW		MHWS	MHWN	MLWN	MLWS
0900	0300	0200	0900	11.1	8.1	4.1	1.3
2100	1500	1400	2100				

Differences ST PETER PORT
| +0012 | 0000 | −0008 | +0002 | −2.1 | −1.4 | −0.6 | −0.3 |

First are shown the average time difference for HW on Dover (GMT) as on the bookmark, Mean Level, Duration and Zone Time. Then the Standard Port, with time and height differences for the Secondary Port(s). If for example HW St Helier is at 0900 or 2100, the time difference for St Peter Port is +0012 (+ 12 mins). At 0300 or 1500 no correction is needed. For intermediate times of HW St Helier, the time difference for St Peter Port must be interpolated — either by eye or by calculator, or by the graphical method in 9.3.4 which is convenient when calculating for a number of tides. Thus for HW St Helier at 1030, the time difference for HW St Peter Port is +0009. The same principle applies to time corrections for LW. Times obtained are in zone time of the Secondary Port, irrespective of the zone time used for the Standard Port predictions. Care is needed in some places (e.g. Area 17) — see 9.1.2. The calculations are best done on the top part of the tidal prediction form in Fig. 9(5), by completing boxes 1, 2, 5, 6, 9 and 10.

9.3.3 Secondary Ports — heights of HW and LW
The height of HW or LW at a Secondary Port is obtained by applying the height difference shown to the height of HW or LW at the Standard Port. Average height differences are shown for Mean Spring and Mean Neap levels. From the St Peter Port data, at MHWS the height at St Helier is 11.1m, and at St Peter Port it is 2.1m less (i.e. 9.0m). At MLWN the height at St Helier is 4.1m, and at St Peter Port 0.6m less (i.e. 3.5m). The calculation can be done by completing boxes 3, 4, 7, 8, 11 and 12 in Fig. 9(5). Resulting heights are referred to chart datum at the Secondary Port. Heights for dates between springs and neaps are obtained by interpolation, in a similar way to time differences above.

9.3.4 Graphical method for interpolating time and height differences

Any suitable squared paper can be used, see Fig. 9(2), the scales being chosen as required. Using the data for St Peter Port in 9.3.2 above, find the time and height differences for HW St Peter Port if HW St Helier is at 1126, height 8.9m.

Select a scale for the time at St Helier on right-hand side covering 0900 and 1500 when the relevant time differences for St Peter Port are known. At the top, the scale for time differences must cover 0000 and +0012, the two which are shown for times 0900 and 1500.

Plot point A, the time difference (+0012) for HW St Helier at 0900; and point B, the time difference (0000) for HW St Helier at 1500. Join AB. Enter the graph on the right at time 1126 (HW St Helier) and mark C where that time meets AB. From C proceed vertically to the time difference scale at the top, +0007. So that morning HW St Peter Port is 7 minutes after HW St Helier, i.e. 1133.

In the bottom of the diagram, select scales which cover the height of HW at St Helier vertically (i.e. 8.1 to 11.1m) and the relevant height differences (−1.4 to −2.1m) horizontally. Plot point D, the height difference (−1.4m) at neaps when the height of HW St Helier is 8.1m; and E, the height difference (−2.1m) at springs when the height of HW St Helier is 11.1m. Join DE. Enter the graph at 8.9m (the height of HW St Helier that morning) and mark F where that height meets DE. From F follow the vertical line to the scale of height differences, −1.6m. So that morning the height of HW St Peter Port is 7.3m.

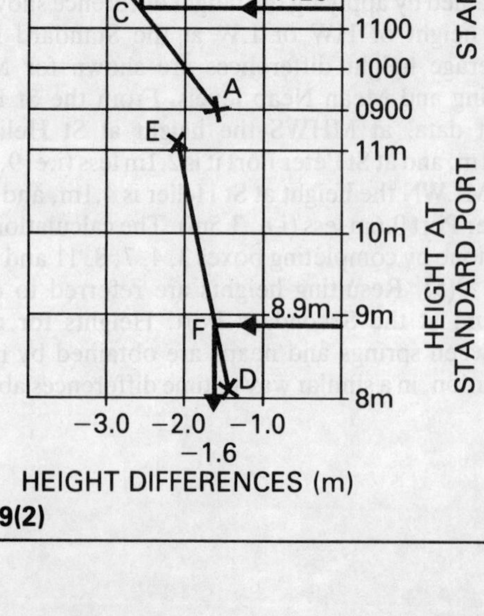

9.4 CALCULATING HEIGHT OF TIDE AT TIMES BETWEEN HIGH AND LOW WATER

9.4.1 Standard Ports

Intermediate times and heights are best predicted from the Mean Spring and Neap Curves for Standard Ports in Chapter 10. Examples below are for Leith, on a day when the predictions are:

22 0202 5.3
    0752 1.0
    1417 5.4
TU 2025 0.5

Example: Find the height at Leith at 1200.
(1) On the Leith tidal diagram plot the heights of HW and LW each side of the required time, and join them by a sloping line, Fig. 9(3).
(2) Enter the HW time and other times as necessary in the boxes below the curves.
(3) From the required time proceed vertically to the curves. Use the heights plotted in (1) to help interpolation between the spring and neap curves. The spring curve is a solid line, and the neap curve (where it differs) is pecked. Do not extrapolate. Here the spring curve applies.
(4) Proceed horizontally to the sloping line plotted in (1), and thence vertically to the height scale, to give 4.2m.

Example: To find the time at which the afternoon tide falls to 3.7m.
(1) On the Leith tidal diagram, plot the heights of HW and LW each side of the required event, and join them by a sloping line, Fig. 9(4).
(2) Enter the HW time and others to cover the required event, in the boxes below.
(3) From the required height, proceed vertically to the sloping line and thence horizontally to the curves, using the heights plotted in (1) to help interpolation between the spring and neap curves. Here the spring curve applies. Do not extrapolate.
(4) Proceed vertically to the time scale, and read off the time required, 1637.

Fig. 9(2)

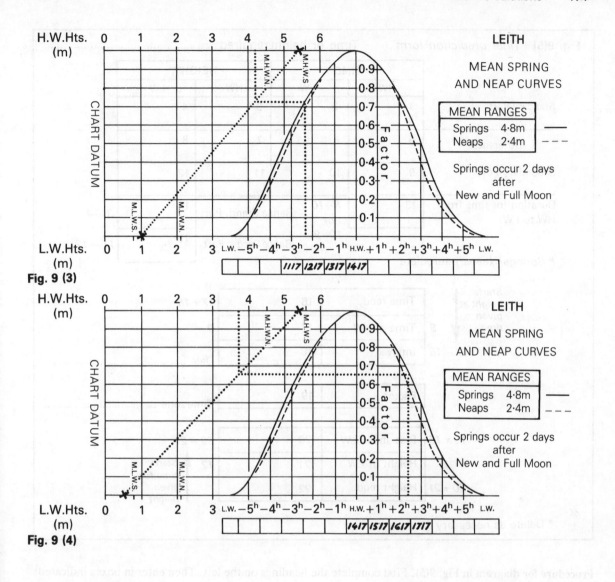

Fig. 9 (3)

Fig. 9 (4)

9.4.2 Secondary Ports

On coasts where there is little change of shape between tidal curves for adjacent Standard Ports, and where the duration of rise or fall at the Secondary Port is like that of the appropriate Standard Port (where the time differences for HW and LW are nearly the same), intermediate times and heights may be predicted from the tidal curves for the Standard Port in a similar manner to 9.4.1 above. The curves are entered with the times and heights of HW and LW at the Secondary Port, calculated as in 9.3.2 and 9.3.3.

Interpolation between the curves can be made by eye, using the range at the Standard Port as argument. Do not extrapolate — use the spring curve for spring ranges or greater, and the neap curve for neap ranges or less. With a large change in duration between springs and neaps the results may have a slight error, greater near LW.

Special curves for places between Swanage and Selsey (where the tide is very complex) are given in 10.2.8.

9.4.3 The use of factors

Tidal curves show the factor of the range attained at times before and after HW. By definition a factor of 1 = HW, and 0 = LW. So the factor represents the proportion of the range (for the day in question) by which the height of tide is above the height of LW (that day) at the interval (time) concerned.

Range × factor = Rise above LW
and   Factor = Rise above LW ÷ range

In determining or using the factor it may be necessary to interpolate between the spring and neap curves as described in 9.4.2.

The procedure is shown on the following page, using the tidal prediction form as in Fig. 9(5). Table 9(1) may be used for factor and range calculations.

Fig. 9(5) *Tidal prediction form* Time or height required ....

	TIME		HEIGHT	
	HW	LW	HW	LW
Standard Port	1	2	3	4
Differences	5	6	7	8
Secondary Port	9	10	11	12

Duration (or time from HW to LW)	13	9–10 or 10–9	Range Stand. Port	14	3 – 4
			Range Secdy. Port	15	11 – 12

* Springs/Neaps/Interpolate

Start: height at given time ↓

	Time reqd.	16	17 + 18
9	Time of HW	17	9
17 – 16	Interval	18	

Date ....

	Factor	19	

Time zone ....

19 x 15	Rise above LW	20	22 – 21
12	Height of LW	21	12
20 + 21	Height reqd.	22	

↑ Start: time for given height

* Delete as necessary

Procedure for diagram in Fig. 9(5). First complete the headings on the left. Then enter in boxes indicated:

1 to 4   Predictions for Standard Port
5 to 8   Differences for Secondary Port (interpolated if necessary). Not required for Standard Port.
9   Sum of 1 and 5 (for Standard Port enter 1)
10   Sum of 2 and 6 (for Standard Port enter 2)
11   Sum of 3 and 7 (for Standard Port enter 3)
12   Sum of 4 and 8 (for Standard Port enter 4)
13   Duration (difference of 9 and 10)
14   Range at Standard Port (difference of 3 and 4)
15   Range at Secondary Port (difference of 11 and 12)

TO FIND HEIGHT AT GIVEN TIME
16   Required time
17   HW time from 9
18   Interval (difference of 16 and 17)
19   Factor, from appropriate tidal curve, entered with Interval (18)
20   Rise above LW = Factor (19) × Range (15), or for Standard Port use Range (14)
21   Height of LW from 12
22   Sum of Rise above LW (20) and Height of LW (21)

TO FIND TIME FOR A GIVEN HEIGHT
(start at bottom of diagram)
22   Required height
21   LW Height from 12
20   Rise = Height (22) − LW Height (21)
19   Factor = Rise (20) ÷ Range (15). (For Standard Port use 14)
18   Interval, from interpolation of appropriate tidal curve, entered with Factor (19)
17   HW time from 9
16   Interval (18) applied to HW time (17)

TABLE 9(1)

MULTIPLICATION TABLE for use with Fig. 9(5)

RANGE (columns) — FACTOR (rows)

1.00	1.2	1.6	2.0	2.4	2.8	3.2	3.6	4.0	4.4	4.8	5.2	5.6	6.0	6.4	6.8	7.2	7.6	8.0	8.4	8.8	9.2	9.6	10.0	10.4	10.8	11.2	11.6	12.0	12.4	1.00
.98	1.2	1.6	2.0	2.4	2.7	3.1	3.5	3.9	4.3	4.7	5.1	5.5	5.9	6.3	6.7	7.1	7.4	7.8	8.2	8.6	9.0	9.4	9.8	10.2	10.6	11.0	11.4	11.8	12.1	.98
.96	1.2	1.5	1.9	2.3	2.7	3.1	3.5	3.8	4.2	4.6	5.0	5.4	5.8	6.1	6.5	6.9	7.3	7.7	8.1	8.4	8.8	9.2	9.6	10.0	10.4	10.8	11.1	11.5	11.9	.96
.94	1.1	1.5	1.9	2.3	2.6	3.0	3.4	3.8	4.1	4.5	4.9	5.3	5.6	6.0	6.4	6.8	7.1	7.5	7.9	8.3	8.6	9.0	9.4	9.8	10.2	10.5	10.9	11.3	11.7	.94
.92	1.1	1.5	1.8	2.2	2.6	2.9	3.3	3.7	4.0	4.4	4.8	5.2	5.5	5.9	6.3	6.6	7.0	7.4	7.7	8.1	8.5	8.8	9.2	9.6	9.9	10.3	10.7	11.0	11.4	.92
.90	1.1	1.4	1.8	2.2	2.5	2.9	3.2	3.6	4.0	4.3	4.7	5.0	5.4	5.8	6.1	6.5	6.8	7.2	7.6	7.9	8.3	8.6	9.0	9.4	9.7	10.1	10.4	10.8	11.1	.90
.88	1.1	1.4	1.8	2.1	2.5	2.8	3.2	3.5	3.9	4.2	4.6	4.9	5.3	5.6	6.0	6.3	6.7	7.0	7.4	7.7	8.1	8.4	8.8	9.2	9.5	9.9	10.2	10.6	10.9	.88
.86	1.0	1.4	1.7	2.1	2.4	2.8	3.1	3.4	3.8	4.1	4.5	4.8	5.2	5.5	5.8	6.2	6.5	6.9	7.2	7.6	7.9	8.3	8.6	8.9	9.3	9.6	10.0	10.3	10.7	.86
.84	1.0	1.3	1.7	2.0	2.4	2.7	3.0	3.4	3.7	4.0	4.4	4.7	5.0	5.4	5.7	6.0	6.4	6.7	7.1	7.4	7.7	8.1	8.4	8.7	9.1	9.4	9.7	10.1	10.4	.84
.82	1.0	1.3	1.6	2.0	2.3	2.6	3.0	3.3	3.6	3.9	4.3	4.6	4.9	5.2	5.6	5.9	6.2	6.6	6.9	7.2	7.5	7.9	8.2	8.5	8.9	9.2	9.5	9.8	10.2	.82
.80	1.0	1.3	1.6	1.9	2.2	2.6	2.9	3.2	3.5	3.8	4.2	4.5	4.8	5.1	5.4	5.8	6.1	6.4	6.7	7.0	7.4	7.7	8.0	8.3	8.6	9.0	9.3	9.6	9.9	.80
.78	0.9	1.2	1.6	1.9	2.2	2.5	2.8	3.1	3.4	3.7	4.1	4.4	4.7	5.0	5.3	5.6	5.9	6.2	6.6	6.9	7.2	7.5	7.8	8.1	8.4	8.7	9.0	9.4	9.7	.78
.76	0.9	1.2	1.5	1.8	2.1	2.4	2.7	3.0	3.3	3.6	4.0	4.3	4.6	4.9	5.2	5.5	5.8	6.1	6.4	6.7	7.0	7.3	7.6	7.9	8.2	8.5	8.8	9.1	9.4	.76
.74	0.9	1.2	1.5	1.8	2.1	2.4	2.7	3.0	3.3	3.6	3.8	4.1	4.4	4.7	5.0	5.3	5.6	5.9	6.2	6.5	6.8	7.1	7.4	7.7	8.0	8.3	8.7	8.9	9.2	.74
.72	0.9	1.2	1.4	1.7	2.0	2.3	2.6	2.9	3.2	3.5	3.7	4.0	4.3	4.6	4.9	5.2	5.5	5.8	6.0	6.3	6.6	6.9	7.2	7.5	7.8	8.1	8.4	8.6	8.9	.72
.70	0.8	1.1	1.4	1.7	2.0	2.2	2.5	2.8	3.1	3.4	3.6	3.9	4.2	4.5	4.8	5.0	5.3	5.6	5.9	6.2	6.4	6.7	7.0	7.3	7.6	7.8	8.1	8.4	8.7	.70
.68	0.8	1.1	1.4	1.6	1.9	2.2	2.4	2.7	3.0	3.3	3.5	3.8	4.1	4.4	4.6	4.9	5.2	5.4	5.7	6.0	6.3	6.5	6.8	7.1	7.3	7.6	7.9	8.2	8.4	.68
.66	0.8	1.1	1.3	1.6	1.8	2.1	2.4	2.6	2.9	3.2	3.4	3.7	4.0	4.2	4.5	4.8	5.0	5.3	5.5	5.8	6.1	6.3	6.6	6.9	7.1	7.4	7.7	7.9	8.2	.66
.64	0.8	1.0	1.3	1.5	1.8	2.0	2.3	2.6	2.8	3.1	3.3	3.6	3.8	4.1	4.4	4.6	4.9	5.1	5.4	5.6	5.9	6.1	6.4	6.7	6.9	7.2	7.4	7.7	7.9	.64
.62	0.7	1.0	1.2	1.5	1.7	2.0	2.2	2.5	2.7	3.0	3.2	3.5	3.7	4.0	4.2	4.5	4.7	5.0	5.2	5.5	5.7	6.0	6.2	6.4	6.7	6.9	7.2	7.4	7.7	.62
.60	0.7	1.0	1.2	1.4	1.7	1.9	2.2	2.4	2.6	2.9	3.1	3.4	3.6	3.8	4.1	4.3	4.6	4.8	5.0	5.3	5.5	5.8	6.0	6.2	6.5	6.7	7.0	7.2	7.4	.60
.58	0.7	0.9	1.2	1.4	1.6	1.9	2.1	2.3	2.6	2.8	3.0	3.2	3.5	3.7	3.9	4.2	4.4	4.6	4.9	5.1	5.3	5.6	5.8	6.0	6.3	6.5	6.7	7.0	7.2	.58
.56	0.7	0.9	1.1	1.3	1.6	1.8	2.0	2.2	2.5	2.7	2.9	3.1	3.4	3.6	3.8	4.0	4.3	4.5	4.7	4.9	5.2	5.4	5.6	5.8	6.0	6.3	6.5	6.7	6.9	.56
.54	0.6	0.9	1.1	1.3	1.5	1.7	1.9	2.2	2.4	2.6	2.8	3.0	3.2	3.5	3.7	3.9	4.1	4.3	4.5	4.8	5.0	5.2	5.4	5.6	5.8	6.0	6.3	6.5	6.7	.54
.52	0.6	0.8	1.0	1.2	1.5	1.7	1.9	2.1	2.3	2.5	2.7	2.9	3.1	3.3	3.5	3.7	4.0	4.2	4.4	4.6	4.8	5.0	5.2	5.4	5.6	5.8	6.0	6.2	6.4	.52
.50	0.6	0.8	1.0	1.2	1.4	1.6	1.8	2.0	2.2	2.4	2.6	2.8	3.0	3.2	3.4	3.6	3.8	4.0	4.2	4.4	4.6	4.8	5.0	5.2	5.4	5.6	5.8	6.0	6.2	.50
.48	0.6	0.8	1.0	1.2	1.3	1.5	1.7	1.9	2.1	2.3	2.5	2.7	2.9	3.1	3.3	3.5	3.6	3.8	4.0	4.2	4.4	4.6	4.8	5.0	5.2	5.4	5.6	5.8	5.9	.48
.46	0.6	0.7	0.9	1.1	1.3	1.5	1.7	1.8	2.0	2.2	2.4	2.6	2.8	2.9	3.1	3.3	3.5	3.7	3.9	4.0	4.2	4.4	4.6	4.8	5.0	5.2	5.3	5.5	5.7	.46
.44	0.5	0.7	0.9	1.1	1.2	1.4	1.6	1.8	1.9	2.1	2.3	2.5	2.6	2.8	3.0	3.2	3.3	3.5	3.7	3.9	4.0	4.2	4.4	4.6	4.8	4.9	5.1	5.3	5.5	.44
.42	0.5	0.7	0.8	1.0	1.2	1.3	1.5	1.7	1.8	2.0	2.2	2.4	2.5	2.7	2.9	3.0	3.2	3.4	3.5	3.7	3.9	4.0	4.2	4.4	4.5	4.7	4.9	5.0	5.2	.42
.40	0.5	0.6	0.8	1.0	1.1	1.3	1.4	1.6	1.8	1.9	2.1	2.2	2.4	2.6	2.7	2.9	3.0	3.2	3.4	3.5	3.7	3.8	4.0	4.2	4.3	4.5	4.6	4.8	5.0	.40
.38	0.5	0.6	0.8	0.9	1.1	1.2	1.4	1.5	1.7	1.8	2.0	2.1	2.3	2.4	2.6	2.7	2.9	3.0	3.2	3.3	3.5	3.6	3.8	4.0	4.1	4.3	4.4	4.6	4.7	.38
.36	0.4	0.6	0.7	0.9	1.0	1.2	1.3	1.4	1.6	1.7	1.9	2.0	2.2	2.3	2.4	2.6	2.7	2.9	3.0	3.2	3.3	3.5	3.6	3.7	3.9	4.0	4.2	4.3	4.5	.36
.34	0.4	0.5	0.7	0.8	1.0	1.1	1.3	1.4	1.5	1.6	1.8	1.9	2.0	2.2	2.3	2.4	2.6	2.7	2.9	3.0	3.1	3.3	3.4	3.5	3.7	3.8	3.9	4.1	4.2	.34
.32	0.4	0.5	0.6	0.8	0.9	1.0	1.2	1.3	1.4	1.5	1.7	1.8	1.9	2.0	2.2	2.3	2.4	2.6	2.7	2.8	2.9	3.1	3.2	3.3	3.5	3.6	3.7	3.8	4.0	.32
.30	0.4	0.5	0.6	0.7	0.8	1.0	1.1	1.2	1.3	1.4	1.6	1.7	1.8	1.9	2.0	2.2	2.3	2.4	2.5	2.6	2.8	2.9	3.0	3.1	3.2	3.4	3.5	3.6	3.7	.30
.28	0.3	0.4	0.6	0.7	0.8	0.9	1.0	1.1	1.2	1.3	1.5	1.6	1.7	1.8	1.9	2.0	2.1	2.2	2.4	2.5	2.6	2.7	2.8	2.9	3.0	3.1	3.2	3.4	3.5	.28
.26	0.3	0.4	0.5	0.6	0.7	0.8	0.9	1.0	1.1	1.2	1.4	1.5	1.6	1.7	1.8	1.9	2.0	2.1	2.2	2.3	2.4	2.5	2.6	2.7	2.8	2.9	3.0	3.1	3.2	.26
.24	0.3	0.4	0.5	0.6	0.7	0.8	0.9	1.0	1.1	1.2	1.2	1.3	1.4	1.5	1.6	1.7	1.8	2.1	2.0	2.1	2.2	2.3	2.4	2.5	2.6	2.7	2.8	2.9	3.0	.24
.22	0.3	0.4	0.4	0.5	0.6	0.7	0.8	0.9	1.0	1.1	1.1	1.2	1.3	1.4	1.5	1.6	1.7	1.8	1.8	1.9	2.0	2.1	2.2	2.3	2.4	2.5	2.6	2.6	2.7	.22
.20	0.2	0.3	0.4	0.5	0.6	0.6	0.7	0.8	0.9	1.0	1.0	1.1	1.2	1.3	1.4	1.4	1.5	1.6	1.7	1.8	1.8	1.9	2.0	2.1	2.2	2.2	2.3	2.4	2.5	.20
.18	0.2	0.3	0.4	0.4	0.5	0.6	0.6	0.7	0.8	0.9	0.9	1.0	1.1	1.2	1.2	1.3	1.4	1.4	1.5	1.6	1.7	1.7	1.8	1.9	1.9	2.0	2.1	2.2	2.2	.18
.16	0.2	0.3	0.3	0.4	0.4	0.5	0.6	0.6	0.7	0.8	0.8	0.9	1.0	1.0	1.1	1.2	1.2	1.3	1.3	1.4	1.5	1.5	1.6	1.7	1.7	1.8	1.9	1.9	2.0	.16
.14	0.2	0.2	0.3	0.3	0.4	0.4	0.5	0.6	0.6	0.7	0.7	0.8	0.8	0.9	1.0	1.0	1.1	1.1	1.2	1.2	1.3	1.3	1.4	1.5	1.5	1.6	1.6	1.8	1.7	.14
.12	0.1	0.2	0.2	0.3	0.3	0.4	0.4	0.5	0.5	0.6	0.6	0.7	0.7	0.8	0.8	0.9	0.9	1.0	1.0	1.1	1.1	1.2	1.2	1.2	1.3	1.3	1.4	1.4	1.5	.12
.10	0.1	0.2	0.2	0.2	0.3	0.3	0.4	0.4	0.4	0.5	0.5	0.6	0.6	0.6	0.7	0.7	0.8	0.8	0.8	0.9	0.9	1.0	1.0	1.0	1.1	1.1	1.2	1.2	1.2	.10
.08	0.1	0.1	0.2	0.2	0.2	0.3	0.3	0.3	0.4	0.4	0.4	0.4	0.5	0.5	0.5	0.6	0.6	0.6	0.7	0.7	0.7	0.8	0.8	0.8	0.9	0.9	0.9	1.0	1.0	.08
.06	0.1	0.1	0.1	0.1	0.2	0.2	0.2	0.2	0.3	0.3	0.3	0.3	0.4	0.4	0.4	0.4	0.5	0.5	0.5	0.5	0.6	0.6	0.6	0.6	0.6	0.7	0.7	0.7	0.7	.06
.04	0.0	0.1	0.1	0.1	0.1	0.1	0.1	0.2	0.2	0.2	0.2	0.2	0.2	0.3	0.3	0.3	0.3	0.3	0.3	0.4	0.4	0.4	0.4	0.4	0.4	0.4	0.5	0.5	0.5	.04
.02	0.0	0.0	0.0	0.0	0.1	0.1	0.1	0.1	0.1	0.1	0.1	0.1	0.1	0.1	0.1	0.1	0.2	0.2	0.2	0.2	0.2	0.2	0.2	0.2	0.2	0.2	0.2	0.2	0.2	.02

FACTOR

9

9.5 CALCULATIONS OF CLEARANCES UNDER BRIDGES ETC

It is sometimes necessary to calculate whether a boat can pass underneath such objects as bridges or power cables. The heights of such objects are shown on the chart above MHWS, so the clearance will nearly always be greater than the figure shown. The height is shown in metres on metric charts, but in feet on older charts. It is sometimes useful to draw a diagram, as shown in Fig. 9(6), which shows how the measurements are related to chart datum.

Clearance = (Elevation of object + height of MHWS) minus (height of tide at the time + height of mast above above water)

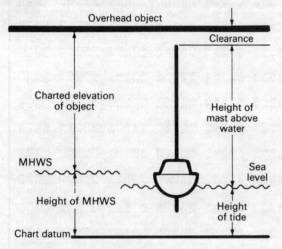

Fig. 9(6)

9.6 TIDAL STREAMS

9.6.1 General

Tidal streams are the horizontal movement of water caused by the vertical rise and fall of the tide. They normally change direction about every six hours and are quite different from ocean currents, such as the Gulf Stream which run for long periods in the same direction.

Tidal streams are important to yachtsmen around the British Isles because they often run at about two knots, and much more strongly in a few areas, and at spring tides. There are a few places where they can attain rates of six to eight knots.

The strength and direction of the tidal stream in the more important areas is shown in *Admiralty Tidal Stream Atlases*, as follows:

NP 209 Edition 4 Orkney and Shetland Islands, 1986
    218 Edition 4 North Coast of Ireland, West Coast of Scotland, 1983
    233 Edition 2 Dover Strait, 1975
    249 Edition 2 Thames Estuary, 1985
    250 Edition 3 English and Bristol Channels, 1973
    251 Edition 3 North Sea, Southern Portion, 1976
    252 Edition 3 North Sea, Northern Portion, 1975
    253 Edition 1 North Sea, Eastern Portion, 1978
    256 Edition 3 Irish Sea, 1974
    257 Edition 3 Approaches to Portland, 1973
    264 Edition 4 Channel Islands & Adjacent Coasts of France, 1984
    265 Edition 1 France, West Coast, 1978
    337 Edition 3 Solent and Adjacent Waters, 1974

Extracts from the above (by permission of the Hydrographer and HMSO) are given in Chapter 10.

The directions of the streams are shown by arrows which are graded in weight and, where possible, in length to indicate the strength of the tidal stream. Thus→ indicates a weak stream and ➡ indicates a strong stream. The figures against the arrows give the mean neap and spring rates in tenths of a knot, thus 19,34 indicates a mean neap rate of 1.9 knots and a mean spring rate of 3.4 knots. The comma indicates the approximate position at which the observations were taken.

9.6.2 Computation of tidal stream rates

Using Table 9(2) it is possible to predict the rate of a tidal stream at intermediate times, assuming that it varies with the range of tide at Dover.

Example
It is required to predict the rate of the tidal stream off the northerly point of the Isle of Skye at 0420 on 19th December 1980. Readings for Dover for that day are

0328   1.4
0819   6.3
1602   1.1
2054   6.4

The range of tide is therefore 6.3 − 1.4 = 4.9m. The appropriate chart in the Tidal Stream Atlas NP 218 or in Chapter 10 Area 8 is that for '4 hours before HW Dover' and this gives a mean neap and a mean spring rate of 09 and 17 respectively (0.9 and 1.7 knots). On Table 9(2), Computation of Rates, on the horizontal line marked Neaps, mark the dot above 09 on the horizontal scale; likewise on the line marked Springs, mark the dot below the figure 17 on the horizontal scale. Join these two dots with a straight line. On the vertical scale

Table 9(2)

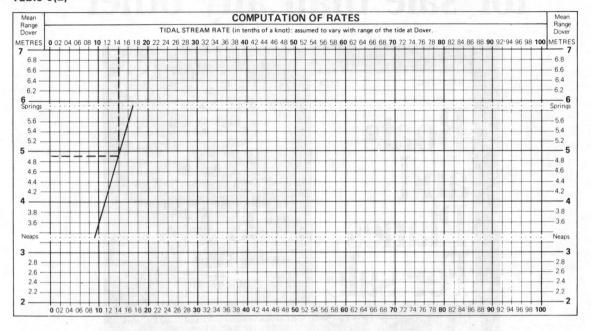

'Mean Range Dover', find the range 4.9. From this point follow across horizontally until the pencil line just drawn is cut; from this intersection follow the vertical line to the scale of Tidal Stream Rates, either top or bottom, and read off the predicted rate — in this example it is 14 or 1.4 knots.

A perspex sheet or a sheet of tracing paper can be used on top of Table 9(2), so as to preserve it for future use.

It should be remembered that tidal atlases cannot show the details of inshore eddies and the tide often sets towards the coast in bays. Along open coasts the turn of the tidal stream does not necessarily occur at High and Low water. It often occurs at about half tide. The tidal stream usually turns earlier inshore than offshore. On modern charts, lettered diamonds give information on the tidal streams by reference to a table showing neap and spring rates at hourly intervals before and after High Water at a Standard Port. Information on tidal streams is also included in *Admiralty Sailing Directions*.

9.7 METEOROLOGICAL CONDITIONS

Meteorological conditions can have a significant effect on tides and tidal streams. Sea level tends to rise in the direction towards which a wind is blowing and be lowered in the other direction. The stronger the wind and the longer it blows, the greater the effect. Also the sudden onset of a gale can set up a wave or 'storm surge' which travels along the coast. Under exceptional conditions this can raise the height of the tide by two or three metres, or a 'negative surge' can lower the height of low water by one or two metres which may be more serious for the yachtsman.

Tidal heights are predicted for average barometric pressure. When the barometer is high, tidal heights are likely to be lower, and vice versa. A change of 34 millibars (one inch of mercury) can cause a change of 0.3 metres in the height of sea level, although it may not be felt immediately. Severe conditions giving rise to a storm surge as described above are likely to be caused by a big depression, and the low barometric pressure tends to raise the sea level still more.

Intense minor depressions can have local effects on the height of water, setting up what is known as a 'seiche' which can raise or lower the sea level a metre or more in the space of a few minutes.

9.8 CONVERSION — FEET TO METRES, AND METRES TO FEET

In the table below the figures in italics are metres.

Feet		6	12	18	24	30	36	42
Fathoms		1	2	3	4	5	6	7
Feet		*1.8*	*3.6*	*5.5*	*7.3*	*9.1*	*10.9*	*12.8*
1	*0.3*	*2.1*	*3.9*	*5.8*	*7.6*	*9.4*	*11.3*	*13.1*
2	*0.6*	*2.4*	*4.2*	*6.1*	*7.9*	*9.7*	*11.6*	*13.4*
3	*0.9*	*2.7*	*4.5*	*6.4*	*8.2*	*10.0*	*11.9*	*13.7*
4	*1.2*	*3.0*	*4.9*	*6.7*	*8.5*	*10.3*	*12.2*	*14.0*
5	*1.5*	*3.3*	*5.2*	*7.0*	*8.8*	*10.6*	*12.5*	*14.3*

The magazine with more craft for sale than any other

FIND IT FAST: If you're buying a boat you'll be sure to find it fast amongst over 1200 for sale every month in **"Boats & Planes For Sale"**.

And the range of power and sailing boats is enormous, from £500 runabouts to luxury craft at over £1,000,000.

SELL IT FAST: Your boat will sell fast and cost you less in **"Boats & Planes For Sale"**.

Your advertisement will appear with a large photograph, black & white or colour, at a price to beat all other boating magazines.

And your advertisement appears in two consecutive monthly issues.

AVAILABLE AT ALL LEADING NEWSAGENTS, ON THE THIRD THURSDAY OF EVERY MONTH

For further details contact:
Freedom House Publishing Co. Ltd,
23 Denmark Street, London WC2H 8NA. Tel: (01) 379 0819
Fax: 01 831 1209

Chapter 10

Harbour, Coastal and Tidal Information

Contents

10.0 INTRODUCTION Page 177
10.0.1 General information
10.0.2 Abbreviations — harbour facilities with foreign translations
10.0.3 Abbreviations — lights, fog signals and waypoints
10.0.4 Glossary of terms used for charts and sailing directions with foreign translations
10.0.5 Map of areas

10.1 AREA 1 185
SW England. Scilly Islands to River Exe

10.2 AREA 2 217
Central S England. Lyme Regis to Chichester

10.3 AREA 3 265
SE England. Littlehampton to Havengore Creek

10.4 AREA 4 307
E England. River Crouch to King's Lynn

10.5 AREA 5 335
NE England. Boston to Berwick-on-Tweed

10.6 AREA 6 363
SE Scotland. Eyemouth to Peterhead

10.7 AREA 7 389
NE Scotland. Fraserburgh to Scrabster including Orkney and Shetland Islands

10.8 AREA 8 417
NW Scotland. Stornoway to Craobh Haven

10.9 AREA 9 443
SW Scotland. Crinan to Kirkcudbright

10.10 AREA 10 465
NW England, Isle of Man and N Wales. Maryport to Pwllheli

10.11 AREA 11 493
S Wales and Bristol Channel. Porthmadog to St Ives

10.12 AREA 12 525
S Ireland. Shannon to Malahide

10.13 AREA 13 555
N Ireland. Carlingford Lough to Galway

10.14 AREA 14 583
S Biscay. Spanish border to River Loire

10.15 AREA 15 615
S Brittany. River Loire to Camaret

10.16 AREA 16 639
N Brittany. Brest to Paimpol

10.17 AREA 17 661
Channel Islands and adjacent coast of France. Portrieux to Cherbourg

10.18 AREA 18 691
NE France. Barfleur to Dunkerque

10.19 AREA 19 725
Belgium and the Netherlands. Nieuwpoort to Delfzijl

10.20 AREA 20 761
Federal Republic of Germany. Borkum to Danish border

Harbour, Coastal and Tidal Information

10.0.1 General information

Harbour, coastal and tidal information is given for the area indicated by the map in 10.0.5, with details of 314 harbours and notes on a further 253 minor harbours and anchorages. The information provided enables a skipper to assess whether he can get into a harbour (tidal height, depth, wind direction etc.), and whether he wants to enter the harbour (shelter, facilities available, early closing days etc.). Each area is arranged as follows:

Index, of the harbours covered in that area.

A sketch of the area showing diagrammatically the positions and characteristics of the harbours covered, principal radiobeacons, coast radio stations, weather information offices, RNLI stations etc.

Tidal stream charts for the area, based on Admiralty tidal stream atlases (by kind permission of the Hydrographer of the Navy and the Controller, HM Stationery Office), showing the strengths and directions of tidal streams for each hour referred to HW Dover and to HW at the nearest Standard Port. For details of the use of tidal stream diagrams see 9.6.2.

A list of principal coastal lights, fog signals and useful waypoints in the area. More powerful lights (ranges 15 miles or more) are in **bold** type; light-vessels and Lanbys are in *CAPITAL ITALICS*; fog signals are in *italics*. Latitude and longitude are shown for more isolated lights and for selected waypoints (underlined). Unless otherwise stated, lights are white. Elevations are in metres (m), and nominal ranges in nautical miles (M). Where appropriate, a brief description is given of the lighthouse or tower. Arcs of visibility, sector limits, and alignment of leading lights etc are true bearings as seen from seaward measured in a clockwise direction. Where the latitude and longitude of a light are given (e.g. Bull Point 51 12.0N/4 12.0W) W stands for West. Elsewhere W means white, and except with longitude the word west is written in full to avoid any possible ambiguity. Abbreviations used in the lists of lights, fog signals and waypoints are given in 10.0.3.

Passage information, briefly calling attention in note form to some of the principal features of the coast, offlying dangers, tide races, better anchorages etc.

Table of distances in nautical miles by the most direct route, avoiding dangers, between selected places in that area and in adjacent areas.

In certain areas (Solent, Ireland, France, Belgium, Netherlands and Germany) there follows a section (in the case of the Solent, two sections) giving **special peculiarities** regarding the area — chart references, double tides etc.

Harbour information includes the following:

a. Chartlets. These are in many cases based upon Admiralty charts (with the kind permission of Hydrographer of the Navy and the Controller, HM Stationery Office). It is also acknowledged with thanks that in certain cases the source information is from Service Hydrographique et Oceanographique de la Marine, France, Deutsches Hydrographisches Institut, Germany, or the Directie Waterhuishouding en Waterbeweging, Netherlands. All depths on the chartlets are in metres. It must be emphasised that these chartlets are not designed or intended for pilotage or navigation, although every effort has been made to ensure that they give an accurate portrayal of the harbour concerned. The publishers and editors disclaim any responsibility for resultant accidents or damage if they are so used. Chartlets do not always cover the whole area referred to in the text (e.g. in E. Anglian rivers). The light tint on the chartlets shows drying area, the dark tint indicates land.

b. After the harbour name, the county (or equivalent abroad) is given, followed by the Admiralty, Stanford (where applicable), Imray Laurie Norie and Wilson (referred to as Imray) chart numbers for the area, and the Ordnance Survey map numbers in the new 1:50,000 series.

c. The tidal information in these sections is also kindly provided with the permission of the Controller, HM Stationery Office, the Hydrographer of the Navy, and the Proudman Oceanographic Laboratory (see 1.1.4). Each Standard Port (see 9.3.1) has the times and heights of High and Low Water for every day of the year. The Secondary Ports (see 9.3.2) have the differences quoted on the most suitable (not necessarily the nearest) Standard Port. Time and height differences from the nominated Standard Port are given for each Secondary Port. Explanations of tidal calculations are given in Chapter 9.

At the top of each port's tidal information, the average time difference on Dover is given, whereby the GMT for High Water can be quickly obtained to an accuracy of 15 minutes or so. The times of High Water Dover are under Dover (10.3.14), and also on the bookmark for quick reference, both being shown in GMT. By subtracting the Duration, quoted for most ports, the time of the previous Low Water can be approximately obtained.

Mean Level is also quoted. Zone times are given for each port, but no account is taken of BST or other daylight saving times (DST).

d. Tidal curves are given for all Standard Ports and for Portland and Dunkerque. For calculations with tidal data (finding the height of tide at a

given time, for example, or the time for a given height), use the tidal curves provided for the appropriate Standard Port.
e. Shelter. Times of lock openings etc are local times (LT), unless stated.
f. Navigation, starting with the lat/long of a waypoint for the approach or entrance.
g. Principal lights, marks and leading lines.
h. Radio telephone. Details of Port Operations and Traffic Management radio services are shown as appropriate for each harbour covered. Other Port Radio stations which may be useful are shown under the nearest harbour covered.

Where significant, the call name of a station is shown in *italics*. Frequencies are indicated by their International Maritime Services Channel (Ch) designator. Ch M refers to the UK Marina Channel (157.85 MHz). MF frequencies are shown in kHz.

Frequencies used for calling and working may be separated by a semi-colon. For example: Ch 16; 12 14 indicates that Ch 16 is used for calling, and that Ch 12 and Ch 14 are working frequencies. Where known, primary frequencies are shown in bold type, thus **14**.

Where there is a choice of calling frequency, always indicate the channel that you are using when calling another station. This avoids confusion when the station being called is listening on more than one channel. As example, 'Dover Port Control. This is NONSUCH, NONSUCH on Channel 74'.

Where local times are stated, the words '(local times)' or 'LT' appear in the heading or the text. H24 means continuous watch. Times of scheduled broadcasts are shown (for example) as H +20, meaning 20 mins past the hour. HW −3 means 3 hours before local HW; HW +2 means 2 hours after local HW.
i. Telephone numbers are either on the exchange whose dialling code is given after 'Telephone', **or** on the exchange quoted. E.g. Portsmouth and Gosport numbers are on 0705, so no exchange is shown. But Fareham numbers are on another exchange, as quoted.

Although the numbers of the local Coastguard and the appropriate MRCC or MRSC (see 8.2.2) are given, in emergency dial 999 and ask for the Coastguard. Equivalent numbers are given in 10.14.7, 10.19.7 and 10.20.7.

Where Customs Freefone applies, dial 100 and ask for Freefone Customs Yachts.
j. At the end of facilities is indicated whether there is a Post Office (PO), Bank, Railway Station (Rly), or Commercial Airport (Air) in or near the port. Where there is not, the nearest one is indicated in brackets. For continental ports, the nearest UK ferry link is shown.

Minor harbours and anchorages. Notes on selected places are given at the end of each area.

10.0.2 Abbreviations — harbour facilities with foreign translations

In the harbour facilities, the following abbreviations are used which are also repeated on the book-mark for convenience:

AB	Alongside berth (omitted for marinas)
AC	220v AC electrical supplies
ACA	Admiralty chart agent
Air	Airport
Bar	Licensed bar
BH	Boat hoist (tons)
BY	Boatyard
C	Crane (tons)
CG	Coastguard
CH	Chandlery
D	Diesel fuel
Dr	Doctor
EC	Early closing day
El	Electrical repairs
FS	Flagstaff
FW	Fresh water supply
Gas	Calor Gas
Gaz	Camping Gaz
HMC	Her Majesty's Customs
Hosp	Hospital
Hr Mr	Harbourmaster
Kos	Kosangas
L	Landing place (omitted for marinas)
Lau	Launderette
LB	Life-boat
M	Moorings available (omitted for marinas)
ME	Marine engineering repairs
MRCC	Maritime Rescue Co-ordination Centre
MRSC	Maritime Rescue Sub-Centre
P	Petrol
PO	Post Office
R	Restaurant
Rly	Railway station
SC	Sailing club
Sh	Shipwright, hull repairs etc
Slip	Slipway for launching, scrubbing etc
SM	Sailmaker
V	Victuals, food stores etc
YC	Yacht club
⚓	Visitors' berth, or where to report

Since yachtsmen may be using local information, these and other harbour terms are given in the following table in French, Dutch and German.

180 Harbour, Coastal and Tidal Information

Abbreviation	English	French	German	Dutch
AB	Alongside berth	Accostage	Stegplatz	Aanlegplaats
Air	Airport	Aéroport	Flughafen	Vliegveld
—	Anchorage	Mouillage	Ankerplatz	Ankerplaats
BH	Boat hoist	Elévateur	Bootlift	Botenlift
BY	Boatyard	Chantier naval	Yachtwerft	Jachtwerf
CH	Chandlery	Accastillage	Beschlag- und schiffsbedarfshändler	Beslag en scheepsleverancier
C.G.	Coast Guard	Garde-Côtière	Küstenwache	Kustwacht
C	Crane	Grue	Kran	Hijskraan
H.M.C.	Customs	Douane	Zoll	Douane
D	Diesel	Gas-oil	Dieselkrafstoff	Dieselolie
Dr	Doctor	Médecin, docteur	Arzt	Huisarts, dokter
E.C.	Early closing	Demi-congé des magasins	Früher Ladenschluss	Tijden winkelsluiting
El	Electrical repairs	Electricien	Bootselektrik	Elektrische
F.S.	Flagstaff	Mât	Flaggenmast	Vlaggestok
FW	Fresh water	Eau potable	Trinkwasser	Drinkwater
Hr Mr	Harbour Master	Chef ou Capitaine du Port	Hafenkapitän, Hafenmeister	Havenmeester
Hosp	Hospital	Hôpital	Krankenhaus	Ziekenhuis
L	Landing	Escalier du quai	Kaitreppe	Haventrappen
Bar	Licensed bar	Bar	Bar	Bar
LB	Lifeboat	Canot de sauvetage	Rettungsboot	Reddingboot
—	Lock	Écluse	Schleuse	Sluis
ME	Marine Engineering	Ingénieur ou Mécanicien	Ingenieur, Bootswerkstatt	Monteur
M	Moorings	Accostage	Festmachebojen	Meerboei
—	Paraffin	Petrole	Petroleum	Petroleum
—	Passport	Passeport	Reisepass	Paspoort
P	Petrol	Essence	Benzin	Benzine
—	Pilot Station	Station de pilotage	Lotsenstation	Loodsstation
P.O.	Post Office	Bureau de poste	Postamt	Postkantoor
Rly Sta	Railway Station	Gare	Bahnhof	Station
R	Restaurant	Restaurant	Restaurant	Restaurant
SM	Sailmaker	Voilier	Segelmacher	Zeilmaker
Sh	Shipwright/hull repairs	Constructeur	Schiffswerft	Bouwmeester
Slip	Slipway	Cale	Slip, Schlipp	Sleephelling
—	Traffic Signals	Signaux de Mouvements de navire	Verkehrssignal	Verkeersseinen
V	Victuals/food supplies	Marché	Lebensmittel	Winkelcentrum
Y.C.	Yacht Club	Yacht Club	Yachtclub, Segelclub	Jacht Club
S.C.	Sailing Club	Club Nautique		
—	Yacht Harbour	Port de plaisance	Yachthafen	Jachthaven

10.0.3 Abbreviations — lights, fog signals and waypoints

Al	Alternating	F	Fixed
Bu	Blue	F.Fl	Fixed and flashing
Bn	Beacon	F.Fl(x)	Fixed and group flashing (no. of flashes)
Brg	Bearing (light)	Fl(x)	Group flashing (no. of flashes)
Dia	Diaphone	G	Green
Dir	Direction light	harb	harbour
Explos	Explosive signal	(hor)	horizontal

incr	increasing
intens	Intensified sector
IQ	Interrupted quick flashing
Irreg	Irregular
Iso	Isophase
Lanby	Large Automatic Navigational Buoy
Ldg Lts	Leading Lights
Lt	Light
Lt F	Light-float
Lt Ho	Lighthouse
Lt V	Light-vessel
M	Sea miles
m	metres
Mo	Morse code light or fog signal
min	minutes
obsc	obscured
Occas	Occasional
Oc	Occulting
Oc(x)	Group Occulting (no. of eclipses)
(P)	Provisional
Q	Quick flashing
R	Red
Racon	Radar responder beacon
Ra refl	Radar reflector
RC	Circular radiobeacon
RD	Directional radiobeacon
RG	Radio direction finding station
s	seconds
SV	Sodium vapour discharge lamp (orange)
(T)	Temporary
TD	Temporarily Discontinued
TE	Temporarily Extinguished
Tr	Tower
ufn	until further notice
unintens	unintensified
(vert)	vertical
Vi	Violet
Vis	Visible
VQ	Very quick flashing
W	White (see 10.0.1)
Whis	Whistle
Y	Yellow, Orange or Amber

10.0.4 Glossary of terms used for charts and sailing directions, with foreign translations

1. Lights — characteristics (examples)

International abbreviations	Older forms	English	French	German	Dutch
F	F	Fixed	Fixe	F.	V
Oc	Occ	Occulting	Occ	Ubr.	O
Oc(2)	Gp Occ (2)	Group occulting (2)	… Occ	Ubr.Grp. (2)	GO
Iso	Isophase	Isophase	Iso	Glt.	Iso
Fl	Fl	Flashing	É, é	Blz.	S
LFl	LFl	Long flashing	Él, él	Blk.	
Fl(3)	GpFl(3)	Group flashing (3)			
Q	QkFl	Quick flashing	Scint, sc	Fkl.	Fl
Q(3)	QkFl(3)	Group quick flashing (3)	… É	Blz (…)	GS
IQ	IntQkFl	Interrupted quick flashing	Scint. dis.	Fkl. unt.	Int Fl
VQ	VQkFl	Very quick flashing	Scint. rapide		
VQ(3)	VQkFl(3)	Group very quick flashing (3)			
UQ	UQ	Ultra quick flashing			
IUQ	IUQ	Interrupted ultra quick flashing			
Mo(K)	Mo(K)	Morse Code (K)	Mo(K)	Mo(K)	
FFl	FFl	Fixed and flashing	Fix É	F. & Blz. Mi.	V & S
Al	Alt.	Alternating	Alt	Wchs	Alt
W	W	White	B, b	w.	w
R	R	Red	R, r	r. (rot)	r
G	G	Green	V, v	gn. (grün)	gn
Y	Y	Yellow	Jaune, J	g. (gelb)	Geel
Y	Or	Orange	org	or. (orange)	gl
Bu	Bl	Blue	bl	bl. (blau)	b
Vi	Vo	Violet	Vio.	viol.	vi
B	B	Black	Noir	s. (schwarz)	Zwart
		Grey	Gris	grau	Grijs

182 Harbour, Coastal and Tidal Information

2. Fog signals

English	French	German	Dutch
Diaphone	Diaphone	Nebelhorn	Diafoon
Horn	Corne de brume	Nautofon	Nautofoon
	Nautophone	Nautophon	Tyfoon
	Typhon	Tyfon	
	Klaxon	Typhon	
Siren	Sirène	Sirene	Mistsirene
			Sirene
Reed	Trompette	Zungenhorn	Mistfluit
Explosive	Explosion	Nebelknallsignal	Knalmistsein
	Canon	Kanone	Mistkanon
Bell	Cloche	Glocke	Mistklok
			Mistbel
Gong	Gong	Gong	Mistgong
Whistle	Sifflet	Heulbaje	Mistfluit

3. Other navigational aids and structures

English	French	German	Dutch
Radiobeacon	Radiophare	Funkfeuer	Radiobaken
Light	Feu	Leuchtfeuer	Licht
Lighthouse	Phare	Leuchtturm	Lichttoren
Light-vessel	Feu flottant	Feurschiff	Lichtschip
	Bateau-feu		
Light-float	Feu flottant	Leuchtfloss	Lichtvlot
Beacon	Balise	Bake	Baken
Column	Colonne	Laternenträger	Lantaarnpaal
Dolphin	Duc d'Albe	Dalben	Ducdalf
			Meerpaal
Framework tower	Charpente Pyl.	Gittermast	Traliemast Geraamte
House	Bâtiment	Haus	Huis
	Maison		
Post	Poteau	Laternenpfahl	Lantaarnpaal
Tower	Tour	Turm	Toren
	Tourelle		
Concrete	Béton	Betonierter	Beton
Metal	de fer	Eiserner	IJzeren
Stone	Maconnerie	Steinerner	Stenen
Wooden	en bois	Hölzerner	Houten
Band	Bande	Waagerecht gestreift	Horizontaal gestreept
Stripe	Raie	Senkrecht gestreift	Vertikaal gestreept
Chequered	à damier	gewürfelt	Geblokt
Topmark	voy.	Toppzeichen	Topteken
Round	cyl.	rundum	Rond
Cone	Cône	Kegel	Kegel
Conical	Conique	kegelförmig	Kegelvormig
Diamond	los.	Raute, rautenförmig	Ruitvormig
Square	Carré	Viereck, quadrat	Vierkant
Triangular	Triang.	dreieckig	Driehoekig
Destroyed	Détruit	zerstört	Vernield
Occasional	Occasionnel	zeitweise	Facultatief
Temporary	Temporaire	vorübergehend	Tijdelijk
Extinguished	Éteint	gelöscht	Gedoofd

10.0.5 Map of areas

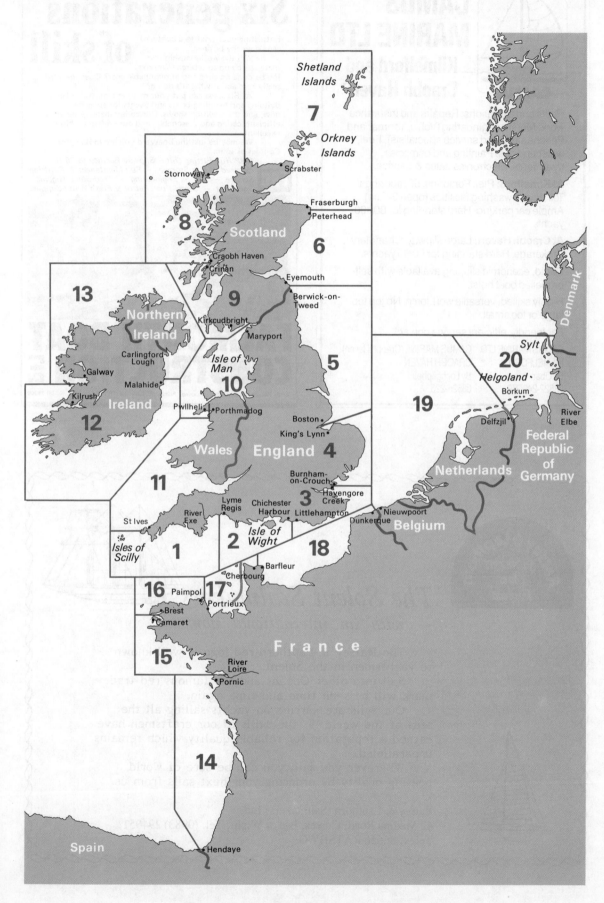

CAMUS MARINE LTD
Kilmelford and Craobh Haven

Refits and alterations: Repairs and insurance work: Marine engineering (Volvo, Yanmar and Perkins sales and service specialists): Fuel, water and gas: Painting and osmosis treatments. Electronics sales & service.

At Kimelford: Pier. Pontoons. 50 moorings. Toilets and washing facilities (open 24 hrs). Ample car parking. Hard standing for 80/100 yachts.

At Craobh Haven: Large slipway. Chandlery Brokerage. Hard standing for 50/60 yachts.

Rapid, economic slipping available with self-propelled boat hoist.

Highly skilled, versatile workforce. No job too large or too small.

For friendly, efficient service contact

CAMUS MARINE LTD	CAMUS MARINE (Craobh Haven)
KILMELFORD	CRAOBH HAVEN
By Oban	By Lochgilphead
08522-248/279	08525-225

Six generations of skill

Entrusting your boat to a boatyard can be a risky business.

Will the workmanship and materials be up to scratch? Will the job be done on time and at reasonable cost? Does the yard really *care* about what it's doing?

At FBC, we can put your mind at rest. We've been building and handling boats and yachts for six generations now, and the tradition shows. Our comprehensive service is efficient, caring and reasonable, and our craftsmanship is excellent.

We may be just the boatyard you are looking for. Contact us soon.

FBC GRP Hull Fitting Out — Slipway Facility to 75 ft
Covered Winter Storage — Full Maintenace & Refitting Service — Marine Engineering — Traditional Rigging Specialist — Diesel Fuel, Petrol & Fresh Water Supplies

FALMOUTH BOAT CONSTRUCTION

Little Falmouth Yacht Yard   Flushing   Falmouth   Cornwall TR11 5TJ   Tel. (0326) 74309

The Solent Sailmakers
— *with an international name*

The Ratsey & Lapthorn red logo is well known to yachtsmen in the Solent.

Sail into other Oceans and the famous red trade mark will be seen time and time again.

Our sails are carried on yachts sailing all the seas of the world - the skills of our craftsmen have earned a reputation for reliable quality which remains unparalleled.

Wherever you sail, you can be sure of world beating quality by ordering your next sails from us.

Ratsey & Lapthorn (Sailmakers) Ltd,
42 Medina Road, Cowes, Isle of Wight, Tel: (0983) 294051
Telex: 86656 RATSEY G.

VOLVO PENTA SERVICE

Sales and service centres in area 1
CORNWALL **Marine Engineering Co. (Looe)** The Quay, EAST LOOE PL13 1AQ Tel (05036) 2887. **Penryn Marine** Freemans Wharf, Falmouth Road, PENRYN TR10 8AD Tel (0326) 76202/1/0. DEVON **J Stone and Son** Cottles Quay, Island Street, SALCOMBE TQ8 8DW Tel (054884) 3655. **Philip & Son Ltd** Noss Works, DARTMOUTH TQ6 0EA Tel (08043) 3351. **Pilkington Marine Engineering** 9 Pottery Units, Forde Road, Brunel Trading Estate, NEWTON ABBOT Tel (0626) 52663. **Retreat Boatyard (Topsham) Ltd** Retreat Boatyard, Topsham, EXETER EX3 0LS Tel (039287) 4720.

VOLVO PENTA

Area 1

South-West England
Isles of Scilly to River Exe

10.1.1	Index	**Page 185**	10.1.20	Dartmouth, Standard Port, Tidal Curves **207**
10.1.2	Diagram of Radiobeacons, Air Beacons, Lifeboat Stations etc **186**		10.1.21	Brixham **212**
			10.1.22	Torquay **212**
10.1.3	Tidal Stream Charts **188**		10.1.23	Teignmouth **213**
10.1.4	List of Lights, Fog Signals and Waypoints **190**		10.1.24	River Exe **214**
10.1.5	Passage Information **191**		10.1.25	Minor Harbours and Anchorages **216**
10.1.6	Distance Tables **192**			Mousehole
10.1.7	English Channel Waypoints **193**			St Michael's Mount Porthleven Port Mullion Coverack
10.1.8	Isles of Scilly **194**			Truro River Porthscatho
10.1.9	St Mary's (Scillies) **195**			Gorran Haven Portmellon
10.1.10	Newlyn **195**			Charlestown Par
10.1.11	Penzance **196**			Polperro Hope Cove
10.1.12	Helford River **196**			Paignton
10.1.13	Falmouth **197**			
10.1.14	Mevagissey **198**			
10.1.15	Fowey **198**			
10.1.16	Looe **199**			
10.1.17	Plymouth (Devonport), Standard Port, Tidal Curves **200**			
10.1.18	Yealm River **206**			
10.1.19	Salcombe **206**			

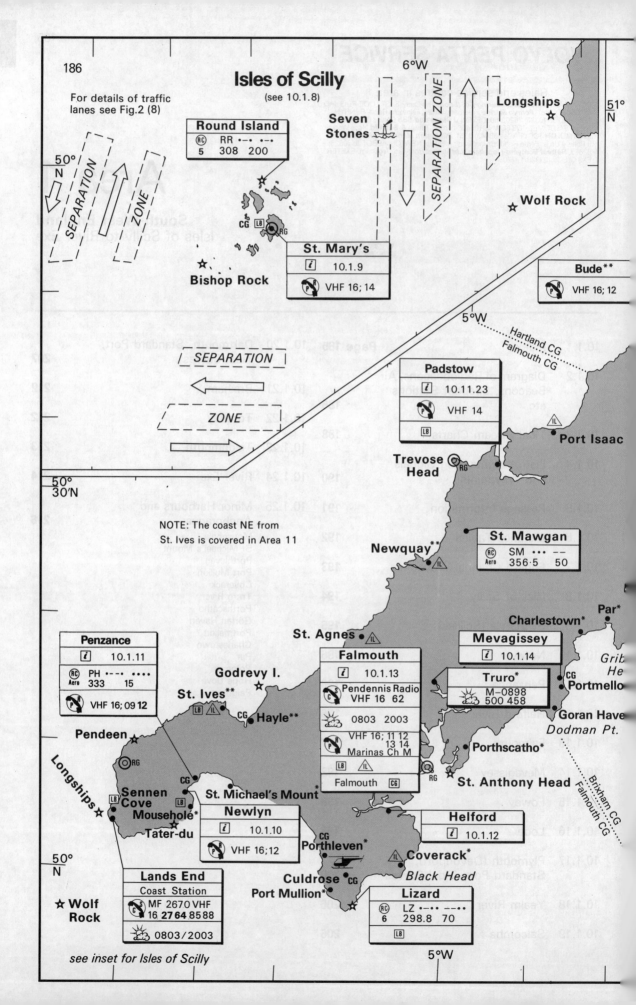

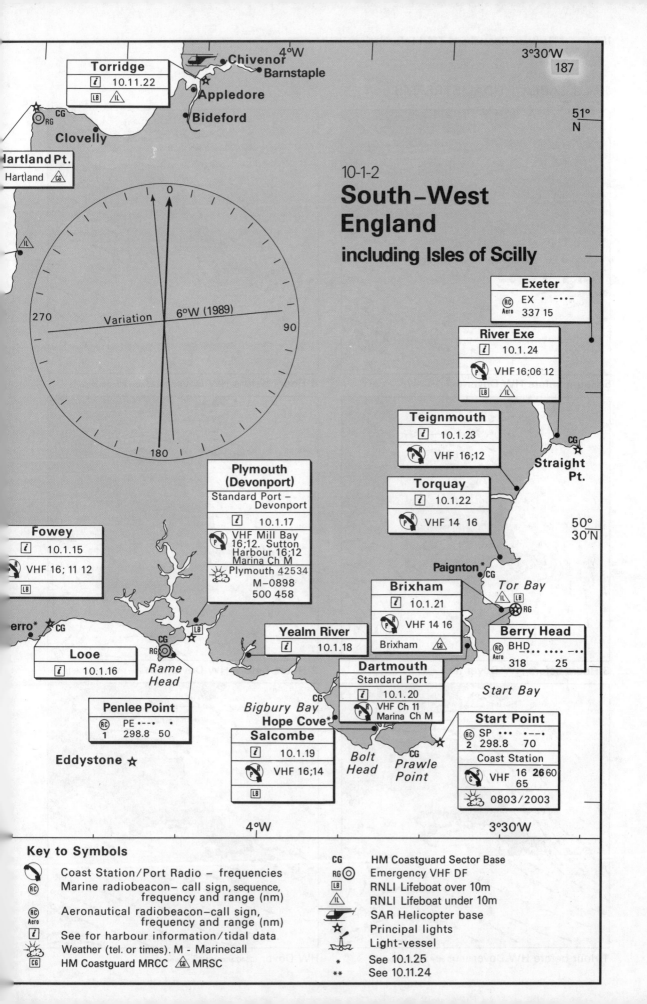

10.1.3 AREA 1 TIDAL STREAMS

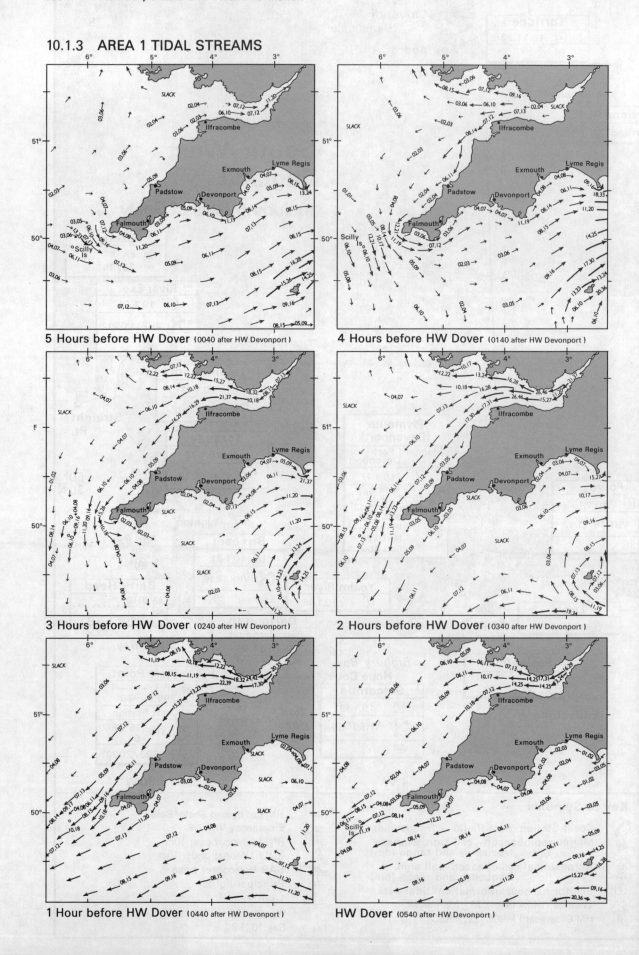

AREA 1 – SW England

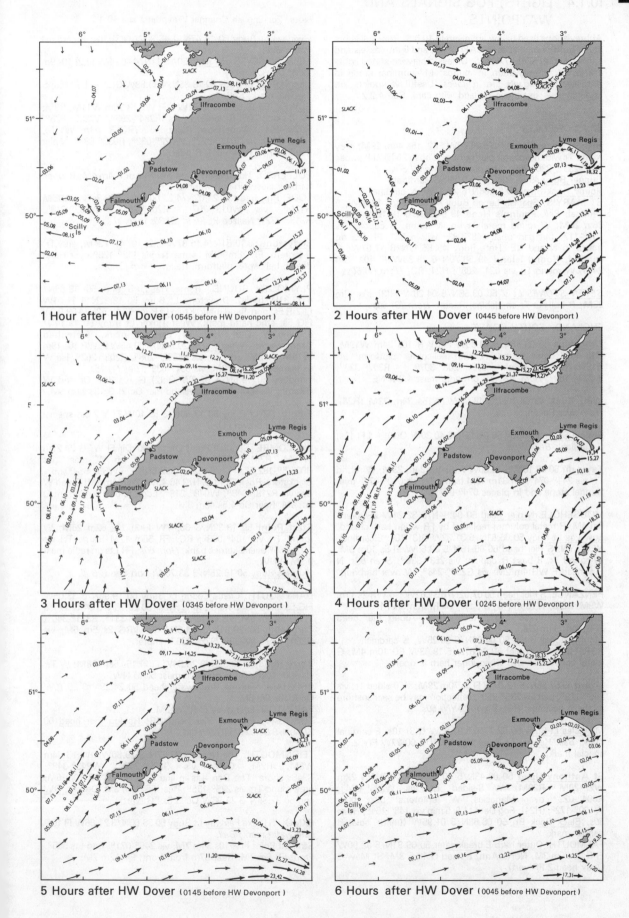

10.1.4 LIGHTS, FOG SIGNALS AND WAYPOINTS

Abbreviations used below are given in 10.0.3. Principal lights are in **bold** print, places in CAPITALS, and light-vessels and Lanbys in *CAPITAL ITALICS*. Unless otherwise stated lights are white. m—elevation in metres; M—nominal range in n. miles. Fog signals are in *italics*. Useful waypoints are underlined — use those on land with care. See 4.2.2.

ISLES OF SCILLY

Bishop Rock 49 52.33N/6 26.68W Fl(2) 15s 44m **29M**; grey round Tr with helicopter platform; part obsc 204°-211°, obsc 211°-233°, 236°-259°; Racon; *Horn Mo(N) 90s*.
Gunner Buoy 49 53.60N/6 25.03W; S cardinal.
Round Rock Buoy 49 53.06N/6 25.14W; N cardinal.
ST MARY'S. Spanish Ledge Buoy 49 53.90N/6 18.80W; E cardinal; *Bell*. **Peninnis Head** 49 54.24N/6 18.15W Fl 20s 36m **20M**; W round metal Tr on W frame, B cupola; vis 231°-117° but part obsc 048°-083° within 5M. FR Lts on masts to N and NE. Hats Buoy 49 56.17N/6 17.07W; S cardinal. **Round Island** 49 58.70N/6 19.33W Fl 10s 55m **24M**; W round Tr; vis 021°-288°; H24; RC; *Horn (4) 60s*.

SEVEN STONES Lt V 50 03.58N/6 04.28W Fl(3) 30s 12m **25M**; R hull, Lt Tr amidships; Racon; *Horn (3) 60s*.

ENGLAND—SOUTH COAST

Longships 50 03.97N/5 44.85W Iso WR 10s 34m **W19M, R18/15M**; grey round Tr with helicopter platform; vis R189°-208°, R(unintens) 208°-307°, R307°-327°, W327°-189°; *Horn 10s*. FR on radio mast 4.9M NE.

Wolf Rock 49 56.70N/5 48.50W Fl 15s 34m **23M** (H24); *Horn 30s*; Racon.

Runnel Stone Lt Buoy 50 01.15N/5 40.30W Q(6) + LFl 15s; S cardinal; *Bell*, *Whis*.

Tater-du 50 03.10N/5 34.60W Fl(3) 15s 34m **23M**; W round Tr; vis 241°-074°. FR 31m 13M (same Tr) vis 060°-074° over Runnel stone and in places 074°-077°.

MOUSEHOLE. N pier head 50 04.94N/5 32.21W 2FG(vert) 6m 4M; R metal column; replaced by FR when harb closed.
Low Lee Lt Buoy 50 05.51N/5 31.32W Q(3) 10s; E cardinal.
NEWLYN. S Pier head 50 06.15N/5 32.50W Fl 5s 10m 9M; W round Tr, R base & cupola; vis 253°-336°; *Siren 60s*. N Pier head F WG 4m 2M; vis G238°-248°, W over harbour.

PENZANCE. S Pier head 50 07.02N/5 31.62W Iso WR 2s 11m W9M, R8M; W round Tr, B base; vis R(unintens) 159°-224°, R224°-268°, W268°-345°, R345°-shore. Albert Pier head 2FG(vert) 11m 2M.
Mountamopus Buoy 50 04.60N/5 26.25W; S cardinal.
PORTHLEVEN. S pier 50 04.87N/5 19.03W FG 10m 4M; G metal column, shown when inner harb is open.

Lizard 49 57.58N/5 12.07W Fl 3s 70m **29M**; W 8-sided Tr; vis 250°-120°, part vis 235°-250°; reflection may be seen inshore of these bearings; RC; *Siren Mo (N) 60s*.

Manacles Lt Buoy 50 02.77N/5 01.85W Q(3) 10s; E cardinal mark; *Bell*. Helston Lt Buoy 50 04.92N/5 00.77W FlY 2.5s; special mark.

St Anthony Head 50 08.43N/5 00.90W Oc WR 15s 22m **W22/20M, R20M**; W 8-sided Tr; vis W295°-004°, R004°-022° over Manacles, W (unintens) 022°-100°, W100°-172°; (H24); Fog Det Lt LFl 5 min vis 148°-151°; *Horn 30s*. Black Rock Bn 50 08.68N/5 01.95W (unlit); isolated danger mark.
FALMOUTH. Outer harb E breakwater 50 09.31N/5 02.90W Fl R 2s 20m 3M. North Arm, E head Q 19m 3M. St Mawes Quay head 2 FR(vert).

Note. For English Channel Waypoints see 10.1.7

Gwineas Lt Buoy 50 14.47N/4 45.30W Q(3) 10s; E cardinal mark; *Bell*.
MEVAGISSEY. S pier head 50 16.11N/4 46.85W Fl(2) 10s 9m 12M; *Dia 30s*.
Cannis Rock Lt Buoy 50 18.35N/4 39.88W Q(6) + L Fl 15s; S cardinal mark; *Bell*.
FOWEY. 50 19.59N/4 38.77W L Fl WR 5s 28m W11M, R9M; W 8-sided Tr, R lantern; vis R284°-295°, W295°-028°, R028°-054°. Whitehouse Point Iso WRG 3s 11m W11M, R8M, G8M; vis G017°-022°, W022°-032°, R032°-037°. N pier head 2 FR(vert) 4m 8M; R post.

POLPERRO. Tidal basin, West pier FW or R 4m 4M; R when harbour closed in bad weather.
LOOE. **Pier head** 50 21.02N/4 27.00W Oc WR 3s 8m **W15M** R12M; vis W013°-207°, R207°-267°, W267°-313°, R313°-332°; Nailzee Pt *Siren (2) 30s (occas)*.

Eddystone 50 10.81N/4 15.87W Fl(2) 10s 41m **24M**; grey Tr, R lantern. FR 28m 13M (same Tr) vis 112°-129° over Hand deeps; helicopter platform, Racon; *Horn (3) 60s*.

PLYMOUTH SOUND. Rame Head, S end 50 18.63N/4 13.31W (unlit). Draystone Lt Buoy 50 18.82N/4 11.01W Fl(2)R 5s; port-hand mark. Knap Lt Buoy 50 19.52N/4 09.94W FIG 5s; stbd-hand mark. West Tinker Lt Buoy 50 19.14N/4 08.62W Q(9) 15s; W cardinal mark. **Detached Breakwater**—West end 50 20.04N/4 09.45W Fl WR 10s 19m **W15M**, R12M; grey Tr; vis W262°-208°, R208°-262°. Iso W 4s (same Tr) 12m 12M; vis 031°-039°; *Bell (1) 15s*.
Whidbey 50 19.5N/4 07.2W Oc(2) G 10s 3M; Or and W column; vis 000°-160°. Bovisand Pier Oc(2) G 15s 17m 3M.

E Rutts DZ Lt Buoy 50 12.60N/3 59.10W Fl Y 2.5s; special mark. Withdrawn (T).

SALCOMBE. Sandhill Point Dir Lt 000°Dir Fl WRG 2s 27m W10M, R7M, G7M; R&W diamond on W mast; vis R002.5°-182.5°, G182.5°-357.5°, W357.5°-002.5°. Blackstone Rock 50 13.57N/3 46.43W QWR 4m 2M; G and W Bn; vis R218°-048°, W048°-218°. Ldg Lts 042°30'. Front Q 5m 8M. Rear Oc 4.5s 45m 8M.

Start Point 50 13.32N/3 38.47W Fl(3) 10s 62m **25M**; W round Tr; vis 184°-068°; RC. FR 55m 12M (same Tr) vis 210°-255° over Skerries bank; *Horn 60s*. FR Lts on radio mast 0.9M WNW.
Skerries Buoy 50 16.25N/3 33.70W; port-hand; *Bell*.

DARTMOUTH. Castle Ledge Lt Buoy 50 19.96N/3 33.05W FIG 5s; stbd-hand mark. Kingswear 50 20.78N/3 34.02W Iso WRG 3s 9m 8M; vis G318°-325°, W325°-331°, R331°-340°. Dartmouth 50 20.81N/3 34.56W Fl WRG 2s 5m 6M; vis G280°-289°, W289°-297°, R297°-shore.

Berry Head 50 23.95N/3 28.94W Fl(2) 15s 58m **18M**; W Tr; vis 100°-023°. R Lts on radio mast 5.7M NW.
BRIXHAM. Victoria Breakwater head 50 24.29N/3 30.70W Oc R 15s 9m 6M.
PAIGNTON. East quay Fl R 7m 3M.
TORQUAY. Princess Pier head QR. Haldon Pier head 50 27.40N/3 31.67W QG 9m 6M.

TEIGNMOUTH. The Ness 50 32.17N/3 29.80W QWRG 44m W8M, R7M, G7M; vis G229°-243°, W243°-348°, R348°-shore. The Den, Lts in line 334°. Front FR 10m 6M; grey round Tr; vis 225°-135°. Rear FR 11m 3M. Den Pt, SW end 50 32.38N/3 29.98W Oc G 5s FG (vert).

EXMOUTH. Exe Fairway Lt Buoy 50 36.00N/3 21.86W Fl 10s; safe water mark; *Bell*.
Straight Point Fl R 10s 34m 7M; vis 246°-071°. Ldg Lts 305°. Front FY 6m 7M. Rear 57m from front, FY 12m 7M.

10.1.5 PASSAGE INFORMATION

NORTH CORNWALL (charts 1149, 1156)

The coast of North Cornwall is covered in Area 11, to which reference should be made. For Padstow see 10.11.23. For Bude, Newquay, Hayle and St Ives see 10.11.24. For ease of reference, certain general information is repeated below.

Approach to Bristol Chan along N coast of Cornwall is very exposed, with little shelter in bad weather. Padstow is a refuge, but in strong NW winds the sea breaks on bar and prevents entry. Shelter is available under lee of Lundy Island; but there are bad races to NE (White Horses), the NW (Hen and Chickens), and to SE; also overfalls over NW Bank. St Ives (dries) is sheltered from E and S, but exposed to N. So in this area yachts need to be sturdy and well equipped, since if bad weather develops no shelter may be at hand. Streams are moderate W of Lundy, but strong round the island. They get much stronger towards Bristol Chan proper.

ISLES OF SCILLY (10.1.8 and chart 34)

The Scillies comprise 48 islands, extending 21-31M WSW of Land's End. There are many rocky outcrops and offlying dangers, and although they are all well charted care is needed particularly in poor visibility. For pilotage details see *Channel Pilot* or *South England Pilot, Vol 5* (Brandon). There is a useful local publication — *A Yachtsman's Guide to Scilly*.

Several transits are shown on chart 34, and these should be followed, because the tidal streams in and between the islands are difficult to predict with any accuracy. They run harder off points and over rocks, where overfalls may occur.

Conspic landmarks are Bishop Rock Lt Ho, Round Island Lt Ho, the disused Lt Ho on St Agnes, the daymark at the E end of St Martin's, Penninis Lt Ho at the S end of St Mary's, and the TV mast and CG signal station (at the old telegraph tower) both in the NW corner of St Mary's. Yachts must expect to lie to their anchors. There is no one anchorage giving shelter in all wind directions, so it may be necessary to move at short notice. All anchorages may be penetrated by swell. The most popular and useful ones are shown in 10.1.8.

From St Ives to Land's End coast is rugged and exposed. There are overfalls SW of Pendeen Pt (Lt, fog sig). Vyneck Rks lie awash about 3 ca NW of C Cornwall. The Brisons are two rocky islets ½M SW of C Cornwall, and rocky ledges extend inshore and to the S and SW. The Longships (Lt, fog sig) are a group of rks about 1M W of Land's End, with ledges 2 ca further seaward. The inshore passage is not recommended.

LAND'S END/SCILLIES

Between Land's End and the Scillies (chart 1148) streams are rotatory, clockwise. They run ENE from HW Devonport −0100 (1 kn at sp), SSE from HW Devonport +0200 (2 kn at sp), WNW from HW Devonport +0600 (1 kn at sp), and N from HW Devonport −0400 (1¾ kn at sp). The Seven Stones (rks) lie 14M W of the Longships and 7M NE of the Scillies; many of them dry, with ledges in between. They are marked by Lt V (Lt, fog sig) on E side. Wolf Rk (Lt, fog sig) is 8M SW of Land's End, and is steep-to. For traffic schemes see Fig. 2(8).

LAND'S END TO LIZARD HEAD (chart 777)

From Land's End to Gwennap Hd, 2½M SE, rks extend up to 1½ ca offshore, and depths are irregular to seaward causing a bad sea in strong W winds with W-going tide. The Runnel Stone (dries) lies 7 ca S of Gwennap Hd, with rks between it and shore. These dangers are in R sectors of Longships and Tater-du Lts. Do not anch off Porth Curno, due to cables.

Entering Mount's B, the Bucks (dry) are 2 ca SE and E of Tater-du Lt Ho. Gull Rk (24m) is 9 ca NE of Tater-du, close off the E point of Lamorna Cove. Little Heaver (dries) is ½ ca SW of Gull Rk, and Kemyel Rk (dries) is 1¾ ca ENE. Mousehole is a small drying harb, sheltered from W and N, but exposed to winds in E or S, when entrance may be closed: approach from S side of St Clement's I. In W winds there is good anch off the harb. See 10.1.25.

Low Lee, a dangerous steep-to rk, is 4 ca NE of Penlee Pt, marked by buoy. Carn Base Rk lies 3 ca NNW of Low Lee. Newlyn (10.1.10) is only harb in Mount's B safe to approach in strong onshore winds, but only near HW. From here to Penzance (10.1.11) beware Dog Rk and Gear Rk.

From Penzance to St Michael's Mount the head of the b is flat, and dries for 4 ca in places. Dangers include Cressar Rks, Long Rk, Hogus Rks, and Outer Penzeath Rk. Venton chy on with pierheads of St Michael's Mount harb at 084° leads S of these dangers. This tiny harb dries, but is well sheltered, with anch about 1 ca W of entrance, see 10.1.25.

Two dangerous rks, Guthen Rk and Maltman Rk, lie within 2 ca of St Michael's Mount. 1M SE is The Greeb (7m), with rks between it and shore. The Bears (dry) lie 1¾ ca E of The Greeb. The Stone (dries) is ½M S of Cudden Pt, while offshore is Mountamopus shoal marked by buoy which should be passed to seaward. Welloe Rk (dries) lies ½M SW of Trewavas Hd.

Porthleven is small tidal harb, entered between pier on S side and Deazle Rks (dry) on N. Dry out alongside in inner harb, closed in bad weather when approach is dangerous. In fair weather there is good anch off Porth Mellin, about 1½ ca NE of Mullion Island; the harb is not recommended. See 10.1.25.

2½M W of Lizard Hd is The Boa, a rky shoal on which sea breaks in SW gales. The Lizard (Lt, fog sig, RC) is a significant headland (chart 2345) with tidal streams up to 3 kn at sp. The outer rks, all of which dry, from W to E are Mulvin, (2½ ca SW of Lizard Pt), Taylor's Rk (2ca SSW of Lizard Pt), Clidgas Rks (5ca SW of Lt Ho), Men Hyr Rk and the Dales or Stags (5ca SSW of Lt Ho), and Enoch Rk (3ca S of Lt Ho). S of the rks the stream turns E at HW Devonport −0500, and W at HW Devonport +0155. Offshore it turns an hour earlier. A dangerous race extends 2-3M S when stream is strong in either direction, worst in W winds against W-going tide. Then keep at least 3M to seaward. There may be a race SE of the hd.

LIZARD HEAD TO STRAIGHT POINT (chart 442)

Vrogue, a dangerous sunken rk, is 4 ca ESE of Bass Pt, NE of Lizard Hd. Craggan Rks are ½M offshore, and 1M N of Vrogue Rk. From Black Hd to beyond Chynnalls Pt, rks extend at least 1 ca offshore. Coverack gives good anch in W winds, see 10.1.25. From Dolor Pt to E of Lowland Pt are drying rks ¼M offshore.

The Manacles (dry), ¾M E and SE of Manacle Pt, are marked by buoy to seaward and are in R sector of St Anthony Hd Lt. Off the Manacles the stream runs NE from HW Devonport −0345, and SW from HW Devonport +0200, sp rates 1¼ kn. From E of the Manacles there are no offshore dangers on courses NNW to Helford entrance (10.1.12) or N to Falmouth (10.1.13).

Sailing NE from St Anthony Hd, Gull Rk lies 6 ca E of Nare Hd, at W side of Veryan B. The Whelps (dry) are ½M SW of Gull Rk. There is a passage between Gull Rk and the shore. In Veryan B beware Lath Rk 1M SE of Portloe.

On E side of Veryan B, Dodman Pt is a 110m cliff, with a stone cross near SW end. Depths are irregular for 1M S, with heavy overfalls in strong winds, when it is best to pass 2M off. Gorran Haven, a sandy cove with L-shaped pier which dries at sp, is a good anch in offshore winds. See 10.1.25.

2.1M NE of Dodman Pt cross, and 1M ENE of Gorran Haven, is Gwineas Rk (8m). 1 ca E of Gwineas Rk is Yaw Rk (dries 0.9m), marked by buoy on E side. Passage inside Gwineas Rk is possible, but not recommended in strong onshore winds or poor vis. Portmellon and Mevagissey B (see 10.1.25 and 10.1.14) are good anchs in offshore winds. For Charlestown and Par see 10.1.25.

Gribbin Hd has a conspic daymark — a square Tr 25m high with R & W stripes. In bad weather the sea breaks on rks round hd. Cannis Rk (dries) is ¼M SE, and marked by buoy to seaward. 3M E of Fowey (10.1.15) is Udder Rk (dries), ½M offshore in E part of Lantivet B. Larrick Rk (dries) is 1½ ca off Nealand Pt.

Polperro harb dries, but the inlet gives good anch in offshore winds. See 10.1.25. Beware E Polca Rk roughly in mid-chan. There are shoals extending ¼M seaward from Downend Pt, E of Polperro. The chan between Hannafore Pt and St George's (or Looe) Island nearly dries. The Rennies (dry) are rks extending ¼M E and SE of the island. (See 10.1.16). There are overfalls S of Looe Island in bad weather.

Eddystone Rks (chart 1613) lie 8M S of Rame Hd. Shoals extend 3 ca E. Close NW of the Lt Ho (Lt, fog sig) is the stump of old one. The sea can break on Hand Deeps — sunken rks 3½M NW of Eddystone.

Rame Hd is on W side of entrance to Plymouth Sound (10.1.17). It is conspic cone shaped, with small chapel on top. Rks extend about 1 ca off. On the E side of Plymouth Sound, the Mewstone (59m) is a conspic rocky islet 4 ca off Wembury Pt. Drying rks extend 1½ ca SW of Little Mewstone, and the Slimers, which dry, (see 10.1.18) lie 2 ca E of Mewstone. E and W Ebb Rks (awash) lie 2½ ca off Gara Pt (chart 30).

Between Gara Pt and Stoke Pt, 2½M to E, offlying dangers extend about 4 ca offshore in places. In Bigbury B beware Wells Rk and other dangers ½M S of Erme Hd. From Bolt Tail to Bolt Hd keep ½M offshore to clear Greystone Ledge, sunken rks near Ham Stone (11m), and Gregory Rks ½M SE of Ham Stone. The Great and Little Mewstones lie just off Bolt Hd, where coast turns N to Salcombe entrance (10.1.19).

Start Pt (Lt, fog sig, RC) is 3M NE of Prawle Pt, and is a long headland with white Lt Ho near end and conspic radio masts in rear. Rks extend 2½ ca offshore, and the stream runs 4 kn at sp, causing a race extending 1M to seaward. 3M S of Start the stream turns ENE at HW Devonport −0150, and WSW at HW Devonport +0420. Inshore it turns ½ hr earlier. In reasonable weather the overfalls can be avoided by passing close to rks, but in bad weather keep at least 2M off.

Start Pt is W end of Lyme B (chart 3315), stretching 50M NE to Portland Bill, and with no secure harb in its E part. NE of Start Pt is Skerries Bank, on which sea breaks in bad weather (chart 1634). Good anch off Hallsands in offshore winds.

Between Dartmouth (10.1.20) and Brixham (10.1.21) rks extend ½M offshore. Berry Hd (Lt) is steep, flat-topped headland (55m). Here the stream turns N at HW Devonport −0105, and S at HW Devonport +0440, sp rates 1½ kn. In Torbay (chart 26) the more obvious dangers are steep-to, but beware the Sunker ½ ca SW of Ore Stone, and Morris Rogue ½M W of Thatcher Rk.

There are good anchs in Babbacombe B and in Anstey's cove in W winds: beware the Three Brothers (drying rks), S side of Anstey's cove. From Long Quarry Pt for 4M N to Teignmouth (10.1.23) there are no offlying dangers. Off Teignmouth the NNE-going stream begins at HW Devonport −0135, and the SSW-going at HW Devonport +0510. In the entrance the flood begins at HW Devonport −0535, and the ebb at HW Devonport +0040. The stream runs hard off Ferry Pt.

Between Teignmouth and Dawlish rks extend 1 ca offshore. Beware Dawlish Rk (depth 2.1m) about ½M off N end of town. Warren Sands and Pole Sands lie W of entrance to Exmouth (10.1.24), and are liable to shift. Along the NE (Exmouth) side of the chan, towards Orcomb Pt and Straight Pt (Lt), drying rks and shoals extend up to ¼M from shore.

10.1.6 DISTANCE TABLE

Approximate distances in nautical miles are by the most direct route while avoiding dangers and allowing for traffic separation schemes etc. Places in *italics* are in adjoining areas.

1 *Fastnet Rock*	1																			
2 *Tuskar Rock*	138	2																		
3 *South Bishop*	162	36	3																	
4 *Lundy Island*	187	85	50	4																
5 *Appledore*	207	103	70	20	5															
6 *Padstow*	186	112	82	40	48	6														
7 Longships	167	130	109	82	89	49	7													
8 Bishop Rock	152	141	132	110	120	76	30	8												
9 Wolf Rock	169	135	117	90	97	57	8	26	9											
10 Lizard Point	190	153	132	105	112	72	23	50	24	10										
11 Falmouth	206	169	148	121	128	88	39	66	40	16	11									
12 Fowey	224	187	166	139	146	106	57	84	58	34	22	12								
13 Plymouth	239	202	181	154	161	121	72	99	73	49	39	22	13							
14 Eddystone	229	192	171	144	151	111	62	89	63	39	31	17	12	14						
15 Start Point	253	216	195	168	175	135	86	113	87	63	55	40	24	24	15					
16 Berry Head	266	229	208	181	188	148	99	126	100	76	68	53	37	37	13	16				
17 Straight Point	280	243	222	195	202	162	113	140	114	90	82	67	51	51	27	14	17			
18 *Portland Bill*	302	265	244	217	224	184	135	162	136	112	104	89	73	73	49	41	36	18		
19 *Le Four*	250	229	208	181	188	148	99	103	95	87	99	109	115	103	113	125	139	153	19	
20 *Casquets*	300	263	142	215	222	182	133	160	134	110	107	96	80	79	57	61	66	48	120	20

AREA 1—SW England 193

10.1.7 ENGLISH CHANNEL WAYPOINTS

Selected waypoints for Channel crossings, listed from West to East in each Area nominated below. Further waypoints in coastal waters are given in section 4 of each relevant Area (i.e. 10.1.4, 10.2.4 and 10.3.4 on the English coast, and 10.16.4, 10.17.4 and 10.18.4 for the French coast and the Channel Islands).

ENGLISH COAST
Area 1

Bishop Rock Lt	49°52'.33N	06°26'.68W
Seven Stones Lt V	50°03'.58N	06°04'.28W
Wolf Rock Lt	49°56'.70N	05°48'.50W
Tater Du Lt	50°03'.10N	05°34'.60W
Lizard Point Lt	49°57'.58N	05°12'.07W
Manacles Lt Buoy	50°02'.77N	05°01'.85W
St Anthony Head Lt	50°08'.43N	05°00'.90W
Eddystone Lt	50°10'.80N	04°15'.87W
Start Point Lt	50°13'.32N	03°38'.47W
Berry Head Lt	50°23'.95N	03°28'.94W

Area 2

Straight Point Lt	50°36'.46N	03°21'.68W
Portland Bill Lt	50°30'.82N	02°27'.32W
E Shambles Lt Buoy	50°30'.74N	02°20'.00W
Anvil Point Lt	50°35'.50N	01°57'.55W
Poole Fairway Lt Buoy	50°38'.97N	01°54'.78W
Needles Fairway Lt Buoy	50°38'.20N	01°38'.90W
Needles Lt	50°39'.70N	01°35'.43W
St Catherine's Point Lt	50°34'.52N	01°17'.80W
Bembridge Ledge Lt Buoy	50°41'.12N	01°02'.74W
Nab Tower Lt	50°40'.05N	00°57'.07W

Area 3

Owers Lanby	50°37'.30N	00°40'.60W
Brighton Marina Lt	50°48'.46N	00°06'.29W
Royal Sovereign Lt	50°43'.38N	00°26'.13E
Dungeness Lt	50°54'.77N	00°58'.67E
S Goodwin Lt Float	51°07'.95N	01°28'.60E

OFFSHORE AIDS
Areas 17 and 3

Channel Lt V	49°54'.42N	02°53'.67W
E Channel Lt Float	49°58'.67N	02°28'.87W
EC1 Lt Buoy	50°05'.90N	01°48'.35W
EC2 Lt Buoy	50°12'.10N	01°12'.40W
EC3 Lt Buoy	50°18'.30N	00°36'.10W
Greenwich Lanby	50°24'.50N	00°00'.00
CS1 Lt Buoy	50°33'.67N	00°03'.83W
CS2 Lt Buoy	50°39'.08N	00°32'.70E
CS3 Lt Buoy	50°52'.00N	01°02'.30E
CS4 Lt Buoy	51°08'.57N	01°34'.02E
Bullock Bank Lt Buoy	50°46'.90N	01°07'.70E
S Varne Lt Buoy	50°55'.57N	01°17'.40E
Varne Lanby	51°01'.25N	01°24'.00E
MPC Lt Buoy	51°06'.10N	01°38'.33E
Sandettie Lt V/Buoy	51°09'.40N	01°47'.20E

OFFSHORE AIDS (continued)
Area 18

Bassurelle Lt Buoy	50°32'.70N	00°57'.80E
Vergoyer SW Lt Buoy	50°26'.90N	01°00'.00E
Vergoyer N Lt Buoy	50°39'.70N	01°22'.30E
ZC1 Lt Buoy	50°44'.85N	01°27'.10E
ZC2 Lt Buoy	50°53'.50N	01°31'.00E

FRENCH COAST
Area 16

Créac'h Lt	48°27'.62N	05°07'.65W
Ouessant NE Lt Buoy	48°45'.90N	05°11'.60W
Ile Vierge Lt	48°38'.38N	04°34'.00W
Ile de Batz Lt	48°44'.78N	04°01'.55W
Les Triagoz Lt	48°52'.35N	03°38'.73W
Les Sept Iles Lt	48°52'.78N	03°29'.33W
Les Heaux Lt	48°54'.53N	03°05'.20W
Rosédo (Bréhat) Lt	48°51'.50N	03°00'.32W
Barnouic Lt	49°01'.70N	02°48'.40W
Roches Douvres Lt	49°06'.47N	02°48'.82W

Area 17

Cap Fréhel Lt	48°41'.10N	02°19'.07W
St Malo Fairway Lt Buoy	48°41'.42N	02°07'.21W
SW Minquiers Lt Buoy	48°54'.40N	02°19'.30W
NW Minquiers Lt Buoy	48°59'.70N	02°20'.50W
Chausey, Grand Ile Lt	48°52'.23N	01°49'.26W
Cap de Carteret Lt	49°22'.46N	01°48'.35W
Cap de la Hague Lt	49°43'.37N	01°57'.19W
CH1 Lt Buoy	49°43'.30N	01°42'.10W
Cherbourg Fort Ouest Lt	49°40'.50N	01°38'.87W
Cap Lévi Lt	49°41'.80N	01°28'.30W

CHANNEL ISLANDS *(Area 17)*

Les Hanois Lt	49°26'.20N	02°42'.10W
Corbière Lt	49°10'.85N	02°14'.90W
Point Robert Lt (Sark)	49°26'.25N	02°20'.67W
Casquets Lt	49°43'.38N	02°22'.55W
Alderney Main Lt	49°43'.81N	02°09'.77W

FRENCH COAST
Area 18

Pointe de Barfleur Lt	49°41'.87N	01°15'.87W
Iles St Marcouf Lt	49°29'.90N	01°08'.90W
Ver-sur-Mer Lt	49°20'.47N	00°31'.15W
Le Havre Lanby	49°31'.67N	00°09'.80W
Cap de la Hève Lt	49°30'.80N	00°04'.24E
Cap d'Antifer Lt	49°41'.07N	00°10'.00E
Pointe d'Ailly Lt	49°55'.13N	00°57'.63E
Ault Lt	50°06'.35N	01°27'.24E
Pointe du Haut Blanc Lt	50°23'.90N	01°33'.75E
Cap d'Alprech Lt	50°41'.95N	01°33'.83E
Cap Gris Nez Lt	50°52'.05N	01°35'.07E
CA4 Lt Buoy	50°58'.94N	01°45'.18E
Sangatte Lt	50°57'.23N	01°46'.57E
Dunkerque Lanby	51°03'.07N	01°51'.83E
Walde Lt	50°59'.57N	01°55'.00E

ISLES OF SCILLY 10-1-8

The Isles of Scilly are made up of 48 islands and numerous rocky outcrops, extending approx 10M by 5M and lying 21-31M WSW of Land's End. Only six islands are inhabited, St Mary's, Gugh, St Agnes, St Martin's, Tresco and Bryher. The islands belong to the Duchy of Cornwall and details of arrangements for visiting uninhabited islands are given in the booklet *Duchy of Cornwall – Information for visiting craft* obtainable from the Harbour Master, Hugh Town Harbour, St Mary's. There is a lifeboat and a HM Coastguard Sector Base at St Mary's.

CHARTS
Admiralty 883, 34; passage from Lands End 1148; Stamford 2; Imray C7.

TIDES
Standard Port is Devonport. Tidal differences for St Mary's are given in 10.1.9 below. The tidal heights, times, speeds and directions round islands are irregular.

SHELTER
There are numerous anchorages, the following being but a selection:
HUGH TOWN HARBOUR (St Mary's). See 10.1.9 below.
PORTH CRESSA (St Mary's S of Hugh Town). Beware of dangers on each side of entrance and the submarine cable. Exposed to swell from SE to SW. Anchor in approx 2m. Facilities: all facilities in Hugh Town to N.
NEW GRIMSBY (between Tresco and Bryher). Approach through New Grimsby Sound, or with sufficient rise of tide across Tresco Flats. Good shelter except in NW winds. Anchor between Hangman Is and the quay in 1.5 to 4.5m. Facilities: Bar, R, FW, V, PO, Ferry to St Mary's.
WATERMILL COVE (NE corner of St Mary's). Excellent shelter in winds NW to S. Anchor in approx 5m.
THE COVE (St Agnes/Gugh). Well sheltered from W and N winds, except when the sand bar between the islands covers near HW sp with a strong NW wind.
PORTH CONGER (St Agnes/Gugh). On the N side of the sand bar, sheltered in winds from E through S to W. May be uncomfortable when the bar is covered. Facilities; L (two quays), ferry to St Mary's; in Middle Town (¼M), St Agnes, V, PO, R, Bar
OLD GRIMSBY (between Tresco and St Helen's). Green Porth and Raven's Porth, divided by a quay, form the W side of Old Grimsby and both dry. Anchor 2 ca NE of quay in 2.5m. Access more difficult than New Grimsby. Well sheltered in SW winds but open to swell if wind veers N of W. Facilities; L(quay), hotel, slip (hotel).
TEAN SOUND (St Martin's, W side). Requires careful pilotage, but attractive anchorage in better weather. More suitable for shoal draught boats which can anchor out of main tidal stream in channel. There are several other anchorages which can be used in settled weather, or by shoal draught boats which can take the ground.

NAVIGATION
See chart 34 (essential), 10.1.5 and *A Yachtsman's Guide to Scilly*. For traffic separation schemes, see Fig. 2(8). The many offlying dangers need care if the leading lines (lettered on chartlet) cannot be identified, when it is best to lie off. Tidal streams are irregular and run hard off points and over shoals, where overfalls may occur. Many of the channels between the islands have dangerous shallows, often rocky ledges. Beware lobster pots and the ferry entering or leaving. Also beware the grave possibility of normal yacht anchors dragging through the fine sand, even with plenty of scope.
St Mary's Sound (line B) is the normal access to St Mary's Road. From the NE beware the Gilstone (dries) 3ca E of Peninnis Head. An E cardinal bell buoy marks Spanish Ledges which extend from the Gugh shore, and a port-hand buoy marks Bartholomew Ledges (least depth 0.6m). The Woolpack Bn (S cardinal) marks SW end of St Mary's. In the approach to Hugh Town, transit E1 clears Woodcock Ledge, which breaks in bad weather.
Crow Sound (line A) is not difficult with sufficient rise of tide, but is rough in strong winds from E or S. From the NE a yacht can pass close to Menawethan and Biggal Rock. Avoid Trinity Rock and the Ridge, which break in bad weather. Hats buoy (S cardinal) marks a drying shoal with an old boiler on it. The bar (dries) lies off Bar Point, but there is slightly deeper water nearer Bar Point. Leave Crow Rock Bn either side before altering course to SW.
North Channel (line D) is about ¾M wide and provides easy access if the leading marks can be seen. Beware cross tide and Steeple Rock (awash) ¾M SW of Mincarlo. In bad weather the sea breaks on Jeffrey Rock and Spencer's Ledge either side of channel.
Broad Sound (line C) is about ¼M wide at narrowest point, and is entered between Bishop Rock and Fleming's Ledge about ¾M to the N. Leading marks are not conspic, but the channel is shown by Round Rock, Gunner and Old Wreck buoys which mark the main dangers other than Jeffrey Rock.

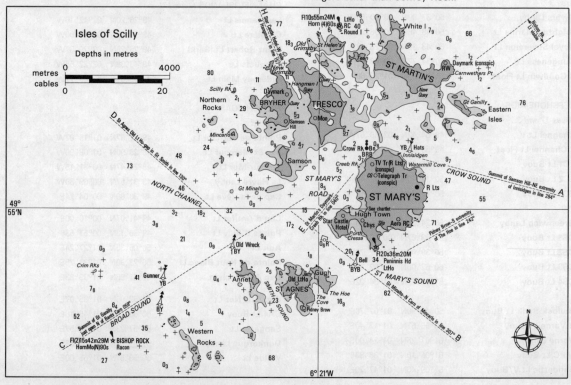

AREA 1—SW England   195

ISLES OF SCILLY continued

LIGHTS AND MARKS
See 10.1.4 for lights. It is wise to study the various marks on the chart before arrival.

RADIO TELEPHONE
For port operations see 10.1.9. Scillies area is served by directional aerial from Land's End Radio on Ch 64. St Mary's Coastguard Ch 16; 67 (manned in bad weather only).

TELEPHONE (0720)
Hr Mr Hugh Town Harbour 22768; CG 22651; HMC 22571; Info Office 22537; MRSC Falmouth 317575; Airport Reservations 22646; Marinecall 0898 500 458; Hosp 22508.

ST MARY'S 10-1-9
Isles of Scilly

CHARTS
Admiralty 34, 883; Stanford 2; Imray C7; OS 203

NOTE: The old Admiralty fathom chart 34 shows more inshore dangers.

TIDES
+0551 Dover; ML 3.1; Duration 0600; Zone 0 (GMT)

Standard Port DEVONPORT (→)

Times				Height (metres)			
HW		LW		MHWS	MHWN	MLWN	MLWS
0000	0600	0000	0600	5.5	4.4	2.2	0.8
1200	1800	1200	1800				

Differences ST MARY'S ROADS
−0030 −0110 −0100 −0020 +0.2 −0.1 −0.2 −0.1

SHELTER
St Mary's Pool gives good shelter in most winds but is exposed to W and NW, when Porth Cressa (see 10.1.8) is better. Anchoring is not permitted S of a line from the

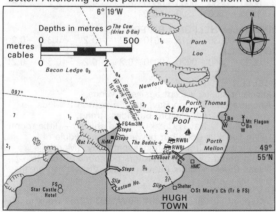

pierhead to the lifeboat slip, off the pierhead where the ferry turns, or in the harbour approaches. Speed limit 3kn.

NAVIGATION
Waypoint St Mary's Sound. Spanish Ledge E cardinal buoy, Bell, 49°53'.90N 06°18'.80W, 128°/308° from/to Great Minalto, 2.4M. See also 10.1.4 and 10.1.8. Other channels into St Mary's Road include Smith Sound between St Agnes and Annet, and with sufficient rise of tide through New Grimsby across Tresco Flats. In the approaches to St Mary's Pool beware Woodcock Ledge (see 10.1.8). Bacon Ledge (depth 0.3m) lies 2½ca W of Newford Island, while further N are The Cow and The Calf (both dry). The 097° transit on the chartlet leads S of Bacon Ledge. If coming from the N, the 151° transit shown leads W of The Cow and E of Bacon Ledge.

LIGHTS AND MARKS
There are two leading marks, rather distant W beacons, bearing 097° for entry into the harbour. Nearest mark is a triangle, the distant mark an X. There are no navigational lights in the harbour except the little lighthouse on the pier showing FG 4m 3M; vis 072°-192°.

RADIO TELEPHONE (local times)
Call: *St Mary's Harbour* VHF Ch 16; 14. (0800-1700).

TELEPHONE (0720)
Hr Mr 22768; CG 22651; MRCC Falmouth 317575; HMC 22571; Marinecall 0898 500 458; Dr 22628.

FACILITIES
EC Wednesday (or Thursday according to boat sailings); Harbour Office will hold mail for visiting yachts if addressed c/o the Harbour Master. **Harbour** Slip, M, Gas, Gaz, P (cans), D, L, FW*, C (6 ton); **T. H. Chudleigh** Tel. 22505, Sh, CH; **Chris Jenkins** Tel. 22321, ME; **H.J. Thomas** Tel. 22710, L, ME, CH; **Scillonian Marine** Tel. 22124, CH, El, ACA; **Isles of Scilly YC** Tel. 22352, Bar, Lau, R; **Island Supply Stores** Tel. 22388, Gas; **Hugh Town** has limited shopping facilities. PO, Bank, Rly (ferry to Penzance), Air (Helicopter service to Penzance). There are PO, R, V, Bar etc. at Tresco, Bryher, St Martins and St Agnes.
*0830-1130 except Sat. — see Hr Mr.

NEWLYN 10-1-10
Cornwall

CHARTS
Admiralty 2345, 777; Stanford 13; Imray C7; OS 203

TIDES
+0550 Dover; ML 3.1; Duration 0555; Zone 0 (GMT)

Standard Port DEVONPORT (→)

Times				Height (metres)			
HW		LW		MHWS	MHWN	MLWN	MLWS
0000	0600	0000	0600	5.5	4.4	2.2	0.8
1200	1800	1200	1800				

Differences NEWLYN
−0040 −0105 −0045 −0020 +0.1 0.0 −0.2 0.0

SHELTER
Good except in SE winds when heavy swell occurs in the harbour. In off-shore winds there are good anchorages outside the harbour in Gwavas Lake. Anchoring inside breakwaters is prohibited. No visitors' moorings; overnight stay only.

NAVIGATION
Waypoint Low Lee E cardinal buoy, Q(3) 10s, 50°05'.52N 05°31'.32W, 130°/310° from/to Newlyn pierhead, 1.0M. Approaching from E beware Mountamopus, from S beware Low Lee and Carn Base. Also beware Dog Rk, 3½ ca NE of harbour entrance. Harbour is available at all states of the tide but entrance is narrow, i.e. 45m between pierheads. Leave Lt Ho to port and keep close to New Pier. Newlyn is a commercial fishing port and fishing boats must take priority.

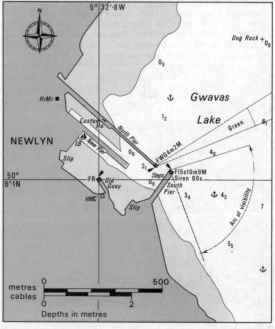

NEWLYN continued

LIGHTS AND MARKS
S Pier, head Fl 5s 10m 9M; W Tr, R base and cupola; Siren 60s. N Pier, head F WG 4m 2M; vis G238°-248°, W over harbour. Old Quay, head FR 3m 1M.
RADIO TELEPHONE (local times)
Call: *Newlyn Harbour* VHF Ch 16; 12 (Mon-Fri 0800-1700, Sat 0800-1200).
TELEPHONE (0736)
Hr Mr 62523; CG 871351; MRCC Falmouth 317575; HMC Freefone Customs Yachts: RT (0752) 669811; Marinecall 0898 500 458; Dr 63866.
FACILITIES
EC Wednesday; **North Pier** Slip, D, L, FW, ME, El, C (6 ton); **South Pier** L; **C. K. Jones** Tel. 63095, ME, SH; **J. H. Bennetts** Tel. 69988, Gas; **Kernow Marine** Tel. 68606, El, Electronics.
There is a BY with all repair facilities but it is usually booked up with work on fishing boats.
Town CH, Lau, V, R, Bar; PO (Penzance), Bank (AM only), Rly (bus to Penzance), Air (Penzance).

PENZANCE 10-1-11
Cornwall

CHARTS
Admiralty 2345, 777; Stanford 13; Imray C7; OS 203
TIDES
−0635 Dover; ML 3.1; Duration 0550; Zone 0 (GMT)

Standard Port DEVONPORT (→)

Times				Height (metres)			
HW		LW		MHWS	MHWN	MLWN	MLWS
0000	0600	0000	0600	5.5	4.4	2.2	0.8
1200	1800	1200	1800				

Differences PENZANCE
−0040 −0105 −0045 −0020 +0.1 0.0 −0.2 0.0
PORTHLEVEN
−0045 −0105 −0035 −0025 0.0 −0.1 −0.2 0.0
LIZARD POINT
−0045 −0055 −0040 −0030 −0.2 −0.2 −0.3 −0.2

SHELTER
Perfectly protected within the wet dock. Dock is open HW−2 to HW+1. Mounts Bay is unsafe anchorage in S or SE winds. Strong S or SE winds make the harbour entrance dangerous.
NAVIGATION
Waypoint 50°06′.70N 05°31′.10W, 135°/315° from/to Penzance S pierhead, 0.48M. Beware Gear Rk coming from S. Coming from the Lizard in wind or swell approach on NNW heading to avoid the Boa and Iron Gates. Cressar (5 ca NE of entrance) and Long Rks are marked by beacons. There is a harbour speed limit of 5 kn.

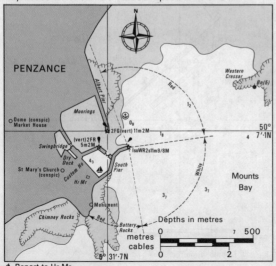

▲ Report to Hr Mr

LIGHTS AND MARKS
There are no Ldg Lts or marks. South pierhead Lt Iso WR 2s 11m 9/8M; vis R159°-268°, W268°-345°, R345°-shore.
Dock entry signals, shown from FS at N side of Dock gate (may not be given for yachts).
2B balls (hor) (2FR (vert) by night) − Dock gates open.
2B balls (vert) (FR over FG by night) − Dock gates shut.
RADIO TELEPHONE
VHF Ch 16; 09 12 (HW−2 to HW+1, and office hours).
TELEPHONE (0736)
Hr Mr 07415; Hr Office 66113; CG 871351; MRCC Falmouth 317575; HMC Freefone Customs Yachts: RT (0752) 669811; Marinecall 0898 500 458; Dr 63866.
FACILITIES
EC Wednesday (except in summer); **Wet Dock** (50 visitors) Tel. 66113, Access HW−2 to HW+1, Slip, M, D (cans), L, FW, AC, C (3 ton); **Holman Marine** Tel. 63838, Slip (dry dock), ME, El, C, CH, AB; **Matthews Sail Loft** Tel. 64004, CH, SM; **Cosalt Newlyn** Tel. 63094, CH; **R. Curnow** Tel. 69800, CH, Sh; **South Pier** Tel. 66113, D, FW, V, R, Bar; **BY** Tel. Germoe 762606; **Penzance YC** Tel. 64989, Bar; **J. H. Bennetts** Tel. 69988, Gas, Gaz; **Town** Lau, CH, V, R, Bar. PO, Bank, Rly, Air.

HELFORD RIVER 10-1-12
Cornwall

CHARTS
Admiralty 154, 147; Stanford 13; Imray C6, Y57; OS 204
TIDES
−0613 Dover; ML 3.0; Duration 0550; Zone 0 (GMT).

Standard Port DEVONPORT (→)

Times				Height (metres)			
HW		LW		MHWS	MHWN	MLWN	MLWS
0000	0600	0000	0600	5.5	4.4	2.2	0.8
1200	1800	1200	1800				

Differences HELFORD RIVER (Ent)
−0030 −0035 −0015 −0010 −0.2 −0.2 −0.3 −0.2
COVERACK
−0030 −0040 −0020 −0010 −0.2 −0.2 −0.3 −0.2

SHELTER
Excellent shelter except against E winds.
Anchorages:- Durgan Bay, good except in E winds.
　　　　　　Off Helford but tides strong.
　　　　　　Navas Creek — good shelter — little room.
Gillan Creek is good but beware rock in middle of entrance, Car Croc E cardinal spar buoy. (Moorings administered by Hr Mr). Visitors pontoons off ent to Navas Creek. Visitors buoys marked 'Visitors' on green pick up buoys.
NAVIGATION
Waypoint 50°05′.70N 05°04′.50W, 093°/273° from/to The Voose N cardinal Bn, 1.5M. If coming from N beware August Rock (or the Gedges). August Rock buoy G conical (seasonal). If from SE keep well clear of Nare Point and Dennis Head. Keep Helford Pt open of Bosahan Pt to clear rocky reef E end of Bosahan Point (The Voose) marked by N Cardinal buoy (seasonal). Avoid mud bank marked by G conical buoy (Bar buoy) on N side of river opposite Helford Creek. Port and Stbd markers mark channel from Mawgan Creek to Gweek. Speed limit 6 kn.
LIGHTS AND MARKS
There is a Lt on August Rk Buoy FlG 5s. A useful natural transit exists as a line from Bosahan Point to Mawnan Shear (259°) keeps one clear of the dangerous rocks, The Gedges. Note: There is a local bye-law stating that there are oyster beds in the river and creeks W of buoy, mid channel SW of Navas Creek, and yachts must not anchor or take the ground in that area.
RADIO TELEPHONE
Helford River SC and Gweek Quay VHF Ch M.
TELEPHONE (0326)
Hr Mr 280422; MRCC 317575; Moorings Officer 280422; HMC Freefone Customs Yachts: RT (0752) 669811; Marinecall 0898 500 458; Hosp Helston 572151.

AREA 1—SW England   197

HELFORD RIVER continued

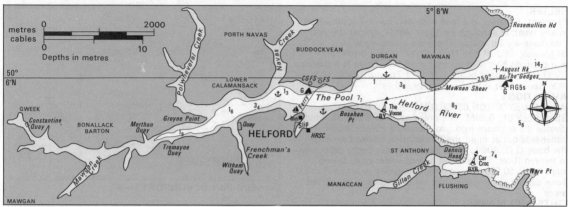

FACILITIES
EC Wednesday; **Helford Passage** Slip, L, FW; **Helford River SC** Tel. Manaccan 460, Slip, M, L, FW, Bar; **Ferry Boat Inn** Tel. Falmouth 250278, M, FW, R, Bar; **J. Badger (Durgan)** Slip, L, FW; **Porth Navas YC** Tel. Falmouth 404419, Slip, P, D, FW, CH, L, C (3 ton), AB, V, R, Bar; **Gweek Quay** Tel. Mawgan 657, AB, BY, C (20 ton), CH, El, FW, L, M, ME, P and D (in cans), R, Sh, V, Gas, Gaz; **Sailaway** Tel. Manaccan 357, M, FW, Sh, CH, L.
PO (Helford, Gweek, Mawnan Smith, Mawgan); Bank (Mawnan Smith June-Sept Mon, Wed, Fri, AM only. Oct-May Tues, Fri, AM only); Rly (bus to Falmouth); Air (Penzance or Newquay).
Note: Ferry will collect yachtsmen from boats.

FALMOUTH 10-1-13
Cornwall
CHARTS
Admiralty 154, 32, 18; Stanford 13; Imray C6, Y58; OS 204
TIDES
−0602 Dover; ML 3.0; Duration 0550; Zone 0 (GMT).
Standard Port DEVONPORT (→)

Times				Height (metres)			
HW		LW		MHWS	MHWN	MLWN	MLWS
0000	0600	0000	0600	5.5	4.4	2.2	0.8
1200	1800	1200	1800				
Differences FALMOUTH							
−0030	−0030	−0010	−0010	−0.2	−0.2	−0.3	−0.2
TRURO							
−0020	−0025	Dries out		−2.0	−2.0	Dries out	

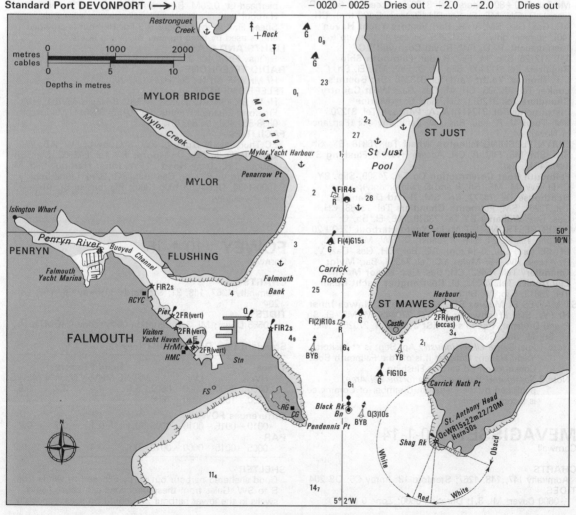

FALMOUTH continued

SHELTER
Excellent. Entrance is a mile wide and can be approached in any weather although on-shore winds against the ebb tide make it a bit rough. There are anchorages at St Mawes, St Just and Mylor Creek. Yacht anchorages and pontoon berths in marinas available. There is a marina for shallow draught boats ½M S of Truro available HW ∓3. Speed limit is 5 kn in upper reaches.

NAVIGATION
Waypoint 50°08'.00N 05°02'.00W, 183°/003° from/to Black Rock Bn, 0.68M. The only hazard is Black Rk which divides the entrance into two. Entrance can be made either side but at night the E channel is advised leaving the buoy Q (3) 10s to port. Beware oyster beds, especially in Penryn River. Falmouth is a deep water port, taking ships up to 90,000 tons; appropriate facilities are available. Take care not to impede shipping or anchor in prohibited areas.

LIGHTS AND MARKS
St Anthony Head light, R sector covers the Manacles Rocks.

RADIO TELEPHONE (local times)
Call: *Falmouth Harbour Radio*, VHF Ch 16; 12 13 14 (Mon-Fri 0900-1700, Sat 0900-1200). Coastguard – Ch 16 10 67 73. Customs launch Ch 16 12 09 06 10 14. Port Health Ch 16 12 06. Mylor Yacht Harbour, Falmouth Yacht Marina, Truro Yacht Marina and Royal Cornwall YC Ch M.

TELEPHONE (0326)
Hr Mr 312285 Penryn 73352; St Mawes 270553; CG & MRCC 317575; HMC Freefone Customs Yachts: RT (0752) 669811; Weather Plymouth 42534; Marinecall 0898 500 458; Dr 312033; Hosp 315522.

FACILITIES
FALMOUTH EC Wednesday (winter only); **Falmouth Yacht Marina** (250+80 visitors) Tel. 316620, Access all tides; P (cans), D, FW, ME, El, Sh, BH (50 ton), C (30 ton), CH, AC, Gas, Gaz, SM, V, R, Bar; **Visitors Yacht Haven** (40), summer only, Tel. 312285, access all tides up to 1.8m draught, P, D, FW, R, Bar; **Royal Cornwall YC** Tel. 312126, Slip, M, FW, R, Bar; **Falmouth Ship Repairers** Tel. 311400, Slip, P, D, FW, ME, El, Sh, C, CH; **Thomas' Yacht Yard** Tel. 313248, Sh; **Bosun's Locker** Tel. 312595, CH, M, Gas, Gaz; **West Country Chandlers** Tel. 312611, CH, Gas, Gaz; **Marine Instruments** Tel. 312414, ACA; **Penrose** Tel. 312705, SM; **Town** V, R, Lau, Bar, PO, Bank, Rly, Air (Penzance or Newquay).
PENRYN/FLUSHING **Islington Wharf** Tel. 76818, BY, Sh, Slip, SM, ME, FW, R, CH, Access HW∓1; **Flushing SC** Tel. 74043, Bar;
Falmouth Boat Construction Co Tel. 74309, Slip, BY, C, El, FW, M, ME, Sh, P and D (winter only); **Boathouse** Tel. 74177, CH, SM; **H and D Daniels** Tel. 73261, SM; **Falmouth Chandlers** Tel. 73988, CH, Gas, Gaz; **Williams BY** Tel. 73819, ME, El, Sh, C;
MYLOR/RESTRONGUET **Mylor Yacht Harbour** (250+20 visitors) Tel. 72121, Slip, P (cans), D, FW, ME, El, Sh, BH (25 ton), Lau, C (4 ton), AC, Bar, CH, Gas, Gaz, V, R, Access HW∓5; **Mylor YC** Tel. 74391, Bar; **Mylor Chandlery** Tel. 75482, CH, Gas, Gaz; **Mylor Marine Electronics** Tel. 74001, El; **Restronguet Yacht Basin** Tel. 73613, Slip, M, Sh;
ST MAWES **St Mawes SC** Tel. 270686; **St Mawes Inner Hr** FW, Slip; **Freshwater BY** Tel. 270443, M, ME, El, Sh, P and D (cans); **Pasco's BY (St Just)** Tel. 270269, ME, El, Sh, M, SM, Gas.
Note 1 Fuel Barge *Ulster Industry* Apl-Sept is at Visitors Yacht Haven; Oct-Mar it is off the Falmouth Boat Construction Co yard at Flushing.
Note 2 *Sailing and Boating Guide to the Fal Area*, published by the Falmouth Chamber of Commerce is recommended.

MEVAGISSEY 10-1-14
Cornwall

CHARTS
Admiralty 147, 148, 1267; Stanford 13; Imray C6; OS 204
TIDES
−0600 Dover; ML 3.1; Duration 0600; Zone 0 (GMT).

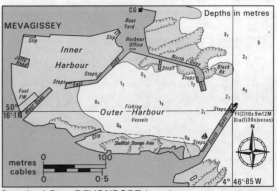

Standard Port DEVONPORT (→)

Times				Height (metres)			
HW		LW		MHWS	MHWN	MLWN	MLWS
0000	0600	0000	0600	5.5	4.4	2.2	0.8
1200	1800	1200	1800				

Differences MEVAGISSEY
−0010 −0015 −0005 +0005 −0.1 −0.1 −0.2 −0.1

SHELTER
Harbour is available at all states of the tide but entrance faces NE and is exposed to E winds. Dangerous to approach when there are strong SE winds. Anchorage off is unsuitable when wind is E of S. Inner harbour is reserved for fishing boats except when taking on fuel or water. Visitors berth S Pier.

NAVIGATION
Waypoint 50°16'.11N 04°46'.54W, 090°/270° from/to pierhead Lt, 0.20M. Beware rock ledges off the N Quay. Harbour entrance between breakwaters is 46m wide. Speed limit in the harbour is 3 kn. Harbour is crowded and is used by large number of fishing boats.

LIGHTS AND MARKS
S Quay Fl (2) 10s 9m 12M; Dia (1) 30s (occas).

RADIO TELEPHONE
Hr Mr VHF Ch 16; 56 (H24).

TELEPHONE (0726)
Hr Mr 842496; CG 842353; MRSC Brixham 882704; HMC Freefone Customs Yachts: RT (0752) 669811; Marinecall 0898 500 458; Dr 843701.

FACILITIES
EC Thursday; **N Quay** L, FW; **S Quay** L, FW, AB; **W Quay** P, D, L, FW; **J. Moores BY** Tel. 842962, Slip, Sh (wood), CH; **Jetty Head** L, FW; **Inner Harbour** Slip, M, L; **Village** V, R, Lau, Gas, Bar. PO; Bank (June-Sept 1000-1430, Oct-June 1000-1300); Rly (bus to St. Austell); Air (Newquay).

FOWEY 10-1-15
Cornwall

CHARTS
Admiralty 1267, 148, 31; Stanford 13; Imray C6, Y52; OS 204
TIDES
−0555 Dover; ML 3.1; Duration 0605; Zone 0 (GMT).

Standard Port DEVONPORT (→)

Times				Height (metres)			
HW		LW		MHWS	MHWN	MLWN	MLWS
0000	0600	0000	0600	5.5	4.4	2.2	0.8
1200	1800	1200	1800				

Differences FOWEY
−0010 −0015 −0010 −0005 −0.1 −0.1 −0.2 −0.2
PAR
−0005 −0015 0000 −0010 −0.4 −0.4 −0.4 −0.2

SHELTER
Good sheltered harbour but partly exposed to winds from S to SW. Gales from these directions can cause heavy swells in the lower harbour and confused seas, especially

AREA 1—SW England 199

FOWEY continued

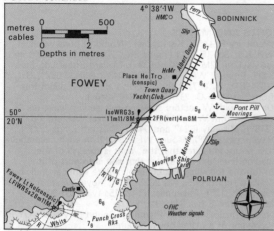

▲ visitors moorings

on the ebb. Can be entered at any tide in any conditions. Boats over 6m are not allowed to anchor or moor on W side of the river. Visitors buoys are yellow or white and marked 'FHC – visitors'. There are two visitors pontoons at the entrance to Pont Pill and one at Albert Quay (May-Sept), the latter for short stays only.

NAVIGATION
Waypoint 50°19'.30N 04°38'.70W, 205°/025° from/to Whitehouse Pt Lt, Iso WRG, 0.72M. From E beware Udder Rk (3 M E of entrance); from SW beware Cannis Rk (4 ca SE of Gribbin Head). Entering, keep well clear of Punch Cross Ledge. Fowey is a busy commercial port, so take necessary precautions. Speed limit 6 kn. Navigable up to Lostwithiel at HW for boats with shallow draught. Channel is unmarked.

LIGHTS AND MARKS
A daymark on Gribbin Head, RW Tower on headland, 76m, can be seen from all sea directions. Fowey Lt Ho, L Fl WR 5s 28m 11M R 284°-295°, W 295°-028°, R 028°-054°.

RADIO TELEPHONE
Call *Fowey Harbour Radio* VHF Ch 16; 12 (office hours). Boat Marshall Patrol (0900-2000 LT) Ch 16 12. Other stations: Charlestown Ch 16; 14 (HW − 1 to HW + 1 and when vessel expected). Par (call: *Par Port Radio*) Ch 16; 12 (Office hours and HW − 2 to HW + 1). Water taxi VHF Ch 06.

TELEPHONE (072 683)
Hr Mr 24712; CG Polruan 228; MRSC Brixham 882704; HMC Freefone Customs Yachts: RT (0752) 669811; Marinecall 0898 500 458; Dr 2451.

FACILITIES
EC Wednesday and Saturday; **Royal Fowey YC** Tel. 2245, M, L, FW, R, Bar; **Upper Deck Marine** Tel. 2287, M, Gas, Gaz, CH; **Troy Chandlery** Tel. 3265, CH, Gas, Gaz, ACA; **Albert Quay Pontoon** L, FW; **Polruan Quay** Slip, P, D, L, FW, C (3 ton); **C. Toms BY** Tel. Polruan 232, Slip (up to 21m by arrangement), FW, ME, El, Sh, C (7 ton), CH; **Fowey Gallants SC** Tel. 2335, AB, Bar; **Winklepicker** Tel. Polruan 296, CH; **Town** Lau, PO; Bank; Rly (bus to Par); Air (Newquay). A port guide is available from harbour office.

LOOE 10-1-16
Cornwall

CHARTS
Admiralty 147, 148, 1267; Stanford 13; Imray C6; OS 201

TIDES
−0545 Dover; ML 3.0; Duration 0610; Zone 0 (GMT).

Standard Port DEVONPORT (→)

Times				Height (metres)			
HW		LW		MHWS	MHWN	MLWN	MLWS
0000	0600	0000	0600	5.5	4.4	2.2	0.8
1200	1800	1200	1800				

Differences LOOE
−0010 −0010 −0005 −0005 −0.1 −0.2 −0.2 −0.2

WHITSAND BAY
0000 0000 0000 0000 0.0 +0.1 −0.1 +0.2

SHELTER
Good except in strong SE winds when the harbour becomes uncomfortable. The whole harbour dries. Visitors berth on W side of Hr marked in Y.

NAVIGATION
Waypoint 50°19'.73N 04°24'.60W, 130°/310° from/to pierhead Lt, 2.0M. Entrance dangerous in strong SE winds, when seas break over the bar heavily. Coming from W, beware Ranneys Rks extending SE and E from Looe Is. From E, beware Longstone Rks extending 1½ ca from shore NE of harbour entrance. At springs, tide runs up to 5 kn. Do not secure to W bank anywhere to S of shops due to rocky outcrops.

LIGHTS AND MARKS
Mid Main Beacon (off Hannafore Pt, halfway between pierhead and Looe Island) Q(3) 10s; E cardinal mark. From West at night keep in W sector (267°-313°) of pierhead Lt Oc WR 3s 8m 15/12M; vis W013°-207°, R207°-267°, W267°-313°, R313°-332°. Nailzee Pt Siren (2) 30s. R flag is displayed from FS on quay when it is too rough for small craft to put to sea.

RADIO TELEPHONE
VHF Ch 16 (no constant watch).

TELEPHONE (050 36)
Hr Mr 2839; CG 2138; MRSC Brixham 882704; HMC Freefone Customs Yachts: RT (0752) 669811; Marinecall 0898 500 458; Dr 3195.

FACILITIES
EC Thursday (winter only); **East Looe Quay** Access HW ∓3, Slip, P and D (cans), L, FW, ME, El, C (2½ ton), CH; **Looe SC** Tel. 2559, L, R, Bar; **West Looe Quay** Slip, M, P and D (cans), L, FW, ME, El, CH, AB; **Curtis Frank & Pope** Tel. 2332, Slip, M, Sh (Wood); **Marine Engineering Looe** Tel. 2887, ME, El; **Pearn Norman & Co** Tel. 2244, Slip, M, ME, El, Sh, CH; **Jack Bray & Son** Tel. 2504, CH; **Millendreath Marine** Tel. 3003, CH; **Looe Harbour Chandlers** Tel. 4760, CH, Gas; **Town** P, FW, V, R, Bar, Lau, PO; Bank; Rly; Air (Plymouth).

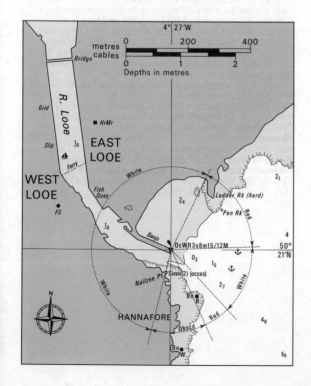

PLYMOUTH 10-1-17
(DEVONPORT)
Devon

CHARTS
Admiralty 30, 1967, 871, 1901, 1902; Stanford 13; Imray C14, OS 201

TIDES
−0540 Dover; ML 3.3; Duration 0610; Zone 0 (GMT).

Standard Port DEVONPORT (→)

Times				Height (metres)			
HW		LW		MHWS	MHWN	MLWN	MLWS
0000	0600	0000	0600	5.5	4.4	2.2	0.8
1200	1800	1200	1800				

Differences JUPITER POINT
+0010 +0005 0000 −0005 0.0 0.0 +0.1 0.0
SALTASH
0000 +0010 0000 −0005 +0.1 +0.1 +0.1 +0.1
CARGREEN
0000 +0010 +0020 +0020 0.0 0.0 −0.1 0.0
COTEHELE QUAY
0000 +0020 +0045 +0045 −0.9 −0.9 −0.8 −0.4
ST GERMANS
0000 0000 +0020 +0020 −0.3 −0.1 0.0 +0.2
BOVISAND PIER
0000 −0020 0000 −0010 −0.2 −0.1 0.0 +0.1

NOTE: Devonport is a Standard Port and times and heights of tides are given below.

SHELTER
Excellent shelter. The entrance to the Sound can be made through either the west or east channel in both of which there is deep water. The Sound itself has no shallow patches with less than 3.7m at MLWS. Winds from SE to W increase the flood and retard the ebb, while winds from the NW to E have the opposite effect. The R Tamar which is joined by the R Tavy about 1¼ M above Saltash offers good shelter, as does the R Lynher. R Tamar is navigable to Cotehele, and beyond by shallow draught boats.

BYE LAWS
Plymouth is a Naval Base. The whole harbour comes under the jurisdiction of the Queen's Harbour Master although specific parts are allowed local control; Cattewater Harbour Commissioners; Sutton Harbour Improvements Co; Associated British Ports run Mill Bay Docks. There is frequent movement of naval vessels. These vessels have right of way in the channels. Signals are hoisted on HM Ships and at the Long Room and Flag Staff Steps and apply to the waters off the dockyard port and 125m either side of the deep water channel out to the western entrance of the breakwater. The Cattewater, Mill Bay Docks and Sutton Harbour are excluded. Plymouth is also a busy commercial port principally using Mill Bay Docks, and a busy fishing port based mainly on the Barbican area.

NAVIGATION
Waypoint 50°19'.70N 04°09'.80W, 213°/033° from/to West breakwater Lt, 0.40M. There are no real hazards entering the harbour as it is very well lit and buoyed. It is not advised to go between Drake's Island and Mount Edgcumbe although this is a short cut to the Hamoaze (known as the Bridge) and has about 1.7m at LWS and is buoyed. As will be seen from the chart, yachts need not keep to the deep water channels.

HMS Cambridge at Wembury has a firing range out to sea. Firing usually occurs between 0830 and 1700 (LT) Tuesday to Friday inclusive and when in progress 3 large red flags are flown at Wembury Point and from Penlee. HMS Cambridge can be called up on VHF Ch 16 (working 12, 14) if you are worried about being near the closed area. To enter the Sound during firings, keep to the W of Eddystone. HMS Cambridge Range Officer Tel. 553740.

LIGHTS AND MARKS
Wembury Pt Oc Y 10s (occas). W end of breakwater Fl WR 10s 19m 15/12M (vis W 262°-208°, R 208°-262°) and Iso 4s 12m 12M (vis 031°-039°). Mallard Shoal Ldg Lts 349°. Front Q WRG (vis G 233°-043°, R 043°-067°, G 067°-087°, W 087°-099°, R 099°-108°). Rear 396m from front (on Hoe) Oc G 1.3s (vis 310°-040°).

Dir Lts, with W sectors showing the leading lines, are established at Withyhedge (070°), Royal Western YC (315°), Mill Bay (324° and 047°), Western Kings (271°), Ravenness (225°), Mount Wise (343°). In each case above a G sector indicates that the yacht is to stbd of the leading line, and a R sector that she is to port of it.

In fog the following Dir Lts show white lights as indicated: Mallard (front) Fl 5s (vis 232°-110°); Royal Western YC F (vis 313°-317°); Ravenness Fl (2) 15s (vis 160°-305°); Mount Wise F (vis 341°-345°).

Note: Principal lights in Plymouth Sound may show QY in the event of a mains power failure.

TRAFFIC SIGNALS
See below.

RADIO TELEPHONE (local times)
Call: *Long Room Port Control* VHF Ch 16; 08 12 14 (H24). Call: *Mill Bay Docks* Ch 16; 12 14 (H24). Call: *Sutton Harbour Radio* Ch 16; 12 M (1 Apr-15 Oct: 0830-2000. 16 Oct-31 Mar: 0900-1800). Mayflower Marina, Queen Anne's Battery Marina, Sutton Harbour Marina, Ballast Pound Yacht Harbour and Royal Western YC Ch M. Call: *Cattewater Harbour Office* Ch 16; 12 (Mon-Fri, 0900-1700).

TELEPHONE (0752)
Hr Mr QHM 663225; Cattewater Hr Mr 665934; Associated British Docks at Mill Bay 662191; CG 822239; MRSC Brixham 882704; Control Centre 53777 Ex 12 or 62; HMC Freefone Customs Yachts: RT (0752) 669811; Marinecall 0898 500 458; Weather Centre 402534; Dr 53533; Hosp 668080.

FACILITIES
EC Wednesday
Mayflower International Marina (220+60 visitors) Tel. 556633, Slip, P, D, FW, ME, El, Sh, AC, C (2 ton), BH (25 ton), CH, V, R, Gas, Gaz, Divers, SM, Bar, BY, YC, Lau; **Sutton Harbour Marina** (310) Tel. 664186, P, D, FW, AC, C (3 ton), Slip, ME, El, Sh, SM; **Queen Anne's Battery Marina** (240+60 visitors) Tel. 671142, Slip, ME, El, Sh, FW, BH (20 ton), C (50 ton), P, D, CH, Gas, Lau, SM, V, R, AC, Bar; **Mill Bay Docks** Tel. 662191; facilities here are limited and yachtsmen should contact the Royal Western YC on arrival. **Mill Bay Village Marina** No visitors; **Ballast Pound Yacht Harbour** (50+10 visitors) Tel. 813658, FW, ME, El, Sh, AC, Diver, SM, Access HW∓2; **Boathaven** (10) Tel. Tavistock 832502, ME, El, SH, C (10 ton); **Royal Dockyard** has unlimited facilities but is expensive; **A S Blagdon** Tel. 28155, ME, El, Sh, C (7 ton), P, D; **Eddystone Marine Repairs** Tel. 509974, El, ME, Sh; **Mashford Bros** Tel. 822232, ME, El, Sh, D, C (10 ton), Slip, Gas; **Sutton Marine Services** Tel. 662129, C (7 ton), CH, P, D; **Ocean Marine Services** Tel. 23922, El, CH; **Saltash Marine** Tel. Saltash 3885, CH, ACA; **A E Monsen** Tel. 664384, CH, ACA; **Fox Haggart** Tel. 662587, ME, Sh; **The Quarterdeck** Tel. 24567, ME, C (12 ton); **Royal Western YC of England** Tel. 660077; **Royal Plymouth Corinthian** Tel. 664327; **RNSA** c/o QHM Tel. 663225; **Weir Quay YC** Tel. Tavistock 840474; **City** P, D, V, R, CH, PO, Bank, Rly, Air.

AREA 1—SW England

PLYMOUTH continued

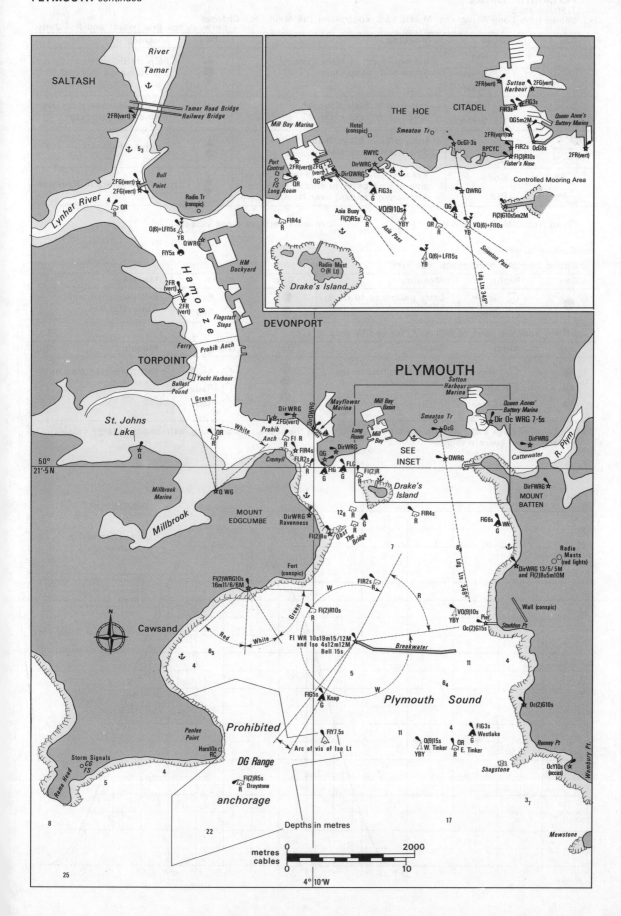

PLYMOUTH continued

Traffic signals
Shown from Long Room and Morice Yard controlling the Main Ship Channel between the Breakwater and No1 Jetty

Day	Night	Meaning	Day	Meaning
▨	Red / Green / Green	No movement in Main Channel. Small craft keep clear	●↕●	Ship traffic between Sound and Hamoaze stopped
▨ / ●	White / Green	Exit only in Main Channel. Small craft keep clear	↕	Caution necessary between Sound and Hamoaze
● / ▨	Green / White	Entry only in Main Channel. Small craft keep clear	▲▲	Caution necessary in Sound
▨ / ▨	–Code / –Pt9	Vessels may enter and leave but must give a wide berth to ships displaying Answering Pendant over Numeral Zero	▼▼	Caution necessary in Hamoaze

Flown from HM ships and tugs   ▨ Code / Zero   Give wide berth to these vessels

Dock signals–Mill Bay
Shown from the head of Mill Bay Pier

Day	Night	Meaning	Day	Night	Meaning
●	Green / Green	Entry permitted to Outer Mill Bay	●	Red / Red	Exit permitted from Outer Mill Bay

Shown from the head of Mill Bay Pier and W side of Lock

Day	Night	Meaning	Day	Night	Meaning
● ● ●	Green / Green / Green	Entry permitted to the Inner Basin	● ● ●	Red / Red / Red	Exit permitted from the Inner Basin

Note. Coloured flags and lights are waved on occasions but these are only berthing instructions to large vessels and do not concern yachts.

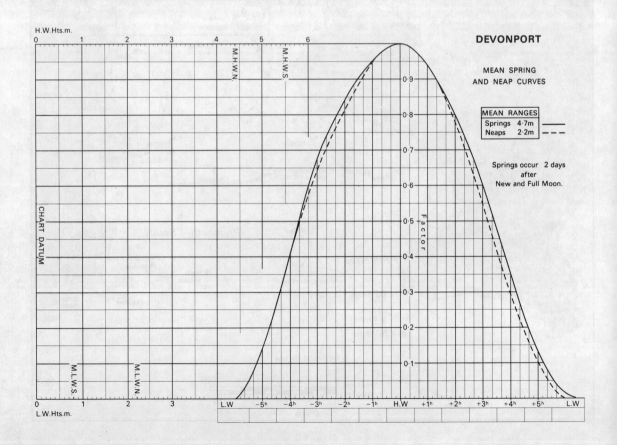

DEVONPORT

MEAN SPRING AND NEAP CURVES

MEAN RANGES
Springs 4·7m
Neaps 2·2m

Springs occur 2 days after New and Full Moon.

ENGLAND, SOUTH COAST - PLYMOUTH (DEVONPORT)

TIME ZONE UT (GMT)
For Summer Time add ONE hour in non-shaded areas

LAT 50°22'N   LONG 4°11'W

TIMES AND HEIGHTS OF HIGH AND LOW WATERS

YEAR **1989**

JANUARY

	TIME	m		TIME	m
1 SU	0512 1112 1745 2356	2.2 4.6 2.2 4.4	**16** M	0605 1220 1840	1.9 4.7 1.9
2 M	0612 1214 1850	2.3 4.5 2.2	**17** TU	0106 0717 1337 1956	4.6 2.0 4.6 2.0
3 TU	0104 0723 1325 2002	4.4 2.3 4.5 2.2	**18** W	0223 0839 1459 2112	4.7 2.0 4.6 1.9
4 W	0216 0837 1441 2111	4.5 2.2 4.5 2.0	**19** TH	0336 0951 1610 2216	4.6 1.8 4.8 1.7
5 TH	0326 0943 1553 2210	4.7 2.0 4.7 1.8	**20** F	0437 1049 1708 2309	5.1 1.5 4.9 1.4
6 F	0427 1038 1655 2302	5.0 1.6 4.9 1.5	**21** SA	0527 1139 1755 2355	5.3 1.3 5.1 1.3
7 SA	0520 1128 1748 ● 2349	1.3 5.1 1.3	**22** SU	0610 1222 1836	1.2 1.1 5.2
8 SU	0607 1214 1835	5.5 1.1 5.3	**23** M	0035 0647 1300 1910	1.2 5.5 1.1 5.2
9 M	0033 0651 1258 1918	1.1 5.6 0.9 5.4	**24** TU	0112 0719 1335 1939	1.1 5.5 1.1 5.1
10 TU	0117 0732 1342 2000	1.0 5.7 0.8 5.4	**25** W	0145 0745 1406 2004	1.2 5.4 1.2 5.0
11 W	0159 0812 1425 2040	0.9 5.7 0.8 5.3	**26** TH	0214 0808 1434 2026	1.3 5.3 1.3 4.9
12 TH	0242 0852 1508 2121	1.0 5.6 0.9 5.2	**27** F	0242 0831 1500 2050	1.4 5.1 1.5 4.8
13 F	0327 0933 1553 2206	1.1 5.4 1.1 5.0	**28** SA	0310 0857 1528 2118	1.6 5.0 1.6 4.7
14 SA ☾	0413 1020 1642 2256	1.4 5.2 1.3 4.9	**29** SU	0340 0928 1601 2155	1.8 4.9 1.8 4.6
15 SU	0505 1114 1736 2356	1.6 4.9 1.6 4.7	**30** M ☾	0417 1009 1641 2246	2.0 4.6 2.0 4.4
			31 TU	0505 1106 1735 2356	2.2 4.4 2.2 4.3

FEBRUARY

	TIME	m		TIME	m
1 W	0610 1225 1852	2.4 4.3 2.4	**16** TH	0202 0829 1452 2104	4.5 2.3 4.3 2.2
2 TH	0122 0744 1359 2034	4.3 2.4 4.3 2.3	**17** F	0324 0948 1605 2210	4.6 2.0 4.5 1.8
3 F	0251 0921 1532 2154	4.5 2.2 4.5 2.0	**18** SA	0427 1042 1659 2258	4.9 1.6 4.8 1.5
4 SA	0407 1027 1643 2250	4.8 1.7 4.8 1.6	**19** SU	0515 1126 1742 2340	5.2 1.2 5.0 1.2
5 SU	0505 1118 1737 2338	5.2 1.2 5.1 1.1	**20** M ○	0555 1204 1818	5.4 1.0 5.2
6 M	0555 1204 1824 ●	5.5 0.8 5.4	**21** TU	0016 0629 1238 1849	1.0 5.5 0.9 5.2
7 TU	0022 0639 1246 1906	0.8 5.8 0.5 5.6	**22** W	0050 0658 1309 1916	0.9 5.4 0.9 5.2
8 W	0104 0720 1328 1947	0.6 5.9 0.4 5.6	**23** TH	0119 0723 1337 1939	0.9 5.4 1.0 5.2
9 TH	0145 0801 1409 2025	0.5 5.9 0.4 5.5	**24** F	0146 0744 1402 1958	1.0 5.3 1.1 5.1
10 F	0226 0839 1449 2103	0.6 5.7 0.6 5.4	**25** SA	0212 0803 1427 2017	1.1 5.1 1.2 4.9
11 SA	0307 0918 1530 2143	0.8 5.3 0.9 5.1	**26** SU	0238 0822 1453 2037	1.4 5.0 1.4 4.8
12 SU	0350 1000 1614 2228 ☽	1.1 5.1 1.3 4.9	**27** M	0306 0845 1522 2104	1.6 4.8 1.6 4.7
13 M	0436 1050 1702 2324	1.5 4.8 1.7 4.6	**28** TU ☾	0339 0917 1557 2147	1.8 4.6 1.9 4.5
14 TU ☾	0531 1155 1802	1.9 4.6 2.1			
15 W	0036 0646 1318 1929	4.5 2.2 4.3 2.3			

MARCH

	TIME	m		TIME	m
1 W	0421 1011 1645 2301	2.1 4.3 2.2 4.3	**16** TH	0010 0627 1305 1912	4.3 2.4 4.1 2.5
2 TH	0520 1143 1758	2.4 4.1 2.5	**17** F	0139 0824 1439 2052	4.3 2.3 4.2 2.3
3 F	0038 0701 1330 2007	4.2 2.5 4.1 2.5	**18** SA	0303 0934 1547 2151	4.5 1.9 4.4 1.9
4 SA	0219 0906 1514 2139	4.4 2.2 4.4 2.0	**19** SU	0403 1021 1636 2235	4.8 1.5 4.7 1.5
5 SU	0344 1012 1626 2234	4.8 1.6 4.8 1.5	**20** M	0449 1100 1715 2314	5.1 1.2 5.0 1.1
6 M	0445 1101 1719 2320	5.2 1.0 5.2 0.9	**21** TU	0527 1136 1749 2348	5.3 0.9 5.2 0.9
7 TU ●	0535 1144 1804	5.6 0.6 5.5	**22** W ○	0600 1208 1820	5.4 0.8 5.3
8 W	0003 0620 1226 1846	0.5 5.9 0.2 5.7	**23** TH	0020 0630 1237 1847	0.8 5.4 0.8 5.3
9 TH	0044 0702 1307 1926	0.3 6.0 0.1 5.8	**24** F	0049 0657 1305 1912	0.8 5.3 0.9 5.2
10 F	0125 0743 1347 2005	0.2 5.9 0.2 5.7	**25** SA	0117 0720 1331 1933	0.9 5.2 1.0 5.1
11 SA	0205 0822 1426 2042	0.4 5.7 0.5 5.4	**26** SU	0143 0739 1357 1951	1.1 5.1 1.2 5.0
12 SU	0245 0902 1506 2120	0.7 5.4 0.9 5.2	**27** M	0210 0756 1424 2008	1.3 4.9 1.4 4.9
13 M	0327 0944 1548 2204	1.1 5.0 1.4 4.8	**28** TU	0239 0815 1454 2032	1.5 4.7 1.6 4.7
14 TU ☾	0412 1034 1634 2258	1.6 4.6 1.9 4.5	**29** W	0313 0848 1530 2116	1.7 4.6 1.9 4.5
15 W	0506 1139 1733	2.0 4.3 2.3	**30** TH ☾	0356 0950 1618 2234	2.0 4.3 2.2 4.3
			31 F	0457 1127 1733	2.2 4.1 2.5

APRIL

	TIME	m		TIME	m
1 SA	0009 0640 1309 1942	4.3 2.4 4.1 2.4	**16** SU	0215 0857 1504 2114	4.4 1.9 4.3 1.9
2 SU	0144 0839 1445 2112	4.4 2.0 4.4 1.9	**17** M	0317 0943 1554 2159	4.6 1.6 4.6 1.6
3 M	0310 0944 1556 2207	4.8 1.4 4.8 1.3	**18** TU	0406 1023 1634 2237	4.9 1.3 4.9 1.3
4 TU	0415 1033 1650 2254	5.2 0.9 5.2 0.8	**19** W	0447 1058 1711 2313	5.1 1.0 5.1 1.0
5 W	0508 1118 1737 2338	5.6 0.5 5.6 0.4	**20** TH	0523 1131 1744 2346	5.2 0.9 5.2 0.9
6 TH	0555 1201 1821	5.8 0.2 5.7	**21** F ○	0557 1202 1816	5.2 0.9 5.3
7 F	0021 0640 1242 1902	0.2 5.9 0.2 5.8	**22** SA	0017 0628 1232 1845	0.9 5.2 0.9 5.2
8 SA	0102 0723 1323 1942	0.2 5.8 0.3 5.7	**23** SU	0048 0657 1302 1911	1.0 5.1 1.1 5.1
9 SU	0143 0805 1404 2021	0.4 5.5 0.7 5.4	**24** M	0118 0722 1332 1934	1.1 4.9 1.2 5.0
10 M	0225 0846 1444 2101	0.8 5.2 1.1 5.1	**25** TU	0149 0746 1403 1957	1.3 4.8 1.4 4.9
11 TU	0307 0930 1526 2144	1.2 4.9 1.6 4.8	**26** W	0223 0812 1438 2028	1.4 4.6 1.6 4.8
12 W	0353 1020 1612 2235 ☽	1.7 4.4 2.0 4.6	**27** TH	0302 0853 1519 2117	1.6 4.5 1.9 4.6
13 TH	0449 1122 1713 2340	2.1 4.2 2.3 4.4	**28** F	0349 0958 1612 2227	1.8 4.3 2.1 4.5
14 F	0611 1238 1847	2.3 4.1 2.5	**29** SA	0453 1118 1727 2346	2.0 4.2 2.3 4.5
15 SA ☾	0057 0752 1359 2016	4.3 2.2 4.1 2.3	**30** SU	0623 1242 1908	2.1 4.3 2.2

Chart Datum: 3.22 metres below Ordnance Datum (Newlyn)

ENGLAND, SOUTH COAST - PLYMOUTH (DEVONPORT)

LAT 50°22'N LONG 4°11'W

TIME ZONE UT (GMT)
For Summer Time add ONE hour in non-shaded areas

TIMES AND HEIGHTS OF HIGH AND LOW WATERS

YEAR **1989**

	MAY				JUNE				JULY				AUGUST			
	TIME	m	TIME	m	TIME	m	TIME	m	TIME	m	TIME	m	TIME	m	TIME	m
1 M	0108 0759 1405 2032	4.6 1.8 4.5 1.8	**16** TU	0207 0850 1451 2108	4.5 1.7 4.5 1.7	**1** TH	0259 0926 1539 2153	5.0 1.1 5.1 1.1	**16** F	0258 0927 1536 2152	4.6 1.6 4.7 1.7	**1** SA	0345 0959 1619 2231	4.9 1.4 5.1 1.3	**16** SU	0322 0947 1600 2218	4.5 1.8 4.7 1.7	**1** TU	0538 1137 1755 ●	5.0 1.2 5.4	**16** W	0510 1116 1729 2341	4.9 1.3 5.3 1.0
2 TU	0228 0906 1516 2132	4.8 1.3 4.9 1.3	**17** W	0304 0934 1540 2152	4.7 1.5 4.7 1.5	**2** F	0405 1020 1638 2247	5.2 0.9 5.3 0.9	**17** SA	0357 1015 1629 2239	4.7 1.5 4.9 1.5	**2** SU	0451 1056 1716 2326	5.0 1.2 5.2 1.1	**17** M	0429 1042 1657 2310	4.6 1.6 5.0 1.4	**2** W	0005 0622 1220 1836	1.0 5.1 5.0 5.5	**17** TH O	0558 1159 1814	5.2 0.9 5.6
3 W	0337 1000 1615 2224	5.1 0.9 5.2 0.9	**18** TH	0355 1014 1624 2233	4.8 1.3 4.9 1.3	**3** SA	0504 1111 1730 2337 ●	5.3 0.8 5.4 0.8	**18** SU	0451 1100 1718 2324	4.8 1.4 5.0 1.3	**3** M	0547 1147 1807 ●	5.1 1.1 5.4	**18** TU O	0526 1130 1747 2356	4.9 1.3 5.2 1.2	**3** TH	0045 0700 1259 1911	0.9 5.2 0.9 5.5	**18** F	0023 0640 1240 1855	0.6 5.4 0.6 5.7
4 TH	0436 1049 1706 2311	5.4 0.6 5.5 0.6	**19** F	0440 1052 1706 2311	4.9 1.2 5.1 1.2	**4** SU	0558 1159 1819 O	5.3 0.8 5.5	**19** M	0540 1143 1802	4.9 1.3 5.1	**4** TU	0015 0636 1234 1852	1.0 5.1 1.0 5.4	**19** W	0614 1215 1831	5.0 1.1 5.4	**4** F	0122 0733 1334 1941	0.9 5.1 1.0 5.4	**19** SA	0103 0720 1321 1934	0.4 5.5 0.5 5.7
5 F ●	0528 1134 1753 2357	5.6 0.4 5.6 0.4	**20** SA O	0522 1128 1744 2347	5.0 1.1 5.2 1.1	**5** M	0025 0648 1245 1905	0.8 5.3 0.9 5.4	**20** TU	0007 0625 1225 1843	1.2 4.9 1.2 5.2	**5** W	0101 0719 1317 1931	1.0 5.1 1.1 5.4	**20** TH	0039 0658 1257 1911	0.9 5.1 0.9 5.5	**5** SA	0155 0800 1406 2005	1.0 5.0 1.1 5.2	**20** SU	0143 0758 1401 2013	0.4 5.5 0.5 5.6
6 SA	0617 1218 1838	5.6 0.4 5.6	**21** SU	0601 1204 1821	5.0 1.1 5.2	**6** TU	0111 0734 1329 1947	0.9 5.1 1.1 5.3	**21** W	0049 0707 1306 1921	1.1 4.9 1.2 5.2	**6** TH	0143 0757 1357 2006	1.0 5.0 1.1 5.3	**21** F	0121 0738 1338 1950	0.8 5.2 0.9 5.5	**6** SU	0225 0822 1435 2028	1.1 4.9 1.2 5.1	**21** M	0223 0836 1441 2052	0.5 5.3 0.7 5.4
7 SU	0041 0703 1302 1921	0.5 5.5 0.6 5.5	**22** M	0023 0638 1239 1855	1.1 5.0 1.2 5.1	**7** W	0156 0816 1412 2026	1.1 4.9 1.3 5.2	**22** TH	0130 0747 1347 1959	1.1 4.9 1.2 5.2	**7** F	0222 0830 1434 2036	1.1 4.9 1.3 5.1	**22** SA	0202 0816 1419 2027	0.7 5.2 0.9 5.4	**7** M	0253 0844 1502 2052	1.3 4.8 1.4 4.9	**22** TU	0304 0915 1523 2134	0.8 5.1 1.0 5.1
8 M	0125 0748 1344 2002	0.6 5.3 0.9 5.4	**23** TU	0100 0713 1315 1927	1.2 4.9 1.3 5.1	**8** TH	0240 0856 1454 2103	1.3 4.7 1.5 5.0	**23** F	0212 0825 1430 2036	1.1 4.9 1.2 5.2	**8** SA	0258 0859 1508 2104	1.3 4.7 1.5 5.0	**23** SU	0243 0854 1501 2106	0.8 5.1 0.9 5.3	**8** TU	0320 0911 1531 2121	1.5 4.7 1.6 4.8	**23** W ☾	0347 0959 1609 2225	1.2 4.9 1.4 4.8
9 TU	0208 0832 1426 2043	0.9 5.0 1.3 5.1	**24** W	0137 0747 1353 1959	1.3 4.8 1.5 5.0	**9** F	0323 0934 1536 2139	1.5 4.5 1.7 4.8	**24** SA	0256 0906 1514 2117	1.1 4.8 1.3 5.1	**9** SU	0331 0927 1541 2135	1.5 4.6 1.6 4.8	**24** M	0326 0934 1544 2148	0.9 5.0 1.1 5.1	**9** W ☾	0349 0945 1604 2159	1.7 4.6 1.8 4.6	**24** TH	0434 1055 1702 2330	1.6 4.7 1.8 4.5
10 W	0253 0916 1509 2125	1.3 4.7 1.6 4.9	**25** TH	0216 0823 1433 2036	1.4 4.7 1.5 4.9	**10** SA	0406 1012 1619 2219	1.7 4.4 1.9 4.7	**25** SU	0342 0950 1602 2204	1.2 4.7 1.4 5.0	**10** M	0405 1000 1616 2211	1.6 4.5 1.8 4.7	**25** TU ☾	0411 1019 1632 2239	1.1 4.8 1.4 4.9	**10** TH	0425 1032 1646 2252	1.9 4.4 2.1 4.4	**25** F	0531 1207 1811	2.0 4.5 2.1
11 TH	0339 1001 1556 2210	1.6 4.4 1.9 4.6	**26** F	0259 0907 1518 2122	1.5 4.6 1.7 4.8	**11** SU ☾	0452 1054 1707 2304	1.9 4.3 2.0 4.5	**26** M	0433 1042 1656 2259	1.3 4.7 1.5 4.9	**11** TU	0441 1042 1656 2256	1.8 4.4 1.9 4.5	**26** W	0501 1115 1726 2340	1.4 4.7 1.6 4.7	**11** F	0512 1136 1742	2.1 4.3 2.3	**26** SA	0051 0652 1332 1953	4.3 2.3 4.5 2.2
12 F ☾	0433 1052 1651 2302	1.9 4.2 2.2 4.5	**27** SA	0349 1000 1611 2218	1.6 4.5 1.8 4.8	**12** M	0543 1144 1801 2356	2.0 4.3 2.1 4.5	**27** TU	0529 1141 1757	1.4 4.6 1.7	**12** W	0524 1133 1746 2351	1.9 4.3 2.1 4.4	**27** TH	0559 1222 1833	1.7 4.6 2.1	**12** SA	0004 0618 1254 1906	4.2 2.4 4.2 2.4	**27** SU	0223 0834 1456 2122	4.3 2.2 4.5 1.9
13 SA	0538 1150 1800	2.1 4.1 2.3	**28** SU ☾	0448 1103 1716 2322	1.7 4.5 1.9 4.7	**13** TU	0639 1239 1902	2.0 4.3 2.1	**28** W	0002 0634 1248 1906	4.8 1.6 4.6 1.7	**13** TH	0618 1234 1848	2.0 4.3 2.2	**28** F	0055 0712 1340 1957	4.5 1.9 4.6 2.0	**13** SU	0129 0757 1418 2052	4.2 2.4 4.4 2.3	**28** M	0340 0945 1602 2219	4.6 1.8 4.9 1.5
14 SU	0001 0652 1253 1915	4.4 2.1 4.1 2.2	**29** M	0558 1211 1831	1.8 4.5 1.9	**14** W	0054 0738 1338 2003	4.5 1.9 4.4 2.0	**29** TH	0114 0745 1400 2020	4.7 1.6 4.7 1.7	**14** F	0055 0726 1342 2003	4.3 2.1 4.4 2.2	**29** SA	0220 0837 1501 2121	4.4 1.9 4.7 1.8	**14** M	0259 0929 1536 2204	4.3 2.1 4.6 1.9	**29** TU	0437 1037 1654 2305	4.8 1.5 5.2 1.2
15 M	0105 0758 1355 2018	4.4 2.0 4.3 2.0	**30** TU	0032 0715 1324 1947	4.7 1.6 4.6 1.7	**15** TH	0155 0835 1439 2100	4.5 1.8 4.5 1.8	**30** F	0231 0855 1513 2130	4.8 1.5 4.9 1.5	**15** SA	0208 0840 1453 2116	4.3 2.0 4.5 2.0	**30** SU	0341 0951 1611 2227	4.6 1.7 4.9 1.5	**15** TU	0413 1029 1638 2256	4.6 1.7 5.0 1.4	**30** W	0522 1120 1736 2344	5.1 1.1 5.4 0.9
			31 W	0146 0825 1434 2054	4.8 1.4 4.8 1.4								**31** M	0446 1049 1708 2319	4.8 1.4 5.2 1.2				**31** TH ●	0600 1158 1812	5.2 0.9 5.5

Chart Datum: 3.22 metres below Ordnance Datum (Newlyn)

ENGLAND, SOUTH COAST - PLYMOUTH (DEVONPORT)

LAT 50°22'N LONG 4°11'W

TIMES AND HEIGHTS OF HIGH AND LOW WATERS

YEAR **1989**

TIME ZONE UT(GMT)
For Summer Time add ONE hour in non-shaded areas

SEPTEMBER

Day	Time	m	Day	Time	m
1 F	0021 / 0633 / 1233 / 1844	0.8 / 5.3 / 0.8 / 5.5	16 SA	0000 / 0616 / 1218 / 1833	0.4 / 5.6 / 0.4 / 5.9
2 SA	0053 / 0702 / 1305 / 1911	0.8 / 5.3 / 0.8 / 5.5	17 SU	0040 / 0657 / 1259 / 1914	0.3 / 5.7 / 0.3 / 5.9
3 SU	0123 / 0726 / 1334 / 1933	0.9 / 5.2 / 1.0 / 5.3	18 M	0120 / 0736 / 1339 / 1955	0.3 / 5.7 / 0.4 / 5.7
4 M	0150 / 0747 / 1400 / 1953	1.0 / 5.1 / 1.1 / 5.2	19 TU	0201 / 0815 / 1420 / 2037	0.5 / 5.5 / 0.7 / 5.4
5 TU	0215 / 0806 / 1425 / 2013	1.2 / 5.0 / 1.3 / 5.0	20 W	0241 / 0856 / 1503 / 2122	0.9 / 5.2 / 1.1 / 5.1
6 W	0239 / 0827 / 1452 / 2036	1.5 / 4.8 / 1.6 / 4.8	21 TH	0324 / 0942 / 1549 / 2216	1.4 / 5.0 / 1.5 / 4.7
7 TH	0307 / 0854 / 1523 / 2107	1.7 / 4.7 / 1.8 / 4.6	22 F ☾	0412 / 1040 / 1644 / 2325	1.8 / 4.7 / 2.0 / 4.4
8 F ☽	0340 / 0935 / 1602 / 2158	1.9 / 4.5 / 2.1 / 4.3	23 SA	0512 / 1153 / 1801	2.3 / 4.5 / 2.3
9 SA	0423 / 1044 / 1655 / 2324	2.2 / 4.3 / 2.4 / 4.1	24 SU	0049 / 0643 / 1319 / 1952	4.2 / 2.5 / 4.5 / 2.3
10 SU	0528 / 1213 / 1822	2.5 / 4.2 / 2.6	25 M	0216 / 0824 / 1439 / 2107	4.3 / 2.3 / 4.6 / 1.9
11 M	0101 / 0724 / 1345 / 2033	4.1 / 2.6 / 4.3 / 2.4	26 TU	0323 / 0926 / 1540 / 2157	4.6 / 1.9 / 4.9 / 1.5
12 TU	0237 / 0911 / 1508 / 2145	4.3 / 2.2 / 4.7 / 1.8	27 W	0413 / 1012 / 1627 / 2238	4.9 / 1.5 / 5.2 / 1.2
13 W	0352 / 1008 / 1613 / 2234	4.7 / 1.7 / 5.1 / 1.3	28 TH	0453 / 1052 / 1706 / 2315	5.1 / 1.1 / 5.4 / 0.9
14 TH	0447 / 1054 / 1704 / 2318	5.1 / 1.1 / 5.5 / 0.8	29 F ●	0528 / 1128 / 1741 / 2349	5.3 / 0.8 / 5.5 / 0.8
15 F ○	0534 / 1137 / 1750	5.4 / 0.7 / 5.7	30 SA	0600 / 1202 / 1811	5.4 / 0.8 / 5.5

OCTOBER

Day	Time	m	Day	Time	m
1 SU	0020 / 0628 / 1232 / 1839	0.8 / 5.4 / 0.9 / 5.5	16 M	0016 / 0632 / 1236 / 1854	0.4 / 5.8 / 0.4 / 5.9
2 M	0048 / 0653 / 1301 / 1903	0.9 / 5.4 / 1.0 / 5.3	17 TU	0057 / 0714 / 1318 / 1938	0.5 / 5.8 / 0.5 / 5.7
3 TU	0115 / 0716 / 1327 / 1924	1.1 / 5.3 / 1.2 / 5.2	18 W	0139 / 0756 / 1401 / 2023	0.7 / 5.6 / 0.8 / 5.4
4 W	0140 / 0736 / 1354 / 1943	1.3 / 5.1 / 1.4 / 5.0	19 TH	0222 / 0839 / 1446 / 2112	1.1 / 5.3 / 1.2 / 5.0
5 TH	0207 / 0755 / 1422 / 2003	1.5 / 4.9 / 1.6 / 4.8	20 F	0306 / 0927 / 1534 / 2206	1.6 / 5.0 / 1.6 / 4.7
6 F	0235 / 0819 / 1454 / 2032	1.8 / 4.8 / 1.8 / 4.6	21 SA ☾	0355 / 1022 / 1631 / 2311	2.0 / 4.8 / 2.0 / 4.4
7 SA	0309 / 0858 / 1534 / 2127	2.1 / 4.6 / 2.1 / 4.3	22 SU	0457 / 1129 / 1748	2.3 / 4.6 / 2.3
8 SU ☽	0354 / 1008 / 1629 / 2257	2.3 / 4.4 / 2.3 / 4.2	23 M	0025 / 0621 / 1245 / 1921	4.3 / 2.5 / 4.5 / 2.2
9 M	0500 / 1137 / 1756	2.5 / 4.3 / 2.5	24 TU	0141 / 0748 / 1358 / 2030	4.4 / 2.3 / 4.6 / 2.0
10 TU	0032 / 0653 / 1307 / 1959	4.2 / 2.6 / 4.4 / 2.3	25 W	0243 / 0849 / 1457 / 2119	4.6 / 2.0 / 4.8 / 1.7
11 W	0204 / 0837 / 1431 / 2112	4.4 / 2.2 / 4.7 / 1.7	26 TH	0332 / 0936 / 1545 / 2201	4.8 / 1.6 / 5.1 / 1.4
12 TH	0318 / 0937 / 1538 / 2204	4.8 / 1.6 / 5.1 / 1.2	27 F	0413 / 1016 / 1626 / 2238	5.1 / 1.3 / 5.2 / 1.1
13 F	0415 / 1025 / 1634 / 2250	5.2 / 1.1 / 5.5 / 0.8	28 SA	0449 / 1053 / 1703 / 2312	5.3 / 1.1 / 5.4 / 1.0
14 SA	0504 / 1110 / 1723 / 2333	5.5 / 0.7 / 5.8 / 0.5	29 SU ●	0523 / 1128 / 1736 / 2345	5.4 / 1.0 / 5.4 / 1.0
15 SU	0549 / 1153 / 1809	5.8 / 0.4 / 5.9	30 M	0554 / 1200 / 1808	5.5 / 1.0 / 5.4
			31 TU	0015 / 0624 / 1231 / 1837	1.1 / 5.4 / 1.1 / 5.3

NOVEMBER

Day	Time	m	Day	Time	m
1 W	0045 / 0652 / 1301 / 1904	1.2 / 5.3 / 1.3 / 5.1	16 TH	0122 / 0741 / 1347 / 2012	0.9 / 5.6 / 0.9 / 5.3
2 TH	0114 / 0717 / 1331 / 1928	1.4 / 5.2 / 1.4 / 4.9	17 F	0207 / 0825 / 1434 / 2059	1.2 / 5.4 / 1.3 / 5.0
3 F	0144 / 0740 / 1403 / 1953	1.6 / 5.0 / 1.6 / 4.8	18 SA	0252 / 0910 / 1522 / 2148	1.6 / 5.2 / 1.6 / 4.8
4 SA	0217 / 0808 / 1439 / 2027	1.8 / 4.9 / 1.8 / 4.6	19 SU	0341 / 0958 / 1615 / 2240	1.9 / 4.9 / 1.9 / 4.5
5 SU	0255 / 0848 / 1522 / 2121	2.0 / 4.8 / 2.0 / 4.5	20 M ☾	0435 / 1050 / 1716 / 2337	2.2 / 4.7 / 2.1 / 4.4
6 M	0342 / 0949 / 1618 / 2236	2.2 / 4.6 / 2.1 / 4.3	21 TU	0539 / 1148 / 1825	2.3 / 4.6 / 2.2
7 TU	0448 / 1105 / 1736 / 2358	2.4 / 4.6 / 2.2 / 4.4	22 W	0038 / 0648 / 1250 / 1931	4.4 / 2.3 / 4.6 / 2.1
8 W	0618 / 1225 / 1912	2.4 / 4.6 / 2.1	23 TH	0138 / 0752 / 1351 / 2027	4.5 / 2.1 / 4.7 / 1.9
9 TH	0120 / 0750 / 1345 / 2029	4.5 / 2.1 / 4.8 / 1.7	24 F	0232 / 0846 / 1447 / 2114	4.7 / 1.9 / 4.8 / 1.7
10 F	0235 / 0858 / 1458 / 2128	4.8 / 1.7 / 5.1 / 1.3	25 SA	0321 / 0933 / 1536 / 2156	4.8 / 1.7 / 5.0 / 1.5
11 SA	0338 / 0951 / 1601 / 2220	5.2 / 1.2 / 5.4 / 0.9	26 SU	0405 / 1015 / 1622 / 2235	5.1 / 1.5 / 5.1 / 1.3
12 SU	0433 / 1043 / 1657 / 2307	5.5 / 0.9 / 5.6 / 0.7	27 M	0446 / 1054 / 1704 / 2312	5.3 / 1.4 / 5.2 / 1.2
13 M	0523 / 1130 / 1748 / 2353	5.7 / 0.7 / 5.7 / 0.6	28 TU	0525 / 1132 / 1743 / 2348	5.4 / 1.3 / 5.2 / 1.3
14 TU	0610 / 1216 / 1837	5.8 / 0.6 / 5.7	29 W ●	0602 / 1207 / 1819	5.4 / 1.3 / 5.2
15 W	0038 / 0656 / 1302 / 1925	0.7 / 5.8 / 0.7 / 5.6	30 TH	0023 / 0636 / 1243 / 1854	1.3 / 5.4 / 1.3 / 5.1

DECEMBER

Day	Time	m	Day	Time	m
1 F	0058 / 0707 / 1318 / 1925	1.4 / 5.3 / 1.4 / 5.0	16 SA	0155 / 0811 / 1423 / 2042	1.2 / 5.5 / 1.1 / 5.1
2 SA	0133 / 0737 / 1355 / 1957	1.5 / 5.2 / 1.5 / 4.9	17 SU	0238 / 0850 / 1506 / 2121	1.4 / 5.3 / 1.4 / 4.9
3 SU	0210 / 0808 / 1434 / 2031	1.6 / 5.1 / 1.6 / 4.8	18 M	0321 / 0926 / 1550 / 2158	1.6 / 5.1 / 1.6 / 4.7
4 M	0251 / 0845 / 1519 / 2116	1.8 / 5.0 / 1.7 / 4.7	19 TU ☾	0403 / 1004 / 1634 / 2238	1.8 / 4.9 / 1.8 / 4.5
5 TU	0338 / 0933 / 1611 / 2213	1.9 / 4.8 / 1.8 / 4.6	20 W	0448 / 1046 / 1721 / 2324	2.0 / 4.8 / 1.9 / 4.5
6 W	0435 / 1034 / 1713 / 2320	2.0 / 4.9 / 1.9 / 4.6	21 TH	0539 / 1135 / 1814	2.2 / 4.6 / 2.1
7 TH	0543 / 1143 / 1826	2.1 / 4.8 / 1.9	22 F	0018 / 0636 / 1232 / 1913	4.4 / 2.2 / 4.6 / 2.1
8 F	0035 / 0701 / 1300 / 1943	4.6 / 2.1 / 4.9 / 1.8	23 SA	0118 / 0740 / 1335 / 2013	4.5 / 2.2 / 4.6 / 2.0
9 SA	0152 / 0816 / 1419 / 2052	4.8 / 1.8 / 5.0 / 1.5	24 SU	0220 / 0842 / 1440 / 2110	4.6 / 2.1 / 4.6 / 1.9
10 SU	0304 / 0922 / 1533 / 2152	5.0 / 1.5 / 5.2 / 1.3	25 M	0320 / 0937 / 1542 / 2201	4.8 / 1.9 / 4.8 / 1.7
11 M	0408 / 1020 / 1637 / 2246	5.3 / 1.2 / 5.4 / 1.1	26 TU	0414 / 1027 / 1637 / 2247	5.0 / 1.7 / 4.9 / 1.6
12 TU	0505 / 1114 / 1735 / 2337	5.5 / 1.0 / 5.5 / 0.9	27 W	0502 / 1112 / 1725 / 2329	5.2 / 1.5 / 5.0 / 1.4
13 W	0557 / 1204 / 1827	5.7 / 0.9 / 5.5	28 TH ●	0546 / 1153 / 1809	5.3 / 1.4 / 5.1
14 TH	0025 / 0645 / 1252 / 1915	0.9 / 5.7 / 0.9 / 5.5	29 F	0010 / 0625 / 1233 / 1848	1.4 / 5.3 / 1.3 / 5.1
15 F	0111 / 0729 / 1338 / 2000	1.0 / 5.7 / 1.0 / 5.3	30 SA	0049 / 0701 / 1312 / 1924	1.3 / 5.4 / 1.2 / 5.1
			31 SU	0127 / 0735 / 1350 / 1957	1.3 / 5.4 / 1.2 / 5.1

Chart Datum: 3.22 metres below Ordnance Datum (Newlyn)

YEALM RIVER 10-1-18
Devon

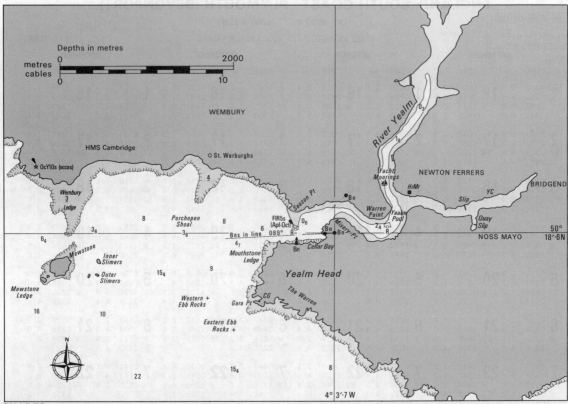

CHARTS
Admiralty 30; Stanford 13; Imray C6, C14; OS 201
TIDES
−0540 Dover; ML 3.2; Duration 0615; Zone 0 (GMT).

Standard Port DEVONPORT (←)

| Times | | | | Height (metres) | | | |
HW		LW		MHWS	MHWN	MLWN	MLWS
0000	0600	0000	0600	5.5	4.4	2.2	0.8
1200	1800	1200	1800				

Differences YEALM RIVER ENTRANCE
+0006 +0006 +0002 +0002 −0.1 −0.1 −0.1 −0.1

SHELTER
Very good shelter and easy entrance except in strong onshore winds. Anchorage in Cellar Bay open to NW winds.
NAVIGATION
Waypoint 50°18′.00N 04°06′.00W, 240°/060° from/to Season Pt, 1.4M. The W and E Ebb rocks and the Inner and Outer Slimers are dangerous, lying awash on either side of Wembury Bay. Beware sand bar running S from Season Pt, marked by port hand buoy Fl R 5s (April-Oct), leaving narrow channel on S side. Also sand spit running S from Warren Pt. You cannot beat in on an ebb tide. Strong SW winds hold up the ebb and increase levels, as does the river when in spate. Speed limit 6 kn.
For information about Wembury firing range, see 10.1.17.
LIGHTS AND MARKS
R can buoy, Fl R 5s, marks end of sand bar (Apl-Oct). Ldg Bns at 089°. When abeam of Bn (G triangle on W back) on S shore turn NE towards Bn on N shore. Leave 'spit' R can buoy off Warren Pt to port. From Sand Bar to Misery Pt is only 1.2m at LWS.
RADIO TELEPHONE
None.
TELEPHONE (0752)
Hr Mr 872533; MRSC Brixham 882704; HMC Freefone Customs Yachts: RT (0752) 669811; Marinecall 0898 500 458; Dr 880392.

FACILITIES
EC Thursday; **Yealm Pool** M, L, FW; **Newton Ferrers Village** L, Slip, FW, V, Gas, Gaz, R, Bar; **Yealm Boat Co** Tel. 872564, ME, El, Sh, Gas, Gaz, CH; **Yealm YC** Tel. 872291, FW, R, Bar; **Bridgend** L, Slip (HW+/to HW−2½), FW; **Bridgend Boat Co** Tel. 872162, ME, El, Sh; **Noss Mayo** Tel. L, Slip, FW, V, R, Bar; **J. Hockaday** Tel. 872369, ME; **A. Hooper** Tel. 872861, SM. PO (Newton Ferrers, Noss Mayo); Bank (Newton Ferrers, Tues and Thurs mornings only). Rly (Plymouth); Air (Plymouth).

SALCOMBE 10-1-19
Devon

CHARTS
Admiralty 28, 1634, 1613; Stanford 13; Imray C6, Y48; OS 202
TIDES
−0535 Dover; ML 3.1; Duration 0615; Zone 0 (GMT).

Standard Port DEVONPORT (←)

| Times | | | | Height (metres) | | | |
HW		LW		MHWS	MHWN	MLWN	MLWS
0100	0600	0100	0600	5.5	4.4	2.2	0.8
1300	1800	1300	1800				

Differences SALCOMBE
0000 +0010 +0005 −0005 −0.2 −0.3 −0.1 −0.1
START POINT
+0005 +0030 −0005 +0005 −0.2 −0.4 −0.1 −0.1

SHELTER
Perfectly protected harbour but entrance is affected by S winds. The estuary is 4 M long and has 8 creeks off it, called lakes, which dry. Limited anchorage between ferry and prohibited area. Plenty of visitors deep water moorings. (Hr Mr's launch will contact − on duty 0700-2000 (LT) in season (0600-2200 (LT) in peak season) all have VHF Ch 14). There can be an uncomfortable swell with S winds in the anchorage off the town. Visitors' pontoon and moorings in the Bag are sheltered.

AREA 1—SW England 207

SALCOMBE continued
NAVIGATION
Waypoint 50°12'.40N 03°46'.60W, 180°/000° from/to Sandhill Pt Lt, 1.3M. The bar can be dangerous at spring tides when there are strong on-shore winds on the ebb tide. Speed limit 8 kn. Channel up to Kingsbridge is navigable for vessels with draughts up to 2m. Channel marked by R and W port hand beacons.

LIGHTS AND MARKS
White House in line with Poundstone beacon at 327°. Sandhill Pt beacon in line with Poundstone beacon at 000°. Sandhill Pt Dir Lt 000° Fl WRG 2s 27m 10/7M. R and W diamond on W mast — R002°-182°, G182°-357°, W357°-002°.
Harbour Ldg Lts 042° as shown.
Blackstone Rock QWR 4m 2M vis W048°-218°, R218°-048°.

RADIO TELEPHONE (local times)
VHF call *Salcombe Harbour* Ch 16; 14 (18 May-18 Sep: 0830-1600, other times of year — not Sat or Sun). Call Harbour Launches: *Salcombe Harbour One, Saltstone, Blackstone,* or *Poundstone.*
Call: *ICC Base* (clubhouse) and *Egremont* (ICC floating HQ) Ch M. Fuel barge Ch 06. Harbour Water taxi service Ch 14.

TELEPHONE (054 884)
Hr Mr 3791; CG Chivelstone 259; MRSC Brixham 882704; HMC Freefone Customs Yachts: RT (0752) 669811; Marinecall 0898 500 458; Dr 2284.

FACILITIES
EC Thursday; **Harbour** (300+50 visitors) Tel. 3791, Slip, P, D, FW, ME, El, M, L, C (15 ton), Sh, CH, SM; **Salcombe YC** Tel. 2593, L, C, R, Bar; **Island Cruising Club** Tel. 3481, Bar; **Winters BY** Tel. 3838, Slip, M, P, D, FW, ME; **Salcombe Houseboats** Tel. 3479, Gas; **J. Stone** Tel. 3470, M, FW, ME, El, Sh, CH; **Salcombe Chandlers** Tel. 2620, CH, ACA; **Tideway Boat Construction** Tel. 2987, Slip, CH; **J. Alsop** Tel. 3702, SM; **J. McKillop** Tel. Kingsbridge 2343, SM; **Burwin Marine Electronics** Tel. 3321, El; **Dawson Rigging** Tel. 3195, ME; **Victory Yard** Tel. 2930, ME; **Sailing** Tel. 2094, ME; **Hudson Thomas Fuel Barge** Tel. 2984 (local operator's number), D, P.
Salcombe is very well equipped with all facilities. FW is available from a boat — if needed, fly a bucket from a halyard. Public slipways at Batson Creek and at Kingsbridge.
Town PO; Bank; Rly (bus to Plymouth or Totnes); Air (Plymouth).
Note: There is a water taxi service run by the Hr Mr up to 2315 (LT).

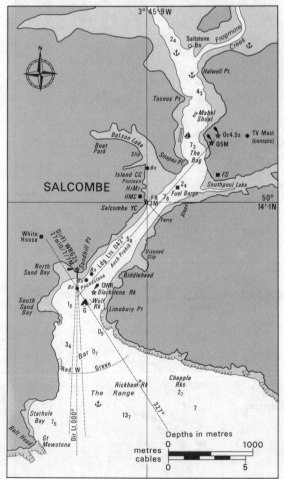

DARTMOUTH 10-1-20
Devon

CHARTS
Admiralty 2253, 1613, 1634; Stanford 12; Imray C5, Y47, Y43; OS 202

TIDES
−0510 Dover; ML 2.8; Duration 0630; Zone 0 (GMT).

NOTE: Dartmouth is a Standard Port and tidal predictions for each day of the year are given below.

SHELTER
Excellent protection inside the harbour but the entrance can be difficult with strong winds from the SE to SW. Anchorages E of fairway opposite Nos 3a to 5 buoys. There are three marinas, as shown on the chartlet, and three sets of pontoons for visiting yachts as follows: − by The Dartmouth YC; near the Hr Mr's office; for short stays only, by the ferry pontoon on the East side.

NAVIGATION
Waypoint 50°19'.50N 03°32'.80W, 148°/328° from/to Kingswear Lt, Iso WRG, 1.5M. To the E of entrance erected on Inner Froward Pt (153m high) is a stone pyramid (24.5m high) which is very conspic. There is no bar and the harbour is always available. The river is navigable on the tide to Totnes. Information from Hr Mr. Speed limit 6 kn.

LIGHTS AND MARKS
Kingswear Main Lt Iso WRG 3s W sector 325°-331°.
Dartmouth Lt Fl WRG 2s W sector 289°-297°.
Entry buoys as on chartlet.

RADIO TELEPHONE
VHF Ch 11 (office hours). Darthaven Marina (Kingswear), Dart Marina, Dart Sailing Centre VHF Ch M; Kingswear Marina Noss Works, Call *Dart Marina Four* Ch M; Fuel barge Ch 16.

TELEPHONE (080 43)
Hr Mr 2337; CG & MRSC Brixham 882704; HMC Freefone Customs Yachts: RT (0752) 669811; Marinecall 0898 500 458; Dr 2212; Hosp 2255.

FACILITIES
EC Wednesday/Saturday; **Darthaven Marina** (230+4 visitors) Tel. Kingswear 545, FW, ME, El, Gas, Gaz, Lau, Sh, CH, Bar, R, Slip, AC; **Kingswear Marina** (110) Tel. 3351, Slip, FW, ME, El, Sh, C (14 ton), CH, BH (16 ton), Gas, Gaz, AC; **Dart Marina** (80+40 visitors) Tel. 3351, Slip, D, FW, AC, ME, El, Sh, C (4 ton), BH (4 ton), Bar, R, Lau, Gas, Gaz, CH; **Dart Harbour and Navigation Authority** (450+90 visitors), Tel. 2337, Slip, D, FW, ME, El, Sh, CH, V, R; **Royal Dart YC** Tel. Kingswear 272, M, L, FW, Bar; **The Dartmouth YC** Tel. 2305, L, FW, Bar, R; **Creekside BY** Tel. 2649, Slip, M, ME, Sh, El, C (1 ton), CH, AB; **Dartmouth Yacht Services** Tel. 2035, FW, ME, El, Sh, BH, CH; **Philip and Son** Tel. 3351, M, ME, El, Sh, BH, CH; **Bosuns Locker** Tel. 2595, CH, ACA; **Dart Sails** Tel. 2185, SM; **Peter Lucas** (Rigging) Tel. 3094, SM; **Electro Marine** Tel. 2817, El; **Torbay Boat Construction** (Dolphin Haven at Galmpton) Tel. Churston 842424, CH, D, FW, L, M, C (27 ton), ME, Sh, Slip, SM, V; **River Taxi** Tel. 3727 (or VHF Ch 16); **Dart Sailing Centre** Tel. 4716, AB, AC, Bar, V; **Misc** Fuel Barge P, D; Dry dock Tel. 2776; Slip at Higher Ferry; **Town** V, P (cans), Lau, R, Bar, PO, Bank, Rly (steam train, in season, or bus to Paignton), Air (Plymouth or Exeter).

DARTMOUTH continued

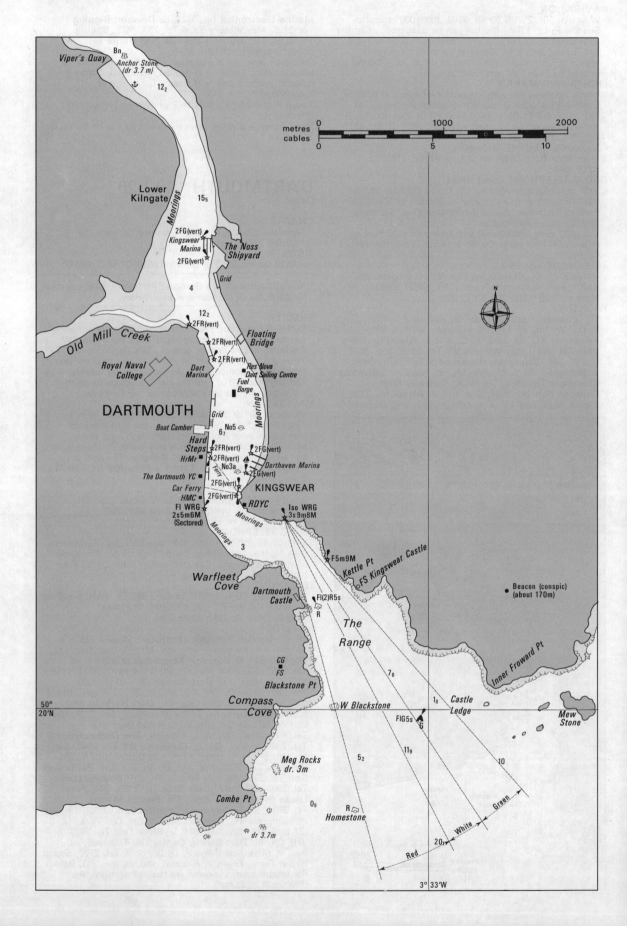

ENGLAND, SOUTH COAST - DARTMOUTH

LAT 50°21'N LONG 3°34'W

TIMES AND HEIGHTS OF HIGH AND LOW WATERS

YEAR **1989**

TIME ZONE UT(GMT)
For Summer Time add ONE hour in non-shaded areas

JANUARY

Day	Time	m	Day	Time	m
1 SU	0514 / 1141 / 1746	2.1 / 4.0 / 2.1	**16** M	0023 / 0605 / 1246 / 1841	4.0 / 1.8 / 4.0 / 1.8
2 M	0023 / 0612 / 1241 / 1852	3.8 / 2.2 / 3.9 / 2.1	**17** TU	0131 / 0720 / 1404 / 2000	4.0 / 1.9 / 4.0 / 1.9
3 TU	0129 / 0726 / 1351 / 2006	3.8 / 2.2 / 3.9 / 2.1	**18** W	0252 / 0844 / 1530 / 2119	4.0 / 1.9 / 4.0 / 1.8
4 W	0245 / 0842 / 1511 / 2118	3.9 / 2.1 / 3.9 / 1.9	**19** TH	0409 / 1000 / 1645 / 2226	4.2 / 1.7 / 4.1 / 1.6
5 TH	0358 / 0951 / 1627 / 2219	4.0 / 1.9 / 4.0 / 1.7	**20** F	0513 / 1100 / 1745 / 2320	4.4 / 1.4 / 4.2 / 1.2
6 F	0502 / 1048 / 1732 / 2313	4.3 / 1.5 / 4.2 / 1.4 ○	**21** SA	0605 / 1151 / 1835	4.5 / 1.1 / 4.4
7 SA	0558 / 1140 / 1827 ●	4.5 / 1.1 / 4.4	**22** SU	0008 / 0650 / 1236 / 1915	1.1 / 4.6 / 0.9 / 4.5
8 SU	0002 / 0647 / 1227 / 1914	1.1 / 4.7 / 0.9 / 4.5	**23** M	0049 / 0725 / 1315 / 1948	1.0 / 4.7 / 0.9 / 4.5
9 M	0047 / 0729 / 1313 / 1955	0.9 / 4.8 / 0.7 / 4.6	**24** TU	0126 / 0756 / 1348 / 2016	0.9 / 4.7 / 1.0 / 4.4
10 TU	0131 / 0809 / 1355 / 2036	0.8 / 4.9 / 0.6 / 4.6	**25** W	0158 / 0822 / 1418 / 2040	1.0 / 4.6 / 1.0 / 4.3
11 W	0211 / 0848 / 1436 / 2115	0.7 / 4.9 / 0.6 / 4.5	**26** TH	0225 / 0844 / 1444 / 2101	1.1 / 4.5 / 1.1 / 4.2
12 TH	0252 / 0926 / 1517 / 2154	0.8 / 4.8 / 0.7 / 4.5	**27** F	0252 / 0906 / 1509 / 2124	1.2 / 4.4 / 1.4 / 4.1
13 F	0335 / 1005 / 1559 / 2237	0.9 / 4.6 / 0.9 / 4.3	**28** SA	0319 / 0931 / 1536 / 2151	1.5 / 4.3 / 1.5 / 4.0
14 SA	0418 / 1050 / 1646 / 2325 ☽	1.2 / 4.5 / 1.1 / 4.2	**29** SU	0347 / 1001 / 1607 / 2226	1.7 / 4.1 / 1.7 / 4.0
15 SU	0508 / 1143 / 1737	1.5 / 4.2 / 1.5	**30** M	0422 / 1040 / 1645 / 2315 ☾	1.9 / 4.0 / 1.9 / 3.8
			31 TU	0508 / 1135 / 1736	2.1 / 3.8 / 2.1

FEBRUARY

Day	Time	m	Day	Time	m
1 W	0023 / 0610 / 1251 / 1854	3.7 / 2.3 / 3.7 / 2.3	**16** TH	0230 / 0834 / 1523 / 2110	3.9 / 2.2 / 3.7 / 2.1
2 TH	0148 / 0747 / 1427 / 2039	3.7 / 2.3 / 3.7 / 2.2	**17** F	0356 / 0956 / 1639 / 2219	4.0 / 1.9 / 3.9 / 1.7
3 F	0322 / 0928 / 1605 / 2203	3.9 / 2.1 / 3.9 / 1.9	**18** SA	0502 / 1052 / 1736 / 2309	4.2 / 1.5 / 4.1 / 1.4
4 SA	0441 / 1037 / 1719 / 2301	4.1 / 1.6 / 4.1 / 1.5	**19** SU	0553 / 1138 / 1821 / 2352	4.5 / 1.0 / 4.3 / 1.0
5 SU	0542 / 1130 / 1816 / 2350 ○	4.5 / 1.0 / 4.4 / 0.9	**20** M	0635 / 1217 / 1857	4.6 / 0.8 / 4.5
6 M	0635 / 1217 / 1903 ●	4.7 / 0.6 / 4.6	**21** TU	0030 / 0708 / 1252 / 1927	0.8 / 4.7 / 0.7 / 4.5
7 TU	0036 / 0718 / 1301 / 1944	0.6 / 4.9 / 0.3 / 4.8	**22** W	0105 / 0736 / 1324 / 1953	0.7 / 4.7 / 0.7 / 4.5
8 W	0119 / 0757 / 1342 / 2023	0.4 / 5.0 / 0.2 / 4.8	**23** TH	0133 / 0800 / 1350 / 2016	0.7 / 4.6 / 0.8 / 4.5
9 TH	0158 / 0837 / 1421 / 2100	0.3 / 5.0 / 0.2 / 4.7	**24** F	0159 / 0821 / 1414 / 2034	0.8 / 4.5 / 0.9 / 4.4
10 F	0237 / 0914 / 1459 / 2137	0.4 / 4.9 / 0.4 / 4.6	**25** SA	0223 / 0839 / 1438 / 2052	1.0 / 4.4 / 1.0 / 4.2
11 SA	0316 / 0951 / 1538 / 2215	0.6 / 4.7 / 0.7 / 4.4	**26** SU	0248 / 0857 / 1502 / 2112	1.2 / 4.3 / 1.2 / 4.1
12 SU	0357 / 1031 / 1619 / 2258 ☽	0.9 / 4.4 / 1.1 / 4.2	**27** M	0315 / 0920 / 1530 / 2138	1.5 / 4.1 / 1.5 / 4.0
13 M	0440 / 1119 / 1705 / 2352	1.4 / 4.1 / 1.6 / 4.0	**28** TU	0346 / 0950 / 1603 / 2219 ☾	1.7 / 4.0 / 1.8 / 3.9
14 TU	0532 / 1222 / 1802	1.8 / 3.9 / 2.0			
15 W	0102 / 0648 / 1344 / 1932	3.9 / 2.1 / 3.7 / 2.2			

MARCH

Day	Time	m	Day	Time	m
1 W	0426 / 1042 / 1649 / 2330	2.0 / 3.7 / 2.1 / 3.7	**16** TH	0037 / 0628 / 1330 / 1914	3.7 / 2.3 / 3.6 / 2.4
2 TH	0522 / 1211 / 1758	2.3 / 3.6 / 2.4	**17** F	0206 / 0829 / 1509 / 2058	3.7 / 2.2 / 3.6 / 2.2
3 F	0104 / 0703 / 1357 / 2011	3.6 / 2.4 / 3.6 / 2.4	**18** SA	0334 / 0942 / 1620 / 2200	3.9 / 1.8 / 3.8 / 1.8
4 SA	0248 / 0912 / 1546 / 2147	3.8 / 2.1 / 3.8 / 1.9	**19** SU	0437 / 1031 / 1712 / 2245	4.1 / 1.4 / 4.0 / 1.4
5 SU	0417 / 1021 / 1701 / 2244	4.1 / 1.5 / 4.1 / 1.4	**20** M	0525 / 1111 / 1753 / 2325	4.4 / 1.0 / 4.3 / 0.9
6 M	0521 / 1112 / 1757 / 2332	4.5 / 0.8 / 4.5 / 0.7	**21** TU	0605 / 1148 / 1828 ○	4.5 / 0.7 / 4.5
7 TU	0614 / 1156 / 1844 ●	4.8 / 0.4 / 4.7	**22** W	0001 / 0640 / 1221 / 1859	0.7 / 4.6 / 0.6 / 4.5
8 W	0016 / 0659 / 1240 / 1924	0.3 / 5.0 / 0.0 / 4.9	**23** TH	0034 / 0709 / 1251 / 1925	0.6 / 4.6 / 0.6 / 4.5
9 TH	0058 / 0740 / 1322 / 2003	0.1 / 5.1 / -0.2 / 4.9	**24** F	0104 / 0735 / 1320 / 1950	0.6 / 4.5 / 0.7 / 4.5
10 F	0139 / 0820 / 1400 / 2041	0.0 / 5.0 / 0.0 / 4.9	**25** SA	0131 / 0757 / 1344 / 2010	0.7 / 4.5 / 0.8 / 4.4
11 SA	0217 / 0857 / 1437 / 2117	0.2 / 4.9 / 0.3 / 4.6	**26** SU	0156 / 0816 / 1409 / 2027	0.9 / 4.4 / 1.0 / 4.3
12 SU	0255 / 0936 / 1515 / 2153	0.5 / 4.6 / 0.7 / 4.5	**27** M	0222 / 0832 / 1435 / 2044	1.1 / 4.2 / 1.2 / 4.2
13 M	0335 / 1016 / 1555 / 2235	0.9 / 4.3 / 1.2 / 4.1	**28** TU	0249 / 0851 / 1503 / 2107	1.4 / 4.0 / 1.5 / 4.0
14 TU	0417 / 1104 / 1638 / 2327 ☽	1.5 / 4.0 / 1.8 / 4.0	**29** W	0321 / 0922 / 1538 / 2149	1.6 / 3.9 / 1.8 / 3.9
15 W	0509 / 1207 / 1734	1.9 / 3.7 / 2.2	**30** TH	0402 / 1022 / 1623 / 2304 ☾	1.9 / 3.7 / 2.1 / 3.7
			31 F	0500 / 1155 / 1734	2.1 / 3.6 / 2.4

APRIL

Day	Time	m	Day	Time	m
1 SA	0036 / 0641 / 1334 / 1945	3.7 / 2.3 / 3.6 / 2.3	**16** SU	0244 / 0903 / 1535 / 2121	3.8 / 1.8 / 3.7 / 1.8
2 SU	0211 / 0844 / 1515 / 2119	3.8 / 1.9 / 3.8 / 1.8	**17** M	0349 / 0951 / 1628 / 2208	4.0 / 1.5 / 3.9 / 1.5
3 M	0342 / 0952 / 1630 / 2216	4.1 / 1.2 / 4.1 / 1.1	**18** TU	0440 / 1033 / 1710 / 2247	4.2 / 1.1 / 4.2 / 1.0
4 TU	0450 / 1043 / 1727 / 2305	4.5 / 0.7 / 4.5 / 0.6	**19** W	0523 / 1109 / 1749 / 2324	4.4 / 0.8 / 4.4 / 0.8
5 W	0545 / 1130 / 1816 / 2350	4.8 / 0.3 / 4.8 / 0.2	**20** TH	0601 / 1143 / 1823 / 2359	4.5 / 0.7 / 4.5 / 0.7
6 TH	0635 / 1214 / 1900 ●	4.9 / 0.0 / 4.9	**21** F	0637 / 1215 / 1855 ○	4.5 / 0.7 / 4.5
7 F	0035 / 0719 / 1256 / 1940	0.0 / 5.0 / 0.0 / 4.9	**22** SA	0031 / 0707 / 1246 / 1924	0.7 / 4.5 / 0.7 / 4.5
8 SA	0117 / 0800 / 1337 / 2019	0.0 / 4.9 / 0.1 / 4.9	**23** SU	0103 / 0735 / 1317 / 1949	0.8 / 4.4 / 0.9 / 4.4
9 SU	0156 / 0841 / 1416 / 2056	0.2 / 4.7 / 0.5 / 4.6	**24** M	0132 / 0759 / 1345 / 2011	0.9 / 4.2 / 1.0 / 4.3
10 M	0236 / 0920 / 1454 / 2135	0.6 / 4.5 / 0.9 / 4.4	**25** TU	0202 / 0822 / 1415 / 2033	1.1 / 4.1 / 1.2 / 4.1
11 TU	0316 / 1003 / 1534 / 2216	1.0 / 4.1 / 1.5 / 4.1	**26** W	0234 / 0848 / 1448 / 2103	1.2 / 4.0 / 1.5 / 4.1
12 W	0359 / 1050 / 1617 / 2305 ☽	1.6 / 3.8 / 1.9 / 3.9	**27** TH	0311 / 0927 / 1527 / 2150	1.5 / 3.9 / 1.8 / 4.0
13 TH	0453 / 1150 / 1715	2.0 / 3.6 / 2.2	**28** F	0356 / 1029 / 1617 / 2257 ☾	1.7 / 3.7 / 2.0 / 3.9
14 F	0008 / 0611 / 1304 / 1849	3.8 / 2.2 / 3.5 / 2.4	**29** SA	0456 / 1146 / 1729	1.9 / 3.6 / 2.2
15 SA	0122 / 0756 / 1427 / 2021	3.7 / 2.1 / 3.6 / 2.2	**30** SU	0013 / 0624 / 1308 / 1910	3.9 / 2.0 / 3.7 / 2.1

Chart Datum: 2.62 metres below Ordnance Datum (Newlyn)

ENGLAND, SOUTH COAST – DARTMOUTH

LAT 50°21′N   LONG 3°34′W

TIMES AND HEIGHTS OF HIGH AND LOW WATERS

YEAR 1989

TIME ZONE UT (GMT)
For Summer Time add ONE hour in non-shaded areas

	MAY				JUNE				JULY				AUGUST		
	TIME m		TIME m		TIME m		TIME m		TIME m		TIME m		TIME m		TIME m
1 M	0133 4.0 0803 1.7 1433 3.9 2037 1.7	**16** TU	0235 3.9 0856 1.6 1522 3.9 2114 1.6	**1** TH	0330 4.3 0933 0.9 1612 4.4 2202 0.9	**16** F	0329 4.0 0934 1.5 1609 4.0 2201 1.6	**1** SA	0418 4.2 1008 1.2 1654 4.4 2241 1.1	**16** SU	0354 3.9 0955 1.7 1634 4.0 2228 1.6	**1** TU	0617 4.3 1149 1.0 1835 4.6 ●	**16** W	0548 4.2 1128 1.1 1807 4.5 2353 0.8
2 TU	0257 4.1 0912 1.1 1548 4.2 2140 1.1	**17** W	0335 4.0 0942 1.4 1613 4.0 2201 1.4	**2** F	0439 4.5 1030 0.7 1714 4.5 2258 0.7	**17** SA	0431 4.0 1025 1.4 1704 4.2 2249 1.4	**2** SU	0528 4.3 1107 1.0 1754 4.5 2338 0.9	**17** M	0504 4.0 1052 1.5 1734 4.3 2321 1.2	**2** W	0018 0.8 0701 4.4 1234 0.8 1915 4.7	**17** TH	0638 4.5 1212 0.7 1854 4.8 ○
3 W	0410 4.4 1009 0.7 1650 4.5 2234 0.7	**18** TH	0429 4.1 1023 1.1 1659 4.3 2243 1.1	**3** SA	0541 4.5 1122 0.6 1756 4.5 ● 2349 0.6	**18** SU	0528 4.1 1111 1.2 1756 4.3 2336 1.1	**3** M	0626 4.4 1200 0.9 1847 4.6	**18** TU	0604 4.2 1142 1.1 1826 4.5 ○	**3** TH	0100 0.7 0738 4.5 1314 0.7 1949 4.7	**18** F	0037 0.4 0719 4.6 1254 0.4 1933 4.9
4 TH	0512 4.6 1100 0.4 1743 4.7 2322 0.4	**19** F	0516 4.2 1103 1.0 1743 4.4 2322 1.0	**4** SU	0638 4.5 1212 0.6 1858 4.7 ○	**19** M	0619 4.2 1155 1.1 1842 4.4	**4** TU	0029 0.8 0715 4.4 1248 0.8 1930 4.6	**19** W	0009 1.0 0654 4.3 1229 0.9 1910 4.6	**4** F	0136 0.7 0810 4.4 1347 0.8 2018 4.6	**19** SA	0118 0.2 0757 4.7 1335 0.3 2011 4.9
5 F ●	0606 4.8 1146 0.2 1833 4.8	**20** SA ○	0600 4.3 1140 0.9 1823 4.5	**5** M	0039 0.6 0726 4.5 1300 0.7 1943 4.6	**20** TU	0020 1.0 0704 4.2 1239 1.0 1922 4.5	**5** W	0116 0.8 0756 4.4 1331 0.9 2008 4.6	**20** TH	0053 0.7 0836 4.4 1312 0.7 1949 4.7	**5** SA	0207 0.8 0836 4.3 1418 0.9 2041 4.5	**20** SU	0156 0.2 0834 4.7 1413 0.3 2049 4.8
6 SA	0010 0.2 0656 4.8 1232 0.2 1917 4.8	**21** SU	0000 0.9 0641 4.3 1217 0.9 1900 4.5	**6** TU	0125 0.7 0811 4.4 1343 0.9 2023 4.5	**21** W	0104 0.9 0745 4.2 1321 1.0 1958 4.5	**6** TH	0156 0.8 0833 4.3 1409 0.9 2042 4.5	**21** F	0135 0.6 0815 4.5 1351 0.7 2026 4.7	**6** SU	0236 0.9 0857 4.2 1445 1.0 2103 4.4	**21** M	0234 0.3 0911 4.5 1451 0.5 2126 4.6
7 SU	0055 0.3 0741 4.7 1317 0.4 1958 4.7	**22** M	0037 0.9 0717 4.3 1253 1.0 1933 4.4	**7** W	0208 0.9 0851 4.2 1423 1.1 2101 4.5	**22** TH	0144 0.9 0823 4.2 1400 1.0 2035 4.5	**7** F	0233 0.9 0905 4.2 1444 1.1 2111 4.4	**22** SA	0214 0.5 0851 4.5 1430 0.7 2102 4.6	**7** M	0302 1.1 0919 4.1 1511 1.2 2126 4.2	**22** TU	0313 0.6 0948 4.4 1531 0.8 2206 4.4
8 M	0139 0.4 0824 4.5 1357 0.7 2038 4.6	**23** TU	0115 1.0 0751 4.2 1329 1.1 2004 4.4	**8** TH	0250 1.1 0930 4.0 1503 1.4 2137 4.3	**23** F	0223 0.9 0900 4.2 1441 1.0 2111 4.5	**8** SA	0307 1.1 0933 4.0 1517 1.4 2138 4.3	**23** SU	0253 0.6 0928 4.4 1510 0.7 2140 4.5	**8** TU	0328 1.4 0944 4.0 1538 1.5 2154 4.1	**23** W ☾	0354 1.0 1030 4.2 1615 1.2 2255 4.1
9 TU	0220 0.7 0907 4.3 1437 1.1 2118 4.4	**24** W	0150 1.1 0823 4.0 1405 1.2 2035 4.3	**9** F	0331 1.4 1006 3.9 1543 1.6 2211 4.1	**24** SA	0305 0.9 0940 4.1 1522 1.1 2150 4.4	**9** SU	0338 1.4 1000 4.0 1548 1.5 2207 4.1	**24** M	0334 0.7 1006 4.3 1551 0.9 2220 4.4	**9** W	0356 1.6 1017 4.0 1610 1.7 2230 4.0	**24** TH ☽	0438 1.5 1124 4.0 1705 1.7 2358 3.9
10 W	0302 1.1 0949 4.0 1518 1.5 2158 4.2	**25** TH	0227 1.2 0858 4.0 1443 1.4 2111 4.2	**10** SA	0412 1.6 1043 3.8 1624 1.8 2249 4.0	**25** SU	0349 1.0 1022 4.1 1608 1.2 2235 4.3	**10** M	0411 1.5 1031 3.9 1621 1.7 2242 4.0	**25** TU ☾	0416 0.9 1049 4.1 1636 1.2 2309 4.2	**10** TH	0430 1.8 1102 3.8 1650 2.0 2321 3.8	**25** F	0532 1.9 1234 3.9 1811 2.0
11 TH	0346 1.5 1032 3.8 1602 1.8 2241 4.0	**26** F	0308 1.4 0941 4.0 1526 1.6 2155 4.1	**11** SU ☽	0455 1.8 1123 3.7 1710 1.9 2333 3.9	**26** M ☾	0437 1.1 1112 4.0 1659 1.4 2328 4.2	**11** TU	0445 1.7 1112 3.8 1659 1.8 2325 3.9	**26** W	0504 1.2 1144 4.0 1728 1.5	**11** F	0514 2.0 1204 3.7 1743 2.2	**26** SA	0116 3.7 0654 2.2 1359 3.9 1957 2.1
12 F ☽	0437 1.8 1121 3.6 1654 2.1 2331 3.9	**27** SA	0356 1.5 1031 3.9 1616 1.7 2248 4.1	**12** M	0544 1.9 1212 3.7 1801 2.0	**27** TU	0531 1.2 1212 4.0 1757 1.6	**12** W	0526 1.8 1201 3.7 1747 2.0	**27** TH	0008 4.0 0559 1.6 1248 4.0 1834 1.8	**12** SA	0031 3.6 0619 2.3 1319 3.6 1908 2.3	**27** SU	0252 3.7 0839 2.1 1527 4.0 2129 1.8
13 SA	0539 2.0 1217 3.6 1800 2.2	**28** SU ☾	0452 1.6 1132 3.9 1718 1.8 2350 4.0	**13** TU	0023 3.9 0640 1.9 1305 3.7 1904 2.0	**28** W	0029 4.1 0635 1.5 1313 4.0 1908 1.6	**13** TH	0018 3.8 0619 1.9 1300 3.7 1850 2.1	**28** F	0120 3.9 0714 1.8 1407 4.0 2001 1.9	**13** SU	0155 3.6 0801 2.3 1447 3.8 2058 2.2	**28** M	0413 4.0 0953 1.7 1636 4.2 2229 1.4
14 SU	0028 3.8 0654 2.0 1318 3.6 1918 2.1	**29** M	0558 1.6 1238 3.9 1832 1.8	**14** W	0119 3.9 0741 1.8 1405 3.8 2007 1.9	**14** TH	0140 3.9 0749 1.5 1428 4.0 2025 1.6	**14** F	0120 3.8 0729 2.0 1409 3.8 2007 2.1	**14** SA	0249 3.9 0842 1.8 1532 4.0 2128 1.7	**14** M	0330 3.7 0936 2.0 1609 4.0 2213 1.8	**29** TU	0513 4.1 1047 1.4 1731 4.5 2316 1.0
15 M	0130 3.8 0802 1.9 1423 3.7 2023 1.9	**30** TU	0058 4.0 0718 1.5 1350 4.0 1951 1.6	**15** TH	0223 3.9 0840 1.7 1509 3.9 2106 1.7	**15** F	0301 4.1 0901 1.4 1545 4.2 2138 1.4	**15** SA	0236 3.7 0845 1.9 1524 3.9 2123 1.9	**15** SU	0414 4.0 1000 1.6 1646 4.2 2237 1.4	**15** TU	0448 4.0 1039 1.6 1714 4.3 2307 1.2	**30** W	0600 4.4 1132 0.9 1815 4.6 2356 0.7
		31 W	0213 4.1 0830 1.2 1504 4.1 2100 1.2							**31** M	0522 4.1 1100 1.2 1745 4.5 2331 1.0			**31** TH ●	0640 4.5 1211 0.7 1852 4.7

Chart Datum: 2.62 metres below Ordnance Datum (Newlyn)

ENGLAND, SOUTH COAST - DARTMOUTH

LAT 50°21'N   LONG 3°34'W

TIMES AND HEIGHTS OF HIGH AND LOW WATERS

YEAR 1989

TIME ZONE UT(GMT)
For Summer Time add ONE hour in non-shaded areas

Chart Datum: 2.62 metres below Ordnance Datum (Newlyn)

SEPTEMBER

	TIME	m		TIME	m
1 F	0035 0712 1247 1923	0.6 4.5 0.6 4.7	**16** SA	0013 0655 1232 1912	0.2 4.8 0.2 5.0
2 SA	0108 0740 1320 1949	0.6 4.5 0.6 4.7	**17** SU	0054 0735 1314 1952	0.1 4.9 0.1 5.0
3 SU	0137 0803 1347 2010	0.7 4.5 0.8 4.5	**18** M	0134 0813 1352 2031	0.1 4.9 0.2 4.9
4 M	0203 0823 1412 2029	0.8 4.4 0.9 4.5	**19** TU	0213 0851 1431 2112	0.5 4.7 0.5 4.6
5 TU	0226 0842 1436 2049	1.0 4.3 1.1 4.3	**20** W	0251 0930 1512 2155	0.7 4.5 0.9 4.4
6 W	0249 0902 1501 2111	1.4 4.1 1.5 4.1	**21** TH	0332 1014 1556 2246	1.2 4.3 1.4 4.0
7 TH	0316 0928 1531 2141	1.6 4.0 1.7 4.0	**22** F	0417 1110 1648 2353 ☾	1.7 4.0 1.9 3.8
8 F	0347 1007 1608 2229 ☾	1.8 3.9 2.0 3.7	**23** SA	0514 1220 1801	2.2 3.9 2.2
9 SA	0428 1114 1658 2352	2.1 3.7 2.3 3.6	**24** SU	0114 0644 1345 1956	3.6 2.4 3.9 2.2
10 SU	0530 1240 1823	2.4 3.6 2.5	**25** M	0245 0829 1509 2113	3.7 2.2 4.0 1.8
11 M	0126 0727 1412 2038	3.6 2.5 3.7 2.3	**26** TU	0355 0933 1613 2206	4.0 1.8 4.2 1.4
12 TU	0307 0918 1539 2153	3.7 2.1 4.0 1.7	**27** W	0448 1021 1702 2248	4.2 1.4 4.5 1.0
13 W	0426 1017 1648 2244	4.0 1.6 4.4 1.1	**28** TH	0530 1103 1743 2327	4.4 0.9 4.6 0.7
14 TH	0523 1105 1741 2330	4.4 0.9 4.7 0.6	**29** F	0606 1140 1820 ●	4.5 0.7 4.7
15 F	0613 1149 1830 ○	4.6 0.5 4.9	**30** SA	0002 0640 1215 1851	0.6 4.6 0.6 4.7

OCTOBER

	TIME	m		TIME	m
1 SU	0034 0707 1246 1918	0.6 4.6 0.7 4.7	**16** M	0030 0711 1250 1932	0.2 4.9 0.2 5.0
2 M	0103 0731 1316 1941	0.7 4.6 0.8 4.5	**17** TU	0112 0752 1332 2015	0.3 4.9 0.3 4.9
3 TU	0129 0753 1341 2001	0.9 4.5 1.0 4.5	**18** W	0152 0832 1413 2058	0.5 4.8 0.6 4.6
4 W	0153 0813 1406 2020	1.1 4.4 1.2 4.3	**19** TH	0233 0914 1456 2145	1.1 4.5 1.0 4.3
5 TH	0219 0831 1433 2039	1.4 4.2 1.5 4.1	**20** F	0315 1000 1541 2237	1.5 4.3 1.5 4.0
6 F	0245 0854 1503 2107	1.7 4.1 1.7 4.0	**21** SA	0401 1052 1635 2340 ☾	1.9 4.1 1.9 3.8
7 SA	0318 0932 1541 2200	1.9 4.0 2.0 3.7	**22** SU	0500 1157 1749	2.2 4.0 2.2
8 SU	0400 1039 1634 2326 ☾	2.2 3.8 2.2 3.6	**23** M	0051 0622 1311 1924	3.7 2.4 3.9 2.1
9 M	0503 1205 1756	2.4 3.7 2.4	**24** TU	0208 0752 1426 2035	3.8 2.2 4.0 1.9
10 TU	0058 0655 1332 2003	3.6 2.5 3.8 2.2	**25** W	0313 0855 1528 2126	4.0 1.9 4.1 1.6
11 W	0232 0842 1501 2119	3.8 2.1 4.0 1.6	**26** TH	0405 0944 1618 2210	4.1 1.5 4.4 1.2
12 TH	0350 0945 1611 2213	4.1 1.5 4.4 1.0	**27** F	0448 1105 1701 2248	4.4 1.1 4.6 0.9
13 F	0450 1035 1710 2301	4.5 0.9 4.7 0.6	**28** SA	0525 1104 1740 2323	4.5 0.9 4.6 0.8
14 SA	0541 1121 1801 2345 ●	4.7 0.5 4.9 0.3	**29** SU	0601 1140 1815 2358 ●	4.6 0.6 4.6 0.8
15 SU	0628 1206 1849	4.9 0.2 5.0	**30** M	0634 1213 1848	4.7 0.8 4.6
			31 TU	0029 0703 1245 1916	0.9 4.6 0.9 4.5

NOVEMBER

	TIME	m		TIME	m
1 W	0100 0730 1316 1942	1.0 4.5 1.1 4.4	**16** TH	0136 0818 1400 2048	0.7 4.8 0.7 4.5
2 TH	0128 0754 1344 2005	1.2 4.4 1.2 4.2	**17** F	0219 0900 1444 2133	1.0 4.6 1.1 4.3
3 F	0157 0817 1415 2029	1.5 4.3 1.5 4.1	**18** SA	0301 0944 1530 2220	1.5 4.5 1.5 4.0
4 SA	0228 0844 1449 2102	1.7 4.2 1.7 4.0	**19** SU	0348 1029 1620 2310	1.8 4.2 1.8 3.9
5 SU	0304 0922 1530 2154	1.9 4.1 1.9 3.9 ☾	**20** M	0439 1119 1718	2.1 4.0 2.0
6 M	0349 1021 1623 2306	2.1 4.0 2.1 3.7	**21** TU	0005 0540 1215 1826	3.8 2.2 4.0 2.1
7 TU	0452 1134 1737	2.3 4.0 2.1	**22** W	0104 0650 1315 1934	3.8 2.2 4.0 2.0
8 W	0025 0619 1251 1914	3.8 2.3 4.0 2.0	**23** TH	0205 0756 1419 2032	3.9 2.0 4.0 1.8
9 TH	0146 0754 1412 2034	3.9 2.0 4.1 1.6	**24** F	0302 0852 1517 2121	4.0 1.8 4.1 1.6
10 F	0305 0904 1529 2135	4.1 1.6 4.4 1.1	**25** SA	0353 0941 1609 2205	4.2 1.6 4.3 1.4
11 SA	0411 1002 1635 2230	4.4 1.0 4.6 0.7	**26** SU	0439 1025 1657 2245	4.4 1.4 4.4 1.1
12 SU	0509 1053 1734 2318	4.7 0.7 4.8 0.5	**27** M	0522 1105 1741 2323	4.5 1.2 4.5 1.1
13 M	0601 1142 1827 ○	4.9 0.5 4.9	**28** TU	0603 1144 1822 ●	4.6 1.1 4.5
14 TU	0006 0650 1230 1916	0.4 4.9 0.4 4.9	**29** W	0001 0642 1220 1858	1.1 4.6 1.1 4.5
15 W	0052 0734 1317 2002	0.5 4.9 0.5 4.8	**30** TH	0037 0715 1257 1932	1.1 4.6 1.1 4.4

DECEMBER

	TIME	m		TIME	m
1 F	0113 0745 1332 2002	1.2 4.5 1.2 4.3	**16** SA	0207 0847 1434 2117	1.0 4.7 0.9 4.4
2 SA	0146 0814 1407 2033	1.4 4.5 1.4 4.2	**17** SU	0248 0924 1515 2154	1.2 4.5 1.2 4.2
3 SU	0222 0844 1444 2106	1.5 4.4 1.5 4.1	**18** M	0329 0959 1557 2229	1.5 4.4 1.5 4.0
4 M	0300 0920 1527 2149	1.7 4.3 1.6 4.0	**19** TU	0409 1035 1638 2308 ☾	1.7 4.2 1.7 3.9
5 TU	0345 1005 1616 2244	1.8 4.2 1.7 4.0	**20** W	0452 1115 1723 2352	1.9 4.1 1.9 3.9
6 W	0439 1104 1715 2348 ☾	1.9 4.2 1.8	**21** TH	0540 1203 1814	2.1 4.0 2.0
7 TH	0544 1211 1827	2.0 4.1 1.8	**22** F	0044 0637 1258 1915	3.8 2.1 4.0 2.0
8 F	0101 0703 1325 1946	4.0 1.9 4.2 1.7	**23** SA	0144 0743 1402 2017	3.9 2.1 4.0 1.9
9 SA	0220 0821 1448 2058	4.1 1.7 4.3 1.4	**24** SU	0249 0847 1510 2117	4.0 2.0 4.0 1.8
10 SU	0335 0929 1606 2201	4.3 1.4 4.5 1.1	**25** M	0352 0945 1615 2210	4.1 1.8 4.1 1.6
11 M	0442 1030 1713 2257	4.4 1.0 4.6 0.9	**26** TU	0449 1037 1713 2258	4.3 1.6 4.2 1.5
12 TU	0542 1125 1814 2349 ○	4.7 0.8 4.7	**27** W	0539 1123 1803 2341	4.5 1.4 4.3 1.2
13 W	0637 1217 1906	4.9 0.7 4.7	**28** TH	0625 1206 1849 ●	4.5 1.2 4.4
14 TH	0039 0724 1307 1953	0.7 4.9 0.7 4.7	**29** F	0023 0704 1247 1926	1.2 4.6 1.1 4.4
15 F	0125 0806 1351 2036	0.8 4.9 0.8 4.5	**30** SA	0104 0739 1326 2001	1.1 4.6 1.0 4.4
			31 SU	0141 0812 1403 2033	1.1 4.6 1.0 4.4

AREA 1—SW England 211

DARTMOUTH continued

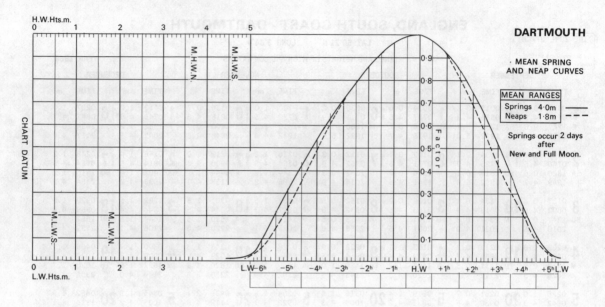

BRIXHAM 10-1-21
Devon

CHARTS
Admiralty 26, 1613; Stanford 12; Imray C5, Y43; OS 202
TIDES
−0505 Dover; ML 2.9; Duration 0635; Zone 0 (GMT).

Standard Port DEVONPORT (←)

Times				Height (metres)			
HW		LW		MHWS	MHWN	MLWN	MLWS
0100	0600	0100	0600	5.5	4.4	2.2	0.8
1300	1800	1300	1800				

Differences BRIXHAM
+0025 +0045 +0010 0000 −0.6 −0.7 −0.2 −0.1

SHELTER
Good, but outer harbour is dangerous in NW winds. Visitors buoys (white) to E of main channel.
NAVIGATION
Waypoint 50°24'.70N 03°30'.00W, 050°/230° from/to Victoria breakwater Lt, 0.60M. No dangers — Little run in tide. Easy access — Inner harbour dries.

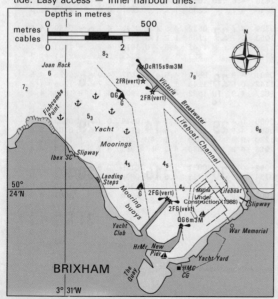

LIGHTS AND MARKS
There are no leading marks or leading lights coming into Brixham. Breakwater head — Oc R 15s. Three R balls or three R Lts (vert) at entrance indicate 'harbour closed'.
RADIO TELEPHONE (local times)
VHF Ch 14 16 (May-Sep): 0900-1300, 1400-1700, 1800-2000. Oct-Apr: Mon-Fri 0900-1300, 1400-1700). Brixham Coastguard: Ch 16 10 67 73.
TELEPHONE (0803) (08045 for 4 and 5 fig numbers)
Hr Mr 3321; CG & MRSC 882704; Pilot 882214; Berthing Master 3211; HMC Freefone Customs Yachts: RT (0752) 669811; Marinecall 0898 500 458; Dr 2731; Hosp 882153.
FACILITIES
EC Wednesday; **Harbour Office New Fish Quay** Slip, M, L, FW, C (4 ton), AB; **Brixham YC** Tel. 3332, M, L, FW, R, Bar; **Upham's Yard** Tel. 882365, Slip, M, L, ME, El, Sh, C, CH, AB; **Brixham Bunkering & Devon Marine** Tel. 882250, D, FW, Sh; **Outer End Victoria Jetty** P, D, FW; **Brixham Yacht Supplies** Tel. 882290, CH, ACA; **R.A. Brewer** Tel. 3465, M; **Inner Harbour** C; **Quay Electrics** Tel. 3030, El; **Lawrence & Rae Marine Electronics** Tel. 51202, El; **Town** PO; Bank; Rly (bus to Paignton); Air (Exeter).

TORQUAY 10-1-22
Devon

CHARTS
Admiralty 26, 1613; Stanford 12; Imray C5, Y43; OS 202
TIDES
−0500 Dover; ML 2.9; Duration 0640; Zone 0 (GMT).

Standard Port DEVONPORT (←)

Times				Height (metres)			
HW		LW		MHWS	MHWN	MLWN	MLWS
0100	0600	0100	0600	5.5	4.4	2.2	0.8
1300	1800	1300	1800				

Differences TORQUAY
+0025 +0045 +0010 0000 −0.6 −0.7 −0.2 −0.1

AREA 1—SW England

TORQUAY continued

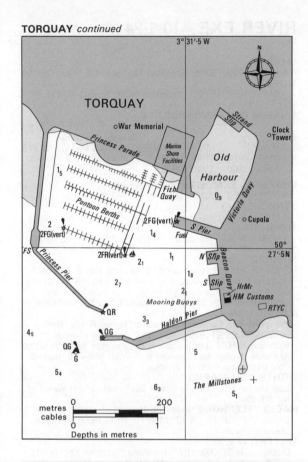

SHELTER
Good shelter but some swell in harbour with strong southerly winds.

NAVIGATION
Waypoint 50°27'.00N 03°31'.50W, 165°/345° from/to Haldon Pier Lt, 0.40M. Inner harbour (known as Old Harbour) dries completely. Three R balls or three R Lts show harbour closed due to navigational hazard.

LIGHTS AND MARKS
No leading marks. Princess Pier QR 9m 6M. Haldon Pier QG 9m 6M. South Pier 2FG (vert) 5M.

RADIO TELEPHONE (local times)
VHF Ch 14 16 (Oct-Mar 0900-1700 Mon-Fri) (Apl-Sept 0900-2000). Marina Ch M (H24).

TELEPHONE (0803)
Hr Mr 22429; CG & MRSC Brixham 882704; HMC Freefone Customs Yachts: RT (0752) 669811; Marinecall 0898 500 458; Hosp 64567; Marina 214624.

FACILITIES
Torquay Marina (440+60 visitors) Tel. 214624, FW, ME, Slip, Gas, Gaz, P, D, Lau, AC, SM, El, Sh, V, R, Bar; **Haldon Pier** L, FW, AB; **Princess Pier** L; **South Pier** P, D, L, FW, C (6 ton); **Beacon Quay** Slip, CH; **Royal Torbay YC** Tel. 22006, R, Bar; **Town** All normal facilities, V, R, Bar, PO, Bank, Rly, Air (Exeter or Plymouth).

TEIGNMOUTH 10-1-23
Devon

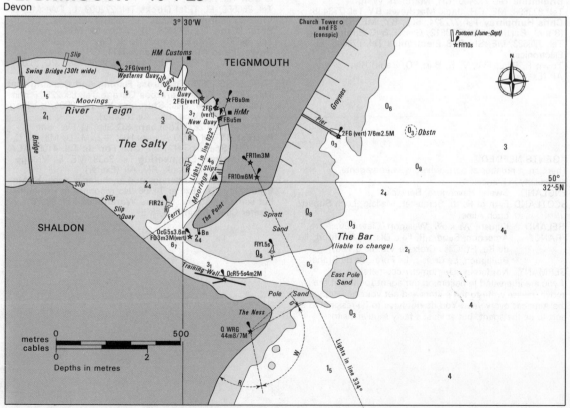

TEIGNMOUTH continued

CHARTS
Admiralty 26, 3315; Stanford 12; Imray C5, Y43; OS 192
TIDES
−0450 Dover; ML 2.7; Duration 0625; Zone 0 (GMT).

Standard Port DEVONPORT (←)

Times				Height (metres)			
HW		LW		MHWS	MHWN	MLWN	MLWS
0100	0600	0100	0600	5.5	4.4	2.2	0.8
1300	1800	1300	1800				

Differences TEIGNMOUTH (Approaches)
+0025 +0040 0000 0000 −0.7 −0.8 −0.3 −0.2

SHELTER
Do not attempt to come in over bar without local knowledge. Harbour completely sheltered but difficult to enter especially with strong winds between NE & S when surf forms on the bar. Access HW∓3.
NAVIGATION
Waypoint 50°32'.40N 03°29'.20W, 076°/256° from/to Training Wall Lt, 0.43M. Bar shifts very frequently.
Beware — rocks off the Ness
— variable extent of Salty flats.
LIGHTS AND MARKS
QWRG 44m 8/7M on The Ness, vis G229°-243°, W243°-348°, R348°-shore. Oc G, FG (vert) on The Point. Two sets of leading lights:- Approaching from SE, lights at 334°. This line keeps clear of the Ness Rocks but does not normally lead through channel through the bar. Once round The Point, lights on the quays at 022°.
RADIO TELEPHONE (local times)
VHF Ch 16; 12 (Mon-Fri: 0900-1230, 1400-1700; Sat 0930-1200).
TELEPHONE (0626)
Hr Mr 773165; CG and MRSC Brixham 882704; HMC Freefone Customs Yachts: RT (0752) 669811; Marinecall 0898 500 458; Dr 774355; Hosp 772161.
FACILITIES
EC Thursday; **E Quay Polly Steps** Slip (up to 10m); **DCS Electrics** Tel. 775960, ME, El; **Teignmouth YC** Tel. 772734, M, FW; **Harbour Master** Tel. 772303, M; **Brigantine** Tel. 772400, CH; **Mariners Weigh** Tel. 873698, ME, CH; **Jack Matthews BY** Tel. 773438; **Chris Humphrey** Tel. 772324, Slip, BY, ME, Sh, D, C (8 ton); **Bartletts** Tel. 773812, Gas; **T & C Engineering** Tel. 778633, ME, Sh; **PFB Electronics** Tel. 776845, El, Electronics.
Town P, D, L, FW, V, R, Bar; PO; Bank; Rly; Air (Exeter).

AGENTS NEEDED
There are a number of ports where we need agents, particularly in France.
ENGLAND   Swale, Havengore, Berwick.
SCOTLAND   Firth of Forth, Scrabster, Mallaig, Loch Sunart, Loch Aline.
IRELAND   Kilrush, Wicklow, Westport/Clew Bay.
FRANCE   Arcachon, Seudre R, Ile d'Oleron, Rochfort, Ile de Re, St. Giles-Croix-de-Vie, Ile d'Yeu, Pouliguen, Le Croisic, La Forêt, Ile de Bréhat.
GERMANY   Norderney, Dornumer-Accumersiel.
If you are interested in becoming our agent for any of the above, please write to the editors and get your free copy of the Almanac every year. You do not have to be resident in a port to be the agent, but at least a fairly regular visitor.

RIVER EXE   10-1-24
Devon

CHARTS
Admiralty 2290, 3315; Stanford 12; Imray C5, Y43; OS 192
TIDES
−0445 Dover; ML 2.1; Duration 0625; Zone 0 (GMT).

Standard Port DEVONPORT (←)

Times				Height (metres)			
HW		LW		MHWS	MHWN	MLWN	MLWS
0100	0600	0100	0600	5.5	4.4	2.2	0.8
1300	1800	1300	1800				

Differences EXMOUTH (Approaches)
+0030 +0050 +0015 +0005 −0.9 −1.0 −0.5 −0.3
STARCROSS
+0040 +0110   No data   −1.4   −1.5   No data
TOPSHAM
+0045 +0105   No data   −1.5   −1.6   No data

SHELTER
Entrance difficult when wind between S and E. Shelter good in Exe River. Access to Exeter is through the Exeter Canal and passage through locks, S of Topsham, by arrangement with Exeter City Council Tel: Exeter 74306.
NAVIGATION
Waypoint Exe Fairway (safe water) buoy, Fl 10s, Bell, 50°36'.00N 03°21'.87W, 102°/282° from/to channel entrance, 0.60M. Long shallow bar extends SE with rocks to the N of channel. Channel is difficult but well marked. Pole sands are liable to change.
LIGHTS AND MARKS
Ldg Lts by The Point in line at 305°; not safe seaward of No 3 Buoy.
RADIO TELEPHONE (local times)
Exeter VHF Ch 16; 06 12 (Mon-Fri: 0730-1630, and when vessel expected).
TELEPHONE (0395)
Dockmaster 272009; HMC Freefone Customs Yachts: RT (0752) 669811; CG 263232; MRSC Brixham 882704; Pilot 264036; Marinecall 0898 500 458; Dr 273001; Hosp 279684.
FACILITIES
EXMOUTH EC Wednesday; **Dixon & Son** Tel. 263063, ME, Sh, CH; **Exe Sailing Club** Tel. 264607, M, L, AB; **Lavis & Son** Tel. 263095, El, Sh, CH; **D.S. Electrics** Tel. 265550, El; **Tidal Docks** Tel. 272009, L, FW, ME, C; **Victoria Marine Services** Tel. 265044, ME, Gaz, CH; **Peter Dixon** Tel. 273248, CH, ACA; **M. McNamara** Tel. 264907, SM; **Exmouth Dock Co.** Tel. 272009, C; **Rowzell and Morrison** Tel. 263911, CH, El, SM; **Pierhead** V, R, Bar, CH; **Town** P, V, R, Bar, PO, Bank, Rly, Air (Exeter).
STARCROSS (0626) **Starcross Fishing and Cruising Club** Tel. 890582; **Wills Starcross Garage** Tel. 890225, P and D (cans), ME, M, Gas; **Starcross YC** Tel. 890470; **Village** P, V, Bar, PO, Bank, Rly, Air (Exeter).
TOPSHAM (039 287) **Topsham SC** Slip, L, FW, Bar; **Retreat BY** Tel. 4720 Access HW ∓2, M, D, ME, El, C, Gas, CH; **Scanes** Tel. 7527, SM; **Foc'sle** Tel. 4105, ACA, CH; **Grimshaw Engineering** Tel. 3539, ME, El; **Village** P, R, V, Bar, PO, Bank, Rly, Air (Exeter).
EXETER (0392) For moorings and anchorages apply Hr Mr Exeter City, Tel. Exeter 74306. No moorings in Exmouth, but some in estuary. **Leisure Warehouse** Tel. 50970 (Exeter Canal Basin) CH, ME.

AREA 1—SW England 215

RIVER EXE continued

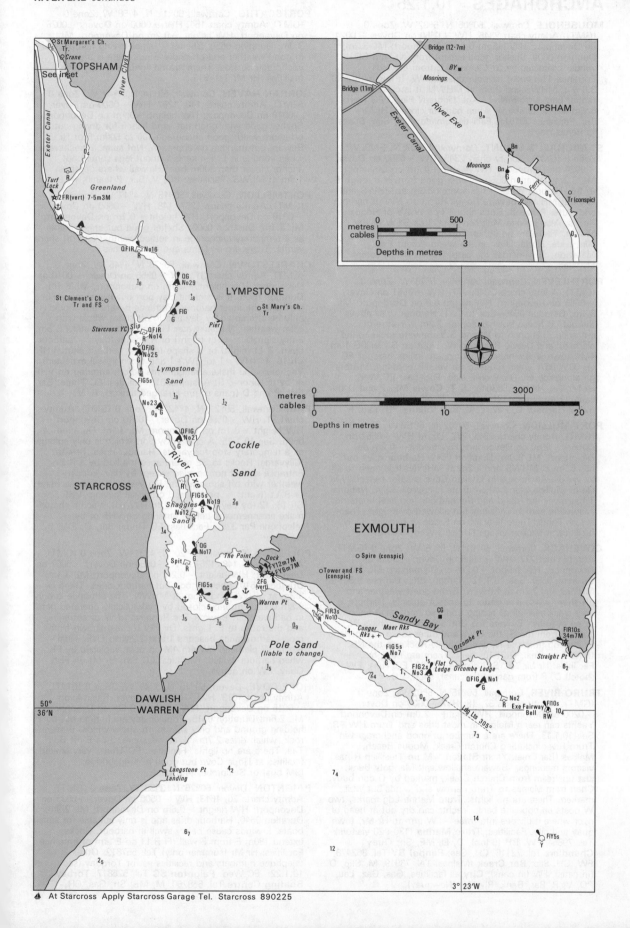

MINOR HARBOURS AND ANCHORAGES 10.1.25

MOUSEHOLE, Cornwall, 50°05′ N, 5°32′ W, Zone 0 (GMT), Admty chart 2345. HW +0550 on Dover, −0024 on Devonport; HW height 0.0 on Devonport; ML 3.2m; Duration 0600. Shelter good except from NE and SE winds; protected by St Clements I. from E winds. Entrance 11m wide; harbour dries at LW. It has depths of 3.8m at MHWS and 2.6m at MHWN. It is closed with timber baulks from Nov – Apl. Lights N Pier 2 FG (vert) 5/6m 4M. R Lt shown when harbour closed. Hr Mr Tel. Penzance 731511. Facilities limited: FW, Slip. Buses to Penzance.

ST MICHAEL'S MOUNT, Cornwall, 50°07′ N, 5°29′ W, Zone 0 (GMT), Admty chart 2345. HW +0550 on Dover, −0024 on Devonport; HW height +0.1m on Devonport; ML 3.2m; Duration 0550. Shelter good from N through E to SE. Harbour dries, it has approx 3.3m at MHWS and 1.4m at MHWN. Beware Hogus Rocks to NW of harbour and Outer Penzeath Rock about 3 ca WSW of Hogus Rocks. Also beware Maltman Rock 1 ca SSW of the Mount. There are no lights. Entrance between piers is 30m wide. No facilities on the island except FW. Marazion, ½ M across the causeway to N has some. EC Wed.

PORTHLEVEN, Cornwall, 50°05′ N, 5°19′ W, Zone 0 (GMT), Admty charts 2345, 777. HW +0551 on Dover, −0055 on Devonport; HW height 0.0 on Devonport; ML 3.1m; Duration 0545. See 10.1.11. Harbour dries above the old LB house but has approx 2.3m in centre of entrance. It is open to W and SW. Beware rocks round pierhead and Deazle Rocks to W. Light on S pier FG 10m 4M shown when inner harbour open. Inside harbour FG vis 033°-067° when required for vessels entering. Visitors go alongside the Quay on E side. Facilities: EC Wed; Hr Mr Tel. Helston 563040; **J. T. Cowis** ME, P and D (in cans); **Handyman** Tel. Helston 562245, CH, Gas, Gaz; **Inner Harbour** C, FW, L, V, Bar; **Village** PO, Lau, R.

PORT MULLION, Cornwall, 50°02′ N, 5°15′ W, Zone 0 (GMT), Admty charts 2345, 777. Lizard HW +0552 on Dover, −0050 on Devonport; HW height −0.2m on Devonport; ML 3.0m; Duration 0545. Harbour dries but has 5.3m at MHWS and 4.2m at MHWN. It is open to all W winds. Anchorage in Mullion Cove is safer especially in lee of Mullion Island where there is approx 3.5m. National Trust owns the island and the harbour. Fin keel boats not allowed in harbour. Visitors not allowed overnight. There are no lights. There is a slip on E side of harbour. Only facilities at Mullion village (1M) EC Wed; Bar, V.

COVERACK, Cornwall, 50°01′ N, 5°05′ W, Zone 0 (GMT), Admty chart 777. HW +0605 on Dover, −0035 on Devonport; HW height −0.2m on Devonport; ML 3.0m; Duration 0550. See 10.1.12. Harbour dries but has 3.3m at MHWS and 2.2m at MHWN. Given good weather and off-shore winds it is better to anchor outside. Harbour is very small and full of fishing boats so berths are unlikely to be available. Beware the Guthens, off Chynhalls Pt, coming from the S, the Dava and other rocks off Lowland Pt coming from the N, and Manacle Rocks to the NE (BYB E cardinal, Q(3) 10s bell). There are no lights. Facilities: Hr Mr Tel. St Keverne 280393; EC Tues; FW (hotel) D, P from garage (in cans), shop, PO.

TRURO RIVER, Cornwall, 50°15′ N, 5°02′ W, Zone 0 (GMT), Admty charts 32, 18; HW −0600 on Dover, −0200 on Devonport; HW height −2.0m on Devonport. Yachts can reach Malpas at most tides and Truro HW∓3. See 10.1.13. There are a number of good anchorages in Truro River including Church Creek, Mopus Reach, Malpas (Bar Creek Yacht Station ¼M up Tresillian R has visitors moorings). Beware shallow spit Maggoty Bank, just upstream from Church Creek, marked by G con buoy. Chan from Malpas to Truro narrow and winds but well marked. There are no lights. Truro Marina Ldg marks, two W posts on opposite bank. Yachts can dry out at head of river where it divides into three — W arm to Hr Mr, town quay and city. Facilities: **Truro Marina** (130+20 visitors) Tel. 79854, FW, BH (5 ton), D, El, ME, Sh; **Quay Chandlers** Tel. 72116, CH, Gas; **Penpol BY** Tel. 862478, FW, CH, Slip; **Bar Creek** (Malpas) Tel. 73919, M, Slip, D (in cans), FW (in cans); **City** all facilities, Gas, Gaz, Lau, PO, V, R, Bar, Bank, Rly, Air (Newquay).

PORTSCATHO, Cornwall, 50°11′ N, 4°58′ W, Zone 0 (GMT), Admty chart 154. HW −0600 on Dover, −0025 on Devonport; HW height −0.2m on Devonport; ML 3.0m; Duration 0550. Shelter − small drying harbour but in good weather and off-shore winds there is a good anchorage outside. There are no lights and very few facilities. Hr Mr Tel. 616.

GORRAN HAVEN, Cornwall, 50°14′ N, 4°47′ W, Zone 0 (GMT), Admty charts 148, 1267. HW −0600 on Dover, −0010 on Devonport; HW height −0.1m on Devonport. Shelter good with good flat sand beach for drying out in off-shore wind; good anchorage 100 to 500m E of Hr. Beware pot markers on approach. Not suitable anchorage when wind is in E. Fin keels without legs should not anchor closer than 300m from Hr wall where depth is 1.8m at MLWS. Facilities: PO, V, Bar, P (cans), R, L.

PORTMELLON, Cornwall, 50°15′ N, 4°47′ W, Zone 0 (GMT), Admty charts 1267, 148. HW −0600 on Dover, −0010 on Devonport; HW height −0.1m on Devonport; ML 3.1m; Duration 0600. Shelter good but only suitable as a temporary anchorage in settled weather and off-shore winds. There are no lights and few facilities.

CHARLESTOWN, Cornwall, 50°20′ N, 4°45′ W, Zone 0 (GMT), Admty chart 31. HW −0555 on Dover, −0010 on Devonport; HW height −0.1m on Devonport; ML 3.1m; Duration 0605. It is a china clay port but has some yacht berths in the inner harbour. Entrance is safe but should only be attempted by day and in off-shore winds with calm weather. N breakwater FG 5m, S breakwater FR 5m. Entry signals − G Lt (night) or R ensign (day) = harbour open. R Lt (night) or B shape (day) = harbour shut. VHF Ch 16; 14 (HW−1 to HW+1 and when vessel expected). Yachts should make arrangements before entering on VHF or by telephone, St Austell 3331. Facilities: EC Thurs; Bar, FW, P and D (cans or pre-arranged tanker), R, V.

PAR, Cornwall, 50°21′ N, 4°42′ W, Zone 0 (GMT), Admty chart 31. HW −0555 on Dover, −0010 on Devonport; HW height −0.1m on Devonport; ML 3.1m; Duration 0605. See 10.1.15. A china clay port which is only suitable as a temporary stop for yachts. Harbour dries. Beware Killyvarder Rocks to SE of entrance marked by a R Bn. Entrance should not be tried except by day, in calm weather with off-shore winds. Entry signals R shape (day) or R Lt (night) = port closed or vessel leaving. VHF Ch 16; 12 (by day and HW−2 to HW+1). Yachts should make arrangements before entering on VHF or by telephone Par 2282. Facilities: EC Thurs; Bar, D, FW, P, R, V.

POLPERRO, Cornwall, 50°20′ N, 4°31′ W, Zone 0 (GMT), Admty charts 148, 1267. HW −0554 on Dover, −0007 on Devonport; HW height −0.2m on Devonport; ML 3.1m; Duration 0610. Shelter good but harbour dries. There is 3.3m at MHWS and 2.5m at MHWN. The entrance is only 9.8m wide. Entrance closed by hydraulically operated gate in bad weather. Beware The Ranneys to W of entrance, and the rocks to E. Lights. Dir FW (occas) shown from measured distance beacons 1 and 2.2M to ENE. Tidal Basin, W pierhead FW 4m 4M on post, replaced by FR when harbour closed. Facilities: EC Sat; Hr Mr on Fish Quay, FW on quays.

HOPE COVE, Devon, 50°15′ N, 3°49′ W, Zone 0 (GMT), Admty chart 1613. HW −0525 on Dover, River Avon +0015 on Devonport; HW height −0.6m on Devonport; ML 2.6m; Duration 0615. Popular day anchorage but poor holding ground and only safe in off shore winds. Beware rock, which dries 2.7m, ½ ca off shore, 3 ca E off Bolt Tail. There are no lights. Facilities: EC Thurs; very limited facilities at Hope Cove but good at Kingsbridge, (9M bus) or Salcombe, (4M bus).

PAIGNTON, Devon, 50°26′ N, 3°33′ W, Zone 0 (GMT), Admty charts 26, 1613; HW −0500 on Dover, +0035 on Devonport; HW height −0.6m on Devonport; ML 2.9m; Duration 0640. Harbour dries and is only suitable for small boats. E winds cause heavy swell in harbour. Rocks extend 180m E from E wall. Fl R Lt on E arm of entrance. Facilities: Hr Mr (summer only) Tel. 557812. Other telephone numbers and facilities as for Torquay (see 10.1.22). EC Wed. **Paignton SC** Tel. 525817; **Torbay Boating Centre** Tel. 558760, M, ME, Sh, Gas, CH.

MAYFLOWER INTERNATIONAL MARINA

OCEAN QUAY
RICHMOND WALK
PLYMOUTH PL1 4LS
Telephone: Plymouth
(0752) 556633 & 567106

Owned & operated by yachtsmen for yachtmen

- Situated in the centre of the SOUTH-WEST cruising area with direct air and ferry routes to France, Spain, Channel & Scilly Isles.
- Motorway and dual carriageway to within ½ mile of Marina (under 2hrs from Bristol — 3½hrs from London & Birmingham).
- Easy access at all states of the tide in sheltered, deepwater berths in beautiful surroundings — yet only minutes from the open sea.
- 24 hour security, safety and radio watch (Marina Channel) maintained at all times.
- 25 ton Hoist available. No restrictions on own maintenance or choice of work personnel.
- Dry berthing for runabouts and yachts up to 30' — no limit on lifts.
- Visitors rates include, free showers, electricity and car parking (one car per boat).

AWARDED NATIONAL YACHT HARBOUR ASSOCIATION'S HIGHEST ACCOLADE FIVE GOLD ANCHORS FOR OUR HIGH STANDARDS SERVICE AND VALUE FOR MONEY

Should you wish to receive further details please write or phone

---

Brixham Yacht Supplies Ltd.

Complete range of sailing and leisure clothing

"Peter Storm — Henri Lloyd — Guy Cotten"

English & continental pure wool traditional knitwear
Admiralty charts
Camping accessories

Visit us for personal service
0803 882290
72 MIDDLE STREET, BRIXHAM

Falmouth Yacht Marina
North Parade, Falmouth
Tel: (0326) 316620 Fax: (0326) 313939

★ Visitor Berths   ★ 24hr Security
★ Electricity   ★ VHF Channel M (37)
★ Water   ★ All tide Access
★ Diesel fuel   ★ Repairs
★ Clubhouse   ★ Cranage

And our Famous Friendly, Service

All Services
DARTHAVEN MARINA
Close on your starboard hand
Call us On V.H.F. Channel 37

---

and
J. W. & A. UPHAM Ltd
(Kingswear Quay)
Engineer — Electricians
Chandlery — Brokerage
Boat Repairs
Kingswear
545/242

Dealing with Moody is just plain sailing.

Moody don't just build boats. The Swanwick Marina boasts every facility a yachtsman could desire...

- New boat sales. Moody yachts from 31' upwards and Princess and Fairline motor cruisers up to 53'
- Custom Yacht Construction
- Brokerage • Repairs and refits
- Marina berthing • Chandlery and provisions • Lifting out and scrubbing-off facilities • Winter storage • Boat bottom treatment
- Agents for Thorneycroft marine engines, Hood sailing systems

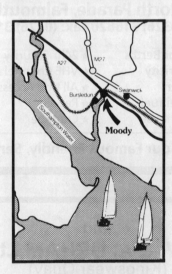

Let Moody keep
a weather eye on all
your marine requirements

A.H. Moody & Son Limited
Swanwick Marina, Southampton, Hampshire
SO3 7ZL, England Tel: Locks Heath (0489) 885000
Telex: 477536 MOODY G Fax: (0489) 885509

Abingdon Boat Centre
The Bridge, Abingdon, Oxon OX14 3HX
Tel: 0235 21125

Oxfordshire's Main Chandlery

TOPPER AGENTS · LASER SPARES
INTERNATIONAL PAINTS
MARLOW ROPES · BRUCE BANKS SAILS
AND COVERS · CREWSAVER · HOLT
R.W.O. · SIMPSON-LAWRENCE
HENRI-LLOYD · MUSTO · GILL
TYPHOON WET AND DRY SUITS
EDEC · SNIPE TRAILERS
S.P. SYSTEMS · PROCTOR
TALURIT and TERMINAL RIGGING
SERVICE · FENDERS · PUMPS

**Open 7 days
APRIL – OCTOBER**

Closed Sundays November – March
Come and visit us

Car Park   Access – Visa – Diners

THE WHEELHOUSE
SCHOOL OF
NAVIGATION

(R.Y.A. & C.A.C.C. approved)

offers certificated courses by correspondence for:

★ **RYA Day Skipper**
★ **RYA/DoT Yachtmaster Offshore**
★ **RYA/DoT Yachtmaster Ocean**

AUDIO TAPES: VHS VIDEO
COMPUTER PROGRAMME

All tuition by qualified RYA/DoT Yachtmaster Offshore Instructors
Brochure:

Rudley Mill, Hambledon,
Hants PO7 6QZ
Tel 070132 467 & 0705 255901

The Berthon Boat Company – Lymington Marina

The Shipyard, Lymington
Hants. Tel: (0590) 73312
Telex 477831

Lymington Marina—the most comfortable in the Solent! All the best facilities at very competitive rates. Contact the Dockmasters for details of permanent or visitors berths.

Repairs and Refits—Berthon's reputation for craftsmanship is unrivalled. Full facilities for all yachts up to 100 ft and 100 tons in heated undercover workshops.

New Boats—Berthon specialise in the construction of large custom built sailing and motor yachts and small commercial craft.

Brokerage—Berthon International is the No. 1 broker in Europe.
Contact us with your requirement.

CALL US FOR FRIENDLY EFFICIENT SERVICE

Ropes and Rigging
Jimmy Green Marine

Britain's Telephone and Mail Order Rope Specialists

Top quality rope splicing.
Genuine Cabco Talurit wire splicing.
Roller swaging for standing rigging.
Guardwires and jackstays made to order.

Comprehensive range of top quality ropes always in stock — list by return post.

Free delivery UK.
Next day delivery available.

VAT free export ★ Tax free shopping.

Telephone orders welcome — Access, Visa/Barclaycard and American Express.

Jimmy Green Marine

The Coachyard, Berryhill, Beer, East Devon EX12 3JP
Telephone: 0297 20744 8 am - 8 pm 7 days
Telex: 42513 Sharet G 24 hr Ansafone 0297 21744

MDL COBB'S QUAY LIMITED

Situated in Poole Harbour
Hamworthy, Poole, Dorset BH15 4EL
Telephone Poole 674299 (STD Code 0202)

Slipping, Craning and Moorings
for all types of craft up to 80 tons

Sheltered pontoon finger berths, quayside and trot moorings, storage ashore outside and undercover, slipping and relaunching, craning (up to 10 tons)

Moorings may also be booked by VHF R/T Marina Band – Channel 37

BOAT SALES DEPARTMENT OPEN 7 DAYS A WEEK

Telephone Poole 672588

There is always a large selection of boats for sale at our Yard, attracting many buyers and offering owners, particularly of the larger and more valuable craft, an excellent chance of effecting an early sale.

Licensed Club and Restaurant

All facilities available including specialist fitting out and repair services, chandlery, water, electricity and fuel supplies, showers and toilets. Brick built lock-up stores with metered electricity supply. All facilities provided, for owners wishing to build, fit out, or maintain their craft.

For further information please telephone or write for Yard brochure.

LM Littlehampton Marina Ltd.

Full Marina facilities available for all craft — Power or Sail

Pontoon Berths — boat park with launching & recovery services. Compressed air for divers — Easy access to town.

Ferry Road, Littlehampton, West Sussex
Tel: 0903 713553

Warsash Nautical Bookshop

Books · Charts · Instruments ·
Shipping · Yachting · Nautical Studies
Callers & Mail Order New & Secondhand Books

**31 Newtown Road, Warsash, Southampton
SO3 6FY Locks Heath (04895) 2384**

Send for your free Nautical Book Catalogue
All standard credit cards accepted

Publishers of the **Bibliography of Nautical Books 1989**
Price includes two **Four-monthly** supplements

BOAT FINANCE

Finance your Dream Boat with the help of experts

Find out how

CALL
07914 81911

South Eastern Credit
206 South Coast Road
Peacehaven, East Sussex BN9 8JP
Evening/weekends (07914) 87655
Fax 07914 88409
The professionals with the personal touch

PORT SOLENT MARINA
The Ultimate Maritime Lifestyle in
PORTSMOUTH HARBOUR

- Visitors Welcome
- Full tidal access — 24 hours a day
- Fuel bunkering
- Full facilities for engineering and repairs
- 30 Tonne travelhoist
- Supermarket and Launderette
- Restaurant and Wine Bars

Port Solent Ltd.
South Lockside
Port Solent,
Portsmouth, England,
Telephone (0705) 210765

STEPHEN RATSEY SAILMAKERS

A family Tradition of Fine Sailmaking Since 1790

The Loft, Building 3,
Shamrock Quay, William Street
Southampton SO1 1QL

Tel: (0703) 221683
Fax: 0703 212077

REPLACE MOST OF YOUR HEADSAILS WITH ONE EFFICIENT SAIL

THE B1- RADIAL FURLING GENOA - THE LIGHTER WEIGHT SOPHISTICATED FURLING HEADSAIL

Radiating head and clew panels for greater strength, lower stretch and a lighter, bettershaped sail.

Two leach and foot panels in U.V. (sunlight) resistant terylene. Replaces sunstrip.

Concave corner patches for increased strength in the high stress areas.

Tape lugs at head and tack for a tight furl.

Multi coloured sails available.

Luff flattener flattens genoa as it is furled.

Furling gears can be fitted at Shamrock Quay Loft

Continous furling and unfurling of the genoa puts added strain on the leach and foot panels. We therefore design these to reduce stretch to a minimum.

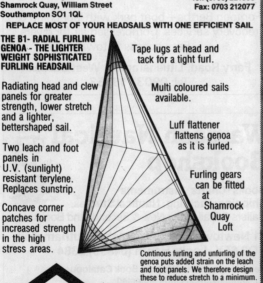

This business has no connection with the business of Ratsey & Lapthorn (Sailmakers) Ltd.

Hoo Marina Medway Ltd

VICARAGE LANE, HOO, ROCHESTER, KENT ME3 9LE.
Telephone No. MEDWAY (0634) 250311

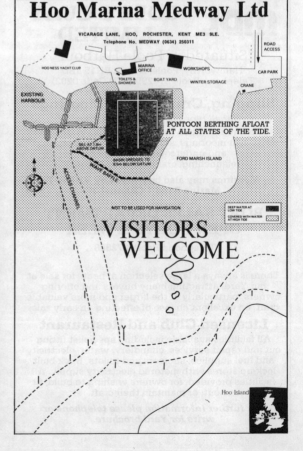

VISITORS WELCOME

VOLVO PENTA SERVICE

Sales and service centres in area 2
BERKSHIRE **D B Marine Engineering Ltd** Cookham Bridge, COOKHAM-ON-THAMES SL6 9SN Tel (06285) 26032. BUCKINGHAMSHIRE **Marlow Marine Services Ltd** Harleyford, MARLOW SL7 2DX Tel (06284) 71368. DORSET **Barretts Marine Service** The Slipway, Nothe Parade, WEYMOUTH, DT4 8TX Tel (0305) 779550. **Poole Marine Services** Sunseeker International Marina, West Quay Road, POOLE BH15 1HX Tel (0202) 679577. **Salterns Boatyard** 38 Salterns Way, Lilliput, POOLE BH14 8JR Tel (0202) 707321/707391. HAMPSHIRE **A R Savage Ltd** Thorney Road EMSWORTH PO10 8BT Tel (0243) 371151/2. **Haven Boatyard Ltd** King's Saltern Road, LYMINGTON SO41 9QD Tel (0590) 77073/4/5. **R K Marine** Hamble River Boatyard, Bridge Road, Swanwick, SOUTHAMPTON SO3 7EB Tel (04895) 83572 or 83585. **S.A.L. Marine** Mill Lane, LYMINGTON SO4 9AZ Tel (0590) 79588. ISLE OF WIGHT **Ancasta Marine Services** Ancasta Marina, COWES PO31 7BD Tel (0983) 294861. **Harold Hayles (Yarmouth IOW) Ltd** The Quay, YARMOUTH PO41 0RS Tel (0983) 760373. MIDDLESEX **Marlow Marine Services Ltd** Shepperton Marina, Felix Lane, SHEPPERTON TW17 8NJ Tel (0932) 247427.SUSSEX **B A Peters and Partners** Birdham Pool, CHICHESTER PO20 7BG Tel (0243) 512831.

VOLVO PENTA

Area 2

Central Southern England
Lyme Regis to Chichester

10.2.1	Index	Page 217
10.2.2	Diagram of Radiobeacons, Air Beacons, Lifeboat Stations etc	218
10.2.3	Tidal Stream Charts	220
10.2.4	List of Lights, Fog Signals and Waypoints	222
10.2.5	Passage Information	224
10.2.6	Distance Tables	225
10.2.7	English Channel Waypoints	See 10.1.7
10.2.8	Special tidal problems between Swanage and Selsey — Tidal Curves	226
10.2.9	The Solent area	232
10.2.10	Lyme Regis	235
10.2.11	Bridport	235
10.2.12	Portland — Tidal Curves and Tidal Streams	236
10.2.13	Weymouth	239
10.2.14	Swanage	239
10.2.15	Poole — Times and heights of LW	240
10.2.16	Christchurch	244
10.2.17	Keyhaven	245
10.2.18	Lymington	245
10.2.19	Yarmouth (I o W)	246
10.2.20	Newtown (I o W)	247
10.2.21	Beaulieu River	247
10.2.22	Cowes (I o W)	248
10.2.23	Southampton, Standard Port, Tidal Curves	249
10.2.24	Hamble River	253
10.2.25	Wootton Creek (I o W)	255
10.2.26	Bembridge (I o W)	255
10.2.27	Portsmouth, Standard Port, Tidal Curves	256
10.2.28	Langstone Harbour	262
10.2.29	Chichester Harbour	262
10.2.30	Minor Harbours and Anchorages Beer River Axe Lulworth Cove Chapman's Pool Studland Bay Wareham Newport (I o W) Titchfield Haven	264

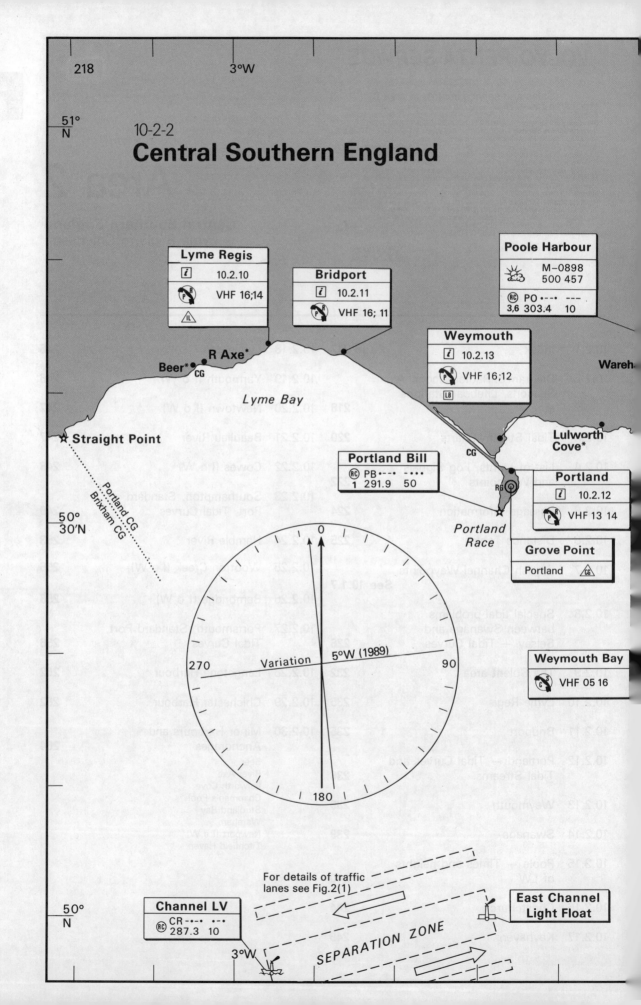

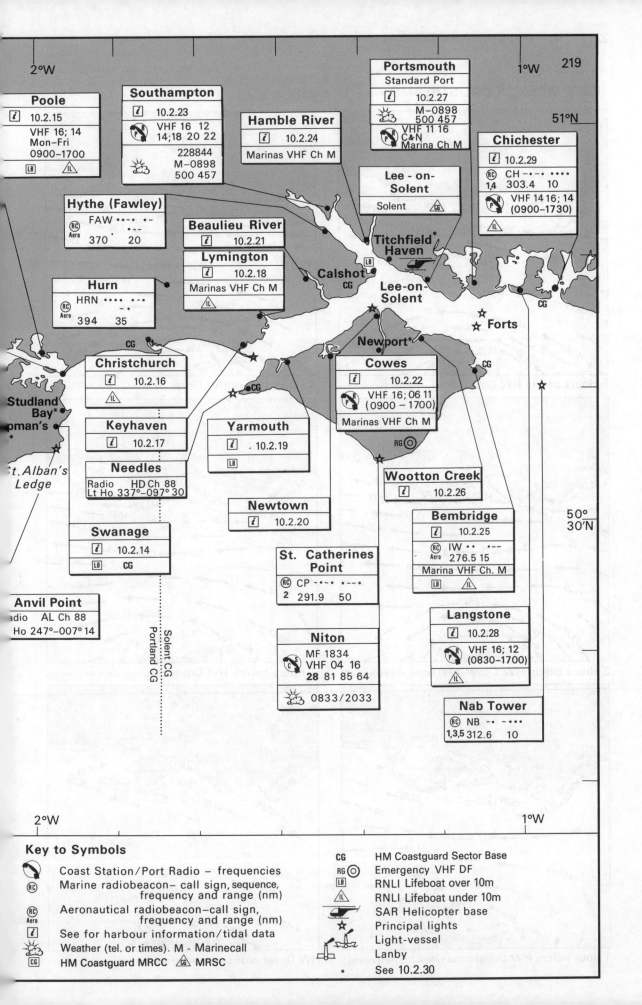

10.2.3 AREA 2 TIDAL STREAMS (see also 10.2.9 and 10.2.12)

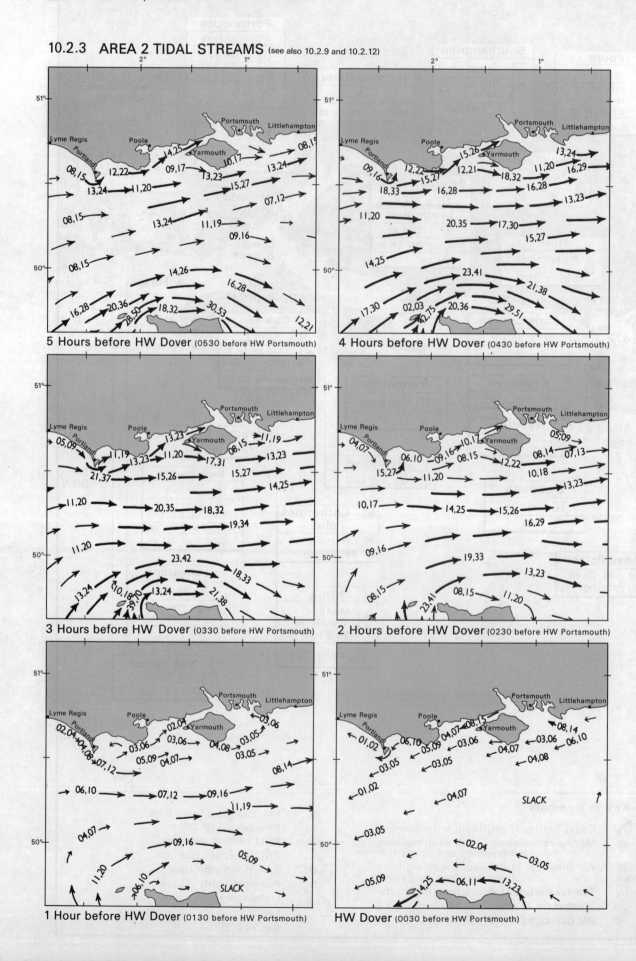

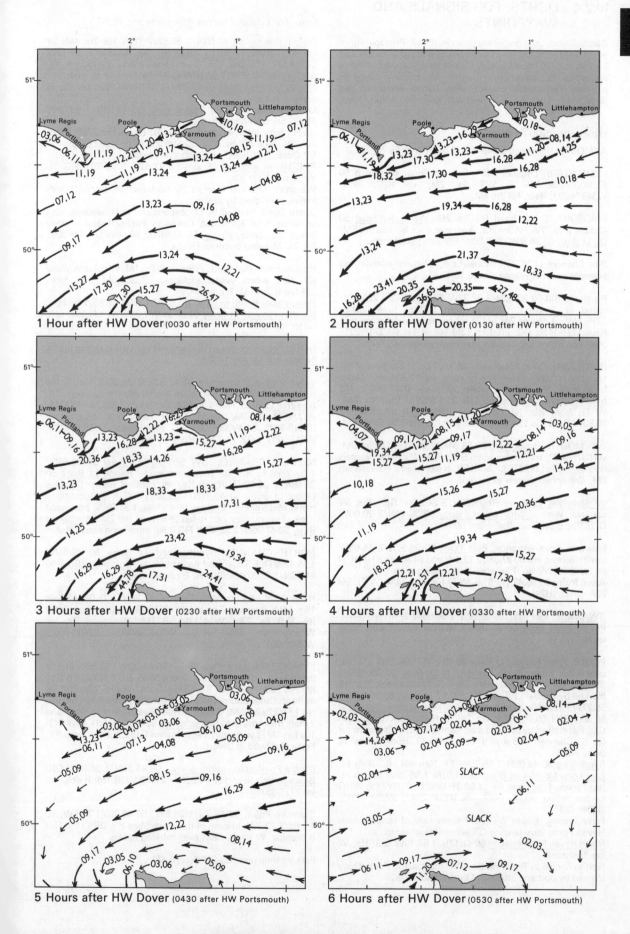

10.2.4 LIGHTS, FOG SIGNALS AND WAYPOINTS

Abbreviations used below are given in 10.0.3. Principal lights are in **bold** print, places in CAPITALS, and light-vessels and Lanbys in *CAPITAL ITALICS*. Unless otherwise stated lights are white. m—elevation in metres; M—nominal range in n. miles. Fog signals are in *italics*. Useful waypoints are underlined — use those on land with care. See 4.2.2

ENGLAND—SOUTH COAST

AXMOUTH. Pier head 50 42.10N/3 03.21W Fl 5s 7m 2M.

LYME REGIS. Ldg Lts 296°. Front at inner pier head 50 43.14N/2 56.09W F WR 6m 2M; vis R208°-343°, W343°-028°. Rear, FR 8m 2M.

BRIDPORT. E pier head FG 3m 2M. West pier head 50 42.52N/2 45.77W FR 3m 2M. West pier, root Iso R2s 9m 5M. DZ Lt Buoy 50 36.50N/2 42.00W FlY 3s; special mark.

(For Channel Lt V, E Channel Lt F, Channel Islands and adjacent coast of France — see 10.17.4).

Portland Bill 50 30.82N/2 27.32W Fl(4) 20s 43m **29M**; W round Tr, R band. Changes from 1 flash to 4 flashes 221°-244°, 4 flashes 244°-117°, changes from 4 flashes to 1 flash 117°-141°; RC. FR 19m 13M (same Tr) vis 271°-291° over The Shambles; *Dia 30s maximum power 005°-085°*.

PORTLAND. Outer breakwater (S end) Oc R 24s 9m 5M.FR on radar Tr 0.68M WSW and on radio mast 2.8M NW. Outer breakwater (N end) QR 14m 2M; vis 013°-268°. **NE breakwater (SE end)** 50 35.11N/2 24.99W Fl 10s 22m **20M**; W Tr; *Horn 10s*. NE breakwater (NW end) 50 35.62N/2 25.80W Oc R 15s 11m 5M. North arm (SE end) Oc G 10s 5M.

WEYMOUTH. S pier head 50 36.54N/2 26.41W Q 10m 9M; vis 037°-300°; traffic sigs 188m SW; *Reed 15s* (when vessels docking). Ldg Lts 237°30' both FR. N pier head 2 FG(vert) 9m 6M. *Bell* (when vessels expected).

W Shambles Lt Buoy 50 29.75N/2 24.35W Q(9) 15s; W cardinal; *Whis*. E Shambles Lt Buoy 50 30.74N/2 20.00W Q(3) 10s; E cardinal; *Bell, Horn (2) 30s*.

Lulworth Cove entrance, East Point 50 36.97N/2 14.69W (unlit). Bindon Hill 50 37.3N/2 13.6W and St Alban's Head 50 34.8N/2 03.4W Iso R 2s when firing taking place.
Anvil Point 50 35.48N/1 57.52W Fl 10s 45m **24M**; W Tr; vis 237°-076°; (H24).

SWANAGE. Pier head 50 36.52N/1 56.88W 2FR(vert) 3M. Peveril Point Buoy 50 36.38N/1 56.01W (unlit); port hand mark.

POOLE. Poole Fairway Lt Buoy 50 38.97N/1 54.78W LFl 10s; safe water mark. Poole Bar (No 1) Lt Buoy 50 39.42N/1 55.14W QG; stbd-hand mark; *Bell* (historic wreck 3ca NE marked by special Lt Buoy, FlY). No 2 Lt Buoy 50 39.33N/1 55.27W Fl(2)R 10s; port-hand mark. Training bank, outer end 50 39.75N/1 55.76W QR 7m 2M; R framework Tr. Swash Channel marked by port and stbd-hand buoys, unlit except for No 12 (Channel) Lt Buoy FlR 3s and No 13 (Hook Sands) Lt Buoy FlG 3s.
Sandbanks 50 40.95N/1 56.79W FY 10m 4M (RC 80m E).
East Looe No 16A Lt Buoy 50 41.02N/1 56.08W QR; port-hand mark. East Looe Dir Lt 50 41.16N/1 56.38W Oc WRG 6s 9m W10M, R6M, G6M; vis R234°-294°, W294°-304°, G304°-024°.
Ferry landing, E side 2FG(vert) either side of ramp. South Haven Point, ferry landing QR either side of ramp.
North Haven Point Beacon 50 41.12N/1 57.10W Q(9) 15s; W cardinal mark.
North channel to Poole Harbour YC marina and Poole Quay marked by port and stbd-hand buoys, mostly lit.

Note. For English Channel Waypoints see 10.1.7

Bullpit Beacon 50 41.69N/1 56.62W Q(9) 15s 7m 4M; W cardinal mark. Salterns Beacon 50 42.18N/1 57.12W Q(6)+LFl 15s 5m 2M; S cardinal mark. Parkstone YC platform 50 42.33N/1 58.00W Q 8m 1M; hut on dolphin.
Stakes No 55 Lt Buoy 50 42.39N/1 58.91W Q(6)+LFl 15s; S cardinal mark.
Little Channel, E side, Oyster Bank 50 42.59N/1 59.02W Fl(3)G 5s; stbd-hand mark.
Ro-Ro ferry terminal, two 2FG(vert). Lts in line 270°. Front FlY 10s. Rear FlY 5s. Mark centre line of dredged basin.
BP Dock. 2 FG(vert) on E and West sides.
PYC Haven entrance, E side 50 42.41N/1 59.70W FlG 5s. West side Fl(3) 10s.
Wareham Channel marked by port and stbd-hand buoys initially, and then by stakes.
South Deep. Marked by Lt beacons and unlit beacons from entrance S of Brownsea Castle to Furzey Island. Furzey Island, SE corner, slipway 50 40.88N/1 58.86W Ldg Lts 305° FlY 2s. Mooring dolphins FlG 5s.
Bournemouth Pier head 2FR(vert) 9/7m 1M; W column. *Reed (2) 120s* when vessel expected. Boscombe Pier head 2FR(vert) 7m 1M; R column.
Hengistbury Head, groyne, Bn 50 42.63N/1 44.85W (unlit).
Needles Fairway Lt Buoy 50 38.20N/1 38.90W LFl 10s; safe water mark; *Whis*.

NOTE: For waypoints of navigational buoys and racing marks in Solent area, see 10.2.9.

Wave Research Structure 50 42.87N/1 38.24W Mo(U) 15s and Mo(U)R 15s 10m 5M; *Horn Mo(U) 30s* (temp inop).
Needles Lt 50 39.70N/1 35.43W Oc(2) WRG 20s 24m **W17M, R17/15M**, G14M; round Tr, R band and lantern; vis R291°-300°, W300°-083°, R(unintens) 083°-212°, W212°-217°, G217°-224°; *Horn (2) 30s*.

NEEDLES CHANNEL. SW Shingles Lt Buoy 50 39.53N/1 37.24W Fl R 2.5s; port-hand mark. Bridge Lt Buoy 50 39.59N/1 36.80W VQ(9) 10s; W cardinal mark. Shingles Elbow Lt Buoy 50 40.31N/1 35.92W Fl(2)R 5s; port-hand mark. Mid Shingles Lt Buoy 50 41.19N/1 34.59W; port-hand mark. Totland Bay Pier, head 2FG(vert) 6m 2M. Warden Lt Buoy 50 41.46N/1 33.48W FlG 2.5s; stbd-hand mark; *Bell*.

NORTH CHANNEL. North Head Lt Buoy 50 42.65N/1 35.42W Fl(3)G 10s; stbd-hand mark. NE Shingles Lt Buoy 50 41.93N/1 33.32W Q(3) 10s; E cardinal mark.

Hurst Point Ldg Lts 042°. Front 50 42.36N/1 33.05W Iso 4s 15m 14M; R square Tr; vis 029°-053°. Rear 215m from front Iso WR 6s 23m W14/13M, R11M; W round Tr; vis W(unintens) 080°-104°, W234°-244°, R244°-250°, W250°-053°.

LYMINGTON. Jack in the Basket 50 44.25N/1 30.50W Fl R 2s 9m; Ra refl. Ldg Lts 319°30'. Front 50 45.2N/1 31.6W FR 12m 8M; vis 309.5°-329.5°. Rear 335m from front FR 17m 8M; vis as front. Cross Boom No 2 Fl R 2s 4m 3M; X and R can on pile. No 1 Fl G 2s 2m 3M; G cone on pile. Piles similar to Nos 1 and 2 mark the channel to Harper's Post 50 45.05N/1 31.40W Q(3) 10s 5m 1M. Lymington Yacht Haven Ldg Lts 244°, both FY. Ferry pier head 2FG(vert).

Durn's Point obstruction, S end, 50 45.37N/1 26.95W QR. Beaulieu Spit, E end 50 46.83N/1 21.67W QR 1M; R dolphin; vis 277°-037°.

Stansore Point. 50 46.7N/1 20.8W, 50 46.8N/1 20.5W, 50 46.9N/1 20.4W. In each of these positions is a QR Lt 4m 1M; R triangle, W band, on R pile, marking cables.

Fort Victoria pier head 2FG(vert) 4M.

AREA 2—Central S England

YARMOUTH. Pier head, centre, 2FR(vert) 2M; G column. High intensity FW (occas). Jetty head 2FR(vert); W mast; second FR shows harbour full. In fog FY 5m from same structure, vis 167°-192°. Ldg Lts 188°. Front 50 42.3N/1 30.0W FG 5m 2M; W diamond on B post. Rear, 63m from front FG 9m 2M; W diamond on B mast.

Egypt Point 50 46.00N/1 18.75W Fl 10s 8m **18M**; R column with W lantern; vis 061°-272°.
West Bramble Lt Buoy 50 47.15N/1 18.55W VQ(9) 10s; W cardinal mark; *Bell*; Racon.
Outfall 50 48.25N/1 18.73W Iso R 10s 6m 5M; column on square structure; Ra refl; FR Lt on each corner; *Horn 20s*.

CALSHOT SPIT Lt F 50 48.32N/1 17.55W Fl 5s 12m 11M; R hull, Lt Tr amidships; *Horn (2) 60s*.
Calshot jetty 50 49.20N/1 18.30W 2FR (vert). Obstn Lts, FR 34m on radar Tr 120m W, and QR on chy 0.9M W.
Fawley Channel No 2 50 49.45N/1 18.75W FIR 3s. Ldg Lts 219°. Front QR (occas). Rear Q (occas). Hamble Point Lt Buoy 50 50.12N/1 18.58W Q(6) + LFl 15s; S cardinal mark.

RIVER HAMBLE. Ldg Lts 345°30'. Front No 6 pile 50 50.58N/1 18.74W Oc(2) R 12s 4m 2M. Rear 820m from front QR 12m; W mast; vis 341.5°-349.5°. No 1 pile 50 50.31N/1 18.57W FlG 3s; stbd-hand mark. No 2 pile 50 50.36N/1 18.68W Q(3) 10s; E cardinal mark. No 3 pile Fl(2)G 5s. No 5 pile Fl(3)G 10s. No 7 pile FlG 3s. No 8 pile FIR 3s. No 9 pile Fl(2)G 5s. No 10 pile Fl(2)R 5s. Ldg Lts 026°, Warsash shore. Front 50 51.0N/1 18.3W QG; B & W chequered pile Bn. Rear, Sailing Club, Iso G 6s; vis 022°-030°. Up river Lts 2FG(vert) or QG are on the E side, and 2FR (vert) on the West side.

SOUTHAMPTON WATER. Esso Marine terminal, SE end 50 50.05N/1 19.33W 2FR(vert) 9m 10M; *Whis (2) 20s*. Shell Mex Jetty 50 50.8N/1 19.4W 2FG(vert) 5/3m 2M (on each side of the 4 dolphins).
Hythe Pier head 50 52.45N/1 23.52W 2FR(vert) 12/5m 5M. Hythe Marina Village 50 52.6N/1 23.8W Q(3) 10s; E cardinal mark. Lock entrance 2FG(vert) and 2FR(vert).
Queen Elizabeth II terminal, S end 50 52.97N/1 23.64W 4FG(vert) 16/14/12/11m 3M.

RIVER ITCHEN. East side. No 1 dolphin 50 53.12N/1 23.32W QG; stbd-hand mark. No 2 dolphin 50 53.26N/1 23.30W FlG 5s; stbd-hand mark. No 3 dolphin 50 53.45N/1 23.20W FlG 7s; stbd-hand mark. No 4 pile 50 53.59N/1 23.08W QG; stbd-hand mark. Itchen Bridge. FW on bridge span each side marks main channel. 2FG(vert) each side on E pier. 2FR(vert) each side on West pier.
Crosshouse Beacon 50 54.01N/1 23.11W OcR 5s; port-hand mark, power cable. Chapel Beacon 50 54.12N/1 23.13W FlG 3s; stbd-hand mark.
Shamrock Quay pontoon, SW and NE ends 2FR (vert). Millstone Point No 5 Bn FlG 3s. No 6 Bn FIR 3s. No 7 Bn Fl(2)G 5s. No 9 Bn Fl(4)G 10s.

RIVER TEST. Lower Foul Ground 50 53.22N/1 24.48W Fl(2)R 10s; port-hand Bn. Upper Foul Ground 50 53.49N/1 24.80W Fl(2)R 10s; port hand Bn. Town Quay Ldg Lts 329°, both FY (occas). Middle Swinging Ground, Lts in line 336°, both OcG 6s.

COWES. No 4 Lt Buoy 50 46.04N/1 17.78W FIR. Ldg Lts 164°. Front 50 45.9N/1 17.8W Iso 2s 3m 6M. Rear 290m from front Iso R 2s 5m 3M; vis 120°-240°. East breakwater head Fl R 3s 3M. Fountain Pontoon 2FG(vert) each end. Yacht Haven 2FG(vert) each end. East Cowes ferry pontoon 2FR(vert); *Horn*. Trinity Wharf 2FR(vert). Above this point Lts 2FG(vert) are shown on the W side, and 2FR(vert) on the E side.

WOOTTON. Beacon 50 44.46N/1 12.08W Q 1M; N cardinal mark. West side of channel marked by pile beacons: No 1 Fl G 3s, No 2 Fl(2) G 5s, No 3 QG.

RYDE. Pier, NW corner, N corner and E corner each marked by 2FR(vert). In fog FY from N corner, vis 045°-165°, 200°-320°.

Fort Gilkicker 50 46.40N/1 08.38W Oc G 10s 7M. Spit Sand Fort, N side 50 46.20N/1 05.85W Fl R 5s 18m 7M; large round stone structure. Horse Sand Fort 50 44.97N/1 04.25W Fl 10s 21m **15M**; large round stone structure. No Man's Land Fort 50 44.37N/1 05.61W Fl 5s 21m **15M**; large round stone structure.
St Helen's Fort (IOW) 50 42.27N/1 04.95W Fl(3) 10s 16m 8M; large round stone structure.
Brading Harbour entrance, West side, tide gauge 50 42.55N/1 05.24W Fl Y 2s.

Sandown Pier head 2FR(vert) 7/5m 2M. Shanklin Pier head 2FR(vert) 6m 4M; TE; *Bell*. Ventnor Pier 2FR(vert) 10m 3M.

St Catherine's Point (IOW) 50 34.52N/1 17.80W Fl 5s 41m **30M**; W 8-sided tower; vis 257°-117°; RC. FR 35m **17M** (same Tr) vis 099°-116°.

Nab Tower. 50 40.05N/0 57.07W Fl(2) 10s 27m **19M**; *Horn(2)30s*; RC; Racon. Fog Det Lt, vis 300°-120°.

PORTSMOUTH. Outer Spit Lt Buoy 50 45.55N/1 05.41W Q(6) + LFl 15s; S cardinal mark. Southsea Castle, N corner 50 46.66N/1 05.25W Iso 2s 16m 11M, W stone Tr, B band; vis 339°-066°. Dir Lt 001°30' (same structure) Dir WRG 11m W13M, R5M, G5M; FG 351.5°-357.5°, AlWG 357.5°-000° (W phase incr with brg), FW 000°-003°, Al WR 003°-005.5° (R phase incr with brg), FR 005.5°-011.5°. Fort Blockhouse 50 47.34N/1 06.65W Dir Lt 320°; Dir WRG 6m W13M, R5M, G5M; OcG 310°-316°, AlWG 316°-318.5° (W phase incr with brg), Oc 318.5°-321.5°, AlWR 321.5°-324° (R phase incr with brg), OcR 324°-330°. 2FR(vert) 20m E. Dolphin, close E of C&N marina, 50 47.8N/1 06.9W, Dir WRG 2m 1M; vis IsoG 2s 322°-330°, AlWG 330°-332°, Iso 2s 332°-335° (main channel), AlWR 335°-337°, IsoR 2s 337°-345° (Small Boat Channel).

Victoria Pile 50 47.31N/1 06.40W OcG 15s; stbd-hand mark. Round Tower 2FG(vert). The Point 50 47.54N/1 06.48W IQG; pile, stbd-hand mark. In Portsmouth Harbour jetties on the E (Portsmouth) shore are marked by 2FG(vert) and those on the West (Gosport) shore by 2FR(vert). Pile 98 (N of North Corner Jetty) 50 48.58N/1 06.69W Q(6) + LFl 15s; S cardinal mark.

LANGSTONE. Roway Wreck beacon 50 46.08N/1 02.20W Fl(2) 5s; isolated danger mark. Langstone Fairway Lt Buoy 50 46.28N/1 01.27W LFl 10s; safe water mark. Eastney Point, drain Bn 50 47.20N/1 01.58W QR 2m 2M. FR Lts combined with Y Lts shown when firing taking place. Water intake 50 47.7N/1 01.7W Fl R 10s; pile. Hayling Island ferry landing 2FG(vert). Langstone Channel, South Lake 50 49.45N/0 59.80W FlG 3s 3m 2M; pile, stbd-hand mark. Binness 50 49.60N/0 59.85W FlR 3s 3m 2M; pile, port-hand mark.

CHICHESTER. Bar beacon 50 45.88N/0 56.37W Fl WR 5s 14m W7M, R5M; vis W322°-080°, R080°-322°; Fl(2)R 10s 7m 2M; same structure; Racon; RC. Eastoke Beacon 50 46.62N/0 56.08W QR. West Winner 50 46.83N/0 55.89W QG.
EMSWORTH CHANNEL Verner 50 48.33N/0 56.62W Fl R 10s; pile, port-hand mark. Marker Point 50 48.87N/0 56.62W Fl(2)G 10s; pile, stbd-hand mark. NE Hayling 50 49.60N/0 56.75W Fl(2)R 10s; pile, port-hand mark. Emsworth 50 49.63N/0 56.67W Q(6) + LFl 15s; pile, S cardinal mark, tide gauge.
CHICHESTER CHANNEL. East Head 50 47.32N/0 54.70W Fl(4)G 10s; pile, stbd-hand mark, tide gauge. Camber Beacon 50 47.84N/0 53.98W Q(6) + LFl 15s; S cardinal mark. Chalkdock Beacon 50 48.46N/0 53.20W Fl(2)G 10s; pile, stbd-hand mark. Itchenor, jetty 50 48.44N/0 51.89W 2FG(vert); tide gauge. Birdham 50 48.30N/0 50.17W Fl(4)G 10s; pile, stbd-hand mark, depth gauge.

10.2.5 PASSAGE INFORMATION

LYME BAY (chart 3315)

Between Torbay and Portland there is no harb accessible in onshore winds, and yachtsmen must take care not to be caught on a lee shore. Tides are weak, seldom reaching more than ¾ kn. There are no dangers offshore. In offshore winds there is a good anch off Beer, NE of Beer Hd, the most W chalk cliff in England, see 10.2.30. 3½M E of Lyme Regis (10.2.10) is Golden Cap (186m and conspic). High Ground and Pollock are rks 7 ca offshore, 2M and 3M ESE of Golden Cape.

E of Bridport (10.2.11) is the start of the Chesil Beach, which runs almost straight to N end of Portland peninsula, with its conspic wedge-shaped appearance. In E winds there is anch in Chesil Cove on NW side of Portland, but beware any shift of wind to the W.

PORTLAND

Portland (chart 2255) is mostly steep-to, but rks extend ¼ ca beyond the extremity of the Bill, where there is a stone Bn (18m). Depths S of Bill are irregular. On E side there is rk awash at HW ½ ca off Grove Pt, the E extremity of Portland. From a distance Portland looks like an island, with the highest land (144m) in N, sloping to the Bill where the Lt Ho is conspic.

Tidal streams run very strongly, with eddies either side of Portland. Stream runs S down each side for about 10 hrs out of 12, causing great turbulence where they meet off the Bill. The direction and strength of stream is shown hour by hour, relative to HW Devonport, in 10.2.12. Note that the sp rate reaches 7 kn, but even stronger streams can occur in vicinity of the race. The diagrams show that the race varies in position and extent — usually SE of the Bill on E-going stream, and SW of the Bill on W-going stream.

There is usually a stretch of relatively smooth water ¼ – ½M offshore, between the Bill and the race, but yachts using this must take care not to be swept S into the race by the strong stream usually running down either side of Portland. Also beware lobster pots — this is no place for a fouled propeller.

On passage E, unless bound for Weymouth (10.2.13) it is safest to pass 5M S of the Bill, to seaward of the race and of the Shambles bank. If bound for Weymouth, and conditions are suitable for the inshore passage, aim for a point well N of the Bill itself, to allow for tide. It is ideal to round the Bill about HW Devonport −0130 (HW Portland −0230) when there will be a favourable tide up the E side of Portland, taking the boat away from the race.

On passage W it is best to round the Bill with the first of the ebb, leaving Weymouth about HW Devonport +0400 (HW Portland +0300), so that a yacht will carry the tide well W of Portland.

PORTLAND TO ISLE OF WIGHT (chart 2615)

The Shambles bank is about 3M E of Portland Bill, and should be avoided at all times. In bad weather the sea breaks heavily on it. It is marked by buoys on its E side and at SW end. E of Weymouth are rky ledges extending 3 ca offshore as far as Lulworth Cove, which provides a reasonable anch in fine, settled weather and offshore winds. Rky ledges extend each side of entrance, more so on the W. See 10.2.30.

A firing range extends 5M offshore between Lulworth and St Alban's Hd. Yachts must pass through this area as quickly as possible, when the range is in use. See 10.2.30. Beware Kimmeridge Ledges, which run over ½M seaward. Warbarrow B and Chapman's Pool provide anchs in calm weather and offshore winds. See 10.2.30.

St Alban's Hd (107m and conspic) is steep-to and has a dangerous race off it which may extend 3M seaward. The race lies to the E on the flood and to the W on the ebb; the latter is the more dangerous. An inshore passage about ½M wide avoids the worst of the overfalls. There is an eddy on W side of St Alban's, where the stream runs almost continuously SE. 1M S of St Alban's the ESE stream begins at HW Portsmouth +0520, and the WNW stream at HW Portsmouth −0030, with sp rates of 4¾ kn.

There is deep water quite close inshore between St Alban's Hd and Anvil Pt (Lt). 1M NE of Durlston Hd is Peveril Ledge running ¼M seaward, and causing quite a bad race which extends nearly 1M eastwards, particularly on W-going stream against a SW wind. Proceeding towards Poole (10.2.15), overfalls may be met off Ballard Pt and Old Harry on the W-going stream. Studland B (chart 2172) is a good anch except in winds between NE and SE. Anch about 4 ca WNW of Handfast Pt. Avoid foul areas on chart. See 10.2.30.

Poole B offers good sailing in waters sheltered from W and N winds, and with no dangers to worry the average yacht. Tidal streams are weak N of a line between Handfast Pt and Hengistbury Hd. The latter is a dark headland, S of Christchurch harb (10.2.16), with a groyne extending 1 ca S and Beerpan Rks a further ½ ca offshore. Christchurch Ledge extends 2¾M SE from Hengistbury Hd. There are often lobster pots in this area. The tide runs hard over the ledge at sp, and there may be overfalls. Within Christchurch B the streams are weak.

APPROACHES TO SOLENT (charts 2219, 2050)

The Needles with their Lt Ho are conspic landmark for W end of Isle of Wight. There are rks ¼ ca (dries) and ½ ca WNW of Lt Ho, with remains of wreck close SW of inner one. In bad weather broken water and overfalls extend along The Bridge — a reef which runs 8 ca of W of the Lt Ho with extremity marked by buoy.

The NW side of Needles Chan is defined by the Shingles bank, parts of which dry and on which the sea breaks very heavily. The SE side of the bank is fairly steep, the NW side much more gradual in slope. On the ebb the stream sets very strongly in a WSW direction across the Shingles. Gale force winds between S and W against the ebb tide cause a very dangerous sea in Needles Chan. In such conditions the eastern approach to the Solent, via the Forts and Spithead, should be used. In strong winds the North Channel, N of the Shingles, is to be preferred to the Needles Chan. The two join S of Hurst Pt, where overfalls may be met. Beware The Trap, a bank close S of Hurst Fort.

The E approach to the Solent, via Spithead, presents few problems. The main chan is well buoyed and easy to follow, but there is plenty of water for the normal yacht to the S of it when approaching No Man's Land Fort and Horse Sand Fort, which must be passed between. Ryde Sand dries extensively and is a trap for the unwary; so too is Hamilton Bank on the W side of the chan to Portsmouth (10.2.27).

THE SOLENT (charts 2040, 394)

Within the Solent there are few dangers in mid-chan. The most significant is Bramble bank (dries) between Cowes and Calshot. The main chan (buoyed) passes S and W of the Brambles, but yachts can use the North Chan running N of the Brambles at any state of tide.

Tidal stream are strong at sp, but principally follow the direction of the main chans. For details see 10.2.8.

For waypoints of navigational buoys and racing marks see 10.2.9.

There are several spits, banks, rks and ledges which a yachtsmen should know. From the W they include: Pennington and Lymington Spits on the N shore; Black Rk 4 ca W of entrance to Yarmouth (10.2.19); Hamstead Ledge 8 ca W of entrance to Newtown River (10.2.20) and Saltmead Ledge 1½M to E; Gurnard Ledge 1½M W of Cowes; Lepe Middle and Beaulieu Spit, S and W of the entrance to Beaulieu River (10.2.21); the shoals off Stone Pt, where three Bns mark cable area; Shrape Mud, which extends N from the breakwater of Cowes harb (10.2.22) and along to Old Castle Pt; the shoals and occasional rks which fringe the Island shore from Old Castle Pt to Ryde, including either side of the entrance to Wootton Creek (10.2.25); and Calshot Spit which extends almost to the Lt F which marks the turn of chan into Southampton Water.

Depending on the direction of wind, there are many good anchs in Solent for yachts which are on passage up or down the coast. For example, in E winds Alum B, close NE of the Needles, is an attractive daytime anch with its coloured cliffs — but beware Long Rk (dries) in middle of b, and Five Fingers Rk 1½ ca SW of Hatherwood Pt on N side. Totland B is good anch in settled weather, but avoid Warden Ledge.

In W winds there is anch on E side of Hurst, as close inshore as depth permits, NE of High Lt. In S winds, or in good weather, anch W of Yarmouth harb entrance, as near shore as possible; reasonably close to town (see 10.2.19).

In winds between W and N there is good anch in Stanswood B, about 1M NE of Stansore Pt. Just N of Calshot Spit there is shelter from SW and W. Osborne B, 2M E of Cowes, is sheltered from winds between S and W. In E winds Gurnard B, the other side of Cowes is preferable. In N winds anch in Stokes B. At E end of IOW there is good anch off Bembridge in winds from S, SW or W; but clear out if wind goes into E.

Within the Solent there are many places which a shoal-draught boat can explore at the top of the tide — such as Ashlett Creek between Fawley and Calshot, Eling up the River Test, and the upper reaches of the Medina.

ISLE OF WIGHT — SOUTH COAST (chart 2045)

From the Needles into Freshwater B the cliffs can be approached to within 1 ca, but beyond the E end of chalk cliffs there are ledges off Brook and again off Atherfield which demand keeping at least ½M offshore. The E-going stream sets towards these dangers. 4M SSW of the Needles the stream turns E x N at HW Portsmouth +0530, and W at HW Portsmouth −0030, sp rate 2 kn. The E-going stream starts running NE, but soon changes to E x N, towards Atherfield ledges. In Freshwater B, 3M E of Needles, the stream turns ESE at HW Portsmouth +0445, and WNW at HW Portsmouth −0130, sp rate 1 kn.

St Catherine's Lt Ho (Lt, RC) is conspic. It is safe to pass 2 ca off, but a race occurs off the Pt and can be very dangerous at or near sp with a strong opposing wind, particularly SE of the Pt on a W-going stream in a W gale — when St Catherine's should be given a berth of at least 2M. 1¼M SE of the Pt the stream turns E x N at HW Portsmouth +0520, and W x S at HW Portsmouth −0055, sp rate 3¾ kn.

From St Catherine's Pt to Dunnose, rks extend about 2½ ca in places. A race occurs at Dunnose. In Sandown B, between Dunnose and Culver Cliff, the streams are weak inshore. Off the centre of the B they turn NE x E at HW Portsmouth +0500, and SW x W at HW Portsmouth −0100, sp rates 2 kn. Whitecliff B provides an anch in winds between W and N. From here to Foreland (Bembridge Pt) the coast is fronted by a ledge of rks (dry) extending up to 3 ca offshore, and it is advisable to keep to seaward (E) of Bembridge Ledge buoy.

4½M E of Foreland is Nab Tr (Lt, fog sig, RC), a conspic steel and concrete structure (28m), marking Nab Shoal for larger vessels and of no direct significance to yachtsmen. It is however a most useful landmark when approaching the E end of Isle of Wight, or when making for the harbs of Langstone (10.2.28) or Chichester (10.2.29).

For notes on cross-Channel passages see 10.3.5.

10.2.6 DISTANCE TABLE

Approximate distances in nautical miles are by the most direct route while avoiding dangers and allowing for traffic separation schemes etc. Places in *italics* are in adjoining areas.

1 *Le Four*	1																			
2 *Start Point*	113	2																		
3 *Berry Head*	126	13	3																	
4 *Casquets*	120	57	61	4																
5 *Cherbourg*	151	85	86	31	5															
6 Portland Bill	153	49	41	48	62	6														
7 Anvil Point	172	69	61	56	58	20	7													
8 Poole Bar	177	74	66	61	63	25	5	8												
9 Needles	180	84	76	64	61	35	15	12	9											
10 St Catherines	188	95	86	67	58	45	25	25	12	10										
11 Nab Tower	206	110	101	81	66	60	40	30	27	15	11									
12 Lymington	187	91	83	71	68	42	22	19	7	19	25	12								
13 Cowes	194	98	90	78	75	49	29	26	14	25	15	10	13							
14 Hamble	198	102	94	82	79	53	33	30	18	29	19	14	6	14						
15 Portsmouth	203	107	99	87	84	58	38	35	23	20	10	19	10	13	15					
16 Chichester Bar	207	112	104	92	89	63	43	40	28	19	6	23	15	18	8	16				
17 Owers Lanby	208	119	110	87	70	69	49	49	36	24	11	34	25	29	20	14	17			
18 *Le Havre*	223	156	151	101	70	118	104	107	98	86	85	105	112	116	95	91	77	18		
19 *Royal Sovereign*	248	163	153	124	102	114	92	92	82	67	54	77	68	72	63	57	43	82	19	
20 *North Foreland*	310	225	215	186	162	176	154	154	144	129	116	139	130	134	125	119	105	137	62	20

SPECIAL TIDAL PROBLEMS BETWEEN SWANAGE AND SELSEY 10-2-8

Owing to the distorted tidal regime in this area, special curves, as shown on the two following pages, are given for certain ports. Since their low water (LW) points are better defined than high water (HW), the times on these curves are referred to LW, but otherwise they are used as described in 9.4. Box 17 of the tidal prediction form shown in Fig.9(5) should be amended to read 'Time of LW'.

Height differences for places between Swanage and Yarmouth always refer to the higher HW (that which reaches a factor of 1.0 on the curves). The time differences, which are not needed for this calculation, also refer to the higher HW.

Since the tides at these places cannot be adequately defined by two curves, a third is shown for the range at Portsmouth (indicated on the right of the graph) at which the two high waters are equal at the port concerned. Interpolation should be between this critical curve and either the spring or neap curve, as appropriate. The higher HW should be used to obtain the range at the Secondary Port.

While the critical curve extends throughout the tidal cycle, the spring and neap curves stop at higher HW. Thus for a range at Portsmouth of 3.5m, the factor for 7 hrs after LW Poole (Town Quay) should be referred to the following LW, whereas had the range at Portsmouth been 2.5m it should be referred to the preceding LW.

For special remarks on Newport, Calshot, Wareham and Tuckton see notes 1, 2 and 3 on the following page.

The procedure to be followed is shown in the example below, referring to the special tidal curve for Swanage, Poole (Entrance) and Bournemouth which is shown at the foot of the page.

Example: Using the tidal data for Swanage below, find the height of tide at Swanage at 0200 on a day when the tidal predictions for Portsmouth are:

19 0100 4.6
0613 1.1
M 1314 4.5
1833 0.8

PORTSMOUTH

HW		LW		MHWS	MHWN	MLWN	MLWS
0000	0600	0500	1100	4.7	3.8	1.8	0.6
1200	1800	1700	2300				

Differences **SWANAGE**
−0250 +0105 −0105 −0105 −2.7 −2.2 −0.7 −0.3

(1) Complete the top part of the tidal prediction form, as in Fig.9(5), omitting the HW time column (boxes 1, 5 and 9).
(2) On the left of the Swanage diagram, plot the Secondary Port HW and LW heights (1.9 and 0.6m from (1) above), and join the points by a sloping line.
(3) The time required (0200) is 3h 8m before the time of LW at the Secondary Port, so from this point draw a vertical line to the curve. It is necessary to interpolate from the range at Portsmouth (3.5m), which is midway between the spring range (4.1m) and the critical curve (2.9m). This should give a point level with a factor of 0.84.

Fig. 9(5) *Tidal prediction form*   Time or height required 0200

	TIME		HEIGHT	
	HW	LW	HW	LW
Standard Port *Portsmouth*	1	2 0613	3 4.6	4 1.1
Differences	5	6 −0105	7 −2.7	8 −0.5
Secondary Port *Swanage*	9	10 0508	11 1.9	12 0.6
Duration (or time from HW to LW)	13	9–10 or 10–9	14 Range Stand. Port 3.5	3–4
			15 Range Secdy. Port 1.3	11–12

*Springs/Neaps/Interpolate

Start height at given time ↓
9
17 − 16

Time reqd.	16 0200	17+18
Time of HW LW	17 0508	9
Interval	18 −0308	

Date 19th Nov
| Factor | 19 0.84 | Time zone 0 (GMT) |

19 × 15 | Rise above LW | 20 1.1 | 22−21
12 | Height of LW | 21 0.6 | 12 Start: time for given height
20+21 | Height reqd. | 22 1.7 |

*Delete as necessary

(4) From this point draw a horizontal line to meet the sloping line constructed in (2) above.
(5) From the point where the horizontal line meets the sloping line, proceed vertically to the height scale at the top, and read off the height required, 1.7m.

The completed lower part of the tidal prediction form above shows how the calculation is done using the factor method instead of the graphical method just described. The procedure follows that given in 9.4.3. Note that box 9 is amended to read 'Time of LW'.

A less accurate but quick method of finding intermediate heights for certain places is by use of Fig 10A on p. 229. It is useful to note that the area enclosed by the dotted lines represents the period during which the tide stands and in which a second high tide may occur.

To use this table, note that the first of the three figures given is for mean spring tides, the second for average and the third for mean neap tides. Work out whether the time required is nearer the HW or LW at Portsmouth and find the interval (between time required and the nearest HW or LW at Portsmouth). Extract the height of the corresponding predicted HW (left hand column) or LW (right hand column). By interpolation between the heights of HW or LW, read off the height of tide under the appropriate column (hrs before or after HW or hrs before or after LW).

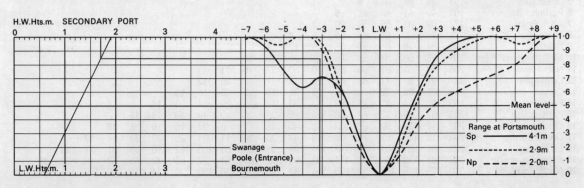

Swanage
Poole (Entrance)
Bournemouth

Range at Portsmouth
Sp ——— 4.1m
-------- 2.9m
Np — — — 2.0m

AREA 2—Central S England

TIDAL CURVES — SWANAGE TO SELSEY

Tidal curves for places between Swanage and Selsey are given below, and their use is explained above. In this area the times of Low Water are defined more sharply than the times of High Water, and the curves are therefore drawn with their times relative to Low Water instead of High Water.

Apart from referring the times to Low Water, the procedure for obtaining intermediate heights with these curves is the same as that used for normal Secondary Ports (see 9.4.2). For most places a third curve is shown, for the range at Portsmouth at which the two High Waters are equal at the port concerned: for interpolation between the curves see the previous page.

Note 1. Owing to constrictions in the River Medina, Newport requires special treatment. The calculation should be made using the time and Low Water height differences for Cowes, and the High Water height differences for Newport. Any height which falls below 1.9m can be disregarded since the tide never falls below this level.

Note 2. Calshot is referred to Southampton as Standard Port. The curves shown below for Calshot should be taken as referring to the Spring and Neap ranges at Southampton (4.0 and 1.9m respectively).

Note 3. Wareham and Tuckton LWs do not fall below 0.7m except under very low river flow conditions.

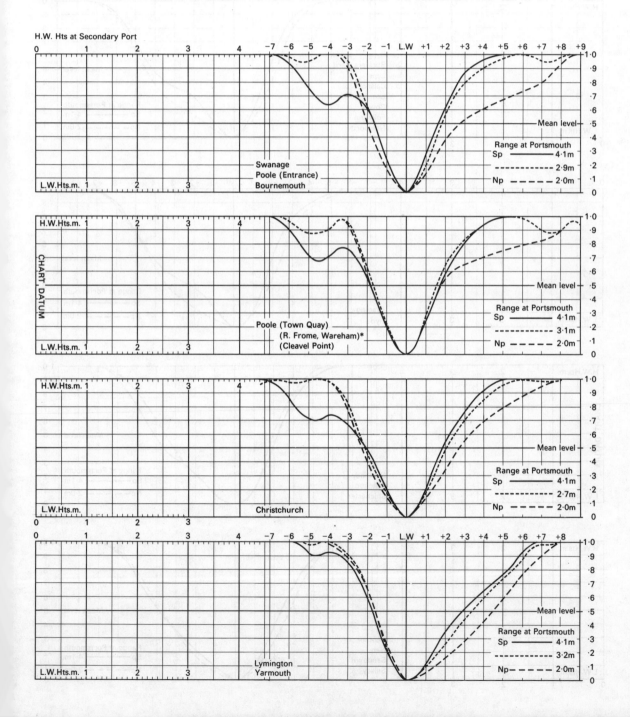

Harbour, Coastal and Tidal Information

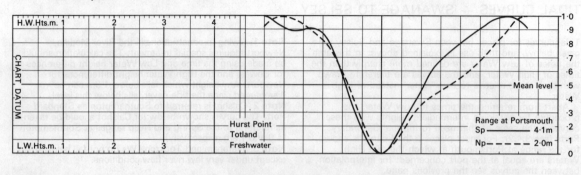

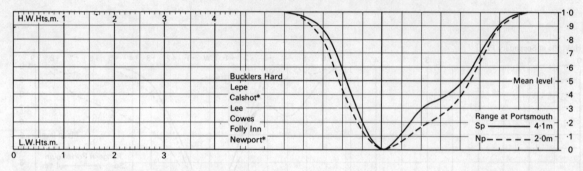

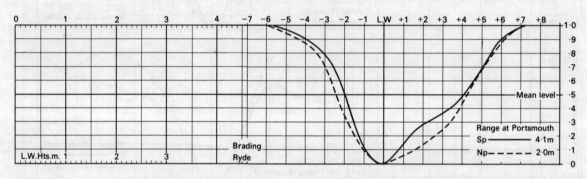

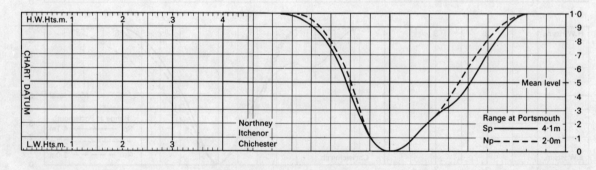

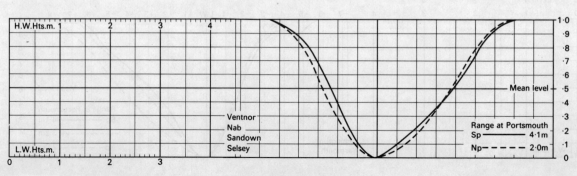

AREA 2—Central S England

SWANAGE TO NAB TOWER

FIG 10A (To be used for finding intermediate heights in metres for places named)

Place	Height of H.W. at Portsmouth	HOURLY HEIGHTS ABOVE CHART DATUM AT THE PLACE													Height of L.W. at Portsmouth
		Hours before or after HIGH WATER AT PORTSMOUTH							Hours before or after LOW WATER AT PORTSMOUTH						
		3b.	2b.	1b.	H.W.	1a.	2a.	3a.	2b.	1b.	L.W.	1a.	2a.	3a.	
	m.	m.	m.	m.	m.	m.	m.	m.	m.	m.	m.	m.	m.	m.	m.
SWANAGE	4·7	1·9	1·8	1·5	1·3	1·4	1·2	0·8	0·6	0·3	0·8	1·3	1·7	1·8	0·6
	4·3	1·6	1·6	1·5	1·5	1·2	1·4	1·1	0·9	0·8	1·0	1·3	1·5	1·6	1·2
	3·8	1·5	1·5	1·5	1·5	1·6	1·5	1·3	1·2	1·1	1·2	1·3	1·4	1·4	1·8
POOLE ENTRANCE	4·7	2·0	1·9	1·6	1·4	1·5	1·4	0·9	0·7	0·3	0·8	1·3	1·6	1·9	0·6
	4·3	1·6	1·6	1·5	1·5	1·5	1·5	1·1	1·0	0·8	0·9	1·2	1·5	1·6	1·2
	3·8	1·4	1·5	1·5	1·5	1·6	1·5	1·3	1·2	1·1	1·1	1·2	1·3	1·4	1·8
POOLE BRIDGE	4·7	2·2	2·2	1·9	1·6	1·6	1·7	1·3	1·1	0·6	0·5	1·0	1·6	1·9	0·6
	4·3	1·8	1·7	1·6	1·7	1·8	1·8	1·5	1·4	1·0	0·8	1·1	1·5	1·7	1·2
	3·8	1·5	1·5	1·6	1·7	1·8	1·8	1·5	1·5	1·3	1·2	1·3	1·5	1·5	1·8
BOURNE-MOUTH	4·7	2·1	2·0	1·7	1·5	1·5	1·5	0·9	0·7	0·3	0·6	1·2	1·7	1·9	0·6
	4·3	1·6	1·7	1·6	1·6	1·6	1·5	1·2	1·0	0·8	0·8	1·2	1·4	1·5	1·2
	3·8	1·5	1·5	1·5	1·5	1·6	1·5	1·4	1·3	1·1	1·1	1·2	1·3	1·4	1·8
CHRIST-CHURCH HARBOUR †	4·7	1·8	1·8	1·5	1·3	1·5	1·1	0·7	0·7	0·5	0·4	0·9	1·3	1·6	0·6
	4·3	1·5	1·5	1·4	1·5	1·5	1·2	0·9	0·8	0·6	0·6	1·0	1·2	1·4	1·2
	3·8	1·2	1·2	1·3	1·4	1·4	1·1	0·9	0·8	0·6	0·7	0·9	1·0	1·1	1·8
FRESHWATER BAY	4·7	2·4	2·5	2·4	2·2	2·2	1·9	1·1	0·9	0·5	0·8	1·5	2·0	2·2	0·6
	4·3	2·2	2·3	2·2	2·2	2·2	1·9	1·4	1·2	0·9	1·1	1·5	1·8	2·0	1·2
	3·8	1·9	2·0	2·2	2·2	2·2	2·0	1·6	1·6	1·3	1·3	1·6	1·7	1·8	1·8
TOTLAND BAY	4·7	2·3	2·5	2·4	2·3	2·3	2·1	1·4	1·1	0·5	0·8	1·4	1·8	2·1	0·6
	4·3	2·1	2·3	2·3	2·3	2·2	2·1	1·6	1·4	0·9	1·1	1·4	1·7	1·9	1·2
	3·8	2·0	2·1	2·3	2·3	2·2	2·1	1·8	1·7	1·4	1·4	1·6	1·8	1·8	1·8
HURST POINT	4·7	2·3	2·6	2·7	2·5	2·5	2·3	1·6	1·2	0·5	0·7	1·3	1·7	2·0	0·6
	4·3	2·0	2·3	2·5	2·5	2·4	2·3	1·7	1·5	1·1	1·0	1·4	1·7	1·9	1·2
	3·8	1·9	2·1	2·3	2·3	2·3	2·2	1·8	1·7	1·4	1·3	1·5	1·7	1·9	1·8
YARMOUTH I.O.W.	4·7	2·4	2·8	3·0	2·8	2·8	2·7	1·8	1·5	0·7	0·8	1·4	1·8	1·8	0·6
	4·3	2·2	2·5	2·7	2·7	2·7	2·6	1·9	1·8	1·1	1·2	1·6	1·8	2·0	1·2
	3·8	2·0	2·3	2·5	2·5	2·5	2·3	1·9	1·7	1·5	1·5	1·6	1·7	1·8	1·8
LYMINGTON	4·7	2·2	2·6	3·0	2·8	2·9	2·8	2·1	1·7	0·7	0·5	1·1	1·6	1·9	0·6
	4·3	2·0	2·3	2·7	2·7	2·7	2·6	2·1	1·8	1·1	1·0	1·3	1·6	1·7	1·2
	3·8	1·9	2·2	2·4	2·5	2·5	2·4	2·0	1·8	1·5	1·4	1·5	1·6	1·7	1·8
SOLENT BANKS	4·7	2·4	2·9	3·4	3·3	3·2	3·0	2·2	1·8	0·7	0·6	1·2	1·8	2·0	0·6
	4·3	2·2	2·6	3·0	3·1	3·0	2·9	2·2	1·9	1·2	1·1	1·4	1·6	1·9	1·2
	3·8	2·1	2·3	2·6	2·7	2·7	2·6	2·2	2·0	1·6	1·5	1·7	1·9	2·0	1·8
COWES ROAD	4·7	2·5	3·4	4·1	4·2	4·1	3·8	3·0	2·5	1·1	0·6	1·2	1·8	2·1	0·6
	4·3	2·4	3·1	3·7	3·8	3·7	3·5	2·8	2·6	1·5	1·2	1·5	1·8	2·1	1·2
	3·8	2·5	3·0	3·3	3·4	3·4	3·2	2·7	2·5	2·0	1·7	1·9	2·0	2·2	1·8
CALSHOT CASTLE	4·7	2·6	3·6	4·3	4·4	4·3	4·1	3·2	2·6	1·2	0·7	1·3	1·9	2·2	0·6
	4·3	2·6	3·3	3·8	4·0	4·0	3·7	3·0	2·6	1·6	1·2	1·6	2·0	2·2	1·2
	3·8	2·7	3·2	3·5	3·6	3·6	3·4	2·8	2·6	2·0	1·9	2·0	2·2	2·3	1·8
LEE-ON-SOLENT	4·7	2·7	3·6	4·4	4·5	4·4	4·2	3·1	2·5	1·1	0·6	1·2	1·8	2·1	0·6
	4·3	2·7	3·4	3·9	4·1	4·0	3·7	2·9	2·5	1·4	1·2	1·6	2·0	2·2	1·2
	3·8	2·7	3·2	3·6	3·7	3·6	3·3	2·8	2·6	2·1	1·9	2·0	2·2	2·4	1·8
RYDE	4·7	2·7	3·7	4·3	4·5	4·3	4·0	2·9	2·4	1·1	0·7	1·2	1·8	2·1	0·6
	4·3	2·7	3·4	4·0	4·1	4·0	3·7	2·9	2·6	1·6	1·3	1·6	1·9	2·2	1·2
	3·8	2·7	3·2	3·6	3·7	3·6	3·4	2·9	2·7	2·1	1·9	2·0	2·2	2·4	1·8
NAB TOWER	4·7	2·9	3·8	4·4	4·5	4·3	3·6	2·4	2·0	0·9	0·6	1·0	1·5	2·1	0·6
	4·3	2·9	3·6	4·1	4·2	4·0	3·4	2·5	2·2	1·4	1·2	1·4	1·8	2·2	1·2
	3·8	2·9	3·3	3·7	3·7	3·6	3·1	2·6	2·3	1·9	1·7	1·8	2·1	2·4	1·8
SANDOWN	4·7	2·6	3·3	3·8	4·0	3·8	3·3	2·1	1·7	0·8	0·8	1·0	1·5	1·9	0·6
	4·3	2·6	3·0	3·5	3·6	3·5	3·1	2·2	1·9	1·3	1·1	1·4	1·7	2·0	1·2
	3·8	2·6	2·9	3·2	3·3	3·1	2·8	2·3	2·1	1·8	1·6	1·7	1·9	2·2	1·8
VENTNOR	4·7	2·8	3·3	3·7	3·8	3·5	2·9	2·0	1·7	1·0	0·9	1·3	1·7	2·1	0·6
	4·3	2·6	3·1	3·4	3·4	3·2	2·8	2·1	1·9	1·4	1·3	1·6	1·9	2·2	1·2
	3·8	2·5	2·8	3·1	3·1	3·0	2·8	2·3	2·2	1·8	1·7	1·9	2·1	2·3	1·8

Note.—Area enclosed by pecked lines represents the period during which the tide stands, or during which a second high water may occur.
† Heights at Christchurch are for inside the bar; outside the bar, L.W. falls about 0·6 metres lower at Springs.

230 Harbour, Coastal and Tidal Information

ISLE OF WIGHT TIDAL STREAMS

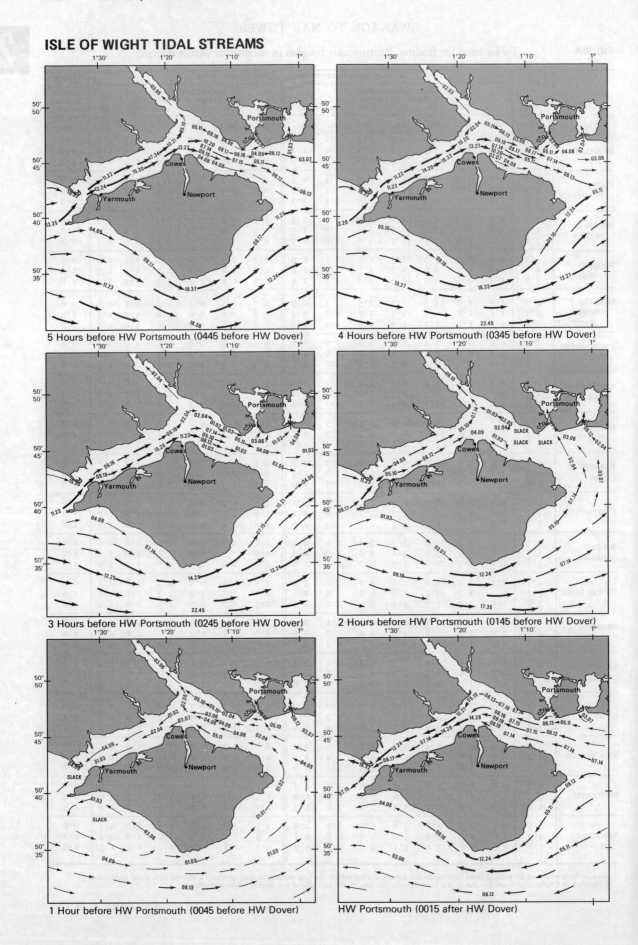

AREA 2—Central S England 231

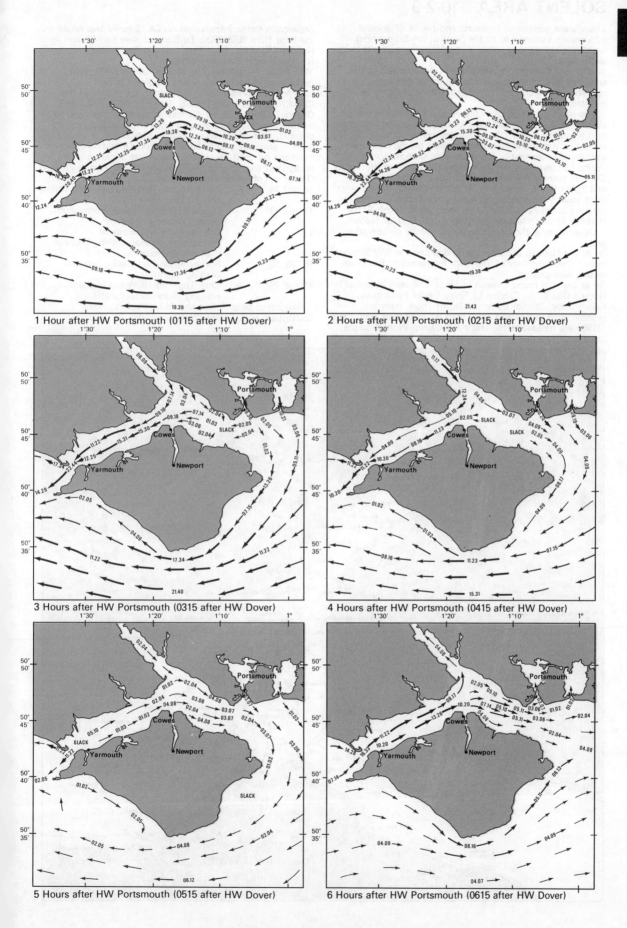

SOLENT AREA 10-2-9

There are a number of problems and pieces of general information common to all the ports in the Solent area which are collected together in the following paragraphs to avoid repetition.

Yachts and Commercial Shipping. In the interests of the safety of all concerned it is important that good co-operation between yachts and commercial shipping be maintained.

The most crucial point in the central Solent is in the vicinity of the NE Gurnard and W Bramble buoys where the turn to and from the Thorn Channel has to be made. It should be noted that large ships inward bound from the east usually turn first to the southward on passing the Prince Consort buoy before starting the turn into the Thorn Channel somewhere between the Royal Yacht Squadron and Egypt Point, depending on tide and wind. When very large vessels are entering or leaving the Thorn Channel, they are usually preceded by an Associated British Ports Harbour Master's launch (see also 'Local Signals' below).

Other areas in the Solent requiring special co-operation are:—
(a) at the Needles between the Bridge and the SW Shingles buoys; there is plenty of sea-room for yachts to the south of the main channel.
(b) at the turn round Calshot Lt Float. The area bounded by Bourne Gap buoy, Calshot Lt Float, Castle Point buoy, Reach buoy, Calshot buoy and North Thorn buoy is prohibited to yachts and small craft when vessels over 100m (329ft) are navigating the channel between West Bramble buoy and Hook buoy. Southampton Port Radio (SPR) broadcasts traffic information on Ch 12 every two hours on the hour from 0600-2200, Fri-Sun inclusive and Bank Holiday Mondays, from Easter to end September.
(c) Portsmouth Harbour entrance.

VHF Radio Telephone. The proliferation of VHF radio telephones in yachts is causing problems in port communication networks. Yachtsmen are reminded that CHANNEL 16 is a DISTRESS, SAFETY AND CALLING Channel. Other than for distress, it must only be used to establish contact before going onto a recognised working channel. If another calling channel is available, it should be used in preference to Channel 16. Port Operation channels must not be used for ship to ship communication. The recognised ship to ship channels are 06, 08, 72 and 77. Port Operations channels are as follows:—

12 and 14	— Southampton Port Radio general calling and working channels
11 and 13	— Queen's Harbour Master Portsmouth general working channels
11	— Commercial Harbourmaster Portsmouth general working channel
09	— Pilot working channel
18, 20, 22	— Southampton Port Radio, selected working channels
71 and 74	— Ship/Tug/Pilot/Berthing Master working channels (keep off)

Reports of oil pollution should be made to HM Coastguard (call *Solent Coastguard* VHF Ch 16 67).

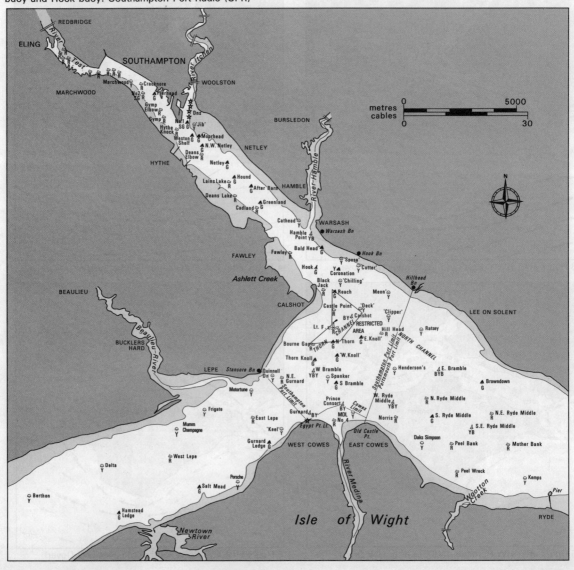

AREA 2—Central S England 233

SOLENT AREA WAYPOINTS

Waypoints marked by an asterisk (*) are special (yellow) racing marks, which may be removed in winter. Other waypoints are navigational buoys, unless otherwise stated.

Name	Lat	Long
Beaulieu Spit Bn	50°46'.83N	01°21'.67W
Bembridge Ledge	50°41'.12N	01°02'.74W
Bembridge tide gauge	50°42'.55N	01°05'.24W
*Berthon	50°44'.18N	01°29'.13W
Boulder (Looe Channel)	50°41'.53N	00°49'.00W
Bourne Gap	50°47'.81N	01°18'.25W
Boyne	50°46'.12N	01°05'.19W
Bramble Bn	50°47'.38N	01°17'.06W
Browndown	50°46'.54N	01°10'.87W
Bridge	50°39'.59N	01°36'.80W
Calshot	50°48'.38N	01°16'.95W
Calshot Spit Lt Float	50°48'.32N	01°17'.55W
*Champagne Mumm	50°45'.60N	01°23'.03W
Chichester Bar Bn	50°45'.88N	00°56'.37W
*Clipper	50°48'.41N	01°15'.65W
Cowes Breakwater Lt	50°45'.84N	01°17'.43W
Cowes No 4	50°46'.04N	01°17'.78W
*Daks-Simpson	50°45'.50N	01°14'.30W
*Deck	50°48'.60N	01°16'.57W
*Delta	50°45'.22N	01°27'.08W
Durns Pt obstn (S end)	50°45'.37N	01°26'.95W
East Bramble	50°47'.20N	01°13'.55W
East Knoll	50°47'.93N	01°16'.75W
East Lepe	50°46'.09N	01°20'.81W
East Winner	50°45'.07N	01°00'.01W
*Frigate	50°46'.10N	01°22'.10W
Greenland	50°51'.08N	01°20'.30W
Gurnard	50°46'.18N	01°18'.76W
Gurnard Ledge	50°45'.48N	01°20'.50W
Hamble Point	50°50'.12N	01°18'.58W
Hamstead Ledge	50°43'.83N	01°26'.10W
*Hendersons	50°47'.23N	01°15'.82W
Hill Head	50°48'.02N	01°15'.91W
Horse Sand	50°45'.49N	01°05'.18W
Horse Sand Fort Lt	50°44'.97N	01°04'.25W
Hound	50°51'.66N	01°21'.45W
*Keel	50°45'.82N	01°19'.60W
*Kemps	50°45'.15N	01°09'.55W
Langstone Fairway	50°46'.28N	01°01'.27W
*MDL	50°46'.12N	01°16'.55W
*Meon	50°49'.12N	01°15'.63W
Mixon Bn	50°42'.35N	00°46'.21W
Mother Bank	50°45'.45N	01°11'.13W
Motortune	50°46'.55N	01°21'.37W
*Mumm Champagne	50°45'.60N	01°23'.03W
Nab 1	50°41'.23N	00°56'.43W
Nab 2	50°41'.70N	00°56'.71W
Nab 3	50°42'.17N	00°57'.05W
Nab Tower	50°40'.05N	00°57'.07W
Needles Fairway	50°38'.20N	01°38'.90W
Netley	50°51'.99N	01°23'.16W
New Grounds	50°41'.96N	00°58'.52W
No Mans Land Fort Lt	50°44'.37N	01°05'.61W
Norris	50°45'.94N	01°15'.42W
North East Gurnard	50°47'.03N	01°19'.34W
North East Mining Ground	50°44'.71N	01°06'.30W
North East Ryde Middle	50°46'.18N	01°11'.80W
North East Shingles	50°41'.93N	01°33'.32W
North Head	50°42'.65N	01°35'.42W
North Ryde Middle	50°46'.57N	01°14'.28W
North Sturbridge	50°45'.31N	01°08'.15W
North Thorn	50°47'.89N	01°17'.75W
No 2 Yarmouth (mooring)	50°42'.73N	01°29'.54W
No 4 Portsmouth	50°46'.98N	01°06'.27W
Outer Spit	50°45'.55N	01°05'.41W
Peel Bank	50°45'.47N	01°13'.25W
Peel Wreck	50°44'.88N	01°13'.34W
Poole Fairway	50°38'.97N	01°54'.78W
*Porsche	50°44'.60N	01°21'.80W
Prince Consort	50°46'.38N	01°17'.48W
*Quinnell	50°47'.03N	01°19'.78W
*Ratsey	50°47'.63N	01°13'.57W
Reach	50°49'.02N	01°17'.57W
Ryde Pier, head	50°44'.35N	01°09'.51W
RYS flagstaff	50°45'.97N	01°17'.97W
Salt Mead	50°44'.49N	01°22'.95W
Sconce	50°42'.50N	01°31'.35W
Shingles Elbow	50°40'.31N	01°35'.92W
South Bramble	50°46'.95N	01°17'.65W
South East Ryde Middle	50°45'.90N	01°12'.00W
South Ryde Middle	50°46'.10N	01°14'.07W
South West Mining Ground	50°44'.63N	01°07'.95W
South West Shingles	50°39'.53N	01°37'.24W
*Spanker	50°47'.08N	01°17'.99W
Spit Sand Fort Lt	50°46'.20N	01°05'.85W
Thorn Knoll	50°47'.47N	01°18'.35W
Warden	50°41'.46N	01°33'.48W
Warner	50°43'.83N	01°03'.92W
West Bramble	50°47'.15N	01°18'.55W
West Knoll	50°47'.52N	01°17'.68W
West Lepe	50°45'.20N	01°24'.00W
West Princessa	50°40'.12N	01°03'.58W
West Ryde Middle	50°46'.45N	01°15'.70W
Weston Shelf	50°52'.68N	01°23'.16W
Wootton Bn	50°44'.46N	01°12'.08W

SOLENT AREA continued

Sailing Information from BBC Radio Solent
999 kHz (300m) and 96.1 MHz VHF.
1359 kHz (221m) in Bournemouth area.
A. Local weather forecasts.
B. Shipping forecasts for Channel sea areas, with general synopsis and Coastal Station reports.
C. Coastguard Station reports.
D. Tidal details.
E. Shipping movements.
F. Gunnery range firing times.

MONDAY TO FRIDAY	SATURDAY
0633 — A B D	0630 — A B D E
0659 — A	0659 — A
0733 — A*	0730 — A* B C D E F
0735 — E	0759 — A
0745 — B C D F	0835 — A* B C D E F
0759 — A	0859, 1000, 1104, 1204,
0833 — A*	1259, 1404, 1533 (approx) } A
0859, 1004, 1104,	1804
1204, 1259, 1404 } A	1925 — A*
1504, 1604, 1704	
1725 — A* D F	**SUNDAY**
1739 — A	0659 — A
1825 (Fridays only) — A*	0730 — A* B C D E F
	0759 — A
	0850 — A*
	0859 — A
*Live forecast from	0930, 1030 (approx),
Southampton Weather	1204, 1304 } A
Centre.	1504 — A*
	1804 — A

The above forecasts can be heard on the telephone by dialling Southampton 8030 or Portsmouth 8030 at the appropriate times.

Local Signals
Outward bound vessels normally hoist the following flag signals during daylight hours.

Signal	Meaning
International 'E' Flag over Answering Pendant	I am bound East (Nab Tower)
Answering Pendant over International 'W' Flag	I am bound West (The Needles)

Southampton Patrol launch by day has Harbour Master painted on after cabin in Black lettering on Yellow background. By night it has a blue all round light above the white masthead light.

HM Coastguard — Solent District. The Maritime Rescue Sub Centre (MRSC) at Lee-on-Solent (Tel. Lee-on-Solent 552100) controls all Search and Rescue (SAR) activities in HM Coastguard Solent District.

Anyone wishing to obtain the services of any rescue authority should contact the Coastguard and not the rescue authority itself. The Coastguard is able to call on lifeboats and also on the fast inshore boats stationed at Hamble, Lepe, Gosport, Ryde, Cowes, Freshwater Bay and Shanklin; he can also call on the police launch *Ashburton*, helicopters, ships at sea, naval vessels and RAF long range aircraft.

There is an auxiliary Coastguard yacht section made up of yachtsmen willing to assist the Coastguards by making their vessels available for lifesaving purposes; they supplement the lifeboat service. Many of these boats have the Coastguard VHF frequency fitted and so are in touch with all the rescue services. They are all registered and are attached to the MRSC. MRCCs and MRSCs broadcast strong wind warnings and local forecasts.

The area of responsibility for MRSC Solent is bounded on the West by a line from Hengistbury Head South to the Anglo/French median line, and on the East by a line from Beachy Head to the Greenwich buoy.

Radio sets under the control of MRSC Solent (call: *Solent Coastguard*) are:

VHF (Ch 0 06 10 16 67 73)	VHF DF	MF (2182 3023 5680 kHz)
Needles	Boniface	Lee-on-Solent
Boniface	Selsey	
Lee-on-Solent	Newhaven	
Selsey	NB. Call	
Newhaven	Portland Coastguard for fix from Highdown, IOW	

To contact Solent Coastguard call on Ch 16; yachtsmen will be transferred to Ch 67 or 73 for traffic.

Sector Stations manned by regular Coastguards and Auxiliaries are sited at Totland (Tel. Isle of Wight 753451), Bembridge (Tel. Isle of Wight 873943), Calshot (Tel. Portsmouth 840200), Hayling (Tel. Portsmouth 464095), Shoreham (Tel. Shoreham 2226), Newhaven (Tel. Newhaven 514008) and Littlehampton (Tel. Littlehampton 715512).

Auxiliary Stations are sited at Needles, Atherfield, Ventnor, Eastney and Selsey.

All Sector and Auxiliary Lookouts are manned for Casualty Risk or heavy traffic periods.

Note: Much useful information is given in the *Solent Year Book* published by the Solent Cruising and Racing Association (SCRA).

AGENTS NEEDED
There are a number of ports where we need agents, particularly in France.
ENGLAND Swale, Havengore, Berwick.
SCOTLAND Firth of Forth, Scrabster, Mallaig, Loch Sunart, Loch Aline.
IRELAND Kilrush, Wicklow, Westport/Clew Bay.
FRANCE Arcachon, Seudre R, Ile d'Oleron, Rochfort, Ile de Re, St. Giles-Croix-de-Vie, Ile d'Yeu, Pouliguen, Le Croisic, La Forêt, Ile de Bréhat.
GERMANY Norderney, Dornumer-Accumersiel.
If you are interested in becoming our agent for any of the above, please write to the editors and get your free copy of the Almanac every year. You do not have to be resident in a port to be the agent, but at least a fairly regular visitor.

AREA 2—Central S England 235

LYME REGIS 10-2-10
Dorset

CHARTS
Admiralty 3315; Stanford 12; Imray C5; OS 193
TIDES
−0455 Dover; ML 2.4; Duration 0700; Zone 0 (GMT).
Standard Port DEVONPORT (←—)

Times				Height (metres)			
HW		LW		MHWS	MHWN	MLWN	MLWS
0100	0600	0100	0600	5.5	4.4	2.2	0.8
1300	1800	1300	1800				

Differences LYME REGIS
+0040 +0100 +0005 −0005 −1.2 −1.3 −0.5 −0.2

NOTE: Rise is relatively fast for the first hour after LW, then a slackening for the next 1½ hours after which the rapid rate is resumed — there is usually a stand of up to 1½ hours at HW.

SHELTER
The harbour, known as The Cobb, offers excellent shelter except in strong winds from E or SE. Harbour dries at LW. Max length allowed in harbour 9m. Keel boats can dry out alongside Victoria Pier or anchor as shown. Six spherical R visitors buoys available. Access about HW∓2½.
NAVIGATION
Waypoint 50°43'.00N 02°55'.60W, 116°/296° from/to front Ldg Lt. Rock breakwater south of entrance marked by R can beacon. Beware numerous fishing floats in all directions. Red flag on flagpole on Victoria Pier indicates 'Gale warning in force'.
LIGHTS AND MARKS
Ldg Lts 296°.
RADIO TELEPHONE (local times)
Call: *Lyme Regis Harbour Radio* Ch 16; 14. (May-Sept: 0900-1200, 1600-1800).
TELEPHONE (029 74)
Hr Mr 2137; MRSC Portland 820441; HMC Freefone Customs Yachts: RT (0305) 71189; Marinecall 0898 500 457; Dr 2696 & 2263.
FACILITIES
EC Thursday; **Harbour (The Cobb)** Tel. 2137, Slip, M, P* and D (cans), L, FW, ME*, El*, Sh*, AB, V, R, Bar; **Lyme Regis Sailing Club** Tel. 3573, FW, R, Bar; **Lyme Regis Power Boat Club** Tel. 2941, R, Bar; **Axminster Chandlery** Tel. Axminster 33980, CH, ACA.
Town V, R, Bar; Gas, Gaz, PO; Bank; Rly (bus to Axminster); Air (Exeter).
* See Hr Mr.

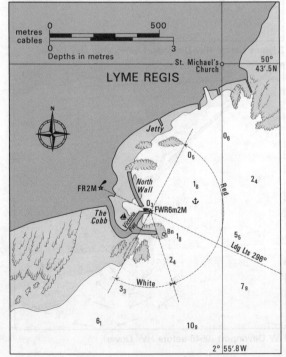

BRIDPORT 10-2-11
Dorset

CHARTS
Admiralty 3315; Stanford 12; Imray C5; OS 193
TIDES
−0500 Dover; ML 2.4; Duration 0650; Zone 0 (GMT).
Standard Port DEVONPORT (←—)

Times				Height (metres)			
HW		LW		MHWS	MHWN	MLWN	MLWS
0100	0600	0100	0600	5.5	4.4	2.2	0.8
1300	1800	1300	1800				

Differences BRIDPORT (West Bay)
+0025 +0040 0000 0000 −1.4 −1.4 −0.6 −0.2

NOTE: Rise is relatively fast for first hour after LW thence a slackening for the next 1½ hours after which the rapid rise is resumed. There is usually a stand of up to 1½ hours at HW.

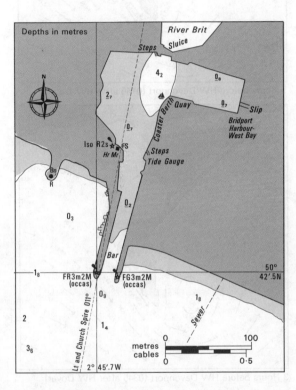

SHELTER
Good once inside the 12m wide, 180m long entrance, but entrance becomes dangerous in even moderate on-shore winds.
NOTE: Bridport town is 1½ M inland of the harbour.
NAVIGATION
Waypoint 50°42'.20N 02°46'.07W, 210°/030° from/to entrance, 0.37M. Concrete block and unlit BY buoy about ¾ M S of entrance indicates end of sewer line. Pier lights are only switched on for commercial vessels. It is safe to enter HW∓3, in favourable weather.
LIGHTS AND MARKS
Harbour can be identified at night by Iso R Lt 2s on roof of Hr Mr office at the shore end of W pier.
Entry signals: Night — No lights on pier — unfit to enter. Day — B ball if port closed.
RADIO TELEPHONE
Call: *Bridport Radio* VHF Ch 16; 11 12 14.
TELEPHONE (0308)
Hr Mr 23222; CG & MRSC Portland 820441; HMC Freefone Customs Yachts: RT (0305) 71189; Marinecall 0898 500 457; Dr 23771.
FACILITIES
Quay Tel. 23222, Access HW∓3, Slip, P, Lau, Sh, M, FW, ME, CH, AB, V; **Lyme Bay Marine** Tel. 22347, CH; **D. Ackerman** Tel. 56815, ME. **Town** P, D, V, R, Bar; PO; Bank; Rly (bus to Axminster); Air (Exeter).

PORTLAND 10-2-12 TIDAL STREAMS
Dorset

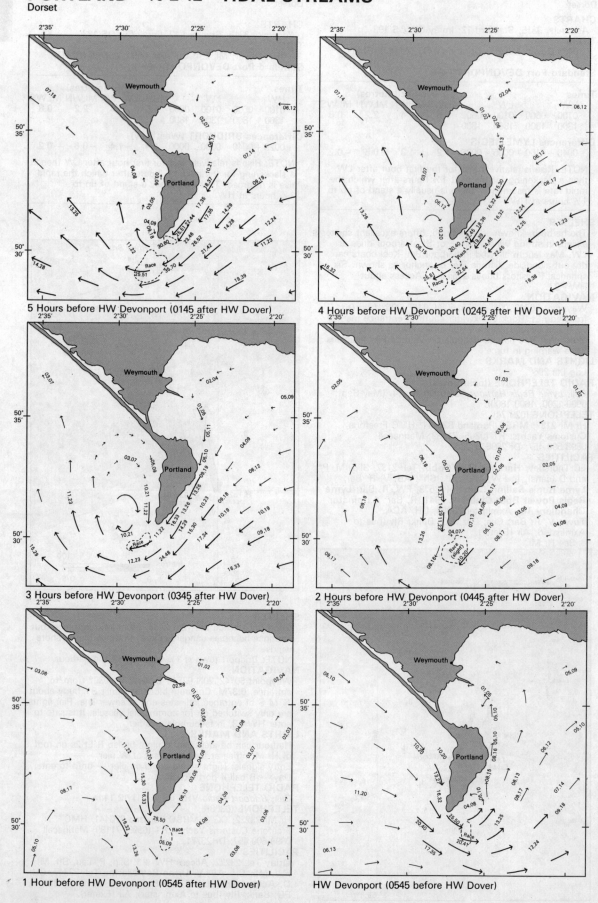

AREA 2 – Central S England

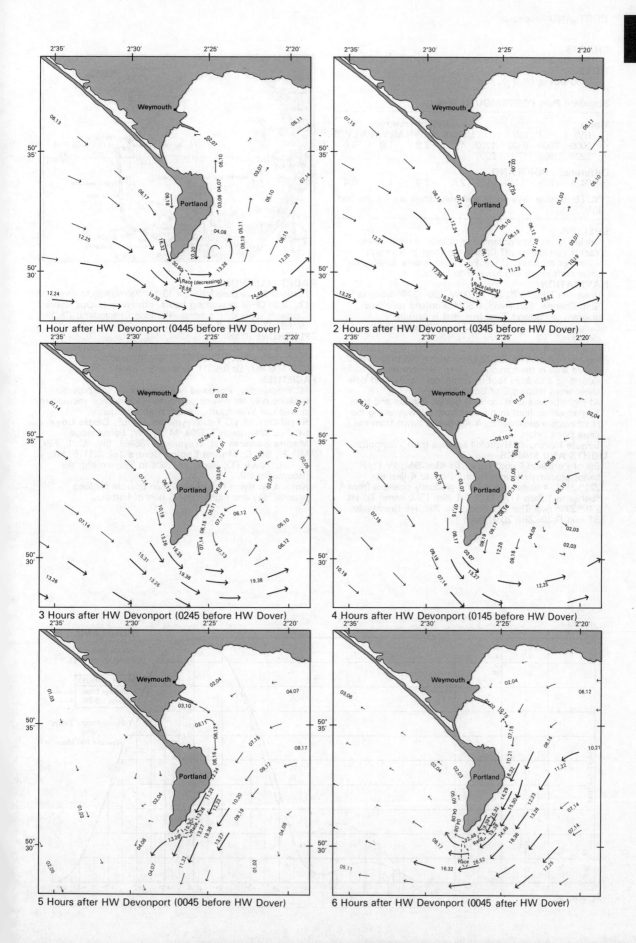

PORTLAND continued

CHARTS
Admiralty 2255, 2268; Stanford 12; Imray C4, C5; OS 194
TIDES
−0430 Dover; ML 1.0; Zone 0 (GMT).

Standard Port PORTSMOUTH (→)

Times				Height (metres)			
HW		LW		MHWS	MHWN	MLWN	MLWS
0000	0600	0500	1100	4.7	3.8	1.8	0.6
1200	1800	1700	2300				
Differences PORTLAND							
−0421	−0525	−0520	−0515	−2.6	−2.4	−1.1	−0.4

NOTE: Double LWs occur and predictions are for the first LW.

SHELTER
Shelter is poor in Portland due to lack of wind breaks. Yachts can anchor off Castletown (best in S or SW winds) or off Castle Cove. East Fleet is only suitable for small craft which can lower their masts.

NAVIGATION
Waypoint 50°35'.07N 02°24'.00W, 090°/270° from/to East Ship Channel, Fort Head, 0.50M. Portland race (see diagram of Portland Tidal Streams) is extremely dangerous. The South Ship Channel is permanently closed.
Due to many naval movements in vicinity, yachts should keep watch on VHF Ch 13 when within 3M of 'A' Head. Speed limit in the harbour is 12 kn. Beware reef of rocks extending 1 ca from foot of Sandsfoot Castle, and new shoal areas forming E of Small Mouth.
At night beware unlit mooring buoys, lighters and rafts. Torpedoes are fired to the east from midway along the Northeastern Breakwater. A red flag is flown from the firing point before firing.
Beware Hovercraft/Hydrofoil services from Weymouth.

LIGHTS AND MARKS
Bill of Portland (S end) Fl (4) 20s 43m 29M; W Tr, R band; gradually changes from 1 flash to 4 flashes 221°-244°, 4 flashes 244°-117°, gradually changes from 4 flashes to 1 flash 117°-141°. FR 19m 13M; same Tr; vis 271°-291° over The Shambles; Dia 30s. NE Breakwater, SE end Fl 10s 22m 20M; Horn 10s.

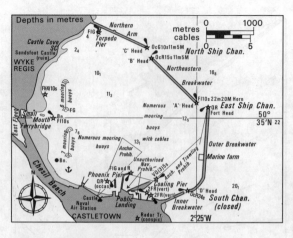

RADIO TELEPHONE
Portland Naval Base VHF Ch 13 14. Permission to enter, Ch 13 to QHM in working hours, otherwise to Ops Room. Casquets separation scheme, Portland Coastguard: Ch 16 69; 69. 67 (for small craft).

TELEPHONE (0305)
QHM 820311; CG & MRSC 820441; HMC Freefone Customs Yachts: RT (0305) 71189; Marinecall 0898 500 457; Dr 820311 (Emergency only).

FACILITIES
EC Wednesday; **Portland** is a Naval Base and provides facilities only for service personnel. Non-service yachtsmen should use Weymouth, except in an emergency.
Royal Dorset YC Tel. Weymouth 771155; **Castle Cove SC** Tel. Weymouth 783708, M, L, FW; **Ferrybridge Marine Services** Tel. Weymouth 786463, Slip, M, L, FW, ME, El, Sh, C; **Marine Engine Centre** Tel. 821175, ME, El, CH. **Town** PO; Bank; Rly (bus to Weymouth); Air (Bournemouth).
Note: Low flying helicopters often operate for long periods, day and night, over S part of harbour.

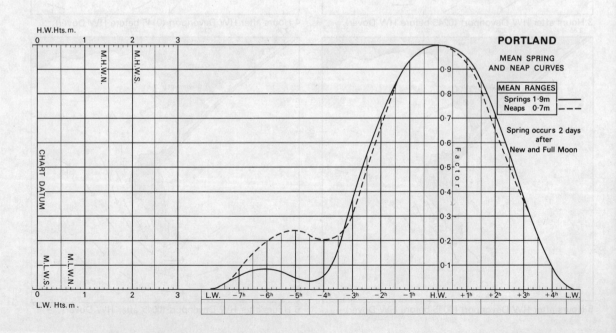

PORTLAND MEAN SPRING AND NEAP CURVES

MEAN RANGES
Springs 1·9m
Neaps 0·7m

Spring occurs 2 days after New and Full Moon

WEYMOUTH 10-2-13
Dorset

CHARTS
Admiralty 2172, 2255, 2268; Stanford 12; Imray C4, C5; OS 194

TIDES
−0438 Dover; ML 1.1; Zone 0 (GMT).

Standard Port PORTSMOUTH (→)

Times				Height (metres)			
HW		LW		MHWS	MHWN	MLWN	MLWS
0000	0600	0500	1100	4.7	3.8	1.8	0.6
1200	1800	1700	2300				

Differences WEYMOUTH
−0421 −0525 −0520 −0515 −2.6 −2.4 −1.1 −0.4

NOTE: Double LWs occur. Predictions are for first LW.

SHELTER
Swell runs up the harbour if strong wind from E, otherwise shelter is good. Visitors' berth in Cove area or as directed.

NAVIGATION
Waypoint 50°36'.74N 02°26'.00W, 059°/239° from/to front Ldg Lt, 0.60M. Principal dangers are ferries; they sometimes put warps across the harbour. Town Bridge will open on request (2 hours notice) 0900-1300, 1400-1700, 1800-2000 (LT). Clearance under is 2.7m to 3.8m at HW, 4.6 to 5.2m at LW.
NOTE: Due to the eddy effect, tide in Weymouth Roads is westerly at all times except HW −0510 to HW −0310. Speed limit in the harbour is 'Dead Slow'.

LIGHTS AND MARKS
Ldg Lts 237°. Traffic signals are shown from position near root of S pier.

3 Fl R	Port closed (emergency)
2 R over 2 G	Entry & departure forbidden
G W G (vert)	Vessels may only proceed with specific orders
3 G	Departure forbidden
3 R	Entry forbidden
No signals	All clear to enter or leave

RADIO TELEPHONE
VHF Ch 16; 12 (when vessel expected). (See Portland).

TELEPHONE (0305)
Hr Mr 760620; CG and MRSC 820441; HMC Freefone Customs Yachts: RT (0752) 669811; Marinecall 0898 500 457; Hosp 772211;

FACILITIES
EC Wednesday; **Weymouth Marina** (580) Tel. 773538; **Outer Harbour** M; **Weymouth SC** Tel. 785481, M, Bar; **Custom House Quay** D, FW, AB; **British Rail Slipway** Tel. 786363 ext. 53, Slip, ME, El, Sh; **Automarine** Tel. 783408, P, D, C (10 ton); **Ladyline** Tel. 771603, CH; **Royal Dorset YC** Tel. 786258, M, Bar; **Small Boat** Tel. 782109, Sh; **R. J. Davis** Tel. 834415, ME, El, CH; **Moto-sails** Tel. 786710, CH; **Weymouth Chandlers** Tel. 771603, CH. **Town** P, D, FW, V, R, Bar. PO; Bank; Rly; Air (Bournemouth).

NOTE: If proceeding eastwards, check Lulworth firing programme — see 10.2.30.

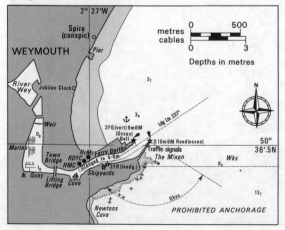

SWANAGE 10-2-14
Dorset

CHARTS
Admiralty 2172; Stanford 12; Imray C4; OS 195

TIDES
HW Spring −0235, Neap −0515 Dover; ML 1.5
+0125 +0120 Zone 0 (GMT)

Standard Port PORTSMOUTH (→)

Times				Height (metres)			
HW		LW		MHWS	MHWN	MLWN	MLWS
0000	0600	0500	1100	4.7	3.8	1.8	0.6
1200	1800	1700	2300				

Differences SWANAGE
−0250 +0105 −0105 −0105 −2.7 −2.2 −0.7 −0.3

NOTE: Double HWs occur except at neaps and predictions are for the higher HW. Near neaps there is a stand, and the predictions shown are for the middle of the stand. See 10.2.8

SHELTER
Shelter is good in winds between SW and N. It is bad in E or SE when winds over Force 4 cause swell. Over Force 6 holding becomes very difficult. Nearest haven Poole. Pier is for commercial traffic, only available for yachts on payment to piermaster.

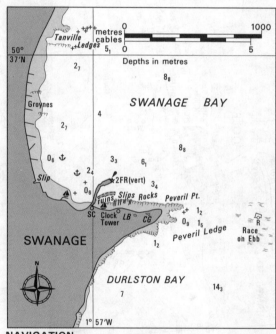

NAVIGATION
Waypoint 50°36'.70N 01°56'.50W, 054°/234° from/to Swanage Pier head, 0.30M. Beware Peveril Ledge coming from S and Tanville Ledges coming from N. On the S side of the pier there are the ruins of an old pier. It is difficult to pick up pierhead Lts at night due to confusing street lights behind.

LIGHTS AND MARKS
The only lights are 2 FR (vert) on the end of the pier. Peveril Ledge buoy, R can unlit (pass to E).

RADIO TELEPHONE
None

TELEPHONE (0929)
Piermaster 423565; CG 422596; MRSC Portland 820441; HMC Freefone Customs Yachts: RT (0703) 229251; Marinecall 0898 500 457; Dr 422676; Hosp 422282.

FACILITIES
EC Thursday; **Boat Park** (Peveril Point), Slip, FW, L; **Town Jetty**, L, AB (HW only); **Swanage SC** Tel. 422987, Slip, L, FW, Bar; **Diving facilities on Pier** Tel. 423565. **Town** P and D (cans), FW, V, R, Bar. PO; Bank; Rly (bus to Wareham); Air (Bournemouth).

POOLE 10-2-15
Dorset

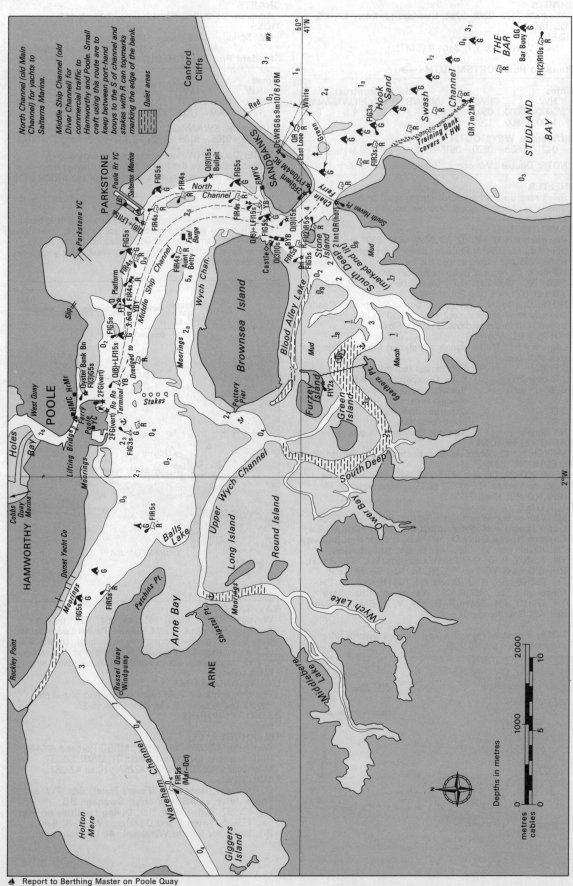

ENGLAND, SOUTH COAST - POOLE (TOWN QUAY)

LAT 50°43′N LONG 1°59′W

TIMES AND HEIGHTS OF HIGH AND LOW WATERS

YEAR 1989

TIME ZONE UT(GMT)
For Summer Time add ONE hour in non-shaded areas

JANUARY				FEBRUARY				MARCH				APRIL			
TIME	m	TIME	m	TIME	m	TIME	m	TIME	m	TIME	m	TIME	m	TIME	m
1 1041 2307 SU	1.7 1.2 1.5 1.2	**16** 1110 2347 M	1.8 0.9 1.6 1.0	**1** 1148 W	1.5 1.2 1.4	**16** 0054 1333 TH	1.2 1.5 1.0 1.6	**1** 0927 2207 W	1.5 1.1 1.4 1.2	**16** 1150 TH	1.5 1.1 1.5	**1** 1155 SA	1.4 1.0 1.5	**16** 0124 1345 SU	1.1 1.5 0.9 1.7
2 1145 M	1.6 1.2 1.5	**17** 1226 TU	1.7 0.9 1.7	**2** 0026 1307 TH	1.2 1.5 1.0 1.5	**17** 0205 1434 F	1.0 1.2 0.9 1.7	**2** 1054 TH	1.4 1.1 1.4 1.2	**17** 0042 1319 F	1.2 1.5 1.0 1.5	**2** 0043 1311 SU	1.0 1.6 0.8 1.8	**17** 0214 1428 M	0.9 1.7 0.8 1.8
3 0014 1249 TU	1.2 1.7 1.1 1.5	**18** 0103 1337 W	1.0 1.7 0.9 1.7	**3** 0137 1408 F	1.0 1.7 0.9 1.7	**18** 0256 1520 SA	0.9 1.8 0.7 1.8	**3** 1230 F	1.5 1.1 1.5	**18** 0155 1419 SA	1.0 1.5 0.9 1.7	**3** 0145 1407 M	0.8 1.8 0.6 1.9	**18** 0250 1502 TU	0.8 1.8 0.6 1.9
4 0114 1344 W	1.1 1.7 1.0 1.7	**19** 0206 1436 TH	1.0 1.8 0.8 1.8	**4** 0233 1459 SA	0.9 1.8 0.7 1.9	**19** 0338 1558 SU	0.8 1.8 0.6 1.9	**4** 0112 1343 SA	1.0 1.6 0.9 1.7	**19** 0243 1501 SU	0.9 1.7 0.7 1.8	**4** 0234 1456 TU	0.7 1.9 0.3 2.1	**19** 0321 1533 W	0.6 1.9 0.6 2.0
5 0205 1432 TH	1.0 1.8 0.9 1.8	**20** 0300 1526 F	0.9 1.8 0.7 1.9	**5** 0319 1544 SU	0.7 1.9 0.5 2.0	**20** 0412 1630 M ○	0.6 1.9 0.5 2.0	**5** 0212 1436 SU	0.9 1.8 0.6 1.9	**20** 0319 1534 M	0.8 1.8 0.6 1.9	**5** 0318 1542 W	0.4 2.1 0.2 2.2	**20** 0350 1604 TH	0.6 1.9 0.5 2.0
6 0252 1516 F	0.9 1.9 0.8 1.9	**21** 0346 1609 SA ○	0.8 1.9 0.6 1.9	**6** 0402 1626 M	0.5 2.0 0.4 2.1	**21** 0444 1701 TU	0.6 1.9 0.6	**6** 0301 1523 M	0.6 1.9 0.5 2.0	**21** 0351 1604 TU	0.6 1.9 0.5 2.0	**6** 0402 1627 TH ●	0.3 2.2 0.2 2.3	**21** 0418 1636 F ○	0.5 1.9 0.5 2.0
7 0336 1559 SA ●	0.8 2.0 0.6 1.9	**22** 0425 1646 SU	0.7 1.9 0.6	**7** 0445 1710 TU	0.5 2.1 0.3	**22** 0516 1733 W	2.0 0.5 1.9 0.5	**7** 0344 1606 TU ●	0.5 2.0 0.2 2.2	**22** 0419 1634 W ○	0.5 1.9 0.5 2.0	**7** 0444 1712 F	0.3 2.2 0.2	**22** 0449 1708 SA	0.5 1.9 0.6
8 0418 1640 SU	0.6 2.0 0.5	**23** 0503 1722 M	1.9 0.6 1.9 0.5	**8** 0530 1754 W	2.2 0.4 2.2 0.2	**23** 0548 1803 TH	2.0 0.5 1.9 0.5	**8** 0426 1649 W	0.3 2.2 0.2	**23** 0449 1706 TH ○	1.9 0.5 1.9 0.5	**8** 0528 1754 SA	2.3 0.3 2.2 0.3	**23** 0520 1738 SU	1.9 0.5 1.9 0.6
9 0501 1724 M	2.0 0.6 2.0 0.5	**24** 0538 1755 TU	2.0 0.6 1.9 0.5	**9** 0614 1840 TH	2.2 0.4 2.2 0.2	**24** 0619 1832 F	2.0 0.6 1.9 0.5	**9** 0509 1735 TH	2.3 0.3 2.2 0.2	**24** 0519 1736 F	2.0 0.5 1.9 0.5	**9** 0611 1835 SU	2.2 0.3 2.1 0.5	**24** 0551 1808 M	1.9 0.5 1.8 0.6
10 0546 1810 TU	2.0 0.6 2.0 0.4	**25** 0611 1828 W	2.0 0.7 1.9 0.6	**10** 0659 1924 F	2.2 0.4 2.1 0.3	**25** 0646 1901 SA	1.9 0.6 1.8 0.6	**10** 0553 1819 F	2.3 0.3 2.2 0.2	**25** 0549 1805 SA	1.9 0.5 1.9 0.5	**10** 0653 1917 M	2.1 0.5 2.0 0.7	**25** 0624 1840 TU	1.8 0.6 1.8 0.7
11 0632 1856 W	2.1 0.6 2.0 0.4	**26** 0645 1900 TH	1.9 0.7 1.9 0.5	**11** 0743 2010 SA	2.2 0.5 2.0 0.5	**26** 0716 1930 SU	1.9 0.7 1.8 0.7	**11** 0636 1900 SA	2.3 0.3 2.2 0.3	**26** 0616 1831 SU	0.5 1.8 0.6	**11** 0741 2007 TU	2.0 0.6 1.9 0.9	**26** 0658 1919 W	1.8 0.7 1.7 0.8
12 0719 1944 TH	2.1 0.6 2.0 0.4	**27** 0719 1936 F	1.9 0.8 1.8 0.7	**12** 0830 2059 SU ☽	2.0 0.6 1.9 0.7	**27** 0748 2007 M	1.8 0.8 1.7 0.9	**12** 0718 1943 SU	2.2 0.4 2.0 0.5	**27** 0645 1859 M	1.9 0.6 1.8 0.7	**12** 0835 2110 W ☽	1.8 0.9 1.7 1.1	**27** 0743 2011 TH	1.7 0.8 1.7 1.0
13 0807 2034 F	2.1 0.6 1.9 0.5	**28** 0755 2013 SA	1.9 0.9 1.7 0.8	**13** 0925 2200 M	1.9 0.8 1.7 0.9	**28** 0829 2055 TU ☾	1.7 0.9 1.5 1.0	**13** 0803 2031 M	2.0 0.6 1.9 0.8	**28** 0716 1933 TU	1.8 0.7 1.7 0.9	**13** 0946 2237 TH	1.6 1.0 1.5 1.2	**28** 0840 2123 F ☾	1.6 0.9 1.6 1.0
14 0859 2129 SA ☽	2.0 0.8 1.8 0.7	**29** 0834 2056 SU	1.8 0.9 1.7 1.0	**14** 1038 2323 TU	1.7 1.0 1.5 1.1			**14** 0858 2133 TU ☽	0.8 1.7 1.0	**29** 0756 2023 W	1.7 0.9 1.6 1.0	**14** 1119 F	1.5 1.1 1.5	**29** 0955 2250 SA	1.5 0.9 1.6 1.0
15 0958 2233 SU	1.9 0.9 1.7 0.9	**30** 0924 2150 M ☾	1.7 1.0 1.5 1.1	**15** 1209 W	1.6 1.0 1.5			**15** 1012 2303 W	1.6 1.0 1.5 1.2	**30** 0853 2137 TH ☾	1.5 1.0 1.5 1.2	**15** 0011 1245 SA	1.2 1.4 1.0 1.6	**30** 1121 SU	1.5 0.9 1.7
		31 1028 2305 TU	1.5 1.2 1.4 1.2					**31** 1018 2317 F	1.4 1.1 1.4 1.2						

SEA LEVEL IS ABOVE MEAN TIDE LEVEL FROM 2.0 HOURS AFTER L.W. TO 2.0 HOURS BEFORE THE NEXT L.W. AND H.W. WILL OCCUR BETWEEN 5.0 HOURS AFTER L.W. AND 3.0 HOURS BEFORE THE NEXT L.W.

Chart Datum: 1.40 metres below Ordnance Datum (Newlyn)

ENGLAND, SOUTH COAST - POOLE (TOWN QUAY)
LAT 50°43′N LONG 1°59′W

TIME ZONE UT(GMT)
For Summer Time add ONE hour in non-shaded areas

TIMES AND HEIGHTS OF HIGH AND LOW WATERS

YEAR 1989

	MAY				JUNE				JULY				AUGUST		
	TIME m		TIME m		TIME m		TIME m		TIME m		TIME m		TIME m		TIME m

1 M 0008 1.0 / 1234 1.6 0.8 1.8 — **16** TU 0129 1.0 / 1344 1.6 0.9 1.8
1 TH 0133 0.7 / 1359 1.9 0.6 2.0 — **16** F 0207 0.9 / 1426 1.7 0.9 1.8
1 SA 0212 0.8 / 1439 1.9 0.8 1.9 — **16** SU 0221 0.9 / 1443 1.7 0.9 1.8
1 TU 0355 0.6 / 1614 1.9 0.7 1.9 ● — **16** W 0329 0.6 / 1551 2.0 0.6 2.0

2 TU 0111 0.8 / 1334 1.8 0.6 2.0 — **17** W 0210 0.9 / 1423 1.7 0.8 1.9
2 F 0226 0.6 / 1451 2.0 0.6 2.1 — **17** SA 0247 0.8 / 1505 1.8 0.9 1.9
2 SU 0309 0.6 / 1532 1.9 0.8 2.0 — **17** M 0306 0.8 / 1526 1.8 0.8 1.9
2 W 0434 0.6 / 1652 2.0 0.6 1.9 ○ — **17** TH 0410 0.5 / 1631 2.1 0.5 2.0

3 W 0203 0.6 / 1425 1.9 0.4 2.1 — **18** TH 0245 0.8 / 1459 1.8 0.7 1.9
3 SA 0317 0.5 / 1543 2.0 0.5 2.1 ● — **18** SU 0325 0.8 / 1544 1.8 0.8 1.9
3 M 0359 0.6 / 1621 1.9 0.7 2.0 ● — **18** TU 0347 0.7 / 1607 1.9 0.7 1.9 ○
3 TH 0511 / 1730 1.9 0.5 2.0 0.6 — **18** F 0450 0.3 / 1713 2.2 0.5

4 TH 0250 0.5 / 1515 2.1 0.3 2.2 — **19** F 0317 0.7 / 1533 1.9 0.7 1.9
4 SU 0405 0.5 / 1630 2.0 0.6 2.0 — **19** M 0402 0.7 / 1622 1.9 0.8 1.9 ○
4 TU 0443 0.5 / 1705 2.0 0.7 2.0 — **19** W 0427 0.6 / 1648 1.9 0.6 2.0
4 F 0546 / 1806 1.9 0.5 2.0 0.7 — **19** SA 0535 0.3 / 1758 2.1 0.5 2.2

5 F 0336 0.3 / 1602 2.2 0.3 2.2 ● — **20** SA 0349 0.6 / 1607 2.0 0.7 1.9 ○
5 M 0450 0.5 / 1715 2.0 0.6 2.0 — **20** TU 0439 0.6 / 1702 1.9 0.7 1.9
5 W 0525 / 1746 1.9 0.5 2.0 0.7 — **20** TH 0508 0.5 / 1732 2.0 0.6 2.0
5 SA 0621 / 1840 1.9 0.6 2.0 0.7 — **20** SU 0620 0.3 / 1842 2.2 0.5 2.2

6 SA 0422 0.3 / 1647 2.2 0.3 — **21** SU 0422 0.6 / 1641 1.9 0.7 1.9
6 TU 0536 / 1758 2.0 0.5 2.0 0.7 — **21** W 0521 0.6 / 1743 1.9 0.7 1.9
6 TH 0604 / 1826 1.9 0.6 2.0 0.8 — **21** F 0551 0.4 / 1814 2.0 0.6 2.0
6 SU 0654 / 1914 1.9 0.6 2.0 0.8 — **21** M 0704 0.3 / 1925 2.2 0.5 2.2

7 SU 0506 / 1732 2.2 0.3 2.1 0.5 — **22** M 0455 0.6 / 1717 1.9 0.7
7 W 0619 / 1840 2.0 0.6 2.0 0.8 — **22** TH 0602 0.5 / 1826 1.9 0.7
7 F 0643 / 1904 1.9 0.6 2.0 0.8 — **22** SA 0636 0.3 / 1859 2.1 0.6
7 M 0728 / 1950 1.9 0.7 1.9 0.9 — **22** TU 0748 0.5 / 2010 2.0 0.6 2.1

8 M 0550 / 1813 2.1 0.4 2.0 0.6 — **23** TU 0534 0.6 / 1753 1.9 0.7 1.8
8 TH 0701 / 1925 1.9 0.7 1.9 0.9 — **23** F 0647 0.5 / 1911 1.9 0.7
8 SA 0722 / 1946 1.9 0.7 1.9 0.9 — **23** SU 0721 0.3 / 1945 2.0 0.6 2.1
8 TU 0807 / 2028 1.8 0.8 1.9 0.9 — **23** W 0835 0.6 / 2101 1.9 0.8 1.9 ☾

9 TU 0634 / 1855 2.0 0.5 2.0 0.8 — **24** W 0610 0.6 / 1832 1.8 0.8
9 F 0749 / 2017 1.8 0.8 1.9 1.0 — **24** SA 0732 0.5 / 1958 1.9 0.8
9 SU 0804 / 2031 1.8 0.8 1.9 1.0 — **24** M 0808 0.4 / 2031 2.0 0.6 2.0
9 W 0849 / 2114 1.7 1.0 1.7 1.1 — **24** TH 0932 0.9 / 2205 1.8 1.0

10 W 0720 / 1945 1.9 0.7 1.9 1.0 — **25** TH 0651 0.6 / 1915 1.8 0.8
10 SA 0842 / 2115 1.7 0.9 1.8 1.1 — **25** SU 0822 0.5 / 2051 1.8 1.0
10 M 0851 / 2120 1.7 0.9 1.8 1.0 ☾ — **25** TU 0857 0.6 / 2123 1.9 0.8
10 TH 0940 / 2215 1.5 1.2 1.6 1.2 — **25** F 1049 1.1 / 2334 1.7 1.1

11 TH 0813 / 2045 1.8 0.9 1.8 1.1 — **26** F 0739 0.6 / 2007 1.8 0.9
11 SU 0940 / 2221 1.7 1.0 1.7 1.2 ☾ — **26** M 0917 0.6 / 2149 1.9 0.9 ☾
11 TU 0944 / 2218 1.6 1.0 1.7 1.1 — **26** W 0954 0.8 / 2228 1.8 0.9
11 F 1052 / 2333 1.4 1.3 1.5 1.2 — **26** SA 1222 1.5 1.2 1.6

12 F 0916 / 2200 1.7 1.0 1.7 1.2 ☽ — **27** SA 0833 0.7 / 2108 1.7 0.9
12 M 1047 / 2328 1.5 1.0 1.7 1.2 — **27** TU 1019 0.8 / 2257 1.7 0.9
12 W 1046 / 2323 1.5 1.1 1.7 1.2 — **27** TH 1105 0.9 / 2344 1.7 1.0
12 SA 1214 1.4 1.3 1.5 — **27** SU 0104 1.0 / 1342 1.6 1.1 1.7

13 SA 1032 / 2321 1.5 1.1 1.7 1.2 — **28** SU 0936 0.8 / 2219 1.7 0.9 ☾
13 TU 1152 1.5 1.0 1.7 — **28** W 1128 0.8 1.8 1.7
13 TH 1152 1.5 1.2 1.6 — **28** F 1223 1.2 1.7
13 SU 0053 1.2 / 1328 1.5 1.5 1.7 1.2 — **28** M 0213 0.9 / 1440 1.8 1.0 1.8

14 SU 1149 1.5 1.0 1.7 — **29** M 1049 0.8 / 2330 1.7 0.9 1.8
14 W 0029 1.0 / 1252 1.5 1.0 1.7 — **29** TH 0004 0.9 / 1236 1.7 0.9 1.9
14 F 0029 1.2 / 1259 1.5 1.2 1.7 — **29** SA 0104 1.0 / 1338 1.7 1.0 1.8
14 M 0158 1.1 / 1424 1.7 1.0 1.8 — **29** TU 0303 0.8 / 1524 1.9 0.8 1.9

15 M 0033 1.1 / 1255 1.5 1.0 1.7 — **30** TU 1158 0.8 1.9
15 TH 0123 1.0 / 1342 1.6 1.0 1.8 — **30** F 0111 0.8 / 1340 1.8 0.8 1.9
15 SA 0131 1.0 / 1355 1.6 1.1 1.7 — **30** SU 0211 0.9 / 1439 1.8 0.9 1.8
15 TU 0247 0.8 / 1510 1.8 0.9 1.9 — **30** W 0343 0.6 / 1602 2.0 0.7 1.9

 31 W 0034 0.8 / 1301 1.8 0.7 1.9
 31 M 0308 0.8 / 1530 1.9 0.8 1.9
 31 TH 0418 0.6 / 1635 2.0 0.6 1.9 ●

SEA LEVEL IS ABOVE MEAN TIDE LEVEL FROM 2.0 HOURS AFTER L.W. TO 2.0 HOURS BEFORE THE NEXT L.W. AND H.W. WILL OCCUR BETWEEN 5.0 HOURS AFTER L.W. AND 3.0 HOURS BEFORE THE NEXT L.W.
Chart Datum: 1.40 metres below Ordnance Datum (Newlyn)

ENGLAND, SOUTH COAST - POOLE (TOWN QUAY)

LAT 50°43'N   LONG 1°59'W

TIMES AND HEIGHTS OF HIGH AND LOW WATERS

YEAR 1989

TIME ZONE UT(GMT)
For Summer Time add ONE hour in non-shaded areas

AREA 2—Central S England

SEPTEMBER

	TIME	m		TIME	m
1 F	0450 1708	0.5 2.0 0.6	**16** SA	0430 1650	0.3 2.3 0.3 2.2
2 SA	0523 1741	2.0 0.5 2.0 0.6	**17** SU	0513 1735	0.3 2.3 0.3
3 SU	0555 1812	1.9 0.5 2.0 0.6	**18** M	0557 1819	2.2 0.3 2.3 0.4
4 M	0626 1843	1.9 0.6 2.0 0.6	**19** TU	0641 1901	2.2 0.4 2.2 0.5
5 TU	0655 1913	1.9 0.7 1.9 0.8	**20** W	0724 1948	2.1 0.6 2.1 0.6
6 W	0726 1945	1.8 0.8 1.9 0.9	**21** TH	0812 2041	1.9 0.8 1.9 0.8
7 TH	0801 2024	1.7 1.0 1.7 1.0	**22** F ☾	0911 2149	1.8 1.0 1.7 1.0
8 F ☽	0848 2119	1.6 1.2 1.6 1.2	**23** SA	1035 2322	1.6 1.2 1.6 1.2
9 SA	0957 2240	1.4 1.3 1.5 1.3	**24** SU	1215	1.5 1.3 1.5
10 SU	1133	1.4 1.3 1.5	**25** M	0055 1334	1.1 1.7 1.2 1.6
11 M	0014 1259	1.2 1.5 1.2 1.5	**26** TU	0200 1427	1.0 1.8 1.0 1.7
12 TU	0127 1400	1.0 1.7 1.0 1.7	**27** W	0245 1507	0.8 1.9 0.8 1.8
13 W	0221 1446	0.8 1.9 0.8 1.9	**28** TH	0320 1540	0.7 2.0 0.7 1.9
14 TH	0305 1528	0.6 2.0 0.6 2.0	**29** F ●	0353 1610	0.6 2.0 0.6 2.0
15 F ○	0347 1608	0.4 2.2 0.5 2.2	**30** SA	0424 1639	0.6 2.0 0.6 2.0

OCTOBER

	TIME	m		TIME	m
1 SU	0455 1711	0.6 2.0 0.6	**16** M	0449 1709	0.3 2.3 0.3
2 M	0526 1743	2.0 0.6 2.0 0.6	**17** TU	0534 1753	2.3 0.4 2.3 0.4
3 TU	0556 1811	1.9 0.6 2.0 0.6	**18** W	0616 1839	2.2 0.5 2.2 0.5
4 W	0624 1840	1.9 0.7 1.9 0.7	**19** TH	0659 1925	2.1 0.7 2.0 0.7
5 TH	0653 1912	1.8 0.8 1.8 0.8	**20** F	0749 2020	1.9 0.9 1.9 0.9
6 F	0727 1950	1.7 1.0 1.7 1.0	**21** SA ☾	0852 2129	1.8 1.1 1.7 1.1
7 SA	0814 2043	1.6 1.1 1.6 1.1	**22** SU	1014 2256	1.7 1.3 1.6 1.2
8 SU ☽	0921 2157	1.5 1.3 1.5 1.2	**23** M	1145	1.7 1.3 1.5
9 M	1055 2331	1.5 1.3 1.5 1.2	**24** TU	0022 1304	1.2 1.7 1.2 1.6
10 TU	1222	1.5 1.2 1.6	**25** W	0126 1357	1.0 1.8 1.0 1.7
11 W	0048 1326	1.0 1.8 1.0 1.8	**26** TH	0213 1437	0.9 1.9 0.9 1.8
12 TH	0145 1414	0.8 1.9 0.8 1.9	**27** F	0248 1510	0.8 2.0 0.8 1.9
13 F	0234 1458	0.5 2.1 0.6 2.1	**28** SA	0321 1540	0.7 2.0 0.7 2.0
14 SA	0318 1541	0.4 2.3 0.5 2.2	**29** SU ●	0353 1610	0.6 2.0 0.6 2.0
15 SU	0403 1625	0.3 2.3 0.4 2.3	**30** M	0425 1639	0.6 2.0 0.6 2.0
			31 TU	0456 1712	0.7 2.0 0.6

NOVEMBER

	TIME	m		TIME	m
1 W	0528 1744	1.9 0.7 1.9 0.6	**16** TH	0555 1819	2.2 0.6 2.1 0.6
2 TH	0557 1815	1.9 0.7 1.9 0.7	**17** F	0639 1906	2.0 0.8 2.0 0.7
3 F	0629 1849	1.8 0.9 1.8 0.8	**18** SA	0727 1958	2.0 1.0 1.9 0.9
4 SA	0707 1929	1.8 0.9 1.7 0.9	**19** SU	0826 2059	1.9 1.1 1.8 1.0
5 SU	0753 2020	1.7 1.0 1.7 1.0	**20** M ☾	0936 2210	1.8 1.2 1.7 1.1
6 M ☽	0856 2127	1.7 1.2 1.6 1.0	**21** TU	1056 2325	1.7 1.2 1.5 1.2
7 TU	1016 2248	1.7 1.2 1.5 1.0	**22** W	1209	1.7 1.2 1.6
8 W	1136	1.7 1.1 1.7	**23** TH	0034 1309	1.1 1.8 1.0 1.7
9 TH	0002 1242	0.9 1.8 0.9 1.8	**24** F	0128 1357	1.0 1.8 0.9 1.8
10 F	0105 1337	0.8 2.0 0.8 1.9	**25** SA	0210 1434	0.9 1.9 0.9 1.8
11 SA	0158 1426	0.6 2.1 0.6 2.1	**26** SU	0248 1509	0.9 1.9 0.8 1.9
12 SU	0249 1514	0.5 2.2 0.5 2.2	**27** M	0322 1542	0.8 2.0 0.8 1.9
13 M ○	0337 1601	0.5 2.3 0.4 2.2	**28** TU	0357 1614	0.8 2.0 0.7 1.9
14 TU	0425 1645	0.5 2.3 0.4	**29** W	0431 1647	0.8 2.0 0.7
15 W	0511 1733	2.2 0.5 2.2 0.5	**30** TH	0504 1722	1.9 0.8 1.9 0.6

DECEMBER

	TIME	m		TIME	m
1 F	0539 1757	1.9 0.8 1.9 0.7	**16** SA	0622 1846	2.0 0.8 2.0 0.6
2 SA	0614 1835	1.8 0.8 1.8 0.7	**17** SU	0706 1931	2.0 0.9 1.9 0.8
3 SU	0654 1917	1.8 0.9 1.8 0.7	**18** M	0755 2021	2.0 1.0 1.8 0.9
4 M	0740 2004	1.8 0.9 1.8 0.8	**19** TU ☾	0850 2116	1.9 1.1 1.7 1.0
5 TU	0833 2100	1.8 1.0 1.7 0.8	**20** W	0951 2219	1.8 1.2 1.7 1.1
6 W	0936 2205	1.8 1.0 1.7 0.9	**21** TH	1100 2327	1.7 1.2 1.5 1.1
7 TH	1048 2317	1.8 1.0 1.7 0.9	**22** F	1207	1.7 1.1 1.5
8 F	1156	1.8 0.9 1.8	**23** SA	0032 1307	1.1 1.7 1.0 1.6
9 SA	0024 1300	0.8 1.9 0.8 1.9	**24** SU	0130 1357	1.1 1.7 1.0 1.7
10 SU	0128 1358	0.8 2.0 0.7 2.0	**25** M	0216 1439	1.0 1.8 0.9 1.8
11 M	0224 1452	0.7 2.1 0.6 2.0	**26** TU	0257 1518	0.9 1.9 0.8 1.8
12 TU	0317 1543	0.6 2.2 0.6 2.1	**27** W ○	0335 1555	0.9 1.9 0.8 1.9
13 W	0407 1631	0.6 2.2 0.5 2.1	**28** TH ●	0410 1630	0.8 1.9 0.7 1.9
14 TH	0453 1719	0.6 2.1 0.5	**29** F	0445 1706	0.8 1.9 0.6
15 F	0538 1802	2.1 0.7 2.0 0.6	**30** SA	0523 1743	1.9 0.7 1.9 0.6
			31 SU	0602 1823	1.9 0.7 1.9 0.5

SEA LEVEL IS ABOVE MEAN TIDE LEVEL FROM 2.0 HOURS AFTER L.W. TO 2.0 HOURS BEFORE THE NEXT L.W. AND H.W. WILL OCCUR BETWEEN 5.0 HOURS AFTER L.W. AND 3.0 HOURS BEFORE THE NEXT L.W.

Chart Datum: 1.40 metres below Ordnance Datum (Newlyn)

POOLE continued

CHARTS
Admiralty 2175, 2611; Stanford 15, 12, 7; Imray C4, Y23; OS 195

TIDES
−0221, +0114 Dover; ML 1.5; Zone 0 (GMT).

Standard Port PORTSMOUTH (→)

Times				Height (metres)			
HW		LW		MHWS	MHWN	MLWN	MLWS
0000	0600	0500	1100	4.7	3.8	1.8	0.6
1200	1800	1700	2300				

Differences POOLE ENTRANCE
−0240 +0105 −0100 −0030 −2.7 −2.2 −0.7 −0.3
TOWN QUAY
−0210 +0140 −0015 −0005 −2.6 −2.2 −0.7 −0.2
WAREHAM (River Frome)
−0140 +0205 +0110 +0035 −2.5 −2.1 −0.7 +0.1

NOTE: Times and heights of LW for each day of the year are given above. Double HWs occur except at neaps and predictions are for the higher HW. Near neaps there is a stand and the predictions shown are for the middle of the stand. See 10.2.8.

SHELTER
An excellent harbour with narrow entrance from sea, accessible in all conditions except very strong E and SE winds. Anchorages may be found in most parts of the harbour, anywhere sheltered from the wind and free from moorings. Speed limit throughout harbour 8 kn; in quiet areas 6 kn.

NAVIGATION
Waypoint Poole Bar (stbd-hand) buoy, QG, 50°39'.42N 01°55'.15W, 147°/327° from/to Haven Hotel, 1.8M. The Bar is dangerous in strong SE-S winds especially on the ebb. Beware number of lobster pots round Studland Bay and close to training bank. There are two channels up to Poole; the Middle Ship Channel (formerly the Diver Channel) and the North Channel (formerly the Main Channel). The former is dredged to 5m for traffic to the ferry terminal at Hamworthy, and is mostly only 60m wide. Recreational traffic must keep out of the Middle Ship Channel but may use the small vessel channel running to the S of the dredged channel between port hand buoys to the S and stakes having R can topmarks marking the edge of the bank. Depth in this channel is 2.0m above CD except close to the stakes where it drops to 1.5m. Alternatively yachts can use the North Channel (formerly the main channel). Both channels are clearly marked by lateral buoys, mostly lit, with divisions marked by cardinal buoys.
Yachts departing westwards should note regulations regarding Lulworth gunnery range (see 10.2.30). Information is shown in Hr Mr's mooring office on Poole Quay and broadcast on Radio Solent at times given in 10.2.9.

LIGHTS AND MARKS
Radio beacon 303.4 kHz PO 10 M located on Haven Hotel at harbour entrance.
Chain Ferry Sandbanks
Moving: Day—B Ball For'd; Night—WGR Lts For'd; Fog—1 long, 2 short blasts every 2 mins.
Stationary: Night—W Lt.; Fog—5 sec bell every min.
Poole Bridge (Lights shown from tower)
FR — Do not approach bridge;
Fl Or — Bridge opening, proceed with caution;
FG — Proceed;
Bridge opened on request (large vessels) and at routine times for small — Mon-Fri 0930, 1130, 1430, 1630, 1830, 2130, 2230. Sat, Sun & Bank holidays 0730, 0930, 1130, 1330, 1530, 1730, 1930, 2130, 2330.
Salterns Marina
Lts 2FG (vert) to stbd, 2FR (vert) to port.
Poole YC Haven
Entrance E side FlG 5s. W side Fl (3) 10s.

RADIO TELEPHONE
Call: *Poole Harbour Control* VHF Ch 16; 14. (Mon-Fri: 0900-1700 LT). Salterns Marina (call: *Gulliver Base*), Rotork Boat Park and Cobbs Quay Marina, Ch M.

TELEPHONE (0202)
Hr Mr 685261; CG (Poole Sector) 670776; MRSC Portland 820441; Pilots 673320; Bridge 674115; HMC Freefone Customs Yachts: RT (0202) 674567; Weather Southampton 228844; Marinecall 0898 500 457; Hosp 675100.

FACILITIES
EC Wednesday. The facilities in Poole Harbour are very extensive and the following is not an exhaustive list.
Salterns Marina (350, some visitors) Tel. 707321, ME, El, Sh, C (5 ton), BH (40 ton), AC, FW, CH, Bar, Gas, Gaz, P, D, R, Lau; **Cobbs Quay Marina** (600, some visitors) Tel. 674299, Slip, P, D, Gas, Lau, SM, FW, AC, ME, El, Sh, C (10 ton), CH, R, Bar; **Sunseekers International Marina** (50) Tel. 685335, Slip, BH (2 ton), C (40 ton), AC, Sh, D, FW, ME, El, CH, V, R, Bar; **Dorset Yacht Co** (120 + 12 visitors) Tel. 674531, Slip, P, D, Gas, Gaz, FW, ME, El, Sh, C (5 ton), Bar, AC;
Harbour Office Tel. 685261 is open during office hours to give berthing directions. There is an office on the quay which stays open until about 2300.
Landing Places: Public landing places on Poole Quay, at Holes Bay and by ferry hards at Sandbanks.
Fresh water is available at Town Quay, at all marinas and boatyards.
Fuel (P and D) and FW is available from Esso barge when on station near Aunt Betty buoy (No 50) and all major yards.
Yacht Clubs
Poole YC Tel. 672687; **Poole Quay YC** Tel. 680746; **Poole Harbour YC** Tel. 707321; **Parkstone YC** Tel. 743610; **Royal Motor YC** Tel. 707227; **Cobb's Quay YC** Tel. 673690.
Note: **Poole Quay** (AB, C (24 ton) FW) is close to town centre and many facilities listed below.
Arthur Bray Tel. 676469, M, L, FW, C (2 ton), AB; **J. Harvey** Tel. 674063, M; **Marine Engine Centre** Tel. 672082, FW, ME, El, Sh, CH, AB, C (4 ton); **Rockley Boating Centre** Tel. 665001, Slip, M, C (5 ton), FW, CH, AB; **Quay Sails** Tel. 681128, SM; **Tab Sails** Tel. 684638, SM; **Crusader Sails** Tel. 698727, SM; **Sandbanks Yacht Co.** Tel. 707500, Slip, M, D, P, L, FW, ME, El, CH, V, Sh; **Mitchell BY** Tel. 747857, Slip, M, FW, ME, El, L, C (18 ton), Sh, CH; **Davis' BY** Tel. 674349, Slip, C (12 ton), AB, CH, FW, ME, El, Sh; **Lilliput Yacht Station** Tel. 707176, Slip, CH, L, M, FW, Sh; **Latham and Sons** Tel. 748029, Slip, L, FW, C (4 ton), CH, ME, El, Sh; **H. Pipler** Tel. 673056, L, CH, ACA; **Marine Power** Tel. 676469 Spares & Silencers; **Greenham Marine** Tel. 676363, El, Radio; **Melmarine** Tel. 680462, CH; **Poole Boat Park** (300) Tel. 681458, Slip, FW, AC, M, ME, Sh, C (10 ton), CH; **Barratts** Tel. 675920, CH, ME, El; **Poole Inflatables** Tel. 677777, Safety Equipment and Inflatables; **Town** PO; Bank; Rly; Air (Bournemouth).

CHRISTCHURCH 10-2-16
Dorset

CHARTS
Admiralty 2172; Stanford 7, 12; Imray C4; OS 195

TIDES
HW Spring −0210 Neap −0140 Dover; ML 1.2
         +0120   +0105   Zone 0 (GMT)

Standard Port PORTSMOUTH (→)

Times				Height (metres)			
HW		LW		MHWS	MHWN	MLWN	MLWS
0000	0600	0500	1100	4.7	3.8	1.8	0.6
1200	1800	1700	2300				

Differences CHRISTCHURCH (Ent)
−0230 +0030 −0035 −0035 −2.9 −2.4 −1.2 −0.2
CHRISTCHURCH (Tuckton)
−0205 +0110 +0110 +0105 −3.0 −2.5 −1.0 +0.1
BOURNEMOUTH
−0240 +0055 −0050 −0030 −2.7 −2.2 −0.8 −0.3

NOTE: Double HWs occur except near neaps and predictions are for the higher HW. Near neaps there is a stand and the predictions shown are for the middle of the stand. Tidal levels are for inside the bar. Outside the bar the tide falls about 0.6m lower at springs. Floods (or drought) in the 2 rivers cause considerable variations from predicted heights. See 10.2.8

CHRISTCHURCH continued

SHELTER
Good shelter in lee of Hengistbury Head but otherwise very exposed to prevailing SW winds. The Stour is navigable at HW up to Tuckton Bridge and the Avon up to the first bridge across to the island, and both give good shelter from all winds. Recommend entry or departure on the stand. All anchorage areas in the harbour dry. Anchoring forbidden in main channel. It is also forbidden to secure alongside ferry terminal on Mudeford sandbank. SC offers limited moorings to monohulls up to 7.9m.

NAVIGATION
Waypoint 50°43'.50N 01°43'.50W, 090°/270° from/to channel entrance, 0.30M. The position of the bar is liable to change after storms. The entrance is difficult on the ebb tide due to very strong streams, 4 to 5 kn in 'The Run'. Beware groynes S of Hengistbury Head and also Beerpan Rks and Yarranton (or Clarendon) Rks.
There is a speed limit in the harbour of 5 kn.

LIGHTS AND MARKS
Buoyage in harbour and entrance is carried out privately by the Christchurch Harbour Association and buoys are often withdrawn in winter. Information can be obtained from Christchurch Marine Tel. 483250.

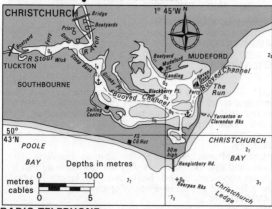

RADIO TELEPHONE
None.
TELEPHONE (0202)
Quay & Moorings Supt. Highcliffe 4933;
CG 425204; MRSC Lee-on-Solent 552100; HMC Freefone Customs Yachts: RT (0703) 229251; Marinecall 0898 500 457; Hosp 486361.
FACILITIES
Mudeford Quay (Apl-Sept), Slip, M, P (cans, ½M), L, FW, C (mobile), AB, V; **Christchurch Marine** Tel. 483250, Slip, M, FW, ME, El, Sh, CH, ACA; **John Lack** Tel. 483191, FW, ME, El; **RIBS Marine** Tel. 477327, M, FW, ME, El, Sh; **Leak Hayes** Tel. 483141, C (8 ton); **Bob Hoare** Tel. 485708, Sh, CH; **Slaters Marine** Tel. 482751, CH; **R. A. Stride** Tel. 485949, CH; **Mudeford YC**.
Town PO; Bank; Rly; Air (Bournemouth).

KEYHAVEN 10-2-17
Hampshire

CHARTS
Admiralty 2219, 2021, 2040; Stanford 11, 7; Imray C4, C3, Y20; OS 196
TIDES
−0020, +0105 Dover; ML 2.0; Zone 0 (GMT).

Standard Port PORTSMOUTH (→)

Times				Height (metres)			
HW		LW		MHWS	MHWN	MLWN	MLWS
0000	0600	0500	1100	4.7	3.8	1.8	0.6
1200	1800	1700	2300				

Differences HURST POINT
−0115 −0005 −0030 −0025 −2.0 −1.5 −0.5 −0.1
NOTE: Double tides occur at or near springs and on other occasions there is a stand which lasts about 2 hours. Predictions refer to the first HW when there are two. At other times they refer to the middle of the stand. See 10.2.8.
River is administered by New Forest District Council aided by the Keyhaven Consultative Committee.

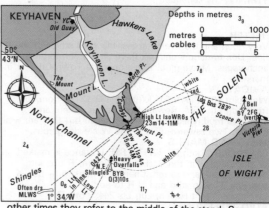

SHELTER
Good, but the river gets extremely congested. All moorings and anchorages are exposed to winds across the marshland. Access HW ∓4½.

NAVIGATION
Waypoint 50°42'.70N 01°32'.50W, 115°/295° from/to channel entrance, 0.40M. Bar is constantly changing. Entrance should not be attempted in strong E winds. Approaching from W, beware Shingles over which seas break and which very occasionally dry. Give 'The Trap' a wide berth.

LIGHTS AND MARKS
When E of the High Lt, two leading beacons in line ('X' topmarks) at 283° lead to entrance of buoyed channel.
NOTE: Ldg Lts 042° do *not* lead into the harbour but keep clear of the Shingles.

RADIO TELEPHONE
None.
TELEPHONE (0590)
River Warden 45695; CG Bournemouth 425204; MRSC Lee-on-Solent 552100; New Forest District Council Lyndhurst 3121; HMC Freefone Customs Yachts: RT (0703) 229251; Marinecall 0898 500 457; Dr 72179.
FACILITIES
EC (Milford-on-Sea) Wednesday; **Keyhaven YC** Tel. 42165, M, L (on beach), FW; **West Solent Boat Builders** Tel. 42080, Slip, ME, El, Sh, C (6 ton), CH; **New Forest District Council** Tel. Lyndhurst 3121, Slip, M; **Milford-on-Sea** P, D, FW, CH, V, R, Bar; **Hurst Castle SC** M, L, FW; **Quay** Slip, L.
Village R, Bar. PO (Milford-on-Sea); Bank (Milford-on-Sea); Rly (bus to New Milton); Air (Bournemouth).

LYMINGTON 10-2-18
Hampshire

CHARTS
Admiralty 2021, 2040; Stanford 7, 11; Imray C3, Y20; OS 196
TIDES
Spring −0040, Neap +0020 Dover; ML 2.0; Zone 0 (GMT) +0100

Standard Port PORTSMOUTH (→)

Times				Height (metres)			
HW		LW		MHWS	MHWN	MLWN	MLWS
0000	0600	0500	1100	4.7	3.8	1.8	0.6
1200	1800	1700	2300				

Differences LYMINGTON
−0110 +0005 −0020 −0020 −1.7 −1.2 −0.5 −0.1
NOTE: Double high waters occur at or near springs and on other occasions there is a stand which lasts about 2 hrs. Predictions refer to the first HW when there are two. At other times they refer to the middle of the stand. See 10.2.8.

LYMINGTON continued

SHELTER
Very good and the river is accessible at all states of the tide. Due to narrow channel there are no comfortable anchorages. Moorings available opposite Town Quay or in marinas.

NAVIGATION
Waypoint 50°44'.04N 01°30'.00W, 139°/319° from/to front Ldg Lt, 1.5M. Beware large ferries. There are extensive mud banks round the entrance but they are well marked. Two sewer outlets to the W are dangerous.

LIGHTS AND MARKS
Ldg Lts 319°, both FR 12/17m 8M. Entrance is marked on W side by Jack in the Basket Fl R 2s 9m; R pile with barrel topmark and ra refl. Channel is marked by buoys and Bns, Fl G to stbd and Fl R to port.
Lymington Yacht Haven, Ldg Lts 244° both FY.

RADIO TELEPHONE
Marinas Ch M (office hours).

TELEPHONE (0590)
Hr Mr 72014; MRSC Lee-on-Solent 552100; HMC Freefone Customs Yachts: RT (0703) 229251; Marinecall 0898 500 457; Dr 77011.

FACILITIES
EC Wednesday; **Lymington Marina** (300 + 100 visitors) Tel. 73312, Slip, P, AC, D, FW, ME, El, Sh, CH, BH (40 ton), C, Gas, Gaz; **Lymington Yacht Haven** (450 + 100

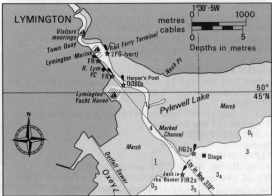

visitors) Tel. 77071, P, D, FW, ME, AC, El, Sh, C (10 ton), BH (45 ton), CH, Gas, Gaz, Lau; **Berthon Boat Co.** Tel. 73312, M, P, FW, ME, El, Sh, C, CH, AB; **Royal Lymington YC** Tel. 72677, Slip, L (pontoon), FW, V, R, Bar; **Lymington Town SC** Tel. 74514, AB, R, Bar; **Town Quay** Slip, M, FW, AB (see Hr Mr); **Bill Smith** Tel. 73876, ME, El, Sh, CH; **Aquaboats** Tel. 74266, ME, Sh, CH; **J.G. Claridge** Tel. 74821, CH; **Shipmates** Tel. 72765, CH; **Greenham Marine** Tel. 75771, El; **Brookes and Gatehouse** Tel. 75808, El; **Hood** Tel. 75011, SM; **Lymington Sail and Tent** Tel. 73139, SM, CH; **Haven BY** Tel. 73489, ACA; **Sanders** Tel. 73981, SM; **Yacht Mail Mariner Electronics** Tel. 76511, El; **S.A.L. Marine** Tel. 79588, ME; **Quayside Marine** Tel. 79582, Sh (Wood and GRP); **Capt OM Watts** Tel. 73186, CH; **Town** has every facility including PO; Bank; Rly, Air (Bournemouth or Southampton).

YARMOUTH 10-2-19
Isle of Wight

CHARTS
Admiralty 2021, 2040; Stanford 11; Imray C3, Y20; OS 196

TIDES
Spring −0050, Neap +0020 Dover; ML 2.0; Zone 0 (GMT) +0150

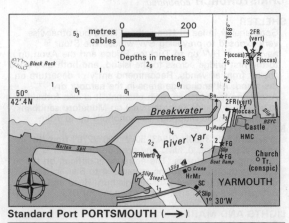

Standard Port PORTSMOUTH (→)

Times				Height (metres)			
HW		LW		MHWS	MHWN	MLWN	MLWS
0000	0600	0500	1100	4.7	3.8	1.8	0.6
1200	1800	1700	2300				

Differences YARMOUTH
−0105 +0005 −0025 −0030 −1.6 −1.3 −0.4 0.0
FRESHWATER
−0210 +0025 −0040 −0020 −2.1 −1.5 −0.4 0.0
TOTLAND BAY
−0130 −0045 −0040 −0040 −2.0 −1.5 −0.5 −0.1

NOTE: Double HWs occur at or near springs; at other times there is a stand which lasts about 2hrs. Predictions refer to the first HW when there are two. At other times they refer to the middle of the stand. See 10.2.8

SHELTER
Harbour affords complete shelter from all directions of wind and sea. Harbour gets very full, so much so that it is, on occasions, closed. Boats over 15m length, 4m beam or 2.4m draft should give notice of arrival by telephone.

NAVIGATION
Waypoint 50°42'.60N 01°29'.93W, 008°/188° from/to front Ldg Lt, 0.28M. Only danger on approach is Black Rock. There are four large Admiralty buoys (two unlit), three E of pier, one W of pier. The most E and the most W are lit. Beware ferries entering and leaving. The road bridge across R Yar opens on request (Oct-May) for boats proceeding up the river to anchor above the bridge.
Bridge opening times (May-Sept) 0805, 0900, 1000, 1200, 1400, 1600, 1700, 1800, 2015 (LT).
Speed limit in harbour 4 kn.

LIGHTS AND MARKS
Leading beacons (two W diamonds) or Ldg Lts (FG), on quay, 188°. When harbour is closed (eg when full in summer at week-ends) a R flag is hoisted at seaward end of Ferry jetty by day and an illuminated board displayed saying 'Harbour Full', by night.

RADIO TELEPHONE
None.

TELEPHONE (0983)
Hr Mr 760300; MRSC Lee-on-Solent 552100; HMC Freefone Customs Yachts: RT (0703) 229251; Marinecall 0898 500 457; Dr 760434.

FACILITIES
EC Wednesday, **South Quay** Tel. 760300, Slip, P, D, L, FW, ME, C; **Harold Hayles Yacht Yard** Tel. 760373, Slip, M, ME, El, Sh, CH; **River Yar Boatyard** Tel. 760520, M, ME, Sh, CH; **Royal Solent YC** Tel. 760270, Bar, R, L, M; **Harwoods Store** Tel. 760258, P, CH; **Yarmouth Outboards** Tel. 760436, ME, Sh, CH; **Brookes & Gatehouse Services** Tel. 760309, El; **Buzzard Engineering and Marine** Tel. 760065, ME, Sh; **Hylton Mortimer** Tel. 760120, SM; **Yarmouth Marine Services** Tel. 760521, ME, El, Sh, Divers;
Town V, R, Bar, PO; Bank (Sept-May AM only, May-Sept 1000-1445); Rly (Ferry to Lymington); Air (Bournemouth, IoW airport or Southampton).

NEWTOWN 10-2-20
Isle of Wight

CHARTS
Admiralty 2021, 2040, 1905; Stanford 11; Imray C3, Y20; OS 196

TIDES
Spring −0108 Dover; ML 2.3; Zone 0 (GMT)
Neap +0058

Standard Port PORTSMOUTH (→)

Times				Height (metres)			
HW		LW		MHWS	MHWN	MLWN	MLWS
0000	0600	0500	1100	4.7	3.8	1.8	0.6
1200	1800	1700	2300				

Differences SOLENT BANK
−0100 0000 −0015 −0020 −1.3 −1.0 −0.3 −0.1

NOTE: Double HWs occur at or near springs; at other times there is a stand which lasts about 2 hrs. Predictions refer to the first HW when there are two. At other times they refer to the middle of the stand. See 10.2.8

SHELTER
Good although exposed to strong N winds. Do not anchor

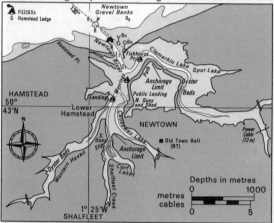

▲ Buoys marked "visitors"

above 'Anchorage limit' boards on account of oyster beds. Fin keel boats can stay afloat from entrance to Hamstead landing or to Clamerkin Limit Boards.

NAVIGATION
Waypoint 50°44'.00N 01°25'.00W, 335°/155° from/to Fishhouse Pt, 0.67M. After entry there are so many perches that confusion may result. Off Clamerkin Lake keep to E to avoid gravel spit off W shore, marked by three perches. Beware numerous oyster beds.
Speed limit in the harbour, 5 kn.

LIGHTS AND MARKS
Entry beacons RW 'Y' Bn and BW 'O' Bn at 130°. Harbour belongs to the National Trust. The whole peninsular ending in Fishhouse Point is a Nature Reserve. Yachtsmen are asked not to land there between April and June inclusive.

RADIO TELEPHONE
None

TELEPHONE (0983)
Hr Mr Calbourne 424; MRSC Lee-on-Solent 552100; HMC Freefone Customs Yachts: RT (0703) 827350; Marinecall 0898 500 457; Dr 760434

FACILITIES
EC Newport Thursday; **Newtown Quay** Tel. Calbourne 424, M, L, FW; **Shalfleet Quay** Slip, M, L, AB; **Lower Hamstead Landing** L, FW.
Shalfleet Village V, Bar; PO (Newport); Bank (Yarmouth or Newport); Rly (bus to Yarmouth, ferry to Lymington); Air (Southampton).

BEAULIEU RIVER 10-2-21
Hampshire

CHARTS
Admiralty 1905, 2021, 2040; Stanford 11; Imray C3, Y20; OS 196

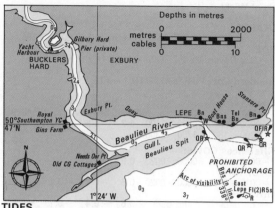

TIDES
−0100 and +0140 Dover; ML 2.4; Zone 0 (GMT)

Standard Port PORTSMOUTH (→)

Times				Height (metres)			
HW		LW		MHWS	MHWN	MLWN	MLWS
0000	0600	0500	1100	4.7	3.8	1.8	0.6
1200	1800	1700	2300				

BUCKLERS HARD
−0040 −0010 +0010 −0010 −1.0 −0.8 −0.2 −0.3
STANSORE POINT
−0050 −0010 −0005 −0010 −0.9 −0.6 −0.2 0.0

NOTE: Double HWs occur at or near springs; on other occasions there is stand which lasts about 2 hours. The predictions refer to the first HW when there are two, or to the middle of the stand.

SHELTER
Very good shelter from all winds. Anchorage possible in reach between Lepe and Needs Oar Point but preferable to proceed up to Bucklers Hard.

NAVIGATION
Waypoint 50°46'.50N 01°21'.47W, 159°/339° from/to Beaulieu Spit Bn, QR, 0.35M. The swatchway at W end of Beaulieu Spit is closed.
A speed limit of 5 knots applies to the whole river.

LIGHTS AND MARKS
Entrance beacons in line at 339°, N beacon is triangle over a square topmark, S beacon is R and carries a W topmark. Beaulieu Spit, E end, R dolphin with QR Lt. River is clearly marked by R and G beacons and perches.

RADIO TELEPHONE
None.

TELEPHONE (059 063 − Bucklers Hard)
Hr Mr 200; CG Portsmouth 840200;
MRSC Lee-on-Solent 552100; HMC Freefone Customs Yachts: RT (0703) 229251; Marinecall 0898 500 457; Dr 612451 or Hythe 845955.

FACILITIES
EC Village Shop opens 7 days a week; **Bucklers Hard Yacht Harbour** (110+20 visitors) Tel. 200, Slip, M, P, D, AC, FW, ME, El, Sh, C (2 ton), BH (26 ton), SM, Gas, Gaz, CH, V, R, Bar; **Palace Quay BY** Tel. Beaulieu 612338, M, ME, El, Sh, CH; **Agamemnon BY** Tel. 214/5, ME, El, Sh; **Village** V, R, Bar. PO (Beaulieu); Bank (Mon, Wed, Fri AM or Hythe); Rly (bus to Brockenhurst); Air (Bournemouth or Southampton).

COWES 10-2-22
Isle of Wight

CHARTS
Admiralty 2793; Stanford 7, 11; Imray C3, Y20; OS 196

TIDES
+0023 Dover; ML 2.7; Zone 0 (GMT)

Standard Port PORTSMOUTH (→)

Times				Height (metres)			
HW		LW		MHWS	MHWN	MLWN	MLWS
0000	0600	0500	1100	4.7	3.8	1.8	0.6
1200	1800	1700	2300				

Differences COWES
−0015 +0015 0000 −0020 −0.5 −0.3 −0.1 0.0
FOLLY INN
−0015 +0015 0000 −0020 −0.6 −0.4 −0.1 +0.2
NEWPORT
 no data   no data −0.6 −0.4 +0.1 +1.3
NOTE: Double high tides occur at or near springs; on other occasions a stand occurs which lasts up to 2 hrs. Times given represent the middle of the stand. See 10.2.8

SHELTER
Good, but outer harbour is exposed to N and NE winds. R Medina is navigable to Newport but the upper reaches dry. Visitors may pick up any mooring so labelled: buoys off the Parade; piles S of Ancasta Marina, opposite Lallows; Thetis Pontoon Landing, between Shepards Wharf and Spencer Thetis Wharf (for short stay and overnight only, dredged to 2m); piles S of the chain ferry, Whitegates area; and 12 pile moorings opposite the Folly Inn. Marina berths at Ancasta Marina (W Cowes), Cowes Marina (E Cowes, above chain ferry), and Island Harbour (on E shore beyond Folly Inn).

NAVIGATION
Waypoint 50°46'.20N 01°17'.90W, 344°/164° from/to front Ldg Lt, 0.35M. On the E side of the entrance, the Shrape (mud flats) extend to Old Castle Point. Yachts are required to use the main channel near W shore. It is forbidden to sail through or anchor in the mooring area. Beware of the floating bridge, ferries and commercial shipping. Speed limit 6 kn in harbour. Board sailing forbidden within fairway and approaches.

LIGHTS AND MARKS
Egypt Point Fl 10s 8m 18M; R column, W lantern; vis 061°-272°; lies 6ca W of entrance which is marked by No 3 buoy (unlit) and No 4 buoy (Fl R). Ldg Lts 164°. Front Iso 2s 3m 6M; post by Customs House. Rear, 290m from front, Iso R 2s 5m 3M; dolphin by Jubilee Pontoon; vis 120°-240°. E breakwater head FlR 3s 3M. The ends of jetties and certain dolphins are marked by 2FR (vert) on E side of harbour, and by 2FG (vert) on W side.

RADIO TELEPHONE
VHF Ch 16; 06 11 (0900-1700, LT). Island Harbour and Cowes Marina Ch M; Ancasta Marina Ch M (office hours).

TELEPHONE (0983)
Hr Mr 293952; MRSC Lee-on-Solent 552100; HMC Freefone Customs Yachts: RT (0703) 229251; Folly Reach Harbour Office 295722; Weather Centre Southampton 228844; Marinecall 0898 500 457; Hosp 524081; Dr 295251.

FACILITIES
EC Wednesday; **Cowes Marina** (120 + 130 visitors) Tel. 293983, FW, BH (10 ton), D, ME, El, Sh, AC, Gas, V; **Ancasta Marina** (80 + 120 visitors) Tel. 293996, D, FW, ME, El, Sh, AC, Gas, Gaz, Kos, R, SM, V, Lau, CH, C (6 ton), BH (30 ton); **Pascall Atkey** Tel. 292381, ME, CH, ACA, Gas, Gaz; **Clare Lallow** Tel. 292112, Gas, Slip, P, D, L, FW, CH, ME, El, Sh; **Souters Shipyard** Tel. 292551, Slip, P, D, FW, AB, ME, El, Sh; **Fairey Marineteknik** Tel. 292561, ME, El, Sh; **Spencer Rigging** Tel. 292022, CH, M, Rigging; **Ancasta Chandlery** Tel. 296315, CH; **Marine Bazaar** Tel. 298869, CH; **The Book Cabin** Tel. 295409, Books and charts; **McWilliam** Tel. 298855, SM; **Ratsey and Lapthorn** Tel. 294051, SM; **RHP Marine** Tel. 290421, ME, El; **Eddie Richards** Tel. 298949, Sh; **Adrian Stone** Tel. 297898, Sh; **Island Boat Repairs** Tel. 298015, Sh; **Martin Marine** Tel. 292892, ME; **Lecmar** Tel. 293996, El; **Martec** Tel. 296913; El; **Mustang Marine** Tel. 290565, ME, El, Sh; **Powerplus Marine** Tel. 200036, ME; **E. Cole & Sons** Tel. 292724, Slip, Sh, ME; **D. Floyd** Tel. 295408, Divers.

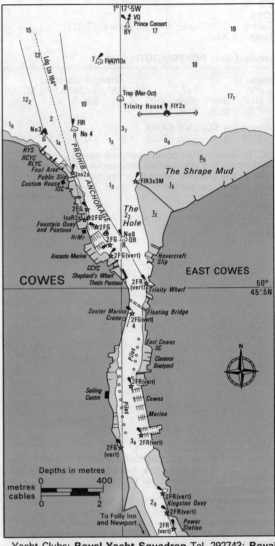

Yacht Clubs: **Royal Yacht Squadron** Tel. 292743; **Royal London YC** Tel. 293153; **Royal Corinthian YC** Tel. 293581; **Cowes Corinthian YC** Tel. 296333; **Cowes Combined Clubs** Tel. 295744; **Island SC** Tel. 293061.

FW is available at Town Quay and Folly Pier. **Do-It-Yourself** at Shepard's Wharf Tel. 297821, CH, C (6 ton), BH (20 ton), Gas, AC, Tools etc; Thetis Public Pontoon Landing
Town P, D, Bar, R, V, Slip, PO, Bank, Rly (Ferry to Southampton), Air (Southampton).
Ferry Services: Red Funnel ferries Tel. 292101; Hydrofoil Tel. 292101.
WHIPPINGHAM: **Folly Inn** Tel. 297171, L, Bar, R, M, Slip, Lau.

SOUTHAMPTON 10-2-23
Hampshire

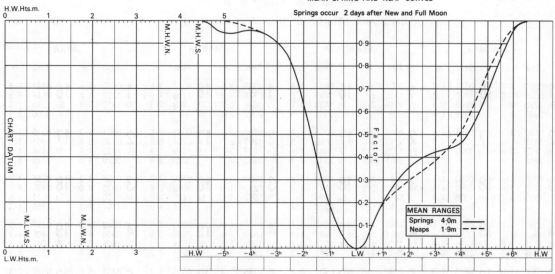

CHARTS
Admiralty 2041, 1905; Stanford 11; Imray C3; OS 196
TIDES
HW (1st) −0001 Dover; ML 2.9; Zone 0 (GMT)

Southampton is a Standard Port and tidal predictions for each day of the year are given below. See Sections 10.2.8 and 10.2.9. A NE gale combined with high barometer may lower sea level by 0.6m. At springs there are two separate HWs about 2 hours apart; at neaps there is a long stand. Predictions are for the first HW where there are two or for the middle of the stand.

SHELTER
Good from most wind directions although a heavy chop develops in SE winds above force 4. It is better to shelter in marinas.
There are no specific yacht anchorages but temporary anchoring is permitted (subject to Hr Mr) off club moorings at Netley, Hythe, Weston & Marchwood in about 2m. Keep clear of main or secondary channels and Hythe Pier. Public moorings opposite Royal Pier near Gymp Shoal in 4m by arrangement with the Hr Mr. The nearest landing is at steps W side of Town Quay adjacent Red Funnel Terminal. Visitors berths available in Hythe Marina (with lock entrance), at Shamrock Quay Marina (R Itchen), at Kemp's Quay Marina (R Itchen) and at Ocean Village Marina.

NAVIGATION
Waypoint Weston Shelf (stbd-hand) buoy, Fl(3)G 15s, 50°52′.68N 01°23′.16W, 138°/318° from/to Port Sig Stn, 0.40M. There are no hidden dangers or navigational hazards. Main channels are well marked. There are several unlit large mooring buoys at Hythe both E and W of the main channel. Other isolated buoys exist and yachtsmen entering the port during darkness should keep just outside the lighted buoyed fairway. Very large tankers operate from Fawley and very large passenger and container ships from Southampton. It is essential to keep clear of ocean-going shipping; see 10.2.9.
R Test   Eling Creek dries. There is foul ground at Marchwood and Royal Pier.
R Itchen   Care is necessary, particularly at night, above Itchen Bridge; the channel bends sharply to port and favours the W side. There are unlit moorings in the centre of the river. Navigation above Northam Bridge (14′ 6″, 4.4m) is not advisable.
There is a speed limit of 6 kn. in both rivers above the line Hythe Pier to Weston Shelf.

LIGHTS AND MARKS
These are too numerous to elaborate and reference should be made to the charts.
The following should be noted:
(1) Southampton Water bifurcates at Dock Head which is clearly marked by a high lattice mast showing traffic signals which are mandatory for commercial vessels but may be disregarded by yachts navigating outside the main channels. The signals are shown on the side of Test or Itchen as appropriate thus:

(a)	A green shape or light	A vessel may enter or leave the docks
(b)	A red shape or light	Entry or departure forbidden
(c)	A red shape or light over a green shape or light	Departure forbidden to facilitate entry of large vessel.
(d)	A green shape or light over a red shape or light	Entry forbidden to facilitate entry of large vessel.

(2) Hythe Marina Village, close NW of Hythe Pier, marked by Q (3) 10s; E cardinal Bn. Lock entrance, N side 2 FG (vert); S side 2 FR (vert).
(3) Dock Head, West side (Queen Elizabeth II Terminal, S end) 4 FG (vert) 3M; framework Tr; mark entrance to R Test.
(4) Entrance to R Itchen marked by Lt Buoy Oc G 4s, beyond which piles with G Lts mark E side of channel leading under road bridge.
(5) Above Itchen Bridge, which is marked by 2 FR (vert) and 2 FG (vert), the principal marks are Crosshouse Bn Oc R 5s, Chapel Bn Fl G 3s, Shamrock Quay pontoons 2 FR (vert) at SW and NE ends, No 5 Bn Fl G 3s, No 7 Bn Fl (2) G 5s, Millstone Pt jetty 2 FR (vert), No 9 Bn Fl (4) G 10s and Kemps Quay Marina 2 FG (vert).

RADIO TELEPHONE
Call: *Southampton Port Radio (SPR)* VHF Ch 12 14 16 (H24)
Southampton Harbour Patrol Call: *Southampton Patrol* VHF Ch 12 16; 01-28, 60-88 (H24) both have Ch 10, 12, 14, 16, 18, 22, 71 & 74.
Marinas VHF Ch M.

TELEPHONE (0703)
Dock and Harbour Master 330022; MRSC Lee-on-Solent 552100; HMC Freefone Customs Yachts: RT (0703) 229251; Southampton Weather Centre 228844; Marinecall 0898 500 457; Doctors 226631 (Port Health); Hosp 634288.

ENGLAND, SOUTH COAST — SOUTHAMPTON

Lat 50°54′ N   Long 1°24′ W

TIMES AND HEIGHTS OF HIGH AND LOW WATERS

YEAR 1989

TIME ZONE UT(GMT)
For Summer Time add ONE hour in non-shaded areas

JANUARY

Day	Time	m	Day	Time	m
1 Su	0501 / 1020 / 1720 / 2248	3.7 / 1.9 / 3.5 / 1.9	16 M	0531 / 1102 / 1803 / 2337	4.1 / 1.5 / 3.9 / 1.6
2 M	0602 / 1128 / 1829 / 2358	3.7 / 1.9 / 3.6 / 1.9	17 Tu	0644 / 1221 / 1925	4.0 / 1.5 / 3.9
3 Tu	0706 / 1239 / 1939	3.8 / 1.9 / 3.7	18 W	0058 / 0756 / 1338 / 2037	1.6 / 4.0 / 1.4 / 4.0
4 W	0107 / 0803 / 1342 / 2039	1.8 / 3.9 / 1.6 / 3.8	19 Th	0212 / 0921 / 1445 / 2136	1.5 / 4.1 / 1.2 / 4.1
5 Th	0208 / 0854 / 1438 / 2130	1.6 / 4.0 / 1.3 / 4.0	20 F	0313 / 0948 / 1540 / 2225	1.3 / 4.2 / 1.0 / 4.3
6 F	0302 / 0940 / 1529 / 2216	1.3 / 4.2 / 1.0 / 4.2	21 Sa ○	0405 / 1034 / 1626 / 2307	1.1 / 4.2 / 0.8 / 4.3
7 Sa ●	0352 / 1024 / 1616 / 2300	1.1 / 4.3 / 0.8 / 4.3	22 Su	0448 / 1112 / 1705 / 2345	1.0 / 4.3 / 0.7 / 4.4
8 Su	0440 / 1105 / 1701 / 2341	0.9 / 4.4 / 0.6 / 4.4	23 M	0524 / 1149 / 1739	0.9 / 4.3 / 0.6
9 M	0525 / 1146 / 1745	0.7 / 4.5 / 0.5	24 Tu	0021 / 0557 / 1222 / 1811	4.4 / 0.9 / 4.3 / 0.7
10 Tu	0022 / 0610 / 1227 / 1828	4.5 / 0.6 / 4.5 / 0.4	25 W	0053 / 0625 / 1253 / 1838	4.3 / 0.9 / 4.2 / 0.7
11 W	0102 / 0653 / 1309 / 1910	4.6 / 0.6 / 4.4 / 0.4	26 Th	0123 / 0654 / 1325 / 1905	4.3 / 1.0 / 4.2 / 0.9
12 Th	0146 / 0734 / 1355 / 1949	4.6 / 0.7 / 4.4 / 0.5	27 F	0153 / 0720 / 1357 / 1932	4.2 / 1.1 / 4.1 / 1.0
13 F	0233 / 0815 / 1445 / 2031	4.5 / 0.8 / 4.3 / 0.8	28 Sa	0227 / 0750 / 1435 / 2003	4.1 / 1.2 / 3.9 / 1.3
14 Sa ☽	0324 / 0901 / 1540 / 2120	4.4 / 1.0 / 4.2 / 1.1	29 Su	0306 / 0826 / 1518 / 2043	4.0 / 1.5 / 3.8 / 1.6
15 Su	0424 / 0956 / 1647 / 2220	4.2 / 1.3 / 4.0 / 1.4	30 M ☾	0352 / 0912 / 1613 / 2136	3.8 / 1.7 / 3.6 / 1.8
			31 Tu	0450 / 1016 / 1722 / 2250	3.7 / 1.9 / 3.5 / 2.0

FEBRUARY

Day	Time	m	Day	Time	m
1 W	0601 / 1138 / 1849	3.7 / 2.0 / 3.5	16 Th	0044 / 0745 / 1329 / 2039	1.9 / 3.8 / 1.7 / 3.9
2 Th	0019 / 0718 / 1305 / 2011	2.0 / 3.7 / 1.8 / 3.7	17 F	0208 / 0855 / 1441 / 2137	1.7 / 3.9 / 1.4 / 4.1
3 F	0141 / 0826 / 1416 / 2113	1.8 / 3.9 / 1.5 / 3.9	18 Sa	0311 / 0947 / 1534 / 2220	1.5 / 4.0 / 1.1 / 4.2
4 Sa	0245 / 0921 / 1513 / 2202	1.4 / 4.1 / 1.1 / 4.1	19 Su	0357 / 1027 / 1615 / 2256	1.2 / 4.1 / 0.8 / 4.3
5 Su	0337 / 1007 / 1601 / 2244	1.1 / 4.2 / 0.7 / 4.4	20 M ○	0432 / 1100 / 1648 / 2327	0.9 / 4.2 / 0.7 / 4.4
6 M ●	0425 / 1049 / 1647 / 2323	0.8 / 4.4 / 0.4 / 4.6	21 Tu	0503 / 1130 / 1717 / 2356	0.7 / 4.3 / 0.6 / 4.4
7 Tu	0511 / 1130 / 1730	0.5 / 4.6 / 0.2	22 W	0530 / 1159 / 1745	0.7 / 4.3 / 0.6
8 W	0002 / 0555 / 1209 / 1813	4.7 / 0.4 / 4.6 / 0.1	23 Th	0022 / 0556 / 1225 / 1809	4.4 / 0.7 / 4.3 / 0.6
9 Th	0042 / 0637 / 1250 / 1853	4.7 / 0.3 / 4.7 / 0.1	24 F	0047 / 0622 / 1252 / 1835	4.4 / 0.7 / 4.3 / 0.7
10 F	0122 / 0716 / 1333 / 1930	4.8 / 0.3 / 4.6 / 0.3	25 Sa	0114 / 0647 / 1323 / 1859	4.3 / 0.8 / 4.2 / 0.9
11 Sa	0206 / 0753 / 1420 / 2007	4.7 / 0.5 / 4.5 / 0.6	26 Su	0144 / 0714 / 1357 / 1925	4.2 / 1.0 / 4.1 / 1.1
12 Su	0253 / 0832 / 1514 / 2049	4.5 / 0.8 / 4.2 / 1.0	27 M ☽	0220 / 0745 / 1437 / 2000	4.1 / 1.2 / 3.9 / 1.4
13 M	0349 / 0919 / 1618 / 2146	4.2 / 1.2 / 4.0 / 1.4	28 Tu ☾	0303 / 0825 / 1528 / 2049	4.0 / 1.6 / 3.7 / 1.8
14 Tu ☾	0456 / 1027 / 1740 / 2308	4.0 / 1.6 / 3.8 / 1.8			
15 W	0618 / 1156 / 1917	3.8 / 1.7 / 3.7			

MARCH

Day	Time	m	Day	Time	m
1 W	0357 / 0924 / 1636 / 2202	3.8 / 1.9 / 3.6 / 2.1	16 Th	0557 / 1136 / 1909	3.6 / 1.9 / 3.7
2 Th	0509 / 1051 / 1812 / 2343	3.6 / 2.0 / 3.5 / 2.1	17 F	0032 / 0734 / 1315 / 2030	2.0 / 3.6 / 1.8 / 3.8
3 F	0639 / 1233 / 1947	3.7 / 1.9 / 3.6	18 Sa	0156 / 0845 / 1425 / 2124	1.8 / 3.8 / 1.5 / 4.0
4 Sa	0117 / 0800 / 1354 / 2051	1.8 / 3.9 / 1.5 / 3.9	19 Su	0253 / 0932 / 1514 / 2201	1.5 / 4.0 / 1.2 / 4.2
5 Su	0226 / 0858 / 1451 / 2138	1.4 / 4.0 / 1.0 / 4.2	20 M	0333 / 1008 / 1550 / 2231	1.2 / 4.1 / 0.9 / 4.3
6 M	0318 / 0944 / 1539 / 2218	1.0 / 4.2 / 0.6 / 4.5	21 Tu	0405 / 1038 / 1619 / 2258	0.9 / 4.2 / 0.7 / 4.3
7 Tu ●	0403 / 1026 / 1623 / 2257	0.6 / 4.4 / 0.3 / 4.7	22 W ○	0432 / 1106 / 1646 / 2325	0.8 / 4.3 / 0.6 / 4.4
8 W	0448 / 1106 / 1707 / 2336	0.3 / 4.6 / 0.1 / 4.8	23 Th	0459 / 1130 / 1713 / 2348	0.6 / 4.3 / 0.6 / 4.4
9 Th	0532 / 1146 / 1751	0.1 / 4.7 / 0.0	24 F	0525 / 1157 / 1740	0.6 / 4.3 / 0.6
10 F	0015 / 0615 / 1227 / 1831	4.9 / 0.0 / 4.7 / 0.0	25 Sa	0014 / 0553 / 1224 / 1805	4.4 / 0.7 / 4.3 / 0.7
11 Sa	0055 / 0653 / 1311 / 1908	4.8 / 0.2 / 4.6 / 0.2	26 Su	0040 / 0619 / 1254 / 1831	4.4 / 0.7 / 4.3 / 0.8
12 Su	0138 / 0728 / 1358 / 1945	4.6 / 0.4 / 4.5 / 0.6	27 M	0110 / 0645 / 1328 / 1900	4.3 / 0.9 / 4.1 / 1.1
13 M	0226 / 0807 / 1453 / 2027	4.2 / 0.7 / 4.2 / 1.1	28 Tu	0146 / 0717 / 1409 / 1936	4.2 / 1.1 / 4.0 / 1.4
14 Tu ☽	0321 / 0853 / 1558 / 2124	4.1 / 1.2 / 3.8 / 1.6	29 W	0230 / 0757 / 1502 / 2026	4.0 / 1.5 / 3.8 / 1.8
15 W	0429 / 1001 / 1725 / 2249	3.8 / 1.6 / 3.7 / 2.0	30 Th ☾	0324 / 0857 / 1613 / 2141	3.8 / 1.8 / 3.6 / 2.0
			31 F	0438 / 1022 / 1749 / 2320	3.6 / 1.9 / 3.6 / 2.0

APRIL

Day	Time	m	Day	Time	m
1 Sa	0610 / 1202 / 1919	3.6 / 1.8 / 3.7	16 Su	0121 / 0814 / 1348 / 2050	1.8 / 3.6 / 1.6 / 3.9
2 Su	0051 / 0730 / 1323 / 2020	1.7 / 3.8 / 1.4 / 4.0	17 M	0215 / 0901 / 1436 / 2126	1.6 / 3.8 / 1.3 / 4.1
3 M	0156 / 0829 / 1420 / 2107	1.3 / 3.9 / 0.9 / 4.3	18 Tu	0255 / 0935 / 1511 / 2156	1.2 / 4.0 / 1.0 / 4.2
4 Tu	0248 / 0916 / 1508 / 2147	0.8 / 4.1 / 0.5 / 4.5	19 W	0327 / 1005 / 1541 / 2223	1.0 / 4.1 / 0.8 / 4.3
5 W	0333 / 0958 / 1553 / 2226	0.5 / 4.3 / 0.3 / 4.7	20 Th	0355 / 1033 / 1609 / 2249	0.8 / 4.2 / 0.7 / 4.4
6 Th ●	0420 / 1040 / 1639 / 2307	0.3 / 4.6 / 0.1 / 4.8	21 F ○	0424 / 1102 / 1640 / 2316	0.7 / 4.3 / 0.7 / 4.4
7 F	0504 / 1123 / 1724 / 2348	0.1 / 4.7 / 0.1 / 4.8	22 Sa	0455 / 1130 / 1710 / 2343	0.6 / 4.3 / 0.7 / 4.4
8 Sa	0549 / 1206 / 1808	0.1 / 4.7 / 0.2	23 Su	0525 / 1201 / 1741	0.6 / 4.3 / 0.8
9 Su	0030 / 0629 / 1252 / 1849	4.7 / 0.2 / 4.6 / 0.4	24 M	0013 / 0556 / 1235 / 1813	4.4 / 0.7 / 4.2 / 0.9
10 M	0115 / 0708 / 1342 / 1927	4.5 / 0.5 / 4.4 / 0.7	25 Tu	0046 / 0628 / 1313 / 1846	4.3 / 0.8 / 4.1 / 1.1
11 Tu	0203 / 0747 / 1440 / 2012	4.2 / 0.8 / 4.1 / 1.2	26 W	0125 / 0702 / 1358 / 1927	4.1 / 1.1 / 3.9 / 1.4
12 W	0300 / 0835 / 1546 / 2109	3.9 / 1.2 / 3.8 / 1.6	27 Th ☽	0211 / 0746 / 1452 / 2020	4.0 / 1.3 / 3.8 / 1.7
13 Th	0407 / 0942 / 1711 / 2231	3.7 / 1.7 / 3.7 / 2.0	28 F ☾	0309 / 0845 / 1604 / 2133	3.8 / 1.7 / 3.7 / 1.8
14 F	0534 / 1112 / 1845	3.5 / 1.8 / 3.7	29 Sa	0420 / 1003 / 1726 / 2258	3.7 / 1.7 / 3.7 / 1.8
15 Sa	0004 / 0705 / 1241 / 1959	2.0 / 3.5 / 1.8 / 3.8	30 Su	0542 / 1129 / 1844	3.7 / 1.6 / 3.9

Chart Datum: 2.74 metres below Ordnance Datum (Newlyn)

ENGLAND, SOUTH COAST — SOUTHAMPTON

Lat 50°54′ N Long 1°24′ W

TIMES AND HEIGHTS OF HIGH AND LOW WATERS

TIME ZONE UT(GMT) — For Summer Time add ONE hour in non-shaded areas

YEAR 1989

MAY

Day	Time	m	Day	Time	m
1 M	0017 / 0656 / 1243 / 1942	1.6 / 3.8 / 1.3 / 4.1	16 Tu	0118 / 0811 / 1340 / 2037	1.6 / 3.7 / 1.5 / 3.9
2 Tu	0120 / 0755 / 1341 / 2031	1.2 / 4.0 / 0.9 / 4.3	17 W	0203 / 0852 / 1421 / 2112	1.4 / 3.8 / 1.3 / 4.1
3 W	0212 / 0846 / 1434 / 2114	0.8 / 4.2 / 0.6 / 4.5	18 Th	0241 / 0927 / 1459 / 2143	1.1 / 4.0 / 1.1 / 4.2
4 Th	0301 / 0931 / 1523 / 2156	0.5 / 4.4 / 0.4 / 4.6	19 F	0316 / 1001 / 1532 / 2215	0.9 / 4.1 / 1.0 / 4.3
5 F ●	0349 / 1016 / 1611 / 2239	0.3 / 4.5 / 0.3 / 4.7	20 Sa ○	0352 / 1036 / 1608 / 2246	0.8 / 4.2 / 0.9 / 4.3
6 Sa	0437 / 1103 / 1659 / 2323	0.2 / 4.6 / 0.3 / 4.6	21 Su	0427 / 1111 / 1645 / 2319	0.7 / 4.2 / 0.9 / 4.3
7 Su	0524 / 1150 / 1747	0.2 / 4.5 / 0.4	22 M	0504 / 1147 / 1724 / 2355	0.7 / 4.2 / 0.9 / 4.3
8 M	0009 / 0608 / 1239 / 1830	4.5 / 0.3 / 4.4 / 0.6	23 Tu	0541 / 1224 / 1803	0.7 / 4.2 / 1.0
9 Tu	0056 / 0650 / 1332 / 1914	4.4 / 0.5 / 4.3 / 0.9	24 W	0032 / 0620 / 1307 / 1842	4.2 / 0.8 / 4.1 / 1.1
10 W	0147 / 0732 / 1429 / 1959	4.1 / 0.8 / 4.1 / 1.3	25 Th	0112 / 0659 / 1353 / 1927	4.1 / 0.9 / 4.1 / 1.2
11 Th	0242 / 0820 / 1532 / 2054	3.9 / 1.2 / 3.9 / 1.6	26 F	0201 / 0745 / 1447 / 2019	4.0 / 1.1 / 4.0 / 1.4
12 F ☽	0344 / 0918 / 1644 / 2200	3.7 / 1.5 / 3.7 / 1.8	27 Sa	0256 / 0837 / 1549 / 2119	3.9 / 1.4 / 4.0 / 1.5
13 Sa	0457 / 1030 / 1758 / 2314	3.5 / 1.7 / 3.7 / 1.9	28 Su	0400 / 0941 / 1656 / 2227	3.9 / 1.4 / 4.0 / 1.5
14 Su	0614 / 1145 / 1904	3.5 / 1.7 / 3.7	29 M	0510 / 1051 / 1804 / 2336	3.8 / 1.3 / 4.1 / 1.4
15 M	0023 / 0720 / 1249 / 1957	1.8 / 3.6 / 1.6 / 3.8	30 Tu	0620 / 1201 / 1904	1.2 / 3.9 / 4.2
			31 W	0040 / 0724 / 1303 / 1958	1.2 / 4.0 / 1.0 / 4.3

JUNE

Day	Time	m	Day	Time	m
1 Th	0138 / 0819 / 1401 / 2047	0.9 / 4.2 / 0.9 / 4.4	16 F	0156 / 0851 / 1416 / 2106	1.4 / 3.8 / 1.1 / 4.1
2 F	0233 / 0911 / 1457 / 2133	0.7 / 4.3 / 0.7 / 4.5	17 Sa	0242 / 0934 / 1501 / 2145	1.2 / 4.0 / 1.2 / 4.2
3 Sa ●	0325 / 1000 / 1550 / 2219	0.5 / 4.4 / 0.6 / 4.5	18 Su	0325 / 1015 / 1544 / 2223	1.0 / 4.1 / 1.1 / 4.2
4 Su	0416 / 1051 / 1642 / 2306	0.4 / 4.4 / 0.6 / 4.4	19 M ○	0408 / 1056 / 1629 / 2302	0.8 / 4.2 / 1.0 / 4.3
5 M	0506 / 1141 / 1731 / 2354	0.4 / 4.4 / 0.7 / 4.3	20 Tu	0449 / 1137 / 1713 / 2342	0.7 / 4.2 / 0.9 / 4.3
6 Tu	0552 / 1231 / 1818	0.5 / 4.3 / 0.8	21 W	0532 / 1218 / 1756	0.7 / 4.3 / 0.9
7 W	0042 / 0637 / 1323 / 1902	4.2 / 0.6 / 4.2 / 1.0	22 Th	0021 / 0615 / 1259 / 1839	4.3 / 0.7 / 4.3 / 0.9
8 Th	0130 / 0718 / 1413 / 1944	4.1 / 0.8 / 4.1 / 1.2	23 F	0103 / 0655 / 1343 / 1923	4.2 / 0.7 / 4.3 / 1.0
9 F	0220 / 0801 / 1507 / 2028	3.9 / 1.1 / 3.9 / 1.4	24 Sa	0148 / 0737 / 1431 / 2008	4.2 / 0.8 / 4.3 / 1.1
10 Sa	0311 / 0846 / 1601 / 2115	3.7 / 1.3 / 3.8 / 1.6	25 Su	0239 / 0823 / 1523 / 2058	4.1 / 0.9 / 4.1 / 1.2
11 Su	0407 / 0936 / 1657 / 2211 ☽	3.6 / 1.5 / 3.7 / 1.8	26 M	0334 / 0914 / 1622 / 2154 ☽	4.0 / 1.1 / 4.2 / 1.3
12 M	0508 / 1036 / 1755 / 2312	3.5 / 1.7 / 3.7 / 1.8	27 Tu	0438 / 1014 / 1725 / 2257	4.0 / 1.2 / 4.2 / 1.4
13 Tu	0611 / 1136 / 1851	3.5 / 1.7 / 3.7	28 W	0548 / 1122 / 1830	3.9 / 1.3 / 4.2
14 W	0011 / 0711 / 1236 / 1940	1.7 / 3.6 / 1.7 / 3.8	29 Th	0004 / 0659 / 1232 / 1932	1.4 / 4.0 / 1.3 / 4.2
15 Th	0106 / 0804 / 1328 / 2025	1.6 / 3.7 / 1.5 / 3.9	30 F	0111 / 0804 / 1340 / 2029	1.4 / 4.1 / 1.2 / 4.1

JULY

Day	Time	m	Day	Time	m
1 Sa	0215 / 0902 / 1442 / 2121	1.0 / 4.2 / 1.1 / 4.3	16 Su	0215 / 0913 / 1439 / 2121	1.4 / 3.9 / 1.4 / 4.1
2 Su	0313 / 0957 / 1541 / 2211	0.8 / 4.3 / 0.9 / 4.3	17 M	0306 / 1001 / 1530 / 2206	1.1 / 4.0 / 1.2 / 4.2
3 M ●	0407 / 1048 / 1633 / 2259	0.7 / 4.3 / 0.9 / 4.3	18 Tu ○	0354 / 1043 / 1616 / 2246	0.9 / 4.2 / 1.0 / 4.3
4 Tu	0456 / 1136 / 1721 / 2345	0.6 / 4.4 / 0.8 / 4.3	19 W	0439 / 1124 / 1702 / 2327	0.7 / 4.3 / 0.8 / 4.4
5 W	0541 / 1223 / 1804	0.6 / 4.3 / 0.9	20 Th	0522 / 1203 / 1747	0.5 / 4.4 / 0.7
6 Th	0027 / 0621 / 1306 / 1842	4.2 / 0.6 / 4.3 / 1.0	21 F	0007 / 0605 / 1242 / 1830	4.4 / 0.4 / 4.5 / 0.6
7 F	0110 / 0658 / 1346 / 1917	4.1 / 0.7 / 4.2 / 1.1	22 Sa	0047 / 0645 / 1323 / 1910	4.4 / 0.4 / 4.5 / 0.7
8 Sa	0149 / 0731 / 1426 / 1951	4.0 / 0.9 / 4.1 / 1.2	23 Su	0129 / 0724 / 1405 / 1951	4.4 / 0.5 / 4.5 / 0.7
9 Su	0230 / 0804 / 1508 / 2027	3.9 / 1.1 / 3.9 / 1.4	24 M ☽	0215 / 0803 / 1453 / 2032	4.3 / 0.6 / 4.4 / 0.9
10 M	0314 / 0841 / 1553 / 2108	3.8 / 1.4 / 3.8 / 1.6	25 Tu ☽	0307 / 0847 / 1547 / 2120	4.2 / 0.9 / 4.3 / 1.2
11 Tu	0403 / 0925 / 1642 / 2159	3.6 / 1.6 / 3.7 / 1.8	26 W	0409 / 0940 / 1651 / 2222	4.0 / 1.2 / 4.1 / 1.4
12 W	0500 / 1022 / 1740 / 2300	3.5 / 1.8 / 3.7 / 1.9	27 Th	0522 / 1051 / 1802 / 2338	3.9 / 1.5 / 4.0 / 1.6
13 Th	0607 / 1128 / 1840	3.5 / 1.9 / 3.7	28 F	0643 / 1213 / 1917	3.9 / 1.6 / 4.0
14 F	0009 / 0717 / 1238 / 1940	1.8 / 3.6 / 1.9 / 3.8	29 Sa	0058 / 0803 / 1334 / 2024	1.5 / 4.0 / 1.6 / 4.1
15 Sa	0116 / 0819 / 1343 / 2033	1.7 / 3.7 / 1.7 / 3.9	30 Su	0211 / 0906 / 1442 / 2122	1.3 / 4.1 / 1.4 / 4.2
			31 M	0312 / 0959 / 1540 / 2209	1.0 / 4.3 / 1.1 / 4.2

AUGUST

Day	Time	m	Day	Time	m
1 Tu ●	0403 / 1046 / 1626 / 2253	0.8 / 4.4 / 0.9 / 4.3	16 W	0340 / 1026 / 1602 / 2229	0.8 / 4.3 / 0.9 / 4.4
2 W	0447 / 1127 / 1707 / 2332	0.6 / 4.4 / 0.8 / 4.3	17 Th ○	0423 / 1103 / 1646 / 2308	0.5 / 4.5 / 0.6 / 4.5
3 Th	0525 / 1204 / 1743	0.6 / 4.4 / 0.8	18 F	0505 / 1140 / 1729 / 2345	0.3 / 4.5 / 0.5 / 4.6
4 F	0008 / 0559 / 1239 / 1815	4.3 / 0.6 / 4.4 / 0.8	19 Sa	0547 / 1217 / 1811	0.2 / 4.7 / 0.4
5 Sa	0042 / 0629 / 1310 / 1845	4.2 / 0.7 / 4.3 / 0.9	20 Su	0025 / 0628 / 1255 / 1852	4.6 / 0.2 / 4.7 / 0.4
6 Su	0114 / 0658 / 1343 / 1911	4.2 / 0.8 / 4.2 / 1.0	21 M	0106 / 0705 / 1337 / 1929	4.6 / 0.3 / 4.6 / 0.5
7 M	0147 / 0724 / 1416 / 1940	4.0 / 1.0 / 4.1 / 1.2	22 Tu	0150 / 0742 / 1423 / 2007	4.5 / 0.5 / 4.5 / 0.8
8 Tu	0224 / 0751 / 1453 / 2013	3.9 / 1.2 / 4.0 / 1.4	23 W ☽	0242 / 0823 / 1517 / 2052	4.3 / 0.9 / 4.3 / 1.1
9 W ☽	0306 / 0828 / 1537 / 2056	3.8 / 1.5 / 3.8 / 1.7	24 Th	0345 / 0917 / 1621 / 2155	4.0 / 1.4 / 4.0 / 1.5
10 Th	0358 / 0917 / 1632 / 2157	3.6 / 1.9 / 3.7 / 1.9	25 F	0504 / 1032 / 1743 / 2321	3.8 / 1.8 / 3.9 / 1.7
11 F	0507 / 1028 / 1741 / 2315	3.5 / 2.1 / 3.7 / 2.0	26 Sa	0642 / 1209 / 1913	3.8 / 1.9 / 3.8
12 Sa	0632 / 1157 / 1859	3.5 / 2.1 / 3.7	27 Su	0056 / 0810 / 1337 / 2029	1.7 / 3.9 / 1.8 / 4.0
13 Su	0042 / 0755 / 1319 / 2008	1.9 / 3.6 / 1.9 / 3.8	28 M	0212 / 0912 / 1445 / 2123	1.4 / 4.1 / 1.5 / 4.1
14 M	0156 / 0856 / 1423 / 2103	1.6 / 3.9 / 1.6 / 4.0	29 Tu	0310 / 0958 / 1535 / 2204	1.1 / 4.3 / 1.2 / 4.2
15 Tu	0252 / 0944 / 1516 / 2147	1.2 / 4.1 / 1.2 / 4.2	30 W	0355 / 1035 / 1614 / 2240	0.8 / 4.4 / 0.9 / 4.3
			31 Th ●	0431 / 1108 / 1647 / 2313	0.6 / 4.5 / 0.8 / 4.3

Chart Datum: 2.74 metres below Ordnance Datum (Newlyn)

AREA 2—Central S England

ENGLAND, SOUTH COAST — SOUTHAMPTON

Lat 50°54′ N Long 1°24′ W

TIMES AND HEIGHTS OF HIGH AND LOW WATERS

YEAR 1989

TIME ZONE UT(GMT)
For Summer Time add ONE hour in non-shaded areas

SEPTEMBER

	Time	m		Time	m
1 F	0502 1138 1717 2342	0.6 4.5 0.7 4.4	**16** Sa	0440 1111 1705 2321	0.2 4.8 0.3 4.7
2 Sa	0531 1206 1745	0.6 4.4 0.7	**17** Su	0523 1149 1748	0.1 4.8 0.2
3 Su	0011 0558 1233 1810	4.3 0.6 4.4 0.8	**18** M	0001 0606 1227 1829	4.7 0.1 4.8 0.3
4 M	0039 0622 1300 1835	4.3 0.8 4.3 0.9	**19** Tu	0044 0645 1310 1906	4.7 0.3 4.7 0.4
5 Tu	0109 0646 1330 1900	4.2 0.9 4.2 1.0	**20** W	0130 0723 1356 1945	4.5 0.6 4.5 0.7
6 W	0141 0712 1403 1929	4.1 1.2 4.1 1.3	**21** Th	0224 0805 1451 2031	4.3 1.0 4.2 1.2
7 Th	0221 0745 1446 2008	3.9 1.5 3.9 1.6	**22** F	0329 0901 1559 2137	4.0 1.5 3.9 1.6
8 F	0311 0832 1538 2106	3.7 1.9 3.8 1.9	**23** Sa	0456 1022 1728 2312	3.8 1.9 3.7 1.8
9 Sa	0420 0944 1650 2231	3.5 2.2 3.6 2.1	**24** Su	0641 1205 1907	3.8 2.0 3.7
10 Su	0556 1124 1822	3.5 2.2 3.6	**25** M	0049 0805 1331 2022	1.8 3.9 1.8 3.9
11 M	0014 0732 1258 1942	2.0 3.6 2.0 3.8	**26** Tu	0202 0902 1431 2113	1.5 4.1 1.5 4.1
12 Tu	0135 0835 1404 2040	1.6 3.9 1.6 4.0	**27** W	0254 0942 1516 2149	1.2 4.3 1.2 4.2
13 W	0232 0919 1456 2124	1.2 4.2 1.1 4.2	**28** Th	0333 1013 1549 2221	0.9 4.4 1.0 4.3
14 Th	0316 0958 1540 2204	0.8 4.4 0.8 4.4	**29** F	0405 1041 1619 2248	0.7 4.4 0.8 4.4
15 F	0358 1034 1621 2242	0.4 4.6 0.5 4.6	**30** Sa	0433 1106 1645 2315	0.7 4.5 0.7 4.4

OCTOBER

	Time	m		Time	m
1 Su	0459 1132 1712 2342	0.6 4.5 0.7 4.4	**16** M	0459 1121 1724 2340	0.2 4.8 0.2 4.7
2 M	0525 1158 1738	0.7 4.5 0.7	**17** Tu	0544 1202 1807	0.3 4.8 0.3
3 Tu	0008 0551 1223 1803	4.4 0.8 4.4 0.8	**18** W	0026 0626 1248 1848	4.6 0.4 4.6 0.4
4 W	0038 0617 1253 1830	4.3 0.9 4.3 1.0	**19** Th	0116 0708 1336 1929	4.5 0.7 4.4 0.8
5 Th	0112 0645 1327 1900	4.1 1.2 4.2 1.2	**20** F	0212 0753 1432 2017	4.2 1.2 4.1 1.2
6 F	0152 0718 1409 1939	4.0 1.5 4.0 1.5	**21** Sa	0319 0850 1541 2122	4.0 1.6 3.8 1.6
7 Sa	0241 0806 1502 2034	3.8 1.9 3.8 1.9	**22** Su	0444 1008 1707 2249	3.8 1.9 3.6 1.8
8 Su	0350 0918 1614 2157	3.6 2.1 3.6 2.0	**23** M	0617 1141 1841	3.8 2.0 3.6
9 M	0526 1055 1746 2339	3.5 2.2 3.6 2.0	**24** Tu	0019 0736 1300 1954	1.8 3.9 1.9 3.8
10 Tu	0657 1228 1908	3.7 2.0 3.8	**25** W	0130 0832 1359 2045	1.6 4.0 1.6 3.9
11 W	0100 0800 1335 2008	1.6 4.0 1.5 4.0	**26** Th	0221 0911 1441 2122	1.4 4.2 1.3 4.1
12 Th	0158 0846 1426 2053	1.2 4.3 1.1 4.2	**27** F	0300 0941 1516 2152	1.1 4.3 1.1 4.2
13 F	0246 0925 1509 2135	0.8 4.5 0.7 4.5	**28** Sa	0331 1009 1545 2221	1.0 4.4 0.9 4.3
14 Sa	0329 1003 1554 2215	0.5 4.6 0.4 4.6	**29** Su	0400 1035 1613 2249	0.8 4.4 0.8 4.3
15 Su	0413 1041 1638 2256	0.3 4.8 0.2 4.7	**30** M	0429 1101 1642 2318	0.8 4.5 0.7 4.4
			31 Tu	0458 1129 1711 2347	0.8 4.5 0.7 4.4

NOVEMBER

	Time	m		Time	m
1 W	0526 1157 1742	0.9 4.4 0.8	**16** Th	0015 0612 1231 1834	4.6 0.6 4.5 0.5
2 Th	0019 0557 1228 1812	4.3 1.0 4.3 0.9	**17** F	0108 0657 1322 1918	4.4 0.9 4.3 0.8
3 F	0055 0630 1305 1845	4.2 1.2 4.2 1.1	**18** Sa	0204 0743 1418 2005	4.3 1.2 4.1 1.1
4 Sa	0136 0708 1348 1926	4.0 1.4 4.0 1.4	**19** Su	0306 0835 1519 2100	4.1 1.5 3.8 1.4
5 Su	0228 0757 1441 2019	3.9 1.7 3.9 1.6	**20** M	0415 0937 1629 2208	3.9 1.8 3.7 1.7
6 M	0332 0901 1546 2129	3.8 1.9 3.7 1.8	**21** Tu	0530 1050 1746 2323	3.8 1.9 3.6 1.8
7 Tu	0451 1023 1705 2254	3.7 2.0 3.7 1.8	**22** W	0641 1202 1901	3.8 1.9 3.6
8 W	0611 1145 1824	3.8 1.8 3.8	**23** Th	0033 0741 1304 1959	1.7 3.8 1.8 3.7
9 Th	0013 0716 1253 1929	1.6 4.0 1.5 4.0	**24** F	0129 0826 1353 2043	1.6 4.0 1.6 3.9
10 F	0115 0807 1348 2021	1.2 4.3 1.1 4.2	**25** Sa	0213 0902 1433 2120	1.4 4.1 1.3 4.0
11 Sa	0209 0851 1438 2107	0.9 4.5 0.8 4.4	**26** Su	0252 0934 1510 2153	1.3 4.2 1.1 4.1
12 Su	0258 0934 1526 2153	0.7 4.6 0.5 4.5	**27** M	0326 1006 1544 2228	1.1 4.3 1.0 4.2
13 M	0346 1016 1613 2237	0.5 4.7 0.4 4.6	**28** Tu	0400 1036 1617 2301	1.0 4.4 0.8 4.3
14 Tu	0437 1059 1701 2326	0.4 4.7 0.3 4.6	**29** W	0435 1107 1652 2335	1.0 4.4 0.8 4.3
15 W	0526 1144 1749	0.5 4.7 0.3	**30** Th	0511 1140 1728	1.0 4.4 0.8

DECEMBER

	Time	m		Time	m
1 F	0010 0547 1215 1804	4.3 1.0 4.3 0.8	**16** Sa	0059 0646 1309 1905	4.5 0.8 4.5 0.7
2 Sa	0048 0625 1254 1841	4.2 1.1 4.3 0.9	**17** Su	0149 0728 1358 1946	4.3 1.0 4.1 0.9
3 Su	0129 0706 1335 1921	4.2 1.2 4.2 1.1	**18** M	0240 0810 1448 2028	4.2 1.2 4.0 1.2
4 M	0216 0750 1424 2007	4.1 1.3 4.0 1.3	**19** Tu	0332 0855 1541 2114	4.0 1.5 3.8 1.5
5 Tu	0310 0843 1520 2102	4.0 1.6 3.9 1.4	**20** W	0427 0944 1639 2209	3.8 1.7 3.6 1.7
6 W	0411 0946 1626 2208	3.9 1.7 3.9 1.5	**21** Th	0526 1044 1744 2312	3.7 1.9 3.5 1.8
7 Th	0520 1055 1738 2320	4.0 1.7 3.9 1.5	**22** F	0626 1148 1851	3.7 1.9 3.5
8 F	0628 1204 1848	4.1 1.5 3.9	**23** Sa	0017 0723 1251 1951	1.9 3.7 1.8 3.6
9 Sa	0030 0729 1309 1951	1.4 4.2 1.3 4.1	**24** Su	0117 0813 1347 2045	1.8 3.8 1.6 3.8
10 Su	0134 0821 1407 2046	1.2 4.3 1.0 4.3	**25** M	0211 0858 1435 2130	1.6 3.9 1.4 3.9
11 M	0232 0910 1503 2138	0.9 4.5 0.7 4.4	**26** Tu	0257 0938 1521 2210	1.4 4.1 1.2 4.1
12 Tu	0328 0958 1556 2230	0.8 4.6 0.5 4.5	**27** W	0340 1016 1602 2249	1.2 4.2 1.0 4.2
13 W	0423 1046 1648 2319	0.7 4.6 0.4 4.6	**28** Th	0421 1053 1641 2327	1.1 4.3 0.8 4.3
14 Th	0514 1133 1736	0.7 4.5 0.4	**29** F	0502 1129 1719	1.0 4.4 0.7
15 F	0010 0602 1222 1822	4.5 0.7 4.4 0.5	**30** Sa	0003 0542 1205 1758	4.4 0.9 4.4 0.7
			31 Su	0040 0620 1243 1836	4.4 0.9 4.4 0.7

Chart Datum: 2.74 metres below Ordnance Datum (Newlyn)

AREA 2—Central S England   253

SOUTHAMPTON continued

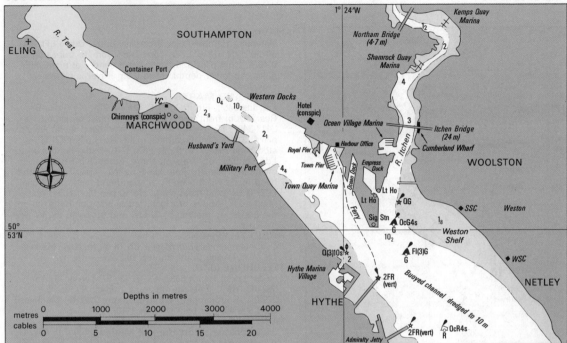

FACILITIES
EC Monday or Wednesday (but few close on weekdays); **Ocean Village Marina** (450+50 visitors) Tel. 229385, FW, V, R, Bar, CH, Slip, Gas, Gaz, Kos, ME, Sh, Lau, P, D, (Access H24); **Shamrock Quay Marina** (220+40 visitors) Tel. 229461, BH (50 ton), C (12 ton), ME, El, Sh, SM, Lau, R, Bar, AC, CH, FW, Gas, Gaz, Kos, V; **Hythe Marina** (180+50 visitors) Tel. 849263, BH (30 ton), C (12 ton), AC, P, D, El, ME, Sh, FW, SM, (Access H24); **Kemp's Quay Marina** (180+5 visitors) Tel. 32323, AC, C (5 ton), D, FW, Gas, ME, (Access HW∓3½); **Town Quay Marina** (400) Tel. 211080, R, Bar; **Itchen Marina** Tel. 631500, D, BY, C (12 ton); **Royal Southampton YC** Tel. 223352; **Hythe SC** Tel. 846563; **Marchwood YC** Tel. 864641; **Netley SC** Tel. 454272; **Southampton SC** Tel. 446575; **Weston SC** Tel. 452527; **Hythe Marina Yacht Services** Tel. 840460, ME, El, Sh; **Kelvin Hughes** Tel. 631286, CH, ACA; **Stephen Ratsey** Tel. 221683, SM; **Pumpkin Marine** Tel. 229713, CH; **Southampton Yacht Services** Tel. 335266, Sh; **Solent Rigging** Tel. 39976, Rigging; **Southern Spar Services** Tel. 331714, Spars; **Alpha Sails** Tel. 553623, SM; **G Sailmakers** Tel. 221453, SM; **Ullman Sails** Tel. 335596, SM; **B. D. Marine** Tel. 220178, ME, El; **Waters Yacht Engineers** Tel. 220144, ME, El; **Boat Shop** Tel. 449338, CH; **Belsize BY** Tel. 671555, CH; **Larry Marks** Tel. 447037, CH; **Shamrock Chandlery** Tel. 632725, CH; **Captain Pumpkin** Tel. 229713, CH; **Bosun's Locker** Tel. 849046, CH; **Hards** at Crackmore, Eling, Mayflower Park (Test), Northam (Itchen). **City** V, R, Bar, Bank, PO, Rly, Air, Ferries/Hydrofoil to IoW Tel. 333042.

HAMBLE 10-2-24
Hampshire

CHARTS
Admiralty 1905, 2022; Stanford 11; Imray C3; OS 196

TIDES
+0130, −0010 Dover; ML 2.9; Zone 0 (GMT)

Standard Port SOUTHAMPTON (←)

Times				Height (metres)			
HW		LW		MHWS	MHWN	MLWN	MLWS
0400	1100	0000	0600	4.5	3.7	1.8	0.5
1600	2300	1200	1800				

Differences WARSASH
+0020 +0010 +0010 0000 0.0 +0.1 +0.1 +0.3
BURSLEDON
+0020 +0020 +0010 +0010 +0.1 +0.1 +0.2 +0.2
CALSHOT CASTLE
+0015 +0030 +0015 +0005 0.0 0.0 +0.2 +0.3

NOTE: Double high waters occur at or near springs — on other occasions there is a stand which lasts about 2 hours. Predictions are for the first HW where there are two or for the middle of the stand. See 10.2.8.

SHELTER
Excellent with many marinas and boat yards. Anchoring prohibited.

HAMBLE continued

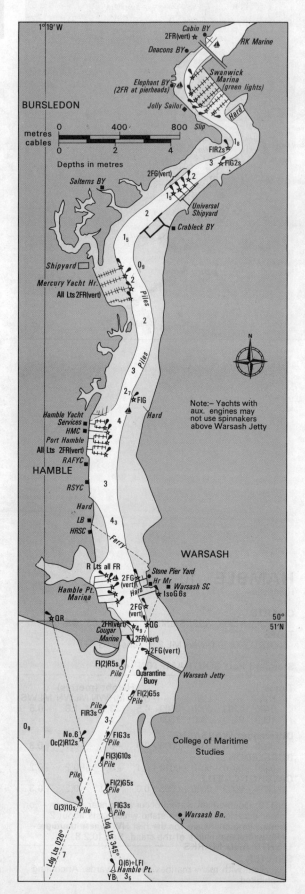

NAVIGATION
Waypoint Hamble Point S cardinal buoy, Q(6)+LFI 15s, 50°50'.12N 01°18'.57W, 168°/348° from/to front Ldg Lt, 0.48M. Unlit piles and buoys are a danger at night. NE gales can lower depths up to 0.6m. River is extremely crowded.

LIGHTS AND MARKS
Ldg Lts 345°. Front Oc(2) R 12s 4m 2M; No 6 pile Bn. Rear, 820m from front, QR 12m; W mast on shore; vis 341°-349°. No 1 pile FlG 3s 3M; stbd-hand mark. No 2 pile Q(3) 10s 3M; E cardinal mark. No 3 pile Fl(2) G 5s; stbd-hand mark. No 5 pile Fl(3) G 10s; stbd-hand mark. Ldg Lts 026° on Warsash shore. Front QG; B&W chequered pile Bn; Rear Iso G 6s. No 7 pile FlG 3s; stbd-hand mark. No 8 pile FlR 3s; port-hand mark. No 9 pile Fl(2) G 5s; stbd-hand mark. No 10 pile Fl(2) R 5s; port-hand mark. Above Warsash, jetties and pontoons etc on the E side are marked by G Lts, and those on the W side by R Lts.

RADIO TELEPHONE
Call: *Hamble Harbour Master* Ch 16 68 (occas). Marinas Ch M. See also Southampton 10.2.23.

TELEPHONE (0703)
Hr Mr Locks Heath 6387; CG & MRSC Lee-on-Solent 552100; HMC Freefone Customs Yachts: RT (0703) 229251; Southampton Weather Centre 228844; Marinecall 0898 500 457; Dr Locks Heath 3110.

FACILITIES
EC Wednesday; **Port Hamble Marina** (369) Tel. 452741, AC, Bar, BH (36 ton), C (7 ton), Slip, CH, D, P, El, FW, Gas, Gaz, ME, Sh, SM, V, (Access H24); **Mercury Yacht Harbour** (323) Tel. 452741, AC, Bar, BH (20 ton), CH, D, P, El, FW, Gas, Gaz, ME, Sh, SM, V, (Access H24); **Hamble Point Marina** (200+10 visitors) Tel. 452464, AC, Bar, BH (40 ton), C (8 ton), CH, D, El, FW, Gas, Gaz, ME, Sh, V, (Access H24); **Moody Marina** (350+50 visitors) Tel. Locks Heath 885000, AC, BH (60 ton), Slip, C (3 ton), CH, ACA, D, El, FW, Gas, Gaz, ME, Sh, V, Lau, P, Sh, SM, (Access H24); **Cougar Marina** Tel. 453513, BY, Sh, Slip, P, D, Access HW∓4; **Royal Southern YC** Tel. 453271; **RAF YC** Tel. 452208; **Hamble River SC** Tel. 452070; **Warsash SC** Tel. Locks Heath 583575; **Aladdin's Cave** Tel. Bursledon 2182, CH; **Cabin BY** Tel. Bursledon 2516, AB, C (3 ton), El, FW, Gas, M, ME, R, Sh, Slip; **Compass Point Chandlery** Tel. 452388, CH; **Crableck BY** Tel. Locks Heath 572570, Sh; **Ditty Box** Tel. 453436, CH; **Elephant BY** Tel. Bursledon 3268, ME, El, Sh, Slip; **Foulkes** Tel. Bursledon 2182, C (10 ton), CH; **Greenham Marine** Tel. Locks Heath 575166, El, Radio; **Hamble River BY** Tel. Locks Heath 572318, ME, Sh, Slip; **Hamble Yacht Services** Tel. 454111, BY, BH (36 ton), El, ME, Sh, Slip (100 ton); **BK Marine** Tel. Locks Heath 572170, El; **Hudson Marine Electronics** Tel. 455129, El; **Victoria Rampart** Tel. Locks Heath 885400, BH (25 ton), BY, C (12 ton), D, El, FW, ME, P, Sh; **Sea Fever** Tel. Locks Heath 582804, CH, Gaz; **Universal Shipyards** (110) Tel. Locks Heath 574272, El, ACA, BH (50 ton), M, CH, FW, BY, C, ME, Sh, Slip; **Stone Pier Yard** Tel. Locks Heath 885400, C, Slip, BY, Sh, D, FW; **Warsash Nautical Bookshop** Tel. Locks Heath 572384, ACA;

Divers
Andark Tel. Locks Heath 581755 (emergency/after hours Tel. Botley 6006); **Warsash Divers** Tel. 0860-627800.

Sailmakers
Alpha Sails Tel. 553623; **Bruce Banks** Tel. Locks Heath 582444; **Hamble Sailing Services** Tel. 455868; **Lucas** Tel. 452247; **Shoe Sails** Tel. Locks Heath 589450; **Relling Sails** Tel. Titchfield 46816; **Richardson Sails** Bursledon 3914; **Sobstad** Tel. 456205; **Ullman Sails** Tel. 454254; **Williams Sails** Tel. 453109.

Hards at Warsash, Hamble and Swanwick. Slips at Warsash, Hamble, Bursledon and Lower Swanwick. PO (Hamble, Bursledon, Warsash and Lower Swanwick); Bank (Hamble, Bursledon, Sarisbury Green, Swanwick, Warsash); Rly (Hamble and Bursledon); Air (Southampton).

'Hamble River Guide' available from Hr Mr.

AREA 2—Central S England   255

WOOTTON CREEK   10-2-25
Isle of Wight

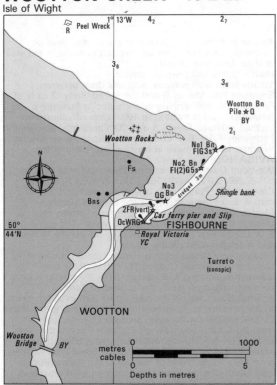

CHARTS
Admiralty 2022, 394; Stanford 11; Imray C3, Y20; OS 196
TIDES
+0023 Dover; ML 2.8; Zone 0 (GMT)

Standard Port PORTSMOUTH (→)

Times				Height (metres)			
HW		LW		MHWS	MHWN	MLWN	MLWS
0000	0600	0500	1100	4.7	3.8	1.8	0.6
1200	1800	1700	2300				

Differences RYDE
−0010 +0010 −0005 −0010 −0.2 −0.1 0.0 +0.1

NOTE: Wootton Creek comes under the authority of the Queen's Harbour Master, Portsmouth but it is exercised through a private association, The Wootton Creek Fairways Committee. Sec. R. Perraton Tel. 882225. For special tidal information see 10.2.8

SHELTER
Good except when stormy winds in N or E. The creek dries at MLWS above ferry terminal. Anchoring in the fairway prohibited.
NAVIGATION
Waypoint Wootton N cardinal Bn, Q, 50°44'.47N 01°12'.08W, 043°/223° from/to ferry slip, 0.59M. Beware large Sealink ferries; ferries leave astern and turn at Wootton Bn. It is difficult to beat in on the ebb. Speed limit 5 kn.
LIGHTS AND MARKS
Entrance to creek due S of SE Ryde Middle buoy and 1¾ M W of Ryde Pier. Once in channel follow four beacons on stbd side. Keep in W sector of Lt at root of ferry pier Oc WRG 10s G221°-224°, W224°-225½°, R225½°-230½°. By ferry terminal, turn on leading marks on W shore (two marks, an upright and an inverted triangle, which together form a diamond when in line).
RADIO TELEPHONE
None.
TELEPHONE (0983)
Royal Victoria YC 882325; CG 872886; MRSC Lee-on-Solent 552100; Fairways Committee 882763; HMC Freefone Customs Yachts: RT (0703) 229251; Marinecall 0898 500 457; Dr 882542.

FACILITIES
EC Wootton Bridge — Wednesday, Wootton Village — Thursday; **Royal Victoria YC** Tel. 882325, Slip, M, FW, V, R, Bar; **Marine Installations** Tel. 883960, Slip, ME, El, Sh;
PO (Ryde, Wootton Bridge); Bank (Ryde); Rly (ferry to Portsmouth); Air (Southampton).

BEMBRIDGE   10-2-26
Isle of Wight

CHARTS
Admiralty 2022, 2050, 2045; Stanford 11; Imray C3, C9; OS 196
TIDES
+0020 Dover; Zone 0 (GMT)

Standard Port PORTSMOUTH (→)

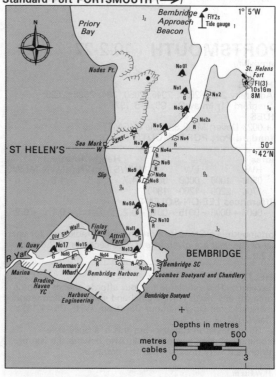

Times				Height (metres)			
HW		LW		MHWS	MHWN	MLWN	MLWS
0000	0600	0500	1100	4.7	3.8	1.8	0.6
1200	1800	1700	2300				

Differences BEMBRIDGE
−0010 +0005 +0020 0000 −1.6 −1.5 −1.4 −0.6
VENTNOR
−0025 −0030 −0025 −0030 −0.8 −0.6 −0.2 +0.2
SANDOWN
0000 +0005 +0010 +0025 −0.6 −0.5 −0.2 0.0

NOTE: For special tidal conditions, see 10.2.8

SHELTER
Good. Very occasional trouble in NNE gale conditions. Entry HW −3 to HW +2.
NAVIGATION
Waypoint Tide gauge, Fl Y 2s, 50°42'.54N 01°05'.23W, close N of entrance to buoyed channel. The bar dries. Avoid the gravel banks between St. Helen's Fort, Nodes Pt and on to Seaview, by keeping to entry times above. Area round St. Helen's Fort is a prohibited anchorage. Harbour speed limit 6 kn.
LIGHTS AND MARKS
Lt on tide gauge at the entrance, Fl Y 2s 1M.
RADIO TELEPHONE
Bembridge Marina VHF Ch 16; M.

BEMBRIDGE continued

TELEPHONE (0983)
Hr Mr 872828; CG 872886; Berthing Master 874436; MRSC Lee-on-Solent 552100; HMC Freefone Customs Yachts: RT (0703) 229251; Marinecall 0898 500 457; Dr 872614

FACILITIES
EC Thursday; **Marina** (40+100 visitors) Tel. 874436, FW, ME, El, D, AC, Lau, V, R, Bar; Access HW−3 to HW+2; **F. Attrill & Sons** Tel. 872319, Slip, M, ME, El, Sh, CH; **Spinnaker Yacht Chandlery** Tel. 874324, CH, Gas; **Alan Coombes** Tel. 872296, Slip, M, FW, ME, D, El, Sh, CH; **St. Helen's Quay** P, D, FW, CH; **Harbour Engineering** Tel. 872306, D, FW, ME; **Bembridge Outboards** Tel. 872817, ME, Sh; **Bembridge BY** Tel. 872423, Slip, M, ME, Sh; **Stratton BY** Tel. 873185, D; **Bembridge SC** Tel. 872683; **Brading Haven YC** Tel. 872289; Various other boatyards with limited facilities. **Town** P, D, CH, V, R, Bar. PO (Bembridge, St. Helens); Bank (Bembridge); Rly (Brading); Air (Southampton).

PORTSMOUTH 10-2-27
Hampshire

CHARTS
Admiralty 2625, 2631, 2045, 2050, 394, 2628, 2629; Stanford 11; Imray C3, C9; OS 197

TIDES
+0023 Dover; ML 2.8; Zone 0 (GMT)
Standard Port PORTSMOUTH (→)

Times				Height (metres)			
HW		LW		MHWS	MHWN	MLWN	MLWS
0500	1000	0000	0600	4.7	3.8	1.8	0.6
1700	2200	1200	1800				

Differences LEE-ON-SOLENT
−0005 +0005 −0015 −0010 −0.2 −0.1 +0.1 +0.2

Portsmouth is a Standard Port and tidal predictions for each day of the year are given below. See sections 10.2.8 and 10.2.9.

SHELTER
Excellent. This very large harbour affords shelter in some area for any wind. Navigational and shipping information available from Duty Harbour Controller (VHF Ch 11 13). Call *QHM*.
Good shelter in Camber, but this is a busy little commercial dock and often full. Also beware the Isle of Wight car ferry docking near the entrance.
Anchor in Portchester Lake near the Castle, clear of moorings. Otherwise pick up a vacant mooring in the area desired, land and consult local YC.
Hard is public but piers are private.
Yachts fitted with engines must use them between Southsea War Memorial and the Ballast Buoy (2 ca N of Fort Blockhouse). Fishing or anchoring in fairways is forbidden. Yachts must keep 100m clear of submarines berthed in Haslar Creek. If over 20m in length, ask QHM's permission (VHF R/T Ch 11) or telephone for permission to enter, leave or move in harbour, especially in fog.

NAVIGATION
Waypoint Outer Spit S cardinal buoy, Q(6)+LFl 15s, 50°45'.55N 01°05'.41W, 185°/005° from/to Southsea Castle, 1.1M. Portsmouth is a major naval port and all comes under the authority of the Queen's Harbour Master. On approaching from E inshore, yachts may go through the gap in the submerged barrier. The gap is 1 M S of Lumps Fort and is marked by a pile on the N side and a dolphin (QR) on the S side.
Beware: very strong tides in Harbour mouth; Commercial shipping and ferries; Gosport ferry; HM Ships and submarines entering and leaving. (HM Ships hoist a Code pendant, ship's pendants and a pendant Zero).
A Small Boat Channel for craft under 20m in length exists at the entrance to Portsmouth Harbour. It lies to the West of (and parallel to) the main dredged channel, and runs from No 4 Bar Buoy (QR, off Clarence Pier) to the north end of Fort Blockhouse, extending about 50m off the latter. Yachts entering harbour by the Small Boat Channel must keep to the West of the line to the Ballast Buoy until North of same. Yachts entering on the Portsmouth side of the main dredged channel must keep well to starboard, clear of the main channel. All yachts leaving harbour must use the Small Boat Channel, as described above.
Yachts crossing the entrance may only do so to the N of Ballast Buoy or to the S of No 4 Bar Buoy. Dir Lt on dolphin S of Cold Harbour jetty, Dir WRG 2m 1M, Iso G 2s 322.5°-330°, Al WG 330°-332.5°, Iso W 2s 332.5°-335° (covering the main channel), Al WR 335°-337.5°, Iso R 2s 337.5°-345° (covering the small boat channel).
Haslar Creek; a reserved area, prohibited to yachts, exists in Haslar Creek near all submarines berthed at HMS Dolphin. Speed limit within the harbour 10 kn.

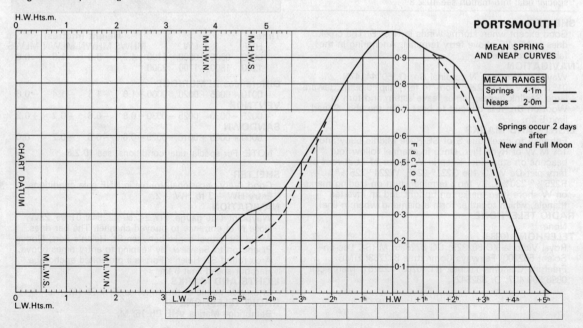

PORTSMOUTH
MEAN SPRING AND NEAP CURVES

MEAN RANGES
Springs 4·1m
Neaps 2·0m

Springs occur 2 days after New and Full Moon

AREA 2 — Central S England 257

PORTSMOUTH continued

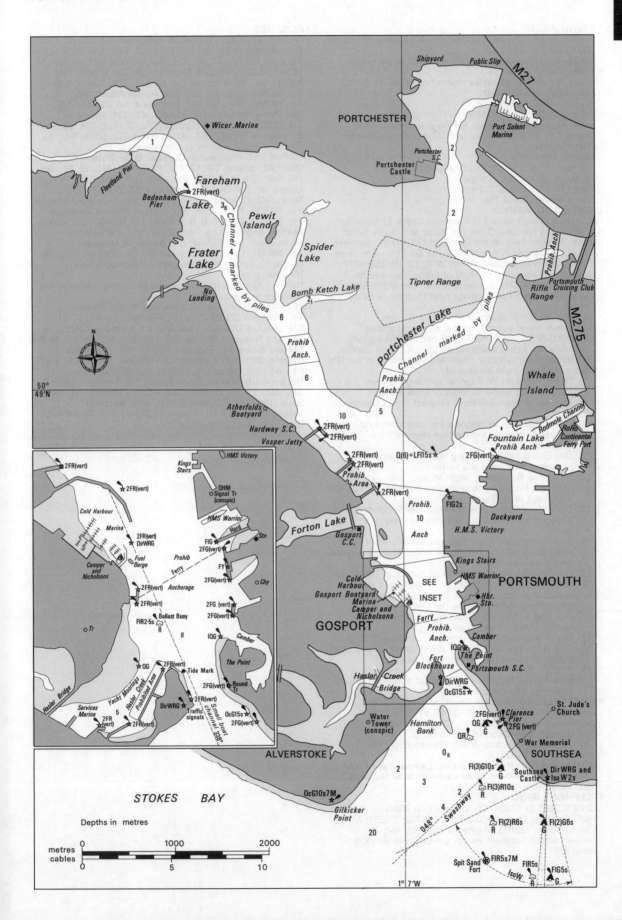

PORTSMOUTH continued

FIRING AREAS
Fraser Gunnery Range, Eastney, in use Monday to Friday, 0900-1600.
Details from Portsmouth 822351 ext 6420 or from Radio Solent (see 10.2.9).
Areas and sectors of circles centred on 50°47′ N, 01°02′ W
Area 1 200 yds 120°-200°
Area 2 6,000 yds 120°-155°
Area 3 18,000 yds 120°-155°
From 30 mins before firing, R Flag with International Numeral (1 for area 1, 2 for area 2, etc) displayed from:
Range Building
Hayling Island YC
Admiralty Trial Station, W of Langstone Harbour
FS by Hr Mr's Office, Langstone
FS by Hr Mr's Office, Itchenor
FS on South Parade Pier

LIGHTS AND MARKS
St Jude's church spire and Southsea Castle Lt Ho in line at 003° lead between Outer Spit buoy and Horse Sand buoy. At night keep in the W sector (000°-003°) between the Al WG and Al WR sectors of the Dir Lt on Southsea Castle, which also shows an Iso W 2s Lt, vis 339°-066° Swashway Channel at the NW of Elbow Spit and No 2 buoy, St Jude's Church spire and the War Memorial in line at 048°. High rise buildings tend to obscure St Jude's Church.
At night the Oc W sector of the Dir Lt on Fort Blockhouse (west side of harbour entrance) shows 318.5°-321.5° between the Al WG and Al WR sectors. The following signals displayed at Central Signal Station, Fort Blockhouse, Gilkicker Point and sometimes in the HM Ship concerned, must be obeyed.

(1) Day-R flag with W diagonal bar; Night-R over 2 G Lts(vert): No vessel to leave or enter Harbour Channel or approach from seaward N of Outer Spit buoy.
(2) Day-R flag with W diagonal bar over one B ball: No vessel to enter the harbour channel from seaward.
(3) Day-One B ball over R flag with W diagonal bar: No vessel may leave the harbour.
(4) Day-Large B pendant; Night-W light over two R Lts(vert): No vessel to anchor in the Man-of-War anchorage at Spithead.
(5) Day-International Code Pendant over Pendant Zero. Keep clear of HM Ships entering, leaving or shifting berth.
(6) Day-International Code Pendant over Pendant 9; Night-3 G Lts(vert): HM Ships under way; all vessels give a wide berth.
(7) Day-International Code Pendant over flags NE; Night-G over R Lt: proceed with great caution at slow speed. It is hoisted at Fort Blockhouse when ships, other than car ferries leave the Camber.
(8) Day-Flag E; Night-R Lt: Submarine entering or leaving Haslar Creek; keep clear.
(9) Day-International Code Pendant over Flag A; Night-Two R Lts horizontal: Have divers down. Diving boat shows all round R Lt in stern and bows.
(10) Traffic Lights.
International Port Traffic Signals at entrance, shown from Fort Blockhouse.

RADIO TELEPHONE (local times)
Queens Harbour Master (call: *QHM*) VHF Ch 11 (commercial vessels & yachts), Ch 13 (RN vessels) (H24).
Fort Gilkicker (call *Gilkicker*) Ch 16 (H24).
Call: *Portsmouth Harbour Radio* (Commercial Port Manager) Ch 11 (H24).
Camper and Nicholson Marina Call: *Camper Base* Ch M (office hours), Fareham Yacht Harbour Ch M (summer 0900-1800, winter Sun 0900-1300).

TELEPHONE (0705)
QHM 822351 Ex 23694; Commercial Port Manager 820436; HMC Freefone Customs Yachts: RT (0703) 229251; MRSC Lee-on-Solent 552100; Marinecall 0898 500 457; Dr Gosport 80922; Fareham Health Centre Fareham 282911; Hosp 822331

FACILITIES
PORTSMOUTH EC Wednesday (Southsea Saturday)
Port Solent (900) Tel. 210765, P, D, FW, ME, El, Sh, BH (30 ton), CH, V, AC, R, Bar, Gas, Access H24 via lock (43m × 12.5m); **Vosper Thornycroft** Tel. 379481, FW, ME, El, Sh; **W G Lucas** Tel. 826629, SM; **Tudor S C** Tel. 662002, Slip, M, FW, Bar; **Royal Albert YC** Tel. 820594, Slip, M, Bar; **Ron Hale** Tel. 732985, ME, CH; **Peter Smailes** Tel. 829386, ME, CH; **Nautech** Tel. 693611, El; **Chris Hornsey** Tel. 734728, CH; **Gieves & Hawkes** Tel. 821351, ACA; **Town** PO, Bank, Rly, Air (Southampton). Ferries to Continent and IoW.

GOSPORT EC Wednesday **Camper & Nicholson Marina** (350 + 40 visitors) Tel. 524811, P, D, FW, ME, El, Sh, BH (40 ton), CH, V, R, AC, Bar, Gas, Gaz, SM, Lau; **W G Lucas** Tel. 504434, CH; **Hardway SC** Tel. 581875, Slip, M, L, FW, C (mast stepping only), AB; **Hardway Marine** Tel. 580420, P, D, ACA, CH; **Solent Marine Services** Tel. 584622, El, CH; **B & H Sails** Tel. 589431, CH; **Ostar Marine Engineering** Tel. 526522, ME, CH; **Allday Aluminium** Tel. 587741, Sh; **Barter & Hugget** Tel. 589431, CH, SM; **Gosport BY** Tel. 586216, Slip, ME, El, Sh, C; **Town** PO, Bank, Rly (Portsmouth), Air (Southampton)
Note:— Approval has been given to build a marina at the entrance to Haslar Creek.

FAREHAM EC Wednesday **Fareham Yacht Harbour** Tel. Fareham 232854, Slip, D, FW, ME, El, Sh, V, Bar, CH; (Access HW∓3); **Ladyline** Tel. Fareham 234297, M, Sh, CH; **Hamper Marine** Tel. Fareham 280203, Sh; **M B Marine Sales** Tel. Fareham 234277, ME, CH: **Wicor Marine** (200) Tel. Fareham 237112, Slip, M, D, ME, Sh, CH, AC, BH (10 ton), C (7 ton), FW, El, Gas, Gaz; **Portsmouth Marine Engineering** Tel. Fareham 232854, Slip, ME, El, Sh, D, FW, Gas, AC, C (10 ton), CH; **North Sails (UK)** Tel. Fareham 231525, SM; **Sobstad Sailmakers (UK)** Tel. Fareham 233242, SM; **Town** PO, Bank, Rly, Air (Southampton).

ENGLAND, SOUTH COAST - PORTSMOUTH
LAT 50°48′N   LONG 1°07′W
TIMES AND HEIGHTS OF HIGH AND LOW WATERS

TIME ZONE UT(GMT)
For Summer Time add ONE hour in non-shaded areas

YEAR 1989

JANUARY

	TIME	m		TIME	m
1 SU	0523 1046 1747 2312	3.9 1.9 3.6 1.9	**16** M	0600 1116 1833 2354	4.1 1.5 3.8 1.6
2 M	0623 1152 1858	3.8 1.9 3.6	**17** TU	0710 1234 1955	4.0 1.5 3.9
3 TU	0021 0726 1257 2005	1.9 3.9 1.8 3.7	**18** W	0111 0817 1347 2105	1.7 4.0 1.4 4.0
4 W	0123 0825 1354 2103	1.8 4.0 1.6 3.9	**19** TH	0217 0918 1448 2204	1.6 4.1 1.3 4.2
5 TH	0216 0918 1443 2153	1.6 4.2 1.4 4.1	**20** F	0312 1012 1539 2252	1.4 4.2 1.1 4.3
6 F	0304 1006 1528 2238	1.4 4.3 1.2 4.3	**21** SA	0359 1100 1623 2335 ○	1.2 4.3 1.0 4.4
7 SA ●	0349 1051 1612 2321	1.2 4.5 1.0 4.4	**22** SU	0439 1142 1701	1.1 4.4 0.9
8 SU	0432 1134 1655	1.0 4.5 0.8	**23** M	0013 0517 1218 1736	4.4 1.0 4.4 0.8
9 M	0004 0515 1216 1738	4.5 0.9 4.6 0.7	**24** TU	0048 0551 1252 1808	4.5 1.0 4.3 0.8
10 TU	0049 0559 1300 1823	4.6 0.9 4.6 0.6	**25** W	0121 0624 1325 1840	4.5 1.1 4.3 0.9
11 W	0136 0644 1345 1908	4.7 0.9 4.6 0.6	**26** TH	0155 0657 1359 1912	4.4 1.1 4.3 1.0
12 TH	0224 0730 1433 1954	4.7 0.9 4.5 0.6	**27** F	0229 0730 1433 1946	4.4 1.2 4.2 1.1
13 F	0314 0816 1522 2043	4.7 1.0 4.4 0.8	**28** SA	0301 0805 1508 2022	4.3 1.4 4.0 1.3
14 SA ☽	0405 0907 1617 2136	4.5 1.2 4.2 1.1	**29** SU	0335 0843 1547 2104	4.1 1.5 3.9 1.6
15 SU	0459 1005 1719 2239	4.3 1.4 4.0 1.4	**30** M ☾	0415 0931 1637 2157	3.9 1.7 3.6 1.8
			31 TU	0508 1034 1744 2310	3.7 1.9 3.5 2.0

FEBRUARY

	TIME	m		TIME	m
1 W	0619 1155 1911	3.7 1.9 3.5	**16** TH	0102 0807 1342 2106	1.9 3.7 1.6 3.8
2 TH	0034 0739 1315 2031	2.0 3.7 1.7 3.7	**17** F	0215 0915 1446 2202	1.7 3.9 1.4 4.0
3 F	0147 0848 1419 2132	1.7 3.9 1.5 4.0	**18** SA	0308 1007 1533 2245	1.5 4.1 1.1 4.2
4 SA	0244 0946 1511 2222	1.4 4.2 1.1 4.3	**19** SU	0351 1050 1611 2320	1.2 4.2 0.9 4.4
5 SU	0332 1035 1557 2306	1.1 4.4 0.8 4.5	**20** M ○	0426 1126 1644 2352	1.0 4.3 0.8 4.5
6 M	0416 1120 1640 2348	0.8 4.6 0.6 4.7	**21** TU	0459 1159 1715	0.9 4.4 0.7
7 TU	0500 1201 1724	0.7 4.7 0.4	**22** W	0022 0530 1230 1746	4.5 0.8 4.4 0.7
8 W	0032 0544 1243 1807	4.8 0.6 4.8 0.3	**23** TH	0053 0601 1302 1816	4.5 0.8 4.3 0.7
9 TH	0117 0627 1328 1852	4.9 0.6 4.8 0.3	**24** F	0124 0631 1334 1844	4.5 0.9 4.3 0.8
10 F	0203 0711 1413 1935	4.9 0.6 4.7 0.4	**25** SA	0155 0658 1405 1913	4.4 0.9 4.2 0.9
11 SA	0249 0753 1501 2019	4.8 0.8 4.5 0.7	**26** SU	0224 0729 1436 1941	4.3 1.1 4.1 1.1
12 SU	0335 0839 1550 2107	4.6 1.0 4.3 1.1	**27** M	0252 0758 1508 2016	4.2 1.3 3.9 1.4
13 M	0423 0932 1650 2207	4.3 1.3 4.0 1.5	**28** TU ☾	0326 0838 1549 2103	4.0 1.5 3.7 1.7
14 TU	0524 1044 1810 2329	4.0 1.6 3.7 1.8			
15 W	0643 1216 1947	3.8 1.7 3.6			

MARCH

	TIME	m		TIME	m
1 W	0412 0934 1650 2214	3.7 1.8 3.5 2.0	**16** TH	0617 1157 1934	3.6 1.8 3.6
2 TH	0524 1059 1826 2354	3.5 1.9 3.4 2.0	**17** F	0050 0753 1328 2053	2.0 3.6 1.7 3.7
3 F	0658 1238 2001	3.6 1.8 3.6	**18** SA	0205 0903 1430 2146	1.7 3.7 1.4 4.0
4 SA	0121 0821 1353 2108	1.7 3.8 1.4 4.0	**19** SU	0255 0953 1513 2224	1.4 4.0 1.1 4.2
5 SU	0223 0923 1448 2200	1.4 4.1 1.0 4.3	**20** M	0332 1031 1547 2255	1.2 4.2 0.9 4.4
6 M	0313 1013 1536 2245	1.0 4.4 0.6 4.6	**21** TU	0404 1103 1618 2324	1.0 4.3 0.7 4.5
7 TU ●	0357 1058 1620 2329	0.7 4.6 0.3 4.8	**22** W ○	0433 1132 1649 2352	0.8 4.4 0.7 4.5
8 W	0440 1141 1704	0.5 4.8 0.2	**23** TH	0504 1203 1720	0.8 4.4 0.7
9 TH	0011 0523 1225 1748	5.0 0.4 4.9 0.2	**24** F	0022 0533 1235 1749	4.5 0.7 4.4 0.7
10 F	0055 0606 1308 1831	5.0 0.4 4.9 0.3	**25** SA	0052 0602 1307 1818	4.4 0.7 4.3 0.8
11 SA	0138 0648 1353 1912	5.0 0.6 4.8 0.5	**26** SU	0122 0629 1339 1843	4.4 0.9 4.2 0.9
12 SU	0220 0729 1439 1953	4.8 0.6 4.5 0.8	**27** M	0150 0657 1411 1911	4.3 0.9 4.1 1.1
13 M	0303 0813 1528 2040	4.6 0.9 4.3 1.3	**28** TU	0220 0727 1444 1944	4.1 1.1 3.9 1.4
14 TU	0349 0906 1627 2140 ☽	4.2 1.3 3.9 1.7	**29** W	0254 0806 1525 2032	3.8 1.4 3.8 1.7
15 W	0450 1018 1750 2308	3.8 1.7 3.6 2.0	**30** TH ☾	0342 0901 1625 2144	3.7 1.6 3.6 1.9
			31 F	0453 1024 1757 2323	3.5 1.8 3.5 1.9

APRIL

	TIME	m		TIME	m
1 SA	0627 1202 1930	3.5 1.7 3.7	**16** SU	0133 0831 1355 2110	1.8 3.7 1.5 4.0
2 SU	0051 0750 1320 2037	1.7 3.8 1.3 4.1	**17** M	0225 0923 1439 2149	1.5 3.9 1.2 4.2
3 M	0155 0852 1418 2131	1.3 4.1 0.9 4.4	**18** TU	0302 1001 1514 2221	1.2 4.1 1.0 4.4
4 TU	0246 0944 1508 2218	1.0 4.4 0.5 4.7	**19** W	0334 1033 1546 2251	1.0 4.3 0.9 4.5
5 W	0331 1030 1555 2304	0.6 4.7 0.3 4.9	**20** TH	0403 1103 1618 2319	0.9 4.4 0.8 4.5
6 TH	0415 1117 1641 2347 ●	0.4 4.8 0.2 5.0	**21** F ○	0432 1134 1651 2349	0.8 4.4 0.8 4.5
7 F	0459 1203 1726	0.4 4.9 0.3	**22** SA	0504 1207 1722	0.8 4.4 0.9
8 SA	0031 0542 1249 1807	5.0 0.5 4.9 0.4	**23** SU	0019 0534 1241 1751	4.4 0.8 4.3 0.9
9 SU	0113 0624 1335 1847	4.9 0.5 4.7 0.7	**24** M	0051 0604 1317 1821	4.3 0.8 4.2 1.0
10 M	0153 0705 1422 1928	4.7 0.7 4.5 1.1	**25** TU	0123 0636 1353 1852	4.2 0.9 4.1 1.1
11 TU	0234 0751 1511 2016	4.5 1.0 4.3 1.5	**26** W	0158 0710 1432 1930	4.1 1.0 4.0 1.3
12 W	0321 0844 1609 2118 ☽	4.2 1.4 4.0 1.8	**27** TH	0239 0753 1519 2020	4.0 1.2 3.9 1.6
13 TH	0422 0953 1728 2243	3.8 1.7 3.7 2.0	**28** F	0329 0848 1619 2130 ☾	3.8 1.4 3.8 1.7
14 F	0545 1125 1900	3.6 1.8 3.7	**29** SA	0437 1002 1737 2255	3.7 1.5 3.8 1.7
15 SA	0018 0718 1253 2016	2.0 3.5 1.7 3.8	**30** SU	0558 1127 1857 —	3.7 1.5 3.9 —

Chart Datum: 2.73 metres below Ordnance Datum (Newlyn)

ENGLAND, SOUTH COAST - PORTSMOUTH

LAT 50°48′N   LONG 1°07′W

TIMES AND HEIGHTS OF HIGH AND LOW WATERS

YEAR **1989**

TIME ZONE UT(GMT)
For Summer Time add ONE hour in non-shaded areas

	MAY				JUNE				JULY				AUGUST		
	TIME m		TIME m		TIME m		TIME m		TIME m		TIME m		TIME m		TIME m
1 M	0015 1.6 0715 3.8 1242 1.2 2002 4.2	**16** TU	0138 1.6 0837 3.8 1354 1.4 2103 4.1	**1** TH	0142 1.1 0844 4.3 1409 0.9 2117 4.6	**16** F	0218 1.5 0924 4.0 1437 1.5 2137 4.2	**1** SA	0223 1.2 0933 4.3 1451 1.3 2154 4.4	**16** SU	0232 1.5 0942 4.0 1455 1.5 2154 4.2	**1** TU	0408 1.0 1118 4.4 1628 1.1 ● 2326 4.4	**16** W	0342 1.0 1049 4.5 1604 1.0 2301 4.5
2 TU	0120 1.2 0818 4.1 1343 0.9 2056 4.5	**17** W	0221 1.4 0921 4.0 1434 1.2 2140 4.3	**2** F	0237 0.9 0941 4.5 1503 0.9 2209 4.7	**17** SA	0259 1.3 1006 4.1 1517 1.4 2218 4.3	**2** SU	0321 1.0 1031 4.4 1545 1.2 2247 4.5	**17** M	0318 1.3 1027 4.2 1539 1.3 2238 4.3	**2** W	0449 0.9 1159 4.5 1707 1.0	**17** TH	0424 0.8 1129 4.7 1645 0.8 ○ 2340 4.6
3 W	0213 0.9 0912 4.4 1436 0.6 2147 4.7	**18** TH	0257 1.2 0959 4.2 1511 1.1 2214 4.4	**3** SA	0329 0.8 1036 4.6 1556 0.8 ● 2300 4.7	**18** SU	0338 1.2 1045 4.2 1557 1.3 2256 4.4	**3** M	0412 0.9 1124 4.4 1635 1.1 ● 2336 4.5	**18** TU	0400 1.1 1109 4.3 1621 1.1 ○ 2319 4.4	**3** TH	0006 4.4 0525 0.8 1237 4.5 1744 1.0	**18** F	0505 0.5 1211 4.8 1727 0.7
4 TH	0302 0.7 1003 4.7 1527 0.5 2235 4.9	**19** F	0329 1.1 1033 4.3 1546 1.1 2246 4.4	**4** SU	0419 0.7 1131 4.6 1644 0.9 2348 4.6	**19** M	0416 1.1 1125 4.3 1636 1.2 ○ 2334 4.4	**4** TU	0458 0.8 1213 4.5 1719 1.1	**19** W	0441 0.9 1149 4.4 1703 1.0 2358 4.5	**4** F	0042 4.4 0559 0.8 1312 4.5 1819 1.1	**19** SA	0020 4.7 0548 0.4 1254 4.9 1811 0.7
5 F	0349 0.5 1053 4.8 1616 0.5 ● 2322 4.9	**20** SA	0402 1.0 1108 4.3 1621 1.1 ○ 2319 4.4	**5** M	0505 0.7 1222 4.6 1729 1.0	**20** TU	0454 1.0 1204 4.3 1716 1.1	**5** W	0021 4.4 0539 0.8 1257 4.5 1759 1.1	**20** TH	0522 0.8 1230 4.5 1745 0.9	**5** SA	0115 4.4 0633 0.9 1346 4.5 1852 1.1	**20** SU	0103 4.8 0632 0.4 1339 4.9 1854 0.7
6 SA	0436 0.5 1144 4.8 1702 0.5	**21** SU	0436 0.9 1143 4.3 1656 1.1 2353 4.4	**6** TU	0033 4.6 0549 0.8 1310 4.5 1811 1.1	**21** W	0013 4.3 0535 0.9 1247 4.3 1756 1.1	**6** TH	0100 4.4 0617 0.9 1337 4.5 1838 1.2	**21** F	0039 4.5 0604 0.6 1315 4.6 1827 0.9	**6** SU	0149 4.3 0706 1.0 1421 4.5 1925 1.2	**21** M	0148 4.8 0715 0.4 1426 4.9 1936 0.8
7 SU	0007 4.9 0520 0.5 1233 4.7 1745 0.7	**22** M	0510 0.9 1221 4.3 1731 1.1	**7** W	0113 4.5 0631 0.9 1355 4.5 1852 1.3	**22** TH	0053 4.3 0615 0.8 1331 4.4 1838 1.1	**7** F	0137 4.3 0655 1.0 1415 4.5 1915 1.3	**22** SA	0122 4.5 0648 0.5 1402 4.7 1911 0.9	**7** M	0223 4.3 0739 1.1 1454 4.4 2000 1.4	**22** TU	0235 4.6 0758 0.7 1511 4.7 2019 1.0
8 M	0050 4.7 0603 0.6 1321 4.6 1826 1.0	**23** TU	0028 4.3 0547 0.9 1300 4.2 1806 1.1	**8** TH	0154 4.3 0713 1.1 1438 4.4 1936 1.5	**23** F	0136 4.3 0659 0.8 1419 4.4 1922 1.1	**8** SA	0214 4.3 0733 1.1 1454 4.4 1956 1.5	**23** SU	0208 4.5 0732 0.5 1450 4.7 1955 0.9	**8** TU	0259 4.1 0816 1.3 1528 4.3 2037 1.5	**23** W	0324 4.4 0844 1.0 1559 4.4 ☾ 2109 1.2
9 TU	0130 4.6 0646 0.8 1408 4.5 1907 1.3	**24** W	0106 4.2 0623 0.9 1342 4.2 1844 1.3	**9** F	0237 4.2 0759 1.3 1525 4.3 2026 1.7	**24** SA	0223 4.3 0743 0.8 1508 4.4 2008 1.2	**9** SU	0253 4.2 0814 1.3 1534 4.3 2040 1.6	**24** M	0255 4.5 0817 0.6 1538 4.6 2040 1.0	**9** W	0337 4.0 0857 1.6 1606 4.0 ☾ 2122 1.8	**24** TH	0419 4.1 0939 1.4 1654 4.1 2212 1.6
10 W	0211 4.4 0731 1.1 1456 4.3 1955 1.6	**25** TH	0147 4.1 0703 0.9 1428 4.1 1926 1.3	**10** SA	0324 4.0 0850 1.5 1615 4.2 2123 1.8	**25** SU	0311 4.2 0831 0.8 1559 4.4 2059 1.3	**10** M	0337 4.0 0859 1.4 1616 4.2 2128 1.7	**25** TU	0344 4.3 0905 0.9 1628 4.4 ☾ 2130 1.2	**10** TH	0424 3.7 0947 1.9 1655 3.8 2221 2.0	**25** F	0531 3.8 1054 1.8 1808 3.9 2340 1.8
11 TH	0257 4.1 0822 1.4 1550 4.1 2053 1.8	**26** F	0232 4.1 0749 1.0 1517 4.1 2016 1.4	**11** SU	0420 3.9 0947 1.6 1711 4.0 ☾ 2227 1.9	**26** M	0404 4.1 0925 1.0 1653 4.3 ☾ 2156 1.4	**11** TU	0426 3.8 0951 1.6 1705 4.0 2224 1.8	**26** W	0439 4.1 1001 1.2 1722 4.2 2234 1.5	**11** F	0527 3.5 1057 2.1 1802 3.7 2339 2.0	**26** SA	0707 3.7 1229 2.0 1936 3.8
12 F	0354 3.9 0924 1.6 1655 3.9 ☾ 2207 2.0	**27** SA	0322 4.0 0842 1.1 1613 4.1 2116 1.5	**12** M	0523 3.7 1052 1.7 1812 3.9 2334 1.9	**27** TU	0503 4.0 1025 1.2 1752 4.3 2302 1.4	**12** W	0524 3.7 1051 1.8 1802 3.9 2329 1.9	**27** TH	0546 3.9 1110 1.5 1828 4.1 2351 1.6	**12** SA	0653 3.5 1221 2.1 1922 3.7	**27** SU	0112 1.7 0835 3.8 1352 1.9 2051 3.9
13 SA	0504 3.7 1038 1.8 1808 3.9 2327 2.0	**28** SU	0423 3.9 0943 1.2 1717 4.1 ☾ 2225 1.5	**13** TU	0632 3.6 1159 1.7 1912 3.9	**28** W	0610 4.0 1134 1.3 1854 4.2	**13** TH	0632 3.6 1159 1.9 1905 3.8	**28** F	0709 3.8 1231 1.7 1941 4.0	**13** SU	0101 1.9 0815 3.6 1337 1.9 2034 3.8	**28** M	0224 1.5 0938 4.1 1452 1.6 2148 4.1
14 SU	0624 3.6 1156 1.7 1919 3.9	**29** M	0530 3.9 1054 1.2 1823 4.1 2336 1.5	**14** W	0037 1.7 0739 3.7 1300 1.7 2007 4.0	**29** TH	0011 1.4 0721 4.0 1244 1.4 1957 4.3	**14** F	0037 1.9 0745 3.6 1307 1.9 2007 3.9	**29** SA	0112 1.6 0829 3.9 1348 1.7 2050 4.1	**14** M	0208 1.7 0917 3.9 1435 1.6 2131 4.1	**29** TU	0315 1.2 1024 4.3 1537 1.3 2233 4.3
15 M	0041 1.8 0739 3.6 1303 1.6 2017 4.0	**30** TU	0640 3.9 1205 1.2 1926 4.3	**15** TH	0132 1.6 0835 3.8 1352 1.6 2055 4.1	**30** F	0120 1.3 0829 4.1 1350 1.3 2056 4.3	**15** SA	0140 1.7 0849 3.8 1405 1.8 2104 4.0	**30** SU	0222 1.4 0936 4.1 1451 1.5 2150 4.2	**15** TU	0259 1.3 1007 4.2 1522 1.3 2219 4.3	**30** W	0356 1.0 1103 4.5 1615 1.1 2311 4.4
		31 W	0042 1.3 0745 4.1 1309 1.1 2024 4.4							**31** M	0320 1.2 1031 4.3 1543 1.3 2241 4.3			**31** TH	0432 0.9 1137 4.6 1650 1.0 ● 2345 4.4

Chart Datum: 2.73 metres below Ordnance Datum (Newlyn)

AREA 2 — Central S England

ENGLAND, SOUTH COAST - PORTSMOUTH
LAT 50°48'N   LONG 1°07'W
TIMES AND HEIGHTS OF HIGH AND LOW WATERS
YEAR 1989

TIME ZONE UT(GMT) — For Summer Time add ONE hour in non-shaded areas

SEPTEMBER

	TIME	m		TIME	m
1 F	0505 1210 1722	0.8 4.6 0.9	**16** SA	0444 1147 1705 2359	0.4 5.0 0.5 4.9
2 SA	0017 0537 1242 1754	4.5 0.8 4.6 0.9	**17** SU	0527 1231 1748	0.4 5.1 0.5
3 SU	0049 0608 1314 1825	4.4 0.8 4.6 1.0	**18** M	0043 0610 1315 1831	4.9 0.4 5.0 0.6
4 M	0122 0638 1346 1855	4.4 0.9 4.5 1.0	**19** TU	0129 0653 1358 1913	4.9 0.6 4.9 0.7
5 TU	0155 0707 1415 1924	4.3 1.1 4.4 1.2	**20** W	0216 0735 1443 1958	4.7 0.9 4.7 1.0
6 W	0227 0737 1446 1955	4.2 1.3 4.3 1.4	**21** TH	0306 0821 1530 2049	4.4 1.3 4.4 1.3
7 TH	0300 0811 1518 2033	4.0 1.6 4.0 1.7	**22** F ☾	0403 0919 1628 2156	4.1 1.7 4.0 1.7
8 F ☾	0339 0856 1602 2127	3.8 1.9 3.8 1.9	**23** SA	0520 1041 1750 2328	3.8 2.0 3.8 1.9
9 SA	0437 1004 1709 2245	3.5 2.1 3.6 2.1	**24** SU	0700 1222 1925	3.7 2.1 3.7
10 SU	0605 1139 1839	3.4 2.2 3.6	**25** M	0103 0824 1343 2041	1.8 3.9 1.9 3.8
11 M	0021 0740 1307 2003	2.0 3.7 2.0 3.7	**26** TU	0210 0921 1438 2134	1.6 4.1 1.6 4.0
12 TU	0136 0847 1410 2103	1.7 3.9 1.6 4.0	**27** W	0257 1004 1519 2214	1.3 4.3 1.3 4.2
13 W	0232 0939 1458 2152	1.3 4.3 1.2 4.4	**28** TH	0333 1037 1553 2248	1.1 4.5 1.1 4.4
14 TH	0317 1022 1541 2236	0.9 4.6 0.9 4.6	**29** F ●	0406 1108 1624 2319	0.9 4.6 1.0 4.5
15 F ○	0400 1138 1622 2317	0.6 4.9 0.7 4.8	**30** SA	0438 1134 1654 2349	0.9 4.6 0.9 4.5

OCTOBER

	TIME	m		TIME	m
1 SU	0510 1209 1725	0.9 4.6 0.9	**16** M	0504 1206 1723	0.5 5.1 0.5
2 M	0022 0540 1239 1756	4.5 0.9 4.6 0.9	**17** TU	0024 0547 1249 1806	5.0 0.6 5.0 0.6
3 TU	0055 0609 1310 1824	4.4 1.0 4.5 1.0	**18** W	0112 0629 1333 1851	4.9 0.8 4.9 0.8
4 W	0128 0636 1340 1852	4.3 1.1 4.4 1.1	**19** TH	0200 0711 1417 1936	4.7 1.1 4.6 1.1
5 TH	0200 0705 1411 1923	4.2 1.3 4.2 1.3	**20** F	0251 0759 1505 2029	4.4 1.5 4.3 1.4
6 F	0233 0738 1444 2000	4.0 1.6 4.0 1.6	**21** SA ☾	0349 0900 1604 2136	4.2 1.8 4.0 1.8
7 SA	0313 0823 1529 2051	3.8 1.8 3.8 1.8	**22** SU	0502 1020 1723 2301	3.9 2.1 3.8 1.9
8 SU ☾	0408 0929 1633 2204	3.7 2.1 3.6 2.0	**23** M	0630 1152 1853	3.9 2.1 3.7
9 M	0530 1100 1801 2337	3.6 2.1 3.6 1.9	**24** TU	0029 0749 1312 2010	1.9 3.9 1.9 3.8
10 TU	0700 1229 1924	3.7 1.9 3.8	**25** W	0135 0846 1407 2104	1.7 4.1 1.6 4.0
11 W	0056 0809 1335 2028	1.6 4.1 1.6 4.1	**26** TH	0224 0929 1449 2145	1.4 4.3 1.4 4.2
12 TH	0155 0903 1425 2118	1.2 4.4 1.2 4.4	**27** F	0300 1003 1522 2219	1.2 4.5 1.2 4.4
13 F	0245 0950 1510 2205	0.8 4.7 0.9 4.6	**28** SA	0334 1035 1553 2250	1.0 4.6 1.1 4.5
14 SA ○	0331 1036 1554 2250	0.6 4.9 0.7 4.9	**29** SU ●	0406 1105 1624 2321	1.0 4.6 1.0 4.5
15 SU	0417 1121 1639 2336	0.5 5.1 0.6 5.0	**30** M	0439 1135 1654 2355	1.0 4.6 1.0 4.5
31 TU	0511 1207 1726	1.1 4.5 1.0			

NOVEMBER

	TIME	m		TIME	m
1 W	0029 0542 1239 1757	4.4 1.1 4.4 1.0	**16** TH	0058 0608 1311 1831	4.8 1.0 4.7 0.9
2 TH	0104 0610 1311 1828	4.3 1.2 4.3 1.1	**17** F	0147 0651 1356 1917	4.6 1.3 4.6 1.1
3 F	0139 0641 1346 1901	4.2 1.4 4.2 1.2	**18** SA	0236 0738 1442 2008	4.5 1.6 4.3 1.4
4 SA	0216 0718 1423 1940	4.1 1.5 4.0 1.4	**19** SU	0329 0835 1537 2107	4.3 1.8 4.1 1.7
5 SU	0300 0803 1510 2029	4.0 1.7 3.9 1.6	**20** M ☾	0431 0943 1642 2216	4.2 2.0 3.9 1.8
6 M	0353 0904 1609 2134	3.9 1.9 3.8 1.7	**21** TU	0541 1101 1758 2331	4.0 2.0 3.7 1.9
7 TU	0502 1022 1724 2253	3.9 1.9 3.7 1.7	**22** W	0652 1216 1916	3.9 1.9 3.8
8 W	0618 1142 1841	4.0 1.8 3.9	**23** TH	0042 0754 1318 2019	1.8 4.1 1.7 3.9
9 TH	0009 0726 1250 1947	1.5 4.2 1.5 4.1	**24** F	0137 0843 1407 2107	1.6 4.2 1.5 4.1
10 F	0113 0823 1347 2042	1.2 4.5 1.2 4.4	**25** SA	0221 0923 1446 2148	1.5 4.3 1.4 4.2
11 SA	0208 0915 1437 2134	0.9 4.7 0.9 4.7	**26** SU	0300 1000 1521 2224	1.4 4.4 1.2 4.4
12 SU	0301 1005 1526 2226	0.7 4.9 0.7 4.8	**27** M	0335 1035 1555 2258	1.3 4.5 1.2 4.4
13 M	0350 1054 1614 2317 ●	0.7 5.0 0.6 4.9	**28** TU	0410 1108 1628 2333	1.2 4.5 1.1 4.4
14 TU	0439 1142 1700	0.7 5.0 0.6	**29** W	0445 1142 1702	1.2 4.5 1.1
15 W	0008 0525 1228 1746	4.9 0.8 4.9 0.7	**30** TH	0008 0518 1216 1736	4.3 1.2 4.4 1.0

DECEMBER

	TIME	m		TIME	m
1 F	0045 0552 1251 1810	4.3 1.2 4.3 1.1	**16** SA	0133 0634 1338 1858	4.6 1.2 4.5 1.0
2 SA	0123 0627 1329 1847	4.2 1.3 4.3 1.1	**17** SU	0218 0717 1420 1942	4.6 1.4 4.4 1.2
3 SU	0204 0706 1410 1928	4.2 1.4 4.1 1.1	**18** M	0302 0805 1506 2030	4.5 1.6 4.2 1.4
4 M	0249 0750 1456 2014	4.2 1.5 4.1 1.2	**19** TU ☾	0350 0858 1558 2124	4.4 1.8 4.0 1.6
5 TU	0339 0842 1549 2108	4.2 1.6 4.0 1.3	**20** W	0443 0958 1658 2225	4.2 1.9 3.9 1.8
6 W	0435 0943 1650 2212	4.1 1.6 3.9 1.4	**21** TH	0543 1105 1807 2333	4.0 1.9 3.7 1.8
7 TH	0538 1053 1758 2323	4.2 1.6 3.9 1.4	**22** F	0645 1214 1919	3.9 1.8 3.7
8 F	0644 1203 1908	4.2 1.5 4.0	**23** SA	0040 0746 1316 2023	1.8 4.0 1.7 3.8
9 SA	0032 0745 1308 2012	1.3 4.4 1.3 4.3	**24** SU	0139 0839 1407 2117	1.8 4.0 1.6 4.0
10 SU	0137 0843 1408 2112	1.1 4.5 1.1 4.5	**25** M	0227 0926 1451 2200	1.6 4.2 1.4 4.1
11 M	0235 0939 1504 2210	1.1 4.6 0.9 4.6	**26** TU	0309 1008 1531 2240	1.5 4.3 1.3 4.2
12 TU	0329 1034 1556 2306 ○	1.0 4.8 0.8 4.7	**27** W	0348 1046 1608 2317	1.4 4.4 1.2 4.3
13 W	0421 1125 1646	0.9 4.8 0.7	**28** TH	0424 1123 1644 2351 ●	1.2 4.4 1.1 4.3
14 TH	0000 0508 1212 1733	4.7 1.0 4.8 0.8	**29** F	0500 1158 1720	1.2 4.4 1.0
15 F	0048 0551 1257 1815	4.7 1.1 4.6 0.9	**30** SA	0029 0537 1236 1756	4.3 1.1 4.3 0.9
31 SU	0107 0615 1314 1835	4.4 1.1 4.3 0.8			

Chart Datum: 2.73 metres below Ordnance Datum (Newlyn)

LANGSTONE 10-2-28
Hampshire

CHARTS
Admiralty 3418; Stanford 10, 11; Imray C3, Y33; OS 196, 197

TIDES
+0022 Dover; Zone 0 (GMT)

Standard Port PORTSMOUTH (←)

Times				Height (metres)			
HW		LW		MHWS	MHWN	MLWN	MLWS
1000	0500	0000	0600	4.7	3.8	1.8	0.6
2200	1700	1200	1800				

Differences LANGSTONE
0000 0000 +0010 +0010 +0.1 +0.1 0.0 0.0

SHELTER
Very good shelter but it must be entered with caution when a heavy sea is running. Shelter in marina to W inside entrance. Anchorage in Russell's Lake. For overnight stay apply Hr Mr at the ferry on Hayling Island. (Tel. 3419). Also anchorage in Langstone Channel out of fairway.

NAVIGATION
Waypoint Langstone Fairway (safe water) buoy, LFI 10s, 50°46'.28N 01°01'.27W, 167°/347° from/to QR Lt at entrance, 0.94M. Entrance between E Winner bank and W Winner bank but apart from these the entrance is straightforward. Tides are strong in the entrance. Harbour speed limit 10 kn.

LIGHTS AND MARKS
Leading marks (concrete dolphins) 344°. Red flag, with numeral pendant below, flown at the Fort, Battery and Hr Mr's flag staff indicates firing from range to W of harbour entrance (frequently across harbour approach). See 10.2.27.

RADIO TELEPHONE (local times)
VHF Ch 16; 12 (Summer 0830-1700. Winter: Mon-Fri 0830-1700; Sat-Sun 0830-1300). Marina Ch M.

TELEPHONE (0705)
Hr Mr 463419; MRSC Lee-on-Solent 552100; HMC Freefone Customs Yachts: RT (0703) 229251; Marinecall 0898 500 457; Dr 465721.

FACILITIES
EC Havant - Wednesday; **Langstone Marina** (300+36 visitors) Tel. 822719, BH (20 ton), D, AC, ME, El, Sh, Gas, FW, Gaz, V, SM, Lau, CH, R, Bar; Access HW∓3½; **Hayling Pontoons** (Entrance to Sinah Lake), Slip, D, FW, L, P, AB; **Langstone SC** Tel. Havant 484577, Slip, M, L, FW, Bar; **Eastney Cruising Ass.** Tel. 734103; **Solartron SC; Locks SC** Tel. 829833; **Tudor SC** (Eastney) Tel. 662002.
PO (Eastney, Hayling); Bank (Havant, Hayling, Emsworth); Rly (bus to Havant); Air (Southampton).

CHICHESTER 10-2-29
W. Sussex

CHARTS
Admiralty 3418; Stanford 10, 11; Imray C3, C9, Y33; OS 197

TIDES
+0020 Dover; ML 2.8; Zone 0 (GMT)

Standard Port PORTSMOUTH (←)

Times				Height (metres)			
HW		LW		MHWS	MHWN	MLWN	MLWS
0500	1000	0000	0600	4.7	3.8	1.8	0.6
1700	2200	1200	1800				

Differences HARBOUR ENTRANCE
−0010 +0005 +0015 +0020 +0.2 +0.2 0.0 +0.1
BOSHAM
0000 +0010 No data +0.2 +0.1 No data
ITCHENOR
−0005 +0005 +0005 +0025 +0.1 0.0 −0.2 −0.2
DELL QUAY
+0005 +0015 No data +0.2 +0.1 No data
NAB TOWER
+0015 0000 +0015 +0015 −0.2 0.0 +0.2 0.0
SELSEY BILL
−0005 −0005 +0035 +0035 +0.6 +0.6 0.0 0.0

SHELTER
Chichester harbour has excellent shelter in all four main arms of the harbour, Emsworth, Thorney, Bosham and Chichester. Beware shallow patches W and S of Bar Bn. There are a large number of yacht harbours, marinas etc. Good anchorages at East Head; in Thorney Channel off Pilsey Is; W of Fairway buoy on S side of channel. No anchoring E of Cobnor Pt. Advisable to make prior arrangement for mooring.

NAVIGATION
Waypoint 50°45'.50N 00°56'.50W, 193°/013° from/to Bar Bn, 0.47M. The entrance is about ½ M wide at HW but this dries exposing a sandbank known as the Winner from the East Head side leaving a deep channel close to Hayling Island. There is a bar 1 M S of the entrance with a channel giving depths of 0.4m close E of the Bar Bn. There is a patch drying 0.3m about 2 ca ESE of the Bn. A port-hand buoy lies 194° 2.2 ca from Bar Bn (Apl-Nov). In winds above force 5 from S quarter and particularly during spring ebbs, seas over the bar are confused. It is advised to cross the bar from HW−3 to HW+1. There is a speed limit throughout the harbour of 8 kn.

LIGHTS AND MARKS
Bar Bn Fl(2) R 10s; vis 020°-080°; and Fl WR 5s; vis W322°-080°, R080°-322°; both on W pylon with tide gauge; RC. All channels well marked.

RADIO TELEPHONE (local times)
Harbour Office, Itchenor VHF Ch **14** 16 (Call *Chichester* on Ch 14). (Apl-Sept 0900-1730. Oct-Mar: Mon-Sat 0900-1300, Mon-Fri 1400-1730). Emsworth and Northney Marinas Ch M.

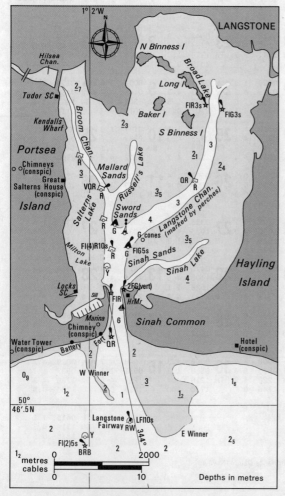

CHICHESTER continued

TELEPHONE (0243)
Harbour Office Itchenor 512301; CG Hayling 464095; Selsey 602274; HMC Freefone Customs Yachts: RT (0703) 229251; MRSC Lee-on-Solent 552100; Weather information Portsmouth 8091; Marinecall 0898 500 457; Hosp 787970.

FACILITIES

EMSWORTH CHANNEL: channel is straight, broad and deep. Good anchorages especially N of Sandy Pt near entrance to channel. Visitors moorings on 5 outer piles. EC Wednesday; **Slips** at South St, Kings St, and Slipper Mill. Contact the Warden Tel. 376422; **Andrews BY** Tel. 373335, FW, C, Slip, C (10 ton); **Emsworth Yacht Harbour** (200 + 4 visitors) Tel. 375211, Slip, Gas, ME, El, AC, Sh, P, D, FW, C (20 ton); (Access HW∓2); **David Still Marine** Tel. 374242, ME, Sh, CH; **Ostar Marine** Tel. 376414; **Greenham Marine Ltd** Tel. 378314, El.

HAYLING ISLAND: EC Wednesday; **Chichester Harbour Yacht Services** Tel. Hayling 465051 ME, El, Sh, Slip, C, CH; **Northney Marina** (260 + 17 visitors) Tel. Hayling 466321, D, C (25 ton), Bar, FW, AC, El, Sh, R, CH, ME; **Hayling Yacht Co.** Tel. Hayling 463592, ME, El, Slip, BH (8 ton), P, D, FW, AB Sh, CH; **Hayling Is Marine Services** Tel. Hayling 464869; **Sparkes Yacht Harbour** (140) Tel. Hayling 463572, Slip, ME, El, FW, P, D, Gas, Gaz, Lau, Sh, C (20 ton), CH.

THORNEY CHANNEL: Strangers are advised to go up at half flood. Channel is marked by perches. After Thorney village, channel splits, Prinsted Chan to port (full of moorings) and Nutbourne Chan to stbd. There is plenty of room to anchor in Thorney Chan and it is well protected from E and SE winds.
Prinsted – **Paynes BY** Tel. 572224, Slip, L, FW, ME; **Andrews BY** Tel. 45335.

BOSHAM CHANNEL: There are some Harbour Authority moorings in this channel and there is a hard at Bosham Quay. Most of the channel dries. Anchoring in channel is prohibited.
EC Wednesday; **Bosham Quay** L, FW, AB; **Capt. Charles Currey** Tel. 573174, CH; **Rockall Sails** Tel. 572149, SM; **Copp Sails** Tel. 573981, SM; **Bosham SC** Tel. 572341.

CHICHESTER CHANNEL: this runs up to Itchenor. From Sandhead buoy proceed 034° to Fl (2) G 10s Chaldock Bn; alter course to 039° to Fairway buoy, Fl (3). Anchoring prohibited in Itchenor Reach.
EC Itchenor Thursday. Hard available at all stages of the tide. There are six visitor's buoys off Itchenor jetty. For single moorings apply Hr Mr. **G. Haines** Tel. 512228, Slip, P, D, CH, Sh; **E. M. Coombes** Tel. 573194, FW, CH, Slip, P and D (cans); **Northshore Yacht Yards** Tel. 512611, Slip, M; **H. C. Darley** Tel. 512243, P and D (cans), CH; **E. Bailey** Tel. 512374, ME, P, D, Sh.

CHICHESTER LAKE AND FISHBOURNE CHANNEL:
Birdham Pool Yacht Basin, Access HW∓3 when lock gates open. At Chichester Yacht Basin, the approach dries at CD but is dredged to about 1m at normal LWS. Secure to pontoons whilst waiting: lock control boards show 'Red' indicating 'wait' or 'Green' for 'enter'. They are floodlit at night. When the gates are open with a free flow in and out, there is a QY Lt on top of the Control Tr.
Birdham – **Chichester Yacht Basin** (900 + 50 visitors) Tel. 512731, Slip, P, D, FW, ME, El, Sh, AC, Gas, Gaz, CH, V, R, BH (20 ton), Access HW∓4½; **Vernons Shipyard** Tel. 512606; **Regis Marine** Tel. 512020, Electronics (7 days); **Greenham Marine** Tel. 512995, El; **Yacht and Sports Gear** Tel. 512720, CH; 784572, CH, ACA.
Birdham Pool – **Birdham Pool Marina** (230 + 10 visitors) Tel. 512310, Slip, P, D, FW, El, AC, Sh, CH, Gas, Gaz, ME, SM, C (3 ton), Access HW∓3; **Birdham Shipyard** Tel. 512310; Slip, M, P, Sh, CH; **Sterndrives** Tel. 512831, ME; **Seahorse Sails** Tel. 512195, SM;
Dell Quay – Launching is possible on the hard, depending on the tide, on payment of harbour dues.
Ted Bailey (Dell Quay) Tel. 781110 Sh; **Wyche Marine** Tel. 782768, M, L, Sh.

Clubs: **Birdham YC** Tel. 512642; **Bosham SC** Tel. 572341; **Chichester YC** Tel. 512918; **Chichester Cruiser and Racing Club; Dell Quay SC** Tel. 785080; **Emsworth SC** Tel. 43065; **Emsworth Slipper SC** Tel. 42523; **Hayling Island SC** Tel. Hayling 463768; **Itchenor SC** Tel. 512400; **Langstone SC** Tel. Havant 484577; **Mengham Rythe SC** Tel. Hayling 463337; **Thorney Island SC; West Wittering SC.**

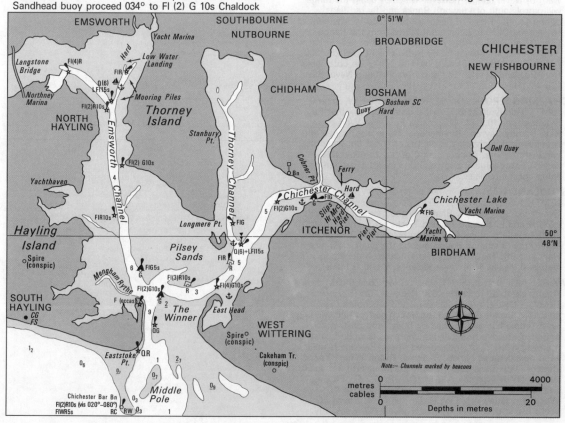

MINOR HARBOURS AND ANCHORAGES 10.2.30

BEER, Devon, 50°41′ N, 3°06′ W, Zone 0 (GMT), Admty chart 3315. HW −0440 on Dover, +0045 on Devonport; HW height −1.1m on Devonport; ML 2.3m; Duration 0640. Beer roads afford a sheltered anchorage in winds from W to N. It is open to prevailing winds. Go ashore by boat over open beach. From E beware rocks at the bottom of headland on E side of cove. There is a FW Lt, 26m near the church in Beer. Facilities: EC Thurs in Beer and Seaton; CG, SC, PO, Bar, BY, R, V.

RIVER AXE, Devon 50°41′ N, 3°06′ W, Zone 0 (GMT), Admty chart 3315. HW −0440 on Dover, +0045 on Devonport; HW height +1.1m on Devonport; ML 2.3m; Duration 0640. Good shelter for smallish boats; boats up to 1m draught can enter at HW. Cross bar, which dries, then turn sharply to W. Entrance is 7m wide. The channel through the bar is unmarked and shifts constantly. Local knowledge is needed. A road bridge (2m clearance) crosses the river 2 ca within the entrance. Facilities: EC Thurs (Seaton), BY; **Axe YC** Tel. Seaton 20966, Slip, Bar; **Seaton** PO, R, V, Gas, Gaz, P and D (cans).

LULWORTH COVE, Dorset, 50°37′ N, 2°15′ W, Zone 0 (GMT), Admty chart 2172. HW −0449 on Dover, −0410 on Portsmouth; HW height −2.4m on Portsmouth; ML 1.2m. Good shelter in offshore winds but S and SW winds cause heavy swell in the cove. The centre of the cove has a depth of about 4m below CD. There is a can buoy in the centre of the cove. Anchor in NE part in 2.5m. Holding ground is poor. Village is on W bank. Facilities: EC Wed and Sat; FW at tap in car park, D and P at garage (cans), Bar, PO, Slip, R.
Note. There are two danger areas to E, one extending 5M seawards, the other 12M. Firing occurs most weekdays 0930-1800 (1500 on Fridays), often on Tues and Thurs nights and at weekends once a month. Normally there is no firing in August. An International Letter Flag U and red flag are flown, and red flashing lights are displayed from the flagstaffs on Bindon Hill and St Albans Head ONLY WHEN ARMY FIRING IS TAKING PLACE. However, further ashore, mariners may notice some red flags which fly whether or not firing is taking place, these mark the range boundary. There are no visual signals ashore to warn of Naval firings, however warships using the range will patrol south of Lulworth Banks and fly red flags during firings. Times of firing are published locally, sent to Hr Mrs and YCs and can be obtained from the Range Officer Tel. Bindon Abbey 462721 ext 819 or 824. Firing programmes are broadcast by Radio Solent (999 kHz-300 AM or 96.1 MHz VHF) daily (see 10.2.9). Also obtainable from Portland CG (Ch 67), and from Portland Naval Base (Ch 13 14).

CHAPMAN'S POOL, Dorset, 50°35′ N, 2°04′ W, Zone 0 (GMT), Admty chart 2172. Tidal data is approx that for Swanage see 10.2.14. Chapman's Pool, like Worbarrow Bay, Brandy Bay and Kimmeridge Bay is picturesque and is convenient anchorage in depths of about 3m when the wind is off-shore. Anchor centre of bay to avoid tidal swirl. Beware large B buoy unlit in centre. There are no lights and no facilities.

STUDLAND BAY, Dorset, 50°39′ N, 1°56′ W, Zone 0 (GMT), Admty chart 2172. Tidal data is approx that for Swanage. See 10.2.14. Shelter is good except from winds from N and E. Beware Redend Rocks off S shore. Best anchorage in depths of about 3m below CD, 3 ca NW of The Yards (3 strange projections on the chalk cliffs near Handfast Pt). Facilities: EC Thurs; shops, hotel etc in Studland village. No marine facilities. P and D (cans).

WAREHAM, Dorset, 50°41′ N, 2°05′ W, Zone 0 (GMT), Admty chart 2611. HW −0030 (Np), +0320 (Sp) on Dover, −0140 (Np), +0205 (Sp) on Portsmouth (See 10.2.15). Shelter very good, approach narrow and winding up R Frome but well marked by buoys and posts at entrance. Beware prohibited anchorages (salmon holes) marked on the chart and moored boats. Passage is unlit; keep outside at all bends. Facilities: **Ridge Wharf Yacht Centre** (180+6 visitors) (½M upstream of Ent) Tel. 2650, Access HW∓2 approx AB, M, FW, P, D, ME, El, Gas, Gaz, AC, BH (20 ton), Slip, Sh, CH; **Redclyffe YC**; **Wareham Quay** AB, FW, R, Dr Tel. 3444; **Town** EC Wed, P and D in cans, V, Gas, R, Bar, PO, Bank, Rly.

NEWPORT, Isle of Wight, 50°42′ N, 1°17′ W, Zone 0 (GMT), Admty chart 2793. Tidal predictions for Newport have special problems and the details are given in Section 10.2.8. See 10.2.22. Good shelter from all winds in Newport and in Island Harbour. There are visitors moorings opposite Folly Pt. S Folly Lt Beacon, QG, beyond this, opposite Pioneer Wharf (2 FG (vert)) lies the **Island Harbour** (100+100 visitors) Tel. 526733; Bar, D, P, Slip, AC, FW, Gas, BH (16 ton), Lau, ME, El, Sh, V. The channel S from Island Harbour dries, but there is 1.3m in channel and at pontoons at Newport from HW−2 to HW+2. Depths alongside in Newport about 2.3m at MHWS and 1.6m at MHWN. On Town Quay there is a pontoon for visiting yachts with FW and other facilities. Contact Harbour Office for details. (VHF Ch 16; 06 (office hours or by arrangement). Ldg Lts into Newport are 2FR (hor) 7m 2M front, 2FR (hor) 11m 2M rear, in line at 192°. Facilities: EC Thurs; Hr Mr Tel. 525994, Bank, Bar, C (5 ton mobile, 12 ton fixed) D, P, PO, R, Slip, V. **Newport Yacht Harbour** (40 visitors) Tel. 525994, AC, C (4 ton), D, FW, ME; Access HW∓2½.

TITCHFIELD HAVEN, Hampshire 50°49′ N, 01°14′ W, Zone 0 (GMT), Admty charts 1905, 2022. HW +0030 on Dover (GMT); 0000 on Portsmouth; HW height −0.2m on Portsmouth. Good protected anchorage for small craft at mouth of R. Meon. Bar dries ¼ M offshore. Ent. dries 1.2m at LWS. Entrance marked by Hillhead SC house (white conspic), beacons mark channel, spherical topmarks to port, triangular ones to stbd. Small harbour to W inside ent where yachts can lie in soft mud alongside bank. Facilities: **Hillhead village** EC Thurs, PO, CH, V, P and D (cans).

Five Gold Anchors with every mooring.

It's not surprising that with all the facilities offered at Brighton Marina Village (easy access, 24 hour security staff, private W.C.s and showers, full boatyard services, Yacht Club to name but a few) we've been awarded Five Gold Anchors by the National Yacht Harbour Association.

Annual berths at Brighton are only £130 per metre (inc. VAT). For more information phone (0273) 693636.

Brighton Marina Village

DMS Seatronics Ltd No:1

For Sales, Service & Installation of all marine-electronic/electrical equipment. Main Agents for all leading brands.

OPEN 7 DAYS A WEEK

Call: (0273) 605166 Tlx: 878210
Unit 14, Brighton Marina, Brighton, BN2 5UD

DEDICATED MARINE POWER
9 to 422hp

VOLVO PENTA

DEDICATED MARINE SERVICE
OVER 1300 LOCATIONS ACROSS EUROPE

DAN · WEBB · FEESEX
MALDON & MAYLANDSEA
TEL (0621) 54280 & 741267

SET SAIL FOR THE FUTURE WITH THE POWER OF THE PAST

ON THE RIVER BLACKWATER
TRADITION AND TECHNOLOGY

★ 600 moorings   ★ VHF (M) watch
★ Excellent shore-based facilities
★ Shipwright repairs and G.R.P.
★ Courtesy launch   ★ Chandlery   ★ Rigging

Shipways, North Street, Maldon, Essex CM9 7HN.
The Shipyard, Marine Parade, Maylandsea, Essex CM3 6AN.

Cruise to the heart of London

(straight up the Thames and we're just by Tower Bridge!)

Finding St. Katharine's Yacht Haven is as easy as driving on a motorway.

And that's what makes St. Katharine's rather special, because you will cruise up the River Thames to the only yacht haven in the heart of London.

When you arrive we'll look after you in one of our 150 berths, with electricity and drinking water. There are shower rooms, water, electricity, chandlery, delicatessen and vintners, as well as the St. Katharine's Yacht Club.

Write for our illustrated brochure and booking form to:-

**The Manager,
St. Katharine Yacht Haven Ltd,
St. Katharine-by-the-Tower Ltd,
Ivory House,
London E1 9AT.
01-488-2400   Telex 884671**

The only Yacht Haven in the heart of London

VOLVO PENTA SERVICE

Sales and service centres in area 3
KENT **John Hawkins Marine** Elmhaven Marina, Rochester Road (A228), Halling, ROCHESTER ME2 1AQ Tel (0634) 242256. SURREY **Marlow Marine Services Ltd** Penton Hook Marina Staines Road, CHERTSEY KT16 8PY Tel (0932) 568772. SUSSEX **Arrow Marine Ltd** Littlehampton Marina, LITTLEHAMPTON BN17 5DE Tel (0903) 721686/7. **Felton Marine Engineering** The Boatyard, Brighton Marina, BRIGHTON BN2 5UF Tel (0273) 601779. **Leonard Marine** The Old Shipyard, Robinson Road, NEWHAVEN BN9 9BL Tel (0273) 515987.

Area 3

South-East England
Littlehampton to Havengore Creek

10.3.1	Index	Page 265	10.3.18	The Swale	290
10.3.2	Diagram of Radiobeacons, Air Beacons, Lifeboat Stations etc	266	10.3.19	Medway River (Sheerness, Standard Port, Tidal Curves)	291
10.3.3	Tidal Stream Charts	268	10.3.20	River Thames (London Bridge, Standard Port, Tidal Curves)	297
10.3.4	List of Lights, Fog Signals and Waypoints	270	10.3.21	Southend-on-Sea/Leigh-on-Sea	303
10.3.5	Passage Information	272	10.3.22	Havengore Creek	304
10.3.6	Distance Tables	273	10.3.23	Minor Harbours and Anchorages	305
10.3.7	English Channel Waypoints	See 10.1.7		Eastbourne, Hastings, Sandwich, Margate, Gravesend, St Katharine Yacht Haven, Holehaven	
10.3.8	Littlehampton	274			
10.3.9	Shoreham, Standard Port, Tidal Curves	274			
10.3.10	Brighton	279			
10.3.11	Newhaven	280			
10.3.12	Rye	281			
10.3.13	Folkestone	282			
10.3.14	Dover, Standard Port, Tidal Curves	282			
10.3.15	Ramsgate	287			
10.3.16	Whitstable	287			
10.3.17	Thames Estuary Tidal Stream charts	288			

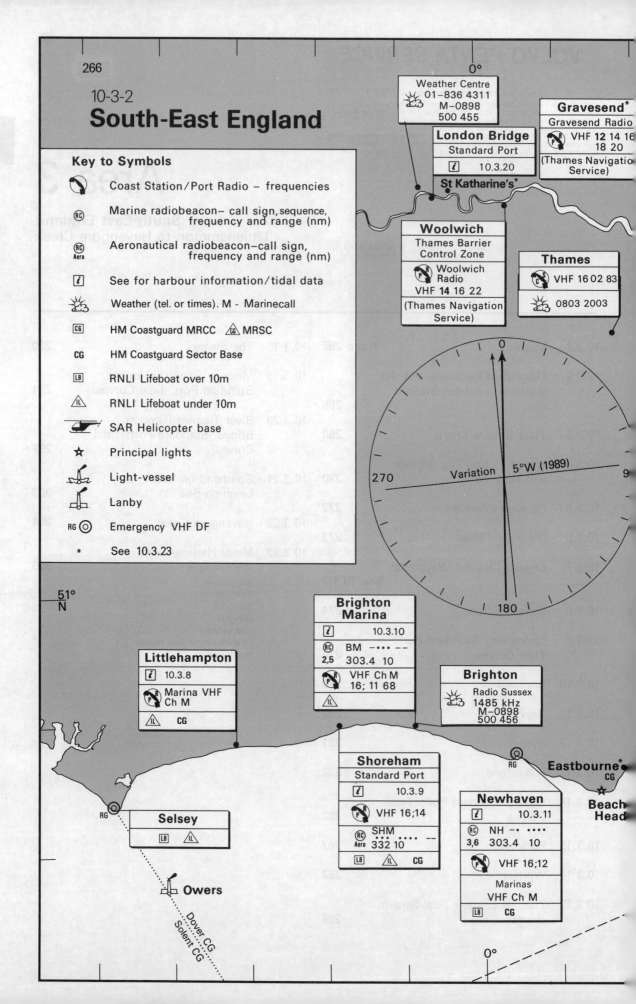

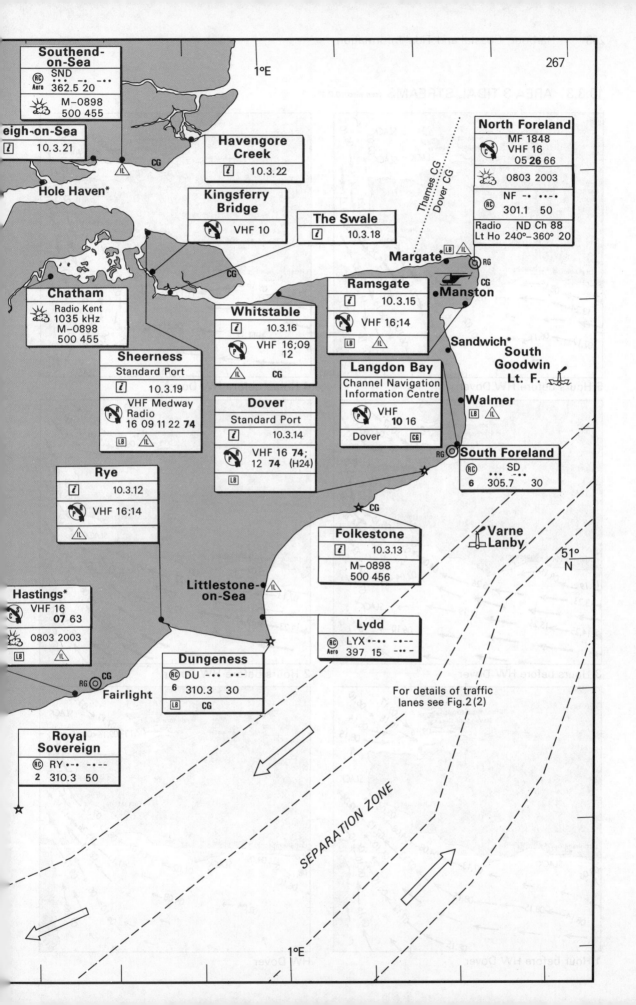

10.3.3 AREA 3 TIDAL STREAMS (see also 10.3.17)

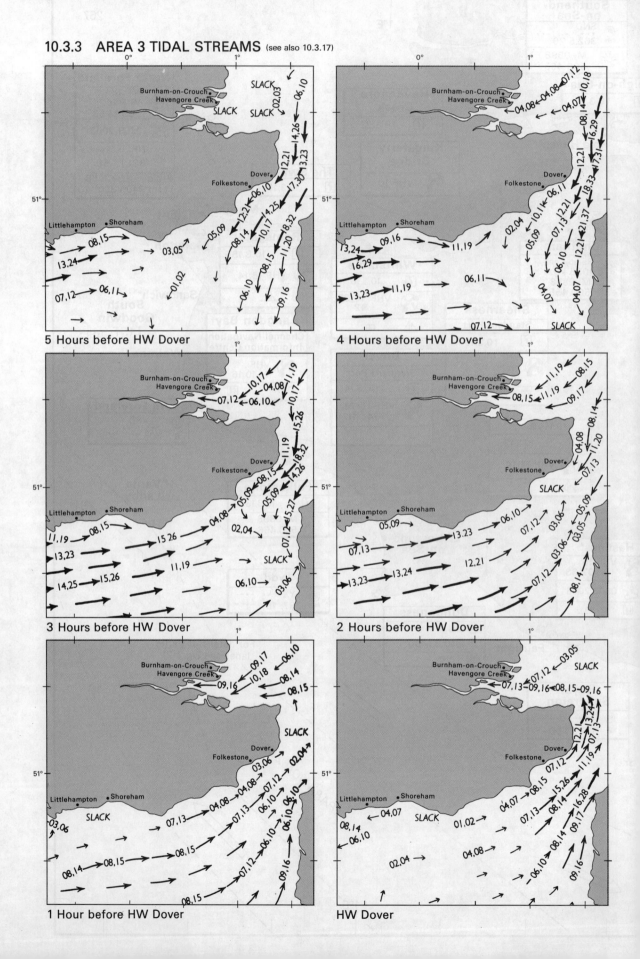

AREA 3—SE England

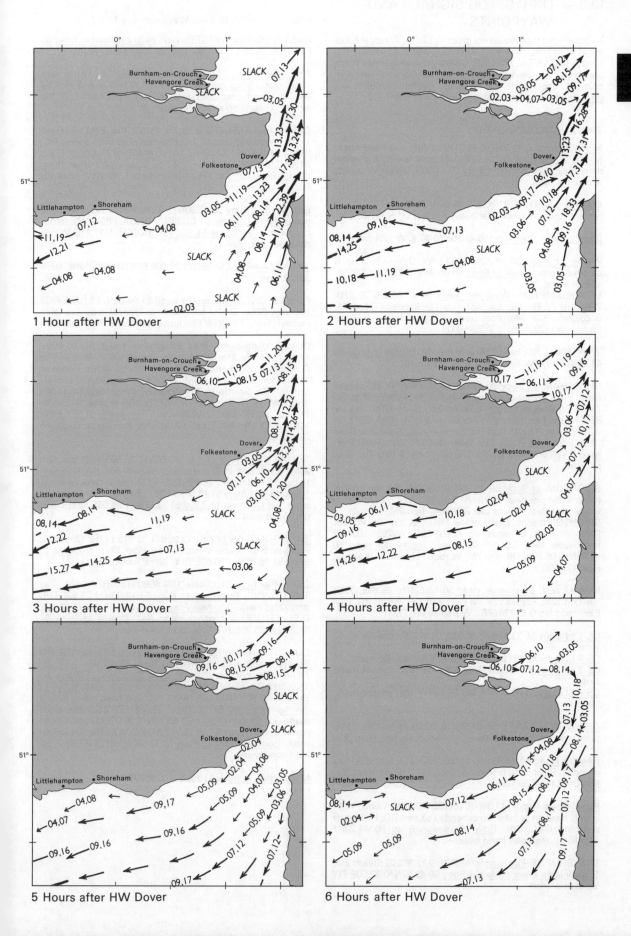

10.3.4 LIGHTS, FOG SIGNALS AND WAYPOINTS

Abbreviations used below are given in 10.0.3. Principal lights are in **bold** print, places in CAPITALS, and light-vessels and Lanbys in *CAPITAL ITALICS*. Unless otherwise stated lights are white. m—elevation in metres; M—nominal range in n. miles. Fog signals are in *italics*. Useful waypoints are underlined – use those on land with care. See 4.2.2.

ENGLAND—SOUTH COAST

Mixon Beacon 50 42.35N/0 46.21W (unlit); port-hand mark.
Middle Owers Buoy 50 40.15N/0 44.90W (unlit); E cardinal mark. East Borough Head Buoy 50 41.50N/0 39.00W (unlit); N cardinal mark.

OWERS LANBY 50 37.30N/0 40.60W Fl(3) 20s 12m **22M**; tubular structure on circular buoy; F riding Lt; *Horn (3) 60s*.

Bognor Regis. Pier, head 50 46.70N/0 40.42W 2FR (vert).

Winter Lt Buoy 50 45.60N/0 33.60W Q(6)+LFl 15s; S cardinal mark. Outfall Lt Buoy 50 46.20N/0 30.45W FlY 5s; special mark.
LITTLEHAMPTON. West Pier head 50 47.85N/0 32.37W 2FR(vert) 7m 6M. Ldg Lts 346°. Front E pierhead FG 6m 7M; B column. Rear 64m from front Oc WY 7.5s 9m 10M; W tower; vis W290°-356°, Y 356°-042°.

WORTHING. Pier head 2FR(vert) 6m 1M. Outfall Lt Buoy 50 48.45N/0 19.40W Fl(2)R 10s; port-hand mark.

SHOREHAM. Lt Buoy 50 47.00N/0 15.33W FlY 5s; special mark. West Breakwater head 50 49.45N/0 14.79W FlR 5s 7m 7M. E Breakwater head Fl G 5s 7m 8M; *Siren 120s*. Ldg Lts 355° Middle Pier. Front F WR or G 8m W10M, R9M, G9M; W watchhouse, R base; tidal Lts, traffic signals; *Horn 20s*. **Rear** 192m from front Fl 10s 13m **15M**; vis 283°-103°. West Pier, head F WR 6m; R to seaward. E pier, head F WG 7m; G to seaward.

BRIGHTON. Lt Buoy 50 46.00N/0 08.30W FlY 3s; special mark. West pier head 2FR(vert) 13m 4M; *Bell (1) 13s* (when vessel expected). Palace pier 2FR(vert) 10m 2M.
BRIGHTON marina. E breakwater Fl(4) WR 20s 16m W10M, R8M; W pillar, G bands; vis R260°-295°, W295°-100°. E breakwater head QG 8m 6M. West breakwater head 50 48.46N/0 06.29W QR 10m 7M; W round structure, R bands; *Horn (2) 30s*.

NEWHAVEN. Breakwater Head 50 46.52N/0 03.60E Oc(2) 10s 17m 12M; *Horn 30s*. West Pier head 2 FR (vert); RC. East Pier, head Iso G 5s 12m 6M; W framework Tr.

CS 1 Lt Buoy 50 33.67N/0 03.83W Fl Y 2.5s; special mark; *Whis*.
CS 2 Lt Buoy 50 39.08N/0 32.70E Fl Y 5s; special mark.
CS 3 Lt Buoy 50 52.00N/1 02.30E Fl Y 10s; special mark; *Bell*.
CS 4 Lt Buoy 51 08.57N/1 34.02E Fl(4)Y 15s; special mark; *Whis*.

GREENWICH LANBY 50 24.50N/0 00.00 Fl 5s 12m **21M**; tubular structure on circular buoy; Racon; Ra refl; *Horn 30s*.

Beachy Head 50 44.00N/0 14.60E Fl(2) 20s 31m **25M**; W round Tr, R band and lantern; vis 248°-101°; (H24); *Horn 30s*. Fog Det Lt vis 085°-265°.

Royal Sovereign 50 43.38N/0 26.13E Fl 20s 28m **28M**; W Tr with R band on W cabin on concrete column; RC; Helicopter platform; *Dia(2) 30s*. Royal Sovereign Buoy 50 44.20N/ 0 25.95E (unlit); port-hand mark.

EASTBOURNE. Pier head 50 45.88N/0 17.85E 2FR(vert) 2M. St Leonard's sewer outfall Lt Buoy 50 49.27N/0 32.00E FlY 5s; special mark.

Note. For English Channel Waypoints see 10.1.7

HASTINGS. Pier head 2FR(vert). West breakwater head 50 51.13N/0 35.70E Fl R 2.5s. Ldg Lts 356° both FR 4M.

RYE. Rye Fairway Lt Buoy 50 54.00N/0 48.13E LFl 10s; safe water mark. West groyne head 50 55.55N/0 46.65E Fl R 5s 9m 5M; Ra refl.; replaced by Buoy(T). E Arm, head Q(9) 15s 9m 5M; G cone on structure; *Horn 7s*.

Dungeness 50 54.77N/0 58.67E Fl 10s 40m **27M**; B round Tr, W bands and lantern, floodlit; Part obsc 078°-shore; RC; (H24). F RG 37m 11M (same Tr); vis R057°-073°, G073°-078°, R196°-216°; FR Lts shown 2.4 and 5.2M WNW when firing taking place. QR and FR on radio mast 1.2M NW; *Horn (3) 60s*.

East Road Lt Buoy 50 58.30N/1 03.8E Fl(4)R 15s; port-hand mark. Hythe Lt Buoy 51 01.30N/1 08.80E Fl(3)G 10s; stbd-hand mark. Sandgate Lt Buoy 51 02.20N/1 11.17E VQ; N cardinal mark; *Bell*.

VARNE LANBY 51 01.25N/1 24.00E Fl R 20s 12m **19M**; *Horn 30s*; Racon.

FOLKESTONE. **Breakwater head** 51 04.53N/1 11.79E Fl(2) 10s 14m **22M**; *Dia(4) 60s; Bell (4) (occas)*. Ldg Lts 267° (occas). Front FR. Rear FG. Inner harbour, E Pier head QG.

DOVER. **Admiralty Pier Extension head** 51 06.65N/ 1 19.77E Fl 7.5s 21m **20M**; W Tr; vis 096°-090°, obsc in The Downs by South Foreland inshore of 226°; *Dia 10s*. Prince of Wales Pier head FG 14m 4M; W Tr; Fl 1.5s (intens, occas); *Bell (2) 15s*. **Southern Breakwater**, West head Oc R 30s 21m **18M**; W Tr. **Knuckle** Oc WR 10s 15m **W15M**, R13M; W Tr; vis R059°-239°, W239°-059°. N head 51 07.17N/ 1 20.72E FY 17m 4M. Eastern Arm head, Port Control Sig Stn; *Dia(2) 30s*.

SOUTH GOODWIN Lt F 51 07.95N/1 28.60E Fl(2) 30s 12m **25M**; R hull with Lt Tr amidships; *Horn(2) 60s*. SW Goodwin Lt Buoy 51 08.55N/1 28.80E Q(6)+LFl 15s; S cardinal mark. S Goodwin Lt Buoy 51 10.56N/1 32.35E Fl(4)R 15s; port-hand mark. SE Goodwin Lt Buoy 51 12.95N/1 34.55E Fl(3)R 10s; port-hand mark.

EAST GOODWIN Lt F 51 13.05N/1 36.31E Fl 15s 12m **26M**; R hull with Lt Tr amidships; Racon; *Horn 30s*. E Goodwin Lt Buoy 51 16.00N/1 35.60E Q(3) 10s; E cardinal mark.

DEAL. Pier head 51 13.40N/1 24.65E 2FR(vert) 7m 5M.
THE DOWNS. Deal Bank Lt Buoy 51 12.90N/1 25.68E QR; port-hand mark. Goodwin Fork Lt Buoy 51 13.25N/1 27.13E Q(6)+LFl 15s; S cardinal mark; *Bell*. Downs Lt Buoy 51 14.31N/1 26.90E Fl(2)R 5s; port-hand mark.

GULL STREAM. W Goodwin Lt Buoy 51 15.28N/1 27.32E FlG 5s; stbd-hand mark. S Brake Lt Buoy 51 15.40N/1 27.00E Fl(3)R 10s; port-hand mark. NW Goodwin Lt Buoy 51 16.55N/1 28.55E Q(9) 15s; W cardinal mark; *Bell*. Brake Lt Buoy 51 16.90N/1 28.40E Fl(4)R 15s; port-hand mark; *Bell*. N Goodwin Lt Buoy 51 17.60N/1 30.03E FlG 2.5s; stbd-hand mark. Gull Stream Lt Buoy 51 18.10N/1 30.00E QR; port-hand mark. Gull Lt Buoy 51 19.55N/1 31.40E VQ(3) 5s; E cardinal mark. Goodwin Knoll Lt Buoy 51 19.55N/1 32.30E Fl(2)G 5s; stbd-hand mark.

NE Goodwin Lt Buoy 51 20.28N/1 34.27E Q(3) 10s; E cardinal mark; Racon.

AREA 3—SE England   271

RIVER STOUR. Channel marked by port and stbd-hand buoys and beacons. Pegwell Bay, Sandwich approach 51 18.72N/1 23.05E FIR 10s 3m 4M; framework Tr; moved to meet changes in channel.

RAMSGATE. East Brake Lt Buoy 51 19.45N/1 29.10E Q(3) 10s; E cardinal mark. Dredged approach channel marked by Lt Buoys, Fl(4)Y 10s on S side and Q Fl on N side. Channel Lt Buoy, S side 51 19.44N/1 26.00E Q; N cardinal mark. N side 51 15.51N/1 26.00E Q(6)+LFl 15s; S cardinal mark. South breakwater head VQR; R+W Bn. North breakwater head QG; G+W Bn. West ferry terminal Dir Lt 270°, Dir Oc WRG 10s; vis G259°-269°, W269°-271°, R271°-281°. Royal Harbour entrance, E side Oc 10s 8m 4M. West side FR or FG 12m 7M.

BROADSTAIRS. Broadstairs Knoll Lt Buoy 51 20.85N/1 29.57E FlR 2.5s; port-hand mark. Pier, SE end 51 21.46N/1 26.83E 2FR(vert) 7m 4M.

North Foreland 51 22.47N/1 26.80E Fl(5) WR 20s 57m **W21M, R18M**; W 8-sided Tr; vis W shore-150°, R150°-200°, W200°-011°; RC.

ENGLAND—EAST COAST

FALLS Lt F 51 18.10N/1 48.50E Fl(2) 10s 12m **24M**; R hull with Lt Tr amidships; RC; Racon; *Horn Mo(N) 60s*.
F3 LANBY 51 23.82N/2 00.62E Fl 10s 12m **22M**; Racon; *Horn 10s*.

Elbow Lt Buoy 51 23.20N/1 31.68E Q; N cardinal mark. Foreness Point, sewer outfall Lt Buoy 51 24.60N/1 26.10E FlY 5s; special mark. Longnose Buoy 51 24.12N/1 26.20E (unlit); port-hand mark. Longnose Spit Buoy 51 23.90N/1 25.85E (unlit); N cardinal mark.

MARGATE. Promenade Pier, N head 51 23.65N/1 22.93E Fl (3)R 10s. Stone Pier head FR 18m 4M.

GORE CHANNEL. SE Margate Lt Buoy 51 24.10N/1 20.50E Q(3) 10s; E cardinal mark. S Margate Lt Buoy 51 23.90N/1 16.75E FlG 2.5s; stbd-hand mark. Hook Spit Buoy 51 24.03N/1 12.68E (unlit); stbd-hand mark. E Last Lt Buoy 51 24.00N/1 12.27E QR; port-hand mark.

HERNE BAY. Pier head (isolated from shore) 51 22.88N/1 07.00E Q 8m 4M (TE 1988). 2FR(vert) near root.

WHITSTABLE. Whitstable Street Lt Buoy 51 23.83N/1 01.70E VQ; N cardinal mark. Whitstable Oyster Lt Buoy 51 22.03N/1 01.16E Fl(2)R 10s; port-hand mark. On E side of harbour F 15m 8M; W mast; FR 10m 5M (same structure) shown when entry/departure prohibited. West Quay, dolphin 51 21.82N/1 01.55E Fl WRG 5s W5M, R3M, G3M; vis W118°-156°, G156°-178°, R178°-201°. E Quay, N end 2FR (vert). Ldg Lts 122°30'. Front 51 21.7N/1 01.8E FR 7m on mast. Rear, 30m from front FR 13m on mast.

THE SWALE. Columbine Buoy 51 24.23N/1 01.45E (unlit); stbd-hand mark. Columbine Spit Buoy 51 23.83N/1 00.12E (unlit); stbd-hand mark. Pollard Spit Lt Buoy 51 22.95N/0 58.66E QR; port-hand mark. Ham Gat Buoy 51 23.05N/0 58.41E (unlit); stbd-hand mark. Sand End Lt Buoy 51 21.40N/0 56.00E FlG 5s; stbd-hand mark. Faversham Spit Buoy 51 20.72N/0 54.28E (unlit); N cardinal mark.

QUEENS CHANNEL/FOUR FATHOMS CHANNEL. E Margate Lt Buoy 51 27.00N/1 26.50E FlR 2.5s; port-hand mark. NE Spit Lt Buoy 51 27.92N/1 30.00E VQ(3) 5s; E cardinal mark. Spaniard Lt Buoy 51 26.20N/1 04.10E Q(3) 10s; E cardinal mark. Spile Lt Buoy 51 26.40N/0 55.85E FlG 2.5s; stbd-hand mark.

PRINCES CHANNEL/EDINBURGH CHANNELS. Outer Tongue Lt Buoy 51 30.70N/1 26.50E LFl 10s; safe water mark; *Whis*; Racon.

Tongue Sand Tower 51 29.55N/1 22.10E (unlit — N and S cardinal Lt Buoys close N and S).
PRINCES CHANNEL. E Tongue Lt Buoy 51 28.73N/1 18.70E Fl(2)R 5s; port-hand mark. S Shingles Lt Buoy 51 29.20N/1 16.15E Q(6)+LFl 15s; S cardinal mark; *Bell*. N Tongue Lt Buoy 51 28.80N/1 13.20E Fl(3)R 10s; port-hand mark. SE Girdler Lt Buoy 51 29.47N/1 10.00E Fl(3)G 10s; stbd-hand mark. W Girdler Lt Buoy 51 29.60N/1 06.90E Q(9) 15s; W cardinal mark; *Bell*. Girdler Lt Buoy 51 29.15N/1 06.50E Fl(4)R 15s; port-hand mark. Shivering Sand Tr 51 29.90N/1 04.90E (unlit — N and S cardinal Lt Buoys close N and S). E Redsand Lt Buoy 51 29.38N/1 04.12E Fl(2)R 5s; port-hand mark.

OAZE DEEP. Red Sand Tr 51 28.60N/0 59.50E (unlit — port and stbd-hand Lt Buoys close NW and E). S Oaze Lt Buoy 51 30.00N/1 00.80E Fl(2)G 5s; stbd-hand mark. SW Oaze Lt Buoy 51 29.03N/0 57.05E Q(6)+LFl 15s; S cardinal mark. W Oaze Lt Buoy 51 29.03N/0 55.52E VQ(9) 10s; W cardinal mark. Cant Bn 51 27.73N/0 55.45E (unlit). E Cant Lt Buoy 51 28.50N/0 55.70E QR; port-hand mark.

Medway Lt Buoy 51 28.80N/0 52.92E Iso 2s; safe water mark.
SHEERNESS. Garrison Pt. 51 26.80N/0 44.73E L Fl(2) R 15s 13m 4M; R diamond, W top, on column on fort; vis 027°-255°; *Horn(3) 30s*. Isle of Grain 51 26.6N/0 43.5E Q WRG 20m W13M, R7M, G8M; R&W diamond on R Tr; vis R220°-234°, G234°-241°, W241°-013°; Ra refl.

SEA REACH. No 1 Lt Buoy 51 29.42N/0 52.67E FlY 2.5s; special mark; Racon.
No 2 Lt Buoy 51 29.37N/0 49.85E Iso 5s; safe water mark.
No 3 Lt Buoy 51 29.30N/0 46.65E LFl 10s; safe water mark.
No 4 Lt Buoy 51 29.58N/0 44.28E FlY 2.5s; special mark.
No 5 Lt Buoy 51 29.92N/0 41.54E Iso 5s; safe water mark.
No 6 Lt Buoy 51 30.00N/0 39.94E Iso 2s; safe water mark.
No 7 Lt Buoy 51 30.83N/0 37.15E FlY 2.5s; special mark; Racon.
Mid Blyth Lt Buoy 51 30.05N/0 32.50E Q; N cardinal mark.

Canvey Island. Jetty head, E end 2FG(vert); *Bell (1) 10s*. West end 2FG(vert). Lts 2FR(vert) to port, and 2FG(vert) to starboard, are shown from wharves etc above this point.

Shornmead 51 27.0N/0 26.6E Fl(2) WRG 10s 12m **W17M**, R13M, G13M; vis G shore-080°, R080°-085°, W085°-088°, G088°-141°, W141°-205°, R205°-213°. **Northfleet Lower** 51 26.9N/0 20.4E Oc WR 5s 15m **W17M**, R14M; vis W164°-271°, R271°-S shore. **Northfleet Upper** 51 26.9N/0 20.2E Oc WRG 10s 30m **W16M**, R12M, G12M; vis R126°-149°, W149°-159°, G159°-269°, W269°-279°.

Broadness. 51 28.0N/0 18.7E Oc R 5s 12m 12M; R metal Tr.

SOUTHEND. Pier, E end 51 30.84N/0 43.51E 2FG(vert) 7/5m. *Horn Mo(N) 30s* (temp inop, 1986). West head 2FG(vert) 13m 8M.

Shoebury Bn 51 30.27N/0 49.40E Fl(3) G 10s. Inner Bn 51 30.96N/0 49.28E Fl Y 2.5s. Maplin Survey Platform 51 30.80N/0 53.16E Fl(2) 5s. Blacktail (West) Bn 51 31.43N/0 55.30E Iso G 10s 10m 5M. Blacktail (East) Bn 51 31.75N/0 56.60E Iso G 5s 10m 5M.

IMPORTANT NOTE. Extensive changes to buoyage in Thames Estuary are in hand, and some details are not yet published.

10.3.5 PASSAGE INFORMATION

This area embraces the greatest concentration of commercial shipping in the world, and it must be recognised that in such waters the greatest danger to a small yacht is being run down by a larger vessel, especially in poor visibility. In addition to the many ships plying up and down the traffic lanes, there are fast ferries, hovercraft and hydrofoils passing to and fro between English and Continental harbs; warships and submarines on exercises; fishing vessels operating both inshore and offshore; many other yachts; and static dangers such as lobster pots and fishing nets which are concentrated in certain places.

Even for coastal cruising it is essential to have knowledge of the traffic separation schemes in force, see Fig. 2(2), observing that the SW-bound lane from the Dover Strait passes only 4M off Dungeness, for example. Radar surveillance of the Dover Strait is maintained continously by the Channel Navigation Information Service (CNIS).

It should be remembered that in the English Chan wind can have a big effect on tidal streams, and on the range of tides. Also N winds, which give smooth water and pleasant sailing off the shores of England, can cause rough seas on the French coast. The rates of tidal streams vary with the locality, and are greatest in the narrowest parts of the Chan and off large headlands: in the Dover Strait sp rates can reach 4 kn, but elsewhere in open water they seldom exceed 2 kn. In strong winds the Dover Strait can become very rough. With strong S winds the English coast between Isle of Wight and Dover is very exposed, and shelter is hard to find.

CROSS-CHANNEL PASSAGES

Various factors need to be considered when planning a passage between, say, England and France in whichever direction. It is an advantage for the average yacht to minimise the length of time out of sight of land, thus reducing not only fatigue (which can soon become apparent in a family crew) but also the risk of navigational error due in part to tidal streams. Thus, when proceeding from Brighton to St Malo for example, it pays to cruise coastwise to Poole or Weymouth before setting off for Cherbourg, Alderney or the Casquets. Such tactics also combine the legal requirement to cross traffic separation schemes at right angles (see 2.1.2). Naturally the coastwise sections of any cruise should be timed to take maximum advantage from favourable tidal streams.

Prevailing winds also need to be considered. If it is accepted (as statistics show) that in the English Channel between Portland and the Channel Islands the wind more commonly blows from a direction between S and W, the likelihood of getting a good slant is improved by sailing from Portland rather than Brighton: departure from Brighton might give a long beat to windward. Such decisions may depend of course on forecast wind directions in the immediate future.

It is also helpful to plan that points of departure and arrival have powerful lights and conspic landmarks, and radio-beacons which can be used in the event of visibility closing down. Do not forget the importance of taking back bearings in the early stages of a Channel crossing, in order to assess the correct course and proper allowance for leeway and tidal stream.

On some passages soundings can be a useful aid to navigation: thus in the example quoted above from Portland to the Casquets, the Hurd Deep is a useful guide to progress made.

THE OWERS (chart 1652)

Selsey Bill is a low headland, and off it lie the Owers — groups of rks and shoals extending 3M to the S, and 5M to the SE. Just W and SW of the Bill, The Streets (awash) extend 1¼ M seaward. 1¼ M SSW of the Bill are The Grounds (or Malt Owers) and The Dries (dry). 1M E of The Dries, and about 1¼ M S of the lifeboat house on E side of Selsey Bill is a group of rks called The Mixon, marked by Bn at E end.

Immediately S of dangers above is the Looe chan, which runs E/W about ¾ M S of Mixon Bn, and is marked by buoys at W end, where it is narrowest between Brake (or Cross) Ledge on N side and Boulder Bank to the S. In daylight and in good vis and moderate weather, the Looe chan is an easy and useful short cut. The E-going stream begins at HW Portsmouth +0430, and the W-going at HW Portsmouth −0135, sp rates 2½ kn. Beware lobster pots in this area.

In poor vis or in bad weather (and always in darkness) keep S of the Owers Lanby, moored 7M SE of Selsey Bill, marking SE end of Owers. Over much of Owers there is less than 3m, and large parts virtually dry: so a combination of tidal stream and strong wind produces heavy breaking seas and overfalls over a large area.

SELSEY BILL TO DUNGENESS (charts 1652, 536)

The coast from Selsey Bill to Brighton is low, faced by a shingle beach, and with few offlying dangers, Bognor Rks (dry in places) extend 1¾ M E from a point 1M W of the pier, and Bognor Spit extends E and S from the end of them. Middleton ledge are rks running 8 ca offshore, about 1½ M E of Bognor pier, with depths of less than 1m. Shelley rks (marked by buoy) lie ½ M S of Middleton ledge, with depths of less than 1m.

Winter Knoll, about 2½ M SSW of Littlehampton (10.3.8) has depths of 2.1m, and is marked by buoy. Kingston rks, with depth of 2.0m lie about 3¼ M ESE of Littlehampton. Grass Banks, an extensive shoal with depths of 3.0m at W end, lie about 1M S of Worthing: Elbow, with depth of 3.1m, lies 1¾ M SE of Worthing pier.

Off Shoreham (10.3.9) Church Rks, with depth of 0.3m, lie 1½ M W of entrance and ¼ M offshore. Jenny Rks, with depth 0.9m, are 1¼ M E of the entrance, 3 ca offshore.

At Brighton (10.3.10) the S Downs join the coastline, and high chalk cliffs are conspic from here to Beachy Hd. There are no dangers more than 3 ca from shore, until Birling Gap, where a rky ledge begins, on which is built Beachy Hd Lt Ho (Lt, fog sig). Below the watch house on cliff top, Head Ledge (dries) extends about 4 ca S. 2M S of Beachy Hd the W-going stream begins at HW Dover +0030, and the E-going at HW Dover −0520, sp rates 2¼ kn. In bad weather there are overfalls off the hd, which should then be given a berth of 2M.

Royal Sovereign Lt Tr (Lt, fog sig, RC) lies about 7M E of Beachy Hd. The extensive Royal Sovereign shoals lie from between 1½ M N of the Tr to 3M NW of it, and have a least depth of 3.5m. There are strong eddies over the shoals at sp, and the sea breaks on them in bad weather.

On the direct course from Royal Sovereign Lt Tr to clear Dungeness there are no dangers. Along the coast in Pevensey B and Rye B there are drying rky ledges or shoals extending ½ M offshore in places. These include Boulder Bank near Wish Tr, S of Eastbourne; Oyster Reef off Cooden; Bexhill Reef off Bexhill-on-Sea; Bopeep Rks off St Leonards; and the shoals at the mouth of R Rother, at entrance to Rye (10.3.12). There are also shoals 2-3M offshore, on which the sea builds in bad weather.

AREA 3—SE England   273

Dungeness (Lt, fog sig, RC) is at SE extremity of Romney Marsh. The nuclear power station is conspic. The pt is steep-to on SE side. Trinity House has an experimental station at Dungeness: disregard any uncharted light, fog signal or radar beacon therefrom. Good anch close NE of Dungeness.

DUNGENESS TO NORTH FORELAND (charts 1892, 1828)

From Dungeness to Folkestone (10.3.13) the coast forms a b. Beware Roar bank, depth 2.7m, E of New Romney: otherwise there are no offlying dangers. Good anch off Sandgate in offshore winds. Off Folkestone harb the E-going stream begins HW Dover −0155, sp rate 2 kn; the W-going begins HW Dover +0320, sp rate 1½ kn.

Past Dover (10.3.14) and S Foreland keep ½M offshore: do not pass too close to Dover harb because ferries etc leave at speed, and also there can be considerable backwash off harb walls. 8M S of Dover is the Varne, a shoal 5M in extent with least depth 3.7m and a heavy sea in bad weather, marked by Lanby. Between S and N Foreland the N-going stream begins at about HW Dover −0150, and the S-going at about HW Dover +0415.

Goodwin Sands are extensive, shifting shoals, running about 10M from N to S, and 5M from E to W at their widest part. Along the E and N sides large areas dry up to 3m. The sands are well marked by Lt Fs and by buoys. Kellett Gut is a chan about ½M wide, running from SW to NE through the middle of the sands, but it is not regularly surveyed and liable to change. The Gull Stream (buoyed) leads inside Goodwin Sands and outside Brake Sands off Ramsgate (10.3.15).

THAMES ESTUARY (chart 1607)

N Foreland (Lt, RC) marks the S entrance to Thames Estuary, an area encumbered by banks many of which dry. N Foreland (chart 1828) is well marked by Lt Ho and buoys offshore. From HW Dover −0120 to HW Dover +0045 the stream runs N and W from the Downs into Thames Estuary. From HW Dover +0045 to HW Dover +0440 the N-going stream from The Downs meets the E-going stream from Thames Estuary, which in strong winds causes a bad sea. From HW Dover −0450 to HW Dover −0120 the streams turn W into Thames Estuary and S towards The Downs.

The sandbanks shift constantly, and it is important to have corrected charts showing recent buoyage changes. With wind against tide a short, steep sea is raised, particularly in E or NE winds. The stream runs 3 kn at sp in places, mostly in the directions of the chans but sometimes across the shoals in between. The main chans carry much commercial shipping and are well buoyed and lit, but this does not apply to lesser chans and swatchways which are convenient for yachtsmen, particularly when crossing the estuary from N to S, or S to N. Note that there have been many changes to buoyage during 1988.

The main chans on the S side of the estuary are as follows. S Chan, which leads W about 1M off the N Kent coast and into Gore Chan, Horse Chan and Four Fathoms Chan across the Kentish Flats (most of this route is sparsely buoyed, and very sparsely lit); Queens Chan; Princes Chan, a main E/W route, passing S of Tongue Sand Tr, where the W-going stream begins at HW Sheerness −0610, and the E-going at HW Sheerness +0030, sp rates 2 kn; the Edinburgh Chans, which run into Knob Chan; and Oaze Deep, a continuation of Knock John Chan and Barrow Deep, which join it from a NE direction.

When crossing the estuary it is essential to study the tides carefully, both to ensure sufficient depth in places and also to make the best use of tidal streams. Good vis is needed to pick out the buoys and marks, and to avoid commercial shipping. Proceeding N from N Foreland to Orford Ness or beyond (or coming in opposite direction) it may be preferable to keep to seaward of the main banks, passing Kentish Knock buoy and thence to Shipwash Lt F 20M further N.

Bound from N Foreland to the Crouch or Blackwater proceed through S Edinburgh Chan, through the swatchway about ½M NE of Knock John Tr into Barrow Deep, and thence round the end of W Barrow Sand. With sufficient rise of tide it is possible to cut across between Maplin Spit and Barrow Sand, but usually it is better to proceed SW to Maplin Bank buoy, and thence into E Swin to the Whitaker buoy. This is just one of many routes which could be followed, depending on wind direction and tidal conditions.

10.3.6 DISTANCE TABLE

Approximate distances in nautical miles are by the most direct route while avoiding dangers and allowing for traffic separation schemes etc. Places in *italics* are in adjoining areas.

	1	2	3	4	5	6	7	8	9	10	11	12	13	14	15	16	17	18	19	20
1 *Casquets*	1																			
2 *Cherbourg*	31	2																		
3 *Nab Tower*	81	66	3																	
4 *Owers Lanby*	87	70	11	4																
5 Shoreham	106	88	32	21	5															
6 Brighton Marina	109	90	35	24	5	6														
7 Newhaven	114	94	40	29	12	7	7													
8 Royal Sovereign	124	102	54	43	27	22	15	8												
9 Folkestone	162	140	92	81	65	60	53	38	9											
10 Dover	167	145	97	86	70	65	58	43	5	10										
11 Ramsgate	182	160	112	101	85	80	73	58	20	15	11									
12 North Foreland	186	164	116	105	89	84	77	62	24	19	4	12								
13 Whitstable	203	179	133	122	106	101	94	79	41	36	22	17	13							
14 Sheerness	216	194	146	135	119	114	107	92	54	49	34	30	14	14						
15 London Bridge	258	236	188	177	161	156	149	134	96	91	76	72	55	45	15					
16 Shoeburyness	215	193	145	134	118	113	106	91	53	48	33	29	14	4	43	16				
17 Orford Ness	231	209	161	150	134	129	122	107	69	64	49	45	50	53	92	49	17			
18 *Dieppe*	139	108	92	81	75	70	65	55	71	75	90	94	111	124	166	123	135	18		
19 *Cap Gris Nez*	175	147	101	90	74	69	62	47	19	19	28	32	49	62	104	61	74	61	19	
20 *Goeree Tower*	278	250	204	193	177	172	165	150	106	101	94	91	104	114	154	111	78	164	103	20

LITTLEHAMPTON 10-3-8
W. Sussex

CHARTS
Admiralty 1991; Stanford 9; Imray C9; OS 197

TIDES
+0015 Dover; ML 2.8; Zone 0 (GMT).

Standard Port SHOREHAM (→)

Times				Height (metres)			
HW		LW		MHWS	MHWN	MLWN	MLWS
0500	1000	0000	0600	6.2	5.0	1.9	0.7
1700	2200	1200	1800				

Differences LITTLEHAMPTON (ENT)
+0010 0000 −0005 −0010 −0.4 −0.4 −0.2 −0.2
LITTLEHAMPTON (NORFOLK WHARF)
+0015 +0005 0000 +0045 −0.7 −0.7 −0.3 +0.2
PAGHAM
+0015 0000 −0015 −0025 −0.7 −0.5 −0.1 −0.1
BOGNOR REGIS
+0010 −0005 −0005 −0020 −0.6 −0.5 −0.2 −0.1

NOTE: Tidal heights inside harbour are affected by flow down River Arun. Tide seldom falls lower than 0.7m above datum.

SHELTER
Good. SE winds cause a certain swell up the harbour; SW winds cause roughness on the bar. Entrance dangerous with strong SE winds.

NAVIGATION
Waypoint 50°47'.50N 00°32'.20W, 166°/346° from/to front Ldg Lt 346°, 0.60M. Bar ½ M offshore. Harbour available from HW−4 to HW+3 for boats with about 2m draught. The ebb stream runs so fast at springs that yachts may have difficulty entering. When Pilot boat with P at the bow displays a RW flag by day or W over R lights at night, all boats keep clear of entrance. Speed limit 6½ kn. Depth over the bar is 0.6m less than depth shown on tide gauges. On departure check tide gauge at East Pier to calculate depth on bar.

LIGHTS AND MARKS
Leading marks for entrance are lighthouse at inshore end of E breakwater and the black steel column for the light at the outer end of E breakwater; Ldg Lts 346°.
Swing bridge Fl G Lt – Open
              Fl R Lt – Closed   from high mast to port.
The central retractable section of the bridge has 2 FR (vert) to port at each end and 2 FG (vert) to stbd at each end. Opening section is 22m wide. Bridge retracted on request provided notice given before 1630 previous day.

RADIO TELEPHONE
Marinas Ch M (office hours).

TELEPHONE (0903)
Hr Mr 721215/6; CG 715512; MRSC Lee-on-Solent 552100; HMC Freefone Customs Yachts: RT (0703) 229251; Marinecall 0898 500 456; Dr 714113.

FACILITIES
EC Wednesday **Littlehampton Marina** (120+30 visitors) Tel. 713553, Slip, BH (16 ton), CH, P, D, V, R, Bar, FW, AC, Sh, ME, SM; **Ship and Anchor Marina** (182, some visitors) Tel. Yapton 551262, Slip, FW, C (6 ton), ME, Sh, CH, V, R, Bar (Access HW∓4); **Arun YC** (90+10 visitors) Tel. 714533, Slip, AC, FW, R, Bar (Access HW∓3); **Arun Sails** Tel. 882933, SM; **E side of Harbour** M, C, FW; **Littlehampton Sailing and Motor Club** Tel. 715859, M, FW, Bar; **Wm Osbornes** Tel. 713996, BY; **Seawolf Yachts** Tel. 717369, BY; **Arrow Marine** Tel. 721686, FW, ME, El, Sh, P, D; **County Wharf** FW, C (5 ton); **Hillyards** Tel. 713327, M, FW, ME, Sh (Wood), C; **Viking Marine Services** Tel. 724873, ME, El, Sh, Electronics.
Town P, D, V, R, Bar. PO; Bank; Rly; Air (Shoreham).

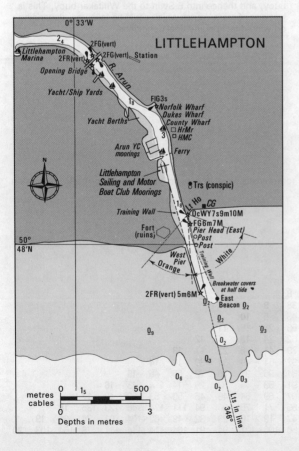

SHOREHAM 10-3-9
W. Sussex

CHARTS
Admiralty 2044; Stanford 9; Imray C9; OS 197/8

TIDES
+0005 Dover; ML 3.4; Duration 0605; Zone 0 (GMT)

Standard Port SHOREHAM (→)

Times				Height (metres)			
HW		LW		MHWS	MHWN	MLWN	MLWS
0500	1000	0000	0600	6.2	5.0	1.9	0.7
1700	2200	1200	1800				

Differences WORTHING
+0010 0000 −0005 −0010 −0.1 −0.2 0.0 0.0

NOTE: Shoreham is a Standard Port and tidal predictions for the year are given below.

SHELTER
Excellent. There is complete shelter once through the locks into Southwick Canal. Locks are manned HW∓4. The shallow water at the entrance can be very rough in strong on-shore winds. Berthing space for yachts is very limited and arrangements should be made in advance. Harbour dues are high.

NAVIGATION
Waypoint 50°49'.20N 00°14'.72W, 175°/355° from/to front Ldg Lt 355°, 0.52M. From E, beware Jenny Rks; from W beware Church Rks. Yachtsmen are advised not to use the W arm if possible. Harbour is mainly commercial, so before entering yachtsmen should familiarise themselves with bye-laws, control signals, lock signals, tidal signals etc. Prince George lock usually used for light craft and yachts.
Prince George Lock opens
Outward HW−3¾, −1¾, +½, +2¾
Inward HW−3¼, −1¼, +1, +3¼
Apl to Oct on Sat, Sun and Bank Holidays
Outward HW−3¾, −2½, −1, +¼, +1½, +3
Inward HW−3¼, −2, −½, +½, +1¾, +3¼

SHOREHAM continued

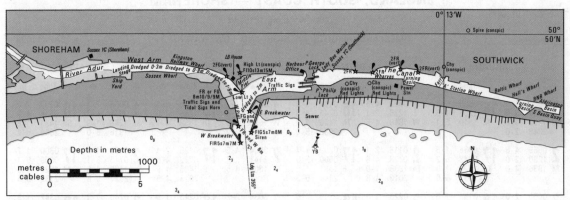

LIGHTS AND MARKS
International Port Traffic Signals (see 6.2.7) are displayed from Middle Pier Control Station. In line with High Lighthouse they form leading line for entrance.
Additional signals are displayed as follows. From Lifeboat House: focussed over Eastern Arm, and controlling movement therein. From Lifeboat House: focussed over Western Arm, and controlling movement therein. From Prince George Lock: controlling entrance thereto. From Prince Philip Lock: controlling entrance thereto.

RADIO TELEPHONE
VHF Ch 16; 14 (H24). Lady Bee Marina Ch M.

TELEPHONE (0273)
Hr Mr 592613; MRSC Lee-on-Solent 552100; HMC Freefone Customs Yachts: RT (0703) 229251; Marinecall 0898 500 456; Hosp 455622; Dr 461101 (Health Centre).

FACILITIES
EC Wednesday; **Lady Bee Marina** (110 + 10 visitors) Tel. 593801, Slip, P, D, FW, ME, El, Sh, SM, AC, R, V, CH; Access HW∓3½; **Surry BY** Tel. 461491, Slip, M, FW, AB, Access HW ∓2½; **Sussex Motor YC** Tel. 453078, M, L, Bar; **Shoreham Ship Services** Tel. 461274, CH; **Riverside Marine** Tel. 464831, ME, FW, El, Sh, SM, Slip, M; **Sussex YC (Shoreham)** Tel. 464868, M, L, FW, ME, El, Sh, AB, R, Bar;
A. O. Muggeridge Tel. 592211, CH, Rigging, ACA;
G. P. Barnes Tel. 591705, Gas.
Town Lau, PO; Bank; Rly; Air.

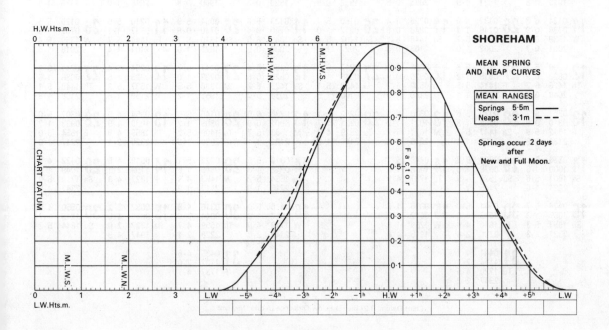

ENGLAND, SOUTH COAST — SHOREHAM

Lat 50°50′ N   Long 0°15′ W

TIMES AND HEIGHTS OF HIGH AND LOW WATERS

YEAR 1989

TIME ZONE UT (GMT)
For Summer Time add ONE hour in non-shaded areas

	JANUARY			FEBRUARY			MARCH			APRIL	
	Time m	Time m		Time m	Time m		Time m	Time m		Time m	Time m
1 Su	0454 4.9 / 1125 2.0 / 1727 4.6 / 2348 2.0	**16** M	0529 5.4 / 1206 1.5 / 1810 5.1	**1** W	0558 4.6 / 1235 2.1 / 1849 4.5	**16** Th	0141 1.9 / 0745 4.8 / 1426 1.7 / 2038 5.0	**1** W	0345 4.7 / 1016 2.0 / 1623 4.5 / 2254 2.2	**16** Th	0546 4.6 / 1233 2.0 / 1858 4.7	**1** Sa	0006 2.1 / 0618 4.7 / 1248 1.9 / 1916 5.0	**16** Su	0207 2.0 / 0814 5.1 / 1431 1.8 / 2046 5.2
2 M	0559 4.8 / 1230 2.0 / 1835 4.6	**17** Tu	0036 1.5 / 0643 5.2 / 1325 1.5 / 1931 5.0	**2** Th	0113 2.1 / 0723 4.8 / 1355 1.9 / 2010 4.9	**17** F	0259 1.7 / 0900 5.0 / 1531 1.5 / 2141 5.3	**2** Th	0504 4.5 / 1141 2.1 / 1804 4.5	**17** F	0125 2.1 / 0729 4.6 / 1405 1.9 / 2023 4.9	**2** Su	0135 1.7 / 0745 5.1 / 1406 1.5 / 2027 5.5	**17** M	0300 1.7 / 0909 5.1 / 1520 1.5 / 2132 5.5
3 Tu	0056 2.0 / 0705 4.9 / 1335 1.9 / 1943 4.7	**18** W	0155 1.5 / 0757 5.2 / 1437 1.4 / 2043 5.2	**3** F	0228 1.8 / 0835 5.1 / 1501 1.5 / 2115 5.3	**18** Sa	0356 1.5 / 0956 5.4 / 1620 1.2 / 2228 5.7	**3** F	0034 2.2 / 0646 4.6 / 1320 1.9 / 1943 4.8	**18** Sa	0243 1.8 / 0846 4.9 / 1510 1.6 / 2123 5.3	**3** M	0240 1.3 / 0847 5.6 / 1505 1.0 / 2119 6.0	**18** Tu	0342 1.3 / 0950 5.4 / 1559 1.2 / 2208 5.8
4 W	0201 1.9 / 0807 5.1 / 1433 1.7 / 2041 5.0	**19** Th	0305 1.4 / 0903 5.4 / 1539 1.3 / 2144 5.5	**4** Sa	0327 1.5 / 0932 5.6 / 1555 1.2 / 2207 5.7	**19** Su	0440 1.2 / 1040 5.7 / 1659 1.0 / 2307 5.9	**4** Sa	0202 1.8 / 0811 5.0 / 1436 1.5 / 2055 5.4	**19** Su	0336 1.6 / 0941 5.2 / 1556 1.3 / 2207 5.7	**4** Tu	0333 0.8 / 0935 6.1 / 1556 0.6 / 2205 6.3	**19** W	0416 1.1 / 1022 5.6 / 1633 0.9 / 2238 5.9
5 Th	0257 1.6 / 0901 5.3 / 1526 1.4 / 2133 5.4	**20** F	0402 1.3 / 0959 5.6 / 1630 1.1 / 2235 5.7	**5** Su	0418 1.1 / 1021 6.0 / 1644 0.9 / 2254 6.1	**20** M	0516 1.0 / 1115 5.9 / 1733 0.8 / 2340 6.1	**5** Su	0307 1.4 / 0912 5.6 / 1535 1.1 / 2148 5.9	**20** M	0417 1.2 / 1021 5.6 / 1635 1.0 / 2243 5.9	**5** W	0421 0.5 / 1019 6.4 / 1643 0.6 / 2248 6.6	**20** Th	0448 0.9 / 1051 5.8 / 1704 0.8 / 2307 6.0
6 F	0347 1.4 / 0949 5.6 / 1612 1.2 / 2221 5.7	**21** Sa	0450 1.2 / 1046 5.8 / 1714 1.0 / 2320 5.9	**6** M	0504 0.9 / 1106 6.3 / 1730 0.7 / 2339 6.4	**21** Tu	0548 0.9 / 1145 6.0 / 1804 0.8	**6** M	0400 1.0 / 1002 6.1 / 1623 0.7 / 2233 6.3	**21** Tu	0450 1.0 / 1052 5.8 / 1707 0.8 / 2312 6.1	**6** Th	0507 0.4 / 1104 6.5 / 1728 0.3 / 2333 6.6	**21** F	0517 0.8 / 1119 5.8 / 1734 0.8 / 2334 5.9
7 Sa	0434 1.2 / 1035 5.9 / 1658 1.0 / 2307 5.9	**22** Su	0531 1.1 / 1126 5.9 / 1751 0.9 / 2358 6.0	**7** Tu	0550 0.8 / 1150 6.5 / 1815 0.6	**22** W	0008 6.1 / 0618 0.9 / 1215 6.0 / 1834 0.7	**7** Tu	0445 0.7 / 1046 6.4 / 1708 0.5 / 2317 6.5	**22** W	0521 0.8 / 1121 5.9 / 1737 0.7 / 2340 6.1	**7** F	0551 0.3 / 1150 6.6 / 1813 0.3	**22** Sa	0546 0.8 / 1149 5.8 / 1802 0.9 / 2359 5.9
8 Su	0518 1.1 / 1119 6.1 / 1742 0.9 / 2353 6.1	**23** M	0607 1.1 / 1201 6.0 / 1825 0.9	**8** W	0025 6.5 / 0635 0.6 / 1237 6.5 / 1901 0.5	**23** Th	0035 6.1 / 0647 0.8 / 1244 5.9 / 1903 0.7	**8** W	0530 0.5 / 1130 6.6 / 1754 0.4	**23** Th	0549 0.7 / 1148 5.9 / 1806 0.7	**8** Sa	0015 6.6 / 0635 0.4 / 1236 6.6 / 1854 0.5	**23** Su	0614 0.8 / 1220 5.8 / 1832 1.0
9 M	0603 1.0 / 1204 6.2 / 1829 0.8	**24** Tu	0030 6.1 / 0640 1.1 / 1235 5.9 / 1858 0.9	**9** Th	0108 6.6 / 0721 0.6 / 1323 6.5 / 1943 0.4	**24** F	0101 6.0 / 0715 0.8 / 1313 5.8 / 1932 0.8	**9** Th	0000 6.7 / 0614 0.4 / 1216 6.6 / 1838 0.3	**24** F	0005 6.1 / 0616 0.7 / 1216 5.9 / 1832 0.7	**9** Su	0054 6.5 / 0715 0.5 / 1320 6.3 / 1934 0.7	**24** M	0027 5.8 / 0645 0.9 / 1250 5.7 / 1902 1.1
10 Tu	0039 6.3 / 0649 1.0 / 1251 6.3 / 1915 0.8	**25** W	0102 6.1 / 0713 1.1 / 1309 5.8 / 1931 0.9	**10** F	0150 6.6 / 0804 0.6 / 1407 6.4 / 2025 0.5	**25** Sa	0125 5.9 / 0744 0.9 / 1340 5.7 / 1959 0.9	**10** F	0042 6.7 / 0658 0.4 / 1301 6.6 / 1920 0.3	**25** Sa	0030 6.0 / 0644 0.7 / 1243 5.8 / 1900 0.8	**10** M	0132 6.2 / 0756 0.7 / 1401 6.1 / 2015 1.0	**25** Tu	0055 5.7 / 0716 1.1 / 1322 5.6 / 1936 1.3
11 W	0124 6.3 / 0736 0.9 / 1339 6.3 / 2001 0.7	**26** Th	0132 6.0 / 0746 1.1 / 1341 5.7 / 2002 1.0	**11** Sa	0228 6.4 / 0846 0.6 / 1448 6.1 / 2105 0.6	**26** Su	0150 5.7 / 0812 1.0 / 1407 5.4 / 2027 1.2	**11** Sa	0123 6.6 / 0740 0.4 / 1343 6.5 / 2000 0.5	**26** Su	0054 5.9 / 0710 0.8 / 1310 5.7 / 1926 1.0	**11** Tu	0210 5.9 / 0837 1.1 / 1444 5.7 / 2101 1.4	**26** W	0129 5.5 / 0752 1.3 / 1359 5.4 / 2018 1.5
12 Th	0209 6.3 / 0823 0.9 / 1426 6.2 / 2046 0.7	**27** F	0200 5.8 / 0818 1.2 / 1414 5.5 / 2033 1.1	**12** Su	0306 6.1 / 0929 0.9 / 1530 5.7 / 2149) 1.0	**27** M	0217 5.4 / 0843 1.3 / 1438 5.1 / 2100 1.5	**12** Su	0159 6.4 / 0820 0.6 / 1423 6.2 / 2038 0.7	**27** M	0117 5.7 / 0739 1.0 / 1338 5.5 / 1955 1.2	**12** W	0254 5.4 / 0925 1.5 / 1537 5.2 / 2159 1.8	**27** Th	0211 5.2 / 0836 1.5 / 1448 5.1 / 2108 1.8
13 F	0251 6.1 / 0910 0.9 / 1512 5.9 / 2131 0.8	**28** Sa	0230 5.6 / 0852 1.3 / 1447 5.2 / 2107 1.4	**13** M	0349 5.7 / 1021 1.2 / 1621 5.2 / 2246 1.4	**28** Tu	0252 5.1 / 0921 1.7 / 1519 4.8 / 2145 (1.9	**13** M	0236 6.0 / 0901 0.9 / 1503 5.7 / 2122 1.1	**28** Tu	0145 5.5 / 0809 1.2 / 1409 5.2 / 2030 1.5	**13** Th	0354 4.9 / 1030 1.9 / 1653 4.9 / 2320 2.1	**28** F	0307 5.0 / 0933 1.7 / 1554 5.0 / 2217 1.9
14 Sa	0336 6.0 / 1000 1.0 / 1601 5.6 / 2221) 1.0	**29** Su	0302 5.3 / 0928 1.5 / 1523 4.9 / 2145 1.7	**14** Tu	0449 5.1 / 1128 1.6 / 1737 4.9			**14** Tu	0318 5.5 / 0948 1.3 / 1554 5.2 / 2218 (1.6	**29** W	0222 5.1 / 0849 1.6 / 1453 4.9 / 2117 1.9	**14** F	0519 4.5 / 1159 2.1 / 1827 4.8	**29** Sa	0422 4.8 / 1049 1.8 / 1719 5.0 / 2342 1.8
15 Su	0426 5.7 / 1057 1.3 / 1657 5.3 / 2321 (1.3	**30** M	0344 5.0 / 1013 1.8 / 1610 4.6 / 2236 (2.0	**15** W	0006 1.8 / 0612 4.8 / 1259 1.8 / 1915 4.8			**15** W	0416 5.0 / 1057 1.8 / 1712 4.8 / 2344 2.0	**30** Th	0316 4.8 / 0944 1.9 / 1559 4.7 / 2228 2.1	**15** Sa	0052 2.2 / 0656 4.5 / 1327 2.0 / 1947 4.9	**30** Su	0550 4.9 / 1215 1.7 / 1844 5.2
		31 Tu	0441 4.7 / 1114 2.1 / 1720 4.5 / 2347 2.2						**31** F	0435 4.6 / 1107 2.0 / 1737 4.7				

Chart Datum: 3.27 metres below Ordnance Datum (Newlyn)

ENGLAND, SOUTH COAST — SHOREHAM

Lat 50°50′ N   Long 0°15′ W

TIMES AND HEIGHTS OF HIGH AND LOW WATERS

TIME ZONE UT (GMT)
For Summer Time add ONE hour in non-shaded areas

YEAR 1989

MAY

	Time	m		Time	m
1 M	0103 0712 1331 1953	1·6 5·2 1·3 5·6	**16** Tu	0211 0821 1431 2044	1·8 4·8 1·7 5·3
2 Tu	0209 0815 1433 2046	1·2 5·6 1·0 6·0	**17** W	0257 0907 1516 2125	1·5 5·1 1·4 5·5
3 W	0304 0906 1527 2134	0·8 5·9 0·7 6·2	**18** Th	0338 0944 1555 2200	1·3 5·3 1·2 5·7
4 Th	0354 0953 1616 2219	0·6 6·2 0·5 6·4	**19** F	0414 1018 1631 2234	1·1 5·5 1·1 5·7
5 F ●	0443 1040 1702 2305	0·5 6·3 0·5 6·4	**20** Sa ○	0448 1052 1705 2304	1·0 5·6 1·1 5·8
6 Sa	0529 1128 1748 2348	0·5 6·3 0·6 6·4	**21** Su	0521 1127 1738 2336	0·9 5·7 1·1 5·8
7 Su	0612 1216 1831	0·6 6·3 0·8	**22** M	0553 1203 1812	1·0 5·7 1·2
8 M	0030 0654 1302 1914	6·3 0·8 6·2 1·0	**23** Tu	0009 0628 1240 1849	5·8 1·1 5·7 1·3
9 Tu	0112 0736 1347 1958	6·1 1·0 6·0 1·2	**24** W	0045 0707 1319 1930	5·7 1·1 5·7 1·4
10 W	0153 0819 1431 2045	5·8 1·2 5·7 1·5	**25** Th	0126 0749 1403 2016	5·6 1·2 5·6 1·5
11 Th	0238 0907 1522 2140	5·4 1·5 5·4 1·8	**26** F	0213 0836 1452 2107	5·5 1·3 5·5 1·5
12 F ☾	0334 1004 1625 2247	5·0 1·8 5·1 2·0	**27** Sa	0306 0930 1550 2208	5·3 1·4 5·4 1·6
13 Sa	0444 1115 1739	4·6 2·0 4·9	**28** Su ☽	0410 1032 1656 2316	5·1 1·5 5·3 1·5
14 Su	0003 0605 1230 1852	2·1 4·5 2·1 4·9	**29** M	0522 1144 1807	5·1 1·4 5·4
15 M	0114 0722 1338 1954	2·0 4·6 1·9 5·1	**30** Tu	0028 0634 1254 1914	1·4 5·2 1·3 5·6
			31 W	0136 0739 1359 2013	1·2 5·4 1·1 5·8

JUNE

	Time	m		Time	m
1 Th	0237 0837 1458 2105	1·0 5·7 0·9 6·0	**16** F	0257 0904 1517 2123	1·6 5·0 1·6 5·3
2 F	0332 0930 1554 2155	0·8 5·9 0·8 6·1	**17** Sa	0342 0949 1601 2204	1·4 5·3 1·4 5·5
3 Sa ●	0423 1022 1644 2244	0·7 6·0 0·8 6·1	**18** Su	0421 1029 1641 2242	1·2 5·5 1·3 5·7
4 Su	0512 1113 1732 2330	0·8 6·1 0·9 6·1	**19** M ○	0500 1111 1720 2320	1·1 5·6 1·3 5·8
5 M	0558 1204 1818	0·9 6·1 1·0	**20** Tu	0539 1152 1800 2359	1·1 5·7 1·3 5·9
6 Tu	0014 0640 1251 1901	6·1 1·0 6·0 1·2	**21** W	0619 1234 1842	1·1 5·8 1·3
7 W	0056 0722 1335 1945	5·9 1·1 5·9 1·4	**22** Th	0040 0702 1318 1925	5·9 1·1 5·9 1·3
8 Th	0138 0804 1418 2029	5·7 1·2 5·8 1·5	**23** F	0125 0747 1402 2012	5·9 1·1 5·9 1·2
9 F	0221 0847 1502 2117	5·5 1·4 5·6 1·7	**24** Sa	0211 0832 1447 2100	5·8 1·2 5·9 1·2
10 Sa	0309 0934 1550 2208	5·2 1·6 5·3 1·8	**25** Su	0300 0920 1536 2151	5·7 1·1 5·8 1·2
11 Su ☽	0403 1027 1644 2307	4·9 1·8 5·1 2·0	**26** M ☾	0352 1012 1628 2249	5·5 1·2 5·6 1·3
12 M	0505 1128 1746	4·6 1·9 5·0	**27** Tu	0449 1112 1729 2354	5·3 1·3 5·5 1·4
13 Tu	0010 0613 1232 1849	2·0 4·5 2·0 4·9	**28** W	0556 1221 1836	5·2 1·3 5·4
14 W	0112 0719 1336 1948	2·0 4·6 1·9 5·0	**29** Th	0105 0706 1332 1943	1·3 5·2 1·3 5·5
15 Th	0209 0816 1429 2039	1·8 4·8 1·8 5·2	**30** F	0214 0815 1440 2046	1·3 5·3 1·3 5·6

JULY

	Time	m		Time	m
1 Sa	0318 0918 1541 2142	1·1 5·5 1·0 5·7	**16** Su	0312 0922 1535 2140	1·7 5·1 1·7 5·4
2 Su	0414 1015 1636 2234	1·0 5·7 1·1 5·9	**17** M	0359 1010 1621 2224	1·4 5·4 1·5 5·7
3 M ●	0503 1108 1724 2320	1·0 5·9 1·1 5·9	**18** Tu ○	0443 1055 1704 2306	1·2 5·7 1·3 5·9
4 Tu	0548 1156 1808	1·0 6·0 1·2	**19** W	0526 1139 1747 2347	1·1 5·9 1·2 6·0
5 W	0003 0629 1239 1849	6·0 1·1 6·0 1·3	**20** Th	0610 1223 1830	1·0 6·1 1·1
6 Th	0044 0708 1319 1928	5·9 1·1 6·0 1·3	**21** F	0031 0653 1306 1915	6·1 0·9 6·2 1·0
7 F	0122 0745 1356 2007	5·8 1·1 5·9 1·3	**22** Sa	0116 0737 1350 1959	6·2 0·8 6·3 0·9
8 Sa	0200 0824 1431 2046	5·6 1·2 5·8 1·4	**23** Su	0200 0820 1431 2045	6·1 0·8 6·2 0·9
9 Su	0239 0901 1509 2126	5·4 1·3 5·5 1·6	**24** M	0243 0903 1513 2129	6·0 0·8 6·1 1·0
10 M	0319 0942 1548 2211	5·1 1·6 5·3 1·8	**25** Tu ☾	0328 0948 1558 2220	5·7 1·0 5·8 1·2
11 Tu	0405 1028 1636 2302	4·8 1·8 5·0 2·0	**26** W	0416 1041 1652 2323	5·4 1·3 5·4 1·4
12 W	0500 1124 1736 2354	4·5 2·1 4·8 1·4	**27** Th	0521 1151 1803	5·1 1·6 5·2
13 Th	0004 0607 1230 1844	2·1 4·4 2·2 4·7	**28** F	0040 0643 1313 1922	1·6 4·9 1·7 5·1
14 F	0113 0721 1339 1951	2·1 4·5 2·1 4·8	**29** Sa	0202 0805 1432 2037	1·6 5·0 1·6 5·3
15 Sa	0216 0826 1442 2050	1·9 4·7 1·9 5·1	**30** Su	0312 0916 1537 2139	1·5 5·3 1·5 5·5
			31 M	0409 1022 1632 2230	1·3 5·6 1·3 5·7

AUGUST

	Time	m		Time	m
1 Tu ●	0457 1102 1717 2313	1·1 5·9 1·2 5·9	**16** W	0424 1035 1644 2247	1·2 5·9 1·1 6·1
2 W	0537 1145 1755 2350	1·0 6·0 1·2 6·0	**17** Th ○	0507 1118 1729 2329	0·9 6·2 0·9 6·3
3 Th	0613 1222 1831	0·9 6·1 1·1	**18** F	0551 1202 1812	0·8 6·4 0·8
4 F	0024 0648 1256 1905	6·0 1·0 6·1 1·1	**19** Sa	0012 0635 1245 1856	6·4 0·7 6·5 0·7
5 Sa	0059 0721 1326 1937	5·9 1·0 6·0 1·1	**20** Su	0056 0718 1327 1940	6·5 0·6 6·5 0·6
6 Su	0131 0754 1355 2010	5·8 1·1 5·9 1·2	**21** M	0140 0800 1407 2023	6·4 0·6 6·4 0·7
7 M	0203 0825 1425 2044	5·6 1·2 5·7 1·3	**22** Tu	0221 0841 1445 2105	6·2 0·7 6·2 0·9
8 Tu	0235 0858 1456 2118	5·3 1·4 5·4 1·6	**23** W ☾	0302 0923 1525 2152	5·8 1·0 5·8 1·2
9 W	0309 0934 1534 2200	5·0 1·7 5·0 1·9	**24** Th	0348 1014 1620 2254	5·4 1·5 5·3 1·6
10 Th	0353 1021 1627 2256	4·6 2·1 4·7 2·2	**25** F	0456 1128 1736	5·0 1·8 4·9
11 F	0457 1127 1740	4·4 2·4 4·5	**26** Sa	0020 0631 1303 1911	1·9 4·8 2·0 4·8
12 Sa	0014 0624 1252 1904	2·3 4·3 2·4 4·6	**27** Su	0152 0803 1429 2034	1·9 4·9 1·9 5·0
13 Su	0136 0751 1410 2020	2·2 4·6 2·2 4·9	**28** M	0306 0912 1533 2136	1·7 5·3 1·6 5·3
14 M	0244 0857 1511 2119	1·8 5·0 1·8 5·3	**29** Tu	0359 1004 1621 2223	1·4 5·6 1·4 5·7
15 Tu	0338 0950 1601 2206	1·5 5·5 1·4 5·8	**30** W	0442 1047 1700 2259	1·1 5·9 1·2 5·9
			31 Th ●	0518 1122 1734 2331	1·0 6·1 1·0 6·0

Chart Datum: 3.27 metres below Ordnance Datum (Newlyn)

ENGLAND, SOUTH COAST — SHOREHAM

Lat 50°50′ N   Long 0°15′ W

TIMES AND HEIGHTS OF HIGH AND LOW WATERS

TIME ZONE UT(GMT)
For Summer Time add ONE hour in non-shaded areas

YEAR 1989

SEPTEMBER

Time	m	Time	m
1 0550 F 1154 1806	0.9 6.1 1.0	**16** 0526 Sa 1133 1749 2346	0.6 6.6 0.6 6.6
2 0000 Sa 0621 1223 1836	6.0 0.9 6.1 0.9	**17** 0610 Su 1216 1833	0.5 6.6 0.5
3 0029 Su 0650 1251 1905	5.9 0.9 6.1 0.9	**18** 0032 M 0653 1259 1916	6.6 0.5 6.6 0.6
4 0059 M 0720 1317 1934	5.8 1.0 5.9 1.0	**19** 0116 Tu 0734 1338 1958	6.5 0.6 6.5 0.7
5 0127 Tu 0747 1342 2002	5.7 1.1 5.8 1.2	**20** 0156 W 0815 1416 2040	6.3 0.9 6.2 1.0
6 0153 W 0816 1409 2032	5.4 1.4 5.5 1.4	**21** 0239 Th 0900 1459 2128	5.9 1.2 5.7 1.4
7 0223 Th 0849 1443 2109	5.1 1.7 5.1 1.8	**22** 0328 F 0954 1555 2230	5.4 1.7 5.2 1.8
8 0303 F 0932 1531 2201	4.8 2.1 4.7 2.2	**23** 0440 Sa 1113 1719	5.0 2.1 4.8
9 0403 Sa 1037 1646 2320	4.5 2.4 4.5 2.4	**24** 0002 Su 0619 1252 1900	2.1 4.8 2.2 4.7
10 0537 Su 1210 1823	4.4 2.5 4.5	**25** 0136 M 0749 1413 2021	2.0 5.0 2.0 4.9
11 0055 M 0717 1339 1951	2.3 4.7 2.2 4.9	**26** 0244 Tu 0853 1513 2120	1.8 5.3 1.7 5.3
12 0212 Tu 0828 1443 2053	1.9 5.2 1.7 5.4	**27** 0335 W 0941 1557 2202	1.5 5.7 1.4 5.6
13 0309 W 0921 1535 2140	1.4 5.7 1.3 5.9	**28** 0415 Th 1019 1633 2236	1.2 5.9 1.1 5.9
14 0357 Th 1007 1621 2222	1.0 6.1 0.9 6.3	**29** 0449 F 1052 1706 2304 ●	1.0 6.1 1.0 6.0
15 0442 F 1049 1704 2303 ○	0.7 6.4 0.7 6.5	**30** 0520 Sa 1121 1735 2331	0.9 6.1 0.9 6.0

OCTOBER

Time	m	Time	m
1 0550 Su 1149 1803	0.9 6.1 0.8	**16** 0543 M 1145 1809	0.5 6.6 0.5
2 0000 M 0619 1215 1832	5.9 0.9 6.0 0.9	**17** 0007 Tu 0628 1229 1852	6.6 0.6 6.6 0.6
3 0028 Tu 0646 1241 1859	5.9 1.0 5.9 1.0	**18** 0052 W 0710 1311 1935	6.4 0.7 6.4 0.8
4 0055 W 0714 1306 1927	5.7 1.2 5.8 1.1	**19** 0137 Th 0753 1353 2019	6.2 1.1 6.1 1.1
5 0121 Th 0743 1333 1957	5.5 1.4 5.5 1.4	**20** 0221 F 0841 1438 2107	5.9 1.4 5.7 1.5
6 0153 F 0817 1410 2035	5.2 1.7 5.2 1.7	**21** 0314 Sa 0938 1536 2211 ☾	5.5 1.8 5.2 1.8
7 0233 Sa 0900 1500 2126	5.0 2.1 4.8 2.0	**22** 0423 Su 1055 1656 2333	5.1 2.1 4.8 2.0
8 0334 Su 1006 1611 2241 ☾	4.7 2.3 4.6 2.2	**23** 0551 M 1223 1830	5.0 2.2 4.7
9 0503 M 1136 1747	4.6 2.3 4.6	**24** 0058 Tu 0713 1340 1949	2.0 5.1 2.1 4.8
10 0015 Tu 0640 1303 1916	2.1 4.9 2.1 5.0	**25** 0208 W 0815 1437 2048	1.8 5.3 1.8 5.1
11 0135 W 0752 1410 2019	1.8 5.3 1.6 5.5	**26** 0259 Th 0904 1523 2130	1.6 5.6 1.5 5.5
12 0235 Th 0847 1503 2109	1.3 5.8 1.1 6.0	**27** 0340 F 0944 1600 2204	1.3 5.8 1.2 5.7
13 0326 F 0934 1551 2151	0.9 6.2 0.8 6.3	**28** 0415 Sa 1016 1633 2234	1.1 6.0 1.0 5.8
14 0412 Sa 1017 1637 ○ 2234	0.6 6.5 0.5 6.5	**29** 0449 Su 1047 1704 2304 ●	1.0 6.0 0.9 5.9
15 0459 Su 1100 1723 2320	0.5 6.6 0.5 6.6	**30** 0520 M 1116 1733 2332	1.0 6.0 0.9 5.8
		31 0549 Tu 1144 1803	1.0 5.9 0.9

NOVEMBER

Time	m	Time	m
1 0003 W 0618 1211 1831	5.8 1.2 5.9 1.0	**16** 0036 Th 0653 1252 1919	6.3 1.0 6.3 1.0
2 0033 Th 0647 1241 1903	5.7 1.3 5.8 1.2	**17** 0124 F 0739 1336 2004	6.2 1.2 6.0 1.2
3 0103 F 0721 1314 1937	5.5 1.5 5.6 1.4	**18** 0211 Sa 0827 1423 2052	5.9 1.5 5.7 1.4
4 0139 Sa 0759 1354 2019	5.4 1.7 5.3 1.6	**19** 0301 Su 0920 1516 2147	5.6 1.7 5.3 1.7
5 0222 Su 0847 1445 2109	5.2 1.9 5.1 1.8	**20** 0359 M 1024 1622 2252 ☾	5.4 2.0 5.0 1.9
6 0320 M 0948 1551 2216	5.1 2.0 4.9 1.9	**21** 0506 Tu 1136 1739	5.2 2.1 4.8
7 0436 Tu 1105 1713 2335	5.0 2.0 4.9 1.8	**22** 0005 W 0619 1247 1856	2.0 5.1 2.1 4.8
8 0557 W 1224 1834	5.2 1.8 5.1	**23** 0113 Th 0723 1349 1959	1.9 5.2 1.9 4.9
9 0053 Th 0709 1332 1940	1.6 5.5 1.4 5.5	**24** 0211 F 0817 1440 2048	1.7 5.3 1.7 5.2
10 0158 F 0809 1431 2033	1.2 5.9 1.1 5.9	**25** 0259 Sa 0902 1523 2129	1.5 5.5 1.4 5.4
11 0254 Sa 0859 1523 2122	0.9 6.2 0.8 6.2	**26** 0341 Su 0941 1601 2206	1.3 5.7 1.3 5.6
12 0344 Su 0946 1613 2209	0.7 6.4 0.6 6.4	**27** 0419 M 1017 1636 2239	1.2 5.8 1.0 5.7
13 0434 M 1033 1702 ● 2257	0.6 6.5 0.6 6.4	**28** 0453 Tu 1105 1710 ● 2313	1.2 5.8 1.0 5.7
14 0522 Tu 1120 1748 2348	0.7 6.4 0.6 6.4	**29** 0526 W 1122 1742 2347	1.2 5.9 1.1 5.8
15 0608 W 1207 1833	0.8 6.2 0.8	**30** 0559 Th 1154 1815	1.3 5.8 1.1

DECEMBER

Time	m	Time	m
1 0021 F 0633 1229 1851	5.8 1.4 5.8 1.2	**16** 0114 Sa 0726 1322 1949	6.2 1.7 6.0 1.1
2 0057 Sa 0711 1306 1929	5.7 1.4 5.7 1.2	**17** 0156 Su 0811 1407 2032	6.0 1.4 5.8 1.2
3 0136 Su 0752 1350 2012	5.6 1.5 5.6 1.3	**18** 0239 M 0856 1453 2116	5.9 1.6 5.5 1.4
4 0219 M 0838 1438 2100	5.6 1.6 5.4 1.4	**19** 0323 Tu 0946 1542 ☾ 2205	5.6 1.7 5.2 1.6
5 0309 Tu 0931 1534 2154	5.5 1.6 5.3 1.4	**20** 0412 W 1039 1639 2303	5.4 1.8 4.9 1.8
6 0408 W 1034 1638 2258	5.2 1.6 5.2 1.5	**21** 0510 Th 1141 1746	5.1 2.0 4.7
7 0515 Th 1143 1748	5.4 1.5 5.2	**22** 0007 F 0615 1247 1855	1.9 5.0 2.0 4.7
8 0010 F 0624 1254 1859	1.4 5.5 1.4 5.4	**23** 0114 Sa 0719 1350 2000	2.0 5.0 1.9 4.8
9 0120 Sa 0730 1400 2002	1.2 5.7 1.1 5.6	**24** 0214 Su 0817 1444 2053	1.8 5.1 1.7 5.0
10 0225 Su 0828 1501 2059	1.0 5.9 0.9 5.9	**25** 0306 M 0908 1532 2139	1.7 5.3 1.4 5.3
11 0323 M 0923 1555 2153	0.9 6.1 0.8 6.1	**26** 0352 Tu 0951 1614 2221	1.5 5.5 1.2 5.5
12 0418 Tu 1014 1648 ○ 2247	0.8 6.2 0.7 6.2	**27** 0432 W 1030 1651 2259	1.3 5.7 1.1 5.7
13 0508 W 1105 1737 2338	0.9 6.3 0.8 6.2	**28** 0508 Th 1107 1727 ● 2336	1.3 5.8 1.1 5.8
14 0555 Th 1153 1821	1.0 6.3 0.9	**29** 0544 F 1143 1804	1.2 5.9 1.2
15 0028 F 0642 1237 1905	6.2 1.1 6.2 1.0	**30** 0013 Sa 0622 1220 1842	5.9 1.2 6.0 1.0
		31 0051 Su 0701 1301 1923	6.0 1.2 6.0 1.0

Chart Datum: 3.27 metres below Ordnance Datum (Newlyn)

AREA 3—SE England 279

BRIGHTON 10-3-10
E. Sussex

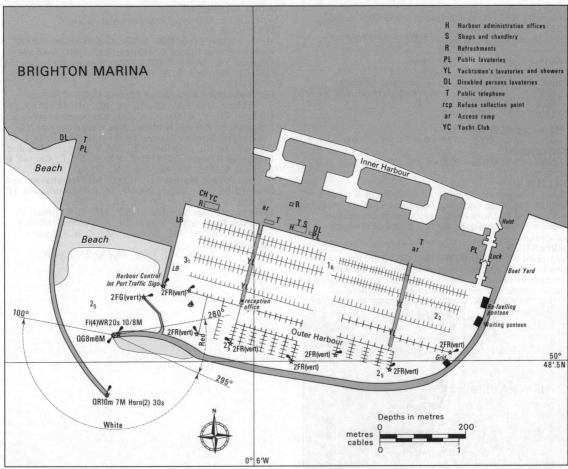

CHARTS
Admiralty 1652, 1991; Stanford 9; Imray C9; OS 198
TIDES
+0001 Dover; ML 3.5; Duration 0605; Zone 0 (GMT)

Standard Port SHOREHAM (←—)

Times				Height (metres)			
HW		LW		MHWS	MHWN	MLWN	MLWS
0500	1000	0000	0600	6.2	5.0	1.9	0.7
1700	2200	1200	1800				

Differences BRIGHTON
−0010 −0005 −0005 −0005 +0.3 +0.1 0.0 −0.1

BRIGHTON MARINA
SHELTER
The marina gives good shelter under all conditions, but the approach in strong S winds could be very rough.
NAVIGATION
Waypoint 50°48'.20N 00°06'.29W, 180°/000° from/to West breakwater Lt, 0.26M. 180° safe sector of approach except shallow water to E of entrance in main Lt R sector. Shoaling may occur at entrance after gales.
LIGHTS AND MARKS
E breakwater Fl(4) WR 20s 16m 10/8M; vis R260°-295°, W295°-100°. E breakwater head QG 8m 6M. W breakwater head QR 10m 7M; Horn (2) 30s. International Port Traffic Signals at Ent.
Inner Harbour Lock controlled by normal R and G traffic control.
Times: In H+15, +45; Out H+00, +30; (0800-2000 LT).
RADIO TELEPHONE
Call: *Brighton Control* VHF Ch 16; 68 11. Marina Ch M.
TELEPHONE (0273)
Hr Mr 693636 Ex 69; CG & MRSC Lee-on-Solent 552100; HMC Freefone Customs Yachts: RT (0703) 229251; Marinecall 0898 500 456; Hosp 606444; Dr 686863.

FACILITIES
Brighton Marina (1600+200 visitors) Tel. 693636, HMC, FW, P, D, AC, Gas, Gaz, Lau, R; **Brighton Marina BY** Tel. 609235, BH (40 ton), C (35 ton); **Brighton Marina YC** Tel. 697049, Bar; **Felton Marine Engineering** Tel. 601779, ME; **Kaymar Marine** Tel. 683348, Sh; **DMS Seatronics** Tel. 605166, El; **Nautical Electronics Services** Tel. 693258, El; **Terry Pachol** Tel. 682724, Sh; **Welch Marine Services** Tel. 675972, SM; **Russell Simpson** Tel. 697161, CH; **Moores Marine Maintenance** Tel. (0860) 525966, Diving; **Gateway Superstore** Tel. 606611, V.
Town V, R, Bar, PO; Bank; Rly; Air (Shoreham).

NEWHAVEN 10-3-11
E. Sussex

CHARTS
Admiralty 1652, 2154; Stanford 9; Imray C9; OS 198
TIDES
0000 Dover; ML 3.6; Duration 0550; Zone 0 (GMT).

Standard Port SHOREHAM (←)

Times				Height (metres)			
HW		LW		MHWS	MHWN	MLWN	MLWS
0500	1000	0000	0600	6.2	5.0	1.9	0.7
1700	2200	1200	1800				

Differences NEWHAVEN
−0015 −0010 0000 0000 +0.4 +0.2 0.0 −0.2
EASTBOURNE
−0010 −0005 +0015 +0020 +1.1 +0.6 +0.2 +0.1

SHELTER
Good, but with strong on-shore winds there is often a difficult sea at the entrance. Accessible in all weathers but with strong on-shore winds pass close to breakwater; there are heavy breaking seas on E side of dredged channel.

NAVIGATION
Waypoint 50°46'.20N 00°03'.70E, 168°/348° from/to West breakwater Lt, 0.32M. Harbour silts up and dredging is in continuous operation. Cargo vessels and ferries often warp off by means of hawsers run across the harbour.

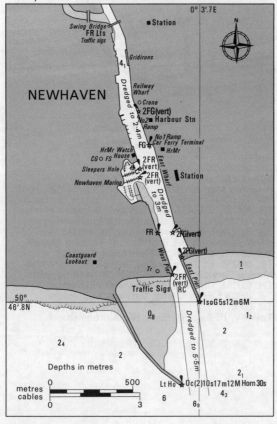

LIGHTS AND MARKS
Traffic signals, displayed from Tr on W side of river.
Triangle over ball, or G Lt — only entry permitted.
Ball over triangle, or R Lt — only departure permitted.
Ball, triangle, ball (vert) or
    RGR Lts (vert) — No entry or departure
Ball or G R Lts (vert) — Entry and departure permitted with care for vessels under 15m

Swing bridge signals
FIG — Bridge opening or closing
FR — Vessels may pass N to S
FG — Vessels may pass S to N

RADIO TELEPHONE
VHF Ch 16; 12 (H24). Newhaven Marina Ch M.
TELEPHONE (0273)
Hr Mr 514131; CG 514008; MRSC Lee-on-Solent 552100; Harbour Signal Station 514131 ext. 247; HMC Freefone Customs Yachts: RT (0703) 229251; Marinecall 0898 500 456; Dr 515076; Hosp 609411.
FACILITIES
EC Wednesday; **Newhaven Marina** (300+50 visitors) Tel. 513881, Slip, FW, ME, El, Sh, AC, BH (18 ton), C (10 ton), CH, V, R, Bar, Gas, Gaz, Lau, (Access HW∓5), fuel pontoon 200 yds N of ent; **Newhaven Marina YC** Tel. 513976; **Sealink Quays** Tel. 514131, Slip, P, D, L, FW, ME, El, Sh, C (3 ton), CH, AB, V, R, Bar; **Ship and Industrial Repairs** Tel. 516298, ME, El, Sh, C; **Cantell & Son** Tel. 514118, ME, Sh, CH, SM, Slip, C, ACA; **Golden Arrow Marine** Tel. 513987, ME; **Newhaven and Seaford SC** Tel. Seaford 890077, M, FW; **Newhaven YC** Tel. 513770, Slip, M, P, D, ME, El, Sh, CH, AB; **Meeching Boats** (80) Tel. 514907, ME, El, Sh; **Russell Simpson Marine** Tel. 513458 CH, El; **Leonard Marine** Tel. 515987 BY, El, ME, Sh, Slip, SM; **C & E Sports** Tel. 515450, Gas.

Town P, V, R, Bar. PO; Bank; Rly; Air (Shoreham).

AREA 3—SE England 281

RYE 10-3-12
E. Sussex

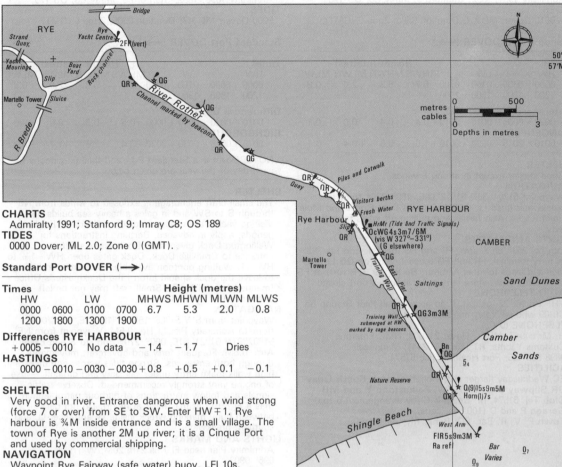

CHARTS
Admiralty 1991; Stanford 9; Imray C8; OS 189
TIDES
0000 Dover; ML 2.0; Zone 0 (GMT).

Standard Port DOVER (→)

Times				Height (metres)			
HW		LW		MHWS	MHWN	MLWN	MLWS
0000	0600	0100	0700	6.7	5.3	2.0	0.8
1200	1800	1300	1900				

Differences RYE HARBOUR
+0005 −0010   No data   −1.4   −1.7   Dries
HASTINGS
0000 −0010 −0030 −0030 +0.8 +0.5 +0.1 −0.1

SHELTER
Very good in river. Entrance dangerous when wind strong (force 7 or over) from SE to SW. Enter HW∓1. Rye harbour is ¾M inside entrance and is a small village. The town of Rye is another 2M up river; it is a Cinque Port and used by commercial shipping.

NAVIGATION
Waypoint Rye Fairway (safe water) buoy, LFl 10s, 50°54′.00N 00°48′.13E, 150°/330° from/to West Arm head Lt, 1.81M. Beware:
(1) Narrow entrance. Limited width of channel (42m).
(2) Bar at entrance.
(3) Strong flood stream in river — up to 3.5 kn (max HW−3 to HW−1).
(4) Shallow water E & W of entrance with ground swell or surf.
Harbour speed limit 6 kn.

LIGHTS AND MARKS
East Arm, head Q(9) 15s 9m 5M; Horn 7s.
Tide Signals on Hr Mr office:—
Night — FG 2.1-3.0m on bar
         FR Over 3.0m on bar
Day — None, but horizontal timbers on tripod beacon at entrance indicate depths 5′, 10′ and 15′.
Traffic signals on Hr Mr office to indicate big ship movements
1 B Ball — Vessel entering
2 B Balls (hor) — Vessel leaving
3 B Balls (triangle) — Vessels entering and leaving.
In addition a Fl Y Lt is exhibited.

RADIO TELEPHONE
VHF Ch 16; 14 (0900–1700 LT).

TELEPHONE (0797)
Hr Mr 225225; CG 813171; MRCC Dover 210008; HMC Freefone Customs Yachts: RT (0703) 229251; Marinecall 0898 500 456; Dr 222031; Hosp 222109.

FACILITIES
EC Tuesday; **Admiralty Jetty** Slip, M*, L, FW, AB*; **Strand Quay** Slip, M*, P and D (50m, cans), L, FW, AB*; **Rye Yacht Centre** Tel. 223336, M, L, FW, ME, El, C (15 ton), AB; **Phillips BY** Tel. 223234, Slip, M, L, FW, ME, El, C (3 ton); **Sandrock Marine** Tel. 222679, Slip, M, D, L, FW, ME, El, CH, AB; **Sea Cruisers** Tel. 222070, ME, Sh, CH, ACA.
Town PO; Bank; Rly; Air (Lydd).
* See Hr Mr.

Harbour, Coastal and Tidal Information

FOLKESTONE 10-3-13
Kent

CHARTS
Admiralty 1991, 1892; Stanford 9; Imray C8; OS 179
TIDES
−0010 Dover; ML 3.7; Duration 0500; Zone 0 (GMT).

Standard Port DOVER (→)

Times				Height (metres)			
HW		LW		MHWS	MHWN	MLWN	MLWS
0000	0600	0100	0700	6.7	5.3	2.0	0.8
1200	1800	1300	1900				

Differences FOLKESTONE
−0020 −0005 −0010 −0010 +0.4 +0.4 0.0 −0.1
DUNGENESS
−0010 −0015 −0020 −0010 +1.0 +0.6 +0.4 +0.1

SHELTER
Good shelter except in strong E winds.
NAVIGATION
Waypoint 51°04'.30N 01°12'.00E, 150°/330° from/to breakwater head Lt, 0.26M. Beware Copt Rocks and Mole Head Rocks. Also ferries and pulling off wires from the jetty.
LIGHTS AND MARKS
Ldg Lts 267° on S arm, FR and FG (occas). Ldg Lts 295° at ferry terminal, FR and FG (occas). 305° on QG Lt at E pierhead leads to inner harbour. Bu flag or 3FR(vert) at FS on S arm, ¼ hr before ferry sails, indicates port closed.
RADIO TELEPHONE
VHF Ch 16; 22 (occas). If no answer call Pilot Station on Ch 09 who will relay message to Hr Mr.
TELEPHONE (0303)
Hr Mr 54947; MRCC Dover 210008; HMC Freefone Customs Yachts: RT (0304) 202441; Marinecall 0898 500 456; Port Health Office 57574.
FACILITIES
EC Wednesday (larger shops open all day); **South Quay BR Slipway** Slip (free), FW; **Folkestone Y and MB Club** Tel. 51574, M; **Sealink** C (by arrangement 5 ton); **Garage** P and D (100 yds, cans).
Town P, V, R, Bar. PO; Bank; Rly; Air (Lydd).

DOVER 10-3-14
Kent

CHARTS
Admiralty 1698, 1892; Stanford 9; Imray C8; OS 179
TIDES
0000 Dover; ML 3.7; Duration 0505; Zone 0 (GMT)

Standard Port DOVER (→)

Times				Height (metres)			
HW		LW		MHWS	MHWN	MLWN	MLWS
0000	0600	0100	0700	6.7	5.3	2.0	0.8
1200	1800	1300	1900				

Differences DEAL
+0010 +0020 +0010 +0005 −0.6 −0.3 0.0 0.0
RICHBOROUGH
+0015 +0015 +0030 +0030 −3.4 −2.6 −1.7 −0.7

NOTE: Dover is a Standard Port and tidal predictions for each day of the year are given below.

SHELTER
The small craft anchorage is exposed to winds from NE through S to SW and in gales a heavy sea builds up. Visiting yachts are welcome for up to 14 days. For longer periods, apply in advance. Berthing instructions for Wellington Dock given from Dockmaster's Office at entrance to Granville Dock. Dock gates open HW−1½ to HW+1. Waiting pontoon available. Yachtsmen intending to leave the dock should inform the Dockmaster's Office (manned from HW−2). Small craft may not be left unattended in Outer Harbour.
NAVIGATION
Waypoint from SW 51°06'.15N 01°19'.77E, 180°/000° from/to Admiralty Pier Lt Ho, 0.5M. Waypoint from NE 51°07'.27N 01°21'.51E, 090°/270° from/to S end Eastern Arm, 0.5M. Frequent ferry and hovercraft movements through both entrances. Strong tides across entrances and high walls make entry under sail slow and difficult — use of engine very strongly recommended. Observe traffic signals and follow instructions of harbour patrol launch (has a Fl Bu Lt at masthead at night). Do not pass between buoy marking wreck inside W entrance, Q, and southern breakwater.
LIGHTS AND MARKS
Admiralty Pier head Fl 7.5s 21m 20M; W Tr; vis 096°-090°.
International Port Traffic Signals are in operation, (see 6.2.7) shown for the Eastern entrance, day and night, on panels near Port Control; for Western entrance, day and night, on panels near Admiralty Pier Signal Station.
N.B. Specific permission to enter or leave Eastern or Western entrance *must* first be obtained from Port Control on VHF Ch 74, Ch 12 or, if not fitted with
VHF/RT, with Aldis Lamp signals:
SV − I wish to enter port
SW − I wish to leave port.
Port Control will reply 'OK' or 'Wait'.
A Q Fl lamp from the Control Tower means keep clear of entrance you are approaching.
Docking signals
International Port Traffic Signals together with small Fl Y light to be shown 5 min before bridge is swung.
RADIO TELEPHONE
Call: *Dover Port Control* VHF Ch 12 74; 12 **74** (H24). Channel Navigation Information Service (CNIS) — call: *Dover Coastguard* Ch 16 **69**; 11 **69** 80. Information broadcasts on Ch 11 at H+40, and also at H+55 when visibility is less than 2M.
TELEPHONE (0304)
Hr Mr 240400; CG 852515; MRCC 210008; HMC Freefone Customs Yachts: RT (0304) 202441; Marinecall 0898 500 456; Hosp 201624.
FACILITIES
EC Wednesday; **Wellington Dock** Tel. 240400, Slip, L, FW, C, AB; **Royal Cinque Ports YC** Tel. 206262, L, Bar, R; **Dover Yacht Co** Tel. 201073, D, FW, ME, El, Sh, C; **Dover Marine Supplies** Tel. 201677, D, FW, ME, El, Sh, Slip, C, SM, ACA, CH; **Smye-Rumsby** Tel. 201187, El.
Town P and D (cans), V, R, Bar. PO; Bank; Rly; Air (Lydd).
Note: *A Yachtsman's Guide* is available from Harbour House or Dockmaster.

AREA 3—SE England 283

DOVER continued

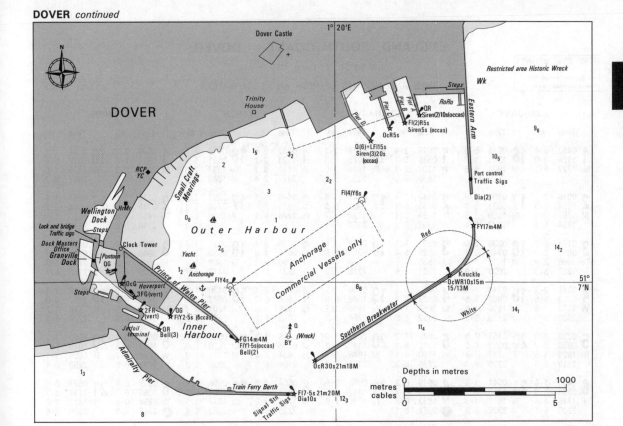

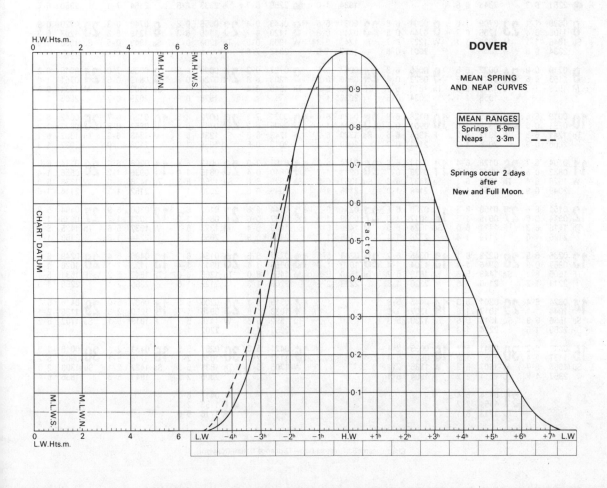

DOVER

MEAN SPRING AND NEAP CURVES

MEAN RANGES	
Springs	5·9m
Neaps	3·3m

Springs occur 2 days after New and Full Moon.

ENGLAND, SOUTH COAST — DOVER

Lat 51°07′ N   Long 1°19′ E

TIMES AND HEIGHTS OF HIGH AND LOW WATERS

YEAR 1989

TIME ZONE UT(GMT)
For Summer Time add ONE hour in non-shaded areas

JANUARY

Day	Time	m	Day	Time	m
1 Su	0452 1156 1727	5.4 2.0 5.1	16 M	0519 1243 1804	5.9 1.7 5.5
2 M	0015 0554 1257 1834	2.3 5.3 2.1 5.1	17 Tu	0109 0631 1359 1923	2.0 5.7 1.8 5.4
3 Tu	0124 0659 1406 1935	2.3 5.3 2.1 5.3	18 W	0230 0748 1519 2039	2.0 5.7 1.7 5.6
4 W	0242 0758 1518 2030	2.2 5.5 1.9 5.5	19 Th	0348 0900 1630 2139	1.8 5.8 1.6 5.8
5 Th	0353 0850 1621 2119	1.9 5.7 1.6 5.8	20 F	0451 0957 1727 2226	1.5 6.0 1.4 6.1
6 F	0451 0938 1715 2206	1.6 5.9 1.4 6.1	21 Sa	0544 1045 1817 2308	1.3 6.1 1.3 6.3 ○
7 Sa ●	0542 1023 1804 2251	1.3 6.2 1.2 6.3	22 Su	0629 1127 1857 2346	1.1 6.2 1.2 6.5
8 Su	0628 1108 1848 2334	1.1 6.4 1.1 6.5	23 M	0709 1205 1930	1.0 6.3 1.1
9 M	0710 1153 1928	0.9 6.5 1.0	24 Tu	0022 0744 1239 1958	6.5 0.9 6.3 1.1
10 Tu	0019 0751 1241 2008	6.6 0.8 6.5 0.9	25 W	0056 0815 1312 2025	6.5 1.0 6.2 1.2
11 W	0104 0832 1328 2049	6.7 0.8 6.4 0.9	26 Th	0128 0846 1341 2051	6.4 1.0 6.1 1.3
12 Th	0151 0914 1416 2129	6.6 0.8 6.3 1.0	27 F	0158 0915 1411 2119	6.3 1.2 6.0 1.4
13 F	0236 0957 1505 2213	6.5 0.9 6.1 1.2	28 Sa	0229 0946 1443 2150	6.1 1.4 5.8 1.6
14 Sa ☽	0324 1044 1556 2259	6.4 1.1 5.9 1.5	29 Su	0301 1017 1519 2224	5.9 1.6 5.6 1.9
15 Su	0417 1137 1655 2357	6.2 1.4 5.7 1.8	30 M ☾	0341 1057 1607 2311	5.7 1.9 5.3 2.2
			31 Tu	0434 1151 1713	5.4 2.2 5.1

FEBRUARY

Day	Time	m	Day	Time	m
1 W	0018 0549 1307 1836	2.4 5.1 2.3 5.0	16 Th	0215 0742 1514 2033	2.1 5.3 2.0 5.3
2 Th	0148 0713 1436 1958	2.4 5.1 2.1 5.2	17 F	0345 0901 1627 2132	1.8 5.5 1.7 5.7
3 F	0321 0825 1555 2101	2.1 5.4 1.8 5.6	18 Sa	0449 0956 1725 2216	1.4 5.8 1.4 6.0
4 Sa	0430 0922 1657 2152	1.6 5.8 1.4 6.0	19 Su	0540 1038 1810 2254	1.2 6.0 1.2 6.3
5 Su	0526 1012 1750 2238	1.2 6.1 1.1 6.3	20 M ○	0621 1115 1845 2327	1.0 6.2 1.1 6.4
6 M	0615 1057 1839 2322	0.9 6.4 0.9 6.6	21 Tu	0655 1147 1912	0.9 6.3 1.0
7 Tu	0702 1142 1923	0.6 6.6 0.8	22 W	0000 0723 1217 1934	6.5 0.8 6.3 1.0
8 W	0004 0744 1227 2001	6.8 0.5 6.7 0.6	23 Th	0031 0749 1242 1957	6.6 0.8 6.3 1.0
9 Th	0046 0823 1310 2034	6.9 0.4 6.7 0.6	24 F	0057 0816 1306 2022	6.5 0.9 6.3 1.0
10 F	0128 0901 1354 2111	6.9 0.5 6.5 0.7	25 Sa	0121 0840 1330 2049	6.4 1.0 6.2 1.2
11 Sa	0211 0938 1437 2148	6.8 0.7 6.3 1.0	26 Su	0144 0911 1357 2115	6.3 1.2 6.0 1.4
12 Su	0256 1017 1524 2228 ☽	6.6 1.0 6.0 1.4	27 M	0211 0939 1429 2145	6.1 1.5 5.8 1.7
13 M	0345 1104 1619 2320	6.2 1.4 5.7 1.8	28 Tu ☾	0244 1010 1510 2223	5.8 1.8 5.5 2.1
14 Tu ☽	0445 1207 1730	5.8 1.9 5.3			
15 W	0035 0605 1335 1904	2.1 5.4 2.1 5.1			

MARCH

Day	Time	m	Day	Time	m
1 W	0332 1101 1610 2327	5.5 2.1 5.1 2.4	16 Th	0014 0551 1317 1849	2.2 5.1 2.3 5.0
2 Th	0449 1219 1746	5.1 2.4 4.9	17 F	0201 0740 1500 2016	2.2 5.1 2.1 5.3
3 F	0103 0639 1359 1931	2.4 4.9 2.3 5.0	18 Sa	0331 0853 1609 2112	1.7 5.4 1.7 5.7
4 Sa	0250 0808 1529 2043	2.1 5.3 1.8 5.5	19 Su	0430 0942 1701 2153	1.3 5.7 1.4 6.0
5 Su	0406 0908 1634 2134	1.5 5.7 1.4 6.0	20 M	0518 1020 1743 2230	1.1 6.0 1.2 6.2
6 M	0504 0955 1729 2219	1.0 6.1 1.0 6.4	21 Tu	0556 1051 1817 2302	1.1 6.1 1.1 6.4
7 Tu ●	0556 1038 1819 2259	0.7 6.5 0.7 6.7	22 W ○	0627 1119 1842 2333	0.9 6.2 1.0 6.5
8 W	0643 1120 1903 2342	0.4 6.7 0.5 7.0	23 Th	0655 1146 1904	0.9 6.3 0.9
9 Th	0727 1203 1940	0.3 6.8 0.4	24 F	0000 0720 1208 1928	6.6 0.8 6.4 0.9
10 F	0021 0805 1243 2013	7.1 0.2 6.8 0.5	25 Sa	0022 0747 1231 1955	6.5 0.9 6.3 1.0
11 Sa	0102 0840 1326 2049	7.0 0.4 6.6 0.6	26 Su	0043 0815 1256 2023	6.4 1.0 6.3 1.1
12 Su	0144 0917 1409 2125	6.8 0.6 6.4 0.9	27 M	0107 0843 1323 2053	6.3 1.2 6.1 1.3
13 M	0227 0955 1456 2206	6.5 1.1 6.0 1.3	28 Tu	0134 0911 1355 2122	6.1 1.4 6.0 1.6
14 Tu ☽	0318 1038 1552 2257	6.1 1.6 5.6 1.8	29 W	0209 0945 1436 2202	5.9 1.7 5.7 1.9
15 W	0421 1142 1705	5.5 2.1 5.2	30 Th ☾	0257 1034 1536 2305	5.5 2.1 5.3 2.2
			31 F	0417 1153 1716	5.1 2.3 5.0

APRIL

Day	Time	m	Day	Time	m
1 Sa	0036 0621 1331 1907	2.2 5.0 2.2 5.1	16 Su	0254 0822 1529 2037	1.7 5.4 1.8 5.7
2 Su	0219 0749 1458 2018	1.9 5.4 1.8 5.6	17 M	0352 0908 1620 2119	1.4 5.7 1.5 5.9
3 M	0334 0846 1603 2108	1.4 5.8 1.3 6.1	18 Tu	0438 0946 1701 2156	1.2 5.9 1.3 6.1
4 Tu	0433 0932 1658 2150	0.9 6.2 0.9 6.5	19 W	0516 1017 1736 2230	1.1 6.0 1.2 6.3
5 W	0526 1013 1749 2233	0.6 6.5 0.7 6.8	20 Th	0550 1045 1804 2259	1.0 6.2 1.1 6.4
6 Th ●	0615 1055 1834 2313	0.4 6.7 0.5 7.0	21 F ○	0621 1111 1832 2326	0.9 6.3 1.0 6.4
7 F	0700 1137 1912 2354	0.3 6.8 0.5 7.1	22 Sa	0650 1137 1902 2350	0.9 6.3 1.0 6.4
8 Sa	0740 1219 1949	0.3 6.8 0.5	23 Su	0720 1203 1933	0.9 6.3 1.0
9 Su	0036 0819 1302 2029	7.0 0.5 6.6 0.7	24 M	0015 0751 1231 2005	6.3 1.0 6.3 1.1
10 M	0120 0857 1348 2108	6.7 0.8 6.3 1.0	25 Tu	0042 0823 1302 2037	6.2 1.2 6.1 1.3
11 Tu	0206 0936 1436 2152	6.4 1.2 6.0 1.3	26 W	0114 0856 1338 2114	6.0 1.4 6.0 1.5
12 W	0300 1023 1532 2244 ☽	5.9 1.7 5.6 1.8	27 Th	0154 0935 1425 2159	5.8 1.7 5.7 1.7
13 Th	0406 1123 1642 2356	5.4 2.1 5.2 2.1	28 F ☾	0249 1026 1531 2258	5.5 1.9 5.4 1.9
14 F	0537 1250 1819	5.0 2.3 5.1	29 Sa	0416 1136 1702	5.2 2.1 5.2
15 Sa	0131 0716 1422 1941	2.1 5.1 2.1 5.3	30 Su	0018 0603 1303 1835	1.9 5.2 2.0 5.4

Chart Datum: 3.67 metres below Ordnance Datum (Newlyn)

ENGLAND, SOUTH COAST — DOVER

Lat 51°07′ N Long 1°19′ E

TIMES AND HEIGHTS OF HIGH AND LOW WATERS

YEAR 1989

TIME ZONE UT (GMT)
For Summer Time add ONE hour in non-shaded areas

MAY	JUNE	JULY	AUGUST
Time m / Time m	Time m / Time m	Time m / Time m	Time m / Time m
1 0145 1.6 / **16** 0257 1.7 0720 5.5 / 0823 5.5 M 1422 1.6 / Tu 1524 1.8 1942 5.8 / 2036 5.8	**1** 0321 1.1 / **16** 0341 1.7 0832 6.1 / 0857 5.6 Th 1545 1.3 / F 1609 1.7 2050 6.4 / 2112 5.8	**1** 0400 1.3 / **16** 0400 1.8 0912 5.9 / 0911 5.6 Sa 1626 1.4 / Su 1633 1.7 2131 6.2 / 2129 5.7	**1** 0601 1.2 / **16** 0529 1.2 1051 6.3 / 1021 6.2 Tu 1817 1.0 / W 1754 1.0 ● 2312 6.2 / 2240 6.3
2 0257 1.3 / **17** 0348 1.5 0815 5.9 / 0904 5.7 Tu 1525 1.3 / W 1610 1.6 2034 6.2 / 2117 5.9	**2** 0420 1.0 / **17** 0433 1.5 0922 6.2 / 0936 5.8 F 1641 1.1 / Sa 1658 1.5 2141 6.5 / 2152 5.9	**2** 0504 1.2 / **17** 0457 1.5 1009 6.1 / 0956 5.9 Su 1725 1.2 / M 1725 1.4 2227 6.3 / 2213 6.0	**2** 0649 1.1 / **17** 0617 1.0 1130 6.5 / 1102 6.6 W 1900 0.9 / Th 1842 0.7 2351 6.3 / ○ 2320 6.5
3 0357 0.9 / **18** 0430 1.4 0901 6.2 / 0938 5.9 W 1620 1.0 / Th 1651 1.4 2119 6.5 / 2152 6.1	**3** 0516 0.9 / **18** 0519 1.3 1013 6.4 / 1014 6.0 Sa 1736 1.0 / Su 1744 1.3 ● 2231 6.6 / 2230 6.0	**3** 0601 1.1 / **18** 0547 1.3 1058 6.3 / 1038 6.1 M 1819 1.0 / Tu 1812 1.1 ● 2318 6.4 / ○ 2257 6.2	**3** 0726 1.0 / **18** 0702 0.8 1207 6.6 / 1142 6.7 Th 1937 0.8 / F 1924 0.5
4 0451 0.7 / **19** 0509 1.2 0946 6.4 / 1009 6.0 Th 1712 0.9 / F 1729 1.3 2204 6.7 / 2224 6.2	**4** 0610 0.8 / **19** 0604 1.2 1104 6.5 / 1054 6.2 Su 1827 0.9 / M 1827 1.2 2322 6.6 / ○ 2309 6.1	**4** 0652 1.0 / **19** 0634 1.1 1144 6.4 / 1120 6.4 Tu 1907 0.9 / W 1856 0.9 2340 6.3	**4** 0028 6.3 / **19** 0003 6.7 0755 1.0 / 0740 0.7 F 1242 6.6 / Sa 1222 6.9 2009 0.9 / 2002 0.5
5 0543 0.6 / **20** 0547 1.1 1030 6.6 / 1040 6.2 F 1800 0.7 / Sa 1805 1.2 ● 2248 6.9 / ○ 2254 6.2	**5** 0700 0.8 / **20** 0645 1.1 1151 6.5 / 1133 6.3 M 1916 0.8 / Tu 1906 1.1 2350 6.2	**5** 0004 6.3 / **20** 0714 1.0 0737 1.0 / 1204 6.5 W 1225 6.5 / Th 1937 0.8 1951 0.9	**5** 0100 6.3 / **20** 0043 6.7 0822 1.1 / 0813 0.7 Sa 1314 6.5 / Su 1303 6.9 2039 0.9 / 2039 0.5
6 0631 0.5 / **21** 0624 1.1 1115 6.7 / 1111 6.2 Sa 1845 0.6 / Su 1842 1.1 2333 6.9 / 2325 6.2	**6** 0011 6.5 / **21** 0723 1.1 0745 0.9 / 1215 6.3 Tu 1238 6.4 / W 1945 1.0 2002 0.9	**6** 0048 6.3 / **21** 0024 6.4 0816 1.1 / 0754 0.9 Th 1304 6.4 / F 1246 6.6 2030 0.9 / 2016 0.7	**6** 0130 6.2 / **21** 0126 6.6 0847 1.2 / 0849 0.8 Su 1347 6.4 / M 1344 6.8 2108 1.1 / 2115 0.7
7 0716 0.6 / **22** 0659 1.1 1201 6.6 / 1144 6.3 Su 1930 0.7 / M 1917 1.1 2357 6.2	**7** 0057 6.3 / **22** 0032 6.2 0829 1.1 / 0801 1.1 W 1321 6.3 / Th 1259 6.3 2046 1.0 / 2025 1.0	**7** 0126 6.2 / **22** 0109 6.4 0850 1.2 / 0830 0.8 F 1344 6.4 / Sa 1330 6.6 2107 1.0 / 2056 0.7	**7** 0159 6.1 / **22** 0208 6.4 0912 1.3 / 0925 1.0 M 1416 6.2 / Tu 1426 6.6 2136 1.3 / 2155 0.9
8 0018 6.7 / **23** 0734 1.1 0759 0.7 / 1219 6.2 M 1248 6.5 / Tu 1952 1.1 2013 0.8	**8** 0144 6.1 / **23** 0119 6.1 0910 1.3 / 0840 1.1 Th 1406 6.2 / F 1345 6.3 2128 1.2 / 2105 1.0	**8** 0205 6.0 / **23** 0155 6.4 0921 1.3 / 0908 0.9 Sa 1422 6.2 / Su 1412 6.6 2142 1.2 / 2135 0.8	**8** 0230 5.9 / **23** 0254 6.2 0941 1.6 / 1006 1.3 Tu 1447 6.0 / W 1515 6.3 2207 1.6 / ☾ 2237 1.3
9 0106 6.5 / **24** 0032 6.1 0842 1.0 / 0809 1.2 Tu 1334 6.3 / W 1259 6.2 2057 1.0 / 2030 1.2	**9** 0230 5.9 / **24** 0209 6.0 0949 1.5 / 0922 1.2 F 1451 6.0 / Sa 1433 6.2 2210 1.4 / 2149 1.0	**9** 0244 5.8 / **24** 0239 6.2 0952 1.5 / 0949 1.0 Su 1501 6.0 / M 1457 6.5 2216 1.4 / ☾ 2217 1.1	**9** 0305 5.6 / **24** 0348 5.8 1013 1.9 / 1054 1.7 W 1524 5.7 / Th 1613 5.9 ☾ 2242 1.9 / 2334 1.8
10 0155 6.2 / **25** 0112 6.0 0924 1.3 / 0847 1.3 W 1422 6.0 / Th 1341 6.0 2141 1.3 / 2111 1.3	**10** 0321 5.6 / **25** 0301 5.9 1030 1.8 / 1007 1.3 Sa 1542 5.8 / Su 1522 6.1 2255 1.6 / 2237 1.2	**10** 0325 5.6 / **25** 0325 6.1 1024 1.7 / 1031 1.3 M 1543 5.8 / Tu 1545 6.3 2252 1.6 / ☾ 2304 1.2	**10** 0349 5.4 / **25** 0454 5.4 1055 2.2 / 1201 2.0 Th 1614 5.4 / F 1729 5.5 2332 2.2
11 0247 5.8 / **26** 0159 5.8 1009 1.7 / 0931 1.4 Th 1514 5.8 / F 1433 5.9 2231 1.6 / 2156 1.4	**11** 0419 5.4 / **26** 0356 5.8 1113 2.0 / 1058 1.4 Su 1638 5.6 / M 1616 6.1 ☾ 2344 1.8 / ☾ 2330 1.3	**11** 0410 5.4 / **26** 0419 5.8 1102 1.9 / 1120 1.6 Tu 1631 5.6 / W 1641 6.0 ☾ 2334 1.9	**11** 0451 5.1 / **26** 0057 2.1 1156 2.4 / 0627 5.2 F 1727 5.0 / Sa 1335 2.1 1912 5.3
12 0349 5.4 / **27** 0301 5.6 1101 2.0 / 1020 1.6 F 1616 5.5 / Sa 1534 5.7 ☾ 2330 1.9 / 2251 1.5	**12** 0522 5.2 / **27** 0454 5.7 1205 2.1 / 1156 1.6 M 1743 5.4 / Tu 1716 6.0	**12** 0505 5.2 / **27** 0000 1.5 1150 2.2 / 0522 5.6 W 1730 5.3 / Th 1225 1.8 1750 5.8	**12** 0041 2.3 / **27** 0236 2.0 0617 4.9 / 0808 5.4 Sa 1320 2.5 / Su 1512 1.8 1859 5.0 / 2040 5.6
13 0505 5.2 / **28** 0416 5.5 1207 2.2 / 1120 1.7 Sa 1733 5.3 / Su 1644 5.7 ☾ 2357 1.5	**13** 0041 1.9 / **28** 0034 1.4 0627 5.2 / 0558 5.7 Tu 1304 2.2 / W 1302 1.7 1846 5.4 / 1821 5.9	**13** 0028 2.0 / **28** 0114 1.7 0608 5.1 / 0639 5.4 Th 1253 2.3 / F 1347 1.9 1836 5.2 / 1910 5.6	**13** 0208 2.3 / **28** 0357 1.7 0747 5.1 / 0911 5.5 Su 1456 2.2 / M 1623 1.5 2016 5.3 / 2139 5.8
14 0042 1.9 / **29** 0532 5.5 0628 5.2 / 1231 1.7 Su 1321 2.2 / M 1756 5.7 1849 5.4	**14** 0141 1.9 / **29** 0142 1.4 0724 5.3 / 0704 5.7 W 1408 2.1 / Th 1411 1.6 1942 5.5 / 1928 6.0	**14** 0135 2.1 / **29** 0237 1.8 0717 5.1 / 0805 5.5 F 1411 2.3 / Sa 1511 1.8 1942 5.3 / 2030 5.7	**14** 0332 1.9 / **29** 0501 1.4 0850 5.5 / 0956 6.1 M 1609 1.7 / Tu 1719 1.1 2112 5.6 / 2221 6.1
15 0155 1.8 / **30** 0110 1.4 0733 5.3 / 0641 5.7 M 1429 2.0 / Tu 1342 1.6 1948 5.6 / 1902 5.9	**15** 0243 1.8 / **30** 0253 1.5 0813 5.4 / 0811 5.8 Th 1511 1.9 / F 1519 1.5 2030 5.6 / 2032 6.0	**15** 0251 2.0 / **30** 0357 1.6 0819 5.3 / 0914 5.8 Sa 1531 2.0 / Su 1624 1.5 2040 5.5 / 2135 5.9	**15** 0435 1.6 / **30** 0551 1.2 0939 5.9 / 1034 6.3 Tu 1705 1.3 / W 1805 1.0 2157 6.0 / 2258 6.2
	31 0219 1.3 0740 5.9 W 1446 1.4 1958 6.1		**31** 0505 1.4 1006 6.0 M 1725 1.2 2227 6.1	**31** 0632 1.1 1109 6.5 Th 1842 0.9 ● 2330 6.3

Chart Datum: 3.67 metres below Ordnance Datum (Newlyn)

AREA 3 — SE England

ENGLAND, SOUTH COAST — DOVER

Lat 51°07′ N Long 1°19′ E

TIMES AND HEIGHTS OF HIGH AND LOW WATERS

YEAR 1989

TIME ZONE UT(GMT)
For Summer Time add ONE hour in non-shaded areas

SEPTEMBER

Day	Time	m	Day	Time	m
1 F	0702 / 1143 / 1912	1.1 / 6.6 / 0.8	16 Sa	0636 / 1115 / 1900 / 2334	0.7 / 7.0 / 0.5 / 6.8
2 Sa	0000 / 0726 / 1214 / 1938	6.4 / 1.1 / 6.7 / 0.9	17 Su	0714 / 1153 / 1940	0.6 / 7.1 / 0.4
3 Su	0028 / 0747 / 1242 / 2005	6.4 / 1.1 / 6.6 / 0.9	18 M	0014 / 0749 / 1232 / 2016	6.8 / 0.7 / 7.1 / 0.5
4 M	0052 / 0812 / 1306 / 2033	6.3 / 1.2 / 6.5 / 1.1	19 Tu	0056 / 0825 / 1314 / 2054	6.7 / 0.8 / 6.9 / 0.7
5 Tu	0116 / 0837 / 1330 / 2100	6.2 / 1.3 / 6.3 / 1.3	20 W	0140 / 0904 / 1359 / 2134	6.5 / 1.0 / 6.6 / 1.1
6 W	0142 / 0905 / 1355 / 2128	6.1 / 1.5 / 6.1 / 1.6	21 Th	0229 / 0946 / 1450 / 2219	6.2 / 1.4 / 6.2 / 1.6
7 Th	0213 / 0935 / 1427 / 2200	5.9 / 1.8 / 5.8 / 1.9	22 F	0324 / 1037 / 1552 / 2319	5.8 / 1.8 / 5.7 / 2.0
8 F	0251 / 1014 / 1510 / 2247	5.6 / 2.2 / 5.4 / 2.2	23 Sa	0434 / 1150 / 1718	5.4 / 2.1 / 5.2
9 Sa	0348 / 1115 / 1624	5.2 / 2.5 / 5.0	24 Su	0048 / 0615 / 1330 / 1917	2.3 / 5.1 / 2.2 / 5.2
10 Su	0000 / 0525 / 1241 / 1825	2.5 / 4.9 / 2.6 / 4.9	25 M	0227 / 0754 / 1501 / 2034	2.1 / 5.4 / 1.8 / 5.5
11 M	0133 / 0716 / 1423 / 1957	2.4 / 5.0 / 2.3 / 5.2	26 Tu	0341 / 0853 / 1606 / 2125	1.7 / 5.8 / 1.4 / 5.9
12 Tu	0301 / 0827 / 1541 / 2053	2.0 / 5.5 / 1.7 / 5.7	27 W	0437 / 0934 / 1655 / 2202	1.4 / 6.1 / 1.1 / 6.1
13 W	0407 / 0915 / 1638 / 2136	1.6 / 6.0 / 1.2 / 6.1	28 Th	0522 / 1010 / 1737 / 2233	1.3 / 6.4 / 1.0 / 6.2
14 Th	0502 / 0956 / 1729 / 2216	1.2 / 6.4 / 0.8 / 6.5	29 F	0558 / 1042 / 1810 / 2301	1.2 / 6.6 / 1.0 / 6.4
15 F	0551 / 1035 / 1817 / 2255	0.9 / 6.7 / 0.6 / 6.7	30 Sa	0625 / 1113 / 1838 / 2329	1.1 / 6.6 / 1.0 / 6.4

OCTOBER

Day	Time	m	Day	Time	m
1 Su	0649 / 1143 / 1904 / 2354	1.1 / 6.6 / 1.0 / 6.5	16 M	0643 / 1143 / 1912 / 2349	0.7 / 7.2 / 0.5 / 6.9
2 M	0713 / 1207 / 1931	1.1 / 6.6 / 1.0	17 Tu	0723 / 1207 / 1952	0.7 / 7.1 / 0.7
3 Tu	0017 / 0740 / 1229 / 1959	6.4 / 1.2 / 6.5 / 1.1	18 W	0034 / 0805 / 1252 / 2034	6.8 / 0.9 / 6.9 / 0.9
4 W	0041 / 0809 / 1252 / 2029	6.3 / 1.3 / 6.3 / 1.3	19 Th	0120 / 0858 / 1340 / 2118	6.5 / 1.1 / 6.5 / 1.3
5 Th	0107 / 0840 / 1319 / 2058	6.2 / 1.6 / 6.1 / 1.6	20 F	0211 / 0934 / 1433 / 2206	6.2 / 1.4 / 6.1 / 1.7
6 F	0138 / 0911 / 1351 / 2132	6.0 / 1.8 / 5.9 / 1.9	21 Sa	0307 / 1027 / 1538 / 2306	5.8 / 1.8 / 5.6 / 2.1
7 Sa	0216 / 0952 / 1433 / 2219	5.7 / 2.1 / 5.5 / 2.2	22 Su	0414 / 1137 / 1705	5.5 / 2.1 / 5.2
8 Su	0310 / 1049 / 1545 / 2329	5.3 / 2.4 / 5.1 / 2.5	23 M	0028 / 0546 / 1306 / 1852	2.3 / 5.3 / 2.1 / 5.2
9 M	0445 / 1212 / 1757	5.0 / 2.4 / 4.9	24 Tu	0154 / 0714 / 1427 / 2001	2.2 / 5.4 / 1.8 / 5.5
10 Tu	0100 / 0641 / 1348 / 1928	2.4 / 5.1 / 2.1 / 5.3	25 W	0303 / 0815 / 1528 / 2050	1.9 / 5.8 / 1.5 / 5.8
11 W	0226 / 0754 / 1504 / 2025	2.0 / 5.6 / 1.6 / 5.8	26 Th	0356 / 0900 / 1617 / 2128	1.6 / 6.0 / 1.3 / 6.0
12 Th	0332 / 0843 / 1603 / 2107	1.6 / 6.1 / 1.2 / 6.1	27 F	0440 / 0938 / 1658 / 2200	1.5 / 6.2 / 1.2 / 6.1
13 F	0427 / 0925 / 1657 / 2146	1.2 / 6.5 / 0.8 / 6.5	28 Sa	0516 / 1012 / 1732 / 2230	1.4 / 6.4 / 1.2 / 6.2
14 Sa	0516 / 1004 / 1746 / 2226	0.9 / 6.8 / 0.6 / 6.8	29 Su	0547 / 1042 / 1803 / 2258	1.3 / 6.5 / 1.1 / 6.4
15 Su	0601 / 1045 / 1831 / 2306	0.8 / 7.1 / 0.4 / 6.9	30 M	0615 / 1111 / 1832 / 2323	1.2 / 6.5 / 1.1 / 6.4
			31 Tu	0645 / 1136 / 1903 / 2350	1.2 / 6.3 / 1.1 / 6.4

NOVEMBER

Day	Time	m	Day	Time	m
1 W	0716 / 1201 / 1934	1.3 / 6.4 / 1.2	16 Th	0019 / 0751 / 1238 / 2020	6.7 / 0.9 / 6.7 / 1.0
2 Th	0017 / 0749 / 1228 / 2005	6.3 / 1.4 / 6.3 / 1.4	17 F	0109 / 0837 / 1328 / 2105	6.5 / 1.1 / 6.4 / 1.3
3 F	0048 / 0823 / 1257 / 2039	6.2 / 1.5 / 6.1 / 1.6	18 Sa	0158 / 0924 / 1422 / 2153	6.3 / 1.3 / 6.0 / 1.7
4 Sa	0121 / 0858 / 1334 / 2117	6.1 / 1.7 / 5.9 / 1.8	19 Su	0250 / 1014 / 1521 / 2245	6.0 / 1.6 / 5.7 / 2.0
5 Su	0202 / 0941 / 1422 / 2203	5.8 / 1.9 / 5.6 / 2.1	20 M	0348 / 1113 / 1630 / 2349	5.7 / 1.9 / 5.4 / 2.2
6 M	0258 / 1035 / 1534 / 2306	5.5 / 2.1 / 5.2 / 2.3	21 Tu	0457 / 1222 / 1754	5.5 / 2.0 / 5.2
7 Tu	0421 / 1146 / 1720	5.3 / 2.1 / 5.1	22 W	0057 / 0615 / 1333 / 1906	2.3 / 5.5 / 1.9 / 5.3
8 W	0025 / 0556 / 1309 / 1845	2.2 / 5.4 / 1.9 / 5.4	23 Th	0205 / 0723 / 1434 / 2001	2.2 / 5.6 / 1.8 / 5.5
9 Th	0145 / 0707 / 1423 / 1944	2.0 / 5.7 / 1.6 / 5.8	24 F	0303 / 0815 / 1528 / 2047	2.0 / 5.8 / 1.7 / 5.7
10 F	0253 / 0804 / 1525 / 2032	1.6 / 6.1 / 1.2 / 6.1	25 Sa	0352 / 0900 / 1614 / 2125	1.8 / 6.0 / 1.5 / 5.9
11 Sa	0349 / 0850 / 1620 / 2117	1.3 / 6.5 / 1.0 / 6.4	26 Su	0434 / 0938 / 1654 / 2159	1.7 / 6.1 / 1.4 / 6.1
12 Su	0440 / 0934 / 1712 / 2200	1.1 / 6.8 / 0.8 / 6.7	27 M	0513 / 1013 / 1730 / 2230	1.5 / 6.2 / 1.3 / 6.2
13 M	0529 / 1019 / 1801 / 2245	0.9 / 7.0 / 0.7 / 6.8	28 Tu	0550 / 1044 / 1807 / 2302	1.4 / 6.3 / 1.3 / 6.3
14 Tu	0617 / 1104 / 1849 / 2333	0.8 / 7.0 / 0.7 / 6.8	29 W	0625 / 1115 / 1842 / 2333	1.3 / 6.3 / 1.2 / 6.3
15 W	0703 / 1150 / 1935	0.8 / 6.9 / 0.8	30 Th	0700 / 1146 / 1916	1.3 / 6.3 / 1.3

DECEMBER

Day	Time	m	Day	Time	m
1 F	0007 / 0735 / 1218 / 1949	6.3 / 1.3 / 6.2 / 1.3	16 Sa	0059 / 0827 / 1320 / 2053	6.6 / 1.0 / 6.3 / 1.2
2 Sa	0042 / 0812 / 1253 / 2026	6.3 / 1.4 / 6.1 / 1.4	17 Su	0142 / 0911 / 1405 / 2134	6.4 / 1.1 / 6.1 / 1.5
3 Su	0120 / 0850 / 1334 / 2105	6.2 / 1.4 / 5.9 / 1.6	18 M	0227 / 0955 / 1453 / 2214	6.2 / 1.3 / 5.8 / 1.7
4 M	0204 / 0932 / 1423 / 2150	6.0 / 1.6 / 5.7 / 1.7	19 Tu	0315 / 1038 / 1545 / 2255	6.0 / 1.6 / 5.6 / 2.0
5 Tu	0254 / 1020 / 1525 / 2242	5.9 / 1.7 / 5.6 / 1.9	20 W	0407 / 1126 / 1645 / 2343	5.8 / 1.8 / 5.4 / 2.2
6 W	0356 / 1118 / 1638 / 2346	5.7 / 1.7 / 5.5 / 2.0	21 Th	0508 / 1219 / 1753	5.6 / 2.0 / 5.2
7 Th	0506 / 1227 / 1751	5.8 / 1.7 / 5.6	22 F	0041 / 0615 / 1320 / 1859	2.3 / 5.4 / 2.1 / 5.2
8 F	0059 / 0617 / 1340 / 1857	1.9 / 5.9 / 1.6 / 5.7	23 Sa	0147 / 0721 / 1426 / 1959	2.3 / 5.4 / 2.0 / 5.3
9 Sa	0208 / 0720 / 1446 / 1957	2.0 / 6.1 / 1.4 / 5.9	24 Su	0256 / 0818 / 1527 / 2050	2.2 / 5.5 / 1.9 / 5.5
10 Su	0311 / 0816 / 1548 / 2051	1.5 / 6.3 / 1.2 / 6.2	25 M	0356 / 0905 / 1620 / 2132	2.0 / 5.7 / 1.7 / 5.7
11 M	0410 / 0910 / 1645 / 2143	1.3 / 6.5 / 1.1 / 6.4	26 Tu	0445 / 0946 / 1705 / 2210	1.7 / 5.8 / 1.5 / 5.9
12 Tu	0508 / 1003 / 1742 / 2235	1.1 / 6.6 / 1.0 / 6.5	27 W	0530 / 1024 / 1747 / 2245	1.4 / 6.0 / 1.4 / 6.1
13 W	0601 / 1054 / 1835 / 2326	1.0 / 6.7 / 0.9 / 6.6	28 Th	0611 / 1101 / 1828 / 2322	1.3 / 6.1 / 1.3 / 6.3
14 Th	0653 / 1144 / 1924	0.9 / 6.7 / 0.9	29 F	0650 / 1137 / 1904	1.2 / 6.2 / 1.2
15 F	0014 / 0741 / 1234 / 2011	6.6 / 0.9 / 6.5 / 1.1	30 Sa	0000 / 0727 / 1214 / 1940	6.4 / 1.1 / 6.2 / 1.2
			31 Su	0038 / 0804 / 1253 / 2015	6.4 / 1.1 / 6.2 / 1.2

Chart Datum: 3.67 metres below Ordnance Datum (Newlyn)

AREA 3—SE England 287

RAMSGATE 10-3-15
Kent

CHARTS
Admiralty 1827, 1828; Stanford 5, 19; Imray C1, C8; OS 179

TIDES
+0020 Dover; ML 2.6; Duration 0530; Zone 0 (GMT).

Standard Port DOVER (←)

Times				Height (metres)			
HW		LW		MHWS	MHWN	MLWN	MLWS
0000	0600	0100	0700	6.7	5.3	2.0	0.8
1200	1800	1300	1900				

Differences RAMSGATE
+0020 +0020 −0007 −0007 −1.8 −1.5 −0.8 −0.4
HW Broadstairs = HW Dover +0040

SHELTER
Good in inner harbour (marina). Access HW−2 to HW+2 approx.

NAVIGATION
Waypoint 51°19'.50N 01°26'.00E, 090°/270° from/to new breakwater entrance, 0.25M. Beware Dyke Bank to the N, and Brake and Cross Ledge to the S; all these dry. Access is by main channel from the E. Contact Harbour Control VHF Ch 16 14. Reception pontoon in West Gully of Royal Harbour.

LIGHTS AND MARKS
Entrance channel has lateral and cardinal buoys and marks. Breakwater entrance N side QG, S side VQR. Ldg Lts 270°. Front Dir Oc WRG 10s; G259°-269°, W269°-271°, R271°-281°. Rear Oc 5s. International Port Traffic Signals control main channel and entry to Royal Harbour and marina. West Pier Lt refers to depth between piers:— FR over 3m, FG less than 3m.

RADIO TELEPHONE
VHF Ch 14 16 (H24). Marina Ch 14.

TELEPHONE (0843)
Hr Mr 592277; CG Dover 210008; MRSC Frinton-on-Sea 5518; HMC Freefone Customs Yachts: RT (0304) 202441; Marinecall 0898 500 456; Dr 595051.

FACILITIES
EC Thursday; **Ramsgate Yacht Marina** (400+100 visitors) Tel. 592277, Slip, Gas, Gaz, Kos, AC, FW, ME, El, Sh, P, D, C (20 ton), CH, V, Lau, (Access HW−2 to HW+2); **Royal Temple YC** Tel. 591766, Bar; **Ramsgate Marine** Tel. 593140, Slip, M, ME, El, Sh, BH; **Marine Centre** Tel. 596772, Slip, ME, El, Sh, SM; **Seagear** Tel. 591733, ACA, CH; **Foy Boat Marine** Tel. 592662, P, D, L, AB; **Bosun's Locker** Tel. 597158, CH, ACA; **Walkers Marine** Tel. 592176, ME; **Davis Marine** Tel. 586172, ME.
Town P, Gas, Gaz, V, R, Bar. PO; Bank; Rly; Air (Manston).

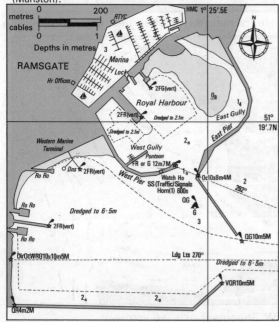

WHITSTABLE 10-3-16
Kent

CHARTS
Admiralty 2571; Stanford 5; Imray Y14; OS 179

TIDES
+0120 Dover; ML 3.0; Duration 0605; Zone 0 (GMT).

Standard Port SHEERNESS (→)

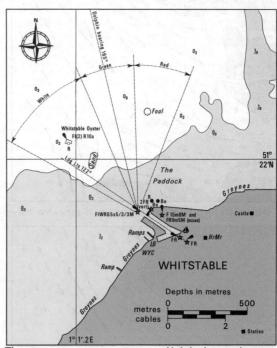

Times				Height (metres)			
HW		LW		MHWS	MHWN	MLWN	MLWS
0200	0800	0200	0700	5.7	4.7	1.5	0.6
1400	2000	1400	1900				

Differences WHITSTABLE
−0008 −0011 +0005 0000 −0.3 −0.3 0.0 −0.1
MARGATE
−0050 −0040 −0020 −0050 −0.9 −0.9 −0.1 0.0
HERNE BAY
−0025 −0015 0000 −0025 −0.5 −0.5 −0.1 −0.1

SHELTER
Good shelter except in strong winds from NNW to NE. Berthing is restricted to genuine refuge seekers since priority is given to commercial shipping. Moorings to NW of harbour controlled by YC, Access HW∓1.

NAVIGATION
Waypoint 51°22'.62N 01°01'.20E, 165°/345° from/to West Quay dolphin, 0.83M. Beware shoals near approaches, which are very shallow. Harbour dries. Oyster beds are numerous. Approach should not be attempted before half tide.

LIGHTS AND MARKS
Ldg Lts 122°, both FR, West Quay, off head Fl WRG 5s; dolphin; vis W118°-156°, G156°-178°, R178°-201°. NE arm FW 15m 8M; indicates harbour open; FR below this Lt indicates harbour closed.

RADIO TELEPHONE (local times)
VHF Ch 16; 09 12 (Mon-Fri: 0800-1700. Other times: HW−3 to HW+1). Tidal information on request.

TELEPHONE (0227)
Hr Mr 274086; MRSC Frinton-on-Sea 5518; HMC Freefone Customs Yachts: RT (0304) 202441; Marinecall 0898 500 455; Dr 263811.

FACILITIES
EC Wednesday; **Harbour** Tel. 274086, L, FW, D, C (15 ton), AB; **Whitstable YC** Tel. 272942, M, P (cans), FW, CH, Bar; **The Dinghy Store** Tel. 274168, ME, Sh, CH, ACA; **Whitstable Marine** Tel. 262525, ME, El, Sh, Slip, CH, C, V, R; **H. Goldfinch** Tel. 272295, SM; **Waldens** Tel. 272098, Gas.
Town Lau, P, V, R, Bar. PO; Bank; Rly (bus to Herne Bay or Canterbury); Air (Lydd or Manston).

10.3.17 THAMES ESTUARY TIDAL STREAMS

CAUTION:— Due to very strong rates of tidal streams in some areas, eddies may occur. Where possible, some indication of these is shown but in many areas there is insufficient information or eddies are unstable

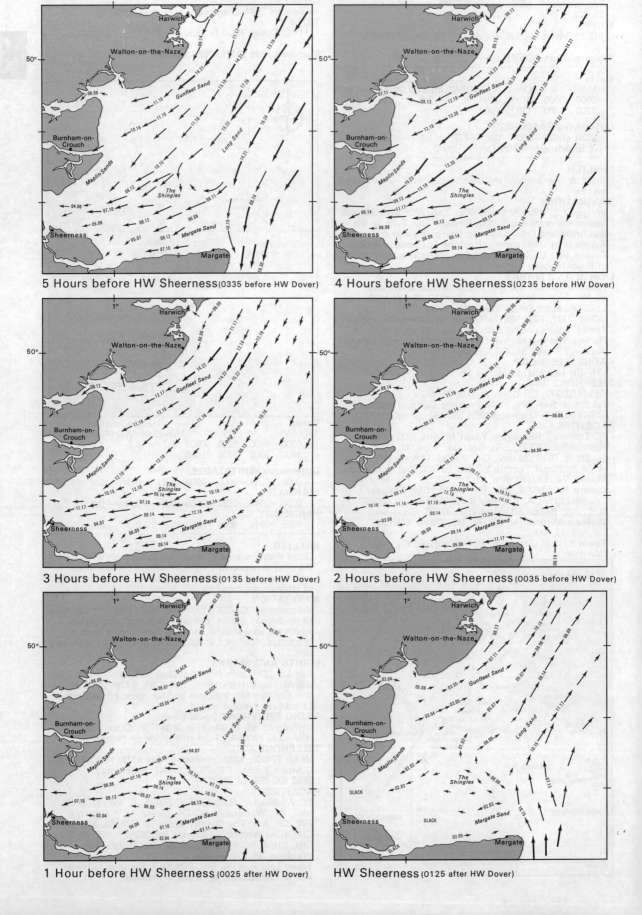

AREA 3—SE England 289

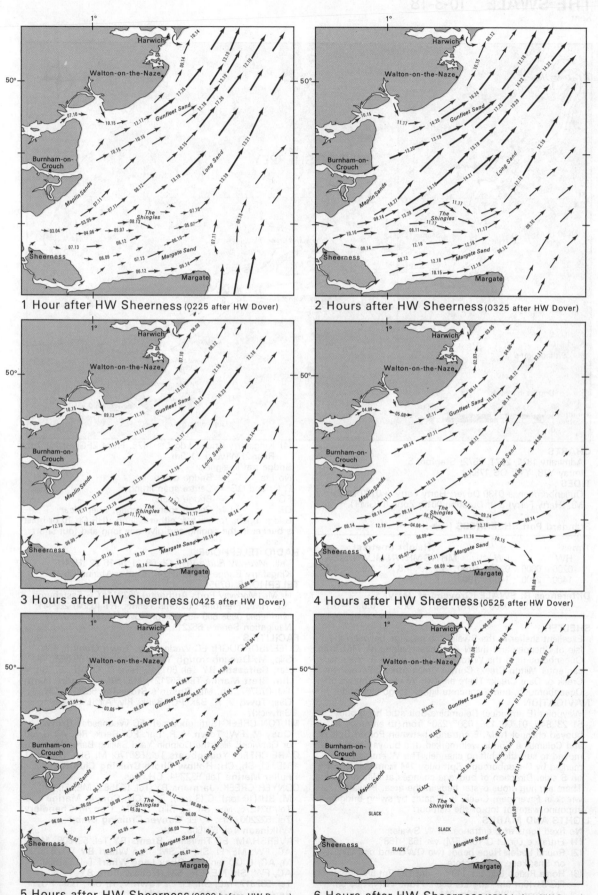

THE SWALE 10-3-18
Kent

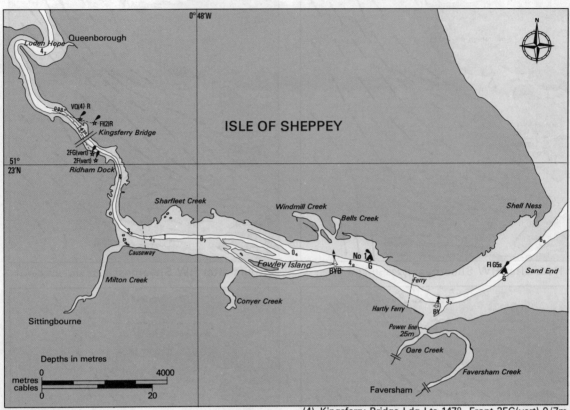

CHARTS
Admiralty 1834, 2571, 2572; Stanford 5; Imray Y18, Y14; OS 178.

TIDES
Queenborough +0130 Dover; Harty Ferry +0120 Dover; ML (Harty Ferry) 3.0; Duration 0610; Zone 0 (GMT).

Standard Port SHEERNESS (→)

Times				Height (metres)			
HW		LW		MHWS	MHWN	MLWN	MLWS
0200	0800	0200	0700	5.7	4.8	1.5	0.6
1400	2000	1400	1900				

Differences R. SWALE (Grovehurst Jetty)
−0007 0000 0000 +0016 0.0 0.0 0.0 −0.1

SHELTER
Excellent shelter in the Swale, the passage between the Isle of Sheppey and the N Kent coast, along its 14M from Queenborough in the W to Shell Ness in the E. Yachts can enter Milton Creek, Conyer Creek (dries), Faversham Creek or Oare Creek. There are two W visitors buoys off Queenborough, and a concrete lighter to go alongside.

NAVIGATION
Waypoint (E entrance) Columbine Spit stbd-hand buoy, 51°23'.84N, 01°00'.13E, 050°/230° from/to entrance to buoyed channel 1.3M. E entrance between Pollard Spit and Columbine Spit is well marked, the buoys being moved to suit the shifting channel. The W entrance is marked by Queenborough Spit buoy 1M from Garrison Pt on S side. Direction of buoyage changes at Milton Creek. There are numerous oyster beds in the area. Beware wreck in Faversham Creek; lock gates by swing bridge impound water at top of the harbour.

LIGHTS AND MARKS
No fixed lights at E entrance. In W Swale:
(1) Entrance Dir Lt Q 16m 5M; vis 163°-168°.
(2) Round Loden Hope bend: two QWG and one QWRG on Bns; keep in G sectors.
(3) Horse Reach Ldg Lts 113°. Front QG 7m 5M. Rear FIG 3s 10m 6M. Ldg Lts 098°. Front VQ(4)R 5s 6m 5M. Rear Fl(2)R 5s 10m 9M.
(4) Kingsferry Bridge Ldg Lts 147°. Front 2FG(vert) 9/7m. Rear 2FW(vert) 11/9m.

Bridge traffic signals: —
No Lts = Bridge down
Q Or and QG = Centre span lifting
FG = Bridge open
QR = Centre span lowering. Keep clear.

Bridge opens for yachts; either call on VHF Ch 10 or hoist a bucket in the rigging and give one long and four short blasts.

RADIO TELEPHONE
Call: *Medway Radio* VHF Ch 09 11 **74** 16 22 (H24); Kingsferry Bridge Ch 10 (H24); Conyer Marine Ch M.

TELEPHONE (0795)
Hr Mr (Medway Ports Authority) 662211; CG Eastchurch 262; HMC Freefone Customs Yachts; RT (0304) 202441; Marinecall 0898 500 455; Dr or Hosp contact Medway Navigation Service 662211.

FACILITIES
QUEENBOROUGH: EC Wednesday. **Town Quay** L, FW, Slip, M; **Queenborough YC** Tel. Sheerness 663955, M, Bar; **Calfcastle BY** Tel. 661141, Slip, C, D, ME, El, Sh; **Jim Brett Marine** Tel. 667121, ME, Sh, El; **André Hardy** Tel. 873767, Sh, ME; **Bosun's Store** Tel. 662674, CH, Gas; **Town** V, R, Bar, PO, Bank, Rly, Air (Lydd or Gatwick).
MILTON CREEK (Sittingbourne): EC Wednesday **Crown Quay** M, FW; **Town** V, R, Bar, PO, Bank, Rly, Air (Lydd or Gatwick); also the Dolphin Yard Sailing Barge Museum.
OARE CREEK: **Youngboats** Tel. 536176, M, Slip, C (8 ton), ME, El, Sh, CH; **Hollow Shore Cruising Club** Bar; **Fuller Marine** Tel. 532294, CH, Gas.
CONYER CREEK: **Jarmans BY** Tel. 521562, ME, El, Sh, M, BH (10 ton), C (3 ton), Slip, CH; **Conyer Marine** Tel. 521276, ME, Sh, El, Slip, D, SM; **Conyer Chandlery** Tel. 522200, CH, V, El; **Conyer Cruising Club**; **Wilkinson** Tel. 521503, SM.
FAVERSHAM: EC Thursday; **Brents BY** Tel. 537809, M, AC, FW, ME, El, Sh, C, SM; **Iron Wharf BY** Tel. 536296, M, AC, C (20 ton), D; **Quay Lane Wharf** Tel. 531660, AC, FW, Sh, ME, CH, SM; **Hollowshore BY** Tel. 532317, ME, Sh, El, C. **Town** V, R, Bar, Gas, PO, Bank, Rly Air (Lydd or Gatwick).

MEDWAY RIVER (SHEERNESS)
10-3-19

Kent

CHARTS
Admiralty 3683, 1834, 1835, 1185; Stanford 5; Imray Y18; OS 178

TIDES
+0130 Dover; ML 3.1; Duration 0610; Zone 0 (GMT).

Standard Port SHEERNESS (→)

Times				Height (metres)			
HW		LW		MHWS	MHWN	MLWN	MLWS
0200	0800	0200	0700	5.7	4.8	1.5	0.6
1400	2000	1400	1900				

Differences UPNOR
+0015 +0015 +0015 +0025 +0.2 +0.2 −0.1 −0.1

ROCHESTER (STROOD PIER)
+0018 +0018 +0018 +0028 +0.2 +0.2 −0.2 −0.3

ALLINGTON LOCK
+0050 +0035 No data −2.1 −2.2 −1.3 −0.4

NOTE: Sheerness is a Standard Port and tidal predictions for each day of the year are given below.

SHELTER
The Medway and Swale give shelter from all directions if the correct anchorage is chosen. Lower reaches are bad in strong NE winds; Stangate Creek is good in all weathers. River is tidal up to Allington Lock (21.6M).

NAVIGATION
Waypoint Medway (safe water) buoy, Iso 2s, 51°28'.80N, 00°52'.92E, 069°/249° from/to Garrison Pt Lt, 5.5M. The wreck of the ammunition ship 'Richard Montgomery' lies 2 M from the harbour mouth, her masts showing above the water. Yachtsmen intending to proceed up the Medway or Swale should obtain a copy of *Medway Ports River Byelaws 1979* from Medway Ports Authority Sheerness Docks, Sheerness, Kent.

Bridge Clearances (MHWS):
- Rochester — 5.9m
- New Hythe (M20) — 11.3m
- Aylesford (Stone) — 2.87m
- Aylesford (Bailey) — 3.26m
- Maidstone Bypass — 9.45m
- Kingsferry (Shut) — 3.35m

Speed Limits: —
- 6 kn — W of Folly Pt Longitude
- 8 kn — S of Kingsferry Bridge
- 8 kn — in Queenborough harbour in area between line Swale Ness — Queenborough Pt and line 270° from Long Pt

For ease of reference the area is split up as follows:-
(1) Lower Medway
(2) Upper Medway

LIGHTS AND MARKS
Garrison Pt L Fl (2) R 15s 13m 4M; vis 027°-255°. Sig Stn, Traffic sigs, Horn (3) 30s.
Isle of Grain Lt Q WRG 20m 13/7M.
Traffic Signals: Powerful Lt, Fl 7s shown from Garrison Pt Signal Station indicates movement of large vessels; shown up river, inward bound; shown to seaward, outward bound.

RADIO TELEPHONE (local times)
Call: *Medway Radio* (at Garrison Point, Sheerness) VHF Ch 09 11 16 22 73 **74** (H24). Yachts should keep watch on Ch 74 underway and Ch 16 at anchor. Kingsferry Bridge, West Swale, Ch 10 (H24). Gillingham Marina, Hoo Marina, Medway Bridge Marina, Ch M (0900-1700).

TELEPHONE (0634)
Hr Mr Sheerness 580003; CG Eastchurch 262; MRSC Frinton-on-Sea 5518; HMC Freefone Customs Yachts: RT (0304) 202441; Marinecall 0898 500 455; Dr Contact Medway Navigation Service Sheerness 666596.

FACILITIES
EC Wednesday; The Medway estuary and the Swale provide a huge area to explore, although much of it dries out to mud. The Medway is well buoyed and perched up to Rochester. *Medway Ports River Byelaws* obtainable from Medway Ports Authority.

LOWER MEDWAY Sheerness is a Standard Port and is the headquarters of Medway Ports Authority. It is a commercial harbour with Ro-Ro ferry berths and no accommodation for yachts. Shelter is good except in the entrance reaches in strong NE winds. In Stangate Creek it is good in all conditions. There are also good anchorages in Sharfleet Creek; from about 4 hrs flood it is possible to go right through into Half Acre Creek. Some minor creeks are buoyed.

Sheppey YC Tel. Sheerness 663052; **W. Hurst** Tel. Sheerness 662356, ACA.

UPPER MEDWAY Admiralty Chart 1835 Folly Point to Maidstone. Harbour Master (Medway Ports Authority) Tel. Sheerness 580003. Speed limit above Folly Point is 6 kn. There are landing facilities (only) at Gillingham Pier, Gillingham Dock steps, Sun Pier (Chatham), Ship Pier, Town Quay steps (Rochester) and Strood Pier. Medway Port Authority have two visitors moorings in Tower Reach just upstream of Rochester Bridge. All other moorings are administered by YCs or marinas. There are slips at Commodore Hard, Gillingham and Upnor Causeway opposite Pier public house. Slips also available at Auto Marine Boatyard, Cuxton, Rochester (HW ∓3).

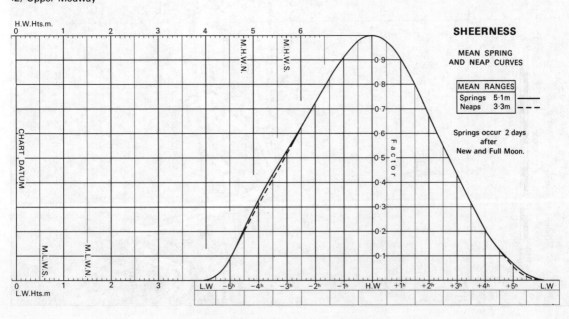

SHEERNESS
MEAN SPRING AND NEAP CURVES

MEAN RANGES
Springs 5·1m
Neaps 3·3m

Springs occur 2 days after New and Full Moon.

MEDWAY RIVER continued

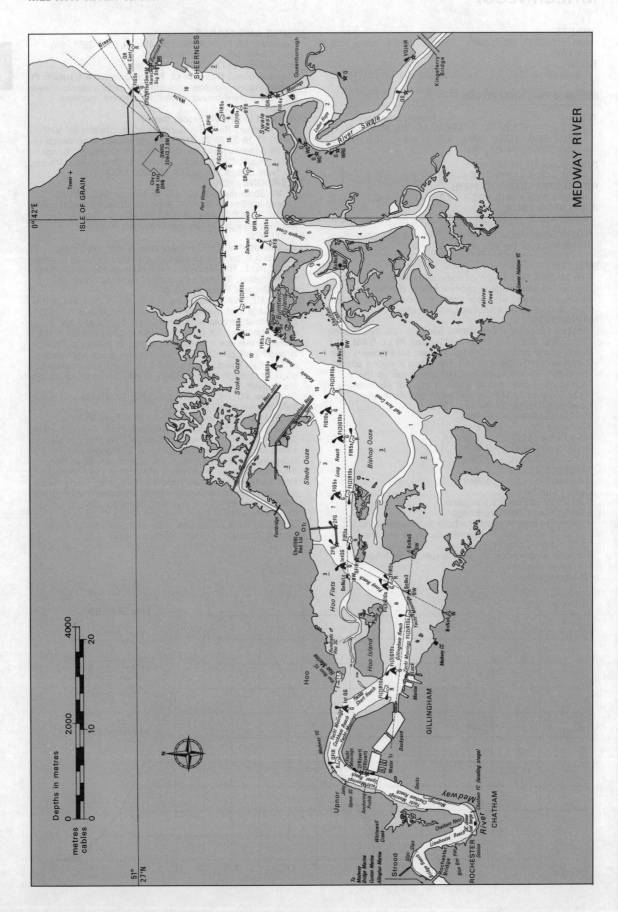

AREA 3—SE England

ENGLAND, EAST COAST — SHEERNESS
Lat 51°27′ N   Long 0°45′ E
TIMES AND HEIGHTS OF HIGH AND LOW WATERS
YEAR 1989

TIME ZONE UT(GMT)
For Summer Time add ONE hour in non-shaded areas

JANUARY

Day	Time	m	Day	Time	m
1 Su	0604 / 1219 / 1848	4.8 / 1.2 / 4.7	16 M	0015 / 0643 / 1310 / 1940	1.4 / 5.1 / 1.1 / 4.9
2 M	0032 / 0704 / 1324 / 1949	1.7 / 4.7 / 1.3 / 4.7	17 Tu	0127 / 0759 / 1429 / 2053	1.5 / 5.0 / 1.3 / 4.9
3 Tu	0144 / 0812 / 1436 / 2054	1.7 / 4.7 / 1.4 / 4.8	18 W	0254 / 0919 / 1550 / 2204	1.6 / 5.0 / 1.3 / 5.0
4 W	0258 / 0919 / 1541 / 2155	1.6 / 4.8 / 1.3 / 4.9	19 Th	0419 / 1033 / 1657 / 2308	1.3 / 5.1 / 1.2 / 5.2
5 Th	0402 / 1020 / 1635 / 2249	1.5 / 4.9 / 1.2 / 5.1	20 F	0527 / 1133 / 1750	1.0 / 5.3 / 1.1
6 F	0458 / 1115 / 1725 / 2340	1.3 / 5.1 / 1.1 / 5.3	21 Sa	0000 / 0622 / 1224 / 1832	5.3 / 0.8 / 5.5 / 1.1 ○
7 Sa ●	0549 / 1205 / 1810	1.1 / 5.3 / 1.1	22 Su	0043 / 0707 / 1306 / 1909	5.4 / 0.7 / 5.6 / 1.0
8 Su	0027 / 0639 / 1250 / 1853	5.4 / 0.9 / 5.5 / 1.0	23 M	0121 / 0747 / 1344 / 1941	5.5 / 0.6 / 5.6 / 1.0
9 M	0109 / 0730 / 1335 / 1937	5.5 / 0.8 / 5.7 / 0.9	24 Tu	0157 / 0820 / 1419 / 2012	5.6 / 0.6 / 5.6 / 0.9
10 Tu	0151 / 0819 / 1420 / 2020	5.6 / 0.6 / 5.8 / 0.9	25 W	0229 / 0851 / 1451 / 2042	5.6 / 0.6 / 5.6 / 0.9
11 W	0233 / 0907 / 1505 / 2105	5.6 / 0.5 / 5.8 / 0.9	26 Th	0300 / 0919 / 1522 / 2110	5.5 / 0.7 / 5.4 / 0.9
12 Th	0317 / 0952 / 1552 / 2148	5.6 / 0.5 / 5.7 / 0.9	27 F	0329 / 0946 / 1555 / 2139	5.5 / 0.7 / 5.4 / 1.0
13 F	0402 / 1035 / 1640 / 2231	5.5 / 0.5 / 5.5 / 1.0	28 Sa	0400 / 1013 / 1627 / 2209	5.4 / 0.8 / 5.2 / 1.1
14 Sa ☽	0448 / 1119 / 1732 / 2319	5.5 / 0.7 / 5.3 / 1.2	29 Su	0434 / 1045 / 1704 / 2242	5.2 / 1.0 / 5.0 / 1.3
15 Su	0540 / 1208 / 1831	5.3 / 0.9 / 5.1	30 M ☾	0511 / 1120 / 1747 / 2326	5.0 / 1.2 / 4.8 / 1.5
			31 Tu	0600 / 1210 / 1843	4.8 / 1.4 / 4.6

FEBRUARY

Day	Time	m	Day	Time	m
1 W	0027 / 0709 / 1324 / 1957	1.7 / 4.5 / 1.6 / 4.5	16 Th	0242 / 0911 / 1535 / 2152	1.5 / 4.8 / 1.6 / 4.8
2 Th	0155 / 0832 / 1456 / 2112	1.8 / 4.5 / 1.6 / 4.6	17 F	0417 / 1028 / 1645 / 2257	1.3 / 5.0 / 1.4 / 5.1
3 F	0325 / 0949 / 1606 / 2223	1.6 / 4.7 / 1.4 / 4.8	18 Sa	0525 / 1126 / 1737 / 2346	1.0 / 5.3 / 1.2 / 5.3
4 Sa	0433 / 1057 / 1704 / 2322	1.3 / 5.0 / 1.3 / 5.1	19 Su	0612 / 1211 / 1815	0.7 / 5.5 / 1.1
5 Su	0534 / 1151 / 1756	1.1 / 5.3 / 1.1	20 M	0027 / 0652 / 1249 / 1849	5.4 / 0.6 / 5.6 / 0.9 ○
6 M ●	0012 / 0634 / 1239 / 1845	5.3 / 0.8 / 5.6 / 0.9	21 Tu	0102 / 0726 / 1323 / 1920	5.6 / 0.5 / 5.7 / 0.7
7 Tu	0057 / 0726 / 1324 / 1930	5.6 / 0.5 / 5.8 / 0.7	22 W	0134 / 0755 / 1354 / 1949	5.7 / 0.5 / 5.7 / 0.7
8 W	0138 / 0813 / 1408 / 2015	5.8 / 0.3 / 6.0 / 0.6	23 Th	0204 / 0823 / 1422 / 2018	5.7 / 0.5 / 5.7 / 0.7
9 Th	0219 / 0857 / 1450 / 2056	5.9 / 0.2 / 6.0 / 0.6	24 F	0232 / 0850 / 1451 / 2046	5.7 / 0.5 / 5.6 / 0.7
10 F	0300 / 0938 / 1534 / 2135	6.0 / 0.2 / 5.9 / 0.7	25 Sa	0258 / 0915 / 1519 / 2111	5.6 / 0.6 / 5.5 / 0.7
11 Sa	0342 / 1016 / 1617 / 2213	5.9 / 0.4 / 5.6 / 0.8	26 Su	0328 / 0941 / 1549 / 2135	5.5 / 0.8 / 5.4 / 1.0
12 Su	0426 / 1054 / 1705 / 2255	5.7 / 0.6 / 5.3 / 1.0	27 M	0357 / 1003 / 1621 / 2200	5.3 / 1.0 / 5.2 / 1.2
13 M	0516 / 1134 / 1758 / 2346	5.4 / 1.0 / 5.0 / 1.3	28 Tu ☾	0431 / 1028 / 1659 / 2234	5.1 / 1.2 / 4.9 / 1.4
14 Tu	0617 / 1231 / 1906	5.1 / 1.3 / 4.7			
15 W	0057 / 0738 / 1358 / 2029	1.5 / 4.8 / 1.6 / 4.6			

MARCH

Day	Time	m	Day	Time	m
1 W	0513 / 1109 / 1749 / 2329	4.8 / 1.5 / 4.6 / 1.6	16 Th	0036 / 0721 / 1327 / 2001	1.5 / 4.7 / 1.8 / 4.5
2 Th	0617 / 1218 / 1900	4.5 / 1.7 / 4.4	17 F	0230 / 0856 / 1511 / 2129	1.5 / 4.7 / 1.7 / 4.7
3 F	0057 / 0748 / 1411 / 2030	1.8 / 4.4 / 1.8 / 4.4	18 Sa	0400 / 1012 / 1621 / 2233	1.2 / 5.0 / 1.5 / 5.0
4 Sa	0251 / 0922 / 1536 / 2155	1.6 / 4.6 / 1.6 / 4.7	19 Su	0502 / 1105 / 1711 / 2320	0.9 / 5.3 / 1.2 / 5.3
5 Su	0410 / 1037 / 1641 / 2259	1.3 / 5.0 / 1.3 / 5.1	20 M	0546 / 1147 / 1747	0.7 / 5.5 / 1.0
6 M	0519 / 1133 / 1737 / 2351	0.9 / 5.4 / 1.0 / 5.4	21 Tu	0000 / 0622 / 1224 / 1821	5.4 / 0.6 / 5.6 / 0.9
7 Tu ●	0619 / 1221 / 1828	0.6 / 5.8 / 0.8	22 W	0034 / 0653 / 1255 / 1852	5.6 / 0.5 / 5.6 / 0.7 ○
8 W	0035 / 0710 / 1304 / 1916	5.7 / 0.3 / 6.0 / 0.6	23 Th	0104 / 0721 / 1323 / 1921	5.7 / 0.5 / 5.7 / 0.6
9 Th	0117 / 0755 / 1347 / 1958	5.9 / 0.1 / 6.1 / 0.5	24 F	0133 / 0749 / 1349 / 1952	5.7 / 0.4 / 5.7 / 0.6
10 F	0158 / 0837 / 1429 / 2039	6.1 / 0.1 / 6.0 / 0.5	25 Sa	0201 / 0818 / 1418 / 2020	5.7 / 0.5 / 5.7 / 0.7
11 Sa	0239 / 0915 / 1510 / 2117	6.1 / 0.2 / 5.9 / 0.6	26 Su	0229 / 0844 / 1446 / 2046	5.6 / 0.7 / 5.6 / 0.9
12 Su	0321 / 0952 / 1553 / 2155	6.0 / 0.4 / 5.6 / 0.7	27 M	0258 / 0908 / 1515 / 2108	5.5 / 0.9 / 5.4 / 1.0
13 M	0406 / 1026 / 1637 / 2234	5.8 / 0.8 / 5.3 / 1.0	28 Tu	0328 / 0929 / 1548 / 2131	5.3 / 1.1 / 5.2 / 1.1
14 Tu ☽	0455 / 1105 / 1729 / 2323	5.4 / 1.2 / 5.0 / 1.2	29 W	0404 / 0953 / 1624 / 2204	5.2 / 1.2 / 5.0 / 1.3
15 W	0558 / 1158 / 1835	5.0 / 1.6 / 4.6	30 Th ☾	0448 / 1034 / 1712 / 2301	4.9 / 1.5 / 4.7 / 1.5
			31 F	0551 / 1144 / 1822	4.6 / 1.7 / 4.5

APRIL

Day	Time	m	Day	Time	m
1 Sa	0032 / 0720 / 1335 / 1954	1.6 / 4.5 / 1.8 / 4.4	16 Su	0318 / 0936 / 1535 / 2155	1.1 / 5.0 / 1.5 / 4.9
2 Su	0223 / 0856 / 1505 / 2122	1.4 / 4.8 / 1.5 / 4.7	17 M	0417 / 1030 / 1627 / 2244	0.9 / 5.2 / 1.3 / 5.2
3 M	0345 / 1010 / 1612 / 2230	1.1 / 5.2 / 1.2 / 5.1	18 Tu	0502 / 1113 / 1708 / 2325	0.8 / 5.4 / 1.1 / 5.3
4 Tu	0455 / 1105 / 1711 / 2322	0.7 / 5.6 / 0.9 / 5.5	19 W	0539 / 1149 / 1744	0.7 / 5.5 / 0.9
5 W	0557 / 1157 / 1805	0.4 / 5.8 / 0.7	20 Th	0000 / 0612 / 1221 / 1819	5.5 / 0.6 / 5.6 / 0.8
6 Th ●	0008 / 0648 / 1242 / 1853	5.8 / 0.2 / 6.0 / 0.5	21 F ○	0031 / 0643 / 1249 / 1852	5.6 / 0.6 / 5.6 / 0.7
7 F	0052 / 0731 / 1324 / 1938	6.0 / 0.1 / 6.1 / 0.4	22 Sa	0100 / 0714 / 1319 / 1926	5.6 / 0.6 / 5.7 / 0.7
8 Sa	0134 / 0813 / 1405 / 2020	6.1 / 0.1 / 6.0 / 0.4	23 Su	0131 / 0745 / 1348 / 1957	5.6 / 0.7 / 5.6 / 0.8
9 Su	0218 / 0850 / 1447 / 2100	6.1 / 0.3 / 5.8 / 0.5	24 M	0202 / 0815 / 1418 / 2026	5.6 / 0.8 / 5.5 / 0.9
10 M	0301 / 0927 / 1529 / 2139	6.0 / 0.6 / 5.5 / 0.7	25 Tu	0234 / 0842 / 1449 / 2053	5.4 / 1.0 / 5.4 / 1.0
11 Tu	0349 / 1000 / 1614 / 2220	5.7 / 1.0 / 5.2 / 1.0	26 W	0310 / 0908 / 1524 / 2121	5.3 / 1.2 / 5.2 / 1.1
12 W	0441 / 1040 / 1704 / 2309	5.3 / 1.4 / 4.9 / 1.2	27 Th	0350 / 0939 / 1603 / 2200	5.2 / 1.3 / 5.0 / 1.2
13 Th	0543 / 1130 / 1805	5.0 / 1.7 / 4.6	28 F ☾	0440 / 1027 / 1655 / 2301	5.0 / 1.5 / 4.8 / 1.3
14 F ☽	0019 / 0657 / 1249 / 1924	1.4 / 4.7 / 1.9 / 4.5	29 Sa	0543 / 1137 / 1803	4.8 / 1.7 / 4.6
15 Sa	0159 / 0823 / 1426 / 2049	1.4 / 4.7 / 1.8 / 4.6	30 Su	0028 / 0703 / 1309 / 1926	1.3 / 4.8 / 1.7 / 4.7

Chart Datum: 2.90 metres below Ordnance Datum (Newlyn)

ENGLAND, EAST COAST — SHEERNESS

Lat 51°27' N   Long 0°45' E

TIME ZONE UT(GMT)
For Summer Time add ONE hour in non-shaded areas

TIMES AND HEIGHTS OF HIGH AND LOW WATERS

YEAR 1989

MAY

Time	m		Time	m
1 0201 M 0829 1432 2047	1.2 5.0 1.4 4.9	**16** Tu	0318 0941 1532 2155	1.0 5.0 1.4 5.0
2 0315 Tu 0939 1538 2155	0.9 5.3 1.2 5.2	**17** W	0407 1027 1621 2241	0.9 5.2 1.2 5.1
3 0424 W 1038 1640 2251	0.6 5.6 0.9 5.5	**18** Th	0451 1108 1705 2320	0.8 5.3 1.0 5.3
4 0527 Th 1130 1739 2342	0.4 5.8 0.7 5.8	**19** F	0530 1143 1746 2357	0.8 5.4 0.9 5.4
5 0619 F 1217 1831 ●	0.3 5.9 0.6	**20** Sa	0607 1217 1824 ○	0.8 5.5 0.9
6 0028 Sa 0706 1302 1919	6.0 0.3 5.9 0.5	**21** Su	0034 0643 1250 1900	5.4 0.8 5.5 0.8
7 0114 Su 0748 1344 2004	6.0 0.4 5.8 0.4	**22** M	0107 0717 1324 1937	5.5 0.9 5.5 0.9
8 0201 M 0827 1427 2047	6.0 0.6 5.7 0.5	**23** Tu	0144 0751 1358 2012	5.4 1.0 5.5 0.9
9 0247 Tu 0904 1511 2129	5.8 0.9 5.4 0.7	**24** W	0220 0825 1433 2050	5.4 1.1 5.4 0.9
10 0336 W 0939 1555 2212	5.6 1.2 5.2 0.9	**25** Th	0301 0900 1512 2129	5.4 1.2 5.3 1.0
11 0427 Th 1019 1642 2257	5.3 1.4 5.0 1.1	**26** F	0346 0941 1557 2216	5.3 1.3 5.1 1.0
12 0522 F 1104 1736 2354 ☾	5.0 1.6 4.8 1.2	**27** Sa	0438 1030 1649 2313	5.2 1.4 5.0 1.0
13 0624 Sa 1204 1838	4.8 1.8 4.7	**28** Su	0537 1132 1750 ☾	5.1 1.4 4.9
14 0109 Su 1323 1948	1.2 4.8 1.7 4.7	**29** M	0022 0646 1243 1859	1.0 5.0 1.5 4.9
15 0219 M 0842 1434 2058	1.1 4.9 1.6 4.8	**30** Tu	0134 0759 1357 2013	0.9 5.1 1.4 5.1
		31 W	0246 0908 1505 2122	0.8 5.3 1.2 5.3

JUNE

Time	m		Time	m
1 0355 Th 1010 1612 2223	0.7 5.5 1.0 5.5	**16** F	0404 1021 1626 2242	1.0 5.1 1.2 5.1
2 0459 F 1105 1715 2319	0.6 5.6 0.9 5.5	**17** Sa	0452 1108 1713 2329	1.0 5.2 1.1 5.2
3 0556 Sa 1157 1814 ●	0.6 5.7 0.7	**18** Su	0536 1150 1758	1.0 5.3 1.0
4 0012 Su 0643 1245 1906	5.7 0.6 5.7 0.6	**19** M	0011 0618 1231 1842 ○	5.3 1.0 5.4 1.0
5 0103 M 0727 1330 1954	5.8 0.7 5.7 0.5	**20** Tu	0053 0656 1309 1926	5.3 1.0 5.4 0.9
6 0151 Tu 0808 1413 2039	5.8 0.8 5.6 0.5	**21** W	0133 0735 1348 2009	5.4 1.0 5.4 0.8
7 0237 W 0846 1456 2121	5.7 1.1 5.4 0.7	**22** Th	0215 0816 1427 2054	5.5 1.0 5.4 0.8
8 0324 Th 0921 1538 2202	5.5 1.2 5.3 0.8	**23** F	0257 0858 1510 2141	5.5 1.0 5.4 0.7
9 0409 F 0956 1620 2240	5.4 1.3 5.2 0.9	**24** Sa	0343 0942 1553 2227	5.5 1.1 5.4 0.7
10 0454 Sa 1035 1704 2320	5.2 1.4 5.1 1.0	**25** Su	0431 1027 1640 2313	5.5 1.1 5.3 0.7
11 0542 Su 1120 1750 ☾	5.0 1.5 4.9	**26** M	0523 1118 1733 ☾	5.3 1.2 5.3
12 0010 M 0634 1215 1845	1.1 4.9 1.6 4.8	**27** Tu	0005 0622 1215 1832	0.8 5.2 1.3 5.2
13 0107 Tu 0731 1321 1948	1.5 4.8 1.6 4.8	**28** W	0104 0730 1321 1942	0.9 5.1 1.4 5.1
14 0212 W 0833 1432 2051	1.1 4.8 1.5 4.8	**29** Th	0215 0839 1436 2056	1.0 5.1 1.3 5.2
15 0311 Th 0931 1532 2150	1.1 4.9 1.4 4.9	**30** F	0329 0946 1552 2206	1.0 5.2 1.2 5.3

JULY

Time	m		Time	m
1 0440 Sa 1048 1704 2311	0.9 5.3 1.0 5.4	**16** Su	0420 1035 1645 2305	1.3 5.3 1.3 5.0
2 0539 Su 1144 1805	0.9 5.4 0.8	**17** M	0511 1127 1739 2356	1.3 5.1 1.2 5.2
3 0007 M 0629 1235 1859 ●	5.5 0.9 5.5 0.6	**18** Tu	0558 1215 1829 ○	1.3 5.3 1.0
4 0057 Tu 0713 1320 1947	5.7 0.9 5.6 0.5	**19** W	0041 0642 1257 1919	5.4 1.1 5.4 0.8
5 0144 W 0751 1401 2029	5.7 0.9 5.6 0.5	**20** Th	0123 0726 1338 2006	5.6 1.0 5.5 0.6
6 0226 Th 0827 1440 2107	5.7 1.0 5.5 0.5	**21** F	0205 0809 1419 2051	5.7 0.9 5.6 0.5
7 0305 F 0901 1517 2142	5.6 1.0 5.5 0.7	**22** Sa	0247 0853 1458 2136	5.8 0.8 5.7 0.4
8 0343 Sa 0932 1552 2213	5.5 1.1 5.4 0.7	**23** Su	0331 0935 1541 2217	5.8 0.8 5.7 0.5
9 0420 Su 1004 1627 2242	5.4 1.1 5.3 0.8	**24** M	0416 1016 1623 2258	5.7 0.9 5.6 0.6
10 0458 M 1040 1705 2318	5.2 1.2 5.2 0.9	**25** Tu	0504 1058 1711 2340 ☾	5.5 1.0 5.5 0.8
11 0539 Tu 1120 1749	5.0 1.4 5.0	**26** W	0557 1147 1807	5.2 1.3 5.3
12 0001 W 0627 1211 1843	1.1 4.9 1.6 4.8	**27** Th	0034 0700 1250 1919	1.0 5.0 1.4 5.0
13 0059 Th 0724 1319 1949	1.3 4.7 1.7 4.6	**28** F	0147 0813 1415 2040	1.3 4.9 1.5 5.0
14 0211 F 0830 1437 2100	1.4 4.7 1.7 4.6	**29** Sa	0312 0929 1545 2202	1.3 5.0 1.4 5.1
15 0322 Sa 0935 1548 2207	1.4 4.8 1.5 4.8	**30** Su	0428 1040 1702 2309	1.3 5.1 1.0 5.3
		31 M	0529 1137 1804 ●	1.2 5.3 0.8

AUGUST

Time	m		Time	m
1 0004 Tu 0617 1225 1852 ●	5.5 1.1 5.5 0.6	**16** W	0539 1156 1815	1.2 5.3 0.9
2 0049 W 0657 1306 1934	5.6 1.0 5.6 0.5	**17** Th	0024 0627 1239 1906 ○	5.6 1.0 5.6 0.6
3 0130 Th 0733 1342 2011	5.7 0.9 5.6 0.5	**18** F	0106 0713 1320 1952	5.8 0.8 5.8 0.4
4 0206 F 0805 1416 2043	5.7 0.8 5.7 0.5	**19** Sa	0148 0757 1359 2036	6.0 0.7 5.9 0.3
5 0240 Sa 0836 1449 2112	5.7 0.9 5.7 0.6	**20** Su	0229 0839 1439 2118	6.0 0.7 6.0 0.3
6 0312 Su 0905 1519 2139	5.6 1.0 5.6 0.7	**21** M	0310 0918 1519 2156	5.9 0.7 6.0 0.4
7 0343 M 0932 1550 2204	5.5 1.0 5.6 0.8	**22** Tu	0353 0956 1602 2233	5.8 0.7 5.9 0.6
8 0416 Tu 1002 1623 2233	5.4 1.1 5.5 0.9	**23** W	0438 1035 1648 2311 ☾	5.5 1.0 5.6 1.0
9 0449 W 1033 1659 2308	5.2 1.3 5.1 1.2	**24** Th	0529 1123 1746	5.2 1.3 5.2
10 0530 Th 1113 1744 2353	4.9 1.5 4.8 1.5	**25** F	0001 0632 1228 1902	1.3 4.7 1.5 4.9
11 0621 F 1210 1848	4.7 1.8 4.5	**26** Sa	0120 0751 1406 2033	1.6 4.7 1.6 4.8
12 0102 Sa 0730 1334 2012	1.7 4.5 1.9 4.4	**27** Su	0300 0917 1545 2159	1.7 4.8 1.3 5.1
13 0237 Su 0850 1510 2135	1.8 4.5 1.7 4.6	**28** M	0419 1028 1658 2302	1.5 5.0 1.0 5.4
14 0350 M 1004 1620 2244	1.6 4.7 1.4 4.9	**29** Tu	0515 1123 1753 2351	1.3 5.2 0.7 5.6
15 0448 Tu 1106 1720 2337	1.4 5.1 1.1 5.3	**30** W	0558 1207 1835	1.1 5.5 0.6
		31 Th	0032 0634 1243 1910 ●	5.7 1.0 5.6 0.6

Chart Datum: 2.90 metres below Ordnance Datum (Newlyn)

AREA 3—SE England

ENGLAND, EAST COAST — SHEERNESS
Lat 51°27′ N Long 0°45′ E
TIMES AND HEIGHTS OF HIGH AND LOW WATERS

YEAR 1989

TIME ZONE UT(GMT)
For Summer Time add ONE hour in non-shaded areas

SEPTEMBER		OCTOBER		NOVEMBER		DECEMBER	
Time m	Time m	Time m	Time m	Time m	Time m	Time m	Time m
1 0107 5·7 / 0706 0·9 / F 1317 5·7 / 1941 0·5	**16** 0042 6·0 / 0650 0·7 / Sa 1253 6·0 / 1930 0·3	**1** 0107 5·7 / 0704 0·8 / Su 1316 5·8 / 1931 0·6	**16** 0057 6·1 / 0712 0·6 / M 1307 6·2 / 1947 0·3	**1** 0130 5·7 / 0737 0·9 / W 1347 5·6 / 1952 0·9	**16** 0204 5·8 / 0825 0·6 / Th 1423 5·9 / 2042 0·9	**1** 0140 5·5 / 0751 1·0 / F 1404 5·4 / 2001 1·1	**16** 0234 5·6 / 0901 0·6 / Sa 1501 5·7 / 2101 1·1
2 0138 5·8 / 0737 0·8 / Sa 1348 5·8 / 2009 0·5	**17** 0124 6·1 / 0735 0·6 / Su 1334 6·1 / 2013 0·2	**2** 0134 5·8 / 0735 0·8 / M 1344 5·8 / 1959 0·6	**17** 0140 6·1 / 0757 0·6 / Tu 1351 6·2 / 2026 0·5	**2** 0159 5·6 / 0806 1·0 / Th 1418 5·4 / 2020 1·1	**17** 0247 5·6 / 0910 0·8 / F 1514 5·7 / 2119 1·2	**2** 0213 5·4 / 0826 1·0 / Sa 1442 5·4 / 2034 1·2	**17** 0317 5·5 / 0943 0·7 / Su 1548 5·5 / 2138 1·3
3 0209 5·8 / 0806 0·8 / Su 1418 5·8 / 2037 0·6	**18** 0205 6·1 / 0818 0·6 / M 1413 6·2 / 2053 0·3	**3** 0202 5·7 / 0805 0·9 / Tu 1413 5·7 / 2026 0·8	**18** 0222 5·9 / 0839 0·7 / W 1436 6·1 / 2104 0·7	**3** 0230 5·4 / 0833 1·2 / F 1453 5·3 / 2047 1·3	**18** 0334 5·3 / 0953 0·9 / Sa 1604 5·5 / 2157 1·4	**3** 0250 5·3 / 0904 1·1 / Su 1524 5·3 / 2112 1·3	**18** 0359 5·3 / 1021 0·9 / M 1633 5·3 / 2213 1·4
4 0237 5·7 / 0834 0·8 / M 1446 5·7 / 2103 0·7	**19** 0246 6·0 / 0857 0·7 / Tu 1456 6·1 / 2129 0·5	**4** 0230 5·6 / 0830 1·0 / W 1443 5·5 / 2050 1·0	**19** 0305 5·7 / 0919 0·8 / Th 1524 5·8 / 2141 1·1	**4** 0303 5·3 / 0901 1·3 / Sa 1531 5·2 / 2118 1·4	**19** 0421 5·1 / 1040 1·1 / Su 1658 5·2 / 2241 1·6	**4** 0332 5·2 / 0946 1·1 / M 1610 5·2 / 2156 1·4	**19** 0442 5·2 / 1101 1·0 / Tu 1718 5·1 / ☾ 2254 1·5
5 0305 5·6 / 0901 1·0 / Tu 1515 5·6 / 2127 0·9	**20** 0328 5·7 / 0936 0·8 / W 1541 5·9 / 2206 0·9	**5** 0258 5·5 / 0854 1·2 / Th 1514 5·3 / 2114 1·2	**20** 0350 5·4 / 1003 1·0 / F 1617 5·4 / 2219 1·4	**5** 0341 5·1 / 0938 1·4 / Su 1617 5·0 / 2200 1·6	**20** 0513 5·0 / 1133 1·2 / M 1757 5·0 / ☾ 2336 1·8	**5** 0419 5·1 / 1037 1·2 / Tu 1704 5·1 / 2249 1·5	**20** 0527 5·1 / 1143 1·1 / W 1808 4·9 / 2343 1·6
6 0335 5·4 / 0925 1·2 / W 1545 5·4 / 2150 1·1	**21** 0413 5·4 / 1017 1·1 / Th 1630 5·5 / 2244 1·2	**6** 0329 5·3 / 0917 1·3 / F 1548 5·1 / 2138 1·4	**21** 0441 5·0 / 1052 1·3 / Sa 1718 5·1 / ☾ 2308 1·7	**6** 0428 4·9 / 1031 1·5 / M 1715 4·8 / ☾ 2301 1·7	**21** 0611 4·9 / 1241 1·3 / Tu 1902 4·9	**6** 0512 5·0 / 1137 1·2 / W 1804 5·0 / ☾ 2356 1·5	**21** 0618 4·9 / 1236 1·2 / Th 1903 4·8
7 0406 5·2 / 0950 1·5 / Th 1619 5·1 / 2217 1·3	**22** 0502 5·1 / 1105 1·3 / F 1730 5·1 / ☾ 2333 1·6	**7** 0404 5·0 / 0948 1·5 / Sa 1630 4·9 / 2216 1·7	**22** 0540 4·8 / 1157 1·4 / Su 1829 4·9	**7** 0529 4·8 / 1150 1·5 / Tu 1827 4·8	**22** 0045 1·8 / 0719 4·8 / W 1351 1·3 / 2012 4·9	**7** 0615 5·0 / 1248 1·1 / Th 1914 5·0	**22** 0045 1·7 / 0719 4·8 / F 1341 1·3 / 2005 4·7
8 0441 5·0 / 1023 1·5 / F 1659 4·8 / ☽ 2255 1·6	**23** 0605 4·8 / 1212 1·5 / Sa 1849 4·8	**8** 0449 4·8 / 1038 1·6 / Su 1727 4·6 / ☽ 2318 1·9	**23** 0017 2·0 / 0653 4·7 / M 1330 1·4 / 1949 4·8	**8** 0025 1·8 / 0645 4·7 / W 1321 1·4 / 1949 4·9	**23** 0201 1·7 / 0829 4·8 / Th 1453 1·2 / 2115 5·0	**8** 0109 1·5 / 0727 5·0 / F 1359 1·0 / 2026 5·1	**23** 0158 1·7 / 0826 4·7 / Sa 1447 1·3 / 2108 4·8
9 0527 4·7 / 1113 1·7 / Sa 1757 4·5 / 2358 1·9	**24** 0052 1·9 / 0726 4·6 / Su 1358 1·5 / 2020 4·8	**9** 0553 4·5 / 1204 1·7 / M 1850 4·5	**24** 0151 1·9 / 0815 4·7 / Tu 1451 1·2 / 2105 5·0	**9** 0152 1·7 / 0806 4·9 / Th 1440 1·1 / 2104 5·2	**24** 0305 1·6 / 0931 5·0 / F 1546 1·0 / 2207 5·2	**9** 0222 1·4 / 0840 5·2 / Sa 1511 0·9 / 2132 5·3	**24** 0310 1·6 / 0932 4·8 / Su 1545 1·2 / 2206 4·9
10 0634 4·4 / 1238 1·9 / Su 1924 4·4	**25** 0237 1·9 / 0854 4·7 / M 1531 1·2 / 2142 5·1	**10** 0059 2·0 / 0720 4·5 / Tu 1357 1·6 / 2025 4·7	**25** 0305 1·7 / 0925 5·0 / W 1553 1·0 / 2204 5·3	**10** 0303 1·4 / 0918 5·2 / F 1546 0·8 / 2204 5·5	**25** 0359 1·4 / 1021 5·2 / Sa 1631 1·0 / 2251 5·3	**10** 0331 1·2 / 0946 5·4 / Su 1619 0·8 / 2233 5·5	**25** 0409 1·4 / 1030 4·9 / M 1635 1·2 / 2254 5·1
11 0145 2·0 / 0802 4·4 / M 1432 1·8 / 2101 4·6	**26** 0353 1·6 / 1004 5·1 / Tu 1637 0·9 / 2240 5·4	**11** 0236 1·8 / 0850 4·8 / W 1518 1·2 / 2141 5·1	**26** 0402 1·4 / 1020 5·2 / Th 1642 0·9 / 2252 5·5	**11** 0403 1·1 / 1017 5·6 / Sa 1649 0·6 / 2259 5·8	**26** 0445 1·2 / 1105 5·3 / Su 1712 0·9 / 2329 5·4	**11** 0438 1·0 / 1048 5·6 / M 1722 0·7 / 2327 5·6	**26** 0459 1·3 / 1118 5·1 / Tu 1720 1·2 / 2337 5·2
12 0317 1·8 / 0931 4·7 / Tu 1552 1·4 / 2216 5·0	**27** 0447 1·3 / 1057 5·4 / W 1726 0·7 / 2326 5·6	**12** 0343 1·4 / 0959 5·2 / Th 1623 0·9 / 2240 5·6	**27** 0447 1·2 / 1104 5·4 / F 1720 0·8 / 2330 5·5	**12** 0502 0·9 / 1111 5·8 / Su 1746 0·5 / 2349 5·9	**27** 0527 1·1 / 1143 5·4 / M 1749 0·9 / ○	**12** 0542 0·8 / 1144 5·8 / Tu 1815 0·7 / ○	**27** 0544 1·2 / 1201 5·2 / W 1800 1·2
13 0419 1·4 / 1037 5·1 / W 1655 1·0 / 2312 5·4	**28** 0527 1·2 / 1139 5·5 / Th 1803 0·7 / ○	**13** 0440 1·1 / 1054 5·6 / F 1723 0·6 / 2329 5·9	**28** 0525 1·1 / 1139 5·5 / Sa 1753 0·8 / ○	**13** 0558 0·7 / 1200 6·0 / M 1836 0·5 / ○	**28** 0003 5·5 / 0605 1·0 / Tu 1219 5·4 / 1822 0·9	**13** 0018 5·7 / 0638 0·7 / W 1236 5·9 / 1902 0·7	**28** 0017 5·3 / 0625 1·2 / Th 1241 5·3 / 1836 1·1
14 0513 1·1 / 1127 5·4 / Th 1753 0·7 / 2358 5·7	**29** 0004 5·7 / 0601 1·0 / F 1215 5·6 / 1835 0·6	**14** 0533 0·8 / 1140 5·8 / Sa 1817 0·4 / ○	**29** 0004 5·6 / 0600 0·9 / Su 1214 5·6 / 1824 0·7	**14** 0034 6·0 / 0650 0·6 / Tu 1248 6·1 / 1921 0·5	**29** 0035 5·5 / 0642 1·0 / W 1253 5·4 / 1856 1·0	**14** 0106 5·7 / 0730 0·6 / Th 1327 5·9 / 1945 0·8	**29** 0055 5·3 / 0706 1·0 / F 1319 5·4 / 1913 1·1
15 0604 0·9 / 1212 5·7 / F 1845 0·4 / ○	**30** 0038 5·7 / 0634 0·9 / Sa 1246 5·7 / 1903 0·6	**15** 0015 6·0 / 0624 0·7 / Su 1225 6·1 / 1903 0·3	**30** 0034 5·6 / 0634 0·9 / M 1245 5·6 / 1855 0·7	**15** 0119 5·9 / 0738 0·6 / W 1335 6·1 / 2002 0·7	**30** 0107 5·5 / 0716 1·0 / Th 1328 5·4 / 1928 1·1	**15** 0151 5·6 / 0816 0·5 / F 1415 5·8 / 2025 1·0	**30** 0131 5·3 / 0747 0·9 / Sa 1357 5·5 / 1951 1·0
			31 0102 5·7 / 0706 0·9 / Tu 1316 5·6 / 1924 0·8				**31** 0208 5·4 / 0829 0·8 / Su 1436 5·5 / 2030 1·0

Chart Datum: 2·90 metres below Ordnance Datum (Newlyn)

MEDWAY RIVER continued

MARINAS (from seaward)
Mariners Farm Boatpark Tel. 33179; **Gillingham Marina** (250+12 visitors), Tel. 54386, Slip, P, D, FW, ME, El, Sh, CH, V, AC, Bar, BH (20 ton), C (1 ton), Gas, Gaz; Access W basin HW∓2, E basin (via lock), HW∓4½; **Medway Pier Marina** Tel. 51113, D, FW, BY, C (6 ton), Slip; **Hoo Marina** (245) Tel. 250311, FW, Sh, C (20 ton), ME, SM, AC, CH, El, Gas, Gaz; Access W basin HW∓1½, E basin (via lock) HW∓2; **Medway Bridge Marina** (160+15 visitors), Tel. 43576, Slip, D, P, FW, ME, El, Sh, C (3 ton), BH (10 ton), Gas, Gaz, R, SM, AC, CH, V, Bar; **Cuxton Marina** (150+some visitors), Tel. 721941, Slip, FW, ME, El, Sh, BH (12 ton), AC, CH; **Elmhaven Marina** (60) Tel. 240489, Slip, FW, ME, El, Sh, C, AC; **Allington Lock Marina** (120), Tel. Maidstone 52864, CH, ME, El, Sh, P, D, Slip, C (10 ton), FW, Gas, Gaz; (non-tidal). Note: Old HM Dockyard basins are being converted to Chatham Maritime Marina, Tel. 815081, due to open 1989.

Yacht Clubs
Chatham YC Tel. 723051; **Hoo Ness YC** Tel. 250052; **Medway Cruising Club**; **Medway Motor Cruising Club** Tel. 827194; **Medway Motor YC** Tel. 389856; **Medway YC** Tel. 718399; **Rochester CC** Tel. 41350; **Strood YC** Tel. 718261.

Other Facilities
Gransden Marine Tel. 826770, CH, ACA; **Letley Moorings** (Sufference Wharf), Tel. 814429, ME, El, Sh, dry dock, AB; **A + B Textiles** Tel. 579686, SM; **Invicta Services** Tel. 574191, Sh; **Cabin Yacht Stores** Tel. 718020, CH; **Dave Elliot** Tel. 408160, SM; **OC Diesel Refuelling Barge** (Ship pier, Rochester), Tel. 813773, D, CH;
Towns — all facilities R, V, Lau, PO, Rly, Air (Lydd or Gatwick).

AGENTS NEEDED
There are a number of ports where we need agents, particularly in France.
ENGLAND   Swale, Havengore, Berwick.
SCOTLAND Firth of Forth, Scrabster, Mallaig, Loch Sunart, Loch Aline.
IRELAND   Kilrush, Wicklow, Westport/Clew Bay.
FRANCE   Arcachon, Seudre R, Ile d'Oleron, Rochfort, Ile de Re, St. Giles-Croix-de-Vie, Ile d'Yeu, Pouliguen, Le Croisic, La Forêt, Ile de Bréhat.
GERMANY   Norderney, Dornumer-Accumersiel.
If you are interested in becoming our agent for any of the above, please write to the editors and get your free copy of the Almanac every year. You do not have to be resident in a port to be the agent, but at least a fairly regular visitor.

AREA 3—SE England 297

RIVER THAMES 10-3-20
London

CHARTS
Admiralty 3319, 2484, 3337, 2151, 1185; Stanford 5; Imray C2, C1; OS 176, 177, 178. More details obtainable from *'Nicholsons Guide to the Thames'*, from PLA publications and *'London Waterways Guide'* (Imray).

TIDES
+0245 Dover; ML 3.7; Duration 0555; Zone 0 (GMT).

Standard Port LONDON BRIDGE (→)

Times				Height (metres)			
HW		LW		MHWS	MHWN	MLWN	MLWS
0300	0900	0400	1100	7.1	5.8	1.6	0.5
1500	2100	1600	2300				

Differences WOOLWICH (GALLIONS POINT)
−0020 −0020 −0025 −0040 −0.1 −0.1 +0.1 0.0
CHELSEA BRIDGE
+0020 +0015 +0055 +0100 −0.8 −0.7 −0.6 −0.3
RICHMOND LOCK
+0100 +0055 +0325 +0305 −2.1 −2.2 −1.4 −0.3

NOTE: London Bridge is a Standard Port and tidal predictions for each day of the year are given below. The river is tidal up to Teddington Lock with half-tide lock at Richmond. When the Thames Barrier is closed, water levels will vary considerably from predictions.

TIDES — TIME DIFFERENCES ON LONDON BRIDGE

Place	MHWS	MHWN	MLWN	MLWS
Teddington Lock	+0106	+0056	—	—
Richmond Lock	+0106	+0056	+0325	+0305
Chiswick Bridge	+0049	+0044	+0235	+0224
Hammersmith Bridge	+0038	+0037	+0200	+0156
Putney Bridge	+0032	+0030	+0138	+0137
Battersea Bridge	+0023	+0020	+0109	+0110
Chelsea Bridge	+0020	+0016	+0056	+0059
Westminster Bridge	+0012	+0011	+0031	+0035
London Bridge	0000	0000	0000	0000
Surrey Comm Dock				
Greenland Entrance	−0010	−0008	−0013	−0015
Millwall Dock Entrance	−0010	−0008	−0014	−0016
Deptford Creek	−0012	−0011	−0018	−0021
Greenwich Pier	−0014	−0012	−0020	−0023
India and Millwall Dock Entrance	−0018	−0015	−0026	−0029
Royal Victoria Dock Entrance	−0021	−0018	−0031	−0025
Woolwich Ferry	−0028	−0024	−0042	−0047
Royal Albert Dock Entrance	−0029	−0024	−0043	−0050
Southern Outfall (below Crossness)	−0030	−0025	−0048	−0056
Coldharbour Point	−0037	−0030	−0053	−0103
Stoneness Lighthouse	−0048	−0037	−0059	−0114
Broadness Lighthouse	−0052	−0040	−0101	−0119
Tilburyness	−0056	−0042	−0103	−0123
Gravesend Town Pier	−0059	−0044	−0106	−0125

SHELTER
Very good shelter in many places above the Thames Barrier. In Bow Creek (Bugsby's Reach), in Deptford Creek (HW∓2), Regent's Canal Entrance (Lower Pool), St Katharine's Yacht Haven (see 10.3.23), Lambeth Pier (PLA), Chelsea Yacht & Boat Co (Cheyne Walk), Cadogan Pier (PLA), Hurlingham YC, Edwin Phelps (BY, Putney), Walter See (Hammersmith), Hammersmith Pier, Chiswick Quay Marina, Auto Marine Services (Chiswick), Grand Union Canal ent (Brentford), Howlett's BY (Twickenham) and Tough's BY (Teddington) as well as numerous Draw Docks belonging to the PLA.

NAVIGATION
River users are advised to read the Port of London River Bye Laws and *Pleasure Users Guide to the Thames*, obtainable from Thames House, St Andrews Road, Tilbury, Essex, RM18 7JH (Tel. Tilbury 3444 Ext 584). When proceeding up or down river, unless impracticable, keep to the starboard side of mid-channel. Boats approaching a bridge against the tide give way to those approaching with the tide. However, vessels over 40m always have priority. Below Wandsworth there is no speed limit. Above, limit is 8 kn.

A triangle of three red discs, or lights, hanging below an arch apex down indicates that that arch is closed. The Thames Tidal Barrier is in the centre of Woolwich Reach and consists of 9 piers between which rotating gates can form a barrier. The spans between piers are designated 'A' to 'K' from South to North. Spans A and from H to K inclusive are not navigable. The main navigational spans are C, D, E and F. Green lights forming arrows, point towards the span in use and clear for traffic. Red lights forming a St Andrews Cross each side of a span indicate that it is closed. Lights may be exhibited from: Thamesmead, False Point, Blackwall Point and Brunswick Wharf. These will indicate: Y flashing — Proceed with extreme caution, barrier about to close; R flashing — all vessels stop, barrier being closed. The Thames Barrier Navigation Centre, call sign *Woolwich Radio*, is the communications centre for that part of the river to the W of Crayfordness. The Traffic Controller regulates all traffic through the barrier, controlling it from Margaret Ness to Blackwall Point. Radio VHF Chs 14 and 22. General shipping information is not available from this centre but from the Thames Navigation Service (Woolwich Radio or Gravesend Radio or Tel. Gravesend 67684). All vessels with VHF (vessels without VHF see below) intending to go through the barrier must inform Woolwich Radio on Ch 14 of their ETA when passing Crayfordness inward bound or Tower Bridge outward bound. Then, when passing Margaret Ness Point inward or Blackwall Point outward obtain permission from the Controller to proceed. Yachts without VHF should proceed with caution, normally using span G (N side) if bound up river or span B (S side) if bound down river, and keeping clear of larger vessels.

Note however that depths in spans B and G may only be about 1.5m at LW springs. If possible ring Barrier Control 855 0315, before arrival. The barrier is completely closed monthly for test. Individual spans are closed weekly. When barrier is closed, water levels will differ considerably from predictions.

The Half tide lock at Richmond is operated to maintain a min depth of 1.7m between Teddington and Richmond bridges.

Name of Bridge	Distance above London Bridge Sea Miles	Headway of Centre Span above Chart Datum (m)	MHWS (m)
Richmond	13.94	7.9	5.3
Richmond Railway	13.64	7.9	5.3
Twickenham	13.61	8.5	5.9
Richmond Footbridge	13.45	9.7	4.8
Kew	11.31	10.6	5.3
Kew Railway	10.95	10.9	5.6
Chiswick	10.20	12.2	6.9
Barnes Railway	9.53	10.9	5.4
Hammersmith	7.95	9.4	3.7
Putney	6.44	11.4	5.5
Fulham Railway	6.30	12.8	5.9
Wandsworth	5.46	11.9	5.8
Battersea Railway	4.85	12.2	6.1
Battersea	4.29	11.7	5.5
Albert	4.06	11.1	4.9
Chelsea	3.41	12.9	6.6
Victoria Railway	3.30	12.3	6.0
Vauxhall	2.52	12.1	5.6
Lambeth	2.11	13.1	6.5
Westminster	1.72	12.2	5.4
Charing Cross Rly	1.38	13.8	7.0
Waterloo	1.13	15.3	8.5
Blackfriars	0.64	14.0	7.1
Southwark	0.25	14.3	7.4
Cannon St. Railway	0.16	14.0	7.1
London Bridge	0	16.0	8.9
Tower	below 0.49	15.7	8.6

RIVER THAMES continued

Note: The tidal Thames is divided by the PLA into two sections
Upper Section – Teddington to Cross Ness
Lower Section – Cross Ness to the Sea

LIGHTS AND MARKS
Margaret Ness or Tripcock Pt Fl(2) W 5s 11m 8M. Tower Bridge sounds horn 20s or gong 30s when bascules open for shipping. When Richmond half-tide barrier is shut, a triangle of red discs (red lights at night) is hung below the centre span of the footbridge.

RADIO TELEPHONE
Call *Woolwich Radio* VHF Ch **14** 16 22 (H24) (for Thames Barrier control)
St Katharines Yacht Haven VHF Ch M (HW−2 to HW+1½; 0600-2030 LT in summer, 0800-1800 LT in winter)
Brentford Dock Marina VHF Ch 14 16
Gravesend Radio VHF Ch **12** 14 16 18 20 (H24)
Information broadcasts every H+15 and H+45 on Ch 12 by Gravesend, and every H+15 and H+45 on Ch 14 by Woolwich. These are repeated by North Foreland Radio Ch 26, Thames Radio Ch 02, Hastings Radio Ch 07 and Orfordness Radio Ch 62 at the end of first silence period after receipt. PLA Patrol Launches (call: *Thames Patrol*) VHF Ch 12 14 16 06.

TELEPHONE (01)
Hr Mr (Upper Section) 481 0720; (Lower Section) Gravesend 67684; River Police 488 5212; Richmond Lock 940 0634; Tower Bridge Master 407 0922; HMC Freefone Customs Yachts: RT 283 8633 Ext 570 (or night 626 3524); Thames Navigation Service 476 7801; Weather 836 4311; Marinecall 0898 500 455; Hosp 987 7011.

FACILITIES (Letters in brackets refer to chartlets)
TEDDINGTON
Tough Shipyards (A) Tel. 977 4494, BY, CH, AC, ME, Gas, C (6 ton), Sh, El, FW; **Swan Island Harbour (B)** Tel. 892 2861, D, M, ME, El, Sh, FW.
TWICKENHAM
Impala Marine Tel. 892 6296, BY, ME, El, Sh; **Francis Newman** Tel. 892 8932, BY, El, ME, Sh; **Richmond Marine** Tel. 898 2771, CH.
RICHMOND
J T Howlett (C) Tel. 892 3183, BY, CH, D, Gas, M, FW; **Richmond Slipways** Tel. 892 5062, BY, Gas, ME, El, Sh.
BRENTFORD
Brentford Dock Marina (D) (80 + 10 visitors) Tel. 568 0287 (VHF Ch 14, 16), AC, Bar, CH, El, FW, ME, R, V; Access HW∓2½; **Thames Locks (No 101)** Tel. 560 8942, M, AB, Entrance to Grand Union Canal; **T Norris** Tel. 560 3453, CH, ME.
KEW
Robbins (Marine) Tel. 837 1392, BY, ME, El, Sh; **Auto Marine Services (E)** Tel. 994 6396, BY, ME, El, Sh; **Bason and Arnold (F)** Tel. 994 2100, CH, D, P, Gas; **Cranfield** Tel. 788 9255; SM; **Kew Pier** Tel. 940 7632, L.
CHISWICK
Chiswick Quay Marina (50) CH, D, Gas, M, ME, El, Sh; (Access HW∓2); **Walter See** Tel. 748 7738, BY, M, ME, El, Sh.
PUTNEY
Edwin Phelps Tel. 788 9391, M; **Putney Pier** Tel. 788 5140, L.
WANDSWORTH
Hurlingham YC Tel. 788 5547, M.
CHELSEA
Chelsea Harbour (G) (75 no visitors) Tel. 351 2300, AC, YC, FW, CH, V; Access HW∓1; **Chelsea Yacht & Boat Co (H)** Tel. 352 1427, M, Gas; **Cadogan Pier (I)** Tel. 352 4604, M, L.
KINGS REACH
Threestokes Marine Tel. 247 4595, CH, Gas.
POOL OF LONDON
St Katharines Yacht Haven (J) Tel. 488 2400 (see 10.3.23); **Crawleys (K)** Tel. 481 1774, (Fuel barge), P, D; **Kelvin Hughes** Tel. 709 9076, CH, ACA; **Pumpkin Marine (J)** Tel. 480 6630, CH.
LIMEHOUSE REACH
Limehouse Ship Lock Tel. 790 3444, Access HW−3½ to HW; AB, M. Entry to Regents Canal.
GREENWICH
East Greenwich Garage Tel. 858 4881, ME, P, D; **Charlton Marine** Tel. 858 1446, CH, Gas; **W H Donovan** Tel. 858 1143, BY; **Caldergate (L)** P, D; **Greenwich YC** Tel. 858 7339; **Greenwich Pier** Tel. 858 0079, L.

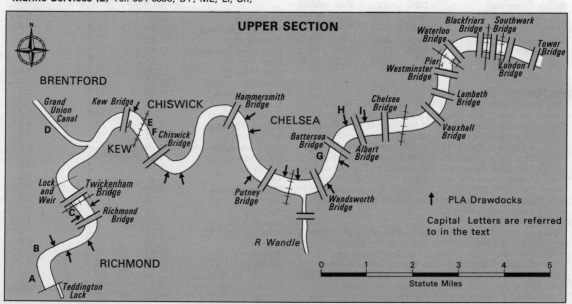

RIVER THAMES continued

AREA 3—SE England

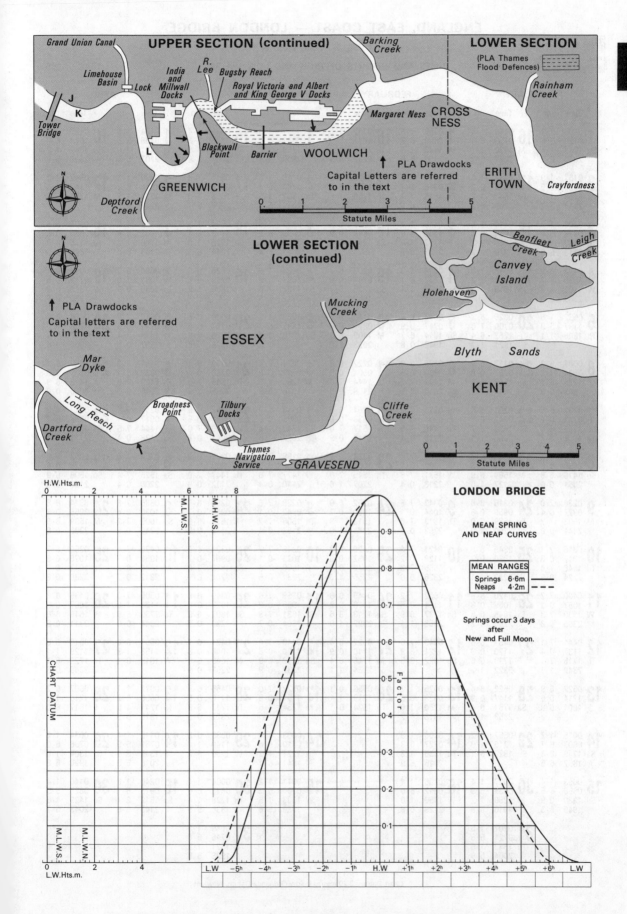

ENGLAND, EAST COAST — LONDON BRIDGE

Lat 51°30′ N Long 0°05′ W

TIMES AND HEIGHTS OF HIGH AND LOW WATERS

YEAR 1989

TIME ZONE UT (GMT) — For Summer Time add ONE hour in non-shaded areas

Chart Datum: 3.20 metres below Ordnance Datum (Newlyn)

JANUARY

Time	m	Time	m
1 Su 0106 / 0730 / 1341 / 2015	1.7 / 5.6 / 1.4 / 5.7	**16** M 0144 / 0758 / 1426 / 2057	1.4 / 6.4 / 1.1 / 6.2
2 M 0157 / 0834 / 1439 / 2114	1.9 / 5.5 / 1.5 / 5.7	**17** Tu 0253 / 0912 / 1541 / 2209	1.4 / 6.2 / 1.2 / 6.2
3 Tu 0300 / 0938 / 1556 / 2213	1.9 / 5.5 / 1.5 / 5.8	**18** W 0414 / 1034 / 1658 / 2320	1.4 / 6.1 / 1.3 / 6.2
4 W 0423 / 1038 / 1706 / 2309	1.8 / 5.7 / 1.3 / 6.0	**19** Th 0540 / 1151 / 1819	1.3 / 6.2 / 1.3
5 Th 0532 / 1137 / 1805	1.5 / 5.9 / 1.1	**20** F 0025 / 0700 / 1252 / 1921	6.3 / 1.0 / 6.4 / 1.2
6 F 0007 / 0634 / 1235 / 1903	6.2 / 1.2 / 6.2 / 0.9	**21** Sa 0117 / 0757 / 1342 / 2011	6.4 / 0.8 / 6.5 / 1.2
7 Sa 0102 / 0733 / 1328 / 1958	6.5 / 0.9 / 6.6 / 0.7	**22** Su 0201 / 0843 / 1425 / 2050	6.5 / 0.7 / 6.7 / 1.1
8 Su 0151 / 0830 / 1416 / 2051	6.8 / 0.6 / 6.9 / 0.6	**23** M 0240 / 0924 / 1504 / 2127	6.7 / 0.6 / 6.9 / 1.0
9 M 0236 / 0922 / 1503 / 2141	7.0 / 0.4 / 7.2 / 0.6	**24** Tu 0315 / 0959 / 1539 / 2200	6.8 / 0.6 / 7.0 / 1.0
10 Tu 0318 / 1012 / 1546 / 2224	7.1 / 0.3 / 7.4 / 0.6	**25** W 0349 / 1030 / 1613 / 2228	6.8 / 0.7 / 6.9 / 1.1
11 W 0400 / 1057 / 1630 / 2305	7.2 / 0.3 / 7.3 / 0.8	**26** Th 0420 / 1055 / 1648 / 2255	6.8 / 0.8 / 6.8 / 1.2
12 Th 0441 / 1137 / 1715 / 2342	7.1 / 0.4 / 7.1 / 1.1	**27** F 0451 / 1120 / 1718 / 2322	6.6 / 0.9 / 6.6 / 1.2
13 F 0522 / 1214 / 1801	6.9 / 0.6 / 6.8	**28** Sa 0522 / 1147 / 1751 / 2353	6.4 / 1.1 / 6.4 / 1.3
14 Sa 0015 / 0607 / 1219 / 1852	1.2 / 6.7 / 0.8 / 6.5	**29** Su 0557 / 1219 / 1828	6.2 / 1.0 / 6.1
15 Su 0053 / 0656 / 1330 / 1949	1.4 / 6.5 / 0.9 / 6.3	**30** M 0027 / 0635 / 1256 / 1910	1.4 / 5.9 / 1.2 / 5.9
		31 Tu 0107 / 0721 / 1342 / 2004	1.6 / 5.5 / 1.5 / 5.6

FEBRUARY

Time	m	Time	m
1 W 0158 / 0829 / 1443 / 2111	1.8 / 5.3 / 1.7 / 5.5	**16** Th 0342 / 1012 / 1624 / 2257	1.5 / 5.9 / 1.7 / 5.9
2 Th 0311 / 0946 / 1616 / 2223	1.9 / 5.4 / 1.7 / 5.6	**17** F 0525 / 1139 / 1756	1.3 / 6.1 / 1.5
3 F 0448 / 1102 / 1732 / 2337	1.7 / 5.6 / 1.4 / 5.9	**18** Sa 0008 / 0649 / 1241 / 1904	6.2 / 0.9 / 6.4 / 1.3
4 Sa 0604 / 1217 / 1838	1.3 / 6.1 / 1.1	**19** Su 0102 / 0744 / 1328 / 1952	6.4 / 0.6 / 6.7 / 1.1
5 Su 0043 / 0717 / 1314 / 1942	6.4 / 0.9 / 6.7 / 0.8	**20** M 0145 / 0827 / 1409 / 2033	6.6 / 0.5 / 6.9 / 1.0
6 M 0135 / 0820 / 1404 / 2040	6.8 / 0.5 / 7.2 / 0.5	**21** Tu 0222 / 0904 / 1444 / 2108	6.8 / 0.5 / 7.0 / 0.9
7 Tu 0220 / 0914 / 1449 / 2129	7.2 / 0.1 / 7.5 / 0.4	**22** W 0254 / 0938 / 1517 / 2141	6.9 / 0.5 / 7.0 / 0.9
8 W 0303 / 1002 / 1531 / 2213	7.4 / -0.1 / 7.6 / 0.4	**23** Th 0324 / 1006 / 1545 / 2207	6.9 / 0.6 / 7.0 / 1.0
9 Th 0342 / 1044 / 1613 / 2252	7.5 / -0.1 / 7.6 / 0.6	**24** F 0352 / 1030 / 1613 / 2231	6.9 / 0.7 / 6.9 / 1.0
10 F 0421 / 1123 / 1655 / 2326	7.4 / 0.2 / 7.3 / 0.8	**25** Sa 0420 / 1052 / 1642 / 2255	6.8 / 0.8 / 6.8 / 1.0
11 Sa 0502 / 1154 / 1737 / 2356	7.2 / 0.5 / 7.0 / 1.1	**26** Su 0449 / 1116 / 1713 / 2322	6.6 / 0.8 / 6.6 / 1.0
12 Su 0543 / 1221 / 1822	7.0 / 0.8 / 6.5	**27** M 0522 / 1146 / 1746 / 2354	6.4 / 0.9 / 6.4 / 1.1
13 M 0025 / 0629 / 1250 / 1913	1.2 / 6.6 / 1.0 / 6.2	**28** Tu 0558 / 1219 / 1824	6.1 / 1.1 / 6.1
14 Tu 0106 / 0727 / 1335 / 2015	1.3 / 6.3 / 1.3 / 6.0		
15 W 0206 / 0842 / 1450 / 2129	1.4 / 6.0 / 1.6 / 5.8		

MARCH

Time	m	Time	m
1 W 0031 / 0641 / 1259 / 1909	1.3 / 5.7 / 1.4 / 5.7	**16** Th 0134 / 0818 / 1408 / 2051	1.4 / 5.9 / 1.8 / 5.6
2 Th 0116 / 0740 / 1352 / 2013	1.6 / 5.4 / 1.8 / 5.4	**17** F 0311 / 0948 / 1555 / 2227	1.6 / 5.8 / 1.9 / 5.7
3 F 0219 / 0903 / 1527 / 2141	1.8 / 5.3 / 1.9 / 5.4	**18** Sa 0504 / 1118 / 1726 / 2344	1.3 / 6.1 / 1.6 / 6.1
4 Sa 0409 / 1037 / 1702 / 2312	1.8 / 5.6 / 1.5 / 5.8	**19** Su 0625 / 1219 / 1836	0.8 / 6.6 / 1.2
5 Su 0539 / 1157 / 1815	1.3 / 6.2 / 1.2	**20** M 0038 / 0717 / 1306 / 1926	6.5 / 0.5 / 6.9 / 1.0
6 M 0021 / 0702 / 1255 / 1926	6.4 / 0.8 / 6.9 / 0.8	**21** Tu 0120 / 0758 / 1345 / 2005	6.7 / 0.4 / 7.0 / 0.9
7 Tu 0113 / 0805 / 1342 / 2022	6.9 / 0.3 / 7.4 / 0.5	**22** W 0157 / 0835 / 1419 / 2042	6.9 / 0.5 / 7.0 / 0.8
8 W 0158 / 0856 / 1427 / 2110	7.3 / 0.0 / 7.6 / 0.4	**23** Th 0227 / 0907 / 1447 / 2112	6.9 / 0.6 / 7.0 / 0.9
9 Th 0239 / 0942 / 1508 / 2153	7.5 / -0.1 / 7.7 / 0.4	**24** F 0256 / 0934 / 1514 / 2139	6.9 / 0.7 / 6.9 / 0.8
10 F 0318 / 1023 / 1550 / 2231	7.7 / -0.1 / 7.6 / 0.5	**25** Sa 0321 / 0957 / 1539 / 2204	6.9 / 0.7 / 6.8 / 0.9
11 Sa 0359 / 1058 / 1631 / 2304	7.6 / 0.2 / 7.3 / 0.7	**26** Su 0349 / 1023 / 1607 / 2230	6.8 / 0.8 / 6.9 / 0.8
12 Su 0440 / 1126 / 1712 / 2332	7.4 / 0.4 / 6.9 / 0.9	**27** M 0421 / 1048 / 1640 / 2257	6.7 / 0.7 / 6.8 / 0.9
13 M 0522 / 1151 / 1754 / 2327	7.0 / 0.9 / 6.6 / 0.9	**28** Tu 0455 / 1110 / 1712 / 2327	6.5 / 0.9 / 6.6 / 0.9
14 Tu 0003 / 0610 / 1221 / 1841	1.0 / 6.6 / 1.2 / 6.1	**29** W 0533 / 1149 / 1750	6.3 / 1.1 / 6.2
15 W 0041 / 0706 / 1303 / 1938	1.2 / 6.2 / 1.5 / 5.8	**30** Th 0004 / 0618 / 1228 / 1836	1.1 / 6.0 / 1.4 / 5.8
		31 F 0049 / 0716 / 1321 / 1938	1.3 / 5.7 / 1.7 / 5.5

APRIL

Time	m	Time	m
1 Sa 0151 / 0839 / 1451 / 2107	1.6 / 5.5 / 2.0 / 5.5	**16** Su 0427 / 1044 / 1647 / 2309	1.4 / 6.0 / 1.7 / 5.9
2 Su 0339 / 1014 / 1631 / 2242	1.6 / 5.8 / 1.7 / 5.8	**17** M 0542 / 1147 / 1754	1.0 / 6.4 / 1.3
3 M 0512 / 1132 / 1746 / 2353	1.2 / 6.4 / 1.2 / 6.4	**18** Tu 0005 / 0635 / 1235 / 1846	6.4 / 0.6 / 6.8 / 1.0
4 Tu 0638 / 1231 / 1859	0.7 / 7.0 / 0.9	**19** W 0049 / 0719 / 1313 / 1930	6.7 / 0.5 / 6.9 / 0.9
5 W 0046 / 0742 / 1319 / 1957	7.0 / 0.3 / 7.4 / 0.6	**20** Th 0126 / 0755 / 1347 / 2006	6.8 / 0.6 / 6.9 / 0.8
6 Th 0131 / 0833 / 1402 / 2046	7.3 / 0.1 / 7.5 / 0.5	**21** F 0157 / 0829 / 1415 / 2040	6.8 / 0.7 / 6.9 / 0.9
7 F 0213 / 0917 / 1444 / 2128	7.5 / 0.1 / 7.5 / 0.4	**22** Sa 0223 / 0858 / 1440 / 2111	6.8 / 0.7 / 6.9 / 0.9
8 Sa 0254 / 0956 / 1525 / 2206	7.6 / 0.2 / 7.4 / 0.4	**23** Su 0253 / 0928 / 1508 / 2139	6.8 / 0.8 / 6.9 / 0.8
9 Su 0336 / 1030 / 1606 / 2240	7.6 / 0.4 / 7.2 / 0.5	**24** M 0324 / 0956 / 1539 / 2209	6.8 / 0.8 / 6.9 / 0.8
10 M 0420 / 1058 / 1648 / 2312	7.4 / 0.7 / 6.9 / 0.7	**25** Tu 0359 / 1026 / 1613 / 2238	6.7 / 0.7 / 6.8 / 0.8
11 Tu 0506 / 1126 / 1730 / 2344	7.0 / 1.0 / 6.5 / 0.9	**26** W 0435 / 1057 / 1649 / 2311	6.6 / 0.9 / 6.6 / 0.9
12 W 0556 / 1158 / 1815	6.6 / 1.3 / 6.1	**27** Th 0518 / 1129 / 1729 / 2349	6.5 / 1.1 / 6.4 / 0.9
13 Th 0022 / 0650 / 1241 / 1910	1.1 / 6.2 / 1.6 / 5.8	**28** F 0607 / 1210 / 1817	6.2 / 1.3 / 6.0
14 F 0113 / 0757 / 1340 / 2018	1.4 / 5.9 / 2.0 / 5.5	**29** Sa 0034 / 0706 / 1304 / 1920	1.2 / 6.0 / 1.6 / 5.7
15 Sa 0240 / 0915 / 1519 / 2143	1.6 / 5.7 / 2.1 / 5.5	**30** Su 0140 / 0823 / 1429 / 2043	1.4 / 5.9 / 1.8 / 5.7

ENGLAND, EAST COAST — LONDON BRIDGE

Lat 51°30′ N Long 0°05′ W

TIMES AND HEIGHTS OF HIGH AND LOW WATERS

TIME ZONE UT(GMT)
For Summer Time add ONE hour in non-shaded areas

YEAR 1989

AREA 3—SE England   301

	MAY				JUNE				JULY				AUGUST		
	Time m		Time m		Time m		Time m		Time m		Time m		Time m		Time m
1 M	0317 1·3 0948 6·1 1559 1·6 2209 6·0	**16** Tu	0445 1·1 1101 6·1 1705 1·5 2320 6·1	**1** Th	0519 0·8 1133 6·7 1744 1·1 2347 6·6	**16** F	0533 1·0 1140 6·1 1756 1·3	**1** Sa	0601 1·1 1208 6·4 1838 1·1	**16** Su	0540 1·2 1143 6·0 1807 1·3	**1** Tu ●	0121 6·5 0754 1·1 1342 6·5 2027 0·6	**16** W	0053 6·4 0716 0·9 1314 6·6 1955 0·7
2 Tu	0440 1·0 1102 6·6 1709 1·2 2319 6·5	**17** W	0542 0·9 1153 6·4 1758 1·2	**2** F	0636 0·8 1229 6·8 1859 1·0	**17** Sa	0003 6·1 0621 0·9 1224 6·3 1845 1·1	**2** Su	0031 6·4 0713 1·1 1304 6·4 1944 1·0	**17** M	0018 5·9 0636 1·0 1242 6·2 1909 1·0	**2** W	0208 6·7 0839 1·0 1425 6·6 2111 0·5	**17** Th ○	0142 7·0 0813 0·6 1359 7·0 2049 0·3
3 W	0601 0·7 1203 7·0 1822 1·0	**18** Th	0008 6·3 0628 0·8 1234 6·6 1845 1·0	**3** Sa ●	0042 6·7 0737 0·9 1319 6·7 1957 0·9	**18** Su	0046 6·2 0709 0·8 1307 6·5 1934 1·0	**3** M ●	0127 6·4 0806 1·2 1354 6·4 2036 0·8	**18** Tu ○	0112 6·3 0734 0·8 1333 6·5 2009 0·8	**3** Th	0249 6·9 0918 0·9 1501 6·8 2149 0·4	**18** F	0226 7·3 0904 0·4 1440 7·3 2138 0·1
4 Th	0015 6·9 0713 0·5 1253 7·2 1927 0·8	**19** F	0048 6·5 0709 0·7 1307 6·6 1927 0·9	**4** Su	0133 6·7 0825 0·9 1404 6·7 2046 0·8	**19** M ○	0130 6·4 0757 0·8 1349 6·6 2023 0·8	**4** Tu	0216 6·6 0851 1·1 1437 6·6 2122 0·6	**19** W	0201 6·7 0830 0·7 1419 6·8 2104 0·5	**4** F	0325 7·0 0952 0·9 1536 6·9 2223 0·5	**19** Sa	0307 7·5 0949 0·4 1518 7·5 2220 0·0
5 F ●	0104 7·1 0805 0·5 1338 7·2 2019 0·7	**20** Sa ○	0121 6·5 0748 0·7 1340 6·7 2006 0·9	**5** M	0222 6·9 0905 0·9 1447 6·8 2129 0·6	**20** Tu	0213 6·6 0844 0·8 1432 6·7 2112 0·7	**5** W	0301 6·8 0932 1·0 1518 6·7 2203 0·5	**20** Th	0244 7·0 0921 0·6 1501 7·0 2152 0·3	**5** Sa	0400 7·0 1023 1·0 1609 6·8 2249 0·7	**20** Su	0348 7·5 1030 0·5 1556 7·5 2259 0·2
6 Sa	0149 7·2 0849 0·5 1422 7·2 2103 0·6	**21** Su	0154 6·6 0825 0·8 1411 6·7 2044 0·8	**6** Tu	0310 7·0 0942 0·9 1531 6·8 2210 0·5	**21** W	0257 6·8 0929 0·7 1512 6·8 2157 0·6	**6** Th	0343 7·0 1009 1·0 1557 6·7 2241 0·5	**21** F	0328 7·2 1006 0·6 1541 7·1 2237 0·3	**6** Su	0433 6·9 1048 1·1 1640 6·7 2311 0·8	**21** M	0428 7·3 1104 0·8 1634 7·3 2332 0·5
7 Su	0233 7·3 0928 0·6 1503 7·1 2143 0·5	**22** M	0229 6·6 0903 0·8 1446 6·8 2122 0·8	**7** W	0356 7·1 1020 0·9 1614 6·8 2249 0·5	**22** Th	0339 6·9 1012 0·8 1553 6·8 2240 0·6	**7** F	0424 7·0 1041 1·0 1634 6·7 2312 0·7	**22** Sa	0409 7·3 1047 0·7 1619 7·1 2318 0·4	**7** M	0505 6·7 1112 1·1 1711 6·6 2334 0·9	**22** Tu	0509 6·9 1132 1·0 1715 7·0 2357 0·8
8 M	0319 7·3 1003 0·7 1546 7·0 2221 0·4	**23** Tu	0307 6·7 0939 0·8 1521 6·8 2159 0·8	**8** Th	0441 7·0 1054 1·1 1655 6·6 2325 0·7	**23** F	0421 7·0 1051 0·9 1633 6·8 2319 0·7	**8** Sa	0502 6·8 1111 1·2 1711 6·6 2340 0·9	**23** Su	0449 7·1 1122 0·9 1657 7·0 2351 0·6	**8** Tu	0537 6·5 1139 1·2 1746 6·3	**23** W	0551 6·5 1158 1·2 1758 6·7
9 Tu	0406 7·2 1035 0·7 1630 6·8 2257 0·6	**24** W	0346 6·8 1014 0·9 1559 6·8 2234 0·8	**9** F	0526 6·8 1129 1·3 1736 6·3 2358 1·0	**24** Sa	0505 6·9 1126 1·1 1712 6·7 2356 0·8	**9** Su	0540 6·6 1140 1·3 1746 6·4	**24** M	0532 6·9 1150 1·2 1736 6·8	**9** W ☾	0004 0·9 0612 6·2 1212 1·2 1825 6·0	**24** Th	0022 1·0 0639 6·2 1236 1·2 1853 6·4
10 W	0454 7·0 1108 1·1 1712 6·5 2333 0·8	**25** Th	0427 6·8 1048 1·0 1638 6·7 2309 0·8	**10** Sa	0610 6·4 1204 1·5 1818 6·1	**25** Su	0550 6·7 1200 1·2 1756 6·6	**10** M	0007 1·0 0618 6·3 1212 1·4 1825 6·1	**25** Tu ☾	0021 0·8 0617 6·6 1221 1·5 1821 6·7	**10** Th	0041 1·1 0655 5·9 1252 1·4 1912 5·6	**25** F	0103 1·2 0738 5·9 1330 1·5 2005 6·1
11 Th	0543 6·6 1143 1·4 1756 6·1	**26** F	0512 6·7 1120 1·1 1720 6·5 2346 0·9	**11** Su ☾	0035 1·2 0657 6·2 1243 1·7 1906 5·9	**26** M ☾	0034 0·8 0641 6·5 1241 1·3 1845 6·5	**11** Tu	0041 1·1 0702 6·1 1249 1·5 1913 5·9	**26** W	0052 0·9 0709 6·3 1302 1·3 1914 6·4	**11** F	0124 1·3 0744 5·6 1338 1·6 2012 5·3	**26** Sa	0206 1·5 0851 5·8 1454 1·5 2129 5·9
12 F ☽	0010 1·1 0634 6·3 1222 1·6 1845 5·8	**27** Sa	0600 6·5 1201 1·3 1808 6·3	**12** M	0119 1·3 0749 5·9 1331 1·8 2005 5·7	**27** Tu	0119 0·9 0738 6·4 1334 1·4 1945 6·4	**12** W	0121 1·2 0752 5·8 1334 1·7 2012 5·7	**27** Th	0135 1·0 0811 6·1 1401 1·3 2025 6·2	**12** Sa	0219 1·6 0846 5·4 1440 1·9 2124 5·3	**27** Su	0343 1·6 1016 5·8 1640 1·3 2301 6·1
13 Sa	0055 1·3 0730 6·0 1313 1·9 1944 5·6	**28** Su ☾	0032 1·0 0657 6·3 1253 1·5 1904 6·1	**13** Tu	0218 1·4 0850 5·8 1437 1·9 2111 5·7	**28** W	0216 0·9 0850 6·3 1442 1·4 2057 6·3	**13** Th	0213 1·4 0849 5·7 1430 1·8 2114 5·5	**28** F	0244 1·2 0922 6·1 1524 1·4 2146 6·1	**13** Su	0343 1·8 0955 5·4 1616 1·8 2238 5·4	**28** M	0516 1·5 1136 6·1 1818 0·9
14 Su	0159 1·5 0834 5·8 1426 2·1 2054 5·6	**29** M	0134 1·1 0804 6·2 1405 1·6 2016 6·1	**14** W	0339 1·4 0952 5·8 1602 1·8 2216 5·7	**29** Th	0328 0·9 0956 6·4 1557 1·3 2212 6·3	**14** F	0322 1·5 0946 5·7 1549 1·8 2214 5·5	**29** Sa	0410 1·3 1040 6·1 1652 1·4 2312 6·1	**14** M	0506 1·5 1111 5·7 1736 1·5 2356 5·9	**29** Tu	0012 6·5 0639 1·2 1238 6·5 1921 0·5
15 M	0338 1·4 0950 5·9 1559 1·9 2214 5·7	**30** Tu	0251 1·0 0919 6·3 1522 1·4 2132 6·3	**15** Th	0442 1·2 1051 6·0 1704 1·5 2313 5·9	**30** F	0442 1·2 1105 6·4 1713 1·2 2325 6·3	**15** Sa	0440 1·4 1044 5·8 1706 1·6 2316 5·7	**30** Su	0536 1·3 1153 6·2 1828 1·1	**15** Tu	0611 1·1 1221 6·1 1849 1·1	**30** W	0107 6·8 0734 0·9 1326 6·7 2011 0·3
		31 W	0406 0·8 1030 6·6 1633 1·2 2244 6·5							**31** M	0024 6·3 0656 1·2 1253 6·4 1935 0·8			**31** Th ●	0151 7·0 0818 0·8 1405 6·9 2050 0·3

Chart Datum: 3.20 metres below Ordnance Datum (Newlyn)

ENGLAND, EAST COAST — LONDON BRIDGE
Lat 51°30′ N Long 0°05′ W

TIMES AND HEIGHTS OF HIGH AND LOW WATERS

YEAR 1989

TIME ZONE UT(GMT)
For Summer Time add ONE hour in non-shaded areas

SEPTEMBER

Time	m		Time	m
1 0229 / 0856 / F 1440 / 2125	7.0 / 0.8 / 6.9 / 0.4	**16**	0201 / 0842 / Sa 1442 / 2114	7.5 / 0.4 / 7.5 / 0.0
2 0301 / 0928 / Sa 1511 / 2153	7.0 / 0.8 / 7.0 / 0.6	**17**	0243 / 0925 / Su 1451 / 2156	7.6 / 0.4 / 7.7 / 0.0
3 0332 / 0955 / Su 1539 / 2217	6.9 / 0.9 / 6.9 / 0.7	**18**	0322 / 1006 / M 1531 / 2234	7.5 / 0.5 / 7.7 / 0.2
4 0359 / 1019 / M 1606 / 2238	6.8 / 1.0 / 6.8 / 0.8	**19**	0403 / 1040 / Tu 1613 / 2305	7.3 / 0.6 / 7.5 / 0.6
5 0427 / 1041 / Tu 1635 / 2301	6.7 / 1.0 / 6.7 / 0.8	**20**	0444 / 1111 / W 1657 / 2332	6.9 / 0.8 / 7.1 / 0.9
6 0457 / 1108 / W 1709 / 2330	6.6 / 1.0 / 6.5 / 0.9	**21**	0527 / 1142 / Th 1744	6.5 / 1.0 / 6.7
7 0530 / 1139 / Th 1746	6.4 / 1.0 / 6.1	**22**	0001 / 0615 / F 1221 / ☾ 1841	1.2 / 6.1 / 1.2 / 6.3
8 0005 / 0607 / F 1217 / ☽ 1829	1.0 / 6.1 / 1.2 / 5.8	**23**	0042 / 0713 / Sa 1312 / 1951	1.5 / 5.8 / 1.4 / 5.9
9 0045 / 0650 / Sa 1300 / 1923	1.3 / 5.7 / 1.5 / 5.4	**24**	0142 / 0823 / Su 1436 / 2111	1.8 / 5.6 / 1.6 / 6.0
10 0135 / 0748 / Su 1357 / 2037	1.7 / 5.4 / 1.8 / 5.2	**25**	0319 / 0949 / M 1627 / 2242	1.9 / 5.7 / 1.3 / 6.1
11 0251 / 0908 / M 1531 / 2209	1.9 / 5.3 / 1.9 / 5.4	**26**	0452 / 1113 / Tu 1757 / 2353	1.6 / 6.1 / 0.9 / 6.6
12 0433 / 1042 / Tu 1708 / 2332	1.7 / 5.6 / 1.5 / 6.0	**27**	0611 / 1214 / W 1856	1.1 / 6.6 / 0.4
13 0544 / 1156 / W 1825	1.3 / 6.0 / 1.0	**28**	0043 / 0704 / Th 1300 / 1941	7.0 / 0.8 / 6.9 / 0.2
14 0029 / 0650 / Th 1253 / 1934	6.7 / 0.9 / 6.8 / 0.5	**29**	0126 / 0748 / F 1340 / ● 2018	7.1 / 0.7 / 7.0 / 0.3
15 0117 / 0751 / F 1333 / ◯ 2027	7.2 / 0.6 / 7.2 / 0.2	**30**	0202 / 0825 / Sa 1412 / 2050	7.3 / 0.7 / 7.0 / 0.4

OCTOBER

Time	m		Time	m
1 0233 / 0857 / Su 1442 / 2118	7.0 / 0.7 / 7.0 / 0.6	**16**	0216 / 0858 / M 1426 / 2128	7.5 / 0.5 / 7.7 / 0.3
2 0300 / 0924 / M 1507 / 2142	6.9 / 0.8 / 6.9 / 0.8	**17**	0257 / 0941 / Tu 1508 / 2206	7.4 / 0.5 / 7.7 / 0.4
3 0324 / 0948 / Tu 1535 / 2204	6.9 / 0.9 / 6.8 / 0.8	**18**	0339 / 1017 / W 1555 / 2238	7.2 / 0.5 / 7.5 / 0.6
4 0350 / 1012 / W 1606 / 2231	6.8 / 0.9 / 6.7 / 0.8	**19**	0424 / 1054 / Th 1642 / 2311	6.9 / 0.6 / 7.2 / 0.9
5 0421 / 1040 / Th 1640 / 2301	6.7 / 0.8 / 6.6 / 0.9	**20**	0509 / 1130 / F 1734 / 2346	6.5 / 0.9 / 6.7 / 1.3
6 0455 / 1112 / F 1718 / 2334	6.5 / 0.8 / 6.3 / 1.0	**21**	0557 / 1211 / Sa 1831 / ☾	6.1 / 1.1 / 6.3
7 0532 / 1149 / Sa 1801	6.2 / 1.0 / 6.0	**22**	0027 / 0652 / Su 1259 / 1934	1.6 / 5.8 / 1.4 / 6.0
8 0014 / 0615 / Su 1231 / ☽ 1855	1.3 / 5.9 / 1.3 / 5.7	**23**	0121 / 0757 / M 1416 / 2046	1.9 / 5.6 / 1.6 / 5.8
9 0102 / 0712 / M 1327 / 2006	1.7 / 5.5 / 1.6 / 5.5	**24**	0249 / 0914 / Tu 1559 / 2210	2.0 / 5.6 / 1.4 / 6.0
10 0215 / 0832 / Tu 1458 / 2139	1.9 / 5.5 / 1.7 / 5.6	**25**	0417 / 1040 / W 1716 / 2322	1.7 / 5.9 / 1.0 / 6.4
11 0357 / 1009 / W 1637 / 2301	1.7 / 5.7 / 1.3 / 6.0	**26**	0529 / 1143 / Th 1815	1.3 / 6.4 / 0.6
12 0511 / 1122 / Th 1754	1.3 / 6.3 / 0.9	**27**	0014 / 0625 / F 1229 / 1859	6.7 / 0.9 / 6.8 / 0.4
13 0001 / 0611 / F 1217 / 1904	6.8 / 0.9 / 6.9 / 0.5	**28**	0056 / 0707 / Sa 1309 / 1937	7.0 / 0.7 / 7.0 / 0.4
14 0050 / 0721 / Sa 1303 / ◯ 2001	7.2 / 0.7 / 7.3 / 0.3	**29**	0131 / 0748 / Su 1342 / ● 2011	7.0 / 0.7 / 7.0 / 0.5
15 0134 / 0813 / Su 1345 / 2047	7.4 / 0.6 / 7.5 / 0.2	**30**	0201 / 0822 / M 1411 / 2040	6.9 / 0.7 / 6.9 / 0.6
		31	0226 / 0853 / Tu 1439 / 2108	6.9 / 0.8 / 6.9 / 0.7

NOVEMBER

Time	m		Time	m
1 0253 / 0921 / W 1510 / 2138	6.9 / 0.8 / 6.8 / 0.8	**16**	0322 / 1002 / Th 1543 / 2219	7.0 / 0.4 / 7.4 / 0.7
2 0322 / 0950 / Th 1543 / 2209	6.8 / 0.8 / 6.7 / 0.8	**17**	0407 / 1042 / F 1634 / 2255	6.9 / 0.5 / 7.2 / 1.0
3 0355 / 1021 / F 1620 / 2240	6.7 / 0.8 / 6.6 / 0.9	**18**	0454 / 1123 / Sa 1725 / 2333	6.6 / 0.7 / 6.9 / 1.3
4 0431 / 1055 / Sa 1659 / 2313	6.6 / 0.9 / 6.5 / 1.0	**19**	0542 / 1203 / Su 1815	6.3 / 1.0 / 6.5
5 0509 / 1130 / Su 1744 / 2351	6.4 / 1.0 / 6.3 / 1.3	**20**	0012 / 0629 / M 1248 / ☾ 1910	1.6 / 6.0 / 1.3 / 6.1
6 0553 / 1212 / M 1838	6.1 / 1.2 / 6.0	**21**	0059 / 0724 / Tu 1347 / 2011	1.8 / 5.8 / 1.5 / 5.9
7 0039 / 0648 / Tu 1309 / 1945	1.5 / 5.8 / 1.4 / 5.9	**22**	0204 / 0829 / W 1512 / 2121	2.0 / 5.7 / 1.5 / 5.9
8 0148 / 0801 / W 1433 / 2108	1.8 / 5.7 / 1.4 / 5.9	**23**	0331 / 0948 / Th 1623 / 2235	1.9 / 5.8 / 1.2 / 6.1
9 0318 / 0929 / Th 1600 / 2226	1.7 / 5.9 / 1.1 / 6.3	**24**	0441 / 1059 / F 1720 / 2333	1.6 / 6.1 / 0.9 / 6.4
10 0433 / 1044 / F 1715 / 2329	1.3 / 6.3 / 0.8 / 6.8	**25**	0537 / 1153 / Sa 1808	1.2 / 6.4 / 0.7
11 0539 / 1143 / Sa 1829	1.0 / 6.8 / 0.6	**26**	0018 / 0625 / Su 1235 / 1850	6.6 / 1.0 / 6.6 / 0.6
12 0022 / 0648 / Su 1235 / 1931	7.1 / 0.9 / 7.1 / 0.5	**27**	0055 / 0707 / M 1310 / 1928	6.7 / 0.8 / 6.7 / 0.6
13 0110 / 0747 / M 1321 / ◯ 2020	7.1 / 0.7 / 7.2 / 0.5	**28**	0127 / 0747 / Tu 1344 / ● 2005	6.7 / 0.8 / 6.7 / 0.7
14 0154 / 0836 / Tu 1406 / 2103	7.1 / 0.6 / 7.3 / 0.6	**29**	0158 / 0825 / W 1416 / 2042	6.8 / 0.7 / 6.7 / 0.7
15 0237 / 0921 / W 1454 / 2142	7.1 / 0.5 / 7.4 / 0.6	**30**	0230 / 0903 / Th 1453 / 2119	6.8 / 0.7 / 6.7 / 0.8

DECEMBER

Time	m		Time	m
1 0305 / 0941 / F 1531 / 2157	6.8 / 0.8 / 6.8 / 0.8	**16**	0356 / 1038 / Sa 1623 / 2245	6.9 / 0.4 / 7.2 / 0.9
2 0342 / 1019 / Sa 1610 / 2231	6.8 / 0.8 / 6.8 / 0.9	**17**	0438 / 1116 / Su 1708 / 2320	6.8 / 0.5 / 7.0 / 1.1
3 0419 / 1054 / Su 1651 / 2305	6.7 / 0.9 / 6.7 / 1.1	**18**	0520 / 1153 / M 1753 / 2354	6.6 / 0.8 / 6.7 / 1.4
4 0458 / 1127 / M 1734 / 2339	6.5 / 1.0 / 6.5 / 1.3	**19**	0603 / 1227 / Tu 1838 / ☾	6.3 / 1.1 / 6.3
5 0540 / 1205 / Tu 1824	6.4 / 1.1 / 6.3	**20**	0031 / 0648 / W 1306 / 1928	1.6 / 6.0 / 1.3 / 6.0
6 0021 / 0628 / W 1255 / 1923	1.4 / 6.2 / 1.1 / 6.1	**21**	0113 / 0741 / Th 1358 / 2025	1.8 / 5.8 / 1.4 / 5.8
7 0120 / 0730 / Th 1402 / 2034	1.6 / 6.1 / 1.2 / 6.1	**22**	0212 / 0846 / F 1514 / 2128	1.9 / 5.7 / 1.4 / 5.8
8 0234 / 0847 / F 1521 / 2149	1.6 / 6.1 / 1.1 / 6.3	**23**	0338 / 0955 / Sa 1623 / 2233	1.9 / 5.7 / 1.3 / 5.9
9 0352 / 1004 / Sa 1635 / 2258	1.4 / 6.3 / 0.9 / 6.6	**24**	0447 / 1101 / Su 1716 / 2329	1.6 / 5.9 / 1.1 / 6.1
10 0502 / 1112 / Su 1750 / 2357	1.2 / 6.6 / 0.8 / 6.7	**25**	0540 / 1154 / M 1804	1.3 / 6.0 / 1.0
11 0617 / 1212 / M 1902	1.0 / 6.7 / 0.8	**26**	0015 / 0629 / Tu 1239 / 1850	6.3 / 1.1 / 6.2 / 0.9
12 0050 / 0727 / Tu 1307 / ◯ 1958	6.8 / 0.9 / 6.8 / 0.9	**27**	0056 / 0717 / W 1321 / 1937	6.5 / 0.9 / 6.4 / 0.8
13 0140 / 0820 / W 1358 / 2046	6.7 / 0.8 / 6.8 / 0.9	**28**	0137 / 0805 / Th 1402 / ● 2025	6.6 / 0.8 / 6.6 / 0.7
14 0226 / 0911 / Th 1447 / 2128	6.8 / 0.6 / 7.1 / 0.9	**29**	0218 / 0853 / F 1443 / 2110	6.8 / 0.7 / 6.8 / 0.7
15 0311 / 0956 / F 1536 / 2207	6.9 / 0.4 / 7.2 / 0.8	**30**	0257 / 0939 / Sa 1524 / 2153	6.9 / 0.6 / 7.0 / 0.7
		31	0335 / 1021 / Su 1603 / 2233	6.9 / 0.6 / 7.0 / 0.9

Chart Datum: 3.20 metres below Ordnance Datum (Newlyn)

SOUTHEND-ON-SEA/
LEIGH-ON-SEA 10-3-21
Essex

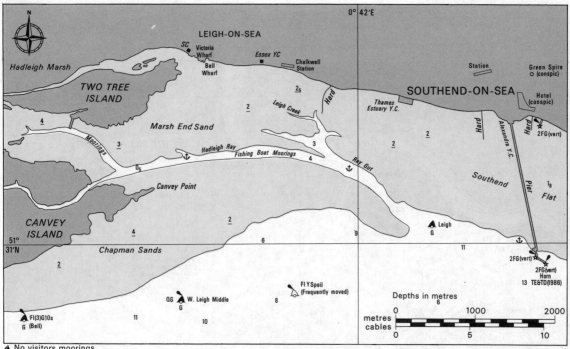

▲ No visitors moorings

CHARTS
Admiralty 1185; Stanford 5; Imray C2, Y6; OS 178
TIDES
+0125 Dover; ML 3.0; Duration 0610; Zone 0 (GMT).

Standard Port SHEERNESS (←)

Times				Height (metres)			
HW		LW		MHWS	MHWN	MLWN	MLWS
0200	0800	0200	0700	5.7	4.8	1.5	0.6
1400	2000	1400	1900				

Differences SOUTHEND-ON-SEA
−0005 −0005 −0005 −0005 0.0 0.0 −0.1 −0.1

SHELTER
The whole area dries soon after half ebb, except Ray Gut. Some moorings are available at Leigh-on-Sea. Yachts can take the ground alongside Bell Wharf or Victoria Wharf.
NAVIGATION
Waypoint Leigh (stbd-hand) buoy, 51°31′.10N 00°42′.43E, at entrance to Ray Gut. Approaching from Shoeburyness, keep outside the two buoys Mid Shoebury (Con G) and W Shoebury (Con G Fl G 2.5s). Beware some 3000 small boat moorings a mile either side of Southend pier. Speed limit in Canvey Island/Hadleigh Ray areas is 8kn.
NOTE: Southend-on-Sea and Leigh-on-Sea are both part of the lower Port of London Authority Area. Southend BC launches 'Alec White' and 'Loway' patrol area (VHF Ch 12), April-September.
LIGHTS AND MARKS
Pier lights as shown in chartlet.
RADIO TELEPHONE
Police launches, Southend BC launches, Thames Navigation Service all VHF Ch 12.
TELEPHONE (0702)
Hr Mr 611889; Hr Mr Leigh-on-Sea 710561; CG Shoeburyness 4998; MRSC Frinton-on-Sea 5518; Essex Police Marine Section Rayleigh 775533; HMC Freefone Customs Yachts: RT (0702) 547141 Ext 26; Marinecall 0898 500 455; Dr 49451; Hosp 348911.

FACILITIES
SOUTHEND-ON-SEA EC Wednesday. **Alexandra YC** Tel. 340363, Bar, FW; **Thames Estuary YC** Tel. 345967; **Thorpe Bay YC** Tel. 587563, Bar, FW; **Southend Pier** L, FW, Bar; **Town** V, R, Bar, PO, Bank, Rly, Air.
LEIGH-ON-SEA EC Wednesday. **Essex YC** Tel. 78404, FW, Bar; **Leigh on Sea SC** Tel. 76788, FW, Bar; **Bell Wharf**, AB; **Victoria Wharf** AB, SM, Slip; **Mikes BY** Tel. 713151, Slip, D (cans), L, FW, ME, El, Sh, C, CH; **Sea King** Tel. 73612, Slip, L, D, ME, El, Sh, C, CH; **Johnson, Sons & Jago** Tel. 76339, ME, El, Sh, D (cans); **W Sails** Tel. 714550, SM; **Two Tree Island** Tel. 711010, Slip, FW (see Hr Mr).
CANVEY ISLAND. **Halcon Marine** Tel. Canvey Island 685001 (250) Slip, D, FW, ME, El, Sh, C, Gas, Access HW∓2.
BENFLEET **Benfleet YC** Tel. South Benfleet 792278, Slip, FW, Bar; **Dauntless** Tel. South Benfleet 793782, Slip, D, FW, ME, El, Sh, C, Access HW∓2½.

HAVENGORE 10-3-22
Essex

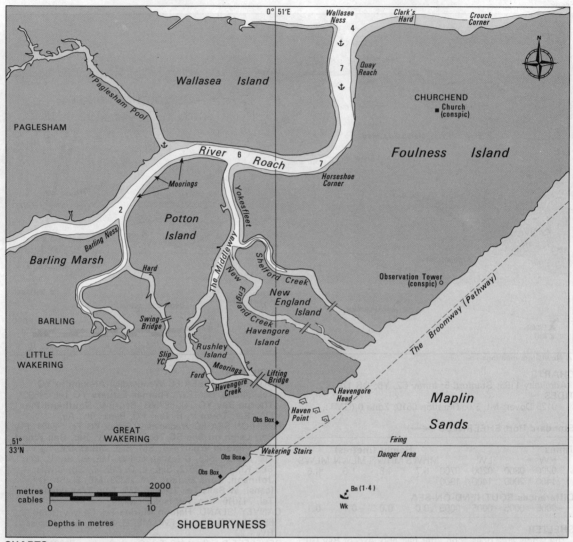

CHARTS
Admiralty 3750, 1185; Stanford 4, 5; Imray C1, Y17; OS 178

TIDES
+0110 Dover; ML 2.7; Duration 0615; Zone 0 (GMT).

Standard Port SHEERNESS (←)

Times				Height (metres)			
HW		LW		MHWS	MHWN	MLWN	MLWS
0200	0700	0100	0700	5.7	4.8	1.5	0.6
1400	1900	1300	1900				

Differences SHIVERING SANDS TOWER
−0025 −0019 −0008 −0026 −0.6 −0.6 −0.1 −0.1

SHELTER
Good shelter and a short cut at spring tides between the Thames and the Crouch for craft drawing less than 1.5m. Recommended to start passage about HW−1.

NAVIGATION
Waypoint 51°30′.78N 00°54′.07E, 146°/326° from/to Havengore Creek entrance, 3.0M. Approach crosses Shoeburyness gunnery range which is in use most weekdays. The rules are laid down in *Statutory Rules & Orders No 714 of 1936* obtainable from HMSO. Key paragraph is 'Any vessel wishing to enter Havengore Creek during such time or times as the whole of the target area is not closed in accordance with Bye Law No 3 must enter the target area not later than half an hour before high water and proceed by the shortest possible course to the Creek'. Red flags are hoisted at many points 1 hour before firing starts. Before setting off to Havengore, ring the Range Planning officer, Tel. Shoeburyness 2271 Ext 211; permission must be obtained preferably 24 hours in advance. Range is clear on bank holidays and at weekends. No passage allowed in dark hours. Lifting bridge across the creek will normally be opened on request. Tel. Shoeburyness 2271 Ext 436 (Bridge manned HW∓2 sunrise to sunset). There is no charge. When tide gauge at Southend pier shows 5m there is 1.5 over the Broomway.

LIGHTS AND MARKS
There are no lights.

RADIO TELEPHONE
Gravesend Radio VHF Ch 12 gives half hourly navigational information.

TELEPHONE (037 08)
CG 4998; HMC Freefone Customs Yachts: RT (01) 626 1515 Ext 5861; MRSC Frinton-on-Sea 5518; Marinecall 0898 500 455; Dr Southend-on-Sea 218678.

FACILITIES
EC Great Wakering Wednesday; **Sutton's BY** Tel. Southend-on-Sea 219422, Slip, P, D, FW, ME, El, Sh, C, CH, AB; **Wakering YC** M, L, Bar; **Great Wakering** V, R, Bar.
PO (Great Wakering & Barling); Bank (Shoeburyness, Thorpe Bay); Rly (Shoeburyness); Air (Southend).

MINOR HARBOURS AND ANCHORAGES 10.3.23

EASTBOURNE, E Sussex, 50°46′ N, 0°17′ E, Zone 0 (GMT), Admty chart 536. HW −0005 on Dover, −0007 on Shoreham; HW height +0.9m on Shoreham; ML 3.8m; Duration 0540. See 10.3.11. Anchorage in Eastbourne Bay sheltered from W winds through N to NE and good holding ground. Beware Boulder Bank, rocks dry 0.2m, extend ¼M seaward by Wish Tower. There is a beacon on shore 1 ca E of Tower. Landing steps on pier (NE side). Pierhead Lts 2 FR (vert) 8m 2M. R Lt occas is shown W of lifeboat station as a guide to local fishermen. There are all shore facilities but few marine facilities. A marina is planned.

HASTINGS, E Sussex, 50°51′ N, 0°35′ E, Zone 0 (GMT), Admty chart 536. HW −0005 on Dover; HW height +0.7m on Dover; ML 3.8m; Duration 0530. See 10.3.12. Anchorage off Hastings is only recommended in fair weather; beware dangerous wreck 3 ca SW of pier head. Lts: Pierhead 2 FR (vert) 8m 5M from white hut; W breakwater head Fl R 2.5s 5m 4M; Ldg Lts 356° front FR 14m 4M on W metal column, rear 357m from front, FR 55m 4M on 5-sided W Tr on West Hill. The stone breakwater is in a state of dis-repair and is only for the protection of fishing vessels. Facilities: EC Wed; HMC Freefone Customs Yachts: RT (0304) 202441; all shore facilities at Hastings and St Leonards. Few marine facilities − **Hastings and St Leonards YC** Tel. Hastings 420656. Landing places on pier.

SANDWICH, Kent, 51°18′ N, 1°21′ E, Zone 0 (GMT), Admty chart 1828. HW +0015 on Dover; HW height −1.0m on Dover; ML 1.4m; Duration 0520. Sandwich is on the S side of the R Stour, Richborough on the N side. Sandwich being 3M above Richborough Wharf. The river is winding but vessels with draught up to 2m can reach Sandwich on spring tides. There are moorings for visitors at Sandwich Quay (ring Quaymaster Tel. 613283). Facilities: EC Wed, Slip; HMC Freefone Customs Yachts: RT (0304) 202441; **Marina** (120+some visitors) Tel. 613335 (max 18m length, 2.1m draught), BH (15 ton), Sh, Slip, FW; Access HW∓2; **A. J. C. White** Tel. 617800, SM; **Richborough Port** has D by tanker. **Sandwich Sailing and Motorboat Club** Tel. 611103 and **Sandwich Bay Sailing and Water Ski Clubs** offer some facilities. Both ports are administered by The Sandwich Port and Haven Commissioners.

MARGATE, Kent, 51°23′ N, 1°23′ E, Zone 0 (GMT), Admty charts 1827, 1607, 1828. HW +0050 on Dover; −0040 on Sheerness; HW height −0.9m on Sheerness; ML 2.6m; Duration 0610. See 10.3.16. There is a small harbour open to NW winds inside Stone Pier − Pierhead Lt FR 4M. Beware remains of old pier, known as Margate Iron Jetty, extending 2½ ca N from root of Stone Pier, Fl(3)R 10s. The harbour dries 2m; alternative anchorages W of Margate or N of pier. Facilities: EC Thurs; Bar, LB, D and P in cans from garage, R, V, all stores. HMC Freefone Customs Yachts: RT (0304) 202441; **Margate YC** Tel. Thanet 292602, R, Bar.

GRAVESEND, Kent, 51°26′ N, 0°22′ E, Zone 0 (GMT), Admty charts 1186, 2151. HW +0150 on Dover, +0020 on Sheerness; HW height +0.7m on Sheerness; ML 3.3m; Duration 0610. See 10.3.20. Anchorage E of the piers just below entrance to Gravesend Canal Basin. There are two visitors buoys (orange) at the Club. Lock gates open HW −1½ to HW+½ on request to Lockkeeper, Tel. 352392. East end of Gravesend Reach marked by Ovens Buoy Fl 1s Bell. Royal Terrace Pierhead FR. Gravesend Radio (Thames Navigation Service) VHF Ch 12 14 16 18 20 (H24). Broadcasts on Ch 12 every H+00, H+30. Facilities: EC Wed; Bar, C (at canal entrance − ask at Club), CH, FW (standpipe near lock), P (from garage), R, V; HMC Freefone Customs Yachts: RT 626 1515 Ext 5861; **Gravesend SC**. Note: Boats can be left unattended in canal basin but not at anchorages.

ST KATHARINE YACHT HAVEN, Greater London, 51°30′ N, 0°05′ W, Zone 0 (GMT), Admty chart 1185. HW +0245 on Dover. HW times and heights as London Bridge. See 10.3.20. Good shelter under all conditions. Lock (42.7m × 9.1m with 2 lifting bridges), access HW−2 to HW+1½, winter 0800-1800, (lock shut Tues and Wed, Nov to Feb); summer 0600-2030. Moor to St Katharine's Pier 183m downstream whilst waiting (no shore access). Radio telephone (call *St Katharines*) VHF Ch M. **Yacht Haven** (100+50 visitors) Tel. 488 2400, AC, Bar, YC, FW, Gas, R, V, HMC; **Robbins Marine** Tel. 987 5884, ME, El, Sh; **Millwall Marine** Tel. 515 9351, ME, El, Sh; **Captain Pumpkin** Tel. 480 6630, CH.

HOLEHAVEN, Essex, 51°30′ N, 0°33′ E, Zone 0 (GMT), Admty chart 1185/6. HW +0140 on Dover, +0010 on Sheerness; height as Leigh-on-Sea; (10.3.21) ML 3.0m; Duration 0610. Shelter is good. Moorings S side of pier − very crowded; see Piermaster for mooring. Beware swell from passing traffic; keep to E (Canvey) side on entrance; once over bar, depths increase to 5m; anchor on W side as E side has long stone groynes running out. Lts Coryton Refinery Jetty No 4 2 FG (vert). ½ M up creek a bridge crosses, 9m at MHWS and opening span of 30m. Facilities: EC Thurs; FW (from 'The Lobster Smack' yard), P and D from village (1M); all other facilities at Canvey Island.

SEEING THE POINT

Clear, unambiguous headings are the hallmark of Rigel compasses. The Scorpio bulkhead compass has electric lighting, correction, covers and no hidden extras.

RIGEL

RIGEL COMPASSES LTD. Shamrock Quay, William Street, Southampton SO1 1QL
Tel: (0703) 32967/38663 Telex: 47669-SHAMOY (ATT: RIGEL)

L.H.Morgan & Sons (Marine) Ltd

★ Large Traditional Chandlery
★ Admiralty Chart Stockist (B)
★ Marine Fashion Clothing Dept.
★ Water ★ Calor Gas ★ Rigging Tools
★ Insurance ★ Sales ★ Yacht Brokers
★ Water ★ Mooring ★ Engineers
★ Mariner Outboard ★ Essex Area Distributors

Open 7 days 020 630 2003
Listening Channel 8.16
Dual Watch

 MARINER OUTBOARDS

32/42, Waterside ★ Brightlingsea ★ Essex ★ CO7 0AY

Bradwell Marina

- 280 Pontoon Berths in Rural Setting
- Access 4 hrs either side H.W.
- VHF monitoring (CH.M, P1, 37)
- Water/Electricity to all Pontoons
- Fuel jetty — Petrol and Diesel
- Chandlery
- Hot Showers
- 1st Class Workshop/Repairs
- Marine Slip to 20 tons
- Boat Hoistage to 16 tons
- Winter Storage
- Licenced Club (Membership Free)
- Yacht Brokerage

 Port Flair Ltd., Waterside, Bradwell-on-Sea Essex. CM0 7RB. Tel: (0621) 76235/76391

Elmhaven Marina Halling, Kent

Offer pleasant location

Good facility inc. Toilets, shower, elect-water etc — pontoon berths, walkway and mud berths. Plus John Hawkins Volvo-Penta Dealer onsite

Phone for More Details
0634-240489
0322-521595

DEDICATED MARINE POWER
9 to 422hp

VOLVO PENTA

DEDICATED MARINE SERVICE
OVER 1300 LOCATIONS ACROSS EUROPE

Rochford Marine Enterprises

Manufacturers, distributors of marine equipment.
★ ROCHFORD gangways, telescopic and ridged in aluminium and stainless steel.
★ Distributors of GUEST (USA) range of spotlights, interior/exterior lighting, battery switches, isolators, fans, synchronixer, marinaspec, navigation lights etc.

GUEST MARINASPEC NAVIGATION LIGHTS

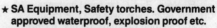
GANGWAYS

★ SA Equipment, Safety torches. Government approved waterproof, explosion proof etc.
★ Custom made switchboards.

Unit 3, Orlando Court, Vicarage Lane,
Walton-on-Naze, Essex, ENGLAND
Tel: (0255) 672036. Telex. 987703 attn. Rochford
Fax: (0473) 225488 attn. Rochford

SA 812 SAFETY TORCH

VOLVO PENTA SERVICE

Sales and service centres in area 4

BEDFORDSHIRE **Harry Kitchener Marine Ltd** Barkers Lane, BEDFORD MK41 9RL Tel (0234) 51931. ESSEX **French Marine Motors Ltd** 61/63 Waterside, BRIGHTLINGSEA CO7 0AX Tel (020630) 2133. **Volspec Ltd** Woodrolfe Road, Tollesbury, MALDON CM9 8SE Tel (0621) 369756 (24-hour Ansafone). HERTFORDSHIRE **The Hayes Allen (Boatyard) Co Ltd** Rye House Quay, HODDESDON EN11 0EH Tel (09924) 60888/64981. NORFOLK **Aquapower Marine Engineering** Easticks Yacht Station Boat Dyke Lane, ACLE Tel (0493) 751845. NORTHAMPTONSHIRE **C V S Pentapower** St. Andrews Road, NORTHAMPTON NN1 2LF Tel (0604) 38537/38409/36173. SUFFOLK **A D Truman Ltd** Old Maltings Boatyard, Oulton Broad, LOWESTOFT NR32 3PH Tel (0502) 65950.

VOLVO PENTA

Area 4

East England
River Crouch to King's Lynn

10.4.1	Index	**Page 307**	
10.4.2	Diagram of Radiobeacons, Air Beacons, Lifeboat Stations etc	308	
10.4.3	Tidal Stream Charts	310	
10.4.4	List of Lights, Fog Signals and Waypoints	312	
10.4.5	Passage Information	314	
10.4.6	Distance Tables	315	
10.4.7	Burnham-on-Crouch	315	
10.4.8	Blackwater River	316	
10.4.9	River Colne	317	
10.4.10	Walton-on-the-Naze	318	
10.4.11	River Stour (Harwich, Standard Port, Tidal Curves)	318	
10.4.12	River Orwell	319	
10.4.13	River Deben	324	
10.4.14	River Ore/Alde	324	
10.4.15	Southwold	325	
10.4.16	Lowestoft, Standard Port, Tidal Curves	325	
10.4.17	Great Yarmouth	329	
10.4.18	Blakeney	330	
10.4.19	Wells-next-the-Sea	331	
10.4.20	King's Lynn	332	
10.4.21	Minor Harbours and Anchorages River Roach Norfolk Broads Burnham Overy Staithe Brancaster Staithe Wisbech River Welland	333	

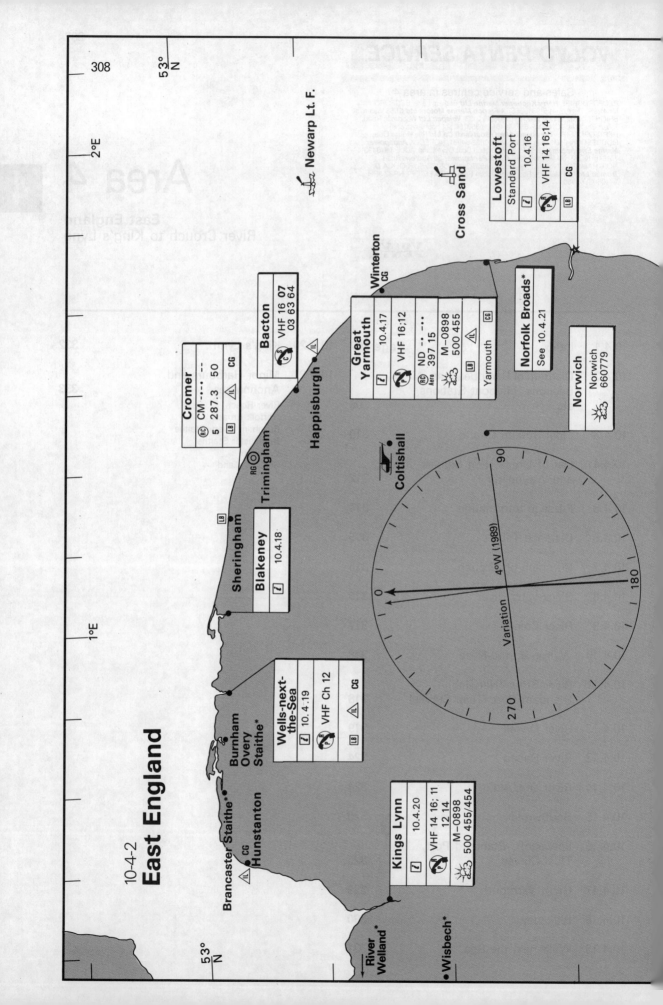

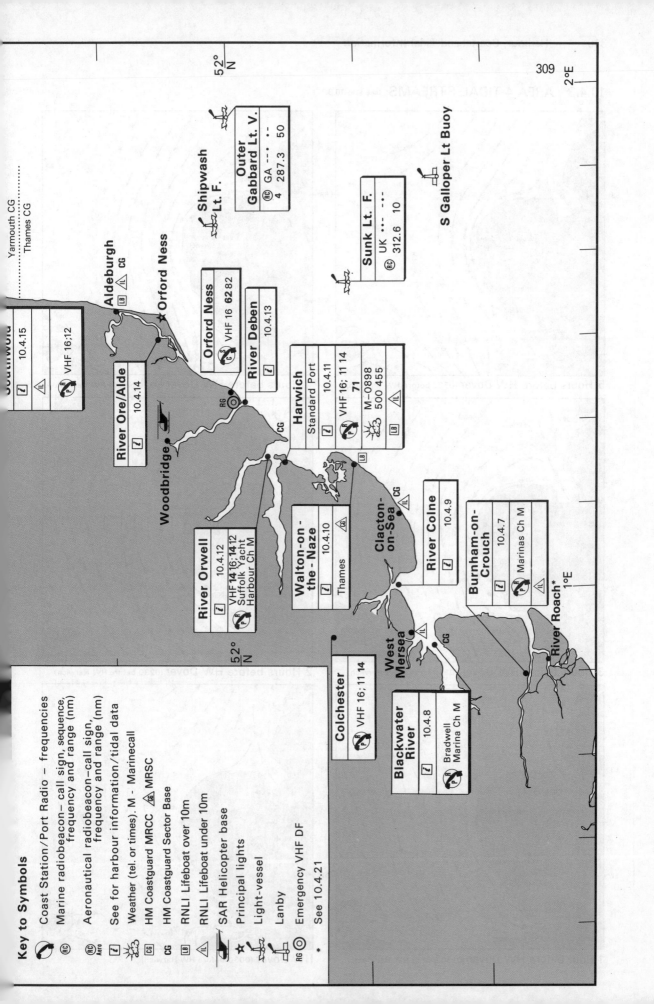

10.4.3 AREA 4 TIDAL STREAMS (see also 10.3.17)

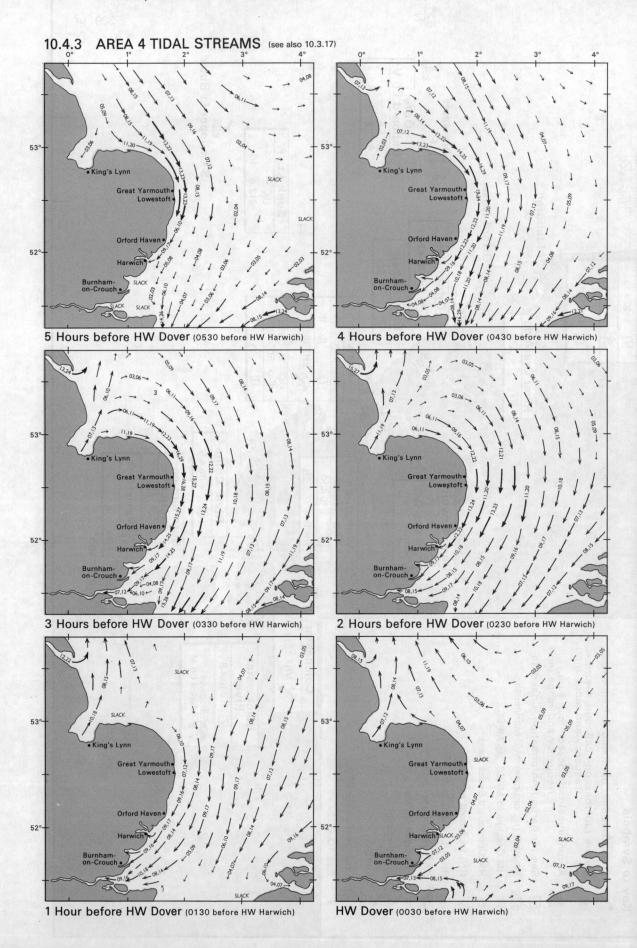

AREA 4 – E England

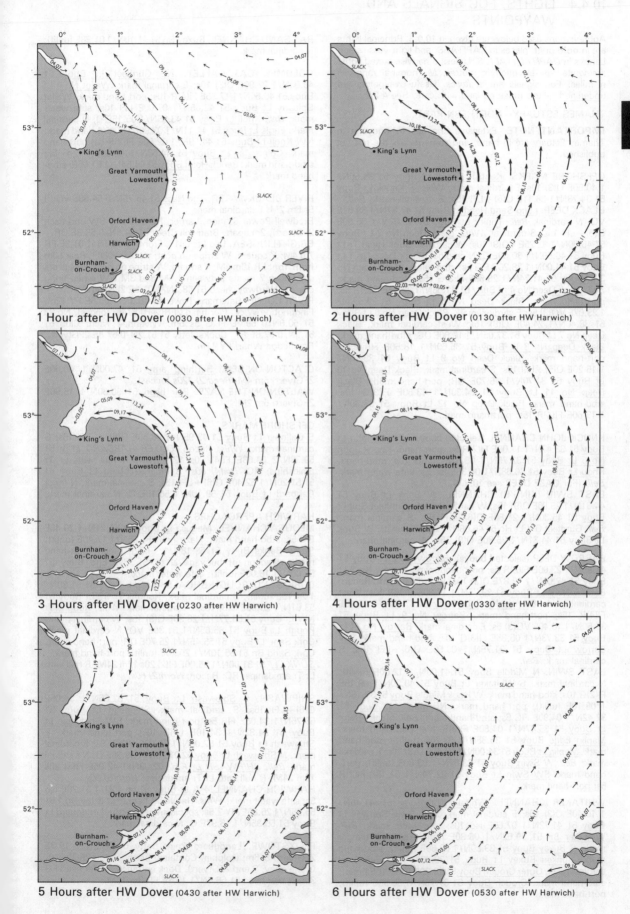

10.4.4 LIGHTS, FOG SIGNALS AND WAYPOINTS

Abbreviations used below are given in 10.0.3. Principal lights are in **bold** print, places in CAPITALS, and light-vessels and Lanbys in *CAPITAL ITALICS*. Unless otherwise stated lights are white. m—elevation in metres; M—nominal range in n. miles. Fog signals are in *italics*. Useful waypoints are underlined — use those on land with care. See 4.2.2.

THAMES ESTUARY—NORTHERN PART

IMPORTANT NOTE. Extensive changes to buoyage in Thames Estuary are in hand, and some details are not yet published.

OFFSHORE MARKS. Kentish Knock Lt Buoy 51 38.50N/1 40.50E Q(3) 10s; E cardinal mark; *Whis*. S Knock Lt Buoy 51 34.73N/1 36.11E Q(6) + LFl 15s; S cardinal mark; *Bell*.
BLACK DEEP. Long Sand Head Lt Buoy 51 47.87N/1 39.51E VQ; N cardinal; *Bell*. Trinity Lt Buoy 51 49.02N/1 36.50E Q(6) + LFl 15s; S cardinal; *Whis*. Black Deep No 2 Lt Buoy 51 45.60N/1 32.25E VQ(9) 10s; W cardinal. Sunk Head Tr Lt Buoy 51 46.59N/1 30.60E Q; N cardinal. Black Deep No 1 Lt Buoy 51 44.00N/1 28.20E FlG 5s; stbd-hand. Black Deep No 3 Lt Buoy 51 41.74N/1 25.65E Fl(3)G 15s; stbd-hand mark; *Bell*. Black Deep No 4 Lt Buoy 51.41.58N/1 28.57E Fl(2)R 5s; port-hand mark. Black Deep No 5 Lt Buoy 51.39.52.N/1 23.07E VQ(3) 5s; E cardinal mark. Black Deep No 6 Lt Buoy 51 38.40N/1 24.40E Q(9) 15s; West cardinal mark. Black Deep No 7 Lt Buoy 51 37.05N/1 17.80E QG; stbd-hand mark. Black Deep No 8 Lt Buoy 51 35.60N/1 18.50E FlR 2.5s; port-hand mark. Black Deep No 9 Lt Buoy 51 35.10N/1 15.20E Q(6) + LFl 15s; S cardinal mark. Black Deep No 10 Lt Buoy 51 34.70N/1 15.70E QR; port-hand mark. Black Deep No 11 Lt Buoy 51 34.30N/1 13.50E Fl(3)G 10s; stbd-hand mark. Black Deep No 12 Lt Buoy 51 33.80N/1 13.60E Fl(4)R 15s; port-hand mark.

KNOCK JOHN CHANNEL (selected buoys). Knock John Lt Buoy 51 33.47N/1 11.08E Fl(2)R 5s; port-hand. Knock John No 5 Lt Buoy 51 32.75N/1 08.67E Fl(3)G 10s; stbd-hand. Knob Lt Buoy 51 30.68N/1 04.35E Iso 5s; safe water mark; *Bell*. (For OAZE DEEP see 10.3.4).
EAST SWIN (KING'S) CHANNEL. W Sunk Lt Buoy 51 44.31N/1 25.86E Q(9) 15s; W cardinal mark. Gunfleet Spit Lt Buoy 51 45.30N/1 21.80E Q(6) + LFl 15s; S cardinal mark; *Bell*. Gunfleet Old Lt Ho 51 46.08N/1 20.52E (unlit). W Sunk Lt Buoy 51 44.31N/1 25.90E Q(9) 15s; W cardinal mark.
BARROW DEEP (selected buoys). Barrow No 2 Lt Buoy 51 41.95N/1 23.00E Fl(2)R 5s; port-hand mark. Barrow No 3 Lt Buoy 51 42.00N/1 19.87E Q(3) 10s; E cardinal mark; Racon. Barrow No 4 Lt Buoy 51 39.75N/1 17.80E VQ(9) 10s; West cardinal mark. Barrow No 6 Lt Buoy 51.37.78N/1 15.18E Fl(4)R 15s; port-hand mark. Barrow No 9 Lt Buoy 51 35.50N/1 10.40E VQ(3) 5s; E cardinal mark. Barrow No 11 Lt Buoy 51 33.73N/1 05.85E Fl(3)G 10s; stbd-hand mark. SW Barrow Lt Buoy 51 31.75N/1 00.50E Q(6) + LFl 15s; S cardinal mark; *Bell*.
EAST SWIN. N Middle Buoy 51.41.00N/1 12.00E (unlit); N cardinal mark. S Whitaker Lt Buoy 51 40.20N/1 09.15E Fl(2)G 10s; stbd-hand mark. W Hook Middle Buoy 51 39.13N/1 08.10E (unlit); port-hand mark. NE Maplin Lt Buoy 51 37.42N/1 04.90E FlG 5s; stbd-hand mark; *Bell*. Maplin Bank Lt Buoy 51 35.47N/1 04.80E Fl(3)R 10s; port-hand mark. Maplin Edge Buoy 51 35.35N/1 03.70E (unlit); stbd-hand mark. Maplin Lt Buoy 51 34.00N/1 02.40E Q(3) 10s; E cardinal mark; *Bell*. W Swin Buoy 51 33.82N/1 03.80E (unlit); port-hand mark. SW Swin Lt Buoy 51 32.74N/1 01.18E Fl(2)R 5s; port-hand mark.

WHITAKER CHANNEL. Whitaker Lt Buoy 51 41.45N/1 10.40E Q(3) 10s; E cardinal mark; *Bell*. Swin Spitway Lt Buoy 51 41.72N/1 07.65E Iso 10s; safe water mark; *Bell*. Whitaker Bn 51 39.62N/1 06.30E (unlit); isolated danger mark. S Buxey Buoy 51 39.80N/1 02.60E (unlit); cardinal mark. Sunken Buxey Lt Buoy 51 39.48N/1 00.75E Q; N cardinal mark. Outer Crouch Buoy 51 38.20N/0 58.10E (unlit); stbd-hand mark. Crouch Buoy 51 37.75N/0 56.95E (unlit); port-hand mark.

RAY SAND CHANNEL. Buxey Bn 51 41.10N/1 01.38E (unlit); N cardinal mark.

GOLDMER GAT/WALLET. NE Gunfleet Lt Buoy 51 49.90N/1 27.90E Q(3) 10s; E cardinal mark. Wallet No 4 Lt Buoy 51 46.50N/1 17.35E Fl(4)R 10s; port-hand mark. Wallet Spitway Lt Buoy 51 42.65N/1 07.00E LFl 10s; safe water mark; *Bell*. Knoll Lt Buoy 51 43.85N/1 05.15E Q; N cardinal mark. Eagle Lt Buoy 51 44.11N/1 03.90E Q; stbd-hand mark. NW Knoll Lt Buoy 51 44.19N/1 02.50E Fl(2)R 5s; port-hand mark. Colne Bar Lt Buoy 51 44.58N/1 02.65E Fl(2)G 5s; stbd-hand mark. Bench Head Buoy 51 44.54N/1 01.05E; stbd-hand mark.

RIVER BLACKWATER, The Nass 51 45.75N/0 54.88E VQ(3) 5s 6m 2M; E cardinal mark.
Bradwell Power Station, barrier, NE end and SW end each 2FR(vert), 2m apart. Bradwell Creek 51 44.6N/0 53.3E QR.
BRIGHTLINGSEA. Ldg Lts 041°. Front 51 48.4N/1 01.3E FR 7m 4M; Y square, W stripe on post; vis 020°-080°. Rear 50m from front FR 10m 4M; Y square, W stripe on post. FR Lts are shown on 7 masts between 1.5M and 3M NW when firing occurs. Hardway, head 51 48.2N/1 01.5E 2FR(vert).
RIVER COLNE. Batemans Tower 51 48.3N/1 00.8E FY 12m. Fingringhoe Wick, Pier head 2FR(vert) 6/4m (occas). No 23 51 50.6N/0 59.0E Fl G 5s 5m. Wivenhoe Yacht Club 51 51.3N/0 57.6E FY. Cooks Jetty 51 51.2N/0 57.8E 2FG(vert). Rowhedge Wharf FY 11m.

CLACTON-ON-SEA. Berthing arm 51 47.00N/1 09.60E 2FG(vert) 5m 4M; *Reed(2) 120s* (occas).
WALTON-ON-THE-NAZE. Pier head 51 50.60N/1 16.90E 2FG(vert) 5m 2M.

OFFSHORE MARKS
S Galloper Lt Buoy 51 43.95N/1 56.50E Q(6) + LFl 15s; S cardinal mark; *Whis*; Racon. *OUTER GABBARD Lt V* 51 59.38N/2 04.63E Fl(4) 20s 12m **24M**; R hull with Lt Tr amidships; RC; *Dia(4) 60s*. S Inner Gabbard Lt Buoy 51 51.20N/1 52.40E Q(6) + LFl 15s; S cardinal mark. N Inner Gabbard Lt Buoy 51 59.10N/1 56.10E Q; N cardinal mark.

HARWICH APPROACHES
MEDUSA CHANNEL. Medusa Lt Buoy 51 51.21N/1 20.45E FlG 5s; stbd-hand mark. Naze Tr 51 51.85N/1 17.40E (unlit). Stone Banks Buoy 51 53.18N/1 19.32E (unlit); port-hand mark, direction of buoyage N to S). Pye End Buoy 51 55.00N/1 18.00E (unlit); safe water mark.
CORK SAND/ROUGH SHOALS. S Cork Buoy 51 51.30N/1 24.15E (unlit); S cardinal mark. Roughs Tr SE Lt Buoy 51 53.61N/1 29.05E Q(3) 10s; E cardinal mark; *Whis*. Roughs Tr NW Lt Buoy 51 53.77N/1 28.90E Q(9) 15s; W cardinal mark. Rough Lt Buoy 51 55.65N/1 31.30E VQ; N cardinal mark. Cork Sand Lt Buoy 51 55.45N/1 25.70E QR; port-hand mark. Cork Sand Bn 51 55.10N/1 25.20E (unlit); port-hand mark. *SUNK Lt F* 51 51.00N/1 35.00E Fl(2) 20s 12m **24M**; R hull with Lt Tr amidships; RC; Racon; *Horn(2) 60s*.

SHIPWASH. S Shipwash Lt Buoy 51 52.65N/1 34.05E Q(6) + LFl 15s; S cardinal mark. Shiphead Lt Buoy 51 53.75N/1 34.05E FlR 5s; port-hand mark. SW Shipwash Lt Buoy 51 54.82N/1 34.10E LFlR 10s; port-hand mark. E Shipwash Lt Buoy 51 57.05N/1 38.00E VQ(3) 5s; E cardinal mark. N Shipwash Lt Buoy 52 01.70N/1 38.38E Q; N cardinal mark; *Bell*. *SHIPWASH Lt F* 52 02.03N/1 42.05E Fl(3) 20s 12m **24M**; R hull with Lt Tr amidships; *Horn(3) 60s*.
HARWICH CHANNEL. HA Lt Buoy 51 56.40N/1 31.20E Iso 5s; safe water mark; *Whis*. Harwich Channel Lt Buoy No 1 51 55.95N/1 26.85E FlY 2.5s; special mark; Racon. Landguard Lt Buoy 51 55.35N/1 18.97E Q; N cardinal mark.

FELIXSTOWE. Landguard Point Jetty 51 56.2N/1 19.2E 2FG(vert) 8/6m; Dolphin. Container berth, S end 2FG(vert). Crane quay, S end 2FG(vert). Dock entrance, S pier head FG 3m. N pier head 51 56.9N/1 19.1E FR 3m 10M.

AREA 4—E England

HARWICH. Wharves, jetties and piers show 2 FR (vert).
Shotley Spit 51 57.23N/1 17.70E Q(6) + L Fl 15s; S cardinal Bn.
Erwarton Ness 51 57.08N/1 13.35E Q(6) + LFl 15s; S cardinal Bn. Holbrook 51 57.19N/1 10.46E VQ(6)+LFl 10s; S cardinal Bn.

Shotley Ganges Pier, head 2FG(vert) 4m 1M; G post.
Mistley, Baltic wharf 51 56.7N/1 05.3E 2FR(vert). Suffolk Yacht Harbour. Ldg Lts FY.
Woolverstone Marina 52 00.4N/1 11.8E 2FR(vert).
Orwell Bridge FY at centre; 2FR(vert) on pier 9 and 2FG(vert) on pier 10.

Felixstowe Town Pier, head 51 57.4N/1 21.0E 2FG(vert) 7m.

WOODBRIDGE HAVEN. Buoy 51 58.36N/1 24.17E (unlit); safe water mark. Ldg Lts FlW or FlY moved as required (on request). Front W triangle on R post. Rear; R line on post. Ferry landing, E side 2FG(vert). W side 2FR(vert). Groyne, outer end QR.
BAWDSEY BANK. S Bawdsey Lt Buoy 51 57.20N/1 30.32E Q(6) + LFl 15s; S cardinal mark; Whis. Mid Bawdsey Lt Buoy 51 58.85N/1 33.70E Fl(3)G 10s; stbd-hand mark. NE Bawdsey Lt Buoy 52 01.72N/1 36.20E FlG 10s; stbd-hand mark.

CUTLER/WHITING BANK. Cutler Buoy 51 58.50N/1 27.60E (unlit); stbd-hand mark. SW Whiting Buoy 52 01.22N/1 30.90E (unlit); S cardinal mark. Whiting Hook Buoy 52 02.95N/1 31.95E (unlit); port-hand mark. NE Whiting Buoy 52 03.75N/1 33.85E (unlit); E cardinal mark.

Orford Ness 52 05.00N/1 34.55E Fl 5s 28m **24M**; W round Tr, R bands. F RG 14m R14M, **G15M** (same Tr); vis R shore-210°, R038°-047°, G047°-shore; Racon.

SIZEWELL. Power station, S pipeline 52 12.5N/1 37.7E 2FR(vert) 12/10m. N pipeline 2FR(vert) 12/10m.

SOUTHWOLD. Lt Ho 52 19.60N/1 41.00E Fl(4) WR 20s 37m **W22M**, **R22M**, **R20M**; W round Tr; vis R(intens) 204°-220°, W220°-001°, R001°-032°. N pier head 52 18.77N/1 40.63E FlG 1.5s and 3 FlR 1.5s (vert); R shown when harb inaccessible. S pier head QR 4m 2M.
E Barnard Lt Buoy 52 24.60N/1 46.18E Q(3) 10s; E cardinal mark. S Newcome Lt Buoy 52 25.70N/1 45.20E FlG 2.5s; stbd-hand mark. Pakefield Lt Buoy 52 27.05N/1 45.29E Fl(2) 5s; stbd-hand mark. E Newcome Lt Buoy 52 28.50N/1 49.31E Fl(2)R 5s; port-hand mark.
Lowestoft Roads, Claremont Pier 52 27.9N/1 45.0E 2FR(vert) 5/4m 4M; W column. 3 FR mark firing range 1.5M SSW.

LOWESTOFT. Fl 15s 37m **28M**; W Tr; part obscd 347°-shore. FR 30m **18M** (same Tr); vis 184°-211°. Outer harb S pier head Oc R 5s 12m 6M; Horn (4) 60s. N pier head 52 28.29N/1 45.50E Oc G 5s 12m 8M.
Corton Lt Buoy 52 31.10N/1 51.50E Q(3) 10s; E cardinal.

GREAT YARMOUTH. Gorleston S pier head Fl R 3s 11m 11M; vis 235°-295°. F (same structure) vis up harbour; Horn (3) 60s. Ldg Lts 264°. Front Oc 3s 6m 10M. Rear Oc 6s 7m 10M. N pier head 52 34.36N/1 44.49E Oc G 8s; vis 215°-035°. Brush lighthouse FR 20m 6M; R round Tr. Groyne, head 52 34.3N/1 44.4E 2FG(vert). 52 34.3N/1 44.3E 2FG(vert). 52 34.3N/1 44.2E 3FG(vert) in form of triangle. 5F mark pile structure close West. Haven Bridge 52 36.4N/1 43.5E marked by pairs of 2FR(vert) and 2FG(vert) showing up and down stream. 1 FR marks centre of channel.

Yarmouth Roads, South Denes, outfall 52 35.10N/1 44.50E QR 5m 2M; B and Y triangle. Wellington Pier, head 52 35.9N/1 44.4E 2FR(vert) 8m 3M. Jetty head 2FR(vert) 7m 2M. Britannia Pier, head 52 36.5N/1 44.6E 2FR(vert) 11m 4M; W column.
North Scroby Lt Buoy 52 42.20N/1 45.79E VQ; N cardinal; Bell, Whis.

Cockle Lt Buoy 52 44.00N/1 43.70E VQ(3) 5s; E cardinal mark.
Cross Sand Lt Buoy 52 37.00N/1 59.25E LFl 10s 6m 5M; W HFP buoy, RWVS; Racon. E Cross Sand Lt Buoy 52 40.00N/1 53.80E Fl(4)R 15s; port-hand mark. NE Cross Sand Lt Buoy 52 43.00N/1 53.80E VQ(3) 5s; E cardinal mark.
SMITH'S KNOLL Lt F 52 43.50N/2 18.00E Fl(3) 20s 12m **24M**; R hull with Lt Tr amidships; RC; Racon; Dia (3) 60s.
S Winterton Ridge Lt Buoy 52 47.20N/2 03.55E Q(6) + LFl 15s; S cardinal mark.

NEWARP Lt F 52 48.35N/1 55.80E Fl 10s 12m **26M** (H24); R hull with Lt Tr amidships; Horn 20s (H24), Racon.
S Haisbro Lt Buoy 52 50.80N/1 48.38E Q(6)+LFl 15s; S cardinal mark; Bell. Mid Haisbro Lt Buoy 52 54.20N/1 41.70E Fl(2)G 5s; stbd-hand mark.
N Haisbro Lt Buoy 53 00.20N/1 32.40E Q; N cardinal mark; Bell; Racon.
CROMER 52 55.5N/1 19.1E Fl 5s 84m **23M**; W 8-sided Tr; RC; Racon. Lifeboat House 2FR(vert) 8m 5M.
E Sheringham Lt Buoy 53 02.20N/1 15.00E Q(3) 10s; E cardinal mark. W Sheringham Lt Buoy 53 02.98N/1 07.70E Q(9) 15s; W cardinal mark. Blakeney Overfalls Lt Buoy 53 03.00N/1 01.50E Fl(2)R 5s; port-hand mark; Bell.

WELLS. Channel, West side 52 59.0N/0 50.2E Fl 5s 2m 3M; metal tripod. East side Fl R 3s; square on metal tripod.
E Docking Lt Buoy 53 09.80N/0 50.50E FlR 2.5s; port-hand mark. N Docking Lt Buoy 53 14.80N/0 41.60E Q; N cardinal mark. S Race Lt Buoy 53 08.65N/0 55.80E Q(6)+LFl 15s; S cardinal mark; Bell. N Race Lt Buoy 53 14.85N/0 44.20E FlG 5s; stbd-hand mark; Bell. (For marks further N, see 10.5.4).
Burnham Flats Lt Buoy 53 07.50N/0 35.00E VQ(9) 10s; W cardinal mark; Bell.
North Well Lt Buoy 53 03.00N/0 28.00E LFl 10s; safe water mark; Horn; Racon. Lynn Knock Lt Buoy 53 04.40N/0 27.30E QG; stbd-hand mark. Woolpack Lt Buoy 53 02.65N/0 31.50E FlR 10s; port-hand mark. *ROARING MIDDLE* Lt F 52 58.50N/0 21.00E Q 5m 8M; N cardinal mark; Bell. Replaced May-June by B&Y pillar and bell buoy, Q.

CORK HOLE. Sunk Lt Buoy 52 56.50N/0 23.85E Q(9) 15s; W cardinal mark. No 1 Lt Buoy 52 55.70N/0 22.10E VQ; N cardinal mark; Bell. No 3 Lt Buoy 52 54.43N/0 24.50E Q(3) 10s; E cardinal mark. No 3A Lt Buoy 52 53.45N/0 24.10E FlG 5s; stbd-hand mark. (Buoyage from Cork Hole to Lynn Cut subject to change).

KING'S LYNN. Beacon B 52 49.1N/0 21.2E Fl Y 2s 3m 2M; cone point up on B post. Beacon E 58 48.2N/0 21.5E Fl Y 6s 3m 2M; cone point up on B post. West Bank 52 47.4N/0 22.1E Fl Y 2s 3m 4M; R pile structure. Marsh Cut Ldg Lts 155°. Front 52 47.0N/0 22.6E QR 11m 3M. Rear 233m from front F 16m 4M. Old Lynn Channel 52 51.3N/0 15.7E Fl G 10s 6m 1M; B cone point up on B column. Trial Bank Beacon 52 50.5N/0 14.7E Fl(2) 5s 13m 3M.
Wisbech Channel (Note: beacons are moved as required) Dale 52 50.8N/0 12.8E Fl G 2s 3M. Double Brush 52 50.1N/0 13.1E QG 3M; X on Bn. Big Tom 52 49.6N/0 13.2E Fl(2) R 10s; R Bn. Walker 52 49.7N/0 13.1E Fl G 5s.
River Nene, Marsh 52 49.0N/0 13.0E QR. West End 52 49.5N/0 13.0E Fl G 5s 3M; B mast, Ra refl. Scottish Sluice, West bank 52 48.5N/0 12.7E FG and QG. Stakes on West side of River Nene to Wisbech carry QW Lts, and those on E side QR Lts.

10.4.5 PASSAGE INFORMATION

On the N side of Thames Estuary (chart 1975) the main chans run seaward in a NE direction (e.g. West Swin, Barrow Deep, Black Deep, Knock Deep) lined by banks which dry in places. The main chans are well buoyed. All the sandbanks in Thames Estuary are liable to change, and particular care is needed when crossing the estuary in a NW/SE direction. Bad visibility and much shipping can add to the hazards. Even with a moderate wind against tide, a short and steep sea is easily raised in these shallow waters. For further notes on Thames Estuary, see 10.3.5. For North Sea crossing see 10.19.5. There have been many buoyage changes in this area during 1988.

FOULNESS TO GREAT YARMOUTH (charts 1975, 2052)

Foulness Sand extends nearly 6M NE from Foulness Pt, the extremity being marked by Whitaker Bn. On N side of Whitaker Chan leading to R Crouch (10.4.7 and chart 3750) lies Buxey Sand, inshore of which is the Ray Sand Chan (dries), a convenient short cut between R Crouch and R Blackwater with sufficient rise of tide. For detailed directions to East Coast harbours (R Swale to Humber) see *East Coast Rivers* (Yachting Monthly). To seaward of Buxey Sand is Wallet Spitway (buoyed), and to NE of this is Gunfleet Sand, marked by buoys and drying in places along much of its 10M length. A conspic disused Lt Tr stands on SE side of Gunfleet Sand, about 6M SSE of Naze Tr, and here the SW- going (flood) streams begins about HW Sheerness +0600, and the NE-going stream at about HW Sheerness −0030, sp rates 2 kn.

Goldmer Gat and the Wallet give access to R Colne (10.4.9) and R Blackwater (10.4.8 and chart 3741). The chan is well marked and always accessible. The final approach 2M S of Colne Pt leads between Knoll (dries) on S side, and Eagle (depth 0.1m) on N side.

11M E of The Naze, the Sunk Lt F (Lt, fog sig, RC) guards the entrance to Harwich (10.4.11), an extensive and well sheltered harb accessible at all times (chart 2693). From the S, approach through Medusa Chan about 1M E of Naze Tr: at N end of this chan, 1M off Dovercourt, is Pye End buoy marking chan to Walton Backwaters (10.4.10). Making Harwich from the E beware Cork Sand (dries), S of which there is a minor chan. The main chan (well buoyed) passes N of Cork Sand and carries much commercial shipping. Approaching from NE beware Wadgate Ledge and the Platters about 1½M ENE of Landguard Pt. S of Landguard Pt the W-going (flood) stream begins at HW Harwich +0600, and the E-going stream at HW Harwich, sp rates about 1½ kn.

From 9M E of Felixstowe to 4M SSE of Orford Ness is the Shipwash shoal, marked by buoys and with a drying patch near its S end. Inshore of this is Bawdsey Bank, marked by buoys, with depths of 1.4m, and on which the sea breaks in E gales. Still further inshore, and slightly S, 2M off the entrance to R Deben (10.4.13) is a rky shoal called Cutler, with least depth of 1.8m, marked by buoy on E side.

Whiting Bank (buoyed) lies close SW of Orford Ness, and has a least depth of 0.9m. Hollesley Chan, about 1M wide, runs inshore W and N of this bank. In the SW part of Hollesley B is the entrance to Orford Haven (10.4.14).

There are overfalls S of Orford Ness on both the ebb and flood streams. 2M E of Orford Ness the SW-going stream begins at HW Harwich +0605, sp rate 2½ kn: the NE-going stream begins at HW Harwich −0010, sp rate 3 kn.

N of Orford Ness the coast is clear of offlying dangers past Aldeburgh and Southwold (10.4.15), as far as Benacre Ness, 5M S of Lowestoft (10.4.16). Sizewell power station is a conspic square building 1½M N of Thorpe Ness.

Lowestoft may be approached from S by buoyed chan to Pakefield Road, Lowestoft South Road (close E of pier heads), W of Lowestoft Bank: beware strong set possibly across harb entrance. From the N, approach through Yarmouth Road and then proceed S through Gorleston, Corton and Lowestoft North Roads (buoyed). From E, approach through Holm Chan. 2M E of harb entrance the S-going stream begins at HW Harwich +0530, and the N-going at HW Harwich −0110, sp rates 3 kn.

In the approaches to Great Yarmouth (10.4.17) from seaward there are frequent changes to the chans. The sea often breaks on North Scroby, Middle Scroby and Caister Shoal (all of which dry), and there are heavy tide rips over parts of Corton and South Scroby Sands, Middle and South Cross Sands, and Winterton Overfalls. 1M E of entrance to Gt Yarmouth Haven the S-going stream begins at HW Harwich +0530, and the N-going at HW Harwich −0050, sp rates 2¼ kn. Breydon Water (tidal) affects streams in the Haven: after heavy rain the out-going stream at Brush Quay may attain over 5 kn. For Norfolk Broads see 10.4.21.

About 12 M NE of Great Yarmouth lie Newarp Banks, on which the sea breaks in bad weather.

NORTH NORFOLK COAST (charts 106, 108)

Haisborough Sand (buoyed) lies 8M off the Norfolk coast, with depths of less than 1.8m in many places, and down to 0.4m. The shoal is steep-to, on its NE side in particular, and there are tidal eddies. Even a moderate sea or swell breaks on the shallower parts. There are dangerous wrecks near the S end. Haisborough Tail and Hammond Knoll (with wreck depth 1.2m) lie to the E of S end of Haisborough Sand.

The streams follow the directions of the coast and of chans between the banks NE of the Norfolk coast. But in the outer chans the stream is somewhat rotatory: when changing from SE-going to NW-going it sets SW, and when changing from NW-going to SE-going it sets NE, across the shoals. Close S of Haisborough Sand the SE-going stream begins at HW Harwich +0530, and the NW-going at HW Harwich −0040, sp rates about 2½ kn.

The coast of N Norfolk is unfriendly in bad weather, with no harb accessible when there is any N in the wind. The harbs all dry, and seas soon build up in the entrances or over the bars, some of which are dangerous even in a moderate breeze and an ebb tide. But in settled weather and moderate offshore winds it is a peaceful area to explore, particularly for boats which can take the ground. See 10.4.21.

If proceeding direct from Cromer to the Humber, pass S of Sheringham Shoal (buoyed) where the ESE-going stream begins at HW Immingham −0225, and the WNW-going at +0430. Proceed to NE of Blakeney Overfalls and Docking Shoal, and to SW of Race Bank, so as to fetch Inner Dowsing Lt Tr (Lt, fog sig). Thence pass E of Protector Overfalls, and steer for Rosse Spit buoy at SE side of entrance to Humber (10.5.8).

THE WASH (charts 108, 1177, 1157)

The Wash is a shallow area of shifting sands, formed by the estuaries of the rivers Great Ouse, Nene, Welland and Witham. Important factors are the strong tidal streams, the low shore line, and frequent poor vis. Keep a careful watch on the echo sounder, and remember that buoys may have been moved to accommodate changes in the chan. In general terms the in-going stream begins at HW Immingham −0520, and the out-going at HW Immingham −0045, sp rates about 2 kn. The in-going stream is usually stronger than the out-going, but its duration is less. Prolonged NE winds cause an in-going current, which can increase the rate and duration of the in-going stream and also the height of sea level at the head of the estuary. Do not attempt entry to the rivers too early on the flood, which runs hard in the rivers.

10.4.6 DISTANCE TABLE

Approximate distances in nautical miles are by the most direct route while avoiding dangers and allowing for traffic separation schemes etc. Places in *italics* are in adjoining areas.

1	*Cap Gris Nez*	**1**																		
2	*North Foreland*	32	**2**																	
3	*Shoeburyness*	61	29	**3**																
4	Burnham-on-Crouch	68	36	27	**4**															
5	West Mersea	67	35	30	22	**5**														
6	Walton-on-Naze	68	36	31	25	18	**6**													
7	Harwich	74	42	37	31	24	6	**7**												
8	River Deben	75	43	41	35	28	10	8	**8**											
9	Orford Ness	74	44	49	43	36	18	15	8	**9**										
10	Southwold	91	59	64	58	51	33	30	23	15	**10**									
11	Lowestoft	101	69	74	68	61	43	40	33	25	10	**11**								
12	Great Yarmouth	109	77	82	76	69	51	48	41	33	18	7	**12**							
13	Wells	157	125	130	124	117	99	96	89	81	66	56	50	**13**						
14	Kings Lynn	184	152	157	151	144	126	123	116	108	93	83	77	33	**14**					
15	Grimsby	201	169	174	168	161	143	140	133	125	110	100	94	54	61	**15**				
16	*Flamborough Head*	224	192	197	191	184	166	163	156	148	133	123	117	77	86	44	**16**			
17	*Oostende*	50	58	80	91	88	74	76	73	72	81	83	93	143	187	194	211	**17**		
18	*Goeree Tower*	103	91	111	108	104	100	99	85	78	79	81	83	125	152	169	186	51	**18**	
19	*IJmuiden*	149	137	155	151	140	125	126	122	110	105	102	102	143	170	182	193	97	46	**19**
20	*Brunsbüttel*	362	337	348	342	337	320	320	312	306	296	288	287	310	343	332	330	314	263	221	**20**

BURNHAM-ON-CROUCH
Essex   10-4-7

CHARTS
Admiralty 1975, 3750, 1183; Stanford 4, 5; Imray C1, Y17; OS 168/178

TIDES
+0110 Dover; ML 3.0; Duration 0610; Zone 0 (GMT).

Standard Port SHEERNESS (←)

Times				Height (metres)			
HW		LW		MHWS	MHWN	MLWN	MLWS
0000	0600	0100	0700	5.7	4.7	1.5	0.6
1200	1800	1300	1900				

Differences BURNHAM-ON-CROUCH
−0020 −0020 −0005 −0005 −0.5 −0.5 −0.5 −0.4

NOTE: Rivers Crouch and Roach, out as far as Foulness Point, are controlled by the Crouch River Authority. The Crouch is navigable up to Battlesbridge, 15 M from Foulness.

SHELTER
Rivers are exposed to all winds. Anchoring prohibited in fairway but is possible just above or below the moorings. Sometimes anchoring is possible to W of moorings by Crouch YC. Cliff Reach is important to yachtsmen as it provides shelter from SW winds. There are many moorings and landing places in the R Roach S of Wallasea Island. See 10.4.21.

NAVIGATION
Waypoint Outer Crouch (stbd-hand) buoy, 51°38'.20N 00°58'.10E, 072°/252° from/to Holliwell Pt Bn, 1.4M. There are few landmarks to assist entering the Crouch. By day, the buoys marking Whitaker Channel are easily found but at night the only lighted buoys are the Whitaker, the Swin Spitway and the Sunken Buxey, some 5 to 6 M away. There is a 'Firing Danger Area' just to S of Foulness Point. Speed limit in river is 8kn.
N.B. Landing on Bridgemarsh Is (up river) is strictly forbidden.

LIGHTS AND MARKS
From Sunken Buxey N cardinal Lt Buoy, St Mary's church spire bearing 233° leads to Outer Crouch stbd-hand buoy and from here 240° leads into river.

RADIO TELEPHONE (local times)
Essex Marina VHF Ch M (0900-1700). West Wick Marina Ch M (1000-1700).

TELEPHONE (0621)
Hr Mr 783602; CG Shoeburyness 4998; MRSC Frinton-on-Sea 5518; HMC Freefone Customs Yachts: RT (0473) 219481; Marinecall 0898 500 455; Dr 782054.

FACILITIES
BURNHAM EC Wednesday; **Burnham Yacht Harbour** (350) Tel. 782150, ME, El, Sh, BH (30 ton), CH, Lau, R, Access H24; **Royal Corinthian YC** Tel. 782105, AB, FW, M, L, R, Bar; **Royal Burnham YC** Tel. 782044, FW, L, R, Bar; **Tucker Brown** Tel. 782150, BY, C (5 ton), CH, El, FW, M, Gas, Gaz, L, ME, Sh, AB, D, Slip, AC, BH; **R. J. Prior** Tel. 782160, AB, BY, C (15 ton), D, FW, L, ME, El, Sh, M, Slip; **Kelvin Aqua** Tel. 782659, L, CH, FW, ACA, Gas, Gaz; **W. Yardley** Tel. 782076, BY, D, FW, M, ME, CH, M; **Newall Petticrow** (200) Tel. 782115, BY, C (4 ton), D, El, FW, M, ME, Sh, Slip; **Rice and Cole** Tel. 782063, BY, C (4½ ton), CH, D, M, Sh, L, Gas, Gaz, M, FW; **Crouch YC** Tel. 782252, L, FW, R, Bar; **Sadler Sails** Tel. 782124, SM; **Cranfield Sails** Tel. 782669, SM; **Crouch Engineering** Tel. 782130, D, El, L, ME, P, AB; **Duerr Engineering** Tel. 782726, El, ME (H24); **Peter Barker** Tel. 782403, Sh; **Tubby Lee Yachting Services** Tel. 783562, Rigging; **Town** V, R, Bar, PO, Bank, Rly, Air (Southend).
WALLASEA **Essex Marina** (400) Tel. Canewdon 531, AB, BY, Gas, Bar, C (13 ton), CH, D, El, FW, L, M, ME, P, R, Sh, Slip, V, YC; **Essex YC**.
FAMBRIDGE **West Wick Marina** (Stow Creek) (180) Tel. 741268, AB, Gas, Gaz, CH, El, Slip, FW, D, C (5 ton), YC, Bar; **North Fambridge Yacht Station** (150) Tel. 740370, CH, Sh, M, ME, BY, C (5 ton), El, FW, L, Slip, Gas, Gaz, Access HW∓5; **Brandy Hole Yacht Station** (120) Tel. Southend-on-Sea 230248, L, M, ME, Sh, Slip, Gas, Gaz, Bar, BY, D, FW, Access HW∓4.

BURNHAM-ON-CROUCH continued

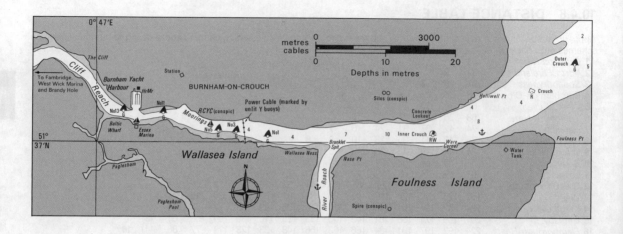

BLACKWATER RIVER 10-4-8
Essex

CHARTS
Admiralty 3741, 1183; Stanford 4, 5; Imray C1, Y17; OS 168

TIDES
+0105 Dover; ML 2.8; Duration 0620; Zone 0 (GMT).

Standard Port SHEERNESS (←)

Times				Height (metres)			
HW		LW		MHWS	MHWN	MLWN	MLWS
0000	0600	0100	0700	5.7	4.7	1.5	0.6
1200	1800	1300	1900				

Differences TOLLESBURY (Mill Creek)
−0027 −0027 −0019 −0019 −0.9 −0.8 −0.6 −0.4
OSEA ISLAND
−0005 −0005 −0016 −0016 −0.4 −0.4 −0.2 −0.1
MALDON
+0005 +0005 No data −2.8 −2.4 No data

SHELTER
By choosing the appropriate area good shelter can always be found. Entrance to the Chelmer and Blackwater canal by Heybridge basin at Maldon. Lock opens HW−1 to HW. For moorings, ring River Bailiff.

NAVIGATION
Waypoint Knoll N cardinal buoy, Q, 51°43′.85N 01°05′.18E, 107°/287° from/to Nass Bn, 6.7M.
WEST MERSEA Beware oyster beds between Cobmarsh Is and Packing Marsh Is and in Salcott Chan.
BRADWELL No dangers but only suitable for small craft and area gets very crowded.
TOLLESBURY FLEET Proceeding up Woodrolfe Creek, a tide mark shows the depth over the yacht harbour entrance sill (approx 3m at MHWS and 2m at MHWN).

⚓ Contact YCs

LIGHTS AND MARKS
Bradwell Yacht Marina entrance W side has two beacons B triangular topmarks; E side has beacon with R square topmark.
MALDON After rounding S of Osea Is, 'The Doctor', Blackwater SC Lt showing Iso G 5s at 300° leads approximately up the channel. Alternatively No 3 and No 8 buoys in line at 305°.

RADIO TELEPHONE
Bradwell Marina VHF Ch M (office hours). Tollesbury Marina Ch M (office hours).

TELEPHONE (0621)
Hr Mr 53646; River Bailiff 54477; CG (MRSC) Frinton-on-Sea 5518; Canal Lockmaster 53506; HMC Freefone Customs Yachts: RT (0473) 219481; Marinecall 0898 500 455; Dr 54118, West Mersea 382015.

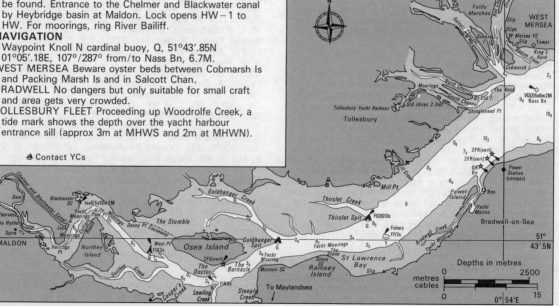

BLACKWATER RIVER continued

FACILITIES
EC W Mersea, Tollesbury, Maldon — Wednesday.
TOLLESBURY: Tollesbury Marina (240+20 visitors) Tel. 868471, Slip, D, AC, BH (10 ton), Gas, Gaz, FW, ME, El, Sh, C (5 ton), CH, V, R, Bar, Lau, Access HW∓1½; **Tollesbury Cruising Club** Tel. 869561, **Pier** FW; **Volspec** Tel. 869756, ME; **Tollesbury Saltings** Tel. 868421, Access HW∓2 Slip, FW, ME, El, Sh; **A.P. Marine** Tel. 869589, CH.
Village P, V, R, Bar. PO; Bank (Tues, Thurs 1000-1430); Rly (bus to Witham); Air (Southend or Cambridge).
WEST MERSEA: (0206) **Clarke & Carter BY** Tel. 382244, Slip (+ dock), M, L (at high water), FW, ME, El, C (10 ton), CH; **William Wyatt** Tel. 382856, Slip, M, ME, El, CH; **Causeway** P and D (A. Clark), L, FW, ME, El, V; **Gowen Sails** Tel. 382922, SM; **West Mersea YC** Tel. 382947, M (see boatman), R, Bar; **Mersea Chandlers** Tel. 384433, CH, ACA;
Town P, D, FW, ME, El, CH, V, R, Bar. PO; Bank; Rly (bus to Colchester); Air (Southend or Cambridge).
BRADWELL: Bradwell Marina (280, some visitors) Tel. 76235, Slip, AC, Gas, Gaz, D, P, FW, ME, El, Sh, BH (16 ton), CH, R, Bar, Access HW∓4½; **Bradwell Quay** Slip, L; **Bradwell Quay Yacht Club** Tel. 76539, M, FW.
Town PO; Bank (Maldon); Rly (bus to Southminster); Air (Southend).
MALDON: Maldon Quay Slip, M, P, D, FW, AB; **Dan Webb & Feesey** Tel. 54280, Slip, D, Sh, CH, M; **Fairways Marine Engineers** Tel. 52866, ME, El; **Heybridge Basin** Tel. 54022, D, ME, El, Sh, CH, Bar; **Blackwater SC** Tel. 53923, L, FW; **Holt and James** Tel. 54022, Slip, D, L, M, FW, CH; **Mantsbrite Marine Electronics** Tel. 53003, El; **A Taylor** Tel. 53456, SM; **Anglian Yacht Services** Tel. 52290 emergencies.
Town PO; Bank; Rly (bus to Chelmsford); Air (Southend, Cambridge or Stanstead).
MAYLANDSEA: Marina Access HW∓2; **Dan Webb and Feesey** (350) Tel. 740264, Slip, D, Sh, CH.

RIVER COLNE 10-4-9
Essex

CHARTS
Admiralty 3741, 1183; Stanford 5, 4; Imray C1, Y17; OS 168
TIDES
+0055 Dover; ML 2.5; Duration 0615; Zone 0 (GMT).
Standard Port SHEERNESS (←—)

Times				Height (metres)			
HW		LW		MHWS	MHWN	MLWN	MLWS
0200	0800	0100	0700	5.7	4.7	1.5	0.6
1400	2000	1300	1900				

Differences BRIGHTLINGSEA
−0035 −0035 −0025 −0025 −0.7 −0.9 −0.3 −0.2
WIVENHOE
−0011 −0011 No data No data

SHELTER
Suitable shelter can be found for most winds. Anchorages in Brightlingsea Creek; to the W of Mersea Stone Pt; in Pyefleet, E of Pewit Is; R Colne navigable up to Wivenhoe (4.5m draught) or Hythe, Colchester (3m draught). Outer harbour is exposed to westerly winds.
NAVIGATION
Waypoint Colne Bar (stbd-hand) buoy, 51°44'.58N 01°02'.66E, 158°/338° from/to Mersea Stone, 3.6M. See also 10.4.8. Extensive mud and sand banks flank the entrance channel. The entrance to Brightlingsea creek is very narrow at LW. River traffic in the Colne is considerable. Large coasters use the Brightlingsea channels. Speed limits — 8kn; Buoy 13 to Fingringhoe (except Buoy 12 to 16) 5kn; Fingringhoe to Colchester 4kn.

LIGHTS AND MARKS
Up as far as Wivenhoe, where the river dries out, it is extremely well buoyed. Brightlingsea has FR Ldg Lts adjusted according to the channel and channel buoys Fl (3) G 8s and Fl (2) R 8s. A Y Lt is exhibited from The Tower by Westmarsh Pt.
RADIO TELEPHONE
Colchester Harbour Radio VHF Ch 16; **14** 11 (Office hrs & HW−2 to HW+1). Tel. Colchester 575858.
TELEPHONE (0621)
Hr Mr 52110; Harbour Office Brightlingsea 2200; CG & MRSC Frinton-on-Sea 5518; Canal Lockmaster 53506; HMC Freefone Customs Yachts: RT (0473) 219481; Marinecall 0898 500 455; Dr 52535, West Mersea 382015.
FACILITIES
BRIGHTLINGSEA: (020 630) **Town Hard** Tel. 3535, L, FW; **Colne YC** Tel. 2594, L, FW, R, Bar; **L.H. Morgan** Tel. 2003, M, Gas, D, L, FW, ME, El, Sh, CH, ACA; **Cox Marine** ME, El, Sh; **French Marine Motors** Tel. 2133, CH, P, D, ME, El; **Brightlingsea SC** Slip, Bar; **James Lawrence** Tel. 2863, SM; **The Boatcentre** Tel. 2003, CH; **St Osyth BY** Tel. St Osyth 820005, M, L, FW, ME, El, Slip; **Town** P, D, FW, ME, El, Sh, C (mobile), CH, V, R, Bar; PO; Bank; Rly (bus to Wivenhoe or Colchester); Air (Southend or Cambridge).

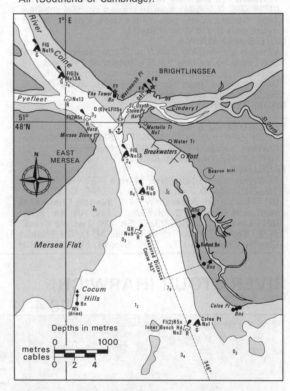

WIVENHOE: (020 622) **Colne Marine Yacht** Tel. 2417, L, ME, El, Sh, C; **Ian Brown** Tel. Rowhedge 867144, Slip, D, L, FW, Sh, CH; **Wivenhoe SC; Village near quay** P, V, Bar; PO; Bank (AM only); Rly; Air (Southend or Cambridge).
Note: A marina is planned at Brightlingsea shipyard.

WALTON-ON-THE-NAZE
Essex   10-4-10

CHARTS
Admiralty 2695, 2052; Stanford 5, 6; Imray C1, Y16; OS 169

TIDES
+0040 Dover; ML 2.3; Duration 0615; Zone 0 (GMT).

Standard Port HARWICH (→)

Times				Height (metres)			
HW		LW		MHWS	MHWN	MLWN	MLWS
0000	0600	0000	0600	4.0	3.4	1.1	0.4
1200	1800	1200	1800				

Differences WALTON-ON-THE-NAZE
−0008 −0011 +0008 +0005 +0.2 0.0 0.0 0.0

SHELTER
The 'Walton Backwaters' give good shelter in any weather. Go alongside in marina in The Twizzle or anchor in Hamford Water, or in Walton Channel. W & F Yacht Trust Basin, Tel. Frinton 5873, entry HW−1 to HW. Keel boats can stay afloat in Walton Channel (N end 7m, S end 5m) or in the Twizzle (E end 5m).

NAVIGATION
Waypoint Pye End (safe water) buoy, 51°55'.00N 01°18'.00E, 054°/234° from/to buoyed channel entrance, 1.0M. Beware lobster pots off the Naze, Pye Sands and oyster beds S and W of Horsey I.

LIGHTS AND MARKS
Whether approaching from N or S it is necessary to find Pye End buoy marking the N extremity of Pye sand. The most conspic landmark is the Naze tower just N of Walton-on-the-Naze. There are no lights so the Backwaters cannot be entered at night.

RADIO TELEPHONE
None.

TELEPHONE (0255 67)
Hr Mr Harwich 6535; CG & MRSC 5518; HMC Freefone Customs Yachts: RT (0473) 219481; Marinecall 0898 500 455; Dr Harwich 6451.

FACILITIES
EC Wednesday; **Titchmarsh Marina** (420+10 visitors), Tel. 2185, Slip, D, FW, Gas, Gaz, AC, ME, C (25 ton), BH (10 ton), CH, Access HW∓4½; **Walton Channel** M; **Walton & Frinton YC** Tel. 5526, L, FW, AB, Bar; **BYs (Mill Lane)** Tel. 5596, 5873, D, Sh, CH; **Twizzle Creek** M, L; **Bedwell & Co** Tel. 5873, Slip, M, D, C (½ ton), Sh, CH; **Frank Halls** Tel. 5596, Slip, M, FW, D, Sh, CH,

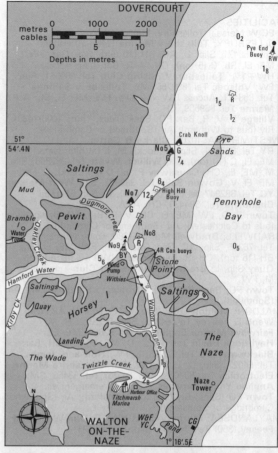

ME, El, Sh; **Fremar Marine** Tel. 78888, ME, CH; **Town** P, V, R, Bar, PO; Bank; Rly; Air (Southend or Cambridge).

RIVER STOUR (HARWICH)
Essex/Suffolk   10-4-11

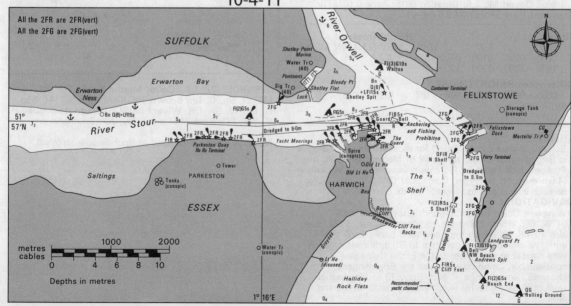

RIVER STOUR (HARWICH) continued

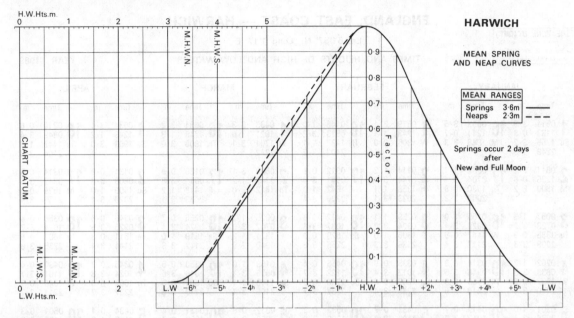

CHARTS
Admiralty 1491, 2693, 1593, 1594; Stanford 5, 6; Imray C1, Y16; OS 169

TIDES
+0040 Dover; ML 2.2; Duration 0630; Zone 0 (GMT)

Standard Port HARWICH (→)

Times				Height (metres)			
HW		LW		MHWS	MHWN	MLWN	MLWS
0000	0600	0000	0600	4.0	3.4	1.1	0.4
1200	1800	1200	1800				

Differences MISTLEY
+0025 +0025 0000 +0020 +0.2   0.0   −0.1   −0.1

NOTE: Harwich is a Standard Port and times and heights of tidal predictions for each day of the year are given below.

SHELTER
Good. Anchor off Erwarton Ness, Wrabness or Holbrook Creek. Alongside berths at Shotley Point Marina, Harwich Quay (dries) for short periods near HW, and at New Mistly and Manningtree (dry; off chartlet). No facilities for yachts at Felixstowe.

NAVIGATION
Waypoint Cork Sand (port-hand) buoy, QR, 01°55'.43N 01°25'.70E, 086°/266° from/to position 0.7M S of Landguard Pt, 4.0M. Beware commercial shipping. Yachts must keep clear of deep water channel: recommended track to S of it from Cork Sand Bn and thence to W of it past Harwich. Entrance to Shotley Point Marina through marked channel, with outer limits lit, to lock gate at all states of tide. The channel up the R Stour is well marked. Beware 'The Horse' 2 ca NW of Stone Pt and also the drying bank 1½ ca NW of Wrabness Point. The channel from Mistley Quay to Manningtree is narrow and tortuous, local knowledge is invaluable.

Special Local Sound Signals
Commercial vessels may use these additional signals:

For short and rapid blasts followed by one short blast } = { I am turning short around to starboard.

Four short and rapid blasts followed by two short blasts) } = { I am turning short around to port.

One prolonged blast = { I am leaving a dock, quay or anchorage.

LIGHTS AND MARKS
The channel West from Harwich past Parkeston Quay to Harkstead Pt (off chartlet) is marked by G conical stbd-hand buoys, Fl G or QG. There is a conspic factory chimney at Cattawade, straight up the river about 8 M.

If kept dead ahead on course 270° it will lead through the best water up to Harkstead Pt.

RADIO TELEPHONE
Call: *Harwich Harbour Control* VHF Ch 16 11 14 **71** (H24). Yachts are requested not to use Ch 71 but it is useful to monitor the channel to obtain information on shipping. Local weather, tidal information etc available on request. The Harwich Harbour Board Patrol launch, *Godwit* keeps regular patrol and listening watch of VHF Ch 11. Shotley Point Marina Ch M.
See also River Orwell. (10.4.12).

TELEPHONE (0255)
Hr Mr 504303; CG & MRSC Frinton-on-Sea 5518; Harwich Harbour Conservancy Board 504303; Harbour Operations Room 506535; HMC Freefone Customs Yachts: RT (0255) 502246; Marinecall 0898 500 455; Dr 506451.

FACILITIES
EC Wednesday
HARWICH: **Town Pier** L, FW, AB (tidal); **Stour & Orwell Engineering (Gas House Creek)** Tel. 502359, Slip, ME, El, Sh, CH, AB; **Royal Harwich YC** Tel. Woolverstone 319; **Dolphin Sails** Tel. 504202, SM; **F.M. Services** Tel. 506808, Gas; **R & J Marine** Tel. 502849, El; **Town** P, D, ME, El, Sh, V, R, Bar. PO; Bank; Rly; Air (Cambridge or Norwich).
SHOTLEY: **Shotley Point Marina** (350 – visitors welcome) Tel. Shotley 348908, FW, AC, D, P, Lau, ME, El, Sh, Slip, BH, V, CH, SM, Bar, R; Access H24 via lock; **Shotley SC** Tel. Shotley 500, Slip, L, FW, Bar; **Harwich and Dovercourt SC** Tel. Harwich 502619, M, L, FW, C (5 ton), AB, V, R, Bar. **Shotley BY** Tel. Shotley 714, Slip, M, ME, El, Sh.
WRABNESS: M, FW, V, Bar.
MISTLEY: AB, M, FW, V, P, D, Bar.
MANNINGTREE: M, AB, FW, V, Bar; **Stour SC** Tel. Colchester 393924, FW, M, Bar; **DN Howells** Tel. 392577, Gas.

RIVER ORWELL 10-4-12
Suffolk

CHARTS
Admiralty 2693, 2052, 1491; Stanford 5, 6; Imray C1, Y16; OS 169

TIDES
Pin Mill +0100 Dover; Ipswich +0115 Dover; ML 2.4; Duration 0555; Zone 0 (GMT)

ENGLAND, EAST COAST — HARWICH

Lat 51°57′ N   Long 1°17′ E

TIMES AND HEIGHTS OF HIGH AND LOW WATERS

YEAR 1989

TIME ZONE UT(GMT)
For Summer Time add ONE hour in non-shaded areas

JANUARY

Time	m	Time	m
1 Su 0511 1133 1756 2349	3.3 0.8 3.2 1.3	**16** M 0553 1211 1848	3.5 0.7 3.3
2 M 0611 1235 1900	3.2 0.9 3.2	**17** Tu 0032 0704 1330 2002	1.1 3.4 0.8 3.3
3 Tu 0057 0723 1338 2008	1.3 3.2 0.9 3.3	**18** W 0155 0829 1447 2117	1.0 3.4 0.9 3.4
4 W 0202 0833 1437 2110	1.2 3.3 0.9 3.4	**19** Th 0314 0946 1550 2219	0.9 3.5 0.9 3.5
5 Th 0301 0936 1531 2204	1.0 3.4 0.9 3.5	**20** F 0417 1047 1640 2311	0.7 3.7 0.8 3.6
6 F 0355 1031 1619 2254	0.9 3.5 0.8 3.6	**21** Sa 0506 1136 1722 2354 ○	0.5 3.8 0.8 3.7
7 Sa 0444 1120 1704 2342 ●	0.7 3.7 0.8 3.7	**22** Su 0549 1218 1758	0.4 3.8 0.8
8 Su 0530 1207 1746	0.6 3.8 0.7	**23** M 0034 0627 1256 1832	3.8 0.3 3.8 0.7
9 M 0025 0615 1253 1828	3.7 0.4 3.9 0.6	**24** Tu 0109 0702 1331 1906	3.8 0.3 3.8 0.7
10 Tu 0109 0659 1337 1910	3.8 0.3 4.0 0.6	**25** W 0141 0735 1404 1937	3.8 0.3 3.8 0.7
11 W 0151 0742 1422 1952	3.8 0.2 3.9 0.6	**26** Th 0212 0808 1434 2008	3.8 0.3 3.7 0.7
12 Th 0233 0827 1507 2036	3.8 0.2 3.9 0.7	**27** F 0242 0839 1505 2039	3.8 0.4 3.6 0.8
13 F 0315 0914 1553 2122	3.8 0.2 3.8 0.8	**28** Sa 0311 0912 1538 2111	3.7 0.5 3.5 0.9
14 Sa 0402 1003 1644 2214 ☽	3.8 0.3 3.6 0.9	**29** Su 0345 0948 1614 2148	3.6 0.6 3.4 1.0
15 Su 0452 1101 1742 2315	3.7 0.5 3.4 1.0	**30** M 0424 1030 1658 2235 ☾	3.5 0.8 3.3 1.1
		31 Tu 0512 1127 1754 2349	3.3 1.0 3.1 1.3

FEBRUARY

Time	m	Time	m
1 W 0619 1243 1906	3.1 1.1 3.0	**16** Th 0148 0823 1437 2103	1.0 3.2 1.1 3.2
2 Th 0114 0744 1358 2023	1.2 3.1 1.1 3.1	**17** F 0312 0941 1542 2206	0.8 3.4 1.0 3.4
3 F 0229 0904 1505 2135	1.1 3.2 1.0 3.3	**18** Sa 0412 1037 1628 2255	0.6 3.6 0.9 3.6
4 Sa 0335 1012 1602 2235	0.9 3.4 0.9 3.4	**19** Su 0455 1122 1706 2337	0.4 3.7 0.8 3.7
5 Su 0431 1106 1651 2326	0.7 3.6 0.7 3.6	**20** M 0533 1200 1739 ○	0.3 3.8 0.7
6 M 0519 1156 1733 ●	0.4 3.8 0.6	**21** Tu 0012 0605 1235 1810	3.8 0.3 3.8 0.6
7 Tu 0012 0604 1241 1814	3.8 0.2 4.0 0.5	**22** W 0046 0636 1306 1839	3.8 0.2 3.8 0.5
8 W 0056 0646 1324 1855	3.9 0.0 4.1 0.4	**23** Th 0116 0706 1335 1909	3.9 0.2 3.8 0.5
9 Th 0137 0727 1406 1934	4.0 0.0 4.1 0.4	**24** F 0144 0735 1404 1937	3.9 0.2 3.8 0.5
10 F 0218 0809 1449 2016	4.1 0.0 4.0 0.5	**25** Sa 0212 0805 1432 2006	3.8 0.3 3.7 0.6
11 Sa 0257 0851 1532 2100	4.0 0.1 3.8 0.6	**26** Su 0240 0833 1501 2034	3.8 0.5 3.6 0.7
12 Su 0339 0936 1617 2148 ☾	3.9 0.3 3.6 0.8	**27** M 0311 0903 1535 2105	3.7 0.6 3.5 0.8
13 M 0428 1030 1709 2245	3.7 0.6 3.4 0.9	**28** Tu 0348 0936 1613 2143 ☽	3.5 0.8 3.3 1.0
14 Tu 0526 1137 1812	3.5 0.9 3.2		
15 W 0008 0645 1309 1935	1.0 3.3 1.1 3.1		

MARCH

Time	m	Time	m
1 W 0433 1021 1704 2242	3.3 1.0 3.1 1.1	**16** Th 0631 1245 1909	3.2 1.3 3.0
2 Th 0533 1142 1812	3.2 1.2 3.0	**17** F 0135 0811 1418 2039	0.9 3.2 1.2 3.1
3 F 0025 0702 1321 1940	1.2 3.0 1.2 2.9	**18** Sa 0256 0924 1519 2142	0.7 3.4 1.0 3.3
4 Sa 0159 0836 1440 2105	1.0 3.1 1.1 3.1	**19** Su 0350 1016 1604 2230	0.5 3.6 0.9 3.5
5 Su 0312 0952 1541 2213	0.8 3.4 0.9 3.4	**20** M 0431 1058 1640 2311	0.3 3.7 0.8 3.6
6 M 0412 1048 1630 2305	0.5 3.7 0.7 3.6	**21** Tu 0505 1134 1712 2344	0.3 3.8 0.6 3.7
7 Tu 0459 1137 1713 2351 ●	0.2 3.9 0.5 3.8	**22** W 0536 1207 1742 ○	0.3 3.8 0.5
8 W 0543 1222 1754	0.0 4.1 0.4	**23** Th 0017 0605 1236 1810	3.8 0.2 3.8 0.5
9 Th 0035 0625 1304 1834	4.0 -0.1 4.1 0.3	**24** F 0046 0634 1304 1839	3.9 0.2 3.8 0.4
10 F 0116 0706 1345 1914	4.1 -0.1 4.1 0.3	**25** Sa 0114 0702 1331 1907	3.9 0.3 3.7 0.5
11 Sa 0155 0747 1426 1955	4.2 0.0 3.9 0.4	**26** Su 0142 0730 1359 1937	3.8 0.4 3.6 0.6
12 Su 0236 0827 1507 2039	4.1 0.2 3.8 0.5	**27** M 0212 0759 1429 2006	3.7 0.6 3.5 0.7
13 M 0318 0911 1549 2127	4.0 0.5 3.6 0.6	**28** Tu 0244 0827 1501 2037	3.6 0.7 3.5 0.7
14 Tu 0407 1002 1638 2226 ☽	3.7 0.8 3.3 0.8	**29** W 0321 0901 1539 2118	3.5 0.8 3.4 0.8
15 W 0506 1109 1740 2351	3.4 1.1 3.1 0.9	**30** Th 0406 0949 1628 2217 ☽	3.3 1.0 3.2 1.0
		31 F 0508 1105 1736 2354	3.1 1.2 3.0 1.0

APRIL

Time	m	Time	m
1 Sa 0634 1249 1903	3.1 1.2 3.0	**16** Su 0219 0849 1442 2104	0.6 3.4 1.1 3.3
2 Su 0131 0808 1409 2029	0.9 3.2 1.0 3.1	**17** M 0314 0942 1529 2155	0.5 3.5 0.9 3.4
3 M 0246 0925 1511 2141	0.6 3.5 0.8 3.4	**18** Tu 0356 1024 1607 2235	0.4 3.7 0.8 3.6
4 Tu 0345 1024 1602 2237	0.3 3.8 0.6 3.7	**19** W 0430 1101 1640 2311	0.4 3.7 0.7 3.7
5 W 0434 1113 1647 2326	0.1 4.0 0.5 3.9	**20** Th 0501 1133 1711 2344	0.3 3.8 0.6 3.8
6 Th 0519 1158 1730 ●	0.0 4.1 0.3	**21** F 0530 1204 1740 ○	0.3 3.8 0.5
7 F 0011 0601 1242 1812	4.1 -0.1 4.1 0.3	**22** Sa 0015 0600 1234 1811	3.8 0.4 3.9 0.5
8 Sa 0053 0642 1323 1855	4.2 0.0 4.1 0.3	**23** Su 0046 0629 1302 1842	3.8 0.4 3.8 0.5
9 Su 0134 0723 1402 1938	4.2 0.2 3.9 0.3	**24** M 0117 0700 1331 1914	3.8 0.5 3.7 0.6
10 M 0216 0806 1442 2023	4.1 0.4 3.7 0.4	**25** Tu 0149 0731 1402 1947	3.7 0.7 3.7 0.7
11 Tu 0301 0850 1524 2114	3.9 0.7 3.5 0.6	**26** W 0225 0804 1437 2025	3.6 0.7 3.6 0.7
12 W 0350 0941 1612 2214 ☾	3.6 1.0 3.3 0.7	**27** Th 0305 0844 1518 2111	3.5 0.9 3.4 0.7
13 Th 0449 1044 1709 2333	3.4 1.2 3.3 0.8	**28** F 0355 0936 1609 2214 ☽	3.4 1.1 3.3 0.8
14 F 0610 1207 1831	3.2 1.3 3.0	**29** Sa 0457 1051 1716 2339	3.3 1.2 3.1 0.8
15 Sa 0104 0738 1335 1958	0.8 3.2 1.3 3.1	**30** Su 0615 1218 1835	3.2 1.2 3.1

Chart Datum: 2.02 metres below Ordnance Datum (Newlyn)

ENGLAND, EAST COAST — HARWICH

Lat 51°57′ N   Long 1°17′ E

TIMES AND HEIGHTS OF HIGH AND LOW WATERS

TIME ZONE UT(GMT)
For Summer Time add ONE hour in non-shaded areas

YEAR 1989

MAY

	Time	m		Time	m
1 M	0103 0738 1334 1952	0.7 3.3 1.0 3.3	**16** Tu	0222 0854 1440 2105	0.6 3.4 1.0 3.4
2 Tu	0215 0853 1437 2103	0.5 3.6 0.8 3.5	**17** W	0310 0941 1525 2153	0.5 3.5 0.9 3.5
3 W	0314 0955 1532 2204	0.3 3.8 0.7 3.7	**18** Th	0349 1021 1603 2234	0.5 3.6 0.8 3.6
4 Th	0406 1047 1621 2258	0.1 3.9 0.5 3.9	**19** F	0424 1058 1638 2312	0.5 3.7 0.7 3.7
5 F ●	0454 1134 1708 2347	0.1 4.0 0.4 4.0	**20** Sa	0457 1132 1713 2349	0.5 3.8 0.6 3.7
6 Sa	0539 1218 1754	0.1 4.0 0.3	**21** Su	0530 1205 1749	0.6 3.8 0.6
7 Su	0032 0621 1300 1839	4.1 0.2 4.0 0.3	**22** M	0024 0604 1239 1824	3.7 0.6 3.8 0.6
8 M	0117 0704 1341 1926	4.1 0.4 3.9 0.3	**23** Tu	0059 0638 1313 1900	3.7 0.7 3.7 0.6
9 Tu	0201 0748 1422 2013	4.0 0.6 3.7 0.4	**24** W	0137 0713 1348 1938	3.7 0.7 3.7 0.6
10 W	0247 0833 1503 2104	3.8 0.8 3.5 0.5	**25** Th	0216 0752 1426 2023	3.6 0.8 3.6 0.6
11 Th	0335 0921 1549 2159	3.6 1.1 3.4 0.6	**26** F	0300 0839 1511 2112	3.6 0.9 3.5 0.6
12 F ☽	0430 1014 1641 2302	3.4 1.2 3.2 0.7	**27** Sa	0350 0932 1602 2213	3.5 1.0 3.4 0.6
13 Sa	0534 1119 1746	3.3 1.3 3.1	**28** Su ☾	0449 1035 1701 2320	3.4 1.1 3.3 0.6
14 Su	0014 0646 1232 1900	0.7 3.2 1.3 3.1	**29** M	0557 1147 1808	3.4 1.1 3.3
15 M	0123 0757 1344 2009	0.6 3.3 1.2 3.2	**30** Tu	0034 0709 1259 1919	0.5 3.5 1.0 3.4
			31 W	0142 0819 1405 2029	0.4 3.6 0.9 3.5

JUNE

	Time	m		Time	m
1 Th	0246 0924 1504 2136	0.4 3.7 0.8 3.7	**16** F	0303 0938 1525 2157	0.7 3.5 0.9 3.4
2 F	0342 1021 1600 2237	0.3 3.8 0.6 3.8	**17** Sa	0349 1023 1610 2244	0.7 3.6 0.8 3.5
3 Sa ●	0434 1113 1654 2330	0.3 3.9 0.5 3.9	**18** Su	0430 1104 1652 2326	0.7 3.7 0.7 3.6
4 Su	0522 1200 1743	0.4 3.9 0.4	**19** M ○	0509 1144 1733	0.8 3.7 0.6
5 M	0019 0607 1243 1831	4.0 0.5 3.9 0.3	**20** Tu	0008 0547 1224 1814	3.7 0.8 3.7 0.6
6 Tu	0104 0650 1326 1919	4.0 0.6 3.8 0.3	**21** W	0049 0625 1303 1855	3.7 0.7 3.7 0.5
7 W	0149 0733 1406 2004	3.9 0.7 3.7 0.3	**22** Th	0130 0704 1342 1937	3.8 0.7 3.7 0.4
8 Th	0233 0815 1446 2049	3.8 0.9 3.6 0.4	**23** F	0212 0747 1423 2022	3.8 0.7 3.7 0.4
9 F	0318 0857 1528 2135	3.7 1.0 3.5 0.5	**24** Sa	0257 0832 1507 2108	3.7 0.8 3.7 0.4
10 Sa	0403 0941 1610 2223	3.5 1.1 3.4 0.5	**25** Su	0345 0919 1552 2159	3.7 0.8 3.6 0.4
11 Su ☽	0452 1030 1658 2316	3.4 1.1 3.3 0.6	**26** M	0435 1012 1644 2257	3.6 0.9 3.6 0.4
12 M	0546 1127 1754	3.3 1.2 3.3	**27** Tu	0533 1113 1743	3.5 1.0 3.5
13 Tu	0015 0646 1232 1859	0.7 3.2 1.2 3.2	**28** W	0003 0638 1224 1849	0.5 3.5 1.0 3.5
14 W	0116 0749 1337 2005	0.7 3.3 1.1 3.3	**29** Th	0114 0748 1337 2002	0.6 3.5 1.0 3.5
15 Th	0212 0847 1436 2105	0.7 3.4 1.0 3.3	**30** F	0223 0858 1446 2118	0.6 3.5 0.9 3.6

JULY

	Time	m		Time	m
1 Sa	0328 1003 1552 2226	0.6 3.6 0.7 3.7	**16** Su	0318 0949 1548 2220	1.0 3.4 0.7 3.4
2 Su	0424 1058 1649 2322	0.6 3.7 0.5 3.8	**17** M	0409 1041 1637 2309	0.9 3.5 0.8 3.6
3 M ●	0513 1147 1739	0.7 3.8 0.4	**18** Tu ○	0454 1127 1722 2354	0.9 3.6 0.6 3.7
4 Tu	0011 0556 1232 1825	3.9 0.7 3.8 0.3	**19** W	0534 1211 1804	0.8 3.7 0.5
5 W	0056 0636 1313 1907	3.9 0.7 3.8 0.2	**20** Th	0038 0615 1253 1846	3.8 0.7 3.8 0.3
6 Th	0137 0716 1351 1948	3.8 0.7 3.8 0.3	**21** F	0121 0655 1334 1927	3.9 0.6 3.9 0.2
7 F	0216 0752 1426 2026	3.8 0.8 3.8 0.3	**22** Sa	0204 0734 1415 2009	4.0 0.6 3.9 0.2
8 Sa	0254 0829 1501 2104	3.7 0.8 3.7 0.4	**23** Su	0246 0815 1454 2051	3.9 0.6 3.9 0.2
9 Su	0331 0905 1536 2142	3.7 0.9 3.7 0.5	**24** M	0329 0858 1536 2136	3.8 0.7 3.9 0.3
10 M	0409 0943 1613 2223	3.5 1.0 3.6 0.6	**25** Tu ☾	0416 0946 1623 2227	3.7 0.8 3.8 0.5
11 Tu ☽	0448 1028 1655 2312	3.4 1.1 3.5 0.7	**26** W	0508 1041 1718 2330	3.6 1.0 3.6 0.7
12 W	0536 1125 1749	3.3 1.2 3.3	**27** Th	0608 1153 1825	3.4 1.1 3.5
13 Th	0010 0635 1234 1859	0.9 3.2 1.2 3.2	**28** F	0049 0720 1319 1947	0.8 3.3 1.1 3.4
14 F	0116 0744 1344 2015	0.9 3.2 1.2 3.3	**29** Sa	0211 0840 1442 2114	0.9 3.4 0.9 3.5
15 Sa	0219 0850 1449 2122	1.0 3.3 1.1 3.3	**30** Su	0324 0952 1553 2223	0.9 3.5 0.7 3.7
			31 M ●	0420 1048 1648 2316	0.8 3.7 0.5 3.8

AUGUST

	Time	m		Time	m
1 Tu ●	0504 1136 1733	0.8 3.8 0.4	**16** W	0435 1108 1704 2337	0.9 3.6 0.5 3.8
2 W	0001 0543 1218 1811	3.9 0.8 3.9 0.3	**17** Th ○	0516 1154 1747	0.7 3.8 0.3
3 Th	0042 0618 1255 1848	3.9 0.7 3.9 0.2	**18** F	0022 0556 1236 1828	4.0 0.6 4.0 0.2
4 F	0119 0652 1330 1923	3.9 0.7 3.9 0.2	**19** Sa	0104 0635 1316 1907	4.1 0.5 4.1 0.1
5 Sa	0152 0726 1401 1957	3.9 0.7 3.9 0.3	**20** Su	0145 0713 1355 1947	4.1 0.5 4.1 0.1
6 Su	0225 0757 1432 2027	3.8 0.7 3.9 0.4	**21** M	0226 0754 1434 2027	4.1 0.5 4.1 0.2
7 M	0256 0829 1501 2100	3.7 0.8 3.8 0.5	**22** Tu	0307 0836 1515 2110	3.9 0.6 4.0 0.4
8 Tu	0327 0900 1534 2134	3.6 0.9 3.7 0.6	**23** W ☾	0350 0921 1600 2159	3.8 0.8 3.9 0.6
9 W	0400 0936 1610 2213	3.5 1.0 3.5 0.8	**24** Th	0440 1016 1655 2302	3.6 1.0 3.7 0.9
10 Th	0440 1020 1654 2305	3.4 1.1 3.3 1.0	**25** F	0539 1132 1808	3.3 1.1 3.4
11 F	0529 1126 1756	3.2 1.3 3.1	**26** Sa	0028 0656 1310 1942	1.1 3.2 1.1 3.3
12 Sa	0018 0638 1253 1921	1.2 3.1 1.3 3.0	**27** Su	0202 0826 1442 2111	1.2 3.3 0.9 3.5
13 Su	0138 0801 1416 2050	1.2 3.1 1.2 3.2	**28** M	0315 0939 1548 2213	1.1 3.5 0.6 3.7
14 M	0251 0918 1525 2159	1.1 3.2 1.0 3.4	**29** Tu	0407 1033 1635 2302	1.0 3.7 0.4 3.9
15 Tu	0349 1019 1619 2251	1.0 3.5 0.7 3.6	**30** W	0447 1118 1715 2343	0.9 3.8 0.3 3.9
			31 Th ●	0520 1156 1749	0.8 3.9 0.3

Chart Datum: 2.02 metres below Ordnance Datum (Newlyn)

ENGLAND, EAST COAST — HARWICH
Lat 51°57′ N Long 1°17′ E
TIMES AND HEIGHTS OF HIGH AND LOW WATERS YEAR 1989

TIME ZONE UT(GMT)
For Summer Time add ONE hour in non-shaded areas

SEPTEMBER

Day	Time	m	Day	Time	m
1 F	0019 / 0553 / 1229 / 1821	3.9 / 0.7 / 4.0 / 0.3	16 Sa	0530 / 1211 / 1803	0.5 / 4.1 / 0.1
2 Sa	0052 / 0624 / 1302 / 1852	3.9 / 0.6 / 4.0 / 0.3	17 Su	0042 / 0611 / 1252 / 1842	4.2 / 0.5 / 4.2 / 0.0
3 Su	0123 / 0655 / 1331 / 1921	3.9 / 0.6 / 4.0 / 0.3	18 M	0123 / 0650 / 1333 / 1921	4.2 / 0.4 / 4.3 / 0.1
4 M	0151 / 0724 / 1359 / 1951	3.9 / 0.6 / 4.0 / 0.4	19 Tu	0202 / 0731 / 1412 / 2002	4.1 / 0.5 / 4.2 / 0.3
5 Tu	0219 / 0754 / 1427 / 2019	3.8 / 0.7 / 3.9 / 0.6	20 W	0243 / 0815 / 1456 / 2046	3.9 / 0.6 / 4.1 / 0.5
6 W	0247 / 0823 / 1458 / 2049	3.7 / 0.9 / 3.7 / 0.7	21 Th	0325 / 0903 / 1542 / 2135	3.8 / 0.8 / 3.9 / 0.9
7 Th	0318 / 0853 / 1532 / 2121	3.6 / 1.0 / 3.6 / 0.9	22 F	0413 / 1000 / 1640 / 2238 ☾	3.5 / 0.9 / 3.6 / 1.2
8 F ☾	0355 / 0929 / 1614 / 2204	3.5 / 1.1 / 3.4 / 1.2	23 Sa	0511 / 1119 / 1756	3.3 / 1.0 / 3.4
9 Sa	0440 / 1023 / 1709 / 2315	3.3 / 1.3 / 3.2 / 1.4	24 Su	0005 / 0631 / 1300 / 1933	1.4 / 3.2 / 1.0 / 3.3
10 Su	0543 / 1201 / 1834	3.1 / 1.3 / 3.0	25 M	0140 / 0802 / 1426 / 2054	1.4 / 3.2 / 0.8 / 3.5
11 M	0059 / 0710 / 1342 / 2015	1.4 / 3.0 / 1.2 / 3.1	26 Tu	0251 / 0914 / 1527 / 2152	1.2 / 3.5 / 0.6 / 3.7
12 Tu	0220 / 0840 / 1456 / 2131	1.2 / 3.2 / 0.9 / 3.4	27 W	0342 / 1007 / 1612 / 2237	1.0 / 3.7 / 0.4 / 3.9
13 W	0321 / 0938 / 1552 / 2227	1.0 / 3.5 / 0.6 / 3.7	28 Th	0420 / 1049 / 1648 / 2316	0.9 / 3.9 / 0.4 / 3.9
14 Th	0409 / 1041 / 1638 / 2315	0.8 / 3.7 / 0.4 / 4.0	29 F	0454 / 1126 / 1719 / 2350	0.8 / 3.9 / 0.4 / 3.9
15 F ○	0451 / 1127 / 1722 / 2358	0.7 / 3.9 / 0.2 / 4.1	30 Sa	0525 / 1200 / 1749	0.7 / 4.0 / 0.4

OCTOBER

Day	Time	m	Day	Time	m
1 Su	0021 / 0554 / 1231 / 1817	3.9 / 0.6 / 4.0 / 0.4	16 M	0017 / 0547 / 1228 / 1817	4.2 / 0.4 / 4.2 / 0.1
2 M	0050 / 0624 / 1259 / 1846	4.0 / 0.6 / 4.0 / 0.4	17 Tu	0057 / 0629 / 1310 / 1859	4.2 / 0.4 / 4.3 / 0.3
3 Tu	0117 / 0653 / 1328 / 1914	3.9 / 0.7 / 3.9 / 0.6	18 W	0138 / 0713 / 1354 / 1941	4.1 / 0.5 / 4.2 / 0.5
4 W	0144 / 0723 / 1357 / 1942	3.9 / 0.8 / 3.8 / 0.7	19 Th	0219 / 0759 / 1439 / 2026	3.9 / 0.6 / 4.1 / 0.8
5 Th	0212 / 0751 / 1429 / 2012	3.8 / 0.8 / 3.7 / 0.9	20 F	0303 / 0850 / 1528 / 2115	3.7 / 0.7 / 3.9 / 1.0
6 F	0244 / 0822 / 1504 / 2043	3.7 / 0.9 / 3.6 / 1.0	21 Sa ☾	0349 / 0949 / 1624 / 2214	3.5 / 0.8 / 3.6 / 1.3
7 Sa	0319 / 0900 / 1548 / 2127	3.5 / 1.0 / 3.4 / 1.2	22 Su	0445 / 1102 / 1737 / 2329	3.3 / 0.9 / 3.4 / 1.4
8 Su ☾	0404 / 0953 / 1642 / 2233	3.3 / 1.1 / 3.2 / 1.4	23 M	0558 / 1229 / 1902	3.2 / 0.9 / 3.4
9 M	0506 / 1123 / 1801	3.2 / 1.2 / 3.1	24 Tu	0055 / 0723 / 1348 / 2018	1.4 / 3.3 / 0.8 / 3.5
10 Tu	0017 / 0629 / 1306 / 1934	1.4 / 3.1 / 1.1 / 3.2	25 W	0209 / 0834 / 1450 / 2117	1.3 / 3.4 / 0.6 / 3.6
11 W	0141 / 0755 / 1419 / 2054	1.3 / 3.2 / 0.8 / 3.5	26 Th	0304 / 0929 / 1535 / 2203	1.1 / 3.6 / 0.5 / 3.7
12 Th	0243 / 0907 / 1517 / 2155	1.0 / 3.5 / 0.5 / 3.8	27 F	0346 / 1014 / 1613 / 2242	1.0 / 3.7 / 0.5 / 3.8
13 F	0335 / 1006 / 1607 / 2245	0.8 / 3.8 / 0.3 / 4.0	28 Sa	0421 / 1052 / 1645 / 2316	0.8 / 3.8 / 0.5 / 3.9
14 Sa	0420 / 1057 / 1652 / 2333 ○	0.7 / 4.0 / 0.2 / 4.2	29 Su ●	0454 / 1127 / 1717 / 2349	0.7 / 3.9 / 0.5 / 3.9
15 Su	0504 / 1143 / 1734	0.5 / 4.2 / 0.1	30 M	0525 / 1200 / 1743	0.7 / 3.9 / 0.5
			31 Tu	0018 / 0556 / 1231 / 1812	3.9 / 0.7 / 3.9 / 0.6

NOVEMBER

Day	Time	m	Day	Time	m
1 W	0046 / 0627 / 1302 / 1842	3.9 / 0.7 / 3.9 / 0.7	16 Th	0119 / 0702 / 1340 / 1924	4.0 / 0.4 / 4.2 / 0.7
2 Th	0116 / 0657 / 1334 / 1913	3.9 / 0.7 / 3.8 / 0.8	17 F	0201 / 0749 / 1426 / 2009	3.9 / 0.5 / 4.0 / 0.9
3 F	0145 / 0730 / 1408 / 1945	3.8 / 0.8 / 3.7 / 0.9	18 Sa	0244 / 0840 / 1514 / 2057	3.7 / 0.5 / 3.8 / 1.1
4 Sa	0219 / 0805 / 1447 / 2022	3.7 / 0.9 / 3.6 / 1.1	19 Su	0329 / 0936 / 1607 / 2148	3.6 / 0.6 / 3.6 / 1.2
5 Su	0257 / 0849 / 1532 / 2110	3.5 / 0.9 / 3.5 / 1.2	20 M ☾	0420 / 1034 / 1706 / 2247	3.4 / 0.7 / 3.5 / 1.3
6 M	0345 / 0945 / 1628 / 2214	3.4 / 1.0 / 3.4 / 1.3	21 Tu	0519 / 1140 / 1814 / 2354	3.3 / 0.8 / 3.4 / 1.4
7 Tu	0444 / 1102 / 1739 / 2337	3.3 / 1.0 / 3.3 / 1.3	22 W	0628 / 1252 / 1923	3.3 / 0.8 / 3.3
8 W	0557 / 1228 / 1857	3.3 / 0.9 / 3.4	23 Th	0107 / 0738 / 1357 / 2027	1.3 / 3.3 / 0.7 / 3.4
9 Th	0057 / 0713 / 1340 / 2012	1.2 / 3.4 / 0.7 / 3.6	24 F	0213 / 0840 / 1450 / 2119	1.2 / 3.4 / 0.7 / 3.5
10 F	0204 / 0823 / 1442 / 2118	1.0 / 3.6 / 0.5 / 3.8	25 Sa	0304 / 0932 / 1532 / 2203	1.1 / 3.6 / 0.7 / 3.6
11 Sa	0300 / 0928 / 1535 / 2216	0.9 / 3.8 / 0.4 / 4.0	26 Su	0348 / 1017 / 1610 / 2241	0.9 / 3.7 / 0.7 / 3.7
12 Su	0352 / 1026 / 1624 / 2306	0.7 / 4.0 / 0.3 / 4.1	27 M	0426 / 1057 / 1642 / 2318	0.8 / 3.7 / 0.7 / 3.8
13 M ○	0440 / 1119 / 1712 / 2353	0.6 / 4.1 / 0.3 / 4.1	28 Tu ●	0459 / 1134 / 1715 / 2350	0.8 / 3.8 / 0.7 / 3.8
14 Tu	0527 / 1207 / 1756	0.5 / 4.2 / 0.3	29 W	0533 / 1210 / 1747	0.7 / 3.8 / 0.8
15 W	0036 / 0614 / 1253 / 1841	4.1 / 0.4 / 4.3 / 0.5	30 Th	0024 / 0608 / 1245 / 1819	3.8 / 0.7 / 3.8 / 0.8

DECEMBER

Day	Time	m	Day	Time	m
1 F	0056 / 0643 / 1320 / 1855	3.8 / 0.7 / 3.8 / 0.9	16 Sa	0147 / 0741 / 1413 / 1952	3.9 / 0.3 / 4.0 / 0.8
2 Sa	0130 / 0720 / 1358 / 1931	3.7 / 0.7 / 3.7 / 0.9	17 Su	0229 / 0826 / 1458 / 2034	3.8 / 0.4 / 3.8 / 0.9
3 Su	0206 / 0801 / 1439 / 2013	3.7 / 0.7 / 3.7 / 1.0	18 M	0310 / 0911 / 1543 / 2117	3.7 / 0.4 / 3.7 / 1.0
4 M	0247 / 0847 / 1525 / 2101	3.6 / 0.7 / 3.6 / 1.0	19 Tu ☾	0352 / 0957 / 1630 / 2203	3.6 / 0.5 / 3.5 / 1.1
5 Tu	0334 / 0939 / 1617 / 2156	3.5 / 0.7 / 3.5 / 1.1	20 W	0437 / 1047 / 1719 / 2254	3.5 / 0.7 / 3.4 / 1.2
6 W	0427 / 1038 / 1716 / 2301	3.5 / 0.7 / 3.5 / 1.2	21 Th	0529 / 1143 / 1817 / 2356	3.4 / 0.8 / 3.2 / 1.3
7 Th	0529 / 1149 / 1824	3.5 / 0.7 / 3.5	22 F	0629 / 1246 / 1920	3.3 / 0.9 / 3.2
8 F	0014 / 0636 / 1300 / 1934	1.2 / 3.5 / 0.6 / 3.5	23 Sa	0106 / 0738 / 1349 / 2023	1.3 / 3.2 / 0.9 / 3.3
9 Sa	0124 / 0747 / 1408 / 2043	1.1 / 3.6 / 0.6 / 3.6	24 Su	0213 / 0846 / 1447 / 2119	1.2 / 3.3 / 0.9 / 3.4
10 Su	0229 / 0856 / 1508 / 2148	0.9 / 3.7 / 0.5 / 3.8	25 M	0312 / 0943 / 1536 / 2207	1.1 / 3.4 / 0.9 / 3.5
11 M	0328 / 1003 / 1604 / 2244	0.8 / 3.9 / 0.5 / 3.9	26 Tu	0400 / 1031 / 1619 / 2251	1.0 / 3.5 / 0.9 / 3.6
12 Tu ○	0424 / 1102 / 1655 / 2334	0.6 / 4.0 / 0.5 / 3.9	27 W	0442 / 1115 / 1657 / 2330	0.8 / 3.6 / 0.9 / 3.6
13 W	0518 / 1154 / 1743	0.5 / 4.1 / 0.5	28 Th ●	0520 / 1154 / 1732	0.7 / 3.7 / 0.9
14 Th	0021 / 0607 / 1243 / 1827	3.9 / 0.4 / 4.1 / 0.6	29 F	0008 / 0558 / 1234 / 1808	3.7 / 0.6 / 3.7 / 0.8
15 F	0104 / 0655 / 1330 / 1910	3.9 / 0.3 / 4.1 / 0.7	30 Sa	0046 / 0636 / 1313 / 1845	3.7 / 0.5 / 3.8 / 0.8
			31 Su	0124 / 0716 / 1352 / 1924	3.7 / 0.5 / 3.8 / 0.7

Chart Datum: 2.02 metres below Ordnance Datum (Newlyn)

AREA 4—E England   323

RIVER ORWELL continued

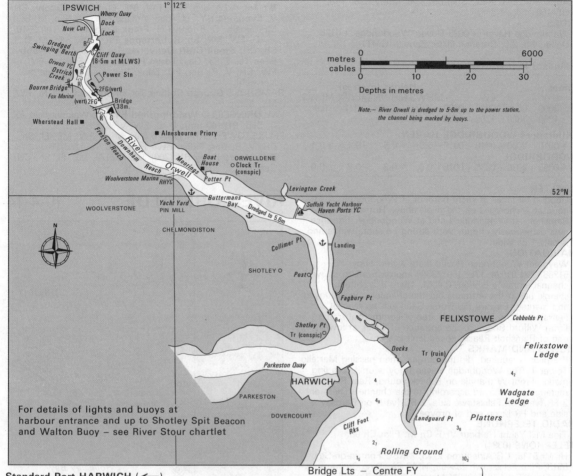

Standard Port HARWICH (←)

Times				Height (metres)			
HW		LW		MHWS	MHWN	MLWN	MLWS
0000	0600	0000	0600	4.0	3.4	1.1	0.4
1200	1800	1200	1800				

Differences IPSWICH
+0015 +0025 0000 +0010 +0.2 0.0 −0.1 −0.1

SHELTER
Good. Entrance and river well marked. Anchorages above Shotley Pt on W side, or in Buttermans Bay. No facilities for yachts at Felixstowe. Berths available by arrangement in Suffolk Yacht Harbour, Woolverstone Marina and Fox Marina (Ipswich). For berths in Ipswich Dock it is essential to contact Ipswich Port Radio before arrival (Tel. 53999) or call *Wherry Quay* Ch 14. Then call Ipswich Port Radio (Ch 14) for lock to open. Keep clear of commercial shipping at all times.

NAVIGATION
Waypoint Shotley Spit S cardinal Bn, Q(6)+LFl 15s, 51°57'.22N 01°17'.70E, at river entrance. Main danger is the large amount of merchant shipping, ferries etc entering and leaving Harwich, Felixstowe and passing up river to Ipswich. Ipswich dock opens from HW−1 to HW but mainly for commercial vessels.
There is a speed limit of 6 knots in the R Orwell.

LIGHTS AND MARKS
Entrance between Shotley Spit Lt Bn Q(6) + L Fl 15s to port and Walton buoy, Fl (3) G 10s to stbd. For detail see R Stour Chartlet. (←) 10.4.11.
Suffolk Yacht Harbour approach marked by four Bns. Ldg Lts both FY. Woolverstone Marina Lts 2 FR (vert). R and G Lts control entry to Ipswich Dock (H24).
New Cut, W of Ipswich Dock, 3 FR (vert) indicates cut closed.

Bridge Lts − Centre FY
No 9 Pier 2 FR (vert) } shown up and
No 10 Pier 2 FG (vert) } down stream.

RADIO TELEPHONE (local times)
Call: *Ipswich Port Radio* VHF Ch **14** 16; 12 **14** (H24). Once above the bridge keep constant listening watch on Ch 14. Fox Marina, Woolverstone Marina, Suffolk Yacht Harbour Ch M. (0800 – 1730)

TELEPHONE (0473)
Hr Mr Orwell Navigation Service 231010; CG & MRSC Frinton-on-Sea 5518; HMC Freefone Customs Yachts: RT (0473) 52837; Marinecall 0898 500 455; Hosp 212477.

FACILITIES
EC Chelmondiston - Thursday; Ipswich - Wednesday;
PIN MILL/WOOLVERSTONE: **Woolverstone Marina** (300+22 visitors) Tel. 84206, Slip, P (cans), D, FW, ME, El, Sh, AC, Gas, Gaz, Lau, C (25 ton mobile), CH, V;
Royal Harwich YC Tel. 84319, R, Bar; **Pin Mill SC** Tel. 84271; **J. Ward** Tel. 84276, Slip, M, (10 visitors), D, L, FW, ME, El, Sh, CH, AB, V, R, Bar; **F. A. Webb BY** Tel. 84291, ME, El, Sh, Slip.
Town PO; Bank (Ipswich); Rly (Ipswich); Air (Cambridge).
LEVINGTON: **Suffolk Yacht Harbour** (390+10 visitors) Tel. Nacton 240, Slip, P, D, FW, ME, El, Sh, C (9 ton), BH (10 ton), CH, V, Gas, Gaz, AC, SM, Bar Access H24;
Town PO; Bank (Felixstowe); Rly (bus to Ipswich); Air (Cambridge or Norwich).
IPSWICH: **Fox's Marina** (100+some visitors) Tel. 689111 FW, AC, D, Gas, Gaz, BH (26 ton), C (7 ton), ME, El, Sh, Bar, CH, ACA; **Neptune Marina** (18 visitors) Tel. 34578, Sh, ME, FW; **Wherry Quay Yacht (Oysterworld)** (30) Tel. 53999, AB, M, FW, AC, R, Access via lock (HW−2 to HW+½; **Orwell YC** Tel. 55288, Slip, L, FW, Bar;
J. Parker Tel. 210181, SM; **Austin Farrar** Tel. 622666, SM; **Bob Spalding** Tel. 79891, ME.
Town PO; Bank; Rly; Air (Cambridge or Norwich).

RIVER DEBEN 10-4-13
Suffolk

CHARTS
Admiralty 2693, 2052; Stanford 3, 6; Imray C28, Y15; OS 169

TIDES
Woodbridge Haven +0025 Dover; Woodbridge +0125 Dover; ML 2.0; Duration 0635; Zone 0 (GMT)

Standard Port HARWICH (←)

Times				Height (metres)			
HW		LW		MHWS	MHWN	MLWN	MLWS
0000	0600	0000	0600	4.0	3.4	1.1	0.4
1200	1800	1200	1800				
Differences WOODBRIDGE HAVEN							
+0010	0000	+0010	+0010	−0.3	−0.5	−0.1	+0.1
WOODBRIDGE							
+0055	+0030	+0055	+0015	0.0	−0.3	−0.2	0.0

SHELTER
Good at all times in yacht harbour at Woodbridge. Also good at anchorages up the river, above Horse Sand, at Ramsholt, at Waldringfield and at Woodbridge. Entrance gets dangerously choppy with strong on-shore winds and is only 1 ca wide.

NAVIGATION
Waypoint Woodbridge Haven (safe water) buoy, 51°58'.36N 01°24'.17E, 150°/330° approx from/to buoyed channel entrance (shifts), 0.40M. There is a shifting shingle bar at the entrance but the channel is well buoyed and marked. Beware the Horse Sand just up river of Felixstowe ferry. It is entirely free of commercial traffic. From Wilford Bridge to the entrance of Créek at lower end of Ramsholt Reach, speed limit is 8 kn.

LIGHTS AND MARKS
If a pilot is required, dip the burgee when passing Martello Tower T. The Woodbridge Haven Buoy is unlit. Leading marks, Front W triangle on R background, Rear R rectangle, are moved according to the channel. There are 2 FR (vert) on Felixstowe side, 2 FG (vert) on Bardsey side and FR Lt on end of groyne.

RADIO TELEPHONE
Tide Mill Yacht Harbour VHF Ch M. Pilot Ch 08.

TELEPHONE (0394)
Hr Mr 5118; CG Aldeburgh 2779; MRSC Frinton-on-Sea 5518; Pilot Felixstowe 283469; HMC Freefone Customs Yachts: RT (0473) 219481; Marinecall 0898 500 455; Dr 2046.

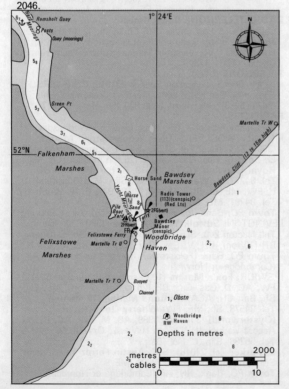

FACILITIES
EC Wednesday.
WOODBRIDGE: **Tide Mill Yacht Harbour** (150+50 visitors) Tel. 385745, Slip, D, L, FW, ME, El, Sh, C (10 ton), AC, V, Access HW−3½ to HW+2½; **Whisstocks BY** Tel. 4222, Slip, MD, FW, ME, El, Sh, C (20 ton), CH, V; **Eversons BY** Tel. 4358, Access HW∓4, Slip, M, D, ME, El, Sh, C (6 ton), CH; **Frank Knights BY** Tel. 2318, M, D, FW, ME, Sh; **Robertsons** Tel. 2305, Slip, M, D, L, Sh, CH; **Small Craft Deliveries** Tel. 2600 ACA; **Webb Bros** Tel. 2183, Gas; **Town** P, D, L (opp Everson BY), FW, CH, V, R, Bar, PO; Bank; Rly; Air (Cambridge or Norwich).
RAMSHOLT **George Collins** Tel. Shottisham 229, M, FW, ME, Bar;
WALDRINGFIELD **Waldringfield BY** Tel. Waldringfield 260, C, Slip, BH (40 ton), D, L, FW, V, Bar;
FELIXSTOWE FERRY QUAY Slip, M, L, FW, ME, El, Sh, CH, V, R, Bar; **Felixstowe Ferry SC** Tel. 283785; **Felixstowe Ferry BY** (200) Tel. 282173, ME, El, Sh, CH, Gas.

RIVER ORE/ALDE 10-4-14
Suffolk

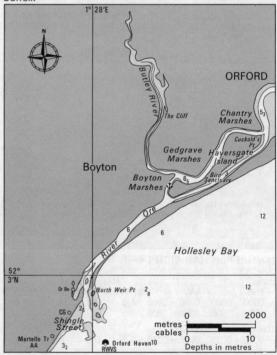

CHARTS
Admiralty 2695, 1543, 2052, 2693; Stanford 6, 3; Imray Y15, C28; OS 169

TIDES
Ent. +0010 Dover Slaughden Quay +0155 Dover; ML 1.5; Duration 0620; Zone 0 (GMT).

Standard Port LOWESTOFT (→)

Times				Height (metres)			
HW		LW		MHWS	MHWN	MLWN	MLWS
0300	0900	0200	0800	2.4	2.1	1.0	0.5
1500	2100	1400	2000				
Differences ALDEBURGH							
+0120	+0120	+0120	+0110	+0.4	+0.6	0.0	0.0
ORFORDNESS							
+0135	+0135	+0135	+0125	+0.4	+0.6	−0.1	0.0

Standard Port HARWICH (←)

Times				Height (metres)			
HW		LW		MHWS	MHWN	MLWN	MLWS
0000	0600	0000	0600	4.0	3.4	1.1	0.4
1200	1800	1200	1800				
Differences ORFORD HAVEN BAR							
−0025	−0025	−0040	−0015	−0.8	−0.8	−0.2	−0.1
ORFORD QUAY							
+0021	+0021	+0053	+0053	−1.4	−1.3	+0.2	0.0

RIVER ORE/ALDE continued

Differences at Slaughden Quay are +0115 on Harwich; Snape Bridge +0315 on Harwich

SHELTER
R Ore changes name to R Alde between Orford and Aldeburgh, and is navigable up to Snape. Good anchorages where shown, also between Slaughden Quay and Martello Tower and at Iken. Upper reaches of river are shallow and winding, and although marked by withies these may be damaged. Landing prohibited on MoD land and on Havergate Island bird sanctuary.

NAVIGATION
Waypoint Orford Haven (safe water) buoy, 52°01'.78N 01°27'.98E. The bar shifts after onshore gales and is dangerous in rough or confused seas. Ebb runs up to 6 kn. Up to date plan of entrance available from Orford Chandlery or Secretary, Aldeburgh YC. In absence of local information do not enter before half flood. Beware shoals off the orange tipped Bn on W shore just inside entrance (keep well to E side), and S to SW of Dove Pt (the most SW point of Havergate Island).

LIGHTS AND MARKS
Shingle Street, about 0.2M S of entrance, is located by a Martello Tower and CG Stn. Buoy and Bn are moved according to channel.

RADIO TELEPHONE
VHF Ch 16; 67 (H24).

TELEPHONE (Aldeburgh, 0728 85; Orford and Shottisham, 0394)
Hr Mr Aldeburgh 2896 and Orford c/o 450210; MRSC Frinton-on-Sea 5518; CG Aldeburgh 2061 and Shingle Street (week ends, bad weather), Shottisham 411558; Unofficial Pilot Mr Emmens Shottisham 411706; HMC Freefone Customs Yachts: RT (0473) 219481; Marinecall 0898 500 455; Dr Orford 450315, 450369, Shottisham 411214, and Aldeburgh 2027.

FACILITIES
ORFORD. **Orford Quay** Slip, AB (short-stay), M, L, FW, C (5 ton), SC, HMC, Bar; **Orford Chandlery** Tel. 450210, CH, D (cans), Gas, Gaz, R, Sh (small craft); **Orford SC** Slip, L, FW. **Village** (¼ M) EC Wed. P and D (cans), PO, V, R, Bar, Rly (occ. bus to Woodbridge).
ALDEBURGH. **Slaughden Quay** L, FW, HMC; **R. F. Upson** Tel. 2896, Slip, M, D (cans), ME, Sh; **D. G. Cable** Tel. Leiston 0119, M, Sh; **Aldeburgh BY** Tel. 2019, Slip, Gas, Gaz, P, D, ME, Sh, CH; **Aldeburgh YC** Tel. 2562, Slip, L, FW. **Town** (¾M), EC Wed, V, R, Bar, PO, Bank; Rly (bus to Wickham Market); Air (Norwich).

SOUTHWOLD 10-4-15
Suffolk

CHARTS
Admiralty 2695, 1543; Stanford 3; Imray C28; OS 156
TIDES
−0105 Dover; ML 1.3; Duration 0620; Zone 0 (GMT).

Standard Port LOWESTOFT (→)

Times				Height (metres)			
HW		LW		MHWS	MHWN	MLWN	MLWS
0300	0900	0200	0800	2.4	2.1	1.0	0.5
1500	2100	1400	2000				

Differences SOUTHWOLD
+0035 +0035 +0040 +0030 +0.1 +0.1 −0.1 −0.1

SHELTER
Good shelter and a stage, ¼ M beyond the ferry on N bank opposite the Harbour Inn, is reserved for visiting yachtsmen. In strong winds from S through E to N entrance is dangerous.

NAVIGATION
Waypoint 52°18'.06N 01°41'.80E, 135°/315° from/to North Pier Lt, 1.0M. Entry should be made on the flood since the ebb runs up to 5 kn. Some shoals are unpredictable. A sand and shingle bar lies off the harbour entrance, whose extent varies. Obtain details of approach channels from Hr Mr before entering or ask for a pilot (Ch 12 or 09 or Tel. 723502).

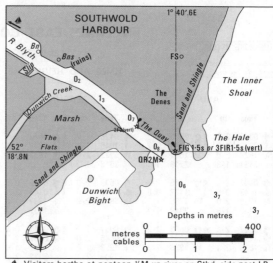

Visitors berths at pontoon ¼M up river on Stbd. side past LB house and Hr Mr office

LIGHTS AND MARKS
N Pier Lt & Lt Ho in line - 010°
N Pier Lt & Water Tr (conspic) - 342°
N Pier Lt & Walberswick Church - 268°
N Pier Lt & Windmill - 239°
Harbour mouth opens on - 300°
Two red flags or 3Fl R 1.5s (vert) at N pier means entry inadvisable or shipping movements in harbour.
NOTE: Lt Ho is in Southwold town, about 1 M N from harbour entrance, Fl (4) WR 20s 37m 22/20M; vis R (intens) 204°-220°, W220°-001°, R001°−032°.

RADIO TELEPHONE
Call: *Southwold Port Radio* VHF Ch 12; 16 (0800-1800 LT). Pilots Ch 12 09.

TELEPHONE (0502)
Hr Mr 723502; CG 722360; MRCC Great Yarmouth 851338; Pilot 723502; HMC Freefone Customs Yachts: RT (0473) 219481; Marinecall 0898 500 455; Weather Norwich 660779; Dr 722326; Hosp 723333.

FACILITIES
EC Wednesday (both Southwold and Walberswick); **Hr Mr** Slip, D (cans), L, FW, Sh, C (10 ton), CH, AB; **Harbour Inn (on quay)** FW, Bar; **Southwold SC** P, D, L. **Jeckells** Tel. 65007, SM; **Town** (½ M) V, R, Gas, Gaz, Kos, PO; Bank; Rly (bus to Brampton or Darsham); Air (Norwich).

LOWESTOFT 10-4-16
Suffolk

CHARTS
Admiralty 1536; Stanford 3; Imray C28; OS 156/134
TIDES
−0145 Dover; ML 1.6; Duration 0620; Zone 0 (GMT).
Lowestoft is a standard port and tidal predictions for each day of the year are given below.

SHELTER
Harbour is always available. Yacht basin in the SW corner. Yachts lie in 3 tiers on the port side (bad in E winds). Yacht basin is controlled by Royal Norfolk & Suffolk YC. Mooring on inner side of S Pier is dangerous due to bad state of the pier. Bridge to inner harbour (and Lake Lothing) opens (local times)
October to April
| Mon−Sat | 0700 | 0930 | 1900 | 2100 | |
| Sunday | 0800 | 0930 | 1400 | 1900 | 2100 |

May to September
| Mon−Sat | 0700 | 0930 | 1900 | 2100 | |
| Sunday | 0730 | 0930 | 1400 | 1900 | 2100 |

and also when ships pass through. Bridge clearance 2.2m at MHWS. Small craft may pass under the bridge at any time but VHF Ch 14 contact advisable (see 10.4.21).

ENGLAND, EAST COAST — LOWESTOFT

Lat 52°28′ N Long 1°45′ E

TIME ZONE UT (GMT)
For Summer Time add ONE hour in non-shaded areas

TIMES AND HEIGHTS OF HIGH AND LOW WATERS   YEAR 1989

Chart Datum: 1.50 metres below Ordnance Datum (Newlyn)

	JANUARY			FEBRUARY			MARCH			APRIL		
	Time m	Time m		Time m	Time m		Time m	Time m		Time m	Time m

JANUARY

1 Su 0300 2.2 / 0939 0.8 / 1628 2.0 / 2131 1.3
16 M 0326 2.3 / 1011 0.7 / 1652 2.0 / 2216 1.1

2 M 0401 2.2 / 1043 0.9 / 1726 2.0 / 2250 1.3
17 Tu 0445 2.3 / 1122 0.8 / 1756 2.1 / 2343 1.1

3 Tu 0507 2.1 / 1143 0.9 / 1816 2.1
18 W 0603 2.2 / 1231 0.9 / 1856 2.1

4 W 0003 1.2 / 0609 2.1 / 1231 0.9 / 1901 2.1
19 Th 0109 0.9 / 0724 2.2 / 1337 0.9 / 1948 2.2

5 Th 0100 1.1 / 0707 2.2 / 1316 0.9 / 1941 2.2
20 F 0216 0.8 / 0831 2.2 / 1431 0.9 / 2033 2.3

6 F 0152 1.0 / 0800 2.2 / 1400 0.9 / 2020 2.3
21 Sa 0307 0.6 / 0922 2.2 / 1516 0.9 / 2115 2.4 ○

7 Sa 0246 0.8 / 0850 2.3 / 1443 0.8 / 2101 2.4 ●
22 Su 0350 0.5 / 1005 2.2 / 1554 0.9 / 2152 2.4

8 Su 0339 0.6 / 0941 2.3 / 1531 0.8 / 2143 2.5
23 M 0430 0.4 / 1045 2.2 / 1628 0.8 / 2228 2.5

9 M 0426 0.5 / 1030 2.3 / 1620 0.7 / 2226 2.5
24 Tu 0507 0.4 / 1120 2.2 / 1658 0.8 / 2303 2.5

10 Tu 0513 0.3 / 1116 2.3 / 1705 0.7 / 2309 2.6
25 W 0541 0.4 / 1154 2.2 / 1724 0.8 / 2335 2.5

11 W 0556 0.2 / 1203 2.3 / 1750 0.7 / 2352 2.6
26 Th 0613 0.4 / 1228 2.1 / 1752 0.8

12 Th 0641 0.2 / 1252 2.2 / 1833 0.8
27 F 0009 2.4 / 0645 0.5 / 1301 2.1 / 1824 0.9

13 F 0035 2.6 / 0726 0.3 / 1343 2.1 / 1916 0.9
28 Sa 0043 2.4 / 0718 0.6 / 1339 2.1 / 1900 0.9

14 Sa 0122 2.5 / 0813 0.4 / 1439 2.0 / 2005 1.0 ☽
29 Su 0122 2.3 / 0754 0.7 / 1422 2.0 / 1941 1.0

15 Su 0218 2.4 / 0907 0.5 / 1545 2.0 / 2101 1.1
30 M 0209 2.2 / 0835 0.8 / 1516 1.9 / 2031 1.1 ☾

31 Tu 0311 2.1 / 0931 1.0 / 1620 1.9 / 2139 1.2

FEBRUARY

1 W 0426 2.0 / 1048 1.0 / 1722 2.0 / 2318 1.2
16 Th 0622 2.1 / 1231 1.1 / 1837 2.1

2 Th 0539 2.0 / 1156 1.0 / 1816 2.0
17 F 0115 0.8 / 0743 2.1 / 1337 1.0 / 1937 2.2

3 F 0037 1.1 / 0645 2.0 / 1252 1.0 / 1905 2.1
18 Sa 0211 0.6 / 0837 2.1 / 1422 1.0 / 2022 2.2

4 Sa 0141 0.9 / 0746 2.1 / 1343 0.9 / 1954 2.2
19 Su 0254 0.5 / 0916 2.2 / 1500 0.9 / 2101 2.3

5 Su 0237 0.6 / 0845 2.2 / 1433 0.8 / 2041 2.3
20 M 0331 0.4 / 0952 2.2 / 1533 0.8 / 2135 2.4 ○

6 M 0326 0.4 / 0933 2.3 / 1524 0.7 / 2126 2.5 ●
21 Tu 0407 0.3 / 1024 2.2 / 1601 0.7 / 2209 2.4

7 Tu 0413 0.2 / 1018 2.3 / 1611 0.6 / 2211 2.6
22 W 0441 0.3 / 1054 2.2 / 1630 0.7 / 2241 2.5

8 W 0456 0.1 / 1101 2.3 / 1654 0.5 / 2254 2.7
23 Th 0511 0.3 / 1124 2.2 / 1656 0.6 / 2311 2.5

9 Th 0537 0.0 / 1145 2.3 / 1733 0.6 / 2335 2.7
24 F 0539 0.4 / 1150 2.1 / 1722 0.7 / 2341 2.4

10 F 0620 0.1 / 1228 2.2 / 1813 0.6
25 Sa 0607 0.5 / 1216 2.1 / 1752 0.7

11 Sa 0018 2.6 / 0701 0.2 / 1313 2.1 / 1854 0.7
26 Su 0011 2.4 / 0633 0.6 / 1245 2.1 / 1826 0.8

12 Su 0105 2.4 / 0746 0.4 / 1401 2.0 / 1939 0.8 ☽
27 M 0046 2.3 / 0705 0.7 / 1320 2.0 / 1903 0.9

13 M 0201 2.4 / 0835 0.7 / 1501 1.9 / 2035 0.9
28 Tu 0130 2.2 / 0758 0.9 / 1407 2.0 / 1948 1.0 ☾

14 Tu 0311 2.2 / 0939 0.9 / 1624 1.9 / 2154 1.0
15 W 0441 2.1 / 1105 1.0 / 1726 2.0 / 2346 1.0

MARCH

1 W 0230 2.0 / 0833 1.0 / 1511 1.9 / 2050 1.1
16 Th 0450 2.0 / 1054 1.2 / 1650 2.0 / 2343 0.8

2 Th 0354 1.9 / 0952 1.1 / 1628 1.9 / 2235 1.1
17 F 0626 2.0 / 1222 1.2 / 1811 2.0

3 F 0516 1.9 / 1126 1.1 / 1733 1.9
18 Sa 0058 0.7 / 0733 2.1 / 1318 1.1 / 1916 2.1

4 Sa 0015 0.9 / 0630 2.0 / 1233 1.0 / 1833 2.0
19 Su 0148 0.5 / 0818 2.1 / 1400 1.0 / 2001 2.2

5 Su 0122 0.7 / 0737 2.1 / 1328 0.9 / 1928 2.2
20 M 0228 0.4 / 0856 2.2 / 1433 0.9 / 2039 2.3

6 M 0216 0.4 / 0831 2.3 / 1418 0.7 / 2018 2.3
21 Tu 0303 0.4 / 0926 2.2 / 1503 0.7 / 2111 2.4

7 Tu 0305 0.2 / 0916 2.3 / 1505 0.6 / 2107 2.5 ●
22 W 0337 0.3 / 0957 2.2 / 1531 0.7 / 2143 2.4 ○

8 W 0348 0.0 / 0958 2.4 / 1550 0.5 / 2150 2.6
23 Th 0407 0.3 / 1024 2.2 / 1558 0.6 / 2213 2.4

9 Th 0431 0.0 / 1039 2.4 / 1631 0.4 / 2235 2.7
24 F 0435 0.4 / 1050 2.2 / 1626 0.6 / 2243 2.4

10 F 0515 0.0 / 1118 2.3 / 1713 0.4 / 2318 2.7
25 Sa 0501 0.5 / 1113 2.1 / 1656 0.6 / 2313 2.4

11 Sa 0554 0.1 / 1200 2.2 / 1752 0.5
26 Su 0526 0.6 / 1137 2.2 / 1726 0.7 / 2345 2.3

12 Su 0001 2.6 / 0635 0.3 / 1241 2.1 / 1835 0.6
27 M 0550 0.7 / 1205 2.2 / 1758 0.7 ☽

13 M 0050 2.5 / 0718 0.6 / 1326 2.1 / 1922 0.7
28 Tu 0020 2.2 / 0620 0.8 / 1239 2.1 / 1835 0.8

14 Tu 0148 2.3 / 0805 0.9 / 1422 2.0 / 2022 0.8 ☽
29 W 0105 2.1 / 0700 0.9 / 1322 2.0 / 1922 0.9

15 W 0305 2.1 / 0911 1.1 / 1531 2.0 / 2146 0.9
30 Th 0207 2.0 / 0752 1.1 / 1416 2.0 / 2028 0.9 ☾

31 F 0333 1.9 / 0911 1.2 / 1531 1.9 / 2207 0.9

APRIL

1 Sa 0500 2.0 / 1054 1.2 / 1652 2.0 / 2345 0.8
16 Su 0020 0.6 / 0701 2.1 / 1243 1.1 / 1835 2.1

2 Su 0611 2.0 / 1207 1.0 / 1800 2.0
17 M 0111 0.5 / 0746 2.1 / 1324 1.0 / 1928 2.2

3 M 0052 0.5 / 0715 2.1 / 1303 0.9 / 1900 2.2
18 Tu 0152 0.5 / 0824 2.2 / 1358 0.9 / 2007 2.2

4 Tu 0146 0.3 / 0807 2.2 / 1352 0.7 / 1952 2.4
19 W 0228 0.5 / 0856 2.2 / 1428 0.8 / 2041 2.3

5 W 0235 0.1 / 0852 2.3 / 1439 0.6 / 2043 2.5
20 Th 0300 0.5 / 0926 2.2 / 1458 0.7 / 2115 2.3

6 Th 0322 0.0 / 0933 2.4 / 1526 0.5 / 2130 2.7 ●
21 F 0330 0.5 / 0950 2.3 / 1528 0.7 / 2145 2.4 ○

7 F 0405 0.0 / 1013 2.4 / 1609 0.4 / 2215 2.7
22 Sa 0356 0.5 / 1013 2.3 / 1600 0.7 / 2216 2.3

8 Sa 0448 0.1 / 1052 2.4 / 1652 0.4 / 2300 2.7
23 Su 0422 0.6 / 1039 2.3 / 1633 0.7 / 2250 2.3

9 Su 0530 0.3 / 1131 2.3 / 1737 0.4 / 2346 2.6
24 M 0448 0.7 / 1105 2.3 / 1707 0.7 / 2324 2.2

10 M 0611 0.6 / 1211 2.2 / 1822 0.5
25 Tu 0516 0.8 / 1137 2.3 / 1743 0.7

11 Tu 0039 2.4 / 0652 0.8 / 1252 2.2 / 1915 0.6
26 W 0005 2.2 / 0550 0.9 / 1215 2.2 / 1824 0.7

12 W 0141 2.2 / 0739 1.1 / 1346 2.1 / 2015 0.7 ☽
27 Th 0052 2.1 / 0631 1.0 / 1256 2.2 / 1918 0.7

13 Th 0303 2.1 / 0843 1.3 / 1450 2.0 / 2133 0.8
28 F 0158 2.0 / 0730 1.1 / 1348 2.1 / 2026 0.7 ☾

14 F 0439 2.0 / 1022 1.3 / 1605 2.0 / 2311 0.7
29 Sa 0320 2.0 / 0846 1.2 / 1452 2.0 / 2148 0.7

15 Sa 0558 2.0 / 1148 1.2 / 1724 2.0
30 Su 0441 2.0 / 1016 1.2 / 1609 2.0 / 2309 0.6

AREA 4—E England

ENGLAND, EAST COAST — LOWESTOFT
Lat 52°28′ N Long 1°45′ E
TIMES AND HEIGHTS OF HIGH AND LOW WATERS
YEAR 1989

TIME ZONE UT(GMT)
For Summer Time add ONE hour in non-shaded areas

	MAY				JUNE				JULY				AUGUST		
	Time m		Time m		Time m		Time m		Time m		Time m		Time m		Time m
1 M	0548 2·1 1131 1·1 1722 2·1	**16** Tu	0022 0·6 0703 2·1 1235 1·1 1833 2·1	**1** Th	0039 0·4 0709 2·2 1252 0·9 1858 2·4	**16** F	0056 0·8 0737 2·2 1318 1·0 1928 2·1	**1** Sa	0111 0·7 0731 2·3 1345 0·8 1950 2·3	**16** Su	0058 0·9 0724 2·1 1348 1·0 1954 2·1	**1** Tu ●	0256 0·9 0852 2·4 1531 0·4 2148 2·3	**16** W	0213 0·9 0820 2·3 1507 0·5 2120 2·3
2 Tu	0015 0·4 0646 2·2 1230 0·9 1826 2·2	**17** W	0107 0·6 0745 2·1 1315 1·0 1924 2·2	**2** F	0133 0·4 0756 2·3 1350 0·8 1958 2·5	**17** Sa	0133 0·8 0809 2·2 1403 0·9 2015 2·1	**2** Su	0209 0·7 0820 2·3 1448 0·6 2052 2·3	**17** M	0143 0·9 0803 2·2 1439 0·9 2046 2·1	**2** W	0337 0·8 0933 2·5 1613 0·3 2228 2·3	**17** Th ○	0301 0·7 0907 2·5 1552 0·3 2201 2·3
3 W	0113 0·3 0739 2·2 1320 0·8 1924 2·4	**18** Th	0145 0·6 0820 2·2 1350 0·9 2005 2·2	**3** Sa ●	0226 0·4 0841 2·4 1448 0·6 2052 2·5	**18** Su	0209 0·8 0837 2·3 1448 0·8 2058 2·2	**3** M ●	0303 0·8 0903 2·4 1541 0·5 2148 2·3	**18** Tu ○	0226 0·9 0845 2·3 1526 0·6 2133 2·2	**3** Th	0415 0·8 1013 2·5 1650 0·2 2305 2·2	**18** F	0346 0·6 0950 2·6 1633 0·1 2241 2·4
4 Th	0203 0·2 0824 2·3 1411 0·7 2016 2·5	**19** F	0216 0·6 0850 2·2 1426 0·8 2043 2·3	**4** Su	0316 0·5 0922 2·4 1545 0·5 2146 2·5	**19** M ○	0245 0·8 0909 2·3 1533 0·7 2141 2·2	**4** Tu	0352 0·8 0946 2·5 1628 0·3 2237 2·3	**19** W	0313 0·8 0926 2·4 1611 0·4 2218 2·3	**4** F	0448 0·8 1048 2·5 1728 0·3 2341 2·2	**19** Sa	0430 0·6 1033 2·7 1715 0·1 2322 2·3
5 F ●	0254 0·2 0905 2·4 1501 0·6 2107 2·6	**20** Sa ○	0246 0·6 0915 2·3 1503 0·8 2118 2·3	**5** M	0405 0·7 1003 2·4 1635 0·4 2239 2·4	**20** Tu	0324 0·8 0946 2·4 1620 0·5 2226 2·2	**5** W	0433 0·8 1028 2·5 1711 0·3 2324 2·3	**20** Th	0401 0·7 1009 2·5 1654 0·3 2301 2·3	**5** Sa	0518 0·8 1126 2·5 1801 0·3	**20** Su	0511 0·6 1115 2·7 1756 0·1
6 Sa	0339 0·2 0946 2·4 1552 0·5 2156 2·6	**21** Su	0316 0·7 0939 2·3 1543 0·7 2154 2·3	**6** Tu	0448 0·8 1045 2·5 1722 0·3 2331 2·3	**21** W	0405 0·9 1024 2·4 1705 0·5 2309 2·3	**6** Th	0513 0·9 1109 2·5 1752 0·3 2345 2·2	**21** F	0446 0·7 1050 2·5 1737 0·2	**6** Su	0016 2·2 0546 0·8 1200 2·5 1833 0·4	**21** M	0003 2·3 0550 0·6 1156 2·7 1835 0·2
7 Su	0424 0·4 1026 2·4 1641 0·4 2245 2·6	**22** M	0346 0·7 1009 2·3 1620 0·7 2233 2·2	**7** W	0531 0·9 1126 2·4 1809 0·3	**22** Th	0448 0·8 1105 2·4 1748 0·4 2356 2·2	**7** F	0007 2·2 0548 0·9 1148 2·5 1830 0·3	**22** Sa	0530 0·7 1131 2·6 1818 0·2	**7** M	0050 2·1 0616 0·8 1233 2·4 1903 0·6	**22** Tu	0045 2·2 0630 0·7 1241 2·6 1918 0·4
8 M	0507 0·6 1105 2·4 1728 0·4 2337 2·4	**23** Tu	0418 0·8 1043 2·4 1701 0·6 2313 2·2	**8** Th	0024 2·2 0611 1·0 1209 2·4 1854 0·4	**23** F	0535 0·8 1146 2·4 1833 0·4	**8** Sa	0050 2·1 0620 0·9 1228 2·4 1909 0·4	**23** Su	0028 2·2 0611 0·7 1215 2·6 1900 0·2	**8** Tu	0126 2·1 0650 0·9 1311 2·3 1937 0·7	**23** W ☾	0131 2·1 0715 0·8 1331 2·5 2005 0·6
9 Tu	0548 0·8 1146 2·4 1816 0·4	**24** W	0454 0·8 1118 2·4 1745 0·6 2358 2·2	**9** F	0118 2·1 0650 1·1 1254 2·3 1939 0·5	**24** Sa	0045 2·2 0622 0·9 1230 2·4 1918 0·4	**9** Su	0133 2·1 0654 1·0 1309 2·4 1946 0·5	**24** M	0113 2·1 0652 0·8 1258 2·5 1945 0·3	**9** W ☾	0207 2·0 0730 1·0 1358 2·2 2016 0·8	**24** Th	0226 2·1 0809 0·9 1437 2·3 2101 0·9
10 W	0031 2·3 0631 1·0 1230 2·3 1907 0·5	**25** Th	0533 0·9 1158 2·3 1833 0·6	**10** Sa	0213 2·1 0731 1·1 1343 2·3 2026 0·5	**25** Su	0137 2·1 0711 1·0 1315 2·4 2007 0·4	**10** M	0218 2·0 0731 1·0 1352 2·3 2026 0·6	**25** Tu ☾	0205 2·1 0737 0·9 1346 2·4 2031 0·5	**10** Th	0256 2·0 0816 1·1 1456 2·1 2105 1·0	**25** F	0331 2·0 0920 1·0 1603 2·2 2220 1·1
11 Th	0135 2·1 0716 1·1 1318 2·2 2001 0·6	**26** F	0050 2·1 0622 0·9 1241 2·3 1926 0·6	**11** Su ☾	0309 2·0 0818 1·2 1433 2·2 2120 0·6	**26** M ☾	0237 2·1 0801 1·0 1405 2·3 2100 0·4	**11** Tu	0307 2·0 0815 1·1 1443 2·2 2115 0·8	**26** W	0303 2·0 0830 1·0 1448 2·3 2130 0·6	**11** F	0356 2·0 0918 1·2 1609 2·0 2216 1·1	**26** Sa	0445 2·1 1101 1·0 1739 2·1 2348 1·1
12 F ☾	0246 2·1 0809 1·3 1415 2·2 2105 0·6	**27** Sa	0152 2·1 0720 1·0 1330 2·2 2022 0·5	**12** M	0409 2·0 0915 1·2 1531 2·2 2222 0·7	**27** Tu	0341 2·0 0900 1·0 1507 2·3 2201 0·5	**12** W	0403 2·0 0909 1·2 1541 2·1 2213 0·8	**27** Th	0409 2·0 0935 1·0 1605 2·2 2239 0·8	**12** Sa	0458 2·0 1100 1·2 1724 2·0 2335 1·1	**27** Su	0556 2·1 1239 0·8 1911 2·2
13 Sa	0401 2·0 0916 1·3 1518 2·1 2218 0·7	**28** Su	0303 2·0 0826 1·1 1426 2·2 2126 0·5	**13** Tu	0507 2·0 1026 1·2 1631 2·1 2322 0·7	**28** W	0445 2·0 1007 1·1 1620 2·3 2307 0·5	**13** Th	0501 2·0 1024 1·2 1648 2·1 2318 0·9	**28** F	0515 2·1 1101 1·0 1726 2·2 2352 0·9	**13** Su	0554 2·0 1228 1·1 1831 2·0	**28** M	0103 1·1 0701 2·2 1343 0·6 2013 2·2
14 Su	0507 2·0 1045 1·3 1624 2·1 2328 0·6	**29** M	0415 2·0 0935 1·1 1533 2·2 2235 0·5	**14** W	0605 2·1 1137 1·2 1733 2·1	**29** Th	0545 2·0 1122 1·0 1733 2·3	**14** F	0556 2·0 1146 1·2 1754 2·0	**29** Sa	0616 2·1 1231 0·9 1848 2·1	**14** M	0033 1·1 0645 2·1 1331 0·9 1941 2·1	**29** Tu	0158 1·0 0754 2·3 1431 0·5 2056 2·3
15 M	0609 2·1 1148 1·2 1731 2·1	**30** Tu	0518 2·1 1048 1·1 1646 2·2 2339 0·4	**15** Th	0013 0·8 0656 2·1 1231 1·1 1833 2·1	**30** F	0011 0·6 0641 2·2 1231 0·9 1843 2·3	**15** Sa	0013 1·0 0643 2·1 1252 1·1 1856 2·0	**30** Su	0103 0·9 0715 2·2 1350 0·7 2005 2·2	**15** Tu	0124 1·0 0733 2·2 1422 0·7 2035 2·2	**30** W	0239 0·9 0837 2·4 1511 0·3 2133 2·3
		31 W	0616 2·1 1154 1·0 1756 2·3							**31** M	0207 0·9 0805 2·3 1446 0·5 2103 2·2			**31** Th ●	0315 0·8 0915 2·5 1548 0·3 2205 2·3

Chart Datum: 1.50 metres below Ordnance Datum (Newlyn)

ENGLAND, EAST COAST — LOWESTOFT
Lat 52°28′ N Long 1°45′ E

TIMES AND HEIGHTS OF HIGH AND LOW WATERS

YEAR 1989

TIME ZONE UT(GMT)
For Summer Time add ONE hour in non-shaded areas

SEPTEMBER

Time	m	Time	m
1 0348 0950 F 1624 2239	0.8 2.5 0.3 2.3	**16** 0324 0928 Sa 1607 2215	0.6 2.7 0.1 2.4
2 0418 1026 Sa 1656 2309	0.7 2.6 0.3 2.3	**17** 0407 1011 Su 1648 2254	0.6 2.8 0.1 2.4
3 0446 1058 Su 1726 2339	0.7 2.5 0.4 2.2	**18** 0448 1054 M 1730 2335	0.5 2.8 0.2 2.4
4 0513 1130 M 1754	0.7 2.5 0.5	**19** 0530 1137 Tu 1811	0.6 2.7 0.4
5 0007 0543 Tu 1200 1820	2.2 0.8 2.4 0.7	**20** 0015 0613 W 1224 1852	2.3 0.6 2.6 0.6
6 0035 0615 W 1235 1848	2.2 0.9 2.3 0.8	**21** 0100 0701 Th 1320 1939	2.3 0.7 2.4 0.9
7 0107 0652 Th 1318 1924	2.1 1.0 2.2 0.9	**22** 0150 0800 F 1433 2037	2.2 1.0 2.2 1.1 ☾
8 0152 0737 F 1416 2011	2.1 1.1 2.1 1.1 ☾	**23** 0256 0915 Sa 1613 2203	2.1 0.9 2.2 1.3
9 0250 0837 Sa 1537 2118	2.0 1.1 2.0 1.2	**24** 0413 1058 Su 1746 2341	2.1 0.9 2.2 1.3
10 0400 1013 Su 1700 2258	2.0 1.2 2.0 1.2	**25** 0530 1222 M 1901	2.2 0.7 2.2
11 0507 1158 M 1813	2.0 1.0 2.0	**26** 0045 0639 Tu 1320 1954	1.2 2.2 0.6 2.2
12 0009 0607 Tu 1301 1918	1.1 2.1 0.8 2.1	**27** 0133 0733 W 1405 2033	1.1 2.3 0.5 2.3
13 0103 0703 W 1352 2011	1.0 2.2 0.6 2.2	**28** 0211 0815 Th 1443 2107	1.0 2.4 0.4 2.3
14 0152 0754 Th 1439 2054	0.8 2.4 0.3 2.3	**29** 0245 0852 F 1518 2137	0.9 2.5 0.4 2.3 ●
15 0239 0843 F 1524 2135 ○	0.7 2.5 0.2 2.4	**30** 0315 0926 Sa 1550 2207	0.8 2.5 0.4 2.3

OCTOBER

Time	m	Time	m
1 0345 0958 Su 1620 2235	0.7 2.5 0.5 2.3	**16** 0343 0948 M 1622 2228	0.6 2.8 0.3 2.5
2 0413 1030 M 1646 2300	0.7 2.5 0.6 2.3	**17** 0428 1033 Tu 1703 2307	0.6 2.8 0.4 2.5
3 0443 1100 Tu 1709 2324	0.8 2.5 0.7 2.3	**18** 0515 1122 W 1746 2346	0.6 2.7 0.6 2.4
4 0513 1130 W 1733 2350	0.8 2.4 0.8 2.3	**19** 0603 1215 Th 1830	0.6 2.5 0.9
5 0546 1205 Th 1801	0.9 2.3 0.9	**20** 0031 0656 F 1315 1918	2.4 0.6 2.4 1.1
6 0024 0624 F 1250 1839	2.2 0.9 2.2 1.0	**21** 0122 0754 Sa 1433 2015 ☾	2.3 0.7 2.2 1.3
7 0105 0711 Sa 1348 1928	2.2 1.0 2.1 1.2	**22** 0222 0903 Su 1605 2133	2.2 0.8 2.2 1.4
8 0158 0813 Su 1513 2037	2.1 1.0 2.0 1.3	**23** 0333 1030 M 1724 2305	2.1 0.8 2.2 1.4
9 0305 0941 M 1637 2215	2.1 1.0 2.0 1.3	**24** 0448 1146 Tu 1830	2.2 0.7 2.2
10 0420 1116 Tu 1746 2337	2.1 0.9 2.1 1.2	**25** 0009 0600 W 1245 1920	1.3 2.3 0.6 2.2
11 0530 1224 W 1848	2.2 0.7 2.2	**26** 0056 0700 Th 1330 2001	1.1 2.3 0.6 2.3
12 0033 0630 Th 1316 1941	1.0 2.3 0.5 2.3	**27** 0135 0745 F 1409 2037	1.0 2.4 0.6 2.3
13 0122 0724 F 1407 2026	0.9 2.4 0.3 2.4	**28** 0211 0822 Sa 1443 2107	0.9 2.5 0.6 2.3
14 0209 0815 Sa 1454 2107 ○	0.8 2.6 0.3 2.4	**29** 0243 0858 Su 1513 2135 ●	0.8 2.5 0.6 2.4
15 0256 0901 Su 1539 2146	0.7 2.7 0.4 2.5	**30** 0313 0930 M 1539 2200	0.8 2.5 0.7 2.4
		31 0346 1001 Tu 1603 2224	0.8 2.4 0.8 2.4

NOVEMBER

Time	m	Time	m
1 0418 1033 W 1628 2248	0.8 2.4 0.8 2.4	**16** 0507 1113 Th 1728 2324	0.5 2.6 0.8 2.5
2 0452 1107 Th 1656 2320	0.7 2.3 0.9 2.4	**17** 0558 1207 F 1811	0.5 2.4 1.0
3 0530 1146 F 1730 2356	0.8 2.3 1.0 2.4	**18** 0009 0648 Sa 1309 1858	2.5 0.5 2.3 1.2
4 0611 1233 Sa 1809	0.9 2.2 1.1	**19** 0058 0741 Su 1416 1946	2.4 0.6 2.2 1.3
5 0037 0701 Su 1333 1901	2.3 0.9 2.1 1.2	**20** 0152 0839 M 1530 2045 ☾	2.3 0.6 2.1 1.3
6 0126 0805 M 1448 2011	2.2 0.9 2.1 1.2	**21** 0252 0946 Tu 1637 2158	2.3 0.7 2.1 1.4
7 0224 0916 Tu 1609 2131	2.2 0.8 2.1 1.3	**22** 0358 1056 W 1741 2313	2.3 0.7 2.1 1.3
8 0333 1035 W 1716 2252	2.2 0.8 2.1 1.2	**23** 0503 1158 Th 1837	2.2 0.7 2.2
9 0446 1143 Th 1815 2356	2.2 0.6 2.2 1.1	**24** 0009 0609 F 1246 1924	1.2 2.3 0.7 2.2
10 0554 1239 F 1907	2.4 0.5 2.3	**25** 0056 0705 Sa 1328 2003	1.1 2.3 0.7 2.3
11 0048 0652 Sa 1331 1954	1.0 2.5 0.4 2.4	**26** 0137 0750 Su 1401 2035	1.1 2.3 0.8 2.3
12 0139 0746 Su 1422 2037	1.0 2.6 0.3 2.5	**27** 0215 0830 M 1430 2101	1.0 2.3 0.8 2.4
13 0231 0839 M 1509 2120 ●	0.7 2.7 0.4 2.5	**28** 0250 0905 Tu 1458 2126	0.9 2.3 0.8 2.4
14 0324 0930 Tu 1556 2201	0.6 2.7 0.5 2.5	**29** 0330 0941 W 1528 2152	0.9 2.3 0.9 2.4
15 0416 1020 W 1643 2243	0.6 2.7 0.6 2.5	**30** 0409 1016 Th 1600 2224	0.8 2.3 0.9 2.5

DECEMBER

Time	m	Time	m
1 0448 1056 F 1633 2300	0.8 2.3 0.9 2.5	**16** 0548 1200 Sa 1754 2348	0.3 2.3 1.0 2.6
2 0531 1137 Sa 1713 2337	0.7 2.2 1.0 2.4	**17** 0633 1252 Su 1833	0.4 2.2 1.0
3 0615 1226 Su 1756	0.7 2.2 1.0	**18** 0033 0718 M 1345 1913	2.5 0.4 2.1 1.1
4 0018 0703 M 1320 1848	2.4 0.7 2.1 1.1	**19** 0122 0805 Tu 1439 1956 ☾	2.4 0.6 2.1 1.2
5 0103 0754 Tu 1422 1946	2.3 0.7 2.1 1.1	**20** 0211 0856 W 1537 2045	2.4 0.7 2.0 1.2
6 0154 0854 W 1533 2050	2.3 0.7 2.1 1.2	**21** 0307 0954 Th 1639 2146	2.3 0.8 2.0 1.3
7 0254 0956 Th 1641 2201	2.3 0.6 2.1 1.2	**22** 0409 1058 F 1741 2307	2.2 0.8 2.1 1.3
8 0405 1103 F 1741 2315	2.3 0.6 2.2 1.1	**23** 0515 1156 Sa 1837	2.2 0.9 2.1
9 0520 1203 Sa 1835	2.4 0.5 2.2	**24** 0015 0618 Su 1243 1924	1.2 2.2 0.9 2.2
10 0016 0626 Su 1300 1926	1.0 2.4 0.5 2.3	**25** 0109 0718 M 1320 2000	1.2 2.2 0.9 2.2
11 0116 0726 M 1354 2013	0.9 2.5 0.6 2.4	**26** 0156 0809 Tu 1356 2028	1.1 2.2 0.9 2.3
12 0218 0824 Tu 1448 2056 ○	0.8 2.6 0.6 2.5	**27** 0241 0852 W 1430 2056	0.9 2.2 0.9 2.3
13 0318 0920 W 1539 2139	0.6 2.6 0.7 2.5	**28** 0324 0931 Th 1505 2130 ●	0.8 2.2 0.9 2.4
14 0413 1015 Th 1628 2222	0.5 2.5 0.8 2.6	**29** 0407 1011 F 1545 2205	0.7 2.2 0.8 2.5
15 0501 1107 F 1711 2305	0.4 2.4 0.9 2.6	**30** 0448 1052 Sa 1626 2245	0.6 2.2 0.8 2.5
		31 0530 1133 Su 1709 2326	0.5 2.2 0.8 2.5

Chart Datum: 1.50 metres below Ordnance Datum (Newlyn)

LOWESTOFT continued

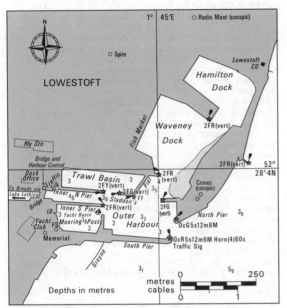

tidal conditions. From S, waypoint East Barnard E cardinal buoy, Q(3) 10s, 52°42'.60N 01°46'.20E; then follow buoyed channel inshore of S Newcome and Pakefield buoys. From N or E, waypoint Corton E cardinal buoy, Q(3) 10s, 52°31'.10N 01°51'.50E; then follow Holm Channel (buoyed) into Corton Road. Speed limit in harbour 4 kn. Passage to Oulton Broad Wednesday p.m., passing two swing bridges and a lock, give Hr Mr at least 48 hours notice.

LIGHTS AND MARKS
Entry Signals: Fl W Lt (obsc 232°-286°) below the Oc R Lt on South Pier Lt Ho — vessels may proceed to sea (with clearance from harbour control on Ch 14 or tel 2286) but entry prohibited. When signal not shown: No vessel to proceed to sea but entry permitted.
Bridge Signals: when bridge is lifting, vessels may not approach within 137m until G Lt is shown on N wall of entrance channel. When G Lt shown: vessels may enter or leave Inner Harbour.

RADIO TELEPHONE
VHF Ch **14** 16; 14 (H24). Pilot Ch 14.

TELEPHONE (0502)
Hr Mr 2286 Ext 41; Hr Mr Oulton Broad 4946; CG 65365; MRCC Great Yarmouth 851338; Bridge Control 2286 Ext 42; Pilot 60277; HMC Freefone Customs Yachts: RT (0473) 219481; Weather Norwich 660779; Marinecall 0898 500 455; Hosp 600611.

FACILITIES
EC Thursday; **Royal Norfolk & Suffolk YC** Tel. 66726, Slip, M, D (cans), L, FW, C (3ton), AB, R, Bar; **Oulton Broad — Jeckells Y. Co.** Tel. 65007, L, ME, El, Sh, CH, SM; **Oulton Broad Calor Centre** Gas, Gaz; **Knights Creek** Tel. 2599, El, CH; **J. E. Fletcher** Tel. 4951, Slip, Sh; **F. Newson** Tel. 4902, Slip, M, L, El, Sh; **A. D. Truman** Tel. 65950, Slip, M, L, ME, El, Sh; **Lowestoft Cruising Club** Tel. 4376, Slip, M, L, FW, AB; **Charity and Taylor** Tel. 81529, ACA. **Town** V, R, Bar. PO; Bank; Rly; Air (Norwich).

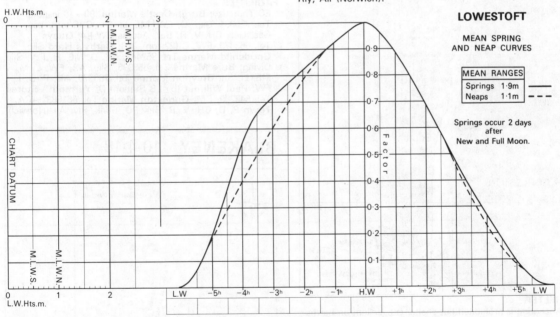

GREAT YARMOUTH 10-4-17
Norfolk

CHARTS
Admiralty 1536, 1543; Stanford 3; Imray C28; OS 134
TIDES
−0210 Dover; ML 1.5; Duration 0620; Zone 0 (GMT).

Standard Port LOWESTOFT (←—)

Times				Height (metres)			
HW		LW		MHWS	MHWN	MLWN	MLWS
0300	0900	0200	0800	2.4	2.1	1.0	0.5
1500	2100	1400	2000				

Differences **GORLESTON**
−0035 −0035 −0030 −0030   0.0 −0.1   0.0   0.0
CAISTER
−0130 −0130 −0100 −0100   0.0 −0.1   0.0   0.0

Rise of tide occurs mainly during 3½ hours after LW. Level is usually within 0.3m of predicted HW height, from 3 hours before HW Lowestoft, until HW Lowestoft. Flood tide runs for about 1½ hrs after predicted HW and Ebb for about 2-2½ hrs after predicted LW.

GREAT YARMOUTH continued

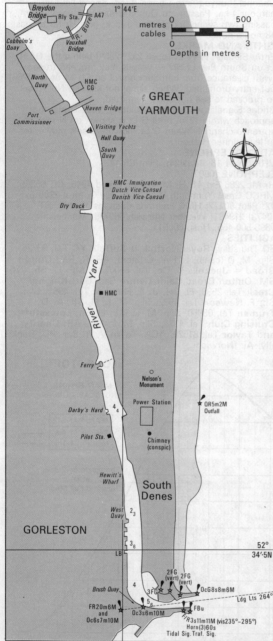

LIGHTS AND MARKS
Ldg Lts 264°. Front Oc 3s 6m 10M. Rear Oc 6s 7m 10M.
S pier head: FY Lt — do not enter. Hr Mr's office (close SE of Brush Quay): 2FR(hor) — do not leave.
Tidal signals (entrance depth) from S pier head
1 Fl every 5 sec   4.3m   (ebb running)
2 Fl every 5 sec   4.6m   (ebb running)
3 Fl every 10 sec   4.9m   (ebb running)
No signal   less than 4.3m   (ebb running)
4 Fl every 10 sec   —   (flood running)
The colour of Lt is altered with the rate of the stream:
less than 1 knot — orange
more than 1 knot — violet.
Haven Bridge:
Northbound — Day: Bu flag on SE buttress. Night: FBu Lt over FW Lt SE buttress (5 min before bridge opening). Southbound — Day: R flag on NW buttress. Night: FR Lt over FW Lt NW buttress (5 min before bridge opening).
Breydon Bridge: (0600-2200 LT in summer)
Air draft 3.5m. Three R Lts (vert) on S buttress control vessels through bridge when raised.
Contact bridge on Ch 12 or hoist International flag V. If possible ring bridge previously giving approx ETA.

RADIO TELEPHONE
Call: *Yarmouth* VHF Ch 12 16; 09 11 12 (H24). Breydon Bridge Ch 12 (0600-2200 LT). Haven Bridge Ch 12 (weekdays — working hours).

TELEPHONE (0493)
Hr Mr 663476; CG & MRCC 851338; HMC Freefone Customs Yachts: RT (0473) 219481; Marinecall 0898 500 455; Breydon Bridge 651275; Port Health Authority 843233.

FACILITIES
EC Thursday; **Burgh Castle Marina** (90 + 10 visitors) Tel. 780331, Slip, D, FW, ME, El, Sh, C (6 ton), Lau, Gas, Gaz, CH, V, R, Bar, Access HW∓4; **Quays** Tel. 663476, FW, C (22 ton), AB; **Darby's Hard** Slip; **Goodchild Marine** Tel. 852442, BY, D, ME, El, FW, Slip, Diving; **Bure Marine** Tel. 656996, Slip, AB, FW, L, M, ME, El, Sh; **Breydon Marine** Tel. 780379, AB, P, D, L, FW; **Paul Willmet** (by LB Station) D; **Yarmouth Stores** Tel. 842289, SM; **Gorleston Marine** Tel. 661883, ACA. **Town** P, D, CH, V, R, Bar. PO; Bank; Rly; Air (Norwich).

SHELTER
Anchoring is prohibited but shelter is excellent on berths on E side, S of Haven Bridge or request Haven Bridge to lift (Ch 12) and pass up to Cobholm's Quay or North Quay.

NAVIGATION
Waypoint 52°34'.40N 01°45'.67E, 084°/264° from/to front Ldg Lt 264°, 1.0M. Harbour is available at all times but small craft should not attempt entrance with strong SE winds when dangerous seas occur especially on the ebb. Temporary shoaling is liable to occur in the harbour entrance during strong easterly winds when depths of 1m less than those charted may be expected. Harbour is a busy commercial port. Beware strong streams sweeping Haven Bridge.

Haven Bridge clearance (closed)   2.7m
A47 bridge (Fixed. R Bure)   2.1m
Vauxhall bridge (Fixed. Old Rly)   2.4m
Breydon bridge (closed)   3.5m
Breydon bridge (side spans)   4.0m

BLAKENEY 10-4-18
Norfolk

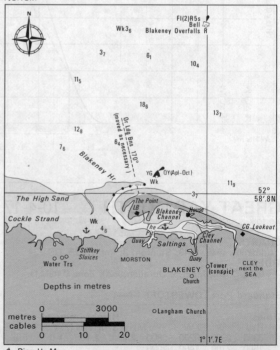

▲ Ring Hr Mr

AREA 4 — E England

BLAKENEY continued

CHARTS
Admiralty 108; Imray C28; OS 133

TIDES
−0445 Dover; Duration 0530; Zone 0 (GMT).

Standard Port IMMINGHAM (→)

Times				Height (metres)			
HW		LW		MHWS	MHWN	MLWN	MLWS
0100	0700	0100	0700	7.3	5.8	2.6	0.9
1300	1900	1300	1900				

Differences BLAKENEY
+0115 +0055 No data −3.9 −3.8 No data
BLAKENEY BAR
+0035 +0025 +0030 +0040 −1.6 −1.3 No data
CROMER
+0050 +0030 +0050 +0130 −2.1 −1.7 −0.5 −0.1

SHELTER
Very good shelter but harbour inaccessible with fresh on-shore winds. Conditions in the entrance deteriorate very quickly with on-shore winds, especially on the ebb. Entry, springs HW ∓2½, neaps HW ∓1. Moorings in The Pit or at Stiffkey Sluices.

NAVIGATION
Waypoint 53°00'.00N 00°58'.24E, 000°/180° from/to Bar Buoy, 0.90M. Large wreck to W of harbour entrance channel; very dangerous, unlit, marked by buoy. The bar is shallow and changes frequently. The channel is marked by B buoys to be left to stbd. Beware mussel lays in the channel off Blakeney spit. Speed limit 8 kn. For pilotage contact Stratton Long Marine (below).

LIGHTS AND MARKS
Orange leading beacons established on dunes at Blakeney Pt, moved according to changes in the channel (not always reliable). Conspic marks are Blakeney Church and Langham Church. There is a conspic chimney on the house on Blakeney Point neck. A Lt buoy, QY, is placed (April-Oct) marking entrance, 1½ ca N of wreck.

RADIO TELEPHONE
None.

TELEPHONE (0263)
Hr Mr 740362; CG Wells 219; MRCC Great Yarmouth 851338; HMC Freefone Customs Yachts: RT (0473) 219481; Marinecall 0898 500 455; Dr 740314.

FACILITIES
EC Wednesday; **Blakeney Quay** Slip, M, D, L, FW, El, C (15 ton), CH, AB; **Blakeney Marine** Tel. 740513, Slip, P, L, FW, Gas, CH, M, ME, El, Sh, C, AB, V. **Stratton Long Marine** Tel. 740362, M, P, D, FW, ME, El, Sh, CH, Slip, L, C, AB, BY, SM, Gas, Gaz, AC;
Village V, R, Bar. PO; Bank; Rly (Sherringham); Air (Norwich).
Note: The areas of Blakeney Point and Dunes are National Trust Property and contain a bird sanctuary.
Note 2: There is a plan to build a marina at Cromer.

WELLS-NEXT-THE-SEA 10-4-19
Norfolk

CHARTS
Admiralty 108; Imray C28, Y9; OS 132

TIDES
−0445 Dover; Duration 0540; Zone 0 (GMT).

Standard Port IMMINGHAM (→)

Times				Height (metres)			
HW		LW		MHWS	MHWN	MLWN	MLWS
0100	0700	0100	0700	7.3	5.8	2.6	0.9
1300	1900	1300	1900				

Differences WELLS-NEXT-THE-SEA
+0035 +0045 +0340 +0310 −3.8 −3.8 No data
WELLS BAR
+0020 +0020 +0020 +0020 −1.3 −1.0 No data
HUNSTANTON
+0010 +0020 +0105 +0025 +0.1 −0.2 −0.1 0.0

SHELTER
The shelter is good except in strong N winds. Max draft 3m at springs. Entrance possible from HW −1½ to HW +1 but recommended on the flood. Entrance impossible to small craft with strong on-shore winds.

NAVIGATION
Waypoint Wells Fairway buoy, 52°59'.80N 00°50'.30E, 320°/140° from/to channel entrance, 0.40M. The bar and entrance vary in depth and position. The position of buoys is altered to suit. Channel is marked with Q buoys to stbd, QR to port, reaching No. 16 (port). Keep to port side of channel from No 12 to quay. Advisable to take a pilot or follow fishing boat.

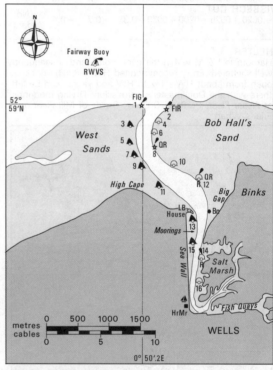

LIGHTS AND MARKS
Harbour ent has a conspic W Lifeboat Ho with R roof on the West point.

RADIO TELEPHONE
VHF Ch 16; 06 08 12 (when vessels expected)

TELEPHONE (0328)
Hr Mr 710655; CG 710587; Pilot 710739; MRCC Gt. Yarmouth 851338; HMC Freefone Customs Yachts: RT (0473) 219481; Marinecall 0898 500 455; Hosp 710218.

FACILITIES
EC Thursday; **East Quay** Slip, M, L; **Main Quay** M (see Hr Mr), L, FW, ME, El, Sh, C (5 ton mobile), CH, AB, V, R, Bar; **Wells SC** Slip, Bar; **Standard House Chandlery** Tel. 710593, M, L, ME, El, Sh, CH, AB, V, ACA;
R. Bicknell Tel. 710593, CH; **M. Walsingham** Tel. 710438, Gas;
Town P and D (cans), CH, V, R, Bar. PO; Bank; Rly (bus to Norwich or King's Lynn); Air (Norwich).

KING'S LYNN 10-4-20
Norfolk

CHARTS
Admiralty 1190, 1200; Imray Y9; OS 132
TIDES
−0450 Dover; ML 3.6; Duration 0340 Sp, 0515 Np; Zone 0 (GMT).

Standard Port IMMINGHAM (→)

Times				Height (metres)			
HW		LW		MHWS	MHWN	MLWN	MLWS
0100	0700	0100	0700	7.3	5.8	2.6	0.9
1300	1900	1300	1900				

Differences KING'S LYNN
+0030 +0030 +0305 +0140 −0.5 −0.8 −0.8 +0.1
WISBECH CUT
+0020 +0025 +0200 +0030 −0.3 −0.7 −0.4 No data

SHELTER
Harbour is 1½ M within the river mouth and consequently well sheltered. Entry recommended HW ∓3. The dock is open from about HW−1½ to HW and yachts can be left there with the Dockmaster's permission. Drying moorings at Fisher Fleet, Purfleet, S Quay, S side of Mill Fleet or Friar's Wharf.

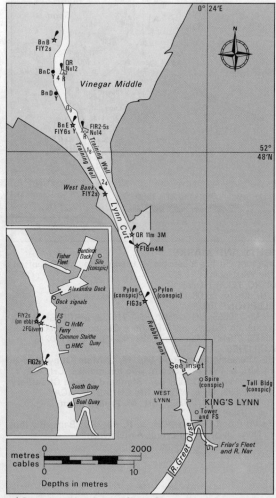

▲ Report to Hr Mr at Common Staithe Quay

NAVIGATION
Waypoint South Well (safe water) buoy, Iso 6s, Whis, Racon, 53°01'.35N 00°25'.79E, 007°/187° from/to Cork Hole entrance, 7.0M. Extensive shifting sand banks extend several miles into the Wash. The best and safest approach is via the Cork Hole route, marked by Lt buoys. Channels are subject to frequent changes particularly between No 7 Light Buoy and Lynn Cut. S of West Stones Bn the deeper water is on E side of channel.
LIGHTS AND MARKS
Ldg Lts 155°. Front QR 11m 3M, Rear FW 16m 4M, both on masts.
Entry signals for Alexander Dock:
Bu flag or R Lt — Vessels can enter
R flag or G Lt — Vessels leaving dock.
RADIO TELEPHONE (local times)
VHF Ch 14 16; 11 12 14 (Mon-Fri: 0900-1700. Sat: 1000-1200. Other times: HW−4 to HW).
King's Lynn Docks Ch **14** 16; 11 (HW−2½ to HW+1).
Other station: Wisbech Ch 16; 09 **14** (HW−4 to HW when vessel expected).
TELEPHONE (0553)
Hr Mr 773411, Dock 691555; CG and MRCC Gt. Yarmouth 851338; HMC Freefone Customs Yachts: RT (0473) 219481; Marinecall 0898 500 455; Dr Ring Hr Mr.
FACILITIES
EC Wednesday; There are virtually no facilities for visiting yachts; **Docks** Tel. 691555, M, L, FW, C (32 ton), AB;
N.C. Shipp Tel. 772831, ME, El; **V. Pratt** Tel. 764058, El, CH;
Town P, D, ME, El, Sh, CH, V, R, Bar. PO; Bank; Rly; Air (Humberside or Norwich).
Note 24 miles up the Great Ouse river **Ely Marina** Tel. Ely 4622, (Lock at Denver Sluice and low bridges above this point), Slip, M, P, D, FW, ME, El, Sh, C (10 ton), CH.

AGENTS NEEDED
There are a number of ports where we need agents, particularly in France.
ENGLAND Swale, Havengore, Berwick.
SCOTLAND Firth of Forth, Scrabster, Mallaig, Loch Sunart, Loch Aline.
IRELAND Kilrush, Wicklow, Westport/Clew Bay.
FRANCE   Arcachon, Seudre R, Ile d'Oleron, Rochfort, Ile de Re, St. Giles-Croix-de-Vie, Ile d'Yeu, Pouliguen, Le Croisic, La Forêt, Ile de Bréhat.
GERMANY Norderney, Dornumer-Accumersiel.
If you are interested in becoming our agent for any of the above, please write to the editors and get your free copy of the Almanac every year. You do not have to be resident in a port to be the agent, but at least a fairly regular visitor.

MINOR HARBOURS AND ANCHORAGES 10.4.21

RIVER ROACH, Essex 51°35′ N, 0°43′ E, Zone 0 (GMT), Admty chart 3750. At Rochford HW +0115 on Dover, −0010 on Sheerness; HW height −2.4m on Sheerness. Excellent shelter. Very good anchorage and shelter about ¼ M up R Roach close to W bank. River runs into R Crouch and is ¼ M wide at the mouth, but mud extends from both banks, the Wallasea Ness bank being marked by Branklet Spit Buoy (Y spher). Few boats go above Barling Ness. For moorings and anchorages, see chartlet 10.3.19 and 10.4.7. A riding light is needed when anchored since R Roach is used by freighters day and night. Facilities: most facilities in **Rochford** EC Wed. **Shuttlewoods** (100) Tel. 226, D, BY, Slip, M; **Paglesham** FW, (from yard), V (at East End), Sh, Bar. **Wakering** El, FW, ME, Sh; Stores at **Great Wakering**; **Wakering YC**.

NORFOLK BROADS. There are about 130 miles of navigable rivers and tributaries in Norfolk and Suffolk together with meres and lakes, known as Broads. The main rivers are the Bure, the Yare and the Waveney which all flow into Breydon Water and thence into the sea at Great Yarmouth (see 10.4.17). This is the principal entrance. Licences are compulsory and Norfolk Broads temporary licences are obtainable from the Haven Commissioners, 21 South Quay, Great Yarmouth, Tel. 855151. Pass up R Yare, under Haven Bridge (headroom when closed 2.4m at MHWS) which is a lifting bridge (contact the Bridge Officer on Yarmouth 663476 or on VHF Ch 12). Bridge signals are Bu Flag, N bound traffic to pass: R flag, S bound traffic to pass. Thence to Breydon Water and to the R Bure, under two bridges (headroom 2m at MHWS.) Once past the two fixed bridges, some other bridges on the broads open – to open swing bridges, give three long blasts. A number of broads are closed to navigation. Tides: LW at Breydon (mouth of R Yare) is 1 hour before LW at Yarmouth Yacht Station (mouth of R Bure). Tide will start to flood on Breydon Water whilst still ebbing from R Bure. LW R Bure mouth = HW at Yarmouth Bar +0700. Tidal differences are quoted on LW at Yarmouth Yacht Station, which is Gorleston (see 10.4.17) +0100. For LW times at the following, add to time of LW at Yarmouth Yacht Station

Acle Bridge +0230	St Olaves +0115
Horning +0300	Beccles +0320
Reedham +0115	Oulton Broad +0300
Cantley +0200	Potter Heigham +0400
Norwich +0430	Burney Arms +0100

The Broads can also be entered at Lowestoft from the sea, via Oulton Dyke and Mutford Lock (opens Wed) into the R Waveney (see 10.4.16).

BURNHAM OVERY STAITHE, Norfolk, 52°59′ N, 0°45′ E, Zone 0 (GMT), Admty chart 108. HW −0420 on Dover, +0021 on Immingham; HW height −2.6m on Immingham. Small harbour which dries; anchorage off the Staithe only suitable in good weather. No lights. Scolt Hd conspic to W and Gun Hill to E. Channel varies constantly and buoys moved to suit. Local knowledge advisable. Facilities: Tel. (0328) **Burnham Overy Staithe SC** Tel. 738348, M, L; **The Boathouse** Tel. 738348, CH, M, ME, Sh, Slip, FW; **Burnham Market** EC Wed; Bar, P and D (cans), R, V.

BRANCASTER STAITHE, Norfolk, 52°58′ N, 0°39′ E, Zone 0 (GMT), Admty chart 108. HW −0425 on Dover, Hunstanton +0020 on Immingham; HW height −0.6m on Immingham. Small unlit harbour which dries; dangerous to enter except in daylight in fine weather. Speed limit 6 kn. Approach from due N. Conspic Golf Club House marks the start of channel buoys. Beware wreck shown on chart. Sandbanks vary continuously and buoys changed to suit. Scolt Hd conspic to E. Pilot advised. Facilities: Tel. (0485) Hr Mr Tel. 210638, Tel. (0485) visitors mooring available occasionally; Bar, BY, CH, D, El, FW, P and D (cans), ME, R, Sh, V; **Brancaster Staithe SC** Tel. 210249, R, Bar; **Scolt Head Nat Trust Warden** Tel. 210330 (Access HW∓3); **Brancaster Sailing Centre** Tel. 210236, CH.

AREA 4—E England 333

WISBECH, Cambridgeshire, 52°40′ N, 0°10′ E, Zone 0 (GMT), Admty charts 1177, 1200. HW −0450 on Dover, Wisbech Cut +0020 on Immingham; HW height −0.5m on Immingham; ML 3.5m; Duration 0520. See 10.4.20. Excellent shelter. Vessels of 4.8m draught at sp can reach Wisbech (3.4m at nps). Entrance to R Nene and inland waterways. Moorings W bank ½ M downstream from Sutton Bridge or at Wisbech Town Quay. Entrance well marked from wreck to Nene Towers with lit stbd Bns. Best entrance HW −3. Wisbech Cut to Sutton Br is 3½ M call on VHF Ch 16 09 14 or Tel. Holbeach 350364. Sutton Br to Wisbech 12M with Fl R Lts to port. Facilities: Tel. (0945) Hr Mr Tel. 582125; Tel. (0945) EC Wed; **Bodger Bros** Tel. 582864 Gas; **Town** P and D (cans), Bar, R, V, L at Western Tower.

WELLAND RIVER, Lincolnshire, 52°56′ N, 0°05′ E, Zone 0 (GMT), Admty chart 1190. At Welland Cut HW −0440 on Dover, +0030 on Immingham; HW height −0.1m on Immingham; ML 0.3m; Duration 0520. Enter by Welland Cut, Iso R 2s to SE and Iso G 2s to NW. Entrance HW ∓3. Beware tides of up to 5 kn on flood at sp. Secure at small quay ½ M NE of bridge on stbd side. Dries. Recommended for short stay only (at Fosdyke). Very limited facilities.

ESSEX MARINA
River Crouch, Essex

★ New Pontoons throughout—complete with finger berths.
★ Deep water Piling completed Spring 1988.
★ Swinging Moorings also available.
★ New Luxury Toilets, Showers and Laundry facilities to be completed Summer 1988.
★ Boats over 28′ GUARANTEED Deep Water.
★ NO LOCKS.
★ Full Marina Services. ★ Bars/Restaurant.
★ Brokerage. ★ Shipwrights.
★ Security. ★ Engineering.
★ Chandlery. ★ Fuel etc.

Essex Marina Ltd.,
Wallasea Island, Nr. Rochford, Essex.
Telephone. Canewdon (03706) 8940
Telex: 995244 POLY G.

EVERSON AND SONS LTD
Est. 1889

Boatbuilders – Marine Engineers
Fitting Out – Chandlery
Brokerage

**Agents for Yamaha Outboards
– Vetus Diesels. Moorings –
Storage – Heavy Cranage**

PHOENIX WORKS
WOODBRIDGE 4358

WOOLVERSTONE MARINA
MDL (Marinas) Ltd
Woolverstone, Ipswich
Tel: 047384 206 & 354

Berths, Moorings, Mobile Crane, Workshops,
Large Chandlery, Pantry/Off Licence,
Launderette, Marine Engineers, Brokerage.

Full facilities
VHF Ch 37
Visitors welcome

RWELL AUTO ELECTRICS LTD

SPECIALISTS IN THE SUPPLY, MAINTENANCE AND REPAIR OF ALL MARINE ELECTRICAL EQUIPMENT

DISTRIBUTORS FOR SU BUTEC, PARIS RHONE, SEV,
CETREX MARINEX & BOSCH ELECTRICAL EQUIPMENT

37 Boss Hall Industrial Estate, Sproughton Road, Ipswich, Suffolk
Telephone: IPSWICH (0473) 49624/5

Granary Yacht Harbour Services
The Granary, Dock Lane, Melton,
Woodbridge, Suffolk IP12 1PE.

● MOORING & LAYING-UP FACILITIES
● CRAFT MAINTENANCE & REPAIRS
● SLIPPING & CRANEAGE
● CHANDLERY
● MARINE ENGINEERING

Telephone Woodbridge (03943) 6327.   Granary Yacht Harbour is a development of Melton Boatyard

VOLVO PENTA SERVICE

Sales and service centres in area 5
HUMBERSIDE **Hall Brothers (Bridlington) Ltd** Bessingby Way, Bessingby Industrial Estate, BRIDLINGTON YO16 4SJ Tel (0262) 673346/676604.
YORKSHIRE **Auto Unit Repairs (Leeds) Ltd** Henshaw Works, Yeadon, LEEDS LS19 7XY Tel (0532) 501222. **Yacht Service Ltd** Naburn, YORK YO1 4RW Tel (0904) 621021.

Area 5

North-East England
Boston to Berwick-on-Tweed

10.5.1	Index	Page 335
10.5.2	Diagram of Radiobeacons, Air Beacons, Lifeboat Stations etc	336
10.5.3	Tidal Stream Charts	338
10.5.4	List of Lights, Fog Signals and Waypoints	340
10.5.5	Passage Information	342
10.5.6	Distance Table	343
10.5.7	Boston	344
10.5.8	River Humber (Immingham, Standard Port, Tidal Curves)	344
10.5.9	Bridlington	350
10.5.10	Scarborough	350
10.5.11	Whitby	351
10.5.12	River Tees, Standard Port, Tidal Curves/Middlesbrough	351
10.5.13	Hartlepool	356
10.5.14	Seaham	356
10.5.15	Sunderland	357
10.5.16	River Tyne/North Shields	358
10.5.17	Blyth	359
10.5.18	Amble	360
10.5.19	Berwick-on-Tweed	361
10.5.20	Minor Harbours and Anchorages Wainfleet Filey Runswick Bay Newton Haven Seahouses/North Sunderland Farne Islands Holy Island	362

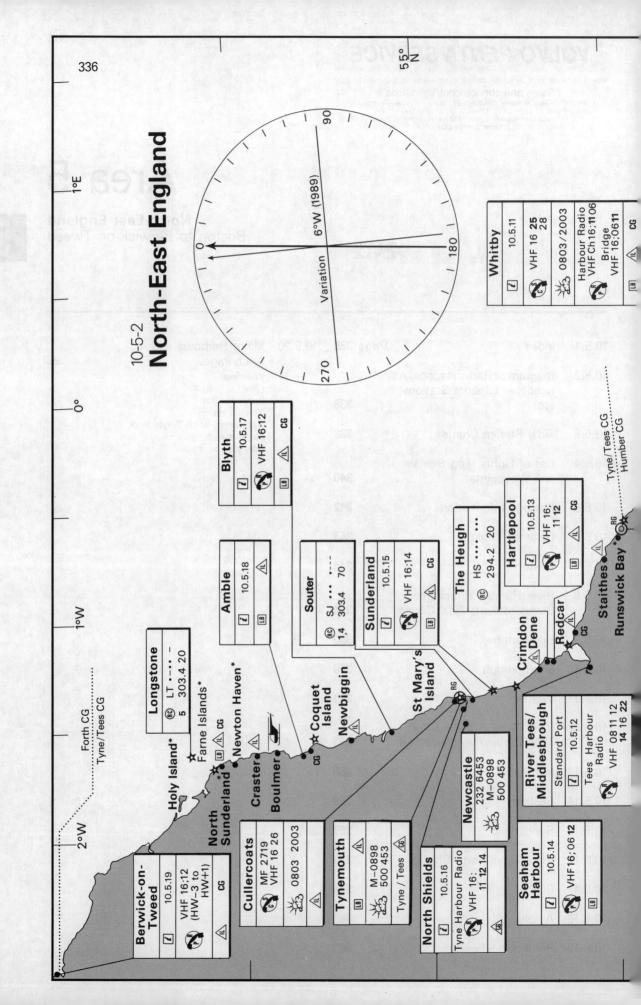

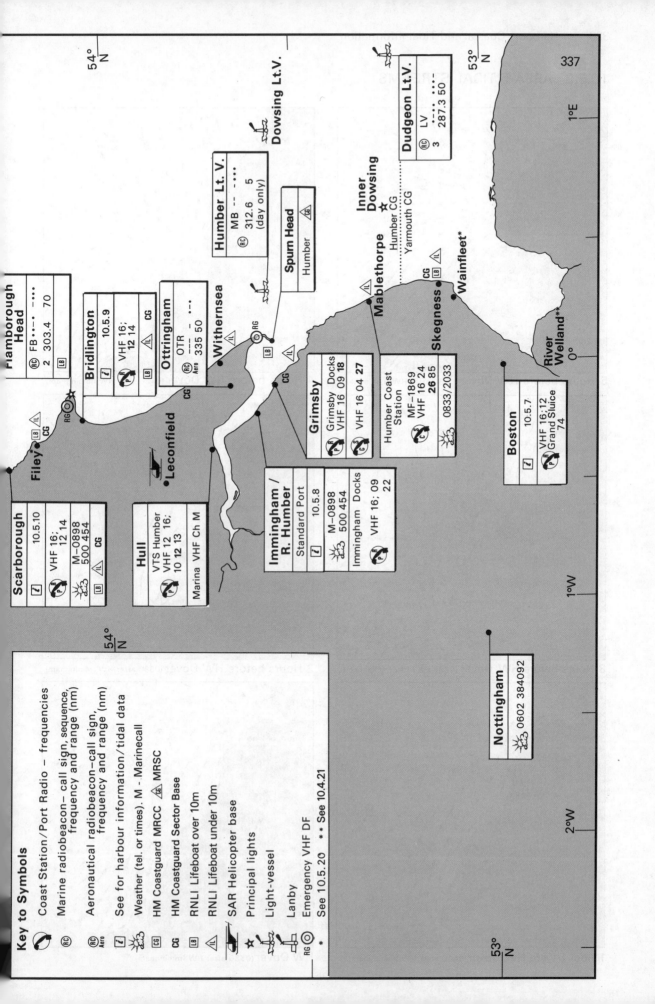

10.5.3 AREA 5 TIDAL STREAMS

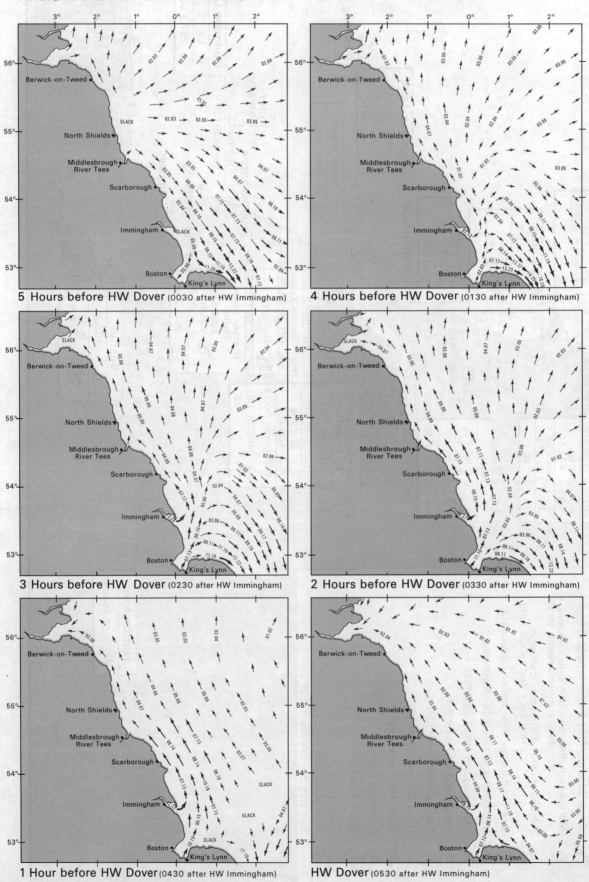

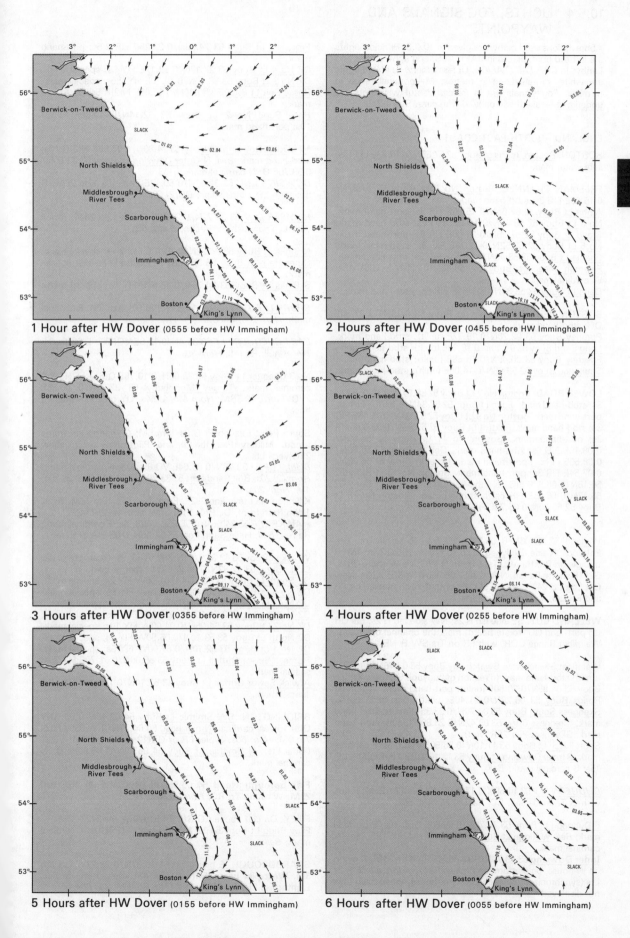

10.5.4 LIGHTS, FOG SIGNALS AND WAYPOINTS

Abbreviations used below are given in 10.0.3. Principal lights are in **bold** print, places in CAPITALS, and light-vessels and Lanbys in *CAPITAL ITALICS*. Unless otherwise stated lights are white. m—elevation in metres; M—nominal range in n. miles. Fog signals are in *italics*. Useful waypoints are underlined — use those on land with care. See 4.2.2.

ENGLAND—NORTH-EAST COAST

BOSTON. Boston Roads Lt Buoy 52 57.55N/0 16.23E LFl 10s; safe water mark.

FREEMAN CHANNEL. Boston No 1 Lt Buoy 52 57.87N/0 15.16E FlG 3s; stbd-hand mark.
Alpha Lt Buoy 52 57.66N/0 15.06E FlR 3s; port-hand mark.
No 3 Lt Buoy 52 58.10N/0 14.15E FlG 6s; stbd-hand mark.
Bravo Lt Buoy 52 57.98N/0 14.00E FlR 6s; port-hand mark.
No 5 Lt Buoy 52 58.52N/0 12.78E FlG 3s; stbd-hand mark.
Charlie Lt Buoy 52 58.43N/0 12.54E FlR 3s; port-hand mark.
Freeman Inner Lt Buoy 52 58.45N/0 11.50E Q(9) 15s; W cardinal mark.
Delta Lt Buoy 52 58.34N/0 11.68E FlR 6s; port-hand mark.

BOSTON DEEP. Lynn Knock Lt Buoy 53 04.35N/0 27.30E QG; stbd-hand mark.
Pompey Buoy 53 02.20N/0 19.37E (unlit) stbd-hand mark.
Long Sand Buoy 53 01.10N/0 18.30E (unlit); stbd-hand mark.
Friskney Buoy 53 00.48N/0 16.68E (unlit); stbd-hand mark.
Scullridge Buoy 52 59.68N/0 14.00E (unlit); stbd-hand mark.

LOWER ROAD. Boston No 7 Lt Buoy 52 58.57N/0 10.05E FlG 3s; stbd-hand mark. Echo Lt Buoy 52 58.34N/0 10.15E FlR 3s; port-hand mark. Boston No 9 Lt Buoy 52 57.58N/0 08.45E FlG 3s; stbd-hand mark. Foxtrot Lt Buoy 52 57.55N/0 09.00E FlR 3s; port-hand mark. Boston No 11 Lt Buoy 52 56.65N/0 08.10E FlG 6s; stbd-hand mark. Golf Lt Buoy 52 56.66N/0 08.50E FlR 3s; port-hand mark. Hotel Lt Buoy 52 56.28N/0 07.54E FlR 3s; port-hand mark. Boston No 13 Lt Buoy 52 56.18N/0 06.93E FlG 3s; stbd-hand mark. India Lt Buoy 52 56.12N/0 06.05E FlR 6s; port-hand mark. Boston No 15 Lt Buoy 52 56.30N/0 05.73E FlG 3s; stbd-hand mark.

Welland 52 56.06N/0 05.37E QR 5m; R square on Bn. Entrance, N side, Dollypeg 52 56.10N/0 05.15E QG 4m 1M; B triangle on Bn; Ra refl. Tabs Head Q WG; vis W shore-251°, G251°-shore; Ra refl. New Cut 52 56.0N/0 04.8E Fl G 3s. New Cut Ldg Lts 240°. Both F 5M. River to Boston marked by FR and FG Lts, and by FW Ldg Lts.

WELLAND CUT. SE side Iso R 2s; NW side Iso G 2s. Lts QR (to port) and QG (to starboard) mark the channel upstream. Wainfleet Range UQR, with FR on Trs SW & NE.

Wainfleet Swatchway. Swatchway Buoy 53 03.76N/0 19.80E (unlit); stbd-hand mark (direction of buoyage S). Inner Knock Buoy 53 04.85N/0 20.50E (unlit); port-hand mark. Wainfleet Roads Buoy 53 06.20N/0 21.40E (unlit); port-hand mark. Skegness South Buoy 53 06.70N/0 23.35E (unlit); stbd-hand mark. Skegness Buoy 53 08.42N/0 23.80E (unlit); stbd-hand mark. Skegness Pier 2FR (vert).
Scott Patch Lt Buoy 53 11.10N/0 36.50E VQ(3) 5s; E cardinal mark. S Inner Dowsing Lt Buoy 53 12.10N/0 33.80E Q(6) + LFl 15s; S cardinal mark; *Bell*.

DUDGEON Lt V 53 16.60N/1 17.00E Fl(3) 30s 12m **25M**; R hull with Lt Tr amidships; RC; Racon; *Horn(4) 60s*.

E Dudgeon Lt Buoy 53 19.70N/0 58.80E Q(3) 10s; E cardinal mark; *Bell*. W Ridge Lt Buoy 53 19.05N/0 44.60E Q(9) 15s; W cardinal mark.

Inner Dowsing 53 19.70N/0 33.96E Fl 10s 41m **21M**; R pylon on W house, on platform on B piles; Racon; *Horn 60s*; FlR 2s Lts (synchronized) on each corner of house, when main Lt is not displayed.

Protector Lt Buoy 53 24.83N/0 25.25E FlR 2.5s; port-hand mark.
DZ No 4 Lt Buoy 53 27.12N/0 19.17E FlY 5s; special mark.
DZ No 3 Lt Buoy 53 28.40N/0 19.20E FlY 2.5s; special mark.
Rosse Spit Lt Buoy 53 30.40N/0 17.05E Fl(2)R 5s; port-hand mark.
Haile Sand No 2 Lt Buoy 53 32.14N/0 12.80E Fl(3)R 10s; port-hand mark.

Mid Outer Dowsing Lt Buoy 53 24.80N/1 07.90E Fl(3)G 10s; stbd-hand mark; *Bell*. N Outer Dowsing Lt Buoy 53 33.50N/0 59.70E Q; N cardinal mark.
DOWSING Lt V 53 34.00N/0 50.17E Fl(2) 10s 12m **26M**; R hull with Lt Tr amidships; *Dia(2) 60s*.

HUMBER Lt V 53 36.72N/0 21.60E Fl 20s 12m **26M**; R hull with Lt Tr amidships; Racon; *Dia(3) 60s*.

N Binks Lt Buoy 53 36.22N/0 18.70E FlY 2.5s; special mark.
Outer Haile Lt Buoy 53 35.25N/0 19.00E Fl(4)Y 15s; special mark.
S Binks Lt Buoy 53 34.70N/0 16.60E FlY 5s; special mark.

SPURN Lt F 53 33.53N/0 14.33E Q(3) 10s 10m 8M; E cardinal mark; *Horn 20s*; Racon.

EAST CHEQUER Lt F 53 33.4N/0 12.6E VQ(6) + L Fl 10s 6m 7M; B hull; Ra refl; *Horn 30s*.

No 3 Chequer Lt Buoy 53 33.05N/0 10.70E Q(6) + LFl 15s; S cardinal mark. Tetney Monobuoy 53 32.34N/0 06.85E 2VQY(vert); Y SBM; *Horn Mo(A) 60s*; QY on 290m floating hose.

Spurn Point Beacon Fl G 3s 11m 5M; G triangle on frame tripod. Military Pier (head) and Pilot Jetty each show 2FG(vert) Lts.
BULL Lt F 53 33.78N/0 05.64E VQ 8m 6M; N cardinal mark; *Horn (2) 20s*. Bull Sand Fort 53 33.70N/0 04.14E Fl(2) 5s 20m 4M; *Horn 20s*.
Killingholme Lts in line 292°. Front Iso R 2s 10m 14M. Rear, 189m from front, Oc R 4s 21m 14M; both vis 287°-297°.
HAWKE Lt F 53 34.65N/0 05.23E VQ(3) 5s; E cardinal mark; Ra refl; *Bell*. Haile Sand Fort 53 32.05N/0 02.14E 2FR(vert).

Wreck Lt Buoy 53 42.35N/0 07.30E VQ(3) 5s; E cardinal mark.
DZ Lt Buoy 53 47.82N/0 00.25E QY; special mark.
DZ South Lt Buoy 53 50.35N/0 01.50W FlY 2s; special mark.
DZ No 5 Lt Buoy 53 50.40N/0 04.00W FlY 5s; special mark.
DZ No 4 Lt Buoy 53 51.50N/0 00.00 FlY 10s; special mark.
DZ No 3 Lt Buoy 53 54.10N/0 02.00W FlY 10s; special mark.
DZ No 2 Lt Buoy 53 54.05N/0 04.00W FlY 10s; special mark.
DZ North Lt Buoy 53 53.10N/0 05.50W FlY 2s; special mark.
DZ No 1 Lt Buoy 53 52.10N/0 07.00W FlY 5s; special mark.
Hornsea Sewer Outfall Lt Buoy 53 55.00N/0 01.70E FlY 20s; special mark.
Atwick Sewer Outfall Lt Buoy 53 57.10N/0 10.20W FlY 10s; special mark.

BRIDLINGTON. SW Smithic Lt Buoy 54 02.40N/0 09.10W Q(9) 15s; W cardinal. N pier head 54 04.77N/0 11.08W Fl 2s 7m 9M; *Horn 60s*. S pier FR or G R12m, G8m, R4M, G3M. (Tidal Lts). N Smithic Lt Buoy 54 06.20N/0 03.80W Q; N cardinal mark.

Flamborough Head 54 06.95N/0 04.87W Fl(4) 15s 65m **29M**; W round Tr; RC; *Horn(2) 90s*.

FILEY. On cliff above CG Stn FR 31m 1M; vis 272°-308°.
Filey Brigg Lt Buoy 54 12.74N/0 14.48W Q(3)10s; E cardinal mark; *Bell*.

SCARBOROUGH. E pier head 54 16.87N/0 23.27W QG 8m 3M. West pier head 2FR(vert) 5m 4M when 1.8m on bar. Lighthouse Pier Iso 5s and FY (tide signals); *Dia 60s*.

AREA 5—NE England

WHITBY. Whitby Lt Buoy 54 30.32N/0 36.49W Q; N cardinal; *Bell*. **High Lt, Ling Hill** 54 28.67N/0 34.00W Iso WR 10s 73m **23M**; W 8-sided Tr and dwellings; vis R128°-143°, W143°-319°. E Pier head FR 14m 3M. West Pier head FG 14m; *Horn 30s*. Ldg Lts 169° Church Steps, Rear FR (between E and W Pier Lts, leads into harbour).

Redcar, The High. Outfall Lt Buoy 54 36.63N/1 00.30W FIY 10s; special mark. Salt Scar Lt Buoy 54 38.15N/1 00.00W Q; N cardinal mark; *Horn(1) 15s; Bell*.

REDCAR. Luffway Ldg Lts 197°. Front, on Esplanade, FR 8m 7M; vis 182°-212°. Rear, 115m from front, FR 12m 7M; vis 182°-212°. High Stone Ldg Lts 247°. Both Oc R 2.5s 7M; vis 232°-262°.

Tees Fairway Lt Buoy 54 40.93N/1 06.37W Iso 4s 9m 8M; safe water mark; *Horn(1) 5s*; Racon.
Tees North (Fairway) Lt Buoy 54 40.28N/1 07.20W FIG 5s; stbd-hand mark. Tees South (Fairway) Lt Buoy 54 40.17N/1 06.95W FIR 5s; port-hand mark.

RIVER TEES. Breakwater head, **South Gare**. 54 38.83N/1 08.16W Fl WR 12s 16m **W20M, R17M**; W round Tr; vis W020°-274°, R274°-357°; *Horn 30s*. Ldg Lts 210°. Both FR 18/20m 13/10M.

HARTLEPOOL. Longscar Lt Buoy 54 40.85N/1 09.80W Q(3) 10s; E cardinal mark; *Bell*. Old Pier head 54 41.59N/1 10.99W QG 13m 7M; B Tr. Middleton Jetty QR 9m 6M. Ldg Lts 329°. Front 54 41.8N/1 11.3W FR. Rear, 100m from front FR.

Pipe jetty head 54 42.80N/1 12.40W 2FR(vert) 1M; *Bell(1) 15s*.

SEAHAM HARBOUR. North Pier head 54 50.25N/1 19.15W Fl G 10s 12m 5M; W column, B bands; *Dia 30s*. S Pier head 2FR(vert).

Hendon Rock Buoy 54 54.21N/1 19.44W (unlit); port-hand mark; *Bell*. Ballast Buoy 54 54.53N/1 18.39W (unlit); special mark.

SUNDERLAND. **Roker Pier** head 54 55.27N/1 21.05W Fl 5s 25m **23M**; W round Tr, R bands and cupola: vis 211°-357°; *Siren 20s*. New South pier head Fl 10s 14m 10M; W Tr. South side Fl R 5s 9m 2M. Old North pier head QG 12m 8M; Y Tr; *Horn 10s*.

Whitburn Steel Buoy 54 56.30N/1 20.80W (unlit); port-hand mark.

WHITBURN FIRING RANGE. 54 57.2N/1 21.3W and 54 57.7N/1 21.2W both FR when firing is taking place. Special buoys 54 57.04N/1 18.81W and 54 58.58N/1 19.80W, both FlY 2.5s.

Lt Buoy 55 00.17N/1 23.60W Fl(3)R 10s; port-hand mark.

TYNEMOUTH. **Entrance**, N pier head 55 00.87N/1 24.08W Fl(3) 10s 26m **26M**; grey round Tr, W lantern; *Horn 10s*. S pier head Oc WRG 10s 15m W13M, R9M, G8M; grey round Tr, R&W lantern; vis W075°-161°, G161°-179° over Bellhues rock, W179°-255°, R255°-075°; *Bell (1) 10s*. Herd Groyne, head Oc WR 10s 13m W13M, R11M, R1M; R pile structure, R&W lantern; vis R(unintens) 080°-224°, W224°-255°, R255°-277°; *Bell (1) 5s*.

NORTH SHIELDS. **Fish Quay** Ldg Lts 258°. **Front** F 25m **20M**; W square Tr. **Rear** 220m from front F 39m **20M**; W square Tr.

CULLERCOATS. Ldg Lts 256°. Front 55 02.1N/1 25.8W FR 27m 3M. Rear 38m from front FR 35m 3M.

BLYTH. Fairway Lt Buoy 55 06.58N/1 28.50W FIG 3s; *Bell*. **E Pier head** Fl(4) 10s 19m **21M**; W Tr. FR 13m 13M (same Tr); vis 152°-249°; *Horn (3) 30s*. West Pier head 2FR(vert) 7m 8M; W Tr. Training wall, S end Fl R 6s 6m 1M. South Harbour, Inner West Pier, N end 2 Fl(2) R 6s (vert) 5m 5M; W metal structure. Ldg Lts 324°. Front 55 07.4N/1 29.8W F Bu 11m 10M; Y diamond on framework Tr. Rear, 180m from front, F Bu 17m 10M; Y diamond on framework Tr. Blyth Snook Ldg Lts 338°. Both F Bu 5M.

Bondicarr Buoy 55 18.58N/1 31.37W (unlit); port-hand mark. Hauxley Buoy 55 19.28N/1 31.54W (unlit); port-hand mark. COQUET CHANNEL. SW Coquet Buoy 55 19.88N/1 32.40W (unlit); stbd-hand mark. Sand Spit Buoy 55 19.95N/1 32.64W (unlit); port-hand mark.

Coquet 55 20.01N/1 32.25W Fl(3) WR 30s 25m **W23M, R19M**; W square Tr, turreted parapet, lower half grey; vis R330°-140°, W140°-163°, R163°-180°, W180°-330°; *Horn 30s*.

NE Coquet Buoy 55 20.55N/1 32.00W (unlit); port-hand mark. NW Coquet Buoy 55 20.41N/1 32.71W (unlit); stbd-hand mark. Pan Bush Buoy 55 20.67N/1 33.22W (unlit); port-hand mark. Outfall Lt Buoy 55 20.32N/1 33.63W FIR 5s; port-hand mark; *Bell*.

WARKWORTH. S breakwater Fl R 5s 9m 5M; W Tr, R bands. N breakwater 55 20.38N/1 34.15W Fl G 6s 11m 6M; W pylon.

ALNMOUTH BAY. Boulmer Stile Buoy 55 23.75N/1 32.60W (unlit); port-hand mark. BEADNALL BAY. Newton Rock Buoy 55 32.08N/1 35.45W (unlit); port-hand mark.

NORTH SUNDERLAND. NW pier head FG 11m 3M; W Tr; vis 159°-294°; *Siren 90s (occas)*. Breakwater head 55 35.05N/1 38.79W Fl R 2.5s 6m.

Black Rock Point, **Bamburgh** 55 37.00N/1 43.35W Oc(2) WRG 15s 12m **W17M**, R13M, G13M; W building; vis G122°-165°, W165°-175°, R175°-191°, W191°-238°, R238°-275°, W275°-289°, G289°-300°.

FARNE ISLANDS, near SW point L Fl(2) WR 15s 27m W13M, R9M; W round Tr; vis R119°-277°, W277°-119°. **Longstone**, West side 55 38.63N/1 36.55W Fl 20s 23m **29M**; R Tr, W band; RC; *Siren (2) 60s*.

Swedman Buoy 55 37.70N/1 41.50W (unlit); stbd-hand mark. HOLY ISLAND. Ridge Buoy 55 39.70N/1 45.87W (unlit); W cardinal mark. Triton Buoy 55 39.61N/1 46.49W (unlit); stbd-hand mark.

BERWICK. Breakwater head 55 45.88N/1 58.95W Fl 5s 15m 10M; W round Tr, R cupola and base; vis 154°-010°. FG (same Tr) 8m 1M; vis 010°-154°; *Reed 60s (occas)*.

10.5.5 PASSAGE INFORMATION

For general notes on The Wash see 10.4.5. The W shore of The Wash (chart 108) is fronted by extensive flats which extend 2–3M offshore and which dry up to 4m, and provide a bombing range. The Wainfleet Swatchway should only be used in good vis: the chan shifts constantly, and beware the various dangers off Gibraltar Pt, 3M S of Skegness. For Wainfleet see 10.5.20. The better route N from Boston (10.5.7) is through Freeman Chan into Lynn Deep which, if bound E, also gives access to Sledway which runs S of Woolpack and Burnham Flats. Near N end of Lynn Deep there are overfalls over Lynn Knock at sp tides.

Inner Dowsing is a narrow N/S sandbank with a least depth of 1.2m, 10M off Mablethorpe. There are overfalls off the W side of the bank at the N end. Inner Dowsing Lt Ho (Lt, fog sig) stands close NE of the bank.

In the outer approaches to R Humber and The Wash there are many offlying banks, but few of them are of direct danger to yachts. The sea however breaks on some of them in bad weather, when they should be avoided. Fishing vessels may be encountered, and there are oil/gas installations offshore (see below).

TIDAL STREAMS IN RIVERS

Tidal streams in rivers are influenced by local weather conditions as well as by the phases of the Moon. At or near springs, in a river which is obstructed, for example, by sandbanks at the entrance, the time of HW gets later going up river: the time of LW also gets later, but more rapidly. So the duration of rise of tide becomes shorter, and duration of fall of tide becomes longer. At the entrance the flood stream starts at an inverval after LW which increases with the degree of obstruction of the channel: this interval between local LW and the start of the flood increases with the distance up river. The ebb begins soon after local HW along the length of the river. Hence the duration of the flood is less than that of the ebb, and the difference increases with distance up river.

The flood stream is normally stronger than the ebb, and runs harder during the first half of the rise of tide.

At neaps the flood and ebb both start soon after local LW and HW respectively, and their durations and rates are roughly equal.

Both at springs and neaps, the streams can be affected by recent precipitation, whereby the duration and rate of the ebb are increased, and of the flood are reduced.

OIL AND GAS INSTALLATIONS

Any yacht crossing the North Sea is likely to encounter oil or gas installations. These are shown on Admiralty charts where the scale permits, and are listed in the *Annual Summary of Admiralty Notices to Mariners* and in the *Admiralty List of Lights and Fog Signals*. Safety zones of radius 500 metres are established round all permanent platforms, mobile exploration rigs, and tanker loading moorings. Yachts must not enter these zones except in emergency or due to stress of weather.

Platforms show a main, white light, flashing Morse U every 15 seconds with a range of 15M. In addition there are secondary red lights, flashing Morse U every 15 seconds with a range of 3M, synchronised with the main light and situated at each corner of the platform if not marked by a white light. Platforms sound a fog signal, Horn Morse U every 30 seconds.

RIVER HUMBER (chart 109)

R Humber is formed by R Ouse and R Trent, which join 15M above Kingston upon Hull, and is important for the access provided to the inland waterways by these rivers. By the time R Humber reaches the sea between Donna Nook and Spurn Hd it is 6M wide, draining as it does most of Yorkshire and the Midlands.

Approaching from the S, a yacht should make for Rosse Spit buoy and then for Haile Sand buoy, before altering course any more W to meet and follow the main buoyed chan, which leads in from a NE direction through New Sand Hole. Beyond Bull Sand Fort (Lt, fog sig) it is advisable to follow one of the buoyed chans, since shoals are liable to change.

If proceeding to or coming from the N, beware The Binks — a shoal (dries in places) extending 3M E from Spurn Hd, with a rough sea when wind is against tide.

10M E of Spurn Hd the tidal streams are not affected by the river; the S-going stream begins at HW Immingham −0455, and the N-going at HW Immingham +0130. Nearer the entrance the direction of the S-going stream becomes more W, and that of the N-going stream more E. ½M S of Spurn Hd the in-going stream runs NW and begins about HW Immingham −0520, sp rate 3½ kn: the out-going stream runs SE and begins about HW Immingham, sp rate 4 kn. In this area the worst seas are experienced in NW gales on a strong flood tide.

HUMBER TO WHITBY (charts 107, 121, 129)

For the coast, harbs, anchs N from R Humber refer to *Sailing Directions Humber Estuary to Rattray Head* (R Northumberland YC). Bridlington B (chart 1882) is clear of dangers apart from Smithic Shoals (buoyed), about 3M off Bridlington (10.5.9): the seas break on these shoals in strong N or E winds.

Flamborough Hd (Lt, fog sig, RC) is a steep, white cliff with conspic Lt Ho on summit. The Lt may be obsc by cliffs when close inshore. An old Lt Ho, also conspic, is 2½ ca WNW. From here the coast runs NW, with no offshore dangers until Filey Brigg where rky ledges extend ½M ESE, marked by buoy. There is anch in Filey B (10.5.20) in N or offshore winds. NW of Filey Brigg beware Horse Rk and foul ground ½M offshore; maintain this offing past Scarborough (10.5.10) to Whitby High Lt. Off Whitby (10.5.11) beware Whitby Rk and The Scar (dry in places) to E of harb, and Upgang Rks (dry in places) 1M to WNW: the sea breaks on all these rks.

WHITBY TO COQUET ISLAND (charts 134, 152, 156)

From Whitby to Hartlepool (10.5.13) there are no dangers more than 1M offshore. Runswick B (10.5.20 and chart 1612), 5M NW of Whitby, provides anch in winds from S and W but is dangerous in onshore winds. The little harb of Staithes is only suitable for yachts which can take the ground, and only in good weather and offshore winds.

Redcliff, dark red and 205m high is a conspic feature of this coast which, along to Hunt Cliff is prone to landslides and is fringed with rky ledges which dry for about 3 ca. There is a conspic radio mast 4 ca SW of Redcliff. Off Redcar and Coatham beware West Scar and Salt Scar — detached rky ledges (dry) lying 1-8 ca offshore. Other ledges lie close SW and S of Salt Scar which is marked by buoy.

Between R Tees (10.5.12) and Hartlepool (10.5.13) beware Long Scar — detached rky ledge (dries) with extremity marked by buoy. From the Heugh an offing of 1M clears all dangers until approaching Sunderland (10.5.15), where White Stones, rky shoals with depth 3.0m, lie 1¾M SSE of Roker Pier Lt Ho, and Hendon Rk, depth 1.5m, lies 1¼M SE of the Lt Ho. N of Sunderland, Whitburn Steel, a rk with less than 2m over it, lies 1M S of Souter Pt, and an obstruction (dries) lies 1 ca SE of Whitburn Steel. Along this stretch of coast industrial smoke haze may reduce vis and obsc Lts.

AREA 5—NE England

The coast N of Tynemouth (10.5.16) is foul, and on passage to Blyth (10.5.17) it should be given an offing of 1M. 3½M N of Tynemouth is St Mary's Island (with disused Lt Ho), connected to land by causeway. The harb of Seaton Sluice, 1M NW of St Mary's Island, is closed.

Proceeding N from Blyth, keep well seaward of The Sow and Pigs buoy, and set course to clear Newbiggin Pt and Beacon Pt by about 1M. Vessels conduct trials on measured distance here. Near Beacon Pt is conspic chy of aluminium smelter. Snab Pt is 2M NNW of Beacon Pt. Cresswell Skeres, rky patches with depth 3.0m, lie 1¼M and 1¾m NNE of Snab Pt. Cresswell Pt is 8 ca NW of Snab Pt, and between the two are rks extending 3 ca seaward. S of Hauxley Hd rks extend 6 ca offshore, with depths of 1.2m, marked by buoy.

COQUET ISLAND AND AMBLE (chart 1627)

Coquet Island (Lt, fog sig) lies about ½M offshore at SE end of Alnmouth B, and nearly 1M NNE of Hauxley Pt — off which drying ledges extend 5½ ca E and 6½ ca NNE towards the island (marked by buoys). On passage, normally pass 1M E of Coquet Island, but Coquet Chan inshore is available in good vis by day. It is very narrow, but buoyed, with least depth of 1.2m near the centre, and the stream runs strongly, S-going from HW Tyne −0515 and N-going from HW Tyne +0045. There are good anchs in Coquet Road, W of the island, in winds from S or W.

Amble (Warkworth) Harb entrance (10.5.18) is about 1M W of Coquet Island, and 1½M SE of Warkworth Castle (conspic). 4 ca NE and ENE of entrance is Pan Bush, rky shoal with least depth of 0.3m on which dangerous seas can build in any swell. The bar off entrance has varying depths, down to less than 1m. Once inside, the harb is safe, but the entrance is dangerous in strong winds from N or E when broken water may extend to Coquet Island.

Between Coquet Island and the Farne Islands, 19M to N, keep at least 1M offshore to avoid various dangers. To seaward, Dicky Shad and Newton Skere lie 1¾M and 4½M E of Baednell Pt, and Craster Skeres lie 5M E of Castle Pt: these are three rky banks on which the sea breaks heavily in bad weather. For Newton Haven and Seahouses (N Sunderland) see 10.5.19.

FARNE ISLANDS AND HOLY ISLAND (chart III)

The coast between N Sunderland Pt and Berwick upon Tweed, 16M NW, is a dangerous area. The Farne Islands and offlying shoals extend 4½M offshore, and are worth visiting in good weather. The area is a bird sanctuary, owned and operated by the National Trust, with large colonies of sea birds and also grey seals. For details of this area refer to R Northumberland YC sailing directions. The group of islands, rks and shoals is divided by Farne Sound running NE/SW, and by Staple Sound running NW/SE. Farne Island (Lt) is the innermost I, separated from mainland by Inner Sound, which in good conditions is a better N/S route than keeping outside the whole group: but the stream runs 3 kn at sp, and with strong wind against tide there is rough water. Beware Swedman reef (dries, marked by buoy) 4 ca WSW of Megstone, which is a rk 5m high 1M NW of Farne Lt Ho. N of Inner Sound, off Holy Island, Goldstone Chan passes between Goldstone Rk (dries) on E side and Plough Seat Reef and Plough Rk (both dry) on W side. See 10.5.20.

On NE side of Farne Island there is anch called The Kettle, sheltered except from NW, but anch out of stream close to The Bridges connecting Knock's Reef and W Wideopen.

If course is set outside Farne Islands, pass 1M E of Longstone (Lt, fog sig, RC) to clear Crumstone Rk 1M to S, and Knivestone (dries) and Whirl Rks (depth 0.6m) respectively 5 ca and 6 ca NE of Longstone Lt Ho. The sea breaks on these rks. Also watch for shipping.

Near the Farne Islands and Holy Island the SE-going stream begins at HW Tyne −0430, and the NW-going at HW Tyne +0130. 1M NE of Longstone the sp rate is about 3½ kn, decreasing seaward. There is an eddy S of Longstone on NW-going stream.

Holy Island (Lindisfarne) lies 6M WNW of Longstone, and is linked to mainland by a causeway covered at HW. There is a good anch on S side (chart 1612). From a point about ½M S of Plough Seat Reef, obelisks on Old Law lead 260° S of Triton Shoal (least depth 0.4m, 4 ca SSE of Castle Pt) to the bar, from where Heugh Bn on with belfry of St Mary's Church leads 310° to anch. The stream runs strongly in and out of harb, W-going from HW Tyne +0510, and E-going from HW Tyne −0045. See 10.5.20.

10.5.6 DISTANCE TABLE

Approximate distances in nautical miles are by the most direct route while avoiding dangers and allowing for traffic separation schemes etc. Places in *italics* are in adjoining areas.

	1	2	3	4	5	6	7	8	9	10	11	12	13	14	15	16	17	18	19	20
1 *North Foreland*	1																			
2 *Goeree Tower*	91	2																		
3 *IJmuiden*	137	46	3																	
4 *Brunsbüttel*	337	263	221	4																
5 *Boston*	160	161	175	344	5															
6 *Grimsby*	169	169	182	332	58	6														
7 Bridlington	194	188	196	329	83	44	7													
8 Flamborough Head	192	186	193	325	84	45	5	8												
9 Scarborough	206	200	207	336	98	59	19	14	9											
10 Whitby	222	116	223	345	114	75	35	30	16	10										
11 River Tees	243	237	244	367	135	96	56	50	37	21	11									
12 Hartlepool	245	239	246	370	137	98	58	53	39	24	4	12								
13 Seaham	253	247	254	373	145	106	66	61	47	33	15	11	13							
14 Sunderland	257	251	258	375	149	110	70	65	51	36	20	16	5	14						
15 Tynemouth	262	256	262	378	154	115	75	70	56	41	27	23	12	7	15					
16 Blyth	270	264	267	382	162	123	83	78	64	49	35	31	20	15	8	16				
17 Longstone (Farne)	298	292	295	388	190	151	111	105	92	76	67	63	52	46	40	32	17			
18 Berwick-on-Tweed	313	307	310	402	205	166	126	120	107	91	82	78	67	61	55	47	15	18		
19 *St Abbs Head*	322	316	322	410	214	175	135	129	116	110	91	87	76	70	64	56	24	12	19	
20 *Peterhead*	404	391	378	439	296	257	217	212	200	186	173	170	161	155	149	143	112	105	97	20

BOSTON 10-5-7
Lincs.

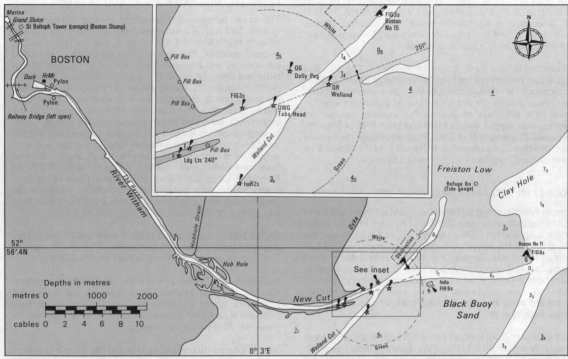

⚓ Visitors berths at River Quay, report to Hr Mr (alongside Lock)

CHARTS
Admiralty 108, 1200; Imray Y9; OS 131
TIDES
−0500 Dover; ML 3.4; Duration Flood 0500, Ebb 0700; Zone 0 (GMT).

Standard Port IMMINGHAM (→)

Times				Height (metres)			
HW		LW		MHWS	MHWN	MLWN	MLWS
0100	0700	0100	0700	7.3	5.8	2.6	0.9
1300	1900	1300	1900				

Differences BOSTON
0000 +0010 +0140 +0050 −0.5 −1.0 −0.9 −0.5
SKEGNESS
+0010 +0015 +0030 +0020 −0.4 −0.5 −0.1 0.0

SHELTER
Very good but mooring in the Dock is not permitted except in emergency. Yachts which can lower masts should pass through the Grand Sluice into fresh water; marina is on the stbd side immediately beyond the sluice. This leads into the R Witham Navigation which goes 31 miles up to Lincoln. The lock is 22.7m x 4.6m and opens approx HW∓1. Alternative is to dry out at pontoon on E bank above the railway bridge (normally left open).
NAVIGATION
Waypoint Boston Roads safe water buoy, LFl 10s, 52°57'.53N 00°16'.23E, 110°/290° from/to Freeman Channel entrance, 0.70M. The channel is liable to change but is well marked. Tabs Head marks the entrance to the river which should be passed not later than HW−3 to enable the Grand Sluice to be reached before the start of the ebb. On reaching Boston Dock, masts should be lowered to negotiate swing bridge (cannot always be opened) and three fixed bridges. Channel through town is narrow and un-navigable at LW.
LIGHTS AND MARKS
New Cut and R Witham are marked by Bns with topmarks. FW Lts mark leading lines — four pairs going upstream and six pairs going downstream. St Boltoph's Church tower, known as the Boston Stump is conspicuous landmark. Entry signals for Boston Dock are not given since yachts are not allowed in. Yachts in difficulty secure just above dock entrance and see Hr Mr. The port is administered by Port of Boston Authority.

RADIO TELEPHONE (local times)
Call: *Boston Dock* VHF Ch 16; 12 (Mon to Fri 0700-1700 and HW−3 to HW+2). Call: *Boston Lock* Ch 74.
TELEPHONE (0205)
Hr Mr 62328; MRCC Great Yarmouth 851338; Port Signal Stn 62328; HMC Freefone Customs Yachts: RT (0205) 63070; Port Authority Boston Docks 65571; Lock Keeper 64864; Marinecall 0898 500 455; Hosp 64801.
FACILITIES
EC Thursday; **Boston Marina** (50 and some visitors) Tel. 64420, Slip, D, ACA, FW, CH; C in dock by arrangement with Hr Mr (emergency only) Access HW∓2; **Boston Diesel** Tel. 63300, Gas; **Town** V, R, Bar, PO, Bank, Rly, Air (Norwich or Humberside).

RIVER HUMBER 10-5-8
Humberside

CHARTS
Admiralty 109, 3497, 1188; Associated British Ports; Imray C29; OS 107
TIDES
−0510 Dover; ML 4.1; Duration 0555; Zone 0 (GMT).

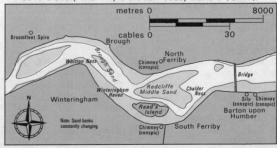

AREA 5—NE England

RIVER HUMBER continued

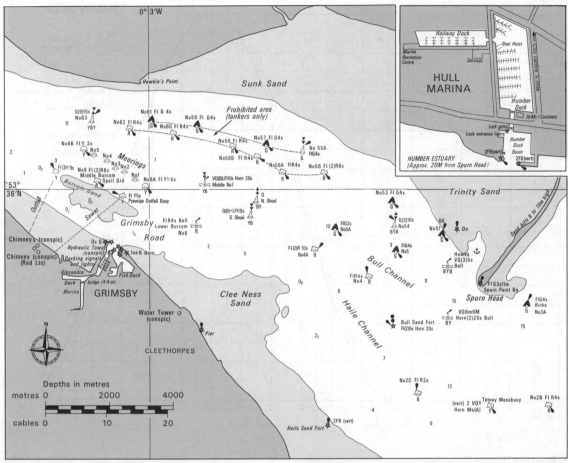

Standard Port IMMINGHAM (→)

Times				Height (metres)			
HW		LW		MHWS	MHWN	MLWN	MLWS
0100	0700	0100	0700	7.3	5.8	2.6	0.9
1300	1900	1300	1900				

Differences HULL
+0005 +0015 +0010 +0020 +0.2 0.0 −0.2 −0.2
GRIMSBY
−0003 −0011 −0015 −0002 −0.3 −0.2 0.0 +0.1
BULL SAND FORT
−0020 −0030 −0035 −0015 −0.4 −0.3 +0.1 +0.2
BURTON STATHER
+0105 +0045 +0335 +0305 −2.1 −2.3 −2.3 Dries
KEADBY
+0135 +0125 +0415 +0355 −2.7 −3.1 Dries Dries
BLACKTOFT
+0055 +0050 +0310 +0255 −1.6 −1.9 −2.0 −0.9
GOOLE
+0130 +0115 +0355 +0350 −1.6 −2.1 −1.9 −0.6

NOTE: Immingham is a standard Port and details of daily HW and LW times and heights are given below.

SHELTER
R Humber is the estuary of R Ouse and R Trent. Yachts can anchor inside Spurn Head except in strong SW to NW winds. There are marinas at Hull, Grimsby, South Ferriby and Naburn. In Grimsby, visiting yachts now use the Fish Dock where lock charges are high. The approach to South Ferriby Marina should not be attempted without reference to up-to-date Associated British Ports Chart. Immingham should be used only in emergency. South Ferriby available HW∓2½, Grimsby Dock HW∓2½, Hull Marina HW∓3. Yachts are welcome at Brough and Winteringham (30M up river from Spurn Head), but contact club first.

NAVIGATION
Waypoint 53°33'.50N 00°08'.00E, 095°/275° from/to Bull Sand Fort, 2.3M. There is a whirlpool effect at confluence of Rivers Hull and Humber. In flood it runs W into Humber and NE into Hull (reaches 2½ kn at springs). N.B. Associated British Ports is the Navigational and Conservancy Authority for the Humber Estuary and are owners of the ports of Hull, Grimsby, Immingham and Goole. They issue up-to-date information about the buoyed channel up river from Hull which is constantly changing.

LIGHTS AND MARKS
The Humber is well marked for its whole length. International Port Traffic Signals control entry to Grimsby (Royal Dock) shown W of entrance; Immingham; Killingholme; Hull; Goole. Entry to Winteringham Haven and Brough Haven should not be attempted without first contacting Humber Yawl Club for up-to-date approach channel details and mooring availability. Havens available HW∓2 approx.

RADIO TELEPHONE (local times)
Humber Vessel Traffic Service – call: *VTS Humber* (Queen Elizabeth Dock, Hull, Tel. Hull 701787) VHF Ch 16; **12** (H24). Weather reports, tidal information and navigational warnings broadcast every odd H+03 on Ch 12. Grimsby Docks call *Royal Dock* Ch 16; 09 **18** (H24). Immingham Docks Ch 16; 09 22 (H24). River Hull Port Operations Service call: *Drypool Radio* Ch 16; 06 **14** (Mon-Fri: 0800–1800; Sat 0830–1300. HW−2 to HW+½). Goole Docks Ch 16; 14 (H24). Booth Ferry Bridge Ch 16; 12 (H24). Selby Railway Bridge Ch 16; **09** 12. Selby Toll Bridge Ch 16; **09** 12. South Ferriby Yacht Harbour, Hull Marina, Ch M (0800–1700). Grimsby Marina Ch 09 18.

RIVER HUMBER continued

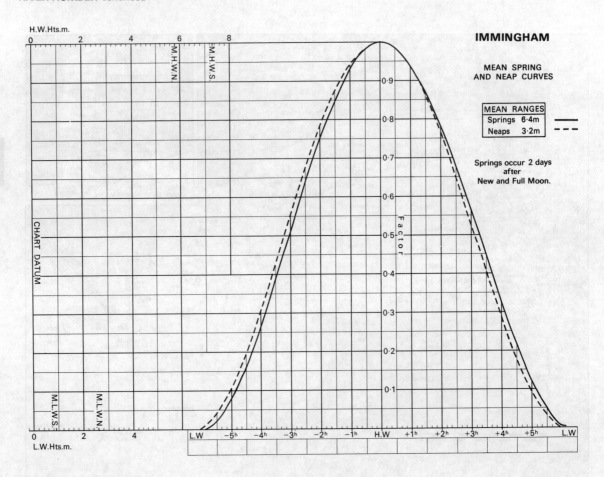

TELEPHONE (Hull 0482 Grimsby 0472)
HULL Hr Mr 783538; MRSC Bridlington 672317; HMC 796161; Marinecall 0898 500 454; Dr – contact Humber Vessel Traffic Service 701787.
GRIMSBY Port Manager 359181; HMC 45441; CG Cleethorpes 698300; Marinecall 0898 500 454; Hosp 74111.

FACILITIES
HULL **Hull Marina** (280+20 visitors) Tel. 25048 (open 0700-2200 LT Fri, Sat, Sun, 0800 – 1800 LT weekdays), P, D, FW, CH, BH (50 ton), C (7 ton), AC, Gas, HMC, ME, El, Sh, Slip, Lau, SM, (Access HW∓3) **B. Cooke** Tel. 223454, ACA.
GRIMSBY EC Thursday; **Grimsby Marina** (150+25 visitors). Tel. 360404, D, FW, ME, AC, Gas, Gaz, El, Sh, BH (30 ton), R, CH, Bar, (Access HW∓3, bridge clearance 4.45m); **Grimsby and Cleethorpes YC** Tel. 356678; **Grahams** Tel. 46673, ACA; **Fish Dock** Tel. 361343, M; **Royal Dock** Tel. 361344.
SOUTH FERRIBY **South Ferriby Marina** (100+20 visitors) Tel. Barton-on-Humber 635620, D, FW, ME, El, Sh, C (30 ton), CH, Gas, Gaz, (Access HW∓3). **Village** V, Bar, P.
WINTERINGHAM HAVEN (belongs to Humber Yawl Club) Tel. Grimsby 734452, PO.
BROUGH HAVEN **Humber Yawl Club** Tel. Hull 667224, Slip, FW, AB*, Bar. *Limited – contact club.
Note. There is a marina at Naburn, up R Ouse near York (approx 80M from Spurn Pt). **Naburn Marina** (450+50 visitors) Tel. York 21021, VHF Ch M, CH, P, D, FW, AC, Sh, ME, BH (16 ton), R.

AGENTS NEEDED
There are a number of ports where we need agents, particularly in France.
ENGLAND   Swale, Havengore, Berwick.
SCOTLAND   Firth of Forth, Scrabster, Mallaig, Loch Sunart, Loch Aline.
IRELAND   Kilrush, Wicklow, Westport/Clew Bay.
FRANCE   Arcachon, Seudre R, Ile d'Oleron, Rochfort, Ile de Re, St. Giles-Croix-de-Vie, Ile d'Yeu, Pouliguen, Le Croisic, La Forêt, Ile de Bréhat.
GERMANY   Norderney, Dornumer-Accumersiel.
If you are interested in becoming our agent for any of the above, please write to the editors and get your free copy of the Almanac every year. You do not have to be resident in a port to be the agent, but at least a fairly regular visitor.

ENGLAND, EAST COAST — IMMINGHAM

Lat 53°38′ N   Long 0°11′ W

TIMES AND HEIGHTS OF HIGH AND LOW WATERS

YEAR 1989

TIME ZONE UT(GMT)
For Summer Time add ONE hour in non-shaded areas

JANUARY

Time	m		Time	m
1 Su	0542 1204 1753	2.4 5.6 3.0	**16** M	0617 1248 1845	2.0 5.9 2.5
2 M	0014 0646 1310 1907	5.7 2.5 5.6 3.0	**17** Tu	0104 0731 1405 2008	6.2 2.2 5.8 2.5
3 Tu	0128 0755 1418 2022	5.7 2.5 5.7 2.8	**18** W	0230 0851 1518 2129	6.1 2.3 6.0 2.2
4 W	0239 0858 1517 2125	5.8 2.3 6.0 2.5	**19** Th	0349 1000 1617 2235	6.2 2.2 6.3 1.9
5 Th	0341 0955 1607 2220	6.0 2.1 6.3 2.1	**20** F	0451 1057 1705 2329	6.4 2.0 6.6 1.5
6 F	0434 1045 1651 2311	6.3 1.9 6.6 1.8	**21** Sa	0540 1143 1746	6.6 1.8 6.9
7 Sa ●	0522 1133 1733 2358	6.6 1.7 6.9 1.4	**22** Su	0015 0622 1224 1822	1.3 6.7 1.7 7.0
8 Su	0608 1218 1814	6.8 1.5 7.1	**23** M	0055 0659 1259 1856	1.2 6.8 1.6 7.1
9 M	0045 0653 1303 1856	1.1 7.0 1.4 7.3	**24** Tu	0128 0733 1331 1930	1.1 6.8 1.6 7.1
10 Tu	0130 0738 1345 1938	0.9 7.1 1.4 7.4	**25** W	0201 0805 1402 2001	1.2 6.7 1.6 7.1
11 W	0215 0823 1427 2022	0.8 7.1 1.4 7.4	**26** Th	0229 0836 1430 2032	1.3 6.6 1.7 6.9
12 Th	0258 0910 1510 2107	0.8 6.9 1.5 7.3	**27** F	0257 0905 1458 2103	1.4 6.4 1.9 6.7
13 F	0341 0956 1552 2153	1.0 6.6 1.7 7.1	**28** Sa	0327 0935 1528 2134	1.6 6.2 2.1 6.5
14 Sa ☽	0426 1045 1638 2245	1.3 6.4 2.0 6.8	**29** Su	0357 1007 1600 2209	1.9 6.0 2.3 6.1
15 Su	0515 1142 1734 2347	1.6 6.1 2.2 6.4	**30** M ☾	0434 1048 1642 2255	2.2 5.7 2.7 5.8
			31 Tu	0525 1146 1746	2.5 5.5 2.9

FEBRUARY

Time	m		Time	m
1 W	0010 0642 1310 1921	5.5 2.8 5.4 3.0	**16** Th	0229 0837 1505 2128	5.7 2.7 5.8 2.3
2 Th	0151 0812 1434 2050	5.5 2.7 5.6 2.7	**17** F	0353 0953 1610 2233	6.0 2.5 6.2 1.8
3 F	0315 0925 1539 2159	5.8 2.4 6.0 2.2	**18** Sa	0449 1047 1657 2322	6.3 2.2 6.5 1.5
4 Sa	0420 1026 1633 2257	6.2 2.1 6.4 1.7	**19** Su	0533 1129 1733	6.5 1.9 6.8
5 Su	0513 1119 1718 2349	6.6 1.7 6.8 1.2	**20** M ○	0001 0608 1205 1807	1.2 6.7 1.6 7.0
6 M ●	0600 1207 1801	7.0 1.4 7.2	**21** Tu	0036 0639 1239 1838	1.1 6.8 1.5 7.1
7 Tu	0036 0643 1252 1842	0.8 7.2 1.1 7.5	**22** W	0106 0709 1309 1907	1.0 6.9 1.4 7.2
8 W	0120 0726 1333 1924	0.5 7.4 1.0 7.7	**23** Th	0134 0737 1337 1935	1.0 6.8 1.3 7.1
9 Th	0202 0808 1413 2006	0.4 7.3 1.0 7.7	**24** F	0201 0804 1404 2004	1.1 6.8 1.4 7.0
10 F	0242 0849 1451 2049	0.5 7.2 1.1 7.6	**25** Sa	0226 0829 1429 2030	1.2 6.6 1.6 6.8
11 Sa	0319 0929 1529 2132	0.8 6.8 1.3 7.3	**26** Su	0250 0856 1454 2057	1.4 6.4 1.8 6.6
12 Su ☾	0357 1012 1610 2219	1.1 6.5 1.7 6.8	**27** M	0317 0921 1522 2128	1.7 6.2 2.0 6.2
13 M	0441 1059 1701 2318	1.7 6.0 2.1 6.3	**28** Tu ☾	0346 0953 1559 2209	2.0 5.9 2.3 5.9
14 Tu	0537 1204 1812	2.3 5.7 2.5			
15 W	0043 0659 1337 1954	5.8 2.7 5.6 2.6			

MARCH

Time	m		Time	m
1 W	0428 1042 1654 2318	2.5 5.6 2.7 5.5	**16** Th	0035 0634 1310 1945	5.5 3.0 5.5 2.5
2 Th	0540 1207 1831	2.8 5.3 2.9	**17** F	0225 0822 1446 2117	5.5 3.0 5.7 2.2
3 F	0117 0731 1355 2022	5.3 2.9 5.5 2.6	**18** Sa	0341 0936 1550 2214	5.9 2.6 6.1 1.7
4 Sa	0256 0900 1512 2139	5.7 2.6 5.9 2.0	**19** Su	0433 1026 1634 2259	6.2 2.2 6.5 1.4
5 Su	0403 1006 1610 2240	6.2 2.1 6.4 1.4	**20** M	0511 1105 1711 2334	6.5 1.9 6.7 1.2
6 M	0457 1101 1658 2332	6.7 1.6 6.9 0.9	**21** Tu	0543 1140 1742	6.6 1.6 6.9
7 Tu ●	0543 1149 1742	7.1 1.2 7.4	**22** W ○	0007 0611 1211 1811	1.1 6.8 1.4 7.1
8 W	0018 0625 1232 1822	0.5 7.4 0.9 7.7	**23** Th	0035 0638 1241 1841	1.0 6.9 1.3 7.1
9 Th	0102 0704 1313 1904	0.2 7.5 0.7 7.9	**24** F	0103 0704 1309 1909	1.0 6.9 1.3 7.1
10 F	0142 0745 1352 1947	0.2 7.4 0.7 7.9	**25** Sa	0130 0731 1337 1935	1.1 6.8 1.3 6.9
11 Sa	0220 0823 1430 2027	0.4 7.2 0.8 7.6	**26** Su	0155 0757 1402 2004	1.2 6.7 1.5 6.8
12 Su	0256 0901 1507 2111	0.8 6.9 1.1 7.2	**27** M	0220 0822 1429 2032	1.5 6.6 1.6 6.6
13 M	0331 0941 1548 2159	1.3 6.5 1.5 6.6	**28** Tu	0246 0849 1457 2104	1.7 6.3 1.9 6.2
14 Tu ☾	0412 1026 1637 2301	1.9 6.0 2.0 6.0	**29** W	0317 0921 1535 2149	2.0 6.1 2.1 5.8
15 W	0505 1129 1753	2.6 5.6 2.5	**30** Th ☾	0359 1010 1631 2304	2.4 5.7 2.5 5.5
			31 F	0509 1132 1807	2.8 5.4 2.6

APRIL

Time	m		Time	m
1 Sa	0059 0700 1320 1955	5.4 2.9 5.6 2.3	**16** Su	0305 0858 1511 2136	5.8 2.7 6.1 1.8
2 Su	0232 0832 1440 2112	5.8 2.5 6.0 1.8	**17** M	0356 0949 1559 2220	6.1 2.3 6.3 1.5
3 M	0339 0938 1541 2213	6.3 2.0 6.5 1.2	**18** Tu	0434 1030 1637 2257	6.3 2.0 6.6 1.4
4 Tu	0433 1034 1631 2306	6.7 1.6 7.0 0.8	**19** W	0506 1106 1711 2330	6.5 1.7 6.7 1.3
5 W	0518 1122 1716 2354	7.1 1.1 7.4 0.4	**20** Th	0537 1139 1742	6.7 1.5 6.9
6 Th ●	0600 1207 1800	7.3 0.8 7.7	**21** F ○	0001 0605 1211 1812	1.2 6.8 1.4 6.9
7 F	0038 0639 1249 1842	0.3 7.4 0.6 7.8	**22** Sa	0032 0634 1241 1842	1.2 6.8 1.3 6.9
8 Sa	0117 0719 1330 1926	0.4 7.4 0.6 7.7	**23** Su	0100 0702 1312 1912	1.2 6.9 1.4 6.8
9 Su	0157 0757 1409 2011	0.6 7.2 0.8 7.4	**24** M	0130 0730 1341 1944	1.4 6.7 1.4 6.6
10 M	0233 0834 1449 2056	1.1 6.9 1.1 6.9	**25** Tu	0158 0759 1412 2019	1.6 6.6 1.5 6.4
11 Tu	0308 0914 1531 2148	1.6 6.5 1.5 6.4	**26** W	0227 0830 1446 2058	1.8 6.4 1.7 6.2
12 W	0349 1000 1623 2251	2.2 6.1 2.0 5.8	**27** Th	0303 0908 1528 2152	2.1 6.2 2.0 5.9
13 Th	0442 1104 1739	2.7 5.7 2.3	**28** F ☽	0349 1000 1627 2306	2.4 5.9 2.2 5.6
14 F	0019 0605 1236 1920	5.5 3.1 5.6 2.4	**29** Sa	0459 1118 1754	2.7 5.8 2.2
15 Sa	0154 0745 1406 2040	5.7 3.0 5.7 2.1	**30** Su	0039 0632 1248 1924	5.7 2.7 5.8 2.0

Chart Datum: 3.90 metres below Ordnance Datum (Newlyn)

AREA 5—NE England   347

ENGLAND, EAST COAST — IMMINGHAM

Lat 53°38′ N Long 0°11′ W

TIMES AND HEIGHTS OF HIGH AND LOW WATERS

YEAR 1989

TIME ZONE UT(GMT)
For Summer Time add ONE hour in non-shaded areas

MAY

	Time	m		Time	m
1 M	0201 0755 1404 2037	5.9 2.4 6.2 1.6	**16** Tu	0303 0858 1508 2129	5.9 2.5 6.1 1.8
2 Tu	0305 0903 1505 2139	6.3 2.0 6.6 1.2	**17** W	0346 0945 1553 2212	6.1 2.2 6.3 1.7
3 W	0400 1000 1600 2235	6.7 1.6 7.0 0.9	**18** Th	0426 1027 1634 2251	6.3 2.0 6.4 1.5
4 Th	0448 1054 1649 2325	6.9 1.3 7.3 0.7	**19** F	0501 1105 1711 2327	6.5 1.8 6.6 1.4
5 F ●	0532 1142 1737	7.1 1.0 7.5	**20** Sa ○	0533 1142 1746	6.7 1.6 6.6
6 Sa	0011 0612 1228 1824	0.6 7.3 0.8 7.5	**21** Su	0003 0605 1217 1821	1.4 6.8 1.5 6.7
7 Su	0055 0653 1312 1912	0.7 7.3 0.8 7.4	**22** M	0036 0636 1252 1857	1.4 6.8 1.4 6.6
8 M	0135 0734 1355 1959	1.0 7.1 0.9 7.1	**23** Tu	0110 0710 1328 1935	1.5 6.8 1.4 6.6
9 Tu	0213 0813 1437 2047	1.4 6.9 1.2 6.7	**24** W	0145 0744 1405 2016	1.7 6.7 1.5 6.4
10 W	0251 0856 1521 2139	1.8 6.6 1.5 6.3	**25** Th	0220 0823 1446 2103	1.8 6.6 1.6 6.3
11 Th	0334 0942 1612 2238	2.3 6.3 1.8 5.9	**26** F	0301 0905 1532 2157	2.0 6.5 1.7 6.1
12 F ☾	0423 1038 1716 2347	2.6 6.0 2.1 5.6	**27** Sa	0350 0959 1630 2301	2.2 6.3 1.8 6.0
13 Sa	0527 1150 1831	2.9 5.8 2.2	**28** Su ☾	0451 1102 1737	2.4 6.2 1.8
14 Su	0100 0648 1307 1942	5.5 2.9 5.8 2.1	**29** M	0012 0603 1214 1849	5.9 2.4 6.2 1.7
15 M	0208 0759 1415 2042	5.7 2.8 5.9 2.0	**30** Tu	0123 0714 1324 1959	6.0 2.3 6.4 1.5
			31 W	0229 0823 1430 2104	6.2 2.1 6.6 1.3

JUNE

	Time	m		Time	m
1 Th	0327 0928 1531 2204	6.5 1.8 6.8 1.2	**16** F	0341 0946 1559 2213	6.0 2.3 6.1 1.9
2 F	0419 1027 1628 2259	6.7 1.5 7.0 1.1	**17** Sa	0424 1033 1644 2257	6.3 2.1 6.3 1.8
3 Sa ●	0506 1120 1722 2349	6.9 1.2 7.1 1.1	**18** Su	0505 1116 1727 2337	6.5 1.8 6.4 1.7
4 Su	0551 1211 1814	7.0 1.0 7.1	**19** M ○	0542 1158 1808	6.7 1.6 6.5
5 M	0035 0634 1300 1903	1.2 7.1 1.0 7.0	**20** Tu	0018 0618 1241 1849	1.6 6.8 1.4 6.6
6 Tu	0119 0716 1345 1952	1.4 7.1 1.0 6.9	**21** W	0057 0656 1323 1931	1.6 6.9 1.3 6.7
7 W	0159 0758 1429 2039	1.6 7.0 1.2 6.6	**22** Th	0138 0735 1405 2016	1.6 6.9 1.2 6.7
8 Th	0237 0839 1511 2125	1.9 6.8 1.4 6.3	**23** F	0219 0816 1447 2103	1.6 6.9 1.2 6.6
9 F	0317 0922 1553 2212	2.1 6.6 1.6 6.0	**24** Sa	0300 0901 1532 2150	1.7 6.9 1.2 6.5
10 Sa	0357 1009 1641 2301	2.3 6.3 1.8 5.8	**25** Su	0345 0949 1620 2242	1.9 6.8 1.3 6.3
11 Su	0445 1101 1734 2357	2.6 6.1 2.0 5.7	**26** M ☾	0434 1041 1713 2340	2.0 6.7 1.5 6.1
12 M	0542 1201 1834	2.7 5.9 2.2	**27** Tu	0530 1142 1814	2.1 6.5 1.6
13 Tu ☾	0056 0648 1307 1935	5.6 2.8 5.8 2.2	**28** W	0043 0636 1249 1921	6.0 2.2 6.4 1.7
14 W	0157 0754 1411 2033	5.7 2.7 5.8 2.1	**29** Th	0152 0749 1402 2033	6.1 2.2 6.4 1.7
15 Th	0251 0854 1507 2125	5.8 2.6 6.0 2.0	**30** F	0258 0903 1514 2141	6.2 2.0 6.5 1.7

JULY

	Time	m		Time	m
1 Sa	0359 1010 1621 2241	6.4 1.7 6.6 1.6	**16** Su	0353 1006 1624 2231	6.0 2.3 6.1 2.0
2 Su	0451 1111 1720 2334	6.6 1.5 6.7 1.5	**17** M	0441 1058 1713 2319	6.3 1.9 6.4 1.8
3 M ●	0539 1204 1812	6.9 1.2 6.8	**18** Tu	0523 1146 1758	6.5 1.5 6.6
4 Tu	0022 0621 1253 1859	1.5 7.0 1.0 6.9	**19** W	0004 0603 1232 1841	1.6 6.9 1.2 6.8
5 W	0106 0703 1337 1942	1.5 7.1 1.0 6.8	**20** Th	0048 0642 1316 1923	1.4 7.1 1.0 7.0
6 Th	0144 0741 1416 2022	1.6 7.1 1.1 6.7	**21** F	0130 0723 1358 2005	1.3 7.3 0.8 7.0
7 F	0220 0819 1451 2100	1.7 7.0 1.2 6.5	**22** Sa	0211 0805 1439 2049	1.3 7.4 0.8 7.0
8 Sa	0253 0856 1525 2136	1.8 6.8 1.4 6.3	**23** Su	0250 0847 1519 2132	1.3 7.3 0.9 6.8
9 Su	0327 0934 1600 2213	2.0 6.6 1.6 6.1	**24** M	0329 0931 1600 2216	1.5 7.2 1.1 6.5
10 M	0402 1013 1638 2254	2.2 6.4 1.9 5.8	**25** Tu ☾	0412 1017 1645 2306	1.7 6.9 1.4 6.2
11 Tu	0442 1058 1725 2343	2.5 6.1 2.2 5.6	**26** W	0501 1113 1740	2.0 6.6 1.8
12 W	0532 1154 1822	2.7 5.8 2.4	**27** Th	0005 0604 1224 1849	6.0 2.2 6.3 2.1
13 Th	0043 0641 1306 1931	5.5 2.9 5.6 2.5	**28** F	0121 0726 1349 2012	5.8 2.3 6.1 2.3
14 F	0154 0759 1422 2040	5.5 2.8 5.6 2.4	**29** Sa	0240 0853 1517 2129	5.9 2.2 6.2 2.2
15 Sa	0258 0908 1528 2139	5.8 2.6 5.8 2.3	**30** Su	0350 1007 1627 2234	6.2 1.9 6.4 2.0
			31 M	0445 1109 1723 2326	6.6 1.5 6.6 1.8

AUGUST

	Time	m		Time	m
1 Tu ●	0530 1200 1808	6.8 1.2 6.8	**16** W	0504 1130 1743 2349	6.7 1.3 6.6 1.5
2 W	0010 0610 1243 1848	1.6 7.1 1.0 6.9	**17** Th ○	0544 1217 1825	7.1 0.9 7.1
3 Th	0049 0646 1320 1923	1.2 7.2 1.0 6.9	**18** F	0031 0624 1300 1904	1.2 7.4 0.7 7.3
4 F	0124 0721 1354 1957	1.4 7.2 1.0 6.8	**19** Sa	0113 0704 1341 1945	1.0 7.6 0.5 7.3
5 Sa	0155 0754 1423 2027	1.5 7.2 1.1 6.7	**20** Su	0152 0745 1420 2026	0.9 7.7 0.6 7.2
6 Su	0225 0826 1451 2057	1.5 7.0 1.3 6.5	**21** M	0230 0826 1458 2105	1.0 7.6 0.8 7.0
7 M	0253 0857 1519 2127	1.7 6.8 1.5 6.3	**22** Tu	0308 0910 1536 2146	1.2 7.4 1.1 6.6
8 Tu	0321 0928 1549 2157	2.0 6.5 1.8 6.1	**23** W ☾	0348 0956 1617 2233	1.5 7.0 1.6 6.1
9 W	0352 1003 1623 2235	2.2 6.2 2.2 5.8	**24** Th	0435 1052 1709 2332	1.9 6.5 2.1 5.9
10 Th	0431 1047 1711 2329	2.6 5.8 2.6 5.5	**25** F	0542 1211 1825	2.3 6.0 2.6
11 F	0529 1158 1824	2.9 5.5 2.9	**26** Sa	0059 0716 1354 2004	5.7 2.5 5.8 2.7
12 Sa ☾	0052 0703 1340 1957	5.3 3.0 5.4 2.9	**27** Su	0233 0856 1524 2127	5.8 2.2 6.1 2.5
13 Su	0219 0845 1504 2111	5.5 2.8 5.7 2.6	**28** M	0345 1007 1627 2226	6.2 1.8 6.4 2.1
14 M	0327 0945 1607 2210	5.9 2.3 6.1 2.2	**29** Tu	0437 1101 1715 2312	6.6 1.4 6.7 1.7
15 Tu	0419 1041 1659 2302	6.3 1.8 6.5 1.8	**30** W	0518 1144 1753 2351	6.9 1.2 6.8 1.6
			31 Th ●	0553 1222 1825	7.1 1.0 6.9

Chart Datum: 3.90 metres below Ordnance Datum (Newlyn)

ENGLAND, EAST COAST — IMMINGHAM

Lat 53°38′ N Long 0°11′ W

TIMES AND HEIGHTS OF HIGH AND LOW WATERS

YEAR 1989

TIME ZONE UT(GMT)
For Summer Time add ONE hour in non-shaded areas

AREA 5—NE England

	SEPTEMBER				OCTOBER				NOVEMBER				DECEMBER			
	Time	m	Time	m	Time	m	Time	m	Time	m	Time	m	Time	m	Time	m
1 F	0025 0625 1255 1856	1.4 7.2 1.0 7.0	**16** Sa	0008 0600 1236 1841	1.0 7.6 0.5 7.5	**1** Su	0027 0627 1249 1849	1.4 7.2 1.2 7.0	**16** M	0024 0617 1252 1852	0.8 7.9 0.6 7.5	**1** W	0056 0659 1314 1913	1.5 6.8 1.6 6.9	**16** Th	0133 0737 1354 1951	0.9 7.3 1.5 7.2	**1** F	0112 0720 1327 1926	1.6 6.6 1.8 6.9	**16** Sa	0211 0819 1422 2019	1.1 6.9 1.9 7.1
2 Sa	0057 0655 1323 1924	1.3 7.3 1.0 6.9	**17** Su	0049 0641 1317 1919	0.8 7.9 0.4 7.5	**2** M	0055 0655 1316 1916	1.4 7.1 1.3 7.0	**17** Tu	0106 0702 1333 1931	0.8 7.8 0.8 7.4	**2** Th	0127 0731 1342 1942	1.6 6.7 1.8 6.8	**17** F	0219 0827 1434 2034	1.1 7.0 1.9 6.9	**2** Sa	0148 0758 1401 2001	1.6 6.6 2.0 6.8	**17** Su	0254 0905 1500 2101	1.3 6.6 2.0 6.9
3 Su	0126 0724 1349 1951	1.4 7.2 1.2 6.8	**18** M	0130 0723 1357 1958	0.8 7.9 0.6 7.3	**3** Tu	0123 0724 1342 1942	1.4 7.0 1.5 6.8	**18** W	0148 0747 1412 2012	0.9 7.6 1.2 7.1	**3** F	0157 0804 1412 2012	1.7 6.5 2.0 6.6	**18** Sa	0304 0919 1517 2121	1.4 6.6 2.3 6.6	**3** Su	0225 0840 1439 2040	1.7 6.4 2.1 6.6	**18** M	0335 0950 1539 2145	1.5 6.3 2.3 6.6
4 M	0154 0754 1416 2019	1.4 7.1 1.3 6.7	**19** Tu	0208 0805 1434 2037	0.9 7.7 0.9 7.1	**4** W	0149 0752 1408 2009	1.6 6.8 1.7 6.6	**19** Th	0229 0836 1450 2053	1.1 7.2 1.7 6.8	**4** Sa	0230 0843 1444 2047	1.9 6.2 2.3 6.3	**19** Su	0355 1017 1604 2216	1.8 6.2 2.6 6.3	**4** M	0305 0927 1521 2125	1.8 6.2 2.3 6.5	**19** Tu ☾	0419 1037 1621 2233	1.8 6.0 2.5 6.3
5 Tu	0220 0822 1442 2044	1.6 6.8 1.6 6.5	**20** W	0247 0850 1511 2118	1.1 7.4 1.4 6.7	**5** Th	0216 0820 1433 2034	1.8 6.5 2.0 6.4	**20** F	0314 0928 1532 2141	1.5 6.6 2.2 6.4	**5** Su	0308 0929 1527 2134	2.1 6.0 2.6 6.1	**20** M ☾	0454 1122 1705 2322	2.1 5.9 2.9 6.0	**5** Tu	0355 1021 1612 2220	2.0 6.1 2.5 6.3	**20** W	0506 1127 1712 2329	2.1 5.8 2.8 6.0
6 W	0246 0850 1507 2111	2.0 6.6 1.9 6.2	**21** Th	0328 0938 1552 2203	1.5 6.8 1.9 6.3	**6** F	0246 0853 1503 2105	2.1 6.2 2.3 6.1	**21** Sa ☾	0406 1031 1626 2241	1.9 6.1 2.7 6.0	**6** M ☽	0400 1035 1626 2240	2.3 5.7 2.9 5.9	**21** Tu	0604 1234 1818	2.3 5.7 3.0	**6** W	0454 1126 1716 2326	2.0 6.0 2.6 6.3	**21** Th	0601 1225 1814	2.3 5.6 2.9
7 Th	0314 0919 1535 2142	2.1 6.2 2.2 5.9	**22** F ☾	0417 1040 1644 2305	1.9 6.2 2.5 5.9	**7** Sa	0319 0936 1541 2149	2.3 5.8 2.7 5.8	**22** Su	0516 1156 1742	2.3 5.8 3.1	**7** Tu	0516 1201 1751	2.4 5.7 3.0	**22** W	0039 0716 1344 1933	5.9 2.3 5.8 2.9	**7** Th	0603 1236 1829	2.0 6.0 2.6	**22** F	0035 0704 1330 1926	5.8 2.4 5.6 3.0
8 F ☽	0348 1000 1614 2227	2.4 5.8 2.7 5.6	**23** Sa	0527 1208 1805	2.4 5.8 3.0	**8** Su ☽	0412 1044 1644 2305	2.6 5.5 3.1 5.5	**23** M	0008 0650 1327 1917	5.8 2.4 5.7 3.1	**8** W	0007 0646 1324 1919	5.9 2.3 5.9 2.8	**23** Th	0151 0819 1442 2036	6.0 2.2 5.9 2.7	**8** F	0039 0714 1347 1942	6.3 1.9 6.1 2.4	**23** Sa	0145 0809 1432 2034	5.8 2.5 5.7 2.8
9 Sa	0438 1108 1722 2349	2.6 5.4 3.1 5.3	**24** Su	0039 0713 1354 1952	5.7 2.5 5.8 3.0	**9** M	0539 1236 1831	2.7 5.4 3.2	**24** Tu	0140 0815 1442 2034	5.9 2.2 6.0 2.8	**9** Th	0127 0802 1432 2029	6.1 1.9 6.3 2.4	**24** F	0250 0911 1529 2128	6.2 2.0 6.1 2.5	**9** Sa	0151 0823 1450 2051	6.5 1.7 6.4 2.1	**24** Su	0251 0907 1525 2132	5.9 2.4 6.0 2.6
10 Su	0612 1306 1914	3.0 5.3 3.1	**25** M	0218 0847 1515 2111	5.9 2.1 6.1 2.6	**10** Tu	0053 0728 1406 2005	5.5 2.6 5.8 2.8	**25** W	0249 0915 1535 2129	6.2 1.9 6.3 2.4	**10** F	0233 0905 1528 2128	6.5 1.5 6.6 1.9	**25** Sa	0338 0955 1609 2212	6.3 1.9 6.4 2.2	**10** Su	0257 0928 1546 2155	6.7 1.5 6.6 1.8	**25** M	0346 0957 1612 2221	6.0 2.2 6.2 2.3
11 M	0140 0805 1440 2043	5.4 2.8 5.7 2.8	**26** Tu	0327 0950 1610 2206	6.3 1.7 6.4 2.2	**11** W	0215 0844 1512 2111	5.9 2.0 6.3 2.3	**26** Th	0339 1002 1616 2213	6.5 1.7 6.5 2.1	**11** Sa	0329 1002 1617 2223	6.9 1.2 6.9 1.5	**26** Su	0420 1034 1644 2251	6.5 1.8 6.6 2.0	**11** M	0359 1027 1638 2252	6.9 1.4 6.9 1.4	**26** Tu	0435 1042 1651 2305	6.2 2.0 6.5 2.0
12 Tu	0256 0919 1545 2145	5.9 2.2 6.2 2.3	**27** W	0416 1038 1652 2248	6.6 1.4 6.7 1.9	**12** Th	0315 0945 1604 2206	6.5 1.5 6.7 1.8	**27** F	0420 1040 1651 2249	6.7 1.5 6.7 1.9	**12** Su	0421 1054 1702 2313	7.3 1.0 7.2 1.0	**27** M	0458 1111 1718 2327	6.6 1.7 6.7 1.7	**12** Tu ○	0455 1122 1725 2346	7.1 1.3 7.1 1.2	**27** W	0518 1123 1727 2346	6.4 1.9 6.7 1.7
13 W	0352 1017 1635 2237	6.4 1.7 6.6 1.8	**28** Th	0454 1113 1726 2323	6.9 1.2 6.8 1.7	**13** F	0404 1043 1651 2255	7.0 1.1 7.1 1.4	**28** Sa	0455 1113 1720 2323	6.8 1.4 6.8 1.7	**13** M ●	0509 1143 1746	7.5 0.9 7.4	**28** Tu	0534 1146 1750	6.7 1.7 6.9	**13** W	0550 1211 1810	7.2 1.3 7.2	**28** Th ●	0557 1203 1803	6.6 1.8 6.9
14 Th	0438 1108 1720 2325	6.8 1.1 7.0 1.4	**29** F ○	0526 1151 1756 2356	7.1 1.2 6.9 1.5	**14** Sa	0449 1125 1732 2340	7.4 0.7 7.4 1.0	**29** Su ●	0526 1146 1732 2356	6.9 1.4 7.0 1.6	**14** Tu	0001 0558 1228 1827	1.0 7.6 0.9 7.4	**29** W	0003 0608 1219 1821	1.7 6.7 1.7 6.9	**14** Th	0038 0641 1257 1853	1.0 7.2 1.4 7.3	**29** F	0025 0635 1241 1838	1.5 6.7 1.7 7.0
15 F	0519 1154 1801	7.3 0.7 7.3	**30** Sa	0557 1221 1822	7.2 1.1 7.0	**15** Su	0533 1210 1812	7.7 0.6 7.5	**30** M	0557 1215 1817	7.0 1.4 7.0	**15** W	0048 0646 1312 1909	0.9 7.6 1.1 7.4	**30** Th	0038 0643 1253 1852	1.6 6.7 1.7 6.9	**15** F	0126 0731 1341 1937	1.0 7.1 1.6 7.2	**30** Sa	0104 0713 1319 1914	1.4 6.7 1.7 7.1
													31 Tu	0027 0628 1245 1845	1.5 6.9 1.5 7.0								**31** Su	0142 0752 1355 1952	1.3 6.8 1.7 7.1

Chart Datum: 3.90 metres below Ordnance Datum (Newlyn)

Harbour, Coastal and Tidal Information

BRIDLINGTON 10-5-9
Humberside

CHARTS
Admiralty 1882, 121, 129, 1190; Imray C29; OS 101
TIDES
+0542 Dover; ML 3.6; Duration 0610; Zone 0 (GMT).

Standard Port RIVER TEES ENT. (→)

Times				Height (metres)			
HW		LW		MHWS	MHWN	MLWN	MLWS
0000	0600	0000	0600	5.5	4.3	2.0	0.9
1200	1800	1200	1800				
Differences BRIDLINGTON							
+0100	+0050	+0055	+0050	+0.6	+0.4	+0.3	+0.2
FILEY BAY							
+0042	+0042	+0047	+0034	+0.3	+0.6	+0.4	+0.1

SHELTER
Good shelter except in winds from E, SE and S. Harbour available from HW∓3 (for draft of 2.7m).
NAVIGATION
Waypoint SW Smithic W cardinal buoy, Q(9) 15s, 54°02'.41N 00°09'.10W, 153°/333° from/to entrance, 2.6M. Beware bar, 1m at MLWN, could dry out at MLWS. Air gunnery and bombing practice is carried out off Cowden 17M south. Range marked by 5 small conical buoys; 3 seaward ones flashing every 10 secs, the 2 shoreward ones every 5 secs.
LIGHTS AND MARKS
Run in keeping the N Pier head light 002°. Tidal signals displayed from Sig Stn 50m from S Pierhead.
Less than 2.7m in harbour:- FG (No signal by day).
More than 2.7m in harbour:- FR (R flag by day).
When harbour is not clear, a W flag with a B circle in it below a R flag.
RADIO TELEPHONE
VHF Ch 16; 12 14 (occas).
TELEPHONE (0262)
Hr Mr 670148; CG 672317; MRSC Bridlington 672317; HMC Scarborough 366631; Marinecall 0898 500 454; Hospital 673451.
FACILITIES
EC Thursday; **S Pier** Tel. 670148, FW, C (7 ton), CH; **Royal Yorkshire YC** Tel. 672041, L, FW, R, Bar; **West end of Harbour** Slip; **C & M Marine** Tel. 672212, Slip, M, Sh, C, CH. See Hr Mr for M, AB.
Town P, D, ME, El, V, R, Bar. PO; Bank; Rly; Air (Humberside).

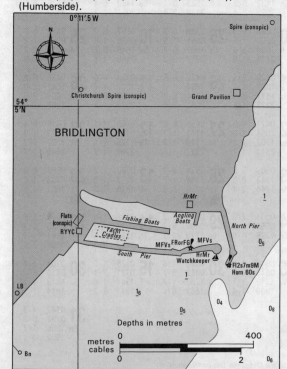

▲ Call to Hr Mr's Watchkeeper

SCARBOROUGH 10-5-10
N. Yorkshire

CHARTS
Admiralty 1612, 129, 1191; Imray C29; OS 101
TIDES
+0527 Dover; ML 3.5; Duration 0615; Zone 0 (GMT).

Standard Port RIVER TEES ENT. (→)

Times				Height (metres)			
HW		LW		MHWS	MHWN	MLWN	MLWS
0000	0600	0000	0600	5.5	4.3	2.0	0.9
1200	1800	1200	1800				
Differences SCARBOROUGH							
+0040	+0040	+0030	+0030	+0.2	+0.3	+0.3	0.0

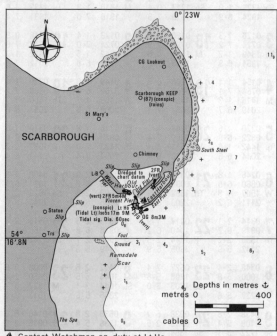

▲ Contact Watchman on duty at Lt Ho

SHELTER
The old harbour is reserved strictly for fishing fleet. E harbour cannot be entered in strong E-SE winds, otherwise available HW∓3.
NAVIGATION
Waypoint 54°16'.50N 00°22'.00W, 122°/302° from/to East Pier Lt, 0.83M. Beware rocks SW of E pier running out for approx 20 m. Keep well clear of Ramsdale Scar. Keep careful watch for salmon nets to E and SE of entrance.
LIGHTS AND MARKS
No leading lights or marks. B ball at Lt Ho masthead (or Iso 5s Lt) — more than 3.7m on the bar. FY — more than 1.8m, less than 3.7m. R flag — do not enter. These depths refer to Old Harbour; East Harbour entrance is about 1.5m less.
RADIO TELEPHONE
Call Scarborough Lt Ho. VHF Ch 16; 12 14 (H24).
TELEPHONE (0723)
Hr Mr Working hrs 373530, non-working hrs 360684; CG 372323; MRSC Bridlington 672317; HMC 366631; Marinecall 0898 500 454; Hosp 368111.
FACILITIES
EC Wednesday; **North Wharf** Slip, C (20 ton); **Scarborough YC** Tel. 373821, Slip, M*, L, FW, ME, El, C (10 ton), AB; **Old Pier** M*, L, FW, ME, El, AB; **Scarborough Marine** Tel. 375199, L, ME, El, Sh, CH, P and D (cans); **Vincent Pier** C (15 ton); **West Pier** Slip, D, L; **Quay Marine** P, D, ME, El, Sh, CH; **Fisherman's Chandlers** Tel. 365266 CH; *Long waiting list — some berths available for visitors.
Town P, D, V, R, Bar, PO; Bank; Rly; Air (Humberside).

AREA 5—NE England 351

WHITBY 10-5-11
N. Yorkshire

CHARTS
Admiralty 1612, 134, 129; Imray C29; OS 94
TIDES
+0500 Dover; ML 3.2; Duration 0605; Zone 0 (GMT).

Standard Port RIVER TEES ENT. (→)

Times				Height (metres)			
HW		LW		MHWS	MHWN	MLWN	MLWS
0000	0600	0000	0600	5.5	4.3	2.0	0.9
1200	1800	1200	1800				

Differences WHITBY
+0014 +0014 +0011 +0011 −0.1 0.0 0.0 −0.1

SHELTER
Good shelter except in lower harbour in strong NW to NE winds. Harbour available from HW∓3 for drafts of approx 2m. In strong winds from NW through N to SE the sea breaks a long way out and entry is difficult. Marina is run by Scarborough Borough Council. Bridge opens at ½ hour intervals HW∓2; and from May to Sept inclusive on Sat and Sun at 0900 and 1800.

NAVIGATION
Waypoint 54°30'.30N 00°36'.90W, 349°/169° from/to entrance, 0.65M. Beware Whitby Rock. Swing bridge is manned HW∓2, shows fixed G Lts when open, fixed R Lts when closed.

LIGHTS AND MARKS
Leading lines
(1) Chapel spire in line with Lt Ho — 176°.
(2) FR Lt, seen between East and West Pier Extension, Lts (FR and FG), leads 169° into harbour. Continue on this line until beacons (W triangle and W circle with B stripe) on East Pier (two FY Lts at night) are abeam.
(3) On course 209° leave marks in line astern.
Entry signals (B ball by day, G Lt by night show from top of W Pier Lt Ho indicating 'Enter') only apply to vessels over 30m in length.

RADIO TELEPHONE (local times)
VHF Ch 16; **11** 06 (0900-1700). Whitby Bridge Ch 16; 06 **11** (listens on Ch 16 HW−2 to HW+2).
TELEPHONE (0947)
Hr Mr 602354; CG 602107; MRSC Humber (0262) 672317; HMC 620107 (working hours);
Marinecall 0898 500 454/453; Dr 602828.
FACILITIES
EC Wednesday; **Whitby Marina** (200+10 visitors) Tel. 600165, Slip, P, AC, D, FW, ME, El, Sh, C, CH, Access HW∓2; **Fish Quay** M, D, L, FW, C (1 ton), CH, AB, R, Bar; **Whitby Marine Services** Tel. 600361, ME, El, Sh, CH; **M.R. Coates Marine** Tel. 604486, ME, El, Sh, CH, BH, ACA; **Whitby Port Services** Tel. 602272, Sh (up to 30 ton); **Whitby YC** Tel. 603623, M, L, Bar; **North East Sails** Tel. 604710 SM; **Endeavour Marine** Tel. 603484, ME, Sh, CH; **Collier** Tel. 602068 Gas, Gaz; **Town** PO; Bank; Rly; Air (Teesside).

RIVER TEES MIDDLESBROUGH 10-5-12
Cleveland

CHARTS
Admiralty 2566, 2567, 134; Imray C29; OS 93
TIDES
+0450 Dover; ML 3.2; Duration 0605; Zone 0 (GMT).

Standard Port RIVER TEES ENT. (→)

Times				Height (metres)			
HW		LW		MHWS	MHWN	MLWN	MLWS
0000	0600	0000	0600	5.5	4.3	2.0	0.9
1200	1800	1200	1800				

Differences MIDDLESBROUGH (Dock Ent)
0000 +0002 0000 −0003 +0.1 +0.2 +0.1 −0.1

NOTE: River Tees Entrance is a Standard Port and detailed tide tables are given below.

SHELTER
Entry is not recommended for small craft during periods of heavy weather especially with NE, E or SE strong winds. Berthing may be arranged at Paddy's Hole (Bran Sands), at Stockton Castlegate Marine Club or elsewhere by arrangement with Hr Mr.
NAVIGATION
Waypoint Tees Fairway safe water buoy, Iso 4s, Horn, Racon, 54°40'.94N 01°06'.39W, 030°/210° from/to S Gare breakwater, 2.4M. Beware Saltscar, Eastscar and Longscar Rocks on approaching. Bridges up river have following clearances:-
(1) ½ mile above Middlesbrough Dock — Transporter Bridge—49m at MHWS;
(2) 2 miles from there, Newport Bridge, 36m at MHWS, or 6.4m if bridge lowered;
(3) ½ mile from there, A 19 Road Bridge 18.3m at MHWS;
(4) 2 miles from there, Stockton-Victoria Bridge 5.6m at MHWS.
LIGHTS AND MARKS
QW Lt at S Gare Lt Ho, or 3FR (vert) at CG Stn, means no entry without approval of Hr Mr. Ldg Lts 210°, FR on framework Trs. Channel well marked to beyond Middlesbrough.
RADIO TELEPHONE
Call: *Tees Harbour Radio* VHF Ch 16; 08 11 12 **14 22** (H24). Information service Ch 14 22.
TELEPHONE (0642)
Hr Mr 452541; CG Redcar 474639; MRSC North Shields 572691; HMC 091 257 2691; Marinecall 0898 500 453; Hosp 813133.

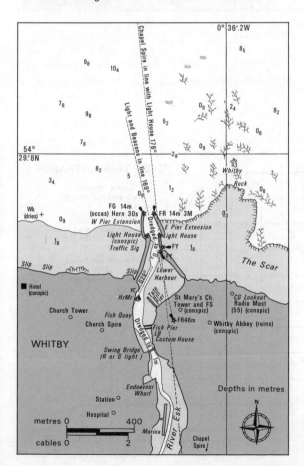

RIVER TEES/MIDDLESBROUGH continued

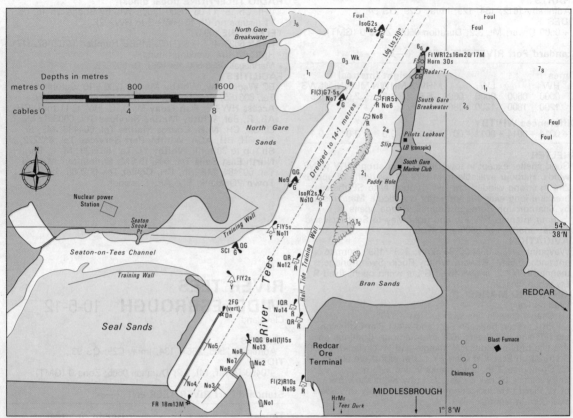

⚠ Contact HrMr

FACILITIES
EC Wednesday; Limited accommodation on River Tees for small pleasure craft. Prior arrangements must be made with Club secretaries or Hr Mr. **South Gare Marine Club** Tel. 482015, Slip, M (check with Warden), L, FW, Bar; **Stockton Castlegate Club** Slip (Public), M (check with secretary), L (at slip), V; **Tees Motor Boat Club** M (check with secretary); **Tees Docks** Tel. 452541, M (check), L, FW, C (30 ton), AB (check); **Eccles Marine Co** Tel. 211756, El, CH, ACA.
Town PO; Bank; Rly; Air.

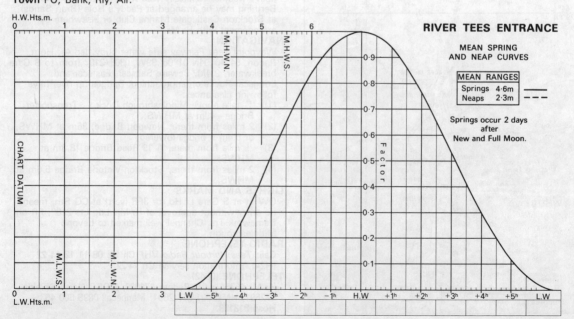

ENGLAND, EAST COAST — RIVER TEES ENTRANCE

Lat 54°38' N Long 1°09' W

TIMES AND HEIGHTS OF HIGH AND LOW WATERS

YEAR 1989

TIME ZONE UT(GMT)
For Summer Time add ONE hour in non-shaded areas

Chart Datum: 2.85 metres below Ordnance Datum (Newlyn)

JANUARY

Day	Time	m	Day	Time	m
1 Su	0356 / 1009 / 1603 / 2214	2.0 / 4.1 / 2.5 / 4.3	16 M	0427 / 1048 / 1652 / 2302	1.6 / 4.4 / 2.1 / 4.6
2 M	0459 / 1115 / 1719 / 2325	2.1 / 4.2 / 2.5 / 4.3	17 Tu	0547 / 1203 / 1814	1.8 / 4.4 / 2.0
3 Tu	0605 / 1218 / 1829	2.1 / 4.3 / 2.3	18 W	0024 / 0702 / 1309 / 1927	4.6 / 1.8 / 4.6 / 1.8
4 W	0032 / 0702 / 1310 / 1928	4.4 / 2.0 / 4.6 / 2.1	19 Th	0133 / 0802 / 1402 / 2026	4.7 / 1.7 / 4.8 / 1.5
5 Th	0128 / 0751 / 1355 / 2019	4.6 / 1.8 / 4.8 / 1.8	20 F	0229 / 0851 / 1449 / 2118	4.9 / 1.6 / 5.0 / 1.3
6 F	0218 / 0836 / 1436 / 2105	4.8 / 1.6 / 5.1 / 1.5	21 Sa	0317 / 0934 / 1529 / 2202 ○	5.0 / 1.6 / 5.2 / 1.1
7 Sa ●	0304 / 0919 / 1517 / 2150	5.0 / 1.4 / 5.3 / 1.2	22 Su	0359 / 1010 / 1606 / 2242	5.1 / 1.5 / 5.3 / 0.9
8 Su	0348 / 1002 / 1557 / 2235	5.2 / 1.3 / 5.5 / 0.9	23 M	0437 / 1045 / 1641 / 2319	5.1 / 1.4 / 5.4 / 0.9
9 M	0433 / 1044 / 1638 / 2318	5.3 / 1.1 / 5.6 / 0.7	24 Tu	0512 / 1116 / 1715 / 2353	5.1 / 1.4 / 5.4 / 0.9
10 Tu	0518 / 1126 / 1720	5.3 / 1.1 / 5.7	25 W	0546 / 1147 / 1749	5.0 / 1.4 / 5.4
11 W	0001 / 0603 / 1208 / 1805	0.6 / 5.3 / 1.1 / 5.6	26 Th	0025 / 0618 / 1218 / 1822	1.0 / 4.9 / 1.5 / 5.3
12 Th	0045 / 0649 / 1252 / 1852	0.6 / 5.2 / 1.2 / 5.5	27 F	0057 / 0652 / 1249 / 1856	1.2 / 4.7 / 1.6 / 5.1
13 F	0131 / 0740 / 1340 / 1942	0.7 / 5.0 / 1.5 / 5.3	28 Sa	0128 / 0728 / 1324 / 1934	1.4 / 4.6 / 1.8 / 4.8
14 Sa ☽	0220 / 0833 / 1432 / 2037	1.0 / 4.7 / 1.9 / 5.0	29 Su	0204 / 0809 / 1404 / 2016	1.7 / 4.4 / 2.1 / 4.6
15 Su	0318 / 0935 / 1535 / 2143	1.3 / 4.5 / 1.9 / 4.8	30 M ☾	0244 / 0900 / 1454 / 2111	1.9 / 4.2 / 2.3 / 4.3
			31 Tu	0339 / 1003 / 1603 / 2224	2.2 / 4.1 / 2.5 / 4.1

FEBRUARY

Day	Time	m	Day	Time	m
1 W	0454 / 1120 / 1734 / 2350	2.3 / 4.1 / 2.4 / 4.1	16 Th	0014 / 0653 / 1255 / 1924	4.3 / 2.1 / 4.3 / 1.8
2 Th	0618 / 1232 / 1902	2.2 / 4.3 / 2.2	17 F	0131 / 0758 / 1352 / 2023	4.5 / 2.0 / 4.6 / 1.5
3 F	0104 / 0727 / 1328 / 2004	4.3 / 2.0 / 4.6 / 1.8	18 Sa	0226 / 0843 / 1437 / 2110	4.7 / 1.8 / 4.9 / 1.2
4 Sa	0202 / 0819 / 1416 / 2053	4.6 / 1.7 / 5.0 / 1.3	19 Su	0308 / 0921 / 1515 / 2149	4.9 / 1.6 / 5.1 / 1.0
5 Su	0250 / 0905 / 1500 / 2138	5.0 / 1.3 / 5.3 / 0.9	20 M ○	0343 / 0955 / 1549 / 2224	5.0 / 1.4 / 5.3 / 0.8
6 M ●	0335 / 0948 / 1541 / 2221	5.2 / 1.1 / 5.6 / 0.5	21 Tu	0416 / 1024 / 1621 / 2257	5.0 / 1.3 / 5.4 / 0.7
7 Tu	0419 / 1030 / 1623 / 2302	5.5 / 0.8 / 5.8 / 0.3	22 W	0445 / 1052 / 1651 / 2325	5.1 / 1.2 / 5.4 / 0.8
8 W	0501 / 1111 / 1705 / 2343	5.5 / 0.7 / 5.9 / 0.2	23 Th	0515 / 1120 / 1720 / 2353	5.1 / 1.2 / 5.4 / 0.9
9 Th	0543 / 1150 / 1747	5.5 / 0.7 / 5.9	24 F	0544 / 1149 / 1750	5.0 / 1.2 / 5.3
10 F	0024 / 0627 / 1232 / 1832	0.3 / 5.4 / 0.9 / 5.7	25 Sa	0019 / 0614 / 1217 / 1821	1.1 / 4.9 / 1.4 / 5.1
11 Sa	0106 / 0710 / 1314 / 1919	0.5 / 5.1 / 1.1 / 5.4	26 Su	0048 / 0645 / 1248 / 1855	1.3 / 4.7 / 1.6 / 4.9
12 Su ☽	0151 / 0759 / 1402 / 2012	0.9 / 4.8 / 1.5 / 5.0	27 M	0117 / 0720 / 1323 / 1934	1.5 / 4.5 / 1.8 / 4.6
13 M	0242 / 0857 / 1501 / 2117	1.4 / 4.4 / 1.8 / 4.6	28 Tu ☾	0154 / 0805 / 1408 / 2026	1.8 / 4.2 / 2.1 / 4.3
14 Tu	0348 / 1009 / 1619 / 2238	1.9 / 4.2 / 2.1 / 4.3			
15 W	0519 / 1136 / 1801	2.2 / 4.1 / 2.1			

MARCH

Day	Time	m	Day	Time	m
1 W	0242 / 0905 / 1511 / 2142	2.1 / 4.1 / 2.3 / 4.0	16 Th	0455 / 1105 / 1750	2.4 / 4.0 / 2.0
2 Th	0355 / 1027 / 1649 / 2318	2.3 / 4.0 / 2.4 / 4.0	17 F	0004 / 0641 / 1234 / 1913	4.1 / 2.3 / 4.2 / 1.7
3 F	0539 / 1153 / 1838	2.3 / 4.2 / 2.1	18 Sa	0120 / 0741 / 1334 / 2006	4.3 / 2.0 / 4.5 / 1.4
4 Sa	0043 / 0703 / 1300 / 1944	4.2 / 2.0 / 4.5 / 1.6	19 Su	0211 / 0822 / 1418 / 2049	4.6 / 1.8 / 4.8 / 1.1
5 Su	0144 / 0759 / 1352 / 2034	4.6 / 1.6 / 4.9 / 1.0	20 M	0247 / 0857 / 1453 / 2124	5.0 / 1.5 / 5.0 / 0.9
6 M	0233 / 0846 / 1439 / 2118	5.0 / 1.2 / 5.3 / 0.6	21 Tu	0319 / 0928 / 1525 / 2156	4.9 / 1.4 / 5.2 / 0.8
7 Tu ●	0315 / 0928 / 1521 / 2159	5.3 / 0.9 / 5.7 / 0.2	22 W ○	0348 / 0956 / 1555 / 2226	5.0 / 1.2 / 5.3 / 0.8
8 W	0357 / 1009 / 1603 / 2240	5.6 / 0.6 / 5.9 / 0.0	23 Th	0416 / 1024 / 1623 / 2252	5.1 / 1.1 / 5.4 / 0.8
9 Th	0438 / 1048 / 1645 / 2320	5.7 / 0.5 / 6.0 / 0.0	24 F	0442 / 1052 / 1652 / 2319	5.1 / 1.1 / 5.3 / 1.0
10 F	0519 / 1129 / 1727	5.6 / 0.5 / 5.8	25 Sa	0511 / 1120 / 1720 / 2346	5.1 / 1.2 / 5.2 / 1.1
11 Sa	0000 / 0600 / 1210 / 1812	0.2 / 5.4 / 0.7 / 5.7	26 Su	0539 / 1150 / 1751	5.0 / 1.3 / 5.1
12 Su	0041 / 0643 / 1252 / 1859	0.6 / 5.2 / 1.0 / 5.4	27 M	0014 / 0610 / 1221 / 1827	1.3 / 4.8 / 1.4 / 4.8
13 M	0123 / 0730 / 1340 / 1952	1.1 / 4.8 / 1.3 / 4.9	28 Tu ☽	0043 / 0645 / 1257 / 1909	1.5 / 4.6 / 1.7 / 4.5
14 Tu ☾	0211 / 0823 / 1436 / 2057	1.6 / 4.4 / 1.7 / 4.5	29 W	0120 / 0728 / 1342 / 2005	1.8 / 4.4 / 1.9 / 4.2
15 W	0314 / 0934 / 1557 / 2221	2.1 / 4.1 / 2.0 / 4.2	30 Th ☾	0209 / 0830 / 1449 / 2122	2.1 / 4.2 / 2.1 / 4.0
			31 F	0324 / 0949 / 1623 / 2254	2.3 / 4.1 / 2.1 / 4.0

APRIL

Day	Time	m	Day	Time	m
1 Sa	0505 / 1113 / 1807	2.3 / 4.2 / 1.8	16 Su	0049 / 0704 / 1257 / 1934	4.2 / 2.1 / 4.4 / 1.4
2 Su	0018 / 0634 / 1227 / 1916	4.3 / 2.0 / 4.5 / 1.3	17 M	0137 / 0747 / 1344 / 2015	4.4 / 1.8 / 4.7 / 1.2
3 M	0120 / 0733 / 1324 / 2006	4.5 / 1.5 / 4.9 / 0.8	18 Tu	0215 / 0822 / 1420 / 2050	4.6 / 1.6 / 4.9 / 1.1
4 Tu	0208 / 0819 / 1412 / 2051	5.0 / 1.1 / 5.3 / 0.5	19 W	0246 / 0854 / 1454 / 2122	4.8 / 1.4 / 5.0 / 1.0
5 W	0251 / 0903 / 1457 / 2134	5.4 / 0.8 / 5.7 / 0.2	20 Th	0315 / 0925 / 1525 / 2152	4.9 / 1.3 / 5.1 / 1.0
6 Th ●	0334 / 0945 / 1541 / 2216	5.6 / 0.6 / 5.9 / 0.1	21 F	0343 / 0955 / 1555 / 2220	5.0 / 1.2 / 5.2 / 1.0
7 F	0413 / 1026 / 1626 / 2257	5.6 / 0.5 / 5.9 / 0.2	22 Sa	0412 / 1026 / 1626 / 2248	5.1 / 1.2 / 5.1 / 1.1
8 Sa	0454 / 1108 / 1709 / 2337	5.6 / 0.5 / 5.8 / 0.5	23 Su	0441 / 1058 / 1658 / 2318	5.1 / 1.2 / 5.1 / 1.2
9 Su	0534 / 1150 / 1756	5.4 / 0.7 / 5.6	24 M	0512 / 1130 / 1733 / 2349	5.0 / 1.3 / 4.9 / 1.4
10 M	0017 / 0618 / 1235 / 1845	0.8 / 5.1 / 0.9 / 5.2	25 Tu	0544 / 1205 / 1812	4.9 / 1.4 / 4.8
11 Tu	0100 / 0704 / 1324 / 1938	1.3 / 4.8 / 1.3 / 4.8	26 W	0022 / 0624 / 1246 / 1900	1.5 / 4.7 / 1.5 / 4.5
12 W ☽	0147 / 0758 / 1422 / 2043	1.8 / 4.5 / 1.6 / 4.4	27 Th	0103 / 0712 / 1337 / 1959	1.7 / 4.5 / 1.6 / 4.3
13 Th	0247 / 0905 / 1542 / 2202	2.2 / 4.2 / 1.9 / 4.1	28 F ☽	0157 / 0812 / 1440 / 2110	2.0 / 4.4 / 1.8 / 4.1
14 F	0421 / 1028 / 1725 / 2334	2.4 / 4.1 / 1.9 / 4.1	29 Sa	0307 / 0922 / 1602 / 2227	2.1 / 4.3 / 1.7 / 4.2
15 Sa	0604 / 1154 / 1841	2.4 / 4.2 / 1.7	30 Su	0434 / 1037 / 1730 / 2346	2.1 / 4.4 / 1.5 / 4.4

ENGLAND, EAST COAST — RIVER TEES ENTRANCE

Lat 54°38′ N   Long 1°09′ W

TIMES AND HEIGHTS OF HIGH AND LOW WATERS

YEAR 1989

TIME ZONE UT (GMT)
For Summer Time add ONE hour in non-shaded areas

MAY

Day	Time	m	Day	Time	m
1 M	0556 / 1150 / 1839	1.9 / 4.6 / 1.2	16 Tu	0050 / 0720 / 1259 / 1931	4.3 / 2.0 / 4.5 / 1.4
2 Tu	0048 / 0657 / 1252 / 1934	4.7 / 1.5 / 4.9 / 0.8	17 W	0131 / 0741 / 1341 / 2009	4.5 / 1.8 / 4.7 / 1.3
3 W	0140 / 0749 / 1345 / 2022	5.0 / 1.2 / 5.3 / 0.6	18 Th	0208 / 0818 / 1419 / 2044	4.7 / 1.6 / 4.8 / 1.3
4 Th	0225 / 0836 / 1434 / 2107	5.3 / 0.9 / 5.5 / 0.5	19 F	0240 / 0853 / 1454 / 2117	4.8 / 1.5 / 4.9 / 1.3
5 F ●	0308 / 0921 / 1521 / 2152	5.4 / 0.7 / 5.7 / 0.5	20 Sa	0312 / 0928 / 1529 / 2149	4.9 / 1.4 / 5.0 / 1.3
6 Sa	0350 / 1006 / 1607 / 2234	5.5 / 0.6 / 5.7 / 0.6	21 Su	0343 / 1004 / 1604 / 2221	5.0 / 1.3 / 5.0 / 1.3
7 Su	0431 / 1051 / 1655 / 2316	5.4 / 0.6 / 5.6 / 0.8	22 M	0417 / 1041 / 1641 / 2257	5.1 / 1.3 / 4.9 / 1.3
8 M	0515 / 1137 / 1743 / 2358	5.3 / 0.7 / 5.4 / 1.1	23 Tu	0451 / 1119 / 1722 / 2332	5.0 / 1.2 / 4.9 / 1.4
9 Tu	0558 / 1224 / 1834	5.1 / 0.9 / 5.0	24 W	0529 / 1200 / 1807	5.0 / 1.2 / 4.8
10 W	0042 / 0646 / 1314 / 1926	1.5 / 4.9 / 1.2 / 4.7	25 Th	0012 / 0611 / 1245 / 1856	1.5 / 4.9 / 1.2 / 4.6
11 Th	0128 / 0737 / 1411 / 2025	1.8 / 4.6 / 1.5 / 4.4	26 F	0056 / 0700 / 1334 / 1951	1.6 / 4.8 / 1.3 / 4.5
12 F ☽	0223 / 0836 / 1518 / 2129	2.1 / 4.4 / 1.7 / 4.2	27 Sa	0148 / 0755 / 1432 / 2051	1.8 / 4.7 / 1.4 / 4.4
13 Sa	0334 / 0943 / 1637 / 2245	2.3 / 4.3 / 1.8 / 4.1	28 Su ☾	0250 / 0857 / 1538 / 2159	1.9 / 4.6 / 1.4 / 4.4
14 Su	0459 / 1058 / 1751 / 2356	2.3 / 4.4 / 1.7 / 4.1	29 M	0402 / 1004 / 1651 / 2309	1.9 / 4.6 / 1.3 / 4.4
15 M	0608 / 1205 / 1848	2.2 / 4.4 / 1.6	30 Tu	0516 / 1113 / 1801	1.8 / 4.7 / 1.2
			31 W	0014 / 0622 / 1221 / 1902	4.6 / 1.6 / 4.9 / 1.0

JUNE

Day	Time	m	Day	Time	m
1 Th	0110 / 0720 / 1321 / 1957	4.9 / 1.3 / 5.1 / 0.9	16 F	0127 / 0741 / 1344 / 2006	4.5 / 1.9 / 4.6 / 1.7
2 F	0201 / 0813 / 1416 / 2046	5.1 / 1.1 / 5.3 / 0.9	17 Sa	0206 / 0825 / 1427 / 2046	4.7 / 1.7 / 4.7 / 1.6
3 Sa ●	0246 / 0903 / 1507 / 2132	5.2 / 0.9 / 5.4 / 0.9	18 Su	0243 / 0907 / 1508 / 2124	4.7 / 1.5 / 4.8 / 1.5
4 Su	0331 / 0952 / 1556 / 2217	5.3 / 0.8 / 5.4 / 1.0	19 M ○	0319 / 0944 / 1548 / 2203	5.0 / 1.4 / 4.9 / 1.4
5 M	0414 / 1041 / 1644 / 2301	5.3 / 0.8 / 5.3 / 1.2	20 Tu	0357 / 1030 / 1630 / 2241	5.1 / 1.2 / 5.0 / 1.3
6 Tu	0458 / 1129 / 1732 / 2344	5.2 / 0.8 / 5.2 / 1.4	21 W	0435 / 1112 / 1712 / 2322	5.1 / 1.0 / 5.0 / 1.3
7 W	0542 / 1215 / 1819	5.1 / 0.9 / 5.0	22 Th	0516 / 1154 / 1757	5.2 / 0.9 / 5.0
8 Th	0025 / 0627 / 1303 / 1907	1.6 / 5.0 / 1.1 / 4.7	23 F	0003 / 0600 / 1238 / 1843	1.3 / 5.2 / 0.9 / 4.9
9 F	0107 / 0714 / 1351 / 1957	1.7 / 4.9 / 1.3 / 4.5	24 Sa	0046 / 0646 / 1323 / 1934	1.4 / 5.1 / 0.9 / 4.8
10 Sa	0154 / 0804 / 1443 / 2049	1.9 / 4.7 / 1.5 / 4.3	25 Su	0134 / 0735 / 1413 / 2027	1.5 / 5.0 / 1.0 / 4.7
11 Su ☽	0244 / 0857 / 1545 / 2146	2.1 / 4.5 / 1.7 / 4.2	26 M ☾	0227 / 0832 / 1510 / 2127	1.6 / 4.9 / 1.2 / 4.5
12 M	0346 / 0956 / 1642 / 2248	2.2 / 4.4 / 1.8 / 4.1	27 Tu	0328 / 0934 / 1614 / 2233	1.7 / 4.8 / 1.3 / 4.5
13 Tu	0454 / 1059 / 1746 / 2350	2.2 / 4.4 / 1.8 / 4.2	28 W	0437 / 1042 / 1725 / 2342	1.6 / 4.7 / 1.4 / 4.5
14 W	0558 / 1203 / 1839	2.2 / 4.4 / 1.8	29 Th	0550 / 1157 / 1835	1.7 / 4.7 / 1.4
15 Th	0042 / 0653 / 1257 / 1926	4.3 / 2.0 / 4.4 / 1.7	30 F	0045 / 0657 / 1306 / 1937	4.6 / 1.6 / 4.8 / 1.4

JULY

Day	Time	m	Day	Time	m
1 Sa	0141 / 0759 / 1405 / 2032	4.8 / 1.4 / 5.0 / 1.4	16 Su	0135 / 0802 / 1405 / 2020	4.6 / 1.8 / 4.5 / 1.8
2 Su	0230 / 0854 / 1500 / 2121	5.0 / 1.2 / 5.1 / 1.3	17 M	0218 / 0850 / 1450 / 2104	4.8 / 1.6 / 4.7 / 1.6
3 M ●	0317 / 0946 / 1549 / 2207	5.1 / 1.0 / 5.2 / 1.3	18 Tu ○	0258 / 0935 / 1532 / 2146	5.0 / 1.3 / 4.9 / 1.4
4 Tu	0400 / 1035 / 1635 / 2248	5.2 / 0.9 / 5.2 / 1.4	19 W	0338 / 1017 / 1614 / 2227	5.2 / 1.0 / 5.1 / 1.2
5 W	0442 / 1120 / 1719 / 2327	5.3 / 0.8 / 5.2 / 1.4	20 Th	0419 / 1058 / 1657 / 2308	5.4 / 0.7 / 5.2 / 1.1
6 Th	0525 / 1201 / 1800	5.3 / 0.8 / 5.0	21 F	0459 / 1139 / 1740 / 2349	5.4 / 0.6 / 5.3 / 1.0
7 F	0005 / 0604 / 1242 / 1841	1.5 / 5.2 / 0.9 / 4.8	22 Sa	0542 / 1221 / 1824	5.5 / 0.5 / 5.2
8 Sa	0041 / 0645 / 1321 / 1921	1.6 / 5.1 / 1.1 / 4.7	23 Su	0029 / 0627 / 1303 / 1909	1.1 / 5.5 / 0.6 / 5.1
9 Su	0119 / 0727 / 1401 / 2002	1.7 / 4.9 / 1.3 / 4.5	24 M	0114 / 0714 / 1348 / 1958	1.2 / 5.3 / 0.8 / 4.9
10 M	0159 / 0811 / 1443 / 2049	1.9 / 4.7 / 1.6 / 4.3	25 Tu ☾	0202 / 0806 / 1439 / 2054	1.4 / 5.0 / 1.2 / 4.6
11 Tu ☽	0244 / 0858 / 1531 / 2141	2.0 / 4.5 / 1.9 / 4.2	26 W	0258 / 0907 / 1539 / 2159	1.6 / 4.8 / 1.5 / 4.4
12 W	0341 / 0955 / 1628 / 2241	2.2 / 4.3 / 2.0 / 4.1	27 Th	0406 / 1019 / 1654 / 2312	1.8 / 4.6 / 1.8 / 4.3
13 Th	0448 / 1102 / 1734 / 2347	2.3 / 4.3 / 2.1 / 4.2	28 F	0527 / 1142 / 1818	1.9 / 4.6 / 1.9
14 F	0601 / 1212 / 1839	2.3 / 4.2 / 2.1	29 Sa	0027 / 0649 / 1300 / 1930	4.4 / 1.8 / 4.6 / 1.8
15 Sa	0045 / 0707 / 1313 / 1934	4.3 / 2.1 / 4.3 / 2.0	30 Su	0130 / 0757 / 1404 / 2027	4.6 / 1.5 / 4.8 / 1.7
			31 M	0222 / 0853 / 1456 / 2114	4.9 / 1.2 / 5.0 / 1.6

AUGUST

Day	Time	m	Day	Time	m
1 Tu ●	0307 / 0941 / 1541 / 2155	5.1 / 1.0 / 5.1 / 1.5	16 W	0236 / 0917 / 1514 / 2127	5.1 / 1.0 / 5.1 / 1.2
2 W	0348 / 1024 / 1620 / 2233	5.2 / 0.8 / 5.2 / 1.4	17 Th ○	0317 / 0957 / 1555 / 2207	5.4 / 0.7 / 5.4 / 1.0
3 Th	0426 / 1102 / 1658 / 2306	5.3 / 0.7 / 5.1 / 1.3	18 F	0357 / 1037 / 1635 / 2247	5.6 / 0.4 / 5.5 / 0.8
4 F	0502 / 1139 / 1732 / 2339	5.4 / 0.8 / 5.1 / 1.3	19 Sa	0438 / 1118 / 1716 / 2326	5.8 / 0.3 / 5.6 / 0.8
5 Sa	0537 / 1212 / 1805	5.3 / 0.9 / 5.0	20 Su	0520 / 1157 / 1758	5.8 / 0.3 / 5.5
6 Su	0010 / 0611 / 1243 / 1839	1.4 / 5.2 / 1.1 / 4.8	21 M	0007 / 0604 / 1238 / 1841	0.8 / 5.7 / 0.5 / 5.3
7 M	0042 / 0648 / 1316 / 1916	1.5 / 5.1 / 1.3 / 4.7	22 Tu	0050 / 0652 / 1321 / 1928	1.0 / 5.5 / 0.9 / 5.0
8 Tu	0116 / 0726 / 1349 / 1955	1.7 / 4.8 / 1.6 / 4.5	23 W ☾	0137 / 0745 / 1409 / 2022	1.3 / 5.1 / 1.3 / 4.6
9 W	0155 / 0808 / 1429 / 2040	1.9 / 4.6 / 1.9 / 4.3	24 Th	0232 / 0847 / 1510 / 2128	1.6 / 4.7 / 1.8 / 4.4
10 Th	0243 / 0901 / 1518 / 2139	2.2 / 4.3 / 2.2 / 4.1	25 F	0343 / 1004 / 1631 / 2249	1.9 / 4.3 / 2.2 / 4.2
11 F	0345 / 1010 / 1626 / 2251	2.3 / 4.1 / 2.4 / 4.1	26 Sa	0518 / 1137 / 1814	2.0 / 4.4 / 2.2
12 Sa	0511 / 1132 / 1753	2.4 / 4.1 / 2.4	27 Su	0014 / 0649 / 1302 / 1928	4.3 / 1.8 / 4.5 / 2.1
13 Su	0004 / 0641 / 1248 / 1906	4.2 / 2.2 / 4.2 / 2.2	28 M	0120 / 0754 / 1401 / 2019	4.6 / 1.5 / 4.7 / 1.8
14 M	0104 / 0746 / 1344 / 1959	4.5 / 1.8 / 4.6 / 1.9	29 Tu	0211 / 0844 / 1447 / 2100	4.9 / 1.1 / 4.9 / 1.6
15 Tu	0154 / 0833 / 1432 / 2044	4.8 / 1.4 / 4.8 / 1.5	30 W	0253 / 0927 / 1525 / 2136	5.1 / 0.9 / 5.1 / 1.5
			31 Th ●	0329 / 1003 / 1559 / 2207	5.3 / 0.8 / 5.2 / 1.3

Chart Datum: 2.85 metres below Ordnance Datum (Newlyn)

AREA 5—NE England 355

ENGLAND, EAST COAST — RIVER TEES ENTRANCE

Lat 54°38′ N   Long 1°09′ W

TIMES AND HEIGHTS OF HIGH AND LOW WATERS   YEAR 1989

TIME ZONE UT(GMT)
For Summer Time add ONE hour in non-shaded areas

SEPTEMBER				OCTOBER				NOVEMBER				DECEMBER			
Time	m	Time	m	Time	m	Time	m	Time	m	Time	m	Time	m	Time	m
1 0403 1037 F 1630 2238	5.4 0.8 5.2 1.3	**16** 0334 1012 Sa 1610 2223	5.8 0.2 5.7 0.7	**1** 0406 1034 Su 1626 2237	5.4 1.0 5.3 1.3	**16** 0356 1026 M 1626 2242	6.0 0.4 5.8 0.7	**1** 0444 1059 W 1655 2318	5.1 1.4 5.2 1.5	**16** 0518 1133 Th 1734	5.5 1.2 5.5	**1** 0506 1113 F 1712 2346	5.0 1.6 5.2 1.4	**16** 0558 1204 Sa 1805	5.2 1.5 5.4
2 0434 1108 Sa 1659 2308	5.4 0.8 5.2 1.2	**17** 0416 1051 Su 1649 2304	5.9 0.2 5.7 0.6	**2** 0435 1101 M 1654 2308	5.4 1.1 5.2 1.3	**17** 0441 1108 Tu 1708 2326	5.9 0.6 5.7 0.8	**2** 0518 1129 Th 1727 2354	5.0 1.6 5.1 1.6	**17** 0004 0610 F 1218 1822	1.0 5.3 1.5 5.2	**2** 0546 1149 Sa 1750	4.9 1.7 5.2	**17** 0043 0646 Su 1246 1850	1.0 5.0 1.7 5.3
3 0506 1137 Su 1729 2337	5.4 0.9 5.1 1.3	**18** 0459 1132 M 1732 2344	5.9 0.4 5.6 0.7	**3** 0506 1129 Tu 1723 2337	5.2 1.3 5.1 1.4	**18** 0529 1150 W 1751	5.7 0.9 5.4	**3** 0557 1203 F 1804	4.8 1.8 5.0	**18** 0056 0704 Sa 1306 1914	1.2 4.9 1.8 5.0	**3** 0025 0631 Su 1229 1832	1.5 4.8 1.8 5.1	**18** 0131 0734 M 1331 1940	1.1 4.7 1.9 5.0
4 0537 1205 M 1758	5.3 1.1 5.0	**19** 0544 1212 Tu 1814	5.8 0.7 5.4	**4** 0539 1156 W 1754	5.1 1.5 5.0	**19** 0012 0619 Th 1235 1839	1.0 5.4 1.4 5.1	**4** 0034 0642 Sa 1241 1849	1.7 4.6 2.0 4.8	**19** 0152 0804 Su 1401 2011	1.4 4.6 2.2 4.7	**4** 0110 0721 M 1316 1921	1.4 4.6 1.9 4.9	**19** 0220 0826 Tu 1419 2030	1.4 4.5 2.1 4.8
5 0007 0610 Tu 1234 1831	1.4 5.1 1.4 4.9	**20** 0029 0634 W 1256 1902	1.0 5.5 1.1 5.0	**5** 0011 0615 Th 1227 1828	1.6 4.8 1.7 4.8	**20** 0104 0717 F 1324 1934	1.3 5.0 1.8 4.8	**5** 0120 0737 Su 1330 1944	1.8 4.4 2.2 4.6	**20** 0256 0908 M 1507 2117	1.6 4.4 2.4 4.6	**5** 0201 0818 Tu 1411 2018	1.5 4.5 2.0 4.8	**20** 0315 0922 W 1517 2128	1.6 4.3 2.3 4.6
6 0039 0645 W 1303 1906	1.6 4.9 1.7 4.7	**21** 0117 0728 Th 1344 1955	1.3 5.1 1.6 4.7	**6** 0048 0657 F 1302 1912	1.8 4.6 2.0 4.6	**21** 0204 0822 Sa 1423 2039	1.6 4.6 2.3 4.5	**6** 0219 0843 M 1433 2049	1.9 4.2 2.3 4.5	**21** 0410 1021 Tu 1628 2228	1.7 4.3 2.5 4.5	**6** 0300 0921 W 1517 2122	1.5 4.4 2.1 4.7	**21** 0417 1026 Th 1626 2234	1.8 4.2 2.4 4.5
7 0116 0727 Th 1338 1949	1.8 4.6 2.0 4.4	**22** 0215 0833 F 1444 2103	1.6 4.6 2.1 4.4	**7** 0133 0754 Sa 1348 2009	2.0 4.3 2.3 4.4	**22** 0318 0939 Su 1549 2157	1.8 4.3 2.5 4.3	**7** 0331 0957 Tu 1555 2200	1.9 4.2 2.4 4.5	**22** 0525 1134 W 1744 2340	1.7 4.3 2.5 4.5	**7** 0409 1030 Th 1631 2231	1.5 4.5 2.1 4.8	**22** 0523 1133 F 1739 2343	1.9 4.3 2.4 4.4
8 0159 0819 F 1425 2047	2.1 4.3 2.3 4.2	**23** 0331 0955 Sa 1614 2227	1.9 4.3 2.5 4.2	**8** 0234 0905 Su 1457 2122	2.2 4.1 2.5 4.2	**23** 0452 1109 M 1732 2322	1.9 4.3 2.5 4.4	**8** 0452 1113 W 1718 2312	1.7 4.4 2.2 4.7	**23** 0625 1234 Th 1841	1.6 4.5 2.2	**8** 0520 1140 F 1744 2343	1.4 4.6 1.9 4.9	**23** 0624 1232 Sa 1842	1.9 4.4 2.3
9 0301 0932 Sa 1532 2202	2.3 4.0 2.5 4.1	**24** 0513 1133 Su 1805 2356	2.0 4.3 2.4 4.3	**9** 0400 1030 M 1631 2241	2.2 4.1 2.5 4.3	**24** 0612 1225 Tu 1839	1.7 4.4 2.3	**9** 0604 1218 Th 1825	1.4 4.7 1.9	**24** 0038 0713 F 1319 1926	1.6 4.6 2.0 4.7	**9** 0627 1241 Sa 1849	1.3 4.9 1.7	**24** 0045 0714 Su 1320 1934	1.5 4.6 1.9 4.7
10 0430 1101 Su 1711 2323	2.4 4.0 2.5 4.2	**25** 0641 1253 M 1913	1.7 4.6 2.2	**10** 0537 1207 Tu 1803 2354	1.9 4.3 2.2 4.6	**25** 0029 0707 W 1317 1926	1.7 4.6 2.0 4.8	**10** 0017 0702 F 1312 1919	5.0 1.1 5.1 1.5	**25** 0126 0754 Sa 1355 2005	1.8 4.9 1.8 4.8	**10** 0049 0726 Su 1334 1947	5.1 1.2 5.1 1.4	**25** 0135 0758 M 1401 2020	1.8 4.6 1.8 4.9
11 0612 1222 M 1838	2.1 4.2 2.2	**26** 0103 0738 Tu 1347 1959	4.6 1.4 4.7 1.9	**11** 0646 1253 W 1903	1.5 4.7 1.8	**26** 0120 0751 Th 1357 2004	4.8 1.2 4.9 1.8	**11** 0113 0751 Sa 1358 2008	5.3 0.8 5.4 1.2	**26** 0205 0829 Su 1429 2042	4.9 1.5 5.0 1.7	**11** 0148 0818 M 1422 2039	5.3 1.1 5.4 1.1	**26** 0219 0836 Tu 1437 2103	4.7 1.7 5.0 1.7
12 0032 0720 Tu 1321 1935	4.5 1.7 4.6 1.9	**27** 0151 0823 W 1426 2036	4.9 1.1 4.9 1.7	**12** 0052 0735 Th 1341 1951	4.9 1.1 5.1 1.4	**27** 0201 0829 F 1429 2037	5.0 1.1 5.0 1.6	**12** 0204 0837 Su 1442 2054	5.6 0.7 5.6 0.9	**27** 0242 0901 M 1500 2118	5.0 1.5 5.1 1.6	**12** 0242 0905 Tu 1508 2131	5.5 1.0 5.5 1.0	**27** 0300 0914 W 1511 2142	4.8 1.6 5.1 1.5
13 0124 0808 W 1408 2020	4.9 1.2 5.0 1.5	**28** 0230 0901 Th 1500 2108	5.1 0.9 5.1 1.5	**13** 0141 0820 F 1423 2034	5.3 0.7 5.4 1.1	**28** 0236 0901 Sa 1458 2110	5.2 1.1 5.2 1.5	**13** 0253 0921 M 1524 2141	5.7 0.6 5.7 0.8	**28** 0317 0934 Tu 1532 2153	5.0 1.5 5.2 1.5	**13** 0332 0953 W 1552 2220	5.5 1.1 5.6 0.8	**28** 0338 0949 Th 1546 2220	4.9 1.5 5.3 1.3
14 0211 0850 Th 1450 2101	5.2 0.8 5.3 1.1	**29** 0304 0934 F 1529 2139	5.3 0.9 5.2 1.3	**14** 0227 0903 Sa 1504 2117	5.7 0.4 5.7 0.8	**29** 0308 0932 Su 1528 2141	5.3 1.1 5.3 1.4	**14** 0341 1006 Tu 1606 2227	5.8 0.7 5.7 0.8	**29** 0352 1006 W 1603 2230	5.1 1.5 5.2 1.5	**14** 0421 1038 Th 1637 2309	5.5 1.2 5.6 0.8	**29** 0416 1024 F 1621 2258	5.0 1.5 5.4 1.2
15 0253 0931 F 1531 2142	5.6 0.4 5.6 0.8	**30** 0335 1004 Sa 1557 2209	5.4 0.9 5.2 1.3	**15** 0311 0945 Su 1545 2159	5.9 0.3 5.8 0.7	**30** 0339 1002 M 1556 2212	5.3 1.2 5.3 1.4	**15** 0428 1049 W 1649 2315	5.7 0.9 5.6 0.8	**30** 0428 1038 Th 1637 2306	5.0 1.5 5.3 1.4	**15** 0511 1122 F 1720 2357	5.4 1.3 5.5 0.8	**30** 0454 1101 Sa 1657 2336	5.1 1.4 5.4 1.0
						31 0410 1030 Tu 1626 2244	5.2 1.3 5.3 1.4							**31** 0533 1139 Su 1734	5.1 1.4 5.4

Chart Datum: 2.85 metres below Ordnance Datum (Newlyn)

HARTLEPOOL 10-5-13
Cleveland

CHARTS
Admiralty 2566; Imray C29; OS 93
TIDES
+0437 Dover; ML 3.0; Duration 0600; Zone 0 (GMT).

Standard Port RIVER TEES ENT. (←)

Times				Height (metres)			
HW		LW		MHWS	MHWN	MLWN	MLWS
0000	0600	0000	0600	5.5	4.3	2.0	0.9
1200	1800	1200	1800				
Differences HARTLEPOOL							
−0010	−0010	−0007	−0007	−0.4	−0.3	−0.2	−0.1

SHELTER
Strong winds from E cause swell to build up making entrance channel hazardous. Once inside there is good shelter. Best berth at Hartlepool YC pontoons, SE corner of Victoria Dock, depth 5m (apply Dock Office Tel. 66127). West Harbour (dries) available for small craft. Access to enclosed docks HW∓1.

NAVIGATION
Waypoint Longscar E cardinal buoy, Q(3) 10s, Bell, 54°40'.85N 01°09'.79W, 137°/317° from/to entrance, 1.0M. Beware from S — Longscar rocks, 6 ca S of bottom of chartlet (Buoy BYB. Q (3) 10s). From N — 'The Stones' NE of breakwater head, 1 ca.

LIGHTS AND MARKS
Leading line 329° into main harbour (FR Lts at night — R dayglow boards by day). Amber Lt near front Ldg Lt means vessels may enter but not leave. When not shown vessels may leave but not enter. Traffic to and from enclosed docks 1G Lt — Vessel may enter N Basin. 2G Lts — Vessel may pass through N Basin. 1R Lt — Vessel may not approach dock.

RADIO TELEPHONE
Call: *Hartlepool Dock Radio* VHF Ch 16; 11 **12** (H24).

TELEPHONE (0429)
Hr Mr 266127; CG & MRSC North Shields 572691; Tees & Hartlepool Port Authority 276771; HMC 861390; Marinecall 0898 500 453; Dr 274570.

FACILITIES
EC Wednesday; **Harbour** (400) Tel. 266127, FW, D, Slip, C (10 ton); **Irvings Quay** Slip, M, L, AB; **Fish Quay** P (in cans), D, FW, CH; **A. M. Marine** Tel. 233820 ME, El, Sh, CH, C (Mobile); **Cliff Reynolds** Tel. 272049, ME, El, CH; **Dock** Tel. 266127, M, FW, ME, El, C (3 ton, 10 ton), AB; **Hartlepool YC** Tel. 274931, M, L, AB, R, Bar; **Tees SC** Tel. 267151, ME, El; **Town** P, V, R, Lau, Bar. Gas, Gaz, PO; Bank; Rly; Air (Teesside).
Note:— South docks are being converted into a marina complex (1988).

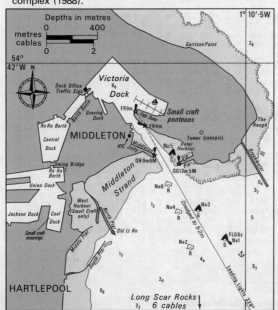

SEAHAM 10-5-14
Durham

CHARTS
Admiralty 1627; Imray C29; OS 88
TIDES
+0435 Dover; ML 3.0; Duration 0600; Zone 0 (GMT).

Standard Port RIVER TEES ENT. (←)

Times				Height (metres)			
HW		LW		MHWS	MHWN	MLWN	MLWS
0000	0600	0000	0600	5.5	4.3	2.0	0.9
1200	1800	1200	1800				
Differences SEAHAM							
−0015	−0015	−0015	−0015	−0.3	−0.2	0.0	−0.2

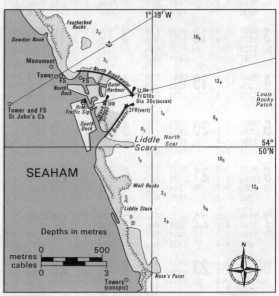

SHELTER
Small boats normally berth in N Dock where shelter is excellent but it dries. Larger boats may enter S Dock; gates open from HW−2 to HW+1.

NAVIGATION
Waypoint 54°50'.20N 01°18'.50W, 094°/274° from/to entrance, 0.40M. Shallow water and rocks to S of S breakwater (Liddle Scars). Entrance should not be attempted during strong on-shore winds. Speed limit 5 kn.

LIGHTS AND MARKS
No leading lights or marks but harbour is easily identified by lighthouse on N Pier (W with B bands). FS at NE corner of S dock in line with N Lt Ho leads in clear of Tangle Rks. Anchorage good ¾ mile off shore with Clock Tower in transit with St John's church tower.
Traffic signals, S Dock:
R flag at half-mast = Half tide, stand by to enter
R flag close up } = Enter
R Lt at night

RADIO TELEPHONE (local times)
VHF Ch 16; 06 **12** (HW−2½ to HW+1½ — office hours 0800−1800 Mon−Fri).

TELEPHONE (091)
Hr Mr 581 3246; MRSC North Shields 257 2691; Seaham Harbour Dock Co 581 3877; HMC Sunderland 565 7113; Marinecall 0898 500 453; Dr 581 2332.

FACILITIES
EC Wednesday; **South Dock** Tel. 581 3877, L, FW, C (40 ton), AB; **North Dock** M; **Town** (½ mile) P, D, FW, ME, El, CH (5 miles), V, R, Bar. PO; Bank; Rly; Air (Teesside or Newcastle).

AREA 5—NE England 357

SUNDERLAND 10-5-15
Tyne and Wear

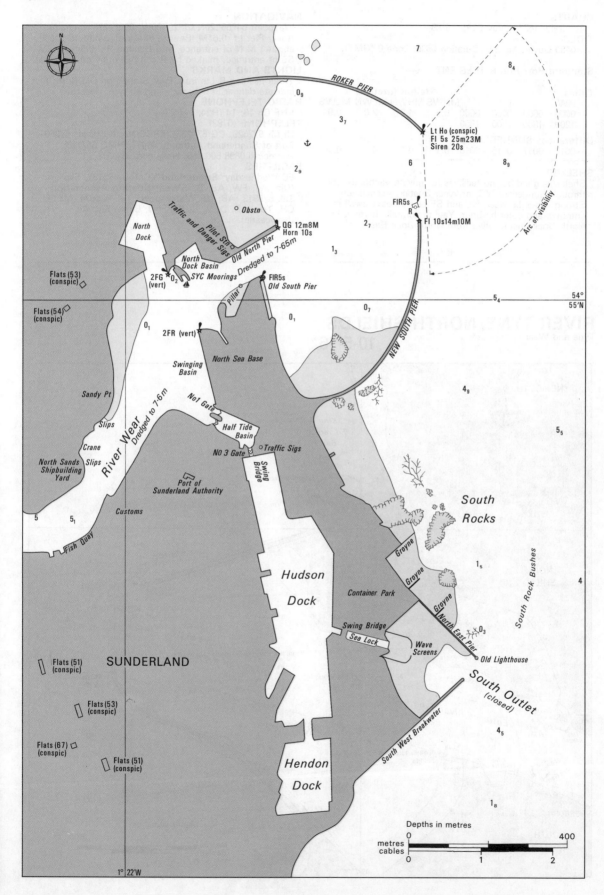

SUNDERLAND continued

CHARTS
Admiralty 1627; Imray C29; OS 88

TIDES
+0430 Dover; ML 2.9; Duration 0600; Zone 0 (GMT).

Standard Port RIVER TEES ENT. (←)

Times				Height (metres)			
HW		LW		MHWS	MHWN	MLWN	MLWS
0000	0600	0000	0600	5.5	4.3	2.0	0.9
1200	1800	1200	1800				
Differences SUNDERLAND							
−0017	−0017	−0016	−0016	−0.3	−0.1	0.0	−0.1

SHELTER
Shelter is good but no facilities for private yachts except through Sunderland YC, or Wear Boating Association. Strong winds between NE and SE cause heavy swell in entrance and outer harbour. Yachts normally berthed in North Dock, just up river from North Dock Basin.

NAVIGATION
Waypoint 54°55′.20N 01°20′.00W, 098°/278° from/to Roker Pier Lt, 0.61M. Beware wreck at Whitburn Steel about 1 M N of entrance, and Hendon Rk (0.9m), 1.2M SE of entrance, marked by R can bell buoy on its E side.

LIGHTS AND MARKS
Three flashing R Lts from Pilot Station on Old North Pier indicate danger in harbour — no entry or departure.

RADIO TELEPHONE
VHF Ch 16; 14 (H24)

TELEPHONE (0783)
Hr Mr 672626; CG 674255; MRSC North Shields 572691; Port of Sunderland Authority 40411; HMC 657113; Marinecall 0898 500 453; Hosp 656256.

FACILITIES
EC Wednesday; **Sunderland YC** Tel. 675133, Slip (Dinghy), FW, AB, Bar; **Wear Boating Association** Tel. 675313, AB; **NE Watersports** Tel. 675874, ME, EI, CH; **Town** P, D, V, R, Bar. PO; Bank; Rly; Air (Newcastle).

RIVER TYNE/NORTH SHIELDS
Tyne and Wear
10-5-16

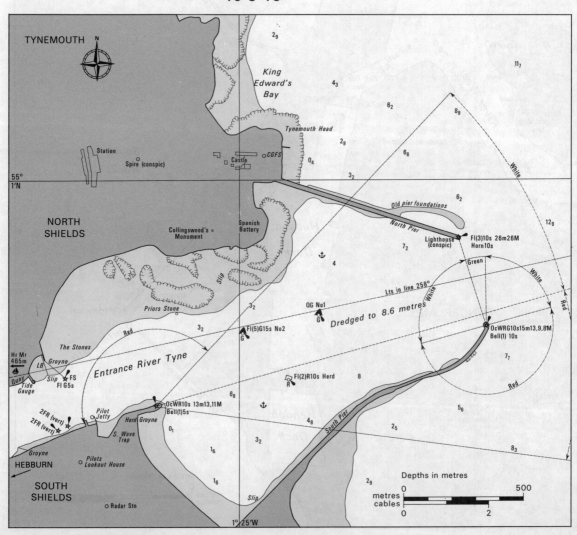

RIVER TYNE/NORTH SHIELDS continued

CHARTS
Admiralty 152, 1934; Imray C29; OS 88
TIDES
+0430 Dover; ML 2.9; Duration 0604; Zone 0 (GMT).

Standard Port RIVER TEES ENT. (←—)

Times				Height (metres)			
HW		LW		MHWS	MHWN	MLWN	MLWS
0000	0600	0000	0600	5.5	4.3	2.0	0.9
1200	1800	1200	1800				

Differences NORTH SHIELDS
−0016 −0018 −0017 −0022 −0.5 −0.4 −0.2 −0.2
NEWCASTLE-UPON-TYNE
−0013 −0015 −0009 −0014 −0.2 −0.2 −0.1 −0.1

SHELTER
Good shelter in all weathers. The approach is difficult for small craft in E and NE strong winds. Yachts may go into Albert Edward Dock giving excellent shelter and protection from heavy river traffic; obtain permission from Hr Mr by VHF or telephone first. Entrance 1M up river from Fl G 5s Lt on No 1 Groyne, on N bank.

NAVIGATION
Waypoint 55°01′.00N 01°22′.22W, 078°/258° from/to front Ldg Lt 258°, 2.2M. From S — No dangers. From N — Beware Bellhues Rock (approx 1 M N of harbour and ¾ M off shore) — Beware old N Pier; give it a wide berth.

LIGHTS AND MARKS
Ldg Lts 258° — two W Trs with W Lts. Buoys mark the dredged channel in Lower Harbour. Tynemouth Castle forms prominent landmark on cliff 26m high on N side of harbour.

RADIO TELEPHONE
Call: *Tyne Harbour Radio* VHF Ch 16; 11 **12** 14 (H24).

TELEPHONE (091)
Hr Mr 257 0407; CG & MRSC 257 2691; HMC 257 9441; Weather: 2326453; Marinecall 0898 500 453; Dr Radio through Tyne Harbour 257 2080.

FACILITIES
EC Wednesday
Limited berthing accommodation on the River Tyne for small pleasure craft. Prior arrangements must be made with club Secretaries or Harbour Master.
Hebburn Marina (50) Tel. 483 2876, FW, Slip; **Friars Goose Marina** (50) Tel. 469 2545, FW, ME, El, CH, Slip; **Robsons Boatyard** Tel. 455 5187, Slip, L, FW, ME, El, AB; **Tyne Slipway** Tel. 456 6209, Slip, M, L, FW; **Jim Marine** Tel. 257 7610, Slip, FW, ME, El, Sh, C, CH, AB, V, R, Bar; **John Lillie and Gillie** Tel. 257 2217 ACA; **Town**, P (cans), D, V, R, Bar. PO: Bank; Rly (Tynemouth & South Shields), Air (Newcastle).

BLYTH 10-5-17
Northumberland

CHARTS
Admiralty 1626; Imray C29; OS 81, 88
TIDES
+0415 Dover; ML 2.8; Duration 0558; Zone 0 (GMT).

Standard Port RIVER TEES ENT. (←—)

Times				Height (metres)			
HW		LW		MHWS	MHWN	MLWN	MLWS
0000	0600	0000	0600	5.5	4.3	2.0	0.9
1200	1800	1200	1800				

Differences BLYTH
−0011 −0025 −0018 +0013 −0.5 −0.4 −0.3 −0.1

SHELTER
Very good — Yachts normally berth in South Harbour, eastern section. At low tide in strong SE winds, seas break across entrance.

NAVIGATION
Waypoint Fairway stbd-hand buoy, Fl G 3s, Bell, 55°06′.58N 01°28′.50W, 140°/320° from/to East Pier Lt, 0.53M. Beware The Pigs, The Sow and Seaton Sea Rocks when approaching from the N. No dangers from the S.

LIGHTS AND MARKS
Ldg Lts 324°, F Bu on framework Trs with Or diamonds 11/17m. Second pair of Ldg Lts 338°, F Bu at 5m and 11m respectively.

RADIO TELEPHONE
Call: *Blyth Harbour Control* VHF Ch 16; 12 (H24).

TELEPHONE (0670)
Hr Mr 352678; CG & MRSC North Shields 572691; HMC 361521; Marinecall 0898 500 453; Dr 353226.

FACILITIES
EC Wednesday; **R Northumberland YC** Tel. 353636, Slip, M, L, FW, C (1½ ton), AB, Bar; **South Harbour** Tel. 352678, M, D, L. FW, C (3 ton 25 ton), CH, Gas, Gaz, AB; **Boat Yard** Tel. 353207, Slip, D, ME, El, Sh; **Town** P, V, R, Bar. PO; Bank; Rly (bus to Cramlington); Air (Newcastle).

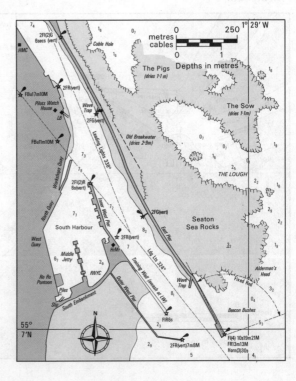

AMBLE 10-5-18
Northumberland

CHARTS
Admiralty 1627; Imray C29; OS 81
TIDES
+0412 Dover; ML 3.0; Duration 0606; Zone 0 (GMT).

Standard Port RIVER TEES ENT. (←)

Times				Height (metres)			
HW		LW		MHWS	MHWN	MLWN	MLWS
0000	0600	0000	0600	5.5	4.3	2.0	0.9
1200	1800	1200	1800				

Differences AMBLE
−0039 −0033 −0040 −0036 −0.5 −0.2 0.0 −0.1
NORTH SUNDERLAND
−0104 −0102 −0115 −0124 −0.7 −0.6 −0.4 −0.2
HOLY ISLAND
−0059 −0057 −0122 −0132 −0.7 −0.6 −0.5 −0.3

SHELTER
The harbour, known as Warkworth Harbour, is safe in all weathers but entrance is dangerous in strong N to E winds or in swell which causes breakers on the bar (Pan Bush). In NE gales, broken water can extend to Coquet Island. Yachts should go to the Braid Yacht Marina on S bank of Coquet River, approx 1000m from ent; or the river above fish dock (ask at Coquet YC). There is a speed limit throughout the harbour of 4 kn.

NAVIGATION
Waypoint 55°21'.00N 01°33'.00W, 045°/225° from/to harbour entrance, 0.9M. Entrance recommended from NE, passing N and W of Pan Bush shoal. A wreck, min depth 1.7m, lies 1.75 ca ENE of breakwater head. Alternative approach in good conditions and with sufficient rise of tide through Coquet channel (buoyed, min depth 0.3m). The S-going stream sets strongly across entrance.

LIGHTS AND MARKS
By day: Ball or Flag on FS = more than 3m over bar
By night: Lt extinguished on S breakwater = less than 3m over bar or bar is dangerous to approach.
Coquet Island Lt Ho, (conspic) W square Tr, turreted parapet, lower half grey; Fl (3) WR 30s 25m 23/19M R330°-140°, W140°-163°, R163°-180°, W180°-330°. Horn 30s.

RADIO TELEPHONE
Call *Amble Harbour* VHF Ch 16 14 (Mon-Fri 0900-1700 LT). Coquet YC Ch 16. Marina, Call *Amble Braid Marina* Ch M (H24).

TELEPHONE (0665)
Hr Mr 710306; MRSC North Shields 572691; CG 710575; HMC Blyth 361521; Marinecall 0898 500 453; Hosp 602661.
FACILITIES
EC Wednesday; **Braid Marina** (250-some visitors) Tel. 712168 AC, BY, C, Gas, FW, V, CH, Lau; Access HW∓4 via sill; **Harbour** D, AB; **J and J Harrison** Tel. 710267 ME, Slip, CH, El, Sh, D; **Coquet YC** Slip, Bar; **Amble Marine** Tel. 711069 SM, El, Electronics; **Marina Filling Station** P (cans); **Town** V, R, Bar, PO, Rly (Alnmouth), Air (Newcastle).

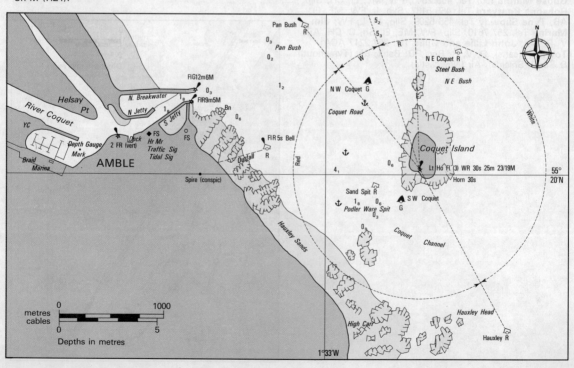

AREA 5—NE England 361

BERWICK-ON-TWEED 10-5-19
Northumberland

CHARTS
Admiralty 1612, 111, 160; OS 75
TIDES
+0330 Dover; ML 2.5; Duration 0620; Zone 0 (GMT).

Standard Port RIVER TEES ENT. (←)

Times				Height (metres)			
HW		LW		MHWS	MHWN	MLWN	MLWS
0000	0600	0000	0600	5.5	4.3	2.0	0.9
1200	1800	1200	1800				

Differences BERWICK
−0109 −0111 −0126 −0131 −0.8 −0.5 −0.7 −0.3

SHELTER
Good shelter or anchorage except in strong winds from E and SE. Yachts lie at W end of Fish Jetty or in Tweed Dock (opens HW−2 to HW). There are also good anchorages at Holy Is. See 10.5.20.

NAVIGATION
Waypoint 55°45'.80N 01°58'.00W, 098°/278° from/to breakwater Lt Ho, 0.54M. On-shore winds and ebb tides cause very confused state over the bar. From HW−2 to HW+1 strong flood tide sets across the entrance — keep well up to breakwater. The sands at the mouth of the Tweed shift so frequently that local knowledge is useful when entering Berwick Harbour.

LIGHTS AND MARKS
Town hall clock tower and lighthouse in line at 294°. When past Crabwater Rock, pick up transit beacons at Spittal in line at 207°. Beacons are B and Y with triangular top marks.

RADIO TELEPHONE (local times)
VHF CG: Ch 16 (0900-1700); Pilot and Hr Mr; Ch 16; 12 (HW−3 to HW+1, when shipping expected).

TELEPHONE (0289)
Hr Mr 7404; CG 6005; MRSC North Shields 572691; HMC 307547; Marinecall 0898 500 453/452; Dr 7484

FACILITIES
EC Thursday; **Dock** Tel. 7404, Slip, M (See Hr Mr), P, D, L, FW, ME, El, Sh, C (Mobile 3 ton), AB; **Town** P, V, R, Bar. PO; Bank; Rly; Air (Newcastle or Edinburgh).

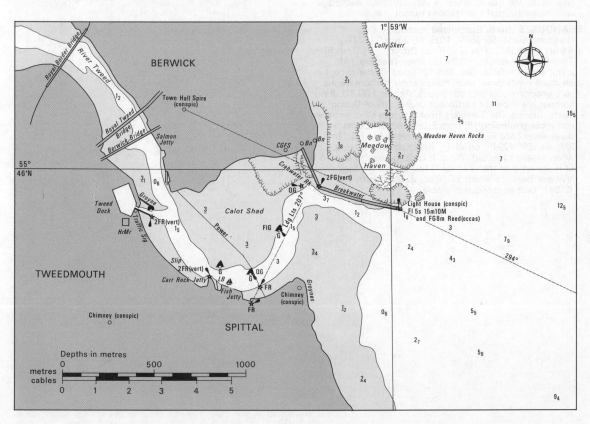

MINOR HARBOURS AND ANCHORAGES 10.5.20

WAINFLEET, Lincolnshire, 53°06′ N, 0°20′ E, Zone 0 (GMT), Admty chart 1190. Skegness HW +0500 on Dover, −0009 on Immingham; HW height −0.4m on Immingham; ML 4.0m; Duration 0600. Shelter good. Swatchway buoyed but not lit; channel through saltings marked with posts with radar reflectors, R can tops to port, G triangular tops to stbd. Enter HW∓1½. No lights. Facilities: EC Thurs, M, AB (larger boats at fishing jetties, smaller at YC), FW at Field Study Centre on stbd side at entrance. All shore facilities at Skegness (3½ miles).

FILEY, N Yorkshire, 54°13′ N, 0°16′ W, Zone 0 (GMT), Admty charts 1882, 129. HW +0532 on Dover, +0042 on River Tees Ent; HW height + 0.4m on River Tees Ent.; ML 3.5m; Duration 0605. See 10.5.9. Good anchorage in winds from NNE through W to S. Anchor in 4 to 5m on hard sandy bottom. Light on cliff above CG Station, G metal column, FR 31m 1M vis 272°−308°. The natural breakwater, Filey Brigg marked by Lt Buoy, Q(3)10s, Bell. Beware the Horse Rock, N of Filey Brigg, foul ground extending ½M from shore. Facilities; EC Wed, V, R, Bar, L, Hosp Tel. 68111, PO, Bank, Rly.

RUNSWICK BAY, N. Yorkshire, 54°32′ N, 0°45′ W; Zone 0 (GMT). Admty chart 1612. HW +0505 on Dover: +0010 on River Tees Ent; HW height −0.1m on River Tees Ent; ML 3.1m; Duration 0605. Good shelter in all winds from NW by W to SSE. Enter bay at 270° keeping clear of many rocks at base of cliffs. Two white posts (2FY by night when required by lifeboat) 18m apart are leading marks to LB house and can be used to lead into anchorage. Good holding in 6m to 9m in middle of bay. Facilities: **Runswick Bay Rescue Boat Station** Tel. Whitby 840965; **Runswick Bay Hotel** Tel. 840210 Bar, R; **Runswick Bay Stores** Tel. 840880 V.

NEWTON HAVEN (St Mary's), Northumberland, 55° 31′ N, 1° 36′.5W, Zone 0 (GMT), Admty chart 156. HW +0342 on Dover, −0102 on River Tees Ent; HW height −0.7m on River Tees Ent.; ML 2.6m; Duration 0625. A safe anchorage in winds from NNW to SE via S but susceptible to swell. Entrance is to S of Newton Point. Beware Fills Rks. Anchor between Fills Rks and Low Newton by the Sea in 4/5m. A very attractive anchorage with no lights, marks or facilities except a pub.

SEAHOUSES (North Sunderland Harbour), Northumberland, 55°35′ N, 1°39′ W, Zone 0 (GMT), Admty chart 1612. HW +0340 on Dover, −0103 on River Tees Ent; HW height −0.7m on River Tees Ent.; ML 2.7m; Duration 0618. See 10.5.18. Good shelter except in on-shore winds when swell makes outer harbour berths very uncomfortable and dangerous. Access HW∓3. Inner harbour has excellent berths but usually full of fishing boats. Beware The Tumblers (rocks) to the W of entrance and rocks protruding NE from E breakwater. Lights — Breakwater head Fl R 2.5s 6m; NW pier head FG 11m 3M; vis 159°−294°, on W Tr; traffic signals; Siren 90s when vessels expected. When it is dangerous to enter a R Lt is shown over the G Lt (or R flag over a Bu flag) on NW pier head. Facilities: EC Wed; **J. Davidson** Tel. 720347, Gas; all facilities available.

FARNE ISLANDS, Northumberland, 55°37′ N, 1°39′ W, Zone 0 (GMT), Admty chart 111. HW +0345 on Dover, −0102 on River Tees Ent.; HW height −0.7m on River Tees Ent; ML 2.6m; Duration 0630. The islands are a nature reserve belonging to the National Trust. Landing is only allowed on Inner Farne, Staple I and Longstone. In the Inner Islands, yachts can anchor in The Kettle on the E side of Inner Farne, near the Bridges (which connect Knox Reef to West Wideopens) or to the S of West Wideopens. In the Outer Islands yachts can anchor in Pinnacle Haven (between Staple I and Brownsman). Beware turbulence over Knivestone and Whirl Rks and eddy S of Longstone during NW tidal streams. Also beware The Bush to E of Farne I which dries 0.6m. Lights Black Rock Point Oc(2) WRG 15s 12m 17/13M; G122°−165°, W165°−175°, R175°−191°, W191°−238°, R238°−275°, W275°−289°, G289°−300°. Farne Is Lt Ho at SW Pt LFl(2) WR 15s 27m 13/9M; W round Tr; R119°−277°, W277°−119°. Longstone Fl 20s 23m 29M, R Tr with W band, RC, Siren (2) 60s. It is a beautiful area with no facilities, and should only be attempted in good weather.

HOLY ISLAND, Northumberland, 55°40′ N, 1°47′ W, Zone 0 (GMT), Admty chart 1612. HW +0344 on Dover, −0058 on River Tees Ent.; HW height −0.7m on River Tees Ent; ML 2.6m; Duration 0630. See 10.5.18. Shelter is complete in The Ooze (or Ouse) aground on soft mud; anchorage good just to E of St Cuthbert's I to the S of The Heugh except in W winds. Beware Plough Rks and the rocks all round them. Keep to Ldg Line 260° with Old Law Bns in line until church and Bn come in line at 310° which leads in to the anchorage S of The Heugh. There are six visitors moorings in the Heugh marked 'RNYC'. There are no lights. Facilities: FW from tap on village green; P and D at Beale (5 miles); limited provisions in village. Hr Mr Tel. Holy Island 207 or 217. Note Lindisfarne is the ancient name for Holy Island; it is linked to the mainland by a causeway which is covered at HW.

VOLVO PENTA SERVICE

Sales and service centres in area 6
LOTHIAN **Port Edgar Marine Services Ltd** Port Edgar Marina. South Queensferry. Nr EDINBURGH EH30 9SQ Tel (031-331) 1233.

Area 6

South-East Scotland
Eyemouth to Peterhead

10.6.1	Index	Page 363	
10.6.2	Diagram of Radiobeacons, Air Beacons, Lifeboat Stations etc	364	
10.6.3	Tidal Stream Charts	366	
10.6.4	List of Lights, Fog Signals and Waypoints	368	
10.6.5	Passage Information	370	
10.6.6	Distance Table	371	
10.6.7	Eyemouth	372	
10.6.8	Dunbar	372	
10.6.9	Firth of Forth (Leith, Standard Port, Tidal Curves)	373	
10.6.10	Burntisland	377	
10.6.11	Methil	378	
10.6.12	Anstruther	378	
10.6.13	River Tay	379	
10.6.14	Arbroath	380	
10.6.15	Montrose	380	
10.6.16	Stonehaven	381	
10.6.17	Aberdeen, Standard Port, Tidal Curves	381	
10.6.18	Peterhead	386	
10.6.19	Minor Harbours and Anchorages	387	

Burnmouth
St Abbs
North Berwick
Fisherrow
Cramond
Inchcolm
Aberdour
Kirkcaldy
Elie
St Monans
Crail
May Island
Gourdon
Boddam

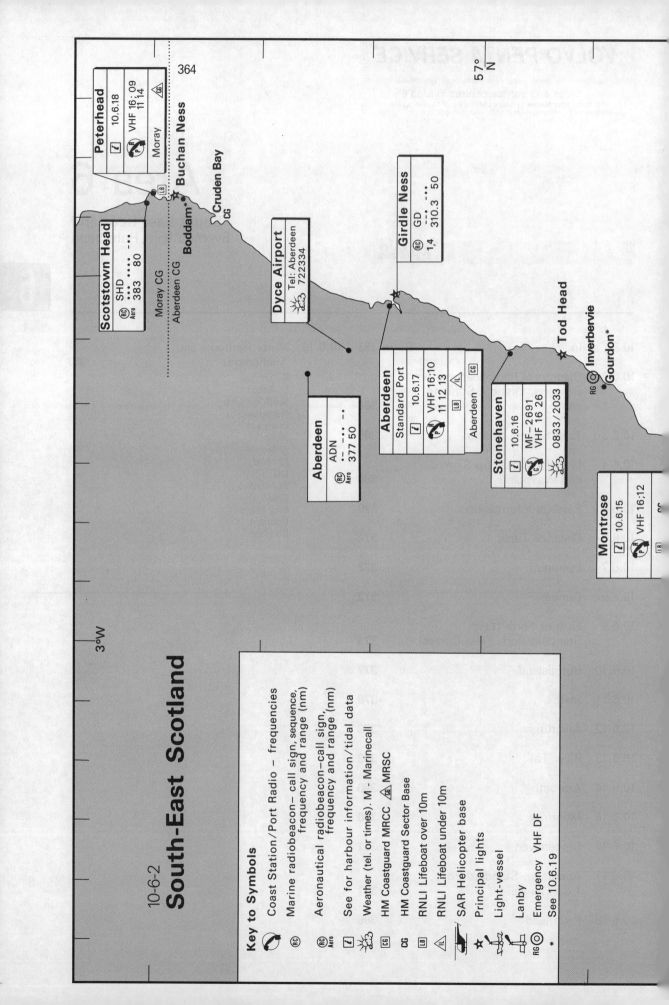

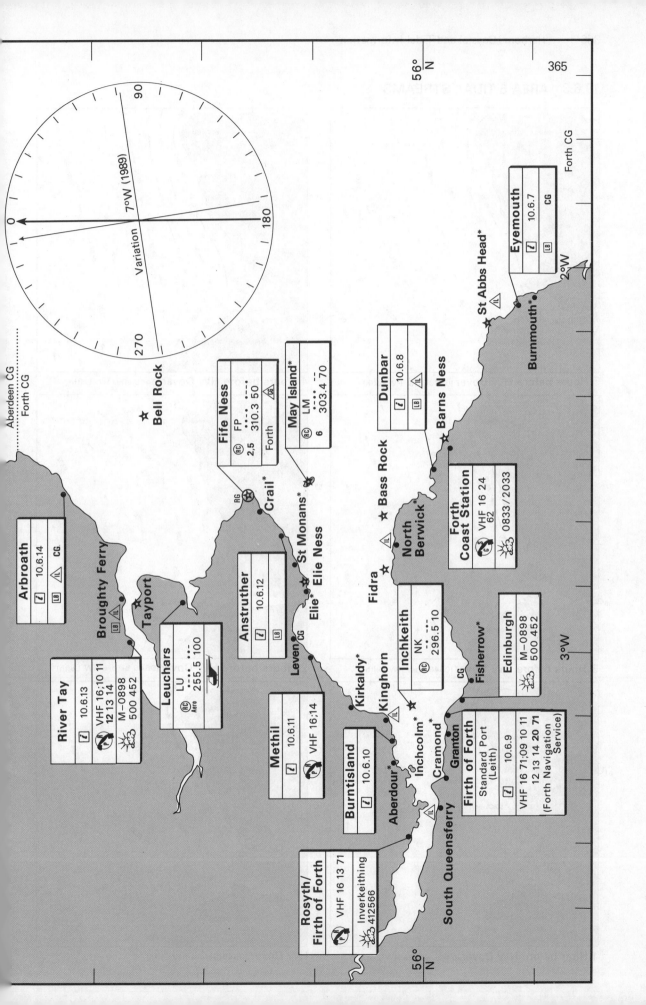

10.6.3 AREA 6 TIDAL STREAMS

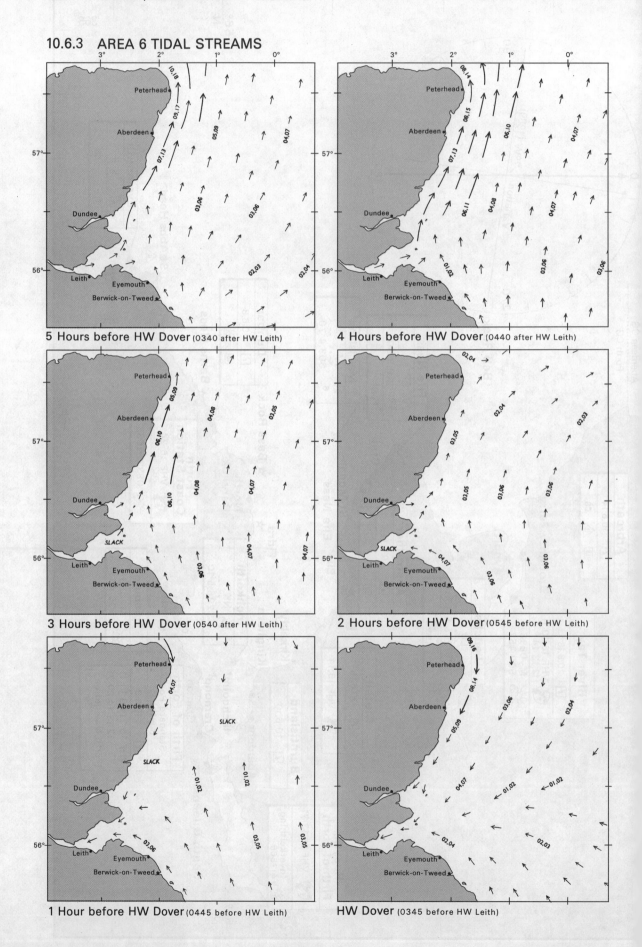

AREA 6 – SE Scotland 367

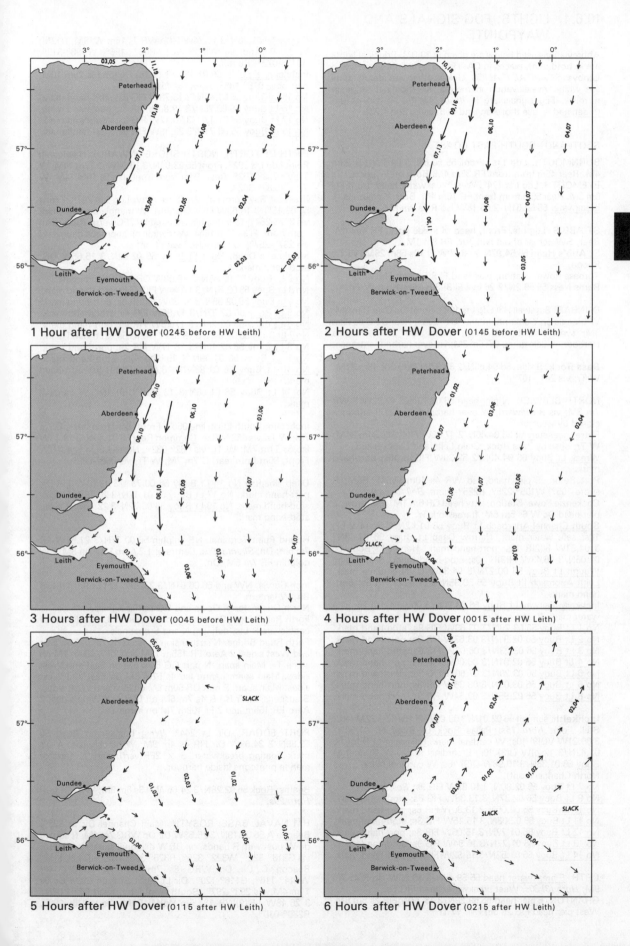

10.6.4 LIGHTS, FOG SIGNALS AND WAYPOINTS

Abbreviations used below are given in 10.0.3. Principal lights are in **bold** print, places in CAPITALS, and light-vessels and Lanbys in *CAPITAL ITALICS*. Unless otherwise stated lights are white. m—elevation in metres; M—nominal range in n. miles. Fog signals are in *italics*. Useful waypoints are underlined — use those on land with care. See 4.2.2.

SCOTLAND—SOUTH-EAST COAST

BURNMOUTH. Ldg Lts. Front 55 50.6N/2 04.1W FR 29m 4M. Rear 45m from front FR 35m 4M. Both on W posts.
EYEMOUTH. Ldg Lts 174°. West breakwater head, front FG 9m 5M. Rear 55m from front FG 10m 4M. Both Y columns. E breakwater 55 52.51N/2 05.18W Iso R 2s 8m 8M.

ST ABB'S. Ldg Lts. Front, head of inside jetty, FR 4m 1M. Rear, SW corner of old harbour, FR 8m 1M.
St Abb's Head 55 54.97N/2 08.20W Fl 10s 68m **29M**; W Tr; Racon.
Torness Power Station, pier head Fl R 5s 10m 5M.
Barns Ness 55 59.2N/2 26.6W Fl(3) 30s 36m 10M; W Tr.

DUNBAR. Bayswell Hill Ldg Lts 198°. Front Oc G 6s 15m 3M; W triangle on Y column; intens 188°-208°. Rear Oc G 6s 22m 3M; synchronised with front, intens 188°-208°. Victoria Harbour, middle quay, QR 6m 3M; vis over harbour entrance.
Bass Rock, S side, 56 04.60N/2 38.37W Fl(6) 30s 46m **21M**; W Tr; vis 241°-107°.

NORTH BERWICK. N pier head 56 03.73N/2 42.92W F WR 7m 3M; vis R seaward, W over harbour. Not lit if entrance closed by weather.
Fidra, near summit 56 04.40N/2 47.00W Fl(4) 30s 34m **24M**; W Tr; obsc by Bass Rock, Craig Leith and Lamb island.
Wreck Lt Buoy 56 04.40N/2 52.30W Fl(2)R 10s; port-hand mark.
Port Seton, E pier head, Iso WR 4s 10m W9M, R6M; R shore-105°, W105°-225°, R225°-shore; *Bell (occas)*.
Cockenzie Power Station, jetty head QR 6m 1M. Fisherrow, E pier head Oc W 6s 5m 6M; framework Tr.
South Channel Approach Lt Buoy 56 01.42N/3 02.14W LFl 10s; safe water mark. Narrow Deep Lt Buoy 56 01.48N/3 04.51W Fl(2)R 10s; port-hand mark. Herwit Lt Buoy 56 01.05N/3 06.43W Fl(3)G 10s; stbd-hand mark; *Bell*. Craigh Waugh Lt Buoy 56 00.27N/3 04.38W Q; N cardinal mark.
Leith Approach Lt Buoy 55 59.95N/3 11.42W FlR 5s; port-hand mark.
Inchkeith Fairway Lt Buoy 56 03.50N/3 00.00W Iso 2s; safe water mark; Racon.
No 1 Lt Buoy 56 03.23N/3 03.63W FlG 9s; stbd-hand mark.
No 2 Lt Buoy 56 02.91N/3 03.63W FlR 9s; port-hand mark.
No 3 Lt Buoy 56 03.23N/3 06.00W FlG 6s; stbd-hand mark.
No 4 Lt Buoy 56 02.91N/3 06.00W FlR 6s; port-hand mark.
No 5 Lt Buoy 56 03.20N/3 07.80W FlG 3s; stbd-hand mark.
No 6 Lt Buoy 56 03.05N/3 08.35W FlR 3s; port-hand mark.
No 8 Lt Buoy 56 02.95N/3 09.54W FlR 9s; port-hand mark.

Inchkeith, summit 56 02.01N/3 08.09W Fl 15s 67m **22M**; RC. Stell Point *Horn 15s*. Pallas Rock Lt Buoy 56 01.50N/3 09.21W VQ(9) 10s; W cardinal mark. E Gunnet Lt Buoy 56 01.42N/3 12.29W Q(3) 10s; E cardinal mark. W Gunnet Lt Buoy 56 01.35N/3 10.98W Q(9) 15s; W cardinal mark.
North Channel (cont.)
No 7 Lt Buoy 56 02.80N/3 10.87W FlG 9s; *Bell*; Racon.
No 9 Lt Buoy 56 02.37N/3 13.38W FlG 6s; stbd-hand mark.
No 10 Lt Buoy 56 02.05N/3 13.30W FlR 6s; port-hand mark.
No 11 Lt Buoy 56 02.08N/3 15.15W FlG 3s; stbd-hand mark.
No 12 Lt Buoy 56 01.77N/3 15.05W FlR 3s; port-hand mark.
No 13 Lt Buoy 56 01.77N/3 16.94W FlG 9s; stbd-hand mark.
No 14 Lt Buoy 56 01.52N/3 16.82W FlR 9s; port-hand mark.

LEITH. E breakwater head 55 59.48N/3 10.85W Iso R 4s 7m 9M; *Horn (3) 30s*. West breakwater head FlG 6s.
GRANTON. E pier head 55 59.28N/3 13.17W FlR 2s 5m 6M. West pier head FlG 2s 5m 7M; W Tr.

Oxcars 56 01.36N/3 16.74W Fl(2) WR 7s 16m W13M, R12M; W Tr, R band; vis W072°-087°, R087°-196°, W196°-313°, R313°-072°; Ra refl.
Inchcolm, E point, 56 01.73N/3 17.75W Fl(3) 15s 20m 10M; part obsc 075°-145°; *Horn (3) 45s*.
No 15 Lt Buoy 56 01.43N/3 18.70W FlG 6s; stbd-hand mark.
No 16 Lt Buoy 56 00.75N/3 19.80W FlR 3s; port-hand mark.
No 17 Lt Buoy 56 01.17N/3 20.12W FlG 3s; stbd-hand mark.
No 19 Lt Buoy 56 00.71N/3 22.38W FlG 9s; stbd-hand mark.

FIRTH OF FORTH, NORTH SHORE (INWARD). Hawkcraig Point Ldg Lts 292°. Front 56 03.04N/3 16.98W Q 12m 14M; W Tr; vis 282°-302°. Rear, 96m from front, Iso 5s 16m 14M; W Tr; vis 282°-302°.
Braefoot Bay Terminal. Western Jetty. Ldg Lts 247°. **Front** 56 02.15N/3 18.63W Fl 3s 6m **15M**; W triangle on E dolphin; vis 237°-257°; four dolphins marked by 2FG(vert). **Rear**, 88m from front, Fl 3s 12m **15M**; W triangle on approach gangway; vis 237°-257°; synchronised with front.
Mortimer's Deep. No 1 Lt Buoy 56 02.82N/3 15.65W QG; stbd-hand mark.
No 2 Lt Buoy 56 02.65N/3 16.08W QR; port-hand mark.
No 3 Lt Buoy 56 02.51N/3 17.44W Fl(2)G 5s; stbd-hand mark.
No 4 Lt Buoy 56 02.38N/3 17.35W Fl(2)R 5s; port-hand mark.
No 5 Lt Buoy 56 02.37N/3 17.86W FlG 4s; stbd-hand mark.
No 6 Lt Buoy 56 02.28N/3 17.78W FlR 4s; port-hand mark.
No 7 Lt Buoy 56 01.94N/3 18.92W Fl(2)G 5s; stbd-hand mark.
No 8 Lt Buoy 56 02.10N/3 18.17W FlR 2s; port-hand mark.
No 9 Lt Buoy 56 01.69N/3 19.08W QG; stbd-hand mark.
No 10 Lt Buoy 56 01.83N/3 18.48W Fl(2)R 5s; port-hand mark.
No 14 Lt Buoy 56 01.56N/3 18.96W Q(9) 15s; W cardinal mark.

Inchcolm. South Lts in line 066°. Front, 84m from rear, Q 7m 7M; W Tr; vis 062°-082°. Common Rear 56 01.80N/3 18.13W Iso 5s 11m 7M; W Tr; vis 062°-082°. North Lts in line 077°. Front, 80m from rear, Q 7m 7M; W Tr; vis 062°-082°.

Deep Channel No 16 Lt Buoy 56 00.75N/3 19.81W FlR 3s; port-hand mark. No 17 Lt Buoy 56 01.17N/3 20.10W FlG 3s; stbd-hand mark. No 19 Lt Buoy 56 00.72N/3 22.40W FlG 9s; stbd-hand mark.

Hound Point Terminal. NE dolphin 56 00.37N/3 21.52W FR 7m 5M; Dn; *Siren(3) 90s*. Centre pier 2 Aero FR 47m 5M. SW dolphin FR 7m 5M, Dn.

Inch Garvie, NW end 56 00.01N/3 23.29W LFl 5s 9m 11M; B Bn, W lantern.
North Queensferry. Oc 5s and QG traffic signals.
Forth Rail Bridge. Centres of spans have W Lts and ends of cantilevers R Lts, defining N and S channels.
Forth Road Bridge. N suspension Tr Iso G 4s 7m 6M on East and West sides; 2 Aero FR 155m 11M and 2 FR 109m 7M on same Tr. Main span, N part QG 50m 6M on East and West sides. Main span, centre Iso 4s 52m 8M on East and West sides. Main span, S part QR 50m 6M on East and West sides. S suspension Tr Iso R 4s 7m 6M on East and West sides; 2 Aero FR 155m and 2 Fr 109m 7M on same Tr.

PORT EDGAR. Dir Lt 244°. West breakwater head 55 59.85N/3 24.69W Dir FlR 4s 4m 8M; W blockhouse; 4 QY mark floating breakwater. 3 × 2FR(vert) mark N ends of marina pontoons inside harbour.

Beamer Rock 56 02.28N/3 24.66W Fl 3s 6m 9M; W Tr, R top; *Horn 20s*.

HM NAVAL BASE, ROSYTH. Main Channel Dir Lt 323°. Beacon A 56 01.19N/3 25.53W Dir Oc WRG 7m 4M; R square on W post with R bands, on B&W diagonal square on W Bn; vis G318°-322°, W322°-325°, R325°-328° (H24). Dir Lt 115°, Beacon C Dir Oc WRG 6s 7m 4M; vis R110°-114°, W114°-116°, G116°-120°. Dir Lt 295°, Beacon E Dir Oc 6s 11m 4M: vis 293°-297°. South Arm Jetty, head 56 01.08N/3 26.48W LFl(2)WR 12s 5m W9M; R6M; vis W010°-280°, R280°-010°.

AREA 6—SE Scotland

Rosyth Main Channel. Whale Bank No 2 Lt Buoy 56 00.70N/3 25.10W Q(3) 10s; E cardinal mark.
No 3 Lt Buoy 56 00.87N/3 24.98W FIG 5s; stbd-hand mark.
No 4 Lt Buoy 56 00.82N/3 25.18W FIR 3s; port-hand mark.
No 5 Lt Buoy 56 01.08N/3 25.80W QG; stbd-hand mark.
No 6 Lt Buoy 56 01.01N/3 25.94W QR; port-hand mark.
Charlestown. Lts in line. Front 56 02.2N/3 30.6W FG 4m 10M; Y triangle on Y pile; vis 017°-037°; marks line of HP gas main. Rear FG 6m 10M; Y triangle on Y pile; vis 017°-037°.
Crombie Pier. S arm, head, E end 2FG(vert) 5m 6M; *Horn 60s*.
Bo'ness. Carriden outfall 56 01.3N/3 33.6W FIY 5s; Y pile Bn.
Torry 56 02.5N/3 35.2W FIG 10s 5m 7M; G structure. Bo'ness Platform 56 01.9N/3 36.1W QR 3m 2M; R pile Bn.

GRANGEMOUTH. No 1 pile 56 02.2N/3 37.9W Fl(3)R 20s 4m 6M. No 2 pile 56 02.4N/3 38.8W FIG 5s 4m 6M; No 3 pile FIR 5s 4m 6M. No 4 pile FIG 2s 4m 5M. No 5 pile FIR 2s 4m 5M. Dock entrance, E jetty *Horn 30s*; docking signals.
Longannet Power Station, intake LFIG 10s 5m 6M.
KINCARDINE. Swing bridge 56 03.9N/3 43.5W FW at centre of each span; FR Lts mark each side of openings.

KIRKCALDY. E pier head 56 06.78N/3 08.81W Fl WG 10s 12m 8M; vis G156°-336°, W336°-156°; S pier head 2FR(vert) 7m 5M.
W Rockheads Buoy 56 07.00N/3 06.90W (unlit); stbd hand mark. E Rockheads Buoy 56 56 07.15N/3 06.35W (unlit); stbd-hand mark. Kirkcaldy Wreck Lt Buoy 56 07.25N/3 05.20W Fl(3)G 18s; stbd-hand mark.

METHIL. Outer pier head 56 10.77N/3 00.39W Oc G 6s 8m 5M; W Tr; vis 280°-100°.
Elie Ness 56 11.05N/2 48.65W Fl 6s 15m **18M**; W Tr.
ST MONANS. Breakwater head 56 12.20N/2 45.80W Oc WRG 6s 5m W7M, R4M, G4M; vis G282°-355°, W355°-026°, R026°-038°. E pier, head 2FG(vert) 6m 4M; Or tripod; *Bell (occas)*. W pier, near head, 2FR(vert) 6m 4M.
PITTENWEEM. Ldg Lts 037° Middle pier head. Front FR 4m 5M. Rear FR 8m 5M. Both W columns, R bands. E pier head extension 56 12.68N/2 43.53W Oc G 6s 5m 5M. Beacon Rock QR 3m 2M. West pier, Elbow, *Horn 90s (occas)*.
ANSTRUTHER. West pier head 2FR(vert) 5m 4M; grey post; *Reed (3) 60s (occas)*. E pier head 56 13.15N/2 41.72W Fl G 3s 6m 4M; R column.

Isle of May, Summit. 56 11.13N/2 33.30W Fl 20s 73m **26M**; square Tr on stone dwelling; RC. Fl(2) 15s 10M (T).

CRAIL. Ldg Lts 295°. Front FR 24m 6M (not lit when harbour closed). Rear 30m from front FR 30m 6M.

Fife Ness 56 16.73N/2 35.10W Iso WR 10s 12m **W21M, R20M**; vis W143°-197°, R197°-217°, W217°-023°; RC.
North Carr Lt Buoy 56 18.05N/2 32.85W Q(3) 10s; E cardinal mark.

Bell Rock 56 26.05N/2 23.07W Fl 5s 32m 10M(T); W round Tr; Racon.

RIVER TAY. Tay Fairway Lt Buoy 56 28.60N/2 37.15W LFl 10s; safe water mark; *Whis*. Middle Lt Buoy (North) 56 28.26N/2 38.87W Fl(3)G 18s; stbd-hand mark. Middle Lt Buoy (South) 56 28.00N/2 38.57W Fl(2)R 12s; port-hand mark. Abertay Lt Buoy 56 27.42N/2 40.63W Q(3) 10s; E cardinal mark; Racon. Abertay (Elbow) Lt Buoy 56 27.15N/2 40.70W FIR 6s; port-hand mark. Inner Lt Buoy 56 27.10N/2 44.25W Fl(2)R 12s; port-hand mark. Lady Lt Buoy 56 27.45N/2 46.58W Fl(3)G 18s; stbd-hand mark. Pool Lt Buoy 56 27.15N/2 48.50W FIR 6s; port-hand mark. Horse Shoe Lt Buoy 56 27.27N/2 50.12W VQ(6) + LFl 10s; S cardinal mark. Scalp Lt Buoy 56 27.18N/2 51.50W Fl(2)R 12s; port-hand mark.

Tentsmuir Point 56 26.6N/2 49.5W FIY 5s; Y Bn; vis 198°-208°; marks gas pipeline. Monifieth 56 28.9N/2 47.8W FIY 5s; Y Bn; vis 018°-028°; marks gas pipeline.
Broughty Castle 56 27.76N/2 52.10W 2FG(vert) 10/8m 4M; FR is shown at foot of old Lt Ho at Buddon Ness, 4M to E, and at other places on firing range when practice is taking place.

TAYPORT. **High Lighthouse**, Dir Lt 269°, 56 27.17N/2 53.85W Dir Iso WRG 3s 24m **W22M, R17M, G16M**; W Tr; vis G267°-268°, W268°-270°, R270°-271°.

ARBROATH. Outfall Lt Buoy 56 32.66N/2 34.96W FIY 3s; special mark. Ldg Lts 299°. Front FR 7m 5M; W column. Rear, 50m from front, FR 13m 5M; W column. West breakwater, E end, VQ(2) 6s 6m 4M; W metal post. E pier, S elbow 56 33.26N/2 34.89W FIG 3s 8m 5M; W Tr; shows FR when harbour closed; *Siren(3) 60s (occas)*.
Scurdie Ness 56 42.12N/2 26.15W Fl(3) 20s 38m **23M**; Racon.
MONTROSE. Scurdie Rocks Lt Buoy 56 42.15N/2 25.43W QR; port-hand mark. Annat Lt Buoy 56 42.37N/2 25.53W QG; stbd-hand mark. Ldg Lts 272°. Front FR 11m 5M; W twin pillars, R bands. Rear, 272m from front, FR 18m 5M; W Tr, R cupola. Inner Ldg Lts 265°. Front FG 21m 5M; Or triangle on pylon. Rear, 180m from front, FG 33m 5M; Or triangle on pylon.

JOHNSHAVEN. Ldg Lts 316°. Front FR 5m; R structure. Rear, 85m from front, FG 20m; shows R when unsafe to enter harbour.

GOURDON HARBOUR. Ldg Lts 358°. Front 56 49.6N/2 17.1W FR 5m 5M; W Tr; shows G when unsafe to enter; *Siren(2) 60s (occas)*. Rear 120m from front FR 30m 5M; W Tr. West pier head Fl WRG 3s 5m W9M, R7M, G7M; vis G180°-344°, W344°-354°, R354°-180°. E breakwater head Q 3m 7M.

Tod Head 56 53.0N/2 12.8W Fl(4) 30s 41m **29M**; W Tr.

STONEHAVEN. Outer pier head 56 57.59N/2 11.89W Iso WRG 4s 7m W11M, R7M, G8M; vis G214°-246°, W246°-268°, R268°-280°. Inner harbour Ldg Lts 273°. Front F 6m 5M. Rear FR 8m 5M.

Girdle Ness 57 08.35N/2 02.82W Fl(2) 20s 56m **22M**; W Tr; obsc by Greg Ness when bearing more than about 020°; RC; Racon.

ABERDEEN. Fairway Lt Buoy 57 09.33N/2 01.85W LFI 10s; safe water mark; Racon. S breakwater head Fl(3) R 8s 23m 7M. N pier head Oc WR 6s 11m 9M; W Tr; vis W145°-055°, R055°-145°. In fog FY (same Tr) vis 136°-336°; *Bell (3) 12s*. Old S breakwater QR 3m 2M; R column. Abercromby Jetty head Oc G 4s 5m 4M; G column; vis 230°-080°. South Jetty head, QR 5m 4M; vis 063°-243°. Torry. Ldg Lts 236°. Front FR or G 14m 5M; W Tr; R when entrance safe, G when dangerous to navigation; vis 195°-279°. Rear 205m from front FR 19m 5M; W Tr; vis 195°-279°.

Buchan Ness 57 28.23N/1 46.37W Fl 5s 40m **28M**; W Tr, R bands; Racon; *Horn (3) 60s*.

PETERHEAD. Kirktown Ldg Lts 314°. Front 57 30.2N/1 47.1W FR 13m 8M; R mast, W triangle point up. Rear 65m from front FR 17m 8M; R mast, W triangle point down. S breakwater head Fl(2) R 12s 24m 7M; W Tr with B base. N breakwater head 57 29.85N/1 46.22W Iso RG 6s 19m 7M; tripod; vis R165°-230°, G230°-165°; *Horn 30s*.

Rattray Head. Ron Rock. 57 36.6N/1 48.9W Fl(3) 30s 28m **24M**; W Tr; *Horn (2) 45s*.

10.6.5 PASSAGE INFORMATION

From Berwick-upon-Tweed to the Firth of Forth there is no good harbour which can be approached with safety in strong onshore winds. So, if on passage with strong winds from N or E, it is advisable to keep well to seaward. Northward from Firth of Forth to Rattray Hd the coast is mostly rky and steep-to, and there are no out-lying dangers within 2M of the coast except those off R Tay, and Bell Rk. Throughout this area reference should be made to *North Sea (West) Pilot* and *Sailing Directions Humber Estuary to Rattray Head* (R Northumberland YC). For the Firth of Forth, refer to the *Forth Yacht Clubs Association Pilot Handbook*, which covers the coast from Berwick-upon-Tweed to Fraserburgh.

BERWICK-UPON-TWEED TO BASS ROCK (chart 175)

The coast N from Berwick is rky with cliffs rising in height to Burnmouth. Keep ½M offshore to avoid outlying rks. Burnmouth has more alongside berths than Eyemouth, which is a busy fishing harb. N of Burnmouth the cliffs fall gradually to Eyemouth Bay.

Close S of St Abb's Hd is St Abb's Harb: the inner harb dries, but outer harb has depth of 2m in places. Do not approach in strong onshore wind or sea, and beware rky ledges close each side of leading line. Temp anchorage in offshore winds in Coldingham Bay. See 10.6.19.

St Abb's Hd (Lt) is a bold, steep headland, 92m high, with no offlying dangers. The stream runs strongly round the hd, causing turbulence with wind against tide; this can be largely avoided by keeping well inshore. The ESE-going stream begins at HW Leith −0345, and the WNW-going at HW Leith +0240. There is a good anch in Pettico Wick, on NW side of hd, in S winds, but dangerous if the wind shifts onshore. There are no off-lying dangers between St Abb's Hd and Fast Castle Hd, 3M WNW. Between Fast Castle Hd and Barns Ness, about 8M NW, is the attractive little harb of Cove, which however dries and should only be approached in very good conditions.

Barns Ness (Lt) lies 2½M ESE of Dunbar (10.6.8) and is fringed with rks: tidal streams as for St Abb's Hd. A conspic chy is ¾M WSW of Barns Ness, and Torness Power Station (conspic) is 1¾M SE of Barns Ness. Between here and Dunbar keep at least ¼M offshore to clear rky patches. Sicar Rk lies about 1¼M ENE of Dunbar, and sea breaks on it in onshore gales.

The direct course from Dunbar to Bass Rk (Lt) is clear of all dangers: inshore of this line beware Wildfire Rks (dry) on NW side of Bellhaven B. In offshore winds there is anch in Scoughall Road. Great Car is ledge of rks, nearly covering at HW, 1M ESE of Gin Hd, with Car Bn (stone Tr surmounted by cross) at its N end. Drying ledges of rks extend 1M SE of Great Car, up to 3 ca offshore. Keep at least ½M off Car Bn in strong onshore winds. Tantallon Castle (ruins) is on cliff edge 1M W of Great Car. Bass Rk lies 1¼M NNE of Gin Hd, and is a steep, conspic rk with no offlying dangers.

BASS ROCK TO INCHKEITH (chart 734)

Westward of Bass Rk Craigleith, Lamb Island and Fidra lie ½M or more offshore, while the coast is generally foul. Craigleith is steep-to, but temporary anchorage can be found on SE and SW sides; if passing inshore of it keep well to N side of chan. N Berwick Harb (dries) lies S of Craigleith, but is unsafe in onshore winds. Lamb Island is 1½M WNW of N Berwick (see 10.6.19) and has a rky ledge extending ¼M SW. Inshore, between Craigleith and Lamb Island, beware drying rks up to 3 ca from land. Fidra Island (Lt) is a bird reserve, nearly connected to the shore by rky ledges, and should be passed to the N. There are anchorages on E, S or W sides, depending on wind, in good weather.

In the B between Fidra and Edinburgh some shelter can be found in SE winds in Aberlady B and Gosford B. The best anch is SW of Craigielaw Pt. Port Seton is ¾M E of the conspic Chys of Cockenzie Power Station, and the E side of this fishing harb (dries) can be entered HW−3 to HW+3, but not advisable in strong on-shore wind or sea. Cockenzie (dries) is close to power station; beware Corsik Rk 400m to E. Access HW−2½ to HW+2½, but no attractions except boatyard. For Fisherrow see 10.6.19.

There are no dangers on the direct course from Fidra to Inchkeith (Lt, fog sig, RC), which stands in the centre of entrance to Leith, Granton (10.6.9), and the higher reaches of Firth of Forth. Rks extend ¾M SE from Inchkeith, and ½M off the SW side. There is a small harb on W side, below the Lt Ho; landing is forbidden without permission. N Craig and Craig Waugh (least depth 0.6m) are shallow patches 2½M SE from Inchkeith Lt Ho, and are buoyed. The deep water chans N and S of Inchkeith are buoyed. In N Chan, close to Inchkeith the W-going (flood) stream begins about HW Leith −0530, and the E-going at HW Leith +0030, sp rates about 1 kn. The streams gather strength towards the Forth bridges, where they reach 2¼ kn and there may be turbulence. For Cramond and Inchcolm see 10.6.19.

INCHKEITH TO FIFE NESS (charts 734, 190)

From Burntisland (10.6.10) the N shore of Firth of Forth leads E to Kinghorn Ness. 1M SSW of Kinghorn Ness Blae Rk (buoyed) has least depth of 4.6m, but the sea breaks on it in E gales. Rost Bank lies halfway between Kinghorn Ness and Inchkeith, with tide rips at sp tides or in strong winds.

Between Kinghorn Ness and Kirkcaldy, drying rks lie up to 3 ca offshore. Kirkcaldy Harb is primarily commercial but by arrangement yachts can enter inner dock near HW. The entrance is dangerous in strong E winds, when the sea breaks a long way out. For Kirkcaldy see 10.6.19.

Between Kirkcaldy and Methil (10.6.11) the only dangers more than 2 ca offshore are The Rockheads, extending 4 ca SE of Dysart, and marked by buoys. Largo B provides anch near E side, well sheltered from N and E. Off Chapel Ness beware W Vows (dries) and E Vows (dries, marked by refuge Bn). There is anch close W of Elie Ness, see 10.6.19. Ox Rk (dries) lies ½M ENE of Elie Ness, and ¼M offshore: otherwise there are no dangers more than 2 ca offshore past St Monans (10.6.19), Pittenweem and Anstruther (10.6.12), but in bad weather the sea breaks on Shield Rk 4 ca off Pittenweem. From Anstruther to Crail and on to Fife Ness keep 3 ca offshore to clear Caiplie Rk and other dangers. For Crail see 10.6.19.

May Island (Lt, RC) lies about 5M S of Fife Ness: its shores are bold except at NW end where rks extend 1 ca off. In good weather it is possible to land. Anch on W side, near N end, in E winds; or on E side, near N end, in W winds. Lt Ho boats use Kirkhaven, close SE of Lt Ho. See 10.6.19.

FIFE NESS TO MONTROSE (chart 190)

Fife Ness is fringed by rky ledges, and a reef extends 1M NE to N Carr Rk (dries, marked by Bn). In strong onshore winds keep to seaward of N Carr Lt buoy. From here keep ½M offshore to clear dangers entering St Andrews B, where there is anch: the little harb dries, and should not be approached in onshore winds.

R Tay (10.6.13) entrance is between Tentsmuir Pt and Buddon Ness (chart 1481), and has many sandbanks which are liable to shift. Abertay Sands extend nearly 4M E of Tentsmuir Pt on S side of chan (buoyed), and Gaa Sands run 1¾M E from Buddon Ness. Passage across Abertay and Gaa Sands is very dangerous. Abertay Lt buoy (Racon) is on N side of chan 1¾M ESE of Buddon Ness. Elbow is an extension of Abertay Sands, to S and SE of Lt buoy. The Bar, NE of Abertay Lt buoy, is dangerous in heavy weather, particularly in strong onshore wind or swell. S of Buddon Ness the W-going (flood) stream begins about HW Aberdeen −0400, and the E-going at about HW Aberdeen +0230, sp rates 2 kn.

Bell Rk (Lt, Racon) lies about 11½M E of Buddon Ness. 2M E of Bell Rk the S-going stream begins HW Aberdeen −0220, and the N-going at HW Aberdeen +0405, sp rates 1 kn. W of Bell Rk the streams begin earlier.

N from Buddon Ness the coast is sandy. 1¼M SW of Arbroath (10.6.14) beware Elliot Houses, rky patches with depth 1.8m, which extend about ½M offshore. Between Whiting Ness and Scurdie Ness, 9½M NNE, the coast is clear of out-lying dangers, but is mostly fringed with drying rks up to 1 ca off. In offshore winds there is anch in Lunan B, off Ethie Haven.

Scurdie Ness (Lt, Racon) stands on S side of entrance to Montrose (10.6.15). Scurdie Rks (dry) extend 2 ca E of the ness. On N side of chan Annat Bank dries up to about ½M E of the shore, opposite Scurdie Ness (chart 1438). The in-going stream begins at HW Aberdeen −0500, and the outgoing at HW Aberdeen +0115; both streams are very strong, up to 7 kn at sp, and there is turbulence off the entrance on the ebb. The entrance is dangerous in strong onshore winds, with breaking seas extending to Scurdie Ness on the ebb. In marginal conditions the last quarter of the flood is best time to enter.

MONTROSE TO RATTRAY HEAD (charts 210, 213)

N from Montrose the coast is sandy for 5M to Milton Ness, where there is anch on S side in N winds. Johnshaven, 2¼M SW of Gourdon, is a small harb (dries), which should not be approached with onshore wind or swell. ½M NE, off Brotherton Castle, drying rks extend 4 ca offshore. Gourdon has a small harb (mostly dries) approached by winding chan marked by Bns: inner harbour has storm gates. Outside Gourdon Harb rks extend ¾M S of entrance, and the sea breaks heavily in strong E winds. See 10.6.19.

N to Inverbervie the coast is fringed with rky ledges up to 2 ca offshore. Just N of Tod Hd (Lt) is Catterline, a small b which forms a natural anch in W winds, but open to E. Downie Pt lies close SE of Stonehaven (10.6.16). The b has a sandy bottom, but is encumbered by rky ledges up to 2 ca from shore. Anch 6 ca E of Bay Hotel. In E gales the b is dangerous, and sea breaks well outside harb entrance.

From Garron Pt to Girdle Ness the coast is mostly steep-to. Fishing nets may be met off headlands during fishing season. Craigmaroinn and Seal Craig (dry) are parts of reef 3 ca offshore SE of Portlethen, a fishing village with landing sheltered by rks. Cove B has a very small fishing harb, off which there is anch in good weather: Cove Rks (dry) lie 1½ ca offshore. From Cove to Girdle Ness keep ½M offshore, avoiding The Hasman, a rk which dries 1 ca off Altens village.

Greg Ness and Girdle Ness (Lt, RC, Racon) are fringed by rks. Girdlestone is a rky patch, depth less than 2m, 2 ca ENE of Lt Ho. A drying patch lies 2 ca SE of Lt Ho. Off Girdle Ness the S-going stream begins at HW Aberdeen −0430, and the N-going at HW Aberdeen +0130, sp rates 2½ kn. A race forms on S-going stream. Girdle Ness lies at SE corner of Aberdeen B (10.6.17).

Northward from Aberdeen there are few offshore dangers to Buchan Ness. R Ythan, 1¾M SSW of Hackley Hd, is navigable by small craft, but chan shifts constantly. 3M N is the very small harb of Collieston (mostly dries), only accessible in fine weather. 4¾M NNE of Hackley Hd lie The Skares, rks (marked by buoy) extending 3½ ca from S point of Cruden B — where there is anch in offshore winds. On N side of Cruden B is Port Erroll (dries).

Buchan Ness (Lt, fog sig, Racon) is a rky peninsula. 2 ca N is the islet Meikle Mackie, close W of which is the small harb of Boddam (dries). 3 ca NE of Meikle Mackie is The Skerry, a rk 6m high on S side of Sandford B; rks on which the sea breaks extend 2 ca NNE. There is a chan between The Skerry and the coast, which can be used when making for Peterhead (10.6.18). See also 10.6.19 for Boddam.

Rattray Hd (Lt, fog sig on Ron Rk, 3 ca E of hd) has rky foreshore, drying for 2 ca. Rattray Briggs is a detached reef, depth 0.6m, 2 ca E of Lt Ho. Rattray Hard is a rky patch, depth 0.2m, 1½M ENE of Lt Ho, which raises a dangerous sea during strong onshore winds. Off Rattray Hd the S-going stream begins at HW Aberdeen −0420, and the N-going at HW Aberdeen +0110, sp rates 3 kn. In normal conditions keep about 1M E of Rattray Hd, but in bad weather pass 5M off, preferably at slack water. Conspic radio masts with R Lts lie 2.5M WNW and 2.2M W of Lt Ho.

For notes on offshore oil and gas installations, see 10.5.5.

10.6.6 DISTANCE TABLE

Approximate distances in nautical miles are by the most direct route while avoiding dangers and allowing for traffic separation schemes etc. Places in *italics* are in adjoining areas.

	1	2	3	4	5	6	7	8	9	10	11	12	13	14	15	16	17	18	19	20
1 *Brunsbüttel*	1																			
2 *Flamborough Head*	325	2																		
3 *Longstone (Farne)*	388	105	3																	
4 Eyemouth	406	126	21	4																
5 St Abbs Head	408	129	24	3	5															
6 Dunbar	422	143	38	17	14	6														
7 Granton	449	170	65	44	41	27	7													
8 Port Edgar	456	177	72	51	48	34	7	8												
9 Burntisland	448	169	64	43	40	26	5	8	9											
10 Methil	441	162	57	36	33	20	14	20	12	10										
11 Anstruther	431	155	50	29	26	14	23	29	22	11	11									
12 Fife Ness	428	155	50	29	26	17	28	34	27	16	5	12								
13 Dundee	446	174	69	49	46	37	48	54	47	36	25	20	13							
14 Arbroath	433	169	64	44	41	34	45	51	44	33	22	17	15	14						
15 Montrose	433	175	69	51	48	43	55	61	54	43	32	27	27	12	15					
16 Stonehaven	434	185	82	66	63	60	72	78	71	60	49	44	45	30	20	16				
17 Aberdeen	436	194	93	78	75	73	84	90	83	72	61	56	57	42	32	13	17			
18 Peterhead	439	212	112	98	95	93	106	112	105	94	83	78	80	64	54	35	25	18		
19 *Duncansby Head*	504	292	192	179	176	174	186	192	185	174	163	158	161	145	135	115	105	82	19	
20 *Lerwick*	514	358	271	256	253	251	264	270	263	252	241	236	238	222	212	193	183	158	110	20

EYEMOUTH 10-6-7
Berwick

CHARTS
Admiralty 1612, 160; OS 67
TIDES
+0330 Dover; ML No data; Duration 0610; Zone 0 (GMT).

Standard Port LEITH (→)

Times				Height (metres)			
HW		LW		MHWS	MHWN	MLWN	MLWS
0300	0900	0300	0900	5.6	4.5	2.1	0.8
1500	2100	1500	2100				

Differences EYEMOUTH
−0015 −0025 −0014 −0004 −0.9 −0.8 No data

SHELTER
Good shelter in harbour in all weathers but entry should not be attempted in strong winds from N through NE to E. There are many fishing boats. Yachts normally berth outer end of E pier in 2m approx. Anchor in bay only in off-shore winds.

NAVIGATION
Waypoint 55°53′.00N 02°05′.28W, 354°/174° from/to E breakwater Lt, 0.50M. Approach can be made N or S of Hurcar Rocks but there are no leading marks to the S. To the N, beware Blind Buss. Entrance and basin dredged to 0.9m.

LIGHTS AND MARKS
Ldg Lts 174° (pillars painted orange) both FG 7/10m 4M on West pier. FR or red flag — unsafe to enter.

RADIO TELEPHONE
VHF Ch 16; 12 (No regular watch).

TELEPHONE (08907)
Hr Mr 50223; CG 50348; MRSC Crail 50666; HMC Edinburgh 554 2421; Marinecall 0898 500 452; Dr 50599.

FACILITIES
EC Wednesday; **Jetty** Slip, D (By delivery), P (cans), FW, AB; **Fishermans Mutual Assoc.** Tel. 50360 CH, Gas; **Eyemouth BY** Tel. 50231, Slip, BH, ME, El, Sh, C (12 ton mobile); **Collins** Tel. 50308, El; **Eyemouth Marine** Tel. 50594, ME; **Leith & Son** Tel. Berwick 307264 SM; **Marconi Marine** Tel. 51252 Electronics; **Coastal Marine** Tel. 50328 ME, Sh; **Town** LB, P, D, CH, V, R, Bar. PO, Lau, Gas, Gaz, Bank, Rly (bus to Berwick-on-Tweed); Air (Edinburgh).
Note: − limited facilities in town on Sundays.

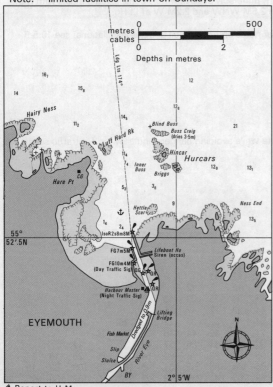
▲ Report to HrMr

DUNBAR 10-6-8
East Lothian

CHARTS
Admiralty 734, 175; Imray 27; OS 67
TIDES
+0330 Dover; ML 3.1; Duration 0600; Zone 0 (GMT).

Standard Port LEITH (→)

Times				Height (metres)			
HW		LW		MHWS	MHWN	MLWN	MLWS
0300	0900	0300	0900	5.6	4.5	2.1	0.8
1500	2100	1500	2100				

Differences DUNBAR
−0005 −0010 +0010 +0017 −0.4 −0.3 −0.1 −0.1
FIDRA
−0005 −0005 −0010 −0010 −0.4 −0.1 0.0 −0.2

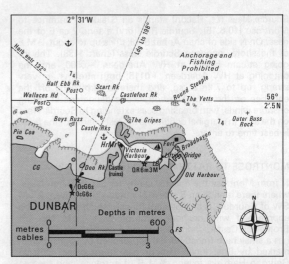

SHELTER
Outer (Victoria) Harbour is dangerous in strong NW to NE winds but Inner (Old or Cromwell) Harbour (dries) is safe at all times: entry through a bridge which is opened on request to Hr Mr.

NAVIGATION
Waypoint 56°00′.70N 02°30′.80W, 018°/198° from/to front Ldg Lt 198°, 0.50M. Min depth at entrance 0.9m. Entrance is unsafe in heavy on-shore swell. N side of Victoria harbour dries but pool 1.25m at LWS in SW corner. Beware Outer Buss Rk (0.6m). Keep to port on entry to avoid rockfall off castle.

LIGHTS AND MARKS
The entrance is to port of the 198° leading line which is marked by two marks to S of Doo Rks, both Oc G 6s, with W triangles on Or columns intens 188°-208° (synchronised). Entrance on Ldg line 132°, QR seen between the cliffs. Port Authority is the Lothian Regional Council.

RADIO TELEPHONE
None.

TELEPHONE (0368)
Hr Mr 63005; CG 63342; MRSC Crail 50666; HMC Edinburgh 554 2421; Marinecall 0898 500 452; Dr 62327.

FACILITIES
EC Wednesday; **Quay** Slip, L, FW, D (delivery), P (cans); **North Wall** M, L, AB; **Inner Harbour** Slip, L, AB; **F & P Blair** Tel. 62371 ME; **Lawrence Turnbull** Tel. 63228 Gas, Gaz; **Town** LB, P, Lau, V, R, Bar. PO; Bank; Rly; Air (Edinburgh).

FIRTH OF FORTH 10-6-9
Lothian/Fife

CHARTS
Admiralty 734, 735, 736; Imray C27; OS 66

TIDES
+0340 Dover; ML 3.3; Duration 0620; Zone 0 (GMT).

Standard Port LEITH (→)

Times				Height (metres)			
HW		LW		MHWS	MHWN	MLWN	MLWS
0300	0900	0300	0900	5.6	4.5	2.1	0.8
1500	2100	1500	2100				

Differences COCKENZIE
−0007 −0015 −0013 −0005 −0.2 0.0 0.0 No data

GRANTON
0000 0000 0000 0000 0.0 0.0 0.0 0.0

Kincardine:— Time difference Leith +0030.
Alloa:— Time difference Leith +0048.
Leith is a Standard Port and tidal predictions for each day of the year are given below.

SHELTER
Rosyth does not offer any facilities to yachts except in emergency. Leith Docks are an impounded dock and accordingly yachts are not normally accepted. Yachts are advised to shelter in Granton or Port Edgar or in Forth Yacht Marina at N Queensferry. Granton — Yachts go to E of Middle Pier. Buoys are run by Royal Forth and Forth Corinthian YCs.

NAVIGATION
Waypoint Granton 56°00'.00N 03°13'.22W, 000°/180° from/to entrance, 0.72M. Beware Forth Railway Bridge — Forth Road Bridge — H.M. Ships entering and leaving Rosyth Dockyard — Hound Pt Oil Terminal — Braefoot Gas Terminal.
Note: Control of Forth Estuary, all commercial impounded docks and Granton Harbour is exercised by Forth Ports Authority.

LIGHTS AND MARKS
Granton — R Flag with W diagonal cross (or G Lt by night) on signal mast at middle pierhead — Entry prohibited.
Leith — R Lt on both walls — Port closed.
— G Lt on both walls — Vessels may proceed.
— G Lt on one wall — Vessels moor on that side.
Port Edgar — On West Pier Dir Lt Fl R 4s 4m 8M 244°; 4 QY Lts mark floating breakwater; 3 x 2 FR (vert) mark N ends of marina pontoons inside harbour.
A Protected Channel 150m wide extends from Nos 13 & 14 light-buoys NNW of Oxcars, under the bridges (N of Inch Garvie and Beamer Rk), to the entrance of Rosyth Dockyard. When the Protected Channel is in operation an Oc 5s Lt and a QR Lt are shown from N Queensferry Naval Sig Stn, and all other vessels must clear the channel for naval traffic.

RADIO TELEPHONE (local times)
North Queensferry Naval Signal Station at Battery Point (call: *Queensferry*) VHF Ch 16; 13 **71** (H24).
Rosyth Naval Base (call: *QHM*) Ch 13 (Mon-Fri: 0730-1700).
Forth Navigation Service (at Leith) Ch 16 **71** (H24); **20** 12 (use 71 within area for calling and for short messages).
Grangemouth Docks Ch 16; 14 (H24).
Port Edgar Marina Ch M (Apl – Sept 0900 – 1900; Oct – Mar 0900 – 1700)

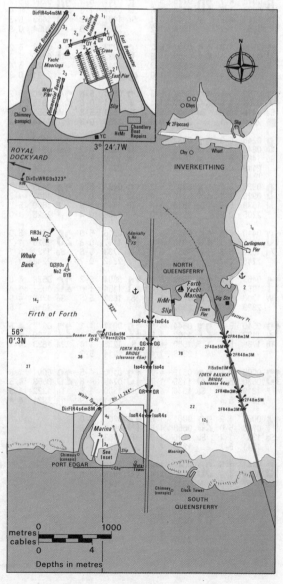

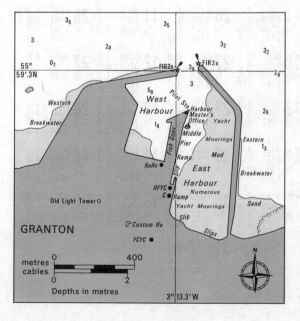

SCOTLAND, EAST COAST — LEITH

TIME ZONE UT(GMT)
For Summer Time add ONE hour in non-shaded areas

Lat 55°59′ N   Long 3°10′ W

TIMES AND HEIGHTS OF HIGH AND LOW WATERS

YEAR 1989

	JANUARY			FEBRUARY			MARCH			APRIL	
	Time m	Time m		Time m	Time m		Time m	Time m		Time m	Time m
1 Su	0158 2.1 / 0845 4.4 / 1423 2.4 / 2104 4.5	**16** M 0250 1.8 / 0948 4.3 / 1525 2.1 / 2146 4.9	**1** W	0321 2.3 / 0948 4.3 / 1608 2.3 / 2228 4.3	**16** Th 0522 2.2 / 1138 4.6 / 1802 1.9	**1** W	0100 2.1 / 0747 4.3 / 1331 2.2 / 2030 4.2	**16** Th 0323 2.5 / 0955 4.3 / 1629 2.1 / 2246 4.5	**1** Sa	0350 2.4 / 1001 4.3 / 1650 1.9 / 2301 4.4	**16** Su 0535 2.3 / 1154 4.6 / 1800 1.6
2 M	0310 2.2 / 0942 4.4 / 1541 2.4 / 2205 4.4	**17** Tu 0410 2.0 / 1036 4.7 / 1652 2.0 / 2302 4.9	**2** Th	0454 2.2 / 1106 4.4 / 1739 2.1 / 2349 4.4	**17** F 0013 4.8 / 0629 2.0 / 1244 4.8 / 1859 1.5	**2** Th	0228 2.4 / 0902 4.2 / 1526 2.3 / 2201 4.2	**17** F 0510 2.4 / 1123 4.5 / 1746 1.8	**2** Su	0514 2.0 / 1120 4.6 / 1754 1.4	**17** M 0019 4.7 / 0618 2.0 / 1242 4.8 / 1838 1.4
3 Tu	0423 2.2 / 1044 4.5 / 1657 2.3 / 2310 4.5	**18** W 0528 2.0 / 1146 4.8 / 1805 1.8	**3** F	0606 2.0 / 1217 4.6 / 1842 1.7	**18** Sa 0112 4.9 / 0719 1.8 / 1336 5.0 / 1945 1.2	**3** F	0423 2.3 / 1033 4.2 / 1719 2.0 / 2329 4.4	**18** Sa 0001 4.7 / 0613 2.1 / 1228 4.7 / 1838 1.5	**3** M	0005 4.8 / 0611 1.6 / 1220 5.0 / 1841 0.9	**18** Tu 0102 4.9 / 0652 1.7 / 1321 5.0 / 1912 1.3
4 W	0527 2.0 / 1145 4.6 / 1800 2.0	**19** Th 0013 5.0 / 0630 1.8 / 1249 5.0 / 1905 1.5	**4** Sa	0054 4.7 / 0700 1.6 / 1313 4.9 / 1933 1.2	**19** Su 0200 5.1 / 0758 1.6 / 1419 5.2 / 2022 1.0	**4** Sa	0545 2.0 / 1152 4.5 / 1825 1.5	**19** Su 0056 4.8 / 0700 1.9 / 1317 4.9 / 1919 1.3	**4** Tu	0056 5.1 / 0655 1.2 / 1307 5.4 / 1926 0.5	**19** W 0137 5.0 / 0720 1.5 / 1353 5.2 / 1941 1.1
5 Th	0012 4.7 / 0623 1.8 / 1241 4.8 / 1853 1.7	**20** F 0115 5.1 / 0724 1.6 / 1342 5.1 / 1954 1.2	**5** Su	0147 5.1 / 0747 1.3 / 1401 5.2 / 2020 0.8	**20** M 0240 5.2 / 0829 1.4 / 1455 5.3 / 2054 0.9	**5** Su	0034 4.7 / 0642 1.6 / 1252 4.9 / 1914 1.0	**20** M 0138 5.0 / 0732 1.6 / 1357 5.1 / 1952 1.1	**5** W	0140 5.5 / 0737 0.8 / 1349 5.7 / 2008 0.2	**20** Th 0206 5.1 / 0746 1.3 / 1422 5.2 / 2007 1.0
6 F	0109 4.9 / 0712 1.6 / 1331 5.0 / 1942 1.4	**21** Sa 0207 5.2 / 0809 1.5 / 1429 5.3 / 2039 1.0	**6** M	0232 5.4 / 0832 1.0 / 1442 5.6 / 2105 0.5	**21** Tu 0313 5.2 / 0855 1.3 / 1525 5.4 / 2121 0.8	**6** M	0127 5.1 / 0726 1.2 / 1337 5.3 / 1959 0.6	**21** Tu 0214 5.1 / 0759 1.4 / 1428 5.2 / 2020 1.0	**6** Th	0220 5.8 / 0821 0.4 / 1431 5.9 / 2050 0.1	**21** F 0233 5.2 / 0812 1.2 / 1450 5.3 / 2034 1.0
7 Sa	0159 5.1 / 0758 1.4 / 1416 5.2 / 2030 1.1	**22** Su 0253 5.3 / 0847 1.4 / 1509 5.3 / 2117 0.9	**7** Tu	0311 5.6 / 0914 0.8 / 1521 5.8 / 2150 0.3	**22** W 0340 5.2 / 0918 1.2 / 1551 5.4 / 2144 0.9	**7** Tu	0210 5.4 / 0808 0.8 / 1419 5.7 / 2041 0.3	**22** W 0243 5.2 / 0822 1.2 / 1456 5.3 / 2046 0.9	**7** F	0259 5.8 / 0905 0.3 / 1513 6.0 / 2132 0.2	**22** Sa 0300 5.2 / 0839 1.1 / 1520 5.3 / 2101 1.0
8 Su	0245 5.3 / 0842 1.2 / 1457 5.4 / 2118 0.9	**23** M 0332 5.3 / 0918 1.4 / 1545 5.4 / 2149 0.9	**8** W	0350 5.7 / 0957 0.7 / 1602 5.9 / 2231 0.3	**23** Th 0407 5.2 / 0940 1.1 / 1620 5.3 / 2204 0.9	**8** W	0248 5.7 / 0850 0.5 / 1458 5.9 / 2123 0.1	**23** Th 0310 5.2 / 0845 1.1 / 1521 5.3 / 2108 0.9	**8** Sa	0341 5.7 / 0948 0.3 / 1559 6.0 / 2209 0.4	**23** Su 0330 5.2 / 0909 1.1 / 1553 5.2 / 2129 1.2
9 M	0328 5.5 / 0928 1.2 / 1539 5.6 / 2205 0.8	**24** Tu 0405 5.2 / 0946 1.4 / 1618 5.3 / 2216 1.0	**9** Th	0431 5.7 / 1038 0.7 / 1646 5.9 / 2308 0.4	**24** F 0435 5.1 / 1003 1.2 / 1650 5.2 / 2225 1.2	**9** Th	0326 5.8 / 0933 0.4 / 1538 6.1 / 2203 0.1	**24** F 0334 5.2 / 0909 1.0 / 1548 5.3 / 2130 0.9	**9** Su	0426 5.6 / 1029 0.5 / 1648 5.7 / 2244 0.8	**24** M 0403 5.1 / 0942 1.2 / 1628 5.1 / 2201 1.3
10 Tu	0410 5.6 / 1012 1.1 / 1622 5.6 / 2248 0.7	**25** W 0438 5.1 / 1010 1.4 / 1651 5.2 / 2238 1.1	**10** F	0516 5.6 / 1117 0.8 / 1733 5.8 / 2342 0.7	**25** Sa 0505 5.0 / 1028 1.2 / 1722 5.1 / 2248 1.2	**10** F	0407 5.8 / 1014 0.4 / 1622 6.0 / 2238 0.3	**25** Sa 0401 5.2 / 0933 1.1 / 1618 5.3 / 2152 1.0	**10** M	0514 5.3 / 1108 1.0 / 1740 5.4 / 2319 1.3	**25** Tu 0437 5.0 / 1020 1.4 / 1709 4.9 / 2236 1.6
11 W	0453 5.5 / 1054 1.2 / 1708 5.6 / 2329 0.8	**26** Th 0509 5.0 / 1035 1.4 / 1725 5.1 / 2302 1.2	**11** Sa	0604 5.3 / 1154 1.0 / 1822 5.5	**26** Su 0538 4.8 / 1057 1.4 / 1756 4.9 / 2319 1.5	**11** Sa	0451 5.6 / 1053 0.5 / 1710 5.8 / 2311 0.6	**26** Su 0430 5.1 / 1000 1.1 / 1651 5.1 / 2218 1.2	**11** Tu	0605 5.0 / 1151 1.2 / 1836 5.0	**26** W 0517 4.8 / 1101 1.5 / 1757 4.7 / 2322 1.9
12 Th	0540 5.4 / 1137 1.3 / 1756 5.5	**27** F 0544 4.8 / 1102 1.5 / 1800 4.9 / 2328 1.4	**12** Su	0017 1.1 / 0654 5.1 / 1236 1.4 / 1914 5.2	**27** M 0612 4.6 / 1133 1.6 / 1833 4.7	**12** Su	0538 5.4 / 1130 0.8 / 1759 5.5 / 2345 1.1	**27** M 0502 4.9 / 1031 1.3 / 1726 4.9 / 2250 1.5	**12** W	0000 1.8 / 0701 4.7 / 1245 1.7 / 1938 4.7	**27** Th 0604 4.6 / 1154 1.7 / 1853 4.5
13 F	0008 0.9 / 0630 5.2 / 1219 1.4 / 1847 5.4	**28** Sa 0619 4.7 / 1135 1.7 / 1836 4.8	**13** M	0100 1.5 / 0748 4.8 / 1330 1.8 / 2011 4.9	**28** Tu 0001 1.8 / 0653 4.5 / 1222 1.9 / 1921 4.4	**13** M	0628 5.0 / 1209 1.2 / 1852 5.1	**28** Tu 0537 4.7 / 1108 1.7 / 1806 4.7 / 2332 1.8	**13** Th	0101 2.3 / 0806 4.4 / 1411 2.0 / 2053 4.4	**28** F 0020 2.1 / 0703 4.4 / 1303 1.9 / 1959 4.4
14 Sa	0050 1.2 / 0723 5.0 / 1307 1.7 / 1941 5.2	**29** Su 0001 1.6 / 0658 4.6 / 1216 1.9 / 1916 4.6	**14** Tu	0207 2.0 / 0853 4.6 / 1453 2.1 / 2126 4.7			**14** Tu	0027 1.7 / 0723 4.7 / 1303 1.7 / 1953 4.8	**29** W 0620 4.5 / 1157 1.8 / 1900 4.4	**14** F	0246 2.5 / 0925 4.3 / 1556 2.0 / 2216 4.4	**29** Sa 0139 2.3 / 0814 4.4 / 1438 1.9 / 2113 4.4
15 Su	0141 1.5 / 0820 4.9 / 1407 1.9 / 2038 5.0	**30** M 0049 1.9 / 0742 4.4 / 1311 2.2 / 2004 4.5	**15** W	0346 2.3 / 1015 4.5 / 1641 2.1 / 2254 4.6			**15** W	0129 2.2 / 0829 4.4 / 1431 2.1 / 2112 4.5	**30** Th 0030 2.1 / 0717 4.3 / 1307 2.1 / 2012 4.2	**15** Sa	0431 2.5 / 1048 4.4 / 1709 1.8 / 2327 4.6	**30** Su 0313 2.3 / 0929 4.5 / 1608 1.7 / 2227 4.6
		31 Tu 0153 2.2 / 0838 4.3 / 1427 2.3 / 2107 4.3						**31** F 0157 2.4 / 0834 4.2 / 1501 2.1 / 2140 4.2			

Chart Datum: 2.90 metres below Ordnance Datum (Newlyn)

AREA 6—SE Scotland 375

SCOTLAND, EAST COAST — LEITH
Lat 55°59′ N   Long 3°10′ W

TIME ZONE UT(GMT)
For Summer Time add ONE hour in non-shaded areas

TIMES AND HEIGHTS OF HIGH AND LOW WATERS   YEAR 1989

	MAY				JUNE				JULY				AUGUST		
	Time m		Time m		Time m		Time m		Time m		Time m		Time m		Time m

MAY

1 0433 2.0 1040 4.7 M 1713 1.4 2328 4.9
16 0525 2.1 1150 4.6 Tu 1750 1.7

2 0531 1.6 1141 5.1 Tu 1804 1.0
17 0012 4.7 0605 1.9 W 1233 4.8 1828 1.5

3 0020 5.1 0620 1.2 W 1234 5.4 1851 0.7
18 0051 4.8 0640 1.7 Th 1311 4.9 1903 1.4

4 0107 5.4 0705 0.8 Th 1321 5.6 1935 0.5
19 0127 5.0 0711 1.5 F 1347 5.0 1936 1.3

5 0150 5.6 0752 0.6 F 1406 5.8 ● 2020 0.4
20 0200 5.1 0745 1.3 Sa 1422 5.1 ○ 2008 1.2

6 0234 5.6 0840 0.4 Sa 1453 5.8 2104 0.5
21 0233 5.1 0820 1.2 Su 1459 5.2 2042 1.2

7 0317 5.6 0927 0.4 Su 1542 5.8 2145 0.7
22 0306 5.2 0859 1.2 M 1538 5.2 2117 1.3

8 0405 5.5 1012 0.6 M 1633 5.6 2223 1.0
23 0344 5.1 0939 1.2 Tu 1618 5.1 2155 1.4

9 0454 5.3 1055 0.9 Tu 1725 5.3 2301 1.5
24 0424 5.1 1022 1.2 W 1702 5.0 2235 1.6

10 0546 5.0 1137 1.2 W 1820 5.0 2342 1.9
25 0507 5.0 1108 1.3 Th 1750 4.9 2322 1.8

11 0640 4.7 1228 1.6 Th 1918 4.7
26 0556 4.9 1159 1.4 F 1843 4.7

12 0035 2.2 0739 4.5 F 1337 1.8 ☽ 2020 4.5
27 0015 1.9 0652 4.8 Sa 1259 1.5 1941 4.6

13 0151 2.4 0843 4.4 Sa 1500 1.9 2128 4.4
28 0119 2.1 0752 4.7 Su 1408 1.6 ☾ 2043 4.6

14 0322 2.5 0952 4.4 Su 1611 1.9 2232 4.3
29 0232 2.1 0855 4.8 M 1524 1.6 2147 4.7

15 0434 2.3 1057 4.5 M 1706 1.8 2327 4.6
30 0346 2.0 1001 4.9 Tu 1631 1.4 2250 4.9

31 0452 1.7 1104 5.1 W 1729 1.2 2345 5.1

JUNE

1 0549 1.4 1202 5.3 Th 1820 1.0
16 0004 4.7 0602 1.9 F 1230 4.7 1830 1.7

2 0038 5.2 0641 1.1 F 1257 5.4 1911 0.9
17 0049 4.8 0646 1.6 Sa 1317 4.8 1912 1.5

3 0127 5.4 0733 0.8 Sa 1349 5.5 ● 2000 0.8
18 0131 4.9 0728 1.4 Su 1403 5.0 1952 1.4

4 0214 5.4 0825 0.6 Su 1441 5.6 2048 0.9
19 0212 5.1 0812 1.2 M 1446 5.1 ○ 2033 1.3

5 0302 5.4 0914 0.6 M 1532 5.5 2131 1.0
20 0251 5.2 0856 1.1 Tu 1528 5.2 2113 1.3

6 0351 5.4 1002 0.7 Tu 1623 5.4 2210 1.2
21 0331 5.3 0940 1.0 W 1610 5.3 2154 1.3

7 0439 5.3 1044 0.9 W 1711 5.2 2246 1.5
22 0412 5.3 1025 1.0 Th 1652 5.2 2236 1.5

8 0527 5.1 1122 1.1 Th 1800 5.0 2322 1.7
23 0456 5.2 1110 1.0 F 1736 5.1 2318 1.5

9 0616 4.9 1204 1.4 F 1848 4.7
24 0544 5.2 1154 1.1 Sa 1824 5.0

10 0003 2.0 0705 4.7 Sa 1249 1.6 1938 4.5
25 0004 1.6 0635 5.1 Su 1241 1.2 1917 4.9

11 0053 2.2 0756 4.5 Su 1346 1.8 ☽ 2030 4.4
26 0055 1.7 0728 5.1 M 1336 1.4 ☾ 2012 4.8

12 0156 2.3 0850 4.5 M 1451 1.9 2126 4.4
27 0154 1.8 0826 5.0 Tu 1440 1.5 2112 4.8

13 0309 2.3 0946 4.4 Tu 1556 1.9 2222 4.4
28 0305 1.9 0927 5.0 W 1550 1.6 2215 4.8

14 0418 2.2 1045 4.5 W 1653 1.9 2315 4.5
29 0419 1.8 1035 5.0 Th 1701 1.6 2320 4.9

15 0515 2.1 1139 4.6 Th 1744 1.8
30 0529 1.6 1143 5.1 F 1804 1.5

JULY

1 0020 5.0 0630 1.3 Sa 1247 5.2 1900 1.3
16 0018 4.6 0630 1.7 Su 1256 4.7 1857 1.6

2 0115 5.2 0727 1.0 Su 1345 5.3 1952 1.2
17 0112 4.7 0719 1.4 M 1349 4.9 1942 1.4

3 0207 5.3 0819 0.8 M 1439 5.4 ● 2041 1.1
18 0157 5.0 0805 1.1 Tu 1435 5.1 ○ 2025 1.2

4 0255 5.3 0908 0.7 Tu 1525 5.4 2122 1.2
19 0238 5.3 0851 0.8 W 1515 5.3 2106 1.1

5 0340 5.3 0951 0.7 W 1612 5.4 2157 1.3
20 0318 5.4 0936 0.7 Th 1555 5.5 2147 1.0

6 0422 5.3 1028 0.8 Th 1654 5.3 2229 1.4
21 0357 5.5 1019 0.6 F 1635 5.6 2228 1.0

7 0504 5.2 1100 0.9 F 1734 5.0 2258 1.5
22 0439 5.6 1100 0.6 Sa 1717 5.4 2308 1.1

8 0545 5.0 1129 1.1 Sa 1814 4.8 2328 1.6
23 0526 5.5 1138 0.7 Su 1804 5.3 2348 1.2

9 0626 4.8 1200 1.4 Su 1854 4.6
24 0613 5.4 1217 0.9 M 1852 5.1

10 0003 1.8 0708 4.7 M 1238 1.6 1936 4.5
25 0031 1.4 0704 5.3 Tu 1302 1.2 ☾ 1944 4.9

11 0048 2.0 0753 4.5 Tu 1329 1.8 2022 4.4
26 0122 1.6 0759 5.1 W 1359 1.6 2042 4.8

12 0146 2.2 0841 4.4 W 1433 2.0 2115 4.3
27 0229 1.8 0902 4.9 Th 1519 1.9 2150 4.7

13 0301 2.3 0937 4.4 Th 1550 2.1 2214 4.4
28 0358 1.9 1019 4.8 F 1646 1.9 2305 4.7

14 0423 2.2 1047 4.4 F 1702 2.0 2318 4.4
29 0523 1.7 1139 4.8 Sa 1802 1.8

15 0533 2.0 1156 4.5 Sa 1805 1.9
30 0014 4.9 0631 1.4 Su 1249 5.0 1902 1.6

31 0114 5.0 0726 1.1 M 1346 5.2 1951 1.4

AUGUST

1 0205 5.2 0813 0.8 Tu 1437 5.3 ● 2033 1.3
16 0139 5.1 0752 0.9 W 1417 5.2 2010 1.1

2 0248 5.3 0855 0.7 W 1518 5.4 2108 1.2
17 0220 5.4 0835 0.5 Th 1456 5.5 ○ 2051 0.9

3 0326 5.4 0932 0.6 Th 1555 5.3 2138 1.2
18 0258 5.7 0918 0.3 F 1534 5.7 2131 0.7

4 0401 5.4 1002 0.7 F 1630 5.2 2204 1.2
19 0336 5.8 0959 0.2 Sa 1611 5.7 2211 0.6

5 0437 5.3 1028 0.8 Sa 1703 5.1 2227 1.2
20 0418 5.9 1038 0.3 Su 1653 5.6 2249 0.7

6 0511 5.1 1050 1.0 Su 1736 4.9 2251 1.3
21 0503 5.8 1114 0.5 M 1739 5.4 2327 0.9

7 0547 5.0 1113 1.2 M 1811 4.7 2318 1.5
22 0552 5.6 1149 0.8 Tu 1828 5.2

8 0624 4.8 1143 1.4 Tu 1847 4.6 2354 1.7
23 0007 1.1 0643 5.3 W 1230 1.3 1921 4.9

9 0703 4.6 1225 1.7 W 1928 4.4
24 0056 1.5 0740 5.0 Th 1327 1.8 2020 4.7

10 0043 2.0 0748 4.3 Th 1324 2.0 2018 4.3
25 0208 1.9 0850 4.7 F 1456 2.2 2135 4.5

11 0154 2.2 0848 4.3 F 1447 2.2 2123 4.2
26 0354 2.0 1017 4.6 Sa 1645 2.2 2301 4.6

12 0335 2.3 1007 4.3 Sa 1626 2.2 2241 4.3
27 0526 1.8 1142 4.7 Su 1804 2.0

13 0512 2.0 1126 4.5 Su 1745 2.0 2354 4.5
28 0013 4.8 0628 1.4 M 1248 5.0 1859 1.8

14 0616 1.7 1239 4.6 M 1843 1.7
29 0109 5.0 0717 1.1 Tu 1341 5.1 1941 1.5

15 0052 4.8 0706 1.3 Tu 1332 4.9 1929 1.4
30 0155 5.2 0757 0.9 W 1424 5.3 2017 1.3

31 0233 5.4 0832 0.7 Th 1500 5.4 ● 2045 1.2

Chart Datum: 2.90 metres below Ordnance Datum (Newlyn)

SCOTLAND, EAST COAST — LEITH

Lat 55°59′ N   Long 3°10′ W

TIMES AND HEIGHTS OF HIGH AND LOW WATERS

TIME ZONE UT(GMT)
For Summer Time add ONE hour in non-shaded areas

YEAR 1989

SEPTEMBER		OCTOBER		NOVEMBER		DECEMBER	
Time m	Time m	Time m	Time m	Time m	Time m	Time m	Time m
1 0305 5.4 F 0902 0.7 1531 5.3 2110 1.1	**16** 0231 5.8 Sa 0850 0.2 1507 5.8 2107 0.5	**1** 0305 5.4 Su 0851 0.9 1525 5.3 2059 1.1	**16** 0244 6.0 M 0902 0.2 1520 5.8 2127 0.4	**1** 0337 5.2 W 0914 1.3 1554 5.2 2132 1.3	**16** 0407 5.7 Th 1008 1.0 1637 5.5 2245 0.9	**1** 0402 5.2 F 0939 1.5 1613 5.2 2208 1.4	**16** 0448 5.5 Sa 1039 1.4 1711 5.4 2318 1.0
2 0336 5.4 Sa 0929 0.7 1559 5.3 2133 1.1	**17** 0310 6.0 Su 0931 0.1 1546 5.8 2149 0.4	**2** 0334 5.4 M 0914 0.9 1552 5.2 2123 1.1	**17** 0331 6.0 Tu 0944 0.4 1605 5.7 2210 0.5	**2** 0414 5.1 Th 0945 1.5 1628 5.1 2206 1.5	**17** 0501 5.5 F 1051 1.4 1729 5.3 2330 1.2	**2** 0443 5.1 Sa 1018 1.7 1653 5.1 2250 1.5	**17** 0537 5.0 Su 1117 1.6 1759 5.2 2355 1.3
3 0405 5.3 Su 0950 0.8 1627 5.2 2153 1.1	**18** 0353 6.0 M 1011 0.2 1629 5.7 2229 0.5	**3** 0404 5.3 Tu 0937 1.1 1621 5.1 2147 1.2	**18** 0421 5.8 W 1024 0.7 1653 5.5 2253 0.8	**3** 0454 5.0 F 1020 1.7 1706 4.9 2246 1.6	**18** 0557 5.1 Sa 1133 1.8 1824 5.0	**3** 0527 5.0 Su 1102 1.8 1737 5.0 2335 1.5	**18** 0626 5.0 M 1155 1.9 1850 5.0
4 0436 5.2 M 1010 0.9 1657 5.0 2216 1.2	**19** 0439 5.9 Tu 1047 0.5 1715 5.5 2308 0.7	**4** 0435 5.1 W 1003 1.3 1653 5.0 2217 1.4	**19** 0514 5.6 Th 1104 1.2 1746 5.2 2337 1.2	**4** 0538 4.8 Sa 1104 1.9 1750 4.8 2334 1.8	**19** 0018 1.5 Su 0656 4.9 1224 2.1 1922 4.8	**4** 0616 4.8 M 1150 2.0 1827 4.9	**19** 0036 1.6 Tu 0717 4.7 1240 2.1 ☾ 1940 4.8
5 0508 5.1 Tu 1033 1.1 1730 4.9 2242 1.4	**20** 0531 5.6 W 1123 1.0 1806 5.2 2348 1.1	**5** 0512 4.9 Th 1034 1.5 1728 4.8 2252 1.6	**20** 0612 5.2 F 1146 1.7 1842 4.9	**5** 0632 4.6 Su 1158 2.2 1844 4.6	**20** 0118 1.8 M 0757 4.7 1330 2.4 ☾ 2025 4.7	**5** 0027 1.6 Tu 0708 4.7 1247 2.1 1922 4.9	**20** 0125 1.8 W 0809 4.6 1335 2.3 2032 4.6
6 0543 4.9 W 1102 1.4 1804 4.7 2316 1.6	**21** 0625 5.3 Th 1205 1.5 1901 4.9	**6** 0553 4.7 F 1114 1.8 1810 4.6 2338 1.9	**21** 0029 1.6 Sa 0715 4.9 1244 2.2 ☾ 1946 4.7	**6** 0035 1.9 M 0734 4.4 1308 2.4 ☾ 1948 4.5	**21** 0233 1.9 Tu 0903 4.6 1455 2.5 2130 4.6	**6** 0126 1.7 W 0808 4.7 1353 2.2 2021 4.9	**21** 0226 2.0 Th 0905 4.5 1448 2.4 2129 4.5
7 0621 4.7 Th 1142 1.7 1843 4.5	**22** 0039 1.5 F 0727 4.9 1303 2.0 ☾ 2004 4.6	**7** 0646 4.4 Sa 1209 2.2 1904 4.4	**22** 0146 1.9 Su 0827 4.6 1414 2.5 2101 4.5	**7** 0157 2.0 Tu 0843 4.4 1435 2.4 2057 4.6	**22** 0345 2.0 W 1009 4.7 1614 2.5 2235 4.7	**7** 0237 1.8 Th 0910 4.7 1508 2.2 2123 4.9	**22** 0334 2.2 F 1004 4.5 1604 2.4 2228 4.5
8 0000 1.9 F 0710 4.4 1236 2.1 ☾ 1934 4.3	**23** 0157 1.9 Sa 0842 4.6 1441 2.4 2123 4.5	**8** 0044 2.1 Su 0753 4.3 1329 2.4 ☾ 2014 4.3	**23** 0324 2.0 M 0948 4.6 1601 2.5 2218 4.6	**8** 0326 1.9 W 0954 4.6 1558 2.2 2205 4.8	**23** 0443 1.9 Th 1109 4.7 1713 2.3 2331 4.8	**8** 0349 1.7 F 1014 4.8 1620 2.0 2227 5.0	**23** 0437 2.2 Sa 1103 4.5 1710 2.3 2327 4.6
9 0107 2.2 Sa 0815 4.2 1400 2.3 2042 4.2	**24** 0351 2.0 Su 1012 4.6 1637 2.4 2249 4.6	**9** 0226 2.2 M 0915 4.2 1516 2.5 2133 4.4	**24** 0439 1.8 Tu 1100 4.7 1714 2.3 2327 4.8	**9** 0437 1.6 Th 1059 4.8 1705 1.9 2308 5.0	**24** 0531 1.8 F 1158 4.8 1757 2.1	**9** 0454 1.5 Sa 1116 5.0 1725 1.7 2330 5.2	**24** 0533 2.1 Su 1156 4.7 1803 2.1
10 0254 2.3 Su 0940 4.2 1553 2.4 2207 4.2	**25** 0514 1.8 M 1132 4.8 1750 2.2 2359 4.8	**10** 0413 2.0 Tu 1035 4.4 1647 2.2 2249 4.6	**25** 0533 1.6 W 1153 4.9 1804 2.1	**10** 0531 1.3 F 1153 5.1 1757 1.6	**25** 0018 4.9 Sa 0613 1.7 1241 4.9 1836 1.9	**10** 0551 1.3 Su 1213 5.2 1823 1.5	**25** 0021 4.7 M 0620 1.9 1244 4.8 1848 1.9
11 0448 2.0 M 1107 4.3 1723 2.1 2325 4.5	**26** 0610 1.5 Tu 1232 5.0 1840 1.9	**11** 0521 1.6 W 1141 4.7 1747 1.8 2349 4.9	**26** 0020 5.0 Th 0616 1.5 1243 5.0 1841 1.9	**11** 0002 5.3 Sa 0619 1.0 1242 5.4 1844 1.1	**26** 0058 5.0 Su 0648 1.6 1319 5.0 1910 1.7	**11** 0028 5.4 M 0644 1.1 1305 5.4 1917 1.0	**26** 0108 4.8 Tu 0703 1.7 1328 4.9 1930 1.6
12 0555 1.6 Tu 1215 4.6 1820 1.7	**27** 0051 5.0 W 0652 1.2 1319 5.1 1917 1.6	**12** 0610 1.2 Th 1233 5.1 1832 1.4	**27** 0101 5.1 F 0651 1.3 1322 5.1 1913 1.6	**12** 0052 5.6 Su 0704 0.7 1328 5.6 1931 0.8	**27** 0134 5.1 M 0723 1.5 1353 5.1 1943 1.5	**12** 0123 5.5 Tu 0736 1.0 1355 5.5 ○ 2011 0.8	**27** 0153 4.9 W 0743 1.6 1408 5.1 2009 1.4
13 0025 4.8 W 0643 1.2 1307 5.0 1905 1.3	**28** 0134 5.2 Th 0728 1.0 1358 5.2 1947 1.4	**13** 0040 5.3 F 0651 0.8 1318 5.4 1913 1.0	**28** 0136 5.2 Sa 0723 1.2 1355 5.2 1939 1.5	**13** 0138 5.8 M 0751 0.6 1412 5.7 ○ 2021 0.6	**28** 0210 5.2 Tu 0756 1.5 1427 5.2 ● 2016 1.4	**13** 0216 5.6 W 0827 0.9 1445 5.6 2103 0.7	**28** 0234 5.1 Th 0820 1.5 1446 5.2 ● 2050 1.3
14 0113 5.2 Th 0726 0.7 1351 5.3 1945 1.0	**29** 0209 5.3 F 0759 0.9 1431 5.3 ● 2013 1.3	**14** 0122 5.6 Sa 0735 0.5 1359 5.7 ○ 1957 0.7	**29** 0206 5.3 Su 0751 1.2 1424 5.3 ● 2006 1.3	**14** 0225 5.9 Tu 0839 0.6 1458 5.7 2111 0.5	**29** 0246 5.2 W 0829 1.5 1500 5.3 2052 1.4	**14** 0307 5.7 Th 0914 1.0 1533 5.6 2152 0.7	**29** 0314 5.1 F 0858 1.4 1523 5.2 2129 1.2
15 0153 5.6 F 0807 0.4 1429 5.6 ○ 2025 0.7	**30** 0239 5.4 Sa 0826 0.9 1500 5.3 2037 1.1	**15** 0203 5.8 Su 0819 0.3 1438 5.8 2041 0.4	**30** 0235 5.3 M 0818 1.1 1452 5.3 2032 1.3	**15** 0315 5.9 W 0924 0.7 1545 5.7 2159 0.6	**30** 0324 5.2 Th 0904 1.4 1535 5.3 2130 1.4	**15** 0358 5.6 F 0959 1.2 1622 5.5 2238 0.8	**30** 0351 5.1 Sa 0937 1.4 1559 5.4 2210 1.1
			31 0306 5.3 Tu 0846 1.2 1522 5.3 2102 1.3				**31** 0429 5.3 Su 1015 1.4 1638 5.4 2249 1.1

Chart Datum: 2.90 metres below Ordnance Datum (Newlyn)

AREA 6—SE Scotland 377

FIRTH OF FORTH continued

TELEPHONE (031)
Hr Mr Leith 554-3661; CG & MRSC Crail 50666; HMC 554-2421; Forth Yacht Clubs Assn 337-5853. Weather, Glasgow 248-3451 or for Port Edgar Inverkeithing 412475; Marinecall 0898 500 452; Dr (Hosps) Dunfermline 35151, Leith 554-3211.

FACILITIES
SOUTH QUEENSFERRY EC Wednesday; **Port Edgar Marina** (300+8 visitors) Tel. 331-3330 Slip, AC, CH, D, El, ME, Sh, SM, Gas, FW, R, C (5 ton); **Forth Area Marine Electronics** Tel. 331 4343 El; **Port Edgar Marine Services** Tel. 331 1233 Sh, ME; **Bosun's Locker** Tel. 331-3875, D, ME, El, Sh, Gas, Gaz, CH; **Town** P, V, R, Bar. PO, Bank, Rly (Dalmeny), Air (Edinburgh).

NORTH QUEENSFERRY **Forth Yacht Marina** (100+some visitors) Tel. Inverkeithing 416101 (Hr Mr 412475) Slip, D (cans), ME, El, Sh, C (20 ton), CH, SM; **Town** D, V, R, Bar, PO, Bank, Rly (Inverkeithing), Air (Edinburgh).
GRANTON Hr Mr Tel. 552-3385 Slip, AB, FW; **Royal Forth YC** Tel. 552-3006 Slip, M, L, FW, C (5 ton), D, ME, El, Sh, Bar; **Forth Corinthian YC** Tel. 552-5959 Slip, M, L, Bar; Access HW∓3½; **Seaspan** Tel. 552-2224, Gas, CH; **Walter and Watson** Tel. 552-6695, ME; **J C Forbes** Tel. 552-8201, Sh; **Town** D, P, V, R, Bar. PO, Bank, Rly, Air (Edinburgh).
EDINBURGH **Chattan Shipping Services** Tel. 557-0988, ACA.

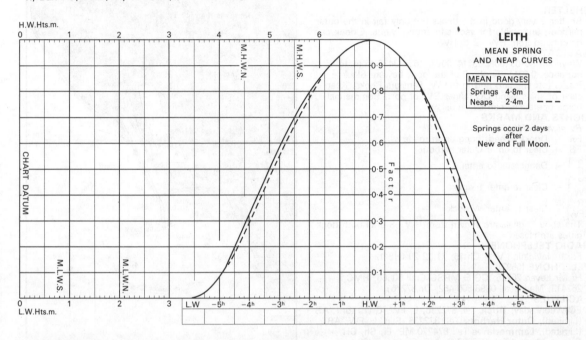

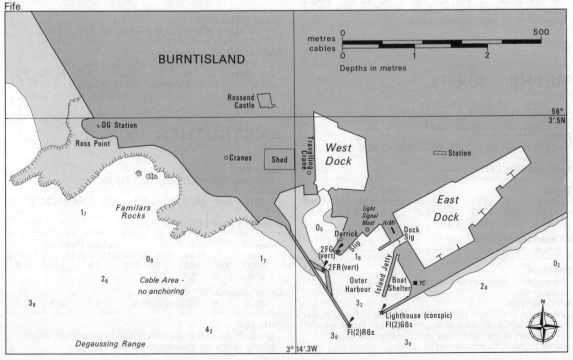

BURNTISLAND 10-6-10
Fife

BURNTISLAND continued

CHARTS
Admiralty 739; Imray C27; OS 66
TIDES
+0340 Dover; ML 2.9; Duration 0625; Zone 0 (GMT).

Standard Port LEITH (←)

Times				Height (metres)			
HW		LW		MHWS	MHWN	MLWN	MLWS
0300	0900	0300	0900	5.6	4.5	2.1	0.8
1500	2100	1500	2100				

Differences BURNTISLAND
+0002 −0002 −0002 −0004 0.0 0.0 0.0 0.0

SHELTER
Shelter is very good in the docks but only fair in the outer harbour; unsuitable for yachts in strong winds. E dock can be entered from HW−2 to HW.
NAVIGATION
Waypoint 56°03′.00N 03°14′.00W, 163°/343° from/to entrance, 0.23M. To the E of the port, beware Black Rocks (off chartlet) and to the W, Familars Rocks. Keep clear of ships using DG ranges SW of port. Commercial barge operations can cause delays.
LIGHTS AND MARKS
By night and day
FR — Docks closed, bring up in roads
FG — Clear to enter Outer harbour
FR / FG } — Dangerous to enter
FR / FW } — Clear to enter E dock
FR / 2FW } — Clear to enter W dock

The above signals are shown from HW−3 until both dock gates are closed.
RADIO TELEPHONE
Forth Navigation VHF Ch 16; 71 12 20 (H24).
TELEPHONE (0592)
Hr Mr Leven 26725; CG & MRSC Crail 50666; HMC 262413; Marinecall 0898 500 452; Dr 872761.
FACILITIES
EC Wednesday; **Dock** Tel. 872236, L, FW, C (10 ton), AB (Limited); **Outer Harbour** Tel. 872236, Slip, L, FW, AB (Limited); **Lammerlaws** Tel. 874270 ME, El, Sh, CH; **Briggs Marine** Tel. 872939 D, ME, El, Sh, C (30 ton); **Burntisland YC** Tel. 873375, M, L, Bar; **Starlyburn BY** Tel. 872939, Slip, L, FW, ME, EL, Sh, C, (25 ton); **Pop's Stores** Tel. 873524, Gas; **Town** P, D, ME, El, C, V, R, Bar. PO; Bank; Rly; Air (Edinburgh).

METHIL 10-6-11
Fife

CHARTS
Admiralty 739; Imray C27; OS 59
TIDES
+0330 Dover; ML 3.2; Duration 0615; Zone 0 (GMT).

Standard Port LEITH (←)

Times				Height (metres)			
HW		LW		MHWS	MHWN	MLWN	MLWS
0300	0900	0300	0900	5.6	4.5	2.1	0.8
1500	2100	1500	2100				

Differences METHIL
−0006 −0012 −0007 −0001 −0.1 −0.1 −0.1 −0.1

SHELTER
Principally a commercial port but good shelter in emergency.
NAVIGATION
Waypoint 56°10′.50N 03°00′.00W, 140°/320° from/to pier head Lt, 0.34M. Beware silting. A sand bar forms rapidly to seaward of the lighthouse and the dredged depth cannot always be maintained.

LIGHTS AND MARKS
By day and night
FR over FG — Dangerous to enter. Bring up in roads.
FR over FW — Clear to enter No 2 dock
FR — Remain in roads until another signal is made.
RADIO TELEPHONE
VHF Ch 16; 14 (HW−3 to HW).

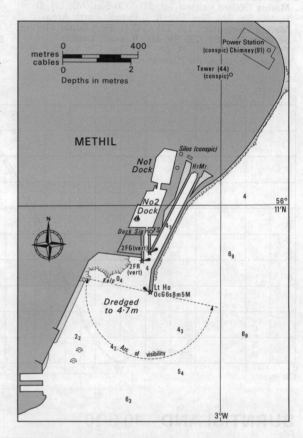

TELEPHONE (0592)
Hr Mr Leven 26725; CG & MRSC Crail 50666; HMC 262413; Marinecall 0898 500 452; Dr Leven 26913.
FACILITIES
EC Thursday; **Harbour** 2 Docks; C (4 ton); **Jas Donaldson** Tel. 26118, Gas; **Town** P, D, V, R, Bar, PO; Bank; Rly (bus to Markinch); Air (Edinburgh).

ANSTRUTHER 10-6-12
Fife

CHARTS
Admiralty 734, 175; Imray C27; OS 59
TIDES
+0315 Dover; ML 3.2; Duration 0620; Zone 0 (GMT).

Standard Port LEITH (←)

Times				Height (metres)			
HW		LW		MHWS	MHWN	MLWN	MLWS
0300	0900	0300	0900	5.6	4.5	2.1	0.8
1500	2100	1500	2100				

Differences ANSTRUTHER
−0010 −0035 −0020 −0020 −0.1 −0.1 −0.1 −0.1

SHELTER
Good shelter but entry should not be attempted when strong winds are blowing from the E and S.

ANSTRUTHER continued

NAVIGATION
Waypoint 56°12'.60N 02°42'.10W, 198°/018° from/to entrance, 0.60M. Harbour dries. Beware lobster pots and fishing vessels during the season (September to May).

LIGHTS AND MARKS
Conspic Lt Tr on W Pier. Ldg Lts 018°, both FG.

RADIO TELEPHONE
Forth Navigation Service. See 10.6.9.

TELEPHONE (0333)
Hr Mr 310836; CG & MRSC Crail 50666; HMC Dundee 22412; Marinecall 0898 500 452; Dr 310352.

FACILITIES
EC Wednesday; **Fisherman's Mutual Association** Tel. Pittenweem 311263, CH, D; **Harbour** Slip, M, FW, CH, AC, AB; **Chandlers** Tel. 311263, CH; **Christie & Co** Tel. 311339, ME; **Gray + Pringle** Tel. 310508, Gas, Gaz; **R. R. Bett** Tel. 31369, El; **Town** PO; Bank; Rly (bus Cupar or Leuchars); Air (Edinburgh, Turnhouse or Dundee Riverside).

RIVER TAY 10-6-13
Fife/Angus

CHARTS
Admiralty 1481, 190; OS 54, 59

TIDES
+ 0350 Dover; ML 3.0; Duration 0610; Zone 0 (GMT).

Standard Port ABERDEEN (→)

Times				Height (metres)			
HW		LW		MHWS	MHWN	MLWN	MLWS
0000	0600	0100	0700	4.3	3.4	1.6	0.6
1200	1800	1300	1900				
Differences BAR							
+0100	+0057	+0100	+0110	+0.9	+0.8	+0.3	+0.1
DUNDEE							
+0132	+0129	+0125	+0150	+1.0	+0.9	+0.3	+0.1
NEWBURGH							
+0212	+0203	+0249	+0337	−0.2	−0.4	−1.1	−0.5

SHELTER
Good shelter in the Tay estuary but the entrance is dangerous in strong winds from SE to E or in an on-shore swell. Tayport is best place for yachts on passage, access HW∓2½. Dundee docks (commercial) gates open HW−2 to HW. Anchorages as shown on chartlet; the one off the city is exposed and landing difficult. There are anchorages up river at Balmerino, Newburgh, Inchyra and Perth.

NAVIGATION
Waypoint Tay Fairway safe water buoy, LFl 10s, Whis, 56°28'.62N 02°37'.15W, 060°/240° from/to Middle Bar buoys, 1.0M. The river is navigable right up to Perth. Entrance to river across Abertay sands is forbidden due to stakes. Channel is well buoyed but caution should be exercised as depths are liable to change. Tayport harbour dries except W side of NE pier; S side is full of yacht moorings.

LIGHTS AND MARKS
There are no leading lights but channel is well marked. N side of entrance marked by Old High Light House on Buddon Ness (conspic) with Old Low Light House below it. Note:- Area is administered by the Dundee Port Authority.

RADIO TELEPHONE
Dundee Harbour Radio VHF Ch 16; 10 11 **12** 13 14 (H24). Perth Harbour Ch 09.

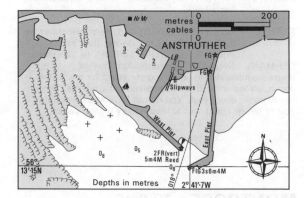

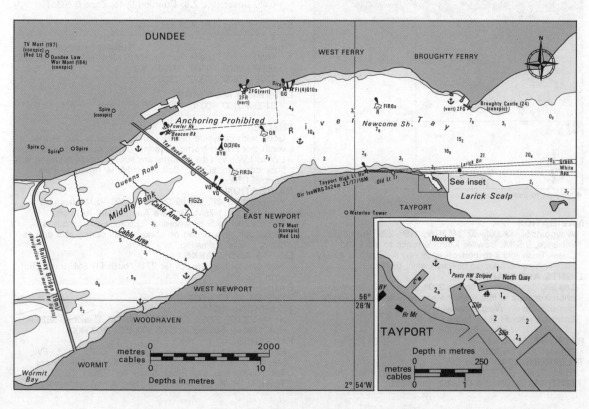

RIVER TAY continued

TELEPHONE (0382)
Hr Mr (Dundee) 24121; Perth Hr Mr Perth 24056; MRSC Crail 50666; HMC 22412; Tayport Boatowners' Assn 553679; Marinecall 0898 500 452; Dr 21953; Hosp 23125.

FACILITIES
NORTH BANK: EC Dundee Wednesday; **Camperdown Dock** FW, ME, El, C (8 ton), AB; **Victoria Dock** FW, ME, C (8 ton), AB; **Royal Tay YC** (Broughty Ferry) Tel. 77516, M, L, R, Bar; **Allison-Gray** 27444, CH, ACA, SM; **Scotcraig Boat Co** Tel. 553230, M, L, ME, El, Sh, C (2 ton), CH; **Broughty Ferry** Slip, M; **Wm C. Scott** Tel. 739192 BY, CH, El, ME, Sh; **Dundee Town** P, D, CH, V, R, Bar, PO, Bank, Rly, Air.
SOUTH BANK: **Tayport Harbour** Tel. 553679 Slip, L, FW, AB, AC; **Wormit Boat Club** Tel. 541400 Slip, L, Bar.

ARBROATH 10-6-14
Angus

CHARTS
Admiralty 1438, 190; OS 54
TIDES
+0335 Dover; ML 2.9; Duration 0620; Zone 0 (GMT).

Standard Port ABERDEEN (→)

Times				Height (metres)			
HW		LW		MHWS	MHWN	MLWN	MLWS
0000	0600	0100	0700	4.3	3.4	1.6	0.6
1200	1800	1300	1900				

Differences ARBROATH
+0056 +0037 +0034 +0055 +0.7 +0.7 +0.2 +0.1

SHELTER
Good, especially in Dock, but entrance can be dangerous in moderate SE swell. Dock gates normally remain open, but will be closed on request. Entry should not be attempted LW ∓ 2½.

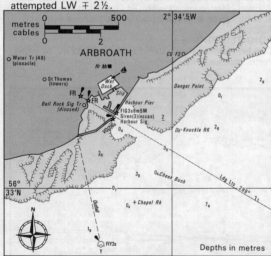

⚠ Report to Senior Harbour Assistant

NAVIGATION
Waypoint 56°33'.00N 02°34'.10W, 119°/299° from/to entrance, 0.50M. Beware Knuckle rocks to stbd and Cheek Bush rocks to port on entering. Arbroath is a very busy fishing harbour so beware heavy traffic.
LIGHTS AND MARKS
Leading lights (FR) in line at 299°. By day St Thomas' church towers in line with the gap between Lt Ho on N pier and Bn on West breakwater.
Entry signals:- Fl G — Entry safe
FR — Entry dangerous
RADIO TELEPHONE
Contact AFA on VHF Ch 16.
TELEPHONE (0241)
Hr Mr 72166; CG 72113; MRSC Crail 50666; HMC Dundee 22412; Marinecall 0898 500 452; Dr 76836.

FACILITIES
EC Wednesday; **Pier** Slip, D, L, FW, AB; **Mackays BY** Tel. 72879, Slip, L, ME, El, Sh, C (8 ton); **Arbroath Fisherman's Association** Tel. 72928, M, D, Gas, CH; **Gerrards** Tel. 73177, Slip, Sh; **W. M. Teviotdale** Tel. 73104, ME; **Elbar Shop** Tel. 73141, Gas; **A. Copland** Tel. 73212 ME; **J. Connely** Tel. 70121 ME; **Town** P, D, V, R, Bar. PO; Bank; Rly; Air (Dundee).

AGENTS NEEDED
There are a number of ports where we need agents, particularly in France.
ENGLAND Swale, Havengore, Berwick.
SCOTLAND Firth of Forth, Scrabster, Mallaig, Loch Sunart, Loch Aline.
IRELAND Kilrush, Wicklow, Westport/Clew Bay.
FRANCE Arcachon, Seudre R, Ile d'Oleron, Rochfort, Ile de Re, St. Giles-Croix-de-Vie, Ile d'Yeu, Pouliguen, Le Croisic, La Forêt, Ile de Bréhat.
GERMANY Norderney, Dornumer-Accumersiel.
If you are interested in becoming our agent for any of the above, please write to the editors and get your free copy of the Almanac every year. You do not have to be resident in a port to be the agent, but at least a fairly regular visitor.

MONTROSE 10-6-15
Angus

CHARTS
Admiralty 1438, 190; OS 54
TIDES
+0320 Dover; ML 2.8; Duration 0645; Zone 0 (GMT).

Standard Port ABERDEEN (→)

Times				Height (metres)			
HW		LW		MHWS	MHWN	MLWN	MLWS
0000	0600	0100	0700	4.3	3.4	1.6	0.6
1200	1800	1300	1900				

Differences MONTROSE
+0100 +0100 +0030 +0040 +0.5 +0.5 +0.3 +0.1

SHELTER
Good with quayside berths usually available but beware wash from other traffic. Channel depth min 5m; best entry LW to LW+1; tidal streams are strong (up to 6 kn). It is a busy commercial port. Call up Hr Mr or report to Port Control Office.
NAVIGATION
Waypoint 56°42'.20N 02°25'.00W, 091°/271° from/to front Ldg Lt, 1.25M. Beware Annat Bank to N and Scurdie Rks to S on entering. Entrance to harbour is dangerous with strong on-shore winds during ebb tide as heavy overfalls develop.
LIGHTS AND MARKS
Two sets of Ldg Lts. Outer 271°, both FR 5M. Inner 265°, both FG, 5M.
RADIO TELEPHONE
VHF Ch 16; 12 (H24).
TELEPHONE (0674)
Hr Mr 72302, 73153; CG 72101; MRCC Aberdeen 592334; HMC 74444; Marinecall 0898 500 452; Dr 73400.
FACILITIES
EC Wednesday; **Riverside Quay** Tel. 72302, D, L, FW, ME, El, C (1½ to 40 ton), CH, AB; **South Esk Boats** Tel. 76580, Slip, Sh; **Elbar Shop** Tel. 73981, Gas; **Town** V, R, P, Bar. PO; Bank; Rly; Air (Aberdeen).

MONTROSE continued

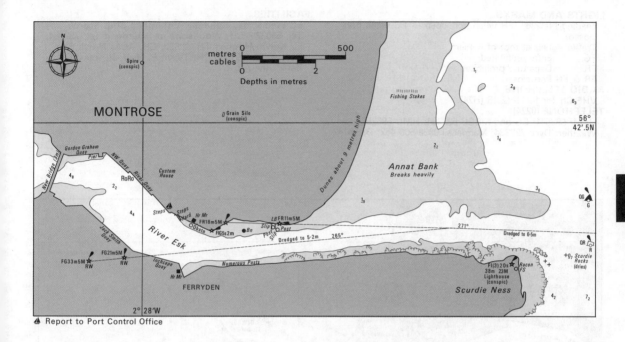

STONEHAVEN 10-6-16
Kincardine

CHARTS
Admiralty 1438, 210; OS 45
TIDES
+0235 Dover; ML 2.7; Duration 0620; Zone 0 (GMT).

Standard Port ABERDEEN (→)

Times				Height (metres)			
HW		LW		MHWS	MHWN	MLWN	MLWS
0000	0600	0100	0700	4.3	3.4	1.6	0.6
1200	1800	1300	1900				

Differences STONEHAVEN
+0013 +0008 +0013 +0009 +0.2 +0.2 +0.1 0.0

SHELTER
Good shelter especially from S through W to N winds. Inner harbour dries; storm gates closed in severe weather. Speed limit in harbour 3 kn.

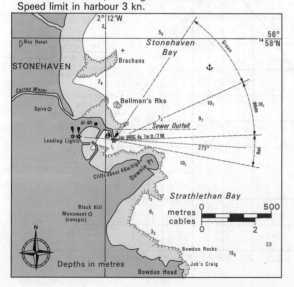

NAVIGATION
Waypoint 56°57'.60N 02°10'.50W, 090°/270° from/to breakwater Lt, 0.77M. Give Downie Pt a wide berth. Do not enter in strong on-shore winds.
LIGHTS AND MARKS
When harbour is closed, a G Lt is displayed.
RADIO TELEPHONE
None.
TELEPHONE (0569)
Hr Mr 65323; CG & MRCC Aberdeen 592334; HMC Aberdeen 586258; Marinecall 0898 500 452; Dr 62945.
FACILITIES
EC Wednesday; **Aberdeen and Stonehaven YC** Slip, Bar; **Harbour** Tel. 65323, M, L, FW, AC, Slip, C (1.5 ton), AB; **Town** LB, V, R, Bar. PO; Bank; Rly; Air (Aberdeen).

ABERDEEN 10-6-17
Aberdeen

CHARTS
Admiralty 210, 1446; OS 38
TIDES
+0220 Dover; ML 2.5; Duration 0620; Zone 0 (GMT).

Note: Aberdeen is a Standard Port and tidal predictions for each day of the year are given below.

SHELTER
Good shelter in harbour. Entry should not be attempted when winds are strong from NE through E to ESE. Yachts are not encouraged and it is very expensive. A busy commercial port and although all facilities are available, they are geared to commercial traffic. There are no special berths for yachts, so using commercial berths entails all the usual obligations and risks. Yachtsmen intending to call should contact the Hr Mr beforehand. Harbour open at all states of tide. Temporary anchorage in Nigg Bay, S of Girdle Ness, but exposed to E and beware salmon nets.
NAVIGATION
Waypoint Fairway safe water buoy, LFl 10s, Racon, 57°09'.34N 02°01'.83W, 058°/238° from/to N pier Lt, 1.1M. Give Girdle Ness a berth of at least ¼M (more in bad weather) and do not pass close round pier heads. Strong tidal streams and, with river in spate, possible overfalls. Channel dredged to 6m on leading line.

ABERDEEN continued

LIGHTS AND MARKS
Leading Lts 236° (FR when port open — FG when port closed).
Traffic signals at root of N Pier:
FG   Entry prohibited
FR   Departure prohibited
FR & FG Port closed

RADIO TELEPHONE
VHF Ch 16; 10 11 12 13 (H24).

TELEPHONE (0224)
Hr Mr 592571; CG and MRCC 592334; HMC 586258;
Weather: Dyce 722334; Marinecall 0898 500 452; Dr 55142.

FACILITIES
EC Wednesday/Saturday; **Enterprise Ship Stores** Tel. 590329, CH; **Woodsons of Aberdeen** Tel. 722884, El; **Kelvin Hughes** Tel. 20823, CH, ACA; **North Sea Stores** Tel. 591896 CH; **Town** P, D, V, R, Bar. PO; Bank; Rly; Air.

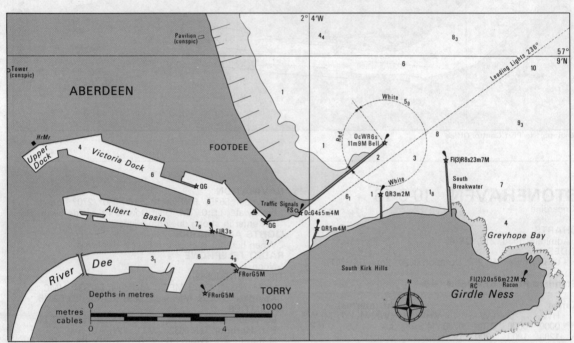

After berthing, report to harbour control at HrMr

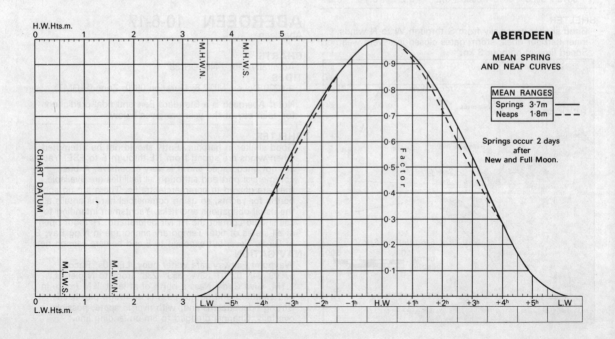

ABERDEEN

MEAN SPRING AND NEAP CURVES

MEAN RANGES
Springs 3·7m
Neaps   1·8m

Springs occur 2 days after New and Full Moon.

SCOTLAND, EAST COAST — ABERDEEN

Lat 57°09′ N Long 2°04′ W

TIMES AND HEIGHTS OF HIGH AND LOW WATERS

YEAR 1989

TIME ZONE UT(GMT)
For Summer Time add ONE hour in non-shaded areas

JANUARY

Day	Time	m	Day	Time	m
1 Su	0114 / 0744 / 1327 / 1947	1·5 / 3·3 / 2·0 / 3·4	16 M	0151 / 0820 / 1408 / 2034	1·3 / 3·4 / 1·6 / 3·7
2 M	0219 / 0850 / 1442 / 2057	1·6 / 3·3 / 2·0 / 3·4	17 Tu	0305 / 0931 / 1531 / 2153	1·4 / 3·5 / 1·6 / 3·6
3 Tu	0324 / 0953 / 1553 / 2203	1·6 / 3·4 / 1·9 / 3·4	18 W	0417 / 1037 / 1645 / 2305	1·4 / 3·6 / 1·4 / 3·7
4 W	0421 / 1047 / 1654 / 2302	1·5 / 3·5 / 1·7 / 3·5	19 Th	0518 / 1132 / 1747	1·4 / 3·8 / 1·2
5 Th	0512 / 1132 / 1743 / 2353	1·4 / 3·8 / 1·5 / 3·7	20 F	0004 / 0608 / 1219 / 1836	3·8 / 1·3 / 3·9 / 1·0
6 F	0557 / 1211 / 1827	1·3 / 4·0 / 1·2	21 Sa ○	0053 / 0650 / 1300 / 1919	3·8 / 1·3 / 4·1 / 0·8
7 Sa ●	0041 / 0639 / 1250 / 1909	3·9 / 1·2 / 4·2 / 1·0	22 Su	0135 / 0727 / 1337 / 1957	4·0 / 1·2 / 4·3 / 0·7
8 Su	0126 / 0721 / 1330 / 1951	4·1 / 1·1 / 4·3 / 0·8	23 M	0212 / 0801 / 1409 / 2030	4·0 / 1·2 / 4·3 / 0·7
9 M	0211 / 0804 / 1411 / 2034	4·2 / 1·1 / 4·4 / 0·6	24 Tu	0247 / 0833 / 1442 / 2103	4·0 / 1·1 / 4·2 / 0·7
10 Tu	0256 / 0846 / 1453 / 2118	4·3 / 1·0 / 4·5 / 0·5	25 W	0319 / 0904 / 1514 / 2135	4·0 / 1·2 / 4·1 / 0·8
11 W	0342 / 0928 / 1538 / 2203	4·3 / 1·0 / 4·5 / 0·6	26 Th	0352 / 0935 / 1546 / 2209	3·9 / 1·2 / 4·1 / 0·9
12 Th	0428 / 1013 / 1626 / 2252	4·1 / 1·1 / 4·4 / 0·7	27 F	0426 / 1009 / 1621 / 2242	3·8 / 1·3 / 4·0 / 1·0
13 F	0519 / 1059 / 1718 / 2344	4·0 / 1·2 / 4·3 / 0·8	28 Sa	0502 / 1044 / 1701 / 2322	3·6 / 1·4 / 3·8 / 1·2
14 Sa ☽	0612 / 1153 / 1815	3·7 / 1·4 / 4·1	29 Su	0544 / 1125 / 1744	3·4 / 1·6 / 3·6
15 Su	0043 / 0712 / 1253 / 1920	1·1 / 3·6 / 1·5 / 3·9	30 M ☾	0005 / 0634 / 1215 / 1841	1·5 / 3·3 / 1·8 / 3·4
			31 Tu	0100 / 0735 / 1323 / 1954	1·7 / 3·2 / 1·9 / 3·2

FEBRUARY

Day	Time	m	Day	Time	m
1 W	0212 / 0847 / 1454 / 2118	1·8 / 3·2 / 1·9 / 3·2	16 Th	0412 / 1020 / 1645 / 2308	1·7 / 3·4 / 1·4 / 3·5
2 Th	0334 / 0959 / 1620 / 2235	1·8 / 3·4 / 1·8 / 3·4	17 F	0516 / 1120 / 1743	1·6 / 3·6 / 1·1
3 F	0444 / 1057 / 1722 / 2336	1·6 / 3·6 / 1·5 / 3·6	18 Sa	0003 / 0601 / 1207 / 1827	3·6 / 1·4 / 3·8 / 0·9
4 Sa	0537 / 1146 / 1810	1·4 / 3·9 / 1·1	19 Su	0045 / 0639 / 1245 / 1902	3·8 / 1·3 / 4·0 / 0·7
5 Su	0025 / 0624 / 1229 / 1853	3·9 / 1·2 / 4·2 / 0·8	20 M ○	0119 / 0712 / 1319 / 1935	3·9 / 1·1 / 4·1 / 0·6
6 M ●	0110 / 0706 / 1312 / 1935	4·1 / 1·0 / 4·4 / 0·5	21 Tu	0149 / 0741 / 1349 / 2005	4·0 / 1·0 / 4·2 / 0·6
7 Tu	0154 / 0747 / 1354 / 2016	4·3 / 0·8 / 4·6 / 0·3	22 W	0220 / 0809 / 1419 / 2034	4·0 / 0·9 / 4·1 / 0·6
8 W	0236 / 0827 / 1436 / 2058	4·4 / 0·7 / 4·7 / 0·2	23 Th	0249 / 0839 / 1447 / 2104	4·0 / 0·9 / 4·2 / 0·6
9 Th	0318 / 0907 / 1519 / 2142	4·4 / 0·7 / 4·7 / 0·3	24 F	0318 / 0907 / 1518 / 2134	3·9 / 0·9 / 4·1 / 0·8
10 F	0402 / 0949 / 1604 / 2227	4·2 / 0·7 / 4·6 / 0·5	25 Sa	0348 / 0936 / 1549 / 2203	3·8 / 1·0 / 3·9 / 0·9
11 Sa	0447 / 1031 / 1654 / 2313	4·0 / 0·9 / 4·4 / 0·8	26 Su	0420 / 1009 / 1624 / 2237	3·7 / 1·2 / 3·8 / 1·2
12 Su	0536 / 1120 / 1749	3·7 / 1·1 / 4·0	27 M	0454 / 1045 / 1705 / 2313	3·5 / 1·4 / 3·6 / 1·4
13 M	0007 / 0632 / 1218 / 1855	1·2 / 3·5 / 1·4 / 3·7	28 Tu ☾	0536 / 1130 / 1758	3·4 / 1·6 / 3·4
14 Tu ☽	0112 / 0741 / 1338 / 2018	1·5 / 3·3 / 1·6 / 3·4			
15 W	0240 / 0903 / 1519 / 2150	1·7 / 3·3 / 1·6 / 3·4			

MARCH

Day	Time	m	Day	Time	m
1 W	0001 / 0632 / 1232 / 1913	1·7 / 3·2 / 1·8 / 3·2	16 Th	0222 / 0833 / 1514 / 2148	2·0 / 3·2 / 1·5 / 3·2
2 Th	0112 / 0748 / 1405 / 2046	1·9 / 3·2 / 1·9 / 3·1	17 F	0404 / 0957 / 1633 / 2259	1·9 / 3·3 / 1·3 / 3·4
3 F	0254 / 0912 / 1548 / 2213	1·9 / 3·3 / 1·7 / 3·3	18 Sa	0502 / 1101 / 1723 / 2346	1·7 / 3·5 / 1·1 / 3·6
4 Sa	0421 / 1024 / 1657 / 2316	1·7 / 3·5 / 1·3 / 3·7	19 Su	0543 / 1146 / 1803	1·4 / 3·7 / 0·9
5 Su	0519 / 1119 / 1747	1·4 / 3·8 / 0·9	20 M	0022 / 0617 / 1222 / 1836	3·7 / 1·2 / 3·8 / 0·7
6 M	0005 / 0604 / 1207 / 1831	4·0 / 1·1 / 4·1 / 0·5	21 Tu	0053 / 0648 / 1255 / 1907	3·8 / 1·0 / 3·9 / 0·6
7 Tu ●	0049 / 0645 / 1249 / 1913	4·2 / 0·8 / 4·4 / 0·2	22 W ○	0123 / 0716 / 1324 / 1937	4·0 / 0·9 / 4·0 / 0·6
8 W	0130 / 0726 / 1331 / 1954	4·4 / 0·6 / 4·7 / 0·1	23 Th	0151 / 0744 / 1354 / 2005	4·0 / 0·8 / 4·1 / 0·6
9 Th	0211 / 0805 / 1413 / 2036	4·4 / 0·4 / 4·8 / 0·1	24 F	0220 / 0812 / 1422 / 2033	4·0 / 0·8 / 4·1 / 0·6
10 F	0253 / 0844 / 1457 / 2117	4·4 / 0·4 / 4·7 / 0·2	25 Sa	0247 / 0840 / 1453 / 2101	3·9 / 0·9 / 4·0 / 0·8
11 Sa	0334 / 0925 / 1543 / 2159	4·2 / 0·5 / 4·6 / 0·5	26 Su	0315 / 0911 / 1524 / 2131	3·9 / 1·0 / 3·9 / 1·0
12 Su	0417 / 1007 / 1633 / 2244	4·0 / 0·7 / 4·2 / 0·9	27 M ☾	0345 / 0943 / 1600 / 2202	3·8 / 1·1 / 3·7 / 1·2
13 M	0504 / 1057 / 1729 / 2333	3·7 / 1·0 / 3·9 / 1·4	28 Tu	0417 / 1020 / 1644 / 2240	3·6 / 1·3 / 3·5 / 1·5
14 Tu ☽	0557 / 1156 / 1839	3·4 / 1·3 / 3·5	29 W	0458 / 1106 / 1740 / 2327	3·5 / 1·4 / 3·2 / 1·7
15 W	0038 / 0706 / 1321 / 2009	1·8 / 3·2 / 1·5 / 3·2	30 Th ☾	0554 / 1208 / 1856	3·3 / 1·6 / 3·2
			31 F	0041 / 0710 / 1340 / 2026	1·9 / 3·2 / 1·7 / 3·2

APRIL

Day	Time	m	Day	Time	m
1 Sa	0226 / 0836 / 1518 / 2149	1·9 / 3·3 / 1·5 / 3·4	16 Su	0428 / 1026 / 1649 / 2312	1·7 / 3·4 / 1·1 / 3·5
2 Su	0353 / 0952 / 1627 / 2249	1·7 / 3·5 / 1·1 / 3·7	17 M	0509 / 1113 / 1729 / 2349	1·4 / 3·5 / 0·9 / 3·6
3 M	0451 / 1051 / 1719 / 2339	1·3 / 3·7 / 0·7 / 4·0	18 Tu	0546 / 1151 / 1804	1·2 / 3·7 / 0·8
4 Tu	0537 / 1140 / 1804	1·0 / 4·1 / 0·4	19 W	0022 / 0618 / 1227 / 1836	3·8 / 1·0 / 3·8 / 0·7
5 W	0024 / 0619 / 1225 / 1848	4·2 / 0·7 / 4·4 / 0·2	20 Th	0053 / 0648 / 1257 / 1906	3·9 / 0·9 / 3·9 / 0·7
6 Th ●	0106 / 0700 / 1309 / 1930	4·3 / 0·5 / 4·6 / 0·1	21 F ○	0124 / 0719 / 1328 / 1935	3·9 / 0·9 / 3·9 / 0·7
7 F	0147 / 0741 / 1354 / 2012	4·3 / 0·4 / 4·7 / 0·1	22 Sa	0152 / 0748 / 1359 / 2005	3·9 / 0·9 / 3·9 / 0·8
8 Sa	0227 / 0823 / 1439 / 2053	4·3 / 0·4 / 4·6 / 0·4	23 Su	0220 / 0819 / 1433 / 2034	3·9 / 0·9 / 3·9 / 0·9
9 Su	0310 / 0905 / 1528 / 2135	4·1 / 0·5 / 4·4 / 0·7	24 M	0251 / 0851 / 1510 / 2105	3·8 / 1·0 / 3·8 / 1·1
10 M	0352 / 0952 / 1620 / 2219	4·0 / 0·7 / 4·0 / 1·1	25 Tu	0319 / 0928 / 1550 / 2141	3·8 / 1·1 / 3·7 / 1·3
11 Tu	0438 / 1042 / 1719 / 2306	3·7 / 1·0 / 3·7 / 1·5	26 W	0355 / 1009 / 1640 / 2223	3·7 / 1·2 / 3·5 / 1·5
12 W	0530 / 1144 / 1828	3·5 / 1·2 / 3·3	27 Th	0440 / 1059 / 1739 / 2316	3·6 / 1·3 / 3·4 / 1·7
13 Th	0010 / 0635 / 1310 / 1954	1·9 / 3·3 / 1·4 / 3·2	28 F ☾	0537 / 1203 / 1849	3·5 / 1·4 / 3·3
14 F	0154 / 0758 / 1449 / 2121	2·1 / 3·2 / 1·4 / 3·3	29 Sa	0029 / 0649 / 1324 / 2005	1·9 / 3·4 / 1·4 / 3·3
15 Sa	0331 / 0922 / 1559 / 2226	1·9 / 3·2 / 1·3 / 3·3	30 Su	0159 / 0806 / 1447 / 2118	1·8 / 3·4 / 1·2 / 3·5

Chart Datum: 2.25 metres below Ordnance Datum (Newlyn)

SCOTLAND, EAST COAST — ABERDEEN

Lat 57°09' N   Long 2°04' W

TIMES AND HEIGHTS OF HIGH AND LOW WATERS

YEAR 1989

TIME ZONE UT (GMT)
For Summer Time add ONE hour in non-shaded areas

MAY

Time	m		Time	m
1 M 0317 0918 1555 2219	1.6 3.6 1.0 3.7	**16** Tu	0423 1028 1648 2308	1.5 3.4 1.1 3.5
2 Tu 0416 1019 1648 2311	1.3 3.8 0.7 3.9	**17** W	0506 1113 1727 2347	1.3 3.5 1.0 3.6
3 W 0506 1112 1737 2357	1.2 4.1 0.4 4.0	**18** Th	0544 1154 1804	1.2 3.6 0.9
4 Th 0553 1201 1824	0.7 4.3 0.3	**19** F	0022 0619 1231 1836	3.7 1.1 3.7 0.9
5 F● 0041 0636 1249 1907	4.1 0.6 4.4 0.3	**20** Sa○	0055 0653 1306 1909	3.8 1.0 3.8 0.9
6 Sa 0124 0721 1337 1951	4.3 0.5 4.4 0.4	**21** Su	0126 0728 1342 1941	3.9 1.0 3.8 1.0
7 Su 0206 0806 1426 2033	4.2 0.5 4.3 0.6	**22** M	0157 0804 1420 2015	3.9 1.0 3.8 1.1
8 M 0249 0853 1518 2115	4.1 0.5 4.1 1.0	**23** Tu	0229 0840 1501 2051	3.9 1.0 3.8 1.2
9 Tu 0332 0942 1612 2159	4.0 0.7 3.9 1.3	**24** W	0304 0919 1548 2132	3.8 1.0 3.7 1.3
10 W 0417 1034 1708 2247	3.8 0.9 3.6 1.6	**25** Th	0345 1004 1637 2219	3.9 1.0 3.7 1.5
11 Th 0508 1133 1810 2344	3.6 1.1 3.4 1.8	**26** F	0431 1055 1732 2312	3.8 1.1 3.6 1.6
12 F☽ 0605 1242 1917	3.4 1.3 3.2	**27** Sa	0527 1154 1832	3.7 1.1 3.5
13 Sa 0100 0714 1359 2026	1.9 3.3 1.4 3.2	**28** Su☾	0015 0631 1303 1937	1.6 3.6 1.1 3.5
14 Su 0226 0829 1508 2131	1.9 3.2 1.3 3.2	**29** M	0126 0738 1413 2043	1.6 3.6 1.1 3.5
15 M 0332 0935 1603 2223	1.7 3.3 1.2 3.3	**30** Tu	0236 0846 1519 2145	1.5 3.7 0.9 3.6
		31 W	0339 0949 1619 2241	1.3 3.9 0.8 3.7

JUNE

Time	m		Time	m
1 Th 0435 1048 1712 2332	1.1 4.0 0.7 3.8	**16** F	0509 1120 1729 2349	1.4 3.5 1.2 3.6
2 F 0529 1143 1803	0.9 4.1 0.6	**17** Sa	0553 1204 1808	1.3 3.6 1.2
3 Sa● 0019 0619 1236 1849	4.0 0.7 4.2 0.6	**18** Su	0025 0634 1248 1846	3.7 1.2 3.7 1.1
4 Su 0104 0710 1328 1934	4.1 0.6 4.2 0.8	**19** M○	0100 0710 1328 1924	3.9 1.2 3.8 1.1
5 M 0148 0758 1420 2018	4.1 0.6 4.1 0.9	**20** Tu	0135 0751 1411 2002	4.0 0.9 3.9 1.1
6 Tu 0232 0846 1510 2100	4.1 0.6 4.0 1.1	**21** W	0212 0830 1453 2042	4.0 0.8 3.9 1.2
7 W 0314 0932 1559 2142	4.0 0.7 3.8 1.3	**22** Th	0250 0911 1538 2124	4.1 0.8 3.9 1.2
8 Th 0356 1019 1647 2226	3.9 0.8 3.7 1.5	**23** F	0332 0956 1624 2209	4.1 0.8 3.9 1.2
9 F 0441 1108 1734 2311	3.8 1.0 3.5 1.6	**24** Sa	0420 1044 1713 2257	4.0 0.8 3.8 1.3
10 Sa 0530 1200 1825	3.6 1.2 3.3	**25** Su	0511 1136 1807 2349	4.0 0.9 3.7 1.3
11 Su☽ 0004 0624 1256 1921	1.7 3.5 1.3 3.2	**26** M☾	0607 1234 1904	3.9 1.0 3.6
12 M 0106 0724 1359 2022	1.8 3.3 1.4 3.2	**27** Tu	0049 0709 1338 2006	1.4 3.8 1.0 3.5
13 Tu 0216 0830 1501 2122	1.7 3.3 1.4 3.2	**28** W	0155 0815 1446 2111	1.4 3.8 1.1 3.5
14 W 0322 0934 1557 2219	1.7 3.3 1.3 3.3	**29** Th	0305 0925 1552 2214	1.4 3.9 1.1 3.6
15 Th 0420 1030 1647 2308	1.6 3.4 1.2 3.4	**30** F	0414 1033 1654 2312	1.2 3.8 1.0 3.7

JULY

Time	m		Time	m
1 Sa 0516 1136 1749	1.0 3.9 1.0	**16** Su	0529 1143 1743 2356	1.4 3.5 1.0 3.7
2 Su 0004 0614 1234 1838	3.9 0.8 4.0 1.0	**17** M	0614 1229 1827	1.2 3.7 1.3
3 M● 0050 0704 1326 1923	4.0 0.7 4.0 1.1	**18** Tu○	0035 0655 1313 1907	3.9 1.0 3.9 1.2
4 Tu 0134 0751 1412 2005	4.1 0.6 4.0 1.1	**19** W	0113 0735 1355 1947	4.1 0.8 4.0 1.0
5 W 0215 0834 1456 2043	4.1 0.6 4.0 1.1	**20** Th	0152 0815 1436 2026	4.2 0.6 4.1 1.0
6 Th 0253 0915 1536 2119	4.1 0.6 3.9 1.2	**21** F	0232 0856 1518 2107	4.3 0.5 4.2 0.9
7 F 0331 0953 1614 2156	4.0 0.7 3.8 1.3	**22** Sa	0314 0938 1602 2148	4.4 0.5 4.1 0.9
8 Sa 0410 1033 1654 2234	3.9 0.8 3.6 1.4	**23** Su	0359 1023 1647 2231	4.4 0.5 4.0 1.0
9 Su 0449 1112 1734 2313	3.8 1.0 3.5 1.5	**24** M	0447 1111 1736 2319	4.3 0.7 3.8 1.1
10 M 0533 1156 1819 2349	3.6 1.2 3.3 1.6	**25** Tu☾	0540 1203 1831	4.1 0.9 3.6
11 Tu 0607 1234 1904	3.5 1.4 3.2	**26** W	0015 0641 1303 1933	1.3 3.9 1.1 3.4
12 W 0057 0721 1345 2015	1.7 3.3 1.5 3.2	**27** Th	0121 0751 1415 2043	1.4 3.7 1.3 3.4
13 Th 0208 0830 1453 2121	1.8 3.2 1.6 3.2	**28** F	0243 0911 1535 2155	1.5 3.6 1.4 3.5
14 F 0327 0942 1559 2221	1.8 3.2 1.6 3.3	**29** Sa	0407 1031 1647 2259	1.4 3.6 1.4 3.6
15 Sa 0434 1048 1655 2312	1.6 3.3 1.5 3.5	**30** Su	0516 1139 1744 2353	1.1 3.7 1.3 3.8
		31 M	0612 1232 1831	0.9 3.8 1.3

AUGUST

Time	m		Time	m
1 Tu● 0038 0657 1317 1912	4.0 0.7 3.9 1.1	**16** W	0008 0635 1252 1846	4.0 0.8 4.0 1.1
2 W 0117 0738 1357 1947	4.1 0.6 4.0 1.1	**17** Th○	0049 0714 1331 1926	4.3 0.6 4.2 0.9
3 Th 0154 0815 1433 2020	4.2 0.5 4.0 1.0	**18** F	0130 0754 1412 2004	4.5 0.4 4.3 0.7
4 F 0229 0849 1505 2051	4.2 0.5 4.0 1.0	**19** Sa	0209 0833 1451 2043	4.6 0.3 4.3 0.7
5 Sa 0301 0921 1538 2124	4.2 0.6 3.9 1.1	**20** Su	0251 0914 1534 2122	4.7 0.3 4.2 0.7
6 Su 0335 0953 1612 2156	4.1 0.8 3.8 1.2	**21** M	0335 0957 1617 2204	4.6 0.4 4.1 0.8
7 M 0409 1027 1647 2231	3.9 0.9 3.6 1.3	**22** Tu	0423 1042 1705 2252	4.4 0.7 3.8 1.0
8 Tu 0447 1104 1726 2309	3.8 1.1 3.5 1.5	**23** W☾	0516 1132 1800 2347	4.1 1.0 3.6 1.3
9 W 0530 1144 1812 2357	3.6 1.4 3.3 1.7	**24** Th	0619 1232 1904	3.8 1.4 3.4
10 Th 0624 1234 1910	3.3 1.6 3.2	**25** F	0100 0740 1354 2022	1.5 3.5 1.7 3.3
11 F 0102 0734 1342 2019	1.9 3.2 1.8 3.2	**26** Sa	0239 0912 1532 2142	1.6 3.4 1.8 3.4
12 Sa 0230 0900 1508 2132	1.9 3.1 1.9 3.3	**27** Su	0412 1037 1645 2249	1.4 3.5 1.7 3.6
13 Su 0403 1020 1626 2235	1.7 3.2 1.8 3.5	**28** M	0516 1137 1737 2340	1.1 3.7 1.5 3.8
14 M 0506 1120 1720 2325	1.5 3.4 1.5 3.7	**29** Tu	0603 1222 1817	0.9 3.8 1.3
15 Tu 0553 1210 1805	1.2 3.6 1.3	**30** W	0022 0642 1300 1852	4.0 0.7 4.0 1.1
		31 Th●	0059 0717 1333 1923	4.1 0.6 4.0 1.0

Chart Datum: 2.25 metres below Ordnance Datum (Newlyn)

AREA 6—SE Scotland

SCOTLAND, EAST COAST — ABERDEEN
Lat 57°09′ N Long 2°04′ W
TIMES AND HEIGHTS OF HIGH AND LOW WATERS
YEAR 1989

TIME ZONE UT(GMT)
For Summer Time add ONE hour in non-shaded areas

SEPTEMBER

Time	m	Time	m
1 F 0131 0748 1404 1952	4.2 0.6 4.1 0.9	**16** Sa 0103 0728 1344 1937	4.7 0.2 4.4 0.6
2 Sa 0202 0818 1433 2022	4.2 0.6 4.0 0.9	**17** Su 0145 0808 1425 2018	4.8 0.2 4.4 0.5
3 Su 0232 0847 1503 2051	4.2 0.6 4.0 1.0	**18** M 0227 0850 1507 2058	4.8 0.3 4.3 0.6
4 M 0301 0917 1534 2122	4.1 0.9 3.9 1.1	**19** Tu 0314 0932 1550 2142	4.7 0.5 4.1 0.8
5 Tu 0334 0948 1604 2155	4.0 1.0 3.7 1.3	**20** W 0403 1016 1637 2231	4.4 0.9 3.9 1.0
6 W 0409 1019 1640 2231	3.8 1.2 3.6 1.4	**21** Th 0501 1105 1732 2330	4.0 1.3 3.6 1.3
7 Th 0451 1055 1720 2315	3.6 1.5 3.5 1.7	**22** F 0611 1207 1839	3.7 1.7 3.4
8 F 0544 1140 1815 ☾	3.4 1.8 3.3	**23** Sa 0052 0738 1342 2004	1.6 3.4 2.0 3.4
9 Sa 0015 0656 1246 1927	1.9 3.2 2.0 3.3	**24** Su 0242 0915 1528 2128	1.6 3.3 2.0 3.4
10 Su 0147 0827 1426 2049	1.9 3.1 2.1 3.3	**25** M 0404 1030 1634 2234	1.4 3.5 1.8 3.6
11 M 0332 0953 1557 2200	1.8 3.3 1.9 3.5	**26** Tu 0459 1122 1718 2322	1.1 3.7 1.3 3.8
12 Tu 0438 1055 1655 2255	1.4 3.6 1.6 3.8	**27** W 0542 1200 1754	0.9 3.8 1.3
13 W 0526 1143 1740 2342	1.1 3.9 1.3 4.1	**28** Th 0001 0617 1234 1827	4.0 0.8 3.9 1.1
14 Th 0608 1224 1819	0.7 4.1 1.0	**29** F 0035 0649 1304 1856 ●	4.1 0.7 4.1 1.0
15 F 0024 0648 1304 1859 ○	4.4 0.4 4.3 0.7	**30** Sa 0106 0719 1334 1926	4.2 0.7 4.1 0.9

OCTOBER

Time	m	Time	m
1 Su 0135 0748 1402 1955	4.2 0.7 4.1 0.9	**16** M 0124 0744 1359 1955	4.8 0.3 4.3 0.5
2 M 0205 0816 1432 2025	4.2 0.8 4.0 1.0	**17** Tu 0211 0826 1442 2040	4.8 0.5 4.3 0.6
3 Tu 0236 0844 1500 2054	4.0 0.9 4.0 1.1	**18** W 0300 0910 1527 2128	4.6 0.8 4.2 0.8
4 W 0308 0914 1529 2128	4.0 1.1 3.9 1.3	**19** Th 0353 0955 1614 2220	4.3 1.1 4.0 1.0
5 Th 0345 0945 1602 2204	3.8 1.3 3.7 1.4	**20** F 0454 1045 1709 2323	3.9 1.6 3.7 1.3
6 F 0428 1021 1642 2249	3.5 1.6 3.6 1.6 ☾	**21** Sa 0605 1149 1815	3.6 1.9 3.5
7 Sa 0525 1108 1736 2350	3.4 1.9 3.5 1.8	**22** Su 0045 0728 1321 1937	1.5 3.4 2.1 3.4
8 Su 0638 1215 1849 ☽	3.2 2.1 3.4	**23** M 0222 0853 1458 2057	1.5 3.4 2.0 3.5
9 M 0119 0804 1354 2011	1.8 3.3 2.1 3.4	**24** Tu 0335 1000 1602 2203	1.5 3.5 1.8 3.6
10 Tu 0256 0924 1524 2125	1.7 3.4 1.9 3.6	**25** W 0428 1049 1647 2252	1.2 3.6 1.6 3.8
11 W 0403 1024 1623 2223	1.3 3.7 1.5 3.9	**26** Th 0511 1125 1725 2333	1.1 3.8 1.3 3.9
12 Th 0454 1112 1709 2312	1.0 3.9 1.2 4.2	**27** F 0547 1203 1758	1.0 3.9 1.2
13 F 0539 1156 1751 2356	0.7 4.1 0.9 4.5	**28** Sa 0008 0619 1235 1831	4.0 0.9 4.0 1.1
14 Sa 0621 1236 1832 ○	0.4 4.3 0.7	**29** Su 0041 0650 1306 1902 ●	4.1 0.8 4.1 1.0
15 Su 0039 0702 1319 1913	4.7 0.3 4.4 0.5	**30** M 0113 0720 1335 1933	4.1 0.9 4.1 1.0
		31 Tu 0145 0748 1405 2004	4.1 1.0 4.1 1.1

NOVEMBER

Time	m	Time	m
1 W 0218 0818 1433 2037	4.0 1.1 4.0 1.2	**16** Th 0253 0853 1508 2119	4.4 1.0 4.3 0.8
2 Th 0254 0850 1503 2112	3.9 1.3 4.0 1.2	**17** F 0348 0939 1555 2213	4.2 1.4 4.1 0.9
3 F 0334 0924 1536 2152	3.8 1.5 3.9 1.4	**18** Sa 0447 1027 1647 2312	3.9 1.7 3.9 1.1
4 Sa 0421 1003 1619 2238	3.7 1.7 3.8 1.5	**19** Su 0549 1123 1746	3.7 1.9 3.7
5 Su 0516 1052 1712 2337	3.5 1.9 3.7 1.6 ☾	**20** M 0019 0655 1234 1853	1.3 3.5 2.0 3.6
6 M 0622 1158 1819	3.4 2.0 3.6	**21** Tu 0134 0804 1357 2006	1.4 3.4 2.0 3.5
7 Tu 0053 0734 1321 1934	1.6 3.4 2.0 3.6	**22** W 0244 0908 1507 2114	1.4 3.4 1.9 3.5
8 W 0215 0846 1442 2046	1.5 3.5 1.8 3.7	**23** Th 0342 1003 1600 2212	1.4 3.5 1.7 3.6
9 Th 0324 0948 1545 2148	1.3 3.7 1.6 3.9	**24** F 0430 1051 1649 2258	1.3 3.6 1.5 3.7
10 F 0419 1041 1637 2241	1.0 3.8 1.3 4.2	**25** Sa 0512 1132 1730 2340	1.2 3.7 1.4 3.8
11 Sa 0508 1127 1723 2332	0.7 4.1 1.0 4.4	**26** Su 0549 1207 1807	1.1 3.8 1.3
12 Su 0554 1212 1810	0.5 4.2 0.8	**27** M 0018 0622 1241 1842	3.9 1.1 4.0 1.2
13 M 0021 0639 1256 1856 ○	4.6 0.5 4.4 0.7	**28** Tu 0053 0655 1312 1917 ●	3.9 1.1 4.1 1.2
14 Tu 0109 0724 1340 1942	4.6 0.6 4.4 0.6	**29** W 0130 0727 1342 1951	4.0 1.1 4.1 1.1
15 W 0201 0808 1423 2030	4.6 0.8 4.4 0.6	**30** Th 0208 0759 1412 2026	4.0 1.3 4.1 1.1

DECEMBER

Time	m	Time	m
1 F 0247 0834 1446 2104	3.9 1.4 4.1 1.1	**16** Sa 0338 0922 1536 2200	4.1 1.3 4.2 0.8
2 Sa 0328 0912 1521 2145	3.9 1.5 4.1 1.1	**17** Su 0427 1006 1621 2248	3.9 1.5 4.1 0.9
3 Su 0413 0953 1603 2230	3.8 1.6 4.0 1.2	**18** M 0515 1049 1708 2337	3.7 1.7 3.9 1.2
4 M 0504 1041 1654 2323	3.7 1.8 3.9 1.3 ☾	**19** Tu 0604 1139 1800	3.5 1.9 3.7
5 Tu 0558 1137 1751	3.6 1.8 3.8	**20** W 0032 0657 1236 1859	1.3 3.4 1.9 3.5
6 W 0024 0700 1242 1857	1.3 3.6 1.8 3.8	**21** Th 0134 0759 1347 2005	1.5 3.3 1.9 3.4
7 Th 0133 0805 1354 2005	1.3 3.6 1.7 3.8	**22** F 0239 0903 1501 2114	1.5 3.3 1.8 3.4
8 F 0240 0910 1503 2112	1.2 3.7 1.6 3.9	**23** Sa 0339 1003 1606 2216	1.5 3.4 1.8 3.4
9 Sa 0343 1010 1604 2214	1.1 3.8 1.4 4.1	**24** Su 0433 1055 1701 2309	1.5 3.6 1.6 3.5
10 Su 0441 1104 1702 2313	0.9 3.9 1.2 4.2	**25** M 0518 1139 1747 2356	1.4 3.7 1.5 3.7
11 M 0533 1153 1756	0.8 4.2 0.9	**26** Tu 0558 1215 1827	1.3 3.9 1.3
12 Tu 0010 0624 1239 1846 ○	4.3 0.8 4.3 0.8	**27** W 0039 0635 1250 1904	3.8 1.3 4.0 1.2
13 W 0103 0710 1324 1937	4.4 0.8 4.4 0.6	**28** Th 0119 0712 1323 1940 ●	3.9 1.3 4.1 1.0
14 Th 0157 0755 1408 2025	4.4 1.0 4.4 0.6	**29** F 0158 0747 1355 2016	4.0 1.3 4.1 0.9
15 F 0249 0840 1451 2112	4.3 1.2 4.4 0.6	**30** Sa 0236 0823 1430 2053	4.0 1.3 4.3 0.9
		31 Su 0315 0901 1508 2134	4.1 1.3 4.3 0.8

Chart Datum: 2.25 metres below Ordnance Datum (Newlyn)

PETERHEAD 10-6-18
Aberdeen

CHARTS
Admiralty 1438, 213; OS 30

TIDES
+0140 Dover; ML 2.3; Duration 0620; Zone 0 (GMT).

Standard Port ABERDEEN (←)

Times				Height (metres)			
HW		LW		MHWS	MHWN	MLWN	MLWS
0000	0600	0100	0700	4.3	3.4	1.6	0.6
1200	1800	1300	1900				

Differences PETERHEAD
−0035 −0045 −0035 −0040 −0.5 −0.3 −0.1 −0.1

SHELTER
Excellent shelter. Can be entered in any weather at any tide. The bay is enclosed by breakwaters behind which yachts can anchor. Alongside berths in the inner harbours are sometimes available. Contact Hr Mr on VHF before entering or leaving.

NAVIGATION
Waypoint 57°29'.14N 01°45'.00W, 134°/314° from/to entrance, 1.0M. Peterhead is a major commercial port with all the implications. No navigational dangers. Yachts should give construction area a wide berth. Buoys are frequently moved.

LIGHTS AND MARKS
Leading lights into Peterhead Bay 314°. Leading lights into harbour 059°
Fishing Harbour signals (on Control Tower over Hr office on West pier)
3 Fl R (hor) – Bay closed to inward traffic
3 FR (hor) – Fishing Hr closed to inward traffic
2 Fl R (hor) – No exit from Bay to sea
2 FR (hor) – No exit from Fishing Hr
4 Fl R (hor) – Bay and harbour closed – No traffic movement permitted.
4 FR (hor) – Fishing harbour closed
When no signals showing, vessels may enter or leave with permission from Control Tower, call on Ch 16.

RADIO TELEPHONE
VHF Ch 16; 09 11 **14** (H24).

TELEPHONE (0779)
Hr Mr 74281/3; CG 74278; MRSC 74278; HMC 74867; Marinecall 0898 500 452; Dr 74841.

FACILITIES
EC Wednesday; **A.J. Buchan** Tel. 72348, Slip, ME, El, Sh, C; **Northern Engineering** Tel. 72406, ME, El, Sh, C; **R. Irvin** Tel. 72044, ME, El, Sh, C; **Murisons** Tel. 72173, Gas; **Town** P, D, V, R, Bar, PO, Bank, Rly (bus to Aberdeen), Air (Aberdeen).

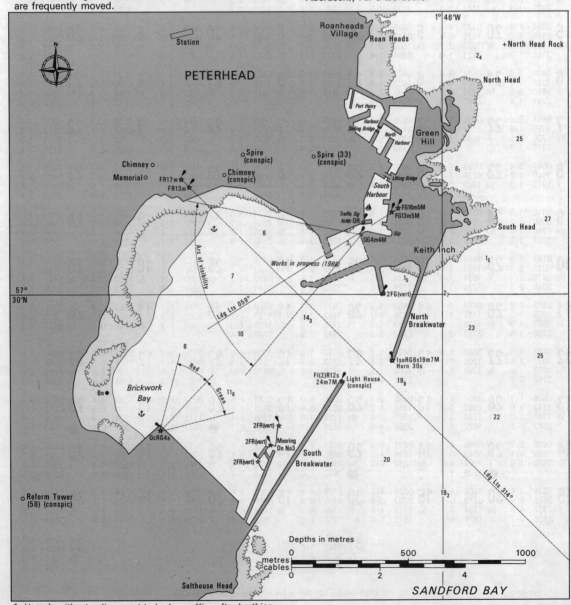

⚓ Vessels without radio, report to harbour office after berthing

MINOR HARBOURS AND ANCHORAGES 10.6.19

BURNMOUTH, Berwick, 55°50′ N, 2°04′ W, Zone O (GMT), Admty charts 1407, 1192. HW +0315 on Dover, −0025 on Leith; Duration 0615. Good shelter especially in inner harbour (dries). With on-shore winds a swell makes outer harbour uncomfortable. Min depth at entrance at LWS is 0.6m. From S beware Quarry Shoal Rocks; from N beware East and West Carrs. Leading line at 274° with two Bns or FR at night 29/35m 4M, 45m apart run in towards the harbour mouth until the Lts in the inner basin at root of W pier (2FG (vert)) come in line with outer harbour entrance. Facilities: FW, limited stores, Bar at top of valley.

ST ABBS, Berwick, 55°54′ N, 2°08′ W, Zone O (GMT), Admty chart 175. HW +0330 on Dover, −0017 on Leith; HW height −0.6m on Leith; Duration 0605; Shelter good. In strong on-shore winds outer harbour suffers from waves breaking over E pier. Inner harbour is best but often full of fishing boats and dries out. Hr Mr will direct visitors. On E side of entrance channel, beware Hog's Nose and on W side the Maw Carr, Ldg Line S face of Maw Carr on village hall (conspic) runs in until the harbour entrance opens to port when the second Ldg Line can be seen, front on head of jetty FR 4m 1M, rear on SW corner of old harbour FR 8m 1M. Access HW∓3. Facilities: FW on quay (tap), V (local grocer), LB, R, more facilities including bar at Coldingham Bay (½ M).

NORTH BERWICK, East Lothian, 56°04′ N, 2°43′ W Zone O (GMT), Admty chart 734. Fidra HW +0344 on Dover, −0005 on Leith; HW height −0.2m on Leith; ML 3.1m; Duration 0625. Shelter good with winds S to W but harbour is dangerous with on-shore winds. Entrance is closed by boom in bad weather. Entrance is 8m wide. Harbour dries out. Beware The Maidenfoot, marked by Bn 1 ca NW of N pierhead. Ldg line − W face of North Berwick Law (conspic) in line with post on N pierhead (160°). Lights; N pierhead FWR 7m 3M R to seaward, W over harbour. Extinguished when vessels cannot enter harbour due to bad weather. Facilities: EC Thurs; **East Lothian YC** Tel. 2698 Bar, P and D (cans), FW on pier, stores etc obtainable in town; **Lawrence Turnbull** Tel. 2134, Gas.

FISHERROW, Midlothian, 55°57′ N, 3°04′ W, Zone O (GMT), Admty chart 734. HW +0345 on Dover, −0005 on Leith; HW height −0.1m on Leith; ML 3.0m; Duration 0620. Shelter good except in winds from NW. Mainly a pleasure craft harbour. Approach dangerous in on-shore winds. Harbour dries; available HW∓2. Light − E pier head, Oc 6s 5m 6M on metal framework tower. Berth on W pier. Facilities: EC Wed; stores and FW. P and D from garage. Hr Mr Tel. Edinburgh 665-5900; CG 665-5639; **Fisherrow YC** 665-2576; **Village** FW, V, R, Bar.

CRAMOND, Midlothian, 55°59′ N, 3°18′ W, Zone O (GMT), Admty chart 736. Tidal details as Granton (see 10.6.9). Enter R Almond, W of Cramond Island, marked by buoys and posts. At shore line is a sill. Access HW∓2. **Cramond Boat Club** M, Bar; Pub.

INCHCOLM, Fife, 56°02′ N, 3°18′ W, Zone O (GMT), Admty chart 736. Tidal details as Aberdour. Best anchorage N of abbey (conspic). Beware Meadulse Rocks (dry). Ends of island foul. Lt Fl(3) 15s (obsc 075°−145°) at SE point, Horn(3) 45s. Land at jetty E of abbey (fee payable). No facilities.

ABERDOUR, Fife, 56°03′ N, 3°18′ W, Zone O (GMT), Admty charts 735. HW +0345 on Dover; +0005 on Leith; HW height 0.0m on Leith; ML 3.3m; Duration 0630. Good shelter except in SE winds when a swell occurs. The anchorage between The Little Craigs and the disused pier is good but exposed to winds from E to SW. Temporary berths are available in harbour (dries) alongside the quay wall. Beware Little Craigs (dries 2.1m) and outfall 2ca N marked by Bn. There are no lights or marks. Facilities: EC Wed; Hr Mr Tel. 860473, FW (tap on pier), P, R, V, Bar in village. **Aberdour Boat Club** Tel. 860029.

KIRKCALDY, Fife, 56°07′ N, 3°09′ W, Zone O (GMT), Admty chart 739. HW +0345 on Dover, −0005 on Leith; HW height −0.1m on Leith; ML 3.2m; Duration 0620. Shelter good except in strong E winds. Yachts can berth on E pier in outer harbour (known as Pier Harbour) or, by arrangement with Hr Mr, enter the Wet Dock (HW−2 to HW) if space allows. If possible make arrangements before arrival (Methil Hr. Radio Ch 16, 14 will relay messages). A special Lt Buoy FlY 5s, lies ½ M E of harbour. If approaching from E, keep S of the Bn 60m SSE of East Pier Lt. Lights; E pier head, Fl WG 10s 10m 8M; G156°−336°, W336°−156°. S pier head 2 FR (vert) 7m, 5M. W pier, inner head FW 5m. Dock entrance W side FR 6m, E side FG 6m. Entry signals, day and night: FR-Port closed, bring up in roads. FG-Entry permitted. Harbour is administered by Forth Ports Authority. Local Office Tel. 260176. Facilities: EC Wed; Bar, D (by road tanker); FW, R, V. Small repairs can be arranged. PO, Bank.

ELIE, Fife, 56°11′ N, 2°46′ W, Zone O (GMT), Admty chart 734. HW +0325 on Dover, −0015 on Leith; HW height −0.1m on Leith; ML 3.2m; Duration 0620; Elie Bay provides good shelter from winds in the N sectors to small craft but local knowledge is needed. Anchor to W of pier or take the mud alongside it. Beware ledge off end of pier which dries. Coming from E beware Ox Rock (dries 1m) ½ M ENE of Elie Ness; coming from W beware rocks off Chapel Ness, West Vows, East Vows (surmounted by cage Bn) and Thill Rock, the E extremity of the latter pair marked by R can buoy. Light; Elie Ness Fl 6s 15m 18M, white Tr. Facilities: In villages of Elie and Earlsferry, FW, R, V, Bar, PO, Bank; **Shell Bay** Tel. Leven 330283, Gas.

ST MONANS, Fife, 56°12′ N, 2°49′ W, Zone O (GMT), Admty chart 734. HW +0335 on Dover, −0020 on Leith; HW height −0.1m on Leith; ML 3.2m; Duration 0620. Shelter good except in strong SE to SW winds when scend occurs in the harbour. Harbour dries. Lie alongside E pier until contact with Hr Mr. Coming from NE keep at least 2½ ca from coast. Lights Breakwater head Oc WRG 6s 5m 7/4M; E pier head 2 FG (vert) 6m 4M. W pier head 2 FR (vert) 6m 4M. Facilities: EC Wed; Bar, D (by road tanker), El, FW, ME, Sh, Slip (by root of E pier), R, V, PO, AC, Bank; **Millers BY** Tel. 864; **Bass Rk Oil** Tel. 501, P, D, Gas.

CRAIL, Fife, 56°15′ N, 2°38′ W, Zone O (GMT), Admty chart 175. HW +0320 on Dover, −0020 on Leith; HW height −0.2m on Leith; ML 3.0m; Duration 0615. Good shelter but only for boats able to take the ground alongside. Harbour is protected by timber booms dropped into grooves at the pier heads in foul weather. Entrance between S pier and Bn on rocks to S following Ldg Line 295° − two W concrete pillars with FR Lts, 24m 6M and 30m 6M. Vessels should contact Forth Coastguard on VHF Ch 16 before entering. Facilities: EC Wed; Bar, C, R, Slip, V, PO, Bank. Coastguard Tel. 50666.

MAY ISLAND, Fife, 56°11′ N, 2°33′ W, Zone O (GMT), Admty chart 734. HW +0325 on Dover, −0025 on Leith. Anchorages at W Tarbert or E Tarbert according to the wind. Landing at Altar Stones. There is a harbour at Kirk Haven. Beware rocks N and S of island, Norman Rock to N and Maiden Hair Rock to S. At summit of island, a square Tr on stone house Fl 20s 73m 26M. RC. There are no facilities. The island is a nature reserve.

GOURDON, Kincardine, 56°50′ N, 2°17′ W, Zone O (GMT), Admty chart 210. HW +0240 on Dover, +0020 on Aberdeen; HW height +0.4m on Aberdeen; ML 2.7m; Duration 0620. Shelter good and inner harbour protected by storm gates. Beware rocks running S from W pier end marked by Bn. Lights − Ldg Lts 358°, both FR. Front Lt shows G when not safe to enter. West pier head Fl WRG 3s 5m 9/7M, vis G180°−344°, W344°−354°, R354°−180°. E breakwater head Q 3m 7M. Facilities: Hr Mr Tel. Inverbervie 61779 − Slip, FW from standpipe, V, R, Bar, D, ME, AC.

BODDAM, Aberdeen, 57°28′ N, 1°46′ W, Zone O (GMT), Admty chart 213. HW +0145 on Dover, −0040 on Aberdeen; HW height −0.4m on Aberdeen; ML 2.3m; Duration 0620. Good shelter, being protected by Meikle Mackie, the island just off Buchan Ness; harbour dries. Beware rocks all round Meikle Mackie and to the SW of it. There are no lights. Yachts lie along N wall. Very limited facilities, but all facilities at Peterhead, 4 miles away.

The magazine for boats and equipment tests
plus - The unique new and secondhand price guide
Available from newsagents
or on subscription.

Ringwood Publishing Ltd
Mansfield House, Ringwood, Hampshire, BH24 1JD
Telephone: (0425) 471035

VOLVO PENTA SERVICE

Sales and service centres in area 7
GRAMPIAN **Henry Fleetwood & Sons Ltd** Baker Street, LOSSIEMOUTH IV31 6NZ Tel (034381) 3015. HIGHLANDS AND ISLANDS **Caley Marina** Canal Road, Muirtown, INVERNESS IV3 6NS Tel (0463) 236539.

Area 7

North-East Scotland
Fraserburgh to Scrabster
including Orkney and Shetland Islands

10.7.1	Index	Page 389
10.7.2	Diagram of Radiobeacons, Air Beacons, Lifeboat Stations etc	390
10.7.3	Tidal Stream Charts	392
10.7.4	List of Lights, Fog Signals and Waypoints	394
10.7.5	Passage Information	397
10.7.6	Distance Table	398
10.7.7	Fraserburgh	399
10.7.8	Macduff/Banff	399
10.7.9	Whitehills	400
10.7.10	Buckie	400
10.7.11	Lossiemouth	400
10.7.12	Hopeman	401
10.7.13	Burghead	402
10.7.14	Findhorn	402
10.7.15	Nairn	403
10.7.16	Inverness	403
10.7.17	Portmahomack	404
10.7.18	Helmsdale	405
10.7.19	Wick	405
10.7.20	Orkney Islands	406
10.7.21	Orkneys — Kirkwall	407
10.7.22	— Stronsay	407
10.7.23	— Stromness	408
10.7.24	Shetland Islands	409
10.7.25	Lerwick, Standard Port, Tidal Curves	410
10.7.26	Scrabster	414
10.7.27	Minor Harbours and Anchorages Cullen Portknockie Findochty Fortrose Cromarty Firth Dornoch Firth Golspie Lybster Houton Bay Shapinsay Pierowall Auskerry Fair Isle Scalloway Balta Sound Kyle of Tongue Loch Eriboll	414

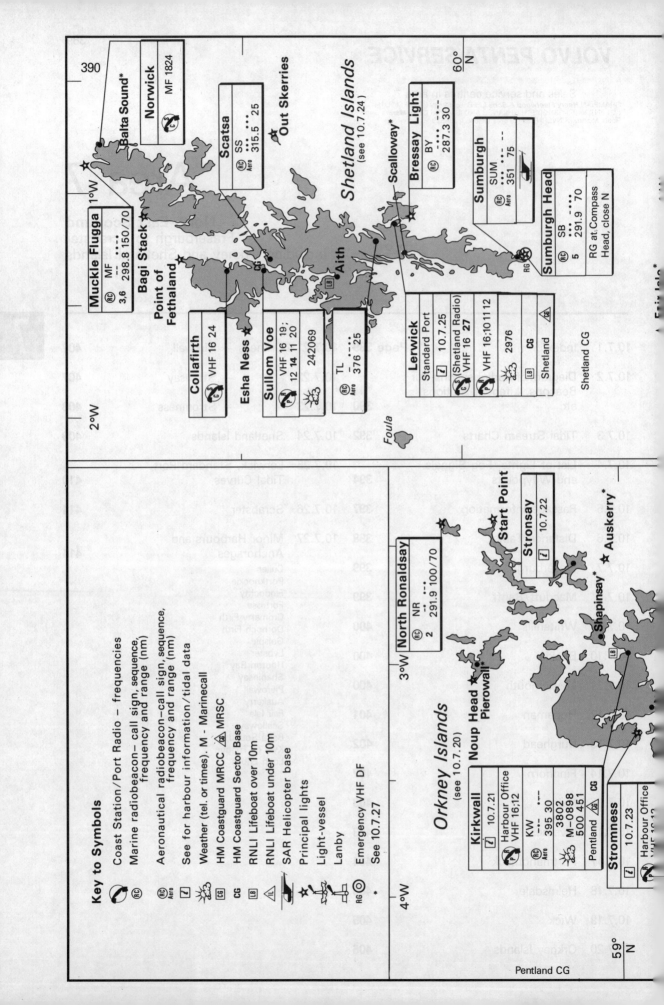

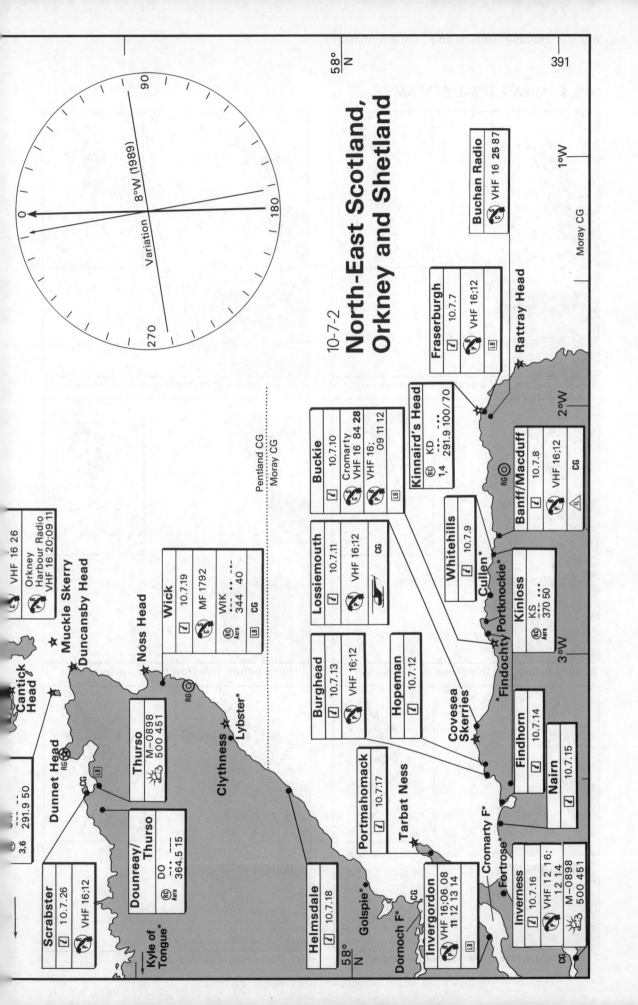

10.7.3 AREA 7 TIDAL STREAMS

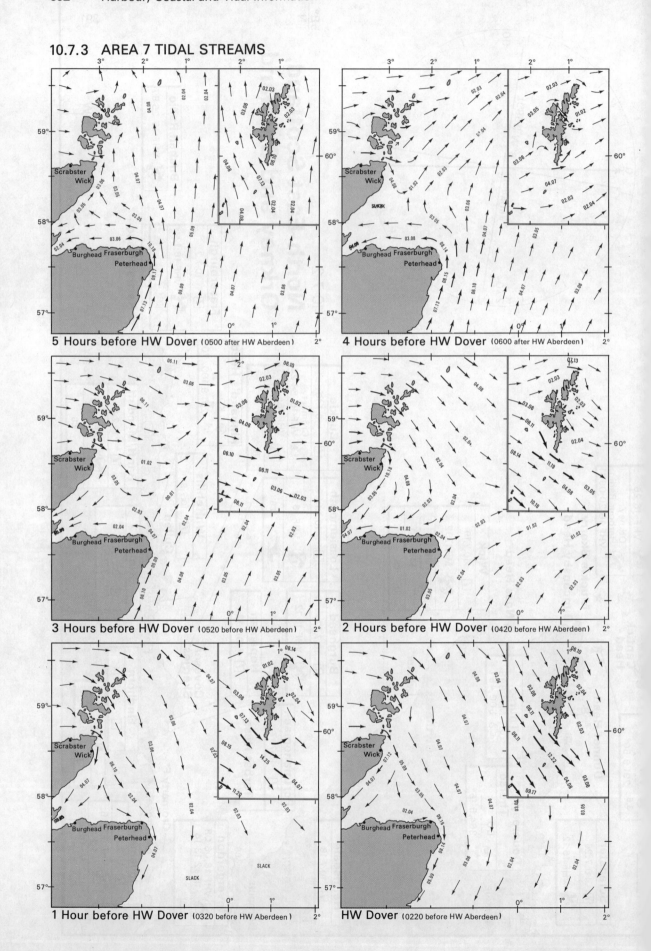

AREA 7—NE Scotland

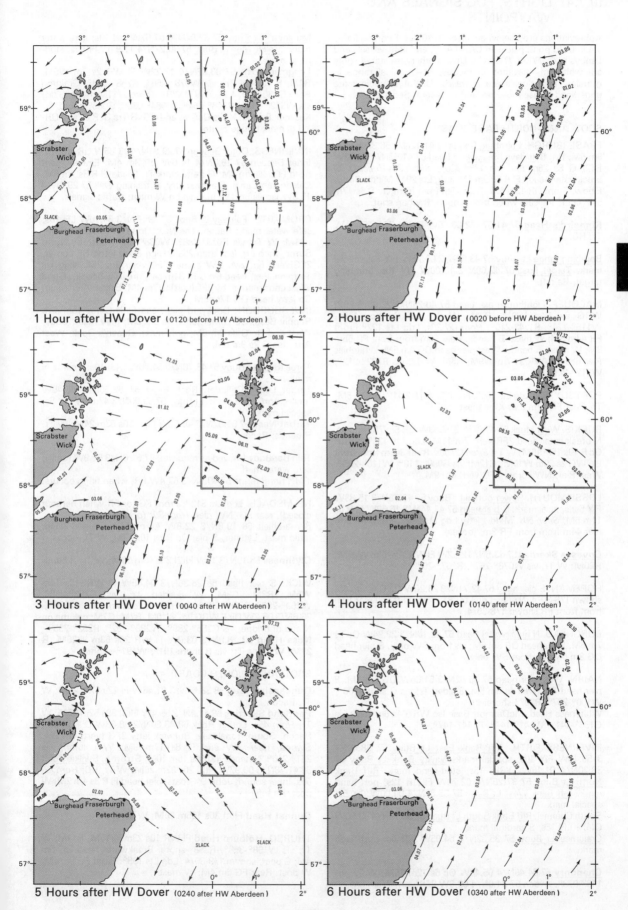

10.7.4 LIGHTS, FOG SIGNALS AND WAYPOINTS

Abbreviations used below are given in 10.0.3. Principal lights are in **bold** print, places in CAPITALS, and light-vessels and Lanbys in *CAPITAL ITALICS*. Unless otherwise stated lights are white. m—elevation in metres; M—nominal range in n. miles. Fog signals are in *italics*. Useful waypoints are underlined — use those on land with care. See 4.2.2.

SCOTLAND—NORTH-EAST COAST

FRASERBURGH. Cairnbulg Briggs 57 41.12N/1 56.37W Fl 3s 9m 5M; Bn. Balaclava breakwater head 57 41.53N/1 59.63W Fl(2) G 8s 26m 6M; vis 178°-326°; *Siren 20s (occas)*. S breakwater head LFl R 6s 9m 5M. Ldg Lts 291° Middle Jetty, Elbow. Front QR 12m 5M. Rear 75m from front Oc R 6s 17m 5M. G Lts indicate when basins open, R when shut.

Kinnaird's Head 57 41.87N/2 00.15W Fl 15s 37m **25M**; W Tr; RC.

Bombing range Lt Buoy 57 43.80N/2 00.75W FlY 5s; special mark. Target float 57 42.00N/2 10.00W FlY 10s; bombing target; Ra refl.

MACDUFF. Lighthouse pier head 57 40.26N/2 29.90W Fl(2) WRG 6s 12m **W7M**, R7M; W Tr; vis G shore-115°, W115°-174°, R174°-210°; *Horn (2) 20s*. Ldg Lts 127° Front FR 44m 3M; W mast, B bands. Rear 60m from front FR 55m 3M; W mast, B bands. West pier head QG 4m 5M. Whitehills Pier head Fl WR 3s 7m W9M, R6M; W Tr; vis R132°-212°, W212°-245°.

PORTSOY. Pier Ldg Lts 160°. Front 57 41N/2 41W F 12m 5M; Tr. Rear FR 17m 5M; X on mast.

BUCKIE. West Muck 57 41.07N/2 57.93W QR 5m 7M; tripod. N breakwater head 2FR(vert) 7m 11M. Ldg Lts 125°. **Front** Oc R 10s 15m **15M**; W Tr; *Siren (2) 60s*. **Rear** 365m from front Iso WG 2s 30m **W16M**, G12M; vis G090°-110°, W110°-225°. West Pier, NW corner 2FG(vert) 4m 9M.

LOSSIEMOUTH. Sewer outfall Lt Buoy 57 43.78N/3 15.83W FlY 5s; special mark. S pier head 57 43.44N/3 16.59W Fl R 6s 11m 5M; *Siren 60s*. Middle jetty Ldg Lts 292°. Front FR 5m; rear 44m from front FR 8m, (occas).

Covesea Skerries 57 43.5N/3 20.2W Fl WR 20s 49m **W24M, R20M**; W Tr; vis W076°-267°, R267°-282°.

HOPEMAN. S pier head 57 42.71N/3 26.20W Oc G 4s 8m 4M shown 1/8-30/4. N quay Ldg Lts 081°. Front FR and rear 10m from front FR shown 1/8-30/4.

BURGHEAD. N breakwater head 57 42.10N/3 29.94W Oc 8s 7m 5M. Spur head QR 3m 5M; vis from SW only. S pier head QG 3m 5M; vis from SW only.

NAIRN. West pier head 57 35.62N/3 51.56W QG 5m 1M. E pier head Oc WRG 4s 6m 5M; 8-sided Tr; vis G shore-100°, W100°-207°, R207°-shore.
Whiteness Head, McDermott Base Iso WRG 4s 6m 10M; vis G138°-141°, W141°-144°, R144°-147°.

INVERNESS FIRTH. Riff Bank East Lt Buoy 57 38.40N/3 58.10W FlY 10s; special mark. Navity Bank Lt Buoy 57 38.20N/4 01.10W Fl(3)G 15s; stbd-hand mark. Riff Bank North Lt Buoy 57 37.25N/4 02.70W Fl(2)R 12s; port-hand mark. Riff Bank West Lt Buoy 57 35.80N/4 04.00W FlY 5s; special mark.
South Channel, Riff Bank South Lt Buoy 57 36.75N/4 00.90W Q(6) + LFl 15s; S cardinal mark.
Craigmee Lt Buoy 57 35.32N/4 04.95W FlR 6s; port-hand mark.

Chanonry 57 34.46N/4 05.48W Oc 6s 12m **15M**; W Tr; vis 148°-073°. Avoch 57 34.0N/4 09.8W 2FR(vert) 7/5m 5M.

Munlochy Lt Buoy 57 32.94N/4 07.57W LFl 10s; safe water mark. Meikle Mee Lt Buoy 57 30.27N/4 11.94W FlG 3s; stbd-hand mark.
Longman Point 57 30.02N/4 13.22W Fl WR 2s 7m W5M, R4M; R conical Bn; vis W078°-258°, R258°-078°. Graigton Point 57 30.07N/4 14.01W Fl WRG 4s 6m W11M, R7M, G7M; vis W312°-048°, R048°-064°, W064°-085°, G085°-shore.
Kessock Bridge. N Trs OcG 6s and QG; S Trs OcR 6s and QR. Y Lts mark bridge centre.

INVERNESS. Outer Beacon 57 29.84N/4 13.85W QR 3m 4M. Inner Beacon 57 29.7N/4 14.0W QR 3m 4M. Embankment head Fl G 2s 8m 4M; G framework Tr. E side Fl R 3s 7m 6M. Caledonian Canal. Clachnaharry, S training wall 57 29.5N/4 15.8W Iso 4s 5m; W triangle on W mast; traffic signals.

CROMARTY. Fairway Lt Buoy 57 39.98N/3 54.10W LFl 10s; safe water mark; Racon. The Ness Oc WR 10s 18m W14M, R11M; W Tr; vis R079°-088°, W088°-275°, obsc by North Sutor when brg less than 253°. Nigg Ferry jetty, SE corner, 2FG(vert) 6m 2M. SW corner 2FG(vert) 6m 2M. Nigg Oil Terminal Pier, head Oc G 5s 31m 5M; grey Tr, floodlit. E and West ends marked by 2FG(vert) 9/7m 4M. British Aluminium Co jetty head QG 17m 5M.
INVERGORDON. Dockyard pier head Fl(3) G 10s 15m 4M. Supply Base, SE corner Iso G 4s 9m 6M; grey mast. Quay, West end Oc G 8s 9m 6M; grey mast. Queen's Dock, West Arm, Iso G 2s 9m 6M.

Three Kings Lt Buoy 57 43.76N/3 54.20W Q(3) 10s; E cardinal mark.
Tarbat Ledge/Culloden Rock Buoy 57 52.50N/3 45.45W (unlit); port-hand mark. Lt Buoy 53.00N/3 47.00W FlY 5s; special mark.
Tarbat Ness 57 51.92N/3 46.52W Fl(4) 30s 53m **24M**; W Tr, R bands; Racon.

Tain Bar Buoy (moved as requisite) 57 51.6N/3 52.9W (unlit); safe water mark.
TAIN firing range 57 49.5N/3 57.4W FR, when firing occurs.

HELMSDALE. Ldg Lts 313°. Front FG (or FR when harbour closed), rear FG. NW pier; *Horn 30s (occas)*.
Ben-a-chielt 58 19.8N/3 22.8W Aero 5FR(vert) 448-265m; radio mast. Lybster, S pier Oc R 6s 10m 3M; W Tr; occas.

Clythness 58 19N/3 13W Fl(2) 30s 45m **16M**; W Tr, R band.

WICK. S pier head 58 26.36N/3 04.65W Fl WRG 3s 12m W5M, R3M, G3M; W 8-sided Tr; vis G253°-269°, W269°-286°, R286°-329°; *Bell (2) 10s (occas)*. Dir Lt 290° 58 26.6N/3 05.2W Dir F WRG 9m W10M, R7M, G7M; column on N end of bridge; vis G285°-289°, W289°-292°, R292°-295°.
Noss Head 58 28.8N/3 03.0W Fl WR 20s 53m **W25M, R 21M**; W stone Tr; vis R shore-191°, W191°-shore.

SCOTLAND—NORTH COAST

Duncansby Head 58 38.67N/3 01.43W Fl 12s 67m **24M**; W Tr; Racon.
Pentland Skerries 58 41.4N/2 55.4W Fl(3) 30s 52m **25M**; W Tr; *Horn 45s*. Lother Rock 58 43.80N/2 58.58W Q 11m 6M; Racon. South Ronaldsay, Burwick jetty 2FG (vert) 6m 2M. Swona, near SW end Fl 8s 17m 9M; W column; vis 261°-210°. N head Fl(3) 10s 16m 10M. **Stroma**, Swilkie Point 58 41.78N/3 06.92W Fl(2) 20s 32m **26M**; W Tr; RC; *Horn (2) 60s*. Inner sound, John O'Groats, pier head Fl R 3s 4m 2M; W post; Ra refl.

Dunnet Head Fl(4) 30s 105m **26M**; W Tr; RG.

THURSO. **Holburn Head** Fl WR 10s 23m **W15M**, R11M; W Tr; vis W198°-358°, R358°-shore. Breakwater head QG 5m 4M; G post; shown 1/9-30/4. Ldg Lts 195°. Front FG 5m 4M; W post. Rear FG 6m 4M; W mast.

AREA 7—NE Scotland

SCRABSTER. Outer pier head 58 36.63N/3 32.48W QG 6m 4M. E pier head 2FG(vert) 6m 4M; shown 1/8-31/5. West pier head 2FR(vert) 6m 4M; shown 1/8-31/5.

Strathy Point 58 36N/4 01W Fl 20s 45m **27M**; W low Tr on dwelling.

Sule Skerry 59 05.10N/4 24.30W Fl(2) 15s 34m **19M**; W Tr; RC; Racon.

North Rona 59 07.3N/5 48.8W Fl(3) 20s 114m **24M**; W Tr. Sula Sgeir 59 05.65N/6 09.50W Fl 15s 74m 11M.

Loch Eriboll, White Head 58 31.1N/4 38.8W Fl WR 3s 18m W13M, R12M; W Tr and building; vis W030°-172°, R172°-191°, W191°-212°.

Cape Wrath 58 37.55N/4 59.87W Fl (4) 30s 122m **24M**; W Tr; RC; *Horn (3) 45s*.

ORKNEY ISLANDS

Tor Ness 58 46.7N/3 17.6W Fl 3s 20m 9M; W Tr.
South Walls, SE end, **Cantick Head** 58 47.2N/3 07.8W Fl 20s 35m **22M**; W Tr.
SCAPA FLOW AND APPROACHES. Ruff Reef, off Cantick Head, Fl(2) 10s 10m 6M; B Bn.
Long Hope, South Ness Pier, head, 58 48.1N/3 12.2W Fl WRG 3s 6m W7M, R5M, G5M; vis G082°-242°, W242°-252°, R252°-082°. Hoxa Head 58 49.3N/3 02.0W Fl WR 3s 15m W9M, R6M; W Tr; vis W026°-163°, R163°-201°, W201°-215°.
Stanger Head 58 49.0N/3 04.6W Fl R 5s 25m 8M.
Roan Head 58 50.8N/3 03.8W Fl(2) R 6s 12m 7M.
Nevi Skerry 58 50.7N/3 02.6W Fl(2) 6s 7m 6M.
Calf of Flotta 58 51.3N/3 03.9W QR 8m 4M.
Flotta Terminal. N end of jetty. East 2FR(vert) 10/8m 3M. West 2FR(vert) 10/8m 3M; *Bell 10s*. Mooring dolphins, East and West, both QR 8m 3M.
Single Point Mooring No 1 58 52.2N/3 07.3W FIY 5s 12m 3M; *Horn Mo(A) 60s*. Single Point Mooring No 2 58 52.3N/3 05.8W Fl(4)Y 15s 12m 3M; *Horn Mo(N) 60s*.
Gibraltar Pier 58 50.3N/3 07.8W 2FG(vert) 7/5m 3M. Golden Wharf, N end 58 50.2N/3 11.4W 2FR(vert) 7/5m 3M. Lyness Wharf, S end 58 50.0N/3 11.3W 2FR(vert) 7/5m 3M.
St Margaret's Hope, Needle Point Reef 58 50.1N/2 57.3W FIG 3s 6m 3M; diamond on post. Pierhead 2FG(vert) 6/4m 2M. Ldg Lts 196°, both FR 7/11m.
Rose Ness 58 52.4N/2 49.9W Fl 6s 24m 8M; W Tr. Scapa Pier 58 57.4N/2 58.3W FIG 3s 6m 8M; W mast.
Barrel of Butter 58 53.4N/3 07.5W Fl(2) 10s 6m 7M.
Cava 58 53.2N/3 10.6W Fl WR 3s 12m W10M, R8M; W octagonal Tr; vis W351°-143°, R143°-196°, W196°-251°, R251°-271°, W271°-298°.
Houton Bay Ldg Lts 316°. Front 58 55.00N/3 11.46W Fl G 3s 8m. Rear 200m from front FG 16m both R triangle on W pole, B bands, vis 312°-320°. Ro-Ro terminal, S end Iso R 4s 5M.

CLESTRAN SOUND. Peter Skerry Lt Buoy 58 55.28N/3 13.42W FIG 6s; stbd-hand mark. Riddock Shoal Lt Buoy 58 55.88N/3 15.06W Fl(2)R 12s; port-hand mark.

HOY SOUND. Ebbing Eddy Rocks Lt Buoy 58 56.62N/3 16.90W Q; N cardinal mark. **Graemsay Island** Ldg Lts 104°. **Front** 58 56.46N/3 18.50W Iso 3s 17m **15M**; W Tr; vis 070°-255°. **Rear** 1.2M from front Oc WR 8s 35m **W20M, R16M**; W Tr; vis R097°-112°, W112°-163°, R163°-178°, W178°-332°; obsc on leading line within 0.5M. Skerry of Ness 58 56.98N/3 17.73W Fl WG 4s 7m 7M, G4M; vis W shore-090°, G 090°-shore.

STROMNESS. Pier, SE corner Iso R 6s 15m 5M. Ldg Lts 317°. Front 58 57.6N/3 18.0W FR 29m 11M; post on W Tr. Rear, 55m from front FR 39m 11M; both vis 307°-327°. N pier, head FIR 3s 8m 5M.

Copinsay 58 53.8N/2 40.2W Fl(5) 30s 79m **21M**; W Tr; *Horn (4) 60s*.
Auskerry 59 01.6N/2 34.2W Fl 20s 34m **18M**; W Tr.

Helliar Holm, S end 59 01.17N/2 53.95W Fl WRG 10s 18m W14M, R10M; W Tr; vis G256°-276° W276°-292° R292°-098° W098°-116° G116°-154°. Balfour, pier 59 01.9N/2 54.4W Q WRG 5m W3M, R2M, G2M; vis G270°-010°, W010°-020°, R020°-090°.

KIRKWALL. Scargun Shoal Buoy 59 00.84N/2 58.57W (unlit); stbd-hand mark. **Pier, N end** 58 59.26N/2 57.58W Iso WRG 5s 8m **W15M**, R13M, G13M; W Tr; vis G153°-183°, W183°-192°, R192°-210°.

WIDE FIRTH. Linga Skerry Lt Buoy 59 02.42N/2 57.46W Q(3) 10s; E cardinal mark. Boray Skerries Lt Buoy 59 03.68N/2 57.55W Q(6) + LFl 15s; S cardinal mark. Skertours Lt Buoy 59 04.15N/2 56.61W Q; N cardinal mark. The Galt Lt Buoy 59 05.25N/2 54.15W Q; N cardinal mark.

Brough of Birsay 59 08.25N/3 20.30W Fl(3) 25s 52m **18M**; white castellated Tr and building. Papa Stronsay, NE end, The Ness 59 09.38N/2 34.80W Iso 4s 8m 9M; W Tr.

STRONSAY, PAPA SOUND. Quiabow Buoy 59 09.85N/2 36.20W (unlit); stbd-hand mark. No 1 Lt Buoy (off Jacks Reef) 59 09.20N/2 36.40W FIG 5s; stbd-hand mark. No 2 Lt Buoy 59 08.95N/2 36.50W FIR 5s; port-hand mark. No 3 Lt Buoy 59 08.73N/2 36.08W Fl(2)G 5s; stbd-hand mark. No 4 Lt Buoy 59 08.80N/2 36.37W Fl(2)R 5s; port-hand mark. Whitehall Pier 50 08.61N/2 35.79W 2FG(vert) 8/6m 4M.

SANDAY. **Start Point** 59 16.70N/2 22.50W Fl(2) 20s 24m **19M**; W Tr, B stripes. Kettletoft, pier Fl WRG 3s 7m W7M, R5M, G5M; W Tr; vis W351°-011°, R011°-180°, G180°-351°.

North Ronaldsay. NE end, 59 23.40N/2 22.80W Fl 10s 43m **19M**; R Tr, W bands; RC; Racon; *Horn 60s*. Nouster, pier head 59 21.4N/2 26.3W QR 5m.

EDAY. Calf of Eday 59 14.2N/2 45.7W Iso WRG 5s 8m W8M, R7M, G6M; W Tr; vis R shore-216°, W216°-223°, G223°-302°, W302°-307°. Backaland pier Fl R 3s 5m 4M; vis 192°-250°.

WESTRAY. **Noup Head** 59 19.90N/3 04.10W Fl 30s 79m **22M**; W Tr; vis 335°-242°, 248°-282°; obsc on easterly bearings within 0.8M, part obsc 240°-275°. Pierowall, E pier head Fl WRG 3s 7m W11M, R7M, G7M; vis G254°-276°, W276°-291°, R291°-308°, G308°-215°. West pier head 2 FR (vert). Papa Westray, Moclett Bay pier head Fl WRG 5s 7m W5M, R3M, G3M; vis G306°-341°, W341°-040°, R040°-074°. Egilsay pier, S end Fl G 3s 4m 4M.

SHETLAND ISLES

FAIR ISLE. **Skadan**, S end 59 30.85N/1 39.08W Fl(4) 30s 32m **24M**; W Tr; vis 260°-146°, obsc inshore 260°-282°; *Horn(2) 60s*. **Skroo**, N end 59 33.16N/1 36.49W Fl(2) 30s 80m **22M**; W Tr; vis 087°-358°; *Horn (3) 45s*.

MAINLAND. **Sumburgh Head** 59 51.30N/1 16.37W Fl(3) 30s 91m **23M**; W Tr; RC.

HOSWICK and SANDWICK. No Ness 59 58.40N/1 12.23W Q(2) 10s 54m 5M. Cumlewick Ness 59 59.00N/1 14.30W Q(2)G 10s 15m 4M. Brownies Taing, pier head Fl G 3s 4m 4M; *Horn 30s (occas)*. Mousa, Perie Bard, Fl 3s 20m 10M; W Tr.

BRESSAY. **Kirkabister Ness**, 60 07.25N/1 07.18W Fl(2) 30s 32m **21M**; W Tr; RC; FR Lts on radio masts 0.95M NE.

LERWICK. Twageos Point 60 08.95N/1 07.83W L Fl 6s 8m 6M. Maryfield, Ferry Terminal Oc WRG 6s 5m 5M; vis W008°-013°, R013°-106°, W106°-111°, G111°-008°. Breakwater, N head 2FR(vert) 5m 4M. Victoria Pier, N head, FIR 3s 5m 1M. Victoria Pier, E Elbow, 2FG(vert) 5m 4M. North Jetty QR 5m 1M. North Ness 60 09.6N/1 08.7W Iso WG 4s 4m 5M; vis G158°-216°, W216°-158°. Loofa Baa 60 09.75N/1 08.67W Q(6) + LFl 15s 4m 5M; S cardinal mark. N entrance Dir Lt 215°. Dir Oc WRG 6s 27m 8M; vis R211°-214°, W214°-216°, G216°-221°. Gremista Marina, S breakwater head Iso R 4s 3m 2M. Greenhead Q(4) R 10s 4m 3M.

Rova Head 60 11.45N/1 08.45W Fl(3) WRG 18s 10m W8M, R7M, G6M; W Tr; vis R shore-180°, W180°-194°, G194°-213°, R213°-241°, W241°-261°, G261°-009°, W009°-shore.
Dales Voe 60 11.8N/1 11.1W Fl(2)WRG 8s W4M, R3M, G3M; vis G220°-227°, W227°-233°, R233°-240°. Dales Voe Quay 2FR(vert) 9/7m.
Hoo Stack 60 14.99N/1 05.25W Fl(4)WRG 12s 40m W7M, R5M, G5M; W pylon; vis R169°-180°, W180°-184°, G184°-193°, W193°-169°. Dir Lt 182°. DirFl(4)WRG 12s 33m W9M, R6M, G6M; same structure; vis R177°-180°, W180°-184°, G184°-187°; synchronised with upper Lt.

Moul of Eswick Fl WRG 3s 50m W9M, R6M, G6M; W Tr; vis R shore-200°, W200°-207°, G207°-227°, R227°-241°, W241°-028°, R028°-shore.

WHALSAY. Symbister Ness 60 20.46N/1 02.15W Fl(2) WG 12s 11m W8M, G6M; vis W shore-203°, G203°-shore.
Symbister Bay, S breakwater head 60 20.6N/1 01.5W QG 4m 2M. N breakwater head Oc G 7s 3m 3M.
Skate of Marrister 60 21.4N/1 01.3W FlG 6s 4m 4M; G mast with platform.
Suther Ness Fl WRG 3s 8m W10M, R8M, G7M; vis W shore-038°, R038°-173°, W173°-206°, G206°-shore.
Mainland. Laxo Voe ferry terminal 60 21.1N/1 10.0W 2FG(vert) 4m 2M.

Out Skerries 60 25.50N/0 43.50W Fl 20s 44m **20M**; W Tr. Bruray ferry berth 60 25.4N/0 45.0W 2FG(vert).

Muckle Skerry 60 26.4N/0 51.7W Fl(2) WRG 10s 13m W7M, R5M, G5M; W Tr; vis W046°-192°, R192°-272°, G272°-348°, W348°-353°, R353°-046°.

YELL SOUND. S entrance, Lunna Holm 60 27.38N/1 02.39W Fl(3)WRG 15s 19m W10M, R7M, G7M; W round Tr; vis R shore-090°, W090°-094°, G094°-209°, W209°-275°, R275°-shore. **Firths Voe**, N shore 60 27.2N/1 10.6W Oc WRG 8s 9m **W15M**, R10M, G10M; W Tr; vis W189°-194°, G194°-257°, W257°-263°, R263°-339°, W339°-066°.
Linga Is. Dir Lt DirQ(4)WRG 8s 10m W9M, R9M, G9M; concrete column; vis R145°-148°, W148°-152°, G152°-155°. Q(4)WRG 8s 10m W7M, R4M, G4M; same structure; vis R052°-146°, G154°-196°, W196°-312°; synchronized with Dir Lt. Yell, Ulsta ferry terminal, breakwater head Oc RG 4s 7m R5M, G5M; vis G shore-354°, R044°-shore. Oc WRG 4s 5m W8M, R5M, G5M; same structure; vis G shore-008°, W008°-036°, R036°-shore.
Toft ferry terminal 60 28.0N/1 12.4W 2FR(vert) 5/3m 2M.
Ness of Sound, West side, Iso WRG 5s 18m W9M, R6M, G6M; vis G shore-345°, W345°-350°, R350°-160°, W160°-165°, G165°-shore. Brother Island, Dir Lt 329°, Dir Fl(4) WRG 8s 16m W10M, R7M, G7M; vis G323°-328°, W328°-330°, R330°-333°.
Mio Ness 60 29.7N/1 13.5W Q(2)WR 10s 12m W7M, R4M; W round Tr; W282°-238°, R238°-282°.
Tinga Skerry 60 30.5N/1 14.7W Q(2)G 10s 9m 5M; W Tr.
YELL SOUND, NORTH ENTRANCE. Bagi Stack 60 43.55N/1 07.40W Fl(4) 20s 45m 10M; W Tr. Gruney Island 60 39.20N/1 18.03W Fl WR 5s 53m W7M, R4M; W Tr; vis R064°-180°, W180°-012°; Racon.

Point of Fethaland Fl(3) WR 15s 65m **W24M, R20M**; W Tr; vis R080°-103°, W103°-160°, R160°-206°, W206°-340°.

Muckle Holm 60 34.85N/1 15.90W Fl(2) 10s 32m 10M; W Tr. Little Holm Iso 4s 12m 6M; W Tr. Outer Skerry Fl 6s 12m 8M; B column, W bands. Quey Firth Oc WRG 6s 22m W12M, R8M, G8M; W Tr; vis W shore (through S and W)-290°, G290°-327°, W327°-334°, R334°-shore. Lamba, S side, Fl WRG 3s 30m W8M, R5M, G5M; W Tr; vis G shore-288°, W288°-293°, R293°-327°, W327°-044°, R044°-140°, W140°-shore.

SULLOM VOE. **Gluss Isle** Ldg Lts 195° (H24). **Front** F 39m **19M. Rear** F 69m **19M**. Little Roe 60 30.05N/1 16.35W Fl(3) WR 10s 16m W5M, R4M; Y and W structure; vis R036°-095°, W095°-036°.
Skaw Taing 60 29.1N/1 16.7W Fl(2)WRG 5s 21m W8M, R5M, G5M; Or and W structure; vis W049°-078°, G078°-147°, W147°-154°, R154°-169°, W169°-288°.
Ness of Bardister 60 28.2N/1 19.5W Oc WRG 8s 20m W9M, R6M, G6M; Or and W structure; vis W180°-240°, R240°-310°, W310°-314°, G314°-030°.
Vats Houllands 60 28.0N/1 17.5W Oc WRGY 3s 73m 6M; grey Tr; vis W343°-029°, Y029°-049°, G049°-074°, R074°-098°, G098°-123°, Y123°-148°, W148°-163°.
Fugla Ness. Lts in line 212°. Rear 60 27.3N/1 19.7W Iso 4s 45m 14M. Common front 60 27.5N/1 19.4W Iso 4s 27m 14M; synchronized with rear Lts. Lts in line 203°. Rear 60 27.3N/1 19.6W Iso 4s 45m 14M.
Sella Ness. Upper Lt 60 26.9N/1 16.5W Q WRG 14m 7M; vis G084°-099°, W099°-100°, W126°-128°, R128°-174°; by day F WRG (occas).
Sella Ness. Lower Lt. Q WRG 10m 7M; vis G084°-106°, W106°-115°, R115°-174°; by day F WRG (occas).
Tug jetty, pier head Iso G 4s 4m 3M.
Garth Pier, N arm, head 60 26.7N/1 16.2W Fl(2)G 5s 4m 3M.
Scatsa Ness. Upper Lt 60 26.5N/1 18.1W Oc WRG 5s 14m 7M; vis G161°-187°, W187°-188°, W207°-208°, R208°-251°; by day F WRG (occas).
Scatsa Ness. Lower Lt Oc WRG 5s 10m 7M; vis G161°-197°, W197°-202°, R202°-251°; by day F WRG (occas).
Ungam Island 60 27.3N/1 18.5W VQ(2) 5s 2m 2M; W column; Ra refl.
Whitehill 60 34.85N/1 00.01W Fl WR 3s 24m W9M, R6M; vis W shore-163°, R163°-211°, W211°-349°, R349°-shore.
Uyea Sound 60 41.2N/0 55.3W Fl(2) 8s 8m 7M; R & W Bn.
Balta Sound 60 44.5N/0 47.6W Iso WR 4s 17m W9M, R6M; vis W249°-010°, R010°-060°, W060°-154°; Q Lt (occas) marks Unst Aero RC 0.7M W.
NORTH UNST. **Muckle Flugga** 60 51.33N/0 53.00W Fl(2) 20s 66m **25M**; W Tr; RC. **Auxiliary Lt** FR 52m **15M** (same Tr); vis 276°-311°.
Yell. Cullivoe breakwater head 60.41.9N/0 59.7W Oc R 7s 5m 2M. Pier head 2FG(vert) 6/4m 1M.
Esha Ness 60 29.35N/1 37.55W Fl 12s 61m **25M**; W square Tr. Hillswick, S end of Ness 60 27.2N/1 29.7W Fl(4) WR 15s 34m W9M, R6M; W house: vis W217°-093°, R093°-114°.

Muckle Roe, Swarbacks Minn, 60 21.05N/1 26.90W Fl WR 3s 30m W9M, R6M; vis W314°-041°, R041°-075°, W075°-137°. Aith breakwater. RNLI berth 60 17.2N/1 22.3W QG 5m 3M. West Burra Firth. Transport Pier, head 60 17.7N/1 32.3W Iso G 4s 4m 4M.
Ve Skerries 60 22.40N/1 48.67W Fl(2) 20s 17m 11M; Racon.
VAILA SOUND. Rams Head 60 12.00N/1 33.40W Fl WG 8s 15m W8M, G5M; W house; vis G265°-355°, W355°-136°, obsc by Vaila I when brg more than 030°. Vaila Pier. 60 13.5N/1 34.0W 2FR(vert).
Skeld Voe. Skeld Pier, head 60 11.2N/1 26.1W 2FR(vert) 4/2m 3M.
SCALLOWAY. Moores slipway, jetty head 60 08.2N/1 16.7W 2FR(vert) 4/3m 1M. Blacksness, West Pier, head 2FG(vert) 6/4m 3M. E Pier, head Oc R 7s 5m 3M.

Fugla Ness 60 06.40N/1 20.75W Fl(2)WRG 10s 20m W10M, R7M, G7M; W Tr; vis G014°-032°, W032°-082°, R082°-134°, W134°-shore.

FOULA. 60 06.78N/2 03.72W Fl(3) 15s 36m **18M**; W Tr.

10.7.5 PASSAGE INFORMATION

RATTRAY HEAD TO DUNCANSBY HEAD (chart 115)

On direct route from Rattray Hd (Lt, fog sig) to Duncansby Hd (Lt) heavy seas may be met in strong W winds. For oil installations see 10.5.5. Tidal streams are moderate off Rattray Hd and strong in inner part of Moray Firth, but weak elsewhere in the firth. 5M NE of Rattray Hd the NE-going stream begins at HW Aberdeen +0140, and the SE-going stream at HW Aberdeen −0440, sp rates 2 kn.

In strong winds the sea breaks over Steraton Rk in the E approach to Fraserburgh (10.7.7), and over Colonel Rk 1¼M E of Kinnairds Hd (Lt). Banff B is shallow: beware Collie Rks N of Macduff (10.7.8). Banff harb dries, and should not be approached in fresh NE-E winds, when sea breaks well offshore.

From Meavie Pt to Scarnose dangers extend up to 3 ca from shore in places. Beware Caple Rk (depth 0.2m) ¾M W of Logie Hd. Spey B is clear of dangers more than ¾M from shore: anch here, but only in offshore winds. Beware E Muck (dries) ½M SW of Craigan Roan, an above-water rky patch ½M SW of Craig Hd, and Middle Muck and W Muck in approach to Buckie (10.7.10). For Portknockie see 10.7.27.

Halliman Skerries (dry) lie 1½M WNW of Stotfield Hd, and are marked by Bn. Covesea Skerries Lt Ho is 2M W of Stotfield Hd, and Covesea Skerries (dry) lie ½M NW of the Lt Ho.

Inverness Firth is approached between Nairn (10.7.15) and S Sutor. In heavy weather there is a confused sea with overfalls on Guillam Bank, 9M S of Tarbat Ness. The sea also breaks on Riff Bank (S of S Sutor) which dries in places. Chans run both N and S of Riff Bank. Off Fort George, on E side of entrance to Inverness Firth (chart 1078), the SW-going stream begins HW Aberdeen +0605, sp rate 2½ kn; the NE- going stream begins at HW Aberdeen −0105, sp rate 3½ kn. There are eddies and turbulence between Fort George and Chanonry Pt when stream is running hard. Most of Inverness Firth is shallow, but a deeper chan skirts the NW shore to within 1M of Craigton Pt, NE of which is a bank called Meikle Mee with depths of less than 1m. There is a clearance of 29m under the bridge, where tidal stream is very strong (see Inverness 10.7.16). For Fortrose see 10.7.27.

Cromarty Firth (chart 1889) is entered between N Sutor and S Sutor, both fringed by rks, some of which dry. Off the entrance the in-going stream begins at HW Aberdeen +0605, and the out-going at HW Aberdeen −0105, sp rates 1½ kn. Good sheltered anchs within the firth. See 10.7.27.

The coast running NE to Tarbat Ness (Lt) is fringed with rks. Beware Three Kings (dries) about 3M NE of N Sutor. Culloden Rk, a shoal with depth of 1.8m, extends ¼M NE of Tarbat Ness, where stream is weak. Beware salmon nets on coast between Tarbat Ness and Portmahomack (10.7.17). Dornoch Firth (10.7.27) is shallow, with shifting banks, and in strong E winds the sea breaks heavily on the bar E of Dornoch Pt.

At Lothbeg Pt, between Brora Pt and Helmsdale (10.7.18), there is a rk ledge extending ½M offshore. Near Berriedale, 7M NE of Helmsdale, is The Pinnacle, a detached rk 61m high, standing close offshore. The Beatrice oil field lies on Smith Bank, 28M NE of Tarbat Ness, and 11M off Caithness coast. Between Dunbeath and Lybster (10.7.27) there are no dangers more than 2 ca offshore. Clyth Ness is fringed by detached and drying rks. From here to Wick (10.7.19) the only dangers are close offshore. There is anch in Sinclair's B in good weather, but Freswick B further N is better while waiting for tide in Pentland Firth (beware wreck in centre of bay). Baxter Rk (depth 2.7m) lies 4 ca S of Duncansby Hd, and Stacks of Duncansby 6 ca further SSW.

PENTLAND FIRTH (chart 2162)

This dangerous chan should only be attempted in moderate winds and good vis, and not at sp tides. At E end the firth is entered between Duncansby Hd and Old Hd (S Ronaldsay), between which lie Pentland Skerries. At W end the firth is entered between Dunnet Hd and Tor Ness (Hoy). Near the centre of firth are the islands of Swona (N side) and Stroma (S side). Outer Sound is between Swona and Stroma; Inner Sound is between Stroma and the Scottish coast E of St John's Pt. Other than as above there are no dangers more than about 2 ca offshore, but the tide runs extremely strongly, and causes very severe eddies and races and a most confused sea at different times and places depending on conditions. See *North Coast of Scotland Pilot*.

In general the E-going stream begins at HW Aberdeen +0500, and the W-going at HW Aberdeen −0105. On both streams eddies form S of Muckle Skerry and around Stroma. Between Pentland Skerries and Duncansby Hd the sp rate is 8-10 kn. On the SE-going stream Duncansby Race extends first towards Muckle Skerry but then swings anti-clockwise until by HW Aberdeen −0440 it extends in a NW direction from Duncansby Hd. At HW Aberdeen −0140 it starts to subside. By HW Aberdeen +0245 the race forms again for about 2h in a ENE direction towards Muckle Skerry. This race is violent in strong E or SE winds against SE going stream.

The stream runs 8-9 kn at sp through Outer Sound, causing a persistent and dangerous race off Swilkie Pt at N end of Stroma, especially with a W-going stream and a strong W wind.

The most dangerous and extensive race in the firth is Merry Men of Mey, which forms off St John's Pt on W-going stream at HW Aberdeen −0150 and for a while extends right across to Tor Ness with heavy breaking seas even in fine weather. By HW Aberdeen +0315 the SE end of race detaches from Men of Mey Rks off St John's Pt, and by HW Aberdeen +0435 the race begins to subside off Tor Ness. With the start of the E-going stream, at about HW Aberdeen +0515, the race subsides in mid-chan.

Any yacht going through Pentland Firth, even in ideal conditions, must avoid Duncansby Race, Swilkie Race, and Merry Men of Mey. A safe passage depends on correct timing and positioning to avoid the worst races mentioned above, regular fixes to detect any dangerous set, and sufficient power to cope with the very strong tidal stream.

ORKNEY ISLANDS (10.7.20 and charts 2249, 2250)

The islands are mostly indented and rky, but with sandy beaches especially on NE sides. Pilotage is easy in good vis, but in other conditions great care is needed since tides run strongly. For details refer to Clyde Cruising Club sailing directions and *North Coast of Scotland Pilot*, which apply also to Shetland Islands.

When cruising in Orkney it is essential to understand and use the tidal streams to the best advantage, while at the same time considering the various tide races and overfalls, particularly near sp. A good engine is needed since, for example, there are many places where it is dangerous to get becalmed. It must also be remembered that swell from the Atlantic or North Sea can contribute to dangerous sea conditions, or penetrate to some of the anchorages. During summer months winds are not normally unduly strong, and can be expected to be Force 7 or more on about two days a month. But in winter the wind reaches this strength for 10-15 days per month, and gales can be very severe in late winter and early spring. Cruising conditions are best near midsummer, when of course the hours of daylight are much extended.

Stronsay Firth and Westray Firth run from SE to NW through the group. Races and/or tide rips, often dangerous in bad weather with wind against tide, occur off Mull Head, over Dowie Sand, between Muckle Green Holm and War Ness (where violent turbulence may extend right across the firth), between Faraclett Hd and Wart Holm, and off Sacquoy Hd. Off War Ness the SE-going stream begins at HW Aberdeen +0435, and the NW-going at HW Aberdeen −0200, sp rates 7 kn.

Tide races and dangerous seas occur at the entrances to most of the firths or sounds when the stream is against strong winds. This applies particularly to Hoy Sound, Eynhallow Sound, Papa Sound (Westray), Lashy Sound, and North Ronaldsay Firth.

There are many good anchs among the islands, including: Deer Sound (W of Deer Ness); B of Firth, B of Isbister, and off Balfour in Elwick B (all leading from Wide Firth); Rysa Sound, B of Houton, Hunda Sound (in Scapa Flow); Rousay Sound; and Pierowall Road (Westray). Plans for some of these are on chart 2622. For Shapinsay, Auskerry, Houton Bay and Pierowall see 10.7.27. There is a major oil terminal and prohibited area at Flotta, on S side of Scapa Flow.

SHETLAND ISLANDS (10.7.24 and charts 3281, 3282, 3283)

These islands mostly have bold cliffs and are relatively high, separated by narrow sounds through which the tide runs strongly, so that in poor vis great care is needed. The Clyde Cruising Club sailing directions are almost indispensable.

As an introduction to these waters, a most violent and dangerous race forms off Sumbrugh Hd (at S end of mainland) on both streams. Other dangerous areas include between Ve Skerries and Papa Stour; the mouth of Yell Sound with strong wind against N-going stream; and off Holm of Skaw (N end of Unst).

Although there are many secluded and attractive anchs, it must be remembered that the weather can change very quickly, with sudden shifts of wind. Also beware salmon fisheries and mussel rafts (unlit) in many Voes, Sounds and harbours. For Scalloway and Balta Sound see 10.7.27.

Fair Isle (North Haven) is a useful port of call when bound to/from Shetland Is. Note that races form off both ends of the island, especially S, during the strength of tidal stream in both directions. See 10.7.27.

SCOTLAND − NORTH COAST (chart 1954)
Dunnet B, S of Dunnet Hd (Lt) gives temp anch in E or S winds, but dangerous seas roll in from NW. On W side of Thurso B is Scrabster (10.7.26).

Between Holborn Hd and Strathy Pt the E-going stream begins at HW Ullapool −0150, and the W-going at HW Ullapool +0420, sp rates 2½ kn. Close to Brims Ness off Ushat Hd the sp rate is 3 kn, and there is often turbulence. SW of Ushat Hd the buildings of Dounreay are conspic, near shore. Dangers extend ¼M seaward off this coast.

Along E side of Strathy Pt an eddy gives almost continuous N-going stream, but there is usually turbulence off the pt where this eddy meets the main E or W stream. Several small bs along this coast give temp anch in offshore winds, but must not be used or approached with wind in a N quarter.

Kyle of Tongue (10.7.27) is entered from E through Caol Raineach, S of Eilean nan Ron, or from N between Eilean Iosal and Cnoc Glass. There is no chan into the kyle W of Rabbit Islands, to which a drying spit extends ½M NNE from the mainland shore. Further S there is a bar across entrance to inner part of kyle. There are anchs on SE side of Eilean nan Ron, SE side of Rabbit Islands, off Skullomie, or S of Eilean Creagach off Talmine. Approach to the latter runs close W of Rabbit Islands, but beware rks to N and NW of them.

Loch Eriboll (see chart 2720 and 10.7.27) provides secure anchs, but in strong winds violent squalls blow down from mountains. Eilean Cluimhrig lies on W side of entrance; the other (E) shore is fringed with rks up to 2 ca offshore. At White Hd (Lt) the loch narrows to 6½ ca. There are chans W and E of Eilean Choraidh. Best anchs in Camas an Duin (S of Ard Neackie) or in Rispond B close to entrance (but not in E winds, and beware Rispond Rk which dries). Westward to C Wrath (see 10.8.5) the coast is indented, with dangers extending 3 ca from the shore and from offlying rks and islets. A firing exercise area extends 8M E of C Wrath, and 4M offshore. When in use, R flags or pairs of R Lts (vert) are shown from E and W limits, and yachts should keep clear.

10.7.6 DISTANCE TABLE

Approximate distances in nautical miles are by the most direct route while avoiding dangers and allowing for traffic separation schemes etc. Places in *italics* are in adjoining areas.

	1	2	3	4	5	6	7	8	9	10	11	12	13	14	15	16	17	18	19	20
1 *Flamborough Head*	1																			
2 *Fife Ness*	155	2																		
3 Peterhead	212	78	3																	
4 Fraserburgh	227	94	16	4																
5 Banff	244	111	33	18	5															
6 Buckie	257	124	46	31	15	6														
7 Lossiemouth	267	134	56	41	25	11	7													
8 Findhorn	280	147	69	54	38	24	13	8												
9 Inverness	301	168	90	75	59	45	34	23	9											
10 Tarbat Ness	283	150	72	57	41	27	18	14	27	10										
11 Helmsdale	285	152	74	59	44	33	26	28	43	16	11									
12 Wick	283	150	72	57	50	46	44	51	69	42	29	12								
13 Duncansby Head	292	158	82	67	62	58	57	63	81	54	41	13	13							
14 Scrabster	310	176	100	85	80	76	75	81	99	72	59	31	18	14						
15 Kirkwall	325	191	115	100	95	91	90	96	114	87	74	46	34	50	15					
16 Stromness	314	180	104	89	84	80	79	85	103	76	63	35	22	25	32	16				
17 Noup Head	342	208	132	117	112	108	107	113	131	104	91	63	50	47	23	28	17			
18 Lerwick	356	238	160	150	156	160	162	170	190	162	148	120	109	124	95	110	82	18		
19 Sullom Voe	385	268	190	180	186	190	192	200	220	192	178	150	139	154	125	136	108	38	19	
20 *Cape Wrath*	355	221	145	130	125	121	120	126	144	117	104	76	63	47	79	58	73	155	176	20

AREA 7—NE Scotland 399

FRASERBURGH 10-7-7
Aberdeen

CHARTS
Admiralty 1462, 222; OS 30
TIDES
+0120 Dover; ML 2.3; Duration 0615; Zone 0 (GMT).

Standard Port ABERDEEN (←)

Times				Height (metres)			
HW		LW		MHWS	MHWN	MLWN	MLWS
0000	0600	0100	0700	4.3	3.4	1.6	0.6
1200	1800	1300	1900				

Differences FRASERBURGH
−0045 −0115 −0110 −0045 −0.4 −0.3 −0.1 0.0

SHELTER
Good shelter but entrance is dangerous in on-shore gales. Yachts normally use South Harbour.
NAVIGATION
Waypoint 57°41'.32N 01°58'.71W, 111°/291° from/to entrance, 0.57M. This is a very busy fishing port and fishing vessels come and go day and night. There are no particular facilities for yachtsmen but it is a safe refuge. Large boulders have been deposited in SW part of harbour ent, reducing depth over an area which extends NE to the 291° ldg line. Take care on entering or leaving.
LIGHTS AND MARKS
Ldg Lts (Front QR, Rear Oc R 6s) lead in between the piers at 291°.
Entry signals at South Pier
2 B Balls or 2 R Lts (vert): No entry. R flag or one R Lt: Harbour open but special care needed.
Docking signals (from head of Burnett Pier, SE end of West Pier, and NE end of head of North Pier). By day, semaphore arm: 45° above horizontal — clear inwards; horizontal — closed to all traffic; 45° below horizontal — clear outwards. By night G Lt shows relevant basin open, R Lt shows relevant basin closed.
RADIO TELEPHONE
VHF Ch 16; 12 (H24).
TELEPHONE (0346)
Hr Mr 23323; CG 24279; MRSC Peterhead 74278; HMC 28033; Marinecall 0898 500 451; Dr 22088.
FACILITIES
EC Wednesday; **Port** Tel. 25858, Slip, P, D, L, FW, ME, El, Sh, C (70 ton mobile), CH, AB, V, R, Bar; **Caleys** Tel. 23241, P, D; **Mitchells** Tel. 22021, ME, El; **Buchan, Hall and Mitchell** Tel. 23336, ME, El, Sh; **May & Bruce** Tel. 25222, ME; **C. Will** Tel. 23364, C (30 ton); **G Walker** Tel. 23211 Slip, SM, Sh, C; **J. Noble** Tel. 3179, El, Sh; **Murisons** Tel. 23376, Gas.
Town PO; Bank; Rly (bus to Aberdeen); Air (Aberdeen).

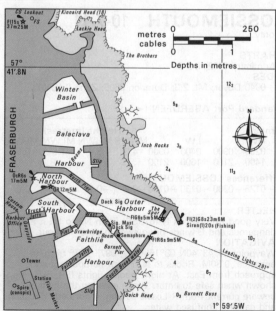

▲ Apply harbour office if not previously directed.

MACDUFF 10-7-8
(BANFF)
Banff

CHARTS
Admiralty 1462, 222; OS 29
TIDES
+0055 Dover; ML 2.0; Duration 0615; Zone 0 (GMT).

Standard Port ABERDEEN (←)

Times				Height (metres)			
HW		LW		MHWS	MHWN	MLWN	MLWS
0200	0900	0400	0900	4.3	3.4	1.6	0.6
1400	2100	1600	2100				

Differences BANFF
−0100 −0150 −0150 −0050 −0.8 −0.6 −0.5 −0.2

SHELTER
Reasonably good shelter but the entrance is open to westerly winds. Entry not recommended in strong NW winds. Harbour is three basins with entrance 17m wide. A busy cargo and fishing port.
NAVIGATION
Waypoint 57°40'.50N 02°30'.50W, 307°/127° from/to Macduff ent, 0.40M. Same waypoint 056°/236° from/to Banff entrance, 0.44M.
Beware rocky coast to the N and S of harbour entrance. There is a slight to moderate surge in the outer harbour with N to NE gales.
LIGHTS AND MARKS
Ldg Lts 127°, both FR.
RADIO TELEPHONE (local times)
VHF Ch 16; 12 (0900-1700 and 1 hr before vessel expected).

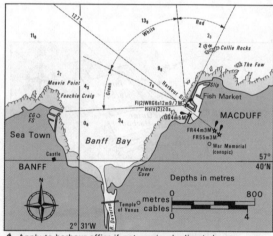

▲ Apply to harbour office if not previously directed

TELEPHONE (Banff 02612, Macduff 0261)
Hr Mr (Macduff) 32236; Hr Mr (Banff) 5093; CG Banff 2415; MRSC Peterhead 74278; HMC 32217; Marinecall 0898 500 451; Dr Banff 2027.
FACILITIES
EC Wednesday; **Harbour** Tel. 32236, Slip, P, D, FW, ME, El, Sh, CH, V, R, Bar; **Macduff Boat Building & Engineering Co** Tel. 32234, ME, El, Sh. **Banff SC**. **W. Thompson** Tel. 32388, Gas; **Town** PO (Macduff, Banff); Bank (Banff); Rly (bus to Keith); Air (Aberdeen).
Note: Banff harbour, controlled by Grampian Region Council, is a popular yachting port. When entrance to Macduff is very rough with winds from SW to N, Banff can be a safe refuge. When winds are strong between N and ENE, Banff is unapproachable. Harbour dries. FW on quays. Town has P, D, Rly, V, R, Bar.

WHITEHILLS 10-7-9
Banff

CHARTS
Admiralty 115, 222; OS 29
TIDES
+0050 Dover; ML 2.4; Duration 0610; Zone 0 (GMT).

Standard Port ABERDEEN (←—)

Times				Height (metres)			
HW		LW		MHWS	MHWN	MLWN	MLWS
0200	0900	0400	0900	4.3	3.4	1.6	0.6
1400	2100	1600	2100				

Differences WHITEHILLS
−0122 −0137 −0117 −0127 −0.4 −0.3 +0.1 +0.1

SHELTER
Safe. Swell in the outer basin in strong N to W winds.
NAVIGATION
Waypoint 57°42'.00N 02°34'.80W, 000°/180° from/to breakwater Lt, 1.2M. Reefs on S side of channel marked by two beacons. Beware numerous fishing floats.
LIGHTS AND MARKS
Fl WR 3s on pier head — approach in R sector vis R132°-212°, W212°-245°; Siren 10s (occas).
RADIO TELEPHONE
Whitehills Harbour Radio VHF Ch 16; 09.
TELEPHONE (026 17)
Hr Mr 229; CG Banff 2415; MRSC Peterhead 74278; HMC Peterhead 74867; Marinecall 0898 500 451; Dr Banff 2027.
FACILITIES
EC Wednesday; **Pier** Tel. 229, P, D, FW, ME, El, CH, V, Bar; **End of Harbour** Slip; **Paterson** Tel. 219, P, D, CH. **Town** V, R, Bar. PO; Bank (AM only); Rly (bus to Keith); Air (Aberdeen).

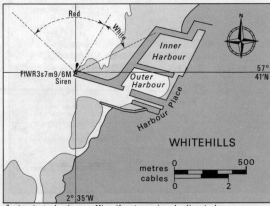

▲ Apply to harbour office if not previously directed

BUCKIE 10-7-10
Banff

CHARTS
Admiralty 1462, 222; OS 28
TIDES
+0040 Dover; ML 2.4; Duration 0550; Zone 0 (GMT).

Standard Port ABERDEEN (←—)

Times				Height (metres)			
HW		LW		MHWS	MHWN	MLWN	MLWS
0200	0900	0400	0900	4.3	3.4	1.6	0.6
1400	2100	1600	2100				

Differences BUCKIE
−0130 −0145 −0125 −0140 −0.2 −0.2 0.0 +0.1

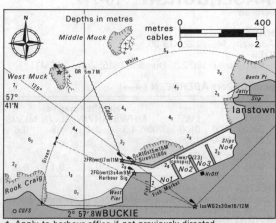

▲ Apply to harbour office if not previously directed

SHELTER
Overnight only. Good shelter and can be entered in all weathers. Hr entrance is 9m wide. Portknockie has good shelter. See 10.7.27.
NAVIGATION
Waypoint 57°41'.32N 02°58'.80W, 306°/126° from/to entrance, 0.80M. Beware West Muck, Middle Muck and East Muck Rocks, 3 ca off shore. There is a dangerous swell over the bar in harbour entrance in strong N to W winds.
LIGHTS AND MARKS
Lights in line at 125° lead SW of West Muck. Entry signals:
W Pier B ball or Fl G Lt — Entrance depth less than 3m
       B ball or FG over Fl G Lts — Entrance dangerous
       2 B balls or 2 FR Lts — Harbour closed.
RADIO TELEPHONE (local times)
VHF Ch 16; 09 11 12 (Mon-Fri: 0900-1700. Other times: Every H+05 to H+15 when vessel expected).
TELEPHONE (0542)
Hr Mr 31700; CG Banff 2415; MRSC Peterhead 74278; HMC 32254; Marinecall 0898 500 451; Dr 31555.
FACILITIES
EC Wednesday; **No 4 Basin (East)** Tel. 31700, Slip, D, (Lorry delivery), AB, FW; **Aberdeen Boat Centre** Sh, CH; **Herd and Mackenzie** Tel. 31245, Slip, ME, El, Sh, C (15 ton); **Jones, Buckie SY** Tel. 32727, ME, El, Sh, CH, ACA, SM, BH; **Moravian Motors** Tel. 33977, Gas. **Town** PO; Bank; Rly (bus to Elgin); Air (Aberdeen or Inverness).

LOSSIEMOUTH 10-7-11
Moray

CHARTS
Admiralty 1462; OS 28
TIDES
+0040 Dover; ML 2.3; Duration 0605; Zone 0 (GMT)

Standard Port ABERDEEN (←—)

Times				Height (metres)			
HW		LW		MHWS	MHWN	MLWN	MLWS
0200	0900	0400	0900	4.3	3.4	1.6	0.6
1400	2100	1600	2100				

Differences LOSSIEMOUTH
−0125 −0200 −0130 −0130 −0.2 −0.2 0.0 0.0

SHELTER
Very good from winds N to SSE. From NE to SE it is dangerous and E winds bring swell in outer harbour.
NAVIGATION
Waypoint 57°43'.40N 03°16'.00W, 097°/277° from/to entrance, 0.30M. Rocks to N & S of harbour entrance — approach from East. At night leading lights (FR 292°) are shown when safe to enter. When entering the harbour, beware current from R Lossie setting in northerly direction and causing confused water.

AREA 7—NE Scotland 401

LOSSIEMOUTH continued

LIGHTS AND MARKS
By day — B ball at S pier } Dangerous
By night — R Lt over G Lt } to enter
R Flag at South Jetty when merchant vessels entering or leaving. Leading lights 292°, both FR.

RADIO TELEPHONE (local times)
VHF Ch 16; 12 (0800-1700, and 1h before vessel expected).

TELEPHONE (034 381)
Hr Mr 3066; CG 2009; MRSC Peterhead 74278; HMC Elgin 547518; Marinecall 0898 500 451; Dr 2277.

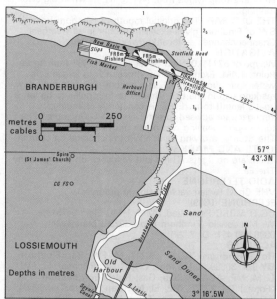

⚠ Apply to harbour office if not previously directed

FACILITIES
EC Thursday; **Harbour** Slip, D, M (See Hr Mr), FW (on quay), ME, El, Sh, C, AB, SM, BH (15 ton); **Jones Buckie Shipyard** Tel. 2029, ME, El, Sh, CH. **Lossiemouth SC** Tel. 2928; **Mallard** Tel. 3001, Gas. **Town** P, V, R, Bar, PO; Bank; Rly (bus to Elgin); Air (Inverness).

HOPEMAN   10-7-12
Moray

CHARTS
Admiralty 1462, 223; OS 28

TIDES
+0050 Dover; ML 2.4; Duration 0610; Zone 0 (GMT)

Standard Port ABERDEEN (←)

Times				Height (metres)			
HW		LW		MHWS	MHWN	MLWN	MLWS
0200	0900	0400	0900	4.3	3.4	1.6	0.6
1400	2100	1600	2100				

Differences HOPEMAN
−0120 −0150 −0135 −0120 −0.2 −0.2 0.0 0.0

SHELTER
Entrance is difficult in winds from NE to SE. Once in SW basin, shelter good from all winds.

NAVIGATION
Waypoint 57°42'.68N, 03°26'.50W, 263°/083° from/to entrance, 0.17M. Dangerous rocks lie off harbour entrance. Harbour dries. Do not attempt entry in heavy weather. Beware salmon stake nets E and W of harbour Mar to Aug and lobster pot floats.

LIGHTS AND MARKS
Ldg Lts 081°, Front FR3m, Rear FR4m (1 August – 30th April).

RADIO TELEPHONE
None.

TELEPHONE (0343)
Hr Mr 830650; CG Lossiemouth 2009; MRSC Peterhead 74278; HMC Elgin 547518; Marinecall 0898 500 451; Doctor Elgin 3141.

FACILITIES
EC Wednesday. **John More** Tel. 830221 CH, ME, Gas; **North East Sailing** Tel. 830889 Sh; **J Sutherland** Tel. 830236 ME, El, D, P (cans); **Harbour** Slip, AB, D, FW; **Town** V, R, Bar, PO, Bank, Rly (bus to Elgin); Air (Inverness).

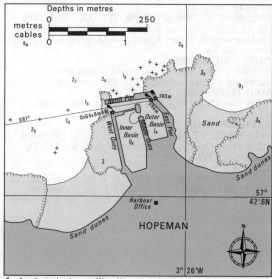

⚠ Apply to harbour office if not previously directed

BURGHEAD 10-7-13
Moray

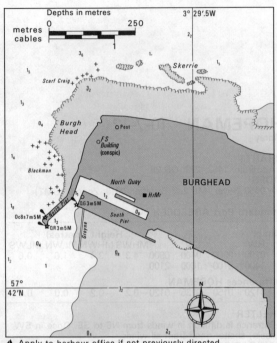

⚠ Apply to harbour office if not previously directed

CHARTS
Admiralty 1462, 223; OS 28
TIDES
+0050 Dover; ML 2.4; Duration 0610; Zone 0 (GMT).

Standard Port ABERDEEN (←)

Times				Height (metres)			
HW		LW		MHWS	MHWN	MLWN	MLWS
0200	0900	0400	0900	4.3	3.4	1.6	0.6
1400	2100	1600	2100				

Differences BURGHEAD
−0120 −0150 −0135 −0120 −0.2 −0.2 0.0 0.0

SHELTER
One of the few harbours open in Moray Firth with strong E winds. NE wall of basin dries. Go alongside where available and contact Hr Mr's office. Can become very congested with fishing vessels.
NAVIGATION
Waypoint 57°42'.30N 03°30'30W, 317°/137° from/to N Pier Lt, 0.28M. Channel is subject to variations due to sand movement. Details from Hr Mr. Approach from SW. Access HW∓4.
LIGHTS AND MARKS
No leading lights but night entrance is safe after identifying the two outer pier Lts, QR and Oc 8s.
RADIO TELEPHONE
Hr Mr VHF Ch 16; 14 (office hours and when vessel expected).
TELEPHONE (0343)
Hr Mr 835337, CG Lossiemouth 2009; MRSC Peterhead 74278; HMC Elgin 547518; Marinecall 0898 500 451; Dr Lossiemouth 2277.
FACILITIES
EC Thursday; **Harbour** D, FW, AB, C (15 ton mobile), L, Slip. **Ernest Colburn** Tel. 835088 BY, Sh; **Town** Bar, PO, V, P (cans), Bank, Rly (Bus to Elgin), Air (Inverness).

FINDHORN 10-7-14
Moray

CHARTS
Admiralty 223; OS 27
TIDES
+0110 Dover; ML 2.5; Duration 0615; Zone 0 (GMT)

Standard Port ABERDEEN (←)

Times				Height (metres)			
HW		LW		MHWS	MHWN	MLWN	MLWS
0200	0900	0400	0900	4.3	3.4	1.6	0.6
1400	2100	1600	2100				

Differences FINDHORN
−0120 −0150 −0135 −0130 0.0 −0.1 0.0 +0.1

SHELTER
Anchor off north pier and enquire at YC. Do not attempt entry in strong winds from NW to NE or when big swell is running.
THE OLD BAR. The original mouth of Findhorn River 4M SW of Findhorn gives excellent shelter in all weathers. Channel changes — local knowledge needed.
NAVIGATION
Waypoint 57°41'.00N 03°38'.80W 328°/148° from/to Ee Point, 1.4M. From waypoint proceed S to locate safe water spar buoy, RWVS, and beyond this two red buoys marking gap in the bar. Thence to four poles to be left a boat's length to port. Bar is constantly changing; strangers are advised to enter near HW. Once past The Ee turn to port, leaving two green buoys to stbd. Beware The Sturdy, a drying bank NW of anchorage on chartlet.
LIGHTS AND MARKS
There are no lights. There is a windsock on a flagpole by The Ee.
RADIO TELEPHONE
VHF Ch M (when yacht racing in progress).
TELEPHONE (0309)
CG & MRSC Peterhead 74278; HMC Elgin 547518; W. Macdonald (Findhorn Pilot) 30236; Marinecall 0898 500 451; Dr 72221.
FACILITIES
Royal Findhorn YC Tel. 30247, M, FW, Bar; **Findhorn BY** Tel. 30099, C (16 ton), CH, D and P (cans), El, FW, L, M, ME, ACA, R, Slip, Gas; **Reids Engineering** (Forres) Tel. 72175, ME, El; **Norman Whyte** Tel. 30504, AC, L, ME, FW, BY, Sh; **Town** V, R, Bar, PO; Bank (Nairn and Forres); Rly (Forres), Air (Inverness).

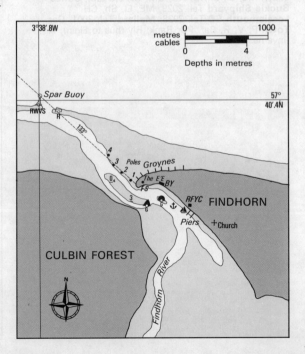

NAIRN 10-7-15
Nairn

CHARTS
Admiralty 1462, 223; OS 27
TIDES
+0110 Dover; ML 2.2; Duration 0615; Zone 0 (GMT).

Standard Port ABERDEEN (←)

Times				Height (metres)			
HW		LW		MHWS	MHWN	MLWN	MLWS
0200	0900	0400	0900	4.3	3.4	1.6	0.6
1400	2100	1600	2100				

Differences NAIRN
−0120 −0150 −0135 −0130 0.0 −0.1 0.0 +0.1
McDERMOTT BASE
−0110 −0140 −0120 −0115 −0.1 −0.1 +0.1 +0.3

SHELTER
Entry channel faces NNE so winds from that direction make entry difficult.
NAVIGATION
Waypoint 57°35′.90N, 03°51′.80W, 335°/155° from/to entrance, 0.30M. Advisable to enter between HW∓1½. Harbour is no longer used by commercial shipping.
LIGHTS AND MARKS
Lt Ho on E pier head, sectored Oc WRG 4s, G shore-100°, W100°-207°, R207°-shore. Keep in W sector.
RADIO TELEPHONE
None. Ch 37 (weekends only).
TELEPHONE (0667)
Hr Mr 54704; MRSC Peterhead 74278; HMC Inverness 231608; Marinecall 0898 500 451; Dr 53421; Clinic 5092.
FACILITIES
EC Wednesday: **Grants** Tel. 52243 D and P (drums); **Nairn Basin** FW (standpipes), Slip, AB, P, D, L; **Nairn SC** Tel. 53897, Bar; **Town** V, R, Bar, PO, Bank, Rly, Air (Inverness).

INVERNESS 10-7-16
Inverness

CHARTS
Admiralty 115, 1077, 1078; OS 26/27
TIDES
+0100 Dover; ML 2.7; Duration 0620; Zone 0 (GMT).

Standard Port ABERDEEN (←)

Times				Height (metres)			
HW		LW		MHWS	MHWN	MLWN	MLWS
0300	1000	0000	0700	4.3	3.4	1.6	0.6
1500	2200	1200	1900				

Differences INVERNESS
−0050 −0150 −0200 −0105 +0.5 +0.3 +0.2 +0.1
FORTROSE
−0125 −0125 −0125 −0125 0.0 0.0 No data
CROMARTY
−0120 −0155 −0155 −0120 0.0 0.0 +0.1 +0.2
INVERGORDON
−0105 −0200 −0200 −0110 +0.1 +0.1 +0.1 +0.1
DINGWALL
−0045 −0145 No data +0.1 +0.2 No data

SHELTER
Good shelter in all weathers. Inverness is the NE entrance to the Caledonian Canal, entrance to which can be difficult in strong tides. Caledonian Canal — See 10.8.15. Fortrose offers some safe anchorages at the entrance to the inner part of the Moray Firth. See 10.7.27.
NAVIGATION
Waypoint Meikle Mee stbd-hand buoy, FlG 3s, 57°30′.28N 04°11′.93W, 070°/250° from/to Longman Pt Bn, 0.74M. Tidal streams strong S of Craigton Pt (East going stream at springs exceeds 5 kn). Beware marine farms S of Avoch.
LIGHTS AND MARKS
Entrance to river is very narrow but deep. Craigton Pt Lt is Fl WRG 4s 6m 11/7M vis W312°-048°, R048°-064°, W064°-085°, G085°-shore.
RADIO TELEPHONE (local times)
Call: *Inverness Harbour Office* VHF Ch 12 16; 12 14 (Mon-Fri: 0800-1500). Caley Marina Ch M. Other station: Cromarty Firth Port Control Ch 16; 06 08 **11** 12 13 14 (H24). Call: *Clachnaharry Sea Lock* Ch 74. For canal office call: *Caledonian Canal* Ch 74.
TELEPHONE (0463)
Hr Mr 233291 (after hours 231725); CG Cromarty 967211; MRSC Peterhead 74278; HMC 231608; Marinecall 0898 500 451; Dr 234151.
FACILITIES
EC Wednesday; **Caley Marina** (25+25 visitors) Tel. 236539 ACA, AC, C (20 ton), CH, D, ME, El, Sh, FW, (Access H24); **Muirtown Marina** (20+20 visitors) Tel. 239745 AC, D, FW, Gas, Gaz, ME, Lau, C (4 ton), CH, Sh, El; Access HW∓4; **Shore Street Quay** Tel. 233291, Slip, P, D, L, FW, ME, El, C (2 mobiles), CH, AB; **Citadel Quay** Tel. 233291, Slip, P, D, L, FW, ME, El, AB; **Loch Ness Marine** Tel. 236539, Slip, M, FW, ME, El, Sh, CH; **Thornbush Slipway & Engineering Co** Tel. 233813, Slip, M, ME, El, Sh; **P. T. McHardy** Tel. 233632, Gas; **Inverness Boat Centre** Tel. 73383, CH. **Town** V, R, Bar. PO; Bank; Rly; Air.

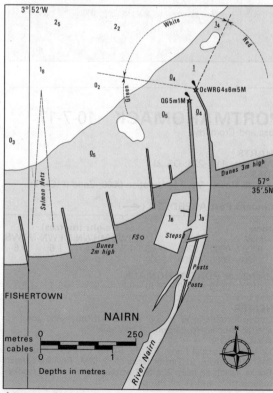

⚓ Nairn, apply to YC

INVERNESS continued

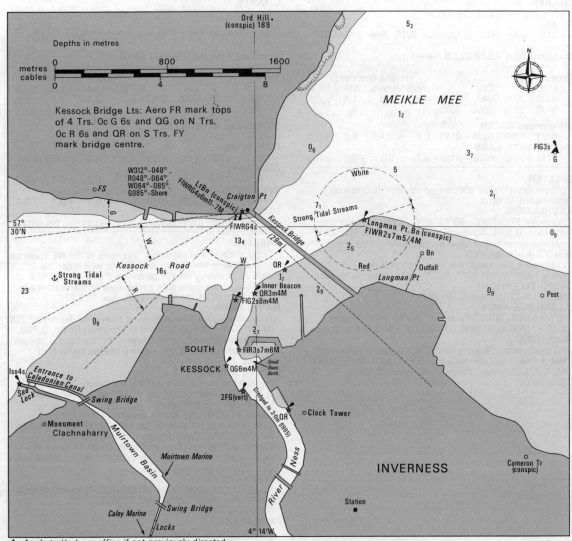

AGENTS NEEDED
There are a number of ports where we need agents, particularly in France.
ENGLAND   Swale, Havengore, Berwick.
SCOTLAND Firth of Forth, Scrabster, Mallaig, Loch Sunart, Loch Aline.
IRELAND   Kilrush, Wicklow, Westport/Clew Bay.
FRANCE   Arcachon, Seudre R, Ile d'Oleron, Rochfort, Ile de Re, St. Giles-Croix-de-Vie, Ile d'Yeu, Pouliguen, Le Croisic, La Forêt, Ile de Bréhat.
GERMANY   Norderney, Dornumer-Accumersiel.
If you are interested in becoming our agent for any of the above, please write to the editors and get your free copy of the Almanac every year. You do not have to be resident in a port to be the agent, but at least a fairly regular visitor.

PORTMAHOMACK 10-7-17
Ross and Cromarty

CHARTS
Admiralty 115, 223; OS 21
TIDES
+0035 Dover; ML 2.5; Duration 0600; Zone 0 (GMT).

Standard Port ABERDEEN (←)

Times				Height (metres)			
HW		LW		MHWS	MHWN	MLWN	MLWS
0300	0800	0200	0800	4.3	3.4	1.6	0.6
1500	2000	1400	2000				
Differences PORTMAHOMACK							
−0120	−0210	−0140	−0110	−0.2	−0.1	+0.1	+0.1

SHELTER
Uncomfortable in W to SW winds, but otherwise good shelter. Access to harbour at HW only.
NAVIGATION
Waypoint Special buoy, FIY 5s, 57°53'.03N 03°47'.02W, 346°/166° from/to Tarbat Ness Lt, 1.13M. Beware Curach Rocks which lie from 2ca SW of pier to the shore. Rocks extend to N and W of the pier.

PORTMAHOMACK continued

LIGHTS AND MARKS
There are no lights or marks.
RADIO TELEPHONE
None.
TELEPHONE (086 287)
CG 564; MRSC Peterhead 74278; HMC Invergordon 852221; Marinecall 0898 500 451; Dr 358 Tain 2759.
FACILITIES
EC Wednesday; **Harbour** C (6 ton), M, P and D (¾M on Tain road in cans), L, FW, CH, AB; **Port Service Stn** Tel. 231, Gas; **Town** PO; Bank (Mobile from Dingwall 1235-1345); Rly (bus to Tain); Air (Inverness).

FACILITIES
EC Wednesday; **NW Pier** Tel. 354, Slip, M (See Hr Mr), L, FW, AB; **E Pier** Tel. 354, M (See Hr Mr), L, FW, AB; **A. R. McLeod** Tel. 234, Gas; **Town** P, D, V, R, Bar. PO; Bank (Brora); Rly; Air (Wick).

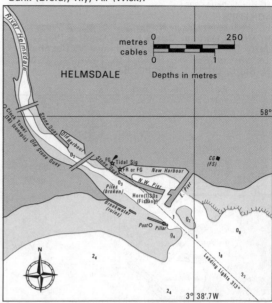

▲ Apply to Harbour office if not previously directed

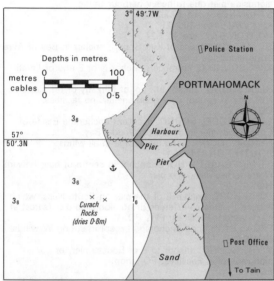

▲ Anchor off and wait

HELMSDALE 10-7-18
Sutherland

CHARTS
Admiralty 115, 1462; OS 17
TIDES
+0035 Dover; ML 2.2; Duration 0615; Zone 0 (GMT).

Standard Port ABERDEEN (←)

Times				Height (metres)			
HW		LW		MHWS	MHWN	MLWN	MLWS
0300	0800	0200	0800	4.3	3.4	1.6	0.6
1500	2000	1400	2000				

Differences HELMSDALE
−0145 −0217 −0202 −0155 −0.6 −0.5 −0.1 −0.1
GOLSPIE
−0130 −0215 −0155 −0130 −0.3 −0.3 −0.1 0.0

SHELTER
Good in all weathers except strong SE winds.
NAVIGATION
Waypoint 58°06′.61N 03°38′.30W, 133°/313° from/to entrance, 0.35M. Beware spate coming down river after heavy rains. Shallow both sides of channel and bar builds up in river when in spate.
LIGHTS AND MARKS
Ldg Lts 313°. Front FG (harbour open) or FR (harbour closed). Rear FG; both on W masts.
RADIO TELEPHONE
VHF Ch 16.
TELEPHONE (043 12)
Hr Mr 347; CG Wick 2332; MRSC Peterhead 74278; HMC Wick 3125; Marinecall 0898 500 451; Dr 225 (Home) or 221 (Surgery).

WICK 10-7-19
Caithness

CHARTS
Admiralty 1462, 115; OS 12
TIDES
+0010 Dover; ML 2.0; Duration 0625; Zone 0 (GMT).

Standard Port ABERDEEN (←)

Times				Height (metres)			
HW		LW		MHWS	MHWN	MLWN	MLWS
0300	0800	0200	0800	4.3	3.4	1.6	0.6
1500	2000	1400	2000				

Differences WICK
−0155 −0220 −0210 −0220 −0.9 −0.7 −0.2 −0.1
DUNCANSBY HEAD
−0320 −0320 −0320 −0320 −1.2 −1.0 No data

SHELTER
Shelter good except in strong E winds. NB. The river harbour is leased and must *not* be entered without prior permission.
NAVIGATION
Waypoint 58°26′.20N 03°03′.30W, 104°/284° from/to South Pier Lt, 0.72M. Harbour entrance dangerous in strong E winds as boats have to make a 90° turn at the end of S Pier. Unlit pole beacon S side of bay marks seaward end of ruined breakwater.
LIGHTS AND MARKS
Harbour inaccessible — a black ball (G light at night) on look-out tower over CG Station on South Head.
South pier Lt — Fl WRG 3s 12m 5/3M G253°-269°, W269°-286°, R286°-329°.
Dir Lt on N end of bridge, W sector 289°-292°.
When harbour temporarily obstructed B ball or R Lt at S Pier head. Leading Lts 234°, both FR.

WICK continued

RADIO TELEPHONE
VHF Ch 16; 14 (when vessel expected).
TELEPHONE (0955)
Hr Mr 2030; CG 2332; MRSC Kirkwall 3268; HMC 3650; Hospitals 2434, 2261.
FACILITIES
EC Wednesday; **Inner N Pier** Slip, L, ME, AB (see Hr Mr), V; **Harbour Quays** L, ME; **Outer S Pier** L; **Jetty** Tel. 2689, D, L, FW, CH; **James McCaughey** Tel. 3701 or 2858, ME, El, Sh, Gas; **Town** Slip, P, ME, El, Sh, C, CH, V, R, Bar. PO; Bank; Rly. Air.

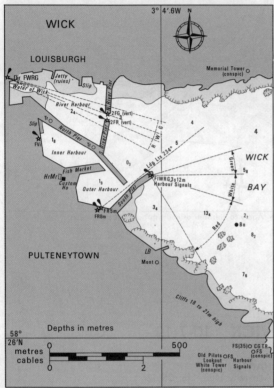

ORKNEY ISLANDS 10-7-20

The Orkneys consist of some 70 islands, of which about 24 are inhabited. They extend 5 to 50M NNE from Duncansby Head, and are mostly low-lying, but Hoy in the SW of the group reaches 475m (1560ft). Coasts are generally rocky and much indented, but there are many sandy beaches. A passage with least width of about 3M runs from NW to SE through the group. The islands are divided from Scotland by Pentland Firth, a very dangerous stretch of water (see 10.7.5). The climate is mild but windy, and very few trees grow. The principal island is Mainland (or Pomona) on which stands Kirkwall, the capital. There is a small naval base at Lyness in Scapa Flow.

Beware the many lobster pots (creels) all round the coast. Severe gales blow in winter and early spring (see 10.7.5). There are lifeboats at Longhope, Stromness and Kirkwall. There is a MRSC at Kirkwall, Tel. (0856) 3268, with Auxiliary (Watch and Rescue) Stations at Longhope (South Walls), Brough Ness (South Ronaldsay), Stromness (Mainland), Deerness (Mainland), Westray, Papa Westray, Sanday, North Ronaldsay and all inhabited islands except Egilsay and Wyre.

CHARTS
Charts 2249 and 2250 cover the islands. For larger scale charts, see under individual harbours.
TIDES
Tidal information is related to Aberdeen (10.6.17) as Standard Port. Tidal streams are strong, particularly in Pentland Firth and in the firths and sounds among the islands (see 10.7.5.).
SHELTER
There are piers at all the main islands, and there are many anchorages, of which the following are a selection.
MAINLAND
SCAPA BAY: good except in S winds. No yacht berths alongside pier due to heavy harbour traffic.
STROMNESS: see 10.7.23
KIRKWALL: see 10.7.21
HOUTON BAY: see 10.7.27
ST MARY'S: N side of Kirk Sound; anchor in Bay of Ayre or go alongside E side of pier.
KIRK SOUND (E entrance): anchor N or E end of Lamb Holt.
DEER SOUND: anchor in Pool of Mirkady or off pier on NW side of sound; very good shelter, no facilities.
BURRAY
EAST WEDDEL SOUND: excellent anchorage E side of Churchill Barrier in centre of Weddel Bay.
HUNDA SOUND: good anchorage in all winds.
SOUTH RONALDSAY
ST MARGARET'S HOPE: anchor in centre of bay; beware Flotta ferries using the pier.
HOY
LONG HOPE: anchor E of the pier on South Ness, which is used by steamers/ferries or go alongside pier (safest at slack water). Facilities: FW, PO, V.
PEGAL BAY: good anchorage except in strong W winds.
ROUSAY
WYRE SOUND: anchor to E of Rousay pier, or go alongside pier. Facilities: PO, shop.
EDAY
FERS NESS BAY: good holding, and shelter from all S winds.
WESTRAY
PIEROWALL: see 10.7.27
BAY OF MOCLETT: Excellent anchorage but open to S winds.
SOUTH WICK: Small vessel anchorage on E of Papa Westray, off the old pier or ESE of pier off Holm of Papa.
SANDAY
OTTERSWICK: good anchorage except in N or E winds.
NORTH BAY: on NW side of island, exposed to NW.
KETTLETOFT BAY: anchor in bay or go alongside pier; very exposed to SE winds. Facilities: PO, shop, hotel.
NORTH RONALDSAY
SOUTH BAY: anchor in middle of bay or go alongside pier; open to S and E, and subject to swell.
LINKLET BAY: not a safe anchorage, open to E.
STRONSAY: see 10.7.22
AUSKERRY: see 10.7.27
NAVIGATION
Apart from the strong tidal streams and associated races, navigation is easy in clear weather; see 10.7.5.
LIGHTS AND MARKS
The main harbours and sounds are well lit; for details see 10.7.4. Powerful lights are shown from Cantick Head, Graemsay Island, Copinsay, Auskerry, Kirkwall, Brough of Birsay, Sanday Island, North Ronaldsay and Noup Head.
RADIO TELEPHONE
Orkney Harbours Navigation Service (call: *Orkney Harbour Radio*, Ch 16 20; 09 11 (H24)) covers Scapa Flow, Wide Firth and Shapinsay Sound; see 10.7.21. For other local stations see individual harbours. The local Coast Radio Station is Orkney Radio (VHF Ch 16 26 (H24)), remotely controlled from Wick Radio.
MEDICAL SERVICES
There are doctors available at Kirkwall, Stromness, Rousay, Shapinsay, Eday, North Ronaldsay, Stronsay, Sanday and Westray (Pierowall). Papa Westray is looked after from Westray. The only hospital is at Kirkwall. Serious cases are flown to Aberdeen (1 hour). The only dental services are at Kirkwall.

AREA 7—NE Scotland 407

ORKNEY ISLANDS continued

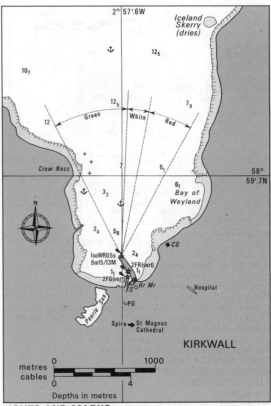

SALMON CAGES
Beware salmon cages in the following areas:—
Rysa Sound
Lyrawa Bay
Ore Bay
Widewall Bay
St Margaret's Hope
Pegal Bay
Hunda Sound
Kirk Sound
Carness Bay
Bay of Ham

OYSTERS AND LONGLINES
Beware Oysters and Longlines located as follows:—
Widenall Bay
Swandister Bay
Water Sound
Hunda Sound
Deer Sound
Inganess Bay
Bay of Firth
Damsay Sound
Gairsay Milldurnday
Pierowall
Bay of Skaill (Westray)
Longhope

LIGHTS AND MARKS
St Magnus Cathedral is very conspic.

RADIO TELEPHONE (local times)
VHF Ch 16; 12 (0900-1700). Orkney Harbours Navigation Service, call: *Orkney Harbour Radio* Ch 16 20; 09 11 (H24). Local Weather on Ch 11 at 0915 and 1715.

TELEPHONE (0856)
Hr Mr 2292; CG & MRSC 3268; HMC 2108; Weather 3802; Marinecall 0898 500 451; Dr 2763 or 3201.

FACILITIES
EC Wednesday; **Pier** P, D, FW CH; **N & E Quays** M; **Ellewick (Pier)** M, D, FW, V; **Orkney SC** Tel. 2331, M, L, C, AB; **Hatston** Slip; **John Scott and Millar** Tel. 3146, Gas; **Town** P, D, ME, El, Sh, CH, V, Gas, R, Bar. PO; Bank; Rly (Ferry to Scrabster, bus to Thurso); Air.

KIRKWALL 10-7-21
Orkney Islands

CHARTS
Admiralty 2250, 1553, 2584; OS 6

TIDES
−0045 Dover; ML 1.7; Duration 0620; Zone 0 (GMT).

Standard Port ABERDEEN (←)

Times				Height (metres)			
HW		LW		MHWS	MHWN	MLWN	MLWS
0300	1100	0200	0900	4.3	3.4	1.6	0.6
1500	2300	1400	2100				
Differences KIRKWALL							
−0305	−0245	−0305	−0250	−1.4	−1.2	−0.5	−0.2
MUCKLE SKERRY							
−0230	−0230	−0230	−0230	−1.7	−1.4	−0.6	−0.2
BURRAYNESS							
−0200	−0200	−0155	−0155	−1.0	−0.9	−0.3	0.0

SHELTER
Harbour well sheltered except in N winds or W gales when there is a surge at the entrance. Safe anchorage for small yachts between W pier head and Crowness Pt.

NAVIGATION
Waypoint 59°01'.40N 02°57'.00W, 008°/188° from/to pierhead Lt, 2.2M. SW of the bay is a shoal.

STRONSAY 10-7-22
Orkney Islands

CHARTS
Admiralty 2622; OS 6

TIDES
−0140 Dover; ML 1.7; Duration 0620; Zone 0 (GMT).

Standard Port ABERDEEN (←)

Times				Height (metres)			
HW		LW		MHWS	MHWN	MLWN	MLWS
0300	1100	0200	0900	4.3	3.4	1.6	0.6
1500	2300	1400	2100				
Differences DEER SOUND							
−0245	−0245	−0245	−0245	−1.1	−0.9	−0.3	0.0

SHELTER
Good from winds in all directions. Good anchorage between seaward end of piers. Many other good sheltered anchorages round the bay.

STRONSAY continued

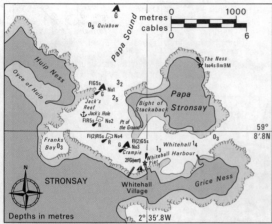

NAVIGATION
Waypoint Quiabow stbd-hand buoy, 59°09'.83N 02°36'.20W, 009°/189° from/to No 1 Lt Buoy, 6.5 ca.
800m NE of Huip Ness is Quiabow, a submerged rock marked by G conical buoy.
Jack's Reef extends 400m E from Huip Ness, and is marked by G conical buoy.
A bank extends 350m SW from Papa Stronsay. Crampie Shoal is in mid-channel, marked by a buoy. The buoyed channel to Whitehall pier is dredged to 3.5 m.
Spit to E of Whitehall pier extends 400m N.
The E entrance is narrow and shallow.

LIGHTS AND MARKS
Pierhead Lts, 2FG (vert).

RADIO TELEPHONE
See Kirkwall.

TELEPHONE (085 76)
Hr Mr 257; CG & MRSC Kirkwall 3268; HMC Kirkwall 2108; Marinecall 0898 500 451; Dr 321.

FACILITIES
EC Thursday; **W. Pier** M, L, AB; **Main Pier** M, L, FW, AB; **Village (Whitehall)** P, D, V, Bar, PO; Bank; Rly (ferry to Scrabster, bus to Thurso); Air.

⚓ Contact Hr Mr

NAVIGATION
Waypoint 58°57'.00N 03°16'.90W, 137°/317° from/to front Ldg Lt, 0.88M. Tides in Hoy sound are very strong, entry should not be attempted in bad weather or when strong winds are against the tide. Spring tides can exceed 7 kn.

LIGHTS AND MARKS
Ldg Lts 317°, both FR 29/39m 11M (H24); W Trs; vis 307°-327°. Skerry of Ness Lt, Fl WG 4s 7m 7/4M; W shore-090°, G090°-shore.

RADIO TELEPHONE (local times)
VHF Ch 16; 12 (0900-1700). (See also Kirkwall).

TELEPHONE (0856)
Hr Mr 850744; CG & MRSC Kirkwall 3268; HMC Kirkwall 2108; Marinecall 0898 500 451; Dr 850205.

FACILITIES
EC Thursday; **I. Richardson** Tel. 850321, Sh;
Town FW, D, V, R, Bar, Gas, PO; Bank, Rly (Ferry to Scrabster, bus to Thurso); Air (Kirkwall).
Note: Yachtsmen may contact, for help, **J. Stout** Tel. 850100, **I. Mackenzie** Tel. 850587 or **S. Mowat** Tel. 850624.

STROMNESS 10-7-23
Orkney Islands

CHARTS
Admiralty 2249, 2568; OS 6

TIDES
−0145 Dover; ML 2.0; Duration 0620; Zone 0 (GMT).

Standard Port ABERDEEN (←)

Times				Height (metres)			
HW		LW		MHWS	MHWN	MLWN	MLWS
0300	1100	0200	0900	4.3	3.4	1.6	0.6
1500	2300	1400	2100				

Differences STROMNESS
−0430 −0355 −0415 −0420 −0.7 −0.8 −0.1 −0.1
OTTERSWICK
−0355 −0355 −0355 −0355 −0.8 −0.7 −0.1 +0.1
PIEROWALL
−0355 −0355 −0355 −0355 −0.6 −0.6 −0.2 0.0
EYNHALLOW SOUND
−0400 −0400 −0355 −0355 −0.6 −0.6 −0.2 −0.1

SHELTER
Very good shelter with no tidal stream in harbour.
Anchorage in harbour or berth alongside.
The southern pier is the Northern Lights Pier and is for no other use. Berthing piers further N.

SHETLAND ISLANDS 10-7-24

The Shetland Islands consist of approx 100 islands, holms and rocks of which fewer than 20 are inhabited. They lie 90 to 150 miles NNE of the Scottish mainland. By far the biggest island is Mainland with Lerwick, the capital, on the E side and Scalloway, the only other town and old capital, on the W side. At the very S there is the airport at Sumbrugh.

Two islands of the Shetland group not shown on the chartlet are Fair Isle, lying 20M SSW of Sumburgh Head (see 10.7.27) and owned by the National Trust for Scotland, and Foula lying 16M W of Mainland.

There are lifeboats at Lerwick and Aith, and there is a MRSC at Lerwick (0595) 2976. There is a Sector Base at Sella Ness (Sullom Voe), and an Auxiliary Station (Watch and Rescue) at Fair Isle.

CHARTS
Admiralty; general 3281, 3282, 3283; also 3291, 3292, 3294, 3295, 3297, 3298.

TIDES
Tidal information is related to Lerwick, Standard Port, see 10.7.25 below. In open waters tidal streams are mostly weak but in some sounds and other places rates of 6 kn or more can be attained and there are some dangerous races and overfalls. Tidal streams run mostly N and S or NW and SE, and cause dangerous disturbances at the N and S extremities of the islands and in the two main sounds (Yell Sound and Bluemull Sound/Colgrave Sound).

SHELTER
Weather conditions are bad in winter, and yachts should restrict visits to April to September. Around mid-summer it is light during all 24 hours. There are numerous anchorages but the following are those which it is safe to enter in all weathers.
LERWICK: see 10.7.25
CAT FIRTH: excellent shelter, anchor in approx 6m. Facilities: PO (Skellister), FW, V (both at Lax Firth).
GRUNNA VOE: off S side of Dury Voe, good shelter, anchor in 5-10m, good holding, beware prohibited anchoring areas. Facilities: V, FW, PO (Lax Firth).
SCALLOWAY: see 10.7.27.
VAILA SOUND: on SW of Mainland, use Easter Sound entrance (Wester Sound entrance is dangerous), very good shelter, anchor N of Salt Ness in 4 to 5m in mud. Facilities: FW, V, PO.
GRUTING VOE: tides −0150 on Lerwick, anchor in main voe or in Browland, Seli or Scutta voes. Facilities: Stores and PO at Bridge of Walls (at head of Browland Voe).
SWARBACKS MINN: a large complex of voes and islands SE of St Magnus Bay. Best anchorages Uyea Sound or Aith Voe, both well sheltered and good holding. Facilities: former none; Aith FW, PO, V, Bar, LB.
OLNA FIRTH: NE of Swarbacks Minn, beware rk 1 ca off S shore which dries. Anchor in firth, 4-8m or in Gon Firth or go alongside pier at Voe. Facilities: (Voe) FW, V, D, PO, Bar.
URA FIRTH: NE of St Magnus Bay, anchor in Hills Wick on W side or in Hamar Voe on E side. The latter has excellent shelter in all weathers and good holding. Facilities: Hamar Voe, none; Hills Wick V, R, FW, V, PO, D, ME, El, Sh, Bar.
HAMNA VOE: tides − 0200 on Lerwick, very good shelter, ldg line 153° old house on S shore with prominent rk on pt of W shore 3 ca within ent. Facilities: PO (½M), Stores, D (1½M), L (at pier).
S OF YELL SOUND: in the complex there are a number of good anchorages; Hamna Voe, Boatsroom Voe, West Lunna Voe, Dales Voe, Colla Firth. Tides −0025 on Lerwick; all well protected and good holding. Facilities: none except at West Lunna Voe where there is a small hotel.
SULLOM VOE: tides −0130 on Lerwick. 6½M long deep water voe which has partly been taken over by the oil industry. Anchor above the narrows. Facilities at Brae, FW, V, PO, D, ME, El, Sh, Bar.
MID YELL VOE: tides −0040 on Lerwick, enter through South Sd or Hascosay Sd, good anchorage in wide part of voe 2½-10m. Facilities: PO, FW at pier on S side, D, Stores, ME, Sh, El.
BASTA VOE: good anchorage above shingle bank in 5-15m; good holding in places. Facilities: FW, Stores, Motel, PO.
BALTA SOUND; see 10.7.27.

NAVIGATION
A careful lookout must be kept for salmon farming cages and low lying mussel rafts (approx 100 in no), mostly marked by plastic yellow buoys and combinations of Y lights. The Clyde Cruising Club *Shetland Sailing Directions and Anchorages* is a most valuable publication for visitors. For general passage information see 10.7.5. Local magnetic anomalies may be experienced.

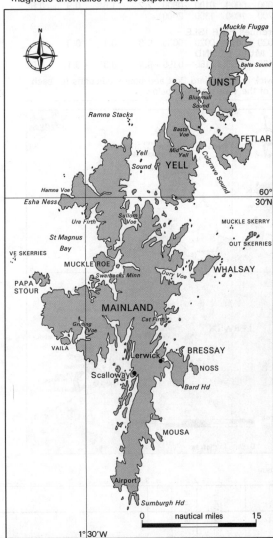

LIGHTS AND MARKS
See 10.7.4.

RADIO TELEPHONE
For Port Radio services see 10.7.25. Local Coast Radio Stations are Shetland Radio (Ch 16 27), and Collafirth Radio (Ch 16 24). Weather messages are broadcast by Shetland Radio (Ch 27) and Collafirth Radio (Ch 24) at 0803 and 2003 GMT. Gale warnings after next silence period and at 0303 0903 1503 2103 GMT.

TELEPHONE (0595)
Hr Mr Lerwick 2828; Sullom Voe 242551; Scalloway 636; Weather 2239.

FACILITIES
See shelter above.

LERWICK 10-7-25
Shetland Islands

CHARTS
Admiralty 3290, 3283, 3291; OS 4

TIDES
−0008 Dover; ML 1.3; Duration 0620; Zone 0 (GMT).

Standard Port LERWICK (→)

Times				Height (metres)			
HW		LW		MHWS	MHWN	MLWN	MLWS
0000	0600	0100	0800	2.2	1.6	0.9	0.5
1200	1800	1300	2000				

Differences FAIR ISLE
−0020 −0025 −0020 −0035 0.0 +0.1 0.0 −0.1

BLUE MULL SOUND
−0135 −0135 −0155 −0155 +0.4 +0.3 +0.1 0.0

Lerwick is a Standard Port and tidal predictions for each day of the year are given below.

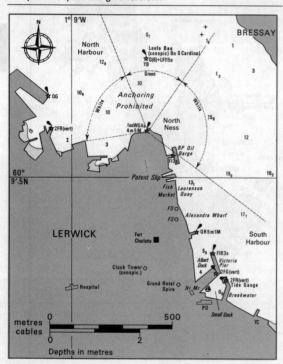

SHELTER
Good. Hr Mr allocates berths in Small Dock or Albert Dock. Fishing boats occupy most alongside space. Anchoring prohibited for about 2 ca off the waterfront. Gremista marina is in North Harbour, mainly for local boats, and is a long way from the town.

NAVIGATION
Waypoint 60°06'.00N 01°08'.45W, 185°/005° from/to Twageos Pt Lt, 3.0M. From S, Bressay Sound is clear of dangers. From N, beware Soldian Rock (dries), Nive Baa (0.6m) and Green Holm (10m high) and the Brethren (two rocks 2m and 1.5m high).

LIGHTS AND MARKS
Kirkabister Ness Lt Fl(2) 30s 32m 21M on Bressay. Twageos Pt Lt LFl 6s 8m 6M. Loofa Baa S Cardinal Bn marks shoal between North and South Harbours. Light-buoys mark Middle Ground in North Harbour.

RADIO TELEPHONE (local times)
VHF Ch 16; 11 **12** (H24).
Other stations: Sullom Voe Ch 19 16; 12 **14** 20 (H24). Information broadcasts on Ch 20 every 4h commencing 0000. Local Fcst Ch 20 at 0830 and 1830.
Scalloway Ch 16; 12 (Mon-Fri: 0800-1700. Sat: 0800-1200).
Balta Sound Ch 16 20 (occas).

TELEPHONE (0595)
Hr Mr 2828; CG Sullom Voe 2561; MRSC 2976; HMC 4040; Dr 3201; Weather 2239.

FACILITIES
EC Wednesday (all day); **Harbour** Slip, M, P, D, L, FW, ME, Sh, Gas; **Malakoff** Tel. 5544, ME, El, Sh, Slip, BY, CH, Sh, SM; **Thulecraft** Tel. 3192, CH, ME, Gas; **Hay** Tel. 3057, ACA; **Lerwick Harbour Trust** Tel. 2991, M, FW, D, P, AB; **Rearo Supplies** Tel. 2636, Gas; **Lerwick Boating Club**; **Town** V, R, Bar, CH, PO; Bank; Rly (ferry to Aberdeen); Air.

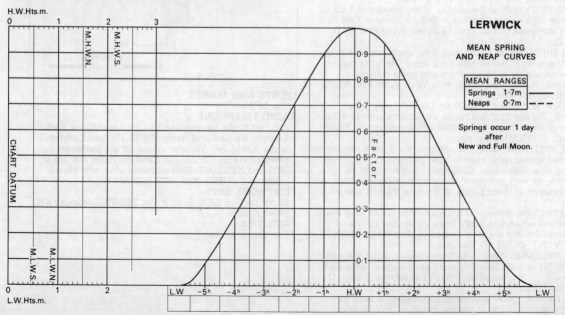

LERWICK MEAN SPRING AND NEAP CURVES

MEAN RANGES
Springs 1.7m
Neaps 0.7m

Springs occur 1 day after New and Full Moon.

SHETLAND ISLANDS - LERWICK
LAT 60°09'N   LONG 1°08'W

TIMES AND HEIGHTS OF HIGH AND LOW WATERS

YEAR 1989

TIME ZONE UT(GMT)
For Summer Time add ONE hour in non-shaded areas

JANUARY

Day	Time	m	Day	Time	m
1 SU	0458 / 1047 / 1703 / 2343	1.6 / 1.1 / 1.7 / 0.9	**16** M	0541 / 1146 / 1802	1.7 / 0.9 / 1.8
2 M	0607 / 1215 / 1823	1.6 / 1.1 / 1.7	**17** TU	0045 / 0658 / 1312 / 1930	0.8 / 1.7 / 0.9 / 1.8
3 TU	0051 / 0718 / 1333 / 1941	1.0 / 1.7 / 1.1 / 1.7	**18** W	0155 / 0805 / 1422 / 2041	0.9 / 1.8 / 0.8 / 1.8
4 W	0149 / 0814 / 1428 / 2041	1.0 / 1.8 / 1.0 / 1.8	**19** TH	0251 / 0859 / 1519 / 2136	0.9 / 1.9 / 0.7 / 1.9
5 TH	0239 / 0900 / 1515 / 2130	0.9 / 1.9 / 0.9 / 1.9	**20** F	0338 / 0945 / 1607 / 2222	0.9 / 2.0 / 0.6 / 1.9
6 F	0324 / 0943 / 1557 / 2215	0.9 / 2.0 / 0.7 / 2.0	**21** SA	0419 / 1027 / 1650 / 2303	0.8 / 2.1 / 0.5 / 1.9
7 SA	0407 / 1024 / 1639 / ●2258	0.8 / 2.2 / 0.6 / 2.1	**22** SU	0456 / 1106 / 1729 / 2340	0.8 / 2.2 / 0.4 / 1.9
8 SU	0449 / 1105 / 1721 / 2341	0.7 / 2.2 / 0.5 / 2.1	**23** M	0531 / 1143 / 1806	0.7 / 2.2 / 0.4
9 M	0532 / 1146 / 1804	0.7 / 2.3 / 0.4	**24** TU	0015 / 0604 / 1219 / 1839	1.9 / 0.7 / 2.2 / 0.4
10 TU	0024 / 0614 / 1227 / 1847	2.1 / 0.7 / 2.3 / 0.3	**25** W	0049 / 0635 / 1253 / 1912	1.9 / 0.7 / 2.1 / 0.5
11 W	0108 / 0657 / 1310 / 1933	2.1 / 0.7 / 2.3 / 0.3	**26** TH	0122 / 0706 / 1327 / 1943	1.9 / 0.7 / 2.1 / 0.6
12 TH	0154 / 0741 / 1355 / 2021	2.0 / 0.7 / 2.2 / 0.4	**27** F	0155 / 0737 / 1400 / 2014	1.8 / 0.8 / 2.0 / 0.7
13 F	0241 / 0828 / 1442 / 2114	1.9 / 0.8 / 2.1 / 0.5	**28** SA	0229 / 0810 / 1435 / 2048	1.7 / 0.9 / 1.9 / 0.8
14 SA ☾	0332 / 0921 / 1536 / 2213	1.8 / 0.8 / 2.0 / 0.6	**29** SU	0306 / 0848 / 1515 / 2128	1.7 / 1.0 / 1.8 / 0.9
15 SU	0430 / 1025 / 1639 / 2326	1.7 / 0.9 / 1.9 / 0.8	**30** M ☽	0348 / 0936 / 1603 / 2218	1.6 / 1.0 / 1.7 / 1.0
			31 TU	0443 / 1045 / 1712 / 2331	1.6 / 1.1 / 1.6 / 1.1

FEBRUARY

Day	Time	m	Day	Time	m
1 W	0603 / 1237 / 1859	1.6 / 1.1 / 1.6	**16** TH	0148 / 0749 / 1421 / 2044	1.0 / 1.7 / 0.8 / 1.7
2 TH	0106 / 0733 / 1403 / 2021	1.1 / 1.7 / 1.0 / 1.7	**17** F	0246 / 0847 / 1516 / 2133	1.0 / 1.8 / 0.6 / 1.8
3 F	0216 / 0834 / 1457 / 2115	1.0 / 1.8 / 0.8 / 1.8	**18** SA	0330 / 0933 / 1559 / 2212	0.9 / 1.9 / 0.5 / 1.8
4 SA	0307 / 0922 / 1541 / 2200	0.9 / 2.0 / 0.6 / 2.0	**19** SU	0407 / 1012 / 1636 / 2246	0.9 / 2.0 / 0.5 / 1.9
5 SU	0352 / 1005 / 1623 / 2243	0.7 / 2.1 / 0.5 / 2.1	**20** M ○	0440 / 1048 / 1709 / 2318	0.7 / 2.1 / 0.4 / 1.9
6 M	0435 / 1046 / 1705 / ●2324	0.6 / 2.2 / 0.3 / 2.1	**21** TU	0511 / 1122 / 1740 / 2349	0.6 / 2.1 / 0.4 / 1.9
7 TU	0516 / 1128 / 1746	0.5 / 2.3 / 0.2	**22** W	0540 / 1155 / 1809	0.5 / 2.1 / 0.4
8 W	0006 / 0557 / 1209 / 1829	2.2 / 0.5 / 2.4 / 0.2	**23** TH	0019 / 0609 / 1226 / 1837	1.9 / 0.6 / 2.1 / 0.4
9 TH	0047 / 0638 / 1250 / 1912	2.1 / 0.5 / 2.4 / 0.2	**24** F	0049 / 0637 / 1257 / 1905	1.9 / 0.6 / 2.1 / 0.5
10 F	0129 / 0720 / 1335 / 1957	2.1 / 0.5 / 2.3 / 0.3	**25** SA	0119 / 0706 / 1329 / 1933	1.9 / 0.7 / 2.0 / 0.6
11 SA	0213 / 0804 / 1422 / 2044	2.0 / 0.6 / 2.2 / 0.5	**26** SU	0150 / 0736 / 1402 / 2004	1.8 / 0.8 / 1.9 / 0.7
12 SU ☾	0258 / 0853 / 1513 / 2138	1.8 / 0.7 / 2.0 / 0.7	**27** M	0224 / 0811 / 1440 / 2039	1.8 / 0.9 / 1.8 / 0.9
13 M	0350 / 0954 / 1615 / 2249	1.7 / 0.8 / 1.8 / 0.9	**28** TU ☽	0302 / 0855 / 1525 / 2124	1.7 / 1.0 / 1.6 / 1.0
14 TU	0457 / 1122 / 1745	1.6 / 0.9 / 1.7			
15 W	0026 / 0628 / 1305 / 1931	1.0 / 1.6 / 0.9 / 1.6			

MARCH

Day	Time	m	Day	Time	m
1 W	0351 / 0959 / 1632 / 2235	1.6 / 1.0 / 1.5 / 1.1	**16** TH	0008 / 0557 / 1259 / 1931	1.1 / 1.6 / 0.8 / 1.5
2 TH	0504 / 1147 / 1827	1.6 / 1.1 / 1.5	**17** F	0136 / 0728 / 1410 / 2036	1.0 / 1.6 / 0.7 / 1.6
3 F	0029 / 0648 / 1336 / 2001	1.1 / 1.6 / 0.9 / 1.6	**18** SA	0231 / 0827 / 1459 / 2117	0.9 / 1.8 / 0.6 / 1.7
4 SA	0154 / 0804 / 1434 / 2056	1.0 / 1.8 / 0.7 / 1.8	**19** SU	0312 / 0912 / 1538 / 2150	0.8 / 1.9 / 0.5 / 1.8
5 SU	0248 / 0856 / 1519 / 2140	0.8 / 1.9 / 0.5 / 1.9	**20** M	0346 / 0950 / 1611 / 2221	0.7 / 2.0 / 0.4 / 1.8
6 M	0332 / 0941 / 1602 / 2221	0.7 / 2.1 / 0.3 / 2.1	**21** TU	0417 / 1024 / 1640 / 2250	0.6 / 2.0 / 0.4 / 1.9
7 TU ●	0414 / 1024 / 1643 / 2302	0.5 / 2.2 / 0.2 / 2.2	**22** W ○	0446 / 1056 / 1709 / 2319	0.6 / 2.1 / 0.4 / 1.9
8 W	0454 / 1107 / 1724 / 2342	0.4 / 2.4 / 0.1 / 2.2	**23** TH	0514 / 1128 / 1736 / 2348	0.5 / 2.1 / 0.4 / 2.0
9 TH	0535 / 1149 / 1806	0.3 / 2.4 / 0.1	**24** F	0542 / 1159 / 1803	0.5 / 2.1 / 0.5
10 F	0022 / 0616 / 1232 / 1848	2.2 / 0.3 / 2.4 / 0.2	**25** SA	0017 / 0610 / 1231 / 1831	2.0 / 0.6 / 2.0 / 0.5
11 SA	0103 / 0658 / 1316 / 1931	2.1 / 0.4 / 2.3 / 0.4	**26** SU	0047 / 0640 / 1304 / 1859	1.9 / 0.6 / 1.9 / 0.6
12 SU	0144 / 0742 / 1403 / 2016	2.0 / 0.5 / 2.1 / 0.6	**27** M	0119 / 0712 / 1339 / 1931	1.9 / 0.7 / 1.8 / 0.8
13 M	0228 / 0832 / 1455 / 2107	1.9 / 0.6 / 1.9 / 0.8	**28** TU	0153 / 0749 / 1419 / 2007	1.8 / 0.8 / 1.7 / 0.9
14 TU ☽	0317 / 0934 / 1558 / 2215	1.7 / 0.7 / 1.7 / 1.0	**29** W	0232 / 0835 / 1509 / 2055	1.7 / 0.9 / 1.6 / 1.0
15 W	0421 / 1111 / 1737	1.6 / 0.8 / 1.5	**30** TH	0322 / 0941 / 1618 / 2209	1.7 / 0.9 / 1.5 / 1.1
			31 F	0431 / 1120 / 1802 / 2358	1.6 / 0.9 / 1.5 / 1.1

APRIL

Day	Time	m	Day	Time	m
1 SA	0604 / 1301 / 1933	1.6 / 0.8 / 1.6	**16** SU	0200 / 0754 / 1427 / 2044	0.9 / 1.7 / 0.6 / 1.6
2 SU	0125 / 0727 / 1403 / 2029	0.9 / 1.7 / 0.6 / 1.8	**17** M	0241 / 0840 / 1504 / 2118	0.8 / 1.8 / 0.5 / 1.7
3 M	0220 / 0825 / 1451 / 2114	0.8 / 1.9 / 0.4 / 1.9	**18** TU	0316 / 0919 / 1536 / 2148	0.7 / 1.8 / 0.5 / 1.8
4 TU	0306 / 0914 / 1536 / 2156	0.6 / 2.1 / 0.2 / 2.0	**19** W	0347 / 0955 / 1606 / 2218	0.6 / 1.9 / 0.4 / 1.9
5 W	0349 / 1000 / 1618 / 2236	0.5 / 2.2 / 0.1 / 2.1	**20** TH	0417 / 1028 / 1634 / 2248	0.6 / 2.0 / 0.5 / 1.9
6 TH	0431 / 1045 / 1701 / ●2317	0.3 / 2.3 / 0.1 / 2.2	**21** F ○	0447 / 1101 / 1703 / 2318	0.5 / 2.0 / 0.5 / 2.0
7 F	0513 / 1129 / 1743 / 2357	0.3 / 2.3 / 0.2 / 2.1	**22** SA	0517 / 1135 / 1732 / 2349	0.5 / 2.0 / 0.5 / 2.0
8 SA	0555 / 1214 / 1825	0.3 / 2.3 / 0.3	**23** SU	0548 / 1209 / 1802	0.6 / 1.9 / 0.6
9 SU	0038 / 0640 / 1300 / 1907	2.1 / 0.3 / 2.2 / 0.5	**24** M	0022 / 0622 / 1246 / 1835	2.0 / 0.6 / 1.9 / 0.7
10 M	0119 / 0727 / 1348 / 1951	2.0 / 0.4 / 2.0 / 0.7	**25** TU	0056 / 0659 / 1325 / 1911	1.9 / 0.6 / 1.8 / 0.8
11 TU	0203 / 0819 / 1441 / 2040	1.9 / 0.5 / 1.8 / 0.9	**26** W	0133 / 0740 / 1407 / 1953	1.9 / 0.7 / 1.7 / 0.9
12 W	0252 / 0925 / 1545 / 2145	1.8 / 0.7 / 1.6 / 1.0	**27** TH	0216 / 0831 / 1504 / 2046	1.8 / 0.7 / 1.6 / 1.0
13 TH	0353 / 1058 / 1720 / 2335	1.6 / 0.8 / 1.5 / 1.1	**28** F ☾	0307 / 0936 / 1610 / 2157	1.7 / 0.8 / 1.5 / 1.0
14 F	0520 / 1236 / 1905	1.6 / 0.7 / 1.5	**29** SA	0410 / 1058 / 1734 / 2326	1.7 / 0.7 / 1.5 / 1.0
15 SA	0105 / 0652 / 1340 / ☾2005	1.0 / 1.6 / 0.7 / 1.5	**30** SU	0527 / 1222 / 1856	1.7 / 0.6 / 1.6

Chart Datum: 1.22 metres below Ordnance Datum (Newlyn)

SHETLAND ISLANDS - LERWICK

LAT 60°09'N   LONG 1°08'W

TIMES AND HEIGHTS OF HIGH AND LOW WATERS

YEAR **1989**

TIME ZONE UT(GMT)
For Summer Time add ONE hour in non-shaded areas

Chart Datum: 1.22 metres below Ordnance Datum (Newlyn)

	MAY				JUNE				JULY				AUGUST		
	TIME m		TIME m		TIME m		TIME m		TIME m		TIME m		TIME m		TIME m
1 M	0047 0.9 0646 1.7 1328 0.5 1955 1.7	**16** TU	0200 0.8 0758 1.7 1421 0.6 2037 1.6	**1** TH	0211 0.6 0823 1.9 1445 0.4 2103 1.9	**16** F	0243 0.8 0852 1.7 1453 0.7 2114 1.8	**1** SA	0252 0.6 0911 1.9 1520 0.7 2132 1.9	**16** SU	0304 0.8 0918 1.7 1509 0.9 2128 1.9	**1** TU	0430 0.4 1044 1.9 1638 0.7 ● 2247 2.1	**16** W	0404 0.5 1023 2.0 1613 0.7 2226 2.2
2 TU	0147 0.7 0752 1.9 1421 0.4 2044 1.9	**17** W	0240 0.8 0843 1.7 1456 0.6 2112 1.7	**2** F	0303 0.5 0920 2.0 1533 0.4 2149 2.0	**17** SA	0324 0.7 0937 1.8 1532 0.7 2152 1.9	**2** SU	0346 0.5 1005 1.9 1607 0.7 2218 2.0	**17** M	0346 0.7 1002 1.8 1552 0.8 2209 2.0	**2** W	0511 0.4 1123 1.9 1715 0.7 2327 2.2	**17** TH	0443 0.4 1102 2.1 1653 0.6 ○ 2306 2.3
3 W	0237 0.6 0847 2.0 1509 0.3 2129 2.0	**18** TH	0315 0.7 0923 1.8 1529 0.6 2145 1.8	**3** SA	0354 0.4 1012 2.1 1620 0.5 ● 2233 2.0	**18** SU	0403 0.7 1019 1.8 1610 0.7 2229 2.0	**3** M	0436 0.4 1054 1.9 1651 0.7 ● 2302 2.1	**18** TU	0426 0.6 1044 1.9 1633 0.7 ○ 2248 2.1	**3** TH	0550 0.4 1159 1.9 1750 0.6	**18** F	0523 0.3 1142 2.2 1733 0.5 2346 2.4
4 TH	0324 0.4 0937 2.1 1554 0.2 2211 2.0	**19** F	0349 0.6 1001 1.8 1601 0.6 2218 1.9	**4** SU	0442 0.4 1102 2.1 1704 0.5 2316 2.1	**19** M	0442 0.6 1059 1.9 1649 0.7 ○ 2307 2.0	**4** TU	0524 0.3 1139 1.9 1732 0.7 2344 2.1	**19** W	0505 0.5 1124 2.0 1713 0.6 2328 2.2	**4** F	0004 2.2 0625 0.4 1234 1.9 1823 0.6	**19** SA	0603 0.2 1222 2.2 1813 0.5
5 F	0409 0.3 1026 2.2 1638 0.3 ● 2253 2.1	**20** SA	0423 0.6 1038 1.9 1633 0.6 ○ 2252 2.0	**5** M	0531 0.3 1149 2.0 1747 0.6	**20** TU	0521 0.5 1140 1.9 1728 0.7 2346 2.1	**5** W	0608 0.3 1221 1.9 1812 0.7	**20** TH	0545 0.4 1205 2.0 1754 0.6	**5** SA	0040 2.1 0658 0.4 1307 1.9 1855 0.7	**20** SU	0028 2.4 0645 0.2 1303 2.1 1854 0.5
6 SA	0454 0.3 1113 2.2 1721 0.3 2334 2.1	**21** SU	0457 0.6 1115 1.9 1707 0.6 2326 2.0	**6** TU	0619 0.3 1236 1.9 1829 0.7	**21** W	0601 0.5 1222 1.9 1809 0.7	**6** TH	0025 2.1 0651 0.4 1302 1.8 1850 0.7	**21** F	0008 2.2 0626 0.3 1246 2.0 1835 0.6	**6** SU	0114 2.1 0730 0.5 1341 1.8 1927 0.7	**21** M	0110 2.3 0728 0.3 1346 2.0 1937 0.6
7 SU	0540 0.3 1200 2.1 1804 0.5	**22** M	0533 0.6 1154 1.9 1742 0.7	**7** W	0043 2.1 0707 0.4 1322 1.8 1912 0.7	**22** TH	0026 2.1 0643 0.4 1305 1.9 1851 0.7	**7** F	0105 2.1 0732 0.4 1341 1.8 1927 0.7	**22** SA	0049 2.2 0708 0.3 1329 2.0 1917 0.7	**7** M	0149 2.0 0801 0.6 1414 1.8 2000 0.8	**22** TU	0156 2.2 0814 0.5 1430 1.9 2025 0.7
8 M	0016 2.1 0627 0.4 1247 2.0 1847 0.6	**23** TU	0002 2.0 0611 0.6 1234 1.9 1820 0.7	**8** TH	0126 2.0 0756 0.4 1409 1.7 1955 0.8	**23** F	0107 2.1 0727 0.4 1350 1.8 1936 0.7	**8** SA	0144 2.0 0811 0.5 1420 1.7 2004 0.8	**23** SU	0131 2.2 0753 0.3 1413 1.9 2001 0.6	**8** TU	0224 1.9 0834 0.7 1450 1.7 2036 0.9	**23** W	0245 2.1 0905 0.7 1519 1.8 ☾ 2121 0.8
9 TU	0059 2.0 0717 0.4 1336 1.9 1931 0.7	**24** W	0040 2.0 0652 0.6 1317 1.8 1901 0.8	**9** F	0211 1.9 0847 0.5 1457 1.6 2041 0.8	**24** SA	0150 2.0 0814 0.4 1438 1.8 2023 0.7	**9** SU	0223 1.9 0850 0.6 1459 1.6 2043 0.8	**24** M	0216 2.1 0840 0.4 1500 1.8 2049 0.7	**9** W	0302 1.8 0910 0.9 1530 1.7 ☽ 2121 1.0	**24** TH	0344 1.9 1008 0.9 1619 1.7 2238 0.9
10 W	0144 1.9 0810 0.5 1428 1.7 2018 0.9	**25** TH	0121 1.9 0737 0.6 1404 1.7 1947 0.8	**10** SA	0257 1.8 0941 0.6 1549 1.5 2134 0.9	**25** SU	0236 2.0 0905 0.4 1529 1.7 2115 0.8	**10** M	0303 1.8 0930 0.7 1541 1.6 2126 0.9	**25** TU	0305 2.0 0934 0.5 1552 1.7 ☾ 2145 0.8	**10** TH	0347 1.7 0956 1.0 1620 1.6 2224 1.1	**25** F	0502 1.8 1137 1.0 1741 1.7
11 TH	0232 1.8 0912 0.6 1527 1.6 2116 1.0	**26** F	0205 1.9 0827 0.7 1455 1.7 2039 0.9	**11** SU	0347 1.7 1040 0.7 1647 1.5 ☽ 2240 0.9	**26** M	0327 1.9 1003 0.5 1627 1.7 ☾ 2216 0.8	**11** TU	0347 1.7 1015 0.8 1630 1.5 ☽ 2221 1.0	**26** W	0402 1.9 1036 0.7 1653 1.7 2256 0.8	**11** F	0451 1.6 1103 1.1 1732 1.6 1911 1.7	**26** SA	0022 0.9 0648 1.7 1312 1.1
12 F	0327 1.7 1027 0.7 1639 1.5 ☽ 2236 1.0	**27** SA	0253 1.8 0925 0.6 1554 1.6 2140 0.9	**12** M	0445 1.6 1142 0.7 1755 1.5 2359 0.9	**27** TU	0426 1.8 1109 0.5 1732 1.6 2326 0.8	**12** W	0439 1.6 1111 0.9 1731 1.5 2340 1.0	**27** TH	0514 1.8 1154 0.8 1808 1.6	**12** SA	0014 1.1 0637 1.6 1242 1.1 1907 1.7	**27** SU	0147 0.8 0813 1.7 1417 1.0 2017 1.8
13 SA	0434 1.6 1148 0.7 1806 1.4	**28** SU	0349 1.8 1032 0.6 1702 1.6 ☾ 2250 0.9	**13** TU	0553 1.6 1239 0.7 1858 1.5	**28** W	0536 1.8 1220 0.6 1842 1.7	**13** TH	0551 1.5 1220 0.9 1847 1.6	**28** F	0024 0.8 0645 1.7 1315 0.9 1926 1.7	**13** SU	0150 1.0 0806 1.6 1358 1.0 2014 1.8	**28** M	0248 0.7 0908 1.8 1506 0.9 2108 2.0
14 SU	0008 1.0 0554 1.6 1252 0.7 1912 1.5	**29** M	0455 1.7 1144 0.5 1814 1.6	**14** W	0106 0.9 0704 1.6 1329 0.7 1950 1.6	**29** TH	0042 0.8 0656 1.8 1328 0.6 1946 1.7	**14** F	0111 1.0 0719 1.6 1328 0.9 1953 1.7	**29** SA	0146 0.8 0807 1.8 1421 0.9 2029 1.8	**14** M	0243 0.9 0859 1.8 1449 0.9 2103 1.9	**29** TU	0335 0.6 0950 1.9 1545 0.8 2150 2.1
15 M	0112 0.9 0705 1.6 1341 0.6 1959 1.5	**30** TU	0005 0.8 0608 1.8 1252 0.5 1919 1.7	**15** TH	0159 0.9 0802 1.6 1413 0.7 2034 1.7	**30** F	0152 0.7 0809 1.8 1428 0.6 2042 1.8	**15** SA	0215 0.9 0826 1.6 1422 0.9 2044 1.8	**30** SU	0250 0.6 0911 1.8 1513 0.9 2120 1.9	**15** TU	0325 0.7 0943 1.9 1533 0.8 2145 2.1	**30** W	0415 0.5 1026 1.9 1621 0.7 2229 2.2
		31 W	0112 0.7 0720 1.8 1352 0.4 2014 1.8					**31** M	0343 0.5 1001 1.9 1558 0.8 2206 2.1					**31** TH	0450 0.4 1100 2.0 1653 0.7 ● 2304 2.2

SHETLAND ISLANDS - LERWICK

LAT 60°09'N   LONG 1°08'W

TIMES AND HEIGHTS OF HIGH AND LOW WATERS

YEAR 1989

TIME ZONE UT(GMT)
For Summer Time add ONE hour in non-shaded areas

SEPTEMBER

Day	Time	m	Day	Time	m
1 F	0522 / 1131 / 1725 / 2339	0.4 / 2.0 / 0.6 / 2.2	16 SA	0457 / 1116 / 1709 / 2323	0.3 / 2.1 / 0.5 / 2.5
2 SA	0553 / 1202 / 1754	0.4 / 2.0 / 0.6	17 SU	0538 / 1156 / 1750	0.2 / 2.3 / 0.4
3 SU	0012 / 0622 / 1233 / 1824	2.2 / 0.5 / 2.0 / 0.7	18 M	0007 / 0620 / 1237 / 1832	2.5 / 0.3 / 2.2 / 0.5
4 M	0044 / 0650 / 1303 / 1853	2.1 / 0.6 / 2.0 / 0.7	19 TU	0051 / 0704 / 1319 / 1917	2.4 / 0.5 / 2.1 / 0.5
5 TU	0116 / 0718 / 1335 / 1924	2.0 / 0.7 / 1.9 / 0.8	20 W	0138 / 0749 / 1403 / 2007	2.3 / 0.7 / 2.0 / 0.7
6 W	0150 / 0748 / 1408 / 1959	1.9 / 0.8 / 1.9 / 0.9	21 TH	0230 / 0840 / 1452 / 2107	2.1 / 0.9 / 1.9 / 0.8
7 TH	0228 / 0823 / 1446 / 2042	1.8 / 1.0 / 1.8 / 1.0	22 F	0332 / 0944 / 1552 / 2232	1.9 / 1.1 / 1.8 / 0.9
8 F	0313 / 0906 / 1533 / 2139	1.7 / 1.1 / 1.7 / 1.1	23 SA	0458 / 1124 / 1716	1.7 / 1.2 / 1.7
9 SA	0417 / 1012 / 1640 / 2326	1.6 / 1.2 / 1.7 / 1.1	24 SU	0021 / 0653 / 1303 / 1852	0.9 / 1.7 / 1.2 / 1.9
10 SU	0604 / 1201 / 1817	1.6 / 1.2 / 1.7	25 M	0139 / 0807 / 1404 / 1959	0.8 / 1.7 / 1.1 / 1.9
11 M	0119 / 0742 / 1332 / 1939	1.0 / 1.7 / 1.1 / 1.8	26 TU	0233 / 0853 / 1448 / 2047	0.7 / 1.8 / 0.9 / 2.0
12 TU	0214 / 0835 / 1425 / 2033	0.8 / 1.8 / 1.0 / 2.0	27 W	0315 / 0929 / 1524 / 2128	0.6 / 1.9 / 0.8 / 2.1
13 W	0257 / 0918 / 1509 / 2117	0.7 / 2.0 / 0.8 / 2.1	28 TH	0350 / 1001 / 1557 / 2205	0.5 / 1.9 / 0.8 / 2.1
14 TH	0337 / 0957 / 1549 / 2200	0.5 / 2.0 / 0.7 / 2.3	29 F	0421 / 1031 / 1628 / 2239	0.5 / 2.0 / 0.7 / 2.2
15 F	0417 / 1037 / 1629 / 2241	0.3 / 2.2 / 0.5 / 2.4	30 SA	0450 / 1101 / 1658 / 2312	0.5 / 2.1 / 0.7 / 2.2

OCTOBER

Day	Time	m	Day	Time	m
1 SU	0519 / 1130 / 1727 / 2344	0.6 / 2.1 / 0.7 / 2.2	16 M	0515 / 1131 / 1730 / 2349	0.3 / 2.3 / 0.4 / 2.4
2 M	0546 / 1200 / 1756	0.6 / 2.1 / 0.7	17 TU	0558 / 1213 / 1815	0.5 / 2.3 / 0.5
3 TU	0016 / 0614 / 1231 / 1826	2.1 / 0.7 / 2.1 / 0.8	18 W	0036 / 0642 / 1255 / 1903	2.4 / 0.6 / 2.2 / 0.5
4 W	0050 / 0643 / 1302 / 1859	2.0 / 0.8 / 2.0 / 0.8	19 TH	0125 / 0727 / 1340 / 1956	2.2 / 0.8 / 2.1 / 0.6
5 TH	0126 / 0714 / 1337 / 1936	1.9 / 0.9 / 2.0 / 0.9	20 F	0219 / 0818 / 1430 / 2100	2.0 / 1.0 / 2.0 / 0.8
6 F	0206 / 0749 / 1416 / 2022	1.8 / 1.0 / 1.9 / 1.0	21 SA	0322 / 0921 / 1529 / 2224	1.8 / 1.1 / 1.9 / 0.9
7 SA	0255 / 0836 / 1504 / 2124	1.7 / 1.1 / 1.8 / 1.1	22 SU	0446 / 1056 / 1647	1.7 / 1.2 / 1.8
8 SU	0400 / 0943 / 1607 / 2254	1.6 / 1.2 / 1.7 / 1.1	23 M	0001 / 0630 / 1233 / 1817	0.8 / 1.7 / 1.2 / 1.8
9 M	0533 / 1124 / 1731	1.6 / 1.2 / 1.7	24 TU	0112 / 0738 / 1334 / 1927	0.8 / 1.7 / 1.1 / 1.8
10 TU	0034 / 0706 / 1256 / 1855	1.0 / 1.7 / 1.1 / 1.8	25 W	0203 / 0822 / 1419 / 2018	0.7 / 1.7 / 1.0 / 1.9
11 W	0136 / 0803 / 1353 / 1956	0.8 / 1.8 / 1.0 / 2.0	26 TH	0243 / 0857 / 1456 / 2100	0.7 / 1.9 / 1.0 / 2.0
12 TH	0224 / 0847 / 1439 / 2047	0.6 / 2.0 / 0.8 / 2.1	27 F	0317 / 0930 / 1530 / 2137	0.6 / 1.9 / 0.8 / 2.1
13 F	0308 / 0929 / 1522 / 2133	0.5 / 2.1 / 0.7 / 2.2	28 SA	0348 / 1000 / 1602 / 2212	0.6 / 2.0 / 0.7 / 2.1
14 SA	0350 / 1010 / 1604 / 2218	0.4 / 2.2 / 0.5 / 2.4	29 SU	0418 / 1031 / 1633 / 2246	0.6 / 2.1 / 0.7 / 2.1
15 SU	0432 / 1050 / 1646 / 2303	0.3 / 2.3 / 0.5 / 2.5	30 M	0447 / 1102 / 1703 / 2320	0.7 / 2.1 / 0.7 / 2.1
			31 TU	0516 / 1133 / 1735 / 2355	0.7 / 2.1 / 0.7 / 2.1

NOVEMBER

Day	Time	m	Day	Time	m
1 W	0546 / 1205 / 1808	0.8 / 2.1 / 0.8	16 TH	0025 / 0624 / 1237 / 1854	2.3 / 0.7 / 2.3 / 0.5
2 TH	0031 / 0618 / 1239 / 1845	2.0 / 0.9 / 2.1 / 0.8	17 F	0114 / 0710 / 1323 / 1948	2.1 / 0.9 / 2.2 / 0.6
3 F	0110 / 0653 / 1316 / 1925	1.9 / 1.0 / 2.0 / 0.9	18 SA	0207 / 0758 / 1412 / 2047	1.9 / 1.0 / 2.1 / 0.7
4 SA	0154 / 0732 / 1357 / 2013	1.8 / 1.0 / 1.9 / 0.9	19 SU	0304 / 0854 / 1505 / 2157	1.8 / 1.1 / 1.9 / 0.7
5 SU	0243 / 0821 / 1445 / 2111	1.8 / 1.1 / 1.9 / 0.9	20 M	0410 / 1005 / 1608 / 2317	1.7 / 1.1 / 1.8 / 0.8
6 M	0344 / 0925 / 1542 / 2223	1.7 / 1.2 / 1.8 / 0.9	21 TU	0531 / 1135 / 1724	1.6 / 1.1 / 1.8
7 TU	0458 / 1045 / 1651 / 2344	1.7 / 1.2 / 1.8 / 0.8	22 W	0026 / 0645 / 1247 / 1839	0.8 / 1.6 / 1.1 / 1.8
8 W	0618 / 1209 / 1808	1.7 / 1.1 / 1.9	23 TH	0120 / 0738 / 1341 / 1939	0.8 / 1.7 / 1.0 / 1.8
9 TH	0054 / 0723 / 1315 / 1918	0.7 / 1.8 / 1.0 / 2.0	24 F	0204 / 0820 / 1424 / 2027	0.8 / 1.8 / 0.9 / 1.9
10 F	0149 / 0815 / 1408 / 2016	0.6 / 2.0 / 0.8 / 2.1	25 SA	0241 / 0856 / 1503 / 2109	0.7 / 1.9 / 0.9 / 1.9
11 SA	0239 / 0900 / 1456 / 2109	0.5 / 2.1 / 0.7 / 2.2	26 SU	0316 / 0931 / 1538 / 2148	0.8 / 2.0 / 0.8 / 1.9
12 SU	0325 / 0944 / 1543 / 2159	0.5 / 2.2 / 0.6 / 2.3	27 M	0348 / 1004 / 1612 / 2226	0.8 / 2.1 / 0.8 / 2.0
13 M	0411 / 1027 / 1629 / 2247	0.5 / 2.3 / 0.5 / 2.4	28 TU	0421 / 1038 / 1646 / 2303	0.8 / 2.1 / 0.7 / 2.0
14 TU	0455 / 1110 / 1716 / 2336	0.5 / 2.3 / 0.5 / 2.3	29 W	0453 / 1112 / 1721 / 2340	0.8 / 2.2 / 0.7 / 2.0
15 W	0539 / 1153 / 1804	0.6 / 2.3 / 0.4	30 TH	0527 / 1147 / 1758	0.8 / 2.2 / 0.7

DECEMBER

Day	Time	m	Day	Time	m
1 F	0019 / 0603 / 1223 / 1836	2.0 / 0.9 / 2.2 / 0.7	16 SA	0101 / 0653 / 1306 / 1934	2.0 / 0.8 / 2.2 / 0.5
2 SA	0059 / 0641 / 1301 / 1917	2.0 / 0.9 / 2.1 / 0.7	17 SU	0148 / 0736 / 1351 / 2024	1.9 / 0.9 / 2.1 / 0.6
3 SU	0142 / 0723 / 1342 / 2002	1.9 / 1.0 / 2.1 / 0.7	18 M	0234 / 0821 / 1436 / 2115	1.8 / 0.9 / 2.0 / 0.7
4 M	0229 / 0809 / 1427 / 2053	1.8 / 1.0 / 2.0 / 0.7	19 TU	0323 / 0910 / 1525 / 2212	1.7 / 1.0 / 1.9 / 0.8
5 TU	0321 / 0903 / 1517 / 2152	1.8 / 1.0 / 1.9 / 0.8	20 W	0418 / 1010 / 1620 / 2314	1.6 / 1.1 / 1.8 / 0.8
6 W	0421 / 1006 / 1616 / 2259	1.7 / 1.1 / 1.9 / 0.7	21 TH	0522 / 1130 / 1728	1.6 / 1.1 / 1.7
7 TH	0529 / 1120 / 1725	1.7 / 1.1 / 1.9	22 F	0018 / 0633 / 1249 / 1845	0.9 / 1.6 / 1.1 / 1.7
8 F	0011 / 0640 / 1234 / 1841	0.7 / 1.8 / 1.0 / 1.9	23 SA	0115 / 0733 / 1349 / 1951	0.9 / 1.7 / 1.0 / 1.7
9 SA	0117 / 0741 / 1339 / 1951	0.7 / 1.9 / 0.8 / 2.0	24 SU	0203 / 0822 / 1437 / 2044	0.9 / 1.8 / 1.0 / 1.8
10 SU	0215 / 0835 / 1436 / 2052	0.6 / 2.0 / 0.7 / 2.1	25 M	0246 / 0903 / 1519 / 2129	0.9 / 1.9 / 0.9 / 1.8
11 M	0307 / 0923 / 1529 / 2147	0.6 / 2.1 / 0.6 / 2.2	26 TU	0324 / 0942 / 1557 / 2210	0.9 / 2.0 / 0.8 / 1.9
12 TU	0355 / 1009 / 1619 / 2238	0.6 / 2.2 / 0.5 / 2.2	27 W	0401 / 1018 / 1634 / 2249	0.9 / 2.1 / 0.7 / 2.0
13 W	0441 / 1054 / 1708 / 2327	0.6 / 2.3 / 0.4 / 2.2	28 TH	0437 / 1054 / 1710 / 2327	0.9 / 2.1 / 0.7 / 2.0
14 TH	0526 / 1138 / 1757	0.7 / 2.3 / 0.4	29 F	0514 / 1131 / 1747	0.8 / 2.2 / 0.6
15 F	0015 / 0610 / 1222 / 1845	2.1 / 0.7 / 2.3 / 0.4	30 SA	0006 / 0551 / 1208 / 1824	2.0 / 0.8 / 2.2 / 0.5
			31 SU	0045 / 0630 / 1246 / 1903	0.8 / 0.8 / 2.2 / 0.5

Chart Datum: 1.22 metres below Ordnance Datum (Newlyn)

SCRABSTER 10-7-26
Caithness

CHARTS
Admiralty 1462, 2162, 1954; OS 12
TIDES
−0240 Dover; ML 3.2; Duration 0615; Zone 0 (GMT).

Standard Port ABERDEEN (←)

Times				Height (metres)			
HW		LW		MHWS	MHWN	MLWN	MLWS
0300	1000	0100	0800	4.3	3.4	1.6	0.6
1500	2200	1300	2000				

Differences SCRABSTER
−0455 −0510 −0500 −0445 +0.7 +0.3 +0.5 +0.2
STROMA
−0320 −0320 −0320 −0320 −1.2 −1.1 −0.3 −0.1

SHELTER
Very good except in strong NE winds. Anchoring is not recommended. Secure alongside NE wall and contact Hr Mr. Yachts normally use Inner Basin.
NAVIGATION
Waypoint 58°36'.60N 03°32'.00W, 098°/278° from/to E pier Lt, 0.25M. Can be entered at all tides in all weathers. Do not confuse harbour lights with those of Thurso. Beware fishing vessels, merchant ships and the Orkney ferries.
LIGHTS AND MARKS
There are no leading marks or lights. Entry is straightforward once the Lt on end of pier QG 6m 4M has been located.
RADIO TELEPHONE
VHF Ch 16; 12 (H24).
TELEPHONE (0847)
Hr Mr 62779; CG Wick 62332; MRSC Kirkwall 3268; HMC 62727; Marinecall 0898 500 451; Dr. 63154.
FACILITIES
EC Thurso — Thursday; **Harbour** Slip, D, L, FW, ME, El, C (30 ton mobile), CH, AB, R, Bar; **Pentland Firth YC** M, R, Bar; **Thurso** V, R, Bar. PO; Bank; Rly; Air (Wick).

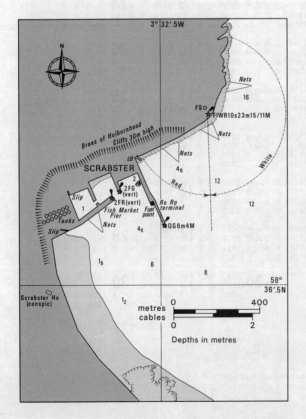

MINOR HARBOURS AND ANCHORAGES 10.7.27

CULLEN, Banff, 57°41' N, 2°49' W, Zone 0 (GMT), Admty Chart 222. HW +0045 on Dover, −0135 on Aberdeen; HW height −0.3m on Aberdeen; ML 2.4m; Duration 0555. Shelter good but entrance hazardous in strong winds from N and W. Harbour, consisting of two basins, dries, access HW∓4 approx. Moor in inner basin. There are no lights or marks. Popular recreational harbour. Facilities: Hr Mr Tel. 41116; **Cullen Bay Hotel** Tel. 40432 R, Bar; **Town** V, R, Bar, PO.

PORTKNOCKIE, Banff, 57°42' N, 2°51' W, Zone 0 (GMT), Admty chart 222. HW +0045 on Dover, −0135 on Aberdeen; HW height −0.3m on Aberdeen; ML 2.3m; Duration 0555. Shelter is good and is one of the safest harbours on S side of Moray Firth but care is needed entering in strong N winds. Scend is often experienced. Most of inner harbour dries. W Tr on S pier head. Yachts use NW jetties. Orange street lights surround the harbour — two conspic white lights are the leading lights of approx 130°, on S pier. Facilities: Hr Mr Tel. Cullen 40705; EC Wed; AB, FW, C (15 ton), CH; **D Reid** Slip, L, Sh. Dr. Tel. Cullen 40272, **Town** Bank, P, PO, V, Bar.

FINDOCHTY, Banff, 57°42' N, 2°54' W, Zone 0 (GMT), Admty chart 222. HW +0045 on Dover, −0140 on Aberdeen; HW height −0.2m on Aberdeen; ML 2.3m; Duration 0550. Good shelter in inner basin which dries in the Western half. Outer basin dries and has numerous rocky outcrops. Entrance faces N and is 20m wide. There are no lights or marks. Facilities: Hr Mr Tel. Cullen 40705. **Town** V, R, Bar, PO.

FORTROSE, Ross and Cromarty, 57°35' N, 4°08' W, Zone 0 (GMT), Admty charts 1078, 223. HW +0055 on Dover, −0125 on Aberdeen; HW height 0.0m on Aberdeen; ML 2.5m; Duration 0620. See 10.7.16. Small Hr which dries, well protected by Chanonry Ness on E. Beware Craigan Rock ESE of Hr entrance. Ldg line 296°, Broomhill House (conspic on hill to NW) in line with school house spire. Follow line until just beyond G buoy Iso G, then turn W to avoid Craigan Rock. There are no Hr Lts. Lt Ho on Chanonry Pt Oc 6s 12m 15M, vis 148°−073°. Facilities: EC Thurs; AB, P, PO, Bank, D, Dr, L, M, R, YC, Slip, V; **Chanonry SC** (near pier); **Fortrose YC**; **Pagwari** Tel. 20356, Gas.

CROMARTY FIRTH, Ross and Cromarty, 57°41' N, 4°02' W, Zone 0 (GMT), Admty charts 1889, 1890. HW +0100 on Dover, −0135 on Aberdeen; HW height 0.0m on Aberdeen; ML 2.5m; Duration 0625. See 10.7.16. Excellent harbour extending 7.5M W and then 9M SW. Good shelter can always be found depending on the direction of the wind. The harbour is run by Cromarty Firth Ports Authority. Visitors should report to the Port Manager (VHF Ch 16). Beware rocks and reefs round N and S Sutor, N and S of entrance. Cromarty harbour is protected by two piers, and dries. Cromarty Lt Ho, on the Ness, Oc WR 10s 18m 14/11M, R079°−088°, W088°−275°, obsc by N Sutor when bearing less than 253°. A Lt is shown 5m from head of S pier 1st Oct to 31st Mar. Anchorage in approx 6m 2 ca W of S pier head. Cromarty Firth Port Control VHF Ch 16; 06 08 11 12 13 14 (H24). Facilities: EC Wed; AB, Bar, C(3 ton) D, FW, PO, P, R, V, L; **Invergordon Boating Club**; **Tomlinson** Tel. 852233, Gas.

DORNOCH FIRTH, Ross and Cromarty/Sutherland 57°51' N, 4°08' W, Zone 0 (GMT), Admty charts 223, 115. HW +0115 on Dover, −0120 on Aberdeen; HW height +0.1m on Aberdeen; ML 2.5m; Duration 0605. Excellent shelter but difficult entrance. Firth extends 15M inland, entered between N edge of Whiteness Sands and S edge of Gizzen Briggs. There are many shifting sandbanks, especially near the entrance. Anchorages in 7m ¾M ESE of Dornoch Pt (sheltered from NE swell by Gizzen Briggs), in 7m 2 ca SSE of Ard na Cailc, in 3.3m 1M below Bonar Bridge. (Admiralty chart coverage ceases ¼M E of Ferry Pt.) Light on Tarbat Ness Fl(4) 30s 53m 24M. R Lt shown on Tain bombing range when firing in progress. Very limited facilities at Ferrytown and Bonar Bridge. Dornoch:- EC Thur; V, P, PO, Dr, Bank, R, Bar. Note: It is planned to build a bridge across Dornoch Firth which will have min clearance of 11m at MHWS under the centre arch.

GOLSPIE, Sutherland, 57°58′ N, 3°59′ W, Zone 0 (GMT), Admty chart 223. HW +0045 on Dover, −0130 sp, −0215 np on Aberdeen; HW height −0.3m on Aberdeen; ML 2.3m; Duration 0610. See 10.7.18. Golspie pier projects 60m across foreshore with arm projecting SW at the head, giving shelter during NE winds. Beware The Bridge, a bank (0.3m to 1.8m) running parallel to the shore ¼ M to seaward of pier head. Seas break heavily over The Bridge in NE winds. There are no lights. To enter, keep Duke of Sutherlands Memorial in line with boathouse SW of pier at 316° until Church spire in middle of village is in line with head of pier at 006°, then keep on those marks. Harbour gets very congested; good anchorage off pier. Facilities: EC Wed; **Lindsay** Tel. 3212, Gas; **Town** Bar, D, Dr, Hosp, L, M, P, PO, R, Rly, V, Bank.

LYBSTER, Caithness, 58°18′ N, 3°17′ W, Zone 0 (GMT), Admty chart 115. HW +0020 on Dover, −0150 sp −0215 np on Aberdeen; HW height −0.6m on Aberdeen; ML 2.1m; Duration 0620. Excellent shelter in Inner Hr. Berth on E side of W pier at N end. Beware rocks on E side of entrance; entrance difficult in strong E winds. Min depth 2.5m in entrance. S pier head Lt Oc R 6s 10m 3M shown during fishing season. Facilities: EC Thurs; FW on S quay, **Town** Bar, D, P, R, V.

HOUTON BAY, Orkney Islands, 58°55′ N, 3°11′ W, Zone 0 (GMT), Admty chart 2568. HW −0140 on Dover, −0400 on Aberdeen; HW height +0.3m on Kirkwall; ML 1.8m; Duration 0615. Anchorage in the bay in approx 5.5m at centre, sheltered from all winds. The entrance is to the E of the island Holm of Houton; entrance channel dredged 3.5m for 15m each side of Ldg line. Care should be taken not to obstruct merchant vessels & ferries plying to Flotta. Ldg Lts in line at 316°, Front Fl G 3s 8m, Rear FG 16m. Ro Ro terminal in NE corner marked by Iso R 4s. There are no facilities, except a slip close E of piers.

SHAPINSAY, Orkney Islands, 59°03′ N, 2°54′ W, Zone 0 (GMT), Admty charts 2249, 2250. HW −0015 on Dover, −0330 on Aberdeen; HW height −1.0m on Aberdeen. Good shelter in Ellwick Bay off Balfour on SW end of island in 2½ to 3m. Enter bay passing W of Helliar Holm which has Lt (Fl WRG 10s) on S end. Keep mid-channel. Balfour Pier Lt Q WRG 5m 3/2M; vis G270°-010°, W010°-020°, R020°-090°. Tides in The String reach 5 kn at springs. Facilities: PO, P (from garage), shop.

PIEROWALL, Orkney Islands, 59°19′ N, 2°58′ W, Zone 0 (GMT), Admty chart 2622. HW −0135 on Dover, −0355 on Aberdeen; HW height +0.7m on Kirkwall; ML 2.2m; Duration 0620. See 10.7.23. The bay is a good anchorage in 2m to 7m and well protected. Deep water berths at Gill Pt. are available – consult Hr Mr (Tel. Westray 273). Beware Skelwick Skerry rocks approaching from S and the rocks off Vest Ness (extent approx 1 ca off shore) coming from the N. The N entrance through Papa Sound should not be used unless local knowledge is available. There is a dangerous tide race off Mull Head at the N of Papa Westray. Lights: E pier head Fl WRG 3s 7m 11/7M; G254°-276°, W276°-291°, R291°-308°, G308°-215°. W pier head 2 FR (vert) 4/6m 3M. Facilities: Hr Mr VHF Ch 16 FW (at Gill Pier) P and D at garage in Pierowall, Bar, PO, R, V.

AUSKERRY, Orkney Islands 59°02′ N, 2°34′ W, Zone 0 (GMT), Admty chart 2250. HW −0010 on Dover, −0315 on Aberdeen; HW height −1.0m on Aberdeen. Small island at entrance to Stronsay Firth has small harbour on W side. Safe entrance and good shelter except in SW winds. Auskerry Sound and Stronsay Firth are dangerous with wind over tide. Ent min 3.5m and alongside pier 1.2m. Yachts can lie secured between ringbolts at entrance and the pier. Auskerry Lt at S end Fl 20s 34m 18M from W Tr. There are no facilities.

FAIR ISLE, Shetland Islands, 59°32′ N, 1°37′ W, Zone 0 (GMT), Admty chart 2622. HW −0030 on Dover, −0020 on Lerwick; HW height −0.1m on Lerwick; ML 1.4m; Duration 0620. See 10.7.25. Good shelter in North Haven although open to NE winds. Lie alongside pier or anchor in approx 2m. Beware rocks all round the island particularly in the 'so called' South Harbour which is not recommended. There is a light at each end; Skadan in the S Fl (4) 30s 32m 24M; vis 260°-146° but obscured close inshore from 260°-282°; Horn(2) 60s. In the N, Skroo Fl (2) 30s 80m 22M (vis about 087°-358°), Horn (3) 45s. Ldg Marks at 199°, the Stack of North Haven in line with top of Sheep Craig leads into North Haven. Facilities: there is a shop and PO at N Shriva, and a bi-weekly ferry to Shetland.

SCALLOWAY, Shetland Islands, 60°08′ N, 1°17′ W, Zone 0 (GMT), Admty chart 3294. HW −0200 on Dover, −0150 on Lerwick; HW height −0.5m on Lerwick; ML 0.9m; Duration 0620. A good sheltered anchorage available in all weathers but care is needed passing through the islands in strong SW winds. Anchor off Scalloway 6 to 10m, in Hamna Voe (W Burra), almost land-locked, in 6m approx, but bottom foul with old moorings. Lights; Scalloway Hr, Moores Slipway jetty head 2 FR (vert) 4/2m 3M; Blacksness W pier head 2 FG (vert) 6/4m 3M; E pier head Oc R 7s 5m 3M; Fugla Ness Fl (2) WRG 10s 20m 10/7M; W Tr; vis G014°-032°, W032°-082°, R082°-134°, W134°-shore. VHF Ch 16; 12 (office hours). Facilities: EC Thurs; Bar, C, D, El, FW, ME, PO, Sh, Slip, R, V, AB, SM, BY, CH, M, P; **Scalloway Boating Club** Tel. 408; **W Moore** Tel. 215, Slip, ME, El, Sh; **H. Williamson** Tel. 645, El.

BALTA SOUND, Shetland Islands, 60°44′ N, 0°48′ W, Zone 0 (GMT), Admty chart 3293. HW −0105 on Dover, −0055 on Lerwick; HW height +0.2m on Lerwick; ML 1.3m; Duration 0640. The Sound and Balta Hr form a large landlocked harbour with good shelter from all winds. Beware kelp which causes bad holding. R can buoys mark 4m patches. Anchor near Sandisons Wharf in approx 6m. Light on S end of Balta I Iso WR 4s 17m 9/6M; W249°-010°, R010°-060°, W060°-154°. Safest entry is between Huney I and Balta I. VHF Ch 16; 20 (office hours) Facilities: at BY FW, D, El, ME, Sh; Hotel by pier; Baltasound village, Bar, R, V, PO.

KYLE OF TONGUE, Sutherland, 58°32′ N, 4°24′ W (Rabbit I.), Zone 0 (GMT), Admty Charts 1954, 2720. HW −0325 on Dover, +0050 on Ullapool; HW height −0.4m on Ullapool. The Kyle runs about 7M inland and contains many sheltered anchorages. Entry (see 10.7.5) should not be attempted in strong N winds. Anchorages at Talmine (W of Rabbit I.) protected from all but NE winds; at Skullomie Hr, protected from E winds; off Mol na Coinne, a small bay on SE side of Eilean nan Ron, protected from W and N winds; off S of Rabbit I., protected from W to N winds. There are no lights or leading marks. Facilities: Limited supplies at Talmine shop (½ M from slip) or at Coldbachie (1½ M from Skullomie).

LOCH ERIBOLL, Sutherland, 58°29′ N, 4°39′ W, Zone 0 (GMT), Admty chart 2076. Tidal figures taken at Rispond: HW −0345 on Dover, +0035 on Ullapool; HW height −0.5m on Ullapool; ML 2.7m; Duration 0610. See 10.8.7. Two good anchorages; in Rispond Bay on W side of loch, entrance good in all but E winds, in approx 5m; in bays to N and S of peninsular at Heilam on E side of loch. Yachts can enter Rispond Hr, access approx HW∓3, where they can dry out alongside. There are no lights or marks. Facilities at Rispond are extremely limited.

APRIL 1988

THE MARKET-PLACE OF THE BOATING WORLD

90p EIRE £1.32 SOR

BOAT MART
INTERNATIONAL
& MARINE ACCESSORIES

- Boating — how to get started
- Know your weather
- Discovering the Canoe
- Navigation Workshop — learning from the basics

THE MONTHLY MAGAZINE FOR ALL YOUR BOATING NEEDS

AVAILABLE AT ALL GOOD NEWSAGENTS

ADVERTISING DEPARTMENT: COLCHESTER (0206) 562939

VOLVO PENTA SERVICE

Sales and service centres in area 8
ARGYLL **Camus Marine Ltd** Kilmelford Yacht Haven, Kilmelford, OBAN PA34 4XD Tel (085 22) 248.

Area 8

North-West Scotland
Stornoway to Craobh Haven

10.8.1	Index	Page 417
10.8.2	Diagram of Radiobeacons, Air Beacons, Lifeboat Stations etc	418
10.8.3	Tidal Stream Charts	420
10.8.4	List of Lights, Fog Signals and Waypoints	422
10.8.5	Passage Information	424
10.8.6	Distance Table	425
10.8.7	Stornoway	426
10.8.8	Ullapool, Standard Port, Tidal Curves	427
10.8.9	Portree	431
10.8.10	Mallaig	431
10.8.11	Loch Sunart	432
10.8.12	Tobermory	432
10.8.13	Loch Aline	433
10.8.14	Fort William/Corpach	433
10.8.15	Caledonian Canal	434
10.8.16	Oban, Standard Port, Tidal Curves	435
10.8.17	Loch Melfort	440
10.8.18	Craobh Haven (Loch Shuna)	440

10.8.19 Minor Harbours and Anchorages 441
Loch Laxford
Loch Inver
Loch Shell (Lewis)
W Loch Tarbert (Harris)
E Loch Tarbert (Harris)
St Kilda
Loch Ewe
Loch Gairloch
Loch Maddy (N Uist)
Loch A'Bhraige (Rona)
Crowlin Islands
Loch Harport (Skye)
Plockton
Loch Alsh
Loch Boisdale (S Uist)
Canna Is
Loch Nevis
Castlebay (Barra)
Eigg Is
Arisaig
Arinagour (Coll)
Dunstaffnage
Loch Lathaich

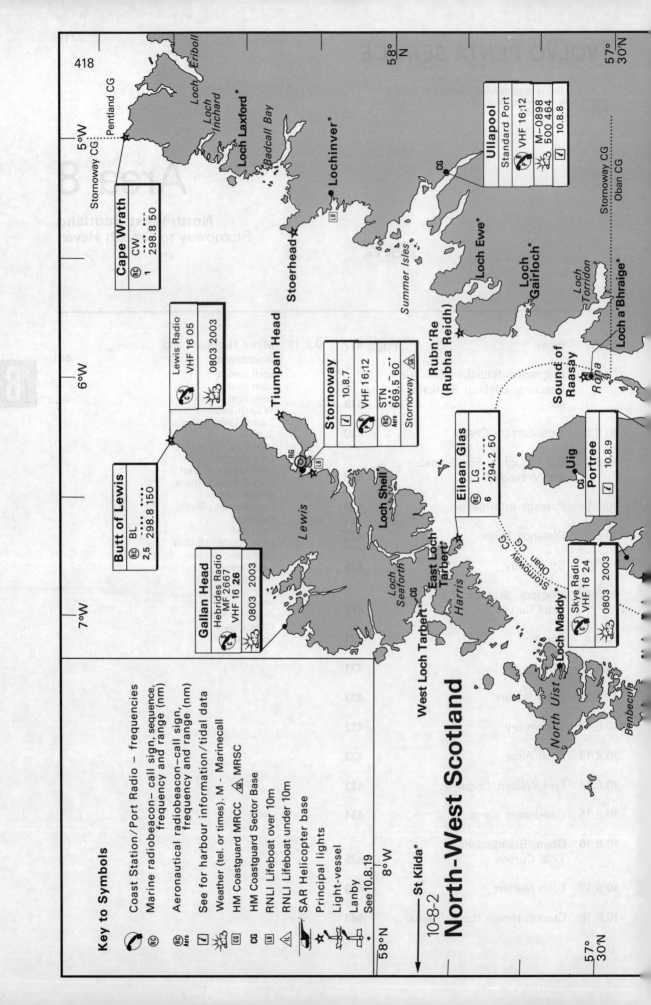

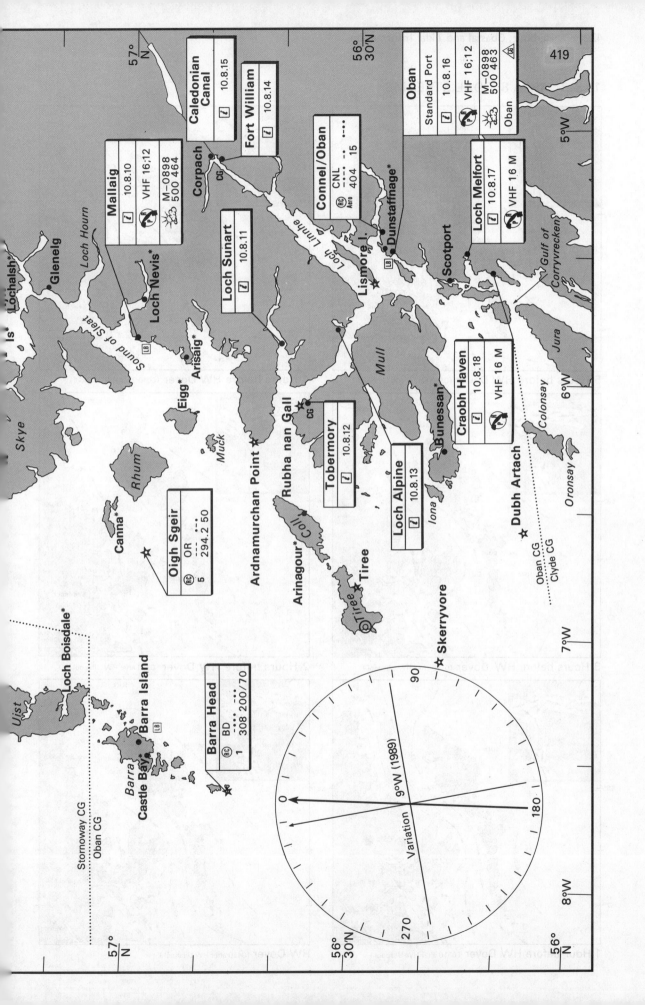

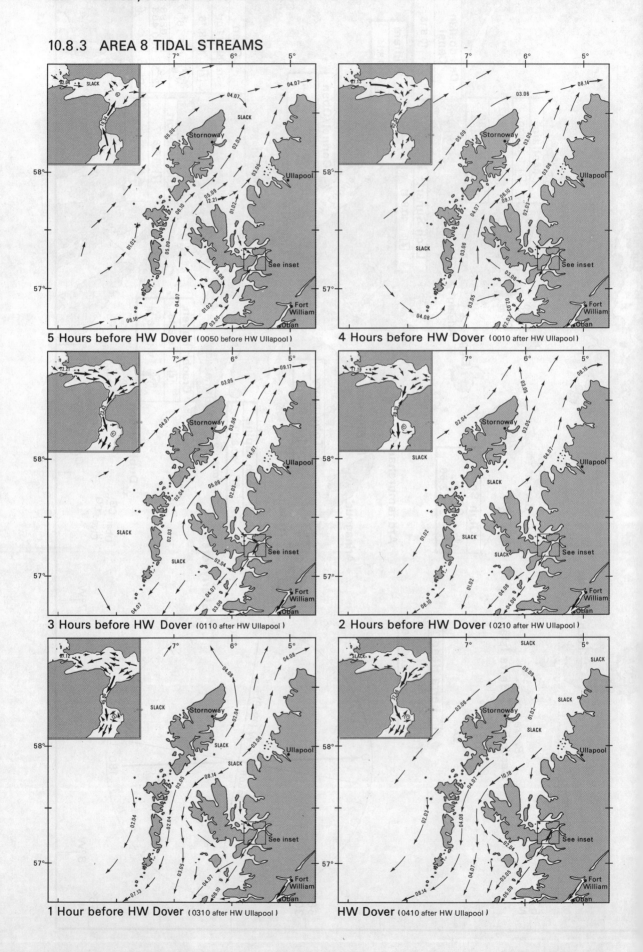

10.8.3 AREA 8 TIDAL STREAMS

AREA 8 – NW Scotland

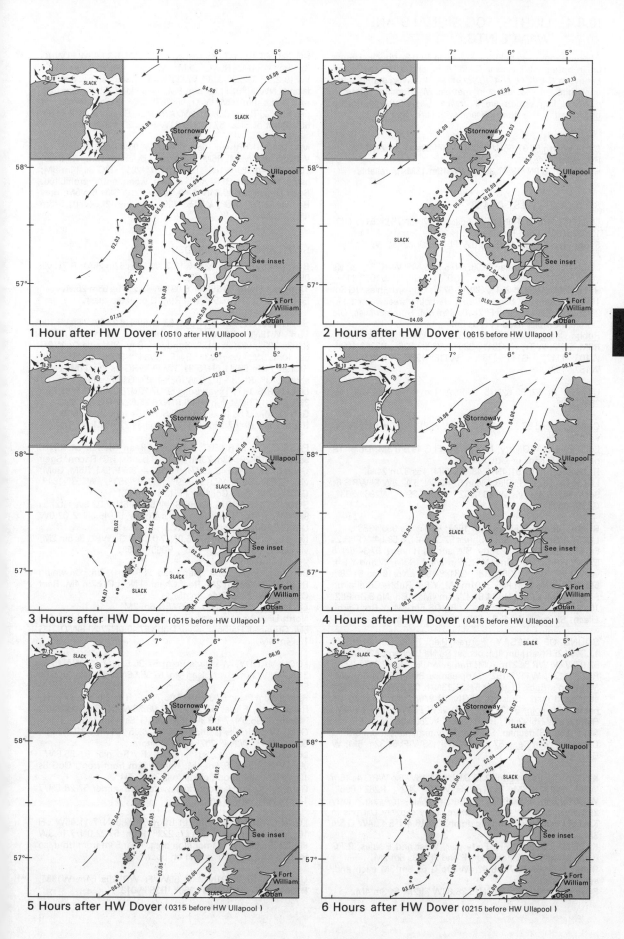

10.8.4 LIGHTS, FOG SIGNALS AND WAYPOINTS

Abbreviations used below are given in 10.0.3. Principal lights are in **bold** print, places in CAPITALS, and light-vessels and Lanbys in *CAPITAL ITALICS*. Unless otherwise stated lights are white. m—elevation in metres; M—nominal range in n. miles. Fog signals are in *italics*. Useful waypoints are underlined — use those on land with care. See 4.2.2.

Cape Wrath 58 37.55N/4 59.87 Fl (4) 30s 122m **24M**; W Tr; RC; *Horn(3) 45s*.
Rockall 57 35.8N/13 41.3W Fl 15s 19m 13M (unreliable).

SCOTLAND—WEST COAST

LOCH INCHARD, Rubha na Lecaig 58 27.43N/5 04.51W Fl(2) 10s 30m 8M. Loch Bervie Ldg Lts 327°. Both FG 8/14m.
Stoer Head 58 14.4N/5 24.0W Fl 15s 59m **24M**; W Tr.

LOCH INVER. Glas Leac 58 08.7N/5 16.3W Fl WRG 3s 7m; vis W071°-080°, R080°-090°, G090°-103°, W103°-111°, R111°-243°, W243°-251°, G251°-071°. Off Aird Ghlas QG 3m 1M; B column, W bands. Culag Harbour, breakwater head 2FG(vert). Soyea Island Fl(2) 10s 34m 6M. Summer Isles, Old Dornie, new pier head Fl G 3s 5m.
Rubha Cadail 57 55.53N/5 13.30W Fl WRG 6s 11m W9M, R6M, G6M; W Tr; vis G311°-320°, W320°-325°, R325°-103°, W103°-111°, G111°-118°, W118°-127°, R127°-157°, W157°-199°.

Ullapool Pt 57 53.62N/5 09.85W Iso R 4s 8m 6M; vis 258°-108°.

Cailleach Head 57 55.83N/5 24.15W Fl(2) 12s 60m 9M; W Tr; vis 015°-236°.
Loch Ewe, NATO jetty, NW corner Fl G 4s 5m 3M. Dolphins off N and S ends of jetty Fl G 4s.
Rubha Reidh 57 51.4N/5 48.6W Fl(4) 15s 37m **24M**.
LOCH GAIRLOCH, Glas Eilean 57 42.8N/5 42.3W Fl WRG 6s 9m W6M, R 4M; vis W080°-102°, R102°-296°, W296°-333°, G333°-080°.

RONA, **NE point** Fl 12s 69m **19M**; W Tr; vis 050°-358°.
LOCH A'BHRAIGE. Sgeir Shuas 57 35.04N/5 58.54W Fl R 2s 6m 3M; vis 070°-199°. Jetty, SW end, 2FR (vert). 57 34.6N/5 57.9W FlR 5s. No 9 Bn Q WRG 3m W4M, R3M; W and Y Bn; vis W135°-138°, R138°-318°, G318°-135°. No 10 Iso 6s 28m 5M; W Bn. No 1 Bn Fl G 3s 91m 3M, Y Bn. Rubha Chuiltairbh Fl 3s 6m 5M; W Bn. No 11 Bn QY 6m 4M; Y Bn. No 3 Bn Fl(2) 10s 9m 4M; W and Y Bn. No 12 Bn QR 5m 3M; Y Bn. Garbh Eilean, SE point, No 8 Bn Fl 3s 8m 5M; W Bn.

SOUND OF RAASAY. Portree Pier, head, 57 24.66N/6 11.34W 2FR(vert) 6m 4M; occas. Ru Na Lachan 57 29.04N/5 52.07W Oc WR 8s 21m 10M; framework Tr; vis W 337°-022°, R022°-117°, W117°-162°. Applecross Pier FlG 3s 3m 3M. McMillan's Rock Lt Buoy 57 21.13N/6 06.25W Fl(2)G 12s; stbd-hand mark. Suisnish 2FG(vert) 8/6m 2M. Raasay, SE point, Eyre Point 57 20.03N/6 01.22W Fl WR 3s 5m W9M, R6M; vis W215°-266°, R266°-288°, W288°-063°.
Skye, Loch Sligachan, Sconser ferry terminal QR 8m 3M. Crowlin, Eilean Beag 57 21.23N/5 51.33W Fl 6s 32m 6M; W Bn.

KYLE AKIN, Eilean Ban 57 16.68N/5 44.48W Iso WRG 4s 16m W9M, R6M, G6M; W Tr; vis W278°-282°, R282°-096°, W096°-132°, G132°-182°; Racon. Allt-an-Avaig, jetty 2FR(vert); vis 075°-270°. S shore, ferry slipway QR (vis in Kyle of Loch Alsh). Mooring dolphin 57 16.4N/5 43.4W Q 5m 3M.
KYLE OF LOCH ALSH. Ferry pier, West and E sides, 2 FG (vert) 6/5m 5M. Fishery Pier, E end Fl G 3s 6m 2M.
Butec Jetty 57 16.7N/5 42.4W Oc G 6s 5m 3M each end, synchronized.
Sgeir-na-Caillich 57 15.63N/5 38.83W Fl(2)R 6s 3m 4M.

SOUND OF SLEAT, Kyle Rhea 57 14.20N/5 39.90W Fl WRG 3s 7m W11M, R9M, G8M; W Bn; vis R shore-219°, W219-228°, G228°-338°, W338°-346°, R346°-shore. Sandaig Island, NW point Fl 6s 12m 8M; W 8-sided Tr. **Ornsay, SE end** Oc 8s 18m **15M**; W Tr; vis 157°-030°. Eilean Iarmain, off pier head 2FR (vert) 3/2m 2M. Armadale Bay, pier Oc R 6s 6m 6M. Point of Sleat 57 01N/6 01W Fl 3s 20m 9M; W Tr.

MALLAIG. Northern Pier, E end 57 00.48N/5 49.45W Iso WRG 4s 6m W9M, R6M, G6M; grey pylon on concrete base; vis G181°-185°, W185°-197°, R197°-201°. FIG 3s 14m 6M; same structure. 3FR(vert) shown when entry prohibited.
Sgeir Dhearg 57 00.64N/5 49.53W Fl(2)WG 8s 6m 5M; grey Bn; vis G190°-055°, W055°-190°. On reef 2FG(vert) 5/3m 4M; B pylon.

SCOTLAND—THE HEBRIDES

Butt of Lewis 58 30.93N/6 15.72W Fl 5s 52m **25M**; R Tr; vis 056°-320°; RC; *Horn(2) 30s*.
Tiumpan Head 58 15.6N/6 08.3W Fl(2) 15s 55m **25M**.
Eitshal 58 10.7N/6 35.0W 4FR(vert) on radio mast.

STORNOWAY, **Arnish Point** 58 11.50N/6 22.17W FlWR 10s 17m **W19M, R15M**; W round Tr; vis W088°-198°, R198°-302°, W302°-013°. Sandwick Bay, NW side Oc WRG 6s 10m 9M; vis G334°-341°, W341°-347°, R347°-354°. Stoney Field 58 11.6N/6 21.3W Fl WRG 3s 8m 11M; vis G shore-073°, R073°-102°, W102°-109°, G109°-shore.
LOCH ERISORT. Tavag Beag 58 07.22N/6 23.15W Fl 3s 13m 3M. Eilean Chalabrigh 58 06.81N/6 26.62W QG 5m 3M.
Gob na Milaid 58 01.0N/6 21.8W Fl 15s 14m 10M; W Tr.
Rubh' Uisenish 57 56.2N/6 28.2W Fl 3s 22m 9M; W Tr.

EAST LOCH TARBERT, Scalpay, **Eilean Glas** 57 51.43N/6 38.45W Fl(3) 20s 43m **23M**; W Tr, R bands; RC; Racon. Sgeir Ghlas 57 52.38N/6 45.18W Iso WRG 4s 9m W9M, R6M, G6M; vis G282°-319°, W319°-329°, R329°-153°, W153°-164°, G164°-171°. Dun Cor Mor Fl R 5s 10m 5M.
SOUND OF HARRIS. Dubh Sgeir 57 45.5N/7 02.6W Q(2) 5s 9m 6M; R Tr, B bands. Jane's Tower 57 45.8N/7 02.0W Q(2)G 5m 4M; obsc 273°-318°.
Leverburgh, pier head 57 46.0N/7 01.6W Oc WRG 8s 5m 2M; vis G305°-059°, W059°-066°, R066°-125°.

Berneray. Bays Loch, pier head 2FR(vert) 7m. Drowning Rock Q(2) G 8s 2m 2M. Reef Channel No 1 QG 2m 4M. Reef Channel No 2 Iso G 4s 2m 4M.
Eilean Fuam 57 41.9N/7 10.6W Q 6m 2M.
North Uist, Newton jetty 57 41.5N/7 11.5W 2 FG (vert) 9m 8M. Griminish Harbour, Sgeir Dubh 57 40.0N/7 26.8W Q(2) G 10s 4m 4M.

LOCH MADDY, Weaver's Point 57 36.61N/7 05.95W Fl 3s 21m 7M; W hut. Glas Eilean Mor 57 35.98N/7 06.64W Fl(2) 6s 8m.
Rudna Nam Pleac 57 35.8N/7 06.7W Fl R 4s 7m 5M; W post. Ruigh Liath, E islet 57 35.73N/7 08.36W QG 6m. Vallaquie Island 57 35.5N/7 09.3W Fl(3)WRG 8s 11m W7M, R5M, G5M; W pillar; vis G shore-205°, W205°-210°, R210°-240°, G240°-254°, W254°-257°, R257°-shore.
Lochmaddy Ldg Lts 301°. Front, Ro Ro pier 57 35.8N/7 09.3W 2FG(vert) 8/6m 4M. Rear, 112m from front, OcG 8s 10m 4M; column on dolphin; vis 284°-304°.
Grimsay. Kellin Harbour breakwater, NE corner 57 28.9N/7 12.3W 2FR(vert) 6/5m.

LOCH CARNAN. Landfall Lt Buoy 57 22.30N/7 11.45W LFl 10s; safe water mark. Ldg Lts 222°. Front 57 22.0N/7 16.3W Fl R 2s 7m 5M; W diamond on post. Rear 58m from front Iso R 10s 11m 5M; W diamond on post.

Ushenish 57 17.91N/7 11.50W Fl WR 20s 54m **W19M, R15M**; W Tr; vis W193°-356°, R356°-013°.

LOCH BOISDALE, Calvay, E End 57 08.5N/7 15.3W Fl(2) WRG 10s 16m W7M, R4M, G4M; W framework Tr; vis G190°-202°, W202°-286°, R286°-111°, W111°-190°. N side QG 3m 3M. Gasay Island FlWR 5s 10m W7M, R4M; W framework Tr; vis R284°-120°, W120°-284°. Jetty, head 57 09.2N/7 18.2W Iso RG 4s 12m 2M; framework Tr; vis G shore-283°, R283°-shore; 2FG(vert) 8m 3M on dolphin 84m W. Ludaig Dir Lt 297° 57 06.2N/7 19.7W Dir Oc WRG 6s 8m W7M, R4M, G4M; vis G287°-296°, W296°-298°, R298°-307°. Ludaig Pier 2FG(vert) 5/3m 5M. Stag Rock 57 05.9N/7 18.3W Fl(2) 8s 7m 4M. Bank Rock 57 05.6N/7 17.5W Q(2) 4s 5m 4M. Haun Dir Lt 236° Dir Oc WRG 3s 9m W7M, R4M, G4M; vis G226°-234°, W234°-237°, R237°-246°.

ERISKAY. Pier 2FG(vert) 5m. Acairseid Mhor Ldg Lts 285°. Front 57 03.9N/7 17.2W Oc R 6s 9m 4M. Rear 24m from front Oc R 6s 10m 4M. Both grey columns.

VATERSAY SOUND, Dubh Sgeir 56 56.4N/7 28.9W Fl(2)WG 6s 6m W7M, G5M; vis W280°180°, G180°-280°. Sgeir Leadh 56 56.7N/7 30.7W Fl 3s 7m 8M; W building. Rudha Glas. Ldg Lts 295°. Front 56 56.8N/7 30.6W FR 9m 4M; R triangle on W Tr. Rear 550m from front FR 22m 4M; R triangle on W Tr.

South Uist. Falconet Tower 57 21.5N/7 23.6W FR 25m 8M (3M by day); shown 1h before firing; changes to Iso R 2s 15min before firing until completion; similar Lts 1.2M NNW and 7.5M SSW.

BERNERAY, West side, **Barra Head** 56 47.13N/7 39.18W Fl 15s 208m **21M**; W Tr; obsc by islands to NE; RC.
Flannan Islands, Eilean Mor 58 17.32N/7 35.23W Fl(2) 30s 101m **20M**; W Tr; obsc in places by islands to west.

EAST LOCH ROAG, Aird Laimishader, Carloway 58 17.1N/6 49.5W L Fl 12s 61m 8M; W hut; obsc on some brgs. Ardvanich Point 58 13.5N/6 47.7W Fl G 3s 2m 2M. Tidal Rock 58 13.5N/6 47.6W Fl R 3s 2m 2M (synchronised with previous Lt). Great Bernera, Kirkibost jetty 2FG(vert) 6/5m 2M. Greinam 58 13.3N/6 46.2W Fl WR 6s 8m W8M, R7M; W Bn; vis R143°-169°, W169°-143°. Rudha Arspaig, jetty 2FR(vert) 10/7m 4M.

ST KILDA Ldg Lts 270°. Front 57 48.36N/8 34.27W Oc 5s 26m 3M. Rear, 100m from front, Oc 5s 38m 3M; synchronised.

SCOTLAND—WEST COAST

Eilean Trodday 57 43.6N/6 17.8W Fl(2) WRG 10s 49m W12M, R9M, G9M; W Bn; vis W062°-088°, R088°-130°, W130°-322°, G322°-062°.

SKYE, Uig. Pier head Iso WRG 4s 9m W7M, R4M, G4M; vis W180°-008°, G008°-052°, W052°-075°, R075°-180°. Waternish Point 57 36.5N/6 38.0W Fl 20s 21m 8M; W Tr. Loch Dunvegan, Uiginish point 57 26.8N/6 36.5W Fl WG 3s 14m W7M, G5M; W hut; vis G040°-128°, W128°-306°, obsc by Fiadhairt point when brg more than 148°. **Neist Point** 57 25.4N/6 47.2W Fl(2) 30s 43m **24M**; W Tr. Loch Harport, Ardtreck Point 57 20.4N/6 25.8W Iso 4s 17m 9M; small W Tr.

CANNA, E end, Sanday Island 57 02.8N/6 27.9W Fl 6s 32m 9M; W Tr; vis 152°-061°.

Oigh Sgeir, near S end, **Hyskeir** 56 58.13N/6 40.80W Fl(3) 30s 41m **24M**; W Tr; RC. N end Horn 30s.

EIGG, SE point of Eilean Chathastail 56 52.25N/6 07.20W Fl 6s 24m 8M; W Tr; vis 181°-shore.

Bo Faskadale Lt Buoy 56 48.18N/6 06.35W Fl(3)G 18s; stbd-hand mark.
Ardnamurchan 56 43.64N/6 13.46W Fl(2) 20s 55m **16M** (T); grey Tr; vis 002°-217°; Siren 60s.
Cairns of Coll, Suil Ghorm 56 42.27N/6 26.70W Fl 12s 23m 11M; W Tr.

COLL, Loch Eatharna, Bogha Mor Lt Buoy 56 36.67N/6 30.90W FlG 6s; stbd-hand mark. Arinagour pier 2FR(vert) 10m.

TIREE, **Scarinish**, S side of entrance 56 30.02N/6 48.20W Fl 3s 11m **16M**; W square Tr; vis 210°-030°. Ldg Lts 286°30'. Front 56 30.6N/6 47.9W FR (when vessel expected). Rear 30m from front FR.

Skerryvore 56 19.40N/7 06.75W Fl 10s 46m **26M**; grey Tr; Racon; Horn 60s.
Dubh Artach 56 08.0N/6 37.9W Fl(2) 30s 44m **20M**; grey Tr, R band; Horn 45s.
Eileanan na Liathanaich, SE end, Bunessan 56 20.5N/6 16.2W Fl WR 6s 12m W8M, R6M; W Bn; vis R088°-108°, W108°-088°.

SOUND OF MULL. Ardmore Point 56 39.4N/6 07.6W Fl(2) 10s 17m 8M; W column. **Rubha nan Gall** 56 38.33N/6 03.91W Fl 3s 17m **15M**, W Tr. Eileanan Glasa, Green Island 56.32.3N/5 54.7W Fl 6s 7m 8M; W Tr. Ardtornish Point 56 31.1N/5 45.1W Fl(2) WRG 10s 7m W8M, R5M, G5M; W Tr; vis G shore-302°, W302°-310°, R310°-342°, W342°-057°, R057°-095°, W095°-108°, G108°-shore. Glas Eileanan, Grey Rocks 56 29.8N/5 42.7W Fl 3s 10m 6M; W round Tr on W base.

LOCH LINNHE. Entrance, west side, Corran Point 56 43.27N/5 14.47W Iso WRG 4s 12m W10M, R7M; W Tr; vis R shore-195°, W195°-215°, G215°-305°, W305°-030°, R030°-shore. Corran Narrows, NE 56 43.62N/5 13.83W Fl 5s 4m 4M; W framework Tr; vis S shore-214°. Jetty 56 43.42N/5 14.56W FlR 5s 7m 3M.

FORT WILLIAM, pier head Fl G 2s 6m 4M. Corpach, Caledonian Canal lock entrance Iso WRG 4s 6m 5M; W Tr; vis G287°-310°, W310°-335°, R335°-030°.

Sgeir Bhuidhe, Appin 56 33.6N/5 24.6W Fl(2) WR 7s 7m 9M; W Bn; vis R184°-220°, W220°-184°.

LOCH CRERAN, off Airds Point Fl WRG 2s 2m W3M, R1M, G1M; vis R196°-246°, W246°-258°, G258°-041°, W041°-058°, R058°-093°. Eriska, NE point, QG 2m 2M; vis 128°-329°.
Lismore, SW end 56 27.4N/5 36.4W Fl 10s 31m **19M**; W Tr; vis 237°-208°. Lady's Rock 56 27.0N/5 37.0W Fl 6s 12m 5M; R structure on W Bn.
Duart Point 56 26.9N/5 38.7W Fl(3) WR 18s 14m W5M, R3M; grey building; vis W162°-261°, R261°-275°, W275°-353°, R353°-shore.

DUNSTAFFNAGE BAY. Pier head, NE end 2FG(vert) 4m 2M.

OBAN, North spit of Kerrera 56 25.50N/5 29.50W FlR 3s 9m 5M; W column. R bands. Dunollie Fl(2) WRG 6s 7m W5M, G4M, R4M; vis G351°-009°, W009°-047°, R047°-120°, W120°-138°, G138°-143°.

Kerrera Sound, Dubh Sgeir 56 22.82N/5 32.20W Fl(2) 12s 7m 5M; W Tr.
Port Lathaich 56 22.8N/5 31.3W OcG 6s; vis 037°-072°.

Fladda 56 14.9N/5 40.8W Fl(3) WR 18s 13m W14M, R11M; W Tr; vis R169°-186°, W186°-001°.
Dubh Sgeir 56 14.8N/5 40.1W Fl 6s 7m 6M; W tank.
The Garvellachs, Eileach an Naoimh, SW end 56 13.1N/5 48.9W Fl 6s 21m 9M; W Bn; vis 240°-215°.

COLONSAY, Scalasaig, Rubha Dubh 56 04.02N/6 10.83W Fl(2) WR 10s 6m W8M, R5M; W building; vis R shore-230°, W230°-337°, R337°-354°. Pier Ldg Lts 262°, FR (occas).

Loch Melfort, Fearnach Bay pier 56 16.2N/5 30.1W 2FR(vert) 6/5m 3M; shown 1/4 to 31/10.

For Sound of Jura and Sound of Islay see 10.9.4.

10.8.5 PASSAGE INFORMATION

SCOTLAND – WEST COAST

This provides splendid, if sometimes boisterous, sailing and unmatched scenery. In summer the long hours of daylight and warmth of Gulf Stream compensate for lower air temp and higher wind speeds experienced when depressions run typically N of Scotland. Inshore the wind is often unpredictable, due to geographical effects of lochs, mountains and islands offshore: calms and squalls can occur in rapid succession. Magnetic anomalies occur.

A yacht must rely on good anchors. Particularly in N of area, facilities are very dispersed. It is essential to carry good large scale charts, and pilotage information as in *West Coast of Scotland Pilot* or as provided by the Clyde Cruising Club. Also useful is *Scottish West Coast Pilot*, by Mark Brackenbury.

It is helpful to know at least some of the more common Gaelic terms, as follows. *Acairseid*: anchorage. *Ailean*: meadow. *Aird, ard*: promontory. *Aisir, aisridh*: passage between rocks. *Beag*: litttle. *Beinn*: mountain. *Bo, boghar, bodha*: rock. *Cala*: harbour. *Camas*: channel, bay. *Caol*: strait. *Cladach*: shore, beach. *Creag*: cliff. *Cumhamn*: narrows. *Dubh, dhubh*: black. *Dun*: castle. *Eilean, eileanan*: island. *Garbh*: rough. *Geal, gheal*: white. *Glas, ghlas*: grey, green. *Inis*: island. *Kyle*: narrow strait. *Linn, Linne*: pool. *Mor, mhor*: large. *Mull*: promontory. *Rinn, roinn*: point. *Ruadh*: red, brown. *Rubha, rhu*: cape. *Sgeir*: rock. *Sruth*: current. *Strath*: river valley. *Tarbert*: isthmus. *Traigh*: beach. *Uig*: bay.

CAPE WRATH TO SKYE (charts 1785, 1794)

C Wrath (Lt, fog sig, RC) is a steep headland (110m). To N of it the E-going stream begins at HW Ullapool – 0350, and W-going at HW Ullapool + 0235, sp rates 3 kn. Eddies close inshore cause almost continuous W-going stream E of cape, and N-going stream SW of it. Where they meet is turbulence, with dangerous sea in bad weather. Duslic Rk, 7 ca NE of Lt Ho, dries. 6M SW of C Wrath is islet of Am Balg (45m), foul for 2 ca offshore.

There are anchs in Loch Inchard (chart 2503), the most sheltered being in Loch Bervie on N shore. Good anchs among Is along S shore of Loch Laxford, entered between Ardmore Pt and Rubha Ruadh (see 10.8.19). Handa I to WSW is bird sanctuary. Handa Sound is navigable with care, but beware Bogha Morair in mid-chan and associated overfalls. Tide turns 2h earlier in the sound than offshore.

With strong wind against tide there is bad sea off Pt of Stoer. S lies Enard B. The best shelter is Loch Inver (10.8.19 and chart 2504), with good anch off hotel near head of loch.

Summer Isles (chart 2501) lie NW of Ullapool (10.8.8) and provide several good anchs, but approaches are often difficult. The best include the b on E side of Tanera Mor; off NE of Tanera Beg (W of Eilean Fada Mor); and in Caolas Eilean Ristol, between the island and mainland.

Gruinard Island, SW of entrance to Little Loch Broom, is dangerously contaminated and landing is prohibited.

Loch Ewe (chart 3146) provides good shelter and is easy of access. Best anchs are in Poolewe B (beware Boor Rks off W shore) and in SW corner of Loch Thuirnaig (entering, keep close to S shore to avoid rks extending from N side). See 10.8.19.

Off Rubha Reidh (Lt) the NE-going stream begins at HW Ullapool – 0335, and the SW-going at HW Ullapool + 0305. Sp rates 3 kn, but streams lose strength SW of pt.

Longa Island lies N of entrance to Loch Gairloch (10.8.19 and chart 2528). The chan N of it is navigable but narrow at E end. Outer loch is free of dangers, but exposed to swell. Best anch is Caolas Bad a Chrotha on W side of Eilean Horrisdale, on S side of loch.

Loch Torridon (chart 2210) has few dangers, and streams are weak except in narrows between Loch Shieldaig and Upper Loch Torridon, where they run 2-3 kn. Entering Loch Torridon from S or W beware Murchadh Breac (dries) 3 ca NNW of Rubha na Fearna. Best anchs are SW of Eilean Mor (to W of Ardheslaig); in Loch a 'Chracaich, on S shore 3M from entrance; E of Shieldaig Island in Loch Shieldaig; and near head of Upper Loch Torridon.

OUTER HEBRIDES (charts 1785, 1794, 1795)

The E sides of these islands have many good, sheltered anchs, but W coasts give little shelter. The Outer Hebrides sailing directions of Clyde Cruising Club are recommended. The Sea of the Hebrides and the Little Minch can be very rough, particularly between Skye and Harris, and around Shiant Islands where tide runs locally 4 kn at sp, and heavy overfalls can be met. Beware fish farms in many inlets.

Between Skerryvore and Neist Pt the stream runs generally N and S, starting N-going at HW Ullapool + 0550, and S-going at HW Ullapool – 0010. Mostly it does not exceed 1 kn; but it is stronger near Skerryvore, around headlands of The Small Isles, and over rks and shoals. Between N point of Skye and S Harris the NE-going stream begins at HW Ullapool – 0335, and the SW-going stream at HW Ullapool + 0250, sp rates 2½ kn.

From N to S, the better harbs in Outer Hebrides include: *Lewis*. Stornoway (see 10.8.7); Loch Grimshader (beware Sgeir a'Chaolais, dries in entrance); Loch Erisort; Loch Odhairn; Loch Shell (10.8.19). Proceeding S from here, or to E Loch Tarbert beware Sgeir Inoe (dries) 3M ESE of Eilean Glas Lt Ho at SE end of Scalpay.

Harris. E Loch Tarbert (10.8.19); Loch Scadaby; Loch Stockinish; Loch Finsby. W Loch Tarbert (10.8.19).

N Uist. Loch Maddy (10.8.19); Loch Eport.

S Uist. Loch Skiport; Loch Eynort; Loch Boisdale (10.8.19).

Barra. Castlebay (10.8.19).

Berneray. On N side, E of Shelter Rk.

SKYE (charts 1795, 2210, 2209, 2208)

Skye and the islands around it provide many good and attractive anchs, of which the most secure are: Acairseid Mhor on the W side of Rona; Portree (see 10.8.9); Isleornsay; Portnalong, near the entrance to Loch Harport, and Carbost at the head: Dunvegan; and Uig B in Loch Snizort.

Anch behind Fladday Island near the N end of Raasay can be squally and uncomfortable except in settled weather, and Loch Scavaig more so, though the latter is so spectacular as to warrant a visit in fair weather. Soay has a small, safe harb on its N side, but the bar at entrance almost dries at LW sp.

Tides are strong off Rubha Hunish at N end of Skye, and heavy overfalls occur with weather-going tide against fresh or strong winds.

Between Skye and the coast there is the choice of Sound of Raasay and Inner Sound. Using the former, coming S from Portree, beware Sgeir Chnapach (3m) and Ebbing Rk (dries 2.4m) NNW of Oskaig Pt. At the Narrows (chart 2534) the S-going stream begins at HW Ullapool – 0605, and the N-going at HW Ullapool – 0040. Sp rate 1½ kn in mid-chan, but more near shoals each side. Beware McMillan's Rk in mid-chan, marked by buoy. The direction of buoyage here is now N-wards.

The chan between Scalpay and Skye narrows to 2½ ca 2M W of Guillamon Island, where a reef has least depth of 0.3m.

There is a Lt Bn on N side of narrows, where N-going stream begins at HW Ullapool +0550, and S-going at HW Ullapool −0010, sp rates about 1 kn.

Inner Sound is a wider, easier chan than Sound of Raasay: the two are connected by Caol Rona and Caol Mor, respectively N and S of Raasay. Dangers extend about 1M N of Rona, and Cow Island lies off the mainland 8M to S; otherwise approach from N is clear to Crowlin Islands, which should be passed to W. There is a platform construction site with prohibited area S of Crowlin Island. There is a good anch between the two S islands, More and Meadhonach. See 10.8.19.

The British Underwater Test and Evaluation Centre (BUTEC) torpedo range exists in the Inner Sound. While passage through the Sound is normally unrestricted, vessels passing through the area may be requested to keep to the eastern side of the Sound while the range is active. This is indicated by RED FLAGS and International Code NE4 flown at the range head building at Applecross, by all range vessels and at the naval pier at Kyle of Lochalsh.

Approaching Kyle Akin from W, beware dangerous rks to N, off Bleat Island (at S side of entrance to Loch Carron); Bogha Beag (dries 0.6m) on S side of chan, 6½ ca W of Kyle Akin Lt Ho; and Black Eye Rk 4 ca W of Lt Ho with least depth 3.8m. For Plockton (Loch Carron) see 10.8.19.

Pass at least ½ ca distant either N or S of Eileanan Dubha off Kyle of Lochalsh. On S side of main chan String Rk (dries) is marked by buoy. For Loch Alsh see 10.8.19.

Kyle Rhea connects Loch Alsh with NE end of Sound of Sleat. The tidal streams are very strong: N-going stream begins HW Ullapool +0600, sp rate 6-7 kn; S-going stream begins at HW Ullapool, sp rate 8 kn. Eddies form both sides of the Kyle and there are heavy overfalls off S end in strong S winds on S-going stream. Temp anch in Sandaig Bay.

THE SMALL ISLES (charts 2207, 2208)

These consist of Rhum, Eigg, Muck and Canna. Dangers extend SSW from Canna: at 1M Jemina Rk (depth 1.5m) and Belle Rk (depth 3.6m); at 2M Humla Rk (5m high), marked by Lt buoy and with offlying shoals close W and ½ ca N of it; at 5M Oigh Sgeir (Lt, fog sig, RC), the largest of a group of small islets; and at 7M Mill Rks (with depths of 1.8m). The tide runs hard in these areas, and in bad weather the sea breaks heavily up to 15M SW of Canna.

On the N side of Muck, 1M offshore, are Godag Rks, some above water but with submerged dangers extending 2 ca further to N. Most other dangers around The Small Isles are closer inshore, but there are banks on which the sea breaks heavily in bad weather. Local magnetic anomalies exist around the islands.

The best harb is Eigg (10.8.19). Others are Loch Scresort (Rhum) and Canna Harbour (between Canna and Sanday), but both are exposed to E. For Canna see 10.8.19.

SOUTH OF ARDNAMURCHAN POINT (chart 2171)

6M NE of Ardnamurchan Pt (Lt, fog sig) is Bo Faskadale, with two heads, the N one awash and the S with depth of 1.2m, marked by buoy. Ardnamurchan Pt has no offlying dangers, but it is exposed, and a heavy sea can build up even in moderate winds. Here the N-going stream begins at HW Oban −0525, and the S-going at HW Oban +0100, sp rates 1½ kn.

Proceeding S (chart 2171) there is the choice of Passage of Tiree (exposed to SW, and with heavy overfalls in bad weather) or the sheltered route via Sound of Mull, Firth of Lorne and Sound of Jura. The latter denies a visit to Coll or Tiree, where best anchs are at Arinagour and Gott B respectively. Beware Cairns of Coll, off the N point.

The Sound of Mull gives access to Tobermory (10.8.12), Dunstaffnage (10.8.19), Oban (10.8.16), and up Loch Linnhe through Corran Narrows (where tide runs strongly) to Fort William (10.8.14) and to Corpach for the Caledonian Canal (10.8.15). But, apart from these places, there are dozens of lovely anchs in the sheltered lochs inside Mull, as for example in Loch Sunart (10.8.11). For Loch Aline see 10.8.13.

The W coast of Mull is rewarding in really calm and settled weather, but careful pilotage is needed. Beware tide rip off Caliach Pt (NW corner) and Torran Rks off SW end of Mull. Buressan, N side of Ross of Mull, is an anch with boatyard etc. (see 10.8.19). Apart from the attractions of Iona and of Staffa (Fingal's Cave), the remote Treshnish Islands are worth visiting. The best anchs in this area are at Ulva, Gometra, and Bull Hole and Tinker's Hole in Iona Sound. Passage through Iona Sound avoids overfalls W of Iona, but care is needed.

On the mainland shore Eilean nam Beathach is a sheltered anch. For passage S through Sound of Jura, and Gulf of Corryvreckan, see 10.9.5. For Loch Melfort see 10.8.17.

10.8.6 DISTANCE TABLE

Approximate distances in nautical miles are by the most direct route while avoiding dangers and allowing for traffic separation schemes etc. Places in *italics* are in adjoining areas.

	1	2	3	4	5	6	7	8	9	10	11	12	13	14	15	16	17	18	19	20
1 *Lerwick*	**1**																			
2 *Kirkwall*	95	**2**																		
3 *Duncansby Head*	109	34	**3**																	
4 *Cape Wrath*	155	79	63	**4**																
5 Butt of Lewis	183	118	103	40	**5**															
6 Stornoway	208	132	116	53	30	**6**														
7 Ullapool	208	132	116	53	52	43	**7**													
8 Portree	238	162	146	83	69	50	53	**8**												
9 Neist Point	247	171	155	92	72	49	65	48	**9**											
10 Mallaig	265	189	173	110	100	83	82	38	41	**10**										
11 Barra Head	295	219	203	140	120	97	113	96	48	63	**11**									
12 Oigh Sgeir	274	198	182	119	99	76	92	67	27	29	33	**12**								
13 Ardnamurchan Pt	284	218	202	139	119	96	104	60	47	22	47	21	**13**							
14 Tobermory	304	228	212	149	129	106	114	70	57	32	56	31	10	**14**						
15 Fort William	347	271	255	192	171	149	157	113	100	75	99	74	53	43	**15**					
16 Oban	328	252	236	173	153	130	138	94	81	56	80	55	34	24	29	**16**				
17 Crinan	344	268	252	189	169	146	154	110	97	72	92	71	50	42	50	24	**17**			
18 *Mull of Kintyre*	387	311	295	232	212	189	203	159	140	111	112	103	99	89	98	70	51	**18**		
19 *Altacarry Head*	379	303	287	224	204	181	197	151	132	103	104	95	91	92	101	73	54	12	**19**	
20 *Tory Island*	385	309	293	230	210	187	203	170	138	132	95	113	110	115	143	114	110	86	74	**20**

STORNOWAY 10-8-7
Lewis (Outer Hebrides)

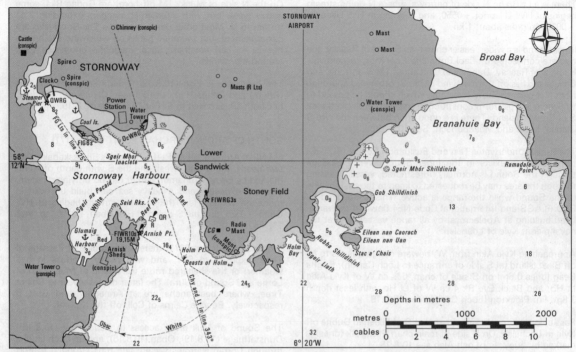

CHARTS
Admiralty 2529, 1794; OS 8

TIDES
−0428 Dover; ML 2.9; Duration 0610; Zone 0 (GMT).
Standard Port ULLAPOOL (→)

Times				Height (metres)			
HW		LW		MHWS	MHWN	MLWN	MLWS
0100	0700	0300	0900	5.2	3.9	2.1	0.7
1300	1900	1500	2100				

Differences STORNOWAY
−0010 −0010 −0010 −0010 −0.4 −0.2 −0.1 0.0
LOCH ERIBOLL (Portnancon)
+0055 +0105 +0055 +0100 0.0 +0.1 +0.1 +0.2
KYLE OF DURNESS
+0030 +0030 +0050 +0050 −0.6 −0.4 −0.3 −0.1
CAPE WRATH
+0025 +0025 +0025 +0025 −0.5 −0.3 No data
LOCH SHELL (Hebrides)
−0023 −0010 −0010 −0027 −0.4 −0.3 −0.2 0.0
EAST LOCH TARBERT (Hebrides)
−0035 −0020 −0020 −0030 −0.2 −0.2 0.0 +0.1
LOCH MADDY (Hebrides)
−0054 −0024 −0026 −0040 −0.4 −0.3 −0.2 0.0
LOCH SKIPORT (Hebrides)
−0110 −0035 −0034 −0034 −0.6 −0.6 −0.4 −0.2
LOCH BOISDALE (Hebrides)
−0105 −0040 −0030 −0050 −1.1 −0.9 −0.4 −0.2
CASTLEBAY
−0125 −0050 −0055 −0110 −0.9 −0.8 −0.4 −0.1
BARRA HEAD (Hebrides)
−0125 −0050 −0055 −0105 −1.2 −0.9 −0.4 +0.1
WEST LOCH TARBERT (Hebrides)
−0103 −0043 −0024 −0044 −1.0 −0.7 −0.8 −0.3
LITTLE BERNERA
−0031 −0021 −0027 −0037 −0.9 −0.8 No data −0.2
CARLOWAY
−0050 +0010 −0045 −0025 −1.0 −0.7 −0.5 −0.1

Standard Port ULLAPOOL (→)

Times				Height (metres)			
HW		LW		MHWS	MHWN	MLWN	MLWS
0000	0600	0300	0900	5.2	3.9	2.1	0.7
1200	1800	1500	2100				

VILLAGE BAY (St Kilda)
−0110 −0040 −0100 −0100 −1.9 −1.4 −0.9 −0.3
FLANNAN ISLES
−0036 −0026 −0026 −0036 −1.3 −0.9 −0.7 −0.2
ROCKALL
−0241 −0231 −0231 −0241 −3.2 −2.4 No data

SHELTER
Good except S swells make the anchorage uncomfortable. The best anchorage is in Glumaig Harbour within W sector. It is possible to berth alongside fishing boats in the inner harbour. Visitors land at steps N of FG Ldg Lts and report to Hr Mr or enter Inner Harbour and berth at Cromwell St quay.

NAVIGATION
Waypoint 58°10'.00N 06°20'.80W, 163°/343° from/to Oc WRG Lt, 2.3M. Reef rocks, W side of entrance marked by lighted R can buoy. The rocky patch off Holm Point is marked by an unlit R beacon. Ferries to Ullapool use the harbour. A local magnetic anomaly exists over a small area 1.75 ca N of Seid Rks.

LIGHTS AND MARKS
Arnish Point Fl WR 10s 17m 19/15M; W Tr; vis W088°−198°, R198°−302°, W302°−013°. Stoney Field Fl WRG 3s 8m 11M; vis G shore−073°, R073°−102°, W102°−109°, G109°−shore. Sandwick Bay, NW side (close E of power station and water Tr) Oc WRG 6s 10m 9M; vis G334°−341°, W341°−347°, R347°−354°; in line with conspic chy bears 343°. Ldg Lts 325° to Ro Ro jetty, both FG.

RADIO TELEPHONE
VHF Ch 16; 12 (H24).

TELEPHONE (0851)
Hr Mr 2688; CG 2013; MRSC 2013; HMC 3626; Marinecall 0898 500 464; Dr 3145

FACILITIES
EC Wednesday; **Pier N end of Bay** FW, C(10 ton), CH, AB; **W Pier** Slip, P, FW; **Steamer Pier** FW; **Duncan MacIver** Tel. 2010 CH, ACA; **Electronic Services** Tel. 2909 El, Electronics; **Stornoway Boatbuilding Co.** Tel. 3488, ME, El, Sh; **Town** P, D, El, V, R, Bar, Gas, PO; Bank; Rly (ferry to Ullapool, bus to Garve); Air.

AREA 8—NW Scotland 427

ULLAPOOL 10-8-8
Ross and Cromarty

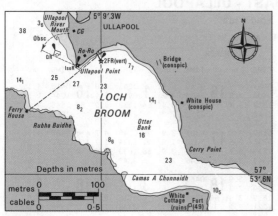

CHARTS
Admiralty 2500; OS 19
TIDES
−0415 Dover; ML 3.0; Duration 0610; Zone 0 (GMT).
Standard Port ULLAPOOL (—→)

Times				Height (metres)			
HW		LW		MHWS	MHWN	MLWN	MLWS
0000	0600	0300	0900	5.2	3.9	2.1	0.7
1200	1800	1500	2100				

Differences LOCH LAXFORD
+0015 +0015 +0005 +0005 −0.3 −0.4 −0.2 0.0
LOCH BERVIE
+0030 +0010 +0010 +0020 −0.3 −0.3 −0.2 0.0
LOCH INVER
−0005 −0005 −0005 −0005 −0.2 0.0 0.0 +0.1
BADCALL BAY
+0005 +0005 +0005 +0005 −0.7 −0.5 −0.5 +0.2
LOCH EWE (Mellon Charles)
−0010 −0010 −0010 −0010 −0.1 −0.1 −0.1 0.0
GAIRLOCH
−0020 −0020 −0010 −0010 0.0 +0.1 −0.3 −0.1
SUMMER ISLES (Tanera Mor)
−0005 −0005 −0010 −0010 −0.1 +0.1 0.0 +0.1

Ullapool is a Standard Port and detailed tidal predictions for each day of the year are given below.

SHELTER
Good shelter in anchorage behind Ullapool Point. Note that Ullapool is primarily a fishing port from Aug to Mar and the pier is extremely congested. Moorings sometimes available (see Hr Mr).
NAVIGATION
Waypoint Loch Broom entrance 57°55'.80N 05°15'.00W, 309°/129° from/to Ullapool Pt Lt, 3.5M. North of Ullapool point is an extensive flat at the mouth of Ullapool River which dries out: deep water marked by QR buoy. Beware fish pens and unlit buoys S of narrows off W shore.
LIGHTS AND MARKS
Ullapool Point Iso R 4s 8m 6M; grey mast; vis 258°–108°. Pier extension 2FR(vert) 6/5m; column on dolphin.
RADIO TELEPHONE
VHF Ch 16; 12 (July-Nov: H24. Dec-June: office hours).
TELEPHONE (0854)
Hr Mr 2091 & 2165; Port Officer 2724; CG 2014; MRSC Stornoway 2013; HMC (Aug–Mar) 2253; Marinecall 0898 500 464; Dr 2015.
FACILITIES
EC Tuesday (but open all day in summer); **Pier** Tel. 2091 & 2165, D, L, FW, CH, AB; **Ardmair Boat Centre** Tel. 2054, Sh, CH; **Ullasport** Tel. 2621, CH, ACA; **Summer Isles Charters** Tel. 2706, ME, El, Sh; **Ullapool YC** Gas; **Highland Coastal Trading Co** Tel. 2488 CH; **Village** P, ME, El, V, R, Bar, PO; Bank; Rly (bus to Garve); Air (Inverness).
Note: daily buses to Inverness and ferries to Stornoway.

AGENTS NEEDED
There are a number of ports where we need agents, particularly in France.
ENGLAND Swale, Havengore, Berwick.
SCOTLAND Firth of Forth, Scrabster, Mallaig, Loch Sunart, Loch Aline.
IRELAND Kilrush, Wicklow, Westport/Clew Bay.
FRANCE Arcachon, Seudre R, Ile d'Oleron, Rochfort, Ile de Re, St. Giles-Croix-de-Vie, Ile d'Yeu, Pouliguen, Le Croisic, La Forêt, Ile de Bréhat.
GERMANY Norderney, Dornumer-Accumersiel.
If you are interested in becoming our agent for any of the above, please write to the editors and get your free copy of the Almanac every year. You do not have to be resident in a port to be the agent, but at least a fairly regular visitor.

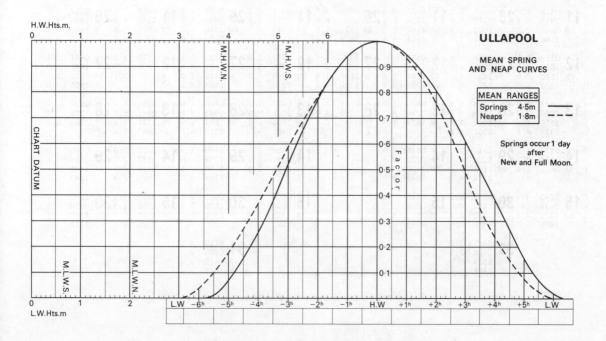

SCOTLAND, WEST COAST - ULLAPOOL
LAT 57°54'N   LONG 5°10'W

TIME ZONE UT(GMT)
For Summer Time add ONE hour in non-shaded areas

TIMES AND HEIGHTS OF HIGH AND LOW WATERS

YEAR **1989**

	JANUARY				FEBRUARY				MARCH				APRIL		
	TIME m		TIME m		TIME m		TIME m		TIME m		TIME m		TIME m		TIME m

1 SU 0120 3.9 / 0704 2.4 / 1317 4.0 / 2004 2.2   **16** M 0159 4.1 / 0749 2.0 / 1417 4.2 / 2028 1.9
1 W 0240 3.9 / 0828 2.5 / 1503 3.8 / 2128 2.3   **16** TH 0415 4.0 / 1037 2.0 / 1642 4.0 / 2258 2.0
1 W 0001 3.9 / 0609 2.3 / 1219 3.7 / 1846 2.4   **16** TH 0240 3.8 / 0900 2.2 / 1522 3.7 / 2138 2.3
1 SA 0241 3.9 / 0912 2.1 / 1533 3.9 / 2153 2.1   **16** SU 0416 4.1 / 1044 1.5 / 1645 4.1 / 2251 1.8

2 M 0229 3.9 / 0816 2.5 / 1432 3.9 / 2111 2.2   **17** TU 0315 4.1 / 0911 2.0 / 1534 4.2 / 2147 1.9
2 TH 0352 4.0 / 1001 2.4 / 1620 4.0 / 2240 2.1   **17** F 0512 4.3 / 1137 1.7 / 1734 4.3 / 2346 1.7
2 TH 0150 3.8 / 0736 2.5 / 1434 3.7 / 2049 2.5   **17** F 0357 3.9 / 1030 1.9 / 1630 3.9 / 2246 2.1
2 SU 0346 4.2 / 1024 1.7 / 1630 4.2 / 2248 1.7   **17** M 0458 4.3 / 1122 1.3 / 1721 4.3 / 2326 1.6

3 TU 0332 4.0 / 0927 2.4 / 1543 4.0 / 2211 2.1   **18** W 0422 4.2 / 1030 1.9 / 1642 4.2 / 2255 1.8
3 F 0448 4.3 / 1110 2.0 / 1715 4.3 / 2334 1.8   **18** SA 0556 4.6 / 1220 1.3 / 1814 4.6
3 F 0318 3.9 / 0939 2.3 / 1559 3.9 / 2221 2.2   **18** SA 0453 4.2 / 1120 1.6 / 1718 4.2 / 2327 1.7
3 M 0438 4.6 / 1113 1.2 / 1715 4.7 / 2332 1.2   **18** TU 0532 4.5 / 1155 1.2 / 1751 4.5 / 2359 1.4

4 W 0425 4.2 / 1030 2.2 / 1641 4.2 / 2302 1.9   **19** TH 0518 4.4 / 1135 1.7 / 1738 4.4 / 2349 1.6
4 SA 0533 4.6 / 1201 1.6 / 1800 4.6   **19** SU 0024 1.5 / 0630 4.8 / 1256 1.1 / 1847 4.7
4 SA 0420 4.2 / 1053 1.9 / 1657 4.2 / 2316 1.8   **19** SU 0534 4.5 / 1157 1.3 / 1753 4.5
4 TU 0523 5.0 / 1156 0.7 / 1755 5.0   **19** W 0601 4.7 / 1226 1.1 / 1818 4.7

5 TH 0511 4.5 / 1124 1.9 / 1730 4.4 / 2348 1.7   **20** F 0604 4.7 / 1226 1.4 / 1823 4.6
5 SU 0018 1.4 / 0614 5.0 / 1244 1.2 / 1840 4.9   **20** M 0057 1.2 / 0700 5.0 / 1328 0.9 / 1916 4.9
5 SU 0509 4.6 / 1142 1.4 / 1741 4.6 / 2359 1.3   **20** M 0001 1.5 / 0606 4.7 / 1230 1.1 / 1822 4.7
5 W 0013 0.8 / 0605 5.4 / 1236 0.3 / 1833 5.3   **20** TH 0030 1.3 / 0628 4.8 / 1255 1.1 / 1844 4.9

6 F 0551 4.7 / 1212 1.6 / 1812 4.6   **21** SA 0033 1.4 / 0643 4.9 / 1309 1.1 / 1902 4.7
6 M 0059 1.1 / 0653 5.3 / 1325 0.8 / 1919 5.1   **21** TU 0128 1.1 / 0727 5.1 / 1358 0.9 / 1943 4.9
6 M 0551 5.0 / 1224 0.9 / 1820 5.0   **21** TU 0031 1.3 / 0633 4.9 / 1300 0.9 / 1848 4.9
6 TH 0053 0.5 / 0646 5.6 / 1315 0.2 / 1911 5.4   **21** F 0100 1.2 / 0654 4.9 / 1324 1.1 / 1910 4.9

7 SA 0031 1.4 / 0630 4.9 / 1256 1.3 / 1853 4.8   **22** SU 0112 1.1 / 0718 5.0 / 1347 1.0 / 1937 4.8
7 TU 0138 0.7 / 0732 5.5 / 1405 0.5 / 1958 5.3   **22** W 0158 1.1 / 0753 5.2 / 1427 0.9 / 2010 4.9
7 TU 0039 0.9 / 0631 5.4 / 1303 0.5 / 1858 5.3   **22** W 0101 1.2 / 0658 5.0 / 1328 0.9 / 1913 5.0
7 F 0133 0.5 / 0727 5.7 / 1355 0.2 / 1948 5.4   **22** SA 0132 1.1 / 0721 4.9 / 1352 1.1 / 1937 5.0

8 SU 0111 1.2 / 0708 5.1 / 1338 1.0 / 1933 4.9   **23** M 0147 1.1 / 0750 5.1 / 1422 0.9 / 2010 4.8
8 W 0218 0.6 / 0812 5.7 / 1444 0.3 / 2038 5.3   **23** TH 0227 1.1 / 0818 5.1 / 1456 1.0 / 2036 4.9
8 W 0118 0.6 / 0710 5.7 / 1342 0.2 / 1935 5.5   **23** TH 0130 1.1 / 0722 5.1 / 1355 0.9 / 1938 5.0
8 SA 0213 0.3 / 0809 5.5 / 1433 0.4 / 2027 5.2   **23** SU 0204 1.1 / 0749 4.8 / 1422 1.2 / 2006 4.9

9 M 0152 1.1 / 0747 5.3 / 1420 0.8 / 2014 4.9   **24** TU 0221 1.1 / 0821 5.1 / 1456 0.9 / 2042 4.9
9 TH 0257 0.6 / 0854 5.6 / 1524 0.4 / 2120 5.1   **24** F 0256 1.2 / 0844 5.0 / 1523 1.2 / 2104 4.8
9 TH 0156 0.4 / 0750 5.8 / 1420 0.1 / 2013 5.4   **24** F 0158 1.1 / 0747 5.1 / 1423 1.0 / 2003 5.0
9 SU 0255 0.5 / 0853 5.2 / 1513 0.7 / 2108 4.9   **24** M 0237 1.2 / 0820 4.6 / 1453 1.3 / 2039 4.7

10 TU 0233 1.0 / 0829 5.3 / 1502 0.7 / 2059 4.9   **25** W 0254 1.2 / 0850 5.0 / 1528 1.1 / 2113 4.7
10 F 0338 0.7 / 0938 5.5 / 1605 0.6 / 2204 4.9   **25** SA 0327 1.3 / 0911 4.9 / 1552 1.4 / 2134 4.6
10 F 0235 0.4 / 0831 5.7 / 1459 0.3 / 2052 5.3   **25** SA 0228 1.1 / 0812 5.0 / 1450 1.2 / 2030 4.9
10 M 0338 0.9 / 0942 4.7 / 1553 1.2 / 2156 4.5   **25** TU 0313 1.3 / 0856 4.4 / 1527 1.5 / 2118 4.5

11 W 0315 1.0 / 0915 5.2 / 1545 0.7 / 2146 4.8   **26** TH 0326 1.3 / 0919 4.9 / 1600 1.3 / 2145 4.5
11 SA 0421 0.9 / 1026 5.1 / 1647 1.0 / 2254 4.5   **26** SU 0359 1.5 / 0941 4.6 / 1622 1.7 / 2208 4.4
11 SA 0315 0.5 / 0913 5.5 / 1538 0.6 / 2133 5.0   **26** SU 0259 1.2 / 0840 4.8 / 1518 1.4 / 2100 4.8
11 TU 0425 1.2 / 1043 4.1 / 1637 1.6 / 2304 4.1   **26** W 0352 1.5 / 0940 4.1 / 1605 1.7 / 2208 4.2

12 TH 0359 1.1 / 1003 5.2 / 1630 0.8 / 2238 4.6   **27** F 0358 1.5 / 0950 4.7 / 1632 1.5 / 2220 4.4
12 SU 0507 1.3 / 1121 4.7 / 1734 1.4 / 2357 4.2   **27** M 0434 1.8 / 1015 4.2 / 1657 1.9 / 2252 4.2
12 SU 0357 0.8 / 1000 5.2 / 1618 1.0 / 2219 4.5   **27** M 0332 1.4 / 0911 4.6 / 1549 1.6 / 2134 4.5
12 W 0521 1.6 / 1206 3.8 / 1731 2.1 / 2319 4.0   **27** TH 0437 1.7 / 1041 3.8 / 1652 2.0 / 2319 4.0

13 F 0446 1.2 / 1056 5.0 / 1718 1.1 / 2336 4.4   **28** SA 0433 1.8 / 1023 4.5 / 1706 1.8 / 2301 4.2
13 M 0600 1.7 / 1230 4.3 / 1829 1.9   **28** TU 0515 2.1 / 1100 4.0 / 1740 2.2
13 M 0442 1.2 / 1056 4.5 / 1701 1.5 / 2321 4.1   **28** TU 0408 1.6 / 0948 4.3 / 1624 1.8 / 2218 4.2
13 TH 0041 3.8 / 0641 1.9 / 1337 3.6 / 1857 2.4   **28** F 0534 1.9 / 1211 3.7 / 1800 2.2

14 SA 0537 1.5 / 1154 4.7 / 1811 1.4   **29** SU 0511 2.0 / 1101 4.3 / 1746 2.0 / 2354 4.0
14 TU 0125 3.9 / 0712 2.1 / 1401 4.0 / 1949 2.2
14 TU 0534 1.7 / 1214 4.0 / 1754 2.1   **29** W 0449 1.9 / 1037 4.0 / 1706 2.1 / 2327 3.9
14 F 0209 3.8 / 0837 2.1 / 1456 3.7 / 2101 2.3   **29** SA 0045 3.9 / 0656 1.9 / 1346 3.7 / 1943 2.2

15 SU 0043 4.2 / 0636 1.8 / 1301 4.4 / 1913 1.7   **30** M 0558 2.3 / 1152 4.0 / 1839 2.2
15 W 0258 3.9 / 0859 2.2 / 1530 3.9 / 2140 2.2
15 W 0102 3.8 / 0650 2.1 / 1353 3.7 / 1920 2.4   **30** TH 0543 2.1 / 1209 3.6 / 1810 2.3
15 SA 0320 3.9 / 0955 1.8 / 1559 3.9 / 2208 2.1   **30** SU 0203 4.0 / 0834 1.8 / 1500 3.9 / 2112 1.9

31 TU 0111 3.9 / 0700 2.5 / 1314 3.8 / 1955 2.4
31 F 0113 3.8 / 0709 2.3 / 1412 3.6 / 2015 2.4

Chart Datum: 2.75 metres below Ordnance Datum (Newlyn)

SCOTLAND, WEST COAST - ULLAPOOL
LAT 57°54'N   LONG 5°10'W
TIMES AND HEIGHTS OF HIGH AND LOW WATERS

TIME ZONE UT(GMT)
For Summer Time add ONE hour in non-shaded areas

YEAR 1989

	MAY				JUNE				JULY				AUGUST		
	TIME m		TIME m		TIME m		TIME m		TIME m		TIME m		TIME m		TIME m

1 0309 4.2 / 0945 1.4 / M 1558 4.3 / 2211 1.6
16 0411 4.1 / 1037 1.5 / TU 1641 4.2 / 2243 1.8
1 0429 4.7 / 1055 1.0 / TH 1706 4.7 / 2320 1.1
16 0456 4.1 / 1115 1.7 / F 1720 4.4 / 2331 1.8
1 0513 4.5 / 1128 1.3 / SA 1743 4.6
16 0521 4.1 / 1134 1.8 / SU 1737 4.5
1 0050 0.9 / 0645 4.6 / TU 1255 1.1 / ● 1859 4.9
16 0026 1.2 / 0622 4.7 / W 1239 1.2 / 1831 5.1

2 0404 4.6 / 1038 1.0 / TU 1646 4.6 / 2300 1.2
17 0452 4.2 / 1115 1.4 / W 1716 4.4 / 2322 1.7
2 0521 4.8 / 1144 0.8 / F 1751 4.9
17 0537 4.3 / 1155 1.5 / SA 1757 4.6
2 0003 1.1 / 0606 4.6 / SU 1219 1.1 / 1828 4.8
17 0000 1.6 / 0603 4.3 / M 1217 1.6 / 1815 4.7
2 0129 0.7 / 0721 4.8 / W 1332 1.0 / 1933 5.1
17 0104 0.8 / 0658 4.9 / TH 1317 0.9 / ○ 1907 5.3

3 0454 4.9 / 1125 0.7 / W 1729 4.9 / 2345 0.9
18 0527 4.4 / 1150 1.4 / TH 1748 4.6 / 2358 1.5
3 0010 0.9 / 0610 4.9 / SA 1230 0.8 / ● 1834 5.0
18 0014 1.6 / 0616 4.4 / SU 1233 1.4 / 1831 4.7
3 0054 0.9 / 0653 4.7 / M 1304 1.0 / ● 1910 4.9
18 0043 1.3 / 0641 4.5 / TU 1257 1.3 / ○ 1852 4.9
3 0205 0.6 / 0755 4.8 / TH 1407 0.9 / 2005 5.1
18 0142 0.5 / 0734 5.1 / F 1355 0.7 / 1945 5.5

4 0540 5.2 / 1208 0.5 / TH 1809 5.1
19 0600 4.5 / 1223 1.3 / F 1818 4.7
4 0058 0.7 / 0657 5.0 / SU 1314 0.8 / 1917 5.0
19 0054 1.4 / 0652 4.5 / M 1311 1.3 / ○ 1906 4.8
4 0140 0.7 / 0736 4.7 / TU 1346 1.0 / 1951 4.9
19 0123 1.0 / 0717 4.7 / W 1335 1.1 / 1928 5.1
4 0239 0.7 / 0827 4.8 / F 1440 1.0 / 2035 5.0
19 0219 0.3 / 0811 5.2 / SA 1433 0.6 / 2025 5.5

5 0028 0.6 / 0624 5.3 / F 1250 0.4 / ● 1849 5.2
20 0034 1.4 / 0631 4.6 / SA 1255 1.2 / ○ 1848 4.8
5 0145 0.7 / 0744 4.9 / M 1357 0.8 / 2000 4.9
20 0134 1.2 / 0729 4.6 / TU 1348 1.2 / 1943 4.9
5 0223 0.7 / 0818 4.7 / W 1426 1.0 / 2030 4.9
20 0202 0.8 / 0755 4.8 / TH 1414 0.9 / 2007 5.2
5 0312 0.8 / 0858 4.7 / SA 1512 1.1 / 2104 4.9
20 0258 0.3 / 0850 5.1 / SU 1513 0.6 / 2107 5.4

6 0112 0.5 / 0708 5.3 / SA 1331 0.4 / 1928 5.2
21 0110 1.3 / 0703 4.6 / SU 1328 1.2 / 1919 4.9
6 0231 0.7 / 0831 4.7 / TU 1439 1.0 / 2045 4.8
21 0214 1.1 / 0808 4.6 / W 1427 1.2 / 2022 4.9
6 0303 0.7 / 0858 4.6 / TH 1504 1.1 / 2108 4.8
21 0241 0.6 / 0834 4.8 / F 1454 0.9 / 2048 5.2
6 0344 1.0 / 0930 4.6 / SU 1545 1.3 / 2134 4.7
21 0337 0.4 / 0932 4.9 / M 1554 0.8 / 2152 5.1

7 0155 0.5 / 0753 5.2 / SU 1412 0.6 / 2010 5.0
22 0146 1.2 / 0736 4.6 / M 1401 1.2 / 1952 4.8
7 0317 0.8 / 0920 4.5 / W 1522 1.2 / 2133 4.6
22 0255 1.0 / 0849 4.5 / TH 1507 1.2 / 2105 4.8
7 0342 0.9 / 0938 4.4 / F 1542 1.2 / 2146 4.6
22 0321 0.6 / 0916 4.7 / SA 1535 0.9 / 2132 5.1
7 0415 1.3 / 1003 4.4 / M 1619 1.6 / 2205 4.4
22 0418 0.8 / 1018 4.6 / TU 1639 1.1 / 2244 4.7

8 0240 0.6 / 0840 4.9 / M 1453 0.9 / 2054 4.8
23 0223 1.2 / 0813 4.5 / TU 1437 1.3 / 2030 4.7
8 0403 1.0 / 1012 4.2 / TH 1605 1.4 / 2225 4.4
23 0337 0.9 / 0936 4.4 / F 1550 1.3 / 2154 4.7
8 0420 1.1 / 1019 4.3 / SA 1620 1.5 / 2224 4.4
23 0402 0.6 / 1002 4.7 / SU 1618 1.0 / 2220 5.0
8 0448 1.6 / 1041 4.2 / TU 1656 1.8 / 2241 4.2
23 0502 1.2 / 1116 4.3 / W 1730 1.5 / ☾ 2351 4.2

9 0326 0.9 / 0932 4.5 / TU 1535 1.2 / 2146 4.5
24 0303 1.2 / 0854 4.3 / W 1515 1.4 / 2113 4.6
9 0451 1.2 / 1108 4.0 / F 1651 1.7 / 2319 4.2
24 0422 1.0 / 1029 4.3 / SA 1638 1.4 / 2247 4.6
9 0459 1.3 / 1102 4.1 / SU 1700 1.7 / 2304 4.2
24 0446 0.8 / 1054 4.5 / M 1705 1.2 / ☾ 2314 4.7
9 0525 1.9 / 1130 4.0 / W 1740 2.1 / 2329 3.9
24 0554 1.7 / 1241 3.9 / TH 1836 1.9

10 0415 1.2 / 1034 4.1 / W 1621 1.6 / 2251 4.2
25 0345 1.3 / 0943 4.2 / TH 1558 1.5 / 2205 4.4
10 0543 1.5 / 1206 3.8 / SA 1743 1.9
25 0511 1.1 / 1129 4.2 / SU 1731 1.5 / 2346 4.5
10 0541 1.6 / 1151 3.9 / M 1744 2.0 / 2350 4.0
25 0534 1.1 / 1154 4.2 / TU 1759 1.5 / ☾
10 0612 2.1 / 1245 3.8 / TH 1838 2.3
25 0124 3.9 / 0705 2.0 / F 1420 3.8 / 2018 2.1

11 0511 1.5 / 1146 3.8 / TH 1714 1.9
26 0433 1.4 / 1044 4.0 / F 1648 1.7 / 2308 4.3
11 0016 4.0 / 0641 1.6 / SU 1306 3.8 / ☾ 1844 2.1
26 0606 1.2 / 1235 4.1 / M 1833 1.6
11 0628 1.9 / 1250 3.8 / TU 1837 2.2
26 0017 4.4 / 0629 1.5 / W 1308 4.0 / 1904 1.8
11 0053 3.6 / 0723 2.3 / F 1418 3.8 / 2005 2.5
26 0258 3.8 / 0856 2.2 / SA 1541 4.0 / 2206 1.9

12 0008 4.0 / 0620 1.7 / F 1300 3.7 / ☾ 1825 2.1
27 0528 1.5 / 1157 3.9 / SA 1751 1.9
12 0114 3.9 / 0745 1.8 / M 1407 3.8 / 1952 2.2
27 0050 4.4 / 0709 1.4 / TU 1344 4.1 / 1942 1.7
12 0049 3.8 / 0727 2.1 / W 1359 3.8 / 1944 2.3
27 0133 4.1 / 0738 1.7 / TH 1430 4.0 / 2027 1.9
12 0250 3.6 / 0904 2.4 / SA 1533 3.9 / 2147 2.3
27 0415 3.9 / 1028 2.0 / SU 1643 4.2 / 2312 1.5

13 0121 3.9 / 0744 1.8 / SA 1409 3.7 / 1954 2.2
28 0017 4.2 / 0636 1.5 / SU 1313 3.9 / ☾ 1908 1.9
13 0214 3.9 / 0847 1.9 / TU 1504 3.8 / 2057 2.2
28 0158 4.3 / 0819 1.5 / W 1453 4.1 / 2055 1.7
13 0206 3.7 / 0838 2.1 / TH 1507 3.9 / 2101 2.3
28 0257 4.0 / 0903 1.9 / F 1546 4.0 / 2156 1.8
13 0409 3.8 / 1023 2.2 / SU 1630 4.1 / 2257 2.0
28 0511 4.2 / 1122 1.6 / M 1731 4.5 / 2358 1.2

14 0226 3.9 / 0858 1.8 / SU 1510 3.8 / 2108 2.1
29 0126 4.2 / 0752 1.5 / M 1422 4.0 / 2026 1.8
14 0314 3.9 / 0943 1.8 / W 1556 4.0 / 2154 2.1
29 0308 4.3 / 0928 1.4 / TH 1556 4.2 / 2204 1.6
14 0314 3.8 / 0947 2.1 / F 1606 4.0 / 2212 2.2
29 0412 4.1 / 1023 1.8 / SA 1648 4.2 / 2309 1.6
14 0503 4.0 / 1117 1.8 / M 1715 4.4 / 2346 1.6
29 0554 4.4 / 1202 1.4 / TU 1808 4.8

15 0323 3.9 / 0953 1.6 / M 1600 4.0 / 2200 2.0
30 0231 4.3 / 0902 1.3 / TU 1524 4.2 / 2132 1.6
15 0408 4.0 / 1031 1.8 / TH 1641 4.2 / 2244 1.9
30 0414 4.3 / 1032 1.4 / F 1653 4.4 / 2307 1.4
15 0430 3.9 / 1046 2.0 / SA 1655 4.2 / 2311 1.9
30 0514 4.2 / 1125 1.6 / SU 1739 4.5
15 0545 4.3 / 1201 1.5 / TU 1754 4.8
30 0035 0.9 / 0628 4.7 / W 1237 1.1 / 1840 5.0

31 0332 4.5 / 1002 1.1 / W 1618 4.5 / 2228 1.3
31 0005 1.2 / 0604 4.5 / M 1214 1.3 / 1822 4.7
31 0108 0.7 / 0658 4.8 / TH 1309 1.0 / ● 1908 5.1

Chart Datum: 2.75 metres below Ordnance Datum (Newlyn)

SCOTLAND, WEST COAST - ULLAPOOL
LAT 57°54'N   LONG 5°10'W

TIMES AND HEIGHTS OF HIGH AND LOW WATERS

YEAR **1989**

TIME ZONE UT(GMT)
For Summer Time add ONE hour in non-shaded areas

SEPTEMBER

Day	Time	m	Day	Time	m
1 F	0140 / 0727 / 1340 / 1935	0.6 / 4.9 / 0.9 / 5.1	16 SA	0116 / 0709 / 1331 / 1921	0.3 / 5.4 / 0.5 / 5.7
2 SA	0209 / 0753 / 1411 / 2000	0.7 / 5.0 / 1.0 / 5.1	17 SU	0154 / 0745 / 1409 / 2001	0.1 / 5.4 / 0.4 / 5.7
3 SU	0238 / 0820 / 1440 / 2026	0.8 / 4.9 / 1.1 / 5.0	18 M	0232 / 0823 / 1450 / 2043	0.2 / 5.3 / 0.5 / 5.5
4 M	0306 / 0847 / 1511 / 2052	1.0 / 4.8 / 1.3 / 4.8	19 TU	0311 / 0903 / 1531 / 2129	0.5 / 5.0 / 0.7 / 5.1
5 TU	0335 / 0916 / 1543 / 2121	1.3 / 4.6 / 1.5 / 4.5	20 W	0351 / 0948 / 1617 / 2223	0.9 / 4.6 / 1.1 / 4.5
6 W	0405 / 0949 / 1618 / 2154	1.6 / 4.4 / 1.7 / 4.2	21 TH	0435 / 1048 / 1709 / 2341	1.4 / 4.2 / 1.5 / 4.0
7 TH	0438 / 1031 / 1659 / 2238	1.9 / 4.1 / 2.0 / 3.9	22 F ☾	0527 / 1227 / 1822	1.8 / 3.9 / 1.9
8 F ☾	0518 / 1140 / 1752 / 2359	2.1 / 3.8 / 2.3 / 3.6	23 SA	0123 / 0644 / 1409 / 2024	3.7 / 2.2 / 3.8 / 2.0
9 SA	0620 / 1331 / 1917	2.4 / 3.7 / 2.5	24 SU	0254 / 0857 / 1527 / 2201	3.7 / 2.3 / 4.0 / 1.8
10 SU	0221 / 0821 / 1458 / 2124	3.5 / 2.5 / 3.8 / 2.3	25 M	0404 / 1017 / 1626 / 2255	3.9 / 2.0 / 4.2 / 1.4
11 M	0344 / 1001 / 1559 / 2236	3.7 / 2.2 / 4.1 / 1.9	26 TU	0455 / 1103 / 1710 / 2335	4.2 / 1.7 / 4.5 / 1.1
12 TU	0439 / 1055 / 1647 / 2322	4.1 / 1.8 / 4.4 / 1.4	27 W	0533 / 1138 / 1745	4.4 / 1.4 / 4.7
13 W	0521 / 1137 / 1727	4.4 / 1.4 / 4.8	28 TH	0008 / 0603 / 1211 / 1814	0.9 / 4.7 / 1.2 / 4.9
14 TH	0001 / 0558 / 1215 / 1805	1.0 / 4.8 / 1.0 / 5.2	29 F ●	0039 / 0631 / 1241 / 1840	0.8 / 4.8 / 1.1 / 5.0
15 F ○	0039 / 0633 / 1253 / 1843	0.6 / 5.1 / 0.7 / 5.5	30 SA	0109 / 0656 / 1311 / 1905	0.8 / 5.0 / 1.0 / 5.1

OCTOBER

Day	Time	m	Day	Time	m
1 SU	0137 / 0721 / 1341 / 1929	0.9 / 5.0 / 1.1 / 5.0	16 M	0128 / 0721 / 1348 / 1940	0.2 / 5.4 / 0.4 / 5.6
2 M	0205 / 0747 / 1411 / 1955	1.0 / 5.0 / 1.1 / 4.9	17 TU	0208 / 0800 / 1430 / 2025	0.3 / 5.3 / 0.5 / 5.3
3 TU	0232 / 0813 / 1442 / 2022	1.1 / 4.9 / 1.3 / 4.8	18 W	0248 / 0842 / 1514 / 2114	0.6 / 5.0 / 0.7 / 4.9
4 W	0301 / 0842 / 1515 / 2052	1.3 / 4.7 / 1.4 / 4.5	19 TH	0330 / 0930 / 1603 / 2215	1.0 / 4.6 / 1.1 / 4.4
5 TH	0331 / 0915 / 1551 / 2127	1.6 / 4.5 / 1.7 / 4.2	20 F	0415 / 1036 / 1659 / 2338	1.5 / 4.2 / 1.5 / 3.9
6 F	0405 / 0958 / 1632 / 2215	1.8 / 4.2 / 1.9 / 3.9	21 SA ☾	0509 / 1213 / 1815	1.9 / 4.0 / 1.8
7 SA	0446 / 1105 / 1725 / 2344	2.1 / 3.9 / 2.2 / 3.6	22 SU	0110 / 0628 / 1342 / 2003	3.7 / 2.2 / 3.9 / 1.9
8 SU ☽	0545 / 1249 / 1846	2.4 / 3.8 / 2.3	23 M	0230 / 0824 / 1454 / 2126	3.8 / 2.3 / 4.0 / 1.7
9 M	0150 / 0740 / 1417 / 2047	3.5 / 2.5 / 3.9 / 2.2	24 TU	0335 / 0940 / 1552 / 2219	3.9 / 2.0 / 4.2 / 1.5
10 TU	0311 / 0925 / 1521 / 2159	3.8 / 2.2 / 4.1 / 1.8	25 W	0424 / 1027 / 1637 / 2300	4.1 / 1.8 / 4.4 / 1.3
11 W	0407 / 1022 / 1612 / 2248	4.1 / 1.8 / 4.5 / 1.3	26 TH	0503 / 1106 / 1714 / 2335	4.4 / 1.6 / 4.5 / 1.2
12 TH	0451 / 1106 / 1656 / 2329	4.5 / 1.4 / 4.9 / 0.9	27 F	0534 / 1139 / 1744	4.6 / 1.5 / 4.7
13 F	0530 / 1146 / 1738	4.8 / 1.0 / 5.3	28 SA	0007 / 0602 / 1211 / 1812	1.1 / 4.8 / 1.3 / 4.8
14 SA	0009 / 0607 / 1226 / 1818	0.5 / 5.2 / 0.7 / 5.5	29 SU ●	0037 / 0629 / 1243 / 1839	1.1 / 4.9 / 1.3 / 4.9
15 SU	0049 / 0644 / 1306 / 1859	0.3 / 5.4 / 0.5 / 5.6	30 M	0106 / 0655 / 1315 / 1906	1.1 / 5.0 / 1.2 / 4.9
			31 TU	0136 / 0722 / 1348 / 1934	1.2 / 5.0 / 1.2 / 4.8

NOVEMBER

Day	Time	m	Day	Time	m
1 W	0205 / 0751 / 1422 / 2005	1.3 / 4.9 / 1.3 / 4.6	16 TH	0232 / 0832 / 1505 / 2108	0.8 / 5.0 / 0.8 / 4.7
2 TH	0236 / 0823 / 1457 / 2039	1.4 / 4.8 / 1.4 / 4.4	17 F	0316 / 0923 / 1555 / 2208	1.1 / 4.7 / 1.0 / 4.4
3 F	0309 / 0900 / 1535 / 2119	1.6 / 4.6 / 1.6 / 4.2	18 SA	0402 / 1026 / 1650 / 2319	1.5 / 4.4 / 1.3 / 4.1
4 SA	0346 / 0946 / 1619 / 2214	1.8 / 4.3 / 1.8 / 3.9	19 SU	0455 / 1141 / 1755	1.8 / 4.2 / 1.6
5 SU	0430 / 1049 / 1712 / 2336	2.0 / 4.1 / 1.9 / 3.7	20 M ☾	0034 / 0600 / 1255 / 1913	3.9 / 2.1 / 4.1 / 1.7
6 M	0530 / 1211 / 1824	2.2 / 4.0 / 2.0	21 TU ☽	0144 / 0721 / 1402 / 2028	3.8 / 2.2 / 4.0 / 1.8
7 TU	0111 / 0701 / 1330 / 1957	3.7 / 2.3 / 4.0 / 1.9	22 W	0247 / 0838 / 1502 / 2128	3.9 / 2.2 / 4.1 / 1.7
8 W	0228 / 0835 / 1437 / 2112	3.9 / 2.1 / 4.2 / 1.6	23 TH	0341 / 0937 / 1554 / 2216	4.0 / 2.1 / 4.2 / 1.6
9 TH	0329 / 0940 / 1534 / 2208	4.2 / 1.8 / 4.5 / 1.3	24 F	0426 / 1024 / 1637 / 2256	4.2 / 1.9 / 4.3 / 1.5
10 F	0419 / 1031 / 1625 / 2256	4.5 / 1.5 / 4.8 / 0.9	25 SA	0503 / 1105 / 1715 / 2333	4.4 / 1.8 / 4.4 / 1.4
11 SA	0503 / 1115 / 1712 / 2341	4.9 / 1.1 / 5.1 / 0.7	26 SU	0536 / 1143 / 1749	4.6 / 1.7 / 4.5
12 SU	0544 / 1202 / 1758	5.1 / 0.8 / 5.3	27 M	0008 / 0607 / 1220 / 1821	1.4 / 4.8 / 1.5 / 4.6
13 M ○	0024 / 0624 / 1247 / 1842	0.5 / 5.3 / 0.6 / 5.4	28 TU ●	0041 / 0637 / 1257 / 1852	1.4 / 4.9 / 1.4 / 4.7
14 TU	0106 / 0705 / 1332 / 1928	0.5 / 5.3 / 0.6 / 5.3	29 W	0114 / 0708 / 1333 / 1925	1.4 / 4.9 / 1.4 / 4.7
15 W	0149 / 0747 / 1418 / 2016	0.6 / 5.2 / 0.6 / 5.1	30 TH	0147 / 0740 / 1410 / 1959	1.4 / 4.9 / 1.3 / 4.6

DECEMBER

Day	Time	m	Day	Time	m
1 F	0221 / 0814 / 1447 / 2036	1.4 / 4.9 / 1.3 / 4.4	16 SA	0304 / 0913 / 1545 / 2149	1.1 / 4.9 / 0.9 / 4.5
2 SA	0257 / 0853 / 1527 / 2119	1.5 / 4.8 / 1.4 / 4.3	17 SU	0348 / 1003 / 1632 / 2242	1.3 / 4.7 / 1.1 / 4.3
3 SU	0337 / 0939 / 1611 / 2210	1.6 / 4.6 / 1.5 / 4.1	18 M	0432 / 1055 / 1720 / 2339	1.6 / 4.5 / 1.4 / 4.1
4 M	0422 / 1034 / 1700 / 2314	1.8 / 4.5 / 1.6 / 4.0	19 TU ☾	0521 / 1150 / 1814	1.8 / 4.3 / 1.7
5 TU	0517 / 1136 / 1758	1.9 / 4.3 / 1.7	20 W	0039 / 0615 / 1248 / 1914	3.9 / 2.1 / 4.1 / 1.9
6 W	0028 / 0624 / 1244 / 1908	4.0 / 2.0 / 4.3 / 1.7	21 TH	0142 / 0719 / 1351 / 2019	3.9 / 2.3 / 4.0 / 2.0
7 TH	0141 / 0742 / 1352 / 2021	4.0 / 2.0 / 4.3 / 1.6	22 F	0244 / 0829 / 1455 / 2121	3.9 / 2.3 / 4.0 / 2.0
8 F	0249 / 0854 / 1457 / 2126	4.2 / 1.9 / 4.5 / 1.4	23 SA	0341 / 0934 / 1556 / 2215	4.0 / 2.3 / 4.0 / 2.0
9 SA	0348 / 0956 / 1558 / 2224	4.4 / 1.7 / 4.6 / 1.2	24 SU	0430 / 1030 / 1647 / 2302	4.2 / 2.2 / 4.1 / 1.9
10 SU	0440 / 1052 / 1654 / 2317	4.7 / 1.4 / 4.8 / 1.0	25 M	0512 / 1120 / 1731 / 2344	4.4 / 2.0 / 4.3 / 1.7
11 M	0528 / 1145 / 1746	4.9 / 1.1 / 5.0	26 TU	0549 / 1205 / 1809	4.6 / 1.8 / 4.4
12 TU	0006 / 0613 / 1236 / 1835	0.9 / 5.1 / 0.9 / 5.1	27 W	0023 / 0623 / 1245 / 1844	1.6 / 4.8 / 1.6 / 4.6
13 W	0052 / 0657 / 1325 / 1923	0.8 / 5.2 / 0.7 / 5.1	28 TH ●	0059 / 0657 / 1324 / 1918	1.5 / 4.9 / 1.4 / 4.6
14 TH	0137 / 0741 / 1412 / 2010	0.8 / 5.2 / 0.7 / 5.0	29 F	0135 / 0730 / 1401 / 1952	1.4 / 5.0 / 1.2 / 4.7
15 F	0221 / 0826 / 1459 / 2059	0.9 / 5.1 / 0.7 / 4.8	30 SA	0211 / 0806 / 1439 / 2029	1.3 / 5.1 / 1.1 / 4.7
			31 SU	0248 / 0844 / 1517 / 2109	1.3 / 5.1 / 1.1 / 4.6

Chart Datum: 2.75 metres below Ordnance Datum (Newlyn)

AREA 8—NW Scotland 431

PORTREE 10-8-9
Skye

CHARTS
Admiralty 2534, 2209; Imray C66; OS 23

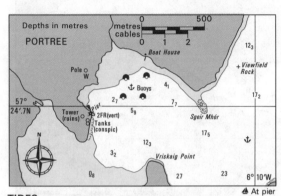

TIDES
−0445 Dover; ML 2.9; Duration 0610; Zone 0 (GMT).
Standard Port ULLAPOOL (←)

Times				Height (metres)			
HW		LW		MHWS	MHWN	MLWN	MLWS
0000	0600	0300	0900	5.2	3.9	2.1	0.7
1200	1800	1500	2100				

Differences PORTREE
−0025 −0025 −0025 −0025 +0.1 −0.2 −0.2 0.0
SHEILDAG
−0020 −0020 −0015 −0015 +0.4 +0.3 +0.1 0.0
PLOCKTON
−0020 −0020 −0010 −0010 +0.3 +0.2 +0.1 +0.1
LOCH SNIZORT-UIG BAY
−0045 −0020 −0005 −0025 +0.1 −0.4 −0.2 0.0
LOCH DUNVEGAN
−0105 −0030 −0020 −0040 0.0 −0.1 0.0 0.0
LOCH HARPORT
−0115 −0035 −0020 −0100 −0.1 −0.1 0.0 +0.1
KYLE OF LOCHALSH
−0040 −0020 −0005 −0025 +0.1 0.0 +0.1 +0.1
GLENELG BAY
−0105 −0034 −0034 −0054 −0.4 −0.4 −0.9 −0.1
LOCH HOURN
−0125 −0050 −0040 −0110 −0.2 −0.1 −0.1 +0.1

SHELTER
Secure shelter in all but strong SW winds but holding ground not reliable. There are four HIDB moorings for visitors as shown. There are a number of anchorages round Skye, so choose one according to wind direction and forecast.

NAVIGATION
Waypoint 57°24'.70N 06°07'.50W, 090°/270° from/to Sgeir Mhor rock, 1.6M. Approaching from S, keep clear of shore round An Tom Pt (off chartlet) where there are rocks.

LIGHTS AND MARKS
The only lights in the harbour are 2 FR (vert) (occas) on the pier.

RADIO TELEPHONE
None.

TELEPHONE (0478)
Hr Mr office on pier (messages to Roads Dept. Tel. 2727); MRSC Oban 63720; CG 2663; HMC Inverness 231608; Marinecall 0898 500 464; Dr 2013; Hosp 2704

FACILITIES
EC Wednesday; **Pier** D (cans), L, FW; **West End Garage** P, Gas; **Town** P, V, Gaz, Lau, R, Bar. PO; Bank; Rly (bus to Kyleakin, ferry to Kyle of Lochalsh); Air (Broadford).

MALLAIG 10-8-10
Inverness

CHARTS
Admiralty 2534, 2208; Imray C65, C66; OS 40
TIDES
−0515 Dover; ML 2.9; Duration 0605; Zone 0 (GMT).
Standard Port OBAN (→)

Times				Height (metres)			
HW		LW		MHWS	MHWN	MLWN	MLWS
0000	0600	0100	0700	4.0	2.9	1.8	0.7
1200	1800	1300	1900				

Differences MALLAIG
+0040 +0020 +0035 +0030 +1.0 +0.9 +0.3 0.0
INVERIE BAY
+0030 +0020 +0035 +0020 +1.0 +0.9 +0.2 0.0
BAY OF LAIG (Eigg)
+0015 +0030 +0040 +0005 +0.7 +0.6 −0.2 −0.2
LOCH MOIDART
+0015 +0015 +0040 +0020 +0.8 +0.6 −0.2 −0.2
LOCH EATHARNA (Coll)
+0025 +0010 +0015 +0025 +0.4 +0.3 No data
GOTT BAY (Tiree)
+0015 0000 0000 +0015 +0.1 +0.1 −0.1 −0.1

SHELTER
Good shelter in SW winds but open to the N. Anchor near the head or go alongside S Pier. Access H24.

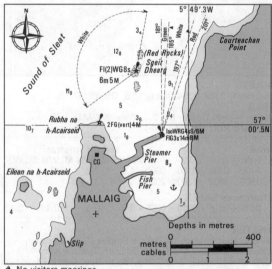

▲ No visitors moorings

NAVIGATION
Waypoint 57°00'.70N 05°49'.80W, 330°/150° from/to Steamer Pier Lt, 0.28M. Red Rocks lie in entrance to the harbour (marked by beacon on centre rock) but can be passed either side, the E side being easier. Harbour is often congested with fishing boats. Also used by car ferry service to Skye.

LIGHTS AND MARKS
Sgeir Dhearg Fl(2)WG 8s 6m 5M; vis G190°−055°, W055°−190° (liable to be obsc by town lights). Steamer Pier, E end Iso WRG 4s 6m 9/6M; vis G181°−185°, W185°−197°, R197°−201°. Int Port Traffic Sigs for ferries (May−Sep).

RADIO TELEPHONE
Call: *Mallaig Harbour Radio* VHF Ch 16; 09 (office hours).
TELEPHONE (0687)
Hr Mr 2154; CG 2336; MRSC Oban 63720; HMC Fort William 2948; Marinecall 0898 500 464; Dr 2202
FACILITIES
EC Wednesday; **Jetty** Slip, M, P, D, L, FW, ME, El, Sh, C (6 ton mobile), CH, AB; **Mallaig Boat Bldg Co.** Tel. 2304, ME, El, Sh, CH; **Western Battery Service** Tel. 2044 El; **Johnson Bros** Tel. 2215 CH, ACA; **Town** V, R, Gas, Bar. PO; Bank; Rly; Air (Broadford or Inverness).

LOCH SUNART (Drumbuie and Salen) 10-8-11
Argyll

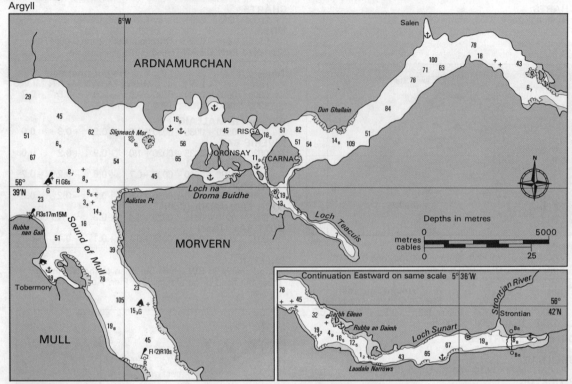

CHARTS
Admiralty 2392, 2394; Imray C65; OS 45, 47, 49
TIDES
Salen −0500 Dover; ML 2.0; Zone 0 (GMT).
Standard Port OBAN (→)

Times				Height (metres)			
HW		LW		MHWS	MHWN	MLWN	MLWS
0100	0700	0100	0800	4.0	2.9	1.8	0.7
1300	1900	1300	2000				

Differences SALEN
−0015 +0015 +0010 +0005 +0.6 +0.5 −0.1 −0.1

SHELTER
Anchorages in Loch na Droma Buidhe (S of Oronsay), between Oronsay and Carna, in Loch Teacuis (very difficult entrance), E of Carna, Salen Bay, Garbh Eilean (NW of Rubha an Daimh), and E of sand spit near head of loch by Strontian R.
NAVIGATION
Waypoints − from W 56°39'.70N 06°03'.00W, 263°/083° from/to Creag nan Sgarbh 3.7M: from S 56°38'00N, 06°00'65W 1M to S of Auliston Pt. Beware Ross Rk S of Rigsa; Broad Rk E of Rigsa; Dun Ghallain Rk; shoals extending 3 ca NNW from Eilean mo Shlinneag off S shore; rk which dries 1 ca W of S end of Garbh Eilean and strong streams at sp in Laudale Narrows.
LIGHTS AND MARKS
There are no Lts. Two transits, 339° and 294° shown on chart.
TELEPHONE (096 785)
MRSC and CG Oban 63720; HMC Oban 2948; Marinecall 0898 500 463; Hosp Fort William 2481; Dr 231.
FACILITIES
SALEN **Jack Walters** Tel. 648, M, V, Gas; **The Jetty Shop** Tel. 648, CH, V, Gas, M, (ME, El, Sh, P, D by arrangement); **Conyer Marine** Tel. 654, M, ME, Sh, El, CH; **Salen Hotel** Tel. 661 R, Bar; **Village** V, FW (at pier), CH, PO, Bank (mobile Tues and Wed), Rly (bus to Loch Ailort), Air (Oban).
ACHARACLE Few facilities.
STRONTIAN V, hotel, PO, Gas, Bar, FW.

TOBERMORY 10-8-12
Mull

CHARTS
Admiralty 2474, 2171; Imray C65; OS 47
TIDES
−0519 Dover; ML 2.5; Duration 0610; Zone 0 (GMT).
Standard Port OBAN (→)

Times				Height (metres)			
HW		LW		MHWS	MHWN	MLWN	MLWS
0100	0700	0100	0800	4.0	2.9	1.8	0.7
1300	1900	1300	2000				

Differences TOBERMORY
+0025 +0010 +0015 +0025 +0.4 +0.4 0.0 0.0
IONA (Mull)
−0010 −0005 −0020 +0015 0.0 +0.1 −0.3 −0.2
LOCH LATHAICH (Bunessan)
−0015 −0015 −0010 −0015 +0.3 +0.1 0.0 −0.1
ULVA SOUND (Mull)
−0010 −0015 0000 −0005 +0.4 +0.3 0.0 −0.1

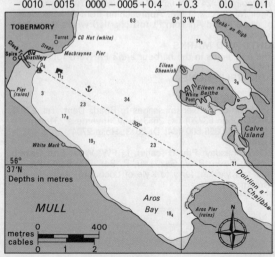

AREA 8—NW Scotland 433

SHELTER
Good except for some swell when strong winds from NW or N. Anchor off Old Distillery pierhead (where there are some Bu spherical mooring buoys marked 'HIDB visitors only'), off Aros pier in SW corner of bay, close to NW side of Calve Is or in bay at NW end of Calve Is.

NAVIGATION
Waypoint 56°38'.00N 06°02'.00W, 058°/238° from/to MacBraynes Pier, 1.1M. South entrance is only 80m wide at HW, with a dangerous wreck, and dries at LW. North entrance is recommended, the only danger being Sgeit Calve on NE of Calve Island.

LIGHTS AND MARKS
Coming up Doirlinn passage Tobermory Free Church spire in line with West entrance Pt to Doirlinn, about 300° leads through the channel.

RADIO TELEPHONE
VHF Ch 16; 12 (office hours. Listens only).

TELEPHONE (0688)
Hr Mr 2017; CG 2200; MRSC Oban 63720; HMC 231608; Marinecall 0898 500 463; Dr 2013.

FACILITIES
EC Wednesday; **Pier** P, FW, AB, V: **Seafare** (listens on Ch 16 office hours) Tel. 2277, M, P, D, Sh, CH, V, ACA; **Archebald Brown** Tel. 2020, FW, AB; **Western Isles YC** Tel. 2097, **Iain MacDonald** Tel. 2023. Gas. **Town** V, R, Bar, PO; Bank; Rly (ferry to Oban); Air (Mull).

LOCH ALINE 10-8-13
Argyll

CHARTS
Admiralty 2390; Imray C65; OS 49

TIDES
−0523 Dover; Duration 0610; Zone 0 (GMT).
Standard Port OBAN (→)

Times				Height (metres)			
HW		LW		MHWS	MHWN	MLWN	MLWS
0100	0700	0100	0800	4.0	2.9	1.8	0.7
1300	1900	1300	2000				

Differences LOCH ALINE
No data +0012 +0012 No data No data +0.5 +0.3 No data

CRAIGNURE
+0030 +0005 +0010 +0015 0.0 +0.1 −0.1 −0.1

SHELTER
Very good shelter but only good anchorage is in S end of loch to E of entrance. There are numerous hazards in N part of loch. Yachts can use the pier outside the loch, 3 ca W of entrance temporarily, or go alongside the old stone pier in the entrance on W side.

NAVIGATION
Waypoint 56°31'.50N, 05°46'30W 356°/176° from/to front Ldg Bn, 0.9M. The entrance is easy and is buoyed. The Ldg Bns 356° lead 0.5 ca W of Bogha Lurcain, a drying rock off Bolorkle Pt on E side of ent. Beyond Bolorkle Pt the ent is narrow with a bar (min depth 2.1m, 1985). Here stream runs 2½ kn at sp. Beware coasters from the sand mine going to/from the jetty and ferries to and from Mull.

LIGHTS AND MARKS
Ardtormish Pt Lt Ho, 1M SE of ent, Fl (2) WRG 10s 7m 8/5M; vis G shore-302°, W302°-310°, R310°-342°, W342°-057°, R057°-095°, W095°-108°, G108°-shore. Three buoys mark entrance channel QR, QG and Fl R 2s. End of ferry slip 2FR (vert). A War Memorial Cross (conspic 9m high) stands on W side of entrance. The Ldg Lts are FW 2m/4m (H24). Half mile up the loch on E side is a Y iron Bn with spherical topmark marking a reef, and ½M further up is a similar Bn marking a larger reef.

RADIO TELEPHONE
None.

TELEPHONE (096 784)
HMC Oban 62892; Marinecall 0898 500 463; MRSC Oban 63720; Hosp Oban 63636; Dr 252.

FACILITIES
K. Masters Tel. 2301, Gas; **J. Hodgson** Tel. 204, M, P, D, FW, Gas, R, Lau; **Lochaline Stores** Tel. 220, V; **Lochaline Hotel** Tel. 657, R, Bar; **Village** Bar, Shop, V, P, PO, Bank (Oban or Fort William), Rly (Oban); Air (Oban).

FORT WILLIAM/CORPACH 10-8-14
Inverness

CHARTS
Admiralty 2372, 2380; Imray C65; OS 41

TIDES
−0505 Dover; ML 2.5; Duration 0610; Zone 0 (GMT).
Standard Port OBAN (→)

Times				Height (metres)			
HW		LW		MHWS	MHWN	MLWN	MLWS
0100	0700	0100	0800	4.0	2.9	1.8	0.7
1300	1900	1300	2000				

Differences CORPACH
+0022 +0022 +0020 +0020 +0.1 +0.2 +0.1 +0.2
CORRAN NARROWS
+0007 +0007 +0004 +0004 +0.4 +0.4 −0.1 0.0
LOCH LEVEN HEAD
+0045 +0045 +0045 +0045 No data No data
LOCH LINNHE (Port Appin)
−0005 −0005 −0030 0000 +0.2 +0.2 +0.1 +0.1
BARCALDINE PIER (Loch Creran)
+0010 +0020 +0040 +0015 +0.1 +0.1 +0.1 +0.1

SHELTER
Exposed to NE through N to SW winds. Anchorages in Bishop's Bay (Loch Leven), Ardgour Bay (beware ferry moorings and fish farm), Camus na Gal Bay (good shelter), Corpach Basin or SW of Eilean A Bhealaidh.

NAVIGATION
Corpach waypoint 56°50'.30N 05°07'.00W, 135°/315° from/to lock entrance, 0.30M. There are no dangers round Fort William but when coming up Loch Linnhe, the Corran Narrows may present a danger. On the port side is Salachan shoal running from Salachan Pt to the narrows and on the starboard is Culchenna spit. These are well marked and buoyed. Spring tide makes up to 6 kn through the narrows.

LIGHTS AND MARKS
There is a Fl G 2s Lt at Fort William pierhead.
NOTE: Corpach is the entrance to the Caledonian Canal (see 10.8.15). There is an Iso WRG 4s Lt at N jetty of lock entrance.

RADIO TELEPHONE
Call: *Corpach Lock* VHF Ch 74.

TELEPHONE (0397)
Hr Mr 7249; CG 7762; MRSC Oban 63720; HMC 2948; Marinecall 0898 500 463; Dr 3136, 2947.

FACILITIES
FORT WILLIAM. EC Wednesday; **Pier** Tel. 3881, L, AB; **Slip** Tel. 3701, L; **Town** P, ME, El, Sh, YC, V, R, Bar; Hosp; PO; Bank; Rly.
CORPACH. **Corpach Basin** Tel. 249, AB, L, FW; **Lochaber Marine** Tel. 861, ME, El, Sh, D (cans), M, CH, C, (18 ton), Slip, Divers; **Lochaber YC** Tel. 473, M; **Village** P, V, R, Bar, PO, Bank, Rly, Air (Oban).

CALEDONIAN CANAL 10.8.15
Inverness

CHARTS
Admiralty 1791. OS 41 34 26

TIDES
Tidal differences: Clachnaharry +0116 on Dover, −0104 on Aberdeen; Corpach −0455 on Dover, +0035 on Oban.

SHELTER
Corpach is the SW end of the Caledonian Canal. It is best to pass through the sea lock and lie between that and the first canal lock. The canal consists of three lochs, Lochs Lochy, Oich and Ness connected by canals with 29 locks totalling in all 60M. There are 15 locks between Loch Linnhe (Corpach), through Lochs Lochy and Oich to the summit (100' above sea level) and 15 locks from the summit through Loch Ness to Inverness. 22 miles are through canals and 38 through lochs. It normally takes two full days to pass through and in the summer it may take longer.

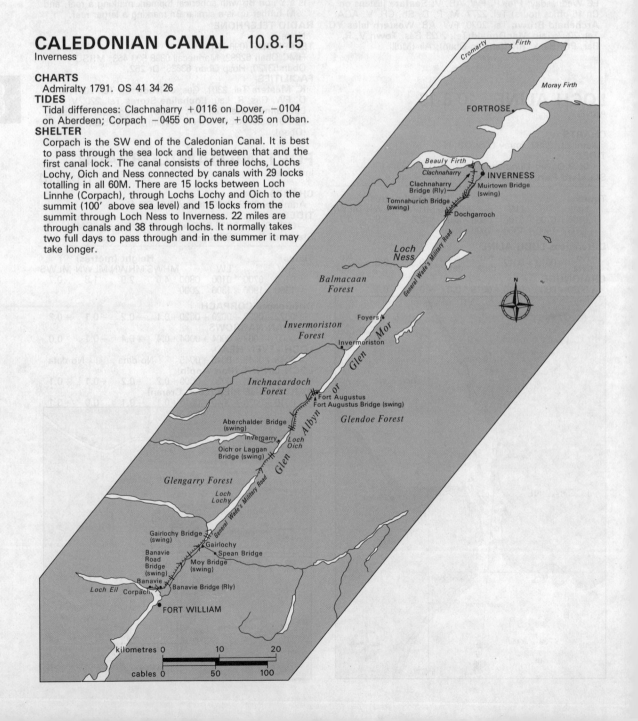

AREA 8—NW Scotland 435

NAVIGATION
There is a speed limit of 5.2 kn in the canal sections. There are eight swing bridges. All locks are manned and operate from 0800-1200 and 1300-1645 LT, May-Oct 7 days a week; winter, Mon-Sat. Sea locks will open outside these hours for extra payment. Dues are payable at the Corpach end. Canal is available for vessels up to 48m LOA, 11m beam and 4m draught, max mast height 36.5m. Entrance is available HW−4 to HW+4 (springs); H24 (neaps). Ring Lockkeeper on Corpach 249. For rules and regulations etc apply to the Engineer and Manager, Caledonian Canal Office, Clachnaharry, Tel. Inverness 233140.

Loch Oich is part of an Hydro-Electric Scheme and for this reason the water level can vary. Do not pass bridges at either end without keeper's instructions.

LIGHTS & MARKS
Channel is marked by posts, cairns and buoys, R on the NW side of the channel and G on the SE side — none are lit. There is an Iso WRG 4s light at canal entrance at Corpach, and an Iso 4s at the sea lock in Inverness.

RADIO TELEPHONE
Call: *Corpach Lock* VHF Ch 74. Call: *Laggan Lock* Ch 74. Call: *Fort Augustus Lock* Ch 74. Call: *Dochgarroch Lock* Ch 74. For Clachnaharry office, call: *Caledonian Canal* Ch 74. Call: *Clachnaharry Sea Lock* Ch 74.

TELEPHONES
Head Office Inverness 233140; Corpach Sea Lock Corpach 249; Clachnaharry Sea Lock Inverness 235439.

FACILITIES
- Corpach see 10.8.14.
- Gairlochy AB.
- NE end of Loch Lochy V, M, AB, R.
- Oich Bridge D, AB, R.
- Invergarry V, L, FW, AB.
- Fort Augustus FW, P, L, ME, El, AB, P, V, PO, Dr, Bar
- Dochgarroch FW, P, V.
- Caley Marina (25+25 visitors) Tel. Inverness 236539 (Muirtown top lock) FW, C, ACA, CH, D, ME, El, Sh, AC.
- Muirtown Marina (20 visitors) (Muirtown basin) Tel. Inverness 239745, D, El, ME, Sh, FW, Gas, Gaz, Lau, C (4 ton), CH, AC.

OBAN 10-8-16
Argyll

CHARTS
Admiralty 1790; Imray C65; OS 49

TIDES
−0530 Dover; ML 2.4; Duration 0610; Zone 0 (GMT).
Standard Port OBAN (→)

Times				Height (metres)			
HW		LW		MHWS	MHWN	MLWN	MLWS
0100	0700	0100	0800	4.0	2.9	1.8	0.7
1300	1900	1300	2000				

Differences DUNSTAFFNAGE BAY
+0005 0000 0000 +0005 +0.1 +0.1 +0.1 +0.1
CONNEL
+0020 +0005 +0010 +0015 −0.3 −0.2 −0.1 +0.1
BONAWE
+0150 +0205 +0240 +0210 −2.0 −1.7 −1.3 −0.5

Oban is a Standard Port and tidal predictions for each day of the year are given below.

SHELTER
Good except in SW-NW strong winds (then Ardantrive Bay is recommended). Anchorages off town, but water is deep. There are no visitors moorings and no alongside berths available.

NAVIGATION
Waypoint 56°25'.80N 05°30'.00W, 306°/126° from/to Dunollie Lt, 0.71M. Beware Sgeir Rathaid in middle of the bay marked by buoys, also MacBraynes ferries run services from Railway Quay.

LIGHTS AND MARKS
North Spit of Kerrera Fl R 3s 9m 5M; W column, R bands. Dunollie Fl(2) WRG 6s 7m 5/4M; vis G351°−009°, W009°−047°, R047°−120°, W120°−138°, G138°−143°. North Pier 2FG(vert). South Quay 2FG(vert). Northern Lights Wharf OcG 6s. FR Lts mark works in progress on Railway Quay (T).

RADIO TELEPHONE (local times)
CG-VHF Ch 16 (H24).

TELEPHONE (0631)
Hr Mr 62892; CG 63720; MRSC 63720; HMC 63079; Marinecall 0898 500 463; Dr 63175.

FACILITIES
EC Thursday; **N. Pier** Tel. 62892, L, FW, C (15 ton mobile); **Cal Mac Pier** Tel. 62285, Slip, D, L, FW, CH; **Curries BY** Tel. 62102, Slip, CH; **Oban Yacht Services BY** Tel. 63666, Slip, M, D, L, FW, ME, Sh, C, CH, AB; **Nancy Black** Tel. 62550, CH, ACA; **Ardantrive Bay** M, P, D, FW (on pier); **Railway Pier** D.
West Highland Gas Services Tel. 64050 Gas; **Ardoran Marine** Tel. 66123 D, Sh, ME, Slip, C, M; **Oban Divers** Tel. 62755 divers; **Vale Engineering** Tel. 64531 ME, El; **Town** P, V, R, Bar. PO; Bank; Rly; Air.
Note: Royal Highland YC, Andentallan, by Oban, Tel. 63309.

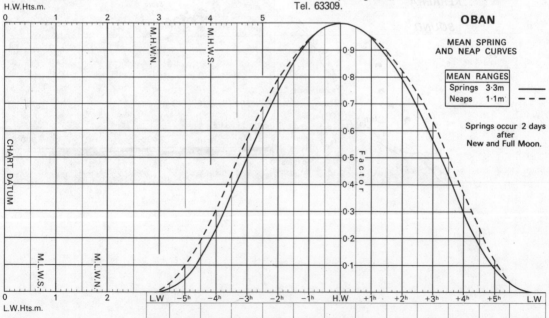

OBAN
MEAN SPRING AND NEAP CURVES

MEAN RANGES	
Springs	3·3m ———
Neaps	1·1m - - -

Springs occur 2 days after New and Full Moon.

OBAN continued

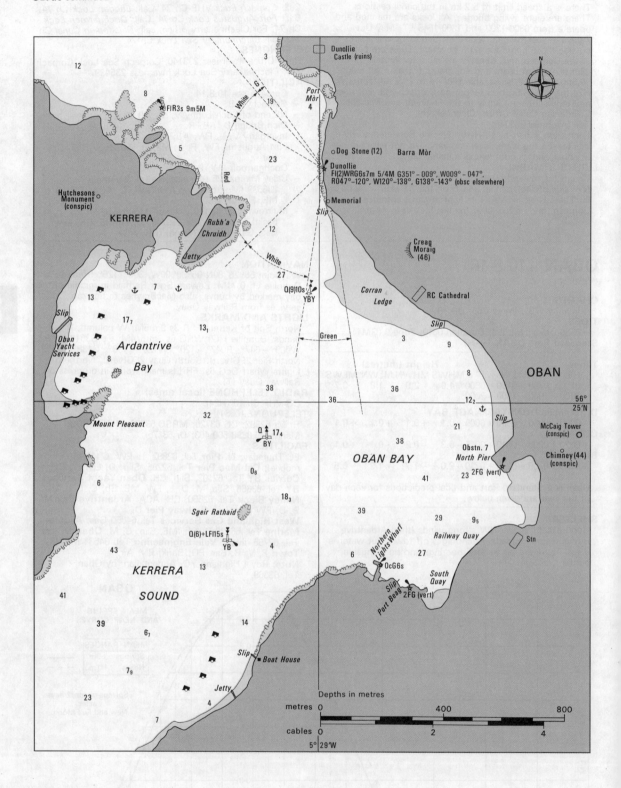

SCOTLAND, WEST COAST - OBAN
LAT 56°25'N LONG 5°29'W

TIMES AND HEIGHTS OF HIGH AND LOW WATERS

YEAR **1989**

TIME ZONE UT (GMT)
For Summer Time add ONE hour in non-shaded areas

JANUARY

#	TIME	m	#	TIME	m
1 SU	0545 1201 1846	1.9 3.1 2.1	16 M	0620 1310 1858	1.6 3.2 1.7
2 M	0101 0650 1314 1954	3.0 2.0 3.1 2.1	17 TU	0114 0747 1441 2016	2.9 1.7 3.2 1.7
3 TU	0212 0757 1419 2057	3.1 2.0 3.2 1.9	18 W	0317 0917 1554 2131	3.1 1.6 3.2 1.6
4 W	0311 0857 1517 2149	3.2 1.9 3.3 1.8	19 TH	0424 1027 1650 2229	3.3 1.5 3.4 1.4
5 TH	0400 0949 1610 2233	3.4 1.7 3.4 1.5	20 F	0509 1119 1732 2315	3.5 1.3 3.5 1.1
6 F	0443 1036 1658 2314	3.6 1.5 3.6 1.3	21 SA	0543 1201 1807 2354	3.7 1.2 3.6 1.0
7 SA	0523 1122 1742 ● 2353	3.8 1.3 3.7 1.0	22 SU	0611 1238 1837	3.8 1.2 3.7
8 SU	0601 1208 1823	4.0 1.1 3.8	23 M	0030 0637 1312 1905	0.8 3.9 1.1 3.8
9 M	0033 0640 1254 1902	0.8 4.1 0.9 3.8	24 TU	0104 0704 1344 1933	0.8 3.9 1.2 3.8
10 TU	0113 0720 1340 1940	0.7 4.2 0.9 3.8	25 W	0137 0732 1415 2001	0.8 3.9 1.3 3.7
11 W	0154 0801 1426 2017	0.7 4.1 0.9 3.7	26 TH	0210 0801 1445 2030	1.0 3.8 1.4 3.6
12 TH	0237 0845 1513 2056	0.7 4.0 1.0 3.6	27 F	0244 0832 1516 2100	1.1 3.7 1.6 3.4
13 F	0323 0933 1601 2138	0.9 3.8 1.2 3.4	28 SA	0318 0904 1549 2132	1.4 3.5 1.7 3.3
14 SA ☽	0412 1029 1653 2227	1.1 3.6 1.4 3.2	29 SU	0356 0940 1626 2211	1.6 3.3 1.9 3.1
15 SU	0510 1140 1750 2332	1.3 3.3 1.6 3.0	30 M ☾	0440 1024 1716 2314	1.8 3.1 2.0 2.9
			31 TU	0540 1132 1829	2.0 3.0 2.1

FEBRUARY

#	TIME	m	#	TIME	m
1 W	0113 0657 1326 2006	2.9 2.0 2.9 2.0	16 TH	0339 0942 1601 2131	3.0 1.7 3.0 1.6
2 TH	0247 0821 1508 2126	3.0 1.9 3.0 1.8	17 F	0432 1039 1647 2225	3.2 1.5 3.2 1.4
3 F	0347 0932 1613 2219	3.2 1.7 3.2 1.5	18 SA	0507 1119 1722 2304	3.5 1.3 3.4 1.1
4 SA	0433 1028 1700 2302	3.5 1.4 3.5 1.2	19 SU	0533 1152 1751 2338	3.7 1.2 3.6 0.9
5 SU	0513 1116 1740 2341	3.8 1.1 3.7 0.8	20 M ○	0554 1221 1817	3.8 1.1 3.8
6 M	0551 1201 1817	4.1 0.8 3.9	21 TU	0010 0615 1249 1842	0.7 3.9 1.0 3.8
7 TU	0019 0629 1243 1851	0.4 4.3 0.6 4.0	22 W	0041 0639 1316 1907	0.7 4.0 1.0 3.9
8 W	0058 0707 1325 1925	0.4 4.4 0.5 4.0	23 TH	0111 0705 1342 1932	0.7 4.0 1.1 3.8
9 TH	0138 0745 1407 1958	0.3 4.4 0.6 4.0	24 F	0141 0732 1408 1957	0.8 3.9 1.2 3.7
10 F	0218 0824 1448 2031	0.4 4.2 0.7 3.8	25 SA	0211 0759 1434 2022	1.0 3.8 1.4 3.6
11 SA	0300 0905 1531 2107	0.6 3.9 1.0 3.5	26 SU	0241 0826 1501 2048	1.2 3.6 1.5 3.4
12 SU	0347 0952 1617 ☾ 2147	1.0 3.5 1.3 3.3	27 M	0314 0855 1533 2118	1.5 3.4 1.7 3.2
13 M	0441 1052 1711 2242	1.4 3.1 1.6 3.0	28 TU ☽	0353 0929 1615 2200	1.7 3.1 1.9 3.0
14 TU	0553 1249 1821	1.7 2.9 1.8			
15 W	0111 0746 1453 1957	2.8 1.8 2.9 1.8			

MARCH

#	TIME	m	#	TIME	m
1 W	0448 1017 1720 2344	1.9 2.9 2.0 2.8	16 TH	0212 0808 1442 1938	2.8 1.9 2.7 1.8
2 TH	0611 1218 1912	2.0 2.7 2.0	17 F	0328 0941 1542 2114	3.0 1.7 2.9 1.6
3 F	0228 0755 1508 2102	2.9 1.9 2.8 1.9	18 SA	0412 1025 1623 2203	3.2 1.5 3.2 1.3
4 SA	0329 0921 1604 2159	3.2 1.6 3.1 1.4	19 SU	0441 1058 1654 2239	3.4 1.3 3.4 1.1
5 SU	0414 1016 1646 2242	3.5 1.3 3.4 1.0	20 M	0502 1125 1721 2311	3.6 1.1 3.6 0.9
6 M	0453 1101 1722 2320	3.9 0.9 3.7 0.6	21 TU	0522 1151 1746 2342	3.7 1.0 3.7 0.8
7 TU ●	0531 1142 1756 2358	4.2 0.6 4.0 0.3	22 W	0544 1217 1812	3.9 1.0 3.9
8 W	0608 1223 1828	4.4 0.3 4.1	23 TH	0012 0609 1243 1837	0.7 3.9 1.0 3.9
9 TH	0037 0645 1302 1900	0.1 4.5 0.3 4.1	24 F	0041 0636 1309 1902	0.8 4.0 1.0 3.9
10 F	0116 0723 1342 1932	0.1 4.4 0.4 4.1	25 SA	0110 0702 1334 1927	0.9 3.9 1.1 3.8
11 SA	0156 0800 1421 2005	0.3 4.1 0.6 3.9	26 SU	0139 0728 1359 1951	1.0 3.8 1.2 3.7
12 SU	0238 0839 1502 2039	0.6 3.8 0.9 3.6	27 M	0209 0755 1427 2017	1.2 3.6 1.4 3.5
13 M	0325 0921 1547 2118	1.0 3.3 1.3 3.3	28 TU	0243 0824 1500 2048	1.4 3.3 1.5 3.2
14 TU ☾	0421 1018 1640 2211	1.4 2.9 1.6 2.9	29 W	0325 0858 1543 2131	1.6 3.1 1.7 3.0
15 W	0541 1252 1753	1.8 2.6 1.8	30 TH ☽	0423 0946 1648 2307	1.8 2.8 1.9 2.8
			31 F	0548 1150 1832	1.9 2.6 1.9

APRIL

#	TIME	m	#	TIME	m
1 SA	0158 0733 1442 2023	2.9 1.8 2.8 1.7	16 SU	0329 0947 1540 2122	3.2 1.5 3.1 1.4
2 SU	0258 0855 1535 2126	3.2 1.5 3.1 1.3	17 M	0355 1019 1613 2202	3.3 1.4 3.3 1.2
3 M	0343 0950 1615 2211	3.6 1.1 3.4 0.9	18 TU	0417 1048 1643 2237	3.5 1.3 3.5 1.1
4 TU	0424 1034 1652 2252	3.9 0.8 3.7 0.5	19 W	0442 1115 1712 2309	3.6 1.1 3.7 1.0
5 W	0503 1116 1726 2332	4.2 0.5 3.9 0.3	20 TH	0509 1143 1740 2341	3.8 1.1 3.8 0.9
6 TH	0543 1156 1800 ●	4.4 0.3 4.1	21 F ○	0538 1210 1807	3.8 1.0 3.8
7 F	0012 0621 1236 1833	0.2 4.4 0.3 4.2	22 SA	0011 0607 1237 1834	0.9 3.8 1.0 3.8
8 SA	0053 0700 1316 1907	0.2 4.2 0.4 4.1	23 SU	0041 0635 1304 1901	1.0 3.8 1.1 3.8
9 SU	0136 0739 1356 1942	0.4 4.0 0.6 3.9	24 M	0113 0704 1333 1928	1.1 3.7 1.1 3.7
10 M	0221 0818 1438 2019	0.7 3.6 0.9 3.6	25 TU	0146 0734 1405 1959	1.2 3.5 1.3 3.5
11 TU	0311 0902 1524 2100	1.1 3.2 1.2 3.2	26 W	0225 0807 1443 2036	1.4 3.3 1.4 3.3
12 W ☽	0412 1002 1618 2201	1.5 2.8 1.5 3.1	27 TH	0314 0848 1531 2128	1.6 3.0 1.6 3.1
13 TH	0537 1220 1728 ☾	1.8 2.6 1.7	28 F	0416 0946 1635 2307	1.7 2.8 1.7 2.9
14 F	0140 0742 1401 1859	2.8 1.8 1.7 1.7	29 SA	0536 1142 1802	1.7 2.7 1.7
15 SA	0250 0904 1458 2027	3.0 1.7 2.9 1.6	30 SU	0114 0702 1350 1933	3.0 1.6 2.8 1.5

Chart Datum: 2.10 metres below Ordnance Datum (Newlyn)

SCOTLAND, WEST COAST - OBAN

LAT 56°25'N LONG 5°29'W

TIMES AND HEIGHTS OF HIGH AND LOW WATERS

YEAR 1989

TIME ZONE UT(GMT)
For Summer Time add ONE hour in non-shaded areas

MAY			JUNE			JULY			AUGUST		
TIME m	TIME m		TIME m	TIME m		TIME m	TIME m		TIME m	TIME m	

MAY

1 0217 3.3 / 0816 1.4 / M 1449 3.0 / 2042 1.3
16 0252 3.2 / 0927 1.6 / TU 1524 3.2 / 2116 1.5

2 0306 3.6 / 0913 1.1 / TU 1535 3.3 / 2135 0.9
17 0326 3.4 / 1003 1.5 / W 1602 3.4 / 2157 1.4

3 0351 3.8 / 1001 0.8 / W 1615 3.6 / 2221 0.7
18 0401 3.5 / 1037 1.3 / TH 1637 3.5 / 2234 1.3

4 0435 4.0 / 1046 0.6 / TH 1655 3.8 / 2306 0.5
19 0435 3.6 / 1109 1.3 / F 1710 3.6 / 2309 1.2

5 0518 4.1 / 1129 0.5 / F 1733 4.0 / ● 2350 0.4
20 0510 3.7 / 1140 1.2 / SA 1741 3.7 / ○ 2343 1.1

6 0601 4.1 / 1211 0.5 / SA 1811 4.0
21 0544 3.7 / 1211 1.1 / SU 1812 3.8

7 0036 0.5 / 0643 4.0 / SU 1254 0.6 / 1849 4.0
22 0018 1.1 / 0617 3.7 / M 1243 1.1 / 1843 3.8

8 0122 0.7 / 0725 3.7 / M 1336 0.7 / 1927 3.8
23 0054 1.2 / 0651 3.6 / TU 1316 1.1 / 1916 3.7

9 0211 1.0 / 0807 3.4 / TU 1420 0.9 / 2007 3.6
24 0134 1.2 / 0727 3.4 / W 1354 1.1 / 1952 3.6

10 0304 1.3 / 0853 3.1 / W 1506 1.1 / 2053 3.3
25 0219 1.3 / 0806 3.3 / TH 1436 1.2 / 2036 3.4

11 0405 1.6 / 0952 2.9 / TH 1558 1.4 / 2154 3.0
26 0312 1.4 / 0852 3.1 / F 1525 1.3 / 2132 3.3

12 0516 1.8 / 1116 2.7 / F 1659 1.6 / ☽ 2353 2.9
27 0411 1.5 / 0950 2.9 / SA 1624 1.4 / 2251 3.2

13 0635 1.8 / 1245 2.7 / SA 1810 1.6
28 0517 1.5 / 1109 2.8 / SU 1732 1.5 / ☾

14 0130 3.0 / 0749 1.8 / SU 1351 2.8 / 1924 1.6
29 0022 3.2 / 0625 1.5 / M 1239 2.9 / 1847 1.4

15 0217 3.1 / 0845 1.7 / M 1441 3.0 / 2026 1.6
30 0131 3.4 / 0732 1.4 / TU 1352 3.0 / 1958 1.3

31 0228 3.5 / 0833 1.2 / W 1450 3.2 / 2100 1.1

JUNE

1 0321 3.6 / 0928 1.1 / TH 1542 3.4 / 2155 0.9
16 0323 3.3 / 1001 1.6 / F 1606 3.3 / 2201 1.5

2 0413 3.7 / 1019 0.9 / F 1631 3.6 / 2247 0.8
17 0408 3.4 / 1040 1.4 / SA 1646 3.5 / 2243 1.4

3 0503 3.8 / 1108 0.8 / SA 1716 3.8 / ● 2338 0.7
18 0452 3.5 / 1117 1.3 / SU 1723 3.6 / 2323 1.3

4 0552 3.8 / 1154 0.7 / SU 1759 3.8 / ○
19 0533 3.5 / 1153 1.2 / M 1758 3.7

5 0028 0.8 / 0638 3.7 / M 1239 0.7 / 1841 3.8
20 0003 1.2 / 0612 3.5 / TU 1229 1.0 / 1834 3.8

6 0117 0.9 / 0720 3.6 / TU 1322 0.7 / 1921 3.8
21 0045 1.1 / 0650 3.6 / W 1306 1.0 / 1911 3.8

7 0207 1.1 / 0801 3.4 / W 1405 0.9 / 2001 3.6
22 0129 1.1 / 0729 3.5 / TH 1346 0.9 / 1950 3.8

8 0256 1.3 / 0843 3.2 / TH 1449 1.0 / 2043 3.4
23 0216 1.1 / 0808 3.4 / F 1428 1.0 / 2034 3.7

9 0345 1.5 / 0927 3.1 / F 1534 1.2 / 2130 3.2
24 0305 1.1 / 0850 3.3 / SA 1514 1.0 / 2124 3.6

10 0436 1.6 / 1019 2.9 / SA 1624 1.4 / 2229 3.1
25 0356 1.2 / 0937 3.2 / SU 1605 1.1 / 2223 3.4

11 0530 1.8 / 1122 2.8 / SU 1719 1.6 / ☽ 2341 3.0
26 0450 1.3 / 1031 3.1 / M 1702 1.3 / ☾ 2333 3.3

12 0627 1.8 / 1230 2.9 / M 1821 1.7
27 0548 1.4 / 1138 3.0 / TU 1808 1.3

13 0049 3.0 / 0727 1.8 / TU 1335 3.0 / 1925 1.7
28 0047 3.3 / 0650 1.4 / W 1255 3.0 / 1920 1.4

14 0146 3.1 / 0825 1.8 / W 1432 3.1 / 2024 1.7
29 0158 3.3 / 0756 1.4 / TH 1415 3.1 / 2035 1.3

15 0236 3.2 / 0916 1.7 / TH 1522 3.2 / 2116 1.6
30 0306 3.3 / 0902 1.3 / F 1528 3.2 / 2144 1.2

JULY

1 0410 3.4 / 1003 1.2 / SA 1629 3.4 / 2245 1.1
16 0357 3.2 / 1019 1.5 / SU 1631 3.4 / 2226 1.5

2 0507 3.5 / 1056 1.0 / SU 1719 3.6 / 2340 1.0
17 0448 3.3 / 1101 1.3 / M 1711 3.6 / 2311 1.3

3 0556 3.5 / 1144 0.8 / M 1801 3.8 / ●
18 0531 3.5 / 1139 1.1 / TU 1748 3.8 / ○ 2354 1.1

4 0030 1.0 / 0637 3.6 / TU 1228 0.7 / 1839 3.8
19 0610 3.6 / 1216 0.9 / W 1825 4.0

5 0115 1.0 / 0714 3.6 / W 1308 0.7 / 1913 3.8
20 0037 0.9 / 0647 3.7 / TH 1254 0.7 / 1902 4.1

6 0156 1.1 / 0747 3.5 / TH 1347 0.7 / 1946 3.7
21 0120 0.8 / 0722 3.7 / F 1333 0.6 / 1940 4.1

7 0235 1.2 / 0820 3.4 / F 1425 0.9 / 2020 3.6
22 0203 0.8 / 0757 3.7 / SA 1413 0.6 / 2020 4.0

8 0312 1.3 / 0853 3.3 / SA 1504 1.1 / 2055 3.5
23 0247 0.8 / 0833 3.6 / SU 1456 0.7 / 2103 3.8

9 0350 1.5 / 0930 3.2 / SU 1545 1.3 / 2135 3.3
24 0332 1.0 / 0911 3.4 / M 1542 0.9 / 2151 3.6

10 0431 1.7 / 1014 3.1 / M 1630 1.5 / 2222 3.2
25 0420 1.2 / 0955 3.2 / TU 1634 1.2 / ☾ 2250 3.3

11 0517 1.8 / 1110 3.0 / TU 1721 1.7 / ☽ 2322 3.0
26 0512 1.4 / 1050 3.0 / W 1737 1.4

12 0612 1.9 / 1224 2.9 / W 1822 1.8
27 0009 3.1 / 0614 1.5 / TH 1210 2.9 / 1857 1.6

13 0034 3.0 / 0718 1.9 / TH 1342 2.9 / 1930 1.9
28 0150 3.0 / 0728 1.6 / F 1421 2.9 / 2034 1.6

14 0148 3.0 / 0828 1.9 / F 1450 3.0 / 2037 1.9
29 0319 3.0 / 0850 1.5 / SA 1553 3.1 / 2159 1.4

15 0257 3.1 / 0930 1.7 / SA 1545 3.2 / 2135 1.7
30 0425 3.2 / 0930 1.3 / SU 1647 3.4 / 2259 1.2

31 0514 3.4 / 1052 1.1 / M 1727 3.6 / 2345 1.1

AUGUST

1 0553 3.5 / 1135 0.8 / TU 1759 3.8 / ●
16 0521 3.5 / 1121 0.9 / W 1731 3.9 / 2340 0.9

2 0024 1.0 / 0625 3.6 / W 1213 0.7 / 1826 3.9
17 0556 3.7 / 1157 0.7 / TH 1807 4.2 / ○

3 0059 1.0 / 0654 3.7 / TH 1249 0.6 / 1853 3.9
18 0020 0.7 / 0629 3.9 / F 1234 0.4 / 1843 4.3

4 0132 1.0 / 0721 3.7 / F 1323 0.7 / 1920 3.9
19 0100 0.5 / 0701 4.0 / SA 1312 0.4 / 1920 4.4

5 0203 1.1 / 0749 3.7 / SA 1356 0.8 / 1948 3.8
20 0141 0.5 / 0734 4.0 / SU 1352 0.4 / 1958 4.2

6 0233 1.2 / 0817 3.6 / SU 1430 1.0 / 2018 3.7
21 0222 0.6 / 0808 3.8 / M 1433 0.6 / 2038 4.0

7 0304 1.4 / 0847 3.4 / M 1505 1.2 / 2049 3.5
22 0304 0.8 / 0843 3.6 / TU 1518 0.9 / 2121 3.6

8 0337 1.6 / 0919 3.3 / TU 1542 1.5 / 2123 3.3
23 0349 1.1 / 0922 3.4 / W 1611 1.2 / ☾ 2215 3.2

9 0414 1.7 / 0958 3.1 / W 1626 1.7 / ☽ 2204 3.1
24 0441 1.4 / 1013 3.1 / TH 1718 1.5 / 2347 2.9

10 0501 1.9 / 1058 2.9 / TH 1723 1.9 / 2304 2.9
25 0546 1.7 / 1148 2.8 / F 1857 1.7

11 0607 2.0 / 1254 2.8 / F 1839 2.0
26 0211 2.8 / 0712 1.7 / SA 1501 3.0 / 2105 1.7

12 0101 2.8 / 0740 2.0 / SA 1430 2.9 / 2005 2.0
27 0329 3.0 / 0850 1.6 / SU 1602 3.2 / 2214 1.5

13 0250 2.9 / 0906 1.8 / SU 1531 3.1 / 2119 1.8
28 0421 3.2 / 0956 1.3 / M 1643 3.5 / 2257 1.3

14 0356 3.1 / 1002 1.6 / M 1616 3.4 / 2214 1.5
29 0501 3.4 / 1040 1.1 / TU 1714 3.7 / 2332 1.1

15 0442 3.3 / 1044 1.3 / TU 1655 3.7 / 2259 1.2
30 0532 3.6 / 1116 0.8 / W 1737 3.8

31 0003 1.0 / 0559 3.7 / TH 1150 0.7 / ● 1759 3.9

Chart Datum: 2.10 metres below Ordnance Datum (Newlyn)

SCOTLAND, WEST COAST - OBAN

LAT 56°25'N LONG 5°29'W

TIMES AND HEIGHTS OF HIGH AND LOW WATERS

YEAR 1989

TIME ZONE UT(GMT)
For Summer Time add ONE hour in non-shaded areas

AREA 8—NW Scotland

Chart Datum: 2.10 metres below Ordnance Datum (Newlyn)

SEPTEMBER

Day	Time	m	Day	Time	m
1 F	0031 / 0625 / 1222 / 1823	0.9 / 3.8 / 0.7 / 4.0	16 SA	0602 / 1209 / 1818	4.1 / 0.3 / 4.5
2 SA	0059 / 0650 / 1254 / 1849	1.0 / 3.9 / 0.7 / 4.0	17 SU	0035 / 0634 / 1248 / 1856	0.4 / 4.2 / 0.2 / 4.5
3 SU	0126 / 0716 / 1325 / 1915	1.0 / 3.9 / 0.8 / 3.9	18 M	0114 / 0707 / 1328 / 1934	0.4 / 4.2 / 0.3 / 4.3
4 M	0154 / 0742 / 1355 / 1943	1.1 / 3.8 / 1.0 / 3.8	19 TU	0155 / 0741 / 1411 / 2013	0.6 / 4.0 / 0.6 / 3.9
5 TU	0222 / 0809 / 1427 / 2010	1.3 / 3.6 / 1.2 / 3.6	20 W	0237 / 0817 / 1459 / 2056	0.8 / 3.7 / 1.0 / 3.5
6 W	0250 / 0836 / 1500 / 2039	1.5 / 3.4 / 1.5 / 3.4	21 TH	0323 / 0857 / 1555 / 2151	1.2 / 3.4 / 1.4 / 3.1
7 TH	0322 / 0907 / 1540 / 2111	1.7 / 3.2 / 1.8 / 3.1	22 F	0416 / 0951 / 1712 / ☾	1.5 / 3.1 / 1.7
8 F ☽	0402 / 0949 / 1634 / 2154	1.9 / 3.0 / 2.0 / 2.9	23 SA	0001 / 0524 / 1326 / 1920	2.7 / 1.7 / 2.9 / 1.9
9 SA	0504 / 1150 / 1756 / 2351	2.0 / 2.8 / 2.1 / 2.7	24 SU	0206 / 0658 / 1455 / 2109	2.8 / 1.8 / 3.1 / 1.7
10 SU	0649 / 1412 / 1940	2.1 / 2.9 / 2.0	25 M	0312 / 0836 / 1545 / 2200	3.0 / 1.6 / 3.3 / 1.5
11 M	0244 / 0838 / 1509 / 2103	2.8 / 1.9 / 3.2 / 1.8	26 TU	0357 / 0935 / 1619 / 2235	3.2 / 1.4 / 3.5 / 1.3
12 TU	0340 / 0936 / 1552 / 2155	3.0 / 1.6 / 3.5 / 1.4	27 W	0432 / 1015 / 1643 / 2304	3.4 / 1.1 / 3.7 / 1.2
13 W	0421 / 1017 / 1629 / 2237	3.3 / 1.2 / 3.8 / 1.1	28 TH	0501 / 1049 / 1703 / 2331	3.6 / 1.0 / 3.8 / 1.1
14 TH	0456 / 1054 / 1705 / 2316	3.6 / 0.8 / 4.1 / 0.7	29 F ●	0527 / 1121 / 1725 / 2358	3.8 / 0.9 / 3.9 / 1.0
15 F ○	0529 / 1131 / 1741 / 2355	3.9 / 0.5 / 4.4 / 0.5	30 SA	0552 / 1152 / 1750	3.9 / 0.8 / 4.0

OCTOBER

Day	Time	m	Day	Time	m
1 SU	0025 / 0618 / 1223 / 1817	1.0 / 4.0 / 0.9 / 4.0	16 M	0008 / 0607 / 1225 / 1834	0.4 / 4.2 / 0.4 / 4.4
2 M	0052 / 0645 / 1253 / 1844	1.1 / 3.9 / 1.0 / 4.0	17 TU	0049 / 0643 / 1308 / 1914	0.5 / 4.2 / 0.5 / 4.2
3 TU	0119 / 0711 / 1321 / 1911	1.2 / 3.9 / 1.1 / 3.8	18 W	0131 / 0720 / 1355 / 1956	0.6 / 4.1 / 0.8 / 3.8
4 W	0146 / 0737 / 1353 / 1939	1.3 / 3.7 / 1.3 / 3.6	19 TH	0215 / 0759 / 1446 / 2041	0.9 / 3.8 / 1.2 / 3.4
5 TH	0214 / 0804 / 1427 / 2007	1.4 / 3.6 / 1.6 / 3.4	20 F	0302 / 0843 / 1548 / 2140	1.2 / 3.5 / 1.5 / 3.0
6 F	0246 / 0835 / 1508 / 2039	1.6 / 3.3 / 1.8 / 3.2	21 SA ☾	0356 / 0944 / 1711 / 2342	1.5 / 3.1 / 1.8 / 2.8
7 SA	0327 / 0917 / 1605 / 2122	1.8 / 3.1 / 2.0 / 2.9	22 SU	0503 / 1304 / 1903	1.7 / 3.0 / 1.9
8 SU ☽	0427 / 1049 / 1727 / 2301	2.0 / 2.9 / 2.1 / 2.7	23 M	0128 / 0627 / 1422 / 2032	2.8 / 1.8 / 3.2 / 1.8
9 M	0602 / 1338 / 1909	2.0 / 3.0 / 2.0	24 TU	0232 / 0753 / 1508 / 2123	3.0 / 1.7 / 3.3 / 1.7
10 TU	0210 / 0750 / 1435 / 2029	2.8 / 1.9 / 3.3 / 1.7	25 W	0317 / 0855 / 1539 / 2158	3.2 / 1.5 / 3.5 / 1.5
11 W	0305 / 0856 / 1518 / 2122	3.1 / 1.5 / 3.6 / 1.4	26 TH	0353 / 0935 / 1601 / 2228	3.4 / 1.3 / 3.6 / 1.4
12 TH	0346 / 0942 / 1556 / 2205	3.4 / 1.2 / 3.9 / 1.0	27 F	0423 / 1015 / 1624 / 2256	3.6 / 1.2 / 3.8 / 1.3
13 F	0422 / 1023 / 1635 / 2247	3.7 / 0.8 / 4.2 / 0.7	28 SA	0452 / 1049 / 1651 / 2325	3.7 / 1.1 / 3.9 / 1.2
14 SA	0457 / 1103 / 1714 / 2327 ●	3.9 / 0.6 / 4.4 / 0.5	29 SU	0521 / 1122 / 1720 / 2353 ●	3.9 / 1.1 / 3.9 / 1.2
15 SU	0532 / 1143 / 1754	4.1 / 0.4 / 4.5	30 M	0550 / 1154 / 1750	3.9 / 1.1 / 4.0
			31 TU	0022 / 0618 / 1225 / 1819	1.2 / 3.9 / 1.2 / 3.9

NOVEMBER

Day	Time	m	Day	Time	m
1 W	0050 / 0646 / 1256 / 1849	1.2 / 3.9 / 1.3 / 3.8	16 TH	0114 / 0709 / 1347 / 1949	0.7 / 4.0 / 1.0 / 3.7
2 TH	0119 / 0715 / 1330 / 1918	1.3 / 3.8 / 1.4 / 3.6	17 F	0159 / 0751 / 1442 / 2036	0.9 / 3.8 / 1.3 / 3.4
3 F	0150 / 0745 / 1407 / 1950	1.4 / 3.7 / 1.6 / 3.4	18 SA	0246 / 0838 / 1542 / 2130	1.1 / 3.6 / 1.6 / 3.1
4 SA	0226 / 0821 / 1452 / 2027	1.5 / 3.5 / 1.8 / 3.2	19 SU	0337 / 0936 / 1650 / 2242	1.3 / 3.3 / 1.8 / 2.9
5 SU	0310 / 0908 / 1551 / 2116	1.7 / 3.3 / 1.9 / 3.0	20 M ☾	0435 / 1117 / 1805	1.5 / 3.1 / 1.9
6 M	0407 / 1028 / 1705 / 2241	1.8 / 3.1 / 1.9 / 2.8	21 TU	0010 / 0541 / 1310 / 1919	2.9 / 1.7 / 3.1 / 1.9
7 TU	0524 / 1240 / 1827	1.9 / 3.2 / 1.9	22 W	0123 / 0653 / 1404 / 2021	3.0 / 1.7 / 3.2 / 1.9
8 W	0100 / 0652 / 1348 / 1941	2.9 / 1.8 / 3.4 / 1.7	23 TH	0220 / 0800 / 1441 / 2108	3.1 / 1.7 / 3.3 / 1.8
9 TH	0212 / 0806 / 1437 / 2040	3.1 / 1.5 / 3.6 / 1.4	24 F	0305 / 0854 / 1514 / 2147	3.3 / 1.6 / 3.5 / 1.7
10 F	0301 / 0902 / 1522 / 2130	3.3 / 1.3 / 3.9 / 1.1	25 SA	0345 / 0939 / 1547 / 2223	3.4 / 1.5 / 3.6 / 1.5
11 SA	0344 / 0950 / 1606 / 2217	3.6 / 1.0 / 4.1 / 0.9	26 SU	0421 / 1018 / 1622 / 2256	3.6 / 1.5 / 3.7 / 1.5
12 SU	0425 / 1036 / 1651 / 2302	3.9 / 0.7 / 4.3 / 0.7	27 M	0456 / 1055 / 1658 / 2329	3.7 / 1.4 / 3.8 / 1.4
13 M ○	0506 / 1122 / 1736 / 2346	4.1 / 0.6 / 4.3 / 0.6	28 TU	0529 / 1130 / 1732 / ●	3.8 / 1.4 / 3.8
14 TU	0547 / 1209 / 1820	4.2 / 0.6 / 4.2	29 W	0000 / 0600 / 1205 / 1806	1.3 / 3.9 / 1.4 / 3.8
15 W	0030 / 0628 / 1257 / 1904	0.6 / 4.2 / 0.7 / 4.0	30 TH	0032 / 0632 / 1240 / 1839	1.3 / 3.9 / 1.4 / 3.7

DECEMBER

Day	Time	m	Day	Time	m
1 F	0104 / 0704 / 1317 / 1912	1.3 / 3.9 / 1.4 / 3.6	16 SA	0147 / 0747 / 1434 / 2025	0.8 / 3.9 / 1.2 / 3.6
2 SA	0138 / 0738 / 1359 / 1948	1.3 / 3.8 / 1.5 / 3.5	17 SU	0230 / 0827 / 1522 / 2105	0.9 / 3.7 / 1.5 / 3.4
3 SU	0216 / 0817 / 1445 / 2027	1.3 / 3.6 / 1.6 / 3.3	18 M	0314 / 0911 / 1610 / 2149	1.1 / 3.5 / 1.7 / 3.2
4 M	0259 / 0903 / 1539 / 2114	1.4 / 3.5 / 1.7 / 3.2	19 TU ☾	0400 / 1001 / 1700 / 2242	1.4 / 3.3 / 1.8 / 3.1
5 TU	0350 / 1006 / 1639 / 2215	1.5 / 3.4 / 1.7 / 3.1	20 W	0452 / 1104 / 1756 / 2349	1.6 / 3.2 / 2.0 / 3.0
6 W	0451 / 1130 / 1744 / 2335 ☾	1.6 / 3.3 / 1.7	21 TH	0551 / 1218 / 1858	1.8 / 3.1 / 2.0
7 TH	0601 / 1250 / 1851	1.7 / 3.4 / 1.6	22 F	0104 / 0657 / 1326 / 2004	3.0 / 1.9 / 3.2 / 2.0
8 F	0100 / 0715 / 1355 / 1956	3.1 / 1.5 / 3.6 / 1.5	23 SA	0212 / 0800 / 1425 / 2104	3.1 / 1.9 / 3.2 / 1.9
9 SA	0211 / 0823 / 1452 / 2057	3.2 / 1.4 / 3.7 / 1.3	24 SU	0310 / 0902 / 1517 / 2154	3.2 / 1.8 / 3.3 / 1.8
10 SU	0311 / 0924 / 1547 / 2152	3.5 / 1.2 / 3.8 / 1.2	25 M	0358 / 0952 / 1605 / 2236	3.4 / 1.8 / 3.4 / 1.6
11 M	0405 / 1020 / 1641 / 2244	3.7 / 1.0 / 3.9 / 1.0	26 TU	0439 / 1036 / 1649 / 2313	3.6 / 1.7 / 3.5 / 1.5
12 TU	0455 / 1113 / 1732 / 2333	3.9 / 0.9 / 4.0 / 0.8	27 W	0516 / 1116 / 1729 / 2347	3.7 / 1.5 / 3.6 / 1.3
13 W	0541 / 1205 / 1820 / ○	4.0 / 0.9 / 3.9	28 TH	0551 / 1154 / 1805 / ●	3.8 / 1.4 / 3.7
14 TH	0019 / 0625 / 1256 / 1904	0.7 / 4.1 / 0.9 / 3.9	29 F	0020 / 0624 / 1232 / 1839	1.2 / 3.9 / 1.3 / 3.7
15 F	0103 / 0706 / 1346 / 1945	0.7 / 4.0 / 1.0 / 3.7	30 SA	0054 / 0658 / 1311 / 1911	1.1 / 4.0 / 1.2 / 3.7
			31 SU	0128 / 0732 / 1352 / 1946	1.0 / 4.0 / 1.2 / 3.7

LOCH MELFORT 10-8-17
Argyll

CHARTS
Admiralty 2326; Imray C65; OS 55
TIDES
Loch Shuna −0615 Dover; ML Seil Sound 1.4; Duration Seil Sound 0615; Zone 0 (GMT).
Standard Port OBAN (←—)

Times				Height (metres)			
HW		LW		MHWS	MHWN	MLWN	MLWS
0100	0700	0100	0800	4.0	2.9	1.8	0.7
1300	1900	1300	2000				
Differences SEIL SOUND							
−0035	−0015	−0040	−0015	−1.3	−0.9	−0.7	−0.3
SCALASAIG (Colonsay)							
−0020	−0005	−0015	+0005	−0.1	−0.2	−0.2	−0.2
GLENGARRISDALE BAY (Jura)							
−0020	0000	−0010	0000	−0.4	−0.2	0.0	−0.2

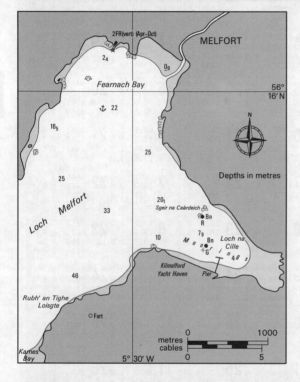

SHELTER
Loch Melfort provides several sheltered anchorages. There is a bay with moorings ½M inside the entrance on S shore, but beware rock which dries 1.5m. 1M further E is Kames Bay with some room to anchor clear of moorings, rocks and fish farm, and sheltered from S and W. Fearnach Bay at NE end of loch is a good anchorage in N winds and moorings are available. Best shelter of all is in Loch na Cille, where Kilmelford Yacht Haven provides swinging and pontoon moorings.
NAVIGATION
Waypoint 56°14'.00N 05°35'.00W, 210°/030° from/to summit Eilean Gamhna, 4ca. Pass either side of Eilean Gamhna. 8ca NE lies Campbell Rock (1.8m). A rock which dries 0.9m lies 1½ca ESE of the FS on Eilean Coltair. The S side of Loch Melfort is mostly steep-to, except in Kames Bay. Approaching Kilmelford Yacht Haven beware drying reef which extends ¾ca from NE shore and is marked by a red and black (port-hand) perch, and also a rock near S shore marked by green (stbd-hand) perch.
LIGHTS AND MARKS
2FR(vert) on the pier in Fearnach Bay.
RADIO TELEPHONE
Camus Marine VHF Ch 16 M (working hours).
TELEPHONE (085 22)
Hr Mr 248; CG and MRSC Oban 63720; HMC Oban 63079; Marinecall 0898 500 463; Hosp Lochgilphead 2323.

FACILITIES
Kilmelford Yacht Haven (Camus Marine) Tel. 248, BH (12 ton), Slip, ME, El, Sh, CH, Gas, Gaz, D, FW. **Melfort Marine** (Fearnach Bay) Tel. 257, M, AB, Gas, D, FW, R, Bar, P, D, AC, Sh, V; **Village** (¾M) V, Bar, PO.

CRAOBH HAVEN (Loch Shuna) 10-8-18
Argyll

CHARTS
Admiralty 2326; Imray C65; OS 55
TIDES
HW Loch Shuna −0615 Dover; ML Seil Sound 1.4; Duration Seil Sound 0615; Zone 0 (GMT). For tidal figures see 10.8.17 opposite.

SHELTER
Craobh (pronounced Creuve) Haven is a yacht harbour on E shore of Loch Shuna. It is enclosed by Eilean Buidhe on the NE side, and by Eilean an Duin and Fraoch Eilean to the NW and W. Each island has a causeway to the shore. The entrance on the N side is formed by breakwaters. Very good shelter within. There are anchorages in Asknish Bay 1M to the N, and in the bays S of Craobh Haven — E of Eilean Arsa and in Bagh an Tight-Stoir.
NAVIGATION
Waypoint 56°12'.84N 05°34'.40W, 270°/090° from/to entrance, 0.5M. Tidal streams in Loch Shuna are weak. Beware un-marked reefs Eich Donna NE of the island Eilean Gamhna. A stbd-hand buoy marks a rock (1m) which lies ¾ca NNW of the N breakwater head. Port-hand buoys mark a shoal area close S of the E breakwater. A pink perch on W corner of harbour marks a spit. Elsewhere there is ample depth in the marina.
LIGHTS AND MARKS
There are no lights or marks.
RADIO TELEPHONE
VHF Ch 16 M (summer 0800–1900 LT; winter 0900–1700 LT).
TELEPHONE (085 25)
Hr Mr 222; CG and MRSC Oban 63720; HMC Oban 63079; Marinecall 0898 500 463; Hosp Lochgilphead 2323.
FACILITIES
Craobh Haven Marina Tel. 222, AC, FW, D, P, SM, V, Bar; **Camus Marine** (200) Tel. 622 Slip, BH (15 ton), C (12 ton), Gas, Gaz, CH, ME, El, Sh. **Craobh Haven** Bank (Wednesday or at Lochgilphead), PO (Kilmelford), Rly (Oban by bus), Air (Glasgow).

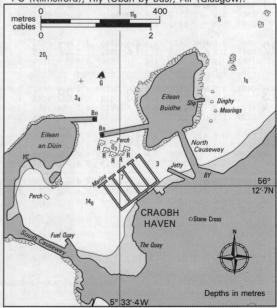

MINOR HARBOURS AND ANCHORAGES 10.8.19

LOCH LAXFORD, Sutherland, 58°25′ N, 5°08′ W, Zone 0 (GMT), Admty chart 2503. HW −0410 on Dover, +0010 on Ullapool. See 10.8.8. Ent between Rubha Ruadh and Ardmore Pt, 1M ENE, clearly identified by three isolated mountains Ben Stack, Ben Arkle and Foinaven inland. There are many anchorages in the loch including: Loch a'Chadh-fi on NE side (John Ridgway's Adventure School has moorings − check with school on Pt on W side of narrows); Bagh nah-Airde Beag, next bay to E, (beware rk ½ ca off SE shore which covers at MHWS); Fanagmore Bay on SW shore (beware head of bay foul with old moorings); Bagh na Fionndalach Mor on SW shore (4-6m); Weaver's Bay on SW shore, 3M from ent (beware drying rock off NW Pt of ent). Beware many fish farming cages. Facilities: none, nearest at Scourie (5M).

LOCH INVER, Sutherland, 58°09′ N, 5°15′ W, Zone 0 (GMT), Admty chart 2504. HW −0433 on Dover, −0005 on Ullapool; HW height −0.1m on Ullapool; ML 3.1m. See 10.8.8. Entrance between Soya I and Kirkaig Pt. Beware rk about 50m off Kirkaig Pt which dries. About 1M up the loch is the islet Glas Leac with a Lt Fl WRG 3s 7m; W071°-080°, R080°-090°, G090°-103°, W103°-111°, R111°-243°, W243°-251°, G251°-071°. Anchor at top of loch in 6m approx in middle of bay off the pier. Perfect shelter from all winds. Facilities: Shop, PO, Hotel, Gas, FW.

LOCH SHELL, Harris, 58°00′ N, 6°25′ W, Zone 0 (GMT), Admty chart 1794. HW −0437 on Dover, −0016 on Ullapool; HW height −0.4m on Ullapool; ML 2.7m. See 10.8.7. Pass S of Eilean Iuvard; beware rks to W of Is. Anchor in Tob Eishken, 2½M up loch on N shore (beware rk awash on E side of entrance), or at head of loch (exposed to E winds, and dries some distance). Facilities: PO/Stores at Lemrevay.

WEST LOCH TARBERT, Harris, 57°55′ N, 6°55′ W, Zone 0 (GMT), Admty chart 2841. HW −0515 on Dover, −0055 on Ullapool; HW height −0.8m on Ullapool; Duration 0550. See 10.8.7. Good shelter once beyond Taransay. Safest entrance to N of Taransay thence up the S side of loch. Beware rks and islets on N shore. Anchor at head of loch in approx 10m off Tarbert village. There are three small lochs off N shore which provide good alternative anchorages − Lochs Leosavay, Meavaig and Bunavoneadar. Facilities: FW only but Tarbert is only ¼M away.

EAST LOCH TARBERT, Harris, 57°54′ N, 6°48′ W, Zone 0 (GMT), Admty chart 2905. HW −0500 on Dover, −0039 on Ullapool; HW height −0.2m on Ullapool; ML 2.8m; Duration 0605. See 10.8.7. Approach through Sound of Scalpay; beware Elliot Rk (depth 2m) 2½ ca SSW of Rubha Crago. Eilean Glas Lt Ho at E end of Scalpay, Fl (3) 20s 43m 23M; W Tr, R bands. In Sound of Scalpay, stream sets W from HW+3, and E from HW−3. Anchor off Tarbert WSW of steamer pier in about 2.5m. Facilities: EC Thurs; Bar, D, Dr, FW at pier, P, PO, R, V, ferry to Uig (Skye).
Alternatively Scalpay North Harbour gives good shelter. Beware rk ½ ca off Aird an Aiseig, E side of entrance. Stbd hand buoy marks wreck off Coddem. ½ ca E of the buoy is a rk, depth 1.1m. Fish pier at S end of harbour has 2FG (vert) lights; anchor ¾ ca N, in about 3m. Facilities FW at pier, PO, V, ferry to Harris.

ST KILDA, 57°50′ N, 8°30′ W, Zone 0 (GMT), Admty chart 2524. Tidal figures taken at Village Bay; HW −0510 on Dover, −0055 on Ullapool; HW height −1.6m on Ullapool; ML 1.9m; Duration 0615. See 10.8.7. A group of four Is and three stacks now taken over by the Army. Anchor in Village Bay, facing SE, in approx 5m about 1.5 ca off the Army pier. If wind is between S and NE big swells come into the bay; poor holding and untenable if winds are strong. Reef lies parallel to and ½ ca off N shore. Ldg Lts 270°, both Oc 5s 26/38m 3M. Yachts with VHF call *Kilda Radio* VHF Ch 16;08 for permission to land. Alternative anchorage is Glen Bay on N side which is only satisfactory in S and E winds. Facilities: FW from wells near landings (by courtesy of the Army); Food, FW, Telephone.

LOCH EWE, Ross and Cromarty, 57°51′ N, 5°38′ W, Zone 0 (GMT), Admty charts 3146, 2509. Tidal figures taken at Mellon Charles; HW −0415 on Dover, −0010 on Ullapool; HW height −0.1m on Ullapool; ML 2.9m; Duration 0610.

See 10.8.8. Excellent shelter in all winds except N. Easy entrance with no dangers in loch except Boor Rks on W shore about 0.7M from loch head. Loch approx 7M long with Isle Ewe with two small Islets about 2M from entrance in centre of loch which can be passed on either side. Beware unlit mooring buoys and also dolphins near MoD (Navy) pier E side of entrance. Loch divides beyond Isle Ewe. MoD fuelling pier, dolphins and marker buoys opposite E end of Isle Ewe. Excellent anchorage in Loch Thurnaig to E. Aultbea Pier, partly derelict, to NE. Beware fish farming cages. Facilities: Dr., P, PO, R, V, Bar and on pier D, L. S of Poolewe Bay (3.5m); FW, Bar, D, L, P (at garage), PO, R, V, Gas.

LOCH GAIRLOCH, Ross and Cromarty, 57°43′ N, 5°45′ W, Zone 0 (GMT), Admty charts 2528, 2210. HW −0440 on Dover, −0020 on Ullapool; HW height +0.1m on Ullapool; ML 2.9m; Duration 0600. See 10.8.8. A wide loch facing W. Entrance clear of dangers. Quite heavy seas roll in bad weather. Yachts can take shelter in Badachro to W of Eilean Horrisdale on S side of loch or in Loch Shieldaig at SE end of the loch. At the head of the loch, Flowerdale Bay, the N branch of the loch, anchor in approx 6m near the head of the pier. Lights: Glas Eilean Fl WRG 6s 9m 6/4M W080°-102°, R102°-296°, W296°-333°, G333°-080°. Pierhead, QR 9m. Facilities: P, D, FW; stores 1M.

LOCH MADDY, North Uist, 57°36′ N, 7°07′ W, Zone 0 (GMT), Admty chart 2825 HW −0500 on Dover, −0039 on Ullapool. See 10.8.7. With strong wind against tide there can be bad seas off ent. Approaches clear but from S inside Madadh Mor beware submerged rk ½ ca N of Leac nam Madadh. Lights: Weaver's Pt Fl 3s 21m 7M; Glas Eilean Mor Fl(2) 6s; Rudna Nam Fl R 4s 7m 5M; inside loch Ruigh Liath QG, Lochmaddy pier Ldg Lts 301°. Front 2FG(vert) 4M. Rear Oc G 8s 10m 4M; vis 284°−304°. Anchorages: off Steamer Pier; in Charles Harbour; Oronsay − go alongside private pier but anchoring not recommended owing to moorings; Sponish Harbour; Loch Portain; Vallaquie. VHF Ch 16; 12. Port Manager Tel. 63337 (day), 63226 (night). Facilities: Lochmaddy, EC Wed, Shop, Gas, PO, P, D, FW; Loch Portain PO, Shop.

LOCH A'BHRAIGE, Rona, 57°34′ N, 5°58′ W, Zone 0 (GMT), Admty chart 2534. HW −0438 on Dover, −0023 on Ullapool; HW height +0.2m on Ullapool; ML 2.9m; Duration 0605. A good safe anchorage in NW of the island except in NNW winds. Beware rocks on the NE side lying up to a ca off shore. Lights Sgeir Shuas, NE of entrance, Fl R 2s 6m 3M; vis 070°−199°. Harbour in NE corner of loch, Lt on SW corner 2 FR (vert). Ldg Lts 137°. Front Q WRG 3m 4/3M, W135°−138°, R138°−318°, G318°−135°. Rear Iso 6s 28m 5M. Facilities very limited. There is a pier, FW and a helipad. Alternative shelter at Acarseid Mhor at SW of Island. (Sketch chart of rocks at entrance is necessary). Lock is used by MoD (Navy) − call on VHF Ch 16 to Rona Naval Establishment before entering.

CROWLIN ISLANDS, Ross and Cromarty, 57°21′ N, 5°51′ W. Zone 0 (GMT), Admty charts 2534, 2209. HW −0435 on Dover, −0020 on Ullapool; HW height +0.3 on Ullapool. Anchor between Eilean Meadhonach and Eilean Mor, approaching from N, keep E of Eilean Beg. Excellent shelter except in strong N winds. There is an inner anchorage with 3½m but entrance channel dries. Eilean Beg Lt Ho FlW 6s 32m 6M. There are no facilities.

LOCH HARPORT, Isle of Skye, 57°20′ N, 6°25′ W, Zone 0 (GMT), Admty chart 1795. HW −0447 (Sp), −0527 (Np) on Dover, −0035 (Sp) −0115 (Np) on Ullapool; HW height −0.1m on Ullapool. See 10.8.9. On E side of Loch Bracadale, entered between Oronsay I and Ardtreck Pt (W Lt Ho, Iso 4s 17m 9M). SW end of Oronsay has conspic rock pillar, called The Castle; keep ¼M off-shore here and off E coast of Oronsay which is joined to Ullinish Pt by drying reef. Anchorages: Oronsay Is, N side of drying reef (4m), or on E side, but beware rk (dries) 0.5ca off N shore of Oronsay. Fiskavaig Bay 1M S of Ardtreck (7m); Loch Beag on N side of loch, exposed to W winds; Port na Long E of Ardtreck, sheltered except from E winds (beware fish farm); Carbost on SW shore. Facilities: (Carbost) EC Wed; V (local shop), P (garage), PO, FW.

PLOCKTON, LOCH CARRON, Ross and Cromarty, 57°01′ N, 5°39′ W, Zone 0 (GMT), Admty charts 2528, 2209. HW −0435 on Dover, −0020 on Ullapool; HW height +0.3m on Ullapool; ML 3.2m; Duration 0600. See 10.8.9. Good anchorage on S side of Outer Loch Carron. Safest entrance between Sgeir Bhuidhe and Sgeir Golach, thence S to pass between Dubgh Sgeir and Hawk Rks off Cat I (with conspic stone tower). Beware Plockton Rks off Yellow Cliff Isle. Lts established 57°21′.93N, 5°37′.32W Dir Fl (3) WRG 10s vis: G060°-063°, W063°-067°, R067°-070°. At 57°20′.33N 5°37′.65 W Dir Fl (3) WRG 10s vis G157°-162°, W162°-166°, R166°-171°. At 57°21′.58N, 5°34′.53W Dir Fl (3) WRG 10s G318°-041°, W041°-044°, R044°-050°. Sgeir Golach marked by G Bn with cone. Anchor in centre of bay in approx 3.5m. Facilities: **Leisure Marine** D, CH; **Village** Bar, FW, PO, R, Rly, Gas, V, airstrip.

LOCH ALSH, Ross and Cromarty, 57°17′ N, 5°34′ W, Zone 0 (GMT), Admty charts 2540, 2541. Tidal figures taken at Dornie Bridge; HW −0450 on Dover, −0025 on Ullapool; HW height 0.0 on Ullapool; ML 3.0m; Duration 0555. See 10.8.9. Almost land-locked between the SE corner of Skye and the mainland. Safe anchorages can be found for most winds; off the hotel at Kyle of Lochalsh in 11m; in Ardintoul Bay on S shore opposite Racoon Rk in 5.5m; in Totaig (or Ob Aonig) opposite Loch Long entrance in 3.5m; in Ratigan Bay in SW corner of the loch in 7m; in Avernish Bay, W of Rubh' an Aisig in 3m (open to SW); inside the pier at Kyleakin. Kyle of Lochalsh is an important ferry port and railhead; yachts can go alongside E end of pier to fuel. Pier head Lts 2 FG (vert) 6/5m 5M on W side, 2 FG (vert) 6/5m 4M on E side. Facilities (at Kyle of Lochalsh): Bar, D, FW, PO, R, Rly, V, airstrip, CH, P. Bus to Glasgow or Inverness.

LOCH BOISDALE, South Uist, 57°09′ N, 7°18′ W, Zone 0 (GMT), Admty chart 2770. HW −0510 on Dover, −0045 on Ullapool; HW height −0.9m on Ullapool; ML 2.4m; Duration 0600. See 10.8.7. Good shelter except in SE gales when swell runs right up the 2M loch. Safest entrance from N passing between Rubha na Cruibe and Calvay I. From S beware Clan Ewan Rk, dries 1m, and McKenzie Rk. Channel up to Boisdale Hr N of Gasay I (beware rks off E end). Anchor off pier in approx 4m, or SW of Gasay I in approx 9m. There are fish cages W of Ru Bhuaitt. Calvay I E end Fl(2) WRG 10s 16m 7/4M; vis G190°−202°, W202°−286°, R286°−111°, W111°−190°. Gasay I Fl WR 5s 10m 7/4M; vis W120°−284°, R284°−120°. Gasay I, N side QG (TE 1985). Ludaig Dir Lt 297°, Dir Oc WRG 6s 8m 7/4M; G287°−296°, W296°−298°, R298°−307°. N pier 2 FG (vert). Ro Ro terminal Iso RG 4s Gshore−283°, R283°−shore. N entrance Ldg line, pier in line with Hollisgeir at 245°. Facilities: EC Tues; Bar, FW (tap on pier); P, PO R, V, ferry to mainland.

CANNA, The Small Islands, 57°03′ N, 6°29′ W, Zone 0 (GMT), Admty chart 1796. HW −0457 (Sp), −0550 (Np) on Dover, −0035 on Ullapool; HW height −0.4m on Ullapool; Duration 0605. Good shelter except in strong E winds. Hr lies between Canna and Sanday I. Most of Hr dries. Anchor in 3m to 4m having entered Hr along Sanday I shore; beware drying patch W of pier on Canna side. Lights: Sanday I Fl 6s 32m 8M, vis 152°−061°. Beware abnormal magnetic variations off NE Canna. Facilities: FW only. **Note:** Canna is a National Trust for Scotland island and yachtsmen are asked to refrain from ditching rubbish.

LOCH NEVIS, Inverness, 57°02′ N, 5°44′ W, Zone 0 (GMT), Admty chart 2208. HW −0515 on Dover, +0030 on Oban. HW height +1.0m on Oban. Anchorage NE of Eilean na Glaschoille, good except in S winds. Beware rocks Bogha cas Sruth (dries 1.8m), Bogha Don and Sgeirean Glasa both marked by Bns. Upper parts of the loch subject to very violent unpredictable squalls. Yachts can pass into the inner loch with care, and anchor N or SE of Eilean Maol, but this passage should be undertaken with caution.

CASTLEBAY, Barra, 56°57′ N, 7°29′ W, Zone 0 (GMT), Admty chart 2769. HW −0525 on Dover, −0105 on Ullapool; HW height −0.8m on Ullapool; ML 2.3m; Duration 0600. See 10.8.7. Very good shelter and holding ground. Best anchorage NW of Kiessimul Castle (on an Island) in approx 8m; NE of castle are rocks. Alternative anchorage in Vattersay Bay in approx 9m. Beware rks NNW of Dubh Sgeir; Dubh Sgeir Lt Fl(2) WG 6s 6m 7/5M; vis W280°-180°, G180°-280°. Sgeir Liath Lt Fl 3s 7m 8M. Ldg Lts 295°, Front on Rudha Glas, FR 9m 4M, R Δ on W framework Tr, rear, 550m from front, FR 22m 4M, R ▽ on W framework Tr. Facilities: Bar, D, FW, P, PO, R, V. Ferry to mainland.

EIGG HARBOUR, Eigg, 56°53′ N, 6°08′ W, Zone 0 (GMT), Admty chart 2207. HW −0523 on Dover, +0025 on Oban, HW height +0.7m on Oban. See 10.8.10. Most of Hr dries but yachts can anchor NE of Galmisdale Pt which is good anchorage, except in NE winds when yachts should go through the narrows and anchor in South Bay in 6/8m. Tide runs strongly in the narrows. Coming from N beware Garbh Sgeir and Flod Sgeir, both marked by Bns. SE point of Eilean Chathastail Fl 6s 24m 8M, vis 181°−shore. VHF, call *Eigg Harbour*, Ch 16; 08. Facilities: FW, D, repairs can be arranged by Hr Mr; **Village** (2M) Bar, PO, R, V, Gas.

ARISAIG, (Loch Nan Ceall), Inverness, 56°54′ N, 5°55′ W, Zone 0 (GMT), Admty chart 2207. HW −0515 on Dover, +0030 on Oban; HW height +0.9m on Oban; ML 2.9m; Duration 0605. Anchorage above Cave Rock, approx 1M within the entrance, is well sheltered. South channel is winding but is marked by perches. Ldg line into loch is S point of Eigg and N point of Muck. There are numerous unmarked rocks. There are no lights. VHF, Call *Arisaig Harbour* Ch 16. Facilities: EC Thurs; FW at hotel; **Sea Leisure Centre** Tel 224 C (10 ton), CH, D, El, M, ME, P, Sh, V. **Village** Bar, PO, R, Rly, V.

ARINAGOUR, Loch Eatharna, Coll, 56°37′ N, 6°31′ W. Zone 0 (GMT). Admty chart 2474. HW −0530 on Dover, +0015 on Oban; HW height +0.4m on Oban; ML 1.4m; Duration 0600; see 10.8.10. Good shelter except with SE swell or strong S winds. Entrance marked by Arinagour pier, NW of G buoy (Fl G 6s) marking Bogha Mor. Continue N towards old pier, pick up a buoy or anchor to S of it. Beware McQuarry's Rock (dries) 1 ca E of Arinagour pier. Lt on Arinagour pier head 2 FR(vert), 10m on metal column. Facilities: six visitors moorings on W side of harbour, N of pier. Alternative anchorage N of Eilean Eatharna. **Highlands and Islands Development Board Trading Post** M, D, FW, Gas, V, CH, R; **Village** Lau, FW, P, Hotel, PO.

DUNSTAFFNAGE, Argyll, 56°27′ N, 5°26′ W, Zone 0 (GMT), Admty chart 2378. HW −0500 on Dover; local tidal figures approx as for Oban. See 10.8.16. Good sheltered anchorage approached between Rubhagarbh and Eilean Mor. No navigational hazards. Anchor near castle in approx 5m or S part of bay or enter yacht harbour. Lts on pier head on NW side of bay 2 FG (vert) 4m 2M. Facilities: Land at pier; **Dunstaffnage Yacht Haven** Tel. Oban 66555 P, D, V, CH, R, BH (20 ton), AB, M, AC, Gas; **Ardtalla Yachts** Tel. Oban 65630 ME, El, Sh, Lau, SM, CH.

LOCH LATHAICH, Mull, 56°19′ N, 6°14′ W, Zone 0 (GMT), Admty chart 2617. HW −0545 on Dover, −0015 on Oban; HW height +0.2m on Oban. ML 2.4. See 10.8.12. Excellent shelter with easy access. Eileanan na Liathanaich (a group of islets) lie off the ent marked by a W Bn at the E end, Fl WR 6s 12m 8/6M, R088°-108°, W108°-088°. Keep to W side of loch and anchor behind White I at the head of loch (approx 5m). Facilities: (Bunessan) Shop, PO, Bar, R, P (cans, 3M), FW (90m).

VOLVO PENTA SERVICE

Sales and service centres in area 9
STRATHCLYDE *J N MacDonald & Co* 47-49 Byron Street, GLASGOW G11 6LP
Tel (041) 334) 6171. *Troon Marine Services Ltd* Harbour Road, TROON KA10 6DJ
Tel: (0292) 316180.

Area 9

South-West Scotland
Crinan to Kirkcudbright

10.9.1	Index	Page 443
10.9.2	Diagram of Radiobeacons, Air Beacons, Lifeboat Stations etc	444
10.9.3	Tidal Stream Charts	446
10.9.4	List of Lights, Fog Signals and Waypoints	448
10.9.5	Passage Information	450
10.9.6	Distance Table	451
10.9.7	Crinan	452
10.9.8	Campbeltown	453
10.9.9	East Loch Tarbert	453
10.9.10	Ardrishaig	453
10.9.11	Lamlash	454
10.9.12	Rothesay	454
10.9.13	Helensburgh/Rhu	455
10.9.14	Greenock, Standard Port, Tidal Curves/Gourock	456
10.9.15	Inverkip (Kip Marina)	460
10.9.16	Troon	461
10.9.17	Portpatrick	461
10.9.18	Kirkcudbright	462
10.9.19	Minor Harbours and Anchorages	462

Loch Craignish
Scalasaig (Colonsay)
Loch Tarbert (Jura)
Loch Sween
Craighouse (Jura)
Gigha Island
Port Askaig (Islay)
Port Ellen (Islay)
Carradale Bay
Brodick (Arran)
Loch Ranza (Arran)
Inverary
Millport (Cumbraes)
Largs
Ayr
Girvan
Loch Ryan
Isle of Whithorn
Wigtown Bay

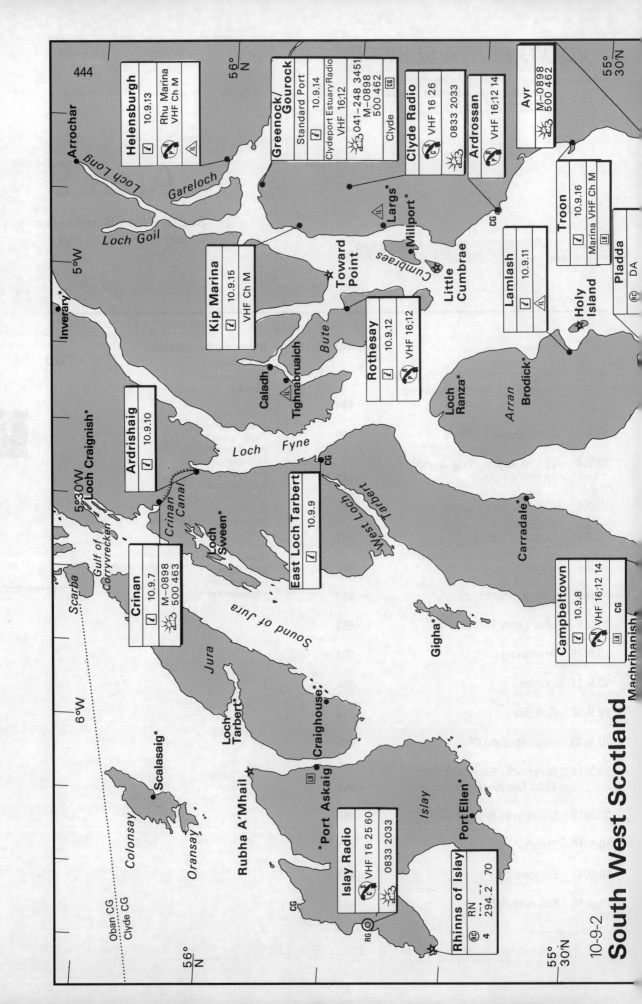

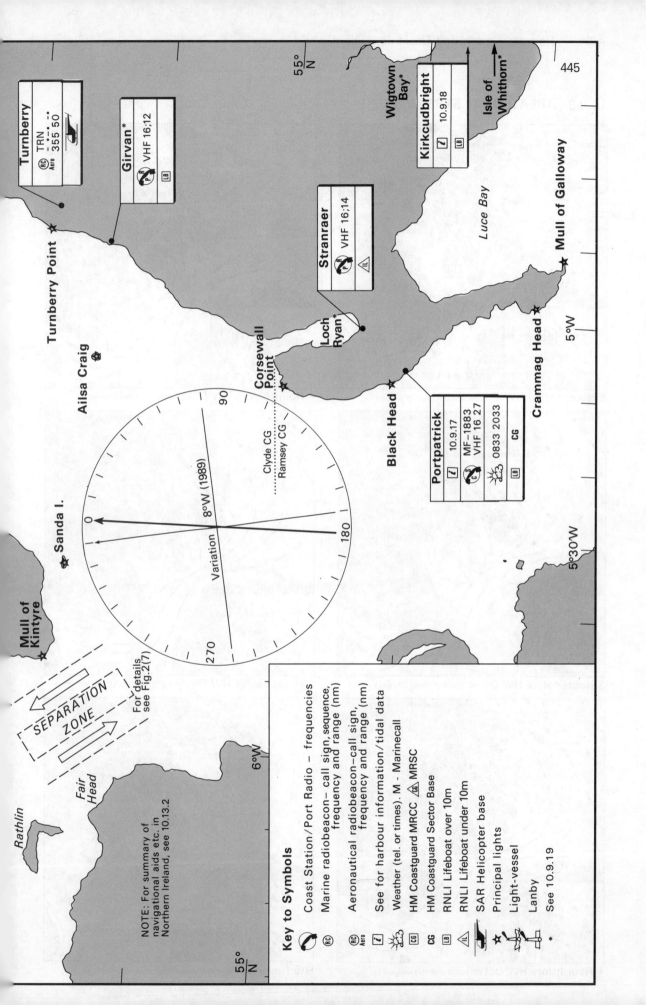

10.9.3. AREA 9 TIDAL STREAMS

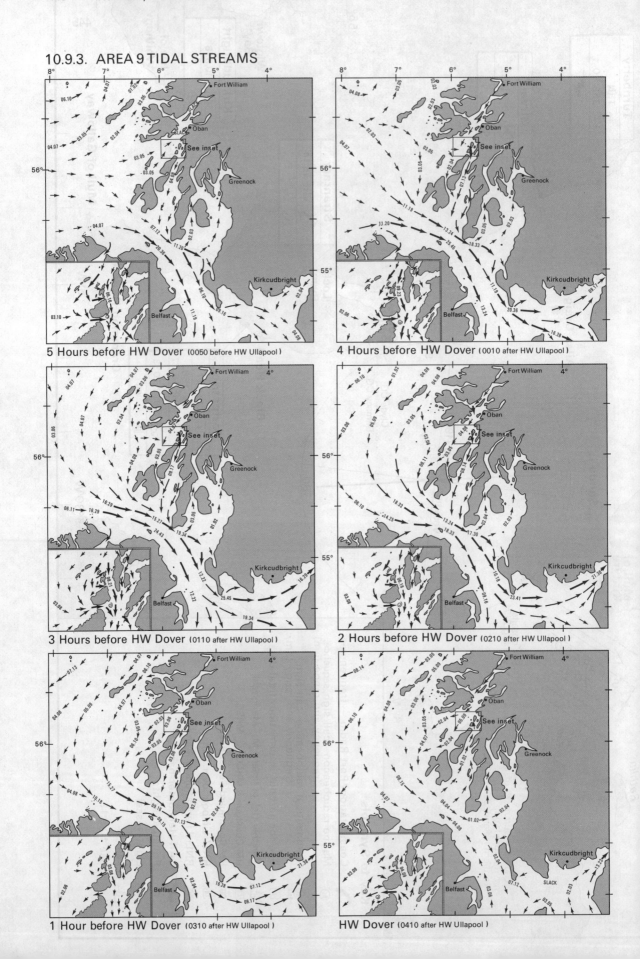

AREA 9—SW Scotland 447

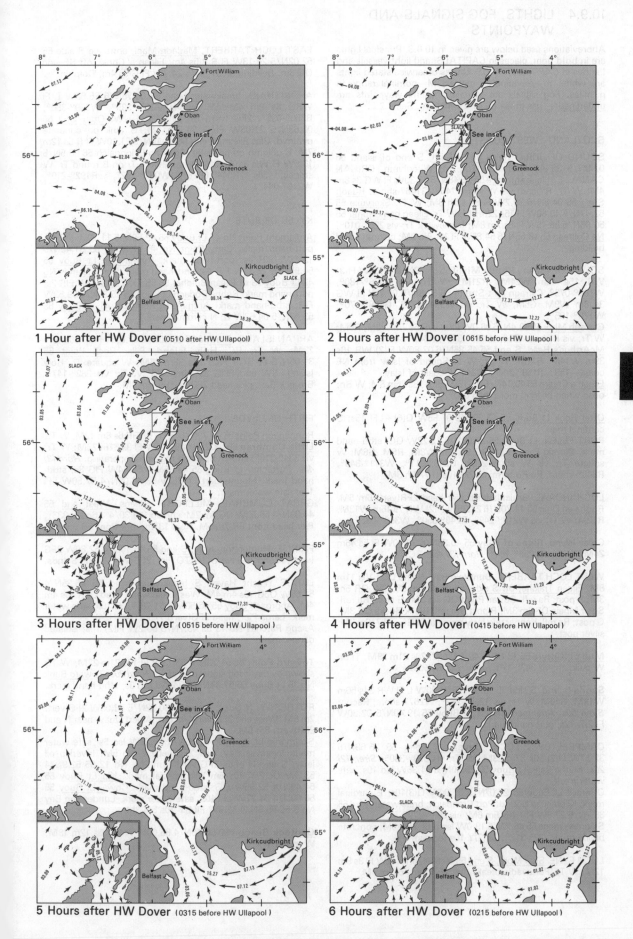

10.9.4 LIGHTS, FOG SIGNALS AND WAYPOINTS

Abbreviations used below are given in 10.0.3. Principal lights are in **bold** print, places in CAPITALS, and light-vessels and Lanbys in *CAPITAL ITALICS*. Unless otherwise stated lights are white. m—elevation in metres; M—nominal range in n. miles. Fog signals are in *italics*. Useful waypoints are underlined — use those on land with care. See 4.2.2.

SCOTLAND—WEST COAST

SOUND OF JURA, Reisa an t-Sruith, S end of island 56 07.8N/5 38.8W Fl(2) 12s 12m 7M; W column. CRINAN CANAL, E of lock entrance 56 05.48N/5 33.30W Fl WG 3s 8m 4M; W Tr, R band; vis W shore-146°, G146°-shore. Ruadh Sgeir 56 04.3N/5 39.7W Fl 6s 13m 8M; W Tr. Skervuile 55 52,47N/5 49.80W Fl 15s 22m 9M; W Tr. Eilean nan Gabhar 55 50.05N/5 56.15W Fl 5s 7m 8M; framework Tr; vis 225°-010°. Na Cuiltean 55 48.65N/5 54.85W Fl 10s 9m 9M; column on W building.

SOUND OF ISLAY, **Rhuda Mhail** 55 56.20N/6 07.35W Fl(3) WR 15s 45m **W24M, R21M**; W Tr; vis R075°-180°, W180°-075°. Carragh an t-Struith 55 52.3N/6 05.7W Fl WG 3s 8m W9M, G6M, W Tr; vis W354°-078°, G078°-170°, W170°-185°.
Carragh Mhor 55 50.4N/6 06.0W Fl(2) WR 6s 7m W8M, R6M; W Tr; vis R shore-175°, W175°-347°, R347°-shore.
McArthur's Head, S end 55 45.85N/6 02.80W Fl(2) WR 10s 39m W14M, R11M; W Tr; W in Sound of Islay from NE coast—159°, R159°-244°, W244°-E coast of Islay.
Eilean a Chuirn 55 40.14N/6 01.15W Fl(3) 18s 26m 8M; W Bn; obsc when bearing more than 040°.

Otter Rock Lt Buoy 55 33.90N/6 07.85W Q(6)+LFl 15s; S cardinal mark.
PORT ELLEN, Lt Buoy 55 37.00N/6 12.22W QG; stbd-hand mark. Carraig Fhada Fl WRG 3s 19m W8M, R6M, G6M; W square Tr; vis W shore-248°, G248°-311°, W311°-340°, R340°-shore. Ro Ro terminal 2FG(vert) 7/6m 3M.

LOCH INDAAL, Bruichladdich pier head 2FR(vert) 6m 5M.
Rudha an Duin 55 44.70N/6 22.35W Fl(2) WR 7s 15m W13M, R12M; W Tr; vis W218°-249°, R249°-350°, W350°-036°.

Orsay Island, **Rinns of Islay** 55 40.38N/6 30.70W Fl 5s 46m **24M**; W Tr; vis 256°-184°; *Horn (3) 45s*. RC.

WEST LOCH TARBERT, Dunskeig Bay, N end Q(2) 10s 11m 8M. Eileen Traighe, off S side, 55 45.40N/5 35.70W QR 5m 3M; R post; Ra refl. Corran Point QG 3m 3M; G post; Ra refl. Sgeir Mhein QR 3m 3M; R post; Ra refl. Black Rocks QG 3M; G post; Ra refl. Kennacraig ferry terminal 2FG(vert) 7/6m 3M; silver post.

Mull of Kintyre 55 18.6N/5 48.1W Fl(2) 20s 91m **29M**; Tr on W building; vis 347°-178°; *Horn Mo(N) 90s*.

Sanda Island, S side 55 16.50N/5 34.90W L Fl WR 24s 50m **W19M, R16M**; W Tr; vis R245°-267°, W267°-shore; Racon; *Siren 60s*. Arranman Barrels Lt Buoy 55 19.40N/5 32.80W Fl(2)R 12s; port-hand mark.

CAMPBELTOWN LOCH, **Davaar** N point 55 25.69N/5 32.37W Fl(2) 10s 37m **23M**; W Tr; vis 073°-330°; *Siren (2) 20s*. Methe Bank Lt Buoy 55 25.30N/5 34.36W Iso 10s; safe water mark.
Otterard Lt Buoy 55 27.07N/5 31.04W Q(3)10s; E cardinal mark. Kilbrannan Sound. Port Crannaich, breakwater head 55 35.7N/5 27.8W Fl R 10s 5m 6M; vis 099°-279°.
Skipness range 55 46.7N/5 19.0W Iso R 8s 7m 10M; concrete building; vis 292°-312°. Oc(2)Y 10s (occas).

LOCH FYNE. Sgat More, S end 55 50.85N/5 18.42W Fl 3s 9m 12M; W Tr. Portavadie breakwater 2FG(vert) 6/4m 4M.

EAST LOCH TARBERT, Madadh Maol, entrance S side 55 52.02N/5 24.18W Fl R 2.5s 4m. Eilean a' Choic, off SE side QG 3m. Beilding 55 51.9N/5 24.7W Fl G 5s 4m; Dolphin.

ARDRISHAIG, breakwater head 56 00.76N/5 26.53W L Fl WRG 6s 9m 4M; W Tr; vis G287°-339°, W339°-350°, R350°-035°. 2FG (vert) on pier 120m NW. Otter Spit 56 00.63N/5 21.03W Fl G 3s 7m 8M; G tank on concrete pyramid. Glas Eilean, S end 56 01.10N/5 21.10W Fl R 5s 12m 7M; R column on pedestal. Sgeir an Eirionnaich 56 06.5N/5 13.5W Fl WR 3s 7m 8M; B framework Tr on B round Tr, W stripes; vis R044°-087°, W087°-192°, R192°-210°, W210°-044°.

KYLES OF BUTE

Ardlamont Point No 47 Lt Buoy 55 49.59N/5 11.70W FlR 4s; port-hand mark. Carry Point No 46 Lt Buoy 55 51.40N/5 12.18W FlR 4s; port-hand mark. Rubha Ban Lt Buoy 55 54.95N/5 12.33W FlR 4s; port-hand mark. Burnt Islands No 42 Lt Buoy (S of Eilean Buidhe) 55 55.77N/5 10.32W FlR 2s; port-hand mark. Rubha Bodach Lt Buoy 55 55.39N/5 09.53W FlG; stbd-hand mark. Ardmaleish Point No 41 Lt Buoy 55 53.03N/5 04.63W Q; N cardinal mark.

ARRAN ISLAND. **Pladda** 55 25.50N/5 07.07W Fl(3) 30s 40m **23M**; W Tr; RC. **Holy Island**, Pillar Rock Point 55 31.05N/5 03.57W Fl(2) 20s 38m **25M**; W square Tr. Holy Island, SW end Fl G 3s 14m 10M; W Tr; vis 282°-147°. Brodick Bay, pier head 2FR(vert) 9/7m 4M.

FIRTH OF CLYDE

Runnaneun Point 55 43.79N/5 00.17W FlR 6s 8m 12M; W Tr. **Little Cumbrae** 55 43.27N/4 57.95W Fl 3s 31m **23M**; W Tr; vis 334°-210°; shown by day if fog signal operating; *Horn (3) 40s*. Portachur Lt Buoy 55 44.35N/4 58.44W FlG 3s; stbd-hand mark. Mountstuart Lt Buoy 55 48.00N/4 57.50W LFl 10s; safe water mark.
GREAT CUMBRAE, MILLPORT. Eileans, West end 55 44.89N/4 55.52W QG 5m 2M; shown 1/9-30/4. Ldg Lts 333°. Pier head front FR 7m 5M. Rear 137m from front FR 9m 5M.

FAIRLIE. Hunterston Jetty, S end 55 45.10N/4 52.80W 2FG (vert); N end 2FG (vert). NATO Pier, 2FG(vert) N and S ends.

LARGS. Yacht Haven S breakwater head 55 46.35N/4 51.67W OcG 10s 4m 4M. West breakwater head OcR 10s 4m 4M. Approach buoy 55 46.50N/4 51.78W LFl 10s; safe water mark. Largs Pier, head 2FG(vert) 7/5m 5M (H24).
Ascog Patches No 13 55 49.7N/5 00.2W Fl(2) 10s; isolated danger mark.

Toward Point 55 51.73N/4 58.73W Fl 10s 21m **22M**; W Tr; shown by day if fog signal operating; *Horn 20s*. Toward Bank No 35 Lt Buoy 55 51.05N/4 59.93W FlG 3s; stbd-hand mark.

ROTHESAY. N Quay, E end 55 50.32N/5 03.03W 2FG(vert) 7m 5M. West end 2FR(vert) 7m 5M. Albert Quay, near N end 2FR(vert) 8m 5M.
Skelmorlie Lt Buoy 55 51.65N/4 56.28W Iso 5s; safe water mark. Wemyss Bay, pier 2FG(vert) 7/5m 5M. Inverkip oil jetty, S and N ends 2FG(vert) 11m 2M. No 12 55 52.9N/4 53.7W Oc(2)Y 10s 5m 3M; special mark. Kip Lt Buoy 55 54.49N/4 52.95W QG; stbd-hand mark. Cowal Lt Buoy 55 56.00N/4 54.77W LFl 10s; safe water mark. Lunderston Bay No 8 55 55.5N/4 52.9W Fl(4)Y 10s 5m 3M.

Gantock Beacon 55 56.46N/4 55.00W Fl R 2.5s 12m **18M**; W round Tr.

AREA 9—SW Scotland

DUNOON, Pier, S end and N end 2FR(vert) 5m 6M.
Cloch Point Fl 3s 24m 8M; W round Tr, B band, W dwellings; *Horn (2) 30s*. McInroy's Point RoRo terminal 2FG(vert) 5/3m 6M. No 5 55 57.0N/4 51.6W Oc(2)Y 10s 5m 3M; special mark. Hunter's Quay RoRo terminal 2FR(vert) 6/4m 6M.

Holy Loch, pier 55 59.0N/4 56.7W 2FR(vert) 6/4m 3M.
LOCH LONG. Loch Long Lt Buoy 55 59.17N/4 52.33W Oc 6s; safe water mark. Baron's Point, No 3 55 59.2N/4 51.0W Oc(2)Y 10s 3M. Ravenrock Point 56 02.1N/4 54.3W Fl 4s 12m 10M; W framework Tr on W column. Dir Lt 204°, Dir WRG 9m (same Tr); vis R201.5°-203°, Al WR203°-203.5° (W phase incr with brg), W203°-204.5°, Al WG204.5°-205° (G phase incr with brg), G205°-206.5°. Portdornaige 56 03.7N/4 53.6W Fl 6s 8m 11M; W column; vis 026°-206°. Carraig nan Ron (Dog Rock) 56 06.0N/4 51.6W Fl 2s 7m 11M; W column. Finnart Oil Terminal, Cnap Point 56 07.4N/4 49.9W Ldg Lts 031°. Front Q 8m 10M; W column. Rear, 87m from front F 13m; R line on W Tr.

Ashton Lt Buoy 55 58.11N/4 50.58W Iso 5s; safe water mark.
GOUROCK, railway pier head 55 57.8N/4 49.0W 2FG(vert) 10/8m 3M; grey framework Tr. Kempock Point No 4 55 57.7N/4 49.3W Oc(2)Y 10s 3M.
Whiteforeland Lt Buoy 55 58.11N/4 47.20W LFl 10s; safe water mark.
Rosneath Patch, S end 55 58.52N/4 47.37W Fl(2) 10s 5m 10M. Ldg Lts 356° Ardencaple Castle, Front 56 00.6N/4 45.3W 2FG(vert) 26/23m 12M; Tr on NW corner of castle; vis 335°-020°. Rear 822m from front FG 10M.

ROSNEATH. Beacon No 1 55 59.12N/4 43.81W VQ(4)Y 5s and Dir Q WRG 7m W3M, R2M, G2M; vis G078°-079°, W079°-080°, G080°-082°. Row Lt Buoy 55 59.85N/4 45.05W FlG 5s; stbd-hand mark. Cairndhu Lt Buoy 56 00.36N/4 45.93W FlG 2.5s; stbd-hand mark. Dir Lt 291°. Dir Oc WRG 3s 7m 14M; vis G288°-290°, W290°-292°, R292°-296°; (H24). Ldg Lts 291°. Front 56 00.6N/4 47.7W FR. Rear 0.66m from front FR. Castle Point 56 00.20N/4 46.43W Fl(2) R 10s 8m 6M; R mast. Rosneath DG Jetty 2FR(vert) 5M; W column; vis 150°-330°. Rhu Narrows 56 00.9N/4 47.1W Q(3) WRG 6s 9m W10M, R7M, G7M; vis G270°-000°, W000°-114°, R114°-188°. Dir Lt 318°, Dir Oc WRG 6s; same Tr; vis G313°-317°, W317°-319°, R319°-320°; (H24).

Rosneath Bay Ldg Lts 163°. Front 56 00.0N/4 47.1W FG 11m 12M; framework Tr; vis 149°-177°. Rear 683m from front FG 24m 12M; stone Tr; vis 149°-177°.

GARELOCH. Rhu South Lt Buoy 56 00.66N/4 47.35W FlG 3s; stbd-hand mark. Rhu Spit 56 00.77N/4 47.38W Fl 2.5s 6m 8M; W Tr, G band. Mambeg Dir Lt 330°, Dir Q(4) WRG 8s 8m 14M; W column; vis G327°-329°, W329°-331°, R331°-332°; (H24). Faslane Base, wharf, S elbow Fl G 5s 11m 5M. Floating dock, E side Fl R, middle 2FG(vert) 14m 5M, N side Q WRG 14m W9M, R6M, G6M; vis G333°-084°, W084°-161°, R161°-196°; (H24).
Garelochhead, S fuel jetty 56 04.23N/4 49.62W 2FG(vert) 10m 5M. N fuel jetty, elbow Iso WRG 4s 10m; vis G351°-356°, W356°-006°, R006°-011°.

GREENOCK. Anchorage Lts in line 196°. Front 55 57.6N/4 46.5W FG 7m 12M. Rear 32m from front FG 9m 12M. Lts in line 194°30'. Front 55 57.4N/4 45.8W FG 18m. Rear 360m from front FG 33m. Clydeport Container Terminal NW corner QG 8m 8M. Victoria Harbour, entrance West side 2FG(vert). Garvel Embankment, West end 55 56.81N/4 43.48W Oc G 10s 9m 4M; E end, Maurice Clark Point 55 56.61N/4 42.78W QG 7m 2M.

PORT GLASGOW, Beacon off entrance 55 56.3N/4 41.2W FG 7m 9M; B&W chequered Tr and cupola. Steamboat Quay, West end FG 12m 12M; B&W chequered column; vis 210°-290°. From here to Glasgow Lts on S bank are Fl G and Lts on N bank are Fl R.

SCOTLAND—WEST COAST

ARDROSSAN, Approach Dir 055°, 55 38.7N/4 49.1W Dir FWRG 15m 9M; vis G048.5°-053.5°, W053.5°-056.5°, R056.5°-061.5°. FR 9m 9M (same Tr); vis 340°-130°. N breakwater head Fl WR 2s 12m 5M; R gantry; vis R041°-126°, W126°-041°. Lighthouse pier head 55 38.47N/4 49.50W Iso WG 4s 11m 9M; W Tr; vis W035°-317°, G317°-035°.

IRVINE. Entrance, N side 55 36.21N/4 42.00W FlR 3s 6m 5M; R mast. S side FlG 3s 6m 5M; G column. Ldg Lts 051°. Front FG 10m 5M. Rear 101m from front FR 15m 5M; both on G masts, vis 019°-120°.

TROON. West pier head 55 33.07N/4 40.95W Oc WR 6s 11m 5M; W Tr; vis R036°-090°, W090°-036°; *Siren 30s*. 14m SE Fl WG 3s 7m 5M; post on dolphin; vis G146°-318°, W318°-146°. E pier head 2FR(vert) 6m 5M; R column; obsc brg more than 199°.

Lady Isle 55 31.63N/4 43.95W Fl(4) 30s 19m 8M; W Bn.

AYR. Bar Lt Buoy 55 28.12N/4 39.38W FlG 2s; stbd-hand mark. N breakwater head 55 28.22N/4 38.71W QR 7m 5M. S pier head Q 7m 7M; vis 012°-161°. FG 5m 5M; same Tr; vis 012°-066° over St Nicholas rocks; *Horn 30s*. Ldg Lts 098°. Front 55 28.2N/4 38.3W FR 10m 5M. Rear 130m from front OcR 10s 18m 9M.

Turnberry Point, near castle ruins 55 19.55N/4 50.60W Fl 15s 29m **24M**; W Tr.

Ailsa Craig 55 15.12N/5 06.42W Fl(6) 30s 18m **17M**; W Tr; vis 145°-028°.

GIRVAN. N groyne head 55 14.75N/4 51.70W Iso 4s 3m 4M. S pier head 2FG(vert) 8m 4M; W Tr.

LOCH RYAN. Cairn Point 54 58.5N/5 01.8W Fl(2) R 10s 14m 12M; W Tr. Cairnryan 54 57.8N/5 00.9W Fl R 5s 5m 5M.

Stranraer approach, No 1 Bn 54 56.6N/5 01.3W Oc G 6s. No 3 Bn 54 55.9N/5 01.6W QG. No 5 54 55.1N/5 01.8W Fl G 3s. STRANRAER centre pier, head 2 F Bu (vert). E pier, head 2FR(vert). Ross pier, head 2 F Vi (vert) 10m 4M; R column.

Corsewall Point 55 00.43N/5 09.50W Al Fl WR 74s 34m **18M**; W Tr; vis 027°-257°.

Killantringan, Black Head 54 51.71N/5 08.75W Fl(2) 15s 49m **25M**; W Tr.
PORTPATRICK. Ldg Lts 050°. Front 54 50.5N/5 06.9W FG (occas). Rear, 68m from front, FG (occas).

Crammag Head 54 39.90N/4 57.80W Fl 10s 35m **18M**; W Tr.
Mull of Galloway, SE end 54 38.05N/4 51.35W Fl 20s 99m **28M**; W Tr; vis 182°-105°.

Port William. Ldg Lts 105°. Front, pier head 54 45.65N/4 35.10W FlG 3s 7m 3M. Rear, 130m from front, FG 10m 2M.

Isle of Whithorn Harbour 54 41.8N/4 21.7W QG 4m 5M. Ldg Lts 335°. Front OcR 8s 7m 7M. Rear, 35m from front, OcR 8s 9m 7M, synchronised.

Little Ross 54 45.93N/4 05.02W Fl 5s 50m 12M; W Tr; obsc in Wigton Bay when brg more than 103°.
KIRKCUDBRIGHT BAY. No 1, Lifeboat House 54 47.68N/4 03.66W Fl 3s 7m 5M. No 12 54 49.1N/4 04.8W FlR 3s 3m; perch. No 14 54 49.25N/4 04.76W Fl 3s 5m; perch. No 22 54 50.1N/4 03.9W FlR 3s 2m. Outfall 54 50.2N/4 03.8W FlY 5s 3m 2M; Y framework Tr.

Hestan Island, E end 54 50.0N/3 48.4W Fl(2) 10s 38m 7M; W house.
Annan, S end of quay 54 59N/3 16W 2FR(vert) 4m 5M.
Barnkirk Point 54 58.0N/3 15.9W Fl 2s 18m 2M.

10.9.5 PASSAGE INFORMATION

SCOTLAND – SOUTH WEST COAST

Although conditions in the South-West of Scotland are in general less rugged than from Mull northwards, some of the remarks at the start of 10.8.5 are equally applicable to this area.

The W coasts of Colonsay and Oronsay (chart 2169) are fringed with dangers up to 2M offshore. The two islands are separated by a narrow chan which dries and has an overhead cable (10m). For Scalasaig see 10.9.19.

Sound of Luing (chart 2326) between Scarba, Lunga, Fiola Meadlonach, Rubha Fiola and Pladda (or Fladda) on the W side, and Luing and Dubh Sgeir on the E side, is the normal chan to or from Sound of Jura – despite dangers at the N end and strong tidal streams. The N-going stream begins at HW Oban +0430, and the S-going at HW Oban –0155. Sp rates are 2½-3 kn at S end of sound, increasing to 6 kn or more in Is off N entrance – where there are eddies, races and overfalls. Good shelter in Ardinmar B, SW of Torsa.

CORRYVRECKAN/CRINAN (charts 2326, 2343)

Between Scarba and Jura is the Gulf of Corryvreckan (chart 2343), with a least width of 6 ca and free of dangers, but noted for its very strong tides which, in conjunction with an uneven bottom, cause extreme turbulence. This is particularly dangerous with strong W winds over a W-going (flood) tide which spews out several miles to seaward of the gulf, with overfalls extending 3M from the W of entrance. The gulf is best avoided, and should never be attempted by yachts except at slack water and in calm conditions. Keep to the S side of the gulf to avoid the whirlpool known as The Hag. The W-going stream in the gulf begins at HW Oban +0410, and the E-going at HW Oban –0210. Sp rate W-going is 8½ kn, and E-going rather less.

The range of tide at sp can vary nearly 2m between the E end of the gulf (1.5m) and the W end (3.4m), with HW ½h earlier at the E end. Slack water lasts about 1h at nps, but only ½h at sps. On the W-going (flood) stream eddies form both sides of the gulf, but the one on the N (Scarba) shore is more important. Where this eddy meets the main stream off Camas nam Bairneach there is violent turbulence, with heavy overfalls extending W at the division of the eddy and the main stream. There are temp anchs with the wind in the right quarter in Bagh Gleann a Mhaoil in the SE corner of Scarba, and in Bagh Gleann nam Muc at N end of Jura but the latter has rocks in approaches E and SW of Eilean Beag.

At N end of Sound of Jura is Loch Craignish (chart 2326). There are very strong tides in Dorus Mor, if approaching from the N. The W-going stream begins at HW Oban +0330, and the E-going at HW Oban –0215, sp rates 8 kn. The main chan up Loch Craignish is clear 1½ ca offshore, but beware charted rks off Eilean Dubh and Eilean Mhic Chrion when approaching head of loch and the Ardfern Yacht Centre, which is behind Eilean Inshaig on W shore. See 10.9.19.

S of Loch Craignish is Loch Crinan, which leads to the Crinan Canal (10.9.7). Beware Black Rk, 2m high and 2 ca N of the canal sea lock, and dangers extending ½ ca from the rk.

SOUND OF ISLAY (chart 2481)

An alternative N/S route is through the Sound of Islay (chart 2481). The chan presents no difficulty: hold to the Islay shore, where all dangers are close in. The N-going stream begins at HW Oban +0440, and the S-going at HW Oban –0140. The sp rates are 2½ kn at N entrance and 1½ kn at S entrance, but increasing in the narrows and reaching 5 kn off Port Askaig. There are overfalls off McArthur's Head (Islay side of S entrance) during the S-going stream. There are anchs in the sound, but mostly holding ground is poor. The best places are alongside at Port Askaig (10.9.19), or at anch off the distillery in Bunnahabhan B, 2½M to N.

The W coast of Islay is very exposed. Off Orsay, Frenchman's Rks and W Bank the NW-going stream begins at HW Oban +0530, and the SE-going at HW Oban –0040. Sp rates are 8 kn off Orsay and Frenchman's Rks, and 6 kn over W bank, but decrease to 3 kn 5M offshore. There are races and overfalls in these areas. In the N of Islay there is anch E of Nave Island at entrance to Loch Gruinart: beware Balach Rks which dry, just to N. Loch Indaal gives some shelter: beware rks extending from Laggan Hd on E side of entrance. Port Ellen (chart 2474) has several dangers in approach, and is exposed to S. See 10.9.19.

SOUND OF JURA TO MULL OF KINTYRE

From Crinan to Gigha the Sound of Jura is safe if a mid-chan course is held. Skervuile (Lt) is a reef roughly in middle of sound. Loch Sween (chart 2397) can be approached N or SE of MacCormaig Islands, where there is an attractive anch on NE side of Eilean Mor, but exposed to NE. Coming from N beware Keills Rk and Danna Rk. Sgeirean a Mhain is a rk in fairway 1½M NE of Castle Sween (conspic on SE shore). Anch at Tayvallich, near head of loch on W side. See 10.9.19.

W Loch Tarbert (chart 2477) is long and narrow, with good anchs but unmarked shoals. On entry give a berth of at least 2½ ca to Eilean Traighe off N shore, E of Ardpatrick Pt. Dun Skeig, an isolated hill, is conspic on S shore. Good anch near head of loch, 1M by road from E Loch Tarbert (see 10.9.9).

On other side of sound, near S end of Jura, are The Small Isles (chart 2396) between Rubh' an Leanachais and Rubha na Caillich. Beware Goat Rk (dries) 1½ ca off southernmost I, Eilean nan Gabhar, behind which is good anch. Also possible to go alongside Craighouse Pier (10.9.19). Another anch exists in Lowlandman's B, about 3M to N, but exposed to S winds: Ninefoot Rks with depth of 2.4m lie off entrance.

S of W Loch Tarbert, and about 2M off the Kintyre shore, is Gigha Island (chart 2475), with Cara Island and Gigalum Island off its S end. See 10.9.19. Dangers extend 1M W off S end of Gigha Island. Outer and Inner Red Rks (least depth 2m) lie 2M SW of N end of Gigha Island.

Gigha Sound need very careful pilotage, since there are several dangerous rks. The N-going stream begins at HW Oban +0430, and S-going at HW Oban –0155, sp rates 1½ kn. Good anchs in Druimyeon B and Ardminish B, respectively N and S of Ardminish Pt on E side of Gigha.

From Crinan to Mull of Kintyre is about 50M, and this long peninsula has great effect on tidal stream in North Channel. Off Mull of Kintyre (Lt, fog sig) the N-going stream begins at HW Oban +0400, and the S-going at HW Oban –0225, sp rates 5 kn. A strong race and overfalls exist S and SW of Mull of Kintyre, dangerous in strong S winds against S-going tide. Careful timing is needed. There is a traffic scheme in N Channel – see Fig. 2(7).

Sanda Sound separates Sanda Island (Lt, fog sig) and its neighbouring rks and islets, from Kintyre. Beware Macosh Rks (dry) forming part of Barley Ridges, 2 ca offshore, E of Rubha McShannuick; Arranman Barrels, drying and submerged, 2½ ca E from Du-na-h-Oighe and marked by buoy; and Blindman rk (depth 2m) 2 ca off Ru Stafnish, where there are 3 radio masts ½M to W. 3 ca off N of Sanda Island is Sheep Island. Paterson's Rk (dries) is 1M E of Sheep Island. There is anch in Sanda Harb. In Sanda Sound the E-going stream begins at HW Greenock +0340, and the W-going at HW Greenock –0230, sp rates 5 kn. Tide races extend N and W from Sheep Island, and in strong S or SW winds the Sound is dangerous. In these conditions pass E of Paterson's Rk, and 2M S of Sanda and Mull of Kintyre.

MULL OF KINTYRE TO MULL OF GALLOWAY (charts 2199, 2198)

Once E of Mull of Kintyre, tidal and pilotage conditions improve greatly. Campbeltown (10.9.8) is entered N of Davaar Island (Lt, fog sig). 1½M N of Lt Ho is Otterard Rk (depth 3m), with Long Rk (dries) 5 ca W of it; both are marked by buoys. E of Davaar Island tide runs 4 kn at sp, and there are overfalls.

Kilbrannan Sound runs 21M from Davaar Island to Skipness Pt, where it joins Inchmarnock Water and Bute Sound. There are few dangers apart from overfalls on Erins Bank, 10M S of Skipness, on S-going stream. Good anch in Carradale B, off Torrisdale Castle. There are overfalls off Carradale Pt on S-going stream. For Carradale see 10.9.19.

Loch Fyne (chart 2381) is relatively clear of dangers to E Loch Tarbert (10.9.9). On E shore beware rks off Ardlamont Pt, and 4M to NW is Skate Island which is best passed to W.

3M from Ardrishaig (10.9.10) beware Big Rk (depth 2.1m). Further N, at entrance to Loch Gilp (much of which dries) note shoals round Gulnare Rk with least depth 1.5m, and Duncuan Island with dangers extending SW to Sgeir Sgalag (depth 0.8m) marked by buoy.

Where Loch Fyne turns NE it is largely obstructed by Otter Spit (dries), extending 8 ca WNW from E shore. Outer end of spit is marked by Bn (lit), but dangers extend somewhat beyond. The stream runs up to 2 kn here. A rk with depth less than 2m lies about ¾M SW of Otter Spit Bn. In Upper Loch Fyne (chart 2382) Minard Narrows are formed by rks and islets in the fairway, but there are charted chans between. For Inveraray see 10.9.19.

Bute Sound leads into Firth of Clyde, and is clear in fairway. Sannox Rk (depth 1.2m) is ¼M off Arran coast 8M N of Lamlash (10.9.11). 1 ca off W side of Inchmarnock is Tra na-h-uil, a rk which dries. In Inchmarnock Sound, Shearwater Rk (depth 0.9m) lies in centre of S entrance.

Arran's mountains tend to cause squalls or calms, but it has good anchs at Lamlash (10.9.11), Brodick (10.9.19) and Loch Ranza (10.9.19).

Kyles of Bute are attractive chan between Inchmarnock Water and Firth of Clyde, and straightforward apart from Burnt Islands. Here it is best to take N chan, narrow but well buoyed, passing S of Eilean Buidhe, and N of Eilean Fraoich and Eilean Mor. Care is needed, since stream may reach 5 kn at sp.

In contrast, the N lochs in Firth of Clyde are less attractive. Loch Goil is worth a visit but Loch Long is squally and has few anchs, while Gareloch has little to attract cruising yachts. For Millport (Great Cumbrae) see 10.9.19. There is anch in Kilchattan Bay (E Bute) sheltered from SSE to WNW.

Navigation in Firth of Clyde presents few problems since chans are well marked, but beware unlit moorings and also the considerable amount of commercial and naval shipping (see 10.9.13). Tidal streams are generally weak, seldom exceeding 1 kn.

S of Largs (10.9.19) the coast is generally uninviting, with mostly commercial harbs until Troon (10.9.16). There are various dangers in Ayr B (10.9.19); beware Troon Rk (depth 5.6m, but sea can break), Lappock Rk (dries, marked by Bn), and Mill Rk (dries, buoyed).

There is a serious race off Bennane Hd (8M SSE of Ailsa Craig) when tide is running strongly. Loch Ryan offers little for yachtsmen but there is anch S of Kirkcolm Pt, inside the drying spit which runs in SE direction 1½M from the point. There is also useful anch in Lady Bay, sheltered except from NE. Between Corsewall Pt and Mull of Galloway the S-going stream begins HW Greenock +0310, and the N-going at HW Greenock −0250. Sp rate off and S of Black Hd is 5 kn; N of Black Hd it decreases, to 2-3 kn off Corsewall Pt. Races occur off Morroch B, Money Hd, Mull of Logan and SE of Crammag Hd.

Mull of Galloway (Lt) is S point of Scotland, a tall headland (83m) and steep-to, but beware dangerous race extending nearly 3M to S. On E-going stream the race extends NNE into Luce B; on W-going stream it extends SW and W.

The Scares are rks in middle of Luce B. Elsewhere b is clear more than ¼M offshore, but there is practice bombing range. Best anch at E Tarbert B, or alongside in Drummore which is sheltered but dries. Off Burrow Hd there is bad race in strong W winds with W-going tide. In Wigtown B (10.9.19) the best anch is in l of Whithorn B, but exposed to S. It is also possible to dry out in Garlieston.

There is a tank firing range extending 14M offshore between Kirkcudbright (10.9.18) and Abbey Head, 4M to E. If unable to avoid the area, cross it at N end close inshore. For information telephone Dundrennan 271 or contact Ramsey Coastguard. Range safety boat *Oaklea* keeps watch on VHF Ch 16.

10.9.6 DISTANCE TABLE

Approximate distances in nautical miles are by the most direct route while avoiding dangers and allowing for traffic separation schemes etc. Places in *italics* are in adjoining areas.

		1	2	3	4	5	6	7	8	9	10	11	12	13	14	15	16	17	18	19	20
1	*Tory Island*	**1**																			
2	*Barra Head*	95	**2**																		
3	*Oigh Sgeir*	113	33	**3**																	
4	*Ardnamurchan Pt*	110	47	21	**4**																
5	Port Askaig	87	77	75	60	**5**															
6	Mull of Kintyre	86	112	103	99	33	**6**														
7	Campbeltown	106	132	123	119	53	20	**7**													
8	E Loch Tarbert	131	157	148	144	78	45	31	**8**												
9	Ardrishaig	140	166	157	153	87	54	39	10	**9**											
10	Toward Point	139	165	156	152	86	53	45	23	31	**10**										
11	Kip Marina	144	170	161	157	91	58	50	28	36	5	**11**									
12	Helensburgh	153	179	170	166	100	67	59	37	45	14	9	**12**								
13	Lamlash	120	146	137	133	67	34	24	25	34	21	25	33	**13**							
14	Troon	130	156	147	143	77	44	33	33	40	24	29	38	16	**14**						
15	Portpatrick	119	148	139	135	69	36	39	66	74	65	68	77	44	49	**15**					
16	Mull of Galloway	134	164	155	151	85	52	56	82	90	81	84	93	60	65	15	**16**				
17	Kirkcudbright	164	196	187	183	117	84	88	114	122	113	116	125	92	97	48	32	**17**			
18	*Mew Island*	116	149	140	136	69	37	44	72	80	74	77	86	54	60	16	22	54	**18**		
19	*Point of Ayre*	160	186	177	173	107	74	78	104	112	103	106	115	82	87	38	22	25	44	**19**	
20	*St Bees Head*	177	207	198	194	128	95	99	125	133	124	127	136	103	108	59	43	24	65	26	**20**

CRINAN 10-9-7
Argyll

CHARTS
Admiralty 2320; Imray C65; OS 55

TIDES
−0608 Dover; ML 2.1; Duration 0605; Zone 0 (GMT)
Standard Port OBAN (←)

Times				Height (metres)			
HW		LW		MHWS	MHWN	MLWN	MLWS
0100	0700	0100	0800	4.0	2.9	1.8	0.7
1300	1900	1300	2000				

Differences GLENGARRISDALE BAY (Jura)
−0020 0000 −0010 0000 −0.4 −0.2 0.0 −0.2
SCALASAIG (Colonsay)
−0020 −0005 −0015 +0005 −0.1 −0.2 −0.2 −0.2
RUBHA A'MHAIL (Islay)
−0020 0000 +0005 −0015 −0.3 −0.1 −0.3 −0.1
ORSAY ISLAND (Islay)
−0110 −0110 −0040 −0040 −1.4 −0.6 −0.5 −0.2
BRUICHLADDICH (Islay)
−0100 −0005 −0110 −0040 −1.7 −1.4 −0.4 +0.1
PORT ELLEN (Islay)
−0530 −0050 −0045 −0530 −3.1 −2.1 −1.3 −0.4
PORT ASKAIG (Islay)
−0110 −0030 −0020 −0020 −1.9 −1.4 −0.8 −0.3
CRAIGHOUSE (Sound of Jura)
−0430 −0130 −0050 −0500 −2.8 −2.0 −1.4 −0.4
LOCH BEAG (Sound of Jura)
−0110 −0045 −0035 −0045 −1.6 −1.2 −0.8 −0.4
GIGHA SOUND (Sound of Jura)
−0450 −0210 −0130 −0410 −2.5 −1.6 −1.0 −0.1
MACHRIHANISH
−0520 −0350 −0340 −0540   Mean range 0.5 metres.
High water in Loch Crinan is 0045 before HW Oban.

SHELTER
Complete shelter in canal basin, but often many fishing boats lie there. Good shelter in Crinan Harbour to E of Eilean da Mheim, but full of moorings. Except in strong winds from W or N there is anchorage off the canal entrance, clear of fairway. R Add is available to small shoal-draught boats in good weather.

NAVIGATION
Waypoint 56°05′.80N 05°33′.69W, 326°/146° from/to Fl WG 3s Lt, 0.38M. Beware the Black Rocks and other nearby rocks in approach and near Ldg line.
CRINAN CANAL. Canal runs from Crinan, 9 miles to Ardrishaig and has 15 locks. All bridges open and canal can be entered at any tide. Max size, 26.5m length, 6m beam, 2.9m draught, mast height 28.9m. Vessels proceeding West have right of way. Canal dues are payable at Ardrishaig or Crinan sea lock. Passage time is between 5 to 6 hours. Locks operate (LT):−
Sea locks 0600-1200, 1230-2130 Mon-Sat.
Inland bridges 0800-1200, 1230-1615 Mon-Sat.
Sunday opening times are flexible (summer months only).

LIGHTS AND MARKS
W wing of entrance 2 FG (vert)
E wing of entrance 2 FR (vert)
Light to E of entrance Fl WG 3s 8m 4M,
W shore-146°, G146°-shore.

RADIO TELEPHONE
None.

TELEPHONE (054 683)
Hr Mr 211; MRSC Oban 63720; HMC 52261; Marinecall 0898 500 463; Dr Lochgilphead 2921

FACILITIES
Canal Office Tel. Lochgilphead 3210, M, L, FW; **Crinan Boats** Tel. 232, Slip, M, L, FW, ME, El, Sh, Gas, C (5 ton), CH, ACA; **Sea Basin** P (cans), D, AB, V, R, Bar; **Crinan Hotel** Tel. 235, D, R, Bar; **Fyneside Service Station** Tel. Lochgilphead 2229, C. **M. Murray** Tel. 238 M; **Village** PO; Bank (Ardrishaig); Rly (Oban); Air (Glasgow or Campbeltown).
Note: There are some Bu spherical buoys marked 'HIDB visitors only'.

AGENTS NEEDED
There are a number of ports where we need agents, particularly in France.
ENGLAND Swale, Havengore, Berwick.
SCOTLAND Firth of Forth, Scrabster, Mallaig, Loch Sunart, Loch Aline.
IRELAND Kilrush, Wicklow, Westport/Clew Bay.
FRANCE Arcachon, Seudre R, Ile d'Oleron, Rochfort, Ile de Re, St. Giles-Croix-de-Vie, Ile d'Yeu, Pouliguen, Le Croisic, La Forêt, Ile de Bréhat.
GERMANY Norderney, Dornumer-Accumersiel.
If you are interested in becoming our agent for any of the above, please write to the editors and get your free copy of the Almanac every year. You do not have to be resident in a port to be the agent, but at least a fairly regular visitor.

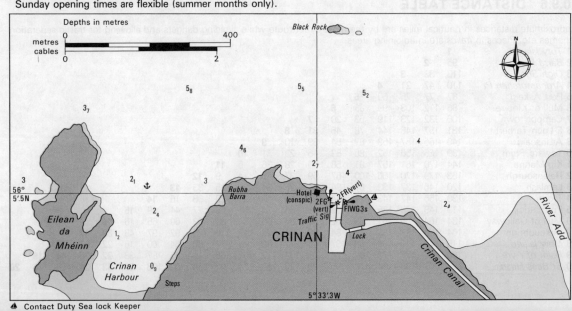

▲ Contact Duty Sea lock Keeper

AREA 9—SW Scotland 453

CAMPBELTOWN 10-9-8
Argyll

CHARTS
Admiralty 1864; Imray C63; OS 68
TIDES
+0045 Dover; ML 1.7; Duration 0630; Zone 0 (GMT).
Standard Port GREENOCK (→)

Times				Height (metres)			
HW		LW		MHWS	MHWN	MLWN	MLWS
0000	0600	0000	0600	3.4	2.9	1.0	0.4
1200	1800	1200	1800				

Differences CAMPBELTOWN
+0010 +0005 +0005 +0020 −0.5 −0.3 +0.1 +0.2
SANDA ISLAND
−0040 −0040 No data −1.0 −0.9 No data
SOUTHEND, KINTYRE
−0020 −0020 −0040 −0035 −1.3 −1.2 −0.5 −0.2
LOCH RANZA
−0015 −0005 −0005 −0010 −0.4 −0.3 −0.1   0.0

SHELTER
Good shelter from all directions. Excellent anchorage.
NAVIGATION
Waypoint 55°26'.30N 05°31'.39W, 060°/240° from/to front Ldg Bn 240°, 2.6M. Strong winds gust over the hills to the SW when wind in that direction.
LIGHTS AND MARKS
Davaar Fl(2) 10s 37m 23M. Ldg Bns 240°.
RADIO TELEPHONE (local times)
VHF Ch 16; 12 14 (0845-1645)
TELEPHONE (0586)
Hr Mr 52552; CG 52770; MRCC Greenock 29988; HMC 52261; Marinecall 0898 500 462; Dr 52105
FACILITIES
EC Wednesday; **Piers** Tel. 52552, Slip, P, D, L, FW, ME, El, Sh, C (4 ton mobile), CH, AB, V, R, Bar; **Town Jetty** L, FW, ME, CH, AB; **New Quay** Slip; **Campbeltown Shipyard** Tel. 52881 L, FW, ME, Sh, CH, AB; **Campbeltown S.C.** Slip (dinghies only), M, L, FW, ME, CH, AB; **Duncan McPhee** Tel. 52115/52772, Slip,

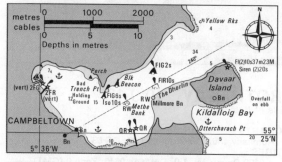

M, ME, El, Sh, C, CH; **County Garage** Tel. 52235 Gas; **Town** P, D, V, R, Bar. PO; Bank; Rly (Air or bus to Glasgow); Air.

EAST LOCH TARBERT 10-9-9
Argyll

CHARTS
Admiralty 2381; Imray C63; OS 62
TIDES
+0120 Dover; ML 1.9; Duration 0640; Zone 0 (GMT).
Standard Port GREENOCK (→)

Times				Height (metres)			
HW		LW		MHWS	MHWN	MLWN	MLWS
0000	0600	0000	0600	3.4	2.9	1.0	0.4
1200	1800	1200	1800				

Differences EAST LOCH TARBERT
+0005 +0005 −0020 +0015 0.0 0.0 +0.1 −0.1

SHELTER
Very good indeed in all weathers but gets very crowded. Access at all times.

NAVIGATION
Waypoint 55°52'.02N 05°22'.96W, 090°/270° from/to FlR 2.5s Lt, 0.70M. Entrance is very narrow. Cock Isle divides the entrance into two. Buteman's Hole to the N, Main harbour to the S. N anchorages are fouled by heavy moorings and lost chains.
LIGHTS AND MARKS
Leading marks, Bn, G column (QG Lt) in line with tower (conspic) at 239°.
RADIO TELEPHONE
VHF Ch 16 (0900-1700)
TELEPHONE (0586)
Hr Mr 344; CG 540; MRCC Greenock 29988; HMC Greenock 28311; Marinecall 0898 500 462; Hosp Lochgilphead 2323.

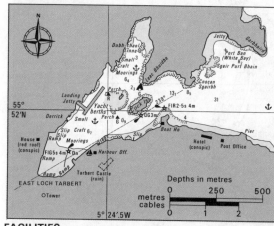

FACILITIES
EC Wednesday; **Old Quay** D, FW; **Tarbert Y.C.** Slip, L; **W. B. Leitch & Son** Tel. 287, SM, Lau, ACA; **M. Elliot** Tel. 560, ME; **A MacCallum** Tel. 209, Sh. **Tarbert (Argyll) Fishermen** Tel. 421, D, V, CH; **Yacht Berthing Facility** AB, FW, L, AC; **A McKay** Tel. 215 Gas; **Town** P and D (cans), Gas, Gaz, L, V, R, Bar. PO (Tarbert); Bank (Tarbert); Rly (bus to Glasgow); Air (Glasgow or Campbeltown).

ARDRISHAIG 10-9-10
Argyll

CHARTS
Admiralty 2381; Imray C63; OS 55
TIDES
+0110 Dover; ML 1.9; Duration 0640; Zone 0 (GMT).
Standard Port GREENOCK (→)

Times				Height (metres)			
HW		LW		MHWS	MHWN	MLWN	MLWS
0000	0600	0000	0600	3.4	2.9	1.0	0.4
1200	1800	1200	1800				

Differences ARDRISHAIG
+0006 +0006 −0015 +0020 0.0 0.0 +0.1 −0.1
INVERARY
+0011 +0011 +0034 +0034 −0.1 +0.1 −0.5 −0.2

SHELTER
Harbour is sheltered except in strong easterly winds. Lock into the Crinan Canal opens at all states of the tide. (See 10.9.7). Shelter can be obtained by going into the canal basin. Canal tolls payable at Ardrishaig or Crinan sea lock.
NAVIGATION
Waypoint No 48 port-hand buoy, FlR 4s, 56°00'.18N 05°26'.24W, 165°/345° from/to breakwater Lt, 0.61M. Tidal rocks to east of approach channel are dangerous.
LIGHTS AND MARKS
Lighted buoy W side of channel Fl R 4s. Unlit B buoy E side of channel. Both to the S of Sgeir Sgalog.
RADIO TELEPHONE
None.

ARDRISHAIG continued

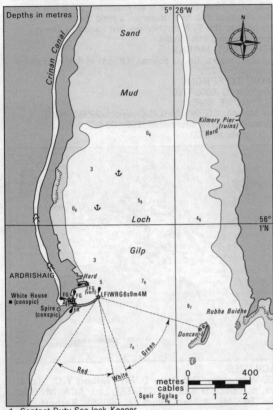

▲ Contact Duty Sea lock Keeper

TELEPHONE (0546)
Hr Mr 3210; CG Oban 3729; MRCC Greenock 29988; HMC 52261; Marinecall 0898 500 462; Dr 2921.

FACILITIES
EC Wednesday; **Ardrishaig Pier/Harbour** Tel. 3210, Slip, L, FW, AB; **Crinan Canal** M, L, FW, AB, R, Bar; **Crinan BY** Tel. 3232, Slip, L, FW, ME, El, Sh, CH, V; **Ardrishaig BY** Tel. 3280 ME, El, Sh, C, D (cans); Cranes up to 20 ton available Lochgilphead (2 M); **Mid-Argyll Caravans** Tel. 2003 Gas; **Village** P and D (cans), CH, V, R, Bar, PO; Bank; Rly (bus to Oban); Air (Glasgow or Campbeltown).

LAMLASH 10-9-11
Isle of Arran, Bute

CHARTS
Admiralty 1864; Imray C63; OS 69

TIDES
+0050 Dover; ML no data; Duration 0635; Zone 0 (GMT).
Standard Port GREENOCK (→)

Times				Height (metres)			
HW		LW		MHWS	MHWN	MLWN	MLWS
0000	0600	0000	0600	3.4	2.9	1.0	0.4
1200	1800	1200	1800				

Differences LAMLASH
−0016 −0036 −0024 −0004 −0.2 −0.2 No data
BRODICK BAY
0000 0000 +0005 +0005 −0.2 −0.2 0.0 0.0

SHELTER
Very good in all weathers. Anchorages — off Lamlash except in E. winds; off Kingscross, good except in N or NW strong winds; off the Farm on NW of Holy Is in E winds. Alternative Loch Ranza to the N (where new marina is planned see 10.9.19).

NAVIGATION
Waypoint 55°32'.63N 05°03'.00W, 090°/270° from/to North Channel buoy (Fl R 5s), 1.0M. Beware submarines which exercise frequently in this area, and also wreck of landing craft (charted) off farmhouse on Holy Island.

LIGHTS AND MARKS
Lights as shown on chartlet. There are two consecutive measured miles marked by poles north of Sannox, courses 322° or 142° (about 12 miles N of Lamlash).

RADIO TELEPHONE
None.

TELEPHONE (077 06)
MRCC Greenock 29988; HMC Ardrossan 63017; Marinecall 0898 500 462; Hosp. 214.

FACILITIES
EC Wednesday (Brodick and Lamlash); **Lamlash Jetty** Slip, M, L, V; **Brodick Pier** L, FW, AB. **Johnston's Marine Stores** Tel. 333, AB, Slip, FW, M, L. **Village** Bar, R, ME, P and D (cans), V, CH, PO; Bank; Rly (bus to Brodick, ferry to Ardrossan); Air (Glasgow or Prestwick).

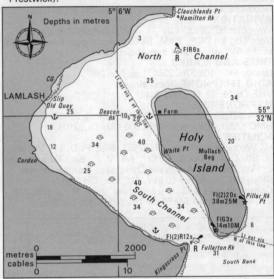

ROTHESAY 10-9-12
Isle of Bute, Bute

CHARTS
Admiralty 1867, 1907; Imray C63; OS 63

TIDES
+0100 Dover; ML 1.9; Duration 0640; Zone 0 (GMT).
Standard Port GREENOCK (→)

Times				Height (metres)			
HW		LW		MHWS	MHWN	MLWN	MLWS
0000	0600	0000	0600	3.4	2.9	1.0	0.4
1200	1800	1200	1800				

Differences ROTHESAY BAY
−0020 −0015 −0010 −0002 +0.2 +0.2 +0.2 +0.2

SHELTER
Good shelter except in strong N to NE winds. Good anchorage in bay. In N to NE winds, yachts advised to go to Kyles of Bute or Kames Bay. Major pier reconstruction due to start in 1989.

NAVIGATION
Waypoint 55°51'.00N 05°02'.69W, 014°/194° from/to Albert Pier Lt, 0.69M. Keep 1 ca off Bogany Pt coming from E, and Ardbeg Pt coming from W. Admiralty mooring buoys, unlit, in Rothesay Bay (painted with Dayglo paint).

LIGHTS AND MARKS
Approaching the harbour, 'A' buoy flashes yellow every 2 seconds. This is the most southerly of the mooring buoys. The bell is sounded on the main pier in fog when a steamer is expected.

AREA 9—SW Scotland

ROTHESAY continued

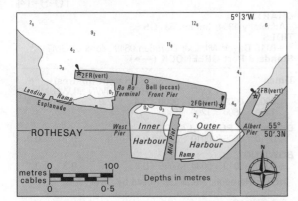

RADIO TELEPHONE (local times)
VHF Ch 16; 12 (1 May-30 Sept; 0600-2100, 1 Oct-30 Apl, 0600-1900).
TELEPHONE (0700)
Hr Mr 3842; CG Ardrossan 61014; MRCC Greenock 29988; HMC Greenock 26331; Marinecall 0898 500 462; Dr 3985; Hosp 3938
FACILITIES
EC Wednesday; **Outer Harbour (E)** Tel. 3842, Slip, D*, L, FW, ME, El, AB; **Outer Harbour (W)** Tel. 3842, D*, L, FW, ME, El, Sh, CH, AB; **Inner Harbour** L, FW, AB; **Main Pier** FW, ME, CH, R, Bar; **Albert Pier** D*, L, FW, C (4 ton mobile); **Scotia Islands** Tel. 2870, ME, Sh; **Town** P, D, CH, V, R, Bar; PO; Bank; Rly (ferry to Wemyss Bay); Air (Glasgow).
*By arrangement (min 200 galls)

HELENSBURGH/RHU 10-9-13
Dumbarton

CHARTS
Admiralty 2000, 1994; Imray C63; OS 56, 63
TIDES
+0110 Dover; ML 1.9; Duration 0640; Zone 0 (GMT).

SHELTER
Excellent in Gareloch — Entrance not difficult. The area is a restricted area. Helensburgh Pier may be used occasionally, but rather exposed.

Standard Port GREENOCK (→)

Times				Height (metres)			
HW		LW		MHWS	MHWN	MLWN	MLWS
0000	0600	0000	0600	3.4	2.9	1.0	0.4
1200	1800	1200	1800				
Differences HELENSBURGH							
0000	0000	0000	0000	0.0	0.0	0.0	0.0
COULPORT							
−0005	−0005	−0005	−0005	0.0	0.0	−0.1	−0.1
ARROCHAR							
−0005	−0005	−0005	−0005	0.0	0.0	−0.1	−0.1
GARELOCHHEAD							
0000	0000	0000	0000	0.0	0.0	0.0	−0.1

NAVIGATION
Waypoint 55°59'.30N 04°45'.19W, 176°/356° from/to front Ldg Lt 356°, 1.3M. Beware submarines from Faslane Base and large unlit MoD mooring buoys and barges off W shore of Gareloch. Gareloch entrance narrows to about 150m. If using Helensburgh Pier keep outside line of pierhead. Beware occasional steamer.
Beaches between Helensburgh Pier and Cairndhu Pt strewn with large boulders above and below MLWS.
NOTE: There are many restricted areas which are closed to all vessels during Nuclear Submarine movements (see W Coast of Scotland Pilot APP 2). Vessels are never allowed within 150m of naval installations; MoD police launches patrol area.

Harbour, Coastal and Tidal Information

HELENSBURGH/RHU continued

LIGHTS AND MARKS
Leading lights into Gareloch 356°, 318°, 291° and 163°.
RADIO TELEPHONE (local times)
VHF Ch 16. Rhu Marina Ch M (office hours). See also Greenock/Gourock.
TELEPHONE (0436)
Queen's Hr Mr 4321; MRCC Greenock 29988 & 29014; HMC Greenock 28311; Marinecall 0898 500 462; Hosp – Alexandria 54121; Dr 2277.
FACILITIES
EC Wednesday; **Rhu Marina** (150) Tel. 820652, D, FW, CH, AC, BH (4½ ton), Gas, Gaz; **Royal Northern & Clyde YC** Tel. 820322, L, FW, Bar; **Helensburgh SC** Tel. 2778 Slip (dinghies, overcrowded) L; **McGruer (Rosneath)** Tel. 831211, Sh, ME, El, M, L, BH (25 ton), Slip; **Timbacraft** Tel. 810391, Sh, ME, El, Slip; **Silvers Marine** Tel. 831222 FW, Sh, ME, El, L, M, BH (18 ton); **Nicholson Hughes** Tel. 831356, SM; **Clyde Corinthian YC** Tel. 820068; **Gareloch Moorings and Engineers** Tel. 820062 ME, El, M; **R. McAlister** Tel. Dumbarton 62396 Sh, ME, El, M, Slip, BH (10 ton), AB, FW, CH, D; **Helensburgh** EC Wednesday, Slip, FW, CH, V, R, Bar, P, D, L. PO; Bank; Rly; Air (Glasgow).

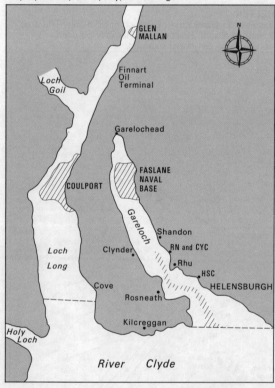

▨ Restricted Area-Max Speed 7Kn
--- Southern Limit of Dockyard Port

Statutory Instruments.

Statutory Instruments.
Clyde Dockyard Port of the Holy Loch Order 1140 of 1967; Clyde Dockyard Port of Gareloch and Loch Long Order 1141 of 1967; Faslane Coulport and Rhu Narrows Order 234 of 1986; Holy Loch Byelaws Order 1708 of 1986.
The S limit of restricted area marked by two Or posts with St Andrews Cross topmarks on Shandon foreshore. The W limit marked by three-sectored Lt at Gareloch Oil fuel depot N jetty. When restrictions are in force
1. Holy Loch, Coulport, Faslane
   by day : 3 G Lts (vert) supplemented by International Code pendant over pendant Nine.
   by night : 3 FG (vert) in conspic position
2. Entrance to Gareloch
   by day : R G G Lts (vert) supplemented by R flag with W diagonal bar.
   by night : R G G Lts (vert)

GREENOCK/GOUROCK
Renfrew
10-9-14

CHARTS
Admiralty 1994; Imray C63; OS 63
TIDES
+0115 Dover; ML 2.0; Duration 0640; Zone 0 (GMT)
Standard Port GREENOCK (→)

Times				Height (metres)			
HW		LW		MHWS	MHWN	MLWN	MLWS
0000	0600	0000	0600	3.4	2.9	1.0	0.4
1200	1800	1200	1800				

Differences PORT GLASGOW
+0010 +0005 +0010 +0020 +0.2 +0.1 0.0 0.0
BOWLING
+0020 +0010 +0030 +0055 +0.6 +0.5 +0.3 +0.1

NOTE: Harbours are all controlled by Clyde Port Authority, 16 Robertson St., Glasgow. Tel. Glasgow 221-8733. Greenock is a Standard Port and tidal predictions for each day of the year are given below.

SHELTER
Greenock — Commercial port
Gourock — very good anchorages on W side of bay but exposed to NE to N winds.
NAVIGATION
Gourock waypoint 55°58'.00N 04°49'.00W, 000°/180° from/to Kempock Pt Lt, 0.22M. No navigational dangers except the continuous shipping using the Clyde estuary and steamers using Railway Pier. Beware the bottom being fouled with lost tackle.

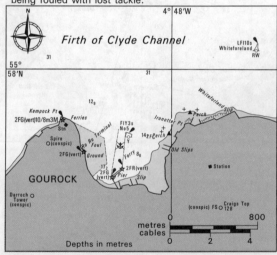

LIGHTS AND MARKS
(Gourock). At the end of the pier 2 FG (vert).
RADIO TELEPHONE
Call: *Clydeport Estuary Radio* VHF Ch 16; 12 (H24). Other station: Dunoon (call: *Dunoon Pier*) Ch 31; 12 16 **31** (0700-2035 LT Mon-Sat; 0900-2015 LT Sun).
TELEPHONE (0475)
Hr Mr 26221; CG 29988; MRCC 29988; HMC 28311; Weather Glasgow 246 8091; Marinecall 0898 500 462; Dr 20000.
FACILITIES
EC Wednesday; **Greenock** All facilities available by prior arrangement with Harbour Dept, but small craft not particularly welcome; **Jas. Adam & Son** Tel. 31346 Slip, P, ME, El, Sh, C, CH; **Mackenzie** Tel. 36196, SM; **G & R McKenzie** Tel. 38811, CH; **Royal Gourock YC** Tel. 32983 M, L, FW, ME, V, R, Bar; **Greenock Sail and Tent Co.**, Tel. 36196, SM; **Ashton Bay** M; **Gourock Bay** M, L; **Maybank** Slip; **Town (Gourock)** P, D, V, R, Bar, PO; Bank; Rly; Air (Glasgow).
Note: Clyde Yacht Clubs Assn is at Anchor House, Blackhall Lane, Paisley 041-887-8296. The Clyde Cruising Club is aboard S.V. Carrick, Glasgow 041-552-2183.

SCOTLAND, WEST COAST - GREENOCK

LAT 55°57'N   LONG 4°46'W

TIME ZONE UT(GMT)
For Summer Time add ONE hour in non-shaded areas

TIMES AND HEIGHTS OF HIGH AND LOW WATERS

YEAR **1989**

JANUARY

	TIME	m		TIME	m
1 SU	0629 1210 1826	2.8 1.4 3.1	**16** M	0013 0630 1239 1916	0.8 3.1 1.2 3.2
2 M	0021 0731 1324 1929	1.2 2.8 1.5 3.0	**17** TU	0124 0741 1405 2041	0.9 3.0 1.2 3.1
3 TU	0126 0840 1434 2034	1.3 2.9 1.5 3.1	**18** W	0235 0902 1513 2157	0.9 3.0 1.0 3.1
4 W	0235 0937 1526 2134	1.3 3.0 1.4 3.2	**19** TH	0334 1006 1606 2300	0.9 3.1 0.9 3.1
5 TH	0332 1023 1610 2229	1.2 3.1 1.1 3.2	**20** F	0426 1059 1653 2355	0.8 3.3 0.7 3.2
6 F	0420 1106 1651 2320	1.0 3.3 0.8 3.2	**21** SA ○	0513 1146 1735	0.7 3.4 0.6
7 SA ●	0505 1147 1731	0.8 3.4 0.6	**22** SU	0046 0558 1230 1817	3.2 0.6 3.6 0.4
8 SU	0011 0549 1231 1813	3.2 0.7 3.5 0.4	**23** M	0131 0640 1311 1855	3.2 0.6 3.7 0.4
9 M	0101 0632 1315 1857	3.3 0.6 3.6 0.3	**24** TU	0211 0718 1349 1931	3.2 0.6 3.7 0.4
10 TU	0151 0717 1401 1943	3.3 0.6 3.7 0.3	**25** W	0247 0754 1424 2003	3.1 0.6 3.7 0.5
11 W	0239 0804 1447 2031	3.3 0.6 3.8 0.4	**26** TH	0318 0827 1457 2035	3.1 0.7 3.6 0.6
12 TH	0325 0852 1533 2122	3.4 0.7 3.8 0.4	**27** F	0348 0857 1532 2106	3.1 0.8 3.5 0.7
13 F	0410 0942 1619 2215	3.4 0.8 3.8 0.5	**28** SA	0416 0928 1608 2141	3.1 1.0 3.4 0.9
14 SA ☽	0452 1032 1708 2311	3.3 0.9 3.6 0.7	**29** SU	0446 1003 1649 2219	3.0 1.2 3.3 1.1
15 SU	0537 1128 1805	3.2 1.0 3.4	**30** M ☾	0522 1047 1735 2309	3.0 1.4 3.2 1.2
			31 TU	0608 1152 1831	2.9 1.5 3.1

FEBRUARY

	TIME	m		TIME	m
1 W	0012 0708 1333 1936	1.4 2.9 1.6 3.0	**16** TH	0223 0826 1503 2201	1.1 2.9 1.0 2.8
2 TH	0145 0833 1454 2053	1.4 2.9 1.4 3.0	**17** F	0327 0958 1559 2308	0.9 3.0 0.8 3.0
3 F	0309 0949 1549 2207	1.3 3.1 1.0 3.1	**18** SA	0420 1055 1645 2358	0.8 3.2 0.6 3.1
4 SA	0406 1045 1634 2308	1.1 3.2 0.7 3.2	**19** SU	0506 1140 1726	1.0 3.4 0.4
5 SU	0452 1133 1716	0.8 3.4 0.4	**20** M ○	0039 0546 1219 1803	3.1 0.5 3.5 0.3
6 M	0003 0536 1220 1758	3.2 0.6 3.6 0.2	**21** TU	0117 0623 1257 1836	3.1 0.5 3.6 0.3
7 TU	0054 0618 1307 1841	3.3 0.5 3.7 0.1	**22** W	0151 0656 1332 1906	3.1 0.4 3.6 0.3
8 W	0142 0659 1353 1924	3.3 0.4 3.9 0.1	**23** TH	0221 0725 1403 1932	3.1 0.5 3.6 0.4
9 TH	0227 0742 1437 2008	3.4 0.5 3.9 0.3	**24** F	0247 0751 1432 1957	3.1 0.6 3.5 0.5
10 F	0308 0826 1520 2054	3.5 0.6 3.9 0.4	**25** SA	0310 0815 1501 2024	3.1 0.7 3.5 0.7
11 SA	0346 0910 1601 2142	3.5 0.7 3.8 0.6	**26** SU	0334 0843 1533 2055	3.2 0.8 3.4 0.8
12 SU	0423 0956 1646 2235	3.5 0.8 3.6 0.8	**27** M ☽	0401 0916 1611 2134	3.2 1.0 3.3 1.0
13 M	0502 1049 1737 2339	3.4 1.0 3.3 1.0	**28** TU ☾	0436 0959 1657 2224	3.1 1.2 3.2 1.2
14 TU ☾	0548 1204 1841	3.2 1.2 3.0			
15 W	0101 0647 1348 2020	1.1 3.0 1.2 2.8			

MARCH

	TIME	m		TIME	m
1 W	0518 1103 1750 2331	3.1 1.4 3.1 1.4	**16** TH	0044 0610 1331 2011	1.1 3.0 1.0 2.6
2 TH	0611 1245 1853	3.0 1.4 3.0	**17** F	0208 0743 1448 2205	1.1 2.8 0.8 2.7
3 F	0114 0729 1424 2020	1.4 2.9 1.2 2.9	**18** SA	0315 0944 1544 2300	0.9 2.9 0.6 2.9
4 SA	0250 0920 1526 2156	1.3 3.0 0.8 3.0	**19** SU	0407 1040 1629 2340	0.7 3.1 0.5 3.0
5 SU	0349 1027 1614 2257	1.0 3.2 0.5 3.1	**20** M	0450 1121 1707	0.6 3.2 0.4
6 M	0435 1117 1656 2348	0.7 3.4 0.2 3.2	**21** TU	0015 0527 1158 1740	3.1 0.5 3.3 0.3
7 TU	0517 1204 1737 ○	0.5 3.6 0.0	**22** W ○	0048 0600 1233 1809	3.1 0.4 3.4 0.3
8 W	0035 0556 1252 1818	3.3 0.3 3.8 0.0	**23** TH	0118 0629 1306 1835	3.1 0.4 3.4 0.4
9 TH	0121 0636 1338 1859	3.4 0.3 3.8 0.1	**24** F	0146 0653 1335 1859	3.1 0.5 3.4 0.4
10 F	0204 0716 1420 1941	3.5 0.4 3.9 0.2	**25** SA	0210 0716 1402 1923	3.1 0.5 3.4 0.5
11 SA	0242 0757 1501 2025	3.5 0.5 3.9 0.5	**26** SU	0231 0741 1430 1951	3.2 0.6 3.4 0.6
12 SU	0318 0810 1541 2113	3.6 0.7 3.8 0.7	**27** M	0255 0810 1503 2025	3.2 0.7 3.3 0.8
13 M	0352 0925 1625 2207	3.4 0.7 3.5 0.9	**28** TU	0325 0847 1543 2108	3.2 0.8 3.3 0.9
14 TU ☽	0430 1020 1713 2314	3.2 0.9 3.2 1.1	**29** W	0401 0937 1630 2203	3.1 1.0 3.2 1.1
15 W	0515 1143 1816	3.2 1.1 2.9	**30** TH ☾	0445 1047 1722 2317	3.2 1.1 3.1 1.3
			31 F	0537 1223 1825	3.0 1.1 2.9

APRIL

	TIME	m		TIME	m
1 SA	0056 0648 1354 1957	1.3 2.9 0.9 2.8	**16** SU	0251 0904 1516 2230	0.9 2.8 0.6 2.8
2 SU	0226 0849 1459 2140	1.1 3.0 0.6 2.9	**17** M	0344 1005 1601 2306	0.7 3.0 0.5 3.0
3 M	0327 1001 1549 2236	0.9 3.2 0.3 3.1	**18** TU	0426 1047 1638 2339	0.6 3.1 0.4 3.0
4 TU	0413 1053 1631 2322	0.6 3.4 0.1 3.2	**19** W	0501 1124 1709	0.6 3.2 0.4
5 W	0453 1141 1712	0.5 3.6 0.0	**20** TH	0011 0531 1159 1737	3.1 0.5 3.2 0.4
6 TH ●	0007 0531 1229 1752	3.3 0.4 3.7 0.1	**21** F ○	0040 0558 1232 1802	3.1 0.5 3.3 0.5
7 F	0051 0611 1316 1833	3.4 0.3 3.8 0.2	**22** SA	0106 0623 1302 1828	3.1 0.5 3.2 0.5
8 SA	0132 0651 1359 1916	3.4 0.4 3.8 0.4	**23** SU	0131 0648 1332 1856	3.1 0.5 3.2 0.5
9 SU	0211 0733 1440 2001	3.6 0.5 3.7 0.6	**24** M	0155 0718 1403 1929	3.2 0.5 3.2 0.6
10 M	0246 0816 1521 2050	3.7 0.5 3.6 0.7	**25** TU	0223 0752 1441 2009	3.1 0.6 3.2 0.7
11 TU	0323 0904 1606 2147	3.6 0.7 3.3 0.9	**26** W	0256 0836 1524 2058	3.3 0.7 3.2 0.9
12 W	0403 1004 1655 2257	3.5 0.8 3.0 1.0	**27** TH ☽	0336 0933 1612 2158	3.3 0.7 3.1 1.0
13 TH	0448 1127 1757	3.3 0.9 2.7	**28** F	0422 1044 1706 2311	3.2 0.8 3.0 1.1
14 F ☾	0023 0543 1304 1957	1.1 3.0 0.8 2.6	**29** SA	0515 1204 1807	3.1 0.8 2.9
15 SA	0144 0703 1419 2140	1.0 2.8 0.7 2.7	**30** SU	0034 0625 1322 1935	1.1 3.0 0.6 2.8

Chart Datum: 1.62 metres below Ordnance Datum (Newlyn)

SCOTLAND, WEST COAST - GREENOCK
LAT 55°57'N LONG 4°46'W
TIMES AND HEIGHTS OF HIGH AND LOW WATERS

YEAR 1989

TIME ZONE UT(GMT) — For Summer Time add ONE hour in non-shaded areas

Chart Datum: 1.62 metres below Ordnance Datum (Newlyn)

MAY

Day	Time	m	Day	Time	m
1 M	0153 / 0812 / 1425 / 2108	1.0 / 3.0 / 0.4 / 2.9	16 TU	0307 / 0911 / 1518 / 2222	0.9 / 2.8 / 0.7 / 2.9
2 TU	0257 / 0928 / 1517 / 2203	0.8 / 3.2 / 0.3 / 3.0	17 W	0352 / 1000 / 1557 / 2256	0.8 / 3.0 / 0.7 / 2.9
3 W	0345 / 1022 / 1601 / 2249	0.7 / 3.4 / 0.2 / 3.2	18 TH	0428 / 1041 / 1631 / 2329	0.8 / 3.0 / 0.7 / 3.0
4 TH	0427 / 1113 / 1644 / 2334	0.5 / 3.5 / 0.2 / 3.3	19 F	0459 / 1119 / 1701 / 2359	0.7 / 3.1 / 0.7 / 3.0
5 F	0507 / 1202 / 1726 / ●	0.5 / 3.6 / 0.2	20 SA	0529 / 1156 / 1731 / ○	0.7 / 3.1 / 0.6
6 SA	0018 / 0548 / 1251 / 1810	3.4 / 0.5 / 3.6 / 0.4	21 SU	0028 / 0558 / 1231 / 1803	3.1 / 0.6 / 3.1 / 0.6
7 SU	0100 / 0630 / 1337 / 1856	3.5 / 0.4 / 3.6 / 0.5	22 M	0057 / 0628 / 1307 / 1838	3.1 / 0.5 / 3.1 / 0.6
8 M	0140 / 0713 / 1421 / 1944	3.6 / 0.4 / 3.5 / 0.6	23 TU	0126 / 0702 / 1345 / 1918	3.2 / 0.4 / 3.1 / 0.6
9 TU	0219 / 0800 / 1505 / 2035	3.7 / 0.5 / 3.3 / 0.8	24 W	0200 / 0743 / 1427 / 2002	3.3 / 0.4 / 3.1 / 0.7
10 W	0258 / 0850 / 1551 / 2131	3.7 / 0.6 / 3.1 / 0.9	25 TH	0238 / 0832 / 1513 / 2054	3.3 / 0.5 / 3.1 / 0.8
11 TH	0340 / 0950 / 1642 / 2237	3.6 / 0.7 / 2.9 / 0.9	26 F	0321 / 0928 / 1603 / 2152	3.3 / 0.5 / 3.1 / 0.8
12 F	0426 / 1102 / 1742 / 2350 ☽	3.3 / 0.7 / 2.7 / 1.0	27 SA	0409 / 1031 / 1655 / 2256	3.4 / 0.5 / 3.0 / 0.9
13 SA	0520 / 1220 / 1909 ☾	3.1 / 0.7 / 2.6	28 SU	0502 / 1138 / 1752	3.3 / 0.5 / 2.9
14 SU	0104 / 0628 / 1332 / 2045	1.0 / 2.9 / 0.7 / 2.6	29 M	0004 / 0606 / 1244 / 1902	0.9 / 3.1 / 0.5 / 2.9
15 M	0211 / 0800 / 1431 / 2141	0.9 / 3.0 / 0.7 / 2.7	30 TU	0113 / 0732 / 1346 / 2024	0.9 / 3.1 / 0.4 / 2.9
			31 W	0220 / 0850 / 1441 / 2124	0.9 / 3.1 / 0.4 / 3.0

JUNE

Day	Time	m	Day	Time	m
1 TH	0315 / 0950 / 1532 / 2215	0.8 / 3.2 / 0.4 / 3.1	16 F	0352 / 0955 / 1551 / 2246	1.0 / 2.9 / 0.9 / 2.9
2 F	0403 / 1046 / 1620 / 2303	0.7 / 3.3 / 0.4 / 3.2	17 SA	0429 / 1040 / 1630 / 2322	0.9 / 3.0 / 0.9 / 3.0
3 SA	0447 / 1139 / 1706 / 2349 ●	0.6 / 3.3 / 0.3 / 3.4	18 SU	0503 / 1124 / 1707 / 2356	0.8 / 3.0 / 0.8 / 3.1
4 SU	0531 / 1231 / 1753	0.5 / 3.3 / 0.5	19 M	0536 / 1206 / 1746	0.6 / 3.0 / 0.7
5 M	0034 / 0615 / 1320 / 1841	3.5 / 0.4 / 3.3 / 0.6	20 TU	0031 / 0613 / 1250 / 1826	3.2 / 0.4 / 2.9 / 0.6
6 TU	0117 / 0701 / 1409 / 1930	3.6 / 0.4 / 3.2 / 0.6	21 W	0108 / 0651 / 1335 / 1907	3.2 / 0.3 / 3.0 / 0.6
7 W	0159 / 0748 / 1455 / 2020	3.7 / 0.4 / 3.1 / 0.7	22 TH	0147 / 0734 / 1421 / 1954	3.3 / 0.3 / 3.0 / 0.6
8 TH	0240 / 0836 / 1541 / 2111	3.6 / 0.4 / 3.0 / 0.7	23 F	0229 / 0822 / 1509 / 2042	3.4 / 0.3 / 3.1 / 0.6
9 F	0322 / 0928 / 1628 / 2206	3.6 / 0.5 / 2.9 / 0.8	24 SA	0313 / 0913 / 1555 / 2135	3.5 / 0.3 / 3.1 / 0.7
10 SA	0406 / 1024 / 1717 / 2303	3.4 / 0.6 / 2.8 / 0.9	25 SU	0359 / 1008 / 1643 / 2230	3.5 / 0.4 / 3.1 / 0.7
11 SU	0455 / 1123 / 1813 ☽	3.2 / 0.7 / 2.7	26 M	0449 / 1105 / 1731 / 2327 ☾	3.4 / 0.4 / 3.1 / 0.8
12 M	0005 / 0549 / 1223 / 1921	1.0 / 3.0 / 0.8 / 2.6	27 TU	0545 / 1205 / 1824	3.3 / 0.5 / 3.0
13 TU	0112 / 0653 / 1323 / 2030	1.1 / 2.8 / 0.9 / 2.7	28 W	0030 / 0654 / 1306 / 1932	0.9 / 3.1 / 0.5 / 2.9
14 W	0216 / 0803 / 1419 / 2125	1.1 / 2.8 / 0.9 / 2.7	29 TH	0143 / 0814 / 1410 / 2045	1.1 / 3.0 / 0.6 / 2.9
15 TH	0310 / 0903 / 1510 / 2208	1.1 / 2.8 / 1.0 / 2.8	30 F	0250 / 0926 / 1510 / 2147	0.9 / 3.0 / 0.6 / 3.0

JULY

Day	Time	m	Day	Time	m
1 SA	0346 / 1028 / 1604 / 2240	0.7 / 3.0 / 0.6 / 3.1	16 SU	0400 / 1006 / 1605 / 2251	1.0 / 2.9 / 1.0 / 3.0
2 SU	0434 / 1126 / 1653 / 2331	0.6 / 3.1 / 0.6 / 3.3	17 M	0439 / 1059 / 1648 / 2332	0.7 / 2.9 / 0.8 / 3.1
3 M	0521 / 1222 / 1742	0.5 / 3.1 / 0.6	18 TU	0518 / 1150 / 1729 ○	0.5 / 2.9 / 0.6
4 TU	0018 / 0606 / 1315 / 1829	3.4 / 0.4 / 3.0 / 0.6	19 W	0014 / 0556 / 1239 / 1811	3.2 / 0.3 / 2.9 / 0.5
5 W	0104 / 0650 / 1403 / 1915	3.5 / 0.3 / 3.0 / 0.6	20 TH	0056 / 0636 / 1328 / 1853	3.3 / 0.1 / 2.9 / 0.4
6 TH	0146 / 0733 / 1447 / 1959	3.6 / 0.3 / 3.0 / 0.6	21 F	0139 / 0718 / 1414 / 1936	3.5 / 0.1 / 3.0 / 0.4
7 F	0227 / 0815 / 1527 / 2043	3.6 / 0.3 / 3.0 / 0.6	22 SA	0222 / 0802 / 1459 / 2021	3.6 / 0.1 / 3.1 / 0.5
8 SA	0306 / 0856 / 1605 / 2126	3.6 / 0.4 / 2.9 / 0.7	23 SU	0305 / 0849 / 1541 / 2107	3.6 / 0.2 / 3.2 / 0.6
9 SU	0344 / 0938 / 1643 / 2208	3.4 / 0.5 / 2.9 / 0.8	24 M	0348 / 0938 / 1622 / 2156	3.6 / 0.3 / 3.2 / 0.6
10 M	0425 / 1020 / 1721 / 2253	3.3 / 0.7 / 2.8 / 1.0	25 TU	0434 / 1030 / 1704 / 2248 ☾	3.5 / 0.5 / 3.2 / 0.8
11 TU	0510 / 1103 / 1804 / 2348 ☽	3.1 / 0.9 / 2.7 / 1.2	26 W	0524 / 1127 / 1750 / 2350	3.3 / 0.6 / 3.0 / 0.9
12 W	0559 / 1153 / 1856	2.9 / 1.0 / 2.7	27 TH	0624 / 1233 / 1846	3.1 / 0.8 / 2.9
13 TH	0102 / 0657 / 1256 / 2003	1.3 / 2.8 / 1.2 / 2.7	28 F	0113 / 0744 / 1350 / 2005	1.0 / 2.9 / 0.8 / 2.8
14 F	0219 / 0801 / 1413 / 2111	1.3 / 3.0 / 1.2 / 2.8	29 SA	0236 / 0914 / 1459 / 2129	0.9 / 2.9 / 0.8 / 2.9
15 SA	0315 / 0907 / 1516 / 2205	1.2 / 2.8 / 1.2 / 2.9	30 SU	0337 / 1024 / 1555 / 2232	0.7 / 2.9 / 0.7 / 3.0
			31 M	0427 / 1130 / 1645 / 2324	0.5 / 2.9 / 0.6 / 3.2

AUGUST

Day	Time	m	Day	Time	m
1 TU	0513 / 1223 / 1732	0.4 / 2.9 / 0.6	16 W	0456 / 1137 / 1710 / 2358	0.3 / 2.9 / 0.6 / 3.3
2 W	0011 / 0555 / 1311 / 1815	3.3 / 0.3 / 3.0 / 0.5	17 TH	0536 / 1226 / 1750 ○	0.1 / 3.0 / 0.4
3 TH	0055 / 0635 / 1354 / 1856	3.5 / 0.2 / 3.0 / 0.4	18 F	0043 / 0615 / 1313 / 1830	3.5 / -0.1 / 3.0 / 0.4
4 F	0135 / 0712 / 1430 / 1933	3.5 / 0.2 / 3.0 / 0.4	19 SA	0128 / 0655 / 1359 / 1911	3.6 / 0.0 / 3.1 / 0.4
5 SA	0212 / 0747 / 1504 / 2008	3.5 / 0.3 / 3.0 / 0.5	20 SU	0211 / 0737 / 1441 / 1953	3.7 / 0.1 / 3.3 / 0.4
6 SU	0245 / 0818 / 1534 / 2040	3.5 / 0.4 / 3.0 / 0.6	21 M	0252 / 0821 / 1519 / 2037	3.8 / 0.2 / 3.3 / 0.5
7 M	0319 / 0849 / 1603 / 2113	3.4 / 0.6 / 2.9 / 0.8	22 TU	0334 / 0907 / 1557 / 2123	3.7 / 0.4 / 3.4 / 0.6
8 TU	0353 / 0921 / 1633 / 2147	3.3 / 0.7 / 2.9 / 1.0	23 W	0416 / 0958 / 1636 / 2214 ☾	3.5 / 0.6 / 3.3 / 0.8
9 W	0431 / 0957 / 1708 / 2230 ☾	3.1 / 0.9 / 2.9 / 1.2	24 TH	0504 / 1056 / 1719 / 2320	3.3 / 0.9 / 3.1 / 1.0
10 TH	0515 / 1042 / 1750 / 2331	3.0 / 1.1 / 2.8 / 1.3	25 F	0602 / 1213 / 1812	3.0 / 1.0 / 2.9
11 F	0606 / 1142 / 1843	2.9 / 1.3 / 2.7	26 SA	0058 / 0726 / 1340 / 1929	1.0 / 2.7 / 1.0 / 2.8
12 SA	0115 / 0707 / 1314 / 1957	1.4 / 2.8 / 1.4 / 2.7	27 SU	0228 / 0925 / 1453 / 2121	0.9 / 2.7 / 0.9 / 2.8
13 SU	0238 / 0821 / 1446 / 2125	1.2 / 2.8 / 1.3 / 2.8	28 M	0330 / 1041 / 1549 / 2230	0.6 / 2.8 / 0.7 / 3.0
14 M	0332 / 0941 / 1544 / 2226	1.0 / 2.8 / 1.1 / 3.0	29 TU	0418 / 1132 / 1636 / 2318	0.4 / 2.9 / 0.6 / 3.2
15 TU	0416 / 1045 / 1630 / 2314	0.6 / 2.9 / 0.7 / 3.1	30 W	0501 / 1215 / 1719	0.3 / 2.9 / 0.5
			31 TH	0001 / 0539 / 1254 / 1757 ●	3.3 / 0.2 / 3.0 / 0.4

SCOTLAND, WEST COAST - GREENOCK
LAT 55°57'N LONG 4°46'W

TIMES AND HEIGHTS OF HIGH AND LOW WATERS

YEAR 1989

TIME ZONE UT(GMT)
For Summer Time add ONE hour in non-shaded areas

AREA 9—SW Scotland

	SEPTEMBER				OCTOBER				NOVEMBER				DECEMBER			
	TIME	m	TIME	m	TIME	m	TIME	m	TIME	m	TIME	m	TIME	m	TIME	m
1 F	0041 0615 1330 1833	3.4 0.2 3.0 0.4	**16** SA	0025 0549 1249 1805	3.6 0.0 3.2 0.3	**1** SU	0052 0618 1328 1834	3.4 0.4 3.1 0.5	**16** M	0048 0605 1302 1820	3.7 0.2 3.5 0.4	**1** W	0122 0643 1344 1902	3.3 0.7 3.2 0.6	**16** TH	0202 0719 1359 1935	3.6 0.7 3.8 0.6	**1** F	0135 0702 1348 1927	3.2 0.7 3.4 0.6	**16** SA	0242 0758 1427 2015	3.4 0.8 3.9 0.6
2 SA	0117 0647 1403 1904	3.5 0.2 3.0 0.4	**17** SU	0110 0630 1333 1845	3.7 0.0 3.3 0.3	**2** M	0122 0643 1354 1900	3.4 0.4 3.1 0.5	**17** TU	0134 0647 1345 1903	3.8 0.4 3.6 0.5	**2** TH	0153 0713 1411 1935	3.2 0.7 3.3 0.7	**17** F	0249 0810 1442 2027	3.5 0.8 3.8 0.7	**2** SA	0214 0743 1424 2010	3.1 0.8 3.4 0.6	**17** SU	0328 0848 1511 2106	3.3 0.8 3.8 0.6
3 SU	0151 0715 1432 1932	3.5 0.3 3.0 0.5	**18** M	0154 0711 1413 1926	3.8 0.2 3.4 0.4	**3** TU	0151 0708 1419 1925	3.3 0.5 3.1 0.6	**18** W	0217 0733 1424 1948	3.7 0.6 3.7 0.6	**3** F	0227 0750 1443 2016	3.2 0.8 3.3 0.8	**18** SA	0337 0906 1526 2126	3.3 0.9 3.7 0.7	**3** SU	0256 0828 1505 2101	3.2 0.9 3.5 0.7	**18** M	0414 0940 1556 2200	3.2 0.9 3.7 0.7
4 M	0220 0741 1457 1959	3.4 0.5 3.0 0.6	**19** TU	0236 0754 1451 2008	3.6 0.4 3.5 0.5	**4** W	0219 0733 1443 1953	3.3 0.7 3.1 0.7	**19** TH	0300 0822 1503 2037	3.6 0.8 3.7 0.7	**4** SA	0308 0833 1522 2109	3.1 1.0 3.3 0.9	**19** SU	0427 1006 1614 2233	3.1 1.0 3.6 0.8	**4** M	0342 0920 1550 2158	3.2 1.0 3.5 0.7	**19** TU ☾	0500 1034 1643 2257	3.1 1.0 3.5 0.8
5 TU	0249 0807 1522 2026	3.4 0.6 3.0 0.7	**20** W	0317 0840 1530 2055	3.7 0.6 3.5 0.7	**5** TH	0250 0805 1512 2029	3.2 0.9 3.2 0.9	**20** F	0346 0918 1545 2137	3.4 0.9 3.6 0.8	**5** SU	0353 0928 1607 2214	3.1 1.1 3.3 0.9	**20** M ☾	0525 1116 1707 2348	2.9 1.1 3.3 0.8	**5** TU	0430 1018 1640 2300	3.2 1.0 3.4 0.7	**20** W	0551 1133 1735 2358	2.9 1.1 3.3 0.9
6 W	0320 0836 1550 2058	3.3 0.8 3.0 0.9	**21** TH	0400 0933 1609 2151	3.5 0.8 3.4 0.9	**6** F	0327 0844 1548 2117	3.1 1.0 3.2 1.0	**21** SA ☾	0436 1025 1632 2254	3.1 1.1 3.4 0.9	**6** M ☾	0445 1034 1658 2328	3.1 1.2 3.2 0.9	**21** TU	0640 1229 1810	2.8 1.1 3.1	**6** W	0521 1119 1737	3.2 1.1 3.4	**21** TH	0650 1241 1836	2.9 1.2 3.1
7 TH	0356 0911 1623 2140	3.2 0.9 3.0 1.1	**22** F ☾	0448 1038 1653 2305	3.2 1.0 3.3 1.0	**7** SA	0411 0936 1631 2223	3.1 1.1 3.1 1.1	**22** SU	0537 1145 1727	2.9 1.1 3.1	**7** TU	0541 1150 1759	3.1 1.2 3.1	**22** W	0101 0813 1339 1933	0.8 3.1 1.1 3.0	**7** TH	0005 0619 1225 1848	0.7 3.1 1.1 3.3	**22** F	0103 0802 1352 1945	1.0 2.8 1.3 3.0
8 F ☾	0438 0957 1705 2242	3.0 1.1 2.9 1.3	**23** SA	0547 1201 1747	2.9 1.1 3.0	**8** SU ☾	0502 1043 1722 2351	3.0 1.2 3.0 1.1	**23** M	0027 0651 1307 1839	0.9 2.7 1.1 2.9	**8** W	0043 0651 1307 1925	0.8 2.9 1.2 3.1	**23** TH	0205 0916 1440 2050	0.8 2.9 1.0 3.0	**8** F	0108 0732 1334 2012	0.7 3.1 1.1 3.3	**23** SA	0207 0905 1452 2050	1.1 2.9 1.3 3.0
9 SA	0529 1101 1755	2.9 1.3 2.9	**24** SU	0049 0722 1330 1902	0.9 2.6 1.1 2.8	**9** M	0601 1213 1826	2.9 1.3 2.9	**24** TU	0147 0911 1418 2032	0.7 2.7 1.0 2.9	**9** TH	0149 0823 1416 2054	0.6 3.0 1.1 3.2	**24** F	0258 1001 1528 2145	0.8 3.0 1.0 3.1	**9** SA	0208 0845 1439 2122	0.6 3.1 1.0 3.4	**24** SU	0304 0955 1540 2147	1.1 3.0 1.2 3.1
10 SU	0019 0628 1236 1901	1.3 2.8 1.4 2.8	**25** M	0214 0934 1441 2109	0.8 2.7 0.9 2.8	**10** TU	0120 0718 1345 2004	1.0 2.8 1.2 2.9	**25** W	0248 1004 1514 2143	0.6 2.9 0.8 2.9	**10** F	0245 0927 1510 2153	0.5 3.1 0.9 3.4	**25** SA	0342 1038 1609 2229	0.8 3.1 0.9 3.2	**10** SU	0304 0942 1532 2220	0.6 3.3 0.9 3.5	**25** M	0350 1037 1621 2235	1.1 3.1 1.1 3.1
11 M	0159 0745 1418 2044	1.1 2.7 1.3 2.8	**26** TU	0315 1034 1537 2216	0.5 2.8 0.7 3.0	**11** W	0228 0902 1451 2132	0.7 2.9 1.0 3.1	**26** TH	0335 1042 1559 2228	0.5 3.0 0.7 3.1	**11** SA	0332 1016 1555 2245	0.3 3.2 0.8 3.6	**26** SU	0419 1112 1644 2309	0.8 3.2 0.9 3.2	**11** M	0355 1034 1620 2315	0.6 3.4 0.8 3.5	**26** TU	0430 1114 1656 2319	1.0 3.2 1.1 3.2
12 TU	0302 0925 1521 2202	0.8 2.8 1.0 3.0	**27** W	0403 1115 1622 2301	0.4 2.9 0.6 3.2	**12** TH	0320 1003 1540 2226	0.4 3.0 0.8 3.4	**27** F	0416 1117 1636 2308	0.5 3.1 0.7 3.3	**12** SU	0416 1101 1637 2335	0.3 3.4 0.6 3.6	**27** M	0453 1145 1717 2347	0.8 3.2 0.8 3.3	**12** TU	0443 1122 1706	0.7 3.5 0.7	**27** W O	0505 1149 1730	0.9 3.3 0.8
13 W	0351 1029 1607 2253	0.5 2.9 0.8 3.2	**28** TH	0442 1150 1700 2340	0.3 3.0 0.5 3.3	**13** F	0403 1049 1621 2314	0.2 3.1 0.6 3.5	**28** SA	0450 1149 1710 2345	0.3 3.1 0.7 3.3	**13** M ○	0459 1147 1719	0.3 3.5 0.6	**28** TU ●	0525 1217 1747	0.8 3.2 0.7	**13** W	0010 0531 1211 1752	3.5 0.8 3.6 0.6	**28** TH ●	0001 0541 1223 1803	3.2 0.9 3.3 0.6
14 TH	0431 1117 1648 2338	0.2 3.0 0.6 3.4	**29** F ●	0518 1224 1735	0.3 3.0 0.5	**14** SA	0443 1133 1700	0.1 3.2 0.5	**29** SU ●	0521 1221 1740	0.5 3.2 0.6	**14** TU	0025 0544 1233 1803	3.7 0.4 3.6 0.5	**29** W	0023 0555 1246 1817	3.2 0.7 3.3 0.7	**14** TH	0103 0620 1258 1839	3.5 0.7 3.8 0.5	**29** F	0042 0617 1257 1838	3.1 0.9 3.4 0.5
15 F ○	0511 1204 1726	0.0 3.1 0.4	**30** SA	0017 0549 1257 1806	3.4 0.3 3.1 0.5	**15** SU	0001 0523 1219 1739	3.7 0.0 3.4 0.4	**30** M	0020 0549 1250 1808	3.3 0.6 3.2 0.6	**15** W	0115 0631 1318 1847	3.6 0.6 3.7 0.5	**30** TH	0058 0628 1316 1849	3.2 0.7 3.3 0.5	**15** F	0153 0708 1343 1927	3.4 0.7 3.8 0.5	**30** SA	0124 0653 1334 1917	3.1 0.7 3.5 0.4
							31 TU	0052 0616 1317 1835	3.3 0.6 3.2 0.6									**31** SU	0206 0733 1413 1958	3.2 0.7 3.6 0.4

Chart Datum: 1.62 metres below Ordnance Datum (Newlyn)

GREENOCK/GOUROCK continued

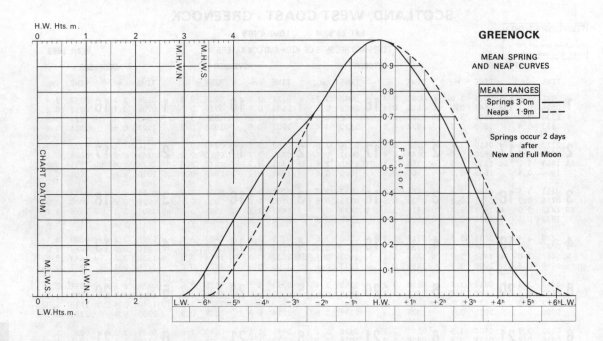

INVERKIP (KIP MARINA)
Renfrew
10-9-15

CHARTS
Admiralty 1907; Imray C63; OS 63

TIDES
+0110 Dover; ML 1.8; Duration 0640; Zone 0 (GMT).
Standard Port GREENOCK (←)

Times				Height (metres)			
HW		LW		MHWS	MHWN	MLWN	MLWS
0000	0600	0000	0600	3.4	2.9	1.0	0.4
1200	1800	1200	1800				
Differences WEMYSS BAY							
−0005	−0005	−0005	−0005	0.0	0.0	+0.1	+0.1

SHELTER
Excellent and is navigable at all states of the tide (2.4m at LWS). Inverkip Bay is exposed to winds NW through W to SW.

NAVIGATION
Waypoint Kip stbd-hand buoy, QG, 55°54'.48N 04°52'.95W, at entrance to buoyed channel. Entrance ½M N of conspic chimney (238m). Beware shifting bank to N of entrance to buoyed channel. No areas in marina dry.

LIGHTS AND MARKS
Buoy Green QG, followed by row of three G buoys to starboard and three R buoys to port, 30m apart form navigable channel 365 metres long.

RADIO TELEPHONE
VHF Ch M (H24).

TELEPHONE (0475)
Hr Mr 521485; CG 29014; MRCC 29988; HMC 28311; Marinecall 0898 500 462; Dr 520248; Hosp 33777.

FACILITIES
EC Wednesday; **Kip Marina** (600+40 visitors) Tel. 521485, D, FW, AC, ME, El, Sh, BH (40 ton), ACA, SM, CH, V, R, Bar, Lau, Gas, Gaz, YC; **Town** PO; Bank (Gourock); Rly; Air (Glasgow).

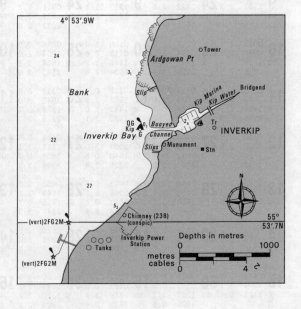

AREA 9—SW Scotland   461

TROON 10-9-16
Ayr

CHARTS
Admiralty 1866; Imray C63; OS 70
TIDES
+0050 Dover; ML 1.9; Duration 0630; Zone 0 (GMT).
Standard Port GREENOCK (←)

Times				Height (metres)			
HW		LW		MHWS	MHWN	MLWN	MLWS
0000	0600	0000	0600	3.4	2.9	1.0	0.4
1200	1800	1200	1800				
Differences TROON							
−0025	−0025	−0020	−0020	−0.2	−0.2	0.0	0.0
ARDROSSAN							
−0020	−0010	−0010	−0010	−0.2	−0.2	+0.1	+0.1
AYR							
−0025	−0025	−0030	−0015	−0.4	−0.3	+0.1	+0.1
GIRVAN							
−0025	−0040	−0035	−0010	−0.3	−0.3	−0.1	0.0

SHELTER
Completely sheltered from all winds. Berth in Yacht Marina.
NAVIGATION
Waypoint 55°33'.20N 04°42'.00W, 283°/103° from/to West Pier Lt, 0.61M. Beware the Mill Rock ½ M NE of West Pier (R buoy on S side). Keep clear of water off Ailsa Shipbuilding yard.
LIGHTS AND MARKS
As on chartlet. No leading lights. Recommended approach Troon on bearing S to SW. Entry signals: Day—2B balls, Night—2RLts (vert) Entry and Exit prohibited: these refer to commercial traffic only.
RADIO TELEPHONE (local times)
Marina VHF Ch M (H24). Other stations: Ardrossan Ch 16; 12 14 (H24). Girvan Ch 16; 12 (Mon-Fri: 0900-1700).
TELEPHONE (0292)
Hr Mr 313412; CG Greenock 29014; MRCC Greenock 29988; HMC 262088; Marinecall 0898 500 462; Hosp 68621; Marina officials will give telephone numbers of doctors on call 315553.

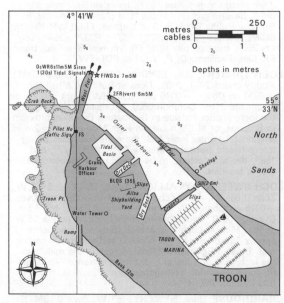

FACILITIES
EC Wednesday; **Marina** (345+40 visitors) Tel. 315553 (Access all tides), Slip, D, FW, ME, El, Sh (on-site repairs Tel. 316180), BH (12 ton), C (2 ton), CH, AC, V, R, Bar; Lau, Gas, Gaz; **Harbour Pier** FW, Sh, AB, V. **Ailsa Shipyard** has all normal 'big ship' repair facilities. **Troon Cruising Club** Tel. 311190; **Marine Mechanical** Tel. 313400 ME; **Boat Electrics and Electronics** Tel. 315355, El.
Town PO; Bank; Rly; Air (Glasgow).

PORTPATRICK 10-9-17
Wigtown

CHARTS
Admiralty 2198; Imray C62; OS 82

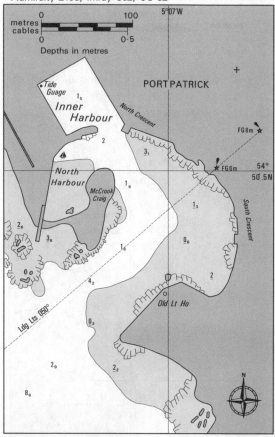

TIDES
+0032 Dover; ML 2.1; Duration 0615; Zone 0 (GMT).
Standard Port LIVERPOOL (→)

Times				Heights (metres)			
HW		LW		MHWS	MHWN	MLWN	MLWS
0000	0600	0200	0800	9.3	7.4	2.9	0.9
1200	1800	1400	2000				
Differences PORTPATRICK							
+0018	+0026	0000	−0035	−5.5	−4.4	−2.0	−0.6

Stranraer Differences Liverpool +0046 (see 10.9.19)

NOTE: Harbour is owned by Mr D Abbott of Millume, Cumbria. No harbour charges at present.

SHELTER
Good but is a difficult entrance in winds from SW to NW when strong. Inner harbour has good shelter and berthing facilities.
NAVIGATION
Waypoint 54°50'.00N 05°08'.00W, 235°/055° from/to entrance, 0.70M. Entrance to outer harbour by way of short narrow entrance with adjacent hazards including rocky shelf submerged at HW.
LIGHTS AND MARKS
Two FG Ldg Lts 050° illuminated on request, 6 and 8m: Front on sea wall, Rear on building; 2 vert orange lines by day.
RADIO TELEPHONE
None. Other station: Stranraer VHF Ch 16; 14 (H24). Portpatrick Coast Radio Station VHF Ch 16; 27.
TELEPHONE (077 681)
Hr Mr Troon 355; CG Greenock 29988; MRSC Greenock 29988; HMC Newton Stewart 2718; Marinecall 0898 500 462; Hosp Stranraer 2323
FACILITIES
EC Thursday. **Harbour** Slip(very small craft), M, FW, L, AB. **Village** P, D, Gas, V, R, Bar. PO; Bank (Stranraer); Rly (bus to Stranraer); Air (Carlisle).

KIRKCUDBRIGHT 10-9-18
Kirkcudbrightshire

CHARTS
Admiralty 1344, 1346, 2094; Imray C62; OS 84

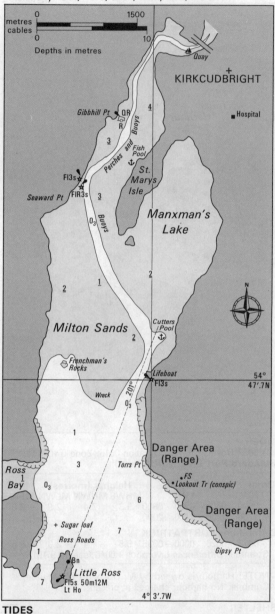

TIDES
+0030 Dover; ML 4.1; Duration 0545; Zone 0 (GMT).
Standard Port LIVERPOOL (→)

Times				Heights (metres)			
HW		LW		MHWS	MHWN	MLWN	MLWS
0000	0600	0200	0800	9.3	7.4	2.9	0.9
1200	1800	1400	2000				

Differences KIRKCUDBRIGHT BAY
+0015 +0015 +0010 0000 −1.8 −1.5 −0.5 −0.1
DRUMMORE
+0030 +0040 +0015 +0020 −3.4 −2.5 −0.9 −0.3
ISLE OF WHITHORN
+0020 +0025 +0025 +0005 −2.4 −2.0 −0.8 −0.2

SHELTER
Very good shelter. Complete shelter at town quays, which dry. Good shelter behind Ross Is and in Flint Bay 1 ca N of Torrs Pt except in winds between W and S which cause heavy swell.
NAVIGATION
Waypoint 54°45'.50N 04°04'.00W, 185°/005° from/to Torrs Pt, 1.4M. The Bar is 7 ca N of Torrs Pt. Note that spring tides run up to 3.5 kn and increase when river is in spate. River is also liable to sudden rise from hydro-electric power station discharge.
LIGHTS AND MARKS
Little Ross Lt Ho West of entrance Fl 5s 50m 12M, (obsc in Wigtown bay when bearing more than 103°).
Note: The harbour authority is the Dumfries & Galloway Regional Council.
RADIO TELEPHONE
VHF Ch 12, 16.
TELEPHONE (0557)
Hr Mr 31135; CG 31133; HMC Newton Stewart 2718; Marinecall 0898 500 461; Dr 30755.
FACILITIES
EC Thursday. **Town Quay** P, D, L, FW, ME, AB, El, CH; **Wooden Pier** AB; **Kirkcudbright YC** Tel. 30963; **SC** Tel. 30032, Slip, M, FW; **Scallop Gear** Tel. 30399, CH; **Town** V, R, Bar. PO; Bank; Rly (bus to Dumfries); Air (Glasgow).

MINOR HARBOURS AND ANCHORAGES 10.9.19

LOCH CRAIGNISH, Argyll, 56°08' N, 5°35' W, Zone 0 (GMT), Admty chart 2326. Tides at Glengarrisdale Bay, Jura, HW −0540 on Dover, −0010 on Oban; HW height −0.3m on Oban; ML 2.1m. Good shelter and once inside the loch there are few navigational dangers. Beware drying reef between Eilean Buidhe and the shore, the drying rocks between Eilean Dubh and Eilean Mhic Chrion, drying rocks N of the E point of Eilean Inshaig and on the E coast of Eilean Righ. Ardfern Yacht Centre entrance just N of Eilean Mhic Chrion, or anchor in 5m by E shore of loch ENE from Eilean Inshaig. There are no lights. Facilities: There are some Bu spherical buoys marked 'HIDB visitors only'. **Ardfern Yacht Centre** (87+20 visitors) Tel. Barbreck 636; Gas, Slip, CH, BH (16 ton), Gaz, D, P, SM, El, FW, ME, Sh; **The Galley of Lorne Hotel** Bar, R, V.

SCALASAIG, Colonsay, 56°04' N, 6°11' W, Zone 0 (GMT), Admty chart 2474. HW −0600 on Dover, −0012 on Oban; HW height −0.1m on Oban; ML 2.2m. See 10.9.7. There is a swell running most of the time so anchor close to the shore, preferably to the S of pier head in approx 2.5m (works in progress 1987). Inner harbour safe, but dries. Beware group of rocks N of pier head marked by Bn. Lts: Rubha Dubh to S of harbour, Fl (2) WR 10s 6m 8/5M; R shore−230°, W230°−337°, R337°−354°. Ldg Lts 262°, front on pier head FR 8m, rear 60m from front, FR 10m. Facilities: D, P, V (all at PO); FW.

LOCH TARBERT, Jura, 55°57' N, 6°00' W, Zone 0 (GMT), Admty charts 2481, 2169. Tides at Rubha A'Mhail HW −0540 on Dover, −0010 on Oban; HW height −0.2m on Oban; ML 2.1m; Duration 0600. Excellent shelter inside the loch; anchor outside in Glenbatrick Bay in approx 6m in S winds, or at Bagh Gleann Righ Mor in approx 2m in N winds. To enter inner loch there are four pairs of ldg marks (white stones) at approx 120°, 150°, 077°, and 188°, the latter astern, to be picked up in sequence. There are no facilities.

LOCH SWEEN, TAYVALLICH, Argyll, 56°01' N, 5°37' W, Zone 0 (GMT), Admty chart 2397. HW +0550 on Dover, −0015 on Oban; HW height −1.1m on Oban; ML 1.5m; Duration 0610. Good shelter; anchor in Port Lunna SSW of Barr Mor, in Loch a Bhealaich (outside Tayvallich in approx 7m) or enter inner harbour and anchor by central reef. Beware Sgeirean a Mhain, a rock in the middle of the loch to S of Taynish I, 3M from entrance. There are no lights. Facilities: **Livingstone** Tel. 226, Gas, Bar, FW (tap by PO), PO, R, V.

CRAIGHOUSE, Jura, 55°50' N, 5°57' W, Zone 0 (GMT), Admty charts 2168, 2396, 2481. HW +0600 on Dover, −0430 np, −0130 sp on Oban; HW height −2.0m on Oban; ML 0.7m; Duration 0640 np 0530 sp. See 10.9.7. Good shelter but squalls occur in W winds. Enter between Lt Bn on SW end of Eilean nan Gabhar, Fl 5s 7m 8M vis 225°−010°, and the beacon to the SW. Yachts may go alongside the pier or anchor in 5m N end of Loch na Mile. Facilities: very limited. There are eight spherical Bu buoys marked 'HIDB visitors only'.Bar, FW, PO, R, V; **Paton** Tel. 242, Gas.

AREA 9—SW Scotland

GIGHA ISLAND, Argyll, 55°41′N, 5°44′W, Zone 0 (GMT), Admty chart 2475. HW +0600 on Dover, −0450 np, −0210 sp on Oban; HW height −2.0m on Oban; ML 0.9m; Duration 0530. See 10.9.7. By careful selection, good shelter can be found round the island in any conditions. Main anchorage is Ardminish Bay but Druimyeon Bay and the bay N of West Tarbert Bay are popular alternatives, the latter being good in NE winds. Beware numerous rocks in Gigha Sound, reefs extending off both points of Ardminish Bay and the Kiln Rk off the jetty. There are no lights but there is a W cardinal buoy Fl (9) 15s marking Gigalum Rks, opposite N end of Gigalum I. There are 12 spherical buoys marked 'HIDB for visitors' in Ardminish Bay. Facilities are limited; Bar, FW in hut at end of jetty, PO, R, V; **Gigha Engineering Services** Tel. 261, ME. **Bannatyne** Tel. 220, Gas.

PORT ASKAIG, Islay, 55°51′N, 6°07′W, Zone 0 (GMT), Admty chart 2481. HW +0610 on Dover, −0050 on Oban; HW height −1.6m on Oban; ML 1.2m. Harbour on W side of Sound of Islay. Anchor close inshore in 4m or secure to ferry pier. Beware strong tides with eddies. Facilities: Hotel, Gas, FW (hose on pier), P, R, Bar, V, PO, ferries to Jura and Kintyre.

PORT ELLEN, Islay, 55°38′N, 6°12′W, Zone 0 (GMT), Admty chart 2474. HW −0620 sp +0130 np on Dover; −0050 sp −0530 np on Oban; the tide seldom exceeds 0.6m and the level of the sea is greatly affected by the weather; at neaps the tide is often not appreciable. See 10.9.7. Good shelter except in S winds when a swell sets into the bay. In these conditions, anchor N of Carraig Fhada. At other times anchor in 3m in NE corner. Beware rocks all along E side of entrance. Carraig Fhada, Lt Ho (conspic) on W side Fl WRG 3s 19m 8/6M; W shore-248°, G248°−311°, W311°−340°, R340°−shore. Keep in W sector until past the G can buoy, QG; Ro Ro terminal shows 2 FG (vert). Facilities: **Bridgend Sawmills** Tel. 598, Gas; **Village** Bar, FW, PO, R, V.

CARRADALE BAY, Argyll, 55°32′N, 5°27′W, Zone 0 (GMT), Admty chart 2131. HW +0115 on Dover, 0000 on Greenock, HW height −0.2m on Greenock. Good anchorage in 7m off Torrisdale Castle in SW corner of bay. With S winds a swell sets into bay and better anchorage to be found N of Carradale Pt, in Port Cranaig. Off Carradale Pt is R buoy Fl(2)R 12s marks outer end of foul ground. Good shelter inside breakwater. Facilities: V, R, Bar, PO.

BRODICK, Arran, 55°35′N, 5°09′W, Zone 0 (GMT), Admty chart 1864. HW +0115 on Dover, 0000 on Greenock; HW height −0.2m on Greenock; ML 1.8m; Duration 0635. See 10.9.11. Shelter is good except in E winds. Anchor W of pier in approx 3m, just below the Castle in 4.5m or further N off Merkland Pt in 3 to 4m. There are no navigational dangers but the bay is in a submarine exercise area. Only Lts are 2FR (vert) 9/7m 4M on pier head. Facilities: EC Wed; Bank, Bar, P and D in cans, FW (at pier head), ME, PO, R, V.

LOCH RANZA, Arran, 55°43′N, 5°17′W, Zone 0 (GMT), Admty charts 2221, 2131. HW +0105 on Dover, −0010 on Greenock; HW height −0.3m on Greenock; ML 1.7m; Duration 0635. Good shelter but swell comes into the harbour with N winds. The 850m mountain 4M to S causes fierce squalls in the loch with S winds. Beware rocky spit extending seaward from S of Newton Pt. Anchor off castle in 5m. Facilities: Bar, FW (tap near pier), PO, R, V. Marina is planned.

INVERARY, Argyll, 56°14′N, 5°04′W, Zone 0 (GMT), Admty chart 2382. HW +0126 on Dover, +0011 on Greenock; HW height −0.1m (sp), +0.1m (np) on Greenock (see 10.9.10). For Upper Loch Fyne see 10.9.5. Beware An Oitir extending 2 ca offshore, ½M S of pier. Anchor NNE of pier in 4m or pick up a buoy (some reserved for visitors). Facilities: FW (on pier), PO, V, R, Bar, Gas, bus to Glasgow.

MILLPORT, Great Cumbrae, Bute, 55°45′N, 4°55′W, Zone 0 (GMT), Admty chart 1867. HW +0100 on Dover, −0015 on Greenock; HW height 0.0m on Greenock; ML 1.9m; Duration 0640. Good shelter except with S winds. Anchor E of the inner of The Eileans in approx 3m, or off pier in approx 3m. Beware The Spoig and The Clach to W of Ldg Line, and the rocks off The Eileans to E of Ldg Line. Ldg Lts 333°, front FR 7m 5M, rear FR 9m 5M. Facilities: EC Wed; Bar, FW, PO, R, Slip, V.

LARGS, Ayr, 55°48′N, 4°52′W, Zone 0 (GMT), Admty charts 1907, 1867. HW +0105 on Dover, −0005 on Greenock; HW height −0.1m on Greenock; ML 1.9m; Duration 0640. Largs Bay exposed to winds from N to W and also from S. Approaching from S, beware Hunterston and Southannan Sands. Lights; N end of pier 2 FG (vert) 7/5m 5M shown when vessels expected. Marina is 1M S of Largs pier. Entrance buoy RW, LFl 10s. S breakwater Oc G 10s, N breakwater Oc R 10s. Facilities: EC Wed; FW, Stores, PO, Rly; **Marina (Largs Yacht Haven)** (250+25 Visitors) Tel. 675333, C, D, FW, Slip, CH, El, ME, AC, Sh, BH (50 ton). VHF Ch M.
Note: N of Largs is the Scottish Sports Council Water Sport Training Centre.

AYR, Ayr, 55°28′N, 4°38′W, Zone 0 (GMT), Admty chart 1866. HW +0050 on Dover, −0025 on Greenock; HW height −0.3m on Greenock; ML 1.8m. Duration 0630. See 10.9.16. Good shelter except in W winds when shelter can be found in the dock. Beware St Nicholas Rock to S of entrance and large amounts of debris being washed down the R Ayr after heavy rains. Entrance is not easy for larger yachts. Ldg Lts 098° Front, by Pilot Stn, FR 10m 5M R Tr, also traffic signals; rear (130m from front) Oc R 10s 18m 10M. N breakwater head QR 7m 5M. S pier head QW 7m 7M; vis 012°−161°, and FG 5m 5M; vis 012°−066°, over St Nicholas Rocks, Horn 30s. Entry signals 2 B balls (2R Lts) = No inward shipping. 1 B ball (1 R Lt or 1 GLt) = proceed with caution. Hr Mr Tel. 67261. VHF Ch 16 14. Facilities: EC Wed; Bank, Bar, CH, Gas, D, El, FW, ME, PO, R, Rly, Sh, V; **Saturn Sails** Tel. 286241 SM; **Ayr Yacht & Cruising Club** (40) Tel. 267963, Bar, R, M.

GIRVAN, Ayr, 55°15′N, 4°52′W, Zone 0 (GMT), Admty chart 2199. HW +0043 on Dover, −0032 on Greenock; HW height −0.3m on Greenock; ML 1.8m; Duration 0630. See 10.9.16. Good shelter. Harbour used by fishing vessels and coasters but yachts may find berths available on quay. Beware Girvan Patch, 2.4m, 4 cables SW of entrance, and Brest Rocks, 3.5M N of harbour extending 6 cables offshore, marked by Bn. Entrance between breakwater to N and pier to S. Lts: Breakwater head Iso 4s 3m 4M; S pier head 2FG (vert) 8m 4M and traffic signals. VHF Ch 16; 12 (office hours). Facilities: EC Wed; Harbour Office Tel. 3048; FW, Slip; **Town** Bank, Bar, PO, R, Rly, V.

LOCH RYAN, Wigtown, 54°55′N, 5°03′W, Zone 0 (GMT), Admty chart 1403. HW (Stranraer) +0055 on Dover, −0020 on Greenock; HW height −0.4m on Greenock; ML 1.6m; Duration 0640. Very good shelter except in strong NW winds. Entrance between Milleur Pt and Finnarts Pt. Anchorages in Lady Bay, 1.3M SSE of Milleur Pt; in The Wig in 3m; off Stranraer NE of steamer pier in 3m or lie alongside quay at Stranraer. Beware The Beef Barrel, rk 1m high 6 cables SSE of Milleur Pt; the sand spit running 1.5M to SE from W shore opposite Cairn Pt Lt Ho Fl (2) R 10s 14m 12M. Lt at Cairnryan ferry terminal Fl R 5s 5m 5M. Lts at Stranraer − centre pier head 2FBu (vert), E pier head 2FR (vert), W Ross pier head 2 FVi (vert) 10m 4M. Facilities (Stranraer): EC Wed; Bank, Bar, D, FW, P, PO, R, Rly, V.

ISLE OF WHITHORN, Wigtown, 54°42′N, 4°22′W, Zone 0 (GMT), Admty chart 2094. HW +0035 on Dover, +0020 on Liverpool; HW height −2.2m on Liverpool; ML 3.7m; Duration 0545. See 10.9.18. Shelter good but Hr dries having approx 2.5m at HW∓3. Beware the Skerries, a ledge on W side of entrance marked by a thin iron stake. E pier head has QG 4m 5M; Ldg Lts 335°, front Oc R 8s 7m 7M; R and W mast; Rear, 35m from front, Oc R 8s 9m 7M; R and W mast; synchronised with front. Berth alongside quay. Facilities: Hr Mr Tel. Whithorn 246, Slip, P, D, FW, ME, Sh, CH, V, Bar, PO, VHF Ch M (occas).

WIGTOWN BAY, Kirkcudbrightshire, 54°47′N, 4°21′W, Zone 0 (GMT), Admty chart 2094. HW (Garlieston) +0100 on Dover, +0030 on Liverpool; HW height −2.0m on Liverpool; ML 3.8m; Duration 0545. Garlieston harbour affords complete shelter but dries. Access (2m) HW∓3. Pier head Lt 2FR (vert) 5m 3M. Beware rocky outcrops in W side of bay marked by a perch; also the rock in the centre which dries 0.6m. Facilities. Hr Mr Tel. Garlieston 259. FW, AC on quay & visitors moorings; stores, P and D in town.

kip marina

SCOTLAND'S PREMIER MARINA

Located in a beautiful rural location on the Clyde with easy access by road, rail and air, just a short sail from the Kyles of Bute and countless other superb cruising anchorages. Over 700 berths with full facilities for the cruising or racing yachtsman or motor yacht owner:

* Beautifully Sheltered
* 40 Ton Boat Hoist
* Service and Repair
* Storage Yard
* Well Stocked Chandlery
* 24 Hr Security and Channel M
* New Boat Sales and Brokerage Team
* Chartroom bar

Distributors: Moody Yachts, MG Yachts
Agents: Fairline Boats, Princess Boats, Cruisers International, Grand Banks

HOLT LEISURE PARKS Ltd

Holt Leisure Parks Ltd Tel: 0475 521485 (Seven Days)
Kip Marina Telex: 777582 TELRAY G
Inverkip, Renfrewshire Attention Kip Marina
Scotland PA16 0AS

**DEDICATED MARINE POWER
9 to 422hp**

VOLVO PENTA

**DEDICATED MARINE SERVICE
OVER 1300 LOCATIONS ACROSS EUROPE**

CRINAN BOATS LTD.

boatbuilders
chandlers
engineers
slipping
repairs
charts
electricians
pontoon

alongside replenishments
Listening Channel 16 VHF

crinan argyll pa31 8sp tel crinan (054 683) 232

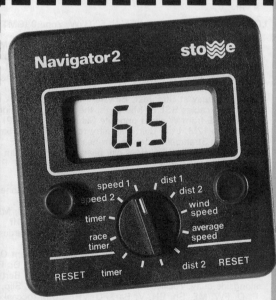

Stowe makes choosing sailing instruments easy, just check out the benefits and our extensive range at your nearest dealer or contact us directly for some friendly, helpful advice.

Navigator Range: ● Navsounder ● Navigator Log ● Wind Monitor System ● Navigator VMG

Micro Range: ● 200 Sounder ● 210 Speed/Log ● 220 Wind Angle/Speed

---

RELIABLE ✓
ACCURATE ✓
PROVEN ✓
RUGGED ✓
STYLISH ✓
LOW COST ✓

stowe ✓

NOW YOU'VE DECIDED ON THE INSTRUMENTATION ....CHOOSE YOUR BOAT!

Stowe Marine Equipment Ltd.
1 Bowes Hill, Rowlands Castle, Hampshire PO9 6BP England
Telephone: Rowlands Castle (0705) 412044 Telex: 869162 Stowe G

VOLVO PENTA SERVICE

Sales and service centres in area 10
CUMBRIA **Waterhead Marine Ltd** AMBLESIDE LA22 0EX Tel (053 94) 32424/5.
Windermere Aquatic Ltd Glebe Road, WINDERMERE Tel (09662) 2121.
MANCHESTER **Pilkington Motor Marine** Watson Street, MANCHESTER M3 4LP Tel (061 834) 5392.

Area 10

North-West England, Isle of Man and North Wales
Maryport to Pwllheli

10.10.1	Index	Page 465
10.10.2	Diagram of Radiobeacon, Air Beacons, Lifeboat Stations	466
10.10.3	Tidal Stream Charts	468
10.10.4	List of Lights, Fog Signals and Waypoints	470
10.10.5	Passage Information	472
10.10.6	Distance Table	473
10.10.7	Maryport	474
10.10.8	Workington	474
10.10.9	Glasson Dock	475
10.10.10	Fleetwood	475
10.10.11	Rivers Mersey and Dee (Liverpool, Standard Port, Tidal Curves)	476
10.10.12	Isle of Man	480
10.10.13	— Port St Mary	481
10.10.14	— Douglas	481
10.10.15	— Ramsey	482
10.10.16	— Peel	482
10.10.17	Conwy	483
10.10.18	Holyhead — Standard Port, Tidal Curves	483
10.10.19	Menai Strait (Bangor, Beaumaris, The Swellies, Port Dinorwic and Caernarfon)	488
10.10.20	Abersoch, St Tudwal's Roads	490
10.10.21	Pwllheli	490
10.10.22	Minor Harbours and Anchorages Silloth Harrington Whitehaven Ravensglass Barrow-in-Furness Heysham River Ribble River Alt Castletown (I o M) Laxey (I o M) Port Erin (I o M) Porth Dinllaen	491

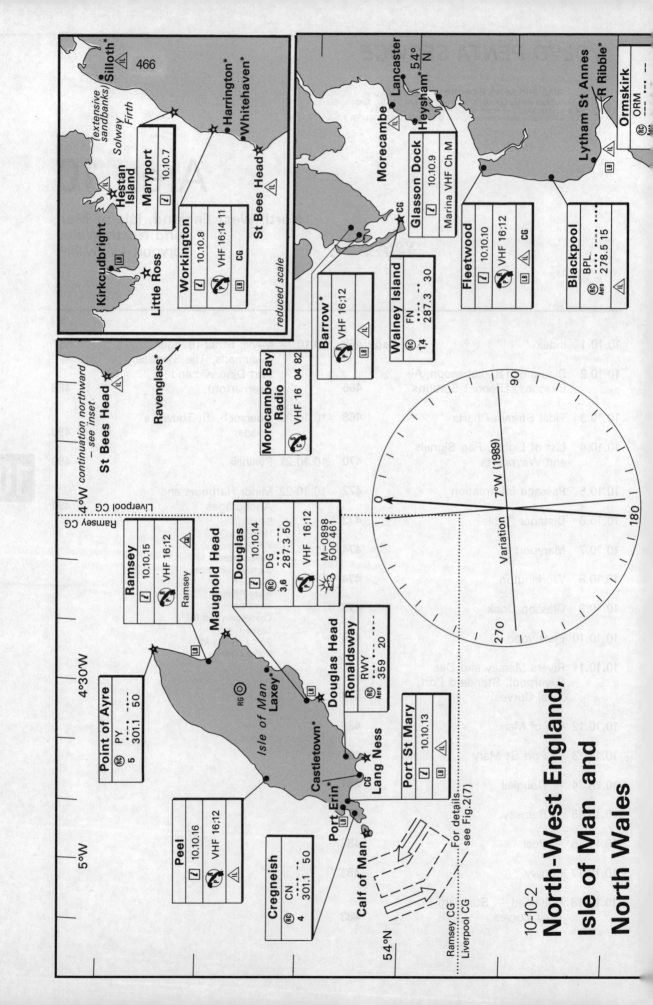

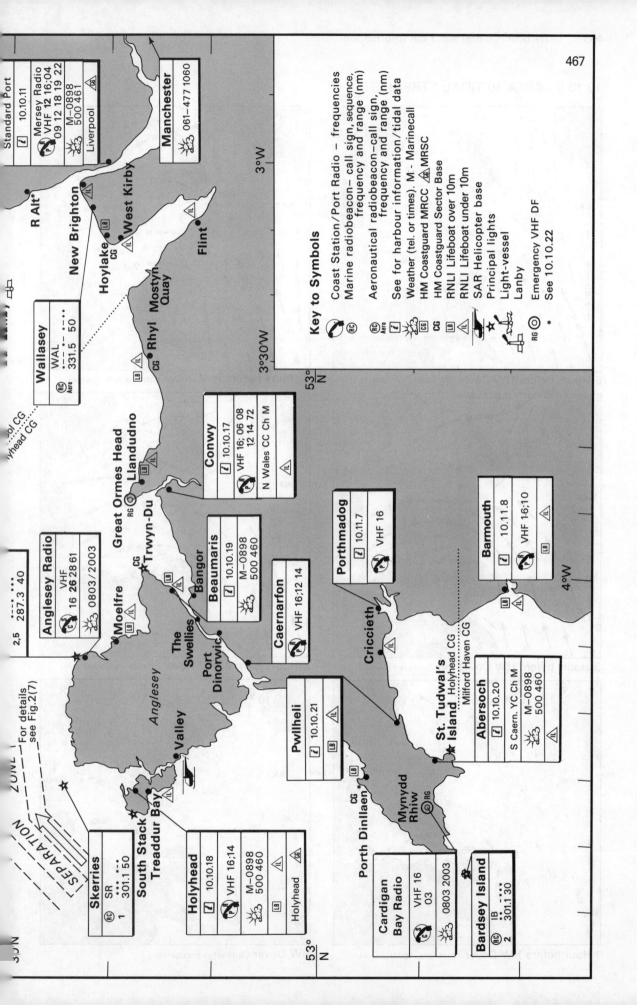

468 Harbour, Coastal and Tidal Information

10.10.3 AREA 10 TIDAL STREAMS

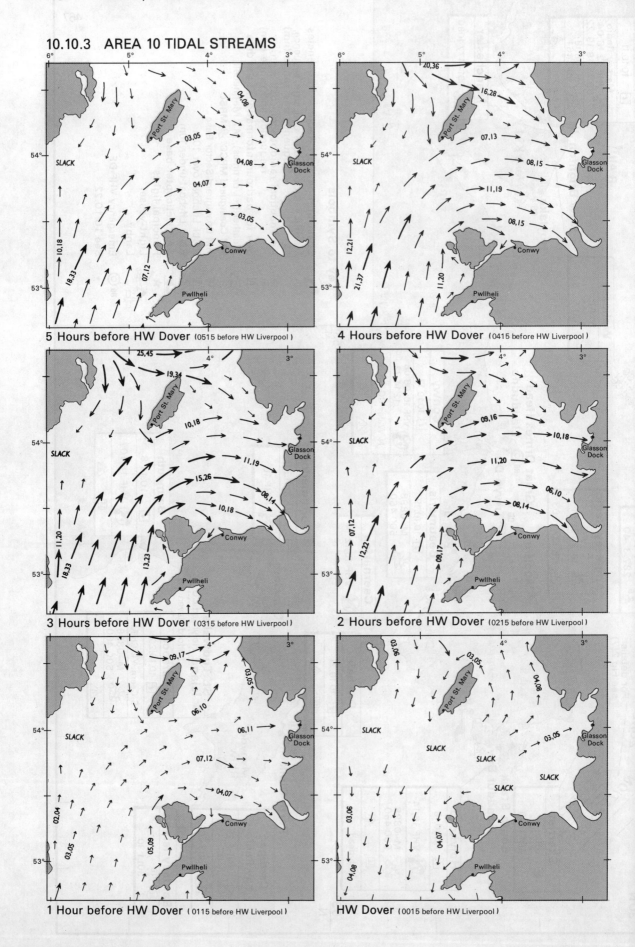

AREA 10—NW England, Isle of Man and N Wales   469

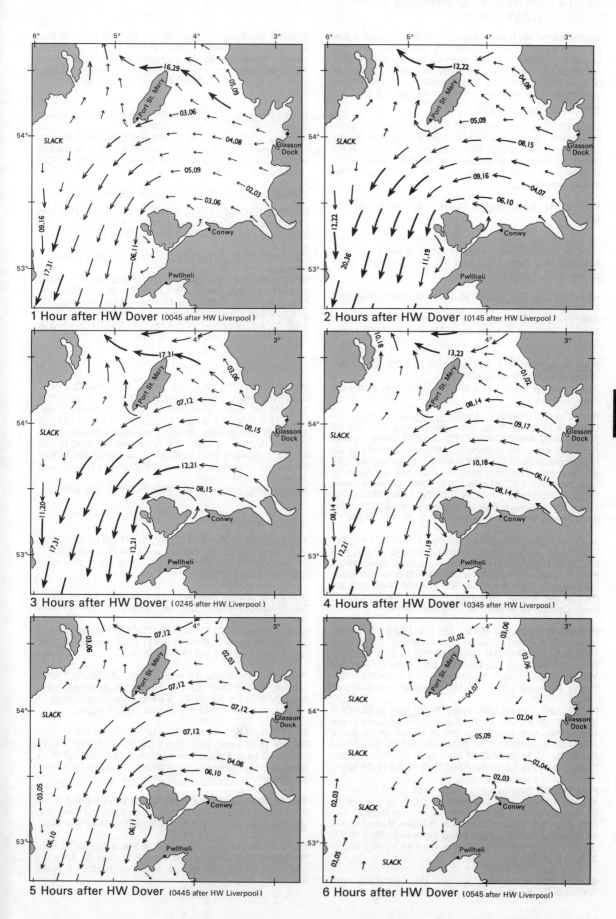

10.10.4 LIGHTS, FOG SIGNALS AND WAYPOINTS

Abbreviations used below are given in 10.0.3. Principal lights are in **bold** print, places in CAPITALS, and light-vessels and Lanbys in *CAPITAL ITALICS*. Unless otherwise stated lights are white. m—elevation in metres; M—nominal range in n. miles. Fog signals are in *italics*. Useful waypoints are underlined — use those on land with care. See 4.2.2.

ENGLAND—WEST COAST

SILLOTH. Corner Lt Buoy 54 49.40N/3 31.30W Fl G; stbd-hand mark. Lees Scar 54 51.80N/3 24.75W QG 11m 8M; W structure on piles; vis 005°-317°. East Cote Dir Lt 052°, Dir FG 15m 12M; W structure on piles; vis 046°-058°, intens 052°. Groyne head 54 52.1N/3 24.0W 2FG(vert); Fl Bu traffic signals close by.

MARYPORT. S pier head 54 43.05N/3 30.60W Fl 1.5s 10m 4M.

WORKINGTON. S pier 54 39.13N/3 34.71W Fl 5s 11m 8M; R building; *Siren 20s*. Ldg Lts 132°. Front 54 38.9N/3 34.1W FR 10m 3M. Rear 134m from front FR 12m 3M; both on W pyramidal Tr, Y bands.

HARRINGTON. Pier head 54 36.67N/3 34.40W Fl G 5s 3M.

WHITEHAVEN. West pier head 54 33.16N/3 35.84W Fl G 5s 16m 13M; W round Tr. N pier head 2FR(vert) 8m 9M; W round Tr.

Saint Bees Head 54 30.80N/3 38.15W Fl(2) 20s 102m **21M**; W round Tr; obsc shore—340°; (H24).
Selker Lt Buoy 54 16.15N/3 29.55W Fl(3)G 10s; stbd-hand mark; *Bell*.

BARROW-IN-FURNESS. Lightning Knoll Lt Buoy 53 59.83N/3 14.20W LFl 10s; safe water mark; *Bell*. **Isle of Walney** 54 02.92N/3 10.65W Fl 15s 21m **23M**; stone Tr; obsc 122°-127° within 3M of shore; RC. Walney Channel Ldg Lts 041°. Front 54 03.3N/3 08.9W Q 6m 6M; B pile structure, W daymark. Rear 640m from front Iso 2s 12m 6M; R column, W face. Haws Point, NE of point QR 8m 6M. Rampside Sands Ldg Lts 006°. Front 54 04.4N/3 09.7W Q 6m 6M; B pile structure, W daymark. Rear 0.77M from front Iso 2s 14m 6M; R column, W face. No 3A 54 04.4N/3 09.7W QG 8m 9M.

MORECAMBE. Morecambe Lt Buoy 53 52.00N/3 24.00W Q(9)15s; W cardinal mark; *Whis*. Lune Deep Lt Buoy 53 55.80N/3 11.00W Q(6)+LFl 15s; S cardinal mark; *Whis*; Racon. Sewer outfall 54 04.3N/2 53.7W Fl G 2s 4m 2M; metal framework Tr. Lts in line about 090°. Front 54 04.4N/2 52.5W FR 10m 2M; G mast. Rear 140m from front FR 14m 2M; G mast. Central Promenade Pier 2 FG (vert) 9m 4M.

HEYSHAM. S outfall 54 01.73N/2 55.73W Fl(2)G 10s 5m 2M; metal post. N outfall FlG 5s 5m 2M; metal post. S jetty head 54 01.90N/2 55.64W 2FG(vert) 9/7m 5M; W pylon; Ra refl; *Siren 30s*. SW Quay Ldg Lts 102°, both F Bu 11/14m 2M; Y&B diamonds on masts. S pier head Oc G 7.5s 9m 6M; W Tr, R base. N pier head 2FR(vert) 11m 2M; obsc from seaward.

RIVER LUNE. Ldg Lts 084°. Front, Plover Scar 53 58.87N/2 52.88W Fl 2s 6m 6M; W Tr, B lantern. Rear, Cockersand Abbey 854m from front F 18m 8M; R framework Tr. Crook Perch, No 7, 53 59.5N/2 52.3W Fl(3) G 5s 3M; G. mast. Bazil Perch, No 16 54 00.2N/2 51.6W Fl(3)R 10s 3M; R mast. Glasson Quay FG 1M. Outfall 54 01.3N/2 49.7W Fl (TE 1985).

FLEETWOOD. Fairway No 1 Lt Buoy 53 57.65N/3 02.15W Q; N cardinal mark; *Bell*. Esplanade Ldg Lts 156°. Front 53 55.7N/3 00.4W Fl Y 2s 14m. Rear 320m from front Fl Y 4s 28m. Both buff-coloured square Trs, B bases, R lanterns and vis on leading line only. Steep Breast Perch 53 55.8N/3 00.5W Iso G 2s 3m 2M. Knott End slipway, head 53 55.7N/3 00.0W 2FR(vert) 3m 2M.

BLACKPOOL. N Pier head 2FG(vert) 3M. Central Pier head 2FG(vert) 4M. South Pier head 2FG(vert) 4M.

RIVER RIBBLE. Gut Lt Buoy 53 41.74N/3 08.91W LFl 10s, safe water mark. Perches show FlR on N side, and FlG on S side of channel. S side, 14¼ mile perch 53 42.75N/3 04.85W FlG 5s 6m 3M; B framework Tr and tripod. Southport Pier, head, 52 39.35N/3 01.25W 2FG(vert) 6m 5M; W post; vis 033°-213°.

El Oso Wreck Lt Buoy 53 37.55N/3 23.50W Q; N cardinal mark.
Jordan's Spit Lt Buoy 53 35.74N/3 19.20W Q(9) 15s; W cardinal mark. FT Lt Buoy 53 34.55N/3 13.12W Q; N cardinal mark. Spoil Ground Lt Buoy 53 34.25N/3 17.30W FlY 3s; special mark.

ENGLAND—WEST COAST, RIVER MERSEY

RIVER MERSEY. *BAR LANBY* 53 32.00N/3 20.90W Fl 5s 12m **21M**; tubular structure on buoy; Racon; *Horn 20s*.
FORMBY Lt F 53 31.10N/3 13.45W Iso 4s 11m 6M; R hull, W stripes.
CROSBY Lt F 53 30.7N/3 06.2W Oc 5s 11m 8M; R hull, W stripes.
SEACOMBE FERRY, N and S corners 3FG 5m 5M; near N corner FY 8m 6M; *Bell (3) 20s*. Cammell Laird slip, SE corner Fl(2) G 6s 5m 5M. BIRKENHEAD Woodside landing stage, N end 3 FG and S end 2FG(vert) with *Bell (4) 15s*.
TRANMERE TERMINAL, N dolphin 53 22.9N/3 00.0W Fl G 3s 7m 7M. N stage, S end 2FG(vert) 10m 5M; *Bell (2) 10s*.

ENGLAND—WEST COAST

RIVER DEE. HE2 Lt Buoy 53 26.20N/3 16.80W QR; port-hand mark. Hilbre Island, N end 53 22.97N/3 13.70W Fl R 3s 14m 5M; W metal framework Tr. MOSTYN, training wall head Fl R 1.3s 8m 4M; B mast. Ldg Lts 216°. Front 53 19.2N/3 16.1W FR 12m; W diamond on B mast. Rear 135m from front FR 22m; W diamond on B mast. N training wall, head Fl R 3s 4m 6M; framework Tr. S training wall, head Fl G 10s 3m 6M; framework Tr. Summersby Wharf FR. Connah's Quay, power station 2FR(vert).
South Hoyle Lt Buoy 53 21.50N/3 24.78W Fl(3)R 10s; port-hand mark.

ISLE OF MAN

Point of Ayre 54 24.95N/4 22.03W Al L Fl WR 60s 32m **19M**; W Tr, two R bands; RC; Racon. Low Lt 54 25.1N/4 21.8W Fl 3s 10m 8M; R Tr, lower part W, on B Base; part obsc 335°-341°; *Siren (3) 90s*.

JURBY, Cronk y Cliwe 54 22.3N/4 31.4W 2 FlR 5s (vert); synchronized, 2m apart. Orrisdale 54 19.3N/4 34.1W 2 FlR 5s (vert); synchronized, 2m apart.
North DZ buoy 54 23.6N/4 36.7W Fl Y 10s (unreliable).
South DZ buoy 54 21.5N/4 38.8W Fl Y 10s (unreliable); with QY on target floats 1.5M ENE and 2.1M E.

PEEL. Pier head, E side of entrance Oc R 7s 8m 5M; W Tr with R band; vis 156°-249°. Groyne, head Iso R 2s. Castle Jetty, head Oc G 7s 5m 4M; W Tr, G bands. Breakwater, head 54 13.67N/4 41.62W Oc 7s 11m 6M; W Tr; *Bell* (occas).

PORT ERIN. Ldg Lts 099°. Front 54 05.3N/4 45.4W FR 10m 5M; W Tr, R band. Rear 39m from front FR 19m 5M; W column, R band. Raglan Pier, head Oc G 5s 8m 5M; W Tr, G band. Thousla Rock 54 03.7N/4 48.0W Fl R 3s 9m 4M; 8-sided tapered pillar.

Calf of Man, West point Fl 15s 93m **28M**; W 8-sided Tr; vis 274°-190°; FR Lts on conspic radio mast 3.2M NE; *Horn 45s*.
Chicken Rock 54 02.3N/4 50.1W Fl 5s 38m 13M; *Horn 60s*.

PORT ST MARY. The Carrick 54 04.30N/4 42.60W Q(2) 5s 6m 3M; isolated danger mark. Alfred Pier, head Oc R 10s 8m 5M; W Tr, R band; *Bell* (occas). Inner Pier, head Oc R 3s 8m 5M; W Tr, R band.

AREA 10—NW England, Isle of Man and N Wales

CASTLETOWN. New Pier, head 54 04.3N/4 38.8W Oc R 15s 8m 5M; W Tr, R band. N side of entrance Oc G 4s 3m; Irish Quay Oc R 4s 5m 5M; W Tr, R band; vis 142°-322°.

Langness, Dreswick Point 54 03.28N/4 37.45W Fl(2) 30s 23m **21M**; W Tr.

Derby Haven, breakwater, SW end Iso G 2s 5m 5M; W Tr, G band (TE 1985). Douglas Head 54 08.58N/4 27.88W Fl 10s 32m **24M**; W Tr; obsc brg more than 037°. FR Lts on radio masts 1 and 3M West.

DOUGLAS. Princess Alexandra Pier, head 54 08.85N/4 27.80W Fl R 5s 16m 8M; R mast; *Whis (2) 40s*. Battery Pier, 140m from breakwater head, QR 12m 1M; W Tr, R band; vis 038°-218°. Ldg Lts 229°, Front Oc 10s 9m 5M; W and R triangle on mast. Rear, 60m from front, Oc 10s 12m 5M; W and R triangle on mast; synchronised with front. Victoria Pier, head Oc G 8s 10m 3M; W column; vis 225°-327°; RC; International Port Traffic Signals; *Bell (1) 2s*. Fort Anne Jetty, head Oc R 4s 6m; W Tr, R band; vis 107°-297°. King Edward VIII Pier, S side, Oc G 4s 6m; W framework Tr, G band; vis 253°-005°. FR each side of swing bridge, 365m West, when closed.

LAXEY. Pier, head Oc R 3s 7m 5M; W Tr, R band; obsc when brg less than 318°. Breakwater, head Oc G 3s 7m; W Tr, G band.

Maughold Head 54 17.70N/4 18.50W Fl(3) 30s 65m **22M**; W Tr.
Bahama Lt Buoy 54 20.00N/4 08.45W Q(6)+LFl 15s; S cardinal mark; *Bell*.

RAMSEY. S Pier, head 54 19.42N/4 22.42W Oc R 5s 8m 4M; W Tr, R band, B base; *Bell* (occas). N Pier, head Oc G 5s 9m 5M; W Tr, B base.

WALES—NORTH COAST

Chester Flat Lt Buoy 53 21.65N/3 27.40W Fl(2)R 5s; port-hand mark.
Middle Patch Spit Lt Buoy 53 21.80N/3 31.55W FlR 5s; port-hand mark.
North Rhyl Lt Buoy 53 22.22N/3 34.45W Q; N cardinal mark.
W Constable Lt Buoy 53 23.10N/3 49.15W Q(9) 15s; W cardinal mark.

Rhyl, River Clwyd, breakwater head 53 19.50N/3 30.30W QR 7m 2M; Bn.
Llanddulas, Llysfaen jetty 53 17.55N/3 39.45W Fl G 10s. Raynes Quarry, jetty head 2FG(vert).
Colwyn Bay, Victoria Pier, head 53 17.90N/3 43.25W 2FG(vert) 5m 3M.
Llandudno Pier, head 53 19.90N/3 49.40W 2FG(vert) 8/6m 4M. Great Ormes Head Lt Ho 53 20.55N/3 52.10W (unlit).

CONWY, entrance, S side 53 18.0N/3 50.9W Fl WR 5s 5m 2M; B column; vis W076°-088°, R088°-171°, W171°-319°, R319°-076°.

MENAI STRAIT. Trwyn-Du 53 18.76N/4 02.38W Fl 5.5s 19m **15M**; W round castellated Tr, B bands; vis 101°-023°; *Bell (1) 30s*, sounded continuously. FR on radio mast 2M SW.
Mountfield 53 16.15N/4 05.10W Fl WG 2s 8m 6M; vis G221°-226°, W226°-341°.
Beaumaris, Pier FWG 5m 6M; vis G212°-286°, W286°-041°, G041°-071°. St George's Pier Fl G 10s. E side of channel 53 13.2N/4 09.5W QR 4m; R mast; vis 064°-222°. Price Point Fl WR 2s 5m 3M; W Bn; vis R059°-239°, W239°-259°. Britannia tubular bridge, S channel Ldg Lts 231° E side. Front FW. West side, Rear 45m from front FW. Centre span of bridge Iso 5s 27m 3M, one either side. S end of bridge, FR 21m 3M either side, N end of bridge FG 21m 3M either side.
Port Dinorwic, pier head F WR 5m 2M; vis R225°-357°, W357°-225°.

Point Lynas 53 24.97N/4 17.30W Oc 10s 39m **20M**; W castellated Tr; vis 109°-315°; (H24); Fog Det Lt vis 212°-214°; *Horn 45s*; RC.
Pilot station 54 24.9N/4 17.1W 2FR(vert).
Amlwch, SBM 53 26.7N/4 19.8W Mo(U) 15s 8M; mooring buoy; Ra refl. Fl Y 5s 3M marks floating hose 285m from buoy; *Horn Mo(U) 30s*.

AMLWCH. Main breakwater 53 25.0N/4 19.8W 2FG(vert) 11/9m 5M; W mast; vis 141°-271°. Inner breakwater 2FR(vert) 12/10m 5M; W mast; vis 164°-239°. Inner harbour FW 9m 8M; W post; vis 233°-257°.

Wylfa power station 2FG(vert) 13m 6M.

The Skerries 53 25.25N/4 36.45W Fl(2) 10s 36m **22M**; W round Tr, R band; RC; Racon. FR 26m **16M**; same Tr; vis 231°-254°; *Horn (2) 20s*.

HOLYHEAD. Clipera Lt Buoy 53 20.08N/4 36.14W Fl(4)R 15s; port-hand mark; *Bell*. Breakwater, head 53 19.83N/4 37.08W Fl(3) 15s 21m 14M; W square Tr, B band; *Siren 20s*. Old Harbour, Admiralty Pier dolphin 53 18.85N/4 37.00W 2FG(vert) 8m 5M; *Bell* (occas).

South Stack 53 18.4N/4 41.9W Fl 10s 60m **28M**; (H24); W Tr; obsc to north by North Stack and part obsc in Penrhos bay. *Horn 30s*. Fog Det Lt vis 145°-325°.
Ynys Meibion 53 11.4N/4 30.2W Fl R 5s 37m 10M (occas) 2FR(vert) shown from flagstaffs 550m NW and 550m SE when firing taking place.

WALES—WEST COAST
CAERNARFON BAY AND MENAI STRAIT (SOUTHERN PART). Llanddwyn Island, S end 53 08.04N/4 24.70W Fl WR 2.5s 12m W7M, R4M; W Tr; vis R280°-015°, W015°-120°.
Cl Lt Buoy 53 07.46N/4 24.90W FlG 5s; stbd-hand mark.
Abermenai Point 53 07.60N/4 19.64W Fl WR 3.5s 6m 3M; W mast; vis R065°-245°, W245°-065°.
Caernarfon Harbour, Tidal Basin, S Pier head 2FG(vert).
Porth Dinllaen. Careg y Chwislen 52 56.96N/4 33.44W isolated danger mark (unlit).

Bardsey Island 52 44.97N/4 47.93W Fl(5) 15s 39m **26M**; W square Tr, R bands; obsc by Bardsey Island 198°-250° and in Tremadoc bay when brg less than 260°; *Horn Mo(N) 45s*; RC.

St Tudwal's, West Island 52 47.90N/4 28.20W Fl WR 20s 46m **W15**/12M, R13M; W round Tr; vis W349°-169°, R169°-221°, W221°-243°, R243°-259°, W259°-293°, R293°-349°; obsc by East Island 211°-231°.

PWLLHELI. Sewer outfall 52 53.18N/4 23.69W QR. F WRG 12m; vis G155°-175°, R175°-245°, W245°-045°.

10.10.5 PASSAGE INFORMATION

ENGLAND – NORTH WEST COAST
On the boundaries of Areas 9 and 10 lies Solway Firth (chart 1346), most of which is encumbered by shifting sandbanks. There are navigable, buoyed channels as far as Annan on the N shore, but buoys are moved as conditions dictate. Particularly in the upper firth, the stream runs very strongly in the chans when the banks are dry, and less strongly over the banks when these are covered. In Powfoot Chan for example the in-going stream begins at HW Liverpool –0300, and the outgoing at HW Liverpool +0100, sp rates up to 6 kn. Off the entrances to Firth of Solway, and in the approaches to Workington (10.10.8) beware shoals over which strong W winds raise a heavy sea. Southwards along the Cumbria coast past St Bees Hd to Walney Island there are no dangers more than 2M offshore, but no shelter either.

For Silloth, Harrington, Whitehaven and Ravenglass see 10.10.22.

BARROW TO RIVER MERSEY (charts 1961, 1981, 1951)
Entrance to Barrow-in-Furness (chart 3164) is about 1M S of Hilpsford Pt at S end of Walney Island. Two chys of Roosecote power station are conspic, 3¼M N of Walney Lt Ho (Lt, RC). Coming from S it is possible to cross the sands between Fleetwood and Barrow with sufficient rise of tide. The stream sets across the chan, which is narrow and shallow but well marked. W winds cause rough sea in the entrance. Moorings and anch off Piel Island, but space is limited and stream runs hard on ebb. See 10.10.22.

Lune Deep, 2M NW of Rossall Pt, is entrance to Morecambe B (chart 2010), gives access to Fleetwood (10.10.10), Glasson Dock (10.10.9), and the commercial port of Heysham (10.10.22). It is well buoyed. Streams runs 3½ kn at sp. Most of b is encumbered with drying sands, intersected by chans which are subject to change. S of Morecambe B, beware shoals extending 4M W of Rossall Pt.

Gut Channel in estuary of R Ribble gives access to the port of Preston. This is now closed to commercial shipping. Many of navigational aids have been withdrawn, dredging of docks has ceased and silting has occurred (10.10.22).

Queen's Chan and Crosby Chan (charts 1951 and 1978) which are well buoyed, dredged and preserved by training banks, give main access to R Mersey, and are entered E of the Bar Lanby. Be careful not to obstruct commercial shipping. From the N the old Formby chan is abandoned, but possible near HW. Towards HW and in moderate winds a yacht can cross the training bank (level of which varies between 2m and 3m above CD) E of Great Burbo Bank, if coming from the W. Rock Chan, parts of which dry and which is unmarked, may also be used but beware wrecks.

In good weather and at nps, the Dee Estuary (chart 1978) is interesting for boats prepared to take the ground. Most of estuary dries and banks extend 6M seaward. Chans shift, and buoys are moved as required. Stream runs hard in chans when banks are dry. Main entrance is Welsh Chan, but if coming from N Hilbre Swash runs W of Hilbre Island (lit).

Sailing W from the Dee on the ebb, it is feasible to take the inshore passage (buoyed) S of West Hoyle Spit, and this enjoys some protection from onshore winds at half tide or below. Rhyl is a tidal harbour, not accessible in strong onshore winds, but gives shelter for yachts able to take the ground. Abergele Road, Colwyn B and Llandudno B are possible anchs in settled weather and S winds.

Between Pt of Air and Great Ormes Hd the E-going stream begins at HW Liverpool +0600, and the W-going at HW Liverpool –0015, sp rates 3 kn.

ISLE OF MAN (charts 2094, 2696)
There are choices when rounding S of IOM. In bad weather or in darkness, keep S of Chicken Rk (Lt, fog sig), noting the traffic scheme offshore – see Fig. 2(7). In good conditions, take chan between Chicken Rk and Calf of Man (Lt, fog sig). Alternatively use Calf Sound (chart 2696) between Calf of Man and coast, passing S of Kitterland Island but N of Thousla Island, which is marked by lit Bn and is close to Calf of Man shore. There is also a minor chan, called Little Sound, N of Kitterland Island.

The stream runs strongly through Calf Sound, starting N-going at HW Liverpool –0145, and S-going at HW Liverpool +0345, sp rates 3½ kn. W of Calf of Man the stream runs N and S, but changes direction off Chicken Rk and runs W and E between Calf of Man and Langness Pt 6M to E. Overfalls extend E from Chicken Rk on E-going stream, which begins at HW Liverpool +0610, and N from the rk on W-going stream, which begins at HW Liverpool.

Off Langness Pt (Lt) the Skerranes (dry) extend 1 ca SW, and tidal stream runs strongly, with eddies and a race. E side of Langness peninsula is foul ground, over which a dangerous sea can build in strong winds. Here the NE-going stream begins at HW Liverpool +0545, and the SW-going at HW Liverpool –0415, sp rates 2¼ kn.

There is anch in Derby Haven, N of St Michael's Island, but exposed to E. Between Derby Island and Douglas (10.10.14) and on to Maughold Hd, there are no dangers more than 4 ca offshore. Near the land the SW-going stream runs for 9 hours and the NE-going for 3 hours, since an eddy forms during the second half of the main NE-going stream. Off Maughold Hd the NE-going stream begins at HW Liverpool +0500, and the SW-going at HW Liverpool –0415.

SE, E and NW of Pt of Ayre are dangerous shoals, over which the sea breaks in bad weather. They are Bahama Bank (least depth 1.5m), Whitestone Bank (0.6m), Ballacash Bank (2.4m), King William Banks (3.7m), and Strunakill Bank (5.5m).

The W coast of IOM has few pilotage features. A spit with depths less than 1.8m runs 4 ca offshore ½M E of Rue Pt. Jurby Rk (depth 2.7m) lies 4 ca off Jurby Hd, and up to 3M offshore there is a target area marked by buoys and Y Lts: yachts are advised to pass to seaward or to keep well inshore by passing close to Jurby Rk (above). Craig Rk (depth 3.7m) and shoals lie 2½M NNE of St Patricks I.

For general pilotage information, tidal streams and harbour details of IOM, see *Isle of Man Sailing Directions, Tidal Streams and Anchorages*, published by the Manx Sailing and Cruising Club.

MENAI STRAIT
The main features of this narrow chan (chart 1464) are: Puffin Island, seaward of NE end; Beaumaris (10.10.19); Garth Pt at Bangor (10.10.19), where NE end of Strait begins; Menai Suspension Bridge (30.5m); The Swellies (10.10.19), a narrow 1M stretch with strong tide and dangers mid-stream; Britannia Rail Bridge, with cables close W at elevation of 24m; Port Dinorwic (10.10.19); Caernarfon (10.10.19); Abermenai Pt, where narrows mark SW end of Strait; and Caernarfon Bar.

The Swellies should be taken near HW slack, and an understanding of tidal streams is essential. The tide is about 1h later, and sp range about 2.7m more, at NE end of Strait than at SW end. Levels differ most about HW +0100 at NE end (when level there is more than 1.8m above level at SW end), and about HW –0445 at NE end (when level there is more than 1.8m below level at SW end). Normally the stream runs as follows (times referred to HW Holyhead). HW –0040 to HW +0420: SW between Garth Pt and Abermenai Pt. HW

AREA 10—NW England, Isle of Man and N Wales

+0420 to HW +0545: outwards from about The Swellies, NE towards Garth Pt and SW towards Abermenai Pt. HW +0545 to HW −0040: NE between Abermenai Pt and Garth Pt. Sp rates generally about 3 kn, but more in narrows, e.g. 5 kn off Abermenai Pt, 6 kn between the bridges, and 8 kn at The Swellies. The timings and rates of streams may be affected by strong winds in either direction.

Note that direction of lateral buoyage changes at Caernarfon.

The following brief notes only cover very basic pilotage. From NE, enter chan W of Puffin Island, taking first of ebb to arrive Swellies at slack HW (HW Holyhead −0100). Slack HW only lasts about 20 mins at sps, a little more at nps. Pass under centre of Suspension Bridge span, and steer to leave Swelly buoy close to stbd, and Platters (dry) on mainland shore to port. From Swelly buoy to Britannia Bridge hold mainland shore, leaving Bn on Price Pt to port, and Gored Goch and Gribbin Rk to stbd. Leave Britannia Rk (centre pier of bridge) to stbd. Thence to SW hold to buoyed chan near mainland shore. Port Dinorwic is useful to await right tidal conditions for onward passage in either direction. Caernarfon Bar is impassable even in moderately strong winds against ebb, but narrows at Abermenai Pt demand a fair tide, or slackish water, since tide runs strongly here. Going seaward on first of ebb, when there is water over the banks, it may not be practicable to return to the Strait if conditions on the bar are bad. Then it is best to anch near Mussel Bank buoy and await slack water, before returning to Caernarfon (say). Leaving Abermenai Pt on last of ebb means banks to seaward are exposed and there is little water in chan or over bar.

Going NE it is safe to arrive at Swellies with last of flood, leaving Caernarfon about HW Holyhead −0230. Do not leave too late, or full force of ebb will be met before reaching Bangor. For detailed instructions see *West Coasts of England and Wales Pilot*, or *Cruising Anglesey* (North West Venturers Yacht Club).

ANGLESEY TO BARDSEY ISLAND (charts 1970, 1971)

On N of Anglesey, off Lynas Pt (Lt, fog sig, RC) a race extends ½M on E-going stream. Off Amlwch is a mooring/unloading buoy for supertankers, which must be given a wide berth. Amlwch is a small commercial harb (mostly dries) 1½M W of Lynas Pt. A drying rk lies ½ ca offshore on W side of approach, which should not be tried in strong onshore winds. From here to Carmel Hd beware E Mouse (and shoals to SE), Middle Mouse, Harry Furlong's Rks (dry), Victoria Bank (least depth 1.6m), Coal Rk (awash), and W Mouse (with dangers to W and SW). The outermost of these dangers is 2M offshore. There are overfalls and races at headlands and over many rks and shoals along this coast.

Between Carmel Hd and the Skerries (Lt, fog sig, Racon, RC) the NE-going stream begins at HW Holyhead +0550, and the SW-going at HW Holyhead −0010, sp rates 5 kn. 1M NW of Skerries the stream turns 1½h later, and runs less strongly. Simplest passage, or at night or in bad weather, is to pass 1M off Skerries, remembering traffic scheme off shore — see Fig. 2(7). In good conditions by day the inshore passage may be taken, close round Carmel Hd, but there is a confused sea here with wind against tide, and good timing is then needed.

Races also occur off N Stack and (more severe) off S Stack (Lt, fog sig), up to 1½M offshore on NNE-going stream which begins at HW Holyhead −0605, sp rate 5 kn. Races do not extend so far on SSW-going stream which begins at HW Holyhead +0020, sp rate 5 kn.

Along W coast of Anglesey, S from Holyhead (10.10.18), coast is rugged with rks, some drying, up to 1½M offshore. There are races off Penrhyn Mawr and Rhoscolyn Hd. Pilot's Cove, E of Llanddwyn Island, is good anch while waiting right conditions for Menai Strait. Porth Dinllaen is good anch, but exposed to N and NE. Beware rks off the pt, and Carreg-y-chad ¾M to SW. Give Carreg-y-Chwislen (a drying rk with Bn, and with shoals extending 2½ ca E) a berth of at least 3½ ca in rough weather. See 10.10.22.

Braich y Pwll is a steep and rky point at end of Lleyn Peninsula (chart 1971). About 1M N of it and about 1M offshore lie The Tripods, a bank on which there are overfalls and a bad sea with wind against tide.

Bardsey Sound can be used in daylight and moderate winds. Stream reaches 6 kn at sp, and passage should be made at slack water, ½h before HW or LW Holyhead. Avoid Carreg Ddu on N side and Maen Bugail on S side of Sound, where there are dangerous races. Passing outside Bardsey Island (Lt, fog sig, RC) make a good offing to avoid overfalls which extend 1½M W of I and 2½M S of it. Turbulence occurs over Bastram Shoal, Devil's Tail and Devil's Ridge, which lie S and E of Bardsey Island. In the approaches to St Tudwal's Island (lit) and Abersoch (10.10.20) there may be overfalls off Trwyn Cilan.

10.10.6 DISTANCE TABLE

Approximate distances in nautical miles are by the most direct route while avoiding dangers and allowing for traffic separation schemes etc. Places in *italics* are in adjoining areas.

#	Place	1	2	3	4	5	6	7	8	9	10	11	12	13	14	15	16	17	18	19	20
1	*Strangford*	**1**																			
2	*Mew Island*	29	**2**																		
3	*Mull of Galloway*	34	23	**3**																	
4	St Bees Head	70	65	43	**4**																
5	Point of Ayre	44	43	21	26	**5**															
6	Douglas	63	62	40	37	19	**6**														
7	Barrow	95	96	74	39	53	52	**7**													
8	Glasson Dock	109	107	85	53	64	63	22	**8**												
9	Fleetwood	103	113	79	47	58	57	17	10	**9**											
10	Liverpool	109	121	101	75	80	70	50	52	46	**10**										
11	Conwy	90	105	94	77	72	59	59	62	56	46	**11**									
12	Beaumaris	89	102	91	77	71	58	62	66	60	49	12	**12**								
13	Holyhead	74	90	79	80	68	50	72	79	73	68	36	32	**13**							
14	Caernarfon	92	111	105	87	81	68	72	76	70	59	22	10	26	**14**						
15	Bardsey Island	103	120	115	116	107	88	103	107	101	90	53	41	38	31	**15**					
16	Abersoch	117	134	129	126	121	102	117	121	115	104	67	55	52	45	14	**16**				
17	Pwllheli	121	138	133	130	125	106	121	125	119	108	71	59	56	49	18	5	**17**			
18	*South Bishop*	151	172	169	174	159	143	163	167	161	150	113	101	98	91	60	70	74	**18**		
19	*Tuskar Rock*	136	155	152	163	152	132	159	170	164	152	115	103	91	93	63	75	79	36	**19**	
20	*Dun Laoghaire*	71	90	93	113	93	78	119	126	120	118	88	84	56	69	61	75	79	90	70	**20**

MARYPORT 10-10-7
Cumbria

CHARTS
Admiralty 1346, 2013; Imray C62; OS 89
TIDES
+0038 Dover; ML 4.7; Duration 0550; Zone 0 (GMT).
Standard Port LIVERPOOL (→)

Times				Height (metres)			
HW		LW		MHWS	MHWN	MLWN	MLWS
0000	0600	0200	0800	9.3	7.4	2.9	0.9
1200	1800	1400	2000				

Differences MARYPORT
+0017 +0032 +0020 +0005 −0.7 −0.8 −0.4 0.0
SILLOTH
+0030 +0040 +0045 +0055 −0.1 −0.3 −0.6 −0.1
TORDUFF POINT
+0105 +0140 +0520 +0410 −4.1 −4.9 — —

SHELTER
Complete shelter in Elizabeth Dock. Harbour dries. Entry HW ∓1½. Nearest anchorage Workington Basin.
NAVIGATION
Waypoint 54°43′.08N 03°32′.39W, 270°/090° from/to South Pier Lt, 1.0M. 1.8m over bar between piers at entrance at half tide and in river channel. Mud banks cover HW−2. Channel not buoyed.
LIGHTS AND MARKS
South Pier. Fl 1.5s.
RADIO TELEPHONE
VHF Ch 16; 12 (occas).
TELEPHONE (0900)
Hr Mr 817440; CG 2238; MRSC Liverpool 931-3341; HMC 4611; Marinecall 0898 500 461; Dr 815544; Hosp 812634.

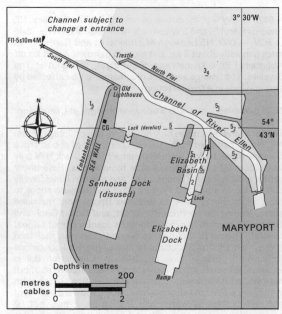

FACILITIES
EC Wednesday. **Elizabeth Dock** Slip (at HW), M, Sh, AB. **Town** P, D, ME, V, R, Bar; PO, Bank, Rly; Air (Newcastle).
Note:— Senhouse Dock is to be converted to a 300 berth marina.

WORKINGTON 10-10-8
Cumbria

CHARTS
Admiralty 1346, 1826, 2013; Imray C62; OS 89
TIDES
+0025 Dover; ML 4.4; Duration 0545; Zone 0 (GMT).
Standard Port LIVERPOOL (→)

Times				Height (metres)			
HW		LW		MHWS	MHWN	MLWN	MLWS
0000	0600	0200	0800	9.3	7.4	2.9	0.9
1200	1800	1400	2000				

Differences WORKINGTON
+0020 +0020 +0020 +0010 −1.1 −1.0 −0.1 +0.3
WHITEHAVEN
+0005 +0015 +0010 +0005 −1.3 −1.1 −0.5 +0.1
TARN POINT
+0005 +0005 +0010 0000 −1.0 −1.0 −0.4 0.0

SHELTER
Good shelter in Turning Basin. Tidal harbour dries and bridge clearance is only 1.8m.
NAVIGATION
Waypoint 54°39′.58N 03°35′.30W, 311°/131° from/to front Ldg Lt, 1.0M. No navigational dangers but in periods of heavy rain a strong freshet may be encountered.
LIGHTS AND MARKS
End of breakwater QG, end of S pier Fl 5s. Leading Lts on W pyramidal Trs with Y bands both FR at 132°.
RADIO TELEPHONE
VHF Ch 16; 11 14 (HW−2½ to HW+2).
TELEPHONE (0900)
Hr Mr 2301; CG 2238; HMC 4611; MRSC Liverpool 931-3341; Dr 64866; Hosp 2244.
FACILITIES
EC Thursday; **Dock** D, FW, ME, El; **Vanguard SC** Tel. 826886, M, FW; **Town** P, V, R, Bar. PO; Bank; Rly; Air (Carlisle or Newcastle).

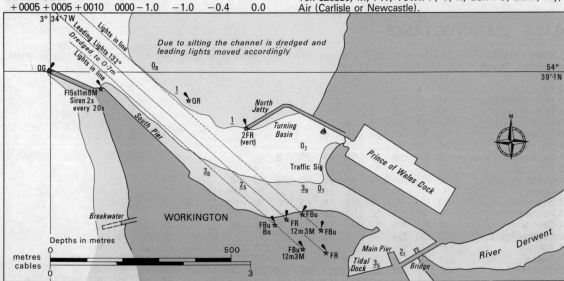

GLASSON DOCK 10-10-9
Lancashire

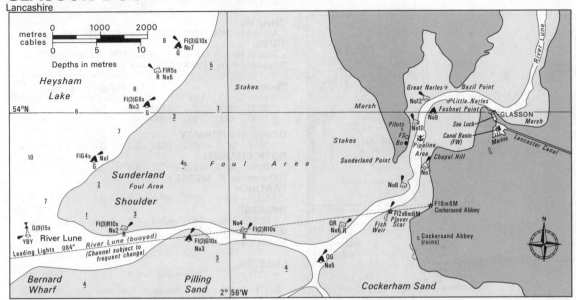

CHARTS
Admiralty 2010, 1552; Imray C62; OS 102, 97
TIDES
+0020 Dover; ML Wyre Lt Ho 11.7, Lancaster 5.2; Duration 0535; Zone 0 (GMT).
Standard Port LIVERPOOL (→)

Times				Height (metres)			
HW		LW		MHWS	MHWN	MLWN	MLWS
0000	0600	0200	0700	9.3	7.4	2.9	0.9
1200	1800	1400	1900				

Differences GLASSON DOCK
+0020 +0030 +0220 +0240 −2.7 −3.0 No data
LANCASTER
+0110 +0030 No data −5.0 −4.9 Dries out
BARROW-IN-FURNESS (Ramsden Dock)
+0015 +0015 +0020 +0020 −0.2 −0.3 −0.1 +0.1
MORECAMBE
+0005 +0010 +0030 +0015 +0.2 0.0 0.0 +0.2
HEYSHAM
+0005 +0005 +0015 0000 +0.1 0.0 0.0 +0.2

SHELTER
Very sheltered anchorage. Even more so up towards Lancaster; navigable for small boats but pilot is advised. Dock connects with British Waterways basin of Lancaster canal. Marina is in canal.
NAVIGATION
Waypoint R Lune entrance 53°58'.40N 03°00'.00W, 264°/084° from/to front Ldg Lt 084°, 4.2M. Dock gates are open from HW−1½ to HW. Wait at R Lune buoy until HW−2 and then go up. Abbey Lt has a tide gauge showing depth over the sill at Glasson Dock. Beware three wrecks on Bernard Wharf within 2½ miles E from Nos 2 and 8 Fleetwood buoys.
LIGHTS AND MARKS
Plover Scar Lt and Cockersand Abbey Lt in line at 084° lead up the channel between Nos 2 and 3 buoys. Lts difficult to pick out at R Lune buoy; steer 085° until Abbey Lt identified. Channel is buoyed up to dock entrance. Buoys no 6, 8, 10, 12, 14, 16 (Bn) and 18 are Fl or QR; no 5, 7 (Bn) and 9 are QG. Harbour authority is the Lancaster Port Commission.
RADIO TELEPHONE
VHF Ch 16; 08 (HW−2 to HW+1). Marina Ch M.
TELEPHONE (0524)
Hr Mr 751724 or 751307; CG & MRSC Liverpool 931-3341; Lancaster Port Commission 751304; HMC Freefone Customs Yachts: RT (051) 933 4292; Marinecall 0898 500 461; Hosp Lancaster 65944.
FACILITIES
EC Lancaster Wednesday; **Glasson Dock Marina** (240+20 visitors) Tel. 751491, Slip, D, FW, ME, El, Sh, C, (50 ton), BH (50 ton), AC, CH; **Glasson Basin** Tel. 751724, M, AB; **Glasson SC** Tel. 751089 Slip, M, C; **Lune Cruising Club** M, Access HW∓2; **Town** P, V, R, Bar, PO; Bank (Lancaster); Rly (bus to Lancaster); Air (Blackpool).

FLEETWOOD 10-10-10
Lancashire

CHARTS
Admiralty 2010, 1552; Imray C62; OS 102
TIDES
+0015 Dover; ML 5.0; Duration 0540; Zone 0 (GMT).
Standard Port LIVERPOOL (→)

Times				Height (metres)			
HW		LW		MHWS	MHWN	MLWN	MLWS
0000	0600	0200	0700	9.3	7.4	2.9	0.9
1200	1800	1400	1900				

Differences FLEETWOOD
0000 0000 +0005 0000 −0.1 −0.1 +0.1 +0.3
ST ANNE'S
−0004 −0004 +0013 +0013 0.0 −0.3 +0.8 +0.5

SHELTER
Very good shelter except in strong winds from N to W. Boats can pass Fleetwood and proceed up to Skipool (5 M) where is HQ of Blackpool & Fleetwood YC. Passage up river not recommended on tides less than 8.0m.
NAVIGATION
Waypoint 53°57'.78N 03°01'.90W, 338°/158° from/to front Ldg Lt 156°, 2.25M. Beware Ro Ro vessels turning in the harbour and dredgers working continuously.
LIGHTS AND MARKS
Channel well buoyed, five G conical buoys and one Bn to starboard, nine R can buoys to port. Fleetwood Ldg Lts 156° (only to be used between No 7 buoy and Black Scar (No 11) perch). Lts are Front Fl Y 2s and Rear Fl Y 4s.
RADIO TELEPHONE
Call: *Fleetwood Harbour Control* VHF Ch 16; 11 **12** (H24). Call: *Fleetwood Docks* Ch 16; 12 (HW−2 to HW+2). Information from Harbour Control on request. Other stations: Ramsden Docks Ch 16; 12 (H24); Heysham Ch 16; 14 (H24).
TELEPHONE (039 17)
Hr Mr 2323; CG 3780; MRSC Liverpool 931-3341; HMC Freefone Customs Yachts: RT (03917) 79211; Marinecall 0898 500 461; Dr Cleveleys 826280.

FLEETWOOD continued

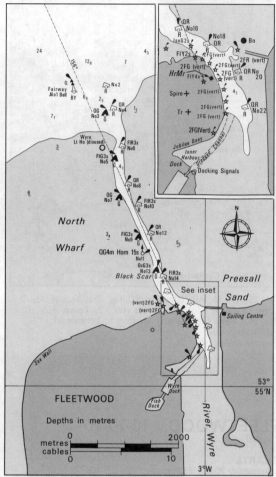

FLEETWOOD
Depths in metres

FACILITIES
EC Wednesday; **Harbour** Tel. 2323, FW, ME, El, CH; **Enclosed dock** M (No deep water moorings), FW, ME, El, Sh, C (by arrangement 25 ton); **Blackpool and Fleetwood YC (Skipool Creek)** Tel. Blackpool 884205, Slip (at Stanah), L, FW, AB (Skipool), Bar; **Fleetwood Trawler Supply Co** Tel. 3476, CH, ACA; **Wardleys Marine** Tel. Hambleton 700117 AB, CH, D, El, FW, L, M (drying), ME, Sh, Slip; **David Moss** Tel. Blackpool 893830, D (cans), FW, ME, El, Sh, C (mobile 50 ton), CH; **Town** P, D, ME, El, Sh, CH, V, R, Bar. PO; Bank; Rly (bus to Poulton-le-Fylde or Blackpool); Air (Blackpool).

Note:— The derelict dock area is to be converted to a Marina Village with berths for 1000 yachts.

POULTON-LE-FYLDE. P, D, ME, El, Sh, V, R, Bar; **Lytham Marine** Tel. Lytham 735531; Slip, D, FW, ME, El, Sh, C (4 ton); CH, AB; **Douglas Boatyard** Tel. Hesketh Bank 2462, Slip, D, FW, ME, El, Sh, C (7 ton), CH, AB; **Fylde Coast Sailmaking Co** Tel. 3476 SM, CH.

AGENTS NEEDED
There are a number of ports where we need agents, particularly in France.
ENGLAND   Swale, Havengore, Berwick.
SCOTLAND   Firth of Forth, Scrabster, Mallaig, Loch Sunart, Loch Aline.
IRELAND   Kilrush, Wicklow, Westport/Clew Bay.
FRANCE   Arcachon, Seudre R, Ile d'Oleron, Rochfort, Ile de Re, St. Giles-Croix-de-Vie, Ile d'Yeu, Pouliguen, Le Croisic, La Forêt, Ile de Bréhat.
GERMANY   Norderney, Dornumer-Accumersiel.

If you are interested in becoming our agent for any of the above, please write to the editors and get your free copy of the Almanac every year. You do not have to be resident in a port to be the agent, but at least a fairly regular visitor.

LIVERPOOL 10-10-11
Merseyside

CHARTS
Admiralty 3490, 1951, 1978; Imray C62; OS 108
TIDES
+0015 Dover; ML 5.2; Duration 0535; Zone 0 (GMT).
Standard Port LIVERPOOL (⟶)

Times				Height (metres)			
HW		LW		MHWS	MHWN	MLWN	MLWS
0000	0600	0200	0700	9.3	7.4	2.9	0.9
1200	1800	1400	1900				

Differences FORMBY
−0015 −0010 −0020 −0020 −0.3 −0.1 0.0 +0.1
ROCK CHANNEL
−0030 −0030 −0030 −0030 −0.4 −0.2 −0.2 0.0
Differences R. MERSEY
EASTHAM
+0003 +0006 +0015 +0030 +0.4 +0.3 −0.1 −0.1
WIDNES
+0040 +0045 +0400 +0345 −4.2 −4.4 −2.5 −0.3
Differences R. DEE
MOSTYN QUAY
−0020 −0015 −0020 −0020 −0.8 −0.7 No data
CHESTER
+0105 +0105 +0500 +0500 −5.3 −5.4 Dries out

NOTE: Liverpool is a Standard Port and tidal predictions for each day of the year are given below.

SHELTER
Yachts should go to marina in Brunswick and Coburg Docks. Shelter good in Stanley, Canning, Albert or Salthouse Docks. There are no other facilities for berthing or anchoring yachts on the NE side. (Anchoring possible on Birkenhead side but only in fair weather).
NAVIGATION
Liverpool waypoint Bar Lanby 53°32'.00N 03°20'.90W, Fl 5s, Horn, Racon, 280°/100° from/to Queen's Channel, 3.0M. River Dee waypoint Hilbre Swash HE2 port-hand buoy, QR, 53°26'.20N 03°16'.80W (channel shifts). Wind against tide causes steep seas in outer reaches of Mersey. Do not go outside buoys marking fairway. Advisable to enter by Bar and Queens Channel. The whole area from

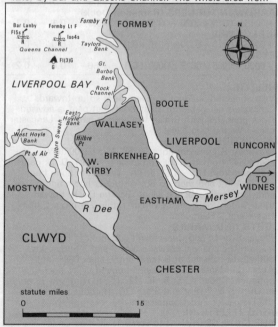

River Dee, River Mersey to River Alt and up to Morecambe Bay is littered with sandbanks. R Mersey is safe within commercial channel, but everywhere else great caution and local knowledge is needed. R Alt is only available HW∓2 and only suitable for shallow draught (1.2m) boats. See 10.10.22.
LIGHTS AND MARKS
The entrance to the Mersey is marked by the Bar Lanby Fl 5s 12m 21M, Horn 20s, Racon.

ENGLAND, WEST COAST — LIVERPOOL

Lat 53°25′ N Long 3°00′ W

TIMES AND HEIGHTS OF HIGH AND LOW WATERS

TIME ZONE UT(GMT) — For Summer Time add ONE hour in non-shaded areas

YEAR 1989

JANUARY

#	Time	m	#	Time	m
1 Su	0504 / 1125 / 1725	7.2 / 3.4 / 7.3	16 M	0539 / 1217 / 1810	7.7 / 2.6 / 8.0
2 M	0018 / 0608 / 1238 / 1832	3.1 / 7.1 / 3.4 / 7.3	17 Tu	0059 / 0656 / 1340 / 1928	2.6 / 7.6 / 2.7 / 7.8
3 Tu	0127 / 0716 / 1349 / 1941	3.1 / 7.3 / 3.2 / 7.4	18 W	0218 / 0811 / 1458 / 2042	2.6 / 7.8 / 2.5 / 7.9
4 W	0230 / 0816 / 1454 / 2042	2.9 / 7.6 / 2.9 / 7.7	19 Th	0328 / 0917 / 1607 / 2145	2.4 / 8.1 / 2.1 / 8.2
5 Th	0327 / 0910 / 1549 / 2134	2.5 / 8.1 / 2.5 / 8.1	20 F	0426 / 1010 / 1702 / 2235	2.1 / 8.5 / 1.7 / 8.5
6 F	0416 / 0956 / 1641 / 2220	2.2 / 8.5 / 2.0 / 8.5	21 Sa	0515 / 1057 / 1750 / 2319	1.9 / 8.9 / 1.5 / 8.7
7 Sa	0502 / 1041 / 1729 / 2305	1.8 / 8.9 / 1.6 / 8.8	22 Su	0556 / 1136 / 1829 / 2358	1.7 / 9.1 / 1.3 / 8.8
8 Su	0547 / 1125 / 1817 / 2350	1.5 / 9.2 / 1.4 / 9.1	23 M	0631 / 1212 / 1906	1.6 / 9.1 / 1.3
9 M	0632 / 1210 / 1903	1.3 / 9.5 / 1.0	24 Tu	0032 / 0703 / 1246 / 1937	8.9 / 1.6 / 9.1 / 1.3
10 Tu	0034 / 0716 / 1253 / 1947	9.2 / 1.2 / 9.6 / 0.8	25 W	0104 / 0733 / 1319 / 2006	8.8 / 1.6 / 9.0 / 1.5
11 W	0119 / 0758 / 1338 / 2030	9.2 / 1.2 / 9.6 / 0.8	26 Th	0135 / 0801 / 1349 / 2033	8.7 / 1.8 / 8.8 / 1.7
12 Th	0204 / 0840 / 1423 / 2114	9.1 / 1.3 / 9.5 / 1.0	27 F	0208 / 0830 / 1420 / 2101	8.5 / 2.0 / 8.6 / 2.1
13 F	0250 / 0924 / 1510 / 2159	8.8 / 1.6 / 9.2 / 1.4	28 Sa	0240 / 0903 / 1453 / 2134	8.2 / 2.3 / 8.3 / 2.3
14 Sa	0338 / 1012 / 1600 / 2248	8.5 / 1.9 / 8.8 / 1.8	29 Su	0314 / 0938 / 1528 / 2210	7.9 / 2.7 / 7.9 / 2.7
15 Su	0433 / 1108 / 1659 / 2347	8.1 / 2.3 / 8.4 / 2.2	30 M	0356 / 1023 / 1614 / 2301	7.5 / 3.1 / 7.5 / 3.2
			31 Tu	0452 / 1125 / 1719	7.1 / 3.4 / 7.1

FEBRUARY

#	Time	m	#	Time	m
1 W	0017 / 0611 / 1253 / 1848	3.4 / 7.0 / 3.5 / 7.0	16 Th	0205 / 0759 / 1456 / 2040	3.1 / 7.4 / 2.7 / 7.5
2 Th	0147 / 0734 / 1418 / 2011	3.3 / 7.2 / 3.2 / 7.3	17 F	0324 / 0910 / 1604 / 2141	2.8 / 7.9 / 2.1 / 8.0
3 F	0258 / 0843 / 1527 / 2115	2.9 / 7.7 / 2.6 / 7.8	18 Sa	0421 / 1002 / 1657 / 2227	2.3 / 8.4 / 1.7 / 8.4
4 Sa	0357 / 0939 / 1626 / 2207	2.3 / 8.4 / 1.9 / 8.4	19 Su	0506 / 1044 / 1737 / 2305	1.9 / 8.8 / 1.4 / 8.7
5 Su	0448 / 1027 / 1718 / 2254	1.8 / 8.9 / 1.3 / 9.0	20 M	0542 / 1119 / 1812 / 2339	1.6 / 9.1 / 1.2 / 8.9
6 M	0536 / 1112 / 1807 / 2337	1.2 / 9.5 / 0.8 / 9.4	21 Tu	0612 / 1151 / 1842	1.4 / 9.2 / 1.1
7 Tu	0621 / 1156 / 1850	0.8 / 9.8 / 0.4	22 W	0008 / 0641 / 1222 / 1909	9.0 / 1.3 / 9.2 / 1.1
8 W	0019 / 0703 / 1238 / 1933	9.6 / 0.6 / 10.1 / 0.2	23 Th	0038 / 0707 / 1250 / 1935	9.0 / 1.3 / 9.1 / 1.2
9 Th	0102 / 0744 / 1320 / 2013	9.7 / 0.5 / 10.1 / 0.3	24 F	0106 / 0734 / 1319 / 1959	8.9 / 1.4 / 9.0 / 1.4
10 F	0144 / 0823 / 1402 / 2051	9.5 / 0.7 / 9.9 / 0.6	25 Sa	0134 / 0802 / 1345 / 2026	8.7 / 1.6 / 8.8 / 1.7
11 Sa	0225 / 0903 / 1444 / 2131	9.2 / 1.1 / 9.5 / 1.2	26 Su	0202 / 0829 / 1413 / 2054	8.5 / 1.9 / 8.5 / 2.1
12 Su	0308 / 0945 / 1531 / 2214	8.7 / 1.6 / 8.9 / 1.8	27 M	0232 / 0904 / 1446 / 2127	8.2 / 2.3 / 8.1 / 2.6
13 M	0357 / 1037 / 1627 / 2309	8.1 / 2.2 / 8.2 / 2.5	28 Tu	0308 / 0943 / 1527 / 2209	7.8 / 2.8 / 7.6 / 3.1
14 Tu	0501 / 1147 / 1742	7.6 / 2.8 / 7.5			
15 W	0027 / 0628 / 1324 / 1916	3.1 / 7.2 / 3.0 / 7.3			

MARCH

#	Time	m	#	Time	m
1 W	0357 / 1038 / 1627 / 2316	7.3 / 2.9 / 7.1 / 3.5	16 Th	0003 / 0605 / 1313 / 1904	3.4 / 7.7 / 3.0 / 7.0
2 Th	0516 / 1204 / 1804	7.0 / 3.5 / 6.8	17 F	0151 / 0742 / 1443 / 2027	3.4 / 7.3 / 2.6 / 7.4
3 F	0104 / 0656 / 1347 / 1945	3.6 / 7.0 / 3.2 / 7.1	18 Sa	0310 / 0851 / 1546 / 2122	2.9 / 7.8 / 2.1 / 7.9
4 Sa	0232 / 0818 / 1505 / 2056	3.0 / 7.6 / 2.5 / 7.8	19 Su	0403 / 0941 / 1633 / 2204	2.3 / 8.3 / 1.7 / 8.4
5 Su	0336 / 0918 / 1607 / 2149	2.3 / 8.4 / 1.7 / 8.5	20 M	0442 / 1019 / 1711 / 2240	1.9 / 8.7 / 1.4 / 8.7
6 M	0431 / 1007 / 1701 / 2234	1.6 / 9.1 / 1.0 / 9.1	21 Tu	0516 / 1052 / 1742 / 2311	1.6 / 8.9 / 1.2 / 8.9
7 Tu	0518 / 1051 / 1747 / 2318	1.0 / 9.7 / 0.4 / 9.6	22 W	0546 / 1123 / 1810 / 2339	1.4 / 9.1 / 1.1 / 9.0
8 W	0603 / 1134 / 1831	0.5 / 10.1 / 0.0	23 Th	0612 / 1153 / 1836	1.3 / 9.1 / 1.1
9 Th	0001 / 0643 / 1217 / 1912	9.9 / 0.2 / 10.3 / -0.1	24 F	0007 / 0639 / 1221 / 1902	9.0 / 1.2 / 9.1 / 1.2
10 F	0039 / 0723 / 1257 / 1949	9.9 / 0.2 / 10.2 / 0.1	25 Sa	0035 / 0707 / 1249 / 1928	9.0 / 1.3 / 9.0 / 1.4
11 Sa	0119 / 0802 / 1338 / 2026	9.7 / 0.4 / 9.9 / 0.6	26 Su	0102 / 0737 / 1316 / 1957	8.9 / 1.6 / 8.8 / 1.7
12 Su	0159 / 0842 / 1420 / 2104	9.3 / 0.9 / 9.4 / 1.3	27 M	0130 / 0806 / 1344 / 2025	8.7 / 1.8 / 8.5 / 2.1
13 M	0240 / 0924 / 1507 / 2145	8.8 / 1.5 / 8.7 / 2.1	28 Tu	0202 / 0840 / 1418 / 2057	8.4 / 2.2 / 8.1 / 2.5
14 Tu	0328 / 1012 / 1603 / 2238	8.1 / 2.3 / 7.8 / 2.8	29 W	0239 / 0921 / 1501 / 2139	8.0 / 2.6 / 7.7 / 3.0
15 W	0433 / 1129 / 1723	7.5 / 2.9 / 7.2	30 Th	0331 / 1014 / 1603 / 2247	7.5 / 3.0 / 7.2 / 3.4
			31 F	0447 / 1137 / 1739	7.2 / 3.2 / 6.9

APRIL

#	Time	m	#	Time	m
1 Sa	0029 / 0624 / 1317 / 1917	3.5 / 7.2 / 2.9 / 7.2	16 Su	0232 / 0815 / 1507 / 2047	3.0 / 7.7 / 2.2 / 7.8
2 Su	0202 / 0747 / 1437 / 2029	2.9 / 7.8 / 2.2 / 7.9	17 M	0324 / 0904 / 1553 / 2129	2.5 / 8.1 / 1.9 / 8.2
3 M	0308 / 0849 / 1541 / 2122	2.2 / 8.5 / 1.5 / 8.6	18 Tu	0406 / 0943 / 1631 / 2204	2.1 / 8.4 / 1.6 / 8.5
4 Tu	0404 / 0939 / 1634 / 2209	1.5 / 9.2 / 0.8 / 9.2	19 W	0440 / 1019 / 1704 / 2237	1.8 / 8.6 / 1.5 / 8.7
5 W	0452 / 1024 / 1720 / 2251	0.9 / 9.7 / 0.3 / 9.6	20 Th	0511 / 1051 / 1733 / 2306	1.6 / 8.8 / 1.3 / 8.9
6 Th	0537 / 1109 / 1804 / 2333	0.4 / 10.1 / 0.1 / 9.8	21 F	0540 / 1122 / 1803 / 2336	1.4 / 8.9 / 1.3 / 9.0
7 F	0619 / 1151 / 1845	0.2 / 10.2 / 0.1	22 Sa	0611 / 1151 / 1831	1.3 / 8.9 / 1.3
8 Sa	0014 / 0702 / 1235 / 1924	9.8 / 0.2 / 10.1 / 0.3	23 Su	0005 / 0642 / 1221 / 1902	9.0 / 1.4 / 8.8 / 1.5
9 Su	0055 / 0742 / 1317 / 2002	9.6 / 0.5 / 9.7 / 0.9	24 M	0036 / 0716 / 1252 / 1933	8.9 / 1.5 / 8.7 / 1.7
10 M	0135 / 0823 / 1401 / 2040	9.2 / 1.0 / 9.1 / 1.5	25 Tu	0109 / 0749 / 1326 / 2005	8.7 / 1.7 / 8.4 / 2.1
11 Tu	0219 / 0908 / 1449 / 2124	8.7 / 1.6 / 8.4 / 2.3	26 W	0144 / 0827 / 1405 / 2042	8.5 / 2.0 / 8.1 / 2.4
12 W	0308 / 1003 / 1548 / 2217	8.1 / 2.3 / 7.7 / 3.0	27 Th	0226 / 0911 / 1453 / 2129	8.2 / 2.3 / 7.7 / 2.8
13 Th	0412 / 1118 / 1704 / 2340	7.5 / 2.8 / 7.1 / 3.5	28 F	0321 / 1009 / 1557 / 2234	7.8 / 2.6 / 7.4 / 3.1
14 F	0537 / 1248 / 1836	7.2 / 2.9 / 7.0	29 Sa	0433 / 1123 / 1720	7.5 / 2.7 / 7.3
15 Su	0119 / 0706 / 1406 / 1952	3.4 / 7.3 / 2.6 / 7.3	30 Su	0001 / 0556 / 1248 / 1845	3.1 / 7.6 / 2.4 / 7.5

Chart Datum: 4.93 metres below Ordnance Datum (Newlyn)

ENGLAND, WEST COAST — LIVERPOOL

Lat 53°25′ N   Long 3°00′ W

TIMES AND HEIGHTS OF HIGH AND LOW WATERS

YEAR 1989

TIME ZONE UT(GMT)
For Summer Time add ONE hour in non-shaded areas

	MAY				JUNE				JULY				AUGUST		
	Time m		Time m		Time m		Time m		Time m		Time m		Time m		Time m
1 M	0126 2·7 0710 8·0 1402 2·0 1954 8·1	**16** Tu	0229 2·8 0815 7·8 1501 2·2 2044 7·9	**1** Th	0257 1·8 0837 8·9 1529 1·4 2111 8·8	**16** F	0312 2·6 0901 7·9 1541 2·3 2125 8·2	**1** Sa	0339 1·9 0919 8·5 1606 1·8 2149 8·7	**16** Su	0332 2·6 0921 7·8 1559 2·4 2143 8·3	**1** Tu ●	0532 1·3 1101 8·8 1740 1·6 2319 9·1	**16** W	0458 1·5 1035 8·7 1715 1·5 2252 9·3
2 Tu	0233 2·1 0813 8·6 1505 1·4 2050 8·6	**17** W	0315 2·5 0900 8·1 1543 2·0 2124 8·2	**2** F	0356 1·4 0932 9·1 1624 1·2 2202 9·0	**17** Sa	0400 2·3 0945 8·1 1624 2·1 2206 8·5	**2** Su	0440 1·6 1016 8·7 1659 1·6 2241 8·9	**17** M	0424 2·2 1009 8·2 1647 2·0 2228 8·7	**2** W	0615 1·1 1142 8·9 1819 1·5 2357 9·3	**17** Th ○	0544 1·0 1116 9·2 1758 1·1 2333 9·7
3 W	0331 1·6 0907 9·1 1602 1·0 2139 9·1	**18** Th	0356 2·2 0939 8·3 1621 1·8 2200 8·5	**3** Sa ●	0451 1·2 1024 9·3 1713 1·1 2249 9·2	**18** Su	0444 2·0 1027 8·3 1705 1·9 2245 8·7	**3** M ●	0534 1·3 1108 8·9 1749 1·5 2329 9·1	**18** Tu ○	0513 1·7 1052 8·5 1732 1·7 2311 9·0	**3** Th	0653 0·9 1218 9·0 1855 1·4	**18** F	0628 0·6 1157 9·5 1841 0·8
4 Th	0423 1·1 0956 9·5 1651 0·6 2224 9·4	**19** F	0434 1·9 1016 8·5 1657 1·7 2234 8·7	**4** Su	0542 1·0 1115 9·3 1800 1·1 2337 9·3	**19** M ○	0527 1·8 1108 8·5 1746 1·7 2325 8·9	**4** Tu	0624 1·1 1154 8·9 1832 1·5	**19** W	0600 1·4 1134 8·9 1815 1·4 2353 9·3	**4** F	0034 9·3 0727 1·1 1253 8·9 1926 1·5	**19** Sa	0014 9·9 0710 0·4 1238 9·6 1921 0·7
5 F ●	0512 0·7 1044 9·8 1737 0·5 2309 9·6	**20** Sa ○	0511 1·7 1051 8·6 1730 1·6 2308 8·8	**5** M	0629 0·9 1203 9·2 1843 1·2	**20** Tu	0610 1·6 1147 8·7 1827 1·6	**5** W	0012 9·2 0709 1·1 1238 8·9 1913 1·5	**20** Th	0645 1·0 1217 9·1 1857 1·2	**5** Sa	0106 9·2 0758 1·2 1324 8·8 1955 1·6	**20** Su	0055 10·0 0749 0·4 1317 9·5 1959 0·8
6 Sa	0557 0·5 1130 9·8 1819 0·5 2351 9·6	**21** Su	0546 1·6 1126 8·7 1805 1·6 2342 8·9	**6** Tu	0022 9·2 0717 1·0 1249 9·0 1926 1·5	**21** W	0005 9·0 0653 1·4 1228 8·7 1907 1·6	**6** Th	0053 9·1 0749 1·2 1317 8·7 1949 1·7	**21** F	0035 9·5 0728 0·8 1259 9·2 1938 1·1	**6** Su	0138 9·0 0826 1·5 1355 8·6 2023 1·8	**21** M	0135 9·9 0827 0·6 1358 9·3 2039 1·1
7 Su	0642 0·5 1215 9·6 1900 0·8	**22** M	0624 1·5 1201 8·7 1841 1·6	**7** W	0106 9·0 0802 1·2 1334 8·7 2008 1·8	**22** Th	0046 9·1 0737 1·3 1310 8·7 1948 1·6	**7** F	0131 9·0 0827 1·4 1355 8·5 2025 1·9	**22** Sa	0116 9·6 0809 0·7 1341 9·1 2019 1·2	**7** M	0209 8·7 0853 1·8 1427 8·3 2053 2·2	**22** Tu	0218 9·6 0905 1·1 1440 8·9 2119 1·5
8 M	0035 9·4 0727 0·8 1300 9·3 1941 1·2	**23** Tu	0017 8·9 0702 1·5 1238 8·6 1916 1·7	**8** Th	0149 8·8 0847 1·5 1419 8·3 2049 2·2	**23** F	0130 9·0 0820 1·3 1358 8·6 2032 1·7	**8** Sa	0209 8·7 0903 1·6 1432 8·3 2058 2·2	**23** Su	0159 9·5 0850 0·9 1423 8·9 2100 1·4	**8** Tu	0242 8·4 0921 2·2 1503 8·0 2127 2·6	**23** W ☾	0303 9·0 0946 1·7 1527 8·4 2207 2·1
9 Tu	0119 9·1 0812 1·1 1347 8·8 2022 1·8	**24** W	0055 8·8 0741 1·6 1317 8·5 1954 1·9	**9** F	0234 8·5 0934 1·9 1505 8·0 2132 2·6	**24** Sa	0215 8·8 0905 1·4 1443 8·5 2117 1·9	**9** Su	0246 8·5 0936 2·0 1510 8·0 2134 2·5	**24** M	0242 9·3 0931 1·1 1508 8·7 2143 1·7	**9** W	0317 7·9 0956 2·7 1542 7·6 2209 3·0	**24** Th	0355 8·4 1037 2·4 1627 7·8 2312 2·7
10 W	0204 8·7 0900 1·6 1436 8·2 2107 2·4	**25** Th	0135 8·7 0823 1·8 1401 8·3 2036 2·2	**10** Sa	0321 8·1 1020 2·2 1553 7·6 2220 2·9	**25** Su	0303 8·8 0952 1·5 1534 8·3 2206 2·1	**10** M	0325 8·1 1013 2·3 1552 7·7 2213 2·8	**25** Tu ☾	0328 9·0 1014 1·6 1557 8·3 2233 2·1	**10** Th	0359 7·5 1042 3·1 1635 7·2 2306 3·4	**25** F	0505 7·7 1147 3·0 1750 7·4
11 Th	0253 8·2 0952 2·1 1531 7·7 2159 2·9	**26** F	0222 8·4 0910 1·9 1451 8·0 2125 2·4	**11** Su ☾	0412 7·8 1111 2·5 1648 7·4 2316 3·1	**26** M	0355 8·6 1042 1·7 1628 8·1 2302 2·3	**11** Tu	0410 7·8 1054 2·7 1641 7·5 2304 3·1	**26** W	0421 8·5 1106 2·0 1657 7·9 2334 2·5	**11** F	0501 7·0 1151 3·5 1750 7·0	**26** Sa	0045 3·0 0638 7·3 1326 3·1 1924 7·5
12 F ☾	0350 7·8 1055 2·5 1634 7·3 2306 3·2	**27** Sa	0315 8·2 1004 2·1 1550 7·8 2224 2·6	**12** M	0508 7·5 1205 2·7 1749 7·2	**27** Tu	0452 8·5 1140 1·9 1732 7·9	**12** W	0502 7·4 1149 3·0 1742 7·2	**27** Th	0527 8·1 1212 2·5 1811 7·7	**12** Sa	0029 3·6 0629 6·8 1321 3·5 1917 7·1	**27** Su	0220 2·7 0809 7·6 1451 2·8 2042 8·0
13 Sa	0458 7·4 1204 2·7 1747 7·1	**28** Su ☾	0417 8·1 1106 2·2 1658 7·7 2333 2·6	**13** Tu	0018 3·3 0610 7·4 1303 2·8 1852 7·3	**28** W	0007 2·4 0558 8·3 1246 2·0 1841 7·9	**13** Th	0008 3·3 0607 7·2 1256 3·1 1850 7·2	**28** F	0053 2·7 0646 7·8 1334 2·6 1931 7·7	**13** Su	0158 3·3 0758 7·1 1437 3·1 2029 7·6	**28** M	0336 2·1 0915 8·1 1556 2·3 2138 8·5
14 Su	0024 3·3 0611 7·4 1312 2·7 1859 7·2	**29** M	0525 8·1 1215 2·0 1810 7·8	**14** W	0121 3·1 0713 7·5 1359 2·7 1949 7·5	**29** Th	0119 2·4 0707 8·3 1357 2·0 1949 8·1	**14** F	0123 3·3 0720 7·2 1405 3·0 1958 7·4	**29** Sa	0220 2·6 0808 7·8 1453 2·5 2046 8·0	**14** M	0308 2·8 0903 7·6 1538 2·6 2124 8·2	**29** Tu	0433 1·6 1004 8·5 1645 1·9 2223 9·0
15 M	0133 3·1 0719 7·5 1411 2·5 1957 7·5	**30** Tu	0045 2·5 0634 8·3 1323 1·9 1917 8·1	**15** Th	0220 2·9 0811 7·6 1453 2·5 2040 7·8	**30** F	0232 2·2 0816 8·4 1504 1·9 2053 8·3	**15** Sa	0232 3·0 0826 7·4 1505 2·7 2054 7·8	**30** Su	0336 2·2 0918 8·1 1600 2·2 2146 8·5	**15** Tu	0407 2·1 0952 8·2 1628 2·0 2210 8·8	**30** W	0518 1·2 1045 8·8 1725 1·6 2301 9·2
		31 W	0154 2·2 0738 8·6 1429 1·6 2018 8·4							**31** M	0440 1·7 1014 8·5 1655 1·9 2235 8·9			**31** Th ●	0554 1·1 1120 9·0 1758 1·4 2334 9·3

Chart Datum: 4·93 metres below Ordnance Datum (Newlyn)

ENGLAND, WEST COAST — LIVERPOOL

Lat 53°25′ N Long 3°00′ W

TIMES AND HEIGHTS OF HIGH AND LOW WATERS

YEAR 1989

TIME ZONE UT(GMT)
For Summer Time add ONE hour in non-shaded areas

Chart Datum: 4·93 metres below Ordnance Datum (Newlyn)

SEPTEMBER

Time	m	Time	m
1 0627 F 1153 1828	1·0 9·1 1·3	**16** 0604 Sa 1132 1818 2349	0·4 9·8 0·6 10·2
2 0007 Sa 0656 1222 1856	9·3 1·1 9·1 1·4	**17** 0645 Su 1212 1857	0·2 9·8 0·5
3 0036 Su 0723 1252 1923	9·0 1·2 9·0 1·5	**18** 0029 M 0724 1252 1938	10·3 0·3 9·7 0·6
4 0104 M 0748 1320 1949	9·1 1·5 8·8 1·7	**19** 0112 Tu 0801 1333 2018	10·0 0·7 9·4 1·0
5 0133 Tu 0813 1348 2019	8·8 1·8 8·6 2·1	**20** 0154 W 0839 1415 2101	9·5 1·3 9·0 1·6
6 0201 W 0840 1419 2051	8·5 2·2 8·2 2·5	**21** 0236 Th 0921 1503 2152	8·9 2·0 8·4 2·2
7 0232 Th 0912 1454 2129	8·0 2·7 7·8 3·0	**22** 0336 F 1014 1606 2304	8·1 2·8 7·7 2·8
8 0311 F 0953 1542 ☽ 2223	7·5 3·2 7·3 3·4	**23** 0452 Sa 1132 1733	7·4 3·3 7·4
9 0407 Sa 1057 1655 2346	7·0 3·7 7·0 3·6	**24** 0042 Su 0631 1317 1910	3·0 7·2 3·4 7·5
10 0543 Su 1241 1836	6·7 3·8 7·0	**25** 0211 M 0758 1439 2025	2·6 7·5 2·9 8·0
11 0127 M 0728 1409 1959	3·4 7·0 2·8 7·6	**26** 0318 Tu 0858 1536 2117	2·1 8·1 2·4 8·5
12 0244 Tu 0839 1512 2057	2·7 7·7 2·6 8·3	**27** 0410 W 0943 1621 2157	1·6 8·5 2·0 8·9
13 0343 W 0928 1606 2143	2·0 8·4 1·9 9·0	**28** 0449 Th 1020 1657 2234	1·3 8·8 1·7 9·2
14 0435 Th 1012 1652 2227	1·3 9·0 1·4 9·6	**29** 0523 F 1052 1727 ● 2306	1·2 9·0 1·5 9·3
15 0520 F 1052 1736 ○ 2308	0·7 9·5 0·9 10·0	**30** 0554 Sa 1123 1757 2336	1·2 9·1 1·4 9·3

OCTOBER

Time	m	Time	m
1 0621 Su 1151 1824	1·2 9·1 1·4	**16** 0618 M 1147 1835	0·3 9·9 0·5
2 0004 M 0646 1219 1852	9·2 1·3 9·1 1·5	**17** 0007 Tu 0657 1228 1917	10·2 0·5 9·8 0·7
3 0032 Tu 0713 1248 1921	9·0 1·6 8·9 1·7	**18** 0050 W 0737 1310 2001	9·9 0·9 9·5 1·1
4 0100 W 0740 1316 1951	8·8 1·9 8·7 2·1	**19** 0135 Th 0818 1355 2049	9·3 1·5 9·0 1·6
5 0130 Th 0809 1347 2025	8·5 2·3 8·4 2·4	**20** 0226 F 0903 1447 2143	8·7 2·2 8·4 2·2
6 0202 F 0842 1422 2104	8·1 2·7 8·0 2·9	**21** 0324 Sa 0957 1549 ☽ 2255	7·9 2·9 7·9 2·7
7 0242 Sa 0922 1510 2157	7·6 3·2 7·6 3·3	**22** 0438 Su 1115 1711	7·4 3·4 7·5
8 0339 Su 1023 1620 ☽ 2316	7·1 3·7 7·2 3·5	**23** 0021 M 0605 1249 1836	2·8 7·2 3·4 7·6
9 0509 M 1200 1756	6·8 3·8 7·2	**24** 0140 Tu 0726 1404 1948	2·2 7·5 3·0 7·9
10 0053 Tu 0650 1333 1920	3·2 7·1 3·3 7·7	**25** 0242 W 0825 1500 2042	2·2 7·9 2·6 8·3
11 0209 W 0802 1440 2020	2·6 7·8 2·6 8·4	**26** 0332 Th 0910 1546 2124	1·9 8·3 2·2 8·6
12 0311 Th 0856 1534 2111	1·8 8·5 1·9 9·1	**27** 0413 F 0948 1621 2200	1·7 8·6 2·0 8·9
13 0403 F 0941 1623 2156	1·2 9·1 1·3 9·7	**28** 0447 Sa 1020 1654 2234	1·6 8·9 1·8 9·0
14 0451 Sa 1023 1708 ● 2240	0·7 9·6 0·7 10·1	**29** 0516 Su 1051 1725 ● 2305	1·5 9·0 1·7 9·0
15 0536 Su 1105 1753 2323	0·4 9·8 0·6 10·2	**30** 0546 M 1120 1754 2336	1·5 9·1 1·6 9·0
		31 0614 Tu 1150 1825	1·6 9·1 1·7

NOVEMBER

Time	m	Time	m
1 0005 W 0643 1221 1857	8·9 1·7 9·0 1·8	**16** 0036 Th 0719 1256 1951	9·6 1·2 9·2 1·5
2 0036 Th 0714 1252 1933	8·7 1·9 8·8 2·0	**17** 0124 F 0802 1342 2042	9·1 1·7 9·0 1·6
3 0109 F 0747 1326 2009	8·4 2·3 8·5 2·3	**18** 0215 Sa 0849 1433 2135	8·6 2·2 8·6 2·0
4 0145 Sa 0822 1405 2051	8·1 2·6 8·2 2·6	**19** 0308 Su 0941 1529 2235	8·0 2·8 8·2 2·4
5 0229 Su 0905 1454 2145	7·8 3·0 7·9 2·9	**20** 0410 M 1044 1633 ☽ 2342	7·6 3·1 7·8 2·6
6 0327 M 1004 1559 ☽ 2254	7·4 3·3 7·6 3·0	**21** 0519 Tu 1157 1743	7·3 3·3 7·6
7 0442 Tu 1123 1718	7·2 3·4 7·6	**22** 0048 W 0631 1307 1853	2·7 7·3 3·2 7·7
8 0014 W 0607 1248 1835	2·8 7·4 3·1 8·0	**23** 0149 Th 0735 1408 1954	2·6 7·6 3·0 7·9
9 0128 Th 0719 1358 1940	2·4 7·9 2·6 8·5	**24** 0243 F 0826 1458 2043	2·4 7·9 2·7 8·1
10 0232 F 0818 1457 2036	1·8 8·5 2·0 9·0	**25** 0328 Sa 0910 1542 2125	2·2 8·2 2·4 8·4
11 0328 Sa 0908 1550 2127	1·3 9·0 1·5 9·5	**26** 0406 Su 0946 1620 2203	2·0 8·5 2·2 8·5
12 0420 Su 0956 1641 2214	1·0 9·4 1·1 9·8	**27** 0441 M 1021 1655 2238	1·9 8·7 2·0 8·7
13 0508 M 1041 1730 ○ 2302	0·8 9·7 0·8 9·9	**28** 0515 Tu 1055 1732 ● 2312	1·8 8·9 1·9 8·7
14 0553 Tu 1126 1817 2349	0·7 9·7 0·8 9·8	**29** 0549 W 1129 1807 2347	1·8 9·0 1·8 8·7
15 0636 W 1211 1903	0·9 9·6 0·9	**30** 0622 Th 1203 1843	1·8 9·0 1·8

DECEMBER

Time	m	Time	m
1 0021 F 0657 1238 1921	8·7 1·9 8·9 1·9	**16** 0114 Sa 0751 1333 2030	9·0 1·6 9·2 1·3
2 0057 Sa 0734 1316 2002	8·5 2·1 8·8 2·0	**17** 0159 Su 0833 1416 2117	8·7 1·9 8·9 1·7
3 0137 Su 0813 1358 2046	8·3 2·3 8·6 2·1	**18** 0244 M 0917 1501 2202	8·3 2·3 8·5 2·1
4 0222 M 0857 1444 2135	8·1 2·5 8·4 2·3	**19** 0331 Tu 1002 1549 ☽ 2249	7·9 2·7 8·1 2·5
5 0314 Tu 0949 1539 2230	7·9 2·7 8·2 2·4	**20** 0421 W 1052 1642 2343	7·6 3·1 7·8 2·8
6 0414 W 1049 1642 2333	7·7 2·9 8·1 2·4	**21** 0519 Th 1153 1743	7·3 3·3 7·5
7 0523 Th 1200 1750	7·7 2·8 8·1	**22** 0041 F 0625 1259 1850	2·9 7·2 3·4 7·4
8 0042 F 0634 1312 1857	2·3 7·9 2·6 8·4	**23** 0142 Sa 0730 1404 1955	3·0 7·4 3·2 7·5
9 0151 Sa 0740 1420 2002	2·0 8·2 2·2 8·6	**24** 0239 Su 0827 1500 2050	2·8 7·7 2·9 7·7
10 0254 Su 0840 1524 2101	1·7 8·6 1·8 9·0	**25** 0328 M 0915 1549 2136	2·6 8·0 2·6 8·0
11 0353 M 0934 1621 2157	1·5 9·0 1·5 9·3	**26** 0412 Tu 0957 1633 2219	2·3 8·4 2·3 8·3
12 0447 Tu 1024 1716 ○ 2249	1·3 9·3 1·2 9·5	**27** 0452 W 1035 1715 2257	2·1 8·7 2·0 8·5
13 0536 W 1111 1807 2340	1·1 9·5 1·0 9·4	**28** 0530 Th 1113 1756 ● 2334	1·9 8·8 1·8 8·7
14 0622 Th 1201 1856	1·2 9·5 0·9	**29** 0608 F 1150 1835	1·7 9·1 1·6
15 0028 F 0707 1248 1944	9·3 1·3 9·4 1·1	**30** 0011 Sa 0648 1228 1916	8·8 1·6 9·1 1·4
		31 0049 Su 0727 1307 1957	8·8 1·6 9·2 1·4

LIVERPOOL continued

RADIO TELEPHONE
Call: *Mersey Radio* (Port of Liverpool Building) VHF Ch **12** 16; 09 12 18 19 22 (H24). Traffic movements, local navigational warnings and weather reports broadcast on Ch 09 at 3h and 2h before HW. Local navigational and gale warnings broadcast on receipt on Ch 12, and on Ch 09 every 4h commencing 0000 local time. Alfred Dock Ch 05 (H24). Tranmere Stages Ch 19 (H24). Garston Dock Ch 20 (H24). Waterloo Dock Ch 20 (H24). Langton Dock Ch 21 (H24). Gladstone Dock Ch 05 (H24). Eastham Locks (Manchester Ship Canal) Ch 07 14 (H24). Latchford Locks (Manchester Ship Canal) Ch 14 20 (H24). Weaver Navigation. Weston Point Dock, call *Weston Point* Ch 74. Yachts entering or leaving Weaver Navigation should call on Ch 74 when approaching locks. For other BWB stations call: *Marsh Lock*, *Dutton Lock*, *Saltisford Lock*, *Anderton Depot* all on Ch 74.

TELEPHONE (051)
Hr Mr 200 4124; CG & MRSC 931 3341; HMC Freefone Customs Yachts: RT (051) 933 4292; Marinecall 0898 500 461; Hosp 709 0141.

FACILITIES
EC Wednesday; **Liverpool Marina** (500) Tel. 708 5228. 0600-2130 Mar to Oct. Min depth 3.5m. Slip, BH, AC, FW, C (50 ton), Bar, Lau, R, CH, V; Ent via Brunswick Dock lock; Access HW∓2; **Albert Dock** Access HW−2 to HW (VHF Ch 37 when entrance manned) AB, FW, AC; **West Kirby SC** Tel. 625 5579, Slip, M (in Dee Est), L, FW, C (30 ton), AB (at HW), Bar; **Blundellsands SC** Tel. 929 2101; Slip, L (at HW), FW, Bar; **Hoylake SC** Tel. 632 2616, Slip, M, FW, Bar; **AMP Marine** Tel. 647 4787 Sh; **Royal Mersey YC** Tel. 645 3204, Slip, M, P, D, L, FW, R, Bar; **Perrys Yacht Centre** Tel. 647 5751, CH; **Davey Jones Marine** Tel. 263 4700, ME, El, CH; **Ship Shape** Tel. 928 4471, Sh, CH; **J. Sewill** Tel. 227 1376 ACA, Nautical Instruments. **Town** PO; Bank; Rly; Air.

Note: Access to East Coast via Leeds and Liverpool Canal (British Waterways Board). Liverpool to Goole 161 miles, 103 locks. Maximum draught 3ft, air draught 8ft, beam 13ft 6in, length 60ft.

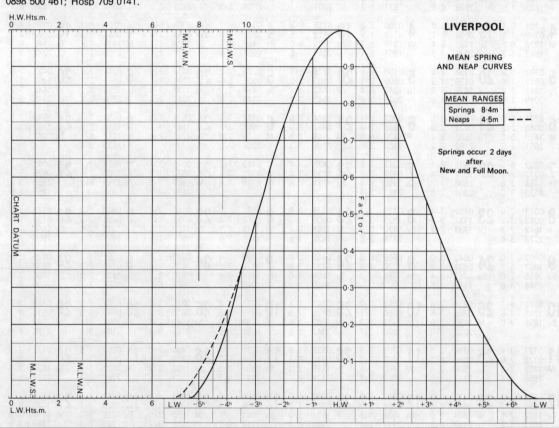

LIVERPOOL MEAN SPRING AND NEAP CURVES. MEAN RANGES Springs 8·4m, Neaps 4·5m. Springs occur 2 days after New and Full Moon.

ISLE OF MAN 10-10-12

CHARTS
Admiralty 2696, 2094

The Isle of Man is one of the British Islands, set in the Irish Sea almost equidistant from England, Scotland and Ireland but it is not part of the United Kingdom. It has a considerable degree of self-government. The island comes under the same customs umbrella as the rest of the United Kingdom, so no customs formalities are necessary on landing or returning to UK ports.

Manx harbours are administered by the Isle of Man Government and the lights are maintained by the Commissioners of Northern Lighthouses in Scotland. Besides the four main harbours given below, there are good anchorages at Castletown in the SE, Laxey Bay in the E and Port Erin in the SW – see 10.10.22. There are also good anchorages in Derby Haven in the SE; this is a rather bleak area and the inner harbour dries. Most of the harbours are on the E and S sides but a visit to the W coast with its characteristic cliffs is worth while. For passage through Calf Sound, see 10.10.5.

When contact with local Harbour Masters cannot be established on VHF, vessels should call *Douglas Radio* Ch 16 12 for urgent messages or other information. Weather forecasts can be obtained from local radio programmes or by telephone from the Met Office Ronaldsway (0624) 823311 between 0630 and 2030 (LT). Visitors are recommended to obtain *Sailing directions, tidal streams and anchorages of the Isle of Man* produced by the Manx Sailing and Cruising Club in Ramsey, Tel. (0624) 813494.

AREA 10—Isle of Man 481

PORT ST MARY 10-10-13
Isle of Man

CHARTS
Admiralty 2696, 2094; Imray C62, Y70; OS 95
TIDES
+0020 Dover; ML 3.2; Duration 0605; Zone 0 (GMT).
Standard Port LIVERPOOL (←)

Times				Height (metres)			
HW		LW		MHWS	MHWN	MLWN	MLWS
0000	0600	0200	0700	9.3	7.4	2.9	0.9
1200	1800	1400	1900				

Differences PORT ST MARY
+0005 +0015 −0010 −0030 −3.4 −2.7 −1.2 −0.3
CALF SOUND
+0005 +0005 −0015 −0025 −3.2 −2.6 −0.9 −0.3

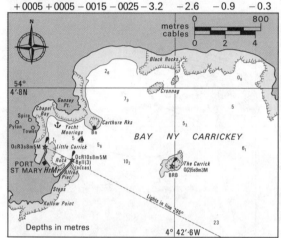

PORT ERIN
−0005 +0015 −0010 −0050 −4.1 −3.2 −1.3 −0.5

SHELTER
Very good shelter except in E or SE winds. Inner harbour dries out. Anchorages off Gansey Pt; poor holding. Visitors moorings.
NAVIGATION
Waypoint 54°03'.60N 04°43'.00W, 150°/330° from/to breakwater Lt, 0.84M. Carrick Rock, 7 ca E of breakwater Q(2) 5s 6m 3M. Rock outcrops to E of breakwater to 2 ca offshore. Beware of lobster and crab pots, especially offshore from Calf Island to Langness Pt.
LIGHTS AND MARKS
Alfred Pier, head Oc R 10s 8m 5M; W Tr, R band. Inner Pier, head Oc R 3s 8m 5M; W Tr, R band. These two Lts in line 295° lead clear S of The Carrick.
RADIO TELEPHONE (local times)
Call: *Port St Mary Harbour* VHF Ch 16; 12 (0830-1630 when manned or through Douglas Harbour Radio).
TELEPHONE (0624)
Hr Mr 833206; CG and MRSC 813255; HMC 74321; Marinecall 0898 500 461; Dr 832281.
FACILITIES
EC Thursday; **Breakwater** Slip, D (by road tanker), L, FW, C (20 ton mobile), AB; **Inner Harbour** Slip, D, L, FW, C (20 ton mobile), AB; **Bay** M, L; **Ballasalla Marine & Auto Eng** Tel. 822715, ME; **Isle of Man YC** Tel. 832088, FW, Bar. **Town** CH, V, R, Bar. PO; Bank; Rly (bus to Douglas, ferry to Heysham); Air (I.o.M.).

DOUGLAS 10-10-14
Isle of Man

CHARTS
Admiralty 2696, 2094; Imray C62, Y70; OS 95
TIDES
+0020 Dover; ML 3.8; Duration 0600; Zone 0 (GMT).
Standard Port LIVERPOOL (←)

Times				Height (metres)			
HW		LW		MHWS	MHWN	MLWN	MLWS
0000	0600	0200	0700	9.3	7.4	2.9	0.9
1200	1800	1400	1900				

Differences DOUGLAS
−0004 −0004 −0022 −0032 −2.4 −2.0 −0.5 −0.1

SHELTER
Shelter is good except with north-easterly winds. Very heavy seas run in during NE gales. Anchorage in outer harbour after obtaining Hr Mr's permission. The Harbour Board provides one B can buoy for visitors, between Lifeboat slip and Fort Anne Jetty, and a mooring pontoon at inner end of Battery Pier.
NAVIGATION
Waypoint 54°09'.00N 04°27'.60W, 049°/229° from/to front Ldg Lt, 0.47M. Inner harbour dries; access HW ∓2. It becomes very full in summer. There is no bar. Beware swing bridge and also concrete step at end of dredged area (diamond mark on King Edward pier). SE and SW corners of linkspan, SE of pier, marked with 2 FG (vert). Concrete dolphin at seaward end of Princess Alexandra Pier 2FR (vert). Approach to N of No 1 buoy and await port entry signal. If available call on VHF Ch 12. Keep clear of larger vessels and ferries.
LIGHTS AND MARKS
Ldg Lts 229°, both Oc 10s 9/12m 5M, synchronised. International Port Traffic Signals Nos 2, 3 and 5 shown from mast at head of Victoria Pier. Signals below are shown from near head of Victoria Pier in panels with FIY Lts in top corners.
X (red)   Unspecified vessels may not proceed
→ (white)   Vessels may proceed in direction shown
RADIO TELEPHONE
VHF Ch 16; 12 (H24). Broadcasts Ch 12 at 0133, 0533, 0733, 0933, 1333, 1733 and 2133 giving navigational warnings for IoM coastal waters.

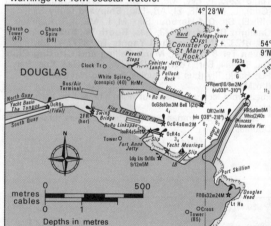

TELEPHONE (0624)
Hr Mr 23813; CG and MRSC 813255; HMC 74321; Marinecall 0898 500 461; Dr 73661 (Hosp).
FACILITIES
EC Thursday; **Outer Harbour** Tel. 23813, Slip, M, P, D, L, FW, ME, El, Sh, C (10, 5 ton), CH; **Inner Harbour (N & S Quays)** Tel. 23813, Slip, M (N & S Quays), P, D, L, FW, ME, El, Sh, C, CH, AB; **Manx Marine** Tel. 74842, CH, ACA; **Manx Marine** Tel. 22995 CH, ACA; **Auto Electric Garage Services** Tel. 28123, El; **Douglas Motor Boat & Sailing Club** Tel. 73965; **Manx Gas Service Centre** Tel. 832102, Gas, Gaz, Kos; **Douglas YC** Tel. 73965.
Town V, R, Bar. PO; Bank; Rly (ferry to Heysham); also in summer to Belfast, Stranraer and Liverpool); Air (I.o.M.).

RAMSEY 10-10-15
Isle of Man

CHARTS
Admiralty 2696, 2094; Imray C62, Y70; OS 95
TIDES
+0020 Dover; ML 4.0; Duration 0545; Zone 0 (GMT).
Standard Port LIVERPOOL (←)

Times				Height (metres)			
HW		LW		MHWS	MHWN	MLWN	MLWS
0000	0600	0200	0700	9.3	7.4	2.9	0.9
1200	1800	1400	1900				

Differences RAMSEY
−0003 +0012 0000 −0015 −2.1 −1.7 −0.3 +0.1

SHELTER
Very good except in strong winds between NE and SE, through E. The harbour dries. Access HW−2½ to HW+2. Berth alongside SW quay — instructions given by Hr Mr as vessels enter between piers. The Manx S.C. has visitors moorings N of Queens Pier.

NAVIGATION
Waypoint 54°19'.40N 04°21'.60W, 095°/275° from/to entrance, 0.48M. The foreshore dries out 100m to seaward of the pier heads. Entrance to the harbour is only permitted from HW−2½ to HW+2.

LIGHTS AND MARKS
There are no leading lights or marks. Lights as shown on chartlet. There is FR Lt in centre of swing bridge.

RADIO TELEPHONE (local times)
VHF Ch 16; 12 (0830-1630 when manned or through Douglas Harbour Radio).

TELEPHONE (0624)
Hr Mr 812245; CG and MRSC 813255; HMC 74321; Marinecall 0898 500 461; Dr 813881.

FACILITIES
EC Wednesday; **Entrance Channel** L, FW, AB; C*;
North Quay L, FW, ME, El, Sh, AB; **East Quay** Tel. 812245, L, FW, ME, El, Sh, AB; **West Quay** Slip (Grid), L, FW, ME, El, AB, Lau; **Shipyard Quay** Slip, ME, El, Sh; **Old Harbour** Slip, M, L, ME, AB; **Ramsay Gaslight Co** Tel. 813143, Gas; **Booth W. Kelly** Tel. 812322, ME; **Freedom Yachts** Tel. 897717, Sh; **Bevan** Tel. 812583, El; **Manx Sailing and Cruising Club** Tel. 813494, Bar; **Town** P, D, V, R, Gas, Gaz, Kos, Bar, PO; Bank; Rly (bus to Douglas, ferry to Heysham); Air (I.o.M.).
Fuels from local suppliers but none situated on harbour estate.
*Various mobile cranes on hire from suppliers through Douglas.

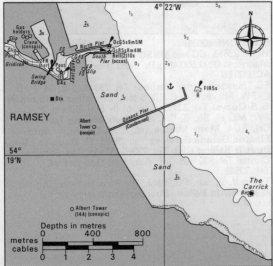

▲ Report to HrMr

PEEL 10-10-16
Isle of Man

CHARTS
Admiralty 2696, 2094; Imray C62, Y70; OS 95
TIDES
+0005 Dover; ML 2.9; Duration 0540; Zone 0 (GMT).

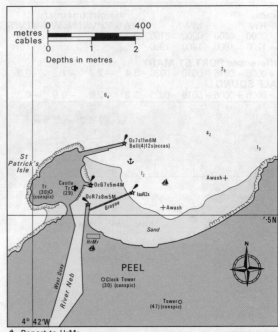

▲ Report to HrMr

Standard Port LIVERPOOL (←)

Times				Height (metres)			
HW		LW		MHWS	MHWN	MLWN	MLWS
0000	0600	0200	0700	9.3	7.4	2.9	0.9
1200	1800	1400	1900				

Differences PEEL
−0015 +0010 0000 −0010 −4.0 −3.2 −1.4 −0.4

SHELTER
Good shelter except in strong NW to NE winds. Inner harbour has water HW∓3; bottom is flat sand.

NAVIGATION
Waypoint 54°14'.00N 04°41'.20W, 037°/217° from/to breakwater Lt, 0.42M. Entry should not be attempted in strong N to NW winds. Beware groyne running ENE from mouth of River Neb submerged at half tide. Harbour dries out and is very crowded from June to October.

LIGHTS AND MARKS
There are no leading marks or lights. Lt on end of groyne, Iso R 2s. Pier head Oc R 7s 8m 5M; vis 156°-249°. Breakwater Oc 7s 11m 6M.

RADIO TELEPHONE (local times)
VHF Ch 16; 12 (0830-1630 when manned or through Douglas Harbour Radio).

TELEPHONE (0624 84)
Hr Mr 2338; CG and MRSC Ramsey 813255; HMC 23813; Marinecall 0898 500 461; Dr 3636.

FACILITIES
EC Thursday; **Breakwater** Slip, L, FW, ME, El, Sh, C (7½ ton mobile), AB; **Inner Pier** Tel. 2338, L, FW, ME, El, Sh, AB; **East Quay** Slip, L, FW, ME, El, C (7½ ton mobile), AB; **West Quay** L, FW, ME, El, Sh, Slip, C (7½ ton mobile), AB; **Bay** M; **J D Faulkner** Tel. 842296, D (cans); **Manx Gas Service Centre** Tel. 842281, Gas; **West Marine** Tel. 842604, BY, CH, ME; **Manx Marine** Tel. 74842, CH, ACA; **Peel Sailing and Cruising Club** Tel. 3357, P and D (cans), R, Lau, Bar; **T and J Autos** Tel. 2096, ME; **Town** P and D (cans), CH, V, R, Bar, PO; Bank; Rly (bus to Douglas, ferry to Heysham); Air (I.o.M.).

AREA 10—N Wales   483

CONWY 10-10-17
Gwynedd

CHARTS
Admiralty 1978, 1977; Imray C61; OS 115
TIDES
−0015 Dover; ML 4.3; Duration 0545; Zone 0 (GMT).
Standard Port HOLYHEAD (→)

Times				Height (metres)			
HW		LW		MHWS	MHWN	MLWN	MLWS
0000	0600	0500	1100	5.7	4.5	2.0	0.7
1200	1800	1700	2300				

Differences CONWY
+0020 +0020 No data +0050 +2.1 +1.6 +0.3 No data

Standard Port LIVERPOOL (←)

Times				Height (metres)			
HW		LW		MHWS	MHWN	MLWN	MLWS
0000	0600	0200	0700	9.3	7.4	2.9	0.9
1200	1800	1400	1900				

LLANDUDNO
−0035 −0025 −0025 −0035 −1.9 −1.5 −0.5 −0.2

NOTE: HW Conwy is −0040 springs and −0020 neaps HW Liverpool approx.

SHELTER
Good except for strong winds from NW. Anchorage off town quay or ask Hr Mr for vacant mooring.
NAVIGATION
Waypoint Fairway (safe water) buoy, 53°17'.90N 03°55'.47W, 290°/110° from/to Penmaen-bach Point, 1.7M. Channel is variable with depths of about 0.5m and strong stream. Keep at least ¼ ca clear of Perch Lt to stbd on entry. Many mussel beds in harbour entrance. Beware unlit moorings. Entry recommended during daylight only and between HW−2 and HW. Entry via 'Inshore Passage' not recommended. Speed limit from the narrows to ½ mile above bridges is 8 kn.
There will be disruption until 1991, due to the building of the Conwy River Tunnel. At times the port may be closed. Vessels must proceed only within the buoyed channel. Vessels must check with harbour authorities before arrival. Beware dredgers in the channel.
LIGHTS AND MARKS
None of the channel marks has either light or bell except the most inshore one, Fl WR 5s 5m 2M; W 076°-088°, R 088°-171°; W 171°-319°, R 319°-076°. The 'Inshore Passage' to E of Perch Lt can be used for vessels up to 2m draught, Neaps HW ∓1, Springs HW ∓3, passing inshore of Or liferaft (on station Apl-Oct).
RADIO TELEPHONE (local times)
VHF Ch 16; 06 08 **12** 14 72 (Apl-Sept 0900-1700. Oct-Mar Mon-Fri 0900-1700). N Wales Cruising Club on Ch M. Llandudno Pier Ch 16; 06 12 (0700-1500 June-mid Sept). Llanddulas Ch 16; 14 (0900−1230, 1400-1700).

TELEPHONE (0492)
Hr Mr 596253; CG Llangoed 224; MRSC Holyhead 2051; HMC Freefone Customs Yachts: RT (0407) 2336; Marinecall 0898 500 460; Dr 593385.
FACILITIES
EC Wednesday; **Conwy Quay Harbour** Tel. 596253, (dries − limited stay for loading, watering, etc.), M, D, L, FW, AB (up to 40 ft); **Conwy Harbour Boat Shop** Tel. 592366, D, ME, Gas, Gaz, CH; **Conwy River Boatyard** Tel. 592489, M, ME, El, Slip, Gas, Gaz, Sh; **Deganwy Dock** (dries), L, FW, C (mobile), AB; **Sailtronic Marine** Tel. Glanconwy 536, El; **Deganwy Yacht Services** Tel. 83869, Slip, D, ME, El, Sh, C, CH; **Deganwy Entrance** Slip, L; **Beacons** Slip, L; **Conwy YC** Tel. 83690, Slip, M, L, FW, R, Bar;
N Wales Cruising Club Tel. 593481, M, L, FW, AB, Bar; **Town** P and D (in cans), V, R, Bar. PO; Bank; Rly; Air (Liverpool).

HOLYHEAD 10-10-18
Gwynedd

CHARTS
Admiralty 2011, 1970; Imray C61; OS 114
TIDES
−0050 Dover; ML 3.2; Duration 0615; Zone 0 (GMT).
Standard Port HOLYHEAD (→)

Times				Height (metres)			
HW		LW		MHWS	MHWN	MLWN	MLWS
0000	0600	0500	1100	5.7	4.5	2.0	0.7
1200	1800	1700	2300				

Differences AMLWCH
+0020 +0010 +0035 +0025 +1.6 +1.3 +0.5 +0.2

PORT TRECASTEL
−0045 −0020 −0005 −0015 −0.6 −0.6 0.0 0.0
Holyhead is a Standard Port and tidal predictions for each day of the year are given below.

SHELTER
Very good. Anchor in New Harbour only. NE winds cause a slight sea.
NAVIGATION
Waypoint 53°20'.00N 04°36'.47W, 019°/199° from/to Admiralty Pier Old Lt Ho, 1.2M. Cliperau Rocks (R can bell buoy) Fl (4) 15s. Beware Sealink ferries, especially crossing the entrance to New Harbour passing down to go to Inner Harbour. Also beware large unlit mooring buoys in NW part of harbour.
LIGHTS AND MARKS
Entrance between Cliperau Rks and breakwater lighthouse. Entry signals:- R Lt or R Flag — Inner harbour closed. 2R Lts or 2R Flags — Old and Inner harbours blocked.
RADIO TELEPHONE
VHF Ch 16 14 (H24). Other station: Anglesey Marine Terminal (Amlwch) Ch 10 12 16 19.
TELEPHONE (0407)
Hr Mr 2304; CG & MRSC 2051; HMC Freefone Customs Yachts: RT (0407) 2714; Marinecall 0898 500 460; Dr via CG.
FACILITIES
EC Tuesday; **Harbour** Tel. 2304, Slip, M, D, L, FW, ME, El, Sh, C (many), CH, AB; **Holyhead SC** Tel. 2526, Slip, M, L, FW, R (Summer only), Bar; **Holyhead BY** Tel. 50111, ACA, FW, CH, D, ME, El, Sh, C (20 ton); **Holyhead Chandlery** Tel. 3632, CH, ACA; **Trearddur Bay BY** Tel. Trearddur Bay 860501, D, FW, Sh, CH; **Trearddur Bay** M (Small craft only), L; **Trearddur Bay Village** P, V, R, Bar.
Town P, V, R, Bar, Lau. PO; Bank; Rly; Air (Liverpool), Ferry to Dun Laoghaire.

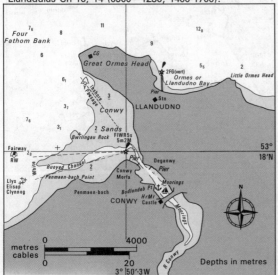

HOLYHEAD continued

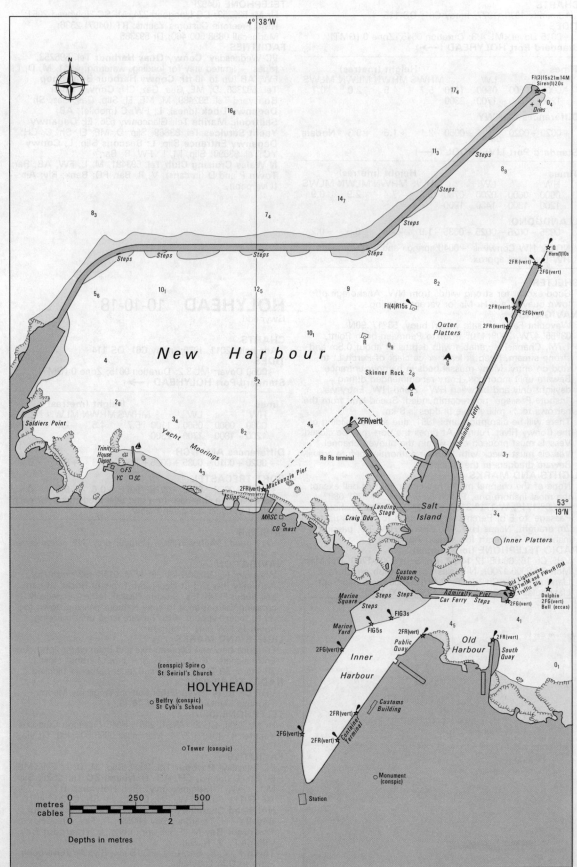

WALES — HOLYHEAD

Lat 53°18′ N   Long 4°38′ W

TIMES AND HEIGHTS OF HIGH AND LOW WATERS

YEAR 1989

TIME ZONE UT(GMT)
For Summer Time add ONE hour in non-shaded areas

JANUARY

Time	m	Time	m
1 0427 1026 Su 1644 2315	4·3 2·3 4·4 2·1	**16** 0459 1106 M 1725 2350	4·6 1·8 4·8 1·7
2 0533 1136 M 1754	4·3 2·3 4·4	**17** 0617 1227 Tu 1845	4·5 1·8 4·7
3 0019 0638 Tu 1243 1900	2·0 4·4 2·2 4·5	**18** 0103 0730 W 1341 1958	1·7 4·7 1·7 4·8
4 0117 0735 W 1341 1958	1·9 4·5 2·0 4·6	**19** 0209 0830 Th 1446 2058	1·6 4·9 1·4 4·9
5 0208 0825 Th 1432 2047	1·8 4·9 1·7 4·8	**20** 0304 0919 F 1536 2146	1·5 5·2 1·0 5·1
6 0253 0907 F 1518 2131	1·5 5·1 1·4 5·0	**21** 0349 1002 Sa 1620 ○ 2227	1·3 5·3 1·0 5·1
7 0336 0948 Sa 1602 ● 2213	1·3 5·4 1·1 5·2	**22** 0428 1040 Su 1659 2304	1·2 5·5 0·9 5·2
8 0417 1028 Su 1645 2255	1·1 5·6 0·9 5·4	**23** 0505 1115 M 1736 2337	1·1 5·5 0·8 5·2
9 0501 1109 M 1729 2337	0·9 5·7 0·7 5·4	**24** 0539 1149 Tu 1810	1·1 5·5 0·9
10 0543 1153 Tu 1814	0·8 5·8 0·6	**25** 0010 0611 W 1219 1842	5·1 1·1 5·4 1·0
11 0022 0628 W 1238 1859	5·4 0·9 5·8 0·6	**26** 0042 0643 Th 1252 1913	5·1 1·2 5·3 1·1
12 0109 0713 Th 1323 1945	5·3 1·0 5·7 0·7	**27** 0114 0716 F 1326 1947	4·9 1·4 5·2 1·3
13 0155 0801 F 1412 2036	5·1 1·1 5·5 1·0	**28** 0149 0751 Sa 1401 2023	4·8 1·6 4·9 1·6
14 0247 0853 Sa 1505 ☽ 2131	4·9 1·3 5·3 1·2	**29** 0229 0832 Su 1442 2105	4·6 1·8 4·7 1·8
15 0348 0953 Su 1609 2235	4·7 1·6 5·0 1·5	**30** 0315 0919 M 1532 2159	4·4 2·1 4·4 2·1
		31 0419 1024 Tu 1647 2313	4·3 2·3 4·2 2·3

FEBRUARY

Time	m	Time	m
1 0539 1150 W 1815	4·2 2·3 4·2	**16** 0055 0724 Th 1341 2005	2·1 4·5 1·7 4·5
2 0035 0657 Th 1310 1934	2·2 4·4 2·1 4·4	**17** 0208 0827 F 1443 2100	1·9 4·8 1·4 4·8
3 0142 0801 F 1412 2033	2·0 4·7 1·7 4·7	**18** 0300 0914 Sa 1529 2139	1·6 5·1 1·1 4·9
4 0236 0853 Sa 1503 2119	1·6 5·0 1·3 5·0	**19** 0339 0950 Su 1607 2213	1·3 5·3 0·9 5·1
5 0322 0934 Su 1548 2200	1·3 5·4 0·9 5·3	**20** 0414 1023 M 1640 ○ 2242	1·1 5·4 0·8 5·2
6 0404 1013 M 1630 ● 2240	0·9 5·7 0·5 5·5	**21** 0445 1052 Tu 1711 2312	1·0 5·5 0·8 5·2
7 0445 1052 Tu 1712 2320	0·6 5·9 0·3 5·7	**22** 0513 1123 W 1740 2340	0·9 5·5 0·9 5·2
8 0526 1134 W 1753	0·5 6·1 0·2	**23** 0543 1151 Th 1808	0·9 5·5 0·9
9 0001 0607 Th 1217 1836	5·7 0·4 6·1 0·2	**24** 0010 0612 F 1221 1836	5·2 1·0 5·4 1·0
10 0043 0650 F 1300 1920	5·6 0·5 5·9 0·5	**25** 0039 0642 Sa 1252 1906	5·1 1·1 5·2 1·2
11 0127 0734 Sa 1345 2005	5·3 0·8 5·7 0·8	**26** 0110 0714 Su 1323 1938	5·0 1·3 5·0 1·4
12 0213 0823 Su 1436 ☾ 2057	5·1 1·1 5·3 1·3	**27** 0144 0751 M 1358 2015	4·8 1·6 4·8 1·7
13 0308 0922 M 1538 2200	4·7 1·5 4·8 1·7	**28** 0225 0834 Tu 1443 ☾ 2103	4·6 1·9 4·5 2·1
14 0420 1041 Tu 1702 2326	4·5 1·9 4·5 2·0		
15 0556 1215 W 1843	4·4 1·9 4·4		

MARCH

Time	m	Time	m
1 0318 0934 W 1550 2216	4·4 2·2 4·2 2·3	**16** 0536 1204 Th 1842	4·3 1·9 4·2
2 0441 1104 Th 1737 2357	4·2 2·3 4·1 2·4	**17** 0042 0710 F 1328 1955	2·2 4·5 1·7 4·4
3 0621 1239 F 1913	4·2 2·1 4·3	**18** 0152 0811 Sa 1425 2043	1·9 4·7 1·4 4·7
4 0117 0735 Sa 1349 2015	2·1 4·6 1·6 4·7	**19** 0242 0853 Su 1507 2118	1·6 5·0 1·1 4·9
5 0216 0827 Su 1442 2100	1·6 5·0 1·1 5·0	**20** 0318 0927 M 1541 2148	1·4 5·2 1·0 5·0
6 0301 0911 M 1525 2139	1·2 5·4 0·7 5·4	**21** 0349 0956 Tu 1612 2214	1·1 5·3 0·8 5·2
7 0342 0950 Tu 1606 ● 2217	0·7 5·8 0·3 5·6	**22** 0417 1024 W 1640 ○ 2241	1·0 5·4 0·8 5·3
8 0421 1030 W 1647 2255	0·4 6·1 0·1 5·8	**23** 0445 1054 Th 1706 2309	0·9 5·4 0·8 5·3
9 0502 1111 Th 1727 2336	0·2 6·2 0·0 5·8	**24** 0513 1122 F 1734 2337	0·9 5·4 0·9 5·3
10 0543 1153 F 1810	0·2 6·2 0·1	**25** 0543 1151 Sa 1803	0·9 5·3 1·0
11 0017 0627 Sa 1236 1852	5·7 0·3 5·9 0·5	**26** 0007 0612 Su 1222 1832	5·2 1·1 5·2 1·2
12 0100 0712 Su 1323 1938	5·4 0·6 5·6 0·9	**27** 0039 0646 M 1255 1904	5·1 1·2 5·0 1·4
13 0145 0802 M 1415 2029	5·1 1·0 5·1 1·4	**28** 0113 0723 Tu 1331 1941	4·9 1·5 4·7 1·7
14 0239 0903 Tu 1519 ☾ 2134	4·8 1·5 4·6 1·9	**29** 0152 0808 W 1418 2030	4·7 1·7 4·4 2·0
15 0350 1026 W 1652 2306	4·4 1·9 4·2 2·3	**30** 0246 0908 Th 1527 ☾ 2143	4·5 2·0 4·2 2·3
		31 0403 1034 F 1712 2325	4·3 2·1 4·1 2·3

APRIL

Time	m	Time	m
1 0544 1208 Sa 1845	4·3 1·8 4·3	**16** 0117 0734 Su 1349 2009	2·0 4·6 1·5 4·6
2 0048 0702 Su 1317 1945	2·0 4·7 1·4 4·7	**17** 0206 0818 M 1432 2044	1·7 4·8 1·3 4·8
3 0145 0757 M 1411 2032	1·5 5·1 0·9 5·1	**18** 0244 0853 Tu 1505 2114	1·5 5·0 1·1 4·9
4 0233 0842 Tu 1457 2111	1·1 5·5 0·5 5·4	**19** 0317 0924 W 1536 2142	1·2 5·1 1·0 5·1
5 0315 0922 W 1538 2150	0·6 5·8 0·2 5·7	**20** 0346 0953 Th 1604 2210	1·1 5·2 0·9 5·2
6 0356 1004 Th 1620 ● 2230	0·3 6·0 0·1 5·8	**21** 0414 1023 F 1633 ○ 2238	1·0 5·3 0·9 5·3
7 0437 1047 F 1702 2311	0·2 6·1 0·1 5·8	**22** 0445 1054 Sa 1702 2309	1·0 5·2 1·0 5·3
8 0520 1132 Sa 1744 2354	0·2 6·0 0·3 5·7	**23** 0516 1126 Su 1733 2342	1·0 5·2 1·1 5·2
9 0607 1218 Su 1829	0·3 5·7 0·7	**24** 0550 1200 M 1805	1·1 5·1 1·2
10 0038 0655 M 1307 1916	5·5 0·6 5·4 1·1	**25** 0015 0627 Tu 1236 1842	5·2 1·2 4·9 1·4
11 0126 0748 Tu 1402 2009	5·2 1·0 4·9 1·6	**26** 0053 0709 W 1319 1924	5·0 1·4 4·7 1·7
12 0220 0851 W 1508 2114	4·8 1·5 4·5 2·0	**27** 0138 0757 Th 1411 2018	4·8 1·6 4·5 1·9
13 0331 1010 Th 1638 2241	4·5 1·8 4·2 2·3	**28** 0233 0858 F 1518 ☾ 2129	4·6 1·7 4·3 2·1
14 0505 1137 F 1814 2254	4·3 1·8 4·2 2·1	**29** 0345 1016 Sa 1648 2254	4·5 1·7 4·3 2·1
15 0008 0634 Sa 1253 1923	2·2 4·4 1·7 4·4	**30** 0509 1134 Su 1810	4·6 1·5 4·5

Chart Datum: 3.05 metres below Ordnance Datum (Newlyn)

WALES — HOLYHEAD

Lat 53°18′ N Long 4°38′ W

TIME ZONE UT(GMT)
For Summer Time add ONE hour in non-shaded areas

TIMES AND HEIGHTS OF HIGH AND LOW WATERS

YEAR 1989

Chart Datum: 3.05 metres below Ordnance Datum (Newlyn)

MAY

Day	Time	m	Day	Time	m
1 M	0010 / 0622 / 1242 / 1910	1.8 / 4.8 / 1.2 / 4.8	16 Tu	0117 / 0730 / 1345 / 2001	1.9 / 4.6 / 1.5 / 4.6
2 Tu	0110 / 0720 / 1337 / 1959	1.4 / 5.1 / 0.9 / 5.1	17 W	0201 / 0811 / 1425 / 2036	1.7 / 4.8 / 1.3 / 4.8
3 W	0201 / 0809 / 1425 / 2043	1.1 / 5.4 / 0.6 / 5.3	18 Th	0239 / 0847 / 1458 / 2108	1.5 / 4.9 / 1.2 / 5.0
4 Th	0247 / 0856 / 1511 / 2124	0.7 / 5.7 / 0.4 / 5.5	19 F	0314 / 0922 / 1531 / 2141	1.3 / 5.0 / 1.2 / 5.1
5 F ●	0332 / 0941 / 1555 / 2206	0.5 / 5.8 / 0.3 / 5.7	20 Sa	0348 / 0956 / 1603 / 2213	1.2 / 5.1 / 1.1 / 5.2
6 Sa	0417 / 1027 / 1640 / 2251	0.3 / 5.8 / 0.4 / 5.7	21 Su	0421 / 1031 / 1637 / 2247	1.1 / 5.1 / 1.1 / 5.3
7 Su	0505 / 1116 / 1725 / 2336	0.4 / 5.7 / 0.6 / 5.6	22 M	0458 / 1106 / 1712 / 2323	1.1 / 5.1 / 1.2 / 5.3
8 M	0554 / 1205 / 1812	0.5 / 5.5 / 0.9	23 Tu	0536 / 1146 / 1750	1.1 / 5.0 / 1.3
9 Tu	0024 / 0645 / 1256 / 1900	5.4 / 0.7 / 5.1 / 1.3	24 W	0001 / 0617 / 1228 / 1831	5.2 / 1.1 / 4.9 / 1.4
10 W	0112 / 0738 / 1351 / 1952	5.2 / 1.0 / 4.8 / 1.6	25 Th	0043 / 0702 / 1314 / 1917	5.2 / 1.2 / 4.8 / 1.5
11 Th	0205 / 0837 / 1453 / 2051	4.9 / 1.3 / 4.5 / 1.9	26 F	0130 / 0752 / 1406 / 2011	5.0 / 1.3 / 4.6 / 1.7
12 F ☾	0305 / 0942 / 1604 / 2202	4.6 / 1.6 / 4.3 / 2.1	27 Sa	0223 / 0849 / 1507 / 2112	4.9 / 1.3 / 4.5 / 1.8
13 Sa	0419 / 1054 / 1720 / 2316	4.5 / 1.7 / 4.2 / 2.2	28 Su ☽	0325 / 0953 / 1619 / 2221	4.8 / 1.4 / 4.5 / 1.8
14 Su	0534 / 1201 / 1828	4.4 / 1.7 / 4.3	29 M	0435 / 1101 / 1730 / 2330	4.8 / 1.3 / 4.7 / 1.7
15 M	0024 / 0639 / 1259 / 1920	2.1 / 4.5 / 1.6 / 4.4	30 Tu	0544 / 1205 / 1832	4.9 / 1.1 / 4.8
			31 W	0034 / 0646 / 1304 / 1927	1.4 / 5.1 / 1.0 / 5.0

JUNE

Day	Time	m	Day	Time	m
1 Th	0131 / 0742 / 1358 / 2018	1.2 / 5.3 / 0.8 / 5.2	16 F	0201 / 0812 / 1422 / 2037	1.8 / 4.6 / 1.5 / 4.8
2 F	0225 / 0834 / 1449 / 2105	1.0 / 5.4 / 0.8 / 5.3	17 Sa	0244 / 0856 / 1503 / 2115	1.6 / 4.8 / 1.4 / 5.0
3 Sa ●	0317 / 0927 / 1538 / 2152	0.8 / 5.5 / 0.7 / 5.5	18 Su	0325 / 0935 / 1541 / 2152	1.4 / 4.9 / 1.3 / 5.1
4 Su	0407 / 1017 / 1626 / 2238	0.6 / 5.5 / 0.8 / 5.5	19 M ○	0404 / 1014 / 1619 / 2230	1.2 / 5.0 / 1.2 / 5.3
5 M	0458 / 1106 / 1713 / 2325	0.6 / 5.4 / 0.9 / 5.4	20 Tu	0445 / 1054 / 1658 / 2309	1.1 / 5.0 / 1.2 / 5.4
6 Tu	0547 / 1157 / 1800	0.6 / 5.3 / 1.0	21 W	0526 / 1136 / 1740 / 2350	1.0 / 5.1 / 1.1 / 5.4
7 W	0011 / 0636 / 1245 / 1846	5.4 / 0.8 / 5.0 / 1.2	22 Th	0608 / 1218 / 1822	0.9 / 5.1 / 1.1
8 Th	0057 / 0724 / 1334 / 1933	5.3 / 1.0 / 4.8 / 1.5	23 F	0034 / 0653 / 1304 / 1907	5.4 / 0.9 / 5.0 / 1.2
9 F	0144 / 0813 / 1422 / 2020	5.1 / 1.2 / 4.6 / 1.7	24 Sa	0119 / 0741 / 1352 / 1957	5.4 / 0.9 / 4.9 / 1.3
10 Sa	0232 / 0904 / 1515 / 2112	4.8 / 1.4 / 4.4 / 1.9	25 Su	0206 / 0830 / 1444 / 2049	5.3 / 1.0 / 4.8 / 1.4
11 Su ☾	0325 / 0959 / 1613 / 2212	4.6 / 1.6 / 4.3 / 2.1	26 M	0300 / 0925 / 1543 / 2148	5.2 / 1.1 / 4.7 / 1.5
12 M	0424 / 1058 / 1715 / 2315	4.5 / 1.7 / 4.2 / 2.1	27 Tu	0400 / 1026 / 1649 / 2254	5.0 / 1.2 / 4.6 / 1.6
13 Tu	0529 / 1157 / 1815	4.4 / 1.8 / 4.3	28 W	0508 / 1132 / 1758	5.0 / 1.3 / 4.7
14 W	0018 / 0629 / 1252 / 1909	2.1 / 4.4 / 1.7 / 4.4	29 Th	0003 / 0618 / 1238 / 1903	1.6 / 4.9 / 1.3 / 4.8
15 Th	0113 / 0724 / 1340 / 1955	1.9 / 4.5 / 1.7 / 4.6	30 F	0110 / 0724 / 1341 / 2002	1.4 / 5.0 / 1.3 / 5.0

JULY

Day	Time	m	Day	Time	m
1 Sa	0213 / 0826 / 1439 / 2057	1.2 / 5.1 / 1.2 / 5.2	16 Su	0220 / 0836 / 1440 / 2056	1.8 / 4.6 / 1.7 / 4.9
2 Su	0312 / 0922 / 1531 / 2145	1.0 / 5.2 / 1.1 / 5.3	17 M	0307 / 0921 / 1524 / 2136	1.5 / 4.8 / 1.4 / 5.2
3 M ●	0404 / 1013 / 1619 / 2231	0.8 / 5.2 / 1.0 / 5.5	18 Tu ○	0350 / 1002 / 1604 / 2214	1.2 / 5.0 / 1.2 / 5.4
4 Tu	0452 / 1101 / 1704 / 2315	0.7 / 5.2 / 1.0 / 5.5	19 W	0431 / 1041 / 1644 / 2254	0.9 / 5.2 / 1.0 / 5.6
5 W	0537 / 1144 / 1746 / 2357	0.7 / 5.2 / 1.0 / 5.5	20 Th	0512 / 1120 / 1725 / 2333	0.7 / 5.3 / 0.8 / 5.7
6 Th	0619 / 1225 / 1825	0.7 / 5.1 / 1.1	21 F	0553 / 1201 / 1805	0.6 / 5.3 / 0.8
7 F	0036 / 0700 / 1304 / 1904	5.4 / 0.9 / 4.9 / 1.3	22 Sa	0015 / 0635 / 1243 / 1849	5.7 / 0.5 / 5.3 / 0.8
8 Sa	0114 / 0740 / 1344 / 1942	5.2 / 1.1 / 4.8 / 1.4	23 Su	0057 / 0719 / 1328 / 1933	5.7 / 0.6 / 5.2 / 1.0
9 Su	0152 / 0819 / 1423 / 2023	5.0 / 1.3 / 4.6 / 1.7	24 M	0142 / 0805 / 1415 / 2020	5.5 / 0.8 / 5.0 / 1.2
10 M	0233 / 0901 / 1508 / 2108	4.8 / 1.5 / 4.4 / 1.9	25 Tu ☾	0232 / 0854 / 1508 / 2115	5.4 / 1.0 / 4.8 / 1.4
11 Tu	0321 / 0949 / 1602 / 2203	4.6 / 1.8 / 4.3 / 2.1	26 W	0328 / 0952 / 1613 / 2221	5.1 / 1.3 / 4.6 / 1.6
12 W	0417 / 1047 / 1704 / 2309	4.4 / 1.9 / 4.2 / 2.2	27 Th	0440 / 1104 / 1730 / 2343	4.8 / 1.6 / 4.5 / 1.8
13 Th	0526 / 1153 / 1811	4.3 / 2.0 / 4.3	28 F	0601 / 1222 / 1850	4.7 / 1.7 / 4.6
14 F	0021 / 0636 / 1256 / 1916	2.2 / 4.3 / 1.9 / 4.4	29 Sa	0104 / 0724 / 1335 / 2001	1.7 / 4.7 / 1.7 / 4.8
15 Sa	0126 / 0742 / 1351 / 2011	2.0 / 4.4 / 1.9 / 4.7	30 Su	0216 / 0832 / 1437 / 2056	1.4 / 4.8 / 1.5 / 5.1
			31 M	0314 / 0925 / 1528 / 2142	1.1 / 5.0 / 1.3 / 5.3

AUGUST

Day	Time	m	Day	Time	m
1 Tu ●	0400 / 1009 / 1610 / 2221	0.9 / 5.1 / 1.1 / 5.5	16 W	0331 / 0943 / 1545 / 2155	1.1 / 5.1 / 1.1 / 5.6
2 W	0442 / 1047 / 1649 / 2259	0.7 / 5.2 / 1.0 / 5.6	17 Th ○	0410 / 1020 / 1623 / 2231	0.7 / 5.4 / 0.8 / 5.8
3 Th	0519 / 1123 / 1725 / 2334	0.7 / 5.2 / 0.9 / 5.6	18 F	0449 / 1058 / 1702 / 2311	0.4 / 5.5 / 0.6 / 6.0
4 F	0556 / 1157 / 1758	0.7 / 5.2 / 1.0	19 Sa	0529 / 1136 / 1742 / 2350	0.3 / 5.6 / 0.5 / 6.1
5 Sa	0008 / 0628 / 1231 / 1831	5.5 / 0.8 / 5.1 / 1.1	20 Su	0610 / 1217 / 1822	0.3 / 5.6 / 0.5
6 Su	0041 / 0702 / 1303 / 1904	5.3 / 1.0 / 5.0 / 1.3	21 M	0032 / 0652 / 1259 / 1906	6.0 / 0.4 / 5.4 / 0.7
7 M	0113 / 0734 / 1337 / 1938	5.2 / 1.2 / 4.8 / 1.5	22 Tu	0117 / 0735 / 1345 / 1954	5.7 / 0.7 / 5.2 / 1.0
8 Tu	0148 / 0809 / 1415 / 2018	5.0 / 1.5 / 4.7 / 1.8	23 W ☾	0206 / 0825 / 1436 / 2050	5.4 / 1.2 / 4.9 / 1.4
9 W	0227 / 0850 / 1500 / 2104	4.7 / 1.8 / 4.5 / 2.0	24 Th	0304 / 0924 / 1542 / 2203	5.0 / 1.6 / 4.6 / 1.8
10 Th	0315 / 0939 / 1557 / 2206	4.4 / 2.0 / 4.3 / 2.3	25 F	0423 / 1044 / 1713 / 2336	4.6 / 2.0 / 4.5 / 1.9
11 F	0424 / 1049 / 1715 / 2329	4.2 / 2.3 / 4.2 / 2.4	26 Sa	0604 / 1217 / 1849	4.4 / 2.1 / 4.6
12 Sa	0553 / 1215 / 1838	4.1 / 2.3 / 4.3	27 Su	0107 / 0733 / 1335 / 1959	1.7 / 4.6 / 1.9 / 4.9
13 Su	0053 / 0719 / 1326 / 1945	2.2 / 4.3 / 2.1 / 4.6	28 M	0216 / 0834 / 1433 / 2050	1.4 / 4.9 / 1.6 / 5.1
14 M	0158 / 0820 / 1419 / 2036	1.9 / 4.5 / 1.8 / 4.9	29 Tu	0305 / 0918 / 1518 / 2129	1.1 / 5.0 / 1.3 / 5.4
15 Tu	0249 / 0905 / 1504 / 2117	1.5 / 4.8 / 1.4 / 5.2	30 W	0346 / 0953 / 1553 / 2203	0.9 / 5.1 / 1.1 / 5.5
			31 Th ●	0421 / 1024 / 1627 / 2235	0.8 / 5.2 / 1.0 / 5.6

AREA 10—N Wales

WALES — HOLYHEAD
Lat 53°18′ N   Long 4°38′ W

TIMES AND HEIGHTS OF HIGH AND LOW WATERS

YEAR 1989

TIME ZONE UT(GMT)
For Summer Time add ONE hour in non-shaded areas

SEPTEMBER		OCTOBER		NOVEMBER		DECEMBER	
Time m	Time m	Time m	Time m	Time m	Time m	Time m	Time m
1 0454 0.7 1055 5.3 F 1657 0.9 2305 5.6	**16** 0420 0.3 1030 5.7 Sa 1634 0.4 2242 6.2	**1** 0449 0.9 1052 5.4 Su 1657 1.0 2305 5.5	**16** 0433 0.2 1042 5.9 M 1657 0.4 2302 6.1	**1** 0518 1.3 1126 5.3 W 1734 1.3 2343 5.1	**16** 0547 0.9 1158 5.7 Th 1821 0.8	**1** 0533 1.4 1144 5.4 F 1800 1.3	**16** 0022 5.3 0624 1.1 Sa 1236 5.6 1902 0.9
2 0523 0.8 1125 5.3 Sa 1727 0.9 2336 5.5	**17** 0459 0.2 1108 5.8 Su 1715 0.3 2325 6.2	**2** 0518 1.0 1120 5.3 M 1726 1.1 2334 5.4	**17** 0516 0.4 1126 5.8 Tu 1739 0.5 2350 5.9	**2** 0550 1.4 1200 5.3 Th 1811 1.4	**17** 0032 5.4 0638 1.2 F 1249 5.5 1916 1.0	**2** 0008 5.0 0612 1.5 Sa 1224 5.3 1842 1.4	**17** 0110 5.1 0710 1.4 Su 1323 5.4 1951 1.1
3 0553 0.9 1154 5.2 Su 1757 1.0	**18** 0542 0.2 1150 5.8 M 1758 0.4	**3** 0546 1.1 1151 5.3 Tu 1757 1.2	**18** 0601 0.7 1212 5.6 W 1829 0.7	**3** 0018 5.0 0625 1.6 F 1236 5.1 1850 1.6	**18** 0127 5.0 0730 1.6 Sa 1342 5.2 2013 1.3	**3** 0050 4.9 0655 1.8 Su 1307 5.2 1927 1.4	**18** 0158 4.8 0758 1.6 M 1409 5.1 2040 1.4
4 0005 5.4 0622 1.0 M 1225 5.1 1828 1.2	**19** 0008 6.0 0624 0.5 Tu 1232 5.8 1843 0.7	**4** 0005 5.2 0615 1.3 W 1224 5.2 1831 1.4	**19** 0041 5.6 0650 1.1 Th 1302 5.4 1923 1.1	**4** 0059 4.8 0704 1.8 Sa 1319 5.0 1937 1.8	**19** 0227 4.7 0827 1.9 Su 1443 4.9 2118 1.6	**4** 0137 4.8 0741 1.8 M 1355 5.1 2018 1.5	**19** 0249 4.6 0847 1.8 Tu 1501 4.9 ☾ 2134 1.6
5 0036 5.2 0652 1.2 Tu 1256 5.0 1900 1.4	**20** 0055 5.7 0710 0.9 W 1320 5.3 1934 1.1	**5** 0038 5.0 0648 1.6 Th 1257 5.0 1907 1.7	**20** 0135 5.1 0744 1.6 F 1358 5.0 2026 1.5	**5** 0148 4.6 0754 2.1 Su 1411 4.8 2034 1.9	**20** 0338 4.5 0934 2.1 M 1553 4.7 2227 1.7	**5** 0232 4.6 0836 1.9 Tu 1450 5.0 2117 1.6	**20** 0345 4.4 0943 2.1 W 1559 4.7 2233 1.8
6 0109 5.0 0724 1.5 W 1331 4.8 1937 1.7	**21** 0147 5.3 0801 1.4 Th 1413 5.0 2034 1.5	**6** 0114 4.8 0724 1.9 F 1338 4.8 1952 1.9	**21** 0243 4.7 0849 2.0 Sa 1507 4.8 ☾ 2143 1.8	**6** 0250 4.4 0858 2.2 M 1515 4.6 2145 1.9	**21** 0452 4.4 1047 2.2 Tu 1709 4.6 2337 1.8	**6** 0335 4.6 0939 1.9 W 1553 4.9 2220 1.5	**21** 0447 4.3 1048 2.2 Th 1704 4.5 2334 1.9
7 0144 4.8 0801 1.8 Th 1411 4.6 2020 2.0	**22** 0250 4.8 0904 1.9 F 1522 4.6 ☾ 2155 1.8	**7** 0201 4.5 0812 2.2 Sa 1429 4.6 2051 2.2	**22** 0410 4.4 1010 2.3 Su 1638 4.6 2309 1.8	**7** 0412 4.3 1017 2.3 Tu 1635 4.7 2302 1.8	**22** 0603 4.4 1157 2.2 W 1817 4.7	**7** 0447 4.6 1048 1.9 Th 1705 5.0 2327 1.4	**22** 0553 4.3 1156 2.2 F 1811 4.5
8 0229 4.5 0847 2.2 F 1504 4.4 ☾ 2119 2.3	**23** 0420 4.4 1031 2.2 Sa 1702 4.5 2332 1.9	**8** 0305 4.2 0919 2.4 Su 1542 4.4 2212 2.3	**23** 0543 4.4 1137 2.3 M 1804 4.7	**8** 0536 4.5 1134 2.1 W 1751 4.9	**23** 0038 1.7 0700 4.5 Th 1256 2.0 1912 4.8	**8** 0556 4.7 1157 1.7 F 1811 5.1	**23** 0035 2.0 0653 4.4 Sa 1257 2.2 1912 4.6
9 0334 4.2 0955 2.4 Sa 1621 4.2 2244 2.4	**24** 0605 4.4 1207 2.2 Su 1838 4.6	**9** 0445 4.1 1055 2.5 M 1719 4.4 2343 2.1	**24** 0027 1.7 0655 4.5 Tu 1248 2.1 1909 4.8	**9** 0011 1.5 0639 4.7 Th 1238 1.7 1850 5.2	**24** 0128 1.6 0745 4.7 F 1344 1.9 1957 4.9	**9** 0031 1.3 0656 4.9 Sa 1259 1.5 1912 5.3	**24** 0128 1.9 0745 4.6 Su 1351 2.0 2005 4.7
10 0516 4.0 1134 2.5 Su 1800 4.3	**25** 0057 1.7 0726 4.6 M 1321 2.0 1942 4.9	**10** 0619 4.3 1219 2.2 Tu 1836 4.7	**25** 0126 1.5 0747 4.7 W 1342 1.8 1957 5.0	**10** 0107 1.2 0731 5.1 F 1331 1.4 1941 5.5	**25** 0209 1.5 0822 4.9 Sa 1425 1.7 2034 5.0	**10** 0128 1.1 0749 5.2 Su 1355 1.2 2008 5.4	**25** 0213 1.8 0829 4.8 M 1437 1.8 2049 4.8
11 0021 2.2 0655 4.2 M 1256 2.2 1916 4.6	**26** 0159 1.4 0819 4.8 Tu 1415 1.7 2029 5.1	**11** 0053 1.7 0721 4.7 W 1319 1.8 1931 5.1	**26** 0212 1.3 0825 4.9 Th 1423 1.6 2033 5.2	**11** 0157 0.9 0815 5.3 Sa 1418 1.0 2027 5.7	**26** 0246 1.4 0856 5.0 Su 1501 1.5 2110 5.1	**11** 0222 1.0 0840 5.4 M 1450 1.0 2100 5.6	**26** 0254 1.6 0908 5.0 Tu 1518 1.6 2129 4.9
12 0130 1.8 0757 4.6 Tu 1354 1.8 2006 5.0	**27** 0244 1.2 0857 5.0 W 1456 1.4 2105 5.3	**12** 0145 1.3 0806 5.1 Th 1405 1.3 2015 5.5	**27** 0249 1.2 0857 5.0 F 1458 1.4 2105 5.3	**12** 0243 0.6 0857 5.6 Su 1504 0.7 2114 5.9	**27** 0318 1.3 0928 5.1 M 1536 1.4 2143 5.2	**12** 0312 0.9 0928 5.6 Tu 1542 0.8 ○ 2152 5.7	**27** 0332 1.5 0943 5.2 W 1556 1.4 2206 5.0
13 0220 1.3 0840 5.0 W 1439 1.4 2049 5.4	**28** 0321 1.0 0928 5.2 Th 1528 1.2 2136 5.4	**13** 0230 0.8 0846 5.4 F 1447 0.9 2056 5.8	**28** 0319 1.1 0925 5.2 Sa 1529 1.3 2136 5.4	**13** 0327 0.5 0941 5.8 M 1550 0.6 ○ 2200 6.0	**28** 0350 1.3 1000 5.3 Tu 1610 1.3 ● 2219 5.2	**13** 0400 0.8 1014 5.7 W 1633 0.6 2242 5.6	**28** 0409 1.4 1020 5.4 Th 1634 1.2 ● 2242 5.1
14 0303 0.9 0917 5.3 Th 1518 0.9 2127 5.7	**29** 0353 0.9 0956 5.2 F 1559 1.1 ● 2206 5.5	**14** 0311 0.5 0924 5.7 Sa 1528 0.6 ○ 2136 6.1	**29** 0349 1.1 0953 5.2 Su 1559 1.2 ● 2207 5.4	**14** 0412 0.5 1024 5.8 Tu 1638 0.5 2248 5.9	**29** 0424 1.3 1034 5.2 W 1645 1.3 2252 5.2	**14** 0449 0.9 1102 5.5 Th 1723 0.6 2333 5.5	**29** 0445 1.3 1055 5.5 F 1711 1.1 2319 5.2
15 0342 0.5 0953 5.6 F 1556 0.6 ○ 2204 6.0	**30** 0421 0.9 1024 5.3 Sa 1627 1.0 2235 5.5	**15** 0352 0.3 1002 5.9 Su 1609 0.4 2219 6.2	**30** 0417 1.0 1023 5.4 M 1630 1.2 2238 5.3	**15** 0459 0.7 1111 5.8 W 1729 0.6 2339 5.7	**30** 0458 1.3 1108 5.4 Th 1722 1.3 2329 5.1	**15** 0537 1.0 1149 5.7 F 1814 0.7	**30** 0522 1.3 1132 5.5 Sa 1750 1.0 2357 5.2
			31 0447 1.1 1054 5.4 Tu 1701 1.2 2309 5.3				**31** 0601 1.2 1211 5.5 Su 1829 1.0

Chart Datum: 3.05 metres below Ordnance Datum (Newlyn)

Harbour, Coastal and Tidal Information

HOLYHEAD continued

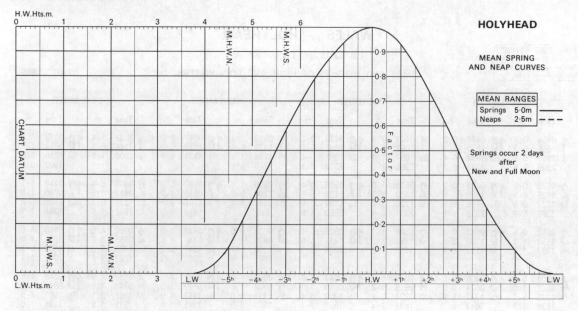

HOLYHEAD

MEAN SPRING AND NEAP CURVES

MEAN RANGES
Springs 5·0m ———
Neaps 2·5m - - -

Springs occur 2 days after New and Full Moon

MENAI STRAIT 10-10-19
Gwynedd

CHARTS
Admiralty 1464; Imray C61; OS 114, 115

TIDES
Beaumaris — 0010, Caernarfon — 0120 Dover;
ML Beaumaris 3.9; Caernarfon 2.9; Duration 0540;
Zone 0 (GMT).

Standard Port HOLYHEAD (←—)

Times				Height (metres)			
HW		LW		MHWS	MHWN	MLWN	MLWS
0000	0600	0500	1100	5.7	4.5	2.0	0.7
1200	1800	1700	2300				

Differences BEAUMARIS
+0025 +0010 +0055 +0035 +2.0 +1.6 +0.5 +0.1
MENAI BRIDGE
+0030 +0010 +0100 +0035 +1.7 +1.4 +0.3 0.0
PORT DINORWIC
−0015 −0025 +0030 0000 0.0 0.0 0.0 +0.1
CAERNARFON
−0030 −0030 +0015 −0005 −0.4 −0.4 −0.1 −0.1
PORTH DINLLAEN
−0120 −0105 −0035 −0025 −1.0 −1.0 −0.2 −0.2
BARDSEY ISLAND
−0220 −0240 −0145 −0140 −1.2 −1.2 −0.5 −0.1

HW in The Swellies −0045 HW Port Dinorwic;
−0200 HW Liverpool; −0115 HW Holyhead.

SHELTER
Very good shelter and facilities at Beaumaris, Bangor, Port Dinorwic (fresh water marina), Caernarfon (only available near HW).

NAVIGATION
NE entrance waypoint 53°20'.00N 04°00'.00W, 045°/225° from/to NE end Puffin Island, 1.0M. Caernarfon Bar waypoint 53°07'.60N 04°26'.00W, 270°/090° from/to Abermenai Pt Lt, 3.8M. From N end of Strait keep to buoyed channel near Anglesey shore. Night pilotage not recommended — many unlit buoys. The bridges and power cables have minimum clearance of 24m at MHWS. Between the bridges is a stretch of water called the Swellies which should only be attempted at slack HW which normally occurs about one hour before HW Holyhead. For further notes see Menai Strait 10.10.5.

LIGHTS AND MARKS
A light FWG exhibited at the end of Beaumaris pier. Menai Strait is marked by lateral marks, lights and light buoys; direction of buoyage changes off Caernarfon. From N approach in G sector of Mountfield Lt. Between the bridges is Price Pt, Fl WR 2s 5m 3M; R059°−239°, W239°−259°. At S end of Strait, Abermenai Pt Lt, Fl WR 3.5s; R065°−245°, W245°−065°.

Sketch showing relationship of Caernarfon, The Swellies, Port Dinorwic, Bangor and Beaumaris in the Menai Strait.

⚓ Apply to Royal Anglesey YC

AREA 10—N Wales 489

MENAI STRAIT continued

BEAUMARIS
RADIO TELEPHONE
None.
TELEPHONE (0248)
Hr Mr 750057 Ex. 212; CG & MRSC Holyhead 2051;
HMC Freefone Customs Yachts: RT (0407) 2336;
Marinecall 0898 500 460; Dr 810501.
FACILITIES
EC Beaumaris – Wednesday. **Pier** FW, L; **Royal Anglesey YC** Tel. 810295, Slip, M, L, R, Bar; **Northwest Venturers YC** Tel. 810023; **Anglesey Boat Co** Tel. 810359, Slip, P and D (in cans), FW, ME, BH (20 ton), Sh, C (2 ton), CH, Gas: **Town** PO; Bank; Rly (bus to Bangor); Air (Liverpool).

BANGOR
RADIO TELEPHONE
None.
TELEPHONE (0248)
Hr Mr 722920 Ex. 212; CG & MRSC Holyhead 2051;
HMC Freefone Customs Yachts: RT (0407) 2336;
Dr 364567.

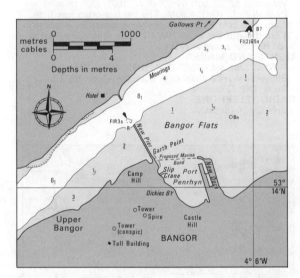

FACILITIES
Dickies Tel. 352775, Slip, D, FW, ME, El, Sh, C, CH, SM, BH (30 ton), Gas, Gaz; **Port Penrhyn**, Slip, D; **Town** PO; Bank; Rly; Air (Chester).

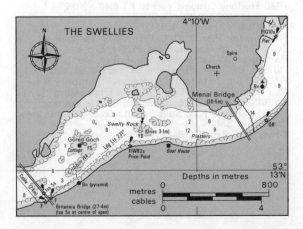

THE SWELLIES
NAVIGATION
For general pilotage notes on Menai Strait, including the Swellies, see 10.10.5. The (Swellies) passage between the bridges is dangerous for yachts and small craft except at or immediately before slack HW. At other times the stream can run up to 8 kn. At slack HW there is 3m over The Platters and over the outcrop off Price Pt, and these can be ignored. Passage is also possible for shallow-draught boats at slack LW at neaps, but there are charted depths of 0.4m close E of Britannia Bridge. Passage at night not recommended.
LIGHTS AND MARKS
St George's Pier FlG 10s.
E side of channel QR 4m; R mast; vis 064°-222°.
Price Pt FlWR 2s 5m 3M; W Bn: vis R059°-239°, W239°-259°. Britannia Bridge, E side, Ldg Lts 231°. Both FW. Centre span of bridge Iso 5s 27m 3M; either side. S end of bridge FR 21m 3M; either side. N end FG 21m 3M; either side.

PORT DINORWIC
RADIO TELEPHONE
Port Dinorwic Yacht Harbour, Call *Dinorwic Marine* VHF Ch M (office hours).
TELEPHONE (0248)
Hr Mr 670441; CG & MRSC Holyhead 2051;
HMC Freefone Customs Yachts: RT (0407) 2336;
Dr 670423.
FACILITIES
Outer harbour dries. **Port Dinorwic Yacht Harbour** (230) Tel. 670559 D, AC, CH, SM; **J. Dawson** Tel. 670103, SM; **P. D. Marine** Tel. 670441, ME, El, Sh, Slip, C; **M. J. Stallard** Tel. 670010 El, CH; **Town** PO (Bangor or Caernarfon); Bank; Rly (Bangor); Air (Liverpool).

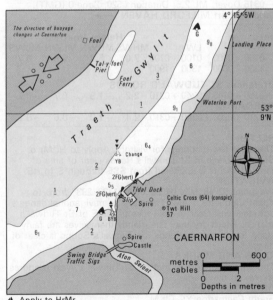

▲ Apply to HrMr

CAERNARFON
RADIO TELEPHONE
VHF Ch 16; 12 14 (day service only).
TELEPHONE (0286)
Hr Mr 2118; CG & MRSC Holyhead 2051; HMC Freefone Customs Yachts: RT (0407) 2336; Dr 2286 (Hospital).
FACILITIES
Harbour Tel. 2118, FW, Slip, L, C (2 ton), V; **Caernarfon Marine** Tel. 4322, P, D, ME, El, Sh, CH; **Caernarfon SC** Tel. 2861, L, Bar; **Royal Welsh YC** Tel. 2599, P, FW, Bar; **Arfon Oceaneering** Tel. 76055, Slip;
Town PO; Bank; Rly (Bangor); Air (Liverpool).
Note: Between Port Dinorwic and Caernarfon is **Plas Menai** Tel. Port Dinorwic 670964, the Sport Council for Wales Sailing and Sports Centre.

ABERSOCH 10-10-20
Gwynedd

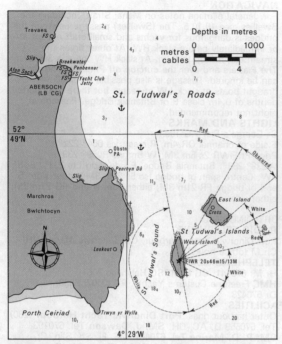

CHARTS
Admiralty 1484, 1512, 1971; Imray C61; OS 123
TIDES
−0315 Dover; ML 2.5; Duration 0520; Zone 0 (GMT).
Standard Port MILFORD HAVEN (→)

Times				Height (metres)			
HW		LW		MHWS	MHWN	MLWN	MLWS
0100	0800	0100	0700	7.0	5.2	2.5	0.7
1300	2000	1300	1900				
Differences ST TUDWAL'S ROADS							
+0155	+0145	+0240	+0310	−2.2	−1.9	−0.7	−0.2
ABERDARON							
+0210	+0200	+0240	+0310	−2.4	−1.9	−0.6	−0.2

SHELTER
There are few moorings for visitors. Apply to Hr Mr or SC. Anchorage in St Tudwal's Roads clear of area of moored yachts is sheltered from SSE through S to NE.
NAVIGATION
Waypoint 52°48'.50N 04°26'.06W, 113°/293° from/to Yacht Club jetty, 2.4M. There are no navigational dangers but steer well clear of the rocks to the E of St Tudwal's Islands, which dry out. The islands themselves are fairly steep to, except at N ends. St Tudwal's Sound is clear of dangers.
LIGHTS AND MARKS
The only major light is that on St Tudwal's West Island as shown on the chartlet.
RADIO TELEPHONE
South Caernarfon YC Ch M.
TELEPHONE (075 881)
Hr Mr 812684; CG & MRSC Holyhead 2051; HMC Freefone Customs Yachts: (RT) (0407) 2336; Marinecall 0898 500 460; Dr Pwllheli 612535.
FACILITIES
EC Wednesday. **S. Caernarvonshire YC** Tel. 2338, Slip, M, L, FW; **Abersoch BY** Tel. 2213, Slip, ME, El, Sh, C (5 ton), ACA, CH; **Abersoch Land & Sea Services** Tel. 3434, ME, El, Sh, FW, P, D, C (12 ton), CH; **Hookes Marine** Tel. 2458, ME, El, Sh; **Abersoch Power Boat Club** Tel. 2027; **Town** CH, V, R, Bar, PO; Bank; Rly (Pwllheli); Air (Chester).

PWLLHELI 10-10-21
Gwynedd

CHARTS
Admiralty 1971, 1512; Imray C61; OS 123
TIDES
−0300 Dover; ML 2.6; Duration 0510; Zone 0 (GMT).
Standard Port MILFORD HAVEN (→)

Times				Height (metres)			
HW		LW		MHWS	MHWN	MLWN	MLWS
0100	0800	0100	0700	7.0	5.2	2.5	0.7
1300	2000	1300	1900				
Differences PWLLHELI							
+0210	+0150	+0245	+0320	−2.0	−1.8	−0.6	−0.2
CRICCIETH							
+0210	+0155	+0255	+0320	−2.0	−1.8	−0.7	−0.3

SHELTER
Small secure harbour with good shelter. Strong winds from SW to E cause breakers off shore. Inner harbour has good shelter but dries. Entry HW∓2. No anchoring in harbour. Mooring enquiries to Hr Mr.
NAVIGATION
Waypoint 52°52'.50N 04°23'.00W, 150°/330° from/to Pwllheli Pt Lt (QR), 0.80M. The bar has often less than 0.3m. It is safe to cross in any wind direction. Boats with 1.5m draught can cross HW∓2. Tide runs up to 3 kn at sp. Harbour speed limit 5 kn.
LIGHTS AND MARKS
F WRG 12m; vis G155°-175°, R175°-245°, W245°-045°. Buoyed entrance channel in G sector. Outer No 1 Lt Buoy, QG, G conical.
RADIO TELEPHONE
VHF Ch 16 (occas).

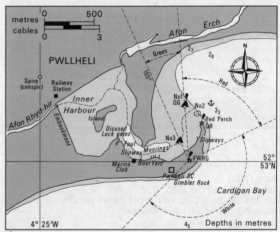

⚓ Secure to mooring and see Hr Mr

TELEPHONE (0758)
Hr Mr 613131 ex 281; MRSC Holyhead 2051; CG 720204; HMC Freefone Customs Yachts: RT (0407) 2336; Marinecall 0898 500 460; Dr 612535
FACILITIES
EC Thursday. **Pier**, P, D, L, FW, AB, V; **Pwllheli SC** Tel. 612219; **Marina Club** Tel. 612611, Slip, L, FW; **Partington BY** Tel. 612808, Slip, L, FW, ME, Gas, Sh, C (10 ton), CH, ACA; **Firmhelm BY** Tel. 612251, Slip, D, L, FW, ME, Sh, C (14 ton), CH; **Harbour Authority** Slip, M, L, FW, AB; **Tony Evans Marine Engineering** Tel. 613219, ME, El; **Rowlands Marine Electronics** Tel. 613193 El; **J.K.A. Sails** Tel. 613266 SM; **Tudor Sails** Tel. 613141 SM.
Town V, R, Bar; PO, Bank, Rly, Air (Chester).

AREA 10—NW England, Isle of Man and N Wales

MINOR HARBOURS AND ANCHORAGES 10.10.22

SILLOTH, Cumbria, 54°52′ N, 3°24′ W, Zone 0 (GMT), Admty charts 1346, 2013. HW −0050 on Dover, +0035 on Liverpool; HW height −0.3m on Liverpool; ML 4.9m; Duration 0520. See 10.10.7. Anchorage off Lees Scar in about 4m, exposed to SW winds; outer harbour dries; or berth in the new wet dock, but this is principally commercial. Beware constantly changing channels and sandbanks. East Cote Dir Lt 052° FG 15m 12M; vis 046°–058°, intens 052°. Lees Scar, S of entrance, QG 11m 8M; vis 005°–317°. Groyne head 2 FG (vert). Entry signals on mast at New Dock — no entry unless Y signal arm raised by day or Q Bu Lt by night. VHF Ch 16; 12 (HW−2½ to HW+1½). Facilities: EC Tues; Hr Mr Tel. 31358; FW on quays, all stores available, Bank, Bar, PO, R, V.

HARRINGTON, Cumbria, 54°37′ N, 3°34′ W, Zone 0 (GMT), Admty chart 1346. HW +0025 on Dover, +0015 on Liverpool; HW height −1.1m on Liverpool; ML 4.6m; Duration 0540. Good shelter in small harbour only used now by local fishermen and yachts. Entrance by stone pier with Lt Fl G 5s 3M. Inner harbour dries. Contact Cockermouth 823741 for moorings. Very limited facilities.

WHITEHAVEN, Cumbria, 54°33′ N, 3°36′ W, Zone 0 (GMT), Admty chart 1346. HW +0015 on Dover, +0010 on Liverpool; HW height −1.2m on Liverpool; ML 4.5m; Duration 0550. See 10.10.8. All harbour dries except Queens Dock, but outer harbour has over 5m at MHWS. Very good shelter and entrance safe in most conditions. Access HW∓2½. W pier head Fl G 5s 16m 13M. N pier head 2 FR (vert) 8m 9M. N Wall Quay 2 FR (vert) 8m 2M. Old Quay head 2 FG (vert) 8m 2M. VHF Ch 16; 12 (HW −2 to HW +2). Facilities: EC Wed; berthing see Hr Mr. Tel. 2435, FW, D, Slip, C (7½ ton), P (on Fish Quay), Bar, Bank, CH, PO, R, V.

RAVENGLASS, Cumbria, 54°21′ N, 3°25′ W, Zone 0 (GMT), Admty chart 1346. HW +0020 on Dover, +0005 on Liverpool; HW height −1.0m on Liverpool; ML 4.6m; Duration 0545. Large harbour formed by estuaries of R Mite, R Irt and R Esk, which dries; there is approx 2.5m in entrance at HW−2. From N beware Drigg Rock and from S Selker Rks. Also beware gun testing range at Eskmeals (R flag when in use). There is a FG Lt (occas). Facilities: FW (in village), but no other facilities.

BARROW-IN-FURNESS, Cumbria, 54°06′ N, 3°12′ W, Zone 0 (GMT), Admty chart 3164. HW +0030 on Dover, +0015 on Liverpool; HW height −0.25m on Liverpool; ML 5.2m; Duration 0530. See 10.10.9. Good shelter. Harbour dries except for Walney Channel which must be kept clear. Landing places at Piel I and Roa I. There are no navigational dangers so long as Ldg Lines are kept to. Lights: Walney I Fl 15s 21m 23M (when within 3M of shore, obsc 122°–127°), RC. Ldg Lts leading in from Lightning Knoll By (RW LFl 10s) at 041° front Q 6m 6M (B structure with W daymark), rear (640m from front) Iso 2s 12m 6M (R column with W face). VHF *Ramsden Dock* Ch 16; 12 (H24). Facilities: EC Thurs; all facilities available in Barrow; **Rawlinson** Tel. 32806 SM.

HEYSHAM, Lancashire, 54°02′ N, 2°55′ W, Zone 0 (GMT), Admty charts 1552, 2010. HW +0015 on Dover, +0001 on Liverpool; HW height +0.1m on Liverpool; ML 5.3m; Duration 0545. See 10.10.9. Good shelter but yachts not normally accepted without special reason. Beware ferries and 'rig' supply vessels. Ldg Lts 102°, front FBu 11m 2M, Y+B diamond on mast; rear (137m from front) FBu 14m 2M, Y+B diamond on mast. S pier head Oc G 7.5s 9m 6M. N pier head 2FR (vert) 11m, obsc from seaward. Entry sigs; R flag or R Lt = no entry; no signal = no departure; 2R flags or 2R Lts = no entry or departure. VHF Ch 16; 14 (H24). Facilities: EC Wed (Heysham and Morecambe); Bar, FW, R, V. All stores at Morecambe (2M).

RIVER RIBBLE, Lancashire, 53°45′ N, 2°47′ W, Zone 0 (GMT), Admty chart 1981. HW +0013 on Dover; Preston 0000, St Anne's −0004 on Liverpool; HW height Preston −3.9m, St Anne's −0.1m on Liverpool; ML Preston 2.4m, St Anne's 5.4m; Duration St Anne's 0520. Preston closed as a commercial port in 1981. Many navigational aids in R Ribble have been withdrawn and silting has occurred. Entrance marked by Gut buoy, RW, lit. Training wall breached and channel now runs through S Gut (unmarked). Local knowledge needed. Contact **Ribble Cruising Club** Tel. Lytham 739983. Yachts berth near BY at Lytham, at Freckleton or near River Douglas entrance. Facilities: AB, C (7 ton), CH, D, FW, Sh, Slip. **Douglas Boatyard**, access HW∓2, Tel. Hesketh Bank 2462. ME, El, Sh, CH. Other facilities very limited except at Preston. Hr Mr (Preston) Tel. Preston 726711. **Preston Marina** Tel. 741031 (30), FW, Slip, Access HW−2 to HW+1 via lock.

RIVER ALT, Merseyside, 53°30′ N, 3°04′ W, Zone 0 (GMT), Admty chart 1951. HW −0008 on Dover, −0021 on Liverpool, see 10.10.11. Good shelter but only available to small craft. Access HW∓1½ for draught of 1.2m. Channel shifts frequently. Entrance channel marked by can buoy N side of groyne between Crosby and Hall Road beach marks. Channel from there marked by perches arranged by Blundellsands SC. Anchor between perches marking channel; local knowledge advised. Facilities very limited. **Blundellsands SC** at Hightown Tel. Liverpool 929 2101 (occas).

CASTLETOWN, Isle of Man, 54°04′ N, 4°39′ W, Zone 0 (GMT), Admty chart 2696. HW +0025 on Dover, +0010 on Liverpool; HW height −2.9m on Liverpool; ML 3.4m; Duration 0555. All 4 harbours dry. Secure in Inner Hr below fixed bridge or in Irish Hr below swing bridge. Anchorage between Lheeah–rio Rks and pier in 3m; at Langness Pt; in Derby Haven (drying). The bay gives good shelter except in SW to SE winds. Beware Lheeah-rio Rocks in W of bay, marked by R can buoy Fl R 3s, Bell. Beware race off Langness Pt, which should be given a good berth. Lt to E of Langness Pt, on Dreswick Pt Fl(2) 30s 23m 21M. N side of entrance Oc G 4s 3m (W metal post on concrete column). S side of entrance, Irish Quay, Oc R 4s 5m 5M vis 142°–322°. 150m NW is swing bridge marked by 2 FR (hor). To SE, New Pier head Oc R 15s 8m 5M. VHF Ch 16; 12 (when vessel expected). Facilities: EC Thurs; Hr Mr Tel. 823549; Dr 823597 **Outer Hr** Slip, L, C (20 ton) AB; **Irish Quay** AB, C, FW; **Inner Hr** AB, C, FW; P and D from garage (500m), **Ballasalla Marine and Auto Eng** Tel. 822715 ME; **J J Clague** Tel. 822525, Gas.

LAXEY, Isle of Man, 54°13′ N, 4°23′ W, Zone 0 (GMT), Admty chart 2094. HW +0025 on Dover, +0010 on Liverpool; HW height −2.0m on Liverpool; ML 4.0m; Duration 0550. The bay gives good shelter in N to SW winds through W. The harbour is only suitable for small yachts; it dries and has approx 2m at MHWN. Access HW∓3. Beware rocks on N side of the narrow entrance. Keep close to pier after entering to avoid training wall to N. Pier head Lt Oc R 3s 7m 5M, obsc when bearing less than 318°. Breakwater head Lt Oc G 3s 7m. Facilities are few other than FW, R, PO, Bank, and Bar.

PORT ERIN, Isle of Man, 54°05′ N, 4°46′ W, Zone 0 (GMT), Admty charts 2696, 2094. HW −0020 on Dover, +0005 on Liverpool; HW height −3.7m on Liverpool; ML 2.9m; Duration 0555. See 10.10.13. The bay has good anchorage in 3m to 8m, and protection from winds except those from SW to NW. There is a small drying harbour on the S side which has 3.6m at MHWN. Beware the ruined breakwater running N from the SW corner, the end being marked with G con buoy. Ldg Lts 099°, front FR 10m 5M, rear FR 19m 5M, lead into middle of the bay. Raglan Pier (E arm of harbour) Oc G 5s 8m 5M. Facilities: EC Thurs; Bar, D, FW, P, R, Slip, V. Two visitors buoys W of Raglan Pier — see Hr Mr.

PORTH DINLLAEN, Gwynedd, 52°57′ N, 4°34′ W, Zone 0 (GMT), Admty charts 1512, 1971. HW −0210 on Dover, −0120 on Holyhead; HW height −1.0m on Holyhead; ML 2.5m; Duration 0535. Shelter good in S to W winds but strong NNW to NNE winds cause heavy seas in the bay. Beware Carreg-y-Chad (1.8m) 0.75M SW of the point, and Carreg-y-Chwislen (dries, with Bn) 2 ca ENE of the point. Best anchorage 0.1M S of LB house in approx 2m. Facilities: EC Wed; Hr Mr Tel. Nefyn 720295; CG Tel. Nefyn 720204, Bar, V by landing stage; Other facilities are at Morfa Nefyn (1M), Bar, P, R, V.

FIBREGLASS

Every fibreglass requirement stocked at lowest prices in Europe. All resins top marine quick-cure quality by the leading U.K. Manufacturers, comply Lloyds, BISS, Admiralty, MOT. Ample hardeners free with every resin. All prices are trade prices and exclude VAT. Enquiries and orders from public welcomed. Please add VAT @ 15% to total of carriage and goods *(except for our Repair Kit which is priced carriage free, U.K. mainland only, and VAT inclusive)*

Prices in pence per lb.	1-9lbs	10-54lbs	55-109lbs	110lbs and over
Lay-Up/Laminating Resin "A"	£1.00	80p	75p	70p
Initial Gel Resin "B", Clear or White pigmented	£1.40	£1.20	90p	85p
Glassfibre C/S mats, all thicknesses	£1.20	£1.00	90p	85p

SPECIAL OFFER: Lay-Up Resin, reprocessed old stock, £25 per 25kg drum; Glassfibre Tissue, 45p sq.m.; Cloth 80p sq.m. Tapes from 10p metre; Pure Acetone £1.80 per ½ gallon; £3.00 gallon; Polyurethane Foam £1.70 lb (min 2 lbs); 10 lbs and over £1.30 lb; Fillers from 20p/lb; Special Purpose Resins; Clear Casting; Fibre Retardant, Heat/Acid Resistant; Release agents, pigments, rollers, brushes, etc. Please add for carriage (U.K. Mainland) 6p lb; Min carr. charge £5; also add VAT at 15% to total of carriage and goods except for our Repair Kit, £13.00 Carr FREE AND VAT inclusive, U.K. MAINLAND ONLY, includes 4 lb Resin, 1 lb Glassmat, 3 lbs Talc Filler, Tissue and Cloth. All orders same day despatch. Write, phone or call for our free 30,000 word instructive literature and price lists. Callers welcome every day; Saturdays, Sundays & Bank Holidays all day.

GLASPLIES
2 Crowland Street, Southport, Lancs
Tel. Southport (0704) 40626

Prop. Joseph Rankin

SHIPSIDES MARINE
YACHT CHANDLERS

5 New Hall Lane
PRESTON, LANCS.
Tel: (0772) 797079

... navigation equipment, Admiralty Charts, marine toilets, bilge pumps, VHF radios, lifejackets, clothing, liferaft hire, books, etc. etc. ...

★ CALL & SEE US OR SEND £1.75 FOR CATALOGUE ★

DUBOIS-PHILLIPS & McCALLUM LIMITED

Admiralty Charts and Publications

Chart Correction Service

Nautical Books and Instruments

**Oriel Chambers
Covent Garden
LIVERPOOL L2 8UD**
Tel: 051-236-2776   Tlx: 627424 DUBOIS G

Yacht Masts & Ancillary Equipment

We specialise in custom built spars together with ancillary equipment for boats 17–60 feet in length.

Aluminium Alloy Toe Rails

Through a comprehensive range of profiles we can supply the answer to your toe rail requirements. Toe rails together with fittings, can be machined and assembled to individual specifications.

S.S. Spars 1982 Ltd

Unit 4 Parsons Hall Ind. Estate
Irchester, Northants
Tel: Rushden (0933) 317143

VOLVO PENTA SERVICE

Sales and service centres in area 11

AVON **Bristol Marine Engineering** Hanover Place, Albion Dockside Estate, BRISTOL BS1 6UT Tel (0272) 262923. DYFED **Burry Port Yacht Services** The Harbour, BURRY PORT SA16 0ER Tel (05546) 2740. **Dale Sailing Co Ltd** DALE SA62 3RB Tel (06465) 349. GLAMORGAN **Swansea Marine Engineering Co** Swansea Yacht Haven, SWANSEA SA1 1WN Tel (0792) 51600. GWYNEDD **Abersoch Land and Sea Ltd** Abersoch, Gwynedd, NORTH WALES LL53 7AG Tel (075 881) 3434/5/6. **Hookes Marine** Sarn Bach Road, ABERSOCH LL53 7ER Tel (075 881) 2458. PEMBROKESHIRE **Dale Sailing Co Ltd** Brunel Quay, Neyland, MILFORD HAVEN SA73 1PY Tel (0646) 601636.

Area 11

South Wales and Bristol Channel
Porthmadog to St Ives

10.11.1	Index	Page 493
10.11.2	Diagram of Radiobeacons, Air Beacons, Lifeboat Stations etc	494
10.11.3	Tidal Stream Charts	496
10.11.4	List of Lights, Fog Signals and Waypoints	498
10.11.5	Passage Information	501
10.11.6	Distance Table	502
10.11.7	Porthmadog	503
10.11.8	Barmouth	503
10.11.9	Aberdovey	504
10.11.10	Aberystwyth	504
10.11.11	Fishguard	505
10.11.12	Milford Haven, Standard Port, Tidal Curves	506
10.11.13	Tenby	511
10.11.14	Burry Port	511
10.11.15	Swansea	512
10.11.16	Barry	512
10.11.17	Cardiff	513
10.11.18	Sharpness	514
10.11.19	Bristol (City Docks). (Avonmouth, Standard Port, Tidal Curves)	515
10.11.20	Burnham-on-Sea	519
10.11.21	Ilfracombe	520
10.11.22	Rivers Taw and Torridge	521
10.11.23	Padstow	522
10.11.24	Minor Harbours and Anchorages	523

Mochras
Aberaeron
New Quay
Cardigan
Solva
St Brides Bay
Skomer
Saundersfoot
Carmarthen
Porthcawl
Newport
Weston-super-Mare
Watchet
Minehead
Porlock Weir
Watermouth
Lundy Is
Bude
Newquay
Hayle
St Ives

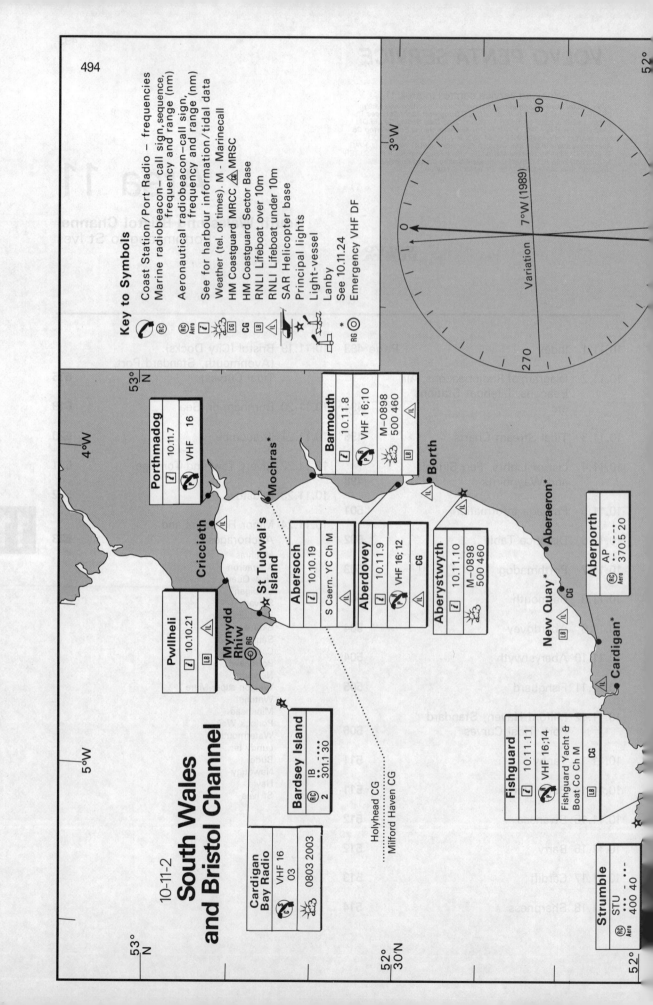

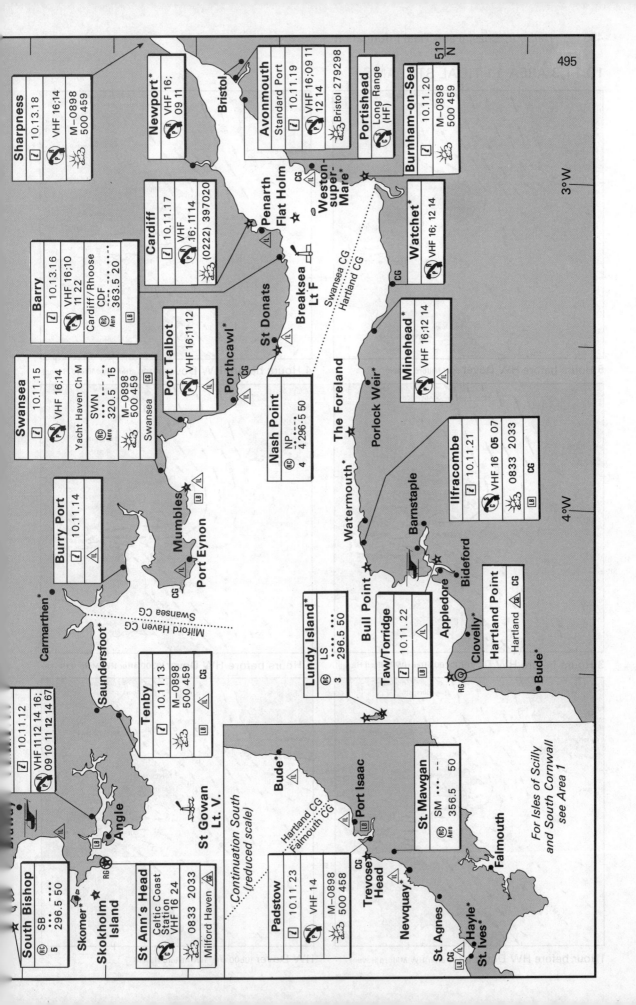

10.11.3 AREA 11 TIDAL STREAMS

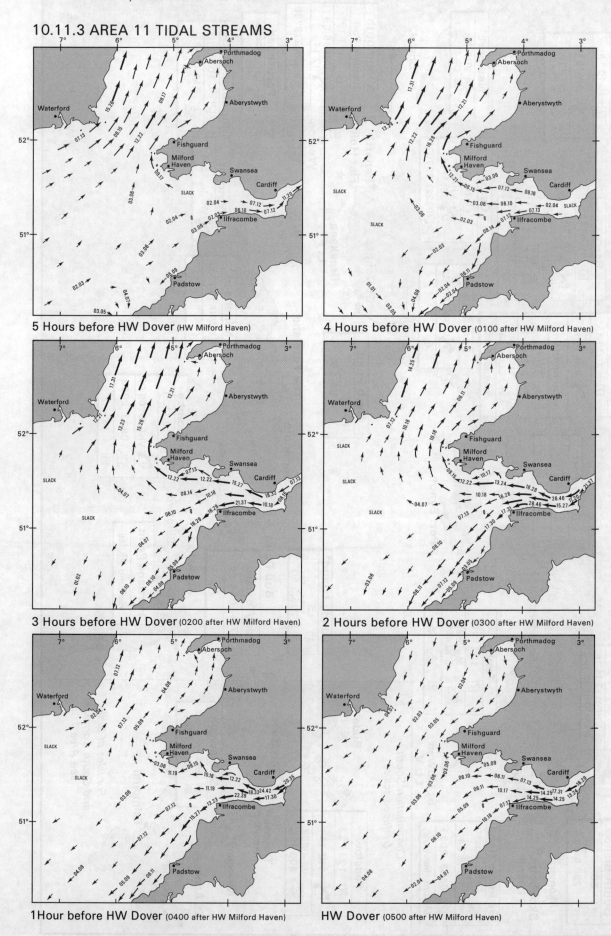

AREA 11 – S Wales and Bristol Channel   497

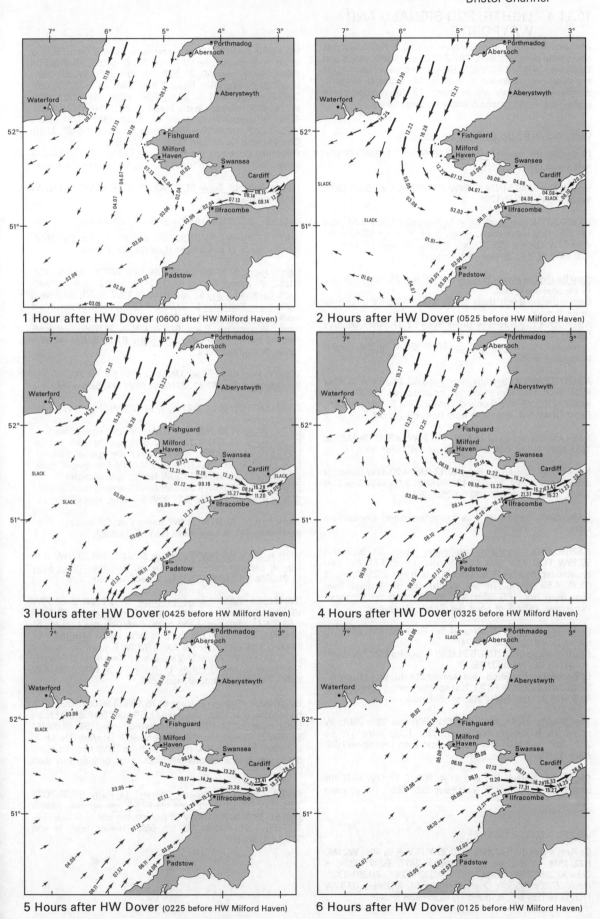

10.11.4 LIGHTS, FOG SIGNALS AND WAYPOINTS

Abbreviations used below are given in 10.0.3. Principal lights are in **bold** print, places in CAPITALS, and light-vessels and Lanbys in *CAPITAL ITALICS*. Unless otherwise stated lights are white. m—elevation in metres; M—nominal range in n. miles. Fog signals are in *italics*. Useful waypoints are underlined — use those on land with care. See 4.2.2.

WALES—WEST COAST (see also 10.10.4)

Porthmadog Fairway Lt Buoy 52 53.58N/4 11.55W LFl 10s; safe water mark.

Shell Island 52 49.54N/4 07.64W Fl WRG 4s; vis G079°-124°, W124°-134°, R134°-179°.

BARMOUTH. Barmouth Outer Lt Buoy 52 42.60N/4 04.76W LFl 10s; safe water mark. North Bank Y perch 52 42.81N/4 03.67W QR 4m 2M. Ynys y Brawd, SE end 52 42.97N/4 03.07W Fl R 5s. Bridge, NW end 2FR (hor).

Sarn Badrig Causeway Lt Buoy 52 41.17N/4 25.30W Q(9) 15s; W cardinal mark; *Bell*.
Sarn-y-Bwch. Bwch Buoy 52 34.80N/4 13.50W (unlit); W cardinal mark.
ABERDOVEY. Aberdovey Outer Buoy 52 31.74N/4 06.20W (unlit); safe water mark.
Cynfelyn Patches. Patches Buoy 52 25.82N/4 16.30W (unlit); W cardinal mark.

ABERYSTWYTH. S breakwater head 52 24.39N/4 05.46W Fl(2) WG 10s 12m 10M; B column; vis G030°-053°, W053°-210°. 4FR(vert) on radio tower 2.8M S. Ldg Lts 138°. Front FR 4m 5M. Rear 52m from front FR 7m 6M.

ABERAERON. S pier 52 14.60N/4 15.87W Fl(3) G 10s 6M; vis 125°-200°. N pier Fl(4) R 15s 6M; vis 104°-178°. FW(T).

NEW QUAY. Carreg Ina Buoy 52 13.09N/4 20.47W (unlit); N cardinal mark. Pier, head 52 12.94N/4 21.27W Fl WG 3s 12m W8M, G5M; vis W135°-252°, G252°-295°.

CARDIGAN. Bridge, Iso Y 2s on upstream and downstream sides.

FISHGUARD. Northern breakwater head 52 00.74N/4 58.15W Fl G 4.5s 18m 13M; 8-sided Tr; *Bell (1) 8s*. East breakwater, head Fl R 3s 10m 5M. Lts in line 282°. Front 52 00.7N/4 59.2W FG 77m 5M; W diamond on W mast. Rear 46m from front FG 89m 5M; W diamond on W mast. Penanglas, 152m S of point *Dia (2) 60s*; W obelisk.

Strumble Head 52 01.8N/5 04.3W Fl(4) 15s 45m **29M**; W round Tr; vis 038°-257°; (H24).
South Bishop 51 51.15N/5 24.65W Fl 5s 44m **24M**; W round Tr; (H24); RC; *Horn (3) 45s*.
St Brides Bay. Research Area, seaward Lt Buoys. Lt Buoy A1 51 49.30N/5 20.00W Fl(4)Y 20s; special mark. Lt Buoy B1 51 48.30N/5 20.00W Fl(4)Y 20s; special mark.

The Smalls 51 43.25N/5 40.15W Fl(3) 15s 36m **25M**; W round Tr, R bands; Racon. FR 33m 13M; same Tr; vis 253°-285° over Hats and Barrels rock; both Lts shown H24; *Horn (2) 60s*.

Skokholm Island, SW end 51 41.60N/5 17.17W Fl R 10s 54m **17M**; W 8-sided Tr; part obsc 226°-258°; (H24); *Horn 15s*.

WALES—SOUTH COAST

St Ann's Head 51 40.85N/5 10.35W Fl WR 5s 48m **W23M, R22/19M**; W 8-sided Tr; vis W233°-247°, R247°-285°, R (intens) 285°-314°, R314°-332°, W332°-124°, W129°-131°; *Horn (2) 60s*. Middle Channel Rocks 51 40.29N/5 09.77W Fl(3) G 7s 18m 8M; B round Tr, aluminium lantern.

MILFORD HAVEN. West Blockhouse Point F 54m 13M; W stripe on B ground, on W Tr; vis 004-5°-040.5°; Q WR 21m W9M, R7M; vis W220°-250°, R250°-020°, W020°-036°, R036°-049°. **Watwick Point** F 80m **15M**; vis 013.5°-031.5°. Dale Fort Fl(2) WR 5s 20m W5M, R3M; vis R222°-276°, W276°-019°. **Great Castle Head** F WRG 27m W5M, R3M, G3M; W square Tr, B stripe; vis R243°-281°, G281°-299°, W299°-029°; Oc 4s 27m **15M**; same Tr; vis 031°-048°. **Little Castle Head** Oc 8s 53m **15M**; vis 031°-048°.
Pembroke Dock Ldg Lts 153°. Front QG 5m; W diamond, B stripe. Rear, 82m from front, QG 9m; W diamond, B stripe.

Turbot Bank Lt Buoy 51 37.40N/5 10.00W VQ(9) 10s; W cardinal mark.

WALES—SOUTH COAST—BRISTOL CHANNEL

SAINT GOWAN LT V 51 30.50N/4 59.80W Fl 20s 12m **26M**; R hull with Lt Tr amidships; *Horn (3) 60s*; Racon.

Caldey Island 51 37.86N/4 41.00W Fl(3) WR 20s 65m W14M, R12M; W round Tr; vis R173°-212°, W212°-088°, R088°-102°. Eel Point Buoy 51 38.84N/4 42.18W (unlit); stbd-hand mark. Giltar Spit Buoy 51 39.00N/4 42.05W (unlit); port-hand mark. Spaniel Buoy 51 38.03N/4 39.67W (unlit); E cardinal mark. Woolhouse Buoy 51 39.32N/4 39.61W (unlit); S cardinal mark. North Highcliff Buoy 51 39.35N/4 40.69W (unlit); N cardinal mark.

TENBY. Pier, head 51 40.37N/4 41.81W FR 7m 7M.
Saundersfoot Pier 51 42.55N/4 41.68W Fl R 5s 6m 7M.

CARMARTHEN BAY. DZ1 Buoy 51 42.05N/4 35.90W (unlit); special mark.
DZ2 Lt Buoy 51 39.95N/4 37.62W FlY 2.5s; special mark.
DZ3 Buoy 51 37.35N/4 37.70W (unlit); special mark.
DZ7 Lt Buoy 51 38.08N/4 30.05W FlY 10s; special mark.
DZ4 Lt Buoy 51 35.70N/4 29.95W FlY 5s; special mark.
DZ8 Buoy 51 41.50N/4 24.30W (unlit); special mark.
DZ6 Buoy 51 38.00N/4 24.30W (unlit); special mark.
DZ5 Lt Buoy 51 36.35N/4 24.30W FlY 2.5s; special mark.

Burry Port 51 40.5N/4 14.9W Barrel Post QR (occas). West breakwater FR. Whiteford Lt Ho Fl 5s (occas).

West Helwick (W.HWK) Lt Buoy 51 31.37N/4 23.58W Q(9) 15s; W cardinal mark; *Whis*; Racon. East Helwick Lt Buoy 51 31.75N/4 12.60W VQ(3) 5s; E cardinal mark; *Bell*.

SWANSEA BAY. Ledge Lt Buoy 51 29.90N/3 58.70W VQ(6)+LFl 10s; S cardinal mark. Mixon Buoy 51 33.10N/3 58.70W (unlit); port-hand mark; *Bell*. Grounds Lt Buoy 51 32.90N/3 53.30W VQ(3) 5s; E cardinal mark. **Mumbles** 51 34.00N/3 58.20W Fl(4) 10s 35m **17M**; W 8-sided Tr; Fog Det Lt vis 331°-336°; *Horn(3) 60s*. Railway Pier, head 2FR(vert). SW Inner Green Grounds Lt Buoy 51 34.04N/3 56.95W Q(6)+LFl 15s; S cardinal mark; *Bell*.

SWANSEA. Channel Lt Buoy 51 35.50N/3 56.06W QG; stbd-hand mark; *Bell*. West pier head Fl(2)R 10s 11m 9M; FR Lts on radio mast 1.3M NNE. E breakwater head 51 36.35N/3 55.55W 2FG(vert) 10m 6M; W framework Tr; *Horn 30s*. Lts in line 020°. Front 2FG(vert) 5m 2M. Rear 250m from front FG 6M (mark E limit of dredged area, obsc bearing less than 020°).

RIVER NEATH. Approach Channel Lt Buoy 51 35.70N/3 52.75W FlG 5s; stbd-hand mark. SE training wall, near S end 51 36.30N/3 51.89W 2FG(vert) 6m 5M (FG 2M (T)). Training wall, middle FG 6m 5M. Training wall, N end 3FG(vert) 6m 5M.

PORT TALBOT. Cabenda Lt Buoy 51 33.43N/3 52.27W VQ(6) + LFl 10s; S cardinal mark. Outer channel buoys: 51 33.67N/3 51.22W FlG 5s; stbd-hand mark. 51 33.76N/3 51.30W FlR 5s; port-hand mark; *Horn*. Ldg Lts 060° (occas). Front Oc R 4s 12m 6M. Rear Oc R 6s 32m 6M; both Y and Or diamonds on pylons. N breakwater head Fl(4)R 10s 11m 3M; metal pylon. S breakwater head 51 34.43N/3 48.95W FlG 3s 11m 3M.
Sker Point. Kenfig Lt Buoy 51 29.71N/3 46.52W Q(3) 10s; E cardinal mark.

BRISTOL CHANNEL—EASTERN PART (NORTH SHORE)

W Scarweather (W.SCAR) Lt Buoy 51 28.28N/3 55.50W Q(9) 15s; W cardinal mark; *Bell*; Racon.
S Scarweather (S.SCAR) Lt Buoy 51 27.58N/3 51.50W Q(6) + LFl 15s; S cardinal mark.
Hugo Buoy 51 28.80N/3 48.30W (unlit); port-hand mark.
E Scarweather Buoy 51 28.10N/3 46.20W (unlit); E cardinal mark.

PORTHCAWL Tusker Lt Buoy 51 26.95N/3 40.60W Fl(2) R 5s; port-hand mark. Porthcawl breakwater, head 51 28.33N/3 41.95W F WRG 10m W6M, R4M, G4M; W 6-sided Tr, B base; vis G302°-036°, W036°-082°, R082°-122°. In line with Saint Hilary radio mast (Aero QR) 094° leads through Shord Channel.
West Nash Lt Buoy 51 25.95N/3 45.88W VQ(9) 10s; W cardinal mark. Middle Nash Buoy 51 25.00N/3 40.00W (unlit); S cardinal mark. East Nash Lt Buoy 51 24.00N/3 34.00W Q(3) 10s; E cardinal mark.

Nash 51 24.00N/3 33.05W Fl(2) WR 10s 56m **W21M, R20/17M**; W round Tr; vis R280°-290°, W290°-097°, R097°-100°, R (intens) 100°-104°, R104°-120°, W120°-128°; *Siren (2) 45s*; RC.
Saint Hilary 51 27.4N/3 24.1W Aero QR 346m 11M; radio mast; 4FR(vert) on same mast 6M.
Breaksea Point, intake 51 22.5N/3 24.5W Fl R 11m. FR Lt on radio mast 3.4M ENE.

BREAKSEA Lt F 51 19.85N/3 19.00W Fl 15s 15M; Racon; F riding Lt; *Horn (2) 30s*.
Wenvoe 51 27.5N/3 16.8W Aero Fl W 364m 12M; radio mast.
Merkur Lt Buoy 51 21.88N/3 16.10W FlR 2.5s; port-hand mark.

BARRY. West breakwater, head Fl 2.5s 12m 10M; W round Tr. E breakwater, head 51 23.50N/3 15.37W QG 7m 8M.

W One Fathom Lt Buoy 51 20.40N/3 14.50W Q(9) 15s; W cardinal mark. N One Fathom Lt Buoy 51 21.11N/3 11.76W Q; N cardinal mark. Mackenzie Lt Buoy 51 21.70N/3 08.15W QR; port-hand mark. Wolves Lt Buoy 51 23.10N/3 08.81W VQ; N cardinal mark.

Flat Holm, SE point 51 22.52N/3 07.05W Fl(3) WR 10s 50m **W16M**, R13M; W round Tr; vis R106°-140°, W140°-151°, R151°-203°, W203°-106°; (H24); *Horn 30s*.

Monkstone Rock 51 24.86N/3 05.93W Fl(2) 10s 13m 5M; R column on round Tr.

CARDIFF/PENARTH ROADS. Ranie Lt Buoy 51 24.22N/3 09.30W Fl(2)R 5s; port-hand mark. S Cardiff Lt Buoy 51 24.15N/3 08.50W Q(6) + LFl 15s; S cardinal mark; *Bell*. Mid Cardiff Lt Buoy 51 25.57N/3 08.02W Fl(3)G 10s; stbd-hand mark. N Cardiff Lt Buoy 51 27.75N/3 05.30W QG; stbd-hand mark.

PENARTH. Promenade Pier, near head 2FR(vert) 8/6m 3M; *Reed Mo(BA) 60s*, sounded 10 min before a steamer is expected. Club pontoon 51 26.8N/3 10.5W Q.
CARDIFF. Outer Wrach Lt Buoy 51 26.17N/3 09.38W Q(9) 15s; W cardinal mark. Ldg Lts 349°. **Front** 51 27.7N/3 09.9W F 4m **17M**. **Rear** 520m from front F 24m **17M**. Queen Alexandra Dock entrance, S jetty head 2FG(vert); *Dia 60s*.
Tail Patch Lt Buoy 51 23.50N/3 03.60W QG; stbd-hand mark.
Hope Lt Buoy 51 24.80N/3 02.60W Q(3) 10s; E cardinal mark.
NW Elbow Lt Buoy 51 26.10N/2 59.95W VQ(9) 10s; W cardinal mark; *Bell*.

ENGLISH AND WELSH GROUNDS Lt F 51 26.90N/3 00.10W L Fl 10s 11m 12M; R hull, W name on sides, Lt Tr; Racon; *Horn (2) 20s*.
NEWPORT DEEP. Newport Deep Lt Buoy 51 29.35N/2 59.05W Fl(3)G 10s; stbd-hand mark; *Bell*.

RIVER USK. **East Usk** 51 32.38N/2 57.93W Fl(2) WRG 10s 11m **W15M**, R11M, G11M; W round Tr; vis W284°-290°, R290°-017°, W017°-037°, G037°-115°, W115°-120°.
Alexandra Dock, South Lock, West pier head 2FR(vert) 9/7m 6M; *Horn 60s*. E pier head 2FG(vert) 9/7m 6M.
Julians Pill Ldg Lts 057°. Rear, 61m from front, FG 8m 4M. Front 51 33.3N/2 57.9W FG 5m 4M (common front). Ldg Lts 149°. Rear, 137m from front FG 9m 4M.
Bellport Jetty 51 33.6N/2 58.0W 2FG(vert). Dallimores Wharf 51 33.8N/2 58.4W 2FG(vert). Transporter Bridge, West side 2FR(vert); 2FY(vert) shown on transporter car; E side 2FG(vert). Centres of George Street and Newport Bridges marked by FY Lts.

BRISTOL DEEP. N Elbow Lt Buoy 51 27.12N/2 57.10W QG; stbd-hand mark; *Bell*. S Mid Grounds Lt Buoy 51 27.80N/2 57.15W Fl(4)R 15s; port-hand mark. E Mid Grounds Lt Buoy 51 27.95N/2 54.60W FlR 5s; port-hand mark. Clevedon Lt Buoy 51 27.34N/2 54.18W VQ; N cardinal mark. Welsh Hook Lt Buoy 51 28.40N/2 52.00W Fl(2)R 5s; port-hand mark; *Bell*.
Avon Lt Buoy 51 27.75N/2 51.65W FlG 2.5s; stbd-hand mark.
Clevedon Pier, detached head FlG 5s 7m 3M.
Walton Bay, Old signal station Fl 2.5s 35m 2M.

Black Nore Point 51 29.05N/2 47.95W Fl(2) 10s 11m **15M**; W round Tr; obsc by Sand point when brg less than 049°; vis 044°-243°.

Newcome Lt Buoy 51 29.94N/2 46.97W Fl(3)R 10s; port-hand mark. Cockburn Lt Buoy 51 30.43N/2 44.00W Fl R 2.5s; port-hand mark; *Bell*.
Portishead Point 51 29.64N/2 46.34W Q(3) 10s 9m **16M**; B framework Tr, W base; vis 060°-262°; *Horn 20s*.

PORTISHEAD. Pier head 51 29.66N/2 45.18W Iso G 2s 5m 3M; W column; *Horn 15s*, sounded from HW−4 to HW+3. Lock, E side 2FR(vert) 7/5m 1M. Lock, West side 2FG(vert) 7/5m 1M. Portbury Wharf. Lts in line 192°. Front 51 29.5N/2 44.1W Oc G 5s 7m 10M; vis 171°-211°. Rear 100m from front Oc G 5s 12m 10M; vis 171°-211°.
Seabank. Lts in line 103°. Front 51 30.0N/2 43.7W Oc(2) 10s 13m 5M; vis 086°-119°. Rear, 150m from front Oc(2) 10s 16m 5M; vis 086°-119°. Royal Portbury Dock 51 30.1N/2 43.6W Fl G 15s 5m 6M. Pier corner Fl G 2s 7m 7M; *Dia 30s*, sounded HW−4 to HW+3.
Knuckle 51 29.92N/2 43.60W Oc G 5s 6m 6M.

AVONMOUTH. Royal Edward Dock, N Pier, head Fl 10s 15m 10M; round Tr; vis 060°-228°. Ldg Lts 184°. Front QG 5m 6M; vis 129°-219°. Rear 220m from front, S pier head 51 30.34N/2 43.02W Oc RG 30s 9m 10M; round Tr; vis R294°-036°, G036°-194°; *Bell 10s*.
King Road Ldg Lts 072°. N pier head. Front 51 30.5N/2 43.0W Oc R 5s 5m 9M; W obelisk, R bands; vis 062°-082°. Rear, 546m from front, QR 15m 10M; B&W striped circle on framework Tr, Y bands; vis 066°-078°. Royal Edward Lock, N side 2FR(vert), S side 2FG(vert). Oil Jetty, head, 51 30.6N/2 42.8W 2FG(vert) 2M. Gypsum effluent pipe Fl Y 3s 3m 2M.

RIVER AVON. Ldg Lts 127°. Front 51 30.05N/2 42.47W FR 7m 3M; vis 010°-160°. Rear 142m from front FR 17m 3M; vis 048°-138°. Monoliths 51 30.2N/2 42.7W FIR 5s 5m 3M; vis 317°-137°. Saint George Ldg Lts 173°, both OcG 5s 1M, on Y columns, synchronised. Nelson Point 51 29.82N/2 42.43W FIR 3s 9m 3M; W mast. Broad Pill 51 29.6N/2 41.8W QY 11m 1M; W framework Tr. Avonmouth Bridge, NE end LFl R 10s, SW end LFl G 10s, both 5m 3M showing up and downstream. From here to City Docks, Oc G Lts are shown on S bank, and R or Y Lts on N bank.

CUMBERLAND BASIN. Entrance, N side 2FR(vert), S side 2FG(vert). Plimsoll Bridge, centre Iso 5s each side.
AVON BRIDGE. N side FR 6m 1M on bridge pier. Centre of span Iso 5s 6m 1M. S side FG 6m 1M on bridge pier.

ENGLAND—WEST COAST—RIVER SEVERN

Bedwin Lt Buoy 51 32.33N/2 43.15W Q(3) 10s; E cardinal mark.
THE SHOOTS. Lower Shoots Bn 51 33.62N/2 42.05W (unlit); stbd-hand mark. Upper Shoots Bn 51 34.20N/2 41.79W (unlit); W cardinal mark. Charston Rock 51 35.32N/2 41.60W Fl 5s 5m 9M; W round Tr, B stripe; vis 203°-049°. Ldg Lts 013° Front F Bu. Rear 320m from front F Bu.

Chapel Rock 51 36.40N/2 39.13W Fl WRG 2.6s 6m W8M, G5M; B framework Tr, W lantern; vis W213°-284°, R284°-049°, W049°-051°, G051°-073°.

RIVER WYE, Wye Bridge, 2F Bu (hor); centre of span.

SEVERN BRIDGE. West Tr 3QR (hor) on upstream and downstream sides; Horn (3) 45s. Centre of span Q Bu, each side. E Tr 3QG (hor) on upstream and downstream sides.

Aust 51 36.1N/2 37.9W 2QG(vert) 11/5m 6M; power cable pylon. Lyde Rock 51 36.9N/2 38.6W QR 5m 5M; B framework Tr, W lantern. Sedbury 2FR(vert) 10m 3M. Slime Road Ldg Lts 210°. Front F Bu 9m 5M; W hut. Rear, 91m from front, F Bu 16m 5M; B framework Tr, W lantern. Inward Rocks Ldg Lts 252°. Front F 6m 6M; B framework Tr. Rear, 183m from front, F 13m 2M; W hut and mast. Sheperdine Ldg Lts 070°. Front F 7m 5M; B framework Tr, W lantern. Rear, 168m from front, F 13m 5M; B framework Tr, W lantern; Bell (26) 60s. Narlwood Rocks Ldg Lts 225°. Front Fl 2s 5m 8M; Y Bn, B lantern. Rear, 198m from front Fl 2s 9m 8M; Y Bn, B lantern. Conigre Ldg Lts 077°. Front F Vi 21m 8M. Rear, 213m from front F Vi 29m 8M. Fishing House Ldg Lts 218°. Front F 5m 2M: W hut and post. Rear F 11m 2M; W hut and mast.

BERKELEY Power Station 3x2FG(vert); Siren (2) 30s. Bull Rock Iso 2s 6m 8M. Lydney Docks, pier head FW or R (tidal); Gong (tidal).
Berkeley Pill Ldg Lts 188°. Front FG 5m 2M. Rear, 152m from front, FG 11m 2M; both B Trs, W lanterns. Panthurst Pill F Bu 6m 1M; W post on concrete hut.

SHARPNESS Docks, S pier head 2FG(vert) 6m 3M; Siren 20s. N Pier 2FR(vert) 6m 3M. Old entrance, S side, Siren 5s (tidal).

ENGLAND—BRISTOL CHANNEL (SOUTH SHORE)

WESTON-SUPER-MARE. Pier head 51 20.85N/2 59.17W 2FG(vert) 6/5m.
South Patches Lt Buoy 51 20.60N/3 03.85W Fl(2) 5s; isolated danger mark; Bell.
East Culver Lt Buoy 51 17.85N/3 11.32W Q(3) 10s; E cardinal mark. West Culver Lt Buoy 51 16.63N/3 19.40W VQ(9) 10s; W cardinal mark. Gore Lt Buoy 51 13.92N/3 09.70W Iso 5s; safe water mark; Bell.

BURNHAM-ON-SEA. Entrance, 51 14.90N/2 59.85W Fl 7.5s 28m 17M; W round Tr, R stripe; vis 074°-164°. Dir Lt 078.5°. Dir F WRG 24m W16M, R12M, G12M; same Tr; vis G073°-077°, W077°-079°, R079°-083°. Seafront Lts in line 112°, moved for changing channel, Front FR 6m 3M. Rear FR 12m 3M; church Tr. Brue 51 13.5N/3 00.2W QR 4m 3M; W mast, R bands. Stert Reach 51 11.3N/3 01.9W Fl 3s 4m 7M; vis 187°-217°.

Hinkley Point, water intake 51 12.9N/3 07.8W 2FG(vert) 7/5m 3M.

WATCHET. West breakwater head 51 11.03N/3 19.67W FG 9m 9M; R 6-sided Tr, W lantern, G cupola. FR Lts on radio masts 1.6M SSW. E pier 2FR(vert) 3M.

MINEHEAD, breakwater head Fl(2)G 5s 4M; vis 127°-262°.

Lynmouth Foreland 51 14.70N/3 47.13W Fl(4) 15s 67m 26M; W round Tr; vis 083°-275° (H24).
LYNMOUTH. River training arm 2FR(vert) 6m 5M. Harbour arm 2FG(vert) 6m 5M.
Sand Ridge Buoy 51 14.95N/3 49.70W (unlit); stbd-hand mark. Copperas Rock Buoy 51 13.75N/4 00.45W (unlit); stbd-hand mark.

ILFRACOMBE. Lantern Hill 51 12.64N/4 06.70W FR; W lantern on chapel. Promenade Pier, N end 3x2FG(vert) (occas).
Horseshoe Lt Buoy 51 15.00N/4 12.90W Q; N cardinal mark.

Bull Point 51 11.95N/4 12.05W Fl(3) 10s 54m 25M; W round Tr. FR 48m 12M; same Tr; vis 058°-096°.

Morte Stone Buoy 51 11.25N/4 14.90W (unlit); stbd-hand mark. Baggy Leap Buoy 51 09.00N/4 16.90W (unlit); stbd-hand mark.

BIDEFORD. Bideford Fairway Lt Buoy 51 05.22N/4 16.20W LFl 10s; safe water mark; Bell. Instow Ldg Lts 118°. Front Oc 6s 22m 15M. Rear 427m from front Oc 10s 38m 15M. Both on W Trs and vis 103°-133°; (H24). Crow Point Fl R 5s 8m 4M; W framework Tr; vis 225°-045°.

LUNDY. Near N point 51 12.07N/4 40.57W Fl(2) 20s 50m 24M; W round Tr; vis 009°-285°.
SE point 51 09.70N/4 39.30W Fl 5s 53m 24M; W round Tr; vis 170°-073°; RC; Horn 25s.

Hartland Point 51 01.3N/4 31.4W Fl(6) 15s 37m 26M; (H24); W round Tr; Horn 60s.
PADSTOW. Stepper Point 50 34.11N/4 56.63W L Fl 10s 12m 4M.

Trevose Head 50 32.93N/5 02.06W Fl 5s 62m 25M; W round Tr.

NEWQUAY. N pier head 2FG(vert) 5m 2M. S pier head 2FR(vert) 4m 2M; round stone Tr.
The Stones Lt Buoy 50 15.60N/5 25.40W Q; N cardinal mark; Bell; Whis. Godrevy Island 50 14.50N/5 23.95W FlWR 10s 37m W12M, R9M; W 8-sided Tr; vis W022°-101°, R101°-145°, W145°-272°. 4 FR(vert) Lts on radio mast 6.5M SE.
HAYLE. Ldg Lts 180°. Front 50 11.5N/5 26.1W 17m 4M. Rear 110m from front. Both F (occas) on pile structures with R&W lanterns.
ST IVES. E pier head 2FG(vert); W pier head 2FR(vert).

Pendeen 50 09.8N/5 40.2W Fl(4) 15s 59m 27M; W round Tr; vis 042°-240°; in bay between Gurnard head and Pendeen it shows to coast; Siren 20s. For Lts further SW see 10.1.4.

10.11.5 PASSAGE INFORMATION

CARDIGAN BAY (charts 1971, 1972, 1973)

Harbs are mostly on a lee shore, and most have bars which make them dangerous to approach in bad weather. In N part of b there are three major dangers to coasting yachts, as are described briefly below. St Patrick's Causeway runs 11M SW from Mochras Pt. It is mostly large loose stones, and dries for much of its length. In strong winds the sea breaks heavily at all states of tide. The outer end is marked by a Lt buoy. At the inner end there is a chan about ½M offshore, which can be taken with care at half tide.

Sarn-y-Bwch runs 4M WSW from Pen Bwch Pt. It is composed of rocky boulders, drying in places 6 ca offshore and with depths of 0.3m extending nearly 3M seaward. There is a buoy off W end. Sarn Cynfelyn and Cynfelyn Patches extend a total of 6½M offshore, with depths of 1.5m in places, from a point about 2M N of Aberystwyth (10.11.10). There is a buoy off outer end. Not quite halfway along the bank is Main Chan, 3 ca wide, running roughly N and S, but not marked.

Firing exercises take place in the S part of Cardigan B (chart 1973). Beware targets and mooring buoys, some unlit.

THE BISHOPS AND THE SMALLS (chart 1478)

If bound N or S along St George's Chan (ie not proceeding to Cardigan B or Milford Haven) the easiest route is W of the Bishops and the Smalls, but note the traffic scheme (see Fig. 2(8)). This is the best route by night but, if bound to/from Milford Haven or S Wales harbours, it is possible to pass inside both the Smalls and Grassholm Island. Even shorter is the route inside the Bishops, passing close W of Ramsey Island, and outside Skomer Island and Skokholm Island. The shortest route of all involves the passage of Ramsey Sound and Jack Sound, but this is not recommended except by daylight, in good weather, and with the right tidal conditions — preferably at neaps. The main features (only) of these various chans are described below, but for full directions refer to the *West Coasts of England and Wales Pilot*, or to *Irish Sea Cruising Guide*.

The Bishops and the Clerks (chart 1482) are islets and rks 2½M W and NW of Ramsey Island, a bird sanctuary SSW of St David's Hd. N Bishop is the N islet of the group, but Bell Rk which has depth of 2.2m lies 3 ca ENE of N Bishop and between them is a ridge with heavy overfalls and tide rips. S Bishop (Lt, fog sig, RC) is the SW islet, with rks extending ½ ca NE and NW. Between S Bishop and Ramsey Island are several dangers, including Daufraich (islet 7 ca NE of S Bishop) with Maen Daufraich (dries) close N of it; Cribog (dries) and Moelyn (dries) respectively 2 and 3½ ca ENE of Daufraich; there are heavy overfalls SE of Moelyn. There is however a chan ¼M W of Ramsey Island, if care is taken to avoid dangers S and close W of Ramsey Island. This chan passes E of Llechau-uchaf (rk 1.5m high) and Llechau-isaf (dries) which are near N and S ends respectively of foul ground with heavy overfalls about 6½ ca W of Ramsey Island. Beware also Carreg Rhoson, with rks extending from it to NE and SW, roughly between N Bishop and S Bishop; Maen Rhoson (rk 9m high) 2 ca NW of Careg Rhoson; and Carreg-trai (dry) 1¾M NE of Carreg Rhoson. Note also Gwahan, a rk 1.5m high, about 4 ca N of Ramsey Island. 2M W of The Bishops the S-going stream begins at HW Milford Haven +0400, and the N-going at HW Milford Haven −0225, sp rates 2 kn. Between The Bishops and Ramsey Island the SW-going stream begins at HW Milford Haven +0330, and the NE-going at HW Milford Haven −0255, sp rates 5 kn.

Ramsey Sound (chart 1482) should be taken at slack water (see also above). The S-going stream begins at HW Milford Haven +0300, and the N-going at HW Milford Haven −0325, sp rates 6 kn at The Bitches where chan is narrowest (2 ca) and decreasing N and S. The Bitches are rks extending 2 ca E from middle of E side of Ramsey Island. Other dangers are Gwahan and Carreg-gafeiliog, respectively on W and E sides of N end of chan: Horse Rk (dries) and associated overfalls about ½M NNE of The Bitches; Shoe Rk (dries) on E side of chan at S end; and rks extending ½M SSE from S end of Ramsey I.

St Brides B provides anch in settled weather or offshore winds, but is a trap in westerlies. Solva is a little harb with shelter for boat capable of taking ground, or anch E of Black Rk off the entrance. See 10.11.24.

If passing outside The Smalls (Lt, fog sig, Racon) beware SW Rk (dries) 3 ca SW of Lt Ho, the only danger W of it. Near The Smalls the S-going stream begins at HW Milford Haven +0515, and the N-going at HW Milford Haven −0045, sp rates 5 kn near the rks but decreasing to 3 kn when 2M S, W or N of them.

As mentioned above, there are various chans inside The Smalls (chart 1478). Hats and Barrels are shallow rky areas respectively about 1¾M and 4M E of The Smalls. They are usually marked by tide rips and overfalls, and by breaking seas in bad weather. The chan between The Smalls and Hats is over 1M wide, but beware E Rk 2½ ca E of Lt Ho. Chan between Hats and Barrels is 2M wide. There are no leading marks for either of these chans.

Grassholm is 7M E of The Smalls, and has no dangers extending more than ½ ca, although there is a race either end and strong tidal eddies so that it is advisable to pass about 1M off. The chan between Barrels and Grassholm is 2½M wide, and here the S-going stream begins at HW Milford Haven +0440, and the N-going at HW Milford Haven −0135, sp rates 5 kn. 5M of clear water lie between Grassholm and Skomer Island/Skokholm Island to the E. But Wildgoose Race, which forms W of Skomer and Skokholm is very dangerous, so it is necessary to keep 2M W of these two islands.

To E of Skomer Island is Midland Island, and between here and Wooltack Pt is Jack Sound (chart 1482) which is only 1 ca wide and should not be attempted without detailed pilotage directions, and only at slack water. Among the dangers which should be identified on the chart are: The Crab Stones, extending E from Midland Island; The Cable, a drying rk on E side of chan; Tusker Rk, steep-to on its W side, off Wooltack Pt; the Black Stones; and The Anvil and other rks off Anvil Pt. In Jack Sound the S-going stream begins at HW Milford Haven +0200, and the N-going at HW Milford Haven −0425, sp rates 6 kn.

BRISTOL CHANNEL (chart 1179, 1076, 1165, 1164)

Sailing E from Milford Haven, along N shore of Bristol Chan, there are few significant dangers to St Gowan's Hd. Crow Rk (dries) is 5½ ca SSE of Linney Hd, and The Toes are dangerous submerged rocks close W and SE of Crow Rk. There is a passage inshore off these dangers. There are overfalls on St Gowan Shoals which extend 4M SW of St Gowan's Hd, and the sea breaks on the shallow patches in bad weather.

Caldy Island (Lt) lies S of Tenby (10.11.13). Off its NW pt is St Margaret's Island connected by a rky reef. Caldy Sound is between St Margaret's Island and Giltar Pt (chart 1482). It is buoyed, but beware Eel Spit near W end of Caldy Island where there can be a nasty sea with wind against tide, and Woolhouse Rks (dry) 1¼ ca NE of Caldy Island.

Carmarthen B (10.11.24) has no offshore dangers for yachts, other than the charted sands at head of b and on its E side off Burry Inlet (10.11.14). There is a harb (dries) at Saundersfoot 2M N of Tenby, with anch off well sheltered from N and W but subject to swell. Streams are weak here. See 10.11.24.

Helwick Sands extend 7M W from Port Eynon Pt. Near their W end are depths of 1.8 m. Stream sets across the sands. There is a narrow chan inshore, close to Port Eynon Pt. Between here and Mumbles Hd the stream runs roughly along coast, sp rates 3 kn off pts, but there are eddies in Port Eynon B and Oxwich B, and overfalls off Oxwich Pt.

Off Mumbles Hd (Lt, fog sig) beware Mixon Shoal (dries), marked by buoy. In good conditions pass N of shoal, 1 ca off Mumbles Hd. Anch N of Mumbles Hd, good holding but exposed to swell. Green Grounds, rky shoals, cover W side of b in approach to Swansea (10.11.15).

Scarweather Sands, much of which dry and where sea breaks heavily, extend 7M W from Porthcawl (10.11.24) and are marked by buoys (chart 1161). There is a chan between the sands and coast to E, but beware Hugo Bank (dries) and rky patches with overfalls up to ¾M offshore between Sker Pt and Porthcawl.

Nash Sands extend 7½M WNW from Nash Pt. Depths vary and are least at inshore end (dries), but there is chan 1 ca wide between sand and rky ledge off Nash Pt. On E-going stream there are heavy overfalls off Nash Pt and at W end of Nash Sands. Between Nash Pt and Breaksea Pt the E-going stream begins at HW Avonmouth +0535, and the W-going at HW Avonmouth −0035, sp rates 3 kn. Off Breaksea Pt there may be overfalls.

From Rhoose Pt to Lavernock Pt the coast if fringed with foul ground. Lavernock Spit extends 1¾M from Lavernock Pt, and E of the spit is main chan to Cardiff (10.11.17), the other side of the chan being Cardiff Grounds − a drying bank which runs parallel with the shore and about 1½M from it. Further offshore is Monkstone Rk (dries), and E of this runs the buoyed chan to Avonmouth and Bristol (10.11.19) S of extensive drying banks along the N shore (chart 1176).

Near the centre of Bristol Chan, and close to main fairway are the Is of Flat Holm (Lt, fog sig) and Steep Holme. 7M SW of Flat Holm lies Culver Sand (dries), 4M in length, marked by buoys. Between Flat Holm and Steep Holme the E-going stream begins at HW Avonmouth −0610, sp rate 3 kn, and the W-going at HW Avonmouth +0015, sp rate 4 kn.

W from Burnham-on-Sea (10.11.20), the S shore of Bristol Chan has fewer obstructions than N shore. There is also less shelter since harbs such as Watchet, Minehead, Porlock Weir or Watermouth dry out. See 10.11.24. In bad weather dangerous overfalls occur NW and NE of Foreland Pt. 5M to W there is a race off Highveer Pt. Between Ilfracombe (10.11.21) and Bull Pt the E-going stream begins at HW Milford Haven +0540, and the W-going at HW Milford Haven −0025, sp rates 3 kn. Overfalls occur up to 1½M N of Bull Point and over Horseshoe Rks, which lie 3M N. There is a dangerous race off Morte Pt, 1½M to W of Bull Pt.

NORTH CORNWALL (charts 1149, 1156)

Approach to Bristol Chan along N coast of Cornwall is very exposed, with little shelter in bad weather. Padstow is a refuge, but in strong NW winds the sea breaks on bar and prevents entry. Shelter is available under lee of Lundy Island; but there are bad races to NE (White Horses), the NW (Hen and Chickens), and to SE; also overfalls over NW Bank. St Ives (dries) is sheltered from E and S, but exposed to N. So in this area yachts need to be sturdy and well equipped, since if bad weather develops no shelter may be at hand. Streams are moderate W of Lundy, but strong round the island. They get much stronger towards Bristol Chan proper.

Proceeding SW from Taw/Torridge (10.11.22), keep 3M off to avoid the race N of Hartland Pt (Lt, fog sig). There is shelter off Clovelly in winds S-SW. Bude Haven dries, and is not approachable in W winds; but it is accessible in calm weather or offshore winds. Near HW a chan leads to canal lock, marked by Ldg marks at 075° and 131°. See 10.11.24.

Boscastle is a tiny harb (dries) 3M NE of Tintagel Hd. Only approach in good weather or offshore winds: anch off or (if room) dry out alongside.

For dangers in Padstow approaches see 10.11.23. Off Trevose Hd (Lt) beware Quies Rks which extend 1M to W. From here S the coast is relatively clear to Godrevy Island, apart from Bawden Rks 1M N of St Agnes Hd. Newquay B (10.11.24) is good anch in offshore winds, and the harb (dries) is sheltered but uncomfortable in N winds. Off Godrevy Island (Lt) are the Stones, extending 1M from island and marked by buoy.

In St Ives B (chart 1168), Hayle (dries) is a commercial port: seas break heavily on bar in strong onshore winds. Stream is strong, so enter just before HW. The bottom is mostly sand. St Ives (dries) gives sheltered anch outside in winds from E to SW, but very exposed to N; there is sometimes room to dry out alongside. See 10.11.24.

From St Ives to Land's End coast is rugged and exposed. There are overfalls SW of Pendeen Pt (Lt, fog sig). Vyneck Rks lie awash about 3 ca NW of C Cornwall. The Brisons are two rocky islets ½M SW of C Cornwall, and rocky ledges extend inshore and to the S and SW. The Longships (Lt, fog sig) are a group of rks about 1M W of Land's End, with ledges 2 ca further seaward. The inshore passage is unwise.

For continuation south, see 10.1.5.

10.11.6 DISTANCE TABLE

Approximate distances in nautical miles are by the most direct route while avoiding dangers and allowing for traffic separation schemes etc. Places in *italics* are in adjoining areas.

1	Bardsey Island	1																		
2	*Abersoch*	14	2																	
3	*Aberdovey*	31	26	3																
4	*Aberystwyth*	33	30	10	4															
5	Fishguard	45	54	47	40	5														
6	South Bishop	60	70	67	61	25	6													
7	Milford Haven	83	93	90	84	48	23	7												
8	Tenby	106	116	113	107	71	46	28	8											
9	Swansea	129	139	136	130	94	69	55	36	9										
10	Barry	151	161	158	152	116	91	77	57	37	10									
11	Newport	168	178	175	169	133	108	94	75	54	17	11								
12	Sharpness	191	201	198	192	156	131	117	106	75	39	32	12							
13	Avonmouth	174	184	181	175	139	114	100	89	58	22	15	17	13						
14	Burnham-on-Sea	168	178	175	169	133	108	94	70	48	18	28	50	33	14					
15	Minehead	148	158	155	149	113	88	76	55	32	14	29	51	34	20	15				
16	Ilfracombe	127	137	134	128	92	67	53	35	25	35	51	74	57	45	25	16			
17	Lundy Island	110	120	117	111	75	50	38	30	37	54	71	95	78	66	46	22	17		
18	*Longships*	168	178	175	169	133	108	105	110	120	130	146	169	152	140	121	95	82	18	
19	*Tuskar Rock*	63	75	84	80	48	36	59	82	105	127	144	167	150	144	124	103	85	130	19
20	*Dun Laoghaire*	61	75	92	94	94	90	113	136	159	181	198	221	204	198	178	157	140	199	70	20

AREA 11—S Wales 503

PORTHMADOG 10-11-7
Gwynedd

CHARTS
Admiralty 1971, 1512; Imray C61; OS 124
TIDES
−0310 Dover; ML 2.8; Duration 0455; Zone 0 (GMT).
Standard Port MILFORD HAVEN (→)

Times				Height (metres)			
HW		LW		MHWS	MHWN	MLWN	MLWS
0100	0800	0100	0700	7.0	5.2	2.5	0.7
1300	2000	1300	1900				

Differences PORTHMADOG
+0235 +0210 No data −1.9 −1.8 No data
Differences on Liverpool −0245

SHELTER
Inner Harbour — Good all year round. Outer Harbour — Summer only and exposed to S winds.
NAVIGATION
Waypoint Fairway (safe water) buoy, LFl 10s, 52°53′.58N 04°11′.55W (channel shifts). Depth at Bar MLWS is only 0.8m. When wind is in SW, waves are steep sided and close, especially on the ebb tide. Bar changes frequently. Up to date situation is supplied by Hr Mr upon request. Advise entering HW∓1½.
LIGHTS AND MARKS
Fairway buoy RW L Fl 10s. Remainder of channel markers (16) have reflective top marks, G to stbd, R to port.
RADIO TELEPHONE
Madoc YC: VHF Ch 16 M. Hr Mr 16; 12 (by prior request).
TELEPHONE (0766)
Hr Mr 512927; CG Nefyn 720204; MRSC Holyhead 2051; Pilot 513545; Harbour Authority Dwyfor District Council, 613131; HMC Freefone Customs Yachts: RT (0407) 2714; Marinecall 0898 500 460; Dr 512239.
FACILITIES
EC Wednesday; **Harbour** (265) Tel. 512927, D, FW, C, Slip, BH (14 ton); **Pen-y-Cei** Slip, V, R, Bar; **Madoc YC** Tel. 512976, M, L, FW, AB, Bar; **Glaslyn Marine Supplies** Tel. 513545, CH, ACA; **Robert Owen** Tel. 513435, D, ME; **W.A. Jones** Tel. 513545, Sh; **Kyffin BY** El, C (2½ ton), CH, AB; **Madog BY** Tel. 513435 AB, C (8 ton mobile); **P and T Sailing Club** Tel. 513546, Slip, M, FW; **Harbour Filling Station** P (45 gall drums). **Town** PO; Bank; Rly; Air (Chester).

BARMOUTH 10-11-8
Gwynedd

CHARTS
Admiralty 1484, 1971; Imray C61; OS 124
TIDES
−0250 Dover; ML 2.8; Duration 0515; Zone 0 (GMT).
Standard Port MILFORD HAVEN (→)

Times				Height (metres)			
HW		LW		MHWS	MHWN	MLWN	MLWS
0100	0800	0100	0700	7.0	5.2	2.5	0.7
1300	2000	1300	1900				

Differences BARMOUTH
+0215 +0205 +0310 +0320 −2.0 −1.7 −0.7 0.0

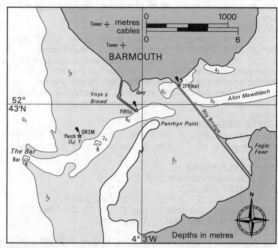

SHELTER
Good shelter. Entry HW∓2½ safe except in strong on-shore winds; entry impossible with strong SW winds. The river (Afon Mawddach) and estuary are tidal and river is navigable for about 7 miles above railway bridge. It is not buoyed and sandbanks move constantly. Local knowledge is essential. Clearance under railway bridge approx 5.5m. Exposed anchorage W of Barmouth Outer buoy in 6 to 10m. In harbour secure to mooring as directed by Hr Mr on account of submarine cables and strong tidal streams. A quay, which dries at half-tide, fronts the town.
NAVIGATION
Waypoint, Barmouth Outer (safe water) buoy, L Fl 10s, 52°42′.60N 04°04′.75W, 253°/073° from/to Y perch Lt, QR, 0.7M. Approach from SW between St Patrick's Causeway (Sarn Badrig) and Sarn-y-Bwch (see 10.11.5). Barmouth can be identified by Cader Idris, a mountain 890m high, 5M ESE. Fegla Fawr, a rounded hill, lies on S side of harbour. Bar lies 0.75M W of Penrhyn Pt, min depth 0.3m but subject to considerable change. Channel marked by two port-hand buoys, moved as necessary. Tidal stream runs 3 to 5 kn on ebb at springs.
LIGHTS AND MARKS
Y perch, QR 4m 2M marks S end of stony ledge extending 3 ca SW from Ynys y Brawd across North Bank. Ynys y Brawd groyne, SE end, marked by Bn with Lt, Fl R 5s. NW end of Rly bridge 2 FR(hor).
RADIO TELEPHONE (local times)
Call *Barmouth Harbour* VHF Ch 16; 10 (0900-1700).
TELEPHONE (0341)
Hr Mr 280671; CG 280176; HMC Freefone Customs Yachts: RT (0407) 2336; Marinecall 0898 500 460; Dr 280521.
FACILITIES
EC Wednesday; There are some visitors' moorings; **Quay** D, FW, El, AC, Slip; **J. Stockford** Tel. 280742, Slip, M, D, L; **Seafarer Chandlery** Tel. 280978, CH, ACA; **Barmouth YC** Tel. 280000; **Marine Stores** Tel. 280742, Slip, M; **Merioneth YC** Tel. 280000; **Town** P, D, V, R, Bar, PO, Bank, Rly, Air (Chester); Ferry across to Penrhyn Pt.

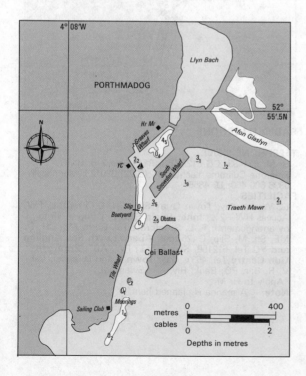

ABERDOVEY 10-11-9
Gwynedd

CHARTS
Admiralty 1484, 1972; Imray C61; OS 135
TIDES
−0320 Dover; ML 2.6; Duration 0535; Zone 0 (GMT).
Standard Port MILFORD HAVEN (→)

Times				Height (metres)			
HW		LW		MHWS	MHWN	MLWN	MLWS
0100	0800	0100	0700	7.0	5.2	2.5	0.7
1300	2000	1300	1900				

Differences ABERDOVEY
+0215 +0200 +0230 +0305 −2.0 −1.7 −0.5 0.0

SHELTER
Good except in strong W to SW winds. Berth alongside jetty.
NAVIGATION
Waypoint Aberdovey Outer (safe water) buoy, 52°31′.75N 04°06′.22W, 250°/070° from/to jetty, 2.3M (channel shifts). Bar is hazardous below ½ tide and is constantly changing position. Visitors are advised to contact Hr Mr before entering, or telephone the Pilot. There are no lights. Submarine cables (prohibited anchorages) marked by beacons with R diamond top marks. There is heavy silting E of jetty.
LIGHTS AND MARKS
No lights or marks. Buoys in the channel are all G conical and are left to stbd — Bar, S Spit and Inner, leading to jetty.
RADIO TELEPHONE
Call *Aberdovey Harbour* VHF Ch 12; 16.
TELEPHONE (065 472)
Hr Mr 626; CG 327; MRSC Dale 218; Pilot 247; HMC Freefone Customs Yachts: RT (0407) 2714 Ext 262; Marinecall 0898 500 460; Dr Tywyn 238.
FACILITIES
EC Wednesday (winter only); **Jetty** L, FW, AB; **Wharf** Slip, L, FW, C, AB; **West Wales Marina BY** Tel. 478, ME, El, Sh, CH; **Frongoch BY** Tel. 644, Slip (Patent 10 ton), ME, El, Sh, CH; **Dovey YC** Slip, L, FW; **Dovey Marine** Tel. 581, CH, ACA. **Town** P and D (in cans), ME, El, CH, V, R, Bar. PO; Bank; Rly; Air (Chester).

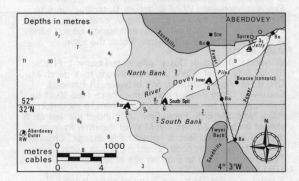

ABERYSTWYTH 10-11-10
Dyfed

CHARTS
Admiralty 1484, 1972; Imray C61; OS 135
TIDES
−0330 Dover; ML 2.7; Duration 0540; Zone 0 (GMT).
Standard Port MILFORD HAVEN (→)

Times				Height (metres)			
HW		LW		MHWS	MHWN	MLWN	MLWS
0100	0800	0100	0700	7.0	5.2	2.5	0.7
1300	2000	1300	1900				

Differences ABERYSTWYTH
+0145 +0130 +0210 +0245 −2.0 −1.7 −0.7 0.0
NEW QUAY
+0150 +0125 +0155 +0230 −2.1 −1.8 −0.6 −0.1
ABERPORTH
+0135 +0120 +0150 +0220 −2.1 −1.8 −0.6 −0.1
PORT CARDIGAN
+0140 +0120 +0220 +0130 −2.3 −1.8 −0.5 0.0

SHELTER
Good but harbour dries. Lie alongside quays. Advise entering between HW∓2; dangerous in strong on-shore winds. Visitors berth halfway up Town Quay, marked 'Visiting Yachts only' — Access HW∓2½.
NAVIGATION
Waypoint 52°24′.80N 04°06′.00W, 318°/138° from/to front Ldg Lt 138°, 0.62M. Beware Castle Rocks when approaching from N. Narrow entrance with right-angle turn inside the pier head.
LIGHTS AND MARKS
Harbour located by Pen-y-Dinas, conspic hill 120m high, with Wellington monument, which lies to S of entrance. The head of the N breakwater in line with Wellington monument bearing 140° leads S of Castle Rock. Leading lights 138°, both FR on Ystwyth Bridge; white markers for daytime. N jetty Q WR 4M, R sector covering Castle Rock. S jetty Fl(2) WG 10s 10M.

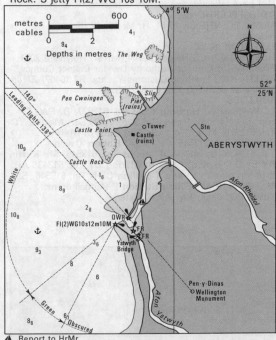

RADIO TELEPHONE
None.
TELEPHONE (0970)
Hr Mr 611433; CG 612220; MRSC Dale 218; HMC Freefone Customs Yachts: RT (0222) 399123; Marinecall 0898 500 460; Dr 4855.
FACILITIES
EC Wednesday; **Town Quay** Slip, CH, C (3 ton), L, FW, Access HW∓2½; **Inner Basin** Slip, L, C (up to 15 ton by arrangement); **F. L. Steelcraft** Tel. Borth 713, CH, D, ME, Sh, M, Slip; C (25 ton); **Aberystwyth Sea Angling and YC** Tel. 612158, Slip, M*, L, FW, AB*; **Primrose Gdn Centre** Tel. 4631, Gas; **Town** P and D (cans), CH, V, R, Bar. PO; Bank; Rly; Air (Swansea).
*Apply to Hr Mr.
Note: − A marina is planned here.

FISHGUARD 10-11-11
Dyfed

CHARTS
Admiralty 1484, 1973; Imray C61/60; OS 157
TIDES
−0400 Dover; ML 2.5; Duration 0550; Zone 0 (GMT).
Standard Port MILFORD HAVEN (→)

Times				Height (metres)			
HW		LW		MHWS	MHWN	MLWN	MLWS
0100	0800	0100	0700	7.0	5.2	2.5	0.7
1300	2000	1300	1900				

Differences FISHGUARD
+0115 +0100 +0110 +0135 −2.2 −1.8 −0.5 +0.1
CARDIGAN TOWN
+0220 +0150 No data −2.2 −1.6 No data
PORTHGAIN
+0055 +0045 +0045 +0100 −2.5 −1.8 −0.6 0.0
RAMSEY SOUND
+0030 +0030 +0030 +0030 −1.9 −1.3 −0.3 0.0
SOLVA
+0015 +0010 +0035 +0015 −1.5 −1.0 −0.2 0.0
LITTLE HAVEN
+0010 +0010 +0025 +0015 −1.1 −0.8 −0.2 0.0
MARTIN'S HAVEN
+0010 +0010 +0015 +0015 −0.8 −0.5 +0.1 +0.1

SHELTER
Good shelter except in strong winds between NE & NW and can be entered at any time.
NAVIGATION
Waypoint 52°01'.00N 04°57'.50W, 057°/237° from/to Northern Breakwater Lt, 0.48M. The harbour can be entered in any weather. The wreck and associated buoys in the entrance have been removed.
Beware large swell against quay near station especially in strong N winds, 340°-010°.
LIGHTS AND MARKS
N Breakwater FlG 4.5s 18m 13M, Bell 8s. E Breakwater FlR 3s 10m 5M. Lts in line 282°, both FG; W diamonds on masts.
RADIO TELEPHONE
VHF Ch 16; 14 (H24). Fishguard Yacht & Boat Co and Goodwick Marina Ch M.
TELEPHONE (0348)
Hr Mr 872247; MRSC Dale 218; HMC Freefone Customs Yachts: RT (0222) 399123; Marinecall 0898 500 460; Dr 872802.
FACILITIES
EC Wednesday; **Goodwick Marina** Tel. 874590 D, AC, CH; **No 3 Berth** L, FW, C (6 ton); **No 4 Berth** L, FW, C (6 ton); **Lower Fishguard** Slip; **Fishguard Marine** Tel. 873377, Slip, ME, Sh, CH; **Harbour** Tel. 872881, Slip (small S end), M (see Hr.Mr.); **Fishguard Bay YC** Tel. 872866, Slip, M, L, FW, Bar; **Fishguard Yacht and Boat Co.** Tel. 873377, BY, ACA; **Fishguard Harbour Garage** Tel. 873814, Gas; **Town** P, D, CH, V, R, Bar. PO; Bank; Rly; Air (Swansea).

AGENTS NEEDED
There are a number of ports where we need agents, particularly in France.
ENGLAND   Swale, Havengore, Berwick.
SCOTLAND Firth of Forth, Scrabster, Mallaig, Loch Sunart, Loch Aline.
IRELAND   Kilrush, Wicklow, Westport/Clew Bay.
FRANCE   Arcachon, Seudre R, Ile d'Oleron, Rochfort, Ile de Re, St. Giles-Croix-de-Vie, Ile d'Yeu, Pouliguen, Le Croisic, La Forêt, Ile de Bréhat.
GERMANY   Norderney, Dornumer-Accumersiel.
If you are interested in becoming our agent for any of the above, please write to the editors and get your free copy of the Almanac every year. You do not have to be resident in a port to be the agent, but at least a fairly regular visitor.

MILFORD HAVEN 10-11-12
Dyfed

CHARTS
Admiralty 1478, 3274, 3275, 2878; Imray C60; OS 157

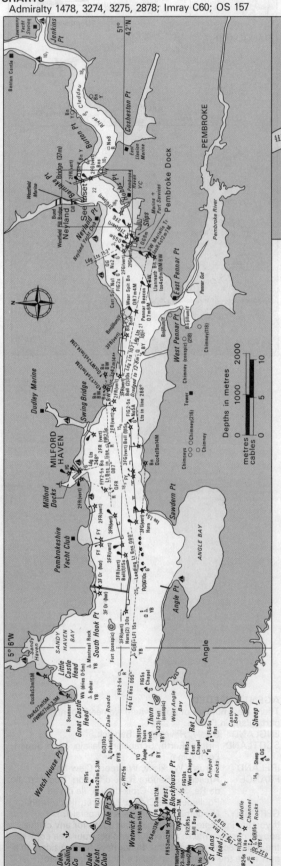

TIDES
−0500 Dover; ML 3.8; Duration 0605; Zone 0 (GMT).
Standard Port MILFORD HAVEN (→)

Times				Height (metres)			
HW		LW		MHWS	MHWN	MLWN	MLWS
0100	0800	0100	0700	7.0	5.2	2.5	0.7
1300	2000	1300	1900				

Differences SKOMER IS
−0005 −0005 +0005 +0005 −0.4 −0.1 0.0 0.0
DALE ROADS
−0005 −0005 −0008 −0008 0.0 0.0 0.0 −0.1
NEYLAND
+0002 +0010 0000 0000 0.0 0.0 0.0 0.0
HAVERFORDWEST
+0010 +0025 No data −4.8 −4.9 Dries out

NOTE: Milford Haven is a Standard Port and all tidal predictions for the year are given below.

SHELTER
Very good shelter in various places round the harbour, especially in Westfield Marina at Neyland. Anchorages – Dale Bay; Off Ellen's Well and Angle Point on S shore; Off Scotch Bay by Milford, above the town; and many others above Pembroke Dock. River Cleddau is navigable up to Haverfordwest for boats with moderate draughts. Clearance under Cleddau Bridge above Neyland is 37m; under power cable 1M upstream, 25m. Check low headroom under bridges and cables approaching Haverfordwest. Contact Milford Haven Signal Station to ascertain most suitable anchorage or berth. Majority of moorings are 'all-tide afloat'. It is possible to dry out safely, depending on weather conditions at inshore areas of Dale, Sandy Haven and Angle Bay.

NAVIGATION
Waypoint Mid Channel Rocks W cardinal buoy, Q(9) 15s, 51°40'.09N 05°10'.04W, 222°/042° from/to Mid Channel Rocks Lt Ho, 0.27M. There is considerable tidal set across the entrance to the Haven particularly at springs. In bad weather avoid passing over Mid Channel Rocks in entrance, where a confused sea and swell will be found. Give St Ann's Head a wide berth, as there is a confused sea there in bad weather. Beware large tankers entering and leaving the haven.
NOTE: – The Port Authority is the Milford Haven Port Authority with jetty, Signal Station and offices at Hubberston Point. Their launches have green hulls and white upperworks and fly a Pilot flag (white over red horizontal) when on pilotage duties or a blue flag with the word 'Harbourmaster' in white letters while on patrol. Their instructions must be obeyed. No vessel may operate within 100 metres of any terminal or any tanker whether at anchor or under way. Milford Dock opens from HW−2 to HW and berths may be had on application to the Dockmaster (Tel. 2275).

⚓ Contact Milford Haven Signal Station

WALES — MILFORD HAVEN

Lat 51°42' N Long 5°01' W

TIMES AND HEIGHTS OF HIGH AND LOW WATERS

YEAR 1989

TIME ZONE UT(GMT)
For Summer Time add ONE hour in non-shaded areas

JANUARY

Time	m	Time	m
1 Su 0601 1212 1842	2.6 5.3 2.6	**16** M 0024 0645 1300 1927	5.7 2.1 5.8 2.2
2 M 0052 0712 1319 1955	5.1 2.7 5.2 2.6	**17** Tu 0137 0809 1416 2049	5.5 2.3 5.6 2.2
3 Tu 0205 0825 1432 2104	5.2 2.6 5.3 2.5	**18** W 0257 0931 1536 2202	5.6 2.1 5.7 2.1
4 W 0311 0931 1535 2202	5.4 2.4 5.5 2.2	**19** Th 0410 1038 1642 2259	5.9 1.9 5.9 1.8
5 Th 0406 1026 1630 2251	5.8 2.1 5.9 1.9	**20** F 0509 1133 1736 2349	6.2 1.6 6.2 1.5
6 F 0455 1115 1718 2337	6.2 1.7 6.2 1.6	**21** Sa 0556 1218 1819	6.5 1.3 6.4
7 Sa 0542 1201 1804 ●	6.5 1.4 6.5	**22** Su 0031 0636 1257 1857	1.3 6.7 1.2 6.6
8 Su 0021 0625 1246 1849	1.3 6.8 1.1 6.7	**23** M 0107 0713 1333 1933	1.2 6.8 1.1 6.6
9 M 0104 0709 1330 1933	1.0 7.1 0.8 6.9	**24** Tu 0141 0747 1405 2005	1.1 6.8 1.1 6.6
10 Tu 0147 0754 1413 2016	0.9 7.2 0.7 6.9	**25** W 0212 0819 1434 2036	1.2 6.7 1.2 6.5
11 W 0229 0837 1458 2100	0.9 7.2 0.7 6.8	**26** Th 0242 0849 1503 2105	1.3 6.6 1.4 6.3
12 Th 0312 0922 1542 2145	1.0 7.1 0.9 6.6	**27** F 0311 0919 1532 2136	1.5 6.4 1.6 6.1
13 F 0357 1009 1627 2231	1.2 6.8 1.2 6.3	**28** Sa 0341 0950 1602 2209	1.7 6.1 1.9 5.8
14 Sa 0444 1058 1718 2323 ☽	1.5 6.5 1.5 6.0	**29** Su 0414 1024 1637 2245	2.0 5.8 2.2 5.5
15 Su 0539 1154 1815	1.8 6.1 1.9	**30** M 0454 1105 1720 2334 ☾	2.4 5.5 2.5 5.2
		31 Tu 0549 1200 1828	2.7 5.1 2.8

FEBRUARY

Time	m	Time	m
1 W 0045 0714 1321 2006	5.0 2.8 5.0 2.8	**16** Th 0246 0927 1534 2156	5.2 2.4 5.3 2.3
2 Th 0216 0849 1456 2129	5.1 2.7 5.2 2.5	**17** F 0407 1035 1638 2254	5.7 2.0 5.7 1.9
3 F 0336 1002 1609 2231	5.5 2.2 5.6 2.0	**18** Sa 0459 1125 1725 2337	6.1 1.6 6.1 1.5
4 Sa 0435 1058 1704 2322	6.0 1.7 6.1 1.5	**19** Su 0542 1204 1803	6.5 1.3 6.4
5 Su 0526 1147 1751	6.6 1.1 6.6	**20** M 0014 0618 1238 1836 ○	1.2 6.7 1.1 6.6
6 M 0007 0611 1234 1835 ●	1.0 7.0 0.7 7.0	**21** Tu 0046 0650 1307 1907	1.0 6.8 1.0 6.7
7 Tu 0050 0655 1316 1917	0.7 7.4 0.4 7.2	**22** W 0116 0720 1335 1935	1.0 6.9 0.9 6.7
8 W 0133 0737 1358 1959	0.4 7.6 0.2 7.3	**23** Th 0144 0749 1402 2004	0.9 6.8 1.0 6.7
9 Th 0213 0819 1439 2040	0.4 7.6 0.3 7.2	**24** F 0211 0816 1429 2032	0.9 6.7 1.1 6.5
10 F 0253 0901 1518 2119	0.5 7.4 0.5 6.9	**25** Sa 0239 0843 1456 2058	1.2 6.5 1.3 6.3
11 Sa 0334 0943 1559 2202	0.8 7.0 1.0 6.5	**26** Su 0307 0911 1522 2127	1.4 6.3 1.6 6.0
12 Su 0414 1027 1641 2247 ☾	1.4 6.5 1.5 6.0	**27** M 0335 0941 1550 2159	1.8 6.0 2.0 5.7
13 M 0504 1118 1733 2344	1.8 5.9 2.1 5.5	**28** Tu 0409 1016 1627 2241 ☾	2.1 5.6 2.3 5.3
14 Tu 0607 1225 1848	2.3 5.4 2.5		
15 W 0104 0749 1357 2033	5.2 2.5 5.1 2.6		

MARCH

Time	m	Time	m
1 W 0455 1106 1722 2347	2.5 5.2 2.7 5.0	**16** Th 0039 0737 1341 2019	5.0 2.7 4.9 2.7
2 Th 0614 1229 1910	2.8 4.9 2.9	**17** F 0232 0915 1521 2139	5.1 2.4 5.2 2.4
3 F 0130 0815 1423 2101	4.9 2.7 5.0 2.6	**18** Sa 0349 1017 1619 2233	5.6 1.9 5.6 1.9
4 Sa 0310 0938 1548 2209	5.4 2.2 5.5 2.0	**19** Su 0437 1102 1701 2312	6.0 1.5 6.1 1.5
5 Su 0414 1038 1644 2301	6.0 1.5 6.2 1.3	**20** M 0516 1137 1736 2347	6.4 1.3 6.4 1.2
6 M 0505 1127 1730 2347	6.7 0.9 6.8 0.7	**21** Tu 0550 1208 1807	6.6 1.1 6.6
7 Tu 0550 1212 1812 ●	7.2 0.4 7.2	**22** W 0017 0621 1236 1836 ○	1.0 6.8 1.0 6.7
8 W 0029 0632 1255 1855	0.4 7.6 0.1 7.5	**23** Th 0046 0649 1304 1904	0.9 6.8 0.9 6.8
9 Th 0110 0714 1334 1934	0.1 7.8 0.0 7.5	**24** F 0114 0717 1331 1933	0.9 6.8 1.0 6.8
10 F 0151 0755 1413 2015	0.1 7.7 0.1 7.4	**25** Sa 0141 0744 1357 1959	1.0 6.7 1.1 6.7
11 Sa 0230 0836 1451 2053	0.3 7.5 0.5 7.1	**26** Su 0209 0812 1423 2026	1.1 6.6 1.3 6.5
12 Su 0310 0917 1531 2134	0.7 7.0 1.0 6.6	**27** M 0237 0839 1450 2056 ☾	1.3 6.3 1.5 6.2
13 M 0350 1000 1612 2217	1.2 6.4 1.6 6.0	**28** Tu 0307 0910 1519 2128	1.7 6.0 1.9 5.9
14 Tu 0437 1049 1701 2313 ☽	1.8 5.7 2.2 5.4	**29** W 0342 0948 1556 2212	2.0 5.7 2.2 5.5
15 W 0543 1158 1817	2.4 5.1 2.7	**30** Th 0431 1040 1654 2320 ☽	2.4 5.2 2.6 5.1
		31 F 0551 1205 1836	2.6 4.9 2.8

APRIL

Time	m	Time	m
1 Sa 0102 0747 1357 2030	5.1 2.5 5.0 2.5	**16** Su 0308 0939 1542 2155	5.5 2.0 5.5 2.0
2 Su 0237 0910 1519 2139	5.5 2.0 5.6 1.9	**17** M 0359 1023 1624 2237	5.9 1.7 5.9 1.7
3 M 0343 1010 1616 2233	6.1 1.3 6.3 1.2	**18** Tu 0440 1059 1701 2312	6.2 1.4 6.2 1.4
4 Tu 0435 1059 1702 2320	6.8 0.7 6.8 0.7	**19** W 0515 1133 1733 2344	6.4 1.2 6.5 1.2
5 W 0522 1144 1746	7.3 0.3 7.2	**20** Th 0546 1203 1804	6.5 1.1 6.6
6 Th 0004 0607 1228 1828 ●	0.3 7.6 0.1 7.5	**21** F 0014 0617 1232 1834 ○	1.1 6.6 1.1 6.7
7 F 0046 0649 1309 1909	0.1 7.7 0.1 7.5	**22** Sa 0045 0646 1300 1902	1.0 6.6 1.1 6.7
8 Sa 0127 0731 1349 1949	0.1 7.6 0.2 7.4	**23** Su 0114 0716 1330 1933	1.1 6.6 1.2 6.6
9 Su 0209 0813 1429 2030	0.4 7.2 0.6 7.0	**24** M 0145 0747 1359 2002	1.2 6.5 1.3 6.5
10 M 0250 0856 1508 2112	0.8 6.7 1.2 6.5	**25** Tu 0218 0819 1430 2036	1.4 6.3 1.5 6.3
11 Tu 0332 0941 1549 2159	1.3 6.1 1.7 6.0	**26** W 0253 0854 1505 2115	1.6 6.0 1.8 6.0
12 W 0421 1031 1640 2257 ☽	1.9 5.5 2.3 5.5	**27** Th 0334 0938 1549 2206	1.9 5.7 2.1 5.7
13 Th 0527 1139 1754	2.4 5.0 2.7	**28** F 0428 1037 1651 2315 ☽	2.1 5.3 2.4 5.4
14 F 0015 0710 1312 1944	5.1 2.6 4.8 2.7	**29** Sa 0546 1157 1819	2.3 5.1 2.5
15 Sa 0154 0839 1443 2103	5.2 2.4 5.1 2.4	**30** Su 0039 0719 1326 1954	5.4 2.2 5.3 2.2

Chart Datum: 3.71 metres below Ordnance Datum (Newlyn)

WALES — MILFORD HAVEN

Lat 51°42′ N   Long 5°01′ W

TIMES AND HEIGHTS OF HIGH AND LOW WATERS

YEAR 1989

TIME ZONE UT(GMT)
For Summer Time add ONE hour in non-shaded areas

	MAY			JUNE			JULY			AUGUST		
	Time m	Time m		Time m	Time m		Time m	Time m		Time m	Time m

MAY

1 M 0201 5.7 0834 1.7 1442 5.7 2103 1.8
16 Tu 0305 5.6 0934 2.0 1536 5.6 2150 2.0
2 Tu 0307 6.2 0935 1.3 1539 6.2 2159 1.3
17 W 0353 5.9 1016 1.7 1619 5.9 2231 1.7
3 W 0403 6.7 1027 0.9 1630 6.7 2249 0.8
18 Th 0434 6.1 1054 1.5 1655 6.2 2308 1.5
4 Th 0452 7.1 1116 0.6 1718 7.0 2337 0.5
19 F 0511 6.2 1129 1.4 1730 6.4 2343 1.4
5 F 0540 7.3 1201 0.4 1803 7.2
20 Sa 0546 6.4 1201 1.3 1804 6.5
6 Sa 0022 0.4 0627 7.3 1245 0.4 1846 7.3
21 Su 0018 1.3 0619 6.4 1235 1.3 1838 6.6
7 Su 0107 0.4 0712 7.2 1328 0.6 1931 7.1
22 M 0053 1.2 0655 6.4 1309 1.3 1913 6.6
8 M 0152 0.7 0757 6.9 1411 0.9 2015 6.9
23 Tu 0128 1.2 0731 6.4 1344 1.3 1949 6.5
9 Tu 0237 1.0 0842 6.5 1451 1.3 2058 6.5
24 W 0206 1.3 0809 6.3 1422 1.5 2029 6.4
10 W 0322 1.4 0927 6.0 1535 1.8 2146 6.1
25 Th 0249 1.5 0851 6.1 1503 1.7 2114 6.2
11 Th 0410 1.9 1016 5.6 1624 2.2 2238 5.7
26 F 0335 1.6 0939 5.9 1552 1.8 2207 6.0
12 F 0508 2.2 1113 5.2 1725 2.5 2342 5.4
27 Sa 0430 1.8 1035 5.7 1649 2.0 2308 5.9
13 Sa 0621 2.4 1224 5.0 1843 2.6
28 Su 0533 1.9 1142 5.5 1758 2.1
14 Su 0056 5.3 0738 2.4 1341 5.1 2002 2.5
29 M 0015 5.8 0645 1.8 1252 5.6 1913 2.0
15 M 0208 5.4 0843 2.2 1446 5.3 2103 2.2
30 Tu 0124 6.0 0754 1.7 1401 5.8 2023 1.7
31 W 0230 6.2 0858 1.4 1503 6.1 2125 1.5

JUNE

1 Th 0329 6.4 0956 1.2 1559 6.4 2221 1.2
16 F 0350 5.7 1013 1.9 1619 5.9 2233 1.9
2 F 0426 6.7 1049 1.0 1652 6.7 2316 1.0
17 Sa 0437 5.9 1057 1.7 1701 6.1 2316 1.6
3 Sa 0520 6.8 1140 0.9 1743 6.8
18 Su 0520 6.1 1137 1.5 1742 6.3 2357 1.4
4 Su 0007 0.8 0611 6.8 1228 0.9 1832 6.9
19 M 0601 6.3 1215 1.4 1821 6.5
5 M 0055 0.8 0700 6.8 1313 0.9 1919 6.9
20 Tu 0038 1.3 0641 6.4 1255 1.3 1900 6.6
6 Tu 0142 0.9 0745 6.6 1357 1.1 2004 6.8
21 W 0119 1.2 0723 6.5 1335 1.2 1942 6.7
7 W 0226 1.1 0829 6.4 1439 1.3 2047 6.5
22 Th 0201 1.1 0804 6.5 1416 1.2 2025 6.7
8 Th 0310 1.4 0911 6.1 1519 1.6 2129 6.2
23 F 0244 1.1 0847 6.4 1500 1.3 2110 6.6
9 F 0352 1.7 0955 5.8 1602 1.9 2213 6.0
24 Sa 0329 1.2 0934 6.3 1545 1.4 2157 6.5
10 Sa 0435 1.9 1040 5.5 1647 2.1 2301 5.7
25 Su 0419 1.3 1023 6.1 1634 1.6 2249 6.3
11 Su 0525 2.2 1130 5.3 1740 2.3 2354 5.5
26 M 0511 1.5 1116 5.9 1730 1.7 2346 6.1
12 M 0622 2.3 1229 5.2 1843 2.4
27 Tu 0608 1.6 1215 5.8 1834 1.9
13 Tu 0056 5.3 0726 2.4 1335 5.2 1951 2.4
28 W 0049 6.0 0713 1.8 1321 5.7 1944 1.9
14 W 0201 5.4 0829 2.3 1439 5.3 2053 2.3
29 Th 0157 6.0 0823 1.8 1430 5.8 2057 1.8
15 Th 0300 5.5 0925 2.1 1532 5.6 2146 2.1
30 F 0304 6.0 0931 1.7 1538 6.0 2204 1.6

JULY

1 Sa 0410 6.1 1033 1.5 1640 6.3 2305 1.4
16 Su 0410 5.6 1031 2.0 1637 5.9 2255 1.8
2 Su 0511 6.3 1127 1.3 1736 6.5 2358 1.2
17 M 0501 5.9 1118 1.7 1723 6.3 2342 1.5
3 M 0604 6.5 1217 1.2 1825 6.7
18 Tu 0546 6.3 1201 1.4 1805 6.6
4 Tu 0048 1.0 0650 6.6 1302 1.1 1910 6.8
19 W 0025 1.1 0628 6.5 1242 1.1 1848 6.9
5 W 0131 1.0 0734 6.6 1342 1.1 1951 6.8
20 Th 0107 0.9 0710 6.7 1324 0.9 1930 7.1
6 Th 0212 1.1 0812 6.5 1420 1.2 2029 6.7
21 F 0149 0.7 0752 6.8 1405 0.8 2012 7.1
7 F 0249 1.2 0849 6.3 1457 1.4 2105 6.5
22 Sa 0232 0.7 0833 6.8 1446 0.8 2054 7.1
8 Sa 0322 1.4 0924 6.1 1531 1.6 2141 6.2
23 Su 0314 0.8 0915 6.7 1528 1.0 2138 6.9
9 Su 0356 1.6 1000 5.9 1606 1.8 2217 6.0
24 M 0356 1.0 0959 6.4 1612 1.2 2224 6.6
10 M 0433 1.9 1038 5.6 1645 2.1 2258 5.7
25 Tu 0441 1.3 1045 6.1 1659 1.6 2315 6.2
11 Tu 0515 2.2 1123 5.4 1733 2.3 2346 5.4
26 W 0532 1.7 1140 5.8 1757 1.9
12 W 0607 2.4 1218 5.1 1835 2.6
27 Th 0015 5.8 0635 2.0 1248 5.5 1914 2.2
13 Th 0045 5.2 0716 2.5 1328 5.1 1949 2.6
28 F 0128 5.6 0758 2.2 1409 5.5 2046 2.2
14 F 0158 5.1 0830 2.5 1443 5.2 2103 2.5
29 Sa 0251 5.5 0921 2.1 1531 5.7 2203 1.9
15 Sa 0311 5.3 0936 2.3 1546 5.5 2203 2.2
30 Su 0409 5.8 1028 1.8 1638 6.1 2304 1.6
31 M 0508 6.1 1122 1.5 1730 6.5 2354 1.3

AUGUST

1 Tu 0556 6.4 1208 1.2 1815 6.7
16 W 0529 6.4 1143 1.2 1747 6.9
2 W 0036 1.0 0638 6.6 1248 1.1 1855 6.9
17 Th 0008 0.9 0611 6.8 1225 0.8 1829 7.2
3 Th 0114 1.0 0714 6.6 1323 1.0 1930 6.9
18 F 0050 0.5 0652 7.1 1306 0.5 1910 7.5
4 F 0148 1.0 0748 6.6 1355 1.0 2002 6.8
19 Sa 0131 0.3 0731 7.2 1345 0.4 1951 7.5
5 Sa 0219 1.0 0819 6.5 1426 1.1 2033 6.7
20 Su 0211 0.3 0811 7.2 1425 0.5 2032 7.4
6 Su 0247 1.2 0850 6.4 1456 1.3 2104 6.5
21 M 0250 0.5 0851 7.0 1504 0.7 2114 7.1
7 M 0317 1.4 0921 6.2 1527 1.6 2135 6.2
22 Tu 0329 0.9 0932 6.7 1545 1.1 2156 6.6
8 Tu 0346 1.7 0952 5.9 1557 1.9 2207 5.8
23 W 0410 1.3 1016 6.2 1631 1.6 2245 6.1
9 W 0419 2.1 1028 5.5 1635 2.3 2245 5.5
24 Th 0459 1.9 1109 5.7 1730 2.1 2347 5.5
10 Th 0459 2.4 1115 5.2 1727 2.6 2337 5.1
25 F 0605 2.4 1224 5.3 1903 2.5
11 F 0601 2.7 1219 5.0 1849 2.8
26 Sa 0114 5.2 0749 2.6 1402 5.2 2050 2.4
12 Sa 0053 4.9 0738 2.8 1352 5.0 2026 2.7
27 Su 0256 5.3 0921 2.3 1534 5.6 2204 2.0
13 Su 0234 5.0 0907 2.6 1518 5.3 2141 2.3
28 M 0409 5.7 1024 1.9 1633 6.1 2259 1.5
14 M 0350 5.4 1010 2.1 1617 5.9 2237 1.8
29 Tu 0459 6.1 1112 1.5 1718 6.5 2342 1.2
15 Tu 0444 5.9 1059 1.6 1705 6.4 2325 1.3
30 W 0540 6.5 1151 1.2 1757 6.8
31 Th 0017 1.0 0615 6.7 1225 1.0 1831 6.9

Chart Datum: 3.71 metres below Ordnance Datum (Newlyn)

WALES — MILFORD HAVEN

Lat 51°42′ N Long 5°01′ W

TIMES AND HEIGHTS OF HIGH AND LOW WATERS

YEAR 1989

TIME ZONE UT(GMT)
For Summer Time add ONE hour in non-shaded areas

SEPTEMBER

Time	m	Time	m
1 F 0049 0648 1256 1902	0.9 6.8 0.9 7.0	**16** Sa 0025 0627 1242 1846	0.3 7.4 0.3 7.7
2 Sa 0117 0719 1326 1931	0.9 6.8 0.9 6.9	**17** Su 0106 0706 1321 1927	0.1 7.5 0.2 7.7
3 Su 0145 0747 1355 1959	1.0 6.7 1.0 6.8	**18** M 0145 0745 1401 2008	0.2 7.4 0.4 7.5
4 M 0213 0815 1422 2027	1.1 6.5 1.2 6.6	**19** Tu 0225 0826 1442 2050	0.5 7.2 0.7 7.1
5 Tu 0239 0843 1450 2056	1.4 6.3 1.5 6.3	**20** W 0304 0907 1524 2134	0.9 6.7 1.2 6.5
6 W 0305 0911 1519 2125	1.7 6.0 1.8 5.9	**21** Th 0346 0952 1612 2224	1.5 6.2 1.8 5.9
7 Th 0335 0943 1553 2159	2.0 5.7 2.2 5.5	**22** F 0437 1049 1716 2330 ☾	2.1 5.6 2.3 5.3
8 F 0410 1023 1638 2245 ☾	2.4 5.3 2.5 5.1	**23** Sa 0549 1211 1904	2.6 5.2 2.6
9 Sa 0502 1126 1756	2.8 5.0 2.9	**24** Su 0106 0747 1358 2046	5.0 2.7 5.3 2.4
10 Su 0004 0646 1307 1957	4.8 3.0 4.9 2.8	**25** M 0250 0911 1521 2152	5.2 2.4 5.7 2.0
11 M 0202 0840 1451 2118	4.9 2.7 5.3 2.3	**26** Tu 0353 1007 1613 2238	5.7 1.9 6.2 1.5
12 Tu 0328 0946 1553 2214	5.4 2.1 5.9 1.7	**27** W 0438 1051 1655 2316	6.2 1.5 6.6 1.3
13 W 0420 1037 1640 2302	6.0 1.5 6.5 1.1	**28** Th 0515 1126 1730 2349	6.5 1.1 6.8 1.1
14 Th 0505 1120 1723 2344	6.6 1.0 6.9 0.6	**29** F 0549 1158 1803	6.7 1.1 6.9
15 F 0546 1201 1804 ○	7.1 0.6 7.5	**30** Sa 0018 0618 1228 1832	1.0 6.8 1.0 6.9

OCTOBER

Time	m	Time	m
1 Su 0048 0648 1257 1900	1.0 6.8 1.0 6.9	**16** M 0041 0641 1259 1904	0.2 7.5 0.3 7.7
2 M 0114 0716 1326 1928	1.1 6.8 1.1 6.8	**17** Tu 0123 0723 1342 1948	0.4 7.5 0.5 7.4
3 Tu 0141 0744 1352 1957	1.2 6.7 1.3 6.6	**18** W 0204 0806 1426 2032	0.7 7.0 0.8 6.9
4 W 0208 0812 1422 2025	1.4 6.4 1.5 6.3	**19** Th 0246 0849 1511 2119	1.1 6.3 1.3 6.4
5 Th 0234 0840 1451 2054	1.7 6.2 1.9 6.0	**20** F 0331 0939 1603 2212	1.7 6.2 1.9 5.8
6 F 0304 0912 1525 2129	2.1 5.8 2.2 5.6	**21** Sa 0423 1038 1711 2318 ☾	2.2 5.7 2.4 5.3
7 Sa 0339 0953 1613 2219	2.4 5.5 2.6 5.2	**22** Su 0536 1154 1848	2.6 5.4 2.6
8 Su 0434 1058 1729 2337 ☾	2.8 5.1 2.8 4.9	**23** M 0045 0720 1327 2015	5.1 2.7 5.4 2.4
9 M 0610 1235 1924	3.0 5.0 2.7	**24** Tu 0216 0839 1446 2118	5.3 2.4 5.7 2.1
10 Tu 0127 0805 1412 2046	5.0 2.7 5.4 2.3	**25** W 0319 0935 1539 2204	5.7 2.0 6.0 1.7
11 W 0253 0914 1518 2143	5.5 2.1 6.0 1.6	**26** Th 0404 1019 1621 2242	6.0 1.7 6.5 1.5
12 Th 0348 1006 1609 2233	6.1 1.5 6.6 1.1	**27** F 0442 1055 1658 2318	6.3 1.5 6.5 1.3
13 F 0434 1048 1654 2316	6.7 1.0 7.2 0.6	**28** Sa 0518 1129 1732 2349	6.6 1.3 6.7 1.2
14 Sa 0518 1134 1737 ○	7.2 0.6 7.5	**29** Su 0549 1200 1803 ●	6.7 1.2 6.7
15 Su 0000 0600 1217 1821	0.3 7.4 0.3 7.7	**30** M 0018 0619 1231 1834	1.2 6.7 1.2 6.7
		31 Tu 0048 0649 1300 1903	1.3 6.7 1.3 6.6

NOVEMBER

Time	m	Time	m
1 W 0116 0719 1331 1934	1.4 6.6 1.4 6.5	**16** Th 0151 0755 1418 2023	0.9 7.1 1.0 6.8
2 Th 0145 0749 1402 2005	1.5 6.5 1.6 6.3	**17** F 0234 0842 1505 2111	1.3 6.8 1.3 6.3
3 F 0216 0822 1437 2039	1.7 6.3 1.8 6.0	**18** Sa 0321 0931 1556 2200	1.7 6.4 1.8 5.9
4 Sa 0250 0858 1517 2119	2.0 6.0 2.1 5.7	**19** Su 0412 1024 1654 2257	2.1 6.0 2.1 5.5
5 Su 0331 0945 1607 2212	2.3 5.7 2.4 5.4 ☾	**20** M 0511 1125 1804	2.4 5.7 2.4
6 M 0427 1047 1716 2323 ☾	2.6 5.4 2.5 5.2	**21** Tu 0003 0627 1235 1919	5.3 2.6 5.5 2.4
7 Tu 0546 1207 1845	2.7 5.4 2.5	**22** W 0117 0742 1347 2025	5.3 2.5 5.5 2.3
8 W 0049 0720 1327 2004	5.2 2.5 5.6 2.1	**23** Th 0226 0846 1449 2119	5.4 2.3 5.7 2.1
9 Th 0208 0832 1436 2105	5.6 2.1 6.1 1.7	**24** F 0321 0938 1541 2204	5.7 2.1 5.9 1.9
10 F 0308 0929 1532 2159	6.1 1.6 6.6 1.2	**25** Sa 0406 1020 1623 2244	6.0 1.9 6.1 1.7
11 Sa 0400 1021 1624 2248	6.6 1.2 7.0 0.9	**26** Su 0445 1059 1702 2320	6.2 1.7 6.3 1.6
12 Su 0448 1109 1712 2334	7.0 0.8 7.3 0.6	**27** M 0522 1136 1737 2354	6.4 1.6 6.4 1.5
13 M 0536 1157 1801 ●	7.3 0.6 7.4	**28** Tu 0556 1210 1812 ●	6.5 1.5 6.4
14 Tu 0021 0622 1243 1848	0.6 7.4 0.6 7.4	**29** W 0027 0629 1243 1846	1.5 6.6 1.5 6.4
15 W 0106 0709 1331 1935	0.7 7.3 0.7 7.1	**30** Th 0059 0703 1317 1920	1.5 6.6 1.5 6.4

DECEMBER

Time	m	Time	m
1 F 0133 0738 1354 1957	1.5 6.6 1.5 6.3	**16** Sa 0225 0833 1457 2058	1.2 6.9 1.2 6.5
2 Sa 0208 0815 1432 2034	1.6 6.4 1.6 6.2	**17** Su 0308 0917 1539 2141	1.4 6.6 1.5 6.2
3 Su 0246 0856 1514 2117	1.6 6.3 1.8 6.0	**18** M 0349 1000 1623 2224	1.7 6.3 1.8 5.8
4 M 0329 0942 1603 2206	2.0 6.1 1.9 5.8	**19** Tu 0434 1045 1709 2312 ☾	2.0 6.0 2.1 5.5
5 Tu 0419 1035 1658 2304	2.1 5.9 2.0 5.6	**20** W 0523 1134 1803	2.3 5.7 2.4
6 W 0519 1137 1804	2.2 5.8 2.1	**21** Th 0007 0622 1234 1906	5.3 2.5 5.4 2.5
7 Th 0010 0631 1245 1914	5.6 2.2 5.9 2.0	**22** F 0113 0733 1341 2016	5.2 2.6 5.3 2.5
8 F 0120 0744 1352 2023	5.7 2.1 6.0 1.8	**23** Sa 0222 0843 1447 2118	5.3 2.5 5.4 2.4
9 Sa 0227 0851 1457 2127	5.9 1.8 6.3 1.6	**24** Su 0324 0942 1546 2210	5.5 2.4 5.6 2.2
10 Su 0329 0953 1557 2223	6.3 1.5 6.6 1.3	**25** M 0414 1031 1634 2255	5.8 2.1 5.8 2.0
11 M 0426 1051 1655 2318	6.6 1.2 6.8 1.1	**26** Tu 0458 1115 1718 2334	6.1 1.9 6.0 1.8
12 Tu 0520 1144 1750 ○	6.9 1.0 7.0	**27** W 0537 1154 1757	6.3 1.6 6.2
13 W 0007 0612 1236 1841	0.9 7.1 0.8 7.0	**28** Th 0011 0615 1231 1835 ●	1.6 6.4 1.5 6.4
14 Th 0056 0702 1326 1928	0.9 7.2 0.9 6.9	**29** F 0048 0652 1309 1912	1.4 6.7 1.3 6.5
15 F 0141 0748 1412 2015	1.0 7.1 0.9 6.8	**30** Sa 0123 0728 1347 1948	1.3 6.8 1.2 6.5
		31 Su 0201 0808 1426 2027	1.3 6.8 1.2 6.5

Chart Datum: 3.71 metres below Ordnance Datum (Newlyn)

MILFORD HAVEN continued

LIGHTS AND MARKS
Middle Channel Rocks Fl(3)G 7s 18m 8M; B Tr, aluminium lantern. Ldg Lts 040°. Front Oc 4s 27m 15M. Rear Oc 8s 53m 15M; both vis 031°-048° (H24). Milford Dock Ldg Lts 348° both FG, with W circular daymarks. Dock entry signals on E side of lock: Blue flag or 2FG(vert) means gates open, vessels may enter. Signal arm lowered or FG at inner end of lock means vessels may leave.

RADIO TELEPHONE
Call: *Milford Haven Radio* (Port Signal Station), VHF Ch 11 12 14 16 (H24); 09 10 11 12 14 16 67. Local weather forecasts broadcast on Ch 12 and 14 at approx. 0300, 0900, 1500 and 2100 GMT. Gale Warnings broadcast on receipt Ch 12 and 14. Shipping movements for next 24 hours on Ch 12 between 0800–0830 and 2000–2030 local time.
Milford Haven Patrol Launches, VHF Ch 11 12 (H24); 06 08 11 12 14 16 67.
Milford Docks Ch 09 12 14 16 (HW−2 to HW).
Kelpie Boat Services, Dale SC and Westfield Marina, Lawrenny Yacht Stn Ch M.

TELEPHONE (0646 — 6 fig. nos; 064 62 4 and 5 fig. nos) Hr Mr Milford Haven Signal Station 2342/3; Dock Master 2275; CG & MRSC Dale 218; HMC Freefone Customs Yachts; RT (0646) 681310; Marinecall 0898 500 459; Dr via Milford Haven Signal Station 2342/3.

FACILITIES
EC Wednesday (Milford Haven, Pembroke Dock and Neyland): Thursday (Haverfordwest); **Westfield Marina** (at Neyland) (260+40 visitors) Tel. 601601; Access Lower basin H24, Upper basin HW∓3½, FW, D, R, Bar, CH, V, AC, Gas, Gaz, Lau, C (15 ton), SM — via Dale SC, ME, El, Sh; **Westfield Chandlery** Tel. 601667, ACA, CH, Gas, Gaz, V; **Dale SC** Tel. Dale 349/Neyland 601636, C (20 ton), BY, ME, El, Sh, CH, D, BH, V, SM, FW, L, M, ACA; **Milford Haven Port Authority Jetty** Tel. 2342/3 (occasional use by visiting yachtsmen with special permission from Harbourmaster — contact Signal Station) AB, FW, L, M; **Milford Docks** (230) Tel. 3091 (contact Pierhead for permission to enter, VHF Ch 14) AB, C (various), CH, El, FW, HMC, D, L, ME, Sh; **Dudley Marine** Tel. 2787, BY, AB, FW, L, ME, CH, Sh, El; **Pennar Park** Tel. 684609, AB, Bar, FW, L, M, R, Slip, V; **Marine and Port Services** Tel. 682271, AB, BY, C, D, El, FW, L, M, ME, Sh, Slip; **Kelpie Boats** Tel. 683661, BY, C, CH, D, El, FW, L, ACA, M, SM, ME, Sh. (Slip and YC adjacent, Bar, PO, P, R & V in town nearby); **Jenkins Boats** Tel. 601836, ME, Sh, BY; **East Llanion Marine** Tel. 2044, BY, D (cans), Slip, C (1 ton), El, FW, Sh; **Lawrenny Yacht Station** (100) Tel. Carew 212, AB, Bar, BY, C (10 ton), CH, D, FW, L, M, ME, P, R, Slip, V (Sub PO in village). **Milford Haven Calor Gas Centre** Tel. 2252, Gas.
Yacht Clubs — **Dale SC** Tel. Dale 362, **Pembrokeshire YC** Tel. 2799, **Neyland YC** Tel. 600267, **Pembroke Haven YC** Tel. 684403;
Towns — Milford Haven, PO; Bank; Rly.
Pembroke Dock, PO; Bank; Rly, Hosp.
Neyland, PO; Bank.
Haverfordwest, PO; Hosp; Bank; Rly; Air (Swansea).

Note: The Castlemartin Range Danger Area extends 12M S from St Govans Head and from there in an arc to a point 12M WNW of Linney Head. The Danger Area operative on any one day however depends on the ranges used. Call Castlemartin Control Tower on Channel 16 VHF, Tel. Castlemartin 321 Ext 336, Range Safety Boats on Ch 16 or 12, or Milford Haven Coastguard. When firing is in progress R flags are flown (R lights at night), along the coast from St Govans Head to Linney Head to Freshwater West. Times of firing are published locally and can be obtained from the Range Officer on Castlemartin 321 Ext 336 or 344, or Ext 364 (the Chief Clerk).

During the firing period (May to December) firing takes place between 0900 and 1630. Night firing takes place from 1830 to 2359 normally on Tuesdays and Thursdays. (These times vary according to the hours of darkness). Firing also takes place for one week in January and February.

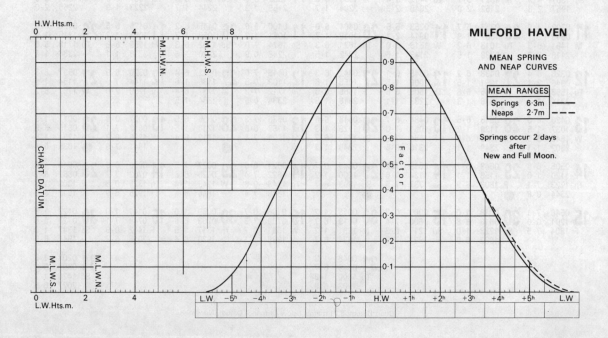

MILFORD HAVEN

MEAN SPRING AND NEAP CURVES

MEAN RANGES	
Springs 6·3m	——
Neaps 2·7m	- - -

Springs occur 2 days after New and Full Moon.

AREA 11—S Wales 511

TENBY 10-11-13
Dyfed

CHARTS
Admiralty 1482; Stanford 14; Imray C60; OS 158
TIDES
−0510 Dover; ML 4.5; Duration 0610; Zone 0 (GMT)
Standard Port MILFORD HAVEN (←)

Times				Height (metres)			
HW		LW		MHWS	MHWN	MLWN	MLWS
0100	0800	0100	0700	7.0	5.2	2.5	0.7
1300	2000	1300	1900				
Differences TENBY							
−0015	−0010	−0015	−0020	+1.4	+1.1	+0.5	+0.2
STACKPOLE QUAY							
−0005	+0025	−0010	−0010	+0.9	+0.7	+0.2	+0.3

SHELTER
Harbour dries out but shelter good. Sheltered anchorages in Tenby Roads or Caldey Roads depending on wind direction.
NAVIGATION
Waypoint 51°40'.00N 04°38'.00W, 099°/279° from/to CGFS on Castle Hill, 2.2M. Beware Woolhouse Rocks (buoy unlit) and Sker Rocks (unbuoyed). It is recommended that boats enter Tenby harbour HW∓2½. On approaching Tenby Roads, keep outside the line of buoys.
LIGHTS AND MARKS
FR 7m 7M on pierhead. Tenby church spire and N side of St Catherine's Is in line at 276°.
RADIO TELEPHONE
None.
TELEPHONE (0834)
Hr Mr 2717; MRSC Dale 218; HMC Freefone Customs Yachts: RT (0222) 399123; Marinecall 0898 500 459; Dr. 2529 or 2284.

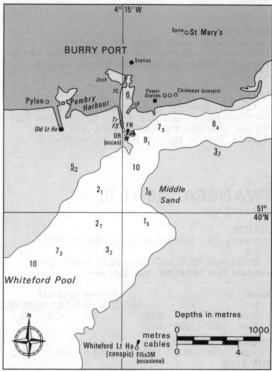
▲ Report to Burry Port Yacht Services

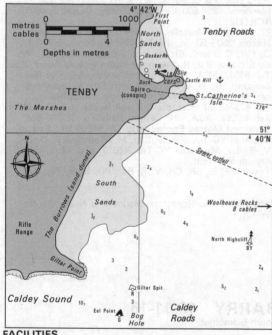

FACILITIES
EC Wednesday; **Harbour** Tel. 2717, Slip (up to 4.2m), L, FW; **Pier** L; **Tenby YC** Tel. 2762, M; **Morris Bros** Tel. 2105, Gas.
Town P, CH, V, R, Bar. PO; Bank; Rly; Air (Swansea).

BURRY PORT 10-11-14
Dyfed

CHARTS
Admiralty 1167; Stanford 14; Imray C59, C60; OS 159
TIDES
−0500 Dover; ML 4.7; Duration 0555; Zone 0 (GMT).
Standard Port MILFORD HAVEN (←)

Times				Height (metres)			
HW		LW		MHWS	MHWN	MLWN	MLWS
0100	0800	0100	0700	7.0	5.2	2.5	0.7
1300	2000	1300	1900				
Differences BURRY PORT							
+0003	+0003	+0007	+0007	+1.6	+1.4	+0.5	+0.4
LLANELLI							
−0003	−0003	+0150	+0020	+0.8	+0.6	No data	
FERRYSIDE							
0000	−0010	+0220	0000	−0.3	−0.7	−1.7	−0.6
CARMARTHEN							
+0010	0000	No data		−4.4	−4.8	dries out	

SHELTER
Shelter in outer harbour is good but it dries out; as it is entirely filled with moorings, entry without prior inspection at LW is not practical. Can be approached HW∓2. The dock is not available to yachtsmen, but gates have been removed and it could be useful in emergency. Port is owned by Llanelli Borough Council and rate payers have first priority.
NAVIGATION
Waypoint 51°37'.74N 04°21'.00W, 250°/070° from/to Whiteford Lt Ho, 4.0M (channel shifts). No dangers in Carmarthen Bay but ground swell develops into rollers on the bar. Bar should not be attempted by small craft in W winds of Force 5 or over. Best time to enter is HW−2 to HW+1, but the channel is not buoyed. Before entry, check which sections of Carmarthen Bay are closed for range practice. (See 10.11.12). Channel through the sandbanks is liable to vary and is unlit.
LIGHTS AND MARKS
A barrel post with a QR light is about 1 ca S of conspic Tr and FS on West breakwater. Entrance to Burry Inlet close N of Burry Holms marked by Y buoy Fl Y 2.5s. Three conspic chimneys east of harbour.
RADIO TELEPHONE
None.

Harbour, Coastal and Tidal Information

BURRY PORT continued

TELEPHONE (055 46)
Superintendent Llanelli 758181; CG and MRCC Swansea 366534; HMC Freefone Customs Yachts: RT (0222) 399123; Marinecall 0898 500 459; Dr 2240.
FACILITIES
EC Tuesday; **Outer Harbour West pier** Tel. 3342, Slip, M (ring Hr Mr); **West Basin** Tel. 3342, Slip (ring Hr Mr), L, CH; **East Pier** Slip, L; **Burry Port Yacht Services** Tel. 2740, Slip, M, D, L, ME, El, Sh, C, CH; **Burry Port YC** Bar; **Shoreline Caravan Park** Tel. 2657, Gas.
Town P, D, V, R, Bar. PO; Bank; Rly; Air (Cardiff).

SWANSEA 10-11-15
West Glamorgan

CHARTS
Admiralty 1161; Stanford 14; Imray C59; OS 159
TIDES
−0515 Dover; ML 5.3; Duration 0620; Zone 0 (GMT).
Standard Port MILFORD HAVEN (←)

Times				Height (metres)			
HW		LW		MHWS	MHWN	MLWN	MLWS
0100	0800	0100	0700	7.0	5.2	2.5	0.7
1300	2000	1300	1900				
Differences SWANSEA							
+0004	+0006	−0006	−0003	+2.6	+2.1	+0.7	+0.3
MUMBLES							
+0005	+0010	−0020	−0015	+2.3	+1.7	+0.6	+0.2
PORTHCAWL							
0000	0000	0000	−0015	+2.9	+2.3	+0.8	+0.3

SHELTER
Very good. Small craft may lie in the river but they dry out on mud at LW. Good anchorage off Mumbles in W wind. It is advised to enter the marina in South Dock, available HW∓3, (Swansea Yacht Haven). There is a pontoon to S of marina entrance and three buoys marked SYH for waiting yachts.
NAVIGATION
Waypoint stbd-hand buoy, QG, Bell, 51°35′.50N 03°56′.06W, 200°/020° from/to Eastern Breakwater Lt, 0.92M. At springs, yachts should not enter river until LW+2. Yachts must be under power in the harbour and approaches, maximum speed 5 kn.
LIGHTS AND MARKS
Ldg Lts 020°. Front 2FG (vert) 5m 2M. Rear FG 6M; mark E side of dredged channel. When approaching from E, keep to seaward side of the inner fairway buoy (G conical, Fl G 2.5s). Marina Master will advise on movements on VHF Ch M. Traffic signals are shown at W side of entrance to King's Lock. Three FR Lts (vert) mean that dock and harbour are closed. Three FR Lts (hor) mean that docks are closed. Other signals consist of three rows of R or G Lts in three columns. Yachts entering are controlled by the light in the middle row of the left-hand column. When red this prohibits movement of yachts in the port or in the approach channel; when green, incoming craft may enter between breakwaters and proceed to river berth or marina, as below:

R R R	Yachts may enter river and proceed
G R R	to Yacht Haven
R R R	

Traffic signals at lock to yacht haven as follows:

Fixed light(s)	Meaning
R (or no R lights)	Lock not operating
R	Lock closed
G	Proceed into lock
R G	Lock gates closing, do not proceed

RADIO TELEPHONE
Call: *Swansea Docks Radio* VHF Ch 16; 14 (H24). Marina Ch M.

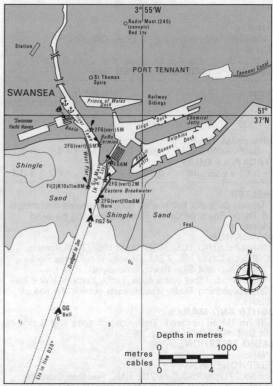

TELEPHONE (0792)
Hr Mr 50855 Ext 260; CG 366534; MRCC 366534; HMC Freefone Customs Yachts: RT (0222) 399123; Marinecall 0898 500 459; Hosp 205666; Dr 53452, 66057.
FACILITIES
EC Thursday (Mumbles Wednesday); **Swansea Yacht Haven** (350+50 visitors) Tel. 470310, Access HW∓3¼, D, FW, C (9 ton), BH (25 ton), AC, Gas, CH, Gaz, ME, El, Sh, Lau, Bar, R; **Swansea Yacht & Sub Aqua Club** Tel. 54863, M, L, FW, C (½ ton static), AB, Bar; **Near Landing** ME, El, CH; **River Neath** Slip, M (entry over bar, after ½ flood); **Westland Engine Supplies** Tel. 873249 ME; **Cambrian Small Boats and Chandlery** Tel. 467263, ACA, CH; **Canard Sails** Tel. 367838 SM; **Swansea Marine Engineering** Tel. 51600 ME; **Bristol Channel YC** Tel. 366000, Slip, M; **Mumbles** P, D, CH, V, R, Bar; **Mumbles YC** Tel. 369321, Slip, M, L, FW, C (by appointment).
Town ME, El, Sh, CH, V, R, Bar. PO; Bank; Rly; Air.

Note: 4M to E is Monkstone Marina in R Neath (access HW∓2).

BARRY 10-11-16
South Glamorgan

CHARTS
Admiralty 1182; Stanford 14; Imray C59; OS 171
TIDES
−0435 Dover; ML 6.1; Duration 0630; Zone 0 (GMT).
Standard Port BRISTOL (AVONMOUTH) (→)

Times				Height (metres)			
HW		LW		MHWS	MHWN	MLWN	MLWS
0600	1100	0300	0800	13.2	10.0	3.5	0.9
1800	2300	1500	2000				
Differences BARRY							
−0030	−0015	−0125	−0030	−1.8	−1.3	+0.2	0.0

AREA 11 – S Wales 513

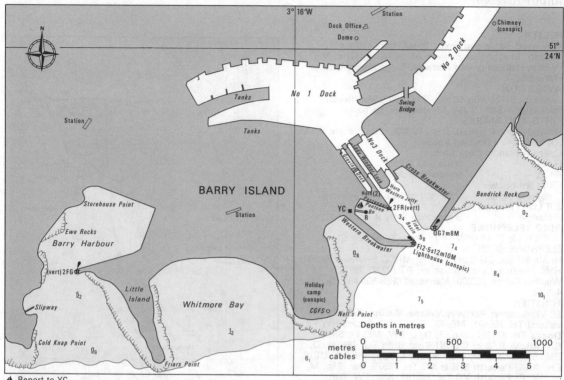

SHELTER
Good shelter except in strong E to SE winds. In these winds yachts are advised to enter the dock (HW∓3) by prior arrangement. Outer harbour always available (free). Old harbour to W of Barry Island dries out and is no longer any use.

NAVIGATION
Waypoint 51°23′.00N 03°15′.00W, 152°/332° from/to entrance, 0.53M. Beware heavy merchant traffic. Approaching from E keep well out from the shore. Strong tidal stream across entrance.

LIGHTS AND MARKS
W Breakwater Fl 2.5s 10M. E Breakwater QG 8M.

RADIO TELEPHONE
Barry Docks VHF Ch 16; 10 11 22 (HW−4 to HW+4).

TELEPHONE (0446)
Hr Mr 732311; MRCC Swansea 366534; HMC Freefone Customs Yachts: RT (0222) 399123; Marinecall 0898 500 459; Dr 739543.

FACILITIES
EC Wednesday; **Windway Marine** Tel. 737312, L, FW, BY, AB, C (30 ton); **Ray Harris Marine** Tel. 740924, Slip, P, D, FW, ME, El, Sh, CH, SM; **Barry YC** (130) Tel. 735511, Slip, M, FW, D, C (20 ton), Access HW∓3½; **ARG Ltd** Gas.
Town P, D, CH, V, R, Bar. PO; Bank; Rly; Air (Cardiff).

CARDIFF 10-11-17
South Glamorgan

CHARTS
Admiralty 1182; Stanford 14; Imray C59; OS 171
TIDES
−0425 Dover; ML 6.4; Duration 0610; Zone 0 (GMT).
Standard Port BRISTOL (AVONMOUTH) (→)

Times				Height (metres)			
HW		LW		MHWS	MHWN	MLWN	MLWS
0600	1100	0300	0800	13.2	10.0	3.5	0.9
1800	2300	1500	2000				

Differences CARDIFF
−0015 −0015 −0100 −0030 −1.0 −0.6 +0.1 0.0
NEWPORT
−0020 −0010 0000 −0020 −1.1 −1.0 −0.6 −0.7

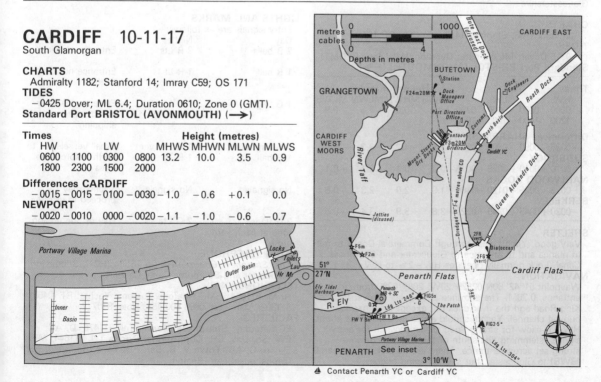

▲ Contact Penarth YC or Cardiff YC

CARDIFF continued

SHELTER
Very good within port limits. Small boats anchor in Penarth. In strong winds from NE to SE, safest to N of Penarth Harbour or into Portway Village Marina (access HW∓3¾).

NAVIGATION
Waypoint 51°24'.00N 03°08'.73W, 169°/349° from/to front Ldg Lt 349°, 3.7M. Shipping can be heavy.

LIGHTS AND MARKS
Buoyed channel lighted. Ldg Lts 349°. Ldg Lts into R Ely 304° and 246°. Docking signals:-
Queen Alexandra Dock Lock;
Fl R Lt — prepare to dock.
FR Lt — vessels may dock.
FR Lt 9m below docking signal — outer half of lock only available.
3 R Lts in inverted triangle — S approach Jetty is occupied.

RADIO TELEPHONE
VHF Ch 16; 11 14 (HW−4 to HW+3). Marina Ch M.

TELEPHONE (0222)
Hr Mr 461083; CG Barry 735016; MRCC Swansea 366534; HMC Freefone Customs Yachts: RT (0222) 399123; Weather Centre 397020; Marinecall 0898 500 459; Dr 415258.

FACILITIES
EC Wednesday; **Portway Village Marina** (320+30 visitors) Tel. 705021, ME, El, Sh, C, CH, Access HW∓3; **Docks** Tel. 471311, Slip, P, D, L, FW, ME, El, Sh, C, CH, AB, V, R, Bar; **Cardiff Boat Bldg** Tel. 488034, D, SM, Sh, C (20 ton); **Blair Nautical Supplies** Tel. 21810, CH, ACA; **Penarth MB & SC** Tel. 26575, Bar, Slip, M, L, FW, C; **Penarth YC** Tel. 708196, Slip, FW, Bar; **Cardiff YC** Tel. 387697, Slip, M, L (Floating Pontoon), FW, Bar; **Barry Marine Centre** Tel. 562000, CH; **EH Hewett** Tel. 480624, ME, El; **Cambrian Marine Services** Tel. 43459, ME, El, Sh; **Barda Yachts** Tel. 705106, Sh; **ARG Ltd** Tel. 481076, Gas. **Town** P D, ME, El, V, R, Bar. PO; Bank; Rly; Air.

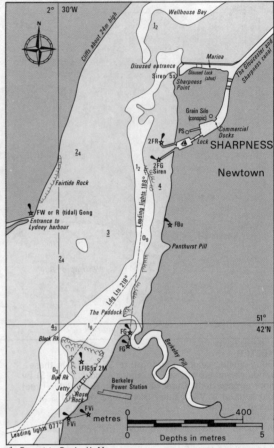

▲ Report to Docks Hr Mr

SHARPNESS 10-11-18
Gloucestershire

CHARTS
Admiralty 1166; Stanford 14; Imray C59; OS 162

TIDES
−0315 Dover; ML No data; Duration 0415; Zone 0 (GMT).
Standard Port BRISTOL (AVONMOUTH) (→)

Times				Height (metres)			
HW		LW		MHWS	MHWN	MLWN	MLWS
0000	0600	0000	0700	13.2	10.0	3.5	0.9
1200	1800	1200	1900				

Differences SHARPNESS DOCK
+0035 +0050 +0305 +0245 −3.9 −4.2 −3.3 −0.4
SUDBROOK
+0010 +0010 +0025 +0015 +0.2 +0.1 −0.1 +0.1
NARLWOOD ROCKS
+0025 +0025 +0120 +0100 −1.9 −2.0 −2.3 −0.8
BERKELEY
+0030 +0045 +0245 +0220 −3.8 −3.9 −3.4 −0.5

SHELTER
Very good. Yachts pass through Commercial Docks to get to marina and to entrance to Gloucester and Sharpness Canal. Lock normally operates HW−2½ to HW+½.

NAVIGATION
Waypoint 51°42'.80N 02°29'.20W, 208°/028° from/to entrance, 0.20M. The entrance is 15 miles up river from King Road and the river dries out except for a narrow buoyed channel. Yachts should arrive off Sharpness about HW−1; allow for strong flood stream. Watch for entry signals, stemming tide south of the F Bu Lt. Beware strong set across entrance. There are no locks in the canal (BWB) to Gloucester, but several swing bridges.

LIGHTS AND MARKS
Entry signals are as follows:—

Day	Night	
2 B balls	2 R Lts	Entrance closed
1 B ball	1 R Lt	Entrance not clear (traffic)
1 G flag	1 G Lt	Entrance clear for commercial shipping
1 G flag over 1 B ball	1 G Lt over 1 R Lt	Small vessels are to berth before large ones
No signals	No signals	Yachts may proceed in or out.
2 G flags (one at each yard arm)	2 G Lts	HW or tide ebbing

RADIO TELEPHONE
Call: *Sharpness Control* VHF Ch 16; 14 (H24). Bridges on Gloucester and Sharpness Canal Ch 74.

TELEPHONE (0453)
Hr Mr 811644; HMC Freefone Customs Yachts: RT (0453) 811302; Marinecall 0898 500 459; Hosp 810777.

FACILITIES
EC Saturday; **Sharpness Marine** Tel. 811476, D, AC, FW, Sh, Gas, CH; **Commercial Docks** Tel. 811644, ME, El, Sh, C, AB. **Town** V, R, Bar. PO; Bank (Berkeley); Rly (Stonehouse); Air (Bristol).

AREA 11—Bristol Channel 515

BRISTOL (CITY DOCKS)
Avon
10-11-19

CHARTS
Admiralty 1859, 1176; Stanford 14; Imray C59; OS 172
TIDES
−0410 Dover; ML 7.0; Duration 0620; Zone 0 (GMT).
Standard Port BRISTOL (AVONMOUTH) (→)

Times				Height (metres)			
HW		LW		MHWS	MHWN	MLWN	MLWS
0200	0800	0300	0800	13.2	10.0	3.5	0.9
1400	2000	1500	2000				

Differences CUMBERLAND BASIN (Ent)
+0010 +0010 Dries out −2.9 −3.0 Dries out
PORTISHEAD
−0002 0000 No data −0.1 −0.1 No data

NOTE: The Port of Bristol (Avonmouth) is a Standard Port and tidal predictions for each day are given below.

SHELTER
Except in emergency, neither Avonmouth, Royal Portbury or Portishead Docks are available for yachts. Some shelter is available in the approach channel to the locks at Portishead. Yachts must not anchor adjoining the pier or stone jetty and they must be prepared to dry out on soft mud 2 hours each side of MLWS.
The regulations for craft entering the Avon, Cumberland Basin, Floating Harbour or City Docks are set out in detail in a pamphlet called *Bristol City Docks — Information for Owners of Pleasurecraft*, obtainable from the Hr Mr's Office, Underfall Yard, Cumberland Road, Bristol BS1 6XG. Tel. 264797.
NAVIGATION
Avonmouth waypoint 51°30'.40N 02°43'.21W, 307°/127° from/to front Ldg Lt 127°, 0.58M. The channel from Flatholm is buoyed. Tidal stream can run at five knots or more. When approaching Walton Bay, contact Avonmouth Radio on Ch 12 and transfer to either Ch 09 or 14 low power, for instructions to pass Royal Portbury Dock and enter the River Avon. If no radio fitted, signal Avonmouth Signal Station with international flag R or flash morse letter R. The signal station will reply by light or loud hailer.

LIGHTS AND MARKS
R Avon is entered south of S Pier Lt Oc RG 30s; vis R 294°-036°, G036°-194°. Ldg Lts 127° both FR. Saint George Ldg Lts 173°, both Oc G 5s synchronised. Above Pill Creek stbd hand Lts are mostly Oc G 5s, and port hand are FY.
Yachts should arrive Cumberland Basin entrance by HW. Entry signals to Bristol City Docks are shown from two positions on E bank, 1½ and 2½ ca above Clifton Suspension Bridge. FG Lt — come ahead with caution. FR Lt — stop and await orders. Prince's Street and Redcliffe bridges are manned 0600-2230 summer, 0900-1645 winter. Other bridges HW−3 to HW+1. R Lts indicate bridges are closed. Inform Bridgemaster Tel. 299338 in advance, or sound one short blast followed by one long and one short.
RADIO TELEPHONE
Avonmouth Signal Station (South Pier, Royal Edward Dock) Call: *Avonmouth Radio*: VHF Ch 12 (VTS), Port Operations Ch 14 09 11 16 (H24): navigational information provided on request. Royal Portbury Dock Ch 16; 12 **14** (HW−4½ to HW+3½).
Portishead Dock Ch 16; 12 **14** (HW−2½ to HW+1½).
City Docks Radio Ch 16; 09 **14** (HW−3 to HW+1).
Bristol Floating Harbour Ch 16; 73 (Office hours).
Other station: Newport Ch 16; 09 11 (HW−4 to HW+4).
TELEPHONE (0272)
Hr Mr 264797; CG Barry 735016; MRCC Swansea 366534; Marina Ch M; Netham Lock Keeper 776590; Princes St & Redcliffe Bridge Masters 299338; HMC Freefone Customs Yachts: RT (0272) 826451; Dock Master, Cumberland Basin 273633; Weather Centre 279298; Marinecall 0898 500 459; Hosp 230000.
FACILITIES
EC Wednesday/Saturday; **Bristol Marina** (60+20 visitors) Tel. 265730, D, FW, ME, El, Sh, AC, SM, Slip, C, BH (30 ton), Access HW−3 to HW+1; **Baltic Wharf Leisure Centre** Tel. 297608, Slip, L, Bar; **Cabot Cruising Club** Tel. 268318, M, L, FW, AB, Bar; **Mud Dock** FW, C (4 ton); **City Dock** Tel. 264797 AB,C (4 ton), M; gridiron available outside Cumberland Basin — contact Dockmaster Tel. 273633; **Saltford Marine** Tel. Saltford 872226, Slip, C (20 ton), M, L, FW, ME, El, Sh, CH, R; **Portavon Marina** Tel. 861626, Slip, M, L, FW, ME, Sh, CH, R; **Bristol Boating Centre** Tel. 294160, CH; **Bristol Marine Engineering** Tel. 262923, BY, Gas, Gaz, Kos, ME, El, CH; **SW Petroleum Co** P, D; **W F Price** Tel. 823888, CH, ACA; **City** PO; Bank; Rly; Air.

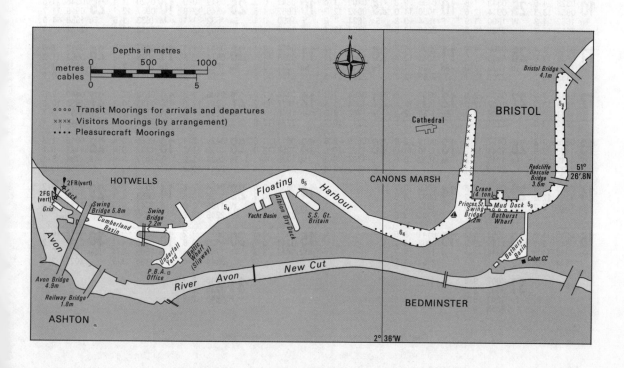

ENGLAND, WEST COAST — PORT OF BRISTOL (AVONMOUTH)

TIME ZONE UT (GMT)
For Summer Time add ONE hour in non-shaded areas

Lat 51°30′ N Long 2°43′ W

TIMES AND HEIGHTS OF HIGH AND LOW WATERS

YEAR 1989

JANUARY

Day	Time	m	Day	Time	m
1 Su	0032 / 0636 / 1257 / 1907	10.0 / 3.6 / 10.2 / 3.8	16 M	0113 / 0740 / 1352 / 2016	11.0 / 3.3 / 10.9 / 3.3
2 M	0137 / 0730 / 1408 / 2013	9.7 / 4.1 / 10.0 / 4.1	17 Tu	0226 / 0856 / 1511 / 2134	10.5 / 3.7 / 10.6 / 3.5
3 Tu	0250 / 0908 / 1515 / 2153	9.8 / 4.2 / 10.2 / 3.8	18 W	0346 / 1021 / 1627 / 2258	10.6 / 3.6 / 10.8 / 3.3
4 W	0355 / 1031 / 1617 / 2259	10.2 / 3.6 / 10.7 / 3.2	19 Th	0458 / 1146 / 1733	11.0 / 3.1 / 11.2
5 Th	0454 / 1132 / 1718 / 2357	10.9 / 2.9 / 11.2 / 2.6	20 F	0012 / 0557 / 1248 / 1828	2.7 / 11.6 / 2.5 / 11.7
6 F	0549 / 1228 / 1812	11.6 / 2.3 / 11.8	21 Sa	0109 / 0648 / 1337 / 1914 ○	2.2 / 12.1 / 2.1 / 12.1
7 Sa ●	0055 / 0639 / 1326 / 1903	2.2 / 12.2 / 1.9 / 12.2	22 Su	0155 / 0730 / 1420 / 1955	1.9 / 12.4 / 1.9 / 12.3
8 Su	0151 / 0727 / 1419 / 1949	1.9 / 12.6 / 1.7 / 12.6	23 M	0236 / 0808 / 1502 / 2032	1.7 / 12.6 / 1.8 / 12.4
9 M	0242 / 0811 / 1507 / 2034	1.7 / 13.0 / 1.5 / 13.0	24 Tu	0312 / 0843 / 1532 / 2105	1.7 / 12.7 / 1.8 / 12.3
10 Tu	0328 / 0854 / 1552 / 2117	1.5 / 13.3 / 1.3 / 13.2	25 W	0343 / 0914 / 1600 / 2135	1.8 / 12.6 / 1.9 / 12.3
11 W	0410 / 0936 / 1633 / 2159	1.4 / 13.5 / 1.3 / 13.2	26 Th	0409 / 0943 / 1626 / 2203	1.9 / 12.5 / 2.0 / 12.1
12 Th	0448 / 1020 / 1711 / 2242	1.5 / 13.3 / 1.4 / 12.9	27 F	0433 / 1012 / 1648 / 2231	2.0 / 12.3 / 2.1 / 11.7
13 F	0526 / 1105 / 1749 / 2327	1.8 / 12.9 / 1.7 / 12.3	28 Sa	0457 / 1041 / 1713 / 2259	2.1 / 11.8 / 2.4 / 11.1
14 Sa ☽	0603 / 1153 / 1828	2.2 / 12.2 / 2.2	29 Su	0523 / 1113 / 1742 / 2329	2.6 / 11.2 / 2.9 / 10.5
15 Su	0015 / 0646 / 1246 / 1914	11.6 / 2.7 / 11.5 / 2.8	30 M	0553 / 1149 / 1814 ☾	3.2 / 10.5 / 3.4
			31 Tu	0007 / 0632 / 1241 / 1859	9.8 / 3.8 / 9.9 / 4.0

FEBRUARY

Day	Time	m	Day	Time	m
1 W	0117 / 0728 / 1404 / 2016	9.3 / 4.4 / 9.5 / 4.5	16 Th	0317 / 0959 / 1610 / 2238	9.8 / 4.1 / 10.0 / 3.8
2 Th	0258 / 0942 / 1532 / 2219	9.4 / 4.4 / 9.8 / 4.3	17 F	0442 / 1129 / 1722 / 2354	10.4 / 3.2 / 10.4 / 2.9
3 F	0419 / 1058 / 1649 / 2327	10.2 / 3.4 / 10.6 / 3.0	18 Sa	0546 / 1229 / 1815	11.3 / 2.4 / 11.5
4 Sa	0526 / 1204 / 1756	11.2 / 2.6 / 11.6	19 Su	0050 / 0634 / 1319 / 1857	2.1 / 12.0 / 1.8 / 12.1
5 Su	0036 / 0622 / 1312 / 1849	2.3 / 12.2 / 1.9 / 12.5	20 M	0135 / 0713 / 1401 / 1935 ○	1.6 / 12.4 / 1.6 / 12.5
6 M	0141 / 0712 / 1409 / 1935	1.7 / 13.0 / 1.3 / 13.1	21 Tu	0215 / 0748 / 1437 / 2009	1.5 / 12.7 / 1.5 / 12.7
7 Tu	0233 / 0755 / 1458 / 2019	1.3 / 13.7 / 0.8 / 13.7	22 W	0250 / 0819 / 1510 / 2039	1.3 / 12.9 / 1.4 / 12.8
8 W	0319 / 0837 / 1542 / 2100	0.8 / 14.1 / 0.5 / 14.0	23 Th	0321 / 0847 / 1536 / 2107	1.3 / 13.0 / 1.4 / 12.8
9 Th	0400 / 0919 / 1620 / 2141	0.6 / 14.3 / 0.5 / 13.9	24 F	0346 / 0915 / 1600 / 2132	1.3 / 13.0 / 1.4 / 12.6
10 F	0435 / 1000 / 1654 / 2220	1.0 / 14.0 / 0.8 / 13.5	25 Sa	0409 / 0942 / 1621 / 2157	1.7 / 12.7 / 1.6 / 12.1
11 Sa	0508 / 1041 / 1725 / 2259	1.2 / 13.4 / 1.4 / 12.7	26 Su	0431 / 1009 / 1642 / 2221	1.7 / 12.1 / 2.0 / 11.5
12 Su	0537 / 1123 / 1756 / 2340 ☽	1.9 / 12.4 / 2.2 / 11.6	27 M	0454 / 1034 / 1706 / 2242	2.3 / 11.4 / 2.5 / 10.7
13 M	0610 / 1210 / 1832	2.9 / 11.2 / 3.0	28 Tu ☾	0519 / 1102 / 1734 / 2312	2.9 / 10.6 / 3.1 / 10.0
14 Tu	0028 / 0652 / 1312 / 1921	10.7 / 3.5 / 10.2 / 3.8			
15 W	0141 / 0802 / 1439 / 2054	9.9 / 4.2 / 9.7 / 4.2			

MARCH

Day	Time	m	Day	Time	m
1 W	0551 / 1144 / 1811	3.5 / 9.9 / 3.7	16 Th	0107 / 0719 / 1411 / 2009	9.4 / 4.3 / 9.1 / 4.5
2 Th	0007 / 0639 / 1300 / 1910	9.4 / 4.1 / 9.3 / 3.9	17 F	0250 / 0938 / 1548 / 2216	9.3 / 4.2 / 9.5 / 3.9
3 F	0157 / 0823 / 1447 / 2134	9.1 / 4.5 / 9.4 / 4.3	18 Sa	0421 / 1104 / 1659 / 2329	10.1 / 3.2 / 10.5 / 2.8
4 Sa	0343 / 1024 / 1623 / 2258	9.8 / 3.6 / 10.3 / 3.2	19 Su	0523 / 1203 / 1750	11.1 / 2.3 / 11.7
5 Su	0502 / 1139 / 1734	11.1 / 2.6 / 11.6	20 M	0022 / 0608 / 1249 / 1832	2.0 / 11.9 / 1.7 / 12.1
6 M	0018 / 0600 / 1255 / 1828	2.3 / 12.4 / 1.7 / 12.7	21 Tu	0107 / 0648 / 1331 / 1907	1.5 / 12.4 / 1.4 / 12.5
7 Tu ●	0126 / 0649 / 1354 / 1914	1.4 / 13.4 / 0.9 / 13.6	22 W ○	0147 / 0721 / 1408 / 1940	1.2 / 12.7 / 1.3 / 12.8
8 W	0218 / 0734 / 1442 / 1958	0.7 / 14.1 / 0.3 / 14.1	23 Th	0222 / 0751 / 1440 / 2009	1.1 / 12.9 / 1.2 / 12.9
9 Th	0303 / 0816 / 1524 / 2039	0.3 / 14.5 / 0.1 / 14.3	24 F	0253 / 0819 / 1508 / 2036	1.0 / 13.0 / 1.1 / 12.9
10 F	0342 / 0857 / 1600 / 2117	0.2 / 14.5 / 0.1 / 14.2	25 Sa	0319 / 0846 / 1532 / 2103	0.9 / 13.0 / 1.1 / 12.7
11 Sa	0414 / 0938 / 1631 / 2156	0.4 / 14.1 / 0.6 / 13.6	26 Su	0343 / 1038 / 1555 / 2128	1.1 / 11.5 / 1.4 / 12.2
12 Su	0444 / 1017 / 1658 / 2234	1.0 / 13.2 / 1.3 / 12.5	27 M ☽	0406 / 0941 / 1616 / 2150	1.5 / 12.0 / 1.8 / 11.5
13 M	0511 / 1058 / 1725 / 2311	1.8 / 12.0 / 2.2 / 11.4	28 Tu	0428 / 1006 / 1638 / 2214	2.1 / 11.2 / 2.3 / 10.8
14 Tu ☽	0539 / 1142 / 1757 / 2354	2.6 / 10.7 / 3.1 / 10.3	29 W	0454 / 1035 / 1706 / 2245	2.6 / 10.4 / 2.9 / 10.2
15 W	0617 / 1241 / 1842	3.5 / 9.7 / 3.9	30 Th ☾	0526 / 1120 / 1744 / 2344	3.2 / 9.9 / 3.4 / 9.6
			31 F	0615 / 1238 / 1843	3.7 / 9.4 / 3.9

APRIL

Day	Time	m	Day	Time	m
1 Sa	0127 / 0747 / 1418 / 2051	9.3 / 4.0 / 9.3 / 3.9	16 Su	0339 / 1019 / 1620 / 2244	10.0 / 3.2 / 10.3 / 2.9
2 Su	0311 / 0948 / 1553 / 2223	10.1 / 3.2 / 10.5 / 3.0	17 M	0444 / 1116 / 1713 / 2340	10.9 / 2.4 / 11.3 / 2.1
3 M	0431 / 1104 / 1705 / 2349	11.3 / 2.3 / 11.8 / 2.1	18 Tu	0532 / 1207 / 1756	11.6 / 1.9 / 11.8
4 Tu	0533 / 1228 / 1801	12.6 / 1.4 / 12.9	19 W	0028 / 0612 / 1252 / 1834	1.7 / 12.1 / 1.6 / 12.3
5 W	0102 / 0624 / 1330 / 1850	1.2 / 13.5 / 0.7 / 13.6	20 Th	0110 / 0648 / 1331 / 1907	1.4 / 12.4 / 1.4 / 12.5
6 Th ●	0155 / 0710 / 1419 / 1934	0.6 / 14.1 / 0.2 / 14.1	21 F ○	0147 / 0719 / 1405 / 1938	1.2 / 12.6 / 1.2 / 12.7
7 F	0240 / 0754 / 1500 / 2015	0.3 / 14.2 / 0.1 / 14.2	22 Sa	0220 / 0749 / 1437 / 2006	1.0 / 12.7 / 1.2 / 12.6
8 Sa	0318 / 0834 / 1536 / 2054	0.2 / 14.2 / 0.3 / 13.9	23 Su	0251 / 0819 / 1505 / 2036	1.0 / 12.6 / 1.2 / 12.4
9 Su	0352 / 0917 / 1606 / 2134	0.5 / 13.7 / 0.8 / 13.2	24 M	0321 / 0849 / 1532 / 2104	1.2 / 12.2 / 1.5 / 12.0
10 M	0420 / 0957 / 1633 / 2212	1.1 / 12.7 / 1.5 / 12.2	25 Tu	0346 / 0919 / 1556 / 2132	1.6 / 11.7 / 1.9 / 11.4
11 Tu	0447 / 1038 / 1701 / 2251	1.9 / 11.5 / 2.3 / 11.1	26 W	0412 / 0952 / 1620 / 2204	2.0 / 11.1 / 2.3 / 10.9
12 W ☽	0518 / 1123 / 1733 / 2337	2.7 / 10.3 / 3.1 / 10.4	27 Th	0440 / 1030 / 1651 / 2248	2.5 / 10.6 / 2.7 / 10.4
13 Th	0556 / 1222 / 1817	3.4 / 9.4 / 3.9	28 F ☾	0516 / 1123 / 1734 / 2354	2.8 / 10.2 / 3.1 / 10.0
14 F	0046 / 1035 / 1341 / 1930	9.3 / 4.1 / 9.0 / 4.4	29 Sa	0612 / 1234 / 1841	3.2 / 10.0 / 3.4
15 Sa	0215 / 0900 / 1505 / 2136	9.3 / 4.1 / 9.4 / 3.9	30 Su	0116 / 0742 / 1357 / 2025	10.1 / 3.2 / 10.2 / 3.3

Chart Datum: 6.50 metres below Ordnance Datum (Newlyn)

ENGLAND, WEST COAST — PORT OF BRISTOL (AVONMOUTH)

Lat 51°30′ N   Long 2°43′ W

TIMES AND HEIGHTS OF HIGH AND LOW WATERS

TIME ZONE UT (GMT)
For Summer Time add ONE hour in non-shaded areas

YEAR 1989

Chart Datum: 6.50 metres below Ordnance Datum (Newlyn)

MAY

Day	Time	m	Day	Time	m
1 M	0242 / 0914 / 1522 / 2149	10.7 / 2.6 / 11.0 / 2.6	16 Tu	0346 / 1021 / 1620 / 2247	10.6 / 2.8 / 10.7 / 2.6
2 Tu	0359 / 1026 / 1634 / 2306	11.6 / 2.0 / 11.9 / 2.1	17 W	0442 / 1113 / 1711 / 2339	11.2 / 2.3 / 11.3 / 2.1
3 W	0502 / 1150 / 1733	12.6 / 1.5 / 12.7	18 Th	0529 / 1203 / 1754	11.6 / 2.0 / 11.8
4 Th	0031 / 0557 / 1300 / 1824	1.5 / 13.2 / 1.0 / 13.3	19 F	0025 / 0610 / 1249 / 1832	1.8 / 12.0 / 1.8 / 12.1
5 F ●	0128 / 0646 / 1351 / 1909	1.0 / 13.5 / 0.7 / 13.6	20 Sa ○	0110 / 0646 / 1330 / 1907	1.5 / 12.1 / 1.6 / 12.2
6 Sa	0215 / 0731 / 1434 / 1952	0.7 / 13.6 / 0.6 / 13.6	21 Su	0149 / 0723 / 1408 / 1941	1.4 / 12.2 / 1.5 / 12.2
7 Su	0256 / 0816 / 1512 / 2034	0.7 / 13.5 / 0.8 / 13.4	22 M	0227 / 0758 / 1444 / 2016	1.4 / 12.1 / 1.6 / 12.1
8 M	0331 / 0900 / 1545 / 2115	1.0 / 13.0 / 1.2 / 12.8	23 Tu	0303 / 0833 / 1517 / 2051	1.5 / 11.9 / 1.8 / 11.9
9 Tu	0403 / 0942 / 1616 / 2157	1.5 / 12.2 / 1.8 / 11.9	24 W	0336 / 0910 / 1548 / 2128	1.8 / 11.7 / 2.0 / 11.7
10 W	0433 / 1026 / 1645 / 2238	2.1 / 11.2 / 2.5 / 11.0	25 Th	0409 / 0950 / 1619 / 2210	2.0 / 11.4 / 2.3 / 11.4
11 Th	0505 / 1111 / 1719 / 2325	2.8 / 10.3 / 3.1 / 10.2	26 F	0442 / 1035 / 1654 / 2259	2.3 / 11.2 / 2.5 / 11.1
12 F ☽	0543 / 1201 / 1800	3.3 / 9.6 / 3.6	27 Sa	0526 / 1127 / 1743 / 2356	2.5 / 11.0 / 2.7 / 11.0
13 Sa	0022 / 0632 / 1303 / 1855	9.8 / 3.7 / 9.4 / 3.9	28 Su ☾	0622 / 1227 / 1846	2.6 / 10.8 / 2.9
14 Su	0131 / 0749 / 1412 / 2033	9.7 / 3.8 / 9.6 / 3.8	29 M	0100 / 0731 / 1334 / 2002	11.0 / 2.6 / 10.9 / 2.8
15 M	0242 / 0921 / 1519 / 2149	10.0 / 3.3 / 10.1 / 3.2	30 Tu	0215 / 0843 / 1450 / 2117	11.2 / 2.4 / 11.2 / 2.6
			31 W	0328 / 0953 / 1602 / 2231	11.7 / 2.1 / 11.7 / 2.4

JUNE

Day	Time	m	Day	Time	m
1 Th	0434 / 1109 / 1704 / 2356	12.1 / 2.0 / 12.2 / 2.1	16 F	0440 / 1115 / 1711 / 2343	11.0 / 2.5 / 11.2 / 2.4
2 F	0532 / 1228 / 1758	12.5 / 1.7 / 12.6	17 Sa	0530 / 1207 / 1758	11.3 / 2.3 / 11.6
3 Sa ●	0100 / 0625 / 1326 / 1849	1.7 / 12.7 / 1.5 / 12.8	18 Su	0034 / 0618 / 1257 / 1842	2.0 / 11.6 / 2.0 / 11.8
4 Su	0152 / 0716 / 1412 / 1935	1.5 / 12.7 / 1.4 / 12.9	19 M ○	0123 / 0702 / 1345 / 1924	1.8 / 11.8 / 1.9 / 12.0
5 M	0236 / 0802 / 1454 / 2019	1.4 / 12.6 / 1.4 / 12.7	20 Tu	0211 / 0744 / 1430 / 2004	1.8 / 11.9 / 1.9 / 12.1
6 Tu	0317 / 0847 / 1531 / 2103	1.5 / 12.4 / 1.7 / 12.4	21 W	0254 / 0825 / 1512 / 2044	1.8 / 12.0 / 1.9 / 12.3
7 W	0352 / 0931 / 1604 / 2145	1.8 / 11.9 / 2.1 / 11.9	22 Th	0336 / 0907 / 1553 / 2125	1.8 / 12.2 / 2.0 / 12.3
8 Th	0426 / 1013 / 1637 / 2224	2.3 / 11.3 / 2.5 / 11.3	23 F	0416 / 0948 / 1631 / 2209	1.8 / 12.2 / 2.0 / 12.3
9 F	0457 / 1052 / 1706 / 2304	2.7 / 10.7 / 2.9 / 10.8	24 Sa	0455 / 1031 / 1709 / 2255	1.8 / 12.2 / 2.1 / 12.2
10 Sa	0529 / 1133 / 1740 / 2347	3.0 / 10.3 / 3.2 / 10.5	25 Su	0534 / 1118 / 1750 / 2343	1.9 / 11.8 / 2.3 / 11.9
11 Su ☽	0605 / 1219 / 1819	3.2 / 10.1 / 3.3	26 M ☾	0618 / 1206 / 1838	2.1 / 11.6 / 2.5
12 M	0041 / 0650 / 1316 / 1910	10.3 / 3.3 / 10.0 / 3.5	27 Tu	0039 / 0709 / 1306 / 1934	11.6 / 2.3 / 11.3 / 2.8
13 Tu	0141 / 0752 / 1418 / 2027	10.4 / 3.5 / 10.0 / 3.6	28 W	0144 / 0811 / 1418 / 2044	11.2 / 2.6 / 11.1 / 3.0
14 W	0244 / 0915 / 1519 / 2150	10.4 / 3.4 / 10.3 / 3.3	29 Th	0257 / 0921 / 1531 / 2202	11.2 / 2.8 / 11.2 / 3.0
15 Th	0343 / 1020 / 1617 / 2249	10.7 / 3.0 / 10.7 / 2.8	30 F	0407 / 1038 / 1640 / 2327	11.3 / 2.8 / 11.4 / 2.8

JULY

Day	Time	m	Day	Time	m
1 Sa	0513 / 1201 / 1740	11.6 / 2.5 / 11.8	16 Su	0458 / 1132 / 1732	10.6 / 2.9 / 11.1
2 Su	0039 / 0614 / 1304 / 1835	2.3 / 11.8 / 2.1 / 12.1	17 M	0004 / 0556 / 1231 / 1822	2.6 / 11.2 / 2.5 / 11.7
3 M ●	0135 / 0706 / 1355 / 1924	2.0 / 12.1 / 1.9 / 12.4	18 Tu ○	0103 / 0646 / 1328 / 1907	2.1 / 11.7 / 2.2 / 12.2
4 Tu	0223 / 0752 / 1440 / 2009	1.8 / 12.2 / 1.8 / 12.5	19 W	0158 / 0731 / 1422 / 1951	1.9 / 12.1 / 1.9 / 12.6
5 W	0305 / 0836 / 1519 / 2050	1.8 / 12.2 / 1.8 / 12.4	20 Th	0247 / 0813 / 1508 / 2032	1.7 / 12.6 / 1.7 / 13.0
6 Th	0342 / 0915 / 1555 / 2128	1.9 / 12.1 / 2.0 / 12.1	21 F	0332 / 0854 / 1550 / 2114	1.4 / 12.9 / 1.5 / 13.3
7 F	0414 / 0952 / 1624 / 2203	2.1 / 11.8 / 2.3 / 11.7	22 Sa	0413 / 0935 / 1630 / 2155	1.3 / 13.1 / 1.4 / 13.3
8 Sa	0442 / 1026 / 1651 / 2235	2.3 / 11.4 / 2.5 / 11.3	23 Su	0451 / 1017 / 1705 / 2238	1.2 / 13.0 / 1.4 / 13.1
9 Su	0508 / 1059 / 1716 / 2311	2.5 / 11.2 / 2.6 / 11.0	24 M	0525 / 1058 / 1739 / 2322	1.5 / 12.6 / 1.9 / 12.5
10 M	0536 / 1134 / 1746 / 2349	2.6 / 10.8 / 2.8 / 10.7	25 Tu ☾	0600 / 1143 / 1815	1.9 / 11.9 / 2.4
11 Tu ☽	0608 / 1209 / 1821	2.9 / 10.4 / 3.2	26 W	0011 / 0639 / 1234 / 1900	11.7 / 2.5 / 11.2 / 3.0
12 W	0038 / 0648 / 1310 / 1907	10.5 / 3.3 / 9.9 / 3.7	27 Th	0110 / 0731 / 1341 / 2006	10.9 / 3.1 / 10.6 / 3.6
13 Th	0138 / 0740 / 1419 / 2022	10.1 / 3.8 / 9.7 / 4.1	28 F	0229 / 0849 / 1504 / 2141	10.3 / 3.6 / 10.4 / 3.7
14 F	0247 / 0915 / 1528 / 2204	9.9 / 3.9 / 9.9 / 3.8	29 Sa	0349 / 1019 / 1623 / 2312	10.4 / 3.5 / 10.7 / 3.3
15 Sa	0353 / 1033 / 1633 / 2308	10.2 / 3.5 / 10.4 / 3.1	30 Su	0505 / 1144 / 1732	10.9 / 3.0 / 11.3
			31 M ●	0025 / 0607 / 1249 / 1828	2.6 / 11.4 / 2.3 / 11.9

AUGUST

Day	Time	m	Day	Time	m
1 Tu ●	0120 / 0657 / 1340 / 1913	2.0 / 11.9 / 1.9 / 12.3	16 W	0045 / 0627 / 1314 / 1848	2.2 / 11.9 / 2.1 / 12.6
2 W	0208 / 0740 / 1425 / 1954	1.7 / 12.3 / 1.7 / 12.6	17 Th ○	0145 / 0713 / 1409 / 1931	1.7 / 12.7 / 1.6 / 13.3
3 Th	0249 / 0819 / 1503 / 2030	1.6 / 12.5 / 1.6 / 12.7	18 F	0234 / 0755 / 1457 / 2013	1.3 / 13.3 / 1.2 / 13.8
4 F	0325 / 0853 / 1536 / 2104	1.6 / 12.5 / 1.7 / 12.7	19 Sa	0319 / 0836 / 1538 / 2054	0.8 / 13.7 / 0.9 / 14.0
5 Sa	0355 / 0925 / 1603 / 2135	1.7 / 12.5 / 1.8 / 12.6	20 Su	0359 / 0915 / 1616 / 2135	0.7 / 13.8 / 0.9 / 14.0
6 Su	0419 / 0955 / 1627 / 2203	1.8 / 12.3 / 1.9 / 12.4	21 M	0434 / 0955 / 1648 / 2216	0.8 / 13.5 / 1.2 / 13.4
7 M	0441 / 1023 / 1648 / 2233	2.0 / 11.9 / 2.1 / 12.0	22 Tu	0505 / 1034 / 1718 / 2258	1.3 / 12.9 / 1.8 / 12.5
8 Tu	0504 / 1051 / 1713 / 2304	2.2 / 11.3 / 2.5 / 11.3	23 W ☾	0534 / 1115 / 1749 / 2343	2.0 / 11.8 / 2.5 / 11.3
9 W	0529 / 1120 / 1742 / 2337	2.6 / 10.6 / 3.0 / 10.5	24 Th	0607 / 1201 / 1827	2.8 / 10.8 / 3.3
10 Th	0600 / 1156 / 1817	3.2 / 9.7 / 3.7	25 F	0041 / 0653 / 1310 / 1927	10.3 / 3.6 / 9.9 / 4.1
11 F	0022 / 0639 / 1252 / 1906	9.7 / 3.9 / 9.2 / 4.4	26 Sa	0205 / 0813 / 1443 / 2127	9.6 / 4.3 / 9.7 / 4.2
12 Sa ☽	0133 / 0740 / 1427 / 2105	9.2 / 4.5 / 9.1 / 4.6	27 Su	0338 / 1007 / 1613 / 2302	9.9 / 3.9 / 10.3 / 3.3
13 Su	0305 / 0950 / 1556 / 2235	9.3 / 4.3 / 9.7 / 3.8	28 M	0457 / 1130 / 1723	10.6 / 3.0 / 11.2
14 M	0428 / 1102 / 1706 / 2339	10.1 / 3.5 / 10.7 / 2.9	29 Tu	0008 / 0554 / 1231 / 1814	2.4 / 11.5 / 2.1 / 12.0
15 Tu	0534 / 1208 / 1801	11.0 / 2.7 / 11.8	30 W	0102 / 0641 / 1320 / 1856	1.7 / 12.1 / 1.6 / 12.5
			31 Th ●	0145 / 0719 / 1401 / 1933	1.4 / 12.5 / 1.4 / 12.8

ENGLAND, WEST COAST — PORT OF BRISTOL (AVONMOUTH)

TIME ZONE UT (GMT)
For Summer Time add ONE hour in non-shaded areas

Lat 51°30′ N Long 2°43′ W

TIMES AND HEIGHTS OF HIGH AND LOW WATERS

YEAR 1989

SEPTEMBER

Time	m		Time	m
1 0225	1·3	**16** 0215	0·8
0754	12·8		0731	13·8
F 1437	1·3	Sa 1436	0·8
2005	13·0		1949	14·2
2 0257	1·3	**17** 0258	0·4
0826	12·9		0812	14·1
Sa 1510	1·3	Su 1518	0·6
2036	13·1		2032	14·4
3 0325	1·4	**18** 0338	0·4
0856	12·9		0851	14·1
Su 1535	1·4	M 1555	0·7
2104	13·0		2112	14·1
4 0349	1·4	**19** 0412	0·7
0922	12·7		0931	13·6
M 1559	1·5	Tu 1626	1·1
2131	12·7		2153	13·3
5 0410	1·6	**20** 0440	1·4
0948	12·2		1012	12·7
Tu 1620	1·8	W 1654	1·8
2157	12·1		2235	12·1
6 0431	2·0	**21** 0508	2·2
1012	11·4		1051	11·5
W 1642	2·3	Th 1723	2·7
2223	11·2		2322	10·8
7 0454	2·5	**22** 0542	3·1
1033	10·6		1139	10·4
Th 1708	3·0	F 1803	3·5
2249	10·4	☾	
8 0520	3·2	**23** 0022	9·7
1057	9·8		0625	3·9
F 1737	3·7	Sa 1250	9·5
2325	9·6		1900	4·3
☽				
9 0554	3·9	**24** 0147	4·5
1143	9·2		0744	4·5
Sa 1819	4·3	Su 1423	4·4
			2114	4·2
10 0031	9·0	**25** 0319	9·5
0643	4·5		0950	3·9
Su 1317	8·8	M 1556	10·2
1935	4·8		2240	3·2
11 0211	8·9	**26** 0437	10·5
0846	4·6		1105	2·8
M 1517	9·4	Tu 1701	11·2
2157	4·0		2340	2·2
12 0356	9·8	**27** 0530	11·5
1027	3·6		1201	1·9
Tu 1635	10·7	W 1750	12·1
2306	2·9			
13 0506	11·1	**28** 0031	1·6
1139	2·6		0614	12·3
W 1733	12·0	Th 1249	1·4
			1829	12·6
14 0019	2·1	**29** 0114	1·3
0601	12·3		0650	12·7
Th 1253	1·9	F 1330	1·2
1822	13·1	● 1904	12·9
15 0124	1·3	**30** 0151	1·2
0648	13·2		0726	12·9
F 1349	1·2	Sa 1405	1·2
○ 1907	13·8		1937	13·1

OCTOBER

Time	m		Time	m
1 0225	1·2	**16** 0234	0·5
0755	13·0		0748	14·1
Su 1437	1·1	M 1454	0·6
2006	13·1		2009	14·2
2 0253	1·2	**17** 0314	0·5
0825	13·0		0830	13·9
M 1505	1·1	Tu 1532	0·8
2033	13·0		2053	13·8
3 0318	1·3	**18** 0348	0·9
0850	12·7		0911	13·4
Tu 1529	1·3	W 1604	1·2
2100	12·6		2136	12·9
4 0341	1·5	**19** 0419	1·6
0915	12·1		0953	12·5
W 1553	1·7	Th 1634	2·0
2127	11·9		2221	11·8
5 0403	2·0	**20** 0449	2·4
0939	11·4		1037	11·3
Th 1616	2·3	F 1708	2·8
2152	11·1		2309	10·6
6 0426	2·6	**21** 0523	3·2
1000	10·6		1127	10·3
F 1641	2·9	Sa 1747	3·6
2219	10·4	☾	
7 0451	3·1	**22** 0008	9·6
1027	10·0		0607	4·0
Sa 1709	3·5	Su 1232	9·6
2257	9·7		1843	4·2
8 0523	3·6	**23** 0120	9·2
1116	9·4		0717	4·4
Su 1753	4·0	M 1351	9·5
☽			2037	4·1
9 0003	9·2	**24** 0240	9·5
0614	4·1		0912	3·9
M 1248	9·1	Tu 1514	10·1
1906	4·4		2155	3·3
10 0137	9·2	**25** 0356	10·4
0755	4·3		1021	3·0
Tu 1434	9·7	W 1623	11·0
2111	3·8		2254	2·4
11 0318	10·1	**26** 0454	11·3
0946	3·5		1118	2·2
W 1559	10·9	Th 1713	11·9
2226	2·7		2346	1·9
12 0433	11·4	**27** 0539	12·0
1059	2·5		1207	1·7
Th 1701	12·3	F 1756	12·4
2342	1·9			
13 0530	12·6	**28** 0032	1·6
1222	1·8		0618	12·5
F 1754	13·3	Sa 1250	1·4
			1834	12·7
14 0055	1·2	**29** 0112	1·4
0623	13·1		0653	12·7
Sa 1323	1·1	Su 1328	1·3
○ 1842	13·9	● 1906	12·8
15 0149	0·7	**30** 0148	1·3
0706	13·9		0726	12·8
Su 1412	0·7	M 1404	1·2
1927	14·2		1937	12·8
		31 0220	1·3
			0755	12·7
		Tu 1436	1·2
			2006	12·6

NOVEMBER

Time	m		Time	m
1 0250	1·4	**16** 0329	1·2
0823	12·5		0857	13·1
W 1505	1·4	Th 1549	1·5
2036	12·3		2125	12·6
2 0317	1·7	**17** 0404	1·8
0851	12·0		0941	12·4
Th 1532	1·8	F 1624	2·1
2104	11·7		2210	11·7
3 0342	2·1	**18** 0438	2·5
0918	11·4		1026	11·5
F 1557	2·4	Sa 1658	2·8
2135	11·2		2257	10·8
4 0406	2·6	**19** 0512	3·2
0948	10·9		1113	10·7
Sa 1624	3·0	Su 1737	3·4
2209	10·6		2347	10·1
5 0433	3·0	**20** 0551	3·7
1024	10·4		1207	10·2
Su 1657	3·2	M 1824	3·8
2255	10·2	☾	
6 0509	3·3	**21** 0043	9·7
1120	10·0		0643	4·0
M 1743	3·6	Tu 1307	10·0
2357	9·9		1933	3·9
7 0605	3·7	**22** 0147	9·8
1235	9·9		0808	3·9
Tu 1859	3·7	W 1415	10·2
			2057	3·5
8 0113	10·0	**23** 0254	10·2
0734	3·8		0927	3·4
W 1401	10·3	Th 1524	10·7
2032	3·3		2159	3·0
9 0242	10·5	**24** 0400	10·8
0907	3·2		1024	2·8
Th 1522	11·3	F 1626	11·3
2146	2·6		2252	2·5
10 0357	11·5	**25** 0455	11·4
1021	2·6		1118	2·3
F 1628	12·3	Sa 1716	11·8
2259	2·0		2343	2·1
11 0459	12·5	**26** 0542	12·0
1144	2·0		1207	1·9
Sa 1726	12·8	Su 1758	12·1
			2346	1·9
12 0021	1·5	**27** 0029	1·8
0553	13·1		0621	12·3
Su 1255	1·5	M 1252	1·6
1818	13·5		1838	12·3
13 0121	1·1	**28** 0113	1·6
0642	13·5		0657	12·5
M 1347	1·1	Tu 1333	1·5
○ 1906	13·7	● 1913	12·3
14 0209	0·9	**29** 0152	1·6
0728	13·7		0731	12·5
Tu 1433	0·9	W 1412	1·5
1952	13·6		1947	12·2
15 0251	0·9	**30** 0227	1·7
0812	13·5		0804	12·2
W 1512	1·1	Th 1447	1·7
2039	13·2		2019	12·0

DECEMBER

Time	m		Time	m
1 0301	1·9	**16** 0356	1·9
0837	12·0		0929	12·6
F 1522	2·0	Sa 1617	2·0
2054	11·8		2157	12·0
2 0334	2·2	**17** 0430	2·3
0911	11·8		1010	12·0
Sa 1555	2·3	Su 1651	2·5
2129	11·6		2238	11·4
3 0403	2·5	**18** 0501	2·8
0948	11·6		1051	11·5
Su 1626	2·6	M 1722	2·9
2210	11·4		2318	10·9
4 0434	2·7	**19** 0532	3·1
1031	11·3		1132	11·1
M 1702	2·8	Tu 1756	3·2
2255	11·1	☾	
5 0513	3·0	**20** 0000	10·6
1120	11·1		0607	3·3
Tu 1749	2·9	W 1218	10·8
2347	10·9		1835	3·4
6 0607	3·2	**21** 0050	10·3
1219	11·0		0650	3·6
W 1848	3·0	Th 1314	10·5
			1927	3·6
7 0049	10·8	**22** 0151	10·1
0713	3·3		0758	3·8
Th 1328	11·0	F 1418	10·4
1957	3·0		2049	3·6
8 0205	10·9	**23** 0256	10·2
0830	3·2		0927	3·6
F 1446	11·3	Sa 1524	10·6
2111	2·8		2200	3·3
9 0322	11·3	**24** 0402	10·6
0948	3·0		1031	3·2
Sa 1557	11·8	Su 1628	10·9
2226	2·5		2258	2·9
10 0430	11·9	**25** 0502	11·1
1111	2·6		1126	2·6
Su 1701	12·3	M 1725	11·3
2349	2·1		2351	2·4
11 0529	12·5	**26** 0553	11·6
1228	2·1		1219	2·2
M 1758	12·7	Tu 1812	11·7
12 0056	1·7	**27** 0042	2·1
0624	12·9		0635	11·9
Tu 1327	1·6	W 1309	2·0
○ 1852	12·9		1853	11·9
13 0154	1·4	**28** 0130	2·0
0714	13·1		0714	12·0
W 1416	1·4	Th 1355	1·9
1942	13·0	● 1933	12·0
14 0236	1·3	**29** 0215	1·9
0803	13·1		0751	12·3
Th 1501	1·4	F 1437	1·9
2029	12·9		2009	12·2
15 0318	1·5	**30** 0256	2·0
0846	13·0		0827	12·4
F 1541	1·6	Sa 1519	1·9
2114	12·5		2047	12·3
		31 0335	2·0
			0904	12·6
		Su 1557	2·0
			2124	12·4

Chart Datum: 6.50 metres below Ordnance Datum (Newlyn)

AREA 11—Bristol Channel 519

BRISTOL continued

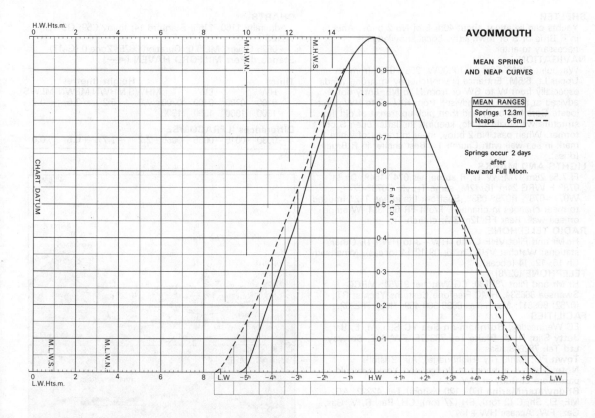

AVONMOUTH

MEAN SPRING
AND NEAP CURVES

MEAN RANGES
Springs 12.3m ———
Neaps 6.5m - - - -

Springs occur 2 days
after
New and Full Moon.

BURNHAM-ON-SEA 10-11-20
Somerset

CHARTS
Admiralty 1152; Stanford 14; Imray C59; OS 182
TIDES
−0435 Dover; ML 5.4; Duration 0620; Zone 0 (GMT).
Standard Port BRISTOL (AVONMOUTH) (←)

Times				Height (metres)			
HW		LW		MHWS	MHWN	MLWN	MLWS
0200	0800	0300	0800	13.2	10.0	3.5	0.9
1400	2000	1500	2000				

Differences BURNHAM
 −0020 −0025 −0030 0000 −2.3 −1.9 −1.4 −1.1
BRIDGWATER
 −0015 −0030 +0305 +0455 −8.6 −8.1 dries out
WESTON-SUPER-MARE
 −0020 −0030 −0130 −0030 −1.2 −1.0 −0.8 −0.2
WATCHET
 −0035 −0050 −0145 −0040 −1.9 −1.5 +0.1 +0.1
MINEHEAD
 −0035 −0045 −0100 −0100 −2.6 −1.9 −0.1 0.0
PORLOCK BAY
 −0045 −0055 −0205*−0050 −3.0 −2.2 −0.1 −0.1

BURNHAM-ON-SEA continued

SHELTER
Yachts can lie afloat about 40m E of No 2 buoy. Anchor in R Brue or S of town jetty. Local knowledge is necessary to enter.

NAVIGATION
Waypoint 51°13'.56N 03°10'.00W, 258°/078° from/to Upper Lt, 6.5M. Entrance is very choppy in strong winds especially from W to SW or from N to NE. Entry not advised at night. Entry advised from HW−3 to HW. First locate Gore bell buoy and then pick up transit of old light structure with main Lt Ho, keeping the latter open S of former. When past No 2 buoy, steer in transit of white mark in sea wall with Church Tr. Best shelter in R Brue (dries).

LIGHTS AND MARKS
Fl 7.5s 28m 17M; W Tr, R stripe; vis 074°-164°. Dir Lt 078° F WRG 24m 16/12M; same Tr; vis G073°-077°, W077°-079°, R079°-083°. Seafront Lts in line 112° (moved to meet changes in channel). Front FR 6m 3M; W stripe on sea wall. Rear FR 12m 3M; church Tr.

RADIO TELEPHONE
Hr Mr and Pilot VHF Ch 16 (HW−3 to HW+1). Other stations: Watchet VHF Ch 16; 09 12 14 (occas). Minehead Ch 16; 12, 14 (occas).

TELEPHONE (0278)
Hr Mr and Pilot 782180; CG Watchet 31751; MRCC Swansea 366534; HMC Freefone Customs Yachts: RT (0752) 669811; Marinecall 0898 500 459; Hosp 782262.

FACILITIES
EC Wednesday; **Burnham-on-Sea YC** Slip, M, L, Bar; **Jetty** Slip; **D. C. Blake** Tel. 783275, FW, ME; **Solway Ltd** Tel. 782908, Gas.
Town PO; Bank; Rly (Highbridge); Air (Bristol).
Note: There is a marina at Bridgwater. Yachts should pass Gore Lt Buoy at HW−3.
Bridgwater Marina (50+120 visitors) Tel. 42222, AC, ME, El, Sh, C (2 ton), BH (27 ton), CH, Bar, R, V, Gas, Gaz, FW, Access HW∓1½.

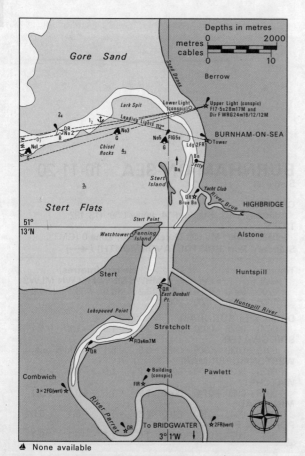

▲ None available

ILFRACOMBE 10-11-21
Devon

CHARTS
Admiralty 1160, 1165; Stanford 14; Imray C59; OS 180
TIDES
−0525 Dover; ML 5.0; Duration 0625; Zone 0 (GMT).
Standard Port MILFORD HAVEN (←)

Times				Height (metres)			
HW		LW		MHWS	MHWN	MLWN	MLWS
0100	0700	0100	0700	7.0	5.2	2.5	0.7
1300	1900	1300	1900				

Differences ILFRACOMBE
−0030 −0015 −0035 −0055 +2.2 +1.7 +0.5 0.0

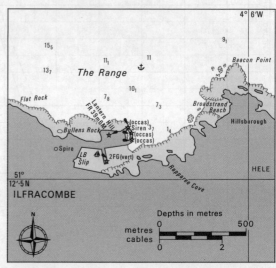

SHELTER
Good except E to NE Winds. On entering obtain directions from pierhead gates on NE side. SW gales cause a very severe surge in the harbour. 12 visitors buoys in outer harbour.

NAVIGATION
Waypoint 51°13'.20N 04°06'.60W, 000°/180° from/to pierhead, 0.55M. Coming from E, beware Copperas Rocks (4M to E), marked by buoy. Do not anchor in line with LB Slip. 1 or 2 balls hoisted on flagpole yard indicates steamer is to berth alongside pier.

LIGHTS AND MARKS
There are no leading marks. Inner Pier Head Lt 2 FG (vert). 3 Lts on Promenade Pier, all 2 FG (vert). Shown 1/9 to 30/4. Siren 30s when vessel expected. Lantern Hill Lt FR, 39m 6M.

RADIO TELEPHONE
Call: *Ilfracombe Harbour* VHF Ch 16; 12 (Apl-Oct 0800-2000 when manned. Nov-Mar occas). Ch M (occas).

TELEPHONE (0271)
Hr Mr 62108; CG 62117; MRSC Hartland 641; HMC Freefone Customs Yachts: RT (0752) 669811; Marinecall 0898 500 459; Dr 63119.

FACILITIES
EC Thursday; **Harbour** Tel. 62108, Slip, M (See Hr.Mr.), D (cans), L, FW, CH, V, R, Bar; **Ilfracombe YC** Tel. 63969; **Pier** FW; **Watermouth Marine** Tel. 62504, M, FW, Slip, D; **Harbour Chandlery** Tel. 62299, CH, Gas; **Ilfracombe Marine Services** Tel. 66815 ME, El, C (5 ton), Sh; **Town** PO; Bank; Rly (bus to Barnstaple); Air (Exeter).
Note: It is planned to turn the harbour into a marina.

AREA 11—Bristol Channel 521

RIVERS TAW & TORRIDGE
Devon
10-11-22

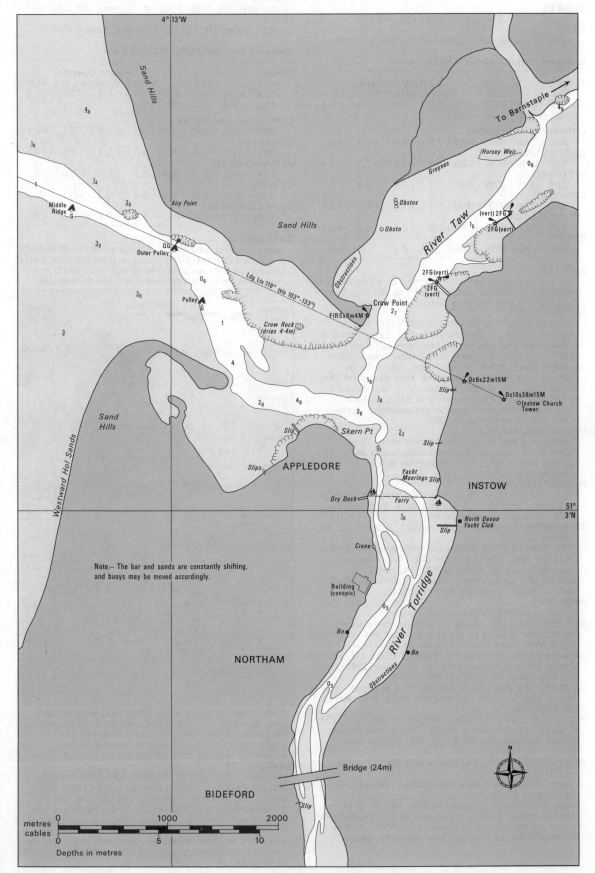

RIVERS TAW & TORRIDGE continued

CHARTS
Admiralty 1160; Stanford 14; Imray C58; OS 180
TIDES
−0520 Dover; ML 3.6; Duration 0600; Zone 0 (GMT)

Standard Port MILFORD HAVEN (←)

Times				Height (metres)			
HW		LW		MHWS	MHWN	MLWN	MLWS
0100	0700	0100	0700	7.0	5.2	2.5	0.7
1300	1900	1300	1900				
Differences APPLEDORE							
−0020	−0025	+0015	−0045	+0.5	0.0	−0.9	−0.5
BARNSTAPLE							
0000	−0015	−0155	−0245	−2.9	−3.8	−2.2	−0.4
BIDEFORD							
−0020	−0025	0000	0000	−1.1	−1.6	−2.5	−0.7
CLOVELLY							
−0030	−0030	−0020	−0040	+1.3	+1.1	+0.2	+0.2
LUNDY ISLAND							
−0030	−0030	−0020	−0040	+1.0	+0.7	+0.2	+0.1
BUDE							
−0040	−0040	−0035	−0045	+0.7	+0.6	No data	

SHELTER
Very well protected, but entry in strong on-shore winds is dangerous. Yachts can anchor or pick up mooring buoy in Appledore Pool. A marina is planned at Northam.
NAVIGATION
Waypoint 51°05'.40N 04°16'.04W, 298°/118° from/to front Ldg Lt 118°, 3.8M (channel shifts). Estuary dries out at low water and entry is only recommended from HW−2 to HW. Once tide is running out, breakers quickly form between Bar Buoy and Middle Ridge. Advice on bar available from Hartland Coastguard. Tidal stream off Skern Pt can reach 5kts at springs. Passage up to Bideford is not difficult but boats proceeding to Barnstaple are advised to take a pilot (available at Appledore). Some buoys may be withdrawn in 1989.
LIGHTS AND MARKS
Apart from jetties etc only the Fairway Buoy, Outer Pulley, Crow Point and the two leading marks are lit. The Torridge has no lights. Entry at night is NOT recommended.
RADIO TELEPHONE
Pilots: VHF Ch 16; 06 09 12. MF 2182; 2241kHz. (Listens on 2182kHz from HW−2). Other station: Bude VHF Ch 16; 12 (when vessel expected.)
TELEPHONE (Barnstaple + Instow 0271, Bideford 023 72, Appledore 023 383)
Hr Mr Bideford 74569, Barnstaple apply Amenities Officer 72511 Ext 7408; MRSC Hartland 641; HMC Freefone Customs Yachts: RT (0752) 669811; Marinecall 0898 500 459; Dr Appledore (Bideford 74994), Bideford 76444 or 76363, Barnstaple 75221 or 73443.
FACILITIES
EC Barnstaple & Bideford — Wednesday;
APPLEDORE is a free port so no authority can charge for use of public facilities eg slip at town quay and alongside berths. **Watts and Allen** Tel. Bideford 74167, CH; **Marine Engine Services** Tel. Bideford 75986, ME, El; **Marine Electronics Systems** Tel. Torrington 22870, El; **The Sea Chest** Tel. Bideford 76191, CH;
BIDEFORD: AB (few), V, R, Bar; Berthing arranged through Capt V. Harris Tel. Bideford 74569; **Blanchards** Tel. 72084, Gas;
INSTOW: **F. Johns** Tel. Instow 860578, M, AB, ACA, CH; **N. Devon YC** Tel. Instow 860367, Slip, L, Bar; **Instow Marine Services** Tel. 861081, D, ME, El, C (4 ton);
Town V, R, FW, Bar;
BARNSTAPLE: AB, V, Bar; **Barnstaple Calor Centre** Tel. 75794, Gas.
Towns PO (all four); Bank (Barnstaple, Bideford); Rly (Barnstaple); Air (Exeter).
Note: There are no alongside facilities for FW, D or P. Small quantities supplied in jerrycans. A larger quantity of D or P can be supplied by bowser from Plymouth (see Hr Mr). FW can be collected from the North Devon YC.

PADSTOW 10-11-23
Cornwall

CHARTS
Admiralty 1156, 1168; Stanford 13; Imray C58; OS 200
TIDES
−0550 Dover; ML 4.0; Duration 0600; Zone 0 (GMT)

Standard Port MILFORD HAVEN (←)

Times				Height (metres)			
HW		LW		MHWS	MHWN	MLWN	MLWS
0100	0700	0100	0700	7.0	5.2	2.5	0.7
1300	1900	1300	1900				
Differences PADSTOW							
−0055	−0050	−0040	−0050	+0.3	+0.4	+0.1	+0.1
NEWQUAY							
−0100	−0110	−0105	−0050	0.0	+0.1	0.0	−0.1
PERRANPORTH							
−0100	−0110	−0110	−0050	−0.1	0.0	0.0	+0.1
ST IVES							
−0050	−0115	−0105	−0040	−0.4	−0.3	−0.1	+0.1
CAPE CORNWALL							
−0130	−0145	−0120	−0120	−1.0	−0.9	−0.5	−0.1

SHELTER
Heavy breaking seas often encountered on Doom Bar and in adjacent channel during strong onshore winds or heavy ground swell. Allow ample rise of tide (least depth in channel at LWS late 1987 was 1.3m). Anchorage just

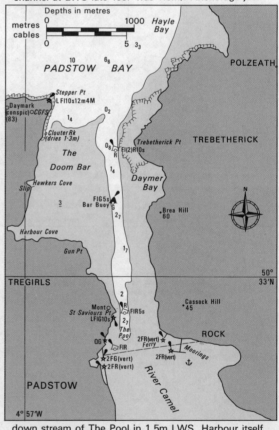

down stream of The Pool in 1.5m LWS. Harbour itself gives alongside berthing in good shelter except in strong SE winds. Harbour dries. Vessels may not be left unattended in harbour. Swinging mooring (drying) sometimes available for small vessels on passage.
NAVIGATION
Waypoint 50°35'.00N 04°58'.50W, 305°/125° from/to Stepper Pt, 1.5M. Coming from S, beware Quies Rks, Gulland Rk, The Hen, Gurley Rk and Chimney Rks and the wreck 5 ca W of Stepper Pt. All are hazardous (off the chartlet). From N, keep well clear of Newland Is and its reefs jutting out. The sandbanks in the estuary change frequently, so local knowledge, care and a rising tide are recommended. Final approach S of St Saviours Pt is close to W shore.

PADSTOW continued

LIGHTS AND MARKS
Entrance marked by the stone tower (daymark) conspic, W of Stepper Point. Stepper Pt Lt L Fl 10s 12m 4M and harbour entrance marked by 2 FG (vert) to stbd and 2FR (vert) to port.

RADIO TELEPHONE
VHF Ch 16 (HW−2½ to HW+1½).

TELEPHONE (0841)
Hr Mr 532239; CG St. Merryn 520407; MRCC Falmouth 317515; HMC Freefone Customs Yachts: RT (0752) 669811; Marinecall 0898 500 458; Dr 532346.

FACILITIES
EC Wednesday; **Harbour** Tel. 532239, Access HW∓2½, Slip, M, FW, El, C, D, AB, V, R, Bar; **Pier** Tel. 532239, D; **Dock** Tel. 532239, ME, C (6 ton), AB; **Rock SC** Tel. Trebetherick 2431, Slip; **Wadebridge Boat Centre** Tel. Wadebridge 3809, M, ME, El, CH; **Padstow SC** Tel. 53281, M; **Padstow Fishing & Chandlery Gear** Tel. 532617 (H24), CH; **Chapman & Hewitt** Tel. Wadebridge 2981, BY; **Westerly Boats** Tel. Trebetherick 3439, Slip, L, Sh; **Cornish Crabbers** Tel. Trebetherick 2666, Slip, ME, Sh, C; **Nettec Marine Chandlers** (on quay), CH, Gas, Gaz.
Town Lau, P, PO, Bank, Rly (bus to Bodmin Road), Air (Newquay or Plymouth).

MINOR HARBOURS AND ANCHORAGES 10.11.24

MOCHRAS, Gwynedd, 52°49′ N, 4°07′ W, Zone 0 (GMT), Admty chart 1512. HW −0245 on Dover, +0205 on Milford Haven. HW height −1.7m on Milford Haven. Small yacht harbour on E end of Shell Island. Mochras lagoon dries. See 10.11.5. Entrance between Shell Is (Lt Fl WRG 4s; G079°-124°, W124°-134°, R134°-179°; shown Apl-Nov) and sea wall. Bar, about 2 ca to seaward. Three R posts mark N side of channel. Tide runs strongly in the narrows on the ebb. Entry advised after HW∓2. Inside, the channel runs NE to Pen-Sarn, marked by posts. To S buoyed channel runs to Shell Island Yacht Harbour. Shallow draught boats can lie afloat below the railway bridge. Facilities (Pen-Sarn) Slip, AB, FW, Rly. **Shell Island** Tel. Llanbedr 453 M, FW, R, Slip.

ABERAERON, Dyfed, 52°19′ N, 4°09′ W, Zone 0 (GMT), Admty Chart 1484. HW −0325 on Dover, +0140 on Milford Haven; HW height −1.9m on Milford Haven; ML 2.7m; Duration 0540. A small drying harbour at the mouth of the R. Aeron (or Ayron). Short piers extend each side of the river ent. Pierheads dry. In strong NW winds there is little shelter. Foul ground with depths of 1.5m extend 3 ca offshore to SW of Aberaeron. Beware Carreg Gloyn (0.3m) 4 ca WSW of harbour; Sarn Cadwgan (1.8m) extending ½M offshore from Cadwgan Pt to NE of Aberaeron. Lights:− N pier Fl(4)R 15s 6m 6M; vis 104°-178°; (this Lt is FW(T)). S pier Fl(3)G 10s 6m 7M; vis 125°-200°. No radio telephone. Facilities: Hr Mr Tel. 570407, **Aberaeron YC** Tel. 570077.

NEW QUAY, Dyfed, 52°13′ N, 4°21′ W, Zone −0100. Admty charts 1484, 1972. HW −0335 on Dover, +0140 on Milford Haven. HW height −2.0m on Milford Haven; ML 2.6m; Duration 0540. The bay gives shelter in offshore winds, but NW winds make it untenable. On E side of bay is Carreg Ina, a rock which dries 1.6m marked by N cardinal buoy. Bns show alignment of sewer pipe extending 7ca NNW from Ina Point. There are moorings in the bay. The harbour (dries) is protected by a pier with a Lt, FlWG 3s 12m 8/5M; vis W135°-252°, G252°-295°. Rocks, marked by a Bn, extend 50m SSE from pierhead. Facilities: Hr Mr Tel. 560368; CG Tel. 560212; Dr Tel. 560203; YC Tel. 560516; D (from fishermen) Tel. 560375. **Town** D, FW, V, P (3M), PO, R, Bar.

CARDIGAN, Dyfed, 52°06′ N, 4°41′ W, Zone 0 (GMT), Admty chart 1484. HW −0350 on Dover, +0130 on Milford Haven; HW height −2.1m on Milford Haven; ML 2.7m; Duration 0550. See 10.11.10. Shelter is good but entrance dangerous in strong N to NW winds. Bar dries but has 2.5m at MHWS. Channel between Pen-y-Ergyd and Bryn Du changes continuously. Anchor in pools in R Teifi near St Dogmaels. There are no lights. Facilities: EC Wed; Hr Mr Tel. 612084; **Ynys Marine** Tel. 613179, ME, Sh; **Town** Bank, Bar, CH, D, FW, ME, P, R, Rly, V.

SOLVA, Dyfed, 51°52′ N, 5°12′ W, Zone 0 (GMT), Admty chart 1478. HW −0450 on Dover, +0012 on Milford Haven, HW height −1.3m on Milford Haven; ML 3.2m; Duration 0555. Good shelter for small boats that can take the ground. Beware Black Rock in centre of entrance (E side entrance recommended) and stone spit at Trwyn Caws on W just inside entrance. Strong S winds make entrance difficult. There are nine visitors buoys painted red and some drying moorings available. Anchor behind the rock in approx 3m. Small craft can go up to the quay. Facilities very limited; stores available in village. FW on quay. Hr Mr St. Davids 721373, M, CH, Access HW∓3; **Solva Boat Owners Assn.** Tel. St Davids 721209.

ST BRIDES BAY, Dyfed, 51°48′ N, 5°10′ W, Zone 0 (GMT), Admty chart 1478. HW (Little Haven) − 0450 on Dover, +0010 on Milford Haven; HW height −1.3m on Milford Haven; ML 3.2m; Duration 0555. Many good anchorages, especially between Little Haven and Borough Head in S or E winds or between Solva and Dinas Fawry in N or E winds, but in W winds boats should shelter in Solva (see above) or Skomer (see below). For approaches from the N or S see 10.11.5. Keep clear of Research Area in the middle of the bay, marked by buoys. Facilities: (Little Haven) CH, V, R, Bar, FW (cans).

SKOMER, Dyfed, 51°44′ N, 5°18′ W, Zone 0 (GMT), Admty charts 2878, 1478. HW −0455 Dover, −0005 on Milford Haven; HW height −0.4m (Sp), −0.1m (Np) on Milford Haven. The island is a nature reserve (fee payable to Warden on landing); anchorage in North Haven or South Haven; in both cases keep close to W shore entering, and land on The Neck (E side). For Jack Sound see 10.11.5. There are no lights or marks and no facilities.

SAUNDERSFOOT, Dyfed, 51°43′ N, 4°42′ W, Zone 0 (GMT), Admty chart 1482. HW −0510 on Dover, −0010 on Milford Haven. HW height +1.4m on Milford Haven; ML 4.4m; Duration 0605. Shelter good except in strong SW to S winds. Hr dries but there is approx 2.5m at MHWS. Alongside berths are sometimes available (see Hr Mr Tel. 812094) as are moorings in the middle. Light on pierhead Fl R 5s 6m 7M on stone cupola. Facilities: EC Wed; Hr Mr Tel. 812094, CH, FW (on SW wall), Slip; **Jones and Teague** Tel. 813429, ME; **Town** D, P, V, R, Bar, PO, Bank, Rly.

CARMARTHEN, Dyfed, 51°51′ N, 4°18′ W, Zone 0 (GMT), Admty chart 1076. HW −0455 on Dover, +0010 on Milford Haven; HW height −4.6m on Milford Haven. R Towy dries out, except for river water. Beware Carmarthen Bar off mouth of Rivers Towy and Taf. Channel into rivers changes frequently and is not buoyed. Access HW∓2. There are six electric cables crossing between the mouth and the railway bridge in Carmarthen, min clearance 15m. Visitors berths at R Towy YC at Ferryside (9M below Carmarthen), access HW∓3. Facilities: **Carmarthen** normal facilities of a market town; **Tawe Works** Tel. 236601, Gas; **River Towy YC** (Ferryside) Tel. Ferryside 366, Bar, FW, M; **Town** Normal facilities Bank, Bar, PO, V, Rly, Air (Swansea).

PORTHCAWL, Mid Glamorgan, 51°24′ N, 3°42′ W; Zone 0 (GMT). Admty chart 1169. HW −0505 on Dover; 0005 on Milford Haven; HW height +2.6m on Milford Haven; ML 5.3m. See 10.11.15. A small tidal harbour protected by breakwater running SE from Porthcawl Pt. Beware rock ledge (dries) to W of breakwater. At end of breakwater Porthcawl Lt Ho, W 6-sided tower with B base; F WRG 10m 6/4M; vis G302°-036°, W036°-082°, R082°-122°. In line 094° with Saint Hilary radio mast leads through Shord channel. Tidal streams can reach 6kn at springs off end of breakwater. Anchor approx 3ca SSE of Lt Ho. Facilities: EC Wed; Hr Mr 2756, 3 visitors moorings (HW∓2); **Porthcawl Marine** Tel. 4785 CH; **Porthcawl Harbour Boating Club** Tel. 2342; **Town** V, R, Bar, P and D (cans), PO, Bank, Rly (Bridgend), Air (Cardiff).

NEWPORT, Gwent, 51°33′ N, 2°59′ W, Zone 0 (GMT), Admty chart 1176. HW −0425 on Dover, −0015 on Avonmouth, HW height −1.0m on Avonmouth; ML 6.0m; Duration 0620. See 10.11.17. A commercial port controlled by Associated British Ports (Tel. 65411) but a safe sheltered port for yachts. Enter R Usk over bar (approx 0.5m) E of West Usk buoy and follow buoyed and lit channel to South Lock entrance; turn NE, yacht moorings on S side between power station pier and YC. W Usk buoy QR Bell, East Usk Lt Ho Fl (2) WRG 10s 11m 15/11M, W284°−290°, R290°−017°, W017°−037°, G037°−115°, W115°−120°. Ldg Lts 057°, both FG. Alexandra Dock, S lock W pier head 2 FR (vert) 9/7m 6M. E pier head 2 FG (vert) 9/7m 6M. VHF Ch 16; 09 11 (HW −4 to HW +4). Facilities: EC Thurs; HMC Tel. 273709 **Beechwood Marine** Tel. 277955 CH, El, ME, Sh; **Leeway Leisure** Tel. 276611 CH, El, ME, Sh; **Town** All facilities.

WESTON-SUPER-MARE, Avon, 51°21′ N, 2°59′ W, Zone 0 (GMT), Admty charts 1152, 1176. HW −0435 on Dover, −0025 on Avonmouth; HW height −1.1m on Avonmouth; ML 6.1m; Duration 0655. See 10.11.20. Good shelter except in S winds in Knightstone Harbour (dries) at N end of bay. Causeway at entrance marked by Bn. Access HW∓2. Grand pierhead 2 FG (vert) 6/5m. Alternative anchorage in good weather in R Axe (dries), entry HW∓2. Facilities: EC Mon, **Passey and Porter** Tel. 28291; CH, El, ME, Sh, Slip; **Uphill Boat Services** Tel. 418617; CH, El, D, FW, Slip, BH (10 ton), ME, Sh; **Weston Bay YC** Tel. 20772; FW, Bar, VHF Ch M. **Town** Bar, Bank, D, FW, P, PO, R, Rly, V.

WATCHET, Somerset, 51°11′ N, 3°20′ W, Zone 0 (GMT), Admty chart 1160. HW −0455 on Dover, −0043 on Avonmouth; HW height −1.7m on Avonmouth; ML 6.2m; Duration 0655. See 10.11.20. Good shelter but harbour dries; there is approx 6m in entrance at MHWS. Available approx HW∓2. Yachts should berth on W breakwater. Beware tidal streams round W pier head. Rocks and mud dry 0.5M to seaward. W breakwater head FG 9m 9M. E pier head 2 FG (vert) 3M. VHF Ch 16; 14 12 (HW−2 to HW+2). On W breakwater head B ball (Fl G at night) and on E pier 2 FR (vert) = at least 2.4m on flood or 3m on ebb. Facilities: EC Wed; Hr Mr Tel. 31264; HMC Tel. 31214, FW, D, P, Slip, CH, V, R, Bar; **Watchet Boat Owners Ass**. Tel. 31625.

MINEHEAD, Somerset, 51°13′ N, 3°28′ W, Zone 0 (GMT), Admty chart 1160. HW −0450 on Dover, −0040 on Avonmouth; HW height −2.6m (Sp), −1.9m (Np) on Avonmouth; ML 5.7m. There is no bar; access HW∓2. Small harbour formed by pier curving E and then SE, over which seas break in gales at MHWS. Best approach from N or NW; beware The Gables, shingle bank (dries 3.3m) about ¾ M ENE of pier. Berth alongside pier. A sewer outfall runs from a position 2ca S of pierhead to 1¾ ca NNE of it, passing ½ ca E of pierhead. Outer portion is protected by rock covering, rising to 2.8m above sea bed. Outer end marked by stbd-hand Bn with Lt. Harbour gets very crowded. Pierhead Lt Fl(2)G 5s 4M; vis 127°-262°. Facilities: EC Wed; Hr Mr Tel. 2566; **Tarr and Foy** Tel. 2029, Gas; **Pier** FW, Slip; **Town** normal facilities, Bar, D, El, Gas, ME, P, R, Sh, V, PO, Bank, Rly (Taunton or Barnstaple), Air (Exeter).

PORLOCK WEIR, Somerset, 51°13′ N, 3°38′ W, Zone 0 (GMT), Admty chart 1160. HW −0500 on Dover, −0050 on Avonmouth; HW height −2.6m on Avonmouth; ML 5.6m; Duration 0655. See 10.11.20. Shelter very good once in harbour but entrance difficult. Channel between pebble bank to NW and wooden wall to SE approx 15m wide. There is a small pool just inside entrance for shallow draft boats — others dry out inside harbour. There are no lights. Hr Mr Tel. 862106. Facilities: very limited; FW and stores on quay; **Porlock Weir SC** Tel. 862028.

WATERMOUTH, Devon, 51°13′ N, 4°05′ S, Zone 0 (GMT), Admty chart 1165. HW −0525 on Dover, −0020 on Milford Haven; HW height +2.0m on Milford Haven; ML 4.9m; Duration 0625. Good shelter, except from NW winds, but harbour dries. Entrance identified by white CG cottage above Rillage Pt. SW gales cause heavy surge in harbour. Access HW∓3 at sp, outside breakwater only at np. Visitors buoys distinguished by Y handles on R buoys. Facilities: Hr Mr Tel. Ilfracombe 883265; M, D (cans), FW (cans), Slip; **Watermouth Caves** Tel. 62504, Gas; **Watermouth YC** Tel. 65048, Bar.

LUNDY ISLAND, Devon, 51°11′ N, 4°40′ W, Zone 0 (GMT), Admty chart 1164. HW −0530 on Dover, −0030 on Milford Haven; HW height +0.9m on Milford Haven; ML 4.3m; Duration 0605. See 10.11.22. Shelter good if correct side of island is selected according to the wind. In winds from NW to SSW anchor N of SE point of island. In winds from E to N some shelter can be obtained on W side of the island but holding ground is poor. Jenny's Cove on W side is safe so long as there is no W ground swell. The high land (145m) gives considerable shelter. Waters round the island are a Marine Nature Reserve. Near N Pt, Lt Fl (2) 20s 50m 24M; vis 009°−285°. On SE Pt, Lt Fl 5s 53m 24M; vis 170°−073°; RC. Facilities: Landing place by the anchorage in the SW corner; **Lundy Co Landmark Trust** Tel. Woolacombe 870870, CH, Gas, bar and hotel.

BUDE, Cornwall, 50°50′ N, 4°33′ W, Zone 0 (GMT), Admty chart 1156. HW −0540 on Dover, −0040 on Milford Haven; HW height +0.7m on Milford Haven; Duration 0605. See 10.11.22. Limited shelter; Hr dries — available for average yacht HW∓2. Provided prior arrangements have been made, yachts can pass lock and berth in canal (approx 5m over CD required). Locking fees are approx £40 entry and £40 exit, and gates will only open if there is no sea running. Ldg marks, front W spar with Y diamond topmark, rear W flagstaff, in line at 075°. Continue on this line until inner ldg marks in line at 131°, front W pile with Y triangular topmark, rear W spar with Y triangular topmark. There are no lights. VHF Ch 16; 12 (when vessel expected). Facilities: EC Thurs; Hr Mr Tel. 3111; very limited facilities; **NE Truscott** Tel. 2423, Gas; **Town** Bank, Bar, PO, R. V.

NEWQUAY, Cornwall, 50°25′ N, 5°05′ W, Zone 0 (GMT), Admty chart 1168. HW −0604 on Dover, −0100 on Milford Haven; HW height +0.1m on Milford Haven; ML 3.7m; see 10.11.23. Entrance to harbour between two walls, rather narrow. Beware Old Dane Rk and Listrey Rk outside harbour towards Towan Hd. Swells cause a surge in the harbour. Enter HW∓2 but not in strong onshore winds. Hr dries; berth as directed by Hr Mr. Lts — North Pier 2 FG (vert) 2M, South Pier 2 FR (vert) 2M. VHF Ch 16 14 (May to Sept). Facilities: EC Wed (winter only); Hr Mr Tel. 872809, FW, Slip, D, V, R, Bar; **H. N. Bennett** Tel. 875900, Gas, Gaz. (Shallow draft boats can take the ground in Gannel Creek, close S of Newquay, but only in settled weather. Beware causeway bridge about half way up the creek.).

HAYLE, Cornwall, 50°11′ N, 5°25′ W, Zone 0 (GMT), Admty chart 1168. HW −0605 on Dover, −0100 on Milford Haven; HW height −0.4m on Milford Haven; ML 3.6m; Duration 0555. Shelter is very good but dangerous sea breaks on bar in strong on-shore winds; harbour dries. Approx 5m at entrance at MHWS. The harbour is divided by long arm (approx 300m) stretching almost to Ldg Lts which should be left to W. Beware training bank W of entrance channel marked by perches. Ldg Lts 180°, front FW 17m 4M, rear FW 23m 4M. E arm of harbour leads to Hayle, the W arm to Lelant Quay. Facilities: EC Thurs; Bank, Bar, FW, PO, R, V, P and D (in cans).

ST IVES, Cornwall, 50°13′ N, 5°28′ W, Zone 0 (GMT), Admty chart 1168. HW −0605 on Dover, −0100 on Milford Haven; HW height −0.4m on Milford Haven; ML 3.6m; Duration 0555. See 10.11.23. Shelter is good except in on-shore winds when heavy swell builds up. Harbour dries but has approx 4.5m at MHWS. Alternatives are anchorages between the harbour and Porthminster Pt to S in 3m, or Hayle in the S of St Ives Bay. Beware Hoe Rock off St Ives Hd coming from NW, and The Carracks off Porthminster Pt coming from SE. Keep E of G conical buoy about 1 ½ ca ENE of Smeaton's pierhead. Lights, E pier head (Smeatons pier) 2 FG (vert) 8m 5M. W pier head 2 FR (vert) 5m 3M. VHF Ch 16. Facilities: EC Thurs; Hr Mr Tel. Penzance 795018; six visitor moorings; **Fisherman's Co-op**, Gas, Gaz; **Smeaton's Pier** FW; **Town** Bank, Lau, Bar, D, FW, P (in cans), PO, R, Rly, V.

VOLVO PENTA SERVICE

Sales and service centres in area 12
Names and addresses of Volvo Penta dealers in this area are available from:
Western Marine, *Bulloch Harbour, Dalkey, Co. Dublin, Ireland, Tel 800321, Telex 24839*

Area 12

South Ireland
Shannon to Malahide

10.12.1	Index	**Page 525**
10.12.2	Diagram of Radiobeacons, Air Beacons, Lifeboat Stations etc	526
10.12.3	Tidal Stream Charts	528
10.12.4	List of Lights, Fog Signals and Waypoints	530
10.12.5	Passage Information	532
10.12.6	Distance Table	534
10.12.7	Special differences in Ireland	534
10.12.8	Kilrush	535
10.12.9	River Shannon	535
10.12.10	Bantry	536
10.12.11	Crookhaven	536
10.12.12	Schull	536
10.12.13	Baltimore	537
10.12.14	Castle Haven	537
10.12.15	Kinsale	538
10.12.16	Cork/Ringaskiddy, Standard Port, Tidal Curves	539
10.12.17	Youghal	543
10.12.18	Dunmore East	544
10.12.19	Waterford	545
10.12.20	Rosslare Harbour	545
10.12.21	Wexford	546
10.12.22	Arklow	547
10.12.23	Wicklow	547
10.12.24	Dublin, Standard Port, Tidal Curves	548
10.12.25	Howth	553
10.12.26	Malahide	553
10.12.27	Minor Harbours and Anchorages Ventry Dingle Port Magee Sneem Castletown Glengariff Glandore Courtmacsherry Oysterhaven Ballycotton Dungarvan	554

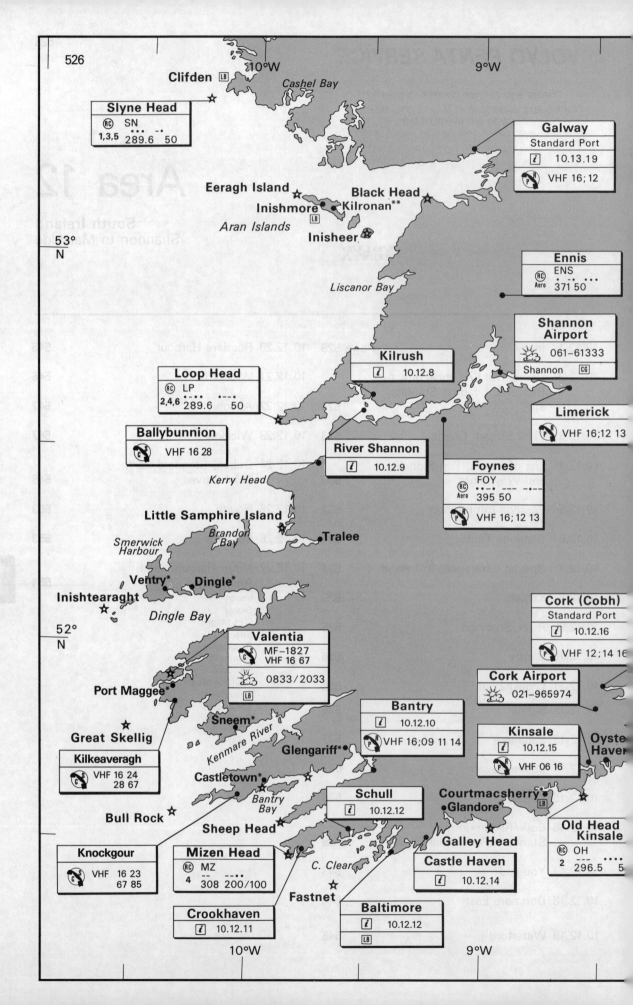

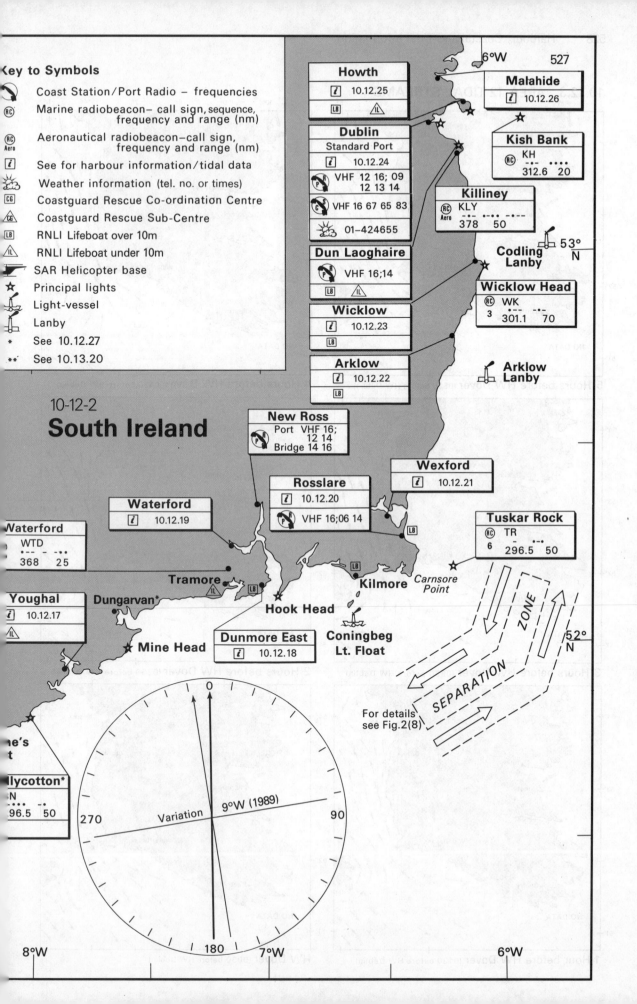

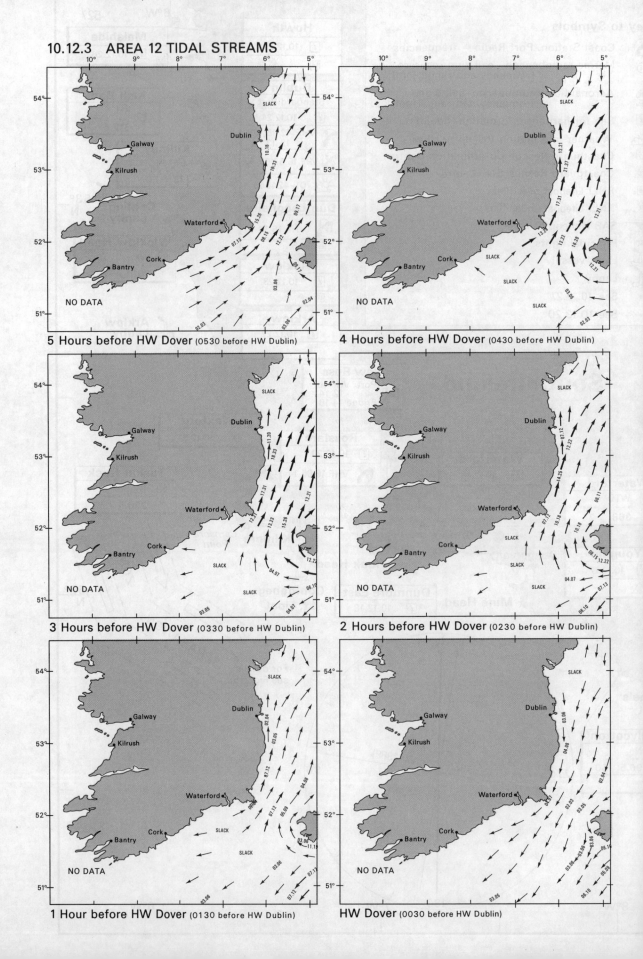

AREA 12—S Ireland 529

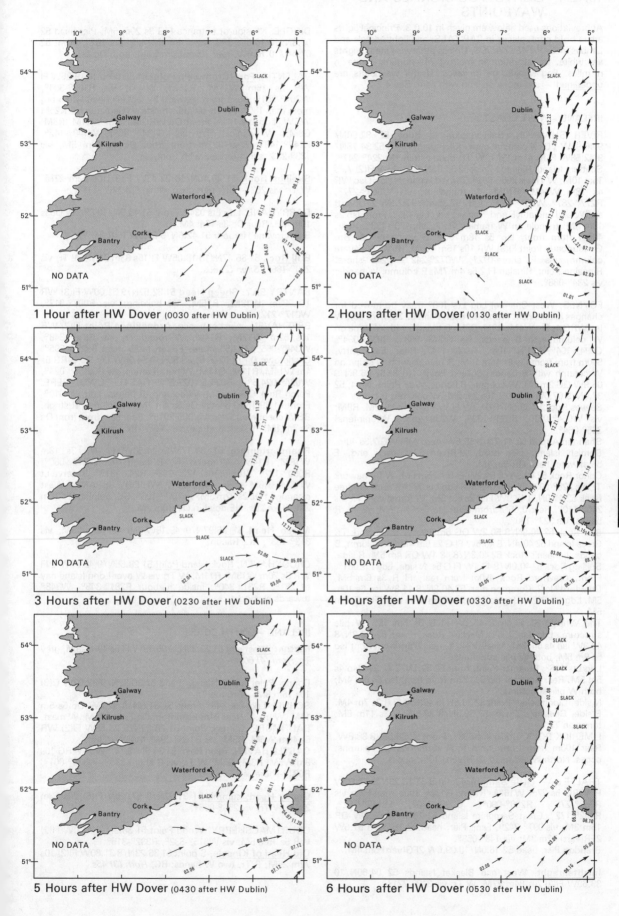

10.12.4 LIGHTS, FOG SIGNALS AND WAYPOINTS

Abbreviations used below are given in 10.0.3. Principal lights are in **bold** print, places in CAPITALS, and light-vessels and Lanbys in *CAPITAL ITALICS*. Unless otherwise stated lights are white. m—elevation in metres; M—nominal range in n. miles. Fog signals are in *italics*. Useful waypoints are underlined — use those on land with care. See 4.2.2.

IRELAND—WEST COAST

RIVER SHANNON. Ballybunnion Lt Buoy 52 32.50N/9 46.92W VQ; N cardinal mark. Kilcredaun Head 52 34.78N/9 42.58W Fl 6s 41m 13M; W Tr; obsc within 1M 224°-247°. Scattery Island, Rineana Point 52 36.33N/9 31.05W Fl(2) 7.5s 15m 10M; W Tr; vis 208°-092°. Tarbert Island, N point Iso WR 4s 18m W14M, R10M; W round Tr; vis W069°-277°, R277°-287°, W287°-339°. Jetty 52 35.4N/9 22.5W 2FG(vert) at SW end and at NE end. Tarbert Ldg Lts 128°. Front Iso 2s 13m 3M; triangle on W framework Tr; vis 123°-133°. Rear, 400m from front, Iso 5s 18m 3M; G stripe on W Bn. Garraunbaun Point Fl(3) WR 10s 16m W8M, R5M; W square column, vis R shore-072°, W072°-242°, R242°-shore. Rinealon Point, Rinalan Fl 2.5s 4m 7M; B column, W bands; vis 234°-088°.

FOYNES, West Channel Ldg Lts 108° (may be moved for changes in channel). Front, Barneen Point 52 36.9N/9 06.5W Iso WRG 4s 3m W4M, R3M, G3M; B triangle with W stripe on W column with B bands; vis W273°-038°, R038°-094°, G094°-104°, W104°-108°, R108°-114°. Rear, East Jetty, 540m from front Oc 4s 16m 10M; B triangle with W stripe on W column with B bands. Colleen Point, No 3, 52 36.9N/9 06.9W QG 2m 2M; W column, B bands. Weir Point, No 4, 52 37.0N/9 07.0W VQ(4)R 10s 2m 2M; W column, B bands. Beeves Rock 52 39.0N/9 01.3W Fl WR 5s 12m W12M, R9M; vis W064°-091°, R091°-238°, W238°-265°, W(unintens) 265°-064°.

Shannon Airport 52 41.7N/8 55.6W Aero AlFl WG 7.5s 40m. Dernish Island, pier head, 2FR(vert) 2M each end. E breakwater head QR 3m 1M.
Conor Rock 52 40.9N/8 54.2W Fl R 4s 6m 6M; W framework Tr; vis 228°-093°. North Channel Ldg Lts 093°. Front, Tradree Rock 52 41.0N/8 49.9W Fl R 2s 6m 5M; W framework Tr; vis 246°-110°. Rear 0.65M from front Iso 6s 14m 5M; W Tr, R bands; vis 327°-190°.
Bird Rock 52 40.9N/8 50.2W QG 6m 5M; W framework Tr. Grass Island 52 40.4N/8 48.5W Fl G 2s 6m 4M; W column, B bands. Laheen's Rock 52 40.3N/8 48.1W QR 4m 5M. S side, Spilling Rock 52 40.0N/8 47.1W Fl G 5s. N side, Ldg Lts 061°. Front, Crawford Rock 490m from rear, Fl R 3s 6m 5M. Crawford No 2, Common rear, 52 40.8N/8 44.8W Iso 6s 10m 5M. Ldg Lts 302°. Flagstaff Rock, 670m from rear, Fl R 2s 7m 5M.
The Whelps 52 40.6N/8 45.1W Fl G 3s 5m 5M; W pile structure. Ldg Lts 106°. Meelick Rock, front 52 40.2N/8 42.3W Iso 4s 6m 3M. Meelick No 2, rear 275m from front Iso 6s 9m 5M; both W pile structures.
Ldg Lts 146°, Braemar Point, front 52 39.1N/8 41.9W Iso 4s 5m 5M. Rear Braemar No 2, 122m from front Iso 6s 6m 4M; both W pile structures.
N side, Clonmacken Point 52 39.5N/8 40.6W Fl R 3s 7m 4M. E side, Spillane's Tower 52 39.3N/8 39.8W Fl 3s 11m 6M; turret on Tr.
LIMERICK DOCK. Lts in line 098°. Front 52 39.5N/8 38.8W. Rear 100m from front; both F; R diamonds on columns; occas. North Wharf head, 2FR(vert) 10m; occas.

TRALEE BAY. **Little Samphire Island** 52 16.23N/9 52.80W Fl WRG 5s 27m **W16M**, R13M; G13M; Blue round Tr; vis R262°-275°, R280°-090°, G090°-140°, W140°-152°, R152°-172°. Great Samphire Island 52 16.1N/9 52.2W QR 15m 3M; vis 097°-242°. Fenit Pier, head 52 16.2N/9 51.5W 2FR(vert) 12m 3M; vis 148°-058°.
Brandon Pier, head 52 16.0N/10 09.6W 2FG(vert) 5m 4M.

Inishtearaght, West end Blasket Islands 52 04.60N/10 39.60W Fl(2) 20s 84m **27M**; W Tr; vis 318°-221°.

DINGLE. NE side of entrance FlG 3s 20m 6M. Pier head 52 08.3N/10 16.5W 2FR (vert) 4m 2M. Ldg Lts 182°. Front 52 07.4N/10 16.6W, rear 100m from front, both Oc 3s.

VALENTIA. **Fort (Cromwell) Point** 51 56.00N/10 19.25W Fl WR 2s 16m **W17M, R15M**; W Tr; vis R102°-304°, W304°-351°; obsc from seaward by Doulus head when brg more than 180°. FR Lts on radio masts on Geokaun hill 1.25M WSW. Ldg Lts 141°. Front Oc WRG 4s 25m W11M, R8M, G8M; W conical Tr; vis G134°-140°, W140°-142°, R142°-148°. Rear 122m from front Oc 4s 43m 5M; vis 133°-233° synchronised with front.

Skelligs Rock 51 46.10N/10 32.43W Fl(3) 10s 53m **27M**; W Tr; vis 262°-115°; part obsc within 6M 110°-115°.

DARRYNANE. Ldg Lts 034°. Front 51 45.9N/10 09.2W Oc3s 10m 4M. Rear Oc 3s 16m 4M.
Ballycrovane Harbour 51 42.6N/9 57.5W Fl R 3s.

Bull Rock 51 35.47N/10 18.05W Fl 15s 83m **31M**; W Tr; vis 220°-186°; *Siren (2) 60s*.

BANTRY BAY. **Sheep Head** 51 32.57N/9 51.00W Fl(3) WR 15s 83m **W18M, R15M**; W building; vis R007°-017°, W017°-212°.
BEREHAVEN, West entrance, **Ardnakinna Point** Fl(2) WR 10s 62m **W17M**, R14M; W round Tr; vis R319°-348°, W348°-066°, R066°-shore. FR on radio mast 3.45M 295°. Castletown Dir Lt 024° 51 38.80N/9 54.30W Dir Oc WRG 5s 4m W14M, R11M, G11M; W hut, R stripe; vis G020.5°-024°, W024°-024.5°, R024.5°-027.5°. CASTLETOWN BERE, Perch Rock QG 4m 1M; W column, B bands. Ldg Lts 010°. Front 51 39.1N/9 54.4W Oc 3s 4m 1M; W column, R stripe, B&W chequered sides; vis 005°-015°. Rear 80m from front Oc 3s 7m 1M; W with R stripe; vis 005°-015°.

Roancarrigmore 51 39.17N/9 44.80W Fl WR 3s 18m **W18M**, R14M; W round Tr, B band; vis W312°-050°, R050°-122°, R(unintens) 122°-242°, R242°-312°. Reserve Lt W10M, R6M obsc 140°-220°. WHIDDY ISLAND. West clearing Lt Oc 2s 22m 3M; vis 073°-106°. SW dolphin QY 10m 2M; *Horn 20s*. NE dolphin QY 10m 2M.

Mizen Head 51 26.97N/9 49.18W Iso 4s 52m **16M**; vis 313°-133°; RC; Racon.

CROOKHAVEN, Rock Island Point 51 28.55N/9 42.23W L Fl WR 8s 20m W13M, R11M; W Tr; vis W over Long Island bay to 281°, R281°-340°; inside harbour R281°-348°, W348° towards N shore.

IRELAND—SOUTH COAST

Fastnet, West end 51 23.33N/9 36.13W Fl 5s 49m **28M**; grey Tr; *Horn (4) 60s*.

Copper Point, Long Island, E end 51 30.22N/9 32.02W Q(3) 10s 16m 8M; W round Tr.
SCHULL Ldg Lts 346° Front 51 31.6N/9 32.5W Oc 5s 5m 11M, W mast. Rear 91m from front Oc 5s 8m 11M; W mast.
BALTIMORE, Barrack Point 51 28.33N/9 23.65W Fl(2) WR 6s 40m W6M, R3M; vis R168°-294°, W294°-038°.
CASTLE HAVEN, Reen Point 51 30.9N/9 10.5W Fl WRG 10s 9m W5M, R3M, G3M; W Tr; vis G shore-338°, W338°-001°, R001°-shore.

Galley Head, summit 51 31.82N/8 57.13W Fl(5) 20s 53m **28M**; W Tr; vis 256°-065°.

COURTMACSHERRY, Wood Point 51 38.3N/8 41.0W Fl(2) WR 5s 15m 5M; vis W315°-332°, R332°-315°.
Old Head of Kinsale, S point 51 36.23N/8 31.80W Fl(2) 10s 72m **25M**; B Tr, two W bands; RC; *Horn (3) 45s*.

AREA 12—S Ireland

KINSALE. Bulman Lt Buoy 51 40.10N/8 29.70W Q(6) + LFl 15s; S cardinal mark. Charles's Fort Fl WRG 5s 18m W9M, R6M, G6M; vis G348°-358°, W358°-004°, R004°-168°. Marina, each end 2FG(vert).

CORK. **Roche's Point** 51 47.56N/8 15.24W Oc WR 20s 30m **W20M**, **R16M**; vis R shore-292°, W292°-016°, R016°-033°, W(unintens) 033°-159°, R159°-shore; *Dia 30s*. White Bay Ldg Lts 035°. Front Oc R 5s 11m 5M; W hut. Rear 113m from front Oc R 5s 21m 5M; W hut; synchronised with front. Fort Davis Ldg Lts 354°. Front Oc 5s 29m 10M. Rear. Dognose Landing Quay, 203m from front Oc 5s 37m 10M; synchronised with front. Curraghbinney Ldg Lts 252°. Front 51 48.6N/8 17.6W F 10m 3M. Rear, 61m from front, F 15m 3M; both white diamonds on columns; vis 229°-274°.
Crosshaven Marina 2FR(vert) at NE and NW corners. Whitegate Marine Terminal, Jetty, 2FG(vert) at S and N heads. East Ferry Marina, East Passage, 2FR(vert) at N and S ends. Spit Bank Pile 51 50.70N/8 16.41W Iso WR 4s 10m W10M, R7M; W house on R piles; vis R087°-196°, W196°-221°, R221°-358°.

Ballycotton 51 49.50N/7 59.00W Fl WR 10s 59m **W22M**, **R18M**; B Tr, within W walls, B lantern; vis W238°-063°, R063°-238°; RC; *Dia(4) 90s*.

YOUGHAL. West side of entrance 51 56.55N/7 50.49W Fl WR 2.5s 24m W12M, R9M; W Tr; vis W183°-273°, R273°-295°, W295°-307°, R307°-351°, W351°-003°.

Mine Head 51 59.57N/7 35.60W Fl(4) 20s 87m **28M**; W Tr, B band; vis 228°-shore.

Ballinacourty Point 52 04.67N/7 33.10W Fl(2) WRG 10s 16m W12M, R9M, G9M; W Tr; vis G245°-274°, W274°-302°, R302°-325°, W325°-117°.

DUNGARVAN. Ballinacourty Ldg Lts 083°. Front F 9m 2M; W column, B bands. Rear 46m from front F 12m 2M; W column, B bands. Esplanade Ldg Lts. Front and Rear FR 2M.

WATERFORD. **Hook Head** 52 07.40N/6 55.72W Fl 3s 46m **24M**; W Tr, two B bands; Racon; *Horn (2) 45s*. Dunmore East, E pier head L Fl WR 8s 13m W12M, R9M; grey Tr, W lantern; vis W225°-310°, R310°-004°.
E breakwater extension 52 08.96N/6 59.32W Fl R 2s 6m 4M; vis 000°-310°. West wharf Fl G 2s 6m 4M; vis 165°-246°.
Duncannon Ldg Lts 002°. **Front** Oc 4s 13m **15M**; W Tr, R stripe on fort; vis 340°-030°. Oc WR 4s 13m W9M, R7M; same Tr; vis R119°-149°, W149°-172°. **Rear** 0.54M from front Oc 4s 39m **15M**; W Tr, R stripe; vis 354°-069°.

Passage Point 52 14.23N/6 57.70W Fl WR 5s 7m W6M, R5M; R pile structure; vis W shore-127°, R127°-302°.
Cheek Point 52 16.1N/6 59.3W Q WR 6m 5M; W mast; vis W007°-289°, R289°-007°. Sheagh 52 16.3N/6 59.4W Fl R 3s 29m 3M; vis 090°-318°. Kilmokea 52 16.4N/6 58.9W Fl 5s. Power station jetty, 4 in No 2FG(vert) 3M. Railway Bridge 8FR; traffic signals.
Snowhill Point Ldg Lts 255°. Front Fl WR 2.5s 5m 3M; vis W222°-020°, R020°-057°, W057°-107°. Rear, Flour Mill, 0.4M from front Q 12m 5M.
Queen's Channel Ldg Lts 098°. Front QR 8m 5M; B Tr, W band; vis 030°-210°. Rear 550m from front Q 15m 5M; W mast.
Giles Quay 52 15.4N/7 04.2W Fl 3s 9m; vis 255°-086°. Cove 52 15.0N/7 05.1W Fl WRG 6s 6m 2M; W Tr; vis R111°-161°, G161°-234°, W234°-111°. Smelting House Point 52 15.1N/7 05.2W Q 8m 3M; W mast.
Ballycar 52 15.0N/7 05.4W Fl RG 3s 5m; vis G127°-212°, R212°-284°.

CONINGBEG Lt F 52 02.38N/6 39.45W Fl(3) 30s 12m **24M**; R hull, and Tr, lantern amidships; *Horn (3) 60s*. Racon. Kilmore, breakwater head Q RG 6m 5M; vis R269°-354°, G354°-003°, R003°-077°.
Carne, pier head Fl R 3s 6m.

IRELAND—EAST COAST

Tuskar 52 12.15N/6 12.40W Q(2) 7.5s 33m **28M**; W Tr; RC; *Horn (4) 45s*.

ROSSLARE. Pier head 52 15.41N/6 20.22W L Fl WRG 5s 15m W13M, R10M, G10M; R Tr; vis G098°-188°, W188°-208°, R208°-246°, G246°-283°, W283°-286°, R286°-320°. Ldg Lts 124°. Front 52 15.3N/6 20.1W. Rear 67m from front. Both FR 2M; vis 079°-169°.
New Ferry Pier, head 52 15.3N/6 20.2W Q 10m 3M. Ldg Lts 146°. Front 52 15.2N/6 20.1W, rear 110m from front, both Oc 3s 3M, synchronised.

ARKLOW LANBY 52 39.50N/5 58.10W Fl(2) 12s 12m **16M**; tubular structure on buoy; Racon; *Horn Mo(A) 30s*. Arklow Head, Pier head 52 46.7N/6 08.4W Oc R 10s 9m 9M.

ARKLOW. S pier head 52 47.59N/6 08.16W Fl WR 6s 10m 13M; framework Tr; vis R shore-223°, W223°-350°; R350°-shore. N pier head LFl G 7s 7m 10M; vis shore-287°.

Wicklow Head 52 57.93N/5 59.83W Fl(3) 15s 37m **26M**; W Tr; RC.

WICKLOW, E pier head 52 58.98N/6 02.01W Fl WR 5s 11m 6M; W Tr, R base and cupola; vis R136°-293°, W293°-136°. West pier head Fl G 1.5s 5m 6M.

CODLING LANBY 53 03.02N/5 40.70W Fl 4s 12m **16M**; tubular structure on buoy; Racon; *Horn 20s*.

Kish Bank 53 18.68N/5 55.38W Fl(2) 30s 29m **28M**; W Tr, R band, helicopter platform; RC; Racon; *Horn (2) 30s*.

DUBLIN BAY. Muglins 53 16.53N/6 04.52W Fl 5s 14m 8M; W conical Tr, R band.
DUN LAOGHAIRE. **East Breakwater, head** 53 18.13N/6 07.55W Fl(2) 15s 16m **22M**; stone Tr, W lantern; *Dia 30s*. West Breakwater, head Fl(3) G 7.5s 11m 7M; stone Tr, W lantern; vis 188°-062°.

PORT OF DUBLIN. Great South Wall, head, **Poolbeg** 53 20.52N/6 09.02W Oc(2) R 20s 20m **15M**; R round Tr; *Horn (2) 60s*. North Bull Wall, **Bull** Fl(3) G 10s 15m **15M**; G round Tr; *Bell (4) 30s*. **North Bank** Oc G 8s 10m **16M**; G square Tr on piles; *Bell (3) 20s*. 53 20.4N/6 11.3W Aero 2QR(vert) 205/85m 11M; 3FR(vert) Lts mark intermediate heights on chimney.

BEN OF HOWTH. **Baily** 53 21.67N/6 03.10W Fl 20s 41m **27M**; stone Tr; *Dia 60s*. HOWTH. Howth Lt Buoy 53 23.72N/6 03.53W FlG 5s; stbd-hand mark. **E pier head** Fl(2) WR 7.5s 13m **W17M**, R 13M; W Tr; vis W256°-295°, R295°-256°. W pier head Fl G 3s 7m 6M.

Dublin Airport 53 25.7N/6 14.7W Aero Al Fl WG 7.5s 95m.

Rockabill 53 35.80N/6 00.30W Fl WR 12s 45m **W23M**, **R19M**; W Tr, B band; vis W178°-329°, R329°-178°; *Horn (4) 60s*.

12

10.12.5 PASSAGE INFORMATION

IRELAND — WEST COAST

This coast gives wonderful cruising, but is exposed to the Atlantic and any swell offshore, but this diminishes midsummer. In bad weather however the sea breaks dangerously on shoals with quite substantial depths. There is usually a refuge close by, but if caught out in deteriorating weather and poor visibility, a stranger may need to make an offing until conditions improve, so a stout yacht and good crew are required. Tidal streams are weak, except round headlands.

There are few Lts, and inshore navigation is not wise after dark. Coastal navigation is feasible at night in good visibility, and fog is less frequent than in Irish Sea. A good watch must be kept for drift nets off the coast, and for lobster pots in inshore waters.

Stores, fuel and water are not readily available. Even in midsummer a yacht may meet at least one gale in a two-week cruise. Listen regularly to the Radio Telefis Eireann forecasts, as described in Table 7(2).

For all Irish waters the Sailing Directions published by the Irish Cruising Club are strongly recommended, and particularly on the W coast, where other information is scant.

The stretch of coast from Black Hd to Loop Hd has no safe anchs, and no Lts S of Inisheer. Take care not to be set inshore, although there are few offlying dangers except in Liscanor Bay and near Mutton Island. Loop Hd (Lt, RC) marks the N side of Shannon estuary, and should be passed 3 ca off. Here the stream runs SW from HW Galway +0300, and NE from HW Galway −0300.

RIVER SHANNON (charts 1547, 1548, 1549, 1540)

R Shannon is the longest river in Ireland, running 100M from Lough Allen to Limerick and thence, in the tidal section, 50M from Limerick dock to its mouth between Loop Hd and Kerry Hd.

In the lower reaches, as far as the junction with R Fergus about 15M below Limerick, the tides and streams are those of a deep-water inlet, with roughly equal durations of rise and fall, and equal rates of flood and ebb streams. In the entrance the flood stream begins at HW Galway −0555, and the ebb at HW Galway +0015. Above the junction with R Fergus the tidal characteristics become more like those of most rivers: the flood stream is stronger than the ebb, but it runs for a shorter time. In the Shannon the stream is much affected by the wind: S and W winds increase the rate and duration of the flood stream, and reduce the ebb. Strong N or E winds have the opposite effect. Prolonged or heavy rain increases the rate and duration of the ebb.

Off Kilcredaun Pt the ebb runs at 4 kn at sp, and in strong winds between S and NW it forms a bad race. This can be mostly avoided by keeping near the N shore, which is free from offlying dangers, and thereby avoiding the worst of the tide. When leaving the Shannon in strong W winds, aim to pass Kilcredaun Pt at slack water, and again keep near the N shore.

In the estuary and lower reaches of the Shannon there are several anchs available for yachts on passage up or down the coast. Killala B (chart 1819) is about 3M E of Loop Hd, and is convenient in good weather or in N winds, but exposed to SE and to any swell. Carrigaholt B (chart 1547), entered about 1M N of Kilcredaun Pt, is well sheltered from W winds and has little tidal stream, but is exposed to E. In N winds there is anch SE of Querrin Pt (chart 1547), 4½M further up river on N shore. Off Kilrush (10.12.8) there are anchs E of Scattery Island and N of Hog Island. (Note that there are overfalls ¾M S of Scattery Island with W winds and ebb tide).

For further information on R Shannon see 10.12.9.

LOOP HEAD TO MIZEN HEAD (charts 2254, 2423)

There is no Lt from Loop Hd to Inishtearaght, apart from Little Samphire Island in Tralee Bay, where Fenit Pier provides the only secure refuge along this bit of coast.

There is an anch, but exposed to N winds and to swell, on the W side of Brandon Bay. From here the scenery is spectacular to Smerwick Harb, entered between the Duncapple islets and the E Sister. It is sheltered except from NW or N winds.

Sybil Pt has steep cliffs, and offlying rks extend 3½ca. There is a race in W or NW winds with N-going tide, and often a nasty sea between Sybil Pt and Blasket Sound.

The Blasket Islands are very exposed, with strong tides and overfalls, but worth a visit in settled weather (chart 2970). Gt Blasket and Inishvickillane each have anch and landing on their NE side. Inishtearaght is the most W island (Lt), but further W lie Tearaght rks, and 3M SSW are Little Foze and Gt Foze rks. Extensive rks and shoals form the W side of Blasket Sound; this is the most convenient N-S route, 1M wide, and easy in daylight and reasonable weather with fair wind or tide. The N-going stream starts at HW Galway +0430, and the S-going at HW Galway −0155, with sp rate 3 kn. Wild Bank (or Three Fathom Pinnacle), a shallow patch with overfalls, lies 2½M SSW of Slea Hd. 3M SW of Wild Bank is Barrack Rk, with breakers in strong winds.

Dingle Bay is wide and deep, with few dangers round its shores. The best anchs are at Ventry, Dingle and Valentia — see 10.12.27.

The SW end of Puffin Island is steep-to, but the sound to the E is rocky and not recommended. Rough water is met between Bray Hd and Bolus Hd with fresh onshore winds or swell. Great Skellig (lit) is 6M, and Little Skellig 5M WSW of Puffin Island. Lemon Rk lies between Puffin Island and Little Skellig. Here the stream turns N at HW Cobh +0500, and S at HW Cobh −0110. There is a rk 3 ca SW of Great Skellig. When very calm it is possible to go alongside at Blind Man's Cove on NE side of Great Skellig, where there are interesting ruins.

Ballinskelligs Bay has an anch N of Horse Island, which has two Rks close off E end. Centre of bay is a prohib anch (cables). Darryname is an attractive, sheltered harb N of Lamb Hd. The entrance has Ldg Lts and marks, but is narrow and dangerous in bad weather.

Great Hog (or Scariff) Island has a rk close N, and a reef extending 2 ca W. Little Hog (or Deenish) Island is rocky 1 ca to NE. Moylaun Island lies 1½M E of Deenish, and has a rk 1½ ca SW of it. Two Head Island, off Lamb Hd, is steep-to. Kenmare R (chart 2495) has attractive harbs and anchs, but its shores are rocky, with no Lts. The best places are Sneem (see 10.12.27), Kilmakilloge and Ardgroom.

Dursey Island is steep-to except for rks ¾M NE and 1½ ca SW of Dursey Hd. Bull Rk (Lt, fog sig) and two rks W of it lie 2½M WNW of Dursey Hd. Cow Rk is midway between Bull Rk and Dursey Hd, with clear water each side. Calf and Heifer Rks are ¾M SW of Dursey Hd, where there is often broken water. 2M W of The Bull the stream turns NW at HW Cobh +0150, and SE at HW Cobh −0420. Dursey Sound (chart 2495) is a good short cut, but the stream runs 4 kn at sp. W-going starts at HW Cobh +0135, and E-going at HW Cobh −0450. A rk lies almost awash in mid-chan at the narrows, where there are also cables 24m above MHWS. Hold very close to the island shore. Beware wind changes in the sound, and broken water at N entrance.

Keep 3 ca off Crow Island to clear dangers. Off Blackball Hd at entrance to Bantry B (10.12.10) there can be a nasty race, particularly on W-going stream against the wind. Bantry B (charts 1838, 1840) has excellent harbs — notably Castletown, Adrigole and Glengariff (see 10.12.27). There are few dangers offshore, except around Bere and Whiddy Islands.

Sheep Hd (Lt) separates Bantry from Dunmanus B (chart 2552) which has three harbs — Dunmanus, Kitchen Cove and Dunbeacon. Carbery, Cold and Furze Islands lie in middle of bay, and it is best to keep N of them. Three Castle Hd at S end of bay has rks 1 ca W, and sea can break on S Bullig 4 ca off hd.

MIZEN HEAD TO TUSKAR ROCK (charts 2424, 2049)

From Mizen Head to Cork are many natural harbours. Only the best are mentioned here. Offshore the stream seldom exceeds 1½ kn, but it is stronger off headlands causing races and overfalls with wind against tide. Prolonged W winds increase the rate/duration of the E-going stream, and strong E winds have a similar effect on the W-going stream.

Off Mizen Hd the W-going stream starts at HW Cobh +0120, and the E-going at HW Cobh −0500. The sp rate is 4 kn, which with wind against tide forms a dangerous race, sometimes extending to Three Castle Hd or Brow Hd, with broken water right to the shore.

Crookhaven (10.12.11) is a well sheltered harb, accessible at all states of tide, entered between Rock Island Lt Ho and Alderman Rks, ENE of Streek Hd. Anchor off the village. The passage from here to Schull (10.12.12) can be made inside Long Island.

Fastnet Rk (Lt, fog sig) is nearly 4M WSW of C Clear. There is one rk ¼ M NE of Fastnet. Long Island Bay can be reached from here or through Gascanane Sound, between C Clear Island and Sherkin Island. Carrigmore Rks lie in the middle of this chan, with Gascanane Rk 1 ca W of them. The chan between Carrigmore Rks and Badger Island is best. If bound for Crookhaven, beware Bullig Reef, N of C Clear Island.

10M E of Baltimore (10.12.13) is Castle Haven (10.12.14), a sheltered and attractive harb, entered between Reen Pt (Lt) and Battery Pt. On passage Toe Hd has foul ground ½ ca S, and ¾ M S is a group of rks called the Stags.

Sailing E to Glandore, pass outside or inside High Island and Low Island, but if inside beware Belly Rk (awash) about 3 ca S of Rabbit Island. Good anch off Glandore (10.12.27), or off Union Hall.

Keep at least ½ M off Galley Hd to clear Dhulic Rk, and further off in fresh winds. Clonakilty B has little to offer. Offshore the W-going stream makes at HW Cobh +0200, and the E-going at HW Cobh −0420, sp rates 1½ kn.

Rks extend ¼ ca from Seven Heads, and beware Cotton Rk and Shoonta Rk close to the E. In middle of Courtmacsherry B are several dangers — from W to E Horse Rk, Black Tom, Barrel Rk and Blueboy, with a patch called Inner Barrels further inshore. These must be avoided going to or from Courtmacsherry, a rather shallow harb with a bar which breaks in strong S or SE winds. See 10.12.27.

Old Head of Kinsale (Lt, fog sig, RC) is quite steep-to, but a race extends 1M to SW on W-going stream, and to SE on E-going stream. There is an inshore passage in light weather, but in strong winds keep 2M off.

½ M S of Hangman Pt at entrance to Kinsale (10.12.15) is Bulman Rk. The Sovereigns are large rks off Oyster Haven — a good harb, clear but for Harbour Rk in mid-chan off Ferry Pt which must be passed on its W side. See 10.12.27.

Daunt Rk is 7 ca SE of Robert's Hd. Little Sovereign on with Reanies Hd 241° leads inshore of it. Ringabella B offers temp anch in good weather, near entrance to Cork (10.12.16).

From Cork to Ballycotton keep 3 ca off as far as Power Hd, and then at least ½M off for dangers including Smiths Rks 1½M WSW of Ballycotton Island (Lt, fog sig, RC). Sound Rk lies between Ballycotton Island and Small Island. Ballycotton Harb (10.12.27) is small and crowded, but usually there is sheltered anch outside. N side of Ballycotton B is foul ½M offshore.

Pass 1 ca S of Capel Island. The sound is not recommended. For Youghal see 10.12.17. To the E, there is a submerged rk ½ ca SE of Ram Hd. Here the W-going stream starts at HW Cobh +0230, and the E-going, at HW Cobh −0215, sp rates 1½ kn. Mine Hd (Lt) has two dangerous rks, The Rogue over ¼M E and The Longship 1M SW.

Helvick is a small sheltered harb on S side of Dungarvan B (10.12.27) and is approached along S shore, S of Helvick Rk, N of which other dangers exist towards Ballinacourty Pt. From here to Tramore there are few offlying rks, until Falskirt — a dangerous rk off Swines Pt in W approach to Dunmore East (10.12.18). Beware salmon nets.

From a point 1M S of Hook Hd (avoiding the overfalls known as Tower Race, which extend about 1M S of the hd at times), there are no obstructions on a direct course for Saltee Sound (chart 2740). Jackeen Rk is 1M NW of the S end of North Saltee, and Sebber Bridge extends ¾M N from the NE point of South Saltee, so care is needed through the sound, where the stream runs 3½ kn at sp. There are several rks round the Saltees, including the Bohurs to the E of the sound. From N of Long Bohur steer to pass N of Black Rk, before altering course eastward to pass 2 ca off Carnsore Pt, and thence to Greenore Pt. Watch for lobster pots in this area. In settled weather the little harb of Kilmore Quay (mostly dries) is available, but there are rocks and shoals in the approaches.

There can be a dangerous race off Carnsore Pt. In bad weather or in poor vis, pass S of Coningbeg Lt F, the Barrels and Tuskar Rk (Lt). Dangerous rks lie up to 6½ ca SSW and 2 ca NW of Tuskar Rk.

TUSKAR ROCK TO LAMBAY ISLAND (charts 1787, 1468)

From Carnsore Pt to Dublin (10.12.24) the shallow offshore banks cause dangerous overfalls and dictate the route. This is not a good cruising area, but for passage making it is sheltered from the W winds. Tidal streams run mainly N and S, but the N-going flood sets across the banks on the inside, and the S-going ebb sets across them on the outside.

From Tuskar or Greenore Pt there is the choice of going inside Lucifer and Blackwater Banks, or seawards to fetch the Arklow Lanby. Again, from here either the inshore or seaward passage can be taken. Approaching Dublin B, yachts normally use Dalkey Sound, but with a foul tide or light wind it is better to use Muglins Sound. Muglins (Lt) is steep-to except for a rk about 1 ca WSW of the Lt. Beware Leac Buidhe (dries) 1 ca E of Clare Rk.

Kish Bank and Burford Bank lie offshore in the approaches to Dublin B, and Rosbeg Bank lies on the N side of it. The N-going stream begins at HW Dublin −0600, and the S-going at HW Dublin, sp rates 3 kn. The sea breaks on Burford Bank in E gales.

Ben of Howth, on N side of Dublin B, is steep-to, with no dangers more than 1 ca offshore. Ireland's Eye, a rky island which rises steeply to a height of 99m, lies about ¾M N of Howth (10.12.25) with reefs running SE and SW from Thulla rk at its SE end.

For notes on crossing the Irish Sea, see 10.13.5.

10.12.6 DISTANCE TABLE

Approximate distances in nautical miles are by the most direct route while avoiding dangers and allowing for traffic separation schemes etc. Places in *italics* are in adjoining areas.

1	Galway	**1**																		
2	*Slyne Hd*	49	**2**																	
3	Loop Hd	55	52	**3**																
4	Valentia	103	92	48	**4**															
5	Bantry	155	144	100	58	**5**														
6	Fastnet Rk	150	139	95	53	34	**6**													
7	Kinsale	199	188	144	102	83	49	**7**												
8	Cork	210	199	155	113	94	60	17	**8**											
9	Youghal	228	217	173	131	112	78	35	25	**9**										
10	Dunmore East	262	251	207	165	146	112	69	59	34	**10**									
11	Tuskar Rk	287	276	232	190	171	137	98	85	63	32	**11**								
12	Arklow	324	313	269	227	208	174	135	122	100	69	37	**12**							
13	Wicklow	339	328	284	242	223	189	150	137	115	84	52	15	**13**						
14	Dun Laoghaire	357	346	302	260	241	207	168	155	133	102	70	36	21	**14**					
15	*Mew Island*	309	260	312	345	326	292	253	240	218	187	155	118	107	90	**15**				
16	*Douglas*	394	345	364	322	303	269	230	217	195	164	132	102	92	80	62	**16**			
17	*South Bishop*	310	299	255	214	194	160	123	110	90	61	36	62	72	90	172	143	**17**		
18	*Lundy Island*	336	325	281	239	220	186	147	140	126	105	85	112	122	140	222	193	50	**18**	
19	*Bishop Rock*	298	287	243	201	182	148	135	136	134	140	141	175	185	215	290	260	125	110	**19**
20	*Ouessant*	396	385	341	299	280	246	235	236	234	234	228	265	275	298	390	361	218	180	100	**20**

10.12.7 SPECIAL DIFFERENCES IN IRELAND

The Ordnance Survey map numbers refer to the Irish Ordnance Survey maps, scale 1:12670 or ½in to 1 mile, which cover the whole island, including Ulster, in 25 sheets.

Irish Customs: Yachts should preferably make their first port of call at one of the following places where there are customs posts Dublin port, Dun Laoghaire, Dunmore East, Waterford, New Ross, Dungarvan, Cobh, Cork, Crosshaven, Kinsale, Baltimore (summer only), Crookhaven, Bantry, Castletownbere, Cahirciveen, Fenit, Kilrush, Foynes, Limerick, Galway, Westport, Sligo and Killybegs. Yachts, in fact, are permitted to make their first call anywhere and if, after the proper procedures have been gone through, no Customs Officer arrives, report to the Civil Guard station. Passports are not required by UK citizens.

Irish Coast Life Saving Service: This is staffed by volunteers who are trained in first aid and equipped with breeches buoys, cliff ladders etc and their telephone numbers are those of the Leader's residence. Where appropriate these telephone numbers are given for each port.

Irish Marine Rescue Co-ordination Centre: The centre for the whole of the Irish Republic is situated at Shannon Airport. Its telephone numbers are (061) 61219 and (061) 61969. Should these be engaged call Air Traffic Control, Shannon (061) 61233 or Cork (021) 965326 or Dublin (01) 3764971. In extreme emergency dial 999. The centre can call on the RNLI, Coast Life Saving Service, Irish Army Air Corps helicopters, search aircraft, Irish naval vessels, civil aircraft, the Irish lighthouse service and the Garda Siochana. The Centre liaises with UK and France and acts as a clearing house for all messages received during the rescue operation within 100 miles of the Irish coast.

For Northern Ireland: The HM Coastguard Maritime Rescue Sub-Centre is in Bangor, Co Down, Tel. Donaghadee (0247) 883184. HM Customs have seven main places where yachtsmen should report in the event of a local Customs Officer not being available;
Larne Customs Freefone Yachts: RT (0232) 752511
Londonderry Customs Freefone Yachts: RT (0504) 262273
Kilkeel Customs Freefone Yachts: RT (06937) 62158
Coleraine Customs Freefone Yachts: RT (0265) 4803
Belfast Customs Freefone Yachts: RT (0232) 752511
Warrenpoint Customs Freefone Yachts: RT (069372) 3288
Bangor No telephone.

Liquefied petroleum gas. In the Republic of Ireland LPG is provided by Kosan, a sister company of Calor Gas Ltd. But the bottles have different connections, and the smallest bottle is taller than the normal Calor one fitted in most yachts. Yachts visiting the Irish Republic are advised to take an ample stock of Calor. Availability of Kosan gas is indicated by symbol Kos.

Telephones To dial the Irish Republic from the UK, dial 010 (International Exchange) followed by 353 (Irish Republic) followed by the area code (given in UK telephone books, or shown below omitting the initial 0) followed by the number. Dublin is an exception: for Dublin dial 0001 followed by the number. To dial UK from the Irish Republic: prefix the UK code with 03 and retain the 0 at the beginning of the UK area code, eg to call Southampton 123456 dial 03 0703 123456. London is the exception: to call London 246 3456, dial 03 1 246 3456.

Weather Weather messages (comprising gale warnings, 24 hr forecast for Irish coastal waters up to 30M offshore and Irish Sea) are broadcast by RTE Radio 1. For details see Table 7(2).

Radio Radio weather bulletins on RTE Radio 1 are at:
0633 Coastal waters
1155 Coastal waters (1255 on Sun)
1755 Coastal waters
2352 Coastal waters

Note 1. There are two books of sailing directions, published by the Irish Cruising Club, one for the South and West coasts of Ireland, the other for the North and East coasts. These are highly recommended.

Note 2. For further information regarding the Republic of Ireland, write to Irish Yachting Association, 4 Haddington Terrace, Dun Laoghaire, Co. Dublin. Tel. Dublin 800239.

AREA 12—S Ireland 535

KILRUSH 10-12-8
Clare

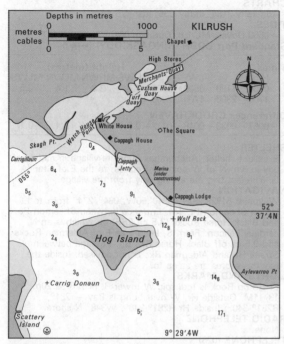

CHARTS
Admiralty 1819, 1547; Irish OS 17
TIDES
−0555 Dover; ML 2.7; Duration 0610; Zone 0 (GMT).

Standard Port GALWAY (→)

Times				Height (metres)			
HW		LW		MHWS	MHWN	MLWN	MLWS
1000	0500	0000	0600	5.1	3.9	2.0	0.6
2200	1700	1200	1800				

Differences KILRUSH
−0006 +0027 +0057 −0016 −0.1 −0.2 −0.3 −0.1

SHELTER
Kilrush harbour dries. The deep water harbour is Cappagh, well sheltered except in SE winds. Pier is in constant use by Shannon pilot boats and tugs.
NAVIGATION
Waypoint 52°37′.00N 09°32′.10W, 235°/055° from/to Watch House Point, 1.5M. See also 10.12.9. Coming between Hog Is and mainland, beware Wolf Rock. Approaching between Scattery Is and mainland, beware Baurnahard Spit and Carrigillaun.
LIGHTS AND MARKS
High store house (Glynn's Mill) in line with conspic white house on Watch House Pt in line at 055° leads into Kilrush Channel until S of Carrigillaun when alter to stbd towards Cappagh Jetty.
RADIO TELEPHONE
VHF Ch 16; 12. Other stations: Foynes Ch 16; 12 13 (occas). Limerick Ch 16; 12 13 (office hours).
TELEPHONE (065)
Hr Mr 51027; MRCC Limerick 61219; Coast Life Saving Service 51004; Customs Limerick 44426; Weather Shannon 61333; Dr. 51275
FACILITIES
EC Thursday; **Cappagh** Tel. 51027; Slip, M, L, FW, CH, AB; **George Brew** Tel. 51028, Kos; **Town** P, D, El, CH, AB, V, R, Bar. PO; Bank; Rly (bus to Limerick); Air (Shannon).
Note: Car ferry across Shannon from Tarbert to Killimer (4M upstream of Kilrush).

RIVER SHANNON 10-12-9
CHARTS
Admiralty 1819, 1547, 1548, 1549, 1540. Upper reaches 5080, 5078.
TIDES

HW at	HW Galway	HW Dover
Kilbaha	−0015	+0605
Carrigaholt	−0015	+0605
Tarbert	+0035	−0530
Foynes	+0050	−0515
Limerick	+0130	−0435

SHELTER
The Shannon Estuary is 50M long, from Loop Head to Limerick. Between Kilconly and Kilcredaun Pts in the W and the entrance to R Fergus there are many sheltered anchorages protected from all but E winds. The most convenient for boats on passage N or S is Carrigaholt Bay which gives good shelter from W winds. Kilbaha Bay, 3M inside Loop Head, is sheltered in winds from W to NE but holding is poor and it is exposed to swell. The best anchorage in the estuary is at Foynes on the S bank opposite R Fergus. Above this point the river narrows and becomes shallower although there is a minimum of 2m at LWS. Yachts may proceed up to Limerick Dock but this has all the drawbacks of a commercial port — frequent shifting of berth, dirt and someone constantly on watch.

NAVIGATION
Waypoint 52°33′.50N 09°43′.00W, 236°/056° from/to Tail of Beal Bar buoy, 1.7M. For notes on entrance and tidal streams see 10.12.5. The ebb can reach 4 kn.
Yachtsmen intending to visit the Shannon are advised to obtain the Irish Coast Pilot and also the Irish Cruising Club's *Sailing Directions for the South and West Coasts of Ireland*. It is the longest navigable river in the UK or Ireland and is controlled by the Limerick Harbour Commissioners. Pilots can be taken from Cappagh Pier to Limerick and from Limerick Dock to Killaloe. The channel is marked by buoys, beacons and perches. There are six

large locks on the upper reaches (above Portumna) and all are manned (toll). Above Lough Ree the ideal draught is less than a metre (1½m with careful navigation). The river is slow with many lakes; Lough Derg is 24M long and Lough Ree is 16M long. Between them at Shannon Harbour, a canal runs across to Dublin, the Grand Canal with 46 locks; max length 18.6m, beam 3.9m, height 2.74m, draught 1m. The Shannon bridges have a clearance of 4.6m except the swing bridge at Portumna which has 2.4m when closed. Lifting spans at Termonbarry and Rooskey are manned during daylight hours and are controlled by traffic lights. There is little commercial traffic.
LIGHTS AND MARKS
Principal lights are shown in 10.12.4
FACILITIES
Fuel, water and stores are obtainable at many villages.
KILBAHA V, FW, D (cans) at Hehir's Pub, P, PO; Bus to Limerick (Sat).
CARRIGAHOLT V, P, PO, R (summer).
KILRUSH and CAPPAGH — see 10.12.8. Pilot Stn Tel. 51027.
TARBERT V, P, PO; Bus to Limerick.
FOYNES V, D, FW, PO, R; Bus to Limerick. **Foynes YC** Tel. Foynes 90, Slip, Bar.
Ballina (Logh Derg) Marina with all facilities.
LIMERICK **J & G Boyd** Tel. 44366, Kos; **Shannon Yacht Fitters** Tel. Portumna 41105, Sh, ME, El; **Peter Lawless** Tel. Limerick 51567, CH, arrangement for pilot, ME, El, Sh; **Limerick Harbour Commission** Tel. (661) 315377.

BANTRY 10-12-10
Cork

CHARTS
Admiralty 1838, 1840; Imray C56; Irish OS 24
TIDES
+0600 Dover; ML 1.8; Duration 0610; Zone 0 (GMT).

Standard Port COBH (RINGASKIDDY) (→)

Times				Height (metres)			
HW		LW		MHWS	MHWN	MLWN	MLWS
0500	1100	0500	1100	4.2	3.3	1.4	0.5
1700	2300	1700	2300				

Differences BANTRY
−0045 −0025 −0040 −0105 −0.7 −0.7 −0.1 0.0
FENIT PIER (TRALEE BAY)
−0057 −0017 −0029 −0109 +0.5 +0.2 +0.3 +0.1
DINGLE HARBOUR
−0111 −0041 −0049 −0119 −0.3 −0.4 0.0 0.0
DUNKERRON HARBOUR (KENMARE RIVER)
−0117 −0027 −0050 −0140 −0.2 −0.3 +0.1 0.0

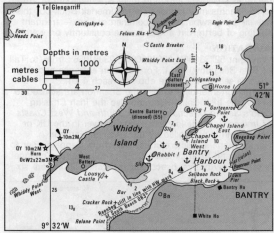

SHELTER
Apart from Bantry there are several good harbours in Bantry Bay — Dunboy Bay on W side of Piper Sound (exposed to E), Castletown (major fishing port, see 10.12.27), Lawrence Cove and Lonehort on Bear Island, Adrigole (attractive anchorage half way along N shore), and Glengariff (see 10.12.27).
NAVIGATION
Waypoint 51°40'.00N 09°36'.00W, 253°/073° from/to Whiddy Point West, 2.5M. Yachts can approach the shore fairly close everywhere except off Bear and Whiddy Islands. Beware Ducalia Rks, awash at LW, 1 M E of Roancarrigmore Lt, and Gerranes Rks 1 M W of Whiddy Is. In Bantry Harbour keep well clear (1½ to 2 ca) of all the islands. Also beware of Carrignafeagh off Whiddy Island and mussel rafts in many parts.
LIGHTS AND MARKS
The main entrance, marked by buoys, is the N channel (10m depth) to anchorage S of Chapel Is. The S channel is unlit (min depth on bar 1.7m over Cracker Rk). Whiddy Is Lt is sectored Oc W 2s 22m 3M 073°-106°.
RADIO TELEPHONE
VHF Ch 16; 14 11 (H24).
TELEPHONE (027)
Lifeboat Valentia 6214; MRCC Limerick 61219; Customs 50061; Dr 50404; Hosp 50133.
FACILITIES
EC Wednesday; **Bantry Pier** L, FW; **Glengarriff** FW, V, Bar; **Carroll Shipping** ME; **Donal and Noreen Casey** Tel. 50342, Kos; **Bantry Bay SC** Slip, L; **Town**, P and D (cans), L, FW, CH, V, R, Bar. PO; Bank; Rly (bus to Cork); Air (Cork or Shannon); Car Ferry (Cork).

CROOKHAVEN 10-12-11
Cork

CHARTS
Admiralty 2184; Imray C56; Irish OS 24
TIDES
+0550 Dover; ML 1.8; Duration 0610; Zone 0 (GMT).
Standard Port COBH (RINGASKIDDY) (→)

Times				Height (metres)			
HW		LW		MHWS	MHWN	MLWN	MLWS
0500	1100	0500	1100	4.2	3.3	1.4	0.5
1700	2300	1700	2300				

Differences CROOKHAVEN
−0057 −0033 −0048 −0112 −0.8 −0.6 −0.4 −0.1

SHELTER
Excellent shelter. Anchorages opposite village in middle of bay in 3m; N of W point of Rock Is; to the E of Granny Is. The last two are a long way from the village.
NAVIGATION
Waypoint 51°28'.50N 09°40'.50W, 094°/274° from/to Lt Ho on Sheemon Pt, 1M. Entrance between Sheemon Pt and Black Horse Rocks (3.5 ca ESE) on which is an N cardinal beacon. From S, keep 1 ca E of Alderman Rocks and 0.5 ca off Black Horse Rks Bn. Passage between Streek Hd and Alderman Rks is not advised. Inside the bay the shores are steep to.
LIGHTS AND MARKS
Lt Ho on Rock Is (conspic W tower) L Fl WR 8s 20m 13/11M. Outside Hr, W over Long Is Bay — 281°, R281°-340°; Inside Hr R281°-348°, W348°-N shore.
RADIO TELEPHONE
None.
TELEPHONE (028)
MRCC (061) 61219; Coast Life Saving Service Goleen 35117; Customs Bantry 50061; Dr 35148.
FACILITIES
Ron Holland Tel. 35116, Sh, ME; **Hotel** Tel. 35309, R, V, FW, Bar; **PO** Tel. 35200; **Village** V, Kos, Bank (Bantry), PO, Rly (Cork), Air (Cork).

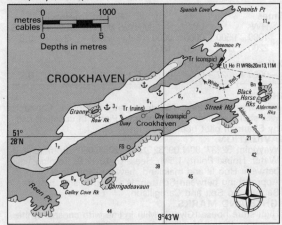

SCHULL 10-12-12
Cork

CHARTS
Admiralty 2184; Imray C56; Irish OS 24
TIDES
+0610 Dover; ML 1.8; Duration 0610; Zone 0 (GMT).
Standard Port COBH (RINGASKIDDY) (→)

Times				Height (metres)			
HW		LW		MHWS	MHWN	MLWN	MLWS
0500	1100	0500	1100	4.2	3.3	1.4	0.5
1700	2300	1700	2300				

Differences SCHULL
−0040 −0015 −0015 −0110 −0.9 −0.6 −0.2 −0.1
DUNMANUS HARBOUR
−0107 −0031 −0044 −0120 −0.7 −0.6 −0.2 0.0

AREA 12—S Ireland 537

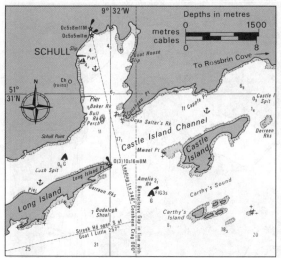

▲ Report to Hr Mr

SHELTER
Good shelter except in strong S winds — in these winds, best shelter is behind Long Island. Schull Harbour is available at all times. Anchor SE of pier, clear of fairway.

NAVIGATION
Waypoint 51°29'.60N 09°31'.60W, 166°/346° from/to front Ldg Lt 346°, 2.1M. Entering between Schull Pt on the W and Coosheen Pt on E, beware Bull Rock in middle of channel, marked by a R iron perch.

LIGHTS AND MARKS
Leading lights NE of village in line at 346°, both Oc 5s. The pier is normally lit by street lights all night.

RADIO TELEPHONE
None.

TELEPHONE (028)
Hr Mr 28136; MRCC Limerick 61219; Coast Life Saving Service 35117; Customs Bantry 50061; Dr 28311; Hosp Bantry 50133.

FACILITIES
EC Tuesday; **Schull Pier** Slip, M, D, L, FW, AC, AB; **Sailing Club** Tel. 28286; **J. O'Reilly** Tel. 28136, Kos; **Rossbrin BY** Tel. 37352 M, Sh, Slip; **Simon Nelson** Tel. 28554 CH.
Village P, ME, El, Sh, CH, V, R, Bar. PO; Bank; Rly (bus to Cork); Air (Cork); Car Ferry (Cork)

BALTIMORE 10-12-13
Cork

CHARTS
Admiralty 3725, 2129; Imray C56; Irish OS 24

TIDES
−0605 Dover; ML 2.1; Duration 0610; Zone 0 (GMT)

Standard Port COBH (RINGASKIDDY) (→)

Times				Height (metres)			
HW		LW		MHWS	MHWN	MLWN	MLWS
0500	1100	0500	1100	4.2	3.3	1.4	0.5
1700	2300	1700	2300				

Differences BALTIMORE
−0025 −0005 −0010 −0050 −0.5 −0.3 +0.1 +0.1
CLONAKILTY BAY
−0033 −0011 −0019 −0041 −0.3 −0.2 No data

SHELTER
Excellent shelter and harbour is always available. Anchor N or W of New Pier, or in Church Strand Bay past lifeboat slip in 2-3m. In strong W winds off ruined abbey on Sherkin I. (off chartlet), clear of ferry.

NAVIGATION
Waypoint 51°27'.80N 09°23'.42W, 180°/000° from/to Loo Rock buoy, 0.62M. On entering beware Loo Rock to starboard, marked by a buoy with Lt and Radar reflector. Beware Lousy Rocks and Wallis Rocks in the middle of the harbour. Harbour can be entered from the N but this is tricky and not recommended.

LIGHTS AND MARKS
Entrance easily identified by conspic W Tr called Lot's Wife on Beacon Pt and W Lt Ho on Barrack Pt.

RADIO TELEPHONE
Call: *Baltimore Harbour Radio* VHF Ch 16 09.

TELEPHONE (028)
Hr Mr 20184; Lifeboat 20125; MRCC Limerick 61219; Coast Life Saving Service 52; Customs Bantry 50061; Dr 21488; Hosp 21677.

FACILITIES
EC None; **New Pier** Slip, AB; **K. Cotter** Tel. 20106, D, CH, V, Gas, Gaz, Kos, FW; **M. Casey** Bar, P; **H. Skinner** Tel. 20114, Slip, Sh; **Old Pier** FW; **R. Bushe** Tel. 20125, ACA; **T. O'Driscoll** Tel. 20344, ME.
Village P, Bar. PO; Rly (bus to Cork); Air (Cork).

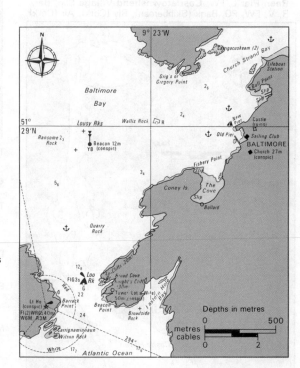

CASTLE HAVEN 10-12-14
Cork

CHARTS
Admiralty 2129, 2092; Imray C56; Irish OS 24

TIDES
+0605 Dover; ML 2.2; Duration 0605; Zone 0 (GMT)

Standard Port COBH (RINGASKIDDY) (→)

Times				Height (metres)			
HW		LW		MHWS	MHWN	MLWN	MLWS
0500	1100	0500	1100	4.2	3.3	1.4	0.5
1700	2300	1700	2300				

Differences CASTLETOWNSHEND
−0020 −0030 −0020 −0050 −0.4 −0.3 0.0 +0.2

KINSALE 10-12-15
Cork

CHARTS
Admiralty 2053, 1765; Imray C56; Irish OS 25

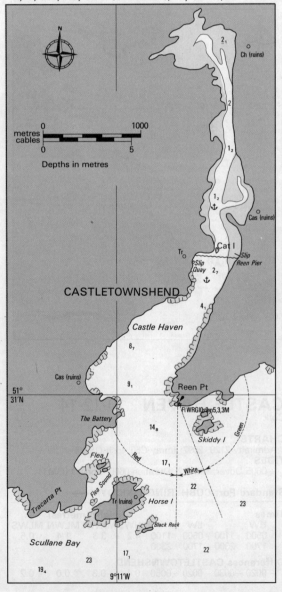

TIDES
−0600 Dover; ML 2.2; Duration 0600; Zone 0 (GMT)

Standard Port COBH (RINGASKIDDY) (→)

Times				Height (metres)			
HW		LW		MHWS	MHWN	MLWN	MLWS
0500	1100	0500	1100	4.2	3.3	1.4	0.5
1700	2300	1700	2300				

Differences KINSALE
−0019 −0005 −0009 −0023 −0.1 −0.1 +0.1 0.0

SHELTER
Excellent sheltered harbour except in very strong SE winds. Anchorage SW of James Fort between moorings and bridge. Moorings at Kinsale YC.

NAVIGATION
Waypoint 51°40'.00N 08°30'.00W, 181°/001° from/to Charles's Fort Lt, 1.7M. Bulman Rock off Hangman Pt marked by S cardinal buoy. Also beware Farmer Rock ¾ ca off shore on W bank and Carrignarone on E bank opposite Money Pt.

LIGHTS AND MARKS
Charles's Fort Fl WRG 5s; 18m 9/6M vis G348°-358°, W358°-004°, R004°-168°. Port-hand Lt buoys mark chan.

RADIO TELEPHONE
VHF Ch 16; 06 14 (H24). Kinsale YC Ch M.

TELEPHONE (021)
Hr Mr 772503; MRCC Limerick 61219; Coast Life Saving Service 773223; Customs 772271; Dr 772253, 772133; Hosp Cork 546400.

FACILITIES
EC Thursday; **Kinsale Marina** Tel. 772196, FW, AC, D, P, AB; **Pier** Tel. 772503, Slip, M, P (cans), D, L, FW, ME, El, Sh, CH, AB, V, R, Bar; **Kinsale YC** Tel. 772196, Slip, M, P, D, L, FW, C, CH, R, Bar; **Atlantic Yacht Co.** Tel. 772167, Slip, Kos, ME, El, Sh, CH; **Kilmacsimon BY** Tel. 775134, Slip, ME, Sh, C; **McWilliam Sails** Tel. 831505, SM; (Several excellent restaurants — the gourmet centre of Ireland).

Town P, D, V, R, Bar. PO; Bank; Rly (bus to Cork); Air (Cork).

SHELTER
Excellent anchorage protected from all weathers and available at all tides, day and night, although the outer part of harbour is subject to swell in S winds. Anchor in midstream SE of the slip at Castletownshend, N of Cat Island, or further upstream as shown on chartlet.

NAVIGATION
Waypoint 51°29'.00N 09°10'.00W, 171°/351° from/to Reen Point Lt, 2M. Enter between Horse I. and Skiddy I. both of which have foul ground all round. Black Rk lies off the SE side of Horse I. and is steep-to along its S side. On the other side of Horse I. is Flea Sound, a narrow boat passage, obstructed by rocks. Colonel Rk (0.5m) lies close to the E shore, 2ca N of Reen Pt. A submarine cable runs E/W across the harbour from the slip close N of Reen Pier to the slip at Castletownshend.

LIGHTS AND MARKS
Reen Point Lt, Fl WRG 10s 9m 5/3M; W Tr; vis G shore-338°, W338°-001°, R001°-shore. A ruined Tr stands on E end of Horse I.

RADIO TELEPHONE
None.

TELEPHONE (028)
MRCC (061) 61219; Coast Life Saving Service 35117; Customs Bantry 50061; Dr 21752; Hosp 21677.

FACILITIES
Reen Pier L, FW; **Castletownshend Village** Slip, Bar, R, V, FW, PO, Bank (Skibbereen), Rly (Cork), Air (Cork).

AREA 12—S Ireland 539

CORK (COBH) 10-12-14
Cork

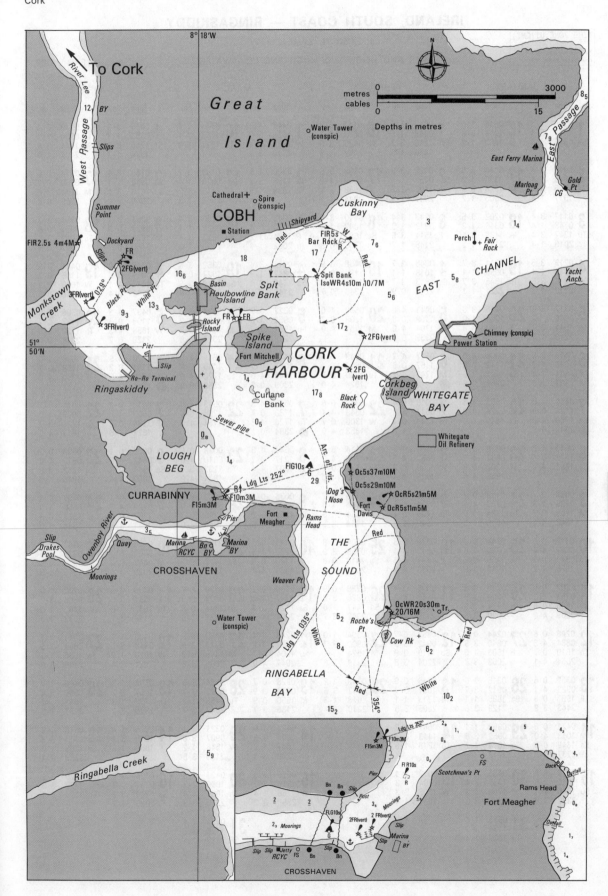

IRELAND, SOUTH COAST — RINGASKIDDY

Lat 51°50′ N   Long 8°19′ W

TIMES AND HEIGHTS OF HIGH AND LOW WATERS

YEAR 1989

TIME ZONE UT(GMT)
For Summer Time add ONE hour in non-shaded areas

JANUARY		FEBRUARY		MARCH		APRIL	
Time m	Time m	Time m	Time m	Time m	Time m	Time m	Time m
1 0546 1.3 1147 3.4 Su 1818 1.4	**16** 0618 1.1 1214 3.6 M 1852 1.2	**1** 0012 3.3 0643 1.5 W 1246 3.2 1920 1.5	**16** 0147 3.3 0833 1.3 Th 1427 3.3 2107 1.3	**1** 0447 1.3 1038 3.3 W 1720 1.4 2319 3.3	**16** 0645 1.4 1243 3.1 Th 1921 1.4	**1** 0015 3.3 0653 1.3 Sa 1255 3.2 1934 1.3	**16** 0219 3.4 0900 1.1 Su 1444 3.4 2122 1.0
2 0015 3.4 0645 1.4 M 1245 3.4 1917 1.4	**17** 0048 3.5 0730 1.2 Tu 1326 3.5 2005 1.2	**2** 0128 3.3 0759 1.4 Th 1402 3.3 2037 1.4	**17** 0308 3.5 0950 1.2 F 1538 3.5 2214 1.1	**2** 0600 1.5 1156 3.2 Th 1839 1.5	**17** 0133 3.2 0820 1.4 F 1413 3.2 2053 1.3	**2** 0140 3.4 0812 1.1 Su 1416 3.4 2049 1.0	**17** 0314 3.6 0950 0.9 M 1534 3.6 2209 0.8
3 0117 3.4 0748 1.4 Tu 1347 3.4 2019 1.4	**18** 0202 3.5 0846 1.2 W 1439 3.6 2118 1.0	**3** 0243 3.4 0912 1.3 F 1514 3.5 2145 1.2	**18** 0410 3.7 1047 1.0 Sa 1631 3.7 2305 0.9	**3** 0045 3.2 0720 1.4 F 1324 3.2 2002 1.4	**18** 0256 3.4 0935 1.1 Sa 1521 3.4 2156 1.0	**3** 0251 3.7 0922 0.9 M 1521 3.7 2150 0.7	**18** 0357 3.7 1031 0.8 Tu 1613 3.8 2247 0.7
4 0219 3.5 0849 1.4 W 1446 3.5 2119 1.3	**19** 0314 3.7 0956 1.1 Th 1545 3.7 2221 1.1	**4** 0348 3.7 1016 1.1 Sa 1613 3.7 2242 0.9	**19** 0458 3.9 1132 0.8 Su 1715 3.9 2346 0.7	**4** 0211 3.4 0842 1.3 Sa 1447 3.4 2118 1.1	**19** 0352 3.6 1027 0.9 Su 1610 3.6 2242 0.8	**4** 0348 4.0 1019 0.6 Tu 1613 4.1 2242 0.4	**19** 0434 3.9 1105 0.7 W 1648 3.9 2318 0.6
5 0317 3.6 0946 1.2 Th 1541 3.7 2213 1.2	**20** 0416 3.8 1054 1.0 F 1640 3.8 2313 1.0	**5** 0441 4.0 1109 0.8 Su 1704 4.0 2332 0.7	**20** 0539 4.0 1208 0.7 M 1751 4.0 ○	**5** 0322 3.7 0950 1.0 Su 1552 3.7 2219 0.8	**20** 0434 3.8 1108 0.7 M 1649 3.8 2320 0.6	**5** 0437 4.3 1108 0.3 W 1658 4.3 2329 0.2	**20** 0506 4.0 1133 0.7 Th 1720 4.0 2346 0.6
6 0410 3.8 1038 1.1 F 1631 3.8 2302 1.0	**21** 0508 4.0 1142 0.9 Sa 1727 3.9 ○ 2358 0.8	**6** 0527 4.2 1157 0.6 M 1747 4.2 ●	**21** 0019 0.6 0612 4.1 Tu 1239 0.7 1824 4.0	**6** 0419 4.0 1047 0.7 M 1642 4.0 2309 0.5	**21** 0511 4.0 1140 0.7 Tu 1725 4.0 2353 0.6	**6** 0520 4.5 1151 0.2 Th 1740 4.5 ●	**21** 0536 4.0 1158 0.7 F 1750 4.0 ○
7 0458 4.0 1126 0.9 Sa 1718 4.0 ● 2349 0.9	**22** 0553 4.1 1224 0.8 Su 1810 4.0	**7** 0018 0.4 0611 4.4 Tu 1241 0.4 1829 4.3	**22** 0049 0.6 0643 4.1 W 1306 0.7 1853 4.0	**7** 0505 4.3 1133 0.4 Tu 1726 4.3 ● 2354 0.2	**22** 0543 4.0 1208 0.6 W 1754 4.0 ○	**7** 0012 0.1 0603 4.5 F 1234 0.2 1821 4.5	**22** 0011 0.7 0603 4.0 Sa 1224 0.7 1817 4.0
8 0543 4.2 1212 0.8 Su 1801 4.1	**23** 0038 0.7 0632 4.1 M 1302 0.8 1846 4.0	**8** 0100 0.3 0652 4.5 W 1323 0.3 1909 4.4	**23** 0114 0.6 0709 4.0 Th 1330 0.7 1919 4.0	**8** 0547 4.5 1217 0.2 W 1807 4.4	**23** 0018 0.6 0610 4.0 Th 1232 0.6 1821 4.0	**8** 0055 0.1 0642 4.5 Sa 1314 0.3 1900 4.4	**23** 0038 0.7 0629 4.0 Su 1252 0.8 1845 4.0
9 0034 0.7 0625 4.3 M 1256 0.7 1843 4.2	**24** 0113 0.7 0707 4.1 Tu 1334 0.8 1919 4.0	**9** 0142 0.3 0731 4.5 Th 1405 0.4 1949 4.3	**24** 0138 0.7 0734 4.0 F 1354 0.8 1944 3.9	**9** 0036 0.1 0628 4.6 Th 1259 0.2 1846 4.5	**24** 0042 0.6 0635 4.0 F 1255 0.7 1846 4.0	**9** 0137 0.3 0723 4.3 Su 1357 0.5 1941 4.3	**24** 0107 0.8 0656 4.0 M 1323 0.8 1914 4.0
10 0117 0.6 0707 4.4 Tu 1341 0.6 1926 4.2	**25** 0144 0.7 0812 4.0 W 1405 0.8 1949 3.9	**10** 0225 0.3 0812 4.4 F 1447 0.5 2030 4.2	**25** 0204 0.8 0758 3.9 Sa 1418 0.8 2011 3.8	**10** 0119 0.1 0707 4.5 F 1340 0.3 1924 4.4	**25** 0104 0.7 0657 4.0 Sa 1317 0.7 1910 4.0	**10** 0220 0.5 0805 4.1 M 1442 0.7 2026 4.0	**25** 0141 0.9 0728 3.9 Tu 1358 0.9 1949 3.9
11 0202 0.6 0751 4.3 W 1425 0.6 2009 4.2	**26** 0215 0.8 0809 3.9 Th 1434 0.9 2020 3.8	**11** 0308 0.4 0856 4.2 Sa 1531 0.7 2115 4.0	**26** 0230 0.9 0826 3.8 Su 1447 1.0 2042 3.7	**11** 0159 0.2 0747 4.4 Sa 1420 0.4 2004 4.3	**26** 0130 0.8 0721 3.9 Su 1344 0.8 1937 3.9	**11** 0308 0.8 0851 3.8 Tu 1531 1.0 2118 3.7	**26** 0222 1.0 0806 3.8 W 1443 1.0 2033 3.8
12 0246 0.6 0834 4.3 Th 1510 0.7 2054 4.1	**27** 0244 0.9 0840 3.8 F 1503 1.0 2053 3.7	**12** 0353 0.7 0942 3.9 Su 1619 0.9 2204 3.8	**27** 0304 1.0 0900 3.7 M 1524 1.1 2119 3.6	**12** 0243 0.4 0827 4.2 Su 1504 0.7 2047 4.0	**27** 0159 0.8 0749 3.8 M 1415 0.9 2008 3.8	**12** 0403 1.1 0948 3.5 W 1628 1.2 2221 3.4	**27** 0311 1.1 0854 3.6 Th 1536 1.1 2129 3.6
13 0332 0.6 0922 4.1 F 1557 0.8 2143 3.9	**28** 0315 1.0 0912 3.7 Sa 1535 1.1 2129 3.6	**13** 0445 0.9 1037 3.7 M 1713 1.1 2305 3.5	**28** 0348 1.2 0942 3.5 Tu 1613 1.3 2210 3.4	**13** 0328 0.7 0914 3.9 M 1550 0.9 2136 3.7	**28** 0234 1.0 0823 3.7 Tu 1453 1.1 2047 3.7	**13** 0508 1.3 1057 3.2 Th 1737 1.3 2340 3.2	**28** 0410 1.2 0956 3.4 F 1641 1.2 2237 3.5
14 0421 0.8 1013 4.0 Sa 1649 0.9) 2237 3.8	**29** 0352 1.1 0952 3.6 Su 1614 1.2 2213 3.5	**14** 0547 1.2 1143 3.4 Tu 1819 1.3		**14** 0421 1.0 1007 3.5 Tu 1647 1.2) 2238 3.4	**29** 0321 1.1 0907 3.6 W 1546 1.2 2139 3.5	**14** 0625 1.4 1219 3.1 F 1859 1.3	**29** 0518 1.2 1111 3.3 Sa 1753 1.2 2353 3.4
15 0516 0.9 1111 3.8 Su 1746 1.1 2337 3.6	**30** 0435 1.2 1037 3.4 M 1704 1.3) 2306 3.3	**15** 0019 3.4 0704 1.3 W 1302 3.3 1941 1.4		**15** 0525 1.3 1116 3.2 W 1756 1.4 2358 3.2	**30** 0421 1.3 1006 3.4 Th 1655 1.4 2249 3.3	**15** 0106 3.2 0751 1.3 Sa 1341 3.2 2020 1.2	**30** 0629 1.2 1228 3.3 Su 1907 1.1
	31 0533 1.4 1134 3.3 Tu 1807 1.4				**31** 0534 1.4 1126 3.2 F 1812 1.4		

Chart Datum: 0.13 metres above Ordnance Datum (Dublin)

AREA 12 — S Ireland

IRELAND, SOUTH COAST — RINGASKIDDY
Lat 51°50' N Long 8°19' W
TIMES AND HEIGHTS OF HIGH AND LOW WATERS

YEAR 1989

TIME ZONE UT (GMT)
For Summer Time add ONE hour in non-shaded areas

MAY		JUNE		JULY		AUGUST	
Time m	Time m	Time m	Time m	Time m	Time m	Time m	Time m
1 0109 3·5 / 0744 1·0 / M 1342 3·5 / 2018 0·9	**16** 0223 3·5 / 0904 1·0 / Tu 1446 3·5 / 2124 0·9	**1** 0246 3·9 / 0922 0·7 / Th 1515 3·9 / 2150 0·6	**16** 0311 3·5 / 0945 1·0 / F 1536 3·6 / 2207 1·0	**1** 0328 3·8 / 1004 0·8 / Sa 1559 3·9 / 2237 0·7	**16** 0331 3·5 / 1002 1·1 / Su 1600 3·7 / 2228 1·0	**1** 0518 3·9 / 1147 0·7 / Tu 1743 4·1 ●	**16** 0454 3·9 / 1119 0·7 / W 1715 4·2 / 2343 0·6
2 0216 3·8 / 0851 0·8 / Tu 1447 3·8 / 2121 0·6	**17** 0311 3·6 / 0948 0·9 / W 1531 3·7 / 2206 0·8	**2** 0343 4·0 / 1017 0·6 / F 1610 4·1 / 2245 0·5	**17** 0357 3·7 / 1028 0·9 / Sa 1621 3·8 / 2249 0·9	**2** 0427 3·9 / 1101 0·7 / Su 1655 4·0 / 2330 0·7	**17** 0423 3·7 / 1051 0·9 / M 1648 3·9 / 2315 0·8	**2** 0014 0·6 / 0601 4·0 / W 1229 0·6 / 1824 4·2	**17** 0536 4·1 / 1203 0·5 / Th 1756 4·4 ○
3 0315 4·0 / 0949 0·6 / W 1542 4·1 / 2214 0·4	**18** 0352 3·7 / 1026 0·8 / Th 1612 3·8 / 2242 0·8	**3** 0435 4·1 / 1108 0·5 / Sa 1701 4·2 / 2334 0·5 ●	**18** 0441 3·8 / 1109 0·9 / Su 1704 3·9 / 2330 0·8	**3** 0520 4·0 / 1150 0·7 / M 1746 4·1 ●	**18** 0509 3·9 / 1137 0·8 / Tu 1732 4·1 ○	**3** 0053 0·6 / 0639 4·0 / Th 1306 0·5 / 1900 4·1	**18** 0024 0·4 / 0615 4·3 / F 1243 0·3 / 1835 4·5
4 0407 4·2 / 1040 0·4 / Th 1631 4·3 / 2304 0·3	**19** 0430 3·8 / 1059 0·8 / F 1648 3·9 / 2315 0·8	**4** 0525 4·2 / 1157 0·5 / Su 1749 4·3	**19** 0522 3·9 / 1150 0·8 / M 1743 4·0 ○	**4** 0019 0·6 / 0607 4·0 / Tu 1236 0·6 / 1832 4·1	**19** 0000 0·7 / 0551 4·0 / W 1219 0·6 / 1814 4·2	**4** 0128 0·6 / 0713 4·0 / F 1341 0·6 / 1933 4·1	**19** 0104 0·3 / 0652 4·4 / Sa 1323 0·3 / 1912 4·5
5 0455 4·4 / 1126 0·3 / F 1716 4·4 / 2350 0·2 ●	**20** 0505 3·9 / 1130 0·8 / Sa 1723 4·0 / 2347 0·8 ○	**5** 0024 0·5 / 0611 4·1 / M 1243 0·6 / 1835 4·2	**20** 0010 0·8 / 0600 3·9 / Tu 1229 0·7 / 1822 4·1	**5** 0104 0·6 / 0650 4·0 / W 1320 0·6 / 1914 4·1	**20** 0042 0·6 / 0631 4·1 / Th 1302 0·5 / 1852 4·3	**5** 0201 0·7 / 0745 3·9 / Sa 1412 0·7 / 2004 3·9	**20** 0144 0·3 / 0728 4·4 / Su 1404 0·3 / 1949 4·4
6 0539 4·4 / 1211 0·4 / Sa 1800 4·4	**21** 0537 3·9 / 1203 0·8 / Su 1757 4·0	**6** 0110 0·6 / 0656 4·1 / Tu 1328 0·6 / 1921 4·1	**21** 0050 0·7 / 0639 4·0 / W 1310 0·7 / 1902 4·1	**6** 0147 0·6 / 0731 3·9 / Th 1402 0·6 / 1955 4·0	**21** 0123 0·5 / 0710 4·2 / F 1342 0·4 / 1931 4·3	**6** 0230 0·8 / 0815 3·8 / Su 1442 0·8 / 2032 3·8	**21** 0223 0·4 / 0808 4·3 / M 1444 0·4 / 2030 4·3
7 0035 0·3 / 0622 4·3 / Su 1255 0·4 / 1843 4·3	**22** 0021 0·8 / 0610 3·9 / M 1236 0·8 / 1829 4·0	**7** 0157 0·7 / 0741 3·9 / W 1415 0·7 / 2006 3·9	**22** 0133 0·7 / 0719 4·0 / Th 1352 0·7 / 1942 4·1	**7** 0227 0·7 / 0811 3·8 / F 1442 0·7 / 2034 3·8	**22** 0204 0·5 / 0749 4·2 / Sa 1423 0·4 / 2012 4·3	**7** 0258 0·9 / 0846 3·7 / M 1511 0·9 / 2103 3·7	**22** 0305 0·6 / 0850 4·1 / Tu 1529 0·7 / 2115 4·0
8 0120 0·4 / 0706 4·2 / M 1340 0·6 / 1927 4·1	**23** 0056 0·8 / 0645 3·9 / Tu 1314 0·8 / 1906 4·0	**8** 0244 0·8 / 0827 3·7 / Th 1503 0·8 / 2054 3·8	**23** 0215 0·7 / 0801 3·9 / F 1436 0·7 / 2026 4·0	**8** 0307 0·8 / 0849 3·7 / Sa 1521 0·8 / 2112 3·7	**23** 0246 0·5 / 0830 4·1 / Su 1507 0·5 / 2054 4·1	**8** 0329 1·0 / 0919 3·6 / Tu 1545 1·1 / 2139 3·5	**23** 0352 0·9 / 0936 3·9 / W 1620 0·9 / 2206 3·7
9 0206 0·6 / 0749 4·0 / Tu 1426 0·8 / 2013 3·9	**24** 0135 0·8 / 0721 3·9 / W 1355 0·9 / 1945 3·9	**9** 0332 0·9 / 0915 3·6 / F 1552 0·9 / 2143 3·6	**24** 0300 0·7 / 0846 3·9 / Sa 1524 0·7 / 2114 4·0	**9** 0345 0·9 / 0928 3·6 / Su 1600 0·9 / 2153 3·5	**24** 0329 0·6 / 0915 4·0 / M 1553 0·6 / 2142 4·0 ☾	**9** 0406 1·2 / 1000 3·5 / W 1627 1·2 / 2223 3·4	**24** 0445 1·1 / 1034 3·6 / Th 1719 1·2 / 2311 3·5
10 0256 0·9 / 0837 3·8 / W 1517 0·9 / 2107 3·7	**25** 0219 0·9 / 0804 3·8 / Th 1440 0·9 / 2032 3·9	**10** 0423 1·0 / 1006 3·4 / Sa 1644 1·0 / 2238 3·4	**25** 0348 0·7 / 0936 3·8 / Su 1614 0·7 / 2206 3·9	**10** 0426 1·0 / 1012 3·5 / M 1644 1·1 / 2237 3·4	**25** 0417 0·8 / 1004 3·8 / Tu 1644 0·8 / 2235 3·8 ☾	**10** 0452 1·3 / 1052 3·4 / Th 1723 1·4 / 2320 3·2	**25** 0550 1·3 / 1147 3·4 / F 1835 1·4
11 0349 1·1 / 0932 3·5 / Th 1612 1·1 / 2206 3·5	**26** 0307 0·9 / 0853 3·7 / F 1532 0·9 / 2125 3·7	**11** 0516 1·1 / 1101 3·3 / Su 1741 1·1 / 2334 3·3	**26** 0441 0·8 / 1031 3·7 / M 1709 0·8 / 2304 3·7 ☾	**11** 0509 1·2 / 1101 3·3 / Tu 1732 1·2 / 2329 3·3	**26** 0511 1·0 / 1102 3·6 / W 1743 1·0 / 2337 3·6	**11** 0554 1·4 / 1158 3·2 / F 1832 1·5	**26** 0032 3·3 / 0712 1·4 / Sa 1319 3·3 / 2008 1·4
12 0449 1·2 / 1035 3·3 / F 1715 1·2 / 2313 3·3 ☾	**27** 0402 1·0 / 0949 3·6 / Sa 1630 1·0 / 2224 3·7	**12** 0612 1·2 / 1200 3·3 / M 1838 1·1	**27** 0539 0·9 / 1132 3·6 / Tu 1811 0·9	**12** 0600 1·3 / 1157 3·3 / W 1828 1·3	**27** 0612 1·1 / 1210 3·5 / Th 1853 1·1	**12** 0032 3·1 / 0707 1·5 / Sa 1317 3·2 / 1948 1·5	**27** 0206 3·3 / 0844 1·3 / Su 1451 3·5 / 2132 1·2
13 0557 1·3 / 1144 3·2 / Sa 1825 1·2	**28** 0501 1·0 / 1054 3·5 / Su 1733 1·0 / 2330 3·6 ☾	**13** 0032 3·3 / 0710 1·3 / Tu 1259 3·3 / 1937 1·1	**28** 0007 3·7 / 0642 1·0 / W 1238 3·6 / 1919 0·9	**13** 0027 3·2 / 0659 1·3 / Th 1259 3·2 / 1931 1·3	**28** 0049 3·4 / 0727 1·2 / F 1328 3·5 / 2013 1·1	**13** 0152 3·2 / 0825 1·3 / Su 1434 3·4 / 2103 1·3	**28** 0325 3·5 / 0957 1·0 / M 1556 3·7 / 2231 1·0
14 0024 3·3 / 0706 1·2 / Su 1253 3·3 / 1934 1·1	**29** 0605 1·0 / 1201 3·5 / M 1839 0·9	**14** 0128 3·3 / 0806 1·1 / W 1355 3·4 / 2032 1·1	**29** 0114 3·6 / 0751 1·0 / Th 1346 3·6 / 2029 0·9	**14** 0130 3·2 / 0802 1·3 / F 1404 3·3 / 2034 1·3	**29** 0211 3·4 / 0849 1·2 / Sa 1450 3·6 / 2132 1·1	**14** 0307 3·4 / 0935 1·2 / M 1539 3·6 / 2206 1·1	**29** 0420 3·7 / 1051 0·8 / Tu 1647 3·8 / 2318 0·8
15 0128 3·3 / 0811 1·1 / M 1354 3·3 / 2034 1·0	**30** 0038 3·6 / 0713 0·9 / Tu 1310 3·6 / 1948 0·8	**15** 0222 3·4 / 0858 1·1 / Th 1449 3·5 / 2122 1·0	**30** 0223 3·7 / 0901 0·9 / F 1456 3·7 / 2136 0·8	**15** 0233 3·3 / 0904 1·2 / Sa 1505 3·5 / 2135 1·2	**30** 0327 3·5 / 1000 1·0 / Su 1559 3·7 / 2237 1·0	**15** 0406 3·6 / 1031 1·0 / Tu 1631 3·9 / 2257 0·8	**30** 0505 3·9 / 1134 0·7 / W 1727 4·1 / 2357 0·6
	31 0144 3·8 / 0820 0·8 / W 1415 3·8 / 2053 0·7				**31** 0427 3·7 / 1058 0·8 / M 1655 3·9 / 2329 0·7		**31** 0543 4·1 / 1211 0·6 / Th 1803 4·2 ●

Chart Datum: 0.13 metres above Ordnance Datum (Dublin)

IRELAND, SOUTH COAST — RINGASKIDDY

Lat 51°50′ N Long 8°19′ W

TIMES AND HEIGHTS OF HIGH AND LOW WATERS

YEAR 1989

TIME ZONE UT(GMT)
For Summer Time add ONE hour in non-shaded areas

SEPTEMBER

Day	Time	m	Day	Time	m
1 F	0032 0617 1243 1835	0.6 4.1 0.6 4.2	16 Sa	0549 1218 1808	4.4 0.3 4.6
2 Sa	0100 0646 1310 1902	0.6 4.1 0.6 4.1	17 Su	0039 0627 1259 1846	0.3 4.5 0.3 4.6
3 Su	0126 0713 1335 1927	0.7 4.1 0.7 4.0	18 M	0119 0704 1340 1924	0.4 4.5 0.4 4.5
4 M	0149 0738 1359 1951	0.8 4.0 0.8 3.9	19 Tu	0159 0744 1422 2005	0.5 4.4 0.5 4.3
5 Tu	0213 0804 1426 2018	0.9 3.9 1.0 3.8	20 W	0243 0825 1508 2050	0.7 4.2 0.8 4.0
6 W	0242 0834 1500 2049	1.0 3.8 1.1 3.7	21 Th	0329 0914 1600 2143	1.0 3.9 1.1 3.7
7 Th	0318 0911 1542 2129	1.2 3.6 1.3 3.5	22 F	0426 1014 1704 2251	1.2 3.6 1.3 3.4
8 F	0406 0959 1638 2224	1.4 3.4 1.5 3.3	23 Sa	0533 1133 1821	1.4 3.4 1.5
9 Sa	0509 1106 1749 2342	1.5 3.3 1.6 3.2	24 Su	0018 0657 1309 1957	3.2 1.5 3.4 1.5
10 Su	0627 1232 1907	1.6 3.2 1.6	25 M	0154 0829 1436 2115	3.3 1.4 3.5 1.2
11 M	0112 0748 1358 2029	3.2 1.5 3.4 1.4	26 Tu	0305 0938 1535 2210	3.5 1.1 3.8 1.0
12 Tu	0234 0903 1508 2136	3.4 1.3 3.6 1.1	27 W	0356 1027 1620 2254	3.8 0.9 4.0 0.8
13 W	0338 1003 1603 2230	3.7 1.0 4.0 0.8	28 Th	0437 1108 1658 2330	4.0 0.7 4.1 0.7
14 Th	0427 1052 1648 2316	4.0 0.7 4.3 0.6	29 F	0512 1143 1732	4.1 0.7 4.2
15 F	0509 1136 1729 2358	4.3 0.4 4.5 0.4	30 Sa	0000 0544 1211 1801	0.7 4.2 0.7 4.2

OCTOBER

Day	Time	m	Day	Time	m
1 Su	0027 0612 1236 1827	0.7 4.2 0.7 4.2	16 M	0012 0600 1234 1819	0.5 4.6 0.4 4.6
2 M	0049 0639 1300 1850	0.8 4.1 0.8 4.1	17 Tu	0055 0641 1317 1902	0.5 4.6 0.5 4.5
3 Tu	0113 0704 1326 1916	0.9 4.1 0.9 4.0	18 W	0138 0723 1402 1944	0.6 4.4 0.7 4.3
4 W	0138 0731 1354 1942	1.0 4.0 1.0 3.9	19 Th	0223 0808 1451 2032	0.8 4.2 0.9 4.0
5 Th	0209 0801 1429 2013	1.1 3.9 1.2 3.8	20 F	0314 0900 1546 2128	1.1 4.0 1.2 3.7
6 F	0247 0839 1514 2056	1.2 3.8 1.3 3.6	21 Sa	0410 1002 1649 2235	1.3 3.7 1.4 3.4
7 Sa	0338 0928 1610 2152	1.4 3.6 1.5 3.4	22 Su	0518 1116 1804 2356	1.4 3.5 1.5 3.3
8 Su	0441 1034 1719 2306	1.5 3.4 1.6 3.3	23 M	0635 1241 1927	1.4 3.4 1.4
9 M	0554 1154 1834	1.6 3.4 1.5	24 Tu	0117 0757 1357 2039	3.3 1.3 3.6 1.3
10 Tu	0032 0712 1316 1951	3.3 1.5 3.5 1.4	25 W	0223 0901 1454 2134	3.5 1.1 3.7 1.1
11 W	0152 0826 1427 2100	3.4 1.2 3.7 1.1	26 Th	0315 0952 1539 2217	3.7 1.0 3.9 0.9
12 Th	0258 0928 1524 2156	3.7 1.0 4.0 0.8	27 F	0357 1033 1619 2254	3.9 0.9 4.0 0.8
13 F	0350 1020 1613 2245	4.0 0.7 4.3 0.6	28 Sa	0435 1108 1654 2325	4.0 0.8 4.1 0.8
14 Sa	0437 1106 1658 2329	4.3 0.5 4.5 0.5	29 Su	0509 1137 1725 2351	4.1 0.7 4.1 0.9
15 Su	0519 1150 1739	4.5 0.3 4.6	30 M	0540 1204 1754	4.2 0.8 4.1
			31 Tu	0018 0610 1232 1822	0.9 4.2 1.0 4.1

NOVEMBER

Day	Time	m	Day	Time	m
1 W	0045 0639 1302 1850	1.0 4.1 1.0 4.1	16 Th	0121 0709 1349 1931	0.7 4.4 0.8 4.2
2 Th	0116 0709 1334 1921	1.0 4.1 1.1 4.0	17 F	0209 0758 1440 2020	0.9 4.2 1.0 4.0
3 F	0151 0742 1413 1957	1.1 4.0 1.2 3.9	18 Sa	0300 0850 1534 2115	1.0 4.0 1.1 3.7
4 Sa	0232 0823 1458 2040	1.2 3.9 1.3 3.7	19 Su	0356 0948 1633 2216	1.1 3.8 1.3 3.6
5 Su	0321 0914 1552 2135	1.3 3.7 1.4 3.6	20 M	0457 1052 1736 2320	1.2 3.6 1.3 3.4
6 M	0420 1014 1654 2241	1.4 3.6 1.4 3.5	21 Tu	0601 1158 1843	1.3 3.5 1.3
7 Tu	0525 1123 1800 2354	1.4 3.6 1.4 3.4	22 W	0028 0710 1303 1947	3.4 1.3 3.5 1.3
8 W	0635 1234 1910	1.3 3.6 1.3	23 Th	0130 0812 1401 2044	3.5 1.2 3.6 1.2
9 Th	0107 0744 1342 2019	3.6 1.2 3.8 1.1	24 F	0225 0905 1451 2132	3.6 1.1 3.7 1.1
10 F	0213 0849 1443 2119	3.8 1.0 4.0 0.9	25 Sa	0314 0950 1536 2213	3.8 1.0 3.8 1.0
11 Sa	0311 0946 1538 2213	4.0 0.8 4.2 0.8	26 Su	0356 1030 1616 2249	3.9 1.0 3.9 1.0
12 Su	0403 1037 1627 2301	4.3 0.6 4.4 0.7	27 M	0437 1106 1654 2322	4.0 1.0 4.0 1.0
13 M	0451 1126 1715 2349	4.4 0.6 4.5 0.6	28 Tu	0513 1139 1729 2354	4.1 1.0 4.1 1.0
14 Tu	0537 1212 1800	4.5 0.6 4.5	29 W	0549 1212 1803	4.1 1.0 4.1
15 W	0035 0622 1300 1845	0.6 4.5 0.6 4.4	30 Th	0028 0622 1246 1835	1.0 4.1 1.0 4.0

DECEMBER

Day	Time	m	Day	Time	m
1 F	0103 0657 1323 1910	1.0 4.1 1.0 4.0	16 Sa	0158 0749 1427 2011	0.7 4.2 0.8 4.0
2 Sa	0141 0734 1404 1948	1.0 4.1 1.1 3.9	17 Su	0246 0837 1515 2057	0.8 4.0 1.0 3.8
3 Su	0223 0815 1447 2032	1.1 4.0 1.1 3.8	18 M	0334 0925 1604 2146	0.9 3.9 1.1 3.7
4 M	0308 0901 1534 2121	1.1 3.9 1.1 3.8	19 Tu	0424 1015 1657 2238	1.0 3.7 1.2 3.5
5 Tu	0359 0953 1627 2217	1.1 3.8 1.2 3.7	20 W	0516 1109 1750 2333	1.2 3.6 1.3 3.4
6 W	0454 1051 1725 2319	1.2 3.8 1.2 3.6	21 Th	0612 1205 1846	1.3 3.5 1.3
7 Th	0556 1154 1828	1.2 3.8 1.2	22 F	0032 0712 1303 1942	3.4 1.3 3.5 1.3
8 F	0024 0702 1259 1935	3.7 1.1 3.8 1.2	23 Sa	0131 0809 1359 2039	3.4 1.3 3.5 1.3
9 Sa	0131 0809 1405 2042	3.6 1.0 3.9 1.1	24 Su	0227 0904 1453 2129	3.5 1.2 3.6 1.2
10 Su	0236 0914 1507 2143	3.9 0.9 4.0 0.9	25 M	0321 0953 1542 2216	3.7 1.2 3.7 1.2
11 M	0336 1013 1604 2240	4.1 0.8 4.1 0.8	26 Tu	0409 1037 1628 2258	3.8 1.1 3.8 1.1
12 Tu	0433 1109 1658 2332	4.2 0.7 4.2 0.7	27 W	0452 1118 1709 2337	3.9 1.1 3.9 1.0
13 W	0525 1201 1749	4.3 0.7 4.3	28 Th	0533 1157 1749	4.0 1.1 4.0
14 Th	0022 0615 1250 1836	0.7 4.4 0.7 4.2	29 F	0015 0611 1235 1825	0.9 4.1 1.0 4.0
15 F	0110 0703 1340 1924	0.8 4.3 0.8 4.1	30 Sa	0053 0646 1313 1902	0.9 4.2 0.9 4.1
			31 Su	0131 0723 1352 1938	0.8 4.2 0.8 4.0

Chart Datum: 0.13 metres above Ordnance Datum (Dublin)

AREA 12—S Ireland 543

CORK continued

CHARTS
Admiralty 1777, 1773, 1765; Imray C56, C57; Irish OS 25
TIDES
−0535 Dover; ML 2.3; Duration 0555; Zone 0 (GMT)

Standard Port COBH (RINGASKIDDY) (—→)

Times				Height (metres)			
HW		LW		MHWS	MHWN	MLWN	MLWS
0500	1100	0500	1100	4.2	3.3	1.4	0.5
1700	2300	1700	2300				

Differences CORK
+0020 +0020 +0020 +0020 +0.3 +0.1 0.0 0.0

NOTE: Cobh is a Standard Port — Times and heights of tides for the year are given below.

SHELTER
Very good shelter under all conditions. Cork itself is largely a commercial port, but berths are available; before proceeding up river contact Port Operations. The main yachting centre is at Crosshaven in Owenboy River, where there are two marinas (Crosshaven Boatyard and Royal Cork YC) and anchorages further upstream, in particular in Drake's Pool. There is also a marina at East Ferry on the E side of Great Island.

NAVIGATION
Waypoint 51°46'.00N 08°15'.30W, 174°/354° from/to front Ldg Lt 354°, 2.8M. There are no navigational dangers for yachtsmen. It is one of the safest harbours to enter in the world. Harbour is deep and well marked. Tidal streams run about 1½ kn in entrance at springs, but more between the forts. Entrance to Owenboy River carries a minimum of 2m at LWS, and the channel is buoyed.

TELEPHONE (021)
Hr Mr 273125; MRCC Limerick 61219; Port Operation & Information Stn. 811380; Coast Life Saving Service (Crosshaven) 831448; Customs 271322; Weather 965974; Dr 831716 (Crosshaven) or 811546 (Cork).
FACILITIES
EC Wednesday; **Royal Cork YC Marina** Tel. 831023, AC, FW, Bar, R; **Crosshaven BY Marina** (100 + some visitors) Tel. 831161, AC, BH (25 ton), C (1.5 ton), CH, D, El, FW, Gas, Gaz, ME, P, Sh, Slip; **East Ferry Marina** (60 + 12 visitors) Tel. 811342, D, AC, FW; **Salve Engineering** Tel. 831145, BY, FW, L, M, ME, Slip; **McWilliam Sailmakers** Tel. 831504, SM; **Union Chandlery** Tel. 271643, ACA, CH; **Glen Marine** Tel. 841163, BY, CH; **Rides Services** Tel. 841176, electronics; **Irish Marine Electronics** Tel. 894155, electronics; **Crosshaven Village** Bar, Dr, L, PO, R, V, YC; **Cork City** ACA, Bar, CH, Lau, V, Bank, PO, Rly, Air.

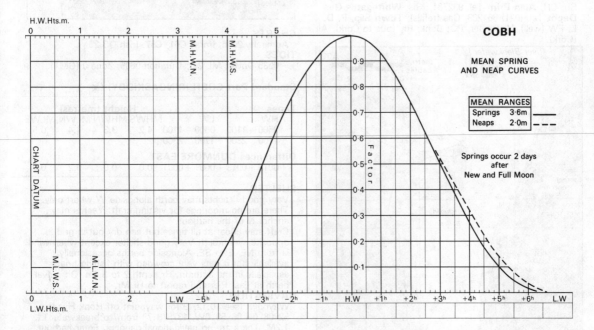

COBH
MEAN SPRING AND NEAP CURVES

MEAN RANGES
Springs 3·6m
Neaps 2·0m

Springs occur 2 days after New and Full Moon

LIGHTS AND MARKS
The hammer-head water tower S of Crosshaven is a very conspic mark, 24.5m high.
Ldg Lts, two sets, both below Fort Davis on E shore;
(1) Lts Oc 5s, front 29m 10M, rear 37m 10M lead E of Harbour Rocks.
(2) Lts Oc R 5s, front 11m 5M, rear 21m 5M lead W of Harbour Rocks.
FW leading lights with W diamond day marks in line 252° lead to Crosshaven, but not easy to see.
RADIO TELEPHONE (local times)
Call: *Cork Harbour Radio* VHF Ch 16; 12 14 (H24); Crosshaven Marina Ch M (Mon-Fri: 0830-1700). Royal Cork YC Marina Ch M (0900-2300). East Ferry Marina Ch M.

YOUGHAL 10-12-17
Cork

CHARTS
Admiralty 2071; Imray C57; Irish OS 22
TIDES
−0545 Dover; ML 2.3; Duration 0555; Zone 0 (GMT).

Standard Port COBH (RINGASKIDDY) (←—)

YOUGHAL continued

Times				Height (metres)			
HW		LW		MHWS	MHWN	MLWN	MLWS
0500	1100	0500	1100	4.2	3.3	1.4	0.5
1700	2300	1700	2300				

Differences YOUGHAL
−0002 +0014 +0016 0000 0.0 0.0 −0.1 0.0
BALLYCOTTON
−0011 +0001 +0003 −0009 0.0 0.0 −0.1 0.0
DUNGARVAN HARBOUR
+0004 +0012 +0007 −0001 0.0 +0.1 −0.2 0.0

SHELTER
Good, but strong SE to S by W winds cause a swell inside the harbour. Good anchorages off Mall Dock, opposite the most northerly warehouse or N of Ferry Point.

NAVIGATION
East Bar waypoint 51°55′.60N 07°48′.00W, 122°/302° from/to Fl WR 2.5s Lt, 1.8M. Beware Blackball Ledge and Bar Rocks outside harbour entrance in R sector of lighthouse Lt. Do not attempt entry in rough seas. Beware salmon nets in entrance and bay during season; also mussel banks off town quays.

LIGHTS AND MARKS
A 15m white tower on W side of entrance shows Lt Fl WR 2.5s, 24m 12/9M, W 183°-273°, R 273°-295°, W 295°-307°, R 307°-351°, W 351°-003°, the W sectors leading over the bars. Convent Tr in line with edge of Town Hall at 178° keeps clear of Red Bank to N.

RADIO TELEPHONE
VHF Ch 16.

TELEPHONE (024)
Hr Mr 92365; MRCC Limerick 61219; Coast Life Saving Service 9 24 27; Customs Cork 21322; Dr 72253.

FACILITIES
EC Wednesday; **Ferry Point Boat Co** Tel. 94232, ME, El, Sh, CH; **Alan Prim** Tel. 932781, Kos; **Whitegates Gas Depot** Tel. (021) 661269, Gas (refills); **Town** Slip, P, D, L, FW (well), V, R, Bar; PO; Bank; Rly (bus to Cork); Air (Cork).

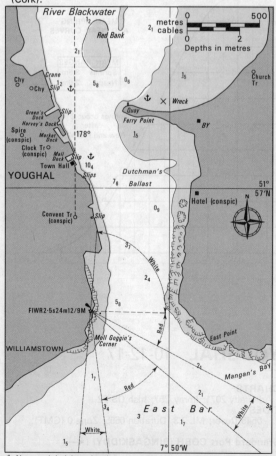
▲ No special visitors berths

DUNMORE EAST 10-12-18
Waterford

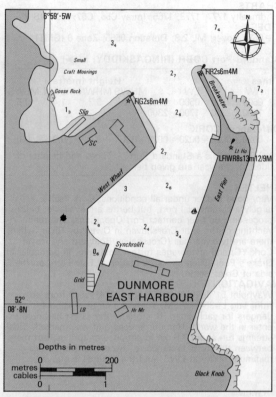

CHARTS
Admiralty 2046; Imray C61, C57; Irish OS 23
TIDES
−0535 Dover; ML 2.3; Duration 0605; Zone 0 (GMT).

Standard Port COBH (RINGASKIDDY) (←)

Times				Height (metres)			
HW		LW		MHWS	MHWN	MLWN	MLWS
0500	1100	0500	1100	4.2	3.3	1.4	0.5
1700	2300	1700	2300				

Differences DUNMORE EAST
+0013 +0013 +0001 +0001 0.0 0.0 −0.2 0.0

SHELTER
Very good. Yachts may berth alongside W wharf only. There are no moorings for visiting craft. Yachts may anchor N of the harbour.
Craft stay afloat at all times but can dry out in grid opposite synchrolift. Very good shelter except when wind is from NE, E to SE. Alongside berths occasionally available but it gets very crowded with fishing boats especially in the autumn. Hr depth 2 to 3.4m. On arrival, berth at East Pier and report to Hr Mr.

NAVIGATION
Waypoint (see 10.12.19 for waypoint off Hook Head) 52°08′.0N 06°58′.0W, 137°/317° from/to breakwater Lt, 1.2M. There are no navigational dangers. Enter harbour under power, if available. Approaching by night from E, give Hook Head a wide berth (½M); when clear of Hook Head alter course for Dunmore East in R sector of breakwater Lt. From W, steer for Hook Head until in R sector and then alter course to N.

LIGHTS AND MARKS
East Pier Lt Ho L Fl WR8s 13m 12/9M, W225°-310°, R310°-004°. E Breakwater head Fl R2s 6m 4M, vis 000°-310°. W wharf Fl G 2s 6m 4M, vis 165°-246°.

RADIO TELEPHONE
VHF Ch 16; 14.

TELEPHONE (051)
Hr Mr 8.31.66; CG Cork 26552; MRCC Limerick (061) 61219; Customs 75391; Coast Life Saving Service 83115; Dr 83194.

AREA 12 — S Ireland 545

FACILITIES
Harbour Tel. 8.31.66, BH (230 ton), D (on NW quay), FW (on East Pier); **Waterford Harbour SC** Tel. 8.33.89, R, Bar; **Dunmore East Garage** Tel. 8.31.24, Kos; **Dunmore East Fishermans Co-operative Society** Tel. 8.33.07, CH; **Dunmore Marine Supply Co** Tel. 8.31.65, CH; **Village** Bar, D, P (cans), Slip, R, V, Bank, PO, Rly (Waterford), Air (Dublin).

WATERFORD 10-12-19
Waterford

CHARTS
Admiralty 2046; Imray C57; Irish OS 23
TIDES
−0520 Dover; ML 2.4; Duration 0605; Zone 0 (GMT).

Standard Port COBH (RINGASKIDDY) (←)

Times				Height (metres)			
HW		LW		MHWS	MHWN	MLWN	MLWS
0500	1100	0500	1100	4.2	3.3	1.4	0.5
1700	2300	1700	2300				

Differences WATERFORD
+0057 +0057 +0046 +0046 +0.4 +0.3 −0.1 −0.1
CARNSORE POINT
+0029 +0019 −0002 +0008 −1.1 −1.0 No data

SHELTER
Very good and there are many excellent anchorages; off the quays just W of Cheek Pt; W of Little Is on S side of main river about 1½ M E of Waterford; in Barrow R on E side just S of New Ross; in Barrow R on the N side of the E/W stretch about 2 M S of New Ross.
NAVIGATION
Waypoint 52°06'.50N 07°56'.60W, 182°/002° from/to front Ldg Lt 002°, 6.7M. Beware of Falskirt Rock (2 ca off Swine Head) and Brecaun reef (20 ca NE of Hook Head). Give Tower Race, extending 1 M to S of Hook Head, a wide berth.
LIGHTS AND MARKS
Hook Head Fl 3s 46m 24M; W Tr, two B bands. Dunmore East LFl WR 8s 13m 12/9M; vis W225°-310°, R310°-004°. Duncannon Ldg Lts 002°, both Oc 4s. Do not confuse Waterford entrance with Tramore Bay.
RADIO TELEPHONE
New Ross VHF Ch 16; 12 14 (H24). River Barrow Railway Bridge Ch 14 16 (H24); contact bridge 1 h before arrival.
TELEPHONE (051)
Hr Mr 74499; CG Cork 26552; MRCC Limerick 61219; Customs 75391; Dr 83194; Hosp 75429.
FACILITIES
EC Thursday; **Carrolls BY (Ballyhack)** ME, El, Sh, C; **Town** AB, M, FW, P, D, Gaz, V, R, Bar, Rly; Air (Cork or Dublin).

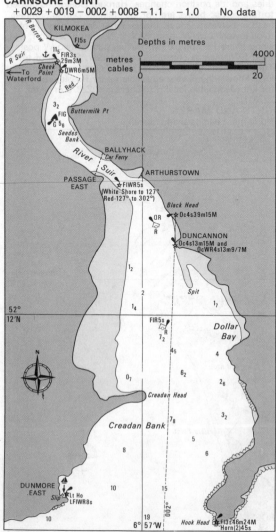

⚓ Dunmore East: Berth W Wharf, report Hr Mr.

ROSSLARE HARBOUR 10-12-20
Dublin

CHARTS
Admiralty 1772; Imray C61; Irish OS 23
TIDES
−0525 Dover; ML 1.1; Duration 0640; Zone 0 (GMT)

Standard Port DUBLIN (NORTH WALL) (→)

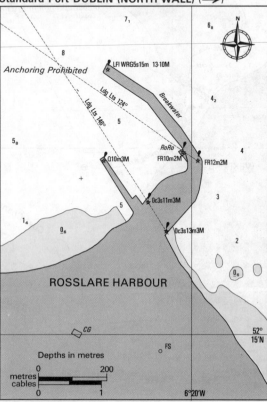

ROSSLARE continued

Times				Height (metres)			
HW		LW		MHWS	MHWN	MLWN	MLWS
0000	0700	0000	0500	4.1	3.4	1.5	0.5
1200	1900	1200	1700				

Differences ROSSLARE HARBOUR
−0440 −0710 −0710 −0440 −2.2 −2.0 −0.7 −0.3

SHELTER
An artificial harbour and busy ferry port which is a useful stopping place. Yachts can anchor in the shelter of the breakwater or go alongside No 1 berth West Pier. Often uncomfortable in winds WNW to NE and dangerous with even moderate winds WNW to NNE.

NAVIGATION
Waypoint 52°14′.60N 06°15′.00W, 105°/285° from/to breakwater Lt, 3.3M. Main approach channel is South Shear. From S, beware Splaugh Rock off Greenore Point, and overfalls here and further S over The Baillies. South Shear runs S of Holdens Bed, a shallow bank with changeable depths, and the tide sets across the channel. Approach from N through North Shear. If wind shifts to dangerous sector leave immediately, through South Shear.

LIGHTS AND MARKS
Two sets of Ldg Lts, 146° both Oc 3s 10/11m, 3M and 124° both FR 10/12m 2M. Lt on the end of breakwater L Fl WRG 5s 15m 13/10M, showing R320°-286° over foul ground from Greenore Pt; W283°-286° over South Shear; G246°-283° over section of Holden's Bed with min depth 5.8m; R208°-246° over N section of Holden's Bed; W188°-208° over North Shear; G098°-188° over South Bay.

RADIO TELEPHONE
Call: *Rosslare Harbour Radio* VHF Ch 16; 06 12 14 (H24).

TELEPHONE (053)
Hr Mr 33114; MRCC Limerick 61219; Coast Life Saving Service 3 21 02; Customs 33116; Dr 31154; Hosp Wexford 22233.

FACILITIES
EC Thursday; **Pier** Tel. 33114, M, P, D, L, FW, ME, C; **Rosslare Ship Repairers** Tel. 3 31 94, ME, Sh, El, Slip, C; **J. Devereaux** Tel. 33104, Kos; **Town** V, R, Bar, PO, Bank. Rly, Air (Dublin).

WEXFORD 10-12-21
Wexford

CHARTS
Admiralty 1772; Imray C61; Irish OS 23

TIDES
Neap −0630 Spring −0330 Dover; ML 1.3; Duration 0630; Zone 0 (GMT)

Standard Port DUBLIN (NORTH WALL) (→)

Times				Height (metres)			
HW		LW		MHWS	MHWN	MLWN	MLWS
0000	0700	0000	0500	4.1	3.4	1.5	0.5
1200	1900	1200	1700				

Differences WEXFORD
−0350 −0720 −0725 −0325 −2.4 −2.0 −1.0 −0.3
TUSKAR ROCK
−0457 −0627 −0601 −0517 −1.5 −1.4 No data

SHELTER
Safe sheltered anchorage off town quays but difficult entrance which should not be attempted in strong winds from E to S when seas break on the bar.

NAVIGATION
Waypoint 52°20′.00N 06°18′.00W, 107°/287° from/to FS on The Raven Point, 2.7M (channel shifts). Bar partly dries and it's position changes. Harbour Board has ceased to function so buoys are not maintained. During summer head in between G local buoys. After 3rd R buoy, N of ruins (awash at HW) head NW × N. About ¼M off; keep along shore for approx 1M. Having passed two W posts, head SW towards conspic factory tower, SE end of town. Turn in when 1ca off training wall. There are no pilots but local knowledge is very advisable — try Rosslare harbour for knowledgeable local. (J. Sherwood 22875 or night 22731).

LIGHTS AND MARKS
No lights. Leading marks (local).

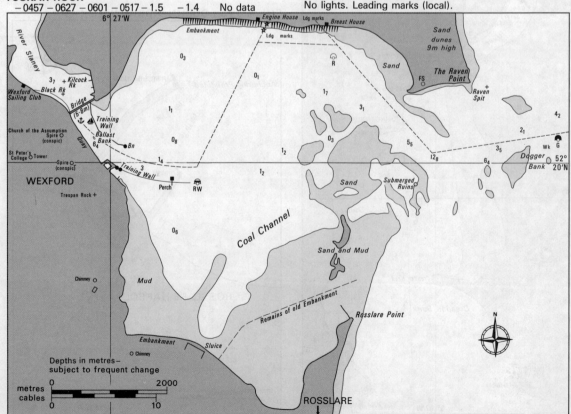

WEXFORD continued
RADIO TELEPHONE
None.
TELEPHONE (053)
Hr Mr 33114; MRCC Limerick 61219; Customs 33116; Dr 31154; Hosp 42233.
FACILITIES
EC Thursday; **Wexford Quays** Slip, P, D, L, FW, ME, El, CH, AB, V; **Tuskar Boats** Tel. 22936, Sh; **J. Jenkins & Son** CH; **Wexford Harbour Boat Club** Tel. 22039, Slip, C (5 ton), Bar; **Barrow Valley Marine** Tel. 21902, CH; **Joyce's Hardware** Tel. 22744, Kos.
Town PO; Bank; Rly; Air (Cork or Dublin).

ARKLOW 10-12-22
Wicklow

CHARTS
Admiralty 633, 1468; Imray C61; Irish OS 19
TIDES
−0200 Dover; ML 1.3; Duration 0640; Zone 0 (GMT).

Standard Port DUBLIN (NORTH WALL) (→)

Times				Height (metres)			
HW		LW		MHWS	MHWN	MLWN	MLWS
0000	0700	0000	0500	4.1	3.4	1.5	0.5
1200	1900	1200	1700				

Differences ARKLOW
−0215 −0255 −0245 −0225 −2.4 −2.0 −0.3 +0.3

SHELTER
Good shelter except in strong SE winds, when seas break across the entrance. Entrance is difficult without power. Once in dock, shelter is perfect (3m). Temporary anchorage 1M S of piers in 4m during SE winds off Roadstone Jetty (Oc R 10s 9m 9M). 2 ca S of jetty a breakwater extends ENE for 3 ca (QY). Best anchorage between jetty and breakwater.
NAVIGATION
Waypoint 52°47'.60N 06°07'.50W, 090°/270° from/to entrance, 0.40M. No navigational dangers. The entrance to the dock is only 13.5m wide. Due to obstructions it is dangerous to proceed up river of dock entrance without local knowledge.
LIGHTS AND MARKS
There are no leading lights or marks. The only two lights are on the piers. North pier LFl G 7s 7m 10M. South pier

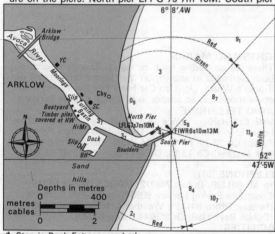

⚠ Stop in Dock Entrance and ask

Fl WR 6s 10m 13M; R shore-223°, W223°-350°, R350°-shore.
RADIO TELEPHONE
VHF Ch 16 (office hours).
TELEPHONE (0402)
Hr Mr 3 24 26; MRCC Limerick 61219; RNLI 3 20 01; Coast Life Saving Service 3 24 30; Customs 3 24 97; Dr 32421.
FACILITIES
EC Wednesday; **Tyrrell's Yard** Tel. 3 20 01, Slip, L, FW, ME, El Sh, C (5 ton mobile), AB; **Dock** Tel. 3 24 26, Slip, M, D, L, FW, ME, El, C (1 ton) CH, AB; **J. Annelsey** Kos; **Town** CH, V, R, P and D (cans), Bar, PO; Bank; Rly; Air (Dublin).

WICKLOW 10-12-23
Wicklow

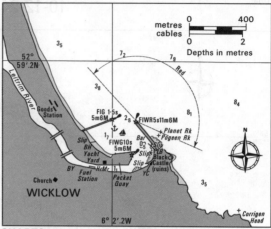

CHARTS
Admiralty 633, 1468; Imray C61; Irish OS 16
TIDES
−0010 Dover; ML 1.5; Duration 0640; Zone 0 (GMT).

Standard Port DUBLIN (NORTH WALL) (→)

Times				Height (metres)			
HW		LW		MHWS	MHWN	MLWN	MLWS
0000	0700	0000	0500	4.1	3.4	1.5	0.5
1200	1900	1200	1700				

Differences WICKLOW
−0035 −0047 −0044 −0038 −1.4 −1.1 −0.6 0.0

SHELTER
Very safe, if not always clean, and always accessible. Outer harbour is open to NE winds which cause a swell. Inner harbour (river) gives excellent shelter. Berthing on W pier is not recommended.
NAVIGATION
Waypoint 52°59'.20N 06°01'.80W, 040°/220° from/to entrance, 0.27M. Entry presents no difficulty so long as one keeps in the R sector of the Lt on E Pier. This avoids Planet Rock and Pogeen Rock. Depth over bar 0.5m.
LIGHTS AND MARKS
There are no leading marks or lights. The only lights are those shown on the chartlet on East Pier, West Pier and Packet Quay.
RADIO TELEPHONE
VHF Ch 16; 02 06 07 08 26 27 28.
TELEPHONE (0404)
Hr Mr 2455; MRCC Limerick 61219; Coast Life Saving Service 2310; Customs 2222; Dr Ring SC 2526.
FACILITIES
EC Thursday; **East Pier** L, FW, AB; **South Quay** Slip, P, D, L, FW, AB; **Wicklow Marine Services** Tel. 3408, M, ME, El, Sh, C, AB; **Neil Watson** Tel. 2492, Slip, BH (16 ton), M, FW, ME, El, Sh, C, CH, AB; **Wicklow SC** Tel. 2526, M, L, FW, Bar; **J. P. Hopkins** Tel. 2413, Kos.
Town CH, V, R, Bar, PO; Bank; Rly; Air (Dublin).

AGENTS NEEDED
There are a number of ports where we need agents, particularly in France.
ENGLAND   Swale, Havengore, Berwick.
SCOTLAND Firth of Forth, Scrabster, Mallaig, Loch Sunart, Loch Aline.
IRELAND   Kilrush, Wicklow, Westport/Clew Bay.
FRANCE   Arcachon, Seudre R, Ile d'Oleron, Rochfort, Ile de Re, St. Giles-Croix-de-Vie, Ile d'Yeu, Pouliguen, Le Croisic, La Forêt, Ile de Bréhat.
GERMANY   Norderney, Dornumer-Accumersiel.
If you are interested in becoming our agent for any of the above, please write to the editors and get your free copy of the Almanac every year. You do not have to be resident in a port to be the agent, but at least a fairly regular visitor.

DUBLIN/DUN LAOGHAIRE
Dublin
10-12-24

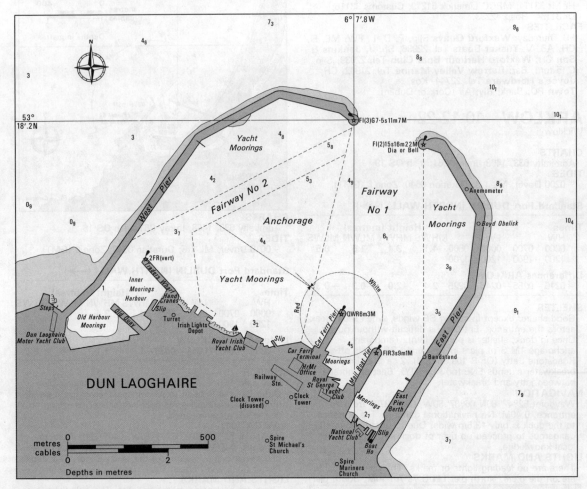

CHARTS
Admiralty 1447, 1468, 1415; Imray C61, C62; Irish OS 16

TIDES
Dublin Bar +0015   Dun Laoghaire +0010 Dover; ML 2.4; Duration 0640; Zone 0 (GMT).

Standard Port DUBLIN (NORTH WALL) (→)

Times				Height (metres)			
HW		LW		MHWS	MHWN	MLWN	MLWS
0000	0700	0000	0500	4.1	3.4	1.5	0.5
1200	1900	1200	1700				

Differences DUBLIN BAR
−0006 −0001 −0002 −0003   0.0   0.0   0.0   +0.1

DUN LAOGHAIRE
−0006 −0001 −0002 −0003   0.0   0.0   0.0   +0.1

NOTE: Dublin is a Standard Port and times and heights of tides for the year are given below.
Since Dun Laoghaire is the principal yachting centre for Dublin, the following all refer to Dun Laoghaire.

SHELTER
The harbour is always available but the harbour is open to NE swell and anchor holding ground is poor. In these conditions yachts should proceed to Dublin unless YCs can provide sheltered moorings.

NAVIGATION
Waypoint 53°18'.40N 06°07'.00W, 060°/240° from/to entrance, 0.47M. Keep clear of the many coasters and ferries. Warning — do not anchor in areas marked 'moorings' or in fairways. Beware — the drying rocks approx 10m from the East Pierhead right in the entrance.

LIGHTS AND MARKS
There are no leading lights or marks. There are lights on both pierheads as shown, on Mail Boat Pierhead and on Trader's Wharf. Light on Car Ferry Pier has R sector to cover moorings, so anchor in W sector outside fairways.

RADIO TELEPHONE
VHF Call *Harbour Office Dun Laoghaire* Ch 16; 14 (office hours Mon-Fri). YCs Ch M. Dublin (Call *Dublin Port Radio*) Ch 12 16; 09 12 13 14 (H24). Dublin lifting bridge (call *Eastlink*) Ch 12 13; Dublin Radio Ch 16 67 65 83 (H24).

TELEPHONE (01)
Hr Mr 801130, Dublin 748772; MRCC Limerick 61219; Coast Life Saving Service 809641 or 376497; Customs Harbour Office 801321; Weather 424655; Local weather (Dublin Bay) 1199; Dr 859244; Hosp 806901.

FACILITIES
EC Wednesday; **National YC** Tel. 801198, Slip, M, L, C (5 ton), FW, D, AB, R, Bar; **Royal St. George YC** Tel. 802953, Slip, M, D, L, FW, C (5 ton), R, Bar; **Royal Irish YC** Tel. 809452, Slip, M, P, D, L, FW, C (5 ton), R, Bar; **Dun Laoghaire Motor YC** Tel. 801371, Slip, FW, AB, Bar; **Cross Berth (Traders Wharf)** FW, C (2, 5, 12 ton), AB; **Downer International Sails (Dun Laoghaire)** Tel. 804286, SM; **B. J. Marine** Tel. 719605, CH; **All Weather Marine** Tel. 713305 CH; **Western Marine** Tel. 800321, CH; **Windmill Leisure & Marine** Tel. 772008, CH, ACA; **Dinghy Supplies** Tel. 322312, CH; **Watson and Jameson** Tel. 326466, SM; **H. Pilsworth** Tel. 822023 ME; **Viking Marine** Tel. 806654, CH, Gas; **Town** P, D, ME, El, Sh, CH, V, R, Bar. PO; Bank; Rly; Air (Dublin).

IRELAND, EAST COAST — DUBLIN (NORTH WALL)

Lat 53°21′ N   Long 6°13′ W

TIMES AND HEIGHTS OF HIGH AND LOW WATERS

TIME ZONE UT(GMT)
For Summer Time add ONE hour in non-shaded areas

YEAR 1989

JANUARY

Time	m		Time	m
1 0558 1129 Su 1815	3.2 1.7 3.3	**16**	0634 1157 M 1855	3.5 1.4 3.7
2 0019 0702 M 1236 1920	1.5 3.2 1.7 3.3	**17**	0050 0744 Tu 1319 2009	1.3 3.5 1.4 3.6
3 0124 0758 Tu 1344 2020	1.5 3.3 1.7 3.4	**18**	0208 0847 W 1434 2117	1.4 3.6 1.3 3.6
4 0220 0849 W 1442 2112	1.4 3.5 1.5 3.5	**19**	0308 0943 Th 1534 2214	1.3 3.7 1.1 3.7
5 0307 0934 Th 1528 2200	1.3 3.7 1.3 3.6	**20**	0356 1033 F 1621 2302	1.2 3.9 0.9 3.7
6 0348 1016 F 1612 2245	1.1 3.8 1.0 3.7	**21**	0435 1116 Sa 1704 2344	1.1 4.0 0.8 3.7 ○
7 0427 1058 Sa 1651 2329 ●	0.9 3.8 0.8 3.8	**22**	0513 1156 Su 1743	1.0 4.0 0.6
8 0505 1139 Su 1732	0.8 4.1 0.6	**23**	0021 0549 M 1232 1821	3.7 0.9 4.1 0.6
9 0012 0544 M 1222 1814	3.9 0.7 4.2 0.4	**24**	0056 0625 Tu 1309 1859	3.6 0.9 4.0 0.6
10 0057 0627 Tu 1306 1859	3.8 0.6 4.3 0.3	**25**	0128 0702 W 1344 1937	3.6 0.9 4.0 0.7
11 0142 0710 W 1352 1945	3.9 0.6 4.3 0.3	**26**	0202 0738 Th 1418 2013	3.5 1.0 3.9 0.8
12 0230 0758 Th 1440 2036	3.8 0.7 4.2 0.4	**27**	0236 0815 F 1451 2050	3.5 1.1 3.8 1.0
13 0321 0849 F 1532 2128	3.7 0.8 4.2 0.6	**28**	0312 0853 Sa 1529 2129	3.6 1.2 3.6 1.2
14 0417 0943 Sa 1630 2226 ☽	3.6 1.0 4.0 0.9	**29**	0355 0934 Su 1613 2213	3.3 1.4 3.4 1.4
15 0522 1045 Su 1739 2332	3.5 1.2 3.8 1.2	**30**	0445 1023 M 1708 2308 ☾	3.2 1.6 3.3 1.5
		31	0551 1127 Tu 1822	3.1 1.7 3.2

FEBRUARY

Time	m		Time	m
1 0022 0704 W 1253 1941	1.6 3.2 1.7 3.2	**16**	0204 0833 Th 1434 2115	1.5 3.5 1.3 3.5
2 0145 0812 Th 1415 2049	1.6 3.3 1.5 3.4	**17**	0307 0935 F 1532 2213	1.4 3.6 1.1 3.6
3 0246 0908 F 1512 2145	1.4 3.6 1.2 3.5	**18**	0350 1024 Sa 1616 2257	1.3 3.8 0.9 3.6
4 0332 0956 Sa 1557 2233	1.1 3.8 0.9 3.7	**19**	0427 1106 Su 1652 2332	1.1 3.9 0.7 3.7
5 0412 1041 Su 1637 2316	0.9 4.0 0.6 3.9	**20**	0459 1142 M 1726 ○	1.0 4.0 0.6
6 0449 1123 M 1716 ●	0.6 4.2 0.3	**21**	0003 0530 Tu 1212 1757	3.7 0.8 4.0 0.5
7 0527 1205 Tu 1757	0.4 4.3 0.1	**22**	0029 0600 W 1243 1829	3.7 0.7 4.0 0.6
8 0039 0608 W 1248 1838	4.0 0.3 4.4 0.1	**23**	0056 0631 Th 1313 1900	3.7 0.7 4.0 0.6
9 0121 0650 Th 1333 1921	4.0 0.3 4.4 0.1	**24**	0124 0703 F 1342 1931	3.6 0.8 3.9 0.8
10 0205 0735 F 1419 2008	4.0 0.4 4.4 0.3	**25**	0155 0735 Sa 1415 2004	3.6 0.9 3.8 0.9
11 0253 0823 Sa 1510 2057	3.9 0.5 4.2 0.6	**26**	0229 0810 Su 1450 2040	3.5 1.0 3.6 1.1
12 0345 0915 Su 1604 2152 ☾	3.7 0.8 4.0 0.9	**27**	0308 0851 M 1531 2121	3.4 1.2 3.4 1.3
13 0447 1016 M 1712 2258	3.5 1.1 3.7 1.3	**28**	0356 0939 Tu 1624 2214 ☽	3.2 1.4 3.2 1.5
14 0600 1132 Tu 1835	3.4 1.4 3.5			
15 0029 0720 W 1310 2001	1.6 3.4 1.4 3.4			

MARCH

Time	m		Time	m
1 0459 1042 W 1740 2332	3.2 1.6 3.1 1.7	**16**	0012 0657 Th 1302 1954	1.7 3.3 1.4 3.3
2 0617 1215 Th 1910	3.1 1.6 3.1	**17**	0151 0816 F 1422 2105	1.6 3.4 1.2 3.4
3 0112 0735 F 1349 2027	1.6 3.3 1.4 3.3	**18**	0251 0917 Sa 1517 2157	1.4 3.6 1.0 3.5
4 0222 0840 Sa 1451 2127	1.4 3.5 1.1 3.5	**19**	0334 1006 Su 1556 2237	1.2 3.8 0.8 3.6
5 0310 0934 Su 1536 2214	1.1 3.8 0.7 3.7	**20**	0406 1044 M 1630 2308	1.0 3.9 0.7 3.7
6 0350 1020 M 1617 2257	0.8 4.0 0.4 3.9	**21**	0437 1116 Tu 1659 2333	0.9 3.9 0.6 3.7
7 0428 1102 Tu 1655 2334 ●	0.5 4.3 0.1 4.1	**22**	0504 1146 W 1727 2358 ○	0.8 3.9 0.6 3.7
8 0506 1144 W 1733	0.3 4.4 0.0	**23**	0532 1214 Th 1756	0.7 3.9 0.6
9 0014 0546 Th 1227 1814	4.1 0.1 4.5 0.0	**24**	0024 0600 F 1242 1824	3.7 0.7 3.9 0.7
10 0054 0627 F 1310 1855	4.1 0.1 4.5 0.1	**25**	0050 0631 Sa 1312 1853	3.7 0.8 3.8 0.8
11 0138 0712 Sa 1358 1940	4.1 0.2 4.4 0.3	**26**	0121 0703 Su 1344 1926	3.7 0.8 3.7 0.9
12 0225 0759 Su 1447 2029	4.0 0.4 4.1 0.7	**27**	0157 0740 M 1422 2002	3.6 0.9 3.6 1.1
13 0315 0853 M 1543 2122	3.8 0.7 3.8 1.1	**28**	0236 0822 Tu 1505 2046	3.5 1.1 3.4 1.3
14 0416 0955 Tu 1652 2231 ☾	3.5 1.0 3.5 1.5	**29**	0325 0914 W 1602 2142	3.4 1.2 3.2 1.5
15 0532 1118 W 1821	3.4 1.3 3.3	**30**	0426 1020 Th 1716 2301 ☽	3.3 1.4 3.1 1.6
		31	0540 1150 F 1845	3.2 1.4 3.1

APRIL

Time	m		Time	m
1 0036 0700 Sa 1319 2004	1.6 3.3 1.2 3.3	**16**	0218 0846 Su 1444 2125	1.5 3.5 1.0 3.4
2 0148 0811 Su 1422 2101	1.4 3.5 0.9 3.6	**17**	0301 0934 M 1525 2203	1.3 3.7 0.9 3.6
3 0240 0905 M 1508 2149	1.1 3.8 0.6 3.8	**18**	0336 1012 Tu 1557 2233	1.1 3.7 0.8 3.6
4 0324 0953 Tu 1550 2230	0.7 4.1 0.3 4.0	**19**	0407 1044 W 1627 2258	1.0 3.8 0.7 3.7
5 0403 1037 W 1630 2309	0.5 4.3 0.1 4.1	**20**	0435 1113 Th 1655 2325	0.9 3.8 0.7 3.8
6 0442 1120 Th 1708 2349 ●	0.3 4.4 0.1 4.2	**21**	0504 1143 F 1722 2351 ○	0.8 3.8 0.7 3.8
7 0523 1204 F 1749	0.1 4.4 0.1	**22**	0532 1212 Sa 1750	0.8 3.8 0.8
8 0029 0605 Sa 1249 1831	4.2 0.1 4.4 0.3	**23**	0021 0603 Su 1246 1821	3.8 0.8 3.7 0.9
9 0113 0650 Su 1338 1914	4.1 0.2 4.2 0.5	**24**	0055 0638 M 1323 1856	3.8 0.9 3.6 1.0
10 0159 0741 M 1430 2005	4.0 0.4 4.0 0.9	**25**	0133 0719 Tu 1405 1937	3.7 0.9 3.5 1.1
11 0253 0837 Tu 1528 2101	3.8 0.7 3.7 1.2	**26**	0216 0805 W 1453 2025	3.6 1.0 3.3 1.3
12 0353 0943 W 1637 2212	3.6 1.0 3.4 1.5	**27**	0305 0901 Th 1549 2125	3.5 1.1 3.3 1.4
13 0505 1105 Th 1804 2347	3.4 1.2 3.2 1.7	**28**	0403 1007 F 1658 2237	3.4 1.2 3.2 1.5
14 0629 1238 F 1931	3.4 1.2 3.2	**29**	0512 1125 Sa 1818 2357	3.4 1.1 3.4 1.5
15 0117 0745 Sa 1352 2037	1.6 3.4 1.1 3.3	**30**	0627 1241 Su 1933	3.5 1.0 3.4

Chart Datum: 0.20 metres above Ordnance Datum (Dublin)

IRELAND, EAST COAST — DUBLIN (NORTH WALL)
Lat 53°21′ N   Long 6°13′ W
TIMES AND HEIGHTS OF HIGH AND LOW WATERS

TIME ZONE UT(GMT) — For Summer Time add ONE hour in non-shaded areas

YEAR 1989

	MAY		JUNE		JULY		AUGUST
	Time m	Time m	Time m	Time m	Time m	Time m	Time m	Time m

	MAY			JUNE			JULY			AUGUST	
1 M	0109 1.3 / 0735 3.6 / 1345 0.8 / 2030 3.6	**16** Tu	0219 1.4 / 0850 3.5 / 1444 1.0 / 2119 3.5	**1** Th	0226 0.9 / 0901 3.9 / 1457 0.7 / 2138 3.9	**16** F	0301 1.4 / 0931 3.5 / 1522 1.2 / 2148 3.6	**1** Sa	0304 1.0 / 0943 3.8 / 1531 1.0 / 2209 3.9	**16** Su	0321 1.3 / 0948 3.5 / 1536 1.2 / 2200 3.7	**1** Tu ●	0441 0.7 / 1125 3.7 / 1652 1.0 / 2334 4.1	**16** W	0421 0.7 / 1057 3.9 / 1631 0.8 / 2301 4.2
2 Tu	0205 1.1 / 0834 3.8 / 1437 0.6 / 2119 3.8	**17** W	0300 1.3 / 0932 3.6 / 1521 1.0 / 2152 3.6	**2** F	0317 0.8 / 0953 4.0 / 1543 0.7 / 2221 4.0	**17** Sa	0341 1.2 / 1012 3.5 / 1557 1.1 / 2224 3.7	**2** Su	0357 0.8 / 1038 3.8 / 1617 0.9 / 2257 4.0	**17** M	0402 1.1 / 1033 3.6 / 1614 1.0 / 2241 3.9	**2** W	0522 0.6 / 1205 3.7 / 1729 0.9	**17** Th ○	0458 0.5 / 1137 4.0 / 1708 0.6 / 2342 4.3
3 W	0254 0.8 / 0927 4.0 / 1522 0.4 / 2203 4.0	**18** Th	0335 1.2 / 1007 3.6 / 1553 0.9 / 2223 3.7	**3** Sa ●	0403 0.6 / 1042 4.0 / 1627 0.6 / 2305 4.0	**18** Su	0417 1.1 / 1051 3.6 / 1631 1.0 / 2301 3.8	**3** M ●	0445 0.7 / 1127 3.8 / 1701 0.9 / 2342 4.0	**18** Tu ○	0440 0.8 / 1116 3.7 / 1651 0.9 / 2322 4.0	**3** Th	0015 4.1 / 0601 0.5 / 1242 3.7 / 1807 0.8	**18** F	0534 0.5 / 1215 4.1 / 1744 0.5
4 Th	0338 0.6 / 1013 4.2 / 1604 0.3 / 2242 4.1	**19** F	0406 1.1 / 1041 3.7 / 1623 0.9 / 2252 3.7	**4** Su	0449 0.5 / 1132 4.0 / 1709 0.7 / 2350 4.1	**19** M ○	0454 0.9 / 1132 3.6 / 1706 0.9 / 2340 3.9	**4** Tu	0532 0.6 / 1215 3.7 / 1743 0.9	**19** W	0518 0.6 / 1157 3.8 / 1727 0.8	**4** F	0053 4.1 / 0641 0.5 / 1317 3.7 / 1845 0.9	**19** Sa	0022 4.4 / 0614 0.2 / 1256 4.1 / 1825 0.4
5 F ●	0420 0.4 / 1058 4.2 / 1644 0.3 / 2323 4.1	**20** Sa ○	0437 1.0 / 1115 3.7 / 1654 1.0 / 2323 3.8	**5** M	0537 0.5 / 1221 3.9 / 1754 0.9	**20** Tu	0532 0.8 / 1214 3.6 / 1743 0.9	**5** W	0028 4.1 / 0618 0.5 / 1300 3.7 / 1827 0.9	**20** Th	0003 4.1 / 0557 0.4 / 1239 3.8 / 1807 0.7	**5** Sa	0130 4.1 / 0719 0.6 / 1351 3.6 / 1923 0.9	**20** Su	0104 4.5 / 0655 0.2 / 1338 4.1 / 1907 0.4
6 Sa	0504 0.3 / 1144 4.2 / 1726 0.4	**21** Su	0509 0.9 / 1150 3.7 / 1725 0.9 / 2358 3.8	**6** Tu	0038 4.1 / 0627 0.5 / 1312 3.8 / 1841 0.9	**21** W	0021 3.9 / 0612 0.7 / 1256 3.6 / 1824 0.9	**6** Th	0113 4.1 / 0703 0.5 / 1344 3.6 / 1910 0.9	**21** F	0045 4.2 / 0638 0.3 / 1321 3.9 / 1848 0.6	**6** Su	0206 4.0 / 0757 0.8 / 1426 3.6 / 2001 1.1	**21** M	0151 4.5 / 0738 0.3 / 1423 4.0 / 1954 0.5
7 Su	0005 4.2 / 0549 0.3 / 1228 4.1 / 1808 0.5	**22** M	0544 0.8 / 1228 3.6 / 1758 0.9	**7** W	0127 4.0 / 0719 0.5 / 1402 3.6 / 1930 1.0	**22** Th	0103 4.0 / 0656 0.6 / 1341 3.6 / 1907 0.8	**7** F	0157 4.0 / 0749 0.6 / 1427 3.5 / 1955 1.0	**22** Sa	0128 4.3 / 0721 0.3 / 1406 3.8 / 1933 0.6	**7** M	0243 3.8 / 0834 1.0 / 1503 3.5 / 2040 1.2	**22** Tu	0239 4.3 / 0825 0.4 / 1512 3.9 / 2044 0.8
8 M	0052 4.1 / 0636 0.4 / 1323 4.0 / 1856 0.7	**23** Tu	0035 3.8 / 0624 0.8 / 1309 3.6 / 1838 1.0	**8** Th	0218 3.9 / 0813 0.7 / 1454 3.5 / 2023 1.2	**23** F	0148 4.0 / 0742 0.5 / 1427 3.6 / 1955 0.9	**8** Sa	0240 3.9 / 0836 0.7 / 1510 3.4 / 2042 1.2	**23** Su	0213 4.3 / 0806 0.3 / 1453 3.8 / 2020 0.7	**8** Tu	0319 3.6 / 0914 1.2 / 1543 3.4 / 2122 1.4	**23** W ☽	0332 4.1 / 0917 0.9 / 1610 3.7 / 2142 1.0
9 Tu	0141 4.0 / 0728 0.5 / 1416 3.8 / 1947 1.0	**24** W	0117 3.8 / 0707 0.8 / 1354 3.5 / 1921 1.0	**9** F	0310 3.8 / 0910 0.8 / 1549 3.3 / 2119 1.3	**24** Sa	0234 4.0 / 0833 0.5 / 1518 3.6 / 2047 0.9	**9** Su	0324 3.8 / 0922 0.9 / 1555 3.3 / 2128 1.3	**24** M	0303 4.2 / 0856 0.5 / 1543 3.7 / 2112 0.8	**9** W	0402 3.5 / 0956 1.4 / 1631 3.3 / 2210 1.6	**24** Th	0435 3.8 / 1017 1.3 / 1719 3.6 / 2252 1.3
10 W	0234 3.9 / 0827 0.7 / 1514 3.6 / 2044 1.2	**25** Th	0202 3.8 / 0757 0.8 / 1443 3.5 / 2012 1.1	**10** Sa	0403 3.7 / 1006 0.9 / 1648 3.2 / 2217 1.4	**25** Su	0325 3.9 / 0925 0.6 / 1613 3.5 / 2142 1.0	**10** M	0409 3.6 / 1009 1.1 / 1642 3.2 / 2216 1.5	**25** Tu ☽	0355 4.0 / 0948 0.7 / 1641 3.6 / 2209 1.0	**10** Th	0455 3.3 / 1048 1.6 / 1732 3.2 / 2312 1.8	**25** F	0553 3.6 / 1137 1.5 / 1838 3.5
11 Th	0332 3.7 / 0931 0.9 / 1619 3.3 / 2150 1.5	**26** F	0251 3.7 / 0851 0.8 / 1536 3.4 / 2108 1.2	**11** Su ☽	0501 3.5 / 1105 1.1 / 1751 3.1 / 2319 1.5	**26** M	0420 3.9 / 1021 0.7 / 1715 3.5 / 2241 1.1	**11** Tu	0458 3.4 / 1058 1.3 / 1739 3.2 / 2311 1.6	**26** W	0457 3.9 / 1048 1.0 / 1747 3.5 / 2315 1.2	**11** F	0604 3.2 / 1200 1.7 / 1843 3.2	**26** Sa	0025 1.5 / 0721 3.5 / 1319 1.7 / 1955 3.6
12 F ☽	0437 3.5 / 1042 1.1 / 1733 3.2 / 2305 1.6	**27** Sa	0345 3.7 / 0950 0.8 / 1638 3.3 / 2212 1.3	**12** M	0603 3.4 / 1205 1.2 / 1852 3.2	**27** Tu	0523 3.8 / 1123 0.9 / 1821 3.5 / 2347 1.2	**12** W	0556 3.3 / 1156 1.5 / 1839 3.2	**27** Th	0608 3.7 / 1200 1.3 / 1859 3.5	**12** Sa	0039 1.8 / 0724 3.2 / 1327 1.7 / 1951 3.4	**27** Su	0158 1.4 / 0842 3.5 / 1433 1.6 / 2101 3.7
13 Sa	0550 3.4 / 1157 1.1 / 1850 3.2	**28** Su ☽	0445 3.7 / 1055 0.9 / 1747 3.4 / 2318 1.3	**13** Tu	0022 1.6 / 0704 3.3 / 1304 1.3 / 1945 3.2	**28** W	0632 3.7 / 1231 1.0 / 1926 3.5	**13** Th	0015 1.7 / 0702 3.2 / 1302 1.5 / 1937 3.3	**28** F	0034 1.4 / 0727 3.6 / 1323 1.4 / 2008 3.6	**13** Su	0205 1.7 / 0832 3.3 / 1430 1.5 / 2047 3.6	**28** M	0303 1.4 / 0945 3.7 / 1524 1.4 / 2156 3.9
14 Su	0022 1.6 / 0702 3.4 / 1304 1.1 / 1952 3.2	**29** M	0553 3.7 / 1201 0.9 / 1856 3.4	**14** W	0124 1.6 / 0759 3.3 / 1358 1.3 / 2030 3.3	**29** Th	0056 1.2 / 0747 3.7 / 1340 1.1 / 2025 3.6	**14** F	0128 1.7 / 0759 3.3 / 1404 1.5 / 2030 3.4	**29** Sa	0157 1.3 / 0840 3.6 / 1434 1.4 / 2110 3.7	**14** M	0301 1.3 / 0928 3.6 / 1517 1.3 / 2136 3.8	**29** Tu	0350 0.9 / 1034 3.8 / 1603 1.2 / 2242 4.1
15 M	0128 1.5 / 0801 3.5 / 1401 1.1 / 2042 3.3	**30** Tu	0027 1.2 / 0702 3.7 / 1307 0.8 / 1958 3.6	**15** Th	0216 1.5 / 0847 3.4 / 1443 1.2 / 2111 3.5	**30** F	0205 1.1 / 0846 3.7 / 1440 1.1 / 2119 3.8	**15** Sa	0232 1.5 / 0858 3.3 / 1454 1.4 / 2119 3.6	**30** Su	0304 1.1 / 0945 3.6 / 1528 1.3 / 2203 3.8	**15** Tu	0343 1.0 / 1014 3.7 / 1555 1.0 / 2220 4.0	**30** W	0430 0.7 / 1113 3.8 / 1638 1.0 / 2320 4.2
		31 W	0130 1.1 / 0805 3.8 / 1406 0.8 / 2050 3.7							**31** M	0356 0.9 / 1040 3.7 / 1613 1.1 / 2251 4.0			**31** Th ●	0505 0.6 / 1146 3.8 / 1711 0.9 / 2356 4.2

Chart Datum: 0.20 metres above Ordnance Datum (Dublin)

AREA 12—S Ireland

IRELAND, EAST COAST — DUBLIN (NORTH WALL)
Lat 53°21′ N   Long 6°13′ W

TIMES AND HEIGHTS OF HIGH AND LOW WATERS   YEAR 1989

TIME ZONE UT(GMT)
For Summer Time add ONE hour in non-shaded areas

	SEPTEMBER				OCTOBER				NOVEMBER				DECEMBER		
	Time	m	Time	m	Time	m	Time	m	Time	m	Time	m	Time	m	Time	m
1 F	0539 1215 1743	0.6 3.8 0.9	**16** Sa	0509 1149 1720 2357	0.2 4.3 0.4 4.6	**1** Su	0539 1208 1746	0.8 3.9 0.9	**16** M	0522 1200 1739	0.3 4.4 0.3	**1** W	0031 0608 1241 1827	3.8 1.1 4.3 1.1	**16** Th	0053 0629 1314 1903	4.2 0.8 4.3 0.6	**1** F	0052 0622 1300 1852	3.7 1.1 4.0 1.0	**16** Sa	0135 0704 1354 1947	3.9 0.9 4.2 0.6
2 Sa	0027 0611 1243 1815	4.2 0.6 3.8 0.9	**17** Su	0547 1227 1801	0.2 4.3 0.3	**2** M	0027 0608 1236 1817	4.0 0.9 3.9 1.0	**17** Tu	0018 0603 1243 1824	4.5 0.4 4.4 0.4	**2** Th	0106 0641 1317 1904	3.8 1.2 3.9 1.2	**17** F	0147 0719 1408 1959	4.0 1.0 4.2 0.8	**2** Sa	0133 0703 1342 1937	3.7 1.2 4.0 1.0	**17** Su	0227 0757 1446 2042	3.7 1.1 4.1 0.8
3 Su	0059 0643 1313 1848	4.1 0.7 3.8 0.9	**18** M	0041 0628 1309 1843	4.6 0.2 4.3 0.4	**3** Tu	0057 0638 1307 1849	3.9 1.0 3.9 1.1	**18** W	0107 0646 1331 1913	4.4 0.6 4.3 0.6	**3** F	0147 0719 1359 1949	3.6 1.3 3.8 1.2	**18** Sa	0244 0815 1505 2101	3.8 1.2 4.0 1.0	**3** Su	0218 0748 1427 2026	3.6 1.3 3.9 1.0	**18** M	0321 0850 1539 2136	3.6 1.2 3.9 0.9
4 M	0130 0716 1344 1923	4.0 0.9 3.7 1.1	**19** Tu	0127 0710 1355 1931	4.5 0.4 4.2 0.5	**4** W	0130 0710 1341 1926	3.8 1.2 3.8 1.2	**19** Th	0159 0735 1423 2009	4.2 0.9 4.1 0.8	**4** Sa	0232 0805 1446 2043	3.5 1.4 3.7 1.3	**19** Su	0348 0917 1610 2210	3.6 1.5 3.9 1.1	**4** M	0307 0840 1517 2121	3.5 1.3 3.9 1.0	**19** Tu ☾	0419 0946 1637 2235	3.4 1.4 3.7 1.1
5 Tu	0204 0749 1418 1958	3.8 1.1 3.6 1.2	**20** W	0218 0758 1446 2023	4.3 0.8 4.0 0.8	**5** Th	0206 0745 1422 2008	3.7 1.3 3.7 1.4	**20** F	0258 0830 1524 2114	3.9 1.3 3.9 1.1	**5** Su	0325 0900 1541 2145	3.4 1.6 3.7 1.3	**20** M ☾	0501 1028 1720 2323	3.4 1.6 3.7 1.2	**5** Tu	0403 0936 1612 2220	3.5 1.5 3.9 1.0	**20** W	0522 1047 1740 2337	3.3 1.6 3.6 1.3
6 W	0239 0825 1456 2039	3.7 1.3 3.5 1.4	**21** Th	0314 0850 1545 2124	4.0 1.1 3.8 1.1	**6** F	0250 0827 1508 2058	3.5 1.5 3.6 1.5	**21** Sa ☾	0406 0936 1634 2231	3.7 1.6 3.8 1.3	**6** M	0430 1006 1642 2255	3.3 1.7 3.6 1.3	**21** Tu	0619 1144 1834	3.4 1.7 3.7	**6** W	0506 1038 1715 2323	3.5 1.4 3.8 1.1	**21** Th	0629 1153 1846	3.3 1.7 3.5
7 Th	0319 0905 1542 2127	3.5 1.5 3.4 1.6	**22** F ☾	0419 0955 1655 2241	3.7 1.5 3.6 1.4	**7** Sa	0345 0922 1606 2204	3.4 1.7 3.5 1.6	**22** Su	0529 1102 1756	3.5 1.8 3.7	**7** Tu	0544 1120 1753	3.4 1.7 3.7	**22** W	0036 0728 1257 1940	1.3 3.4 1.7 3.7	**7** Th	0617 1144 1822	3.5 1.4 3.8	**22** F ☾	0043 0730 1303 1947	1.4 3.3 1.7 3.5
8 F ☾	0412 0956 1641 2230	3.3 1.7 3.3 1.7	**23** Sa	0543 1123 1819	3.5 1.8 3.6	**8** Su	0454 1035 1716 2327	3.2 1.9 3.4 1.6	**23** M	0001 0657 1236 1914	1.4 3.4 1.8 3.7	**8** W	0008 0659 1232 1902	1.3 3.5 1.6 3.8	**23** Th	0140 0823 1358 2033	1.3 3.5 1.6 3.7	**8** F	0029 0723 1252 1930	1.1 3.6 1.3 3.9	**23** Sa	0145 0820 1406 2040	1.5 3.4 1.7 3.5
9 Sa	0523 1111 1754	3.2 1.9 3.3	**24** Su	0021 0717 1307 1940	1.5 3.5 1.8 3.7	**9** M	0618 1205 1832	3.3 1.8 3.5	**24** Tu	0121 0809 1347 2019	1.3 3.6 1.7 3.8	**9** Th	0114 0801 1334 2004	1.1 3.7 1.3 4.0	**24** F	0229 0907 1444 2118	1.3 3.6 1.5 3.8	**9** Sa	0133 0819 1354 2030	1.0 3.7 1.2 4.0	**24** Su	0237 0904 1457 2125	1.4 3.5 1.6 3.5
10 Su	0000 0650 1249 1912	1.8 3.2 1.8 3.4	**25** M	0149 0834 1419 2047	1.3 3.6 1.6 3.8	**10** Tu	0053 0737 1321 1941	1.4 3.4 1.6 3.7	**25** W	0220 0903 1437 2111	1.2 3.7 1.5 3.9	**10** F	0209 0851 1425 2056	0.9 3.9 1.1 4.2	**25** Sa	0310 0942 1524 2157	1.2 3.7 1.4 3.8	**10** Su	0229 0910 1449 2125	0.9 3.9 1.0 4.1	**25** M	0318 0943 1539 2206	1.3 3.6 1.4 3.6
11 M	0134 0808 1401 2016	1.6 3.4 1.6 3.6	**26** Tu	0249 0932 1507 2139	1.1 3.7 1.4 4.0	**11** W	0157 0836 1415 2037	1.2 3.7 1.3 4.0	**26** Th	0305 0945 1518 2153	1.0 3.8 1.4 4.0	**11** Sa	0256 0935 1511 2143	0.7 4.1 0.9 4.3	**26** Su	0345 1014 1559 2231	1.1 3.8 1.3 3.8	**11** M	0318 0956 1539 2216	0.8 4.1 0.8 4.1	**26** Tu	0355 1019 1616 2245	1.2 3.8 1.2 3.6
12 Tu	0233 0905 1450 2110	1.3 3.6 1.3 3.9	**27** W	0334 1014 1545 2221	0.9 3.8 1.2 4.1	**12** Th	0244 0922 1458 2125	0.9 3.9 1.0 4.2	**27** F	0341 1017 1550 2228	1.0 3.9 1.2 4.0	**12** Su	0339 1016 1555 2228	0.6 4.2 0.7 4.4	**27** M	0416 1044 1630 2305	1.1 3.9 1.2 3.8	**12** Tu ○	0404 1041 1627 2305	0.8 4.2 0.7 4.1	**27** W	0427 1055 1649 2322	1.1 3.9 1.1 3.7
13 W	0317 0952 1529 2155	0.9 3.8 1.0 4.1	**28** Th	0409 1049 1617 2258	0.8 3.9 1.1 4.1	**13** F	0325 1003 1538 2209	0.6 4.1 0.8 4.4	**28** Sa	0413 1045 1621 2259	0.9 3.9 1.1 4.0	**13** M ●	0420 1057 1638 2315	0.5 4.3 0.5 4.4	**28** Tu ●	0445 1115 1702 2337	1.1 4.0 1.1 3.8	**13** W	0448 1126 1715 2354	0.7 4.3 0.5 4.1	**28** Th ●	0459 1130 1725	1.0 4.0 0.9
14 Th	0355 1033 1606 2235	0.6 4.0 0.7 4.4	**29** F ●	0441 1118 1647 2329	0.8 3.9 1.0 4.2	**14** Sa	0404 1042 1617 2251	0.4 4.3 0.5 4.5	**29** Su ●	0441 1112 1651 2329	0.9 4.0 1.1 4.0	**14** Tu	0501 1139 1723	0.5 4.4 0.5	**29** W	0515 1147 1736	1.1 4.0 1.1	**14** Th	0532 1214 1804	0.8 4.3 0.7	**29** F	0000 0533 1207 1800	3.7 1.0 4.0 0.8
15 F ○	0431 1111 1642 2316	0.3 4.2 0.5 4.5	**30** Sa	0511 1143 1716 2358	0.7 3.9 0.9 4.1	**15** Su	0442 1120 1657 2333	0.3 4.4 0.4 4.6	**30** M	0509 1139 1720 2358	1.0 4.0 1.1 3.9	**15** W	0003 0543 1224 1811	4.3 0.6 4.4 0.5	**30** Th	0014 0547 1222 1812	3.8 1.1 4.0 1.0	**15** F	0045 0617 1303 1853	4.0 0.8 4.3 0.5	**30** Sa	0038 0608 1246 1839	3.7 1.1 4.1 0.7
							31 Tu	0537 1208 1751	1.0 4.1 1.1											**31** Su	0119 0648 1326 1920	3.7 1.1 4.1 0.6

Chart Datum: 0.20 metres above Ordnance Datum (Dublin)

DUBLIN continued

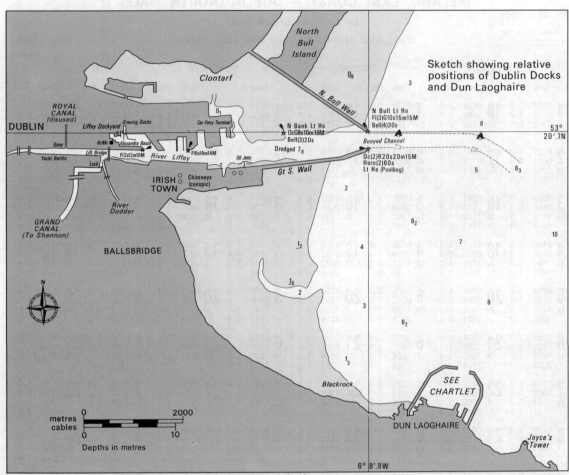

▲ Call Hailing Station (conspic) on N side of Dock area verbally or on Ch 16

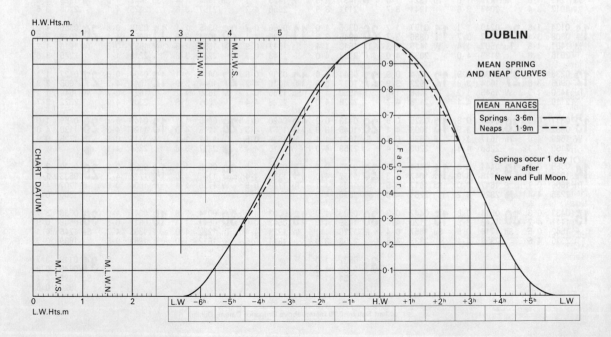

DUBLIN

MEAN SPRING AND NEAP CURVES

MEAN RANGES	
Springs	3·6m
Neaps	1·9m

Springs occur 1 day after New and Full Moon.

HOWTH 10-12-25
Dublin

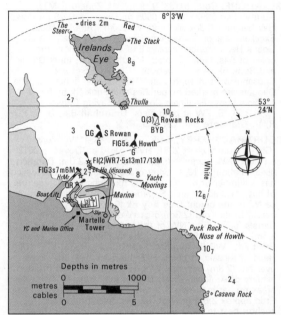
▲ Call Marina on Ch M

CHARTS
Admiralty 1415; Imray C61, C62; Irish OS 16
TIDES
+0025 Dover; ML 2.4; Duration 0625; Zone 0 (GMT)

Standard Port DUBLIN (NORTH WALL) (←)

Times				Height (metres)			
HW		LW		MHWS	MHWN	MLWN	MLWS
0000	0700	0000	0500	4.1	3.4	1.5	0.5
1200	1900	1200	1700				

Differences HOWTH
−0005 −0015 −0005 +0005 0.0 0.0 −0.3 0.0

SHELTER
Good shelter, and available at all tides and in almost any conditions. Anchoring in the harbour is prohibited. See Hr Mr on arrival. Marina dredged to 3m.
NAVIGATION
Waypoint Howth stbd-hand buoy, FlG 5s, 53°23'.72N 06°03'.53W, 071°/251° from/to breakwater Lt, 0.30M. Beware Casana Rk on the E side of the Ben of Howth. Also beware Puck Rk on the NE corner, and a rock about 50m off Puck Rk which dries. A mile N is Ireland's Eye with rocks running out SE and SW from Thulla, their ends marked by Rowan Rocks and South Rowan buoys. There are also drying rocks off the Stack and the Steer on the N. The best approach is S of Ireland's Eye. Between the Nose and the harbour, watch out for lobster pots. Beware rocks off both pierheads.
LIGHTS AND MARKS
E Pierhead Lt — Fl(2) WR 7.5s 13m 17/13M; W 256°-295°, R elsewhere. W sector leads safely to NE corner of harbour.
RADIO TELEPHONE
Hr Mr VHF Ch 16; 08. Marina Ch M 16.
TELEPHONE (01)
Hr Mr 32.22.52; MRCC Limerick 61219; Customs 742961; Coast Life Saving Service 77.34.81; Dr 32.36.36; Hosp 47.84.33
FACILITIES
EC Saturday. **Howth Marina** (220) Tel. 322141, Slip, FW, AC (110 volts), D (cans); **Howth YC** Tel. 32.21.41, R, Bar; **W Pier** D, FW; **Sailmakers** Tel. 326466, SM; **Summit Stores** Tel. 322136, Kos; **Fish Dock** Tel. 32.22.52 D; **Town** P and D (cans), PO; Bank; Rly (Dublin); Air (Dublin).

MALAHIDE 10-12-26
Dublin

CHARTS
Admiralty 1468, 633; Imray C61, C62; Irish OS 13, 16
TIDES
+0030 Dover; ML 2.4; Duration 0615; Zone 0 (GMT)

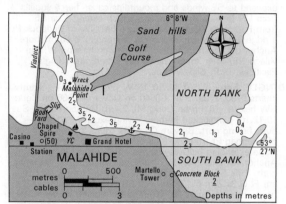

Standard Port DUBLIN (NORTH WALL) (←)

Times				Height (metres)			
HW		LW		MHWS	MHWN	MLWN	MLWS
0000	0700	0000	0500	4.1	3.4	1.5	0.5
1200	1900	1200	1700				

Differences MALAHIDE
−0019 −0013 +0014 +0006 +0.1 +0.1 0.0 0.0
RIVER BOYNE BAR
−0025 −0015 +0110 0000 +0.4 +0.3 No data
DUNANY POINT
−0028 −0018 −0008 −0006 +0.7 +0.9 No data
DUNDALK − SOLDIERS POINT
−0010 −0010 0000 +0045 +1.0 +0.8 +0.1 −0.1

SHELTER
A safe anchorage for yachts of up to 2m draught. Entry possible after half flood.
NAVIGATION
Waypoint 53°27'.20N 06°06'.00W, 085°/265° from/to Grand Hotel, 1.6M. The bar off the entrance is 0.3m in depth and there are sandbanks on either side. Do not attempt entry in thick weather or after dark, or in strong onshore winds. The flood stream reaches 3 kn sp, and the ebb 3½ kn. Beware wreck of trawler off the shipyard.
LIGHTS AND MARKS
During the summer only, the S bank is marked with a R buoy, the N bank with a G buoy by the bar. (Laid and altered as necessary by the Malahide YC). There are no navigational Lts. The entry bearing was 267° (1986) and marked by a RW Fairway buoy ½M seaward of the bar.
RADIO TELEPHONE (01)
Malahide YC. Call *Yacht Base* Ch M (occas).
TELEPHONE (01)
MRCC Limerick 61219; Coast Lifesaving Service Howth 77.34.81; Customs Dublin 74.04.59; Dr 45.19.53; Hosp 30.11.22
FACILITIES
Malahide YC Tel. 45.33.72, Slip, C; **Malahide Hardware** Tel. 45.09.44, Kos; **Town** Lau, PO; Bank; Rly; Air (Dublin).

MINOR HARBOURS AND ANCHORAGES 10.12.27

VENTRY, Kerry, 52°06′ N, 10°19′ W, Zone 0 (GMT), Admty charts 2790, 2789. HW +0540 on Dover, −0056 on Cork; HW height −0.3m on Cork; ML 2.1m; Duration 0605. Note: there are magnetic anomalies near the Blaskets. Hr exposed to SE winds and mountains cause fierce squalls from W. Beware Reenvare Rks which extend 1 ca SE from Parkmore Pt; also NE shore 0.5M NW of Paddock Pt a ridge 3m on which heavy seas break, extends 2.5 ca offshore. Anchor off beach in approx 4m (church bearing W, the village NE) or in 3m S side of bay. There are no lights. Facilities: L, P, Slip, V, Kos.

DINGLE, Kerry, 52°07′ N, 10°15′ W, Zone 0 (GMT), Admty charts 2790, 2789. HW +0540 on Dover, −0056 on Cork; HW height −0.3m on Cork; ML 2.1m; Duration 0605. See 10.12.10. Landlocked harbour giving excellent shelter. A busy fishing port. Beware Crow Rk (dries 3.7m) 0.8M SW of Reenbeg Pt, running 3 ca off shore; also the rocky ledge extending half way across entrance at Black Pt. Anchor 1M S of pier, W of dredged channel; or, for small craft, 1 ca SSW of pier in 1.5m. Lights: Tr on E side of ent Fl G 3s 20m 6M; Pier head FR 4m 2M; Ldg Lts 182° front and rear Oc W 3s, 100m apart, leading down dredged channel. E side of S end of channel, G cone buoy. Facilities: Coast Life Saving Service (066) 5 12 78; **Pier** D, FW; **Town** Bar, P, PO, R, V, Kos.

PORTMAGEE, Kerry, 51°53′ N, 10°21′ W, Zone 0 (GMT), Admty chart 2125. HW +0550 on Dover, −0120 sp −0040 np on Cork; HW height −0.4m on Cork; ML 2.3m; Duration 0610. A good safe anchorage but entrance dangerous in heavy seas. There are no hidden dangers but navigation from Reencaragh Pt up to the pier and bridge needs to be very accurate. Anchor off the pier in approx 6m and beware of strong tides; going alongside pier is not recommended. Bridge opens on request − ring Hr Mr Tel. 6101. Facilities: limited to stores in the village, Kos.

SNEEM, Kerry, 51°49′ N, 9°09′ W, Zone 0 (GMT), Admty chart 2495. HW +0600 on Dover, −0115 sp −0035 np on Cork; HW height −0.5m on Cork; ML 2.1m; Duration 0615. Good shelter except for a swell coming in from the Kenmare R. Ent between Sherky I on W and Rossdohan I on E or N of Sherky I, but then beware Cottoner Rk 1.5 ca from middle of NW side of Sherky I. Also beware Seal Rk in NE part of Hr, marked by Bn. Anchor in bight in NE of Garnish I in approx 4m; N of Rossdohan I in approx 7m (bad in SW winds); near hotel (the Parknasilla) which has moorings. Facilities: L (Hotel Parknasilla, Oystershed House pier or near town); **Town** (2M from Hr.), Bar, PO, R, Slip, V, Kos.

CASTLETOWN, (Bearhaven), Cork, 51°39′ N, 9°54′ W, Zone 0 (GMT), Admty chart 1840. HW +0605 on Dover, −0035 on Cork; HW height −0.6m on Cork; ML 1.9m; Duration 0610. Good shelter; entrance chan only 50m wide abreast Came Pt. Anchor to E of Ldg line NW of Dinish I in 3/5m. Perch Rk in ent. chan. marked by Bn with QG Lt. Ldg Lts 010°, both Oc W 3s, 80m apart, 4/7m 1M, vis 005°−015°. There is also a Dir Oc WRG 5s 4m 14/11M with W sector 024°−024.5°. Beware Walter Scott Rk (2.7m) 2 ca S of Dinish I, marked by S Card Lt Buoy Q(6) + LFl 15s; also Carrigaglos (0.6m high) 1.5 ca from SW point of Dinish I. VHF Ch 16; 08. Facilities: FW (on pier and on quay); Synchrolift on Dinish I; Hr Mr on quay, Tel. 70220; **Town** El, ME, Sh, D and P at garage, Bar, Bank, PO, V, R, Kos.

GLENGARIFF, Cork, 51°44′ N, 9°33′ W, Zone 0 (GMT), Admty chart 1838. HW +0600 on Dover, −0030 on Cork; HW height −0.4m on Cork; ML 2.0m; Duration 0610. Excellent anchorage; ent. between Big Pt and Gun Pt. Keep well to E of Illnacullen and small I to E of it. Chan to W of I is strewn with rks and has power line overhead. Beware Tinker Rks by Big Pt; Yellow Rks S of Illnacullen; Ship I with rks all round in W of ent., E of Illnacullen. Anchor; S of Bark I in 7m; in centre of Hr. in 8/10m. Facilities: **Eccles Hotel** Tel. 63003, Bar, FW, R; **Town** Bar, D, P, PO, R, V, Kos.

GLANDORE, Cork, 51°33′ N, 9°07′ W, Zone 0 (GMT), Admty chart 2092. HW +0605 on Dover, −0025 on Cork; HW height −0.3m on Cork; ML 2.0m; Duration 0605. Excellent shelter. Safest entrance between Adam I and Goat's Head thence keeping E of Eve I and W of the chain of Rks, Outer Danger, Middle Danger, Inner Danger and Sunk Rk. Anchor between Glandore pier and Coosaneigh Pt in approx 3m. There are no lights, but The Dangers are marked by perches and Sunk Rk by BY N Cardinal buoy. Facilities: Coast Life Saving Service (028) 3 31 15; FW at both piers; **Unionhall** Bar, D, P, PO, R, V. **Glandore** Bar, CH, PO, R, V, Kos.

COURTMACSHERRY, Cork, 51°38′ N, 8°41′ W; Zone 0 (GMT), Admty chart 2081. HW −0610 on Dover, −0012 on Cork; HW height −0.4m on Cork; ML 2.0m; Duration 0545. Harbour in NW corner of Courtmacsherry Bay which has a number of hazards; Barrel Rk (dries 2.6m) in centre of bay, marked by unlit S cardinal perch; sunken rocks 2 ca NE of the perch; Blue Boy Rk (0.2m) 4 ca E of the perch; Black Tom (2.3m) 6 ca W of the perch; Inner Barrels (0.5m) 5 ca NW of the perch; Horse Rk (dries 3.6m) 4½ ca off Barry Pt on the W shore. Lights: Wood Point Fl(2) WR 5s 15m 5M, W315°−332°, R332°−315°. Old Head of Kinsale, S point, Fl(2) 10s 72m 25M; RC. Enter over bar (2.3m) between Wood Pt and Coolmain Pt into estuary of Argideen River. Depths may vary. Beware seas break on bar in strong S or SE winds. Anchor NE of Ferry Pt in approx 2.5m or N of the pier. Weed may foul anchor; best to moor with anchors up and down stream. Facilities: Coast Life Saving Service (053) 2.96.37. **Courtmacsherry Hotel** Tel. 4.61.98, Bar, R, P, V; **Quay** FW, D; **Village** V, PO.

OYSTER HAVEN, Cork, 51°41′ N, 8°27′ W, Zone 0 (GMT), Admty charts 2053, 1765. HW −0600 on Dover, −0018 on Cork; ML 2.2m; Duration 0600. Good shelter but subject to swell in S winds. Enter 0.5M N of Big Sovereign, a steep islet divided into two. Keep to S of Little Sovereign on E side of entrance. There is foul ground off Ballymacus Pt on W side, and off Kinure Pt on E side. Pass W of Harbour Rock (0.9m) off Ferry Pt, the only danger within harbour. Anchor NNW of Ferry Pt or up N arm of harbour off the W shore. Weed in higher reaches may foul anchor. No lights, marks or radio telephone. Coast Life Saving Service (021) 7.37.11. Facilities at Kinsale.

BALLYCOTTON, Cork, 51°50′ N, 8°01′ W, Zone 0 (GMT), Admty chart 2424. HW −0555 on Dover, +0006 on Cork; HW height 0.0m on Cork; ML 2.3m; Duration 0550. See 10.12.17. Small Hr. at W end of bay; suffers from scend in strong SE winds. Numerous fishing boats alongside piers, but yachts should go alongside and not anchor in harbour. Good anchorage in offshore winds in 6m NE of breakwater, protected by Ballycotton I. Ballycotton Lt Fl WR 10s 59m 22/18M, B Tr in W walls, on Ballycotton I; W238°−063°, R063°−238°; RC: Dia(4) 90s. Facilities: FW on pier. **Village** Bar, PO, R, V, LB, Kos.

DUNGARVAN, Waterford, 52°05′ N, 7°34′ W, Zone 0 (GMT), Admty chart 2017. HW −0540 on Dover, +0008 on Cork; HW height +0.1m on Cork; ML 2.3m; Duration 0600. See 10.12.17. A large bay, the W side of which dries. Across the entrance there are Carrickapane Rk, Helvick Rk and The Gainers, all marked. Yachts can anchor off Helvick Hr. in approx 4m, or pass up to Dungarvan Town Hr. Beware salmon nets. Channel from Ballynacourty Pt marked by buoys. Ballynacourty Pt Lt Fl(2) WRG 10s 16m 12/9M, G245°−274°, W274°−302°, R302°−325°, W325°−117°. Ballynacourty Ldg Lts 083°, both FW 9/12m 2M on W columns with B bands, 46m apart; Esplanade Ldg Lts 298°, both FR 8/9m 2M. Facilities: EC Thurs; AB, Bar, Bank, D (from garage), P (on quay), PO, R, V, Kos.

VOLVO PENTA SERVICE

Sales and service centres in area 13
NORTHERN IRELAND **Robert Craig & Sons Ltd** 15-21 Great Georges Street, BELFAST BT15 1BW Tel (0232) 232971.

Area 13

North Ireland
Carlingford Lough to Galway

10.13.1	Index	**Page 555**	
10.13.2	Diagram of Radiobeacons, Air Beacons, Lifeboat Stations etc	**556**	
10.13.3	Tidal Stream Charts	**558**	
10.13.4	List of Lights, Fog Signals and Waypoints	**560**	
10.13.5	Passage Information	**563**	
10.13.6	Distance Table	**565**	
10.13.7	Special differences in Ireland	**See 10.12.7**	
10.13.8	Carlingford Lough	**565**	
10.13.9	Strangford Lough	**566**	
10.13.10	Belfast Lough, Standard Port, Tidal Curves	**568**	
10.13.11	Larne	**572**	
10.13.12	Portrush	**573**	
10.13.13	River Bann/Coleraine	**574**	
10.13.14	Lough Foyle (Londonderry, Moville)	**574**	
10.13.15	Lough Swilly	**575**	
10.13.16	Killybegs	**576**	
10.13.17	Sligo	**576**	
10.13.18	Westport/Clew Bay	**577**	
10.13.19	Galway, Standard Port, Tidal Curves	**577**	
10.13.20	Minor Harbours and Anchorages Kilkeel Ardglass Portavogie Donaghadee Carnlough Red Bay Mulroy Bay Sheep Haven Burtonport Blacksod Bay Clifden Bay Kilronan	**582**	

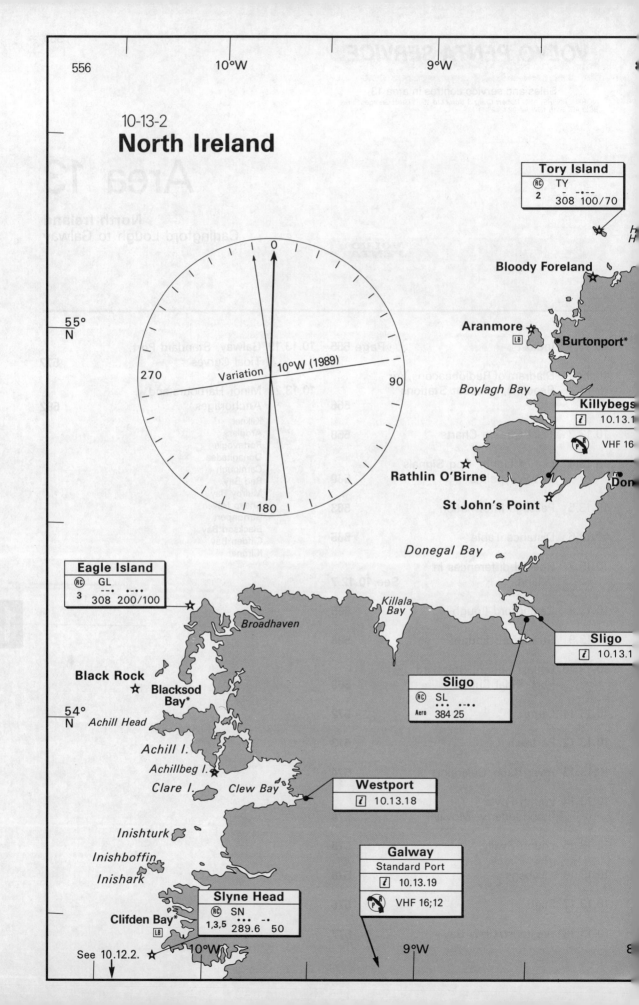

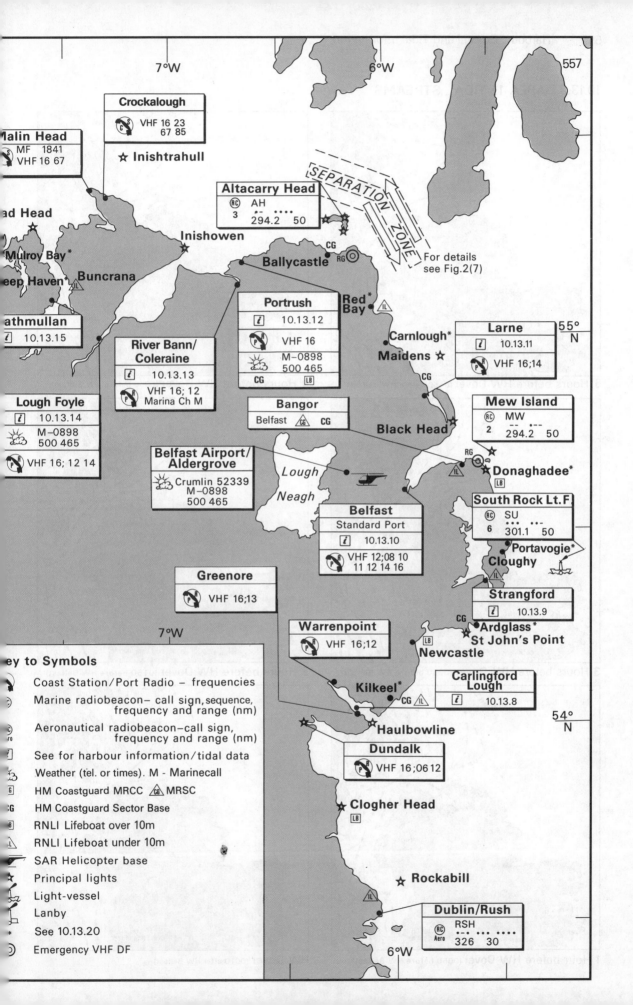

558 Harbour, Coastal and Tidal Information

10.13.3 AREA 13 TIDAL STREAMS

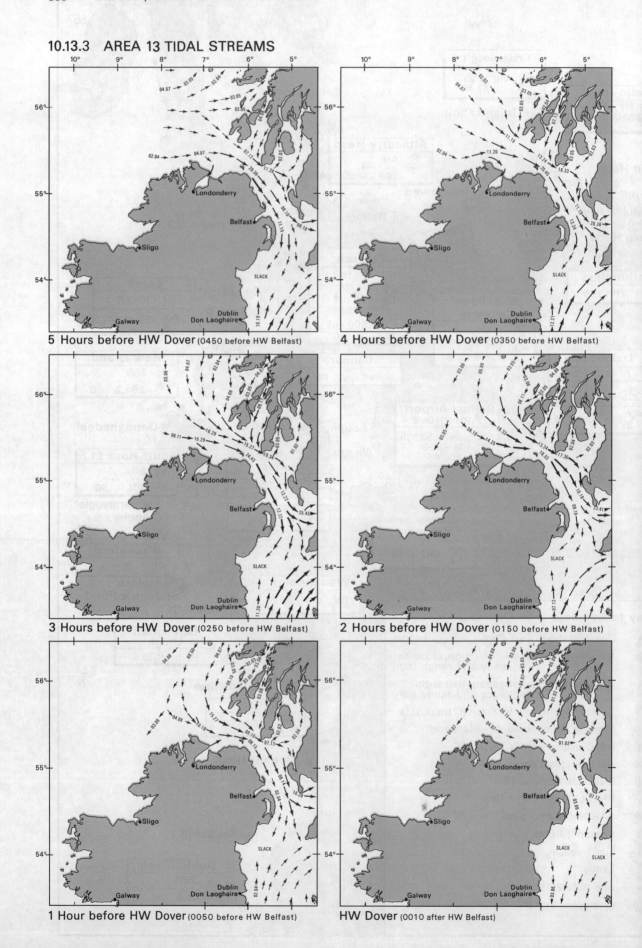

AREA 13—N Ireland 559

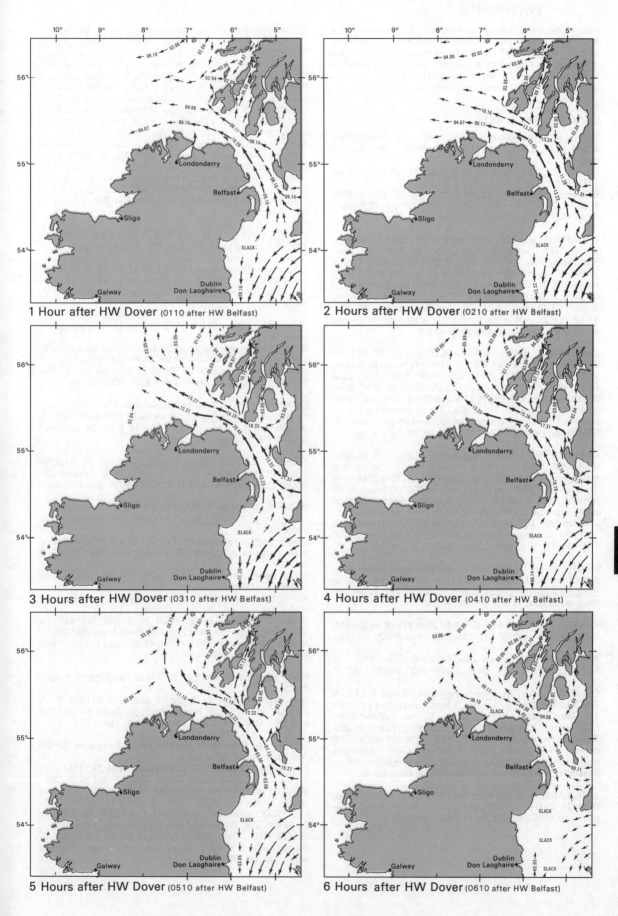

10.13.4 LIGHTS, FOG SIGNALS AND WAYPOINTS

Abbreviations used below are given in 10.0.3. Principal lights are in **bold** print, places in CAPITALS, and light-vessels and Lanbys in *CAPITAL ITALICS*. Unless otherwise stated lights are white. m—elevation in metres; M—nominal range in n. miles. Fog signals are in *italics*. Useful waypoints are underlined — use those on land with care. See 4.2.2.

IRELAND—EAST COAST

Rockabill 53 35.80N/6 00.30W Fl WR 12s 45m **W23M, R19M**; W Tr, B band; vis W178°-329°, R329°-178°; *Horn (4) 60s*.
Skerries Bay, pier head Oc R 6s 7m 7M; W column; vis 103°-154°.
Balbriggan 53 36.7N/6 10.7W Fl(3) WRG 20s 12m W13M, R10M, G10M; W Tr; vis G159°-193°, W193°-288°, R288°-305°.

DROGHEDA. Ldg Lts 248°. **Front** 53 43.13N/6 14.82W Oc 12s 8m **15M**; framework Tr, W lantern; vis 203°-293°. **Rear** 85m from front Oc 12s 12m **17M**; framework Tr; vis 246°-252°. **North light** 53 43.4N/6 15.2W Fl R 4s 7m **15M**; framework Tr, W lantern; vis 282°-288°. Drogheda Bar 53 43.3N/6 13.7W Fl(3) R 5s 6m 3M; R column; Ra refl. Aleria Beacon 53 43.3N/6 14.2W QG 11m 3M. Lyons 53 43.2N/6 41.2W Fl R 2s. Above this point Lts on starboard hand when entering are green, and on port hand red.

DUNDALK. N training wall, head, **Pile light** 53 58.50N/6 17.68W Fl WR 15s 10m **W21M, R18M**; W house on R piles; vis W124°-151°, R151°-284°, W284°-313°, R313°-124°. Oc G 5s 8m; same Tr; vis 352°-355°; Fog Det Lt, Q, vis 358°; *Horn (3) 60s*. No 2 Bn 53 58.3N/6 17.8W Fl(2)R 5s; pile. No 8 Bn 3 59.4N/6 19.0W Fl R 3s; R pile. Above this point Lts on starboard hand when entering are QG, and on port hand QR. Giles Quay, pier head 53 59.1N/6 14.3W Fl G 3s; occas.

CARLINGFORD LOUGH. Hellyhunter Lt Buoy 54 00.34N/6 01.99W Q(6)+LFl 15s; S cardinal mark; *Horn(2) 20s*.
Haulbowline 54 01.19N/6 04.68W Fl(3) 10s 32m **20M**; grey Tr; reserve Lt 15M; Fog Det Lt, VQ, vis 330°. Turning Lt FR 21m 9M; same Tr; vis 196°-208°; *Horn 30s*. Ldg Lts 310° Vidal Bank, Front 54 01.8N/6 05.4W Oc 3s 7m 11M; G house on piles; vis 295°-325°. Rear, Green Island 457m from front Oc 3s 12m 11M; G house on piles; vis 295°-325°. Greenore Pier FlR 7.5s 10m 5M. Carlingford Quay, head Fl 3s 5m 2M.

NEWRY RIVER. Ldg Lts 310°. Front 54 06.4N/6 16.5W. Rear 274m from front. Both Iso 4s 2M; stone columns. Warren Point breakwater, head 54 05.78N/6 15.20W Fl G 3s 6m 3M. Deep water quay Fl G 5s 5m 3M; R post.

KILKEEL. Pier head 54 03.45N/5 59.27W Fl WR 2s 8m 8M; vis R296°-313°, W313°-017°. Meeney's Pier FlG 3s 6m 2M.

ANNALONG. E breakwater head Oc WRG 5s 8m 9M; framework Tr; vis G204°-249°, W249°-309°, R309°-024°.

DUNDRUM BAY. **St John's Point** 54 13.58N/5 39.23W Q(2) 7.5s 37m **23M**; B round Tr, Y bands. **Auxiliary Lt** Fl WR 3s 14m **W15M**, R11M; same Tr, vis W064°-078°, R078°-shore; H24 when fog sig operating; *Horn (2) 60s*. Dundrum Harbour Ldg Lts 330°. Both Oc 8s 4M (occas). FR on west side of channel outside harbour and 3FR on west side of channel inside harbour when local vessels expected. FR on flagstaffs S and E of entrance when firing takes place.

ARDGLASS. Inner pier head Iso WRG 4s 10m W8M, R7M, G5M; Tr; vis G shore-310°, W310°-318°, R318°-shore. Outer pier head 54 15.6N/5 36.0W Fl R 3s 10m 5M.

STRANGFORD LOUGH. Strangford Lt Buoy 54 18.61N/5 28.63W LFl 10s; safe water mark; *Whis*. Bar Pladdy Lt Buoy 54 19.33N/5 30.45W Q(6) + LFl 15s; S cardinal mark. Angus Rock 54 19.83N/5 31.45W Fl R 5s 15m 6M. Ldg Lts 341°. Front, Dogtail Point, Oc(4) 10s 2m 5M. Rear, Gowland Rock 0.8M from front, Oc(2) 10s 6m 5M. Salt Rock FlR 3s 8m 3M. Swan Island Fl(2) WR 6s 5m; W column; vis W115°-334°, R334°-115°. S Pladdy Fl(3) 10s. N Pladdy Q. Church Point Fl(4) R 10s. Portaferry Pier, head Oc WR 10s 9m W9M, R6M; vis W335°-005°, R005°-017°, W017°-128°. North Ldg Lts 181°. Rear, 70m from front, Oc R 10s 12m 6M; Y mast; vis 178°-185°. Pier head, N front Oc WRG 10s 8m W9M, R6M, G6M; Y mast; vis G173°-180°, W180°-183°, R183°-190° (common front); E Front, same mast, Oc WRG 5s 6m; vis R190°-244°, G244°-252°, W252°-260°, R260°-294°. E Ldg Lts 256°, Rear 46m from front Oc R 5s 10m 6M; Y mast; vis 250°-264°.

SOUTH ROCK Lt F 54 24.47N/5 21.92W Fl(3) R 30s 12m **20M**; R hull and Lt Tr, W Mast; RC; Racon; *Horn (3) 45s*.

Portavogie. Plough Rock Lt Buoy 54 27.37N/5 25.07W QR; port-hand mark; *Bell*. S pier head 54 27.44N/5 26.08W, Iso WRG 5s 9m 9M; vis G shore-258°, W258°-275°, R275°-348°.

Skulmartin Lt Buoy 54 31.83N/5 24.85W LFl 10s; safe water mark; *Whis*.
Ballywalter, breakwater head 54 32.67N/5 28.75W Fl WRG 1.5s 5m 9M; vis G240°-267°, W267°-277°, R277°-314°.
Donaghadee, S pier 54 38.70N/5 31.81W Iso WR 4s 17m **W18M**, R14M; W Tr; vis W shore-326°, R326°-shore.

Governor Rocks Lt Buoy 54 39.36N/5 31.94W FlR 3s; port-hand mark. Deputy Reefs Lt Buoy 54 39.50N/5 31.90W FlG 2s; stbd-hand mark. Foreland Spit Lt Buoy 54 39.63N/5 32.25W FlR 6s; port-hand mark.

BELFAST LOUGH. **Mew Island**, NE end 54 41.92N/5 30.75W Fl(4) 30s 37m **30M**; B Tr, W band; RC; *Dia (4) 30s*.
Briggs Lt Buoy 54 41.18N/5 35.67W Fl(2)R 10s; port-hand mark.
BANGOR. N pier 54 40.02N/5 40.30W Iso R 12s 9m 9M.
Belfast No 1 Channel Lt Buoy 54 41.67N/5 46.30W IQG; stbd-hand mark; *Bell*.
Cloghan Jetty Lt Buoy 54 44.12N/5 41.52W FlG 3s; stbd hand mark. Cloghan Jetty, N end Fl G 3s 2M; *Horn 15s*. S end Fl G 3s.
Kilroot power station, intake 54 43.2N/5 45.9W OcG 4s; 2 QR on chimney 500m N.
Kilroot Point. Jetty head Oc G 10s 6m 9M; G framework Tr.

CARRICKFERGUS, E pier head Fl G 7.5s 5m 4M; vis 050°-255°. West pier head Fl R 7.5s 5m 4M; vis 068°-256°. Marina E breakwater 54 42.71N/5 48.63W QG 8m 3M. West breakwater QR 7m 3M; vis 125°-065°.

Black Head 54 46.00N/5 41.27W Fl 3s 45m **27M**; W 8-sided Tr.
North Hunter Rock Lt Buoy 54 53.04N/5 45.06W Q; N cardinal mark. South Hunter Rock Lt Buoy 54 52.68N/5 45.22W Q(6) + LFl 15s; S cardinal mark; *Horn(3) 30s*.

LARNE. Barr Point 54 51.5N/5 46.7W; R framework Tr; *Dia 30s*. Reserve fog signal *Horn*.
Chaine Tower 54 51.27N/5 47.82W Iso WR 5s 23m 11M; grey Tr; vis W230°-240°, R240°-shore. **Ferris Point** 54 51.08N/5 47.34W Iso WRG 10s 18m **W17M**, R13M, G13M; lantern on square W Tr; vis W345°-119°, G119°-154°, W154°-201°, R201°-223°. Entrance Ldg Lts 184°. Front 54 49.6N/5 47.7W Oc 4s 6m 12M; W diamond with R stripe on R pile structure; vis 179°-189°. Rear 610m from front Oc 4s 14m 12M; W diamond with R stripe on aluminium round Tr; synchronised with front, vis 179°-189°.

AREA 13—N Ireland

Maidens 54 55.73N/5 43.60W Fl(3) 20s 29m **23M**; W Tr, B band. Auxiliary Lt Fl R 5s 15m 8M; same Tr; vis 142°-182° over Russel and Highland Rocks.

Carnlough Harbour, N pier 54 59.6N/5 59.2W Fl G 3s 5M; W column, B bands. S pier Fl R 3s 5M; W column, B bands.

Red Bay, pier 55 03.9N/6 03.1W Fl 3s 10m 5M.

RATHLIN ISLAND. Rue Point 55 15.53N/6 11.40W Fl(2) 5s 16m 14M; W 8-sided Tr, B bands. Altacarry Head, **Rathlin East** 55 18.07N/6 10.20W Fl(4) 20s 74m **26M**; W Tr, B band; vis 110°-006° and 036°-058°; RC. Rathlin West, 0.5M NE of Bull point 55 18.05N/6 16.75W Fl R 5s 62m **22M**; W Tr, lantern at base; vis 015°-225°; H24 when fog sig operating; *Horn (4) 60s*. Manor House Pier 55 17.5N/6 11.6W Fl(2) R 6s 5m 4M.

IRELAND—NORTH COAST

Ballycastle pier, head 55 12.5N/6 14.3W L Fl WR 9s 5m; vis R110°-212°, W212°-000°.

PORTRUSH. N pier head Fl R 3s 6m 3M; vis 220°-160°. S pier head Fl G 3s 6m 3M; vis 220°-100°.
Portstewart Point 55 11.3N/6 43.2W Oc R 10s 21m 5M; R square hut; vis 040°-220°.

COLERAINE. River Bann, Ldg Lts 165°. Front 55 09.9N/6 46.2W Oc 5s 6m 2M; W Tr. Rear 245m from front Oc 5s 14m 2M; W Tr. River marked by Fl G on starboard hand, and Fl R on port. West pier, near head Fl G 5s 2M; grey mast. East pier head Fl R 5s 6m 2M; W Tr.

Lough Foyle Lt Buoy 55 15.30N/6 52.50W LFl 10s; safe water mark; *Whis*.
Tuns Lt Buoy 55 14.01N/6 53.38W Fl R 3s; port-hand mark.

Inishowen 55 13.56N/6 55.70W Fl(2) WRG 10s 28m **W18M**, R14M, G14M; W Tr, 2 B bands; vis G197°-211°, W211°-249°, R249°-000°; *Horn (2) 30s*.

LOUGH FOYLE. Warren Point 55 12.58N/6 57.06W Fl 1.5s 9m 10M; W round Tr, G abutment. G; vis 232°-061°.
Magilligan Point 55 11.74N/6 57.97W Fl R 2s 7m 4M; R pile structure. McKinney's Bank 55 10.92N/7 00.50W Fl R 5s 6m 4M; R pile structure. Moville 55 11.00N/7 02.06W Fl WR 2.5s 11m 4M; vis W240°-064°, R064°-240°.
Above this point the channel to River Foyle is marked by Lts Fl G, when entering, on starboard hand, and Fl R on port hand. G Lts are shown from W structures on G or B piles; R Lts from W structures on R piles.
Kilderry 55 04.09N/7 13.95W Fl G 2s 6m 6M; W structure on B piles; Muff 55 03.63N/7 14.21W Fl G 2s 5m 3M; G structure. Coneyburrow 55 03.32N/7 14.42W Fl G 2.5s 5m 3M; G mast on G piles. Faughan 55 03.12N/7 14.42W Fl R 4s 8m 3M; W lantern on R piles. Culmore Point 55 02.78N/7 15.20W Q 6m 5M; G Tr on B base. Culmore Bay 55 02.72N/7 15.65W Fl G 5s 4m 2M; W lantern on G piles. Ballynagard 55 02.28N/ 7 16.37W Fl 3s 6m 3M; W lantern on G structure. Otter Bank 55 01.95N/7 16.65W Fl R 4s 4m 3M; W structure on R Tr. Brook Hall 55 01.70N/7 17.07W QG 4m 3M; W structure on G base. Mountjoy 55 01.25N/7 17.49W QR 5m 3M; W lantern on R piles. Demolished, FW(T).

Inishtrahull 55 25.85N/7 14.60W Fl(3) 15s 59m **25M**; W Tr; Racon.

LOUGH SWILLY. **Fanad Head** 55 16.58N/7 37.85W Fl(5) WR 20s 39m **W18M**, R14M; W Tr; vis R100°-110°, W110°-313°, R313°-345°, W345°-100°. FR on radio mast 3.08M 200°. Swilly More Lt Buoy 55 15.15N/7 35.73W Fl G 3s; stbd-hand mark. Dunree 55 11.85N/7 33.20W Fl(2) WR 5s 46m W12M, R9M; vis R320°-328°, W328°-183°, R183°-196°. Buncrana pier, near head Iso WR 4s 8m W14M, R11M; vis R shore-052° over Inch spit, W052°-139°, R139°-shore over White Strand rock. Rathmullan, pier head Fl G 3s 5M; vis 206°-345°.

MULROY LOUGH. Limeburner Buoy 55 18.55N/7 48.40W (unlit); N cardinal mark; *Whis*. Ravedy island 55 15.1N/7 46.7W Fl 3s 9m 3M; concrete Tr; vis 177°-357°. Dundooan Rocks 55 13N/7 48W QG 4m 1M; G concrete Tr. Crannoge Point 55 12N/7 48W Fl G 5s 5m 2M; G concrete Tr.

IRELAND—WEST COAST

SHEEPHAVEN. Downies Bay, Pier head Fl R 3s 5m 2M; vis 283° through N till obsc by Downies Point.
Ldg Lts 125°. Front 55 10.8N/7 55.6W Oc 6s 7m 2M; B column, W bands. Rear, 81m from front, Oc 6s 12m 2M; B column, W bands.

Tory Island, NW point 55 16.35N/8 14.92W Fl(4) 30s 40m **30M**; B Tr, W band; vis 302°-277°; RC; *Dia 60s*.

Inishbofin Pier Fl 8s 3m 3M; part obsc.
Ballyness Harbour. Ldg Lts 119°. Front 55 09.0N/8 06.9W Iso 4s 24m 1M. Rear, 61m from front, Iso 4s 26m 1M.
Bloody Foreland 55 09.5N/8 17.0W Fl WG 7.5s 14m W6M, G4M; vis W062°-232°, G232°-062°.
Inishsirrer, NW end 55 07.4N/8 20.9W Fl 3.7s 20m 4M; vis 083°-263°.

BUNBEG. Gola Island Ldg Lts 171°. Front 55 05.1N/8 21.0W Oc 3s 9m 2M; W Bn, B band. Rear, 86m from front, Oc 3s 13m 2M; B Bn, W band; synchronised with front. Bo Island, E point 55 04.8N/8 20.1W Fl G 3s 5m; Bn. Inishinny No 1 55 04.5N/8 19.8W QG 3m 1M; square column with steps. Carrickbullog No 2 QR. Inishcoole No 4 QR 12m 1M; square column on base, with steps; Neon. Yellow Rocks No 6 QR 3m 1M; square column with steps; Neon. Magheralosk No 5 QG 4m 1M; G square column with steps.

Cruit Island. Owey Sound Ldg Lts 068°. Front Oc 10s. Rear, 107m from front, Oc 10s (TE 1983).
Rinnalea Point 55 02.5N/8 23.7W Fl 7.5s 19m 9M; square Tr; vis 132°-167°.
Mullaghdoo Ldg Lts 184°, Front 55 02.4N/8 21.6W Iso 8s 19m 2M; W mast. Rear, 358m from front Iso 8s 29m 2M; W mast (TE 1983).

Aranmore, Rinrawros Point 55 00.9N/8 33.6W Fl(2) 20s 71m **29M**; W Tr; obsc by land about 234°-007° and about 013°. Auxiliary Lt Fl R 3s 61m 13M, same Tr; vis 203°-234°.

NORTH SOUND OF ARAN. Ldg Lts 186°. Front 54 58.9N/8 29.2W Oc 8s 8m 3M; B Bn, W band. Rear 395m from front Oc 8s 17m 3M; B Bn. Ballagh Rocks 54 59.97N/8 28.80W Fl 2.5s 13m 5M; W structure, B band.
Black Rocks 54 59.4N/8 29.6W Fl R 3s 3m 1M; R column.

RUTLAND NORTH CHANNEL. Inishcoo Ldg Lts 119°. Front 54 59.1N/8 27.7W Iso 6s 6m 1M; W Bn, B band. Rear 248m from front Iso 6s 11m 1M; B Bn, Y band. Carrickatine 54 59.2N/8 28.0W QR 3m 1M; R beacon with steps. Rutland Island Ldg Lts 138°. Front 54 58.9N/8 27.6W Oc 6s 8m 1M; W Bn, B band. Rear 330m from front Oc 6s 14m 1M; B Bn, Y band. Inishcoo No 4 QR 3m 1M. Nancy's Rock No 1 QG 3m 1M. No 6 QR 3m 1M.

BURTONPORT Ldg Lts 068°. Front 54 58.9N/8 26.4W FG 17m 1M; grey Bn, W band. Rear 355m from front FG 23m 1M; grey Bn, Y band.

SOUTH SOUND OF ARAN. Illancrone Island 54 56.28N/ 8 28.53W Fl 5s 7m 6M; square Tr. Wyon Point 54 56.50N/ 8 27.50W Fl(2) WRG 10s 8m W6M, R3M; W square Tr; vis G shore-021°, W021°-042°, R042°-121°, W121°-150°, R 150°-shore. Turk Rocks 54 57.30N/8 28.15W Fl G 5s 3m 2M; G square Tr. Aileen Reef 54 58.2N/8 28.8W QR 3m 1M. Carrickbealatroha, Upper 54 58.64N/8 28.58W Fl 5s 3m 2M; square brickwork Tr.

RUTLAND SOUTH CHANNEL. Corren's Rock 54 58.12N/ 8 26.68W Fl R 3s 4m 2M; R square Tr. Teige's Rock 54 58.61N/8 26.75W Fl 3s 4m 2M; W round Tr, square base.

Dawros Head 54 49.6N/8 33.6W L Fl 10s 39m 4M; W square column.

Rathlin O'Birne. West side 54 39.77N/8 49.90W Fl WR 20s 35m **W22M, R18M**; W Tr; vis R195°-307°, W307°-195°.

Teelin Point 54 37.32N/8 37.72W Fl R 10s; R structure.

Donegal Bay, St John's Point 54 34.15N/8 27.60W Fl 6s 30m 14M; W Tr.

KILLYBEGS. **Rotten Island** 54 36.87N/8 26.39W Fl WR 4s 20m **W15M**, R11M; W Tr; vis W255°-008°, R008°-039°, W039°-208°. Ldg Lts 338°. Pier, Front Oc R 8s 5m 2M. Rear 65m from front Oc R 8s 7m 2M. Both Y diamonds, on buildings.

SLIGO. Wheat Rock Lt Buoy 54 18.82N/8 39.03W Q(6) + LFl 15s; S cardinal mark. Blackrock 54 18.47N/8 37.00W Fl 5s 24m 13M; W Tr, B band. Auxiliary Lt Fl R 3s 12m 5M; same Tr; vis 107°-130° over Wheat and Seal rocks. Lower Rosses, N of point (Cullaun Bwee) 54 19.72N/8 34.37W Fl(2) WRG 10s 8m W10M, R8M, G8M; W hut on piles; vis G over Bungar bank-066°, W066°-070°, R070° over Drumcliff bar. Ldg Lts 125°. Front Metal Man 54 18.22N/8 34.50W Fl 4s 3m 7M. Rear Oyster Island, 365m from front Oc 4s 13m 10M.

KILLALA. Inishcrone Pier Fl WRG 1.5s 8m 2M; vis W098°-116°, G116°-136°, R136°-187°. Ldg Lts 230°. Rinnaun Point, Front 54 13.5N/9 12.2W Oc 10s 7m 5M. Rear 150m from front Oc 10s 12m 5M.
Dir Lt 215°, Inch Island, 54 13.3N/9 12.3W Fl WRG 2s 6m 3M; square Tr; vis G205°-213°, W213°-217°, R217°-225°. Ldg Lts 196°. Kilroe, Front 54 12.6N/9 12.2W Oc 4s 5m 2M; square Tr. Rear 120m from front Oc 4s 10m 2M; square Tr. Ldg Lts 236°. Pier, Front Iso 2s 5m 2M; W diamond on Tr. Rear, 200m from front, Iso 2s 7m 2M; W diamond on pole.
Killala Bay. Bone Rock, NE end 54 15.8N/9 11.2W Q 7m; N cardinal mark.

Broadhaven, Gubacashel Point 54 16.05N/9 53.28W Iso WR 4s 27m W12M, R9M; W Tr; vis W shore (S side of bay)-355°, R355°-shore. Ballyglass 54 15.3N/9 53.4W Fl G 3s.

Eagle Island, West end 54 16.98N/10 05.52W Fl(3) 10s 67m **26M**; W Tr; RC.

Blackrock 54 04.0N/10 19.2W Fl WR 12s 86m **W22M, R16M**; W Tr; vis W276°-212°, R212°-276°.

Blacksod pier 54 05.90N/10 03.63W Fl(2) WR 7.5s 13m W12M, R9M; W Tr on dwelling; vis R189°-210°, W210°-018°.
Achill Island, Ridge Pt. 54 01.8N/9 58.5W Fl 5s 21m 5M.

ACHILL SOUND. Innish Biggle QR. Carrigeenfushta Fl G 3s. Achill Sound 54 06.0N/9 55.4W QG. Ldg Lts 330° Whitestone, Front and rear both Oc 4s. Saulia Pier 53 57.1N/9 55.5W Fl G 3s 12m. 53 52.1N/9 56.5W Fl R 2s 5m; R square Tr.
Carrigin-a-tShrutha 53 52.3N/9 56.7W Q (2) R 5s .

CLEW BAY. **Achillbeg Island**, S point 53 51.48N/9 56.80W Fl WR 5s 56m **W18M, R18M, R15M**; W round Tr on square building; vis R262°-281°, W281°-342°, R342°-060°, W060°-092°, R(intens) 092°-099°, W099°-118°. Clare Island, E pier Fl R 3s 5m 3M. Cloghcormick Buoy 53 50.54N/9 43.27W (unlit); W cardinal mark.

WESTPORT BAY. Dorinish Lt Buoy 53 49.46N/9 40.61W Fl G 3s; stbd-hand mark. Inishgort, S point 53 49.58N/9 40.17W L Fl 10s 11m 10M; W Tr.
Westport approach 53 47.97N/9 34.30W Fl 3s; G box on conical Bn. Roonagh Quay Ldg Lts 144°, both Iso 10s.

INISHBOFIN. Inishlyon, Lyon Head 53 36.7N/10 09.6W Fl WR 7.5s 13m W7M, R4M; W post; vis W036°-058°, R058°-184°, W184°-325°, R325°-036°. Gun Rock Fl(2) 6s 8m 4M; W column; vis 296°-253°.

Cleggan Point 52 34.5N/10 07.7W Fl(3) WRG 15s 20m W6M, R3M, G3M; W column on W hut; vis W shore-091°, R091°-124°, G124°-221°.

Slyne Head, N Tr, Illaunamid 53 23.97N/10 14.00W Fl(2) 15s 35m **28M**; B Tr; RC.
Inishnee 53 22.7N/9 54.4W Fl(2) WRG 10s 9m W5M, R3M, G3M; W column on W square base; vis G314°-017°, W017°-030°, R030°-080°, W080°-194°.
Croaghnakeela Island 53 19.4N/9 58.3W Fl 3.7s 7m 5M; W column; vis 034°-045°, 218°-286°, 311°-325°.

GALWAY BAY. **Eeragh**, E side 53 08.90N/9 51.78W Fl 15s 35m **23M**; W Tr, two B bands; vis 297°-262°.
Straw Island 53 07.05N/9 37.80W Fl(2) 5s 11m **17M**; W Tr. Ldg Lts 192°. Front 53 06.3N/9 39.7W Oc 5s 6m 3M; W column on W square base; vis 142°-197°. Rear 43m from front Oc 5s 8m 2M; W column on W square base; vis 142°-197°. Kilronan Pier, head Fl WG 1.5s 5m 3M; W column; vis G240°-326°, W326°-000°. Kiggaul Bay Fl WR 3s 5m W5M, R3M; vis W329°-359°, R359°-059°, part obsc by West shore of bay.

CASHLA BAY, entrance, West side Fl(3) WR 10s 8m W6M, R3M; W column on concrete structure; vis W216°-000°, R000°-069°. Rossaveel Pier Ldg Lts 120° Front Oc 3s 4m; W mast. Rear 90m from front Oc 3s 8m; W mast. Spiddle Pier, head Fl WRG 7.5s 11m W6M, R4M, G4M; Y column; vis G102°-282°, W282°-024°, R024°-066°.
Lion Pt Dir Lt 53 15.83N/9 33.97W Dir Iso WRG 4s 6m W8M, R6M, G6M; W square Tr on column; vis G357°-008°, W008°-011°, R011°-017°.

GALWAY, Mutton Island Lt Buoy 53 15.05N/9 02.88W Fl(2)R 6s; port-hand mark. Leverets 53 15.32N/9 01.87W Q WRG 9m 10M; B round Tr, W bands; vis G015°-058°, W058°-065°, R065°-103°, G103°-143.5°, W143.5°-146.5°, R146.5°-015°. Rinmore Iso WRG 4s 7m 5M; W square Tr; vis G359°-008°, W008°-018°, R018°-027°. Approach Channel Ldg Lts 325°. Front 53 16.1N/9 02.8W Fl R 1.5s 12m 7M; R diamond, Y diagonal stripes on mast; vis 315°-345°. Rear 310m from front Oc R 10s 19m 7M; R diamond, Y diagonal stripes on framework Tr; vis 315°-345°. Nimmo's Pier 53 15.99N/9 02.77W Iso Y 6s 7m 6M.

Black Head 53 09.25N/9 15.78W Fl WR 5s 20m W11M, R8M; W square Tr; vis W045°-268°, R268°-276°.

Inisheer 53 02.77N/9 31.60W Iso WR 20s 34m **W20M, R16M**; W Tr, B band; vis W (part vis beyond 7M) 225°-231°, W231°-245°, R245°-269°, W269°-115°.

Loop Head 52 33.65N/9 55.90W Fl(4) 20s 84m **28M**; W Tr; vis 280°-218°; RC.
For Lts further S see 10.12.4.

10.13.5 PASSAGE INFORMATION

LAMBAY ISLAND TO FAIR HEAD (charts 44, 2093, 2198)

In general this coast is fairly steep-to except in larger bs, particularly Dundalk. Streams offshore run up to 2½ kn as far as Rockabill, but are weaker further N until approaching Belfast Lough.

Lambay Island is private, and steep-to except on W side, where there can be overfalls. Skerries Islands (Colt, Shenicks and St Patrick's) are 1M E and SE of Red Island, to E of Skerries Harbour. Colt Island and Shenicks Island are connected to shore at LW. Pass between them and St Patrick's Island, but the latter has off-liers 3 ca to S. Rockabill, two steep-to rks with Lt Ho, is 2½M E of St Patrick's Island. Going NE from Carlingford (10.13.8), after rounding Hellyhunter buoy there are no real offshore dangers until Strangford Lough (10.13.9). For Kilkeel and Ardglass see 10.13.20.

N from Strangford keep ½M off Ballyquintin Pt. 3M to NE are Butter Pladdy rks; keep to E of these. 2M further N is South Rock, with disused Lt Ho, part of cluster of rks to be avoided in poor visibility or bad weather by closing South Rock Lt F. In good weather pass inshore of South Rk, and between it and North Rk (chart 2156).

There are three routes for Belfast Lough (10.13.10). E of Mew Island is Ram Race (to the N on the ebb, and the S on the flood). Copeland Sound, between Mew Island and Copeland Island, is passable but not recommended. Donaghadee Sound is buoyed and a good short cut for yachts: the stream runs SSE from HW Belfast +0530 and NW from HW Belfast −0030, and can attain 4½ kn. An eddy extends S to Ballyferris Pt, and about 1M offshore. For Donaghadee see 10.13.20.

N from Belfast Lough, Black Hd is straightforward. Pass E of Muck Island, which is steep-to. Hunter Rk, 2½M off Larne, is marked by buoys. Further offshore are the Maidens, two dangerous groups of rks extending 2M N and S; E Maiden in the S group is lit.

The very small harb of Carnlough (10.13.20) provides shelter for small yachts, but should not be approached in strong onshore winds. There is anch here in offshore winds, and also in Red B 5M further N. Either can provide a useful anch on passage to or from the Western Isles. There is temp anch in good weather in Cushendun B, 5M NW of Garron Pt.

Fair Hd is a bold headland, steep-to all round, and marks the NE corner of Ireland.

CROSSING THE IRISH SEA

Passages across the Irish Sea can range from the fairly long haul from Land's End to Cork (140M), to the relatively short hop from Mull of Kintyre to Torr Pt (11M). But such distances are deceptive, because the average cruising yacht needs to depart from and arrive at a reasonably secure harbour; also in the North Chan strong tidal streams can cause heavy overfalls: so each passage needs to be treated on its merits.

Many yachts use the Land's End/Cork route on their way to (and from) the delightful cruising ground along the S coast of Ireland (see 10.12.5). Penzance B, or one of the Scilly Is anchs, make a convenient place from which to leave, with good Lts and Round Island radiobeacon to take departure. Although the Celtic Sea is exposed to the Atlantic, there are no dangers on passage and the tidal streams are weak. A landfall between Ballycotton and Old Hd of Kinsale (both have good Lts and radiobeacons) presents no offlying dangers, and in poor vis decreasing soundings indicate approach to land. There is likelihood, outward bound under sail, that the boat will be on the wind − a possible benefit on the return passage. If however the wind serves, and if it is intended to cruise along the southern coast, a landfall at the Fastnet with arrival at (say) Baltimore will place the yacht more to windward, with little extra distance.

From Land's End the other likely destination is Dun Laoghaire. A stop at (say) Milford Haven enables the skipper to select the best time for passing the Bishops and Smalls (see 10.11.5), and roughly divides the total passage into two equal parts. From S Bishop onwards there are the options of making the short crossing to Tuskar Rk and going N inside the banks (theoretically a good idea in strong W winds), or of keeping to seaward. But in bad weather the area off Tuskar is best avoided; apart from the traffic scheme, the tide is strong at sp and the sea can be very rough.

The ferry route Holyhead/Dun Laoghaire is another typical crossing, and is relatively straightforward with easy landfalls either end. The tide runs hard round Anglesey at sp, so departure just before slack water minimises the set N or S. Beware also the traffic scheme off The Skerries.

The IOM (see 10.10.5) is a good centre for cruising in the Irish Sea, and its harbs provide convenient staging points whether bound N/S or E/W.

Between Scotland and Northern Ireland there are several possible routes, but much depends on weather and tide. Time of departure must take full advantage of the stream, and avoid tide races and overfalls (see 10.9.5). Conditions can change quickly, so a flexible plan is needed.

FAIR HEAD TO BLOODY FORELAND (chart 2723)

This is a good cruising area, under the lee of land in SW wind, but very exposed to NW or N. Beware fishing boats and nets in many places. Tidal streams are complex, up to 6 kn at sp in Rathlin Sound but weaker further W. The flood runs W from HW Galway −0500, but at HW Galway −0300 an E eddy starts inshore between Bloody Foreland and Malin Hd. The ebb runs E from HW Galway +0130, but at HW Galway +0330 a W eddy starts inshore from Malin Hd to Bloody Foreland. These eddies initially run only 1M or so offshore but gradually extend seawards, and if timed correctly can be used when sailing in either direction.

A fair tide is essential through Rathlin Sound where the stream runs 6 kn at sp, and causes dangerous overfalls. The main stream runs NW from HW Galway −0600 for five hrs, and SE from HW Galway +0100 for four hrs. A W-going eddy runs from Fair Hd close inshore towards Carrickmannanon Rk from HW Galway +0100, and an E-going eddy runs from HW Galway −0500 to −0100. The worst overfalls in the chan are from HW Galway −0500 to −0300, and it is best to enter W-bound at the end of this period, on the last of favourable tide. Keep seaward of Carrickmannanon Rk and Sheep Island. A harb has been built in Church Bay, on the W side of Rathlin Island.

Proceeding to Portrush (10.13.12), use Skerries Sound in good weather. For Lough Foyle use either the main chan W of The Tuns, or S chan passing 2 ca N of Magilligan Pt and allowing for set towards The Tuns on the ebb (up to 3½ kn).

Torr Rks, Inishtrahull and Garvan Islands lie N and E of Malin Hd. Inishtrahull is lit and is about 1M long, with rks extending N about 3 ca into Torr Sound. Inishtrahull Sound, between Inishtrahull and Garvan Islands is exposed, and tidal stream up to 4 kn at sp can cause a dangerous sea with no warning. Stream also sets hard through Garvan Islands, S of which Garvan Sound can be passed safely in daylight avoiding two sunken rks − one 1½ ca NE of Rossnabartan, and the other 5 ca NW. The main stream runs W for only 3 hrs, from HW Galway −0500 to −0200; otherwise it runs E. W of Malin Hd a W-going eddy starts at HW Galway +0400, and an E-going one at HW Galway −0300. In bad weather it is best to pass at least 3M N of Torr Rks.

From Malin Hd to Dunaff Hd, at entrance to Lough Swilly (10.13.15), keep ½M offshore. Trawbreaga Lough gives shelter, but is shallow, and sea can break on bar: only approach when no swell, and at half flood. Entrance to Lough Swilly is clear except for Swilly Rks off the W shore, SSE of Fanad Head.

W from Lough Swilly the coast is very foul. Beware Limeburner Rk 3M N of Melmore Hd. Mulroy B has good anchs but needs accurate pilotage, as in Irish Cruising Club Sailing Directions. See 10.13.20.

Between Mulroy B and Sheephaven there is inshore passage S of Frenchman Rk, and between Guill Rks and Carnabollion, safe in good weather. Otherwise keep 1M offshore. Sheephaven B has good anchs except in strong NW or N winds, and is easy to enter between Rinnaflagla Pt and Horn Hd. Beware Wherryman Rks, which dry, 1 ca off E shore.

Between Horn Hd and Bloody Foreland are three low islands — Inishbofin, Inishdooey and Inishbeg. The first is almost part of the mainland, but there is a temp anch on S side. 6M offshore is Tory Island (Lt, fog sig, RC) with rks for 5 ca on SW side. Temp anch in good weather in Camusmore B. In Tory Sound the stream runs W from HW Galway +0230, and E from HW Galway −0530, sp rates 2 kn.

BLOODY FORELAND TO ARAN IS. (charts 2725, 2420)

Off Bloody Foreland there is often heavy swell. The coast and islands S to Aranmore give good cruising. Offlying dangers are Buniver and Brinlack shoals, which can break; Bullogconnell 1M NW of Gola Island; and Stag Rks 2M NNW of Owey Island. Anchs include Bunbeg and Gweedore Harb (chart 1883), and Cruit B which is easier access. Behind Aranmore are several good anchs. Use N entrance, since S one is shallow (chart 2792). Rutland N Channel is main approach to Burtonport — general facilities, see 10.13.20.

Boylagh B has shoals and rks N of Roaninish Island. Connell Rk is 1M N of Church Pool — a good anch, best approached from Dawros Hd 4½M to W. There is temp anch (but not in W or NW winds) just E of Rinmeasa Pt, on S side of Glen B. Rathlin O'Birne has steps E side. Anch SE of them ¾ ca offshore. The sound is ½M wide; keep island side to avoid rks off Malin Beg Head.

In Donegal B beware uncharted rks W of Teelin, a good natural harb but exposed to S and W. Killybegs (10.13.16) has better shelter and is always accessible. Good shelter with fair access in Donegal Harb (chart 2792). Moderate anch at Mullaghmore in good weather with winds from SE through S to NW. Inishmurray is worth a visit in good weather: anch off S side. There are shoals close E and NE of the island, and rks 1½M to N. Keep well clear of coast S to Sligo (10.13.17) in onshore winds, and watch for lobster pots.

Killala B has temp anch 1M S of Kilcummin Pt, on W side. Proceeding to Killala beware St Patrick's Rks. Entrance has Ldg Lts and marks, but bar is dangerous in strong NE winds. The coast W to Broadhaven is inhospitable. Only Belderg and Portacloy give a little shelter. Stag Rks are steep-to. Broadhaven (chart 2703) is good anch and refuge, but in N/NW gales sea can break in entrance. In approaches beware Slugga Rk on E side with off-lier, and Monastery Rk on S side.

The coast from Eagle Island to Slyne Hd has many inlets, some sheltered. Streams are weak offshore. There are few Lts. Keep ½M off Erris Hd, and further in bad weather. Unless calm, keep seaward of Eagle Island (Lt, RC) where there is race to N. Frenchport (chart 2703) is good temp anch except in strong W winds. Inishkea Is (chart 2704) can be visited in good weather: anch N or S of Rusheen Island. The sound off Mullett Peninsula is clear, but for Pluddany Rks 6 ca E of Inishkea N.

Blacksod B (chart 2704) has easy entrance, but in W gales beware breakers 1M SE of Duvillaunmore. From N, in good weather, there is chan between Duvillaunbeg and Gaghty Island. Black Rk (Lt) has rks up to 1¼M SW, but entry at night is possible. See 10.13.20.

Rough water is likely off Achil Hd. Achill Sound (chart 2667) is obstructed by cables 11m high at swing bridge. Anchs each end of Sound, but the stream runs strongly.

Clare Island has Two Fathom Rk 1½M off NW coast, and Deace's Rk ½M to the N. In Clew B both Newport and Westport (see chart 2057 and 10.13.18) need detailed pilotage instructions. S of Clare Island beware Meemore Shoal 1½M W of Roonagh Hd. 2M further W is the isolated rk Mweelaun. The offshore islands Coher, Bullybeg, Inishturk (with anch on E side) and Inishdalla have few hidden dangers, but the coast to the E must be given a berth of 1½M even in calm weather, while in strong winds breakers extend much further seaward.

Killary B (chart 2706) and Little Killary both have good anchs in magnificent scenery. Consult sailing directions, and only approach in reasonable weather and good visibility.

Ballynakill (chart 2706), easily entered either side of Freaghillaun South, has excellent shelter. Beware Mullaghadrina and Ship Rk in N chan. Anch in Fahy, Derryinver or Barnaderg B. Rks and breakers exist E of Inishbofin and S of Inishshark: see chart 2707 for clearing lines. Carrickmahoga is dangerous rk between Inishbofin and Cleggan. Lecky Rks lie 1M SSE of Davillaun. There is anch on S side of Inishbofin (Lt), but difficult in strong SW wind or swell.

Cleggan B is moderate anch, exposed to NW but easy access. High Island sound can be used, but not Friar I Sound or Aughrus Passage. Clifden B (chart 2708) has offlying dangers with breakers: enter 3 ca S of Seal Rks Bn and anch between Drinagh Pt and Larner's Rk. See 10.13.20.

Slyne Hd (Lt and RC) marks SW end of rks and islets stretching 2M WSW from coast. Here the stream turns N at HW Galway −0320, and S at HW Galway +0300. It runs 3 kn at sp, and in bad weather causes a dangerous race. The sea may break on Barret Shoal, 3M NW of Slyne Hd.

The Connemara coast (charts 2709, 2096) and Aran Islands (chart 3339) gives excellent cruising in good vis. But there are many rks, and few navigational marks. Between Slyne Hd and Roundstone B are many offlying dangers. Further E the better harbs are Roundstone B, Cashel B, Kilkieran B, Greatman B and Cashla B. Kilronan (10.13.20) on Inishmore is only reasonable harb in Aran Islands, but is exposed in E winds. Cruising coastwise keep well offshore of Skird Rks.

Normal approach to Galway B is through N Sound or S Sound. N Sound is 3M wide between Eagle Rk and other dangers off Lettermullan shore, and banks on S side which break in strong winds. S Sound is wider, with no dangers except Finnis Rk ½M SE of Inisheer. The other chans are Gregory Sound, 1M wide between Inishmore and Inishmaan, and Foul Sound between Inishmaan and Inisheer. The latter has one danger — Pipe Rk and the reef inshore of it, extending 3 ca NW of Inisheer.

The N side of Galway B is exposed, with no shelter. Galway (10.13.19) is a commercial port, and New Harb (chart 1984) is a more pleasant anch.

For all Irish waters the Sailing Directions published by the Irish Cruising Club are strongly recommended, and particularly on the W coast, where other information is scant.

AREA 13—N Ireland

10.13.6 DISTANCE TABLE

Approximate distances in nautical miles are by the most direct route while avoiding dangers and allowing for traffic separation schemes etc. Places in *italics* are in adjoining areas.

1	*Dun Laoghaire*	**1**																		
2	*Pt of Ayre*	93	**2**																	
3	*Mull of Galloway*	93	21	**3**																
4	*Barra Hd*	238	184	163	**4**															
5	Rockabill	20	75	73	218	**5**														
6	Carlingford	50	70	60	203	30	**6**													
7	Strangford	71	44	34	175	51	36	**7**												
8	Mew Island	92	43	23	146	72	57	29	**8**											
9	Bangor	96	48	29	149	76	61	34	6	**9**										
10	Larne	108	59	37	135	88	73	45	16	16	**10**									
11	Altacarry Hd	135	85	64	104	117	102	74	45	45	31	**11**								
12	Portrush	150	98	79	103	130	115	87	58	58	48	19	**12**							
13	Londonderry	177	127	106	120	157	142	114	85	85	75	45	30	**13**						
14	Rathmullan	200	149	128	105	181	166	138	109	109	96	65	51	65	**14**					
15	Tory Island	209	156	135	95	189	174	146	117	117	105	74	61	74	35	**15**				
16	Killybegs	267	214	193	153	247	232	204	175	175	163	132	119	132	93	58	**16**			
17	Sligo	281	228	207	167	261	246	218	189	189	177	146	133	146	107	72	30	**17**		
18	Eagle Island	297	244	223	183	277	262	234	205	205	193	162	149	162	123	88	62	59	**18**	
19	Slyne Hd	352	299	278	238	332	317	289	260	260	248	217	204	217	178	143	117	114	55	**19**
20	Galway	348	348	327	287	381	366	338	309	309	297	266	253	266	227	192	166	163	104	49	**20**

CARLINGFORD LOUGH
Louth/Newry and Mourne

10-13-8

CHARTS
Admiralty 44, 2800; Imray C62; Irish OS 9

TIDES
Cranfield Point +0025 and Warrenpoint +0035 Dover; ML 2.9; Duration Cranfield Point 0615, Warrenpoint 0540; Zone 0 (GMT).

Standard Port DUBLIN (NORTH WALL) (←)

Times				Height (metres)			
HW		LW		MHWS	MHWN	MLWN	MLWS
0000	0700	0000	0500	4.1	3.4	1.5	0.5
1200	1900	1200	1700				

Differences CRANFIELD POINT
−0027 −0011 +0017 −0007 +0.7 +0.9 +0.3 +0.2
WARRENPOINT
−0020 −0010 +0040 +0040 +1.0 +0.9 +0.1 +0.2

Note: NE coast is Ulster, SW coast the Republic of Ireland.

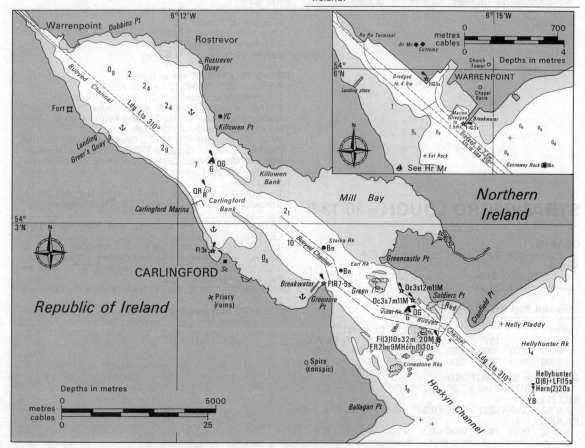

CARLINGFORD LOUGH continued

SHELTER
Good shelter. Anchorages at Greenore between quay and breakwater; off Greer's Quay; between Killowen Pt and Rostrevor; N of Carlingford Hr. Alongside berth at pier in Hr (dries).

NAVIGATION
Waypoint 54°00'.23N 06°02'.20W, 130°/310° from/to front Ldg Lt 310°, 2.4M. The lough becomes choppy in S winds and with on-shore winds the bar becomes impassable. Beware sudden squalls. Tides run up to 5 kn off Greenore. Drying rocks and shoals extend nearly across the entrance. The main channel is Carlingford Cut, about 3 ca SW of Cranfield Point, and passing 2 ca N of Haulbowline Lt Ho.

LIGHTS AND MARKS
Haulbowline Fl(3) 10s 32m 20M; granite Tr. Turning Lt FR 21m 9M; same Tr; vis 196°-208°; Horn 30s. Ldg Lts 310° both Oc 3s 7/12m 11M; G houses on piles, vis 295°-325°. Greenore Pier FIR 7.5s 10m 5M. Newry River Ldg Lts 310° both Iso 4s 5/15m 2M, stone columns.

RADIO TELEPHONE (local times)
Warrenpoint VHF Ch 16; 12 (H24). Greenore (call: *Ferry Greenore*) Ch 16; 13 (H24). Other stations: Dundalk Ch 16; 06 12. Kilkeel Ch 16; **12** 14 (Mon-Fri: 0900-2000).

TELEPHONE (042) (Warrenpoint 06937)
Hr Mr Warrenpoint 73381; CG Kilkeel 62232; MRCC Limerick 61219 or Donaghadee 883184; Coast Life Saving Service Clogerhead 2 22 63; HMC Freefone Customs Yachts: RT (0232) 752511; Irish Customs Dundalk 3 41 14; Marinecall 0898 500 465; Dr Dundalk 3 14 70; Hosp Newry 2543, Dundalk 3 47 01

FACILITIES
EC, Warrenpoint Wednesday. **Carlingford Marina** (50), AC, D, FW, P, Slip; **Harbour** Slip; **Carlingford YC** FW, Slip; **Dundalk SC** FW, Slip; **Greenore Pt** FW; **Warrenpoint** FW, P, D, AB; **D.A. Stukins** Tel. 72322, ME, El, Sh; **T. McArdle** Tel. 73100 Kos; **P. A. McGill** Tel. 73278, Gas.
PO (Warrenpoint, Rostrevor, Carlingford); Bank, (Dundalk, Warrenpoint); Rly (Dundalk, Newry); Air (Dublin).

SHELTER
Excellent — largest inlet on E coast. Good anchorages in the Narrows at Cross Roads, Strangford Creek, Audley Roads and in Ballyhenry Bay. Berthing at piers in Strangford or Portaferry dependent on state of tide. Good anchorages up the lough in Quoile, Ringhaddy Sound and Whiterock. Some visitors moorings available.

NAVIGATION
Waypoint Strangford (safe water) buoy, LFl 10s, Whis, 54°18'.62N 05°28'.62W, 126°/306° from/to Angus Rock Lt, 2.05M. Beware overfalls in the SE approaches and at the bar, which can be dangerous when ebb from narrows is running against strong SSW to E winds. Tidal flow in narrows, and hence the overfalls, relates to HW Strangford Quay (+0200 Dover) not to HW Killard Pt. During flood the bar presents no special problem but for preference enter when tide in the narrows is slack or on the young flood. Visitors should use the East Channel. Strong tidal streams flow through the Narrows, up to 7 kn at springs. Beware St Patricks Rk, Bar Pladdy and Pladdy Lug and whirlpool at Routen Wheel. Beware car ferry between Strangford and Portaferry. Up the lough, beware drying patches, known as pladdies and often un-marked. Swan Is, seen as grassy mound at HW, is edged with rocks and a reef extends 32m E ending at W Bn (Fl (2) WR 6s).

LIGHTS AND MARKS
Entrance identified by Strangford safe water buoy, W Tr on Angus Rk, Pladdy Lug Bn (W) and St Patrick's Rk perch. Leading marks to clear St Quintin Rks are perch on St Patrick's Rks and obelisk (off chartlet) in line at 224°. Leading marks to Cross Roads anchorage are stone Bns ½M inland in line at 260°.

Bar Pladdy Buoy	Q (6) + one long ev 15s
Angus Rk	Fl R 5s (on Tower)
Dogtail Pt	Oc (4) 10s Ldg Lts 341°
Salt Rk Bn	Fl R 3s
Gowland Rk Bn	Oc (2) 10s Ldg Lts 341°
Swan Is Bn	Fl (2) WR 6s
SW outlier (S Pladdy)	Fl (3) 10s
N outlier (N Pladdy)	Q
Church Pt Bn	Fl (4) R 10s
Portaferry Quay	Oc WR 10s
Ballyhenry Is Bn	Q G
Limestone Rk	Q R

RADIO TELEPHONE (local times)
Killyleagh VHF Ch 16; 12 (occas). Hr Mr/Ferry Superintendent Strangford Ch 16 (office hours). In Strangford Lough most YCs and Seaquip Ch M. Other station: Portavogie Ch 16; 14 (Mon-Fri: 0900-2000).

TELEPHONE (Strangford 039 686)
Hr Mr (Strangford) 637; MRSC Donaghadee 883184; HMC Freefone Customs Yachts: RT (0232) 752511; Marinecall 0898 500 465; Medical Clinic Downpatrick 2971; Casualty Downpatrick 3311.

FACILITIES
STRANGFORD M, L, FW, CH, AB, V, R, Bar, PO; **Seaquip** Tel. 303, D (cans), El, Sh, CH; **Milligan** Tel. 233, Gas.
PORTAFERRY M, P and D (cans), V, R, Bar, PO, Bank, L, LB, Gas.
KILLYLEAGH M, P and D (cans), L, C (mobile), CH, V, R, Bar, PO, YC; **Seaboard Sailing Centre** Tel. Killyleagh 828511; **East Down YC** Tel. Killyleagh 828375; **Maguire** Tel. Killyleagh 828439, Gas.
SKETRICK ISLAND R, Bar, L; **Irwin Yachts** Tel. Killinchy 541592, CH, SM.
KIRCUBBIN **Maxwell Bros** Tel. Kircubbin 211, Gas.

STRANGFORD LOUGH   10-13-9
Co. Down

CHARTS
Admiralty 2156, 2159; Imray C62; Irish OS 5, 9

TIDES
Killard Pt. 0000, Strangford Quay +0200 Dover; ML 2.0; Duration 0610; Zone 0 (GMT).

Standard Port BELFAST (→)

Times				Height (metres)			
HW		LW		MHWS	MHWN	MLWN	MLWS
0100	0700	0000	0600	3.5	3.0	1.1	0.4
1300	1900	1200	1800				

Differences STRANGFORD
+0147 +0157 +0148 +0208 +0.1 +0.1 −0.2 0.0
KILLYLEAGH
+0157 +0207 +0211 +0231 +0.3 +0.3 No data
KILKEEL
+0010 +0010 0000 0000 +1.8 +1.4 +0.8 +0.3

AREA 13—N Ireland

STRANGFORD LOUGH continued

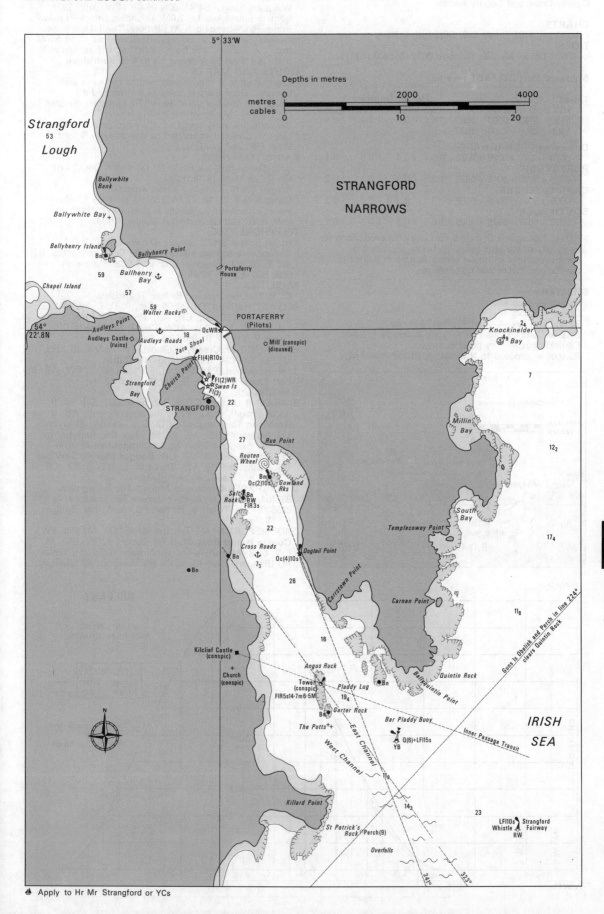

BELFAST LOUGH 10-13-10
County Down and County Antrim

CHARTS
Admiralty 1753; Imray C62, C64; Irish OS 5
TIDES
+0001 Dover; ML 2.0; Duration 0620; Zone 0 (GMT).

Standard Port BELFAST (→)

Times				Height (metres)			
HW		LW		MHWS	MHWN	MLWN	MLWS
0100	0700	0000	0600	3.5	3.0	1.1	0.4
1300	1900	1200	1800				

Differences DONAGHADEE
+0020 +0020 +0023 +0023 +0.5 +0.4 0.0 +0.1
PORTAVOGIE
+0010 +0020 +0010 +0020 +1.2 +0.9 +0.3 +0.2
CARRICKFERGUS
+0005 +0005 +0005 +0005 −0.3 −0.3 −0.2 −0.1
SOUTH ROCK
+0023 +0023 +0025 +0025 +1.0 +0.8 +0.1 +0.1

NOTE: Belfast is a Standard Port and tidal predictions for every day of the year are given below. Belfast is a commercial port and main sailing centres in Belfast Lough are Bangor, Cultra and Carrickfergus.

SHELTER
Belfast — excellent.
Cultra — good.
Ballyholme Bay — good in offshore winds.
SSE of Carrickfergus Pier — good except in E winds.
Carrickfergus — very good in marina.
Bangor — exposed to N winds; good for small boats.

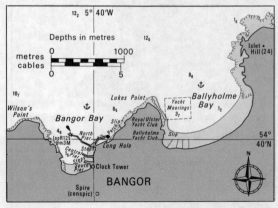

NAVIGATION
Waypoint Bangor 54°41'.00N 05°40'.00W, 010°/190° from/to breakwater Lt, 1.0M. Rounding Orlock Pt beware Briggs Rks extending ¾ M offshore. The Lough is well marked. Beware Carrickfergus Bank extends 15 ca SSW from the harbour, and from Carrickfergus E to Kilroot Pt there is a sand bank drying up to 4 ca from shore.

LIGHTS AND MARKS
Channel up to Belfast is well marked by buoys and beacons. Once the pile beacons are reached it is dangerous to leave the channel. Tie up where directed by the Berthing Master at the Hailing House.
Carrickfergus entry signal — when sufficient water to enter — Square flag hoisted by day or extra R light on West Pierhead by night.

RADIO TELEPHONE
Call: *Belfast Harbour Radio* (at Clarendon Dock) VHF Ch 12; 08 11 **12** 14 16 (H24).
Bangor Ch 16; 11 (when vessel expected).
Carrickfergus Ch 16; 12 14 (HW−3 to HW+1, when vessel expected). Copelands Marina, Royal Northern Ireland YC and Carrickfergus Marina, Ch M.

TELEPHONE (0232)
Hr Mr Belfast 34422, Bangor 472596; MRSC & CG Donaghadee 883184; HMC Freefone Customs Yachts: RT (0232) 752511; Weather: Crumlin 52339, Marinecall 0898 500 465; Dr 38506.

FACILITIES
EC Belfast Wednesday; Bangor Thursday;
BELFAST Harbour Tel. 38506, Slip, P, D, L, FW, ME, El, Sh, C (up to 20 ton), CH, AB; **James Tedford** Tel. 226763, SM, Gas, ACA, CH; **J. McCready** CH; **Town** P, CH, V, R, Bar; PO; Bank; Rly; Air.
HOLYWOOD McCready (Sailboat) Tel. Holywood 2888, ME, El, Sh, CH; **R N Ireland YC,** Cultra Tel. Holywood 2041, Slip, M, P (½ M), D, L, FW, AB, R, Bar; **Town** P, CH, V, R, Bar; PO, Bank, Rly, Air (Belfast).
BANGOR R Ulster YC Tel. Bangor 465002, Slip, M, P (local garage), D, L, FW, C (by arrangement), AB, R, Bar; **Hamilton** Tel. Bangor 461013, Gas; **Bangor Shipyard** Tel. Bangor 463939, ME, El, Sh, Slip, CH; **Ballyholme YC** Tel. Bangor 462467, M, L; **Clandeboye Marine** Tel. Bangor 460763, ME, CH; **Bangor Marine Services** Tel. Bangor 469916, ME; **Town** M, L, ME, El, Sh, CH, V, R, Bar; PO, Bank, Rly, Air (Belfast).
DONAGHADEE Copelands Marina Tel. Donaghadee 882184, D, FW, C (20 ton); **James** Tel. Donaghadee 883300, Gas. **Town** V, R, Bar, PO, Bank, Rly (Bangor), Air (Belfast).
GROOMSPORT Hr.Mr. Tel. Bangor 454371, M, Slip, FW; **Cockle Island Boat Club** Tel. Bangor 42662, Slip, M, FW, L, R. **Murphy's** Tel. Groomsport 337, Gas. **Town** PO (Bangor), Bank (Bangor), Rly (Bangor), Air (Belfast).

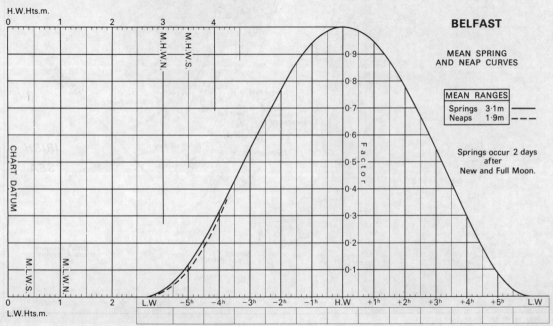

BELFAST
MEAN SPRING AND NEAP CURVES

MEAN RANGES
Springs 3·1m
Neaps 1·9m

Springs occur 2 days after New and Full Moon.

IRELAND, EAST COAST — BELFAST

Lat 54°36′ N   Long 5°55′ W

TIMES AND HEIGHTS OF HIGH AND LOW WATERS

YEAR 1989

TIME ZONE UT(GMT)
For Summer Time add ONE hour in non-shaded areas

	JANUARY				FEBRUARY				MARCH				APRIL		
	Time	m	Time	m	Time	m	Time	m	Time	m	Time	m	Time	m	Time	m
1 Su	0520 1126 1743 2358	2·8 1·4 3·0 1·1	**16** M	0544 1146 1819	3·0 1·2 3·2	**1** W	0000 0624 1246 1903	1·2 2·7 1·3 2·8	**16** Th	0134 0747 1408 2037	1·2 3·0 1·1 2·9	**1** W	0413 1031 1658 2306	2·7 1·2 2·7 1·2	**16** Th	0556 1228 1909	2·9 1·1 2·7	**1** Sa	0611 1252 1912	2·8 0·9 2·7	**16** Su	0148 0752 1411 2046	1·1 3·0 0·7 2·8
2 M	0628 1235 1849	2·7 1·4 3·0	**17** Tu	0042 0700 1307 1938	1·0 3·0 1·2 3·1	**2** Th	0112 0741 1358 2009	1·2 2·8 1·2 2·9	**17** F	0239 0849 1505 2132	1·1 3·1 0·9 3·0	**2** Th	0523 1157 1821	2·6 1·2 2·7	**17** F	0114 0720 1348 2023	1·3 2·9 1·0 2·8	**2** Su	0114 0733 1358 2015	1·1 2·9 0·6 2·9	**17** M	0240 0844 1453 2125	0·9 3·1 0·6 2·9
3 Tu	0059 0734 1340 1949	1·1 2·8 1·3 3·0	**18** W	0151 0808 1418 2043	1·0 3·1 1·1 3·1	**3** F	0213 0839 1451 2101	1·1 3·0 1·0 3·0	**18** Sa	0328 0938 1549 2214	1·1 3·3 0·8 3·0	**3** F	0029 0652 1324 1941	1·2 2·7 1·1 2·8	**18** Sa	0220 0826 1444 2117	1·1 3·0 0·8 2·9	**3** M	0215 0832 1450 2105	0·8 3·2 0·3 3·1	**18** Tu	0321 0927 1529 2159	0·8 3·2 0·5 3·0
4 W	0154 0826 1432 2040	1·0 2·9 1·2 3·1	**19** Th	0249 0904 1514 2136	1·0 3·2 1·0 3·2	**4** Sa	0303 0925 1538 2148	0·9 3·2 0·7 3·2	**19** Su	0409 1019 1626 2251	0·9 3·4 0·6 3·0	**4** Sa	0145 0806 1426 2040	1·1 2·9 0·8 2·9	**19** Su	0310 0917 1527 2156	1·0 3·2 0·6 2·9	**4** Tu	0304 0921 1534 2149	0·6 3·4 0·1 3·3	**19** W	0357 1003 1602 2230	0·7 3·2 0·4 3·0
5 Th	0242 0910 1515 2124	1·0 3·1 1·0 3·1	**20** F	0336 0950 1559 2221	1·0 3·4 0·9 3·1	**5** Su	0348 1007 1620 2230	0·8 3·4 0·5 3·3	**20** M	0444 1057 1659 2325	0·8 3·5 0·6 3·0	**5** Su	0242 0900 1515 2128	0·9 3·2 0·5 3·1	**20** M	0349 0956 1602 2230	0·8 3·3 0·5 2·9	**5** W	0348 1006 1616 2231	0·4 3·6 0·0 3·4	**20** Th	0428 1038 1633 2259	0·6 3·2 0·4 3·0
6 F	0324 0948 1556 2204	0·9 3·2 0·9 3·2	**21** Sa	0419 1033 1640 2302	1·0 3·5 0·8 3·1	**6** M	0430 1048 1701 2313	0·6 3·6 0·3 3·4	**21** Tu	0519 1132 1733 2356	0·7 3·5 0·5 3·0	**6** M	0328 0945 1559 2212	0·7 3·4 0·2 3·3	**21** Tu	0424 1033 1634 2259	0·7 3·3 0·4 3·0	**6** Th	0430 1049 1658 2313	0·2 3·7 0·0 3·4	**21** F	0459 1111 1704 2327	0·6 3·1 0·5 3·0
7 Sa	0403 1026 1635 2245	0·8 3·4 0·7 3·3	**22** Su	0457 1112 1719 2340	0·9 3·5 0·7 3·1	**7** Tu	0512 1132 1744 2358	0·5 3·8 0·2 3·4	**22** W	0551 1207 1807	0·6 3·5 0·5	**7** Tu	0410 1028 1640 2254	0·5 3·6 0·1 3·4	**22** W	0455 1106 1705 2329	0·6 3·3 0·4 3·0	**7** F	0512 1133 1742 2357	0·1 3·7 0·0 3·5	**22** Sa	0529 1143 1734 2357	0·6 3·1 0·5 3·0
8 Su	0444 1105 1718 2329	0·8 3·5 0·6 3·3	**23** M	0534 1151 1756	0·9 3·6 0·6	**8** W	0554 1217 1829	0·4 3·9 0·1	**23** Th	0028 0625 1243 1841	3·0 0·6 3·4 0·5	**8** W	0452 1111 1723 2337	0·3 3·8 0·0 3·4	**23** Th	0526 1140 1734 2358	0·6 3·3 0·4 3·0	**8** Sa	0556 1219 1825	0·1 3·6 0·2	**23** Su	0600 1217 1808	0·6 3·0 0·6
9 M	0526 1147 1801	0·7 3·7 0·5	**24** Tu	0018 0612 1231 1835	3·1 0·8 3·6 0·6	**9** Th	0045 0639 1302 1917	3·5 0·4 3·9 0·1	**24** F	0100 0657 1317 1916	3·0 0·7 3·4 0·5	**9** Th	0534 1156 1807	0·2 3·8 0·0	**24** F	0557 1212 1807	0·6 3·2 0·4	**9** Su	0042 0642 1309 1913	3·5 0·2 3·5 0·4	**24** M	0029 0634 1253 1845	3·0 0·6 3·0 0·7
10 Tu	0015 0611 1234 1849	3·4 0·7 3·8 0·4	**25** W	0056 0652 1310 1913	3·1 0·8 3·6 0·6	**10** F	0133 0726 1349 2006	3·5 0·4 3·9 0·2	**25** Sa	0133 0731 1352 1951	3·0 0·7 3·3 0·6	**10** F	0021 0618 1242 1852	3·5 0·2 3·8 0·1	**25** Sa	0028 0628 1246 1839	3·0 0·6 3·2 0·5	**10** M	0130 0731 1359 2004	3·4 0·3 3·3 0·6	**25** Tu	0104 0712 1333 1924	3·0 0·7 2·9 0·8
11 W	0104 0657 1320 1938	3·4 0·7 3·9 0·3	**26** Th	0134 0730 1349 1952	3·1 0·8 3·5 0·6	**11** Sa	0220 0813 1440 2057	3·4 0·5 3·7 0·4	**26** Su	0206 0805 1429 2029	3·0 0·8 3·2 0·7	**11** Sa	0107 0703 1328 1938	3·5 0·2 3·7 0·2	**26** Su	0059 0659 1320 1914	3·0 0·6 3·1 0·6	**11** Tu	0219 0825 1454 2100	3·4 0·5 3·1 0·9	**26** W	0144 0757 1418 2011	3·0 0·7 2·9 0·9
12 Th	0154 0747 1409 2030	3·4 0·7 3·9 0·4	**27** F	0211 0806 1427 2032	3·0 0·9 3·4 0·7	**12** Su	0310 0905 1532 2153	3·3 0·7 3·5 0·7	**27** M	0242 0844 1508 2111	2·9 0·9 3·0 0·9	**12** Su	0154 0751 1419 2029	3·4 0·3 3·6 0·5	**27** M	0131 0734 1357 1952	2·9 0·7 3·0 0·7	**12** W	0312 0925 1556 2203	3·2 0·7 2·9 1·1	**27** Th	0229 0849 1510 2104	2·9 0·7 2·9 1·0
13 F	0244 0837 1500 2125	3·4 0·7 3·8 0·5	**28** Sa	0247 0846 1507 2114	3·0 1·0 3·3 0·8	**13** M	0403 1003 1634 2258	3·2 0·9 3·2 1·0	**28** Tu	0322 0929 1556 2202	2·8 1·1 2·9 1·1	**13** M	0243 0843 1512 2124	3·3 0·5 3·3 0·8	**28** Tu	0208 0813 1437 2034	2·9 0·8 2·9 0·9	**13** Th	0412 1035 1711 2320	3·1 0·9 2·7 1·3	**28** F	0321 0950 1612 2207	3·0 0·8 2·8 1·1
14 Sa	0338 0931 1556 2224	3·3 0·9 3·6 0·7	**29** Su	0327 0928 1550 2200	2·9 1·1 3·1 1·0	**14** Tu	0508 1116 1754	3·0 1·1 3·0				**14** Tu	0335 0942 1613 2228	3·2 0·8 3·0 1·1	**29** W	0249 0903 1527 2127	2·9 0·9 2·8 1·0	**14** F	0523 1158 1839	3·0 1·0 2·7	**29** Sa	0423 1105 1722 2322	3·0 0·8 2·7 1·1
15 Su	0435 1033 1701 2330	3·1 1·0 3·4 0·9	**30** M	0412 1017 1641 2254	2·8 1·3 2·9 1·1	**15** W	0015 0628 1248 1924	1·2 2·9 1·2 2·9				**15** W	0437 1055 1733 2349	3·0 1·0 2·8 1·2	**30** Th	0339 1004 1630 2231	2·8 1·0 2·7 1·2	**15** Sa	0041 0643 1314 1952	1·2 2·9 0·9 2·7	**30** Su	0539 1221 1839	3·0 0·7 2·8
			31 Tu	0509 1122 1747	2·7 1·4 2·8									**31** F	0445 1126 1749 2353	2·8 1·0 2·7 1·2						

Chart Datum: 2.01 metres below Ordnance Datum (Newlyn)

IRELAND, EAST COAST — BELFAST

Lat 54°36′ N Long 5°55′ W

TIME ZONE UT(GMT)
For Summer Time add ONE hour in non-shaded areas

TIMES AND HEIGHTS OF HIGH AND LOW WATERS

YEAR 1989

	MAY		JUNE		JULY		AUGUST	
	Time m	Time m	Time m	Time m	Time m	Time m	Time m	Time m
	1 0039 1·0 0657 3·1 M 1327 0·5 1945 3·0	**16** 0201 1·0 0804 3·0 Tu 1412 0·7 2046 2·9	**1** 0212 0·7 0833 3·3 Th 1444 0·4 2101 3·3	**16** 0253 1·0 0857 3·0 F 1456 0·8 2125 3·0	**1** 0251 0·8 0912 3·2 Sa 1517 0·8 2132 3·3	**16** 0304 1·0 0911 2·9 Su 1508 0·9 2134 3·1	**1** 0420 0·7 1044 3·1 Tu 1637 0·9 ● 2249 3·5	**16** 0400 0·5 1012 3·2 W 1609 0·7 2226 3·5
	2 0144 0·8 0802 3·2 Tu 1420 0·3 2039 3·1	**17** 0247 0·9 0851 3·1 W 1453 0·6 2124 2·9	**2** 0304 0·6 0924 3·4 F 1531 0·4 2146 3·4	**17** 0332 0·9 0938 3·0 Sa 1534 0·8 2200 3·0	**2** 0342 0·7 1002 3·2 Su 1602 0·8 2217 3·4	**17** 0345 0·8 0952 3·0 M 1549 0·9 2210 3·2	**2** 0459 0·6 1123 3·1 W 1716 0·9 2330 3·6	**17** 0440 0·3 1051 3·3 Th 1648 0·6 ○ 2305 3·6
	3 0237 0·6 0856 3·4 W 1508 0·2 2124 3·3	**18** 0325 0·8 0931 3·1 Th 1528 0·6 2157 3·0	**3** 0352 0·5 1012 3·4 Sa 1614 0·5 ● 2231 3·4	**18** 0407 0·8 1016 3·0 Su 1609 0·8 2234 3·1	**3** 0427 0·6 1048 3·2 M 1647 0·8 ● 2302 3·5	**18** 0421 0·7 1033 3·1 Tu 1627 0·8 ○ 2248 3·3	**3** 0537 0·5 1203 3·1 Th 1756 0·8	**18** 0520 0·2 1133 3·3 F 1729 0·5 2347 3·7
	4 0325 0·4 0943 3·5 Th 1552 0·1 2207 3·4	**19** 0400 0·7 1007 3·1 F 1602 0·6 2228 3·0	**4** 0437 0·5 1058 3·3 Su 1658 0·6 2316 3·5	**19** 0442 0·8 1052 3·0 M 1645 0·8 ○ 2309 3·2	**4** 0512 0·6 1134 3·1 Tu 1730 0·9 2347 3·6	**19** 0501 0·5 1113 3·1 W 1708 0·8 2327 3·5	**4** 0011 3·6 0617 0·5 F 1242 3·1 1835 0·8	**19** 0603 0·1 1217 3·4 Sa 1811 0·4
	5 0409 0·3 1027 3·5 F 1634 0·2 ● 2249 3·4	**20** 0433 0·7 1041 3·0 Sa 1634 0·6 ○ 2259 3·0	**5** 0522 0·5 1146 3·2 M 1743 0·7	**20** 0519 0·7 1132 3·0 Tu 1725 0·9 2346 3·3	**5** 0556 0·6 1221 3·1 W 1814 0·9	**20** 0542 0·4 1156 3·2 Th 1749 0·7	**5** 0052 3·5 0656 0·5 Sa 1320 3·1 1914 0·8	**20** 0032 3·8 0648 0·1 Su 1303 3·4 1857 0·4
	6 0452 0·2 1113 3·5 Sa 1718 0·3 2334 3·5	**21** 0504 0·7 1116 3·0 Su 1708 0·7 2330 3·0	**6** 0003 3·5 0610 0·5 Tu 1236 3·1 1831 0·8	**21** 0600 0·6 1215 3·1 W 1805 0·8	**6** 0032 3·6 0642 0·6 Th 1309 3·0 1900 0·9	**21** 0010 3·6 0625 0·3 F 1242 3·3 1834 0·6	**6** 0133 3·5 0735 0·6 Su 1359 3·0 1955 0·9	**21** 0119 3·8 0734 0·2 M 1351 3·4 1945 0·5
	7 0537 0·3 1200 3·4 Su 1803 0·5	**22** 0537 0·7 1153 3·0 M 1743 0·8	**7** 0050 3·5 0700 0·5 W 1328 3·1 1921 0·9	**22** 0028 3·4 0643 0·5 Th 1302 3·1 1852 0·8	**7** 0119 3·6 0727 0·6 F 1354 3·0 1947 0·9	**22** 0055 3·7 0712 0·3 Sa 1328 3·3 1920 0·6	**7** 0213 3·4 0816 0·7 M 1437 3·0 2036 0·9	**22** 0209 3·7 0825 0·4 Tu 1440 3·4 2036 0·6
	8 0021 3·5 0624 0·3 M 1250 3·3 1850 0·6	**23** 0005 3·1 0615 0·7 Tu 1232 3·0 1822 0·8	**8** 0140 3·5 0752 0·6 Th 1419 3·0 2015 1·0	**23** 0113 3·5 0731 0·5 F 1349 3·1 1940 0·8	**8** 0204 3·5 0813 0·6 Sa 1437 3·0 2033 1·0	**23** 0141 3·7 0801 0·3 Su 1418 3·3 2009 0·6	**8** 0254 3·2 0858 0·8 Tu 1518 2·9 2119 1·1	**23** 0301 3·5 0919 0·6 W 1534 3·3 ☾ 2134 0·8
	9 0109 3·5 0714 0·4 Tu 1344 3·2 1941 0·8	**24** 0045 3·2 0657 0·6 W 1316 3·0 1906 0·9	**9** 0229 3·5 0844 0·7 F 1511 2·9 2108 1·1	**24** 0159 3·5 0823 0·4 Sa 1440 3·2 2030 0·8	**9** 0249 3·4 0858 0·7 Su 1521 3·0 2119 1·0	**24** 0230 3·7 0853 0·4 M 1507 3·3 2100 0·7	**9** 0339 3·1 0943 0·9 W 1602 2·8 ☾ 2209 1·2	**24** 0402 3·3 1020 0·9 Th 1634 3·1 2242 1·0
	10 0159 3·4 0809 0·6 W 1439 3·0 2037 1·0	**25** 0127 3·2 0745 0·6 Th 1405 3·0 1955 0·9	**10** 0319 3·4 0938 0·7 Sa 1603 2·9 2203 1·1	**25** 0250 3·5 0917 0·4 Su 1532 3·1 2125 0·9	**10** 0334 3·3 0946 0·8 M 1607 2·9 2210 1·1	**25** 0322 3·6 0948 0·5 Tu 1600 3·2 ☾ 2157 0·8	**10** 0430 2·9 1035 1·1 Th 1657 2·8 2313 1·3	**25** 0515 3·0 1134 1·1 F 1747 3·0
	11 0251 3·4 0908 0·7 Th 1536 2·9 2138 1·1	**26** 0215 3·3 0839 0·6 F 1457 3·0 2049 1·0	**11** 0412 3·2 1037 0·8 Su 1659 2·8 ☾ 2304 1·2	**26** 0343 3·5 1016 0·5 M 1628 3·1 ☾ 2224 0·9	**11** 0424 3·1 1037 0·9 Tu 1659 2·8 ☽ 2308 1·2	**26** 0421 3·4 1048 0·7 W 1701 3·1 2304 1·0	**11** 0533 2·8 1139 1·2 F 1807 2·7	**26** 0010 1·1 0648 2·9 Sa 1257 1·2 1909 3·1
	12 0346 3·2 1012 0·8 F 1641 2·8 ☽ 2245 1·2	**27** 0305 3·3 0938 0·6 Sa 1553 3·0 2148 1·0	**12** 0511 3·1 1132 0·8 M 1803 2·8	**27** 0445 3·4 1118 0·6 Tu 1732 3·0 2332 1·0	**12** 0520 3·0 1133 1·0 W 1801 2·8	**27** 0533 3·2 1157 0·9 Th 1812 3·1	**12** 0034 1·3 0648 2·7 Sa 1250 1·2 1921 2·8	**27** 0135 1·1 0809 2·9 Su 1408 1·2 2018 3·2
	13 0448 3·1 1119 0·9 Sa 1753 2·7 2356 1·2	**28** 0403 3·2 1041 0·6 Su 1657 2·9 ☾ 2252 1·0	**13** 0008 1·2 0615 3·0 Tu 1231 0·9 1906 2·8	**28** 0556 3·3 1224 0·7 W 1841 3·1	**13** 0015 1·3 0627 2·9 Th 1234 1·0 1907 2·8	**28** 0022 1·1 0655 3·1 F 1310 1·0 1927 3·1	**13** 0145 1·2 0754 2·8 Su 1354 1·2 2020 2·9	**28** 0239 0·9 0910 3·1 M 1501 1·1 2111 3·3
	14 0558 3·0 1227 0·8 Su 1904 2·7	**29** 0511 3·2 1149 0·6 M 1805 2·8	**14** 0112 1·1 0719 3·0 W 1326 0·8 2001 2·8	**29** 0043 1·0 0712 3·2 Th 1328 0·7 1947 3·1	**14** 0123 1·2 0730 2·9 F 1333 1·0 2005 2·8	**29** 0141 1·0 0811 3·0 Sa 1415 1·0 2030 3·2	**14** 0237 1·0 0847 3·0 M 1446 1·0 2105 3·1	**29** 0327 0·7 0955 3·1 Tu 1546 1·0 2155 3·4
	15 0103 1·1 0707 3·0 M 1324 0·8 2001 2·8	**30** 0004 1·0 0625 3·2 Tu 1255 0·5 1913 2·8	**15** 0206 1·1 0812 3·0 Th 1413 0·8 2046 2·8	**30** 0152 0·9 0818 3·2 F 1426 0·8 2043 3·1	**15** 0219 1·1 0825 2·9 Sa 1425 1·0 2053 3·0	**30** 0246 0·9 0910 3·1 Su 1510 1·0 2122 3·3	**15** 0321 0·7 0931 3·0 Tu 1528 0·9 2146 3·3	**30** 0406 0·6 1033 3·1 W 1623 0·9 2233 3·5
		31 0113 0·9 0734 3·3 W 1354 0·5 2011 3·1				**31** 0336 0·8 1000 3·1 M 1555 1·0 2207 3·4		**31** 0440 0·5 1106 3·1 Th 1658 0·8 ● 2309 3·5

Chart Datum: 2.01 metres below Ordnance Datum (Newlyn)

IRELAND, EAST COAST — BELFAST

Lat 54°36′ N Long 5°55′ W

TIMES AND HEIGHTS OF HIGH AND LOW WATERS

YEAR 1989

TIME ZONE UT(GMT)
For Summer Time add ONE hour in non-shaded areas

SEPTEMBER

Time	m	Time	m
1 F 0513 1139 1733 2346	0.5 3.1 0.7 3.5	**16** Sa 0455 1109 1706 2325	0.0 3.5 0.3 3.8
2 Sa 0547 1211 1807	0.5 3.1 0.7	**17** Su 0537 1151 1749	0.0 3.5 0.3
3 Su 0022 0621 1245 1842	3.4 0.5 3.1 0.7	**18** M 0010 0621 1238 1834	3.8 0.1 3.6 0.3
4 M 0100 0657 1320 1917	3.4 0.6 3.1 0.8	**19** Tu 0057 0707 1326 1923	3.8 0.3 3.6 0.4
5 Tu 0137 0734 1355 1955	3.3 0.7 3.0 0.9	**20** W 0148 0758 1416 2015	3.6 0.5 3.4 0.6
6 W 0216 0813 1433 2036	3.1 0.8 3.0 1.0	**21** Th 0243 0853 1510 2115	3.4 0.8 3.4 0.8
7 Th 0257 0856 1515 2122	3.0 1.0 2.9 1.1	**22** F 0345 0956 1610 2227 ☾	3.2 1.1 3.2 1.0
8 F 0345 0945 1604 2224 ☽	2.9 1.2 2.8 1.3	**23** Sa 0502 1113 1723 2357	2.9 1.3 3.1 1.1
9 Sa 0447 1047 1709 2347	2.7 1.3 2.8 1.3	**24** Su 0638 1241 1848	2.8 1.4 3.1
10 Su 0605 1207 1831	2.7 1.3 2.8	**25** M 0121 0759 1352 1958	1.0 2.9 1.3 3.2
11 M 0109 0723 1323 1944	1.1 2.8 1.2 3.0	**26** Tu 0222 0856 1446 2051	0.9 2.9 1.1 3.3
12 Tu 0208 0820 1419 2036	0.9 2.9 1.0 3.2	**27** W 0307 0938 1528 2134	0.7 3.1 0.9 3.4
13 W 0254 0907 1504 2119	0.6 3.1 0.8 3.4	**28** Th 0342 1012 1603 2212	0.5 3.1 0.8 3.5
14 Th 0335 0948 1545 2200	0.3 3.3 0.6 3.6	**29** F 0414 1042 1637 2245 ●	0.5 3.1 0.7 3.5
15 F 0414 1028 1626 2241 ○	0.1 3.4 0.4 3.8	**30** Sa 0445 1111 1708 2319	0.5 3.1 0.7 3.4

OCTOBER

Time	m	Time	m
1 Su 0516 1142 1739 2353	0.5 3.1 0.7 3.3	**16** M 0512 1127 1727 2347	0.1 3.7 0.3 3.8
2 M 0549 1212 1811	0.5 3.1 0.7	**17** Tu 0556 1214 1814	0.3 3.7 0.3
3 Tu 0028 0621 1245 1845	3.2 0.6 3.1 0.8	**18** W 0036 0643 1302 1903	3.7 0.5 3.7 0.5
4 W 0103 0656 1319 1921	3.2 0.8 3.1 0.9	**19** Th 0130 0737 1354 1958	3.5 0.7 3.6 0.6
5 Th 0141 0734 1357 2002	3.1 0.9 3.0 1.0	**20** F 0226 0829 1449 2100	3.3 1.0 3.5 0.9
6 F 0223 0816 1437 2050	3.0 1.1 3.0 1.1	**21** Sa 0329 0934 1548 2212 ☾	3.1 1.2 3.4 1.0
7 Sa 0311 0905 1527 2150	2.9 1.2 3.0 1.2	**22** Su 0442 1048 1657 2333	2.9 1.4 3.2 1.1
8 Su 0412 1007 1627 2308 ☽	2.7 1.3 2.9 1.2	**23** M 0612 1210 1815	2.8 1.4 3.2
9 M 0526 1123 1744	2.7 1.4 2.9	**24** Tu 0050 0730 1321 1927	1.0 2.9 1.3 3.2
10 Tu 0029 0646 1243 1903	1.0 2.8 1.3 3.1	**25** W 0149 0826 1418 2022	0.9 3.0 1.1 3.3
11 W 0134 0749 1347 2004	0.8 3.0 1.1 3.3	**26** Th 0236 0910 1501 2107	0.7 3.1 1.0 3.4
12 Th 0225 0840 1436 2053	0.5 3.2 0.8 3.5	**27** F 0314 0943 1539 2145	0.6 3.2 0.8 3.4
13 F 0308 0922 1521 2136	0.3 3.3 0.6 3.7	**28** Sa 0346 1014 1613 2220	0.6 3.2 0.8 3.4
14 Sa 0349 1003 1602 2219 ○	0.1 3.4 0.4 3.8	**29** Su 0417 1044 1644 2252 ●	0.6 3.2 0.8 3.3
15 Su 0430 1045 1644 2302	0.1 3.6 0.3 3.8	**30** M 0448 1113 1715 2326	0.6 3.2 0.8 3.2
		31 Tu 0519 1144 1747	0.7 3.2 0.8

NOVEMBER

Time	m	Time	m
1 W 0000 0551 1217 1821	3.2 0.8 3.2 0.9	**16** Th 0021 0624 1243 1849	3.5 0.7 3.7 0.6
2 Th 0035 0627 1252 1857	3.1 0.9 3.2 0.9	**17** F 0114 0714 1335 1944	3.4 0.9 3.7 0.7
3 F 0114 0704 1328 1940	3.0 1.0 3.2 0.9	**18** Sa 0211 0809 1427 2044	3.2 1.1 3.6 0.8
4 Sa 0158 0748 1411 2030	3.0 1.1 3.2 1.0	**19** Su 0310 0910 1524 2148	3.1 1.2 3.5 0.9
5 Su 0247 0839 1500 2128	2.9 1.2 3.2 1.1	**20** M 0414 1014 1624 2255 ☾	3.0 1.3 3.4 1.0
6 M 0343 0936 1556 2237 ☽	2.9 1.3 3.2 1.0	**21** Tu 0526 1125 1732	2.9 1.4 3.3
7 Tu 0451 1045 1704 2350	2.8 1.3 3.2 0.9	**22** W 0004 0641 1236 1843	1.0 2.9 1.3 3.2
8 W 0604 1200 1821	2.9 1.2 3.2	**23** Th 0106 0742 1338 1944	0.9 2.9 1.2 3.2
9 Th 0056 0713 1309 1930	0.7 3.0 1.1 3.4	**24** F 0157 0832 1429 2034	0.9 3.0 1.1 3.3
10 F 0152 0809 1406 2025	0.5 3.2 0.9 3.5	**25** Sa 0240 0912 1511 2117	0.8 3.1 0.9 3.3
11 Sa 0242 0857 1456 2114	0.4 3.4 0.7 3.6	**26** Su 0317 0948 1549 2155	0.7 3.2 0.9 3.3
12 Su 0327 0941 1542 2159	0.3 3.5 0.5 3.7	**27** M 0352 1020 1621 2230	0.7 3.3 0.9 3.2
13 M 0409 1024 1626 2244 ○	0.3 3.6 0.4 3.7	**28** Tu 0424 1051 1654 2304 ●	0.8 3.3 0.9 3.2
14 Tu 0451 1107 1711 2330	0.4 3.7 0.4 3.6	**29** W 0457 1122 1727 2337	0.8 3.3 0.9 3.1
15 W 0536 1154 1758	0.5 3.7 0.5	**30** Th 0530 1156 1803	0.9 3.3 0.9

DECEMBER

Time	m	Time	m
1 F 0015 0605 1231 1841	3.1 1.0 3.4 0.9	**16** Sa 0100 0657 1317 1928	3.3 0.9 3.8 0.7
2 Sa 0055 0646 1310 1924	3.1 1.0 3.4 0.8	**17** Su 0152 0748 1406 2020	3.2 1.0 3.7 0.7
3 Su 0140 0730 1352 2012	3.0 1.0 3.4 0.8	**18** M 0243 0840 1457 2114	3.1 1.1 3.6 0.8
4 M 0227 0818 1439 2107	3.1 1.1 3.5 0.8	**19** Tu 0335 0934 1549 2209 ☾	3.0 1.2 3.5 0.9
5 Tu 0319 0911 1531 2206	3.1 1.1 3.4 0.8	**20** W 0431 1033 1645 2308	2.9 1.3 3.3 0.9
6 W 0417 1010 1631 2311	3.0 1.2 3.4 0.8	**21** Th 0534 1139 1750	2.9 1.3 3.2
7 Th 0523 1118 1742	3.0 1.2 3.3	**22** F 0010 0643 1249 1857	1.0 2.9 1.3 3.1
8 F 0018 0634 1231 1856	0.8 3.1 1.1 3.4	**23** Sa 0110 0747 1351 1958	1.0 2.9 1.3 3.1
9 Sa 0121 0738 1338 2001	0.7 3.2 1.0 3.4	**24** Su 0204 0837 1443 2047	1.0 3.0 1.2 3.1
10 Su 0218 0834 1436 2056	0.6 3.4 0.8 3.5	**25** M 0249 0919 1525 2129	0.9 3.2 1.1 3.1
11 M 0307 0924 1527 2145	0.6 3.5 0.7 3.5	**26** Tu 0328 0957 1602 2207	0.9 3.2 1.0 3.1
12 Tu 0353 1009 1614 2233 ○	0.6 3.6 0.6 3.5	**27** W 0404 1031 1635 2244	0.9 3.2 0.9 3.1
13 W 0437 1054 1701 2320	0.7 3.6 0.6 3.5	**28** Th 0438 1104 1711 2320 ●	0.9 3.2 0.8 3.1
14 Th 0522 1142 1749	0.7 3.8 0.6	**29** F 0513 1137 1746 2357	0.9 3.4 0.8 3.2
15 F 0010 0608 1229 1838	3.4 0.8 3.8 0.6	**30** Sa 0550 1212 1825	0.9 3.5 0.7
		31 Su 0038 0629 1252 1907	3.2 0.8 3.6 0.6

Chart Datum: 2.01 metres below Ordnance Datum (Newlyn)

BELFAST LOUGH continued

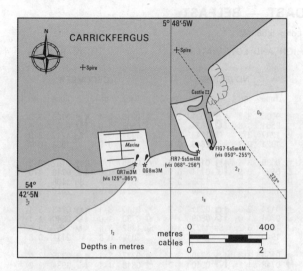

CARRICKFERGUS. Easily recognised by conspic castle to E of commercial port. Marina depths 1.9m to 2.4m.
Carrickfergus Marina (270 + 30 visitors)
Tel. Carrickfergus 66666 AC, BH (10½ ton), CH, D, El, FW, Sh; **Carrickfergus S.C.** Tel. Carrickfergus 63402 M, L, FW, C, AB; **Castle Stores** Tel. Carrickfergus 63322 Gas; **Carrick Marine** Tel. Carrickfergus 63019 ME, El, Sh, CH; **Belfast Lough Marine Electronics**
Tel. Carrickfergus 78243 El; **Clandeboye Marine**
Tel. Carrickfergus 63658 ME, CH; **Town** V, R, Bank, PO, Rly, Air (Belfast).

LARNE 10-13-11
Antrim

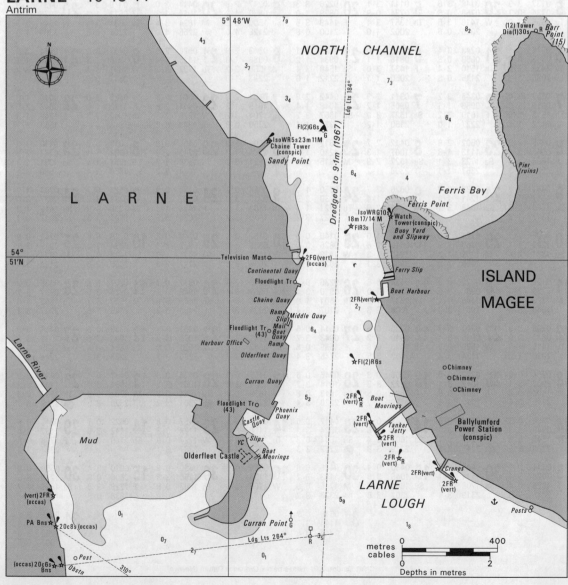

AREA 13—N Ireland 573

LARNE continued

CHARTS
Admiralty 1237; Imray C62, C64; Irish OS 5
TIDES
0000 Dover; ML 1.6; Duration 0620; Zone 0 (GMT).

Standard Port BELFAST (←)

Times				Height (metres)			
HW		LW		MHWS	MHWN	MLWN	MLWS
0100	0700	0000	0600	3.5	3.0	1.1	0.4
1300	1900	1200	1800				
Differences LARNE							
+0005	0000	+0010	−0005	−0.7	−0.5	−0.3	0.0
RED BAY							
+0022	−0010	+0007	−0017	−1.9	−1.5	−0.8	−0.2

SHELTER
Secure shelter in Larne Lough. Anchorage S of Ballylumford Power Stn. Harbour can be entered day or night in any conditions. W side is commercial until Curran Pt where there are two YCs with moorings.
NAVIGATION
Waypoint 54°51′.70N 05°47′.47W, 004°/184° from/to front Ldg Lt 004°, 2.1M. Beware Hunter Rock 2 miles NE of harbour entrance. N.B. Abnormal magnetic variation exists near Hunter Rock and between it and the mainland. Larne is a busy commercial port. Tide in the entrance runs at up to 3½ kn. Beyond the narrow entrance, the only recommended channel is close along the E shore.
LIGHTS AND MARKS
Entrance Ldg Lts 184°, Oc 4s, synchronised and vis 179°-189°; W diamonds with R stripes.
RADIO TELEPHONE
VHF Ch 16.
TELEPHONE (0574)
Hr Mr 2604; CG Island Magee 227; MRSC Donaghadee 883184; Ferrans Pilotage Co 4085; HMC Freefone Customs Yachts: RT (0232) 752511; Marinecall 0898 500 465; Dr 5331; Hosp (Casualty Dept) 5431.
FACILITIES
EC None; **Pier (HW only)** Tel. 77204, M, L, FW, C (up to 20 ton); **Larne Harbour** M, L, FW; **E.A. BC** Tel. 77204, Slip, M, L, FW, V, Bar; **Plant & Marine** Tel. 3012, ME, El, Sh; **AC Mitcheson** Tel. 3207, El; **Adams Bros** Slip, ME, El, Sh; **Barnhill Service Station** P; **Alan Lyttle** Tel. 2850, D; **F Watson Glynn** Sh; **Curran Saw Mills**, CH; **T R Fulton** Tel. 72288, Gas. **Town** P and D (delivered), CH, V, R, Bar. PO; Bank; Rly; Air (Belfast).

AGENTS NEEDED

There are a number of ports where we need agents, particularly in France.
ENGLAND Swale, Havengore, Berwick.
SCOTLAND Firth of Forth, Scrabster, Mallaig, Loch Sunart, Loch Aline.
IRELAND Kilrush, Wicklow, Westport/Clew Bay.
FRANCE Arcachon, Seudre R, Ile d'Oleron, Rochfort, Ile de Re, St. Giles-Croix-de-Vie, Ile d'Yeu, Pouliguen, Le Croisic, La Forêt, Ile de Bréhat.
GERMANY Norderney, Dornumer-Accumersiel.
If you are interested in becoming our agent for any of the above, please write to the editors and get your free copy of the Almanac every year. You do not have to be resident in a port to be the agent, but at least a fairly regular visitor.

PORTRUSH 10-13-12
Antrim

CHARTS
Admiralty 49, 2499, 2798; Imray C64; Irish OS 2
TIDES
−0410 Dover; ML 1.1; Duration 0610; Zone 0 (GMT).

Tidal figures based on Londonderry (→)

Times		Height (metres)			
HW	LW	MHWS	MHWN	MLWN	MLWS
−0105	−0105	−0.8	−0.6	−0.2	0.0

SHELTER
Anchorage on E side of Ramore Head in Skerries Roads gives good shelter in most conditions, but exposed to sea/swell from N. Harbour is sheltered except in strong NW-N winds.
NAVIGATION
Waypoint 55°13′.00N 06°41′.00W, 308°/128° from/to N Pier Lt, 1.1M. Entrance with on-shore winds over force 4 is difficult. Beware submerged breakwater projecting 20m SW from N pier.
LIGHTS AND MARKS
Ldg Lts 028° (occas, for lifeboat use) both FR 6/8 1M; R triangles on metal Bn and metal mast. N pier Fl R 3s 6m 3M; vis 220°-160°. S pier Fl G 3s 6m 3M; vis 220°-100°. Fixed R aero obstruction Lt on CG look-out, Ramore Hd.
RADIO TELEPHONE
VHF Ch 16; 12 (0900-1730 LT).
TELEPHONE (0265)
Hr Mr 822307; CG 823356; MRSC (0247) 883184; Customs Coleraine 4803; Marinecall 0898 500 465; Dr 823767; Hosp Coleraine 4177.
FACILITIES
EC Wednesday; **Harbour** AB, D, FW, M, Slip; **Portrush YC** Tel. 823932, Bar, FW; **A. Doherty** Tel. 824735 Electronics; **C. Trolan** Tel. Portstewart 2221 Gas; **J. Mullan** Tel. 822209, Gas, Gaz; **Town** V, R, Bar, Lau, P, PO, Bank, Rly, Air (Belfast).

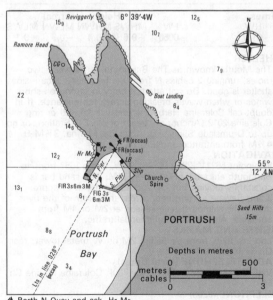

⚓ Berth N Quay and ask Hr Mr

RIVER BANN 10-13-13
Londonderry/Antrim

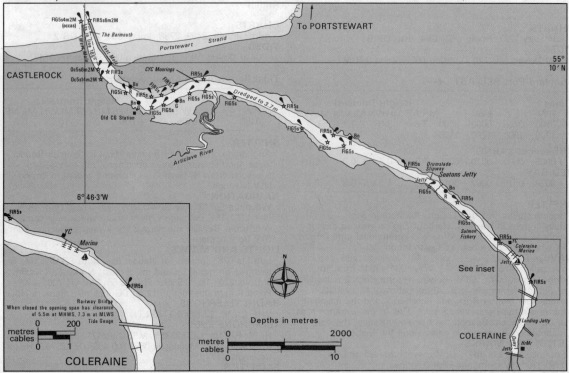

CHARTS
Admiralty 2499, 2723, 2798; Imray C64; Irish OS 2

TIDES
−0345 Dover (Coleraine); ML 1.1; Duration 0540; Zone 0 (GMT).

Tidal figures based on Londonderry (→)

Times		Height (metres)			
HW	LW	MHWS	MHWN	MLWN	MLWS
−0117	−0050	−0.6	−0.4	−0.3	−0.1

SHELTER
The Mouth, known as The Barmouth, is between two moles, running 2 cables N from the beaches. Once inside shelter is good. Do not attempt entry in strong on-shore winds or when waves breaking across the entrance. If in doubt call Coleraine Harbour Radio (Ch 16, 12) or ring Coleraine 2012. Anchor upstream of old CG Station, or go up to Drumslade Slipway or Coleraine Marina, 3½M & 4½M from entrance on NE bank.

NAVIGATION
Waypoint 55°11′.28N, 06°46′.77W, 345°/165° from/to Barmouth entrance breakwaters, 1M. The sand bar is constantly moving but has a dredged depth of approx 3.5m. Beware salmon nets across the width of the river during the salmon fishing season at 2M or 4M from entrance. Also beware commercial traffic.

LIGHTS AND MARKS
Ldg Lts 165°, front Oc 5s 6m 2M on W metal tower; rear Oc 5s 14m 2M.

RADIO TELEPHONE (local times)
VHF Ch 16; 12 (Mon-Fri: 0900-1700). Coleraine Marina Ch M.

TELEPHONE (0265)
Hr Mr 2012; HMC Freefone Customs Yachts: RT (0232) 752511; CG Portrush 823356; MRSC Donaghadee 883184; Rly Bridge 2403; Marinecall 0898 500 465; Hosp 4177; Dr 4831.

FACILITIES
COLERAINE EC Thursday. **Coleraine (Borough Council) Marina** (45+15 visitors), Tel. 4768, Slip, BY, AB, P, D, L, FW, CH, R, BH (15 ton), AC, ME, El, Sh; **Coleraine Yacht Supplies** Tel. 52525, CH; **Coleraine YC** Tel. 4503, Bar; **Seaton Sail & Power (Drumslade Slipway)** Tel. 832086, Slip, CH; **Calor Kosangas Gas Centre** Tel. 57057, Gas, Kos. **Town** P, D, V, R, PO, Bank, Rly, Air (Belfast).

LOUGH FOYLE 10-13-14
Londonderry/Donegal

CHARTS
Admiralty 2486, 2499; Imray C64; Irish OS 2

TIDES
Culmore Point −0025 Londonderry
Moville (−0055 Londonderry
(−0300 Dover
(−0400 Belfast
Warren Point −0400 Dover
Londonderry −0300 Dover
ML 1.5; Duration 0615; Zone 0 (GMT)

Standard Port GALWAY (→)

Times				Height (metres)			
HW		LW		MHWS	MHWN	MLWN	MLWS
0200	0900	0200	0800	5.1	3.9	2.0	0.6
1400	2100	1400	2000				

Differences LONDONDERRY
+0254 +0319 +0322 +0321 −2.4 −1.9 −1.0 −0.2

SHELTER
The SE side of the Lough is low lying and shallow. The NW rises steeply and has a number of village harbours between the entrance and Londonderry (often referred to as Derry).
GREENCASTLE — a fishing harbour, safe in NNW to WSW winds. Anchor off or go alongside fishing boats.
MOVILLE — the pier is close to the village and has 1.5m at the end (shops closed all day Wed).

AREA 13—N Ireland 575

LOUGH FOYLE continued

CARRICKARORY — has a good pier/quay, the end 25m of which has a depth of 2m. Good shelter with winds NNW to SW.
CULMORE BAY — Complete shelter (anchor 1½ cables W of Culmore Pt). 4M from Londonderry.
LONDONDERRY — Good shelter — anchor close below Craigavon Bridge. There is a bridge at Rosses Pt about 2M downstream, clearance 32m.

NAVIGATION
Waypoint Tuns (port-hand) buoy, FlR 3s, 55°14'.01N 06°53'.38W, 055°/235° from/to Warren Point Lt, 2.5M. Outside entrance beware The Tuns which run 3M NE from E side of entrance. The main channel ¾M wide is the North Channel NW of The Tuns but there is a channel, min depth 4m, 3 cables off shore along NE side of Magilligan Pt. Beware commercial traffic; in June and July the channel is at times obstructed by salmon nets at night. N channel tides reach 3½ kn, and up in the river the ebb runs up to 6 kn.

LIGHTS AND MARKS
Inishowen Fl(2) WRG 10s 28m 18/14M; W Tr, two B bands; vis G197°-211°, W211°-249°, R249°-000°; Horn(2) 30s. Warren Point Fl 1.5s 9m 10M; W Tr, G abutment; vis 232°-061°. Magilligan Point FlR 2s 7m 4M; R structure. The main channel up to Londonderry is very well lit. Foyle Bridge centre FW each side; VQG on W pier; VQR on E pier.

RADIO TELEPHONE
VHF Ch 16; 12 14 (H24).

TELEPHONE (0504)
Hr Mr 263680; CG Portrush 823356; MRSC Donaghadee 883184; HMC Freefone Customs Yachts: RT (0235) 752511; Marinecall 0898 500 465; Dr 264868; Hosp 45171.

FACILITIES (Londonderry)
EC Thursday; **Harbour Office** Tel. 263680, M, FW, C (10 ton mobile), AB; **8 & 9 Sheds** M, FW, C (2 x 1½ ton elec), AB; **7 Shed** M, FW, AB; **14/15 Berth** M, FW, AB; **Dolphins** M, FW, El, AB; **NR Quay** P, D, ME, El; **Prehen Boat Club** Tel. 43405;
Town P, D, ME, El, CH, V, R, Bar, PO; Bank; Rly; Air.

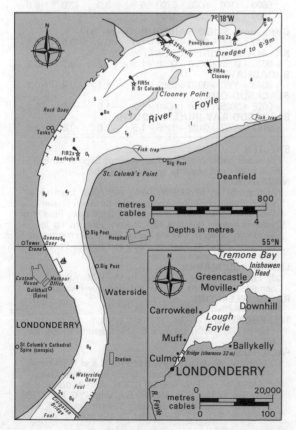

LOUGH SWILLY 10-13-15
Donegal

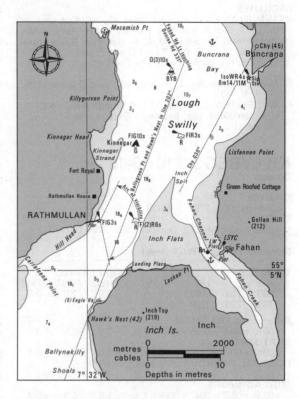

CHARTS
Admiralty 2697; Irish OS 1

TIDES
−0500 Dover; ML 2.5; Duration 0605; Zone 0 (GMT)

Standard Port GALWAY (→)

Times				Height (metres)			
HW		LW		MHWS	MHWN	MLWN	MLWS
0200	0900	0200	0800	5.1	3.9	2.0	0.6
1400	2100	1400	2000				

Differences RATHMULLAN
+0125 +0050 +0126 +0118 −0.8 −0.7 −0.1 −0.1
FANAD HEAD
+0115 +0040 +0125 +0120 −1.1 −0.9 −0.5 −0.1
MULROY BAY BAR
+0108 +0052 +0102 +0118 −1.2 −1.0 No data
SHEEPHAVEN (DOWNIES BAY)
+0057 +0043 +0053 +0107 −1.1 −0.9 No data

SHELTER
Entrance easy in all weathers but swell can render anchorages below Inch Island uncomfortable. Anchorages — Port Salon Bay when wind not in E, but liable to swell (off chartlet to NW); Fahan Creek E of Inch Is entered at HW; W of Macamish Pt sheltered from SE to N through W; Rathmullan Road N of pier off town.

NAVIGATION
Lough Swilly waypoint 55°17'.50N 07°34'.50W, 352°/172° from/to Dunree Head Lt, 5.7M. Six lit lateral buoys and one lit cardinal buoy mark main channel. Beware Swilly More Rks, Kinnegar Spit, Colpagh rks off E shore, Kinnegar Strand, Inch Flats and fish farms.

LIGHTS AND MARKS
Leading lines into Lough Swilly
(1) Fanad Hd Lt touching Dunree Head at 331°.
(2) Ballygreen Pt and Hawk's Nest in line at 202°.

RADIO TELEPHONE
None.

TELEPHONE (074)
Hr Mr 58177; MRSC Limerick 61219; Customs 21935; Dr 58135.

LOUGH SWILLY continued

FACILITIES
RATHMULLAN EC Wednesday; **Pier** AB, AC, C (5 ton), FW, L, M, Slip; **E. Toomey** Tel. 58125, CH; **Rathmullan House Hotel** Tel. 58117, M, L, R, Bar; **Fort Royal Hotel** Tel. 58100, M, L, R, Bar; **Town** D and P (cans), Kos, Bar, R, V, PO, Bank, Rly (bus to Londonderry), Air (Londonderry).
RAMELTON **Quay** AB, L; **Town** Bar, P and D (cans), FW, Kos, R, V, PO.
FAHAN Slip, FW, L, M (ask YC), R; **Lough Swilly YC** Tel. Fahan 60189, Bar; **Bradley's Garage** (1M SE), Kos, V, P and D (cans).

KILLYBEGS 10-13-16
Donegal

CHARTS
Admiralty 2702, 2792; Irish OS 3
TIDES
−0520 Dover; ML 2.2; Duration 0620; Zone 0 (GMT).

Standard Port GALWAY (→)

Times				Height (metres)			
HW		LW		MHWS	MHWN	MLWN	MLWS
0600	1100	0000	0700	5.1	3.9	2.0	0.6
1800	2300	1200	1900				

Differences KILLYBEGS
+0040 +0050 +0055 +0035 −1.0 −0.9 −0.5 0.0
BURTONPORT
+0042 +0055 +0115 +0055 −1.2 −1.0 −0.6 −0.1
DONEGAL HARBOUR (SALTHILL QUAY)
+0038 +0050 +0052 +0104 −1.2 −0.9 No data
MULLAGHMORE
+0036 +0048 +0047 +0059 −1.4 −1.0 −0.4 −0.2

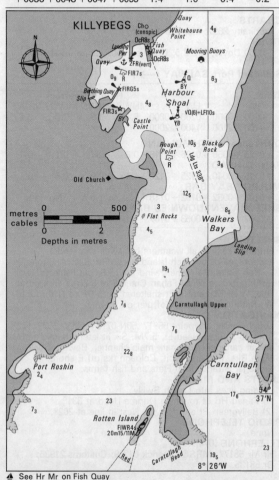

SHELTER
A secure harbour although SSW winds cause a little swell. Accessible in all weathers day and night.
NAVIGATION
Waypoint 54°36'.00N 08°27'.00W, 202°/022° from/to Rotten Island Lt, 0.94M. Beware of Fintragh. Note that Manister Rock is covered at HW and dries at LW.
LIGHTS AND MARKS
Keep mid channel until off Rough Pt then follow the leading lights into harbour by Fish Quay. Ldg Lts 338°, both Oc R 8s.
RADIO TELEPHONE
VHF Ch 16.
TELEPHONE (073)
Hr Mr 31032; MRCC Limerick 61219; Customs 31070; Dr 31181 (Home) 31148 (Surgery).
FACILITIES
EC Wednesday; **Landing Pier** Tel. 31032, Slip, M, D, P (Pier Bar), L, ME, El, CH, AB, V, R, Bar; **Berthing Quay** Slip, M, D, AB; **Mooney Boats** Tel. 31152, Sh, C (12 ton), ME, El, Sh; **Gallaher Bros** Tel. 31004, Kos. **Town** PO; Bank; Rly (bus to Sligo); Air (Strandhill).

SLIGO 10-13-17
Sligo

CHARTS
Admiralty 2852, 2767; Irish OS 7
TIDES
−0511 Dover; ML 2.3; Duration 0620; Zone 0 (GMT).

Standard Port GALWAY (→)

Times				Height (metres)			
HW		LW		MHWS	MHWN	MLWN	MLWS
0600	1100	0000	0700	5.1	3.9	2.0	0.6
1800	2300	1200	1900				

Differences SLIGO HARBOUR (Oyster Is)
+0043 +0055 +0042 +0054 −1.0 −0.9 −0.5 −0.1
KILLALA BAY (INISHCRONE)
+0035 +0055 +0030 +0050 −1.3 −1.2 −0.7 −0.2
BROADHAVEN
+0040 +0050 +0040 +0050 −1.4 −1.1 −0.4 −0.1
BLACKSOD QUAY
+0025 +0035 +0040 +0040 −1.2 −1.0 −0.6 −0.2
BLACKSOD BAY (BULL'S MOUTH)
+0101 +0057 +0109 +0105 −1.5 −1.0 −0.6 −0.1

SHELTER
The lower harbour is fairly exposed but Sligo town is 5 miles from open sea and gives good shelter.
NAVIGATION
Waypoint 54°18'.62N 08°39'.00W, 248°/068° from/to Cullaun Bwee Lt, 3.0M. The passage between Oyster Island and Coney Island is marked 'Dangerous'. Pass N of Oyster Is leaving Blennick Rks to Port. Passage up to Sligo town between training walls. Some perches are in bad repair. Pilots at Raghley Head and Rosses Pt. Channel up to quays dredged to 2.4m.
LIGHTS AND MARKS
Ldg Lts into harbour 125°, lead to Metal Man Rocks. Channel lights up the harbour are presently discontinued.
RADIO TELEPHONE
Pilots VHF Ch 12; 16.
TELEPHONE (071)
Hr Mr 61197; MRCC Limerick 61219; Customs 61064; Dr 2746.
FACILITIES
EC Monday, all day; **Deepwater Pier** P and D (in cans), L, FW, ME, El, Sh, C (15 ton), CH, AB; **Ballast Quay** Tel. 61197, M, L, FW, ME, El, C, CH, AB; **Rodney Lomax** Tel. 66124, Slip, ME, El, Sh; **Sligo YC** Tel. 77168, M, FW, Bar, Slip; **Sligo Bedding Centre** Tel. 2303, Kos. **Town** V, R, Bar. PO; Bank; Rly; Air (Strandhill).

SLIGO HARBOUR continued

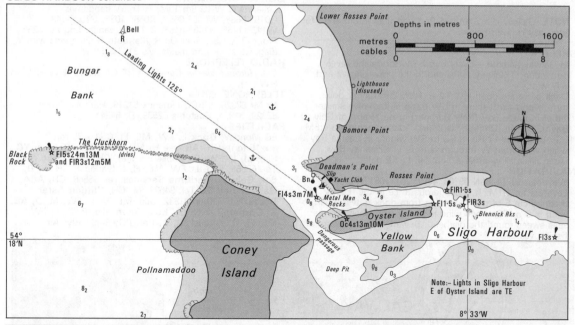

WESTPORT (CLEW BAY)   10-13-18
Mayo

CHARTS
Admiralty 2057; Irish OS 10 and 11

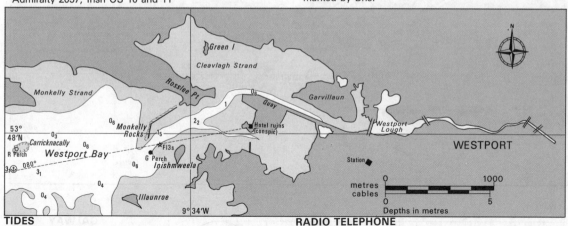

LIGHTS AND MARKS
Westport Bay entrance — Inishgort Lt L Fl 10s 11m 10M. North of hotel ruins in line with South of Lt Bn (Fl 3s) at 080°. Passage from Westport Bay to Westport, 5 miles, is marked by Bns.

TIDES
−0545 Dover; ML 2.5; Duration 0610; Zone 0 (GMT)

Standard Port GALWAY (→)

Times				Height (metres)			
HW		LW		MHWS	MHWN	MLWN	MLWS
0600	1100	0000	0700	5.1	3.9	2.0	0.6
1800	2300	1200	1900				

Differences INISHRAHER
+0030 +0012 +0058 +0026 −0.6 −0.5 −0.3 −0.1
CLARE ISLAND
+0019 +0013 +0029 +0023 −1.0 −0.7 −0.4 −0.1
KILLARY HARBOUR
+0021 +0015 +0035 +0029 −1.0 −0.8 −0.4 −0.1
CLIFDEN BAY
+0005 +0005 +0016 +0016 −0.7 −0.5 No data

SHELTER
Westport Bay affords secure anchorage amongst the islands and is available at all states of tide.

NAVIGATION
Waypoint 53°49′.20N 09°42′.10W, 251°/071° from/to Inishgort Lt Ho, 1.2M. Pilot for Westport lives on Inishlyre. Beware of the Spit and Mondelly (or Monkelly) Rks which are unmarked.

RADIO TELEPHONE
None.

TELEPHONE (098)
MRCC Limerick 61219; Customs 142; Hosp 109.

FACILITIES
EC Wednesday; **Quays** M, L, AB, V, R, Bar; **M O'Grady** Tel. 25072, Kos. **Town** PO; Bank; Rly; Air (Enniskillen).
Note: — Between Westport and Galway, repair facilities at **Aster Boats**, Drimagh Harbour, Errislannan Tel. Clifden 166. See 10.13.20.

GALWAY   10-13-19
Galway

CHARTS
Admiralty 1903, 1984, 3339; Irish OS 14
TIDES
−0605 Dover; ML 2.9; Duration 0620; Zone 0 (GMT)

Standard Port GALWAY (→)

Times				Height (metres)			
HW		LW		MHWS	MHWN	MLWN	MLWS
0600	1100	0000	0700	5.1	3.9	2.0	0.6
1800	2300	1200	1900				

Differences KILLEANY BAY (Aran Islands)
−0008 −0008 +0003 +0003 −0.4 −0.3 −0.2 −0.1

NOTE: Galway is a Standard Port and tidal information for each day is given below.

SHELTER
In Galway harbour — very good. Mutton Island gives protection to Galway roadstead in the prevailing W and SW winds.
Dock gates are open HW−2 to HW. Enter inner dock, turn to port and secure in SW basin. Alternative shelter either in Riville Bay (New Harbour) or on N arm of Galway Bay at Clifden (Drinagh Harbour). A safe anchorage 1.5M ESE Blackhead Lt Ho in SE to SW winds; in Cashla Bay when wind from W through N to E; in Killeaney Bay (Aran Is) in all weather except NE to SE gales.

NAVIGATION
Waypoint 53°14′.80N 09°03′.40W, 241°/061° from/to Leverets Lt, 1.1M. It is dangerous to lie in the 'Layby' (a dredged cut E of pier extending from dock gates) when wind is S or SE; when wind is strong from these points seas sweep round the pierhead. Beware large numbers of fishing vessels.

LIGHTS AND MARKS
Leverets Q WRG 9m 10M; B Tr, W bands; vis G015°-058°, W058°-065°, R065°-103°, G103°-143°, W143°-146°, R146°-015°. B Tr, W bands. Ldg Lts 325°. Front Fl R 1.5s. Rear Oc R 10s; both R diamonds with Y diagonal stripes on masts, vis 315°-345°.

RADIO TELEPHONE
Call *Harbour Master Galway*. VHF Ch 16; 12 (HW−2½ to HW+1).

TELEPHONE (091)
Hr Mr 62329; MRCC Limerick 61219; Harbour Office 62329, 61874; Customs 62539; Dr 64241.

FACILITIES
EC Monday; **Dock** L, FW, ME, El, C (1 × 35 ton 4 × 15 ton), CH, AB (see Hr Mr), V, R, Bar; **Galway YC** Slip, M, L, FW, C (mobile on hire), CH, Bar; **Boat Yard** Tel. 62568, Slip, L, FW, ME, El, C (various), CH, V, R, Bar; **Galway Maritime Services** Tel. 66568, CH, ACA; **Galway Bay SC** Tel. 94527, M, CH; **Clifden Aster Boats** Tel. Clifden 21332, Slip, ME, El, L, FW, M, D, AB, C; **Corbett** Tel. 65151, Kos. **Town** Slip, P, D, L, FW, ME, El, C (various), V, R, Bar. PO; Bank; Rly; Air (Shannon).

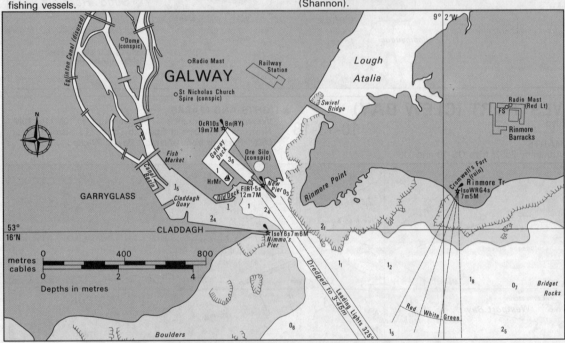

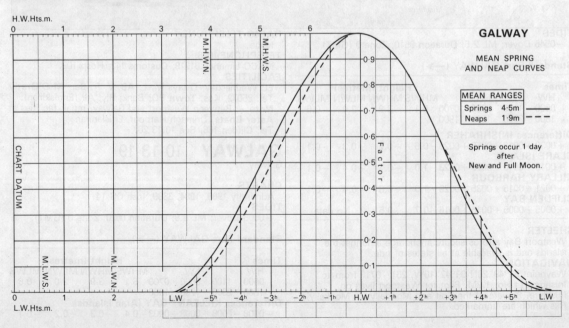

GALWAY

MEAN SPRING AND NEAP CURVES

MEAN RANGES
Springs 4·5m
Neaps 1·9m

Springs occur 1 day after New and Full Moon.

TIME ZONE UT (GMT)
For Summer Time add ONE hour in non-shaded areas

IRELAND, WEST COAST — GALWAY
Lat 53°16′ N   Long 9°03′ W
TIMES AND HEIGHTS OF HIGH AND LOW WATERS

YEAR 1989

AREA 13—N Ireland

JANUARY

Time	m	Time	m
1 0506 / 1125 / Su 1749	2.2 / 3.9 / 2.0	**16** 0537 / 1201 / M 1810	1.8 / 4.2 / 1.8
2 0015 / 0617 / M 1235 / 1853	3.9 / 2.3 / 3.8 / 2.0	**17** 0045 / 0702 / Tu 1321 / 1928	4.1 / 1.8 / 4.1 / 1.9
3 0120 / 0726 / Tu 1342 / 1952	4.0 / 2.2 / 3.9 / 2.0	**18** 0158 / 0816 / W 1432 / 2034	4.2 / 1.7 / 4.2 / 1.8
4 0215 / 0822 / W 1439 / 2043	4.1 / 2.0 / 4.0 / 1.9	**19** 0300 / 0917 / Th 1531 / 2131	4.4 / 1.5 / 4.4 / 1.6
5 0301 / 0911 / Th 1528 / 2129	4.3 / 1.7 / 4.2 / 1.7	**20** 0350 / 1007 / F 1620 / 2216	4.6 / 1.2 / 4.6 / 1.4
6 0345 / 0956 / F 1613 / 2213	4.6 / 1.4 / 4.5 / 1.5	**21** 0435 / 1049 / Sa 1704 / ○ 2257	4.8 / 1.0 / 4.7 / 1.2
7 0426 / 1038 / Sa 1657 / ● 2254	4.7 / 1.1 / 4.7 / 1.2	**22** 0518 / 1129 / Su 1744 / 2334	4.9 / 0.8 / 4.8 / 1.1
8 0508 / 1120 / Su 1740 / 2336	5.0 / 0.8 / 4.9 / 1.0	**23** 0556 / 1205 / M 1822	5.0 / 0.8 / 4.8
9 0551 / 1203 / M 1824	5.2 / 0.7 / 5.0	**24** 0011 / 0634 / Tu 1242 / 1859	1.0 / 5.0 / 0.8 / 4.7
10 0017 / 0636 / Tu 1245 / 1907	0.9 / 5.3 / 0.6 / 5.0	**25** 0046 / 0710 / W 1317 / 1935	1.0 / 4.9 / 0.8 / 4.6
11 0100 / 0721 / W 1328 / 1951	0.9 / 5.3 / 0.6 / 5.0	**26** 0121 / 0745 / Th 1351 / 2011	1.1 / 4.8 / 1.0 / 4.5
12 0144 / 0806 / Th 1413 / 2037	1.0 / 5.2 / 0.7 / 4.8	**27** 0157 / 0819 / F 1425 / 2046	1.3 / 4.6 / 1.2 / 4.3
13 0232 / 0856 / F 1501 / 2127	1.1 / 5.0 / 1.0 / 4.6	**28** 0232 / 0854 / Sa 1501 / 2122	1.5 / 4.3 / 1.5 / 4.1
14 0322 / 0948 / Sa 1553 / ☽ 2221	1.4 / 4.7 / 1.3 / 4.4	**29** 0311 / 0931 / Su 1541 / 2204	1.8 / 4.1 / 1.7 / 3.9
15 0423 / 1048 / Su 1655 / 2327	1.6 / 4.4 / 1.6 / 4.2	**30** 0356 / 1016 / M 1630 / 2257	2.0 / 3.8 / 2.0 / 3.7
		31 0458 / 1119 / Tu 1740	2.2 / 3.6 / 2.2

FEBRUARY

Time	m	Time	m
1 0010 / 0624 / W 1248 / 1909	3.7 / 2.2 / 3.6 / 2.2	**16** 0152 / 0823 / Th 1437 / 2043	3.9 / 1.7 / 3.9 / 1.9
2 0130 / 0751 / Th 1413 / 2025	3.7 / 2.1 / 3.7 / 2.0	**17** 0300 / 0921 / F 1541 / 2132	4.1 / 1.4 / 4.1 / 1.6
3 0236 / 0857 / F 1515 / 2119	4.0 / 1.7 / 4.0 / 1.7	**18** 0349 / 1002 / Sa 1616 / 2210	4.4 / 1.2 / 4.4 / 1.4
4 0328 / 0946 / Sa 1604 / 2204	4.3 / 1.3 / 4.4 / 1.3	**19** 0428 / 1038 / Su 1652 / 2244	4.6 / 0.9 / 4.6 / 1.1
5 0414 / 1028 / Su 1647 / 2244	4.7 / 0.8 / 4.7 / 0.9	**20** 0504 / 1111 / M 1726 / ○ 2316	4.8 / 0.7 / 4.7 / 0.9
6 0457 / 1109 / M 1727 / ● 2323	5.1 / 0.5 / 5.0 / 0.6	**21** 0537 / 1143 / Tu 1800 / 2349	4.9 / 0.6 / 4.8 / 0.7
7 0539 / 1147 / Tu 1807	5.3 / 0.3 / 5.2	**22** 0611 / 1214 / W 1832	4.9 / 0.5 / 4.8
8 0001 / 0621 / W 1227 / 1848	0.4 / 5.5 / 0.1 / 5.2	**23** 0021 / 0643 / Th 1245 / 1904	0.7 / 4.9 / 0.6 / 4.7
9 0042 / 0703 / Th 1306 / 1928	0.3 / 5.5 / 0.2 / 5.1	**24** 0052 / 0714 / F 1314 / 1935	0.8 / 4.7 / 0.8 / 4.6
10 0123 / 0747 / F 1347 / 2011	0.4 / 5.3 / 0.4 / 5.0	**25** 0123 / 0745 / Sa 1344 / 2006	1.0 / 4.6 / 1.0 / 4.4
11 0205 / 0832 / Sa 1430 / 2056	0.7 / 5.0 / 0.8 / 4.7	**26** 0154 / 0816 / Su 1415 / 2037	1.2 / 4.3 / 1.3 / 4.2
12 0253 / 0921 / Su 1519 / ☽ 2146	1.1 / 4.6 / 1.3 / 4.3	**27** 0227 / 0849 / M 1447 / 2114	1.5 / 4.1 / 1.6 / 4.0
13 0349 / 1019 / M 1617 / 2249	1.5 / 4.2 / 1.7 / 4.0	**28** 0305 / 0929 / Tu 1531 / ☾ 2159	1.8 / 3.8 / 1.9 / 3.7
14 0505 / 1134 / Tu 1742	1.8 / 3.9 / 2.1		
15 0017 / 0653 / W 1314 / 1927	3.8 / 1.9 / 3.8 / 2.1		

MARCH

Time	m	Time	m
1 0400 / 1027 / W 1638 / 2309	2.0 / 3.4 / 2.2 / 3.5	**16** 0652 / 1312 / Th 1930	1.9 / 3.6 / 2.2
2 0529 / 1201 / Th 1834	2.2 / 3.4 / 2.3	**17** 0144 / 0813 / F 1432 / 2034	3.7 / 1.7 / 3.8 / 1.9
3 0046 / 0726 / F 1354 / 2009	3.6 / 2.0 / 3.6 / 2.0	**18** 0249 / 0903 / Sa 1519 / 2115	4.0 / 1.4 / 4.1 / 1.6
4 0212 / 0839 / Sa 1501 / 2104	3.8 / 1.6 / 3.9 / 1.6	**19** 0331 / 0941 / Su 1556 / 2149	4.2 / 1.1 / 4.3 / 1.3
5 0310 / 0928 / Su 1549 / 2145	4.3 / 1.1 / 4.4 / 1.1	**20** 0406 / 1013 / M 1638 / 2220	4.4 / 0.9 / 4.5 / 1.0
6 0355 / 1009 / M 1627 / 2224	4.7 / 0.6 / 4.8 / 0.6	**21** 0440 / 1044 / Tu 1659 / 2251	4.6 / 0.7 / 4.7 / 0.8
7 0437 / 1045 / Tu 1705 / ● 2301	5.2 / 0.2 / 5.1 / 0.2	**22** 0512 / 1113 / W 1732 / ○ 2322	4.7 / 0.6 / 4.7 / 0.7
8 0518 / 1123 / W 1744 / 2339	5.5 / -0.1 / 5.3 / 0.0	**23** 0543 / 1143 / Th 1801 / 2353	4.8 / 0.6 / 4.7 / 0.6
9 0558 / 1201 / Th 1824	5.6 / -0.2 / 5.3	**24** 0614 / 1211 / F 1832	4.8 / 0.6 / 4.8
10 0018 / 0641 / F 1241 / 1903	0.0 / 5.6 / 0.0 / 5.2	**25** 0023 / 0643 / Sa 1239 / 1900	0.7 / 4.7 / 0.8 / 4.7
11 0059 / 0723 / Sa 1320 / 1945	0.2 / 5.3 / 0.4 / 5.0	**26** 0052 / 0711 / Su 1307 / 1930	0.9 / 4.5 / 1.0 / 4.5
12 0141 / 0808 / Su 1402 / 2029	0.5 / 5.0 / 0.8 / 4.7	**27** 0121 / 0744 / M 1337 / 2002	1.1 / 4.3 / 1.3 / 4.3
13 0229 / 0857 / M 1450 / 2118	1.0 / 4.5 / 1.4 / 4.3	**28** 0155 / 0819 / Tu 1412 / 2039	1.4 / 4.1 / 1.6 / 4.1
14 0325 / 0956 / Tu 1550 / ☾ 2221	1.5 / 4.1 / 1.9 / 3.9	**29** 0236 / 0904 / W 1457 / 2127	1.6 / 3.8 / 1.9 / 3.8
15 0447 / 1118 / W 1725 / 2356	1.9 / 3.7 / 2.2 / 3.6	**30** 0332 / 1003 / Th 1607 / ☾ 2235	1.9 / 3.6 / 2.2 / 3.6
		31 0502 / 1136 / F 1805	2.0 / 3.5 / 2.2

APRIL

Time	m	Time	m
1 0012 / 0655 / Sa 1326 / 1938	3.6 / 1.8 / 3.7 / 1.9	**16** 0213 / 0825 / Su 1444 / 2039	3.9 / 1.5 / 4.0 / 1.7
2 0140 / 0806 / Su 1432 / 2033	3.9 / 1.4 / 4.0 / 1.5	**17** 0257 / 0903 / M 1522 / 2115	4.1 / 1.3 / 4.2 / 1.4
3 0240 / 0856 / M 1518 / 2117	4.2 / 0.9 / 4.5 / 1.0	**18** 0334 / 0938 / Tu 1556 / 2149	4.3 / 1.1 / 4.4 / 1.1
4 0327 / 0938 / Tu 1559 / 2156	4.8 / 0.5 / 4.8 / 0.5	**19** 0409 / 1009 / W 1628 / 2221	4.4 / 0.9 / 4.6 / 0.9
5 0410 / 1017 / W 1638 / 2235	5.2 / 0.1 / 5.1 / 0.2	**20** 0441 / 1040 / Th 1659 / 2254	4.6 / 0.8 / 4.7 / 0.8
6 0452 / 1055 / Th 1718 / ● 2315	5.4 / -0.1 / 5.3 / 0.0	**21** 0513 / 1109 / F 1730 / ○ 2323	4.6 / 0.8 / 4.8 / 0.8
7 0534 / 1134 / F 1757 / 2356	5.5 / 0.0 / 5.4 / 0.0	**22** 0544 / 1139 / Sa 1800 / 2354	4.6 / 0.9 / 4.8 / 0.8
8 0617 / 1214 / Sa 1838	5.4 / 0.2 / 5.3	**23** 0615 / 1207 / Su 1829	4.6 / 1.0 / 4.7
9 0036 / 0702 / Su 1256 / 1921	0.2 / 5.2 / 0.6 / 5.0	**24** 0025 / 0646 / M 1238 / 1900	0.9 / 4.5 / 1.2 / 4.6
10 0121 / 0748 / M 1340 / 2006	0.6 / 4.8 / 1.1 / 4.7	**25** 0059 / 0721 / Tu 1313 / 1937	1.1 / 4.4 / 1.4 / 4.4
11 0212 / 0840 / Tu 1429 / 2058	1.1 / 4.4 / 1.6 / 4.3	**26** 0137 / 0802 / W 1352 / 2019	1.3 / 4.2 / 1.7 / 4.2
12 0311 / 0941 / W 1532 / 2200	1.5 / 4.0 / 2.0 / 3.9	**27** 0223 / 0851 / Th 1446 / 2111	1.5 / 3.9 / 1.9 / 3.9
13 0434 / 1059 / Th 1704 / 2327	1.8 / 3.7 / 2.3 / 3.7	**28** 0325 / 0953 / F 1559 / ☾ 2219	1.7 / 3.8 / 2.1 / 3.9
14 0619 / 1242 / F 1850	1.9 / 3.6 / 2.2	**29** 0447 / 1115 / Sa 1732 / 2340	1.8 / 3.7 / 2.0 / 3.9
15 0109 / 0735 / Sa 1358 / 1957	3.7 / 1.7 / 3.8 / 2.0	**30** 0614 / 1245 / Su 1855	1.6 / 3.8 / 1.8

Chart Datum: 0.20 metres below Ordnance Datum (Dublin)

IRELAND, WEST COAST — GALWAY

Lat 53°16′ N Long 9°03′ W

TIMES AND HEIGHTS OF HIGH AND LOW WATERS

YEAR 1989

TIME ZONE UT(GMT)
For Summer Time add ONE hour in non-shaded areas

MAY		JUNE		JULY		AUGUST	
Time m	Time m	Time m	Time m	Time m	Time m	Time m	Time m
1 0100 4.1 / 0724 1.3 / M 1352 4.2 / 1954 1.4	**16** 0211 3.9 / 0832 1.3 / Tu 1439 4.1 / 2036 1.6	**1** 0225 4.6 / 0857 1.6 / Th 1457 4.7 / 2103 0.9	**16** 0257 4.0 / 0904 1.6 / F 1517 4.3 / 2122 1.5	**1** 0304 4.5 / 0918 1.7 / Sa 1529 4.7 / 2142 1.0	**16** 0318 3.9 / 0918 1.7 / Su 1531 4.2 / 2148 1.4	**1** 0444 4.6 / 1037 1.1 / Tu 1657 4.9 / ● 2309 0.6	**16** 0430 4.5 / 1026 1.0 / W 1635 4.9 / 2249 0.5
2 0204 4.4 / 0818 0.9 / Tu 1443 4.5 / 2043 1.0	**17** 0254 4.1 / 0857 1.3 / W 1518 4.3 / 2115 1.4	**2** 0318 4.8 / 0921 0.8 / F 1545 4.9 / 2150 0.7	**17** 0339 4.1 / 0938 1.5 / Sa 1555 4.4 / 2202 1.3	**2** 0359 4.6 / 0956 1.1 / Su 1619 4.8 / 2233 0.8	**17** 0404 4.1 / 1002 1.5 / M 1613 4.5 / 2228 1.1	**2** 0525 4.7 / 1116 0.9 / W 1736 4.9 / 2347 0.5	**17** 0508 4.7 / 1102 0.7 / Th 1716 5.1 / ○ 2326 0.2
3 0254 4.8 / 0904 0.6 / W 1528 4.8 / 2127 0.7	**18** 0334 4.2 / 0932 1.2 / Th 1553 4.5 / 2150 1.2	**3** 0407 4.9 / 1007 0.8 / Sa 1630 5.1 / ● 2238 0.6	**18** 0420 4.2 / 1016 1.4 / Su 1631 4.5 / 2240 1.1	**3** 0449 4.7 / 1044 1.1 / M 1705 4.9 / ● 2319 0.7	**18** 0447 4.3 / 1041 1.2 / Tu 1654 4.7 / ○ 2308 0.8	**3** 0604 4.7 / 1153 0.9 / Th 1815 5.0	**18** 0546 5.0 / 1139 0.4 / F 1756 5.3
4 0342 5.1 / 0946 0.4 / Th 1610 5.1 / 2210 0.4	**19** 0409 4.3 / 1006 1.2 / F 1626 4.6 / 2226 1.1	**4** 0457 5.0 / 1052 0.8 / Su 1716 5.1 / 2325 0.6	**19** 0459 4.4 / 1054 1.3 / M 1709 4.7 / ○ 2319 1.0	**4** 0537 4.7 / 1129 1.0 / Tu 1751 5.0	**19** 0526 4.5 / 1120 1.0 / W 1734 4.9 / 2346 0.6	**4** 0024 0.5 / 0642 4.7 / F 1229 0.9 / 1853 4.9	**19** 0003 0.1 / 0624 5.1 / Sa 1217 0.3 / 1836 5.4
5 0427 5.2 / 1028 0.3 / F 1651 5.3 / ● 2252 0.3	**20** 0444 4.4 / 1038 1.1 / Sa 1658 4.7 / ○ 2258 1.0	**5** 0546 4.9 / 1139 0.9 / M 1803 5.1	**20** 0539 4.5 / 1132 1.2 / Tu 1749 4.7 / 2358 0.9	**5** 0004 0.6 / 0622 4.7 / W 1211 1.0 / 1835 4.9	**20** 0607 4.7 / 1200 0.8 / Th 1817 5.0	**5** 0059 0.6 / 0719 4.6 / Sa 1304 0.9 / 1928 4.7	**20** 0041 0.1 / 0703 5.1 / Su 1256 0.2 / 1919 5.3
6 0512 5.3 / 1111 0.4 / Sa 1733 5.3 / 2336 0.3	**21** 0518 4.5 / 1111 1.2 / Su 1730 4.7 / 2332 1.0	**6** 0012 0.6 / 0634 4.8 / Tu 1224 1.1 / 1849 4.9	**21** 0619 4.5 / 1212 1.2 / W 1829 4.8	**6** 0046 0.7 / 0706 4.6 / Th 1253 1.1 / 1917 4.8	**21** 0025 0.4 / 0648 4.8 / F 1239 0.7 / 1859 5.1	**6** 0134 0.8 / 0755 4.5 / Su 1341 1.1 / 2005 4.5	**21** 0120 0.3 / 0744 5.0 / M 1338 0.6 / 2004 5.0
7 0558 5.2 / 1153 0.6 / Su 1817 5.2	**22** 0553 4.5 / 1144 1.2 / M 1804 4.7	**7** 0100 0.8 / 0723 4.6 / W 1310 1.3 / 1935 4.7	**22** 0039 0.8 / 0702 4.6 / Th 1255 1.2 / 1913 4.8	**7** 0128 0.8 / 0748 4.5 / F 1334 1.2 / 1959 4.6	**22** 0106 0.4 / 0728 4.8 / Sa 1320 0.7 / 1942 5.1	**7** 0209 1.0 / 0832 4.3 / M 1418 1.4 / 2040 4.3	**22** 0202 0.6 / 0827 4.8 / Tu 1425 0.9 / 2051 4.7
8 0021 0.5 / 0645 5.0 / M 1236 0.9 / 1903 5.0	**23** 0007 1.0 / 0629 4.5 / Tu 1221 1.3 / 1841 4.7	**8** 0149 1.0 / 0812 4.4 / Th 1358 1.5 / 2025 4.5	**23** 0123 0.8 / 0747 4.5 / F 1340 1.2 / 1959 4.7	**8** 0209 1.0 / 0830 4.3 / Sa 1416 1.4 / 2040 4.4	**23** 0147 0.5 / 0811 4.7 / Su 1405 0.9 / 2027 4.9	**8** 0244 1.3 / 0910 4.1 / Tu 1457 1.6 / 2118 4.0	**23** 0249 1.1 / 0917 4.4 / W 1518 1.3 / ☾ 2146 4.3
9 0109 0.8 / 0734 4.7 / Tu 1323 1.3 / 1951 4.7	**24** 0046 1.1 / 0710 4.4 / W 1302 1.4 / 1923 4.6	**9** 0240 1.2 / 0901 4.2 / F 1449 1.7 / 2114 4.3	**24** 0209 0.9 / 0833 4.5 / Sa 1427 1.3 / 2049 4.6	**9** 0251 1.2 / 0912 4.2 / Su 1458 1.6 / 2122 4.2	**24** 0230 0.7 / 0857 4.6 / M 1451 1.1 / 2115 4.7	**9** 0324 1.6 / 0950 3.9 / W 1542 1.9 / 2202 3.7	**24** 0345 1.6 / 1017 4.1 / Th 1630 1.7 / 2258 3.9
10 0201 1.1 / 0827 4.4 / W 1415 1.7 / 2042 4.4	**25** 0130 1.2 / 0755 4.3 / Th 1347 1.6 / 2009 4.4	**10** 0332 1.5 / 0952 4.0 / Sa 1542 1.9 / 2204 4.1	**25** 0258 1.0 / 0922 4.4 / Su 1519 1.4 / 2141 4.5	**10** 0335 1.4 / 0956 4.0 / M 1546 1.8 / 2207 3.9	**25** 0319 1.0 / 0946 4.4 / Tu 1546 1.3 / ☾ 2210 4.4	**10** 0412 1.9 / 1041 3.7 / Th 1642 2.1 / 2302 3.5	**25** 0501 1.9 / 1137 3.9 / F 1815 1.9
11 0301 1.4 / 0924 4.1 / Th 1514 2.0 / 2141 4.1	**26** 0220 1.3 / 0846 4.2 / F 1442 1.7 / 2103 4.3	**11** 0427 1.6 / 1047 3.9 / Su 1641 2.0 / ☾ 2302 3.9	**26** 0352 1.1 / 1019 4.3 / M 1620 1.5 / ☾ 2238 4.3	**11** 0423 1.6 / 1045 3.9 / Tu 1642 2.0 / ☾ 2301 3.7	**26** 0414 1.3 / 1045 4.2 / W 1654 1.6 / 2318 4.1	**11** 0520 2.1 / 1150 3.6 / F 1810 2.2	**26** 0034 3.8 / 0648 2.0 / Sa 1314 3.9 / 1952 1.7
12 0409 1.7 / 1030 3.8 / F 1624 2.1 / ☾ 2248 3.9	**27** 0318 1.4 / 0943 4.1 / Sa 1545 1.8 / 2202 4.2	**12** 0526 1.7 / 1150 3.8 / M 1747 2.0	**27** 0451 1.3 / 1120 4.2 / Tu 1727 1.6 / 2346 4.2	**12** 0518 1.8 / 1144 3.8 / W 1749 2.1	**27** 0523 1.6 / 1158 4.1 / Th 1818 1.7	**12** 0031 3.4 / 0652 2.2 / Sa 1312 3.7 / 1942 2.1	**27** 0205 3.9 / 0811 1.9 / Su 1429 4.2 / 2054 1.4
13 0523 1.8 / 1144 3.7 / Sa 1744 2.2	**28** 0423 1.4 / 1049 4.0 / Su 1655 1.8 / ☾ 2309 4.2	**13** 0008 3.8 / 0627 1.8 / Tu 1253 3.8 / 1853 2.0	**28** 0557 1.4 / 1229 4.2 / W 1841 1.5	**13** 0007 3.6 / 0624 1.9 / Th 1252 3.8 / 1902 2.0	**28** 0038 4.0 / 0645 1.7 / F 1317 4.1 / 1941 1.6	**13** 0201 3.6 / 0811 2.0 / Su 1420 3.9 / 2047 1.7	**28** 0307 4.2 / 0904 1.6 / M 1521 4.4 / 2138 1.1
14 0007 3.8 / 0634 1.8 / Su 1259 3.8 / 1856 2.0	**29** 0532 1.4 / 1201 4.1 / M 1810 1.7	**14** 0113 3.8 / 0723 1.7 / W 1348 4.0 / 1951 1.8	**29** 0057 4.2 / 0704 1.4 / Th 1337 4.3 / 1948 1.4	**14** 0120 3.6 / 0730 1.9 / F 1354 3.9 / 2006 1.9	**29** 0158 4.0 / 0802 1.7 / Sa 1427 4.3 / 2050 1.3	**14** 0304 3.8 / 0905 1.7 / M 1512 4.2 / 2134 1.3	**29** 0352 4.4 / 0945 1.3 / Tu 1602 4.7 / 2214 0.8
15 0117 3.8 / 0730 1.6 / M 1355 3.9 / 1951 1.8	**30** 0021 4.2 / 0639 1.3 / Tu 1310 4.2 / 1914 1.5	**15** 0209 3.8 / 0812 1.7 / Th 1436 4.1 / 2040 1.7	**30** 0204 4.3 / 0808 1.3 / F 1436 4.5 / 2049 1.2	**15** 0225 3.7 / 0827 1.8 / Sa 1446 4.0 / 2101 1.7	**30** 0304 4.2 / 0904 1.5 / Su 1524 4.5 / 2143 1.1	**15** 0350 4.1 / 0948 1.4 / Tu 1556 4.5 / 2213 0.9	**30** 0430 4.6 / 1020 1.1 / W 1640 4.8 / 2248 0.6
	31 0127 4.4 / 0738 1.1 / W 1408 4.5 / 2012 1.2				**31** 0357 4.4 / 0953 1.3 / M 1613 4.7 / 2228 0.8		**31** 0505 4.7 / 1055 0.9 / Th 1715 4.9 / ● 2322 0.5

Chart Datum: 0.20 metres below Ordnance Datum (Dublin)

IRELAND, WEST COAST — GALWAY

Lat 53°16′ N Long 9°03′ W

TIMES AND HEIGHTS OF HIGH AND LOW WATERS

TIME ZONE UT (GMT)
For Summer Time add ONE hour in non-shaded areas

YEAR 1989

	SEPTEMBER			OCTOBER			NOVEMBER			DECEMBER		
	Time m	Time m		Time m	Time m		Time m	Time m		Time m	Time m
1 F	0539 4.8 1127 0.7 1749 5.0 2354 0.5	**16** Sa	0518 5.2 1113 0.2 1730 5.6 2334 0.0	**1** Su	0542 4.9 1132 0.8 1753 4.9 2351 0.8	**16** M	0529 5.5 1129 0.3 1749 5.6 2347 0.3	**1** W	0612 4.9 1210 1.2 1829 4.6	**16** Th	0012 1.0 0638 5.3 1246 0.9 1910 5.0	**1** F	0005 1.5 0625 4.9 1232 1.3 1853 4.6	**16** Sa	0049 1.3 0714 5.2 1327 1.0 1948 4.8
2 Sa	0612 4.8 1201 0.7 1824 4.9	**17** Su	0556 5.3 1151 0.2 1812 5.6	**2** M	0612 4.9 1204 0.9 1824 4.8	**17** Tu	0611 5.5 1211 0.4 1834 5.4	**2** Th	0021 1.4 0645 4.8 1243 1.4 1904 4.4	**17** F	0059 1.3 0727 5.1 1338 1.2 2004 4.7	**2** Sa	0043 1.6 0704 4.8 1313 1.4 1935 4.5	**17** Su	0135 1.5 0802 4.9 1416 1.2 2037 4.6
3 Su	0025 0.6 0646 4.8 1234 0.8 1856 4.8	**18** M	0012 0.1 0636 5.3 1232 0.3 1855 5.4	**3** Tu	0021 1.0 0643 4.8 1235 1.1 1856 4.6	**18** W	0028 0.7 0655 5.3 1256 0.7 1921 5.1	**3** F	0055 1.7 0720 4.6 1321 1.6 1944 4.3	**18** Sa	0151 1.7 0819 4.8 1437 1.5 2101 4.4	**3** Su	0127 1.7 0748 4.7 1358 1.5 2022 4.4	**18** M	0225 1.7 0851 4.7 1505 1.5 2127 4.3
4 M	0056 0.8 0719 4.8 1306 1.0 1928 4.6	**19** Tu	0052 0.5 0717 5.1 1314 0.6 1940 5.1	**4** W	0049 1.2 0714 4.6 1306 1.3 1927 4.4	**19** Th	0113 1.1 0741 5.0 1347 1.1 2015 4.7	**4** Sa	0134 1.9 0800 4.4 1406 1.8 2032 4.1	**19** Su	0249 2.0 0918 4.5 1542 1.9 2204 4.2	**4** M	0215 1.9 0837 4.6 1450 1.6 2114 4.3	**19** Tu	0315 1.9 0942 4.4 1559 1.7 ☾ 2220 4.1
5 Tu	0127 1.0 0751 4.5 1338 1.3 2001 4.3	**20** W	0135 0.8 0802 4.9 1402 1.0 2030 4.7	**5** Th	0120 1.5 0747 4.4 1340 1.6 2004 4.1	**20** F	0204 1.6 0834 4.6 1447 1.6 2115 4.3	**5** Su	0223 2.1 0851 4.2 1505 2.0 2131 3.9	**20** M	0356 2.3 1024 4.2 1657 1.9 ☾ 2318 4.0	**5** Tu	0311 2.0 0931 4.5 1548 1.6 2213 4.2	**20** W	0412 2.1 1037 4.2 1657 1.9 2320 4.0
6 W	0158 1.4 0825 4.2 1413 1.6 2036 4.0	**21** Th	0222 1.3 0853 4.5 1458 1.5 2129 4.2	**6** F	0154 1.8 0823 4.2 1422 1.9 2046 3.9	**21** Sa	0305 2.1 0938 4.3 1609 1.9 ☾ 2234 4.0	**6** M	0331 2.3 0953 4.1 1620 2.0 2247 3.9	**21** Tu	0513 2.3 1139 4.1 1808 1.9	**6** W	0414 2.0 1033 4.4 1652 1.6 2320 4.2	**21** Th	0516 2.2 1142 4.0 1801 2.0
7 Th	0232 1.7 0901 4.0 1454 1.9 2117 3.8	**22** F	0321 1.9 0955 4.1 1619 1.9 ☾ 2247 3.9	**7** Sa	0239 2.1 0912 3.9 1519 2.1 2146 3.7	**22** Su	0433 2.4 1102 4.0 1753 2.0	**7** Tu	0458 2.3 1109 4.1 1743 1.9	**22** W	0032 4.1 0629 2.3 1250 4.1 1909 1.8	**7** Th	0527 2.0 1140 4.4 1800 1.6	**22** F	0028 4.0 0628 2.2 1250 3.9 1903 2.0
8 F ☾	0315 2.0 0949 3.8 1550 2.2 2214 3.5	**23** Sa	0449 2.2 1123 3.9 1822 1.9	**8** Su ☾	0349 2.4 1019 3.8 1651 2.2 2316 3.6	**23** M	0012 3.9 0619 2.3 1238 4.1 1910 1.8	**8** W	0012 4.0 0621 2.1 1227 4.3 1853 1.6	**23** Th	0133 4.2 0728 2.1 1348 4.2 1957 1.7	**8** F	0031 4.3 0638 1.8 1250 4.5 1904 1.4	**23** Sa	0130 4.1 0734 2.1 1354 4.0 1958 2.0
9 Sa	0423 2.3 1057 3.6 1723 2.3 2349 3.4	**24** Su	0038 3.8 0655 2.2 1310 4.0 1947 1.7	**9** M	0543 2.4 1150 3.8 1836 2.0	**24** Tu	0131 4.1 0730 2.1 1345 4.2 2001 1.6	**9** Th	0123 4.3 0724 1.8 1331 4.6 1948 1.3	**24** F	0219 4.4 0816 1.9 1434 4.3 2037 1.6	**9** Sa	0134 4.5 0740 1.6 1354 4.7 2002 1.3	**24** Su	0222 4.2 0827 2.0 1446 4.1 2047 1.9
10 Su	0621 2.4 1232 3.6 1919 2.1	**25** M	0202 4.0 0804 2.0 1418 4.2 2037 1.4	**10** Tu	0104 3.7 0713 2.2 1314 4.1 1944 1.6	**25** W	0220 4.3 0815 1.9 1432 4.4 2040 1.4	**10** F	0215 4.6 0815 1.4 1425 4.9 2034 0.9	**25** Sa	0300 4.5 0857 1.7 1515 4.4 2115 1.5	**10** Su	0229 4.8 0834 1.3 1450 4.9 2054 1.1	**25** M	0307 4.4 0915 1.8 1532 4.3 2129 1.8
11 M	0141 3.6 0751 2.1 1354 3.9 2023 1.7	**26** Tu	0253 4.2 0847 1.7 1503 4.4 2115 1.2	**11** W	0209 4.1 0808 1.8 1412 4.4 2030 1.2	**26** Th	0300 4.5 0853 1.6 1510 4.5 2115 1.2	**11** Sa	0300 5.0 0902 1.1 1512 5.2 2118 0.7	**26** Su	0336 4.7 0935 1.5 1553 4.5 2150 1.4	**11** M	0318 5.1 0925 1.0 1542 5.1 2142 1.0	**26** Tu	0348 4.5 0956 1.6 1613 4.4 2209 1.6
12 Tu	0244 3.9 0843 1.7 1447 4.3 2108 1.2	**27** W	0331 4.5 0924 1.4 1541 4.6 2148 0.9	**12** Th	0253 4.5 0850 1.3 1458 4.9 2111 0.7	**27** F	0334 4.7 0928 1.4 1545 4.7 2148 1.1	**12** Su	0342 5.3 0943 0.7 1559 5.4 2202 0.5	**27** M	0410 4.8 1012 1.4 1628 4.6 2224 1.4	**12** Tu	0406 5.3 1014 0.8 1633 5.2 ○ 2230 0.9	**27** W	0424 4.7 1034 1.4 1652 4.5 2247 1.5
13 W	0327 4.3 0922 1.3 1531 4.7 2145 0.7	**28** Th	0404 4.7 0956 1.1 1614 4.8 2220 0.8	**13** F	0332 4.9 0929 0.9 1541 5.2 2150 0.4	**28** Sa	0407 4.8 1002 1.2 1619 4.8 2219 1.0	**13** M ○	0424 5.5 1027 0.5 1644 5.5 2244 0.6	**28** Tu ●	0444 4.9 1047 1.3 1704 4.6 2258 1.4	**13** W	0452 5.4 1104 0.7 1722 5.2 2316 1.0	**28** Th ●	0501 4.8 1111 1.2 1729 4.6 2322 1.4
14 Th	0404 4.7 0959 0.9 1610 5.1 2221 0.3	**29** F ●	0437 4.8 1028 0.9 1648 4.9 2251 0.7	**14** Sa	0410 5.2 1009 0.5 1623 5.5 ○ 2228 0.2	**29** Su ●	0438 4.9 1034 1.1 1652 4.9 2249 1.0	**14** Tu	0506 5.6 1110 0.5 1732 5.5 2327 0.7	**29** W	0516 4.9 1123 1.3 1739 4.7 2330 1.5	**14** Th	0540 5.4 1151 0.7 1811 5.1 2358 1.3	**29** F	0537 4.9 1147 1.1 1805 4.7 2358 1.3
15 F ○	0441 5.0 1035 0.5 1651 5.4 2258 0.1	**30** Sa	0509 4.9 1101 0.8 1720 4.9 2322 0.7	**15** Su	0449 5.4 1048 0.3 1705 5.6 2306 0.2	**30** M ●	0511 5.0 1106 1.0 1819 4.8 2320 1.1	**15** W	0551 5.5 1157 0.6 1819 5.3	**30** Th	0550 4.9 1156 1.3 1815 4.6	**15** F	0003 1.1 0627 5.3 1239 0.8 1900 5.0	**30** Sa	0615 5.0 1224 1.0 1845 4.7
				31 Tu	0542 5.0 1137 1.1 1757 4.7 2350 1.3							**31** Su	0036 1.3 0655 5.0 1303 1.0 1924 4.7

Chart Datum: 0.20 metres below Ordnance Datum (Dublin)

MINOR HARBOURS AND ANCHORAGES 10.13.20

KILKEEL, Down, 54°04' N, 5°59' W, Zone 0 (GMT), Admty chart 2800. HW +0015 on Dover, +0025 on Belfast; HW height +1.6m on Belfast; ML 2.9m; Duration 0620. See 10.13.9. Shelter is complete in inner basin, but it becomes very crowded. Depth off quays approx 1m. There are drying banks both sides of entrance channel and SE gales cause sand bank right across entrance. This is removed by dredging or is slowly washed away in E winds. CG stn is conspic mark for entering, red brick with W flagstaff to W of entrance. Secure in inner basin and see Hr Mr. Breakwater Lt is Fl WR 2s 8m 8M, R296°–313°, W313°–017°, storm signals. Meeney's pier Fl G 3s. VHF Ch 16 12 14 (Mon-Fri: 0900–2000). Facilities: EC Thurs; Hr Mr Tel. 62287 FW on quay; **Shipyard** (between fish market and dock) El, ME, Sh, Slip; **Town** (¾M) Bar, PO, R, V, Gas.

ARDGLASS, Down, 54°16' N, 5°36' W, Zone 0 (GMT), Admty chart 633. HW +0025 on Dover, +0030 on Belfast; HW height +1.1m on Belfast; ML 2.6m; Duration 0620. A rocky bay partly sheltered by breakwater with quays on inside. Further up hr on SW side is old tidal dock giving excellent shelter. Safe harbour except in strong winds from E to S. Yachts should consult Hr Mr who will allocate berth or anchorage clear of fishing vessels. Yachts can go alongside E wall of inner hr. VHF Ch 16 14 12. Lts on inner pier head E of Tidal Dock, Iso WRG 4s 10m 8/5M G shore–310°, W310°–318°, R318°–shore. Breakwater Fl R 3s 10m 5M. Facilities: Hr Mr Tel. 841291; FW (on quay); **Town** P and D nearby; Bar, PO, R, V, Gas.

PORTAVOGIE, Down, 54°27' N, 5°26' W, Zone 0 (GMT), Admty chart 2156. HW +0016 on Dover, +0015 on Belfast; HW height +1.1m on Belfast; ML 2.6m; Duration 0620. See 10.13.10. Good shelter; harbour very full of fishing boats. Beware Plough Rks to S, and McCammon Rks to N of ent. Keep in W sector of Lt on Outer Breakwater Iso WRG 5s 12m 9M G shore-258°, W258°-275°, R275°-348°. Inner Breakwater 2 FG (vert) 6m 4M. About 3 ca NE of Plough Rk is a R can buoy, Bell, Fl(2) R 10s. VHF Ch 16; 14 12 (Mon-Fri: 0900–2000 LT). Facilities: EC Thurs; Hr Mr Tel. 71470; Slip, FW (on central quay) Sh, ME, El; **Town** CH, D, P, PO, R, Gas, V. No licenced premises.

DONAGHADEE, Down, 54°38' N, 5°32' W, Zone 0 (GMT), Admty chart 3709. HW +0025 on Dover, +0020 on Belfast; HW height +0.5m on Belfast; ML 2.2m; Duration 0615. See 10.13.10. Hr is small and very full; scend often sets in. Best berth alongside SE quay. Depth in harbour approx 3m. Beware ledge with less than 2m extends 1.5 ca ENE from S pier head. Alternatively go to the Marina 3 ca S of hr., excellent shelter and facilities but tricky entrance (pilots available). South Pier Lt, Iso WR 4s 17m 18/14M W shore–326°, R326°–shore. No lights on marina. Facilities: Hr Mr Tel. 882377; **Copeland's Marina** Tel. 882184 Access HW∓4; all facilities. **Town** Bar, D, Gas, FW, P, PO, R, V.

CARNLOUGH HARBOUR, Antrim, 54°59' N, 05°58°W, Zone 0 (GMT), Admty chart 2198. HW +0006 on Dover, +0005 on Belfast; HW height –1.6m on Belfast; ML 0.9m; Duration 0625. Small Hr accommodating yachts & small fishing boats, visitors welcome. Good shelter except in SE gales. Entrance dredged annually in May to 1.6m, harbour 2.0m. Entrance difficult in strong onshore winds. Beware rocks which cover at HW on either side of entrance. N pier Lt, Fl G 3s; S Pier Fl R 3s. VHF Ch 16 (occas.). Facilities: EC Wed, Hr Mr Tel. 0574 72313; **Quay** AB, AC (see Hr Mr), FW, L, Slip; **Town** Bar, P and D (cans), Gas, Gaz, PO, R, V.

RED BAY, Antrim, 55°04' N, 6°03' W, Zone 0 (GMT), Admty chart 2199. HW +0006 on Dover, +0022 sp –0010 np on Belfast; HW height –1.7m on Belfast; ML 0.9m; Duration 0625. See 10.13.11. Good anchorage but open to N and E winds. Beware rks and two ruined piers W of Garron Pt. In S and E winds anchor 2 ca off W stone arch near head of bay in approx 3.5m; in N or W winds anchor S of small pier in 2/5m 0.5M NE of Waterfoot village. Facilities: **Cushendall** (1M N of pier) Bar, D, Hosp, P, PO, R, V, Gas.

MULROY BAY, Donegal, 55°15' N, 7°47' W, Zone 0 (GMT), Admty chart 2699; HW (bar) –0455 on Dover, +0100 on Galway; HW height –1.1m on Galway. Beware Limeburner Rk, (marked by buoy) 2½M N of ent and the bar which is dangerous in swell or onshore winds. Channel runs between Black Rks and Sessiagh Rks thence through First, Second and Third Narrows (with strong tides) to Broad Water. HW at the head of the lough is 2¼ hrs later than at the bar. Anchorages: Close SW of Ravedy Is (Fl 3s); Fanny's Bay (2m), excellent; Rosnakill Bay (3.5m) in SE side; Cranford Bay; Milford Port (3 to 4m). Beware electric cable 6m, over Moross channel, barring North Water to high masted boats. Facilities: **Milford Port** AB, FW, V; **Fanny's Bay** PO, Shop at Downings village (1M), hotel at Rosepenna (¾M).

SHEEP HAVEN, Donegal, 55°11' N, 7°51' W, Zone 0 (GMT), Admty chart 2699. HW –0515 on Dover, +0050 on Galway; HW height –1.0m on Galway. See 10.13.15. Bay is 4M wide with numerous anchorages, easily accessible in daylight, but exposed to N winds. Beware rocks for 3ca off Rinnafaghla Pt; also Wherryman Rocks, which dry, 1ca off E shore 2¼M S of Rinnafaghla Pt. Anchor in Downies (or Downings) Bay to SW of pier; in Pollcormick Inlet close W in 3m; in Ards Bay for excellent shelter, but beware the bar in strong winds. Lights: Portnablahy Ldg Lts 125°, both Oc 6s 7/12m 2M; Downies pier head Fl R 3s 5m 2M. Facilities: (Downies) EC Wed, V, FW, P(cans, 300m), R, Bar; (Portnablahy Bay) V, P (cans), R, Bar.

BURTONPORT, Donegal, 54°59' N, 8°26' W, Zone 0 (GMT), Admty charts 2792, 1879. HW –0525 on Dover, +0050 on Galway; HW height –0.9m on Galway; ML 2.2m; Duration 0605. See 10.13.16. Normal ent. via North Channel. Hr very full. Berth on local boat on pier or go to Rutland Hr or Aran I. Ent. safe in all weathers except NW gales. N Channel Ldg Lts 119° on Inishcoo, Front Iso 6s 6m 1M. Rear, 248m from front, Iso 6s 11m 1M. Rutland I Ldg Lts 138° Front Oc 6s 8m 1M. Rear, 330m from front, Oc 6s 14m 1M. Burtonport Ldg Lts 068°, Front FG, 17m 1M, rear, 355m from front, FG 23m 1M. Facilities: D (just inside pier), FW (root of pier), P (0.5M inland), **Village** Bar, PO, R, V, Kos.

BLACKSOD BAY, Mayo, 54°06' N, 10°04' W, Zone 0 (GMT), Admty chart 2704. HW –0525 on Dover, +0030 on Galway; HW height –1.1m on Galway; ML 2.2m; Duration 0610. See 10.13.17. Safe anchorage with no hidden dangers, accessible by day or night. Beware Rk which dries 3.5 ca SSE of Ardmore Pt. Good anchorages at Elly Bay (1.8m); NW of Blacksod Quay (3m); Saleen Bay; Elly Hr; N of Claggan Pt. Blacksod Pier Lt, Fl(2) WR 7.5s 13m 12/9M, R189°–, W210°–018°. There are no facilities but normal supplies can be obtained at Belmullet 2.5M N of Claggan Pt.

CLIFDEN BAY, Galway, 53°29' N, 10°01' W, Zone 0 (GMT), Admty chart 2708. HW –0600 on Dover, +0005 on Galway; HW height –0.6m on Galway; ML 2.3m; Duration 0610. See 10.13.18. It is essential to identify the high W beacon on Carrickarana Rks, 2.8M SW of Fishing Pt before entering. Ldg Marks 080° W Bn on Fishing Pt in line with Clifden Castle. Beware bar at ent by Fishing Pt and another SE of creek going up to Clifden; also Doolick Rks, Coghan Rks and Rks off Errislannon Pt. Anchor between Larner Rks and Drinagh Pt, dry out alongside Clifden Quay or anchor beyond Yellow Slate Rks in 3.4m in Ardbear Bay. Keep clear of fish farming cages. Facilities: EC Thurs; **A. O'Connell** Tel. 166, El, FW, M, ME, Sh, Slip; **Aster Boats** Tel. 21332 Slip, L, FW, D, AB, C, CH, M, ME, El, Sh; **Town** Bar, Bank, CH, D, P, PO, R, V, Kos, FW, V, R, Dr, Hosp.

KILRONAN, Aran Islands, 53°07' N, 9°39' W, Zone 0 (GMT), Admty chart 3339 HW –0555 on Dover, –0008 on Galway, HW height –0.4m on Galway; ML 2.6m; Duration 0610. Good shelter except in E or NE winds but harbour very crowded with fishing boats. Entrance well marked with Straw Is Lt Ho and Stbd hand buoy marking E end of Bar of Aran. Beware foul ground N and W of Straw Is. Anchor near LB off Kilronan Pier. Lts – N side of Straw Is, Fl(2) 5s, 11m, 17M. Pierhead, Fl WG 1.5s, G240°-326°, W326°-000°. Facilities: D, FW, Ferry to Galway.

VOLVO PENTA SERVICE

Sales and service centres in area 14
Names and addresses of Volvo Penta dealers in this area are available from:

FRANCE **Volvo Penta France SA**, BP45, F78130 Les Mureaux Tel (01) 3 474 72 01, Telex 695221 F.
SPAIN **Volvo Concesionarios SA**, Paeso De La Castellana 130, 28046 Madrid 16, Tel 1-262 2207, Telex 23296 VOLCO E.

VOLVO PENTA

Area 14

South Biscay
Spanish Border to Pornic

10.14.1	Index	Page 583
10.14.2	Diagram of Radiobeacons, Air Beacons, Lifeboat Stations etc	584
10.14.3	Tidal Stream Charts	586
10.14.4	List of Lights, Fog Signals and Waypoints	588
10.14.5	Passage Information	590
10.14.6	Distance Table	591
10.14.7	Special notes for French areas	592
10.14.8	Tidal information for North Spain	594
10.14.9	Arcachon	594
10.14.10	Port Bloc/La Gironde (Pointe de Grave, Standard Port, Tidal Curves)	596
10.14.11	Canal connections	600
10.14.12	Royan	601
10.14.13	Seudre River (Marennes, La Tremblade)	601
10.14.14	Ile d'Oléron (Le Chateau, Boyardville)	603
10.14.15	Rochefort	603
10.14.16	La Rochelle	604
10.14.17	St Martin, Ile de Ré	605
10.14.18	Bourgenay	606
10.14.19	Les Sables d'Olonne	607
10.14.20	St Giles-Croix-de-Vie	608
10.14.21	Ile d'Yeu (Port Joinville)	609
10.14.22	Ile de Noirmoutier (L'Herbaudière)	610
10.14.23	Pornic	611
10.14.24	Minor Harbours and Anchorages Hendaye St Jean-de-Luz Anglet Bayonne Capbreton La Vigne Pauillac Bordeaux Blaye Douhet Ile d'Aix La Flotte, Ile de Ré Ars-en-Ré L'Aiguillon-la Faute-sur-Mer Jard-sur-Mer Fromentine	612

14

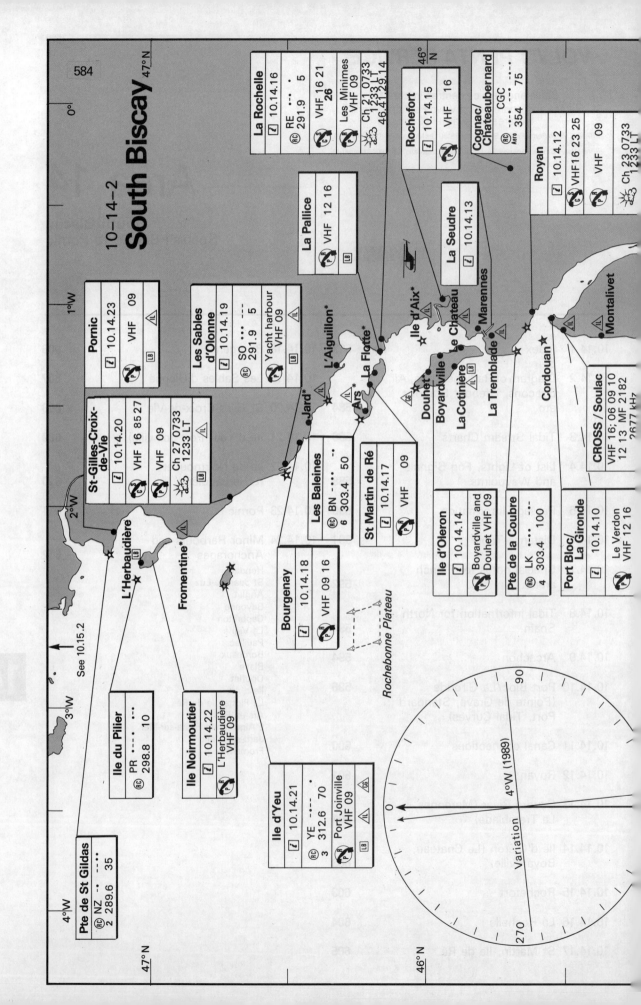

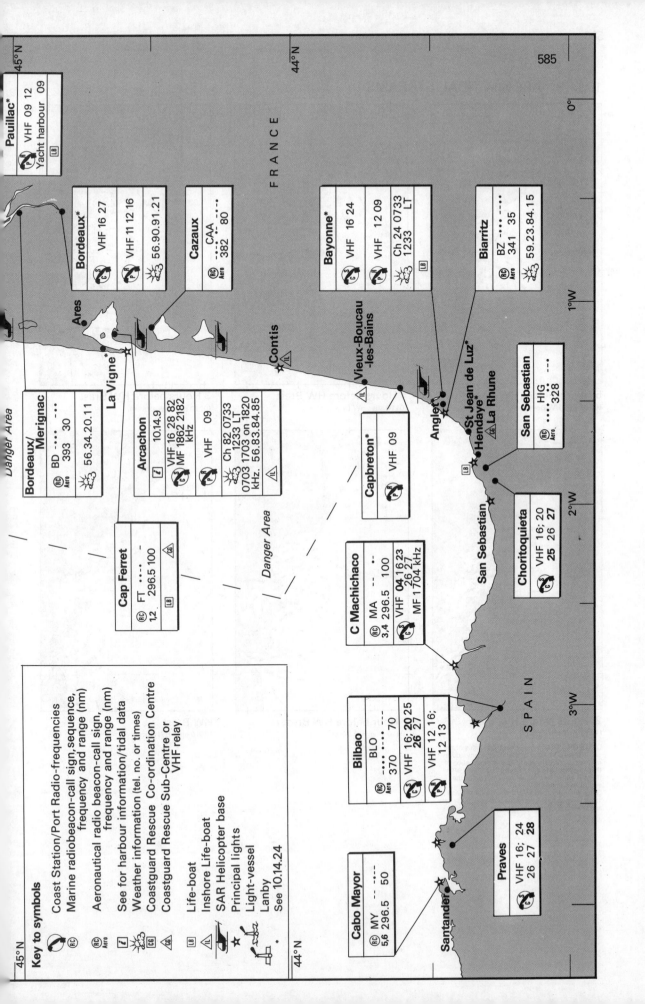

10.14.3 AREA 14 TIDAL STREAMS

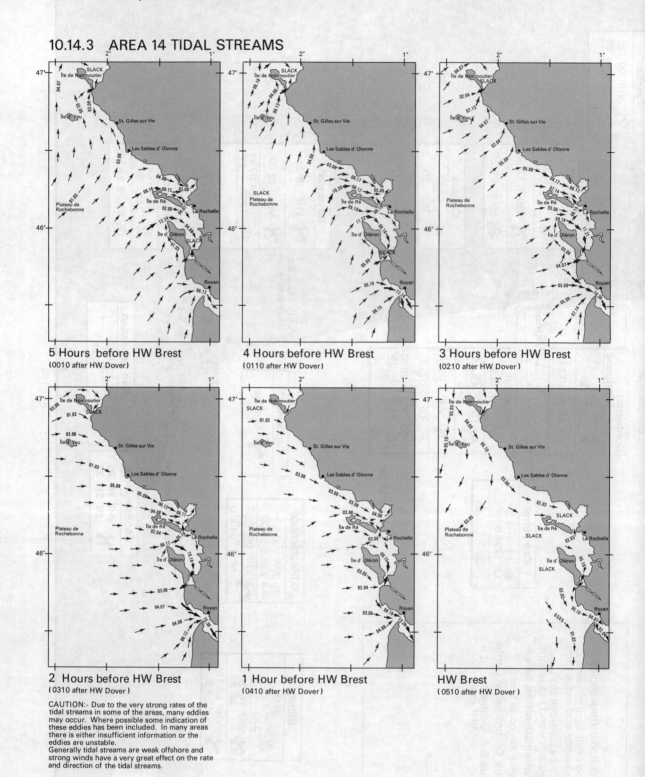

CAUTION:- Due to the very strong rates of the tidal streams in some of the areas, many eddies may occur. Where possible some indication of these eddies has been included. In many areas there is either insufficient information or the eddies are unstable.
Generally tidal streams are weak offshore and strong winds have a very great effect on the rate and direction of the tidal streams.

AREA 14—S Biscay 587

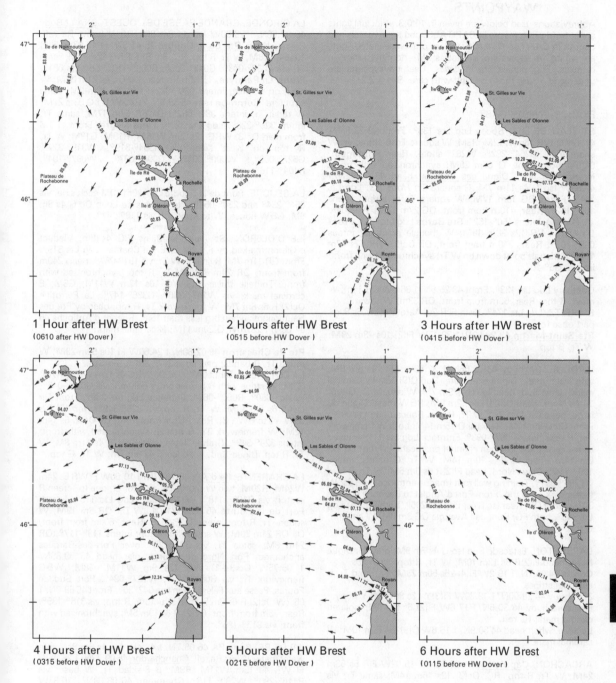

10.14.4 LIGHTS, FOG SIGNALS AND WAYPOINTS

Abbreviations used below are given in 10.0.3. Principal lights are in **bold** print, places in CAPITALS, and light-vessels and Lanbys in *CAPITAL ITALICS*. Unless otherwise stated lights are white. m—elevation in metres; M—nominal range in n. miles. Fog signals are in *italics*. Useful waypoints are underlined — use those on land with care. See 4.2.2.

FRANCE—WEST COAST

ST JEAN DE LUZ, Socoa Ldg Lts 138°. Front 43.23.8N/1 41.1W Q WR 36m W12M, R8M; W square Tr, B stripe; vis W shore-264°, R264°-282°, W282°-shore. Rear **Bordagain**, 0.79M from front, Q 67m **20M**; synchronised with front; intens 135°-141°. Digue des Criquas, head 43 23.92N/1 40.59W Iso G 4s 11m 7M; G square Tr; *Horn 15s*. Ldg Lts 151°, **Front** QG 18m **17M**; W square Tr, R stripe; intens 150°-152°. **Rear**, 410m from front, QG 27m **17M**; W square Tr, G stripe; intens 150°-152°. **Ste Barbe** Ldg Lts 101°. **Front** Dir Oc(3+1)R 12s 30m **18M**; W triangle point up; intens 095°-107°. **Rear**, 340m from front, Dir Oc(3+1)R 12s 47m **18M**; B triangle point down on W Tr; synchronised with front; intens 095°-107°.

Guethary Ldg Lts 133°. Front 43 25.6N/1 36.5W QR 11m; W mast, R top. Rear, 66m from front, QR 33m; W tower.
BIARRITZ. Ldg Lts 174°. Both Fl R 2s. Aero Mo(L)7.5s 80m; part obsc (occas).
Pte Saint-Martin 43 29.69N/1 33.17W Fl(2) 10s 73m **29M**; W Tr, B top.

L'ADOUR. BA buoy 43 32.66N/1 32.68W LFl 10s 8m 8M; safe water mark. Digue exterieure, Sud Q(9) 15s; W cardinal mark. Jetée Sud, head Iso G 4s 9m 7M; W square Tr, G top. Jetée Nord, head Oc(2)R 6s 12m 8M; R&W mast. Digue du large, head QR 11m 7M; W Tr, R top. Boucau Ldg Lts 090°. Front Dir Q 9m 14M. Rear Dir Q 15m 14M; both W framework Trs, R tops, intens 087°-093°. Entrance Ldg Lts 111° (moved as necessary and lit when channel practicable). Front Dir FG 6m 14M. Rear Dir FG 10m 14M; W Tr, G bands; intens 109°-114°. Digue Nord, head Fl(2)R 6s 9m 6M; W Tr, R top; vis 296°-091°. Training wall root (marina entrance) Fl G 2s; W framework Tr, G top. From Port d'Anglet to Bayonne Ldg Lts 322°, Front QR, Rear Iso R 4s; Ldg Lts 205°, both Dir FG; Ldg Lts 345°, both FG; Pont de l'Aveugle QG 8m 8M; W column, G top.

CAPBRETON. Estacade Sud Iso G 4s 7m 9M; grey Tr. Digue Nord, head Fl(2)R 6s 13m 10M; W Tr, R top; *Horn 30s*.
Contis 44 05.7N/1 19.2W Fl(4) 25s 50m **23M**; W round Tr, B diagonal stripes.
ZDS buoy 44 28.00N/1 19.30W Fl(3)Y 12s 8m 7M; Y buoy.
Emissaire buoy 44 30.5N/1 17.6W Fl(2) 6s 8m 5M; isolated danger mark; Ra refl.
La Salie, wharf head 44 30.9N/1 15.6W Q(9)15s 19m 10M; W cardinal mark.

ARCACHON. **Cap Ferret** 44 38.83N/1 15.02W FlR 5s 53m **24M**; W Tr, R top; RC. Oc(3) 12s 46m 14M; same Tr; vis 045°-135°. Arcachon, West breakwater head QG. E breakwater head QR. Port de la Vigne Iso R 4s 7m 4M.
Hourtin 45 08.5N/1 09.7W Fl 5s 55m **24M**; R square Tr.

LA GIRONDE, PASSE SUD. **Cordouan** 45 35.25N/1 10.34W Oc(2+1) WRG 12s 60m **W21M**, **R17M**, **G17M**; W conical Tr, dark grey band and top; vis W014°-126°, G126°-178°, W178°-250°, W(unintens)250°-267°, R(unintens)267°-294°, W294°-014°; obsc in estuary when brg more than 285°. Ldg Lts 063°. **St Nicolas Front** 45 33.80N/1 04.93W Dir QG 22m **17M**; W square Tr; intens 060°-066°. **Rear Pte de Grave**, 0.84M from front, Oc WRG 4s 26m **W19M**, R15M, G15M; W square Tr, B corners and top; vis W(unintens)033°-054°, W054°-234°, R234°-303°, W303°-312°, G312°-330°, W330°-341°, W(unintens) 341°-025°. Ldg Lts 041°, **Le Chay Front**, 45 37.35N/1 02.40W Dir QR 33m **18M**; W Tr, R top; intens 039°-042°. **Rear St Pierre**, 0.97M from front, Dir QR 61m **18M**; R water Tr; intens 039°-043°.

LA GIRONDE, GRANDE PASSE DE L'OUEST. BXA Lt Buoy 45 37.60N/1 28.60W Iso 4s 8m 8M; R&W buoy; Ra refl; Racon; *Whis*. **Pte de la Coubre** 45 41.87N/1 13.93W Fl(2) 10s 64m **28M**; W Tr, R top; RC; Sig Stn. F RG 42m 12M; same Tr; vis R030°-043°, G043°-060°, R060°-110°. Ldg Lts 081°. **Front, La Palmyre**, 1.1M from rear, Dir Oc 4s 21m **22M**; W mast on dolphin; intens 080°-082°; Q(2) 5s 10m 3M; same structure. **Common rear** 45 39.8N/1 07.2W Dir Q 53m **27M**; W radar Tr; intens 080°-082°. Dir FR 57m **17M**; same Tr; intens 326°-328°. Ldg Lts 327°. **Terre-Nègre**, Front, 1.1M from rear, Oc(3) WRG 12s 39m **W16M**, R13M, G13M; W Tr, R top on west side; vis R304°-319°, W319°-327°, G327°-000°, W000°-004°, G004°-097°, W097°-104°, R104°-116°.

LA SEUDRE. Pont de la Seudre QW 20m 10M each side, vis 054°-234° and 234°-054°. Pte de Mus de Loup Oc G 4s 8m 6M; G&W house, W to seaward; vis 118°-147°.

ILE D'OLERON. St Trojan-les-Bains FlG 4s 8m; Viaduct d'Oleron marked by F 1.1M ENE. Le Chateau Ldg Lts 319°. Front QR 11m 7M; R line on W Tr; vis 191°-087°. Rear, 240m from front, QR 24m 7M; W Tr, R top; synchronised with front. Tourelle Juliar Q(3)WG 10s 12m W11M; G8M; E cardinal mark; vis W147°-336°, G336°-147°. La Pérrotine Oc(2) R 6s 8m 7M; W framework Tr, R top; obsc by Pte des Saumonards when brg less than 150°. Rocher d'Antioche 46 04.00N/1 23.70W Q 20m 11M; N cardinal mark.

Pte de Chassiron 46 02.80N/1 24.60W Fl 10s 50m **28M**; W round Tr, B bands; part obsc 297°-351°; Sig Stn.
LA COTINIERE. Dir Lt 048°. Dir Oc WRG 4s 13m W9M, R7M, G7M; W stripe with B border on W column; vis G033°-046°, W046°-050°, R050°-063°. Entrance Ldg Lts 339°. Front Dir Oc(2) 6s 6m 13M; W Tr, R top; vis 329°-349°; *Horn(2) 20s* (HW−3 to HW+3). Rear, 425m from front, Dir Oc(2) 6s 14m 12M; W framework Tr, R bands; synchronised with front; intens 329°-349°. Grande Jetée, elbow Oc R 4s 11m 8M; W Tr, R top. Digue Sud, head Iso G 4s 9m 7M; W Tr, G top.

LA CHARENTE. **Ile d'Aix** 46 00.67N/1 10.60W FlWR 5s 24m **W24M**, **R20M**, two W round Trs, one for Lt, one to screen R sector; vis R103°-118°, W118°-103°. Ldg Lts 115°. **Front, Fort de la Pointe** 45 58.0N/1 04.3W Dir QR 8m **19M**; W square Tr, R top; intens 113°-117°. **Rear**, 600m from front, Dir QR 21m **20M**; W square Tr, R top; intens 113°-117°. QR 21m 8M; same Tr; vis 322°-067° over Port-des-Barques anchorage. Port Nord de Fouras, pier head 45 59.88N/1 05.75W Oc(3+1) WG 12s 9m W11M, G8M; W&G framework Tr; vis G084°-127°, W127°-084°. Port Sud de Fouras, Passe aux Filles, Lts in line 042°30'. Front 45 59.0N/1 05.7W Oc(2) R 6s 8m 5M; W structure, R top; vis 315°-135°. Rear, 75m from front, Oc(2) R 6s 19m 5M; synchronised with front; vis 315°-135°.

ILE DE RÉ. Buoy PA 46 05.7N/1 42.4W Iso 4s 8m 8M; safe water mark; *Whis*; Ra refl. Chanchardon 46 09.72N/1 28.45W Fl WR 4s 15m W11M, R9M; B 8-sided Tr, W base; vis R118°-290°, W290°-118°. Chauveau 46 08.09N/1 16.33W Oc(2+1) WR 12s 23m **W15M**, R11M; W Tr, R top; vis W057°-094°, R094°-104°, W104°-342°, R342°-057°. Ferry landing 46 10.0N/1 15.4W F Vi 6m 1M; Vi column on dolphin. Pte de Sablanceaux 46 09.82N/1 15.08W Q Vi 7m 1M; W mast and hut, G top. Rivedoux-Plage Ldg Lts 200°. Front QG 6m 6M; W Tr, G top. Rear, 100m from front, QG 9m 7M; W and G chequered column. La Flotte 46 11.3N/1 19.3W Fl WG 4s 10m W12M, G9M; W Tr, G top; vis G130°-205°, W205°-220°, G220°-257°; *Horn(3) 30s* (by day HW−2 to HW+2).

St MARTIN DE RÉ 46 12.5N/1 21.9W on ramparts E of entrance Oc(2) WR 6s 18m W10M, R7M; W Tr, R top; vis W shore-245°, R245°-281°, W281°-shore. Mole head, 90m 328°, Iso G 4s 10m 6M; W tripod, G top; obsc by Pte de Loix when brg less than 124°. Le Fier d'Ars Ldg Lts 265°. Front 46 14.0N/1 28.8W Iso 4s 8m 10M; W stripe on grey framework Tr; vis 141°-025°. Rear, 237m from front, Dir Iso G 4s 12m 13M; G square Tr on dwelling; synchronised with front, intens 263°-267°. ARS-EN-RÉ Ldg Lts 232°. Front 46 12.8N/1 30.5W Q 5m 9M; W hut, R lantern. Rear, 370m from front, Q 13m 11M; B stripe on W framework Tr, G top; vis 142°-322°. <u>Les Baleines</u> 46 14.70N/1 33.60W Fl(4) 15s 53m **27M**; grey 8-sided Tr, R lantern; RC. <u>Les Baleineaux</u> 46 15.85N/ 35.20W Oc(2) 6s 23m 11M; pink Tr, R top.

LA ROCHELLE. <u>Chauveau Lt Buoy</u> 46 06.62N/1 15.98W VQ(6) + LFl 10s; S cardinal mark; *Whis.* <u>Roche du Sud Lt Buoy</u> 46 06.43N/1 15.15W Q(9) 15s; W cardinal mark. <u>Le Lavardin</u> 46 08.15N/1 14.45W Fl(2) WG 6s 14m W11M, G8M; B Tr, R band; vis G160°-169°, W169°-160°. <u>Tour Richelieu</u> 46 08.95N/1 10.27W Fl(4)R 12s 10m 9M; R Tr; RC; *Siren(4) 60s* (HW − 1 to HW + 1). Ldg Lts 059°. Front 46 09.4N/1 09.1W Dir Q 15m 14M; R Tr, W bands; intens 056°-062°; by day Fl 4s. Rear, 235m from front, Q 25m 14M; W 8-sided Tr, G top; synchronised with front, vis 350°-125°, obsc 061°-065° by St Nicolas Tr; by day Fl 4s.

LA PALLICE. <u>Mole d'Escale, head,</u> 46 09.42N/1 14.43W Dir Lt 016°. DirQ WRG 33m W14M, R13M, G13M; grey Tr; vis G009°-015°, W015°-017°, R017°-031°. Mole, SE corner Oc(2)R 6s 7m 6M; W framework Tr, R top. Mole, NW corner Fl G 4s 5m 6M; G structure. Oil jetty, head, Q(6) + LFl 15s 6m 9M; S cardinal mark. Bassin Chef-de-Baie, head Fl(3)G 12s 13m 9M; W Tr, G top. Jetée Sud *Reed(3) 30s.* Ldg Lts 126°, both QG. Avant Port, Jetée Nord, head Oc(2)R 6s 14m 5M; pylon, R top. Basin Ldg Lts 085°, both QR 7M; synchronised.

PERTUIS BRETON, NORTH SHORE. Port du Plomb, West mole 46 12.4N/1 12.1W Fl R 4s 9m 7M; W column R top. Sèvre Niortaise, entrance, Port du Pavé 46 18.1N/1 08.0W Fl G 4s 9m 7M; W column, G top. La Tranche-sur-Mer, pier head 46 20.7N/1 25.5W Fl(2) R 6s 6m 7M; R column. <u>Pte du Grouin-du-Cou</u> 46 20.68N/1 27.80W Fl WRG 5s 29m **W22M, R18M, G18M**; W 8-sided Tr, B top; vis R034°-061°, W061°-117°, G117°-138°, W138°-034°.

PLATEAU DE ROCHEBONNE. <u>Rochebonne NW buoy</u> 46 12.9N/2 31.9W Q(9) 15s 8m 8M; W cardinal mark; *Whis*; Ra refl. <u>Rochebonne SW buoy</u> 46 10.1N/2 27.0W Fl(2)R 6s 9m 5M; R buoy; Ra refl. <u>Rochebonne SE buoy</u> 46 09.2N/2 21.2W Q(3) 10s 8m 8M; E cardinal mark; *Bell*; Ra refl. <u>Rochebonne NE buoy</u> 46 12.7N/2 25.0W Iso G 4s 8m 5M; G buoy; Ra refl. BOURGENAY. Ldg Lts 040°. Front 46 26.4N/1 40.5W QG 9M. Rear QG 9M. Digue West, head Fl R 4s 9M.

LES SABLES D'OLONNE. <u>Nouch Sud Lt Buoy</u> 46 28.63N/ 1 47.43W Q(6) + LFl 15s; S cardinal mark. PASSE DU SW, Ldg Lts 033°. **Front** 46 29.5N/1 46.3W Iso R 4s 14m **16M**; mast; H24. **Rear**, 330m from front, Iso R 4s 33m **16M**; W square Tr; H24. Ldg Lts 320°, <u>Jetée des Sables, Front</u> 46 29.45N/1 47.52W QG 11m 10M; W Tr, G top. Rear, Tour de la Chaume Oc(2 + 1) 12s 33m 13M; large grey Tr, W turret; RC. Ldg Lts 327°, Front FR 6m 5M; R line on W hut. Rear, 65m from front, FR 9m 11M; R line on W Tr; intens 324°-330°. Jetée St Nicolas, head UQ(2)R 1s 16m 10M; W Tr, R top; vis 094°-043°. **L'Armandèche** Fl(2 + 1) 15s 42m **23M**; W 6-sided Tr, R top; vis 295°-130°. *LA PETITE BARGE LANBY* 46 28.9N/1 50.6W Q(6) + LFl 15s 8m 7M; S cardinal mark; *Whis*; Ra refl. **Les Barges** 46 29.7N/1 50.4W Fl(2) R 10s 25m **17M**; grey Tr, helicopter platform; vis 265°-205°.

St GILLES-SUR-VIE. <u>Pill'Hours Lt Buoy</u> 46 41.1N/1 58.2W Q(6) + LFl 15s; S cardinal mark; *Bell*. Ldg Lts 043°. Front Oc(3 + 1)R 12s 7m **15M**; W square Tr, R top; intens 033°-053°. **Rear**, 260m from front, Oc(3 + 1)R 12s 28m **15M**; W square Tr, R top; synchronised with front, intens 033°-053°. Jetée de la Garenne, head Iso WG 4s 7m W8M, G5M; G Tr; vis G220°-335°, W335°-220°. Jetée de Boisvinet Fl(2) WR 6s 8m W10M, R7M; R column; vis R045°-225°,

W225°-045°; *Reed(2) 20s.* **Pte de Grosse Terre** Fl(4) WR 12s 25m **W18M**, R14M; W truncated conical Tr; vis W290°-125°, R125°-145°.
St Jean de Monts, jetty head 46 47.1N/2 05.1W Q(2) R 5s 10m 3M; W mast, R top.

ILE D'YEU. **Pte des Corbeaux** 46 41.4N/2 17.1W Fl(2 + 1)R 15s 25m **18M**; W square Tr, R top; obsc 083°-143°. PORT JOINVILLE, jetty NW head Oc(3) WG 12s 6m W11M, G9M; W 8-sided Tr, G top; vis G shore-150°, W150°-232°, G232°-279°, W279°-285°, G285°-shore; *Horn(3) 30s.* Quai de Canada. Ldg Lts 219° both QR; vis 169°-269°. Quai de Canada, head Iso G 4s 7m 6M. <u>Les Chiens Perrins</u> 46 43.6N/ 2 24.6W Q(9)WG 15s 16m W8M, G4M; W cardinal mark; vis G330°-350°, W350°-200°. Pte du Butte *Horn 60s.* **Petite Foule** 46 43.1N/2 22.9W Fl 5s 56m **24M**; W square Tr, G lantern; RC. La Meule 46 41.7N/2 20.6W Oc WRG 4s 9m W9M, R6M, G5M; grey square column, R top; vis G007°-018°, W018°-027°, R027°-041°.

FROMENTINE. Pte de Notre Dame-de-Monts 46 53.3N/2 08.5W Oc(2) WRG 6s 21m W13M, R10M, G10M; W Tr, B top; vis G000°-043°, W043°-063°, R063°-073°, W073°-094°, G094°-113°, W113°-116°, R116°-175°, G175°-196°, R196°-230°. **Bridge**, each side on centre span Iso 4s 32m **18M**; H24. Tourelle Milieu 46 53.6N/2 09.6W Fl(4)R 12s 6m 5M; R Tr.

ILE DE NOIRMOUTIER. Passage du Gois, E shore Fl R 4s 6m 6M; R hut; vis 038°-218°. E turning point Fl 2s 5m 6M; grey pyramid structure. W turning point Fl 2s 5m 3M; grey pyramid structure; Bassotière Fl G 2s 7m 2M; W tripod, G lantern; vis 180°-000°. Noirmoutier, jetty Oc(2) R 6s 6m 7M. **Pte des Dames** 47 00.7N/2 13.3W Oc(3) WRG 12s 34m **W19M, R15M, G15M**; W square Tr; vis G016°-057°, R057°-124°, G124°-165°, W165°-191°, R191°-267°, W267°-357°, R357°-016°. <u>Pierre Moine</u> 47 03.43N/2 12.30W Fl(2) 6s 14m 9M; isolated danger mark. <u>Basse du Martroger</u> 47 02.65N/2 17.05W Q WRG 10m W9M, R6M, G6M; N cardinal mark; vis G033°-055°, W055°-060°, R060°-095°, G095°-124°, W124°-153°, R153°-201°, W201°-240°, R240°-033°. Port de l'Herbaudière Jetée Ouest, head Oc(2 + 1) WG 12s 9m W10M, G7M; W column and hut, G top; vis W187°-190°, G190°-187°. Jetée Est, head Fl(2) R 6s 8m 5M; R tripod.
Ile du Pilier 47 02.62N/2 21.53W Fl(3) 20s 33m **29M**; grey square Tr. Auxiliary Lt QR 10m 11M, same Tr; vis 321°-034; *Reed(3) 60s.* Pte de Devin 46 59.1N/2 17.6W Oc(4) WRG 12s 10m W11M, R8M, G8M; W column and hut, G top; vis G314°-028°, W028°-035°, R035°-134°.

BAIE DE BOURGNEUF. Bec de l'Épois 46 56.4N/2 04.5W Dir Iso WRG 4s 6m W12M, R9M, G8M; W square Tr, R top; vis G106°-113°, R113°-122°, G122°-157°, W157°-158°, R158°-171°, W171°-176°. Étier des Brochets 46 59.9N/2 01.9W Oc(2 + 1) WRG 12s 8m W10M, R7M, G7M; G Tr, W band; vis G071°-091°, W091°-102°, R102°-116°, W116°-119°, R119°-164°. Le Collet 47 01.8N/1 59.0W Oc(2) WR 6s 7m W9M, R6M; vis W shore-093°, R093°-shore. Ldg Lts 118° both QG 6M; W pylons, G tops. La Bernerie-en-Retz 47 04.6N/2 02.4W Fl R 2s 3m 2M; W structure, R top.

PORNIC Jetée SW, elbow Fl(2 + 1) 7s 4m 3M. <u>Head</u> 46 06.53N/2 06.61W Fl(2) R 6s 4m 2M; B&R column. Jetée Est, head Fl G 2s 4m 2M; B&G column. Pte de Noveillard Oc(3 + 1) WRG 12s 22m W14M, R10M, G10M; W square Tr, G top, W dwelling; vis G shore-051°, W051°-079°, R079°-shore. Pte de Gourmalon, breakwater head Fl(2).G 4m 8M; W mast, G top.

Pointe de St-Gildas 47 08.10N/2 14.67W Q WRG 23m **W15M**, R12M, G10M; framework Tr on W house; vis R264°-308°, G308°-078°, W078°-088°, R088°-174°, W174°-180°, G180°-264°; RC.

10.14.5 PASSAGE INFORMATION

For general notes on French waters see 10.14.7. Further passage information appears in 10.15.5. A glossary of French terms used on charts or in sailing directions is given in 10.16.5.

BAY OF BISCAY (chart 1104)

Despite its reputation, weather in B of Biscay is no worse than in English Chan, and the S part generally enjoys a warmer and more settled climate. Bigger seas and longer swells from the Atlantic may be met. Although W winds mostly prevail, NE winds are often experienced with anticyclones over the continent. In summer the wind is seldom from SE or S. Often the wind varies in speed and direction from day to day. Gales may be expected once a month in summer. S of La Gironde, sea and land breezes are well developed in summer months. Coastal rainfall is moderate, increasing in the SE corner, where thunder is more frequent. Sea fog may be met from May to October, but is less common in winter.

The general direction of the surface current in summer is SE, towards the SE corner of B of Biscay, where it swings W along N coast of Spain. Rate and direction much depend on wind. With W gales in winter, the current runs E along N coast of Spain, sometimes at 3kn or more. When crossing B of Biscay, allow for a likely set to the E, particularly after strong W winds. Tidal streams are weak offshore, but can be strong in estuaries and channels, and around headlands. It is wise to study the prevailing current and stream from fish floats etc.

Larger scale French charts are often more suitable for inshore waters. The *Bay of Biscay Pilot* is recommended, as are *South Biscay Pilot* and *North Biscay Pilot* (Adlard Coles Ltd) and *French Pilot Volume Four* (Nautical Books).

BAIE DE FONTARABIE (chart 1343)

Known to the Spanish as Rada de Higuer, this lies on the border of Spain and France, and is entered between Cabo Higuer (a bare, rugged cape with Lt Ho, connected by drying reefs to Isla Amuitz, 24m high) and Pte Ste Anne 1¾M ESE. Les Briquets (dry) lie 1M N of Pte Ste Anne. In strong W winds pass to W of Banc Chicharvel in entrance to b.

In the middle of the roadstead is a neutral area, marked by Bns and shown on chart. Outside this area the boundary line (approximately 1°46′.2W) runs N from a white pyramid on the S shore, about 1M SW of Pte Ste Anne. Ria Fuenterrabia is entered between breakwaters in SW corner of the b, giving access to Hendaye-Plage (see 10.14.24). Entry should not be attempted with strong onshore winds or heavy swell.

POINTE SAINT ANNE TO L'ADOUR (chart 2665)

In bad weather the sea breaks on Plateau de St Jean de Luz, a chain of rky shoals lying 1-4M offshore along this coast, which as far as Pte St Martin has mostly a sandy beach and rky cliffs, with mountains inland.

St Jean de Luz (chart 1343 and 10.14.23) may be approached through four passes. Passe d'Illarguita (between Illarguita and Belhara Perdun) is recommended in bad weather: follow the 138° transit (Le Socoa in line with Bordagain) until the Ste Barbe Ldg Lts (101°) are in line, and thence by Passe de l'Ouest on the 151° transit of the St Jean de Luz Ldg Lts.

N of Pte St Martin the coast changes to sand dunes. The sea breaks over Loutrou shoal in strong W winds. At L'Adour entrance the flood runs E and SE, sp rate 2-4 kn; the ebb runs W, sp rate 3-5 kn.

L'ADOUR TO LA GIRONDE (charts 2665, 2664)

Apart from the entrance to Arcachon (10.14.9), this is a featureless coast bordered by sand dunes and fir trees, with no shelter from W winds. 5M offshore a current usually sets N at about ½ kn, particularly with a S wind: in winter this may be stronger after W winds. Within 1M of the coast there may be a counter-current setting S.

A submerged canyon, the Fosse (or Gouf) de Capbreton, runs at right angles to and within 2M of the coast. In strong W winds a dangerous sea develops along the N and S sides of this. A firing area lies offshore, bounded to the S by a line 295° from Capbreton Bn, and to the N by a line 245° from Pte de la Négade Bn, and extending to 2°25′ W. The inshore limit is 3M off, but connected to the shore at three places, two between Biscarosse and Pte d'Arcachon, and one at Hourtin. The area is divided into 31S and 31N, which are S and N of a clear chan 8M wide running 270° from Arcachon. 31S and 31N are further divided into areas indicated by distance from coast. Thus, 31S 27.45 means the S section, 27-45M offshore. Areas in use are bcst by Bordeaux-Arcachon Radio at 0803 and 1803 (zone −0100). Sectors 31S 12.27 and 31N 12.27 are used on working days from 0800-1800 (zone −0100).

LA GIRONDE − APPROACHES (chart 2910)

La Gironde estuary is formed by the Garonne and the Dordogne, which join at Bec d'Ambes, 38M above Pte de Grave. BXA Lt buoy is moored about 11M WSW of Pte de la Coubre, off which lies Banc de la Mauvaise, the S end of which dries. Cordouan Lt Ho is on a large rky bank in the middle of the estuary. Grande Passe de l'Ouest starts about 4M E of BXA buoy and is dredged through Grand Banc − the outer bar of La Gironde. Enter to seaward of Buoys Nos 1 and 2, and keep in buoyed channel with Ldg Lts. Off Terre-Nègre Lt the SE-going stream begins at HW −0500 (sp rate 1½ kn), and the NW-going at HW +0130 (sp rate 2½ kn).

Passe Sud, a lesser chan, is entered 5½M W of Pte de la Négade, and runs NE past Pte de Grave. There are two sets of Ldg Lts; the second lead over Platin de Grave, but it is better to pass NW of this shoal. Both entrance chans are dangerous in strong on-shore winds, due to breakers and also the mascaret (bore) on the outgoing stream. Westerly swell breaks on La Mauvaise and around Cordouan, and sandbanks shift constantly. In places tidal stream runs 4 kn or more, and with wind against tide a dangerous sea can build up.

ILE D'OLERON (chart 2648, 2746)

From Pte de l'Epinette 13M N to Pte de Chassiron, the W coast is bounded by drying rks and shoals. In bad weather the sea breaks 4 or 5M offshore, where tidal streams are weak, sp rate 1 kn, starting NW at +0300 HW Pte de Grave and SE at −0505 HW Pte de Grave, but often overcome by current due to prevailing wind. The rate however increases towards Pte de Gatseau, where Pertuis de Maumusson separates the island from mainland. Its entrance is marked by buoy (safe water mark) about 2¾M W of Pte d'Arvert. Banc des Mattes and Banc de Gatseau, both of which dry in places, lie S and N of the chan, and are joined by a sand bar which usually has a depth of about 1.5m. Depth and position vary, and buoys are moved accordingly. Any swell forms breakers, and the chan is very dangerous then or in even moderately strong W winds, especially on the ebb (sp rate 3½ kn). In good weather and with no swell, enter about HW−1.

On N shore, Pte du Grouin du Crou has steep cliffs: with fresh NW winds against NW-going stream a bad sea builds on the bank which extends 8M W. 1M S of the point is Roche de l'Aunis (depth 0.3m). From the point sand dunes run 8M ESE to the entrance to Rivière Le Lay, which is fronted by a bar (dries 1m), dangerous in bad weather. The chan to L'Aiguillon (10.14.24) is marked by Bns and buoys. 4M further E is entrance to Anse de l'Aiguillon, in which are extensive mussel beds. In NE corner is entrance to Sèvre Niortaise which, after 3½M, gives access to the canal leading to the port of Marans. The lock operates at HW.

BAIE DE BOURGNEUF (charts 2646, 2647)

B de Bourgneuf is entered between Pte de l'Herbaudière, the NW extremity of Ile de Noirmoutier, and Pte de St Gildas about 7M NNE. The S entrance through Goulet de Fromentine is obstructed by a causeway, Route du Gois, which dries 3m. The b is sheltered except in W winds, which can raise a heavy sea on the ebb stream. There are good anchs but the S and E sides of the b are encumbered with shoals, rocks and oyster or mussel fisheries.

From the S, beware Chaussée des Boeufs, marked by buoys, extending 3½M W and 7½M SW of Ile de Noirmoutier. Some of these dry, and the tide sets on to them. Chenal de la Grise (chart 3216) between Ile du Pilier and Pte de l'Herbaudière has 2.7m in the W sector of Basse du Martroger Lt, and gives access to L'Herbaudière marina (10.14.22). If proceeding E to Pornic (10.14.23), pass N of Basse du Martroger, and beware Roches des Pères about 1M ENE.

From the N (chart 3216) the approach is simple, but beware La Couronnée, dries 1.8m and marked by buoys, on a rky bank about 2M WSW of Pte de St Gildas. 2M S of Pte de St Gildas, Banc de Kerouars extends 3M in an E/W direction with depths of 1m, where the sea breaks.

From westward, approach Pornic in the W sector of Pte de Noveillard Lt – S of Banc de Kerouars and N of Notre Dame Bn Tr, which lies 2M SW of Pornic and marks end of a line of rks extending WNW from La Bernerie. Pierre du Chenal is an isolated rk about 1M SSE of Notre Dame. Within the b are Noirmoutier (10.14.22) and minor drying harbs – Bec de l'Epoids, Port des Brochets, Le Collet and La Bernerie-en-Retz – with a good anch at Pte des Dames.

Coureau d'Oleron, joined by La Seudre (10.14.13) near its S end, leads between ledges and oyster beds, and constantly changes, with buoys moved to conform. It is crossed by a bridge, clearance 15m, about 2M SE of Le Chateau (10.14.14): chan (prohibited anch) is marked by boards at road level, triangular W&G to be left to stbd, rectangular R&W to be left to port, illuminated at night. S-going stream starts – 0230 HW Pte de Grave, N-going at +0500 HW Pte de Grave, sp rates 2 kn. Just N of bridge is Fort du Chapus, connected to mainland by causeway.

The N end of Coureau d'Oleron leads into Grande Rade, where good anch is found except in fresh NW winds. At N end of Grande Rade is entrance to La Charente (10.14.15). From the N, Grande Rade is entered from Pertuis d'Antioche through Passage de l'Est close to Ile d'Aix (10.14.24) or Passage de l'Ouest, which run each side of La Longe du Boyard – an extensive sandbank on which stands a fort.

PERTUIS D'ANTIOCHE (chart 2746)

A buoy (safe water mark) is moored in W approach to Pertuis d'Antioche which runs between Ile d'Oleron and Ile de Ré, giving access to Rochefort (10.14.15) and La Rochelle (10.14.16). Its shores are low. Off Pte de Chassiron (Lt Ho, Sig Stn) reefs extend ½M W, 1½M N to Rocher d'Antioche (lit), and 1½M E, and there is often a nasty sea. Ile de Ré forms the N shore, and its coast is fringed by rky ledges extending 2½M SE from Pte de Chanchardon (Lt) and nearly 1M from Pte de Chauveau (marked by Bn and Lt Tr).

Plateau de Rochebonne, with depths of less than 4m, is a large rky flat 35-40M W of Ile de Ré, marked by buoys. It is steep-to on all sides, and the sea breaks dangerously on it.

PERTUIS BRETON (chart 2641)

Pertuis Breton is entered between Pte des Baleines (Ile de Ré) and Pte du Grouin du Cou (both lit). It gives access to the harbs on N shore of Ile de Ré, to L'Aiguillon (10.14.24) and Marans on the mainland shore, and provides (except in NW winds) a sheltered route to La Pallice and La Rochelle, through Coureau de la Pallice. Beware rocky ledges (dry) extending 2½M NW from Les Baleines. There are shallows and drying areas all along the N coast of Ile de Ré. Near St Martin and La Flotte there are extensive fisheries, with seaward limits marked by buoys (special marks).

10.14.6 DISTANCE TABLE

Approximate distances in nautical miles are by the most direct route while avoiding dangers and allowing for traffic separation schemes etc. Places in *italics* are in adjoining areas.

1 *Pointe du Raz*	**1**																		
2 *Le Palais (Belle Ile)*	81	**2**																	
3 Pornic	125	45	**3**																
4 Port Joinville (Ile d'Yeu)	126	50	30	**4**															
5 St-Gilles-Croix-de-Vie	141	64	40	18	**5**														
6 Les Sables d'Olonne	156	79	55	31	20	**6**													
7 St Martin (Ile de Ré)	177	105	75	55	44	27	**7**												
8 La Rochelle	186	110	92	66	51	36	12	**8**											
9 Pointe de Chassiron	185	107	84	58	46	32	17	13	**9**										
10 Royan	224	146	115	97	85	71	56	52	39	**10**									
11 Bordeaux	278	200	169	151	139	125	110	106	93	54	**11**								
12 Castets (canal lock)	308	230	199	181	169	155	140	136	123	84	30	**12**							
13 Cap Ferret (Arcachon)	253	185	160	138	130	113	102	98	85	68	118	152	**13**						
14 Capbreton	298	235	215	192	186	171	162	158	145	124	174	208	58	**14**					
15 Bayonne	307	242	223	200	195	178	149	166	153	132	182	216	70	12	**15**				
16 St Jean de Luz	310	250	230	206	200	186	153	174	161	136	186	220	76	19	16	**16**			
17 *San Sebastian*	307	249	233	210	204	191	159	178	165	142	192	226	85	31	29	17	**17**		
18 *Santander*	280	239	241	212	210	201	190	198	185	173	223	257	130	104	102	95	83	**18**	
19 *La Coruna*	321	334	357	335	340	342	365	354	341	348	398	432	327	315	317	314	298	223	**19**
20 *Cabo Finisterre*	369	385	408	386	391	393	416	405	392	399	449	483	378	366	368	365	349	274	59	**20**

SPECIAL NOTES FOR FRENCH AREAS (Areas 14, 15, 16, 17 and 18) 10-14-7

There are certain differences in the harbour information where French ports are concerned. Instead of 'County' the 'Department' is given. The word Customs is used instead of HMC. It should also be remembered that the Time Zone for France is — 0100 (i.e. 1300 Standard Time in France is 1200 GMT), daylight saving schemes not having been taken into account (see 9.1.2.)

For special information about passports, visas, triptyques, carnets, Permis de Circulation, etc. see the relevant sections in earlier chapters or, for greater detail apply to the French Government Tourist Office, 178 Piccadilly, London, W1V 0AL. Tel. 01-499-6911.

AFFAIRES MARITIMES

In every French port there is a representative of the central government, L'Administration Maritime, known as Affaires Maritimes. This organisation watches over all maritime activities and helps them develop harmoniously — commercial, fishing, pleasure, etc. All information about navigation and any other maritime problems can be supplied by the local representative. If in doubt whom to consult, the Affaires Maritimes is the best place to start. Their headquarters is at

Bureau de la Plaisance — 3 Place Fonteroy, 75007 Paris.
Tel: (1) 45.67.55.05

and the telephone number of their local representative is given under each port in the Almanac.

CHARTS

Two French chart numbers are given; SHOM are those issued by the 'Service Hydrographique et Oceanographique de la Marine', the Hydrographic Service of the French Navy; also ECM 'Éditions Cartographiques Maritimes', which are charts for sea and river navigation (Carte-Guide de navigation maritime et fluviale).

The letters SHOM after entries in the 'Facilities' indicates an agent for SHOM charts.

SIGNALS

There are standard sets of signals for Traffic, Storm Warnings, Distress and Tidal, and these apply in all French ports unless otherwise stated.

Traffic Signals

Day		Night		
Full Code	Simplified Code	Full Code	Simplified Code	
▲	or R Flag	R / W / R	or ● R	ENTRY PROHIBITED
▼ over ▲	R Flag over G Flag	G / W / R	or ● R / ● G	ENTRY AND DEPARTURE PROHIBITED
▼	or G Flag	G / W / G	or ● G	DEPARTURE PROHIBITED
R R Balls R		R R R		EMERGENCY — ENTRY PROHIBITED
INTERNATIONAL CODE SIGNAL		G G G		PORT OPEN

A Black Flag displayed indicates a shipping casualty in the area.

Flag 'P' is sometimes displayed to indicate that lock or dock gates are open.

Distress Signals (from Lighthouse) or Danger Signals

Most isolated French lighthouses have radiotelephones, but the following signals may be displayed:

A ball above or below a square flag — Require immediate assistance

A black flag at the masthead — Shipwreck in vicinity of lighthouse

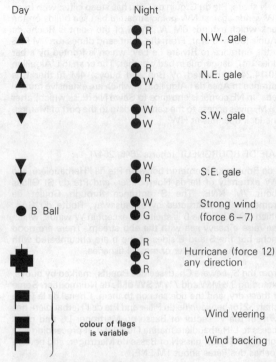

Storm Signals (International System)

Day	Night	
▲	R / R	N.W. gale
▲▲	R / W	N.E. gale
▼	W / W	S.W. gale
▼▼	W / R	S.E. gale
● B Ball	W / G	Strong wind (force 6–7)
✚	R / G / R	Hurricane (force 12) any direction
▮ ▬	colour of flags is variable	Wind veering / Wind backing

In France, flashing white lights, by day only, indicate wind over Force 6, as follows: — Quick flashing — within three hours; Interrupted quick flashing — within six hours.

Tidal Signals

There are two sets of signals, one showing the state of the tide and the other showing the depth of water.

State of the tide is shown by: —

	Day	Night	
High Water...	⊠ W Flag / B Cross	W — W	
Tide falling...	▼	● / G	
Low Water...	▭ Bu	● — ● G G	
Tide rising...	▲	G / ● / W	

The depth of water signals show the depth above Chart Datum and to ascertain the figure the various shapes have to be added together;
by day: a cone point down = 0.2 m, a cylinder = 1.0 m, a ball = 5.0 m.
by night: G light = 0.2 m, R light = 1.0 m, W light = 5.0 m.

The three different shapes are shown horizontally, with cones to the left of cylinders and balls to the right, viewed from seaward. Lights are disposed similarly.

The following examples will help to explain.

BY DAY

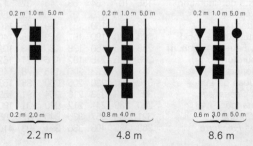

BY NIGHT

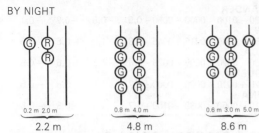

2.2 m   4.8 m   8.6 m

The following French equivalents may be useful for understanding tidal data published on notice boards etc in French harbours:

High water (HW)	Pleine mer (PM)
Low water (LW)	Basse mer (BM)
Springs (Sp)	Vive eau (VE)
Neaps (Np)	Morte eau (ME)
Mean high water springs (MHWS)	Pleine mer moyenne de VE
Mean high water neaps (MHWN)	Pleine mer moyenne de ME
Mean low water neaps (MLWN)	Basse mer moyenne de ME
Mean low water springs (MLWS)	Basse mer moyenne de VE
Mean level	Niveau moyen (ou niveau de mi-marée)
Chart datum (CD)	Zero des cartes
Lowest astronomical tide (LAT)	Plus basse mer connue (PBM)
Flood (tide)	Flot (ou flux)
Ebb (tide)	Reflux (ou jusant)
Range	Marnage (ou amplitude)
Stand (at high water)	Etale
Greenwich Mean Time (GMT)	Temps universel (TU)

TIDAL COEFFICIENTS
The French use tidal coefficients which vary from 45 at mean neaps (Morte Eau) to 95 at mean springs (Vive Eau), but which with exceptional tides can be as little as 20 or as much as 120. For an average tide the coefficient is 70. The coefficients, which are published in France for every tide of the year, therefore indicate the size of each tide. The ratio of the coefficients of different tides equals the ratio of their ranges. For example, from the figures given above it can be seen that the range of the largest spring tide (coefficient 120) is six times the range of the smallest neap tide (coefficient 20).

SAFETY AT SEA
Control of Search and Rescue is exercised by the CROSS organisation (Centres Régionaux Operationnels de Surveillance et de Sauvetage) available 24 hrs a day. The organisation is divided as follows:-
Spanish border to Pointe du Raz: CROSS Etel, Tel. 97.55.35.35, (2182 kHz and VHF Ch 16).
Pointe du Raz to Mont St Michel: CROSS Corsen, Tel. 98.89.31.31, (2182 kHz and VHF Ch 11 16).
Mont St Michel to Antifer: CROSS Jobourg, Tel. 33.52.72.13, (2182 kHz and VHF Ch 11 16).
Antifer to Belgian border: CROSS Gris Nez; Tel. 21.87.21.87, (2182kHz and VHF Ch 16 69).
Note: CROSS Soulac, near Pointe du Grave, is a subsidiary station of CROSS Etel, Tel. 56.59.82.00, (2182 kHz and VHF Ch 16, 0600-2100).
Besides the Search and Rescue function CROSS will provide urgent up-to-date navigational information such as buoys adrift, lights not working etc.
The following CROSS VHF relay stations keep watch on Ch 16; La Rhune (near Spanish border), Cap Ferret, Chassiron (Ile de Oléron), Ile d'Yeu, Belle Ile and Penmarc'h. CROSS can be contacted by radiotelephone, by telephone, through coast radio stations, via the semaphore system (French Navy on Ch 16) as shown below, or via the National Gendarmerie or Affaires Maritimes.
Naval 'Semaphore' stations keep watch on Ch 16 (working channel Ch 10) or by telephone as follows:

SEMAPHORE	TELEPHONE
● SOCOA	59.47.18.54
CAP FERRET	56.60.60.03
POINTE-DE-GRAVE	56.09.60.03
LA COUBRE	46.22.41.73
ILE D'AIX	46.88.66.07
CHASSIRON (Oléron)	46.47.85.43
LES BALEINES (Ré)	46.29.42.06
ST-SAUVEUR (Yeu)	51.58.31.01
● CHEMOULIN	40.91.99.00
PIRIAC-SUR-MER	40.23.59.87
ST-JULIEN	97.50.09.35
TALUT (Belle-Ile)	97.31.85.07
TAILLEFER (Belle-Ile)	97.31.83.18
BEG MELEN (Groix) (PEN MEN)	97.05.80.13
ETEL	97.55.35.59
● PORT-LOUIS	97.85.52.10
BEG MEIL	98.94.98.92
● PENMARCH	98.58.61.00
● POINTE-DU-RAZ	98.70.66.57
CAP-DE-LA-CHEVRE	98.27.09.55
TOULINGUET	98.27.90.02
● MINOU (Vigie)	98.22.10.43
● ST-MATHIEU	98.89.01.59
OUESSANT CREACH	98.48.80.49
● OUESSANT STIFF	98.48.81.50
● BRIGNOGAN	98.83.50.84
● ST QUAY-PORTRIEUX	96.70.42.18
BATZ	98.61.76.06
● PLOUMANACH	96.23.21.50
BREHAT	96.20.00.12
ST-CAST	96.41.85.30
LE GROUIN	99.89.60.12
LE ROC	33.50.05.85
CARTERET	33.53.85.08
LA HAGUE	33.52.71.87
● LE HOMET	33.92.60.08
LEVY	33.54.31.17
● BARFLEUR	33.54.04.37
ST-VAAST	33.54.44.50
● PORT-EN-BESSIN	31.21.81.51
VILLERVILLE	31.88.11.13
● LA HEVE	35.46.07.81
FECAMP	35.28.00.91
DIEPPE	35.84.23.82
AULT	22.60.47.33
BOULOGNE	21.31.32.10
● DUNKERQUE	20.66.86.18

● H24. Remainder sunrise to sunset.

Whereas in UK, lifeboats should be contacted only through HM Coastguard, in France, although the lifeboat service comes under the CROSS organisation, it is encouraged to contact the lifeboat stations (Societe Nationale de Sauvetage en Mer, SNSM) direct. Where appropriate, therefore, the telephone numbers of local SNSMs are given.

MEDICAL
Telephone numbers of doctors and/or hospitals are given for each port. There is also the Services d'Aide Médicale Urgente (SAMU), which liaises closely with CROSS, which can be contacted as follows: –

PYRÉNÉES-ATLANTIQUE (Bayonne)	59.63.33.33
LANDES (Mont-de-Marsan)	58.75.44.44
GIRONDE (Bordeaux)	56.96.70.70
CHARENTE-MARITIME (La Rochelle)	46.27.15.15
VENDEE (La Roche-sur-Yon)	51.62.62.15
LOIRE-ATLANTIQUE (Nantes)	40.48.35.35
MORBIHAN (Vannes)	97.54.22.11
FINISTERE (Brest)	98.46.11.33
CÔTES-DU-NORD (Saint Brieuc)	96.94.40.15
CALVADOS (Caen)	31.44.88.88
SEINE-MARITIME (Le Havre)	35.21.11.00
SOMME (Amiens)	22.44.33.33
PAS-DE-CALAIS (Arras)	21.71.51.51
NORD (Lille)	20.54.22.22

SPECIAL NOTES FOR FRENCH AREAS continued

PUBLIC HOLIDAYS
New Year's Day, Easter Sunday and Monday, Labour Day (1 May), Ascension Day, Armistice Day 1945 (8 May), Whit Sunday and Monday, National (Bastille) Day (14 July), Feast of the Assumption (15 August), All Saints' Day (1 November), Remembrance Day (11 November), Christmas Day.

METEO (Weather)
The BQR (Bulletin Quotidien des Renseignements) is a daily information bulletin displayed in Harbour Masters' Offices and in yacht clubs, and is very informative.

Telephone numbers of the principal coastal Bureaux Meteorologiques (Met. Offices) are given in 7.2.4. This also shows numbers for Repondeurs Automatiques (recorded messages) which are sufficiently detailed for short coastal passages.

For details of weather information from French radio stations see — Table 7 (5).

TELEPHONES
To telephone France from UK, dial 010 33 followed by the number. For all exchanges outside Paris, area codes have been added on to individual numbers. All numbers therefore have eight digits. From inside France, dial the eight figures shown in the Almanac. From UK dial 010 33 followed by the eight digit number shown. There are two sorts of ringing tone: one consists of three second bursts of tone separated by three seconds silence; the other consists of one and a half second bursts of tone followed by three and a half seconds silence. Engaged is similar to that in UK. Number unobtainable is indicated by a recorded announcement. A rapid series of pips indicates your call is being connected. For Police dial 17. For Fire or Hospital dial 18.

To telephone UK from France dial 19; wait for the second dialling tone then dial 44 followed by the area dialling code (omitting the first 0) and then the number.

Phonecards for public telephones may be bought at the PTT or at some cafés and tabacs. Cheap rates are from 2130 to 0800 Mon-Sat; and all day Sun.

Note 1.
See 2.5.3 regarding French Customs regulations.
Note 2.
Speed under power is limited in France to 5 kn. when within 300 m of the shore.
Note 3.
There are strict penalties for infringing the International Regulations for Prevention of Collision at Sea and for French yachts not carrying the correct documentation or safety equipment.
Note 4.
The following abbreviations are used in telephone numbers:

Capitaine du Port	Hr Mr
Affaires Maritimes	Aff Mar
Centre régional opérationnel de surveillance et sauvetage	CROSS
Societe Nationale de Sauvetage en Mer (Lifeboats)	SNSM
Meteorologie/Weather	Meteo
Meteo Repondeur Automatique	Auto

Note 5.
Foreign vessels over 25m must get permission to navigate or anchor in French internal waters and keep watch on Ch 16 or other nominated frequency.

TIDAL INFORMATION FOR NORTH COAST OF SPAIN 10-14-8
Standard Port POINTE DE GRAVE (→)

Times				Height (metres)			
HW		LW		MHWS	MHWN	MLWN	MLWS
0000	0600	0500	1200	5.3	4.3	2.1	1.0
1200	1800	1700	2400				

Differences SAN SEBASTIAN
−0055 −0040 −0020 −0035 −1.2 −1.2 −0.6 −0.5
LEQUEITO
−0100 −0050 −0025 −0045 −1.2 −1.2 −0.6 −0.5
BILBAO
−0045 −0030 −0010 −0030 −1.3 −1.2 −0.7 −0.5
SANTANDER
−0020 −0100 0000 −0050 −0.9 −0.8 −0.2 −0.1
GIJON
−0050 −0040 −0015 −0045 −1.3 −1.2 −0.6 −0.5
LUARCA
−0030 −0020 +0005 −0025 −1.4 −1.3 −0.6 −0.5
ORTIGUEIRA
−0020 +0010 +0020 0000 −1.7 −1.5 −0.8 −0.5
CORUNA
−0110 −0050 −0030 −0100 −1.7 −1.5 −0.8 −0.5

ARCACHON 10-14-9
Gironde

CHARTS
Admiralty 2664; SHOM 6766, 6787; ECM 255
TIDES
+0620 Dover; ML 2.2; Zone −0100
Standard Port POINTE DE GRAVE (→)

Times				Height (metres)			
HW		LW		MHWS	MHWN	MLWN	MLWS
0000	0600	0500	1200	5.3	4.3	2.1	1.0
1200	1800	1700	2400				

Differences ARCACHON
+0005 +0030 +0015 +0040 −1.1 −1.1 −1.0 −0.8
ST JEAN DE LUZ
−0050 −0045 −0025 −0040 −1.0 −1.0 −0.5 −0.5
L'ADOUR (BOUCAU)
−0035 −0030 −0010 −0030 −1.0 −0.9 −0.3 −0.2
CAP FERRET
−0015 0000 +0005 +0015 −1.2 −1.0 −0.7 −0.6

SHELTER
Good shelter in marina in Bassin d'Arcachon. Entry impossible at night or in strong winds from SW to N. Max length 15m. Besides the Port d'Arcachon there are numerous other small drying harbours round the basin. On the West, La Vigne, Le Canon, Piquey and Claouey; on the South La Teste and Gujan; on the NE Port de Lège, Ares, Andernos, Port de Fontaine Vielle, Lanton (Cassy) and Audenge.

NAVIGATION
Waypoint ATT ARC (safe water) buoy, LFl 10s, Whis, 44°31′.76N 01°18′.27W, approx 260°/080° from/to Nos 1 and 2 channel buoys, 1.0M. Beware shifting sand banks in entrance channel, the sea breaks on these banks in any wind but the entrance channel can be seen through the breakers. With a strong on-shore wind the bar is impassable from HW+1 until LW, and it is best to wait until LW+3. Buoys are adjusted each spring but channel can change later. Latest information from Service de la Marine Gironde at Arcachon 56.82.32.97 or at Bordeaux 56.83.03.00. Average depth over bar is 4.5m. Beware firing ranges to seaward and S of Arcachon. See 10.14.5.

LIGHTS AND MARKS
Cap Ferret Fl R 5s 53m 26M and Oc (3) 12s vis 045°-135°.
Arcachon harbour W breakwater QG, E breakwater QR.
Port de la Vigne Iso R 4s 7m 4M.
La Salie Q (9) 15s 19m 10M.
RADIO TELEPHONE
VHF Ch 09.
TELEPHONE
Hr Mr 56.83.22.44; Aff Mar 56.83.03.00; Customs 56.83.08.27; SNSM 56.83.22.44; CROSS 56.09.82.00; Auto 56.83.84.85; Dr 56.83.04.72; Hosp 56.83.39.50; Brit Consul 56.52.28.35.
FACILITIES
Marina (1900+57 visitors), FW, AC, Slip, C (10 ton), BH (45 ton), D; **Station Elf** Tel. 56.83.29.72, P, D; **C V d'Arcachon** Tel. 56.83.05.92, P, D, FW, ME, El, Sh; **Arcachon Plaisance** Tel. 56.83.31.31, P, D, L, FW, ME, El, Sh, CH; **Arcachon Ship Chandler** Tel. 56.83.23.69 CH: **Nautic Service** Tel. 56.83.84.10, ME, El, Sh, CH; **Voilerie Ries** Tel. 56.83.12.96 SM; **Service Nautique Coop Maritime** Tel. 56.83.27.45 SHOM; **Voiles Assistance** Tel. 56.83.58.55 SM; **Yachting Arcachon** Tel. 56.54.61.48, ME, El, Sh, CH; **N Jetty** Tel. 56.83.22.44, Slip, L, FW, C (10 ton), AB; **Town** AB, V, R, Gaz, PO, Bank, Rly, Air (Bordeaux).
Ferry UK — Roscoff–Plymouth; St Malo–Portsmouth.

AREA 14—S Biscay

ARCACHON *continued*

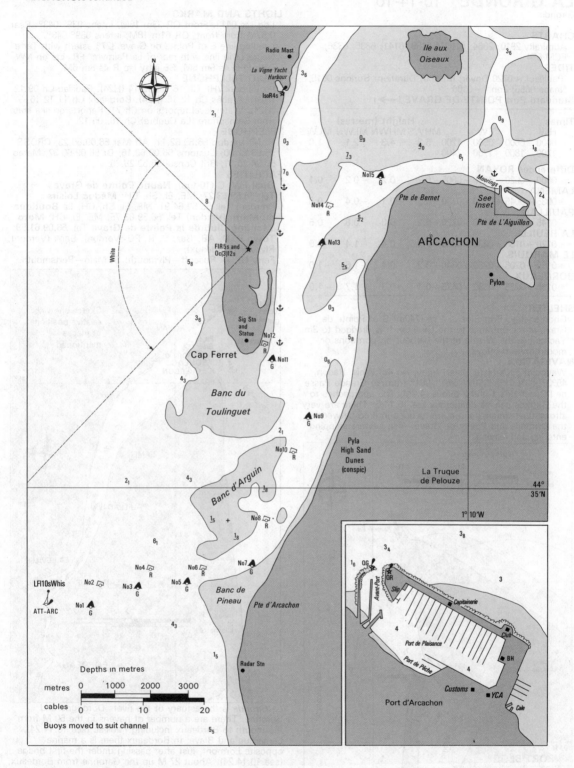

PORT BLOC/ LA GIRONDE 10-14-10
Gironde

CHARTS
Admiralty 2910, 2664; SHOM 6139, 6141, 6335, 6336; ECM 554

TIDES
(Pauillac) +0720 Dover; ML 3.0; Duration: Springs 0615; Neaps 0655; Zone −0100
Standard Port POINTE DE GRAVE (→)

Times				Height (metres)			
HW		LW		MHWS	MHWN	MLWN	MLWS
0000	0600	0500	1200	5.3	4.3	2.1	1.0
1200	1800	1700	2400				

Differences ROYAN
0000 −0020 −0010 −0005 −0.2 −0.2 −0.2 −0.1
LAMENA
+0035 +0045 +0100 +0130 +0.2 +0.1 −0.4 −0.3
PAUILLAC
+0045 +0110 +0140 +0220 +0.2 0.0 −0.8 −0.5
LA REUILLE
+0120 +0155 +0230 +0320 −0.2 −0.4 −1.4 −0.9
LE MARQUIS
+0130 +0205 +0250 +0340 −0.2 −0.4 −1.6 −1.0
BORDEAUX
+0155 +0235 +0330 +0425 −0.1 −0.3 −1.7 −1.0

SHELTER
Good shelter. Port Bloc is 4 ca (740m) S of Pointe de Grave at the entrance to the Gironde. It is dredged to 3m. Yachts use the W side of the harbour on pontoons or moor between buoys.

NAVIGATION
Waypoint BXA (safe water) buoy, Iso 4s, Whis, Racon, 45°37'.60N 01°28'.60W, 261°/081° from/to Grande Passe de l'Ouest Nos 1 and 2 buoys, 4.8M. The approaches to the Gironde can be dangerous see 10.14.5. There are very strong tidal streams and currents encountered between the entrance and Pointe de Grave. Also beware shipping entering and leaving.

LIGHTS AND MARKS
Ldg Lts 041°. Front QR 33m 16M, intens 039°-043°. Rear, 0.97M from front, QR 61m 18M, intens 039°-043°. Leading line E of Pointe de Grave 327° astern with Tèrre Negre Lt in line with rear of La Palmyre, FR. Lts on NW jetty Fl G 4s 8m 6M, SE jetty Iso R 4s 8m 6M.

RADIO TELEPHONE
Le Verdon VHF Ch 16; 11 12 14 (H24). Pauillac Ch 09 12 (H24). Ambès Ch 12 16 (H24). Bordeaux Ch 11 12 16 (H24). Water level reports on Ch 71. Reports on sea state from Semaphore (La Coubre) Ch 16; 06 12.

TELEPHONE
Hr Mr Verdon 56.59.62.11; Aff Mar 56.09.60.23; CROSS 56.09.82.00; Customs 56.09.60.16; Dr 56.09.60.37; Meteo 56.34.20.11; Brit Consul 56.52.28.35.

FACILITIES
Quai FW, C (10 ton); **Nautic Pointe de Grave** Tel. 56.59.63.27; ME, El, Sh, CH; **Médoc Loisirs** (Verdon) Tel. 56.59.64.91, ME, El, Sh, CH; **La Boutique du Marin** (Verdon) Tel. 56.59.63.75, ME, El, CH; **Moto Yachting Club de la Pointe de Grave** Tel. 56.09.61.58; **Town** P, D, AB, Gaz, V, R, PO (Verdon), Bank (Verdon), Rly, Air (Bordeaux);
Ferry UK — Roscoff—Plymouth; St Malo—Portsmouth.

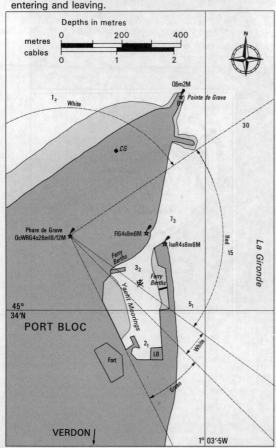

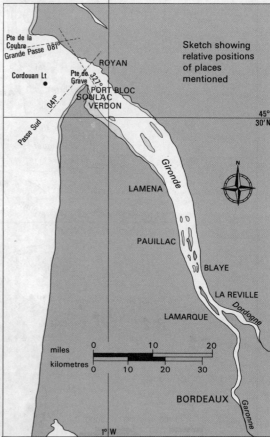

Sketch showing relative positions of places mentioned

La Gironde is the estuary of the rivers Dordogne and Garonne. There are a number of havens in the 50 M from the mouth to Bordeaux including Pauillac (see 10.14.24), Lamarque and Blaye. In Bordeaux there is a marina opposite Lormont, just after passing under the first bridge (see 10.14.24). About 27 M up the Garonne from Bordeaux, at Castets, there is the first lock into the Canal Lateral à la Garonne which leads into the Canal du Midi. In the Garonne at springs, the flood stream starts with a small bore and then runs at about 3 kn, while the ebb runs at 5 kn. In the other rivers the spring rate does not exceed 3 kn on the flood and 4 kn, on the ebb.
NOTE: A major marina is due to be built at Verdon starting 1989.

FRANCE, WEST COAST - POINTE DE GRAVE
LAT 45°34′N   LONG 1°04′W
TIMES AND HEIGHTS OF HIGH AND LOW WATERS

YEAR 1989

TIME ZONE −0100 (French Standard Time)
For French Summer Time add ONE hour in non-shaded areas

JANUARY

	TIME	m		TIME	m
1 SU	0457 / 1119 / 1736	2.2 / 4.2 / 2.2	**16** M	0531 / 1221 / 1809	1.8 / 4.5 / 2.0
2 M	0010 / 0600 / 1236 / 1842	4.1 / 2.2 / 4.2 / 2.2	**17** TU	0105 / 0650 / 1343 / 1928	4.3 / 1.9 / 4.4 / 2.0
3 TU	0117 / 0705 / 1344 / 1944	4.2 / 2.1 / 4.2 / 2.1	**18** W	0218 / 0807 / 1454 / 2037	4.5 / 1.8 / 4.5 / 1.8
4 W	0214 / 0805 / 1441 / 2039	4.4 / 2.0 / 4.4 / 1.9	**19** TH	0318 / 0912 / 1549 / 2132	4.7 / 1.6 / 4.7 / 1.6
5 TH	0302 / 0859 / 1530 / 2126	4.6 / 1.7 / 4.6 / 1.7	**20** F	0407 / 1004 / 1633 / 2219	4.9 / 1.4 / 4.8 / 1.4
6 F	0347 / 0948 / 1614 / 2211	4.8 / 1.5 / 4.8 / 1.5	**21** SA	0447 / 1049 / 1710 / 2300 ○	5.0 / 1.3 / 4.9 / 1.3
7 SA	0429 / 1033 / 1656 / 2254 ●	5.0 / 1.3 / 4.9 / 1.3	**22** SU	0523 / 1129 / 1744 / 2338	5.1 / 1.2 / 5.0 / 1.2
8 SU	0511 / 1118 / 1738 / 2336	5.2 / 1.1 / 5.1 / 1.2	**23** M	0557 / 1206 / 1815	5.2 / 1.2 / 5.0
9 M	0554 / 1202 / 1820	5.3 / 1.0 / 5.1	**24** TU	0013 / 0628 / 1240 / 1844	1.2 / 5.1 / 1.2 / 5.0
10 TU	0020 / 0638 / 1246 / 1903	1.1 / 5.4 / 0.9 / 5.1	**25** W	0047 / 0657 / 1313 / 1913	1.3 / 5.1 / 1.3 / 4.9
11 W	0104 / 0724 / 1331 / 1947	1.1 / 5.4 / 1.0 / 5.0	**26** TH	0119 / 0726 / 1344 / 1943	1.3 / 5.0 / 1.4 / 4.8
12 TH	0149 / 0812 / 1416 / 2034	1.1 / 5.3 / 1.1 / 4.9	**27** F	0153 / 0757 / 1417 / 2015	1.4 / 4.9 / 1.5 / 4.6
13 F	0236 / 0902 / 1503 / 2125	1.2 / 5.2 / 1.3 / 4.7	**28** SA	0228 / 0831 / 1452 / 2053	1.6 / 4.7 / 1.7 / 4.5
14 SA	0326 / 0958 / 1555 / 2226 ☾	1.4 / 4.9 / 1.5 / 4.5	**29** SU	0307 / 0911 / 1533 / 2141	1.8 / 4.5 / 1.9 / 4.3
15 SU	0423 / 1103 / 1656 / 2342	1.6 / 4.7 / 1.8 / 4.3	**30** M	0354 / 1003 / 1624 / 2249 ☾	2.0 / 4.2 / 2.2 / 4.1
			31 TU	0454 / 1119 / 1734	2.2 / 4.0 / 2.3

FEBRUARY

	TIME	m		TIME	m
1 W	0018 / 0611 / 1253 / 1857	4.0 / 2.3 / 4.0 / 2.3	**16** TH	0214 / 0800 / 1452 / 2027	4.3 / 2.0 / 4.3 / 2.0
2 TH	0135 / 0728 / 1412 / 2008	4.2 / 2.1 / 4.2 / 2.1	**17** F	0315 / 0903 / 1544 / 2120	4.6 / 1.7 / 4.6 / 1.7
3 F	0237 / 0834 / 1512 / 2105	4.5 / 1.9 / 4.5 / 1.8	**18** SA	0400 / 0951 / 1622 / 2204	4.8 / 1.5 / 4.8 / 1.5
4 SA	0329 / 0930 / 1601 / 2154	4.8 / 1.5 / 4.8 / 1.5	**19** SU	0436 / 1032 / 1654 / 2242	5.0 / 1.3 / 4.9 / 1.3
5 SU	0415 / 1019 / 1645 / 2240	5.1 / 1.2 / 5.0 / 1.2	**20** M	0507 / 1108 / 1722 / 2318 ○	5.1 / 1.2 / 5.0 / 1.2
6 M	0459 / 1105 / 1726 / 2324	5.4 / 0.9 / 5.2 / 1.0	**21** TU	0535 / 1142 / 1750 / 2351	5.2 / 1.1 / 5.1 / 1.1
7 TU	0543 / 1149 / 1807	5.6 / 0.7 / 5.3	**22** W	0603 / 1214 / 1816	5.2 / 1.1 / 5.1
8 W	0007 / 0627 / 1232 / 1848	0.8 / 5.7 / 0.7 / 5.4	**23** TH	0021 / 0629 / 1243 / 1843	1.1 / 5.2 / 1.1 / 5.0
9 TH	0050 / 0710 / 1314 / 1930	0.7 / 5.7 / 0.7 / 5.3	**24** F	0051 / 0656 / 1312 / 1910	1.1 / 5.1 / 1.2 / 5.0
10 F	0132 / 0754 / 1356 / 2011	0.8 / 5.5 / 0.8 / 5.1	**25** SA	0121 / 0723 / 1341 / 1939	1.2 / 5.0 / 1.3 / 4.8
11 SA	0215 / 0839 / 1439 / 2055	0.9 / 5.3 / 1.1 / 4.8	**26** SU	0152 / 0753 / 1412 / 2011	1.4 / 4.8 / 1.5 / 4.6
12 SU	0301 / 0928 / 1526 / 2148 ☾	1.2 / 4.9 / 1.5 / 4.5	**27** M	0227 / 0827 / 1449 / 2051	1.6 / 4.6 / 1.8 / 4.4
13 M	0355 / 1029 / 1623 / 2304	1.6 / 4.5 / 1.9 / 4.3	**28** TU ☾	0307 / 0911 / 1528 / 2150	1.8 / 4.3 / 2.0 / 4.1
14 TU	0505 / 1155 / 1741	1.9 / 4.2 / 2.1			
15 W	0047 / 0636 / 1338 / 1913	4.2 / 2.1 / 4.2 / 2.2			

MARCH

	TIME	m		TIME	m
1 W	0400 / 1024 / 1631 / 2324	2.1 / 4.0 / 2.3 / 4.0	**16** TH	0030 / 0621 / 1328 / 1854	4.1 / 2.1 / 4.0 / 2.3
2 TH	0519 / 1211 / 1809	2.3 / 3.9 / 2.4	**17** F	0159 / 0744 / 1436 / 2006	4.3 / 2.0 / 4.3 / 2.0
3 F	0057 / 0654 / 1344 / 1938	4.1 / 2.2 / 4.1 / 2.2	**18** SA	0258 / 0843 / 1523 / 2057	4.5 / 1.7 / 4.5 / 1.7
4 SA	0210 / 0810 / 1451 / 2042	4.4 / 1.8 / 4.4 / 1.8	**19** SU	0340 / 0927 / 1558 / 2139	4.7 / 1.5 / 4.7 / 1.5
5 SU	0313 / 0909 / 1541 / 2133	4.8 / 1.3 / 4.8 / 1.4	**20** M	0413 / 1005 / 1627 / 2216	4.9 / 1.3 / 4.9 / 1.3
6 M	0359 / 0959 / 1625 / 2220	5.2 / 1.1 / 5.1 / 1.1	**21** TU	0441 / 1040 / 1653 / 2251	5.0 / 1.2 / 5.0 / 1.1
7 TU ●	0441 / 1045 / 1706 / 2304	5.5 / 0.8 / 5.3 / 0.7	**22** W ○	0508 / 1113 / 1720 / 2323	5.1 / 1.1 / 5.1 / 1.1
8 W	0525 / 1129 / 1747 / 2348	5.7 / 0.6 / 5.5 / 0.6	**23** TH	0535 / 1144 / 1748 / 2354	5.2 / 1.1 / 5.1 / 1.0
9 TH	0608 / 1212 / 1827	5.8 / 0.5 / 5.5	**24** F	0601 / 1213 / 1815	5.1 / 1.1 / 5.1
10 F	0030 / 0650 / 1253 / 1907	0.5 / 5.8 / 0.6 / 5.4	**25** SA	0024 / 0628 / 1241 / 1842	1.1 / 5.1 / 1.1 / 5.0
11 SA	0112 / 0733 / 1333 / 1948	0.6 / 5.6 / 0.8 / 5.2	**26** SU	0053 / 0655 / 1310 / 1911	1.1 / 5.0 / 1.3 / 4.9
12 SU	0155 / 0816 / 1415 / 2031	0.8 / 5.2 / 1.1 / 4.7	**27** M	0124 / 0723 / 1340 / 1943	1.3 / 4.8 / 1.4 / 4.7
13 M	0240 / 0904 / 1500 / 2121	1.2 / 4.8 / 1.5 / 4.5	**28** TU	0157 / 0757 / 1413 / 2022	1.5 / 4.6 / 1.7 / 4.5
14 TU ☾	0333 / 1004 / 1557 / 2235	1.6 / 4.3 / 2.0 / 4.2	**29** W	0236 / 0843 / 1456 / 2121	1.7 / 4.3 / 2.0 / 4.2
15 W	0446 / 1135 / 1720	2.0 / 4.0 / 2.3	**30** TH ☾	0328 / 1000 / 1557 / 2250	2.0 / 4.0 / 2.2 / 4.1
			31 F	0445 / 1145 / 1733	2.2 / 3.9 / 2.3

APRIL

	TIME	m		TIME	m
1 SA	0023 / 0623 / 1316 / 1904	4.2 / 2.1 / 4.1 / 2.1	**16** SU	0222 / 0808 / 1446 / 2022	4.4 / 1.8 / 4.4 / 1.8
2 SU	0140 / 0741 / 1422 / 2010	4.5 / 1.7 / 4.5 / 1.7	**17** M	0306 / 0854 / 1521 / 2106	4.6 / 1.6 / 4.6 / 1.6
3 M	0246 / 0840 / 1513 / 2105	4.9 / 1.3 / 4.8 / 1.3	**18** TU	0340 / 0932 / 1552 / 2144	4.8 / 1.4 / 4.8 / 1.4
4 TU	0335 / 0932 / 1559 / 2154	5.2 / 1.0 / 5.1 / 0.9	**19** W	0410 / 1008 / 1621 / 2220	4.9 / 1.3 / 4.9 / 1.2
5 W	0419 / 1020 / 1642 / 2241	5.5 / 0.7 / 5.4 / 0.7	**20** TH	0438 / 1041 / 1650 / 2254	5.0 / 1.2 / 5.0 / 1.1
6 TH	0503 / 1105 / 1723 / 2326 ●	5.7 / 0.6 / 5.5 / 0.5	**21** F ○	0507 / 1113 / 1720 / 2327	5.0 / 1.1 / 5.0 / 1.1
7 F	0547 / 1148 / 1805	5.7 / 0.5 / 5.5	**22** SA	0536 / 1144 / 1750 / 2359	5.0 / 1.1 / 5.0 / 1.1
8 SA	0010 / 0630 / 1230 / 1847	0.5 / 5.6 / 0.6 / 5.4	**23** SU	0604 / 1214 / 1819	4.9 / 1.2 / 5.0
9 SU	0053 / 0713 / 1311 / 1929	0.6 / 5.4 / 0.9 / 5.2	**24** M	0030 / 0633 / 1244 / 1850	1.1 / 4.8 / 1.3 / 4.9
10 M	0137 / 0758 / 1354 / 2013	1.0 / 5.0 / 1.2 / 4.9	**25** TU	0103 / 0705 / 1317 / 1926	1.3 / 4.7 / 1.5 / 4.7
11 TU	0223 / 0846 / 1440 / 2105	1.2 / 4.6 / 1.6 / 4.5	**26** W	0138 / 0744 / 1354 / 2011	1.4 / 4.5 / 1.7 / 4.6
12 W ☾	0317 / 0947 / 1539 / 2215	1.6 / 4.2 / 2.0 / 4.3	**27** TH	0220 / 0837 / 1440 / 2112	1.6 / 4.3 / 1.9 / 4.4
13 TH	0428 / 1117 / 1657 / 2356	2.0 / 4.0 / 2.2 / 4.1	**28** F	0314 / 0952 / 1544 / 2230	1.8 / 4.1 / 2.1 / 4.3
14 F	0553 / 1254 / 1820	2.1 / 4.0 / 2.2	**29** SA	0428 / 1122 / 1708 / 2353	1.9 / 4.1 / 2.1 / 4.4
15 SA	0123 / 0710 / 1359 / 1929	4.2 / 2.0 / 4.2 / 2.0	**30** SU	0553 / 1244 / 1828	1.8 / 4.3 / 1.9

Chart Datum: 2.93 metres below Lallemand System (Mean Sea level, Marseilles)

AREA 14 — S Biscay

FRANCE, WEST COAST - POINTE DE GRAVE

LAT 45°34'N LONG 1°04'W

TIMES AND HEIGHTS OF HIGH AND LOW WATERS

YEAR 1989

TIME ZONE –0100 (French Standard Time). For French Summer Time add ONE hour in non-shaded areas.

MAY

	TIME	m		TIME	m
1 M	0107 / 0706 / 1348 / 1934	4.6 / 1.6 / 4.5 / 1.6	**16** TU	0216 / 0810 / 1436 / 2023	4.4 / 1.7 / 4.4 / 1.7
2 TU	0212 / 0808 / 1443 / 2033	4.9 / 1.3 / 4.8 / 1.3	**17** W	0257 / 0853 / 1512 / 2106	4.5 / 1.6 / 4.6 / 1.5
3 W	0307 / 0902 / 1531 / 2126	5.2 / 1.0 / 5.1 / 1.0	**18** TH	0334 / 0932 / 1549 / 2146	4.7 / 1.4 / 4.8 / 1.4
4 TH	0356 / 0953 / 1617 / 2217	5.4 / 0.8 / 5.2 / 0.7	**19** F	0408 / 1008 / 1622 / 2224	4.8 / 1.3 / 4.9 / 1.3
5 F	0443 / 1040 / 1702 / 2305 ●	5.5 / 0.7 / 5.4 / 0.6	**20** SA	0442 / 1043 / 1655 / 2301 ○	4.8 / 1.3 / 4.9 / 1.2
6 SA	0529 / 1126 / 1745 / 2352	5.5 / 0.7 / 5.4 / 0.6	**21** SU	0515 / 1117 / 1729 / 2337	4.8 / 1.2 / 5.0 / 1.2
7 SU	0613 / 1209 / 1829	5.3 / 0.8 / 5.3	**22** M	0547 / 1151 / 1802	4.8 / 1.3 / 5.0
8 M	0037 / 0658 / 1253 / 1914	0.7 / 5.1 / 1.0 / 5.1	**23** TU	0012 / 0620 / 1226 / 1837	1.2 / 4.7 / 1.3 / 4.9
9 TU	0123 / 0743 / 1336 / 2000	1.0 / 4.8 / 1.3 / 4.9	**24** W	0049 / 0657 / 1303 / 1918	1.2 / 4.7 / 1.5 / 4.8
10 W	0210 / 0832 / 1424 / 2051	1.3 / 4.5 / 1.6 / 4.6	**25** TH	0128 / 0741 / 1345 / 2007	1.3 / 4.5 / 1.6 / 4.7
11 TH	0302 / 0928 / 1519 / 2151	1.6 / 4.3 / 1.9 / 4.4	**26** F	0214 / 0835 / 1435 / 2105	1.5 / 4.4 / 1.7 / 4.6
12 F	0402 / 1038 / 1624 / 2305 ☽	1.8 / 4.1 / 2.0 / 4.2	**27** SA	0308 / 0941 / 1535 / 2212	1.6 / 4.3 / 1.8 / 4.6
13 SA	0510 / 1156 / 1732	2.0 / 4.0 / 2.1	**28** SU	0412 / 1055 / 1643 / 2323 ☾	1.6 / 4.3 / 1.9 / 4.6
14 SU	0022 / 0618 / 1303 / 1837	4.2 / 2.0 / 4.1 / 2.0	**29** M	0521 / 1208 / 1752	1.6 / 4.4 / 1.7
15 M	0126 / 0719 / 1354 / 1934	4.3 / 1.9 / 4.3 / 1.9	**30** TU	0034 / 0629 / 1314 / 1858	4.7 / 1.5 / 4.5 / 1.5
			31 W	0140 / 0734 / 1413 / 2002	4.8 / 1.4 / 4.7 / 1.3

JUNE

	TIME	m		TIME	m
1 TH	0241 / 0834 / 1507 / 2101	5.0 / 1.2 / 4.9 / 1.1	**16** F	0257 / 0855 / 1518 / 2113	4.4 / 1.7 / 4.6 / 1.6
2 F	0336 / 0928 / 1557 / 2157	5.1 / 1.1 / 5.1 / 1.0	**17** SA	0340 / 0938 / 1555 / 2157	4.5 / 1.5 / 4.7 / 1.4
3 SA	0427 / 1019 / 1645 / 2249 ●	5.1 / 1.0 / 5.2 / 0.9	**18** SU	0420 / 1017 / 1634 / 2238	4.6 / 1.4 / 4.9 / 1.3
4 SU	0515 / 1107 / 1731 / 2338	5.2 / 1.0 / 5.2 / 0.8	**19** M	0458 / 1056 / 1711 / 2318 ○	4.7 / 1.3 / 5.0 / 1.2
5 M	0600 / 1153 / 1816	5.1 / 1.0 / 5.2	**20** TU	0535 / 1134 / 1749 / 2358	4.8 / 1.3 / 5.0
6 TU	0024 / 0644 / 1237 / 1901	0.9 / 5.0 / 1.1 / 5.1	**21** W	0613 / 1214 / 1829	4.8 / 1.3 / 5.0
7 W	0109 / 0728 / 1320 / 1945	1.0 / 4.8 / 1.3 / 4.9	**22** TH	0039 / 0653 / 1255 / 1913	1.2 / 4.8 / 1.3 / 5.0
8 TH	0154 / 0812 / 1405 / 2029	1.2 / 4.6 / 1.5 / 4.7	**23** F	0122 / 0737 / 1339 / 2001	1.2 / 4.7 / 1.4 / 5.0
9 F	0239 / 0857 / 1452 / 2116	1.5 / 4.4 / 1.7 / 4.5	**24** SA	0207 / 0826 / 1426 / 2053	1.2 / 4.6 / 1.4 / 4.9
10 SA	0327 / 0948 / 1543 / 2208	1.7 / 4.2 / 1.8 / 4.4	**25** SU	0256 / 0921 / 1518 / 2150	1.3 / 4.5 / 1.5 / 4.8
11 SU	0419 / 1046 / 1638 / 2308 ☽	1.8 / 4.1 / 2.0 / 4.2	**26** M	0349 / 1023 / 1616 / 2253 ☾	1.4 / 4.5 / 1.6 / 4.7
12 M	0516 / 1152 / 1737	1.9 / 4.1 / 2.0	**27** TU	0449 / 1132 / 1719	1.5 / 4.4 / 1.6
13 TU	0014 / 0617 / 1254 / 1836	4.2 / 2.0 / 4.1 / 2.0	**28** W	0002 / 0555 / 1243 / 1828	4.6 / 1.6 / 4.4 / 1.6
14 W	0116 / 0715 / 1347 / 1934	4.2 / 1.9 / 4.3 / 1.9	**29** TH	0113 / 0704 / 1350 / 1938	4.6 / 1.6 / 4.5 / 1.5
15 TH	0210 / 0808 / 1435 / 2026	4.3 / 1.8 / 4.4 / 1.8	**30** F	0222 / 0812 / 1451 / 2045	4.7 / 1.5 / 4.7 / 1.4

JULY

	TIME	m		TIME	m
1 SA	0323 / 0912 / 1545 / 2145	4.8 / 1.4 / 4.9 / 1.2	**16** SU	0315 / 0911 / 1531 / 2133	4.3 / 1.7 / 4.6 / 1.6
2 SU	0417 / 1005 / 1634 / 2238	4.9 / 1.3 / 5.0 / 1.1	**17** M	0400 / 0955 / 1614 / 2218	4.5 / 1.5 / 4.9 / 1.4
3 M	0504 / 1053 / 1720 / 2326 ●	4.9 / 1.2 / 5.1 / 1.0	**18** TU	0442 / 1038 / 1654 / 2302 ○	4.7 / 1.4 / 5.0 / 1.2
4 TU	0548 / 1138 / 1802	5.0 / 1.1 / 5.2	**19** W	0521 / 1119 / 1735 / 2345	4.8 / 1.2 / 5.2 / 1.0
5 W	0010 / 0627 / 1220 / 1842	1.0 / 4.9 / 1.2 / 5.1	**20** TH	0601 / 1201 / 1818	5.0 / 1.1 / 5.3
6 TH	0052 / 0705 / 1300 / 1921	1.1 / 4.8 / 1.2 / 5.0	**21** F	0027 / 0642 / 1243 / 1901	0.9 / 5.0 / 1.1 / 5.3
7 F	0131 / 0741 / 1339 / 1957	1.2 / 4.7 / 1.4 / 4.9	**22** SA	0109 / 0724 / 1325 / 1946	0.9 / 5.0 / 1.1 / 5.3
8 SA	0209 / 0817 / 1418 / 2033	1.3 / 4.6 / 1.5 / 4.7	**23** SU	0152 / 0807 / 1409 / 2033	0.9 / 4.9 / 1.1 / 5.2
9 SU	0247 / 0854 / 1459 / 2111	1.5 / 4.4 / 1.7 / 4.5	**24** M	0236 / 0854 / 1456 / 2124	1.1 / 4.7 / 1.2 / 5.0
10 M	0328 / 0937 / 1544 / 2157	1.7 / 4.3 / 1.8 / 4.3	**25** TU	0323 / 0948 / 1548 / 2223 ☾	1.3 / 4.6 / 1.4 / 4.7
11 TU	0415 / 1032 / 1636 / 2254 ☽	1.9 / 4.1 / 2.0 / 4.1	**26** W	0417 / 1056 / 1650 / 2334	1.6 / 4.4 / 1.7 / 4.5
12 W	0511 / 1141 / 1737	2.1 / 4.0 / 2.1	**27** TH	0524 / 1218 / 1806	1.8 / 4.3 / 1.8
13 TH	0006 / 0616 / 1253 / 1843	4.0 / 2.1 / 4.1 / 2.1	**28** F	0056 / 0643 / 1338 / 1928	4.3 / 1.9 / 4.4 / 1.8
14 F	0119 / 0722 / 1357 / 1946	4.1 / 2.1 / 4.2 / 2.0	**29** SA	0216 / 0800 / 1446 / 2040	4.4 / 1.8 / 4.6 / 1.6
15 SA	0222 / 0820 / 1448 / 2043	4.2 / 1.9 / 4.4 / 1.8	**30** SU	0320 / 0901 / 1541 / 2138	4.5 / 1.6 / 4.8 / 1.4
			31 M	0411 / 0954 / 1627 / 2226	4.7 / 1.4 / 5.0 / 1.2

AUGUST

	TIME	m		TIME	m
1 TU	0452 / 1039 / 1706 / 2310 ●	4.9 / 1.2 / 5.1 / 1.1	**16** W	0421 / 1018 / 1635 / 2242	4.8 / 1.3 / 5.2 / 1.1
2 W	0529 / 1121 / 1743 / 2350	4.9 / 1.1 / 5.2 / 1.0	**17** TH	0501 / 1100 / 1717 / 2325 ○	5.0 / 1.1 / 5.4 / 0.9
3 TH	0603 / 1159 / 1816	5.0 / 1.1 / 5.1	**18** F	0541 / 1143 / 1759	5.2 / 0.9 / 5.5
4 F	0027 / 0634 / 1235 / 1847	1.1 / 4.9 / 1.2 / 5.1	**19** SA	0007 / 0621 / 1224 / 1842	0.7 / 5.2 / 0.8 / 5.6
5 SA	0101 / 0704 / 1308 / 1917	1.2 / 4.9 / 1.2 / 5.0	**20** SU	0049 / 0702 / 1306 / 1925	0.7 / 5.2 / 0.8 / 5.5
6 SU	0133 / 0733 / 1341 / 1947	1.3 / 4.7 / 1.3 / 4.8	**21** M	0130 / 0743 / 1348 / 2010	0.8 / 5.1 / 0.9 / 5.3
7 M	0205 / 0805 / 1415 / 2020	1.4 / 4.6 / 1.5 / 4.7	**22** TU	0212 / 0827 / 1433 / 2059	1.0 / 4.9 / 1.1 / 4.9
8 TU	0239 / 0841 / 1453 / 2058	1.6 / 4.4 / 1.7 / 4.4	**23** W	0257 / 0918 / 1524 / 2157 ☾	1.4 / 4.6 / 1.5 / 4.6
9 W	0317 / 0926 / 1537 / 2146 ☾	1.8 / 4.2 / 2.0 / 4.2	**24** TH	0350 / 1027 / 1629 / 2316	1.7 / 4.3 / 1.8 / 4.2
10 TH	0404 / 1029 / 1635 / 2257	2.1 / 4.0 / 2.2 / 4.0	**25** F	0501 / 1203 / 1756	2.1 / 4.2 / 2.0
11 F	0510 / 1154 / 1751	2.3 / 4.0 / 2.3	**26** SA	0056 / 0633 / 1337 / 1925	4.1 / 2.2 / 4.3 / 1.9
12 SA	0029 / 0633 / 1314 / 1910	3.9 / 2.2 / 4.1 / 2.2	**27** SU	0218 / 0753 / 1445 / 2034	4.3 / 2.0 / 4.6 / 1.7
13 SU	0150 / 0748 / 1420 / 2016	4.0 / 2.2 / 4.3 / 1.9	**28** M	0316 / 0852 / 1535 / 2125	4.5 / 1.7 / 4.8 / 1.4
14 M	0251 / 0845 / 1510 / 2110	4.3 / 1.9 / 4.6 / 1.6	**29** TU	0359 / 0939 / 1614 / 2208	4.7 / 1.4 / 5.0 / 1.3
15 TU	0339 / 0933 / 1553 / 2157	4.6 / 1.6 / 4.9 / 1.3	**30** W	0433 / 1020 / 1647 / 2247	4.9 / 1.3 / 5.1 / 1.1
			31 TH	0503 / 1058 / 1717 / 2323 ●	5.0 / 1.1 / 5.2 / 1.1

Chart Datum: 2.93 metres below Lallemand System (Mean Sea level, Marseilles)

FRANCE, WEST COAST - POINTE DE GRAVE
LAT 45°34'N   LONG 1°04'W
TIMES AND HEIGHTS OF HIGH AND LOW WATERS

YEAR 1989

TIME ZONE –0100 (French Standard Time) For French Summer Time add ONE hour in non-shaded areas

SEPTEMBER

	TIME	m		TIME	m
1 F	0532 1133 1745 2356	5.0 1.1 5.2 1.1	**16** SA	0516 1120 1736 2343	5.3 0.7 5.7 0.6
2 SA	0559 1205 1812	5.0 1.1 5.1	**17** SU	0556 1203 1819	5.4 0.6 5.7
3 SU	0027 0627 1236 1839	1.2 5.0 1.2 5.0	**18** M	0025 0638 1245 1903	0.7 5.4 0.7 5.5
4 M	0057 0655 1306 1907	1.2 4.9 1.3 4.9	**19** TU	0106 0720 1328 1948	0.8 5.2 0.9 5.2
5 TU	0126 0724 1337 1936	1.4 4.7 1.4 4.7	**20** W	0148 0804 1413 2037	1.1 4.9 1.2 4.8
6 W	0156 0756 1410 2009	1.6 4.6 1.6 4.5	**21** TH	0233 0857 1505 2139	1.5 4.6 1.6 4.4
7 TH	0229 0834 1449 2051	1.8 4.3 1.9 4.2	**22** F ☾	0329 1009 1615 2306	1.9 4.3 1.9 4.1
8 F ☽	0310 0931 1540 2202	2.1 4.1 2.2 3.9	**23** SA	0446 1154 1748	2.2 4.2 2.1
9 SA	0408 1101 1657 2348	2.4 4.0 2.4 3.8	**24** SU	0054 0619 1328 1913	4.1 2.5 4.3 2.0
10 SU	0545 1235 1835	2.5 4.0 2.3	**25** M	0207 0735 1431 2016	4.3 2.0 4.6 1.7
11 M	0120 0715 1347 1948	4.0 2.3 4.3 2.0	**26** TU	0257 0830 1516 2103	4.5 1.7 4.8 1.5
12 TU	0224 0816 1445 2043	4.3 1.9 4.7 1.6	**27** W	0335 0915 1552 2142	4.7 1.5 5.0 1.3
13 W	0313 0906 1528 2131	4.6 1.6 5.0 1.2	**28** TH	0405 0954 1621 2219	4.9 1.3 5.1 1.2
14 TH	0355 0952 1611 2217	4.9 1.2 5.3 0.9	**29** F ●	0432 1030 1647 2252	5.0 1.2 5.1 1.2
15 F ○	0436 1036 1654 2301	5.2 0.9 5.6 0.7	**30** SA	0459 1103 1713 2324	5.0 1.2 5.1 1.2

OCTOBER

	TIME	m		TIME	m
1 SU	0526 1135 1740 2354	5.0 1.2 5.1 1.2	**16** M	0533 1141 1759	5.5 0.6 5.6
2 M	0554 1205 1807	5.0 1.2 5.0	**17** TU	0002 0616 1225 1844	0.8 5.4 0.7 5.4
3 TU	0022 0622 1235 1834	1.3 5.0 1.3 4.9	**18** W	0044 0700 1310 1931	1.0 5.2 0.9 5.1
4 W	0051 0651 1306 1902	1.4 4.8 1.4 4.7	**19** TH	0128 0748 1358 2022	1.3 5.0 1.2 4.7
5 TH	0121 0722 1339 1934	1.6 4.7 1.6 4.5	**20** F	0216 0843 1452 2125	1.6 4.7 1.6 4.3
6 F	0154 0800 1416 2016	1.8 4.5 1.9 4.2	**21** SA ☾	0313 0954 1601 2250	2.0 4.4 2.0 4.1
7 SA	0234 0855 1505 2129	2.1 4.2 2.1 4.0	**22** SU	0428 1129 1724	2.2 4.3 2.1
8 SU ☽	0331 1022 1618 2315	2.3 4.1 2.3 3.9	**23** M	0026 0550 1256 1843	4.1 2.2 4.3 2.0
9 M	0501 1155 1755	2.4 4.1 2.2	**24** TU	0134 0702 1359 1944	4.3 2.1 4.5 1.8
10 TU	0046 0634 1311 1912	4.1 2.3 4.4 1.9	**25** W	0224 0758 1446 2032	4.5 1.8 4.7 1.7
11 W	0151 0739 1410 ★ 2010	4.4 1.9 4.8 1.5	**26** TH	0302 0843 1522 2112	4.6 1.6 4.8 1.5
12 TH	0242 0833 1500 2101	4.7 1.5 5.1 1.2	**27** F	0333 0923 1551 2147	4.8 1.5 4.9 1.4
13 F	0326 0922 1546 2148	5.0 1.2 5.4 0.9	**28** SA	0401 1000 1618 2221	4.9 1.4 5.0 1.3
14 SA	0409 1010 1631 ○ 2234	5.3 0.9 5.6 0.7	**29** SU ●	0430 1034 1646 2253	5.0 1.3 5.0 1.3
15 SU	0450 1056 1715 2318	5.4 0.7 5.7 0.7	**30** M	0459 1107 1714 2324	5.0 1.3 5.0 1.3
31 TU	0528 1139 1742 2354	5.0 1.3 4.9 1.3			

NOVEMBER

	TIME	m		TIME	m
1 W	0558 1211 1810	5.0 1.3 4.8	**16** TH	0027 0648 1257 1918	1.1 5.2 1.0 5.0
2 TH	0024 0627 1243 1840	1.5 4.9 1.4 4.7	**17** F	0113 0737 1346 2009	1.3 5.0 1.3 4.7
3 F	0056 0701 1318 1916	1.6 4.8 1.6 4.5	**18** SA	0201 0830 1438 2105	1.6 4.8 1.6 4.4
4 SA	0132 0743 1357 2003	1.8 4.6 1.8 4.3	**19** SU	0255 0930 1536 2212	1.8 4.6 1.8 4.2
5 SU	0215 0838 1447 2111	2.0 4.4 2.0 4.1	**20** M ☾	0357 1041 1642 2330	2.0 4.4 2.0 4.1
6 M	0313 0953 1554 2241	2.2 4.3 2.1 4.1	**21** TU	0504 1159 1751	2.1 4.3 2.1
7 TU	0430 1117 1716	2.3 4.2 2.1	**22** W	0040 0611 1307 1856	2.3 2.1 4.4 2.0
8 W	0007 0551 1233 1830	4.2 2.1 4.5 1.8	**23** TH	0136 0711 1401 1950	4.3 2.0 4.5 1.9
9 TH	0114 0658 1337 1933	4.5 1.8 4.8 1.6	**24** F	0221 0803 1444 2035	4.5 1.9 4.6 1.8
10 F	0209 0758 1432 2029	4.7 1.5 5.1 1.3	**25** SA	0258 0849 1519 2115	4.7 1.7 4.7 1.6
11 SA	0259 0843 1523 2121	5.0 1.2 5.3 1.0	**26** SU	0332 0929 1552 2152	4.8 1.6 4.8 1.5
12 SU	0345 0945 1611 2210	5.2 1.0 5.5 0.9	**27** M	0405 1007 1625 2226	4.9 1.5 4.8 1.4
13 M	0430 1035 1658 ○ 2256	5.4 0.8 5.5 0.9	**28** TU ●	0438 1043 1657 2300	5.0 1.4 4.9 1.4
14 TU	0515 1123 1744 2342	5.4 0.7 5.4 0.9	**29** W	0510 1118 1728 2333	5.0 1.4 4.8 1.4
15 W	0601 1210 1831	5.4 0.8 5.2	**30** TH	0542 1153 1758	5.0 1.4 4.8

DECEMBER

	TIME	m		TIME	m
1 F	0007 0615 1229 1831	1.5 5.0 1.4 4.7	**16** SA	0059 0723 1332 1949	1.2 5.2 1.2 4.8
2 SA	0042 0651 1306 1910	1.6 4.9 1.5 4.6	**17** SU	0144 0808 1417 2033	1.4 5.0 1.4 4.6
3 SU	0121 0735 1348 1957	1.7 4.8 1.6 4.5	**18** M	0229 0854 1503 2120	1.6 4.8 1.6 4.4
4 M	0206 0827 1437 2054	1.8 4.7 1.7 4.4	**19** TU ☾	0318 0943 1552 2214	1.8 4.5 1.9 4.2
5 TU	0259 0929 1534 2205	1.9 4.6 1.8 4.3	**20** W	0410 1039 1647 2320	2.0 4.4 2.1 4.1
6 W	0402 1039 1639 2322	1.9 4.6 1.8 4.3	**21** TH	0508 1148 1749	2.1 4.2 2.2
7 TH	0509 1152 1747	1.9 4.6 1.8	**22** F	0031 0611 1258 1854	4.1 2.2 4.2 2.2
8 F	0034 0618 1302 1855	4.5 1.8 4.8 1.7	**23** SA	0131 0714 1358 1953	4.2 2.1 4.3 2.1
9 SA	0138 0724 1406 1959	4.6 1.6 4.9 1.5	**24** SU	0221 0811 1448 2043	4.4 2.0 4.4 1.9
10 SU	0235 0827 1506 2057	4.9 1.4 5.1 1.3	**25** M	0306 0900 1530 2126	4.6 1.8 4.5 1.7
11 M	0328 0926 1600 2151	5.1 1.2 5.2 1.2	**26** TU	0347 0945 1609 2205	4.7 1.6 4.7 1.6
12 TU	0418 1020 1649 ○ 2241	5.2 1.0 5.3 1.1	**27** W	0422 1025 1645 2243	4.9 1.5 4.8 1.5
13 W	0506 1111 1736 2328	5.4 0.9 5.2 1.1	**28** TH ●	0457 1103 1719 2319	5.0 1.4 4.8 1.4
14 TH	0552 1159 1821	5.4 0.9 5.2	**29** F	0532 1141 1752 2356	5.1 1.3 4.9 1.4
15 F	0014 0638 1247 1906	1.1 5.3 1.0 5.0	**30** SA	0608 1219 1827	5.1 1.3 4.9
31 SU	0034 0646 1258 1905	1.4 5.1 1.3 4.9			

AREA 14 – S Biscay   599

Chart Datum: 2.93 metres below Lallemand System (Mean Sea level, Marseilles)

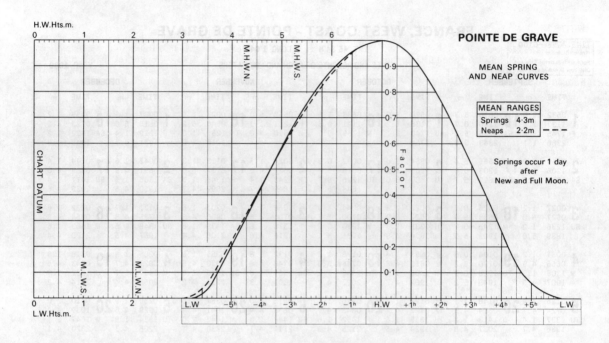

POINTE DE GRAVE

MEAN SPRING AND NEAP CURVES

MEAN RANGES
Springs 4·3m
Neaps 2·2m

Springs occur 1 day after New and Full Moon.

CANAL CONNECTIONS 10-14-11

The Canal Latéral à la Garonne and the Canal du Midi provide a popular route to the Mediterranean, despite the large number of locks to be negotiated. If necessary the transit can be done in about a week, but this is not recommended.

Masts can be unstepped at Royan, Pauillac or Bordeaux. Leave Bordeaux at LW Pointe de Grave for the passage up river to the first lock at Castets-en-Dorthe. If necessary a tow can usually be arranged.

Plenty of stout fenders (for example car tyres in strong canvas bags) are needed. Commercial traffic and boats bound west have right of way. On the Canal Latéral à la Garonne most of the locks are automatic. On the Canal du Midi there are many hire cruisers in the summer months. Fuel is available alongside at Agen, Castelnaudary and Port de la Robine, and at other places by can. Food and water are readily obtained along the route. Further information from Service de la Navigation de Toulouse, 2 Port St Etienne, Toulouse. Tel. 61.80.79.91.

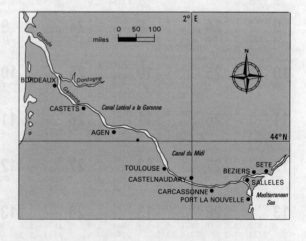

Canal	From	To	Km/Locks	Min Depth (m)	Min Height (m)
Latéral à la Garonne	Castets	Toulouse	193/53	2.2	3.5
Du Midi	Toulouse	Sete	240/65	1.8	3.0
De la Nouvelle	Salleles	Port la Nouvelle	37/14	1.8	3.1

Notes: Max length 30m. Max beam 5.5m.
Headroom of 3.0m is to top of arch. Over a width of 4m, clearance is about 2.50m. Speed limit 8 km/hour (about 4½ kn), but 3 km/hour over aquaducts and under bridges.

AREA 14—S Biscay 601

ROYAN 10-14-12
Charente Maritime

CHARTS
Admiralty 2910, 2916, 2664; SHOM 6141, 6336; ECM 553
TIDES
+0615 Dover; ML 3.0; Duration 0615 Springs, 0655 Neaps; Zone −0100
Standard Port POINTE DE GRAVE (←)

Times				Height (metres)			
HW		LW		MHWS	MHWN	MLWN	MLWS
0000	0600	0500	1200	5.3	4.3	2.1	1.0
1200	1800	1700	2400				

Differences ROYAN
| 0000 | −0020 | −0010 | −0005 | −0.2 | −0.2 | −0.2 | −0.1 |

SHELTER
Good shelter and easy access day and night except with strong winds from the W or NW. The entrance channel is dredged to 1.0m. Yachts use N basin. A good port of call for Canal du Midi with crane for masts.
NAVIGATION
Waypoint No 12 (port-hand) buoy, OcR 4s, 45°36′.45N 01°03′.00W, 237°/057° from/to Jetée Sud Lt, 1.1M. The approaches to La Gironde can be dangerous — see 10.14.5 and 10.14.10. The sandbanks off Royan shift and buoys are consequently altered. Give the Mauvaise bank and shoals round Cordouan a wide berth. Off harbour entrance, there is an eddy, running S at about 1 kn on the flood and 3 kn on the ebb. Beware ferries.
LIGHTS AND MARKS
La Coubre light — Fl (2) W 10s 64m 31M RC, also FRG 42m 12M R030°-043°, G043°-060°, R060°-110°. Channel buoyed past Courdouan Lt Oc(2+1) WRG 12s 60m 21/17M, W014°-126°, G126°-178.5°, W178.5°-250°, W (unintens) 250°-267°, R (unintens) 267°-294.5°, R294.5-014°. Obscured in estuary when bearing over 285°.
RADIO TELEPHONE
VHF Ch 09. Le Verdon Ch 12 16.
TELEPHONE
Hr Mr 46.38.72.22; Aff Mar 46.38.32.75; CROSS 56.09.82.00; SNSM 46.38.75.79; Customs 46.38.51.27; Meteo 56.34.20.11; Dr 46.05.68.69; Hosp 46.38.01.77; Brit Consul 56.52.28.35.

FACILITIES
Marina (620) Slip, P, D, FW, ME, AC, El, Sh, C (6 ton), BH (26 ton); **Les Regates de Royan** Tel. 46.38.59.64;
Mole Nord C, P, D; **Radio Ocean** Tel. 46.38.35.34 El;
Royan Sports Nautique Tel. 46.38.61.35, ME, El, Sh, CH; **Winckel** (la Voilerie) Tel. 46.38.36.95 SM;
Royan Marine Service Tel. 46.38.54.00, CH, SHOM;
Atlas Marine Tel. 46.38.47.99, Sh; **Bernard Luquiau** Tel. 46.39.94.34, Sh, ME; **Meneau Marine** Tel. 46.39.81.27, ME; **Depan' Elec' Marina** Tel. 46.38.35.34, El.
Town Slip, P, D, FW, V, R, Bar, Gaz, PO, Bank, Rly, Air (Bordeaux).
Ferry UK — Roscoff—Plymouth, St Malo—Portsmouth.

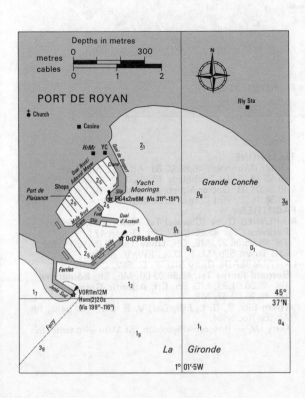

SEUDRE RIVER 10-14-13
Charente Maritime

CHARTS
Admiralty 2648; SHOM 6335, 6912, 6037; ECM 552
TIDES
+0545 Dover; ML 3.5; Duration Springs 0545, Neaps 0700; Zone −0100
Standard Port POINTE DE GRAVE (←)

Times				Height (metres)			
HW		LW		MHWS	MHWN	MLWN	MLWS
0000	0600	0500	1200	5.3	4.3	2.1	1.0
1200	1800	1700	2400				

Differences LA CAYENNE
| −0015 | −0035 | −0020 | 0000 | +0.5 | +0.3 | +0.3 | +0.1 |

SHELTER
La Seudre is navigable to lock at Riberou. There are secure anchorages at La Cayenne, La Grève (½M upstream), and off entrance to Chenal de la Tremblade; or yachts can lock into the basin near Marennes, 2M from the river.

SEUDRE RIVER continued

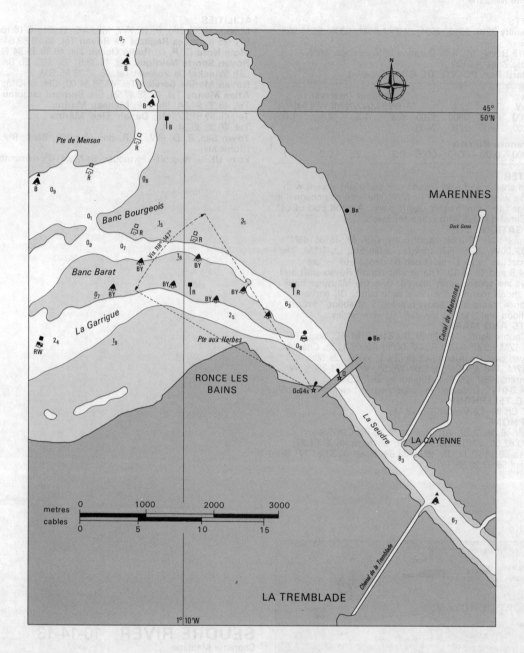

NAVIGATION
Waypoint Pertuis de Maumusson (safe water) buoy, Whis, 45°47′.70N 01°17′.85W, 260°/080° from/to Pte de Gatseau, 2.7M. Pertuis de Maumusson can be used only in good weather, at about HW−1, see 10.14.5. In even moderate weather it is extremely dangerous, particularly with out-going stream or any swell. It is advisable to approach La Seudre from N, through Coreau d'Oleron, and thence through Chenal de la Soumaille (dries about 0.7m). Chenal de la Garrigue carries slightly more water. Both are marked by Bns and buoys. Beware oyster beds. Pont de Seudre has clearance of 18m. An overhead cable (16m) spans Canal de Marennes.

LIGHTS AND MARKS
There are no leading lights or marks. Lights: Pte de Mus de Loup Oc G 4s 8m 6M vis 118°-147°. On bridge between piers 6 and 7; downstream Q 20m 10M vis 054°-234°; upstream Q 20m 10M vis 234°-054°. Channel is marked by W boards.

RADIO TELEPHONE
VHF Ch 09.

TELEPHONE
Hr Mr (Marennes) 46.85.02.68; Aff Mar, Marennes 46.85.14.33, La Tremblade 46.36.00.22; Customs 46.36.03.03; Doctor Marennes 46.85.23.06, La Tremblade 46.36.16.35; Brit Consul 56.36.16.35.

FACILITIES
MARENNES **Quay (Claude)** Tel. 46.85.15.11, ME, CH; **Paraveau** Tel. 46.85.10.77, Sh; **Ocean 17** Tel. 46.85.36.27, ME, CH; **Quay** Tel. 46.85.15.11, ME, CH; **Town** Slip, M, P, D, L, FW, V, R, Bar.
LA TREMBLADE **Quay** Slip, P, D, FW, C (5 ton); **Bernard Frères** Tel. 46.36.09.00, ME, Sh; **Boulanger** Tel. 46.36.01.34, ME, Sh, CH; **Atlantic Garage** Tel. 46.36.00.67, ME;
Town Slip, P, D, L, FW, Gaz, V, R, Bar, PO, Bank, Rly, Air (La Rochelle).
Ferry UK — Roscoff—Plymouth; St Malo—Portsmouth.

AREA 14—S Biscay 603

ILE D'OLÉRON 10-14-14
Charente Maritime

CHARTS
Admiralty 2648, 2746; SHOM 3711, 6334, 6335, 6913, 6914; ECM 552

TIDES
+0545 Dover; ML 3.8; Duration 0540; Zone −0100
Standard Port POINTE DE GRAVE (←)

Times				Height (metres)			
HW		LW		MHWS	MHWN	MLWN	MLWS
0000	0600	0500	1200	5.3	4.3	2.1	1.0
1200	1800	1700	2400				

Differences ROCHEFORT
+0015 −0020 +0045 +0120 +1.1 +0.9 +0.1 +0.4

There are two harbours, Le Chateau d'Oléron and Boyardville. A third on W coast, La Cotiniere suffers almost constantly from Atlantic swells.

SHELTER
LE CHATEAU: Good shelter, but very full of oyster boats which use inner harbour. Yachts lie on quay SW side (dries 3m) — check with Hr Mr. Access ∓3.
BOYARDVILLE: Good anchorage, sheltered from S and W, in 3m (sand) ½M N of La Perrotine Lt. Landing jetty inshore. Good shelter alongside quays (dry) on N side of harbour. Access HW∓2. Boyardville yacht harbour, non-tidal, entered through automatic gates which open when height of tide is 2.4m. (Approx HW∓2½.)

NAVIGATION
Waypoint Pertuis d'Antioche PA (safe water) buoy, Iso 4s,

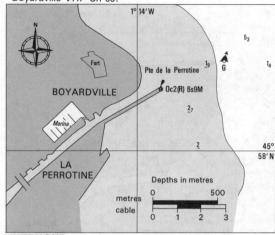

TELEPHONE
Hr Mr (Le Chateau) 46.47.00.01; Hr Mr (Boyardville) 46.47.23.71; Customs 46.47.62.53; Aff Mar 46.47.60.01; CROSS 56.09.82.00; Meteo 46.41.29.14; Auto 46.41.11.11; Dr 46.47.60.68; Brit Consul 56.52.28.35.

FACILITIES
LE CHATEAU **Jetty**, Slip, L, FW, C (2.5 ton); **Blondel** Tel. 46.47.61.89, ME, El; **Sorlut Marine** Tel. 46.47.54.08, ME, El, Sh, CH; **Oléron-Motors-Cycles** Tel. 46.47.62.78, ME, El, Ch; **Dubois** Tel. 46.47.62.57, ME, Sh, El, CH; **Magasin 2000** Tel. 46.47.61.43, CH;
BOYARDVILLE **Marina** (150+50 visitors), AC, FW, C (7 ton), Slip; **Môle chenal de la Perrotine** P, D; **Sodinautic** Tel. 46.47.25.65, ME, Sh, CH; **Marine Oléron** Tel. 46.47.01.36, ME, El, Sh, Ch; **Barbaud** Tel. 46.47.01.10, ME.
Town (Le Chateau) V, Gaz, R, Bar, PO; Bank; Rly (Marennes); Air (La Rochelle).
Ferry UK — Roscoff—Plymouth, St Malo—Portsmouth

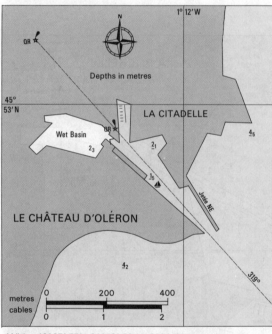

Whis, 46°05'.55N 01°42'.60W, 283°/103° from/to Pte de Chassiron Lt, 13M.
LE CHATEAU: Bridge to mainland lies 2M S, 15m clearance under spans 20-24, channel marked by boards — W&G to stbd, W&R to port, illuminated at night. Beware Grande Mortanne, a drying ledge, marked by Bns. From N, beware Rocher du Doux (dries) and Rocher Juliar (dries). Mortanne Sud Bn marks entrance to channel (depth 0.8m) indicated by Ldg Lts.
BOYARDVILLE: Bar dries 2m and extends 3 ca (560m) E of breakwater head. Inside bar the channel mostly dries. Beware ferries.

LIGHTS AND MARKS
LE CHATEAU: Ldg Lts 319°, Both QR 11/24m 7M; W Tr with R top; synchronised. Tourelle Juliar Q(3) WG 10s 12m 11/8M; E cardinal mark; vis W147°-336°, G336°-147°.
BOYARDVILLE: Breakwater head Oc(2) R 6s 8m 6M; W Tr with R top; obsc by Pte des Saumonards when brg less than 150°.

ROCHEFORT 10-14-15
Charente Maritime

CHARTS
Admiralty 2748, 2746; SHOM 4333; ECM 552

TIDES
+0610 Dover; Zone −0100
Standard Port POINTE DE GRAVE (←)

Times				Height (metres)			
HW		LW		MHWS	MHWN	MLWN	MLWS
0000	0600	0500	1200	5.3	4.3	2.1	1.0
1200	1800	1700	2400				

Differences ROCHEFORT
+0015 −0020 +0045 +0120 +1.1 +0.9 +0.1 +0.4
ILE D'AIX
−0005 −0035 −0025 −0015 +0.9 +0.7 +0.4 0.0

SHELTER
Rochefort is on N bank of Charente, about 10M from the mouth. Entry advised on late flood, as bar breaks on ebb. Anchor out of channel at Martrou or Soubise, or enter Port de Plaisance in Basin No1 (access HW∓1), first basin on W bank: pontoon N side of entrance for waiting. No 3 Basin, 2½ ca (400m) N, is for commercial craft only. The river is navigable to Tonnay-Charente.

NAVIGATION
Waypoint Les Palles N cardinal buoy 45°59'.58N 01°09'.53W, 293°/113° from/to front Ldg Lt 115°, 4.0M. Stream in river runs about 2 kn (4 kn in narrows), and at springs there is a bore. Beware wreck just S of first Ldg Line (115°). When WSW of Fouras pick up second (Port-des-Barques) Ldg Line (135°). The bar at Fouras carries about 0.5m. From Port-des-Barques follow the alignment of lettered pairs of Bns. There is a lifting bridge at Martrou, with moorings above and below it.

ROCHEFORT continued

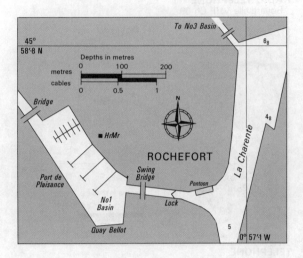

LIGHTS AND MARKS
Ile d'Aix Fl WR 5s 24m 24/20M; twin W Trs with R tops; vis R103°-118°, W118°-103°. Ldg Lts 115°, both QR; W square Trs with R tops; intens 113°-117°. Port Sud de Fouras, pierhead, Fl WR 4s 6m, 9/6M; vis R117°-177°, W177°-117°. Port-des-Barques Ldg Lts 135°, both Iso G 4s; W square Trs; synchronised and intens 125°-145°.

RADIO TELEPHONE
VHF Ch 09 16

TELEPHONE
Hr Mr 46.84.30.30; Aff Mar 46.84.22.67; CROSS 56.09.82.00; Customs 46.99.03.90; Meteo 46.41.29.14; Dr 46.99.61.11; Brit Consul 56.52.28.35.

FACILITIES
Marina (180+20 visitors) FW, ME, El, Sh, AC, C (30 ton); **Port Neuf** Slip, FW; **Club Nautique Rochefortais** Slip; **Saco** Tel. 46.99.06.02, ME, El, Sh, CH; **Rochefort Marine** Tel. 46.87.52.08, ME, El, Sh, CH; **Town** P, D, V, Gaz, R, Bar, PO, Bank, Rly, Air (La Rochelle).
Ferry UK — Roscoff—Plymouth; St Malo—Portsmouth.

LA ROCHELLE 10-14-16
Charente Maritime

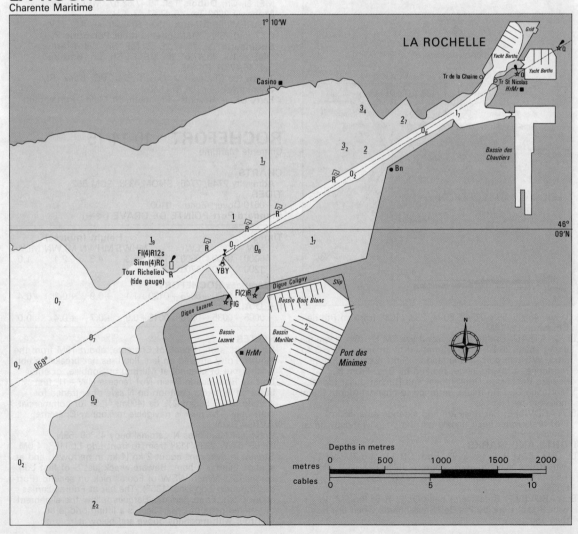

AREA 14—S Biscay 605

LA ROCHELLE continued

CHARTS
Admiralty 2743, 2746, 2648, 2641; SHOM 6334, 6468; ECM 551, 1022
TIDES
+0555 Dover; ML 3.6; Zone −0100
Standard Port POINTE DE GRAVE (←)

Times				Height (metres)			
HW		LW		MHWS	MHWN	MLWN	MLWS
0000	0600	0500	1200	5.3	4.3	2.1	1.0
1200	1800	1700	2400				

Differences LA ROCHELLE
+0005 −0035 −0020 −0015 +0.8 +0.6 +0.4 0.0
LA PALLICE
+0005 −0035 −0020 −0015 +0.8 +0.6 +0.4 0.0

NOTE: La Pallice is the commercial port of La Rochelle and yachts are not welcome there, nor are any facilities provided. La Rochelle has two ports for yachts, La Rochelle-Ville and the Port des Minimes (a large marina).

SHELTER
La Rochelle-Ville harbour, in the old town, is entered between the two towers of St Nicolas and La Chaine. It has an outer basin (tidal), and an inner basin (non-tidal with depth of 3m) on the E side which is entered by a gate (open HW−2 to HW+½). In outer basin (100 berths), the first pontoon is reserved for fishing boats. Port des Minimes is a marina with a depth of 2m (max length 14m). All the above give excellent shelter. There is a water bus from Minimes to the town.

NAVIGATION
Waypoint Chauveau S cardinal buoy, VQ(6)+LFl 10s, Whis, 46°06′.63N 01°15′.98W, 240°/060° from/to Tour Richelieu, 4.6M. Approach from about 1M S of Le Lavardin Lt Tr on Ldg Line 059°. Drying rocks off Pte des Minimes (SW of marina) extend ¼M offshore. Tour Richelieu marks S extremity of rocks extending from N shore. Least depth in channel 0.2m. Just past Tour Richelieu (tide gauge for channel), entrance to Port des Minimes is marked by W cardinal buoy, and buoys mark NE side of channel to marina. For La Rochelle-Ville keep on 059° Ldg Line in channel marked by buoys on N side, least depth 0.2m.

LIGHTS AND MARKS
Ldg Lts 059°. Front Q 15m 14M (Fl 4s by day); R round Tr, W bands. Rear Q 25m 14M (Fl 4s by day); W octagonal Tr, G top; synchronised with front, obsc 061°-065° by St Nicolas Tr. Le Lavardin Fl(2) WG 6s 14m 11/8M; B Tr, R band; vis G160°-169°, W169°-160°. Tourelle Richelieu Fl(4)R 12s 10m 9M; R Tr; RC; Siren(4) 60s. Port des Minimes W Mole head Fl G 4s, E Mole head Fl(2)R 6s.

RADIO TELEPHONE
La Pallice VHF Ch 12 16 (HW−2 to HW+1). Port des Minimes Ch 09 (H24).

TELEPHONE
Port des Minimes Hr Mr 46.44.41.20, Aff Mar 46.41.43.91, CROSS 56.09.82.00; Customs 46.42.64.64; Meteo 46.41.29.14; Auto 46.41.11.11; Dr 46.42.19.22; Hosp 46.27.33.33; Brit Consul 56.52.28.35.

FACILITIES
PORT DES MINIMES **Marina** (2,800+250 visitors pontoons 14 and 15), Gaz, AC, Slip, C (10 ton), P, D, FW, BH (15 ton), R: (also control berths in wet dock in town);
Société des Regates Rochellaises Tel. 46.44.62.44;
Atlantic Loisirs Tel. 46.44.21.35, ME, Sh, CH;
Atlantique-Plaisance Tel. 46.44.32.80, ME, Sh, CH;
Atlantique Voile Tel. 46.44.20.68, SM, ME, El, Sh, CH;
Comptoir Maritime Rochelais Tel. 46.44.34.97, ME, Sh, CH; **Fabre Marine** Tel. 46.41.33.25, El; **Proust** Tel. 46.44.13.66, ME, El, Sh, CH; **La Rochelle Yachting** Tel. 46.44.13.66, Sh, El, ME, CH; **Pochon** Tel. 46.41.30.53, El; **Accastillage Diffusion** Tel. 46.41.33.08, ME, El, Electronics, Sh, CH, SHOM; **Chantier Pinta** Tel. 46.44.40.70, SM; **Chéret** Tel. 46.44.17.99, SM; **Voile Système** Tel. 46.41.83.52, SM;

LA ROCHELLE-VILLE **Quay** Slip, FW, C (10 ton), SM, ME, Sh, CH; **Outer Basin** (100)
Town P, D, V, Gaz, R, Bar, PO, Bank, Rly, Air.
Ferry to the Ile de Ré; internal air services from Laleu airport (2½km N of port).
Ferry UK — Roscoff—Plymouth; St Malo—Portsmouth.
Note: Bridge under construction between Pte de La Repentie (on mainland) and Pte de Sablanceaux (on Ile de Ré). Passage to/from Rade de La Pallice must be made N-bound between piles Nos 13 and 14, and S-bound between piles Nos 11 and 12. The passages are buoyed.

ST MARTIN, Ile de Ré 10-14-17
Charente Maritime

CHARTS
Admiralty 2641; SHOM 6521, 6668; ECM 551
TIDES
+0535 Dover; ML 3.4; Zone −0100
Standard Port POINTE DE GRAVE (←)

Times				Height (metres)			
HW		LW		MHWS	MHWN	MLWN	MLWS
0000	0600	0500	1200	5.3	4.3	2.1	1.0
1200	1800	1700	2400				

Differences ST. MARTIN, Ile de Ré
−0025 −0045 −0005 −0005 +0.8 +0.4 +0.1 −0.3

SHELTER
Approach channel dries 1.2m; access HW−3 to HW+1. Avant port protected by mole on W side, but exposed to N and NE, leads to drying basin where fishing boats lie. Wet dock (with marina) entered by gates which open about HW−1 to HW+1, with sill 0.8m above CD. Complete shelter in wet dock (depth 3m), but often very crowded. Berthing instructions from Hr Mr.

NAVIGATION
Waypoint Rocha N cardinal buoy, Q, 46°14′.75N 01°20′.80W, 020°/200° from/to St Martin mole head, 2.4M. Time arrival for dock gates opening. From the NW, pass N and E of Le Rocha, a rocky bank running 2½M ENE from Pte du Grouin. Approach with church Tr in transit with mole head Lt (202°). From SE, pass well N of Le Couronneau N cardinal Bn, about ¾M NE of entrance, marking a drying ledge in R sector of St Martin Lt (245°-281°).

LIGHTS AND MARKS
Ldg Line, church and Lt Ho in line 210°. NW of entrance Iso G 4s 10m 7M obscured by Pte de Loix when bearing less than 124°. SE of entrance, Oc(2)WR 6s 18m 10/7M W shore-245°, R245°-281°, W281°-shore.

RADIO TELEPHONE
VHF Ch 09 (0800-1900 local time) in summer season.

TELEPHONE
Hr Mr 46.09.26.69; Aff Mar 46.09.68.89; Customs 46.09.21.78; Meteo 46.41.29.14; Auto 46.41.11.11; CROSS 56.09.82.00; Dr 46.09.20.08; Hosp 46.09.20.01; Brit Consul 56.52.28.35.

FACILITIES
Marina (135+50 visitors), P, D, FW, ME, El, Sh; **Garage du Port** Tel. 46.09.20.41, ME; **YC St Martin** Tel. 46.09.22.07; **Chantiers Naval** Tel. 46.29.41.94, ME, El, Sh, CH, SHOM; **Quay** FW, C (4 ton); **Ré-Multiservices** Tel. 46.09.40.74, ME, El, Sh, CH.
Town P, D, V, Gaz, R, Bar, PO, Bank, Rly (La Rochelle), Air (La Rochelle).
Ferry UK — Roscoff-Plymouth; St Malo-Portsmouth.

ST MARTIN continued

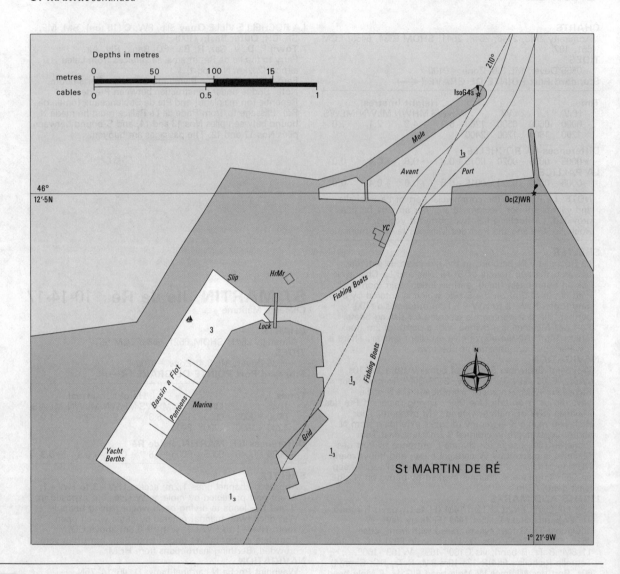

BOURGENAY 10-14-18
Vendee

CHARTS
Admiralty 2648; SHOM 6522, 6337; ECM 1022
TIDES
+0600 Dover; ML 3.1; Duration 0640; Zone −0100
Standard Port BREST (→)

Times				Height (metres)			
HW		LW		MHWS	MHWN	MLWN	MLWS
0500	1100	0500	1100	7.5	5.9	3.0	1.4
1700	2300	1700	2300				

Differences BOURGENAY
−0030 +0015 −0035 −0030 −2.2 −1.7 −0.9 −0.6

SHELTER
Good shelter in the marina, access H24. In bad weather, especially with SW winds, a big swell can break at the entrance, Yachts can anchor off in good weather with offshore winds.
NAVIGATION
Waypoint 46°25'.60N, 1°41'.50W, 220°/040° from/to pierhead, 0.9M. Entrance channel leads 040° from safe water Lt Buoy to seaward of Roches de Joanne (which are dangerous in bad weather). Beware shallow patch to W of entrance, marked by stbd-hand Bn.
LIGHTS AND MARKS
Ldg Lts 040°, both QG 9M; G panels. Digue Ouest head Fl R 4s 9M; R structure. Mole E head Iso G 4s 5M; not vis to seaward. Breakwater elbow Fl (2) R 6s 5M, not vis to seaward.

BOURGENAY continued

RADIO TELEPHONE
VHF Ch 09 16 (office hours; in summer 0800-2100).
TELEPHONE
Hr Mr 51.22.20.36; Port Office 51.22.20.36; SNSM 51.32.26.69; CROSS 97.55.35.35; Auto 51.62.45.99; Customs 51.32.02.33; Aff Mar 51.21.01.80; Dr 51.90.62.68.
FACILITIES
Marina (390 + 110 visitors) Tel. 51.22.20.36, Slip, FW, AC, P, D; **Quai 85** Tel. 51.22.24.36, Slip, C (15 ton), ME, Sh, CH; **Aquatic Loisiers** Tel. 51.22.29.47, CH, ME; **Super 2000** Gaz; **Association Nautique de Bourgenay** Tel. 51.22.20.36; **Town** R, V, Bar, PO, Bank, Rly (Les Sables d'Olonne) Air (La Lande, Chateau d'Olonne). Ferry UK — Roscoff—Plymouth; St. Malo—Portsmouth.

AGENTS NEEDED
There are a number of ports where we need agents, particularly in France.
ENGLAND   Swale, Havengore, Berwick.
SCOTLAND Firth of Forth, Scrabster, Mallaig, Loch Sunart, Loch Aline.
IRELAND   Kilrush, Wicklow, Westport/Clew Bay.
FRANCE   Arcachon, Seudre R, Ile d'Oleron, Rochfort, Ile de Re, St. Giles-Croix-de-Vie, Ile d'Yeu, Pouliguen, Le Croisic, La Forêt, Ile de Bréhat.
GERMANY   Norderney, Dornumer-Accumersiel.
If you are interested in becoming our agent for any of the above, please write to the editors and get your free copy of the Almanac every year. You do not have to be resident in a port to be the agent, but at least a fairly regular visitor.

LES SABLES D'OLONNE 10-14-19
Vendee

CHARTS
Admiralty 2648; SHOM 6523, 6552, 6551; ECM 1022
TIDES
+ 0600 Dover; ML 3.1; Duration 0640; Zone − 0100
Standard Port BREST (→)

Times				Height (metres)			
HW		LW		MHWS	MHWN	MLWN	MLWS
0500	1100	0500	1100	7.5	5.9	3.0	1.4
1700	2300	1700	2300				

Differences LES SABLES D'OLONNE
−0030 +0015 −0035 −0030 −2.2 −1.7 −0.9 −0.6

SHELTER
Available at all states of tide, and entry is good except in winds from SE to SW when approaches get rough. There are two main channels − the navigation controlled (SW) channel with La Potence Lt bearing 033°, which leads into E channel on Ldg Line 320°. In bad weather use the E channel. Sailing is forbidden in the access channel to harbour. Visitors marina pontoons A, C, E and F. Access to basin forbidden to yachts. Access to Port Olona Marina H24. There are also 100 berths in harbour for yachts on passage.

NAVIGATION
Waypoint Nouch Sud S cardinal buoy, Q(6)+LFl 15s, 46°28'.63N 01°47'.43W, 220°/040° from/to front Ldg Lt 033°, 1.2M. To the W, beware Les Barges d'Olonne, extending 3M W from Pte de Aiguille. Le Noura and Le Nouch are two isolated rocks on shallow patches SSE of Jetée St Nicolas. Further SE, Barre Marine breaks, even in moderate weather. A buoyed wreck (dries) lies off harbour entrance, to E of 320° Ldg Line. From port to marina, dredged channel (2m) is on W side.
LIGHTS AND MARKS
SW passage Ldg Lts 033°. Front Iso R 4s 14m 14M. Rear, 330m from front, Iso R 4s 33m 14M; (both H24). E passage Ldg Lts 320°. Front QG 11m 10M. Rear, 465m from front, Oc (2+1) 12s 33m 13M; RC. Port signals St Nicolas jetty head UQ(2)R 1s 16m 10M vis 094°-043°; Horn(2) 30s.
RADIO TELEPHONE
Port VHF Ch 16, (HW−2 to HW+2). Marina Ch 09 16 (0600-2400 LT in season; 0800-2000 LT out of season and occasionally outside these times).
TELEPHONE
Hr Mr Port de Plaisance 51.32.51.16; Aff Mar 51.21.01.80; CROSS 56.09.82.00; Meteo 51.36.10.78; Auto 51.62.45.99; Customs 51.32.02.33; SNSM 51.32.26.69; Dr 51.95.14.47; Hosp 51.21.06.33; Brit Consul 56.52.28.35.

LA SABLES D'OLONNE continued

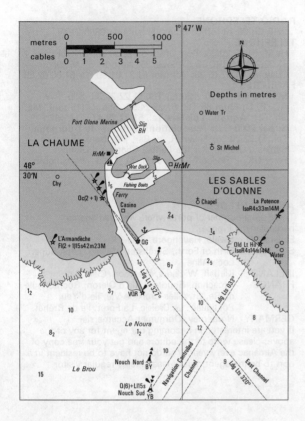

FACILITIES
Port Olona Marina (600+66 visitors), Tel. 51.32.51.16, Slip, P, D, FW, ME, El, AC, Sh, BH (27 ton), Access H24; **Plaisance 85** Tel. 51.95.07.38, ME, El, Sh, CH, BH (27 ton); **Heriaud** Tel. 51.21.06.88, ME, El, Sh, CH; **Parisot Marine** Tel. 51.32.59.97, ME, CH; **Sablais Nautique** Tel. 51.32.62.16, ME, CH, El, Sh, SHOM; **Radio Ocean** Tel. 51.32.01.07, Electronics; **Martineau** Tel. 51.32.30.74, ME; **Compagnie Radio Marine** Tel. 51.32.07.73, El; **Town** P, D, V, Gaz, R, Bar, PO, Bank, Rly, Air.
Ferry UK — Roscoff—Plymouth; St Malo—Portsmouth.

ST GILLES-CROIX-DE-VIE 10-14-20
Vendee

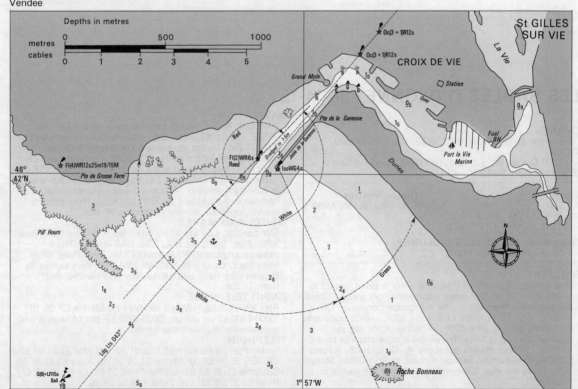

ST GILLES-CROIX-DE-VIE continued

CHARTS
Admiralty 2647; SHOM 6613, 6853; ECM 1022
TIDES
+0555 Dover; ML 3.1; Duration 0600; Zone −0100
Standard Port BREST (→)

Times				Height (metres)			
HW		LW		MHWS	MHWN	MLWN	MLWS
0500	1100	0500	1100	7.5	5.9	3.0	1.4
1700	2300	1700	2300				

Differences ST GILLES-SUR-VIE
−0030 −0015 −0030 −0035 −2.2 −1.7 −0.9 −0.6

SHELTER
Good shelter, and easy access except in strong SW winds or swell when breakers form off entrance. Channel is dredged, but very shallow near breakwater heads. Access HW−2 to HW. In harbour, channel 1.5m deep. On N bank lie:— small yacht basin inside Grand Môle; two tidal fishing boat basins; beyond them the marina nominally dredged to 1.5m. Up river on E bank at St-Gilles-sur-Vie yachts can lie alongside quay (dries). In good weather anchor off entrance, close SE of leading line. Anchoring prohibited in channel.

NAVIGATION
Waypoint Pill'Hours S cardinal buoy, Q(6)+LFl 15s, Bell, 46°41'.10N 01°58'.33W, 233°/053° from/to Jetée de la Garenne Lt, 0.90M. Landmarks are Pte de Grosse-Terre (rocky headland) with Lt Ho, the rear Ldg Lt structure, two spires, and water towers round town. Beware Rocher Pill'Hours and drying reefs extending 1 ca (180m) SE. Tide runs strongly in entrance, particularly on ebb.

LIGHTS AND MARKS
Ldg Lts at 043°. Both Oc(3+1)R 12s 7/28m 15M; intens 033.5°-053.5° synchronized. Pointe de Grosse Terre Fl(4)WR 12s 25m 19/15M; vis W290°-125°, R125°-145°. SE Mole head Iso WG 4s 7m 8/5M; G220°-335°, W335°-220°. NW Mole head Fl(2)WR 6s 8m 9/6M; R045°-225°, W225°-045°, Reed(2) 20s.

RADIO TELEPHONE
VHF Ch 09 (season 0600-2200; out of season 0800-1200, 1400-1800).

TELEPHONE
Hr Mr Port de Plaisance 51.55.30.83; Aff Mar 51.55.10.58; Customs 51.55.10.58; CROSS 97.55.35.35; SNSM 51.55.01.19; Meteo 51.36.10.78; Auto 51.62.45.99; Dr 51.55.11.93; Brit Consul 40.63.16.02.

FACILITIES
Port la Vie Marina (750+60 visitors) Tel. 51.55.30.83, Bar, CH, Gaz, R, SM, V, BH (26 ton), P, D, FW, ME, El, AC, Slip (Access H24); **Quay** Slip, FW, C (6 ton); **Cecillon Yachting** Tel. 51.55.52.90, ME, El, Sh, CH; **Radio Marine** Tel. 51.55.02.84, Electronics; **Massif Marine** Tel. 51.55.42.07, ME, El, Sh, CH, C (15 ton); **Cooperative Maritime des Marins-Pecheurs** Tel. 51.55.31.39, CH, SHOM; **Meca Marine** Tel. 51.55.42.93, ME, El, Sh, CH; **CN de Havre de Vie** Tel. 51.55.87.91; **BM Sails** Tel. 51.55.50.48, SM; **Transnav** Tel. 51.55.31.60 Electronics;
Town V, Gaz, R, Bar, PO, Bank, Rly. There is a ferry to Ile d'Yeu.
Ferry UK — Roscoff—Plymouth; St Malo—Portsmouth.

ILE D'YEU 10-14-21
Vendee

CHARTS
Admiralty 2647; SHOM 6853, 6513, 6890; ECM 549
TIDES
+0550 Dover; ML 3.1; Duration 0600; Zone −0100
Standard Port BREST (→)

Times				Height (metres)			
HW		LW		MHWS	MHWN	MLWN	MLWS
0500	1100	0500	1100	7.5	5.9	3.0	1.4
1700	2300	1700	2300				

Differences PORT JOINVILLE
−0035 −0010 −0035 −0035 −2.2 −1.8 −0.9 −0.6

Although Port de la Meule is a fine weather small port (dries) on the S coast all information following refers to Port Joinville unless otherwise stated.

SHELTER
Swell runs in the entrance in winds from N to NE but shelter is good once in the marina in the NE part of the harbour. Marina gets very crowded mid-season. Yachts can anchor fore and aft (rafted together) between Gare Maritime and the ice factory, leaving room for commercial traffic to west.

NAVIGATION
Waypoint Basse Mayence N cardinal buoy, 46°44'.65N 02°19'.10W, 055°/235° from/to breakwater Lt, 1.4M. Harbour gets very full in summer. Beware Basse du Bouet 3ca (500m) NW, La Sablaire shoal to the E and rocks along the coast both sides of harbour entrance. Anchoring in outer harbour is prohibited. In the main harbour a forbidden zone is marked by R&W paint. At night keep in W sector of Dir Q WRG Lt.

LIGHTS AND MARKS
There is a conspic water tower behind the town; NW jetty head Oc(3)WG 12s 9m 11/9M; G shore-150°, W150°-232°, G232°-279°, W279°-285°, G285°-shore. Horn (3) 30s; tidal signals. Quai de Canada Ldg Lts 219°, Front QR 11m 6M; pylon; Rear (85m from front) QR 16m 6M; mast, both vis 169°-269°. Quai de Canada head Iso G 4s 7m 6M unintens 337°-067°. Galiote jetty root Fl(2)R 5s 1M.

RADIO TELEPHONE
VHF Ch 09 (office hours).

TELEPHONE
Hr Mr 51.58.38.11; Harbour Office 51.58.51.10; Aff Mar 51.58.32.45; Customs 51.58.37.88, CROSS 97.55.35.35; Meteo 51.36.10.78; Auto 51.62.45.99; SNSM 51.58.35.39; Dr 51.58.31.70; Hosp 51.68.30.23; Brit Consul 40.63.16.02.

FACILITIES
Marina (100+40 visitors), FW, AB, P, D, ME, El, Sh; **2nd Tidal Basin** Tel. 51.58.38.11, Slip, L, FW, C (5 ton); **Berlivet** Tel. 51.58.33.11, CH; **CN Ile d'Yeu** Tel. 51.58.31.50; **Mollé Gilbert** Tel. 51.58.36.40, ME, El, CH; **Oya Nautique** Tel. 51.58.35.54, ME, El, CH; **Poiraud** Tel. 51.58.50.69, ME; **Radio Maritime** Tel. 51.58.32.83, Electronics; **M. Naud Gilbert** Tel. 51.58.32.44, Gaz; **Town** V, Gaz, R, Bar, PO, Bank, Rly (St-Gilles-Croix-de-Vie), Air (Nantes). Flights to Nantes and (summers only) to Les Sables d'Olonne. (Airfield 2M westward).
Ferry UK — Roscoff—Plymouth; St Malo—Portsmouth.
Note: There are no cash telephones —
   Phone-cards available from PTT or at some tabacs and cafés.

ILE D'YEU continued

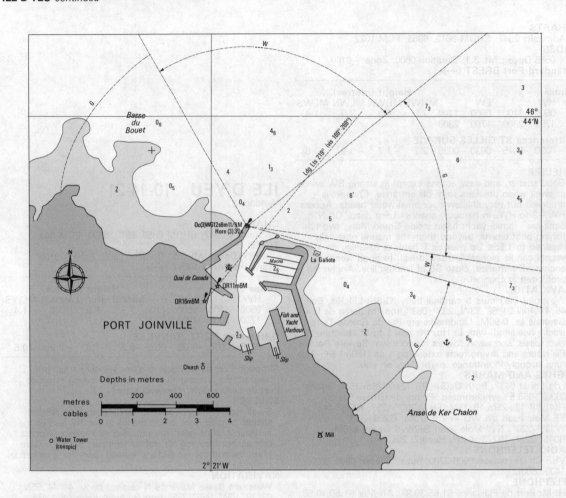

ILE DE NOIRMOUTIER 10-14-22
Vendee

CHARTS
Admiralty 2647, 3216; SHOM 5039; ECM 549

TIDES
+0555 Dover; ML 2.8; Zone −0100
Standard Port BREST (→)

Times				Height (metres)			
HW		LW		MHWS	MHWN	MLWN	MLWS
0500	1100	0500	1100	7.5	5.9	3.0	1.4
1700	2300	1700	2300				

Differences BOIS DE LA CHAISE
−0030 −0020 0000 −0005 −2.1 −1.9 −1.4 −1.0
FROMENTINE
−0025 −0020 −0005 +0015 −2.2 −1.9 −1.3 −0.9

SHELTER
Anchor in Bois de la Chaise, exposed to N and E winds. Good shelter alongside in Noirmoutier-en-l'Ile but harbour dries 1.8m-2.4m, access HW∓1. L'Herbaudière comprises fishing harbour (W side) and marina (E side), dredged 2m-3m with good shelter. Visitors report to Pontoon F.

NAVIGATION
Waypoint Baie de Bourgneuf SN3 (safe water) buoy, LFl 10s, 47°06'.00N 02°21'.50W, 318°/138° from/to Basse du Martroger Lt Bn, 4.5M. Chaussée des Boeufs extends 3½M W and 6M SW of island, which is fringed by rocks and banks on N and E sides. W sector (187.5°-190°) of L'Herbaudière West breakwater Lt leads into entrance channel, dredged to 1.3m and passing close W of two 0.8m patches. W breakwater obscures vessels leaving. For Noirmoutier-en-l'Ile, a long channel marked by beacons leads through rocks. La Vendette is one of the highest parts of rocky ledges, E of entrance to Noirmoutier-en-l'Ile.
For Fosse de Fromentine see 10.14.24.

LIGHTS AND MARKS
Noirmoutier jetty Oc(2)R 6s 6m 6M; L'Herbaudière E jetty Fl(2)R 6s 8m 5M; W jetty Oc(2+1)WG 12s 9m 10/7M; W187.5°-190°, G elsewhere.

RADIO TELEPHONE
L'Herbaudière VHF Ch 09. (Hr Mr in working hours).

ILE DE NOIRMOUTIER continued

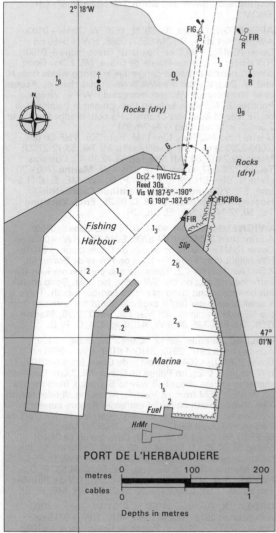

PORT DE L'HERBAUDIERE

TELEPHONE
Hr Mr (L'Herbaudière) 51.39.05.05; Aff Mar 51.39.01.64; SNSM 51.39.33.90; CROSS 97.55.35.35; Meteo 40.75.80.07; Auto 40.04.15.15; Customs 51.39.06.80; Dr 51.39.05.64; Brit Consul 40.63.16.02.

FACILITIES
L'HERBAUDIÈRE. **Marina** (442 + 50 visitors),
Tel. 51.39.05.05, P, D, AC, C (25 ton), Slip, FW, ME, V, SM, Lau, R, Gas, Gaz, Bar (July, Aug), Sh, SC; **Quay** P, D, Bar, R, Gaz, ME, Sh, SM, V; **Gendron Plaisance** Tel. 51.39.18.69, ME, C, El, Sh, CH; **Gendron Yvon** Tel. 51.39.12.06, ME; **Massif Marine** Tel. 51.39.05.86, ME, C, El, Sh, CH, Gaz; **Voilerie Burgaud** Tel. 51.39.23.89, SM; **Voilerie Simonin** Tel. 51.39.41.87, SM.

NOIRMOUTIER. **Quay** FW, C (4 ton); **Nautique 85** Tel. 51.39.05.78, ME, El, Sh, CH; **Quincaillerie de la Mer** Tel. 51.39.47.67, CH; **Voilerie Burgaud** Tel. 51.39.12.66, SM.

Town P, D, V, Gaz, R, Bar, Lau, PO, Bank, Rly (ferry to Pornic, bus to Nantes), Air (Nantes).
Ferry UK — Roscoff—Plymouth; St Malo—Portsmouth.

PORNIC 10-14-23
Loire Atlantique

CHARTS
Admiralty 2646, 2985, 3216; SHOM 5039, 6854; ECM 549
TIDES
+0540 Dover; ML 2.9; Duration 0540; Zone −0100
Standard Port BREST (→)

Times				Height (metres)			
HW		LW		MHWS	MHWN	MLWN	MLWS
0500	1100	0500	1100	7.5	5.9	3.0	1.4
1700	2300	1700	2300				

Differences PORNIC
−0035 −0015 +0005 +0005 −2.1 −1.9 −1.4 −1.1

SHELTER
Very good shelter in the large Pornic-Noëveilland marina, P1, P2 and P3 pontoons reserved for visitors. Old harbour dries (1.8m); access HW∓2½. Entry dangerous in strong winds from SE to W. Marina Access HW∓5.
NAVIGATION
Waypoint Notre Dame Bn, isolated danger mark, 47°05'.49N 02°08'.20W, 227°/047° from/to jetty head, 1.5M. Beware Banc de Kerouars in N of bay; all other hazards are well marked. There is an exit to the S (possible HW∓1 near springs) but a drying causeway joins Ile de Noirmoutier to the mainland. The S end of bay is devoted to oyster beds, and there are many obstructions.

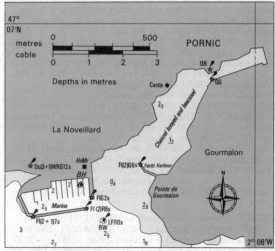

LIGHTS AND MARKS
Marina SW elbow Fl(2+1) 7s 4m 3M; SW jetty head Fl(2)R 6s 4m 2M; E jetty head FlG 2s 4m 2M; Pointe de Noëveilland Oc(3+1)WRG 12s 22m 14/10M; G shore-051°, W051°-079°, R079°-shore. Entry signals (simplified).
RADIO TELEPHONE
VHF Ch 09 (H24).
TELEPHONE
Hr Mr 40.82.05,40; Aff Mar 40.82.01.69; Customs 40.82.03.17; SNSM 40.82.01.54; Meteo 40.04.15.15; CROSS 97.55.35.35; Dr 40.82.01.80; Brit Consul 40.63.16.02.
FACILITIES
Marina (750 + 165 visitors) Tel. 40.82.05.40, P, D (on pontoon), FW, ME, AC, El, Sh, BH (20 ton), C (6 ton) CH; **CN de Pornic** Tel. 40.82.42.26; **Pornic Nautic** Tel. 40.82.04.64, ME, El, Sh, CH; **Petit Breton Nautique** Tel. 40.82.11.82, ME, El, Sh, CH; **Ouest Voile** Tel. 40.82.26.57, SM;

Town V, Gaz, R, Bar, PO, Bank, Rly, Air (Nantes).
Ferry UK — Roscoff—Plymouth; St Malo—Portsmouth.

MINOR HARBOURS AND ANCHORAGES 10.14.24

HENDAYE, Pyrénées Atlantique, 43°22′ N, 1°47′ W, Zone −0100; Admty charts 1343, 2665, SHOM 6556, 6558. HW +0450 on Dover (GMT), −0030 on Pointe de Grave (zone −0100). HW height −1.1m on Pointe de Grave, ML 2.3m. Hendaye lies on the French bank of the Rio Bidasoa, the other bank being Fuenterrabia in Spain. Access to river HW∓3. Beware Les Briquets lying to the E of the bay. Lt Ho on W end of bay, Cabo Higuer, Fl(2) 60s 63m 23M. River ent breakwater heads, East L Fl R 10s 7m 9M, West FG 9m 4M. River dredged to 2m. Yacht anchorage 1 ca (185m) SW of landing place at Hendaye Plage in 3.5m, or in about 3m off Roca Punta at root of W breakwater, or S of Hendaye quay. Much of bay dries. Facilities: Hr Mr Tel. 59.20.16.97; Customs Tel. 59.20.01.98; Aff Mar 59.55.06.68; **Quay** C (20 ton), D, FW; **Club Maritime Hendaye** Tel. 59.20.03.02, VHF Ch 09, P, C (20 ton); **Jabin** Tel. 59.20.08.96, ME, CH; **Town** V, R, Bar, PO, Bank, Rly, Air.

ST JEAN-DE-LUZ, Pyrénées Atlantique, 43°23′ N, 1°40′ W, Zone −0100; Admty chart 1343, SHOM 6526, 6558. HW +0435 on Dover (GMT), −0045 on Pointe de Grave (zone −0100); HW height −1.0m on Pointe de Grave; ML 2.5m. There are two harbours, St Jean-de-Luz on SE of bay and Socoa to the W of the bay. Approaching from N, keep W of Les Esquilletac, a rock ½M off Pointe St Barbe. The bay can be entered at all times except in strong NW winds, and good anchorage found in approx 4m. There is a small yacht harbour in SW corner of St Jean-de-Luz. Socoa harbour dries. Enter bay via W passage between Digue des Criquas and Digue d'Artha. Sailing is forbidden in the port. Beware a submerged obstruction in SE corner of the bay. Ldg Lts at 101° both Oc(3+1)R 12s 30/47m 18M lead in from W, S of Belhara Perdun. Ldg Lts 138° Front QWR 12/8M, Rear Q 20M lead between Illarguita and Belhara Perdun. Inner Ldg Lts 151° both Oc(2)G 6s. Le Socoa Lt Ho QWR 36m 12/8M, W shore-264°, R264°-282°, W282°-264°. Socoa breakwater head QG 11m 9M. Facilities: ST JEAN-DE-LUZ Hr Mr Tel. 59.26.26.81; Aff Mar Tel. 59.47.14.55; **Quay** FW, P, C; **Assistance Technique Marine** Tel. 59.26.45.81, ME, El, Sh, CH; **Haize-Egoa** Tel. 59.26.82.84, CH; Meteo 59.23.84.15. SOCOA Hr Mr Tel. 59.47.18.44; Aff Mar Tel. 59.47.14.55; Customs Tel. 59.47.18.61; **Jetty** C (1 ton), FW, P, D, AC; **Yacht Club Basque** Tel. 59.47.26.81; **Arrantzalat** Tel. 59.47.02.02, ME, CH; **Larmanou-Marine** Tel. 59.47.99.97, ME, El, Sh, CH: **Town** V, R, Bar, PO, Bank, Rly.

ANGLET, Pyrénées Atlantique, 43°32′ N, 1°31′ W, Zone −0100; Admty charts 1343, 2665, SHOM 6571. HW at Boucau (1½M up river) +0450 on Dover (GMT), −0035 on Pointe de Grave (zone −0100); HW height −1.0m on Pointe de Grave; ML 2.6m. Access good except in strong winds from SW to NW when entry may be impracticable. Tidal stream is strong, up to 5 kn on ebb at springs. The marina is on the S side ¾M from ent to R Adour. There are yacht moorings W of marina ent. Pointe St Martin Fl(2) 10s 73m 29M is approx 2½M SSW of ent. BA buoy (safe water mark) L Fl 10s is moored 9 ca (1650m) NW of breakwater head which has QR Lt. Ldg Lts 090°, both Q intens 087°-093°, W Trs with R tops. N jetty head Oc(2)R 6s 12m 8M. S jetty head Iso G 4s 9m 7M. Inner Ldg Lts approx 111°, both FG moved as necessary. Traffic signals (full code, see 10.14.7) shown from Tr S of entrance. Facilities: Hr Mr Tel. 59.63.05.45; Aff Mar and Customs at Bayonne; Meteo 59.23.84.15; **Marina** (390+10 visitors), Tel. 59.63.05.45; P, D, FW, ME, El, AC, C (1.3 ton), BH (13 ton), Slip, Sh; **YC Adour Atlantique** Tel. 59.63.16.22, **BAB Marine** Tel. 59.63.50.19, ME, El, Sh, CH; **Nivadour Nautique** Tel. 59.63.16.85, ME, El, Sh, CH.

BAYONNE, Pyrénées Atlantique, 43°30′ N, 1°29′ W, Zone −0100; Admty chart 1343, SHOM 6536, 6557/8. Details of tides, entry lights etc are as shown under Anglet above. Bayonne is 3M upstream from Anglet, the channel being well marked with Ldg Lts, buoys and beacons. There are numerous commercial wharves. Yachts berth on S bank below the bridge near town hall where R Nive joins the R Adour. VHF Ch 09 12 (H24); Facilities: Hr Mr Tel. 59.63.11.57; Aff Mar Tel. 59.55.06.68; Customs Tel. 59.59.08.29; **Port** C (30 ton), Slip, FW, P, D; **Club Nautique de Bayonne**; **Sorin** Tel. 59.59.16.87, CH, SHOM.

CAPBRETON, Landes, 43°39′ N, 1°26′ W, Zone −0100; Admty chart 2665, SHOM 6557, 6586. HW +0450 on Dover (GMT), −0030 on Pointe de Grave (zone −0100); HW height −1.1m on Pointe de Grave, ML 2.3m. Good shelter, narrow entrance dangerous in strong winds from N and W. Do not enter if waves break in mid-channel. Access at all times except LW; recommended HW∓4. Visitors Pontoon B. Marina is entered through gap in training wall on SE side of Boucaret Channel. Harbour dredged to 1.5 to 2.5m. S jetty head Iso G 4s 7m 9M. N jetty head Fl(2)R 6s 13m 11M, Horn 30s. VHF Ch 09 (0600-2200 in season). Facilities: Hr Mr Tel. 58.72.21.23; Meteo 89.24.58.80; Aff Mar Tel. 58.72.10.43; Customs Tel. 58.72.07.47; **Quay** P, D, Slip, FW; **Marina** (700+70 visitors), Slip, BH (28 ton), AC, P, D, FW, ME, El, C (1.5 ton), Sh; **YC Landais**; **Club Nautique Capbreton-Hossegor-Seignone** Tel. 58.72.05.25; **Erick Yachting** Tel. 58.72.14.32, ME, El, Sh, CH.

LA VIGNE, Gironde, 44°40′ N, 01°14′ W; Zone −0100. Admty chart 2664, SHOM 6766, ECM 255. HW +0620 on Dover (GMT); +0030 on Pointe de Grave (zone −0100); HW height (Cap Ferret) −1.1m on Pointe de Grave; ML 2.3m. Access HW∓2. There are two perches marking the entrance and a Lt on the SW point, Iso R 4s. Good shelter but beware strong currents across harbour mouth. There is a small breakwater (unlit) protruding into the entrance from the NE side. Facilities: Aff Mar Tel. 56.60.52.76, **Marina** (268) Tel. 56.60.85.80, FW, AC, Slip, CH, C, P, D.

PAUILLAC, Gironde, 45°12′ N, 0°45′ W, Zone −0100; Admty charts 2916, 2910; SHOM 6139; HW +0620 on Dover (GMT); +0045 on Pointe de Grave (zone −0100); HW height +0.2m on Pointe de Grave; ML 3.0m. Excellent shelter in marina about half way to Bordeaux from the sea on W bank, 25M from Le Verdon. Access at all tides (depth 3m). Visitors berth on left of entrance. Beware current in the river when entering or leaving. Facilities: Hr Mr (Port de Plaisance) Tel. 56.59.12.16, VHF Ch 09 Aff Mar 56.59.01.58; Cross 56.09.82.00; Customs 56.59.04.01; Meteo 56.34.20.11; SNSM 56.09.82.00. There are facilities for stepping and un-stepping masts; **Quay** FW, D, P, C (15 ton); **Marina** (185+15 visitors), Tel. 56.59.04.53, P, D, FW, AC, Slip, ME, El, Sh; fuel (on quay); **CN de Pauillac** Tel. 56.59.12.58; **Guiet** Tel. 56.59.02.66, CH, ME, El, Sh.

BORDEAUX, Gironde, 44°50′ N, 0°34′ W, Zone −0100; Admty chart 2916, SHOM 6140, 4610. HW +0715 on Dover (GMT), +0155 on Pointe de Grave (zone −0100); HW height −0.1m on Pointe de Grave; ML 2.4m. Bordeaux is about 55M up the Gironde estuary and River Garonne. Beware large merchant vessels, strong currents (up to 5 kn when river in spate) and large bits of flotsam. Point du Jour marina is 2M from Bordeaux, on W bank just S of suspension bridge (Pont d'Aquitaine, clearance 51m), with visitors berths and crane for masts. Moorings may be available just upstream. No 1 Basin, 1½M above bridge, access HW−2 to HW+½, may be dirty; crane available. Or go alongside wharves between No 1 Basin and Pont de Pierre, but stream is strong. The channel up to Bordeaux is well marked and lit. VHF Ch 12. Facilities: Hr Mr Tel. 56.90.91.21; Aff Mar Tel. 56.52.26.23; Customs Tel. 56.44.47.10; Meteo 56.90.91.21; **Sport Nautique de la Gironde** Tel. 56.50.84.14; **Cercle de la Voile de Bordeaux**; **Agence Nautique du Sud Ouest** Tel. 56.91.23.83, ME, El, Sh, CH; **Le Compas** Tel. 56.50.60.02, ME, El, Sh, CH, Electronics; **Poitevin-Duault** Tel. 56.52.55.50, CH, SHOM agent; **Quay** Slip, C (5 ton).

BLAYE, Charente Maritime, 45°07′ N, 0°40′ W, Zone −0100, Admty chart 2916, SHOM 6140. HW +0175 on Dover (GMT), +0145 on Pointe de Grave (zone −0100); HW height −0.3m on Pointe de Grave; ML 2.4m. Good shelter; access good except in winds S to SW. Entrance to NE of Ile du Paté, marked by Lt Fl G on S bank. Max stay 24 hours. VHF Ch 12 (office hours). Facilities: Hr Mr Tel. 57.42.13.49; Customs Tel. 57.42.01.11. **Quay** FW, AC, P, D, C (25 ton), Slip, Access HW∓2½; **Auxemerry** Tel. 57.42.13.43, ME; **A. Moulinier** Tel. 57.42.11.94, ME.

DOUHET, Ile d'Oleron, Charente Maritime, 46°00′ N, 1°20′ W, Zone −0100; Admty charts 2648, 2746, SHOM 6334. HW +0545 on Dover (GMT) −0020 on Pointe de Grave (zone −0100); HW height +0.8m on Pointe de Grave; ML 3.7m. In the NE part of the island, access approx HW∓4. There are no lights. Beware rocks E and W of ent channel. Yacht harbour in wet dock, 1.5m. Max length 7m. Facilities: Hr Mr 46.76.71.13, Auto 46.41.11.11; Aff Mar and Customs − Le Chateau; **Quay** FW, AB, Slip. Correspondence with Hr Mr Boyardville.

ILE D'AIX, Charente Maritime, 46°01′ N, 1°10′ W, Zone −0100; Admty charts 2746, 2748, SHOM 6334, 6468. HW +0545 on Dover (GMT), −0020 on Pointe de Grave (zone −0100); HW height +0.8m on Pointe de Grave; ML 3.7m. Only a fine weather anchorage; anchor or pick up a buoy off the landing jetty at St Catherine's Point, the most southerly tip of the island. Lt on the point is FlWR 5s 24m 23/19M. Mooring available E of St Catherine's Pt (shallow), four W buoys to West of St Catherine's Pt (deeper) and another four SE of the point. Facilities: C (1.5 ton), Slip, AB (SE jetty), shop and restaurant.

LA FLOTTE, Ile de Ré, Charente Maritime, 46°11′ N, 1°19′ W, Zone −0100; Admty charts 2648, 2641, 2746, SHOM 6521, 6668. HW +0535 on Dover (GMT), −0025 on Pointe de Grave (zone −0100); HW height +0.6m on Pointe de Grave; ML 3.4m. La Flotte lies 2M SE of St Martin (see 10.14.17). From NW keep clear of Le Couronneau; from E, keep N of Bn off Pointe des Barres. Approach with La Flotte Lt Ho (W Tr with G top) bearing 215° in W sector. La Flotte Lt Fl WG 4s 10m 12/7M; vis G130°-205°, W205°-220°, G220°-257°, Horn (3)30s by day HW−2 to HW+2. Avant port is sheltered by mole and dries (2.4m); inner harbour dries (2.7m). Access HW∓3. There is an anchorage off La Flotte in 3m, sheltered from S and W. Facilities: Visitors use North Jetty; Hr Mr on Quai Senac Ouest; Aff Mar and customs at St Martin; **Quay** Slip, FW; **Cercle Nautique de la Flotte-en-Ré** (CNLF) (open July-15 Sept); **Chauffour** Tel. 46.09.60.25, ME.

ARS-EN-RÉ, Ile de Ré, Charente Maritime, 46°12′ N, 1°31′ W, Zone −0100; Admty chart 2648, SHOM 6333, 6334 HW +0540 on Dover (GMT), −0045 on Pointe de Grave (zone −0100); HW height +0.6m; ML 3.4m. Port d'Ars is at the head of a creek in the SW corner of the bay Mer du Fier, the entrance obstructed by rocks which dry. Channel through Mer du Fier marked by buoys and bns; Access HW∓3. There are two quays (dry) and a wet dock with sill gate, 2.9m above CD. Ldg Lts to Fiers d'Ars 265°, Front Iso 4s 8m 10M on grey metal framework Tr; Rear 237m from front, Iso G 4s 12m 13M, G square Tr on dwelling (synchronised with front and intens 263°-267°). Port d'Ars Ldg Lts 232°, front Q 5m 9M, W rectangle with R lantern, rear 370m from front, Q 13m 11M, B rectangle on W framework Tr, G top; vis 142°-322°. There is an anchorage close S of Pte du Fier with min depth of 2.4m; most of the bay dries. VHF Ch 09. Facilities: Hr Mr Tel. 46.29.40.19, Aff Mar and customs at St Martin (see 10.14.17); SNSM Tel. 46.29.41.49; **Cercle Nautique d'Ars-en-Ré** Tel. 46.29.41.13 (open Apl to Nov); **Blanchard** Tel. 46.29.40.43, ME, El, Sh; **Blondeau Marine** Tel. 46.29.40.39, ME; **Chantiers Navals de l'Ile de Ré** Tel. 46.29.41.94, ME, El, Sh, CH; **Jetties** Slip, FW, C (6 ton).

L'AIGUILLON-LA-FAUTE-SUR-MER, Vendée, 46°20′ N, 1°18′ W, Zone −0100; Admty charts 2641, 2648, SHOM 6521. HW +0535 on Dover (GMT), −0030 on Pointe de Grave (zone −0100); HW height +0.6m on Pointe de Grave; ML 3.4m. Shelter good except in strong winds from W or S; entry only safe in fine weather with off-shore winds. Access HW∓2½. Beware mussel bed timber piers which cover at HW. Also beware oyster beds. The area is very flat and, being shallow, waves build up quickly in any wind. The bar to seaward dries and is dangerous in bad weather. Entrance can be identified by a hill, La Dive, opposite side of entrance to Pointe d'Arcay, with a conspic transformer on it. Enter with transformer bearing 033°. Anchor in R Lay or go alongside on YC pontoons, (40). Facilities: Hr Mr Tel. 51.56.45.02 Aff Mar Tel. 51.56.45.35; Cross 56.09.82.00; Dr Tel. 51.56.46.17; **Club Nautique Aiguillonais et Fautais** (CNAF) open July-Sept; Tel. 51.56.44.42; **Arrivé** Tel. 51.56.44.42, ME; **Atlantic Garage** Tel. 51.56.40.15, ME, El, Sh, CH; **Co-operative Maritimes** Tel. 51.56.44.89, CH.

JARD-SUR-MER, Vendee, 46°24′ N, 1°35′ W, Zone −0100; Admty chart 2648, SHOM 6522. HW +0600 on Dover (GMT), −0010 on Brest (zone −0100); HW height −2.0m on Brest; ML 3.1m; Duration 0640. White daymarks 4 ca (740m) E of harbour lead 038° between Roches de l'Islatte (dry) and Roches de la Brunette (dry), marked by buoys. Then pick up 293° transit of RW marks on W side of harbour, leading to entrance. There are no lights. Facilities: Hr Mr Tel. 51.33.40.17; Customs Tel. 51.95.11.33; **Jetty** FW, C (5 ton); **Jard Marine** Tel. 51.33.46.84, ME, El, Sh, CH.

FROMENTINE, Vendée, 46°54′ N, 2°08′ W, Zone −0100; Admty chart 2647, SHOM 5039. HW +0550 on Dover (GMT), −0025 on Brest (zone −0100); HW height −2.1m on Brest; ML 2.9m; Duration 0540. Do not approach from Baie de Bourgneuf as there is a road causeway which dries 3m, connecting Ile de Noirmoutier with the mainland. Enter through Goulet de Fromentine, under the bridge (clearance 27m). The channel is marked by buoys, moved as necessary but it is very shallow, so dangerous in bad weather. The ebb stream runs at more than 5 kn at springs. Do not attempt night entry. Anchor W of pier. Coming from N, beware Les Boeufs. Tourelle Milieu lies on the N side of Le Goulet de Fromentine, Fl(4)R 12s 6m 5M, R Tr. Pointe de Notre Dames-de-Monts Lt Oc(2)WRG 6s 21m 13/10M, G000°-043°, W043°-063°, R063°-073°, W073°-094°, G094°-113°, W113°-116°, R116°-175°, G175°-196°, R196°-230°. Fromentine light-buoy (safe water mark) is about 1.5M WSW of Pointe du Notre Dames-de-Monts. Bridge W end Iso W 4s 32m 18M; E end Iso W 4s 32m 18M. Facilities: Aff Mar at Noirmoutier; Customs at Beauvois-sur-Mer; **Quay** Slip, C (3 ton), FW; **Cercle Nautique de Fromentine-Barfatre** (CNFB); **Marine Motoculture** Tel. 51.68.53.77, ME, El, Sh, CH; **Roblin** Tel. 51.68.50.15, ME.

SAILING IN SCOTLAND?

MAKE SURE OF Your Yachting Life.

FOR ALL THE LATEST YACHTING AND BOARDSAILING NEWS NORTH OF THE BORDER, PLUS MONTHLY REPORTS ON MAJOR SCOTTISH RACING EVENTS

AVAILABLE FROM CHANDLERS AND NEWSAGENTS AT ALL PORTS OF CALL.

ONLY 80p

VOLVO PENTA SERVICE

Sales and service centres in area 15
Names and addresses of Volvo Penta dealers in this area are available from:
FRANCE **Volvo Penta France SA,** BP45 F78130 Les Mureaux Tel (01) 3 474 72 01, Telex 695221 F.

Area 15

South Brittany
River Loire to Camaret

10.15.1	Index	Page 615
10.15.2	Diagram of Radiobeacons, Air Beacons, Lifeboat Stations etc	616
10.15.3	Tidal Stream Charts	618
10.15.4	List of Lights, Fog Signals and Waypoints	620
10.15.5	Passage Information	623
10.15.6	Distance Table	624
10.15.7	Special notes for French areas	See 10.14.7
10.15.8	Canal connections in Brittany	625
10.15.9	River Loire/St Nazaire	625
10.15.10	Le Pouliguen (Port of La Baule)	626
10.15.11	Le Croisic	627
10.15.12	La Roche Bernard	627
10.15.13	Crouesty-en-Arzon	628
10.15.14	Vannes	629
10.15.15	Morbihan	629
10.15.16	La Trinité-sur-Mer	630
10.15.17	Port Haliguen — Quiberon	631
10.15.18	Le Palais (Belle Ile)	631
10.15.19	Lorient	632
10.15.20	Concarneau	633
10.15.21	La Forêt	634
10.15.22	Bénodet	634
10.15.23	Audierne	635
10.15.24	Douarnenez	636
10.15.25	Morgat	637
10.15.26	Camaret	637
10.15.27	Minor Harbours and Anchorages	637
	Pornichet	
	Piriac	
	Port d'Arzal	
	Penerf	
	Ile de Houat	
	Port Maria	
	Sauzon	
	River Etel	
	Port Tudy	
	Doëlan	
	Pont-Aven	
	Loctudy	
	Guilvinec	

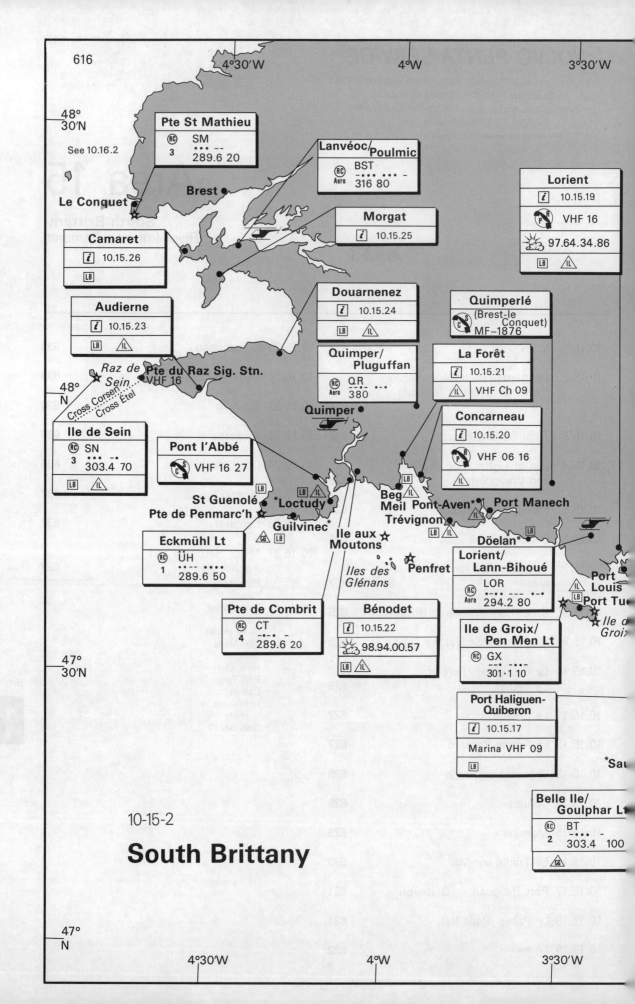

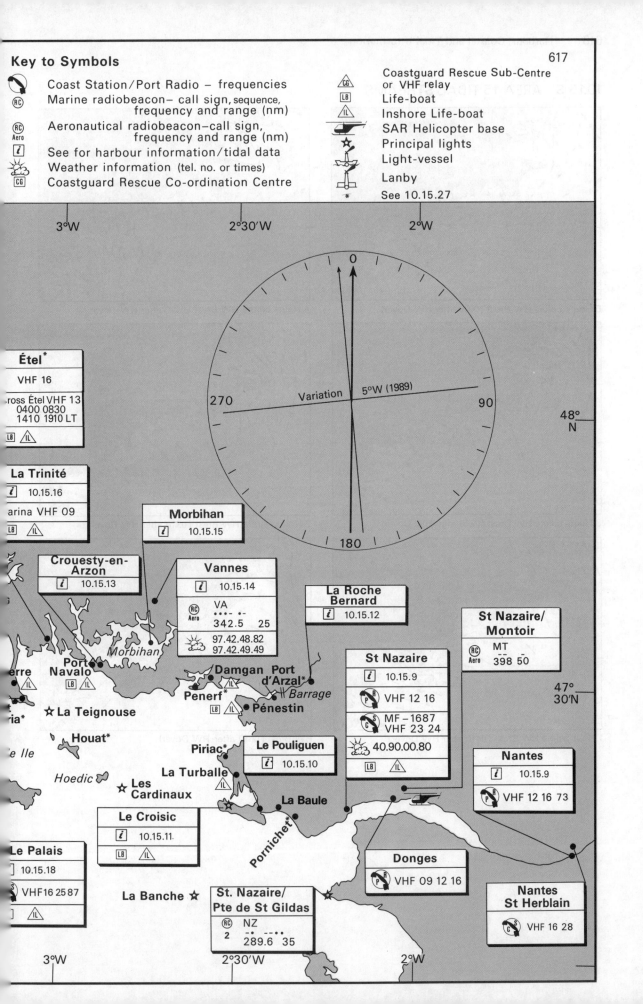

10.15.3 AREA 15 TIDAL STREAMS

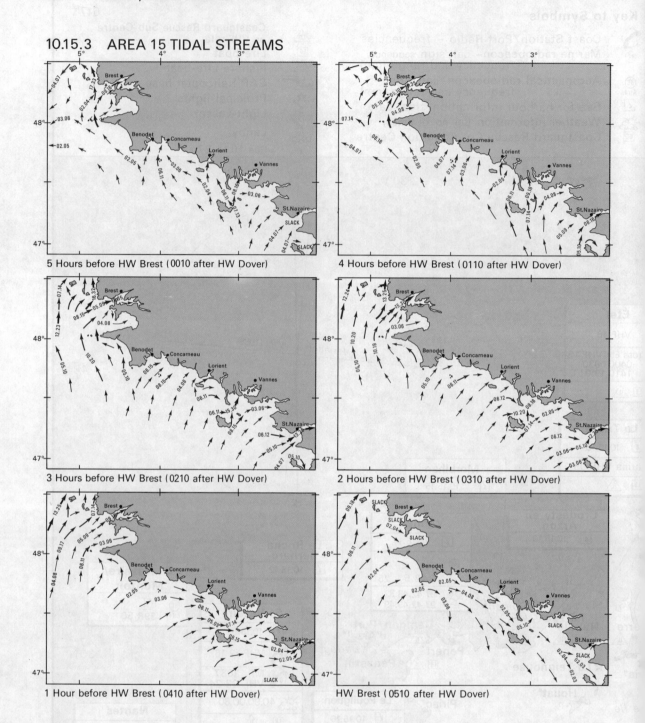

AREA 15 – S Brittany

1 Hour after HW Brest (0610 after HW Dover)

2 Hours after HW Brest (0515 before HW Dover)

3 Hours after HW Brest (0415 before HW Dover)

4 Hours after HW Brest (0315 before HW Dover)

5 Hours after HW Brest (0215 before HW Dover)

6 Hours after HW Brest (0115 before HW Dover)

10.15.4 LIGHTS, FOG SIGNALS AND WAYPOINTS

Abbreviations used below are given in 10.0.3. Principal lights are in **bold** print, places in CAPITALS, and light-vessels and Lanbys in *CAPITAL ITALICS*. Unless otherwise stated lights are white. m—elevation in metres; M—nominal range in n. miles. Fog signals are in *italics*. Useful waypoints are underlined — use those on land with care. See 4.2.2.

FRANCE—WEST COAST

<u>*LOIRE APPROACH LANBY SN1*</u> 47 00.05N/2 39.95W Iso 4s 8m 9M; safe water mark; *Whis*; Racon.
<u>*LOIRE APPROACH LANBY SN2*</u> 47 03.00N/2 30.70W Oc 4s 8m 8M; safe water mark; Ra refl.
<u>*NW BANCHE LANBY*</u> 47 12.90N/2 30.95W Q 8m 8M; N cardinal; *Bell*; Ra refl.
<u>*LA COURONNÉE LANBY*</u> 47 07.65N/2 20.00W QG 8m 6M; stbd hand mark; Ra refl.

Pointe de Saint Gildas 47 08.10N/2 14.67W Q WRG 23m **W15M**, R12M, G10M; framework Tr on W house; vis R264°-308°, G308°-078°, W078°-088°, R088°-174°, W174°-180°, G180°-264°. RC.
Port de La Gravette, jetty head 47 09.80N/2 12.60W Fl(3) WG 12s 7m W8M, G5M; W structure, G top; vis G224°-124°, W124°-224°.
Port de Comberge, S jetty 47 10.60N/2 09.95W Oc WG 4s 7m W9M, G5M; W Tr, G top; vis W123°-140°, G140°-123°.
Le Pointeau 47 14.05N/2 10.90W Fl WG 4s 4m W10M, G6M; W round Tr, G top; vis G050°-074°, W074°-149°, G149°-345°, W345°-050°.

Les Morées 47 15.05N/2 12.95W Oc(2) WR 6s 12m W9M, R6M; G Tr; vis W058°-224°, R300°-058°.
Point de Minden, West mole and E mole, both Fl G 2s.
DONGES, West dolphin 47 18.1N/2 05.0W Iso G 4s 12m 7M. E dolphin (close ENE) Iso G 4s; G structure.
Paimboeuf, mole head 47 17.4N/2 02.0W Oc(3) WG 12s 9m **W16M**, G11M; W round Tr, G top; vis G shore-123°, W123°-shore. Île du Petit Carnet 47 17.3N/2 00.3W Iso G 4s 9m 6M; W framework Tr, G top.
From Paimboeuf to Nantes lights on S side are G, and on N side R.

Portcé Ldg Lts 025°. **Front** 47 14.6N/2 15.4W Q 6m **23M**; W mast; intens 024°-026°. **Rear**, 0.75M from front, 47 15.3N/2 14.9W Q 36m **27M**; W Tr; intens 024°-026° (H24).
Pointe d'Aiguillon 47 14.60N/2 15.70W Oc(4) WR 12s 27m 14M; W Tr; vis W233°-293°, W297°-300°, R300°-327°, W327°-023°, W027°-089°.
Villèz-Martin, jetty head, 47 15.3N/2 13.7W Fl(2) 6s 10m 12M; W Tr, R top.

SAINT-NAZAIRE, West jetty Oc(4) R 12s 11m 10M; W Tr, R top. E jetty Oc(4) G 12s 11m 11M; W Tr, G top. Old Mole head Oc (2+1) 12s 18m 12M; W Tr, R top; weather signals.

Bridge, Iso 4s each side mark channel. *Horn (2) 20s*.
Gron 47 17.7N/2 09.6W Iso R 4s 5m 7M; R pedestal.
Le Village No1 47 18.1N/2 08.8W QR 7m 8M; vis 225°-135°.
Oiling jetty Oc R 4s 10m 9M; W Tr, R top.

Le Grand Charpentier 47 12.90N/2 19.05W Q WRG 22m W14M, R10M, G 10M; grey Tr, G lantern; vis G020°-049°, W049°-111°, R111°-310°, W310°-020°; helicopter platform; sig stn 1.5M NE.
PORNICHET, <u>S breakwater head</u> 47 15.55N/2 21.10W Iso WG 4s 11m W12M, G9M; vis G084°-081°, W081°-084°. S head Fl G 2s 2m 3M. N breakwater head Fl R 2s 2m 3M.
LE POULIGUEN. South jetty QR 13m 9M. Les Petits Impairs Fl(2) G 6s 6m 4M.

<u>**La Banche**</u> 47 10.70N/2 28.00W Fl(2+1) WR 15s 22m **W17M**, R12M; B Tr, W bands; vis R266°-280°, W280°-266°.

LE CROISIC. Jetée de Tréhic, head 47 18.5N/2 31.4W Iso WG 4s 12m W13M, G10M; grey Tr, G top; vis G042°-093°, W093°-137°, G137°-345°. *Horn 15s*. Ldg Lts 156°. **Front** 47 18.0N/2 31.0W Oc(2+1) 12s 10m **18M**; Or topmark on W pylon; intens 154°-158°. **Rear**, 116m from front, Oc(2+1) 12s 14m **18M**; Or topmark on W pylon, G top; synchronised with front, intens 154°-158°. Ldg Lts 174° both QG on G and W structures, vis on leading line. Ldg Lts 134° both Oc R 4s, 11M; RW pylons. Le Grand Mabon 47 18.1N/2 31.0W Oc(2) R 6s 6m 5M, R pylon and pedestal on base.

Le Four 47 17.9N/2 38.0W Fl 5s 23m **19M**; W Tr, B stripes.

ILE DE HOEDIC. Port de l'Argol 47 20.7N/2 52.5W Fl WG 4s 10m W10M, G7M; W Tr, G top; vis W143°-163°, G163°-183°, W183°-203°, G203°-143°.

Les Grands Cardinaux 47 19.3N/2 50.1W Fl(4) 15s 28m 13M; R and W Tr.

ILE DE HOUAT. Port de Saint-Gildas, Môle Nord 47 23.5N/2 57.4W Oc(2) WG 6s 8m W8M, G5M; W Tr, G top; vis W168°-198°, G198°-210°, W210°-240°, G240°-168°.

LA TURBALLE. Ldg Lts 006° both Dir F Vi 3M, intens 004°-009°. Jetée Garlahy Fl(4) WR 12s 13m W10M, R7M; W pylon, R top; vis R060°-315°, W315°-060°. Digue Tourlandroux Fl G 4s 7m 6M; W pedestal, G top; *Siren 10 min*.

PIRIAC-SUR-MER. Pipeline 47 22.1N/2 32.8W Oc(2+1) WRG 12s 14m W12M, R9M, G9M; W square, R stripe on R framework Tr; vis G300°-036°, W036°-068°, R068°-120°. Inner mole, head 47 22.9N/2 32.7W Oc(2) WRG 6s 8m W10M, R7M, G6M; W column; vis R066°-185°, W185°-201°, G201°-224°; *Siren 120s*. Breakwater heads Fl G 4s and Fl R 4s.

Mesquer, jetty head 47 25.3N/2 28.1W Oc(3+1) WRG 12s 7m W12M, R8M, G7M; W column and building; vis W067°-072°, R072°-102°, W102°-118°, R118°-293°, W293°-325°, G325°-067°.

Île Dumet 47 24.7N/2 37.2W Fl(2+1) WRG 15s 14m W8M, R6M, G6M; W column, G top on fort; vis G090°-272°, W272°-285°, R285°-325°, W325°-090°.

LA VILAINE. Basse Bertrand 47 31.1N/2 30.7W Iso WG 4s 6m W9M, G6M; G Tr; vis W040°-054°, G054°-227°, W227°-234°, G234°-040°.
Penlan 47 31.0N/2 30.2W Oc(2) WRG 6s 26m **W16M**, R12M, G11M; W Tr, R bands; vis R292°-025°, G025°-052°, W052°-060°, R060°-138°, G138°-180°.
Pointe du Scal, 47 29.7N/2 26.8W Oc(3)G 12s 8m 6M; W square Tr, G top. Tréhiguier Oc(4)R 12s 21m 11M; W square Tr and dwelling; intens 105°-118°.

PÉNERF. Le Pignon 47 30.1N/2 38.9W Fl(3) WR 12s 6m W9M, R6M; R Tr; vis R028°-167°, W167°-175°, R175°-349°, W349°-028°.
Saint-Jacques-en-Sarzeau 47 29.2N/2 47.4W Oc(2) R 6s 5m 6M; W 8-sided Tr, R top.

CROUESTY EN ARZON. Ldg Lts 058°. **Front** 47 32.6N/2 53.9W Q 10m **19M**; framework Tr; intens 057°-059°. **Rear**, 315m from front, Q 27m **19M**; intens 057°-059°. N jetty head Oc(2) R 6s 9m 7M; R and W Tr. S jetty head Fl G 4s 9m 7M; G and W Tr.

Port-Navalo 47 32.90N/2 55.12W Oc(3) WRG 12s 32m **W16M**, R12M, G11M; W Tr and dwelling; vis W155°-220°, G317°-359°, W359°-015°, R015°-105°.
Rivière D'Auray. Le Gregan Q(6) + LFl 15s 3m 8M.

RIVIÈRE DE CRAC'H. Ldg Lts 347°. Front 47 34.1N/3 00.4W Q WRG 10m W10M, R7M, G7M; W Tr, G top; vis G321°-345°, W345°-013°, R013°-080°. **Rear**, 560m from front, Q 21m **15M**; W Tr, G top; synchronised with front, intens 337°-357°.
Dir Lt 347° 47 35.0N/3 01.0W Dir Oc WRG 4s 9m W14M, R11M, G11M; W Tr; vis G345°-346°, W346°-348°, R348°-349°. La Trinité-sur-Mer S pierhead Oc(2) WR 6s 6m W10M, R7M; W Tr, R top; vis R090°-293°, W293°-300°, R300°-329°. Jetty head Iso R 4s 8m 5M; W Tr, R top.

BELLE ILE. <u>Goulphar</u> 47 18.67N/3 13.67W Fl(2) 10s 87m **24M**; grey Tr; RC; 0.7M SW *Siren(2) 60s*. Pointe de Kerdonis 47 18.6N/3 03.6W Fl(3) R 15s 35m 11M; W square Tr and dwelling; obsc by Pointes d'Arzic and de Taillefer 025°-129°. LE PALAIS Jetée Sud Oc(2) R 6s 11m 11M; W Tr; obsc 298°-170°. Jetée Nord Fl(2+1) G 12s 11m 7M; W Tr, G top; obsc 298°-168°. Sauzon QG 9m 5M; W Tr, G top; vis 194°-045°. <u>Pointe des Poulains</u> 47 23.3N/3 15.1W Fl 5s 34m **24M**; W square Tr and dwelling; vis 023°-291°.

PASSAGE DE LA TEIGNOUSE. <u>La Teignouse</u> 47 27.5N/3 02.8W Fl R 5s 19m 13M; W Tr, R top; auxiliary Lt Dir Q(7) 12s 14m 9M; same Tr; intens 033°-039°. *BASSE DU MILIEU LANBY* 47 25.9N/3 04.2W Fl(2) G 6s 9m 5M; stbd-hand mark.

PORT HALIGUEN. Marina, old breakwater head Fl R 4s 10m 5M; W Tr, R top. New breakwater head 47 29.4N/3 06.0W Oc(2) WR 6s 10m W12M, R9M; W Tr, R top; vis W233°-240°, R240°-299°, W299°-306°, R306°-233°.

PORT MARIA. Ldg Lts 006°. Front 47 28.6N/3 07.2W QG 5m 14M; W Tr, B band; intens 005°-008°. Rear, 230m from front, QG 13m 14M; W Tr, B band; intens 005°-008°. **Main light** 47 28.8N/3 07.5W Q WRG 28m **W15M**, R11M, G11M; W Tr; vis W246°-252°, W291°-297°, G297°-340°, W340°-017°, R017°-051°, W051°-081°, G081°-098°, W098°-143°. Brise-lames Sud, head Oc(2) R 6s 9m 7M; W Tr, R top. Môle Est, head Iso G 4s 9m 7M; W Tr, G top.

<u>Plateau des Birvideaux</u> 47 29.20N/3 17.45W Fl(2) 6s 24m 9M; B 8-sided Tr, R bands.

Rivière d'Étel, <u>West side, entrance</u> 47 38.7N/3 12.8W Oc(2) WRG 6s 13m W9M, R6M, G6M; R framework Tr; vis W022°-064°, R064°-123°, W123°-330°, G330°-022°; 2 FR on radio mast 2.3M NW; FR and F on radio mast 2.4M NW.

LORIENT. Ldg Lts Passe Sud 008°. **Front, Fish market** 47 43.8N/3 21.7W QR 16m **15M**; W square on grey framework Tr; intens 006°-011°. Rear, **Kergroise-La Perrière** 515m from front QR 28m **16M**; R square, W stripe on grey framework Tr; synchronised with front, intens 006°-011°.
Ldg Lts Passe Ouest 060°. **Front, Lohic** 47 42.2N/3 21.0W Q 7m **18M**; W wall of house; intens 058°-063°. **Rear, Kerbel** 0.65M from front Q 30m **18M**; W Tr; intens 058°-063°. **Port Louis** Dir QY 37m **16M**; occas; intens 058°-060°; in line with Lohic Lt 059°.
<u>Les Trois Pierres</u> 47 41.58N/3 22.40W Q RG 11m R7M, G7M; W Tr, B bands; vis G060°-196°, R196°-002°.

Passe de la Citadelle Ldg Lts 016°. **Front** Oc(3) G 12s 8m **16M**; W Tr, G top. **Rear**, 306m from front, Oc(3) G 12s 12m **16M**; W Tr, G top; synchronised with front.
West side, La Jument Oc R 4s 5m 6M; R Tr; vis 182°-024°. E side, Tourelle de la Citadelle Oc G 4s 6m 6M; G Tr; vis 009°-193°. Port-Louis jetty Iso G 4s 7m 6M; W Tr, G top; vis 043°-301°. West side, Le Cochon Fl R 4s 5m 6M; R Tr.

Kéroman, submarine base Ldg Lts 349°. **Front** Oc(2) R 6s 25m **17M**. **Rear**, 92m from front, Oc(2) R 6s 31m **17M**; grey pylon, R & W bands; Lts synchronised and intens 348°-353°. Fishing harbour, SE side of entrance Fl RG 4s 7m 6M; W Tr, G top; vis G000°-235°, R235°-360°.

Kernevel Ldg Lts 217°. Front Oc(4) R 12s 10m 14M; R square on R and W framework Tr; intens 216°-218°. Rear, 290m from front Oc(4) R 12s 18m 14M; W square Tr, R top; synchronised with front, intens 216°-218°.
Pengarne Fl G 2.5s 3m 4M; G Tr.
Pointe de l'Espérance Dir Lt 037°. Dir Q WRG 8m W9M, R7M, G7M; W column; vis G034°-037°, W037°-037°, R037°-047°. Landing stage Fl(4) WR 12s 7m W10M, R7M; W dolphin, R top; vis W110°-347°, R347°-355°, W355°-035°.

ILE DE GROIX. <u>Pointe des Chats</u> 47 37.30N/3 25.25W Fl R 5s 16m **19M**; W square Tr and dwelling. Pointe de la Croix Oc WR 4s 16m W12M, R9M; W pedestal, R lantern; vis W169°-336°, R336°-345°, W345°-353°. Port Tudy, Môle Est, head Fl(2) R 6s 11m 6M; W Tr, R top; vis 110°-226°. Port Tudy, Môle Nord, head Iso G 4s 12m 6M; W Tr, G top.
<u>Pen Men</u> 47 38.87N/3 30.48W Fl(4) 25s 59m **29M**; W square Tr, B top; vis 309°-275°; RC; *Siren (4) 60s*.

Kerroc'h 47 42.0N/3 27.7W Oc(2) WRG 6s 22m W11M, R8M, G8M; W Tr, R top; vis R096°-112°, G112°-132°, R132°-302°, W302°-096°.

DOELAN. Ldg Lts 014°. **Front** 47 46.3N/3 36.5W Oc(2+1) WG 12s 20m W13M, G10M; W Tr, G band and lantern; vis W shore-305°, G305°-314°, W314°-shore. Rear 326m from front, QR 27m 9M; W Tr, R band and lantern.

MERRIEN 47 47.1N/3 39.0W QR 26m 7M; W square Tr, R top; vis 004°-009°.

BRIGNEAU, <u>mole head</u> 47 46.9N/3 40.2W Oc(2) WRG 6s 7m W12M, R9M, G9M; W column, R top: vis G280°-329°, W329°-339°, R339°-034°.

PORT MANECH. <u>Pointe de Beg-ar-Vechen</u> 47 48.0N/3 44.4W Oc(4) WRG 12s 38m W10M, R7M, G7M; W & R Tr; vis W(unintens) 050°-140°, W140°-296°, G296°-303°, W303°-311°, R311°-328° over Les Verres, W328°-050°; obsc by Pointe de Beg-Morg when brg less than 299°.

TRÉVIGNON. <u>Breakwater</u> 47 47.6N/3 51.3W Oc(3+1) WRG 12s 11m W14M, R11M, G11M; W square Tr, G top; vis W004°-051°, G051°-085°, W085°-092°, R092°-127°, R322°-351°. Mole head Fl G 4s 5M; W column, G top.

Baie de Pouldohan 47 51.0N/3 53.7W Fl G 4s 7m 9M; G square Tr, W band; vis 053°-065°.

CONCARNEAU. Ldg Lts 028°. Front <u>La Croix</u> 47 52.22N/3 55.00W Oc(3) 12s 14m 13M; R and W Tr; vis 006°-093°. **Rear Beuzec**, 1.34M from front, Q 87m **23M**; Belfry; intens 026°-030°. Lanriec QG 13m 8M; G stripe on W gable; vis 063°-078°. La Medée Fl R 2.5s 6m 4M; R Tr. Passage de Lanriec 47 52.3N/3 54.8W Oc(2) WR 6s 4m W8M, R6M; R Tr; vis R209°-354°, W354°-007°, R007°-018°; Fl R 4s and Fl(2) R 6s shown on West side, Fl G 4s and Fl(2) G 6s on E side of passage. Le Cochon 47 51.53N/3 55.47W Fl(3) WRG 12s 5m W9M, R6M, G6M; G Tr; vis G048°-205°, R205°-352°, W352°-048°. Basse du Chenal QR 6m 6M; R Tr; vis 180°-163°.

LA FORÊT. <u>Cap Coz, mole head</u> 47 53.55N/3 58.20W Fl(2) R 6s 5m 6M. Kerleven, mole head Fl G 4s 8m 6M. Inner mole head Iso G 4s 5m 5M.

ÎLES DE GLÉNAN. <u>Penfret</u> 47 43.32N/3 57.10W Fl R 5s 36m **21M**; W square Tr, R top; auxiliary Lt Dir Q 34m 12M; same Tr; vis 295°-315°. Fort Cigogne Q(2) RG 5s 2M; vis G106°-108°, R108°-262°, G262°-268°, obsc 268°-106°; shown in summer.
<u>Île-aux-Moutons</u> 47 46.5N/4 01.7W Oc(2) WRG 6s 18m **W15M**, R11M, G11M; W square Tr and dwelling; vis W035°-050°, G050°-063°, W063°-081°, R081°-141°, W141°-292°, R292°-035°; **auxiliary Lt** DirOc(2) 6s 17m **24M**; same Tr; synchronised with main Lt, intens 278°-283°.

Jaune de Glenan Lt Buoy 47 42.6N/3 49.8W Q(3) 10s; E cardinal mark; *Whis*.
JUMENT DE GLENAN LANBY 47 38.8N/4 01.3W Q(6) + LFl 15s 10m 8M; S cardinal; *Whis*; Ra refl.
BASSE PERENNES LANBY 47 41.1N/4 06.3W Q(9) 15s 8m 8M; W cardinal; *Whis*; Ra refl.
ROUGE DE GLENAN LANBY 47 45.5N/4 03.9W VQ(9) 10s 8m 8M; W cardinal; *Whis*; Ra refl.

Beg-Meil, quay head 47 51.72N/3 58.85W Fl R 2s 6m 2M; R and W column.

BENODET. Ldg Lts 346°. Front **Pointe du Coq**, 336m from rear, Oc(2+1) G 12s 11m **17M**; W Tr, G stripe; intens 345°-347°. Common rear Pyramide Oc(2+1) 12s 48m 11M; W Tr, G top; vis 338°-016°, synchronised with previous Lt. Ldg Lts 000°. Pointe de Combrit, 47 51.92N/4 06.70W, front 0.63M from rear, Oc(3+1) WR 12s 19m W12M, R9M; W Tr, grey corners; vis W325°-017°, R017°-325°; RC. Pointe du Toulgoet Fl R 2s 2m 2M. Pont de Cornouaille, E 2FG 3M, West 2FR 3M.

LOCTUDY. S side 47 49.94N/4 09.48W Fl(4) WRG 12s 12m **W15M**, R11M, G11M; W Tr, R top; vis W115°-257°, G257°-284°, W284°-295°, R295°-318°, W318°-328°, R328°-025°. Les Perdrix 47 50.3N/4 10.0W Fl WRG 4s 15m W12M, R9M, G9M; vis G090°-285°, W285°-295°, R295°-090°. Karek-Saoz 47 50.08N/4 09.30W QR 3m 1M; R truncated Tr. Le Blas 47 50.3N/4 10.1W Q(3)G 6s 7m 2M; G truncated column.

LESCONIL. Men-ar-Groas 47 47.8N/4 12.6W Fl(3) WRG 12s 14m W13M, R10M, G10M; W Tr, G top; vis G268°-313°, W313°-333°, R333°-050°. E breakwater head 47 47.77N/4 12.56W QG 5m 9M; G Tr. West breakwater Oc R 4s.

GUILVINEC. Ldg Lts 053°. Mole de Léchiagat, spur, front 47 47.5N/4 17.0W Q 13m 10M; W pylon; vis 233°-066°. Rocher Le Faoute's, middle, 210m from front Q WG 17m W14M, G11M; R circle on W pylon; vis W006°-293°, G293°-006°; synchronised with front. **Rear**, 0.58M from front, Dir Q 31m **15M**; R circle on W pylon; vis 051°-055°; synchronised with front. Lost Moan 47 47.07N/4 16.69W Fl(3) WRG 12s 7m W9M, R6M, G6M; W Tr, R top; vis R327°-014°, G014°-065°, R065°-140°, W140°-160°, R160°-268°, W268°-273°, G273°-317°, W317°-327°. Môle de Lechiagat Fl G 4s 5m 7M; W hut, G top. Môle Ouest Fl R 4s 11m 9M; W Tr, R top. Pier Fl(2) R 6s 4m 5M; R structure.
Kérity. Men Hir 47 47.3N/4 20.6W FlR 2.5s 6m 2M. Detached breakwater LFlG 10s 5m 2M.

Locarec 47 47.3N/4 20.3W Iso WRG 4s 11m W9M, R6M, G6M; W pedestal on rock; vis G063°-068°, R068°-271°, W271°-285°, R285°-298°, G298°-340°, R340°-063°.

POINTE DE PENMARC'H. **Eckmühl** 47 47.95N/4 22.35W Fl 5s 60m **24M**; grey 8-sided Tr; RC; *Siren 60s*. Le Menhir 47 47.8N/4 23.9W Oc(2) WG 6s 19m W8M, G5M; W Tr, B band; vis G135°-315°, W315°-135°. Scoedec 47 48.5N/4 23.1W Fl G 2.5s 6m 3M; G Tr.

SAINT GUÉNOLÉ. Ldg Lts 026°. Front 47 49.1N/4 22.6W QR 8m 4M; R mast. Rear, 51m from front, QR 12m 4M; mast, R and W bands; synchronised with front. Roches de Groumilli Ldg Lts 123°. Front FG 9m 9M; W Tr, B bands. Rear, 300m from front, FG 13m 9M; W Tr, B bands. Ldg Lts 051°. Front 47 48.8N/4 22.6W QG 5m 1M; G and W column. Rear, 330m from front F Vi 12m 1M; G and W mast; vis 036°-066°.

Pors Poulhan, West side of entrance 47 59.1N/4 28.0W QR 14m 9M; W square Tr, R lantern.

AUDIERNE. Gamelle Ouest Lt Buoy 47 59.53N/4 32.76W VQ(9) 10s; W cardinal mark; *Whis*. Passe de l'Est Ldg Lts 331°. Front Jetée de Raoulic 48 00.60N/4 32.37W Oc(2+1) WG 12s 11m W14M, G9M; W Tr; vis W shore-034°, G034°-shore, but may show W037°-055°. Rear, 0.5M from front, FR 9M; W 8-sided Tr, R top; intens 321°-341°. Kergadec Dir Lt 006° 48 01.0N/4 32.8W Dir Q WRG 43m W12M, R9M, G9M; same Tr as previous Lt; vis G000°-005°, W005°-007°, R007°-017°. Jetée de Sainte-Evette Oc(2) R 6s 2m 6M; R lantern. Pointe de Lervily 48 00.1N/4 34.0W Fl(2+1) WR 12s 20m W14M, R11M; W Tr, R top; vis W211°-269°, R269°-294°, W294°-087°, R087°-121°.

RAZ DE SEIN. Le Chat 48 01.44N/4 48.80W Fl(2) WRG 6s 27m W9M, R6M, G6M; South cardinal mark; vis G096°-215°, W215°-230°, R230°-271°, G271°-286°, R286°-096°; Ra refl. La Plate 48 02.36N/4 45.50W VQ(9) 10s 19m 8M; West cardinal mark. La Vieille 48 02.5N/4 45.4W Oc(2+1) WRG 12s 33m **W17M**, R14M, G13M; grey square Tr; vis W290°-298°, R298°-325°, W325°-355°, G355°-017°, W017°-035°, G035°-105°, W105°-123°, R123°-158°, W158°-205°; R Lt on radio mast 3.4M ENE; *Siren (2+1) 60s*. Tévennec 48 04.3N/4 47.6W Q WR 28m W9M, R6M; W square Tr and dwelling; vis W090°-345°, R345°-090°; Dir Lt Iso 4s 24m 12M; same Tr; intens 324°-332°.

CHAUSSÉE DE SEIN. **Île de Sein** 48 02.70N/4 51.95W Fl(4) 25s 49m **29M**; W Tr, B top; RC. Ar Guéveur 48 02.0N/4 51.4W *Dia 60s*; W Tr. Men-Brial, 0.8M 115° from main Lt, Oc(2) WRG 6s 16m W12M, R9M, G7M; W Tr, G top; vis G149°-186°, W186°-192°, R192°-221°, W221°-227°, G227°-254°. **Ar-Men** 48 03.0N/4 59.9W Fl(3) 20s 29m **23M**; W Tr, B top. *Siren (3) 60s*.

CHAUSSÉE DE SEIN LANBY 48 03.80N/5 07.70W VQ(9) 10s 9m 8M; W cardinal mark; *Whis*; Racon; Ra refl.

Pointe du Millier 48 05.9N/4 27.9W Oc(2) WRG 6s 34m **W16M**, R12M, G11M; W house; vis G080°-087°, W087°-113°, R113°-120°, W120°-129°, G129°-148°, W148°-251°, R251°-258°; part obsc 255°-082°.

TRÉBOUL. Épi de Biron, head 48 06.1N/4 20.4W QG 7m 6M; W column, G top.
Île Tristan 48 06.2N/4 20.3W Oc(3) WR 12s 35m W13M, R10M; grey Tr, W band, B top; vis W shore-138°, R138°-153°, W153°-shore; obsc by Pointe de Leidé when brg less than 111°.
DOUARNENEZ. Bassin Nord, N mole head 48 06.0N/4 19.3W Iso G 4s 9m 4M; W & G framework Tr. S mole head Oc(2) R 6s 6m 6M; W & R framework Tr. Elbow, Môle de Rosmeur, head Oc G 4s 6m 6M; W framework Tr, G top; vis 170°-097°.

BASSE VIEILLE LANBY 48 08.40N/4 35.58W Fl(2) 6s 8m 8M; isolated danger mark; *Whis*; Ra refl.
Basse du Bouc Lt Buoy 48 11.50N/4 37.34W Q(9) 15s; W cardinal mark; *Whis*.

Pointe de Morgat 48 13.2N/4 29.9W Oc(4) WRG 12s 77m **W15M**, R11M, G10M; W square Tr, R top, W dwelling; vis W shore-281°, G281°-301°, W301°-021°, R021°-043°; obsc by Pointe du Rostudel when brg more than 027°. Morgat mole head 48 13.5N/4 30.0W Oc(2) WR 6s 8m W9M, R6M; W & R framework Tr; vis W015°-265°, R265°-015°.

LANBY, BASSE DU LIS 48 13.05N/4 44.46W Q(6) + LFl 15s 9m 8M; S cardinal mark; *Whis*.

La Parquette 48 15.91N/4 44.25W Fl RG 4s 17m R6M, G5M; W 8-sided Tr, B diagonal stripes; vis R244°-285°, G285°-244°.

Pointe du Toulinguet 48 16.87N/4 37.64W Oc(3) WR 12s 49m **W15M**, R11M; W square Tr on building; vis W shore-028°, R028°-090°, W090°-shore.

CAMARET. Môle Nord, head 48 16.92N/4 35.20W Iso WG 4s 7m W12M, G8M; W framework Tr, G top; vis W135°-182°, G182°-027°; obsc by Pointe des Capucins when brg more than 187°. Môle Sud, head Fl(2) R 6s 9m 5M; R framework Tr; obsc by Pointe du Grande Gouin when brg less than 143°, and by Pointe des Capucins when brg more than 185°.

AREA 15—S Brittany 623

10.15.5 PASSAGE INFORMATION

BAY OF BISCAY – NORTH PART (chart 20)

In crossing the B of Biscay, allowance should be made for possible set to E. Onshore winds carry cloud which may cause mist or fog rising over land, sometimes obscuring high Lts. Swell usually comes from W or NW. In summer the sea is usually calm or slight. Wind may be changeable in speed and direction, predominantly from SW to NW in summer. Offshore the tidal stream is weak, but gets stronger in N part of B of Biscay towards English Channel, and may run strongly near the coast, off headlands. French charts are often on larger scale than Admiralty charts, and hence are more suitable for inshore waters. Yachts are recommended to carry the *Bay of Biscay Pilot* and *North Biscay Pilot* (Adlard Coles Ltd). Malcolm Robson's *French Pilot Volume Four* (Nautical Books) covers the Biscay coast from the Gironde to the Morbihan, while his *French Pilot Volume Three* continues from Belle Ile to the Raz de Sein. For French terms see 10.16.5.

Many of the more attractive French harbours dry completely, so it is best to have a boat which can take the ground. For this purpose bilge keels are ideal – or otherwise a boat with a long, straight keel, and fitted with legs.

La Loire (chart 2985) carries much commercial traffic. The estuary is divided by two shoals, Plateau de la Banche and Plateau de La Lambarde, which lie about 10M and 6M WNW of Pte de St Gildas. Chenal du Nord runs between these two shoals on S side and the coast on N side. Chenal du Sud, the main chan, leads between Plateau de La Lambarde and Pte de St Gildas. In the approaches to St Nazaire (10.15.9) beware Le Vert, Les Jardinets and La Truie (all dry) which lie SW of Le Pointeau. In Grande Rade the in-going stream begins at HW Brest – 0500, and the out-going at HW Brest + 0050, sp rates about 2¾ kn.

From Chenal du Nord, B de Pouliguen (10.15.10) is entered between Pte de Penchâteau and Pte du Bec, 3M to E. In SE corner of b is the yacht harb of Pornichet (10.15.27). The B is part sheltered from S by rks and shoals extending SE from Pte de Penchâteau, but a heavy sea develops in strong S-SW winds. The chan through these rks runs between Basse Martineau and Les Guérandaises.

Between Pte de Penchâteau and Pte du Croisic, Basse Lovre is a rky shoal with depths of 0.9m, ½M offshore. Off Pte du Croisic dangers extend 1M to W and to N. Plateau du Four, a dangerous drying bank of rks, lies about 4M W and WSW of Pte du Croisic, marked by buoys and Lt Ho near N end. Off Pte du Castelli, Les Bayonelles (dry) extend ½M W, and Plateau de Piriac extends about 1¾M NW with depths of 3m and drying rks closer inshore. There is chan between Plateau de Piriac and Île Dumet (Lt), which is fringed by drying rks and shoals particularly on N and E sides.

In the approaches to La Vilaine (10.15.12) beware Basse du Bile, a rky shoal depth 0.6m, 1M W of Île du Bile; Basse de Loscolo, depth 0.6m, about 1M to N; and La Grande Accroche, an extensive shoal with depths of less than 1.8m, astride the entrance. The main chan is lit and runs NW of La Grande Accroche, to the bar on N side thereof. Here the flood stream begins at HW Brest – 0515, and the ebb at HW Brest + 0035, sp rates 2½ kn. In SW winds against the stream the sea breaks heavily in this area, and then the chan ½M off Pte du Halguen is better, but beware La Varlingue (dries).

S of Penerf (10.15.27), which provides good anch, Plateau des Mats is an extensive rky bank, drying in places, up to 1¾M offshore. Dangers extend 1M seaward of Pte de St Jacques, and 3M offshore lies Plateau de la Recherche with depths of 1.8m.

BAIE DE QUIBERON (chart 2353)

The S side of B is enclosed by a long chain of islands, islets, rks and shoals from Presqu'île de Quiberon to Les Grands Cardinaux 13M SE. This chain includes the attractive islands of Houat (10.15.27) and Hoedik, well worth visiting, preferably mid-week, and through it the main chan is Passage de la Teignouse which is well marked. In this chan the NE-going (flood) stream begins at HW Brest – 0610, and the SW-going at HW Brest – 0005, sp rates 3¾ kn: in strong winds it is best to pass at slack water. A good alternative chan is Passage du Béniguet, NW of Houat.

B de Quiberon is an important and attractive yachting area, with major centres at Crouesty (10.15.13), the Morbihan (10.15.15), La Trinite (10.15.16), Port Haliguen (10.15.17) and Port Maria (10.15.27).

BELLE ILE (chart 2353)

Apart from rks which extend ¾ M W of Pte des Poulains, and La Truie (dries) marked by Bn Tr ½M off the S coast, there are no dangers more than ¼M offshore. The S coast is much indented and exposed to swell from W: in good settled weather (only) and in absence of swell there is an attractive anch in Port du Vieux Chateau, 1M S of Pte des Poulains – see *North Biscay Pilot*. On the NE coast lie Le Palais (10.15.18) and Sauzon (10.15.27), which dries but has good anch off and is sheltered from S and W. Off Le Palais the ESE-going (flood) stream begins at HW Brest – 0610, and the WNW- going at HW Brest + 0125, sp rates 1½ kn. 6M NNW of Pte des Poulains lies Plateau des Birvideaux (Lt), a rky bank (depth 2m) on which the sea breaks in bad weather. Between here and Lorient, Etel River (chart 2352) is an attractive harb, but must only be approached in good weather and on the last of the flood. The chan over the bar is shallow and shifts; local semaphore signals indicate the best water, or when entry is impossible. See 10.15.27.

ILE DE GROIX (chart 2352)

Île de Groix lies in SW approach to Lorient (10.15.19). The main offlying dangers are shoals off Pte de la Croix, Les Chats which extend 1M SE from Pte des Chats, and shoals extending ¾M seaward off Port Loc Maria.

Port Tudy, on N coast, is the main harb, and is easy of access and well sheltered except from NE see 10.15.27. The inner harb has a sill and pontoon berths – entry HW∓2. Port Loc Maria is ¾ M W of Pte des Chats. The unspoiled little harb is open to the S, but provides an attractive anch in offshore winds, and particularly near nps.

LORIENT TO BENODET (chart 2352)

Between Pte du Talat and Pte de Trévignon sandy beaches give way to rky cliffs. Grand Cochon and Petit Cochon lie about 1M offshore 2M NW of Pte de Keroch. There are many shoals in approaches to Rivière de Bélon and L'Aven. To the W, Ile Verte lies 6 ca S of Ile de Raguénès, with foul ground another 2 ca to S; Men an Tréas, a rk which dries, is 1M WSW of Ile Verte and marked by buoy; Corn Vas, depth 1.8m, is 1¾M SE of Pte de Trévignon, marked by a buoy; Men Du, a rk 0.3m high, lies about 1¼M SE of the same pt, and is marked by a Bn. From Pte de Trévignon to Pte de Cabellou (1M S of Concarneau, 10.15.20) rks extend nearly 1½M offshore in places. Chaussée de Beg Meil extends 8½ ca SE, where Linuen rk (dries) is marked by Bn. The coast W from Beg Meil is fringed by rks, many of which dry, extending 1M offshore, and 1¾M S of Pte de Mousterlin.

Along this coast are several nice harbs and anchs, but most are dangerous to approach in strong onshore winds. They include Le Pouldu (Rivière de Quimperlé); Doëlan (but most of harb dries); Merrien (also mostly dries); Brigneau; and Rivières de Bélon and L'Aven. All are described in *North Biscay Pilot*. For Doëlan and Pont-Aven see 10.15.27.

15

ILES DE GLÉNAN (chart 3640, French chart 6648)

Îles de Glénan lie off entrance to Benodet (10.15.22). The main island is Penfret (Lt), well known for its Centre Nautique, at NE corner of the group which, with its offlying dangers, stretches 5M from E to W and 4M from N to S. There are buoys on SE, S and SW sides. The islands are interesting to explore, but anchs are rather exposed. Easiest approach is from N, to W side of Penfret: anch there or proceed W to La Chambre, S of Île de St Nicolas. 3½M E of Penfret lies Basse Jaune, a shallow bank part of which dries, marked by buoy. Between Îles de Glénan and Benodet lie Les Pourceaux, reefs which dry, and Île aux Moutons which has dangers extending SW and NW.

BENODET TO RAZ DE SEIN (chart 2351)

Between Benodet and Raz de Sein lies Pte de Penmarc'h (Lt, fog sig, RC) off which dangers extend 3M to SE, and 1M to S, W and NW, and breaking seas occur in strong winds. The harbs of Loctudy, Lesconil and Le Guilvinec all provide sheltered anchs. See 10.15.27.

Chaussée de Sein (chart 2351) is a chain of islands, rks and shoals extending 12M W from a position 3M WSW of Pte du Raz. The outer end is marked by a Lt buoy, and near inner end is Île de Sein (Lt, RC). 5M W of Île de Sein is Ar-Men Tr (Lt, fog sig), ½M E of which is a narrow N/S chan. Île de Sein is interesting to visit in fair weather and good vis, preferably near nps. Best approach is through Chenal d'Ezaudi, with Men-Brial Lt Ho, Oc(2) WRG 6s, on with third house (W with B stripe) from left, close S of Lt Ho, at 187°. Drying rocks lie each side, and tide sets across the chan. Anchor off or inside mole, but exposed to N and E. Harb partly dries. For directions on Île de Sein and Raz de Sein see *North Biscay Pilot* (Adlard Coles Ltd).

RAZ DE SEIN (chart 798)

Raz de Sein is the chan between Le Chat Bn Tr at E end of Chaussée de Sein and the dangers extending 8 ca off Pte du Raz, the extremity of which is marked by La Plate Lt Tr. 2M N of Raz de Sein is the Plateau de Tévennec, consisting of islets, rks and shoals which extend ½M in all directions from the Lt Ho thereon. Other dangers on the N side of the Raz are rks and shoals extending nearly 1M W and 8 ca WSW from Pointe du Van, and Basse Jaune (dries) 1M to N. On the S side the main dangers are: Cornoc Bras, a rk with depth of 3m, 1½M SW of La Vieille Lt Tr; Masclougreiz, rky shoals on which sea can break heavily, 1½M S of La Vieille; and Roche Moullec 1½M SE of La Vieille.

In the middle of Raz de Sein the NE-going (flood) stream begins at HW Brest +0550, sp rate 6½ kn; the SW-going (ebb) stream begins at HW Brest −0030, sp rate 5½ kn. There are eddies near La Vieille on both streams. In good weather, near neaps, and with wind and tide together, the Raz presents no difficulty, but in moderately strong winds it should be taken at slack water, which lasts for about ½ hour at end of flood stream. In strong winds the chan must not be used with wind against tide, when there are overfalls with a steep breaking sea.

L'IROISE

Entering B de Douarnenez (chart 798) from Raz de Sein beware Basse Jaune, an isolated rk (dries) about 1M N of Pte du Van. 4½M to E lies Duello Rk (4m high) ½M offshore, with other rks extending 1M to E. Otherwise the S shore of the b is clear of dangers more than 2 ca offshore. Approaching Douarnenez (10.15.24) beware Basse Veur and Basse Neuve. On N side of b beware group of drying rks including La Pierre-Profonde and Le Taureau close SSW of Les Verrès (rk 9m high) which lies nearly 2½M ESE of Morgat (10.15.25).

Off C de la Chèvre various dangers, on which the sea can break, extend SW for 2¼M to Basse Vieille (dries) which is marked by Lanby. 2½M NW of C de la Chèvre is Le Chevreau (dries), with La Chèvre ½M to NE of it (1¼M WSW of Pte de Dinan). 7M W of Pte de Dinan lies Basse du Lis, rky shoals with depth of 1.8m, marked by Lanby.

On the NE side of L'Iroise (chart 2643) a chain of rks extends 7M W from Pte du Toulinguet. There are several chans through these rks, of which the most convenient for Camaret (10.15.26) and for Brest (10.16.9) is the inshore one — Chenal du Toulinguet, which runs between La Louve Bn Tr (1 ca W of Pte du Toulinguet) on E side and a rk called Le Pohen on the W side. Keep in the middle of the chan, which is about 3 ca wide. Here the N-going stream begins at HW Brest −0550, and the S-going at HW Brest +0015, sp rates 2¾ kn.

10.15.6 DISTANCE TABLE

Approximate distances in nautical miles are by the most direct route while avoiding dangers and allowing for traffic separation schemes etc. Places in *italics* are in adjoining areas.

1 *Le Four*	1																			
2 *Brest*	25	2																		
3 Camaret	21	8	3																	
4 Morgat	32	23	16	4																
5 Douarnenez	42	27	21	11	5															
6 Pointe du Raz	30	24	18	17	20	6														
7 Audierne	40	34	28	27	30	10	7													
8 Pte de Penmarc'h	52	46	40	39	42	22	15	8												
9 Benodet	70	64	58	57	60	40	33	18	9											
10 Concarneau	74	68	62	61	64	44	37	22	11	10										
11 Lorient	98	92	86	85	88	68	61	46	36	32	11									
12 Port Haliguen	117	111	105	104	107	87	80	65	57	53	32	12								
13 Le Palais	111	105	99	98	101	81	74	59	52	47	26	11	13							
14 La Trinité	122	116	110	109	112	92	85	70	62	58	37	8	16	14						
15 Crouesty	122	116	110	109	112	92	85	70	62	58	37	9	16	8	15					
16 Roche Bernard	146	140	134	133	136	116	109	94	86	82	61	35	40	35	30	16				
17 Le Croisic	136	130	124	123	126	106	99	84	76	72	51	27	27	28	23	25	17			
18 St Nazaire	150	144	138	137	140	120	113	98	91	87	66	42	41	45	40	43	20	18		
19 *Ile d'Yeu*	156	150	144	143	146	126	119	104	98	95	76	56	50	60	56	61	38	34	19	
20 *La Rochelle*	216	210	204	203	206	186	179	164	158	155	136	116	110	120	116	121	98	94	60	20

CANAL CONNECTIONS IN BRITTANY 10.15.8

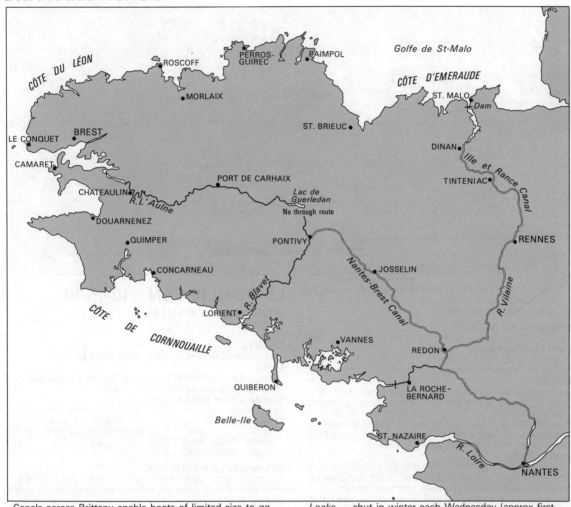

Canals across Brittany enable boats of limited size to go from the Channel to the Bay of Biscay avoiding the passage around Finistere.

SUMMARY

Channel — Biscay (Ille et Rance Canal and Vilaine)

Section	No of locks	Length nm	Length km	Speed limit kn
St. Malo—Dinan	2	20.5	38	7.5
Dinan—Tinteniac	14	17.3	32	4.3
Tinteniac—Rennes	33	25.4	47	4.3
Rennes—Redon	12	48	89	4.3
Redon—La Roche-Bernard	0	21.1	39	—

Nantes — Lorient Link (Nantes — Brest Canal and Blavet)

Nantes—Redon	16	51.3	95	4.5
Redon—Josselin	17	34	63	4.3
Josselin—Pontivy	72	25.9	48	4.3
Pontivy—Lorient	28	39.2	72	4.3

L'Aulne

Chateaulin—Port de Carhaix	34	41	76	4.3

Max length 25 m
Max draught 1.1 m (Ille et Rance Canal)
             0.8 m (Josselin—Pontivy)
Max beam 4.6m
Max height 2.4m

Contact: Channel—Biscay — Rennes 99.59.20.60
                          Redon 99.71.10.66
         Nantes—Lorient (Brest) — Nantes 40.71.70.70
                          Hennebont 97.65.20.82
                          Lorient 97.21.21.54

Locks — shut in winter each Wednesday (approx first week in November to last week in March) for repairs. July and August, in order to conserve water, open on the hour only (and at half hour if traffic demands).

Channel—Biscay Assembly —Northbound— Lengager Lock.
Southbound— Madeleine Lock.

For information and guide books write (with addressed envelope) to
Comité des Canaux Bretons,
Service de Documentation du Comité,
12 rue de Jemmapes,
44000 Nantes (Tel. 40.47.42.94)

RIVER LOIRE/ST NAZAIRE
Loire Atlantique   10-15-9

CHARTS
Admiralty 2985, 3216, 2989; SHOM 6854, 6493, 6797, 6260, 6261, 5992; ECM 248

TIDES
St Nazaire-Dover: Springs +0445, Neaps −0540; ML 3.0; Duration: Springs 0640; Neaps 0445; Zone −0100
Standard Port BREST (→)

RIVER LOIRE/ST. NAZAIRE continued

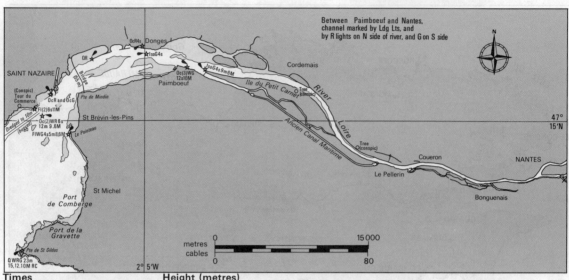

Times				Height (metres)			
HW		LW		MHWS	MHWN	MLWN	MLWS
0000	0600	0000	0600	7.5	5.9	3.0	1.4
1200	1800	1200	1800				

Differences ST NAZAIRE
+0030 −0025 −0005 −0010 −2.0 −1.7 −1.1 −0.8
NANTES (Chantenay)
+0155 +0140 +0330 +0245 −1.7 −1.3 −1.1 +0.2

SHELTER
In strong winds from West the bar is only safe HW−3 to HW. St Nazaire is principally commercial although yachts can stop temporarily. Yachts enter the Bassin de St Nazaire through E lock and berth at S end of Bassin Penhoët. Port Control and Hr Mr NW side of Avant Port. Anchorages in Bonne Anse and Villez Martin.
NAVIGATION
Waypoint Chenal du Sud 47°07'.95N 02°20'.00W, 205°/025° from/to front Ldg Lt 025°, 7.3M. Entrance to Loire divided by La Banche bank with well marked channels N and S. River is shallow beyond Nantes but is navigable in optimum conditions for small craft West of Angers. At Nantes the Brittany canal system can be entered through canal Saint-Felix by Malakoff Lock (see 10.15.8); also there are places reserved for visiting yachts at Trentemoult near the Bureau du Port, accessible at all times.
LIGHTS AND MARKS
Ldg Lts over bar, 025°; both Q 27M, intens 022°-028°.
RADIO TELEPHONE
St Nazaire Port VHF Ch 12 16 (H24). Other stations: Donges VHF Ch 12 16 (H24). Pte de Chémoulin Sig Stn Ch 16.
Nantes VHF Ch 12 16 (H24); water level reports bcst on Ch 73 every 15 min commencing H+00.
TELEPHONE
ST NAZAIRE Hr Mr 40.22.08.46; Aff Mar 40.22.46.32; CROSS 97.55.35.35; SNSM 40.61.03.20; Customs 40.66.82.65; Meteo 40.90.00.80; Auto 40.90.19.19; Dr 40.22.15.32; Hosp 40.90.60.60.
NANTES Hr Mr 40.44.20.54; Aff Mar 40.73.18.70; Customs 40.73.39.55; Meteo 40.84.80.19; Auto 40.04.15.15; Dr 40.47.03.19; Hosp 40.48.33.33; Brit Consul 40.63.16.02.
FACILITIES
ST NAZAIRE **La Société Nautique de St Nazaire** Tel. 40.70.10.47, M; **Genet-Loisirs** Tel. 40.22.44.52, M, ME, El, Sh, CH; **Quai** P, D, L, FW, C; **Perraud** Tel. 40.22.51.39, ME, El; **Burodess** Tel. 40.22.11.12; SHOM.
Town V, Gaz, R, Bar. PO; Bank; Rly; Air.
NANTES **Quai** FW, C; **Librairie Beaumont** Tel. 40.48.24.21, CH, SHOM; **Anore Marine** Tel. 40.73.86.22, ME, CH, SHOM, El; **Town** all facilities; PO; Bank; Rly; Air.
Ferry UK — St Malo−Portsmouth; Roscoff−Plymouth.

LE POULIGUEN 10-15-10
(PORT OF LA BAULE)
Loire Atlantique

CHARTS
Admiralty 3216; SHOM 6825, 4902; ECM 547
TIDES
Dover: Springs +0435 Neaps +0530; ML 2.9; Duration: Springs 0530 Neaps 0645; Zone −0100
Standard Port BREST (→)

Times				Height (metres)			
HW		LW		MHWS	MHWN	MLWN	MLWS
0000	0600	0000	0600	7.5	5.9	3.0	1.4
1200	1800	1200	1800				

Differences LE POULIGUEN
+0015 −0040 0000 −0020 −2.1 −1.8 −1.2 −0.8

SHELTER
Very good shelter, protected from all winds except SE. Approach from West between Pte de Penchateau and the Grand Charpentier. Best approach HW−1. Visitors berths (30) on Quai Rageot de la Touche.
NAVIGATION
Waypoint 47°15'.00N 02°25'.00W, 232°/052° from/to Basse Martineau buoy, 0.58M. Beware rocks extending 4 M to the Grand Charpentier Lt Ho and breakers in shallow water when wind in S. It is not advisable to enter at night. Beware strong ebb tide.
LIGHTS AND MARKS
From Basse Martineau leave La Vieille and Les Impairs as on chartlet to stbd, and two Bns off Penchateau to port. S Jetty QR 13m 9M; vis 171°-081°.
RADIO TELEPHONE
Pouliguen VHF Ch 09. Pornichet Ch 09.
TELEPHONE
Hr Mr 40.60.37.40; Aff Mar 40.42.32.55; CROSS 97.55.35.35; SNSM 40.61.03.20; Customs 40.61.32.04; Meteo 40.90.00.80; Auto 40.90.19.19; Dr Le Pouliguen 40.60.51.73; Dr La Baule 40.60.17.20; Brit Consul 40.63.16.02.
FACILITIES
Quai (Pontoons 620) Tel. 40.60.03.50, Slip, P, D, AC, L, FW, C (18 ton); **Callaghan Naval** Tel. 40.60.78.43, M, ME, Sh, CH; **Petit Breton Nautique** Tel. 40.42.10.14, M, ME, El, Sh, CH; **Prat** Tel. 40.60.31.30, Sh, Divers; **SOS Atlantique** Tel. 40.42.33.29, ME, El, CH; **L'Ancre Marine** Tel. 40.42.33.73, SHOM; **Voilerie X Voiles** Tel. 40.60.47.52, SM; **Le Baule YC** Tel. 40.60.20.90 (allocates berths for visitors); **Town** V, Gaz, R, Bar. PO (Le Pouliguen and La Baule); Bank (Le Pouliguen and La Baule); Rly (La Baule); Air (St Nazaire, La Baule).
Ferry UK — St Malo−Portsmouth; Roscoff−Plymouth.

AREA 15—S Brittany

LE POULIGUEN continued

FACILITIES
Marina (220 + 15 visitors), P, D, FW, C, ME, El, Sh; **Quai** Slip, M, FW, ME, El, Sh, C (10 ton), CH, V, R, Bar; **Bollore** Tel. 40.23.00.85, ME; **Lemerie** Tel. 40.42.92.12, M, ME, El, Sh, CH; **Meca Navale** Tel. 40.23.04.10, ME, El, Sh. **Town** V, Gaz, R, Bar. PO; Bank; Rly; Air (St Nazaire, La Baule).
Ferry UK — St Malo—Portsmouth; Roscoff—Plymouth.

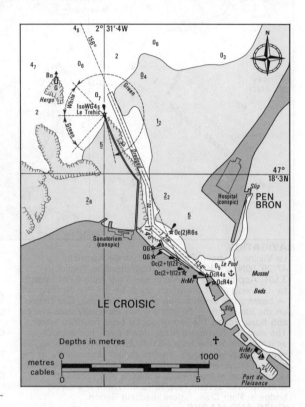

LE CROISIC 10-15-11
Loire Atlantique

CHARTS
Admiralty 2353; SHOM 6826, 5482; ECM 546, 547

TIDES
+0450 Dover; ML 3.0; Duration 0605; Zone −0100
Standard Port BREST (→)

Times				Height (metres)			
HW		LW		MHWS	MHWN	MLWN	MLWS
0000	0600	0000	0600	7.5	5.9	3.0	1.4
1200	1800	1200	1800				

Differences LE CROISIC
+0030 −0030 −0015 −0010 −2.3 −1.8 −1.1 −0.8
ILE DE HÖEDIC
−0005 −0025 −0030 −0020 −2.3 −1.9 −1.1 −0.8

SHELTER
Safest anchorage is up Pen Bron creek. Alongside berths, called *Chambres* formed by islands called *Jonchères*, dry out. Access HW ∓ 1. Yachts use the fifth *Chambre*. A mooring may be available in Le Poul.

NAVIGATION
Waypoint 47°19'.00N 02°31'.80W, 336°/156° from/to front Ldg Lt 156°, 1.2M. At springs, tides run up to 4 kn. Safest entry then is HW−1. Beware the rocks at Hergo tower. Beware that the Tréhic Lt, which leads clear of distant dangers, will lead onto close dangers.

LIGHTS AND MARKS
Ldg Lts 156°. Front Dir Oc(2+1) 12s 10m 18M; Y topmark on W pylon. Rear Dir Oc(2+1) 12s 14m 18M; Y topmark on W pylon, G top; both intens 154°-158°, synchronized. Ldg Lts 174° both QG 5/8m 12M; G & W structures; vis 170°-177°. Ldg Lts 134° both Oc R 4s, synchronized. The sanatorium and hospital are conspic.

RADIO TELEPHONE
VHF Ch 16.

TELEPHONE
Hr Mr 40.23.05.38; Port de Plaisance 40.23.10.95; Aff Mar 40.23.06.56; CROSS 97.55.35.35; SNSM 40.23.01.17; Customs 40.23.05.38; Meteo 40.90.08.80; Dr 40.23.01.70; Hosp 40.23.01.12; Brit Consul 40.63.16.02.

LA ROCHE BERNARD 10-15-12
Morbihan

CHARTS
Admiralty 2353; SHOM 5418, 6418.

TIDES
+0500 Dover; ML (Pénerf) 3.1; Duration 0610; Zone −0100
Standard Port BREST (→)

Times				Height (metres)			
HW		LW		MHWS	MHWN	MLWN	MLWS
0000	0600	0000	0600	7.5	5.9	3.0	1.4
1200	1800	1200	1800				

Differences PÉNERF
−0010 −0020 −0025 −0020 −2.0 −1.7 −1.0 −0.7

SHELTER
Good shelter, on S bank of La Vilaine, about 7M E of Tréhiguier. There are two marinas — Nouveau Port close to bridge (clearance 30m) and Vieux Port (or Quai de Sainte-Antoine) ¼M downstream. Anchor at Tréhiguier or at Vieille Roche while waiting for lock at Arzal (see 10.15.27).

LA ROCHE BERNARD continued

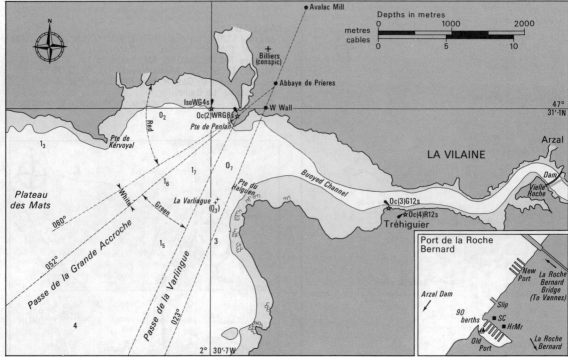

NAVIGATION
La Vilaine waypoint Passe de la Grande Accroche 47°29'.00N 02°34'.28W, 234°/054° from/to Penlan Lt, 3.5M. Approach to La Vilaine has a bar (min 0.5m except for La Varlingue rocks which dry 0.3m off Pte du Halguen — beware this when sailing channel 1 below). Sea breaks on the bar in strong on-shore winds especially with low or ebb tides. Follow the line of cockle beds for deep water. The dam, Le Barrage d'Arzal, down river, has a lock which opens on the hour 0700 to 2200 (LT) in summer; in winter in daylight hours. Yachts should keep strictly to the channel above and below the dam. Yachts with masts can reach Redon where there is a marina and access to Brittany canal system (see 10.15.8). There is a swing bridge at Pont Cran, before reaching Redon.

LIGHTS AND MARKS
There are two approaches to river entrance:
(1) Passe de la Varlingue (W wall, Abbey de Prieres and Avalac Mill in line at 023°), unlit.
(2) Passe de la Grande Accroche (Lt Ho and Abbey de Prieres in line at 052°).
Two principal lights
(1) Basse Bertrand G Tr, Iso WG 4s 6m 9/6M W040°-054°, G054°-227°, W227°-234°, G234°-040°.
(2) Penlan W Tr with R bands Oc (2) WRG 6s 26m 16/11M, R292°-025°, G025°-052°, W052°-060°, R060°-138°, G138°-180°.
Channel from Pte du Halguen to Tréhiguier marked by R and G numbered buoys.

RADIO TELEPHONE
None.

TELEPHONE
Hr Mr 99.90.62.17; Aff Mar 97.41.12.43; CROSS 97.55.35.35; Customs (Vannes) 97.63.18.71; Meteo 40.75.80.07; Dr 99.90.61.25; Hosp 99.90.61.20; Brit Consul 40.63.16.02.

FACILITIES
Marina (New Port) (70), Tel. 99.90.65.91, FW, P and D (Tel. 99.90.66.77), AC, C, CH; **Marina (Old Port)** (200), Lau, FW, Slip; **Chantier Naval de la Couronne** Tel. 99.90.66.77, M, ME, El, Sh, CH. **Arvor Marine** Tel. 99.90.64.98, ME, El, Sh, CH; **Town** V, Gaz, R, Bar. PO; Bank; Rly (Pontchateau); Air (Nantes or Rennes).
Ferry UK — St Malo—Portsmouth; Roscoff—Plymouth.

CROUESTY-EN-ARZON
Morbihan

10-15-13

CHARTS
Admiralty 2359, 2358; SHOM 5420, 3165, 5554; ECM 546

TIDES
+0505 Dover; ML 3.0; Duration 0555; Zone −0100
Standard Port BREST (→)

Times				Height (metres)			
HW		LW		MHWS	MHWN	MLWN	MLWS
0000	0600	0000	0600	7.5	5.9	3.0	1.4
1200	1800	1200	1800				

Differences PORT NAVALO
+0025 −0005 −0010 −0005 −2.5 −2.0 −1.1 −0.7

SHELTER
Good shelter, protected from all winds by Quiberon peninsular. A very large marina gives added protection.

NAVIGATION
Waypoint 47°32'.12N 02°55'.00W, 238°/058° from/to front Ldg Lt, 0.85M. There are no navigational dangers, the entrance being well marked.

LIGHTS AND MARKS
Ldg Lts 058°, Front Q 10m 19M on metal tower, Rear Q on W column: both intens 056°-060°. Marina breakwater heads: N Oc (2) R 6s, S Fl G 4s.

RADIO TELEPHONE
VHF Ch 09.

TELEPHONE
Hr Mr 97.53.73.33; Aff Mar 97.41.84.10; CROSS 97.55.35.35; Meteo 97.64.34.86; Auto 97.84.82.83; SNSM 97.41.27.40; Customs 97.53.25.66; Dr 97.51.60.05; Dr (Arzon) 97.26.23.23; Brit Consul 40.63.16.02.

FACILITIES
Marina (1000+120 visitors) Tel. 97.53.73.33, BH (45 ton), P, D, L, FW, ME, El, Sh, Slip, CH, AB, Bar, C (3 ton); **Armour Yachting** Tel. 97.53.78.42, M, ME, El, Sh, CH; **Ateliers Maritimes du Crouesty** Tel. 97.53.71.30, M, ME, El, Sh, CH; **Chantier Naval du Redo** Tel. 97.53.78.70, BY, CH; **Technique Voile** Tel. 97.53.78.58, SM; **Town** V, Gaz, R, Bar. PO (Arzon); Bank (Arzon); Rly (Vannes); Air (Lorient—Nantes).
Ferry UK — St Malo—Portsmouth; Roscoff—Plymouth.

CROUESTY-EN-ARZON continued

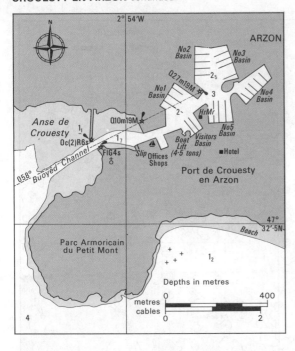

LIGHTS AND MARKS
There are buoys and perches marking the channel from Pointe de Bararac up to Vannes.
G Lt over R Lt = Port closed
No lights = Port open
Lts shown from the lock, but boats must stop and wait at the S end of the approach canal.

RADIO TELEPHONE
VHF Ch 09. (Summer 0800-2100; winter 0900-1200, 1330-1800 LT).

TELEPHONE
Hr Mr 97.54.16.08; Aff Mar 97.63.40.95; CROSS 97.55.35.35; SNSM 97.26.00.56; Customs 97.63.18.71; Meteo 97.42.48.82; Auto 97.42.49.49; Dr 97.47.47.25; Hosp 97.42.66.42; Brit Consul 40.63.16.02.

FACILITIES
Marina (190 + 80 visitors), Tel. 97.54.16.08, Slip, Lau, FW, AC, ME, Sh, C (12 ton), V, R; **La Corderie** Tel. 97.47.15.52, CH; **Vannes Plaisance** Tel. 97.47.32.09, M, ME, El, Sh, CH, AB; **Vannes Nautique** Tel. 97.63.20.17, ME, Sh, El, CH; **Voilerie Daniel** Tel. 97.47.14.41, Gas, CH, SM, SHOM; **Voilerie Le Pors** Tel. 97.42.54.01, CH, SM, Gaz, Electronics; **Quai visiteurs** Slip, FW; **Le Pennec** Tel. 97.47.32.09, P, D, ME, El, Sh (on E side of approach canal).
Town all facilities. PO; Bank; Rly; Air (Lorient).
Ferry UK — St Malo—Portsmouth; Roscoff—Plymouth.

VANNES 10-15-14
Morbihan

CHARTS
Admiralty 2358; SHOM 5420, 3165; ECM 546

TIDES
−0515 Dover; ML 2.7; Duration —; Zone −0100
Standard Port BREST (→)

Times				Height (metres)			
HW		LW		MHWS	MHWN	MLWN	MLWS
0000	0600	0000	0600	7.5	5.9	3.0	1.4
1200	1800	1200	1800				

Differences VANNES
+0145 +0155 +0200 +0105 −4.1 −3.2 −2.0 −1.0

SHELTER
Very good shelter, protected from all winds. Access to the port by day only. Marina boat will meet and indicate pontoon. Lock into wet basin opens HW∓2½ (0800-2000 (LT) in season; 0830-1900 (LT) other periods). Lock sill 1.3m above CD.

NAVIGATION
Waypoint — see 10.15.15. Beware very strong streams off the Réchauds, after passing between the Ile aux Moines and the mainland — Réchauds rocks marked by two beacons to stbd. Thereafter passage to Conleau is straightforward and well marked. Passage on to Vannes only advisable near HW.

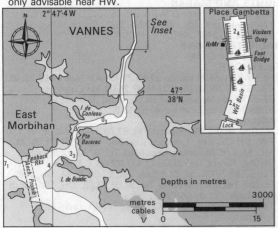

MORBIHAN 10-15-15
Morbihan

CHARTS
Admiralty 2358, 2359; SHOM 3165, 5554, 5420; ECM 546.

TIDES
−0515 Dover; ML 3.1; Zone −0100
Standard Port BREST (→)

Times				Height (metres)			
HW		LW		MHWS	MHWN	MLWN	MLWS
0000	0600	0000	0600	7.5	5.9	3.0	1.4
1200	1800	1200	1800				

Differences AURAY
+0025 −0005 +0025 −0005 −2.6 −2.0 −1.1 −0.6
PORT NAVALO
+0025 −0005 −0010 −0005 −2.5 −2.0 −1.1 −0.7

SHELTER
Golfe du Morbihan is an inland sea with deep approaches and entrance. It is about 50 sq miles and thick with islands, most of which are privately owned. There are numerous anchorages, some of which are shown on the chartlet below and some are as follows (see chart 2358):
PORT NAVALO — anchor in bay but exposed to W to NW winds. All facilities.
LE ROCHER — in Auray river. Good shelter.
AURAY — anchor in middle of river (3.5m) or go alongside at St Goustan.
LARMOR BADEN — perfect anchorage to N (2 to 4m).
ILE AUX MOINES — one of the public islands. Anchor off N end, landing at Pt. de Drech or Pt. de Réchauds, or secure to yacht moorings (see Hr Mr).
ILE PIREN — good anchorage but in tidal stream.
ILE D'ARS — one of the public islands. Anchor NE of Pt de Beluré where there is a landing.
VANNES — see 10.15.14.
PORT BLANC MARINA 1¼ M WSW of Pt d'Aradon.
NOYALLO RIVER — anchor in Anse de Truscatte or between Le Passage and Ile de Pechit.
KERNES — off the Anse de Kernes in 3 to 6m.
AR GAZEK — in Ile de Jument in perfect shelter.

NAVIGATION
Waypoint 47°32'.00N 02°55'.13W, 180°/000° from/to Port-Navalo Lt, 0.90M. Beware very strong tides, max 8 kn, which run in the entrance and between some of the islands. Channels and dangers are well marked.

MORBIHAN continued

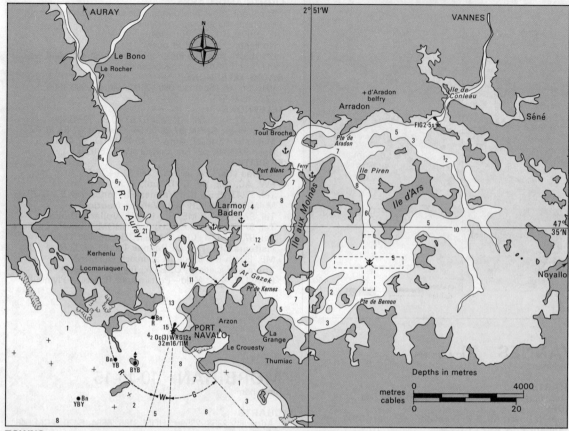

TOWNS
PORT NAVALO. all facilities. Customs Tel. 97.41.21.53
AURAY. All facilities. Aff Mar Tel. 97.24.01.43;
Accastillage Diffusion Tel. 97.24.18.89, ME, El, Sh, CH, SHOM; **Miran** Tel. 97.24.07.19, ME, Sh, CH;
Chatenay et Cie 'La Chateyne' Tel. 97.24.91.52, CH.
LE BONO — Anchorage. **Boursicot** Tel. 97.57.82.57, ME.
LARMOR BADEN. Aff Mar Tel. 97.57.05.66.
ILE AUX MOINES. Bureau du Port Tel. 97.26.30.57, M, FW. Yachts can secure to moorings — see Hr Mr.
LOCMARIAQUER. Protected anchorage. **Madec** Tel. 97.24.92.56, ME.

ARRADON — good anchorage. Slip; Hr Mr 97.26.01.23; **Mechanique Marine** (Pramer) Tel. 97.26.01.16, ME.
CONLEAU — Anchorage in bight just S of village. **Le Mée (Michelet)**, ME, El, Sh.
VANNES (see 10.15.14).
SÉNÉ — **Le Gal** Tel. 97.66.96.90, ME.
ILE D'ARS — Rudevent. Visitors anchorage, slip; exposed to winds from S and E. **Toussaint Guy** Tel. 97.26.30.01, ME, El, Sh.
PORT BLANC — Hr Mr 97.57.01.00. Limited facilities.

LA TRINITÉ-SUR-MER
Morbihan   10-15-16

CHARTS
Admiralty 2358; SHOM 5352, 5420; ECM 546, 545
TIDES
+0455 Dover; ML 3.1; Duration 0610; Zone −0100
Standard Port BREST (→)

Times				Height (metres)			
HW		LW		MHWS	MHWN	MLWN	MLWS
0000	0600	0000	0600	7.5	5.9	3.0	1.4
1200	1800	1200	1800				

Differences LA TRINITÉ
+0010 −0020 −0025 −0015 −2.1 −1.7 −0.9 −0.7

SHELTER
Very good shelter except in strong SE to S winds when the sandbank La Vanererse is covered. Anchorage in the river is forbidden. Marina boat will meet and indicate pontoon. Access day and night at all tides.
NAVIGATION
Waypoint 47°31'.90N 02°59'.63W, 167°/347° from/to front Ldg Lt 347°, 2.3M. Beware many oyster beds, marked with perches. There are no navigational dangers and the river is well marked by buoys and perches. Deepest water close to E shore. Anchoring and fishing prohibited. Speed limit 5 kn.
LIGHTS AND MARKS
Ldg Lts 347°. Front Q WRG (W345°-013°). Rear QW 21m 17M; synchronised; intens 337°-357°. Dir Lt 347° Oc WRG 4s (W346°-348°).
RADIO TELEPHONE
Marina VHF Ch 09.
TELEPHONE
Hr Mr Port de Plaisance 97.55.71.49; Aff Mar 97.55.73.46; CROSS 97.55.35.35; SNSM 97.24.00.14; Customs 97.55.73.46; Meteo 97.64.34.86; Auto 97.84.82.83; Dr 97.55.74.03; Brit Consul 40.63.16.02.
FACILITIES
Marina (900+100 visitors) Tel. 97.55.71.27, AC, P, D, FW, BH (36 ton), Gridiron; **Kervilor** Tel. 97.55.71.69, M, ME, El, CH; **Madame Lecuillier** Tel. 97.55.72.59, CH, SHOM; **Rioux** Tel. 97.55.72.43, ME; **S.E.E.M.A.** Tel. 97.55.78.06, El; **St Mizar** Tel. 97.55.78.44, CH, ME, El, Sh; **Technique Voile** Tel. 97.55.78.08, CH, SM; **Club Nautique** Tel. 97.55.73.48.
Town V, Gaz, R, Bar, PO; Bank; Rly (Auray); Air (Lorient).
Ferry UK — Roscoff—Plymouth; St Malo—Portsmouth.

AREA 15—S Brittany   631

LA TRINITE-SUR-MER continued

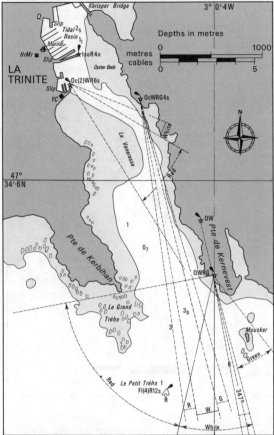

▲ Marina boat will indicate

PORT HALIGUEN-QUIBERON
Morbihan
10-15-17

CHARTS
Admiralty 2353; Imray C38; SHOM 5420, 5352, 135, 5439; ECM 545

TIDES
+0500 Dover; ML 3.0; Duration 0615; Zone −0100
Standard Port BREST (→)

Times				Height (metres)			
HW		LW		MHWS	MHWN	MLWN	MLWS
0000	0600	0000	0600	7.5	5.9	3.0	1.4
1200	1800	1200	1800				

Differences PORT HALIGUEN-QUIBERON
0000 −0020 −0025 −0015 −2.3 −1.9 −1.0 −0.8

SHELTER
Good shelter but strong winds from NW to NE make it uncomfortable. Yachts usually moor stern to pontoons. Secure to visitors' pontoon and ask the Hr Mr. Access at all tides, day and night.

NAVIGATION
Waypoint 47°29'.80N 03°05'.00W, 060°/240° from/to breakwater Lt, 0.75M. Approach from E or SE. Beware shoal (1.8m) ENE of breakwater Lt in R sector 240°-299°.

LIGHTS AND MARKS
W sector 246°-252° of Port Maria Lt (Q WRG) leads N of Banc de Quiberon; W sector 299°-306° of Haliguen Lt (Oc(2)WR 6s) leads S of it.

RADIO TELEPHONE
VHF Ch 09.

TELEPHONE
Hr Mr 97.50.20.56; Aff Mar 97.50.08.71; CROSS 97.55.35.35; SNSM 97.50.14.39; Customs 97.40.14.81; Meteo 97.64.34.86; Auto 97.84.82.83; Dr 97.50.13.94 (Quiberon); Brit Consul 40.63.16.02.

FACILITIES
Marina (640 + 80 visitors) Tel. 97.50.20.56, Slip, P, D, C (2 ton), BH (13 ton), ME, El, Sh, Bar, R, AC, CH, FW, Lau, SM, V; Société Nautique de Quiberon Tel. 97.50.17.45; Loc'Haliguen Tel. 97.50.25.03, CH, ME, El, Sh; La Patelle Tel. 97.30.42.03, El, SHOM, Diving, CH, ME, Electronics; Librairie de Port Maria Tel. 97.50.01.43; Quiberon V, Gaz, R, Bar. PO; Bank; Rly; Air. Ferry UK — Roscoff—Plymouth.

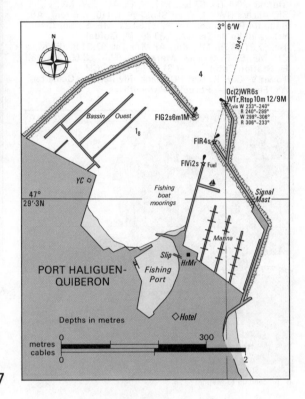

LE PALAIS (BELLE ÎLE)
Morbihan
10-15-18

CHARTS
Admiralty 2353; Imray C38; SHOM 5911, 135; ECM 545

TIDES
+0450 Dover; ML 3.0; Duration 0615; Zone −0100
Standard Port BREST (→)

Times				Height (metres)			
HW		LW		MHWS	MHWN	MLWN	MLWS
0000	0600	0000	0600	7.5	5.9	3.0	1.4
1200	1800	1200	1800				

Differences LE PALAIS
−0015 −0030 −0030 −0025 −2.3 −1.9 −1.0 −0.6

SHELTER
Good shelter except during strong E winds when a marked swell builds up. Yachts can lock into the Bassin à Flot HW −1½ to HW +1 (0600-2200 LT) and thence through lifting bridge into marina. Inner harbour partly dredged. Usual berth for deep keel yachts on three lines of mooring buoys inside Mole Bourdelle. Gets very crowded in summer. As an alternative, Sauzon has space in outer harbour with approx 2m over CD (see 10.15.27).

NAVIGATION
Waypoint 47°21'.20N 03°08'.00W, 065°/245° from/to Jetée Nord Lt, 0.80M. There are no navigational dangers but beware ferries which leave and enter at high speed. Anchorage between Sauzon and Le Palais is prohibited.

LE PALAIS continued

LIGHTS AND MARKS
Mole Bourdelle (N) Fl (2+1) G 12s 11m 7M
Mole Bonnelle (S) Oc (2) R 6s 11m 8M.

RADIO TELEPHONE
VHF Ch 09.

TELEPHONE
Harbour Office 97.31.42.90; Aff Mar 97.31.83.17; CROSS 97.55.35.35; SNSM 97.47.48.49; Customs 97.31.85.95; Dr 97.31.40.90; Hosp 97.31.81.82; Brit Consul 40.63.16.02.

FACILITIES
Marina (200) Tel. 97.52.83.17, P, D, AC, FW, ME, El, Sh, Access HW∓1; **Quai** Slip, FW, C (10 + 5 ton), AB; **Belle-Ile Plaisances** Tel. 97.31.80.23, ME, El, Sh, CH; **YC de Belle Ile** Tel. 97.31.80.46, M; **Guidal** Tel. 97.31.83.75, ME; **Kan Ar Mor** Tel. 97.31.82.05, M, CH, ME, El, Electronics; **Marec** Tel. 97.31.83.60, M, ME, El; **Station Total** Tel. 97.31.80.68, P, D, CH.
Town V, Gaz, R, Bar. PO; Bank; Rly (ferry to Quiberon); Air. Ferry UK — Roscoff—Plymouth.

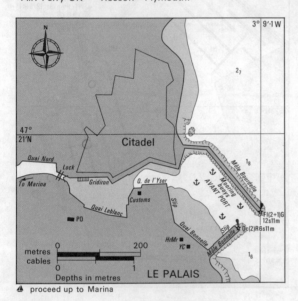

⚓ proceed up to Marina

LORIENT 10-15-19
Morbihan

CHARTS
Admiralty 304, 2352; Imray C38; SHOM 5470, 5560, 6470; ECM 544.

TIDES
+0455 Dover; ML 3.0; Duration 0620; Zone −0100
Standard Port BREST (→)

Times				Height (metres)			
HW		LW		MHWS	MHWN	MLWN	MLWS
0000	0600	0000	0600	7.5	5.9	3.0	1.4
1200	1800	1200	1800				

Differences LORIENT
+0005 −0025 −0025 −0015 −2.4 −1.9 −1.0 −0.5
PORT LOUIS
−0010 −0025 −0020 −0020 −2.4 −1.9 −0.9 −0.5
PORT TUDY (Ile de Groix)
−0005 −0035 −0030 −0025 −2.4 −1.9 −0.9 −0.5
PENFRET (Iles de Glenan)
−0010 −0025 −0030 −0020 −2.6 −2.0 −1.2 −0.6

SHELTER
The Ile de Groix shelters the entrance from SW winds. Good shelter in all weathers. Anchorage N of Kernével. Lock into Marina HW∓1. Waiting berths are provided outside lock. Anchorage very full of mooring buoys. An alternative is to pick up vacant mooring N of Port Louis for the night. In both cases passing traffic makes it uncomfortable. Access day and night tat all tides.

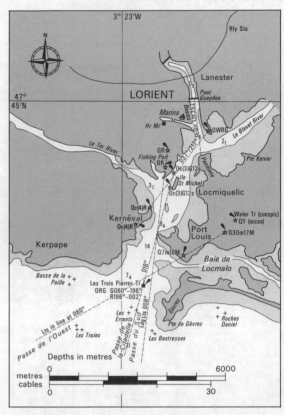

NAVIGATION
Waypoint Passe du Sud 47°40'.50N 03°22'.40W, 189°/009° from/to front Ldg Lt 008°, 3.4M.
Waypoint Passe de l'Ouest 47°40'.80N 03°24'.80W, 240°/060° from/to front Ldg Lt 060°, 3.0M. There are few navigational dangers entering Lorient so long as leading lines are kept to, but yachts must keep clear of shipping in the main channel.

LIGHTS AND MARKS
There are the following leading or directional lights:
(1) Passe Ouest: Ldg Lts 060°; front Q 7m 18M; rear, 0.65M from front, Q 30m 17M; both intens 058°-062°.
(2) Passe Sud: Ldg Lts 008°; front QR 16m 15M; rear, 515m from front, QR 28m 16M; both intens 006°-011°, synchronised.
(3) Passe de la Citadelle: Ldg Lts 016° (N of Trois Pierres). Both Dir Oc (3) G 12s 8/12m 16M; both W Trs, G tops, vis 015°-018°, synchronised.
(4) Les Trois Pierres: Q RG; G 060°-196°, R 196°-002°.
(5) Kernével: Ldg Lts 217°; front Oc (4) R 12s 10m 14M; rear, 290m from front, Oc (4) R 12s 18m 14M; both intens 216°-218°, synchronised.
(6) Pointe de l'Esperance: Dir Lt 037°: Dir Q WRG 8m 9/7M; G 034°-037°, W 037°-037.2°, R 037.2°-047°.
(7) Pont Gueydon: Dir Lt 352°: Dir Iso WRG 4s 6m 9/7M; G 350°-351°, W 351°-353°, R 353°-356°.

RADIO TELEPHONE
Call: *Vigie Port Louis* VHF Ch 16 (H24). Marina Ch 09.

TELEPHONE
Hr Mr Port de Plaisance 97.21.10.14; Aff Mar 97.37.16.22; CROSS 97.55.35.35; SNSM 97.64.32.42; Customs 97.37.29.57; Meteo 97.64.34.86; Auto 97.84.82.83; Hosp 97.37.51.33; Brit Consul 40.63.16.02.

FACILITIES
Marina (230 + 10 visitors) Tel. 97.21.10.14, AC, Slip, ME, El, Sh, M, FW, C (2 ton); **Co-opérative Maritimes Lorient** Tel. 97.37.07.91, ME, El, Sh, CH, SHOM; **Guillet** Tel. 97.37.26.76, Sh; **Intership** Tel. 97.37.00.97, ME, CH, Electronics, SHOM; **Lorient Marine** Tel. 97.37.34.78, M, ME, El, CH, Sh; **V. V. Tonnerre** Tel. 97.37.23.55, CH, SM; **Club Nautique de Lorient** Tel. 97.21.50.85. **Town** All facilities, PO, Bank, Rly, Air.
Ferry UK — Roscoff—Plymouth.

AREA 15—S Brittany 633

CONCARNEAU 10-15-20
Finistere

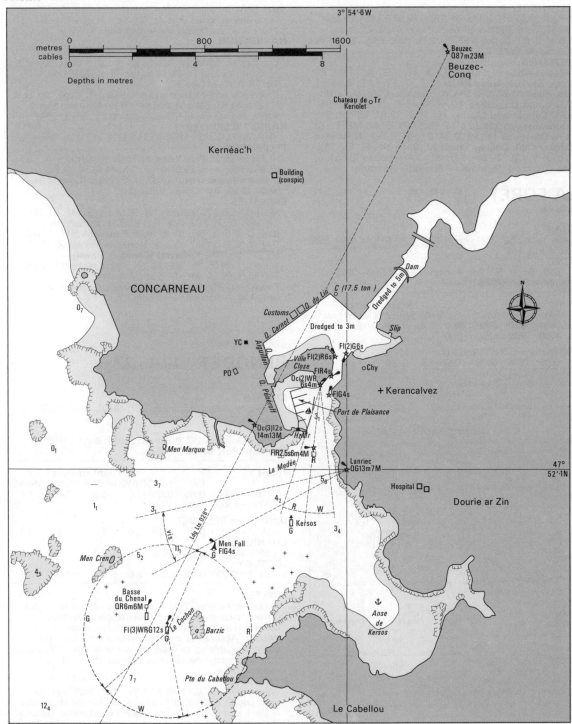

CHARTS
Admiralty 2352, 3641; Imray C38; SHOM 6650, 5368; ECM 544

TIDES
+0455 Dover; ML 2.9; Duration 0615; Zone −0100
Standard Port BREST (→)

Times				Height (metres)			
HW		LW		MHWS	MHWN	MLWN	MLWS
0000	0600	0000	0600	7.5	5.9	3.0	1.4
1200	1800	1200	1800				

Differences CONCARNEAU
−0005 −0035 −0030 −0020 −2.5 −2.0 −1.0 −0.6

SHELTER
Very good shelter in the marina which has the added protection of an anti-wash barrier but exposed to strong S winds. The inner quays are reserved for fishing boats.

NAVIGATION
Waypoint 47°50′.00N 03°56′.80W, 208°/028° from/to front Ldg Lt 028°, 2.52M. Following the leading lights, beware rocks round Men Cren and Le Cochon.

LIGHTS AND MARKS
Ldg Lts 028°, front Oc (3) 12s 14m 13M rear, 1.35M from front, Q 87m 23M. Lanriec Lt, QG 13m 7M vis 063°-078°.

RADIO TELEPHONE
VHF Ch 06 16 (H 24). Marina Ch 09.

CONCARNEAU continued

TELEPHONE
Hr Mr 98.97.01.24; Marina Office 98.97.57.96; Aff Mar 98.97.53.45; CROSS 97.55.35.35; 97.64.32.42; Auto 98.94.00.69; Customs 98.97.01.73; Hosp 98.97.10.60; Brit Consul 40.63.16.02.

FACILITIES
Marina (267+40 visitors), P, D, FW, ME, AC, El, C (17 ton), Slip, Sh; **Quay Pénéroff** P, D, FW, AB; **Barzic** Tel. 98.97.01.57, CH, SHOM; **Chantiers Nautique du Minaouet Grignallou** Tel. 98.97.46.79, M, ME, El, Sh, CH; **Concarneau Marine** Tel. 98.97.19.63, ME, El, Sh; **Concarneau Plaisance** Tel. 98.97.00.50, M, ME, El, Sh, CH; **Gloaguen** Tel. 98.97.04.08, ME, El, Sh; **Nautisme-Motor Club Dolliou** Tel. 98.97.12.05, ME, El, Sh, CH. **Town** V, Gaz, R, Bar. PO; Bank; Rly (bus to Quimper); Air (Quimper). Ferry UK — Roscoff—Plymouth.

LA FORÊT 10-15-21
Finistere

CHARTS
Admiralty 2352, 3641; Imray C37, C38; SHOM 6650, 5368; ECM 543, 544

TIDES
+0450 Dover; ML 2.9; Duration 0615; Zone −0100
Standard Port BREST (→)

Times				Height (metres)			
HW		LW		MHWS	MHWN	MLWN	MLWS
0000	0600	0000	0600	7.5	5.9	3.0	1.4
1200	1800	1200	1800				

Differences CONCARNEAU
−0005 −0035 −0030 −0020 −2.5 −2.0 −1.0 −0.6

SHELTER
Good shelter in the Port de Plaisance in all weathers. There is an anchorage inside Cap Coz.

NAVIGATION
Waypoint 47°52'.70N 03°57'.80W, 162°/342° from/to Cap-Coz Lt, 0.9M. Beware Basse Rouge 6.5ca (1200m) S of Cape Coz and Le Scoré to the SE of Cape Coz.

LIGHTS AND MARKS
Channel is marked by buoys and beacons. Cape Coz Lt is Fl (2) R 6s 5m 6M and that on the breakwater to the NNW, Fl G 4s 8m 6M. Breakwater Lts in line at 334°.

RADIO TELEPHONE
VHF Ch 09.

TELEPHONE
Hr Mr 98.56.98.45; Aff Mar 98.56.01.98; CROSS 97.55.35.35; Auto 98.94.00.57; SNSM 98.56.98.45; Customs (Concarneau) 98.97.01.73; Hosp (Concarneau) 98.97.10.60; Brit Consul 40.63.16.02.

FACILITIES
Marina (690+70 visitors) Tel. 98.56.98.45, ME, El, Sh, CH, AC, D, P, FW, Gaz, R, Lau, SM, V, Bar, BH (16 ton), C (2 ton), Slip (multi hull); **Pontoons** SM, P, AC, D, FW, C (2.5 ton), V, Bar; **Horizons** Tel. 98.56.99.72, ME, El, Sh, CH; **Kerleven (Chantier Naval)** Tel. 98.56.96.85, M, ME, El, Sh, Electronics; **PLF Marine** Tel. 98.56.96.04, ME, El, Sh.
Town Gaz, PO; Bank; Rly (Quimper); Air (Quimper). Ferry UK — Roscoff—Plymouth.

BÉNODET 10-15-22
Finistere

CHARTS
Admiralty 3640, 2352; Imray C37; SHOM 6649, 6679, 5368; ECM 545.

TIDES
+0450 Dover; ML 2.7; Duration 0610; Zone −0100
Standard Port BREST (→)

Times				Height (metres)			
HW		LW		MHWS	MHWN	MLWN	MLWS
0000	0600	0000	0600	7.5	5.9	3.0	1.4
1200	1800	1200	1800				

Differences BÉNODET
−0010 −0025 −0040 −0015 −2.6 −2.2 −1.2 −0.8
CORNIGUEL
+0015 +0010 −0015 −0010 −2.6 −2.1 −1.4 −1.1
LOCTUDY
−0010 −0030 −0030 −0025 −2.6 −2.1 −1.2 −0.9

SHELTER
Anchor in Anse du Trez in offshore winds; or go to one of two marinas; Anse de Penfoul (apply to berthing master on quay); or go to Sainte Marine marina on the W bank. Access day and night at any tide. River Odet easily navigable up to Quimper, but bridge at Poulguinan (just before city) necessitates lowering the mast, or mooring there.

NAVIGATION
Waypoint 47°51'.00N 04°06'.10W, 166°/346° from/to front Ldg Lt 346°, 1.43M. Beware Roches de Mousterlin at the E end of the bay and various rocks round Loctudy at the West end of this bay. In the centre it is clear for small craft. There is a tanker channel past the Ile aux Moutons off the West horn of the bay. Speed limit 3 kn in harbour.

LIGHTS AND MARKS
Ile-aux-Moutons Oc (2) WRG 6s 18m 15/11M and Dir Oc (2) 6s 17m 24M; intens 278°-283°. Ldg Lts 346°. Front Pte du Coq Oc (2+1) G 12s 11m 17M. Rear Oc (2+1) 12s 48m 11M; both W Trs, G tops, synchronized.

RADIO TELEPHONE (local times)
VHF Ch 09 (0800-2000 in season); both marinas.

TELEPHONE
Hr Mr 98.57.05.78; Aff Mar 98.57.03.82; Auto 98.94.00.57; CROSS 97.55.35.35; SNSM 97.57.02.00; Customs 98.55.04.19; Dr 98.57.02.55; Brit Consul 40.63.16.02.

BENODET continued

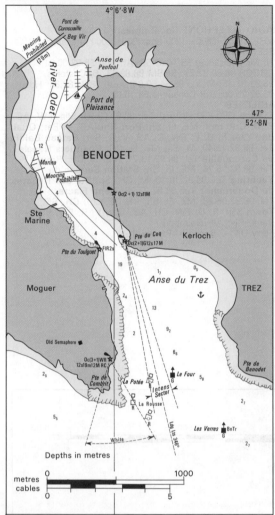

FACILITIES
BENODET **Anse de Penfoul Marina** (200+50 visitors) Tel. 98.57.05.78, AC, FW, Lau, R, CH, ME, V, P, D; **Accastilage 29** Tel. 98.57.20.83, M, ME, El, Sh, CH; **Le Gai Matelot** Tel. 98.57.19.73, C, M, ME, El, Divers; **Town Quay** C (10 ton); **Le Bihan Voiles** Tel. 98.57.18.03, SM; **Navi Composites** Tel. 98.57.02.74, Sh, ME, El; **Town** All facilities, Gaz, PO, Bank, Rly (bus to Quimper), Air (Quimper).
SAINTE MARINE **Ste Marine Marina** (250+40 visitors) has few facilities. **Village** V, R, Bár.

AUDIERNE 10-15-23
Finistere

CHARTS
Admiralty 3640; SHOM 6594, 6377, 7147 CA; Imray C36, C37; ECM 541

TIDES
+0440 Dover; ML 3.1; Duration 0605; Zone −0100
Standard Port BREST (→)

Times				Height (metres)			
HW		LW		MHWS	MHWN	MLWN	MLWS
0000	0600	0000	0600	7.5	5.9	3.0	1.4
1200	1800	1200	1800				

Differences AUDIERNE
−0020 −0040 −0040 −0020 −2.3 −1.8 −0.9 −0.5
GUILVINEC
−0010 −0035 −0035 −0020 −2.4 −1.9 −1.0 −0.5
ILE DE SEIN
−0010 −0010 −0010 −0015 −1.1 −0.9 −0.5 −0.3

SHELTER
River is accessible above half tides in all but strong S-SW winds. Breakwater by Ste Evette gives shelter in all but SE winds although there may be a swell there when wind is S or SW. Keep clear of slip area as vedettes enter with much verve. Visitors buoys are white and marked 'Payant-Visiteurs'.

NAVIGATION
Waypoint 47°59'.00N 04°31'.03W, 151°/331° from/to front Ldg Lt, 1.85M. Access is difficult in strong S-SW winds. There are two rock outcrops, Le Sillon de Galets to the W and La Gamelle in the middle of the bay. A good drying anchorage opposite quays (E side of river). Fishermen resent yachts on quays except in NW corner, which is foul.

LIGHTS AND MARKS
Ldg Lts at 331°, Front Oc (2+1) WG 12s 11m 14/9M, Rear FR 44m 9M. Dir Q WRG 43m 12/9M G 000°-005°, W 005°-007°, R 007°-017°. This latter Lt in line with Old Lt Ho gives Ldg line at 006°.

RADIO TELEPHONE
None. Other station Pointe du Raz Ch 16.

TELEPHONE
Hr Mr 98.70.07.91; Aff Mar 98.70.03.33; CROSS 97.55.35.35; SNSM 98.70.10.85; Customs 98.70.70.97; Meteo 98.94.03.43; Auto 98.94.00.57; Hosp 98.70.00.18; Brit Consul 40.63.16.02.

FACILITIES
Pontoons (70) AC, FW; **Poulgoazec** C (15 ton), Slip, P and D (cans); **Marine Service** Tel. 98.70.22.85, ME, El, Sh; **Bosser** Tel. 98.70.10.52, El, Electronics; **Club Nautique de la Baie d'Audierne** Tel. 98.70.21.69.

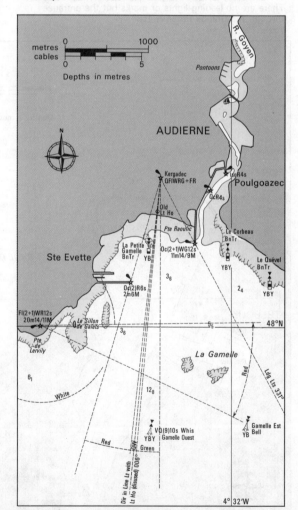

Town V, Gaz, R, Bar. PO; Bank; Rly (bus to Quimper); Air (Quimper).
Ferry UK — Roscoff–Plymouth.

DOUARNENEZ 10-15-24
Finistere

CHARTS
Admiralty 798; SHOM 5316, 6099; Imray C36; ECM 542.
TIDES
+0500 Dover; ML 4.2; Duration 0615; Zone −0100
Standard Port BREST (→)

Times				Height (metres)			
HW		LW		MHWS	MHWN	MLWN	MLWS
0000	0600	0000	0600	7.5	5.9	3.0	1.4
1200	1800	1200	1800				

Differences DOUARNENEZ
−0015 −0010 −0015 −0020 −0.5 −0.5 −0.2 0.0

SHELTER
Very good shelter and open in all weathers and tides. Yachts are not allowed in the Fishing Harbour but can use the marina (Port de Plaisance) in Treboul, Port Rhu, further up river or Port de Rosmeur. Visitors berths are scarce in the marina at Quai de l'Yser. Access day and night at all tides.

NAVIGATION
Waypoint 48°07′.00N 04°20′.30W, 000°/180° from/to Ile Tristan Lt, 0.83M. There are three rocks to the NW of Ile Tristan, Basse Veur, Petite Basse Neuve and Basse Neuve from 3 to 9 ca (555 to 1665m). All are in the R sector of Ile Tristan Lt.

LIGHTS AND MARKS
Approx 5 M to W is Pte du Millier Lt, Oc (2) WRG 6s 34m 15/10M G 080°-087°, W 087°-113°, R 113°-120°, W 120°-129°, G 129°-148°, W 148°-251°, R 251°-258°. There are no leading lights or marks but the entrance through the Grande Passe is 158°.

RADIO TELEPHONE (local times)
VHF Ch 09 (0800 – 1200 and 1330 – 2000 in season).
TELEPHONE
Hr Mr (Plaisance) 98.74.02.56; Aff Mar 98.92.00.91; CROSS 98.89.31.31; SNSM 98.89.63.16; Customs 98.92.01.45; Meteo 98.84.60.64; Auto 98.84.82.83; Hosp 98.92.25.00; Brit Consul 40.63.16.02.
FACILITIES
Marina (380+30 visitors) Tel. 98.92.09.99, Slip, FW, AC, P, D, C (6 ton), ME, El, Sh; **Port Rhu** AB, R, Bar; **Quai de l'Yser** P, D, L, FW, C, AB, V, R, Bar; **Bateau Baloin** Tel. 98.92.10.40, M, Sh, ME, El, CH; **Chantiers Navals de Cornouaille** Tel. 98.92.13.71, ME, El, CH; **Riou** Tel. 98.92.15.76, ME, El, Sh; **Station Service de Yachting** Tel. 98.92.10.10, M, CH, Sh; **Ste des Regates de Douarnenez** Tel. 98.92.27.28; **Henri Fiacre** Tel. 98.92.06.28, ME, El, Sh, BH (12 ton), CH, SM; **Town** Slip, P, D, FW, CH, AB, Gaz, V, R, Bar, PO; Bank; Rly; Air (Quimper).
Ferry UK — Roscoff—Plymouth.

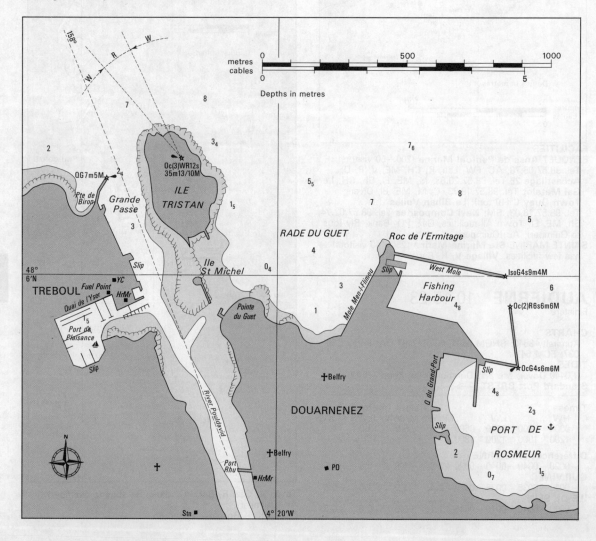

AREA 15 — S Brittany

MORGAT 10-15-25
Finistere

CHARTS
Admiralty 798, 2690; SHOM 6676, 6678, 6099; Stanford 17; Imray C36; ECM 541, 542.

TIDES
+0500 Dover; ML 4.2; Duration No data; Zone −0100
Standard Port BREST (→)

Times				Height (metres)			
HW		LW		MHWS	MHWN	MLWN	MLWS
0000	0600	0000	0600	7.5	5.9	3.0	1.4
1200	1800	1200	1800				

Differences MORGAT
−0010 −0010 −0015 −0015 −0.5 −0.5 −0.2 0.0

SHELTER
The port is exposed to winds from the W and N but the marina is protected by concrete pontoons.

NAVIGATION
Waypoint 48°12'.00N 04°28'.00W, 147°/327° from/to E breakwater head, 1.9M. Few dangers in approaches. There are rocks under the cliffs S of Pte de Morgat, and Les Verres 2M ESE of entrance.

LIGHTS AND MARKS
Morgat Lt Oc (4) WRG 12s 77m 15/10M G sector covers Les Verres.

RADIO TELEPHONE
VHF Ch 25 87 Marina 09 16.

TELEPHONE
Hr Mr 98.27.01.97; Aff Mar 98.27.09.95; CROSS 98.89.31.31; Meteo 98.84.60.64; SNSM 98.27.00 41; Customs 98.27.93.02; Hosp 98.27.02.79; Brit Consul 40.63.16.02.

FACILITIES
Marina (475+50 visitors), AC, FW, C (8 ton), Slip, CH, D, P, ME, Access H24; **Alemany** Tel. 98.27.01.97, Slip, M, L, FW, ME, C (6 ton), AB, V, R, Bar; **Quay by Hr Mr** P, D; **Service-Plaisance** Tel. 98.27.95.90, M, ME, El, Sh, CH, Electronics; **YC du Crozon-Morgat** Tel. 88.27.01.98.
Town (Crozon), V, Gaz, R, Bar. PO; Bank; Rly; Air (Brest or Quimper).
Ferry UK — Roscoff—Plymouth.

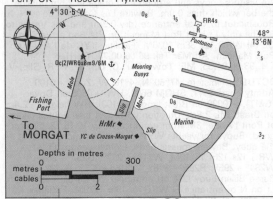

CAMARET 10-15-26
Finistere

CHARTS
Admiralty 2690, 3427; SHOM 6678, 6099; Imray C36; Stanford 17; ECM 542.

TIDES
+0500 Dover; ML 4.1; Duration 0610; Zone −0100
Standard Port BREST (→)

Times				Height (metres)			
HW		LW		MHWS	MHWN	MLWN	MLWS
0000	0600	0000	0600	7.5	5.9	3.0	1.4
1200	1800	1200	1800				

Differences CAMARET
−0015 −0015 −0015 −0020 −0.5 −0.5 −0.2 −0.1

SHELTER
Harbour gives good shelter except from strong E winds, but much of it dries and there is a restricted area: see chartlet. Shelter in both marinas, Plaisance La Pointe or Plaisance Styvel, is always good.

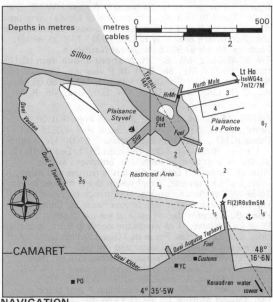

NAVIGATION
Waypoint 48°18'.00N 04°36'.00W, 335°/155° from/to North Mole Lt, 1.2M. Beware rocks W of Pointe du Grand Gouin. Anchoring in harbour forbidden.

LIGHTS AND MARKS
Lt Ho at end of N Mole Iso WG 4s W135°-182°, G182°-027°; Ldg line at 148° with front — top of the old fort, back — Keraudren water tower on hill behind.

RADIO TELEPHONE
None. For Ouessant Traffic see Fig.2(1).

TELEPHONE
Hr Mr 98.27.95.99; Bureau de Port 98.27.93.30; Aff Mar 98.27.93.28; CROSS 98.89.31.31; SNSM 98.27.94.76, Customs 98.27.93.02; Meteo 98.84.60.64; Dr 98.27.91.35; Brit Consul 40.63.16.02.

FACILITIES
Plaisance 'La Pointe (30+120 visitors), Tel. 98.27.95.99, FW, AC, C (8 ton), D, Access H24; **Plaisance 'Styvel'** (180+30 visitors), FW, AC, Access HW∓3; **Service Plaisance** Tel. 98.27.95.90, M, L, FW, ME, El, Sh, CH, AB; **Boennec Frères** Tel. 98.27.90.61, ME; **Le Roy** Tel. 98.27.94.32, ME; **Quai Tephany** P, D, L, C (5 ton); **Voileire Lastennet** 98.27.92.32, SM; **Hugot** Tel. 98.27.90.88, SHOM;
Town V, Gaz, R, Bar. PO; Bank; Rly (Brest); Air (Brest or Quimper).
Ferry UK — Roscoff—Plymouth.

MINOR HARBOURS AND ANCHORAGES 10.15.27

PORNICHET, Loire-Atlantique, 47°15'N, 2°20'W, Zone −0100, Admty chart 2989, SHOM 6797. HW +0500 on Dover (GMT), −0040 sp, +0015 np on Brest (zone −0100); HW height −1.9m on Brest; ML 2.9m; Duration 0530 sp 0645 np. Excellent shelter in artificial harbour. No navigational dangers; available at all tides (up to 3.5m draught). Lt on S breakwater head Iso WG 4s 11m 12/9M, G084°−081°, W081°−084°. S breakwater FIG 2s. N breakwater FlR 2s. VHF Ch 09. Facilities are extensive; Hr Mr Tel. 40.47.23.71. Customs Tel. 40.61.32.04. Meteo Tel. 40.90.08.80; Aff Mar Tel. 40.60.56.13; CROSS Tel. 97.55.35.35. **Marina** (1100+150 visitors), Tel. 40.61.03.20, AC, Slip, FW, P, D, BH (24 ton), V, R, Bar; **Agence Maritime de Pornichet** Tel. 40.61.20.37, CH, El, D, FW, ME, P, Sh. **Town** Bar, Dr, PO, R, Rly, V.

PIRIAC, Loire-Atlantique, 47°23'N, 2°32'W, Zone −0100, Admty chart 2353, SHOM 5482. HW +0505 on Dover (GMT), −0015 on Brest (zone −0100); HW height −1.9m on Brest; ML 3.1m; Duration 0605. A small resort and fishing village with good small harbour. Available HW∓3, 30 berths for visitors, ask at Cercle Nautique Tel. 40.23.52.32 (July-Aug only). Lt Oc(2) WRG 6s 8m 10/6M, vis R066°−185°, W185°−201°, G201°−224°. Facilities: D, P on quay. **Piriac Marine Service** Tel. 40.23.50.86, CH, El, ME, Sh; **Town** Bar, Dr, R, V.

PORT D'ARZAL, Morbihan, 47°30' N, 2°24' W, Zone −0100, Admty chart 2353, SHOM 2381, 5482. Tidal figures below dam as for St Nazaire. Access HW∓3. See 10.15.12 and 10.15.9. Keep to outside of bends going up river. Lock on N side of dam opens as boats approach (no charges); 0700−2000 May to Oct, 0800−1900 in winter. Facilities: Hr Mr 97.90.05.06; **Arzal Nautique** Tel. 97.45.03.52, ME, El, Sh; **N shore above dam**. C (15 ton) D, El, FW, ME, Sh. **Port de Plaisance, Camoel** (on S shore above dam) (660+50 visitors), Tel. 99.90.05.86 AC, C (15 ton), CH, P, D, El, FW, Gas, Gaz, ME, R, Sh, V.

PENERF, Morbihan, 47°31' N, 2°37' W, Zone −0100, Admty chart 2353, SHOM 5418, 5420. HW +0515 on Dover (GMT), −0010 on Brest (zone −0100); HW height −1.9m on Brest; ML 3.0m; Duration 0610. See 10.15.12. Shelter very good in all weathers. Entrances are all difficult. Passe de l'Est has 4m but is narrow and rocky; it leads E of La Traverse, a reef which dries, marked by Bn 1¼M ENE of Pte de Penvins, and thence between Le Pignon and Petite Bayonelle to join Passe du Centre. Passe du Centre is the widest and easiest, leading close W of La Traverse, N of which the depth is 0.1m, and then close E of Le Pignon. Passe de l'Ouest, which leads W of Plateau des Passes, is shallow and not well marked. Beware oyster beds in the river. Anchor in 3.5m near quay. Light, Le Pignon, in the river entrance Fl (3) WR 12s 6m 9/6M on R Tr: R028°−167°, W167°−175°, R175°−349°, W349°−028°. Facilities: P and D (on quay), Slip. **Penerf Marine** Tel. 97.41.13.86, Ch, El, ME, Sh; **Town** Bar, Dr, R, V.

ÎLE DE HOUAT, 47°24' N, 2°57' W, Zone −0100, Admty chart 2353, SHOM 135. HW +0505 on Dover (GMT), −0017 on Brest (zone −0100); HW height −1.9m on Brest; ML 2.9m; Duration 0605. See 10.15.11 (Île de Höedic). Good shelter at Port St Gildas, towards E end of the N coast, protected by wave-break. Beware rock in the middle (1m) and rocks protruding 6m from breakwater. Anchor inside breakwater in approx 2m. S part of harbour dries. Lt on N mole Oc (2) WG 6s 8m 8/5M from W Tr with G top. W168°−198°, G198°−210°, W210°−240°, G240°−168°. Facilities very limited. Ferries to Quiberon (Port Maria) and Île de Höedic.

PORT MARIA, Morbihan, 47°28' N, 3°08' W, Zone −0100, Admty chart 2353, SHOM 5420, 5352, 135. HW +0505 on Dover (GMT), −0010 on Brest (zone −0100); HW height −2.1m on Brest; ML 2.9m; Duration 0615. Shelter good in all winds. Access dangerous in strong winds from SE to SW. Half harbour dries. E mole reserved for ferries. Anchor in S of harbour in approx 2m. Beware rocks at base of S mole. Ldg Lts 006°, both QG intens 005°−008°. E mole head Iso G 4s 9m 6M. S end of wave-break Oc (2) R 6s 9m 8M. Main Lt Ho Q WRG 28m 15/10M, W246°−252°, W291°−297°, G297°−340°, W340°−017°, R017°−051°, W051°−081°, G081°−098°, W098°−143°. Facilities: Hr Mr and Aff Mar Tel. 97.50.08.71. Customs Tel. 97.50.14.81. C (6 ton), FW at E quay. **Monvoisin** Tel. 97.50.09.67 SHOM; **Ateliers Normand** Tel. 97.50.26.17 El, ME, Sh; **Town** (and at Quiberon 0.5M) Bar, R, V.

SAUZON, Belle Ile, 47°22' N, 3°13' W, Zone −0100, Admty Chart 2353, SHOM 135, ECM 545; HW +0450 on Dover (GMT), −0020 on Brest (zone −0100); HW height −2.0m on Brest; ML 3.0m; Duration 0615. Small attractive harbour which dries, 4M N of Le Palais. See 10.15.18. Good shelter except in E winds. Moorings available (17 deep water buoys) just inside E and W moles, or go into inner harbour and dry out. Main Lt QG 9m 5M. NW Jetée FlG 4s. SE Jetée FlR 4s. Facilities: FW on quay; P and D at YC; **Société Nautique de Sauzon** Tel. 97.31.84.56. V, R, Bar in village.

RIVER ÉTEL, Morbihan, 47°39' N, 3°12' W, Zone −0100, Admty chart 2352, SHOM 5560. HW +0505 on Dover (GMT), −0020 on Brest (zone −0100); ML 2.8m; Duration 0617. Shelter excellent but approach should only be made on flood tide, by day in good weather at about HW−1½. Bar dries. There is a small marina inside the town quay. Ldg line at 042° is marked by two water towers. Hoist ensign to mast-head or call Mat Fenoux Ch 16; 12. Semaphore arms will be rotated, one turn to acknowledge and then as follows: Arrow vertical = remain on course. Arrow inclined = alter course in direction indicated. Arrow horizontal with ball over = no entry. R flag = insufficient depth at bar. Light W side of ent Oc (2) WRG 6s 13m 9/6M W022°−064°, R064°−123°, W123°−330°, G330°−022°. VHF Ch 16; 13. In Etel town, anchor S of LB house SW of quay, above town (beware strong currents) or go alongside quay (end of quay reserved for ferries) or alongside in yacht harbour beyond quay. Facilities: Hr Mr Tel. 97.55.46.62; Aff Mar Tel. 97.55.90.32; Semaphore tower Tel. 97.55.35.59; Sig.Stn. Tel. 97.52.35.29; Auto Tel. 97.64.54.43; **Kenkiz Marine** Tel. 97.55.34.58. CH, El, ME, Sh; **Town** Bar, Dr, FW, R, V.

PORT TUDY, Ile de Groix, 47°39' N, 3°27' W, Zone −0100, Admty chart 2352, SHOM 5479, 5912. HW +0505 on Dover (GMT), −0035 on Brest (zone −0100); HW height −2.2m on Brest; ML 3.1m; Duration 0610. See 10.15.19. Very good shelter in marina in inner harbour, but some swell in NE winds. Access HW∓2 (0600-2200). Beware large unlit mooring buoys NE of entrance and rocks both sides of approach. Ldg line at 220°, spire of St Tudy church in line with W end of N jetty. Lights; E mole head Fl (2) R 6s 11m 6M vis 112°−226°. N mole head Iso G 4s 12m 6M. Facilities Limited. Hr Mr Tel. 98.05.80.90; Customs Tel. 97.05.80.93; **Marina** (150), FW, P, D, ME, El, Sh; **F. Mugger** Tel. 97.05.81.86, ME; **Coopérative Maritime** Tel. 97.05.80.03, ME, CH; **Town** V, R, Bar, ferry to Lorient.

DOËLAN, Finistere, 47°46' N, 3°36' W, Zone −0100, Admty chart 2352, SHOM 5479. HW +0450 on Dover (GMT), −0035 on Brest (zone −0100); HW height −2.2m on Brest; ML 3.1m; Duration 0607. Shelter is good except in bad onshore weather. Harbour dries. Anchorage in outer part of creek or just inside breakwater. Ldg Lts 014°, front Oc (2 + 1) WG 12s 20m 13/10M, W shore−305°, G305°−314°, W314°−shore. Rear QR 27m 9M. Facilities: Aff Mar Tel. 98.96.62.38. **Doëlan-Nautic** Tel. 98.71.50.76 CH, El, Electronics, ME, Sh; **Town** Bar, D, Dr, FW, P, PO, R, V.

PONT-AVEN, Finistere, 47°49' N, 3°44' W, Zone −0100, Admty chart 2352, SHOM 5479. HW +0450 on Dover (GMT), −0030 on Brest (zone −0100). HW height −1.9m on Brest; ML 2.8m; Duration 0610. Very good shelter at Pont-Aven. Moorings in 2.5m; alongside the quay dries 2.5m. Pont-Aven is 3.6M from Port Manech where there are visitors buoys and a good anchorage in 2.5m outside the bar which dries 0.9m. Beware Les Cochons de Rousbicout to W of entrance (dry 0.3m) and Les Verres to E. The only Lt is at Port Manech on Pte Beg ar Vechen Oc (4) WRG 12s 38m 10/6M. Facilities: **YC de l'Aven** at Port Manech. Aff Mar Tel. 98.06.00.73; **Derrien** Tel. 98.06.01.35. ME; **Town** Bar, C, D, FW, P, R, Slip, V.

LOCTUDY, Finistere, 47°50' N, 4°10' W, Zone −0100, Admty chart 3641, SHOM 6649. HW +0505 on Dover (GMT), −0020 on Brest (zone −0100). HW height −2.4m on Brest; ML 2.8m; Duration 0615. (All tidal figures taken at Pont l'Abbé, 3M from entrance). See 10.15.22. Shelter good. Buoys for yachts W of Ile Tudy or anchor NW of the quay. Beware mussel beds. Lights: Langoz Fl (4) WRG 12s 12m 15/10M, W115°−257°, G257°−284°, W284°−295°, R295°−318°, W318°−328°, R328°−025°. Basse Bilien buoy (2M off entrance) VQ (3) 5s. Les Perdrix Lt on N of entrance Fl WRG 4s 15m 12/9M. Facilities: Aff Mar Tel. 98.87.41.79; **wet basin** C (10 ton), D, FW, P; **Club Nautique Loctudy** Tel. 98.87.42.84; **SO.ME.CO**. Tel. 98.87.50.00; CH, El, ME, Sh.

GUILVINEC, Finistere, 47°48' N, 4°17' W, Zone −0100, Admty chart 3640, SHOM 6646, 5368. HW +0447 on Dover (GMT), −0023 on Brest (zone −0100); HW height −2.1m on Brest; ML 3.0m. See 10.15.23. Good shelter; harbour always available (unless over 2.5m draught). Good anchorage off entrance protected by reef but keep clear of fairway. Beware Lost Moan Rks SE of entrance marked by RW Bn Tr, Fl (3) WRG 12s 7m 9/6M. Ldg Lts 053°. Front, Mole de Lechiagat 47°47'.5N/4°17'.0W Q 13m 10M; W pylon; vis 233°-066°. Middle, Rocher Le Faoute's, 210m from front QWG 17m W14M, G11M; R circle on W pylon; vis W006°-293°, G293°-006°; synchronised with front. Rear, 0.58M from front, DirQW 31m 15M; R circle on W pylon; vis 052°-054°; synchronised with front. VHF Ch 12. Secure to buoy marked 'Port de Plaisance' at N end of harbour. Facilities: Hr Mr Tel. 98.58.05.67; Aff Mar Tel. 98.58.13.13; Customs; C, D, FW, P; **Glehen** Tel. 98.58.12.00. El, ME, Sh; **Biguais** Tel. 98.58.27.43. CH. **S.E.A.** Tel. 98.58.23.52. El.

VOLVO PENTA SERVICE

Sales and service centres in area 16
Names and addresses of Volvo Penta dealers in
this area are available from:
FRANCE **Volvo Penta France SA,** BP45, F78130 Les Mureaux Tel (01) 3474 72 01, Telex 695221 F.

Area 16

North Brittany
Brest to Paimpol

VOLVO PENTA

10.16.1	Index	**Page 639**	
10.16.2	Diagram of Radiobeacons, Air Beacons, Lifeboat Stations etc	**640**	
10.16.3	Tidal Stream Charts	**642**	
10.16.4	List of Lights, Fog Signals and Waypoints	**644**	
10.16.5	Passage Information	**646**	
10.16.6	Distance Tables	**647**	
10.16.7	English Channel Waypoints	**See 10.1.7**	
10.16.8	Special notes for French areas	**See 10.14.7**	
10.16.9	Brest, Standard Port, Tidal Curves	**648**	
10.16.10	Le Rade de Brest	**648**	
10.16.11	Le Conquet	**653**	
10.16.12	Baie de Lampaul	**653**	
10.16.13	L'Aberbenoit	**654**	
10.16.14	L'Aberwrac'h	**654**	
10.16.15	Roscoff	**654**	
10.16.16	Morlaix	**655**	
10.16.17	Lannion	**656**	
10.16.18	Perros-Guirec/Ploumanac'h	**657**	
10.16.19	Tréguier	**657**	
10.16.20	Lézardrieux	**658**	
10.16.21	Ile de Bréhat	**658**	
10.16.22	Paimpol	**659**	
10.16.23	Minor Harbours and Anchorages	**660**	

L'Aberildut
Argenton
Portsall
Pontusval
Ile de Batz
Carantec
Primel
Locquemeau
Trébeurden
Tregastel
Port Blanc
Pontrieux

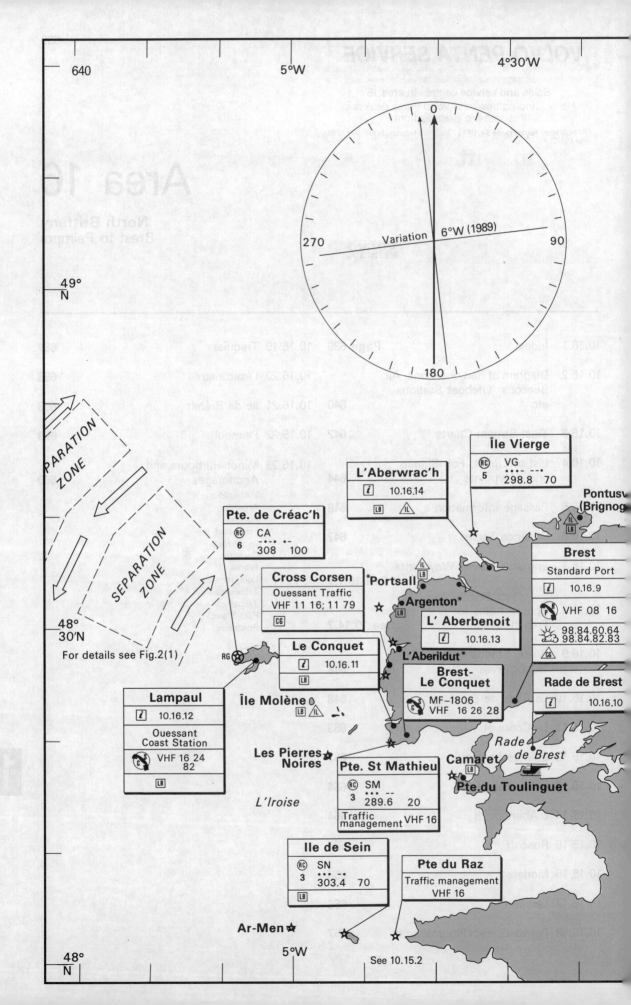

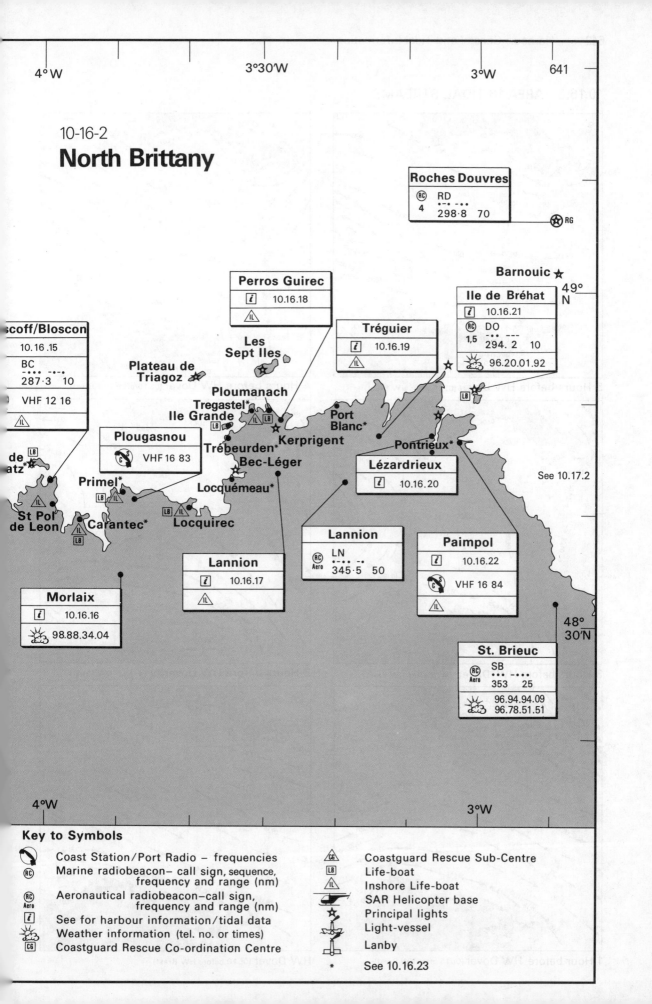

10.16.3 AREA 16 TIDAL STREAMS

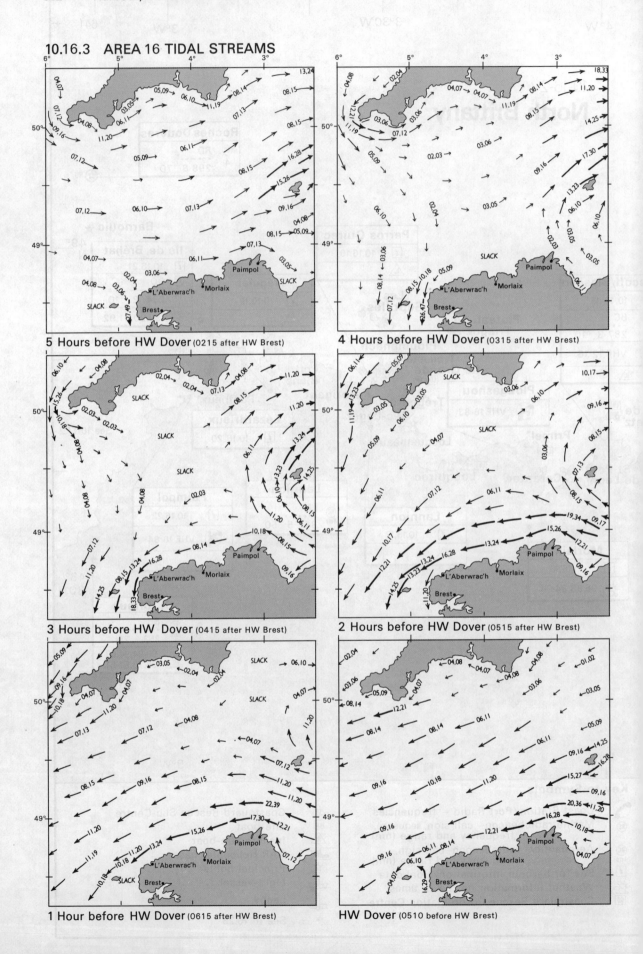

AREA 16 – N Brittany 643

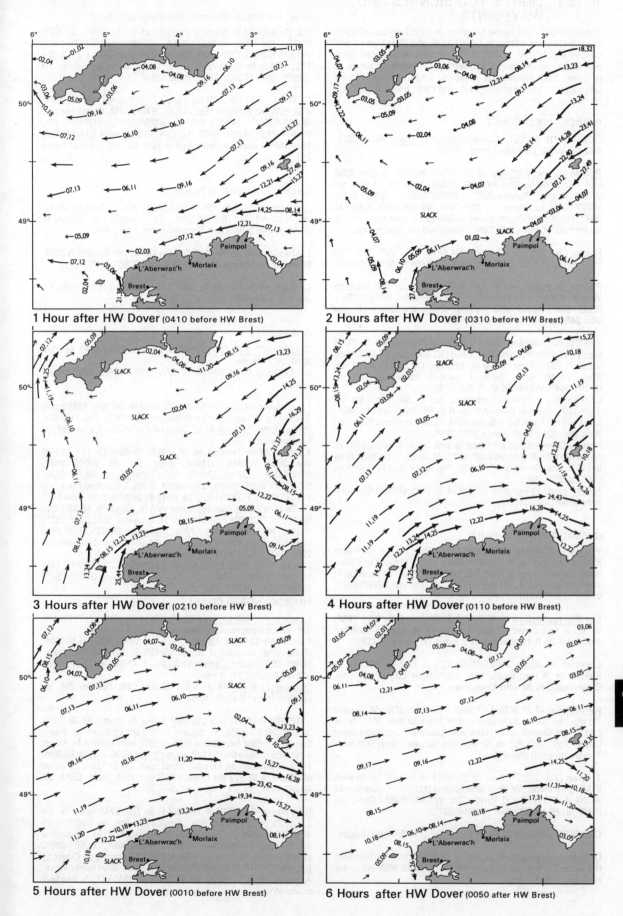

10.16.4 LIGHTS, FOG SIGNALS AND WAYPOINTS

Abbreviations used below are given in 10.0.3. Principal lights are in **bold** print, places in CAPITALS, and light-vessels and Lanbys in *CAPITAL ITALICS*. Unless otherwise stated lights are white. m—elevation in metres; M—nominal range in n. miles. Fog signals are in *italics*. Useful waypoints are underlined — use those on land with care. See 4.2.2.

FRANCE—WEST COAST

GOULET DE BREST. Charles Martel Lt Buoy 48 18.90N/ 4 42.10W QR; port-hand mark; *Whis*.

Pointe du Petit-Minou 48 20.26N/4 36.80W Q 30m **23M** and Fl(2) WR 6s 32m **W19M, R15M**; grey Tr, R top; vis W070°-shore, Rshore-252°, W252°-260°, R260°-307°, W307°-015° (unintens), W015°-065°; *Siren (1) 60s*. Ldg Lts **068°**. Front from Pointe du Petit-Minou Tr, rear from Pointe du Portzic Tr, both QW **23M**, intensified on leading line.

Roche Mengam 48 20.40N/4 34.48W Fl(3) WR 12s 11m W11M, R8M; R Tr, B bands; vis R034°-054°, W054°-034°.

Pointe du Portzic 48 21.56N/4 31.97W Oc(2) WR 12s 56m **W19M, R15M**; grey 8-sided Tr; vis R219°-259°, W259°-338°, R338°-000°, W000°-065°, W070°-219°. Dir Q(6)+LFl 15s 54m **24M**; same Tr; intens 045°-050°.

BREST. Port Militaire, Jetée Sud, head QR 10m 5M; W Tr, R top; vis 094°-048°. Jetée Est, head QG 10m 5M; W Tr, G top; vis 299°-163°. Terre-pleine du Château Oc (3+1) WRG 12s 20m W10M, R7M, G6M; Y over B framework Tr; vis G306°-344°, W344°-351°, R351°-019°.
La Penfeld, E side of entrance, Iso G 3s 11m 7M; vis 316°-180°. Quai de l'Artillerie Iso R 3s 8m 7M; vis 144°-350°. Ldg Lts 314° Front Iso R 5s 8m 10M. Rear, 17m from front, Iso R 5s 15m 12M; both intens 309°-319°.
Port de Commerce. E entrance, S side Oc(2) R 6s 8m 5M; W framework Tr, R top; vis 018°-301°. Mole de l'Est Oc(2) G 6s 8m 7M. Jetée Ouest Iso R 4s 10m 7M. Jetee Sud Fl G 4s 10m 6M; vis 022°-257°.
Moulin Blanc Lt Buoy 48 22.80N/4 26.07W Oc(2)R 6s; port-hand mark.

CHENAL DU FOUR (S PART)

Pte de St Mathieu 48 19.85N/4 46.17W. Ldg Lts 158° with Kermorvan (below) Fl 15s 56m **29M**; W Tr, R top. Dir F 54m **28M**; same Tr; intens 158°; RC. 54m 291° from St Mathieu Q WRG 26m W14M, R11M, G11M; W Tr; vis G086°-107°, W107°-116°, R116°-134°.

Les Pierres Noires 48 18.73N/4 54.80W Fl R 5s 27m **18M**; W Tr, R top; *Siren (2) 60s*. Les Vieux-Moines 48 19.40N/ 4 46.55W Oc R 4s 16m 5M; R 8-sided Tr; vis 280°-133°.

Lochrist 48 20.6N/4 45.9W Oc (2+1) 12s 49m **22M**; W 8-sided Tr, R top; intens 136°-140°; Rear Ldg Lt 138° for Chenal de la Helle with Kermorvan.

Kermorvan 48 21.80N/4 47.30W Fl 5s 20m **22M**; W square Tr; obsc by Pte de St Mathieu when brg less than 341°; front Ldg Lt for Chenal de la Helle with Lochrist. Common front Ldg Lt 158° with Pte de St Mathieu (above), and 007° with Trézien (below); *Reed 60s*.

Trézien Ldg Lt 007°; Rear 48 25.4N/4 46.8W Oc (2) 6s 84m **21M**; grey tower, W on S side; intens 003°-011°. Corsen 48 24.9N/4 47.7W Dir Q WRG 33m W12M, R8M, G8M; vis R008°-012°, W012°-015°, G015°-021°.

CONQUET, Mole 48 21.6N/4 47.1W Oc G 4s 3m 6M; G mast.

La Grande Vinotière, 48 22.00N/4 48.30W Oc R 6s 15m 5M; R 8-sided Tr. Rouget Lt Buoy 48 22.00N/4 48.80W Iso G 4s; stbd-hand mark; *Whis*.

Note. For English Channel Waypoints see 10.1.7

ILE DE MOLÈNE. Molène mole head Dir Lt 191°, Dir Fl(3) WRG 12s 6m W9M, R7M, G7M; vis G183°-190°, W190°-192°, R192°-203°. Chenal des Las, Dir Lt 261°, Dir Fl(2) WRG 6s 9m W9M, R7M, G7M; same structure; vis G252°-259°, W259°-262°, R262°-269°.

CHENAL DE LA HELLE. Les Trois-Pierres 48 24.75N/ 4 56.75W Iso WRG 4s 15m W10M, R7M, G7M; W column; vis G070°-147°, W147°-185°, R185°-191°, G191°-197°, W197°-213°, R213°-070°. Le Faix 48 25.80N/4 53.82W VQ; 16m 9M; N cardinal mark; in line with Le Stiff (below) bears 293°.

CHENAL DU FOUR (N PART)

Les Plâtresses 48 26.35N/4 50.85W Fl RG 4s 17m 6M; W Tr; vis R343°-153°, G153°-333°. La Valbelle Lt Buoy 48 26.55N/4 49.90W Oc R 6s; port-hand mark; *Whis*.

L'Aberildut 48 28.3N/4 45.6W Dir Oc (2) WR 6s 12m **W20M, R16M**; W buildings; vis W081°-085°, R085°-087°.

Le Four 48 31.45N/4 48.23W Fl(5) 15s 28m **20M**; grey Tr; *Siren (3+2) 75s*.

OUESSANT

OUESSANT SW LANBY 48 31.68N/5 49.10W Fl 4s 12m **20M**; RC; Racon.
NE Lt Buoy 48 45.90N/5 11.60W L Fl 10s 9m 12M; *Whis*; Racon.

La Jument 48 25.40N/5 07.95W Fl(3) R 15s 36m **18M**; grey 8-sided Tr, R Top; obsc by Ouessant 199°-241°; *Reed (3) 60s*.
Pierre-Vertes Lt Buoy 48 22.2N/5 04.7W VQ(9) 10s 9m 8M; W cardinal mark; *Whis*.
Kéréon (Men-Tensel) 48 26.30N/5 01.45W Oc (2+1) WR 24s 38m **W18M, R15M**; grey Tr; vis W019°-248°, R248°-019°. *Siren (2+1) 120s*. Men-Korn 48 27.95N/ 5 01.22W VQ(3) WR 5s 21m W8M, R8M; E Cardinal mark; vis W145°-040°, R040°-145°. **Le Stiff** 48 28.60N/5 03.10W Fl(2) R 20s 85m **24M**; two adjoining W Trs. **Créac'h** 48 27.62N/ 5 07.72W Fl(2) 10s 70m **34M**; W Tr, B bands; obsc 247°-255°; Racon, RC; RG; *Horn(2) 120s*. Nividic 48 26.80N/5 08.95W VQ(9) 10s 28m 9M; W 8-sided Tr, R bands; obsc by Ouessant 225°-290°. Helicopter platform.
Port du Stiff. Mole Est, head 48 28.2N/5 03.2W QWRG 11m W10M, R7M, G7M; W Tr, G top; vis G251°-254°, W254°-264°, R264°-267°.

FRANCE—NORTH COAST

Basse de Portsall Lt Buoy 48 36.78N/4 46.05W VQ(9)10s 9m 8M; W cardinal mark; *Whis*.
Portsall 48 33.9N/4 42.3W Oc (3+1) WRG 12s 9m W10M, R7M, G6M; W column, R top; vis G058°-084°, W084°-088°, R088°-058°. Corn-Carhai 48 35.2N/4 43.9W Fl(3) 12s 19m 9M; W 8-sided Tr, B top.
Libenter Lt Buoy 48 37.5N/4 38.4W Q(9) 15s 8m 8M; W cardinal mark; Ra refl; *Whis*.

L'ABERWRAC'H. Ldg Lts 100° Front Ile Vrac'h 48 36.9N/4 34.6W QR 20m 8M; W square Tr, R top, W dwelling. Rear, Lanvaon 1.63M from front Q 55m 10M; grey square Tr, W on west side; intens 090°-110°. Breac'h Ver 48 36.70N/4 35.30W FlG 2.5s 6m 2M; Bn Tr, stbd-hand mark. Dir Lt 128°, La Palue 48 35.9N/4 33.9W Dir Oc(2) WRG 6s W8M, R6M, G6M; vis G126°-127°, W127°-129°, R129°-130°.

Île-Vierge 48 38.38N/4 34.00W Fl 5s 77m **27M**; grey Tr; vis 337°-325°; RC; *Siren 60s*. Pontusval, Pte de Beg-Pol 48 40.7N/4 20.8W Oc(3) WR 12s 16m W10M, R7M; W square Tr, B top; W dwelling; W shore-056°, R056°-096°, W096°-shore. QY Lts on towers 2.4M S. Lizen Ven Ouest Lt Buoy 48 40.57N/4 33.68W VQ(9) 10s; W cardinal mark; *Whis*.
Aman-ar-Ross Lt Buoy 48 41.9N/4 27.0W VQ; N cardinal mark; *Whis*.

PORT DE MOGUÉRIEC. Ldg Lts 162°. Front, jetty head Iso WG 4s 9m W11M, G6M; vis W158°-166°, G166°-158°. Rear, 440m from front FG 7M; W column, G top; vis 142°-182°.

Île de Batz 48 44.78N/4 01.55W Fl(4) 25s 69m **23M**; auxiliary Lt FR 67m 7M; same Tr; vis 024°-059°.

ROSCOFF. Astan Lt Buoy 48 44.95N/3 57.55W VQ(3) 5s; E cardinal mark; *Whis*. Ar-Chaden 48 43.99N/3 58.15W Q(6)+LFl WR 15s 14m W9M, R6M; S cardinal mark; vis R262°-288°, W288°-294°, R294°-326°, W326°-110°. Men-Guen-Bras 48 43.81N/3 57.95W Q WRG 14m W9M, R6M, G6M; N cardinal mark; vis W068°-073°, R073°-197°, W197°-257°, G257°-068°. Basse de Bloscon Lt Buoy 48 43.77N/3 57.48W VQ; N cardinal mark. Ldg Lts 209°. Front, NW mole Oc(2+1) G 12s 7m 7M; W column, G top; vis 078°-318°. **Rear** 430m from front Oc(2+1) 12s 24m **15M**; grey square Tr, W on NE side; vis 062°-242°. Jetty head 48· 44.0N/3 59.0W F Vi 5m 1M. Slip 48 44.3N/4 00.5W VQ(6) + LFl 10s 12m 6M; S cardinal mark.

BLOSCON. Jetty Head 48 43.30N/3 57.65W Fl WG 4s 9m W10M, G7M; W Tr, G top, vis W210°-220°, G220°-210°; RC.

BAIE DE MORLAIX. Ldg Lts 190°. Front, Île Noire 48 40.39N/3 52.58W Oc(2) WRG 6s 15m W11M, R8M, G8M; W square Tr, R top; vis G051°-135°, R135°-211°, W211°-051°; obsc in places. Rear (common with Île Louet 176°) **La Lande** 48 38.2N/3 53.1W Fl 5s 85m **23M**; W square Tr, B top; obsc by Pointe Annelouesten when brg more than 204°. Ldg Lts 176°. Front **Ile Louet** 48 40.5N/3 53.4W Oc(3) WG 12s 17m **W15M**; G10M; W square Tr, B top; vis W305°-244°, G244°-305°, vis 139°-223° from offshore, except when obsc by islands. Rear, La Lande above.

ANSE DE PRIMEL. Ldg Lts 152°. Front FR 35m 6M; W square, R stripe, on framework Tr; vis 134°-168°. Rear, 202m from front FR 56m 6M; W square, red stripe. Jetty head 48 42.82N/3 49.53W Fl G 4s 6m 7M; W column, G top.

LOCQUEMEAU. Ldg Lts 121°. Front FR 21m 6M; W framework Tr, R top; vis 068°-228°. Rear, 484m from front Oc(2+1) R 12s 39m 7M; W gabled house.

Beg-Léguer 48 44.40N/3 32.83W Oc(4) WRG 12s 60m W13M, R10M, G10M; west face of W house, R lantern; vis G007°-084°, W084°-098°, R098°-129°.
Les Triagoz 48 52.35N/3 38.73W Oc(2) WR 6s 31m W13M, R10M; grey square Tr, R lantern; vis W010°-339°, R339°-010°; obsc in places 258°-268° by Les Sept-Îles.
Les Sept-Îles, Île-aux-Moines 48 52.78N/3 29.33W Fl(3) 15s 59m **24M**; grey Tr and dwelling; obsc by Ilot Rouzic and E end of Île Bono 237°-241°, and in Baie de Lannion when brg less than 039°.

Ploumanac'h Méan-Ruz 48 50.32N/3 28.90W Oc WR 4s 26m W13M, R10M; pink square Tr; vis W226°-242°, R242°-226°; obsc by Pte de Trégastel when brg less than 080°, and part obsc by Sept-Îles 156°-207°, and by Île Tomé 264°-278°.

PERROS-GUIREC. Passe de l'Ouest. Kerjean Dir Lt 144°. Dir Oc(2+1) WRG 12s 78m **W15M**, R13M, G13M; W Tr, B top; vis G134°-143°, W143°-144°, R144°-154°. **Passe de l'Est** Ldg Lts 225°. Front **Le Colombier** Oc(4) 12s 28m **18M**; W house; intens 219°-229°. Rear, **Kerprigent** 1.5M from front Q 79m **20M**; W Tr; intens 221°-228°. Jetée Est (Linkin) 48 48.26N/3 26.23W Fl(2) G 6s 4m 7M; W pile, G top. Mole Ouest Fl(2) R 6s 4m 8M; W pile, R top.

Port Blanc, Le Voleur Fl WRG 4s 17m W14M, R11M, G11M; W Tr; vis G140°-148°, W148°-152°, R152°-160°.

TRÉGUIER. La Corne 48 51.40N/3 10.53W Fl(3) WRG 12s 14m W11M, R8M, G8M; W Tr, R base; vis W052°-059°, R059°-173°, G173°-213°, W213°-220°, R220°-052°. Grande Passe Ldg Lts 137°. Front, Port de la Chaîne Oc 4s 12m 12M; W house. Rear **St Antoine** 0.75M from front OcR 4s 34m **15M**; W house, R roof; synchronised with front, intens 134°-140°.

Les Heaux 48 54.53N/3 05.20W Oc(3) WRG 12s 48m **W17M**, R12M, G12M; grey Tr; vis R227°-247°, W247°-270°, G270°-302°, W302°-227°.

LE TRIEUX. Moisie Rk Bn Tr 48 53.85N/3 02.22W (unlit); E cardinal mark. Noguejou Bihan Bn 48 53.43N/3 01.93W (unlit); E cardinal mark. Les Sirlots Buoy 48 52.97N/2 59.60W (unlit); stbd-hand mark; *Whis*. Vieille du Tréou Bn Tr 48 52.02N/3 01.12W (unlit); stbd-hand mark. Rocher Men-Grenn 48 51.3N/3 03.9W Q(9) 15s 7m 8M; W cardinal mark. Ldg Lts 225°. Front **La Croix** 48 50.25N/3 03.25W Oc 4s 15m **19M**; two Trs joined, W on NE side, R tops; intens 215°-235°. Rear **Bodic**, 2.1M from front Q 55m **21M**; W house with G gable; intens 221°-229°. Coatmer Ldg Lts 219°. Front 48 48.30N/3 05.78W F RG 16m R9M, G8M; W gable; vis R200°-250°, G250°-053°. Rear, 660m from front, FR 50m 9M; W gable; vis 197°-242°. Les Perdrix 48 47.77N/3 05.83W Iso WG 4s 5m W6M, G6M; G Tr; vis G165°-197°, W197°-202°, G202°-040°. 3 FBu Lts mark marina pontoons, 0.4M SSW.

ILE DE BRÉHAT. **Rosédo** 48 51.5N/3 00.3W Fl 5s 29m **20M**; W Tr; RC. Le Paon 48 51.95N/2 59.17W F WRG 22m W12M, R9M, G8M; Y square concrete tower; vis W033°-078°, G078°-181°, W181°-196°, R196°-307°, W307°-316°, R316°-348°.
Men-Joliguet 48 50.14N/3 00.23W Iso WRG 4s 6m W13M, R10M, G8M; Y Tr, B band; vis R255°-279°, W279°-283°, G283°-175°. Chenal de Ferlas, Roche Quinonec, Dir Lt 257°, Dir Q WRG 12m W11M, R9M, G9M; vis G254°-257°, W257°-258°, R258°-261°. Embouchure du Trieux, Dir Lt 271°, Dir Fl WRG 2s 16m W11M, R9M, G9M; W column; vis G267°-270°, W270°-272°, R272°-274°.

La Horaine 48 53.55N/2 55.30W Fl(3) 12s 13m 11M; grey 8-sided Tr on B hut.

PAIMPOL. **Pte de Porz-Don** Oc(2) WR 6s 13m **W15M**, R11M; W house; vis W269°-272°, R272°-279°. Ldg Lts 264°. Jetée de Kernoa, front FR 5m 7M; W hut, R top. Rear 360m from front FR 12m 14M; W pylon, R top; intens 261°-267°.

L'Ost Pic 48 46.75N/2 56.45W Oc WR 4s 20m W11M, R8M; 2 W Trs, R tops; vis W105°-116°, R116°-221°, W221°-253°, R253°-291°, W291°-329°; obsc by islets near Bréhat when brg less than 162°.

Barnouic 49 01.70N/2 48.40W VQ(3) 5s 15m 9M; East cardinal mark, B Tr, Y band, W base.

Roches Douvres 49 06.47N/2 48.82W Fl 5s 60m **28M**; pink Tr on dwelling with G roof; RC; RG; *Siren 60s*.

Grand Lejon 48 44.95N/2 39.90W Fl(5) WR 20s 17m **W18M**, R14M; R Tr, W bands; vis R015°-058°, W058°-283°, R283°-350°, W350°-015°.

AREA 16—N Brittany 645

16

10.16.5 PASSAGE INFORMATION

FRENCH TERMS

When cruising in French waters it is essential to know some French words commonly used on charts or in sailing directions: *Anse*: bay. *Arrière port*: inner harbour. *Avant port*: outer harbour. *Baie*: bay. *Balise*: beacon. *Banc*: bank. *Barre*: bar. *Basse*: shoal. *Bassin*: basin. *Canal*: canal, channel. *Cap*: cape. *Chaussée*: bank. *Chenal*: channel. *Clocher*: steeple. *Digue*: breakwater, mole. *Ecluse*: lock. *Est*: east. *Étang*: lagoon, lake. *Falaise*: cliff. *Fleuve*: river. *Golfe*: gulf. *Goulet*: narrows. *Grand*: great. *Île*: island. *Jetée*: jetty. *Maison*: house. *Menhir*: large stone. *Môle*: mole. *Mont*: mountain. *Mouillage*: anchorage. *Moulin*: mill. *Nord*: north. *Ouest*: west. *Passe*: channel. *Pertuis*: strait. *Petit*: small. *Pierre*: stone. *Pointe*: point. *Pont*: bridge. *Port*: harbour, port. *Presquîle*: peninsula. *Quai*: quay. *Rade*: roadstead. *Récif*: reef. *Rivière*: river. *Roche(r)*: rock. *Sable*: sand. *Sud*: south. *Tour*: tower. *Ville*: town.

OUESSANT (USHANT) (chart 2694)

Ouessant is 10M off the W end of coast of France (chart 2694). It is rky island, with dangers extending ½ M to NE, ¾ M to SE, 1½ M to SW and 1M to NW (where Chaussée de Keleren is a dangerous chain of drying and submerged rks running 1M W of Île de Keleren).

Tidal streams are strong close to the island, and in chans between it and mainland. Off Créach Pt (Lt, fog sig, RC) the stream turns NNE at HW Brest − 0550, and SSW at HW Brest + 0045, sp rate 5½ kn.

Apart from Lampaul (10.16.12) the only other anch is Baie du Stiff which gives some shelter in moderate winds between S and NW: beware the rk awash about 3½ ca SSW of the rk with isolated danger Bn in middle of b.

Ouessant is something of a barrier between W and N coasts of France, but in fair weather and with reasonable vis the pilotage in the chans between it and the mainland is not very demanding. They are well buoyed and marked, but the tide runs hard in places, causing overfalls when against wind of over force 5. The route outside Ouessant has little to commend it: unless bound to or from Spain or Portugal it adds much to the distance, it is exposed to sea and swell, and the traffic scheme − see Fig.2(1) − restricts the free passage of yachts. In thick weather this is an unhealthy area, and it is prudent to stay in harbour until the vis improves.

There are three main chans between the island and mainland. The inshore one, Chenal du Four, is most direct and most popular. Chenal de la Helle, partly used for access to Ile Moléne, is not so direct but better in bad weather. Passage du Fromveur, immediately SE of Ouessant, is easiest but can become extremely rough.

The tide runs strongest at S end of Chenal du Four, entered 1M W of Pte St Mathieu, with Kermorvan and Trézien Ldg Lts on at 007°. Here the N-going stream begins at HW Brest − 0550 and reaches 5½ kn at sp. The S-going stream starts at HW Brest + 0015, reaching 4¾ kn. Further N, off Pte de Corsen, the stream is weaker, less than 3 kn at sp. Bound homeward, or along the N coast of France, enter the S end of Chenal du Four at LW Brest: then the strongest tide off Kermorvan will be avoided, since it runs at about HW Brest − 0300, while full use can be made of the NE stream past Ile Vierge.

Compared to larger vessels, there are several routes for a yacht through Chenal du Four. The following can be used day or night, if necessary from buoy to buoy if marks are obscured. The 007° transit above is followed until the Q WRG St Mathieu Lt is abeam (in G sector). Then alter course to make good 325° (Le Faix Tr on with Grande Courleau Bn) for about 1M until La Grande Vinotière and Pte de Corsen Lts are in transit at 009°. At this point steer to make good 005° to pass between La Grande Vinotière and Roche du Rouget buoy which are 1¼ M away. Aim to pass 1 ca W of La Grande Vinotiére, and continue on this course until the transit of Kermorvan and Pte St Mathieu is picked up at 158°. Then alter to port to follow this stern transit. There are buoys marking chan if difficult to follow the marks (or powerful Lts) astern. Do not confuse St Mathieu with Lochrist; the latter is rear Ldg Lt for outer part of Chenal de la Helle. Follow the 158° transit until Le Four is abeam, although in fact open water is reached once past L'Aberildut.

Coming S, the reverse procedure is followed. It may be difficult to identify the Kermorvan/St Mathieu 158° transit, but assuming moderate vis locate Le Four Tr E of the approach, and then Les Plâtresses slightly W of the transit, and only 5M from Kermorvan.

For Chenal de la Helle from S, follow the above route until Trézien is in transit with Corsen, bearing 050°. At this point pick up transit of Le Faix Tr with Stiff (NE corner of Ouessant) at 293°. Follow this transit for 2M until Kermorvan comes on with Lochrist astern at 138°. Take this stern transit out into open water, passing 6 ca NE of Le Faix. From the N, if difficult to see the 138° Ldg marks, the positions of Les Plâtresses to E of Ldg Line, and of Le Faix Tr and La Helle rk (12m elevation, and 7 ca W of Le Faix) to the W of it, will help locate them. At N end of Chenal de la Helle the ENE stream starts at HW Brest − 0520 (sp rate 2¾ kn), and the SW stream at HW Brest − 0045 (sp rate 3¾ kn).

Passage du Fromveur is 1M wide, between Ouessant and the dangers round Kéréon Lt Tr. But the stream runs 7 kn at sp, and with wind against tide the chan can be very dangerous. In good conditions it is safe and convenient with a fair tide.

In fair weather access to Molène is easy from Chenal de la Helle. The key transit is Les Trois Pierres on with North Mill at 215°, picked up ½ M W of La Helle Rk. North Mill may be obsc by Tr of Les Trois Pierres: as the latter is approached alter to stbd to skirt it at 1 ca. When 1 ca W of Les Trois Pierres, alter to 190° to pass close E of Bazou Real Bn Tr, which brings South Mill over the bow and in line with white patch on mole. Do not borrow to port, where there are rks, and anch about 1 ca N of mole.

LE FOUR TO ROCHES DOUVRES (charts 2644, 2668)

Proceeding NE from Le Four, and E towards Roscoff (10.16.15) there are many off-lying dangers, in places 3M offshore. Swell may break on shoals even further to seaward. The tide runs strongly, and in poor vis or bad weather it is a coast to avoid. 1M W of Le Four the E-going stream begins at HW Brest − 0545, sp rate 3½ kn; the W-going stream begins at HW Brest + 0100, sp rate 4 kn. Off Le Libenter, at N side of L'Aberwrac'h entrance the E-going stream starts at HW Brest − 0500, sp rate 3¾ kn, and the W-going stream at HW Brest + 0110. But in good conditions this is an admirable cruising ground with delightful harbours such as Argenton and Portsall (10.16.23), L'Aberbenoit (10.16.13), L'Aberwrac'h (10.16.14), Correjou, Pontusval and L'Aberildut (10.16.23), and Mogueriec.

E of Le Four is an inshore passage leading to Portsall, and thence to L'Aberwrac'h entrance. This is convenient and sheltered, but must only be used by day and in good conditions. French chart 5772 and full pilotage directions, as in *North Brittany Pilot*, are needed. N of L'Aberwrac'h is Ile Vierge Lt Ho, the tallest in the world, and a conspic landmark.

In daylight and above half tide there is a useful short cut to Roscoff through the Canal de l'Ile de Batz, between the island and the mainland at half tide or above. See chart 2745 or French chart 5828. Coming from the W, steer just N of Basse Plate Bn Tr, with Le Loup (a rk with white patch at N end) in transit with St Barbe (a white pyramid beyond Roscoff harb) brg 106°. This transit leads to L'Oignon N cardinal mark, where course is altered to 083° for Pen ar Cleguer at S end of Ile de Batz. Short of this point, when Per Roch N cardinal Bn bears 45° on starboard bow, alter course to leave it ½ ca to starboard. When Per Roch is passed, alter course slightly to starboard to leave An Oan S cardinal Bn to port, and steer for the violet Bn at end of the conspic Roscoff ferry pier. Pass 30m N of this Bn, and at this point alter to 095° which leads to Ar Chaden. For Ile de Batz see 10.16.23.

Approaching from NE, pass E of Basse Astan (marked by buoy) and steer with Men Guen Bras (lit) in transit with Chapelle St Barbe 213° until Ar Chaden is nearly abeam, when course is altered to pass S of Ar Chaden and N of Roc'h Zu N cardinal Bn. Beware rather isolated rk (dries) about 0.8 ca S by W from Ar Chaden. Then pass S of Duslen S cardinal Bn, and N of the ferry pier, and follow the route previously described from W. There are several fair-weather anchs in Canal de l'Ile de Batz, but anch is prohib in the narrows immediately W of the Roscoff ferry pier.

N of Ile de Batz the E-going stream begins at HW Brest −0435, and the W-going stream at HW Brest +0105, sp rates both 3¾ kn.

There are many offlying dangers fringing approaches to Rade de Morlaix (10.16.16 and chart 2745). The more important ones by Grand Chenal are Plateau des Duons (with Tour de Duon about 1½ M N of Ile de Callot), La Vieille (with Bn Tr 1M NE of N end of Île de Callot), Stolvezen (dries) ½M E of La Vieille, and La Fourche and Pierre a l'Évêque (both dry) ½M S of La Vieille. S of this point the 176° Ldg Line passes close E of Bn Trs 1 ca NE and I ca SE of Île Ricard. When Calhic Bn Tr is abeam, alter slightly to port to pass ½ ca E of Le Corbeau Bn Tr, and ½ ca W of Taureau Bn Tr. Pass 1 ca E of Île Louet, and enter the buoyed chan ¼M beyond.

Sailing E from Roscoff, Bloscon or Morlaix pass inside or outside Plateau de la Méloine; the former is obvious choice if bound for Lannion (10.16.17). The radome NE of Trébeurden is conspic. For Trébeurden see 10.16.23.

Plateau des Triagoz has offlying dangers WSW and NE of the Lt, where the sea breaks heavily. Here the stream turns ENE at HW Brest −0325, and WSW at HW Brest +0245, sp rates both 3¾ kn.

Les Sept Îles consist of four main islands and several islets, through which the tide runs strongly. Île aux Moines is lit, and there is an open anch between it and Île de Bono, which is a bird sanctuary.

Between Ploumanac'h and Perros-Guirec (10.16.18) there is a nice anch in Anse Trestraou, sheltered from S and W winds.

A useful inshore passage, avoiding detour round Les Heaux, is Passage de la Gaine between Tréguier River (10.16.19) and Lézardrieux (10.16.20). It should only be used by day, and in good vis. Front Ldg mark is Men Noblance Bn Tr (white, with B stripe). The rear mark (in transit at 242°) is less conspic, and is a white mark with vertical B stripe on a wall just below the skyline, to right of Plougrescant church (conspic). This mark is easier to see from W end of transit, than from Les Heaux. However, there are three Bns which mark the SE side of Basses des Heaux, and should be passed close on required side. If the passage is made at above half tide it presents no problem in fair weather.

Le Ferlas chan (chart 2557) runs S of Île de Brehat, and is useful if entering or leaving Pontrieux R from or to the E. It is well marked and not difficult, but there are unmarked rks in chan almost awash at or near LW, so is best tackled at half tide.

La Horaine (lit) marks NE part of rks extending seawards from Île de Brehat, but there are rky patches ½M N and E, and also up to 2M SE, of the Lt Ho.

NNE of Île de Brehat are Plateau de Barnouic (lit) and Plateau des Roches Douvres (Lt, fog sig, RC), both with drying and submerged rks, to be avoided particularly in poor vis.

10.16.6 DISTANCE TABLE

Approximate distances in nautical miles are by the most direct route while avoiding dangers and allowing for traffic separation schemes etc. Places in *italics* are in adjoining areas.

1 *Le Palais*	**1**																			
2 *Pointe du Raz*	81	**2**																		
3 *Brest*	105	24	**3**																	
4 *Pointe St Mathieu*	99	18	13	**4**																
5 *Le Four*	111	30	25	12	**5**															
6 L'Aberwrac'h	127	46	41	28	15	**6**														
7 Ile Vierge	124	43	38	25	13	6	**7**													
8 Roscoff	152	71	66	53	41	34	28	**8**												
9 Morlaix	163	82	77	64	52	45	39	12	**9**											
10 Les Sept Iles	171	90	85	72	60	53	47	21	26	**10**										
11 Perros Guirec	175	94	89	76	64	57	51	25	30	5	**11**									
12 Tréguier	191	110	105	92	80	73	61	41	46	20	20	**12**								
13 Lézardrieux	199	118	113	100	88	81	75	49	54	28	28	22	**13**							
14 *Roches Douvres*	200	119	114	101	89	82	76	50	55	29	31	26	22	**14**						
15 *St Malo*	233	152	147	134	122	115	109	83	88	62	62	54	49	42	**15**					
16 *Les Hanois*	216	135	130	117	105	98	92	66	71	45	48	45	42	20	56	**16**				
17 *Casquets*	231	150	145	132	120	113	107	89	94	68	71	68	65	43	70	23	**17**			
18 *Bishop Rock*	212	131	127	114	102	109	105	119	131	129	134	150	155	148	187	147	160	**18**		
19 *Eddystone*	214	133	128	115	103	99	93	88	98	84	89	96	109	86	128	74	79	89	**19**	
20 *Portland Bill*	264	183	178	165	153	146	140	125	129	108	115	1130	113	87	118	67	48	162	73	**20**

BREST 10-16-9
Finistere

CHARTS
Admiralty 3427, 3428; SHOM 6427, 5316; ECM 542; Stanford 17; Imray C36

TIDES
Dover +0510; ML 4.4; Duration 0605; Zone −0100

NOTE: Brest is a Standard Port and the tidal predictions for each day of the year are given below.

SHELTER
Excellent shelter in Brest, and in many anchorages round the Rade du Brest (50 sq miles). Yachts should not use the Port du Commerce but the Port de Plaisance in the Anse du Moulin Blanc. Access at any tide day or night.

NAVIGATION
Waypoint 48°18′.30N 04°44′.00W, 248°/068° from/to front Ldg Lt 068° (Pte du Petit Minou), 5.3M. Moulin Blanc port-hand buoy, Oc(2)R 6s, 48°22′.82N 04°26′.06W, 207°/027° from/to MB1 and MB2 channel buoys, 0.55M. There are two forbidden areas — the Port Militaire in Brest and the zone round the Ile Longue. Fast tidal streams run in the Goulet de Brest. Brest is a busy commercial, naval and fishing port.

LIGHTS AND MARKS
Approaches marked by Lts at Pte St Mathieu, Pte du Petit Minou and Pte du Portzic. Marina 2 M E of the Port de Commerce, marked by buoys.

RADIO TELEPHONE
Call: *Brest Port* (at Pointe du Petit Minou) VHF Ch 08 16 (controls approaches to Brest). Moulin Blanc Marina Ch 09 16.

Saint Mathieu Ch 16 is part of Ouessant traffic management system, for vessels proceeding in Chenal de la Helle or Chenal du Four. PC Rade (Tour César) Ch 08 16. All vessels entering keep watch on Ch 16.

TELEPHONE
Hr Mr Brest 98.44.13.44; Moulin Blanc 98.02.20.02; Aff Mar 98.80.62.25; CROSS 98.89.63.16; Customs 98.44.35.20; Meteo 98.84.60.64; Auto 98.84.82.83; Dr 98.44.38.70; Hosp 98.22.33.33; Brit Consul 99.46.26.64.

FACILITIES
Marina (1100+100 visitors), Tel. 98.02.20.02, Slip, P (in cans), D, FW, ME, El, Sh, BH (14 ton), AC, Bar, Gaz, R, Lau, SM, V, C (4 ton), CH, Access H24; **Quai** Slip, P, D, C, CH; **Société des Régates de Brest** (at Moulin Blanc) Tel. 98.02.11.93 R, all facilities; **Occase Mer** Tel. 98.42.03.31, Sh, CH, ME; **Comptoirs Maritime** Tel. 98.02.30.04, CH, ME, El; **Service Plaisance** Tel. 98.02.60.07, ME, El, Sh, CH, Electronics; **Voiles Océan** Tel. 98.80.28.32, SM; **Service Electronique de Navigation** Tel. 98.42.10.35, El, Electronics; **Jouanneau** Tel. 98.44.36.14, SHOM;
Town all facilities, Gaz, PO; Bank; Rly; Air (Brest-Guipavas).
Ferry UK — Roscoff—Plymouth.

RADE DE BREST 10-16-10

CHARTS
Admiralty 3427, 3428; SHOM 6542, 6427, 6678; Imray C36; Stanford 17

The Rade de Brest provides a sheltered cruising ground, useful in bad weather, with many attractive anchorages. Beyond Pont Albert-Louppe (28m) it is possible to explore L'Elorn (or R de Landerneau) for about 6M. There is a good anchorage at Le Passage, ½M above the bridge. The port of Landerneau dries.

L'Aulne (or R de Chateaulin) is a lovely river with steep and wooded banks. Near the mouth of the estuary, on the N shore, is a good anchorage in Anse de l'Auberlach. Other anchorages near Landévennec, below Térénez bridge (27m), and also 1½M above the bridge. At Port Launay, 14M above Landévennec, there is a lock, open HW−2 to shortly after HW, leading into a basin where yachts can lie afloat.

Three small rivers run into the Aulne from the N — R Daoulas, R de l'Hôpital and R du Faou. They mostly dry, but provide sheltered berths for boats which can take the ground.

Along the S side of Rade de Brest are various naval installations, with prohibited anchorage around Ile Longue. There are however anchorages at Le Fret, on SE side of Ile Longue, and also at Roscanvel on E side of the Quelern peninsula.

In most of the bays in Rade de Brest the tidal streams are weak, but in the main rivers they can attain 2 kn or more at springs.

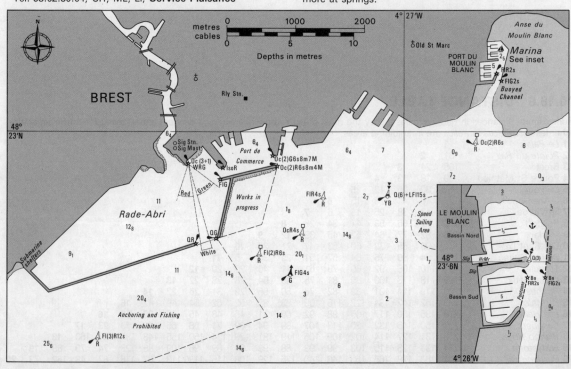

AREA 16—N Brittany 649

RADE DE BREST continued

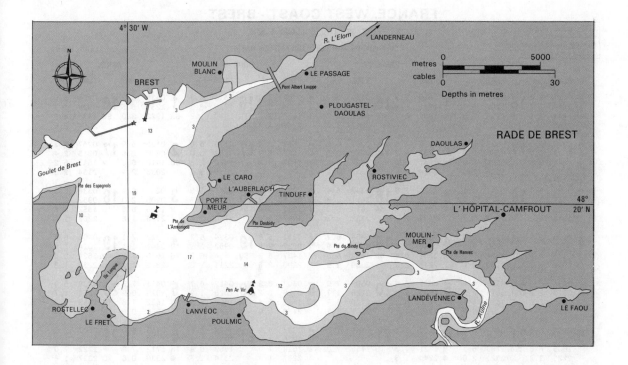

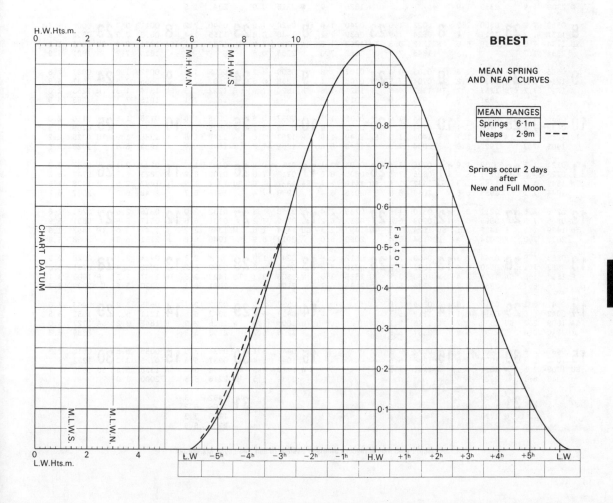

FRANCE, WEST COAST - BREST

LAT 48°23'N   LONG 4°29'W

TIME ZONE −0100 (French Standard Time)
For French Summer Time add ONE hour in non-shaded areas

TIMES AND HEIGHTS OF HIGH AND LOW WATERS

YEAR **1989**

Chart Datum: 4.45 metres below Lallemand System (Mean Sea Level, Marseilles)

JANUARY

	TIME	m		TIME	m
1 SU	0511 1114 1741 2346	3.1 5.8 3.1 5.7	**16** M	0541 1149 1818	2.7 6.1 2.8
2 M	0612 1219 1844	3.1 5.7 3.1	**17** TU	0027 0656 1308 1935	6.0 2.8 6.0 2.9
3 TU	0052 0717 1325 1948	5.7 3.1 5.8 3.0	**18** W	0147 0813 1424 2048	6.1 2.7 6.1 2.8
4 W	0156 0818 1425 2046	5.9 2.9 6.0 2.8	**19** TH	0257 0920 1528 2149	6.3 2.5 6.3 2.5
5 TH	0252 0913 1518 2138	6.2 2.6 6.3 2.5	**20** F	0355 1035 1620 2239	6.6 2.2 6.5 2.2
6 F	0342 1002 1605 2225	6.6 2.3 6.6 2.2	**21** SA ○	0442 1101 1703 2321	6.9 1.9 6.7 2.0
7 SA ●	0428 1047 1650 2309	6.9 1.9 6.8 1.9	**22** SU	0523 1141 1742 2359	7.1 1.8 6.8 1.9
8 SU	0512 1131 1734 2352	7.2 1.7 7.0 1.7	**23** M	0600 1217 1818	7.2 1.7 6.9
9 M	0555 1214 1817	7.4 1.4 7.2	**24** TU	0034 0634 1251 1851	1.8 7.2 1.7 6.9
10 TU	0036 0639 1258 1901	1.5 7.5 1.3 7.2	**25** W	0107 0706 1323 1922	1.8 7.1 1.7 6.8
11 W	0121 0723 1343 1944	1.5 7.3 1.4 7.1	**26** TH	0139 0737 1355 1953	1.9 6.9 1.9 6.6
12 TH	0206 0806 1429 2030	1.6 7.3 1.5 7.0	**27** F	0211 0808 1427 2024	2.1 6.7 2.1 6.4
13 F	0253 0853 1517 2116	1.8 7.1 1.8 6.7	**28** SA	0244 0840 1501 2057	2.3 6.4 2.4 6.2
14 SA ☽	0342 0942 1609 2210	2.1 6.8 2.2 6.4	**29** SU	0319 0916 1539 2137	2.6 6.1 2.7 5.9
15 SU ☾	0437 1040 1708 2313	2.4 6.4 2.5 6.1	**30** M ☾	0401 1000 1627 2227	2.9 5.8 3.1 5.6
			31 TU	0456 1101 1730 2339	3.2 5.5 3.3 5.5

FEBRUARY

	TIME	m		TIME	m
1 W	0610 1221 1852	3.3 5.5 3.3	**16** TH	0137 0807 1421 2047	5.7 3.0 5.7 3.0
2 TH	0106 0735 1347 2014	5.6 3.2 5.6 3.1	**17** F	0257 0920 1527 2147	6.0 2.6 6.0 2.6
3 F	0224 0849 1457 2119	5.9 2.8 6.0 2.7	**18** SA	0352 1011 1614 2232	6.4 2.3 6.3 2.2
4 SA	0326 0946 1551 2211	6.3 2.3 6.4 2.2	**19** SU	0434 1051 1652 2309	6.7 1.9 6.6 1.9
5 SU	0415 1035 1638 2257	6.8 1.8 6.9 1.7	**20** M ○	0509 1126 1726 2341	7.0 1.7 6.9 1.7
6 M	0500 1119 1721 2340	7.3 1.3 7.3 1.3	**21** TU	0541 1157 1756	7.2 1.5 7.0
7 TU	0542 1201 1803	7.7 1.0 7.5	**22** W	0011 0610 1226 1825	1.6 7.2 1.4 7.0
8 W	0022 0624 1242 1845	1.0 7.9 0.8 7.7	**23** TH	0040 0639 1253 1852	1.6 7.2 1.5 7.0
9 TH	0103 0704 1323 1925	0.9 7.9 0.8 7.6	**24** F	0107 0705 1321 1919	1.6 7.1 1.6 6.9
10 F	0144 0744 1405 2004	1.0 7.7 1.1 7.3	**25** SA	0135 0732 1349 1946	1.8 6.9 1.9 6.6
11 SA	0226 0825 1447 2046	1.4 7.3 1.5 6.9	**26** SU	0204 0800 1419 2016	2.1 6.6 2.2 6.4
12 SU ☽	0310 0909 1534 2134	1.9 6.8 2.1 6.4	**27** M	0235 0832 1453 2050	2.4 6.2 2.6 6.0
13 M	0400 1001 1630 2234	2.4 6.2 2.7 5.9	**28** TU ☾	0312 0910 1535 2134	2.8 5.8 3.0 5.7
14 TU	0505 1112 1745 2357	2.9 5.7 3.1 5.6			
15 W	0632 1248 1921	3.1 5.5 3.2			

MARCH

	TIME	m		TIME	m
1 W	0402 1004 1635 2242	3.1 5.5 3.3 5.4	**16** TH	0616 1235 1910	3.3 5.3 3.3
2 TH	0517 1129 1807	3.4 5.3 3.4	**17** F	0128 0757 1410 2035	5.6 3.0 5.5 3.0
3 F	0024 0659 1317 1947	5.4 3.3 5.4 3.2	**18** SA	0244 0905 1511 2130	5.9 2.7 5.9 2.6
4 SA	0201 0827 1437 2100	5.8 2.8 5.9 2.6	**19** SU	0334 0952 1554 2211	6.3 2.3 6.3 2.2
5 SU	0307 0928 1533 2153	6.3 2.2 6.4 2.0	**20** M	0412 1028 1628 2244	6.7 1.9 6.6 1.9
6 M	0357 1016 1619 2237	6.9 1.6 7.0 1.4	**21** TU	0444 1059 1659 2314	6.9 1.7 6.9 1.7
7 TU ●	0440 1058 1700 2318	7.5 1.0 7.5 0.9	**22** W	0513 1128 1727 2341	7.1 1.5 7.0 1.5
8 W	0520 1138 1740 2358	7.9 0.6 7.8 0.6	**23** TH	0540 1155 1754	7.2 1.4 7.1
9 TH	0600 1218 1820	8.2 0.5 7.9	**24** F	0008 0607 1222 1820	1.5 7.2 1.5 7.1
10 F	0037 0639 1257 1859	0.6 8.1 0.6 7.8	**25** SA	0035 0634 1249 1848	1.6 7.1 1.6 6.9
11 SA	0117 0718 1337 1938	0.9 7.8 1.0 7.4	**26** SU	0103 0702 1317 1915	1.7 6.9 1.9 6.7
12 SU	0158 0759 1419 2019	1.3 7.2 1.6 6.9	**27** M	0132 0731 1348 1946	2.0 6.6 2.2 6.5
13 M	0242 0841 1506 2105	1.9 6.6 2.3 6.3	**28** TU ☾	0204 0802 1423 2021	2.3 6.2 2.5 6.1
14 TU ☽	0332 0933 1603 2206	2.6 5.9 2.9 5.7	**29** W	0243 0843 1506 2106	2.7 5.9 2.9 5.8
15 W	0440 1048 1724 2339	3.1 5.4 3.3 5.5	**30** TH	0334 0937 1609 2216	3.0 5.5 3.3 5.5
			31 F	0452 1105 1743 2359	3.3 5.3 3.3 5.5

APRIL

	TIME	m		TIME	m
1 SA	0635 1251 1922	3.1 5.5 3.0	**16** SU	0208 0829 1435 2054	5.9 2.7 5.9 2.6
2 SU	0135 0800 1410 2033	5.9 2.6 6.0 2.4	**17** M	0257 0915 1517 2134	6.2 2.4 6.2 2.3
3 M	0239 0900 1511 2125	6.4 2.0 6.6 1.8	**18** TU	0336 0952 1552 2208	6.5 2.1 6.5 2.0
4 TU	0328 0947 1550 2209	7.1 1.4 7.2 1.2	**19** W	0408 1024 1624 2239	6.8 1.9 6.8 1.8
5 W	0411 1029 1632 2250	7.6 0.9 7.6 0.8	**20** TH	0438 1054 1653 2308	6.9 1.7 6.9 1.7
6 TH	0452 1110 1711 2330	8.0 0.6 7.9 0.6	**21** F	0507 1122 1722 2337	7.0 1.7 6.9 1.6
7 F	0532 1150 1753	8.1 0.6 7.9	**22** SA	0536 1151 1751	7.0 1.7 7.0
8 SA	0011 0613 1231 1834	0.7 7.9 0.8 7.7	**23** SU	0006 0606 1221 1821	1.7 6.9 1.8 6.9
9 SU	0052 0655 1314 1915	1.0 7.5 1.3 7.3	**24** M	0037 0637 1253 1853	1.8 6.8 2.0 6.8
10 M	0135 0736 1358 1959	1.5 7.0 1.9 6.8	**25** TU	0110 0709 1328 1928	2.0 6.5 2.2 6.5
11 TU	0222 0823 1447 2048	2.1 6.3 2.5 6.2	**26** W	0147 0747 1408 2007	2.3 6.2 2.5 6.2
12 W ☽	0315 0917 1543 2151	2.6 5.8 3.0 5.8	**27** TH	0231 0831 1457 2058	2.6 5.9 2.8 5.9
13 TH	0424 1032 1707 2320	2.9 5.4 3.1 5.5	**28** F ☾	0328 0931 1603 2209	2.9 5.6 3.1 5.7
14 F	0555 1211 1842	3.0 5.3 3.0	**29** SA	0443 1053 1727 2339	3.0 5.5 3.0 5.8
15 SA	0056 0725 1336 2000	5.6 3.0 5.5 3.0	**30** SU	0610 1222 1850	2.8 5.7 2.7

FRANCE, WEST COAST - BREST

LAT 48°23′N   LONG 4°29′W

TIMES AND HEIGHTS OF HIGH AND LOW WATERS

YEAR 1989

TIME ZONE −0100
(French Standard Time)
For French Summer Time add
ONE hour in non-shaded areas

MAY

Day	Time	m	Day	Time	m
1 M	0100 / 0725 / 1334 / 1957	6.1 / 2.4 / 6.2 / 2.3	16 TU	0205 / 0825 / 1428 / 2046	6.1 / 2.6 / 6.1 / 2.5
2 TU	0203 / 0824 / 1430 / 2050	6.6 / 1.9 / 6.7 / 1.7	17 W	0249 / 0907 / 1508 / 2126	6.3 / 2.4 / 6.3 / 2.3
3 W	0254 / 0914 / 1518 / 2137	7.1 / 1.5 / 7.1 / 1.3	18 TH	0327 / 0944 / 1545 / 2201	6.5 / 2.2 / 6.5 / 2.1
4 TH	0340 / 0959 / 1602 / 2221	7.5 / 1.1 / 7.5 / 1.1	19 F	0402 / 1019 / 1619 / 2236	6.6 / 2.1 / 6.7 / 2.0
5 F	0424 / 1043 / 1646 / ●2305	7.7 / 1.0 / 7.7 / 1.0	20 SA	0436 / 1052 / 1653 / ○2309	6.7 / 2.0 / 6.8 / 1.9
6 SA	0508 / 1127 / 1730 / 2349	7.7 / 1.0 / 7.6 / 1.1	21 SU	0510 / 1126 / 1727 / 2343	6.8 / 1.9 / 6.9 / 1.9
7 SU	0553 / 1212 / 1815	7.5 / 1.3 / 7.4	22 M	0544 / 1201 / 1802	6.8 / 2.0 / 6.9
8 M	0034 / 0638 / 1257 / 1901	1.4 / 7.2 / 1.6 / 7.1	23 TU	0019 / 0620 / 1238 / 1839	1.9 / 6.7 / 2.1 / 6.8
9 TU	0120 / 0723 / 1344 / 1946	1.7 / 6.7 / 2.1 / 6.7	24 W	0057 / 0659 / 1318 / 1919	2.0 / 6.5 / 2.2 / 6.7
10 W	0209 / 0811 / 1435 / 2036	2.2 / 6.3 / 2.5 / 6.3	25 TH	0139 / 0741 / 1403 / 2004	2.2 / 6.3 / 2.4 / 6.5
11 TH	0303 / 0904 / 1532 / 2135	2.6 / 5.9 / 2.9 / 6.0	26 F	0228 / 0829 / 1455 / 2057	2.3 / 6.1 / 2.6 / 6.3
12 F☽	0405 / 1009 / 1640 / 2246	2.9 / 5.6 / 3.1 / 5.8	27 SA	0324 / 0926 / 1556 / 2159	2.5 / 6.0 / 2.7 / 6.2
13 SA	0517 / 1125 / 1755	3.0 / 5.5 / 3.1	28 SU☾	0429 / 1035 / 1704 / 2311	2.5 / 5.9 / 2.6 / 6.2
14 SU	0004 / 0631 / 1239 / 1904	5.7 / 3.0 / 5.6 / 3.0	29 M	0539 / 1146 / 1813	2.5 / 6.1 / 2.5
15 M	0112 / 0734 / 1340 / 2000	5.9 / 2.8 / 5.8 / 2.7	30 TU	0021 / 0646 / 1253 / 1917	6.4 / 2.3 / 6.3 / 2.2
			31 W	0124 / 0747 / 1353 / 2015	6.6 / 2.0 / 6.6 / 1.9

JUNE

Day	Time	m	Day	Time	m
1 TH	0221 / 0842 / 1448 / 2109	6.9 / 1.8 / 6.9 / 1.7	16 F	0244 / 0903 / 1507 / 2126	6.1 / 2.6 / 6.2 / 2.5
2 F	0314 / 0934 / 1539 / 2200	7.1 / 1.6 / 7.1 / 1.5	17 SA	0328 / 0947 / 1549 / 2208	6.3 / 2.4 / 6.5 / 2.3
3 SA	0404 / 1024 / 1629 / ●2249	7.2 / 1.6 / 7.3 / 1.5	18 SU	0410 / 1028 / 1630 / 2248	6.4 / 2.3 / 6.7 / 2.1
4 SU	0453 / 1113 / 1717 / 2337	7.2 / 1.6 / 7.3 / 1.5	19 M	0450 / 1108 / 1710 / ○2328	6.6 / 2.1 / 6.8 / 1.9
5 M	0541 / 1201 / 1804	7.1 / 1.7 / 7.2	20 TU	0530 / 1148 / 1750	6.7 / 2.0 / 6.9
6 TU	0024 / 0628 / 1247 / 1851	1.6 / 6.9 / 1.8 / 7.0	21 W	0008 / 0610 / 1228 / 1831	1.8 / 6.7 / 1.9 / 7.0
7 W	0110 / 0713 / 1333 / 1935	1.8 / 6.6 / 2.1 / 6.8	22 TH	0049 / 0652 / 1311 / 1913	1.8 / 6.7 / 1.9 / 7.0
8 TH	0157 / 0758 / 1420 / 2021	2.0 / 6.4 / 2.3 / 6.5	23 F	0133 / 0735 / 1356 / 1958	1.8 / 6.7 / 2.0 / 6.9
9 F	0244 / 0844 / 1509 / 2108	2.3 / 6.1 / 2.6 / 6.3	24 SA	0220 / 0821 / 1445 / 2045	1.9 / 6.6 / 2.1 / 6.8
10 SA	0334 / 0934 / 1601 / 2201	2.6 / 5.9 / 2.8 / 6.0	25 SU	0311 / 0911 / 1537 / 2138	2.0 / 6.4 / 2.2 / 6.6
11 SU☽	0428 / 1029 / 1657 / 2300	2.8 / 5.7 / 2.9 / 5.9	26 M	0405 / 1006 / 1634 / ☾2237	2.1 / 6.3 / 2.3 / 6.5
12 M	0526 / 1130 / 1756	2.9 / 5.7 / 3.0	27 TU	0504 / 1106 / 1735 / 2341	2.2 / 6.3 / 2.4 / 6.4
13 TU	0002 / 0626 / 1232 / 1855	5.8 / 2.9 / 5.7 / 2.9	28 W	0608 / 1215 / 1841	2.3 / 6.3 / 2.4
14 W	0100 / 0723 / 1329 / 1950	5.9 / 2.9 / 5.8 / 2.8	29 TH	0048 / 0714 / 1323 / 1947	6.4 / 2.4 / 6.3 / 2.3
15 TH	0155 / 0816 / 1420 / 2040	6.0 / 2.7 / 6.0 / 2.6	30 F	0156 / 0820 / 1428 / 2051	6.4 / 2.3 / 6.5 / 2.2

JULY

Day	Time	m	Day	Time	m
1 SA	0259 / 0921 / 1528 / 2150	6.6 / 2.2 / 6.7 / 2.0	16 SU	0301 / 0922 / 1527 / 2147	6.0 / 2.7 / 6.3 / 2.4
2 SU	0356 / 1017 / 1622 / 2243	6.7 / 2.0 / 6.9 / 1.8	17 M	0351 / 1010 / 1613 / 2232	6.3 / 2.4 / 6.6 / 2.1
3 M	0447 / 1107 / 1711 / ●2331	6.8 / 1.9 / 7.1 / 1.7	18 TU	0435 / 1054 / 1656 / ○2315	6.5 / 2.1 / 6.9 / 1.8
4 TU	0535 / 1154 / 1757	6.9 / 1.8 / 7.1	19 W	0517 / 1136 / 1738 / 2356	6.8 / 1.8 / 7.2 / 1.5
5 W	0015 / 0618 / 1237 / 1839	1.6 / 6.9 / 1.8 / 7.1	20 TH	0559 / 1217 / 1819	7.0 / 1.6 / 7.3
6 TH	0057 / 0659 / 1317 / 1918	1.7 / 6.8 / 1.9 / 7.0	21 F	0038 / 0640 / 1258 / 1901	1.3 / 7.1 / 1.5 / 7.4
7 F	0137 / 0737 / 1357 / 1956	1.8 / 6.6 / 2.1 / 6.8	22 SA	0120 / 0721 / 1341 / 1941	1.3 / 7.1 / 1.4 / 7.4
8 SA	0216 / 0815 / 1436 / 2033	2.0 / 6.4 / 2.3 / 6.6	23 SU	0202 / 0802 / 1424 / 2025	1.5 / 7.0 / 1.6 / 7.2
9 SU	0255 / 0853 / 1516 / 2113	2.3 / 6.2 / 2.5 / 6.3	24 M	0247 / 0846 / 1510 / 2109	1.6 / 6.8 / 1.8 / 6.9
10 M	0336 / 0934 / 1558 / 2157	2.5 / 6.0 / 2.7 / 6.0	25 TU	0335 / 0935 / 1601 / 2202	1.9 / 6.5 / 2.2 / 6.5
11 TU	0422 / 1021 / 1647 / 2248	2.7 / 5.8 / 2.9 / 5.8	26 W	0429 / 1032 / 1700 / 2305	2.3 / 6.2 / 2.5 / 6.2
12 W	0514 / 1118 / 1744 / ☽2349	3.0 / 5.6 / 3.1 / 5.6	27 TH	0534 / 1142 / 1812	2.6 / 6.0 / 2.7
13 TH	0616 / 1223 / 1849	3.1 / 5.6 / 3.1	28 F	0022 / 0652 / 1304 / 1932	6.0 / 2.8 / 5.8 / 2.7
14 F	0057 / 0723 / 1331 / 1955	5.6 / 3.1 / 5.7 / 3.0	29 SA	0145 / 0812 / 1423 / 2048	6.0 / 2.7 / 6.2 / 2.5
15 SA	0203 / 0826 / 1433 / 2055	5.8 / 2.9 / 6.0 / 2.7	30 SU	0257 / 0926 / 1528 / 2149	6.2 / 2.4 / 6.5 / 2.2
			31 M	0355 / 1016 / 1620 / 2240	6.4 / 2.2 / 6.8 / 1.9

AUGUST

Day	Time	m	Day	Time	m
1 TU	0443 / 1102 / 1705 / ●2323	6.7 / 1.9 / 7.1 / 1.7	16 W	0419 / 1037 / 1640 / 2258	6.7 / 1.9 / 7.1 / 1.5
2 W	0525 / 1142 / 1744	6.9 / 1.7 / 7.2	17 TH	0500 / 1118 / 1720 / ○2338	7.1 / 1.4 / 7.5 / 1.1
3 TH	0001 / 0602 / 1219 / 1820	1.5 / 7.0 / 1.7 / 7.2	18 F	0540 / 1158 / 1800	7.4 / 1.1 / 7.8
4 F	0036 / 0637 / 1253 / 1853	1.5 / 6.9 / 1.7 / 7.2	19 SA	0018 / 0619 / 1237 / 1839	0.9 / 7.6 / 1.0 / 7.9
5 SA	0109 / 0708 / 1325 / 1924	1.6 / 6.9 / 1.8 / 7.0	20 SU	0057 / 0659 / 1317 / 1918	0.8 / 7.6 / 1.0 / 7.7
6 SU	0141 / 0739 / 1357 / 1955	1.8 / 6.7 / 2.0 / 6.8	21 M	0137 / 0737 / 1358 / 1958	1.0 / 7.4 / 1.3 / 7.4
7 M	0213 / 0811 / 1430 / 2027	2.0 / 6.5 / 2.3 / 6.5	22 TU	0219 / 0819 / 1441 / 2041	1.4 / 7.0 / 1.8 / 6.9
8 TU	0247 / 0844 / 1505 / 2101	2.4 / 6.2 / 2.6 / 6.1	23 W	0305 / 0904 / 1531 / ☾2131	2.0 / 6.6 / 2.3 / 6.4
9 W☽	0324 / 0921 / 1546 / 2144	2.7 / 5.9 / 2.9 / 5.8	24 TH	0400 / 1002 / 1633 / 2238	2.6 / 6.1 / 2.8 / 5.9
10 TH	0410 / 1010 / 1638 / 2241	3.0 / 5.6 / 3.2 / 5.5	25 F	0511 / 1121 / 1756	3.0 / 5.8 / 3.1
11 F	0511 / 1118 / 1750	3.3 / 5.5 / 3.3	26 SA	0011 / 0644 / 1300 / 1932	5.6 / 3.2 / 5.8 / 3.0
12 SA	0002 / 0632 / 1245 / 1916	5.4 / 3.4 / 5.5 / 3.2	27 SU	0147 / 0814 / 1425 / 2050	5.7 / 2.9 / 6.1 / 2.6
13 SU	0129 / 0756 / 1407 / 2031	5.5 / 3.2 / 5.8 / 2.9	28 M	0258 / 0920 / 1526 / 2146	6.1 / 2.6 / 6.5 / 2.2
14 M	0240 / 0902 / 1508 / 2129	5.8 / 2.8 / 6.2 / 2.5	29 TU	0350 / 1008 / 1611 / 2229	6.4 / 2.2 / 6.8 / 1.9
15 TU	0334 / 0953 / 1557 / 2216	6.2 / 2.3 / 6.7 / 2.0	30 W	0431 / 1048 / 1649 / 2305	6.7 / 1.9 / 7.1 / 1.6
			31 TH	0506 / 1122 / 1722 / ●2338	7.0 / 1.7 / 7.3 / 1.5

Chart Datum: 4.45 metres below Lallemand System (Mean Sea Level, Marseilles)

FRANCE, WEST COAST - BREST

LAT 48°23'N LONG 4°29'W

TIMES AND HEIGHTS OF HIGH AND LOW WATERS

YEAR **1989**

TIME ZONE −0100
(French Standard Time)
For French Summer Time add
ONE hour in non-shaded areas

SEPTEMBER

	TIME	m		TIME	m
1 F	0537 1153 1752	7.1 1.6 7.3	**16** SA	0514 1132 1733 2351	7.8 0.8 8.1 0.6
2 SA	0008 0607 1222 1822	1.4 7.1 1.6 7.3	**17** SU	0553 1210 1812	7.8 0.7 8.1
3 SU	0037 0636 1251 1850	1.5 7.0 1.7 7.1	**18** M	0030 0632 1250 1852	0.7 7.9 0.9 7.9
4 M	0105 0704 1319 1917	1.7 6.9 1.9 6.9	**19** TU	0111 0712 1332 1932	1.1 7.5 1.3 7.4
5 TU	0134 0731 1348 1946	2.0 6.7 2.1 6.6	**20** W	0154 0755 1416 2017	1.6 7.0 1.9 6.7
6 W	0204 0801 1420 2018	2.3 6.4 2.5 6.2	**21** TH	0241 0842 1508 2110	2.3 6.5 2.5 6.1
7 TH	0238 0835 1458 2056	2.7 6.0 2.9 5.8	**22** F	0339 0943 1616 2223	2.9 5.9 3.0 5.6
8 F	0320 0919 1547 2149	3.1 5.7 3.2 5.5	**23** SA	0500 1113 1750	3.3 5.7 3.2
9 SA	0421 1026 1702 2315	3.4 5.4 3.4 5.3	**24** SU	0007 0641 1258 1928	5.5 3.3 5.8 3.0
10 SU	0552 1209 1843	3.5 5.4 3.3	**25** M	0142 0807 1417 2039	5.7 3.0 6.1 2.6
11 M	0059 0729 1343 2008	5.4 3.3 5.8 2.9	**26** TU	0245 0905 1509 2128	6.1 2.6 6.5 2.3
12 TU	0217 0840 1446 2107	5.8 2.8 6.3 2.4	**27** W	0330 0948 1549 2206	6.5 2.2 6.8 1.9
13 W	0311 0931 1534 2153	6.4 2.2 6.9 1.7	**28** TH	0406 1023 1623 2238	6.8 1.9 7.1 1.7
14 TH	0355 1013 1615 2233	6.9 1.6 7.4 1.2	**29** F	0438 1054 1653 2308	7.1 1.7 7.2 1.6
15 F	0435 1053 1655 2312	7.4 1.1 7.5 0.8	**30** SA	0507 1122 1721 2336	7.1 1.6 7.3 1.5

OCTOBER

	TIME	m		TIME	m
1 SU	0535 1150 1749	7.2 1.6 7.2	**16** M	0526 1144 1747	8.0 0.8 8.0
2 M	0004 0603 1218 1817	1.6 7.1 1.7 7.1	**17** TU	0005 0608 1227 1830	0.9 7.9 1.1 7.7
3 TU	0032 0631 1246 1845	1.8 7.0 1.9 6.9	**18** W	0049 0652 1312 1914	1.3 7.5 1.5 7.2
4 W	0100 0659 1315 1914	2.0 6.8 2.2 6.6	**19** TH	0135 0737 1400 2001	1.9 7.0 2.1 6.6
5 TH	0131 0730 1347 1946	2.4 6.5 2.5 6.2	**20** F	0226 0829 1455 2057	2.5 6.5 2.6 6.0
6 F	0206 0804 1426 2026	2.7 6.2 2.8 5.9	**21** SA	0328 0932 1605 2212	3.0 6.0 3.0 5.6
7 SA	0249 0850 1517 2119	3.1 5.8 3.2 5.5	**22** SU	0447 1058 1733 2347	3.3 5.7 3.2 5.5
8 SU	0351 0957 1633 2245	3.4 5.5 3.4 5.3	**23** M	0619 1232 1901	3.2 5.8 3.0
9 M	0522 1138 1813	3.5 5.6 3.3	**24** TU	0112 0736 1345 2006	5.7 3.0 6.1 2.7
10 TU	0028 0658 1311 1936	5.5 3.2 5.9 2.8	**25** W	0212 0832 1436 2054	6.1 2.7 6.4 2.4
11 W	0145 0808 1414 2035	6.0 2.6 6.4 2.2	**26** TH	0257 0914 1516 2132	6.4 2.3 6.7 2.2
12 TH	0239 0859 1502 2121	6.6 2.0 7.0 1.6	**27** F	0333 0950 1550 2206	6.7 2.1 6.9 2.0
13 F	0324 0942 1545 2203	7.1 1.4 7.6 1.1	**28** SA	0406 1021 1621 2237	6.9 1.9 7.0 1.9
14 SA	0405 1023 1625 2243	7.6 1.0 8.0 0.8	**29** SU	0436 1051 1651 2306	7.0 1.8 7.1 1.8
15 SU	0445 1103 1705 2324	7.9 0.8 8.1 0.7	**30** M	0505 1121 1720 2335	7.1 1.8 7.1 1.9
31 TU	0535 1150 1750	7.1 1.9 7.0			

NOVEMBER

	TIME	m		TIME	m
1 W	0005 0605 1221 1821	2.0 7.0 2.0 6.8	**16** TH	0035 0639 1259 1904	1.6 7.4 1.7 7.0
2 TH	0037 0637 1253 1853	2.2 6.8 2.2 6.6	**17** F	0124 0728 1350 1953	2.0 7.0 2.1 6.6
3 F	0110 0710 1329 1929	2.4 6.6 2.4 6.3	**18** SA	0216 0819 1444 2046	2.4 6.6 2.5 6.1
4 SA	0149 0749 1411 2011	2.7 6.3 2.7 6.0	**19** SU	0313 0916 1545 2149	2.8 6.2 2.8 5.8
5 SU	0236 0837 1504 2106	3.0 6.0 3.0 5.7	**20** M	0419 1023 1655 2302	3.1 6.0 3.0 5.7
6 M	0337 0941 1614 2221	3.2 5.8 3.1 5.6	**21** TU	0531 1139 1807	3.1 5.9 3.0
7 TU	0455 1105 1737 2348	3.2 5.9 3.0 5.8	**22** W	0016 0641 1248 1912	5.8 3.0 6.0 2.9
8 W	0617 1227 1853	3.0 6.1 2.7	**23** TH	0119 0740 1346 2006	5.9 2.9 6.2 2.7
9 TH	0101 0725 1332 1954	6.2 2.5 6.6 2.2	**24** F	0210 0829 1433 2051	6.2 2.7 6.3 2.6
10 F	0200 0820 1425 2045	6.6 2.0 7.0 1.7	**25** SA	0253 0911 1513 2130	6.4 2.5 6.5 2.4
11 SA	0249 0908 1512 2131	7.1 1.6 7.4 1.4	**26** SU	0332 0949 1550 2206	6.6 2.3 6.7 2.3
12 SU	0335 0954 1557 2216	7.5 1.3 7.7 1.2	**27** M	0407 1024 1624 2241	6.8 2.1 6.8 2.2
13 M	0420 1039 1642 2302	7.8 1.1 7.8 1.1	**28** TU	0441 1058 1658 2314	6.9 2.1 6.8 2.1
14 TU	0505 1125 1728 2348	7.8 1.1 7.7 1.3	**29** W	0515 1131 1732 2348	7.0 2.0 6.8 2.1
15 W	0552 1211 1815	7.7 1.3 7.4	**30** TH	0549 1206 1806	7.0 2.0 6.8

DECEMBER

	TIME	m		TIME	m
1 F	0023 0624 1242 1843	2.2 6.9 2.1 6.6	**16** SA	0115 0718 1338 1940	1.9 7.2 1.8 6.7
2 SA	0101 0702 1321 1922	2.3 6.8 2.2 6.5	**17** SU	0202 0802 1425 2026	2.1 6.9 2.1 6.4
3 SU	0142 0742 1404 2005	2.4 6.6 2.4 6.3	**18** M	0249 0849 1513 2112	2.4 6.6 2.3 6.2
4 M	0228 0829 1454 2056	2.6 6.4 2.6 6.1	**19** TU	0338 0937 1604 2204	2.7 6.3 2.7 5.9
5 TU	0322 0924 1552 2155	2.7 6.3 2.7 6.0	**20** W	0431 1032 1659 2302	2.9 6.0 2.9 5.8
6 W	0424 1028 1658 2304	2.8 6.2 2.7 6.1	**21** TH	0529 1134 1759	3.1 5.9 3.1
7 TH	0532 1139 1805	2.7 6.3 2.5	**22** F	0006 0631 1238 1902	5.7 3.1 5.8 3.1
8 F	0013 0638 1245 1910	6.2 2.5 6.5 2.3	**23** SA	0110 0733 1340 2002	5.8 3.2 5.9 3.0
9 SA	0118 0740 1347 2010	6.5 2.3 6.7 2.1	**24** SU	0208 0829 1434 2054	6.0 2.9 6.0 2.8
10 SU	0216 0838 1444 2106	6.8 2.0 7.0 1.9	**25** M	0258 0918 1521 2140	6.2 2.7 6.2 2.7
11 M	0311 0932 1538 2159	7.1 1.7 7.2 1.7	**26** TU	0343 1001 1603 2221	6.4 2.4 6.4 2.4
12 TU	0404 1025 1629 2250	7.4 1.5 7.3 1.6	**27** W	0423 1041 1642 2300	6.7 2.2 6.6 2.3
13 W	0455 1115 1719 2340	7.5 1.4 7.3 1.6	**28** TH	0501 1119 1720 2337	6.9 2.0 6.7 2.1
14 TH	0544 1204 1808	7.5 1.5 7.2	**29** F	0538 1156 1757	7.0 1.9 6.8
15 F	0028 0632 1252 1855	1.7 7.4 1.6 7.0	**30** SA	0015 0616 1234 1835	2.0 7.1 1.8 6.9
			31 SU	0053 0655 1313 1914	1.9 7.1 1.8 6.8

Chart Datum: 4.45 metres below Lallemand System (Mean Sea Level, Marseilles)

AREA 16—N Brittany 653

LE CONQUET 10-16-11
Finistere

CHARTS
Admiralty 2694, 3345; SHOM 5159, 5287; ECM 540; Stanford 17; Imray C36

TIDES
Dover +0535; ML 4.3; Duration 0600; Zone −0100

Standard Port BREST (←)

Times				Height (metres)			
HW		LW		MHWS	MHWN	MLWN	MLWS
0000	0600	0000	0600	7.5	5.9	3.0	1.4
1200	1800	1200	1800				
Differences LE CONQUET							
0000	0000	+0010	0000	−0.3	−0.3	−0.1	0.0

SHELTER
Good shelter except in strong winds from the W. Anchor inside the mole, Pierre Glissant. Yachts can go further up harbour but it dries. Six visitors buoys on N side of harbour.

NAVIGATION
Waypoint 48°21'.50N 04°48'.50W, 263°/083° from/to Mole Pierre Glissante Lt, 1.0M. Coming from the NW, beware the Grande Vinotière rocks. Also note strong cross streams in the Chenal du Four.

LIGHTS AND MARKS
Leading line — Lt on end of mole, Pierre Glissant, Oc G 4s in line with spire of Le Conquet church at 095°. Tr beacon and end of Mole St Christophe in line at 079°.

RADIO TELEPHONE
St Mathieu Sig Stn (call: *St Mathieu*) VHF Ch 16 for Chenal du Four and Chenal de la Helle (see 10.16.12).

TELEPHONE
Hr Mr 98.89.00.05; Aff Mar 98.89.00.05; Meteo 98.84.60.64; CROSS 98.89.31.31; SNSM 98.89.02.07; Dr 98.89.01.86; Brit Consul 99.46.26.64.

FACILITIES
Bay Slip, M, L; **Garage Tanniou** Tel. 98.89.00.29, ME, El; **Coopérative Maritime** Tel. 98.89.01.85, CH; **Agmar** Tel. 98.89.12.97, ME; **Town** P, D, FW, ME, El, V, Gaz, R, Bar, PO; Bank; Rly (bus to Brest); Air (Brest). Ferry UK — Roscoff—Plymouth.

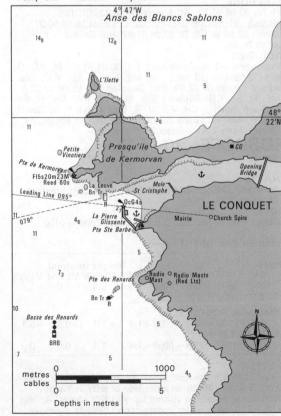

LAMPAUL (ILE D'OUESSANT/ USHANT) 10-16-12
Finistere

CHARTS
Admiralty 2694; SHOM 5567, 5287; ECM 540; Stanford 17; Imray C36

TIDES
Dover +0505; ML 4.4; Duration 0555; Zone −0100

Standard Port BREST (←)

Times				Height (metres)			
HW		LW		MHWS	MHWN	MLWN	MLWS
0000	0600	0000	0600	7.5	5.9	3.0	1.4
1200	1800	1200	1800				
Differences BAIE DE LAMPAUL							
0000	+0005	−0005	−0005	0.0	−0.1	0.0	+0.1
ILE MOLENE							
+0010	+0010	+0015	+0015	0.0	+0.1	−0.1	−0.2

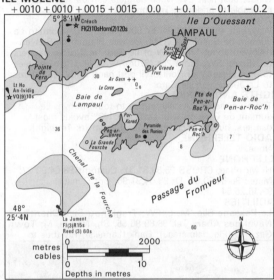

SHELTER
Lampaul Bay is open to strong SW winds but a certain amount of shelter can be found. More secure shelter can be found by entering the harbour to E of piers: entrance 15 m wide.

NAVIGATION
Waypoint 48°26'.30N 05°09'.00W, 250°/070° from/to Le Corce Rk, 1.4M. Bring Le Stiff Lt Ho 055° open N of Le Corce, which may be passed on either side; Bn Trs mark Me-ar-Blanh and Kargroe Rks further inshore. Beware ferries whch use the T shaped new quay. Ferries also use the Chenal de la Fourche but this should be attempted with great caution.

LIGHTS AND MARKS
La Jument Fl (3) R 15s 36m 18M; grey 8-sided Tr, R top; obsc 199°-241°; Reed (3) 60s. An-Ividig (Nividic) VQ(9) 10s 28m 9M; W 8-sided Tr, R bands; obsc 225°-290°; helicopter platform. Creac'h Fl (2) 10s 70m 34M; W Tr, B bands; obsc 247°-255°; Horn (2) 120s; RC; Racon.

RADIO TELEPHONE
The Ushant Control Centre (call: *Ouessant Traffic*) is at Corsen, on mainland combined with CROSS Corsen, Ch 11 16 79 (H24). Traffic information broadcast in English (French on request) on Ch 11 every H+10 and H+40. Weather bcsts in English every 3h from 0150 on Ch 11. Le Stiff Signal and Radar Station Ch 16.

TELEPHONE
Hr Mr 98.89 20 05; Aff Mar 98.89.70.27; CROSS 98.89.31.31; Meteo 98.84.60.64; SNSM 98.89.70.04; Dr 98.89.92.70; Brit Consul 99.46.26.64.

FACILITIES
There are few facilities in Lampaul; L, AB, FW, P and D (in cans), Slip.
Town Gaz, PO; Bank; Rly (ferry to Brest); Air.
Ferry UK — Roscoff—Plymouth.

L'ABERBENOIT 10-16-13
Finistere

CHARTS
Admiralty 1432; SHOM 5772, 964; ECM 539; Stanford 17; Imray C35

TIDES
Dover +0535; ML 4.7; Duration 0555; Zone −0100

Standard Port BREST (←)

Times				Height (metres)			
HW		LW		MHWS	MHWN	MLWN	MLWS
0000	0600	0000	0600	7.5	5.9	3.0	1.4
1200	1800	1200	1800				

Differences L'ABERBENOIT
+0020 +0020 +0035 +0035 +0.6 +0.5 +0.1 −0.2

SHELTER
Good shelter, but do not enter at night or in strong WNW winds; best near LW when dangers can be seen. River navigable to bridge at Treglonou.

NAVIGATION
Waypoint 48°37'.10N 04°39'.00W, 332°/152° from/to La Jument de Guénioc, 1.15M. Beware Plateau de Rusven (buoyed) about 1.7M NW of Ile Garo, and other dangers to NE and SW. From 2ca (370m) W of Ile Guénioc steer to pass close W of Poul Orvil port-hand buoy, and thence to La Jument off Ile Garo.

LIGHTS AND MARKS
There are no Lts. As alternative to route above, La Jument de Garo on with Le Chien 143° avoids most dangers, but leads close to Men Renéat (rk 1m high).

RADIO TELEPHONE
None.

TELEPHONE
Hr Mr none; CROSS 98.89.31.31; Customs 98.04.90.27; Meteo 98.84.60.64; Dr 98.04.91.96; Brit Consul 99.46.26.64.

FACILITIES
Le Passage Slip, M, L, FW; **Treglonou** Slip; **Chantier Naval des Abers** Tel. 98.89.86.55, Sh, ME, El, M; **Town** Gaz, PO (Ploudalméreau); Bank (Lannilis); Rly (bus to

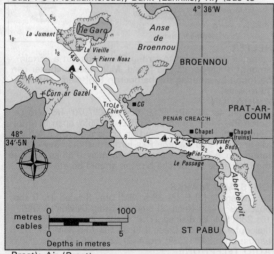

Brest); Air (Brest).
Ferry UK — Roscoff—Plymouth.

L'ABERWRAC'H 10-16-14
Finistere

CHARTS
Admiralty 1432; SHOM 964, 5772; ECM 539, 540; Stanford 17; Imray C35

TIDES
Dover +0540; ML 4.5; Duration 0600; Zone −0100

Standard Port BREST (←)

Times				Height (metres)			
HW		LW		MHWS	MHWN	MLWN	MLWS
0000	0600	0000	0600	7.5	5.9	3.0	1.4
1200	1800	1200	1800				

Differences ABERWRAC'H, FORT CÉZON
+0020 +0030 +0035 +0020 +0.5 +0.2 −0.1 −0.3

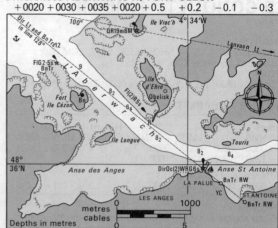

SHELTER
Good except in strong NW winds when it is better to proceed up river to Paluden where there are fore and aft moorings for visitors. At La Palue, pick up a buoy or secure bows to pontoon, stern lines to buoys. Max length 10 m. If anchoring, beware oyster beds.

NAVIGATION
Waypoint Grand Chenal 48°37'.40N 04°38'.40W, 280°/100° from/to front Ldg Lt 100°, 2.6M. W cardinal Lt buoy (Whis), 2.6M WxN from Ile Vrac'h, marks Le Libenter shoal. Also beware Basse Trousquennou to S. There are three channels to inner Ldg line 128°. (1) From N, Chenal de la Malouine. (2) From NW, Chenal de la Pendante (only by day in good weather). (3) From W and best for strangers, Grand Chenal running S of Le Libenter, Grand Pot de Beurre and Petit Pot de Beurre.

LIGHTS AND MARKS
For the Grand Chenal the first pair of Ldg Lts at 100° are Ile Vrac'h (QR) and Lanvaon (QW, 1.63M to the rear). Then Dir Lt 128° takes one to the jetty at La Palue. There are no lights further up river but there is a jetty at Paluden (to the E, off chartlet).

RADIO TELEPHONE (local times)
VHF Ch 09 16 (0700-2100).

TELEPHONE
Hr Mr 98.04.91.62; Aff Mar 98.04.90.13; CROSS 98.89.31.31; SNSM 98.04.92.03; Customs 98.04.90.27; Meteo 98.84.60.64; Dr 98.04.91.87; Brit Consul 99.46.26.64.

FACILITIES
Pontoons (80, some visitors) Tel. 98.04.91.62, M, AC, D, BH (12 ton), C (13 ton), FW, ME, El, CH, AB; **Y.C. des Abers** Tel. 98.04.92.60, Bar; **Segalen** Tel. 98.04.02.32 ME, El, Sh, CH; **Slipway** Slip, M, D, L, FW, ME, El, Sh, C (3 ton mobile); CH, AB, V, R, Bar; **Town** P, V, Gaz, R, Lau, P, Bar. PO; Bank (Lannilis); Rly (bus to Brest); Air (Brest). Ferry UK — Roscoff—Plymouth.

ROSCOFF 10-16-15
Finistere

CHARTS
Admiralty 2745; SHOM 5828, 5827; ECM 538; Stanford 17; Imray C35

TIDES
Dover −0605; ML 5.2; Duration 0600; Zone −0100

Standard Port BREST (←)

Times				Height (metres)			
HW		LW		MHWS	MHWN	MLWN	MLWS
0000	0600	0000	0600	7.5	5.9	3.0	1.4
1200	1800	1200	1800				

Differences ROSCOFF
+0055 +0105 +0115 +0050 +1.4 +1.1 +0.5 −0.1
ILE DE BATZ
+0045 +0100 +0105 +0050 +1.4 +1.1 +0.5 0.0

SHELTER
Good shelter.in Roscoff (dries) except with strong winds from N and E. Berth on inner side of jetty in Yacht Harbour or secure to visitors buoy in SW corner. Access HW∓3. A sill is under construction (1988) to maintain deep water in the harbour. Entry to ferry harbour at

AREA 16—N Brittany 655

ROSCOFF continued

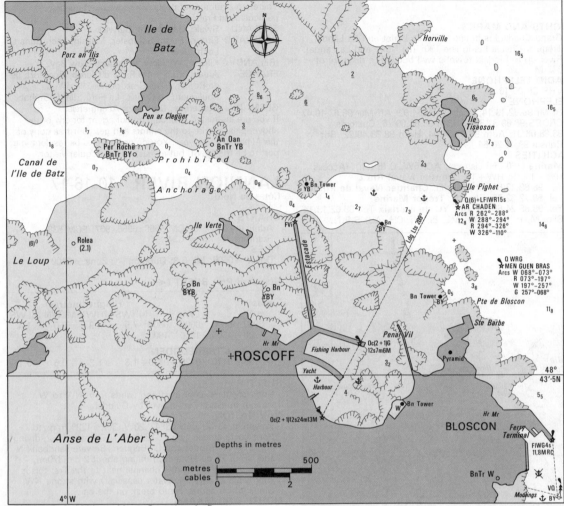

Bloscon is forbidden without Capitaine du Port's permission. Yachts moor or anchor to the S and well clear of ferry jetty.

NAVIGATION
Waypoint 48°46'.00N 03°55'.80W, 033°/213° from/to Men Guen Bras Lt, 2.6M. The entrance to Roscoff is difficult due to the numerous large rocks in the area. Yachtsmen are advised to enter near high tide. The channels are well marked and must be kept to. Bloscon is easier to approach.

LIGHTS AND MARKS
Ile de Batz Lt Ho (conspic) Fl(4) 25s 69m 23M. Off Roscoff are Ar Chaden Q(6)+LFIWR 15s, S cardinal mark, and Men-Guen-Bras Q WRG, N cardinal mark. Ldg Lts 209°. Front on NW mole Oc(2+1)G 12s. Rear Oc(2+1) 12s. Bloscon jetty head FlWG 4s; W sector 210°-220° shows correct approach.

RADIO TELEPHONE (local times)
VHF Ch 09 12 16 (0700-1100, 1300-2200).

TELEPHONE
Hr Mr (Roscoff) 98.69.76.37; Hr Mr (Bloscon) 98.61.27.84; Harbour Office (Port de Plaisance) 98.69.76.37; Aff Mar 98.69.70.15; CROSS 98.89.31.31; SNSM 98.69.74.08; Customs (Roscoff) 98.69.19.67; Customs (Bloscon) 98.61.27.86; Meteo 98.84.60.64; Auto 98.88.34.04; Dr 98.69.71.18; Brit Consul 99.46.26.64.

FACILITIES
Vieux Port (220 + 30 visitors) Tel. 98.69.76.37, FW, AC, ME, BY; **Quai Neuf** Reserved for fishing vessels, C (5 ton); **Harbour** M; **Centre Nautique** Tel. 98.61.20.57, Slip, M, L, FW, CH; **Club Nautique de Roscoff** Tel. 98.69.72.79, Bar; **Comptoirs Maritime** Tel. 98.69.70.47, CH, ME, El; **Le Got** Tel. 98.69.71.86, BY.
Town P, D, ME, El, Sh, CH, Gaz, V, R, Bar. PO; Bank; Rly; Air (Morlaix).
Ferry UK — Plymouth.

MORLAIX 10-16-16
Finistere

CHARTS
Admiralty 2745; SHOM 5827, 5950; ECM 538; Stanford 17; Imray C34, C35

TIDES
Dover −0610; ML 5.2; Duration 0610; Zone −0100

Standard Port BREST (←)

Times				Height (metres)			
HW		LW		MHWS	MHWN	MLWN	MLWS
0000	0600	0000	0600	7.5	5.9	3.0	1.4
1200	1800	1200	1800				

Differences MORLAIX (CHÂTEAU DU TAUREAU)
+0100 +0115 +0115 +0050 +1.5 +1.1 +0.5 −0.1

SHELTER
Good shelter in the bay and in Dourduff although the latter dries. Yachts can proceed up to Morlaix town and enter the lock into the Bassin à Flot where there is a marina and shelter is complete. Lock opens HW−1½ to HW+1 during daytime.

NAVIGATION
Waypoint Grand Chenal 48°43'.00N 03°53'.50W, 356°/176° from/to Ile Louet Lt, 2.5M. Morlaix Bay is divided by the Ile de Callot on E side of which lies Morlaix River. There are three entrances and all have rocky dangers
(1) Grand Chenal E of Ricard Is (shallower but lit 176°).
(2) Chenal Ouest, W of Ricard Is. Big ship channel, not lit.
(3) Chenal de Tréguier (190° and best at night).

MORLAIX continued

LIGHTS AND MARKS
Grand Chenal Lts in line 176° (Ile Louet and La Lande). Treguier Chenal Lts in line 190° (Ile Noire and La Lande). River up to Morlaix town is well buoyed, a distance of 3¾ M.

RADIO TELEPHONE
VHF Ch 09 16.

TELEPHONE
Hr Mr 98.62.13.14; Lock 98.88.54.92; Aff Mar 98.62.10.47; CROSS 98.89.31.31; SNSM 98.88.00.76; Customs 98.88.06.31; Auto 98.88.34.04; Hosp 98.88.40.22; Brit Consul 99.46.26.64.

FACILITIES
Marina (160 + 60 visitors), AC, FW, C (8 ton), (Access HW – 1½ to HW + 1); **Chantier Naval Rio L** Tel. 98.88.13.54, Sh, ME, El; **Chantier Naval de Primel** Tel. 98.72.32.76, Sh, CH; **Trégor Marine** Tel. 98.88.24.68, El, CH; **YC de Morlaix** Tel. 98.62.13.14, Slip, M, P, D, L, FW, ME, El, Sh, CH, AB; **S.E.N.**

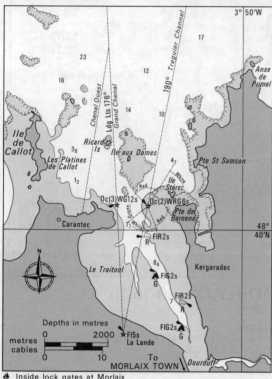

⚓ Inside lock gates at Morlaix

Tel. 98.88.42.42, El, Electronics; **Loisirs Nautique** Tel. 98.88.27.30, ME, CH, El; **Valomer Accastillage** Tel. 98.88.25.85, CH, El, Electronics; **Jegou Ritz** Tel. 98.88.04.15, SHOM. **Comptoir Co-operatif de Pêcheurs** Tel. 98.88.24.68, ME, El, CH, Charts; **Town** P, D, V, Gaz, R, Bar. PO; Bank; Rly; Air.
Ferry UK — Roscoff—Plymouth.

AGENTS NEEDED
There are a number of ports where we need agents, particularly in France.
ENGLAND   Swale, Havengore, Berwick.
SCOTLAND Firth of Forth, Scrabster, Mallaig, Loch Sunart, Loch Aline.
IRELAND   Kilrush, Wicklow, Westport/Clew Bay.
FRANCE   Arcachon, Seudre R, Ile d'Oleron, Rochfort, Ile de Re, St. Giles-Croix-de-Vie, Ile d'Yeu, Pouliguen, Le Croisic, La Forêt, Ile de Bréhat.
GERMANY   Norderney, Dornumer-Accumersiel.
If you are interested in becoming our agent for any of the above, please write to the editors and get your free copy of the Almanac every year. You do not have to be resident in a port to be the agent, but at least a fairly regular visitor.

LANNION RIVER 10-16-17
Côtes du Nord

CHARTS
Admiralty 2644; SHOM 5950, 6056, 967; ECM 538; Stanford 17; Imray C34

TIDES
Dover –0605; ML 5.3; Duration –; Zone –0100

Standard Port BREST (←)

Times				Height (metres)			
HW		LW		MHWS	MHWN	MLWN	MLWS
0000	0600	0000	0600	7.5	5.9	3.0	1.4
1200	1800	1200	1800				

Differences TRÉBEURDEN
+0105 +0110 +0120 +0100 +1.6 +1.3 +0.5 –0.1

SHELTER
Good shelter except in strong winds from NW to W. Anchor in estuary or in river by Guiodel.

NAVIGATION
Waypoint 48°46'.65N 03°42'.20W, 302°/122° from/to front Locquemeau Ldg Lt 122°, 6M. There are non-drying pools off Le Yaudet and Le Beguen. Beware sandbank N and NE of Pt de Dourvin, running about 2 ca (370m), which dries. Pass up the channel near to the Trs; this is impossible at very low water especially with strong NW winds which cause sea to break on the bar.

LIGHTS AND MARKS
The large radome 3.5 M NNE of entrance is conspic landmark. Lt on Pt Beg-Léguer Oc (4) WRG 12s 60m 13/10M; vis W084°-098°, R098°-129°, G007°-084°. Channel to Lannion is narrow and marked by towers and beacons.

RADIO TELEPHONE
None.

TELEPHONE
Hr Mr 96.37.06.52; Aff Mar 96.37.06.52; CROSS 98.89.31.31; SNSM 96.23.52.07; Customs 96.37.45.32; Dr 96.37.42.52; Brit Consul 99.46.26.64.

FACILITIES
Quai de Loguivy Slip, L, FW, C (1 ton mobile), AB; **Coopérative Maritime** Tel. 96.37.03.18, CH; **Lesbleiz** Tel. 96.37.09.12, M, ME, El, Sh, SHOM.
Town M, CH, V, Gaz, R, Bar. PO; Bank; Rly; Air (Morlaix, Lannion). Ferry UK — Roscoff—Plymouth.

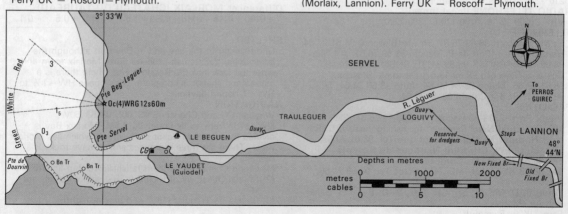

PERROS-GUIREC 10-16-18
(PLOUMANAC'H)
Côtes du Nord

CHARTS
Admiralty 3672, 2668; SHOM 974, 967; ECM 537, 538; Stanford 17; Imray C34

TIDES
Dover −0550; ML 5.1; Duration 0605; Zone −0100

Standard Port ST HELIER (→) (Note 9.1.2 last para)

Times				Height (metres)			
HW		LW		MHWS	MHWN	MLWN	MLWS
0100	0800	0200	0700	11.1	8.1	4.1	1.3
1300	2000	1400	1900				

Differences PLOUMANAC'H
0000 +0005 −0025 0000 −2.1 −1.0 −0.7 −0.2

SHELTER
Anchorage with good shelter but exposed to NE winds. Shelter in the marina is very good. Access HW∓1. Shelter is good in Ploumanac'h and sill is 4.75m lower than in Perros-Guirec; channel is well marked. Entrance and exit difficult in strong NW winds.

NAVIGATION
Waypoint 48°52'.40N 03°20'.00W, 045°/225° from/to front Ldg Lt 225° (Le Colombier), 6.4M. Beware Ile Tomé in the entrance to Anse de Perros. Rocks extend 7 ca (1300m) off the W side and 6 ca (1110m) E of the N side. Gate into marina opens at HW, day and night, for 1 to 4 hours according to the height of tide. May be closed for 3 days at nps. The sill is 7m above CD. Sill is marked at high water by R and W poles. No anchoring allowed in non-tidal harbour.

LIGHTS AND MARKS
From E of Ile Tomé, Ldg Lts 225° — Front Le Colombier Oc (4) 12s 28m 18M, intens 220°-230°. Rear Kerprigent (1.5M from front) Q 79m 20M, intens 221°-228°.
Passe de l'Ouest. Kerjean Dir Lt 144° Oc (2+1) WRG 12s 78m 15/13M vis G134°-143°, W143°-144°, R144°-154°.

RADIO TELEPHONE
VHF Ch 09 16.

TELEPHONE
Hr Mr 96.23.37.82; Gate 96.23.19.03; Aff Mar 96.23.13.78; CROSS 98.89.31.31; SNSM 96.23.20.16; Customs 96.23.18.12; Auto 96.20.01.92; Dr 96.23.20.01; Brit Consul 99.46.26.64.

FACILITIES
Marina (550+50 visitors) Tel. 96.23.19.03, P, D, FW, ME, AC, El, Sh, C (7 ton), CH, V, R, SM, Gas, Gaz, Kos, Lau, Bar; **Loc Armor** Tel. 96.23.05.08, ME, El, Sh, CH; **Ponant Loisirs** Tel. 96.23.18.38, ME, El, Sh, CH, SHOM; **Ship Marine** Tel. 96.91.11.88, D, P, CH, ME, El, Sh; **Town** P, FW, CH, V, Gaz, R, Bar. PO; Bank; Rly (Lannion); Air (Morlaix/Lannion).

Booking for pontoons and moorings — at Town Hall Tel. 96.23.22.64;
PLOUMANAC'H (230+20 visitors) Sill 2.25m above CD — min depth 1.5m. **Quai Bellevue** FW, AC, Slip.
YC Société Nautique de Perros Guirec.
Ferry UK — Roscoff-Plymouth.

TRÉGUIER 10-16-19
Côtes du Nord

CHARTS
Admiralty 3672, 2668; SHOM 973, 972, 967; ECM 537, 538; Stanford 17; Imray C34

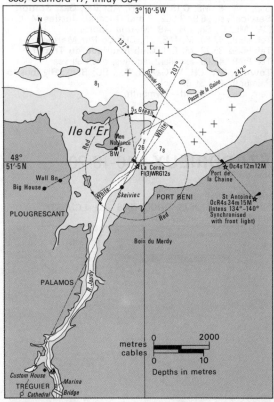

TIDES
Dover −0540; ML 5.5; Duration 0600; Zone −0100

Standard Port ST HELIER (→) (Note 9.1.2 last para)

Times				Height (metres)			
HW		LW		MHWS	MHWN	MLWN	MLWS
0100	0800	0200	0700	11.1	8.1	4.1	1.3
1300	2000	1400	1900				

Differences TRÉGUIER
−0004 +0012 −0025 +0015 −1.3 −0.6 −0.8 −0.2

SHELTER
Good shelter can be found from most winds; SW of La Corne Lt Tr, exposed to NE to NW; off Roche Jaune village; up in Tréguier by the quay; by Palamos.

NAVIGATION
Waypoint 48°55'.00N 03°12'.80W, 317°/137° from/to front Ldg Lt 137°, 5M.
There are three entrance channels:
(1) Grande Passe. Well marked, but strong tidal streams across the channel.
(2) Passe de la Gaine. Navigable with care by day in good visibility.
(3) Passe du Nord-Est. Dangerous with winds from W and NW as sea breaks across the channel.

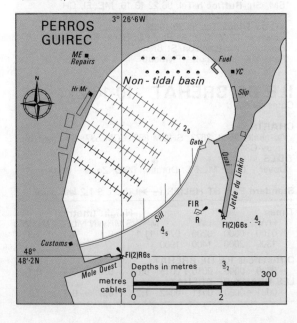

TRÉGUIER continued

LIGHTS AND MARKS
Important marks are Men Noblance Bn Tr (BW) on SE corner of Ile d'Er, Skeiviec Bn Tr (W), and La Corne Lt Tr (WR) Fl(3) WRG 12s. Leading marks:
(1) For Grande Passe. Port de la Chaine and St Antoine Lts at 137°.
(2) For Passe de la Gaine. Men Noblance and Plougrescant wall Bn at 242°.
(3) For approach to La Corne. Tréguier cathedral spire and Skeiviec (from SW of La Jument) at 207°.

RADIO TELEPHONE
None.

TELEPHONE
Hr Mr 96.92.42.37; Aff Mar 96.92.30.38; CROSS 98.89.31.31; Customs 96.92.31.44; Auto 96.20.01.92; Dr 96.92.32.14; Hosp 96.92.30.72; Brit Consul 99.46.26.64.

FACILITIES
EC Sunday; **Marina** (200 + 130 visitors), Tel. 96.92.42.37, Slip, FW, P, ME, C (8 ton), CH, AC, Bar, Gaz; **Station Service** Tel. 96.92.30.52, P and D (cans); **Jetties** M, L, FW, ME, El, Sh, CH, AB; **Marina Sports** Tel. 96.92.47.60, M, ME, Sh, CH; **Co-Per Marine** Tel. 96.92.35.72, M. CH; **Club Nautique du Tregor** Tel. 96.92.42.08, excellent facilities, open all year.
Town P, FW, CH, V, Gaz, R, Bar. PO; Bank; Rly (bus to Paimpol); Air (St Brieuc).
Ferry UK — Roscoff—Plymouth.

LÉZARDRIEUX 10-16-20
Côtes du Nord

CHARTS
Admiralty 2668, 3673; SHOM 2845, 832; ECM 537; Stanford 17, 16; Imray C34

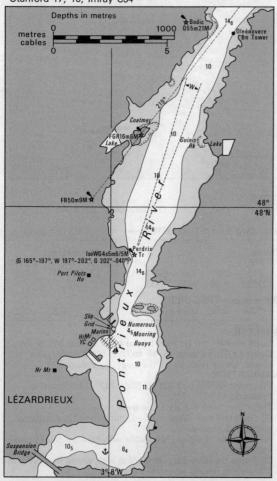

TIDES
Dover −0510; ML 5.6; Duration 0610; Zone −0100

Standard Port ST HELIER (→) (Note 9.1.2 last para)

Times				Height (metres)			
HW		LW		MHWS	MHWN	MLWN	MLWS
0100	0800	0200	0700	11.1	8.1	4.1	1.3
1300	2000	1400	1900				

Differences LÉZARDRIEUX
+0026 +0038 +0015 +0035 −0.9 −0.5 −0.6 −0.3

SHELTER
Very good in all weathers. Harbour and marina can be entered at all tides day or night. Multi-hulls and boats over 12.5 m are not welcome at pontoons, so should moor on visitors buoys in front of marina. Yachts can proceed up river to Pontrieux, about 12 km beyond Lézardrieux, and go into the basin.

NAVIGATION
Waypoint 48°55'.00N 02°56'.20W, 045°/225° from/to front Ldg Lt 225° (La Croix), 6.7M. Outlying dangers are Barnouic Rks and Roches Douvres. There are four entrance channels, Moisie channel, Men du Castric channel (very dangerous at LW), Grand Chenal which is recommended for strangers and the Ferlas Channel from the E. Bridge clearance 18m (above MHWS).

LIGHTS AND MARKS
(1) Moisie channel — Amer du Rosédo W pyramid in line with St Michael's chapel at 160°.
(2) Men du Castric channel — Vieille du Tréou Tr (difficult to see) in line with Arcouest Tr at 174°.
(3) Grand Chenal Lts in line at 225°.
   Front — La Croix Oc 4s 15m 19M; two Trs joined, W on NE side with R tops. Intens 215°-235°.
   Rear — Bodic (2.1 M from front) Q 55m 21M (intens 221°-229°).
(4) Ferlas channel.
   W sector of Joliguet Lt Ho
   W sector of Loguivy
   W sector of Ile à Bois lead to Coatmer line
(5) Coatmer Ldg Lts 219°. Front F RG 16m 9/8M; vis R200°-250°, G250°-053°. Rear, 660m from front, FR 50m 9M.

RADIO TELEPHONE (local times)
VHF Ch 09 (0800-2000 July Aug: 0800-1200 and 1400-1800 rest of year).

TELEPHONE
Hr Mr 96.20.14.22; Aff Mar at Paimpol 96.20.84.30; CROSS 98.89.31.31; Customs at Paimpol 96.20.81.87; Auto 96.20.01.92; Dr 96.20.10.30; Brit Consul 99.46.26.64.

FACILITIES
EC Sunday; **Marina** (404 + 120 visitors), Tel. 96.20.14.22, Slip, P, D, FW, ME, El, C (6 ton), CH, AC, Bar, Gaz, R, SM, Sh; **Ruffloc'h** Tel. 96.22.13.16, ME, El, CH, Electronics; **Trieux Marine** Tel. 96.20.14.71, ME, CH; **YC de Trieux** Tel. 96.20.10.39.
Town P, D, V, Gaz, R, Bar. PO; Bank; Rly (bus to Paimpol); Air (Lannion/St Brieuc).
Ferry UK — Roscoff—Plymouth.

ILE DE BRÉHAT 10-16-21
Côtes du Nord

CHARTS
Admiralty 2668, 3673; SHOM 882; ECM 537; Stanford 16; Imray C34

TIDES
Dover −0525; ML 5.8; Duration 0605; Zone −0100

Standard Port ST HELIER (→) (Note 9.1.2 last para)

Times				Height (metres)			
HW		LW		MHWS	MHWN	MLWN	MLWS
0100	0800	0200	0700	11.1	8.1	4.1	1.3
1300	2000	1400	1900				

Differences ILE DE BRÉHAT
+0020 +0040 +0010 +0015 −0.6 −0.1 −0.4 −0.1
LES HEAUX DE BRÉHAT
+0031 +0030 −0011 +0042 −1.2 −0.5 −0.7 −0.2

AREA 16—N Brittany 659

ILE DE BREHAT continued

SHELTER
Good shelter in Port Clos although harbour dries. Also good shelter in Port de la Corderie at neaps when it is possible to anchor out of the strong tides, and in the Chambre de Bréhat in the SE, each depending on the direction of the wind. There are no alongside berths.

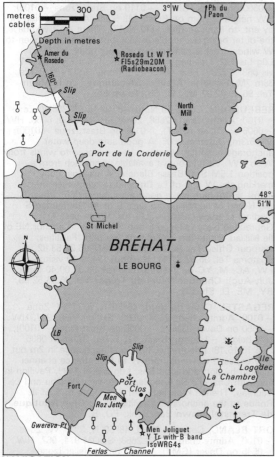

NAVIGATION
Waypoint Ferlas channel 48°49'.45N 02°55'.00W, 098°/278° from/to La Croix Lt, 5.5M. See also 10.16.22. On approach, beware the Barnouic and Roches Douvres (12 and 17 M NNE) and La Horaine, Men March and Ringue Bras closer in. La Chambre and Port Clos best approached from seaward in Ferlas channel (see 10.16.20). There is a prohibited anchorage SW of Port Clos, between Brehat and the mainland.

LIGHTS AND MARKS
There are three principal lights on Ile de Bréhat:
(1) In the N, Phare du Paon F WRG 22m 12/9M; Y Tr; vis W033°-078°, G078°-181°, W181°-196°, R196°-307°, W307°-316°, R316°-348°.
(2) In NW part, Rosédo Fl 5s 29m 20M; W Tr and G gallery; RC.
(3) In the S, on E side of Port Clos entrance, Men-Joliguet Iso WRG 4s 6m 13/10M, Y Tr with B band; vis R255°-279°, W279°-283°, G283°-175°.

RADIO TELEPHONE
None.

TELEPHONE
Hr Mr none; CROSS 98.89.31.31; SNSM 96.20.00.14; Auto 96.20.01.92; Customs 96.20.81.87; Dr 96.20.00.99; Brit Consul 99.46.26.64.

FACILITIES
Club Nautique de Bréhat Tel. 96.20.00.69, M, FW, Bar; **Mervel** Tel. 96.20.01.39, ME; **Harbours** Slip, M, FW; **Village** V, Gaz, Bar. PO; Bank (Paimpol); Rly (ferry to Pointe de l'Arcouest, bus to Paimpol); Air (St Brieuc).
Ferry UK — Roscoff—Plymouth.

PAIMPOL 10-16-22
Côtes du Nord

CHARTS
Admiralty 2668, 3673; SHOM 3670, 832; ECM 537; Stanford 16; Imray C34

TIDES
Dover −0525; ML 5.5; Duration 0600; Zone −0100

Standard Port ST HELIER (→) (Note 9.1.2 last para)

Times				Height (metres)			
HW		LW		MHWS	MHWN	MLWN	MLWS
0100	0800	0200	0700	11.1	8.1	4.1	1.3
1300	2000	1400	1900				

Differences PAIMPOL
+0025 +0038 +0025 +0040 −0.6 −0.3 −0.9 −0.7

SHELTER
There is good shelter in Paimpol from all winds but the whole Anse de Paimpol dries. Lock opens HW−1½ to HW springs, HW−1 to HW neaps, (when sufficient rise of tide, 8m). Visitors' berths are at Pontoon A.

NAVIGATION
Waypoint 48°47'.40N 02°51'.50W, 090°/270° from/to Pte de Porz-Don Lt, 6.6M. There are rocks which dry very close to the leading lines. The Anse de Paimpol is divided by rocks (El Paimpol and El Bras) down the centre. If intending to anchor in Port de Texier, keep to the N of rocks on entering.

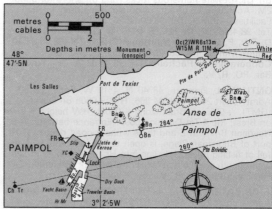

LIGHTS AND MARKS
Leading marks — Paimpol church tower and the top of Pt Brividic 260°. Second leading line, lights on Jetée de Kernoa and Paimpol front (both FR) 264°. Pointe de Porz-Don Lt Oc (2) WR 6s 13m 15/11M; vis W269°-272°, R272°-279°

RADIO TELEPHONE
None.

TELEPHONE
Hr Mr 96.20.80.77; Port de Plaisance 96.20.47.65; Aff Mar 96.20.84.30; CROSS 98.89.31.31; Customs 96.20.81.87; Meteo 96.20.01.92; Dr 96.20.80.04; Brit Consul 99.46.26.64.

FACILITIES
EC Monday all day; **Bassin No 2** (marina 230+10 visitors), FW, AB, D, P, ME, El, C (6 ton, 3 ton); **Quai de Kernoa** P, ME; **Quai neuf** Slip, M, FW, AB; **Dauphin** Tel. 96.20.81.38, Sh, CH; **Despretz** Tel. 96.20.82.00, M, ME, Sh, CH; **Le Lionnaise Marine** Tel. 96.20.85.18, ME, El, Sh, CH; **Comptoir Nautique Paimpolais** Tel. 96.20.88.59, M, ME, Sh, CH; **Centre Nautique des Glenans** Tel. 96.20.84.33; **Le Corre L** Tel. 96.20.85.17, SHOM.
Town P, D, CH, V, Gaz, R, Bar. PO; Bank; Rly; Air (St Brieuc).
Ferry UK — Roscoff—Plymouth.

MINOR HARBOURS AND ANCHORAGES 10.16.23

L'ABERILDUT, Finistere, 48°28' N, 4°45' W, Zone −0100, Admty charts 3345, 2694; SHOM 5721. HW +0520 on Dover (GMT), +0007 on Brest (zone −0100). From N beware Les Liniou Rks 2M S of Le Four Lt (Fl(5)15s). Good shelter but exposed to W winds and harbour partly dries. Beware strong cross tides in approach. For Chenal du Four see 10.16.5. Aberildut Lt Dir Oc(2) WR 6s, vis W081°-085°, R085°-087°. Ldg marks Brélès spire and Lanildut spire at 079° lead into fairway between rks which dry, marked by Bn Trs. Facilities: several slips, village shop; **Cariou** Tel. 98.04.30.44, ME.

ARGENTON, Finistere, 48°31' N, 4°46' W, Zone −0100, Admty chart 2694, SHOM 5772, 5287. HW +0535 on Dover (GMT), +0020 on Brest (zone −0100); HW height +0.5m on Brest; ML 4.6m; Duration 0600. Small drying Hr giving good shelter except in W winds when a swell comes up the bay into the Hr. Access HW∓3. Approach is well marked. Beware strong E−W tides. Anchorage in deep water off Hr ent. There are no lights. Facilities: FW, P on quay; **Town** Bar, R, V.

PORTSALL, Finistere, 48°33' N, 4°42' W, Zone −0100, Admty charts 1432, 20, SHOM 5772, 5287. HW +0535 on Dover (GMT), +0020 on Brest (zone −0100); HW height +0.5m on Brest; ML 4.6m; Duration 0600. Small drying harbour at head of bay. Access HW∓3. Good shelter except in strong N winds. Entrance to bay marked by two Bn Trs. Beware numerous rocks lying off-shore for about 2M. Anchor to W of ent. or go alongside quay in Hr. Facilities: Aff.Mar. Tel. 98.80.62.25; SNSM Tel. 98.48.65.92; Dr Tel. 98.48.10.46. **Quay** C (0.5 ton), D, FW, P, Slip; **Club Nautique** (CNPK) Tel. 98.48.63.10. **Helies Marine** Tel. 98.48.14.56. CH, El, ME, Sh; **Coopérative de Pêcheurs** Tel. 98.48.63.26 CH; **Town** Bar, PO, R, V.

PONTUSVAL, Finistere, 48°40' N, 4°19' W, Zone −0100, Admty chart 2643, SHOM 966. HW +0605 on Dover (GMT), +0055 on Brest (zone −0100); HW height +0.7m on Brest; ML 4.7m Duration 0600. Entrance between An Neudenn Bn Tr to E and three white topped rks to W. Hr, port of Brignogan-Plage, is open to N winds and is often full of fishing boats. Ldg line into Hr, Bn is on shore in line with church steeple at 178°. Entry at night forbidden. Facilities: Bar, FW, R, V.

ILE DE BATZ, Finistere, 48°44' N, 4°00' W, Zone −0100, Admty chart 2745; SHOM 5828. HW +0610 Dover (GMT), +0055 on Brest (zone −0100); HW height +1.3m on Brest. For Canal de l'Ile de Batz see 10.16.5. Porz-Kernoc'h gives good shelter but dries. E slip is reserved for ferries. Anchor in E or W parts of the channel depending on wind, but holding ground poor. Anchorage prohibited in area W of Roscoff landing slip. Ile de Batz Lt Fl(4) 25s. Facilities: there are a few shops.

CARANTEC, Finistere, 48°40' N, 3°55' W, Zone −0100, Admty chart 2745, SHOM 5827. HW +0610 on Dover (GMT), +0105 on Brest (zone −0100); HW height +1.2m on Brest; ML 5.0m; Duration 0605. Approach as for Morlaix, (10.16.16) but pass W of Ile Callot. La Penzé Rivière is narrow and tortuous. Yachts can dry out in shelter off Penpoul on the W shore. The anchorage off Carentec is exposed, especially to NW; landing stage dries about 4.6m. Further S the river is not marked but shelter is better SW of Pte de Lingos, or off the old ferry slipways which provide landing places. Beyond the bridge (15m) the river is navigable on the tide for 3M to Penzé. No lights. Facilities: Aff.Mar. Tel. 98.67.03.80; **Centre Nautique Carantec-Henvic** Tel. 98.67.01.12; **Elies** Tel. 98.67.03.61. El, ME, Sh; **Marine Service** Tel. 98.67.00.04; CH; **Comptoirs Maritime** Tel. 98.67.01.85, CH, P, D; **Lebras** Tel. 98.67.04.50, ME, El, Sh; **Town** Bank, Bar, PO, R, V.

PRIMEL, Finistere, 48°42' N, 3°50' W, Zone −0100, Admty chart 2644, SHOM 5827, 5950. HW −0610 on Dover (GMT), +0100 on Brest (zone −0100); HW height +0.9m on Brest; ML 4.8m; Duration 0600. Good deep anchorage well protected; seas break across ent in strong winds. Beware rks off Pte de Primel to E of ent. and Le Zameques to W. Ldg Lts 152°, front FR 35m 6M; vis 134°−168°, rear, 202m from front, FR 56m 6M: both have framework Tr with W square with R vert stripe. Jetty head Lt Fl G 4s 6m 7M. Anchor in channel in 2 to 9m or go alongside quay (1m). VHF Ch 09 16 (summer). Facilities: Bar, C (12 ton), FW, R. Slip, V; **Rolland Marine** Tel. 98.72.32.76. CH, El, ME, Sh, Electronics.

LOCQUEMEAU, Côtes du Nord, 48°43' N, 3°34' W, Zone −0100, Admty charts 2644, 2668, SHOM 5950. HW −0600 on Dover (GMT), +0110 on Brest (zone −0100); HW height +1.5m on Brest; ML 5.3m. A small drying Hr by ent. to Lannion River. (10.16.17). There are two quays, the outer being accessible at LW, but this quay is open to W winds. Yachts can dry out at inner quay, on S side. Ldg Lts 122°, leading to outer quay, Front FR 21m 6M; vis 068°−228°. Rear 484m from front, Oc (2+1) R 12s 39m 7M. Facilities: **Ateliers Mecaniques de Lannion** Tel. 96.48.72.32 ME, El, Sh; **Town** Bar.

TREBEURDEN, Côtes du Nord, 48°46' N, 03°35' W; Zone −0100; Admty charts 2668, 2644; SHOM 5950, 6056. HW −0605 on Dover (GMT), +0105 on Brest (zone −0100); ML 5.3m; Duration 0605. A good safe deep water anchorage in settled weather. Very exposed to winds from W to NW. Visitors buoys available. Approach from a position 1.5M S of Basse Blanche W cardinal buoy, marking the W end of Le Crapaud (a reef which dries 3.9m). From here make good a course of 065° towards the N end of Ile Milliau, to pass S of Ar Goureudeg S cardinal Lt Buoy VQ(6) + LFl 10s. When past Ar Goureudeg buoy, alter to E and SE, so as to anchor NE of Ile Milliau. There are no lights in the port. Facilities: Harbour Office Tel. 96.23.66.93; SNSM 96.23.53.82, Customs Tel. 96.92.31.44; **Harbour** (340+10 visitors), FW, AC, M; **YC de Trébeurden** Tel. 96.37.00.40 (open July-Aug); **Chantier Naval du Trégor** Tel. 96.23.52.09, BY, ME, El, Sh, CH.

TRÉGASTEL, Côtes du Nord, 48°50' N, 3°31' W, Zone −0100, Admty charts 2644, 2668, SHOM 5950, 967. HW −0550 on Dover (GMT), +0005 on Brest (zone −0100); HW height −1.8m on Brest; ML 5.1m; Duration 0605. See 10.16.18. (Ploumanac'h). Good anchorage in 2m but very exposed to winds from W to N. Entrance channel marked by port and stbd Bns. Ldg line at 149°, Pavilion in line with Chapelle St Anne leads between Ile Dhu and Le Taureau. Thence keep between Bns round to S of Ile Ronde to the anchorage. Facilities: Slip; **Club Nautique de Tregastel**; **Town** Bar, Bank, PO, R, V.

PORT BLANC, Côtes du Nord, 48°50' N, 3°19' W, Zone −0100, Admty charts 3672, 2668, SHOM 974, 967. HW −0545 on Dover (GMT), +0007 on St Helier (zone −0100); HW height −1.3m on St Helier; ML 5.3m; Duration 0600. Good natural Hr but exposed to winds between NW and NNE. Ldg marks in line at 150°, Moulin de la Comtesse and Le Voleur Lt; Le Voleur Dir Lt 150°, Fl WRG 4s 17m 14/11M, G140°−148°, W148°−152°, R152°−160°. Yachts can anchor off or dry out alongside quays (known as Port Bago). Facilities: AB, FW on quay, Slip; **Gelgon Nautisme** Tel. 96.92.67.00. CH, El, ME, Sh; **Town** Bar, R, V.

PONTRIEUX, Côtes du Nord, 48°41' N, 3°08' W, Zone −0100, Admty chart 2668, SHOM 970. Pontrieux is above the lock; for tides below the lock see 10.16.20 Lezardrieux. Complete shelter above lock. Pass under Lezardrieux suspension bridge (18m clearance) and proceed the 7M up river, not well marked. Access up river from mid-tide; lock opens HW∓1 (but does not open unless rise is more than 8.8m). Yachts can go alongside quay in approx 4m. Facilities: **Town** AB, Bar, FW, R, Slip, V, P, D.

VOLVO PENTA SERVICE

Sales and service centres in area 17

CHANNEL ISLANDS **Chicks Marine Ltd**, Collings Road, St. Peter Port, GUERNSEY Tel (0481) 23716/24536. **DK Collins Ltd,** South Pier, St. Helier, JERSEY, C.I. Tel (0534) 32415.
FRANCE For details of local service dealers in France contact **Volvo Penta France SA**, BP45. F78130 Les Mureaux Tel (01) 3 474 72 01, Telex 695221 F.

VOLVO PENTA

Area 17

Channel Islands and adjacent coast of France
Portrieux to Cherbourg

10.17.1	Index	Page 661
10.17.2	Diagram of Radiobeacons, Air Beacons, Lifeboat Stations etc	662
10.17.3	Tidal Stream Charts	664
10.17.4	List of Lights, Fog Signals and Waypoints	666
10.17.5	Passage Information	668
10.17.6	Distance Tables	669
10.17.7	English Channel Waypoints	See 10.1.7
10.17.8	Special notes for French areas	See 10.14.7
10.17.9	Channel Islands	670
10.17.10	Jersey — Gorey	671
10.17.11	— St Helier, Standard Port, Tidal Curves	671
10.17.12	Sark	676
10.17.13	Guernsey — Beaucette	677
10.17.14	— St Peter Port	677
10.17.15	Alderney — Braye	678
10.17.16	St Quay—Portrieux	679
10.17.17	Binic	680
10.17.18	Le Légué/St Brieuc	680
10.17.19	St Malo/Dinard	681
10.17.20	Granville	682
10.17.21	Carteret	683
10.17.22	Iles Chausey	683
10.17.23	Omonville	684
10.17.24	Cherbourg, Standard Port, Tidal Curves	684
10.17.25	Minor Harbours and Anchorages St Aubin (Jersey) St Sampson (Guernsey) Dahouet Val André Erquy Saint Cast Richardais Rotheneuf Cancale Portbail	689

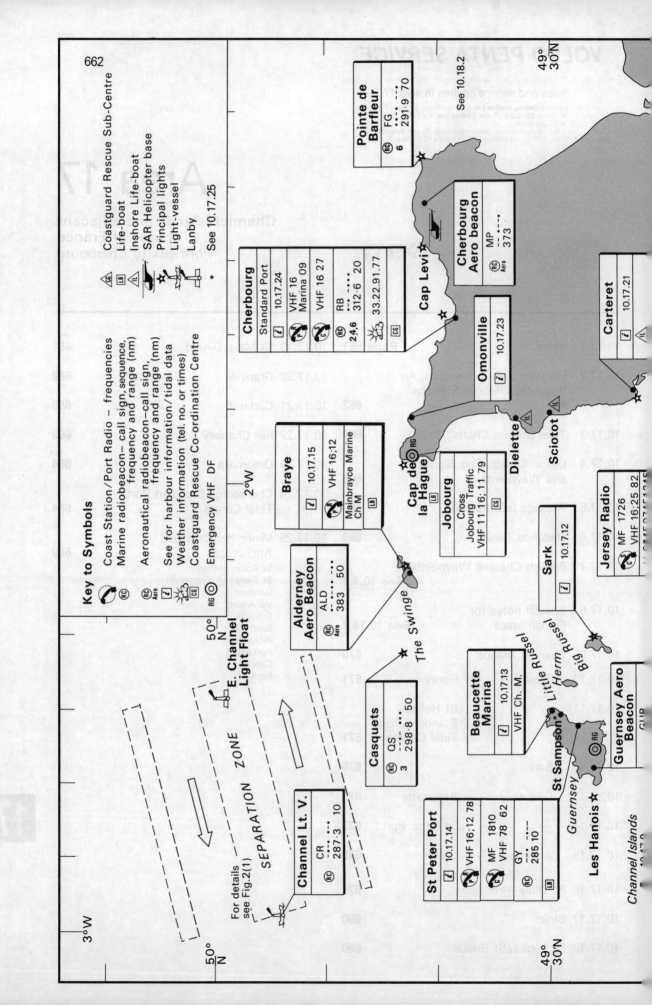

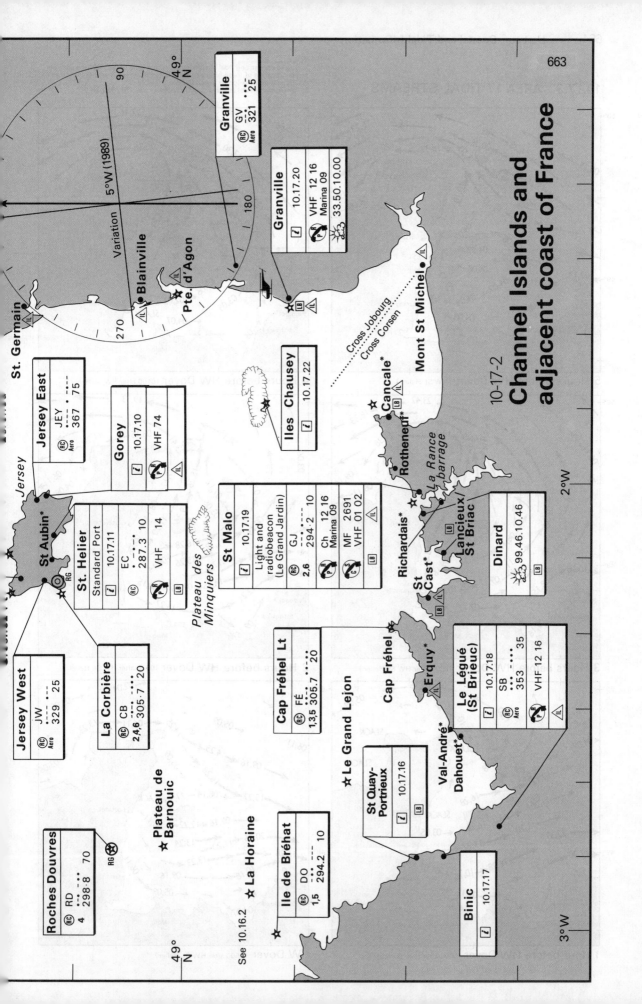

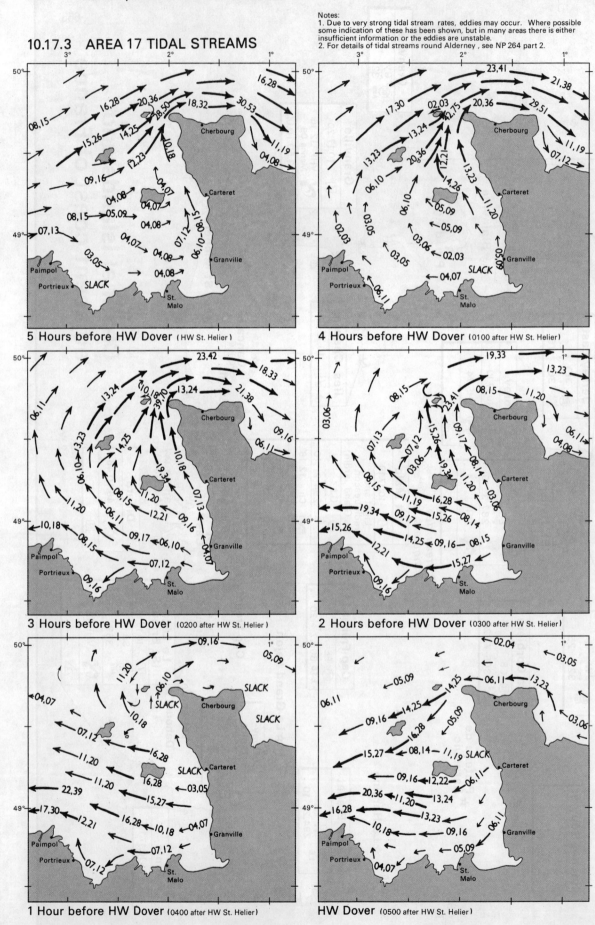

AREA 17 — Channel Islands and French Coast

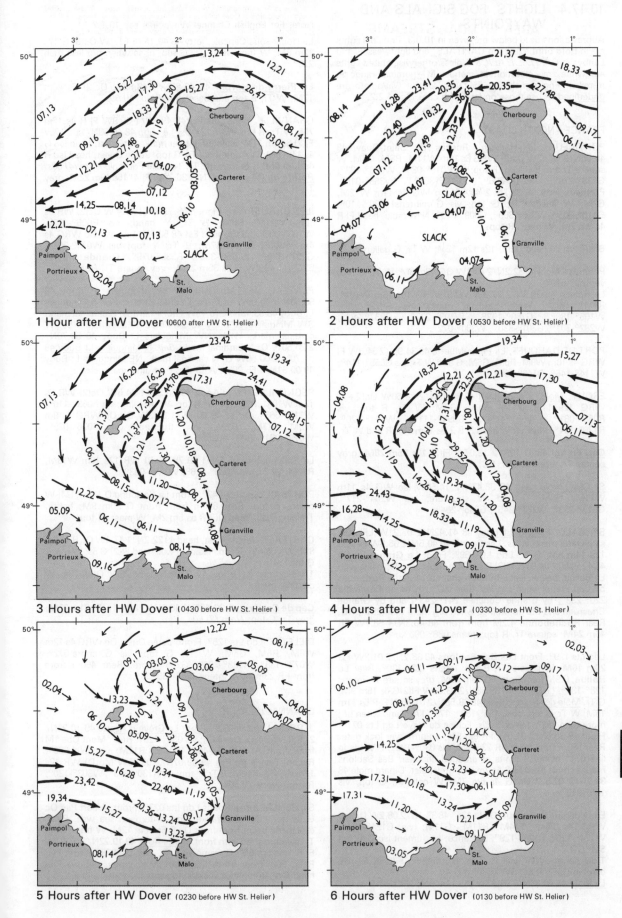

10.17.4 LIGHTS, FOG SIGNALS AND WAYPOINTS

Abbreviations used below are given in 10.0.3. Principal lights are in **bold** print, places in CAPITALS, and light-vessels and Lanbys in *CAPITAL ITALICS*. Unless otherwise stated lights are white. m—elevation in metres; M—nominal range in n. miles. Fog signals are in *italics*. Useful waypoints are underlined — use those on land with care. See 4.2.2.

FRANCE—NORTH COAST

PORTRIEUX. Île Harbour, Roches de Saint-Quay 48 40.05N/2 48.42W Oc(2) WRG 6s 16m W11M, R8M, G8M; W square Tr and dwelling, R top; vis R015°-080°, W080°-133°, G133°-270°, R270°-306°, G306°-358°, W358°-015°. Portrieux jetty 48 38.7N/2 49.4W Iso WG 4s 11m W12M, G7M; W 8-sided Tr, G top; vis G(unintens) 020°-110°, G110°-305°, W305°-309°, G309°-020°. New mole, head Fl R 4s 9m 7M; W mast, R top.

BINIC, mole head Oc(3) 12s 12m 12M; W Tr, G gallery.

Pointe à l'Aigle 48 32.2N/2 43.1W QG 13m 8M; W Tr, G top; vis 160°-070°.
Le Rohein 48 38.9N/2 37.8W VQ(9) WRG 10s 13m W10M, R8M, G8M; West cardinal mark; vis R072°-105°, W105°-180°, G180°-193°, W193°-237°, G237°-282°, W282°-301°, G301°-330°, W330°-072°.

PORT DE DAHOUET, La Petite-Muette 48 34.9N/2 34.3W Fl WRG 4s 10m W9M, R6M, G6M; G and W Tr; vis G055°-114°, W114°-146°, R146°-196°.

ERQUY. Mole, S end, head 48 38.13N/2 28.60W Oc(2+1) WRG 12s 11m W11M, R9M, G9M; W Tr, R top; vis R055°-081°, W081°-094°, G094°-111°, W111°-120°, R120°-134°. Inner jetty, head Fl R 2.5s 10m 3M; R and W Tr.

Cap Fréhel 48 41.10N/2 19.07W Fl(2) 10s 85m **28M**; grey square Tr; *Horn (2) 60s*; RC.

St Cast, mole head 48 38.47N/2 14.50W Iso WG 4s 11m W11M, G8M; G and W structure; vis W204°-217°, G217°-233°, W233°-245°, G245°-204°.

ST MALO. Fairway Lt Buoy 48 41.42N/2 07.21W LFl 10s; safe water mark; *Whis*. Les Courtis 48 40.5N/2 05.8W Fl(3) G 12s 14m 9M; G Tr. Ldg Lts 089°. Front **Le Grand Jardin** 48 40.27N/2 04.90W Fl(2) R 10s 24m **15M**; grey Tr; in line with La Ballue, below, leads through channel of Petite Port; obsc by Cap Fréhel when brg less than 097°, by Île de Cézembre 220°-233°, by Grande Conchée 241°-243°, and by Grande Chevruen and Pte du Meinga when brg more than 251°; RC. Rear **Rochebonne** 4.2M from front 48 40.3N/1 58.7W FR 40m **24M**; square Tr, R top; intens 088°-090°.

Ldg Lts 129°. Front **Les Bas-Sablons** 48 38.2N/2 01.2W FG 20m **16M**; W square Tr, B top; intens 127°-130°. Rear **La Ballue**, 0.9M from front FG 69m **25M**; grey square Tr; intens 128°-130°. Le Buron 48 39.38N/2 03.60W Fl(2)G 6s 15m 9M; G Tr. Môle des Noires, head 48 38.58N/2 01.85W Fl R 5s 11m 13M; W Tr, R top; obsc 155°-159°, 171°-178°, and when brg more than 192°; *Horn(2) 20s*. Ecluse du Naye Ldg Lts 071°. Front 48 38.6N/2 01.5W FR 6m 5M; on arm over lock gate. Rear 580m from front FR 23m 8M; vis 030°-120°. F Vi Lts in line 071° marks S limits of dredged channel. Bas-Sablons marina, mole head Fl G 4s 7m 5M; grey mast. Dinard mole 48 38.2N/2 02.8W Oc G 4s 6m 7M; W column, G top; vis 195°-015°.

Embouchure du Fremur. Dir Lt 125° 48 37.1N/2 08.2W Dir Iso WRG 4s 20m W14M, R11M, G11M; vis G121°-124°, W124°-126°, R126°-129°. La Rance, La Jument Iso G 4s 6m 5M; G Tr.
Tidal barrage NW wall Fl G 4s, NE dolphin Fl(2) R 6s, SE dolphin Oc(2) R 6s, SW wall Iso G 4s.
La Plate 48 40.83N/2 01.83W VQ 11m 9M; N cardinal mark.

Note. For English Channel Waypoints see 10.1.7

La Houle-sous-Cancale, jetty 48 40.1N/1 51.1W Oc(3) G 12s 12m 8M; W framework Tr, G top; obsc when brg less than 223°.

La Pierre-de-Herpin 48 43.83N/1 48.83W Oc(2) 6s 20m **17M**; W Tr, B top and base; *Siren Mo (N) 60s*.

GRANVILLE. Hérel, marina 48 49.9N/1 35.9W Fl R 4s 12m 7M; W Tr, R top; *Horn (2) 40s*. Le Loup 48 49.63N/1 36.17W Fl(2) 6s 8m 11M; isolated danger mark. Jetée Est, head Iso G 4s 11m 6M; W framework Tr, G top. Jetée Ouest, head Iso R 4s 12m 6M; R framework Tr. Tourelle Fourchie *Horn (4) 60s*. **Pointe du Roc** 48 50.11N/1 36.70W Fl(4) 15s 49m **22M**; grey Tr, R top.

ILES CHAUSEY. Le Pignon 48 53.5N/1 43.4 W Oc(2) WR 6s 10m W11M, R8M; B Tr, Y band; vis R005°-150°, W150°-005°. La Crabière Est 48 52.5N/1 49.4W Oc WRG 4s 4m W10M, R7M, G7M; B Tr, Y top; vis W079°-291°, G291°-329°, W329°-335°, R335°-079°. **Grande Ile** 48 52.2N/1 49.3W Fl 5s 39m **23M**; grey square Tr; *Horn 30s*.

PLATEAU DES MINQUIERS. SE Minquiers Lt Buoy 48 53.50N/2 00.00W Q(3) 10s; E cardinal mark; *Bell*. S Minquiers Lt Buoy 48 53.13N/2 10.13W Q(6) + LFl 15s; S cardinal mark. SW Minquiers Lt Buoy 48 54.40N/2 19.30W Q(9) 15s; W cardinal mark; *Whis*. NW Minquiers Lt Buoy 48 59.70N/2 20.50W Q; N cardinal mark; *Bell*. N Minquiers Lt Buoy 49 01.70N/2 00.50W Q; N cardinal mark. NE Minquiers Lt Buoy 49 00.91N/1 55.20W VQ(3) 5s; E cardinal mark; *Bell*.

REGNÉVILLE, Pte d'Agon 49 00.2N/1 34.6W Oc(2) WR 6s 12m W10M, R7M; W Tr, R top, W dwelling; vis R063°-110°, W110°-063°. Dir Lt 49 00.7N/1 33.3W Dir Oc WRG 4s 9m W9M, R7M, G7M; W structure; vis G026°-029°, W029°-031°, R031°-035°.

Le Sénéquet 49 05.5N/1 39.7W Fl(3) WR 12s 18m W13M, R10M; W Tr; vis R083°-116°, W116°-083°.

PORTBAIL. Ldg Lts 042°. Front La Caillourie Q 14m 11M; W pylon, R top. Rear 870m from front Q 20m 9M; belfry. Training wall, head Q(2)R 5s 5m 2M; W mast, R top.

CARTERET. Jetée Ouest, head 49 22.2N/1 47.4W Oc R 4s 6m 8M; W column, R top. Training wall head Fl(2) G 5s; W mast, G top.
Trois-Grunes Lt Buoy 49 21.85N/1 54.76W Q(9) 15s; W cardinal mark.

Cap de Carteret 49 22.46N/1 48.35W Fl(2+1) 15s 81m **26M**; grey Tr, G top; *Horn (3) 60s*.

DIELETTE. Ldg Lts 125°. Front, Jetée Ouest Oc WRG 4s 12m W8M, R6M, G6M; W Tr, G top; vis G shore-072°, W072°-138°, R138°-206°, G206°-shore. Rear 460m from front FR 23m 11M; intens 121°-130°.

CHANNEL ISLANDS

JERSEY. Ldg Lts 082°. Front La Grève d'Azette 49 10.21N/2 05.00W Oc 5s 23m 14M; vis 034°-129°. Rear, Mont Ubé 1M from front Al WR 6s 46m W14M, R12M; vis 250°-095°; Racon. Demie de Pas 49 09.07N/2 06.05W Mo(D) WR 12s 11m W14M, R10M; B Tr, Y top; vis R130°-303°, W303°-130°; Racon; *Horn (3) 60s*.

ST HELIER. Small Roads Ldg Lts 023°. Front, Albert Pier Oc G 5s 8m 11M; W bracket and lantern on sea wall. Rear, esplanade 576m from front Oc R 5s 20m 12M; W framework Tr; synchronised with front. Platte Rock 49 10.22N/2 07.27W Fl R 1.5s 6m 5M; R framework Tr. Ldg Lts 078°. Front FG. Rear, 90m from front, FG; both W columns. Victoria Pier head *Bell*; in reply to vessels' fog signals; traffic signals.

AREA 17 — Channel Islands and French Coast

Canger Rock Lt Buoy 49 07.41N/2 00.30W Q(9) 15s; W cardinal mark. Violet Lt Buoy 49 07.87N/1 57.05W LFl 10s; safe water mark.

GOREY. Ldg Lts 298°. Front, pier head 49 11.9N/2 01.3W Oc RG 5s 8m 12M; W framework Tr; vis R304°-352°, G352°-304°. Rear 490m from front Oc R 5s 24m 8M.
Verclut breakwater 49 13.39N/2 00.57W Fl 1.5s 18m 13M; framework Tr.
Bonne Nuit Bay Ldg Lts 223°. Front, Pierhead FG 7m 6M. Rear, 170m from front FG 34m 6M. Rozel Bay Dir Lt 245° Dir FWRG 11m 5M; vis G240°-244°, W244°-246°, R246°-250°.
Sorel Point 49 15.64N/2 09.45W L Fl WR 7.5s 50m **15M**; B and W chequered Tr; vis W095°-112°, R112°-173°, W173°-230°, R230°-269°, W269°-273°.
Grosnez Point 49 15.55N/2 14.75W Fl(2) WR 15s 50m **W19M**, **R17M**; W concrete hut; vis W081°-188°, R188°-241°.
La Corbière 49 10.85N/2 14.90W Iso WR 10s 36m **W18M**, **R16M**; Round stone Tr; vis W shore-294°, R294°-328°, W328°-148°, R148°-shore; RC; *Horn Mo (C) 60s*.
Noirmont Point Fl(4) 12s 18m 13M; B Tr, W band.
Saint Aubin Harbour, N pierhead Iso R 4s 12m 10M; and Dir Lt 252° F WRG, vis G246°-251°, W251°-253°, R253°-258°. Fort, pierhead Fl(2) Y 5s 8m 1M.

SARK. **Point Robert** 49 26.25N/2 20.67W Fl(2) 5s 65m **18M**; W 8-sided Tr; vis 138°-353°; *Horn (2) 60s*. Blanchard Lt Buoy 49 25.42N/2 17.35W Q(3) 10s; E cardinal mark; *Bell*.

Courbée du Nez 49 27.15N/2 22.08W Fl(4)WR 15s 14m 8M; vis W057°-230°, R230°-057°.
Big Russel, Noire Pute 49 28.27N/2 24.93W Fl(2) WR 15s 8m 6M; vis W220°-040°, R040°-220°. Destroyed (T).

HERM. Fourquies Lt Buoy 49 27.34N/2 26.47W Q; N cardinal mark. Lower Heads Lt Buoy 49 25.91N/2 28.48W Q(6) + LFl 15s; S cardinal mark; *Bell*. Alligande 49 27.9N/2 28.8W Fl(3) G 5s; Orange A on black mast; shown 1/4-1/11; Ra refl. Épec Beacon 49 28.0N/2 27.9W Fl G 3s; Black E on G mast; shown 1/4-1/11. Vermerette Beacon 49 28.3N/2 27.8W Fl(2) Y 5s; Orange V on Bn; shown 1/4-1/11. Percée Pass, Gate Rock 49 27.9N/2 27.4W Q(9)15s; West cardinal Bn.

GUERNSEY. St Peter Port Ldg Lts 220°. **Front**, Castle breakwater Al WR 10s 14m **16M**; dark round Tr, W on NE side; vis 187°-007°; RC; *Horn 15s*. Rear Belvedere Oc 10s 61m 14M; W square with Y stripe on W Tr; vis 179°-269°. White Rock Pier, head 49 27.43N/2 31.60W Oc G 5s 11m 14M; round stone Tr; intens 174°-354°. South pier head Oc R 5s 10m 14M; W framework Tr, R lantern. North Beach Marina. Dir Lt Oc WRG 10s 5m 6M; vis G264°-268°, W268°-272°, R272°-276°.

St Sampson. Crocq Pier FR 11m 5M; R column; vis 250°-340°. N pier head FG 3m 5M; vis 230°-340°. Ldg Lts 286°. Front S pier head 49 28.9N/2 30.8W FR 3m 5M; vis 230°-340°. Rear, 390m from front FG 13m; clock Tr. 2 FR (vert) on chimneys 300m N.
Brehon 49 28.3N/2 29.2W Iso 4s 19m 9M; Bn. Platte 49 29.1N/2 29.5W Fl WR 3s 6m W7M, R5M; G Tr; vis R024°-219°, W219°-024°. Roustel, S end, 49 29.28N/2 28.71W Q 8m 7M; B and W chequered Tr, G lantern. Tautenay 49 30.2N/2 26.7W Q(3) WR 6s 7m W7M, R6M; B and W Bn; vis W050°-215°, R215°-050°.

Beaucette Marina. Ldg Lts 277°. Front FR; W board with R stripe. Rear FR; R board with W stripe.

Platte Fougère 49 30.88N/2 29.05W LFl WR 10s 15m **16M**; W 8-sided Tr, B band; vis W155°-085°, R085°-155°; Racon; *Horn 45s*.

Les Hanois 49 26.2N/2 42.1W Q(2) 5s 33m **23M**; grey round Tr, B lantern, helicopter platform; vis 294°-237°; 4 FR on masts 1.27M ESE; *Horn (2) 60s*.

St Martin's Point 49 25.37N/2 31.61W Fl(3) WR 10s 15m 14M; flat-roofed, W building; vis R185°-191°, W191°-011°, R011°-081°. *Horn (3) 30s*.

ALDERNEY. **Alderney** 49 43.81N/2 09.77W Fl(4) 15s 37m **18M**; W round Tr, B band; vis 085°-027°; *Siren (4) 60s*. Braye Harbour Ldg Lts 215°. **Front**, elbow of old pier Q 8m **17M**; vis 210°-220°. **Rear** 335m from front Iso 10s 17m **18M**; vis 210°-220°. Both metal posts on W columns. Chateau a l'Etoc Point 49 44.00N/2 10.55W Iso WR 4s 20m W10M, R7M; W column; vis R071°-111°, W111°-151°; in line 111° with main Lt.

Casquets 49 43.38N/2 22.55W Fl(5) 30s 37m **28M**; W Tr, the highest and NW of three; RC; Racon; *Dia (2) 60s*.

CHANNEL LT V 49 54.42N/2 53.67W Fl 15s 12m **25M**; R hull with Lt Tr amidships; RC; Racon; *Dia(1) 20s*.
E. CHANNEL LT FLOAT 49 58.67N/2 28.87W Fl(2) 10s 11m **16M**; F riding Lt; Racon; *Horn (2) 30s*.
Lt Buoy EC 1 50 05.90N/1 48.35W FlY 2.5s; X on Y HFP buoy; Racon; *Whis*.
Lt Buoy EC 2 50 12.10N/1 12.40W Fl(4)Y 15s; X on Y HFP buoy; Racon; *Whis*.
Lt Buoy EC 3 50 18.30N/0 36.10W FlY 5s; X on Y HFP buoy; Racon; *Whis*.

FRANCE — NORTH COAST

Cap de la Hague 49 43.37N/1 57.19W Fl 5s 48m **23M**; grey Tr, W top; FR Lts on chimney and radio mast 3.8M and 4.5M SE; *Horn 30s*.
La Plate 49 44.03N/1 55.64W Fl(2+1) WR 12s 11m W9M, R6M; 8-sided Tr, Y with B top; vis W115°-272°, R272°-115°.
Basse Bréfort Lt Buoy 49 43.70N/1 51.05W VQ; N cardinal mark; *Whis*. Omonville, La Rogue 49 42.3N/1 50.2W Iso WRG 4s 13m W11M, R8M, G8M; W framework Tr, R top; vis G180°-252°, W252°-262°, R262°-287°.

CHERBOURG. **CH1 Lt Buoy** 49 43.30N/1 42.10W Oc 4s; safe water mark; *Whis*. Digue de Querqueville, head Oc(3) WG 12s 8m W11M, G8M; W column, G top; vis W120°-290°, G290°-120°.
Passe de l'Ouest Ldg Lts 140° and 142°. Front, Jetée du Homet 2 Q(hor) 5m **15M**; W triangles on parapet at root of jetty; 63m apart; intens 136°-146°. Rear, **Gare Maritime** 0.98M from front DirQ 23m **23M**; grey framework with W triangle point down on building; intens 137°-145°.
Fort de l'Ouest 49 40.50N/1 38.87W Fl(3) WR 15s 19m **W22M, R18M**; grey Tr, R top, on fort; vis W122°-355°, R355°-122°; RC; *Reed (3) 60s*.
Fort Central VQ (6) + LFl 10s 5m 8M; column; vis 322°-032°.
Fort de l'Est 49 40.33N/1 35.93W Iso WG 4s 19m W12M, G9M; W framework Tr, G top; vis W008°-229°, G229°-008°. Passe Est, Forte d'Île Pelée Oc (2) WR 6s 19m W11M, R8M; W and R pedestal on fort; vis W055°-120°, R120°-055°.
Passe Cabart-Danneville, Pierhead Fl(2) R 6s 5m 6M.
Fort des Flamands 49 39.1N/1 35.6W Dir Q WRG 13m W12M, R11M, G11M; vis G173°-176°, W176°-183°, R183°-193°.

LE BECQUET. Ldg Lts 187°. **Front** Oc(2+1) 12s 8m **16M**; W 8-sided Tr; intens 183°-190°. Rear 50m from front Oc (2+1) R 12s 13m 11M; W 8-sided Tr, R top; synchronised with front, intens 183°-190°.

Port de Lévi 49 41.3N/1 28.3W F WRG 7m W11M, R8M, G8M; vis G050°-109°, R109°-140°, W140°-184°.
Cap Lévi 49 41.80N/1 28.30W Fl R 5s 36m **22M**; grey square Tr, W top.
La Pierre Noire Lt Buoy 49 43.57N/1 28.98W Q(9) 15s 8m 9M; W cardinal mark.
Basse du Renier Lt Buoy 49 44.90N/1 22.00W VQ 8m 8M; N cardinal mark; *Whis*.

10.17.5 PASSAGE INFORMATION

Sailing directions for this popular cruising area are listed in Chapter 15 (15.5) of the *The Macmillan & Silk Cut Yachtsman's Handbook*. Particular features are a rugged shoreline, but with sandy bays, numerous offlying rks, strong tidal streams, and a big range of tide. It is important to have large scale charts, and to use recognised leading marks (of which there are plenty) when entering or leaving many of the harbs and anchs. Neap tides are best, particularly for a first visit, and tidal streams need to be worked carefully. Boats which can take the ground have an advantage for exploring the quieter harbs. Be careful to avoid lobster pots, and oyster beds in some rivers. For French terms see 10.16.5.

ST BRIEUC TO ST MALO (chart 2669)

To seaward of B de St Brieuc is Grand Léjon (Lt), a rky shoal 10½M E of Pte de Minard. Rks extend 2½ ca W and 8 ca NNE of the Lt Ho. Petit Léjon (dries) lies 3½M SSE of Lt Ho. B de St Brieuc is encumbered by two rky shoals: on the W side, Roches de St Quay and offlying patches extend 4M E from Portrieux (10.17.16), and on the E side the Plateau du Rohein, at W end of long ridge, is 6M W of C d'Erquy. Between these two shoals is a chan 3M wide leading to the harbs of Binic (10.17.17), Le Légué (10.17.18), Port de Dahouet and Erquy (10.17.25). On the E side of the b, Chenal d'Erquy runs in a WSW/ENE direction about 3 ca off C d'Erquy and inshore of the various rky patches close to seaward. There are several rky shoals within the b itself, some extending nearly 2M from shore. Erquy is a pleasant, drying harb, but often crowded with fishing boats.

From C d'Erquy to C Fréhel and St Malo (10.17.19) there are no worthwhile harbs, but there are anchs in B de la Fresnaie and in B de l'Arguenon. Beware Le Vieux-Banc (dries) 6M ENE of C Fréhel. Here the E-going stream begins at HW St Helier −0555, and the W-going at HW St Helier −0015, sp rates 2½ kn. There are W-going eddies very close inshore on the E-going stream.

In the approaches to St Malo (chart 2700) are many islets, rks and shoals, between which are several chans that can be used in good vis. Tidal streams reach 4 kn at sp, and can set across chans. Chenal de la Grande Porte and Chenal de la Petite Porte are the easiest routes and are well marked. With sufficient rise of tide, and good vis, Chenal du Décollé, Chenal de la Petite Conchée, Chenal de la Grande Conchée, Chenal des Petis Pointus or Chenal de la Bigne can be used, but they all pass over or near to drying patches.

By passing through the lock at W end of the Rance barrage it is possible to cruise up river to Dinan, via the lock at Châtelier (French chart 4233). Beware very rapid changes in level caused by the hydro-electric scheme, explained in pamphlet available at lock, or from harb office St Malo. At Dinan is entrance to Canal d'Ille et Rance (see 10.15.8).

ST MALO TO ALDERNEY RACE

4M E of St Malo is the large drying harb of Rothéneuf (10.17.25 and chart 2700), with anch off in good weather. Proceeding E, beware many dangers off Pte du Grouin, 3M S of which is the drying harb of Cancale, with many oyster beds. There is anch SE of Île des Rimains in good weather. To E of Cancale is the drying expanse of B du Mont St Michel. See 10.17.25.

The W coast of the Cherbourg peninsula is exposed to the W, and is mostly rky and inhospitable. Granville (10.17.20) is the only harb where a yacht can lie afloat. The two main chans to/from the Alderney Race are Passage de la Déroute and Déroute de Terre. The former passes W of Îles Chausey, SE and E of Les Minquiers and Les Ardentes, between Chaussée de Boeufs and Plateau de l'Arconie, between Basses de Taillepied and Les Écrehou, and W of Plateau des Trois Grunes. Parts of this chan are not well marked, and it is advisable not to use it at night.

Déroute de Terre leads between Pte du Roc (off Granville) and Îles Chausey, between Le Sénéquet Lt Tr and Les Boeufs, between Basses de Portbail and Bancs Félés, and E of Plateau des Trois Grunes. For detailed directions see *Channel Pilot*. The S end of this chan, E of Îles Chausey, is very shallow, and along much of this coast as far N as Carteret there is little depth of water, so that a nasty sea can build up. There are drying harbs at Portbail, Carteret (10.17.21) and at Dielette, 11M S of C de la Hague. In approaches to Alderney Race (see below) beware Les Huquets de Jobourg, an extensive bank of drying and submerged rks 5M S of C de la Hague, and Les Huquets de Vauville (dry) close SE of them.

CHANNEL ISLANDS − GENERAL (10.17.9 and chart 2669)

For detailed instructions, refer to sailing directions. In the following paragraphs brief mention is made only of some of the most important navigational features.

Approaching the islands from N, note the traffic scheme off the Casquets − see Fig. 2(1); soundings of Hurd Deep can help navigation. Coming from the W, Guernsey slopes down from S to N, and Jersey from N to S. Individually and collectively the islands are fringed by many rky dangers: there is considerable range of tide (most in Jersey, and least in Alderney), and the streams run very hard through the chans and around headlands; strong W winds cause a heavy sea, usually worst from local HW −0300 to local HW +0300. In bad vis it is prudent to stay in harb.

Tidal streams are rotatory anti-clockwise, particularly in open water and in wider chans. The E-going (flood) stream is of less duration than the W-going, but is stronger. The islands lie across the main direction of the streams, so eddies are common along the shores.

Apart from the main harbs described in 10.17.10 − 10.17.15, there are many delightful anchs in offshore winds. For St Aubin and St Sampson see 10.17.25.

ALDERNEY AND THE CASQUETS (chart 60)

Casquets Lt Ho (Lt, fog sig, RC) is on the largest islet of this group of rocks 5½M W of Alderney (10.17.15). Off-lying dangers extend 4 ca W and WSW (The Ledge and Noire Roque) and 4 ca E (Pte Colotte). The tide runs very hard round and between these various obstructions. A shallow bank, on which are situated Fourquie and l'Equet rks (dry), lies from ½M to 1M E of Casquets, and should not be approached.

3½M E of Casquets is Ortac rk (24m). Ortac Chan runs N/S ½M W of Ortac: here the stream begins to run NE at HW St Helier −0230, and SW at HW St Helier +0355, with sp rates 7 kn. Ortac Chan should not be used in bad weather, when there are very heavy overfalls in this area.

In the approaches to Casquets, overfalls may also be met over Eight-fathom Ledge (8½ ca W of Casquets), Casquets SW bank, Casquets SSW bank, and Casquets SSE bank.

The Swinge (chart 60) lies between Burhou and its bordering rks, and the NW coast of Alderney. It can be a dangerous chan, and should only be used in reasonable vis and in fair weather. The tide runs very hard, and in strong or gale force winds from S or W there are very heavy overfalls on the SW-going stream between Ortac and Les Etacs (off W end of Alderney). In strong E winds, on the NE-going stream, overfalls occur between Burhou and Braye breakwater. These overfalls can mostly be avoided by choosing the best time and route (see below), but due to the uneven bottom and strong tides broken water may be met even in calm conditions.

AREA 17—Channel Islands and French Coast

The NE-going stream begins at HW St Helier −0245, and the SW stream at HW St Helier +0340, sp rates 7 or 8 kn. On the NE-going stream, beware the very strong northerly component of the set in vicinity of Ortac.

On N side of the Swinge the main dangers are Boues des Kaines, almost awash at LW about ¾M ESE of Ortac, and North Rk ¼M SE of Burhou. On S side of the Swinge beware Pierre au Vraic (dries) almost in the fairway 1¾M S of Ortac, Barsier Rk (dries) 3½ ca NNW of Clonque Rk, and Corbet Rk with outliers ½M N of Clonque Rk.

Heading NE, avoid the worst overfalls in W-going tide by keeping near SE side of chan. Great Nannel on with E end of Burhou clears Pierre au Vraic to the E, but passes close W of Les Etacs. When on this transit Roque Tourgis fort is abeam, alter slightly to stbd to pass 1 ca W of Corbet Rk.

The best time to pass SW through the Swinge is when the NE-going stream slackens at about HW St Helier +0400, when the reverse route to that described above should be followed. But after HW St Helier +0500 keep close to Burhou and Ortac, avoiding North Rk and Boues des Kaines, to clear the worst of the overfalls.

Alderney Race (chart 3653) lies between Alderney and C de la Hague (Lt, fog sig), and is so called due to very strong tidal streams. In mid-chan the NNE-going stream starts at HW St Helier −0210, and the SSW stream at HW St Helier +0405, sp rates both 5½ kn. The times at which stream turns do not vary much for various places, but the rates do. For example, 1M W of C de la Hague the sp rates are 7 or 8 kn. In strong winds against the tide the sea breaks almost everywhere, and there are heavy overfalls over submerged rks and banks as shown on chart. Going S through the race, the best time is HW St Helier +0500, before the overfalls get established.

APPROACHES TO GUERNSEY (chart 3654)

The Little Russel between Guernsey and Herm (chart 808) gives the most convenient access to/from N for St Peter Port (10.17.14), and also leads to Beaucette Marina (10.17.13). But it needs care, due to rks which fringe the chan and approaches, and the strong tide which also sets across the entrance. The Big Russel is wider and easier. Doyle Passage, which runs on a line 146°/326° off Doyle Point, can be used with local knowledge by day.

In mid chan, S of Platte and NW of Brehon, the NE-going stream begins at HW St Helier −0250, and the SW stream at HW St Helier +0325, sp rates both 5¼ kn which can cause a very steep sea with wind against tide.

With Lts on Platte Fougère, Tautenay, Roustel, Platte and Bréhon, plus the Ldg Lts (220°) for St Peter Port, the Little Russel can be navigated day or night in reasonable vis, even at LW.

In bad weather or in poor vis it is better to approach St Peter Port from the S, round St Martin's Pt, and this is the natural route when coming from S or W. In onshore winds keep well clear of the W coast of Guernsey, where the sea breaks on dangers up to 4M from land in bad weather.

JERSEY (charts 3655, 1136, 1137, 1138)

To N and NE of Jersey, Les Pierres de Lecq, Les Dirouilles and Les Ecrehou are groups of drying rks, 2-4M offshore. Coming from N, a convenient landfall is Desormes W cardinal Lt Buoy, 4M NNW of Grosnez Pt. Keep ¾M clear of La Corbière Lt, SE of which begins the Western Passage (buoyed) leading E past Noirmont Pt towards St Helier.

Around the SE point of Jersey runs the Violet Channel (chart 1138). Although marked it is not easy, and should be avoided in bad weather or poor vis. There are dangers to seaward of this chan.

CAP DE LA HAGUE TO POINTE DE BARFLEUR

The N coast of the Cotentin peninsula (chart 1106) runs for 26M, mostly bordered by rks which reach 1M offshore between C de la Hague and Pte de Jardeheu, and 2½M offshore from C Lévi to Pte de Barfleur. Tidal streams run 5 kn at sp, and raise a steep sea with wind against tide. Between C de la Hague and Cherbourg (10.17.24) an eddy runs W close inshore during the last of the E-going stream.

There is anch in Anse de St Martin, about 2M E of C de la Hague, exposed to N, but useful while waiting tide for Alderney Race, as alternative to Omonville (10.17.23).

E of Cherbourg Port de Becquet and Port Lévi are two small drying harbs. Off C Lévi a race develops with wind against tide, and extends nearly 2M to N. Pte de Barfleur has dangers up to 2M offshore, and a race in which the sea breaks heavily, extending 3-4M NE and E from Lt Ho: in bad weather, particularly with winds from NW or SE against the tide, it is necessary to keep at least 6M to seaward to avoid the worst effects.

10.17.6 DISTANCE TABLE

Approximate distances in nautical miles are by the most direct route while avoiding dangers and allowing for traffic separation schemes etc. Places in *italics* are in adjoining areas.

1	*Le Four*	**1**																		
2	*Isle de Batz*	37	**2**																	
3	*Roches Douvres*	89	52	**3**																
4	*Portrieux*	100	63	28	**4**															
5	*St Malo*	122	83	42	35	**5**														
6	Granville	134	97	51	54	23	**6**													
7	Carteret	132	95	44	64	50	38	**7**												
8	St Helier	119	82	30	44	39	30	26	**8**											
9	St Peter Port	114	74	25	51	54	55	31	29	**9**										
10	Les Hanois	105	67	20	48	56	58	37	32	10	**10**									
11	Casquets	120	90	43	71	70	63	32	43	18	23	**11**								
12	Braye, Alderney	133	96	46	73	73	66	28	46	23	29	8	**12**							
13	Cap de la Hague	139	102	50	74	73	61	23	45	28	35	17	9	**13**						
14	Cherbourg	153	116	64	88	87	75	37	59	42	49	31	23	14	**14**					
15	*Le Havre*	223	186	134	158	157	145	107	129	112	119	101	93	84	70	**15**				
16	*Cap Gris Nez*	297	260	208	232	231	219	181	203	186	193	175	167	158	147	108	**16**			
17	*Start Point*	113	90	76	103	117	120	89	93	70	61	57	67	77	91	156	210	**17**		
18	*Portland Bill*	153	123	87	115	118	111	75	91	66	67	48	49	52	62	118	161	49	**18**	
19	*Needles*	178	149	108	138	131	119	81	103	82	86	64	62	59	61	98	126	84	35	**19**
20	*Royal Sovereign*	248	212	154	184	183	171	133	155	138	145	124	119	110	102	82	47	163	114	82	**20**

THE CHANNEL ISLANDS
10-17-9

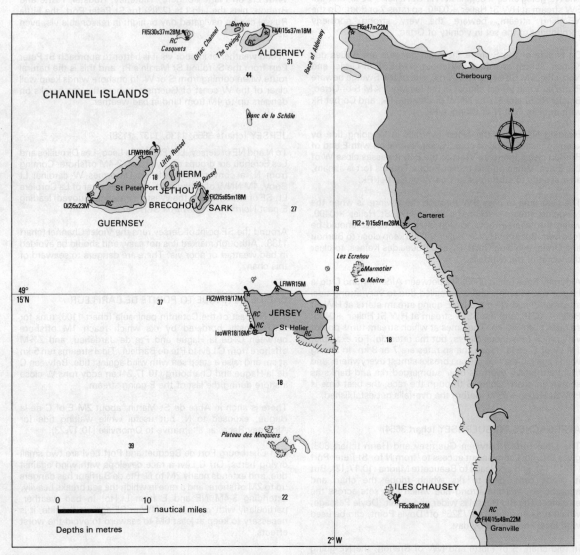

The Channel Islands (Jersey, Guernsey, Alderney, Sark and other small islands) lie, not in the Channel, but in the Bay of St. Malo. Alderney is part of the Bailiwick of Guernsey and the States of Alderney have seats in the States of Guernsey. Brecqhou belongs to Sark which in turn with Herm and Jethou are part of Guernsey. Although the eastern ends of Jersey (15 miles) and Alderney (8½ miles) are close to France, the Channel Islands have never been French. They were part of Normandy and became British with William of Normandy, William the Conqueror in 1066. The French call them the Iles Anglo-Normandes (Jersey, Guernsey, Aurigny and Seroq).

The Islands are British but are not part of the United Kingdom. They are self governing and have their own customs regulations and laws. No special documentation is needed for British yachts entering Channel Island ports but they will be subject to customs formalities on return to U.K. Yachts going to France need the normal documentation (passports etc.) and British yachts returning to the Channel Islands from France, must, like all French yachts, wear the Q flag. (It is advisable to do so when arriving from U.K., but not mandatory, except in Alderney). All Channel Islands have a reciprocal medical arrangement with U.K.

Customs and Immigration clearance formalities are carried out at St. Helier and Gorey (Jersey), St Peter Port, St Sampson and Beaucette (Guernsey) and Braye (Alderney). Unlike U.K., there are no coastguards but lifeboats are stationed at St Helier, St Catherines (Jersey), St Peter Port (Guernsey) and Braye (Alderney).

The main problems around the Channel Islands are steep seas which get up very quickly, overfalls, fog and thick weather and the very large tidal range.

GOREY 10-17-10
Jersey (Channel Islands)

CHARTS
Admiralty 3655, 1138; SHOM 6939; ECM 1014; Stanford 16; Imray C33A

TIDES
−0454 Dover; ML 6.1; Duration 0545; Zone 0 (GMT)

Standard Port ST HELIER (→)

Times				Height (metres)			
HW		LW		MHWS	MHWN	MLWN	MLWS
0300	0900	0200	0900	11.1	8.1	4.1	1.3
1500	2100	1400	2100				

Differences ST. CATHERINE'S BAY
0000 +0010 +0010 +0010 0.0 −0.1 0.0 +0.1

SHELTER
Good shelter in the harbour which dries out completely. Access HW∓3. Good anchorage in Gorey Roads and in St Catherine's Bay to the N of Gorey, except in S to SE winds; deep water.

NAVIGATION
Waypoint 49°10'.50N 01°57'.33W, 118°/298° from/to front Ldg Lt 298°, 2.9M. Beware the very large tidal range. On approaching harbour, keep well outside all local beacons until the leading marks are picked up. Beware Banc du Chateau 1M offshore to N of Ldg Line and Azicot Rk (dries 2.2m) just S of Ldg Line 2 ca from ent.

LIGHTS AND MARKS
(1) Ldg Lts 298°. Front Gorey pier Oc RG 5s 8m 12M; W framework Tr; vis R304°-352°, G352°-304°. Rear, 490m from front, OcR 5s 24m 8M; W square with R border.
(2) Gorey pier Lt on with church spire 304° leads over Road Rk (3.3m), and Azicot Rk.
(3) Gorey pier Lt on with Fort William 250° leads close to Pacquet Rk (0.6m).

RADIO TELEPHONE
Gorey Harbour Ch 74 (HW −3 to HW +3 summer only).

TELEPHONE (0534)
Hr Mr 53616; Customs 32302; Marinecall 0898 500 457; Dr Contact Hr Mr.

FACILITIES
EC Thursday; **Jetty** FW; **Gorey Yacht Services** Tel. 53958, M, P, D, ME, El, Sh, Gas, CH; **Pier** Tel. 53616, M, P, D, L, FW, C (7 ton), AB.
Town P, CH, V, R, Bar. PO; Bank; Rly (ferry to Weymouth, St Malo or Granville); Air (Jersey Airport).
Ferry UK — St Helier—Portsmouth/Weymouth.

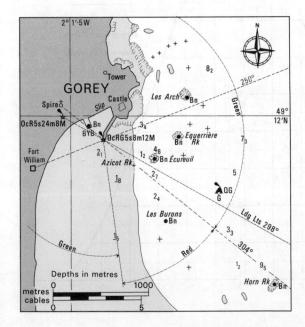

AREA 17 — Channel Islands and French Coast

ST HELIER 10-17-11
Jersey (Channel Islands)

CHARTS
Admiralty 3655, 3278, 1137, 1136, 1138; SHOM 6938, 6937; ECM 1014; Stanford 16; Imray C33B

TIDES
−0455 Dover; ML 6.5; Duration 0545; Zone 0 (GMT)

Standard Port ST HELIER (→)

Times				Height (metres)			
HW		LW		MHWS	MHWN	MLWN	MLWS
0300	0900	0200	0900	11.1	8.1	4.1	1.3
1500	2100	1400	2100				

Differences LES MINQUIERS
+0007 0000 −0008 +0013 +0.5 +0.8 −0.1 +0.1
LES ECREHOU
+0004 +0012 +0010 +0020 −0.2 +0.3 −0.3 0.0

NOTE: St Helier is a Standard Port and the tidal predictions for each day of the year are shown below.

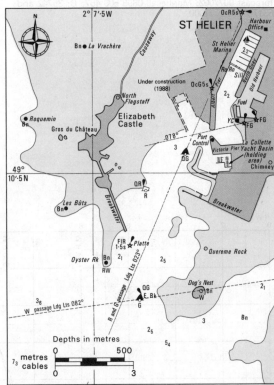

SHELTER
Excellent shelter in marina and good in La Collette. For St Aubin's Bay see 10.17.25. Anchoring prohibited in St Helier Roads due to shipping.

NAVIGATION
Waypoint 49°10'.01N 02°07'.30W, 203°/023° from/to front Ldg Lt 023°, 0.87M. Note the very large tidal range, and many offlying rks. Four main channels are Western Passage, Danger Rk Passage, Red and Green Passage, and South Passage. In Western Passage beware race off Noirmont Pt. Entering harbour note Oyster Rks (RW Bn) to West of Red and Green Passage, and Dog's Nest Rks (W Bn with globe topmark) to E. Speed limit 10 kn N of Platte Rk, and 5 kn N of La Collette. Ldg Lts difficult to see against town lights.

LIGHTS AND MARKS
Western Passage Ldg Lts and Dog's Nest Bn 082° lead N of Les Fours and Ruaudière Lt Buoys, to position close to East Rk Lt Buoy, where course is altered to 023° on the Red and Green Passage Ldg Line (Front Oc G 5s. Rear Oc R 5s, synchronised). Port-hand buoy beyond Platte Rk (Fl R 1.5s) marks channel. Main harbour Ldg Lts 078°, both

ST. HELIER continued

FG on W columns. Bell at Victoria Pierhead is rung in answer to vessel's fog signals.

Entry Signals (Port Control Stn and Marina)
FG or Fl G Lt at Port Control Stn. — Vessels may enter but not leave
FR or Fl R Lt at Port Control Stn. — Vessels may leave but not enter
R and G Lts together at Port Control Stn. — No vessel may enter or leave
QY Lts on NW and NE of Port Control Stn indicate power driven craft of under 25m length may enter or leave against the signals displayed (keeping to stbd at entrance). Marina sill fixed (CD+5m) and hinged flap which lifts 1.4m over sill to maintain 5.0m.

RADIO TELEPHONE
Call: *St Helier Port Control* VHF Ch 14 (H24). Messages can be routed through Jersey Radio, the Coast Radio Station, VHF Ch 16; 25 82 (H24).

TELEPHONE (0534)
Hr Mr 34451; Marina Office 79549; Customs 30232 and 73561; Duty Forecaster 0077007; Recorded weather forecast 0077002; Marinecall 0898 500 457; Dr 35742 and 53178; Hosp 71000.

FACILITIES
EC Thursday; **St Helier Marina** (180+200 visitors), Tel. 79549, Grid, CH, ME, El, Sh, FW, V, AC, Gas, Gaz, Kos, (Visitors area — fingers E, F and G. Vessels over 12m use N side of A finger); Access HW∓3; **La Collette Yacht Basin** Tel. 74434 (holding area when marina is inaccessible), FW, AC, V, R; **St Helier Yacht Club** Tel. 32229, L, FW, R, Bar; **Boat Yard** Tel. 31907, M, P, D, FW, ME, El, Sh, C (15 ton), CH, AB; **South Pier** Slip, M, P, D, L, FW, C (1 ton & 2 ton), AB; **Channel Islands Yacht Services** Tel. 71511, CH, Gas; **Battrick BY** Tel. 43412, CH; **Bernard Amy** Tel. 42071, CH; **Silva Yates** Tel. 36980, CH; **D.K. Collins** Tel. 32415, ME; **South Pier Shipyard** Tel. 31907, ME, El, Sh, CH, ACA, Gas; **Raffray** Tel. 23151, ME; **IPCO Marine** Tel. 44651, ME, El, Sh. Tel. 77044, CH, ACA; **Fox Marine Services** Tel. 21312, ME; **Jersey Marine Electronics** Tel. 21603, El. **Town** P, D, CH, V, R, Bar. PO; Bank; Rly (ferry to Portsmouth, Weymouth, St Malo, Granville), Air (Jersey Airport).
Ferry UK (Weymouth and Portsmouth daily; Torbay twice weekly in summer, weekly in winter).
Hydrofoil to Weymouth, Guernsey and St Malo.

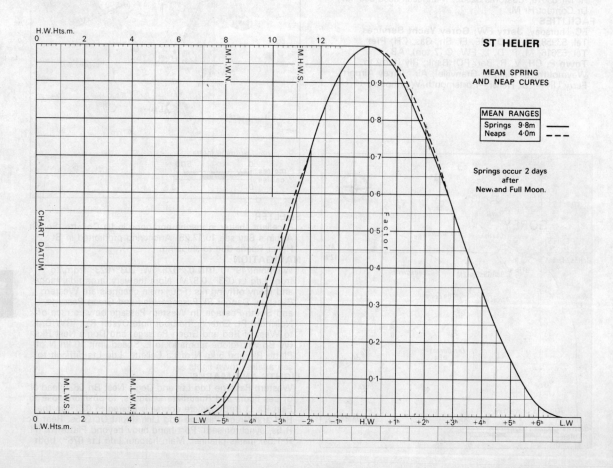

ST HELIER
MEAN SPRING AND NEAP CURVES

MEAN RANGES
Springs 9·8m ———
Neaps 4·0m - - - -

Springs occur 2 days after New and Full Moon.

AREA 17 — Channel Islands and French Coast   673

CHANNEL ISLANDS — ST. HELIER

Lat 49°11′ N Long 2°07′ W

TIMES AND HEIGHTS OF HIGH AND LOW WATERS

YEAR 1989

TIME ZONE UT(GMT)
For Summer Time add ONE hour in non-shaded areas

	JANUARY			FEBRUARY			MARCH			APRIL	
	Time m	Time m		Time m	Time m		Time m	Time m		Time m	Time m
1 Su	0619 4.0 / 1221 8.2 / 1853 4.1	**16** M	0028 8.7 / 0719 3.7 / 1306 8.6 / 1958 3.6	**1** W	0050 7.8 / 0735 4.4 / 1330 7.6 / 2015 4.5	**16** Th	0301 7.8 / 0948 4.1 / 1555 8.0 / 2224 4.0	**1** W	0519 4.1 / 1105 8.0 / 1737 4.4 / 2342 7.8	**16** Th	0035 7.5 / 0740 4.7 / 1406 7.3 / 2036 4.9	**1** Sa	0120 7.7 / 0758 4.4 / 1420 7.8 / 2049 4.3	**16** Su	0318 8.2 / 1009 3.8 / 1555 8.4 / 2224 3.7
2 M	0102 8.1 / 0731 4.2 / 1328 8.1 / 2006 4.1	**17** Tu	0149 8.4 / 0840 3.8 / 1433 8.4 / 2118 3.6	**2** Th	0230 7.8 / 0904 4.3 / 1510 7.8 / 2142 4.1	**17** F	0419 8.4 / 1104 3.5 / 1657 8.6 / 2326 3.4	**2** Th	0635 4.6 / 1228 7.4 / 1919 4.8	**17** F	0247 7.6 / 0941 4.4 / 1543 7.8 / 2213 4.3	**2** Su	0258 8.4 / 0931 3.6 / 1535 8.8 / 2206 3.4	**17** M	0406 8.9 / 1052 3.1 / 1635 9.1 / 2305 3.0
3 Tu	0213 8.1 / 0844 4.0 / 1440 8.1 / 2115 3.9	**18** W	0312 8.5 / 0956 3.5 / 1555 8.6 / 2230 3.4	**3** F	0350 8.4 / 1020 3.7 / 1623 8.5 / 2251 3.5	**18** Sa	0511 9.1 / 1157 2.8 / 1742 9.3	**3** F	0147 7.6 / 0826 4.5 / 1446 7.6 / 2114 4.4	**18** Sa	0400 8.2 / 1049 3.6 / 1637 8.5 / 2306 3.6	**3** M	0359 9.5 / 1035 2.6 / 1628 9.9 / 2302 2.3	**18** Tu	0447 9.5 / 1129 2.5 / 1711 9.7 / 2340 2.4
4 W	0318 8.4 / 0948 3.7 / 1543 8.4 / 2216 3.6	**19** Th	0423 8.9 / 1105 3.1 / 1659 9.1 / 2332 3.0	**4** Sa	0449 9.2 / 1123 2.9 / 1718 9.3 / 2349 2.7	**19** Su	0012 2.8 / 0553 9.7 / 1241 2.2 / 1819 9.8	**4** Sa	0328 8.2 / 0957 3.8 / 1603 8.5 / 2233 3.5	**19** Su	0448 9.0 / 1134 2.9 / 1716 9.2 / 2347 2.9	**4** Tu	0449 10.5 / 1129 1.5 / 1715 10.9 / 2353 1.4	**19** W	0523 10.0 / 1203 2.1 / 1744 10.1
5 Th	0416 8.9 / 1045 3.3 / 1640 8.9 / 2312 3.1	**20** F	0519 9.4 / 1204 2.6 / 1753 9.5	**5** Su	0539 10.1 / 1218 2.1 / 1804 10.2	**20** M	0052 2.3 / 0629 10.2 / 1316 1.8 / 1853 10.2	**5** Su	0428 9.2 / 1104 2.8 / 1657 9.6 / 2330 2.5	**20** M	0526 9.7 / 1211 2.3 / 1750 9.8	**5** W	0534 11.4 / 1218 0.8 / 1758 11.6	**20** Th	0015 2.0 / 0558 10.3 / 1236 1.8 / 1817 10.4
6 F	0506 9.4 / 1140 2.8 / 1730 9.4	**21** Sa	0024 2.7 / 0607 9.8 / 1253 2.2 / 1836 9.8	**6** M	0041 1.9 / 0624 10.9 / 1307 1.3 / 1848 11.0	**21** Tu	0126 1.9 / 0703 10.6 / 1348 1.5 / 1924 10.4	**6** M	0518 10.3 / 1157 1.8 / 1743 10.6	**21** Tu	0021 2.3 / 0601 10.2 / 1245 1.8 / 1822 10.2	**6** Th	0039 0.7 / 0619 11.9 / 1304 0.3 / 1842 12.0	**21** F	0049 1.7 / 0631 10.4 / 1309 1.7 / 1848 10.5
7 Sa	0004 2.7 / 0553 10.0 / 1232 2.2 / 1818 10.0	**22** Su	0109 2.4 / 0648 10.2 / 1335 1.9 / 1916 10.0	**7** Tu	0127 1.2 / 0706 11.5 / 1352 0.7 / 1930 11.5	**22** W	0155 1.7 / 0734 10.7 / 1416 1.5 / 1952 10.5	**7** Tu	0019 1.6 / 0601 11.2 / 1245 0.9 / 1825 11.4	**22** W	0053 1.8 / 0632 10.6 / 1314 1.5 / 1852 10.5	**7** F	0126 0.4 / 0702 12.1 / 1348 0.2 / 1923 12.0	**22** Sa	0123 1.7 / 0702 10.4 / 1341 1.8 / 1917 10.5
8 Su	0055 2.2 / 0638 10.5 / 1321 1.7 / 1903 10.5	**23** M	0148 2.2 / 0726 10.4 / 1412 1.8 / 1949 10.1	**8** W	0211 0.8 / 0747 11.9 / 1434 0.4 / 2009 11.7	**23** Th	0223 1.6 / 0802 10.8 / 1442 1.5 / 2019 10.5	**8** W	0106 0.8 / 0645 11.9 / 1330 0.3 / 1906 12.0	**23** Th	0124 1.5 / 0703 10.7 / 1344 1.4 / 1920 10.7	**8** Sa	0209 0.4 / 0744 11.9 / 1430 0.5 / 2002 11.6	**23** Su	0157 1.8 / 0731 10.3 / 1412 2.0 / 1947 10.4
9 M	0142 1.8 / 0721 10.9 / 1408 1.3 / 1945 10.8	**24** Tu	0220 2.1 / 0758 10.5 / 1443 1.8 / 2020 10.2	**9** Th	0251 0.7 / 0826 11.9 / 1514 0.5 / 2047 11.5	**24** F	0249 1.7 / 0829 10.6 / 1505 1.7 / 2043 10.4	**9** Th	0149 0.2 / 0726 12.2 / 1412 0.0 / 1947 12.1	**24** F	0154 1.5 / 0731 10.8 / 1412 1.5 / 1947 10.7	**9** Su	0250 0.8 / 0823 11.3 / 1510 1.2 / 2040 10.9	**24** M	0229 2.0 / 0801 10.1 / 1442 2.4 / 2015 10.1
10 Tu	0226 1.5 / 0802 11.2 / 1450 1.2 / 2026 11.0	**25** W	0249 2.1 / 0829 10.5 / 1510 1.9 / 2049 10.1	**10** F	0331 1.0 / 0905 11.5 / 1552 0.9 / 2125 11.1	**25** Sa	0314 1.9 / 0854 10.3 / 1528 2.1 / 2108 10.1	**10** F	0230 0.3 / 0805 12.2 / 1453 0.2 / 2025 11.8	**25** Sa	0222 1.7 / 0758 10.6 / 1437 1.7 / 2012 10.5	**10** M	0328 1.6 / 0903 10.5 / 1546 2.1 / 2117 10.1	**25** Tu	0258 2.4 / 0830 9.7 / 1510 2.8 / 2044 9.7
11 W	0307 1.5 / 0843 11.2 / 1531 1.2 / 2107 10.9	**26** Th	0315 2.2 / 0858 10.3 / 1534 2.1 / 2117 9.9	**11** Sa	0407 1.5 / 0945 10.9 / 1630 1.7 / 2204 10.3	**26** Su	0339 2.3 / 0919 9.9 / 1550 2.6 / 2134 9.6	**11** Sa	0310 0.7 / 0843 11.7 / 1529 0.8 / 2101 11.2	**26** Su	0249 1.8 / 0823 10.3 / 1501 2.1 / 2037 10.2	**11** Tu	0406 2.4 / 0942 9.5 / 1623 3.1 / 2156 9.1	**26** W	0329 2.9 / 0903 9.3 / 1538 3.3 / 2119 9.3
12 Th	0348 1.7 / 0925 10.9 / 1612 1.5 / 2148 10.5	**27** F	0341 2.4 / 0927 10.0 / 1559 2.4 / 2145 9.6	**12** Su	0447 2.3 / 1026 10.0 / 1711 2.6 / 2248 9.4	**27** M	0404 2.9 / 0948 9.3 / 1614 3.2 / 2203 9.1	**12** Su	0346 1.4 / 0921 10.9 / 1606 1.7 / 2138 10.3	**27** M	0315 2.3 / 0850 9.9 / 1524 2.6 / 2103 9.8	**12** W	0445 3.4 / 1028 8.5 / 1705 4.0 / 2247 8.2	**27** Th	0404 3.4 / 0942 8.8 / 1614 3.8 / 2204 8.7
13 F	0428 2.1 / 1007 10.5 / 1654 2.1 / 2233 10.0	**28** Sa	0409 2.7 / 0956 9.6 / 1626 2.9 / 2214 9.2	**13** M	0532 3.2 / 1115 9.0 / 1801 3.5 / 2342 8.5	**28** Tu	0435 3.5 / 1020 8.7 / 1645 3.8 / 2241 8.4	**13** M	0423 2.3 / 1000 9.9 / 1644 2.8 / 2217 9.3	**28** Tu	0341 2.6 / 0917 9.4 / 1548 3.2 / 2132 9.2	**13** Th	0539 4.3 / 1137 7.6 / 1807 4.7 / 2353 7.5	**28** F	0449 3.8 / 1037 8.3 / 1709 4.2 / 2313 8.3
14 Sa	0513 2.7 / 1057 9.8 / 1742 2.7 / 2325 9.3	**29** Su	0440 3.2 / 1028 9.1 / 1657 3.4 / 2251 8.7	**14** Tu	0635 3.9 / 1224 8.1 / 1916 4.2			**14** Tu	0504 3.2 / 1044 8.7 / 1727 3.8 / 2306 8.3	**29** W	0410 3.4 / 0950 8.8 / 1619 3.8 / 2210 8.6	**14** F	0012 7.6 / 0709 4.5 / 1335 7.3 / 1957 5.0	**29** Sa	0558 4.1 / 1204 8.0 / 1836 4.4
15 Su	0607 3.3 / 1153 9.2 / 1842 3.3	**30** M	0519 3.7 / 1109 8.5 / 1737 3.9 / 2337 8.2	**15** W	0109 7.8 / 0808 4.3 / 1416 7.7 / 2056 4.4			**15** W	0600 4.1 / 1151 7.7 / 1838 4.7	**30** Th	0452 4.0 / 1037 8.1 / 1711 4.4 / 2312 7.9	**15** Sa	0206 7.6 / 0903 4.4 / 1501 7.8 / 2132 4.5	**30** Su	0055 8.2 / 0733 4.0 / 1345 8.3 / 2018 4.0
		31 Tu	0615 4.2 / 1204 7.9 / 1843 4.4						**31** F	0605 4.5 / 1207 7.5 / 1849 4.8				

Chart Datum: 5.88 metres below Ordnance Datum (Local)

CHANNEL ISLANDS — ST. HELIER

Lat 49°11' N Long 2°07' W

TIMES AND HEIGHTS OF HIGH AND LOW WATERS

YEAR 1989

TIME ZONE UT(GMT)
For Summer Time add ONE hour in non-shaded areas

MAY

	Time	m		Time	m
1 M	0220 0857 1458 2132	8.8 3.3 9.1 3.2	**16** Tu	0314 0955 1545 2212	8.7 3.4 8.8 3.3
2 Tu	0324 1002 1555 2231	9.6 2.5 10.0 2.3	**17** W	0400 1038 1627 2254	9.1 3.0 9.3 2.8
3 W	0417 1058 1645 2323	10.5 1.7 10.8 1.5	**18** Th	0442 1118 1705 2334	9.5 2.6 9.7 2.4
4 Th	0508 1149 1732	11.1 1.1 11.3	**19** F	0522 1157 1743	9.7 2.3 10.0
5 F ●	0014 0556 1239 1818	1.0 11.5 0.8 11.5	**20** Sa ○	0015 0600 1236 1818	2.2 9.9 2.2 10.0
6 Sa	0102 0642 1327 1902	0.8 11.5 0.8 11.4	**21** Su	0056 0636 1314 1853	2.1 9.9 2.2 10.2
7 Su	0149 0727 1412 1944	0.9 11.2 1.1 11.1	**22** M	0135 0712 1352 1928	2.1 9.9 2.3 10.2
8 M	0234 0809 1453 2025	1.3 10.7 1.7 10.5	**23** Tu	0213 0747 1429 2004	2.2 9.8 2.5 10.1
9 Tu	0315 0851 1532 2104	1.9 10.0 2.4 9.9	**24** W	0251 0822 1504 2039	2.4 9.7 2.7 9.8
10 W	0355 0934 1609 2146	2.5 9.3 3.2 9.2	**25** Th	0328 0901 1541 2119	2.6 9.5 3.1 9.6
11 Th	0434 1020 1648 2235	3.3 8.5 3.9 8.5	**26** F	0406 0945 1620 2207	2.9 9.2 3.4 9.2
12 F ☽	0520 1120 1739 2343	3.9 7.9 4.4 8.0	**27** Sa	0452 1038 1712 2308	3.2 8.8 3.7 8.9
13 Sa	0625 1241 1855	4.3 7.7 4.6	**28** Su ☾	0550 1147 1821	3.4 8.6 3.8
14 Su	0106 0749 1358 2020	7.9 4.3 7.9 4.4	**29** M	0022 0703 1306 1942	8.8 3.1 8.8 3.6
15 M	0218 0903 1457 2122	8.2 3.9 8.3 3.9	**30** Tu	0140 0820 1418 2056	9.1 3.1 9.2 3.1
			31 W	0247 0928 1521 2159	9.5 2.6 9.7 2.5

JUNE

	Time	m		Time	m
1 Th	0348 1027 1617 2255	10.0 2.1 10.2 2.0	**16** F	0400 1031 1628 2255	8.8 3.2 9.2 3.0
2 F	0444 1123 1709 2351	10.4 1.7 10.6 1.6	**17** Sa	0449 1120 1713 2344	9.1 2.9 9.5 2.7
3 Sa ●	0537 1218 1800	10.7 1.6 10.8	**18** Su	0534 1207 1756	9.4 2.7 9.8
4 Su	0046 0628 1310 1848	1.5 10.7 1.6 10.8	**19** M ○	0032 0618 1253 1836	2.4 9.6 2.5 10.0
5 M	0137 0717 1358 1933	1.5 10.5 1.8 10.6	**20** Tu	0120 0659 1338 1917	2.2 9.8 2.3 10.2
6 Tu	0225 0802 1442 2016	1.6 10.2 2.1 10.3	**21** W	0204 0738 1420 1957	2.0 10.0 2.2 10.3
7 W	0308 0844 1521 2056	2.0 9.8 2.5 9.9	**22** Th	0244 0818 1500 2036	2.0 10.1 2.3 10.3
8 Th	0346 0925 1556 2135	2.4 9.4 3.0 9.5	**23** F	0324 0858 1538 2115	2.0 10.1 2.4 10.2
9 F	0420 1004 1628 2217	2.9 8.9 3.4 9.1	**24** Sa	0402 0939 1617 2200	2.2 9.9 2.7 10.0
10 Sa	0457 1049 1708 2305	3.5 8.5 3.8 8.7	**25** Su	0444 1026 1702 2249	2.6 9.6 3.0 9.7
11 Su	0539 1142 1757	3.7 8.2 4.0	**26** M ☾	0532 1133 1757 2349	2.8 9.3 3.2 9.3
12 M	0001 0634 1243 1902	8.4 3.9 8.1 4.1	**27** Tu	0632 1224 1904	3.0 9.0 3.4
13 Tu	0106 0741 1348 2011	8.3 4.0 8.2 4.0	**28** W	0057 0742 1337 2020	9.1 3.1 9.0 3.3
14 W	0211 0846 1447 2111	8.4 3.8 8.5 3.6	**29** Th	0212 0856 1450 2131	9.1 3.0 9.1 3.0
15 Th	0308 0941 1541 2206	8.6 3.5 8.8 3.3	**30** F	0324 1004 1556 2237	9.3 2.8 9.5 2.6

JULY

	Time	m		Time	m
1 Sa	0430 1108 1657 2340	9.6 2.5 9.8 2.3	**16** Su	0424 1049 1651 2320	8.5 3.4 9.1 3.1
2 Su	0530 1207 1751	9.9 2.3 10.1	**17** M	0516 1146 1739	9.0 3.0 9.6
3 M ●	0038 0624 1300 1841	2.0 10.1 2.1 10.3	**18** Tu ○	0014 0603 1236 1822	2.6 9.5 2.5 10.1
4 Tu	0131 0712 1349 1926	1.8 10.2 2.1 10.4	**19** W	0103 0646 1323 1903	2.0 10.1 2.0 10.6
5 W	0216 0754 1430 2005	1.7 10.1 2.1 10.3	**20** Th	0149 0726 1406 1944	1.6 10.5 1.7 11.0
6 Th	0256 0832 1505 2040	1.8 10.0 2.3 10.2	**21** F	0230 0805 1447 2022	1.3 10.8 1.5 11.1
7 F	0328 0905 1535 2114	2.1 9.8 2.5 10.0	**22** Sa	0310 0843 1525 2101	1.2 10.9 1.6 11.0
8 Sa	0356 0938 1603 2148	2.4 9.5 2.8 9.7	**23** Su	0348 0922 1603 2141	1.4 10.7 1.9 10.7
9 Su	0424 1010 1633 2224	2.7 9.2 3.1 9.3	**24** M	0426 1003 1642 2224	1.8 10.3 2.3 10.2
10 M	0455 1048 1709 2305	3.0 8.8 3.4 8.9	**25** Tu ☾	0509 1048 1730 2315	2.3 9.7 2.9 9.5
11 Tu	0533 1133 1756 2354	3.5 8.3 3.7 8.4	**26** W	0600 1144 1831	3.0 9.1 3.4
12 W	0624 1231 1859	3.9 8.2 4.0	**27** Th	0018 0709 1257 1949	8.8 3.5 8.5 3.7
13 Th	0056 0731 1341 2011	8.1 4.1 8.1 4.0	**28** F	0144 0832 1429 2114	8.1 3.7 8.4 3.6
14 F	0211 0844 1453 2119	8.0 4.1 8.2 3.9	**29** Sa	0315 0952 1549 2231	8.5 3.5 8.8 3.2
15 Sa	0322 0950 1556 2223	8.1 3.8 8.6 3.5	**30** Su	0430 1102 1654 2339	8.9 3.1 9.3 2.7
			31 M	0529 1201 1746	9.4 2.7 9.8

AUGUST

	Time	m		Time	m
1 Tu ●	0034 0617 1252 1831	2.2 9.8 2.3 10.2	**16** W	0543 1217 1801	9.9 2.3 10.5
2 W	0120 0659 1334 1910	1.8 10.1 2.0 10.5	**17** Th ○	0042 0625 1303 1843	1.7 10.6 1.6 11.2
3 Th	0159 0734 1409 1944	1.6 10.3 1.9 10.6	**18** F	0127 0704 1345 1923	1.0 11.2 1.1 11.7
4 F	0233 0806 1440 2016	1.6 10.3 1.9 10.6	**19** Sa	0209 0742 1426 2001	0.6 11.6 0.9 11.8
5 Sa	0301 0836 1507 2046	1.7 10.2 2.0 10.4	**20** Su	0249 0820 1505 2039	0.6 11.6 1.0 11.6
6 Su	0325 0904 1532 2114	1.9 10.0 2.2 10.2	**21** M	0327 0858 1542 2118	0.9 11.3 1.4 11.1
7 M	0349 0931 1557 2143	2.3 9.8 2.5 9.7	**22** Tu	0404 0936 1621 2157	1.5 10.7 2.0 10.4
8 Tu	0414 1000 1627 2214	2.7 9.4 3.0 9.2	**23** W ☾	0444 1019 1705 2244	2.3 9.8 2.8 9.4
9 W	0444 1034 1705 2251	3.2 8.9 3.5 8.6	**24** Th	0532 1109 1801 2346	3.2 8.9 3.7 8.4
10 Th	0522 1118 1756 2342	3.8 8.3 4.0 8.0	**25** F	0639 1225 1927	4.0 8.1 4.2
11 F	0619 1222 1909	4.3 7.8 4.4	**26** Sa	0131 0816 1423 2111	7.8 4.4 7.9 4.1
12 Sa	0103 0745 1406 2037	7.5 4.6 7.7 4.4	**27** Su	0322 0952 1549 2234	8.0 4.1 8.4 3.5
13 Su	0250 0915 1532 2156	7.6 4.3 8.2 3.9	**28** M	0431 1101 1647 2333	8.7 3.5 9.1 2.8
14 M	0404 1027 1631 2301	8.2 3.8 8.9 3.2	**29** Tu	0519 1151 1732	9.4 2.8 9.8
15 Tu	0458 1126 1719 2354	9.0 3.0 9.7 2.4	**30** W	0019 0558 1232 1810	2.2 9.9 2.3 10.3
			31 Th ●	0057 0634 1309 1845	1.8 10.3 1.9 10.6

Chart Datum: 5.88 metres below Ordnance Datum (Local)

AREA 17 — Channel Islands and French Coast

CHANNEL ISLANDS — ST. HELIER

Lat 49°11′ N   Long 2°07′ W

TIMES AND HEIGHTS OF HIGH AND LOW WATERS   YEAR 1989

TIME ZONE UT(GMT)
For Summer Time add ONE hour in non-shaded areas

SEPTEMBER		OCTOBER		NOVEMBER		DECEMBER	
Time m	Time m	Time m	Time m	Time m	Time m	Time m	Time m
1 0131 1·5 / 0706 10·5 / F 1340 1·7 / 1916 10·8	**16** 0100 0·7 / 0638 11·8 / Sa 1320 0·7 / 1857 12·1	**1** 0126 1·6 / 0702 10·7 / Su 1335 1·6 / 1914 10·8	**16** 0117 0·4 / 0655 12·1 / M 1340 0·6 / 1916 12·0	**1** 0155 2·1 / 0733 10·4 / W 1412 2·1 / 1948 10·1	**16** 0232 1·5 / 0805 11·0 / Th 1457 1·6 / 2033 10·5	**1** 0213 2·5 / 0751 10·1 / F 1436 2·3 / 2011 9·8	**16** 0307 2·1 / 0842 10·4 / Sa 1532 1·9 / 2110 9·9
2 0201 1·5 / 0735 10·6 / Sa 1409 1·6 / 1945 10·7	**17** 0142 0·3 / 0719 12·1 / Su 1402 0·5 / 1938 12·2	**2** 0154 1·6 / 0730 10·7 / M 1405 1·6 / 1942 10·6	**17** 0202 0·6 / 0735 11·8 / Tu 1423 0·8 / 1958 11·6	**2** 0225 2·4 / 0802 10·1 / Th 1444 2·4 / 2018 9·7	**17** 0315 2·1 / 0849 10·3 / F 1541 2·2 / 2118 9·8	**2** 0249 2·7 / 0825 10·0 / Sa 1511 2·6 / 2044 9·6	**17** 0345 2·6 / 0921 10·0 / Su 1610 2·5 / 2149 9·4
3 0227 1·6 / 0802 10·6 / Su 1434 1·7 / 2013 10·7	**18** 0225 0·4 / 0757 12·0 / M 1443 0·7 / 2016 11·8	**3** 0220 1·9 / 0757 10·5 / Tu 1433 1·9 / 2009 10·3	**18** 0246 1·1 / 0816 11·3 / W 1507 1·4 / 2040 10·8	**3** 0256 2·9 / 0832 9·8 / F 1515 2·9 / 2049 9·3	**18** 0357 2·7 / 0932 9·6 / Sa 1624 2·9 / 2206 9·0	**3** 0322 3·0 / 0901 9·7 / Su 1546 2·9 / 2124 9·3	**18** 0419 3·1 / 1000 9·5 / M 1644 3·0 / 2228 8·9
4 0251 1·8 / 0829 10·6 / M 1500 1·9 / 2040 10·4	**19** 0304 0·8 / 0836 11·5 / Tu 1522 1·2 / 2056 11·2	**4** 0246 2·2 / 0823 10·2 / W 1500 2·3 / 2036 9·9	**19** 0327 1·9 / 0857 10·5 / Th 1549 2·2 / 2124 9·9	**4** 0325 3·3 / 0904 9·4 / Sa 1549 3·4 / 2125 8·9	**19** 0440 3·6 / 1021 8·9 / Su 1712 3·6 / 2301 8·3	**4** 0357 3·3 / 0942 9·4 / M 1624 3·2 / 2209 9·1	**19** 0454 3·5 / 1042 9·0 / Tu 1722 3·5 / ☽ 2315 8·5
5 0315 2·1 / 0854 10·1 / Tu 1525 2·1 / 2105 9·9	**20** 0343 1·6 / 0914 10·7 / W 1602 2·0 / 2136 10·2	**5** 0311 2·7 / 0849 9·8 / Th 1528 2·9 / 2104 9·4	**20** 0407 2·8 / 0939 9·6 / F 1631 3·1 / 2212 9·0	**5** 0359 3·8 / 0945 8·9 / Su 1630 3·8 / 2213 8·4	**20** 0530 4·2 / 1120 8·4 / M 1811 4·1 / ☽	**5** 0438 3·6 / 1033 9·1 / Tu 1712 3·4 / 2305 8·8	**20** 0536 3·9 / 1132 8·6 / W 1808 3·9
6 0338 2·6 / 0919 9·7 / W 1553 2·8 / 2134 9·4	**21** 0423 2·6 / 0955 9·7 / Th 1645 3·0 / 2223 9·1	**6** 0336 3·3 / 0918 9·3 / F 1559 3·5 / 2135 8·8	**21** 0454 3·8 / 1031 8·6 / Sa 1729 3·9 / ☽ 2319 8·0	**6** 0444 4·2 / 1042 8·4 / M 1727 4·1 / ☽ 2325 8·0	**21** 0012 7·9 / 0638 4·5 / Tu 1236 8·1 / 1928 4·2	**6** 0534 3·8 / 1136 8·9 / W 1814 3·6	**21** 0010 8·2 / 0632 4·1 / Th 1234 8·3 / 1910 4·1
7 0403 3·2 / 0949 9·1 / Th 1624 3·4 / 2204 8·7	**22** 0509 3·6 / 1044 8·7 / F 1742 3·9 / ☽ 2327 8·0	**7** 0407 3·8 / 0953 8·6 / Sa 1640 4·1 / 2219 8·1	**22** 0557 4·6 / 1150 7·9 / Su 1852 4·4	**7** 0556 4·5 / 1210 8·2 / Tu 1849 4·2	**22** 0130 7·9 / 0758 4·1 / W 1351 8·2 / 2042 4·0	**7** 0015 8·7 / 0649 3·9 / Th 1250 8·9 / 1931 3·5	**22** 0116 8·1 / 0742 4·2 / F 1341 8·2 / 2020 4·0
8 0435 3·8 / 1024 8·5 / F 1708 4·1 / ☽ 2247 8·0	**23** 0617 4·5 / 1205 7·8 / Sa 1913 4·5	**8** 0454 4·5 / 1048 8·0 / Su 1744 4·6 / ☽ 2336 7·5	**23** 0107 7·6 / 0737 4·6 / M 1340 7·8 / 2036 4·3	**8** 0102 8·1 / 0733 4·3 / W 1341 8·5 / 2018 3·7	**23** 0233 8·3 / 0903 4·0 / Th 1451 8·6 / 2136 3·6	**8** 0131 8·9 / 0811 3·6 / F 1405 9·2 / 2046 3·1	**23** 0223 8·2 / 0850 3·9 / Sa 1447 8·6 / 2122 3·8
9 0523 4·5 / 1119 7·8 / Sa 1817 4·6	**24** 0133 7·5 / 0806 4·8 / Su 1416 7·7 / 2108 4·3	**9** 0618 4·9 / 1243 7·7 / M 1924 4·6	**24** 0237 7·9 / 0908 4·4 / Tu 1454 8·3 / 2146 3·8	**9** 0220 8·8 / 0854 3·6 / Th 1449 9·3 / 2127 2·9	**24** 0325 8·7 / 0953 3·5 / F 1542 9·0 / 2220 3·2	**9** 0242 9·4 / 0921 3·0 / Sa 1512 9·7 / 2152 2·6	**24** 0324 8·6 / 0949 3·6 / Su 1548 8·6 / 2217 3·5
10 0001 7·4 / 0653 4·9 / Su 1321 7·5 / 1959 4·7	**25** 0315 8·0 / 0945 4·3 / M 1535 8·4 / 2223 3·6	**10** 0148 7·7 / 0813 4·6 / Tu 1429 8·2 / 2058 4·0	**25** 0334 8·5 / 1004 3·7 / W 1546 8·9 / 2233 3·1	**10** 0321 9·6 / 0956 2·8 / F 1545 10·1 / 2223 2·1	**25** 0409 9·2 / 1037 3·0 / Sa 1626 9·4 / 2301 2·8	**10** 0345 9·9 / 1023 2·4 / Su 1613 10·2 / 2252 2·1	**25** 0417 9·0 / 1042 3·2 / M 1640 8·9 / 2308 3·2
11 0225 7·5 / 0847 4·6 / M 1507 8·1 / 2131 4·1	**26** 0413 8·7 / 1042 3·5 / Tu 1624 9·1 / 2312 2·9	**11** 0305 8·5 / 0934 3·8 / W 1531 9·2 / 2203 3·0	**26** 0416 9·2 / 1045 3·1 / Th 1627 9·5 / 2309 2·6	**11** 0413 10·5 / 1049 1·9 / Sa 1637 10·8 / 2315 1·5	**26** 0449 9·7 / 1118 2·6 / Su 1708 9·7 / 2340 2·5	**11** 0441 10·4 / 1122 1·9 / M 1711 10·6 / 2350 1·7	**26** 0504 9·3 / 1132 2·8 / Tu 1727 9·2 / 2356 2·9
12 0341 8·3 / 1004 3·8 / Tu 1604 9·0 / 2235 3·2	**27** 0454 9·3 / 1125 2·9 / W 1705 9·8 / 2350 2·3	**12** 0359 9·6 / 1030 2·7 / Th 1620 10·2 / 2257 2·0	**27** 0452 9·7 / 1120 2·5 / F 1705 10·0 / 2344 2·2	**12** 0502 11·1 / 1140 1·4 / Su 1726 11·3 / ○	**27** 0529 10·0 / 1158 2·3 / M 1747 9·8	**12** 0534 10·8 / 1219 1·5 / Tu 1805 10·8 ○	**27** 0547 9·7 / 1221 2·5 / W 1811 9·5
13 0433 9·3 / 1102 2·9 / W 1652 10·0 / 2327 2·2	**28** 0529 9·9 / 1201 2·3 / Th 1740 10·3	**13** 0445 10·6 / 1119 1·8 / F 1705 11·1 / 2344 1·2	**28** 0526 10·1 / 1156 2·1 / Sa 1740 10·3	**13** 0007 1·1 / 0550 11·5 / M 1232 1·0 / 1814 11·5 ●	**28** 0019 2·4 / 0605 10·2 / Tu 1239 2·2 / 1825 9·9 ●	**13** 0045 1·5 / 0627 11·0 / W 1314 1·3 / 1856 10·8 ●	**28** 0041 2·6 / 0628 10·0 / Th 1306 2·2 / 1850 9·8 ●
14 0516 10·3 / 1150 1·9 / Th 1734 10·9	**29** 0024 1·9 / 0603 10·3 / F 1234 1·9 / ● 1814 10·6	**14** 0529 11·4 / 1207 1·1 / Sa 1750 11·8 ●	**29** 0017 2·0 / 0601 10·4 / Su 1229 1·9 / 1814 10·4	**14** 0057 0·9 / 0636 11·6 / Tu 1321 1·0 / 1902 11·5	**29** 0057 2·3 / 0642 10·2 / W 1320 2·1 / 1900 9·9	**14** 0137 1·5 / 0714 11·0 / Th 1405 1·3 / 1944 10·7	**29** 0124 2·4 / 0706 10·1 / F 1348 2·0 / 1927 10·0
15 0015 1·3 / 0558 11·2 / F 1235 1·2 / ○ 1817 11·7	**30** 0056 1·6 / 0632 10·6 / Sa 1304 1·6 / 1845 10·8	**15** 0031 0·6 / 0612 11·9 / Su 1253 0·7 / 1834 12·1	**30** 0050 1·9 / 0631 10·6 / M 1304 1·8 / 1846 10·4	**15** 0145 1·1 / 0721 11·4 / W 1411 1·1 / 1948 11·1	**30** 0137 2·4 / 0717 10·2 / Th 1358 2·2 / 1935 9·9	**15** 0225 1·7 / 0759 10·8 / F 1451 1·6 / 2027 10·4	**30** 0204 2·2 / 0742 10·4 / Sa 1427 1·9 / 2002 10·2
			31 0123 1·9 / 0702 10·5 / Tu 1338 1·9 / 1917 10·3				**31** 0242 2·2 / 0818 10·5 / Su 1503 1·9 / 2037 10·2

Chart Datum: 5·88 metres below Ordnance Datum (Local)

SARK 10-17-12
Sark (Channel Islands)

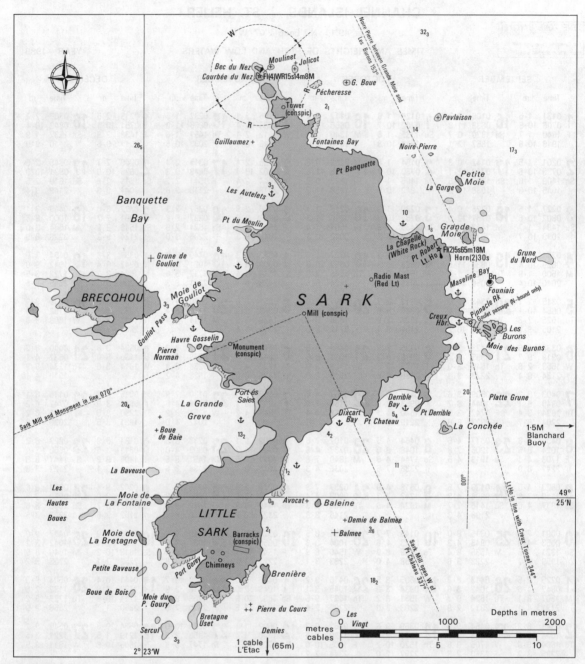

CHARTS
Admiralty 3654, 808; SHOM 6904; ECM 1014; Stanford 16; Imray C33A

TIDES
−0450 Dover; ML 4.9; Duration 0550; Zone 0 (GMT).

Standard Port ST HELIER (←)

Times				Height (metres)			
HW		LW		MHWS	MHWN	MLWN	MLWS
0300	0900	0200	0900	11.1	8.1	4.1	1.3
1500	2100	1400	2100				

Differences SARK (MASELINE PIER)
+0005 +0015 +0005 +0010 −2.1 −1.5 −0.6 −0.3

SHELTER
Many sheltered anchorages, shown on chartlet, depending on wind direction. Main harbours are Creux (dries) and Maseline. Goulet passage for N-bound traffic only. Berthing at jetties prohibited when steamers due, and always when boat is unattended.

NAVIGATION
Creux waypoint 49°25'.30N 02°20'.30W, 164°/344° from/to entrance, 0.57M. Beware large tidal range and strong tidal streams. There are numerous rocks all round Sark but the centres of the bays are clear of dangers for the most part. Sark is not a comfortable place to lie except in settled weather. Beware numerous lobster pots.

LIGHTS AND MARKS
From S, Pinnacle Rock on with E side of Grand Moie at 001°; or W patch on Creux Hr pierhead in line with Pt Robert Lt Ho at 350°. Point Robert Fl (2) 5s 65m 18M; W 8-sided Tr; vis 138°-353°; Horn(2) 30s.

RADIO TELEPHONE
None.

TELEPHONE (0481 83)
Lighthouse Keeper 2021; Customs 2021; Marinecall 0898 500 457; Dr 2045.

AREA 17—Channel Islands and French Coast

SARK continued

FACILITIES
EC Thursday and Saturday — Winter only; **Harbour (east)** Tel. 2025, Slip, M, L, FW, C (1 ton), AB; **Maseline Jetty** Tel. 2070, M, C (3 ton); **Landing (west)** M, L; **Gallery Stores** Tel. 2078, Gas, Gaz, Kos; **Village** P and D (cans), V, R, Bar.
PO; Bank; Rly (ferry to Guernsey-Weymouth, St Malo and Granville); Air (Guernsey).
Ferry UK — Guernsey (St. Peter Port)—Weymouth/Portsmouth.
Hydrofoil — Guernsey, Alderney, Jersey, St Malo, Weymouth.

BEAUCETTE 10-17-13
Guernsey (Channel Islands)

CHARTS
Admiralty 807, 808, 3654; SHOM 6903, 6904; ECM 1014; Stanford 16; Imray C33A

TIDES
−0450 Dover; ML 5.0; Duration 0500; Zone 0 (GMT)

Standard Port ST HELIER (←)

Times				Height (metres)			
HW		LW		MHWS	MHWN	MLWN	MLWS
0300	0900	0200	0900	11.1	8.1	4.1	1.3
1500	2100	1400	2100				

Differences ST PETER PORT
0000 +0012 −0008 +0002 −2.1 −1.4 −0.6 −0.3

NOTE: To ascertain the depth of water over the sill at the Channel Island Yacht Marina at Beaucette
1. Calculate the predicted *Height* of HW at St Peter Port for the day.
2. Calculate the *Time* of LW for the day.
3. On following table, select the line corresponding to the predicted Height of HW. The seven columns at the right of the table show the depth of water over the sill at hourly intervals either side of LW.

Predicted height of HW at St Peter Port	Depth of Water over the Sill in metres (sill dries 2.1m)						
HW metres	L.W.	∓1hr	∓2hr	∓3hr	∓4hr	∓5hr	∓6hr
9.75	—	—	0.30	2.74	5.18	6.79	7.62
9.44	—	—	0.15	2.74	5.03	6.55	7.31
9.14	—	—	0.60	2.74	4.87	6.31	7.01
8.83	—	—	0.76	2.74	4.72	6.03	6.70
8.53	—	—	0.91	2.74	4.57	5.79	6.40
8.22	—	—	1.06	2.74	4.42	5.54	6.09
7.92	—	0.18	1.21	2.74	4.26	5.27	5.79
7.61	—	0.42	1.37	2.74	4.11	5.03	5.48
7.31	0.30	0.70	1.52	2.74	3.96	4.78	5.18
7.00	0.60	0.94	1.67	2.74	3.81	4.51	4.87
6.70	0.91	1.21	1.82	2.74	3.65	4.27	4.57
6.39	1.21	1.46	1.98	2.74	3.50	4.02	4.27
6.09	1.52	1.70	2.13	2.74	3.35	3.75	3.96

SHELTER
Excellent shelter but entry restricted as above, and not advised in strong onshore winds or heavy swell. Secure to mooring buoy (see chartlet) while waiting. Space inside is very limited, so booking advisable.

NAVIGATION
Waypoint 49°30'.15N 02°28'.50W, 096°/276° from/to entrance, 1.1M. Approach either from N or S along Little Russel until between Platte Fougère Lt Ho (W with B band, 25m) and Roustel Lt Tr (BW checked). Half way between, pick up the leading marks, but beware the Petit Canupe Rks (buoyed) to the N. Running in, there are rocks and drying areas to the S. Entrance channel buoyed with R and G buoys. The entrance is very narrow indeed.

LIGHTS AND MARKS
Ldg Lts 276°. Front FR; R stripe on W background at N side of entrance. Rear FR; W stripe on R background, on roof of building, with windsock.

RADIO TELEPHONE
VHF Ch 16; M. Other station: St Sampson Ch 16; 12 (H24).

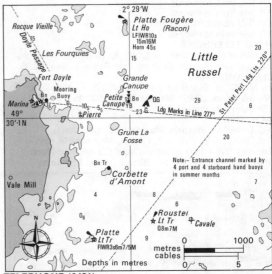

TELEPHONE (0481)
Hr Mr 45000; Customs 45000; Marinecall 0898 500 457; Dr 25211 (St Johns Ambulance St Peter Port).

FACILITIES
EC Thursday; **Channel Island Yacht Marina** (200+50 visitors), Tel. 45000, P, D, L, FW, C, AC, Gas, AB, V, R, Lau, BH (30 ton), Slip, Bar; PO; Bank (St Sampson), Air (Guernsey).
Ferry UK — St Peter Port—Weymouth/Portsmouth.

ST PETER PORT 10-17-14
Guernsey (Channel Islands)

CHARTS
Admiralty 808, 3140, 3654, 807; SHOM 6903, 6904; ECM 1014; Stanford 16; Imray C33A

TIDES
−0450 Dover; ML 5.0; Duration 0500; Zone 0 (GMT)

Standard Port ST HELIER (←)

Times				Height (metres)			
HW		LW		MHWS	MHWN	MLWN	MLWS
0300	0900	0200	0900	11.1	8.1	4.1	1.3
1500	2100	1400	2100				

Differences ST PETER PORT
0000 +0012 −0008 +0002 −2.1 −1.4 −0.6 −0.3

SHELTER
Good shelter especially in the Victoria Marina which has a sill 4.2 m above CD. Access approx HW∓3. Approach via channel along S side of harbour. Marina waiting pontoon and some visitors' moorings to N of marina entrance. Local boat moorings occupy centre of harbour, but there is also a fairway to N of them. Albert Dock Marina and North Beach Marina are for local boats only. Alternative anchorages in Guernsey are Havelet Bay, Icart Pt, Fermain Bay, St Sampsons (see 10.17.25) and Bordeaux.
Customs/immigration at St Peter Port, St Sampson and Beaucette (10.17.13).

NAVIGATION
Waypoint 49°27'.88N 02°30'.70W, 040°/220° from/to front Ldg Lt 220°, 0.68M. Offlying dangers, big tidal range and strong tidal streams demand careful navigation. Easiest approach from N via Big Russel between Herm and Sark passing S of Lower Heads Lt buoy: slightly more direct, the Little Russel which needs care — see 10.17.5. Coming round W and S of Guernsey, give Les Hanois a wide berth. Beware ferries and shipping.

ST. PETER PORT continued

LIGHTS AND MARKS
Ldg Lts: Front Castle Breakwater head Al WR 10s 14m 16M (vis 187°-007°) Horn 15s: Rear Belvedere Oc 10s 61m 14M in line at 220°.
On White Rock Pierhead
FR (vis from seaward) — No entry
FR (vis from landward) — No exit
On SW corner of New Pier
FR (vis from landward) — No exit

RADIO TELEPHONE
Call: *Port Control* VHF Ch **12** 16; 12 78 (H24). Messages may be sent through St Peter Port Radio — See 6.3.15.

TELEPHONE (0481)
Hr Mr 20229; Signal Station (CG) 20085; Customs 26911; Marina Office 25987; Marinecall 0898 500 457; Dr 25211 (St John's Ambulance).

FACILITIES
EC Thursday; **Victoria Marina** (280, all visitors) Tel. 25987, Slip, FW, AC, Lau, R, Max stay 14 days (Access HW∓3 sill maintains level CD+4.2m); **North Pier** Tel. 20085, C (32 ton 20 ton 7 ton), AB; **Jetty**

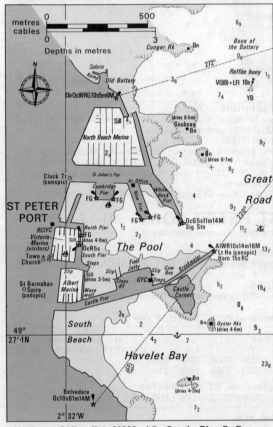

Harbour Office Tel. 20229, AB; **Castle Pier** P, D; **Marquand Bros.** Tel. 20962, CH, Gas, Gaz; **Guernsey Marine Services** Tel. 47158, ME, El, BY; **Channel Is Yacht Services** Tel. 23228, ME, El, CH; **Navigation and Marine Suppliers** Tel. 25838, CH, ACA; **G. Scott** Tel. 28989, SM; **Herm Seaway** Tel. 26829, ME, El, Gas, Gaz; **Seaward Marine** Tel. 45353, ME, El, BY, CH; **John Webster (Marine Management)** Tel. 53755; **Royal Channel Islands YC** Tel. 25500; **Guernsey YC** Tel. 22838; **Auto Electrical & Diesel Service** Tel. 26811, El; **Boatwork+** Tel. 26071, Electronics, CH, ME, El, ACA.
Town P, D, V, CH, R, Lau, Bar. PO; Bank; Rly (ferry to Weymouth, Portsmouth, Cherbourg, St. Malo); Hydrofoil to Sark, Weymouth, Jersey, St Malo; Air (Guernsey Airport).
Ferry UK—Weymouth/Portsmouth.

HERM. Herm belongs to the States of Guernsey and landing is permitted on payment of a landing fee. The best anchorages are: off Rosière steps in the SW (almost landlocked but poor holding); Herm Harbour just N — dries; Belvoir Bay on the E side.
Herm harbour office Tel. 22377.

BRAYE (ALDERNEY) 10-17-15
Alderney (Channel Islands)

CHARTS
Admiralty 3653, 2669, 60, 2845; SHOM 6934; ECM 1014; Stanford 16; Imray C33A

TIDES
−0410 Dover; ML 3.6; Duration 0545; Zone 0 (GMT).

Standard Port ST HELIER (←)

Times				Height (metres)			
HW		LW		MHWS	MHWN	MLWN	MLWS
0300	0900	0200	0900	11.1	8.1	4.1	1.3
1500	2100	1400	2100				

Differences BRAYE
+0050 +0040 +0025 +0105 −4.8 −3.4 −1.5 −0.5

SHELTER
Good shelter in Braye (Alderney) Harbour except in strong N and NE winds. There are 80 visitors' yellow mooring buoys in the western part of the harbour and the area near Fort Albert. Anchorage in harbour is good, but keep clear of the jetty because of steamer traffic. Landing on Admiralty Pier is forbidden. The old derelict length of pier extending in NE direction has been removed. There is a FY Lt on end of remaining quay, as shown. Harbour speed limit 4 kn.

NAVIGATION
Waypoint 49°44′.32N 02°10′.90W, 035°/215° from/to front Ldg Lt 215°, 1.05M. The principal dangers are the strong tidal streams, but the N coast of Alderney is covered with detached rocks. The safest approach is from the NE and even if coming from the S, it may be worth going the long way round, giving the Casquets a wide berth and coming down from the N. Beware the Swinge and Alderney Race (see 10.17.5). In harbour entrance, beware the breakwater submerged extension.

LIGHTS AND MARKS
Clearing line for the Nannels:
N side of Fort Albert and end of Admiralty Pier 115°.
Ldg Lts 215° lead into harbour. Front Q 8m 17M vis 210°-220° Rear Iso 10s 17m 18M vis 210°-220°
Daymarks — St Anne's church spire in line with W beacon on Douglas Quay at 210°
There is a sector Lt at Chateau à l'Etoc Pt E of the harbour, Iso WR 4s; vis R071°-111°, W111°-151°. This is in line 111° with the main Lt Ho Fl (4) 15s.

RADIO TELEPHONE (local times)
Call: *Alderney Radio* VHF Ch 16; 74 (Oct-Mar, Mon-Fri 0800-1700, Apl-Sept 0800 to 1800 daily). Outside these hours call St Peter Port. Mainbrayce Marine Ch M (Apl-mid Sept: 0800-2000). For Casquets traffic scheme see Cherbourg (10.17.24).

TELEPHONE (048 182)
Hr Mr 2620; Customs 2620; Marinecall 0898 500 457; Dr 2077; Hosp 2822.

FACILITIES
EC Wednesday; **Marine Engine Services** Tel. 2531, ME, El, Sh; **Mainbrayce** Tel. 2772, Slip, D, FW, ME, El, Gas, SM, ACA, Sh, CH, P, D (inner harbour HW∓2); (For Water Taxi service, call Mainbrayce on Ch 16 M);
Harbour M, FW, C, Lau; **Jetty** FW; **Sapper Slip** Slip, FW; **Alderney SC** Tel. 2758, Bar; **Riduna Garage** Tel. 2919, P, D (cans), Gaz, Gas, Kos.
Town V, R, Lau, Bar. PO; Bank;
Air — to Guernsey, Jersey, Hurn, Southampton or Cherbourg
Ferry UK — via Guernsey—Weymouth/Portsmouth (Direct to Torquay, Tuesdays mid-June − early Sept).

BRAYE (ALDERNEY) continued

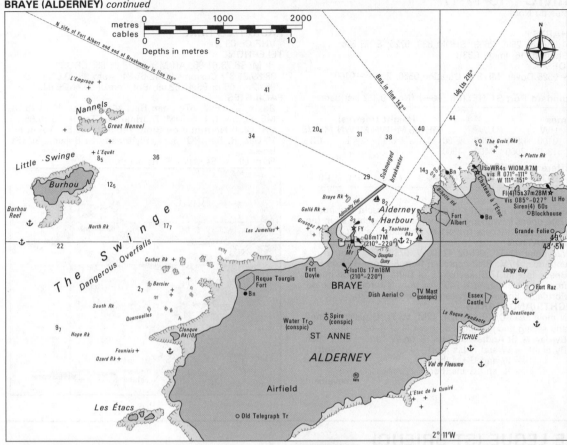

SAINT QUAY-PORTRIEUX
Côtes du Nord
10-17-16

CHARTS
Admiralty 3672, 2668, 2669; SHOM 5725, 833; ECM 536; Stanford 16; Imray C33B, C34

TIDES
−0520 Dover; ML 6.4; Duration 0550; Zone −0100

Standard Port ST HELIER (←) (Note 9.1.2 last para)

Times				Height (metres)			
HW		LW		MHWS	MHWN	MLWN	MLWS
0100	0800	0200	0700	11.1	8.1	4.1	1.3
1300	2000	1400	1900				

Differences ST QUAY-PORTRIEUX
+0030 +0045 +0030 +0030 +0.3 +0.5 0.0 +0.1

SHELTER
There is reasonable shelter in the harbour but in winds from SE to N, a surge occurs. Anchorage in the Rade de Portrieux is good but affected by winds from N to SE. Harbour dries. Max stay three days in season.

NAVIGATION
Waypoint 48°41'.00N 02°49'.70W, 347°/167° from/to Moulières de Portrieux E cardinal Bn, 1.7M. Portrieux lies inside the Roches de St Quay the channel between being about ½ M wide. N end of the Roches de St Quay is the Ile Harbour Lt Ho. Beware the Moulières de Portrieux.

LIGHTS AND MARKS
Ile Harbour (off chartlet) Oc (2) WRG 6s 16m 11/8M; W square Tr and dwelling, R top. Jetty head Iso WG 4s 11m 12/7M; W 8-sided Tr, G top; vis G (unintens) 020°-110°, G110°-305°, W305°-309°, G309°-020°; in line with building (conspic) leads 260°. Mole head FlR 4s 9m 7M; W mast, R top.

RADIO TELEPHONE
None.

TELEPHONE
Hr Mr 96.70.52.04; Aff Mar 96.70.42.27; CROSS 98.89.31.31; SNSM 96.70.56.62; Customs 96.33.33.03; Meteo 99.46.10.46; Dr 96.70.41.31; Brit Consul 99.46.26.64.

FACILITIES
Quay Slip, M, P, D, L, FW, Sh, C (1.5 ton), ME, El, Sh, CH, AB, R, Bar; **Cercle de la Voile de Portrieux** Tel. 96.70.41.76, M, FW, C (1 ton), Bar; **Mathurin Cras** Tel. 96.70.57.83, ME, El, C (5 ton).
Town V, Gaz, R, Bar. PO; Bank; Rly (bus to St Brieuc); Air (St Brieuc—Armor).
Ferry UK — St Malo—Portsmouth.

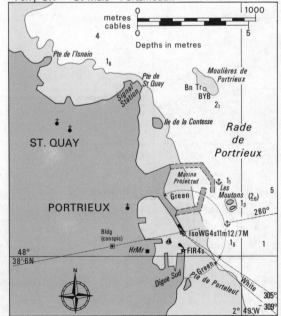

BINIC 10-17-17
Côtes du Nord

CHARTS
Admiralty 2668, 2669; SHOM 833, 5725; ECM 536; Stanford 16; Imray C33B

TIDES
−0525 Dover; ML 5.6; Duration 0550; Zone −0100

Standard Port ST HELIER (←) (Note 9.1.2 last para)

Times				Height (metres)			
HW		LW		MHWS	MHWN	MLWN	MLWS
0100	0800	0200	0700	11.1	8.1	4.1	1.3
1300	2000	1400	1900				

Differences BINIC
+0030 +0045 +0035 +0031 +0.3 +0.6 0.0 +0.1

SHELTER
Good shelter especially in Bassin à Flot (yacht harbour). Gate opens in working hours when tide reaches 9.5m (not near neaps). Access to Port de Penthièvre about HW∓3.

NAVIGATION
Waypoint 48°37'.00N 02°42'.00W, 078°/258° from/to entrance, 4.7M. Best approach from E, from Baie de St Brieuc (see 10.17.5) keeping E of Caffa W cardinal buoy, from which entrance is 246°; or from N through Rade de Portrieux. Entrance between moles, dries 4.2m.

LIGHTS AND MARKS
N mole head Oc(3) 12s 12m 12M; W Tr, G gallery. Gate and sliding bridge signals on mast N of gate:
By day — St Andrew's Cross, B on W flag — Gate open
By night — W and R Lts (hor)   No entry
— W and G Lts (hor)   No exit
— R and G Lts (hor)   No movements

RADIO TELEPHONE
VHF Ch 09.

TELEPHONE
Hr Mr 96.73.61.86; Aff Mar 96.73.61.86; CROSS 98.89.31.31; Customs 96.42.61.84; Auto 96.20.01.92; Dr 96.42.61.05 or 96.42.62.09; Brit Consul 99.46.26.64.

FACILITIES
Basin Slip, FW, AB; **Jean Bart Marine** Tel. 96.73.75.28, ME, El, Sh, CH, SHOM; **Marinarmor** Tel. 96.73.60.55, CH; **Club Nautique de Binic** Tel. 96.61.93.55; **Town** P, V, Gaz, R, Bar. PO; Bank; Rly (bus to St Brieuc); Air (St Brieuc).
Ferry UK — St Malo—Portsmouth.

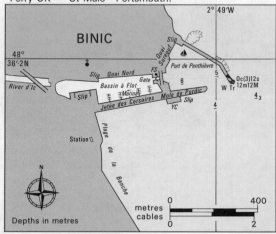

LE LÉGUÉ (ST BRIEUC) 10-17-18
Côtes du Nord

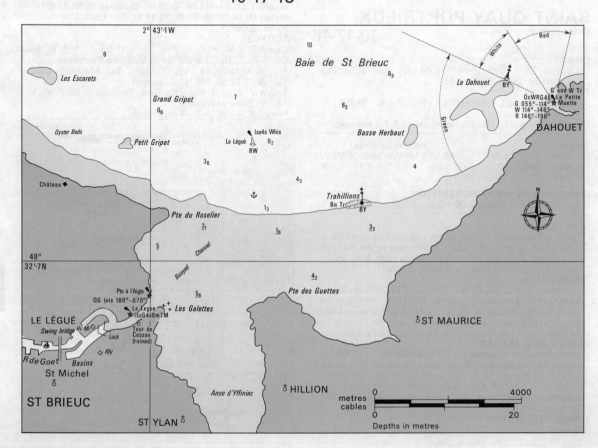

AREA 17—Channel Islands and 681
French Coast

LE LÉGUÉ (ST. BRIEUC) continued

CHARTS
Admiralty 2668, 2669; SHOM 833, 5725; ECM 536;
Stanford 16; Imray C34, C33B

TIDES
−0520 Dover; ML 5.6; Duration 0550; Zone −0100

Standard Port ST HELIER (←) (Note 9.1.2 last para)

Times				Height (metres)			
HW		LW		MHWS	MHWN	MLWN	MLWS
0100	0800	0200	0700	11.1	8.1	4.1	1.3
1300	2000	1400	1900				
Differences LE LÉGUÉ							
+0030	+0045	+0035	+0031	+0.3	+0.6	0.0	+0.1
DAHOUET							
+0031	+0038	+0027	+0036	+0.2	+0.6	−0.2	0.0

SHELTER
Le Légué is the port for St Brieuc and offers very good shelter especially in the wet basin. Le Quai Gilette is a commercial quay but yachts can lie there whilst waiting for lock to open (HW−2 to HW+1 springs; HW ∓1 neaps). Yachts use Bassin No 2 (min 3m) near viaduct.

NAVIGATION
Waypoint Le Légué (safe water) buoy, Iso 4s, 48°34'.40N 02°41'.20W, 032°/212° from/to Pte à l'Aigle Lt, 2.6M. The area dries beyond Pte du Roselier. See 10.17.5. Keep close to Pte à l'Aigle to avoid the Galettes Rks. Channel buoyed (with some gaps). The lock sill is 5.0m above CD.

LIGHTS AND MARKS
There are no leading lines, but there are two Lts on the N side of the river de Gouet ent.
Pte à l'Aigle QG vis 160°-070° 13m 7M. Jetée de la Douane (W column with G top) Iso G 4s 6m 7M.

RADIO TELEPHONE
Call: *Légué Port* VHF Ch 12 16 (occas).

TELEPHONE
Hr Mr 96.33.35.41; Aff Mar 96.61.22.61; CROSS 98.89.31.31; Meteo 96.94.94.09; Auto 96.78.51.51; SNSM 96.88.35.47; Customs 96.33.33.03; Dr St Brieuc 96.61.49.07; Brit Consul 99.46.26.64.

FACILITIES
Quai AB, C (30 ton); **Roger Nautique** Tel. 96.33.81.38, ME; **22 Voiles** Tel. 96.33.14.04, SM; **L'Habitat et la Mer** Tel. 96.33.71.68, El, Electronics; **Inter-Nautic** Tel. 96.33.28.71, ME, El, CH; **Lecoq** Tel. 96.33.16.68, Sh; **Travadon** Tel. 96.33.38.54, ME, Sh. **Town (St Brieuc)** P, D, FW, ME, El, CH, Gaz, V, R, Bar, PO; Bank; Rly; Air.
Ferry UK — St Malo—Portsmouth.

ST MALO/DINARD 10-17-19
Ille et Vilaine

CHARTS
Admiralty 2669, 2700; SHOM 5645, 844; ECM 535;
Stanford 16; Imray C33B

TIDES
−0515 Dover; ML 6.8; Duration 0535; Zone −0100

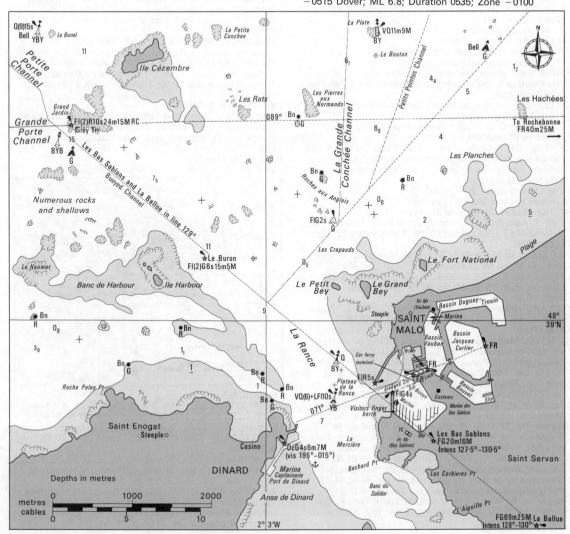

ST. MALO/DINARD continued

Standard Port ST HELIER (←) (Note 9.1.2 last para)

Times				Height (metres)			
HW		LW		MHWS	MHWN	MLWN	MLWS
0100	0800	0200	0700	11.1	8.1	4.1	1.3
1300	2000	1400	1900				
Differences ST MALO							
+0034	+0044	+0105	+0050	+1.1	+1.1	+0.3	+0.2
ERQUY							
+0030	+0040	+0035	+0032	+0.3	+0.6	0.0	+0.1

SHELTER
Excellent shelter in Bassin Vauban or Bassin Duguay-Trouin, min depth 6m. Lock opens HW−2½ to HW+1½. Also good shelter in Port des Bas Sablons Marina entered over sill 2m above CD. There is a yacht anchorage at Dinard.

NAVIGATION
Waypoint Fairway (safe water) buoy, LFl, Whis, 48°41'.42N 02°07'.20W, 307°/127° from/to Grand Jardin Lt, 1.9M. There are numerous dangerous rocks round the entrance channel and there are strong tidal streams. There are three principal entrance channels, Grande Porte channel, Petite Porte channel and the Grande Conchée channel. Care needs to be exercised on all three. Lights at night are very good for the first two. Speed limit in basins is 5 kn. Anchoring in basins is forbidden.

LIGHTS AND MARKS
Chenal de la Grande Porte; Outer Ldg Lts at 089° are:
Front — Le Grand Jardin Fl (2) R 10s 24m 15M grey Tr.
Rear — Rochebonne (off chartlet) 4.2M from Front, FR 40m 25M leads into Chenal de la Petite Porte.
Leading Lts (Les Bas Sablons FG 20m 16M in front and La Ballue FG 69m 25M at rear) at 129° lead in from the Grand Jardin Lt Ho.
St Malo Lock signals
R Lt — Entrance prohibited
R and G Lts — Entrance and departure prohibited
G Lt — Departure prohibited
When these signals are accompanied by a W Lt on the other yard-arm, the meaning is the same, but it indicates both gates open.
R Lt over 2 G Lts — all movements prohibited except entrance of large ships
2 R Lts over G Lt — all movements prohibited except departure of large ships
Port des Bas Sablons — there is a sill and the depth over the sill is shown on a neon indicator; white figures indicate metres, red figures indicate decimetres and zero indicates 'No entry'.
Ldg Lts into basin, 2 FR, at 071°.
Les Bas Sablons Marina Lt Fl G 4s.

RADIO TELEPHONE (local times)
Call: *St Malo Port* or *Grand Jardin* VHF Ch 12 (H24). Bas Sablons Marina, St Servan Ch 09. Port Vauban Ch 12.

TELEPHONE
ST MALO Hr Mr 99.81.62.86; Hr Mr (Sablons) 99.81.71.34; Hr Mr (Vauban) 99.56.51.91; Aff Mar 99.56.87.00; CROSS 98.89.31.31; SNSM 98.89.31.31; Customs 99.81.65.90; Meteo 99.46.10.46; Hosp 99.81.60.40; Brit Consul 99.46.26.64
DINARD Hr Mr 99.46.65.55; Customs 99.46.12.42; Meteo 99.46.10.46; Auto 99.46.18.77; Hosp 99.46.18.68.

FACILITIES
ST MALO: **Marina des Bas-Sablons** (1216+60 visitors, visitors Pontoon A) Tel. 99.81.71.34, (sill 2m above CD), Slip, P and D (in cans), AC, BH (10 ton), FW, CH, Gaz, R, YC, Bar; **Port Vauban** Tel. 99.56.51.91, FW, C (1 ton), AC; **Chantier Naval de la Ville Audrain** Tel. 99.56.48.06, ME, CH, Sh; **Chantier Labbé** Tel. 99.56.31.72, ME, El, C, Sh, CH; **Electrotechnique Malouine** Tel. 99.81.52.01, El; **Chantier Naval de la Rance** Tel. 99.40.00.36, Sh; **Chantier Naval de la Plaisance** Tel. 99.82.62.97, M, ME, El, Sh, CH; **Robert Hus** Tel. 99.81.84.90, El, Sh; **Chatelais et le Gall** Tel. 99.56.14.67, El, Sh, CH; **Richard** Tel. 99.81.63.81, SM, CH; **Armor Voile** Tel. 99.81.49.30, SM; **Société Nautique de la Baie de Saint-Malo** Tel. 99.40.84.42, Bar (visitors welcome); **F. Lessard** Tel. 99.56.13.61, P, D; **Back** Tel. 99.40.91.73, CH, SHOM; **Town** Slip, P, Gaz, D, ME, El, Sh, C, V, R, Bar, PO; Bank; Rly; Air (Dinard); Ferry UK—Portsmouth; or Jersey.

DINARD: **Port de Dinard** Tel. 99.46.65.55, Slip, M, P, D, L, FW, AB; **Dinard Marine** Tel. 99.46.22.49, ME, El, CH; **Duval** Tel. 99.46.19.63, El; **YC de Dinard** Tel. 99.46.14,32, Bar; **G. L. Voile** Tel. 99.46.48,99, SM; **Town** P, D, ME, El, CH, V, Gaz, R, Bar.
PO; Bank; Rly, Air.

GRANVILLE 10-17-20
Manche

CHARTS
Admiralty 3672, 2669; SHOM 5897, 824; ECM 535; Stanford 16; Imray C33B

TIDES
−0510 Dover; ML 7.2; Duration 0525; Zone −0100

Standard Port ST HELIER (←) (Note 9.1.2 last para)

Times				Height (metres)			
HW		LW		MHWS	MHWN	MLWN	MLWS
0100	0800	0200	0700	11.1	8.1	4.1	1.3
1300	2000	1400	1900				
Differences GRANVILLE							
+0040	+0049	+0115	+0053	+1.9	+1.7	+0.5	+0.1
CANCALE							
+0035	+0050	+0115	+0100	+2.4	+2.2	+1.0	+0.8

SHELTER
Shelter is good in the marina, Port de Hérel. Approach is rough in strong W winds. The old harbour is only for commercial and fishing vessels: yachts use the marina. Enter and leave under power. Access HW−2½ to HW+3½. Speed limit 4 kn; 2 kn between pontoons.

NAVIGATION
Waypoint 48°49'.40N 01°37'.00W, 235°/055° from/to Digue Principale Lt (Fl R 4s), 0.95M. From W, Le Videcoq rks (dry 0.8m) lie 3¼M W of Pte du Roc, marked by W cardinal whis buoy. Beware rks off Pte du Roc and La Fourchie Bn, and Banc de Tombelaine 1M SSW of Le Loup Lt Tr.

LIGHTS AND MARKS
Port de Hérel marina is a half-tide harbour and depths at the entrance are shown by illuminated figures. W numbers are metres, Or numbers decimetres. Zero means no entry. Ldg Lts as on chartlet.

RADIO TELEPHONE
VHF Ch 12 16 (HW−1½ to HW+1). Marina Ch 09.

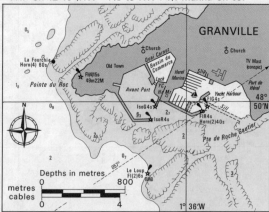

TELEPHONE
Hr Mr 33.50.12.45; Hr Mr (Hérel) 33.50.20.06; Aff Mar 33.50.00.59; CROSS 33.52.72.13; SNSM 33.50.05.85; Customs 33.50.19.90; Meteo 33.22.91.17; Auto 33.50.10.00; Dr 33.50.00.07; Brit Consul 33.44.20.13.

FACILITIES
Hérel Marina (850+150 visitors) Tel. 33.50.20.06, Slip, P, D, FW, ME, AC, BH (12 ton), C (10 ton), CH, Gaz, R, Lau, V, Bar, SM, El, Sh; **YC de Granville** Tel. 33.50.04.25, L, FW, AB, Bar; **Comptoir Maritime** Tel. 33.50.18.71, CH; **Ayello et fils** Tel. 33.50.11.54, CH; **Granville Plaisance** Tel. 33.50.23.82, ME, El, CH; **Lecoulant Marine** Tel. 33.50.20.34, M, ME, El, Sh, CH;

AREA 17—Channel Islands and French Coast 683

GRANVILLE continued

La Marine Tel. 33.50.71.31, SHOM; **Voiles Mora** Tel. 33.50.38.69, SM; **Town** P, D, ME, V, Gaz, R, Bar. PO; Bank; Rly; Air (Dinard). Ferry UK via Jersey or Cherbourg—Weymouth/Portsmouth.

CARTERET 10-17-21
Manche

CHARTS
Admiralty 2669; SHOM 827; ECM 535; Stanford 16; Imray C33A

TIDES
−0440 Dover; ML 6.3; Duration 0545; Zone −0100

Standard Port ST HELIER (←) (Note 9.1.2 last para)

Times				Height (metres)			
HW		LW		MHWS	MHWN	MLWN	MLWS
0100	0800	0200	0700	11.1	8.1	4.1	1.3
1300	2000	1400	1900				

Differences CARTERET
+0100 +0110 +0120 +0115 +0.1 +0.4 0.0 +0.2

SHELTER
Shelter good but harbour dries completely — entry only possible at HW∓2. Exposed to winds from W and SW which make entrance rough. There are no safe anchorages off shore. It dries ½ M to seaward. Bar at right angles to pier-head rises up to 3m and 15m long. The Petit Port provides the best shelter but is only suitable for shallow draught yachts able to take the bottom. Bar forms after NW winds but is removed by local sand company.

NAVIGATION
Waypoint 49°21'.00N 01°52'.25W, 250°/070° from/to

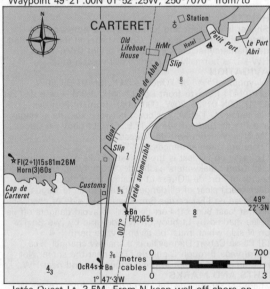

Jetée Ouest Lt, 3.5M. From N keep well off shore on approach to avoid rocks 1 M N of Cap de Carteret extending about 1M from coast. From W, about 4 M off shore, beware Trois Grune Rks (dry 1.6m) marked by S cardinal buoy.

LIGHTS AND MARKS
Cap de Carteret Fl (2+1) 15s 81m 26M; grey Tr, G top; Horn (3) 60s. Training wall, head Fl (2) G 5s; in line with lifeboat house leads 007°. Jetée Ouest, head Oc R 4s 6m 8M; W column, R top.

RADIO TELEPHONE
None.

TELEPHONE
Hr Mr 33.04.70.84; CROSS 33.52.72.13; Customs 33.04.90.08; Meteo 33.22.91.17; Auto 33.43.20.40; Hosp (Valognes) 33.40.14.39; Brit Consul 33.44.20.13.

FACILITIES
West Jetty Slip, FW, AB, R, Bar; **Club Nautique de Carteret — Barneville** Slip, M, Bar; **Port de Plaisance** Tel. 33.04.70.84, AB; **Garage Tollemer** Tel. 33.53.85.62, ME, P and D (cans). **Town** V, Gaz, R, Bar, PO; Bank (Barneville); Rly (Valognes); Air (Cherbourg). Ferry UK — Cherbourg—Portsmouth, Poole; Weymouth (summer only).
Note: 5M to S, Portbail is a safe harbour although choppy in E winds. See 10.17.25.
Note: A marina is planned.

ILES CHAUSEY 10-17-22
Manche

CHARTS
Admiralty 2669, 3659; SHOM 829, 824, 830, 4599; ECM 534; Stanford 16; Imray C33B

TIDES
−0500 Dover; ML 7.5; Duration 0530; Zone −0100

Standard Port ST HELIER (←) (Note 9.1.2 last para)

Times				Height (metres)			
HW		LW		MHWS	MHWN	MLWN	MLWS
0100	0800	0200	0700	11.1	8.1	4.1	1.3
1300	2000	1400	1900				

Differences ILES CHAUSEY
+0044 +0048 +0104 +0058 +1.9 +1.8 +0.8 +0.7

SHELTER
Shelter good except in strong NW or SE winds. When anchoring remember the big tidal range.

NAVIGATION
Waypoint 48°51'.50N 01°48'.48W, 152°/332° from/to La Crabière Est Lt, 1.2M. Although very large range of tide (see above), tidal streams are not excessive. The most direct route into Sound of Chausey is from the S, but beware rocks extending from Pte de la Tour. Alternative route from N with sufficient height of tide but refer to detailed sailing directions for transits to be followed. Local knowledge is needed. Dangerous wreck reported N of La

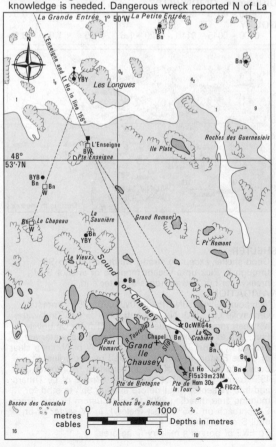

ILES CHAUSEY continued

Petite Entrée (off chartlet).
Note:— Yachts may not visit the Iles Chausey without having first made official entry elsewhere, such as Granville.
Harbour gets very crowded in summer weekends.

LIGHTS AND MARKS
Grande Ile, Pte de la Tour Fl 5s 39m 23M; Horn 30s. La Crabière Est Oc WRG 4s 5m 10/7M; B pylon, Y. top; vis W079°-291°, G291°-329°, W329°-335°, R335°-079°.
From N, L'Enseigne Bn Tr (19m) in line with Pte de la Tour Lt Ho leads 156°. From S, La Crabière in line with L'Enseigne leads 333°.
Note:— The island is private property.

RADIO TELEPHONE
None.

FACILITIES
R. Tourelle L; **Town** FW, Gaz, V, R, Bar.
PO (Granville); Bank (Granville); Rly (Granville); Air (Dinard).
Ferry UK via Granville and Jersey—Weymouth/Portsmouth.

OMONVILLE 10-17-23
Manche

CHARTS
Admiralty 2669, 1106; SHOM 5636, 5631, 828; ECM 528, 1014; Stanford 16, 7; Imray C33A

TIDES
−0330 Dover; ML 3.8; Duration 0545; Zone −0100

Standard Port CHERBOURG (→)

Times				Height (metres)			
HW		LW		MHWS	MHWN	MLWN	MLWS
0300	1000	0400	1000	6.3	5.0	2.5	1.1
1500	2200	1600	2200				

Differences OMONVILLE
−0015 −0010 −0020 −0025 −0.1 −0.1 +0.1 0.0
GOURY
−0100 −0045 −0110 −0120 +1.6 +1.5 +1.0 +0.1

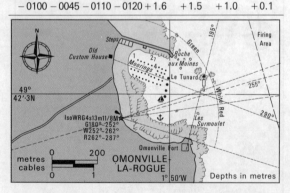

SHELTER
Good shelter except in strong winds from N to SE through E. Pick up a vacant mooring or anchor S of breakwater but beware rocks at outer end.

NAVIGATION
Waypoint 49°42'.50N 01°48'.60W, 075°/255° from/to Omonville Lt, 1.0M. Narrow entrance, ½ ca (93m) wide, between rocks. On S side rocks run N from Omonville Fort, and to N rocks running ESE from breakwater are marked by G Bn Tr, Le Tunard. From W or N, keep clear of Basse Bréfort (depth 1m, marked by N cardinal Lt buoy) 6ca (1100m) N of Pte de Jardeheu. Approach on 195° transit (below), but pass ½ca (93m) E of Le Tunard and into W sector of Lt before altering 95° to stbd for mooring area, heading 290° for old Custom House. From E, approach on 255° transit (below), in W sector of Lt, until S of Le Tunard.
To ENE of port is a military firing area.

LIGHTS AND MARKS
Omonville Lt Iso WRG 4s 13m 11/8M on W framework Tr with R top; vis G180°-252°, W252°-262°, R262°-287°. Lt in transit with church steeple, 650m beyond, leads 255° S of Le Tunard. Le Tunard Bn Tr in line with centre of Omonville Fort leads 195°, but pass ½ca (93m) E of Le Tunard.

RADIO TELEPHONE
None. For Casquets traffic scheme see Cherbourg (10.17.24).

TELEPHONE
Hr Mr Cherbourg 33.53.05.60; Aff Mar Cherbourg 33.53.21.76; Customs Cherbourg 33.53.05.60; CROSS 33.52.72.13; Dr 33.53.08.69; Assn. of Users of Port of Omonville 33.52.71.33; Meteo 33.22.91.17; Brit Consul 33.44.20.13.

FACILITIES
Jetty M, L, FW, AB, V, R, Bar; **Village** V, Gaz, R, Bar.
PO (Beaumont Hague); Bank (Beaumont Hague); Rly (bus to Cherbourg); Air (Cherbourg).
Ferry UK — Cherbourg—Portsmouth, Poole; Weymouth (summer only).

CHERBOURG 10-17-24
Manche

CHARTS
Admiralty 1106, 2602; SHOM 5628, 5627; ECM 528; Stanford 16, 7; Imray C32, C33A

TIDES
−0320 Dover; ML 3.8; Duration 0535; Zone −0100

Cherbourg is a Standard Port and tidal predictions for every day of the year are given below.

SHELTER
Shelter is excellent and harbour can be entered in all states of tide and weather. Good shelter in marina in Port de Chantereyne. Lock into Bassin à Flot (HW−1 to HW+1) is normally for commercial vessels only.

NAVIGATION
Waypoints Passe de l'Ouest 49°41'.10N 01°39'.80W, 321°/141° from/to front Ldg Lt 141°, 1.9M. Passe de l'Est 49°41'.00N 01°35'.70W, 000°/180° from/to Fort des Flamands Lt, 1.85M. For coast between Cap de la Hague and Pte de Barfleur see 10.17.5. CH1 Lt buoy lies about 3M N × W from Fort de l'Ouest.
There are three entrances:
(1) Passe de l'Ouest is the easiest. Rocks extend about 80m from breakwaters each side. From West, the W sector of Fort de l'Ouest Lt (bearing more than 122° by day) keeps clear of all dangers E of Cap de la Hague.
(2) Passe de l'Est carries 6m. Keep to W side of channel (but at least 80m off Fort de l'Est) to avoid dangers off Ile Pelée marked by Lt buoy on W side, and by two Bn Trs on N side which must be given a wide berth.
(3) Passe Cabart Danneville is a shallow channel ½ca (93m) wide through Digue de l'Est, near the shore. Not recommended except in good conditions and near HW.

LIGHTS AND MARKS
There are three powerful Lts near Cherbourg. To the W Cap de la Hague, Fl 5s 48m 23M; to the E Cap Levi, Fl R 5s 36m 22M, and Pte de Barfleur, Fl(2) 10s 72m 29M. For details see 10.17.4 and 10.18.4.
Fort de l'Ouest Lt, at W end of Digue Centrale, Fl(3)WR 15s 19m 22/18M; vis W122°-355°, R355°-122°; RC; Reed(3) 60s.
Passe de l'Ouest Ldg Lts 141°. Gare Maritime Lt (Q 35m 23M) in line with centre of 2Q(hor) 5m 15M at base of Digue du Homet.
Grande Rade Ldg Lts 124°. Front FG 10m 10M; W pylon, G top, on blockhouse; Reed(2+1) 60s. Rear, 0.75M from front, Iso G 4s 16m 11M; W column, B bands, W top; intens 114°-134°.
Passe de l'Est is indicated by W sector (176°-183°) of Dir Q WRG Lt at Fort des Flamands.

AREA 17—Channel Islands and French Coast   685

CHERBOURG continued

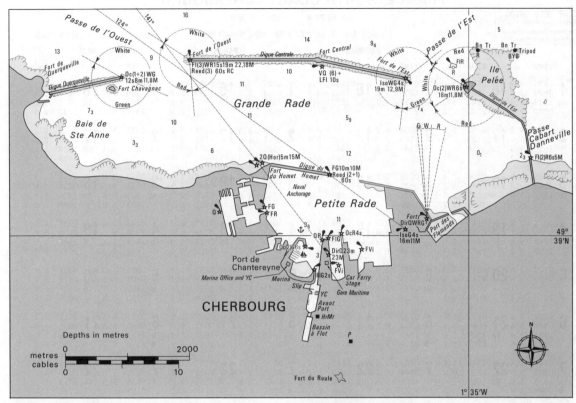

RADIO TELEPHONE (local times)
Marina Ch 09. (0800-2000).
Call: *COM Cherbourg* VHF Ch 16 (H24). Casquets traffic scheme, Ship Movement Report System (MAREP), call *Jobourg Traffic* (near Cap de la Hague) Ch **11**; 16 11 79 (H24). Information broadcasts in English and French on Ch 11 at H+20 and H+50, also at H+05 and H+35 when visibility is less than 2M.

TELEPHONE
Hr Mr 33.53.10.42; Hr Mr (Port de Plaisance) 33.53.75.16; Aff Mar 33.44.00.13; CROSS 33.52.72.13; Customs 33.44.16.00; Marina 33.53.79.65; Meteo 33.22.91.77; Auto 33.43.20.40; Dr 33.53.05.68; Brit Consul 33.44.20.13.

FACILITIES
Marina (620+100 visitors), Slip, P, D, FW, ME, El, Sh, BH (27 ton), CH; **YC de Cherbourg** Tel. 33.53.02.83, FW, Bar; **Cherbourg Marine** Tel. 33.93.11.36, P, D, ME, El, Sh, CH; **Cherbourg Plaisance** Tel. 33.53.27.34, CH, ME, El; **Cotentin Marine** Tel. 33.53.07.30, CH; **Cherbourg Général Yachting** Tel. 33.53.63.73, ME, El, SM, CH; **Nordie** Tel. 33.53.02.67, V; **Daily Tanker** D; **Ergelin** Tel. 33.53.20.26, El; **Nicollet** Tel. 33.53.11.74, SHOM.
Town P, D, Gaz, V, R, Bar, PO; Bank; Rly; Air.
Ferry UK—Portsmouth, Poole; Weymouth (summer only).

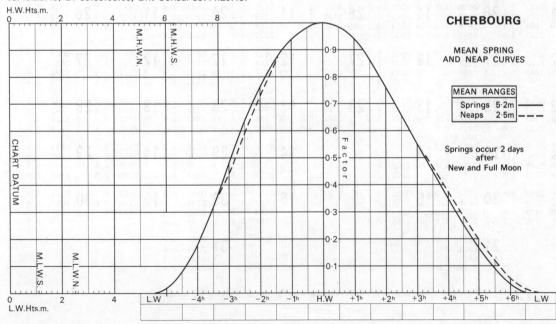

CHERBOURG
MEAN SPRING AND NEAP CURVES

MEAN RANGES	
Springs	5·2m
Neaps	2·5m

Springs occur 2 days after New and Full Moon

FRANCE, NORTH COAST - CHERBOURG

LAT 49°39'N LONG 1°38'W

TIMES AND HEIGHTS OF HIGH AND LOW WATERS

YEAR **1989**

TIME ZONE −0100 (French Standard Time)
For French Summer Time add ONE hour in non-shaded areas

JANUARY

	TIME	m		TIME	m
1 SU	0235 / 0929 / 1505 / 2201	4.9 / 2.7 / 4.9 / 2.7	**16** M	0304 / 1001 / 1542 / 2241	5.2 / 2.2 / 5.2 / 2.3
2 M	0339 / 1035 / 1613 / 2308	4.8 / 2.7 / 4.9 / 2.7	**17** TU	0422 / 1120 / 1704	5.1 / 2.4 / 5.1
3 TU	0448 / 1142 / 1721	4.9 / 2.6 / 5.0	**18** W	0000 / 0544 / 1237 / 1821	2.4 / 5.2 / 2.3 / 5.2
4 W	0013 / 0553 / 1242 / 1822	2.6 / 5.1 / 2.4 / 5.2	**19** TH	0111 / 0655 / 1343 / 1926	2.3 / 5.4 / 2.0 / 5.4
5 TH	0109 / 0650 / 1336 / 1916	1.5 / 5.3 / 2.1 / 5.4	**20** F	0211 / 0753 / 1437 / 2018	2.0 / 5.6 / 1.8 / 5.5
6 F	0200 / 0740 / 1424 / 2003	2.1 / 5.6 / 1.9 / 5.6	**21** SA ○	0302 / 0840 / 1524 / 2101	1.8 / 5.9 / 1.6 / 5.7
7 SA ●	0247 / 0826 / 1510 / 2048	1.8 / 5.9 / 1.6 / 5.8	**22** SU	0344 / 0921 / 1605 / 2141	1.6 / 6.0 / 1.4 / 5.8
8 SU	0332 / 0910 / 1554 / 2133	1.5 / 6.1 / 1.3 / 6.0	**23** M	0423 / 0959 / 1641 / 2216	1.5 / 6.1 / 1.3 / 5.9
9 M	0416 / 0954 / 1638 / 2215	1.4 / 6.3 / 1.2 / 6.1	**24** TU	0458 / 1032 / 1714 / 2249	1.4 / 6.1 / 1.3 / 5.9
10 TU	0459 / 1037 / 1721 / 2258	1.2 / 6.4 / 1.1 / 6.2	**25** W	0529 / 1104 / 1745 / 2319	1.5 / 6.1 / 1.4 / 5.8
11 W	0543 / 1120 / 1804 / 2340	1.2 / 6.4 / 1.1 / 6.1	**26** TH	0600 / 1134 / 1816 / 2349	1.5 / 5.9 / 1.5 / 5.6
12 TH	0626 / 1202 / 1848	1.3 / 6.3 / 1.2	**27** F	0630 / 1203 / 1846	1.7 / 5.7 / 1.7
13 F	0024 / 0711 / 1247 / 1933	6.0 / 1.4 / 6.1 / 1.4	**28** SA	0018 / 0702 / 1234 / 1918	5.5 / 1.9 / 5.5 / 2.0
14 SA ☽	0110 / 0757 / 1335 / 2025	5.7 / 1.7 / 5.8 / 1.8	**29** SU	0051 / 0735 / 1310 / 1954	5.3 / 2.2 / 5.2 / 2.3
15 SU	0202 / 0854 / 1431 / 2126	5.5 / 2.0 / 5.5 / 2.1	**30** M ☾	0130 / 0816 / 1352 / 2043	5.0 / 2.5 / 5.0 / 2.6
			31 TU	0219 / 0913 / 1451 / 2149	4.8 / 2.7 / 4.7 / 2.8

FEBRUARY

	TIME	m		TIME	m
1 W	0331 / 1033 / 1615 / 2316	4.7 / 2.8 / 4.7 / 2.9	**16** TH	0534 / 1231 / 1818	4.9 / 2.6 / 4.8
2 TH	0502 / 1200 / 1744	4.8 / 2.7 / 4.8	**17** F	0110 / 0655 / 1343 / 1925	2.5 / 5.2 / 2.2 / 5.2
3 F	0038 / 0621 / 1312 / 1855	2.6 / 5.0 / 2.4 / 5.1	**18** SA	0209 / 0750 / 1433 / 2012	2.1 / 5.5 / 1.9 / 5.4
4 SA	0142 / 0724 / 1408 / 1949	2.2 / 5.4 / 1.9 / 5.5	**19** SU	0254 / 0832 / 1514 / 2050	1.8 / 5.7 / 1.5 / 5.7
5 SU	0233 / 0813 / 1458 / 2036	1.8 / 5.8 / 1.4 / 5.9	**20** M ○	0332 / 0907 / 1549 / 2124	1.5 / 6.0 / 1.3 / 5.9
6 M	0320 / 0858 / 1542 / 2119 ●	1.4 / 6.2 / 1.1 / 6.2	**21** TU	0405 / 0940 / 1621 / 2155	1.4 / 6.1 / 1.2 / 6.0
7 TU	0404 / 0941 / 1625 / 2202	1.0 / 6.5 / 0.8 / 6.4	**22** W	0435 / 1008 / 1650 / 2223	1.3 / 6.2 / 1.2 / 6.0
8 W	0446 / 1022 / 1705 / 2243	0.8 / 6.7 / 0.6 / 6.5	**23** TH	0503 / 1037 / 1716 / 2250	1.2 / 6.2 / 1.2 / 6.0
9 TH	0525 / 1102 / 1745 / 2321	0.7 / 6.7 / 0.7 / 6.5	**24** F	0529 / 1103 / 1743 / 2316	1.3 / 6.1 / 1.3 / 5.9
10 F	0605 / 1140 / 1825	0.8 / 6.6 / 0.9	**25** SA	0556 / 1129 / 1810 / 2342	1.4 / 5.9 / 1.5 / 5.7
11 SA	0000 / 0645 / 1219 / 1905	6.2 / 1.1 / 6.2 / 1.2	**26** SU	0624 / 1156 / 1838	1.7 / 5.6 / 1.8
12 SU ☽	0040 / 0727 / 1303 / 1949	5.9 / 1.5 / 5.8 / 1.7	**27** M	0011 / 0653 / 1227 / 1911	5.4 / 2.0 / 5.3 / 2.1
13 M	0127 / 0815 / 1353 / 2046	5.4 / 2.0 / 5.3 / 2.3	**28** TU ☾	0044 / 0728 / 1304 / 1950	5.2 / 2.3 / 5.0 / 2.5
14 TU	0225 / 0922 / 1503 / 2206	5.1 / 2.5 / 4.9 / 2.7			
15 W	0351 / 1056 / 1644 / 2346	4.8 / 2.7 / 4.7 / 2.8			

MARCH

	TIME	m		TIME	m
1 W	0127 / 0817 / 1356 / 2052	4.8 / 2.7 / 4.7 / 2.9	**16** TH	0331 / 1039 / 1630 / 2335	4.7 / 2.8 / 4.5 / 2.9
2 TH	0233 / 0935 / 1521 / 2229	4.6 / 2.9 / 4.5 / 3.0	**17** F	0524 / 1222 / 1807	4.8 / 2.6 / 4.7
3 F	0419 / 1123 / 1713	4.6 / 2.8 / 4.6	**18** SA	0058 / 0642 / 1328 / 1909	2.5 / 5.1 / 2.2 / 5.1
4 SA	0012 / 0558 / 1251 / 1835	2.7 / 4.9 / 2.4 / 5.0	**19** SU	0153 / 0732 / 1414 / 1952	2.1 / 5.4 / 1.9 / 5.4
5 SU	0123 / 0705 / 1351 / 1931	2.2 / 5.4 / 1.8 / 5.5	**20** M	0233 / 0810 / 1450 / 2026	1.8 / 5.7 / 1.5 / 5.6
6 M	0215 / 0755 / 1438 / 2017	1.6 / 5.9 / 1.2 / 6.0	**21** TU	0307 / 0842 / 1522 / 2057	1.5 / 5.9 / 1.3 / 5.9
7 TU ●	0300 / 0838 / 1521 / 2058	1.1 / 6.4 / 0.8 / 6.4	**22** W ○	0337 / 0911 / 1551 / 2125	1.3 / 6.1 / 1.2 / 6.0
8 W	0341 / 0918 / 1602 / 2139	0.7 / 6.7 / 0.5 / 6.7	**23** TH	0405 / 0939 / 1619 / 2153	1.2 / 6.2 / 1.2 / 6.1
9 TH	0422 / 0959 / 1642 / 2218	0.5 / 6.9 / 0.4 / 6.7	**24** F	0432 / 1006 / 1646 / 2218	1.2 / 6.2 / 1.3 / 6.1
10 F	0500 / 1037 / 1720 / 2256	0.5 / 6.9 / 0.5 / 6.6	**25** SA	0458 / 1032 / 1712 / 2246	1.2 / 6.1 / 1.3 / 5.9
11 SA	0539 / 1115 / 1758 / 2335	0.7 / 6.6 / 0.8 / 6.3	**26** SU	0525 / 1059 / 1739 / 2312	1.4 / 5.9 / 1.5 / 5.7
12 SU	0618 / 1154 / 1838	1.1 / 6.2 / 1.3	**27** M	0554 / 1127 / 1809 / 2342	1.6 / 5.6 / 1.8 / 5.5
13 M	0014 / 0700 / 1235 / 1923	5.9 / 1.5 / 5.6 / 1.9	**28** TU	0624 / 1158 / 1842	1.9 / 5.3 / 2.1
14 TU ☽	0059 / 0747 / 1326 / 2018	5.4 / 2.1 / 5.1 / 2.5	**29** W	0016 / 0701 / 1236 / 1923	5.2 / 2.2 / 5.0 / 2.5
15 W	0158 / 0857 / 1439 / 2143	4.9 / 2.6 / 4.6 / 2.9	**30** TH ☾	0100 / 0749 / 1330 / 2025	4.9 / 2.6 / 4.7 / 2.8
			31 F	0208 / 0909 / 1455 / 2203	4.7 / 2.8 / 4.5 / 2.9

APRIL

	TIME	m		TIME	m
1 SA	0353 / 1059 / 1647 / 2347	4.7 / 2.7 / 4.7 / 2.6	**16** SU	0025 / 0605 / 1253 / 1833	2.5 / 5.1 / 2.2 / 5.0
2 SU	0532 / 1225 / 1807	5.0 / 2.2 / 5.1	**17** M	0117 / 0655 / 1338 / 1915	2.2 / 5.3 / 1.9 / 5.3
3 M	0056 / 0637 / 1321 / 1903	2.0 / 5.5 / 1.6 / 5.6	**18** TU	0156 / 0734 / 1414 / 1950	1.9 / 5.6 / 1.7 / 5.6
4 TU	0148 / 0726 / 1409 / 1948	1.4 / 6.0 / 1.1 / 6.1	**19** W	0230 / 0806 / 1446 / 2022	1.6 / 5.8 / 1.5 / 5.8
5 W	0231 / 0809 / 1451 / 2030	1.0 / 6.5 / 0.7 / 6.5	**20** TH	0302 / 0836 / 1517 / 2051	1.4 / 5.9 / 1.4 / 5.9
6 TH	0313 / 0850 / 1533 / 2110 ●	0.7 / 6.8 / 0.5 / 6.7	**21** F ○	0331 / 0905 / 1545 / 2120	1.3 / 6.0 / 1.3 / 6.0
7 F	0353 / 0931 / 1614 / 2152	0.5 / 6.9 / 0.5 / 6.7	**22** SA	0401 / 0935 / 1615 / 2150	1.3 / 6.0 / 1.3 / 6.0
8 SA	0435 / 1011 / 1655 / 2232	0.6 / 6.8 / 0.7 / 6.6	**23** SU	0430 / 1004 / 1645 / 2219	1.3 / 5.9 / 1.4 / 5.9
9 SU	0515 / 1053 / 1736 / 2312	0.8 / 6.4 / 1.0 / 6.2	**24** M	0500 / 1035 / 1716 / 2251	1.4 / 5.8 / 1.6 / 5.8
10 M	0556 / 1133 / 1818 / 2354	1.2 / 6.0 / 1.5 / 5.8	**25** TU	0532 / 1107 / 1750 / 2324	1.6 / 5.6 / 1.8 / 5.6
11 TU	0641 / 1217 / 1905	1.7 / 5.4 / 2.0	**26** W	0608 / 1143 / 1828	1.9 / 5.3 / 2.1
12 W ☽	0042 / 0731 / 1311 / 2002	5.3 / 2.2 / 4.9 / 2.5	**27** TH	0003 / 0650 / 1226 / 1914	5.3 / 2.1 / 5.0 / 2.4
13 TH	0144 / 0840 / 1423 / 2125	4.9 / 2.6 / 4.6 / 2.8	**28** F	0052 / 0743 / 1324 / 2018	5.1 / 2.4 / 4.8 / 2.6
14 F	0311 / 1017 / 1605 / 2306	4.7 / 2.7 / 4.6 / 2.8	**29** SA	0201 / 0900 / 1444 / 2146	4.9 / 2.5 / 4.7 / 2.6
15 SA	0452 / 1150 / 1733	4.8 / 2.6 / 4.7	**30** SU	0331 / 1033 / 1616 / 2314	5.0 / 2.4 / 4.9 / 2.3

Chart Datum: 3.70 metres below Lallemand System (Mean Sea Level, Marseilles)

AREA 17 — Channel Islands and French Coast

FRANCE, NORTH COAST - CHERBOURG
LAT 49°39'N LONG 1°38'W
TIMES AND HEIGHTS OF HIGH AND LOW WATERS
YEAR 1989

TIME ZONE −0100 (French Standard Time)
For French Summer Time add ONE hour in non-shaded areas

	MAY				JUNE				JULY				AUGUST		
	TIME m		TIME m		TIME m		TIME m		TIME m		TIME m		TIME m		TIME m

1 M 0456 5.3 / 1150 2.0 / 1731 5.3 **16** TU 0025 2.3 / 0602 5.2 / 1249 2.1 / 1825 5.2
1 TH 0039 1.5 / 0618 5.9 / 1305 1.4 / 1846 5.9 **16** F 0103 2.2 / 0642 5.2 / 1326 2.1 / 1905 5.3
1 SA 0114 1.8 / 0657 5.6 / 1344 1.8 / 1926 5.7 **16** SU 0118 2.3 / 0659 5.1 / 1345 2.2 / 1925 5.4
1 TU 0303 1.5 / 0841 5.7 / 1525 1.5 / ● 2103 6.0 **16** W 0238 1.6 / 0817 5.7 / 1500 1.5 / 2038 6.1

2 TU 0022 1.8 / 0600 5.6 / 1248 1.5 / 1827 5.7 **17** W 0109 2.1 / 0647 5.4 / 1330 1.9 / 1906 5.4
2 F 0132 1.4 / 0712 6.0 / 1356 1.3 / 1937 6.1 **17** SA 0149 2.0 / 0726 5.4 / 1409 2.0 / 1947 5.5
2 SU 0212 1.6 / 0754 5.7 / 1439 1.6 / 2020 5.9 **17** M 0209 2.0 / 0749 5.3 / 1432 2.0 / 2011 5.6
2 W 0346 1.3 / 0923 5.9 / 1606 1.4 / 2143 6.2 **17** TH 0321 1.2 / 0858 6.1 / 1541 1.1 / ○ 2118 6.4

3 W 0113 1.4 / 0652 6.1 / 1337 1.2 / 1916 6.1 **18** TH 0149 1.9 / 0725 5.5 / 1406 1.8 / 1943 5.6
3 SA 0222 1.2 / 0802 6.1 / 1446 1.2 / ● 2027 6.2 **18** SU 0230 1.8 / 0808 5.5 / 1450 1.8 / 2028 5.7
3 M 0306 1.4 / 0845 5.8 / 1530 1.5 / 2109 6.1 **18** TU 0254 1.7 / 0833 5.6 / 1517 1.7 / ○ 2054 5.9
3 TH 0425 1.2 / 1001 6.0 / 1643 1.3 / 2218 6.2 **18** F 0402 0.9 / 0939 6.3 / 1622 0.9 / 2159 6.6

4 TH 0159 1.1 / 0738 6.4 / 1421 0.9 / 2000 6.4 **19** F 0223 1.7 / 0800 5.6 / 1441 1.7 / 2017 5.7
4 SU 0312 1.2 / 0851 6.1 / 1536 1.2 / 2115 6.2 **19** M 0311 1.7 / 0848 5.6 / 1531 1.7 / ○ 2108 5.8
4 TU 0354 1.3 / 0934 5.9 / 1618 1.4 / 2156 6.1 **19** W 0338 1.4 / 0915 5.8 / 1600 1.4 / 2137 6.1
4 F 0459 1.3 / 1035 5.9 / 1716 1.3 / 2251 6.1 **19** SA 0442 0.7 / 1017 6.5 / 1700 0.8 / 2237 6.7

5 F 0243 0.8 / 0822 6.5 / 1506 0.8 / ● 2044 6.5 **20** SA 0259 1.6 / 0834 5.7 / 1515 1.6 / ○ 2051 5.8
5 M 0401 1.2 / 0940 6.1 / 1625 1.3 / 2203 6.2 **20** TU 0351 1.5 / 0929 5.7 / 1612 1.6 / 2149 5.9
5 W 0439 1.3 / 1016 5.8 / 1700 1.4 / 2237 6.1 **20** TH 0420 1.2 / 0958 6.0 / 1641 1.3 / 2217 6.3
5 SA 0531 1.3 / 1106 5.9 / 1747 1.4 / 2321 6.0 **20** SU 0520 0.7 / 1056 6.5 / 1739 0.8 / 2315 6.6

6 SA 0328 0.8 / 0906 6.5 / 1550 0.8 / 2128 6.5 **21** SU 0332 1.5 / 0908 5.8 / 1549 1.6 / 2125 5.9
6 TU 0448 1.3 / 1026 5.9 / 1710 1.5 / 2249 6.0 **21** W 0432 1.5 / 1008 5.7 / 1652 1.5 / 2229 6.0
6 TH 0520 1.3 / 1056 5.8 / 1739 1.5 / 2315 6.0 **21** F 0501 1.1 / 1038 6.1 / 1721 1.2 / 2258 6.3
6 SU 0602 1.4 / 1135 5.7 / 1817 1.6 / 2350 5.8 **21** M 0558 0.8 / 1134 6.3 / 1818 1.0 / 2353 6.3

7 SU 0413 0.9 / 0952 6.4 / 1636 1.0 / 2213 6.4 **22** M 0407 1.5 / 0943 5.8 / 1625 1.6 / 2201 5.9
7 W 0532 1.4 / 1110 5.7 / 1755 1.7 / 2332 5.8 **22** TH 0512 1.4 / 1050 5.7 / 1733 1.5 / 2310 6.0
7 F 0558 1.4 / 1134 5.6 / 1817 1.7 / 2351 5.8 **22** SA 0542 1.0 / 1118 6.1 / 1802 1.2 / 2337 6.3
7 M 0632 1.6 / 1206 5.5 / 1849 1.8 **22** TU 0638 1.0 / 1214 6.0 / 1859 1.4

8 M 0458 1.1 / 1036 6.1 / 1720 1.3 / 2258 6.1 **23** TU 0443 1.5 / 1018 5.7 / 1701 1.7 / 2237 5.8
8 TH 0617 1.6 / 1153 5.4 / 1839 1.9 **23** F 0555 1.4 / 1132 5.7 / 1816 1.6 / 2353 5.9
8 SA 0635 1.6 / 1210 5.5 / 1854 1.8 **23** SU 0622 1.1 / 1158 6.0 / 1843 1.3
8 TU 0021 5.5 / 0705 1.9 / 1238 5.3 / 1922 2.1 **23** W 0035 5.9 / 0722 1.6 / 1258 5.6 / ☾ 1946 1.9

9 TU 0542 1.4 / 1120 5.7 / 1805 1.7 / 2342 5.7 **24** W 0520 1.6 / 1056 5.6 / 1740 1.8 / 2316 5.7
9 F 0016 5.6 / 0702 1.9 / 1238 5.2 / 1926 2.1 **24** SA 0639 1.5 / 1216 5.6 / 1903 1.7
9 SU 0028 5.6 / 0713 1.8 / 1247 5.3 / 1932 2.0 **24** M 0019 6.2 / 0705 1.3 / 1240 5.8 / 1927 1.5
9 W 0055 5.3 / 0740 2.3 / 1315 5.1 / 2001 2.4 **24** TH 0124 5.4 / 0815 2.1 / 1354 5.2 / 2049 2.3

10 W 0629 1.8 / 1206 5.3 / 1853 2.1 **25** TH 0600 1.7 / 1137 5.4 / 1823 2.0
10 SA 0102 5.3 / 0749 2.1 / 1327 5.0 / 2016 2.3 **25** SU 0039 5.8 / 0727 1.6 / 1305 5.5 / 1952 1.8
10 M 0107 5.3 / 0751 2.1 / 1327 5.1 / 2013 2.3 **25** TU 0103 5.9 / 0750 1.5 / ☾ 1328 5.6 / 2016 1.7
10 TH 0137 4.9 / 0826 2.6 / 1402 4.8 / 2055 2.7 **25** F 0229 5.0 / 0929 2.6 / 1513 4.9 / 2218 2.6

11 TH 0030 5.4 / 0720 2.1 / 1258 5.0 / 1947 2.4 **26** F 0000 5.5 / 0647 1.9 / 1224 5.2 / 1913 2.1
11 SU 0153 5.2 / 0844 2.3 / 1421 4.9 / ☽ 2114 2.5 **26** M 0131 5.6 / 0820 1.7 / ☾ 1358 5.4 / 2051 1.9
11 TU 0149 5.2 / 0838 2.2 / 1413 4.9 / 2104 2.5 **26** W 0154 5.6 / 0845 1.9 / 1423 5.3 / 2117 2.1
11 F 0232 4.7 / 0929 2.9 / 1509 4.7 / 2211 2.9 **26** SA 0405 4.8 / 1108 2.7 / 1656 4.9 / 2357 2.5

12 F 0128 5.1 / 0820 2.4 / ☾ 1401 4.8 / 2057 2.6 **27** SA 0051 5.4 / 0740 2.0 / 1320 5.1 / 2011 2.2
12 M 0250 5.0 / 0945 2.4 / 1522 4.8 / 2218 2.5 **27** TU 0228 5.5 / 0921 1.8 / 1459 5.4 / 2155 1.9
12 W 0239 4.9 / 0932 2.5 / 1509 4.8 / 2205 2.6 **27** TH 0255 5.3 / 0954 2.2 / 1535 5.2 / 2235 2.3
12 SA 0355 4.6 / 1056 2.9 / 1641 4.7 / 2341 2.8 **27** SU 0544 4.9 / 1238 2.5 / 1822 5.2

13 SA 0237 4.9 / 0935 2.5 / 1517 4.7 / 2217 2.7 **28** SU 0151 5.3 / 0845 2.1 / ☾ 1426 5.1 / 2121 2.2
13 TU 0355 5.0 / 1049 2.5 / 1627 4.9 / 2319 2.5 **28** W 0333 5.4 / 1030 1.9 / 1609 5.4 / 2305 2.0
13 TH 0342 4.8 / 1039 2.7 / 1617 4.8 / 2313 2.7 **28** F 0416 5.2 / 1116 2.4 / 1700 5.2 / 2357 2.3
13 SU 0525 4.7 / 1221 2.7 / 1804 4.9 **28** M 0113 2.2 / 0656 5.2 / 1343 2.1 / 1924 5.5

14 SU 0357 4.9 / 1054 2.5 / 1635 4.8 / 2329 2.5 **29** M 0302 5.3 / 0959 2.0 / 1539 5.2 / 2236 2.0
14 W 0456 5.0 / 1148 2.4 / 1725 5.0 **29** TH 0444 5.4 / 1139 1.9 / 1719 5.4
14 F 0453 4.8 / 1148 2.6 / 1727 4.9 **29** SA 0542 5.1 / 1236 2.3 / 1820 5.3
14 M 0055 2.5 / 0638 5.0 / 1325 2.4 / 1906 5.3 **29** TU 0208 1.8 / 0748 5.5 / 1430 1.8 / 2009 5.8

15 M 0508 5.0 / 1159 2.3 / 1737 5.0 **30** TU 0415 5.4 / 1110 1.9 / 1649 5.4 / 2342 1.8
15 TH 0015 2.4 / 0552 5.1 / 1240 2.3 / 1817 5.2 **30** F 0012 1.9 / 0553 5.5 / 1244 1.9 / 1825 5.6
15 SA 0015 2.4 / 0552 5.0 / 1240 2.3 / 1817 5.1 **30** SU 0020 2.5 / 0600 4.9 / 1250 2.5 / 1830 5.1
15 TU 0152 2.0 / 0732 5.3 / 1415 1.9 / 1955 5.7 **30** W 0251 1.5 / 0829 5.7 / 1511 1.5 / 2047 6.1

15 (additional) 0111 2.1 / 0655 5.3 / 1343 2.0 / 1926 5.6 **30** 0152 2.0 / 0732 5.3 / 1415 1.9 / 1955 5.7

31 W 0520 5.6 / 1212 1.6 / 1750 5.6
31 M 0211 1.8 / 0753 5.5 / 1438 1.8 / 2018 5.8
31 TH 0328 1.3 / 0904 6.0 / 1545 1.3 / ● 2120 6.2

Chart Datum: 3.70 metres below Lallemand System (Mean Sea Level, Marseilles)

FRANCE, NORTH COAST - CHERBOURG

LAT 49°39′N LONG 1°38′W

TIMES AND HEIGHTS OF HIGH AND LOW WATERS

YEAR **1989**

TIME ZONE −0100 (French Standard Time)
For French Summer Time add ONE hour in non-shaded areas

SEPTEMBER

	TIME	m		TIME	m
1 F	0402 0936 1617 2151	1.2 6.1 1.2 6.3	**16** SA	0335 0912 1555 2132	0.7 6.6 0.7 6.9
2 SA	0432 1005 1646 2220	1.1 6.1 1.2 6.2	**17** SU	0415 0952 1634 2210	0.5 6.8 0.6 6.9
3 SU	0500 1034 1714 2248	1.2 6.0 1.3 6.1	**18** M	0454 1030 1713 2250	0.6 6.7 0.7 6.7
4 M	0527 1101 1741 2314	1.4 5.9 1.5 5.9	**19** TU	0533 1109 1754 2329	0.8 6.1 1.1 6.3
5 TU	0556 1128 1809 2342	1.6 5.7 1.7 5.6	**20** W	0615 1150 1835	1.3 6.0 1.5
6 W	0624 1157 1839	1.9 5.4 2.0	**21** TH	0012 0659 1236 1925	5.7 1.8 5.5 2.1
7 TH	0013 0656 1230 1915	5.3 2.2 5.2 2.4	**22** F ☾	0104 0754 1336 2032	5.2 2.4 5.1 2.6
8 F ☽	0050 0736 1313 2002	5.0 2.6 4.8 2.8	**23** SA	0215 0917 1504 2211	4.8 3.0 4.8 2.8
9 SA	0142 0837 1418 2119	4.7 3.0 4.6 3.0	**24** SU	0400 1105 1654 2353	4.7 2.8 4.9 2.6
10 SU	0306 1013 1603 2307	4.5 3.1 4.6 2.9	**25** M	0539 1231 1814	4.9 2.5 5.2
11 M	0455 1154 1740	4.6 2.8 4.9	**26** TU	0102 0643 1328 1907	2.2 5.2 2.1 5.5
12 TU	0032 0614 1303 1844	2.5 5.0 2.3 5.4	**27** W	0151 0728 1410 1947	1.8 5.5 1.8 5.8
13 W	0130 0709 1354 1932	1.9 5.5 1.8 5.9	**28** TH	0228 0804 1445 2021	1.5 5.8 1.5 6.1
14 TH	0215 0753 1435 2013	1.4 5.9 1.3 6.3	**29** F ●	0301 0836 1517 2051	1.4 6.0 1.4 6.2
15 F ○	0255 0833 1516 2053	1.0 6.4 0.9 6.7	**30** SA	0331 0905 1545 2119	1.2 6.1 1.3 6.2

OCTOBER

	TIME	m		TIME	m
1 SU	0400 0934 1614 2148	1.2 6.1 1.3 6.2	**16** M	0347 0924 1608 2146	0.6 6.8 0.7 6.8
2 M	0428 1002 1642 2215	1.3 6.1 1.4 6.1	**17** TU	0429 1006 1651 2228	0.7 6.7 0.8 6.6
3 TU	0456 1029 1709 2243	1.4 6.0 1.5 5.9	**18** W	0512 1050 1734 2311	1.1 6.4 1.2 6.1
4 W	0523 1056 1737 2311	1.6 5.8 1.8 5.6	**19** TH	0556 1134 1820 2357	1.5 6.0 1.7 5.6
5 TH	0553 1126 1808 2342	1.9 5.5 2.0 5.3	**20** F	0645 1223 1913	2.0 5.5 2.2
6 F	0626 1200 1845	2.3 5.3 2.4	**21** SA ☾	0051 0743 1325 2020	5.2 2.5 5.1 2.6
7 SA	0020 0707 1244 1933	5.0 2.6 5.0 2.7	**22** SU	0204 0904 1449 2152	4.8 2.8 4.9 2.7
8 SU ☽	0113 0806 1349 2049	4.7 3.0 4.7 2.9	**23** M	0340 1042 1627 2325	4.7 2.8 5.0 2.6
9 M	0236 0940 1530 2236	4.6 3.0 4.7 2.8	**24** TU	0508 1201 1742	4.9 2.5 5.2
10 TU	0423 1122 1707	4.7 2.7 5.1	**25** W	0030 0609 1255 1834	2.3 5.2 2.2 5.5
11 W	0001 0542 1232 1811	2.4 5.1 2.2 5.5	**26** TH	0117 0655 1337 1914	2.0 5.5 1.9 5.7
12 TH	0058 0637 1322 1900	1.8 5.6 1.6 6.0	**27** F	0155 0731 1412 1948	1.8 5.7 1.7 5.9
13 F	0144 0722 1404 1943	1.3 6.1 1.2 6.4	**28** SA	0228 0804 1443 2019	1.6 5.9 1.5 6.0
14 SA	0225 0803 1445 2023	0.9 6.5 0.8 6.8	**29** SU ●	0300 0834 1514 2049	1.5 6.0 1.5 6.1
15 SU	0306 0843 1526 2103	0.7 6.8 0.7 6.9	**30** M	0329 0903 1544 2118	1.5 6.1 1.4 6.0
			31 TU	0359 0934 1614 2149	1.5 6.1 1.5 6.0

NOVEMBER

	TIME	m		TIME	m
1 W	0429 1003 1645 2219	1.6 6.0 1.6 5.8	**16** TH	0458 1037 1722 2301	1.3 6.3 1.3 6.0
2 TH	0500 1035 1716 2251	1.8 5.8 1.8 5.6	**17** F	0546 1124 1811 2349	1.6 6.0 1.7 5.6
3 F	0532 1108 1751 2325	2.0 5.6 2.0 5.4	**18** SA	0635 1214 1902	2.0 5.6 2.0
4 SA	0610 1145 1830	2.2 5.4 2.3	**19** SU	0040 0729 1310 2000	5.3 2.4 5.3 2.4
5 SU	0006 0654 1231 1921	5.1 2.5 5.2 2.5	**20** M ☾	0142 0835 1415 2112	5.0 2.6 5.2 2.5
6 M	0100 0752 1334 2030	4.9 2.7 5.0 2.7	**21** TU	0252 0950 1531 2229	4.9 2.7 5.1 2.6
7 TU	0213 0912 1455 2157	4.8 2.8 5.0 2.6	**22** W	0410 1105 1644 2337	4.9 2.6 5.2 2.4
8 W	0341 1040 1622 2317	4.9 2.5 5.2 2.2	**23** TH	0515 1205 1743	5.1 2.4 5.3
9 TH	0458 1150 1729	5.3 2.1 5.6	**24** F	0030 0607 1253 1830	2.3 5.3 2.2 5.4
10 F	0019 0557 1244 1822	1.8 5.7 1.6 6.0	**25** SA	0114 0651 1334 1911	2.1 5.5 2.0 5.5
11 SA	0108 0647 1331 1910	1.4 6.1 1.3 6.3	**26** SU	0153 0730 1411 1948	2.0 5.6 1.9 5.7
12 SU	0154 0733 1416 1955	1.1 6.4 1.0 6.6	**27** M	0228 0805 1446 2022	1.9 5.8 1.7 5.8
13 M	0238 0818 1502 2040 ○	0.9 6.6 0.9 6.6	**28** TU	0304 0839 1521 2056 ●	1.8 5.9 1.7 5.8
14 TU	0325 0903 1548 2126	0.9 6.7 0.9 6.6	**29** W	0337 0913 1554 2131	1.7 6.0 1.6 5.8
15 W	0412 0951 1635 2213	1.0 6.6 1.1 6.3	**30** TH	0412 0948 1630 2204	1.7 6.0 1.6 5.8

DECEMBER

	TIME	m		TIME	m
1 F	0447 1022 1705 2241	1.8 5.9 1.7 5.7	**16** SA	0537 1115 1759 2336	1.5 6.1 1.5 5.7
2 SA	0524 1059 1743 2319	1.9 5.8 1.7 5.5	**17** SU	0622 1158 1844	1.7 5.9 1.7
3 SU	0603 1138 1824	2.0 5.6 2.0	**18** M	0020 0707 1243 1929	5.5 2.0 5.6 2.0
4 M	0001 0647 1224 1912	5.4 2.2 5.5 2.1	**19** TU ☾	0106 0753 1330 2019	5.3 2.3 5.4 2.2
5 TU	0050 0738 1318 2007	5.2 2.3 5.4 2.2	**20** W	0156 0847 1423 2116	5.1 2.5 5.2 2.5
6 W	0148 0840 1420 2115	5.2 2.3 5.3 2.2	**21** TH	0252 0948 1526 2221	4.9 2.6 5.0 2.6
7 TH	0254 0951 1531 2227	5.3 2.3 5.4 2.1	**22** F	0359 1054 1633 2327	4.9 2.7 5.0 2.6
8 F	0407 1102 1641 2335	5.3 2.1 5.5 1.9	**23** SA	0506 1158 1737	5.0 2.6 5.0
9 SA	0514 1205 1744	5.6 1.8 5.7	**24** SU	0027 0605 1253 1832	2.6 5.1 2.4 5.2
10 SU	0034 0613 1301 1842	1.7 5.8 1.6 6.0	**25** M	0117 0656 1341 1919	2.4 5.3 2.2 5.3
11 M	0129 0709 1355 1936	1.5 6.1 1.4 6.2	**26** TU	0202 0741 1423 2001	2.2 5.4 2.0 5.5
12 TU	0221 0802 1447 2027	1.4 6.3 1.2 6.2	**27** W	0243 0821 1504 2040	2.0 5.7 1.8 5.6
13 W	0313 0853 1538 2117	1.3 6.4 1.2 6.2	**28** TH ●	0323 0859 1542 2118	1.9 5.9 1.6 5.7
14 TH	0404 0943 1628 2206	1.3 6.4 1.2 6.2	**29** F	0401 0937 1620 2156	1.7 6.0 1.5 5.8
15 F	0452 1030 1715 2253	1.4 6.3 1.3 6.0	**30** SA	0439 1014 1658 2233	1.6 6.1 1.4 5.8
			31 SU	0516 1053 1735 2311	1.5 6.1 1.4 5.8

Chart Datum: 3.70 metres below Lallemand System (Mean Sea Level, Marseilles)

AREA 17—Channel Islands and French Coast 689

MINOR HARBOURS AND ANCHORAGES 10.17.25

ST AUBIN, Jersey, 49°11′N, 2°10′W, Zone 0 (GMT), Admty chart 1137. HW −0455 on Dover; local times and heights, as St Helier; ML 6.1m; Duration 0545. See 10.17.11. Good quiet anchorage in bay with off-shore winds, or yachts can go alongside N quay and dry. Alternative Belcroute Bay on E of Noirmont Pt — steep to, deep water, excellent shelter from W to SW winds; landing by dinghy. Ent to N of St Aubin Fort (on I) in W sector of Lt. N pier head Iso R 4s 12m 10M and Dir Lt 252°, Dir F WRG 5m, G246°−251°, W251°−253°, R253°−258°. St Aubin Fort I pier head Fl (2) Y 5s 8m 1M. Facilities: FW on N quay, Slip; **Royal Channel Islands YC** Tel. 45783; **Battricks** Tel. 43412, BY, Gas; **Jacksons Yacht Services** Tel. 43819. CH, El, ME, Sh; **St Aubin BY** Tel. 45499 ME, El, Sh, CH; **J. Fox** Tel. 44877 ME; **M. Paddock** ME, Sh; **Town** Bar, D, FW, Lau, P, R, V, buses to St Helier.

ST SAMPSON, Guernsey, 49°29′N, 2°31′W, Zone 0 (GMT), Admty chart 808. HW −0450 on Dover. See 10.17.14. Harbour dries. Guernsey's second harbour and very commercial. Official port of entry. Good shelter but the disadvantages of a commercial port. Lt on Crocq pier head FR 11m 5M; vis 250°−340°, and traffic signals. N pier head FG 3m 5M; vis 230°−340°. Ldg Lts 286°, Front on S pier head FR 3m 5M; vis 230°−340°; Rear, 390m from front, on clocktower, FG 13m. VHF Ch 16; 12 (H24). Facilities: See Hr Mr for AB, C, FW. All facilities at BY for commercial shipping. There are no facilities for yachts. **Marine and General** Tel. 45808, Slip, ME, El, Sh, C. Buses to St Peter Port.

DAHOUET, Côtes du Nord, 48°35′N, 2°34′W, Zone −0100, Admty charts 2668, 2669, SHOM 833. HW −0520 on Dover (GMT), +0035 on St Helier (zone −0100); HW height +0.2m on St Helier; ML 6.1m; Duration 0550. Good shelter except in strong NW winds. Hr dries, accessible HW∓2. Approach channel is marked and there is a Lt Tr, La Petite-Muette, Fl WRG 4s 10m 9/6M, G055°−114°, W114°−146°, R146°−196°. Fishing boats use outer Hr, yachts inner Hr on Vieux Quai. Facilities: **Quay** Hr Mr, Customs, C (5 ton), P, D; **YC du Val-André** Tel. 96.72.21.68. **Bouguet** Tel. 96.72.97.00. CH, El, ME, Sh; **Town** Bar, R, V.

VAL-ANDRÉ, Côtes du Nord, 48°36′N, 2°33′W, Zone −0100, Admty charts 2668, 2669, SHOM 833. HW −0520 on Dover (GMT), +0035 on St Helier (zone −0100); HW height +0.2m on St Helier; ML 6.1m; Duration 0550. A small drying harbour exposed to S and SW winds. Beware Verdelet coming from E and Platier des Trois Têtes from W. Yachts can go alongside the quay, ask YC for mooring off Le Piegu or anchor off. Hr accessible HW∓3. Facilities: Hr Mr Tel. 96.72.83.20, FW, Slip; **YC du Val-André** Tel. 96.72.21.68. Bar, R; **Troalen** Tel. 96.72.20.20. CH, El, ME, Sh; **Town** Bar, R, V.

ERQUY, Côtes du Nord, 48°38′N, 2°28′W, Zone −0100, Admty charts 3672, 2668, 2669, SHOM 833, 5724. HW −0515 on Dover (GMT), +0040 on St Helier (zone −0100); HW height +0.2m on St Helier; ML 6.1m; Duration 0550. See 10.17.19. Sheltered from E, but exposed to SW or W winds. Hr dries and is usually full of fishing boats. Beware Plateau des Portes d'Erquy (dry) about 2M to West. Coming from S, beware of rks off Pte de la Houssaye. Lt on mole-head Oc(2+1) WRG 12s 11m 11/6M R055°−081°, W081°−094°, G094°−111°, W111°−120°, R120°−134°. Inner jetty, head Fl R 2.5s 10m 3M. Facilities: Hr Mr and Customs Tel. 96.72.19.32; **Quay** C (1 ton), D, FW, P; **Cercle de la Voile d'Erquy** Tel. 96.72.32.40; **Régina Plaisance** Tel. 96.72.13.70. CH, El, ME, Sh; **Town** Bar, R, V.

SAINT-CAST, Côtes du Nord, 48°38′N, 2°15′W, Zone −0100, Admty charts 3659, 2669, SHOM 5646. HW −0515 on Dover (GMT), +0040 on St Helier (zone −0100); HW height +0.2m on St Helier; ML 6.3m; Duration 0550. Good shelter from S through W to NW, and moorings are available in 1.8m. Beware Les Bourdinots (dry 2m) with E cardinal buoy ¾M NE of Pte de Saint-Cast, and La Feuillâtre (marked by Bn) and Bec Rond off harbour. Mole-head Lt Iso WG 4s 11m 11/8M, W204°−217°, G217°−233°, W233°−245°, G245°−204°. Facilities: Hr Mr Tel. 96.41.88.34. SNSM Tel. 96.41.88.34. **YC de Saint-Cast** Tel. 96.41.91.77. **La Maison Blanche** Tel. 96.41.81.40. CH, El, ME, Sh; **L.M.B. Marine** Tel. 96.41.80.23. CH, El, ME, Sh; **Town** Bank, Bar, D, P, PO, R, V.

RICHARDAIS, Ille-et-Vilaine, 48°34′N, 2°02′W, Zone −0100, Admty chart 2669, SHOM 4233. HW (at lock) −0515 on Dover (GMT). Pass through the lock in the dam above St Servan, on W side, opens every hour on the hour when tide is over CD+4m. Port of La Richardais approx 1M beyond lock to W. Complete shelter except from NE winds. R Rance navigable up to Dinan with ease. For water levels in Rance basin and times of operation of power station Tel. 99.46.14.46. Facilities: Hr Mr Tel. 99.46.24.20. **Naviga Voile** Tel. 99.46.99.25. CH, El, ME, Sh; **Town** Bar, D, P, PO, R, V, Bank.

ROTHENEUF, Ille-et-Vilaine, 48°42′N, 1°56′W, Zone −0100, Admty charts 2700, 3659, SHOM 5644. HW −0510 on Dover (GMT), +0040 on St Helier (zone −0100); HW height +0.2m on St Helier; ML 7.0m; Duration 0540. Complete shelter in Hr which dries completely. Anchor outside in 4m just N of spar Bn marking ent. Rocks on both sides of ent which is less than 170m wide. Safest to enter when rks uncovered. There are no lights; ldg line at 163° W side of Pte Benard and old converted windmill. Facilities: FW, Slip; **Village** Bar, D, P, R, V.

CANCALE, Ille-et-Vilaine, 48°40′N, 1°51′W, Zone −0100, Admty chart 3659, SHOM 5644, 824. HW −0510 on Dover (GMT), +0045 on St Helier (zone −0100); HW height +0.2m on St Helier; ML 7.5m; Duration 0535. See 10.17.20. A drying Hr just inside Bay of Mont St Michel, 1M SW of Point de la Chaine. Area dries to about 1M offshore; anchor off end of Pte de la Chaine in deep water. Drying berths usually available in La Houle, the Hr in Cancale. Exposed to winds SW to SE. Jetty-head Lt Oc(3)G 12s 12m 8M, obsc when bearing less than 223°. Facilities: **Quay** C (1.5 ton) D, FW, P; **Club Nautique de Cancale** Tel. 99.89.90.22; **Froc** Tel. 99.89.61.74. El, M, ME, Sh; **Town** (famous for oysters), Bank, Bar, D, P, PO, R, V.

PORTBAIL, Manche, 49°19′N, 1°43′W, Zone −0100, Admty chart 2669, SHOM 827. HW −0440 on Dover (GMT), +0110 on St Helier (zone −0100); HW height −0.2m on St Helier; ML 6.3m; Duration 0545. See 10.17.21. Good shelter. Hr dries but access HW∓½ at np, HW∓2½ at sp for a draught of 1m. Beware very strong tide over bar. Passage through sand banks buoyed to Training Wall (covers at HW) which is marked by R spar Bns. Drying harbour E of jetty — 1st line of buoys parallel to jetty for visitors; slip to W of jetty. Training Wall head Q(2)R 5s 5m 2M. Ldg Lts 042°, Front (La Caillourie) QW 14m 11M. Rear, 870m from front, QW 20m 9M. VHF Ch 09. Facilities: Hr Mr Tel. 33.04.33.48; **Quay** C (5 ton) D, FW, P; **Cercle Nautique de Portbail-Denneville** Tel. 33.04.86.15. Bar, R; **YC de Portbail** Tel. 33.04.83.48. AB, C, Slip; **Gérard** Tel. 33.04.80.07. El, ME, Sh; **Le Cornec** Sh (wood); **Fleury** Sh, ME, El; **Flambard** Sh, ME, El; **Town** Bar, Bank, PO, R, Rly (Valognes), V.

BOATWORKS+
CASTLE EMPLACEMENT
ST PETER PORT, GUERNSEY C.I
Telephone (0481) 26071
Telex 4191576 POSTIV G
Fax (0481) 710254

CHANNEL ISLANDS
LEADING CHANDLERS

ELECTRONICS SALES
& SERVICE

FULL BOAT YARD FACILITIES

ADMIRALTY CHART AGENTS
& DISTRIBUTORS

BOATWEAR & LEISURE WEAR

BOATCARE & GARDIENNAGE

FUELING BERTH IN
ST. PETER PORT

BROKERAGE &
NEW BOAT SALES INCLUDING
TAX FREE EXPORT SALES

C.I. DEALERS FOR BENETEAU

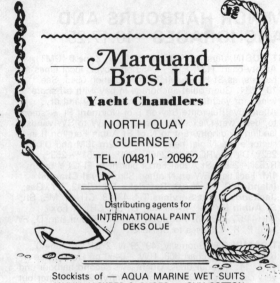

Marquand Bros. Ltd.
Yacht Chandlers

NORTH QUAY
GUERNSEY
TEL. (0481) - 20962

Distributing agents for
INTERNATIONAL PAINT
DEKS OLJE

Stockists of — AQUA MARINE WET SUITS — JAVLIN JACKETS & SHOES — GUY COTTON FOUL WEATHER GEAR — ROMIKA AND NOKIA BOOTS — HELLY HANSEN CLOTHING — OPTIMUS COOKERS — DUBARRY SHOES — CREWSAVER — BLAKES TOILETS — AVON — SEAFARER — THAI TEAK — GIBB — MAIN MARINE — SIMPSON LAWRENCE — PLASTIMO — IMRAY & STANFORD CHARTS — MARLOW ROPES — ANODES — BRETONS — FENDERS — GALLEY EQUIPMENT — POTS — PANS — PARAFFIN — GAS — METHYLATED SPIRITS — FISHING TACKLE — IRONMONGERY — CARPENTERS & MECHANICS TOOLS — TBS YACHTING SHOES, etc., etc., etc.

Channel Yacht Brokers Ltd

Practical offshore yacht brokers specialising in VAT free boats. Conditions and commissions as per BMIF Code of Practice but with no VAT added. No sale no commission.

**Phone George Llewellin
on (0481) 22282**

Concordia, Les Vardes,
St. Peter Port, Guernsey.

Synthetic Lubricants
Marine Diesel Engine Oil
Waterproof Greases
Marine Gear Box Oils

Require Distributors
For FREE information for
Individual and Trade Enquiries

John Woodington (AMSOIL)
184 Watford Road, St. Albans,
Herts., AL2 3EB 0727 66971

**DEDICATED MARINE POWER
9 to 422hp**

VOLVO PENTA

**DEDICATED MARINE SERVICE
OVER 1300 LOCATIONS ACROSS EUROPE**

VOLVO PENTA SERVICE

Sales and service centres in area 18
Names and addresses of Volvo Penta dealers in this area are available from:

FRANCE **Volvo Penta France SA**, BP45, F78130 Les Mureaux
Tel (01) 3 474 72 01, Telex 695221 F.
BELGIUM **Volvo Penta Belgium**, Weiveldlaan 37-G, B-1930 Zaventem
Tel (02) 721-20-62. Telex 65249, Volvo BMB.

VOLVO PENTA

Area 18

North-East France
Barfleur to Dunkerque

10.18.1	Index	Page 691
10.18.2	Diagram of Radiobeacons, Air Beacons, Lifeboat Stations etc	692
10.18.3	Tidal Stream Charts	694
10.18.4	List of Lights, Fog Signals and Waypoints	696
10.18.5	Passage Information	698
10.18.6	Distance Tables	699
10.18.7	English Channel Waypoints	See 10.1.7
10.18.8	Special notes for French areas	See 10.14.7
10.18.9	Barfleur	700
10.18.10	St Vaast-la-Hougue	700
10.18.11	Carentan	701
10.18.12	Grandcamp	702
10.18.13	Port-en-Bessin	702
10.18.14	Courseulles-sur-Mer	703
10.18.15	Ouistreham	704
10.18.16	Deauville/Trouville	704
10.18.17	Honfleur	706
10.18.18	River Seine	706
10.18.19	Le Havre, Standard Port, Tidal Curves	707
10.18.20	Fécamp	711
10.18.21	St Valéry-en-Caux	712
10.18.22	Dieppe, Standard Port, Tidal Curves	713
10.18.23	Le Tréport	717
10.18.24	St Valéry-sur-Somme	718
10.18.25	Le Touquet	719
10.18.26	Boulogne	719
10.18.27	Calais	720
10.18.28	Dunkerque — Tidal Curves	721
10.18.27	Minor Harbours and Anchorages Iles St Marcouf Isigny-sur-Mer Caen Cabourg Rouen Le Crotoy Étaples Gravelines	723

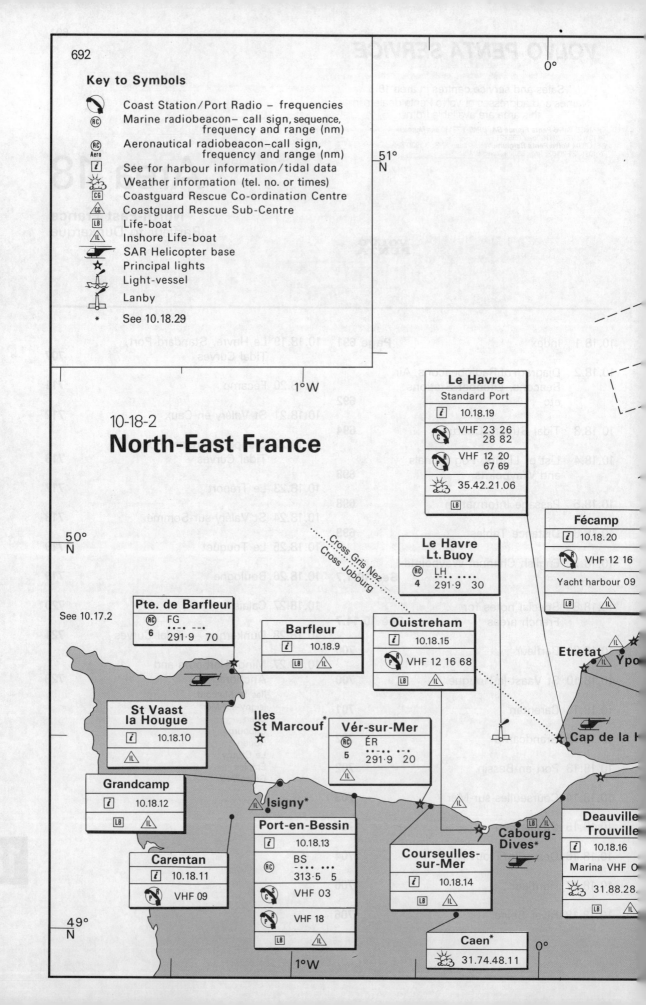

10-18-2
North-East France

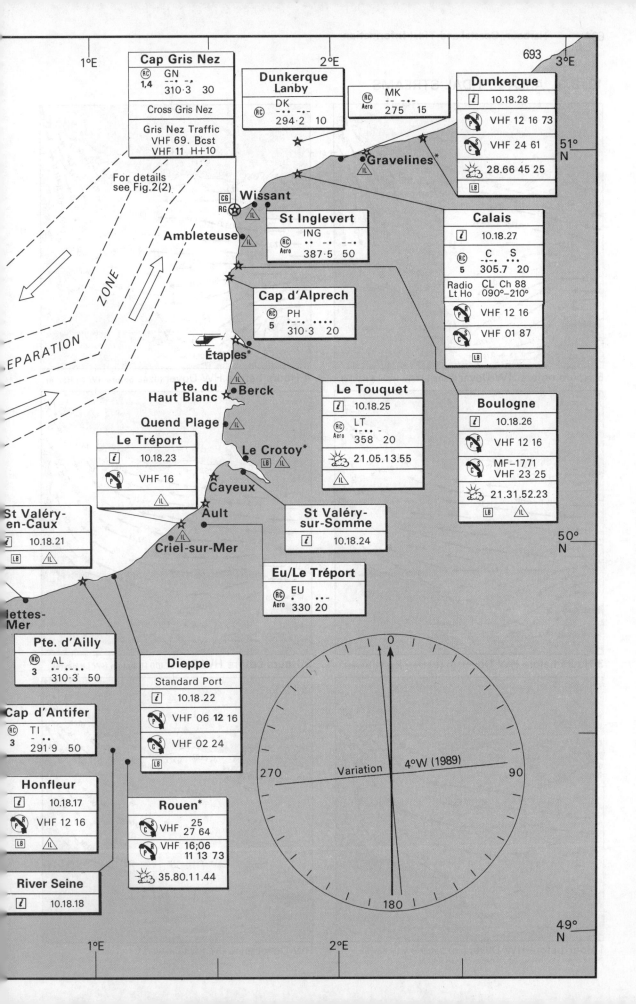

10.18.3 AREA 18 TIDAL STREAMS

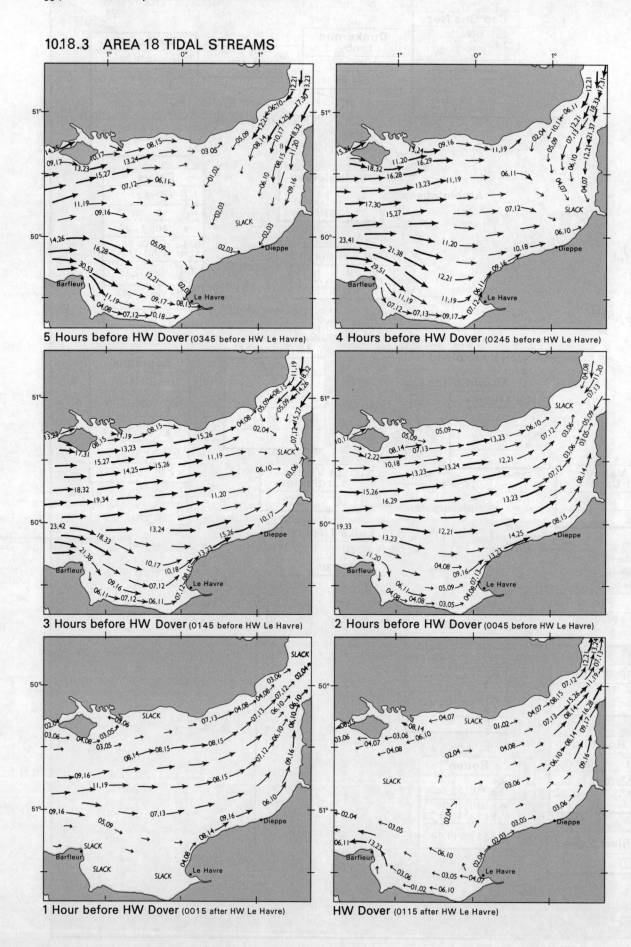

AREA 18 — NE France 695

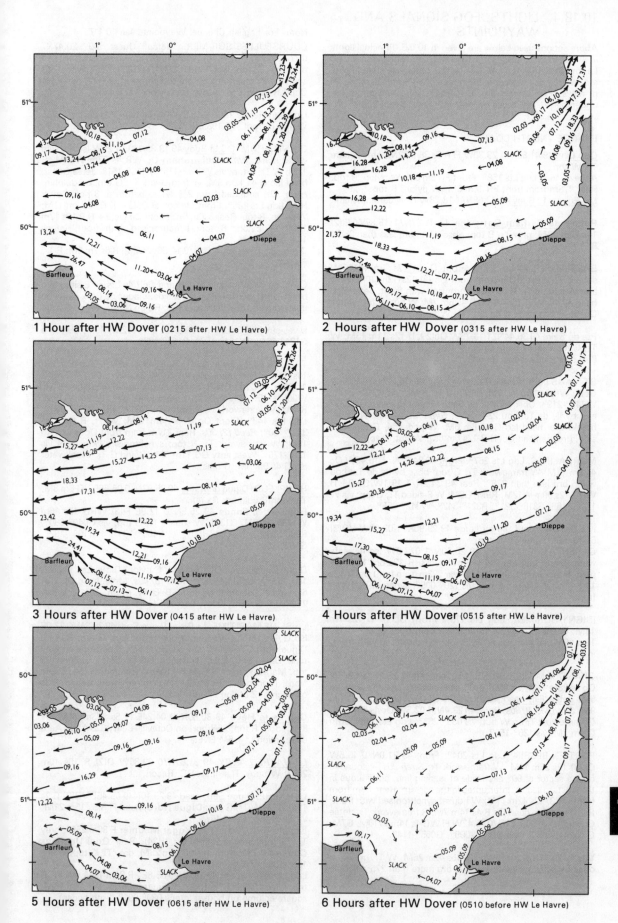

10.18.4 LIGHTS, FOG SIGNALS AND WAYPOINTS

Abbreviations used below are given in 10.0.3. Principal lights are in **bold** print, places in CAPITALS, and light-vessels and Lanbys in *CAPITAL ITALICS*. Unless otherwise stated lights are white. m—elevation in metres; M—nominal range in n. miles. Fog signals are in *italics*. Useful waypoints are underlined — use those on land with care. See 4.2.2.

FRANCE—NORTH COAST

<u>Cap Lévi</u> 49 41.80N/1 28.30W Fl R 5s 36m **22M**; grey square Tr, W top.
Anse de Vicq Ldg Lts 158°. Front FR 8m 6M; W pylon, R top. Rear 428m from front FR 14m 6M; W pylon, R top.
<u>Les Equets Lt Buoy</u> 49 43.68N/1 18.28W Q; N cardinal mark.

<u>Pointe de Barfleur-Gatteville</u> 49 41.87N/1 15.87W Fl(2) 10s 72m **29M**; grey Tr, B top; obsc when brg less than 088°; RC; *Reed (2) 60s*.

BARFLEUR. Ldg Lts 219°. **Front** Oc(3) 12s 7m 10M; W square Tr. **Rear** 283m from front Oc (3) 12s 13m 10M; grey and W square Tr, G top; synchronised with front. Jetée Est, head 49 40.4N/1 15.4W Oc R 4s 5m 7M; W hut, R top. Jetée Ouest, near root Fl G 4s 8m 7M; W framework Tr, G top.

Pte de Saire 49 36.44N/1 13.75W Oc(2+1) 12s 11m 13M; W Tr, G top. <u>Val de Saire Lt Buoy</u> 49 43.7N/1 04.5W Iso 4s 9m 8M; safe water mark; Ra refl.

ST VAAST-LA-HOUGUE. <u>Jetty head</u> 49 35.22N/1 15.33W Oc(2) WR 6s 12m W11M, R8M; W Tr, R top; vis R219°-247°, W247°-219°. Obsc by Pte de Saire when brg less than 223° and by Île de Tatihou 230°-249°. *Siren Mo(N) 30s*. NE side, breakwater head Iso G 4s 6m 6M; W pedestal, G top. SW side, groyne head Oc(4) R 12s 6m 6M; W hut, R top.

MORSALINES. Ldg Lts 267°. <u>Front</u> 49 34.28N/1 16.45W Oc 4s 9m 11M; W framework Tr, G top; obsc by Île de Tatihou when brg less than 228°. Rear 49 34.2N/1 19.1W Oc(3+1) WRG 12s 90m W12M, R9M, G8M; W 8-sided Tr, G top; vis W 171°-316°, G316°-321°, R321°-342°, W342°-355°.

Îles Saint-Marcouf, <u>Île du Large</u> 49 29.90N/1 08.90W VQ(3) 5s 18m 9M; grey Tr, G top.
CARENTAN. <u>Cl Lt Buoy</u> 49 25.50N/1 07.08W Iso 4s; safe water mark. Ldg Lts 210°. **Front** Oc(3) R 12s 6m **17M**; W post, R top; intens 209°-211°. Rear 723m from front Oc(3) 12s 14m 11M; W gantry, G top; vis 120°-005°; synchronised with front.

ISIGNY-SUR-MER. Ldg Lts 173°. **Front** Oc(2+1) 12s 7m **18M**; intens 171°-175°. **Rear** 600m from front Oc(2+1) 12s 19m **18M**; W framework Tr, B top; synchronised with front, intens 171°-175°.

GRANDCAMP-LES-BAINS. Maresquerie 49 23.2N/1 02.7W Oc 4s 28m 12M; vis 090°-270°. Jetée Ouest, head Fl G 4s 9m 6M. <u>Jetée Est, head</u> 49 23.50N/1 03.00W Oc(2) R 6s 10m 9M; *Siren Mo(N) 30s*. Perré 49 23.4N/1 02.5W Oc 4s 8m 13M; G framework Tr on W hut; vis 083°-263°. Ldg Lts 146°, both Q 4m 7M; vis 136°-156°.

PORT-EN-BESSIN. Ldg Lts 204°. Front 49 21.0N/0 45.5W Oc(3) 12s 25m 10M; W framework Tr; *Siren 20s* — sounded over a sector of 90° each side of leading line, continuous in the western sector, interrupted in the eastern. Rear, 93m from front Oc(3) 12s 42m 11M; W house; synchronised with front; RC. Jetée Est, head Oc R 4s 14m 7M; R framework Tr. <u>Jetée Ouest, head</u> 49 21.21N/0 45.42W Fl WG 4s 14m W10M, G7M; G framework Tr; vis W114°-065°, G065°-114°.

<u>Ver</u> 49 20.47N/0 31.15W Fl(3) 15s 42m **26M**; W Tr, grey top; obsc by cliffs of St Aubin when brg more than 275°; RC.

Note. For English Channel Waypoints see 10.1.7

COURSEULLES-SUR-MER. <u>Jetée Ouest</u> 49 20.47N/ 0 27.28W Iso WG 4s 7m W9M, G5M; brown framework Tr on dolphin, G top; vis W135°-235°, G235°-135°; *Horn 30s* sounded from 2 hours before to 2 hours after HW. Jetée Est Oc(2) R 6s 9m 7M; brown framework Tr, R top.

Lion-sur-Mer 49 18.2/0 18.8W Iso 4s 16m 7M; building with W and R front. TE 1985.

OUISTREHAM. <u>OC Lt Buoy</u> 49 19.89N/0 14.25W Iso 4s; safe water mark; *Whis*. **Ouistreham** Oc WR 4s 37m **W17M**, R 13M; W Tr, R top; vis W151°-115°, R115°-151°. Jetée Ouest, head Iso G 4s 12m 9M; W framework Tr, G top, on dolphin; *Horn 10s*, sounded from 2½ hours before to 3 hours after HW. Ldg Lts 186°. Front, Jetée Est Oc(3+1) R 12s 8m 11M; W pylon, R top. Rear 530m from front Dir Oc(3+1) R 12s 17m 14M; W pylon, R top; synchronised with front; intens 184°-188°.
<u>Digue Est, head</u> 49 18.08N/0 14.56W Oc(2) R 6s 7m 8M; R Tr; Ra refl.

Canal de Caen marked by QR and OcR 4s Lts on E side, and by QG and IsoG 4s Lts on West side. Viaduc de Calix Iso 4s on upstream and downstream sides; FG on N side, FR on S side.

Dives-sur-Mer Oc(2+1) WRG 12s 6m W12M, R9M, G9M; vis G124°-154°, W154°-157°, R157°-193°.

TROUVILLE. <u>Trouville SW Lt Buoy</u> 49 22.68N/0 02.64E VQ(9) 10s; W cardinal mark. <u>West jetty</u> 49 22.44N/0 04.17E Fl WG 4s 10m W12M, G9M; B framework mast on dolphin; vis W005°-176°, G176°-005°. E jetty Fl(4) WR 12s 8m W10M, R7M; W framework Tr, R top; vis W131°-175°, R175°-131°. Ldg Lts 148°. Front Oc R 4s 11m 12M; W Tr, R top; vis 330°-150°; *Reed (2) 30s*. Rear 217m from front Oc R 4s 17m 10M; W framework Tr, R top; synchronised with front vis 120°-170°. West jetty QG 11m 9M; W and G Tr. Breakwater Iso G 4s 9m 5M; G mast.

HONFLEUR. <u>Digue du Ratier</u> 49 25.97N/0 06.66E VQ 8m 8M; Tank on B column; Ra refl.
Falaise des Fonds 49 25.5N/0 12.9E Fl(3) WRG 12s 15m **W17M**, R13M, G13M; W square Tr, G top; vis G040°-080°, R080°-084°, G084°-100°, W100°-109°, R109°-162°, G162°-260°.

Digue Ouest, head QG 10m 6M; G pylon. Digue Est, head Q 10m 9M; N cardinal mark; *Reed (5) 40s*. Mole, head Oc(2) R 6s. Jetée Transit F Vi 10m; B & W pylon. Quay, West, 2FG(vert) 6m 5M. Quay, East 2FG(vert) 6m 5M.

LA SEINE MARITIME. La Risle, Digue Sud, 49 26.3N/0 22.0E Iso G 4s 11m 7M; W framework Tr and hut, G top; Ra refl. Digue Nord, Tourelle Ygou, VQR 7m 5M; R pedestal on grey tank. Digue Sud, Épi de la Roque, QG 8m 6M; W column, G top. Marais-Vernier Fl G 4s 8m 5M; W column, G top. Digue Nord, Tancarville, QR 9m 6M; W column, R top. Aero Fl R 3s on each of 2 bridge pillars.

<u>Cap de la Hève</u> 49 30.80N/0 04.24E Fl 5s 123m **24M**; W 8-sided Tr, R top; 2FR(hor) on Octeville belfry 3.2M NE; Aero obstruction Lt 1.6M ENE.

LE HAVRE LANBY 49 31.67N/0 09.80W Q(2) R 10s 10m **20M**; W buoy, R stripes; RC; Racon.

LE HAVRE. <u>Digue Sud, head</u> 49 29.10N/0 05.45E VQ(3) G 2s 15m 12M; W Tr, G top. **Digue Nord**, head Fl R 5s 15m **21M**; W Tr, R top; *Reed 15s*.
Ldg Lts 107°. Front **Quai Roger Meunier** F 36m **25M**; grey Tr, G top; intens 106°-108°; (H24). Rear **Quai Joannes Couvert** 0.73M from front F 78m **25M**; grey Tr, G top; intens 106°-108°; (H24).
Ldg Lts 090°. **Front** 49 29.5N/0 05.8E FR 21m **18M** ; W Tr, R top; intens 089°-091°. **Rear** , 620m from front, FR 43m **18M**; square column on house; intens 089°-091° (both occas).

Yacht harbour, Digue Augustin Normand, Q(2) G 5s 5m 2M. Quai des Abeilles QR 9m 6M; W and R mast; *Bell (1) 2.5s*.

PORT D'ANTIFER. Bassin de Caux, Mole Ouest, head Fl R 4s 13m 5M; W mast, R top. Mole Est Fl G 4s 13m 5M; W mast, G top. Antifer 49 39.5N/0 09.2E Oc WRG 4s 24m W14M, R13M, G13M; W Tr; vis G068°-078°, W078°-088°, R088°-098°. Jetty head 49 39.8N/0 07.1E QR 20m 10M; W Tr, R top.
Port D'Antifer Ldg Lts 127°. **Front** 49 38.3N/0 09.2E Oc 4s 113m **22M**; W mast, G top. **Rear**, 430m from front, Oc 4s 131m **22M**; W mast, G top. By day both show FW Lts **33M**.

Cap d'Antifer 49 41.07N/0 10.00E Fl 20s 128m **29M**; grey 8-sided Tr, G top; vis 021°-222°; RC.

YPORT. Ldg Lts 165°. Front Oc 4s 11m. Rear 30m from front Oc 4s 14m.

FÉCAMP. **Jetée Nord** 49 46.00N/0 21.87E Fl(2) 10s 15m **16M**; grey Tr, R top; *Reed (2) 30s*. Lts in line 085°. Front Jetée Sud, head, QG 14m 9M; grey Tr, G top; vis 072°-217°. Rear, Jetée Nord, root, QR; R circle on W mast.

SAINT VALERY-EN-CAUX. Jetée Ouest 49 52.45N/0 42.50E Oc(2+1) G 12s 13m 14M; G Tr. Jetée Est Fl(2) R 6s 8m 4M; W mast, R top.

Pointe d'Ailly 49 55.13N/0 57.55E Fl(3) 20s 95m **31M**; W square Tr, G top; RC; *Reed (3) 60s* (TD).
D1 Lt Buoy 49 57.10N/1 01.26E VQ(3) 5s; E cardinal mark.

DIEPPE. Jetée Est, head 49 56.18N/1 04.98E Oc(4) R 12s 10m 8M; W Tr, R top. Jetée Ouest, head 49 56.2N/1 05.0E Iso WG 4s 10m W12M, G8M; W Tr, G top on building; vis W095°-164°, G164°-095°; *Reed 30s*. Ldg Lts 138°. Front QR 19m 9M; W framework Tr, R top. Rear 200m from front QR 35m 9M; W hut, R top; vis 106°-170°.
Daffodils Lt Buoy 50 02.50N/1 04.15E VQ(9) 10s; W cardinal mark.

LE TRÉPORT. **Jetée Ouest** 50 03.94N/1 22.22E Fl(2) G 10s 15m **20M**; W Tr, G top; *Reed Mo(N) 30s*. Jetée Est Oc R 4s 8m 6M; W column, R top.

Ault 50 06.35N/1 27.24E Oc(3) WR 12s 95m **W18M**, R14M, W Tr, R top; vis W040°-175°, R175°-220°.

Cayeux-sur-Mer 50 11.75N/1 30.70E Fl R 5s 32m **22M**; W Tr, R top.
Pte du Hourdel 50 12.95N/1 34.00E Oc(3) WG 12s 19m W12M, G9M; W Tr, G top; vis W053°-248°, G248°-323°; *Reed (3) 30s*.

PORT DU CROTOY. Le Crotoy 50 12.9N/1 37.4E Oc(2) R 6s 19m 9M; W pylon; vis 285°-135°. Yacht harbour, West side jetty Fl R 2s 4m 2M. Yacht harbour, E side Fl G 2s 4m 2M.

SAINT VALERY-SUR-SOMME 50 11.5N/1 37.6E Iso G 4s 9m 9M; W pylon, G top; vis 347°-222°. Embankment head 50 12.3N/1 35.9E Q(3) G 6s 2m 2M; W pylon; Ra refl. Mole head 50 11.2N/1 38.6E Fl R 4s 9m 9M; W column, R top; vis 000°-250°.

Pointe du Haut-Blanc 50 23.90N/1 33.75E Fl 5s 44m **23M**; W Tr, R bands, G top.

Pointe du Touquet 50 31.4N/1 35.6E Fl(2) 10s 54m **25M**; Y Tr, brown band; W and G top.

Camiers, Riviere Canche entrance, N side Oc(2) WRG 6s 17m W9M, R7M, G6M; R framework Tr; vis G015°-090°, W090°-105°, R105°-141°.

Bassurelle Lt Buoy 50 32.70N/0 57.80E Fl(4) R 15s 6M; Racon; *Whis*.

Cap d'Alprech 50 41.95N/1 33.83E Fl(3) 15s 62m **24M**; W Tr, B top; RC; FR Lts on radio mast 600m NE.

Hoverport Ldg Lts 119° both FR 14M; W columns, grey tops; occas.

BOULOGNE. Approaches Lt Buoy 50 45.25N/1 31.15E VQ(6) + LFl 10s; S cardinal mark; *Whis*. **Digue Sud (Carnot)** 50 44.48N/1 34.13E Fl(2+1) 15s 25m **19M**; W Tr, G top; *Horn(2+1) 60s*. Digue Nord QR 10m 6M; R Tr.
Ro Ro berth Ldg Lts 197°. Front 50 43.7N/1 34.1E FG 16m; dolphin; vis 107°-287°. Rear, 480m from front FR 23m; R mast, W band; vis 187°-207°.
Darse Sarraz-Bournet, entrance E side Oc(2) R 6s 8m 6M; W framework Tr, R top. West side Iso G 4s 8m 5M; W framework Tr, G top. Jetée NE, head FR 11m 9M; R Tr. Jetée SW, FG 17m 5M; W column, G top; *Horn 30s*.

ZC1 Lt Buoy 50 44.85N/1 27.10E Fl(4)Y 15s; special mark.
ZC2 Lt Buoy 50 53.50N/1 31.00E Fl(2+1)Y 15s; special mark.

Cap Gris-Nez 50 52.17N/1 35.07E Fl 5s 72m **29M**; W Tr, B top; obsc 232°-005°; RC; RG; *Siren 60s*.

Sangatte 50 57.23N/1 46.57E Oc WG 4s 12m W9M, G6M; W column, B top; vis G065°-089°, W089°-152°, G152°-245°; Racon.

CALAIS. CA4 Lt Buoy 50 58.90N/1 45.15E VQ(9) 10s; W cardinal mark; *Whis*. Jetée Ouest, head 50 58.30N/1 50.48E Iso G 3s 12m 9M; W Tr, G top; *Bell (1) 5s*.
Jetée Est, head Fl(2)R 6s 12m **17M**; grey Tr, R top; *Reed(2) 40s*.
Calais 50 57.7N/1 51.2E Fl(4) 15s 59m **23M**; W 8-sided Tr, B top; obsc by Cap Blanc-Nez when brg less than 073°; RC.

SANDETTIE LT V 51 09.40N/1 47.20E Fl 5s 12m **25M**; R hull, Lt Tr amidships; Racon. MPC Lanby 51 06.09N/1 38.34E FlY 4s 10m 6M; special mark.
DUNKERQUE LANBY 51 03.00N/1 51.83E Fl 3s 10m **25M**; R tubular structure on circular buoy; Racon.

Walde 50 59.7N/1 54.9E Fl(3) 12s 13m 5M; B pylon.
DKA Lt Buoy 51 02.59N/1 57.06E LFl 10s; safe water mark.

GRAVELINES. **Petite Fort Phillipe** 51 00.3N/2 06.6E Oc(4) WG 12s 29m **W27M, G22M**; W Tr, B diagonal stripes; intens G186°-193°, W193°-200°. Jetée Ouest 51 00.93N/2 05.68E Fl(2)WG 6s 14m W9M, G6M; vis W317°-327°, G078°-085°, W085°-244°. Jetée Est Q(3) R 6s 5m 2M.

DUNKERQUE – PORT OUEST. Ldg Lts 120°. **Front** Dir FG 16m **19M**; W column, G top; intens 119°-121°. **Rear**, 600m from front, Dir FG 30m **22M**; W column, G top; intens 119°-121°. By day both show FW **28M**. Jetée du Dyck, head 51 02.3N/2 09.9E FlG 4s 24m 10M; W column, G top. Jetée Clipon, head Fl(4) 12s 24m 13M; W column, R top; vis 278°-243°; *Siren (4) 60s*.

DUNKERQUE. **Dunkerque** 51 03.0N/2 21.9E Fl(2) 10s 59m **29M**; W Tr, B top. Ldg Lts 185°. Front F Vi 10m 3M; W column, R top; intens 183°-187°. Common rear F Vi 24m 4M; grey framework Tr; intens 184°-186°, 178°-180°. Front Ldg Lt 179° F Vi 10m 3M; W column, G top; intens 177°-181°. Jetée Est, head Oc(3) R 12s 11m 10M; W framework Tr, R top. *Horn (3) 30s*.
Jetée Ouest, head 51 03.68N/2 21.05E Oc(2+1) WG 12s 35m **W17M**, G13M; W Tr, brown top; vis G252°-310°, W310°-252°; *Dia(2+1) 60s*.

10.18.5 PASSAGE INFORMATION

The coasts of Normandy and Picardy are convenient to harbours along the South Coast of England — the distance from (say) Brighton to Fécamp being hardly more than an overnight passage. It should be noted however that many of the harbs dry, so that a boat which can take the ground is an advantage. For details of traffic schemes in Dover Strait see Fig. 2(2). Notes on the English Channel and on cross Channel passages appear in 10.3.5. For detailed sailing directions, refer to *Normandy Harbours and Pilotage* (Adlard Coles Ltd), which conveniently covers this area. *The Shell Pilot to the English Channel* (Faber and Faber) covers the French coast from Dunkerque to Brest, and the Channel Islands. For French terms see 10.16.5.

POINTE DE BARFLEUR TO DEAUVILLE (chart 2613)

Raz de Barfleur (10.17.5) must be avoided in bad weather. Pte de Barfleur marks the W end of B de Seine, which stretches 53M E to C de la Heve, close NW of Le Havre. There are no obstructions on a direct course across the b, but a transhipment area for large tankers is centred about 10M ESE of Pte de Barfleur. A feature of B de Seine is the stand of tide at HW.

S from Barfleur (10.18.9) the coast runs SSE 4M to Pte de Saire, with rks and shoals up to 1¼M offshore. 2M S of Pte de Saire is St Vaast-la-Hougue (10.18.10): approach S of Île de Tatihou, but beware La Tourelle (rk which dries) 4½ ca E of island, and Le Gavendest (dries) and La Dent (dries) which lie 6 ca SE and 5 ca SSE of island and are marked by buoys. There is a chan inshore of Île de Tatihou which can be used near HW. Anch in Grande Rade in offshore winds, exposed to E and S.

Îles St Marcouf (chart 2073) lie 7M SE of St Vaast-la-Hougue, about 4M offshore, and consist of Île du Large (Lt) and Île de Terre about ¼M apart. (See 10.18.29). Banc de St Marcouf with depths of 2.4m extends 2½M NW from the islands, and the sea breaks over this in strong N or NE winds. There is anch, rather exposed, SW of Lt Ho. Landing is possible by dinghy in the small harb on W side of Île du Large from about local HW −0200 to HW +0200. There is a bird sanctuary on Île de Terre.

At the head of B du Grand Vey, about 10M S of Îles St Marcouf, are the (very) tidal harbs of Carentan (10.18.11) and Isigny (10.18.29). Entry is only possible near HW. The Carentan chan is well buoyed and adequately lit. It trends SSW across sandbanks for about 4M, beyond which it runs between two breakwaters leading to a lock gate, and thence into canal to town of Carentan. The Isigny chan is deeper, but the harb dries. Neither chan should be attempted in strong onshore winds.

On E side of B du Grand Vey, Roches de Grandcamp (dry) extend more than 1M offshore, N and W of Grandcamp (10.18.12), but they are flat and can be crossed from the N in normal conditions from about HW −0130 to HW +0130. Heavy kelp can give false echo soundings.

Between Pt de la Percée and Port-en-Bessin (10.18.13) a bank lies offshore, with drying ledges extending 3 ca. A race forms over this bank with wind against tide. Off Port-en-Bessin the E-going stream begins about HW Le Havre −0500, and the W-going at about HW Le Havre +0050, sp rates 1¼ kn.

From Port-en-Bessin the coast runs E 4M to C Manvieux, E of which lies the wartime harb of Arromanches, where there is anch. Between C Manvieux and Langrune, 10M E, Plateau du Calvados lies offshore. Rocher du Calvados (dries) lies on the W part of this bank, and there are several wrecks in this area.

Roches de Ver (dry) lie near centre of Plateau du Calvados, extending 8 ca offshore about 1M W of Courseulles-sur-Mer (10.18.14). The approach to this harb is dangerous in strong onshore winds. Les Essarts de Langrune (dry) lie E of Courseulles-sur-Mer, and extend up to 2¼M seaward off Langrune: at their E end lie Roches de Lion (dry), which reach up to 1½M offshore in places and extend to a point 2½M W of the harb of Ouistreham (10.18.15).

6M E of Ouistreham is R Dives, where a yacht can dry out (if there is room) alongside the jetty at Cabourg. The banks dry for 1M to seaward, and entry is only possible near HW and in reasonable conditions. See 10.18.29.

Deauville, 8M ENE of Dives, is an important yachting harb (see 10.18.16). Beware Banc de Trouville (dries) which extends ENE from a point 2M N of entrance. The sands dry more than ½M offshore, and in strong winds from W or N the entrance is dangerous. In marginal conditions it is best attempted within 15 mins of HW, when the stream is slack. At other times, or in worse weather, the sea breaks between the jetties.

ESTUAIRE DE LA SEINE/LE HAVRE (chart 2990)

The Seine estuary is entered between Dives and Le Havre, and is encumbered by shallow and shifting banks which extend seawards to Banc de Seine, 15M W of Le Havre. With wind against tide there is a heavy sea on this bank. Here the SW-going stream begins at HW Le Havre +0400, and the NE-going at HW Le Havre −0300, sp rates 1½ kn. Between Deauville and Le Havre the sea can be rough in W winds.

Chenal du Rouen is the main chan into R Seine, and carries a great deal of commercial traffic. The S side of the chan is contained by Digue du Ratier, a training wall which extends E to Honfleur (10.18.17). Note that for tidal reasons Honfleur is not a useful staging port when bound up-river. For notes on R Seine see 10.18.18, and for Rouen see 10.18.29.

Le Havre (10.18.19) is a large commercial port, as well as a yachting centre. Approaching from the NW, the most useful mark is the Lanby moored about 9M W of C de la Hève (Lt). The approach chan, which runs 6M WNW from the harb entrance, is well buoyed and lit. Strong W winds cause rough water over shoal patches either side of the chan. Coming from the N or NE, there is deep water close off C de la Hève, but from here steer S to join the main entrance chan. Beware Banc de l'Éclat (depth 0.1m), which lies on N side of main chan and about 1½M from harb entrance.

CAP D'ANTIFER TO Pte DU HAUT BLANC (chart 2612)

Immediately S of C d'Antifer (Lt, RC) is the large oil tanker harb of Port d'Antifer. A breakwater extends about 1½M seaward, and should be given a berth of about 1M, or more in heavy weather when there may be a race with wind against tide. Off C d'Antifer the NE-going stream begins about HW Le Havre −0430, and the SW-going at about HW Le Havre +0140. There are eddies close inshore E of C d'Antifer on both streams.

From C d'Antifer to Fécamp (10.18.20) drying rks extend up to ¼M offshore. At Fécamp pierheads the E-going stream begins about HW Le Havre −0500, and the W-going at about HW Le Havre +0025, sp rates 2¾ kn. Off Pte Fagnet, close NE of Fécamp, lie Les Charpentiers (rks which dry, about 1½ ca offshore).

From Fécamp to St Valéry-en-Caux (10.18.21), which lies 15M ENE, the coast consists of chalk cliffs broken by valleys. There are rky ledges, extending 4 ca offshore in places. The nuclear power station at Paluel 3M W of St Valéry-en-Caux is conspic. Immediately E of St Valéry-en-Caux shallow sandbanks, Les Ridins de St Valéry, with a least depth of 0.6m, extend about 6½ ca offshore. At St Valéry-en-Caux entrance the E-going stream begins about HW Dieppe −0550, and the W-going stream begins about HW Dieppe −0015, sp rates 2¾ kn. E of the entrance a small eddy runs W on the E-going stream.

Between St Valéry-en-Caux and Pte d'Ailly (Lt, fog sig, RC) there are drying rks 4 ca offshore in places. About 1½M E of Pte de Sotteville a rky bank (depth 4.2m) extends about 1M NNW; a strong eddy causes a race over this bank.

From Pte d'Ailly to Dieppe (10.18.22) the coast is fringed by a bank, drying in places, up to 4 ca offshore. Off Pte d'Ailly are Roches d'Ailly, which dry and extend ½M: on this reef is La Galère, a rk which dries, about 3 ca N of Lt Ho. About 6M N of Pte d'Ailly, Les Ecamias are banks with depths of 11m, dangerous in a heavy sea. E of Pte d'Ailly an eddy runs W close inshore on first half of E-going stream. Off Dieppe the ENE-going stream begins about HW Dieppe −0505, and the WSW-going at about HW Dieppe +0030, sp rates 2 kn.

Between Dieppe and Le Tréport, 14M NE, rky banks, drying in places, extend ½M offshore. A prohibited area extends ¾M offshore of Penly nuclear power station and is marked by Lt Buoys. There are no dangers further to seaward, nor in outer approaches to Le Tréport except Ridins du Tréport (depth 4.9m) about 3M NW of entrance, which should be avoided in bad weather.

Between Le Tréport and B de Somme, Banc Franc-Marqué (depth 3.1m) lies 2M offshore, and about 3M N of Le Tréport. B de Somme, entered between Cayeux-sur-Mer and Pte de St Quentin 6M NNE, is a shallow and drying area of shifting sands. Offshore there are two shoals, Bassurelle de la Somme and Quémer, on parts of which the sea breaks in bad weather.

4½M NW of Cayeux-sur-Mer the stream is rotatory anticlockwise. The E-going stream begins about HW Dieppe −0200, and reaches 2½ kn at sp in a direction 070°: the W-going stream begins about HW Dieppe +0600, and reaches 1½ kn at sp in a direction 240°. The chan, which runs close to Pte du Hourdel, is buoyed, but the whole estuary dries out 3M to seaward, and should not be approached in strong W or NW winds. For St Valéry-sur-Somme see 10.18.24, and for Le Crotoy see 10.18.29.

From Pte de St Quentin the coast runs 7M N to Pte du Haut Blanc, with a shallow coastal bank which dries for about ½M offshore.

LE TOUQUET TO DUNKERQUE (charts 2451, 323)

Le Touquet (10.18.25) lies in the Embouchure de la Canche, entered between Pte du Touquet and Pte de Lornel, and with a drying bank which extends 1M seaward of a line joining these two points. Le Touquet-Paris-Plage Lt is shown from a conspic Tr, 1M S of Pte du Touquet. Off the entrance the N-going stream begins about HW Dieppe −0335, sp rate 1¾ kn; and the S-going stream begins about HW Dieppe +0240, sp rate 1¾ kn.

In the approaches to Pas de Calais a number of shoals lie offshore — La Bassurelle, Le Vergoyer, Bassure de Baas, Le Battur, Les Ridens, and The Ridge (or Le Colbart). In bad weather, and particularly with wind against tide in most cases, the sea breaks heavily on all these shoals.

From Pte de Lornel to Boulogne (10.18.26) the coast dries up to ½M offshore. Beware hovercraft traffic S of the harb (chart 438). Off Digue Carnot the N-going stream begins HW Dieppe −0130, and the S-going at HW Dieppe +0350, sp rates 1¾ kn.

Between Boulogne and C Gris Nez (Lt, fog sig, RC) the coastal bank dries about 4 ca offshore. The NE-bound traffic lane lies only 3M off C Gris Nez. 1M NW of C Gris Nez the NE-going stream begins at HW Dieppe −0150, and the SW-going at HW Dieppe +0355, sp rates 4 kn.

In bad weather the sea breaks heavily on Ridens de Calais, 3M N of Calais (10.18.27), and also on Ridens de la Rade about 1½M NE of the harb. Midway between Calais and Dunkerque (10.18.28) is the drying harb of Gravelines which should not be used in strong onshore winds.

Offshore lie Sandettié bank (about 14M to N), Outer Ruytingen midway between Sandettié and the coast, and the Dyck banks which extend in a NE direction for 30M from a point 5M NE of Calais. There are chans between these banks, and they are well buoyed, but great care is needed in poor vis. In general the banks are steep-to on the inshore side, and slope seaward. In bad weather the sea breaks on the shallower parts.

10.18.6 DISTANCE TABLE

Approximate distances in nautical miles are by the most direct route while avoiding dangers and allowing for traffic separation schemes etc. Places in *italics* are in adjoining areas.

		1	2	3	4	5	6	7	8	9	10	11	12	13	14	15	16	17	18	19	20
1	*Casquets*	1																			
2	*Cherbourg*	31	2																		
3	*Portland Bill*	48	62	3																	
4	*Needles*	64	60	35	4																
5	*Nab Tower*	81	66	60	27	5															
6	*Royal Sovereign*	124	102	114	82	54	6														
7	*Dover*	167	145	157	125	97	43	7													
8	Barfleur	48	20	70	60	62	91	138	8												
9	Port en Bessin	76	48	98	88	80	98	143	28	9											
10	Ouistreham	94	66	112	99	88	95	134	46	22	10										
11	Deauville	104	76	126	102	89	90	127	56	36	14	11									
12	Honfleur	110	82	132	107	91	91	126	62	40	24	10	12								
13	Rouen	179	151	201	176	160	160	195	131	109	93	79	69	13							
14	Le Havre	101	70	118	98	85	85	120	56	36	19	8	9	75	14						
15	Fécamp	111	80	120	93	75	60	95	64	52	39	32	34	100	25	15					
16	Dieppe	139	108	143	115	92	55	75	94	81	68	61	63	129	54	29	16				
17	Boulogne	167	142	160	128	100	46	25	128	125	115	108	110	176	101	76	54	17			
18	Cap Gris Nez	175	147	157	126	101	47	19	135	132	122	115	117	183	108	83	61	7	18		
19	Calais	188	160	174	142	114	60	22	148	145	135	128	130	196	121	96	74	20	13	19	
20	Dunkerque	210	182	196	164	136	82	43	170	167	157	150	152	218	143	118	96	42	35	22	20

BARFLEUR 10-18-9
Manche

CHARTS
Admiralty 1349, 2073, 1106; SHOM 5618, 5609, 847; ECM 528; Stanford 7; Imray C32

TIDES
Dover −0250; ML 3.9; Duration 0550; Zone −0100

Standard Port CHERBOURG (←)

Times				Height (metres)			
HW		LW		MHWS	MHWN	MLWN	MLWS
0300	1000	0400	1000	6.3	5.0	2.5	1.1
1500	2200	1600	2200				

Differences BARFLEUR
+0100 +0100 +0050 +0040 +0.2 +0.3 0.0 +0.1

SHELTER
Excellent shelter although entrance difficult in E to NE winds. Harbour dries. Normally yachts lie along NW wall. Beware rocks in SE of harbour. Anchorage outside harbour is safe in off-shore winds. Access HW∓3.

NAVIGATION
Waypoint 49°41'.30N 01°14'.21W, 039°/219° from/to front Ldg Lt, 1.35M. Beware Barfleur Race, about 3½ M E and NE of Barfleur Pt, in rough weather. There are numerous rocks between Barfleur Pt and the harbour, and also cross currents. Keep ¼ M E of La Grotte Rks (buoyed). Beware Le Hintar Rks (buoyed) and La Raie (Bn) to E of Ldg Line.

LIGHTS AND MARKS
Pte de Barfleur Fl (2) 10s 72m 29M; grey Tr, B top (conspic); Reed (2) 60s. Ldg Lts 219°. Front Oc (3) 12s 7m 10M; W square Tr. Rear Oc (3) 12s 13m 10M; grey and W square Tr, G top; synchronized, not easy to see by day. Jetée Est Oc R 4s 5m 7M. Jetée Ouest Fl G 4s 8m 7M.

RADIO TELEPHONE
None.

TELEPHONE
Hr Mr 33.54.02.68; Aff Mar 33.44.00.13; CROSS 33.52.72.13; SNSM 33.54.04.62; Customs 33.54.48.81; Meteo 33.22.91.17; Auto 33.43.20.40; Dr 33.54.00.02; Brit Consul 33.44.20.13.

FACILITIES
NW Quay Slip, P, L, FW, AB; **Harbour** M; **Bouly** Tel. 33.54.02.66 P, ME, El; **Chantier Bellot** Tel. 33.54.04.29, Sh; **Garage Marine** Tel. 33.54.00.23, D. **Town** P, D, Sh, CH, V, Gaz, R, Bar. PO; Bank; Rly (bus to Cherbourg); Air (Cherbourg).
Ferry UK — Cherbourg—Portsmouth/Weymouth (latter in summer only).

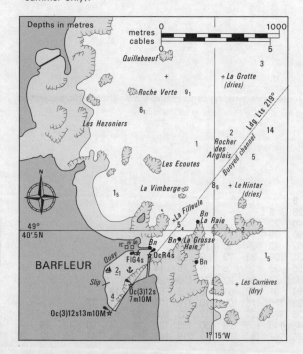

ST VAAST-LA-HOUGUE 10-18-10
Manche

CHARTS
Admiralty 1349, 2073, 2613; SHOM 5522, 847; ECM 527; Stanford 2; Imray C32

TIDES
Dover −0220; ML 3.8; Duration 0530; Zone −0100

Standard Port CHERBOURG (←)

Times				Height (metres)			
HW		LW		MHWS	MHWN	MLWN	MLWS
0300	1000	0400	1000	6.3	5.0	2.5	1.1
1500	2200	1600	2200				

Differences ST VAAST-LA-HOUGUE
+0105 +0055 +0120 +0100 +0.3 +0.3 −0.2 −0.2

SHELTER
The harbour is protected from all sides. If harbour full, anchorage in the approach channel is satisfactory except in strong winds from E to S. Marina lock gates open HW−2¼ to HW+3. Visitors pontoons A, B and C.

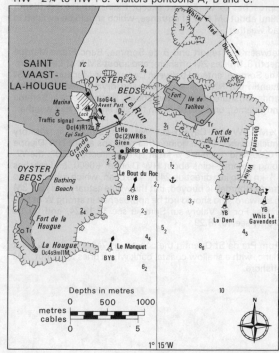

NAVIGATION
Waypoint 49°34'.40N 01°13'.50W, 087°/267° from/to La Hougue Lt (Oc 4s), 1.8M. Do not confuse with Lt on jetty. From waypoint head 267° until Reville Point de Saire Lt is just concealed by the Fort de l'Ilet, turn on to 346°. Head for St Vaast la Hougue jetty Lt leaving the Bout du Roc & Balise de Creux buoys to port. Watch out for cross currents.
"Le Run" approach is not recommended and should not be attempted with a boat of over 1.2m draft. The entrance is wide and well marked. Harbour gets very full in summer. Beware oyster beds.

LIGHTS AND MARKS
There are two prominent towers to indicate position—one on Ile de Tatihou and an identical one on La Hougue Pt. Prominent light structure on the end of the outer jetty, Oc (2) WR 6s, R219°-247°, W elsewhere. Siren Mo (N) 30s. Control of traffic in and out of marina by R & G lights.

RADIO TELEPHONE
VHF Ch 09.

TELEPHONE
Hr Mr 33.54.48.81; Aff Mar 33.54.43.61; CROSS 33.52.72.13; SNSM 33.54.42.52; Customs 33.54.48.81; Meteo 33.43.20.40; Auto 33.43.20.40; Dr 33.54.43.42; Brit Consul 33.44.20.13.

AREA 18—NE France

FACILITIES
Marina (730) Tel. 33.54.48.81, FW, C (15 ton), ME, El, Sh, Gaz, D, P, CH, Bar, R, Lau, V, AC, Slip;
G. Bernhard Tel. 33.54.43.47, ME, El, Sh; **Agence du Port** Tel. 33.54.53.01, ME, El, CH; **E Massieu** Tel. 33.54.41.00, ME, El; **Cercle Nautique de la Hogue** Tel. 33.54.48.81, Bar; **YC** Tel. 33.54.53.73. **Town** P, D, V, Gaz, R, Bar, Lau. PO; Bank; Rly (bus to Valognes); Air (Cherbourg).
Ferry UK — Cherbourg—Portsmouth/Weymouth (summer only).

CARENTAN 10-18-11
Manche

CHARTS
Admiralty 2073; SHOM 6614D, 847; ECM 527; Stanford 1; Imray C32

TIDES
Dover −0025; ML (Port-en-Bessin) 4.2; Duration (Port-en-Bessin) 0520; Zone −0100

Standard Port LE HAVRE (→)

Times				Height (metres)			
HW		LW		MHWS	MHWN	MLWN	MLWS
0000	0500	0000	0700	7.9	6.6	3.0	1.2
1200	1700	1200	1900				

Differences PORT-EN-BESSIN
−0045 −0040 −0040 −0045 −0.7 −0.6 −0.3 −0.1
HW Carentan is HW Cherbourg +0100

SHELTER
Safe anchorage N of buoyed channel in winds up to force 5. Complete shelter in the canal/river and in the marina. Entrance protected from prevailing winds W to S. In emergency the Iles St. Marcouf close by provide excellent shelter see 10.18.29. Depth in locked marina basin max/min 4.0/2.5m.

NAVIGATION
Waypoint CI safe water buoy, Iso 4s, 49°25'.50N 01°07'.00W, 034°/214° from/to entrance to buoyed channel, 1.6M. Iles St Marcouf are a good landmark, 4.5M to N. Channel shifts and buoys (six of which are lit, Fl R or G(3) 12s) are moved as required. Start from CI buoy between HW−2 and HW−1½. After about 4M the channel enters between two breakwaters, marked by Bns. 3.5M further on the channel divides into three, forming a small pool. Enter the lock ahead (opens HW−2 to HW+3, traffic sigs) and proceed up the canal to marina. There are waiting pontoons on E side above and below the lock.

LIGHTS AND MARKS
Breakwaters are marked by Bns, the second of which inwards are lit port and stbd, FlR(3) 12s and FlG(3) 12s respectively. Ldg Lts 210°, Front Oc(3)R 12s and Rear Oc(3) 12s, lead through the part of the channel 1.6M inwards from the ends of the breakwaters and immediately outside (not for the buoyed channel). Lock signals: FG = lock open, FR = lock closed.

RADIO TELEPHONE
VHF Ch 09 (0600−2000 in season; 0800−1800 out of season).

TELEPHONE
Hr Mr 33.42.24.44; Aff Mar 33.44.00.13; Customs 33.44.16.00; CROSS 33.52.28.72; Meteo 33.44.45.00; Auto 33.43.20.40; Lockmaster 33.71.10.85; Hosp 33.42.08.23; Dr 33.42.33.21.

FACILITIES
Marina (495+55 visitors) FW, AC, C (25 ton), BH (16 ton), P and D (Tel. 33.42.24.44), Access HW−2 to HW+3; **YC Croiseurs Côtiers de Carentan** Tel. 33.42.35.94, Bar; **Carentan Sport** 33.42.18.77, CH; **Fleury** Tel. 33.42.12.21, ME; **Hardy** 33.42.02.87, ME, CH; **Itelec Nautic** Tel. 33.42.07.83, ME, El, Sh, CH, divers; **Le Guen Hemidy Marine** Tel. 33.42.26.55, Sh, CH; **G.A.M. Marine** Tel. 33,71.17.02, ME, BY, CH, El; **Town** Bar, Bank, D, P, PO, Rly, R, V, Air (Cherbourg).

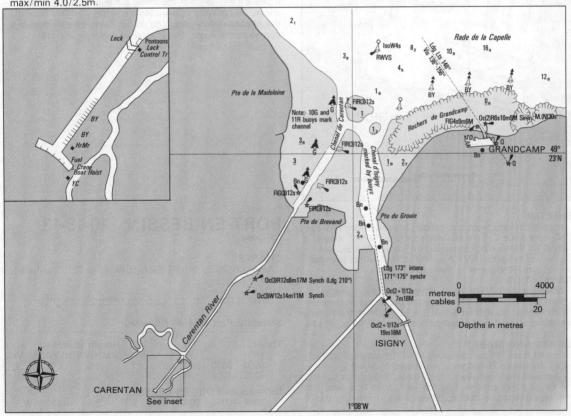

GRANDCAMP 10-18-12
Calvados

CHARTS
Admiralty 2073, 2613; SHOM 847; ECM 527; Stanford 1; Imray C32

TIDES
Dover −0220; ML Port-en-Bessin 4.2; Duration Port-en-Bessin 0520; Zone −0100

Standard Port LE HAVRE (→)

Times				Height (metres)			
HW		LW		MHWS	MHWN	MLWN	MLWS
0000	0500	0000	0700	7.9	6.6	3.0	1.2
1200	1700	1200	1900				

Differences PORT-en-BESSIN
−0045 −0040 −0040 −0045 −0.7 −0.6 −0.3 −0.1

RADIO TELEPHONE
VHF Ch 09.

TELEPHONE
Hr Mr 31.22.63.16; Aff Mar 31.22.60.65; CROSS 33.52.72.13; SNSM 31.22.64.25; Meteo 31.74.48.11; Auto 31.75.14.14; Dr Isigny-sur-Mer 31.22.01.07; Brit Consul 35.42.27.47.

FACILITIES
Marina (268+25 visitors) Tel. 31.22.63.16, El, FW, BH (4 ton), Bar, V, AC; **Conin** Tel. 31.22.62.79, ME; **Galliot Marine** Tel. 31.22.61.95, ME, El, Sh, CH; **Yachting 14** Tel. 31.22.67.02, ME, El, Sh, CH, Gaz; **Cercle Nautique de Grandcamp les Bains** Tel. 31.22.67.37, Bar; **Town** Gaz, PO, Bank, Rly (Carentan), Air (Caen). Ferry UK — Riva Bella (Ouistreham)—Portsmouth.

AGENTS NEEDED
There are a number of ports where we need agents, particularly in France.
ENGLAND Swale, Havengore, Berwick.
SCOTLAND Firth of Forth, Scrabster, Mallaig, Loch Sunart, Loch Aline.
IRELAND Kilrush, Wicklow, Westport/Clew Bay.
FRANCE Arcachon, Seudre R, Ile d'Oleron, Rochfort, Ile de Re, St. Giles-Croix-de-Vie, Ile d'Yeu, Pouliguen, Le Croisic, La Forêt, Ile de Bréhat.
GERMANY Norderney, Dornumer-Accumersiel.
If you are interested in becoming our agent for any of the above, please write to the editors and get your free copy of the Almanac every year. You do not have to be resident in a port to be the agent, but at least a fairly regular visitor.

SHELTER
Access day and night but difficult in winds from NW to NE over force 6. Safe approach springs HW∓2, neaps HW↑½. Wet basin, containing marina on W side, has gate which opens HW−2 to HW+2½. Visitors pontoon near gate. SW corner is very dirty.

NAVIGATION
Waypoint 49°25'.00N 01°04'.60W, 326°/146° from/to front Ldg Lt 146°, 2.0M. Large flat rocks, les Roches de Grandcamp, stretch for about 1½ M out from the harbour and dry about 1.5m. Entrance between HW∓2 is safe for most yachts.

LIGHTS AND MARKS
Line of three N cardinal buoys mark the seaward limit of Les Roches de Grandcamp, numbered 1, 3 and 5. Approach from between any of these. E pierhead Lt Oc(2) R 6s on a RW column. Ldg Lts 146°, both Q. E jetty due to be extended in 1988 as shown and Lt moved to the end.

PORT-EN-BESSIN 10-18-13
Calvados

CHARTS
Admiralty 2073; SHOM 6927, 5515; ECM 527; Stanford 1; Imray C32

TIDES
Dover −0215; ML 4.2; Duration 0520; Zone −0100

Standard Port LE HAVRE (→)

Times				Height (metres)			
HW		LW		MHWS	MHWN	MLWN	MLWS
0000	0500	0000	0700	7.9	6.6	3.0	1.2
1200	1700	1200	1900				

Differences PORT-EN-BESSIN
−0045 −0040 −0040 −0045 −0.7 −0.6 −0.3 −0.1

PORT-EN-BESSIN continued

SHELTER
Good shelter although harbour dries completely. There is little room for yachts. Basins accessible from HW∓2. Yachtsmen should contact Hr Mr on arrival.

NAVIGATION
Waypoint 49°22'.00N 00°44'.90W, 024°/204° from/to front Ldg Lt 204°, 1.1M. Busy fishing port. Yachts are admitted for short stays (up to 24 hrs). Entry is difficult with strong winds from N and NE and may become impossible. Beware submerged jetty from end of E Wharf towards Mole Ouest marked by Bn. Keep out of G sector of Mole Ouest Lt.

LIGHTS AND MARKS
Ldg Lts 204°. Front Oc (3) 12s 25m 10M; W pylon, G top; vis 069°-339°; Siren 20s (sounded over 90° each side of leading line, continuous in W sector, interrupted in E sector). Rear Oc (3) 12s 42m 11M; W and grey house; vis 114°-294°, synchronized with front; RC.
Entry signals: FR over FG, or R flag over G flag means basins closed. Bridge has FR Lt each side, and one in the middle, lit when bridge shut. Bridge opens H+00 and H+30 from HW−2 to HW+2.

RADIO TELEPHONE
VHF Ch 03 and Ch 18 (HW∓2) for lock opening.

TELEPHONE
Hr Mr 31.21.70.49; Aff Mar 31.21.71.52; CROSS 33.52.72.13; SNSM 31.21.71.52; Customs 31.21.71.09; Meteo 31.74.48.11; Auto 31.75.14.14; Lock 31.21.71.77; Dr 31.21.74.26, 31.21.70.84; Brit Consul 35.42.27.47.

FACILITIES
EC: Open every day — including Sunday in summer; **Outer Harbour** Slip, M, L; **Bassin II** Slip, M, L, FW, AB, C (4 ton); **Ayello** Tel. 31.21.72.24, Sh, C (4 ton mobile, 6 ton), CH; **Hutrel** Tel. 31.21.72.36, CH; **Bellot** Tel. 31.21.71.88, Sh; **Sominex** Tel. 31.21.70.53, El; **M. Marie** El; **Digne Francoise** Tel. 31.21.72.16, D, ME, El. **Town** P, AB, V, Gaz, R, Bar. PO; Bank (Bayeux); Rly (bus to Bayeux); Air (Caen).
Ferry UK — Riva Bella (Ouistreham)—Portsmouth.

COURSEULLES-SUR-MER
10-18-14
Calvados

CHARTS
Admiralty 1349, 2073, 1821; SHOM 5515, 5598, 6927; ECM 527, 528; Stanford 1; Imray C32

TIDES
−0200 Dover; ML 3.9; Duration No data; Zone −0100

Standard Port LE HAVRE (→)

Times				Height (metres)			
HW		LW		MHWS	MHWN	MLWN	MLWS
0000	0500	0000	0700	7.9	6.6	3.0	1.2
1200	1700	1200	1900				

Differences COURSEULLES-SUR-MER
−0030   No data   No data   −0020   −0.8   −1.0   −0.7   −0.3

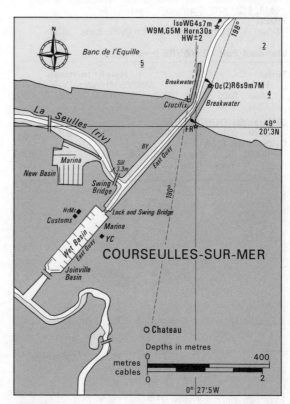

SHELTER
Good except in strong winds from NE through N to W. Best entry is at HW−1. Yachts enter basin (lock opens HW∓2) or go to marina in New Basin.

NAVIGATION
Waypoint 49°22'.50N 00°29'.50W, 313°/133° from/to Bernières-sur-Mer church Tr, 3.7M. Plateau du Calvados extends 2M offshore. Banks dry for ⅔M. RW buoy, Iso 4s, lies 1M N by W from entrance. To S of this beware rks awash at CD and sunken breakwater on W side extending 2 ca (370m) from entrance.

LIGHTS AND MARKS
Large crucifix at root of W pier is a conspic mark when entering. Ldg Lts (Oc (2) R 6s on end of E jetty in line with FR at root of E jetty) in line at 198°. Day — two Ldg marks to seaward on this line.

RADIO TELEPHONE
VHF Ch 09.

TELEPHONE
Hr Mr 31.37.51.69; Lock Keeper 31.37.46.03; Aff Mar 31.37.46.19; CROSS 33.52.72.13; SNSM 31.37.45.24; Customs West Quai, no tel.; Meteo 31.74.48.11; Auto 31.75.14.14; Dr 31.37.45.24; Brit Consul 35.42.27.47.

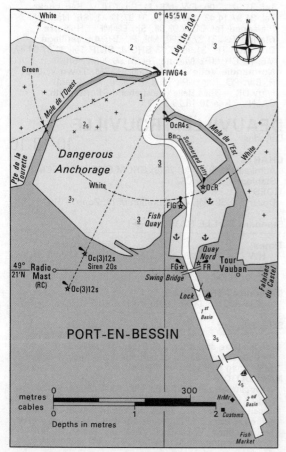

COURSEULLES-SUR-MER continued

FACILITIES
Marina Tel. 31.97.46.03, P, D, FW, ME, El, Sh; **Outer Harbour** Tel. 31.37.46.03, M, FW, C (8 ton), AB; **Wet Basin** Slip, P, FW, Sh, C (10 ton), CH, AB; **Serra Marine** Tel. 31.37.42.34, ME, El, Sh, CH; **Chantier Caullet** Tel. 31.37.90.19, Sh; **Chantiers Navals de la Côte de Nacre** Tel. 31.37.45.08, M, ME, El, Sh, CH; **Courseulles Marine** Tel. 31.37.43.17, M, ME, El, Sh, CH; **Ste des Regates de Courseulles** Tel. 31.37.47.42, Bar. **Town** P, D, V, Gaz, R, Bar. PO; Bank; Rly (bus to Caen); Air (Caen).
Ferry UK — Riva-Bella (Ouistreham) — Portsmouth.

OUISTREHAM 10-18-15
Calvados

CHARTS
Admiralty 1349, 1821, 2613; SHOM 7055, 6928, 6927, 6614; ECM 526; Stanford 1; Imray C32

TIDES
Dover −0150; ML 4.4; Duration 0525; Zone −0100

Standard Port LE HAVRE (→)

Times				Height (metres)			
HW		LW		MHWS	MHWN	MLWN	MLWS
0000	0500	0000	0700	7.9	6.6	3.0	1.2
1200	1700	1200	1900				

Differences OUISTREHAM
−0020 −0010 −0005 −0010 −0.2 −0.2 −0.2 −0.2
DIVES-SUR-MER
−0055 No   No   −0115 −0.4 −0.5 −0.6 −0.3
       data data

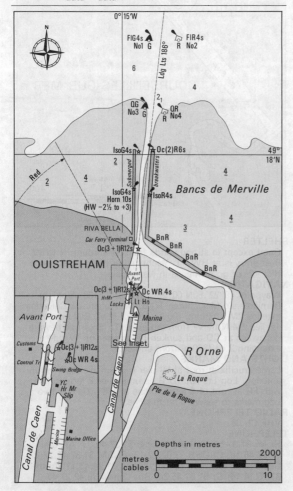

SHELTER
Very good inside locks. Temporary berths on pontoons outside, and moorings in R Orne. Enter canal by locks each side of control tower. Yachts normally use E lock. Locks open for departure HW−2 and HW+1¾; and for entry HW−1½ and HW+2¼. From 15 June−15 Sep, and at weekends and public holidays from 1 Apr−15 June and 15 Sep−31 Oct, extra openings between 0700 and 2000 local time as follows: for departure at HW−3 and HW+2¾, and for entry at HW−2½ and HW+3¼. Marina (depth 3.5m) is 1 ca (185m) inside locks on E side (Access HW∓3). Canal to Caen, 7.5 M, depth 10m to 2.5m, max speed 7 kts, overtaking prohibited. Good marina at Caen. See 10.18.29.

NAVIGATION
Waypoint OC (safe water) buoy, Iso 4s, Whis, 49°19′.89N 00°14′.25W, 006°/186° from/to front Ldg Lt 186°, 2.7M. Training walls run each side of entrance channel, marked by beacons. Beware turbulence in locks. Passage of Caen canal (four lifting bridges) is arranged at Ouistreham or Caen.

LIGHTS AND MARKS
From Ouistreham buoy (RW, Iso 4s, 3M from Lt Ho) follow Ldg Lts 186°. Main Lt Ho, conspic, W with R top, Oc WR 4s. Traffic signals (full code — see 10.14.7) shown from two panels on control tower between locks. The main signals do not apply to yachts or fishing craft, which must only enter lock when white light shows alongside lowest light of signal panel referring to lock concerned.

RADIO TELEPHONE
Call: *Ouistreham Port* VHF Ch 12 16 68 (HW−2 to HW+3). Call: *Caen Port* Ch 12 68; Ste des Regates Ch 09 (office hours).

TELEPHONE
Hr Mr 31.97.14.43; Port de Plaisance/Ste des Regates 31.97.13.05; Aff Mar 31.97.18.65; CROSS 33.52.72.13; SNSM 31.97.17.47; Lock Tower 31.97.14.43; Ferry terminal 31.96.80.80; Customs 31.86.61.50; Meteo 31.74.48.11; Auto 31.75.14.14; Dr 31.97.18.45; Brit Consul 35.42.27.47.

FACILITIES
Marina (600+65 visitors) Tel. 31.97.13.05, Slip, P, D, FW, ME, El, Sh, CH, AC, BH (8 ton), Gas, Gaz, Kos, SM, Bar; **Société des Régates de Caen — Ouistreham** Tel. 31.97.13.05, FW, ME, El, Sh, CH, V, Bar; **Lock** Tel. 31.97.14.43; **Serra** Tel. 31.97.17.41, Sh; **Nauti Plaisance** Tel. 31.97.03.08, Sh, El, ME, CH; **Serra Marine** Tel. 31.97.03.60, ME, El, CH; **Accastillage Diffusion** 31.96.07.75 SHOM; **SNIP** Tel. 31.97.34.47, ME, El, CH; **JPL Marine** Tel. 31.96.29.92, Sh; **Normandie Voile** Tel. 31.97.06.29, SM. **Town** V, Gaz, R, Bar. PO; Bank; Rly (bus to Caen); Air (Caen).
Ferry UK — Riva-Bella (Ouistreham) — Portsmouth.
CAEN — See 10.18.29.

DEAUVILLE/TROUVILLE 10-18-16
Calvados

CHARTS
Admiralty 1349, 2146, 2613; SHOM 5530, 6928; ECM 526; Stanford 1; Imray C32

TIDES
Dover −0130; ML 4.5; Duration 0510; Zone −0100

Standard Port LE HAVRE (→)

Times				Height (metres)			
HW		LW		MHWS	MHWN	MLWN	MLWS
0000	0500	0000	0700	7.9	6.6	3.0	1.2
1200	1700	1200	1900				

Differences TROUVILLE
−0035 −0015 0000 −0010 −0.1 −0.2 −0.2 −0.1

SHELTER
Good shelter — entry to channel difficult in NW to N winds above about force 6. Excellent marina and Yacht Harbour but remainder of harbour dries. Yacht Harbour gates open HW−2 to HW+2½.

NAVIGATION
Waypoint 49°23′.10N 00°03′.68E, 330°/150° from/to front Ldg Lt 150°, 1.2M. About 2½ M from the harbour, on the transit line, there is a wreck marked by a buoy. Do not approach from E of N because of Banc de Trouville. SW Trouville buoy, W cardinal, VQ(9) 10s, lies about 1M WNW of entrance. Access from seaward about HW∓2.

AREA 18—NE France 705

DEAUVILLE/TROUVILLE continued

LIGHTS AND MARKS
The casino is conspic building on Trouville side. Ldg Lts 148° (both Oc R 4s synchronised).

RADIO TELEPHONE
Deauville Yacht Harbour VHF Ch 11; Port Deauville Marina VHF Ch 09.

TELEPHONE
Hr Mr Marina 31.88.56.16; Hr Mr Yacht Harbour 31.88.28.71; Aff Mar 31.88.36.21; SNSM 31.88.13.07; CROSS 33.52.72.13; Customs 31.88.35.29; Meteo 31.88.28.62; Auto 31.88.84.22; Dr 31.88.23.57; Hosp 31.88.14.00; Brit Consul 35.42.27.47.

FACILITIES
Port Deauville (Marina) (738 + 100 visitors) Tel. 31.88.56.16, Slip, P, D, AC, FW, ME, El, Sh, C (6 ton), BH (50 ton), CH, R, Bar; **Ancien Yacht Bassins** Access HW∓2, M, L, FW, ME, El, AB; **Bassin Morny** M, P, D, L, FW, ME, El, Sh, C (6 ton), AB; **Touques Nautisme** Tel. 31.88.45.99, Slip, ME, El, Sh, P, D, C (50 ton); **Marina Ouest** Tel. 31.88.98.37, ME, El, Sh, CH, Electronics; **Voilerie Hervé Lepault** Tel. 31.98.58.61, ME, El, SM; **Deauville YC** Tel. 31.88.38.19, M, L, FW, C (6 ton), AB, R, Bar; **Outer Harbour (quai de la Cahotte)** Slip, M (see Hr Mr), P, L, CH, AB, V, Bar; **Callac** Tel. 31.88.55.94, ME, El, CH; **Voile Normande** Tel. 31.88.31.33, CH, SM; **Manche Electronique** Tel. 31.88.63.07, El. **Town** P, D, FW, CH, V, Gaz, R, Bar. PO; Bank; Rly; Air (Deauville).
Ferry UK — Le Havre—Portsmouth or Riva-Bella (Ouistreham)—Portsmouth.

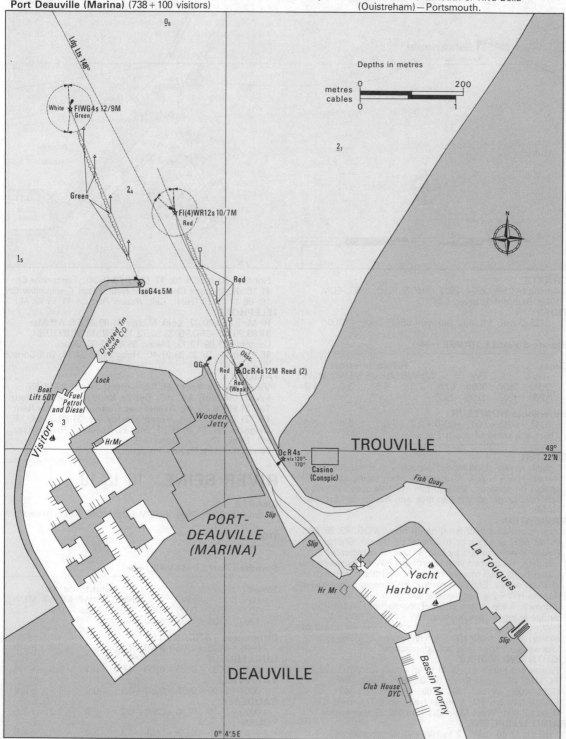

HONFLEUR   10-18-17
Calvados

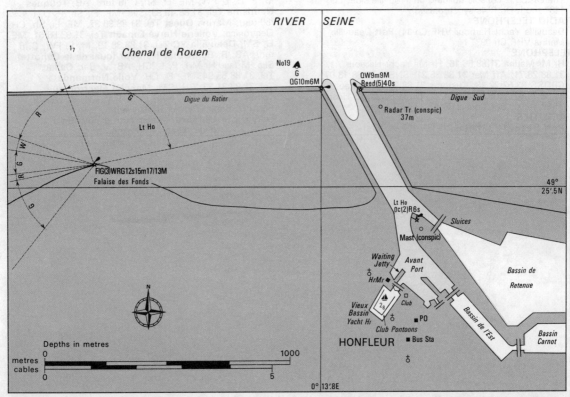

CHARTS
Admiralty 2146, 2994, 2613; SHOM 6796, 6683; ECM 1012; Stanford 1; Imray C31
TIDES
−0135 Dover; ML 4.6; Duration 0540; Zone −0100

Standard Port LE HAVRE (→)

Times				Height (metres)			
HW		LW		MHWS	MHWN	MLWN	MLWS
0000	0500	0000	0700	7.9	6.6	3.0	1.2
1200	1700	1200	1900				

Differences HONFLEUR
−0140 −0015 +0005 +0040 +0.2 +0.1 0.0 0.0
There is a stand of about 2 hours.

SHELTER
Shelter is very good indeed especially in the Vieux Bassin. Access HW∓1. Ask YC member when taking vacant spaces on pontoon. Visitors can use NW quay with no facilities. Larger yachts can lock into Bassin de l'Est; see Hr Mr.
NAVIGATION
Waypoint Ratier NW (stbd-hand) buoy, VQG, 49°26′.58N, 00°03′.56E, at entrance to Chenal de Rouen. Chenal de Rouen is well marked by buoys. Due to commercial shipping, yachtsmen should proceed HW∓3 between the Digue Nord, well marked by posts, and the port hand channel buoys. Entrance marked by conspic Radar tower. Access HW∓2½ but beware strong currents across entrance. Keep in mid channel approaching Avant Port. Bridge lifting & lock opening HW−1, HW, HW+1, HW+2 during day in high season and week-ends: out of season and at night HW−1, HW, HW+1. Make sure lock-keeper knows you are approaching.
LIGHTS AND MARKS
Falaise des Fonds Lt, ¾ M W of ent Fl (3) WRG 12s 15m, 14/11M, W square Tr vis G040°-080°, R080°-084°, G084°-100°, W100°-109° R109°-162°, G162°-260° E Mole QW 9m 9M; Y metal framework Tr with B top, Reed (5) 40s. West Mole QG 10m 6M; G framework Tr.
RADIO TELEPHONE
VHF Ch 11 16 (HW−2 to HW+4). Other stations: Honfleur Radar Ch 16; 11 13 71 73 (H24); Tancarville Ch 11 16 (H24). Lock Ch 18. Call: *Rouen Port Capitainerie* Ch 16; 06 11 13 73 (H24); Call: *Rouen Port* Ch 11 13 73 74.
TELEPHONE
Hr Mr 31.89.20.02; Lock Master 31.89.22.57; Aff Mar 31.89.20.67; CROSS 33.52.72.13; SNSM 31.89.07.84; Customs 31.89.12.13; Meteo 35.42.21.06; Auto 35.21.16.11; Dr 31.89.07.40; Hosp 31.89.04.74; Brit Consul 35.42.27.47.
FACILITIES
Vieux Bassin Yacht Hr (20 visitors) Tel. 31.89.01.85, FW, C (16 ton), AB, AC; **Cercle Nautique de Honfleur** Tel. 31.89.00.29, M; **Ateliers et Chantiers de Honfleur** Tel. 31.89.13.71, Sh; **Volvo Marine** Tel. 31.89.38.54, ME; **Grignon** Tel. 31.89.18.67, ME, El. **Town** V, R, Bar, Gaz, PO, Bank, Rly, Air (Deauville).
Ferry UK — Le Havre—Portsmouth.

RIVER SEINE   10-18-18

CHARTS
Admiralty 2146, 2994; SHOM 6796, 6683, 6117; Stanford 1; Imray C31
TIDES
Zone −0100

Standard Port LE HAVRE (→)

Times				Height (metres)			
HW		LW		MHWS	MHWN	MLWN	MLWS
0000	0500	0000	0700	7.9	6.6	3.0	1.2
1200	1700	1200	1900				

Differences TANCARVILLE
−0105 +0025 +0105 +0140 0.0 0.0 +0.3 +1.0
QUILLEBOEUF
−0045 +0030 +0120 +0200 0.0 +0.1 +0.5 +1.4
VATTEVILLE
+0004 +0100 +0225 +0250 +0.1 0.0 +1.1 +2.4
CAUDEBEC
+0020 +0115 +0230 +0300 −0.2 0.0 +1.2 +2.5
ROUEN
+0440 +0415 +0525 +0525 −0.1 0.0 +1.9 +3.6

AREA 18—NE France

RIVER SEINE continued

The Seine is a beautiful river giving access to Paris and central France, and to the Mediterranean via the canals (see below). But there is constant traffic of *péniches* (barges), and in the tidal section the strong stream and the wash from ships make it dangerous to moor alongside and uncomfortable to anchor. There are locks at Amfreville (202km from Paris), Notre Dame de la Garenne (161km), Mericourt (121km), Andresy (73km), Bougival (49km), Le Chatou (44km) and Suresnes (17km). Above Amfreville the current is about 1kt in summer, but more in winter.

Masts can be unstepped at Le Havre (Société des Régates) and Deauville. Yacht navigation is prohibited at night, and anchorages are scarce. A good anchor light is needed, and even with the mast lowered it is useful to have a radar reflector and a VHF aerial.

Entrance is through the Chenal de Rouen, which is dredged and buoyed but rough in a strong westerly wind and ebb tide, or by the Canal de Tancarville (see 10.18.19 for details). The canal has two locks and nine bridges which cause delay, and is best avoided unless sea conditions are bad.

Care is needed, especially near the mouth where there are shifting banks and often early morning fog. Between Quilleboeuf (335km) and La Mailleraye (303km) the *mascaret* (Seine bore) still runs at HWS when the river is in flood. It is important to study the tides. The flood stream starts progressively later as a boat proceeds upriver. So even a 4-kn boat leaving the estuary at LW can carry the flood for the 78M (123km) to Rouen (see 10.18.29). Coming down river on the ebb however a boat will meet the flood: in a fast boat it is worth carrying on, rather than anchoring for about four hours, because the further downstream the sooner the ebb starts.

The most likely mooring places below Rouen are Quilleboeuf, Villequier, Caudebec, Le Trait and Duclair. In Rouen (a major port) yachts use the Bassin St Gervais, some way from the city centre and rather squalid, but there is also a marina on the north side of Ile Lacroix above Pont Corneille. In the non-tidal section there are alongside berths at Les Andelys, Vernon, Port Maria (near Mantes) and Les Mureaux. In Paris the Touring Club de France (Pont de la Concorde) is helpful and cheap, but noisy and uncomfortable. Fuel is available alongside at Rouen, Rolleboise, Conflans and Paris.

For detailed information see *A Cruising Guide to the Lower Seine* (Imray, Laurie, Norie and Wilson Ltd).

CANALS TO MEDITERRANEAN

The quickest route to the Mediterranean is Le Havre/Paris/St Mammes/canal du Loing/canal du Briare/canal Latéral à la Loire/canal du Centre/Saône/Rhône, a distance of 1319km (824 miles) with 182 locks. Maximum dimensions: length 38.5m, beam 5.0m, draught 1.8m, air draught 3.5m. Further details and list of chomages (stoppages) for current year (available from March), can be obtained from French Tourist Office, 178 Piccadilly, London, W1V 0AL.

LE HAVRE 10-18-19
Seine Maritime

CHARTS
Admiralty 2990, 2146; SHOM 6683, 6736, 6796; ECM 526, 1012; Stanford 1; Imray C31

TIDES
Dover −0120; ML La Roque 5.3 Le Havre 4.6; Duration Le Havre 0543; Zone −0100

Standard Port LE HAVRE (→)

Times				Height (metres)			
HW		LW		MHWS	MHWN	MLWN	MLWS
0000	0500	0000	0700	7.9	6.6	3.0	1.2
1200	1700	1200	1900				

Differences LA ROQUE

| −0120 | −0010 | +0045 | +0145 | +0.3 | +0.2 | +0.1 | +0.9 |
| +0100 | | | | | +0.1 | | |

NOTE: Le Havre is a Standard Port and the tidal predictions for each day of the year are given below.

SHELTER
Excellent shelter in the marina. Visitors go to Pontoon O. If this is full, the Bassin du Commerce can be used but this can only be entered at HW when gates and three bridges are opened. Advance request to the Hr Mr. Harbour accessible at all tides.

NAVIGATION
Waypoint 49°31'.05N 00°04'.00W, 287°/107° from/to front Ldg Lt 107°, 7M. Shingle banks, awash at LW, lie to the N of approach channel. Keep in channel to avoid lobster pots close each side. Beware boulders inside at the base of Digue Nord. Anchoring is forbidden in the fairway and over a large area to the N of the fairway. Crossing the fairway the harbour side of buoys LH7 and LH8 is also forbidden. Obey traffic signals given from end of Digue Nord. Le Havre is a busy commercial port so yachts must not obstruct shipping. The Canal de Tancarville runs eastwards from Le Havre to the River Seine at Tancarville (see 10.18.18). To enter the canal at Le Havre proceed through the small craft lock (Écluse de la Citadelle) in the NE corner of the Arrière Port, across Bassin de la Citadelle and under a bridge into Bassin de l'Eure. Turn south, and after 400m turn east into Bassin Bellot. At the far end alter course to port under two bridges into Bassin Vétillart. From here proceed straight through Garage de Graville, Bassin de Despujois, and across the north end of Bassin de Lancement, which leads under two bridges into Canal de Tancarville.

LIGHTS AND MARKS
Ldg Lts 107° both Dir FW 36/78m 25M; grey Trs, G tops; intens 106°-108° (H24).

RADIO TELEPHONE
Call: *Havre Port* Sig Stn VHF Ch 12 20 (or 2182 kHz). Port Operations Ch 67 69 (H24). Marina Ch 09. Other station: Havre-Antifer Ch 14 16 **22** (H24).

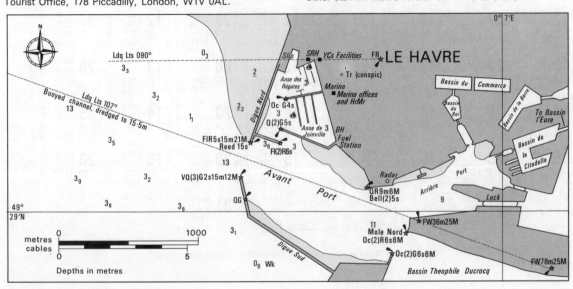

FRANCE, NORTH COAST - LE HAVRE

LAT 49°29′N LONG 0°07′E

TIMES AND HEIGHTS OF HIGH AND LOW WATERS

YEAR **1989**

TIME ZONE −0100 (French Standard Time)
For French Summer Time add ONE hour in non-shaded areas

	JANUARY				FEBRUARY				MARCH				APRIL		
	TIME m		TIME m		TIME m		TIME m		TIME m		TIME m		TIME m		TIME m
1 SU	0431 6.5 1108 3.2 1646 6.4 2333 3.1	**16** M	0511 6.9 1200 2.7 1745 6.7	**1** W	0527 6.3 1209 3.4 1814 6.1	**16** TH	0133 3.3 0736 6.4 1429 2.9 2025 6.5	**1** W	0318 6.5 1011 3.0 1551 6.2 2236 3.3	**16** TH	0535 6.2 1231 3.2 1851 6.1	**1** SA	0555 6.1 1243 3.1 1854 6.3	**16** SU	0207 3.0 0753 6.4 1438 2.6 2028 6.7
2 M	0537 6.5 1207 3.3 1800 6.3	**17** TU	0031 2.8 0626 6.8 1323 2.8 1909 6.7	**2** TH	0050 3.4 0703 6.3 1345 3.2 1946 6.3	**17** F	0307 3.0 0846 6.7 1551 2.4 2122 6.9	**2** TH	0430 6.1 1120 3.3 1728 6.0	**17** F	0117 3.4 0721 6.2 1413 3.0 2011 6.4	**2** SU	0138 3.1 0721 6.5 1416 2.5 2003 6.8	**17** M	0309 2.5 0841 6.8 1533 2.2 2106 7.1
3 TU	0037 3.2 0646 6.5 1318 3.2 1914 6.4	**18** W	0156 2.8 0745 6.8 1441 2.6 2027 6.8	**3** F	0230 3.1 0815 6.6 1509 2.7 2052 6.7	**18** SA	0419 2.5 0936 7.0 1650 2.0 2205 7.2	**3** F	0004 3.6 0624 6.1 1308 3.3 1920 6.2	**18** SA	0258 3.0 0829 6.5 1535 2.5 2103 6.8	**3** M	0254 2.4 0822 7.0 1523 1.8 2056 7.3	**18** TU	0357 2.1 0917 7.1 1617 1.9 2138 7.3
4 W	0154 3.1 0748 6.7 1435 2.9 2017 6.6	**19** TH	0310 2.7 0852 7.0 1549 2.3 2128 7.1	**4** SA	0343 2.6 0911 7.0 1612 2.1 2144 7.1	**19** SU	0508 2.1 1016 7.3 1730 1.7 2241 7.4	**4** SA	0204 3.2 0750 6.5 1444 2.7 2031 6.7	**19** SU	0402 2.4 0916 6.9 1626 2.0 2141 7.1	**4** TU	0355 1.7 0913 7.5 1623 1.3 2143 7.7	**19** W	0436 1.8 0949 7.3 1654 1.7 2208 7.5
5 TH	0307 2.8 0841 6.9 1537 2.5 2110 6.9	**20** F	0413 2.4 0944 7.3 1648 2.0 2216 7.3	**5** SU	0440 2.1 0959 7.4 1709 1.6 2230 7.5	**20** M	0544 1.8 1050 7.5 1803 1.4 ○ 2313 7.6	**5** SU	0322 2.5 0851 7.0 1552 2.0 2125 7.2	**20** M	0444 2.0 0952 7.2 1703 1.7 2213 7.4	**5** W	0452 1.2 0959 7.9 1717 0.8 2226 8.0	**20** TH	0512 1.5 1020 7.4 1728 1.6 2238 7.5
6 F	0403 2.5 0928 7.2 1629 2.1 2158 7.2	**21** SA	0507 2.1 1029 7.5 1737 1.7 ○ 2258 7.5	**6** M	0535 1.6 1044 7.8 1802 1.1 ● 2314 7.8	**21** TU	0615 1.6 1121 7.7 1833 1.3 2342 7.6	**6** M	0423 1.9 0940 7.5 1652 1.4 2210 7.6	**21** TU	0517 1.7 1023 7.5 1734 1.5 2242 7.5	**6** TH	0542 0.8 1044 8.1 1805 0.6 ● 2309 8.1	**21** F	0544 1.6 1052 7.5 1759 1.5 ○ 2308 7.6
7 SA	0453 2.1 1012 7.5 1719 1.7 ● 2242 7.4	**22** SU	0552 1.9 1108 7.6 1817 1.5 2335 7.6	**7** TU	0625 1.2 1128 8.0 1850 0.8 2357 7.9	**22** W	0645 1.5 1151 7.7 1902 1.4	**7** TU	0518 1.3 1024 7.9 1745 0.9 ● 2253 7.9	**22** W	0548 1.5 1052 7.6 1804 1.3 ○ 2311 7.6	**7** F	0627 0.6 1128 8.2 1847 0.6 2351 8.1	**22** SA	0612 1.5 1124 7.5 1826 1.6 2338 7.6
8 SU	0542 1.8 1056 7.7 1809 1.4 2326 7.6	**23** M	0630 1.8 1143 7.7 1852 1.4	**8** W	0710 0.9 1211 8.1 1933 0.6	**23** TH	0012 7.6 0713 1.5 1220 7.7 1929 1.4	**8** W	0608 0.9 1108 8.1 1831 0.5 2335 8.1	**23** TH	0617 1.4 1121 7.7 1832 1.3 2340 7.7	**8** SA	0708 0.6 1212 8.1 1926 0.7	**23** SU	0640 1.5 1155 7.5 1853 1.7
9 M	0630 1.6 1139 7.8 1857 1.2	**24** TU	0009 7.6 0704 1.7 1216 7.7 1925 1.4	**9** TH	0040 8.0 0753 0.9 1254 8.2 2013 0.7	**24** F	0041 7.6 0738 1.5 1249 7.6 1952 1.5	**9** TH	0651 0.6 1151 8.3 1913 0.4	**24** F	0643 1.4 1151 7.7 1857 1.4	**9** SU	0032 8.0 0746 0.8 1255 7.9 2003 1.1	**24** M	0008 7.5 0709 1.6 1227 7.4 1922 1.8
10 TU	0010 7.7 0717 1.5 1224 7.9 1943 1.1	**25** W	0041 7.6 0737 1.7 1248 7.6 1956 1.5	**10** F	0122 8.0 0831 1.0 1337 8.0 2050 0.9	**25** SA	0108 7.5 0802 1.7 1316 7.5 2015 1.8	**10** F	0017 8.2 0732 0.6 1233 8.3 1951 0.5	**25** SA	0008 7.6 0708 1.5 1210 7.6 1921 1.5	**10** M	0113 7.7 0824 1.2 1339 7.5 2037 1.7	**25** TU	0040 7.4 0740 1.7 1303 7.2 1953 2.1
11 W	0055 7.7 0802 1.4 1309 7.9 2026 1.1	**26** TH	0113 7.5 0806 1.8 1320 7.5 2023 1.7	**11** SA	0203 7.8 0908 1.3 1419 7.7 2125 1.4	**26** SU	0134 7.5 0825 1.9 1342 7.3 2038 2.1	**11** SA	0057 8.1 0809 0.8 1315 8.1 2026 0.9	**26** SU	0036 7.6 0733 1.6 1248 7.5 1946 1.7	**11** TU	0153 7.4 0900 1.7 1424 7.1 2113 2.3	**26** W	0115 7.2 0812 2.0 1343 6.9 2026 2.4
12 TH	0141 7.7 0846 1.5 1355 7.7 2107 1.3	**27** F	0143 7.3 0833 2.0 1349 7.3 2048 2.0	**12** SU	0244 7.5 0945 1.8 1503 7.3 ☽ 2201 2.0	**27** M	0200 7.1 0852 2.2 1413 7.0 2105 2.4	**12** SU	0137 7.8 0846 1.2 1357 7.7 2100 1.5	**27** M	0102 7.4 0759 1.8 1317 7.2 2011 2.0	**12** W	0237 6.9 0940 2.3 1520 6.6 ☽ 2159 2.9	**27** TH	0156 6.9 0848 2.3 1430 6.7 2106 2.7
13 F	0227 7.5 0927 1.7 1441 7.6 2148 1.6	**28** SA	0211 7.2 0858 2.3 1417 7.1 2113 2.3	**13** M	0329 7.1 1027 2.3 1557 6.9 2247 2.6	**28** TU	0233 6.8 0924 2.6 1453 6.6 2141 2.9	**13** M	0215 7.5 0921 1.7 1440 7.2 2134 2.1	**28** TU	0131 7.2 0827 2.1 1351 6.9 2039 2.4	**13** TH	0337 6.5 1037 2.8 1642 6.2 2311 3.3	**28** F	0247 6.6 0935 2.6 1530 6.4 ☾ 2204 3.1
14 SA	0314 7.3 1010 2.0 1532 7.3 ☽ 2231 2.0	**29** SU	0241 7.0 0928 2.6 1451 6.9 2144 2.6	**14** TU	0428 6.7 1123 2.8 1716 6.4 2354 3.1			**14** TU	0258 7.0 1000 2.3 1534 6.7 ☽ 2218 2.8	**29** W	0205 6.9 0859 2.4 1433 6.6 2115 2.8	**14** F	0508 6.1 1205 3.1 1818 6.2 2332 3.1	**29** SA	0355 6.4 1044 2.8 1655 6.3
15 SU	0407 7.1 1059 2.4 1631 7.0 2323 2.4	**30** M	0319 6.7 1006 2.9 1535 6.5 ☾ 2226 3.0	**15** W	0558 6.4 1255 3.1 1901 6.3			**15** W	0357 6.5 1055 2.9 1659 6.2 2328 3.4	**30** TH	0251 6.5 0942 2.8 1533 6.2 ☾ 2210 3.3	**15** SA	0046 3.4 0642 6.2 1327 2.9 1936 6.4	**30** SU	0528 6.4 1219 2.7 1822 6.5
		31 TU	0410 6.5 1058 3.2 1638 6.2 2326 3.3							**31** F	0404 6.2 1053 3.2 1708 6.0 2341 3.5				

Chart Datum: 4.72 metres below Lallemand System (Mean Sea Level, Marseilles)

FRANCE, NORTH COAST - LE HAVRE

LAT 49°29'N LONG 0°07'E

TIMES AND HEIGHTS OF HIGH AND LOW WATERS

YEAR **1989**

TIME ZONE −0100
(French Standard Time)
For French Summer Time add
ONE hour in non-shaded areas

MAY

	TIME	m		TIME	m
1 M	0106 0646 1340 1927	2.8 6.7 2.3 7.0	**16** TU	0205 0748 1430 2017	2.7 6.6 2.4 6.9
2 TU	0221 0747 1447 2023	2.2 7.1 1.8 7.4	**17** W	0259 0833 1523 2056	2.4 6.9 2.2 7.1
3 W	0321 0842 1548 2112	1.7 7.5 1.4 7.7	**18** TH	0348 0912 1609 2131	2.2 7.0 2.0 7.3
4 TH	0420 0932 1645 2159	1.3 7.7 1.1 7.9	**19** F	0431 0949 1649 2205	2.0 7.2 1.9 7.4
5 F ●	0513 1020 1735 2244	1.0 7.9 0.9 8.0	**20** SA ○	0508 1025 1725 2240	1.8 7.3 1.9 7.4
6 SA	0601 1107 1821 2328	0.9 8.0 1.0 8.0	**21** SU	0542 1100 1757 2313	1.7 7.3 1.8 7.5
7 SU	0645 1154 1902	0.9 7.9 1.1	**22** M	0616 1136 1831 2348	1.6 7.3 1.8 7.5
8 M	0011 0725 1240 1941	7.8 1.0 7.7 1.4	**23** TU	0651 1214 1907	1.6 7.3 1.9
9 TU	0054 0804 1326 2019	7.6 1.3 7.4 1.9	**24** W	0025 0728 1254 1944	7.4 1.6 7.2 2.0
10 W	0137 0844 1413 2058	7.3 1.7 7.0 2.3	**25** TH	0107 0807 1338 2024	7.3 1.8 7.1 2.2
11 TH	0223 0927 1507 2147	7.0 2.2 6.7 2.8	**26** F	0152 0848 1427 2110	7.1 2.0 6.9 2.5
12 F ☽	0319 1019 1612 2249	6.6 2.6 6.4 3.1	**27** SA	0243 0938 1523 2207	6.9 2.2 6.8 2.6
13 SA	0428 1124 1724 2359	6.3 2.8 6.3 3.1	**28** SU ☾	0344 1040 1632 2318	6.8 2.3 6.7 2.7
14 SU	0543 1233 1833	6.3 2.8 6.5	**29** M	0456 1151 1744	6.8 2.3 6.8
15 M	0105 0652 1333 1932	3.0 6.4 2.7 6.7	**30** TU	0030 0608 1301 1849	2.5 6.9 2.0 7.0
			31 W	0140 0712 1409 1948	2.2 7.1 1.9 7.3

JUNE

	TIME	m		TIME	m
1 TH	0247 0813 1514 2043	1.9 7.3 1.7 7.5	**16** F	0259 0834 1525 2056	2.6 6.7 2.5 7.0
2 F	0348 0910 1613 2135	1.6 7.5 1.5 7.6	**17** SA	0350 0920 1612 2137	2.3 6.9 2.3 7.2
3 SA	0445 1003 1655 2217	1.4 7.6 1.5 7.6	**18** SU	0435 1003 1655 2217	2.1 7.1 2.1 7.3
4 SU	0537 1054 1756 2311	1.2 7.7 1.4 7.8	**19** M ○	0517 1043 1736 2255	1.8 7.2 2.0 7.4
5 M	0624 1142 1842 2356	1.2 7.7 1.5 7.7	**20** TU	0559 1123 1818 2335	1.6 7.3 1.9 7.5
6 TU	0708 1229 1924	1.2 7.6 1.7	**21** W	0642 1204 1900	1.5 7.4 1.8
7 W	0040 0749 1314 2005	7.6 1.4 7.4 1.9	**22** TH	0016 0725 1247 1944	7.5 1.4 7.4 1.8
8 TH	0123 0829 1358 2046	7.4 1.7 7.2 2.2	**23** F	0100 0809 1332 2028	7.5 1.5 7.3 1.9
9 F	0207 0911 1443 2128	7.1 2.0 6.9 2.5	**24** SA	0145 0852 1419 2113	7.4 1.6 7.3 2.0
10 SA	0252 0953 1531 2214	6.9 2.3 6.7 2.7	**25** SU	0233 0937 1508 2201	7.3 1.7 7.2 2.1
11 SU ☽	0342 1038 1625 2304	6.6 2.6 6.6 2.9	**26** M	0325 1025 1603 2254	7.1 1.9 7.1 2.3
12 M	0440 1129 1724 2359	6.4 2.8 6.5 3.0	**27** TU	0424 1119 1705 2354	7.0 2.1 7.0 2.4
13 TU	0542 1226 1823	6.4 2.8 6.5	**28** W	0531 1223 1812	6.9 2.2 7.0
14 W	0059 0645 1326 1920	2.9 6.5 2.8 6.7	**29** TH	0103 0642 1335 1919	2.4 6.9 2.3 7.1
15 TH	0201 0743 1431 2011	2.8 6.5 2.7 6.8	**30** F	0219 0753 1446 2024	2.2 7.0 2.2 7.2

JULY

	TIME	m		TIME	m
1 SA	0324 0859 1550 2123	2.0 7.2 2.0 7.4	**16** SU	0315 0858 1545 2115	2.6 6.7 2.6 7.0
2 SU	0425 0957 1648 2215	1.7 7.4 1.9 7.5	**17** M	0410 0946 1635 2200	2.2 7.0 2.3 7.3
3 M	0521 1048 1741 2302	1.5 7.5 1.8 7.6	**18** TU ○	0500 1030 1723 2242	1.8 7.2 1.9 7.5
4 TU	0612 1134 1829 2345	1.4 7.6 1.7 7.6	**19** W	0550 1111 1813 2323	1.5 7.4 1.7 7.7
5 W	0656 1216 1911	1.3 7.6 1.7	**20** TH	0637 1153 1858	1.3 7.6 1.5
6 TH	0025 0736 1256 1950	7.6 1.4 7.5 1.7	**21** F	0005 0722 1235 1941	7.8 1.1 7.7 1.4
7 F	0104 0813 1334 2026	7.5 1.5 7.4 1.9	**22** SA	0048 0804 1318 2023	7.8 1.0 7.7 1.3
8 SA	0141 0846 1411 2059	7.4 1.7 7.2 2.1	**23** SU	0131 0844 1401 2103	7.8 1.1 7.6 1.5
9 SU	0218 0917 1448 2131	7.1 2.0 7.0 2.4	**24** M	0215 0922 1445 2143	7.6 1.4 7.5 1.7
10 M	0255 0949 1526 2206	6.9 2.3 6.8 2.7	**25** TU ☾	0301 1002 1531 2227	7.4 1.7 7.3 2.0
11 TU	0335 1024 1610 2248	6.7 2.6 6.6 2.9	**26** W	0353 1048 1627 2321	7.1 2.1 7.0 2.4
12 W ☽	0424 1110 1707 2342	6.4 2.9 6.5 3.1	**27** TH	0500 1148 1739	6.8 2.5 6.8
13 TH	0531 1210 1817	6.2 3.1 6.4	**28** F	0033 0624 1308 1902	2.8 6.6 2.8 6.8
14 F	0051 0650 1326 1926	3.2 6.2 2.9 6.5	**29** SA	0159 0750 1430 2019	2.6 6.6 2.7 7.0
15 SA	0209 0800 1444 2025	3.0 6.4 2.9 6.7	**30** SU	0312 0901 1540 2120	2.3 7.0 2.4 7.2
			31 M	0419 0955 1644 2209	1.9 7.3 2.1 7.4

AUGUST

	TIME	m		TIME	m
1 TU ●	0519 1040 1738 2251	1.6 7.5 1.8 7.6	**16** W	0445 1012 1711 2224	1.7 7.4 1.7 7.6
2 W	0605 1119 1820 2329	1.4 7.6 1.7 7.7	**17** TH ○	0537 1053 1800 2306	1.3 7.7 1.4 7.9
3 TH	0643 1156 1856	1.3 7.6 1.6	**18** F	0625 1134 1846 2347	1.0 7.8 1.1 8.1
4 F	0004 0716 1230 1928	7.7 1.3 7.6 1.6	**19** SA	0708 1216 1928	0.8 8.0 1.0
5 SA	0037 0747 1302 1957	7.7 1.4 7.6 1.7	**20** SU	0030 0747 1257 2006	8.1 0.7 8.0 1.0
6 SU	0110 0814 1334 2024	7.6 1.6 7.4 1.9	**21** M	0112 0825 1338 2043	8.0 0.9 7.9 1.2
7 M	0141 0838 1404 2049	7.4 1.9 7.2 2.1	**22** TU	0154 0900 1418 2121	7.8 1.3 7.6 1.6
8 TU	0210 0902 1432 2116	7.1 2.2 7.0 2.5	**23** W ☾	0237 0936 1502 2202	7.5 1.8 7.3 2.1
9 W	0240 0930 1505 2150	6.9 2.5 6.8 2.8	**24** TH	0328 1020 1556 2254	7.0 2.4 6.9 2.6
10 TH	0320 1008 1551 2238	6.5 3.0 6.5 3.2	**25** F	0440 1121 1717	6.6 2.9 6.5
11 F	0418 1103 1703 2346	6.2 3.3 6.2 3.4	**26** SA	0013 0621 1256 1859	2.9 6.4 3.2 6.5
12 SA	0553 1226 1843	6.0 3.5 6.2	**27** SU	0153 0755 1430 2018	2.8 6.6 2.9 6.8
13 SU	0125 0732 1411 1959	3.3 6.2 3.3 6.5	**28** M	0315 0858 1547 2113	2.4 7.0 2.5 7.1
14 M	0249 0838 1523 2055	2.8 6.6 2.7 6.9	**29** TU	0422 0944 1646 2156	1.9 7.3 2.0 7.4
15 W	0350 0928 1618 2142	2.2 7.0 2.2 7.3	**30** W	0511 1022 1728 2232	1.6 7.5 1.7 7.5
			31 TH ●	0548 1057 1801 2305	1.4 7.6 1.6 7.8

Chart Datum: 4.72 metres below Lallemand System (Mean Sea Level, Marseilles)

FRANCE, NORTH COAST - LE HAVRE

LAT 49°29′N   LONG 0°07′E

TIMES AND HEIGHTS OF HIGH AND LOW WATERS

YEAR **1989**

TIME ZONE –0100 (French Standard Time)
For French Summer Time add ONE hour in non-shaded areas

SEPTEMBER

	TIME	m		TIME	m
1 F	0619 1128 1830 2336	1.3 7.7 1.5 7.8	**16** SA	0601 1109 1823 2324	0.8 8.1 0.9 8.2
2 SA	0647 1158 1859	1.3 7.7 1.5	**17** SU	0644 1150 1906	0.6 8.1 0.8
3 SU	0006 0717 1227 1925	7.8 1.4 7.7 1.6	**18** M	0007 0724 1231 1944	8.3 0.7 8.1 0.9
4 M	0036 0739 1256 1949	7.6 1.6 7.5 1.8	**19** TU	0049 0802 1312 2022	8.1 1.0 7.9 1.2
5 TU	0104 0802 1322 2013	7.5 1.8 7.4 2.0	**20** W	0132 0837 1353 2059	7.8 1.4 7.6 1.7
6 W	0130 0824 1348 2038	7.2 2.2 7.1 2.3	**21** TH	0217 0913 1437 2140	7.4 2.0 7.2 2.2
7 TH	0159 0850 1419 2109	6.9 2.6 6.8 2.7	**22** F ☾	0311 0957 1534 2234	6.8 2.7 6.7 2.8
8 F ☾	0238 0924 1502 2153	6.5 3.0 6.5 3.1	**23** SA	0431 1104 1705	6.4 3.3 6.4
9 SA	0335 1016 1610 2258	6.1 3.5 6.1 3.5	**24** SU	0008 0619 1253 1849	3.1 6.3 3.4 6.4
10 SU	0508 1138 1804	5.9 3.7 6.1	**25** M	0147 0745 1426 2004	2.9 6.6 3.0 6.7
11 M	0048 0704 1345 1932	3.4 6.1 3.4 6.4	**26** TU	0303 0841 1533 2054	2.4 7.0 2.4 7.1
12 TU	0225 0813 1500 2030	2.8 6.6 2.7 6.9	**27** W	0400 0922 1621 2133	2.0 7.3 2.0 7.4
13 W	0327 0903 1555 2117	2.2 7.1 2.1 7.4	**28** TH	0442 0956 1658 2206	1.7 7.5 1.7 7.6
14 TH	0422 0947 1648 2200	1.6 7.6 1.5 7.8	**29** F ●	0516 1026 1729 2236	1.5 7.7 1.6 7.7
15 F ○	0514 1028 1737 2242	1.1 7.9 1.1 8.1	**30** SA	0545 1055 1759 2305	1.4 7.7 1.5 7.8

OCTOBER

	TIME	m		TIME	m
1 SU	0614 1124 1827 2335	1.4 7.7 1.5 7.7	**16** M	0617 1123 1841 2344	0.8 8.2 0.8 8.2
2 M	0642 1153 1854	1.5 7.7 1.6	**17** TU	0700 1206 1922	0.9 8.1 0.9
3 TU	0005 0706 1220 1918	7.6 1.7 7.6 1.8	**18** W	0030 0739 1249 2002	8.0 1.2 7.9 1.3
4 W	0033 0730 1246 1944	7.4 1.9 7.4 2.0	**19** TH	0115 0817 1332 2041	7.7 1.7 7.5 1.7
5 TH	0101 0755 1314 2011	7.2 2.2 7.2 2.3	**20** F	0204 0856 1419 2124	7.3 2.3 7.1 2.3
6 F	0133 0823 1347 2043	6.9 2.6 6.9 2.6	**21** SA ☾	0301 0943 1518 2221	6.8 2.9 6.7 2.8
7 SA	0215 0858 1434 2125	6.5 3.0 6.5 3.0	**22** SU	0420 1054 1644 2347	6.4 3.3 6.4 3.1
8 SU ☾	0313 0948 1542 2227	6.2 3.4 6.2 3.3	**23** M	0552 1229 1815	6.4 3.3 6.3
9 M	0440 1109 1726	6.1 3.7 6.1	**24** TU	0112 0711 1347 1929	2.9 6.6 3.0 6.6
10 TU	0013 0630 1311 1856	3.3 6.2 3.4 6.5	**25** W	0219 0807 1448 2022	2.6 6.9 2.6 6.9
11 W	0150 0738 1429 1956	2.8 6.7 2.7 7.0	**26** TH	0313 0848 1536 2101	2.2 7.2 2.2 7.2
12 TH	0254 0830 1524 2046	2.1 7.3 2.0 7.5	**27** F	0357 0922 1617 2135	2.0 7.4 2.0 7.4
13 F	0351 0915 1618 2131	1.5 7.7 1.5 7.9	**28** SA	0435 0953 1653 2206	1.8 7.6 1.8 7.5
14 SA	0443 0958 1709 2215	1.1 8.0 1.1 8.1	**29** SU ●	0509 1022 1726 2237	1.7 7.6 1.7 7.6
15 SU	0532 1040 1757 2300	0.9 8.1 0.9 8.2	**30** M	0541 1052 1757 2308	1.7 7.7 1.7 7.6
31 TU	0610 1122 1825 2339	1.8 7.6 1.7 7.5			

NOVEMBER

	TIME	m		TIME	m
1 W	0638 1151 1853	1.9 7.6 1.8	**16** TH	0016 0720 1232 1945	7.9 1.5 7.8 1.3
2 TH	0010 0706 1221 1923	7.4 2.1 7.4 2.0	**17** F	0104 0802 1318 2029	7.6 1.8 7.5 1.7
3 F	0043 0736 1254 1954	7.2 2.3 7.2 2.2	**18** SA	0154 0845 1405 2113	7.3 2.3 7.2 2.1
4 SA	0122 0808 1334 2029	7.0 2.6 7.0 2.5	**19** SU	0247 0933 1500 2204	6.9 2.7 6.8 2.6
5 SU	0207 0846 1423 2112	6.7 2.9 6.7 2.8	**20** M ☾	0350 1032 1605 2306	6.6 3.1 6.5 2.9
6 M	0302 0937 1525 2211	6.5 3.2 6.5 3.0	**21** TU	0500 1139 1718	6.5 3.2 6.4
7 TU	0416 1053 1648 2339	6.4 3.4 6.4 3.0	**22** W	0013 0609 1246 1829	2.9 6.6 3.1 6.5
8 W	0547 1228 1812	6.5 3.1 6.6	**23** TH	0116 0707 1348 1931	2.9 6.7 2.9 6.7
9 TH	0105 0656 1334 1916	2.7 6.9 2.6 7.0	**24** F	0215 0802 1444 2020	2.7 7.0 2.6 7.1
10 F	0214 0751 1449 2011	2.2 7.3 2.1 7.4	**25** SA	0308 0842 1534 2101	2.4 7.2 2.4 7.1
11 SA	0315 0841 1546 2102	1.7 7.7 1.6 7.7	**26** SU	0354 0918 1617 2137	2.2 7.3 2.1 7.2
12 SU	0411 0929 1641 2151	1.4 7.9 1.3 7.9	**27** M	0435 0952 1655 2212	2.1 7.4 2.0 7.3
13 M	0504 1015 1731 2240	1.2 8.0 1.1 8.1	**28** TU ●	0511 1026 1730 2247	2.0 7.5 1.9 7.4
14 TU	0552 1101 1818 2328	1.1 8.1 1.0 8.0	**29** W	0544 1059 1803 2322	1.8 7.5 1.8 7.4
15 W	0637 1146 1903	1.2 8.0 1.1	**30** TH	0617 1132 1836 2357	2.0 7.5 1.8 7.4

DECEMBER

	TIME	m		TIME	m
1 F	0651 1206 1911	2.1 7.5 1.8	**16** SA	0054 0751 1305 2016	7.7 1.8 7.6 1.5
2 SA	0034 0726 1245 1948	7.3 2.2 7.4 2.0	**17** SU	0139 0833 1348 2056	7.5 2.0 7.4 1.8
3 SU	0115 0804 1327 2026	7.2 2.4 7.2 2.1	**18** M	0223 0914 1432 2137	7.2 2.3 7.1 2.2
4 M	0200 0846 1414 2110	7.0 2.6 7.1 2.3	**19** TU ☾	0309 0955 1519 2218	6.9 2.6 6.8 2.5
5 TU	0250 0935 1507 2203	6.9 2.7 6.9 2.5	**20** W	0359 1040 1612 2304	6.7 2.9 6.6 2.8
6 W	0349 1036 1610 2306	6.8 2.8 6.8 2.6	**21** TH	0455 1131 1713 2358	6.6 3.1 6.4 3.1
7 TH	0459 1144 1723	6.7 2.8 6.8	**22** F	0558 1232 1821	6.5 3.2 6.4
8 F	0016 0609 1256 1833	2.5 6.9 2.6 7.0	**23** SA	0102 0701 1342 1928	3.1 6.6 3.1 6.4
9 SA	0129 0712 1409 1938	2.3 7.2 2.3 7.2	**24** SU	0215 0757 1448 2025	3.0 6.7 2.9 6.6
10 SU	0239 0810 1517 2039	2.1 7.4 1.9 7.4	**25** M	0314 0845 1542 2112	2.8 6.9 2.6 6.9
11 M	0342 0905 1616 2135	1.8 7.6 1.6 7.7	**26** TU	0403 0927 1628 2154	2.5 7.1 2.3 7.1
12 TU	0439 0957 1710 2229	1.6 7.8 1.3 7.8	**27** W ○	0447 1006 1709 2233	2.3 7.3 2.0 7.2
13 W	0532 1047 1802 2319	1.5 7.9 1.2 7.9	**28** TH	0527 1043 1748 2310	2.1 7.5 1.8 7.4
14 TH	0621 1134 1850	1.7 7.9 1.2	**29** F	0605 1120 1827 2347	2.0 7.6 1.6 7.4
15 F	0007 0708 1220 1935	7.8 1.6 7.8 1.3	**30** SA	0645 1157 1908	1.9 7.6 1.5
31 SU	0026 0725 1237 1948	7.5 1.8 7.7 1.5			

Chart Datum: 4.72 metres below Lallemand System (Mean Sea Level, Marseilles)

LE HAVRE continued

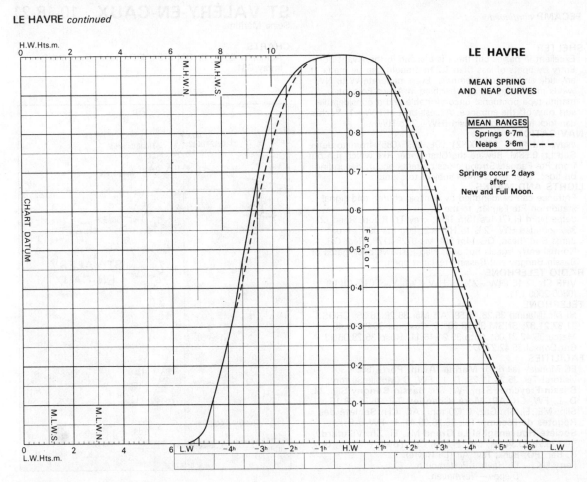

TELEPHONE
Hr Mr 35.22.72.72; Hr Mr Port de Plaisance 35.21.23.95; Aff Mar 35.22.41.03; CROSS 33.52.72.13; SNSM 35.22.41.03; Customs 35.41.33.51; Meteo 35.42.21.06; Auto 35.21.16.11; Dr 35.41.23.61; Hosp 35.22.81.23; Brit Consul 35.42.27.47.

FACILITIES
Le Havre-Plaisance Marina (850+80 visitors) Tel. 35.21.23.95, Slip, AC, BH (16 ton), C (6 ton), CH, D, P, El, FW, Gas, Gaz, R, Lau, V; **Accastillage Diffusion** Tel. 35.43.43.62, ME, CH; **Coloma** Tel. 35.25.30.51, Sh, ME, CH; **Manche Yachting** Tel. 35.21.08.06, ME, CH; **Nautic Channel** Tel. 35.42.18.37, ME; **YC — Societe des Regates du Havre** Tel. 35.42.41.21, R, Bar; **Nautic-Service** Tel. 35.51.75.30 SHOM; **Heilmann** Tel. 35.42.42.49, El, CH, SHOM, ACA; **Havre Voiles** Tel. 35.42.58.15, SM; **Voilerie de l'Estuaire** Tel. 35.43.33.34, SM. **Town** ME, El, CH, V, Gaz, R, Bar. PO; Bank; Rly; Air.
Ferry UK—Portsmouth.

FÉCAMP 10-18-20
Seine Maritime

CHARTS
Admiralty 2612, 1352; SHOM 932, 6765; ECM 1012; Stanford 1; Imray C31
TIDES
Dover −0100; ML 4.5; Duration 0550; Zone −0100

Standard Port DIEPPE (→)

Times				Height (metres)			
HW		LW		MHWS	MHWN	MLWN	MLWS
0100	0600	0000	0700	9.3	7.2	2.6	0.7
1300	1800	1200	1900				

Differences FÉCAMP
  −0022 −0018 −0034 −0043 −1.4 −0.7 0.0 +0.1
ANTIFER
  −0046 −0039 −0051 −0100 −1.3 −0.6 +0.4 +0.5

FECAMP continued

SHELTER
Excellent in basins but there is a'scend in the Avant Port. Entry by boats of less than 1.2 m draught can be made at any tide and in most weathers. Even moderate W to NW winds make entrance and berthing in Avant Port (at marina-type pontoons) uncomfortable and a considerable surf runs off the entrance. Access best at HW+1. Yachts can lock into Berigny Basin (HW−2 to HW).

NAVIGATION
Waypoint 49°45'.90N 00°21'.00E, 265°/085° from/to Jetée Sud Lt, 0.58M. Beware the Charpentier Rks which run out from Pte Fagnet. Strong cross currents occur, depending on tides. Dredged channel subject to silting.

LIGHTS AND MARKS
Entrance can be identified by conspic church and signal station on Pte Fagnet, to the N.
Jetee Nord Fl (2) 10s 15m 16M; grey Tr, R top; Reed (2) 30s sounded HW−2½ to HW+2. Ldg Lts 085°. Front Jetée Sud, head, QG 14m 9M; vis 072°-217°. Rear QR. Normal entry signals but add: Flag P shows dock gates of Bassin Bérigny and Bassin Freycinet open.

RADIO TELEPHONE
VHF Ch 12 16 (HW−2½ to HW+½). Yacht Hr Ch 09 (0800-2000 LT).

TELEPHONE
Hr Mr (Marina) 35.28.13.58; Aff Mar 35.28.16.35; CROSS 21.87.21.87; SNSM 35.28.28.15; Customs 35.28.19.40; Meteo 35.42.21.06; Auto 35.21.16.11; Hosp 35.28.05.13; Brit Consul 35.42.27.47.

FACILITIES
EC Monday (all day); **Marina (Avant Port)** (500+30 visitors) Tel. 35.28.11.25, D, L, FW, C (Mobile 36 ton); **Bassin Freycinet** Slip, FW, Sh; **Bassin Bérigny** Slip, M, D, L, FW, CH, AB; **Moré (Chantier)** Tel. 35.28.28.15, Slip, ME, El, Sh, Gaz, C (30 ton), AC, CH; **Société des Régates de Fécamp** Tel. 35.28.08.44, L, FW, AB; **Société Houvenaghel Le Grand** ME, El, Sh; **Villetard** Tel. 35.28.30.34, M, ME, El, Sh, CH. **Town** P, D, V, Gaz, R, Bar. PO; Bank; Rly; Air (Le Havre).
Ferry UK — Le Havre—Portsmouth.
           Dieppe—Newhaven.

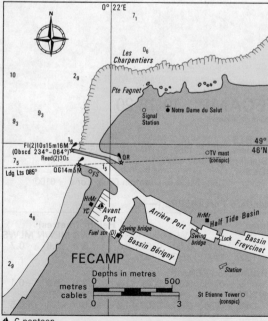
▲ C pontoon

ST VALÉRY-EN-CAUX 10-18-21
Seine Maritime

CHARTS
Admiralty 2612; SHOM 6794; ECM 1012; Stanford 1; Imray C31

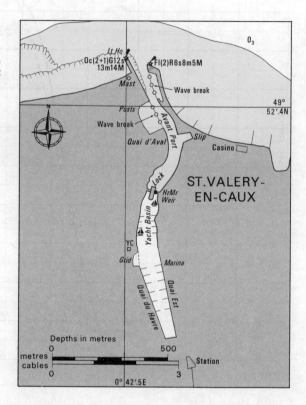

TIDES
Dover −0044; ML 4.9; Duration 0530; Zone −0100

Standard Port DIEPPE (→)

Times				Height (metres)			
HW		LW		MHWS	MHWN	MLWN	MLWS
0100	0600	0000	0700	9.3	7.2	2.6	0.7
1300	1800	1200	1900				

Differences ST VALÉRY-EN-CAUX
−0018 −0016 −0007 −0013 −0.4 −0.1 −0.1 +0.3

SHELTER
Good but outer harbour dries. Entry into outer harbour from HW−3. Marina access HW∓2. Bridge opens every half hour during the period, by day; at HW only at night.

NAVIGATION
Waypoint 49°53'.00N 00°42'.50E, 000°/180° from/to Jetée Ouest Lt, 0.50M. Entry is easy except in strong winds from E through N when seas break across entrance. Coast dries to approx 150 m off the pier heads. Inside the pier heads there are half-tide ramps to dampen the swell marked by posts each side. Shingle tends to build up against W wall.

LIGHTS AND MARKS
Harbour is difficult to identify between high chalk cliffs. Approaching from N or W the nuclear power station at Paluel is conspic 3 M W of entrance.
W pier head Lt Ho Oc (2+1) G 12s 13m 14M. E pier head light structure Fl (2)R 6s. Bu flag indicates sluicing (at about LW−1). Traffic signals are shown on both sides of the bridge/lock
G Lt — Vessels may enter but not leave
R Lt — Vessels may leave but not enter
R+G — Vessels may not enter or leave

RADIO TELEPHONE
VHF Ch 09.

ST VALERY-EN-CAUX continued

TELEPHONE
Hr Mr 35.97.01.30; Aff Mar 35.28.16.35; CROSS 21.87.21.87; SNSM 35.87.05.27; Customs 35.82.46.49; Meteo 35.80.11.44; Dr 35.97.20.13; Hosp 35.97.06.21; Brit Consul 35.42.27.47.

FACILITIES
EC Monday (all day); **Marina** (550+ some visitors) Tel. 35.97.01.30, AC, C (6 ton), V, CH, El, FW, Gaz, ME, Sh, SM, Bar; **S.A. Bondois** Tel. 35.97.04.22, CH, ME, El; **Prigent** Tel. 35.97.17.66, ME, El; **Club Nautique Valeriquais** Tel. 35.97.10.88. **Town** P, D, FW, CH, V, Gaz, R, Bar. PO; Bank; Rly; Air (Dieppe).
Ferry UK — Dieppe—Newhaven.

DIEPPE 10-18-22
Seine Maritime

CHARTS
Admiralty 2147; SHOM 5927, 934; ECM 1011, 1012; Stanford 1; Imray C31

TIDES
Dover −0035; ML 5.0; Duration 0535; Zone −0100

NOTE: Dieppe is a Standard Port. Tidal predictions for each day of the year are given below.

SHELTER
Very good and the harbour can be entered at any state of tide. Channel is exposed to winds from NW to NE, when a heavy scend renders. Beware work in progress in the Avant Port. Port de Plaisance is not comfortable for yachts. The SW part is reserved for pleasure craft. Bassin Duquesne is Yacht Basin.

NAVIGATION
Waypoint 49°56′.48N 01°04′.60E, 318°/138° from/to front Ldg Lt 138°, 0.58M. Dieppe is a busy commercial and fishing port and both entry and exit are forbidden during shipping movements. Entry to harbour is simple and yachts normally lie in SW corner of Avant Port at pontoons and wait to pass through lock into Bassin Duquesne. Access HW−2 to HW+1. The bridge opens two or three times during each period. Beware strong stream setting across entrance. Yachts are forbidden in the Arrière Port.

LIGHTS AND MARKS
Ldg Lts in line at 138° (both QR).
W Jetty — Iso WG 4s 10m 12/8M W095°-164°, G164°-095°
E Jetty — Oc (4) R 12s 10m 8M
Entry signals shown from Station at root of W jetty and from station on W side of harbour, 2½ ca (480m) S of former, (full code). Also repeated from repeating stations at entrances to Bassin Duquesne, Arriere Port and Bassin du Canada (simplified code). In low visibility a fog signal, 1 blast of 1 sec every 5 secs, is sounded from the head of Jettée Ouest to indicate 'Entry Prohibited'.
International signals combined with:−
FG to right   − Ferry entering
FR to right   − Ferry leaving
FW to left   − lock gates open
2 FR (hor) to right − dredger in channel.
Signals are repeated with flags from the W sig stn on the cliff top near church of Notre Dame de Bon-Secours.
Signals for Duquesne Basin:−
Bridge and lock gates to open − Sound 2 blasts
Vessels may enter but not leave − G Lt
Vessels may leave but not enter − R Lt
Request will not be met unless Flag P (W Lt at night) is flying from pier head and dock.

RADIO TELEPHONE
Call: *Dieppe Port* VHF Ch 06 **12** 16 (H24).

TELEPHONE
Hr Mr 35.82.23.85; Hr Mr (Port de Plaisance) 35.84.32.99; Aff Mar 35.82.59.40; CROSS 21.87.21.87; SNSM 35.84.30.76; Customs 35.82.24.47; Meteo 35.80.11.44; Auto 35.21.16.11; Hosp 35.84.20.75; Brit Consul 35.42.27.47.

FACILITIES
Quai Duquesne M, L, FW, C (by arrangement with Hr Mr up to 30 ton), AB; **Chantier de Normandie** Tel. 35.84.51.40, ME, El, Sh; **Coopérative Maritime** Tel. 35.82.26.52, ME, CH; **Cercle de la Voile de Dieppe** Tel. 35.84.32.99, P, D, FW, C (3 ton), AB, R, Bar; **S.A. Dimex (Cauach)** Tel. 35.84.33.75, ME; **Thalassa** Tel. 35.82.16.08, CH; **Dieppe Nautique** Tel. 35.84.13.80, ME; **Plaisance** Tel. 35.84.34.95, CH. **Town** P, D, FW, V, Gaz, R, Bar. PO; Bank; Rly; Air.
Ferry UK—Newhaven.

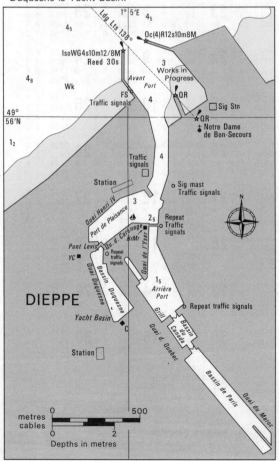

FRANCE, NORTH COAST - DIEPPE

LAT 49°56'N   LONG. 1°05'E

TIMES AND HEIGHTS OF HIGH AND LOW WATERS

YEAR **1989**

TIME ZONE −0100 (French Standard Time)
For French Summer Time add ONE hour in non-shaded areas

	JANUARY				FEBRUARY				MARCH				APRIL			
	TIME	m	TIME	m	TIME	m	TIME	m	TIME	m	TIME	m	TIME	m	TIME	m
1 SU	0528 1215 1758	7.0 2.7 7.1	**16** M	0012 0557 1246 1833	1.9 7.6 2.1 7.6	**1** W	0034 0623 1316 1906	3.0 6.6 3.0 6.5	**16** TH	0234 0826 1525 2112	2.9 7.0 2.6 7.0	**1** W	0422 1109 1650 2340	7.0 2.7 6.6 3.1	**16** TH	0028 0623 1323 1921	3.1 6.6 3.0 6.3	**1** SA	0048 0644 1344 1938	3.1 6.6 2.8 6.6	**16** SU	0317 0859 1549 2127	2.5 7.3 2.1 7.3
2 M	0046 0630 1319 1904	2.7 7.0 2.8 7.1	**17** TU	0125 0713 1407 1955	2.3 7.5 2.3 7.5	**2** TH	0202 0753 1450 2037	3.1 6.8 2.8 6.8	**17** F	0409 0950 1645 2222	2.5 7.5 2.1 7.5	**2** TH	0526 1221 1813	6.5 3.2 6.2	**17** F	0222 0817 1514 2101	3.1 6.8 2.6 6.7	**2** SU	0235 0824 1517 2101	2.6 7.3 2.1 7.4	**17** M	0417 0950 1640 2211	2.0 7.8 1.7 7.8
3 TU	0154 0739 1430 2014	2.8 7.1 2.7 7.2	**18** W	0250 0837 1531 2115	2.4 7.6 2.2 7.6	**3** F	0333 0915 1611 2150	2.7 7.3 2.3 7.4	**18** SA	0515 1048 1741 2311	2.0 8.0 1.6 7.9	**3** F	0113 0709 1410 2005	3.3 6.5 3.0 6.5	**18** SA	0356 0937 1629 2205	2.5 7.4 2.1 7.3	**3** M	0353 0931 1623 2159	1.8 8.1 1.3 8.3	**18** TU	0501 1031 1720 2248	1.6 8.2 1.4 8.2
4 W	0304 0846 1537 2116	2.6 7.4 2.4 7.5	**19** TH	0410 0950 1645 2223	2.2 7.9 1.9 7.8	**4** SA	0444 1021 1714 2247	2.1 7.9 1.6 8.0	**19** SU	0604 1132 1824 2350	1.5 8.5 1.2 8.0	**4** SA	0303 0851 1547 2129	2.8 7.1 2.3 7.2	**19** SU	0456 1029 1720 2250	1.6 7.9 1.6 7.9	**4** TU	0451 1023 1715 2246	1.1 8.9 0.8 9.0	**19** W	0538 1105 1755 2321	1.3 8.5 1.2 8.5
5 TH	0408 0945 1637 2212	2.3 7.8 2.0 7.8	**20** F	0517 1112 1746 2317	1.8 8.6 1.5 8.2	**5** SU	0541 1112 1807 2336	1.5 8.6 1.1 8.6	**20** M	0643 1208 1900	1.2 8.8 1.0	**5** SU	0423 1001 1654 2228	2.0 7.9 1.5 8.1	**20** M	0541 1109 1800 2325	1.5 8.4 1.2 8.3	**5** W	0539 1108 1801 2329	0.6 9.4 0.4 9.4	**20** TH	0612 1136 1827 2351	1.1 8.7 1.1 8.7
6 F	0505 1038 1731 2302	1.9 8.2 1.6 8.2	**21** SA	0612 1140 1835	1.5 8.6 1.2	**6** M	0630 1158 1853	1.0 9.1 0.7	**21** TU	0025 0715 1241 1931	8.6 1.0 9.0 0.8	**6** M	0522 1053 1747 2316	1.3 8.7 0.9 8.8	**21** TU	0617 1142 1832 2357	1.2 8.7 1.0 8.6	**6** TH	0623 1150 1844	0.3 9.7 0.2	**21** F	0642 1206 1856	1.0 8.8 1.0
7 SA	0557 1125 1820 2348	1.5 8.7 1.2 8.6	**22** SU	0001 0655 1222 1915	8.4 1.3 8.9 1.1	**7** TU	0020 0714 1242 1935	9.1 0.7 9.5 0.4	**22** W	0056 0744 1310 1959	8.8 1.0 9.0 0.8	**7** TU	0609 1138 1831 2358	0.8 9.3 0.4 9.3	**22** W	0648 1212 1902	1.0 8.9 0.8	**7** F	0011 0704 1232 1924	9.7 0.2 9.8 0.2	**22** SA	0021 0711 1236 1925	8.8 1.0 8.8 1.0
8 SU	0643 1211 1905	1.2 9.0 1.0	**23** M	0042 0733 1300 1950	8.6 1.2 9.0 1.0	**8** W	0103 0755 1324 2014	9.3 0.4 9.7 0.3	**23** TH	0125 0812 1338 2025	8.9 0.9 9.0 0.8	**8** W	0652 1219 1912	0.4 9.7 0.2	**23** TH	0026 0715 1240 1929	8.8 0.9 9.0 0.8	**8** SA	0053 0744 1313 2004	9.7 0.3 9.7 0.3	**23** SU	0051 0739 1306 1954	8.8 1.0 8.7 1.1
9 M	0034 0726 1255 1947	8.8 1.0 9.2 0.8	**24** TU	0118 0807 1334 2023	8.7 1.1 9.0 1.0	**9** TH	0144 0834 1403 2051	9.5 0.4 9.6 0.3	**24** F	0151 0837 1404 2050	8.8 1.0 8.9 1.0	**9** TH	0040 0732 1300 1951	9.6 0.2 9.9 0.2	**24** F	0054 0741 1307 1955	8.9 0.8 9.0 0.8	**9** SU	0134 0824 1354 2043	9.5 0.5 9.3 0.7	**24** M	0121 0809 1336 2025	8.7 1.1 8.5 1.3
10 TU	0117 0809 1338 2029	9.0 0.9 9.3 0.7	**25** W	0150 0837 1405 2051	8.6 1.1 8.9 1.0	**10** F	0223 0910 1441 2128	9.4 0.5 9.5 0.5	**25** SA	0217 0902 1430 2114	8.6 1.1 8.7 1.2	**10** F	0120 0809 1338 2028	9.7 0.2 9.8 0.3	**25** SA	0120 0808 1334 2021	8.9 0.9 8.9 0.9	**10** M	0213 0902 1434 2122	9.1 0.9 8.8 1.2	**25** TU	0152 0840 1408 2056	8.5 1.3 8.2 1.5
11 W	0159 0850 1421 2109	9.0 0.9 9.3 0.7	**26** TH	0220 0905 1435 2119	8.5 1.2 8.7 1.2	**11** SA	0300 0946 1518 2204	9.1 0.8 9.1 0.9	**26** SU	0243 0927 1456 2140	8.4 1.4 8.3 1.5	**11** SA	0157 0846 1416 2104	9.6 0.3 9.6 0.5	**26** SU	0147 0834 1400 2046	8.7 1.0 8.7 1.2	**11** TU	0254 0942 1516 2204	8.5 1.4 7.9 1.8	**26** W	0225 0912 1444 2130	8.2 1.6 7.8 1.9
12 TH	0241 0929 1501 2149	8.9 0.9 9.1 0.9	**27** F	0249 0933 1503 2147	8.3 1.4 8.5 1.4	**12** SU	0338 1024 1559 2245	8.7 1.2 8.5 1.4	**27** M	0310 0954 1525 2209	8.0 1.7 7.8 2.0	**12** SU	0236 0922 1454 2140	9.2 0.7 9.0 1.0	**27** M	0213 0859 1427 2113	8.5 1.3 8.3 1.5	**12** W	0340 1028 1607 2256	7.8 2.0 7.1 2.6	**27** TH	0302 0950 1524 2213	7.8 2.0 7.3 2.6
13 F	0322 1009 1544 2230	8.8 1.1 8.9 1.1	**28** SA	0317 1002 1532 2216	8.0 1.7 8.1 1.8	**13** M	0422 1108 1647 2335	8.0 1.8 7.7 2.2	**28** TU	0341 1026 1600 2246	7.5 2.2 7.2 2.5	**13** M	0313 1000 1533 2221	8.6 1.2 8.3 1.6	**28** TU	0243 0927 1458 2143	8.1 1.6 7.8 1.9	**13** TH	0438 1130 1716	7.1 2.6 6.4	**28** F	0349 1040 1619 2310	7.3 2.4 6.8 2.7
14 SA	0406 1052 1629 2316	8.4 1.4 8.5 1.5	**29** SU	0349 1032 1606 2249	7.7 2.0 7.6 2.2	**14** TU	0518 1209 1756	7.3 2.5 7.0				**14** TU	0356 1043 1621 2310	7.8 1.9 7.3 2.5	**29** W	0314 1001 1534 2221	7.6 2.1 7.2 2.5	**14** F	0011 0604 1300 1856	3.0 6.7 2.9 6.3	**29** SA	0455 1148 1737	7.0 2.5 6.7
15 SU	0456 1142 1724	8.0 1.7 8.0	**30** M	0425 1109 1646 2333	7.3 2.4 7.2 2.6	**15** W	0050 0642 1340 1935	2.8 6.9 2.8 6.7				**15** W	0452 1145 1732	7.1 2.7 6.5	**30** TH	0357 1045 1626 2316	7.1 2.6 6.6 3.0	**15** SA	0151 0744 1439 2025	3.0 6.8 2.6 6.7	**30** SU	0031 0623 1316 1907	2.6 7.2 2.3 7.1
			31 TU	0512 1201 1744	6.9 2.8 6.7								**31** F	0502 1157 1749	6.6 3.0 6.2					

Chart Datum: 4.89 metres below Lallemand System (Mean Sea Level, Marseilles)

FRANCE, NORTH COAST - DIEPPE

LAT 49°56'N LONG 1°05'E

TIMES AND HEIGHTS OF HIGH AND LOW WATERS

YEAR **1989**

TIME ZONE −0100
(French Standard Time)
For French Summer Time add
ONE hour in non-shaded areas

	MAY				JUNE				JULY				AUGUST		
	TIME m		TIME m		TIME m		TIME m		TIME m		TIME m		TIME m		TIME m
1 M	0200 2.2 0748 7.7 1439 1.7 2023 7.7	**16** TU	0317 2.2 0856 7.6 1545 2.0 2120 7.6	**1** TH	0334 1.2 0912 8.6 1603 1.1 2141 8.7	**16** F	0401 2.1 0937 7.6 1627 2.0 2201 7.8	**1** SA	0413 1.5 0952 8.2 1646 1.5 2223 8.5	**16** SU	0418 2.2 0954 7.4 1647 2.1 2222 7.8	**1** TU●	0613 1.2 1141 8.4 1836 1.2	**16** W	0547 1.3 1116 8.4 1809 1.2 2338 8.9
2 TU	0314 1.6 0853 8.3 1544 1.2 2122 8.4	**17** W	0408 1.9 0942 7.8 1631 1.7 2202 7.9	**2** F	0433 1.0 1008 8.9 1701 1.0 2234 8.9	**17** SA	0452 1.8 1023 7.8 1715 1.8 2245 8.1	**2** SU	0518 1.3 1052 8.4 1748 1.3 2319 8.7	**17** M	0515 1.8 1047 7.8 1740 1.7 2310 8.3	**2** W	0004 8.9 0657 1.0 1224 8.6 1916 1.0	**17** TH○	0631 0.8 1158 8.9 1852 0.8
3 W	0412 1.0 0947 8.9 1639 0.8 2212 8.9	**18** TH	0452 1.6 1022 8.1 1712 1.5 2241 8.2	**3** SA●	0529 0.9 1101 9.0 1755 0.9 2326 9.1	**18** SU	0538 1.6 1107 8.1 1800 1.6 2327 8.4	**3** M●	0616 1.1 1145 8.6 1841 1.2	**18** TU○	0604 1.4 1133 8.2 1827 1.4 2354 8.7	**3** TH	0044 9.0 0735 0.9 1302 8.8 1952 1.0	**18** F	0019 9.3 0712 0.5 1240 9.2 1932 0.5
4 TH	0504 0.7 1035 9.3 1728 0.5 2259 9.3	**19** F	0530 1.4 1059 8.3 1750 1.4 2316 8.5	**4** SU	0622 0.8 1151 9.0 1847 0.9	**19** M○	0621 1.4 1148 8.3 1842 1.4	**4** TU	0010 8.9 0705 1.0 1235 8.6 1928 1.1	**19** W	0649 1.1 1216 8.6 1910 1.1	**4** F	0120 9.0 0809 0.9 1336 8.7 2025 1.0	**19** SA	0100 9.5 0751 0.4 1319 9.4 2009 0.4
5 F●	0552 0.5 1121 9.5 1816 0.4 2344 9.5	**20** SA○	0608 1.3 1134 8.5 1825 1.3 2351 8.6	**5** M	0016 9.1 0711 0.8 1241 8.9 1935 1.0	**20** TU	0009 8.6 0702 1.2 1229 8.4 1922 1.3	**5** W	0057 8.9 0748 1.0 1318 8.6 2009 1.1	**20** TH	0038 9.0 0730 0.9 1259 8.8 1950 0.9	**5** SA	0152 9.0 0839 0.9 1407 8.6 2053 1.1	**20** SU	0138 9.6 0828 0.3 1357 9.4 2046 0.5
6 SA	0639 0.4 1207 9.5 1901 0.5	**21** SU	0643 1.2 1209 8.5 1900 1.2	**6** TU	0104 9.0 0757 0.9 1328 8.7 2019 1.1	**21** W	0050 8.7 0741 1.1 1310 8.5 2001 1.2	**6** TH	0138 8.9 0828 1.0 1357 8.5 2046 1.2	**21** F	0119 9.1 0810 0.7 1339 8.9 2029 0.8	**6** SU	0222 8.8 0907 1.1 1437 8.4 2121 1.3	**21** M	0216 9.5 0904 0.5 1435 9.2 2122 0.7
7 SU	0029 9.4 0723 0.5 1253 9.3 1945 0.7	**22** M	0026 8.7 0717 1.2 1244 8.5 1935 1.3	**7** W	0150 8.8 0840 1.1 1411 8.4 2100 1.4	**22** TH	0131 8.8 0821 1.1 1351 8.5 2041 1.2	**7** F	0216 8.8 0904 1.1 1435 8.3 2121 1.4	**22** SA	0159 9.2 0849 0.7 1419 8.9 2107 0.8	**7** M	0251 8.5 0935 1.3 1506 8.1 2149 1.6	**22** TU	0253 9.2 0940 0.8 1513 8.8 2159 1.1
8 M	0115 9.2 0807 0.7 1337 9.0 2028 1.0	**23** TU	0102 8.7 0752 1.2 1320 8.4 2010 1.4	**8** TH	0233 8.6 0921 1.3 1453 8.0 2141 1.7	**23** F	0211 8.8 0900 1.1 1433 8.4 2120 1.3	**8** SA	0252 8.6 0937 1.3 1509 8.0 2155 1.6	**23** SU	0238 9.2 0925 0.7 1458 8.8 2144 0.9	**8** TU	0320 8.1 1004 1.7 1536 7.7 2220 1.9	**23** W☾	0333 8.7 1020 1.3 1555 8.2 2242 1.6
9 TU	0159 8.9 0849 1.0 1421 8.5 2110 1.4	**24** W	0138 8.6 0828 1.3 1357 8.2 2047 1.5	**9** F	0314 8.2 1002 1.6 1536 7.6 2223 2.0	**24** SA	0253 8.7 0941 1.2 1514 8.2 2202 1.4	**9** SU	0326 8.2 1011 1.6 1544 7.7 2229 1.8	**24** M	0318 9.0 1004 0.9 1538 8.6 2221 1.1	**9** W	0352 7.7 1036 2.1 1610 7.3 2255 2.4	**24** TH	0419 8.0 1108 2.0 1648 7.6 2338 2.3
10 W	0243 8.5 0931 1.5 1506 7.8 2154 1.9	**25** TH	0217 8.4 0905 1.5 1438 7.9 2126 1.7	**10** SA	0358 7.8 1045 1.9 1622 7.3 2309 2.3	**25** SU	0337 8.5 1025 1.3 1601 8.1 2247 1.5	**10** M	0403 7.9 1047 1.9 1622 7.4 2306 2.2	**25** TU☾	0359 8.7 1046 1.2 1623 8.2 2309 1.5	**10** TH	0431 7.1 1117 2.6 1656 6.9 2343 2.8	**25** F	0522 7.2 1215 2.6 1805 7.1
11 TH	0329 7.9 1018 1.9 1555 7.2 2243 2.4	**26** F	0300 8.1 0948 1.7 1522 7.6 2211 1.9	**11** SU☾	0447 7.5 1134 2.2 1714 7.0	**26** M	0425 8.3 1112 1.5 1652 7.9 2339 1.6	**11** TU	0443 7.5 1128 2.2 1707 7.1 2352 2.5	**26** W	0448 8.2 1135 1.6 1716 7.8	**11** F	0525 6.6 1215 3.1 1802 6.6	**26** SA	0101 2.7 0656 6.8 1354 2.8 1948 7.1
12 F☾	0423 7.4 1112 2.4 1655 6.8 2345 2.7	**27** SA	0348 7.8 1036 1.9 1615 7.4 2304 2.1	**12** M	0002 2.4 0543 7.3 1230 2.4 1752 7.8	**27** TU	0521 8.1 1208 1.5 1752 7.8	**12** W	0532 7.1 1218 2.6 1802 6.9	**27** TH	0004 1.9 0549 7.7 1238 2.1 1826 7.5	**12** SA☾	0055 3.1 0646 6.4 1340 3.2 1932 6.6	**27** SU	0246 2.5 0837 7.0 1533 2.5 2116 7.5
13 SA	0530 7.1 1221 2.6 1809 6.6	**28** SU☾	0445 7.7 1135 1.9 1719 7.4	**13** TU	0101 2.4 0646 7.2 1334 2.4 1918 7.0	**28** W	0039 1.7 0625 8.0 1314 1.7 1900 7.8	**13** TH	0049 2.7 0633 6.9 1323 2.7 1908 6.8	**28** F	0119 2.2 0707 7.4 1402 2.3 1951 7.5	**13** SU	0229 2.9 0818 6.6 1513 2.9 2058 7.1	**28** M	0412 2.0 0951 7.5 1645 1.9 2221 8.1
14 SU	0100 2.7 0648 7.1 1339 2.5 1925 6.8	**29** M	0008 2.0 0555 7.7 1244 1.8 1830 7.6	**14** W	0206 2.5 0748 7.2 1436 2.4 2018 7.2	**29** TH	0150 1.7 0735 8.0 1427 1.7 2011 7.9	**14** F	0159 2.7 0745 6.9 1427 2.7 2020 7.0	**29** SA	0246 2.2 0835 7.2 1530 2.2 2114 7.7	**14** M	0351 2.4 0932 7.2 1625 2.3 2202 7.7	**29** TU	0514 1.5 1046 8.0 1738 1.5 2308 8.6
15 M	0216 2.5 0800 7.3 1449 2.3 2030 7.2	**30** TU	0120 1.8 0706 8.0 1356 1.6 1940 7.9	**15** TH	0306 2.3 0845 7.4 1535 2.2 2111 7.5	**30** F	0303 1.6 0846 8.1 1539 1.6 2120 8.2	**15** SA	0306 2.5 0853 7.1 1546 2.5 2125 7.4	**30** SU	0410 1.9 0950 7.7 1645 1.8 2223 8.1	**15** TU	0455 1.8 1029 7.8 1722 1.7 2253 8.4	**30** W	0601 1.2 1128 8.5 1821 1.2 2347 8.9
		31 W	0230 1.5 0812 8.3 1503 1.3 2043 8.3							**31** M	0517 1.5 1051 8.1 1747 1.5 2317 8.6			**31** TH●	0639 0.9 1205 8.8 1856 1.0

Chart Datum: 4.89 metres below Lallemand System (Mean Sea Level, Marseilles)

FRANCE, NORTH COAST - DIEPPE

LAT 49°56′N   LONG 1°05′E

TIMES AND HEIGHTS OF HIGH AND LOW WATERS

YEAR **1989**

TIME ZONE −0100
(French Standard Time)
For French Summer Time add
ONE hour in non-shaded areas

	SEPTEMBER				OCTOBER				NOVEMBER				DECEMBER		
	TIME m		TIME m		TIME m		TIME m		TIME m		TIME m		TIME m		TIME m
1 F	0021 9.1 0712 0.8 1237 8.9 1927 0.9	**16** SA	0646 0.3 1213 9.6 1906 0.3	**1** SU	0020 9.1 0710 0.9 1235 9.0 1924 0.9	**16** M	0004 9.9 0658 0.3 1225 9.8 1918 0.3	**1** W	0050 8.8 0738 1.3 1305 8.8 1954 1.3	**16** TH	0115 9.2 0808 0.9 1338 9.2 2030 1.0	**1** F	0106 8.5 0756 1.5 1324 8.7 2014 1.4	**16** SA	0154 8.8 0844 1.2 1416 9.0 2104 1.1
2 SA	0052 9.1 0741 0.8 1307 8.9 1955 0.9	**17** SU	0033 9.9 0725 0.2 1253 9.7 1943 0.3	**2** M	0049 9.0 0738 0.9 1303 8.9 1951 1.0	**17** TU	0047 9.8 0738 0.4 1308 9.6 2000 0.5	**2** TH	0121 8.6 0809 1.5 1336 8.6 2025 1.5	**17** F	0202 8.8 0852 1.3 1425 8.8 2115 1.4	**2** SA	0142 8.4 0832 1.6 1400 8.6 2050 1.5	**17** SU	0237 8.5 0925 1.4 1458 8.7 2145 1.4
3 SU	0122 9.1 0809 0.9 1336 8.8 2023 1.0	**18** M	0112 9.9 0803 0.3 1332 9.6 2022 0.4	**3** TU	0117 8.9 0805 1.1 1331 8.8 2018 1.2	**18** W	0130 9.5 0821 0.7 1351 9.3 2042 0.9	**3** F	0152 8.3 0840 1.7 1409 8.3 2057 1.8	**18** SA	0249 8.2 0937 1.8 1513 8.3 2202 1.8	**3** SU	0220 8.1 0908 1.8 1439 8.3 2127 1.7	**18** M	0319 8.1 1006 1.7 1540 8.3 2227 1.7
4 M	0149 8.9 0835 1.0 1402 8.7 2048 1.2	**19** TU	0151 9.6 0841 0.6 1410 9.3 2059 0.7	**4** W	0144 8.7 0831 1.3 1357 8.5 2044 1.5	**19** TH	0212 9.0 0902 1.2 1435 8.8 2123 1.4	**4** SA	0226 7.9 0914 2.1 1446 7.9 2133 2.2	**19** SU	0338 7.7 1027 2.3 1606 7.8 2254 2.3	**4** M	0300 7.9 0948 2.0 1522 8.1 2210 1.9	**19** TU	0402 7.7 1048 2.1 1625 7.9 2311 2.1
5 TU	0215 8.7 0901 1.3 1429 8.4 2113 1.4	**20** W	0230 9.2 0918 0.9 1451 8.8 2137 1.2	**5** TH	0212 8.3 0858 1.7 1427 8.1 2112 1.8	**20** F	0257 8.3 0946 1.8 1522 8.1 2211 2.0	**5** SU	0306 7.4 0955 2.5 1530 7.5 2219 2.6	**20** M	0436 7.2 1125 2.6 1709 7.5	**5** TU	0347 7.6 1034 2.2 1613 7.9 2301 2.1	**20** W	0450 7.4 1136 2.4 1716 7.5
6 W	0243 8.3 0927 1.6 1457 8.0 2141 1.8	**21** TH	0311 8.5 0959 1.6 1534 8.1 2222 1.9	**6** F	0243 7.8 0929 2.2 1500 7.7 2146 2.3	**21** SA	0349 7.5 1040 2.5 1620 7.5 2312 2.6	**6** M	0357 7.0 1047 2.9 1628 7.2 2320 2.8	**21** TU	0000 2.5 0545 7.0 1235 2.8 1823 7.4	**6** W	0441 7.5 1130 2.2 1713 7.8	**21** TH	0004 2.5 0545 7.1 1233 2.7 1818 7.3
7 TH	0312 7.7 0956 2.1 1528 7.5 2214 2.4	**22** F	0400 7.6 1049 2.4 1630 7.4 2322 2.6	**7** SA	0319 7.2 1006 2.7 1541 7.2 2230 2.9	**22** SU	0458 6.8 1152 3.0 1742 7.1	**7** TU	0507 6.8 1200 2.9 1749 7.2	**22** W	0113 2.6 0701 7.1 1350 2.6 1935 7.5	**7** TH	0003 2.1 0548 7.6 1236 2.1 1823 7.9	**22** F	0105 2.6 0650 7.0 1339 2.7 1924 7.2
8 F	0347 7.2 1033 2.7 1608 7.0 2256 2.9	**23** SA	0509 6.8 1204 3.0 1757 6.9	**8** SU	0408 6.6 1100 3.3 1643 6.7 2338 3.2	**23** M	0037 2.8 0631 6.7 1326 3.0 1918 7.2	**8** W	0041 2.6 0632 7.1 1324 2.5 1913 7.6	**23** TH	0224 2.4 0807 7.4 1455 2.4 2036 7.7	**8** F	0111 1.9 0658 7.8 1347 1.9 1932 8.2	**23** SA	0213 2.7 0757 7.2 1447 2.6 2030 7.3
9 SA	0436 6.5 1127 3.3 1711 6.5	**24** SU	0055 2.9 0651 6.6 1350 3.0 1946 7.1	**9** M	0529 6.3 1226 3.4 1822 6.7	**24** TU	0212 2.6 0800 7.1 1451 2.5 2035 7.6	**9** TH	0203 2.1 0749 7.7 1439 1.9 2021 8.2	**24** F	0324 2.2 0901 7.7 1549 2.1 2125 7.9	**9** SA	0222 1.7 0806 8.2 1455 1.6 2037 8.5	**24** SU	0319 2.6 0859 7.4 1549 2.4 2126 7.5
10 SU	0006 3.3 0559 6.2 1257 3.5 1853 6.5	**25** M	0242 2.6 0832 7.0 1525 2.5 2108 7.6	**10** TU	0120 3.0 0714 6.7 1409 2.8 1959 7.3	**25** W	0324 2.2 0903 7.6 1552 2.1 2128 8.0	**10** F	0311 1.5 0850 8.4 1539 1.3 2116 8.8	**25** SA	0413 1.9 0946 8.0 1635 1.8 2207 8.2	**10** SU	0328 1.4 0907 8.6 1559 1.3 2137 8.8	**25** M	0417 2.1 0951 7.7 1643 2.1 2215 7.7
11 M	0152 3.1 0747 6.5 1443 3.0 2033 7.1	**26** TU	0400 2.1 0938 7.6 1629 1.9 2203 8.1	**11** W	0251 2.3 0835 7.5 1526 2.0 2105 8.1	**26** TH	0417 1.7 0950 8.0 1639 1.7 2210 8.4	**11** SA	0407 1.0 0942 8.9 1632 0.9 2206 9.2	**26** SU	0456 1.7 1027 8.3 1717 1.6 2246 8.4	**11** M	0430 1.2 1005 8.8 1658 1.0 2233 9.0	**26** TU	0507 2.1 1039 8.1 1730 1.8 2300 8.0
12 TU	0326 2.5 0908 7.2 1601 2.2 2139 7.9	**27** W	0454 1.6 1025 8.1 1716 1.5 2245 8.6	**12** TH	0356 1.5 0931 8.3 1622 1.3 2155 8.8	**27** F	0458 1.5 1028 8.4 1718 1.4 2246 8.7	**12** SU	0457 0.7 1030 9.3 1723 0.6 2253 9.5	**27** M	0536 1.6 1104 8.5 1755 1.5 2321 8.5	**12** TU	0528 1.0 1101 9.2 1757 0.9 2326 9.1	**27** W	0552 1.8 1120 8.4 1814 1.5 2340 8.3
13 W	0431 1.7 1005 8.0 1657 1.5 2229 8.6	**28** TH	0536 1.2 1103 8.5 1754 1.2 2320 8.9	**13** F	0446 0.8 1018 8.9 1709 0.8 2241 9.4	**28** SA	0536 1.1 1103 8.7 1752 1.2 2318 8.8	**13** M	0547 0.6 1117 9.6 1812 0.5 ○ 2340 9.6	**28** TU	0614 1.4 1139 8.7 1831 1.4 ● 2356 8.6	**13** W	0623 0.9 1153 9.3 1849 0.8	**28** TH	0634 1.6 1159 8.7 1853 1.3
14 TH	0522 1.0 1051 8.7 1744 0.9 2312 9.2	**29** F	0610 1.0 1136 8.8 1827 1.0 ● 2351 9.0	**14** SA	0533 0.5 1102 9.3 1754 0.4 ○ 2322 9.7	**29** SU	0609 1.2 1134 8.8 1824 1.1 ● 2349 8.9	**14** TU	0636 0.6 1204 9.6 1859 0.6	**29** W	0648 1.4 1214 8.7 1905 1.3	**14** TH	0018 9.1 0714 0.9 1244 9.3 1938 0.8	**29** F	0019 8.5 0711 1.4 1238 8.8 1930 1.2
15 F	0605 0.6 1133 9.2 1826 0.5 ○ 2353 9.6	**30** SA	0642 0.9 1206 8.9 1856 0.9	**15** SU	0616 0.3 1143 9.7 1837 0.3	**30** M	0640 1.1 1204 8.9 1855 1.1	**15** W	0027 9.5 0722 0.7 1252 9.5 1944 0.7	**30** TH	0032 8.6 0722 1.4 1249 8.8 1939 1.3	**15** F	0108 9.0 0801 1.0 1332 9.2 2024 0.9	**30** SA	0057 8.6 0748 1.3 1316 8.9 2007 1.1
						31 TU	0019 8.9 0709 1.2 1235 8.9 1924 1.2							**31** SU	0135 8.6 0825 1.2 1354 8.9 2043 1.1

Chart Datum: 4.89 metres below Lallemand System (Mean Sea Level, Marseilles)

DIEPPE continued

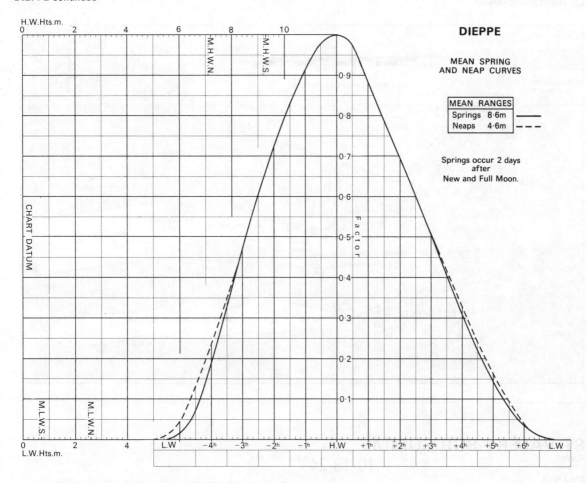

LE TRÉPORT 10-18-23
Seine Maritime

CHARTS
Admiralty 1351, 2147; SHOM 5928, 934; ECM 1011; Stanford 1; Imray C31

TIDES
Dover −0025; ML 5.0; Duration 0530; Zone −0100

Standard Port DIEPPE (←)

Times				Height (metres)			
HW		LW		MHWS	MHWN	MLWN	MLWS
0100	0600	0000	0700	9.3	7.2	2.6	0.7
1300	1800	1200	1900				

Differences LE TRÉPORT
+0001 +0005 +0005 +0011 +0.1 +0.2 −0.1 0.0

SHELTER
Shelter good but harbour dries. Access difficult in strong on-shore winds. Entry channel dries — only usable from HW∓1½. Wet basin opens from HW−1½ to HW. Signal for opening of swing bridge — three blasts on passing heads of jetties. Yachts lie on pontoons in NW corner of Arrière Port.

NAVIGATION
Waypoint 50°04′.30N 01°21′.70E, 315°/135° from/to entrance, 0.52M. Coast dries to approx 300m off the pier heads. No navigational dangers. Entrance difficult in strong winds: scend is caused in the Avant Port.

LIGHTS AND MARKS
Le Treport can be difficult to identify. "P" flag or W Lt indicates entrance open. No Ldg Lts or marks.
Traffic signals shown from E jetty and entrance to Arrière-Port.

RADIO TELEPHONE
Call: *Ovaro* VHF Ch 16; 12 (HW−2 to HW).

TELEPHONE
Hr Mr 35.86.17.91; Aff Mar 35.86.08.88; CROSS 21.87.21.87; SNSM 35.86.33.71; Customs 35.86.15.34; Meteo 21.05.13.55; Dr 35.86.16.23; Brit Consul 21.96.33.76.

FACILITIES
Outer Harbour M; **Arrière Port** AB, FW; **Chantier Naval du Tréport** Tel. 35.86.12.33, ME, El, Sh, C, CH; **Nautic Côte d'Opale** Tel. 22.26.85.85, ME, El, CH; **Yacht Club de la Bresle** Tel. 35.86.19.93, C (3 ton), AB, Bar; **Europ-Yachting** Tel. 35.86.60.52, M, CH. **Town** P, D, V, Gaz, R, Bar. PO; Bank; Rly; Air (Dieppe).
Ferry UK — Dieppe−Newhaven.

LE TRÉPORT continued

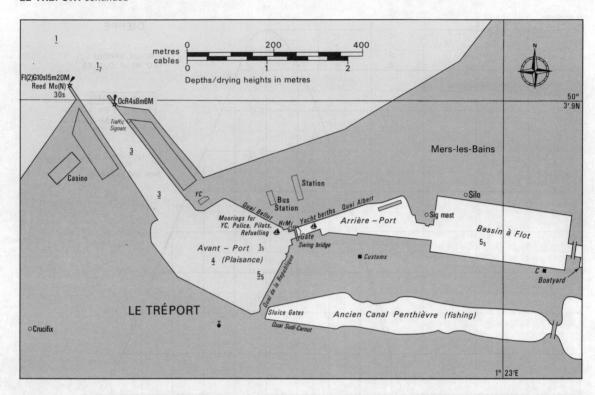

ST VALÉRY-SUR-SOMME
Somme   10-18-24

CHARTS
Admiralty 2612; SHOM 3800; ECM 1011; Stanford 1; Imray C31

TIDES
Dover −0020; ML —; Duration —; Zone −0100

Standard Port DIEPPE (←)

Times				Height (metres)			
HW		LW		MHWS	MHWN	MLWN	MLWS
0100	0600	0000	0700	9.3	7.2	2.6	0.7
1300	1800	1200	1900				

Differences ST VALÉRY-SUR-SOMME
+0028 +0040   No data   +0.7   +0.8   No data
LE HOURDEL and LE CROTOY
+0021 +0026   No data   +0.7   +0.7   No data

SHELTER
The bay is open to W and can be dangerous in onshore winds over force 6.
The three harbours, St Valéry-sur-Somme, Le Crotoy and Le Hourdel, are very well protected. Yachts can
(1) dry out on hard sand or moor alongside at Le Hourdel
(2) go into marina at St Valéry (max. draught 2.5 m). Access HW∓2
(3) go into marina at Le Crotoy when tidal range at Dover exceeds 4.4 m (max. draught 1.6 m). See 10.18.29.
(4) go into Abbeville Canal at St. Valéry (max. draught 3.3 m). Entry by prior arrangement and in daylight only.

NAVIGATION
Waypoint Baie de Somme AT-SO N cardinal buoy, VQ, 50°14'.00N 01°28'.50E, about 1M W of buoyed channel (shifts). The whole estuary dries up to 3 M seaward from Le Hourdel. The landfall buoy N cardinal VQ 'Baie de Somme' should be closed and entrance commenced at HW−2. Departure at HW−1. Channels are constantly changing but buoys are moved as necessary. The sands build up in ridges offshore but inside Pt du Hourdel are generally flat except where the River Somme and minor streams scour their way to the sea. If range of tide exceeds 4.1 m Dover there is sufficient water over the sands inside Pt du Hourdel from HW−1 to HW+½ for vessels with less than 1.5 m draught. At all other times buoys marking the channels must be followed in strict sequence. There are fishing fleets based at all three harbours and small cargo vessels occasionally navigate to the commercial quay inside the lock to the Abbeville Canal.

LIGHTS AND MARKS
Lt Ho Cayeux-sur-Mer Fl R 6s 32m 22M (off chartlet). Thirty-nine numbered buoys mark twisting channel up to Division Buoy YBY. Starboard hand green cone buoys (odd numbers) port hand red can buoys (even numbers). Buoys 15 and 23 are lit. From division buoy, buoys C1 to C10 run N and E to Le Crotoy; buoys 39 to 50 S and SE to St Valery-sur-Somme.
LE HOURDEL Only landmark is Lt Ho Oc(3) WG 12s 19m. Channel unmarked. Follow St Valéry channel buoys until Lt Ho bears 270°, then head in.
ST VALERY-SUR-SOMME From Division Buoy, buoyed channel numbered 39 to 50. From 43, starboard hand marks are beacons on submerged training wall then four on edge of beach.
Hut on lattice tower Iso G 4s marks beginning of tree lined promenade with port hand beacons marking submerged training wall.
W tower on beginning of Digue du Large Fl R 4s.
LE CROTOY — See 10.18.29.

RADIO TELEPHONE
VHF Ch 09 HW∓2.

TELEPHONE
St Valery Hr Mr 22.27.52.07; Aff Mar 22.27.52.57; Customs 22.27.50.36; Lock (canal) 22.27.50.23; Dr 22.27.52.25; CROSS 21.87.21.87; Meteo 21.05.13.55; Brit Consul 21.96.33.76.

AREA 18—NE France   719

ST VALÉRY-SUR-SOMME continued

FACILITIES
EC Monday. **Marina** (280+30 visitors) Tel. 22.26.91.64, C (6 ton), Slip, FW, Bar; **CNMF** Tel. 22.26.82.20, Sh, ME, El, Electronics; **Latitude 50** Tel. 22.26.82.06, ME, El, CH, charts, Electronics; **Nautic Côte d'Opale** Tel. 22.26.85.85, ME, El, CH. **Town** P, D, CH, V, Gaz, R, Bar, PO; Bank; Rly (Noyelles-sur-Mer); Air (Le Touquet). Ferry UK — Boulogne—Dover/Folkestone.

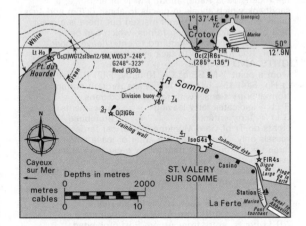

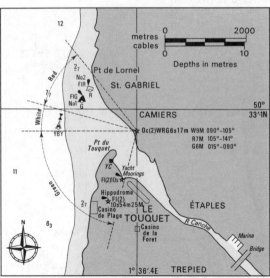

Sh, SM; **Technical Composite** Tel. 21.05.62.97, Sh. **Town** P, D, V, Gaz, R, Bar. PO; Bank; Rly; Air. Ferry UK — Boulogne—Dover/Folkestone.

LE TOUQUET 10-18-25
Pas de Calais

CHARTS
Admiralty 2451; SHOM 6795; ECM 1011; Stanford 1, 9; Imray C31

TIDES
Dover −0010; ML 4.8; Duration 0520; Zone −0100

Standard Port DIEPPE (←)

Times				Height (metres)			
HW		LW		MHWS	MHWN	MLWN	MLWS
0100	0600	0000	0700	9.3	7.2	2.6	0.7
1300	1800	1200	1900				

Differences LE TOUQUET
+0012 No data −0.3 0.0 +0.2 +0.3

SHELTER
Good except in strong W winds. Yacht moorings, which dry, are to stbd after passing Pt de Touquet. There are plans to build a dam across the mouth of the R Canche, with locks. Marina at Étaples — see 10.18.29.

NAVIGATION
Waypoint 50°35'.00N 01°31'.80E, 308°/128° from/to Camiers Lt, 3.8M. Entrance is not easy — local pilots or fishermen are available. Best entrance is at HW−1. In W winds the sea breaks heavily a long way out and entry should not be attempted. The channel is always changing and buoys are moved accordingly. Beware stranded wreck 2M NW of Le Touquet-Paris Plage Lt Ho, marked by Lt buoy.

LIGHTS AND MARKS
Good landmarks are the Terres de Tourmont, a conspic range 175m high, visible for 25 miles. Le Touquet is at the S end of this range. Entrance between Pt de Lornel and Pt de Touquet. Keep in the W sector of Camiers Lt. Pt du Touquet light is Y tower, Brown band, W & G top — Fl (2) 10s 54m 25M.

RADIO TELEPHONE
VHF Ch 21; 16.

TELEPHONE
Hr Mr 21.05.12.77; Aff Mar Etaples 21.94.61.50; Meteo 21.05.13.55; Customs 21.05.01.72; CROSS 21.87.21.87; Dr 21.05.14.42; Brit Consul 21.96.33.76.

FACILITIES
Cercle Nautique du Touquet Tel. 21.05.12.77, Slip, M, P, ME, D, FW, C, CH, R, Bar; **Marina** (Etaples — see 10.18.29), AB, FW, P, D, AC; **Demoury** Tel. 21.84.51.76,

BOULOGNE-SUR-MER 10-18-26
Pas de Calais

CHARTS
Admiralty 1892, 438; SHOM 6436, 6682, 6795; ECM 1010, 1011; Stanford 1, 9; Imray C31, C8

TIDES
Dover 0000; ML 5.0; Duration 0515; Zone −0100

Standard Port DIEPPE (←)

Times				Height (metres)			
HW		LW		MHWS	MHWN	MLWN	MLWS
0100	0600	0000	0700	9.3	7.2	2.6	0.7
1300	1800	1200	1900				

Differences BOULOGNE
+0015 +0026 +0037 +0036 −1.8 −0.9 −0.5 +0.5

SHELTER
Good except in strong NE winds. Turbulent water when R Liane in spate. Entrance possible in most weathers and at any state of tide. Yachts secure to pontoons in SW side of tidal basin alongside Quai Chanzy. Max length 10 m (over 10 m apply before arrival). Yacht berths tend to silt up.

NAVIGATION
Waypoint 50°44'.50N 01°33'.00E, 270°/090° from/to Digue Carnot Lt, 0.72M. Very busy commercial passenger and fishing port. Respect warnings and instructions given by harbour lights. There are no navigational dangers and entrance is easily identified and is well marked. Beware heavy wash from fishing boats. Also beware Digue Nord which partially covers at HW. Keep W and S of N light tower, QR.

LIGHTS AND MARKS
Ldg Lts 123° lead towards R Liane and marina. Besides normal traffic signals, special signals apply as follows: —

*G	Movement suspended except for vessels with
W G	special permission to enter
R	

*G	Movement suspended except for vessels with
W R	special permission to leave inner and
R	outer harbour

*G	Movement suspended except for vessels with
W R	special permission to leave inner harbour
R R	

*Normal light signal prohibiting entry and departure. Two Bu Lts (hor) mean sluicing from R Liane.

RADIO TELEPHONE
VHF Ch 12 16 (H24).

BOULOGNE continued

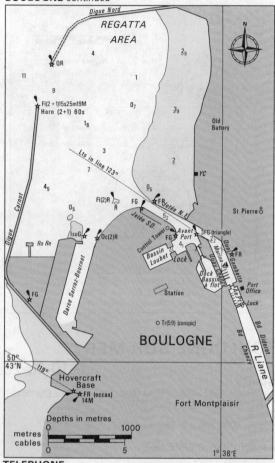

SHELTER
Shelter is very good especially in the marina. Bar is dredged to 4.5m but is liable to build up in heavy weather. Entrance is difficult in strong NW to NE winds. Enter Bassin Carnot if headed for the French canals otherwise use Bassin de l'Ouest, the marina.

NAVIGATION
Waypoint 50°58'.50N 01°49'.90E, 298°/118° from/to Jetée Ouest Lt, 0.43M. Entrance is relatively easy but beware the Ridens de la Rade, about 1 M N of entrance, a sandbank which shoals in places to about 0.75m and seas break on it. Entrance is well marked but there is a great deal of shipping, including ferries, which can be hazardous. Yachts are required to enter/leave under power. It is forbidden to tack in entrance channel.

LIGHTS AND MARKS
From a point ¼M SE of CA 10 buoy, Fl(2) R 6s, the main Lt Ho, Fl(4) 15s, bearing 140° leads through entrance. Normal entry signals with following additions.
R Lt = indicates ferry leaving ⎫ movements
G Lt = ferry entering ⎭ prohibited
Lights on Tr of Gare Maritime are for car ferries and cargo vessels only. Yachts may not enter or leave when at least three Lts are on. They may follow a car ferry entering or leaving but must keep to the stbd side of the fairway. One R Lt denotes the presence of a dredger and does not mean movement is forbidden.

Bassin Carnot
Lock signals, gates open HW – 1½ to HW + ¾.
2 G Lt hor = entry from Arrière Port permitted.
2 R Lt hor = entry from Arrière Port prohibited.
1 G Lt = entry from basin to Arrière Port permitted.
1 R Lt = entry from basin to Arrière Port prohibited.
2 blasts = request permission to enter Bassin Carnot.
4 blasts = request permission to enter Bassin de l'Ouest.

Bassin de l'Ouest
Orange Lt = 10 mins before opening of lock.
Red Lt = All movement prohibited.
Green Lt = Movement authorised.
Dock gates and bridge open HW – 1½, HW and HW + ½. (Sat and Sun HW – 2 and HW + 1).

TELEPHONE
Hr Mr 21.30.10.00; Hr Mr Plaisance 21.31.70.01; Harbour office 21.30.90.46; Aff Mar 21.30.53.23; CROSS 21.87.21.87; SNSM 21.31.42.58; Customs 21.30.14.24; Meteo 21.31.52.23; Auto 21.33.82.55; Hosp 21.31.62.07; Brit Consul 21.30.25.11.

FACILITIES
Marina (350 + 50 visitors) Tel. 21.31.70.01, M, L, FW, C (1 ton); **Bassin F. Sauvage** Tel. 21.31.70.01, M, L, AB; **Quai Gambetta** M, P*, D*, L, FW, AB, V, R, Bar; **Baude Electronique** Tel. 21.30.01.15, Electronics; **Opale Marine** Tel. 21.30.36.19, ME, El, Sh, CH, Divers; **Angelo** Tel. 21.31.37.61, CH; **Librairie Duminy** Tel. 21.30.06.75, SHOM; **YC Boulonnais** Tel. 21.31.80.68, C, Bar.
Town P, D, FW, ME, El, Sh, CH, V, Gaz, R, Bar. PO; Bank; Rly; Air (Le Touquet/Calais).
Ferry UK – Dover/Folkestone.
*Obtainable from Société Maritime Carburante Liquide.

CALAIS 10-18-27
Pas de Calais

CHARTS
Admiralty 1892, 1352; SHOM 6474, 6651, 6681; ECM 1010; Stanford 1, 19; Imray C8

TIDES
Dover +0025; ML 4.1; Duration 0525; Zone –0100

Standard Port DIEPPE (←)

Times				Height (metres)			
HW		LW		MHWS	MHWN	MLWN	MLWS
0100	0600	0000	0700	9.3	7.2	2.5	0.7
1300	1800	1200	1900				

Differences CALAIS
+0043 +0057 +0105 +0054 –2.2 –1.3 –0.5 +0.2

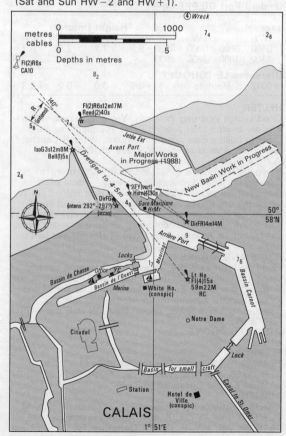

CALAIS continued

RADIO TELEPHONE
VHF Ch 12 16 (H24). Hoverport Ch 20 (occas). Carnot Lock Ch 12 (occas). Cap Gris Nez, Channel Navigation Information Service (CNIS), call: *Gris Nez Traffic* Ch 69 (H24). Information broadcasts in English and French on Ch 11, at H+10, and also at H+25 when visibility is less than 2 M.

TELEPHONE
Hr Mr 21.96.31.20; Aff Mar 21.34.52.70; Pilot (Quai de la Gare Maritime) 21.96.40.18; CROSS 21.87.21.87; SNSM 21.96.40.18; Port de Plaisance 21.34.55.23; Customs 21.34.75.40; Meteo 21.31.52.23; Auto 21.33.82.55; Dr 21.96.31.20; Hosp 21.97.99.60; Brit Consul 21.96.33.76.

FACILITIES
Marina (135+100 visitors) Tel. 21.34.55.23, D, FW, C (6 ton), BH (3 ton), AC, CH, Gaz, R, Lau, Sh, SM, V, Bar; Access HW−1½ to HW+½. **Bassin du Paradis** AB, Grid; **The Calais YC** Tel. 21.34.60.00, M, P, Bar; **Chantiers Navals** Tel. 21.34.30.40, Sh; **Nautic Sport** Tel. 21.36.41.03, ME, El, CH; **Godin Moteurs** Tel. 21.96.29.97, ME; **Marinerie** Tel. 21.34.47.83, CH. **Town** CH, V, Gaz, R, Bar. PO; Bank; Rly, Air. Ferry UK−Dover.

DUNKERQUE 10-18-28
Nord

CHARTS
Admiralty 1350, 323; SHOM 6500, 6501, 6651; ECM 1010; Stanford 1, 19; Imray C30

TIDES
Dover +0050; ML 3.2; Duration 0530; Zone −0100

Standard Port DIEPPE (←)

Times				Height (metres)			
HW		LW		MHWS	MHWN	MLWN	MLWS
0100	0600	0000	0700	9.3	7.2	2.5	0.7
1300	1800	1200	1900				

Differences DUNKERQUE
+0059 +0122 +0121 +0108 −3.5 −2.4 −1.1 −0.1

SHELTER
Good shelter and harbour is available at all states of tide and weather. Yachts must use E of harbour; all pleasure craft are prohibited from entering W harbour. YC pontoons are ¾ M down E side of harbour, or proceed through Ecluse Trystram and under two opening bridges to municipal marina in Bassin du Commerce.

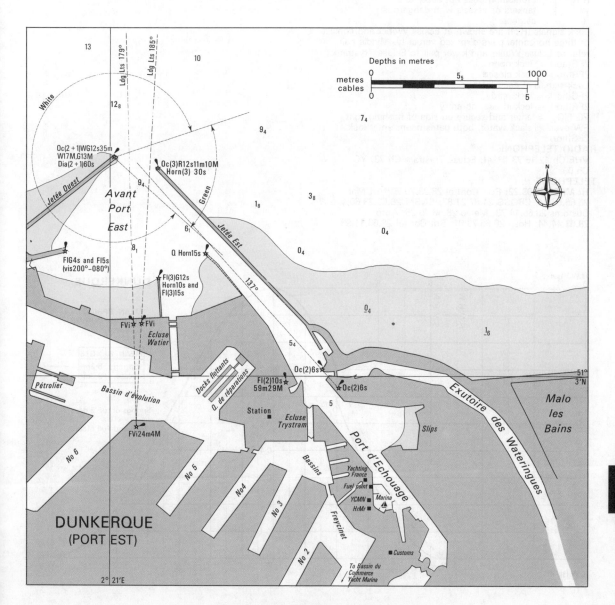

DUNKERQUE continued

NAVIGATION
Waypoint Port Est 51°03'.90N 02°21'.00E, 355°/175° from/to Jetée Ouest Lt, 0.20M. From the W, fetch the Dunkerque Lanby 5M N of Calais, thence to DKA safe water pillar Lt Buoy, and then follow series of DW Lt buoys past Dunkerque Ouest to buoy DW 29, ½M WNW of ent. From the E, the series of E Lt buoys lead S of Banc Hills from the Nieuwpoort Bank W Cardinal Lt buoy. There is no bar. Strong currents reach about 3½ kn. This is a busy commercial port. Entrance is easy in most weathers, but fresh winds from NW to NE cause heavy seas at entrance to Avant Port and scend in harbour.

LIGHTS AND MARKS
Two sets of Ldg Lts (there are others in the Commercial Port):—
(1) Two Lts 179° (F Vi) giving entrance line between the E and W jetties. A second pair, with common rear, lead 185°.
(2) Two Lts in line at 137° (Oc(2) 6s) lead down to Port d'Echouage.
Normal entry signals with following additions, shown from head of Jetée Ouest

R   } Prohibition does not apply to tugs,
W   } warships or fishing vessels
R W   }

R R   } Prohibition does not apply to
W   } tankers or vessels with dangerous
R   } cargoes.

Lock signals (H24) are shown at Ecluse Watier and consist of three horizontal pairs disposed vertically. Middle pair refer to Ecluse Watier and lower pair to Ecluse Trystram:
2FG(hor) = lock open
2FR(hor) = lock closed
Lock signals at each lock:
2FG(hor) = lock ready
2FR(hor) = lock in use, no entry
FG FlG = enter and secure on side of flashing light
FW over = slack water, both gates open; enter lock.
2FG(hor)

RADIO TELEPHONE
VHF Ch 12 16 **73** (H24). Ecluse Trystram Ch 73. YC Ch 09.

TELEPHONE
Hr Mr 28.65.99.22; Port Control 28.29.70.70; Aff Mar 28.66.56.14; CROSS 21.87.21.87; SNSM 28.63.23.60; Customs 28.65.14.73; Meteo 28.66.45.25; Auto 28.63.44.44; Hosp 28.66.70.01; Brit Consul 28.66.11.98.

FACILITIES
YC de la Mer du Nord Marina (250+40 visitors) Tel. 28.66.79.90, Slip, D, P, FW, C (4½ ton), BH (15 ton), AC, ME, SM, R, Bar, Access H24. **Bassin du Commerce** Tel. 28.66.11.06, M, D, L, FW, CH, AB; **Port de Pêche** Slip, P, D, FW, ME, El, Sh, CH, V, R, Bar; **Ayello** Tel. 28.65.02.11, ME, CH; **Weizsaeker et Carrere** Tel. 28.66.64.00, SHOM; **Leroy Garage** Tel. 28.66.82.37, P, D; **Flandre-Chantier** Tel. 28.66.49.62, M, ME, El, Sh, CH; **Norbert-Peche** Tel. 28.24.35.14, ME, CH. **Town** P, D, V, Gaz, R, Bar. PO; Bank; Rly, Air (Calais).
Ferry UK—Ramsgate/Dover.

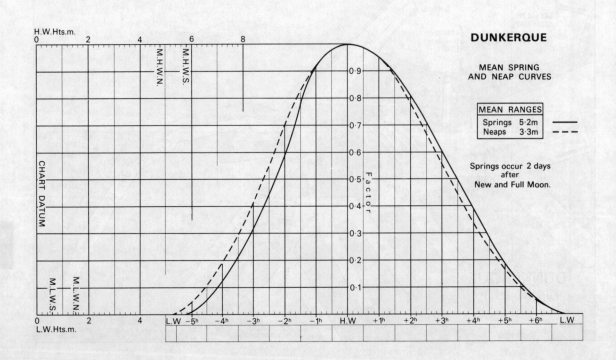

DUNKERQUE

MEAN SPRING AND NEAP CURVES

MEAN RANGES
Springs 5·2m
Neaps 3·3m

Springs occur 2 days after New and Full Moon.

MINOR HARBOURS AND ANCHORAGES 10.18.29

ILES St MARCOUF, Manche, 49°30′ N, 1°09′ W, Zone −0100, Admty chart 2073; SHOM 847. For tides, see 10.18.11. There are two islands, Ile de Terre and to the N, Ile du Large. The former is a bird sanctuary, the latter has a small dinghy harbour. Anchor SW or SE of Ile du Large or SE of the Ile de Terre. Both islands surrounded by drying rocks. Ile du Large Lt, VQ(3) 5s 18m 9M. Both islands are uninhabited and there are no facilities.

ISIGNY-SUR-MER, Calvados, 49°19′ N, 1°06′ W, Zone −0100, Admty chart 2073, SHOM 847. HW −0230 on Dover (GMT), −0025 on Le Havre (zone −0100). See 10.18.11. The channel is buoyed from the RWVS buoy NW of the Roches de Grandcamp to breakwaters (marked by Bns); best to start from outer buoy HW −2½. Where channel divides, take E leg up to Isigny. Ldg Lts 173° lead up between the breakwaters, front Oc(2+1) 12s 7m 18M, intens 171°-175°. Rear, 600m from front, Oc(2+1) 12s 19m 18M, intens 171°-175°, synchronised with front. Access difficult with strong NW winds. Berth on SW side ¼ M N of town. Facilities: Hr Mr 31.22.00.40; Aff.Mar. Tel. 31.22.03.11. **Quay** D, FW, P, Slip; **Club Nautique d'Isigny** AB, C (2, 5, 7 ton) Bar, R; **Isigny Garage** Tel. 31.22.02.33. El, ME, Sh; **Town** Bank, Bar, PO, R, V.

CAEN, Calvados, 49°10′ N, 0°19′ W, Zone −0100, Admty chart 1349, SHOM 7055. HW depths sp 7.2m, np 5.8m. Proceed up Caen Canal from Ouistreham, 8M see 10.18.15. Report to lock control Tr at Ouistreham to pay and arrange. Speed limit 7 kn; passage only possible by day. Three bridges, at Bénouville (2½ M from locks), Hérouville (5M) and Calix (6½ M), each show G Lt when passage clear. Seaward bound vessels have right of way. Marina at Bassin St Pierre near city centre. VHF Ch 12 68. Facilities: Hr Mr Tel. 31.52.12.88. Aff Mar. Tel. 31.85.40.55. Customs Tel. 31.86.61.50; **Marina** (64 visitors) Tel. 31.93.74.47, P, D, ME, Sh, El, AB, FW; **Clinique du Bateau** Tel. 31.84.48.63 ME, El, Sh, CH; **Caen Marine** Tel. 31.86.24.89. CH, El, M, ME, Sh, SHOM; **Quay** FW; **City** Bank, Bar, Hosp, PO, R, Rly, V.

CABOURG, Calvados, 49°19′ N, 0°07′ W, Zone −0100, Admty charts 2146, 1892, SHOM 6928, 890, 6614. Tides for Dives-sur-Mer. See 10.18.15. HW −0135 on Dover (GMT), −0032 on Le Havre (zone −0100); HW height −0.4m on Le Havre; ML 4.1m. Shelter good but entrance rough in winds from NW to NE. Entrance Lt, between Dives and Houlgate, Oc(2+1) WRG 12s 6m 12/9M G124°-150°, W150°-157°, R157°-193°. Entrance buoys, R can and G conical are moved according to shifting channel. There is a landing place at Dives at Société des Régates de la Dive. Channel from Dives to Cabourg is buoyed and accessible approx HW∓2. Yachts can go alongside jetty in front of Cabourg YC. Facilities: Aff Mar 31.91.23.55; **SNIP** Tel. 31.97.34.47, ME. Harbour run by **Cabourg YC** Tel. 31.91.23.55, AB, Bar, M, Slip; **Delanoé Marine** Tel. 31.91.28.49, El, M, ME, CH; **Cabourg Marina** Tel. 31.91.69.08, CH, ME, El, BY; **Town** Bar, Bank, PO, R, V.

ROUEN, Seine Maritime, 49°29′ N, 01°05′ E, Zone −0100. Admty charts 2880, 2994; SHOM 6117. HW +0330 on Dover (GMT), +0430 on Le Havre (zone −0100); HW height −0.1m on Le Havre; ML 6.2m; Duration 0400. See 10.18.18 for Seine navigation. Yacht navigation forbidden from ½ hr after sunset to ½ hr before sunrise. Yachts use Bassin St Gervais, on N bank, berth SE side for mast removal or replacements (max stay 48 hrs); berth in La Halte de Plaisance de Rouen NE side of Ile Lacroix. VHF 11 13 73 74 call *Rouen Port Capitainerie*. Facilities: Hr Mr 35.88.81.55; Aff Mar 35.98.53.98; Customs 35.98.27.60; Meteo 35.80.11.44; **Bassin St Gervais** C (3 to 25 ton), FW; **La Halte de Plaisance de Rouen** (50) Tel. 35.07.33.94, FW, AC, Slip, BH (4 ton), C (30 ton); **Villetard** Tel. 35.88.00.00, ME, El, Sh, P, D, CH; **Rouen YC** Tel. 35.66.52.52; **Eponville Nautic** Tel. 35.72.28.24, ME, El, Sh, CH.

LE CROTOY, Somme, 50°13′ N, 1°37′ E, Zone −0100, Admty chart 2612, SHOM 3800. HW −0020 on Dover (GMT), −0035 on Dieppe (zone −0100); HW height +0.7m on Dieppe. See 10.18.24. It is on the N side of R Somme estuary, which dries. Access when tidal range at Dover exceeds 4.4m. Max draught 1.6m. Follow buoyed channel for St Valery-sur-Somme until the Division Buoy YBY, from where Le Crotoy channel runs N and E marked by small R and G buoys numbered C1 to C10. Moor to the S side of S pontoon and check with YC. Access to non-tidal basin HW −1½ to HW +2. Le Crotoy Lt, Oc(2)R 6s 19m 9M, vis 285°-135°. Yacht harbour W side, Fl R 2s 4m 2M. E side Fl G 2s 4m 2M. Facilities: Hr Mr Tel. 22.27.81.59. Customs Tel. 22.27.50.36. **Quay** FW, C (3.5 ton); **YC Nautique de la Baie de Somme** Tel. 22.27.83.11, M, FW, Slip, Bar; **Marina Plaisance Baie de Somme** (280) Tel. 22.27.86.05, FW, CH, El, M, ME, Sh; **Phil Nautique** Tel. 22.27.86.47 CH, ME; **Town** Bank, Bar, D, P, PO, R, Rly, V.

ÉTAPLES, Pas de Calais, 50°32′ N, 1°38′ E, Zone −0100, Admty chart 2451, SHOM 6795. HW 0000 on Dover (GMT), +0012 on Dieppe (zone −0100); HW height −0.3m on Dieppe; ML 4.8m; Duration 0520. See 10.18.25. Very good shelter but entry to R Canche should not be attempted in strong on-shore winds. Access HW∓2. Channel in river marked by posts between sunken training walls to Étaples, where shoal draft boats can lie afloat on pontoons at small marina just below the bridge. Alternatively yachts which can take the ground can dry out on sand opposite the YC at Le Touquet. Whole estuary dries. Camiers Lt Oc(2) WRG 6s 17m 9/6M. Facilities: Hr Mr Tel. 21.94.74.26; Aff Mar. Tel. 21.94.61.50. **Marina** (60 + 15 visitors), FW, AC; **Quay** C (3 ton), FW, Slip; **Centre Nautique YC** Tel. 21.94.74.26, Bar, Slip; **Agence Nautique du Nord** Tel. 21.94.26.55, CH, El, M, ME, Sh; **ETS Lamour** Tel. 21.94.61.21, D; **Town** Bank, Bar, D, P, PO, R, Rly, V.

GRAVELINES, Nord, 51°01′ N, 2°06′ E, Zone −0100, Admty chart 323, SHOM 6501, 6651. HW +0045 on Dover (GMT), +0100 on Dieppe (zone −0100); HW height −2.8m on Dieppe; ML 3.1m; Duration 0520. See 10.18.27. Good shelter but bar dries; entry should not be attempted in strong on-shore winds. Entrance between Grand Fort Philippe and Petit Fort Philippe. Beware strong tidal stream across ent to the E, at HW. Safest entry HW −1. Keep to W entering and to E when inside. Yachts may take the bottom or go into the yacht harbour to W of town, available HW∓1. Harbour dries but is soft mud. Petit Fort Philippe Lt Oc(4) WG 12s 29m 27/22M intens G186°-193°, W193°-200°. E Jetty Lt Q(3)R 6s 5m 2M. W jetty Fl(2)WG 6s 14m 9/6M; vis W317°-327°, G078°-085°, W085°-244°. Facilities: Aff Mar. Tel. 28.23.06.12. **Bassin Vaubin** Tel. 28.23.06.12, C (1.5 ton to 15 ton); **Yacht Club Gravelines**; **MER (Marine et Reparations)** Tel. 28.23.14.68. CH, El, M, ME, Sh; **Town** Bank, Bar, D, P, PO, R, Rly, V.

Best for power cruising!

STILL ONLY £1

Practical features...
Equipment Reviews... Boat Reports...
Navigation Courses... Family Cruise Reports...

Motor cruiser
— for the discerning boating enthusiast

For further information contact:
Editorial: Chris Cattrall and Norman Alborough. Tel: (03727) 41411
Advertising: Barry Hollamby. Tel: (03727) 41411

Published by A.E. Morgan Publications Ltd, Stanley House, 9 West Street, Epsom, Surrey KT18 7RL

VOLVO PENTA SERVICE

Sales and service centres in area 19
Names and addresses of Volvo Penta dealers in this area are available from:

BELGIUM **Volvo Penta Belgium,** Weiveldlaan 37-G, B-1930 Zaventem
Tel (02) 721-20-62, Telex 65249, Volvo BMB.
NETHERLANDS **Nebim Handelmaatschappij BV,** Postbus 195, 3640
Ad Mijdrecht Tel 02979-84411, Telex 15505 NEHANL.

Area 19

Belgium and the Netherlands
Nieuwpoort to Delfzijl

10.19.1	Index	Page 725
10.19.2	Diagram of Radiobeacons, Air Beacons, Lifeboat Stations etc	726
10.19.3	Tidal Stream Charts	728
10.19.4	List of Lights, Fog Signals and Waypoints	730
10.19.5	Passage Information	734
10.19.6	Distance Table	735
10.19.7	Special factors affecting Belgium and the Netherlands	736
10.19.8	Nieuwpoort (Nieuport)	737
10.19.9	Oostende (Ostend)	738
10.19.10	Blankenberge	738
10.19.11	Zeebrugge	739
10.19.12	Breskens	739
10.19.13	Terneuzen	740
10.19.14	Westerschelde	740
10.19.15	Antwerpen (Antwerp)	741
10.19.16	Vlissingen (Flushing), Standard Port, Tidal Curves	742
10.19.17	Oosterschelde	746
10.19.18	Stellendam	748
10.19.19	Rotterdam	748
10.19.20	Hoek van Holland (Hook of Holland), Standard Port, Tidal Curves	749
10.19.21	Scheveningen	754
10.19.22	IJmuiden	754
10.19.23	Amsterdam	755
10.19.24	IJsselmeer	756
10.19.25	Den Helder	757
10.19.26	Harlingen	758
10.19.27	Vlieland	758
10.19.28	Terschelling	759
10.19.29	Delfzijl	759
10.19.30	Minor Harbours and Anchorages Oudeschild Nes (Ameland) Oostmahorn Zoutkamp Lauwersoog Schiermonnikoog	760

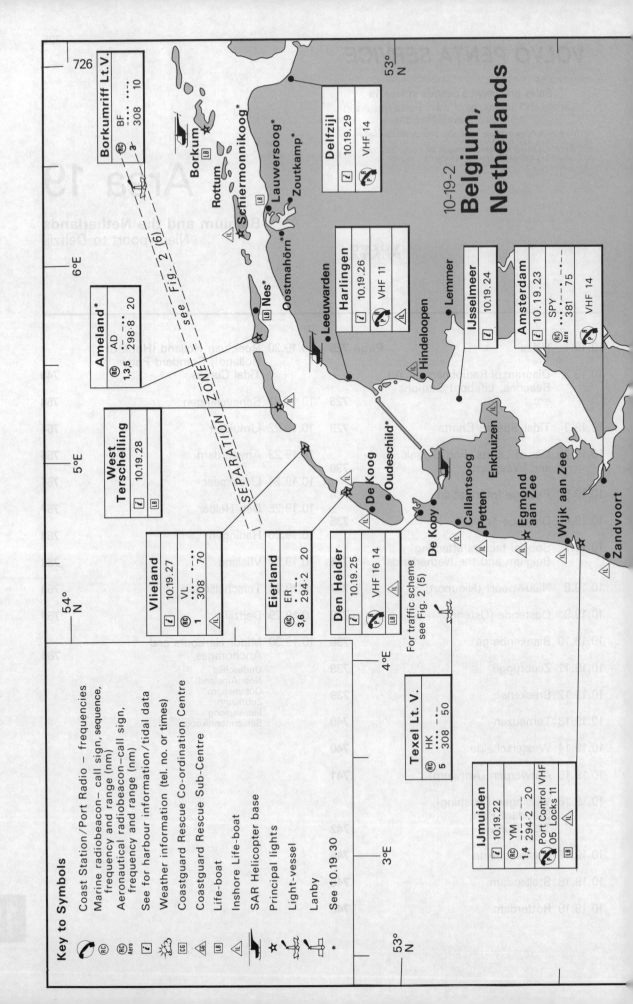

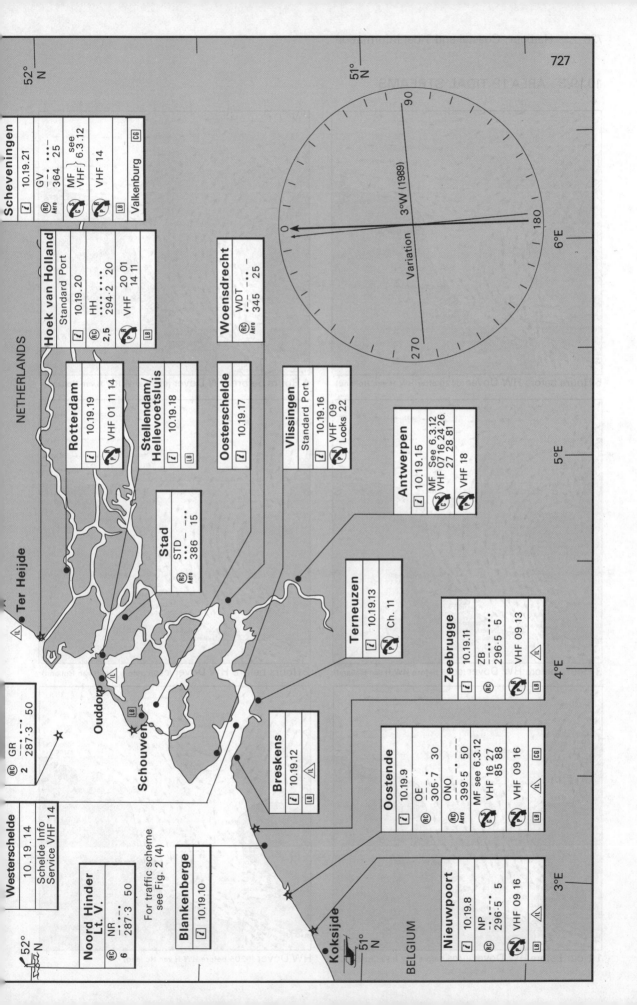

10.19.3 AREA 19 TIDAL STREAMS

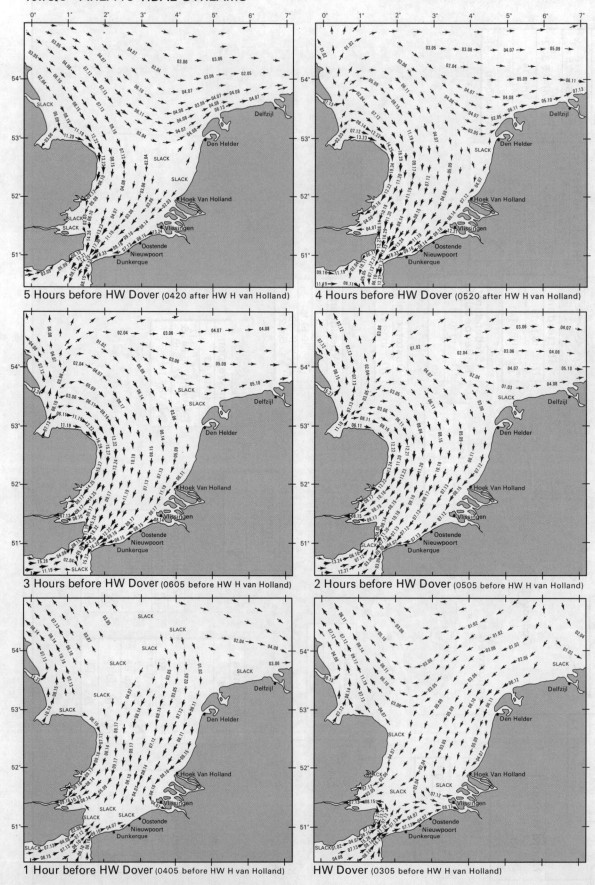

AREA 19—Belgium and the Netherlands   729

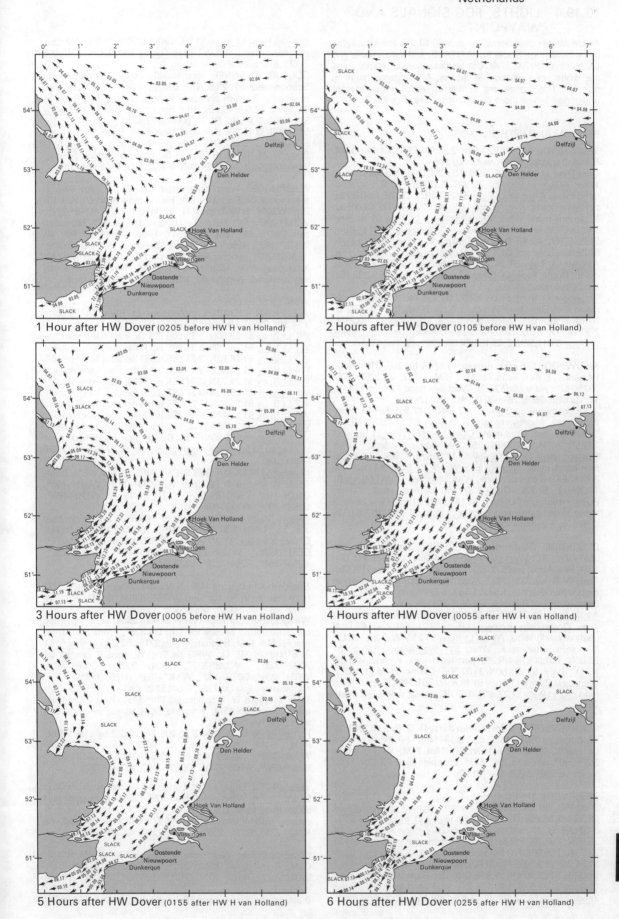

10.19.4 LIGHTS, FOG SIGNALS AND WAYPOINTS

Abbreviations used below are given in 10.0.3. Principal lights are in **bold** print, places in CAPITALS, and light-vessels and Lanbys in *CAPITAL ITALICS*. Unless otherwise stated lights are white. m—elevation in metres; M—nominal range in n. miles. Fog signals are in *italics*. Useful waypoints are underlined — use those on land with care. See 4.2.2.

BELGIUM AND NETHERLANDS

WEST HINDER LT V 51 23.08N/2 26.40E Fl(4) 30s 14m **17M**; R hull, 2 masts; RC; *Horn Mo(U) 30s*.
Oost-Dyck Lt Buoy 51 21.55N/2 31.20E QG; stbd-hand mark; *Whis*. A-Zuid Lt Buoy 51 21.50N/2 37.00E Fl(3)G 10s; stbd-hand mark. A-Noord Lt Buoy 51 23.50N/2 37.00E Fl(4)R 20s; port-hand mark. Kwintebank Lt Buoy 51 21.75N/2 43.00E Q; N cardinal mark; *Whis*. Middelkirkebank Lt Buoy 51 18.25N/2 42.80E FlG 5s; stbd-hand mark. Middelkirkebank N Lt Buoy 51 20.87N/2 46.40E Q; N cardinal mark. Akkaert SW Lt Buoy 51 22.30N/2 46.40E Q(9) 15s; W cardinal mark; *Whis*. Midden Akkaert Lt Buoy 51 24.23N/2 53.50E VQ(3) 5s; E cardinal mark. Goote Bank Lt Buoy 51 27.00N/2 52.70E Q(3) 10s; E cardinal mark; *Whis*.
Trapegeer Lt Buoy 51 08.45N/2 34.45E FlG 10s; stbd-hand mark; *Bell*. Den Oever Wreck Lt Buoy 51 09.20N/2 39.50E Q; N cardinal mark. Nieuwpoortbank Lt Buoy 51 10.21N/2 36.16E Q(9) 15s; W cardinal mark; *Whis*. Weststroombank Lt Buoy 51 11.39N/2 43.15E Fl(4)R 20s; port-hand mark.

NIEUWPOORT. **E pier**, near root Fl(2) R 14s 26m **21M**; R Tr, W bands. E pier, head FR 11m 10M; W Tr; vis 025°-250°, 307°-347°; *Horn Mo(K) 30s*. West pier head FG 11m 9M; W Tr; vis 025°-250°, 284°-324°; RC; *Bell (2) 10s*.
Zuidstroombank Lt Buoy 51 12.33N/2 47.50E FlR 5s; port-hand mark. Middelkirkebank S Lt Buoy 51 14.78N/2 42.00E Q(9)R 15s; port-hand mark. Oostendebank West Lt Buoy 51 16.25N/2 44.85E Q(9) 15s; W cardinal mark; *Whis*. Oostendebank Oost Lt Buoy 51 17.36N/2 52.00E Fl(4)R 20s; port-hand mark. Nautica Ena Wreck Lt Buoy 51 18.12N/2 52.85E Q; N cardinal mark. Wenduinebank West Lt Buoy 51 17.30N/2 52.87E Q(9) 15s; W cardinal mark; *Whis*. Buitenstroombank Lt Buoy 51 15.20N/2 51.80E Q; N cardinal mark; *Whis*. Binnenstroombank Lt Buoy 51 14.50N/2 53.73E Q(3) 10s; E cardinal mark.

OOSTENDE. **Oostende** Fl(3) 10s 63m **27m**; W Tr; obsc 069°-071°; RC. West pier head 51 14.36N/2 55.12E FG 12m 10M; W column; vis 057°-327°; *Bell (1) 4s*. E pier head FR 13m 12M; W Tr; vis 333°-243°; *Horn Mo(OE) 30s*. QY Lt at signal mast when channel closed. Ldg Lts 128°, both FR 4M on W framework Trs, R bands; vis 051°-201°.
A1 Lt Buoy 51 21.70N/2 58.10E LFl 10s; safe water mark; *Whis*. Oostendebank Noord Lt Buoy 51 21.25N/2 53.00E Q; N cardinal mark; *Whis*. A1 bis Lt Buoy 51 21.70N/2 58.10E LFl 10s; safe water mark; *Whis*. SW Wandelaar Lt Buoy 51 22.00N/3 01.00E Fl(4)R 20s; port-hand mark. Wenduine Bank N Lt Buoy 51 21.50N/3 02.70E QG; stbd-hand mark. Wenduine Bank E Lt Buoy 51 18.85N/3 01.70E QR; port-hand mark. A2 Lt Buoy 51 22.50N/3 07.05E Iso 8s; safe water mark; *Whis*.

BLANKENBERGE. **Comte Jean Jetty** Oc(2) 8s 30m **20M**; W Tr, B top; vis 065°-245°. Ldg Lts 134° both FR 3M; R crosses on masts. West mole head FG 14m 11M; W Tr; intens 065°-290°, unintens 290°-335°. East pier head FR 12m 11M; W Tr; vis 290°-245°; *Bell (2) 15s*.

ZEEBRUGGE APPROACHES. Scheur 1 Lt Buoy 51 23.18N/3 00.15E FlG 5s; stbd-hand mark. Scheur 2 Lt Buoy 51 23.38N/2 58.20E Fl(4)R 15s. Scheur 3 Lt Buoy 51 24.35N/3 02.90E Q; N cardinal mark; *Whis*. Scheur 4 Lt Buoy 51 25.05N/3 03.10E FlR 5s; port-hand mark. Scheur 5 Lt Buoy 51 23.73N/3 05.90E FlG 5s; stbd-hand mark. Scheur 6 Lt Buoy 51 24.25N/3 05.90E Fl(4)R 20s; port-hand mark. Scheur-Zand Lt Buoy 51 23.70N/3 07.68E Q(3) 10s; E cardinal mark. Zand Lt Buoy 51 22.50N/3 10.12E QG; stbd-hand mark. Wielingen Zand Lt Buoy 51 22.60N/3 10.80E Q(9) 15s; W cardinal mark.

Bol van Heist Lt Buoy 51 23.15N/3 12.05E Q(6) + LFlR 15s; port-hand mark. MOW3 Tide Gauge 51 23.45N/3 12.00E Fl(5)Y 20s; special mark; *Whis*.
SCHEUR CHANNEL (continued). HAAZ, Droogte van Schooneveld 51 25.50N/3 09.00E Fl(4)Y 20s 12m 2M; measuring Bn with platform; Ra refl. Scheur 7 Lt Buoy 51 24.00N/3 10.50E FlG 5s; stbd-hand mark. Scheur 8 Lt Buoy 51 24.45N/3 10.45E FlR 5s; port-hand mark. Scheur 9 Lt Buoy 51 24.45N/3 15.05E QG; stbd-hand mark. Scheur 10 Lt Buoy 51 24.90N/3 15.05E Fl(4)R 15s; port-hand mark. Scheur 12 Lt Buoy 51 24.70N/3 18.50E FlR 5s; port-hand mark. Scheur-Wielingen Lt Buoy 51 24.26N/3 18.00E Q; N cardinal mark.

ZEEBRUGGE. West breakwater Oc G 7s 33m 7M. E breakwater Oc R 7s 33m 7M. **Heist**, mole head Oc WR 15s 22m **W20M, R18M**; grey Tr; vis W068°-145°, R145°-212°, W212°-296°; RC; *Horn (3 + 1) 90s*. Ldg Lts 136°. Front and rear Oc 5s 8M. Ldg Lts 154°. Front OcWR 6s 3M; vis W135°-160°, R160°-169°. Rear, 520m from front, Oc 6s 3M; vis 150°-162°, synchronised. Ldg Lts 235°. Front QR 10m 5M; X on mast. Rear, 155m from front, QR 14m 5M; X on mast, synchronised. Westhoofd, NE corner FG, NW corner FR.

WESTERSCHELDE

WIELINGEN CHANNEL. Wielingen Lt Buoy 51 23.00N/3 14.10E Fl(3)G 15s; stbd-hand mark.
W1 Lt Buoy 51 23.50N/3 18.00E FlG 5s; stbd-hand mark.
W2 Lt Buoy 51 24.64N/3 21.58E Iso R 8s; port-hand mark.
W3 Lt Buoy 51 24.03N/3 21.58E Iso G 8s; stbd-hand mark.
W4 Lt Buoy 51 24.92N/3 24.48E Iso R 4s; port-hand mark.
W5 Lt Buoy 51 24.33N/3 24.48E Iso G 4s; stbd-hand mark.
W6 Lt Buoy 51 25.15N/3 27.25E Iso R 8s; port-hand mark.
W7 Lt Buoy 51 24.65N/3 27.30E Iso G 8s; stbd-hand mark.
W8 Lt Buoy 51 25.48N/3 30.15E Iso R 4s; port-hand mark.
W9 Lt Buoy 51 24.97N/3 30.13E Iso G 4s; stbd-hand mark.
W10 Lt Buoy 51 25.80N/3 33.00E QR; port-hand mark.
Wielingen Sluis. Wave observation post 51 22.84N/3 22.82E Fl(5)Y 20s.
Kruishoofd 51 23.73N/3 28.36E Iso WRG 8s 14m W8M, R6M, G5M; W square Tr, B post; vis R074°-091°, W091°-100°, G100°-118°, W118°-153°, R153°-179°, W179°-198°, G198°-205°, W205°-074°.
Nieuwe Sluis, on embankment 51 24.49N/3 30.33E *Horn (3) 30s*. 51 24.48N/3 31.38E Oc WRG 10s 27m W14M, R11M, G10M; B 8-sided Tr, W bands; vis R055°-084°, W084°-091°, G091°-132°, W132°-238°, G238°-244°, W244°-258°, G258°-264°, R264°-292°, W292°-055°.

Noorderhoofd Ldg Lts 149°, NW head of dyke. Front, 0.73M from rear Oc WRG 10s 18m W13M, R10M, G10M; R Tr, W band; vis R353°-008°, G008°-029°, W029°-169°.
Westkapelle, Common rear Fl 3s 48m **28M**; Tr, R top; obsc on certain brgs. Zoutelande Ldg Lts 326°, Front 1.8M from rear FR 21m 12M; R square Tr; vis 321°-352°. Molenhoofd 51 31.61N/3 26.07E Oc WRG 6s 9m; W mast R bands; vis R306°-328°, W328°-347°, R347°-008°, G008°-031°, W031°-035°, G035°-140°, W140°-169°, R169°-198°.
Kaapduinen, Ldg Lts 130°. Front 51 28.5N/3 31.0E Oc 5s 26m 13M; Y square Tr, R bands; vis 115°-145°. Rear 220m from front Oc 5s 36m 13M; Y square Tr, R bands; synchronised with front; vis 108°-152°.
Fort de Nolle 51 27.00N/3 33.20E Fl WRG 2.5s 11m W6M, R4M, G4M; W column, R bands; vis R293°-309°, W309°-333°, G333°-351°, R351°-013°, G013°-062°, R062°-086°, W086°-093°, G093°-110°, W110°-130°.

VLISSINGEN. Ldg Lts 117°. Leugenaar causeway, front Oc R 5s 5m 11/7M; W&R pile; intens 108°-126°. Sardijngeul, rear 550m from front Oc WRG 5s 8m W12M, R9M, G8M; R triangle with W bands on R and W mast; synchronised; vis R245°-271°, G271°-285°, W285°-123°, R123°-147°.

Koopmanshaven, West mole Iso WRG 3s 15m W12M, R10M, G9M; R pylon; vis R253°-270°, W270°-059°, G059°-071°, W071°-077°, R077°-101°, G101°-110°, W110°-114°. E mole head FG.
Buitenhaven, E mole head FG 7m 4M; grey mast on Dn; in fog

AREA 19—Belgium and the Netherlands

FY. W mast; West mole head 51 26.44N/3 36.12E Iso WRG 4s 10m; vis W072°-021°, G021°-042°, W042°-056°, R056°-072°; traffic signals. FR (same structure) 10m 5M; *Horn 15s.*
Schone Waardin 51 26.60N/3 37.95E Oc WRG 9s 10m W13M, R10M, G9M; R mast, W bands; vis R248°-260°, G260°-270°, W270°-282°, G282°-325°, W325°-341°, G341°-023°, W023°-024°, G024°-054°, R054°-066°, W066°-076°, .G076°-094°, W094°-248°.

VLISSINGEN EAST (SLOEHAVEN), west mole head FR 8m 5M; W column; in fog FY; *Horn (2) 20s.* East mole head FG 8m 4M; W column. Ldg Lts 023°. Front Oc R 8s 7m 8M; G post. Rear 100m from front Oc R 8s 12m 8M; G mast; synchronised with front, both vis 015°-031°.

BRESKENS. Ferry harbour, West mole head FG 8m 4M; B and W mast; in fog FY. E mole head FR 8m 5M; B and W mast; in fog FY. Breskens, West mole head 51 24.09N/ 3 34.12E FWRG 6m; grey mast; vis R090°-128°, W128°-157°, R157°-172°, W172°-175°, G175°-296°, W296°-300°, R300°-008°, G008°-090°; *Horn Mo(U) 30s.* E mole head FR.

BRAAKMANHAVEN. Ldg Lts 191°. Front 51 20.3N/3 45.8E Iso 4s 10m; B pile, W bands. Rear, 60m from front, Iso 4s 14m; B pile, W bands; synchronised with front, showing over harbour mouth. Ldg Lts 211°, both Oc G 4s, synchronised, showing middle of turning basin. West side 51 21.0N/3 45.9E FG; G pile, W bands; Ra refl. E side 51 21.1N/3 46.3E FR; R mast, W bands; Ra refl, traffic signals.
Braakman 51 21.03N/3 46.31E Oc WRG 8s 7m W7M, R5M, G4M; B pedestal, W band; vis R116°-132°, W132°-140°, G140°-202°, W202°-116°.

TERNEUZEN. Nieuw Neuzenpolder Ldg Lts 125°. Front Oc 5s 5m 13/9M; W column, B bands; intens 117°-133°. Rear 365m from front Oc 5s 16m 13/9M; B and W framework mast; synchronised with front, intens 117°-133°. Dow Chemical jetty, 4 dolphins showing Fl 3s and Fl R 3s; *Horn 15s.* Veer Haven, West jetty Oc WRG 5s 13m W9M, R7M, G6M; B and W framework Tr; vis R092°-115°, W115°-238°, G238°-248°, W248°-277°, R277°-309°, W309°-003°. West mole head 51 20.62N/3 49.71E FG 6m. E mole head FR 7m.
Borssele-Noordnol, pier head 51 25.55N/3 42.80E Oc WRG 5s; 9m; R mast, W bands; vis R305°-331°, W331°-341°, G341°-000°, W000°-007°, R007°-023°, G023°-054°, W054°-057°, G057°-113°, W113°-128°, R128°-155°, W155°-305°; FR on chimney 0.4M NNE.
Borssele, Total jetty, NW end 51 24.85N/3 43.61E Oc WR 10s; vis R135°-160°, W160°-135°.
Borssele-Everingen 51 24.73N/3 44.20E Iso WRG 4s 9m; W structure, R band; vis R021°-026°, G026°-080°, W080°-100°, R100°-137°, W137°-293°, R293°-308°, W308°-344°, G344°-357°, W357°-021°.
Wave observation post 51 30.35N/3 14.53E Fl Y 5s.

OOSTERSCHELDE
OUTER APPROACHES. SW Thornton Lt Buoy 51 31.01N/ 2 51.00E Iso 8s; safe water mark. TB Lt Buoy 51 34.45N/ 2 59.15E Q; N cardinal mark. Westpit Lt Buoy 51 33.70N/ 3 10.00E Iso 8s; safe water mark. Rabsbank Lt Buoy 51 38.30N/3 10.05E Iso 4s; safe water mark. Middelbank Lt Buoy 51 40.90N/3 18.30E Iso 8s; safe water mark. Schouwenbank Lt Buoy 51 45.00N/3 14.40E Mo(A) 8s; safe water mark; Racon. Buitenbank Lt Buoy 51 51.20N/3 25.80E Iso 4s; safe water mark.
WESTGAT/OUDE ROOMPOT (selected marks). WG1 Lt Buoy 51 38.00N/3 26.30E QG; stbd-hand mark. WG Lt Buoy 51 38.25N/3 28.90E Q; N cardinal mark. WG4 Lt Buoy 51 38.62N/3 27.80E Iso R 8s; port-hand mark. Wave observation post OS11 51 38.63N/3 28.95E FlY 5s; Y pile. WG7 Lt Buoy 51 39.45N/3 32.75E Iso G 4s; stbd-hand mark. WG OR Lt Buoy 51 39.75N/3 32.60E VQ(6)+LFl 10s; S cardinal mark. OR2 Lt Buoy 51 39.50N/3 33.70E LFlR 8s; port-hand mark. OR5 Lt Buoy 51 38.55N/3 35.50E QG; stbd-hand mark. OR11 Lt Buoy 51 37.00N/3 38.50E Iso G 2s; stbd-hand mark.

Roompotsluis Ldg Lts 073°. Front Oc G 5s. Rear, 280m from front, Oc G 5s; synchronised. N breakwater head FR 6m; *Horn(2) 30s.* S breakwater head FG 6m. Inner mole head QR.
COLIJNSPLAAT. E jetty head 51 36.25N/3 51.15E FR 3m 3M; *Horn.* W jetty head FG 5m 3M.
Zeeland Bridge. N and S passages marked by FY Lts, 14m.
KATS. N jetty head FG 5m 5M. S jetty head 51 34.44N/ 3 53.72E Oc WRG 8s 5m 5M; vis W344°-153°, R153°-165°, G165°-200°, W200°-214°, G214°-258°, W258°-260°, G260°-313°, W313°-331°.
SAS VAN GOES. S mole head FR. N mole head FG.
WEMELDINGE. W jetty head 51 31.34N/4 00.25E Oc WRG 5s 7m W9M, R7M, G6M; B mast, W band; vis R shore-116°, W116°-123°, G123°-151°, W151°-153°, R153°-262°, W262°-266°, R266°-shore; in fog Oc Y 4s; traffic signals. E jetty head FR; *Horn(4) 30s.*
YERSEKE. Ldg Lts 155° (through Schaar van Yerseke). Front Iso 4s 8m; in fog FY. Rear, 180m from front, Iso 4s 13m; synchronised; in fog 2FY. FG and FR mark mole heads.
BERGEN OP ZOOM. Molenplaat Ldg Lts 119°, both Oc 5s. W breakwater FG. E breakwater FR. Bergsche Diep Ldg Lts 065°. Front Iso 6s 6m; W daymark, B band. Rear, 100m from front, Oc 5s 9m; W daymark, B stripe. Theodorushaven W mole head FR, E mole head FG. Ldg Lts 034° both Oc 5s; synchronised. Lock Ldg Lts 057° both Iso G 6s; synchronised.

Tholensche Gat. Strijenham Oc WRG 5s 9m W8M, R5M, G5M; R square, W bands, on mast; vis W shore-298°, R298°-320°, W320°-052°, G052°-069°, W069°-085°, R085°-095°, W095°-shore.
Gorishoek Iso WRG 8s 7m W6M, R4M, G4M; R pedestal, W bands; vis R260°-278°, W278°-021°, G021°-025°, W025°-071°, G071°-085°, W085°-103°, R103°-120°, W120°-260°.

STAVENISSE. E mole head 51 35.73N/4 00.35E Oc WRG 5s 10m W12M, R9M, G8M; B pylon; vis W075°-087°, R087°-106°, W106°-108°, G108°-115°, W115°-124°, G124°-157°, W157°-162°, G162°-231°, W231°-238°, R238°-253°, W253°-350°.
ST ANNALAND. Entrance, West side FG. E side FR.

ZIJPE. ANNA JACOBAPOLDER S mole head FG. N mole head FR; in fog FY; *Horn 2s (occas).* St Philipsland, on dyke Oc WRG 4s 9m W8M, R5M, G4M; pylon on B round column; vis W051°-100°, R100°-144°, W144°-146°, G146°-173°. Zijpsche Bout Oc WRG 10s 9m W12M, R9M, G8M; mast on R column; vis R208°-211°, W211°-025°, G025°-030°, W030°-040°, R040°-066°. Tramweghaven S mole FR; in fog FY; *Siren 2s (occas).* N mole FG. Stoofpolder Iso WRG 4s 10m W12M, R9M, G8M; B Tr, W bands; vis R153°-229°, W229°-235°, G245°-256°, W256°-026°.

Hoek Van Ouwerkerk 51 36.92N/3 58.27E Iso WRG 6s; vis R268°-305°, W305°-313°, G313°-008°, W008°-011°, G011°-059°, W059°-065°, R065°-088°, W088°-098°, G098°-112°, R112°-125°, W125°-268°.

DE VAL. Engelsche Vaarwater Ldg Lts 019°. Front Iso WRG 3s 7m W6M, R4M, G4M; R pedestal, W band; vis R290°-306°, W306°-317°, G317°-334°, W334°-336°, G336°-017°, W017°-026°, G026°-090°, R090°-108°, W108°-290°. Rear, 300m from front, Iso 3s 15m 6M; R square on W mast, R bands.
ZIERISKEE. W jetty head 51 37.95N/3 53.45E Oc WRG 6s 10m W6M, R4M, G4M; R pedestal, W band; vis G060°-107°, W107°-133°, R133°-156°, W156°-278°, R278°-304°, G304°-314°, W314°-331°, R331°-354°, W354°-060°. West mole head FR. E mole head FG.

FLAUWERSPOLDER. W mole head 51 40.70N/3 50.86E Iso WRG 4s 7m W6M, R4M, G4M; W daymark, B band on pylon; vis R303°-344°, W344°-347°, G347°-083°, W083°-086°, G086°-103°, W103°-110°, R110°-128°, W128°-303°.
HAMMEN. Schelphoek Ldg Lts 357°. Front, E mole head FG. Rear, 400m from front, Iso G 4s; vis 342°-012°. E breakwater head Fl(2) 10s 8m. Burghsluis, S mole head F WRG 9m W8M, R5M, G4M; mast on R column; vis W218°-227°, R227°-246°, W246°-250°, G250°-277°, W277°-009°, G009°-050°, W050°-058°, R058°-070°, W070°-081°.

APPROACHES TO EUROPOORT
West Schouwen 51 42.58N/3 41.60E Fl(2+1) 15s 57m **30M**; grey Tr, R diagonal stripes on upper part.
Verklikker 51 43.58N/3 42.39E F WR 13m W9M, R7M; Tr with R lantern; vis R115°-127°, W127°-169°, R169°-175°, W175°-115°.
Wave observation posts. OS13 51 44.00N/3 33.40E FlY 5s; Y pile. OS14 51 43.30N/3 40.60E FlY 5s; Y pile. BG2 51 46.10N/3 37.20E FlY 5s; Y pile. BG5 51 49.50N/3 45.70E FlY 5s; Y pile. Ha10 51 51.80N/3 51.70E FlY 5s; Y pile.
Buitenbank Lt Buoy 51 51.20N/3 25.80E Iso 4s; safe water mark.

Westhoofd 51 48.83N/3 51.90E Fl(3) 15s 55m **30M**; R square Tr. Kwade Hock 51 50.3N/3 59.1E Iso WRG 4s 8m W12M, R9M, G8M; B mast, W bands; vis W 235°-068°, R068°-088°, G088°-107°, W107°-113°, R113°-142°, W142°-228°, R228°-235°. FR on radio mast 4.2M NE.
Slijkgat SG Lt Buoy 51 52.00N/3 51.50E Iso 4s; safe water mark. Hinder Lt Buoy 51 54.60N/3 55.50E Q(9) 15s; W cardinal mark.

HARINGVLIET. Heliushaven, West jetty 51 49.3N/4 07.2E FR 7m 4M. E jetty FG 7m 3M; *Horn (3) 20s*.
HELLEVOETSLUIS. 51 49.2N/4 07.7E Iso WRG 10s 16m W11M, R8M, G7M; W Tr, R cupola; vis G shore-275°, W275°-294°, R294°-316°, W316°-036°, G036°-058°, W058°-095°, R095°-shore. West mole head FR 6m. E mole head FG 6m; in fog FY.
Hoornsche Hoofden, watchhouse on dyke, 51 48.3N/4 11.0E Oc WRG 5s 7m W7M, R5M, G4M; vis W288°-297°, G297°-313°, W313°-325°, R325°-335°, G335°-345°, W345°-045°, G045°-055°, W055°-131°, R131°-shore.

MIDDELHARNIS. West pier head F WRG 5m W8M, R5M, G4M; vis W144°-164°, R164°-176°, G176°-144°. E pier head FR 5m 5M; in fog FY.
Nieuwendijk. Ldg Lts 303°. Front 51 45.1N/4 19.5E Iso WRG 6s 8m W9M, R7M, G6M; B framework Tr; vis G093°-100°, W100°-103°, R103°-113°, W113°-093°. Rear, 450m from front F 11m 9M; B framework Tr.
VOLKERAK. Noorder Voorhaven. West mole head FG 6m 4M; R lantern on pedestal; in fog FY.

NOORD HINDER LT V 52 00.15N/2 51.20E Fl(2) 10s 16m **27M**; R hull, W upperworks; RC; Racon; *Horn(2) 30s*. Euro Platform 51 59.9N/3 16.6E Mo(U) 15s; W structure, R bands; helicopter platform; *Horn Mo(U) 30s*.
Goeree 51 55.53N/3 40.18E Fl(4) 20s 31m **28M**; R and W chequered Tr on platform; RC; helicopter platform; Racon; *Horn (4) 30s*.
Adriana Lt Buoy 51 56.13N/3 50.65E VQ(9) 10s; W cardinal mark. Maas Center Lt Buoy 52 01.18N/3 53.57E Iso 4s; safe water mark; Racon. Indusbank N Lt Buoy 52 02.92N/4 03.73E Q; N cardinal mark.

HOEK VAN HOLLAND. **Maasvlakte** 51 58.2N/4 00.9E Fl(5) 20s 66m **28M**; B 8-sided Tr, Y bands. Nieuwe Noorderdam, head 51 59.71N/4 02.92E FR 24m 10M; Y Tr, B bands; Helicopter platform; in fog Al Fl WR 6s, vis 278°-255°.
Nieuwe Zuiderdam, head 51 59.19N/4 02.58E FG 24m 10M; Y Tr, B bands; Helicopter platform; in fog Al Fl WG 6s, vis 330°-307°; *Horn 10s*. Norderpier Oc R 10s 14m 6M. South pier Oc G 10s 14m 6M. Maasmond Ldg Lts 112° for very deep draught vessels. **Front** 51 58.9N/4 04.9E Iso 4s 28m **21M**; W Tr, B bands; vis 101°-123°. **Rear**, 0.6M from front Iso 4s 45m **21M**; W Tr, B bands; synchronised with front. Ldg Lts 107° for other vessels. **Front** 51 58.6N/4 07.6E Iso R 6s 29m **18M**; R Tr, W bands; vis 100°-115°. **Rear** 450m from front Iso R 6s 43m **18M**; vis 100°-115°, synchronised.

EUROPOORT. Calandkanaal Entrance Ldg Lts 116°. **Front** Oc G 6s 29m **16M**; W Tr, R bands; vis 108°-124°, synchronised with **Rear** Lt 550m from front Oc G 6s 43m **16M**; W Tr, R bands; vis 108°-124°. Beerkanaal Ldg Lts 192°. Front Iso G 3s. Rear, 50m from front, Iso G 3s. Rotterdamsche Waterweg, S side 20 G Lts 7M; B circle on groynes and dolphins. N side 20 R Lts 5M; R circle on groynes and dolphins.

SCHEVENINGEN. SCH Lt Buoy 52 07.80N/4 14.20E Iso 4s; safe water mark. **Scheveningen** 52 06.3N/4 16.2E Fl(2) 10s 48m **29M**; brown Tr; vis 014°-244°. Ldg Lts 156°. Front Iso 4s 17m 14M. Rear Iso 4s 21m 14M; synchronised with front. S mole head 52 06.28N/4 15.22E FG 11m 9M; B 6-sided Tr, Y bands, R lantern; *Horn (3) 30s*. N mole head FR 11m 9M; B 6-sided Tr, Y bands, R lantern.
Noordwijk-aan-Zee 52 15.00N/4 26.10E Oc(3) 20s 32m **17M**; W square Tr. Survey platform 52 16.4N/4 17.9E FR and Mo(U) 15s; *Horn Mo(U) 20s*.

IJMUIDEN. IJmuiden Lt Buoy 52 28.70N/4 23.93E Mo(A) 8s; safe water mark; Racon. Ldg Lts 100°. **Front** F WR 31m **W16M**, R13M; dark R Tr; vis W050°-122°, R122°-145°, W145°-160°; RC. **Rear** 560m from front Fl 5s 52m **29M**; dark R Tr; vis 019°-199°. S breakwater head 52 27.86N/4 32.00E FG 14m 10M; in fog Fl 3s, *Horn (2) 30s*. N breakwater head FR 14m 10M. N pier head QR 11m 9M; vis 263°-096°; in fog FW. S pier head QG 11m 9M; vis 096°-295°; in fog FW.

Egmond-aan-Zee 52 37.20N/4 37.40E Iso WR 10s 36m **W18M**, R14M; W Tr; vis W010°-175°, R175°-188°. FR Lts on chimney 10.2M N.

ZEEGAT VAN TEXEL
TEXEL LT V. 52 47.10N/4 06.60E Fl(3+1) 20s 16m **26M**; R hull, W band; RC; Racon; *Horn (3) 30s*. During maintenance replaced by RW buoy Oc 10s; Racon.

OUTER GROUNDS. ZH (Zuider Haaks) Lt Buoy 52 54.70N/4 34.84E VQ(6)+LFl 10s; S cardinal mark. MR (Middelrug) Lt Buoy 52 56.80N/4 33.90E Q(9) 15s; W cardinal mark. NH (Noorder Haaks) Lt Buoy 53 00.30N/4 35.45E VQ; N cardinal mark.
Grote Kaap Oc WRG 10s 31m W11M, R8M, G8M; vis G041°-088°, W088°-094°, R094°-131°.
Schulpengat Ldg Lts 026°. **Front** 53 00.9N/4 44.5E Iso 4s **18M**; vis 024°-028°. Rear, **Den Hoorn** 0.83M from front Oc 8s **18M**; church spire; vis 024°-028°.
Huisduinen 52 57.20N/4 43.37E F WR 27m W14M, R11M; square Tr; vis W070°-113°, R113°-158°, W158°-208°.
Kijkduin, Rear 52 57.35N/4 43.60E Fl(4) 20s 56m **30M**; brown Tr; vis except where obsc by dunes on Texel. Ldg Lt 253° with Den Helder, Harssens Island (QG).

SCHULPENGAT. SG Lt Buoy 52 52.95N/4 38.00E Mo(A) 8s; safe water mark.
S1 Lt Buoy 52 53.74N/4 39.25E Iso G 4s; stbd-hand mark.
S2 Lt Buoy 52 54.05N/4 38.20E Iso R 4s; port-hand mark.
S3 Lt Buoy 52 54.60N/4 39.80E Iso G 4s; stbd-hand mark.
S4 Lt Buoy 52 54.70N/4 39.30E Iso R 8s; port-hand mark.
S5 Lt Buoy 52 55.40N/4 40.30E Iso G 4s; stbd-hand mark.
S6 Lt Buoy 52 55.50N/4 39.95E Iso R 4s; port-hand mark.
S7 Lt Buoy 52 56.20N/4 40.80E Iso G 8s; stbd-hand mark.
S6A Lt Buoy 52 56.55N/4 40.50E QR; port-hand mark.
S8 Buoy 52 57.10N/4 41.14E (unlit); port-hand mark.
S9 Buoy 52 56.90N/4 42.15E (unlit); stbd-hand mark.
S10 Lt Buoy 52 57.65N/4 41.65E Iso R 8s; port-hand mark.
S11 Lt Buoy 52 57.60N/4 43.35E Iso G 4s; stbd-hand mark.

AREA 19—Belgium

MOLENGAT. MG Lt Buoy 53 03.42N/4 39.10E Mo(A) 8s; safe water mark. MG1 Lt Buoy 53 01.75N/4 41.20E Iso G 8s; stbd-hand mark. MG2 Lt Buoy 53 02.30N/4 41.42E Iso R 8s; port-hand mark. MG2A Buoy 53 01.75N/4 41.57E (unlit); port-hand mark. MG1A Buoy 53 01.15N/4 41.20E (unlit); stbd-hand mark. MG4 Lt Buoy 53 01.20N/4 41.65E Iso R 4s; port-hand mark. MG3 Buoy 53 00.06N/4 41.25E (unlit); stbd-hand mark. MG6 Buoy 53 00.63N/4 41.75E (unlit); port-hand mark. MG5 Lt Buoy 53 00.02N/4 41.30E Iso G 4s; stbd-hand mark. MG8 Lt Buoy 53 00.00N/4 41.85E Iso R 8s; port-hand mark. MG7 Buoy 52 59.60N/4 41.74E (unlit); stbd-hand mark. MG9 Lt Buoy 52 59.20N/4 42.30E Iso G 8s; stbd-hand mark. MG10 Buoy 52 59.52N/4 42.50E (unlit); port-hand mark. MG11 Buoy 52 58.80N/4 42.75E (unlit); stbd-hand mark. MG 12 Buoy 52 59.00N/4 43.10E (unlit); port-hand mark. S14/MG13 Lt Buoy 52 58.40N/4 43.30E VQ(3) 5s; E cardinal mark. MG14 Lt Buoy 52 58.50N/4 43.68E Iso R 4s; port-hand mark.

MARSDIEP. DEN HELDER, Marinehaven, W breakwater head (Harssens Island) QG 12m 8M; Horn 20s. MH4, E side of entrance, Iso R 4s; R pile; Ra refl. Entrance, W side, FlG 5s 9m 4M; vis 180°-067° (H24). Entrance, E side, QR 9m 4M; (H24). Ldg Lts 191°. Front Oc G 5s 15m 14M; B triangle on building; vis 161°-221°. Rear, 275m from front, Oc G 5s 25m 14M; B triangle on building; vis 161°-247°, synchronised. Schilbolsnol 53 00.6N/4 45.8E F WRG 27m **W15M**, R12M, G11M; G Tr; vis W338°-002°, G002°-035°, W035°-038° (leading sector for Schulpengat), R038°-051°, W051°-068°.

MOK 53 00.25N/4 46.85E Oc WRG 10s 10m W10M, R7M, G6M; vis R229°-317°, W317°-337°, G337°-112°. Ldg Lts 284°. Front Iso 2s 7m 6M. Rear, 245m from front, Iso 8s 10m 6M; both vis 224°-344°.

WADDENZEE
Malzwin KM/RA1 52 58.6N/4 49.2E Fl(5)Y 20s; post. Malzwin M5 52 58.3N/4 49.9E Iso G 4s; G pile; Ra refl. Wierbalg W3A 52 58.1N/4 57.1E QG; G pile. Pile 01 52 57.0N/5 00.6E Iso G 8s; G pile. Pile 05 52 56.9N/5 01.9E Iso G 4s; G pile.
DEN OEVER. Ldg Lts 132°. Front and rear both Oc 10s; vis 127°-137°. Detached breakwater, N head, LFl R 10s. Stevinsluizen, E wall, Iso WRG 5s; vis G226°-231°, W231°-235°, R235°-290°, G290°-327°, W327°-335°, R335°-345°. West wall, 80m from head, Iso WRG 2s; vis G195°-213°, W213°-227°, R227°-245°.

TEXELSTROOM. T3/MH2 Lt Buoy 52 58.38N/4 47.80E Fl(2+1)G 12s; preferred channel to port. T6A/Mk1 Lt Buoy 52 59.82N/4 47.55E Fl(2+1)R 12s; preferred channel to stbd. T5A/GvS2 Lt Buoy 52 59.95N/4 49.20E Fl(2+1)G 12s; preferred channel to port. T9 Lt Buoy 53 01.20N/4 51.50E Iso G 8s; stbd-hand mark. T12 Lt Buoy 53 02.27N/4 51.50E Iso R 8s; port-hand mark. Note: Buoys being renumbered.
OUDESCHILD. S mole head FR; Horn(2) 30s (sounded 0600-2300). N mole head FG. Oc 6s Lt seen between FR and FG leads into harbour. T13 Lt Buoy 53 03.45N/4 55.70E FlG 4s; stbd-hand mark.
DOOVE BALG. D4 53 02.7N/5 04.1E Iso R 8s; R pile; Ra refl. D3A/J2 53 02.1N/5 07.0E Fl(2+1)G 12s; G post, R band. D14 53 02.5N/5 09.2E Iso R 4s; R pile; Ra refl. D11 53 02.9N/5 12.0E Iso G 8s; G pile; Ra refl.

KORNWERDERZAND. West side, 53 04.0N/5 17.6E Iso R 4s 9m 4M; grey pedestal, R lantern; vis 049°-229°. Buitenhaven, West mole head FG; Horn Mo(N) 30s. E Mole head FR. Spuihaven Noord, W mole head LFlG 10s.
BOONTJES. BO11/K2/2 53 05.0N/5 20.3E Q; N cardinal mark. BO15 53 05.7N/5 22.0E FlG 2s. BO21 53 06.8N/5 22.3E Iso G 4s; G pile. BO22 53 06.8N/5 22.2E Iso R 4s; R pile. BO27 53 07.8N/5 22.7E Iso G 8s; G pile. BO28 53 07.9N/5 22.6E Iso R 8s; R pile. BO33 53 08.9N/5 23.2E Iso G 2s; G pile. BO34 53 08.9N/5 23.0E Iso R 2s; R pile. BO39 53 10.0N/5 23.4E Iso G 4s; G pile. BO40 53 10.0N/5 23.3E Iso R 4s; R pile.

Eierland, N point of Texel 53 10.97N/4 51.40E Fl(2) 10s 52m **29M**; R Tr; RC. Tide gauge 53 11.33N/4 48.05E Fl(5)Y 20s.
Off Vlieland TSS. VL Center Lanby 53 27.00N/4 40.00E Fl 5s; Racon; Horn(2) 30s.
Vlieland 53 17.8N/5 03.6E Iso 4s 53m **20M**; brown Tr; RC.

TERSCHELLING
VSM Lt Buoy 53 19.05N/4 55.73E Iso 4s; safe water mark.
TG Lt Buoy 53 24.22N/5 02.40E Q(9) 15s; W cardinal mark.
VNG Lt Buoy 53 24.82N/5 11.41E Iso 4s; safe water mark.
ZUIDER STORTEMELK (selected marks). ZS-bank Lt Buoy 53 18.73N/4 57.90E VQ; N cardinal mark.
ZS1 Lt Buoy 53 18.75N/4 59.56E FlG 4s; stbd-hand mark.
ZS2 Lt Buoy 53 19.03N/4 59.70E FlR 4s; port-hand mark.
ZS5 Lt Buoy 53 18.66N/5 01.68E LFlG 7s; stbd-hand mark.
ZS6 Lt Buoy 53 18.92N/5 01.68E LFlR 7s; port-hand mark.
ZS9 Lt Buoy 53 18.65N/5 03.70E FlG 4s; stbd-hand mark.
ZS10 Lt Buoy 53 18.85N/5 03.70E FlR 4s; port-hand mark.
ZS13/VS2 Lt Buoy 53 18.80N/5 05.93E Fl(2+1)G 12s; preferred channel to port.
ZS14 Lt Buoy 53 19.00N/5 05.55E LFlR 7s; port-hand mark.
ZS15 Lt Buoy 53 18.97N/5 07.10E LFlG 5s; stbd-hand mark.
ZS18 Lt Buoy 53 19.36N/5 06.95E LFlR 5s; port-hand mark.

TERSCHELLING. **Brandaris Tr** Fl 5s 55m **29M**; Y square Tr; vis except where obsc by dunes on Vlieland and Terschelling. Ldg Lts 053°, West harbour mole head, front 53 21.3N/5 13.1E F WR 5m W8M, R5M; R post, W bands, vis W049°-055°, R055°-252°, W252°-263°, R263°-049°; Horn 15s. **Rear**, on dyke, 1.1M from front Iso 5s 14m **19M**; vis 045°-061°, intens 045°-052°. E pier head FG 5m 4M.

WADDENZEE
VLIESTROOM (selected marks). VL1 Lt Buoy 53 19.00N/5 08.80E QG; stbd-hand mark. VL2/SG1 Lt Buoy 53 19.30N/5 09.80E Fl(2+1)R 12s; preferred channel to stbd.
VL5 Lt Buoy 53 18.60N/5 09.55E LFlG 8s; stbd-hand mark.
VL6 Lt Buoy 53 18.60N/5 11.00E LFlR 8s; port-hand mark.
VL9 Lt Buoy 53 17.65N/5 10.10E IsoG 4s; stbd-hand mark.
VL10/WM1 Lt Buoy 53 17.20N/5 11.26E VQ(9) 10s; W cardinal mark. VL14 Lt Buoy 53 15.90N/5 10.72E LFlR 8s; port-hand mark. VL15 Lt Buoy 53 16.05N/5 09.80E LFlG 8s; stbd-hand mark.
BLAUWE SLENK (selected marks). BS1/IN2 Lt Buoy 53 14.74N/5 10.05E Fl(2+1)G 12s; preferred channel to port.
BS2 Lt Buoy 53 14.70N/5 10.20E QR; port-hand mark.
BS3 Lt Buoy 53 14.40N/5 10.20E LFlG 5s; stbd-hand mark.
BS4 Lt Buoy 53 14.50N/5 10.40E LFlR 5s; port-hand mark.
BS7 Lt Buoy 53 13.85N/5 11.45E LFlG 8s; stbd-hand mark.
BS8 Lt Buoy 53 14.05N/5 11.54E LFlR 8s; port-hand mark.
BS11 Lt Buoy 53 13.63N/5 13.30E IsoG 4s; stbd-hand mark.
BS12 Lt Buoy 53 13.80N/5 13.30E IsoR 4s; port-hand mark.
BS19 Lt Buoy 53 13.40N/5 17.00E QG; stbd-hand mark.
BS20 Lt Buoy 53 13.55N/5 17.10E QR; port-hand mark.
BS27 Lt Buoy 53 11.95N/5 18.25E QG; stbd-hand mark.
BS28 Lt Buoy 53 12.10N/5 18.42E QR; port-hand mark.
BS31 Beacon 53 11.60N/5 19.70E LFlG 8s; G pile, stbd-hand mark.
BS32 Lt Buoy 53 11.72N/5 19.75E LFlR 8s; port-hand mark.
Pollendam. Stbd-hand marks: P1 FlG 2s; P3 IsoG 4s; P5 FlG 2s. Port-hand marks: P2 FlR 2s; P4 IsoR 4s; P6 FlR 2s; BS54 Beacon 53 10.80N/5 23.50E VQR; R pile, port-hand mark.

HARLINGEN. Ldg Lts 112°. Front Iso 4s 8m 4M. Rear Oc 6s 19m 14M; both on B masts, W bands; vis 097°-127° (H24). N mole IsoR 5s 8m 4M. S mole; Horn(3) 30s.

Ameland, West end 53 27.02N/5 37.60E Fl(3) 15s 57m **30M**; brown Tr, W bands; RC.

Schiermonnikoog 53 29.20N/6 08.90E Fl(4) 20s 43m **28M**; round Tr, dark R Tr. F WR 28m **W15M**, R12M (same Tr); vis W210°-221°, R221°-230°.

(For Die Ems see 10.20.4).

10.19.5 PASSAGE INFORMATION

CROSSING SOUTHERN NORTH SEA

From Crouch, Blackwater or Orwell the first stage is to make Long Sand Head Lt Buoy — from Crouch via Whitaker Chan and East Swin; from Blackwater via the Wallet and N of Gunfleet Bn to Wallet No 2 Lt Buoy; and from Orwell via Roughs Tower and Sunk Lt F. From Long Sand Head Lt Buoy proceed to Gallop Lt Buoy and thence to W Hinder Lt V, crossing traffic scheme at right angles. Care must be taken throughout with tidal streams, which may be setting across the yacht's track. The area is relatively shallow, and in bad weather seas are steep and short.

BELGIUM

Features of this coast (chart 1872) are the long shoals lying roughly parallel to it. Mostly the deeper, buoyed chans run within 3M of shore, where the outer shoals give some protection in strong onshore winds. Approaching from seaward it is essential to fix position from one of the many marks, so that the required chan is identified before shoal water is reached. Shipping is another hazard, but it helps to identify the main routes.

From the W, the natural entry to the buoyed chans is at Dunkerque Lanby: from the Thames, bound for Oostende (10.19.9) or the Schelde (10.19.14), identify W Hinder Lt V, or the N Hinder Lt V if bound from the N. For traffic schemes see Fig. 2(2). *North Sea Harbours and Pilotage* (Adlard Coles Ltd) is recommended for the coast of Belgium and of the Netherlands to Den Helder.

Off the Belgian coast the E-going stream begins at HW Vlissingen −0320 (HW Dover −0120), and the W-going at HW Vlissingen +0240 (HW Dover +0440), sp rates 2 kn. Mostly the streams run parallel with the coast. Nieuwpoort lies 8M from the French border. From the W (Dunkerque), approach through Passe de Zuydcoote (buoyed with least depth 3.3m) and West Diep. From ENE approach through Kleine Rede, the inner road off Oostende which carries a depth of 6m. There are other approaches through the channels and over the banks offshore, but they need care in bad weather.

Off Oostende the E-going stream begins at HW Vlissingen −0245 (HW Dover −0045), sp rate 2½ kn; the W-going stream begins at HW Vlissingen +0245 (HW Dover +0445), sp rate 1½ kn. So sailing E from Oostende, leave about HW Vlissingen −0300 to catch the E-going stream. If bound for Blankenberge (10.19.10) it is only necessary to keep a mile or two offshore, but if heading E of Zeebrugge (10.19.11) it is advisable to clear the new harbour extension by a mile or more. The main route to Zeebrugge for commercial shipping is through Scheur (the deep water channel of the Westerschelde) as far as Scheur-Zand Lt Buoy, about 3M NW of harbour entrance. There is much commercial traffic, and yachts should keep clear (S of) the buoyed channel so far as possible. Beware strong tidal stream and possibly dangerous seas in approaches to Zeebrugge.

NETHERLANDS

The main approach chans to Westerschelde (chart 325) are Wielingen and Oostgat, but yachts are required to keep clear of these. Coming from Zeebrugge keep close to S side of estuary until past Breskens (10.19.12) when, if proceeding to Vlissingen (10.19.16), cross close W of buoy H-SS. From N, use Deurloo/Spleet chans to S side of estuary. The tide runs hard in the estuary, causing a bad sea in chans and overfalls on some banks in strong winds. Vessels under 20m must give way to larger craft; and yachts under 12m are requested to stay clear of main buoyed chan, and any buoyed chan between Walsoorden and Antwerpen (10.19.15) if navigation permits.

The final stages of the barrier across the Oosterschelde (chart 192) are now completed, and entry must be made through Roompotsluis in the S part of the barrage. Several banks (e.g. Schaar, Schouwenbank, Middelbank, Steenbank) lie in the W approaches to Oosterschelde, and the main channel runs through Westgat and Oude Roompot, which are well marked. Oude Roompot would also be used if approaching from the N.

Coming from the S, Oostgat runs close to the Walcheren shore. Westkapelle Lt Ho is conspic near the W end of Walcheren. There are two lesser Lts nearby — Molenhoofd ½M WSW and Noorderhoofd ¾M NNW. Having passed the latter the coast runs NE past Domburg but becomes shallower as Roompot is approached, so it is necessary to keep near the Roompot channel which here is marked by buoys (unlit). It is important to have updated information on the buoyage and the channels.

From here to Hoek van Holland (10.19.20) banks extend 7M W of Schouwen and WSW of Goeree. Shipping is very concentrated at the entrance to Europoort and Rotterdam (10.19.19). Traffic schemes must be noted and regulations for yachts (see 10.19.20) obeyed.

The coast N to Den Helder (10.19.25) is low, and not easily visible from seaward, like most of the Dutch coast. Conspic landmarks include Noordwijk aan Zee Lt, chys of steelworks N of IJmuiden, Egmond aan Zee Lt, and chys of nuclear power station 1½M NNE of Petten. For traffic schemes see Figs. 2(3)-2(5). 3M W of IJmuiden (10.19.22) the N-going stream begins at HW Hoek van Holland −0120, and the S-going at HW Hoek van Holland +0430, sp rates about 1½ kn. Off entrance to IJmuiden the stream turns about 1h earlier and is stronger, and in heavy weather there may be a dangerous sea.

FRISIAN ISLANDS (charts 2593, 3761)

N from Den Helder (10.19.25), and then running E for nearly 150M along the N coasts of Netherlands and Federal Republic of Germany, is the chain of Frisian Islands. The Dutch islands are in general larger and further offshore, but the great majority of Frisian Islands have similar characteristics — being low, long and narrow, with the major axis parallel to the coast.

Between the islands, narrow chans (zeegat in Dutch, seegat in German) give access to/from the North Sea. Most of these chans are shallow for at least part of their length, and in these shoal areas a dangerous sea builds up in strong winds between W and N on the outgoing (ebb) tide. In onshore winds there are occasions when safe entry is possible on the flood tide, but departure is dangerous.

The flood stream along this coast is E-going, so it starts to run in through the zeegaten progressively from W to E. Where the tide meets behind each island, as it flows in first at the W end and a little later at the E end, is formed a bank called a wad (Dutch) or watt (German). These banks between the islands and the coast are major obstacles to E/W progress inside the islands. The chans are narrow and winding, marked by buoys or by withies in the shallower parts, and they mostly dry — so that it is essential to time the tide correctly.

This is an area most suited to shallow-draught yachts, particularly with bilge keels or centreboards, that can take the ground easily. While the main zeegaten are described very briefly below, no attempt has been made to mention the many channels inside the islands. Reference should be made to *Frisian Pilot* (Stanford Maritime).

While British Admiralty charts are adequate for through passages, coastal navigation, and entry to the main ports, foreign charts are essential for any yacht exploring the cruising grounds along this coast or using the smaller harbours. In non-tidal waters chart datum usually refers to the level at which the water is kept. In the Netherlands this may be Kanaalpeil, which in turn is related to Normaal Amsterdams Peil (NAP), which is about Mean Sea Level (MSL). In the North Sea shoal waters are liable to change due to gales and tidal streams. Changes in sea level due to special meteorological conditions may also be encountered.

AREA 19—Belgium and the Netherlands

It should be realised that there are numerous wrecks and obstructions which lie offshore and in coastal areas. Some of these are marked, but many are not. The Texel traffic scheme extends NNE from the Texel Lt V — see Fig.2(5): the separation zone incorporates the Helder gas field. Some 20–30M N lie the Placid and Petroland fields with production platforms. For general notes on N Sea oil and gas installations see 10.5.5.

Zeegat van Texel (chart 191) lies between Den Helder and the island of Texel, and gives access to the Waddenzee, the tidal part of the former Zuider Zee. Haaksgronden shoals extend 5M seaward, with three chans: Schulpengat on S side, leading into Breewijd; Westgat through centre of shoals, where the stream sets across the chan but suitable for passage in good weather and in daylight; and Molengat near the Texel shore. Schulpengat is the main chan, and is well marked, buoys being prefixed with letter 'S'; but strong SW winds cause rough sea against the SW-going (ebb) stream which begins at HW Helgoland −0330, while the NE-going (flood) stream begins at HW Helgoland +0325, sp rates 1½ kn. In Molengat the NW-going (ebb) stream begins at HW Helgoland −0145, and the SE-going (flood) stream at HW Helgoland +0425, sp rates 1¼ kn. Molengat is marked by buoys prefixed by letters 'MG', but strong winds between W and N cause a bad sea, and in such conditions Schulpengat should be used.

E of Zeegat van Texel the flood stream runs in three main directions: to E and SE through Malzwin, Wierbalg and Den Oever (where there is a lock into IJsselmeer), NE along the Afsluitdijk, and then N towards Harlingen (10.19.26); to NE and E through Texelstroom, Doove Balg towards the Pollen flats; and to NE through Texelstroom,. Scheurrak, Omdraai and Oude Vlie, where it meets the flood stream from Zeegat van Terschelling. The ebb stream runs the other way. The locks at Kornwerderzand, near NE end of Afsluitdijk, also give access to the IJsselmeer (10.19.24). For Oudeschild see 10.19.30.

Eierlandsche Gat, between Texel and Vlieland, is a dangerous unmarked chan, only used by fishermen.

Zeegat van Terschelling (chart 112) is the chan between Vlieland (10.19.27) and Terschelling, leading to the harbs of Oost Vlieland, West Terschelling (10.19.28) and Harlingen (10.19.26), and also to the locks at Kornwerderzand. Terschellinger Gronden are shallow banks extending more than 3M seaward; the main chan (buoyed) through them is Zuider Stortemelk passing close N of Vlieland. In this chan the buoys are prefixed by letters 'ZS', and the E-going (flood) stream begins at HW Helgoland +0325, while the W-going (ebb) stream begins at HW Helgoland −0230, sp rates 2½ kn. Zuider Stortemelk leads into Vliesloot, and thence to Oost Vlieland where there is a yacht harbour, and on to Vliestroom. Vliestroom is a deep, well buoyed chan (buoy numbers prefixed by letters 'VL') which runs S about 4M until its junction with Blauwe Slenk and Inschot. The latter leads to Kornwerderzand and the IJsselmeer, while Blauwe Slenk runs SE to Harlingen. At the N end of Vliestroom, Schuitengat is a buoyed chan leading to West Terschelling.

Other chans through Terschellinger Gronden are, from W to E: Stortemelk, Thomas Smit Gat and Noordgat. The first two are unmarked, and are dangerous with an ebb tide and strong wind from W or N.

Zeegat van Ameland is the chan between Terschelling and Ameland, fronted by the sandbank of Bornrif extending 3M seaward. Westgat is the main entrance, with buoys prefixed by letters 'WG'. In Westgat the flood stream begins at HW Helgoland +0425, and the ebb stream at HW Helgoland −0150, sp rates 2 kn. The chan runs close to the Terschelling shore: a dangerous sea develops in strong onshore winds. For Ameland see 10.19.30.

Between Ameland and Schiermonnikoog (10.19.30) is Friesche Zeegat, with the main chan also called Westgat and buoys also marked 'WG'. In strong winds the sea breaks across the whole passage. Westgat leads SE through Wierumer Gronden, and then past Englesmanplaat (a prominent sandbank in mid-channel) into Zoutkamperlaag which is the main chan (marked by buoys prefixed 'Z') to Lauwersoog (10.19.30), where locks give access to the inland waterways.

Going E, the next main chan through the Is is Westerems (chart 3509), which runs SW of the German island of Borkum (10.20.8) and leads to Delfzijl (10.19.29) and Emden (10.20.21). Hubertgat, which runs S of the main Westerems chan, is more useful when bound to or from the W, but in both these chans there is a dangerous sea on the ebb stream in strong NW winds. The E-going (flood) stream begins at HW Helgoland +0530, and the W-going (ebb) stream begins at HW Helgoland −0030, sp rates 1½ kn.

10.19.6 DISTANCE TABLE

Approximate distances in nautical miles are by the most direct route while avoiding dangers and allowing for traffic separation schemes etc. Places in *italics* are in adjoining areas.
Note: In this table all distances along and between the Frisian Islands are for routes passing north of same.

	1	2	3	4	5	6	7	8	9	10	11	12	13	14	15	16	17	18	19	20
1 *North Foreland*	**1**																			
2 *Dunkerque*	40	**2**																		
3 *Burnham-on-Crouch*	36	75	**3**																	
4 *Great Yarmouth*	77	101	76	**4**																
5 *Grimsby*	169	194	168	94	**5**															
6 Oostende	58	26	91	93	194	**6**														
7 Breskens	80	54	103	101	191	28	**7**													
8 Vlissingen	82	56	104	101	191	30	3	**8**												
9 Goeree Tower	91	73	108	83	169	51	40	40	**9**											
10 IJmuiden	137	119	151	102	182	97	86	86	46	**10**										
11 *Amsterdam*	150	132	164	115	195	110	99	99	59	13	**11**									
12 Den Helder	171	153	169	115	180	131	120	120	80	38	51	**12**								
13 West Terschelling	187	185	196	135	193	163	152	152	112	70	83	39	**13**							
14 Westerems Lt Buoy	226	224	235	174	232	202	191	191	151	109	122	78	47	**14**						
15 Delzijl	263	261	272	211	269	239	228	228	188	146	159	115	84	33	**15**					
16 *Norderney*	259	257	268	207	265	235	224	224	184	142	155	111	80	34	41	**16**				
17 *Wangerooge*	283	281	292	231	289	259	248	248	208	166	179	135	104	58	65	29	**17**			
18 *Wilhelmshaven*	307	305	316	255	313	283	272	272	232	190	203	159	128	82	89	53	27	**18**		
19 *Helgoland*	300	298	309	241	285	276	265	265	225	183	196	152	121	68	81	44	24	43	**19**	
20 *Brunsbüttel*	337	336	342	287	332	314	303	303	263	221	234	190	159	113	120	84	55	70	51	**20**

SPECIAL FACTORS AFFECTING BELGIUM AND THE NETHERLANDS 10-19-7

BELGIUM
TIME ZONE is −0100, which is allowed for in tidal predictions but no provision is made for daylight saving schemes which are indicated by the non-shaded areas on the tide tables (see 9.1.2).

SIGNALS International Port Traffic Signals (see page 122) are shown at Nieuwpoort, Oostende and Zeebrugge. Small craft warnings (shown when wind from seaward is Force 4 or over) remain in force at Nieuwpoort, Oostende, Blankenberge and Zeebrugge as follows: −
By day: two black cones, points together
By night: blue flashing light.

CHARTS The chart most widely used is the 'Vlaamse Banken' issued by the Hydrografische Dienst der Kust. Admiralty, Stanford and Imray chart numbers are quoted.

PROVINCES In place of 'counties' in the UK, Provinces are given for Belgian ports.

MARINE RESCUE CO-ORDINATION CENTRES Sea Rescue Co-ordination Centre, Oostende, Tel. 70.10.00, 70.11.00, 70.77.01 or 70.77.02. Coast station, Oostende Radio Tel. 70.24.38.
In emergency Tel. 900, or call *Oostende Radio* VHF Ch 16. For medical advice call *Radiomédical Oostende* on Ch 16.

PUBLIC HOLIDAYS New Year's Day, Easter Monday, Labour Day (1 May), Ascension Day, Whit Monday, National Day (21 July), Feast of the Assumption (15 August), All Saints' Day (1 November), Armistice Day (11 November), King's Birthday or Fete de la Dynastie (15 November), Christmas Day.

TELEPHONE To call UK from Belgium, dial 00 (pause for new tone to be heard) 44. Then dial the full UK area code but omitting the prefix 0, followed by the number required. To call Belgium from UK, dial 010-32 followed by the code and number.

Emergencies Police − dial 101
Fire − dial 100

NOTE 1: Although the Hr Mr's tel no is given for Belgian ports, he is not the key figure for yachtsmen that he is in British ports. Berths and moorings are always administered by the local yacht clubs. British yachtsmen are very welcome in Belgian ports and YCs.

NOTE 2: Belgian harbour police are very strict about yachts using their engines entering harbour. If sails are used as well, display a black cone. Yachts may not navigate within 200m of shore (MLWS).

NOTE 3: Ports of entry are Nieuwpoort, Oostende and Zeebrugge.

NOTE 4: Further information can be obtained from The Belgian National Tourist Office, 38 Dover St., London, W1X 3RB, Tel. 01 499 5379.

NETHERLANDS
TIME ZONE is −0100, which is allowed for in tidal predictions but no provision is made for daylight saving schemes which are indicated by the non-shaded areas (see 9.1.2).

SIGNALS Traffic signals The standard French/Belgian system is not used in the Netherlands. Where possible the local system is given.

Sluicing signals In the Netherlands the following sluicing signals may be shown:
By day: A blue board, with the word 'SPUIEN' on it, often in addition to the night signal, of three red lights in a triangle, point up.

Visual storm signals Light signals only, shown day and night, in accordance with the International System (see 10.14.7) are shown at the following harbours: Vlissingen, Hoek van Holland, Amsterdam, IJmuiden, Den Helder, West Terschelling, Texel Lt V, Harlingen, Eierland, Ameland, Oostmahorn, Schiermonnikoog, Zoutkamp and Delfzijl.

Inland waterways
Bridge signals
Y − You may pass under this arch
R each side − Bridge closed (opens on request)
R one side
G other side } Bridge about to open
G each side − Bridge open
2 R (vert) − Bridge out of use
2 G (vert)
each side − Bridge open but not in use (you may pass)
To request bridges to open sound 'long, short, long'.

Railway bridges
Opening times of railway bridges in the Netherlands are given in a leaflet 'Openingstijden Spoorwegbruggen' published annually by ANWB and available free (send A5 self addressed envelope with international reply coupon) from Aan de Chef der Hydrografie, Postbus 90704, 2509 LS 's-Gravenhage.

Buoyage
In certain inland waters, including the whole of the IJsselmeer, a buoyage system known as SIGNI is used. The main features are:
(1) Channel buoyage as for IALA (Region A).
(2) Supplementary port and starboard hand buoys may be red and white or green and white respectively.
(3) At division of channel, a spherical buoy with following characteristics:
 a. Channels of equal importance − red and green bands; topmark red and green sphere;
 b. Main channel to port − green above red; topmark green cone or green cone above a green sphere;
 c. Main channel to starboard − red above green; topmark red can or red can above a red sphere.
Offshore buoys are often marked with abbreviations of the banks or channels which they mark (e.g. ZS − Zuider Stortemelk). A separation buoy has the abbreviations of both channels meeting there.
Some channels are marked with perches, starboard hand bound, port hand unbound. On tidal flats (e.g. Friesland) where the direction of main flood stream is uncertain, bound perches are on the S side of a channel and unbound on the N side.
In minor channels the buoyage may be moved without notice to accommodate changes.

CHARTS Chart numbers are printed in the following order: − Admiralty; Zeekaarten (equivalent to Admiralty, issued by the Hydrographer, Royal Netherlands Navy and up-dated by Dutch Notices to Mariners); Dutch Yacht Charts (Kaarten voor Zeil en Motorjachten, also issued by the Hydrographer of the Royal Netherlands Navy. A new edition comes out each year in March and is updated in Dutch Notices to Mariners); the letters of the ANWB Waterkaarten are quoted, lettered from A in the north to O in the South; Stanford (where applicable); Imray (where applicable). Visiting yachts are required to carry the ANWB publication '*Almanak voor Watertoerisme*'.

PROVINCES In place of 'counties' in the UK, Provinces are given for Netherlands ports.

MARINE RESCUE CO-ORDINATION CENTRES Coast station Scheveningen Radio Tel. 550.19104. In emergency call *Scheveningen Radio* VHF Ch 16. For medical advice call *Radiomédical Scheveningen* on Ch 16.

PUBLIC HOLIDAYS New Year's Day, Easter Monday, Queen's Birthday (30 April), Liberation Day (5 May), Ascension Day, Whit Monday, Christmas Day and Boxing Day.

TELEPHONE To call UK from the Netherlands, dial 09 (pause for new tone to be heard) 44. Then dial the full UK area code but omitting the prefix 0, followed by the number required.
To call the Netherlands from UK dial 010-31 followed by the area code and number.

Emergencies − Police dial 101; Ambulance and Fire 100.

CUSTOMS Main customs ports are Breskens, Vlissingen, Roompotsluis, Hoek van Holland, Scheveningen, IJmuiden, Den Helder, Harlingen, Oost Vlieland, West Terschelling, Lauwersoog and Delfzijl. No entry/customs facilities at Stellendam.

Note Further information, including a useful publication '*Watersports Paradise*' can be obtained from the Netherlands Board of Tourism, 25 Buckingham Gate, London, SW1E 6LD, Tel. (01) 630 0451.

AREA 19—Belgium   737

NIEUWPOORT (NIEUPORT)   10-19-8
West Flanders

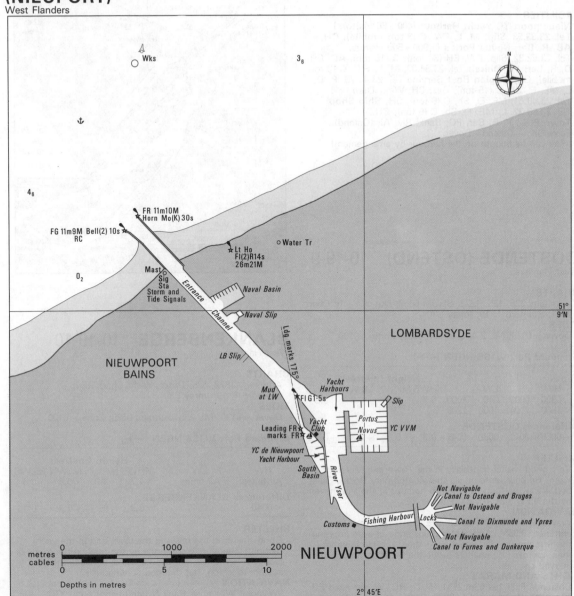

CHARTS
Admiralty 125, 1872; Belgian D102; SHOM 6652; Stanford 1, 19; Imray C30

TIDES
Dover +0105; ML 2.4; Duration 0515; Zone −0100

Standard Port VLISSINGEN (→)

Times				Height (metres)			
HW		LW		MHWS	MHWN	MLWN	MLWS
0300	0900	0400	1000	4.7	3.9	0.8	0.3
1500	2100	1600	2200				

Differences NIEUWPOORT
−0110 −0050 −0035 −0045 +0.6 +0.4 +0.4 +0.1

SHELTER
Shelter is good except for strong winds from the NW. There are two yacht harbours, both with good shelter. The YC of Nieuwpoort gets very full but there is always ample room in the Portus Novus.

NAVIGATION
Waypoint 51°10′.00N 02°42′.00E, 308°/128° from/to entrance, 0.90M. The bar is liable to silt up but there is usually sufficient water for yachts. Entrance and channels to both yacht harbours are dredged, although channel to both yacht harbours can drop to about 2m. At springs there is a strong stream across the entrance. The firing range E of Nieuwpoort is not used mid-June to end of September. At other times, contact range officer on Ch 67.

LIGHTS AND MARKS
Near root, E pier Fl(2)R 14s 26m 21M; R Tr, W bands. E pier, head FR 11m 10M; W Tr; vis 025°-250°, 307°-347°; Horn Mo(K) 30s. W pier, head FG 11m 9M; W Tr; vis 025°-250°, 284°-324°; Bell(2) 10s; RC. International Port Traffic Signals from root of W pier, with addition of:
⊻ or Fl Bu Lt   Departure prohibited for craft under 6m length.

RADIO TELEPHONE
VHF Ch 09 19 16 (H24)

TELEPHONE (51 or 58)
Hr Mr 23.30.00; CG 23.30.45; Sea Rescue Helicopter 31.17.14; Customs 23.34.51; Duty Free Store 23.34.33; Hosp (Oostende) 70.76.31; Dr 23.30.89; Brit Consul (02) 217.90.00.

NIEUWPOORT continued

FACILITIES
Nieuwpoort YC Yacht Harbour (400+80 visitors) Tel. 23.33.53, Slip, M, L, FW, C (3 ton mobile), CH, AB, R, Bar; **Novus Portus** (1,900+500 visitors) Tel. 23.52.32, Slip, FW, BH (30 ton), C (15 ton), AC, CH, D, R, Bar; **YC Militair** Tel. 23.34.33, M, L, FW, C (2 ton mobile), CH; **Belgian Boat Service** Tel. 23.44.73, P, D, L, ME, El, Sh, C (15 ton), Gaz, CH; **West Diep** Tel. 23.40.61, ME, El, Sh, C (6 ton), CH; **Ship Shop** Tel. 23.50.32, L, ME, El, Sh, C (6 ton), CH.
Town P, D, V, R, Bar. PO; Bank; Rly; Air (Ostend).
Ferry UK — Ostend—Dover.
(Fuel can be bought on the harbour by arrangement).

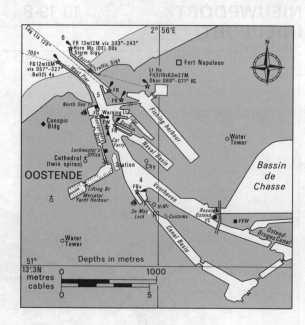

OOSTENDE (OSTEND)   10-19-9
West Flanders

CHARTS
Admiralty 325, 125, 1872; SHOM 6652; Dutch Yacht Chart 1801; Stanford 1, 19; Imray C30
TIDES
Dover +0110; ML 2.4; Duration 0530; Zone −0100

Standard Port VLISSINGEN (→)

Times				Height (metres)			
HW		LW		MHWS	MHWN	MLWN	MLWS
0300	0900	0400	1000	4.7	3.9	0.8	0.3
1500	2100	1600	2200				

Differences OOSTENDE
−0055 −0040 −0030 −0045 +0.3 +0.3 +0.3 +0.1

SHELTER
Very good shelter especially in the Town and Mercator Yacht Harbour (entered via the Montgomery Dock). Quiet berths may be found further up harbour in the Voorhaven by the VVW and Royal Oostende YCs.
NAVIGATION
Waypoint 51°15′.00N 02°53′.97E, 308°/128° from/to entrance, 0.98M. Approaching Oostende beware crossing the Stroombank and avoid altogether in rough weather; then approach by channel inside the bank or through the buoyed channel.
LIGHTS AND MARKS
Oostende Fl(3) 10s 63m 27M; W Tr; RC. W pier head FG 12m 10M; W Tr; vis 057°-327°; Bell(1) 4s. E pier head FR 13m 12M; W Tr; vis 333°-243°; Horn Mo(OE) 30s. Ldg Lts 128° both FR 12/18m 4M; both X on W pylons, R bands; vis 051°-201°.
International Port Traffic Signals are shown from head of E pier, with addition of following signals from S side of entrance to Montgomery dock:—

⊥ or Fl Bu Lt — Departure of craft under 6m length prohibited.

RADIO TELEPHONE
VHF Ch 09 16 (H24). Mercator Marina Ch 14 (H24).
TELEPHONE (59)
Hr Mr 70.57.62; Life Saving Service 70.11.00; Customs 70.20.09; Hosp 70.76.37; Brit Consul (02) 217.90.00.
FACILITIES
North Sea YC Tel. 70.27.54, L, FW, AB, R, Bar; **Royal YC of Oostende** (160+40 visitors) Tel. 70.36.07, Slip, P and D (cans), FW, AC, AB, R, Bar; **Mercator Yacht Harbour** (450+50 visitors) Tel. 70.57.62, FW, AC, P, D, C, Slip; **Montgomery Dock** (70+50 visitors) YC, Bar, R, FW, Slip; **Compas** Tel. 70.00.57, ME, CH; **Maritime Bureau Hindery CKT** Tel. 70.72.61, ME, Sh, CH; **N end of Fishing harbour** D; **North Sea Marine** Tel. 32.06.88, ME, El, Sh; **Marina Yachting Centre** Tel. 32.00.28, CH.
Town P and D (cans), CH, V, R, Bar. PO; Bank; Rly; Air.
Ferry UK—Dover.

BLANKENBERGE   10-19-10
West Flanders

CHARTS
Admiralty 1872, 325; Dutch Yacht Chart 1801; Stanford 1, 19; Imray C30
TIDES
Dover +0130; ML 2.5; Duration 0535; Zone −0100

Standard Port VLISSINGEN (→)

Times		Height (metres)			
HW	LW	MHWS	MHWN	MLWN	MLWS
All times	All times	4.7	3.9	0.8	0.3

Differences BLANKENBERGE
−0040   −0040   −0.3 −0.1 +0.3 +0.1

SHELTER
Good shelter in the basin in the town and in the yacht harbour. There are also berths by the floating pontoons or alongside the quay. Entrance channel between moles subject to silting, but dredged continuously.
NAVIGATION
Waypoint Wenduine Bank N (stbd-hand) buoy, QG, 51°21′.55N 03°02′.67E, 317°/137° from/to entrance, 3.6M. Entry should not be attempted when strong winds are from the NW. Beware strong tides across entrance.
LIGHTS AND MARKS
Ldg Lts 134°, Front (Red X on column) FR 3M, Rear (Red X on concrete post) FR, 3M, show the best water.
FS by Lt Ho shows

⊥ or Fl Bu Lt — Departure of craft under 6m length prohibited.

RADIO TELEPHONE
Private volunteer VHF service, Ch 08 (or relay via Ostend or Zeebrugge).
TELEPHONE (50)
Hr Mr 41.14.20; Customs Zeebrugge 54.42.23; Dr 33.36.68; Hosp 41.37.01; Brit Consul (02) 217.90.00.
FACILITIES
Marine Centre Tel. 41.35.60, ME, El, Sh, C (15 ton), CH; **Agemex** Tel. 41.49.38, Slip, ME, El, Sh, C (5 ton), CH; **Yacht Harbour** Slip, FW, C (10 ton); **Scarphout YC** Tel. 41.33.48, CH; **Wittervrongel Sails** Tel. 41.18.63, CH; **N.V. Internautic on Sea** Tel. 41.31.78, Slip, P, D, L, ME, El, Sh, C (22 ton), CH; **Marine Yachting Centre** Tel. 41.57.12, CH; **YC Vrije Noordzeezeilers (VNZ)** Tel. 41.64.58, R, Bar; **YC VVW** Tel. 41.75.36.
Town V, R, Gaz, Bar. PO; Bank; Rly; Air (Ostend).
Ferry UK — Zeebrugge—Dover/Hull/Felixstowe.

BLANKENBERGE continued

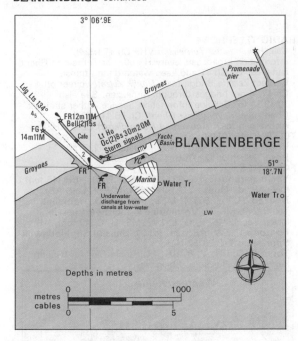

ZEEBRUGGE 10-19-11
West Flanders

CHARTS
Admiralty 325, 97; Zeekaart 1442; Dutch Yacht Chart 1803; Stanford 1, 19; Imray C30

TIDES
Dover +0110; ML 2.4; Duration 0535; Zone −0100

Standard Port VLISSINGEN (→)

Times				Height (metres)			
HW		LW		MHWS	MHWN	MLWN	MLWS
0300	0900	0400	1000	4.7	3.9	0.8	0.3
1500	2100	1600	2200				

Differences ZEEBRUGGE
−0035 −0015 −0020 −0035 +0.1 0.0 +0.3 +0.1

SHELTER
Shelter in Zeebrugge is very good and harbour is always available. Zeebrugge is the port of Brugge (Zee Brugge) and the canal up to Brugge (or Bruges) is 6 M long. The yacht harbour is very well protected except in strong NE winds. New marina is projected.

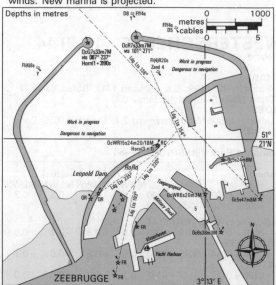

AREA 19—Belgium and the Netherlands

NAVIGATION
Waypoint Scheur-Zand E cardinal buoy, Q(3) 10s, 51°23′.70N 03°07′.68E, 309°/129° from/to entrance, 3.1M. There are no hazards approaching the harbour mouth except harbour works which are still in progress (1986), marked by buoys. Beware strong currents in harbour approaches (up to 4 kn at HW−1). Zeebrugge is the main fishing port of Belgium and is a ferry terminal, so keep clear of fishing boats and ferries. Call Harbour Control Ch 71 before entry.

LIGHTS AND MARKS
(1) Ldg Lts 136°. Oc 5s 8M, vis 131°-141° synchronised (H24).
(2) Ldg Lts 154°. Front Oc WR 6s. Rear Oc 6s synchronised (H24).
(3) Ldg Lts 235°. Both QR 5M, synchronised.
(4) Ldg Lts 220°, Front 2FW(vert) 30/22m. Rear FW 30m. Both W concrete columns, B bands.
(5) Ldg Lts 193°. Front 2FR(vert) 30/22m. Rear FR 29m. Both W concrete columns, R bands.
International Port Traffic Signals are shown at NE end of Leopold II Dam (old Zeebrugge mole). A QY Lt at S side of Visserhaven prohibits entry/departure to/from Visserhaven.

RADIO TELEPHONE
VHF 13 71 (H24).

TELEPHONE (50)
Hr Mr Bruges 54.42.68; Hr Mr 54.32.40; Port Control 54.40.07; CG 54.50.72; Sea Saving Service 54.40.07; Customs 54.42.23; Dr 54.45.90; Hosp 32.08.32; Brit Consul (02) 217.90.00.

FACILITIES
Yacht Harbour Slip, Sh, CH, D, AB, R, Bar; **Pontoon L**, FW; **Alberta (bar/restaurant of Royal Belgian SC)** Tel. 54.41.97, R, Bar; **Royal Belgian SC** Tel. 54.49.03, M, AB; **Brugge Marine** Tel. 35.33.75, ME, El; **Hennion Luc Yachting** Tel. 54.48.29, CH.
Town P, D, FW, ME, El, Sh, CH, Gaz, V, R, Bar. PO; Bank; Rly; Tram to Oostende; Air (Ostend).
Ferry UK—Dover/Hull/Folkestone.

BRESKENS 10-19-12
Zeeland

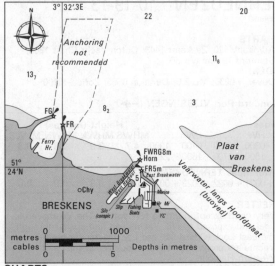

CHARTS
Admiralty 325; Zeekaart 1533; Dutch Yacht Chart 1801, 1803; Stanford 1,19; Imray C30

TIDES
Dover +0210; ML 2.5; Duration 0600; Zone −0100

Standard Port VLISSINGEN (→)

Times				Height (metres)			
HW		LW		MHWS	MHWN	MLWN	MLWS
0300	0900	0400	1000	4.7	3.9	0.8	0.3
1500	2100	1600	2200				

Differences BRESKENS
−0005 −0005 −0002 −0002 +0.1 0.0 0.0 0.0

BRESKENS continued

SHELTER
Good shelter in all winds. The North finger of the marina is for visiting yachts. The marina is operated by Oostburg municipality. Anchor off Plaat van Breskens, not in harbour.

NAVIGATION
Waypoint SS-VH N cardinal buoy, Q, 51°24'.75N 03°34'.00E, 357°/177° from/to W breakwater Lt, 0.69M. Beware strong tides across the entrance. There is a shoal just to port after passing the entrance. Make sure the entrance is not confused with that of the ferry port 1000m WNW, where yachts are forbidden.

LIGHTS AND MARKS
West mole head F WRG 8m; vis R090°-128°, W128°-157°, R157°-172°, W172°-175°, G175°-296°, W296°-300°, R300°-008°, G008°-090°; Horn Mo(U) 30s. E mole head FR 5m. Nieuwe Sluis Lt Oc WRG 10s is 1.2M W of Ferry Hr.

RADIO TELEPHONE
VHF Ch 14.

TELEPHONE (1172)
Hr Mr 1902; Customs 2610; Dr 1566; Hosp (01170) 3355; Brit Consul (020) 76.43.43.

FACILITIES
Marina Tel. 1902, FW, Lau; **YC Breskens** Tel. 3278, R, Bar; **Neil Pryde** Tel. 2101, SM; **C. Kosten** Tel. 1257, CH, D, Gaz, P, chart agent; **Standfast** Tel. 1797, BY, C (15 ton), El, ME, Sh, Slip; **Proctor** Tel. 1397 masts; **Jachtwerf Delta** Tel. 2440, BY, C (18 ton), El, ME, Sh; **M D Meeusen** Tel. 1996, Sh; **Tronik Shop** Tel. 3031, El, Electronics; **Town** V, R, Bar, PO, Bank, Gas, Rly (Flushing), Air (Ghent or Brussels).
Ferry UK — Flushing—Sheerness.

TERNEUZEN 10-19-13
Zeeland

CHARTS
Admiralty 120; Zeekaart 1443; Dutch Yacht Chart 1803; Stanford 19; Imray C30

TIDES
Dover +0230; ML 2.6; Duration 0555; Zone −0100

Standard Port VLISSINGEN (→)

Times				Height (metres)			
HW		LW		MHWS	MHWN	MLWN	MLWS
0300	0900	0400	1000	4.7	3.9	0.8	0.3
1500	2100	1600	2200				

Differences TERNEUZEN
+0021 +0022 +0022 +0033 +0.3 +0.3 0.0 0.0

SHELTER
Very good except in strong NW winds. The Veerhaven is tidal and exposed to NE. Yachts can find shelter through the East Lock (Oostsluis); see Lockmaster for berth. Yachts are prohibited in West Lock (Westsluis) and Western Harbour.

NAVIGATION
Waypoint No 23A (stbd-hand) buoy, IsoG 4s, 51°21'.37N 03°45'.66E, 285°/105° from/to Veerhaven entrance, 2.6M. Westerschelde (see 10.19.14) is a mass of sand-banks but well marked. It is the waterway to Antwerpen and Ghent, very full of shipping and also of barges (which do not normally conform to the rules). The Ghent canal is 17M long with three bridges, minimum clearance 6.5m when closed; VHF Ch 11 is compulsory.

LIGHTS AND MARKS
For the Veerhaven, the water Tr to SE and the Oc WRG Lt on W mole are conspic. When entry prohibited, a R flag is shown at W mole head by day, or a second R Lt below FR on E mole by night.
Signals for Oostsluis: R Lts each side of lock — entry prohibited; G Lts each side — entry permitted.

RADIO TELEPHONE
Call: *Havendienst Terneuzen* VHF Ch 11 (H24). Information broadcasts every H+00. For Terneuzen-Ghent canal call on Ch 11 and keep watch during transit. Contact Zelzate Bridge (call: *Uitkijk Zelzate*) direct on Ch 11, other bridges through Terneuzen. See also 10.19.14 for Schelde Information Service. Other stations: Ghent (call *Havendienst Ghent*) Ch 05 11 (H24); Antwerpen (call: *Antwerpen Havendienst*) Ch 18 (H24).

TELEPHONE (01150)
Hr Mr 95551; CG 13017 (H24); Customs 12377; Hosp 12851; Dr 12200; Brit Consul (20) 76.43.43.

FACILITIES
Yacht harbour Tel. 96331, Slip, ME, El, Sh, BH (15 ton), FW, AB; **de Honte YC** Tel. 17633, L, FW, Lau, AB; **Neuzen YC** Tel. 96331, M, L, FW, AB; **D. Hamelink** Tel. 97240, ME, El; **Vermeulen's Jachtwerf** (inside locks) Tel. 12716, P, D, L, FW, ME, El, C (40 ton), AB; **Pontoon** L, FW, AB (see Hr Mr); **Sluiskil** Sh; **Flodewijk Sportshop** Tel. 96157 Gaz.
Town P, D, CH, V, R, Bar. PO; Bank; Rly; Air (Ghent).
Ferry UK — Flushing—Sheerness.

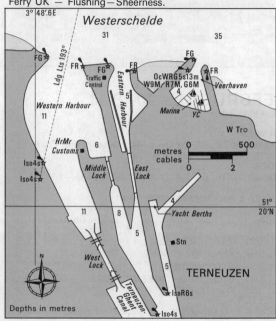

WESTERSCHELDE 10-19-14
Zeeland

CHARTS
Admiralty 120, 139, 325; Zeekaart 1443; Dutch Yacht Chart 1803; Stanford 19; Imray C30

TIDES
Dover +0200; ML (Hansweert) 2.7; Duration 0555; Zone −0100

Standard Port VLISSINGEN (→)

Times				Height (metres)			
HW		LW		MHWS	MHWN	MLWN	MLWS
0300	0900	0400	1000	4.7	3.9	0.8	0.3
1500	2100	1600	2200				

Differences WESTKAPELLE
−0024 −0014 −0012 −0023 −0.6 −0.5 −0.1 0.0
HANSWEERT
+0114 +0054 +0040 +0100 +0.5 +0.6 0.0 −0.1
BATH
+0126 +0117 +0117 +0144 +0.7 +0.9 0.0 0.0

WESTERSCHELDE continued

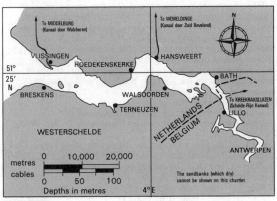

SHELTER
There are few harbours in the 38M from Terneuzen to Antwerpen. The following are possible for yachts:
HOEDEKENSKERKE — on N bank, disused ferry harbour (dries).
HANSWEERT — can be used temporarily, but it is the busy entrance to Zuid Beveland canal.
WALSOORDEN — on S bank with yacht harbour (Tijhaven, dries) on N side, and Diepe Haven on S side (depth 3m, but exposed to swell), both approached through same entrance.
ZANDVLIET — in Belgium, on N bank, is a commercial harbour; use only in emergency.
LILLO — customs post, for shallow-draught boats only.

NAVIGATION
Waypoints — see Breskens (10.19.12) and Terneuzen (10.19.13). Note traffic scheme off Vlissingen (see 10.19.5). The winding channel through the shifting banks is well marked, but full of commercial traffic. It is necessary to work the tide, which runs 2½ kn on average and more at springs. Yachts should keep at edge of main channel, and beware unpredictable actions of barges. Alternative channels must be used with caution, particularly going downstream on the ebb.

LIGHTS AND MARKS
The approaches to Westerschelde are well lit — on the S shore by Lt Trs at Kruishoofd and Nieuwe Sluis, and on the N shore at Westkapelle and Kaapduinen. Details of these and other important Lts are in 10.19.4. The main fairway is, for the most part, covered by the narrow white sectors of the various lights along the shore.

RADIO TELEPHONE
The following shore stations comprise the Schelde Information Service (all H24):
(1) Vlissingen.
A. Scheldemonding area, seaward of Kaapduinen and Kruishoofd Lts, call *Vlissingen Radio* VHF Ch 14.
B. Vlissingen area call *Post Vlissingen* Ch 21.
(2) Terneuzen area, call *Post Terneuzen* Ch **03** 14
(3) Hansweert area, call *Hansweert Radio* Ch 71
(4) Zandvliet area, call *Zandvliet Radio* Ch 12 14
Information broadcasts in Dutch and English every H+35 on Ch 12 by Zandvliet and on Ch 14 by Vlissingen, every H+55 on Ch 03 by Radar Terneuzen.
For other stations see harbour concerned.

FACILITIES
TERNEUZEN — see 10.19.13
HOEDEKENSKERKE (01193) — visitors berths, P, D, Gaz, FW; **W. Bek** Tel. 309 ME; **YC WV Hoedekenskerke** Tel. 259. PO, Bank, Rly (Goes), Air (Antwerpen).
HANSWEERT (01130) — visitors berths, ME, C(17 ton) R, P, D; **Ribens** Tel. 1371 CH; **Scheepswerf Reimerswaal** Tel. 3021, BY. PO, Bank, Rly (Goes), Air (Antwerpen)
WALSOORDEN — visitors berths; **Havenmeester** Tel. 1235, P, D, Gaz; **Werf Gebn De Klerk** Tel. 1614, BY. PO, Rly, Air (Antwerpen).

AGENTS NEEDED
There are a number of ports where we need agents, particularly in France.
ENGLAND   Swale, Havengore, Berwick.
SCOTLAND   Firth of Forth, Scrabster, Mallaig, Loch Sunart, Loch Aline.
IRELAND   Kilrush, Wicklow, Westport/Clew Bay.
FRANCE   Arcachon, Seudre R, Ile d'Oleron, Rochfort, Ile de Re, St. Giles-Croix-de-Vie, Ile d'Yeu, Pouliguen, Le Croisic, La Forêt, Ile de Bréhat.
GERMANY   Norderney, Dornumer-Accumersiel.
If you are interested in becoming our agent for any of the above, please write to the editors and get your free copy of the Almanac every year. You do not have to be resident in a port to be the agent, but at least a fairly regular visitor.

ANTWERPEN (ANTWERP) 10-19-15
Antwerpen

CHARTS
Admiralty 139; Stanford 19; Zeekaart 1443; Dutch Yacht Chart 1803;

TIDES
Dover +0334; ML 2.7; Duration 0605; Zone −0100

Standard Port VLISSINGEN (→)

Times				Height (metres)			
HW		LW		MHWS	MHWN	MLWN	MLWS
0300	0900	0400	1000	4.7	3.9	0.8	0.3
1500	2100	1600	2200				

Differences ANTWERPEN
+0128 +0116 +0121 +0144 +1.1 +0.9 0.0 0.0

SHELTER
Excellent in Imalso marina on W bank, 4ca (750m) SW of Kattendijksluis and ½M from city centre which can be reached through two tunnels. Access by lock HW∓1 (0600-2000). Lock closed in winter except by arrangement.

NAVIGATION
For Westerschelde see 10.19.12 and 10.19.13. On river bend before marina is a windmill, and G conical buoy Iso G 8s. Two unlit Y buoys N and S of lock can be used if entrance is closed.

LIGHTS AND MARKS
(marina signals)
R flag or R Lt   — Entrance prohibited
G flag or G Lt   — Departure prohibited
Bu cone/flag or Bu Lt — Approach channel closed
B ball or FW Lt   — Depth over sill 2.5−3.0m
2 B balls or 2 FW Lts — Depth over sill 3.0−3.5m
3 B balls or 3 FW Lts — Depth over sill 3.5−3.8m
Yachts should make following Int Code signals:
UH — I wish to enter marina under power
UP — I have an emergency; request priority entry
Z — Request tug assistance for entry
P — I wish to leave harbour

RADIO TELEPHONE
Call: *Antwerpen Havendienst* VHF Ch 18 (H24). Imalso Marina, (call *Bolleke*) Ch 72.

TELEPHONE (3)
Marina Lock Keeper 219.08.95; Customs 234.08.40; Hosp 217.71.11.

FACILITIES
Imalso Marina Tel. 219.08.95, FW, P, D, R, V, Gaz, El, Sh, AC, C (1.5 ton), Slip; **Royal YC van België** Tel. 219.27.84 (Secretariat), Tel. 219.26.82 (Clubhouse); **Kon. Liberty YC** Tel. 219.11.47; **Martin & Co** Tel. 7994, ACA; **Bogerd Navtec** Tel. 8476 ACA; **Landtmeters** Tel. 233.31.31; **City** All facilities including PO, Bank, Rly, Air, Ferry UK — Flushing-Sheerness.

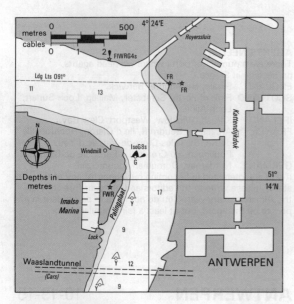

VLISSINGEN (FLUSHING) 10-19-16
Zeeland

CHARTS
Admiralty 325, 120, 139; Zeekaart 1442, 1443, 1533; Dutch Yacht Chart 1803; Stanford 19; Imray C30

TIDES
Dover +0200; ML 2.4; Duration 0555; Zone −0100

NOTE: Flushing is a Standard Port and the tidal predictions for each day of the year are given below.

SHELTER
Very good shelter, the yacht harbour being near the entrance to the lock to Walcheren canal. The old harbour to the W (fishing harbour) has had the lock gates removed and can be used by yachts. It is nearer the town but suffers from constant wash from passing traffic.

NAVIGATION
Waypoint H-SS N cardinal buoy, Q, 51°25'.97N 03°37'.54E, 125°/305° from/to Buitenhaven entrance, 0.95M. Approaching from the SW there are no dangers. Yachts are forbidden to sail in the main fairways. From Zeebrugge, keep S of Wielingen channel past Breskens, and cross to Flushing near buoy H-SS. From N use Deurloo/Spleet channels to S side of estuary, then as above. Heavy ocean-going traffic, often taking on or dropping pilots, off the town. In the harbour, frequent ferries berth at terminals near the locks and enter and leave at speed.

LIGHTS AND MARKS
The Lt Ho at the root of the W breakwater of Koopmanshaven is brown metal framework Tr, Iso WRG 3s 15m 12/9M.
Ldg Lts 117°. Front Leugenaar causeway, near head, OcR 5s 5m 11/7M. Rear Sardijngeul, 550m from front Oc WRG 5s 10m 12/8M, R Δ, W bands on R and W mast.
Entry signals from pier W side of Buitenhaven entrance:

| R flag | or | Additional R Lt near R harbour Lt | Entry prohibited |

| R flag over G flag | or | Additional R and G Lt near R harbour Lt | Entry prohibited if over 6m draught |

Two R Lts (vert) from lock indicate 'lock closed'.

RADIO TELEPHONE
Vlissingen Port Ch 09. Lock information Ch 22. Yacht Harbour Ch 14.
See also Westerschelde (10.19.14).

TELEPHONE (1184)
Hr Mr 68080; East Harbour Port Authority 15045; Yacht Hr Mr 65912; Buitenhaven Lock Keeper 12372; Customs 60000; Hosp 15000; Dr 12233; Brit Consul (20) 76.43.43.

FACILITIES
Jachthaven 'VVW Schelde' (85+60 visitors) Tel. 65912 AC, Bar, BH (10 ton), C (1 ton), D, FW, R, Lau (Access H24); **Royal Schelde Repair Yard** Sh; **Bureau Kramer** Tel. 16364, ME, Sh; **Gerb van de Gruiter** Tel. 65961, SM; **F. J. Roovers Sport** Tel. 13361, Gaz; **Town** P, D, CH, V, R, Bar. PO; Bank; Rly; Air (Antwerpen).
Ferry UK—Sheerness.

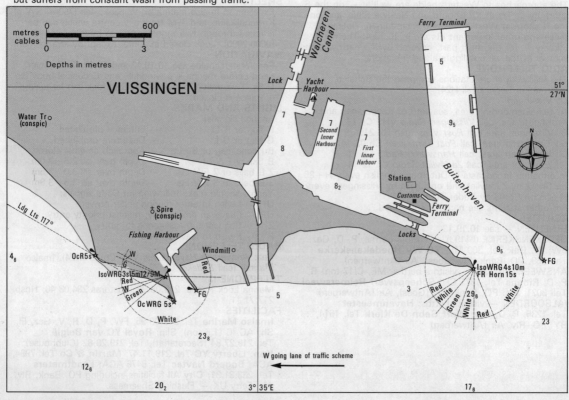

NETHERLANDS - VLISSINGEN (FLUSHING)
LAT 51°27'N   LONG 3°36'E
TIMES AND HEIGHTS OF HIGH AND LOW WATERS

TIME ZONE –0100 (Dutch Standard Time) For Dutch Summer Time add ONE hour in non-shaded areas

YEAR 1989

Chart Datum: 2.32 metres below Normaal Amsterdams Peil

	JANUARY				FEBRUARY				MARCH				APRIL			
	TIME	m	TIME	m	TIME	m	TIME	m	TIME	m	TIME	m	TIME	m	TIME	m
1 SU	0156 0806 1430 2042	1.1 3.8 0.8 3.8	**16** M	0230 0838 1515 2126	0.9 4.1 0.5 4.0	**1** W	0244 0920 1535 2200	1.1 3.6 1.0 3.6	**16** TH	0445 1059 1736 2346	0.9 3.8 0.9 3.8	**1** W	0045 0655 1326 1945	0.8 3.9 0.8 3.7	**16** TH	0256 0916 1545 2216	0.9 3.7 1.0 3.4	**1** SA	0316 0936 1554 2215	0.9 3.6 1.0 3.5	**16** SU	0525 1125 1755 2356	0.6 4.0 0.8 3.9
2 M	0306 0905 1535 2145	1.1 3.7 0.9 3.8	**17** TU	0334 0949 1636 2235	1.0 4.0 0.7 3.9	**2** TH	0425 1045 1705 2325	1.1 3.6 1.0 3.7	**17** F	0605 1221 1846	0.7 4.0 0.8	**2** TH	0149 0814 1444 2116	1.0 3.7 1.1 3.5	**17** F	0436 1044 1727 2324	0.9 3.7 1.0 3.6	**2** SU	0447 1055 1715 2326	0.8 3.9 0.9 3.8	**17** M	0616 1216 1836	0.5 4.2 0.7
3 TU	0405 1021 1634 2301	1.1 3.7 0.9 3.9	**18** W	0454 1106 1740 2349	0.9 4.0 0.7 4.0	**3** F	0547 1156 1810	1.0 3.8 0.9	**18** SA	0046 0705 1309 1930	4.0 0.5 4.3 0.7	**3** F	0334 1005 1636 2256	1.1 3.5 1.1 3.5	**18** SA	0555 1154 1826	0.7 4.0 0.8	**3** M	0544 1155 1812	0.5 4.3 0.6	**18** TU	0031 0655 1255 1905	4.1 0.3 4.3 0.6
4 W	0504 1126 1745 2356	1.1 3.8 0.9 4.0	**19** TH	0612 1215 1839	0.8 4.2 0.7	**4** SA	0026 0646 1248 1906	4.0 0.7 4.2 0.7	**19** SU	0132 0756 1356 2002	4.2 0.4 4.4 0.7	**4** SA	0515 1125 1745	0.9 3.8 0.9	**19** SU	0028 0650 1244 1910	4.0 0.4 4.3 0.7	**4** TU	0018 0646 1239 1906	4.2 0.2 4.6 0.4	**19** W	0110 0725 1326 1936	0.3 4.4 0.5
5 TH	0604 1215 1836	0.9 4.0 0.8	**20** F	0049 0716 1311 1936	4.1 0.6 4.3 0.7	**5** SU	0112 0729 1329 1946	4.2 0.5 4.5 0.6	**20** M O	0206 0828 1426 2029	4.3 0.3 4.5 0.6	**5** SU	0000 0615 1225 1840	3.8 0.6 4.2 0.7	**20** M ☾	0105 0730 1325 1940	4.2 0.3 4.4 0.6	**5** W	0059 0736 1320 1946	4.5 0.2 4.8 0.3	**20** TH	0137 0758 1355 2006	4.4 0.2 4.5 0.4
6 F	0046 0700 1306 1915	4.2 0.8 4.2 0.7	**21** SA O	0145 0759 1406 2016	4.3 0.5 4.5 0.7	**6** M	0151 0815 1409 2030	4.5 0.2 4.7 0.4	**21** TU	0240 0900 1455 2100	4.4 0.2 4.6 0.6	**6** M	0048 0715 1305 1925	4.2 0.3 4.5 0.5	**21** TU	0141 0759 1358 2005	4.3 0.2 4.5 0.5	**6** TH ●	0139 0815 1403 2028	4.8 -0.2 5.0 0.2	**21** F O	0207 0831 1426 2040	4.5 0.2 4.6 0.3
7 SA ●	0126 0746 1342 1959	4.3 0.6 4.5 0.6	**22** SU	0221 0842 1439 2045	4.4 0.4 4.5 0.7	**7** TU	0230 0902 1451 2113	4.6 0.0 4.9 0.2	**22** W	0305 0930 1520 2136	4.5 0.2 4.6 0.5	**7** TU ●	0127 0755 1345 2010	4.6 0.1 4.8 0.3	**22** W O	0208 0830 1425 2036	4.5 0.2 4.6 0.4	**7** F	0221 0858 1446 2116	4.9 -0.2 5.0 0.1	**22** SA	0237 0900 1456 2116	4.6 0.2 4.6 0.3
8 SU	0207 0829 1426 2045	4.5 0.4 4.6 0.5	**23** M	0256 0920 1516 2119	4.4 0.3 4.6 0.7	**8** W	0310 0947 1532 2155	4.8 -0.1 5.0 0.3	**23** TH	0338 1002 1555 2206	4.6 0.2 4.7 0.5	**8** W	0206 0839 1425 2052	4.7 -0.1 5.0 0.2	**23** TH	0237 0902 1455 2108	4.6 0.2 4.6 0.4	**8** SA	0302 0942 1528 2155	5.0 -0.2 4.9 0.1	**23** SU	0305 0936 1525 2146	4.6 0.3 4.6 0.4
9 M	0249 0915 1505 2131	4.6 0.3 4.8 0.5	**24** TU	0331 0955 1547 2156	4.5 0.3 4.6 0.7	**9** TH	0352 1029 1613 2240	4.8 -0.2 5.0 0.4	**24** F	0407 1036 1626 2236	4.7 0.2 4.6 0.5	**9** TH	0246 0923 1505 2136	4.9 -0.2 5.1 0.2	**24** F	0307 0932 1525 2135	4.7 0.1 4.7 0.4	**9** SU	0345 1026 1612 2240	5.0 -0.1 4.8 0.2	**24** M	0337 1002 1557 2215	4.6 0.4 4.5 0.4
10 TU	0329 1002 1548 2216	4.6 0.1 4.9 0.5	**25** W	0405 1030 1622 2230	4.5 0.3 4.6 0.6	**10** F	0436 1116 1659 2326	4.9 -0.2 4.9 0.4	**25** SA	0438 1106 1656 2259	4.6 0.3 4.5 0.5	**10** F	0326 1006 1550 2218	5.0 -0.3 5.0 0.2	**25** SA	0335 1002 1552 2211	4.7 0.2 4.6 0.4	**10** M	0429 1106 1658 2326	4.9 0.1 4.5 0.3	**25** TU	0409 1036 1627 2251	4.5 0.5 4.4 0.4
11 W	0412 1051 1632 2301	4.6 0.1 4.9 0.6	**26** TH	0440 1106 1656 2259	4.5 0.3 4.5 0.7	**11** SA	0518 1200 1746	4.8 -0.1 4.7	**26** SU	0505 1128 1721 2331	4.5 0.3 4.5 0.5	**11** SA	0409 1050 1636 2259	5.0 -0.2 4.9 0.2	**26** SU	0406 1030 1621 2235	4.6 0.3 4.5 0.4	**11** TU	0517 1148 1748	4.6 0.2 4.2	**26** W	0446 1108 1705 2326	4.5 0.5 4.3 0.4
12 TH	0457 1135 1719 2345	4.6 0.1 4.8 0.6	**27** F	0511 1135 1731 2331	4.5 0.3 4.4 0.7	**12** SU	0008 0606 1246 1839	0.5 4.7 0.1 4.4	**27** M	0536 1156 1756	4.4 0.4 4.3	**12** SU	0452 1136 1721 2345	4.9 -0.1 4.7 0.3	**27** M	0436 1058 1648 2302	4.6 0.4 4.4 0.4	**12** W ☽	0009 0615 1238 1844	0.4 4.3 0.6 3.8	**27** TH	0522 1150 1749	4.3 0.7 4.0
13 F	0542 1226 1816	4.6 0.1 4.6	**28** SA	0545 1205 1808	4.4 0.4 4.3	**13** M	0056 0659 1336 1939	0.6 4.4 0.4 4.1	**28** TU ☾	0002 0609 1235 1835	0.6 4.2 0.6 4.0	**13** M	0537 1215 1808	4.7 0.2 4.3	**28** TU	0506 1128 1723 2336	4.5 0.5 4.3 0.5	**13** TH	0104 0725 1334 1959	0.5 3.9 0.9 3.5	**28** F	0016 0611 1246 1856	0.5 4.1 0.8 3.7
14 SA ☾	0035 0635 1316 1909	0.7 4.4 0.2 4.4	**29** SU	0008 0622 1240 1846	0.7 4.2 0.5 4.1	**14** TU	0156 0806 1446 2056	0.8 4.1 0.7 3.8				**14** TU ☽	0029 0631 1306 1909	0.5 4.4 0.5 3.9	**29** W	0540 1206 1801	4.3 0.6 4.1	**14** F	0224 0845 1504 2140	0.8 3.7 1.1 3.4	**29** SA	0114 0735 1355 2020	0.6 3.8 1.0 3.6
15 SU	0125 0731 1405 2016	0.8 4.3 0.4 4.2	**30** M	0045 0706 1320 1935	0.8 4.0 0.7 3.9	**15** W	0304 0925 1606 2225	1.0 3.8 0.9 3.6				**15** W	0130 0746 1410 2031	0.7 4.0 0.8 3.6	**30** TH ☽	0026 0626 1306 1854	0.6 4.0 0.7 3.7	**15** SA	0400 1026 1656 2258	0.8 3.7 1.0 3.6	**30** SU	0245 0905 1536 2135	0.7 3.9 1.0 3.6
			31 TU	0136 0806 1426 2045	1.0 3.8 0.9 3.7							**31** F	0130 0735 1415 2046	0.8 3.7 1.0 3.4							

NETHERLANDS - VLISSINGEN (FLUSHING)

LAT 51°27′N   LONG 3°36′E

TIMES AND HEIGHTS OF HIGH AND LOW WATERS

YEAR 1989

TIME ZONE −0100 (Dutch Standard Time)
For Dutch Summer Time add ONE hour in non-shaded areas

MAY

Day	Time	m	Day	Time	m
1 M	0410 / 1022 / 1640 / 2250	0.6 / 4.1 / 0.8 / 3.9	16 TU	0525 / 1132 / 1745 / 2349	0.6 / 4.1 / 0.8 / 4.0
2 TU	0526 / 1121 / 1740 / 2345	0.4 / 4.4 / 0.6 / 4.2	17 W	0616 / 1216 / 1825	0.5 / 4.2 / 0.6
3 W	0618 / 1211 / 1832	0.2 / 4.6 / 0.4	18 TH	0029 / 0650 / 1252 / 1906	4.2 / 0.4 / 4.3 / 0.5
4 TH	0031 / 0706 / 1257 / 1926	4.5 / 0.0 / 4.8 / 0.3	19 F	0106 / 0725 / 1326 / 1935	4.3 / 0.2 / 4.4 / 0.5
5 F	0115 / 0749 / 1338 / ● 2008	4.7 / −0.1 / 4.9 / 0.2	20 SA	0137 / 0758 / 1358 / ○ 2016	4.4 / 0.2 / 4.5 / 0.4
6 SA	0158 / 0835 / 1425 / 2053	4.9 / −0.1 / 4.8 / 0.1	21 SU	0211 / 0829 / 1432 / 2050	4.5 / 0.4 / 4.5 / 0.4
7 SU	0243 / 0918 / 1509 / 2138	4.9 / 0.0 / 4.7 / 0.1	22 M	0246 / 0905 / 1506 / 2128	4.6 / 0.4 / 4.5 / 0.4
8 M	0328 / 1002 / 1556 / 2226	4.9 / 0.2 / 4.6 / 0.2	23 TU	0317 / 0942 / 1538 / 2206	4.5 / 0.5 / 4.4 / 0.4
9 TU	0415 / 1045 / 1642 / 2310	4.7 / 0.4 / 4.4 / 0.2	24 W	0356 / 1021 / 1617 / 2246	4.5 / 0.5 / 4.3 / 0.4
10 W	0508 / 1128 / 1738	4.5 / 0.6 / 4.1	25 TH	0435 / 1056 / 1657 / 2326	4.5 / 0.6 / 4.2 / 0.4
11 TH	0000 / 0606 / 1216 / 1836	0.4 / 4.3 / 0.8 / 3.9	26 F	0517 / 1139 / 1748	4.4 / 0.7 / 4.1
12 F ☽	0056 / 0716 / 1309 / 1929	0.5 / 4.0 / 1.0 / 3.7	27 SA	0016 / 0616 / 1236 / 1845	0.4 / 4.2 / 0.8 / 3.9
13 SA	0200 / 0816 / 1425 / 2056	0.6 / 3.8 / 1.1 / 3.5	28 SU ☾	0116 / 0726 / 1334 / 1956	0.4 / 4.1 / 0.9 / 3.9
14 SU	0303 / 0935 / 1534 / 2206	0.7 / 3.8 / 1.0 / 3.6	29 M	0230 / 0840 / 1456 / 2105	0.4 / 4.1 / 0.9 / 3.9
15 M	0415 / 1040 / 1655 / 2305	0.7 / 3.9 / 0.9 / 3.8	30 TU	0335 / 0945 / 1606 / 2211	0.4 / 4.2 / 0.8 / 4.0
			31 W	0445 / 1052 / 1710 / 2316	0.3 / 4.4 / 0.7 / 4.2

JUNE

Day	Time	m	Day	Time	m
1 TH	0551 / 1145 / 1811	0.2 / 4.5 / 0.5	16 F	0606 / 1215 / 1830	0.6 / 4.1 / 0.7
2 F	0007 / 0639 / 1237 / 1902	4.5 / 0.1 / 4.6 / 0.4	17 SA	0036 / 0656 / 1256 / 1915	4.1 / 0.6 / 4.3 / 0.6
3 SA	0056 / 0730 / 1326 / ● 1949	4.6 / 0.1 / 4.6 / 0.3	18 SU	0115 / 0730 / 1335 / 1951	4.2 / 0.6 / 4.3 / 0.6
4 SU	0143 / 0816 / 1412 / 2040	4.7 / 0.2 / 4.6 / 0.2	19 M	0156 / 0805 / 1416 / ○ 2036	4.4 / 0.5 / 4.4 / 0.4
5 M	0229 / 0900 / 1459 / 2128	4.8 / 0.3 / 4.5 / 0.2	20 TU	0227 / 0845 / 1448 / 2111	4.5 / 0.5 / 4.4 / 0.3
6 TU	0317 / 0941 / 1546 / 2216	4.7 / 0.4 / 4.4 / 0.2	21 W	0307 / 0925 / 1527 / 2158	4.5 / 0.5 / 4.4 / 0.3
7 W	0406 / 1026 / 1636 / 2300	4.6 / 0.6 / 4.3 / 0.2	22 TH	0345 / 1006 / 1607 / 2246	4.6 / 0.6 / 4.4 / 0.2
8 TH	0456 / 1106 / 1722 / 2346	4.5 / 0.7 / 4.2 / 0.3	23 F	0427 / 1049 / 1651 / 2326	4.6 / 0.6 / 4.4 / 0.2
9 F	0550 / 1144 / 1810	4.4 / 0.8 / 4.1	24 SA	0510 / 1135 / 1737	4.6 / 0.7 / 4.3
10 SA	0024 / 0635 / 1234 / 1856	0.4 / 4.2 / 0.9 / 4.0	25 SU	0015 / 0602 / 1226 / 1829	0.2 / 4.5 / 0.7 / 4.2
11 SU ☽	0120 / 0724 / 1340 / 1944	0.5 / 4.0 / 0.9 / 3.8	26 M ☾	0106 / 0706 / 1320 / 1925	0.2 / 4.4 / 0.8 / 4.2
12 M	0216 / 0825 / 1440 / 2044	0.6 / 3.9 / 1.0 / 3.7	27 TU	0201 / 0805 / 1425 / 2032	0.3 / 4.3 / 0.8 / 4.1
13 TU	0316 / 0936 / 1534 / 2200	0.6 / 3.8 / 0.9 / 3.7	28 W	0306 / 0915 / 1530 / 2140	0.3 / 4.2 / 0.8 / 4.1
14 W	0416 / 1036 / 1646 / 2255	0.7 / 3.9 / 0.9 / 3.8	29 TH	0416 / 1021 / 1646 / 2245	0.4 / 4.2 / 0.7 / 4.2
15 TH	0516 / 1130 / 1740 / 2345	0.6 / 4.0 / 0.8 / 3.9	30 F	0522 / 1126 / 1751 / 2349	0.4 / 4.3 / 0.6 / 4.3

JULY

Day	Time	m	Day	Time	m
1 SA	0626 / 1225 / 1850	0.4 / 4.3 / 0.5	16 SU	0006 / 0619 / 1231 / 1855	3.9 / 0.8 / 4.1 / 0.7
2 SU	0046 / 0716 / 1319 / 1941	4.4 / 0.4 / 4.4 / 0.3	17 M	0055 / 0710 / 1315 / 1936	4.1 / 0.7 / 4.3 / 0.6
3 M	0139 / 0801 / 1407 / ● 2032	4.6 / 0.4 / 4.4 / 0.3	18 TU	0135 / 0750 / 1355 / ○ 2015	4.3 / 0.6 / 4.4 / 0.4
4 TU	0228 / 0848 / 1455 / 2115	4.6 / 0.5 / 4.4 / 0.2	19 W	0216 / 0830 / 1435 / 2102	4.5 / 0.6 / 4.5 / 0.3
5 W	0315 / 0925 / 1536 / 2159	4.6 / 0.6 / 4.4 / 0.2	20 TH	0252 / 0916 / 1516 / 2145	4.7 / 0.5 / 4.6 / 0.1
6 TH	0356 / 1006 / 1618 / 2246	4.6 / 0.7 / 4.4 / 0.2	21 F	0330 / 0955 / 1552 / 2230	4.8 / 0.5 / 4.6 / 0.1
7 F	0436 / 1046 / 1658 / 2319	4.6 / 0.7 / 4.4 / 0.2	22 SA	0412 / 1036 / 1636 / 2316	4.8 / 0.5 / 4.6 / 0.0
8 SA	0517 / 1126 / 1735	4.5 / 0.7 / 4.3	23 SU	0455 / 1119 / 1717	4.8 / 0.6 / 4.6
9 SU	0006 / 0559 / 1206 / 1818	0.3 / 4.4 / 0.8 / 4.2	24 M	0001 / 0542 / 1205 / 1806	0.0 / 4.7 / 0.6 / 4.5
10 M	0046 / 0646 / 1250 / 1900	0.4 / 4.2 / 0.8 / 4.1	25 TU ☾	0045 / 0635 / 1255 / 1855	0.1 / 4.6 / 0.7 / 4.4
11 TU ☽	0114 / 0730 / 1340 / 1951	0.5 / 4.1 / 0.9 / 3.9	26 W	0136 / 0736 / 1349 / 2000	0.2 / 4.6 / 0.8 / 4.2
12 W	0204 / 0819 / 1446 / 2045	0.6 / 3.9 / 1.0 / 3.8	27 TH	0236 / 0846 / 1454 / 2112	0.4 / 4.5 / 0.8 / 4.1
13 TH	0321 / 0919 / 1550 / 2156	0.6 / 3.8 / 1.0 / 3.7	28 F	0345 / 0955 / 1620 / 2230	0.6 / 4.0 / 0.9 / 4.0
14 F	0425 / 1021 / 1701 / 2308	0.7 / 4.0 / 1.0 / 3.7	29 SA	0506 / 1118 / 1740 / 2341	0.7 / 4.0 / 0.7 / 4.1
15 SA	0526 / 1146 / 1755	0.8 / 3.9 / 0.9	30 SU	0616 / 1218 / 1845	0.6 / 4.1 / 0.6
			31 M	0049 / 0704 / 1318 / 1940	4.3 / 0.6 / 4.3 / 0.4

AUGUST

Day	Time	m	Day	Time	m
1 TU	0146 / 0756 / 1401 / ● 2026	4.5 / 0.6 / 4.4 / 0.3	16 W	0115 / 0730 / 1336 / 2000	4.4 / 0.7 / 4.4 / 0.4
2 W	0226 / 0835 / 1438 / 2105	4.6 / 0.7 / 4.5 / 0.2	17 TH	0155 / 0810 / 1412 / ○ 2039	4.7 / 0.6 / 4.6 / 0.2
3 TH	0259 / 0910 / 1516 / 2146	4.6 / 0.7 / 4.5 / 0.2	18 F	0229 / 0849 / 1449 / 2123	4.9 / 0.5 / 4.8 / 0.0
4 F	0335 / 0945 / 1548 / 2215	4.7 / 0.7 / 4.6 / 0.2	19 SA	0309 / 0936 / 1529 / 2205	5.0 / 0.4 / 4.9 / −0.1
5 SA	0408 / 1018 / 1625 / 2252	4.7 / 0.6 / 4.6 / 0.2	20 SU	0350 / 1016 / 1609 / 2252	5.0 / 0.4 / 4.9 / −0.1
6 SU	0446 / 1056 / 1659 / 2325	4.6 / 0.7 / 4.6 / 0.3	21 M	0433 / 1101 / 1650 / 2333	5.0 / 0.5 / 4.9 / 0.0
7 M	0522 / 1126 / 1735	4.6 / 0.7 / 4.5	22 TU	0516 / 1146 / 1737	4.8 / 0.5 / 4.8
8 TU	0001 / 0555 / 1155 / 1808	0.4 / 4.5 / 0.7 / 4.3	23 W ☾	0016 / 0608 / 1231 / 1825	0.1 / 4.6 / 0.6 / 4.6
9 W	0025 / 0636 / 1236 / 1850	0.5 / 4.2 / 0.8 / 4.1	24 TH	0106 / 0700 / 1326 / 1928	0.4 / 4.3 / 0.8 / 4.3
10 TH	0105 / 0720 / 1315 / 1934	0.7 / 4.0 / 1.0 / 3.8	25 F	0206 / 0815 / 1436 / 2045	0.6 / 3.9 / 0.9 / 4.0
11 F	0200 / 0820 / 1430 / 2056	0.9 / 3.7 / 1.2 / 3.6	26 SA	0319 / 0939 / 1616 / 2219	0.9 / 3.7 / 1.0 / 3.9
12 SA	0330 / 0936 / 1616 / 2214	1.1 / 3.6 / 1.2 / 3.5	27 SU	0455 / 1104 / 1736 / 2346	0.9 / 3.8 / 0.8 / 4.1
13 SU	0457 / 1106 / 1726 / 2340	1.1 / 3.6 / 1.0 / 3.8	28 M	0604 / 1215 / 1846	0.7 / 4.1 / 0.6
14 M	0556 / 1210 / 1825	1.0 / 3.9 / 0.8	29 TU	0046 / 0705 / 1305 / 1935	4.4 / 0.5 / 4.3 / 0.4
15 TU	0036 / 0645 / 1255 / 1916	4.1 / 0.8 / 4.2 / 0.6	30 W	0136 / 0745 / 1348 / 2010	4.5 / 0.7 / 4.4 / 0.3
			31 TH	0205 / 0816 / 1418 / ● 2041	4.6 / 0.7 / 4.5 / 0.3

Chart Datum: 2.32 metres below Normaal Amsterdams Peil

NETHERLANDS - VLISSINGEN (FLUSHING)

LAT 51°27'N LONG 3°36'E

TIMES AND HEIGHTS OF HIGH AND LOW WATERS

YEAR 1989

TIME ZONE −0100 (Dutch Standard Time)
For Dutch Summer Time add ONE hour in non-shaded areas

Chart Datum: 2.32 metres below Normaal Amsterdams Peil

SEPTEMBER

	TIME	m		TIME	m
1 F	0236 0848 1448 2116	4.7 0.7 4.6 0.3	**16** SA	0205 0827 1423 2058	5.0 0.4 4.9 0.0
2 SA	0308 0918 1521 2148	4.7 0.6 4.7 0.3	**17** SU	0245 0910 1502 2142	5.1 0.4 5.1 −0.1
3 SU	0337 0952 1552 2215	4.7 0.6 4.7 0.3	**18** M	0326 0956 1543 2226	5.1 0.4 5.1 0.0
4 M	0408 1026 1622 2251	4.7 0.6 4.7 0.4	**19** TU	0408 1035 1626 2306	5.0 0.4 5.1 0.1
5 TU	0442 1050 1656 2316	4.6 0.6 4.6 0.5	**20** W	0452 1121 1710 2350	4.8 0.5 4.9 0.3
6 W	0509 1116 1721 2346	4.5 0.7 4.5 0.6	**21** TH	0541 1207 1801	4.5 0.6 4.6
7 TH	0542 1151 1756	4.3 0.8 4.3	**22** F	0036 0640 1300 1908 ☾	0.6 4.1 0.8 4.2
8 F ☾	0018 0618 1230 1835	0.7 4.1 0.9 4.0	**23** SA	0136 0744 1426 2036	0.9 3.8 1.0 3.9
9 SA	0105 0716 1336 1935	1.0 3.8 1.1 3.6	**24** SU	0255 0930 1606 2216	1.1 3.6 1.0 3.8
10 SU	0214 0846 1515 2134	1.2 3.5 1.2 3.5	**25** M	0457 1106 1725 2329	1.1 3.7 0.8 4.1
11 M	0415 1020 1650 2310	1.3 3.5 1.1 3.8	**26** TU	0555 1205 1825	0.9 4.0 0.5
12 TU	0525 1140 1806	1.1 3.8 0.8	**27** W	0026 0645 1246 1905	4.4 0.8 4.3 0.4
13 W	0006 0626 1225 1850	4.2 0.9 4.2 0.5	**28** TH	0105 0726 1321 1946	4.5 0.7 4.5 0.4
14 TH	0049 0706 1307 1932	4.5 0.7 4.5 0.3	**29** F ●	0139 0752 1351 2016	4.8 0.7 4.6 0.4
15 F ○	0126 0742 1346 2015	4.8 0.5 4.7 0.1	**30** SA	0207 0821 1421 2046	4.7 0.6 4.7 0.3

OCTOBER

	TIME	m		TIME	m
1 SU	0237 0850 1449 2115	4.7 0.6 4.8 0.3	**16** M	0219 0845 1437 2117	5.1 0.3 5.1 0.0
2 M	0307 0926 1519 2149	4.7 0.5 4.8 0.4	**17** TU	0303 0932 1520 2158	5.1 0.3 5.1 0.1
3 TU	0337 0956 1552 2216	4.7 0.6 4.7 0.5	**18** W	0346 1015 1605 2239	4.9 0.4 5.0 0.3
4 W	0406 1020 1620 2240	4.6 0.6 4.6 0.6	**19** TH	0436 1100 1651 2326	4.7 0.4 4.8 0.5
5 TH	0437 1046 1650 2306	4.5 0.7 4.5 0.7	**20** F	0525 1151 1747	4.6 0.6 4.5
6 F	0507 1115 1718 2346	4.4 0.7 4.4 0.8	**21** SA ☾	0015 0626 1245 1856	0.8 4.0 0.7 4.2
7 SA	0541 1206 1801	4.2 0.8 4.1	**22** SU	0115 0736 1354 2026	1.1 3.7 0.9 3.9
8 SU ☾	0036 0635 1255 1859	1.0 3.9 1.0 3.8	**23** M	0234 0900 1523 2146	1.3 3.6 1.0 3.9
9 M	0140 0811 1424 2105	1.3 3.6 1.1 3.6	**24** TU	0425 1025 1701 2300	1.2 3.7 0.8 4.1
10 TU	0324 0935 1616 2230	1.3 3.5 1.0 3.9	**25** W	0524 1126 1755 2356	1.1 4.0 0.6 4.3
11 W	0445 1055 1720 2335	1.1 3.8 0.8 4.2	**26** TH	0621 1211 1835	0.9 4.2 0.4
12 TH	0546 1152 1816	0.9 4.2 0.5	**27** F	0035 0656 1249 1912	4.4 0.8 4.4 0.5
13 F	0016 0636 1235 1906	4.6 0.7 4.5 0.3	**28** SA	0105 0719 1321 1946	4.5 0.7 4.5 0.5
14 SA ●	0057 0715 1316 1948	4.7 0.5 4.8 0.1	**29** SU	0139 0756 1355 2012	4.6 0.6 4.6 0.5
15 SU	0138 0802 1356 2030	5.1 0.4 5.0 0.0	**30** M	0209 0826 1425 2046	4.7 0.6 4.7 0.5
31 TU	0238 0856 1455 2116	4.7 0.6 4.7 0.5			

NOVEMBER

	TIME	m		TIME	m
1 W	0309 0931 1526 2146	4.7 0.6 4.7 0.6	**16** TH	0330 0959 1549 2222	4.7 0.3 4.9 0.5
2 TH	0341 0955 1556 2216	4.6 0.6 4.6 0.7	**17** F	0421 1045 1641 2306	4.5 0.4 4.7 0.7
3 F	0415 1036 1627 2245	4.5 0.6 4.5 0.8	**18** SA	0511 1136 1739 2344	4.3 0.5 4.5 0.9
4 SA	0446 1106 1706 2322	4.3 0.7 4.4 0.9	**19** SU	0610 1224 1839	4.1 0.6 4.2
5 SU	0528 1156 1750	4.2 0.7 4.2	**20** M ☾	0046 0710 1336 1951	1.1 3.9 0.8 4.0
6 M	0004 0626 1246 1856	1.0 3.9 0.8 4.0	**21** TU	0156 0816 1434 2055	1.2 3.7 0.8 3.9
7 TU	0116 0746 1353 2036	1.2 3.7 0.9 3.9	**22** W	0304 0936 1606 2210	1.3 3.7 0.9 3.9
8 W	0240 0900 1536 2146	1.2 3.7 0.9 4.0	**23** TH	0424 1038 1716 2310	1.2 3.8 0.8 4.1
9 TH	0355 1012 1646 2256	1.1 3.9 0.7 4.3	**24** F	0536 1130 1758 2355	1.0 4.0 0.7 4.2
10 F	0506 1111 1740 2345	0.9 4.2 0.5 4.6	**25** SA	0604 1211 1836	0.9 4.2 0.6
11 SA	0554 1205 1836	0.8 4.5 0.3	**26** SU	0035 0656 1249 1905	4.3 0.8 4.3 0.6
12 SU	0032 0650 1250 1922	4.8 0.6 4.8 0.2	**27** M	0111 0725 1325 1946	4.4 0.7 4.4 0.6
13 M ○	0115 0740 1333 2008	4.9 0.4 5.0 0.1	**28** TU	0145 0800 1357 2016	4.5 0.6 4.5 0.6
14 TU	0159 0829 1418 2053	4.9 0.4 5.0 0.2	**29** W	0216 0838 1432 2050	4.5 0.5 4.6 0.6
15 W	0246 0916 1503 2135	4.9 0.3 5.0 0.3	**30** TH	0251 0909 1506 2126	4.5 0.6 4.6 0.7

DECEMBER

	TIME	m		TIME	m
1 F	0325 0946 1542 2156	4.5 0.6 4.6 0.7	**16** SA	0411 1040 1636 2248	4.5 0.5 4.7 0.8
2 SA	0401 1022 1615 2236	4.4 0.6 4.5 0.8	**17** SU	0459 1125 1726 2330	4.4 0.5 4.6 0.9
3 SU	0439 1106 1656 2316	4.3 0.6 4.5 0.9	**18** M	0548 1210 1816	4.3 0.4 4.4
4 M	0521 1146 1741 ☾	4.2 0.6 4.4	**19** TU	0016 0636 1256 1906	1.0 4.2 0.6 4.2
5 TU	0005 0616 1246 1839	1.0 4.1 0.6 4.2	**20** W	0105 0725 1345 1954	1.1 4.0 0.7 4.0
6 W	0056 0716 1334 1956	1.0 4.0 0.6 4.1	**21** TH	0216 0820 1448 2105	1.1 3.8 0.8 3.9
7 TH	0206 0822 1450 2105	1.1 4.0 0.6 4.2	**22** F	0304 0931 1545 2216	1.2 3.7 0.9 3.8
8 F	0304 0929 1600 2216	1.0 4.0 0.6 4.3	**23** SA	0426 1036 1656 2310	1.1 3.9 0.9 3.9
9 SA	0426 1038 1705 2312	0.9 4.2 0.4 4.4	**24** SU	0520 1136 1750	1.0 3.9 0.8
10 SU	0530 1137 1805	0.8 4.4 0.4	**25** M	0006 0616 1219 1835	4.0 0.9 4.0 0.8
11 M	0010 0630 1227 1902	4.6 0.6 4.6 0.3	**26** TU	0048 0659 1305 1915	4.2 0.8 4.2 0.8
12 TU	0059 0726 1317 1952	4.7 0.5 4.7 0.3	**27** W	0126 0745 1341 1956	4.3 0.7 4.3 0.7
13 W	0150 0817 1406 2035	4.7 0.4 4.8 0.4	**28** TH ●	0201 0826 1419 2035	4.4 0.6 4.4 0.7
14 TH	0236 0906 1455 2126	4.6 0.5 4.8 0.5	**29** F	0236 0855 1452 2111	4.5 0.5 4.5 0.7
15 F	0326 0951 1545 2206	4.6 0.3 4.6 0.6	**30** SA	0312 0929 1529 2145	4.5 0.5 4.6 0.7
31 SU	0349 1018 1606 2226	4.5 0.3 4.6 0.7			

VLISSINGEN (FLUSHING) continued

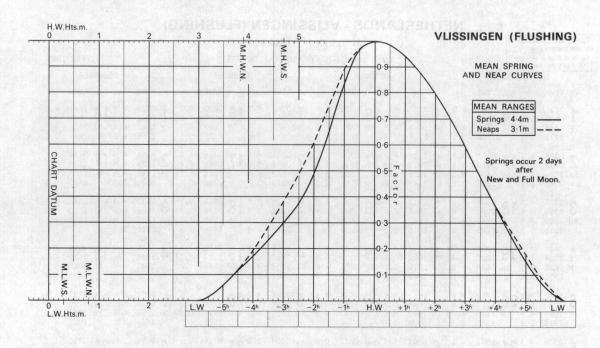

OOSTERSCHELDE 10-19-17
Zeeland

CHARTS
Admiralty 110, 120, 192; Zeekaart 1448; Dutch Yacht Charts 1805, 1807; Stanford 19; Imray C30

TIDES
Dover +0230; ML Sas van Goes 2.0; Zierikzee 1.8
Duration Sas van Goes 0615; Zierikzee 0640; Zone −0100

Standard Port VLISSINGEN (←)

Times				Height (metres)			
HW		LW		MHWS	MHWN	MLWN	MLWS
0300	0900	0400	1000	4.7	3.9	0.8	0.6
1500	2100	1600	2200				

Differences ROOMPOT
+0115 +0115 +0115 +0115 No data   No data
WEMELDINGE
+0145 +0145 +0125 +0125 No data   No data
LODIJKSE GAT
+0145 +0145 +0125 +0125 No data   No data

SHELTER
Inside the barrier good shelter is found in several harbours — for example Colijnsplaat and Zieriezee, which are both west of Zeelandbrug (bridge connecting Noord Beveland and Duiveland) and accessible at any tide. Veerse Meer is a non-tidal waterway with mooring places for yachts: access from Oosterschelde at Zandkreekdam lock, or through Kanaal door Walcheren via Vlissingen and Middelburg from Westerschelde.

NAVIGATION
Waypoint WG1 (stbd-hand) buoy, QG, 51°38′.00N 03°26′.30E at entrance to Westgat/Oude Roompot buoyed channel. Not advisable to enter except in settled weather, and only from HW−6 to HW+1½ Zierikzee. There are several offshore banks, see 10.19.5. All vessels must use the Roompotsluis (lock). The areas each side of the barrier are very dangerous due to strong tidal streams and many obstructions. Passage is prohibited west of Roggenplaat. Zeelandbrug clearance 11.5m at centre of arch in buoyed channels. Bascule bridge near N end lifts weekdays at odd hours 0700-2100, but not if wind in excess of Force 7.

AREA 19—The Netherlands

OOSTERSCHELDE continued

LIGHTS AND MARKS
There are few prominent marks in the approaches. For principal lights and waypoints see 10.19.4.

RADIO TELEPHONE
Zeelandbrug VHF Ch 18. Kreekraksluizen (Schelde-Rijnkanaal) Ch 20. Roompotsluis Ch 18. Krammer Locks. Call *Krammersluizen* Ch 22 (H24).

There are numerous ports to visit and the following is a selection of the more important ones (working anti-clockwise, starting from the Roompotsluis).

COLIJNSPLAAT Hr Mr Tel. 201; YC Tel. 806; Dr Tel. 304; **Marina WV Noord YC Beveland** AC, Slip, FW, berth N side of mole or W side of basin; **Tanker** (see YC) P, D; **De Koster** ME; **Town** V, R, Gaz, PO, Bank, Rly, (Goes), Air (Antwerpen).

GOES The canal up to Goes starts at Sas van Goes where there are a few facilities; there are also a few at Wilhelminadorp half way up the canal. At Goes Hr Mr Tel. 27257; Dr Tel. 27451; Hosp Tel. 27000; **Marina WV 'De Werf'** Tel. 16372, FW, C (3½ ton), P, D; **Beekman** Tel. 27383 ME, El, Sh; **Jachtwerf Goes** Tel. 23944, Slip; **Sporthuis Olympia** Tel. 16008, Gaz; **Delta** Tel. 27338 Lau; **Sakko** Tel. 15100, P, D; **Town** V, R, PO, Bank, Rly, Air (Antwerpen).

WEMELDINGE Hr Mr Tel. 1463; Dr Tel. 1438; Yachts berth W side of Voorhaven or pass through Westsluis locks and berth W side of Binnenhaven. **Florusse** Tel. 1253, ME, SM, AC, C (6.5 ton) ME; **Town** P, D, V, R, PO, Bank, Rly, (Goes) Air (Antwerpen).

YERSEKE Hr Mr Tel. 1726; Dr Tel. 1444; **Marina Prins Willem-Alexanderhaven** FW, P, D, C; **Prins Beatrixhaven** Slip, FW, P, D; **J. Zoetewey** Tel. 1390, Gaz, SM; **W. Bakker** Tel. 1521, ME; **Town** V, BY, R, PO, Bank, Rly (Kruiningen), Air (Antwerpen).

BERGEN OP ZOOM Hr Mr Tel. 37472; Dr Tel. 35120; Hosp Tel. 37980; **Marina WV 'de Schelde'** (115+6 visitors) Tel. 37472, AC, BH (10 ton), D, P, FW, Gas, Gaz, R, Lau, Slip; **L. Ribens** Tel. 35395, SM; **Gebr Westerduin** Tel. 41759, BY; **Town** V, R, Bar, PO, Bank, Rly, Air (Antwerpen).

THOLEN Hr Mr Tel. 2236; Dr Tel. 2542; Yachts secure NW corner of inner basin or anchor N end of outer harbour, off the Schelde – Rijnkanaal. **Marina WV 'de Kogge'** P, D, ME, El, Sh, CH, Slip, FW, C; **M Schot** Tel. 2476, P, D, FW, AC, C (1 ton); **G. P. Contant** Tel. 2581, Gaz; **M. Oerlemans** Tel. 2325; ME; **Town** V, R, PO, Bank, Rly (Bergen op Zoom), Air (Antwerp).

ST ANNALAND Dr Tel. 2400; **Marina** Hr Mr Tel. 2463 Slip, FW; **YC** Tel. 2634; **J. Keur** Tel. 2454, P, D; **Garage Dekker** ME; **Town** Gaz, V, R, Bank, PO, Rly (Bergen op Zoom), Air (Antwerpen).

ST PHILIPSLAND Dr Tel. 500; **Marina** Hr Mr Tel. 416, FW; **P. Wolse** Tel. 363, P, D, ME; **Town** Gaz, V, R, Bar, Rly (Roosendal), Air (Rotterdam).

BRUINISSE Hr Mr Tel. 1444; Dr Tel. 1280 **Marina WV 'Bru'** Tel. 1506, FW, C (1.5 ton). Enter Visserhaven Ldg Lts 281°, berth at NW corner. **Aquadelta** Tel. 1485, Slip, P, D, FW, ME, C, AC, Lau; **J.D. de Korte** SM, Gaz; **Town** V, R, Bar, PO, Bank, Rly (Roosendal), Air (Rotterdam).

ZIERIKZEE Hr Mr Tel. 3151; Dr Tel. 2080; Hosp Tel. 6900. Yachts either go to Marina 't Luitje', NW side of Havenbaraal before reaching Zierikzee or go up to Nieuwe Haven. **'t Luitje' Marina** FW, C (10 ton);

NIEUWE HAVEN (dries) AB, FW; **W Bouwman** Tel. 2966, P, D, Gaz, AC; **Town** V, R, PO, Bank, Rly, Air (Rotterdam).

BURGHSLUIS Hr Mr Tel. 1302; **Marina** FW, C (1 ton), **Blom** Tel. 2020, ME, AC; Facilities at Burg-Haamstede P, D, Gaz, PO, Bank, Rly, Air (Rotterdam)

TOWNS IN THE VEERSE MEER (non-tidal)
MIDDELBURG Dr Tel. 12637; Hosp Tel. 25555; **Marina WV 'Arne'** in the Dockhaven, Tel. 13852; **J. Boone** Tel. 29913, D, Gaz, FW; **Zeeuwse Motoren Revisie** Tel. 13242, SM; **Jachtwerf Jansen** Tel. 13925, AC, Slip; **Wassalon Miele** Tel. 13830 Lau; **Town** V, R, Bar, PO, Bank, Rly, Air (Antwerpen)

VEERE Dr Tel. 271; Yachts can berth in the Stadshaven or in the Buitenhaven. **Jachtwerf Veere** Tel. 551, Slip, P, D, Gaz, FW, ME, AC, C (3½ ton), BH (12 ton), Lau; **Jachtclub Veere** Tel. 246, FW, M; **Jachthaven Oostwatering** Tel. 484, Slip, FW; **Town** V, R, Bar, PO, Bank, Rly (Middelburg), Air (Rotterdam).

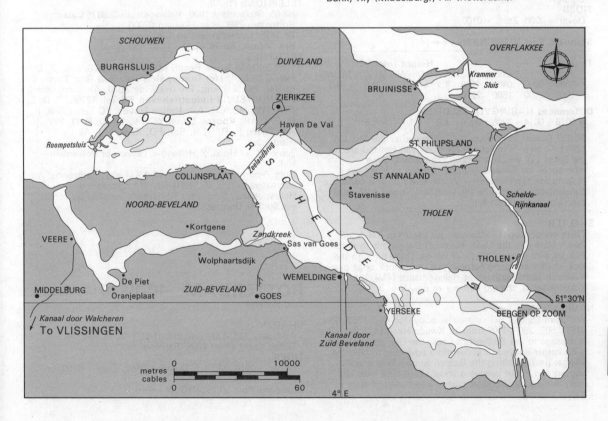

STELLENDAM (HELLEVOETSLUIS) 10-19-18
Zuid Holland

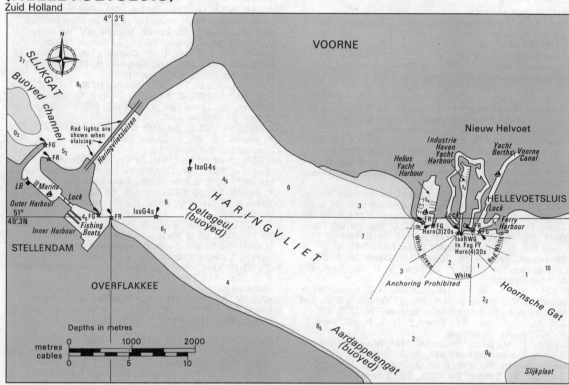

CHARTS
Admiralty 2322; Zeekaart 1448; Dutch Yacht Chart: 1807; Stanford 19; Imray C30

TIDES
Dover +0300; Zone −0100

Standard Port VLISSINGEN (←)

Times				Height (metres)			
HW		LW		MHWS	MHWN	MLWN	MLWS
0300	0900	0400	1000	4.7	3.9	0.8	0.3
1500	2100	1600	2200				

Differences HARINGVLIETSLUIZEN
+0016 +0014 +0006 −0026 −1.8 −1.7 −0.5 0.0

NOTE: Double LWs occur — The rise after the first LW is called Agger — Predictions for Hoek van Holland are for the lower LW which is usually the second. Time differences for Secondary Ports referred to Hoek van Holland are approximate — Water levels on this coast are considerably affected by the weather. Strong NW gales can raise levels up to 3m.

SHELTER
There are pontoons exclusively for yachts in the Aqua Pesch Marina in the Outer Harbour. Dangerous to enter Slijkgat with strong NW to W winds. Once through the lock, shelter in fishing harbour just inside lock or at one of the three yacht harbours at Hellevoetsluis (4M), the Industriehaven, the old harbour to the W or in the Voorne canal.

NAVIGATION
Waypoint Slijkgat SG safe water buoy, Iso 4s, 51°52'.00N 03°51'.50E, 290°/110° from/to Kwade Hoek Lt, 5.0M. The N passage to the Stellendam Lock, the Gat van de Hawk is no longer buoyed. It is therefore advisable to take the Slijkgat passage along the Goeree Shore. This is well buoyed and lit but beware a shoal about 5 ca from the lock, marked by G conical buoys. Keep clear of dam during sluicing.

LIGHTS AND MARKS
Hellevoetsluis Lt Iso WRG 10s 16m 11/7M, W stone Tr, R cupola. G Shore-275°, W275°-294°, R294°-316°, W316°-036°, G036°-058°, W058°-095°, R096°-shore Sluicing signals — 3 R Lts in triangle shown from pillar heads on dam. Danger area marked with small R buoys.

RADIO TELEPHONE
Call: *Goereesesluis* VHF Ch 13.

TELEPHONE (1879)
Hr Mr Stellendam 1000; Hellevoetsluis 30911; Customs Rotterdam 298088 or Flushing 60000
Dr 1425; Dr Hellevoetsluis 12435;
Brit Consul (20) 76.43.43.

FACILITIES
STELLENDAM **Aqua Pesch Marina** Tel. 2600, Slip, P, D, FW, C (8 ton); **Het Dumppaleis** Tel. 1529 Gaz; **Town** V, R, Bar, PO (2½ km), Rly (Hellevoetsluis); Air (Rotterdam).
HELLEVOETSLUIS **Industriehaven YC** Tel. 12166, P, D, L, Gaz, FW, ME, CH, AB; **Helius Haven YC** Tel. 15868, P, D, FW, AB; **Voorne Canal YC** Tel. 12870, P, D, FW, AB; **Town** P, D, V, R, Bar, PO, Bank, Rly, Air (Rotterdam).
Ferry UK — Hook of Holland—Harwich or Rotterdam—Hull.
Note. Voorne canal is dammed 4½M above entrance. An alternative port is Middelharnis, 4M upstream of Hellevoetsluis. Bound for Rotterdam, proceed by canals via Spui or Dordrecht.

ROTTERDAM 10-19-19
Zuid Holland

CHARTS
Admiralty 133, 132; Zeekaart 1540/1/2; Dutch Yacht Chart 1809; Stanford 19

TIDES
Dover +0510; ML 1.1; Duration 0440; Zone −0100

Standard Port VLISSINGEN (←)

AREA 19—The Netherlands

ROTTERDAM continued

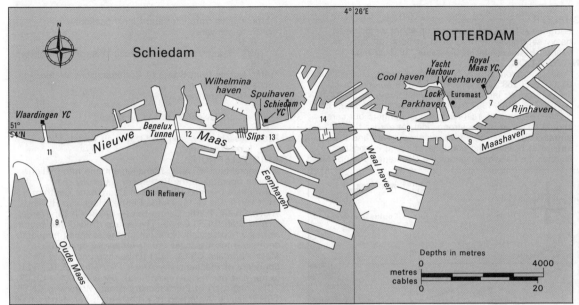

Times				Height (metres)			
HW		LW		MHWS	MHWN	MLWN	MLWS
0300	0900	0400	1000	4.7	3.9	0.8	0.3
1500	2100	1600	2200				

Differences ROTTERDAM
+0202 +0156 +0313 +0400 −2.9 −2.0 −0.4 +0.1

NOTE: Double LWs occur — The rise after the first LW is called the Agger — Predictions for Hoek van Holland are for the lower LW which is usually the second. Time differences for Secondary Ports referred to Hoek van Holland are approximate — Water levels on this coast are considerably affected by the weather. Strong NW gales can raise levels by up to 3m.

SHELTER
Shelter is good in the various yacht harbours but in the river there is always a very considerable sea and swell due to constant heavy traffic (Rotterdam/Europoort is the world's largest port complex). From Berghaven to Rotterdam is about 19 M. All but local yachts are actively discouraged from using the Nieuwe Maas.

NAVIGATION
See 10.19.20. There are no navigational dangers other than the amount of heavy sea-going traffic. 7 M before the centre of Rotterdam, the Oude Maas joins; from here to Dordrecht, sailing is prohibited. Speed limit — 6 kn.

LIGHTS AND MARKS
Marks to locate Yacht Harbours:
(1) Where Oude Maas enters from S, on N bank by Delta Hotel, between kilometre posts 1010-1011 entrance to Vlaardingen YC.
(2) 2 M further up, just above the entrance to Wilhelmina Haven at Schiedam is the Spuihaven.
(3) Parkhaven and the lock through to Coolhaven are clearly marked on the N bank by the huge Euromast in the park.
(4) The Royal Maas YC at the Veerhaven is less than one M further up on the N bank. The Veerhaven is a normal yacht harbour.

RADIO TELEPHONE
There is a comprehensive Traffic Management and Information Service. See details under Hook of Holland.

TELEPHONE (10)
Hr Mr 13.76.81; Local Rijn Pilot 65.91.10; Port Authority 89.40.62; Customs 29.80.88; Hosp 11.28.00; Brit Consul (20) 76.43.43.

FACILITIES
Yacht Harbour Tel. 138514, Slip, M, P, D, L, FW, ME, El, Sh, C, CH, AB, V, R, Bar; **Royal Maas YC** Tel. 137681, D, L, FW, ME, El, Sh, CH, AB; **Schiedam YC** Tel. 267765, D, L, FW, ME, El, Sh, CH, AB; **A.L. Valkhof** Tel. 131226, P, D; **Watersport BV** Tel. 841937, ME; **Vlaardingen YC** M, FW; **Observator** Tel. 130060 ACA; **Handelsmig** Tel. 181860 Gaz; **Town** all facilities. PO; Bank; Rly; Air.
Ferry UK — Hull.
Note: Special regulations apply to yachts in the Rhine. These are given in a French booklet 'Service de la Navigation du Rhin' obtainable from 25 Rue de la Nuée Bleu, 6700, Strasbourg.

HOEK VAN HOLLAND 10-19-20 (HOOK OF HOLLAND)
Zuid Holland

CHARTS
Admiralty 122, 132; Zeekaart 1540/1/2; Dutch Yacht Chart 1809; Stanford 19; Imray C30, Y5

TIDES
Dover +0310; ML 1.1; Duration 0505; Zone −0100

Standard Port VLISSINGEN (←)

Times				Height (metres)			
HW		LW		MHWS	MHWN	MLWN	MLWS
0300	0900	0400	1000	4.7	3.9	0.8	0.3
1500	2100	1600	2200				

Differences HOEK VAN HOLLAND
+0033 +0038 +0118 +0107 −2.4 −2.1 −0.5 +0.1
MAASSLUIS
+0201 +0136 +0040 0000 −2.6 −2.0 −0.4 +0.1

NOTE: Hook of Holland is a Standard Port and tidal predictions for each day of the year are given below. Double LWs occur — The rise after the first LW is called The Agger — Predictions for Hoek van Holland are for the lower LW which is usually the second. Time differences for Secondary Ports referred to Hoek van Holland are approximate — Water levels on this coast are considerably affected by the weather. Strong NW gales can raise the levels by up to 3m.

HOOK OF HOLLAND continued

SHELTER
Entrance safe except in strong on-shore winds when heavy seas develop. Berghaven harbour not open to yachts. Better shelter at Maassluis, 10M up river.

NAVIGATION
Waypoints From S: MV-N N cardinal buoy, Q, 51°59'.65N 04°00'.30E, 288°/108° from/to Nieuwe Zuiderdam Lt, 1.5M. From N: Indusbank N cardinal buoy, Q, 52°02'.93N 04°03'.72E, 010°/190° from/to Nieuwe Noorderdam Lt, 3.2M. There are no real navigational dangers but care should be taken of the strong tidal set across the entrance. The river is a very busy waterway and the stream of ocean-going and local shipping is constant. Neither the Calandkanaal nor the Beerkanaal may be used

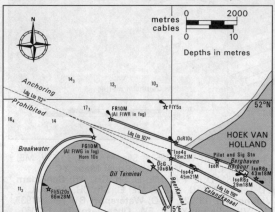

by pleasure craft. Yachts crossing the Hook of Holland roadstead are recommended to call Maas Entrance on Ch 03, reporting yacht's name, position and course, and maintain watch on Ch 03. Yachts should cross under power on a track close W of line joining buoys MV (51°57'.5N, 3°58'.5E), MVN (51°59'.7N, 4°00'.3E) and Indusbank N (52°03'.0N, 4°03'.7E).

LIGHTS AND MARKS
Maasvlakte Fl(5) 20s 66m 28M; B 8-sided Tr, Or bands; vis 340°-267° (H24). Nieuwe Noorderdam, head FR 24m 10M; Or Tr, B bands, helicopter platform; in fog Al Fl WR 6s. Nieuwe Zuiderdam, head FG 24m 10M; Or Tr, B bands, helicopter platform; in fog Al Fl WG 6s; Horn 10s. Ldg Lts 107° (for Rotterdamsche Waterweg). Front Iso R 6s 29m 18M; R Tr, W bands; vis 100°-114° (H24). Rear Iso R 6s 43m 18M; R Tr, W bands; vis 100°-114° (H24), synchronized with front.

Traffic signals from Pilot and Signal Station:
Visible to seaward
R R
W   Entry to Rotterdamsche Waterweg prohibited
R R

Traffic signals from S side of Rotterdamsche Waterweg, opposite Berghaven:
Visible from landward
R R
W   Navigation to sea prohibited
R R

Patrol vessels show a Fl Bu Lt. If such vessels show a Fl R Lt, it means 'Stop'.

RADIO TELEPHONE
A comprehensive Traffic Management and Information Service operates in the Rotterdam waterway. Four Traffic Centres are manned H24: Haven Coordinatie Centrum (HCC) VHF Ch 11 14; Traffic Centre, Hoek van Holland (TCH) Ch 13; Traffic Centre Botlek (VCB) Ch 13; Traffic Centre Stad (VCS) Ch 13. The area is divided into sectors as below; vessels should listen on the assigned frequency and use it for messages unless otherwise directed.
Maas Approach Ch 01, TCH (Outer approaches to W boundary of Precautionary Area); *Pilot Mass* Ch 02, TCH (outer part of Precautionary Area); *Maas Entrance* Ch 03, TCH (inner part of Precautionary Area); *Waterweg* Ch 65, TCH (Nieuwe Waterweg to Kruitsteiger); *Europoort* Ch 66, TCH (Calandkanaal); *Maassluis* Ch 80, VCB (kp 1023 to kp 1017); *Botlek* Ch 61, VCB (kp 1017 to kp 1011); *Eemhaven* Ch 63, VCS (kp 1011 to kp 1007); *Waalhaven* Ch 60, VCS (kp 1007 to kp 1003); *Massbruggen* Ch 81, VCS (kp 1003 to kp 998); *Brieneoord* Ch 21, VCS (kp 998 to kp 993). Yachts should report to Maas Approach or Pilot Maas (or to Maas Entrance if using the Inshore Traffic Zone) and follow instructions. Information broadcasts by Maas Approach and HCC on Ch 01. Weather on Ch 14 every H+00.
Other stations: Oude Maas (call: *RHD Post Hartel*) Ch 10; 13. Bridges at Botlekbrug and Spijkenisserbrug Ch 13. Dordrecht (call: *Post Dordrecht*) Ch 08 10 13 14 71 (H24).

TELEPHONE (10)
Hr Mr Rotterdam 89.69.11; Rotterdam Port Authority 89.40.62; Pilot Hook of Holland 4840; CG 2579; Customs 2418; Hosp Rotterdam 11.28.00;
Brit Consul (20) 76.43.43.

FACILITIES
Schoemeyer Tel. 3464, ME, Sh; **N.C. Ruygers** Tel. 2487., FW; **Maassluis** AB, BY, CH, El, ME, P, PO, R, Sh, V, Rly.
Town P, D, V, R, Bar. PO; Bank; Rly; Air (Rotterdam). Ferry UK—Harwich.

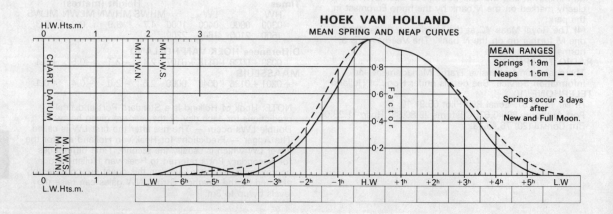

HOEK VAN HOLLAND
MEAN SPRING AND NEAP CURVES

MEAN RANGES
Springs 1·9m
Neaps 1·5m

Springs occur 3 days after New and Full Moon.

AREA 19 — The Netherlands

NETHERLANDS - HOEK VAN HOLLAND
LAT 51°59′N LONG 4°07′E
TIMES AND HEIGHTS OF HIGH AND LOW WATERS

TIME ZONE -0100 (Dutch Standard Time)
For Dutch Summer Time add ONE hour in non-shaded areas

YEAR 1989

JANUARY

Day	TIME	m	Day	TIME	m
1 SU	0245 / 0844 / 1450 / 2125	0.4 / 1.7 / 0.2 / 1.8	16 M	0336 / 0915 / 1525 / 2154	0.5 / 1.8 / 0.1 / 1.8
2 M	0350 / 0933 / 1634 / 2224	0.4 / 1.6 / 0.3 / 1.8	17 TU	0438 / 1029 / 1648 / 2303	0.5 / 1.6 / 0.2 / 1.7
3 TU	0500 / 1054 / 1730 / 2323	0.4 / 1.6 / 0.3 / 1.8	18 W	0607 / 1145 / 1847	0.4 / 1.8 / 0.3
4 W	0617 / 1159 / 1857	0.4 / 1.7 / 0.4	19 TH	0024 / 0728 / 1245 / 2110	1.7 / 0.4 / 1.9 / 0.3
5 TH	0023 / 0835 / 1244 / 2050	1.8 / 1.8 / 1.8 / 0.4	20 F	0123 / 0935 / 1333 / 2200	1.8 / 0.3 / 2.0 / 0.3
6 F	0119 / 0936 / 1334 / 2150	1.9 / 0.3 / 1.9 / 0.4	21 SA	0223 / 1015 / 1428 / 2234 ○	1.8 / 0.2 / 2.0 / 0.4
7 SA	0158 / 1015 / 1414 / 2235 ●	1.9 / 0.3 / 2.0 / 0.4	22 SU	0315 / 1105 / 1503 / 2315	1.9 / 0.2 / 2.0 / 0.4
8 SU	0240 / 0803 / 1455 / 2300	1.9 / 0.3 / 2.1 / 0.4	23 M	0355 / 1140 / 1545 / 2355	1.9 / 0.2 / 2.1 / 0.5
9 M	0325 / 0839 / 1537 / 2335	2.0 / 0.2 / 2.2 / 0.4	24 TU	0414 / 0914 / 1613	1.9 / 0.1 / 2.1
10 TU	0408 / 0918 / 1617	2.0 / 0.1 / 2.2	25 W	0035 / 0444 / 0953 / 1654	0.5 / 2.0 / 0.1 / 2.1
11 W	0020 / 0454 / 1000 / 1701	0.4 / 1.9 / 0.0 / 2.3	26 TH	0100 / 0513 / 1040 / 1735	0.5 / 2.0 / 0.1 / 2.1
12 TH	0105 / 0534 / 1045 / 1754	0.4 / 1.9 / 0.0 / 2.2	27 F	0057 / 0548 / 1125 / 1809	0.4 / 2.0 / 0.1 / 2.0
13 F	0150 / 0618 / 1136 / 1840	0.4 / 1.9 / 0.0 / 2.2	28 SA	0106 / 0620 / 1210 / 1844	0.4 / 1.9 / 0.1 / 2.0
14 SA ☾	0234 / 0708 / 1234 / 1945	0.4 / 1.9 / 0.0 / 2.1	29 SU	0115 / 0658 / 1255 / 1935	0.3 / 1.9 / 0.1 / 1.9
15 SU	0307 / 0804 / 1405 / 2044	0.5 / 1.9 / 0.1 / 1.9	30 M ☾	0154 / 0744 / 1345 / 2024	0.3 / 1.8 / 0.1 / 1.8
			31 TU	0234 / 0855 / 1444 / 2129	0.3 / 1.7 / 0.2 / 1.7

FEBRUARY

Day	TIME	m	Day	TIME	m
1 W	0340 / 1004 / 1628 / 2239	0.4 / 1.6 / 0.3 / 1.6	16 TH	0504 / 1123 / 1850	0.3 / 1.7 / 0.3
2 TH	0540 / 1118 / 1757	0.4 / 1.6 / 0.4	17 F	0024 / 0620 / 1244 / 2057	1.6 / 0.3 / 1.8 / 0.3
3 F	0005 / 0650 / 1224 / 2020	1.6 / 0.3 / 1.7 / 0.4	18 SA	0124 / 0915 / 1332 / 2145	1.7 / 0.2 / 1.9 / 0.3
4 SA	0058 / 0906 / 1314 / 2135	1.7 / 0.3 / 1.9 / 0.3	19 SU	0214 / 1014 / 1418 / 2215	1.8 / 0.1 / 2.0 / 0.3
5 SU	0142 / 1045 / 1353 / 2205	1.8 / 0.1 / 2.0 / 0.4	20 M ○	0254 / 1045 / 1455 / 2255	1.8 / 0.1 / 2.0 / 0.4
6 M	0228 / 0745 / 1425 / 2237 ●	1.9 / 0.2 / 2.1 / 0.4	21 TU	0324 / 0823 / 1517 / 2317	1.9 / 0.1 / 2.0 / 0.4
7 TU	0304 / 0815 / 1517 / 2326	1.9 / 0.1 / 2.2 / 0.4	22 W	0343 / 0853 / 1549	1.9 / 0.1 / 2.1
8 W	0344 / 0851 / 1559	2.0 / 0.0 / 2.3	23 TH	0005 / 0419 / 0929 / 1628	0.4 / 2.0 / 0.1 / 2.1
9 TH	0006 / 0427 / 1237 / 1645	0.4 / 2.0 / -0.1 / 2.3	24 F	0017 / 0444 / 1237 / 1658	0.3 / 2.0 / 0.1 / 2.1
10 F	0051 / 0510 / 1015 / 1727	0.3 / 2.0 / -0.1 / 2.2	25 SA	0035 / 0515 / 1305 / 1728	0.3 / 2.0 / 0.1 / 2.0
11 SA	0124 / 0554 / 1109 / 1815	0.4 / 2.0 / 0.0 / 2.1	26 SU	0115 / 0544 / 1114 / 1754	0.3 / 2.0 / 0.1 / 2.0
12 SU	0215 / 0638 / 1430 / 1915	0.4 / 2.0 / 0.1 / 2.0	27 M	0104 / 0615 / 1205 / 1828	0.3 / 2.0 / 0.1 / 1.9
13 M	0205 / 0730 / 1420 / 2008	0.4 / 2.0 / 0.0 / 1.8	28 TU ☾	0120 / 0648 / 1330 / 1914	0.2 / 1.9 / 0.1 / 1.8
14 TU	0235 / 0839 / 1520 / 2124	0.3 / 1.8 / 0.1 / 1.6			
15 W	0344 / 1004 / 1630 / 2259	0.3 / 1.7 / 0.2 / 1.5			

MARCH

Day	TIME	m	Day	TIME	m
1 W	0154 / 0733 / 1424 / 2023	0.2 / 1.8 / 0.2 / 1.6	16 TH	0314 / 0955 / 1610 / 2243	0.2 / 1.7 / 0.3 / 1.4
2 TH	0244 / 0904 / 1535 / 2159	0.3 / 1.6 / 0.3 / 1.5	17 F	0435 / 1113 / 1840	0.2 / 1.7 / 0.3
3 F	0457 / 1049 / 1720 / 2324	0.3 / 1.6 / 0.3 / 1.5	18 SA	0003 / 0607 / 1235 / 2027	1.5 / 0.2 / 1.8 / 0.3
4 SA	0610 / 1153 / 2015	0.3 / 1.7 / 0.3	19 SU	0058 / 0910 / 1314 / 2124	1.7 / 0.1 / 1.9 / 0.3
5 SU	0039 / 0835 / 1252 / 2115	1.6 / 0.2 / 1.9 / 0.3	20 M	0143 / 0954 / 1354 / 2155	1.8 / 0.1 / 1.9 / 0.3
6 M	0129 / 0936 / 1338 / 2155	1.7 / 0.2 / 2.0 / 0.3	21 TU	0224 / 1037 / 1424 / 2230	1.8 / 0.1 / 2.0 / 0.3
7 TU ●	0205 / 0715 / 1415 / 2215	1.8 / 0.1 / 2.2 / 0.3	22 W ○	0255 / 0804 / 1455 / 2250	1.9 / 0.1 / 2.0 / 0.3
8 W	0244 / 0748 / 1456 / 2306	1.9 / 0.0 / 2.2 / 0.4	23 TH	0314 / 0829 / 1524 / 2314	1.9 / 0.1 / 2.1 / 0.3
9 TH	0324 / 0826 / 1538 / 2335	2.0 / -0.1 / 2.3 / 0.3	24 F	0345 / 1130 / 1554 / 2350	2.0 / 0.1 / 2.1 / 0.2
10 F	0401 / 0907 / 1620	2.1 / -0.1 / 2.2	25 SA	0414 / 1226 / 1624	2.0 / 0.1 / 2.0
11 SA	0036 / 0441 / 0955 / 1705	0.3 / 2.1 / -0.1 / 2.1	26 SU	0015 / 0440 / 1255 / 1657	0.3 / 2.0 / 0.1 / 2.0
12 SU	0104 / 0526 / 1355 / 1754	0.3 / 2.1 / 0.0 / 2.0	27 M	0045 / 0508 / 1314 / 1724	0.2 / 2.0 / 0.1 / 1.9
13 M	0150 / 0610 / 1347 / 1840	0.3 / 2.1 / 0.0 / 1.9	28 TU	0104 / 0541 / 1330 / 1759	0.2 / 2.0 / 0.2 / 1.9
14 TU ☾	0124 / 0705 / 1355 / 1944	0.2 / 2.0 / 0.1 / 1.7	29 W	0008 / 0614 / 1330 / 1838	0.2 / 2.0 / 0.2 / 1.8
15 W	0204 / 0814 / 1454 / 2053	0.2 / 1.8 / 0.2 / 1.5	30 TH	0114 / 0700 / 1420 / 1938	0.1 / 1.9 / 0.2 / 1.6
			31 F	0204 / 0814 / 1520 / 2124	0.2 / 1.6 / 0.3 / 1.4

APRIL

Day	TIME	m	Day	TIME	m
1 SA	0314 / 1019 / 1656 / 2255	0.2 / 1.6 / 0.3 / 1.4	16 SU	0504 / 1155 / 1944	0.1 / 1.8 / 0.2
2 SU	0440 / 1123 / 1937	0.2 / 1.8 / 0.3	17 M	0028 / 0554 / 1245 / 2045	1.6 / 0.1 / 1.9 / 0.2
3 M	0004 / 0524 / 1224 / 2055	1.5 / 0.2 / 1.9 / 0.2	18 TU	0115 / 0925 / 1318 / 2130	1.7 / 0.1 / 1.9 / 0.2
4 TU	0059 / 0606 / 1307 / 2125	1.7 / 0.1 / 2.1 / 0.3	19 W	0143 / 0715 / 1355 / 2157	1.8 / 0.1 / 1.9 / 0.2
5 W	0137 / 0645 / 1346 / 2205	1.8 / 0.0 / 2.2 / 0.3	20 TH	0213 / 0744 / 1424 / 2230	1.8 / 0.1 / 2.0 / 0.2
6 TH	0217 / 0725 / 1431 / 2240	1.9 / -0.1 / 2.2 / 0.3	21 F	0245 / 0804 / 1454 / 2240	1.9 / 0.2 / 2.0 / 0.2
7 F ●	0254 / 0801 / 1514 / 2325	2.0 / -0.1 / 2.2 / 0.3	22 SA	0314 / 1057 / 1524 / 2320	2.0 / 0.1 / 2.0 / 0.2
8 SA	0336 / 0848 / 1556	2.1 / 0.0 / 2.1	23 SU	0344 / 1150 / 1558 / 2357	2.0 / 0.1 / 2.0 / 0.2
9 SU	0016 / 0417 / 1255 / 1639	0.2 / 2.2 / 0.0 / 2.0	24 M	0411 / 1236 / 1628	2.0 / 0.2 / 1.9
10 M	0105 / 0459 / 1335 / 1727	0.2 / 2.2 / 0.0 / 1.9	25 TU	0025 / 0445 / 1245 / 1700	0.2 / 2.0 / 0.2 / 1.9
11 TU	0145 / 0544 / 1400 / 1818	0.2 / 2.1 / 0.1 / 1.7	26 W	0110 / 0519 / 1325 / 1738	0.2 / 2.0 / 0.2 / 1.8
12 W ☾	0034 / 0640 / 1344 / 1913	0.1 / 2.0 / 0.2 / 1.5	27 TH	0000 / 0557 / 1340 / 1825	0.1 / 2.0 / 0.2 / 1.7
13 TH	0135 / 0753 / 1435 / 2044	0.1 / 1.8 / 0.3 / 1.4	28 F	0025 / 0644 / 1418 / 1923	0.1 / 1.9 / 0.3 / 1.5
14 F	0250 / 0923 / 1607 / 2224	0.1 / 1.7 / 0.3 / 1.4	29 SA	0146 / 0809 / 1527 / 2105	0.1 / 1.7 / 0.3 / 1.4
15 SA	0405 / 1055 / 1814 / 2334	0.2 / 1.7 / 0.3 / 1.5	30 SU	0246 / 0955 / 1710 / 2219	0.1 / 1.7 / 0.3 / 1.4

LOW WATERS – IMPORTANT NOTE. DOUBLE LOW WATERS OFTEN OCCUR. PREDICTIONS ARE FOR THE LOWER LOW WATER WHICH IS USUALLY THE SECOND.

Chart Datum: 0.84 metres below Normaal Amsterdams Peil

NETHERLANDS - HOEK VAN HOLLAND

LAT 51°59'N   LONG 4°07'E

TIMES AND HEIGHTS OF HIGH AND LOW WATERS

YEAR **1989**

TIME ZONE –0100 (Dutch Standard Time)
For Dutch Summer Time add ONE hour in non-shaded areas

MAY

Day	TIME	m	Day	TIME	m
1 M	0345 / 1054 / 1915 / 2335	0.1 / 1.8 / 0.3 / 1.5	16 TU	0525 / 1205 / 1955	0.1 / 1.8 / 0.2
2 TU	0439 / 1155 / 2020	0.1 / 2.0 / 0.2	17 W	0028 / 0609 / 1245 / 2035	1.6 / 0.1 / 1.9 / 0.2
3 W	0025 / 0536 / 1240 / 2045	1.7 / 0.0 / 2.1 / 0.3	18 TH	0109 / 0644 / 1318 / 2130	1.7 / 0.1 / 1.9 / 0.2
4 TH	0108 / 0615 / 1325 / 2125	1.8 / 0.0 / 2.1 / 0.3	19 F	0144 / 0740 / 1354 / 2200	1.8 / 0.2 / 1.9 / 0.2
5 F ●	0147 / 0706 / 1410 / 2226	2.0 / 0.0 / 2.1 / 0.3	20 SA ○	0214 / 0950 / 1428 / 2230	1.9 / 0.2 / 1.9 / 0.2
6 SA	0231 / 0745 / 1455 / 2310	2.1 / 0.0 / 2.0 / 0.2	21 SU	0244 / 1040 / 1504 / 2315	1.9 / 0.2 / 1.9 / 0.2
7 SU	0314 / 1135 / 1540	2.1 / 0.1 / 2.0	22 M	0318 / 1115 / 1534 / 2340	2.0 / 0.2 / 1.9 / 0.1
8 M	0006 / 0356 / 1224 / 1624	0.2 / 2.2 / 0.1 / 1.9	23 TU	0355 / 1200 / 1614 / 2145	2.0 / 0.1 / 1.9 / 0.1
9 TU	0057 / 0445 / 1326 / 1713	0.1 / 2.1 / 0.1 / 1.7	24 W	0427 / 1225 / 1648 / 2225	2.0 / 0.3 / 1.8 / 0.1
10 W	0146 / 0535 / 1354 / 1808	0.1 / 2.0 / 0.2 / 1.7	25 TH	0504 / 1315 / 1735 / 2305	2.0 / 0.3 / 1.7 / 0.0
11 TH	0000 / 0623 / 1407 / 1914	— / 1.9 / 0.3 / 1.6	26 F	0544 / 1354 / 1818	2.0 / 0.3 / 1.6
12 F ☽	0054 / 0744 / 1415 / 2009	0.0 / 1.8 / 0.3 / 1.5	27 SA	0005 / 0644 / 1435 / 1924	0.0 / 1.9 / 0.3 / 1.6
13 SA	0210 / 0854 / 1520 / 2124	0.1 / 1.7 / 0.4 / 1.4	28 SU ☾	0053 / 0753 / 1537 / 2035	0.0 / 1.9 / 0.3 / 1.5
14 SU	0325 / 1003 / 1737 / 2249	0.1 / 1.7 / 0.3 / 1.5	29 M	0215 / 0914 / 1700 / 2144	0.0 / 1.9 / 0.3 / 1.5
15 M	0425 / 1115 / 1844 / 2355	0.1 / 1.8 / 0.3 / 1.5	30 TU	0315 / 1025 / 1820 / 2255	0.0 / 1.9 / 0.3 / 1.6

JUNE

Day	TIME	m	Day	TIME	m
1 TH	0504 / 1219 / 2007	0.0 / 2.0 / 0.3	16 F	0022 / 0640 / 1244 / 2037	1.7 / 0.2 / 1.8 / 0.2
2 F	0044 / 0606 / 1306 / 2104	1.9 / 0.0 / 2.0 / 0.3	17 SA	0103 / 0757 / 1329 / 2134	1.7 / 0.3 / 1.8 / 0.2
3 SA ●	0128 / 0656 / 1357 / 2145	2.0 / 0.1 / 2.0 / 0.2	18 SU	0144 / 0940 / 1402 / 2221	1.8 / 0.3 / 1.9 / 0.2
4 SU	0215 / 1025 / 1439 / 2244	2.1 / 0.2 / 1.9 / 0.2	19 M ○	0220 / 1030 / 1445 / 2250	1.9 / 0.3 / 1.9 / 0.2
5 M	0300 / 1135 / 1534 / 2346	2.0 / 0.2 / 1.8 / 0.1	20 TU	0258 / 1110 / 1520 / 2049	2.0 / 0.3 / 1.9 / 0.2
6 TU	0345 / 1205 / 1624	2.1 / 0.2 / 1.8	21 W	0337 / 1205 / 1605 / 2125	2.0 / 0.4 / 1.8 / 0.1
7 W	0036 / 0434 / 1300 / 1713	0.1 / 2.1 / 0.3 / 1.7	22 TH	0417 / 1220 / 1645 / 2205	2.1 / 0.3 / 1.8 / 0.1
8 TH	0127 / 0524 / 1356 / 1758	0.1 / 2.0 / 0.3 / 1.7	23 F	0457 / 1305 / 1724 / 2246	2.1 / 0.3 / 1.8 / 0.0
9 F	0205 / 0612 / 1417 / 1844	0.0 / 2.0 / 0.4 / 1.7	24 SA	0544 / 1356 / 1814 / 2324	2.1 / 0.3 / 1.7 / 0.0
10 SA	0014 / 0703 / 1356 / 1923	0.0 / 1.9 / 0.4 / 1.6	25 SU	0630 / 1435 / 1904	2.1 / 0.3 / 1.7
11 SU ☽	0120 / 0804 / 1437 / 2012	0.0 / 1.8 / 0.4 / 1.6	26 M ☾	0023 / 0735 / 1515 / 2005	0.0 / 2.0 / 0.4 / 1.7
12 M	0250 / 0905 / 1534 / 2113	0.1 / 1.8 / 0.3 / 1.5	27 TU	0135 / 0844 / 1620 / 2116	0.0 / 2.0 / 0.4 / 1.7
13 TU	0345 / 1003 / 1710 / 2244	0.1 / 1.7 / 0.4 / 1.5	28 W	0244 / 0944 / 1710 / 2225	0.0 / 1.9 / 0.4 / 1.7
14 W	0444 / 1115 / 1730 / 2333	0.1 / 1.7 / 0.3 / 1.6	29 TH	0404 / 1055 / 1820 / 2325	0.0 / 1.9 / 0.4 / 1.8
15 TH	0544 / 1158 / 1950	0.2 / 1.8 / 0.2	30 F	0504 / 1153 / 1910	0.1 / 1.9 / 0.2

JULY

Day	TIME	m	Day	TIME	m
1 SA	0020 / 0604 / 1254 / 2014	1.9 / 0.2 / 1.8 / 0.3	16 SU	0039 / 0727 / 1304 / 2008	1.7 / 0.3 / 1.7 / 0.3
2 SU	0114 / 0930 / 1343 / 2150	2.0 / 0.2 / 1.8 / 0.2	17 M	0124 / 0924 / 1348 / 2155	1.8 / 0.4 / 1.8 / 0.2
3 M ●	0204 / 1014 / 1444 / 2234	2.0 / 0.3 / 1.8 / 0.2	18 TU ○	0204 / 1014 / 1434 / 1954	1.9 / 0.4 / 1.8 / 0.2
4 TU	0250 / 1116 / 1523 / 2335	2.1 / 0.3 / 1.8 / 0.1	19 W	0240 / 1037 / 1503 / 2025	2.0 / 0.4 / 1.9 / 0.1
5 W	0338 / 1155 / 1612 / 2122	2.1 / 0.4 / 1.8 / 0.1	20 TH	0324 / 1125 / 1548 / 2055	2.1 / 0.4 / 1.9 / 0.1
6 TH	0425 / 1245 / 1654 / 2205	2.1 / 0.4 / 1.8 / 0.1	21 F	0357 / 1154 / 1627 / 2135	2.2 / 0.5 / 1.9 / 0.0
7 F	0504 / 1325 / 1745 / 2249	2.1 / 0.4 / 1.8 / 0.0	22 SA	0440 / 1246 / 1710 / 2215	2.2 / 0.4 / 1.9 / 0.0
8 SA	0548 / 1410 / 1814 / 2334	2.1 / 0.4 / 1.8 / 0.0	23 SU	0524 / 1336 / 1754 / 2306	2.2 / 0.4 / 1.9 / 0.0
9 SU	0635 / 1327 / 1843	2.0 / 0.3 / 1.8	24 M	0614 / 1405 / 1839 / 2353	2.2 / 0.4 / 1.9 / 0.0
10 M	0024 / 0724 / 1345 / 1935	0.0 / 1.9 / 0.4 / 1.8	25 TU ☾	0705 / 1445 / 1928	2.1 / 0.4 / 1.9
11 TU	0129 / 0804 / 1445 / 2025	0.1 / 1.8 / 0.3 / 1.7	26 W	0120 / 0804 / 1507 / 2035	0.0 / 2.0 / 0.4 / 1.8
12 W	0246 / 0854 / 1527 / 2112	0.1 / 1.8 / 0.4 / 1.6	27 TH	0245 / 0914 / 1535 / 2144	0.1 / 1.8 / 0.4 / 1.8
13 TH	0357 / 0959 / 1640 / 2229	0.2 / 1.7 / 0.3 / 1.6	28 F	0404 / 1023 / 1650 / 2305	0.1 / 1.7 / 0.4 / 1.8
14 F	0510 / 1054 / 1750 / 2334	0.2 / 1.7 / 0.3 / 1.6	29 SA	0515 / 1149 / 1807	0.2 / 1.7 / 0.3
15 SA	0610 / 1209 / 1850	0.3 / 1.7 / 0.3	30 SU	0015 / 0727 / 1254 / 1844	1.8 / 0.3 / 1.7 / 0.3
			31 M	0114 / 0934 / 1353 / 2155	1.9 / 0.3 / 1.8 / 0.2

AUGUST

Day	TIME	m	Day	TIME	m
1 TU ●	0203 / 1005 / 1455 / 2246	2.0 / 0.4 / 1.8 / 0.2	16 W	0142 / 0945 / 1408 / 1925	2.0 / 0.4 / 1.8 / 0.2
2 W	0244 / 1050 / 1535 / 2025	2.1 / 0.4 / 1.9 / 0.2	17 TH ○	0220 / 0744 / 1444 / 1955	2.1 / 0.5 / 1.9 / 0.1
3 TH	0324 / 1124 / 1559 / 2054	2.1 / 0.5 / 1.9 / 0.1	18 F	0257 / 0819 / 1525 / 2032	2.2 / 0.5 / 2.0 / 0.0
4 F	0404 / 1220 / 1628 / 2135	2.1 / 0.5 / 1.9 / 0.1	19 SA	0338 / 1146 / 1604 / 2110	2.3 / 0.5 / 2.0 / 0.0
5 SA	0438 / 1254 / 1657 / 2215	2.1 / 0.5 / 2.0 / 0.1	20 SU	0420 / 1226 / 1644 / 2149	2.3 / 0.5 / 2.1 / 0.0
6 SU	0519 / 1327 / 1734 / 2254	2.1 / 0.5 / 2.0 / 0.1	21 M	0501 / 1315 / 1727 / 2235	2.3 / 0.5 / 2.1 / 0.0
7 M	0554 / 1310 / 1809 / 2350	2.0 / 0.4 / 1.9 / 0.1	22 TU	0546 / 1335 / 1808 / 2329	2.2 / 0.5 / 2.1 / 0.1
8 TU	0623 / 1310 / 1845	2.0 / 0.4 / 1.9	23 W ☽	0634 / 1410 / 1857	2.1 / 0.4 / 2.0
9 W ☾	0034 / 0716 / 1340 / 1930	0.1 / 1.9 / 0.3 / 1.8	24 TH	0200 / 0735 / 1420 / 1953	0.1 / 1.9 / 0.4 / 1.9
10 TH	0140 / 0805 / 1415 / 2025	0.2 / 1.8 / 0.3 / 1.7	25 F	0244 / 0845 / 1504 / 2113	0.2 / 1.7 / 0.4 / 1.8
11 F	0229 / 0905 / 1520 / 2134	0.3 / 1.7 / 0.4 / 1.6	26 SA	0344 / 1025 / 1615 / 2255	0.3 / 1.6 / 0.4 / 1.8
12 SA	0420 / 1004 / 1720 / 2305	0.4 / 1.6 / 0.4 / 1.6	27 SU	0537 / 1155 / 1746	0.4 / 1.6 / 0.4
13 SU	0540 / 1123 / 1805	0.4 / 1.6 / 0.3	28 M	0004 / 0830 / 1305 / 2047	1.9 / 0.4 / 1.7 / 0.3
14 M	0015 / 0640 / 1244 / 1855	1.7 / 0.4 / 1.7 / 0.3	29 TU	0103 / 0930 / 1342 / 2145	2.0 / 0.4 / 1.8 / 0.2
15 TU	0104 / 0915 / 1323 / 1904	1.8 / 0.4 / 1.8 / 0.3	30 W	0153 / 1037 / 1429 / 2234	2.0 / 0.4 / 1.9 / 0.2
			31 TH ●	0228 / 1025 / 1505 / 1954	2.1 / 0.5 / 1.9 / 0.2

LOW WATERS – IMPORTANT NOTE. DOUBLE LOW WATERS OFTEN OCCUR. PREDICTIONS ARE FOR THE LOWER LOW WATER WHICH IS USUALLY THE SECOND.

Chart Datum: 0.84 metres below Normaal Amsterdams Peil

AREA 19 — The Netherlands

NETHERLANDS - HOEK VAN HOLLAND
LAT 51°59′N LONG 4°07′E
TIMES AND HEIGHTS OF HIGH AND LOW WATERS
YEAR 1989

TIME ZONE −0100 (Dutch Standard Time)
For Dutch Summer Time add ONE hour in non-shaded areas

SEPTEMBER

Day	Time	m	Day	Time	m
1 F	0300 / 1055 / 1528 / 2036	2.1 / 0.5 / 2.0 / 0.2	16 SA	0231 / 0756 / 1456 / 2005	2.3 / 0.5 / 2.1 / 0.0
2 SA	0334 / 1137 / 1554 / 2105	2.1 / 0.5 / 2.0 / 0.2	17 SU	0311 / 0828 / 1536 / 2043	2.4 / 0.5 / 2.2 / 0.0
3 SU	0408 / 1220 / 1628 / 2139	2.2 / 0.5 / 2.1 / 0.2	18 M	0354 / 0905 / 1616 / 2126	2.4 / 0.5 / 2.2 / 0.1
4 M	0444 / 1237 / 1654 / 2215	2.1 / 0.5 / 2.1 / 0.2	19 TU	0438 / 0945 / 1657 / 2215	2.3 / 0.5 / 2.2 / 0.1
5 TU	0514 / 1300 / 1724 / 2300	2.1 / 0.5 / 2.1 / 0.2	20 W	0521 / 1330 / 1740	2.2 / 0.5 / 2.2
6 W	0544 / 1217 / 1754 / 2345	2.0 / 0.4 / 2.0 / 0.3	21 TH	0140 / 0610 / 1238 / 1830	0.2 / 2.0 / 0.4 / 2.1
7 TH	0615 / 1237 / 1828	2.0 / 0.3 / 2.0	22 F	0137 / 0702 / 1325 / 1935	0.3 / 1.8 / 0.4 / 2.0 ☾
8 F	0100 / 0643 / 1324 / 1908 ☽	0.3 / 1.9 / 0.3 / 1.9	23 SA	0230 / 0824 / 1434 / 2115	0.3 / 1.6 / 0.4 / 1.8
9 SA	0155 / 0743 / 1420 / 2029	0.3 / 1.7 / 0.4 / 1.7	24 SU	0340 / 1003 / 1610 / 2244	0.5 / 1.5 / 0.5 / 1.8
10 SU	0310 / 0924 / 1530 / 2224	0.4 / 1.6 / 0.4 / 1.6	25 M	0620 / 1134 / 1806 / 2354	0.5 / 1.6 / 0.4 / 1.9
11 M	0500 / 1044 / 1750 / 2344	0.5 / 1.5 / 0.4 / 1.7	26 TU	0807 / 1234 / 2046	0.6 / 1.7 / 0.3
12 TU	0628 / 1214 / 1846	0.5 / 1.6 / 0.4	27 W	0054 / 0914 / 1325 / 2140	2.0 / 0.4 / 1.8 / 0.2
13 W	0033 / 0844 / 1304 / 1819	1.9 / 0.4 / 1.7 / 0.3	28 TH	0134 / 0950 / 1358 / 2224	2.1 / 0.4 / 1.9 / 0.2
14 TH	0113 / 0940 / 1345 / 1855	2.1 / 0.4 / 1.9 / 0.2	29 F	0204 / 1007 / 1428 / 1939 ●	2.2 / 0.5 / 2.0 / 0.3
15 F ○	0154 / 1424 / 1925	2.2 / 2.0 / 0.1 (0714 0.5, 1030 0.5)	30 SA	0237 / 1030 / 1458 / 2004	2.1 / 0.5 / 2.0 / 0.3

OCTOBER

Day	Time	m	Day	Time	m
1 SU	0308 / 1110 / 1524 / 2039	2.2 / 0.5 / 2.1 / 0.3	16 M	0250 / 0801 / 1515 / 2022	2.4 / 0.4 / 2.3 / 0.1
2 M	0337 / 1130 / 1557 / 2104	2.2 / 0.5 / 2.1 / 0.3	17 TU	0335 / 0841 / 1555 / 2108	2.3 / 0.4 / 2.3 / 0.1
3 TU	0410 / 1210 / 1627 / 2146	2.2 / 0.4 / 2.1 / 0.3	18 W	0416 / 0930 / 1634	2.2 / 0.4 / 2.3
4 W	0437 / 1240 / 1654 / 2215	2.1 / 0.4 / 2.1 / 0.4	19 TH	0055 / 0504 / 1026 / 1720	0.3 / 2.1 / 0.4 / 2.3
5 TH	0509 / 1034 / 1725 / 2255	2.1 / 0.4 / 2.1 / 0.4	20 F	0137 / 0555 / 1124 / 1814	0.4 / 1.9 / 0.3 / 2.1
6 F	0538 / 1140 / 1754	2.0 / 0.3 / 2.0	21 SA	0120 / 0643 / 1245 / 1924 ☾	0.4 / 1.7 / 0.3 / 2.0
7 SA	0027 / 0614 / 1245 / 1834	0.4 / 1.9 / 0.3 / 2.0	22 SU	0210 / 0803 / 1400 / 2044	0.5 / 1.6 / 0.3 / 1.9
8 SU ☽	0135 / 0702 / 1345 / 1934	0.4 / 1.8 / 0.3 / 1.8	23 M	0437 / 0944 / 1540 / 2213	0.6 / 1.5 / 0.4 / 1.8
9 M	0235 / 0844 / 1434 / 2144	0.5 / 1.6 / 0.4 / 1.7	24 TU	0555 / 1105 / 1800 / 2322	0.5 / 1.6 / 0.3 / 1.9
10 TU	0418 / 1009 / 1600 / 2304	0.6 / 1.5 / 0.4 / 1.8	25 W	0730 / 1205 / 1937	0.5 / 1.7 / 0.4
11 W	0647 / 1124 / 1655	0.6 / 1.6 / 0.4	26 TH	0019 / 0835 / 1248 / 2100	2.0 / 0.4 / 1.8 / 0.4
12 TH	0004 / 0830 / 1228 / 1739	2.0 / 0.5 / 1.8 / 0.3	27 F	0053 / 0905 / 1328 / 2145	2.1 / 0.4 / 1.9 / 0.3
13 F	0044 / 0915 / 1310 / 1819	2.2 / 0.5 / 1.9 / 0.2	28 SA	0134 / 0935 / 1358 / 1915	2.2 / 0.5 / 2.0 / 0.3
14 SA ○	0128 / 0649 / 1350 / 1900	2.3 / 0.5 / 2.0 / 0.2	29 SU	0208 / 1025 / 1424 / 1944 ●	2.1 / 0.4 / 2.1 / 0.3
15 SU	0207 / 0725 / 1429 / 1938	2.3 / 0.4 / 2.2 / 0.1	30 M	0245 / 0957 / 1457 / 2015	2.1 / 0.5 / 2.1 / 0.4
31 TU	0310 / 1100 / 1526 / 2310	2.1 / 0.4 / 2.1 / 0.4			

NOVEMBER

Day	Time	m	Day	Time	m
1 W	0345 / 1140 / 1558	2.1 / 0.4 / 2.1	16 TH	0005 / 0359 / 0920 / 1620	0.3 / 2.1 / 0.3 / 2.3
2 TH	0000 / 0417 / 1225 / 1628	0.4 / 2.1 / 0.4 / 2.1	17 F	0044 / 0454 / 1009 / 1707	0.4 / 1.9 / 0.3 / 2.2
3 F	0024 / 0447 / 1025 / 1700	0.5 / 2.0 / 0.3 / 2.1	18 SA	0134 / 0542 / 1115 / 1804	0.4 / 1.8 / 0.2 / 2.1
4 SA	0055 / 0525 / 1104 / 1739	0.5 / 2.0 / 0.3 / 2.1	19 SU	0220 / 0643 / 1220 / 1903	0.5 / 1.8 / 0.2 / 2.0
5 SU	0050 / 0558 / 1205 / 1825	0.5 / 1.9 / 0.2 / 2.0	20 M	0150 / 0744 / 1320 / 2025 ☾	0.6 / 1.7 / 0.2 / 1.9
6 M	0134 / 0654 / 1305 / 1924	0.5 / 1.7 / 0.2 / 1.9	21 TU	0427 / 0843 / 1450 / 2134	0.6 / 1.6 / 0.3 / 1.9
7 TU	0227 / 0825 / 1355 / 2104	0.5 / 1.6 / 0.3 / 1.9	22 W	0546 / 1025 / 1605 / 2244	0.6 / 1.6 / 0.3 / 1.9
8 W	0357 / 0944 / 1505 / 2224	0.6 / 1.6 / 0.3 / 1.9	23 TH	0640 / 1119 / 1654 / 2339	0.6 / 1.7 / 0.3 / 1.9
9 TH	0617 / 1055 / 1604 / 2329	0.6 / 1.6 / 0.3 / 2.1	24 F	0735 / 1209 / 1744	0.4 / 1.8 / 0.3
10 F	0745 / 1155 / 1716	0.5 / 1.8 / 0.2	25 SA	0024 / 0830 / 1244 / 1835	2.0 / 0.4 / 1.8 / 0.3
11 SA	0014 / 0845 / 1245 / 1755	2.2 / 0.5 / 1.9 / 0.2	26 SU	0102 / 0905 / 1324 / 1920	2.0 / 0.5 / 1.9 / 0.4
12 SU	0100 / 0920 / 1324 / 1835	2.2 / 0.5 / 2.1 / 0.3	27 M	0144 / 0954 / 1358 / 2026	2.0 / 0.4 / 2.0 / 0.4
13 M	0146 / 0705 / 1406 / 1922 ○	2.3 / 0.5 / 2.2 / 0.2	28 TU	0218 / 1035 / 1437 / 2157	2.0 / 0.4 / 2.0 / 0.4
14 TU	0230 / 0748 / 1449 / 2005	2.2 / 0.4 / 2.3 / 0.4	29 W	0254 / 1037 / 1504 / 2250	2.0 / 0.3 / 2.1 / 0.4
15 W	0316 / 0829 / 1534 / 2115	2.2 / 0.4 / 2.3 / 0.4	30 TH	0324 / 1117 / 1538 / 2325	2.0 / 0.3 / 2.1 / 0.5

DECEMBER

Day	Time	m	Day	Time	m
1 F	0358 / 1157 / 1615	2.0 / 0.3 / 2.1	16 SA	0035 / 0443 / 0959 / 1704	0.4 / 1.9 / 0.2 / 2.2
2 SA	0010 / 0434 / 1004 / 1648	0.5 / 2.0 / 0.3 / 2.1	17 SU	0115 / 0544 / 1044 / 1754	0.5 / 1.9 / 0.1 / 2.1
3 SU	0050 / 0514 / 1050 / 1728	0.5 / 1.9 / 0.2 / 2.1	18 M	0205 / 0624 / 1144 / 1844	0.5 / 1.9 / 0.1 / 2.1
4 M	0130 / 0559 / 1124 / 1815	0.5 / 1.8 / 0.2 / 2.1	19 TU	0300 / 0708 / 1244 / 1938 ☾	0.6 / 1.8 / 0.1 / 2.0
5 TU	0210 / 0649 / 1235 / 1915	0.5 / 1.8 / 0.1 / 2.0	20 W	0155 / 0752 / 1400 / 2045	0.5 / 1.8 / 0.2 / 1.9
6 W	0257 / 0743 / 1325 / 2023	0.5 / 1.7 / 0.2 / 1.9	21 TH	0254 / 0848 / 1505 / 2134	0.5 / 1.7 / 0.2 / 1.8
7 TH	0357 / 0853 / 1436 / 2138	0.6 / 1.7 / 0.2 / 1.9	22 F	0344 / 1003 / 1615 / 2244	0.5 / 1.6 / 0.3 / 1.8
8 F	0530 / 1015 / 1540 / 2249	0.6 / 1.7 / 0.2 / 2.0	23 SA	0500 / 1125 / 1715 / 2354	0.5 / 1.6 / 0.3 / 1.8
9 SA	0650 / 1119 / 1634 / 2349	0.5 / 1.8 / 0.2 / 2.1	24 SU	0717 / 1203 / 1820	0.4 / 1.7 / 0.3
10 SU	0750 / 1214 / 1740	0.5 / 1.9 / 0.2	25 M	0034 / 0830 / 1259 / 1936	1.8 / 0.4 / 1.8 / 0.4
11 M	0044 / 0850 / 1304 / 1824	2.0 / 0.5 / 2.1 / 0.2	26 TU	0113 / 0917 / 1333 / 2106	2.0 / 0.3 / 1.9 / 0.4
12 TU	0130 / 0940 / 1354 / 2210 ○	2.1 / 0.4 / 2.2 / 0.3	27 W	0158 / 1005 / 1414 / 2207	1.9 / 0.3 / 1.9 / 0.4
13 W	0217 / 1024 / 1437 / 2306	2.0 / 0.4 / 2.2 / 0.3	28 TH	0233 / 1041 / 1443 / 2240 ●	1.9 / 0.3 / 2.0 / 0.4
14 TH	0307 / 1126 / 1525 / 2345	2.0 / 0.3 / 2.2 / 0.4	29 F	0314 / 1110 / 1524 / 2320	1.9 / 0.3 / 2.1 / 0.5
15 F	0353 / 1216 / 1610 / 2325	1.9 / 0.2 / 2.2 / 0.5	30 SA	0354 / 0915 / 1557 / 2345	2.0 / 0.2 / 2.1 / 0.5
31 SU	0423 / 0945 / 1638	1.9 / 0.2 / 2.2			

LOW WATERS — IMPORTANT NOTE. DOUBLE LOW WATERS OFTEN OCCUR. PREDICTIONS ARE FOR THE LOWER LOW WATER WHICH IS USUALLY THE SECOND.

Chart Datum: 0.84 metres below Normaal Amsterdams Peil

Harbour, Coastal and Tidal Information

SCHEVENINGEN 10-19-21
Zuid Holland

CHARTS
Admiralty 2322; Zeekaart 1450; Dutch Yacht Chart 1801; Dutch Waterkaarts ANWB H/J; Stanford 19; Imray Y5

TIDES
Dover +0320; ML No data; Duration 0445; Zone −0100

Standard Port VLISSINGEN (←)

Times				Height (metres)			
HW		LW		MHWS	MHWN	MLWN	MLWS
0300	0900	0400	1000	4.7	3.9	0.8	0.3
1500	2100	1600	2200				

Differences SCHEVENINGEN
+0105 +0102 +0226 +0246 −2.6 −2.2 −0.6 −0.1

NOTE: Double LWs occur — The rise after the first LW is called The Agger — Predictions for Hoek van Holland are for the lower LW which is usually the second. Time differences for Secondary Ports referred to Hoek van Holland are approximate — Water levels on this coast are considerably affected by the weather. Strong NW gales can raise levels by up to 3m, whereas strong E winds can lower levels by 1m.

SHELTER
Entrance can be difficult in on-shore NW winds over force 6 but once in the Buitenhaven, shelter is good. The yacht marina in the Second Harbour gives very good shelter. Access H24.

NAVIGATION
Waypoint SCH safe water buoy, Iso 4s, 52°07'.80N 04°14'.20E, 336°/156° from/to entrance, 1.6M. Strong tidal streams running NE/SW across the entrance can cause problems. Strong winds from SW through NW to N cause a scend in the outer harbour. Beware large ships entering and leaving.

LIGHTS AND MARKS
Ldg Lts 156° to outer basin, 131° to inner basin.
Entry signals (from Semaphore Mast)
R over W Lts — Entry prohibited.
W over R Lts — Exit prohibited.
Fl Y Lt — entry difficult due to vessels leaving.
A QY Lt is shown by entrance to First Harbour (W side) when vessels are entering or leaving port.
Tide signals:
G over W Lts — tide rising.
W over G Lts — tide falling.
R Lt — less than 5m in entry channel.

RADIO TELEPHONE
Call: *Scheveningen Haven* VHF Ch 14 (H24).
Yachts must get permission from Traffic Centre before entering.

TELEPHONE (70)
Hr Mr 527.711; Traffic Centre 527.721; Pilot Hook of Holland 38809; Customs 51.44.81; Dr 85.87.00; Ambulance 22.21.11; Brit Consul (20) 76.43.43.

FACILITIES
Marina Scheveningen (223+100 visitors) Tel. 52.00.17, AC, Bar, C (15 ton), CH, D, El, FW, ME, R, Lau, Sh, SM; **Harbour** Slip, FW, ME, Sh, C (60 ton), CH, R, Bar; **Hoogenraad & Kuyt**, ME, El, Sh, CH; **Yachtclub Scheveningen** Tel. 52.03.08; **Vrolyk** Tel. 554.957 SM; **Town** P, D, V, R, Bar. PO; Bank; Rly; Air (Rotterdam).
Ferry UK — Rotterdam—Hull;
Scheveningen—Great Yarmouth;
Hook of Holland—Harwich.

IJMUIDEN 10-19-22
Noord Holland

CHARTS
Admiralty 2322, 124; Zeekaart 1450, 1543; Dutch Yacht Chart 1801; Stanford 19; Imray Y5

TIDES
Dover +0400; ML 1.1; Zone −0100

Standard Port VLISSINGEN (←)

Times				Height (metres)			
HW		LW		MHWS	MHWN	MLWN	MLWS
0300	0900	0400	1000	4.7	3.9	0.8	0.3
1500	2100	1600	2200				

Differences IJMUIDEN
+0145 +0143 +0304 +0321 −2.7 −2.2 −0.6 −0.1

NOTE: Water levels on this coast can be considerably affected by the weather. Strong NW gales can raise the levels by up to 3m.

SHELTER
Good shelter especially on the S quay on the canal side of the small S locks. It is noteworthy that the canal level may be above or below the sea level. Yachts can stay in the Haringhaven for short periods but are not advised to stay at the sea side of the locks longer than necessary. There are marinas at IJmond (under lift bridge in ZiJkanaal C) and at Nauerna (in ZiJkanaal D) 6M and 7M respectively from IJmuiden.

NAVIGATION
Waypoint IJmuiden Lt buoy, Mo(A) 8s, Racon, 52°28'.70N 04°23'.93E, 278°/098° from/to entrance, 5.0M. Beware strong tidal streams across the harbour entrance and heavy merchant traffic. Yachts normally use S lock.

LIGHTS AND MARKS
Ldg Lts 100° (front is the Lt Ho showing FWR 31m 16/13M, W050°-122°, R122°-145°, W145°-160°. RC, storm, tidal & traffic signals; rear Fl 5s 53m 29M). Both display a FW Lt by day. This line leads into the Buitenhaven and the Zuider Buitenkanaal.

RADIO TELEPHONE
IJmuiden Port Operations Centre. (1) IJmuiden. Call *IJmuiden Port Control* (or *IJmuiden Port Control Locks* for direct contact with locks) Ch 09, in area from North Sea to North Sea Locks. (2) Hemtunnel. Call *IJmuiden Locks* Ch 11, in area from North Sea Locks to Hemtunnel. (3) Amsterdam Ch 14, in area of Amsterdam port basins, E and W of Hemtunnel. (All H24).
Reports on visibility, when less than 1000m, every H+00 on Ch 12 by Coastguard IJmuiden, every H+30 on Ch 05 and 11 by IJmuiden, every H+00 on Ch 14 by Amsterdam.
Vessels in the roads area, the Buitenhaven and the Noordzeekanaal should keep watch on the appropriate Ch, and report to Port Operations Centre on departure from or arrival in a port in Noordzeekanaal.

TELEPHONE (2550)
Hr Mr 19027; Traffic Centre IJmuiden (DGSM) 34542; Pilot 19027; Customs 23309; Hosp 19.1.00; Brit Consul (20) 76.43.43.

FACILITIES
North Sea Canal Marina Tel. 384.457, FW, AB, Bar, BY, C (20 ton), Lau, PO, V, R, D, AC; **Hermans** Tel. 12963 Gaz; **Town** P, D, V, R, Bar. PO; Bank; Rly (bus to Beverwijsk); Air (Amsterdam).
Ferry UK — Hook of Holland—Harwich.

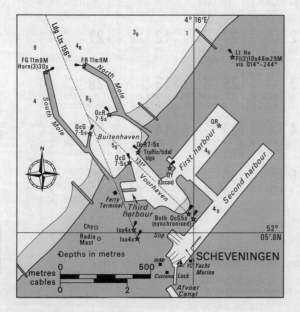

IJMUIDEN continued

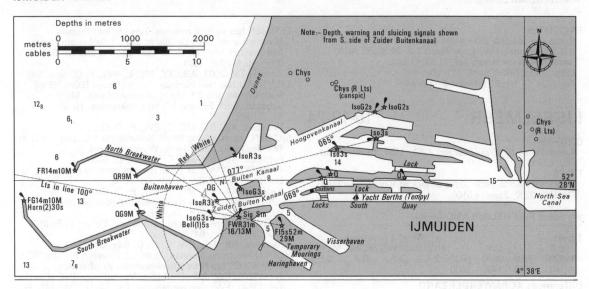

AMSTERDAM 10-19-23
Noord Holland

CHARTS
Admiralty 124; Zeekaart 1543; Dutch Yacht Chart 1801, 1810; ANWB G, I
TIDES
Amsterdam is between the Noordzeekanaal and the IJsselmeer, both of which are non-tidal; Zone −0100.
SHELTER
Shelter is complete. There are several yacht harbours/marinas. The principal one is Sixhaven, on the N bank, NE of the railway station (conspic). Another is on the S bank, close NW of the Harbour Building (conspic), but beware of swell from passing ships. Amsterdam gives access to canals running to N and S and also, via the Oranjesluizen, to the IJsselmeer.

NAVIGATION
See 10.19.22. The transit of the Noordzeekanaal is straightforward, apart from the volume of commercial traffic. The speed limit for yachts is 9 kn. There is a slight set to the W during sluicing.
LIGHTS AND MARKS
Lights are shown along both banks of the canal at the entrances to branch canals and basins. The continuation E towards the Oranjesluizen and the Amsterdam-Rijn canal is marked by Lt buoys.
RADIO TELEPHONE
VHF Ch 14. See also 10.19.22.
TELEPHONE (20)
Hr Mr 94.45.44; Sixhaven Yacht Harbour 32.94.29; Customs 25.53.83; Dr 59.99.111.

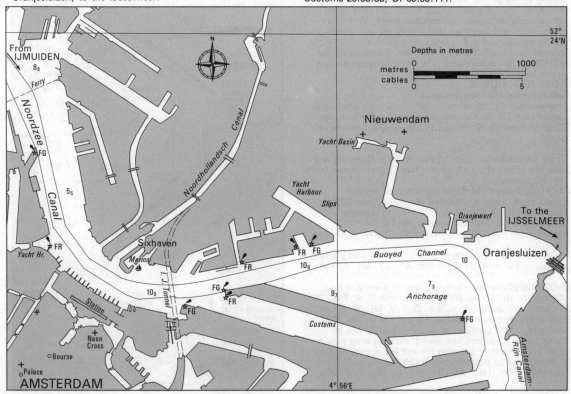

AMSTERDAM continued

FACILITIES
Sixhaven Marina (60+some visitors) Tel. 37.08.92, AC, Bar (weekends), FW; **L. J. Harri** Tel. 24.80.35 ACA; **Jachthaven de Hoop** Tel. 71.83.49 Gaz.
City: All facilities, Bank, PO, Rly, Air.
Ferry UK — Hook of Holland—Harwich.

IJSSELMEER 10-19-24

CHARTS
Admiralty 1408, 2593; Zeekaart 1351, 1454; Dutch Yacht Chart 1810

TIDES
IJsselmeer is non-tidal. Tides at lock at Kornwerderzand Dover −0230; Zone −0100.

Standard Port HELGOLAND (→)

Times				Height (metres)			
HW		LW		MHWS	MHWN	MLWN	MLWS
0200	0700	0200	0800	2.7	2.3	0.4	0.0
1400	1900	1400	2000				

Differences KORNWERDERZAND
−0200 −0305 −0250 −0210 −0.5 −0.5 0.0 +0.2

SHELTER
IJsselmeer is the name for the un-reclaimed part of the Zuiderzee. Shelter is excellent in the many marinas, some of which are listed below. It is divided into two parts, the SW part being called the Markerwaard (20M × 15M, 2 to 4.5m deep) divided from the rest by a dyke with two locks, Enkhuizen in the N and Houtribhaven in the S. The rest of the IJsselmeer is 30M × 20M.

NAVIGATION
Three entrances
(1) via IJmuiden and Noordzeekanaal.
(2) via Stevinsluizen (Den Oever) at the W end of the Afsluitdijk which operates during working hrs.
(3) via the Kornwerderzand lock at the E end of the Afsluitdijk which operates 24 hrs.
Yacht charts are essential to avoid shoals, traps, fishing areas, nets etc which are well marked on the charts. Strong winds get up very quickly and short seas are frequently encountered. Strong winds can raise water level on lee shore, or lower it on weather shore, by 1m or more. Most harbours have water level gauges which should be checked in bad weather. Speed restrictions: in buoyed channels 8.5kn; outside 4.8kn. Between sunset and 0800 4.8kn everywhere. The Signi buoyage system is in use (see 10.19.7).

RADIO TELEPHONES
The following channels are used
Oranjesluizen Ch 18; den Oeversluizen Ch 20; Kornwerder Zandsluizen Ch 18; Krabbersgatsluizen Ch 22; Lemmer Ch 20; Enkhuizen Ch 22; Houtribsluizen Ch 20; Emergency Ch 16.

There are numerous ports to visit and the following is a selection of the more important: (working clockwise from the Waddenzee).
MAKKUM. Approx 2M SE of the Kornwerderzand breakwaters, there is a buoyed channel approx 2000m long. Ldg Lts FR and FG at 092° leading between breakwaters into Makkumharbour. Hr Mr Tel. 1450. Visserijhaven Tel. 1450, AB, BY, ME, C, FW, P, D, SM, Gaz.
WORKUM, 2.5M, N of Hindeloopen, with buoyed channel leading to two marinas. Facilities: AB, BY, ME, C, FW, P, D, Gaz.
HINDELOOPEN has two marinas, one in town and one 180m to N. Hr Mr Tel. 2009. Old Harbour D, P, FW, ME, El, Sh; Jachthaven Tel. 1238 P, D, FW, ME, El, Sh.
STAVOREN. Entrance between FR and FG on line 048° Iso 4s Dir Lt. Yacht berths S of entrance by Rly Stn in Buitenhaven. Stavoren has several marinas after passing lock. Lock entrance between FR and FG approx 1000m S of entrance to Buitenhaven. Hr Mr Tel. 1216.
Jachthaven Tel. 1469, AB, BY, ME, C, FW, P, D, SM, Gaz.

LEMMER has several marinas. Ldg Lts into Prinses Margrietkanaal both Oc 10s at 038°. Ldg Lts into town and marinas both Iso 8s at 083° and FG and Iso G 4s at 065°. Hr Mr Tel. 1604.
Tacozijl Tel. 2003, AB, BY, ME, C, FW, P, D, Gaz, SM.
LELYSTAD has two marinas. Houtribhaven 500m NE of Houtriblocks and 3.5 km N of Houtriblocks, Marina Lelystad. Both FR and FG on breakwaters. Hr Mr Tel. 21048.
Facilities: AB, BY, ME, C, FW, P, D, Gaz.
MUIDEN, home of the Royal Netherlands YC. Ldg Lts Q at 181° lead into Yacht Hr to W of entrance.
Facilities: AB, BY, ME, C, FW, P, D, Gaz.
MONNICKENDAM has three marinas with all facilities. Ldg Lts Iso R 4s at 236°. Hr Mr Tel. 1616.
Facilities: AB, BY, ME, C, FW, P, D, Gaz.
MARKEN (picturesque show piece). Approach from N. Harbour, AB, FW, P, D, Gaz.
VOLENDAM; several yacht berths. Ldg Dir Lts Fl 5s at 313°. Hr Mr Tel. 64122.
Facilities: AB, ME, FW, C, P, D, Gaz, SM.
HOORN has three marinas. Entrance between FR and FG on line 352° Iso R 8s. Grashaven (to W of ent), Tel. 15208, FW, ME, El, D, Sh; Vluchthaven (NE of ent) Tel. 13540, FW; Gemeentehaven Tel. 14012, FW, ME, El, Sh, D. Hr Mr Tel. 14012.
Facilities: AB, ME, C, FW, P, D, Gaz.
ENKHUIZEN. The Krabbersgast FR and FG at breakwaters gives access to the Krabbersgatlock and also to the harbours of Enkhuizen. The town has three marinas Gemeentehaven Tel. 12444, P, D, ME, Sh, El, FW; Buyshaven Tel. 15660, FW; Compagnieshaven Tel. 13353, D, P, ME, El, Sh, FW. Hr Mr Tel. 13122.
Facilities: AB, BY, ME, C, FW, P, D, Gaz, SM.
ANDIJK. Has two marinas. Ent between FR and FG Lts on 248°.
Facilities: ME, C, SM, Gaz.
MEDEMBLIK. Entrance between FR and FG on line 232° Oc 5s Dir Lt. Hr Mr Tel. 1666. Westerhaven Tel. 1861 (3m) FW; Oosterhaven Tel. 1686 (3.5m), P, D, FW, ME, El, Sh.

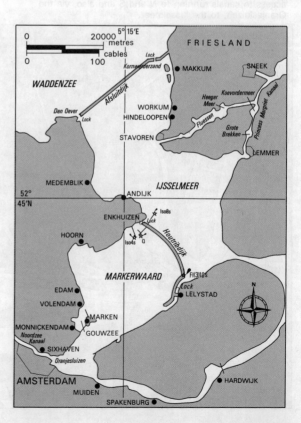

DEN HELDER 10-19-25
Noord Holland

CHARTS
Admiralty 2322, 191; Zeekaart 1454, 1546;
Dutch Yacht Chart 1811; ANWB F; Imray Y5

TIDES
Dover −0430; ML 1.1; Duration No data; Zone −0100

Standard Port HELGOLAND (→)

Times				Height (metres)			
HW		LW		MHWS	MHWN	MLWN	MLWS
0200	0700	0200	0800	2.7	2.3	0.4	0.0
1400	1900	1400	2000				

Differences DEN HELDER
−0350 −0500 −0515 −0425 −0.9 −0.8 0.0 +0.2
OUDESCHILD
−0300 −0430 −0435 −0355 −0.9 −0.8 0.0 +0.2
DEN OEVER
−0245 −0420 −0355 −0300 −0.8 −0.7 0.0 +0.2

SHELTER
There is good shelter in the Naval Harbour, the YC and yacht harbour being immediately to stbd on entering. More secure shelter can be obtained by locking into the Koopvaarders Binnenhaven Yacht Harbour or the Den Helder Yacht Haven.

NAVIGATION
Waypoint Schulpengat SG safe water buoy, Mo(A) 8s, 52°52'.95N 04°38'.00E (on leading line), 216°/036° from/to Kaap Hoofd, 6.0M.
There are two channels to the harbour entrance:—
(1) Molengat is good except in strong winds from NW when heavy breakers occur at this entrance.
(2) Schulpengat is well marked and well lit. Beware very strong tidal streams through the Marsdiep across the harbour entrance.
Note: the port mainly belongs to the Royal Netherlands Navy and, like the YC, is run by them. There is a large fishing fleet.

LIGHTS AND MARKS
Schulpengat Ldg Lts 026° on Texel. Front Iso 4s 18M. Rear Oc 8s 18M. Both vis 025°-028°.
Molengat Ldg Lts 142°. Front Iso 5s 13m 8M; vis 124°-157°. Rear, 650m from front, F 22m 8M; Tr on hosp; vis 124°-157°. Ldg Lts 191°. Front Oc G 5s 16m 14M; B triangle on building. Rear (synchronised) Oc G 5s 25m 14M; B triangle on B framework Tr.
Entry signals, shown from Harssens on W side of entrance:—

R
W } Entry to Marinehaven Willemsoord and
R   Rijkszeehaven Nieuwe Diep prohibited

R } Entry only for ships with permission of RNN
W   Harbour Officer

R   ) All movement prohibited in Marinehaven
W W) Willemsoord and Rijkszeehaven Nieuwe Diep.

Moormanbridge operates 7 days a week H24.
Kinsbergenbridge operates 0500-2300 Mon-Fri; 0700-1400 Sat. Burgemeester Vissersbrug operates 0500-2300 Mon-Fri; 0700-1300, 1400-1900 Sat; in summer 0730-1230, 1400-2100 Sun. All bridges remain closed 0715-0810, 1200-1215, 1245-1300, Mon-Fri. Also 0830-0910 on Mon, 1545-1645 on Fri.

RADIO TELEPHONE
VHF Ch 16; 14 (H24). Other stations: Den Oever Lock Ch 20. Kornwerderzand Locks Ch 18. Coastguard Kijkduin Ch 12 (H24). Koopvardersschutsluis (Locks) Ch 22 (H24). Moorman Bridge Ch 18 (H24).

TELEPHONE (2230)
Hr Mr 53000-5-6822 CG 12732; Municipal Port Control 13955; Naval Commander 53000-5-6929; Customs 15181; Hosp 11414; Water Police 16767; Brit Consul (20) 76.43.43.

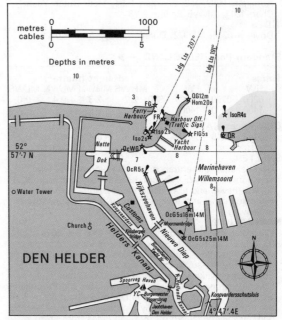

FACILITIES
Yacht Harbours KMYC Tel. 11366, Ext 2645; **MWY YC** (Binnenhaven) Tel. 17076, P (at garage), D, L, FW, AB; **YC WSOV** 53000-5-2173; **YC HWN** Tel. 24422; **W. Visser & Son** ME, El, Sh, Floating dock; **Binnenhaven** Tel. 16641, L, FW, CH, AB, SM; **Yacht Haven** Tel. 37444, AB, ME, El, Sh, C, CH, Slip, FW, R, Bar; **W. Bakker** Tel. 17356 Gaz.
Town P, CH, V, R, Bar. PO; Bank; Rly; Air (Amsterdam).
Ferry UK — Hook of Holland—Harwich.

HARLINGEN 10-19-26
Friesland

CHARTS
Admiralty 2593, 112; Zeekaart 1454, 1456; Dutch Yacht Chart 1811; ANWB B; Imray Y5

TIDES
Dover −0210; ML 1.2; Duration 0520; Zone −0100

Standard Port HELGOLAND (→)

Times				Height (metres)			
HW		LW		MHWS	MHWN	MLWN	MLWS
0200	0700	0200	0800	2.7	2.3	0.4	0.0
1400	1900	1400	2000				

Differences HARLINGEN
−0145 −0245 −0155 −0115 −0.4 −0.4 −0.1 +0.2

SHELTER
Good shelter, but entrance can be rough at HW with winds in W or SW. Very good shelter in yacht harbour in Noorderhaven and also inside the lock into the Van Harinxma canal in the yacht harbour (immediately to stbd on leaving the locks).

NAVIGATION
Waypoint BS1-IN2 buoy, GRG, Fl(2+1)G 10s, 53°14'.74N 05°10'.05E, at Blauwe Slenk/Inschot junction. Approach via the Vliestroom and Blauwe Slenk, both buoyed and lit, but channel is narrow for the last 2½ M. Beware the Pollendam, marked with port and stbd beacons; when covered, strong tidal stream sweeps across the Pollendam. Hanerak, a second buoyed channel, runs about 600m S of the Pollendam. At particularly high tides, the outer yacht harbour may be closed off by flood gates from up to HW−1½ to HW+1½.

LIGHTS AND MARKS
Ldg Lts 112° Front Iso 4s, rear Oc 6s. Entry signals for locks in Nieuwe Voorhaven (at signal station)
2 FR (vert) = Arrival or departure of large vessel – all other movement prohibited.
Bu Flag = Sluicing, entry prohibited.

RADIO TELEPHONE (local times)
VHF Ch 11 (Mon 0000 to Sat 2200).

TELEPHONE (5178)
Hr Mr 3041; CG (Brandaris) 2341; Port Authority 3041; Customs 5241; Hosp 5441; Brit Consul (20) 76.43.43.

FACILITIES
EC Monday a.m. and Wednesday p.m.; **Noorderhaven Yacht Harbour** Tel. 5666, M, P (see Hr Mr), D (by barge), L, FW, ME, El, C (up to 40 ton); CH, AB, V, R, Bar; **Leeuwenbrug Watersport** P (see Hr Mr), ME (by arrangement), El (by arrangement), CH; **Harinxma Canal Yacht Harbour** FW, AB, R, Bar; **R. Bakker** Tel. 6491 Gaz; **Welgelegen Scheepsw** Tel. 2744; big ship firm — emergency only.
Town P, D, V, R, Bar. PO; Bank; Rly; Air (Groningen).
Ferry UK — Hook of Holland—Harwich.

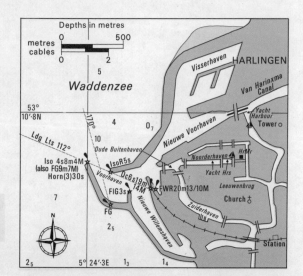

VLIELAND 10-19-27
WEST FRISIAN ISLANDS
Friesland

CHARTS
Admiralty 2593, 112; Zeekaart 1456; Dutch Yacht Chart 1811; Imray Y5

TIDES
Dover −0300; ML and Duration No data; Zone −0100

Standard Port HELGOLAND (→)

Times				Height (metres)			
HW		LW		MHWS	MHWN	MLWN	MLWS
0200	0700	0200	0800	2.7	2.3	0.4	0.0
1400	1900	1400	2000				

Differences VLIELAND
−0255 −0320 −0350 −0330 −0.4 −0.3 +0.1 +0.2

SHELTER
Shelter is good, but it becomes very crowded in high season. Anchorage about 1600m W of the harbour is good except in S to SW winds, and it is near the village. Yachts are forbidden to anchor in buoyed channel or berth at pier ½M W of harbour.

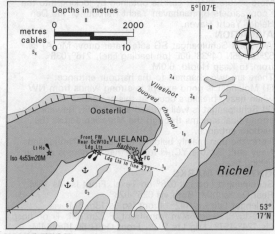

NAVIGATION
Waypoint VSM safe water buoy, Iso 4s, 53°19'.08N 04°55'.74E, 274°/094° from/to ZS1 and ZS2 buoys at Zuider Stortemelk entrance, 2.3M. Best approach is through the Zuider Stortemelk which is deep and well marked and lit. This leads to Vliesloot. There is an alternative entrance from the S via Fransche Gaatje but it is very much shallower and can even dry out at very LW.

LIGHTS AND MARKS
Ldg Lts 277°. Front FW 8M — Rear Oc W 10s.
Main Lt Ho Iso 4s 53m 20M; brown Tr, W lantern; RC.
Entry signals: R Flag or two R Lts on the pier indicates harbour closed by a chain across the entrance.

RADIO TELEPHONE
None.

TELEPHONE (5621)
Hr Mr 20.15.63; Hr Mr (Yacht Hr) 20.17.29; CG 20.13.26; Customs 20.15.22; Dr 20.13.07; Brit Consul (20) 76.43.43.

FACILITIES
Yacht Harbour Tel. 20.17.29, Slip, M, P, D, L, FW, AB; **Harbour** C (12 ton); **North Shore W of Ferry Pier** M; **B. H. Iedema** Tel. 1336 Gaz; **C. D. Hoogland** Tel. 1352, P, D; **Town** P, D, ME, El, CH, V, R, Bar.
PO; Bank; Rly (ferry to Harlingen); Air (Amsterdam).
Ferry UK — Hook of Holland—Harwich.

AREA 19—The Netherlands 759

TERSCHELLING 10-19-28
WEST FRISIAN ISLANDS
Friesland

CHARTS
Admiralty 2593, 112; Zeekaart 1456;
Dutch Yacht Chart 1811; Imray Y5

TIDES
Dover −0300; ML 1.3; Duration No data; Zone −0100

Standard Port HELGOLAND (→)

Times				Height (metres)			
HW		LW		MHWS	MHWN	MLWN	MLWS
0200	0700	0200	0800	2.7	2.3	0.4	0.0
1400	1900	1400	2000				

Differences WEST TERSCHELLING
−0220 −0250 −0330 −0305 −0.4 −0.3 +0.1 +0.2

SHELTER
Good shelter except in strong E winds, and harbour is available at all times. Yachts moor at the N of the harbour, which is very crowded in the season, or go alongside near Hr Mr's office. Eemshaven is for emergency shelter only.

NAVIGATION
See 10-19-27. Waypoint VL2-SG1 buoy, RGR, Fl(2+1)R 12s, 53°19'.30N 05°09'.80E, 225°/045° from/to front Ldg Lt 053°, 2.9M. Entrance via the Schuitengat; the channel is narrow and shifts but it is well buoyed.
The harbour has considerable commercial traffic.

LIGHTS AND MARKS
Ldg Lts 053°. Front (West mole) FWR 5m 8/5M; R post, W bands; vis W049°-055°, R055°-252°, W252°-263°, R263°-049°; Horn 15s. Rear (on dyke 1.1M from front, off chartlet) Iso 5s 14m 19M; metal mast, Y lantern; vis 045°-061° (intens 045°-052°).

RADIO TELEPHONE
None.

TELEPHONE (5620)
Hr Mr 2235; CG 2341; Customs Harlingen 5241; Dr 2181; Brit Consul (20) 76.43.43.

FACILITIES
EC Wednesday; **Harbour** Slip, M, FW (in cans), ME, AB; **W. Bloem** Tel. 2076, ME; **L. G. Schiepstra** Tel. 2694 ME; **C Bloem** Tel. 2178 SM; **Village** P, D, FW, CH, V, R, Bar. PO; Bank (W. Terschelling); Rly (ferry to Harlingen); Air (Groningen).
Ferry UK — Hook of Holland—Harwich.
Note:— A marina is under construction (1988).

DELFZIJL 10-19-29
Groningen

CHARTS
Admiralty 3510; Zeekaart 1555; Dutch Yacht Chart 1812; Dutch ANWB A

TIDES
Dover −0025; ML 2.0; Duration 0605; Zone −0100

Standard Port HELGOLAND (→)

Times				Height (metres)			
HW		LW		MHWS	MHWN	MLWN	MLWS
0200	0700	0200	0800	2.7	2.3	0.4	0.0
1400	1900	1400	2000				

Differences DELFZIJL
+0020 −0005 −0035 +0005 +0.8 +0.9 +0.1 +0.2
EEMSHAVEN
−0025 −0045 −0115 −0035 +0.3 +0.4 +0.2 +0.2
LAUWERSOOG
−0130 −0150 −0230 −0225 +0.1 +0.2 +0.2 +0.2

SHELTER
Good shelter. Yacht harbour in Balkenhaven gives good protection. Yachts can lock through into the Eemskanaal and berth in Farmsumerhaven (1 day only) or at the end of the old Eemskanaal near the obsolete lock.

NAVIGATION
Waypoint Hubertgat safe water buoy, Iso 8s, Whis, 53°34'.90N 06°14'.40E, 270°/090° from/to Borkum Kleiner Lt, 15.5M. Westerems safe water buoy, Iso 4s, Racon, 53°37'.10N 06°19'.50E, 272°/092° from/to Nos 1 and 2 Westerems channel buoys, 1.9M. See also 10.20.8. Channels in the Ems are well buoyed and lit. The entrance is 3 M ESE of the town of Delfzijl. Beware strong tides across the entrance, as these can be dangerous.

LIGHTS AND MARKS
Harbour entrance, FG on W arm and FR on E arm with Horn 15s. The channel from the entrance has W Fl Lts to N and R Fl Lts to S.
Entry signals:—
2 R Flags or 2 R Lts — All movement prohibited except as directed by Hr Mr.
Sluicing signals:—
R Flag or 3 R Lts in triangle —Sluicing at Eemskanaal Lock through gates.
Bu Flag or 3 R Lts in triangle—Sluicing at Eemskanaal over G Lt Lock through draining sluices.

RADIO TELEPHONE (local times)
Call: *Port Office Delfzijl/Eemshaven* VHF Ch 14 (H24). Traffic reports on Ch 14 every 10 min from H+00 when visibility less than 500m. Locks (Eemskanaalsluizen) Ch 11 (Mon-Sat, H24). Die Ems (Ems Revier) Ch 18 20 21 (H24) — information broadcasts in German every H+50 with weather and tidal information, including storm warnings for coastal waters between Die Ems and Die Weser.
Locks: Grosse Seeschleuse (call: *Emden Lock*) Ch 13 16 (H24). Nesserlander Seeschleuse (call: *Nesserland Lock*) Ch 13 16 (H24). Leer Lock Ch 16 13 (0700-2300). Papenburg Lock Ch 16; 13 (H24).
Jann-Berghaus Bridge (call: *Leer Bridge*) Ch 15.
Friesen Bridge (call: *Weener Bridge*) Ch 15.

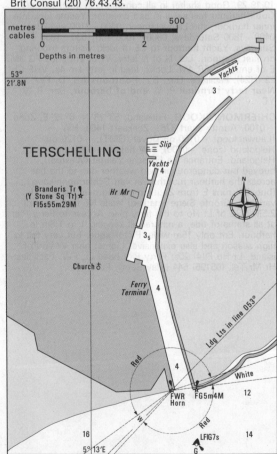

DELFZIJL continued

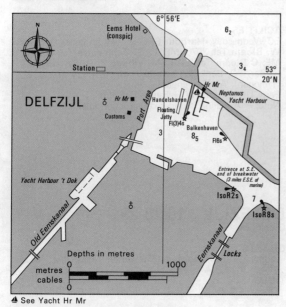

▲ See Yacht Hr Mr

TELEPHONE (5960)
Hr Mr (Delfzijl Port Authorities) 14966; Neptunus Yacht Hr Mr 15004; Sea locks 13293; CG (Police) 13831; Customs 15060; Hosp 14944; Brit Consul (20) 76.43.43.

FACILITIES
Neptunus Yacht Harbour Tel. 15004, D, P, FW; **Yacht Harbour 't Dok** D, FW, AB; **Datema** Tel. 13810, CH, ACA, chart agent; **P. Dinges** Tel. 15010, ME, El; **H. S. Hunfeld** Tel. 13446, ME, El; **Ems Canal** L, FW, AB; **IJzerhandel Delken** Tel. 30200 Gaz;
Town P, D, V, R, Bar. PO; Bank; Rly; Air (Groningen). Ferry UK — Hook of Holland — Harwich.

MINOR HARBOURS AND ANCHORAGES 10.19.30

OUDESCHILD, Texel, Noord Holland, 53°02' N, 4°51' E, Zone −0100, Admty charts 191, 2593, Zeekaart 1546, 1454, Dutch Yacht 1811. HW −0355 on Dover (GMT), −0425 on Helgoland (zone −0100); HW height −0.7m on Helgoland; ML 1.1m; Duration 0625. On SE side of Texel, Hr gives good shelter and has Yacht Harbour in the Werkhaven, NNE of ent. Beware strong cross tides across ent. Keep Ldg Lt, Oc 6s 14m between harbour entrance Lts, FR and FG on course of 291°. VHF Ch 09, 0730–1500. Facilities: Hr Mr Tel. 2710; **Marina** AB, Bar, D, Dr, FW, M, R; **YC W. V. Texel**; **Boom Shiprepairs** Tel. 2661, Sh, Dry dock.

NES, AMELAND, Friesland, 53°26' N, 5°45' E, Zone −0100, Admty charts 2593, 1405, Zeekaart 1458, Dutch Yacht 1811, 1812. HW −0055 on Dover (GMT), −0108 on Helgoland (zone −0100); HW height +0.4m on Helgoland; ML 1.3m; Duration 0625. Harbour at Nes, mid S of the island; no proper yacht hr but sheltered from all but winds from E to S. Harbour dries but fin keels sink into mud. Beware sandbanks in the Zeegat van Ameland. Lt at W end of Ameland Fl(3) 15s 57m 30M, RC. Nes Nieuwe Veerdam head Iso 6s 2m 8M. Facilities are very limited; Gaz.

OOSTMAHORN, Friesland, 53°23' N, 6°09' E, Zone −0100, Admty chart 3761, Zeekaarten 1458, Dutch Yacht Chart 1812; non tidal. Harbour on W side of Lauwersmeer; lock in from the Zoutkamperlaag to W of Lauwersoog. Main harbour with FR and FG Lts at entrance has yacht haven (2.5m to 3.4m). Approx 450m to the SSE there is another yacht haven, the Voorm Veerhaven (1.5m). Floating beacons with Y flags mark fishing areas. Facilities: Hr Mr (05193) 13 31; most normal facilities in the Yacht Haven.

ZOUTKAMP, Friesland, 53°20' N, 6°18' E, Zone −0100. Admty chart 3761, Zeekaarten 1458, Dutch Yacht Chart 1812; non tidal. Approach down the Zoutkamperril (3 to 4.5m); Approx 400m before the bridge, the yacht haven 'Hunzegat' (1.4 to 2.2m) is on the NE side. There is another yacht haven, the Oude Binnenhaven (2m), further SE. Pass through lock, Provinciale Sluis, and yacht haven is immediately to NE. FW and FR Lts shown from lock. Facilities: **Gruno BY** Tel. 2057, ME, El, Sh, C (20 ton), BH; **Oude-Binnenhaven** AC, C, FW; **Hunzegat** Tel. 2588, AC, FW, SC, Slip; **Town** BY, C, D, Dr, El, Gaz, ME, P, PO, R, Sh, SM, V.

LAUWERSOOG, Friesland, 53°25' N, 6°12' E, Zone −0100, Admty chart 3761, Zeekaart 1458, Dutch Yacht 1812. HW −0150 on Dover (GMT), −0120 on Helgoland (zone −0100); HW height +0.5m on Helgoland; ML 1.3m. See 10.19.29. Good shelter in all conditions in inner harbour. Outer harbour has swell in bad weather. Yachts lock into inner harbour, the Lauwersmeer, 0400–2100 weekdays, 0400–1800 Saturdays, 0900–1000 and 1630–1830 Sundays. Yacht harbour to SE of lock; visitors mooring on first pontoon. Lts, N of N jetty. W of entrance FG 3M. E of entrance FR 4M. Locks, lead-in jetty Iso 4s. VHF Ch 22 (H24). Facilities: Gaz, BY, **Pontoons** AB, D, FW; **Near ferry terminal** P; **W end of harbour**, Bar, R, V, YC.

SCHIERMONNIKOOG, Friesland, 53°28' N, 6°12' E, Zone −0100. Admty chart 3761, Zeekaart 1460, HW (Lauwersoog) −0150 on Dover (GMT), −0120 on Helgoland (zone −0100); HW height +0.5m on Helgoland. Entrance via Friesche Zeegat (Westgat), buoyed but dangerous in bad weather due to the bar across the harbour mouth. Gat van Schiermonnikoog (buoyed) runs E from Zoutkamperlaag over bar (depth variable). Groote Siege (marked) leads NE from a point 2.5M SSE of Lt Ho to the ferry pier. Access to Ferry Pier at all states of tide, a natural sill keeping 1 to 1.5m in harbour. Ent only 5M wide. Picturesque, but very full in high season and also expensive. Lights: near W end of island, Lt Ho Fl(4) 20s; Ferry Pier head Lt, FW. Facilities: Hr Mr Tel. (05195) 544 (May-Sept), FW.

VOLVO PENTA SERVICE

Sales and service centres in area 20
Names and addresses of Volvo Penta dealers in
this area are available from:
GERMANY BRD **Volvo Penta Deutschland GmbH**, Redderkoppel 5. Postfach 9069. D-2300 Kiel-Friedrichsort Tel 0431 3948-0. Telex 292764.

Area 20

Federal Republic of Germany
Borkum to Danish border

10.20.1	Index	Page 761	
10.20.2	Diagram of Radiobeacons, Air Beacons, Lifeboat Stations etc	762	
10.20.3	Tidal Stream Charts	764	
10.20.4	List of Lights, Fog Signals and Waypoints	766	
10.20.5	Passage Information	768	
10.20.6	Distance Table	769	
10.20.7	Special factors affecting the Federal Republic of Germany	770	
10.20.8	Borkum	770	
10.20.9	Norderney	771	
10.20.10	Dornumer-Accumersiel	771	
10.20.11	Langeoog	772	
10.20.12	Wangerooge	772	
10.20.13	Hooksiel	773	
10.20.14	Wilhelmshaven	774	
10.20.15	Bremerhaven	774	
10.20.16	Cuxhaven	775	
10.20.17	River Elbe/Hamburg	776	
10.20.18	Brunsbüttel/Nord-Ostsee Kanal	777	
10.20.19	Helgoland, Standard Port, Tidal Curves	778	
10.20.20	Sylt	782	
10.20.21	Minor Harbours and Anchorages Emden Juist Norddeich Baltrum Nessmersiel Bensersiel Neuharlingersiel Spiekeroog Wangersiel Dangastersiel Büsum Tönning Husum Wittdün (Amrun) Wyk (Föhr) Schlüttsiel Dagebüll	782	
10.20.22	Tidal information for the west coast of Denmark	784	

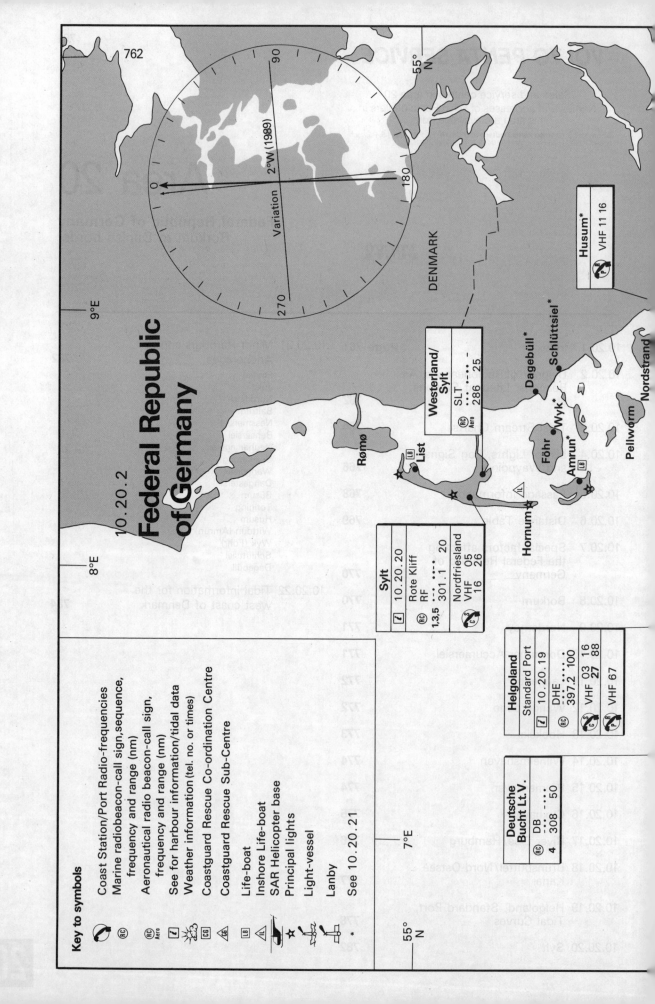

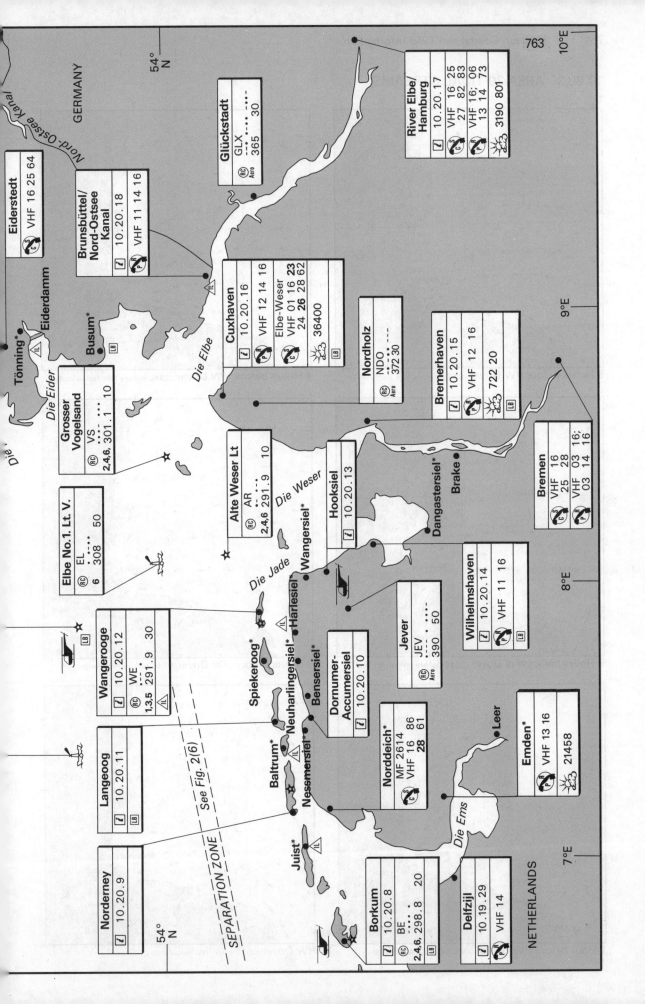

10.20.3 AREA 20 TIDAL STREAMS

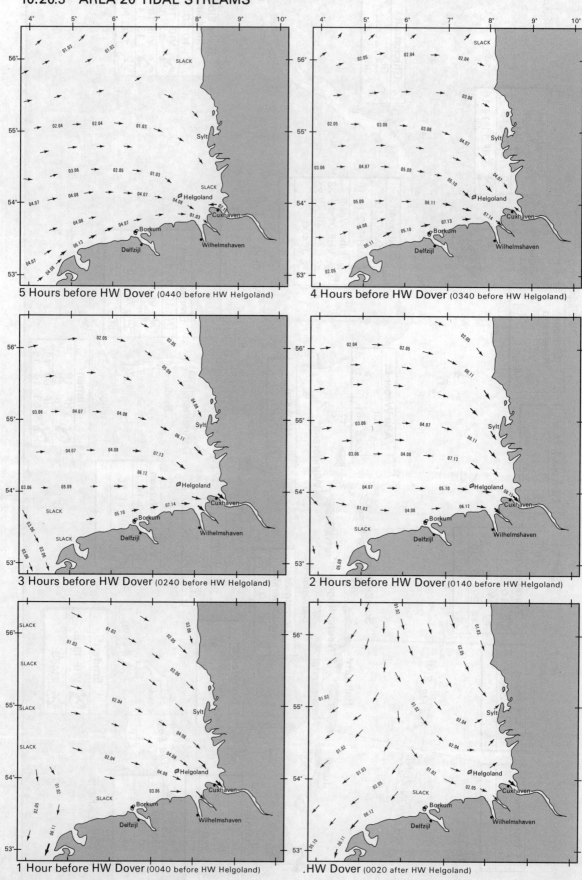

AREA 20—Federal Republic of Germany

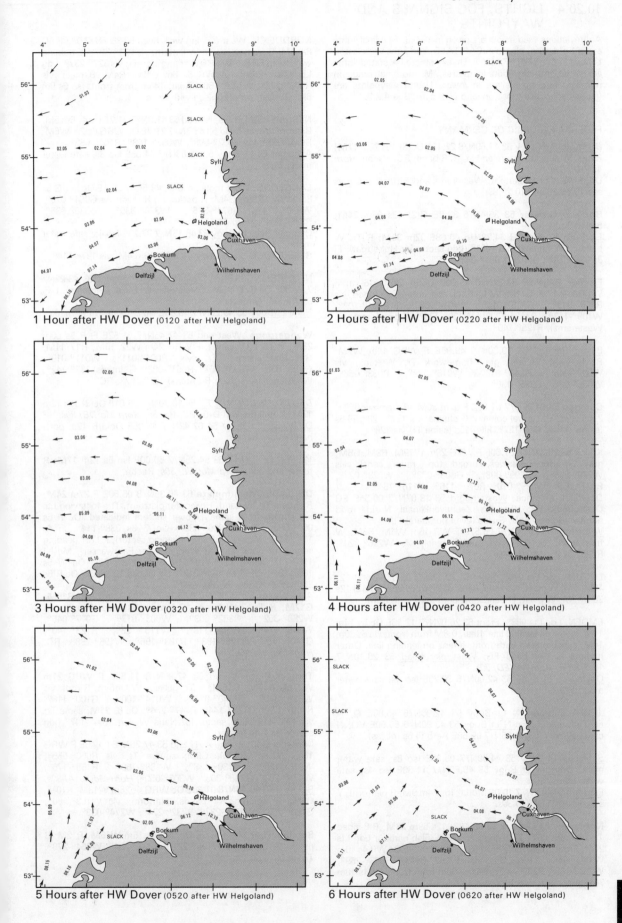

10.20.4 LIGHTS, FOG SIGNALS AND WAYPOINTS

Abbreviations used below are given in 10.0.3. Principal lights are in **bold** print, places in CAPITALS, and light-vessels and Lanbys in *CAPITAL ITALICS*. Unless otherwise stated lights are white. m—elevation in metres; M—nominal range in n. miles. Fog signals are in *italics*. Useful waypoints are underlined — use those on land with care. See 4.2.2.

FEDERAL REPUBLIC OF GERMANY

BORKUMRIFF LT V 53 47.50N/6 22.13E Oc(3) 15s 20m **17M**; R hull, W superstructure and Lt Tr, R band; RC; Racon; *Horn Mo(BF) 30s*.
DIE EMS. For Hubertgat, Westerems and Riffgat Lt Buoys see 10.19.29 and 10.20.8.

Borkum Grosser 53 35.4N/6 39.8E Fl(2) 12s 63m **24M**; brown Tr.
Borkum Kleiner 53 34.78N/6 40.08E 32m **30M**; R Tr, W bands; FW (intens) 090°-091° (Ldg sector for Hubertgat); Fl 3s 088°-090°; Q(4) 10s 091°-093°; RC.

Fischer Balje 53 33.21N/6 43.00E Oc(2) WRG 16s 15m **W16M**, R 12M, G11M; W Tr, R top and lantern, on tripod; vis R256°-313°, G313°-014°, W014°-081°, R081°-109°, W109°-113°, R113°-120°, W120°-125° (Ldg sector to Westerems), R125°-140°.

Binnen-Randzel 53 30.20N/6 49.95E F WRG 14m W7M, R5M, G4M; B and grey framework Tr, W tank; vis W302°-345°, R345°-016°, W016°-033°, R033°-056°, W056°-085°, G085°-130°.

Campen 53 24.39N/7 01.00E F 62m **30M**; R framework Tr, 2 galleries, W central column, G cupola; vis 127°-128°. Fl 5s (same Tr) vis 126°-127°. Fl(4) 15s (same Tr) vis 128°.

Knock 53 20.37N/7 01.50E F WRG 29m W12M, R9M, G8M; grey Tr, four galleries, broad top, radar aerial; vis W270°-299°, R299°-008°, G008°-023°, W023°-027°, R027°-039°, W039°-070°, R070°-116°, W116°-154°.
DELFZIJL entrance. Westerhoofd 53 19.07N/7 00.35E FG Oosterhoofd FR; *Horn 15s*. Zeehavenkanaal, N side (odd numbered piles) FlW. S side (even numbers) FlR.
Wybelsum 53 20.20N/7 06.57E F WR 16m W6M, R5M; W framework Tr, R bands, radar aerial; vis W295°-320°, R320°-024°, W024°-049°.

LOGUM. Ldg Lts 075°. Front 53 20.17N/7 08.05E Oc(2) 12s 16m 12M; W mast, R bands. Rear, 630m from front, Oc(2) 12s 28m 12M; W mast, R bands; synchronised with front.

EMDEN. Ldg Lts 088°. Front 53 20.07N/7 12.15E Oc 5s 14m 14M; intens on leading line. Rear, 0.8M from front, Oc 5s 30m 14M; synchronised with front, intens on leading line. Outer harbour E pier head FG; West pier head 53 20.10N/7 10.57E FR, *Horn Mo(ED) 30s*.
Oosterems Lt Buoy 53 42.00N/6 36.20E Iso 4s; safe water mark; *Whis*.

JUIST. Juisterriff-N Lt Buoy 53 42.90N/6 45.80E Q; N cardinal mark. Juist-N Lt Buoy 53 43.90N/6 55.50E VQ; N cardinal mark. 53 40.95N/7 03.50E Aero Fl 5s (occas).

Schlucter Lt Buoy 53 44.90N/7 05.30E Iso 8s; safe water mark. Dovetief Lt Buoy 53 45.50N/7 11.30E Iso 4s; safe water mark.
Platform 54 42.1N/7 10.0E Mo(U) 15s 24m 9M; R platform, Y stripes; *Horn Mo(U) 30s*.

Norderney 53 42.6N/7 13.8E Fl(3) 12s 59m **23M**; R 8-sided Tr; unintens 067°-077° and 270°-280°. Fish harbour Ldg Lts 274°. Front, West mole head 53 41.9N/7 09.9E Oc WR 4s 10m W7M, R4M; RW Tr; vis W062°-093°, R093°-259°, W259°-289°, R289°-062°. Rear, 460m from front, Oc 4s 18m 7M; grey mast; synchronised with front.

NORDDEICH. West training wall, head 53 38.7N/7 09.0E FG 8m 4M; G framework Tr, W lantern; vis 021°-327°. E training wall, head FR 8m 4M; R&W triangle on Tr; vis 327°-237°. Ldg Lts 144°. Front Iso WR 6s 6m W6M, R4M; B mast; vis R078°-122°, W122°-150°. Rear, 140m from front, Iso 6s 9m 6M; B mast; synchronised with front.

NESSMERSIEL mole, N head 53 41.9N/7 21.7E Oc 4s 6m 5M. Baltrum groyne head 53 43.3N/7 21.7E Oc WRG 6s 7m W6M, R4M, G3M; vis G082°-098°, W098°-103°, R103°-082°.
Accumer Ee Lt Buoy 53 46.83N/7 24.65E Iso 8s; safe water mark; *Bell*.

LANGEOOG. West mole head 53 43.5N/7 30.1E Oc WRG 6s 8m W7M, R5M, G4M; R basket on R mast; vis G064°-070°, W070°-074°, R074°-326°, W326°-330°, G330°-335°, R335°-064°; *Horn Mo(L) 30s*.
Otzumer Balje Lt Buoy 53 48.13N/7 37.23E Iso 4s; safe water mark.
SPIEKEROOG 53 45.0N/7 41.3E FR 6m 4M; vis 197°-114°.

NEUHARLINGERSIEL, training wall, head 53 43.22N/7 42.30E Oc 6s 6m 5M. Carolinensieler Balje. Leitdamm. L Fl 8s 7m 6M; G mast.
Harle Lt Buoy 53 49.28N/7 48.00E Iso 8s; safe water mark.

Wangerooge, West end 53 47.45N/7 51.52E Fl R 5s 60m **23M**; R Tr, 2 W bands. F WRG 24m **W22/15M**, **R17**/11M, **G18**/10M; same Tr; vis R004°-011°, W011°-019°, G019°-038°, W038°-061°, R061°-083°, G (intens) 137°-142°, W (intens) 142°-152°, R (intens) 152°-157°; RC.

DEUTSCHE BUCHT LT F 54 10.70N/7 26.01E Oc(3) 15s 12m **17M**; R hull marked D-B; RC; Racon; *Horn Mo(DB) 30s*.
DB/Weser Lt Buoy 54 02.42N/7 43.05E Oc(3)R 12s; port-hand mark; Racon.

WESER Lt FLOAT 53 54.25N/7 50.00E Iso 5s 12m **17M**; R tower and hull; *Horn Mo(WE) 30s*; Racon.

DIE JADE. **Mellumplate** 53 46.35N/8 05.60E F 27m **24M**; R square Tr, W band, Helicopter platform. The following Lts are shown from same Tr over sectors indicated. Oc R 6s 000°-006°, Oc 6s 006°-038°, Oc G 6s 038°-114°, Fl 4s 114°-115°, Mo(A) 115°-116° (leading sector), F 116° (leading sector for outer part of Wangerooger Fahrwasser), Mo(N) 116°-117°, Fl(4) 117°-118°, Oc R 6s 118°-168°, Oc W 6s 168°-184°, Oc R 6s 184°-212°, Oc W 6s 212°-266°, Oc R 6s 266°-280°, Oc W 6s 280°-000°.
Alte Weser 53 51.85N/8 07.72E F WRG 33m **22M**, **R19M**, **G17M**; R Tr, 2 W bands, G lantern, B base, floodlit; vis W288°-352°, R352°-003°, W003°-017°, G017°-045°, W045°-074°, G074°-118°, W118°-123°, R123°-140°, G140°-175°, W175°-183°, R183°-196°, W196°-238°. RC. *Horn Mo(AL) 60s*.

Tegeler Plate N end 53 47.90N/8 11.50E F WRG 21m **W27M**, **R17M**, **G16M**; R Tr, gallery, W lantern, R roof; vis W329°-340°, R340°-014°, W014°-100°, G100°-116°, R119°-123°, G123°-144°, R147°-264°. Oc 6s **21M**; same Tr; vis 116°-119° (ldg sector for Neue Weser), 144°-147° (ldg sector for Alte Weser).
Minsener Oog, Buhne A, N End 53 47.30N/8 00.45E F WRG 16m W13M, R10M, G9M; square Tr; vis R050°-055°, W055°-130°, G130°-138°, W138°-158°, R158°-176°, W176°-268°, G268°-303°, W303°-050°; *Horn Mo(AAA) 30s*.
Buhne C 53 45.40N/8 01.35E Oc WRG 4s 25m W13M, R10M, G9M; B column, W bands, platform; vis W153°-180°, G180°-203°, W203°-232°, R232°-274°, W274°-033°.

Schilling 53 41.8N/8 01.7E Oc WR 6s 15m **W15M**, R12M; B pylon, W band; vis W196°-221°, R221°-254°, W254°-278°.
Tossens. Ldg Lts 146°. Both Oc 6s 15/51m **20M**.

AREA 20 — Federal Republic of Germany   767

Hooksielplate Cross Lt 53 40.20N/8 09.00E Oc WRG 3s 25m W7M, R5M, G4M; W Tr, R bands; vis R345°-359°, W359°-002°, G002°-012°, W012°-020°, R020°-047°, W047°-062°, G062°-080°, W080°-092°, R092°-110°.

DIE WESER. **Hohe Weg**, NE part, 53 42.80N/8 14.65E F WRG 29m **W19M**, **R16M**, **G15M**; R 8-sided Tr, 2 galleries, G lantern; vis W102°-138°, G138°-142°, W142°-145°, R145°-184°, W184°-278°.
Robbennordsteert 53 42.20N/8 20.45E F WR 11m W10M, R7M; R column on tripod; vis W324°-356°, R356°-089°, W089°-121°. Robbenplate Ldg Lts 122°, both Oc 6s, synchronised. Wremerloch Ldg Lts 141°. Front Iso WRG 6s. Rear Iso 6s, synchronised. Dwarsgat Ldg Lts 321°. **Front** Iso 6s 16m **15M**. **Rear** Iso 6s 35m **17M**, synchronised. Langlutjen Ldg Lts 304°. Front Oc WRG 6s. Rear Oc 6s, synchronised. Imsum Ldg Lts 125°, both Oc 6s, synchronised.

HELGOLAND. Restricted area Helgoland — W Buoy 54 10.65N/7 48.25E (unlit); W cardinal mark. Helgoland Lt Buoy 54 09.00N/7 53.55E Q(3) 10s; E cardinal mark; *Whis*. **Lt Ho** Fl 5s 82m **28M**; brown square Tr, B lantern, W balcony; RC. FR on radio masts 180m SSE and 0.4M NNW. Cable area Oc(3) WRG 8s 17m W6M, R4M, G4M; W mast, R bands; vis W179°-185°, R185°-190°, W190°-196°, W239°-244°, G244°-280°, W280°-285°. Vorhafen. Ostmole, S elbow Oc WG 6s 5m W7M, G4M; vis W203°-250°, G250°-109°. Ostmole head FG 7m 4M; *Horn(3) 30s*. Sudmole, head Oc(2)R 12s 7m 4M. Binnenhaven Ldg Lts 302°, both Oc R 6s.
DÜNE. Ldg Lts 020°. Front Iso 4s 11m 11M; intens on Ldg line. Rear, 120m from front, Iso WRG 4s 17m W14M, R11M, G10M; synchronised, vis G010°-018°, W018°-021°, R021°-030°, G106°-125°, W125°-130°, R130°-144°.
Sellebrunn Lt Buoy 54 14.43N/7 49.83E Q(9) 15s; W cardinal mark; *Whis*.

ELBE 1 LT V 54 00.00N/8 06.58E Iso 10s 15m **17M**; R hull and Lt Tr; RC; Racon; *Horn Mo(EL) 30s*.

Grosser Vogelsand 53 59.78N/8 28.68E Fl(3) 12s 39m 26/9M; helicopter platform on R Tr, W bands; vis 085°-087°. Iso 3s, same Tr, vis 087°-091°. Oc 6s, same Tr, vis 091°-095°. Fl(4) 15s same Tr, vis 095°-102°. Fl(4) R 15s, same Tr, vis 102°-105°. Fl R 3s, same Tr, vis 113°-270°. Oc(4) R 18s same Tr, vis 322°-012°; *Horn Mo(VS) 30s*; RC.

Neuwerk, S side 53 55.0N/8 29.8E L Fl(3) WRG 20s 39m **W16M**, R12M, G11M; vis G165°-215°, W215°-239°, R239°-321°, R343°-100°.

CUXHAVEN. Cuxhaven 53 52.36N/8 42.56E F WR 24m W8M, R6M; dark R Tr, copper cupola; vis W200°-245°, R245°-285°, W285°-290°. Fl(4) 12s, same Tr, vis 290°-301°. Oc 6s, same Tr, vis 301°-310°. Fl(5) 12s, same Tr, vis 310°-318°. Ldg Lts 151°. **Baumrönne**, front, 1.55M from rear Iso 4s 25m **17M**; W Tr, B band; vis 149°-154°. Same Tr Fl 3s vis 144°-149°; Fl(2) 9s vis 154°-157°; Oc R 6s 283°-286°, Oc 6s 286°-293°, Oc G 6s 293°-298°. **Altenbruch**, Common rear, Iso 4s 58m **21M**; B Tr, W bands; intens on leading line, synchronised with front above. Iso 8s 51m **22M**; same Tr; synchronised with front below. Crosslight; same Tr; Oc WR 3s 44m 7/5M; vis W202°-233°, R233°-247°, W247°-255°. **Altenbruch** Ldg Lts 261°. Common front Iso 8s 19m **19M**; W Tr, B bands. Iso WRG 8s, same Tr; vis G117°-124°, W124°-135°, R135°-140°. Rear, Wehldorf Iso 8s 31m 11M; W Tr, B bands; synchronised with front, leads 131°.

Balje Ldg Lts 081°. **Front** 53 51.4N/9 02.7E Iso WG 8s 24m **W17M, G15M**; W round Tr, R bands; vis G shore-080°, W080°-shore. **Rear** 1.35M from front Iso 8s 54m **21M**; W round Tr, R bands; Lts synchronised and intens on ldg line. Same Tr as Rear Lt Oc WR 3s 54m W5M, R3M; vis W180°-195°, R195°-215°, W215°-223°. Zweidorf 53 53.5N/9 05.7E Oc R 5s 9m 3M; R square on W pylon; vis 287°-107°.

BRUNSBÜTTEL. Ldg Lts 065°. Front **Schleuseninsel** Iso 3s and Fl 3s 24m **16M**; R Tr, W bands; Iso to north of leading line and Fl to south. **Industriegebiet**, Rear Iso 3s 46m **21M**; R Tr, W bands; synchronised with front.

BÜSUM. **Büsum** 54 07.65N/8 51.55E Oc(2) WRG 16s 22m **W16M**, R13M, G12M; vis W248°-317°, R317°-024°, W024°-084°, G084°-091°, W091°-093° (ldg sector for Süder Piep), R093°-097°, W097°-148°. West mole Oc(3) R 12s. E mole Oc(3) G 12s. Ldg Lts 355° both Iso 4s 13M, synchronised.
St Peter 54 17.30N/8 39.15E LFl(2) WRG 15s 23m **W15M**, R13M, G11M (W115°-116° Ldg sector for Mittelhever).

Eiderdamm lock. N mole Oc(2)R 12s. S mole OcG 6s.
TÖNNING. West mole FR 5m 2M. Quay FG 5m 2M.
Westerheversand Oc(3)WR 15s 41m **W21M**, **R16M**; R Tr, W bands; vis W012°-145°, R145°-155°, W155°-169°, R169°-199°, W199°-230°, R230°-234°, W234°-252°. Süderoogsand Cross Lt Iso WR 6s 16m W9M, R7M; vis R240°-244°, W244°-246°, G246°-260°, W260°-326°, R326°-345°, W345°-017°, R017°-053°, W053°-099°, R099°-125°, W125°-150°.

PELLWORM. S side Ldg Lts 041°. **Front** Oc WR 5s 14m **W24M**/11M, R8M; vis W303°-313°, R313°-316°. Cross light, same Tr as Rear Ldg Lt, Oc WR 5s 38m 14/6M; vis R123°-140°, W140°-161°, R161°-179°, W179°-210°, R255°-265°, W265°-276°, R276°-297°, W297°-307°. **Rear** Oc 5s 38m **25M**, synchronised.

NORDSTRAND. Strucklahnungshörn, West mole FG 2M.
HUSUM. Ldg Lts 106° both Iso R 8s. Ldg Lts 090° both Iso G 8s (both sets synchronised).

AMRUN. Ldg Lts 273°. Front FR 8m 10M. Rear Fl(3) 30s 63m **23M**. Wriakhorn Cross Lt LFl(2) WR 15s 26m W9M, R7M; vis R287°-297°, W297°-325°, R325°-343°, W343°-014°, R014°-034°. **Witt dun** Dir F WRG 17m **W21M**, **R16M**, **G17M**; vis G000°-004°, W004°-006°, R006°-008° (between buoys 6 and 9 only). **Nebel** Oc WRG 5s 16m **W20M**, **R16M**, **G16M**; vis R255°-258°, W258°-261°, G261°-264°. **Norddorf** 54 40.20N/8 18.60E Oc WRG 6s 22m **W15M**, R12M, G11M; W Tr, R lantern; vis W009°-035°, G035°-038°, W038°-040°, R040°-090°, W090°-156°, R156°-176°, W176°-178°, G178°-188° (unintens) 188°-202°, W(part obsc) 202°-230°.

LANGENESS. Nordmarsch F WRG 13m W14M, R11M, G10M; W268°-279°, R279°-311°, W311°-350°, G350°-033°, W033°-043°, R043°-064°, W064°-123°, R123°-127°, W127°-218°.

FÖHR. **Nieblum** Oc(2) WRG 10s 11m **W19M**, **R15M**, **G15M**; vis G028°-031°, W031°-032°, R032°-035°. Oldenhörn Oc(4) WRG 15s 10m W13M, R10M, G9M. Wyk, S mole Oc(2) WR 12s 12m W7M, R4M; vis W220°-322°, R322°-220°.

OLAND, near West point F WRG 7m W13M, R10M, G9M.

DAGEBÜLL. **Ldg Lts** 053° both Iso 4s (synchronised) **17M**. Ldg Lts 055° both Iso R 4s (synchronised).

SYLT. HÖRNUM **Ldg Lts** 012° both Iso 8s (synchronised). **Hörnum** Fl(2) 9s 48m **20M**. Odde Cross Lt Oc(2) WRG 12s. N pier head FG 6m 3M. Schutzmole FR 6m 3M.

Kampen, Rote Kliff 54 56.87N/8 20.50E Oc(4) WR 15s 62m **W20M**, **R16M**; W Tr, B band; vis W193°-260°, W (unintens) 260°-339°, W339°-165°, R165°-193°; RC. N end, List West 55 03.25N/8 24.19E Oc WRG 6s. List Ost Iso WRG 6s. List Land Oc WRG 3s. List Hafen, N mole FG 5m 3M. S mole FR 5m 3M.

10.20.5 PASSAGE INFORMATION

DIE EMS TO DIE JADE (chart 3761)

The coastal route from the Hubertgat entrance of Die Ems to the Wangerooger Fahrwasser at the entrance to Die Jade (see below) is about 60 miles. The route leads along the Inshore Traffic Zone, S of the Terschellinger-Deutsche Bucht traffic scheme, the E-going lane of which is marked on its S side by stbd-hand buoys with the letters DB (DB1, DB3, DB5 etc). About 4M S of this line of buoys the landward side of the Inshore Traffic Zone is marked by buoys showing the approaches to the various zeegaten between the German Frisian Islands (see below). There are major Lts on Borkum, Norderney and Wangerooge. Near the traffic scheme the E-going stream begins at HW Helgoland −0500, and the W-going at HW Helgoland +0100, sp rates 1.2 kn. Inshore the stream is influenced by the flow through the zeegaten.

DIE EMS (charts 3509, 3510)

Die Ems forms the boundary between Netherlands and Germany, and is a major thoroughfare leading to ports such as Eemshaven, Delfzijl (10.19.29), Emden (10.20.21), Leer and Papenburg. Approach is made through Hubertgat or Westerems, which join W of Borkum. Both are well buoyed but can be dangerous with an ebb stream and a strong W or NW wind. The flood begins at HW Helgoland +0530, and the ebb at HW Helgoland −0030, sp rates 1½ kn.

From close SW of Borkum the chan divides into two — Randzel Gat to the N and Alte Ems running parallel to it to the S — as far as Eemshaven on the S bank. About 3M further on the chan again divides. Bocht van Watum is always varying so keep in main chan, Ostfriesisches Gatje, for Delfzijl (10.19.29) and beyond. Off Eemshaven the stream runs 2-3 kn.

Osterems is the E branch of the estuary of Die Ems, passing between Borkum and Memmert. It is not lit, but is a useful chan in daylight and good weather if bound to/from the E, although much shallower than Westerems. It also gives access to the Ley chan, leading to the harb of Greetsiel.

FRISIAN ISLANDS (chart 3761)

The German islands to the E of Die Ems have fewer facilities for yachtsmen than the Dutch islands to the W, and most of them are closer to the mainland. For general notes on the Frisian Islands see 10.19.5. For navigating inside the islands, in the so-called watt channels, it is essential to have the appropriate German charts and to understand the system of channel marking. Withies, unbound (with twigs pointing up) are used as port-hand markers, and withies which are bound (with twigs pointing down) as stbd-hand marks. Inshore of the islands the conventional direction of buoyage is always from west to east, even though this may conflict with the actual direction of the flood stream in places.

Borkum (10.20.8 and chart 3509) is the first of the German islands. It lies between Westerems and Osterems, with high dunes each end so that at a distance it looks like two separate islands. Round the W end of Borkum are unmarked groynes, some extending ¼M offshore. Conspic landmarks include Grosse Beacon and Neue Beacon, a water Tr, Borkum Grosser Lt Ho, Borkum Kleiner Lt Ho, and a disused Lt Ho — all of which are near the W end of the island.

Memmert on the E side of Osterems is a bird sanctuary, and landing is prohibited. Juist (10.20.21) is the first of the chain of similar, long and narrow islands which lie in an E/W line between the estuaries of Die Ems and Die Jade. Most have groynes and sea defences on their W and NW sides. Their bare sand dunes are not easy to identify, and shoals extend seaward for 2M or more in places. The zeegaten between the islands vary in position and depth, and all of them are dangerous on the ebb tide — even in a moderate onshore wind. There are Nature Reserves (entry prohibited) inshore of Baltrum, Langeoog and the W end of Spiekeroog (chart 1875).

Norderneyer Seegat is a deep chan close W of Norderney, but there are dangerous shoals offshore, through which lead two shallow chans, Dovetief and Schluchter. Both are buoyed. The chans vary considerably and on occasions are silted up. Dovetief is the main chan, but Schluchter is more protected from the NE. Neither should be used in strong winds. Further inward Norderneyer Seegat leads round the W end of the island to Hafen von Norderney (10.20.9). Beware groynes and other obstructions along the shore.

SW of Norderney, Busetief leads in a general S direction to the tidal harb of Norddeich (see 10.20.21), which can also be approached with sufficient rise of tide from Osterems through the Norddeich Wattfahrwasser with a depth of about 2m at HW. Busetief is deeper than Dovetief and Schluchter, and is marked by buoys and Lt buoys. The flood begins at HW Helgoland −0605, and the ebb at HW Helgoland −0040, sp rates 1 kn.

Baltrum (10.20.21) is an island about 2½M long, very low in the E and only rising to dunes about 15m high in the W. With local knowledge and only in good weather it can be approached through Wichter Ee, a narrow chan obstructed by a bar between Norderney and Baltrum. There is a small pier alongside a groyne which extends about 2ca from the SW corner of Baltrum. The little harb dries, and is exposed to the S and SW. Continuing southwards from Wichter Ee is Nessmersiel Balje, with buoys on its W side prefixed by the letter 'N', leading to the sheltered harb of Nessmersiel (see 10.20.21).

Proceeding E, the next major chan through the islands is Acummer Ee between Baltrum and Langeoog. Shoals extend 2M to N, depths in chan vary considerably and it sometimes silts up. In onshore winds the sea breaks on the bar. Chan is marked by buoys prefixed with letter 'A', moved as necessary. Apart from Langeoog (10.20.11), Accumer Ee gives access to the mainland harbs of Dornumer-Accumersiel (10.20.10) and Bensersiel (10.20.21).

Westerbalje and Otzumer Balje are both buoyed and lead inward between Langeoog and Spiekeroog. Otzumer Balje is normally the deeper (0.7m-2.0m) but both chans may silt up and should be used only in good weather and on a rising tide.

Harle chan (buoys prefixed by letter 'H') leads W of Wangerooge (10.20.12), but beware Buhne H groyne which extends 7½ ca WSW from end of island to edge of fairway. Dove Harle (buoys prefixed by letter 'D') leads to the harb. In bad weather Harlesiel, 4M S on the mainland shore, is a more comfortable berth. See 10.20.21.

Blau Balje leads between the E end of Wangerooge and Minsener Oog. Although marked by buoys (lettered 'B' and moved as requisite) this chan is dangerous in N winds and there is a prohibited area (15 May − 31 Aug) S of the E end of Wangerooge for the protection of seals.

DIE JADE (charts 3368, 3369)

To the E of Wangerooge and Minsener Oog lie the estuaries of Die Jade and Die Weser. Die Jade is entered through the Wangerooger Fahrwasser (buoyed) and then leads SSE past Wangersiel and the yachting centre of Hooksiel (10.20.13) to Wilhelmshaven (10.20.14). S of Wilhelmshaven is a large but shallow area of water called Die Jadebusen, through which channels run to the small harbours of Dangastersiel (10.20.21) and Vareler Siel. Although not so dangerous as Die Elbe (see below), the outer parts of both Die Jade and Die Weser become very rough with wind against tide, and are dangerous on the ebb in strong NW winds.

AREA 20—Federal Republic of Germany   769

DIE WESER (charts 3368, 3405, 3406, 3407)

Die Weser is an important waterway leading to the ports of Bremerhaven, Nordenham, Brake and Bremen which in turn connect with the inland waterways. From Bremen to Bremerhaven (10.20.15) the river is called Die Unterweser, and below Bremerhaven it is Die Aussenweser flowing into a wide estuary through which pass two main channels — Neue Weser and Alte Weser. The position and extent of sandbanks vary: on the W side they tend to be steep-to, but on the E side there are extensive shoals (e.g. Tegeler Plate).

Weser Lt Float marks the approach from NW to Neue Weser (the main fairway) and Alte Weser which are separated by Roter Sand and Roter Grund, marked by the disused Roter Sand Lt Tr (conspic). Both channels are well marked and they join about 3M S of Alte Weser Lt Tr (conspic). From this junction Hohewegrinne (buoyed) leads inward in a SE direction past Tegeler Plate Lt Tr (conspic) on the E side to the Fedderwarder Fahrwasser, which passes N of Hohe Weg Lt Tr (conspic). In Die Aussenweser the stream, which runs over 3 kn at sp, often sets towards the banks and the branch channels which pass through them.

Das Weser-Elbe Wattfahrwasser is a useful inshore passage between Die Weser and Die Elbe. It leads in a general NE direction from the Wurster Arm, part of the Weser which runs N of the N leitdamm, and keeps about 3M offshore. The normal yacht will take two tides to traverse it, but there are suitable anchs. Like other inshore passages behind the islands, it is well described in *Frisian Pilot* by Mark Brackenbury (Stanford Maritime).

DIE ELBE (charts 3262, 3266, 3268)

Yachts using the Elbe (see 10.20.17) are probably bound for the Nord-Ostee Kanal entrance at Brunsbüttel (10.20.18). Commercial traffic is very heavy. At Elbe 1 Lt V the E-going (flood) stream begins at HW Helgoland −0500, and the W-going (ebb) stream at HW Helgoland +0500, sp rates 2 kn. The stream runs harder N of Scharnhörn, up to 3½ kn on the ebb, when the Elbe estuary is dangerous in strong W or NW winds. From E end of traffic scheme the channel is well marked by buoys and Bn Trs.

DIE ELBE TO LISTER TIEF (charts 1875, 3767)

The W coast of Schleswig-Holstein is low and marshy, with extensive offshore banks which partly dry and on which are low offshore islands (Die Nordfriesischen Inseln). Between the banks and islands are many chans which change frequently. Süderpiep and Norderpiep are two chans (both buoyed, but the latter unlit) S of Die Eider, which join S of Blauort and lead to Büsum (10.20.21) and Meldorfer Hafen. Norderpiep has a bar (depth 3m) while Süderpiep does not and is preferable in W winds. Landmarks from seaward are Tertius Bn and Blauwortsand Bn, and a conspic silo at Büsum.

Die Eider is merged with much of its length with Der Nord-Ostsee Kanal, with which it connects at Gieselau. Below Tönning (10.20.21) it winds into an estuary, separated from Die Hever further N by the Eiderstedt Peninsula. About 5M below Tönning the estuary is closed by Eiderdamm — a storm barrage with a lock and sluices. The lock operates H24. Approaching from seaward, locate the Ausseneider Lt buoy, about 6M W of the buoyed entrance chan. St Peter Lt Ho is conspic on N shore. The entrance is rough in W winds, and dangerous in onshore gales.

Die Hever consists of several chans on the N side of Eiderstedt Peninsula, and S of the islands of Süderoogsand and Pellworm. Mittelhever is the most important of the three buoyed chans through the outer grounds, all of which meet SE of Süderoogsand before they separate once more into Heverstrom leading to Husum (10.20.21), and Noderhever which runs NE between Pellworm and Nordstrand into a number of watt channels. Schmaltief leads S of Amrun to Amrun Hafen, to Wyk on E side of Fohr, and to Dagebull (see 10.20.21).

Sylt (10.20.20) is the largest of Die Nordfriesischen Inseln, and is some 20M from N to S. It has a straight coast facing the sea, and a peninsula on the E side connected to the mainland by Hindenburgerdamm. Vortrapptief is the chan inward between Amrum and Sylt, leading to Hörnum-Reede and Hafen von Hörnum. It has a depth of about 4m (subject to frequent change) and is marked by buoys and Lt buoys. The area should not be approached in strong W winds. The flood (ESE-going) stream begins at HW Helgoland −0350, and the ebb (WNW-going) at HW Helgoland +0110, sp rates 2½ kn.

Lister Tief, which marks the German-Danish border, leads between the N end of Sylt and Romo, and gives access to List Roadstead and Hafen von List as well as to Danish harbours. Lister Tief is well marked by buoys and Lt buoys, with a least depth over the bar of about 4m. After Süderpiep it is the safest chan on this coast, available for yachts seeking anch under the lee of Sylt in strong W winds (when however there will be a big swell over the bar on the ebb). Beware obstructions (ODAS) marked by Lt buoys, 18M WSW of List West Lt Ho.

10.20.6 DISTANCE TABLE

Approximate distances in nautical miles are by the most direct route while avoiding dangers and allowing for traffic separation schemes etc. Places in *italics* are in adjoining areas.
Note: In this table all distances along and between the Frisian Islands are for routes passing north of same.

1	*North Foreland*	**1**																		
2	*Dunkerque*	40	**2**																	
3	*Burnham-on-Crouch*	36	75	**3**																
4	*Grimsby*	169	194	168	**4**															
5	*Goeree Tower*	91	73	108	169	**5**														
6	*Den Helder*	171	153	169	180	80	**6**													
7	Borkum	243	241	252	249	168	95	**7**												
8	Norderney	259	257	268	265	184	111	31	**8**											
9	Wangerooge	283	281	292	289	208	135	55	29	**9**										
10	Bremerhaven	316	314	325	322	241	168	88	62	38	**10**									
11	Helgoland	300	298	309	285	225	152	67	44	24	44	**11**								
12	Brunsbüttel	337	335	347	332	263	190	110	84	55	78	51	**12**							
13	Holtenau	390	388	400	385	316	243	163	137	108	131	104	53	**13**						
14	Hamburg	374	372	384	369	300	189	104	81	61	88	37	90	**14**						
15	Husum	339	337	348	327	265	190	105	85	52	82	47	76	129	113	**15**				
16	Hörnum Lt (Sylt)	325	323	334	301	251	177	97	77	60	80	38	75	128	112	48	**16**			
17	*Esbjerg*	355	353	360	322	281	207	133	123	109	127	83	126	179	163	95	47	**17**		
18	*Thyboron*	393	390	400	348	329	256	200	186	178	201	154	196	249	233	163	120	90	**18**	
19	*Skagen*	501	499	507	442	437	364	305	291	283	306	259	301	—	338	268	225	195	104	**19**
20	*Oksoy Lt (Norway)*	466	468	470	391	398	332	272	268	260	283	236	278	—	315	245	203	172	83	84	**20**

SPECIAL FACTORS AFFECTING THE FEDERAL REPUBLIC OF GERMANY 10-20-7

TIME ZONE is −0100, which is allowed for in the tidal predictions but no provision is made for daylight saving schemes which are indicated by the non-shaded areas on the tide tables (see 9.1.2).

SIGNALS Local port, tidal, distress and traffic signals are given where possible.

Light signals

R	R	— Passage or entry forbidden.
	R	— Be prepared to pass or enter.
W R	R	Bridge closed or down; vessels which can pass under the available clearance may proceed, but beware of oncoming traffic which have right of way.
W R	W R	Lift bridge will remain at first step; vessels which can pass under the available vertical clearance may proceed.
G	G	Passage or entry permitted; oncoming traffic stopped.
W G	W G	Passage permitted, but beware of oncoming traffic which may have right of way.
R R		Bridge, lock or flood barrage closed to navigation.
R		— Exit from lock forbidden.
G		— Exit from lock permitted.

Visual storm signals in accordance with the International System (see 10.14.7) are shown at: Borkum, Norderney, Norddeich, Accumersiel, Bensersiel, Bremerhaven, Brunsbüttel, Die Oste, Glückstadt, Stadersand, Hamburg, Büsum, Tönning, Husum, Wyk, List and Helgoland.

Lights In coastal waters particular use is made of light sectors. A leading sector (usually white, and often intensified) may be flanked by warning sectors to show the side on which the vessel has deviated. If to port — a red fixed light or a white group flashing light with an even number of flashes. If to starboard — a green fixed light or a white group flashing light with an odd number of flashes. Cross lights with red, white and green sectors may indicate the limits of roadsteads, turning points in channels etc.

Signals hoisted at masts

By day	By night	Meaning
Red cylinder	W R W	Reduce speed to minimize wash
Two black balls over cone point down	R R G	Fairway obstructed
Black ball over two cones points together (or red board with white band)	R G W	Channel permanently closed

CHARTS Where possible the Admiralty, Stanford, Imray and Dutch chart numbers are quoted. The German chart numbers are those of the charts issued by the Deutsches Hydrografisches Institut in Hamburg.

LANDS In place of the 'counties' in the UK, Lands are given.

MARINE RESCUE CO-ORDINATION CENTRES Coast station, Elbe- Weser Radio Tel. 4721-22066 or 22067. Kiel Radio, Tel. 431- 39011. Norddeich Radio, Tel. 4931-1831.

PUBLIC HOLIDAYS New Year's Day, Good Friday, Easter Monday, Labour Day (1 May), Ascension Day, Whit Monday, Day of German Unity (17 June), Prayer Day (18 November), Christmas Day and Boxing Day.

TELEPHONES To call UK from West Germany, dial 0044. Then dial the full UK area code, but omitting the prefix 0, followed by the number required.
To call Germany from UK, dial 010-49.

EMERGENCIES Police – dial 110
Fire, Ambulance – dial 112

Note: Two National Water Parks have been declared; one the Watten Sea area of Lower Saxony excluding the Jade and Weser rivers and the Ems-Dollard estuary; the other the west coast of Schleswig-Holstein. The rules for yachting in these areas have not yet been published other than a ban on leaving grounded boats unattended except in harbours. It is hoped that more detailed rules, due to be published by the Federal Ministry of Transport, will be available for the Supplements.

BORKUM 10-20-8
EAST FRISIAN ISLANDS
Niedersachsen

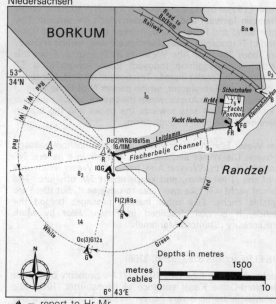

⚓ – report to Hr Mr

CHARTS
Admiralty 3761, 3509; German D90, D87;
Dutch Yacht Chart 1812; ANWB A

TIDES
Dover −0135; ML 1.3; Duration 0610; Zone −0100

Standard Port HELGOLAND (→)

Times				Height (metres)			
HW		LW		MHWS	MHWN	MLWN	MLWS
0200	0700	0200	0800	2.7	2.3	0.4	0.0
1400	1900	1400	2000				

Differences BORKUM (FISCHERBALJE)
−0050 −0050 −0125 −0110 0.0 0.0 0.0 0.0
NORDDEICH HAFEN
−0010 −0020 −0040 −0025 +0.2 +0.2 +0.1 0.0

SHELTER
The shelter is reasonable in the Schutzhafen and it is available at all states of the tide. To the W there is a yacht harbour; access H24 with good shelter and facilities. There is another small harbour to the NE, the triangular Kleinbahnhafen (Light Railway Harbour) whose entrance is 65m wide. None give good protection in really bad weather, but the Yacht Harbour is best.

NAVIGATION
Waypoint Riffgat (safe water) buoy, Iso 8s, 53°38'.90N 06°27'.10E, 302°/122° from/to Fischerbalje Lt, 11M. See also 10.19.29. There are many groynes extending up to 2¾ ca (500 m) off shore. Approaching from any direction, pick up the Fischerbalje Lt at the end of the Leitdamm. Beware strong cross currents across the Fischerbalje channel. Speed limit in harbours 5 kn.
Note: — The S part of the island is a Nature Reserve. Borkum town is 7 km away.

LIGHTS AND MARKS
Landmarks — Water Tower, Grosse Beacon and 2 ca to the SW, Nene Beacon.
From the Fischerbalje to the channel up to the harbour is well buoyed.

AREA 20—Federal Republic of Germany

BORKUM continued

Schutzhafen harbour ent shows FR on W and FG on E moles.
Fischerbalje Lt, Oc (2) WRG 16s 15m 16/11M, W Tr with R top and lantern on tripod; R256°-313°, G313°-014°, W014°-081°, R081°-109°, W109°-113°, R113°-120°, W120°-125° (Ldg Sector to Westerems), R125°-140°.

RADIO TELEPHONE
Hr Mr VHF Ch 16. See also Delfzijl, 10.19.29.

TELEPHONE
Hr Mr 3440; CG Borkum Kleiner Lt Tr; Customs 2287; Hosp 813; Brit Consul (040) 446071.

FACILITIES
Yacht Harbour (200) Tel. 3880 AB, FW, Slip, D, L, FW, C (1½ ton), AB, V, R; **Kleinbahnhof** C (hand), AB; **F. Hellman** Tel. 2576 Gaz.
Town P, ME, El, V, R, Bar. PO; Bank; Rly (ferry to Emden); Air (services to Emden and Bremen).
Ferry UK — Hamburg—Harwich.

NORDERNEY 10-20-9
EAST FRISIAN ISLANDS
Niedersachsen

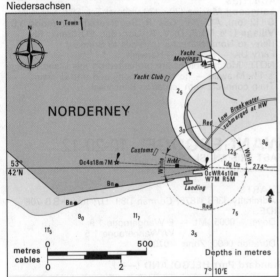

CHARTS
Admiralty 3761; Dutch Yacht Chart 1812; German D89
TIDES
Dover −0050; ML 1.4; Duration 0605; Zone −0100

Standard Port HELGOLAND (→)

Times				Height (metres)			
HW		LW		MHWS	MHWN	MLWN	MLWS
0200	0700	0200	0800	2.7	2.3	0.4	0.0
1400	1900	1400	2000				

Differences NORDERNEY (RIFFGAT)
−0025 −0030 −0055 −0040 +0.1 +0.1 +0.1 0.0

SHELTER
Good shelter and the harbour is always available. Yacht harbour is at the NE of the harbour, where yachts lie bow to pontoon, stern to posts.

NAVIGATION
Waypoint Schlucter (safe water) buoy, Iso 8s, 53°44'.90N 07°05'.30E, 325°/145° from/to W point of Norderney, 3.0M. Entrance through the Dovetief (see 10.20.5) is well buoyed but the bar is dangerous in on-shore winds and seas break on it, especially on an ebb tide. Beware also that tidal streams run across the Dovetief and not up and down it. Beware merchant shipping. Harbour speed limit 3kn.

LIGHTS AND MARKS
Land marks — Norderney Light in centre of island — tall octagonal red brick tower, Fl (3) 12s 59m 23M. Water tower in town (conspic). Storm signal mast 4 ca NW of water tower.
Ldg Lts in line at 274°, synchronised.

RADIO TELEPHONE
None.

TELEPHONE
Hr Mr 793; CG 2293; Customs 3386; Weather Emden 21458; Hosp 477 & 416; Brit Consul (040) 446071.

FACILITIES
Segler-Verein (SVN) Tel. 2850, M, L, FW, C (10 ton), V, R; **W. Visser** Tel. 2207, CH, SM; **Dübbel & Jesse** Tel. 2928, Slip, ME, El, Sh; **Aral** Tel. 2913, P, D; **Shell** (in harbour) Tel. 2468, P; **Yacht Club** Bar; **A. Berghaus** Tel. 1689 Gaz; **Pontoon/Quay** P, D, FW;
Town V, R, Bar. PO; Bank; Rly (ferry to Norddeich); Air (services to Bremen).
Ferry UK — Hamburg—Harwich.

DORNUMER-ACCUMERSIEL 10-20-10
Niedersachsen

CHARTS
Admiralty 3761; German D89
TIDES
−0040 Dover; ML 1.4; Duration 0600; Zone −0100

Standard Port HELGOLAND (→)

Times				Height (metres)			
HW		LW		MHWS	MHWN	MLWN	MLWS
0200	0700	0200	0800	2.7	2.3	0.4	0.0
1400	1900	1400	2000				

Differences LANGEOOG OSTMOLE
+0005 +0005 −0030 −0020 +0.4 +0.4 +0.1 0.0

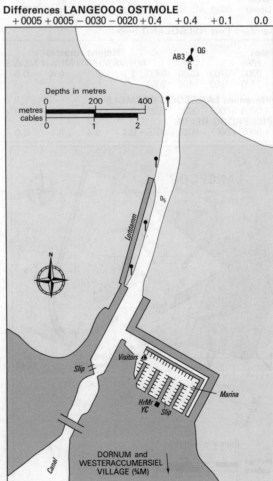

DORNUMER-ACCUMERSIEL continued

SHELTER
The marina provides complete shelter in all winds. Marina depth 3m, access HW∓4. Marina entrance is narrow; keep to W of channel on entering. Marina is fenced so obtain a key before leaving.

NAVIGATION
Approach through Accumer Ee (see 10.20.5 and 10.20.11) leading into Accumersieler Balje and to AB3 buoy, (stbd-hand), QG. From here keep four withies (downturned brooms) to stbd, clear of the Leitdamm. Note warnings on German chart D89.

LIGHTS AND MARKS
None.

RADIO TELEPHONE
None.

TELEPHONE
Hr Mr 1732; Deputy Hr Mr 441; Customs (04971) 7184; Lifeboat (04972) 247; Weather Emden 21458; Hosp 01 15 02; Brit Consul (040) 446071.

FACILITIES
Marina FW, R, Slip, Gaz, D (ME and P by arrangement); **YC Dornumersiel** Tel. 622; **Town** V, R, Bar, PO, Bank, Rly (Harlesiel), Air (Bremen or Hamburg).
Ferry UK — Hamburg—Harwich.

LANGEOOG 10-20-11
EAST FRISIAN ISLANDS
Niedersachsen

CHARTS
Admiralty 3761, 1875; Dutch Yacht Chart 1812; ANWB A; German D89

TIDES
Dover −0030; ML No data; Duration 0600; Zone −0100

Standard Port HELGOLAND (→)

Times				Height (metres)			
HW		LW		MHWS	MHWN	MLWN	MLWS
0200	0700	0200	0800	2.7	2.3	0.4	0.0
1400	1900	1400	2000				
Differences LANGEOOG OSTMOLE							
+0005	+0005	−0030	−0020	+0.4	+0.4	+0.1	0.0
SPIEKEROOG REEDE							
+0007	+0007	−0030	−0015	+0.4	+0.4	+0.1	0.0

SHELTER
At SW corner of the island, the harbour gives good shelter but it is open to the S. Sand dunes, 20m high, give considerable protection from winds.

NAVIGATION
Waypoint Accumer Ee (see 10.20.5) (safe water) buoy, Iso 8s, Bell, 53°46′.73N 07°21′.20E, 310°/130° from/to SW point of Langeoog, 5.3M. The E side of harbour dries and channel to yacht pontoons is marked by withies. Ice protectors extend about 40m E of E mole, awash at HW, marked by withies.

LIGHTS AND MARKS
Lt Bn on W pier Oc WRG 6s 8m 7/3M.
G064°-070°, W070°-074°, R074°-326°, W326°-330°, G330°-335°, R335°-064°, Horn Mo (L) 30s.
Landmarks:— Langeoog church, Esens church spire and Water Tower 2 ca WNW of Langeoog church.

RADIO TELEPHONE
None.

TELEPHONE
Hr Mr 301; Lifeboat CG 247; Customs 275; Weather Emden 21458; Dr 589; Brit Consul (040) 446071.

FACILITIES
Langeoog Marina (70+130 visitors) Tel. 552, Slip, FW, C (12 ton), AC, Bar, Gaz, R; **Segelverein Langeoog YC**; **Village** (1½ M) P, D, V, R, Gaz, Bar. PO; Bank; Rly (ferry to Norddeich); Air (services to Bremen).
Ferry UK — Hamburg—Harwich.
NOTE: Motor vehicles are prohibited on this island. Village is 1½ M away — go by foot, pony and trap or train. Train connects with ferries to Bensersiel.

WANGEROOGE 10-20-12
EAST FRISIAN ISLANDS
Niedersachsen

CHARTS
Admiralty 3368, 1875; German D89, D7; Imray B B 70B

TIDES
Dover −0005; ML — E Wangerooge 1.9
W Wangerooge 1.5
Duration 0600; Zone −0100

Standard Port HELGOLAND (→)

Times		Height (metres)			
HW	LW	MHWS	MHWN	MLWN	MLWS
All Times	All Times	2.7	2.3	0.4	0.0
Differences WEST WANGEROOGE					
+0003	−0019	+0.5	+0.4	+0.1	0.0

SHELTER
Not good shelter except in N winds. Winds from all other directions make the harbour uncomfortable and, if strong, dangerous. Yachts lie against the E pier. In bad weather, yachts should make for Harlesiel, 4 M SSW.

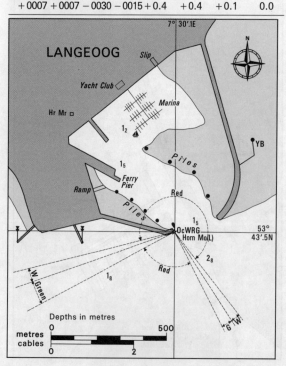

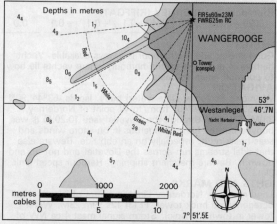

AREA 20—Federal Republic of Germany

WANGEROOGE continued

NAVIGATION
Waypoint Harle (safe water) buoy, Iso 8s, 53°49'.25N 07°49'.00E, 321°/141° from/to Wangerooge Lt, 2.3M. From seaward the Harle channel (see 10.20.5) leads inward between Spiekeroog and Wangerooge: it varies in depth and position, and care is needed. Beware Buhne H groyne, extending 7½ ca (1390 m).
Yachts are forbidden to use the W jetty which is reserved for ferries. Beware protrusions on the E jetty.

LIGHTS AND MARKS
Wangerooge Lt Ho Fl R 5s 60m 23M; R Tr with two W bands. Also FWRG 24m W22/15M, R17/11M, G18/10M R004°-013°, W013°-019°, G019°-038°, W038°-061°, R061°-083°, G (intens)137°-142°, Ldg sector W (intens)142°-152°, R (intens)152°-157°; RC.

RADIO TELEPHONE
None.

TELEPHONE
Hr Mr 630; Customs 223; Weather Bremerhaven 72220; Ambulance 588; Brit Consul (040) 446071.

FACILITIES
Wangerooge YC Tel. 1868, FW; Note: There are virtually no facilities in the harbour; **Grunemann** Tel. 258 Gaz.
Village P, El, V, R. PO; Bank; Rly (ferry to Harlesiel); Air (services to Bremen, Bremerhaven and Helgoland).
Ferry UK — Hamburg—Harwich.

HOOKSIEL 10-20-13
Niedersachsen

CHARTS
Admiralty 3369; German D3011, D7

TIDES
Dover +0025; ML 1.8; Duration 0605; Zone −0100

Standard Port HELGOLAND (→)

Times		Height (metres)			
HW	LW	MHWS	MHWN	MLWN	MLWS
All Times	All Times	2.7	2.3	0.4	0.0
Differences HOOKSIEL					
+0039	+0012	+1.0	+1.0	No data	

SHELTER
Yachts can lie alongside in the Vorhafen (approx 1m at MLWS) but it is very commercial and uncomfortable in E winds. Inside the lock there is complete shelter in a lake 2M long and approx 2.8m deep. Lock opens hourly. Best berths for yachts are in Visitor Yacht Harbour at the far end, in the town. (Max draught 2m — bigger yachts go to YCs — see Lockmaster).

NAVIGATION
Waypoint No 35/Hooksiel 1 (stbd-hand) buoy, QG, 53°39'.43N, 08°05'.52E, 010°/190° from/to entrance, 0.80M. See also 10.20.14. Entrance is marked by H3 stbd-hand buoy, approx 1M W of main Innenjade channel. Approach to lock through Vorhafen, enclosed by two moles. Depth in channel 2.5m. Lock opens weekdays 0800-1900 (LT), Sundays 0900-2000 (LT), actual times on board at lock. Report to lock office. Secure well in lock. Beware tanker pier and restricted area to SE of entrance.

LIGHTS AND MARKS
Conspic chimneys of oil refinery 1.7M S of lock, which is in the W047°-062° sector of the Hooksielplate Cross Lt, Oc WRG 3s. There is a street lamp on the N mole and a pile with R dayglo paint on S mole. Ordinary traffic signals at lock.

RADIO TELEPHONE
VHF Ch 63.

TELEPHONE
Hr Mr 565; Lockmaster 430; Customs (0441) 42031; CG (0421) 5550555; Weather Bremerhaven 72220; Hosp (04421) 2080; Dr 1080; Brit Consul (040) 446071.

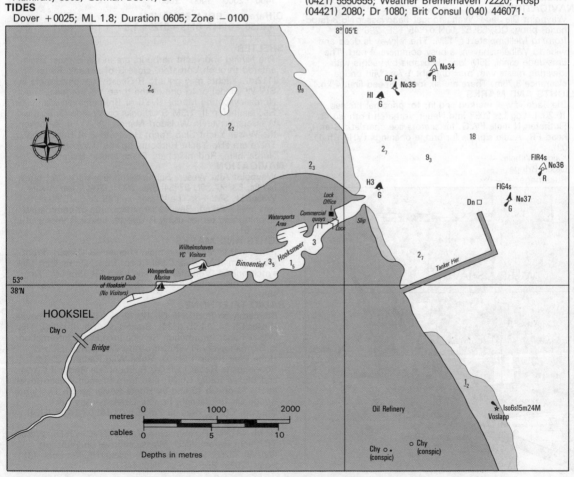

HOOKSIEL continued

FACILITIES
Visitors Yacht Hr (50) AC, C (30 ton), ME, El, Sh, CH, FW, BH (25 ton), R, Slip, SM, V, Gaz, Bar, P and D (cans); **Alter Hafen** AB, FW, V, R, Bar; **Wilhelmshaven YC** Tel. 285; **Boot-Stove-North** Gaz; **Town** V, R, Bar, Bank, PO, Rly (Wilhelmshaven), Air (Wilhelmshaven or Bremen). Ferry UK — Hamburg—Harwich.

WILHELMSHAVEN 10-20-14
Niedersachsen

CHARTS
Admiralty 3369; German D3011, D8
TIDES
Dover +0050; ML 2.0; Duration 0615; Zone −0100

Standard Port HELGOLAND (→)

Times		Height (metres)			
HW	LW	MHWS	MHWN	MLWN	MLWS
All Times	All Times	2.7	2.3	0.4	0.0

Differences WILHELMSHAVEN
+0109 +0034 +1.6 +1.4 +0.2 0.0

SHELTER
The yacht harbour, on E side of Ausrustungshafen inside the locks is sheltered, but unattractive and remote. Locks operate Mon-Fri: 0600-1830, Sat: 0630-1600, Sun and holidays: 0800-1600 (all LT). The tidal yacht harbour in Fluthafen is better for a short stay.
NAVIGATION
Waypoint Die Jade. Wangerooger Fahrwasser No 3 (stbd-hand) buoy, QG, 53°52'.00N 07°45'.58E, 296°/116° from/to Mellumplate Lt, 13M. The fairway is deep and wide to Wilhelmshaven, a busy commercial port. The Ems-Jade canal, 35M long, is usable by yachts with lowering masts with max draught 1.7m. Min bridge clearance 3.75m. There are six locks. Speed limit 4 kn.
LIGHTS AND MARKS
Die Jade is well marked and lit; for principal Lts see 10.20.4. Ldg Lts 208° into Neuer Vorhafen both Iso 4s. Fluthafen N mole FWG (for sectors see chartlet); S arm, head FR. Audio signals for bridge openings (VHF Ch 11 09).
Kaiser Wilhelm — ···
Deich Bridge — — ···
Ruestringer — — ····

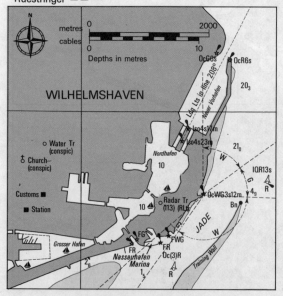

RADIO TELEPHONE (local times)
VHF Ch 11 16 (H24). Wilhelmshaven Lock Ch 13 16 (Mon-Fri: 0600-1830. Sat: 0630-1600. Sun and Public Holidays: 0800-1600). Bridges Ch 11. Other stations: Die Jade (call: *Jade Revier*) Ch 20 63 (H24) gives information broadcasts in German every H + 10.
TELEPHONE (04421)
Hr Mr 61552; Police 43031; Port Authority 26311; Lock 3.34.04; Customs 43136; Weather Bremerhaven 72220; Hosp 8011; Brit Consul (040) 446071.
FACILITIES
Nassauhafen Marina (28 + 100 visitors) Tel. 41439 Slip, AC, Bar, CH, FW, SM, R; **Wiking Sportsboothafen** (30) Tel. 41301 AC, Bar, CH, El, FW, Gaz, ME, R, SM, V; **Turbo-Technik** Tel. 44061, ME, El, Sh; **Farymann** Tel. 23058, ME; **Hochsee YC Germania** Tel. 44121; **Ship-Shop** Tel. 26011 Gaz.
Town PO; Bank; Rly; Air.
Ferry UK — Hamburg—Harwich.

BREMERHAVEN 10-20-15
Federal State of Bremen

CHARTS
Admiralty 3406; German D3011, D4
TIDES
Dover +0030; ML 2.0; Duration 0600; Zone −0100.

Standard Port HELGOLAND (→)

Times				Height (metres)			
HW		LW		MHWS	MHWN	MLWN	MLWS
0200	0800	0200	0800	2.7	2.3	0.4	0.0
1400	2000	1400	2000				

Differences BREMERHAVEN
+0136 +0144 +0119 +0129 +1.5 +1.4 +0.1 0.0

SHELTER
The fishing and yacht harbours are in the S-most basin, entered through Vorhafen, close S of conspic Radar Tr (112m). Yachts may go into R Geeste (uncomfortable in SW-W winds) or to one of the three YC/marinas. The Nordseeyachting Marina (NYC or Bremerhaven Marina) in Fischereihafen II, 1.3M S of locks on W side (3m); the Wassersportverein Wulsdorf Marina (WVW) 0.5M beyond; the Weser Yacht Club Yacht Haven to E of Handelshafen. There are also Yacht Harbours up the R Weser at Nordenham, Rodenkirchen, Brake and at Elsfleth.
NAVIGATION
Waypoint Alte Weser. Schlüsseltonne (safe water) buoy, Iso 8s, 53°56'.30N 07°54'.87E, 300°/120° from/to Alte Weser Lt, 9M. For The Weser, see 10.20.5. R Weser is very well marked. Beware large amounts of commercial shipping and ferries using R Geeste. Leave R Weser at Lt Buoy 61.
LIGHTS AND MARKS
Ldg Lts 151° lead down main channel (R Weser). Front Oc 6s 17m 18M. Rear 0.68M from front, Oc 6s 45m 18M synchronised. Vorhafen N mole head FR 15m 5M. To enter Schleusenhafen sound Q and enter when G Lt is shown.
RADIO TELEPHONE
Bremerhaven Port VHF Ch 12; 12 16 (H24). Bremerhaven Weser Ch 14; **14** 16 (H24). Brake Lock Ch 10 (H24). The Weser Radar Information Service broadcasts information in German every H + 20 from Alte Weser Radar Ch 22, Hohe Weg Radar I Ch 02, Hohe Weg Radar II Ch 02, Robbenplate Radar I Ch 04, Robbenplate Radar II Ch 04 and Blexen Radar Ch 07; also from Bremen-Weser-Revier at H + 30 on Ch 19 and by Hunte-Revier on Ch 17. Hunte Bridge Ch 10 (daylight hours). Oslebshausen Lock, Bremen, Ch 12. Bremen Port Ch 03 16; 03 14 16 (H24).
TELEPHONE (471)
Hr Mr 481260; Bremerhaven Marina (NYC) 77555; Weser YC Yacht Hafen 23531; Wassersportverein Wulsdorf (WVW) 73268; Weather 72220; Fischereihafen Lock 4811; Hosp 42028; Brit Consul (040) 446071.

AREA 20—Federal Republic of Germany

BREMERHAVEN continued

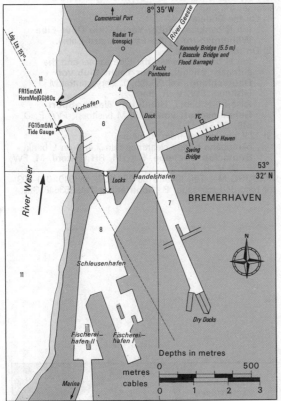

FACILITIES
Bremerhaven Marina (Nordsee Yachting) (200+60 visitors) Tel. 77555, Slip, P, D, AC, Gaz, SM, V, Bar, FW, ME, El, BH (13 ton), Sh; **Weser YC** Tel. 23531, R, Bar, AB; **Yachting Club Wulsdorf** Tel. 23268, AB, FW, Slip; **W. G. Janssen** Tel. 25095 Gaz.
Town all facilities, V, R, Bar, PO; Bank; Rly; Air.
Ferry UK — Hamburg—Harwich.

CUXHAVEN 10-20-16
Niedersachsen

CHARTS
Admiralty 3261; German D87, D44
TIDES
Dover +0050; ML 1.5; Duration 0535; Zone −0100

Standard Port HELGOLAND (→)

Times		Height (metres)			
HW	LW	MHWS	MHWN	MLWN	MLWS
All Times	All Times	2.7	2.3	0.4	0.0
Differences CUXHAVEN					
+0119	+0022	+0.6	+0.5	−0.1	−0.1
SCHARNHÖRN					
+0024	−0050	+0.7	+0.6	+0.1	0.0

SHELTER
Good shelter both in the Alter Hafen and in the Yacht Harbours to the NW. Apply to Cuxhaven YC for moorings. Yachts over 20m may be allowed in Alter Fischereihafen.
NAVIGATION
Waypoint Elbe 1 Lt V, Iso 10s, Horn, RC, Racon, 54°00′.00N 08°06′.58E, 271°/091° from/to Grosser Vogelsand Lt, 13M. Channel well marked by buoys and beacons, see 10.20.4 and 10.20.5. Tide is strong, up to 5 kn off Cuxhaven on the ebb. Much commercial shipping.

LIGHTS AND MARKS
Entrance to S Yacht Harbour is close NW of conspic radar Tr, and N of main Cuxhaven Lt Tr (dark R with copper cupola) FWR, Fl(4) 12s, Oc 6s and Fl(5) 12s (for sectors see chartlet).
Luminous tide signals (G by day, Y at night) from radar Tr. Top panel: chevron point up — tide rising; chevron point down — tide falling; horizontal line below chevron — level below CD. Lower panel: two figures show difference in decimetres between level and CD. S Yacht Harbour entrance, N side FWG, G108°-340°, W340°-108°. S side FWR, W056°-120°, R120°-272°, W272°-295°.
RADIO TELEPHONE
Cuxhaven Elbe. Call: *Cuxhaven Report (Radio)* VHF Ch 12 16; 12 14 16 (H24). Die Elbe information broadcasts in English and German every odd H+00 by Cuxhaven Radar Centre on Ch 21 (see also Brunsbüttel 10.20.18). Cuxhaven Radar is part of Die Elbe Radar and Information Service which includes River Elbe Approach Ch 19, Neuwerk Radar I Ch 18, Neuwerk Radar II Ch 05, Belum Radar Ch 03. See also Brunsbüttel. Information broadcasts every H+55 in English and German by Revierzentrale Cuxhaven on Ch 19, for E part of Deutsche Bucht.
TELEPHONE
Hr Mr 3.41.11; CG 3.80.11; Lifeboat 3.46.22; Weather 3.64.00; Port Authority 2.01.21; Customs 2.10.85; Cuxhaven YC 3.54.04; Brit Cruising Assn 3.58.20; Hosp 181; Brit Consul (040) 446071.
FACILITIES
Yacht Harbour Tel. 3.41.11 (summer only), Slip, M, L, FW, C, AB, R, Bar; **Alter Hafen** Slip, D; **Cuxhavener Bootswerft** Tel. 2.21.32, Slip, ME, El, Sh; **Glüsing** Tel. 2.40.17, P, D, Gaz; **A. Rickmers** Tel. 2.60.26, SM, CH; **Herbert Krause** Tel. 26655, CH, El; **Behrend Hein** Tel. 2.40.69, SM, CH, Chart agent; **Georg Bening** Tel. 2.60.11, CH, Chart agent, ME; **Town** All facilities, R, V, Bar, PO, Bank, Rly, Air (Bremen, Hamburg, Hannover).
Ferry UK — Hamburg—Harwich.

RIVER ELBE/ HAMBURG 10-20-17
Niedersachsen/Schleswig Holstein

CHARTS
Admiralty 3262, 3266, 3268; German D3010, D46, D47, D48

TIDES

	ML	Duration
Glückstadt	1.2	0515
Brunshausen	1.1	0510
Hamburg	1.3	0435
Bunthaus	1.4	0435

Zone −0100

Standard Port HELGOLAND (→)

HW All Times	LW All Times	MHWS	MHWN	MLWN	MLWS
		2.7	2.3	0.4	0.0
Differences GLÜCKSTADT					
+0319	+0234	+0.2	+0.2	−0.4	−0.2
BRUNSHAUSEN					
+0402	+0322	+0.2	+0.2	−0.4	−0.2
HAMBURG					
+0414	+0452	+0.4	+0.4	−0.6	−0.2
BUNTHAUS					
+0538	+0532	+0.5	+0.5	−0.3	+0.2

The river is tidal up to Geesthacht, 24 M above Hamburg, and Hamburg is 78 M from the Elbe 1 Light Vessel. It is a very busy waterway and there are so many lights that at night it can be very confusing. Yachts should keep to starboard, preferably just outside the marked channel. The river is 13m deep all the way to Hamburg. Strong W winds can raise the level in the river by as much as 4m.

It is not a particularly salubrious yachting area but there are places with reasonable shelter:

FREIBURG (7 M above Brunsbüttel) on the SW bank; the river here is ½ M wide. Small craft can enter Freiburg harbour at HW and there is excellent shelter at Freiburg Reede.

GLÜCKSTADT, on the NE bank 2½ M up stream from the entrance of the Stör River, has good harbours, both the Inner and Outer having a min depth of 2.0m.

WEDEL on E bank 12M downstream from Hamburger Jachthafen Wedel (1800) with two entrances (23m wide). Depth 2.5-4.0m. Facilities: Hr Mr 5632, C, FW, P, D, Slip, most facilities available.

STADE, on SW bank, 12M upstream from Glückstadt, Access HW∓2. Yachts lock into inner basin. Most facilities in town. Customs 3014, Hr Mr 10 12 75.

BRUNSHAUSEN Reede is only for commercial vessels. Yachts should go to Brunshausen Yacht Harbour on SW bank. There are several leading lights above Freiburg, shown on the Admiralty chart. Above Brunshausen the channels are subject to change.

There are three pairs of measured mile posts, marked with W diamonds on posts.

HAMBURG There are three harbours, Hamburg, Altona and Harburg-Wilhelmsburg but all are principally commercial. The river divides into several parts passing through Hamburg.

In the Hamburg Altona area there are: (W to E)

Wedel, Hamburger Yachthafen (1800) C (16 ton), D, El, FW, ME, P, Sh, Slip, SM, 20km W of the centre of Hamburg. Hr Mr Tel. 5632; **Bülow** Tel. 8 84 92 El; **Yachtelektrik Wedel** Tel. 8 72 73 El; **Bomotag** Tel. 8 44 02 ME.

Schulau, 3km E of Wedel, a small yacht harbour, Hr Mr Tel. 2784, Customs Tel. 2688, CH, D, El, ME, P, Sh, SC.

Neuenschleuse/Borstel on the S bank a small yacht harbour, 1.5m, with Slip, AB.

Muhlenberg, in the Nienstedten area, a very small yacht harbour, 1.1m, D, FW, P, SC.

Nesskanal on the S bank in Finkenwerden, a yacht harbour with AC, C, CH, El, FW, ME, Sh, Slip, four SCs.

Ruschkanal on the S bank to E of Nesskanal, a largish yacht basin with a SC but no facilities.

Teufelsbrücke, small yacht harbour, 1.3m, Bar, FW, ME, P, SC, bus or ferry to city centre.

Steendiekkanal on S bank to E of Ruschkanal, a yacht basin with AC, CH, FW.

Hamburg Holzhafen Marina, Hr Mr Tel. (040) 789 88 49 P, ME, C, Slip, D, FW, Lau; **Interboat** Tel. 78 44 41 P, D, ME.

There are also a number of YCs where visitors are welcome, the most useful being:
(1) On the Süderelbe —
  (on N bank) — **Motor-Yacht-Club Dove-Elbe Wilhelmsburg** by the Ernst-August Canal.
(2) On the Norderelbe —
  (on N bank, between the Elbe tunnel and the autobahn bridge), **Motor-Yacht Club von Deutschland**, Tel. 36.62.70. Very central M (visitors), FW, V, R, Bar.
  (also on N bank), **Hamburg- Wasser-Sport- Geimeinschaft von 1973**, M (visitors), FW, P, D, ME, El, Sh.
(3) 12½ km up river from Hamburg
  (Autobahn bridge clearance min 7.4m) on L bank. **Bootsclub Oberelbe**, Slip, M, BH (15 ton), M, FW, P, D, V, R, Bar.

HAMBURG Met Tel. (040) 3190 226/827, or (040) 011 509 for recorded message; **B. Schmeding** Tel. 353646 Gaz.

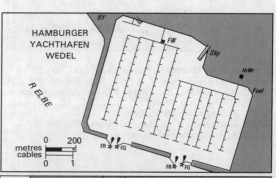

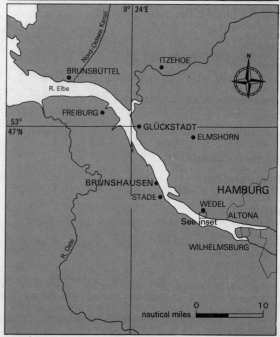

BRUNSBÜTTEL/ NORD-OSTSEE KANAL
10-20-18
Schleswig-Holstein

CHARTS
Admiralty 2469; German D3010, D46, D42
TIDES
Dover +0200; ML 1.4; Duration 0520; Zone −0100

Standard Port HELGOLAND (→)

Times		Height (metres)			
HW	LW	MHWS	MHWN	MLWN	MLWS
All Times	All Times	2.7	2.3	0.4	0.0
Differences BRUNSBÜTTEL					
+0221	+0139	+0.3	+0.3	−0.1	0.0

SHELTER
There is very good shelter in the yacht harbour on the N bank, just inside the locks into the Kiel canal. There is also good shelter in the yacht harbour outside the locks in the Alter Hafen, available HW∓3.

NAVIGATION
For Die Elbe see 10.20.16 and 10.20.5. There are no navigational dangers in the immediate approaches, but commercial traffic is heavy. The stream sets strongly across entrances to locks.

LIGHTS AND MARKS
Ldg Lts 065°, both Iso 3s 16M, synchronised. Alter Hafen Ldg Lts 012°, both Oc(3)R 10s synchronised.

RADIO TELEPHONE (local times)
Brunsbüttel Elbe. Call: *Brunsbüttel Report (Radio)* VHF Ch 14 16; 11 **14** 16 (H24). Die Elbe information broadcasts in English and German every odd H + 05 by Brunsbüttel Radar Centre on Ch 04 (H24) (see also Cuxhaven 10.20.16). Nord-Ostsee Kanal: Ports — Ostermoor (Brunsbüttel to Burg) Ch 73 (H24), Breiholz (Breiholz to Nübbel) Ch 73 (H24). Canal — Kiel Kanal I (Brunsbüttel entrance and locks) Ch 13 (H24), Kiel Kanal II (Brunsbüttel to Breiholz) Ch 02 (H24), Kiel Kanal III (Breiholz to Holtenau) Ch 03 (H24), Kiel Kanal IV (Holtenau entrance and locks) Ch 12 (H24). Information broadcasts by Kiel Kanal II on Ch 02 at H + 15, H + 45 and by Kiel Kanal III on Ch 03 at H + 20, H + 50. Vessels should monitor these broadcasts and not call the station if this can be avoided. Other stations: Oste Bridge Ch 16 69. Este Lock Ch 10 16. Stör Bridge Ch 09 16. Stadersand Elbe. Call: *Stadersand Report (Radio)* Ch 11 16; **11** 12 16 (H24). Hamburg Elbe Port Ch **12** 16; 12 14 (H24). Hamburg Port. Call: *Hamburg Report (Radio)* Ch 16; 06 13 **14** 73 (H24). Hamburg Control Vessel Ch 14 16. Rethe Revier Ch 16 13 (Mon-Sat: 0600–2100, Sun and Public Holidays on request). Süderelbe Revier (at Kattwyk bridge) Ch 16 13 (H24). Harburg Lock Ch 13 16 (H24). See also Cuxhaven 10.20.16.

TELEPHONE
Hr Mr 8011 ex 360; CG 8444; Customs 8.72.41; Weather Cuxhaven 36400; Hosp 601; Brit Consul (040) 446071.

FACILITIES
Alter hafen Tel. 3340, Slip, L, FW, AB, R; **Yachthafen** Tel. 3256, Slip, M, D, FW, C (20 ton), P (cans), AB, Bar; **P. Lützen** Tel. 51016 Gaz;
Town P, D, ME, El, Sh, V, R, Bar. PO; Bank; Rly; Air (Hamburg).
Ferry UK — Hamburg–Harwich.
Note: If transiting the canal without visiting Germany, fly International Code Flag 3rd substitute.

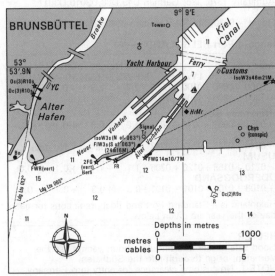

NORD-OSTSEE KANAL (Kiel Canal)

Nord-Ostsee Kanal is 53.3M long (98.7 km) running from Brunsbüttel to Kiel-Holtenau. Width 103 to 162m on the surface, 44 to 90m on the bottom, with a depth of 11m. Kilometre marks start at Brunsbüttel; there are ten passing places (sidings or weichen) and seven bridges with 40m clearance. The speed limit is 8 kn. Sailing, except motor sailing is forbidden. Yachts should report before entering as follows: at Brunsbüttel through Kiel Kanal I on VHF Ch 13; at Holtenau through Kiel Kanal IV on VHF Ch 12.
Canal lock entry signals:—
Iso W — safe for yachts to enter
Other signals — no entry for yachts
In canal:—
3FR (vert) — all movement prohibited
Other signals — not relevant to yachts.
Other ports in the canal, Ostermoor Ch 73 (H24) and Breiholz Ch 73 (H24). The traffic signal of most concern to yachts is 3FR(vert) meaning 'STOP' and usually means the approach of a big ship. Yachts keep to starboard side.
  The principal port along the canal is Rensburg, situated between kilometre posts 60 and 67. There are all facilities at Rensburg including dry docks and several shipyards. At the Baltic end of the canal is Kiel-Holtenau, a major port with all facilities. Here two pairs of locks lead into the Kieler Förde which is practically tideless and almost the same level as the canal.

AGENTS NEEDED
There are a number of ports where we need agents, particularly in France.
ENGLAND Swale, Havengore, Berwick.
SCOTLAND Firth of Forth, Scrabster, Mallaig, Loch Sunart, Loch Aline.
IRELAND Kilrush, Wicklow, Westport/Clew Bay.
FRANCE Arcachon, Seudre R, Ile d'Oleron, Rochfort, Ile de Re, St. Giles-Croix-de-Vie, Ile d'Yeu, Pouliguen, Le Croisic, La Forêt, Ile de Bréhat.
GERMANY Norderney, Dornumer-Accumersiel.
If you are interested in becoming our agent for any of the above, please write to the editors and get your free copy of the Almanac every year. You do not have to be resident in a port to be the agent, but at least a fairly regular visitor.

HELGOLAND 10-20-19
Schleswig-Holstein

CHARTS
Admiralty 1875, 126; German D49, D88, D3010, D3011
TIDES
Dover −0030; ML 1.4; Duration 0540; Zone −0100

Standard Port HELGOLAND (→)

Times				Height (metres)			
HW		LW		MHWS	MHWN	MLWN	MLWS
0100	0600	0100	0800	2.7	2.3	0.4	0.0
1300	1800	1300	2000				

Differences BÜSUM
+0051 +0051 +0014 +0014 +1.0 +0.9 +0.1 0.0
TÖNNING
+0246 +0246 +0242 +0242 No data No data
HUSUM
+0210 +0158 +0120 +0205 +1.1 +1.1 +0.1 0.0
SUDEROOGSAND
+0108 +0108 +0101 +0101 +0.3 +0.3 +0.1 0.0

Helgoland is a Standard Port and tidal predictions for each day of the year are given below.

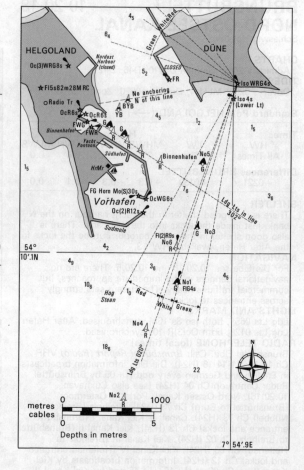

SHELTER
Good safe shelter in artificial harbour; yachts use the Vorhafen, or go through into the Südhafen.
NOTE 1. No customs clearance for entry into Germany. Customs in Helgoland is for passport control.
NOTE 2. It is forbidden to land from yachts on neighbouring Düne Island.
NAVIGATION
Waypoint Helgoland E cardinal buoy, Q(3) 10s, Whis, 54°09'.00N 07°53'.56E, 202°/022° from/to Düne front Ldg Lt, 2.1M. Beware the Hog Stean shoal, 4 ca (740 m) S of Sudmole head Lt. The S entrance is dangerous due to submerged rocks either side of channel. Beware lobster pots round Düne Is.
LIGHTS AND MARKS
Helgoland Lt Ho, Fl 5s 82m 28M; brown sq Tr, B lantern, W balcony; RC.
Binnenhafen Ldg Lts 302°, both Oc R 6s, synchronised.
RADIO TELEPHONE (local times)
Helgoland Port Radio VHF Ch 16 17 (Mon-Sat: 0700-2100): Coast Radio VHF Ch 03 16 27 88 (H24). Information broadcasts in English and German every H+00 and H+30 on Ch 80 by Deutsche Bucht Revier (at Helgoland Lt Ho) for area within 26M of Helgoland.
TELEPHONE
Hr Mr 504; CG 210; Harbour Police 110; Customs 304; Met (04725) 606; Dr 7345; Hosp 8030; Brit Consul (040) 446071.
FACILITIES
Südhafen Tel. 504, D, FW, ME, C (mobile 12 ton), CH, AB, V, R, Bar; **Binnenhafen** Tel. 504, P, D, FW, ME, C (mobile 12 ton), CH, AB, V, R, Bar; **Vorhafen** L, AB; **Rickmers** Tel. 585 Gaz;
Town V, R, Bar. PO; Bank; Rly (ferry Cuxhaven or Hamburg); Air (services to Bremen or Hamburg).
Ferry UK — Hamburg—Harwich.

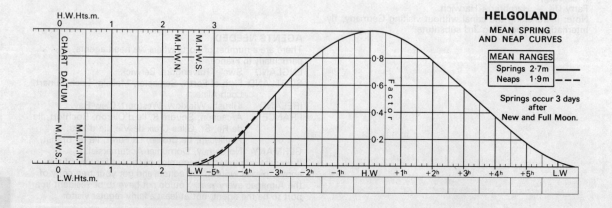

AREA 20 — Federal Republic of Germany

GERMANY - HELGOLAND
LAT 54°11'N LONG 7°53'E

TIMES AND HEIGHTS OF HIGH AND LOW WATERS

YEAR **1989**

TIME ZONE –0100 (German Standard Time)
For German Summer Time add ONE hour in non-shaded areas

	JANUARY				FEBRUARY				MARCH				APRIL		
	TIME m		TIME m		TIME m		TIME m		TIME m		TIME m		TIME m		TIME m
1 SU	0516 2.4 / 1201 0.4 / 1754 2.2	**16** M	0009 0.2 / 0557 2.6 / 1247 0.3 / 1833 2.3	**1** W	0018 0.5 / 0613 2.2 / 1256 0.5 / 1853 2.2	**16** TH	0212 0.2 / 0803 2.1 / 1450 0.3 / 2037 2.2	**1** W	0417 2.3 / 1050 0.3 / 1642 2.2 / 2326 0.3	**16** TH	0023 0.1 / 0613 2.1 / 1252 0.3 / 1842 2.2	**1** SA	0026 0.3 / 0623 2.1 / 1306 0.4 / 1858 2.3	**16** SU	0254 0.1 / 0840 2.0 / 1521 0.2 / 2106 2.4
2 M	0023 0.5 / 0614 2.4 / 1301 0.5 / 1856 2.2	**17** TU	0116 0.2 / 0706 2.4 / 1356 0.3 / 1944 2.3	**2** TH	0141 0.5 / 0736 2.3 / 1420 0.5 / 2014 2.3	**17** F	0344 0.1 / 0930 2.1 / 1615 0.2 / 2157 2.3	**2** TH	0520 2.1 / 1200 0.5 / 1757 2.1	**17** F	0150 0.1 / 0741 1.9 / 1426 0.3 / 2015 2.2	**2** SU	0207 0.2 / 0801 2.1 / 1445 0.3 / 2029 2.4	**17** M	0411 0.0 / 0951 2.2 / 1628 0.1 / 2207 2.5
3 TU	0131 0.6 / 0723 2.4 / 1410 0.5 / 2003 2.3	**18** W	0235 0.2 / 0824 2.4 / 1514 0.3 / 2058 2.3	**3** F	0308 0.4 / 0859 2.3 / 1541 0.4 / 2129 2.4	**18** SA	0501 -0.1 / 1041 2.1 / 1719 0.1 / 2258 2.5	**3** F	0053 0.4 / 0651 2.1 / 1336 0.5 / 1930 2.2	**18** SA	0327 0.0 / 0913 2.0 / 1557 0.2 / 2140 2.3	**3** M	0338 0.0 / 0926 2.2 / 1606 0.1 / 2142 2.5	**18** TU	0502 -0.1 / 1038 2.3 / 1711 0.1 / 2248 2.6
4 W	0244 0.6 / 0833 2.5 / 1517 0.5 / 2107 2.5	**19** TH	0356 0.1 / 0941 2.3 / 1626 0.2 / 2208 2.4	**4** SA	0424 0.2 / 1011 2.4 / 1651 0.3 / 2232 2.5	**19** SU	0556 -0.1 / 1133 2.2 / 1805 0.0 / 2345 2.6	**4** SA	0234 0.3 / 0828 2.2 / 1512 0.4 / 2058 2.3	**19** SU	0447 -0.1 / 1026 2.1 / 1703 0.1 / 2242 2.5	**4** TU	0448 -0.2 / 1029 2.3 / 1708 0.0 / 2240 2.6	**19** W	0536 -0.1 / 1110 2.4 / 1746 0.0 / 2322 2.6
5 TH	0351 0.5 / 0937 2.5 / 1616 0.4 / 2202 2.6	**20** F	0506 0.0 / 1047 2.3 / 1727 0.1 / 2307 2.5	**5** SU	0529 0.1 / 1111 2.5 / 1749 0.2 / 2326 2.7 ○	**20** M	0637 -0.1 / 1212 2.3 / 1844 0.0	**5** SU	0402 0.1 / 0951 2.3 / 1631 0.2 / 2210 2.5	**20** M	0538 -0.1 / 1114 2.2 / 1745 0.0 / 2324 2.6	**5** W	0543 -0.3 / 1118 2.4 / 1800 -0.2 / 2329 2.7	**20** TH	0607 -0.1 / 1142 2.5 / 1823 0.0 / 2356 2.6
6 F	0450 0.4 / 1033 2.6 / 1710 0.3 / 2253 2.6	**21** SA	0602 0.0 / 1141 2.3 / 1816 0.1 / 2355 2.6 ○	**6** M	0624 -0.1 / 1202 2.5 / 1841 0.0	**21** TU	0022 2.6 / 0713 -0.1 / 1245 2.4 / 1920 0.0	**6** M	0512 -0.1 / 1055 2.4 / 1734 0.0 / 2307 2.6	**21** TU	0613 -0.1 / 1148 2.4 / 1821 0.0 / 2358 2.6	**6** TH	0631 -0.4 / 1200 2.4 / 1846 -0.4 ●	**21** F	0640 -0.1 / 1212 2.5 / 1856 -0.1 ○
7 SA	0544 0.2 / 1123 2.6 / 1801 0.3 / 2341 2.7 ●	**22** SU	0648 -0.1 / 1225 2.4 / 1858 0.1	**7** TU	0014 2.7 / 0715 -0.2 / 1248 2.5 / 1928 -0.1	**22** W	0055 2.7 / 0745 -0.1 / 1315 2.4 / 1952 -0.1	**7** TU	0609 -0.2 / 1145 2.4 / 1825 -0.1 ●	**22** W	0645 -0.1 / 1218 2.4 / 1856 -0.1 ○	**7** F	0012 2.7 / 0713 -0.5 / 1239 2.4 / 1928 -0.5	**22** SA	0026 2.6 / 0709 -0.1 / 1240 2.6 / 1924 -0.1
8 SU	0635 0.1 / 1211 2.6 / 1851 0.2	**23** M	0035 2.7 / 0728 0.0 / 1302 2.4 / 1935 0.1	**8** W	0058 2.8 / 0801 -0.3 / 1333 2.5 / 2011 -0.2	**23** TH	0125 2.6 / 0813 -0.1 / 1343 2.4 / 2019 -0.1	**8** W	0657 -0.4 / 1227 2.5 / 1910 -0.3	**23** TH	0029 2.6 / 0716 -0.1 / 1245 2.4 / 1927 -0.2	**8** SA	0054 2.7 / 0752 -0.4 / 1319 2.5 / 2009 -0.5	**23** SU	0055 2.6 / 0735 0.0 / 1307 2.6 / 1952 -0.1
9 M	0027 2.8 / 0724 0.0 / 1301 2.6 / 1938 0.1	**24** TU	0112 2.7 / 0805 0.0 / 1338 2.4 / 2010 0.0	**9** TH	0140 2.8 / 0844 -0.3 / 1412 2.5 / 2049 -0.2	**24** F	0153 2.6 / 0838 -0.1 / 1409 2.4 / 2044 -0.1	**9** TH	0036 2.7 / 0741 -0.4 / 1308 2.4 / 1952 -0.4	**24** F	0057 2.6 / 0743 -0.1 / 1311 2.5 / 1953 -0.2	**9** SU	0138 2.6 / 0832 -0.3 / 1400 2.5 / 2051 -0.4	**24** M	0125 2.6 / 0803 0.0 / 1337 2.6 / 2023 0.0
10 TU	0113 2.8 / 0813 -0.1 / 1348 2.6 / 2022 0.1	**25** W	0148 2.7 / 0838 0.0 / 1411 2.4 / 2042 0.0	**10** F	0220 2.8 / 0922 -0.3 / 1450 2.4 / 2128 -0.2	**25** SA	0219 2.6 / 0903 0.0 / 1436 2.4 / 2111 0.0	**10** F	0117 2.7 / 0821 -0.5 / 1347 2.4 / 2031 -0.4	**25** SA	0124 2.6 / 0806 -0.1 / 1337 2.5 / 2017 -0.1	**10** M	0224 2.6 / 0910 -0.1 / 1440 2.5 / 2134 -0.3	**25** TU	0157 2.6 / 0833 0.1 / 1408 2.6 / 2057 0.0
11 W	0156 2.9 / 0856 -0.1 / 1429 2.5 / 2101 0.1	**26** TH	0219 2.7 / 0908 0.0 / 1442 2.4 / 2110 0.0	**11** SA	0301 2.8 / 0959 -0.2 / 1529 2.4 / 2208 -0.2	**26** SU	0247 2.5 / 0929 0.0 / 1503 2.4 / 2138 0.0	**11** SA	0159 2.7 / 0858 -0.5 / 1425 2.4 / 2110 -0.4	**26** SU	0151 2.6 / 0831 0.0 / 1404 2.5 / 2044 0.0	**11** TU	0309 2.5 / 0949 0.0 / 1522 2.5 / 2217 -0.2	**26** W	0231 2.5 / 0905 0.1 / 1441 2.6 / 2132 0.0
12 TH	0236 2.9 / 0936 -0.1 / 1509 2.5 / 2141 0.1	**27** F	0248 2.7 / 0936 0.1 / 1511 2.4 / 2140 0.1	**12** SU	0346 2.7 / 1038 -0.1 / 1612 2.4 / 2251 -0.1	**27** M	0314 2.5 / 0952 0.1 / 1529 2.4 / 2202 0.1	**12** SU	0242 2.7 / 0934 -0.2 / 1503 2.4 / 2150 -0.3	**27** M	0219 2.6 / 0857 0.1 / 1431 2.5 / 2113 0.0	**12** W	0356 2.3 / 1030 0.1 / 1607 2.4 / 2305 0.0	**27** TH	0308 2.4 / 0939 0.1 / 1519 2.5 / 2211 0.0
13 F	0318 2.9 / 1019 0.0 / 1554 2.4 / 2226 0.1	**28** SA	0318 2.6 / 1004 0.1 / 1541 2.3 / 2209 0.1	**13** M	0434 2.6 / 1118 0.1 / 1658 2.3 / 2339 0.0	**28** TU	0341 2.4 / 1013 0.2 / 1557 2.3 / 2231 0.2	**13** M	0326 2.6 / 1011 -0.1 / 1544 2.4 / 2231 -0.2	**28** TU	0248 2.5 / 0923 0.1 / 1459 2.4 / 2140 0.0	**13** TH	0448 2.2 / 1119 0.2 / 1703 2.3 / 2303 0.1	**28** F	0351 2.3 / 1021 0.2 / 1607 2.4
14 SA	0407 2.8 / 1105 0.0 / 1642 2.4 / 2315 0.1	**29** SU	0348 2.5 / 1030 0.1 / 1611 2.3 / 2238 0.2	**14** TU	0527 2.4 / 1208 0.2 / 1754 2.3			**14** TU	0412 2.4 / 1050 0.1 / 1628 2.4 / 2318 0.0	**29** W	0317 2.4 / 0947 0.1 / 1529 2.4 / 2212 0.1	**14** F	0005 0.1 / 0552 2.1 / 1227 0.3 / 1816 2.3	**29** SA	0449 2.2 / 1123 0.3 / 1712 2.4
15 SU	0500 2.7 / 1152 0.2 / 1734 2.3	**30** M	0421 2.4 / 1058 0.3 / 1646 2.2 / 2316 0.4	**15** W	0045 0.2 / 0636 2.2 / 1320 0.3 / 1910 2.2			**15** W	0505 2.2 / 1140 0.2 / 1725 2.3	**30** TH	0354 2.3 / 1024 0.2 / 1613 2.3 / 2303 0.2	**15** SA	0124 0.1 / 0714 2.0 / 1353 0.3 / 1943 2.3	**30** SU	0016 0.2 / 0607 2.2 / 1247 0.4 / 1834 2.4
		31 TU	0506 2.3 / 1144 0.4 / 1739 2.2							**31** F	0455 2.1 / 1131 0.4 / 1725 2.2				

Chart Datum: 1·76 metres below Normal Null (German reference level)

GERMANY - HELGOLAND

LAT 54°11′N   LONG 7°53′E

TIMES AND HEIGHTS OF HIGH AND LOW WATERS

YEAR **1989**

TIME ZONE −0100 (German Standard Time)
For German Summer Time add ONE hour in non-shaded areas

Chart Datum: 1·76 metres below Normal Null (German reference level)

MAY

	Time	m		Time	m
1 M	0144 0734 1416 1958	0.1 2.2 0.3 2.5	**16** TU	0312 0854 1533 2112	0.1 2.2 0.2 2.5
2 TU	0308 0852 1532 2108	−0.1 2.2 0.1 2.6	**17** W	0407 0945 1623 2200	0.1 2.3 0.2 2.5
3 W	0414 0953 1634 2207	−0.2 2.3 −0.1 2.6	**18** TH	0447 1024 1704 2240	0.0 2.5 0.1 2.6
4 TH	0509 1045 1729 2300	−0.3 2.4 −0.2 2.6	**19** F	0523 1101 1744 2319	0.1 2.5 0.1 2.6
5 F ●	0559 1131 1820 2349	−0.4 2.5 −0.3 2.6	**20** SA ○	0600 1137 1823 2355	0.0 2.6 0.0 2.6
6 SA	0645 1213 1906	−0.3 2.5 −0.4	**21** SU	0635 1210 1858	0.0 2.6 0.0
7 SU	0035 0726 1255 1950	2.6 −0.3 2.6 −0.4	**22** M	0029 0709 1243 1933	2.6 0.1 2.7 0.0
8 M	0121 0808 1338 2035	2.6 −0.2 2.6 −0.3	**23** TU	0105 0744 1319 2011	2.6 0.1 2.7 0.0
9 TU	0209 0850 1422 2121	2.5 −0.1 2.6 −0.2	**24** W	0144 0820 1356 2050	2.6 0.1 2.7 0.0
10 W	0255 0932 1505 2206	2.4 0.1 2.6 −0.1	**25** TH	0223 0858 1433 2131	2.5 0.1 2.7 0.0
11 TH	0341 1013 1551 2252	2.3 0.1 2.6 0.0	**26** F	0306 0938 1516 2216	2.4 0.1 2.7 0.0
12 F ☽	0431 1100 1643 2344	2.2 0.2 2.5 0.1	**27** SA	0353 1025 1606 2308	2.3 0.2 2.6 0.0
13 SA	0528 1157 1744	2.1 0.3 2.4	**28** SU ☾	0449 1121 1705	2.3 0.3 2.6
14 SU	0047 0634 1308 1857	0.2 2.1 0.3 2.4	**29** M	0009 0553 1229 1812	0.1 2.3 0.3 2.6
15 M	0201 0704 1425 2010	0.2 2.1 0.3 2.4	**30** TU	0120 0704 1344 1925	0.0 2.3 0.2 2.6
			31 W	0231 0812 1455 2033	−0.1 2.3 0.1 2.6

JUNE

	Time	m		Time	m
1 TH	0336 0914 1559 2135	−0.1 2.4 −0.1 2.6	**16** F	0352 0933 1618 2157	0.2 2.5 0.3 2.6
2 F	0433 1011 1659 2235	−0.2 2.4 −0.2 2.6	**17** SA	0439 1019 1707 2245	0.2 2.6 0.2 2.6
3 SA ●	0529 1105 1757 2331	−0.2 2.5 −0.2 2.6	**18** SU	0524 1103 1753 2328	0.2 2.6 0.1 2.6
4 SU	0621 1153 1850	−0.1 2.6 −0.2	**19** M ○	0608 1145 1837	0.1 2.7 0.1
5 M	0021 0707 1237 1937	2.6 −0.1 2.7 −0.2	**20** TU	0010 0651 1226 1921	2.6 0.1 2.7 0.0
6 TU	0108 0750 1322 2023	2.5 −0.1 2.7 −0.2	**21** W	0053 0733 1308 2005	2.6 0.1 2.8 0.0
7 W	0155 0834 1407 2108	2.5 0.0 2.7 −0.1	**22** TH	0137 0814 1348 2046	2.6 0.1 2.8 −0.1
8 TH	0241 0916 1451 2152	2.6 0.1 2.8 −0.1	**23** F	0218 0853 1427 2127	2.6 0.1 2.8 −0.1
9 F	0324 0955 1533 2234	2.3 0.1 2.7 0.0	**24** SA	0300 0934 1509 2213	2.5 0.1 2.8 −0.1
10 SA	0408 1037 1617 2316	2.3 0.2 2.6 0.1	**25** SU	0347 1021 1558 2303	2.4 0.1 2.8 0.0
11 SU ☽	0454 1122 1705	2.2 0.2 2.6	**26** M ☾	0438 1112 1652 2354	2.4 0.1 2.8 0.0
12 M	0001 0545 1214 1759	0.2 2.2 0.3 2.5	**27** TU	0532 1208 1749	2.4 0.2 2.7
13 TU	0055 0641 1315 1901	0.2 2.2 0.3 2.4	**28** W	0049 0630 1311 1853	0.1 2.4 0.2 2.6
14 W	0157 0743 1422 2005	0.3 2.3 0.3 2.4	**29** TH	0152 0734 1421 2002	0.1 2.4 0.1 2.6
15 TH	0258 0842 1524 2104	0.2 2.4 0.2 2.5	**30** F	0259 0842 1532 2112	0.1 2.4 0.0 2.6

JULY

	Time	m		Time	m
1 SA	0406 0945 1640 2219	0.1 2.5 0.0 2.5	**16** SU	0358 0941 1635 2215	0.4 2.5 0.3 2.5
2 SU	0509 1047 1744 2320	0.0 2.6 −0.1 2.5	**17** M	0456 1035 1730 2309	0.3 2.6 0.2 2.6
3 M ●	0606 1140 1839	0.0 2.7 −0.2	**18** TU ○	0548 1125 1822 2357	0.2 2.7 0.1 2.6
4 TU	0013 0654 1226 1927	2.5 0.0 2.8 −0.1	**19** W	0637 1211 1910	0.1 2.8 0.0
5 W	0058 0737 1309 2011	2.5 0.0 2.8 −0.1	**20** TH	0043 0724 1256 1955	2.6 0.1 2.9 −0.1
6 TH	0141 0818 1353 2053	2.5 0.0 2.8 −0.1	**21** F	0127 0806 1337 2037	2.7 0.1 2.9 −0.1
7 F	0223 0857 1432 2131	2.5 0.0 2.8 0.0	**22** SA	0207 0843 1414 2116	2.6 0.0 2.9 −0.1
8 SA	0301 0932 1509 2205	2.4 0.1 2.8 0.1	**23** SU	0245 0922 1454 2157	2.6 0.0 2.9 −0.1
9 SU	0336 1007 1545 2239	2.4 0.1 2.7 0.1	**24** M	0327 1006 1541 2243	2.5 0.0 2.9 0.0
10 M	0413 1043 1622 2312	2.4 0.2 2.6 0.2	**25** TU ☾	0415 1053 1632 2327	2.5 0.0 2.8 0.1
11 TU	0452 1120 1702 2349	2.3 0.3 2.5 0.3	**26** W	0504 1143 1724	2.5 0.1 2.7
12 W	0534 1206 1752	2.3 0.4 2.4	**27** TH	0014 0556 1241 1826	0.2 2.5 0.2 2.6
13 TH	0039 0628 1307 1855	0.4 2.3 0.4 2.4	**28** F	0116 0701 1356 1942	0.3 2.4 0.2 2.5
14 F	0144 0733 1420 2006	0.4 2.4 0.4 2.4	**29** SA	0233 0816 1519 2102	0.3 2.4 0.2 2.4
15 SA	0254 0841 1532 2114	0.4 2.4 0.4 2.5	**30** SU	0352 0931 1636 2214	0.2 2.5 0.0 2.5
			31 M ●	0500 1037 1739 2315	0.1 2.6 0.0 2.4

AUGUST

	Time	m		Time	m
1 TU	0556 1132 1831	0.1 2.7 −0.1	**16** W	0531 1106 1805 2343	0.3 2.8 0.1 2.6
2 W	0004 0642 1217 1915	2.5 0.1 2.8 −0.1	**17** TH ○	0622 1154 1853	0.2 2.9 0.0
3 TH	0046 0723 1257 1955	2.5 0.1 2.9 0.0	**18** F	0027 0708 1237 1938	2.7 0.1 3.0 −0.1
4 F	0124 0800 1334 2030	2.6 0.2 2.9 0.0	**19** SA	0108 0749 1317 2018	2.7 0.1 3.0 −0.1
5 SA	0159 0834 1408 2101	2.6 0.1 2.8 0.0	**20** SU	0146 0826 1355 2055	2.7 0.1 3.0 −0.1
6 SU	0230 0904 1438 2129	2.5 0.0 2.8 0.1	**21** M	0222 0903 1434 2133	2.6 −0.1 2.9 −0.1
7 M	0300 0932 1508 2157	2.5 0.1 2.7 0.1	**22** TU	0301 0945 1519 2213	2.6 0.0 2.8 0.1
8 TU	0330 1003 1540 2223	2.5 0.1 2.6 0.2	**23** W ☾	0345 1029 1608 2255	2.6 0.0 2.7 0.2
9 W ☽	0401 1032 1613 2250	2.5 0.3 2.6 0.4	**24** TH	0432 1117 1701 2342	2.5 0.2 2.6 0.4
10 TH	0434 1106 1652 2329	2.5 0.4 2.5 0.5	**25** F	0526 1217 1806	2.5 0.3 2.4
11 F	0520 1200 1752	2.4 0.5 2.4	**26** SA	0048 0636 1339 1928	0.5 2.5 0.4 2.3
12 SA	0034 0629 1320 1912	0.6 2.4 0.6 2.4	**27** SU	0215 0801 1513 2058	0.5 2.4 0.3 2.4
13 SU	0158 0750 1448 2037	0.6 2.4 0.5 2.4	**28** M	0345 0925 1635 2214	0.4 2.5 0.1 2.5
14 M	0322 0907 1607 2152	0.5 2.5 0.4 2.5	**29** TU	0456 1032 1734 2309	0.3 2.7 0.0 2.4
15 TU	0433 1012 1711 2253	0.4 2.6 0.2 2.5	**30** W	0545 1122 1817 2351	0.2 2.8 0.1 2.5
			31 TH ●	0625 1202 1854	0.2 2.8 0.1

AREA 20 — Federal Republic of Germany

GERMANY - HELGOLAND

LAT 54°11′N   LONG 7°53′E

TIMES AND HEIGHTS OF HIGH AND LOW WATERS

YEAR 1989

TIME ZONE −0100 (German Standard Time)
For German Summer Time add ONE hour in non-shaded areas

	SEPTEMBER				OCTOBER				NOVEMBER				DECEMBER		
	TIME m		TIME m		TIME m		TIME m		TIME m		TIME m		TIME m		TIME m
1 F	0027 2.6 0703 0.1 1238 2.9 1929 0.1	**16** SA	0003 2.7 0644 0.1 1212 3.0 1911 −0.1	**1** SU	0028 2.7 0708 0.2 1242 2.8 1925 0.2	**16** M	0012 2.7 0658 0.0 1228 2.9 1920 0.0	**1** W	0054 2.8 0736 0.3 1312 2.8 1944 0.4	**16** TH	0115 2.8 0809 0.0 1347 2.7 2025 0.2	**1** F	0107 2.8 0754 0.3 1331 2.7 2001 0.4	**16** SA	0149 2.9 0847 0.0 1424 2.5 2056 0.2
2 SA	0059 2.6 0737 0.1 1310 2.9 2000 0.1	**17** SU	0041 2.7 0724 0.0 1252 3.0 1950 −0.1	**2** M	0056 2.7 0740 0.2 1310 2.8 1950 0.3	**17** TU	0052 2.7 0740 0.0 1313 2.9 2001 0.1	**2** TH	0124 2.8 0805 0.4 1343 2.7 2013 0.5	**17** F	0201 2.9 0858 0.1 1436 2.6 2109 0.3	**2** SA	0143 2.8 0830 0.3 1407 2.6 2035 0.4	**17** SU	0234 2.9 0932 0.1 1508 2.5 2136 0.3
3 SU	0129 2.6 0807 0.1 1339 2.8 2026 0.1	**18** M	0119 2.7 0803 −0.1 1333 2.9 2028 −0.1	**3** TU	0123 2.7 0801 0.2 1337 2.8 2013 0.3	**18** W	0134 2.8 0824 0.0 1400 2.8 2043 0.3	**3** F	0155 2.8 0837 0.4 1416 2.7 2044 0.5	**18** SA	0246 2.9 0944 0.2 1523 2.5 2151 0.4	**3** SU	0217 2.8 0905 0.3 1444 2.6 2111 0.4	**18** M	0315 2.9 1013 0.2 1549 2.4 2215 0.3
4 M	0156 2.6 0832 0.1 1406 2.8 2050 0.2	**19** TU	0157 2.7 0843 −0.1 1416 2.9 2106 0.1	**4** W	0150 2.8 0827 0.3 1405 2.8 2039 0.4	**19** TH	0216 2.8 0909 0.1 1447 2.7 2124 0.3	**4** SA	0226 2.8 0910 0.4 1451 2.6 2117 0.5	**19** SU	0331 2.8 1030 0.3 1611 2.4 2237 0.5	**4** M	0254 2.8 0945 0.3 1526 2.5 2151 0.4	**19** TU	0357 2.8 1052 0.3 1633 2.3 (2258 0.4
5 TU	0223 2.6 0857 0.2 1434 2.8 2115 0.3	**20** W	0236 2.7 0924 0.0 1500 2.8 2145 0.2	**5** TH	0217 2.8 0855 0.4 1434 2.7 2106 0.5	**20** F	0259 2.8 0954 0.2 1534 2.6 2206 0.4	**5** SU	0302 2.7 0948 0.5 1532 2.5 2156 0.6	**20** M	0422 2.7 1120 0.4 1706 2.3 (2332 0.6	**5** TU	0338 2.8 1031 0.3 1615 2.4 2240 0.5	**20** W	0443 2.6 1134 0.3 1720 2.3 2346 0.5
6 W	0250 2.6 0925 0.3 1504 2.7 2140 0.4	**21** TH	0318 2.7 1008 0.1 1548 2.7 2227 0.3	**6** F	0246 2.7 0925 0.4 1506 2.7 2133 0.5	**21** SA	0346 2.7 1043 0.3 1627 2.4 (2256 0.5	**6** M	0346 2.6 1035 0.5 1624 2.4) 2250 0.7	**21** TU	0521 2.6 1220 0.5 1810 2.2	**6** W	0429 2.7 1125 0.4 1712 2.4) 2341 0.6	**21** TH	0534 2.5 1223 0.4 1814 2.2
7 TH	0318 2.6 0953 0.3 1533 2.6 2204 0.5	**22** F	0405 2.6 1058 0.2 1643 2.5 (2317 0.5	**7** SA	0317 2.6 0958 0.5 1543 2.5 2208 0.6	**22** SU	0441 2.6 1142 0.4 1731 2.3	**7** TU	0444 2.6 1139 0.5 1734 2.3	**22** W	0038 0.7 0631 2.6 1330 0.5 1921 2.3	**7** TH	0531 2.7 1229 0.4 1819 2.4	**22** F	0042 0.5 0634 2.5 1322 0.5 1916 2.2
8 F)	0348 2.5 1023 0.5 1609 2.5 2238 0.6	**23** SA	0501 2.5 1159 0.4 1749 2.3	**8** SU)	0359 2.5 1045 0.6 1637 2.4 2307 0.8	**23** M	0001 0.7 0552 2.6 1257 0.5 1849 2.2	**8** W	0006 0.7 0600 2.6 1302 0.5 1857 2.3	**23** TH	0153 0.6 0745 2.6 1442 0.5 2032 2.3	**8** F	0052 0.6 0642 2.7 1340 0.3 1930 2.4	**23** SA	0149 0.5 0741 2.4 1428 0.5 2021 2.3
9 SA	0429 2.5 1111 0.6 1705 2.4 2340 0.7	**24** SU	0025 0.6 0615 2.5 1322 0.4 1913 2.2	**9** M	0504 2.5 1159 0.6 1757 2.3	**24** TU	0123 0.7 0717 2.6 1425 0.5 2015 2.2	**9** TH	0133 0.7 0723 2.7 1427 0.4 2017 2.4	**24** F	0305 0.6 0852 2.6 1543 0.4 2129 2.5	**9** SA	0207 0.4 0754 2.7 1451 0.3 2037 2.4	**24** SU	0258 0.5 0846 2.5 1530 0.5 2120 2.5
10 SU	0537 2.4 1231 0.6 1828 2.3	**25** M	0155 0.6 0745 2.5 1459 0.4 2046 2.2	**10** TU	0035 0.8 0631 2.5 1335 0.6 1932 2.4	**25** W	0252 0.6 0842 2.6 1546 0.4 2131 2.4	**10** F	0254 0.5 0837 2.7 1537 0.2 2122 2.5	**25** SA	0401 0.5 0944 2.7 1628 0.4 2211 2.6	**10** SU	0318 0.3 0902 2.7 1554 0.2 2139 2.5	**25** M	0400 0.5 0945 2.5 1623 0.5 2210 2.6
11 M	0111 0.7 0706 2.4 1410 0.6 2004 2.3	**26** TU	0329 0.5 0913 2.6 1623 0.3 2204 2.3	**11** W	0213 0.7 0801 2.6 1507 0.4 2057 2.4	**26** TH	0405 0.5 0947 2.7 1642 0.3 2222 2.5	**11** SA	0358 0.3 0937 2.8 1633 0.1 2215 2.6	**26** SU	0445 0.4 1026 2.7 1705 0.4 2247 2.7	**11** M	0423 0.2 1005 2.7 1654 0.2 2236 2.6	**26** TU	0453 0.4 1035 2.6 1710 0.4 2254 2.6
12 TU	0248 0.6 0835 2.5 1540 0.4 2128 2.4	**27** W	0441 0.4 1020 2.7 1719 0.2 2255 2.5	**12** TH	0336 0.6 0916 2.7 1618 0.2 2202 2.5	**27** F	0452 0.4 1031 2.8 1716 0.2 2254 2.6	**12** SU	0454 0.2 1031 2.8 1724 0.1 2303 2.6	**27** M	0526 0.4 1106 2.7 1741 0.4 2323 2.7	**12** TU	0525 0.1 1105 2.7 1750 0.1 2328 2.7	**27** W	0541 0.3 1120 2.6 1754 0.4 2336 2.7
13 W	0409 0.5 0947 2.7 1649 0.2 2233 2.6	**28** TH	0526 0.3 1104 2.8 1753 0.2 2328 2.6	**13** F	0438 0.4 1013 2.8 1712 0.1 2251 2.6	**28** SA	0525 0.3 1105 2.8 1745 0.3 2324 2.7	**13** M	0546 0.1 1122 2.8 1811 0.1 O 2347 2.7	**28** TU	0606 0.4 1144 2.7 1818 0.4 ● 2358 2.8	**13** W	0621 0.0 1158 2.7 1840 0.1	**28** TH	0625 0.3 1200 2.6 1835 0.3 ●
14 TH	0509 0.3 1043 2.8 1743 0.1 2322 2.6	**29** F ●	0600 0.3 1138 2.9 1823 0.2 2358 2.7	**14** SA	0528 0.2 1101 2.9 1758 0.0 O 2333 2.7	**29** SU ●	0601 0.2 1139 2.8 1817 0.3 2355 2.8	**14** TU	0635 0.0 1209 2.8 1855 0.1	**29** W	0643 0.3 1218 2.7 1852 0.4	**14** TH	0015 2.8 0711 0.0 1247 2.6 1926 0.2	**29** F	0015 2.7 0707 0.2 1241 2.6 1916 0.2
15 F O	0559 0.2 1130 2.9 1829 0.0	**30** SA	0635 0.2 1210 2.9 1855 0.2	**15** SU	0615 0.1 1145 2.9 1840 0.0	**30** M	0636 0.3 1212 2.8 1849 0.3	**15** W	0030 2.8 0722 0.0 1257 2.8 1939 0.2	**30** TH	0031 2.8 0718 0.3 1254 2.7 1926 0.4	**15** F	0101 2.8 0759 0.0 1336 2.6 2012 0.2	**30** SA	0055 2.8 0747 0.1 1323 2.6 1955 0.2
						31 TU	0025 2.8 0707 0.3 1242 2.8 1917 0.4							**31** SU	0134 2.8 0826 0.1 1401 2.6 2030 0.2

Chart Datum: 1·76 metres below Normal Null (German reference level)

SYLT 10-20-20
Schleswig-Holstein

CHARTS
Admiralty 3767; German D108, D107
TIDES
Dover +0110; ML 1.3; Duration 0540; Zone −0100

Standard Port HELGOLAND (←)

Times				Height (metres)			
HW		LW		MHWS	MHWN	MLWN	MLWS
0100	0600	0100	0800	2.7	2.3	0.5	0.0
1300	1800	1300	2000				

Differences LIST
+0246 +0246 +0206 +0206 −0.7 −0.6 −0.2 0.0
HÖRNUM
+0220 +0220 +0137 +0137 −0.5 −0.4 −0.2 0.0
AMRUM-HAFEN
+0137 +0137 +0129 +0129 +0.2 +0.2 0.0 0.0

The island of Sylt is about 20M long, has its capital Westerland in the centre, Hornum in the S and List in the N. It is connected to the mainland by the Hindenburgdamm.

SHELTER
HÖRNUM Good in the small harbour, approx 370m × 90m, protected by outer mole on the S side and by two inner moles. Yachts secure on the N side. Good anchorage in Hörnum Reede in W and N winds and in Hörnumtief in E and S winds.
LIST The small harbour is sheltered except in NE to E winds. Entrance is 25m wide and harbour 3m deep.

NAVIGATION
HÖRNUM Waypoint Westvortrapptief (safe water) buoy, Oc 4s, 54°35'.25N 08°10'.85E, 222°/042° from/to Norddorf Lt (Amrun), 6.7M. Access through Vortrapptief buoyed channel (see below) inside drying banks. In strong W winds the sea breaks on off-lying banks and in the channel. Keep mid-channel in harbour entrance, approached from NE. Access at all tides.
LIST Waypoint Lister Tief safe water buoy, Iso 8s, 55°05'.37N 08°16'.87E, 298°/118° from/to List West Lt, 4.7M. Lister Tief (see 10.20.5) is well marked. In strong W or NW winds expect a big swell on the bar (depth 4m) on the ebb, which sets on to Salzsand. Channel buoys through roadstead lead to harbour. Access at all tides, but beware strong tidal streams across entrance.

LIGHTS AND MARKS
HÖRNUM Conspic radio mast 3M N of harbour, and Hörnum Lt Ho close S of harbour. Follow W041°-043° sector of Norddorf Lt (Oc WRG 6s) to line of Hörnum Ldg Lts 012°, both Iso 8s. N pier head FG, vis 024°-260°. Schutzmole FR (mole floodlit).
LIST Kampen Lt, List West Lt, List Ost Lt and Romo church are prominent. For Lts see 10.20.4. N mole head FG, vis 218°-038°. S mole head FR, vis 218°-353°.

RADIO TELEPHONE
None.
TELEPHONE
HÖRNUM Hr Mr 1027, Dr 1016, Hosp (04651) 841.
LIST Hr Mr 374; Hosp (04651) 841, Customs at foot of jetty.
FACILITIES
HÖRNUM **Quay** Tel. 85199, V, FW, AB, CG; **Sylter YC** Tel. 274, AB, Bar, AC, FW, Lau; **Town** V, R, Bar, ME, PO, Bank.
LIST **Harbour** SC, AB, FW, Customs, C (3½ ton), Slip; CG Tel. 85199; **Village** V, R, Bar, ME, P and D (cans); Ferry to Römö (Denmark).
WESTERLAND **HB Jensen** Tel. 7017, Gaz; **Town** V, R, Bar, P and D (cans), Customs, PO, Bank, Rly, Air.

MINOR HARBOURS AND ANCHORAGES 10.20.21

EMDEN, Ostfriesland, 53°21' N, 7°11' E, Zone −0100. Admty chart 3510, German D91. HW +0100 on Dover (GMT), +0040 on Helgoland (zone −0100); HW height +0.8m on Helgoland; ML 1.9m. Ldg Lts 088°, both Oc 5s 12M, lead to Vorhafen entrance, marked by FR and FG Lts. Proceed up Aussenhafen to yacht harbour on E side before Nesserland Lock. Good shelter except in SW winds. If no room there, go through lock into inner harbour, and round to Jarssumer Hafen in SE corner of port. Locks operate 0600-2200 (0800-2000 Sundays). Sound NV (−· ···−) for opening. Jarssumer Hafen is also approached through Grosse Seeschleuse (H24). Both lock-keepers on VHF Ch 13 (H24). Facilities: Hr Mr 2 90 34 Weather 2.14.58; Yacht Hafen Tel. 26020 FW; Customs Tel. 20371; all facilities in city (1M).

JUIST, Niedersachsen, 53°41' N, 7°00' E, Zone −0100. Admty chart 3509, German D90. HW −0105 on Dover (GMT), −0035 on Helgoland (zone −0100); HW height +0.1 on Helgoland (latter two readings taken at Memmert). Entrance through narrow channel running N in the centre of the S side of island, marked by withies to the W. To the W of these is the long (5 ca) landing pier. West Beacon, on Haakdünen at W end of island, Juist water tower (conspic) in centre and East Beacon, 1M from E end of island. Aero Lt Fl W 5s 14m (occas) at the airfield. Facilities: Hr Mr Tel. 724; Customs Tel. 351; Harbour FW, Slip, C; villages of Oosdorp, Westdorp and Loog in centre of island have limited facilities, V, R, Bar, Gaz. Train runs from Juist pier to Oosdorp.

NORDDEICH, Niedersachsen, 53°37' N, 7°09' E, Zone −0100, Admty charts 3761, 1405, German D3012. HW −0030 on Dover (GMT), −0020 on Helgoland (zone −0100); HW height +0.2m on Helgoland. See 10.20.8. Very good shelter in harbour, reached by channel 50m wide, 2m deep and over 1M long between two training walls which cover at HW and are marked by stakes. Hr divided by central mole; yachts berth in West Hr. Lts on

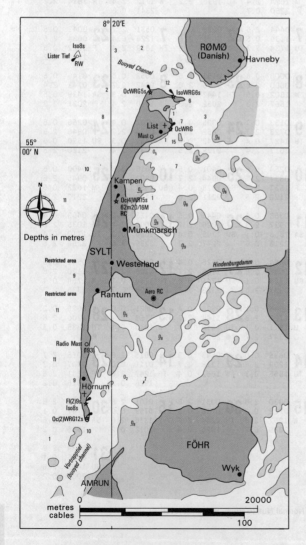

W training wall head FG 8m 4M, vis 021°–327°. On E training wall head FR 8m 4M, vis 327°–237°. Ldg Lts 144°, Front, Iso WR 6s 6m 6/4M, Rear, 140m from front, Iso W 6s 9m 6M synchronised with front. Ldg Lts 350°, Front Iso WR 3s 5m 6/4M, Rear, 95m from front, Iso W 3s 8m 6M. Ldg Lts 170°, Front Iso W 3s 12m 10M, Rear, 420m from front, Iso W 3s 23m 10M, synchronised with front. Facilities: Hr Mr Tel. 8060; **Fuel Stn** Tel. 2721; **Customs** Tel. 2735; **YC** Tel. 3560; **Fritz H. Venske** CH, German Chart Agent; **M. Wagner** Gaz; **Town** AB, Bank, Bar, C (5 ton), D, Dr, FW, PO, R, Rly, Slip, V, Gaz.

BALTRUM, Niedersachsen, 53°43′ N, 7°23′ E, Zone −0100. Admty chart 3761, German D3012. HW −0040 on Dover (GMT), +0005 on Helgoland (zone −0100); HW height +0.4m on Helgoland (latter two readings taken at Langeoog Ostmole). Shelter is good except in SW winds. Without local knowledge, the approach via Wichter Ee Seegat is not recommended. Entrance channel marked by withies to be left approx 10m to the W. Harbour partly dries. Yacht moorings in the Bootshafen at the E end of harbour. Lt at the groyne head Oc WRG 6s 7m 6/3M G082.5°-098°, W098°-103°, R103°-082.5°. Facilities; Hr Mr Tel. 241; **Harbour** AB, FW; Baltrumer Bootsclub YC; **Village** (¼M NNE) V, R, Bar. There is no fuel.

NESSMERSIEL, Niedersachsen, 53°41′ N, 7°22′ E, Zone −0100. Admty chart 3761, German D3012; HW −0040 on Dover (GMT), −0020 on Helgoland (zone −0100); HW height −0.2m on Helgoland. Good shelter in all weathers. Approach down the Nessmersieler Balje, channel marked by stbd buoys and beacons leading to end of Leitdamm. End of Leitdamm has Bn, Oc W, which, together with other unlit Bns mark the course of the Leitdamm on W; withies mark the E side of channel. Leitdamm covers at HW. Beyond ferry berth is a Yachthafen (1½m); secure to catamarans S of Ferry Quay. Harbour dries (2m at HW). Facilities: Hr Mr Tel. 2981; Customs Tel. 2735; **Nordsee Yachtclub Nessmersiel**; no supplies except FW; **Village** (1M to S) has limited facilities.

BENSERSIEL, Niedersachsen, 53°41′ N, 7°35′ E, Zone −0100, Admty chart 1875, German D3012. HW −0024 on Dover (GMT), +0005 on Helgoland (zone −0100); HW height +0.4m on Helgoland. Very good shelter in the yacht harbour (dries). Entrance channel from Rute 1.5M between training walls with depth of 1.5m. Walls submerge at HW. Yacht Hr to SW just before harbour ent (2m). Yachts can also berth in the W side of main hr. Lights: E training wall head Oc WRG 6s 6m 5/2M G110°–119°, W119°–121°, R121°–110°. W mole head FG, E mole head FR. Ldg Lts 138° Front Iso W 6s 7m 9M (intens on line), Rear, 167m from front, Iso W 6s 11m 9M, synchronised with front. Facilities: Slip, C (8 ton), D (on E pier), FW, P (from garage); **Harle-Yachtbau** Tel. (04971) 1760, El, ME, Sh; **Kerkau** Tel. 2401, Gaz; **Town** Bank, Bar, PO, R, Rly, V, Gaz.

NEUHARLINGERSIEL, Niedersachsen, 53°42′ N, 7°42′ E, Zone −0100. Admty chart 1875, German D3012. HW −0100 on Dover (GMT), 0000 on Helgoland (zone −0100); HW height +0.5m on Helgoland. Approach channel well marked from Baklegde N end of Leitdamm, Oc 6s. Beware strong tidal streams across the entrance. Channel runs close E of Leitdamm which is marked by stakes with downturned brooms. It covers at HW. Yachts lie in NE corner of harbour; visitors berths very limited. Facilities: **Quay** FW, D; **Village** V, R, Bar. Very picturesque place.

SPIEKEROOG, Niedersachsen, 53°45′ N, 7°41′ E, Zone −0100, Admty chart 3761, 1875, German D3012. HW −0100 on Dover (GMT), −0005 on Helgoland (zone −0100); HW height +0.5m on Helgoland; ML 1.3m; Duration 0555. See 10.20.11. Good shelter except in winds from S to SW when heavy swell builds up. Small harbour which dries. There is a protective dam to the W of the pier and yachts berth inside the dam. There is also a yacht harbour near the town, up the dredged channel from the Seegat. Lt on rectangular structure FR 6m 4M, vis 197°–114°. There are no facilities except a 5 ton crane but the town can be reached by light railway. Hr Mr Tel. 217.

WANGERSIEL, Niedersachsen, (also known as Horumersiel) 53°41′ N, 8°01′ E, Zone −0100. Admty chart 3369, German D3010. HW −0100 on Dover (GMT), −0030 on Wilhelmshaven (zone −0100); HW height −0.6m on Wilhelmshaven; ML 3.3m. From buoy W3 (G conical) besom perches mark the N side of the channel to the harbour. Best water 10 to 20m from perches. Depth at ent, 1.3m. Channel keeps shifting especially at E end. Boats drawing 1.5m can cross the bar HW∓2½. Most of harbour dries. Secure on N quay. In NW corner is the YC and FW. Facilities: Hr Mr Tel. 238. No fuel. At Horumersiel (¼M) V, R, Bar.

DANGAST, Niedersachsen, 53°27′ N, 8°07′ E, Zone −0100, Admty chart 3369, German D3011. HW +0055 on Dover (GMT), −0007 on Helgoland (zone −0100). S of Wilhelmshaven in the wide bay of Jadebusen, which mostly dries: the Hr dries. Ent via Dangaster Aussentief (0.5m at LWS) leading SW from Stenkentief, marked on NW side by stakes. There are no lights other than Arngast Lt in the middle of Jadebusen. Yacht Hr on W side just before lock into Dangaster Tief, access HW∓2. Facilities: Very limited but usual facilities can be found in the seaside town of Dangast.

BÜSUM, Schleswig Holstein, 54°07′ N, 8°51′ E, Zone −0100, Admty charts 1875, 3767, German D105. HW −0105 on Dover (GMT), +0051 on Helgoland (zone −0100); HW height +0.1m on Helgoland; ML 1.8m; Duration 0625. Very good shelter. Two fairways, Suderpiep and Norderpiep, marked by R and G buoys, each approx 17M, lead into the port. Beware strong tidal streams across ent and sudden winds over the moles. Yachts pass through lock at N of Vorhafen, turn to SE into Segler Hafen. Lock controlled by R and G Lts and manned H24. Signal for opening, one long blast on entering approach chan; two long blasts on reaching lock if not already open. VHF Ch 11 16; Call *Büsum Port*. Facilities: Hr Mr Tel. 3607; CG tel. 2025; Customs Tel. 2376; Dr Tel. 2088; **Büsum YC** Tel. 2997; **Seglerhafen** AB, AC, El, FW, M, ME, D; **Town** CH, V, R, Bar, PO, Bank, Rly, Air (Hamburg).

TÖNNING, Schleswig Holstein, 54°19′ N, 8°57′ E, Zone −0100; Admty chart 3767, German D104. HW +0216 on Dover (GMT), +0246 on Helgoland (zone −0100); HW height +0.5m on Helgoland; ML 2.7m. Eiderdamm lock, 5M seawards, operates H24. VHF Ch 14 16; N mole Oc(2)R 12s, S mole Oc G 6s. Above the dam, Die Eider is tidal, depending on operation of sluices HW+3 to HW+½ (beware strong currents). Secure on S quay of middle section of harbour. If proceeding up river, bridge clearance 5.6m, opened on request Mon-Sat 0600 to sunset. Facilities: Hr Mr Tel. 1400; **Tönning YC** Tel. 754 (all welcome); Dr Tel. 389; Hosp Tel. 706; **Jetty** FW, P, D; **BY** C (5 ton), D, FW, Slip; **Hamkens** Gaz; **Town** Bar, Bank, Hosp (Tel. 706), PO, R, Rly, V.

HUSUM, Schleswig Holstein, 54°29′ N, 9°03′ E, Zone −0100. Admty chart 3767, German D105; 3013 Sheet 13. HW +0130 on Dover (GMT), +205 on Helgoland (zone −0100); ML 1.9m; Duration 0555. See 10.20.19. Yachts can either anchor W of Rly bridge near YC on S bank or pass through bridge (clearance 5m when unopened) to Inner Harbour by SC Nordsee. Both dry. There is a sluice/lock 650m downstream from YC which is shut when level exceeds 0.5m over MHWS. A R Lt is shown when closed. Beware the canal effect when passing big ships in the narrows. VHF call *Husum Port* on Ch 11. Traffic reports on Ch 11 every H+00 from HW−4 to HW+2. Outer Ldg Lts 106° both Iso R 8s, synchronised and intens on leading line. Inner Ldg Lts 090° both Iso G 8s, synchronised and intens on leading line. Facilities: Most facilities available in town. P, D, in harbour. Hr Mr Tel. 667217; Sluice Tel. 2565; **Husum YC** Tel. 65670; **SC Nordsee** Tel. 3436; Customs Tel. 61759.

WITTDÜN, Amrun, 54°38′ N, 8°23′ E; Zone −0100. Admty chart 3767, 3760; German D107; 3013 Sheet 7. HW +0107 on Dover (GMT), +0137 on Helgoland (zone −0100); ML 1.3m; Duration 0540. See 10.20.20. Wittdün is the main harbour of Amrun; there are two others at Nebel and Risum. Yachts berth on yacht bridge S of stone quay, lying bows to pontoon, stern to posts. Good shelter except in E winds. The quay 800m to the E is reserved for ferries only. Channel is 2m. There are two Ldg lines 251° Front (0.55M from rear) FW 15m 14M, Grey mast, Intens on leading line. 273° Front (0.9M from rear) FR 8m 10M, W metal hut B stripes, Intens on leading line. Common rear Fl(3) W 30s 63m 23M R tower with W bands. It is not advisable to enter at night. Channel into harbour marked by withies left to port. Wriakhörn Cross Lt to W of town LFl(2) WR 15s R287°-297°, W297°-325°, R325°-343°, W343°-014°, R014°-034°, 26m 9/7M. To S of the town is a Dir FWRG 17m 21/16M G000.5°-004.5°, W004.5°-005.5°, R005.5°-007.5° (use between Lt Buoys 6 and 9 only). Facilities: Hr Mr Tel. 2294; Customs Tel. 2026, Dr Tel. 2612. P and D (in cans) at Nebel. **Amrun YC** Tel. 2054.

SCHLÜTTSIEL, Schleswig Holstein, 54°41′ N, 8°45′ E; Zone −0100. Admty charts 3760, 3767; German D107; 3013 sheets 7, 9. HW +0156 on Dover (GMT), +0026 on Helgoland (zone −0100). Approach between Langeness/Oland and Gröde-Appelland. Channel 2-3m deep, harbour 2m. Turn into harbour at buoy SCHL 20, proceed between groynes 300m long, ends marked by S and N cardinal beacons. Small harbour mainly used by ferries with no special places for yachts. Moor alongside walls of the quay. In W winds a swell runs into the harbour. Beware warps across harbour from local boats. There are no lights. Facilities: there are very few facilities. Hr Mr Tel. 3301. **Dagebüll YC** Tel. 1463.

WYK, Island of Föhr, 54°42′ N, 8°35′ E, Zone −0100. Admty chart 3760, German D107; 3013 sheet 7. HW +0107 on Dover (GMT), +0137 on Helgoland (zone −0100); ML 2.8. Good yacht harbour to N of entrance, sheltered in all winds, or yachts can berth on E quay of Old Harbour. Yacht harbour (1.5m) Access H24; Commercial harbour (4m). Lights − Oldenhorn (SE point of Föhr), R tower, Oc(4) WRG 15s 10m 13/9M; W208°-250°, R250°-281°, W281°-290°, G290°-333°, W333°-058°, R058°-080°. S mole head Oc(2) WR 12s 12m 7/4M; W220°-322°, R322°-220°. Facilities: Yacht harbour Hr Mr Tel. 3030; Commercial Hr Mr Tel. 2852; **YC** Tel. 2594; Dr Tel. 8998; Customs Tel. 2594; **Harbour road** (500m), CH, Lau, P and D (cans); **Yacht Harbour** R, V. **W Quay** D.

DAGEBÜLL, Schleswig Holstein, 54°44′ N, 8°42′ E, Zone −0100. Admty chart 3760; German D107; 3013 Sheet 7. HW +0156 on Dover (GMT), +0026 on Helgoland (zone −0100). Small harbour mainly for ferries; there are no special yacht berths. Yachts berth on outer side of N mole. Outer Ldg Lts 053°. Front Iso W 4s 10m 17M; red tower; intens on Ldg line. Rear, 400m from front, Iso W 4s 21m 17M; synchronised with front. Inner Ldg Lts 055°, Front Iso R 4s 7m 9M; R mast, W bands; vis 346°-121°. Rear, 50m from front, Iso R 4s 10m 9M; R mast, W bands; synchronised with front. FW Lts on N and S moles. Depth in basin 2.8m. Facilities: Hr Mr Tel. 233; Slip Tel. 209; **Village** R, V, P and D (cans), Rly.

TIDAL INFORMATION FOR WEST COAST OF DENMARK 10-20-22

The following figures are average values and so are approximate and should be used with caution
Standard Port HELGOLAND (←)

Times				Height (metres)			
HW		LW		MHWS	MHWN	MLWN	MLWS
0300	0700	0100	0800	2.7	2.3	0.4	0.0
1500	1900	1300	2000				
Differences HOJER SLUICE							
+0247	+0322	No data		−0.3	−0.2	0.0	0.0
ROMO HAVN							
+0227	+0302	+0221	+0201	−0.8	−0.7	−0.1	0.0
GRADYB BAR							
+0137	+0152	No data		−1.2	−1.1	−0.1	0.0
ESBJERG							
+0307	+0307	+0221	+0221	−1.1	−0.9	−0.2	−0.1
BLAAVANDS HUK							
+0147	+0157	+0131	+0121	−0.9	−0.9	−0.1	0.0
TORSMINDE							
+0337	+0357	+0301	+0231	−1.8	−1.6	−0.3	0.0
THYBORON							
+0427	+0537	+0631	+0431	−2.3	−2.0	−0.3	0.0
HANSTHOLM							
+0407	+0647	+0601	+0351	−2.4	−2.0	−0.3	0.0
HIRTSHALS							
+0402	+0627	+0601	+0321	−2.4	−2.0	−0.3	0.0

Advertisers' Index

Abingdon Boat Centre 216/2
Amsoil UK 690
Aqua-Marine Manufacturing (UK) Ltd 6/1
Berthon Boat Company Ltd 216/3
Boat Mart International 416
Boating Business & Marine Trade News 144/2
Boats and Planes for Sale 176
Boatworks + Ltd 690
Bosun's Locker/Topgear x
Brighton Marina 264/1
Brixham Yacht Supplies Ltd 216/1
Camper & Nicholsons Marinas Ltd 36/1
Camus Marine Ltd 184
Channel Yacht Brokers Ltd 690
Chichester Yacht Basin 6/2
W & H China xviii
Coastguard 788
Cobb's Quay Ltd 216/3
Communication Aerials Ltd 66/1
Crinan Boats Ltd 464
Darthaven Marina 216/1
DMS Seatronics Ltd 264/1
Dubois-Phillips & McCallum Ltd 492
Elmhaven Marina 306
Essex Marina Ltd 334
Everson and Sons Ltd 334
Falmouth Boat Construction Ltd 184
Falmouth Yacht Marina 216/1
Glasplies 492
Granary Yacht Harbour Services 334
Jimmy Green Marine 216/3

Greenham Marine Ltd 36/1
C T Harwood Ltd 66/1
Hoo Marina Medway Ltd 216/4
Housemans Insurance Consultants Ltd 66/2
Hydrographer of The Navy 66/3
Iain Kerr Hunter Yacht Deliveries 66/2
Icom (UK) Ltd 66/2
Kip Marina 464
Largs/Lymington/Swansea Yacht Havens 66/6
Littlehampton Marina Ltd 216/3
Marquand Brothers Ltd 690
Mayflower International Marina 216/1
Ministry of Agriculture Fisheries and Food 22/2
A H Moody & Son 216/2
L H Morgan & Sons (Marine) Ltd 306
Motor Boat & Yachting 801
Motor Cruiser 724
Multihull International 785
Nautech Ltd iv
Navico i
Orwell Auto Electrics Ltd 334
Port Flair Ltd 306
Port Solent Ltd 216/4
Ratsey & Lapthorn (Sailmakers) Ltd 184
Stephen Ratsey Sailmakers 216/4
Rigel Compasses Ltd 306
Rochford Marine Enterprises 306
Royal Ocean Racing Club 22/1
St Katherine Yacht Haven 264/2
Seafarer International Ltd 66/5
Sell's Marine Market 144/1

Shipsides Marine Ltd 492
E C Smith & Sons (Marine Factors) Ltd 6/1
Solent Trading Company, The *inside back cover*
South Eastern Credit 216/4
South Western Marine Factors Ltd *inside front cover*
Southern Marine 6/1
SS Spars (1982) Ltd 492
Stowe Marine Equipment Ltd 464
Telephone Information Services vi
Volvo Penta UK Ltd 6/1, 36/2, 264/1, 306, 464, 690
Warsash Nautical Bookshop 216/3
Dan Webb & Feesey 264/1
What Boat 388
Wheelhouse School of Navigation 216/2
Woolverstone Marina 334
Wyn Ltd 66/1
Yachting Life 614
Yachting Monthly 66/4
Yachts and Yachting 802

Enquiries about advertising in this book should be addressed to:

Communications Management International
Chiltern House
120 Eskdale Avenue
Chesham
Bucks
HP5 3BD

multihull international

The only monthly Journal in the world devoted entirely to the

News and views on multihulls

Annual Subscription £13.00

Five Specimen Copies £5.00

published by
CHANDLER PUBLICATIONS LTD.
53, High Street, Totnes, Devon, TQ9 5NP
Tel: TOTNES (0803) 864668

Late corrections

(up to and including Admiralty Notices to Mariners, Weekly Edition No. 16/88)

Important navigational information in the body of the Almanac is corrected up to Weekly Edition No. 13/1988 of *Admiralty Notices to Mariners*. For ease of reference, each correction is numbered, and its page reference is also given.

Chapter 4
1. 55 Aero Radiobeacon No A620, Eu/Le Treport, delete station.
2. 64 List of Radar Beacons. Insert: No 50 3 & 10cm N Haisbro' Lt Buoy 53°00'.2N 1°32'.4E 30s 360° 10nm T.

Chapter 6
3. 139 Irish Coast Radio Stations. Malin Head Radio and Valentia Radio. Delete VHF services.

Chapter 10
Area 1
4. 190 Longships Lt Ho. Lt now shown continuously.
5. 198 FALMOUTH – RADIO TELEPHONE. Delete: Sat 0900–1200.

Area 2
6. 219 Delete the SAR helicopter symbol at Lee-on-Solent, and insert it at Portland.

Area 3
7. 271 Oaze Deep. E Cant Lt Buoy discontinued.

Area 4
8. 312 Barrow Deep. No 3 Barrow Lt Buoy. Add: *Bell*.
9. 332 KING'S LYNN – RADIO TELEPHONE. Delete lines 1 and 2, and insert: VHF Ch 16; 11 14 (Mon–Fri: 0800–1730. Other times HW–4 to HW+1).

Area 5
10. 351 WHITBY – RADIO TELEPHONE. Delete: VHF Ch 16; **11** 06 (0900–1700). Insert: VHF Ch 16; **11** 12 (0830–1730).

Area 6
11. 386 PETERHEAD – LIGHTS AND MARKS. Delete: Leading lights into harbour 059°.
12. 386 PETERHEAD – CHARTLET. See Correction 11. Delete Ldg Lts and arc of visibility. In place of front Ldg Lt substitute Oc WRG 6s 16m 12M; vis G049°–056°, W056°–060°, R060°–068°.

Area 7
13. 395 ORKNEY. Cava. Lt characteristic changed from Fl WR 3s to Fl 3s 16m 8M.
14. 399 MACDUFF – RADIO TELEPHONE. Delete: (0900–1700 and 1 hr before vessel expected). Insert: (H24).
15. 414 SCRABSTER – RADIO TELEPHONE. Delete: (H24). Insert: (0800–2200 LT).

Area 9
16. 462 KIRKCUDBRIGHT – RADIO-TELEPHONE. Add: (HW−2 to HW+2).

Area 10
17. 483 CONWY – RADIO-TELEPHONE. Delete: Llandudno Pier Ch 16; 06 12 (0700–1500) June–mid Sept).

Area 11
18. 499 Bristol Channel. Merkur Lt Buoy moved to 51 21.85N/3 15.87W.
19. 513 BARRY – RADIO-TELEPHONE. Delete entry and insert: Call: *Barry Radio* VHF Ch 16; 10 (HW−4 to HW+3).
20. 520 BURNHAM-ON-SEA – RADIO-TELEPHONE. Delete: VHF Ch 16. Insert: VHF Ch 16; 08.

Area 14
21. 596 LA GIRONDE – RADIO-TELEPHONE. Add: Blaye Ch 12 (H24). Bordeaux, delete: Ch 11.
22. 612 BORDEAUX. Hr Mr Tel. Delete: 56.90.91.21. Insert: 56.90.58.00.

Area 15
23. 620 LE CROISIC. Ldg Lts 134°. Delete: both Oc R 4s. Insert: both QR.
24. 627 LE CROISIC – LIGHTS AND MARKS. Ldg Lts 134°. Delete: both Oc R 4s, synchronized. Insert: both QR.
25. 627 LE CROISIC – CHARLET. See Correction 24.

Area 19
26. 733 TERSCHELLING. Zuider Stortemelk ZS-Bank Lt Buoy moved about 200m N, to position 53 18.85N/4 57.95E.
27. 741 WESTERSCHELDE – RADIO-TELEPHONE. B. Vlissingen area. Post Vlissingen. Delete: Ch 21. Insert: Ch 61.

Area 20
28. 767 Westerheversand, Oc(3) WR 15s. Amend sectors to read: W012°–089°, R089°–W107°, W107°–155°, R155°–169°, W169°–206° R206°–218°, W218–233°, R233°–248°.
29. 767 Suderoogsand cross light, Iso WRG 6s. Amend sectors to read: R240°–244°, W244°–246°, G246°–263°, W263°–318°, R318°–338°, W338°–017°, R017°–053°, W053°–099°, R099°–125°, W125°–150°.
30. 767 SYLT. Delete: Odde Cross Light Oc(2) WRG 12s.

COASTGUARD

The Magazine of Her Majesty's Coastguard

Index

Explanation

Place names are shown in **bold** type, and without the definite article.
Abbreviations are entered under the full word and not the abbreviation.
  For example, St Abb's comes under Saint, and not St.
Page numbers in **bold** type indicate harbour and tidal information.
Page numbers in *italics* indicate tidal information (only).
The prefix C before a page number indicates Coast Radio Station.
The prefix L before a page number indicates an entry in the lists of Lights, Fog Signals and Waypoints.
The prefix R before a page number indicates Radio Navigational Aid.

A

Abandon ship, 166
Abbreviations, 3, 34–35, 179
Aberaeron, 523
Aberbenoit, L', 654
Aberdeen, R 46, **381**
Aberdeen, fairway buoy, R 64, L 369
Aberdour, 387
Aberdovey, 504
Aberildut, L', L 644, **660**
Aberporth, R 49, *504*
Abersoch, 490
Abertay Lt Buoy, R 64, L 369
Abertay Sands, 370
Aberwrac'h, L', L 644, **654**
Aberystwyth, 506
A'Bhraige, Loch, L 422, **441**
Achilbeg Island, L 562
Achill Sound, L 562, 564
Acknowledgments, 2
Addresses, 22
Adjectives, abbreviations, 34
Adjusting compass, 24
Admiralty chart symbols and abbreviations, 34–35, 116–17
Admiralty Notices to Mariners, 4
Adour, L', L 588, 590, **612**
Advertisers' Index, 785
Aero radiobeacons, 38, 42–61
Affaires Maritimes, 592
Aground, vessels, 17, 118
Aigle, Pointe a l', L 666, 680
Aiguillon-La-Faute-sur-Mer, L', 591, **613**
Ailly, Pointe d', R 55, L 697
Ailsa Craig, L 449
Aircraft directing signals/procedures, 128
Air traffic control centres, SAR, 164
Aix, Ile d', L 588, *603*, **613**
Alarm signal, radiotelephone, 17, 132, 163
Aldeburgh, 324
Alderney, R 50, L 667, 668, **678**
Alderney Race, 669
Aline, Loch, 433
Alprech, Cap d', R 55, **697**
Alsh, Loch, Kyle of, 425, **442**
Altacarry Head, R 51, **561**
Altenbruch, L 767

Alt River, 491
Alte Weser, R 58, L 766
Altitude correction tables, 70, 76
Amble, *359*, **360**
Ameland, R 58, L 733, 735, **760**
Amendments, record of, 4
Amplitudes, 68
Amrun, L 767, *782*, **784**
Amsterdam, R 57, **755**
Anchor lights, 16, 118
Anglesey, 14, C 138, 473, **483**
Anglet, 612
Annalong, L 560
Annan, L 449
Anniversaries, 75
Anstruther, L 369, 370, **378**
Antifer, Cap d', R 55, L 697, 698, *711*
Antwerpen (Antwerp), C 141, **741**
Anvil Point, R 62, L 222, 224
Appledore, 521
Aran Islands, 564, *577*
Aranmore, L 561
Arbroath, L 369, **380**
Arcachon, C 139, L 588, 590, **594**
Arc to time, 114
Ardglass, L 560, **582**
Armandèche, L', L 589, 608
Ardnakinna Point, L 530
Ardnamurchan Point, L 423, 425
Ardrishaig, L 448, 451, **453**
Ardrossan, L 449, *461*
Areas, shipping forecasts, 152, 159
Aries, GHA and Dec, 77, 86–96
Argenton, 660
Arinagour, 442
Arisaig, 442
Arklow, L 531, **547**
Arklow, Lanby, R 65, L 531
Arnish Point, L 422, 426
Arran Isle, L 448, 451, **454, 463**
Arrochar, 455
Ars-en-Ré, 613
Askaig, Port (Islay), *452*, **463**
Astro-navigation, 67–114
Astronomical data, 75–114
Audierne, L 622, **635**
Aut, L 697
Auray, *629*
Auskerry, L 395, 406, **415**
Automatic direction finding, 38
Automatic telephone weather services (Marinecall), 149
Auxiliary planning data, 70, 78–79, 108–113

Aven River, 638
Avonmouth, L 499, **515**
Axmouth, 264
Axe River, 264
Ayr, L 449, *461*, **463**
Ayre, Point of, R 47, R 65, L 470, 472
Azimuth diagrams, 111–113

B

Bacton Radio, C 137
Badcall Bay, *427*
Bad visibility, 16, 17
Baie = bay, see proper name
Baily, L 531
Balbriggan, L 560
Baleines, Les, R 52, L 589
Balje, L 767
Ballinacourty Point, L 531
Ballue, La, L 666
Ballybunnion, C 139
Ballycastle, L 561
Ballycotton, R 50, L 531, 533, **554**
Ballycrovane, L 530
Ballywalter, L 560
Balta Sound, L 396, 409, **415**
Baltimore, L 530, **537**
Baltrum, 768, **783**
Bamburgh, L 341
Banche, La, L 620
Banff, 399
Bangor (N Ireland), L 560, **568**
Bangor (Wales), 488
Bank holidays, 75
Bann, River, L 561, **574**
Bantry, L 530, **536**
Barcaldine Pier (Loch Creran), *433*
Bardsey, R 49, L 471
Bardsey Sound, 473
Barfleur, L 696, **700**
Barfleur, Pointe du, R 55, L 696, 698
Barges, Les, L 589
Bar Lanby, R Mersey, R 65, L 470, 472
Barmouth, L 498, **503**
Barnouic, L 645
Barnstaple, 521
Barometer conversion scale, 147
Barra Head (Hebrides), R 47, L 423

For key to symbols and entries see explanation on page 789.

Barrow-in-Furness, *475*, **491**
Barry, L 499, **512**
Bas-Sablons, Les, L 666, **681**
Basse du Lis Lanby, L 622
Basse du Millieu Lanby, L 621
Basse Perennes Lanby, L 622
Basse Vieille Lanby, L 622
Bass Rock, L 368, 370
Bassurelle Lt Buoy, R 66, L 697
Basta Voe, **409**
Batz, Île de, L 645, *654*, **660**
Baule, La, **626**
Baumronne, L 767
Bawdsey, R 40
Bay of Laig (Eigg), 425, *431*, **442**
Bay of Moclett, **406**
Bayonne, C 139, **612**
Beachy Head, L 270, 272
Beacons sequence numbers, 38, 43
Bearing of sun, rising and setting, 24, 25
Beat frequency oscillator (BFO), 38
Beaucette (Guernsey), **677**
Beaufort scale, 146
Beaulieu River, **247**
Beaumaris, **488**
Becquet, Le, L 667
Beer, **264**
Beg-ar-Vechen, Pointe de, L 621
Beg-Léguer, L 645
Beg-Meil, L 622
Belfast, R 51, L 560, 563, **568**
Belgium
 Charts, 736
 Coast, 734
 Customs, 21
 Lifeboat stations, 727
 Maritime Rescue Centre, 736
 Passage information, 734
 Storm signals, 736
 Weather information, 160
Belle Île, Le Palais, C 140, L 621, 623, **631**
Belle Île, Goulphar Lt, R 52, L 621
Belle Île Radio, C 140
Bell Rock, R 64, L 369
Bembridge, R 42, **255**
Benodet, R 54, L 622, **634**
Bensersiel, 768, **783**
Berehaven, L 530
Bergen op Zoom, **746**
Berneray, L 423, 424
Berry Head, R 40, R 42, L 190
Berwick-on-Tweed, L 341, **361**
Bhraige, Loch a', L 422, **441**
Biarritz, R 52, L 588
Bideford, L 500, **521**
Bilbao, R 51, 594
Bill of Portland, R 42, L 222, 224, 236
Binic, L 666, **680**
Birvideaux, Plateau des, L 621, 623
Biscay, Bay of, 590, 623
Bishop Rock, R 64, L 190, 191
Bishops, The, 501
Blaavands Huk, *784*
Black Head (Belfast Lough), L 560
Black Head (Galway), L 562
Black Head (Killantringan), L 449
Black Nore Point, L 499
Blackpool, R 49
Blackrock, L 562
Blacksod Bay, *576*, **582**
Blacksod Quay, L 562, *576*, **582**

Blackwater River, L 312, **316**
Blakeney, 330
Blankenberge, L 730, 734, **738**
Blasket Islands, 532
Blaye, **612**
Bloody Foreland, **561**, 564
Bloscon, R 54, L 645, **654**
Blyth, L 341, **359**
Boddam, **384**
Bodic (Le Trieux), L 645, 658
Bognor Regis, *274*
Boisdale, Loch, L 423, *426*, **442**
Bois de la Chaise, *610*
Bol van Heist, R 66
Bordagain, **588**
Bordeaux, R 52, C 139, *596*, **612**
Bordeaux-Arcachon Radio, C 139
Borkum, L 766, **770**
Borkum Grosser, L 766
Borkum Kleiner, R 58, L 766
Borkumriff Lt V, R 58, R 66, L 766
Bosham, **262**
Boston, L 340, **344**
Bottom, quality, abbreviations, 34
Boulogne, C 141, L 697, **719**
Bourgenay, L 589, **506**
Bourgneuf, Baie de, 591
Bournemouth, **244**
Bournemouth/Hurn, R 42
Bowling, *456*
Boyne, River, *553*, L 560
Braakmanhaven, L 731
Brading, L 223, **255**
Bradwell, L 312, **316**
Braefoot Bay (Firth of Forth), L 368
Brancaster Staithe, **333**
Brandaris Tower, L 733, 759
Brawdy, R 49
Braye (Alderney), L 667, 668, **678**
Breaksea Lt Float, R 65, L 499
Bréhat, Île, de, **658**
Bréhat, Île, de, Rosédo, R 54, L 645
Bremen, C 143
Bremerhaven, R 58, L 767, **774**
Breskens, L 731, **739**
Bressay, R 46, L 395
Brest, L 644, **648**
Brest-le-Conquet Radio, C 140
Brest, Rade de, 648
Bridges, clearance under, 174
Bridgwater, *519*, 520
Bridlington, L 340, **350**
Bridport, L 222, **235**
Brightlingsea, L 312, **317**
Brighton Marina, R 43, L 270, **279**
Brigneau, L 621
Bristol, L 499, **515**
Bristol Channel, 501
BBC general forecasts, 148
BBC inshore waters forecasts, 148, 154
BBC shipping forecasts, 148, 152
British Marine Industries Federation, 22
British Summer Time, 74, 168
British Telecom Coast Stations, 130
Brittany — canals, 625
Brixham, L 190, **212**
Broadhaven (Gubacashel Point), L 562
Broadstairs, L 271
Brodick, **463**
Brodick Bay, *454*
Brough Haven, 345

Brough of Birsay, L 395
Bruichladdich (Islay), *452*
Bruinisse, **746**
Brunsbüttel, L 767, **777**
Buchan Ness, R 64
Buchan Radio, C 137
Buckie, L 394, **400**
Buckler's Hard, **247**
Bude, *522*, **524**
Buildings, abbreviations, 34
Bull Lt Float, L 340, 345
Bull Point, L 500
Bull Rock, L 530
Bull Sand Fort, L 340, 345
Bunbeg, L 561
Bunratty, R 51
Buoyage, IALA Region A, 36, 119
Burgee, 144
Burghead, L 394, **402**
Burghsluis, **746**
Burnham-on-Crouch, **315**
Burnham-on-Sea, L 500, **519**
Burnham Overy Staithe, **333**
Burnmouth, L 368, **387**
Burntisland, **374**
Burray, **406**
Burrayness, *407*
Burry Port, L 498, **511**
Bursledon, **253**
Burtonport, L 561, *576*, **582**
Büsum, L 767, *778*, **783**
Bute, Isle of, 454
Bute, Kyles of, 451
Butt of Lewis, R 47, L 422
BXA Lanby, R 66, L 588, 590, 596

Cabo = cape, see proper name
Cabourg, **723**
Caen, 704, **723**
Caernarfon, 473, **489**
Caernarfon Bar, 473
Cailleach Head, L 422
Caister, 329
Calais, R 55, R 62, C 141, L 697, **720**
Calais, Pas de, 699
Calculator for interpolation, use of, 74
Calculators, coastal navigation, 24, 33
Calculators for sight reduction, 68
Caldey Island, L 498
Caledonian Canal, **434**
Calendar, festivals 1989, 75
Calibration, DF sets, 39
Call signs/numbers, radiobeacons, 59–61
Calshot Castle, *253*
Calshot Spit Lt Float, L 223
Camaret, L 622, **637**
Camiers, L 697, **715**
Campbeltown, L 448, **453**
Campen, L 766
Canal connections
 Brittany, 625
 Mediterranean, 600
Cancale, *682*, **689**
Canna, L 423, **442**
Cantick Head, L 395
Cap = cape, see proper name
Capbreton, L 588, **612**

Index 791

For key to symbols and entries see explanation on page 789.

Carantec, 656, **660**
Cardiff, L 499, **513**
Cardiff/Rhoose, R 49
Cardigan, 504, **523**
Cardigan Bay, 501
Cardigan Bay Radio, C 138
Cardinal marks, 36, 119
Carentan, L 696, **701**
Cargreen, 200
Carlingford Lough, L 560, **565**
Carmarthen, 523
Carnane, R 47
Carnlough harbour, L 561, **582**
Carnmore, R 51
Carradale Bay, 463
Carrickfergus, L 560, **568**
Carron, Loch, 442
Carteret, L 666, **683**
Carteret, Cap de, L 666
Cashla Bay, L 562, 564
Casquets, The, 9, R 50, R 65, L 667, 668
Castlebay, 442
Castlebay S Buoy, R 65
Castle Breakwater, St Peter Port, R 50
Castle Haven, 538
Castletown (I o M), L 471, **491**
Castletown (S Ireland), 554
Cat Firth (Shetland), 409
Caudebec, 706
Cayenne, La, 601
Cayeux-sur-Mer, L 697, 699
Cazaux, R 52
Celsius (centigrade) to Fahrenheit, 147
Celtic Radio, C 138
CROSS, 593
Certificate of Competence, Helmsman's, 19
Chaine Tower, Larne, L 560, 572
Channel Islands, 670
Channel Lt V, R 50, R 65, L 667
Ch M, radiotelephone, VHF, 130
Channels, VHF radio, 130
Chanonry, L 394
Chapel Rock (R Severn), L 500
Chapman's Pool, 264
Charlestown, 216
Chart corrections, 4
Chart datum, 168
Charted depth, 168
Chart symbols and abbreviations, 34, 116–117
Chassiron, Pointe de, L 588, 590
Chatham, 291
Chats, Pointe des, L 621
Chaume, Tour de la, R 52
Chausey, Iles, 683
Chaussée de Sein Lt Buoy, R 66, L 622
Chauveau, L588
Cherbourg, R 55, C 141, L 667, **684**
Chester, 476
Chichester, R 42, L 223, **262**
Christchurch, 244
Civil calendar, 75
Civil twilight, 82–83
Clacton-on-Sea, L 312
Clare Island, L 562, 577
Classification, 19
Clearance under bridges, 174
Clew Bay, L 562, **577**
Clifden Bay, 577, **582**

Cloch Point, R 47, L 449
Clonakilty Bay, 537
Clovelly, 522
Clyde, Firth of, L 448, 451, **452**
Clyde Radio, C 138
Clythness, L 394, 397
Coastal features, 34
Coastal navigation, 23–36
Coastal refraction, 39
Coastguard, HM, 164
  Weather information, 151
Coast radio stations, 130–143
  SAR, 130, 164
  Weather bulletins, 150
Cobh, 539
Cockenzie, L 368, 373
Cocq, Pointe du, L 622, 635
Code flags, International Code of Signals, 120–121
Codling Lanby, R 65, L 531
Cognac/Chateaubernard, R 52
Coleraine, L 561, **574**
Colijnsplaat, L 731, **746**
Coll, L 423, **442**
Collision, action to avoid, 8
Collision avoidance, radar, 63
Collision regulations, 8, 16–17
Colne, River, L 312, **317**
Colombier, Le, L 645, 657
Colonsay, L 423
Colwyn Bay, L 471, 472
Combrit, Pointe de, R 54, L 622
Communications, 123–144
Compass checks, 24–25
Compass, deviation and variation, 24
Compass Head (Shetland), R 40
Competence, Certificate of Helmsman's, 19
Complements, International Code of Signals, 124
Concarneau, L 621, **633**
Conduct of vessels in sight of one another, 8
Cone, motor sailing, 16
Coningbeg Lt Float, R 65, L 531
Connel (Loch Etive), R 47, **435**
Connemara, 564
Conquet, Le, C 140, L 644, **653**
Continental coast radio stations, 139
Contis, L 588
Convergency, half, 33, 39
Conversion factors, 18
Conversion tables, arc to time, 114
Conversion tables, decimals of a degree to minutes of arc, 114
Conversion tables, hours, minutes and seconds to decimals of a day, 114
Conwy, L 471, **483**
Copinsay, L 395, 407
Coquet, L 341, 343
Corbeaux, Pointe des (Île d' Yeu), L 589
Corbière, La, R 50, L 667
Cordouan, L 588
Corne, La, L 645, 657
Cork, L 531, **538**
Cork Lt Buoy, R 65
Corniguel, 632
Cornwall, North Coast, L 500, **522, 524**
Corpach, L 423, **433**
Correction for Moon's parallax, 76

Corrections, late, 786
Correction tables, altitude, 76
Corryvreckan, Gulf of, 450
Corsen, CROSS, 9, 593
Corsewall Point, L 449
Coruna, La, 594
COSPAS/SARSAT, 164
Coubre, Pointe de la, R 52, L 588, 596
Coulport, 455
Couronnée, La, Lt Buoy, L 620
Courseulles-sur-Mer, L 696, **703**
Courtesy flag, 144
Courtmacsherry, 554
Cove, The (Scilly), 194
Coverack, 196, **216**
Covesea Skerries, L 394
Cowes, L 223, **248**
Crac'h, Rivière de, L 621, **630**
Craobh Haven, 440
Craighouse (Sound of Jura), 452, **462**
Craignish, Loch, 450, **462**
Craigton Point, L 394, 404
Crail, L 369, **387**
Crammag Head, L 449
Cramond, 387
Cranfield Point, 565
Creac'h, Pointe de, R 40, R 54, R 66, L 644, 653
Cregneish (I o M), R 49
Creran, Loch, L 423, 433
Crockalough Radio, C 139
Croisic, Le, L 620, **627**
Croix (Le Trieux), L 645, 658
Cromarty, L 394, 397, 403, **414**
Cromarty Firth Fairway Buoy, R 64, L 394
Cromarty Radio, C 138
Cromer, R 45, L 313, 331
Crookhaven, L 530, **536**
Crosby Lt Float, L 470
Cross channel passages, 272
Cross Sand Lanby, R 64, L 313
Crotoy, Le, L 697, **718, 723**
Crouesty-en-Arzon, L 620, **628**
Crowlin Islands, 441
Cruising formalities, 19
CS1, CS2, CS3, CS4 Lt Buoys, 193, L 270
Cullen, 414
Cullercoats Radio, C 137
Currents, ocean, 174
Customs, HM, procedures, 20–21
  European countries, 21–22
Cuxhaven, L 767, **775**
Cynfelyn Patches, 501

D

Dagebull, L 767, **784**
Dahouet, L 666, 681, **689**
Dames, Pointe des, L 589
Dangastersiel, 768, **783**
Dangers, abbreviations, 34
Darryname, L 530
Dartmouth, L 190, **207**
Davaar (Campbeltown), L 448, 453
Day of the year, 75
Deal, L 270, 282

For key to symbols and entries see explanation on page 789.

Deauville (Trouville), L 696, **704**
Deben, River (Woodbridge Haven), L 313, **324**
Decca warnings, 135–143
Decca Yacht Navigator, 40–41
Dee Estuary, 472, 776
Deer Sound (Orkneys), 406, 407
Definitions, tides, 168
Delfzijl, 759, L 766
Dell Quay, 262
Demie de Pas, L 666
Den Helder, L 733, **757**
Denmark (West), tidal information, 784
Den Oever, L 733, 756, 757
Depth, 168
Derby Haven, L 471
Déroute de Terr, 668
Deutsche Bucht Lt V, R 58, L 766
Deviation, deviation card, 24
Devonport, 200
Diélette, L 666
Dieppe, C 141, L 697, **713**
Dinard (St Malo), L 666, **681**
Dingle, L 530, **554**
Dingle Bay, 532
Dingwall, 403
Dip of the horizon, 68, 70, 76
Directional radiobeacon (RD), 39
Direction finding, VHF, 40
Direction signals, aircraft, 128
Distance finding, beacons, 38
Distance off, vertical sextant angle, 24, 26–28, 33
Distance of lights, rising-dipping, 29
Distance of the horizon, 24, 29
Distress messages and procedures (RT), 132, 163
Distress signals, 16–17, 163
Divers, signals, 118
Dives-sur-Mer, L 696, 704
Documentation of yachts, 19–20
Doëlan, L 621, **638**
Donaghadee, L 560, 563, **682**
Donegal Bay, L 561, 564
Donegal Harbour, 576
Dornoch Firth, 397, **414**
Dornumer-Accumersiel 768, **771**
Douglas (I o M), R 47, L 471, **481**
Douglas Head, L 471, 481
Douarnenez, L 622, 624, **636**
Douhet, 613
Dounreay/Thurso, R 46
Dover, L 270, 273, **282**
Dover Strait, 11, 272
Dowsing Lt V, L 340
Dredgers, lights, 118
Drogheda, L 560
Drummore, 451, **461**
Drying heights, 168
Dubh Artach, L 423
Dublin, C 139, L 531, **548**
Dublin/Rush, R 50
Dudgeon Lt V, R 45, L 340
Dumet, Île, L 620
Dunany Point, 553
Dunbar, L 368, **372**
Duncannon, L 531, 545
Duncansby Head, R 64, L 394, 405
Duncansby Race, 397
Dundalk, 553
Dundee, R 45, **379**
Dundrum Bay, L 560
Düne, L 767, **778**

Dungarvan, L 531, **544**, **554**
Dungeness, R 43, L 270, 272, 282
Dunkerque, C 141, L 697, **721**
Dunkerque/Calais, R 55
Dunkerque Lanby, R 55, R 66, L 697
Dunkerron Harbour (Kenmare R), 536
Dun Laoghaire, L 531, **548**
Dunmanus Harbour, 536
Dunmore East, L 531, **544**
Dunnet Head, R 40, L 394, 397
Dunoon, L 449
Dunstaffnage, 442
Dunvegan, Loch, 431
Duration, 168
Dursey Sound, 532
DW Lt Buoy, R 66

E

Eagle Island, R 51, L 562
Easington, R 40
Eastbourne, L 270, 279, **305**
East Channel Lt Float, R 65, L 667
East Channel, passage information, 272
East Chequer Lt Float, L 340
East Goodwin Lt Float, R 64, L 270
East Loch Roag, L 423
East Loch Tarbert (Harris), L 422, 426, **441**
East Loch Talbert (Mull of Kintyre), L 448, **453**
East Usk, L 499
East Weddel Sound, 406
Eatharna, Loch (Coll), 431
EC1, EC2 and EC3 Lt Buoys, R 65, L 667
Eckmühl (Pointe de Penmarc'h), R 54, L 622
Eclipses, 79
Ecrehou, Les, 668, 671
Eday, L 395, 406
Eddystone, R 64, L 190, 192
Edinburgh, R 45
Eeragh (Galway Bay), L 562
Eglinton/Londonderry, R 51
Egmond-aan-Zee, L 732
Egypt Point, L 223
Eiderdamm, L 767, 769
Eider, Die, 769
Eiderstedt Radio, C 143
Eierland, R 57, L 733
Eigg Island, 431, **442**
Eilean Glas (Scalpay), R 47, R 65, L 422
Eilean Trodday, L 423
Elbe, Die, 769, **776**
Elbe No 1 Lt V, R 58, L 767
Elbe/Weser Radio, C 143
Electronic calculators, 24, 33, 68, 74
Elie, 387
Elie Ness, L 369
Emden, L 766, **782**
Emergency position indicating radio beacons (EPIRBs), 164
Emergency signals, 17, 125–126, 127–128
Emission, radio, types of, 38
Ems, Die, 768

Emsworth, 262
English and Welsh Grounds Lt Float, R 65, L 499
English Channel, 10–11, 272
English Channel waypoints, 193
Enkhuizen, 756
Ennis, R 51
Ensign, 144
Ephemeris — Aries, planets, Sun and Moon, 86–97
Ephemeris, explanation, 69–73
Eriboll, Loch, L 395, **415**
Eriskay, L 423
Erisort, Loch, 424
Erquy, L 666, 682, **689**
Error, quadrantal, 39
Errors in radio bearings, 39
Errors, notification of, 4
Esbjerg, 784
Esha Ness, L 396
Étaples, 733
Etel, River, R 40, L 621, **638**
Eu/Le Treport, R 55
Europoort, L 732, **749**
Ewe, Loch, L 422, 427, **441**
Exe, River, L 190, **214**
Exeter, R 42
Exmouth, L 190, **214**
Eyemouth, L 368, **372**

F

F3 Lanby, R 64, L 271
Facsimile broadcasts, 149
Factors, tidal, 171
Fahrenheit to celsius (centigrade), 147
Fair Isle, L 395, 410, **415**
Fairlight, R 40
Faix, Le, L 644, 646
Falls Lt Float, R 43, R 64, L 271
Falmouth, L 190, **197**
Falmouth (Pendennis), R 40, C 135
Fanad Head, L 561
Fareham, 256
Farne Islands, L 341, 343, **361**
Fastnet, 15, L 530, 533
Fathoms and feet to metres, 175
Fawley/Hythe, L 223
Fécamp, L 697, **711**
Federal Republic of Germany — see **Germany**
Feet to metres, metres to feet, 18, 175
Felixstowe, L 312, **318**
Fenit Pier, L 530, 532, 536
Ferret, Cap, R 52, L 588, 595
Ferris Point, L 560, 572
Fers Ness Bay, 406
Festivals and calendar, 75
Fethaland, Point of, L 396
Fidra, L 368, 370, 372
Fife Ness, R 40, R 45, L 369
Filey, L 340, 342, 350, **361**
Findhorn, 402
Findochty, 414
Firths Voe, L 396
Fisherrow, 387
Fishguard, L 498, **505**
Fishing vessels, lights, 118

Fishing vessels, signals, 120–121
Fladda, L 423
Flag etiquette, 144
Flag signalling, International Code of Signals, 120–121, 124–126
Flags, International Code, 120–121
Flag officers' flags, 144
Flamborough, R 40
Flamborough Head, R 45, L 340
Flannan Islands (Eilean Mor), L 423, *426*
Flat Holm, L 499, 502
Fleetwood, L 470, **475**
Flotte, La, L 588, **613**
Flushing (Vlissingen), L 730, **742**
Fog, 16, 147
Fog signals, 17, 182
   Abbreviations, 180
   Foreign terms, 182
Föhr, L 767, **783**
Folkestone, L 270, **282**
Fontarabie, Baie de, 590
Forecasts by personal telephone call, 148
Forecasts, coast radio stations, 150
Forecasts, inshore waters, 148, 154
Forecasts, local radio, 156
Forecasts, HM Coastguard, VHF, 151
Forecasts, shipping, 148, 152
Fôret, La, L 621, **634**
Formby Lt Float, L 470
Forth, Firth of, L 368, 370, **373**
Forth, North Channel Lt Buoy, R 64, L 368
Forth Radio, C 137
Fortrose, *403*, **414**
Fort William, L 423, **433**
Foula, L 396
Four, Chenal de, L 644, 646
Four, Le (Quiberon), L 620
Fowey, L 190, 192, **198**
Foyle, Lough, L 561, 563, **574**
Foynes, R 51, L 530, **535**
France
   Charts, 592
   Customs, 21
   Forecast areas, 159
   Lifeboat (SNSM) stations, 584, 616, 640, 662, 692
   Maritime rescue, CROSS, 593
   Storm signals, 592
   Tidal terms, 593
   Weather information, 158
Fraserburgh, L 394, **399**
Fréhel, Cap, R 54, L 666
French areas, special notes, 592
Freshwater, *246*
Frequencies, radio, 129–130
Frisian Islands, 734, 758, 768, 770
Fromentine, L 589, **610**, **613**
Fromveur, Passage du, 646, 653
Fyne, Loch, L 448, 451

G

Gaelic terms, 424
Gaine, Passage de la, 647
Gairloch, *427*, **441**
Gale warnings, 147–155
Galley Head, L 530

Galloper (S) Lt Buoy, R 64, L 312
Galloway, Mull of, L 449, 451
Galway, L 562, 564, **577**
Galway Bay, L 562, 564
Gantock Beacon, L 448
Gareloch, L 449, **455**
Garelochhead, *455*
Garraunbaun Point, L 530
Garvellachs, The, L 423
Gas systems, 162
General information, 7–22
General procedures, radiotelephone, 130–132
General provisions, radiotelephone, 129–130
Germany, Federal Republic of,
   Charts, 770
   Customs, 22
   Lifeboat stations, 762
   Maritime Rescue Centres, 770
   Storm signals, 770
   Weather information, 160
Gigha Island, 450, *452*, **463**
Gigha Sound, 450, *452*
Gijon, *594*
Girdle Ness, R 45, R 64, L 369
Gironde, La, L 588, 590, **596**
Girvan, L 449, 461, **463**
Glandore, 533, **554**
Glas, Eilean (Scalpay), R 47, R 65, L 422
Glasgow, Port, *456*
Glasson Dock, *475*
Glasson (Torduff Point), *474*
Glénan, Îles de, L 621, 624
Glenelg Bay, *431*
Glengariff, *554*
Glengarrisdale Bay (Jura), *452*
Glückstadt, R 58
Gluss Isle (Sullom Voe), L 396
Goeree, R 57, R 66, L 732
Goes, Sas van, L 731, **746**
Golspie, *405*, **415**
Goodwin Sands, L 270, 273
Goole, *345*
Gorey (Jersey), L 667, **671**
Gorleston, *329*
Gorran Haven, *216*
Gosport, *256*
Gott Bay (Tiree), *431*
Goulphar (Belle Île), R 52, L 621
Gourdon, L 369, **387**
Gourock, *456*
Goury, *684*
Gradyb Bar, *784*
Graemsay Island, L 395, 407
Grandcamp, L 696, **702**
Grand Jardin, Le (St Malo), R 54, L 666, 681
Grand Lejon, L 645
Grande Île (Chausey), L 666, **683**
Grande Vinotière, La, L 644, 646
Grands Cardinaux, Les, L 620
Granton, L 368, 370, **373**
Granville, R 54, L 666, **682**
Grassholm, 501
Gravelines, L 697, **723**
Grave, Pointe de, L 588, **596**
Gravesend, *298*, **305**
Great circles, 39
Great Cumbrae, L 448, **463**
Great Ormes Head, R 40
Great Yarmouth, R 45, L 313, **329**
Greenock, L 449, **456**

Greenwich Hour Angle, 69, 86–97
Greenwich Lanby, R 64, L 270
Greenwich Mean Time, 69, 74, 168
Greve d'Azette, La, L 666
Gribbin Head, *192*
Grimsby, C 137, **345**
Gris Nez, Cap, 11, R 40, R 55, L 697, 699
Groix, Île de, L 621, 623, **638**
Groix, Île de, Pen Men, R 54
Groomsport, *568*
Grosnez Point, L 667, 669
Grosser Vogelsand, R 58, L 767
Grosse Terre, Pointe de, L 589, 608
Grouin-du-Cou, Pointe du, L 589, 591
Group, signals, 124–126
Grove Point, R 40
Gruney Island, R 65, L 396
Grunna Voe (Shetland), *409*
Gruting Voe (Shetland), *409*
Guernsey, R 40, R 50, L 667, **677**
Guilvinec, L 622, 624, **638**

H

Hague, Cap de la, L 667, 669
Hague, Cap de la, Jobourg, 9–10, R 40, 593
Hamble, L 223, **254**
Hamburg, C 143, **776**
Hamna Voe, *409*
Hanois, Les, L 667
Hanstholm, *784*
Hansweert, *740*
Harbour facilities, abbreviations, 179
Haringvlietsluizen, *748*
Harlesiel, L 766, 768
Harlingen, L 733, 735, **758**
Harport, Loch, L 423, *431*, **441**
Harrington, L 470, **491**
Hartland Point, R 40, L 500
Hartlepool, L 341, **356**
Harwich, L 312, **318**
Harwich Channel Lt Buoy, R 64, L 312
Hastings, C 135, L 270, **305**
Haulbowline, L 560, 565
Haut-Blanc, Pointe du, L 697
Havengore Creek, *304*
Haverfordwest, *506*
Havre, Le, C 141, L 696, 698, **707**
Havre, Le/Octeville, R 55
Havre, Le, Lanby, R 55, R 66, L 696
Hayle, L 500, **524**
Hayling, Island, *262*
Health clearance messages, 127
Health regulations, 20
Heaux, Les, L 645, 647
Hebrides, Outer, L 422, 424, **426**, **441**
Hebrides Radio, C 137
Height of eye and distance of horizon, 24, 29
Height of tide, 168
Helensburgh, L 449, **455**
Helford River, 191, **196**
Helgoland, R 58, C 143, L 767, **778**
Helicopter rescues, 165
Helle, Chenal de la, **646**

794 Index

For key to symbols and entries see explanation on page 789.

Hellevoetsluis, L 732, **748**
Helmsdale, L 394, **405**
Helmsman's Overseas Certificate of Competence, 19
Hendaye, 590, **612**
Herm, L 667, **678**
HM Coastguard, 164
HM Coastguard, VHF (Ch 67), 131, 151, 165
HM Coastguard, VHF DF, 40
HM Customs, 20–21
Herne Bay, L 271, *287*
Hestan Island, L 449
Hève, Cap de la, L 696
Hever, Die, 769
Heysham, L 470, *475*, **491**
High Down, Scratchells Bay, R 62
High Water Dover, 178, 284
Hindeloopen, 756
Hinder, Noord, Lt V, 12, R 57, L 732
Hinder, West, Lt V, 11, R 57, L 730
Hinkley Point, L 500
Hirtshals, *784*
Hoëdic, Ile de, L 620, 623
Hoek van Holland (Hook of Holland), R 57, L 732, **749**
Hohe Weg, L 767, 769
Hoist, signals, 126
Hojer Sluice, *784*
Holburn Head, L 394, 414
Holehaven, 298, **305**
Holland, see **Netherlands**
Hollerwettern, R 58
Holyhead, L 471, 473, **483**
Holy Island (Arran), L 448, 454
Holy Island (Northumberland), *360*, **362**
Holy Loch, L 449, 456
Honfleur, L 696, 698, **706**
Hook Head, R 65, L 531, 533, 545
Hooksiel, 768, **773**
Hooksielplate, L 767
Hoorn, 756
Hope Cove, 216
Hopeman, L 394, **401**
Horaine, La, L 645, 647
Horizon, distance of, 24, 29
Horizontal sextant angles, 33
Hörnum, L 767, **782**
Horse Sand Fort, L 223, 224
Hoswick, L 395
Houat, Ile de, L 620, 623, **638**
Houle-sous-Cancale, La, L 666
Hourdel, Le, 718
Hourdel, Pointe du, L 697, 719
Hourn, Loch, *431*
Hourtin, L 588
Houton Bay, L 395, **406, 415**
House flag, 144
Howth, L 531, 533, **553**
Hoy, 406
Hoy Sound, L 395, 398
Hubertgat Lt Buoy, R 66
Hugh Town (Scilly), 194, 195
Huisduinen, L 732
Hull, Kingston-upon-, 345
Humber Lt V, R 45, R 64, L 340
Humber Radio, C 137
Humber, River, L 340, **344**
Hunda Sound (Orkney), 406
Hurn/Bournemouth, R 42
Hurst Point, L 222, 224, 245
Husum, L 767, **783**
Hyskeir (Oigh Sgeir), R 47, L 423

Hythe/Fawley, L 223
Hythe (Southampton), L 223, 249

I

IALA, 36, 119
Identity, signals, 126
IJmuiden, R 57, L 732, **754**
IJmuiden Lt Buoy, R 66
IJsselmeer, 756
Île = island, see proper name
Ilfracombe, C 138, L 500, **520**
Immigration, 20
Immingham, 344
Improvements, suggestions, 4
Inchard, Loch, L 422, 424
Inchcolm, L 368, 370, **387**
Inchkeith, R 45, L 368, 370
Inchkeith Fairway Lt Buoy, R 64
Indaal, Loch, L 448
Information, general, 7–22
Inishbofin, L 561
Inishbofin (Inishlyon), L 562
Inisheer, L 562
Inishgort, L 562
Inishlyon (Inishbofin), L 562
Inishnee, L 562
Inishowen, L 561
Inishraher Islands, *577*
Inishtearaght, L 530
Inishtrahull, R 65, L 561
Inishtrahull Sound, 563
Inmarsat, 133
Inner Dowsing, R 64, L 340
Inshore waters forecasts, 148, 154
Instow, L 500, **521**
Insurance, 19
IALA Buoyage, Region A, 36, 119
International Certificate for Pleasure Navigation, 19
International Code, 120–121, 124–127
  Flags, 120–121
  Flag signalling, 126
  Morse code, 120–121, 124, 126
  Radio telephony, 126
  Selected groups, 125–126
  Single letter signals, 120–121
International Port Traffic Signals, 122, 128
International Regulations for Preventing Collisions at Sea, 1972, 8, 16–17
Interpolation tables, ephemeris, 98–107
Interpolation by calculator, formulae, 74
Interval, tidal, 168
Inver, Loch, L 422, **441**
Inveraray, *453*, **463**
Inverbervie, R 40
Invergordon, L 394, *403*, **414**
Inverkip, 460
Inverness, L 394, 397, **403**
Inverness Firth, 397
Iona (Mull), 425, *433*
Ipswich, 323
Ireland
  Coast Lifesaving Service, 534
  Coast Radio Stations, 139

  Customs, 21, 534
  East Coast, 533, 563
  Maritime Rescue Centres, 534
  North Coast, 563
  Ordnance Survey, 534
  South Coast, 533
  West Coast, 532, 564
Irish Sea, crossing, 563
Iroise, L', 624
Irvine, L 449
Isigny-sur-Mer, L 696, **723**
Islay, C 138, *452*
Islay, Sound of, L 448, 450
Isle of Man, L 470, 472, **480**
Isle of Man, traffic scheme, 14
Isle of Wight, L 222, 224, **246, 255, 264**
Isolated danger marks, 36, 119
Itchenor, L 223, **262**

J

Jack Sound, 501
Jade, Die, L 766, 768
Jard-sur-Mer, 613
Jaune de Glenan Lt Buoy, L 622
Jersey, R 40, R 50, C 135, L 666, 669, **671, 689**
Jersey East, R 50
Jersey West, R 50
Jever, R 58
Jobourg, CROSS, 9–10, R 40, 593
Juist, 782
Jument de Glénan Lanby, L 622
Jument, La (La Rance), L 666
Jument, La (Ouessant), L 644
Jura, Sound of, L 448, 450, *452*
Jurby, L 470, 472

K

Kampen, Rote Kliff, R 58, L 767, 782
Kéréon (Men-Tensel), L 644
Kerjean (Perros-Guirec), L 645, 657
Kermorvan, L 644, 646
Kerprigent (Perros-Guirec), L 645, 657
Kerrera Sound, L 423, 436
Kettletoft Bay, 406
Keyhaven, 245
Kiel Canal (Nord-Ostsee Kanal), 777
Kijkduin, L 732
Kilchiaran, R 40
Kilcredaun Head, L 530, 532
Kilkeaveragh Radio, C 139
Kilkeel, L 560, **582**
Killala Bay, L 562, 564, *576*
Killantringan (Black Head), L 449
Killary Harbour, 564, *577*
Killeany Bay (Aran Islands), *577*
Killiney, R 50
Killybegs, L 562, **576**
Killyleagh, *566*
Kilronan, 564, **582**

Kilrush, 532, **535**
King's Lynn, L 313, **332**
Kingston-upon-Hull, **345**
Kingswear, L 190, **207**
Kinloss, R 46
Kinnairds Head, R 46, L 394
Kinsale, L 531, 533, **538**
Kintyre, Mull of, L 448, 450
Kirbabister Ness, L 395
Kilkcaldy, L 369, 370, **387**
Kirkcudbright, L 449, 451, **462**
Kirk Sound (Orkney), **406**
Kirkwall, R 46, L 395, **407**
Kish Bank, R 50, R 65, L 531
Knock, L 766
Knockgour Radio, C 139
Kornwerderzand, L 733, *756*
Krautsand, R 58
Kruishoofd, L 730
Kyle Akin (Eilean Ban), R 65, L 422, 425
Kyle of Durness, *426*
Kyle of Loch Alsh, L 422, 425, *431*, **441**
Kyle of Tongue, 415
Kyles of Bute, L 448, 451

L

Lamena, *596*
Lamlash, **454**
Lampaul (Ouessant), 646, **653**
Lancaster, *475*
Land effect, DF bearings, 39
Lande, La, L 645, 656
Landing signals, boats, 127–128
Landivisiau, R 54
Land's End, 15, C 135, L 190, 191
Langeness (Nordmarsch), L 767
Langeoog, L 766, 768, **772**
Langness (Dreswick Point), L 471
Langstone Harbour, L 223, **262**
Lann-Bihoué/Lorient, R 52
Lannion, R 54, **656**
Lanvéoc/Poulmic, R 54
Largs, **463**
Larne, L 560, **572**
Late corrections, 786
Lateral marks, 36, 119
Lathaich, Loch, **442**
Lauwersoog, *759*, **760**
Laxey (I o M), L 471, **491**
Laxford, Loch, 424, *427*, **441**
Lee-on-Solent, R 42, *256*
Legal requirements for safety equipment, 162
Légué, Le, L 666, **680**
Leigh-on-Sea, **303**
Leith, L 368, **373**
Lelystad, **756**
Lemmer, **756**
Lequeito, *594*
Lerwick, R 46, L 396, **410**
Lesconil, L 622, 624
Leuchars, R 45
Lévi, Cap, L 667, 669
Lewis Radio, C 137
Lézardrieux, L 645, 647, **658**
Licences, 19

Licences, radiotelephone, 129
Light characters, 35
Lights, 35, 178
Lights, abbreviations, 35
Lights and shapes, 16, 115, 118
Lights, distance off rising and dipping, 24, 29
Lights, foreign terms, 181
Lights, navigation, 16, 115, 118
Lights, visibility, 29, 178
Lights (navigation), visibility, 115
Limerick, L 530, **535**
Link calls, radiotelephone, 130–131
Linklet Bay, **406**
Linnhe, Loch, L 423, 433
Lion-sur-Mer, L 696
Lismore, L 423
List, L 767, 769, **782**
Lister Tief, 769, 782
Little Cumbrae, L 448
Littlehampton, L 270, **274**
Little Ross, L 449
Little Russel, L 667, 669, 677
Little Samphire Island, L 530, 532
Liverpool, L 470, 472, **476**
Lizard Point, R 42, L 190, 191
Llanddulas, L 471
Llandudno, L 471, 472, *483*
Lloyd's Register of Shipping, 22
Local radio stations, forecasts, 156
Locarec, L 622
Loch, see proper name
Lochrist, L 644, 646
Locquemeau, 660
Loctudy, L 622, 624, *634*, **638**
Lodijkse Gat, *746*
Logum, L 766
Loire, River, L 620, 623, **625**
Loire Approach Lanby SN1/SN2, L 620
London Bridge, 297
Londonderry, L 561, **574**
Long Hope, **406**
Long, Loch, L 449, 451
Longman Point, L 394, 404
Longships, 15, L 190, 191
Longstone, R 45, L 341, 343
Looe, L 190, 192, **199**
Looe Channel (Selsey), 272
Lookout, 8
Loop Head, R 51, 532, L 562
Loran-C, radio aid, 76
Lorient, L 621, 623, **632**
Lorient/Lann-Bihoué, R 52
Lossiemouth, L 394, **400**
Lost Moan, L 622, 638
Lother Rock, R 64, L 394
Louet, Ile, L 645, 647
Lowest astronomical tide (LAT), 168
Lowestoft, L 313, 314, **325**
Luarca, *594*
Lulworth Cove, 224, **264**
Lundy Island, R 49, L 500, 502, *522*, **524**
Lune Deep Lt Buoy, R 65, L 470
Lune, River, L 470, 472, **475**
Lybster, L 394, 397, **415**
Lydd, R 43
Lydney, L 500
Lyme Bay, 224
Lyme Regis, L 222, **235**
Lymington, L 222, **245**
Lynas, Point, R 49, L 471
Lynmouth Foreland, R 49, L 500

M

Maas Center Lt Buoy, R 66, L 732
Maas River, L 732, **749**
Macduff, L 394, **399**
Machichaco, Cabo, R 51
Macrihanish, *452*
Maddy, Loch, L 422, *426*, **441**
Magnetic variation and deviation, 24
Maidens, L 561, 563
Mainland (Orkney), 406
Makkum, **756**
Malahide, **553**
Maldon, **316**
Malin Head, C 139
Mallaig, **431**
Man, Calf of, 14, L 470, 472
Man, Isle of, 14, L 470, 472, **480**
Manacles, L 190, 191
Manech, Port, L 621, **638**
Manoeuvring signals (Rule 34), 17
Map of areas (Chap 10), 183
Marcouf, Iles St, L 696, **723**
Margate, L 271, **305**
Marina frequency, VHF Ch M, 130
Marinecall, 149
Marine radiobeacons, 38, 42–61
Maritime mobile service (RT), 129–132
Marquis, Le, *596*
Maryport, L 470, **474**
Massvlakte, 732
Maughold Head, L 471, 472
May, Isle of, R 45, L 369, 370, **387**
Mayor, Cabo, R 51
Mean level, 169
Meanings of terms, weather bulletins, 147
Measured mile table, 32
Medembilk, **756**
Medical help by radiotelephone, 131
Medway, River, L 271, **291**
Melfort, Loch, **440**
Mellumplate, L 766
Memmert, 768
Menai Strait, L 471, 472, **488**
Mengam, Roche, L 644
Men-Joliquet, L 645, 659
Meridian passage (Mer Pass), Sun and planets, 78–79
Merrien L 621
Merry Men of Mey, 397
Mersea, West, **316**
Mersey, River, L 470, 472, **476**
Mesquer, L 620
Meteo, 593
Meteorology, 145–160
Methil, L 369, **378**
Metres to feet, feet to metres, 18, 175
Mevagissey, **198**
Mew Island, R 51, L 560, 563
Middleburg, **746**
Middelharnis, L 732
Middlesbrough, **351**
Mid Yell Voe, **409**
Milford Haven, L 498, **506**
Millier, Pointe du, L 622
Millport, L 448, **463**
Mine Head, L 531, 533
Minehead, L 500, *519*, **524**

For key to symbols and entries see explanation on page 789.

Minquiers, Les, L 666, 668
Minsener Oog, L 766, 768
Mizen Head, R 50, R 65, L 530
Mochras, 523
Moguerec, L 645, 646
Moidart, Loch, *431*
Moines, Ile-aux- (Les Sept Iles), L 645, 647
Molene, Ile, L 644, 646, **653**
Molengat, L 733, 735
Monach Lt Ho, R 65
Monnickendam, 756
Montoir/St Nazaire, R 52
Montrose, 371, **380**
Moon, GHA and Dec, 87–97
Moon, phases of, 79
Moonrise and moonset, 84–85
Moon's parallax, 70, 73, 76
Morbihan, 629
Morecambe, L 470, 472, *475*
Morecambe Bay Radio, C 138
Morgat, L 622, 624, **637**
Morlaix, L 645, 647, **655**
Morsalines, L 696
Morse code, 120–121, 126
Morse code by hand flags, or arms, 126
Morse code by light, 124–126
Mostyn Quay, L 470, *476*
Moul of Eswick, L 396
Mount's Bay, 191
Mousehole, L 190, 191, **216**
Moutons, Ile-aux-, L 621, 624
Moville, L 561, **574**
MPC Lanby, L 697
Muckle Flugga, R 46, L 396
Muckle Skerry (Orkney), *407*
Muckle Skerry (Shetland), L 396
Muglin's Sound, L 531, 533
Muiden, 756
Mull, L 423, 425, **433**
Mullaghmore, *576*
Mull of Galloway, L 449, 451
Mull of Kintyre, L 448, 450
Mull, Sound of, L 423, 425
Mulroy Lough, L 561, 564, *575*, **582**
Multiplication tables, ranges and factors, 173
Mumbles, L 498, 502, 512
Mynydd Rhiw, R 40

N

Nab Tower, R 43, R 64, L 223, 225
Nairn, L 394, **403**
Nantes, 626
Nantes St Herblain Radio, C 140
Narrow channels, 8, 17
Nash, R 49, L 499
Nautical twilight, 82–83
Navigation by calculator, 24, 33, 68, 74
Navigation lights, 16, 115, 118
Navigational warnings, 133
Navstar GPS, 40
NAVTEX, 134, 149
Neath, River, L 498
Needles Channel, L 222, 224
Neist Point, L 423

Nes (Ameland), L 733, **760**
Nessmersiel, 768, **783**
Netherlands
 Charts, 736
 Customs, 22
 Lifeboat stations, 726
 Maritime Rescue Centre, 736
 Passage Information, 734
 Storm signals, 736
 Weather information, 160
Neuharlingersiel, L 766, **783**
Neuwerk, L 767
Nevis, Loch, **442**
Newarp Lt Float, R 64, L 313
Newburgh, *379*
Newcastle-upon-Tyne, L 341, 342, **358**
New Galloway, R 47
New Grimsby, 194
Newhaven, R 40, R 43, L 270, **280**
Newlyn, L 190, 191, **195**
Newport (Gwent), L 499, *513*, **524**
Newport (I o W), *248*, **264**
New Quay, L 498, **523**
Newquay, *522*, **524**
Newry, L 560, **565**
Newton Ferrers, 206
Newton Haven, 361
Newtown (I o W), 247
Nieuwe Sluis (Westerschelde), L 730
Nieuwpoort (Nieuport), R 57, L 730, 734, **737**
Night effect, DF bearings, 39
Niton Radio, C 135
Nividic (An-Ividig), L 644, 653
Noirmoutier, Île de, L 589, 591, **610**
No Man's Land Fort, L 223
Nominal range, 178
Noorderhoofd (Walcheren), L 730
Noord Hinder Lt V, R 57, R 66, L 732
Noord Hinder, traffic scheme, 12
Noordwijk-aan-Zee, L 732
Norddeich, C 143, 768, **782**
Norderney, L 766, 768, **771**
Nordfriesland Radio, C 143
Nordholz, R 58
Nord-Ostsee Kanal, **777**
Nordstrand, L 767, 769
Norfolk Broads, **333**
Norfolk, North, 314
North bay, 406
North Berwick, L 368, 370, **387**
North Carr Lt Buoy, L 369
North Channel, traffic scheme, 14
North Denes/Gt Yarmouth, R 45
NE Goodwin Lt Buoy, L 270
North Foreland, R 40, R 43, C 135, L 271
North Rona, 395
North Ronaldsay, R 46, R 64, **406**
North Sea, French forecast areas, 159
North Shields, L 341, **359**
North Sunderland, L 341, *359*, **361**
North Well Lt Buoy, R 64, L 313
North Unst (Muckle Flugga), R 46, L 396
Noss Head, L 394
Noss Mayo, 206
Not under command, vessels, 16, 118
Numbering system, 2
Numeral group, signals, 124, 126

O

Oaze Deep, L 271, 273
Oban, L 423, **435**
Octeville/Le Havre, R 55
Oigh Sgeir (Hyskeir), R 47, L 423
Oil and gas installations, 342
Oland, L 767
Old Grimsby, **194**
Old Head of Kinsale, R 50, L 530, 533
Oleron, Île d', L 588, 590, **603**
Olna Firth, **409**
Omega radio aid, 40
Omonville, L 667, 669, **684**
Oostende (Ostend), R 57, C 142, L 730, 734, **738**
Oosterschelde, L 731, 734, **746**
Oostmahorn, **760**
Orford Haven, 324
Orford Ness, C 135, L 313
Organisations, yachting, 22
Orkney, C 138, L 395, **406, 415**
Orlock Point, R 40
Orsay (Rinns of Islay), R 47, L 448
Ortigueira, *594*
Orwell, River, L 312, **319**
Osea Island, *316*
Ost Pic, L', L 645
Otterswick, **406**
Ottringham, R 45
Oudeschild, L 733, **760**
Ouessant, Ile d' (Ushant), 9, R 54, R 66, C 140, L 644, 646, **653**
Ouessant NE Lt Buoy, R 66, L 644
Ouessant SW Lanby, R 54, R 66, L 644
Ouistreham, L 696, 698, **704**
Outer Gabbard Lt V, R 43, L 312
Out Skerries, L 396
Overtaking, 8
Oysterhaven, **554**
Owers, L 270
Oxcars, L 368

P

Padstow, L 500, 502, **522**
Pagham, *274*
Paignton, L 190, **216**
Paimboeuf, L 620, 626
Paimpol, C 140, L 645, **659**
Palais, Le, C 140, L 621, **631**
Pallice, La, L 589
Papa Sound (Stronsay), L 395, 398, **407**
Par, *198*, **216**
Parallax in altitude, 70, 73, 76
Parquette, La, L 622
Pauillac, *596*, **612**
Peel, L 470, **482**
Pegal Bay, **406**
Pellworm, L 767, **769**
Penarth, L 499
Pendeen, R 40, L 500
Pendennis, R 40, C 135
Penerf, L 620, 623, *627*, **638**

Penfret (Iles de Glénan), L 621, 624, *632*
Peninnis Head, L 190, 194
Penlan, L 620
Penlee Point, R 42, 201
Penmarc'h, Pointe de, R 54, L 622, 624
Pen Men (Île de Groix), R 54, L 621
Pentland Firth, L 394, 397
Pentland Skerries, L 394, 397
Penzance, R 42, L 190, **197**
Perranporth, *522*
Perros-Guirec, L 645, 647, **657**
Personal locator beacons, 164
Personal safety equipment, 162
Pertuis Breton, L 589, 591
Pertuis d'Antioche, 591
Peterhead, L 369, 371, **386**
Petite Barge Lanby, La, L 589
Petite Foule, L 589
Petits Impairs, Les, L 620
Petite-Muette, La (Dahouet), L 666
Petit-Minou, Pointe du, L 644
Petroland Platform, R 66
Phases of the Moon, 79
Phonetic tables, figure/letter spelling, 120–121
Pierowall, L 395, **415**
Pierre-de-Herpin, La, L 666
Pierre-Moine, La, L 589
Pierre Noire Lt Buoy, La, L 667
Pilier, Ile du, R 52, L 589
Pillar Rock Point, L 448
Pilot vessels, lights, 118
Piriac, L 620, **637**
Pittenweem, L 369, 370
Pladda, R 47, L 448
Planet planning data, 78–79
Planets, GHA and Dec, 86–96
Platresses, Les, L 644
Platte Fougère, R 65, L 667
Plockton, *431*, **442**
Plougasnou Radio, C 140
Ploumanac'h, L 645, **657**
Plymouth, R 42, L 190, 192, **200**
Point, see proper name
Pointe = point, see proper name
Polaris table, 70, 73, 81
Polperro, L 190, **216**
Pont l'Abbé Radio, C 140
Pontusval, L 644, **660**
Pont-Aven, 638
Pontrieux, 658, 660
Poole, R 42, L 222, **240**
Porlock Bay, *519*
Porlock Weir, *524*
Pornic, L 589, 591, **611**
Pornichet, L 620, 623, **637**
Pors Poulhan, L 622
Port Askaig (Islay), 450, *452*, **463**
Portavogie, L 560, *568*, **580**
Portbail, L 666, 668, **689**
Port Blanc (Côtes du Nord), **660**
Port Blanc (Morbihan), 630
Port Bloc/La Gironde, 596
Port Cardigan, L 498, *504*, **523**
Portcé, L 620
Port d'Arzal, 627, **638**
Port Dinorwic, 473, **488**
Port Edgar, L 368, **373**
Port Ellen (Islay), L 448, 450, *452*, **463**
Port-en-Bessin, R 55, C 141, L 696, 698, **702**

Port Erin (I o M), L 470, *481*, **491**
Port Glasgow, *456*
Port Haliguen-Quiberon, L 621, 623, **631**
Porthcawl, *523*
Porth Conger, **194**
Porth Cressa, **194**
Porth Dinllaen, 473, **491**
Porthleven, L 190, 191, *196*, **216**
Porthmadog, L 498, **503**
Porthscatho, **216**
Portishead, L 499, **515**
Portishead Point, L 499
Port Joinville, **609**
Portknockie 414
Portland, R 42, L 222, 224, **236**
Portland, tidal streams, 236
Port Louis, **632**
Portmagee, **554**
Portmahomack, 397, **404**
Port Maria, L 621, 623, **638**
Portmellon, **216**
Port Mullion, **216**
Port Navalo, L 620, **629**
Port operations, radiotelephone, 131
Portpatrick, C 138, L 449, **461**
Portree, **431**
Portrieux, L 666, 668, **679**
Portrush, L 561, 563, **573**
Portsall, L 644, 646, **660**
Port St Mary, L 470, **481**
Port signals, 122, 128
Portsmouth, L 223, 224, **256**
Portsoy, L 394
Port Tudy, L 621, 623, **638**
Portzic, Pointe de, L 644
Porz-Don, Pointe de, L 645, **659**
Position fixing systems, 40
Poulains, Pointe de, L 621, 623
Pouldohan, Baie de, L 621
Poulmic/Lanvéoc, R 54
Pratique messages, 127
Present weather reports, 151
Primel, L 645, **660**
Prince's Channel, L 271
Procedure signals, 124–125
Procedures, radiotelephone, 130–132
Prowords, radiotelephones, 130
Pwllheli, L 471, **490**

Q

Q flag, quarantine, 20, 121
Quadrantal error, 39
Quality of the bottom, 34
Quarantine regulations, 20
Queenborough, **291**
Queen's Channel, L 271
Querqueville, Digue de, L 667, 685
Quiberon, Baie de, 623, **631**
Quilleboeuf, *706*, 707
Quimperlé Radio, C 140
Quimper/Pluguffan, R 54

R

Raasay, Sound of, L 422, 424
Racons (Radar beacons), 63–66

Radar, 63
Radar beacons (Racons), 63–66
Radar reflectors, 162
Rade de Brest, Le, **648**
Radiobeacons, Aero, 38, 42–61
Radiobeacons, errors, 39
Radiobeacons, Marine, 38, 42–61
Radiobeacons, operating procedure, 38–39
Radio direction finding, 40
Radio emission, types, 38
Radio forecasts, Western Europe, 158
Radio lighthouses, VHF, 62
Radio, port operations, 131
Radio Solent, 234
Radiotelephone, link calls, 130–131
Radiotelephones
 General procedures, 130–131
 General provisions, 129
 Licences, 129
Radiotelephony, 129–143
Radiotelephony, International Code of Signals, 126
Radio time signals, 127
Radio weather reports, 154–160
Rame Head, R 40, 192
Ramsey, L 471, **482**
Ramsey Sound, 501, *505*
Ramsgate, L 271, **287**
Range, of tide, 169
Ranges, of lights, 178
Ranza, Loch, 461, *463*
Rathlin Island, L 561, 563
Rathlin O'Birne, L 562
Rathlin Sound, 563
Rathmullan, L 561, **575**
Rattray Head, L 369, 371, 397
Ravenglass, **491**
Ré, Ile de, L 588, 591, **605, 613**
Recording and interpreting shipping forecasts, 147, 153
Record of amendments, 4
Red Bay, 563, **582**
Redcar, L 341, 342
Refraction, 68, 70
Registrar of Ships, 19
Registration, 19
Regnéville, L 666
Regulations for Prevention of Collisions at Sea, 8, 16–17
Regulations in European countries, 21–22
Religious calendars, 75
Reports of present weather 151
Response to a distress call, 132, 165
Restricted visibility, conduct in, 16
Restricted visibility, sound signals, 17
Reuille, La, *596*
Rhea, Kyle, L 422, 425
Rhinns of Islay, R 47, L 448
Rhoose/Cardiff, R 49
Rhu, **455**
Rhuba Cadail, L 422
Rhuda Mhail, L 448
Rhyl, L 471
Ribble, River, L 470, 472, **491**
Richardais, **689**
Richborough, *282*
Richelieu, Tour (La Rochelle), R 52, L 589
Richmond, *297*, 298
Rineana Point, L 530
Rising and setting phenomena, 70, 82–85

For key to symbols and entries see explanation on page 789.

Risk of collision, 8
Roach, River, 304, 315, 333
Roancarrigmore, L 530
Roaring Middle Lt Float, L 313
Robbennordsteert, L 767
Robert, Point, L 667, 676
Roc, Pointe du, L 666, 682
Roche Bernard, La (La Vilaine),
  L 620, 623, **627**
Rochebonne, L 666
Rochebonne, Plateau de, L 589, 591
Rochefort (La Charente), L 588, **603**
Rochelle, La, R 52, C 139, L 589,
  591, **604**
Roches Douvres, Les, R 40, R 54,
  L 645, 647
Roche's Point, L 531, 539
Rochester, 291
Rockabill, L 531, L 560, 563
Rockall, L 422
Rohein, Le, L 666
Roker Pier, L 341, 357
Romo Havn, *783*
Rona, North, L 395
Rona, South, L 422, 425
Ronaldsway (I o M), R 47
Roompot, L 731, **746**
Roque, La, *707*
Roscoff (Bloscon), R 54, L 645, **654**
Rosédo (Ile de Bréhat), R 54, L 645,
  659
Rosneath, L 449, **455**
Rosslare Harbour, L 531, **545**
Rosyth, L 368
Rote Kliff (Kampen), R 58, L 767,
  782
Rothéneuf, 689
Rothesay, L 448, **454**
Rotten Island, L 562, 576
Rotterdam, 748
Rouen, C 141, *706*, 707
Rouge de Glénan Lanby, L 622
Round Island, R 42, L 190, 194
Rousay, 406
Rova Head, L 396
Royal Air Force, SAR, 164
RNLI, 165
Royal Navy, SAR, 164
Royal Sovereign, R 43, L 270, 272
Royal Yachting Association, 22
Royan, C 139, **601**
Rubha nan Gall, L 423
Rubha Reidh, L 422
Rule of the road, 8, 16–17
Rumble Rock, R 64
Runnaneun Point, L 448
Runswick Bay, 362
Rush/Dublin, R 50
Russel, Big, 669
Russel, Little, 669, 677
Rutland North Channel, L 561
Ryan, Loch, L 449, **463**
Ryde, L 223, *255*
Ryde Sand, 224
Rye, L 270, 272, **281**

S

Sables D'Olonne, Les, R 52, L 589,
  607

Safe speed, 8
Safety, 161–166
Safety equipment, 162
Safety signal, RT, 132
Safe water marks (IALA), 36, 119
Sailing directions, abbreviations, 181
Sailing vessels, rule of the road, 8,
  16–17
St Abb's, R 64, L 368, 370, **387**
St Alban's Head, 224
St Annaland, 746
St Anne's, *475*
St Ann's Head, R 40, C 138 (Celtic
  Radio), L 498, 506
St Anthony Head, L 190, 191, 197
St Aubin (Jersey), L 667, **671, 689**
St Barbe, L 588
St Bees Head, L 470
St Brides Bay, 501, **523**
St Brieuc, R 54, **680**
St Cast, L 666, **689**
St Catherine's Point, R 42, L 223,
  225
St Germans, *200*
St Gildas, Pointe de, R 52, L 620,
  623, 626
St Giles-Croix-de-Vie, C 140, L 589,
  608
St Gowan Lt V, R 65, L 498
St Guenole, L 622
St Helen's Fort, L 223, 255
St Helier, R 50, R 65, L 666, 669,
  671
St Inglevert, R 55
St Ives, L 500, 502, **524**
St Jean-de-Luz, L 588, 590, **612**
St John's Point, L 560
St Katharine's Yacht Haven, 297,
  305
St Kilda, L 423, *426*, **441**
St Malo, C 140, L 666, 668, **681**
St Marcouf, Îles, L 696, 688, **723**
St Margaret's Hope, 406
St Martin, Ile de Ré, L 588, **605**
St Martin, Pointe, L 588, 590
St Martin's Point, L 667, 669
St Mary's (Orkney), 406
St Mary's (Scilly), L 190, 191, 194,
  195
St Mathieu, Pointe de, R 54, L 644,
  646
St Mawgan, R 50
St Michael's Mount, 216
St Monans, L 369, **387**
St Nazaire, R 52, C 140, L 620, 623,
  625
St Nazaire Lt Buoy, R 66
St Nazaire/Montoir, R 52
St Patrick's Causeway, 501
St Peter (Die Eider), L 767
St Peter Port, R 50, C 135, L 667,
  669, **677**
St Philipsland, 746
St Quay-Portrieux, L 666, 668, **679**
St Sampson (Guernsey), L 667, **689**
St Tudwal's, L 471, 473, **490**
St Vaast-la-Hougue, L 696, 698, **700**
St Valéry-en-Caux, L 697, **712**
St Valéry-sur-Somme, L 697, 699,
  718
Saire, Pointe de, L 696, 698
Salcombe, L 190, **206**
Salen (Loch Sunart), 432
Saltash, *200*

Saltee Sound, 533
Salutes, 144
Sanda Island, R 65, L 448, *453*
Sanda Sound, 450
Sanday, 406
Sandettie Lt V, R 66, L 697
Sandown, L 223, 225, *255*
Sandwich, 305
Sandwick Bay, R 40, L 422
Sangatte, R 66, L 697
San Sebastian, R 51, **594**
Santander, *582*
Sark, L 667, **676**
**Sarn Badrig Causeway (St
  Patrick's),** L 498, 501
Sarn-y-Bwch, 501
Sas van Goes, L 731, **746**
Satellite navigation, 40
Saundersfoot, L 498, 501, **523**
Sauzon, 638
Scalasaig (Colonsay), L 423, *452*,
  466
Scalloway, L 396, **415**
Scapa Bay, 406
Scarborough, L 340, 342, **350**
Scarinish, L 423
Scatsa, R 46
Scharnhörn, *775*
Schelde, River, L 730, 734, **740**
Scheveningen, R 57, C 142, L 732,
  754
Scheveningen/Valkenberg, R 57
Schiermonnikoog, L 733, 735, **760**
Schilbolsnol, L 733
Schillig, L 716
Schleuseninsel, L 767
Schluttsiel, 784
Schone Waardin, L 731
Schouwenbank Lt Buoy, R 66,
  L 731
Schull, L 530, 533, **537**
Schulpengat, L 732, 735
Scilly, Isles of, L 190, 191, **194, 195**
Scotland, North Coast, L 394, 398
Scotland, West Coast, L 422, 424
Scotstown Head, R 46
Scrabster, L 395, **414**
Scratchell's Bay, R 62
Scurdie Ness, L 369, 370
Sea areas, shipping forecasts, 152,
  159
Seaham, L 341, **356**
Seahouses (North Sunderland),
  L 341, 343, **361**
Search and rescue (SAR)
  Aircraft, signals 128
  Organisation, 164–166
  Search procedures, aircraft, 128
Sea Reach (Thames Estuary), L 271
Sea Reach Lt Buoy, R 64, L 271
Secondary ports, 169
Seegat, 734
Sein, Chaussée de, L 622, 624
Sein, Chaussée de, Lt Buoy, R 66,
  L 622
Sein, Île de, R 54, L 622, 624
Sein, Raz de, 624
Seine, Baie de, 698
Seine Estuary, 698
Seine River, 706
Selected groups from the
  International Code, 125–126
Selected stars and Sun, for calculator
  use, 77

Selsey Bill, R 40, 272
Semaphore (French), 593
Semi-diameter, Sun, 69, 87–97
Sénéquet, Le, L 666
Separation schemes, traffic, 9–15
Sept Îles, Les, L 645, 647
Sequence numbers, radiobeacons, 38, 43, 45, 47, 49, 51, 52, 55, 58
Seudre, River, L 588, **601**
Seven Stones Lt V, R 64, L 190
Severn, River, L 500, **514**
Sextant angles, horizontal, 33
Sextant angles, vertical, 26–28, 33
Shaldon, 213
Shambles, The, L 222, 224
Shannon, River, L 530, 532, **535**
Shapes, 16, 118
Shapinsay (Orkney), 415
Sharpness, L 500, **514**
Sheep Haven, 564, **582**
Sheep Head, L 530, 532
Sheerness, L 271, **291**
Sheildag, *431*
Shell Island (Mochras), L 498, **523**
Shell, Loch, *426*, **441**
Shetland Islands, L 395, 398, **409, 415**
Shetland Radio, C 138
Shipping forecasts, 148, 152–154
Shipwash Lt Float, L 312
Shivering Sands Tower, L 271, *304*
Shoots, The, L 500
Shoreham, R 43, L 270, 272, **274**
Shornmead, L 271
Signals, 123–143
Signals, aircraft on SAR operations, 128
Signals, methods of transmission, 124
Signals, shore lifesaving equipment, 127–128
Signals, vessel standing into danger, 128
Signals, when towing, 16, 118
Silloth, L 470, *474*, **491**
Single letter signals, 120–121
Skadan (Fair Isle), L 395
Skegness, L 340, 342, *344*
Skelligs Rock, L 530, 532
Skerries Bank (Start Bay), 192
Skerries Bay (Meath), L 560, 563
Skerries, Covesea, L 394
Skerries, Out (Shetland), L 396
Skerries, Pentland, L 394, 397
Skerries, The (Anglesey), R 49, R 65, L 471, 473
Skerries, Ve (Shetland), R 65, L 396
Skerry of Ness, L 395, 408
Skerry, Outer (Yell Sound), L 396
Skerry, Sule L 395
Skerryvore, R 65, L 423
Skiport, Loch, *426*
Skokholm Island, L 498, 501
Skomer Island, *506*, **523**
Skroo (Fair Isle), L 395
Skye, Isle of, L 423, 424, **431**
Skye Radio, C 137
Sky wave effect, DF bearings, 39
Sleat, Sound of, L 422, 425
Sligo, R 51, L 562, **576**
Slyne Head, R 51, L 562, 564
Small Craft Edition, Admiralty Notices to Mariners, 4
Small Craft Warning Service (strong winds), 151, 156

Small Isles, L 423, 425, **442**
Small Ships Register, 19
Smalls, The, 15, R 65, L 498, 501
Smiths Knoll Lt V, R 43, R 64, L 313
Snaefell, R 40
Sneem, 554
Snizort, Loch, 424, *431*
Soldiers Point (Dundalk), *553*
Solent Bank, *247*
Solent, L 222, 224, 226, **245, 264**
  Special factors, 232, 234
  Tidal problems, 226
  Tidal streams, 230
  Waypoints, 233
Solva, 523
Solway Firth, 472
Somme, Baie de, 699, **718**, 723
Sophiapolder, L 731
Sorel Point, L 667
Sound signals, 17
Sound signals for distress purposes, 163
Sound signals, International Code, 126
Sources of weather information, 148–160
Southampton, L 223, **249**
South Bay, 406
South Bishop, R 49, L 498, 501
Southend-on-Sea, R 43, L 271, **303**
South Gare (River Tees), L 341, 352
South Goodwin Lt Float, L 270
South Rock Lt Float, R 51, R 65, L 560, 563
South Ronaldsay, 406
South Shields, 358
South Stack, L 471, 473
South Wick, 406
Southwold, L 313, 314, **325**
Spain (North), tidal information, *594*
Special marks (IALA), 36, 119
Spiekeroog, 768, **783**
Spijkerboor/Amsterdam, R 57
Spring tides, 168
Spurn Head, L 340, 345
Spurn Lt Float, R 64, L 340
Stad, R 57
Standard ports, 169
Standard terms, 2
Standard times, 74, 168
Starcross, *214*
Star diagrams, 70, 108–110
Stars, SHA and Dec, 77, 80–81
Start Point, R 42, C 135, L 190, 192
Start Point (Sanday), L 395
Stavoren, 756
Stellendam, 748
Stenbury Down, R 40
Stiff, Le (Ouessant), L 644
Stoer Head, L 422, 424
Stonehaven, C 137, L 369, 371, **381**
Storm signals, visual, 129, 592
Stornoway, R 47, L 422, **426**
Stour, River, 318
Strangford Lough, L 560, **566**
Stranraer, L 449, **463**
Strathy Point, L 395
Straw Island, L 562
Stroma (Swilkie Point), R 46, L 394, 397, *414*
Stromness, L 395, **408**
Strong breeze warnings, 151, 156
Stronsay, L 395, **407**
Strumble Head, R 49, L 498

Studland Bay, 264
Suderpiep, 769
Suggestions for improvements, 4
Sule Skerry, R 46, R 65, L 395
Sullom Voe, L 396, **409**
Sumburgh, R 46
Sumburgh Head, R 46, L 395, 398
Summary of important sound signals, International Regulations for Preventing Collisions at Sea, 17
Summer Islands (Tanera Mor), 424, *427*
Sun, altitude correction, 76
Sunart, Loch, 432
Sun, bearing, rising and setting, 24–25
Sun, GHA and Dec, 77, 87–97
Sun, semi-diameter (SD), 69, 87–97
Sunderland, L 341, 342, **357**
Sunk Lt Float, R 43, R 64, L 312, 314
Sunrise/sunset, 82–83
Supplement, 4
Supplements, application form, 5
Swale, The, L 271, **290**
Swanage, L 222, **239**
Swansea, R 49, L 498, **512**
Swarbacks Minn, L 396, **409**
Sween, Loch, 450, **462**
Swellies, The, 472, **489**
Swilkie Race (Pentland Firth), 397
Swilly, Lough, L 561, 563, **575**
Swinge, The, 668, 679
Sylt, L 767, 769, **782**
Sylt/Westerland, R 58
Symbols, Admiralty chart, 34–35, 116–117
Synchronised transmissions, 39

T

Tancarville, *706*, 707
Tanera Mor (Summer Islands), 424, *427*
Tarbat, Ness, R 64, L 394, 397
Tarbert, Loch (Jura), 462
Tarn Point, *474*
Taw, River, 521
Tater-du, L 190
Tay, River, L 369, 370, **379**
Tayport, 379
Tayvallich (Loch Sween), 450, **462**
Tean Sound, 194
Tees Fairway Buoy, R 64, L 341
Teeside, R 45
Tees, River, L 341, 342, **351**
Tegeler Plate, L 766
Teignmouth, L 190, 192, **213**
Teignouse, Passage de la, L 621, 623
Telephone, weather forecasts, 148
Television forecasts, 150
Temperature conversion, 147
Tenby, L 498, 501, **511**
Terms, standard, 2
Terms used in weather bulletins, 147
Terneuzen, L 731, **740**
Terre-Nègre, L 588, 590
Terschelling, L 733, 735, **759**
Texel, 12, L 732, **760**

800 Index

For key to symbols and entries see explanation on page 789.

Texel, Lt V, R 57, R 66, L 732
Texel, Zeegat van, L 732, 735
Thames Estuary, L 271, 273, 288, L 312, 314
Thames Estuary, tidal streams, 288
Thames Radio, C 135
Thames, River, L 271, **296**
Thermometer conversion scale, 147
Tholen, 746
Thurso/Dounreay, R 46
Thyboron, *784*
Tidal calculations, 169
Tidal coefficients, 593
Tidal curves, 170
Tidal information, 167–175
Tidal stream atlases, 174
Tidal stream charts, 174
Tidal stream rates, 174
Tidal streams, 174
Tidal streams in rivers, 342
Tides, 167–175
Tides and currents, abbreviations, 34
Tides, French terms, 593
Tide signals, 129, 592
Tides, meteorological conditions, 175
Times, 74, 168
Time signals, 127
Time, speed and distance tables, 30–32
Time zones, 74, 168
Tiree, L 423, 425, *431*
Titchfield Haven, 264
Tiumpan Head, L 422
Tobermory, 425, **432**
Tod Head, L 369, 371
Tollesbury, 316
Tönning, L 767, **783**
Topsham, L 190, **214**
Torquay, 212
Torridge, River, 521
Torr Sound, 563
Torsminde, *784*
Tory Island, R 51, L 561
Tory Sound, 564
Totland Bay, *246*
Toulinguet, Chenal du, 624
Toulinguet, Pointe du, L 622
Touquet, Le, R 55, L 699, **719**
Touquet, Pointe du, L 697
Toward Point, L 448
Towing and pushing, 16
Towing signals, 16, 118
Traffic separation schemes, 9–15
Traffic signals, 122, 128, 592, 736, 770
Tralee Bay, L 530, *536*
Transmitting frequencies/channels, VHF radio, 130
Trébeurden, 660
Tréboul, L 622, **636**
Trégastel, 660
Tréguier, L 645, **657**
Tremblade, La, 602
Tréport, Le, R 55, L 697, 699, **717**
Trévignon, L 621, 623
Trevose Head, R 40, L 500, 502
Trézien, L 644, 646
Triagoz, Les, L 645, 647
Trieux, Le (Lézardrieux), L 645, 647, **658**
Trimingham, R 40
Trinité-sur-Mer, La (Rivière de Crac'h), L 621, 623, **630**
Trinity House, 22

Tristan, Île, L 622, 636
Trois Pierres, Les (Chenal de la Helle), L 644, 646
Trois Pierres, Les (Lorient), L 621, 632
Trouville, L 696, 698, **704**
Troon, L 449, 451, **461**
Truro, *197*, **216**
Trwyn-du, L 471
Turballe, La, L 620
Turnberry, R 47
Turnberry Point, L 449
Tuskar Rock, 15, R 50, R 65, L 531, 533
Twilight, 82–83
Tyne, River (Tynemouth), L 341, 342, **358**

U

Uig Bay, L 423, *431*
Ullapool, L 422, **427**
UK Coast Radio Stations, 135–138
Unst, R 46
Upnor, 291
Ura Firth, 409
Urgency signal, 132
Useful addresses, 22
Ushant (Île d'Ouessant), 9, R 54, R 66, C 140, L 644, 646, **653**
Ushenish, L 422
Usk, River, L 499, **523**

V

Vaila Sound, L 396, **409**
Val-André, 689
Valentia, C 139, L 530
Valkenberg/Scheveningen, R 57
Vannes, R 52, **629**
Variation, 24
Varne Lanby, R 64, L 270, 273
Vatersay Sound, L 423
Vatteville, 706
Veere, 746
Ventnor, L 223, *255*
Ventry, 554
Ver, R 55, L 696
Verclut breakwater, L 667
Vergoyer Lt Buoy, R 66
Vertical sextant angles, 24, 26–28, 33
VHF direction finding, 40
VHF frequencies, 2, 130
VHF Radio, 129–132
VHF Radio Lighthouses, 61–62
Ve Skerries, R 65, L 396
Vessels aground lights, 16, 118
Vessels constrained by their draught, lights, 118
Vessels not under command, 16, 118
Vieille, La (Raz de Sein), L 622
Vierge, Île, R 54, L 644, 646
Vigne, La, 612
Vilaine, La, L 620, 623, **627, 638**
Visibility of navigation lights, 115

Visual signals between UK shore stations and ships in distress, 127–128
Visual storm signals, 129, 592
Vlieland, R 57, L 733, 735, **758**
Vlissingen (Flushing), L 730, 734, **742**
Voice communication, radiotelephone, 126, 130–132
Volendam, 756
Volkerak, L 732
VOLMET, 149
Vortrapptief, 769

W

Wainfleet, L 340, **361**
Wallasea, 315
Wallasey, R 49
Walney, Isle of, R 49, L 470, 472, **491**
Walton-on-the-Naze, L 312, **318**
Wandelaar, R 66
Wangerooge, R 58, L 766, 768, 772
Wangersiel, 768, **783**
Wareham, *244*, **264**
Warning signals (Rule 34), 17
Warkworth Harbour, L 341, 343, **360**
Warrenpoint (Newry River), L 560, **565**
Warsash, L 223, **254**
Wash, The, 314, **332,** 342, **344**
Watchet, L 500, *519*, **524**
Waterford, R 50, L 531, **545**
Watermill Cove, 194
Watermouth, 524
Wavelength-frequencies, 129–130
Waypoints, 40 (and Section 4 of each Area, e.g. 10.1.4), 193, 233
Weather, 145–160
Weather bulletins, coast radio stations, 150
Weather forecasts by telephone call, 148
Weather information by radiotelephone, 148, 151
Weather information from radio stations in Western Europe, 158
Welland, River, 333
Wells-next-the-Sea, L 313, **331**
Wemeldinge, 746
Wemyss Bay, *460*
Weser, Die, L 767, 769, **774**
Weser Lt Float, R 58, R 66, L 766
West Bramble Lt Buoy, R 64, L 223
Westerems, 735
West Hinder Lt V, 11, R 57, L 730
Westerland/Sylt, R 58
Westerschelde, L 730, 734, **740**
Westhoofd, L 732
Westkapelle, L 730, 734, *740*
West Loch Tarbert (Harris), *426*, **441**
Weston, R 49
Weston-super-mare, L 500, *519*, **524**
Westport (Clew Bay), L 562, **577**
Westray, L 395, **406**
West Schouwen, L 732

West Torr, R 40
Wexford, 546
Weymouth, L 222, 224, **239**
Weymouth Bay Radio, C 135
Whalsay, L 396
Whitby, C 137, L 341, **351**
Whitehaven, *474*, **491**
Whitehills, 400
Whithorn, Isle of, 463
Whitsand Bay, *199*
Whitstable, L 271, **287**
Wick, R 46, C 138, L 394, 397, **405**
Wicklow, L 531, **547**
Wicklow Head, R 50, L 531
Widnes, *476*
Wight, Isle of, L 222, 224, 230, **246**, 255, 264
Wigtown Bay, 451, **463**
Wilhelmshaven, R 58, **774**
Windyheads Hill, R 40
Winteringham, 344
Winterton Old Lt Ho, R 64
W/T transmissions, weather, 149
Wisbech, L 313, **333**
Wisbech Cut, *332*

Wittdun, 784
Wivenhoe, L 312, **317**
Woensdrecht, R 57
Wolf Rock, R 64, L 190, 191
Woodbridge, L 313, **324**
Wootton Creek, L 223, **255**
Workington, L 470, 472, **474**
Workum, 756
Worthing, *274*
Wrath, Cape, R 46, L 395, 398, L 422, 424
Wyk, 784
Wyre Lighthouse, 476
Wyre Sound, 406

Y

Yacht and boat safety schemes, 165
Yachting and marine organisations, 22
Yarmouth, Great, L 313, 314, **329**

Yarmouth (I o W), L 223, **246**
Yealm, River, 206
Yell Sound, L 396, **409**
Yerseke, 746
Yeu, Île d', R 52, L 589, **609**
Youghal, L 531, **544**
Yport, L 697

Z

Zeebrugge, R 57, L 730, 734, **739**
Zeegat, 734
Zeegat van Ameland, 735
Zeegat van Terschelling, 735
Zeegat van Texel, 735
Zierikzee, L 731, **746**
Zone time, 74, 168
Zoutkamp, 760
Zuider Stortemelk, 735

SELLING? BUYING?

USE Yachts and Yachting

- 26 issues a year.
- Up to date list of boats for sale.
- Minimum delay in advertising your boat.
- The largest number of private sailing boats for sale of any UK Yachting Magazine.

Yachts and Yachting

196 Eastern Esplanade, Southend-on-Sea, Essex

 Tel: 0702 582245